AA002404

CIDs and ASME Conference Proceedings

Beginning with 2013 conference proceedings, ASME transitioned to e-first publication of conference Proceedings. As a result, instead of traditional page numbers, the online pagination has been by 11-digit citation identifiers (CIDs); For instance, using the CID V###T###A### as an example, the structure of the basic 11-digit citation is defined as follows:

- First four digits (V00#): indicate the **Volume Number**
- Middle three digits (T##): indicate the **Track Number**
- Last four digits (A###): indicate the **Article Order** in which the paper is published within the track.

In the print version, the CID appears on every page of the paper followed sequentially by the page number of each article starting with page one. For example, the pagination for a three page article would use the following sequence:

V###T##A###-1
V###T##A###-2
V###T##A###-3

In the ASME Digital Collection version, the CID appears on the Table of Contents where the paper is published. The CID also appears on the abstract page, followed by the total number of pages in the paper. For example: V###T##A###; 3 pages

CIDs and Citations

When citing an article that has been published wit a CID, enter the CID where the page range would have gone previously; do not include the page numbers used on the PDF or print.

Errata

Errata submitted by authors after publication of this volume are published with the relevant paper on the ASME Digital Collection

http://asmedigitalcollection.asme.org/

This page left blank intentionally.

CONTENTS

45TH DESIGN AUTOMATION CONFERENCE

ARTIFICIAL INTELLIGENCE AND MACHINE LEARNING

DETC2019-97399.. **V02AT03A001**
Learning to Design From Humans: Imitating Human Designers Through Deep
Learning
Ayush Raina, Christopher McComb, Jonathan Cagan

DETC2019-97642.. **V02AT03A002**
Analyzing Customer Needs of Product Ecosystems Using Online Product Reviews
Jackie Ayoub, Feng Zhou, Qianli Xu, Jessie Yang

DETC2019-97689.. **V02AT03A003**
Deep Reinforcement Learning for Transfer of Control Policies
*James D. Cunningham, Simon W. Miller, Michael A. Yukish, Timothy W. Simpson, Conrad S.
Tucker*

DETC2019-97830.. **V02AT03A004**
Machine Learning-Augmented Stochastic Search for the Automated Synthesis and
Optimization of Cooling Channels
Jonas Schwarz, Kristina Shea

DETC2019-97975.. **V02AT03A005**
A Neuroevolution-Based Learning of Reciprocal Maneuver for Collision Avoidance in
Quadcopters Under Pose Uncertainties
Amir Behjat, Krushang Gabani, Souma Chowdhury

DETC2019-98027.. **V02AT03A006**
Gaussian Process Emulation for Big Data in Data-Driven Metamaterials Design
Ramin Bostanabad, Yu-Chin Chan, Liwei Wang, Ping Zhu, Wei Chen

DETC2019-98115.. **V02AT03A007**
Multi-Fidelity Physics-Constrained Neural Network and its Application in Materials
Modeling
Dehao Liu, Yan Wang

COMPUTATIONAL DESIGN FOR BIOMEDICAL APPLICATIONS

DETC2019-97777.. **V02AT03A008**
Computational Design of a Personalized Artificial Spinal Disc With a Data-Driven
Design Variable Linking Heuristic
Zhiyang Yu, Kristina Shea, Tino Stankovic

DETC2019-98190.. **V02AT03A009**
Design and Biological Simulation of 3D Printed Lattices for Biomedical Applications
Paul F. Egan

DATA-DRIVEN DESIGN

DETC2019-97437.. **V02AT03A010**
A Digital Twin-Driven Improved Design Approach of Drawing Bench for Brazing
Material
Bingtao Hu, Yixiong Feng, Yicong Gao, Hao Zheng, Jianrong Tan

DETC2019-97587.. **V02AT03A011**
Computer-Aided Design Ideation Using InnoGPS
Jianxi Luo, Serhad Sarica, Kristin L. Wood

DETC2019-97691.. **V02AT03A012**
Product Service System Design in New Situations: Prediction of Demand Surfaces
From Environment
Bryan C. Watson, Cassandra Telenko

DETC2019-97881.. **V02AT03A013**
Pseudo-Rigid Body Dynamic Modeling of Compliant Members for Design
Vedant, James T. Allison

DETC2019-98000.. **V02AT03A014**
Checking the Automated Construction of Finite Element Simulations From Dirichlet
Boundary Conditions
Kevin N. Chiu, Mark D. Fuge

DETC2019-98111.. **V02AT03A015**
Using Bayesian Optimization With Knowledge Transfer for High Computational Cost
Design: A Case Study in Photovoltaics
Mine Kaya, Shima Hajimirza

DETC2019-98385.. **V02AT03A016**
Data-Driven Dynamic Network Modeling for Analyzing the Evolution of Product
Competitions
*Jian Xie, Youyi Bi, Zhenghui Sha, Mingxian Wang, Yan Fu, Noshir Contractor, Lin Gong, Wei
Chen*

DETC2019-98525.. **V02AT03A017**
3D Shape Synthesis for Conceptual Design and Optimization Using Variational
Autoencoders
*Wentai Zhang, Zhangsihao Yang, Haoliang Jiang, Suyash Nigam, Soji Yamakawa, Tomotake
Furuhata, Kenji Shimada, Levent Burak Kara*

DECISION MAKING IN ENGINEERING DESIGN

DETC2019-97301.. **V02AT03A018**
Investigating Optimal Communication Frequency in Multi-Disciplinary Engineering
Teams Using Multi-Agent Simulation
Mojtaba Arezoomand, Jesse Austin-Breneman

Proceedings of the ASME

INTERNATIONAL DESIGN ENGINEERING TECHNICAL CONFERENCES & COMPUTERS AND INFORMATION IN ENGINEERING CONFERENCE

- 2019 -

VOLUME 2A

45TH DESIGN AUTOMATION CONFERENCE

presented at

ASME 2019 INTERNATIONAL DESIGN ENGINEERING TECHNICAL CONFERENCES &

COMPUTERS AND INFORMATION IN ENGINEERING CONFERENCE

AUGUST 18-21, 2019

ANAHEIM, CALIFORNIA, USA

sponsored by

DESIGN ENGINEERING DIVISION, ASME

COMUTERS AND INFORMATION IN ENGINEERING DIVISION, AMSE

THE AMERICAN SOCIETY OF MECHANICAL ENGINEERS
Two Park Avenue ✴ New York, NY. 10016

Printed from e-media with permission by:

Curran Associates, Inc.
57 Morehouse Lane
Red Hook, NY 12571

Some format issues inherent in the e-media version may also appear in this print version.

Statement from By-Laws: The Society shall not be responsible for statements or opinions
Advanced in papers. . .or printed in its publications (7.1.3)

INFORMATION CONTAINED IN THIS WORK HAS BEEN OBTAINED BY ASME FROM SOURCES
BELIEVED TO BE RELIABLE. HOWEVER, NEITHER ASME NOR ITS AUTHORS OR EDITORS
GUARANTEE THE ACCURACY OR COMPLETENESS OF ANY INFORMATION PUBLISHED IN
THIS WORK. NEITHER ASME NOR ITS AUTHORS AND EDITORS SHALL BE RESPONSIBLE
FOR ANY ERRORS, OMISSIONS, OR DAMAGES ARISING OUT OF THE USE OF THIS
INFORMATION. THE WORK IS PUBLISHED WITH THE UNDERSTANDING THAT ASME AND ITS
AUTHORS AND EDITORS ARE SUPPLYING INFORMATION BUT ARE NOT ATTEMPTING TO
RENDER ENGINEERING OR OTHER PROFESSIONAL SERVICES. IF SUCH ENGINEERING OR
PROFESSIONAL SERVICES ARE REQUIRED, THE ASSISTANCE OF AN APPROPRIATE
PROFESSIONAL SHOULD BE SOUGHT.

For authorization to photocopy material for internal or personal use under circumstances not falling
within the fair use provisions of the Copyright Act, contact the Copyright Clearance Center (CCC),
222 Rosewood Drive, Danvers, MA 01923, Tel: 978-750-8400

Requests for special permission or bulk reproduction should be addressed to
permissions@asme.org.

ISBN NO. 978-0-7918-5918-6

© 2019 ASME
All rights reserved.
Printed in U.S.A with permission by Curran Associates, Inc. (2020)

Additional copies of this publication are available from:

Curran Associates, Inc.
57 Morehouse Lane
Red Hook, NY 12571 USA
Phone: 845-758-0400
Fax: 845-758-2633
Email: curran@proceedings.com
Web: www.proceedings.com

DETC2019-97372.. **V02AT03A019**

Classification and Execution of Coupled Decision Problems in Engineering Design for Exploration of Robust Design Solutions
Gehendra Sharma, Janet K. Allen, Farrokh Mistree

DETC2019-97894.. **V02AT03A020**

Design of Composite Structures Through Decision Support Problem and Multiscale Design Approach
Rizwan Khan Pathan, Soban Babu Beemaraj, Amit Salvi, Gehendra Sharma, Janet K. Allen, Farrokh Mistree

DETC2019-98028.. **V02AT03A021**

Quantum Mechanical Perspectives in Reliability Engineering and System Design
Vijitashwa Pandey

DETC2019-98035.. **V02AT03A022**

A Proposal for a Decision Support Framework to Solve Design Problems in the Automotive Industry
Timothe M. Sissoko, Marija Jankovic, Christiaan J. J. Paredis, Eric Landel

DESIGN AND OPTIMIZATION OF SUSTAINABLE ENERGY SYSTEMS

DETC2019-97171.. **V02AT03A023**

Sustainable Design of Residential Net-Zero Energy Buildings: A Multi-Phase and Multi-Objective Optimization Approach
Lan Lan, Kristin L. Wood, Chau Yuen

DETC2019-97190.. **V02AT03A024**

Designing Optimal Arbitrage Policies for Distributed Energy Systems in Building Clusters Using Reinforcement Learning
Philip Odonkor, Kemper Lewis

DETC2019-97964.. **V02AT03A025**

Optimization Model for Owner-Based Microgrids Using LSTM Predicted Demand for Rural Development
Anosh P. Amaria, Ryan Nguyen, Joshua A. Davison, Souma Chowdhury, John F. Hall

DETC2019-98205.. **V02AT03A026**

Self-Adapting Intelligent Battery Thermal Management System via Artificial Neural Network Based Model Predictive Control
Yuanzhi Liu, Jie Zhang

DESIGN FOR ADDITIVE MANUFACTURING

DETC2019-97248.. **V02AT03A027**

Stress Field Guided Lattice Structure Design Based on Hexahedral Mesh
Lingyun Liu, Yizhou Liao, Shuming Gao

DETC2019-97478.. **V02AT03A028**
But Will it Print?: Assessing Student Use of Design for Additive Manufacturing and
Exploring its Effect on Design Performance and Manufacturability
Rohan Prabhu, Scarlett R. Miller, Timothy W. Simpson, Nicholas A. Meisel

DETC2019-97480.. **V02AT03A029**
A Comparative Study of Virtual Reality and Computer-Aided Design to Evaluate Parts
for Additive Manufacturing
*John K. Ostrander, Lauren Ryan, Snehal Dhengre, Christopher McComb, Timothy W.
Simpson, Nicholas A. Meisel*

DETC2019-97492.. **V02AT03A030**
Simulation-Based Process Optimization of Metallic Additive Manufacturing Under
Uncertainty
Zhuo Wang, Pengwei Liu, Zhen Hu, Lei Chen

DETC2019-97607.. **V02AT03A031**
Digital Design Automation to Support In-Situ Embedding of Functional Components in
Additive Manufacturing
Manoj Malviya, Swapnil Sinha, Nicholas A. Meisel

DETC2019-97649.. **V02AT03A032**
Optimization of Parts Consolidation for Minimum Production Costs and Time Using
Additive Manufacturing
Zhenguo Nie, Sangjin Jung, Levent Burak Kara, Kate S. Whitefoot

DETC2019-97775.. **V02AT03A033**
Computational Design of Active Lattice Structures for 4D Printed Pneumatic Shape
Morphing
Cosima du Pasquier, Pascal Koller, Tino Stankovic, Kristina Shea

DETC2019-97840.. **V02AT03A034**
A Design Modification System for Additive Manufacturing: Towards Feasible
Geometry Development
Seyedeh Elaheh Ghiasian, Prakhar Jaiswal, Rahul Rai, Kemper Lewis

DETC2019-97863.. **V02AT03A035**
Lattice Structure Design for Additive Manufacturing: Unit Cell Topology Optimization
Bradley Hanks, Mary Frecker

DETC2019-97865.. **V02AT03A036**
Evaluating the Potential of Design for Additive Manufacturing Heuristic Cards to
Stimulate Novel Product Redesigns
Alexandra Blosch-Paidosh, Saeema Ahmed-Kristensen, Kristina Shea

DETC2019-97913.. **V02AT03A037**
An Optimal Quantity of Scheduling Model for Mass Customization-Based Additive
Manufacturing
Yosep Oh, Sara Behdad

DETC2019-97915.. **V02AT03A038**
Design for Additive Manufacturing Using a Master Model Approach
Anton Wiberg, Johan Persson, Johan Olvander

DETC2019-98024.. **V02AT03A039**
Rule of Mixtures Model to Determine Elastic Modulus and Tensile Strength of 3D
Printed Carbon Fiber Reinforced Nylon
*Kaiyue Deng, Hamid Khakpour Nejadkhaki, Felipe M. Pasquali, Anosh P. Amaria, Jason N.
Armstrong, John F. Hall*

DETC2019-98068.. **V02AT03A040**
Substrate Optimization for Hybrid Manufacturing
Brandon R. Massoni, Matthew I. Campbell

DETC2019-98103.. **V02AT03A041**
Voxel-Based CAD Framework for Planning Functionally Graded and Multi-Step Rapid
Fabrication Processes
Cole Brauer, Daniel M. Aukes

DESIGN FOR MARKET SYSTEMS

DETC2019-97379.. **V02AT03A042**
Designing the Customer Order Decoupling Point to Facilitate Mass Customization
Lin Guo, Suhao Chen, Janet K. Allen, Farrokh Mistree

DETC2019-97680.. **V02AT03A043**
Word-of-Mouth Recommendations in an Automobile Market System
Amineh Zadbood, Nicholas Russo, Steven Hoffenson

DETC2019-97835.. **V02AT03A044**
Influence of Omitted Variables in Consumer Choice Models on Engineering Design
Optimization Solutions
Waleed Gowharji, Kate S. Whitefoot

DETC2019-98114.. **V02AT03A045**
Implications of Competitor Representation on Optimal Design
Arthur H. C. Yip, Jeremy J. Michalek, Kate S. Whitefoot

DETC2019-98219.. **V02AT03A046**
The Impact of Consumer Preference Distributions on Dynamic Electricity Pricing for
Residential Demand Response
Samuel Dunbar, Scott Ferguson

DESIGN OF COMPLEX SYSTEMS

DETC2019-97456.. **V02AT03A047**
Integrated System Design and Control Optimization of Hybrid Electric Propulsion
System Using a Bi-Level, Nested Approach
Li Chen, Huachao Dong, Zuomin Dong

DETC2019-98393.. **V02AT03A048**
Structural Consequence Analysis: Towards the Quantification of Component
Consequential Importance in System Architecture Design
Hannah S. Walsh, Mohammad Hejase, Daniel Hulse, Guillaume Brat, Irem Y. Tumer

DETC2019-98404.. **V02AT03A049**
An Excess Based Approach to Change Propagation
Daniel Long, Scott Ferguson

DETC2019-98428.. **V02AT03A050**
Estimating the Value of Excess: A Case Study of Gaming Computers, Consoles and
the Video Game Industry
Darshan Yadav, Daniel Long, Beshoy Morkos, Scott Ferguson

DETC2019-98429.. **V02AT03A051**
Using Semantic Fluency Models Improves Network Reconstruction Accuracy of Tacit
Engineering Knowledge
Thurston Sexton, Mark Fuge

DESIGN OF ENGINEERING MATERIALS AND STRUCTURES

DETC2019-97370.. **V02AT03A052**
Topology Optimization of Multi-Material Lattices for Maximal Bulk Modulus
Hesaneh Kazemi, Ashkan Vaziri, Julian Norato

DETC2019-97390.. **V02AT03A053**
Inverse Thermo-Mechanical Processing (ITMP) Design of a Steel Rod During Hot
Rolling Process
Anand Balu Nellippallil, Pranav Mohan, Janet K. Allen, Farrokh Mistree

DETC2019-97547.. **V02AT03A054**
Gaussian Process Based Crack Initiation Modeling for Design of Battery Anode
Materials
Zhuoyuan Zheng, Yanwen Xu, Bo Chen, Pingfeng Wang

DETC2019-97617.. **V02AT03A055**
Generative Design of Multi-Material Hierarchical Structures via Concurrent Topology
Optimization and Conformal Geometry Method
Long Jiang, Shikui Chen, Xianfeng David Gu

DETC2019-97628.. **V02AT03A056**
Thermomechanical Topology Optimization of Lattice Heat Transfer Structure Including
Natural Convection and Design Dependent Heat Source
Tong Wu, Joel C. Najmon, Andres Tovar

DETC2019-97659.. **V02AT03A057**
Design of Gradient Nanotwinned Metal Materials Using Adaptive Gaussian Process
Based Surrogate Models
Haofei Zhou, Xin Chen, Yumeng Li

DETC2019-97672.. **V02AT03A058**

Distributed Design of Two-Scale Structures With Unit Cells
Xingchen Liu

DETC2019-97675.. **V02AT03A059**

A Framework of Multi-Fidelity Topology Design and its Application to Optimum Design
of Flow Fields in Battery Systems
Kentaro Yaji, Shintaro Yamasaki, Shohji Tsushima, Kikuo Fujita

DETC2019-97722.. **V02AT03A060**

Visualizing and Evaluating High-Dimensional Mappings of Sets of High Performance
Designs
Clinton B. Morris, Michael R. Haberman, Carolyn C. Seepersad

DETC2019-97774.. **V02AT03A061**

Computational Design of 4D Printed Shape Morphing Multi-State Lattice Structures
Thomas Lumpe, Kristina Shea

DETC2019-97833.. **V02AT03A062**

Topology Design With Conditional Generative Adversarial Networks
Conner Sharpe, Carolyn Conner Seepersad

DETC2019-97905.. **V02AT03A063**

Stress-Constrained Design of Functionally Graded Lattice Structures With Spline-
Based Dimensionality Reduction
Jenmy Zimi Zhang, Conner Sharpe, Carolyn Conner Seepersad

DETC2019-97934.. **V02AT03A064**

Multimaterial Topology Optimization of Thermoelectric Generators
Xiaoqiang Xu, Yongjia Wu, Lei Zuo, Shikui Chen

DETC2019-98105.. **V02AT03A065**

Bayesian Optimization of Equilibrium States in Elastomeric Beams
David Yoo, Carson Wiley, Andrew Gillman, Vincent Chen, Abigail Juhl, Philip Buskohl

DETC2019-98222.. **V02AT03A066**

Data-Centric Mixed-Variable Bayesian Optimization for Materials Design
*Akshay Iyer, Yichi Zhang, Aditya Prasad, Siyu Tao, Yixing Wang, Linda Schadler, L.
Catherine Brinson, Wei Chen*

DETC2019-98341.. **V02AT03A067**

Multi-Scale Design of Meta-Materials With Offset Periodicity
Rushabh Sadiwala, Georges Fadel

DETC2019-98386.. **V02AT03A068**

Optimizing Topology and Fiber Orientations With Minimum Length Scale Control in
Laminated Composites
Chuan Luo, James K. Guest

DETC2019-98463... **V02AT03A069**

An Adaptive and Efficient Boundary Approach for Density-Based Topology
Optimization
Reza Behrou, Reza Lotfi, Josephine V. Carstensen, James K. Guest

Proceedings of the ASME 2019
International Design Engineering Technical Conferences
and Computers and Information in Engineering Conference
IDETC/CIE2019
August 18-21, 2019, Anaheim, CA, USA

DETC2019-97399

LEARNING TO DESIGN FROM HUMANS: IMITATING HUMAN DESIGNERS THROUGH DEEP LEARNING

Ayush Raina
Department of Mechanical Engineering
Carnegie Mellon University
Pittsburgh, PA 15213 USA
araina@andrew.cmu.edu

Christopher McComb
School of Engineering Design, Technology, and Professional Programs
The Pennsylvania State University
University Park, PA, 16802 USA
mccomb@psu.edu

Jonathan Cagan
Department of Mechanical Engineering
Carnegie Mellon University
Pittsburgh, PA 15213 USA
cagan@cmu.edu

ABSTRACT

Humans as designers have quite versatile problem-solving strategies. Computer agents on the other hand can access large scale computational resources to solve certain design problems. Hence, if agents can learn from human behavior, a synergetic human-agent problem solving team can be created. This paper presents an approach to extract human design strategies and implicit rules, purely from historical human data, and use that for design generation. A two-step framework that learns to imitate human design strategies from observation is proposed and implemented. This framework makes use of deep learning constructs to learn to generate designs without any explicit information about objective and performance metrics. The framework is designed to interact with the problem through a visual interface as humans did when solving the problem. It is trained to imitate a set of human designers by observing their design state sequences without inducing problem-specific modelling bias or extra information about the problem. Furthermore, an end-to-end agent is developed that uses this deep learning framework as its core in conjunction with image processing to map pixel-to-design moves as a mechanism to generate designs. Finally, the designs generated by a computational team of these agents are then compared to actual human data for teams solving a truss design problem. Results demonstrates that these agents are able to create feasible and efficient truss designs without guidance, showing that this methodology allows agents to learn effective design strategies.

INTRODUCTION

Advancements in machine learning and computational modelling have brought us closer to the goal of developing intelligent agents that can solve a variety of problems. Machine intelligence has surpassed human levels in several problem-solving milestones [1–5]. One common feature among all these problems is that they are well-defined with a set of rules and actions that are universally followed. Humans are still seen to perform better than machines for tasks which require skills like strategic reasoning, abstract decision making, creativity, and explainability. All of these skills are characteristic of the field of design. In design, the sheer complexity of most problems and the loosely-bound definitions and requirements necessitate a combination of the above-mentioned skills to generate solutions. This motivates the need to combine the diverse qualities of both humans and machines to facilitate solving challenges in design. This work focuses on developing methods to learn essential design strategies from human data and use that to help improve the performance of computational design agents.

Design is an iterative process; in order to create something, humans interact with an environment by making sequential decisions. Expert designers apply efficient search strategies to navigate massive design spaces [6]. The ability to navigate maze-like design problem spaces [7,8] by making relevant decisions is of great importance and is a crucial part of learning to emulate human design behavior. Based on this understanding, Figure 1 schematically represents a human solving a design problem, in this case a truss design problem as used in the example study in this paper. The problem is decomposed into two steps: perception and problem solving. The designer perceives the current design state and takes an action to modify it based on previous knowledge and understanding of the problem. This new design is then fed back into the current design state and the process is repeated. Designer behavior guides the decisions regarding what actions to take and is shown as a black box, the input and output of which are observable. However, what happens inside is hidden and not entirely known.

V02AT03A001-1

Copyright © 2019 ASME

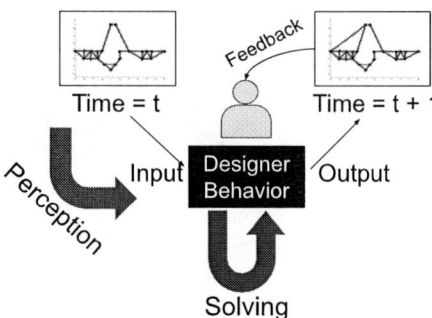

Figure 1: Schematic representation of a human interacting with a design interface

The motivation for this research is primarily based on understanding what goes on within that black box and uncovering the implicit rules and strategies. We present a data-driven approach that learns human design and problem-solving strategies and also learns to generate new designs only from observing the visual states of an evolving design, without explicitly being provided with any information about the performance metrics or meaning behind design modifications. Since the presented deep learning framework does not encode any specific information about allowable actions and instead represents the whole process using images, this research has promise to provide a domain-agnostic method of representing a generative design process. Every design process has a different set of actions or operations, preventing a common algorithm or framework to be used across problems. Our domain-agnostic method uses an image-based common representation to define and track the evolution of design problems.

This work illustrates how sequential images from historical designer data can be used to emulate designer. Specifically, an independent agent algorithm is developed that utilizes methods from unsupervised representation learning, imitation learning and image processing to model the entire design process as shown in Figure 1. The agent uses a convolutional autoencoder to map the design states to low dimensional embeddings, then it uses a neural network-based transition function to predict a design embedding based on the current state and then maps the embedding back to the original image dimensions. This process produces an image of how the next design should look like. Finally, a control algorithm driven by image processing constructs outputs the parameters to execute the operation. This agent is trained offline, doesn't require carefully labelled state-action data and is agnostic to objectives and other parameters. Even though the current framework utilizes images (a *2*-dimensional array), the framework can be extended to other problems where the raw state is representable using an *N* dimensional array. The contributions made by this research can be listed as follows:

a)   A generalized methodology is presented that can be used across various design problems to learn implicit design strategies organically from data, using a pixel-based representation.

b)   An agent framework is developed that generates and follows design suggestions and automates the design process in an end-to-end manner.

c)   The agent maintains a process similar to a human designer and hence provides a platform to use agent in computational teams or along with humans in a hybrid AI/human team

For the entirety of the paper the term *framework* is used to refer to the combined deep leaning model and the term *agent* is used to refer to the end-to-end algorithm that is the deep learning framework and the automated inference algorithm both combined The paper is organized as follows: Section 1 contains the review of the relevant studies and builds up a foundation for the different methods used in the framework. Section 2 begins by explaining the setup of the design study and the nature of the datasets used, then it explains the network architectures, training strategies and some intermediate results with the (image-based, or imaginal) deep learning part. Section 3 explains the experimental setup that was required to evaluate the performance of the agents and their computational teams. The results and discussions from the experiments are found in Section 4. The final conclusions and future work are mentioned at the end in Section 5.

1.   BACKGROUND:

This section is divided into four subsections, each representing a different area of research this paper relates to. A review is conducted to discuss ideas and show how this work is related to the existing research in each of these fields and in what ways it is new and different from them.

1.1 Design as a sequential decision-making process:

The idea of problem solving (or design) as a sequential decision-making process is bolstered by similar representations used in robotics [9], game AI research[10], behavioral economics [11], decision-making [12], and also design research [13–15]. Here, the basic idea is that an agent is placed in a design environment and the goal of the agent is to interact and create a design maximizing the given performance metrics. Formalizing design as a problem-solving process opens up possibilities of applying logic, inference and different mathematical models to often abstractly-defined design processes. Modern-day design processes work in a computer-in-the-loop setting, where designers leverage computational tools along with their learned knowledge to make informed decisions [16]. Data collected from these processes can be utilized to extract key features and insights about them, while some of the previous studies have found that certain probabilistic models can be used to represent design strategies and describe skilled behavior [14,15], generate designs with similar trends to humans [17] and finally generalize them to transfer across certain problems [18]. Modeling design processes in an environment is challenging since, limiting the designer to a small set of operations can limit designer skills and these operations further have multiple parameters associated with them. Also, these variables and actions are specific to the design problem that is being modelled. This makes it difficult to develop an agent or a framework that can be used across different domain of problems since a new learning architecture is required every time. This problem is discussed and solved in the next subsections.

1.2 Pixel based representation of design processes

Humans are shown to think and solve problems visually [19], creating mental imagery [20] and incrementally making changes to it until reaching a solution [21]. Humans are able to abstract and store information as images and then recall them when solving a problem; this process is collectively called imaginal thinking. Some previous works have also shown its utility in helping humans solve design problems using an interface [22,23]. This motivates the use of an image (pixel based) common representation for defining a design problem. It is especially beneficial for design due to the great diversity of design variables in different problems leading to algorithms to be problem specific. An empty image encompasses the complete design space since any design can be represented within that image (if the problem is 2D), hence a design process can be represented as a designer changing the pixels of an image to achieve the best configuration of pixels. This form of representation can be extended to any problem that shows the current problem state as a set of pixels (or even voxels for 3D data). This method leads to a very high dimensional problem that can be very difficult to manipulate directly (e.g., for design). The research presented in the subsequent sections utilize image-based representations and formulates a learning problem based on it.

1.3 Using deep learning for design feature extraction – dimensionality reduction:

Dimensionality reduction of design spaces is a critical problem and numerous techniques have been discussed in the literature. Such reduction is done in order to use computation techniques efficiently, however there exists a tradeoff between geometric variability (possibility of creating novel designs) and dimensionality (complexity) that must be addressed [24]. Techniques like Karhunen-Loève expansion [10,11] and principal component analysis [12] are linear methods that determine a set of dimensions that can maximally explain the variance in the data. Higher accuracy non-linear methods have also been utilized by researchers like kernel PCA and multi-dimensional scaling [25,26] however these methods are non-invertible, i.e., one can go from high dimensional space to a low dimension embedding, however, the embedding cannot be converted to the original design. This is an essential requirement of our framework. Recent advances in deep learning has led to development of methods for highly non-linear dimension reduction that are invertible and accurate. Autoencoders are one such family of methods which can determine lower dimensional embeddings [25,27–30] and is an efficient data-driven way of learning representations [31]. Autoencoders are a type of neural network that are trained to reconstruct an output that is identical to the input. Often a density constraint limits the dimensions of the embedding and forces the autoencoder to generate a lower dimensional representation. Autoencoders have proven to perform much better than other available methods, especially for design tasks [25,32]. A variant of these autoencoders, a deep convolutional autoencoder, combines a convolutional architecture [33] with the encoder-decoder structure of an autoencoder [30]. It has the ability to extract high-level semantic information from image data and encode it to a lower dimension embedding. It is used in this paper as a part of a larger framework to encode design information to a lower dimensional embedding and back to an image which is human readable. Adding further constraints on the distribution of the embedding elements leads to a variational autoencoder (VAE) architecture [34], which has certain advantages in design generation tasks. However, that requires higher amount of data hence it may be explored in future works. Other recent works in design have also used convolution deep learning architectures to solve different tasks with both 2D and 3D data with interesting results [35–37].

1.4 Mimicry – learning to imitate from raw pixels:

Understanding what happens inside the human behavior "black box" has been approached differently in different fields of research. Historically, researchers from psychology consider learning human behavior as *learning a function* that generates the same output given a specific input. These methods were aimed at learning rule-based relationships [38,39], or learning similarities [40,41] in problem solving behavior. Studies in cognitive science have put forth several core ingredients of human intelligence, including intuitive physics [42–45], problem decomposition skills [46,47], ability in learning-to-learn [48], and others [49]. More recently learning from imitation or *imitation learning* has been actively utilized to solve various problems since it doesn't require significant state-space exploration, making it a faster approach in practice. Researchers have taught robots to imitate and learn how to drive [50], reach [51], manipulate [52], play Atari games [53] and also fly and control a helicopter [54]. However, all these require properly labelled state-action pairs, and such datasets are very few in number and very problem-specific. Recent work by Liu et. al [55], proposes a framework that learns only by observing state-space trajectories. This allows a network to learn in third person (or use historical data) which is closer to how humans learn by just observing without explicit action information. The research presented builds off some of the ideas from that work and adapts it to work with design data. Some other works develop on the two step problem representation idea by having an encoded representation of the environment and a control network that can navigate it [56], however they do not utilize any human knowledge and use active learning methods like deep reinforcement learning to solve the problem, which is different from our approach.

Significant research has been done in deriving qualitative explanations for human design strategies [8,57–60] and using them to further design optimization algorithms [61,62]. Also, methods have been developed to extract and verify design knowledge from design data, like inter part dependencies from design databases [63], learning a machine learning based design method recommender [64] and also developing more efficient optimization algorithms [65] from crowdsourced design solutions. These previous efforts, however, do not address the problem of making a generalized parameter-independent algorithm and solving a sequential decision-making design problem which is an essential factor for our framework in order to make it similar to how humans solve it. In conclusion, prior research relates to the methods that we've used; however, our work integrates concepts from different fields and proposes a unique two step deep learning framework that decomposes the solving process of design similar to a human, and hence makes it compatible for future hybrid teaming possibilities.

2. METHODS: A Deep Learning Framework for Visual Design Evolution

In this section the methods involved in this research are explained. The first subsection discusses the original design study and the source of the dataset. The second subsection is a formulation of the problem. The third subsection explains the model architecture and the training strategies used. The final subsection illustrates how the deep learning framework is integrated with an inference algorithm to develop an agent.

2.1 Design study and database

The nature of the developed framework is such that it can utilize historical data for training and is generalizable to any problem with a pixel-based state representation. For the purpose of the work addressed in this paper the database used for learning comes from a truss design study (conducted by McComb et al [66]). The study involved senior undergraduate students in mechanical engineering who were tasked to design a truss structure given a certain set of objectives to fulfil. The students were randomly divided in sixteen teams (three members each). An interface was designed for the study which allowed students to apply one of the nine operations shown in Figure 2 and also interact with the rest of the team members to complete the truss structure. Two metrics were provided to participants for feedback on the quality of these trusses: factor of safety (FOS) and total mass. The former is a metric to represent how strong and resistant to failure the design is, where a higher value represents higher strength. A value above 1 illustrates sufficient strength and deems the design feasible. Mass shows the amount of material that has been used in the structure. These two metrics together are used to evaluate the strength to weight ratio (SWR) i.e. the ratio of the FOS and mass. This represents how optimized the structure is and is also used to evaluate the quality of a structure; a high FOS with low mass will lead to a high SWR showing that high strength was achieved with minimal mass. During the study, the design interface was also used to capture data whenever a subject made any change in the design. The study was conducted in a controlled environment where different constraints were provided to the subjects in gradual time steps.

The resulting database contains the design state information for all the subjects under different design scenarios. Previous work used this sequential design data to develop a method of representing design strategies and contrast high and low performing designers [14,15,17]. For the purpose of our research we are using only the first portion of the study which is unconstrained. The first image in Figure 2 shows the initial state of the design problem, with the pink arrows representing the load bearing nodes while the other nodes are the support nodes. The database contains only the design state information with nodes and members, excluding any loading arrows or water. The subjects were tasked to create feasible designs (>1 FOS) with minimal mass of the truss structure. This consists of a total of 12850 design states that appear in design sequences generated by humans. These states include parametric specifications of nodes and members that exist in the design structure. As the designs progress, the number of nodes and members vary greatly in the dataset, changing the number of variables to completely define a design. In the current work, the design state is represented by an image with a fixed number of pixels instead of the variable number of parametric values. This representation leads to a more visually interpretable representation which is what the subjects in the study actually observed. Every design in the solution space can be fully represented with an image. Most real-life design problems can be visualized similarly using software tools which provide state-space information to the human users along with a list of actions to modify/create the design. Decomposing design as a sequential decision-making process simplifies it to the schematic shown in Figure 1. Having an image as the primary representation of a state can allow developing a framework that can be used across multiple problems since there is no problem specific modeling in terms of actions or design variables. Even though, the truss design problem is visually simpler in comparison to real-life mechanical design problems, it still captures the basic challenges of perception followed by selecting actions and relevant parameters. The focus of the current work is on developing this method for truss design problem, however extending to other problems is a tangible next step and the effectiveness shall be tested and discussed in future work.

2.2 Problem Formulation

The problem that this work deals with is formulated as a prediction problem, where the framework is expected to predict the next design state given previous state information. The dataset used contains a set of sequence demonstrations $\{D_1, D_2, \dots D_n\}$ of human generated designs from the truss study. Each demonstration is sequence of state observation $D_i = [\ o^i_1, o^i_2, \dots o^i_T\ ]$ where, o^i_t is the observation at time (analogous to iteration number, since one iteration is one time unit) t in the ith demonstration sequence and is an image that shows the current design (truss structure). Each designer creates multiple demonstrations and all demonstrations begin from the same initial state. Every demonstration is a trajectory through different

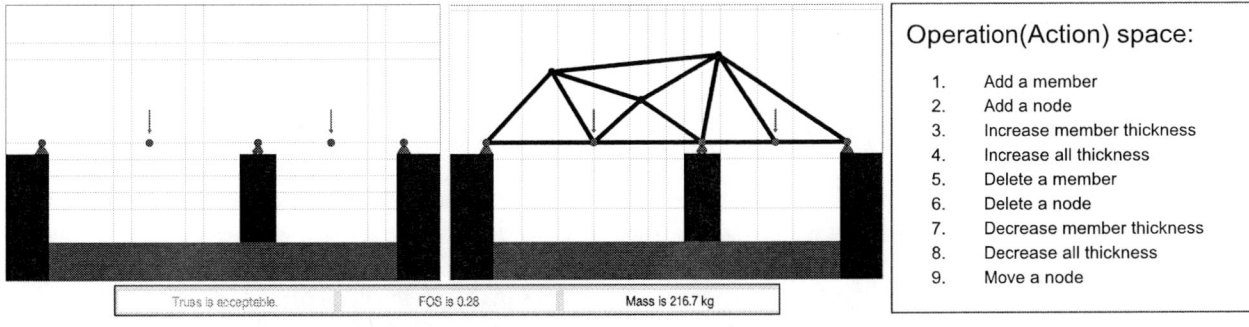

Figure 2: Depicting the interface for the truss design software along with the visible metric and possible operations [66].

intermediate design states that eventually leads to a final design, representing a search through the design space. This dataset does not contain any parameters for the metrics for node and members or information about the loading parameters. However, there are implicit relationships between consecutive states corresponding to design operations which lead to changes in pixel intensities. The framework (in Section 3.3) aims to learn those implicit relationships among these states from the dataset since they correspond to designer strategies for the given problem and boundary conditions.

This task to replicate human strategies can be formulated as learning a function that converts o^i_t to o^i_{t+1} ($\forall t + 1 < T_i$; where T_i is the total time for the ith demonstration). In order to develop new designs in the study humans were provided with a set of nine operations (A) that change the existing designs; the set A = {a_1, a_2, ... , a_9 }, and each of the operations have a parameter set associated with them (w). Hence, $o^i_{t+1} = a_j (o^i_t , w)$ represents generating a new design by applying operation j with parameters w. The whole design process can be seen as a designer analyzing the current state o^i_t, selecting and applying one of these operation functions with relevant parameters, and then repeating the process all over again. The designer looks at the current state and makes the decision about the operations and the parameters using implicit design knowledge and is able to eventually create a well performing truss structure. The database captures that information as it preserves the sequence of the progression of design. The desired framework must satisfy the following requirements:

a) Ability to semantically interpret the design features of state o^i_t.
b) Create an embedding that can represent large variability in the designs with less features.
c) Map consecutive design states to learn how design operations function.
d) Learn how designs progress iteratively to generate design trajectories.

The first two requirements relate to the perception part of the problem and the next two relate to problem solving. In order to fulfil them two different neural network architectures are used. They are explained in the next section.

2.3 Model framework:

To learn the low dimensional embeddings of the design data, a convolutional auto-encoder is designed. It has two parts, encoder ($e(.)$) and decoder ($d(.)$). The encoder is used to encode the raw pixel data into a semantically meaningful embedding that can capture all the important features of the structure at a reduced size. The decoder on the other hand is used to convert this low dimensional embedding back into the original size image. This step helps visually realize the image. Another neural network is used as a transition network ($t(.)$), that takes the embedding from the encoder as the input and converts it into a new design embedding that helps in progressing the design, and which is regained by the decoder. Figure 3 shows the complete deep learning framework.

2.3.1 Perception - dimension reduction:

This part of the framework deals with perception of the design. The raw designs are high-dimensional and using them to predict other high dimensional designs is computationally intensive. Thus, identifying semantically meaningful embeddings that are low dimensional will make later operations less computationally intensive. This was achieved with a nine-layer autoencoder (as shown in Figure 4). Dimension reduction was achieved in the encoder by using a combination of 3 layers of stride 2 convolutions [67] and a max-pool layer. The kernel sizes for the convolution layers were in decreasing order (12x12), (9x9) and (5,5) with increasing number of channels- 32, 64 and 128, respectively, in order to capture large features in the first layer (like nodes and member thicknesses). For the decoder, 1 un-pooling layer and 3 de-convolution (or convolution transpose [68]) layers with stride 2 were used for up-sampling the dimensions. The kernel sizes were in increasing order (7,7), (9,9) and (12,12). The number of channels were same as the encoder but in a reverse order (128, 64, 32). Non-linear activations were used for all the layers, all convolution layers except for the final layer used ReLU activations [69] because of their efficiency in deep-networks, while the linear layer at the center and the final convolutional layers uses *sigmoid* activation to avoid grey value and forcing the network to make a decision for every pixel (since *sigmoid* is biased towards 0 or 1 values). The dataset was split into test and train images with a 80-20 split (80% training, 20% test data) and the autoencoder was trained in an un-supervised

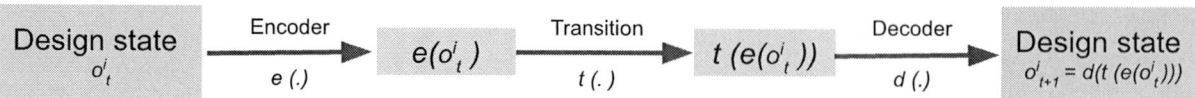

Figure 3: A block (functional) diagram of the deep learning framework

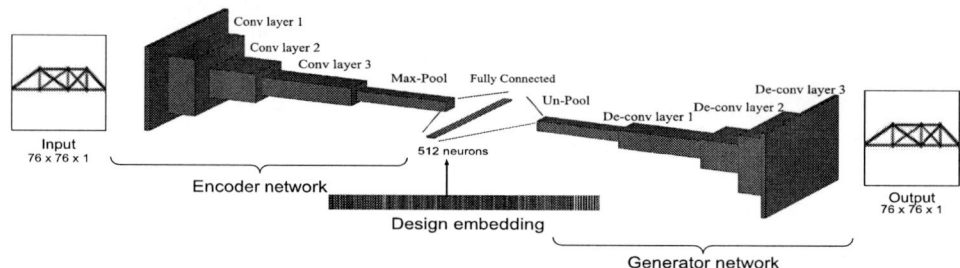

Figure 4: Network architecture for the convolutional autoencoder

manner since there are no labels and the network automatically identifies the optimal latent representations. The main objective for training an autoencoder is to reconstruct the same input, so that the output is the same as the input. Also, an information bottleneck is designed in the hidden layers to reduce the dimensions. The auto-encoder reduces the initial 76x76 image (5776 dimensions) down to 512 dimensions in the embedding. A mean square error (MSE) was used as the loss function to train the autoencoder. Adam optimizer [70] was used to train the weights of the network. A sufficient low value of 0.0012 MSE was achieved with a binary accuracy of over 91% over the test set. An R-squared value of 0.9684 is achieved in the test case, indicating that the network is capable of explaining nearly all of the variance in the data.

2.3.2 Prediction - the transition network:

This part of the framework is a non-linear function that maps the embedded representations of designs in consecutive time steps. This function basically manipulates the current designs to generate a new design that shall eventually lead to a complete final design. This manipulation of the current designs to generate a new design is similar to how a human designer approaches a design problem, in a step-by-step manner taking feedback and applying relevant operations. A neural network is used to approximate this function. The network takes 5 previous design embeddings as input (5 x 512 units), concatenates them and then down-samples it in the two linear layers (1024, 512 units) back to 512 units. A non-linear activation function Leaky-ReLU [71] was used for all the linear layers. Multiple design states are concatenated in order to make sequential information accessible to the network for better predictions [3].The weights of the neural network implicitly represent the design operations that are applied to manipulate the design states. These operations modify the design embeddings by adding, deleting or manipulating their values (matrix operations) in order to generate new designs. Since this step manipulates an abstract representation of a visual image it is similar to how humans build designs [21].

This network was trained after the autoencoder. Training data for the transition network consist of embedding sequences from the design study. $S = \{ e(o_t^i), e(o_t^i), ... e(o_t^i)\} \; \forall \; i \in H$ *(all sequences)* is the total set of embedding sequences, for the purpose of this network with 5 inputs, the dataset was converted to $S = \{ [e(o_1^i), e(o_2^i), ... e(o_5^i); e(o_6^i)], [e(o_2^i), e(o_3^i), ... e(o_6^i); e(o_7^i)], ...[e(o_{Ti-5}^i), e(o_{Ti-4}^i), ... e(o_{Ti-1}^i); e(o_{Ti}^i)] \} \; \forall \; i \in H.$ The semi-colon separates the input data from the labels i.e. the next design state in the sequence. For training, test and training sequences were separated with a 80-20 split (80% training data, 20% test) and the network was evaluated on how well it performed on the test set. This converts the imitation problem into a supervised learning problem where based on the previous 5 design embeddings the next embedding needs to be predicted. The training data used is only the state sequences from the design study, so the network is only trying to learn the implicit relationships between the states; it has no explicit understanding of design goals, constraints or intent. The transition network was combined with the encoder and decoder networks as shown in Figure 5 to complete the deep learning framework. For the training of the transition network weights for the encoder-decoder networks were kept fixed and MSE loss was calculated. Adam optimizer [70] was then used to backpropagate the loss values to train the weights.

The training was accomplished in a 2-step process where the first step pre-trains the network and the second step fine tunes to predict meaningful designs. In the first step, the network must learn to recreate the current design itself. This pre-training step is important since the weights of the network are initialized randomly and training may not lead to meaningful predictions [72]; pre-training helps in finding better weights by providing good initialization to the network weights. In this fine-tuning step the learning rate is lowered, and the network is trained to recreate the next design in the sequence. Through experimentation the mentioned architecture was finalized for the transition network, a final MSE of 0.0072 (or binary accuracy of 90.05%) is achieved for the test case and an R-squared value of

0.8105 was achieved by the final network. This indicates that the network was able to explain a large majority of the variance in the data. The final network was arrived at through an iterative hyper-parameter search, where numerous architectures were compared, preference was given to smaller networks as larger networks tend to over-fit the data. A major boost in performance was seen when past design states were concatenated in the input, showing the importance of the relation between design trajectories and next state prediction.

2.3.3 Framework results

After training both the autoencoder and the transition network, updated weights were put together in the overall framework as shown in Figure 5. The input side of the framework consists of 5 instances of the encoder networks. The embeddings generated by those networks are then concatenated together and a 3-layer neural network down-samples the embeddings before the decoder network and converts the embedding back to the full image size. The final generated image from the framework within an iteration is the suggestion how the next design should look, hence creating a mental image of the design solution. However, these are still pixel-based information and specific parameters need to be identified to create an actual design.

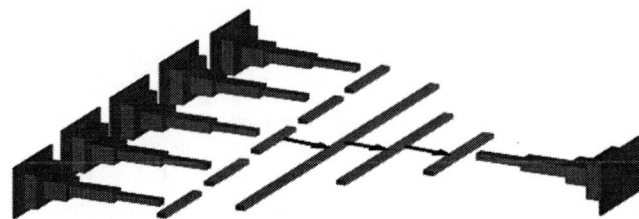

Figure 5: Network architecture for the complete framework

To visualize the suggested move better, a difference of the generated and input image is taken and then overlaid as a heatmap on the original image. Figure 6 shows 3 images: the current design, the suggestion heatmap and the ground truth (i.e., the actual subsequent design from data). The suggestion heatmap shows the prediction of design change as a color gradient on the current design state. The color gradient spans across pink to green with 0 (no change) values being represented as black. Pink regions represent the regions where the network adds materials in the prediction while at the green regions it deletes or reduces the material. Black implies no change is predicted in those regions. The ground truth is the next design state from the original dataset, showing the actual design created by the designer in the study. A heatmap that matches the ground truth shows perfect prediction. Figure 7 shows some more sequences that have been taken from the test set which contain designs that were not seen in the training phase. From both Figure 6 and Figure 7 it can be observed that the network successfully learns to predict semantically meaningful pixel regions (that appear like discrete operations relating to nodes and members) and also correspond to the operations undertaken by the human designers in the test dataset. This shows that the framework was able to extract design features from the problem space along with learning to developing truss designs to match human operations using supervised imitation learning. However, it can be seen that there is some level of noise in the heatmaps which can be improved in future work by using more advanced architectures and other methods of training. Currently the framework is able to provide exceedingly well results using very basic architectures, utilizing more complex architectures and methods may be needed for more complicated visual setting (more than nodes and lines) and also may lead to better results in terms of noise and efficiency however, for the scope of current work the framework appears to be sufficient. The network is also seen to produce several suggestions at once, showing that it is unsure about which move to take. That may be because of several reasons: at a given design state there may not be a single best operation to take that mimics a human strategy; or different sequences of operations can be used to reach the same final state hence which operation to do first could be arbitrary; or at a particular state the design can go in various possible trajectories and if the network is unsure it produces more noise and less discrete features; also the network could possibly be learning an 'average' of multiple designers and their strategies, since it is not trained to only follow a particular style of design generation (although that would be feasible with the current network). In future work, further analysis needs to be done that can identify the primary factor for multiple operation suggestions. Since the heatmaps do seem to follow a singular 'design idea' and the suggestions are coherent, the current performance is sufficient and promising for it to act as a design suggestion engine. Now, in order to evaluate how well the suggestions perform in an actual design process, an inference algorithm (similar to having a control system) needs to be designed that can map these suggestion heatmaps (ideas) to the operator parameters (actions) and then select and implement a single operation from the list required to sequentially generate a parametric truss design.

2.4 Developing a deep learning agent

The framework generates the design as a picture and the suggestions/changes are shown as the highlighted pixels in the heatmaps. These images are basically how the deep learning framework imagines/suggests that the design should progress in the next few steps and is a pictorial visualization with no detailed information about which parts are members, their size, or where the nodes are. However, in order to evaluate a design, it needs to have parametric information for the nodes and members involved in the change/suggestion. Interpreting this heatmap information to make changes to the current design is a trivial activity for humans; however, it is very complex to map the process algorithmically. A rule-based algorithm was developed to act like the control system of the framework which can automatically inference from the heatmaps and map pixel-based suggestions to operator parameters. When combined with the framework, this completes the process of generation of a new design state given the current design states. In brief, the algorithm uses image processing techniques to pre-

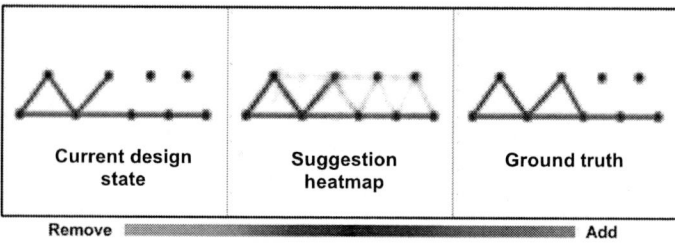

Figure 6: Suggestion heatmaps from the DL framework. The color gradient shows the suggestions where pink means "Add", green means "Delete" and black means no change

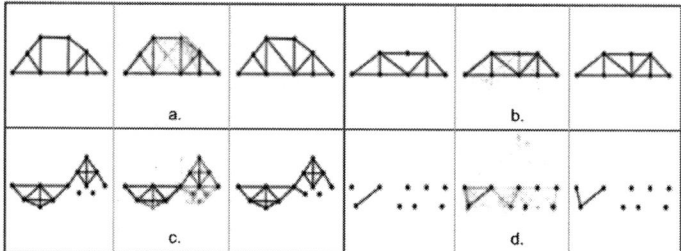

Figure 7: (a-d) represent different test dataset pairs and the heatmaps from the best performing framework.

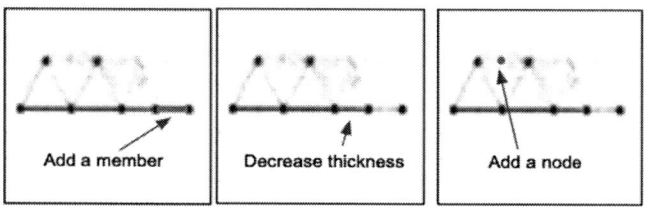

Figure 8: An example of a candidate list with labelled operations generated in one of the applications of the image processing algorithm.

process and isolate different regions that may correspond to distinct operations from the heatmap. Further, the regions are analyzed based on the different parameters to check and classify it as a particular operation and finally a candidate list is prepared based on the different regions; these candidate lists can appear similar to Figure 8. The final step in the algorithm is to select one operation from the list to finally apply on the design. This is achieved by comparing the similarity between the suggestion and the candidates and the one with the highest similarity is probabilistically selected. Hence, the algorithm goes through these different steps and outputs the particular operation parameters to change the design of the truss structure. The framework is shown in Figure 9, it can be considered an independent agent as it senses an environment, makes decisions, and then acts on the environment generating new designs [73]. The design state is the sequence of the previous 5 designs which is used as an input for the framework.

This engine generates a suggestion heatmap, which is fed into the inference algorithm. This algorithm processes the suggestions and classifies the pixel representation into a design operation. This move is then executed, and a new design is generated.

3. EXPERIMENTAL SETUP

This section details how the experiments were carried out to evaluate the quality of the framework by creating a team of identical deep learning-based agents. In order to evaluate the performance of the agent algorithm we compare the generated designs with human data from the original design study. The design study was conducted in a team (of 3 designers) and they interacted at predefined intervals. When interacting, members of a team shared their designs with the option for each designer to select any one of the 3 designs in the team to continue working on (typically the best design of the triplet). In order to make a fair comparison, 3 different agent instances of the algorithm were combined into teams and they were made to interact periodically at the same average rate as the humans, at which point the best design from the agent team is accepted by all agents. Every agent is given a random starting point and then iteratively refines their design. Each agent works independently to generate new designs that are fed back into the deep learning framework to complete the loop. The agents in a team interact every 48th move (mean value from data), when the interaction occurs the current design from every agent is shared in a common pool and the highest performing feasible design (based on SWR and FOS) is adopted by all the agents. When the agents don't interact then the design states of individual agents are updated only with their own designs. This greedy method of interaction was chosen as a simple approximation to the process of human interaction in a team, which is complicated and modeling that is beyond the scope of the current work. This is the only step where the team of agents are implicitly aware of the metrics. However, this step is separate from the design generation steps of the agents (heatmap generation and operation selection) and occurs in < 2% operations; hence the agents individually still continue to work in a metric agnostic manner for generating trusses. The complete loop is shown in Figure 9 and is repeated for 250 iterations per agent (equivalent to the human average). This whole setup was repeated to produce 16 team solutions in order to compare with the 16 design teams in the study.

4. RESULTS AND DISCUSSION

Data from the experiments, generated by deep learning agents, and human designers is analyzed in this section. Figure 10, Figure 11 and Figure 12 show the plots for performance metrics: Factor of Safety (FOS) and Strength to Weight Ratio (SWR) with respect to number of iterations. The lines in the background connect the data-points (designs) from the same designer showing trajectories of how the design progressed. The comparison is made between blue (solid) and green (dashed) trajectories where blue color (solid) represents the Deep Learning Agents (DLAgent) and the green color (dashed) is for Human data. The plot presents the data for all 48 humans and 48 DLAgents. The graph limits to only 250 iterations for the purpose of better visualization since the majority of the data can be shown within this range; for calculating the performance values all iterations are considered. Due to the nature of truss design, whenever a new node is added it is virtually hanging in

V02AT03A001-8   Copyright © 2019 ASME

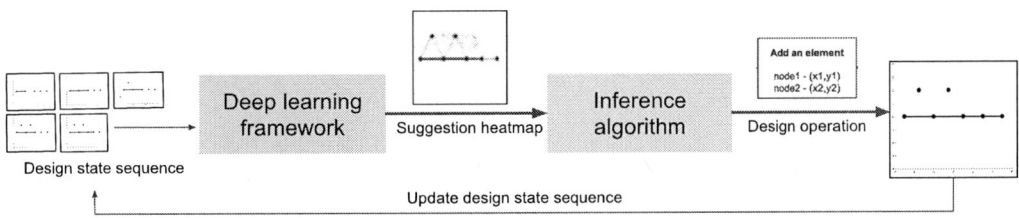

Figure 9: Schematic representation of the complete agent

the air and hence the overall structure is infeasible. This leads to zero values of FOS and SWR, which is not a good representation of the quality of the truss structure since, minus that extra node, the structure could be feasible. This leads to significant amount of noise in the data where the quality metrics unnecessarily dive to zero. In order to avoid that, a denoising operation is carried out once a design becomes acceptable in the process, the most recent non-zero value is adopted as the current value. This way the data is denoised while also maintaining the trends. The solid blue and green lines show the mean values at the given iteration for the respective class. Mean values are shown in order to provide a sense of the average trend in the data. Figure 10 shows a plot of the FOS over time. The red dotted line shows the threshold for a feasible design (FOS = 1); all designs above that line are feasible. It can be observed that although a few human designers reach high FOS values very early in the process, most of the humans struggle to create a feasible design as they remain below the red line, showing that there is a mix of designers throughout. On the other hand, DLAgent builds slowly but more consistently since the majority of the designs beyond ~175 iterations reach levels higher than humans. Also, the mean lines further justify the performance by DLAgents since there is a gradual but consistent increase in FOS and it exceeds the human level and eventually the feasible level. Also, for the human data there is some noise in the mean with immediate ups and downs which smooths out with increased iterations, but there is less noise with the agents. This could be indicative of learning/searching behavior for humans since they had access to immediate metric feedback and could try different things and evaluate their design whereas agents solve the problem in mostly a metric agnostic manner.

Figure 11 shows the plot for comparing the SWR over time. Humans appear to perform better than the DLAgents especially in the early part of the process since the dashed green mean line is higher than solid blue. Select individual human designers also seem to have reached much higher values throughout the process. However, it should be noted that a high SWR can be achieved with infeasible FOS (< 1.0) but very low mass values. This motivated us to calculate a refined SWR value as shown in Figure 12.

For Figure 12, only designs that are feasible (with FOS ≥ 1.0) are allocated a SWR value unlike in Figure 11 where an infeasible design could also contribute in representing the quality of the design. Here a majority of the human data-points are removed,

showing that most of the designs in the early part of the process with high SWR were actually incomplete or infeasible designs. We observe that even though designs by DLAgents don't reach very high values like some over-performing humans, there still are a considerable number of designs that are feasible and have good SWR, and the agents are much more consistent with the quality of their designs. Observing the mean lines, DLAgents seem to perform quite close to the humans throughout and slightly exceed them in the later part of the process. The gradual increase in SWR throughout the process shows that the deep learning framework was able to learn certain design strategies from human data and is able to implement them in the generation process.

Figure 13 shows the mean value across the best designs created by individual designers for the two performance metrics. It shows that the DLAgent performs better in terms of FOS and comes considerably close in terms of SWR. This implies that the DLAgents have learnt to mimic certain aspects of the human strategies and learn to generate design sequences even after they started from mere random initial points. It must be noted that the DLAgents are generating designs only on the basis of transition features learnt from data. On the other hand, the humans had access to various metrics like FOS, mass, and also a colored grading showing which members were under stress at every iteration. All of this information guided them to produce better designs in the process. Agents receive no information about mass, and hence cannot get direct feedback for how heavy the design is which makes limiting it difficult, possibly leading to achieving higher mass in their structures. This helps the designs with strength (FOS) but adversely affects the SWR. However, as seen later in the SWR result plots the agents maintain good SWR values meaning that the agents learn to efficiently utilize the extra mass. It can be concluded that the agents have learned to create feasible good performing designs, however, humans were still more efficient in the early parts of the process indicating a room for improvement.

5. CONCLUSION

Human designers possess great problem-solving skills. Essential insights can be extracted from them to develop computer agents that can work with humans in unison as a team. A novel methodology is presented in this paper that is used to develop agents that learn purely from human data and generate high performing designs. The methodology involves training a deep

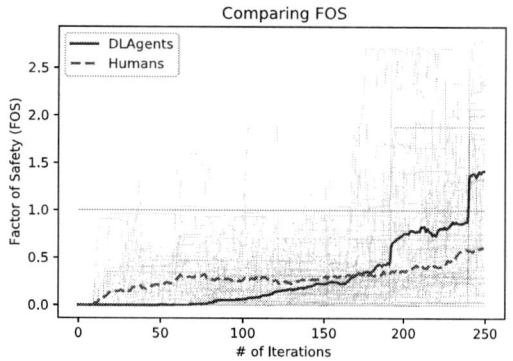

Figure 10 Comparing DLAgents and Humans on FOS

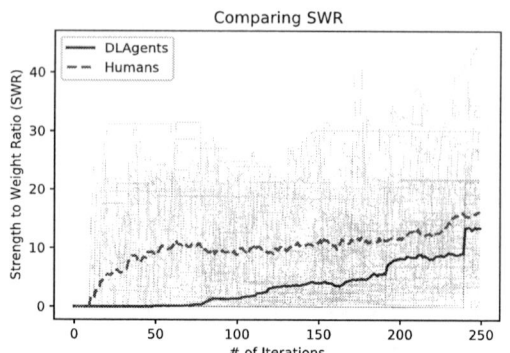

Figure 11 Comparing DLAgents and Humans on SWR

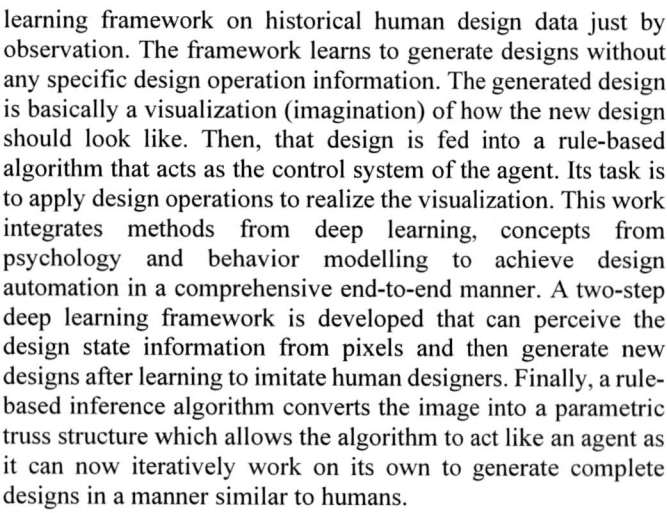

Figure 12 Comparing DLAgents and Humans on RSWR

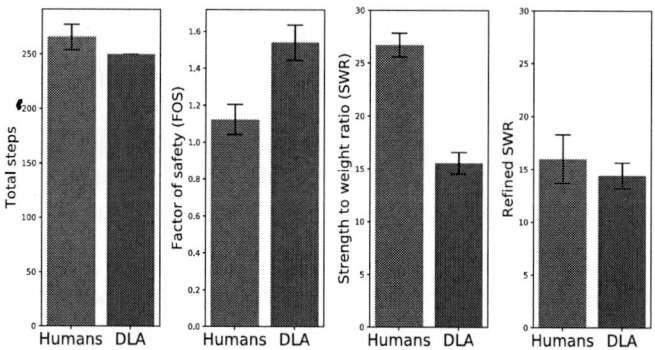

Figure 13: Compiled results. Error bars represent the standard error.

learning framework on historical human design data just by observation. The framework learns to generate designs without any specific design operation information. The generated design is basically a visualization (imagination) of how the new design should look like. Then, that design is fed into a rule-based algorithm that acts as the control system of the agent. Its task is to apply design operations to realize the visualization. This work integrates methods from deep learning, concepts from psychology and behavior modelling to achieve design automation in a comprehensive end-to-end manner. A two-step deep learning framework is developed that can perceive the design state information from pixels and then generate new designs after learning to imitate human designers. Finally, a rule-based inference algorithm converts the image into a parametric truss structure which allows the algorithm to act like an agent as it can now iteratively work on its own to generate complete designs in a manner similar to humans.

The comparative results show that the agent learns to create designs that are comparable to humans designs from the study. While making this comparison it must also be considered that the agent did not have access to real-time feedback on the quality of the design or the objectives of the process (only the team interaction was implicitly guided by it). Humans, on the other hand, could get feedback at every step helping them in the process. Moreover, the agents start the design process from randomly generated initial states, while humans begin on their own. However, irrespective of the differences the agents were seen to perform exceedingly well and achieve feasible designs that are even comparable to human level. This implies that the agents were able to extract implicit design strategies from data and use them to generate designs. The framework has the ability to learn the relationships between consecutive design states and hence can be applied to consecutive decision-making design processes whose sequential states have incremental changes and can be represented as an N dimensional array.

Learning from historical human data helps the agent identify the important regions of a design space and implicitly learn to navigate it, helping reach good performing designs. This imitation learning, however, has its limitations since the agent can at best do as well as the teacher, i.e., the human in this case, and also it may only learn to work with designs similar to shown in the training data in case of overfitting. In order to overcome this, the networks are trained separately to first create design embeddings that can learn general features of a truss design and hence can represent even new unseen truss designs and then learn generic design operations on them. This can allow the agents to explore new designs by possible interpolations between design embeddings. The agent might still learn bad strategies in case the quality of the subject is bad. The aim for the current research was to illustrate an ability to extract design strategies from data however in future work knowledge about metrics could be infused to ignore low performing strategies along with careful selection of training data. Further experimentation also needs to be carried out to test the novelty in designs and how different they are from the training data to evaluate generality of the learnt operations. This motivates us to utilize and test other methods of learning in the agents as well as identify the effect of training data on the final performance. For the current study the agents were only trained to imitate humans using metric agnostic design data. In the future, embedding these agents in a goal driven environment where real-time rewards are provided for optimal designs can lead to interesting results. Also, creating hybrid approaches of imitation learning with other methods of active learning can also be explored since it can significantly enhance the agent performance.

ACKNOWLEDGEMENTS

This material is based upon work supported by the Defense Advanced Research Projects Agency through cooperative agreement No. N66001-17-1-4064. Any opinions, findings, and conclusions or recommendations expressed in this paper are those of the authors and do not necessarily reflect the views of the sponsors.

REFERENCES

[1] Campbell, M., Hoane, A. J., and Hsu, F., 2002, "Deep Blue," Artificial Intelligence, 134(1), pp. 57–83.

[2] Mnih, V., Kavukcuoglu, K., Silver, D., Rusu, A. A., Veness, J., Bellemare, M. G., Graves, A., Riedmiller, M., Fidjeland, A. K., Ostrovski, G., Petersen, S., Beattie, C., Sadik, A., Antonoglou, I., King, H., Kumaran, D., Wierstra, D., Legg, S., and Hassabis, D., 2015, "Human-Level Control through Deep Reinforcement Learning," Nature, 518, p. 529.

[3] Silver, D., Huang, A., Maddison, C. J., Guez, A., Sifre, L., van den Driessche, G., Schrittwieser, J., Antonoglou, I., Panneershelvam, V., Lanctot, M., Dieleman, S., Grewe, D., Nham, J., Kalchbrenner, N., Sutskever, I., Lillicrap, T., Leach, M., Kavukcuoglu, K., Graepel, T., and Hassabis, D., 2016, "Mastering the Game of Go with Deep Neural Networks and Tree Search," Nature, 529, p. 484.

[4] Brown, N., and Sandholm, T., 2018, "Superhuman AI for Heads-up No-Limit Poker: Libratus Beats Top Professionals," Science, 359(6374), pp. 418 LP – 424.

[5] Vinyals, O., Babuschkin, I., Chung, J., Mathieu, M., Jaderberg, M., Czarnecki, W. M., Dudzik, A., Huang, A., Georgiev, P., Powell, R., Ewalds, T., Horgan, D., Kroiss, M., Danihelka, I., Agapiou, J., Oh, J., Dalibard, V., Choi, D., Sifre, L., Sulsky, Y., Vezhnevets, S., Molloy, J., Cai, T., Budden, D., Paine, T., Gulcehre, C., Wang, Z., Pfaff, T., Pohlen, T., Yogatama, D., Cohen, J., McKinney, K., Smith, O., Schaul, T., Lillicrap, T., Apps, C.,

Kavukcuoglu, K., Hassabis, D., and Silver, D., 2019, *AlphaStar: Mastering the Real-Time Strategy Game StarCraft II.*

[6] Cross, N., 2004, "Expertise in Design: An Overview," Design Studies, 25(5), pp. 427–441.

[7] Newell, A., and Simon, H. A., 1972, *Human Problem Solving*, Prentice-Hall, Inc., Upper Saddle River, NJ, USA.

[8] Daly, S., McKilligan, S., Christian, J., Seifert, C., and Gonzalez, R., 2012, "Design Heuristics in Engineering Concept Generation," Journal of Engineering Education, 101.

[9] Ross, S., 2013, "Interactive Learning for Sequential Decisions and Predictions," PhD Thesis, Carnegie Mellon University.

[10] Yannakakis, G. N., and Togelius, J., 2018, *Artificial Intelligence and Games*, Springer Publishing Company, Incorporated.

[11] Payne, J. W., Bettman, J. R., and Johnson, E. J., 1993, *The Adaptive Decision Maker.*, Cambridge University Press, New York, NY, US.

[12] Busemeyer, J. R., and Townsend, J. T., 1993, "Decision Field Theory: A Dynamic-Cognitive Approach to Decision Making in an Uncertain Environment.," Psychol Rev, 100(3), pp. 432–459.

[13] Singer, D. J., Doerry, N., And Buckley, M. E., 2009, "What Is Set-Based Design?," Naval Engineers Journal, 121(4), pp. 31–43.

[14] McComb, C., Cagan, J., and Kotovsky, K., 2017, "Capturing Human Sequence-Learning Abilities in Configuration Design Tasks Through Markov Chains," Journal of Mechanical Design, 139(9), pp. 91101–91112.

[15] McComb, C., Cagan, J., and Kotovsky, K., 2017, "Mining Process Heuristics From Designer Action Data via Hidden Markov Models," Journal of Mechanical Design, 139(11), p. 111412.

[16] Finger, S., and R. Dixon, J., 1989, "A Review of Research in Mechanical Engineering Design. Part II: Representations, Analysis, and Design for the Life Cycle," Research in Engineering Design, 1, pp. 121–137.

[17] McComb, C., Cagan, J., and Kotovsky, K., 2017, "Utilizing Markov Chains to Understand Operation Sequencing in Design Tasks," *Design Computing and Cognition '16*, pp. 401–418.

[18] Raina, A., McComb, C., and Cagan, J., 2018, "Design Strategy Transfer in Cognitively-Inspired Agents," *44th Design Automation Conference.*

[19] Brooks, R. A., 1991, "New Approaches to Robotics.," Science, 253(5025), pp. 1227–1232.

[20] Athavankar, Uday A . 1997, "Mental Imagery As A Design Tool," *Cybernetics and Systems,* 28(1), pp. 25–42.

[21] Goldschmidt, Gabriela 1992, "Serial Sketching: Visual Problem Solving In Designing -," *Cybernetics and Systems,* 23(2), pp. 191–219.

[22] Yin, Y. H., Xie, J. Y., Xu, L. D., and Chen, H., 2012, "Imaginal Thinking-Based Human-Machine Design Methodology for the Configuration of Reconfigurable Machine Tools," IEEE Transactions on Industrial Informatics, 8(3), pp. 659–668.

[23] Yin, Y. H., Zhou, C., and Zhu, J. Y., 2010, "A Pipe Route Design Methodology by Imitating Human Imaginal Thinking," CIRP Annals, 59(1), pp. 167–170.

[24] Diez, M., Campana, E. F., and Stern, F., 2015, "Design-Space Dimensionality Reduction in Shape Optimization by Karhunen–Loève Expansion," Computer Methods in Applied Mechanics and Engineering, 283, pp. 1525–1544.

[25] D'Agostino, D., Serani, A., Campana, E. F., and Diez, M., 2018, "Nonlinear Methods for Design-Space Dimensionality Reduction in Shape Optimization BT - Machine Learning, Optimization, and Big Data," G. Nicosia, P. Pardalos, G. Giuffrida, and R. Umeton, eds., Springer International Publishing, Cham, pp. 121–132.

[26] Chen, W., Fuge, M., and Chazan, J., 2017, "Design Manifolds Capture the Intrinsic Complexity and Dimension of Design Spaces," Journal of Mechanical Design, 139(5), pp. 51102–51110.

[27] Yumer, M. E., Asente, P., Mech, R., and Kara, L. B., 2015, "Procedural Modeling Using Autoencoder Networks," *Proceedings of the 28th Annual ACM Symposium on User Interface Software & Technology*, ACM, New York, NY, USA, pp. 109–118.

[28] Guo, T., Lohan, D., Cang, R., Ren, Y., and Allison, J., 2018, *An Indirect Design Representation for Topology Optimization Using Variational Autoencoder and Style Transfer.*

[29] D'Agostino, D., Serani, A., Campana, E., and Diez, M., 2018, *Deep Autoencoder for Off-Line Design-Space Dimensionality Reduction in Shape Optimization.*

[30] Hinton, G. E., and Salakhutdinov, R. R., 2006, "Reducing the Dimensionality of Data with Neural Networks," Science, 313(5786), p. 504.

[31] Bengio, Y., Courville, A. C., and Vincent, P., 2012, "Unsupervised Feature Learning and Deep Learning: {A} Review and New Perspectives," CoRR, abs/1206.5538.

[32] McComb, C., 2018, *Towards the Rapid Design of Engineered Systems Through Deep Neural Networks.*

[33] LeCun, Y., Bottou, L., and Haffner, P., 1998, "Gradient-Based Learning Applied to Document Recognition."

[34] Kingma, D. P., and Welling, M., 2013, "Auto-Encoding Variational Bayes," eprint arXiv:1312.6114, p. arXiv:1312.6114.

[35] Zhang, Y., Chen, A., Peng, B., Zhou, X., and Wang, D., 2019, "A Deep Convolutional Neural Network for Topology Optimization with Strong Generalization Ability," arXiv:1901.07761 [cs, stat].

[36] Banga, S., Gehani, H., Bhilare, S., Patel, S., and Kara, L., 2018, "3D Topology Optimization Using Convolutional Neural Networks," arXiv:1808.07440 [physics, stat].

[37] Burnap, A., Liu, Y., Pan, Y., Lee, H., Gonzalez, R., and Papalambros, P. Y., 2016, "Estimating and Exploring the Product Form Design Space Using Deep Generative Models," p. V02AT03A013.

[38] Carroll, J. D., 1963, "Functional Learning: The Learning Of Continuous Functional Mappings Relating Stimulus And Response Continua," *ETS Research Bulletin Series*, 1963(2), pp. i–144.

[39] Koh, K., and Meyer, D. E., 1991, "Function Learning: Induction of Continuous Stimulus-Response Relations.," J Exp Psychol Learn Mem Cogn, 17(5), pp. 811–836.

[40] DeLosh, E. L., Busemeyer, J. R., and McDaniel, M. A., 1997, "Extrapolation: The Sine qua Non for Abstraction in Function Learning.," J Exp Psychol Learn Mem Cogn, 23(4), pp. 968–986.

[41] Busemeyer, J. R., Byun, E., Delosh, E. L., and McDaniel, M. A., 1997, "Learning Functional Relations Based on Experience with Input-Output Pairs by Humans and Artificial Neural Networks.," Knowledge, concepts and categories., pp. 408–437.

[42] S. Spelke, E., Gutheil, G., and Van de Walle, G., 2019, *The Development of Object Perception.*

[43] Baillargeon, R., Li, J., Ng, W., and Yuan, S., 2008, *An Account of Infants' Physical Reasoning.*

[44] Bates, C., Yildirim, I., B Tenenbaum, J., and W Battaglia, P., 2015, *Humans Predict Liquid Dynamics Using Probabilistic Simulation.*

[45] Gershman, S. J., Horvitz, E. J., and Tenenbaum, J. B., 2015, "Computational Rationality: A Converging Paradigm for Intelligence in Brains, Minds, and Machines," Science, 349(6245), pp. 273 LP – 278.

[46] Kulkarni, T. D., Narasimhan, K., Saeedi, A., and Tenenbaum, J. B., 2016, "Hierarchical Deep Reinforcement Learning: Integrating Temporal Abstraction and Intrinsic Motivation," CoRR, abs/1604.06057.

[47] Biederman, I., 1987, "Recognition-by-Components: A Theory of Human Image Understanding.," Psychological Review, 94(2), pp. 115–147.

[48] Thrun, S., and Pratt, L., 1998, *Learning to Learn: Introduction and Overview.*

[49] Lake, B. M., Ullman, T. D., Tenenbaum, J. B., and Gershman, S. J., 2016, "Building Machines That Learn and Think Like People," CoRR, abs/1604.00289.

[50] Pomerleau, D. A., 1989, "ALVINN: An Autonomous Land Vehicle in a Neural Network," *Advances in Neural Information Processing Systems 1*, D.S. Touretzky, ed., Morgan-Kaufmann, pp. 305–313.

[51] Billard, A., and Matarić, M. J., 2001, "Learning Human Arm Movements by Imitation:: Evaluation of a Biologically Inspired Connectionist Architecture," Robotics and Autonomous Systems, 37(2), pp. 145–160.

[52] Finn, C., Yu, T., Zhang, T., Abbeel, P., and Levine, S., 2017, "One-Shot Visual Imitation Learning via Meta-Learning," CoRR, abs/1709.04905.

[53] Hester, T., Vecerik, M., Pietquin, O., Lanctot, M., Schaul, T., Piot, B., Sendonaris, A., Dulac-Arnold, G., Osband, I., Agapiou, J., Leibo, J. Z., and Gruslys, A., 2017, "Learning from Demonstrations for Real World Reinforcement Learning," CoRR, abs/1704.03732.

[54] Abbeel, P., Coates, A., and Ng, A. Y., 2010, "Autonomous Helicopter Aerobatics through Apprenticeship Learning," The International Journal of Robotics Research, 29(13), pp. 1608–1639.

[55] Liu, Y., Gupta, A., Abbeel, P., and Levine, S., 2017, "Imitation from Observation: Learning to Imitate Behaviors from Raw Video via Context Translation," CoRR, abs/1707.03374.

[56] Ha, D., and Schmidhuber, J., 2018, "World Models," CoRR, abs/1803.10122.

[57] Pretz, J. E., 2008, "Intuition versus Analysis: Strategy and Experience in Complex Everyday Problem Solving.," Memory & cognition, 36(3), pp. 554–566.

[58] Cagan, J., Dinar, M., J. Shah, J., Leifer, L., Linsey, J., Smith, S., and Vargas Hernandez, N., 2013, *Empirical Studies of Design Thinking: Past, Present, Future.*

[59] Björklund, T. A., 2013, "Initial Mental Representations of Design Problems: Differences between Experts and Novices," Design Studies, 34(2), pp. 135–160.

[60] Egan, P., and Cagan, J., 2016, "Human and Computational Approaches for Design Problem-Solving BT - Experimental Design Research: Approaches, Perspectives, Applications," P. Cash, T. Stanković, and M. Štorga, eds., Springer International Publishing, Cham, pp. 187–205.

[61] Cagan, J., and Kotovsky, K., 1997, "Simulated Annealing and the Generation of the Objective Function: A Model of Learning During Problem Solving," Computational Intelligence, 13(4), pp. 534–581.

[62] McComb, C., Cagan, J., and Kotovsky, K., 2018, *Drawing Inspiration From Human Design Teams For Better Search And Optimization: The Heterogeneous Simulated Annealing Teams Algorithm.*

[63] Matthews, P., Blessing, L., and Wallace, K. M., 2002, *The Introduction of a Design Heuristics Extraction Method.*

[64] Fuge, M., Peters, B., and Agogino, A., 2014, *Machine Learning Algorithms for Recommending Design Methods.*

[65] Sexton, T., and Ren, M. Y., 2017, "Learning an Optimization Algorithm Through Human Design Iterations," Journal of Mechanical Design, 139(10), pp. 101404–101410.

[66] McComb, C., Cagan, J., and Kotovsky, K., 2018, "Data on the Design of Truss Structures by Teams of Engineering Students," Data Brief, 18, pp. 160–163.

[67] Springenberg, J. T., Dosovitskiy, A., Brox, T., and Riedmiller, M. A., 2014, "Striving for Simplicity: The All Convolutional Net," CoRR, abs/1412.6806.

[68] Fergus, R., Zeiler, M. D., Taylor, G. W., and Krishnan, D., 2010, "Deconvolutional Networks," *2010 IEEE Computer Society Conference on Computer Vision and Pattern Recognition(CVPR)*, pp. 2528–2535.

[69] Nair, V., and E. Hinton, G., 2010, *Rectified Linear Units Improve Restricted Boltzmann Machines Vinod Nair.*

[70] Kingma, D. P., and Ba, J., 2014, "Adam: A Method for Stochastic Optimization," CoRR, abs/1412.6980.

[71] Maas, A. L., Hannun, A. Y., and Ng, A. Y., 2013, "Rectifier Nonlinearities Improve Neural Network Acoustic Models," *In ICML Workshop on Deep Learning for Audio, Speech and Language Processing.*

[72] Bengio, Y., Lamblin, P., Popovici, D., and Larochelle, H., 2006, "Greedy Layer-Wise Training of Deep Networks," *Proceedings of the 19th International Conference on Neural Information Processing Systems*, MIT Press, Cambridge, MA, USA, pp. 153–160.

[73] Franklin, S., and Graesser, A., 1997, "Is It an Agent, or Just a Program?: A Taxonomy for Autonomous Agents," *Intelligent Agents III Agent Theories, Architectures, and Languages*, J.P. Müller, M.J. Wooldridge, and N.R. Jennings, eds., Springer Berlin Heidelberg, pp. 21–35.

Proceedings of the ASME 2019
International Design Engineering Technical Conferences
and Computers and Information in Engineering Conference
IDETC/CIE2019
August 18-21, 2019, Anaheim, CA, USA

DETC2019-97642

ANALYZING CUSTOMER NEEDS OF PRODUCT ECOSYSTEMS
USING ONLINE PRODUCT REVIEWS

Jackie Ayoub
Department of Industrial and Manufacturing
Systems Engineering,
University of Michigan, Dearborn
Dearborn, Michigan, 48128
Email: jyayoub@umich.edu

Feng Zhou
Department of Industrial and Manufacturing
Systems Engineering,
University of Michigan, Dearborn
Dearborn, Michigan, 48128
Email: fezhou@umich.edu

Qianli Xu
Image and Video Analytics Department
Institute for Infocomm Research,
1 Fusionopolis Way, #21-01 Connexis,
Singapore, 138632
Email: qxu@i2r.a-star.edu.sg

Jessie Yang
Department of Industrial and Operations
Engineering, University of Michigan, Ann Arbor
Ann Arbor, Michigan, 48109
Email: xijyang@umich.edu

ABSTRACT

It is necessary to analyze customer needs of a product ecosystem in order to increase customer satisfaction and user experience, which will, in turn, enhance its business strategy and profits. However, it is often time-consuming and challenging to identify and analyze customer needs of product ecosystems using traditional methods due to numerous products and services as well as their interdependence within the product ecosystem. In this paper, we analyzed customer needs of a product ecosystem by capitalizing on online product reviews of multiple products and services of the Amazon product ecosystem with machine learning techniques. First, we filtered the noise involved in the reviews using a fastText method to categorize the reviews into informative and uninformative regarding customer needs. Second, we extracted various customer needs related topics using a latent Dirichlet allocation technique. Third, we conducted sentiment analysis using a valence aware dictionary and sentiment reasoner method, which not only predicted the sentiment of the reviews, but also its intensity. Based on the first three steps, we classified customer needs using an analytical Kano model dynamically. The case study of Amazon product ecosystem showed the potential of the proposed method.

1 INTRODUCTION

1.1 Background

Product ecosystem includes multiple related products manufactured by the same company to achieve superior customer satisfaction and user experience. Examples of product ecosystems include what Apple, Amazon, or Samsung offers. The success of these product ecosystems depends more on the overall user experience within the product ecosystem rather than the performance of single products or services. This is probably due to the fact that, within a product ecosystem, a main product is supported by other products and services to create customer satisfaction and user experience that other single offerings cannot beat [1]. In addition, once a customer enters such an ecosystem, it will be hard to exit due to the high cost of transferring the applications to other devices [2].

Although we have witnessed the success of product ecosystems, it is often challenging to create a successful product ecosystem from the perspective of design. Previous researchers have attempted to address different issues of product ecosystem design. For example, Levin et al. [3] created a framework for designing user experience of multiple devices within a product ecosystem, including smartphones, tablets, TVs, and computers, based on consistent, complementary, and continuous techniques. Gawer and Cusumano [4] discussed the impacts of internal (or company-specific) and external (or industry-wide) platforms on product innovation. Internal platform promoted innovation by considering the derivative of products and services while the external platform considered complementary products and services for an innovative ecosystem. Jiao et al. [5] incorporated the notion of ambience or context of human-product interaction in product ecosystem design in order to enhance user experience. Oh et al. [6] presented a product-service system design framework within a business ecosystem, including manufacturers, suppliers, and content providers, to identify design factors for products and services. Zhou et al. [1] presented a three-stage product ecosystem design process for user experience. The design stages included affective-cognitive need acquisition, affective-cognitive analysis, and affective-cognitive fulfillment. Their results showed that high-level needs, including affective and cognitive needs, can improve user experience. However, the framework was mainly conceptual and no concrete research methods were provided for each step. Later, in order to show such a design process, Zhou et al. [7] described the relations between user experience and the components of a product ecosystem using a simulation technique named fuzzy reasoning Petri nets. However, such a simulation method may be restricted in terms of its generalizability. In this paper, we attempted to analyze various customer needs associated with different products and services within a product ecosystem with a concrete case study. While it is relatively easy to elicit and analyze customer needs of single products, it is challenging to do for a whole product ecosystem.

1.2 Research Challenges and Suggested Solutions

1) Collecting Customer Need Data: There are various products and services involved in the product ecosystem. Hence, it is time-consuming to collect data in order to elicit customer needs using traditional methods, such as interviews, focus groups, and surveys [8], compared to the short product development lead time [9]. Recently, many researchers have made use of online-user generated data (e.g., online product reviews) to identify customer needs. For example, Zhou et al. [10] extracted customers' latent and explicit needs based on online product reviews using case analogical reasoning and sentiment analysis. For one thing, users describe the product performance in various use situations from the users' perspectives in these online product reviews, which provide a good channel for customer needs elicitation [11]. For another, the scale of the online user-generated data is often so large that it can extract the customer needs associated with the products and services. For example, Amazon Echo Dot 2 has more than 120,000 product reviews on Amazon.com. Therefore, in this research, we will make use of the online user-generated data to identify customer needs.

2) Analyzing Text Data: Customer needs identification and analysis are ambiguous since they are written in the form of natural languages [12]. It often needs domain experts to analyze customer needs with specific skills. For example, experts who can conduct ethnographic studies can have a deep understanding of customer needs. However, it is extremely time-consuming to analyze such a large amount of text data, if not impossible [13]. In order to reduce the ambiguity of customer needs involved in the text data and analyze text data efficiently and effectively at the same time, we propose to make use of machine learning methods in this research.

First, we cleaned the data by filtering out the noise involved in the textual review data. In order to do that, we proposed to employ a supervised machine learning technique, fastText [14], to distinguish informative reviews from uninformative reviews (i.e., noise) with regard to customer needs. For example, the review, "I love this e-reader" is not informative in terms of customer needs analysis since it does not show what specific customer need is satisfied, although it shows a positive attitude towards the product. fastText is a supervised machine learning technique, and it is reported to be an order of magnitude faster in performance than the state-of-the-art deep learning classifiers, but is as accurate as them [14]. Such a step improves data quality greatly and reduces the ambiguity of customer needs embedded in noisy text data.

Second, in order to identify different topics of customer needs within the product ecosystem, we proposed to apply a topic modeling technique, latent Dirichlet allocation (LDA). LDA is an unsupervised machine learning technique and it assumes that each review is a mixture of a small number of topics related to customer needs and that each word's presence is attributable to one of the review's topics [15]. Therefore, LDA can identify the hidden topics, indicating the voices of customers from a large volume of unstructured online product reviews [16]. Each topic identified is related to a specific customer need for individual products and/or services within the product ecosystem expressed in product reviews. The topics identified can categorize customer needs into different groups, which are then combined with the quantitative customer preferences (i.e., satisfaction and dissatisfaction score of each topic) as input for categorizing customer needs for the product ecosystem so that the superiorities and weaknesses of product ecosystem can be discovered.

Third, in order to quantitatively understand to what extent customers are satisfied with the products and services, we conducted sentiment analysis of each product review associated with each group of customer needs to pave the way for customer needs categorization in the next step. Sentiment analysis is a computational method to predict whether a product review is positively, neutrally, or negatively evaluated [17]. Researchers have proposed various machine learning methods for the purpose of sentiment analysis, including both supervised and unsupervised methods. For example, Hu and Liu [18] compiled a list of lexicon seeds (both positive and negative ones), which were expanded using synonym and antonym relations in WordNet. Ding et al. [19] created a holistic lexicon list by exploiting external evidence and linguistic rules in the language expressions, which outperformed the one in [18]. Chen et al. [20] built a sentiment analysis model with conditional random fields using various linguistic features, including part-of-speech tags and lexicons, and obtained accuracy over 75% in terms of the F_1 measure. Kim [21] proposed a deep learning model based on word embeddings and convolutional neural networks, and the prediction accuracy was between 81.5% and 93.4% across various datasets. Li et al. [22] trained an adversarial memory network for sentiment analysis across different domains (e.g., product reviews vs. movie reviews) and visualized the pivotal words in the review to improve the interpretability of their deep model. Although these sentiment analysis methods classified whether the review belonged to the positive, the negative, or the neutral category, they failed to give an intensity score, which is necessary in this research to quantitatively identify the satisfaction and dissatisfaction scores that are used to categorize customer needs in the next step. In order to do this, we proposed to employ VADER (Valence Aware Dictionary and sEntiment Reasoner) as an unsupervised method to predict customer satisfaction quantitatively. Hutto and Gilbert [23] has shown that this method was able to outperform human raters in terms of the F_1 measure, especially for social media text data.

3) Categorizing Customer Needs: It is necessary to categorize customer needs into different groups in terms of their priority and importance in creating great user experience and improving customer satisfaction. Many techniques have been proposed in the literature in this aspect [24]. For example, the Kano model is widely used to identify four basic drivers of customer satisfaction, i.e., must-be, one-dimensional, attractive, and indifferent [25]. Kim and Han [26] identified three usability dimensions, including product, product-user, and product-user-task, in order to evaluate customer satisfaction in terms of usability. Delin et al. [27] classified customer needs into three groups, i.e., highest, medium, and least concept coverage, in order to identify the most appropriate emotional adjectives in describing different customers. However, the methods mentioned above only offered a qualitative evaluation of customer satisfaction regarding product performance level [28]. In order to make use of quantitative measures to evaluate user experience and satisfaction, it is important to assign customer satisfaction and dissatisfaction scores. In this research, we proposed to use an analytical Kano model [30]. It is consistent with the Kano principles, but incorporates quantitative measures produced from sentiment analysis in the previous step into customer satisfaction, which can be used as the tangible criteria for categorizing customer needs.

2 PROPOSED METHOD AND CASE STUDY

In this work, we presented a method to analyze customer needs of the Amazon product ecosystem based on its online customer reviews to understand to what degree Amazon products and services are satisfying their customers. The Amazon product ecosystem (services and products) is used to illustrate the proposed method. We collected a total number of 91738 review sentences between 2011 and 2018 for the following products: Amazon Kindle Fire tablets, Kindle E-reader, Fire TV, Echo and Alexa devices, and other accessories (e.g., Kindle keyboard, Kindle leather cover, Fire TV power adapter, USB chargers). There are four important steps involved in the proposed method as illustrated in Figure 1, including 1) noise filtering, 2) topic extraction, 3) sentiment analysis, and 4) categorizing customer needs. The noise filtering step applies the fastText method to remove uninformative reviews from the informative reviews in order to reduce the ambiguity involved in customer needs. The topic extraction step uses the LDA method

to identify the underlying topics associated with the customer needs of individual products and services within the Amazon product ecosystem. The sentiment analysis step makes use of the VADER method to not only predict sentiment polarity, but also sentiment intensity scores of product reviews in order to quantitatively evaluate customer satisfaction and dissatisfaction regarding each customer need produced in the previous step. Finally, the analytical Kano model is used to classify customer needs based on the satisfaction and dissatisfaction scores produced in the previous step as the tangible criteria. According to the obtained results from this study, possible design recommendations can be made to improve user experience and customer satisfaction. Such a process is iterative in order to continuously improve the performance of the product ecosystem.

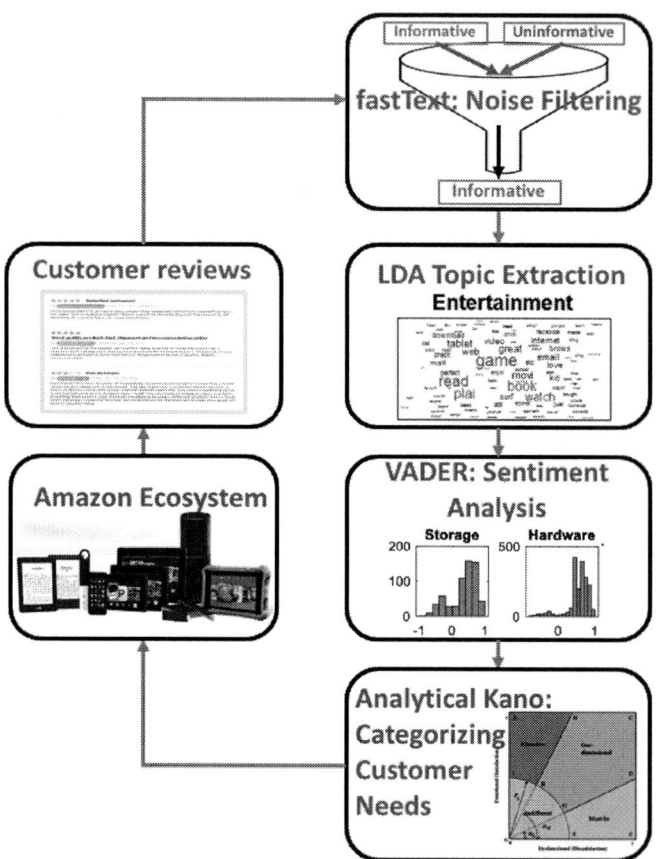

FIGURE 1: THE STEPS INVOLVED IN THE PROPOSED METHOD

3 FILTERING NOISE USING FASTTEXT

The fastText method is a supervised machine learning technique created by Facebook for text classification and word vector representation. fastText assumes that words are formed by a n-grams of characters where n can range from 1 to the length of the word. Compared to word2vec and glove, fastText can find a vector representation for rare words not present in the dictionary by breaking the words into chunks of vectors and combining them to create the final vector, which is particularly useful for text data in social media [19].

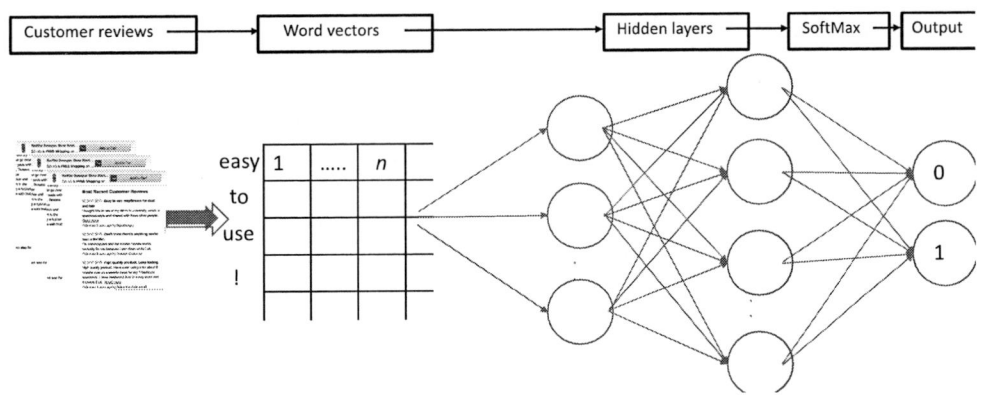

FIGURE 2: THE MODEL OF FASTTEXT

The procedure of noise filtering using fastText is shown in Figure 2. In order to train and test the model, 10,000 reviews were manually labeled into informative (7453 labeled as 1) and uninformative (2543 labeled as 0) reviews and 4 were removed as they were not in English or blank. By uninformative reviews we mean that they do not describe specific features of the product or provide only a general opinion on the product. Thus, they are not helpful to elicit customer needs (e.g., "I have 3 echoes and 2 dots", "Very pleased", and "This was my first tablet…"). The next step was to preprocess the reviews, including removing punctuations, converting the letters into lowercase, and stemming.

Furthermore, by varying the fastText parameters, including the learning rate, number of epochs, dimension of word embeddings, and number of n-grams and using a 5-fold cross-validation, we obtained the following results: precision = 0.91, recall = 0.93, and F_1 measure = 0.92, where precision is defined as tp/(tp + fp), recall = tp/(tp + fn), where tp, fp, and fn mean true positive, false positive, and false negative, respectively. F_1 measure is defined as the harmonic mean of precision and recall, i.e., 2×precision×recall/(precision + recall) [31]. Due to the imbalance between the informative and uninformative reviews in the dataset, we reported precision, recall, and F_1 measure and the fastText model performed reasonably well despite the manual work involved in labeling reviews. Then we trained the model with the 9,996 labeled review data to filter the uninformative reviews from the rest of the data, and this process removed 19,800 (21.58%) reviews from the original dataset.

4 CUSTOMER NEEDS TOPIC MODELING USING LDA

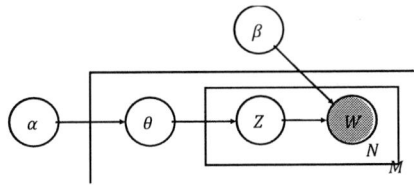

FIGURE 3: REPRESENTATION OF THE LDA MODEL

LDA is an unsupervised topic modeling technique used to cluster informative customer reviews into separate groups [20]. The LDA model is illustrated in Figure 3, where the boxes represent repeated entities. Those in white circles represent latent variables and only words in the shaded circle are the observed variables. The outer box shows the number (i.e., $M = 71938$) of the informative customer reviews extracted using fastText and the inner box represents the number of words (i.e., N) in the customer reviews. Thus, each review is represented as a sequence of N word, $\mathbf{w} = (w_1, w_2, \ldots, w_N)$ and each word is represented as a one-hot vector in V dimension, where V is the total number of the words in all the reviews, and only the $v - th$ word is 1 if it shows up in the review. A description of the generated process is shown below:

for each document

draw a Dirichlet prior topic distribution from $\boldsymbol{\theta} \sim Dir(\boldsymbol{\alpha})$

for $n \in \{1, \ldots, N\}$

draw a topic distribution from $z_n \sim Multinomial(\boldsymbol{\theta})$

draw a word distribution from $w_n \sim Multinomial(w_n|z_n, \boldsymbol{\beta})$ conditioned on the topic z_n and word probability matrix $\boldsymbol{\beta}_{K \times V}$, where $\beta_{kv} = p(w^v = 1|z^k = 1)$

end for

end for

Based on the above process, $\boldsymbol{\theta}$ is a K-dimensional Dirichlet random vector of probabilities, which must sum to 1. The probability density

$$p(\boldsymbol{\theta}|\boldsymbol{\alpha}) = \frac{\Gamma(\sum_{i=1}^{K} \alpha_i)}{\prod_{i=1}^{K} \alpha_i} \theta_1^{\alpha_1 - 1} \ldots \theta_K^{\alpha_K - 1}, \quad (1)$$

where $\boldsymbol{\alpha}$ is a K dimensional vector of positive reals and $\Gamma(x)$ is the Gamma function. We can write the joint distribution of a topic mixture $\boldsymbol{\theta}$, N topics $\boldsymbol{z}$, and N words $\boldsymbol{w}$, conditioned on the given parameters $\boldsymbol{\alpha}$ and $\boldsymbol{\beta}$

$$p(\boldsymbol{\theta}, \boldsymbol{z}, \boldsymbol{w}|\boldsymbol{\alpha}, \boldsymbol{\beta}) = p(\boldsymbol{\theta}|\boldsymbol{\alpha}) \prod_{n=1}^{N} p(z_n|\boldsymbol{\theta}) p(w_n|z_n, \boldsymbol{\beta}). \quad (2)$$

TABLE 1: EXTRACTED TOPICS AND THEIR ASSOCIATED STEMMED TOPIC WORDS USING LDA

Topics	Entertainment	Great Value	Interaction	Music-related	Parental control	Cost-effective	Storage	Hardware	Gift	Smart Home	Turning Page
Topic words	game	great	time	music	kid	tablet	memori	good	bought	echo	page
	read	price	work	alexa	great	recommend	card	great	love	home	turn
	plai	tablet	get	plai	tablet	good	storag	qualiti	gift	light	screen
	book	good	everi	voic	parent	worth	expand	sound	christma	control	get
	watch	product	app	great	love	monei	fast	speaker	purchas	smart	like
	movi	work	devic	amazon	control	well	app	pictur	got	devic	time
	great	bui	need	echo	easi	get	space	screen	tablet	hous	button
	tablet	kid	dai	listen	user	great	add	batteri	daughter	alexa	back
	email	valu	just	command	entertain	price	ad	life	wife	room	read
	web	best	wifi	love	friendli	look	need	camera	son	work	down

Topics	Reading	Buying Experience	For Kids	Streaming	Apps	Usability	Battery	Alexa	Charging	Size	Amazon Prime
Topic words	kindl	purchas	old	fire	app	easi	read	music	list	size	amazon
	read	bui	year	stick	amazon	set	light	ask	thing	easi	prime
	fire	best	love	amazon	store	setup	batteri	plai	charg	read	cabl
	book	kindl	bought	stream	game	learn	easi	weather	just	perfect	watch
	like	fire	tablet	box	download	fun	great	alexa	like	small	movi
	screen	amazon	daughter	fast	googl	navig	life	question	need	book	show
	new	get	purchas	faster	plai	simpl	screen	new	charger	light	stream
	paperwhit	tablet	son	better	access	great	kindl	answer	shop	carri	fire
	better	bought	kindl	work	free	work	long	get	time	travel	member
	version	first	got	devic	avail	user	last	love	make	take	great

The likelihood of an informative review document can be obtained by integrating over $\boldsymbol{\theta}$ and summing over $\boldsymbol{z}$

$$p(\boldsymbol{w}|\boldsymbol{\alpha},\boldsymbol{\beta}) = \int p(\boldsymbol{\theta}|\boldsymbol{\alpha})\left(\prod_{n=1}^{N}\sum_{z_n}p(z_n|\boldsymbol{\theta})p(w_n|z_n,\boldsymbol{\beta})\right)d\boldsymbol{\theta}. \tag{3}$$

Finally, the topic mixture $\boldsymbol{\theta}$ can be obtained using the posterior distribution of the hidden variables of a review document as follows:

$$p(\boldsymbol{\theta},\boldsymbol{z}|\boldsymbol{w},\boldsymbol{\alpha},\boldsymbol{\beta}) = \frac{p(\boldsymbol{\theta},\boldsymbol{z},\boldsymbol{w}|\boldsymbol{\alpha},\boldsymbol{\beta})}{p(\boldsymbol{w}|\boldsymbol{\alpha},\boldsymbol{\beta})}. \tag{4}$$

However, it is intractable to obtain the analytical solution of the parameter inference of the LDA model. Various estimate methods have been proposed, including Gibbs sampling [32] and variational methods [15]. In this research, we made use of the Text Analytics Toolbox in Matlab 2017b for parameter estimation.

In order to evaluate the LDA model and to determine the number of topics, a perplexity measure was used. After training the LDA model, the perplexity of the test dataset was calculated to measure how well the model can predict the test data. A low perplexity score indicates a better prediction performance [19]. The perplexity is defined as follows:

$$Perplexity(D_{test}) = exp\left(-\frac{\sum_{m=1}^{M}logp(\boldsymbol{w_m})}{\sum_{m=1}^{M}N_m}\right), \tag{5}$$

where N_m is the total number of words in the $m-th$ review document in the test dataset D_{test} with M review documents and 10% of the data were randomly selected as the test dataset. Figure 4 shows how the perplexity measure changes when the number of topics varies from 8 to 30 with a step size of 2. The minimum value of perplexity was reached when the number of topics was 22 as shown in Figure 4. After these topics were generated, we manually assign a name to each group.

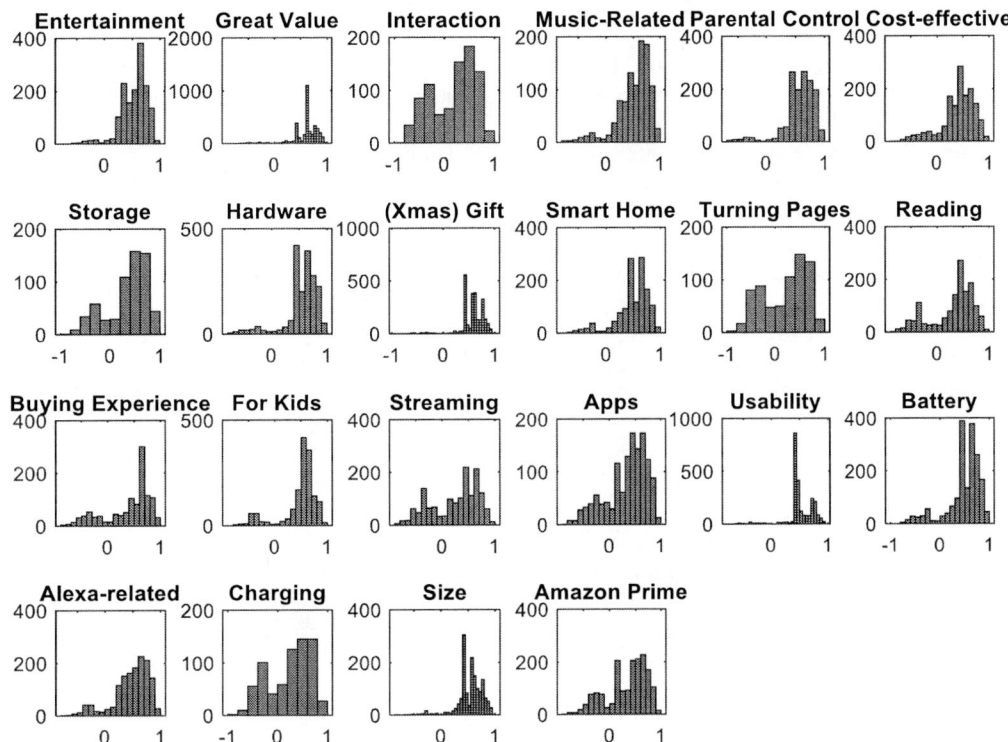

FIGURE 5: HISTOGRAMS OF THE SENTIMENT INTENSITY OF THE DIFFERENT CUSTOMER REVIEWS TOPICS

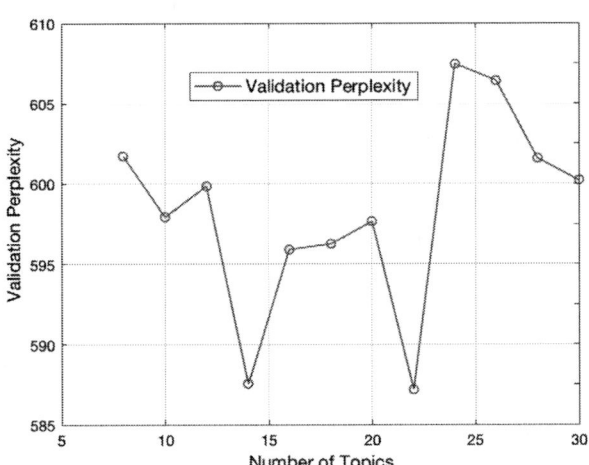

FIGURE 4: HOW THE PERPLEXITY MEASURE CHANGES WITH THE NUMBER OF TOPICS

Furthermore, we set up a threshold of topic probability in order to exclude ambiguous reviews and it was set as one standard deviation below the mean among the maximum probabilities of all the reviews predicted by the LDA model. Then we ended up with 60565 informative reviews and 22 topics as shown in Table 1 (Note the words were stemmed). The extracted topics cover various products and services within the Amazon product ecosystems. The topic "Entertainment" is most relevant to Kindle Fire HD tablets, the topic "Reading" is most relevant to Kindle E-reader, and the topic "Alexa" is most relevant to Amazon Echo and Echo Dot products. Some topics cover more than one product or service. For example, the topic "Amazon Prime" covers all the topics related to the prime service, the topics "Storage" and "Battery" are relevant to both Kindle Fire HD tablets and Kindle E-readers, and the topic "Streaming" is relevant to Fire TV, Kindle Fire HD tablets, and the Amazon Prime service. In terms of interpreting the topics, we examined not only the most relevant topic words associated with these topics as shown in Table 1, but also example reviews with high probabilities. Some topics are easier to understand while others tend to be difficult to interpret. For example, the topic "Storage" is easy to interpret while the topic "Interaction" tends to be difficult without examining review examples (e.g., "trying to type an email and the keyboard would just stop working").

5 CUSTOMER SATISFACTION ANALYSIS WITH VADER

VADER is a lexicon and rule-based sentiment analysis technique used to predict the sentiment and its intensity of the customer review. Hutto and Gilbert [9] showed that VADER performance was similar to human raters based on the correlation coefficient ($r = 0.88$), but regarding the classification accuracy of social media data, the performance was better than human raters. In VADER, over 9000 affective lexicons were used, including word banks and social media slangs. In order to predict the intensity of the lexical features, a systematic control process was used. First, ten Amazon Mechanical Turk workers, who went through a rigorous training and selection process, rated the lexicons in a range from 4 (most positive valence) to -4 (most negative valence). Then the intensity of the review was modified using five generalizable heuristics, including punctuation emphasis, capitalization differentiation, degree intensifiers, contrastive conjunction, and tri-grams before affective lexicons in the data. By averaging and normalizing the affective lexicon scores between -1 and 1, the overall intensity of the reviews was calculated, where those smaller than 0 are negative and those larger than 0 are positive. The larger the absolute value, the more intense the sentiment is. For example, the review "this product so far has not disappointed" was predicted as 0.3724, and the review "not easy for elderly users cease of ads that pop up" was predicted as -0.3412.

Of all the 60565 informative reviews, 68.3% were predicted to be positive, 21.8% neutral, and 9.9% negative. All the 13199 neutral reviews were removed, and the rest was used for the following analysis. The histograms of the normalized sentiment intensity of the customer review topics are shown in Figure 5, representing the distribution of satisfaction (between 0 and 1) and dissatisfaction (between -1 and 0) of customers with the Amazon product ecosystem. Each histogram shows how well the customer needs are satisfied by their corresponding products and/or services in the Amazon product ecosystem. For example, the histograms of "Entertainment", "Music-Related", "Parental-Control", "Cost-effective", "Hardware", "(Xmas) Gift", "Smart Home", "For Kids", "Battery", and "Size" show that most of the reviews were positively evaluated. While almost an equal number of positive and negative reviews were seen for the following topics: "Interaction", "Turning Pages", "Streaming", and "Charging".

6 CUSTOMER NEEDS CATEGORIZATION WITH THE ANALYTIC KANO MODEL

An extension to the traditional Kano model was used in this paper, i.e., the analytical Kano model, to categorize customer needs using the extracted topics of product reviews and the sentiment intensity values, indicating quantitative customer preferences. The categorization process is based on the level of customer satisfaction and dissatisfaction derived from the sentiment analysis. The results are shown in Figure 6, where the x-axis and y-axis represent the normalized dissatisfaction and satisfaction levels of the 22 topics, respectively. The levels of satisfaction and dissatisfaction are calculated using the sentiment intensity and the number of positive and negative reviews as shown below [20]:

$$X_i = \lambda \times \frac{NR_i}{TR_i} \times INR_i, \qquad (6)$$
$$Y_i = \frac{PR_i}{TR_i} \times IPR_i, \qquad (7)$$

where λ indicates the degree of dissatisfaction, $NR_i, PR_i,$ and TR_i represent the total numbers of negative, positive, and total reviews and $INR_i, IPR_i,$ are the mean intensity values of the negative and positive reviews, respectively. The customer need is represented as a vector $\boldsymbol{r_i}$ of magnitude r_i and direction $\alpha_i,$

where $\qquad 0 \leq r_i = |\boldsymbol{r_i}| = \sqrt{X_i^2 + Y_i^2} \leq \sqrt{2} \qquad$ and

$0 \leq \alpha_i = arctan(Y_i/X_i) \leq \pi/2$ is the angle between the horizontal axis and $\boldsymbol{r_i}$. Four types of customer needs were identified based on the Kano model: (1) attractive needs that are unexpected by the customer and can increase his/her satisfaction substantially, (2) one-dimensional needs that linearly increase customer satisfaction, (3) must-be needs that are expected by the customer and lead to dissatisfaction if they are not satisfied, and (4) indifferent needs that the customer is not intrested in. The attractive needs are shown in the area AIHB in Figure 6(a) when $r_i \geq r_0$ and $\alpha_H \leq \alpha_i \leq \pi/2$; the one-dimensional needs lie in the area HBCDG in Figure 6(a) when $r_i \geq r_0$ and $\alpha_L \leq \alpha_i \leq \alpha_H$; the must-be needs are shown in the area GDEF in Figure 6(a) when $r_i \geq r_0$ and $0 \leq \alpha_i \leq \alpha_L$; the indifferent needs lie in the area OIF in Figure 6(a) when $r_i < r_0$. In this paper, the following parameters are assumed based on a sensitivety analysis to illustrate the performance of the Amazon product ecosystem: $\lambda = 3,$ $r_0 = 0.4, \alpha_L = \pi/6,$ and $\alpha_H = \pi/3.$

Figure 6(a) and Figure 6(b) represent the classification results in terms of the different topics of customer needs for the collected reviews before January 1, 2015 and for all the reviews collected, respectively. The blue plus signs represent attractive needs, the red crosses represent one-dimensional needs, the purple circles represent indifferent needs, and the green asterisks represent must-be needs. It is noticed that the customer needs change with time. For instance, the topic "Cost-effective" has changed from a one-dimensional customer need in Figure 6(a) to an attractive need in Figure 6(b). The topic "Smart Home" has changed from an indifferent customer need in Figure 6(a) to an attractive customer need in Figure 6(b). The topic "Charging" has changed from a must-be customer need in Figure 6(a) to a one-dimensional customer need in Figure 6(b). From Figure 6(b), we can tell that all the customer needs are either one-dimensional or attractive. By examining the review data at different times, i.e., different stages of the product ecosystem, it shows the dynamic evolution of the product ecosystem, which is helpful to understand the evolution of different customer needs within the.

The paramteres involved in the analytical Kano model can also be adjusted to show how they influence the results. The aversion of customer dissatisfaction is often around 2.5 to 3.5. The increase of λ was shown to increase the degree of customer dissatisfaction. The range of r_0 is often between 0.3 to 0.5, it was noticed that if r_0 is 0.5, some of the topics (i.e., "Interaction", "Cost-effective", "Storage", "Reading", "Streaming", "Apps", "Charging", and "Amazon Prime") will be considered indifferent which is not reasonable. As long as r_0 is less than 0.4437 (minimum value of r_0 for topic 22) the results in Figure 6 (b) will not change. To study the influence of α_L and α_H, we varied their values between $\pi/12$ to $\pi/4$ and $\pi/4$ to $5\pi/12$ respectively [30]. By increasing the value of α_L, the topic "Turning pages" has changed from being a one-dimensional to a must-be need. And by increasing α_H, some topics (i.e., "Interaction", "Cost-effective", "Storage", "Reading", "Buying Experience", "Charging", and "Amazon Prime") have changed from being attractive to one-dimensional needs, which makes more sense.

7 DISCUSSION

1) Reducing ambiguity: In this study, we analyzed customer needs of the Amazon product ecosystem using online product reviews. The first step was to remove uninformative reviews using fastText, where we manually labeled 9996 reviews into informative and uninformative reviews. During the topic modeling, 11373 reviews were removed, since they had low probabilities for the 22 topics. Then, for the purpose of evaluating the level of satisfaction and dissatisfaction, all the 13199 neutral reviews were removed. Hence, 47366 reviews were used to classify customer needs in the analytical Kano model. Such a multi-step filtering process greatly improved data quality, reliability, and validity of the results.

2) Improving efficiency and effectiveness of data analysis: In order to analyze a large volume of text data, we made use of three machine learning techniques, including fastText, LDA, and VADER. The fastText is a supervised machine learning technique used to remove uninformative reviews. To improve the performance of noise filtering, we had to manually label customer reviews to train and validate the model. Compared to traditional coding in text data analysis in the domain of customer needs elicitation and analysis, it was much easier and more efficient to label the text data here in order to build the fastText model, as it only involved two categories, i.e., informative or not. Second, both LDA and VADER are unsupervised machine learning techniques, and thus no human manual work was involved. Compared with traditional customer needs analysis methods, machine learning methods not only improve the efficiency of the data analysis, but also are likely to produce more valuable customer needs from online product reviews [13].

3) Categorizing customer needs: In order to analyze customer needs within the Amazon product ecosystem, we first made use of the satisfaction and dissatisfaction scores produced from the sentiment analysis step, which gave us a tangible criterion in categorizing different types of customer needs and in measuring the performance of the product ecosystem. Specific products and services associated with these customer needs can be further examined in order to further improve user experience and satisfaction of the product ecosystem. Second, we examined the customer needs along the temporal dimension using the analytical Kano model, which showed the evolution of the customer needs within the Amazon product ecosystem. However, it was also observed that certain customer needs (and their associated product attributes) did not follow the product life cycle suggested by Kano [22] where the product state changes from indifferent to attractive, to one-dimensional, and finally to must-be. Two possible reasons can be identified. First, we only had a small amount of data (493 reviews) before Jan. 1, 2015 and fewer products and services were involved in the product ecosystem, which might produce less reliable results. Second, it is still too short a time (from 2011 to 2018) to show the full life cycle of the product ecosystem. For example, it took 18 years from the remote control to change from an attractive customer need (i.e., 1983) to a must-be one (i.e., 1998) [33].

Furthermore, in order to understand the customer needs within the product ecosystem, we need to study different

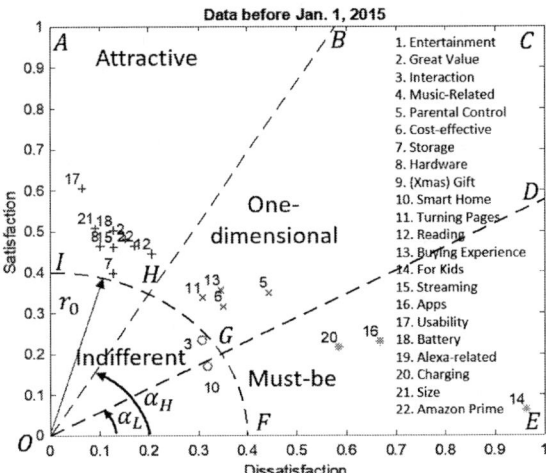

(a)

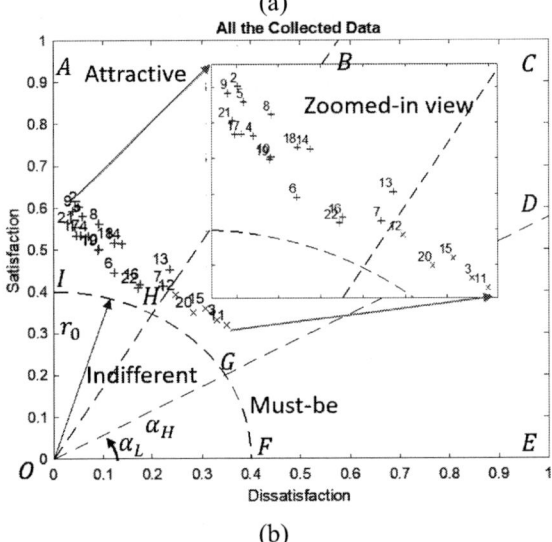

(b)

FIGURE 6: CUSTOMER NEEDS CATEGORIZATION USING ANALYTICAL KANO MODEL (a) USING DATA BEFORE JAN. 1, 2015 AND (b) USING ALL THE COLLECTED DATA

characteristics and relations of the products and services. By making use of the analytical Kano model, we can see that "Amazon Prime" was considered as an attractive need since it can provide many services (i.e. product discounts, free 2 days shipping, free video streaming, and free eBooks...). And these services can positively affect other topics, such as "Cost Effective", "Entertainment", "Music-Related", "Reading", "Streaming", "Alexa-related" and "Buying Experience". Hence, understanding the relationships between the products and services of the ecosystem is essential for creating desired user experience.

4) Limitation and future work: The first limitation of this study is related to the unbalanced distribution of data over the time. The data before 2015 can be unreliable as compared to after 2015 since we only had 493 reviews. Therefore, more data should be collected before 2015 in order to show a complete picture of the product ecosystem. Second, showing the full life cycle of a product ecosystem can provide more reliable and precise analysis. For example, understanding the evolution of "Smart Home" over the time, from the day it was created till now, can help us understand why the customer need was changed from being indifferent before 2015 to be attractive in 2018 (see Figure 6). The topic "Smart Home" is most relevant to Alexa products and services and it was first launched in November 2014. Therefore, it seems that the evolution of this customer need makes sense. However, at the current stage, more data in the future are needed in order to show the full life cycle of the product ecosystem. Third, further investigations should be conducted to understand the needs for individual products and services and their roles and interrelationships in the whole product ecosystem.

8 CONCLUSIONS

In this work, we analyzed the customer needs of the Amazon product ecosystem by making use of online product reviews. Using both supervised (fastText) and unsupervised (LDA and VADER) machine learning techniques, we were able to reduce ambiguity involved in the text data and analyze customer needs for products and services involved in the product ecosystem effectively and efficiently. fastText was used to filter the noise from the online customer reviews. LDA was used to cluster the customer needs into different topics. VADER helped in recognizing customer preferences quantitatively regarding different products and services. Finally, the analytical Kano model was used to categorize customer needs based on satisfaction and dissatisfaction scores. The proposed method was demonstrated using the Amazon product ecosystem and the process and the results showed the potential of the proposed method.

REFERENCES

[1] Zhou, F., Xu, Q., and Jiao, R. J., 2011, "Fundamentals of Product Ecosystem Design for User Experience," Res. Eng. Des., **22**(1), pp. 43–61.

[2] Miguel, J. C., and Casado, M. Á., 2016, "GAFAnomy (Google, Amazon, Facebook and Apple): The Big Four and the B-Ecosystem," *Dynamics of Big Internet Industry Groups and Future Trends*, pp. 127–148.

[3] Levin, M., 2014, *Designing Multi-Device Experiences: An Ecosystem Approach to User Experiences Across Devices*, "O'Reilly Media, Inc."

[4] Gawer, A., and Cusumano, M. A., 2014, "Industry Platforms and Ecosystem Innovation," Journal of Product Innovation Management, **31**(3), pp. 417–433.

[5] Jiao, R. J., Xu, Q., Du, J., Zhang, Y., Helander, M., Khalid, H. M., Helo, P., and Ni, C., 2007, "Analytical Affective Design with Ambient Intelligence for Mass Customization and Personalization," Int. J. Flexible Manuf. Syst., **19**(4), pp. 570–595.

[6] Oh, H. S., Moon, S. K., and Kim, W., 2012, "A Product-Service System Design Framework Based on a Business Ecosystem," *ASME 2012 International Design Engineering Technical Conferences and Computers and Information in Engineering Conference*, American Society of Mechanical Engineers, pp. 1033–1042.

[7] Zhou, F., Jiao, R. J., Xu, Q., and Takahashi, K., 2012, "User Experience Modeling and Simulation for Product Ecosystem Design Based on Fuzzy Reasoning Petri Nets," IEEE Transactions on Systems, Man, and Cybernetics - Part A: Systems and Humans, **42**(1), pp. 201–212.

[8] Crandall, B., Klein, G. A., and Hoffman, R. R., 2006, *Working Minds: A Practitioner's Guide to Cognitive Task Analysis*, MIT Press.

[9] Wang, Y., Mo, D. Y., and Tseng, M. M., 2018, "Mapping Customer Needs to Design Parameters in the Front End of Product Design by Applying Deep Learning," CIRP Ann., **67**(1), pp. 145–148.

[10] Zhou, F., Jiao, R. J., and Linsey, J. S., 2015, "Latent Customer Needs Elicitation by Use Case Analogical Reasoning from Sentiment Analysis of Online Product Reviews," J. Mech. Des., **137**(7), p. 071401.

[11] Bickart, B., and Schindler, R. M., 2001, "Internet Forums as Influential Sources of Consumer Information," Journal of Interactive Marketing, **15**(3), p. 31.

[12] Tseng, M. M., and Jiao, J., 1998, "Computer-Aided Requirement Management for Product Definition: A Methodology and Implementation," Concurrent Eng.: Res. Appl., **6**(2), pp. 145–160.

[13] Timoshenko, A., and Hauser, J. R., 2017, "Identifying Customer Needs from User-Generated Content," SSRN Electronic Journal.

[14] Joulin, A., Grave, E., Bojanowski, P., and Mikolov, T., 2017, "Bag of Tricks for Efficient Text Classification," *Proceedings of the 15th Conference of the European Chapter of the Association for Computational Linguistics: Volume 2, Short Papers*.

[15] Blei, D. M., Ng, A. Y., and Jordan, M. I., 2003, "Latent Dirichlet Allocation," J. Mach. Learn. Res., **3**(Jan), pp. 993–1022.

[16] Wang, W., Feng, Y., and Dai, W., 2018, "Topic Analysis of Online Reviews for Two Competitive Products Using Latent Dirichlet Allocation," Electron. Commer. Res. Appl., **29**, pp. 142–156.

[17] Liu, B., 2010, "Sentiment Analysis and Subjectivity," Handbook of natural language processing, **2**, pp. 627–666.

[18] Hu, M., and Liu, B., 2004, "Mining and Summarizing Customer Reviews," *Proceedings of the 2004 ACM SIGKDD International Conference on Knowledge Discovery and Data Mining - KDD '04*.

[19] Ding, X., Liu, B., and Yu, P. S., 2008, "A Holistic Lexicon-Based Approach to Opinion Mining," *Proceedings of the International Conference on Web Search and Web Data Mining - WSDM '08*.

[20] Chen, L., Qi, L., and Wang, F., 2012, "Comparison of Feature-Level Learning Methods for Mining Online Consumer Reviews," Expert Syst. Appl., **39**(10), pp. 9588–9601.

[21] Kim, Y., 2014, "Convolutional Neural Networks for Sentence Classification," *Proceedings of the 2014 Conference on Empirical Methods in Natural Language Processing (EMNLP)*, Association for Computational Linguistics, Stroudsburg, PA, USA, pp. 1746–1751.

[22] Li, Z., Zhang, Y., Wei, Y., Wu, Y., and Yang, Q., 2017, "End-to-End Adversarial Memory Network for Cross-Domain Sentiment Classification," *Proceedings of the Twenty-Sixth International Joint Conference on Artificial Intelligence.*

[23] Hutto, C. J., and Gilbert, E., 2014, "Vader: A Parsimonious Rule-Based Model for Sentiment Analysis of Social Media Text," *Eighth International Conference on Weblogs and Social Media (ICWSM-14).*

[24] Zhou, F., Ji, Y., and Jiao, R. J., 2012, "Affective and Cognitive Design for Mass Personalization: Status and Prospect," J. Intell. Manuf., **24**(5), pp. 1047–1069.

[25] Kano, N., 1984, "Attractive Quality and Must-Be Quality," Hinshitsu (Quality, The Journal of Japanese Society for Quality Control), **14**, pp. 39–48.

[26] Kim, J., and Han, S. H., 2008, "A Methodology for Developing a Usability Index of Consumer Electronic Products," Int. J. Ind. Ergon., **38**(3-4), pp. 333–345.

[27] Delin, J., Sharoff, S., Lillford, S., and Barnes, C., 2007, "Linguistic Support for Concept Selection Decisions," Artif. Intell. Eng. Des. Anal. Manuf., **21**(02).

[28] Wassenaar, H. J., Chen, W., Cheng, J., and Sudjianto, A., 2005, "Enhancing Discrete Choice Demand Modeling for Decision-Based Design," J. Mech. Des., **127**(4), p. 514.

[29] Zhou, F., Ji, Y., & Jiao, R. J. (2015). "Prospect Theoretic Modeling of Customer Affective - Cognitive Decisions under Uncertainty for User Experience Design".

[30] Xu, Q., Jiao, R. J., Yang, X., Helander, M., Khalid, H. M., and Opperud, A., 2009, "An Analytical Kano Model for Customer Need Analysis," Design Studies, **30**(1), pp. 87–110.

[31] Zhou, F., Qu, X., Helander, M. G., and Jiao, J. (roger), 2011, "Affect Prediction from Physiological Measures via Visual Stimuli," Int. J. Hum. Comput. Stud., **69**(12), pp. 801–819.

[32] Griffiths, T. L., and Steyvers, M., 2004, "Finding Scientific Topics," Proceedings of the National Academy of Sciences, **101**(Supplement 1), pp. 5228–5235.

[33] Kano, S., 2001, "Life Cycle and Creation of Attractive Quality," *Proceedings of the 4th International Quality Management and Organisational Development Conference*, pp. 18–36.

Proceedings of the ASME 2019
International Design Engineering Technical Conferences
and Computers and Information in Engineering Conference
IDETC/CIE2019
August 18-21, 2019, Anaheim, CA, USA

DETC2019-97689

DEEP REINFORCEMENT LEARNING FOR TRANSFER OF CONTROL POLICIES

James D. Cunningham
Computer Science and Engineering
Penn State University
University Park, PA 16802
Email: jdc5549@psu.edu

Simon W. Miller
Applied Research Lab
Penn State University
University Park, PA 16802
Email: swm154@arl.psu.edu

Michael A. Yukish
Applied Research Lab
Penn State University
University Park, PA 16802
Email: may106@arl.psu.edu

Timothy W. Simpson
Engineering Design
and Industrial and Manufacturing
Engineering
Penn State University
University Park, PA 16802
Email: tws8@psu.edu

Conrad S. Tucker
Engineering Design and
and Industrial and Manufacturing
Engineering
Penn State University
University Park, PA 16802
Email: ctucker4@psu.edu

ABSTRACT

We present a form-aware reinforcement learning (RL) method to extend control knowledge from one design form to another, without losing the ability to control the original design. A major challenge in developing control knowledge is the creation of generalized control policies across designs of varying form. Our presented RL policy is form-aware because in addition to receiving dynamic state information about the environment, it also receives states that encode information about the form of the design that is being controlled. In this paper, we investigate the impact of this mixed state space on transfer learning. We present a transfer learning method for extending a control policy to a different design form, while continuing to expose the agent to the original design during the training of the new design. To demonstrate this concept, we present a case study of a multi-rotor aircraft simulation, wherein the designated task is to achieve a stable hover. We show that by introducing form states, an RL agent is able to learn a control policy to achieve the hovering task with both a four rotor and three rotor design at once, whereas without the form states it can only hover with the four rotor design. We also benchmark our method against a test case

that removes the transfer learning component, as well as a test case that removes the continued exposure to the original design to show the value of each of these components. We find that form states, transfer learning, and parallel learning all contribute to a more robust control policy for the new design, and that parallel learning is especially important for maintaining control knowledge of the original design.

1 INTRODUCTION

The design-test cycle is a fundamental building block of how effective problem solving in engineering is understood [1,2]. The testing component of the loop often becomes complicated when a nontrivial control algorithm is required to use the design effectively. Unless a control algorithm for an existing design can be completely transferred to a new design with little to no modifications, time will need to be dedicated to optimizing a control algorithm, which is a costly process [3, 4]. Additionally, while computer aided design methods such as neural networks have demonstrated the ability to automatically generate designs [5], a method for learning control systems to utilize the generated designs is lacking.

Recently, Deep Reinforcement Learning (DRL), which combines traditional Reinforcement Learning (RL) techniques with deep neural networks, has achieved impressive results in complex continuous control tasks, from robotics [6–9] to flight [10–12]. DRL approaches model a control problem as learning a policy for selecting actions, given some state information, in order to maximize some scalar reward signal. Additionally both DRL and deep neural networks in general, have shown the ability to transfer knowledge learned from one task to a new task [13–17]. This motivates the development of a method that employs DRL as a control policy to be able to transfer knowledge to a new design and thus reduce the amount of time spent discovering the optimal control algorithm, including considerations for the particular challenges of design that have been lacking in transfer DRL to this point in the literature.

In RL continuous control applications, the state variables include no information about the form of the object the agent is controlling. When the form changes, the same control actions will map to different state transitions and rewards. Without any information about how the form of the design has changed, we hypothesize that the ability of the RL control policy to generalize to designs of different form is limited. We propose encoding information about the form of the design into the RL state information in addition to the dynamic environment states that change from time-step to time-step, then using the learned control policy for one design as a starting point for a different design. We also propose training the original design and the new design *in parallel* to ensure that the agent does not lose the ability to control the original design, and thus extends its knowledge instead of purely transferring it. This contributions from this work are summarized as follows:

- Introduction of form information into the state space of an RL agent.
- A method of learning a single RL policy that is robust to multiple design configurations, as opposed to learning a separate policy for each.
- A transfer reinforcement learning method that prioritizes preventing loss of control knowledge in the original task using parallel learning.

The remainder of this paper is organized as follows. Section 2 reviews related literature, and Section 3 details the novelty of our method. Section 4 then applies the proposed method to a case study involving multi-rotor aircraft. Section 5 discusses the results before offering conclusions and discussion of future work in Section 6.

2 Related Work

In this section we review related literature in the domain of transfer reinforcement learning. Table 1 compares the features

TABLE 1. Features of the presented method and most closely related literature.

Authors	Form-Aware	Policy Extension	Reward Shaping	Transfer Belief Parameter
Parisotto et al. (2015)				
Barreto et al. (2017)				
Liu et al. (2018)				
This work				

Key: Presence of Feature / Lack of Feature

of the proposed method with features of the most closely related literature that is reviewed in the remainder of this section.

In their survey of transfer learning for reinforcement learning domains, Taylor and Stone [13] define transfer learning as the transfer of knowledge from a source task to a target task. The most similar work in their survey to the proposed method involves cases wherein the state transition function changes in the source task to the target task. This would be analogous to learning a policy to hover with a quadcopter and then transitioning to the task of flying a tricopter without any design configuration states. The reward, state space, and possibly action space (although one could make the choice to reduce the action space to only 3) would stay the same, but the main difference would be how actions taken in a given state map to the next state. The drawback of this approach is a lack of transfer when the change in design is not accounted for in the state space. We use this as a benchmark in our experiments.

Liu and Jin [18] examine the effect of transfer belief and transfer period on transfer learning. Their work frames transfer learning as an expert in one task teaching a student to perform a related task. In their framing, the transfer belief parameter determines the probability that the student will perform the expert's suggested action, and this value linearly decreases over time. The transfer period dictates the time period over which the expert suggests actions to the student. They show that in tasks that are dissimilar, a low transfer belief parameter with a longer transfer period is best. In our work, we do not make any assumptions about the similarity between tasks, and therefore we do not introduce tunable parameters for transfer belief. Instead, we adopt the policy from the original task completely for the new task and adjust the policy from there.

Baretto et al. [19] propose a method for generalized policy improvement between tasks with identical environment dynamics but different rewards. Their method suggests learning *successor features* in the state and action space which correspond to success with respect to multiple reward functions. By learning these features for multiple example reward functions, policies for new reward functions can be inferred. While this is a solution for generalizing a policy to many different tasks, the tasks can only differ in reward function. Our method considers a case in which the reward function is fixed, but the design form changes, enabling us to use a traditional RL framework and only modify the state space.

Parisotto et al. [14] propose a method of transfer learning to generalize to unseen tasks by providing expert guidance in a set of initial tasks (Atari games). They show that these initial policies learn much faster for new tasks. Their goal of generalization is similar to ours; however, we do not provide any expert guidance. Instead, we transfer an initial learned task to a new task while still maintaining performance in the original task. Additionally, because we are presenting a general control structure to designs of varying configuration, there are cases where certain actions cease to be meaningful when transferring to a new task. For example, in the case of transitioning from quadcopter control to tricopter control, manipulating the thrust of the 4th rotor does not have any effect on the tricopter's behavior. Our approach of introducing design states provides robustness in these cases which do not come up in traditional DRL environments such as Atari games used in their study.

Another category of transfer learning is reward shaping [13] which refers to constructing an artificial reward signal to supplement the real reward signal. This is typically a way to inject human knowledge into the training process to provide the agent with intermediate steps in sparse reward environments, which can then be transferred to the desired task. Konidaris and Barto propose a method for automatically shaping the artificial reward signal to avoid having to carefully engineer it manually, by having agents learn which sensory patterns predict reward across tasks [20]. Reward shaping is closely tied to curriculum learning, wherein the agent learns progressively more difficult tasks to work its way up to the ultimately desired task [21–23]. This form of transfer learning has a considerably different purpose than in our proposed method. In our method, we want to generalize policies to different tasks, whereas in curriculum learning and reward shaping, transfer learning is a means-to-an-end of learning a single difficult task. Nonetheless, we still see the benefits of reward shaping in our multi-rotor case study, as the agent is unable to learn the "hard" policy of controlling a tricopter without first learning the "easy" policy of controlling a quadcopter.

Transfer learning has been shown to provide benefits in a wide variety of RL contexts, but as of yet, it has not been constructed to deal with the particular challenges of transferring knowledge between different designs for the same task, which is the problem addressed by our method. As shown in Table 1, the proposed method is form-aware, preserves knowledge between different designs, and benefits from reward shaping.

3 Method

In this section, our method for generalizing a DRL control policy to two different designs is presented. Section 3.1 formalizes the problem and introduces the concept of discrete design states. Section 3.2 describes the policy optimization method and defines Transfer Learning. Section 3.3 discusses how parallel agents are leveraged to extend control knowledge to multiple designs.

3.1 Form-Aware Reinforcement Learning

We consider the standard Markov Decision Process (MDP) formalism [24] of a set of states, $\mathbf{s} \in \mathbf{S}$, a set of possible actions, $\mathbf{a} \in \mathbf{A}$, and a transition function, $T(\mathbf{s}, \mathbf{a}) = PD(\mathbf{S})$ where $PD(\mathbf{S})$ represents a probability distribution over the set $\mathbf{S}$. An agent's task is quantified by a reward function $R(\mathbf{a}, \mathbf{s})$, and the agent's goal is to learn an optimal policy $P(\mathbf{a}) = \pi(\mathbf{a}|\mathbf{s})$, which determines the probability of the agent taking action $\mathbf{a}$ while in state $\mathbf{s}$. In DRL approaches, this policy function is implemented as a deep neural network and is optimized with respect to maximizing the expected sum of discounted reward $E[\sum_{j=0}^{\infty} \gamma^j r_{t+j}]$, where r_{t+j} is the reward for j steps in the future, and $0 \le \gamma < 1$ is a discount factor that controls how much more heavily to weigh near-term rewards than long-term rewards.

We introduce the concept of a design form state, $\mathbf{d} \in \mathbf{D}$, where $\mathbf{D}$ represents the space of design forms that are being explored, and each element of $\mathbf{d}$ is a design variable. In this paper, we limit our discussion to a design space of a discrete configuration of parts with a fixed upper bound of possible configurations. We define a form-aware RL policy as one that considers both the state space $\mathbf{S}$ and the design form space $\mathbf{D}$, and we denote a new state vector $\tilde{\mathbf{s}} = [\mathbf{s}, \mathbf{d}]$. While $\mathbf{s}$ changes over time for a given episode, $\mathbf{d}$ remains fixed, which is an atypical configuration of a state space for an RL agent. This is one of the novel aspects of our proposed method.

3.2 Transfer Learning

We employ Proximal Policy Optimization (PPO) [25], a deep neural network approach to estimate the gradient of the current policy:

$$\hat{g} = E[\nabla_\theta \log \pi_\theta(\mathbf{a}|\tilde{\mathbf{s}})\hat{A}] \qquad (1)$$

where θ denotes the weights of the neural network which defines the policy, and $\hat{A}$ denotes an estimator of the policy advantage function, which is a ratio of the expected sum of discounted rewards of the current policy to that of an older policy. An optimal policy is learned via stochastic gradient ascent of the policy gradient.

When we say that we *transfer* knowledge of one task to another, this means that the policy weights for the new task θ_{new} are initialized to the learned policy weights of the original task θ_{old}. Next, we describe how training is accomplished with multiple agents in parallel.

3.3 Exploiting Parallelism for Generalization

Introducing parallelism into DRL has been shown to generate more robust policies [26, 27]. Every time an agent takes an action, it stores information about what happened into an experience vector $[\tilde{s}, \mathbf{a}, \tilde{s}', r]$ which contains the state it was in, the set of actions it took, the state it transitioned to, and the reward it received as defined in Section 3.1. This experience vector is added to an experience buffer, as shown in Figure 1. When multiple agents are operating in an environment, this experience buffer becomes filled with the most recent experiences agents exploring different state trajectories. A batch of experiences are randomly sampled for every learning update, and having an experience buffer that contains experiences from multiple parallel agents reduces the statistical correlation between the experiences in the batch, and thus increases the robustness of the learned policy.

In this design context, parallel training has additional benefits. Specifically, we hypothesize that in order for effective generalization of a control policy to occur, the agent must continue to be exposed to the original design. Parallel training enables both the new and old designs to be trained at once, thus preventing a loss of control knowledge for the original design. Figure 1 illustrates this concept in terms of the transfer of knowledge from one design concept to another.

Next, we present a case study to demonstrate the proposed method.

4 Case Study

In this section we describe the details of the multi-rotor case study. Section 4.1 specifies how multirotors are modeled in our experiments, Section 4.2 formulates the RL problem with respect to this model and the hovering at a waypoint task, Section 4.3 describes the simulated environment built using Unity real-time engine, and Section 4.4 describes each of the four experiments performed.

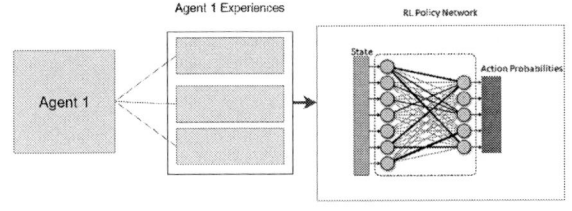

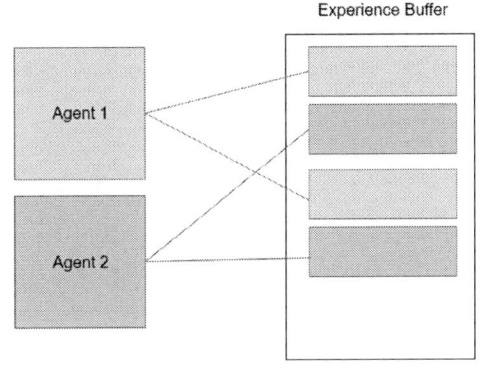

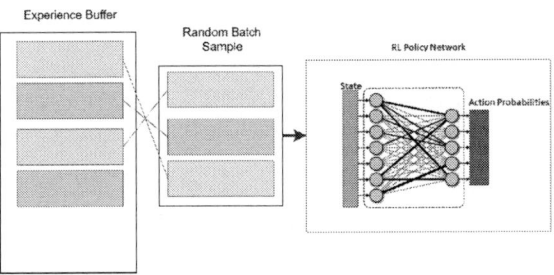

FIGURE 1. (Top) Agent 1 trains on a design configuration. (Bottom) Introduction of Agent 2 learning a new design configuration in parallel.

4.1 Multi-Rotor Model

We consider a multi-rotor unmanned air vehicle as a central body with thrust-providing rotors placed in a subset of up to eight discrete points as shown in Figure 2. All positions are located 0.85m radially from the central body; however, this could also be varied in future implementations. Each rotor is designated as either spinning clockwise or counter-clockwise. Thus, we can completely describe the design form space $\mathbf{D}$ of our multirotor model using an 8-element vector, where each element position corresponds to the rotor position number in Figure 4, and each rotor can take one of the discrete set of values $[-1, 0, 1]$. A value of 1 indicates a clockwise-spinning rotor, a value of -1 indicates a counterclockwise-spinning rotor, and a value of 0 indicates that no rotor exists at that position.

We consider each rotor's orientation to be fixed at a 45 degree tilt towards the body and thus the multi-rotor can only be controlled by varying the thrusts on each of its rotors. Again, this could be allowed to vary in future studies. This simplifying assumption allows us to represent the action space as an 8-dimensional vector, where the value of each position represents the pulse width of an Electronic Speed Controller (ESC) signal, and a higher pulse width corresponds to more thrust being applied to the rotor. The exact thrust and torque from the rotor depends on the choice of the motor and propeller. For our experiments, we used an ESC signal to thrust/torque function based on experiment data collected using 8x6 propellers for a Turnigy 2836 1000kv with HobbyKing 30A ESC 3A UBEC motor [28]. This function can be seen in Figure 3. The ESC pulse width is bounded between 1000 and 2000 μs.

We consider two different designs in our case study: (1) a four-rotor aircraft, which we will refer to as a "quadcopter", and (2) a three-rotor aircraft, which we will refer to as a "tricopter". Figure 4 shows the specific realization of these designs in terms of the design space we have outlined. Translating these configurations into form state vectors yields $\mathbf{d}_{\text{tri}} = [0, 1, 0, -1, 0, 0, -1, 0]$ and $\mathbf{d}_{\text{quad}} = [0, 1, 0, -1, 0, 1, 0, -1]$.

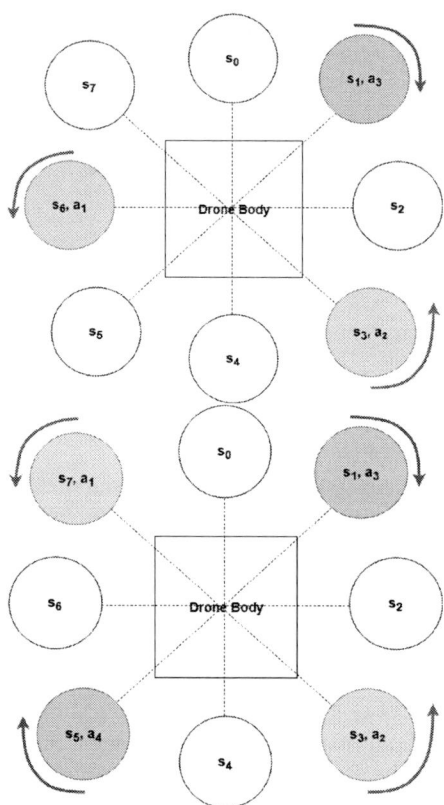

FIGURE 4. (Top) Configuration of tricopter form states and action indexing. (Bottom) Configuration of quadcopter form states and action indexing.

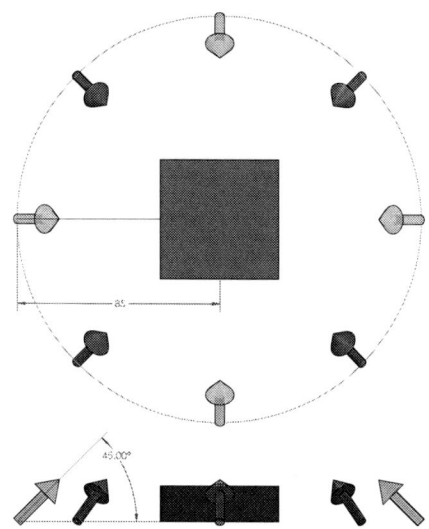

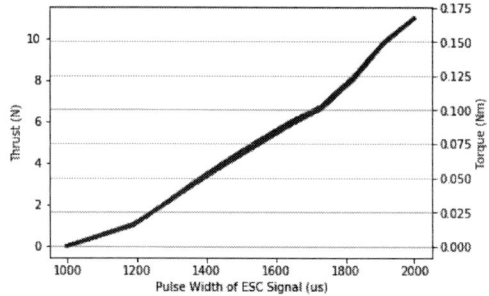

FIGURE 2. (Top) Top-down view of discrete possible rotor positions. (Bottom) Side view of discrete possible rotor positions.

4.2 Reinforcement Learning Formulation

In this section we relate the variables of the DRL formulation discussed in Section 3.1 to the case study of the multi-rotor stable hover at waypoint task.

4.2.1 Action Space As mentioned in Section 4.1, we describe the action space **A** as an 8 dimensional vector corresponding to the amount of thrust to apply to each rotor. However, in order to promote a high degree of transfer between designs,

FIGURE 3. Relationship between output thrust and torque with respect to pulse width of ESC signal used for our experiments. Data provided by [28].

we encode the action to rotor mapping differently than the states. We ensure that for a design with n rotors, the first n actions correspond to existing rotors. Using this paradigm, the first 4 actions for any 4-rotor configuration would correspond to meaningful events in the environment, and when this policy is transitioned to any 3-rotor configuration, then only one of these actions would cease to be meaningful. Figure 4 shows how the action vector maps to the tricopter and quadcopter rotors. Note that while only actions corresponding to existing rotors are shown on the figure, the rest of the actions are still being chosen (actions 4-8 for the tricopter and 5-8 for the quadcopter); however, they have no effect on the resulting behavior and thus we do not map them to any particular empty positions. We choose action indexing such that two of the action-rotor mappings stay consistent between the tricopter and the quadcopter.

4.2.2 Dynamic States

In the previous section, we defined the design configuration state vectors $\mathbf{d}_{\text{tri}}$ and $\mathbf{d}_{\text{quad}}$ for the case study. To form the complete state vector $\tilde{\mathbf{s}}$ we must also define the dynamic system state vector $\mathbf{s}$. We describe the dynamic system for this test case with the following dynamic properties of the multi-rotor body with respect to an inertial frame:

- Waypoint position relative to starting position
- Current position relative to starting position
- Current pitch, roll, and yaw.
- Current velocity
- Current angular velocity
- Current acceleration

These six 3-dimensional vectors are encoded to one 18-dimensional vector that is the dynamic state vector $\mathbf{s}$.

4.2.3 Reward Function

The DRL agent understands the task via the reward function, which maps desirable states to a scalar value. For the task of achieving a stable hover at a waypoint, we consider the following variables:

- g the distance from the body to the waypoint
- g_0 the body's starting distance from the waypoint
- u the projection of the drone's vertical axis onto the world's vertical axis
- ω the angular velocity

In terms of these variables, the reward function for a given timestep is as follows:

$$r_t = \begin{cases} \alpha \tanh(1 - \frac{g}{\lambda}) + \beta u & \text{if } g > 0.5 \\ 10\beta u - \gamma \omega & \text{if } g \leq 0.5 \end{cases} \quad (2)$$

where λ, α, β, and γ are the weights for each of the variables. We choose $\lambda = g_0$ so that when the multi-rotor is at its starting position, the distance component of the reward is 0. We choose all other weights to be 0.001, so that each component of the reward is weighted equally. When the multi-rotor is greater than 0.5m away from the waypoint, then the agent is rewarded for reducing the distance between itself and the waypoint and staying oriented upright. When it is within a radius of 0.5m of the waypoint, then we consider it "at the waypoint" and reward it for maintaining an upright orientation and minimizing its angular velocity. We enclose the distance reward in a *tanh* function to avoid very large negative rewards in cases where the agent flies far away from the waypoint. The reward function is evaluated at every time-step in the simulation environment. Theoretically, the maximum possible reward would be achieved by the agent being within 0.5m of the waypoint, oriented exactly upright, and with zero angular velocity for every time step of the simulation. While this is not possible in practice, it sets an upper bound of the cumulative reward for a single episode of $r \leq 10\beta T$, where T is the total number of time steps of the simulation. Similarly, the lower bound would be the case that $\frac{g}{\lambda} > 2$, and the body is oriented exactly upside-down for every time step. This provides the lower bound of $r \geq -T(\alpha + \beta)$.

4.3 Unity Environment

An RL agent sees the environment as a "black box" to which it sends actions and from which it receives state transitions and rewards. We employ the Unity ML-Agents Toolkit [29], an open source toolkit, for developing RL agents that use simulation environments developed in Unity real-time engine. Unity has grown from an engine focused on videogame development to a flexible general-purpose physics engine, which is used by a large community of developers for a variety of interactive simulations, including high-budget console games and AR/VR experiences [29]. Because of its combination of flexibility and fidelity, Unity is being used as a simulation environment development platform for many DRL research studies [30–32].

The Unity simulation environment only considers rigid body physics; it neglects aerodynamic phenomena. Thrusts on each rotor are calculated from the ESC signal sent from the DRL policy as described in Section 4.1, and these thrusts are oriented directly perpendicular to the rotation plane. The total forces and torques that act on the multi-rotor body are given by:

$$\mathbf{F} = \sum_{i=0}^{N} F_i \hat{\mathbf{n}}_\mathbf{i}$$

$$\mathbf{T} = \sum_{i=0}^{N} \rho_i T_i \hat{\mathbf{n}}_\mathbf{i}$$

where,

- **F** Is the total force vector that acts on the multi-rotor
- **T** Is the total torque vector that acts on the multi-rotor
- F_i Is the scalar force magnitude from a particular rotor's ESC signal
- T_i Is the scalar torque magnitude from a particular rotor's ESC signal
- N Is the number of rotors
- $\hat{\mathbf{n}}_i$ Is the unit vector of the force direction relative to the multi-rotor
- ρ_i Is either -1 for a counter-clockwise rotating rotor, or 1 for a clockwise rotating rotor

It should be noted that given our definition of a multi-rotor, in the case of an odd number of rotors, such as a tricopter, in order for the magnitude of the forces from each rotor to be equal, the magnitude of the torque must be non-zero. This means that while a multi-rotor with an even number of rotors can hover without spinning about its vertical axis, a multi-rotor with an odd number of rotors cannot.

In all experiments, we designate the waypoint starting position to be 2m directly above the multi-rotor starting position.

4.4 Test Cases

Our method combines three components in order to successfully generalize from one design form to another:

- Form information encoded in the state space.
- Using control policy of one design as starting point for another (Transfer Learning).
- Continued exposure to the original design while training the new design (Parallel Training).

In order to demonstrate the necessity of each of these components, we benchmark the proposed method (Test Case 1) against training to fly the tricopter alone (Test Case 2), and against three cases that each remove one of the above components but keep the other two. Specifically, Test Case 3 removes form information from the state space. Test Case 4 removes transfer learning by starting with the task of controlling the quadcopter and the tricopter in parallel. Finally, Test Case 5 removes the continued exposure to the original design by transferring to the task of controlling only the tricopter. The five total experiments we conduct are summarized in Table 2. The "Starting Task" column represents the task that the agent learned with no prior knowledge, and the "Transfer Task" column represents the task that the agent learned when initialized with the learned policy for the starting task.

TABLE 2. The five experiments performed, including the presented method: Test Case 1. The tasks included in the experiments are: Quadcopter Only (QO), Tricopter Only (TO), and Quadcopter and Tricopter in Parallel (QTP).

	Starting Task	Transfer Task	Form States
Test Case 1	QO	QTP	Yes
Test Case 2	TO	N/A	Yes
Test Case 3	QO	QTP	No
Test Case 4	QTP	N/A	Yes
Test Case 5	QO	TO	Yes

5 Results and Discussion

In this section we discuss the results of the our method's ability to generalize from hovering with a quadcopter to hovering with a tricopter. We compare results for each of the four test cases outlined in Section 4.4. Table 4 shows the cumulative reward achieved per episode from the learned policy of each test case. These results were gathered from 80 inference episodes for each configuration and test case. We trained the quadcopter only task for 240,000 steps at which point it was successfully able to hover. For each transfer task, we further trained the agent until step 690,000, at which point the reward of all test cases had converged. In order to directly compare Test Cases 2 and 4, which do not employ transfer learning, we train them for 690,000 steps as well. All experiments were run using a GeForce GTX 1080 GPU and a 12-core Intel i7-8700 CPU. The experiments employ 16 agents training in parallel, and for test cases that train both the tricopter and quadcopter, there are 8 agents training with each. On average, an experiment of 690,000 steps took about 1 hour to complete training.

Table 3 shows hyperparameters used for RL training. The learning rate started at 1e-03 and linearly decayed to 1e-05 over the course of the training. The buffer size indicates how many of the most recent experiences are stored. In cases where quadcopters and tricopters train in parallel, the buffer will consist of 50% quadcopter and 50% tricopter experiences at any given time, because each time step introduces 8 new experiences from each configuration. The batch size determines how many experiences are randomly sampled at once to do an iteration of training. This batch is not guaranteed to have an equal number of quadcopter and tricopter experiences, but because experiences are uniformly sampled, on average an equal number of experiences from each configuration will be sampled. The γ parameter is discussed in Section 3.1 and controls how much more to weigh near-term rewards than long-term rewards.

Given the discussion of the physics of tricopter flight in Section 4.3, they must have a nonzero angular velocity even when hovering at the waypoint, which incurs a reward penalty. This

TABLE 3. Hyperparameters for RL training. Learning rate linearly decayed to $1e-05$ over the course of the training.

Learning Rate	1e-03
Buffer Size	20240
Batch Size	1012
γ	0.99

TABLE 4. Mean and standard deviations of cumulative reward per episode for quadcopters and tricopters in each test case.

	QC Mean	QC SD	TC Mean	TC SD
Test Case 1	1.87	0.0151	0.931	0.0109
Test Case 2	N/A	N/A	0.179	7.06e-05
Test Case 3	1.85	0.0146	0.702	0.117
Test Case 4	1.88	0.00427	0.173	0.0326
Test Case 5	1.225	0.528	0.491	0.347

explains why even in cases where the tricopter learns to hover, it does not achieve as large of a reward as the quadcopter.

We can see that our proposed method achieved the highest reward for the generalization task in Test Case 1, and Figures 5.a and 5.b show that both the quadcopters and tricopters come to a hover at the waypoint. Furthermore, Figure 6 shows consistent differences in the ESC signals sent to rotors 1-3 for tricopters and rotors 1-4 for quadcopters, confirming that the agent is making different decisions based on the form of the design. By comparison, Test Case 2 shows that without the starting point of the quadcopter control policy, the agent fails to hover with the tricopter as shown in Figure 5.c. This result indicates that the transfer learning component is necessary for the agent to learn to fly tricopters at all, but not quadcopters.

TABLE 5. Average Pearson correlation coefficients between the ESC signal time series for quadcopters and tricopters shown in Figures 6-9

	Correlation Coefficient
Test Case 1	0.488
Test Case 2	N/A
Test Case 3	0.224
Test Case 4	0.433
Test Case 5	0.567

Surprisingly, Test Case 3 was the second most successful in terms of reward for the tricopter, and even more successful than Test Case 4. We had expected that without access to states that indicated the form of the design, it would have to learn a "one-size-fits-all" behavior. However, this is not what happened. As

can be seen in Figures 5.d and 5.e, while the quadcopters come to a stable hover at the waypoint, the tricopters display an odd flight pattern of oscillating between the waypoint and the ground. Figure 7 confirms that the control policy is sending different signals to the quadcopters and tricopters, and Table 5 shows that this test case has the lowest correlation between tricopter and quadcopter ESC signals. While this result was surprising to us, we hypothesize that the tricopters are exposed to angular velocities, accelerations, and orientations that the quadcopters are less likely to be exposed to, and because these designs occupied distinct state spaces, a single control policy could be learned that affected each design differently without access to explicit information about the design. The beginning of the episode is the only time when the tricopter and quadcopter must have the same state, and we can see that the ESC signals are very similar for this case. This explanation is consistent with the observed low correlation between quadcopter and tricopter signals, as the agent would need to learn a policy that ensures that the two designs behave very differently to be able to distinguish them without form states.

In Test Case 4, we see a complete failure to hover with the tricopter, the same as Test Case 2, but we find equally good quadcopter results as the methods that started with a policy trained on only quadcopters as shown in Figures 5.f and 5.g. Since this approach saw tricopters as separate from quadcopters from the start, it never applied the quadcopter control algorithm to the tricopter and thus missed out on this vital launching point for tricopter training.

Test Case 5 was the second most successful in learning to hover with tricopters as can be seen in Figures 5.h and 5.i. We can see from Table 4 that it also had the highest standard deviation in reward, and further investigation shows that only 50% of inference experiments ended with the tricopter at the waypoint, which indicates that the policies results in a failure to hover just as often as it results in stable hover. Ultimately, this leads to a lesser reward than the relatively consistent policy of Test Case 3, even though Test Case 3 does not learn to hover. We can also see that the quadcopter behavior became less stable after the tricopter training phase compared to the tricopter performance in every other test case. This was the expected drawback of not continuing to expose the agent to the original design, namely, the exclusive exposure to the new design resulted in changes in the control policy for the new design "leaking" into the policy for the old design. This is confirmed in Figure 9, as we can see that the ESC signals sent to quadcopter rotors for the learned policy resemble a mixture of the signals sent to the tricopter rotors and the signals sent to quadcopter rotors in every other test case. This explains the high correlation between quadcopter and tricopter signals shown in Table 5, because this test case results in the most "tricopter-like" control policy for quadcopters. We can also see that the stability of the tricopter altitude is worse than with the proposed method, which means not only does parallel learning offer the benefit of preserving the knowledge of the old policy, but it also allows the agent to learn a better policy for the new design. This suggests that while the DRL agent was exploring new policies for the tricopter, it lost some information embedded in the quadcopter policy that was valuable for both tasks.

6 Conclusion and Future Work

We have shown a single form-aware DRL controller is able to learn to control multiple design configurations if form state information is provided. We saw that without this form state information, the DRL controller was still able to learn to control multiple configurations, albeit with reduced performance. We have also shown that for a configuration that the controller was not able to learn with no prior knowledge (tricopter), training the controller on an easier configuration for it to control (quadcopter), gave it a starting point from which it was able to learn the more difficult configuration. Finally, we showed that training the new design concept while continuing to expose the agent to the old design produces better outcomes both in terms of preservation of knowledge of the old policy and performance of the new policy.

One limitation of the proposed method is the simplified design space. A discrete configuration of parts in eight possible positions does not capture the typical complexity of a design application. However, this work demonstrates the benefits of control knowledge transfer between designs in a case with several simplifying assumptions. We will relax some of these assumptions in future work. Another limitation is that in this work we chose the starting task to be the design with the simpler control policy and the transfer task to be the difficult control policy in order to benefit from reward shaping. In a design-test loop, the complexity of a design's control policy is not known *a priori*. Therefore, we cannot conclude from the experiments presented in this work that reward shaping would occur naturally in a design-test loop. In future work, we plan to address this limitation by introducing design actions in addition to design states, so that this full design-test loop can be optimized over by the DRL agent. The agent would have the ability to modify the design configuration (e.g., location and orientation of rotors), and spend time optimizing the control policy for this configuration before modifying the design again and generalizing that knowledge to the next design. Finally, we plan to embed our learned control policy onto a chip on a real multi-rotor in order to validate the method in a real-world environment.

ACKNOWLEDGMENT

This research is funded in part by DARPA HR0011-18- 2- 0008. Any opinions, findings, or conclusions found in this paper are those of the authors and do not necessarily reflect the views of the sponsors.

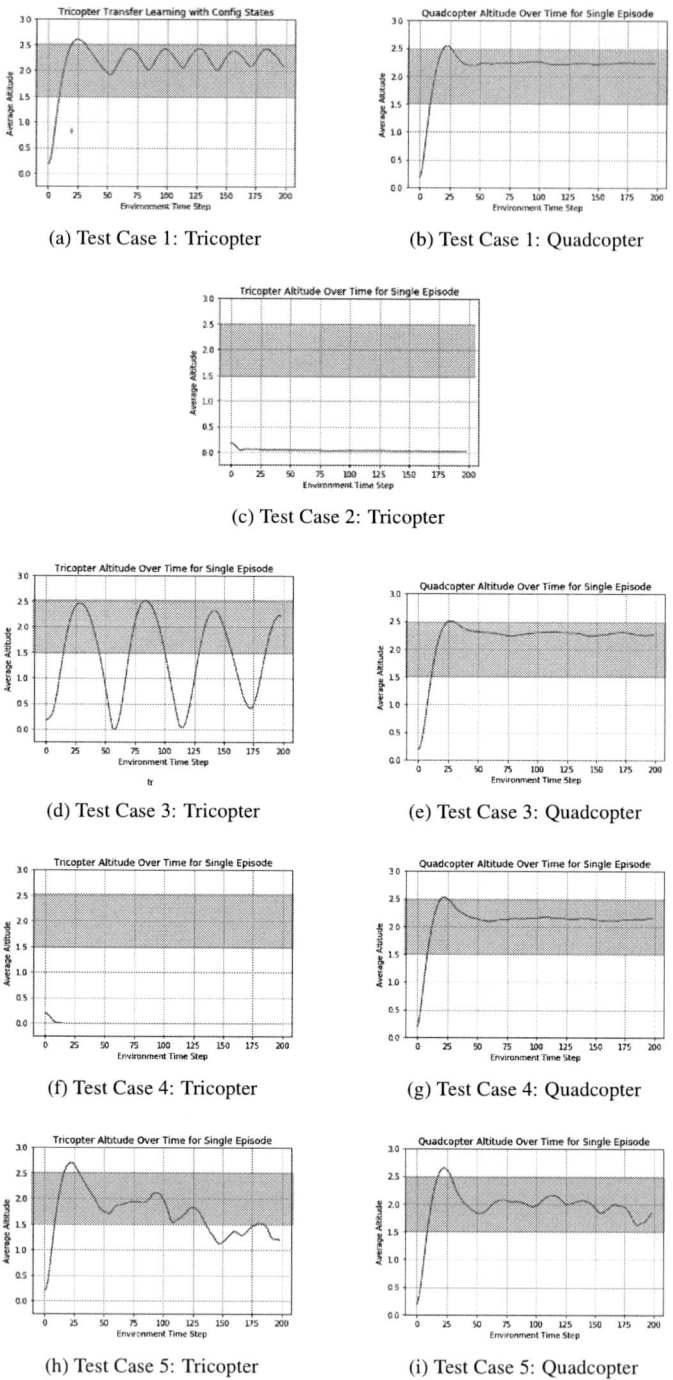

FIGURE 5. The left column shows the mean altitude per time step of all tricopters in the episode for the final learned behavior in each of the four test cases. The right column shows the same for the quadcopters. The range of altitudes that are considered "at the waypoint" are highlighted in green.

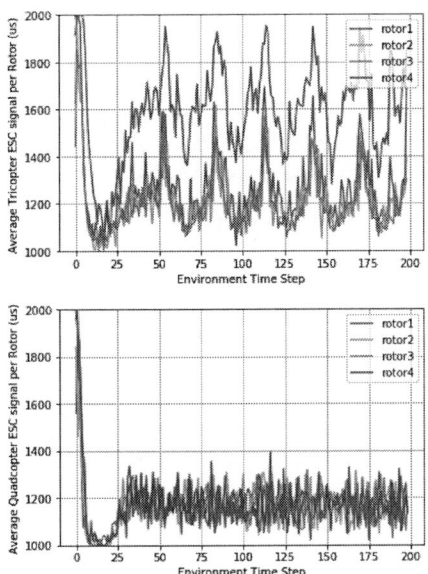

FIGURE 6. Average of ESC signals sent to (Top) tricopters and (Bottom) quadcopters in the episode to rotors 1-4 for test case 1.

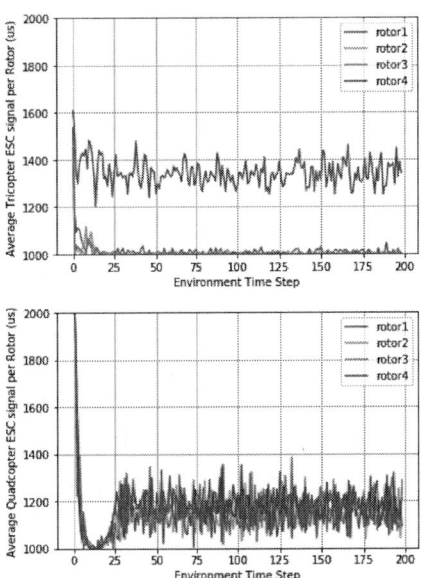

FIGURE 8. Average of ESC signals sent to (Top) tricopters and (Bottom) quadcopters in the episode to rotors 1-4 for test case 4.

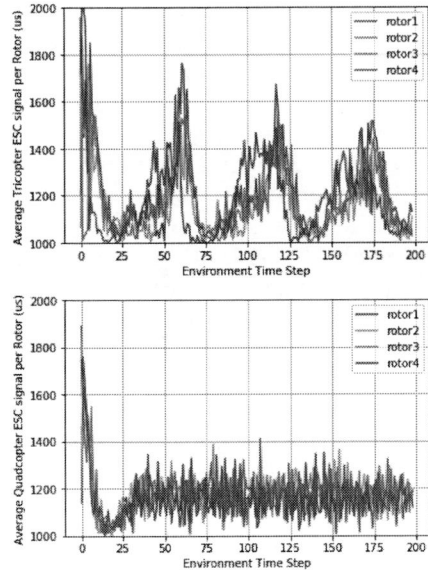

FIGURE 7. Average of ESC signals sent to (Top) tricopters and (Bottom) quadcopters in the episode to rotors 1-4 for test case 3.

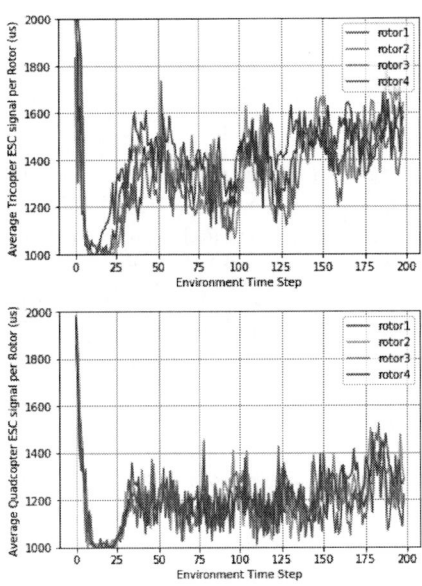

FIGURE 9. Average of ESC signals sent to (Top) tricopters and (Bottom) quadcopters in the episode to rotors 1-4 for test case 5.

REFERENCES

[1] Wheelwright, S. C., and Clark, K. B., 1994. "Accelerating the design-build-test cycle for effective product development". *International Marketing Review,* *11*(1), pp. 32–46.

[2] Hevner, A. R., 2007. "A three cycle view of design science research". *Scandinavian journal of information systems,* *19*(2), p. 4.

[3] Isermann, R., Schaffnit, J., and Sinsel, S., 1999. "Hardware-in-the-loop simulation for the design and testing of engine-control systems". *Control Engineering Practice,* *7*(5), pp. 643–653.

[4] Hanselmann, H., 1996. "Hardware-in-the-loop simulation testing and its integration into a cacsd toolset". In Proceedings of Joint Conference on Control Applications Intelligent Control and Computer Aided Control System Design, IEEE, pp. 152–156.

[5] Dering, M., Cunningham, J., Desai, R., Yukish, M. A., Simpson, T. W., and Tucker, C. S., 2018. "A physics-based virtual environment for enhancing the quality of deep generative designs". In ASME 2018 International Design Engineering Technical Conferences and Computers and Information in Engineering Conference, American Society of Mechanical Engineers, pp. V02AT03A015–V02AT03A015.

[6] Lillicrap, T. P., Hunt, J. J., Pritzel, A., Heess, N., Erez, T., Tassa, Y., Silver, D., and Wierstra, D., 2015. "Continuous control with deep reinforcement learning". *arXiv preprint arXiv:1509.02971.*

[7] Kober, J., Bagnell, J. A., and Peters, J., 2013. "Reinforcement learning in robotics: A survey". *The International Journal of Robotics Research,* *32*(11), pp. 1238–1274.

[8] Gu, S., Holly, E., Lillicrap, T., and Levine, S., 2017. "Deep reinforcement learning for robotic manipulation with asynchronous off-policy updates". In 2017 IEEE International Conference on Robotics and Automation (ICRA), IEEE, pp. 3389–3396.

[9] Zhang, F., Leitner, J., Milford, M., Upcroft, B., and Corke, P., 2015. "Towards vision-based deep reinforcement learning for robotic motion control". *arXiv preprint arXiv:1511.03791.*

[10] Reddy, G., Wong-Ng, J., Celani, A., Sejnowski, T. J., and Vergassola, M., 2018. "Glider soaring via reinforcement learning in the field". *Nature,* *562*(7726), p. 236.

[11] Bansal, S., Akametalu, A. K., Jiang, F. J., Laine, F., and Tomlin, C. J., 2016. "Learning quadrotor dynamics using neural network for flight control". In 2016 IEEE 55th Conference on Decision and Control (CDC), IEEE, pp. 4653–4660.

[12] Zhang, T., Kahn, G., Levine, S., and Abbeel, P., 2016. "Learning deep control policies for autonomous aerial vehicles with mpc-guided policy search". In 2016 IEEE international conference on robotics and automation (ICRA), IEEE, pp. 528–535.

[13] Taylor, M. E., and Stone, P., 2009. "Transfer learning for reinforcement learning domains: A survey". *Journal of Machine Learning Research,* *10*(Jul), pp. 1633–1685.

[14] Parisotto, E., Ba, J. L., and Salakhutdinov, R., 2015. "Actor-mimic: Deep multitask and transfer reinforcement learning". *arXiv preprint arXiv:1511.06342.*

[15] Torrey, L., and Shavlik, J., 2010. "Transfer learning". In *Handbook of Research on Machine Learning Applications and Trends: Algorithms, Methods, and Techniques.* IGI Global, pp. 242–264.

[16] Weiss, K., Khoshgoftaar, T. M., and Wang, D., 2016. "A survey of transfer learning". *Journal of Big Data,* *3*(1), p. 9.

[17] Shin, H.-C., Roth, H. R., Gao, M., Lu, L., Xu, Z., Nogues, I., Yao, J., Mollura, D., and Summers, R. M., 2016. "Deep convolutional neural networks for computer-aided detection: Cnn architectures, dataset characteristics and transfer learning". *IEEE transactions on medical imaging,* *35*(5), pp. 1285–1298.

[18] Liu, X., and Jin, Y., 2018. "Design of transfer reinforcement learning under low task similarity". In ASME 2018 International Design Engineering Technical Conferences and Computers and Information in Engineering Conference, American Society of Mechanical Engineers, pp. V02AT03A061–V02AT03A061.

[19] Barreto, A., Dabney, W., Munos, R., Hunt, J. J., Schaul, T., van Hasselt, H. P., and Silver, D., 2017. "Successor features for transfer in reinforcement learning". In Advances in neural information processing systems, pp. 4055–4065.

[20] Konidaris, G., and Barto, A., 2006. "Autonomous shaping: Knowledge transfer in reinforcement learning". In Proceedings of the 23rd international conference on Machine learning, ACM, pp. 489–496.

[21] Andreas, J., Klein, D., and Levine, S., 2017. "Modular multitask reinforcement learning with policy sketches". In Proceedings of the 34th International Conference on Machine Learning-Volume 70, JMLR. org, pp. 166–175.

[22] Wu, Y., and Tian, Y., 2016. "Training agent for first-person shooter game with actor-critic curriculum learning".

[23] Zaremba, W., and Sutskever, I., 2015. "Reinforcement learning neural turing machines-revised". *arXiv preprint arXiv:1505.00521.*

[24] Littman, M. L., 1994. "Markov games as a framework for multi-agent reinforcement learning". In *Machine learning proceedings 1994.* Elsevier, pp. 157–163.

[25] Schulman, J., Wolski, F., Dhariwal, P., Radford, A., and Klimov, O., 2017. "Proximal policy optimization algorithms". *arXiv preprint arXiv:1707.06347.*

[26] Mnih, V., Badia, A. P., Mirza, M., Graves, A., Lillicrap, T., Harley, T., Silver, D., and Kavukcuoglu, K., 2016. "Asyn-

chronous methods for deep reinforcement learning". In International conference on machine learning, pp. 1928–1937.

[27] Yahya, A., Li, A., Kalakrishnan, M., Chebotar, Y., and Levine, S., 2017. "Collective robot reinforcement learning with distributed asynchronous guided policy search". In 2017 IEEE/RSJ International Conference on Intelligent Robots and Systems (IROS), IEEE, pp. 79–86.

[28] Lariviere-Chartier, Julien , 2019. Turnigy 2836 1000kv with hobbyking 30a esc 3a ubec.

[29] Juliani, A., Berges, V.-P., Vckay, E., Gao, Y., Henry, H., Mattar, M., and Lange, D., 2018. "Unity: A general platform for intelligent agents". *arXiv preprint arXiv:1809.02627.*

[30] Jang, S., and Han, M., 2018. "Combining reward shaping and curriculum learning for training agents with high dimensional continuous action spaces". In 2018 International Conference on Information and Communication Technology Convergence (ICTC), IEEE, pp. 1391–1393.

[31] Namatēvs, I., 2018. "Deep reinforcement learning on hvac control.". *Information Technology & Management Science,* ***21***.

[32] Burda, Y., Edwards, H., Pathak, D., Storkey, A., Darrell, T., and Efros, A. A., 2018. "Large-scale study of curiosity-driven learning". *arXiv preprint arXiv:1808.04355.*

Proceedings of the ASME 2019
International Design Engineering Technical Conferences
and Computers and Information in Engineering Conference
IDETC/CIE2019
August 18-21, 2019, Anaheim, CA, USA

DETC2019-97830

MACHINE LEARNING-AUGMENTED STOCHASTIC SEARCH FOR THE AUTOMATED SYNTHESIS AND OPTIMIZATION OF COOLING CHANNELS

Jonas Schwarz, Kristina Shea[*]
Engineering Design and Computing Laboratory
Department of Mechanical and Process Engineering
ETH Zurich
Switzerland

ABSTRACT

Stochastic search methods are widely used when it comes to design synthesis and optimization of response-based objective functions. In engineering applications, the objective function is typically expensive to evaluate, and stochastic search methods lack efficiency, resulting in the necessity of extensive design evaluations. In order to improve stochastic search methods, we propose a Machine Learning (ML)-based augmentation, consisting of three modules: a design archiver, a data modeler, and a modification advisor. These three modules cooperatively work together to store the gathered data during the design process, build up a representative model of the observations made, and advise the search for further sequences of modifications to apply. The proposed method is benchmarked against its unaugmented parent method in placing cooling channels in a die casting mold. The results show that the efficiency of the method is significantly improved when augmented with ML, i.e. similar results are obtained with 25-50 % fewer evaluations. Additionally, the robustness and reliability of the optimization process is improved with a standard deviation of the obtained results that is 60-85 % smaller. It is shown that the search strategy can be significantly improved with the proposed method, resulting in shorter running times and more reliable convergence behavior.

---

[*]Contact author: kshea@ethz.ch

1 INTRODUCTION

Engineering design is increasingly influenced by the emerging possibilities of computer tools in the design process. CAD and CAE tools are being extended with synthesis and optimization algorithms that automate various design tasks. Additive Manufacturing (AM) intensifies the need for computer tools that aid designer finding and generating optimal shapes and geometries. AM has a unique capability for customization and producing almost any shape at no additional cost [1, 2] but also poses challenging design automation and optimization tasks [3]. In some fields, systems are emerging of such complexity that Computational Design Synthesis (CDS) is essential for success [4]. With the advancements in CDS, it is applied in more diverse areas, such as conceptual design of office buildings [5] and performance-based synthesis of discrete mechanical systems [6]. CDS is a process which enables the computer to automatically generate inventive and optimized solutions. The steps required for CDS are: representation, generation, evaluation and guidance [7–10]. All CDS systems and applications include to some extent all of these four tasks. For engineering design, the computationally expensive part is the evaluation of a candidate design. Complexity increases rapidly with exact modeling and inclusion of all needed constraints making it difficult to quickly move through the design space. It is thus crucial to propose the best possible candidate design in the guidance part of the CDS framework to minimize the required number of expensive function evaluations. This is where the proposed method of this paper pursues improvements.

1.1 CONTRIBUTION AND OUTLINE

The contribution of this work is a machine learning based augmentation for rule-based design synthesis that guides the search in favourable directions based on the knowledge from past rule applications. The augmentation can be used in conjunction with any stochastic search method and consists of three modules (see Fig. 2):

- A *design archiver* keeps track of the rules and sequences of rules applied to the design and their effect on the objective value.
- A *data modeler* uses this information to predict the influence on the objective by building a decision tree to a desired depth.
- A *modification advisor* utilizes the predictions from the data modeler and suggest a new sequence of rules to be applied next.

These components try to improve the search through the use of past knowledge. The general aim of the three components is to identify and exploit search strategies that makes the search more efficient. The technique used is based on a decision tree model of the sequence of rules, where each leaf node is assigned a score based on its probability of improvement and merit of the sequence. The modification advisor then uses these scores to advise the search with a sequence of rules to apply. The work in this paper builds on previous work described in [11] and benchmarks it on the design synthesis problem of cooling channels for die casting molds where it compares a simulated annealing and a greedy search each against its augmented counterpart on two different example geometries.

The remainder of this paper is organized as follows. Section 2 reviews relevant background literature considering rule-based optimization and decision-tree modelling both used in this work as well as the conformal cooling channel problem to which it is applied. An explanation of die casting operation and design potential regarding its cooling channel design is given. In Section 3, the implementation of the ML augmented stochastic search is described generally and later specifically to its application within design generation and optimization of conformal cooling channels. Section 4 presents the results of the application and Section 5 discusses findings suggested by the obtained data. Comparisons between unaugmented and augmented simulated annealing and greedy search with two different design problems are examined. Section 6 concludes this paper.

2 RELATED WORK

This section presents the relevant background of the proposed method focusing on rule-based optimization and decision trees (see Sec. 2.1 and 2.2). An overview of the die casting process and the importance of conformal cooling channel design, to which the proposed method is applied, is given in Sec. 2.3.

2.1 RULE-BASED OPTIMIZATION

Rule-based optimization describes a process in which an initial design, i.e. the starting point of the optimization, is gradually refined by application of rules from a given set. The rules act on a specific design representation, which can have any form, e.g. a set of points, spline control points or geometric features [7]. The current design is modified by changing the value of current design variables, adding additional variables or removing existing variables. Thereafter, the dimensionality of the problem can change its size during the optimization process. The rules are formulated as generally as possible to allow a broad and unbiased exploration of the solution space. Typically, the optimization algorithm seeks to conduct modifications that lead to improved designs after a number of modifications have been made. Various optimization strategies can be applied to rule-based systems. Genetic Algorithms synthesize multiple solutions in an evolutionary manner by genetic modifications such as cross-over and mutation [12]. Another widely used optimization method in CDS is simulated annealing [13], which mimics the annealing of metals where inferior designs are accepted based on a probability that is gradually decreased during the process. Other methods include bursts of sequences of rules [14] or greedy selection [15]. Recent work investigated graph grammars as a subsystem of rule-based optimization for energy-wave converters [16], gearbox design [17] and brachiating robots [18].

2.2 DECISION TREES

Decision making under uncertainty is a broadly investigated field in machine learning. Methods such as Bayesian modelling, influence diagrams and decision trees are widely known and applied to various applications, such as material selection [19], fault diagnosis of engineering applications [20], medical diagnosis [21], energy price classification [22], and design synthesis [11, 23, 24]. In a decision tree, the data is partitioned using a recursive classification procedure. Starting from a root node, the data is partitioned into several split nodes by applying a conditional test on each node to further refine the data until a leaf node is reached. Each node in a decision tree has only one parent and at least two child-nodes. Decision trees have several advantages over other machine learning techniques. In particular decision trees do not have any tuning parameters and are not based on any assumptions about the input data. Additionally, they can handle categorical and numeric data, non-linear relations, and can deal with missing values [25, 26]. Finally, decision trees have an intuitive structure as the classification is carried out in explicit steps. Decision trees can be examined by the designer and easily revised if the outcome is not as desired [27].

2.3 CONFORMAL COOLING CHANNELS

The methods presented in this paper are applied to the problem of the automated design of conformal cooling channels in Die Casting (DC). Complex and lightweight parts made of aluminium or magnesium are often produced in a high-pressure DC process. The mold of DC machines contains the negative of the part to be produced as well as a cooling structure and ejector pins. Typically, the mold is split in half for ease of extraction of the cast part (see Fig. 1). The cost of the produced parts and the quality of the cast aluminium depends on the solidification time and homogeneity. A long known way of improving the solidification time and quality is adding cooling channels that follow the contour of the part. These so-called conformal cooling channels are already being applied in plastic injection molding [28–31]. Current advances in metal AM technologies enable the fabrication of complex cooling channels inside the mold, which can endure the high pressures and temperatures of the DC process. These channels can sufficiently cool the mold omitting the need for spraying coolant directly on the surface of the mold, resulting in drastically reduced cycle times.

2.4 MOTIVATION

The geometric modelling as well as the exact placement of conformal cooling channels pose a challenging design task to the engineer as well as to the CAD system. DC parts consist of many free-form surfaces, holes, ribs and pins, which are difficult to handle and costly to simulate. A single transient heat FEM simulation with cooling circuits takes several hours to solve on state of the art computers and expert knowledge is required to set up such a simulation. This makes it difficult for an engineer to come up with a satisfactory solution. Additionally, the cooling channels need to avoid intersecting the ejector pins and the pressure loss in the tube cannot be too high.

This complex design task with its many requirements gives great opportunity for application of CDS to generate and optimize solutions. For this reason, along with the potential for lightweight parts, the design of conformal cooling channels is selected as a case study for the implementation of this paper's machine learning augmentation for stochastic search.

3 METHOD

In this work, we propose a novel search augmentation with machine learning for rule-based design optimization. According to the CDS framework in [7], this augmentation belongs to the guidance of the search. The proposed augmentation is generic and can be used in conjunction with any rule-based optimization

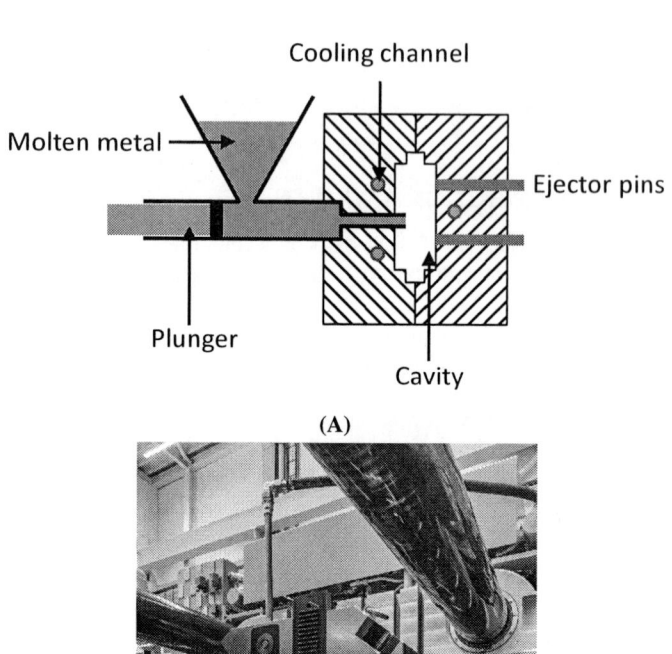

(A)

(B)

FIGURE 1: A SCHEMATIC OVERVIEW (A) AND A REAL (B) DC DIE.

method. In this work it is used to augment Simulated Annealing (SA) and Greedy Search (GS) within a design problem for the generation and placement of conformal cooling channels for die casting molds. After presenting the proposed framework in Sec. 3.1, the three machine learning modules are introduced in detail in Sec. 3.2-3.4. These sections are kept generic. In Sec. 3.5 the specific application to the synthesis and optimization problem for cooling channels in a die casting mold is explained.

3.1 SYNTHESIS AND OPTIMIZATION FRAMEWORK

In this rule-based synthesis framework solutions develop through the successive application of pre-defined modification operators (rules) that act to change the generated design. The search is carried out by generating a new design and evaluating its objective function. This framework can be applied to a wide range of practical scenarios while requiring only minimal effort to define the optimization problem. Within the search, the modification operators are applied to the current design and result in a candidate design. While this process is repeatable and predictable, the effect on the objective value is often unpredictable and hence, a re-evaluation is needed after each design modification.

The flowchart of the proposed method is shown in Fig. 2 and builds on previous work described in [11,14]. The optimization starts with the initial design, which is evaluated and modified through sequential rule applications and re-evaluation. Two different criteria for accepting solutions are used: greedy search and simulated annealing. If the modification is discarded, the best solution from the archive is retrieved. The machine learning augmentation consists of the three modules shown in the darker area and are further explained in the forthcoming subsections.

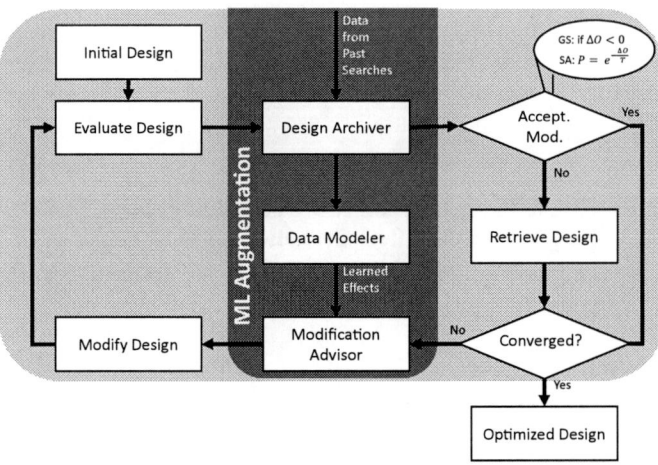

FIGURE 2: FLOW CHART OF THE PROPOSED METHOD BASED ON [11].

3.2 THE DESIGN ARCHIVER

The design archiver keeps track of the changes made and their influence on the objective value. Therefore, it stores the course of the objective value and the rule applied during the process. A snippet of a sample of a typical search is shown in Tab. 1. Referring to the table, the current iteration is 89, the search has just applied rule number 2 to the outcome of iteration 88. It has completed a modification sequence with length two, $< 3 - 2 >$, and a modification sequence with length three, $< 1 - 3 - 2 >$. The change in the objective value for the level three sequence

can be calculated as:

$$\Delta O_{<1-3-2>} = O_{89} - O_{86} \tag{1}$$

where $\Delta O_{<1-3-2>}$ represents the change in the objective value for the sequence $< 1 - 3 - 2 >$ and O_{89} represents the objective value at iteration 89. Depending on the size of the archive, $\Delta O_{<1-3-2>}$ might not be unique, but can occur several times or never. The problem at hand has four rules and a sequence length of two is studied. This results in 16 combinations. To capture the effects of these sequences, a large archive with many occurences of each combination is desired. By keeping a large archive in memory, new trends can not be detected rapidly. The convergence behaviours of the studied problem (see Figs. 6 and 7) show a change in trends roughly around 100 iterations. Thus, all occurrences of each sequence in the last 50 iterations are considered to calculate the merit values.

Arbitrarily long sequences of modifications can be observed with this approach. However, the number of possible sequences increases with r^n, where r is the number of available rules and n the combinatorial level, e.g. the sequence length. With a deeper combinatorial level, there will be fewer observed effects per sequence to support inference of each sequence.

The archive is typically problem specific and built up during the optimization. However, for similar problems, the archive of a past search can be re-used, already providing a wide base of knowledge (exogenous input in Fig. 2). In this work, we focus on building up the archive from scratch and compare combinatorial levels up to two.

3.3 THE DATA MODELER

The data modeler observes the available data and manipulates it in such a way as to extract knowledge to be later used by the modification advisor. This module builds a decision-tree based on the available data of the design archiver.

Each fixed-length sequence is assigned a figure of merit, a score according to the previously observed performance of the sequence. Knowing which sequences are promising and which are not, the modification advisor can guide the search in an efficient manner.

The calculations of the scores is carried out every few iterations as the database of observations is continuously growing throughout the search. In particular, it is of interest to assess which sequences often tend to result in a large improvement of the objective value. Therefore, a score Z containing the probability and merit of improvement concerning the objective value is calculated for each sequence. Z ranges from zero to infinity, where zero indicates complete inability to improve the objective value. The scoring favors distributions that promise a high probability of large, negative change in the objective value. The score corresponds to the maximum possible value of the product of

TABLE 1: SAMPLE OUTPUT OF A DESIGN ARCHIVER.

Iteration	Obj. Value	Applied Rule
0	100	[-]
⋮	⋮	⋮
21	89	1
22	85	3
23	80	2
24	78	2
⋮	⋮	⋮
47	70	2
48	60	1
49	80	3
50	50	2

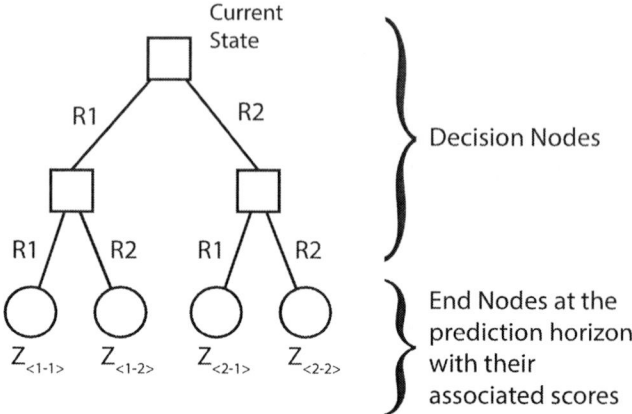

FIGURE 3: DECISION TREE REPRESENTATION OF A SYSTEM WITH TWO RULES AND A PREDICTION HORIZON OF LEVEL TWO.

probability of improvement and *absolute improvement* as shown in Eq. 2.

$$Z = \max\left(I\,P(\Delta O \leq -I)\right) \quad \forall\, I > 0 \qquad (2)$$

Where $P(\Delta O \leq -I)$ is the probability that the sequence in question leads to a decrease of the objective by at least the amount of I, which is calculated as in Eq. 3.

$$P(\Delta O \leq -I) = \frac{n_{\Delta O \leq -I}}{n_{Total}} \qquad (3)$$

The different sequences can be modeled in a decision tree as shown in Fig. 3. It starts at the current state where a series of decision points follow, represented with squares. At each decision point, one of the rules at hand can be applied. The number of these decision nodes correspond to the sequence length. At the end of the sequence are the end nodes, represented with circles. The scores of each sequence are attached to these end nodes. The sequence to apply next is chosen based on the scores at the end nodes and the sequence to apply is recursively assembled backwards from the end node to the root node. The scores can change in the course of the optimization, therefore only the last 50 design points from the design archiver are taken into consideration in this work.

3.4 THE MODIFICATION ADVISOR

Given the knowledge of the design archiver and the model of the data modeler, the modification advisor can advise the search strategy with a sequence of rules to apply next. Therefore, an ordered list of all the scores to a desired sequence depths is generated. The values on this list act as weights for a roulette wheel to choose the next sequence of rules to suggest.

3.5 RULE-BASED DESIGN OPTIMIZATION OF CONFORMAL COOLING CHANNELS FOR DC

The proposed framework is applied to the design generation and optimization of cooling channels for die casting molds. A design space of a square overlapping 20 cm each side of the part to be cooled is defined. Inside, a regular grid of points is defined that act as control points for the cooling channels. That is, cooling channels can be added between two neighboring control points and moved to another control point. An initial design is derived from the temperature distribution of an initial simulation. Four rules are defined to change the set of control points (see Fig. 4):

1. adding a point,
2. removing a point,
3. moving a point away from the part, and
4. moving a point closer to the part.

The first two of these rules change the number of design variables whereas the second two only change one value of a design variable. The rules directly manipulate which control points are active and thus are defined as convective cooled nodes in the FEM formulation. The actual cooling channel design is manually derived from the set of points in a later step.

The generated design is tested by running a thermal FEM simulation and extracting the temperature of the part surface after

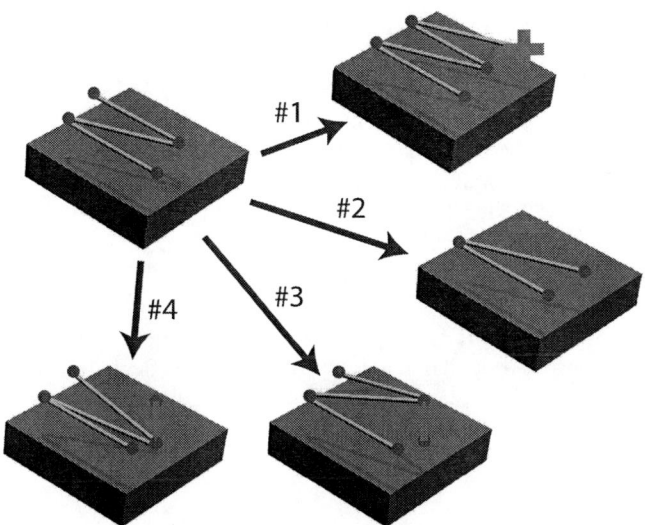

FIGURE 4: SCHEMATIC OVERVIEW OF THE USED RULES FOR CHANGING THE COOLING CHANNEL LAYOUT.

solidification. The objective function is defined as in Eq. 4.

$$O = \int_S \left(T(x) - T^* \right)^2 dS \quad (4)$$

Where S denotes the surface of the part and $T(x)$ the temperature at location x. T^* represents the desired surface temperature of the part right before extraction. In this formulation, the cycle time is considered as a constant which must be provided prior to the optimization. The optimization goal is thus to homogenize the solidification of the part given the cycle time, where the square acts as penalty of too high and too low values.

4 RESULTS

In this section, the performance of the augmentation is benchmarked on two different optimization methods (simulated annealing and a greedy search). For the simulated annealing, we use Ingber's adaptive simulated annealing [32], where an initial temperature of 100 °C, a temperature reduction coefficient of 0.2 and a re-heat iteration limit of 25 is used. Solutions are accepted based on the probability $P = e^{-\frac{\Delta O}{T}}$. In the greedy search, only design changes are accepted that lead to an improvement of the objective function ($\Delta O < 0$), everything else is discarded. The two approaches have a different acceptance criteria (see Fig. 2) but the same set of rules for changing the design. Both, the simulated annealing and the greedy search are augmented by the three modules described in Sec. 3 and are compared to their unaugmented method using two example geometries. The first problem is a simplified geometry with one rib and two pins (further

referred to as geometry A, see Fig. 5A) with a grid of 10x10x10 control points. The second one is a typical die casting part where the geometry is simplified due to confidentiality reasons (further referred to as geometry B, see Fig. 5B) with 10x20x10 control points. The dimensions and geometries of both examples are given in Fig. 5. In order to account for the random nature of the method, each optimization problem was solved 20 times and the mean and Standard Deviations (STD) are reported in the graphs. To give an overview, all results of this section are summarized in Tab. 2.

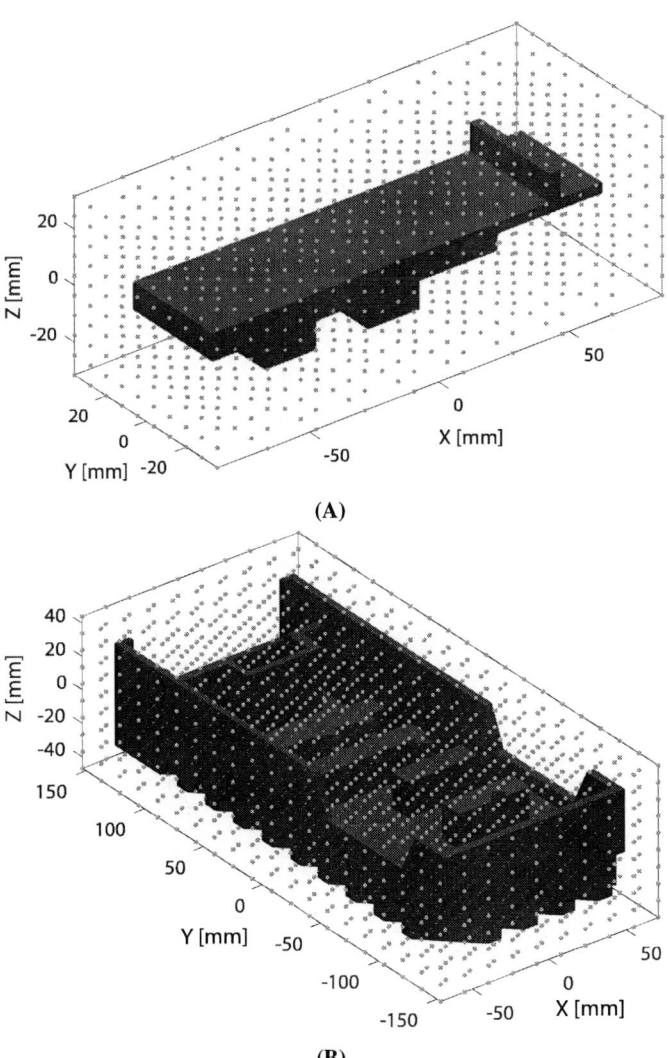

FIGURE 5: DIMENSIONS AND CONTROL POINT LAYOUT OF THE TWO EXAMPLE GEOMETRIES.

The software workflow is as follows. A MATLAB 2018a

script is used to load and generate the model, design space and control points. Given the geometry and control point layout, a FEM mesh is generated using TetGen [33]. The mesh is used for the whole optimization procedure, i.e. no re-meshing is conducted. An initial FEM-Model is generated with ANSYS APDL 19.2 by communication through aaS. The initial simulation is solved and a starting point is derived from the initial temperature distribution. The optimization is then carried out by a generate and test scheme, where the cooling power of the proposed channels is directly mapped to the control points of the FEM model by convection link elements. The cooling channel layout is derived afterwards from the calculated optimal positions. All calculations are carried out on a Xerox E5-1620v3 processor with 32 GB RAM.

Figure 6 shows the convergence history of the unaugmented and augmented greedy search for geometry A. The starting point's objective value is $80.32\,\mathrm{m^2\,^\circ C^2}$. The thick lines show the current mean objective value of the 20 optimization runs while the error bars indicate the standard deviation. The unaugmented greedy search achieves a mean objective value of $4.52\,\mathrm{m^2\,^\circ C^2}$ with a standard deviation of $0.59\,\mathrm{m^2\,^\circ C^2}$, whereas the augmented search achieves a mean objective value of $4.47\,\mathrm{m^2\,^\circ C^2}$ with a standard deviation of $0.96\,\mathrm{m^2\,^\circ C^2}$. The unaugmented greedy search needs 75 iterations to reach an objective value of $6\,\mathrm{m^2\,^\circ C^2}$, whereas the augmented method needs 39 iterations. After 50 iterations, the standard deviation of the unaugmented search was $6.42\,\mathrm{m^2\,^\circ C^2}$ whereas the augmented search results in a standard deviation of $0.96\,\mathrm{m^2\,^\circ C^2}$. Table 2 summarizes all the results of the two example geometries with the different optimization techniques.

Included in Fig. 6 is the temperature distribution on the part surface of the starting point as well as of the final solution. Both temperature distributions are based on the same color-axis displayed on the right. The green color represents the target surface temperature of $T^* = 350\,^\circ\mathrm{C}$.

Figure 7 shows the convergence history of the unaugmented and augmented simulated annealing optimization for geometry B. The starting point's objective value is $2.32 \times 10^4\,\mathrm{m^2\,^\circ C^2}$. The thick lines show the current mean objective value of the 20 runs while the error bars indicate the standard deviation. The unaugmented simulated annealing achieves a mean objective value of $122.3\,\mathrm{m^2\,^\circ C^2}$ with a standard deviation of $33.9\,\mathrm{m^2\,^\circ C^2}$, whereas the augmented SA achieves a mean objective value of $115.2\,\mathrm{m^2\,^\circ C^2}$ with a standard deviation of $37.2\,\mathrm{m^2\,^\circ C^2}$. The unaugmented SA needs 266 iterations to achieve an objective value of $1000\,\mathrm{m^2\,^\circ C^2}$, whereas the augmented method needs 182 iterations. After 200 iterations, the standard deviation of the unaugmented simulated annealing was $525.2\,\mathrm{m^2\,^\circ C^2}$ whereas the augmented search results in a standard deviation of $201.6\,\mathrm{m^2\,^\circ C^2}$.

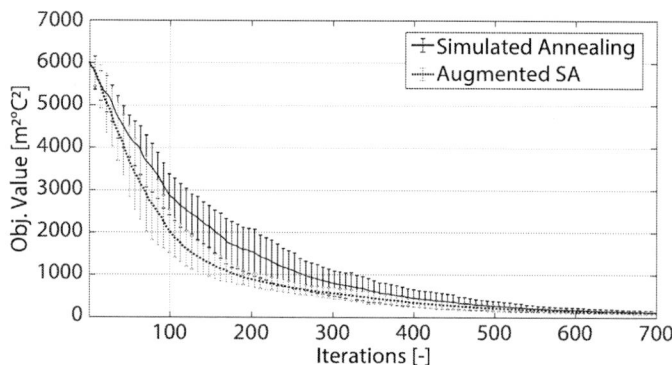

FIGURE 7: CONVERGENCE HISTORY OF THE UNAUGMENTED AND AUGMENTED SIMULATED ANNEALING OPTIMIZATION OF GEOMETRY B

Figure 8 shows the temperature distribution on the part surface of the starting point (Fig. 8A) as well as of the final solution (Fig. 8B). Both temperature distributions are based on the same color-axis displayed below Fig. 8D. The green color represents the target surface temperature of $T^* = 350\,^\circ\mathrm{C}$. Figure 8C shows the resulting control point location and Fig. 8D shows the corresponding cooling channel layout.

The scores of each rule and each sequence of rules up to the prediction horizon two for the geometry A are shown in Fig. 9. Figure 9A shows the scores of the prediction horizon one of the first 50 iterations and the last 50 iterations of the optimization run. Figure 9B shows the scores of the prediction horizon two of the same simulation run again of the first and last 50 iterations. The scores from geometry B have slightly different values but are overall in the same range.

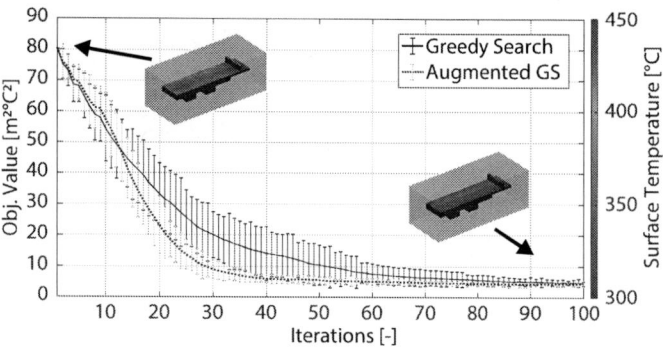

FIGURE 6: CONVERGENCE HISTORY OF THE UNAUGMENTED AND AUGMENTED GREEDY SEARCH OF GEOMETRY A

TABLE 2: OVERVIEW OF STARTING POINTS AND OPTIMIZATION RESULTS OF THE DIFFERENT EXAMPLE GEOMETRIES. ALL LISTED VALUES ARE IN $[M^2{}^\circ C^2]$.

Geometry	Starting Point $[m^2{}^\circ C^2]$	Measure	SA	Aug. SA	GS	Aug. GS
A	80.32	Mean Objective Value $[m^2{}^\circ C^2]$	4.83	4.38	4.52	4.47
		STD $[m^2{}^\circ C^2]$	1.10	0.76	0.59	0.96
		$6\,m^2{}^\circ C^2$ reached after [iter.]	66	47	75	39
		STD after 50 Iterations $[m^2{}^\circ C^2]$	4.00	1.31	6.42	0.96
B	2.32E+04	Mean Objective Value $[m^2{}^\circ C^2]$	122.3	115.2	104.4	101.5
		STD $[m^2{}^\circ C^2]$	33.9	37.2	25.2	29.9
		$1000\,m^2{}^\circ C^2$ reached after [iter.]	266	182	250	184
		STD after 50 Iterations $[m^2{}^\circ C^2]$	525.2	201.6	523.9	164.4

5 DISCUSSION

The discussion of the examples focuses on the results interpretation, the improvement of the computational efficiency and the limitations of the proposed method.

5.1 COMPUTATIONAL EFFICIENCY

The convergence histories of Fig. 6 and 7 show that the number of iterations needed to converge to an optimized solution can be significantly decreased with the proposed method. The unaugmented greedy search needs 75 iterations (mean value of 20 optimization runs) to converge to an objective value of $6\,m^2{}^\circ C^2$ of geometry A, whereas the augmented one only needed 39 iterations (36 % less). Simulated annealing needs 66 iterations to achieve the same objective value whereas the augmented simulated annealing only needs 47 iterations (29 % less). Geometry B shows similar improvements. Here, the unaugmented search reached an objective value of $1000\,m^2{}^\circ C^2$ within 266 iterations with SA and 250 with the greedy search whereas the augmented search managed to reach the same objective value in 182 iterations with the augmented SA (32 % less) and 184 iterations with the augmented greedy search (26 % less).

In the beginning of the optimization, the convergence behavior is similar. Due to an empty or small design archive, no reliable predictions can be made. After around ten iterations, the effect of the augmentation starts to be effective and a much better convergence can be observed for the augmented methods afterwards.

These major improvements of the overall efficiency of the method show, that the learned scores of the rules and sequence of rules turn out to be effective on the convergence speed. Figures 9A and 9B show that in the beginning of the optimization, rule number one and the rule sequence $< 1 - 1 >$ are most effective. This rule and this sequence decreases the overall temperature level as it is too high at the beginning (compare Fig. 6 and 8A). Less dominant but still powerful are the sequences containing rules number one or four, which still reduce the overall temperature level but not as significantly as the sequence $< 1 - 1 >$.

In the course of the optimization, the distributions of the scores change. At the end of the optimization, the scores of the prediction horizon one (see Fig. 9A, left side) are on a more even level. The scores of the prediction horizon two (see Fig. 9B) have changed more drastically. The previously strongest sequence of rules $< 1 - 1 >$ has become one of the weakest, whereas the previous weakest rules have become much more powerful these are: $< 2 - 3 >$, $< 2 - 4 >$, $< 3 - 4 >$, $< 1 - 2 >$ (in descending order). These sequences of rules are a combination between adding/moving closer and removing/moving away. This combination of rules is likely to preserve the overall temperature level, which at this point in the optimization generally fits the target temperature, but might locally improve the temperature distribution. The rules in the sequences $< 1 - 2 >$ and $< 3 - 4 >$ are counteracting. However, the rules in a sequence are most likely not applied to the same control point as the location of the rule application is chosen randomly for each rule. These sequences therefore still result in a new design.

Both convergence behaviors of the greedy search and the simulated annealing compared to their augmented methods show that for both cases, the optimization method convergences faster and more reliable with the augmentation. This shows the importance of rule selection on the search convergence and that the proposed method can be applied to almost any stochastic rule-

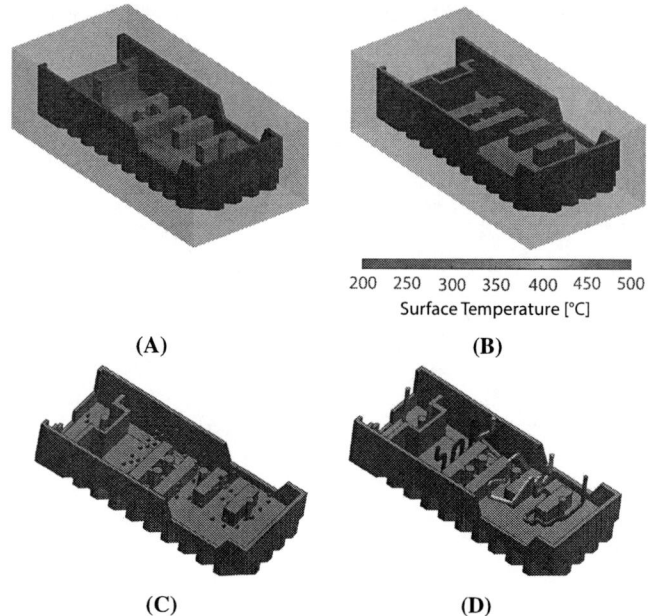

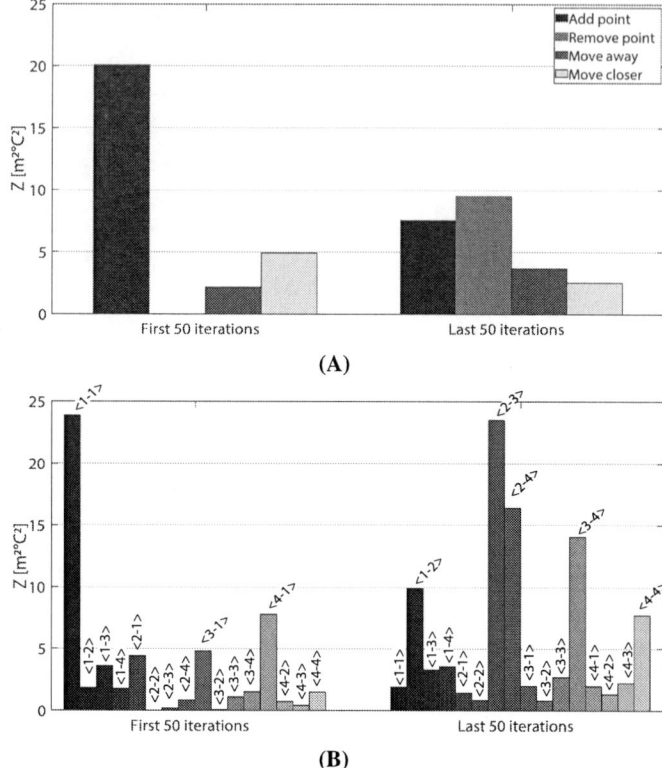

FIGURE 8: SURFACE TEMPERATURE DISTRIBUTION OF GEOMETRY B AT THE STARTING POINT (A) AND AFTER THE OPTIMIZATION (B) AS WELL AS THE RESULTING CONTROL POINTS (C) AND A POSSIBLE COOLING CHANNEL LAYOUT FOLLOWING THE CONTROL POINTS (D)

FIGURE 9: FIRST (A) AND SECOND LEVEL (B) OF RESULTING SCORES DURING THE OPTIMIZATION OF GEOMETRY A.

base optimization method.

5.2 OPTIMALITY

The results show that the method is capable of reducing the iterations needed to converge to an optimized solution. The convergence history of Fig. 6 also shows that the overall deviation of the obtained solution lies within a tighter band than with an unaugmented search. At iteration 50, the standard deviation of the unaugmented greedy search is $6.42\,\mathrm{m^2 {}^\circ C^2}$ whereas that of the augmented search is $0.96\,\mathrm{m^2 {}^\circ C^2}$ (85 % less). These numbers suggest that not only the efficiency but also the repeatability of the method is improved with the proposed method.

Although the results show promising convergence behavior it is not guaranteed that a global optimum is found by either the unaugmented and augmented methods. The stochastic nature of the search methods and the response-based objective function make it almost impossible to judge whether the found solution fulfils the optimality conditions. The comparison of the solutions found by the 20 optimization runs show that the achieved objective function values are similar but the obtained design slightly differs in some areas. It suggests, that the studied problem is non-convex and has many local optima.

The augmentation seems to learn correct effects of rule se-

quences of the shown example for a sequence length up to two even though the model is built up on a small set of observations. This might not be the case for other CDS problems, where a larger rule set is used. A larger rule set also results in exponentially more possible combinations of the rules. Learning all effects of these combinations could require more design evaluations than a traditional meta heuristic approach. Further studies are needed.

5.3 LIMITATIONS AND FUTURE WORK

The results of this work are promising, however further work is needed to study the behavior of the proposed method in other fields and to overcome certain limitations.

The impact of different sequence lengths on the performance of the proposed method has not been deeply studied. Further experiments are needed to tune the best sequence length to a specific problem. An adaptive approach where the sequence length is varied through the search may be beneficial. Additionally, only the last 50 entities of the design archiver are used. This is based on a problem specific trade-off between agility to react to a different behavior and accuracy of the predicted scores. However, its effect is not assessed in detail and so far acts as an input parameter that needs to be tuned for each problem at hand. These studies are beyond the scope of this work and will be addressed in future work.

The magnitude of the application variables of the rules, such as the location of application or the direction of the movement, are assigned randomly in this work. The effect of rule application still has a random effect and thus is hard to predict. Therefore, an extension for intelligent selection of application variables is intended to follow this work.

An optional, external input to the method is the archive of past searches. This might speed up the optimization of similar problems. When the problem differs from the one of which the archive was generated, the additional data is useless. In this work, the effects of re-using an archive are not studied.

The proposed method is studied on the problem of placing cooling channels in a die casting mold. In this case study, it showed promising results. The authors intend to apply the proposed method to other fields and case studies to further generalize and improve it.

6 CONCLUSIONS

Stochastic search is a widespread design synthesis and optimization method used in many fields of application. Typically, engineering design synthesis problems are expensive to evaluate and thus the guidance throughout the search needs to be as efficient as possible. To improve the stochastic search strategy of rule-based optimization, a machine learning-based augmentation of rule selection is proposed in this work. Three modules: a design archiver, a data modeler, and a modification advisor store the arising knowledge, utilize it to build up a representative model and use this model to advise future rule selections and thus design changes. A decision tree is used to capture the model behavior where each leaf node is assigned a score consisting of the probability of success and the amount of merit. The proposed method is applied to the problem of placing cooling channels in a die casting mold, where two different optimization methods and two different example geometries are studied. The results show that the proposed augmentation is capable of not only improving the efficiency of the overall method, i.e. the augmented mehtods converge 26-48% faster, but also in increasing the repeatability of the process, i.e. the standard deviation during the search is 62-85% smaller.

ACKNOWLEDGMENT

The authors acknowledge the support and funding from Bühler AG, Uzwil.

REFERENCES

[1] Gao, W., Zhang, Y., Ramanujan, D., Ramani, K., Chen, Y., Williams, C. B., Wang, C. C., Shin, Y. C., Zhang, S., and Zavattieri, P. D., 2015. "The status, challenges, and future of additive manufacturing in engineering". *Computer-Aided Design, 69*, pp. 65–89.

[2] Thompson, M. K., Moroni, G., Vaneker, T., Fadel, G., Campbell, R. I., Gibson, I., Bernard, A., Schulz, J., Graf, P., Ahuja, B., et al., 2016. "Design for additive manufacturing: Trends, opportunities, considerations, and constraints". *CIRP Annals, 65*(2), pp. 737–760.

[3] Schwarz, J., Chen, T., Shea, K., and Stanković, T., 2018. "Efficient size and shape optimization of truss structures subject to stress and local buckling constraints using sequential linear programming". *Structural and Multidisciplinary Optimization, 58*(1), pp. 171–184.

[4] Gebhardt, A., Schmidt, F.-M., Hötter, J.-S., Sokalla, W., and Sokalla, P., 2010. "Additive manufacturing by selective laser melting the realizer desktop machine and its application for the dental industry". *Physics Procedia, 5*, pp. 543–549.

[5] Grierson, D., and Khajehpour, S., 2002. "Method for conceptual design applied to office buildings". *Journal of Computing in Civil Engineering, 16*(2), pp. 83–103.

[6] Shea, K., and Starling, A. C., 2003. "From discrete structures to mechanical systems: a framework for creating performance-based parametric synthesis tools". In Proceedings of the AAAI 2003 Symposium on Computational Synthesis: From Basic Building Blocks to High Level Functionality, pp. 210–217.

[7] Cagan, J., Campbell, M. I., Finger, S., and Tomiyama, T., 2005. "A framework for computational design synthesis: model and applications". *Journal of Computing and Information Science in Engineering, 5*(3), pp. 171–181.

[8] Chakrabarti, A., Shea, K., Stone, R., Cagan, J., Campbell, M., Hernandez, N. V., and Wood, K. L., 2011. "Computer-based design synthesis research: an overview". *Journal of Computing and Information Science in Engineering, 11*(2), p. 021003.

[9] Stanković, T., Mueller, J., Egan, P., and Shea, K., 2015. "A generalized optimality criteria method for optimization of additively manufactured multimaterial lattice structures". *Journal of Mechanical Design, 137*(11), p. 111405.

[10] Rosen, D. W., 2016. "A review of synthesis methods for additive manufacturing". *Virtual and Physical Prototyping, 11*(4), pp. 305–317.

[11] Vale, C. A., Shea, K., et al., 2003. "A machine learning-based approach to accelerating computational design synthesis". In DS 31: Proceedings of ICED 03, the 14th International Conference on Engineering Design, Stockholm, pp. 183–184.

[12] Bäck, T., Fogel, D., and Michalewicz, Z., 1997. *The Hand-*

book of Evolutionary Computation. CRC Press.

[13] Kirkpatrick, S., Gelatt, C. D., and Vecchi, M. P., 1983. "Optimization by simulated annealing". *Science, 220*(4598), pp. 671–680.

[14] Vale, C., 2002. "Multiobjective dynamic synthesis via machine learning". PhD thesis, University of Cambridge.

[15] Feo, T. A., and Resende, M. G., 1995. "Greedy randomized adaptive search procedures". *Journal of Global Optimization, 6*(2), pp. 109–133.

[16] Puentes, L., McComb, C., and Cagan, J., 2018. "A two-tired grammatical approach for agent-based computational design". In IDETC/CIE, Quebec City, Quebec, Canada, August 26-29, 2017, ASME Paper No. DETC2018-85648.

[17] Königseder, C., and Shea, K., 2016. "Comparing strategies for topologic and parametric rule application in automated computational design synthesis". *Journal of Mechanical Design, 138*(1), pp. 011102–011102–12.

[18] Stöckli, F., and Shea, K., 2017. "Automated synthesis of passive dynamic brachiating robots using a simulation-driven graph grammar method". *Journal of Mechanical Design, 139*(9), p. 092301.

[19] Bickel, B., Bächer, M., Otaduy, M. A., Lee, H. R., Pfister, H., Gross, M., and Matusik, W., 2010. "Design and fabrication of materials with desired deformation behavior". *ACM Trans. on Graphics (Proc. SIGGRAPH), 29*(3).

[20] Abdallah, I., Dertimanis, V., Mylonas, H., Tatsis, K., Chatzi, E., Dervilis, N., Worden, K., and Maguire, E., 2018. "Fault diagnosis of wind turbine structures using decision tree learning algorithms with big data". *Safety and Reliability–Safe Societies in a Changing World*, pp. 3053–3061.

[21] Goodman, K. E., Lessler, J., Cosgrove, S. E., Harris, A. D., Lautenbach, E., Han, J. H., Milstone, A. M., Massey, C. J., and Tamma, P. D., 2016. "A clinical decision tree to predict whether a bacteremic patient is infected with an extended-spectrum β-lactamase–producing organism". *Clinical Infectious Diseases, 63*(7), pp. 896–903.

[22] Reston Filho, J., Affonso, C., and de Oliveira, R., 2015. "Energy price classification in north brazilian market using decision tree". In European Energy Market (EEM), 2015 12th International Conference on the, IEEE, pp. 1–5.

[23] Chen, W., Chiu, K., and Fuge, M., 2019. "Aerodynamic design optimization and shape exploration using generative adversarial networks". In AIAA Scitech 2019 Forum, p. 2351.

[24] McComb, C., 2018. "Toward the rapid design of engineered systems through deep neural networks". In International Conference on-Design Computing and Cognition, Springer, pp. 3–20.

[25] Fayyad, U. M., and Irani, K. B., 1992. "On the handling of continuous-valued attributes in decision tree generation". *Machine Learning, 8*(1), pp. 87–102.

[26] Fayyad, U. M., and Irani, K. B., 1992. "The attribute selection problem in decision tree generation". In AAAI, pp. 104–110.

[27] Friedl, M. A., and Brodley, C. E., 1997. "Decision tree classification of land cover from remotely sensed data". *Remote Sensing of Environment, 61*(3), pp. 399–409.

[28] Park, S., and Kwon, T., 1998. "Optimal cooling system design for the injection molding process". *Polymer Engineering & Science, 38*(9), pp. 1450–1462.

[29] Park, S. J., and Kwon, T. H., 1998. "Optimization method for steady conduction in special geometry using a boundary element method". *International Journal for Numerical Methods in Engineering, 43*(6), pp. 1109–1126.

[30] Sachs, E., Wylonis, E., Allen, S., Cima, M., and Guo, H., 2000. "Production of injection molding tooling with conformal cooling channels using the three dimensional printing process". *Polymer Engineering & Science, 40*(5), pp. 1232–1247.

[31] Huang, J., and Fadel, G. M., 2001. "Bi-objective optimization design of heterogeneous injection mold cooling systems". *Journal of Mechanical Design, 123*(2), pp. 226–239.

[32] Ingber, L., Petraglia, A., Petraglia, M. R., Machado, M. A. S., et al., 2012. "Adaptive simulated annealing". In *Stochastic Global Optimization and its Applications with Fuzzy Adaptive Simulated Annealing*. Springer, pp. 33–62.

[33] Si, H., 2015. "Tetgen, a delaunay-based quality tetrahedral mesh generator". *ACM Transactions on Mathematical Software (TOMS), 41*(2), p. 11.

Proceedings of the ASME 2019
International Design Engineering Technical Conferences
and Computers and Information in Engineering Conference
IDETC/CIE2019
August 18-21, 2019, Anaheim, CA, USA

DETC2019-97975

A NEUROEVOLUTION-BASED LEARNING OF RECIPROCAL MANEUVER FOR COLLISION AVOIDANCE IN QUADCOPTERS UNDER POSE UNCERTAINTIES

Amir Behjat[*]
University at buffalo
Buffalo, New York 14260
Email: amirbehj@buffalo.edu

Krushang Gabani[†]
University at Buffalo
Buffalo, New York, 14260
Email: krushang@buffalo.edu

Souma Chowdhury[‡]
University at Buffalo
Buffalo, New York, 14260
Email: soumacho@buffalo.edu

ABSTRACT

This paper focuses on the idea of energy efficient cooperative collision avoidance between two quadcopters. Two strategies for reciprocal online collision-avoiding actions (i.e., coherent maneuvers without requiring any real-time consensus) are proposed. In the first strategy, UAVs change their speed, while in the second strategy they change their heading to avoid a collision. The avoidance actions are parameterized in terms of the time difference between detecting the collision and starting the maneuver and the amount of speed/heading change. These action parameters are used to generate intermediate way-points, subsequently translated into a minimum snap trajectory, to be executed by a PD controller. For realism, the relative pose of the other UAV, estimated by each UAV (at the point of detection), is considered to be uncertain – thereby presenting substantial challenges to undertaking reciprocal actions. Performing supervised learning based on optimization derived labels (as done in prior work) becomes computationally burden-some under these uncertainties. Instead, an (unsupervised) neuroevolution algorithm, called AGENT, is employed to learn a neural network (NN) model that takes the initial (uncertain) pose as state inputs and maps it to a robust optimal action. In neuroevolution, the NN topology and weights are simultaneously optimized using a special evolutionary process, where the fitness of candidate NNs are evaluated over a set of sample (in this case, various collision)

scenarios. For further computational tractability, a surrogate model is used to estimate the energy consumption and a classifier is used to identify trajectories where the controller fails. The trained neural network shows encouraging performance for collision avoidance over a large variety of unseen scenarios.

NOMENCLATURE

$H(t)$ Unit step function
$\overline{V}$ Average Velocity
$d(t)$ Distance between UAVs over time
ϕ Approach angle
E_A & E_B Energy consumption by UAV A & B
$E_{battery}$ Total Battery Charge
N_m Number of Monte Carlo samples
N_s Total samples of Design of Experiment
$P_{A,1}$ & $P_{B,1}$ Position of UAVs A and B respectively at time t_1
$V_{A,1}$ & $V_{B,1}$ Velocity of UAVs A and B respectively at time t_1
$q_{A,i}, q_{B,i}$ Seventh order polynomials between each consecutive way points in the path
d_{col} Threshold distance of collision
d_{min} Minimum Distance between both UAV in their maneuver
f_D Distance term in objective function
f_E Energy term in objective function
t_1 Time when collision is detected
t_1 Time when decision is taken to avoid collision
t_{col} Time of expected collision
t_{DC} Time to start deviation
t_H Time Horizon

---

[*]Ph.D. Student, Department of Mechanical and Aeronautical Engineering
[†]M.S. Student, Department of Mechanical and Aeronautical Engineering
[‡]Assistant Professor, Department of Mechanical and Aeronautical Engineering. ASME Professional Member. Corresponding Author.

V02AT03A005-1

Copyright © 2019 ASME

t_{min} Time at both UAV will come closest to each other
t_{SC} Time to change velocity
δ_V Velocity change
$\theta_{A,1}$ & $\theta_{B,1}$ Heading angle of UAVs A and B respectively
θ_{DC} Angle to deviation
θ_{col} Angle between the UAVs when they enter into collision distance

INTRODUCTION

Successful introduction of small low-altitude and low-speed unmanned aerial vehicles (UAVs) into the airspace is dependent on the maturity of technologies associated with their operational safety. Avoiding collisions with tall infrastructure, and other small UAVs and manned aircraft sharing the airspace is central to these technologies. It is also important to note that increasing market growth will cause more UAVs to operate in close proximity to each other, thereby increasing the likelihood of UAV-UAV encounters. Thus, unsurprisingly, there is a growing body of work in UAV collision avoidance, mostly in the context of static obstacles or non-cooperative UAV-UAV encounters. However, there is limited understanding of whether mutually reciprocal behavior, e.g., flexible traffic rules, could play a crucial role in guiding the online maneuver decisions in the case of friendly encounters (most commercial/civilian operations arguably falls in this category). This paper focuses on a special kind of cooperative collision avoidance scheme, wherein both UAVs implement the same action model, which ensures mutually reciprocal actions. Accounting for uncertainties within this scheme is particularly difficult, and the primary contribution of the paper lies in developing a novel neuroevolution-derived neural network model for providing robust energy-optimal actions under pose estimation uncertainties. The remainder of this section briefly reviews the literature of UAV collision avoidance and converges on the objectives of the research presented here.

Survey of Collision Avoidance Methods

Existing collision avoidance methods can be broadly classified [1] into: geometric, optimized trajectory, bearing angle, force field and Markov decision process approaches.

Geometric Approach alters the trajectory of the UAVs based on the pose of the ownship and intruder UAV [2]. When a collision is detected [3]), each UAV is given a priority number, and the one with lower priority chooses a way-point that is perpendicular to its velocity. While this method can seek energy-optimal paths, it is not guaranteed that changing the path is the optimal strategy in the first place. Graph search approaches include A* and Dijkstra's algorithms [4], which have been extended from typically dealing with stationary obstacles, to UAV-UAV collision scenarios [5, 6, 7] with limited applicability.

Some of the other online maneuver planning approaches

tend to be computationally expensive. For instance, the Bearing Angle Approach, which uses cameras to estimate the relative angle of the obstacle concerning the UAV and collision is prevented by keeping the obstacle image at a desired safe position in the camera's field of view [8] [9]. Then, there are Force-field approaches, which use attraction/repulsion principles from classical robotics. In some implementations, proximity to another UAV incites a braking force [10], while in other [11], the current states of all UAVs are used to predict the trajectory over a time horizon. Another approach for collision avoidance between UAV is using Markov decision process (MDP) and Partially Observable MDP (POMDP) [12, 13] methods. While POMDPs are potent in providing system-aware optimal and safe actions under uncertainties, the online computing burden can become intractable, and performance is highly sensitive to tunable parameters [12].

Another approach for collision avoidance between UAVs is using reciprocal maneuvers for cooperative UAVs to increase the optimality. Some of the UAV collision avoidance researches are conducted based on this idea [14, 15], but they have limited their maneuvers to acceleration which is not an optimal strategy for many collision scenarios especially for head-on collisions, and it is not necessarily the most energy efficient maneuver.

Research Objectives: Reciprocal Collision Avoidance

In this paper, we build on a new (partly bio-inspired) collision avoidance method. This method uses two different collision avoidance strategies where UAVs can temporarily change its heading (thereby deviating from its original path) or change its speed to avoid a collision. Figure 1 shows the schematic path of a UAV when *speed change* or *direction change* maneuvers are applied. As can be seen from Fig. 1, the actions to be determined by the action model (a neural network system) is encoded in terms of the amount of change in speed or heading angle, and the time of maneuver initiation w.r.t. the collision detection time point. These actions are to be taken in response to estimated pose of the other UAV, where the relative state of the two UAVs can be completely represented with a total of five degrees of freedom (that can serve as inputs to the decision model). The two main quantities of interest that drive the training of the optimum action model include energy consumption of the maneuver, and more importantly the minimum separation experienced during maneuver. The two collision avoidance strategies, introduced previously in [16] with the notion of "complementary maneuvers", applies to encounters between identical multi-rotor UAVs. Here, we aim to expand the applicability of this concept to more realistic collision scenarios, one where uncertainties, primarily pose estimation uncertainties, are ubiquitous.

The pose of a UAV includes (at a minimum) both the location and velocity of the UAV. A standard small multirotor UAV typically gets it pose information from IMU and GPS data. The pose of the other (technically termed "intruder") is typically estimated using sensors such as LIDAR and stereo vision. Due to sensor noise, there is significant uncertainty associated with the estimated pose of the intruder UAV. Rich literature exists on quantifying uncertainty in localization through the use Bayesian principles [17] that predict the future obstacle localization probabilities. A handful of methods exist to deal with uncertainties in the context of UAV collision avoidance.Rathbun et al. [18] used a safe radius approach to deal with uncertain environments. All of these approaches require knowledge of the model unlike the Monte Carlo [19] sampling methods that can handle uncertainty propagation in localization without the explicit knowledge of the model.

In this paper, the Monte Carlo sampling approach is used since the explicit uncertainty in minimum distance and energy is not readily computable. Since such an approach calls for simulating over a large number of scenarios, two notable novel model choices are made in this research – I) a neuroevolution approach is taken to implicitly adapt to uncertainties, as opposed to the original supervised learning approach [16] that would have necessitated prohibitively expensive robust optimizations to generate the labels under uncertainties; II) surrogate models of minimum separation is trained and used to substitute performing the entire control/dynamics in every collision scenario (main computational cost of training) and a classifier is trained to identify and discard state/action pairs where the control would like fail. These, along with the implementation of the design of experiments framework supporting the model training process and subsequent case studies performed to test the framework, collectively constitutes the primary contributions of this paper.Note that neuroevolution, particularly simultaneous evolution of neural network topology and weights, is an emerging machine learning technique used for designing state→action mapping models otherwise designed using reinforcement learning. Further description of the state of the art in neuroevolution and its effectiveness in particularly handling problems of the nature being pursued here can be respectively found in [20, 21, 22].

The remaining portion of this paper is organized as follows: The next section discusses detection of a collision, uncertainty in UAV localization and design of experiment of different uncertain collision scenarios. The section after formulates the framework to avoid a collision which generates a trajectory that is followed by the UAVs using the PD controller. The following three sections respectively describe the surrogate/classifier model training, the evolutionary process of optimizing the neural network topology and weights, and the numerical experiment results. The last two sections present further discussion of results and concluding remarks.

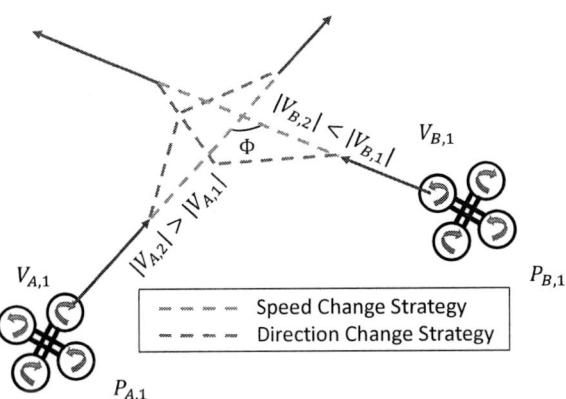

FIGURE 1. Collision avoidance scenario illustrating the two types of avoidance actions (The UAVs actually take smooth trajectories guided by these actions)

SCENARIO DESIGN
The UAV System:

The UAV-UAV encounter scenarios and the autonomous UAV system considered here are in part motivated by the broader research in the area of multiple performing collaborative search [23] and mapping [24] applications (and thus closely sharing the airspace). Specifically, identical quadcopter UAVs are assumed with the key specifications summarized in Table 4. Similar to most other reported work in DSA: i) the motion of the UAVs are assumed to be in a 2D plane; ii) a calm flying environment with no gusts is considered, and iii) only one-on-one collisions are considered. We also assume that each UAV can either communicate its current state (location, speed, heading direction) with its neighboring UAVs or can accurately sense the state of neighboring UAVs (where, practically, sensing range $>>$ minimum separation required). Hence, only deterministic scenarios are considered; and each UAV, based on its current state and the information it has about its peer's state, can predict the possibility and time of the collision with the other UAV.

Inter-UAV Collision: Scenarios and Detection

If the separation between two UAVs becomes lower than a threshold distant d_{col}, then it is termed as a collision. The value of d_{col} can be regulated based on the desired level of safety and can be defined in terms of the UAV size (Here, $3 \times$ diameter of UAV).

As shown in Fig . 1, let $P_{A,1}$ and $P_{B,1}$ be the current phase of UAVs A and B respectively at a given time point (t_1), and $V_{A,1}$ and $V_{B,1}$ are their respective velocity vectors.

The time point at which the two UAVs will come closest to each other can then be estimated as:

$$t_{\min} = \arg \min_{t \in [t_1, t_H)} d(t) \qquad (1)$$

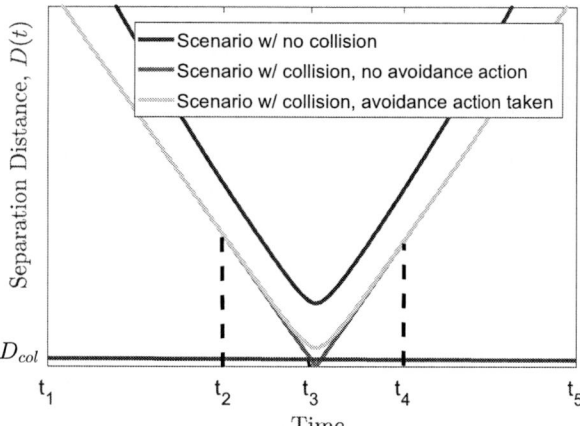

FIGURE 2. Separation distance for illustrative cases with no collision, with collision, and successful collision avoidance. Time points are t_1: collision-evasion action decided; t_2: evasion maneuver starts; t_3: symmetry point of evasion maneuver ($t_3 = t_{col}$); t_4: evasion maneuver ends; and $t_5 = t_H$.

If a solution, $t_H > t_{min} > t_1$, exists within the time horizon, and $d(t_{min}) < d_{col}$, then a collision event is said to have been detected within the time horizon. In that case, at the time point of collision is given by:

$$t_{col} = \arg\min_{t \in (t_1, t_{min})} t, \quad \text{s.t. } d(t) \leq d_{col} \quad (2)$$

Since the maneuver cannot start before t_1, each UAV is then expected to start a collision avoidance maneuver at a designed time, $t_2 : t_1 < t_2 < t_1 + \mu(t_{col} - t_1), \ 0 < \mu \leq 1$. The value of μ can be decided based on desired safety tolerance.

The variation of the inter-UAV separation for generic collision/avoidance scenarios is illustrated In Fig 2, where $t_1 < t_i < t_H, \ \forall \ i = 1, \ldots, 5$. In this figure, the blue curve shows the separation distance between two UAVs for a scenario where no collision is predicted within the time horizon. The red curve shows the separation distance for a scenario where collision is predicted, and the separation distance curve intersects the collision threshold, d_{col} (yellow line), at time $t_3 = t_{col}$ – i.e., collision occurs. The green curve shows the separation distance for the scenario where a collision is predicted (originally similar to the red scenario), but an avoidance action is taken to evade collision successfully. Here 't_2' and 't_4' respectively represent the decided time points at which the collision avoidance maneuver begins and ends.

Uncertainty in UAV Localization

In this paper, uncertainty is considered in the local reference frame to take decision for collision avoidance. The GPS and IMU estimate the position of ownship UAV with uncertainty. The position of intruder UAV is estimated with respect to ownship UAV position using vision sensors. Therefore the error in ownship UAV position will be added to intruder UAV position error. Because of the only relative position is important, it is possible to remove this common error term from the position of ownship and intruder UAVs. There will be no uncertainty in ownship UAV position and for the intruder UAV, only the error due to vision sensor remains. The uncertainty in the position and velocity of the intruder are used as a normal distribution with zero mean and 0.05 [25] and 0.01 [25] standard deviation respectively while the uncertainty in velocity of own-ship UAV is considered as a normal distribution with zero mean and 0.005 [18] standard deviation.

Design of Experiment

In order to train the system to work in different conditions. The DoE includes the initial position and velocity of UAVs. The first step in DoE is to define the time and distance for collision. Instead of using the initial poses of the two UAVs to represent the input space of the DoE, the following parameters are used to provide a closed form representation of the relative state of the two UAVs (UAV-A and UAV-B) at time point t_1:

Angle of the vector P_{AB} when the separation distance becomes smaller than the threshold $\rightarrow \theta_{col}$;
Heading angle of UAV-A $\rightarrow \theta_{A,1}$;
Heading angle of UAV-B $\rightarrow \theta_{B,1}$;
Speed of UAV-A $\rightarrow |V_{A,1}|$
Speed of UAV-B $\rightarrow |V_{B,1}|$.

The heading angle and velocity inputs are all specified in terms of a global coordinate system. Table 1 lists the upper and lower bounds for these inputs.

An inverse approach is taken to perform the DoE, starting with points where the UAVs are separated by a distance of d_{col}, the specified separation threshold. The points $P_{A,col}$ and $P_{B,col}$ are respectively the locations of UAV-A and UAV-B at $t = t_{col}$, if the UAVs had continued on their original path with no avoidance maneuver. The angles $\theta_{A,1}$ and $\theta_{B,1}$ are chosen in a manner that guarantees the separation distance, $d(t < t_{col}) > d_{col}$. The initial speeds are then chosen randomly in a range that is practical for UAVs. Figure 3 illustrates the DoE procedure.

The origin (0,0) of the global coordinate system is considered to be at the mid-point of the vector P_{AB} connecting UAV-A and UAV-B at $t = t_{col}$. Then, for each sample scenario, using the input parameter θ_{col} and the prescribed minimum separation threshold of d_{col}, the position of the two UAVs at t_{col} can be expressed as:

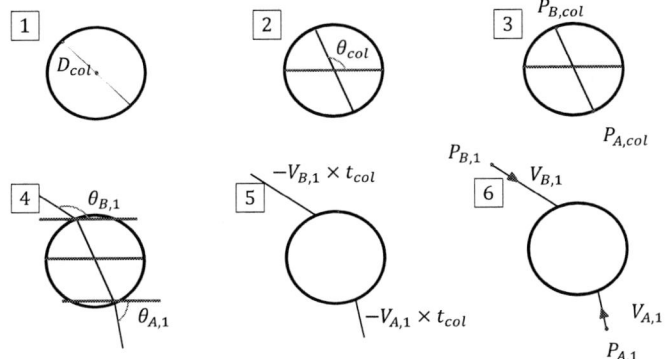

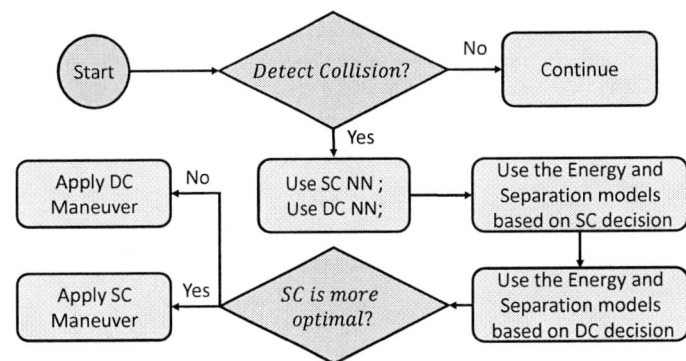

FIGURE 3. DoE process; 1: set global origin: midway between uavs at $t = t_{col}$. 2: find separation vector P_{ab} making angle θ_{col} with global x-axis. 3: determine the uav positions at t_{col}. 4: identify UAV heading directions over the time t_1 to t_{col}. 5: identify the UAV velocities $(V_{A,1}, V_{B,1})$ over this time. 6: estimate the UAVs' initial positions $(P_{A,1}, P_{B,1})$.

FIGURE 4. Collision avoidance framework

where

$$V_{A,1} = (|V_A|\cos(\theta_{A,1}), |V_A|\sin(\theta_{A,1})) \qquad (7)$$
$$V_{B,1} = (|V_B|\cos(\theta_{B,1}), |V_B|\sin(\theta_{B,1})) \qquad (8)$$

and where $\theta_A, \theta_B, |V_A|$, and $|V_B|$ are input parameters determined by the sampling.

COLLISION AVOIDANCE FRAMEWORK

The total decision framework includes two prediction models (which are neural networks in this research) which will be used later to find the optimal decision.

After training each prediction model, we use the validation tests to find the optimal strategy for each input.

In final application, the trained neural networks will be used to find the optimal strategy for each scenarios. Then the minimum distance model and surrogate models of energy consumption which are explained in "Low-Fidelity Modeling" section will be used to find the energy consumption and minimum distance for each model. These models are so fast to use. Therefore, the best decision between only these two choices can be taken online.

Figure 4 explains the complete framework.

Collision avoidance strategies

Two different strategies to avoid a collision are utilized in this paper.

Speed Change (SC) Strategy : In this strategy, two UAVs accelerate and decelerate with the same amount but with the reverse order. While the faster UAV initially decelerates from t_2 to $t_3 = t_{col}$, the slower UAV initially accelerates. After reaching the collision point they follow reverse actions (faster UAV accelerates t_3 to t_4, and slower UAV decelerates during this time period) in a way that at t_5 they are in their original desired position and speed as illustrated in figure 1. The SC strategy is defined in

TABLE 1. Range of the input parameters in the DoE

Parameter	Lower bound	Upper bound
d_{col}	2 m	5 m
t_{col}	2 s	5 s
θ_{col}	0°	360°
$\theta_{A,1}$	$\theta_{col} - 90°$	$\theta_{col} + 90°$
$\theta_{B,1}$	$\theta_{col} + 90°$	$\theta_{col} + 270°$
$|V_{A,1}|$	0.1 m/s	V_{max}
$|V_{B,1}|$	0.1 m/s	V_{max}

$$P_{A,col} = \left(-\frac{d_{col}}{2}\cos(\theta_{col}),\ -\frac{d_{col}}{2}\sin(\theta_{col}) \right) \qquad (3)$$

$$P_{B,col} = \left(+\frac{d_{col}}{2}\cos(\theta_{col}),\ +\frac{d_{col}}{2}\sin(\theta_{col}) \right) \qquad (4)$$

Then, assuming the situation where no maneuver action is taken (i.e., the UAVs had continued on their original paths), the initial position of the two UAVs at t_1 are given by:

$$P_{A,1} = P_{A,col} - V_{A,1}.t_{col} \qquad (5)$$
$$P_{B,1} = P_{B,col} - V_{B,1}.t_{col} \qquad (6)$$

terms of the average change in speed, δ_V, and the time point, t_2 when the SC maneuver initiates. The time point is the same for both UAVs, and the magnitude of acceleration and deceleration is also equal. The average velocity of the UAVs during the avoidance maneuver, i.e., between time points t_2 to t_3 and t_3 to t_4 are given by:

$$\overline{V}_{A,2-3} = (|V_{A,1}| - \delta_V) \cdot \frac{V_{A,1}}{|V_{A,1}|}$$
$$\overline{V}_{B,2-3} = (|V_{B,1}| + \delta_V) \cdot \frac{V_{B,1}}{|V_{B,1}|} \qquad (9)$$

$$\overline{V}_{A,3-4} = \frac{V_{A,1}(t_4 - t_2) - V_{A,2-3}(t_3 - t_2)}{t_4 - t_3}$$
$$\overline{V}_{B,3-4} = \frac{V_{B,1}(t_4 - t_2) - V_{B,2-3}(t_3 - t_2)}{t_4 - t_3} \qquad (10)$$

The maximum time to initiate the strategy is 3 seconds, and the maximum change of speed is 5 m/s.

Direction Change (DC) Strategy : In this strategy, both UAVs deviate from their original path, and then they return to their original path. In order to have a mutually coherent trajectory planning for UAVs, we make both UAVs deviate to their left side (Counter clockwise turn) at t_2. At time point $t_3 = t_{col}$, UAVs start to return to their original path and they plan to be in their original path at t_4 (clockwise turn). A smooth trajectory is planned to execute the maneuver and UAVs plan to return to their original desired position and velocity at t_5. Figure 5 provides a representative illustration of a DC maneuver, showing both the planned and the controller executed trajectory of the UAVs.

Both strategies are designed in a way that the maneuvers do not delay (or advance) entire operation after these maneuvers are finished which is vital for sensitive missions like [24] where entire operation schedule does not have flexibility.

Because the deviated path is longer than the original path, both UAVs need to increase their average speed to preserve their pre-planned (time encoded) mission paths following the maneuver. Their average velocity during the DC maneuver can be expressed as:

$$\overline{V}_{A,2-3} = \frac{1}{\cos\phi} R \, V_{A,1}$$
$$\overline{V}_{B,2-3} = \frac{1}{\cos\phi} R \, V_{B,1} \qquad (11)$$

where R is the standard rotation matrix in 2D space. The maximum allowed time to initiate the strategy (from the point of collision detection) is 3 seconds, and the maximum change of speed is $30°$.

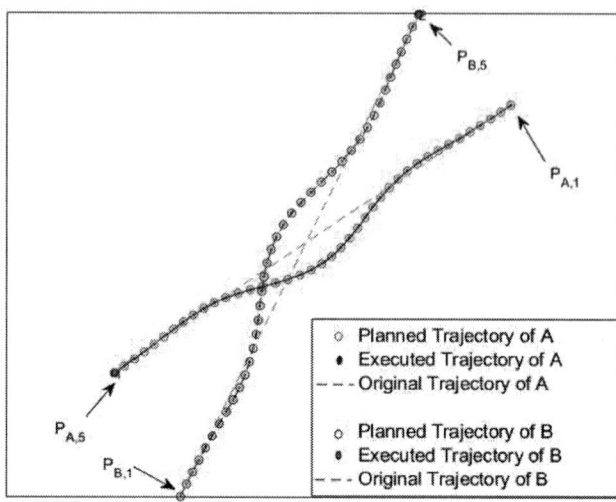

FIGURE 5. Direction change (DC) strategy: illustrating planned and executed trajectories under this strategy

Trajectory Planning and Control

The minimum snap trajectory planning approach [26] is used here. This method uses a seventh order polynomial (spline) to fit the sequence of setpoints or waypoints estimated from the parameters of maneuver decided by the neural network (e.g., deviation angle in DC). These polynomials produce the time stamped position command, which is to be followed by the PD controller.

LOW FIDELITY MODELING

The proposed problem needs a simulation that is relatively expensive to run. In order to decrease the computational expense, a prediction model is trained for each strategy to get the energy instead of running the simulation. Also instead of using the simulation minimum distance is calculated using the geometrical model.

Minimum Distance Model

The minimum distance between two UAVs can be calculated using the simulation model, but running the dynamics simulation is expensive (in the context of the large number of function evaluations needed to train). In order to decrease the computational cost, the trajectory points are used to estimate the minimum dis-

TABLE 2. Parameters of classifier for controllable decision

Parameter	Direction Change	Speed Change
Input	V_1, t_{DC}, θ_{DC}	V_1, t_{SC}, δ_V
Total Samples	1000	1000
Training Algorithm	SVM (Cubic)	Bagged Trees
Cross Validation	10-fold	10-fold
Misclassification	0.132	0.139

tance between the UAVs. Equation 12 explains this process

$$P_A(t) = \sum_{i=1}^{4} q_{A,i}(t) H(t - t_i) H(t_{i+1} - t)$$

$$P_B(t) = \sum_{i=1}^{4} q_{B,i}(t) H(t - t_i) H(t_{i+1} - t) \qquad (12)$$

$$D(t) = |P_A(t) - P_B(t)|$$

Here $q_{A,i}, q_{B,i}$ are the $7^t h$ order polynomials between each consecutive waypoints in the path and $H(t)$ is the unit step function. The minimum point is estimated by computing the roots of $\frac{\partial D(t)}{\partial t}$.

Classification of Actions based on Controller Failure

Using trajectory to find the minimum distance decreases the computational cost because generating trajectories is less expensive compared to simulating the dynamics. But it can cause an error if the controller is unable to keep the UAV in its desired path. Therefore it is essential to find the cases where the controller is unable to keep the UAV in the desired path. In order to avoid such cases, a classifier will be trained to find these cases. The minimum distance is calculated using the trajectories will have negligible error compared to the error due to uncertainty and therefore this method can be applied in optimization.

The properties of trained classifier are summarized in table 2.

Energy Surrogate Modeling

Similar to the minimum distance, energy consumption can be calculated using the simulation. In order to decrease the computation cost, a prediction model is used for each strategy to estimate the energy consumption of each UAV in each strategy. To increase the accuracy of the model, 1500 samples for Speed change strategy and 1500 samples for Direction change strategy were used to train these models. The properties of trained surrogate models are summarized in table 3.

TABLE 3. Parameters of surrogate model for energy consumption of UAV

Parameter	Direction Change	Speed Change
Input	V_1, t_{DC}, θ_{DC}	V_1, t_{SC}, δ_V
Mean of output	858.49 J	845.77 J
Total Samples	1500	1500
Test Samples	225	225
Training Algorithm	Bayesian Regularization	
Hidden Layers	10	10
RMSE	8.54	4.89

OPTIMIZATION
Optimization Problem Definition

In order to find the optimal decision, a neural network is trained to generate the optimal maneuver decision.

The input to the model includes the initial state parameters of the UAVs, and the output will be the best decision. Equation 13 indicates the required input and the expected output of this model. In order to make the neural network training easier, two models are trained for both Speed Change (SC) and Direction Change (DC) strategies.

$$[t_{SC}, \delta_V] = f_{SC}(P_{A,1}, P_{B,1}, V_{A,1}, V_{B,1})$$
$$[t_{DC}, \theta_{DC}] = f_{DC}(P_{A,1}, P_{B,1}, V_{A,1}, V_{B,1}) \qquad (13)$$

In equation 13, f indicates the neural network model. In this equation each term is associated with a 2×1 vector. Although using all 8 inputs is possible in order to decrease the dimension of the input, the following 5 variables, based on 14, were used as input to the model. Not that because the DoE can be explained with 5 variables, these variables are sufficient to define every condition.

$$[t_{SC}, \delta_V] = f_{SC}(\Delta P_{X,1}, \Delta P_{Y,1}, \Delta V_{X,1}, V_{A,Y,1}, V_{B,Y,1})$$
$$[t_{DC}, \theta_{DC}] = f_{DC}(\Delta P_{X,1}, \Delta P_{Y,1}, \Delta V_{X,1}, V_{A,Y,1}, V_{B,Y,1}) \qquad (14)$$

Where $\Delta P_{X,1}$ is $P_{A,X,1} - P_{B,X,1}$. Also $\Delta P_{Y,1}$ and $\Delta V_{X,1}$ are initial relative position in Y and initial velocity in X directions.

In order to have performance robust to uncertainty effects, the UAVs are trained with a Monte Carlo sampling to face different samples with noise in the measurement. The optimal decision is the decision is not only optimal but also robust to the noise.

TABLE 4. UAV Specifiactions

Parameters	Value
Quadcopter UAV structure	Plus Shape
UAV Weight	1 kg
UAV Max speed	15 m/s
UAV Max Thrust	8×Weight
UAV size (diameter)	0.5 m
Safe separation distance (d_{col})	3×size (1.5 m)
Detection-to-collision time (t_{col})	5 seconds
UAV Battery Capacity	5100 (mAh)
UAV Battery Voltage	11.1 (V)

The objective function, which is the same for both strategies, is defined as below:

Maximize:

$$f = 1 - (f_E + f_D)$$
$$f_E = \frac{\sum_{i=1}^{N_s} \sum_{j=1}^{N_M} (E_{A,i,j} + E_{B,i,j})}{2 \times E_{Battery} \times N_s \times N_M} \qquad (15)$$
$$f_D = \frac{\sum_{i=1}^{N_s} \sum_{j=1}^{N_M} max((d_{col} - d_{min,i,j}), 0) \times 2 \times E_{Battery}}{2 \times E_{Battery} \times N_s \times N_M}$$

Where, N_s is the total test scenarios which come from the DoE and N_M is the number of Monte Carlo samples that are used to ensure the robustness of the decision for uncertainty, for the current study, $n_s = 50$ and $n_m = 20$. Here, E_A, E_B are the energy used in the maneuver and the $E_{Battery}$ is the total Battery charge of each UAV. Finally, the term d_{col} is the minimum required distance, which is $1.5m$ and d_{min} is the minimum distance between UAVs in their maneuver. Because the value of $E_{Battery}$ is usually so great, this formulation leads to decisions to avoid collision more than energy optimal decision. Also, note that the maximum achievable objective function is 1 and the minimum will be 0.

Table 4 lists the UAV specifications and parameters for the optimization model.

Based on this voltage and Capacity of Batteries, the total Battery charge will be calculated as equation 16:

$$E_{Battery} = 5.100 \times 11.1 \times 3600 = 230,796J \qquad (16)$$

Handling Special Cases Due To Uncertainty

The optimal decision is affected by the noise in the measurement. Although the noise can affect the control , in this research, we focus on the effect of uncertainty in decisions and not control. The uncertainty can affect the decision in different ways:

1. The uncertainty can cause the system to take an asymmetric decision. The UAVs are trained to have optimal coherent decisions, therefore having asymmetric decisions, may lead to deviating from energy optimal behavior and more seriously leading to collision.

2. The UAVs may start the maneuver in a different time, or even do not start the maneuver at all. This is more hazardous, compared to asymmetric decisions.

3. The Direction change decision relies on self identifying if you are UAV-A or UAV-B, based on the relative pose. Therefore if the noise in measurement/estimation is large, the UAVs may have a problem in finding the UAV on the right and the left side and lead to the wrong decision.

While using the Monte Carlo sampling might help have robust decisions, the last issues need a more thoughtful solution. To avoid these issues, we use the idea of some minor UAV-UAV communication. Instead of sending the information the UAVs give a light signal of starting the maneuver. After receiving a signal like this, the second UAV starts its predicted coherent maneuver. The second UAV cannot mimic the same maneuver as the first UAV, but it still can start the maneuver based on its own estimation of the position and velocity. The key idea here is to train the UAVs for a bigger range of collision but only start the collision if either extreme collision condition is met or the second UAV started the maneuver. Figure 6 explains this procedure.

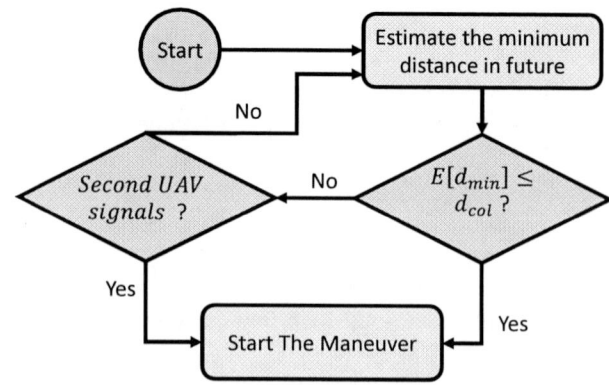

FIGURE 6. Starting the maneuver condition

By having especial signals for turning right and left, the

UAVs can also avoid the problem of wrong UAV position detection.

Optimizing The Decision Making System Using Neuroevolution

The optimization method cannot use the labeled data because the optimal answer need s the asymmetric decision of the second UAV. Therefore instead of supervised learning, we follow an unsupervised method to train the neural network. One of the main approaches to train the neural network is using neuroevolution [27]. Neuroevolution uses an evolutionary algorithm to optimize the topology and weights of a neural network. Different researches compared the effectiveness of neuroevolution with alternative methods like reinforcement learning [28]. One of the derivations of using neuroevolution is Adaptive Genomic Evolution of Neural Network Topologies (AGENT) [22] which was proposed by Behjat et al. This variation uses a similar architecture to NeuroEvolution of Augmented topology (NEAT) [29] and Spectrum-diverse Unified Neuroevolution Architecture (SUNA) [30], which was first published by Chidambaran et al. [21], but AGENT can make adaptive mutation and selection pressure in order to improve the total performance of the algorithm.

By using neuroevolution for each strategy two optimal models can be achieved. Later we use the classification to find the best strategy based on the initial state parameters.

As mentioned Neuroevolution uses a genetic algorithm to optimize the neural network for the current problem a population of 200 networks were trained for 60 generations.

Strategy Choice During Application

After training one neural network for each strategy, they can find the optimal decision with running neural network trained by AGENT. Using the suggest decision parameters, it is possible to run the surrogate model with a classifier for energy and minimum distance model in 0.01 seconds and 0.03 seconds respectively on Intel Xeon CPU E5-1620 @3.50 GHz with 16 GB RAM for each strategy. Therefore it is possible to find the best strategy online. After calculating expected energy consumption and minimum distance for each strategy, the optimal strategy between these strategies is chosen using the conditions below.

If $d_{min,SC}, d_{min,DC} \geq d_{col}$ Choose the strategy with less energy consumption.
If $d_{min,SC} \geq d_{col} > d_{min,DC}$ Choose the Speed Change strategy.
If $d_{min,DC} \geq d_{col} > d_{min,SC}$ Choose the Direction Change strategy.
If $d_{min,SC}, d_{min,DC} \leq d_{col}$ Choose the strategy with more distance.

RESULTS
Speed Change Results

Figure 7 shows the convergence history using Speed Change strategy. Here, the initial value of the objective function is high because collision avoidance accounts for a significant part of the objective function as compared to energy consumption. This bias for collision avoidance is reasonable due to the energy consumption to avoid the collision has negligible impact on the total energy consumption for the whole maneuver. This distribution of objective function is not a problem because the AGENT uses a tournament selection and it is not sensitive to the value of objective function.

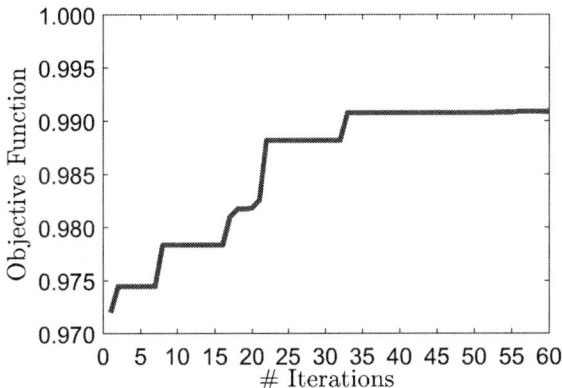

FIGURE 7. Convergence history for speed change strategy

Figure 8 shows the structure of an optimized neural network using AGENT.

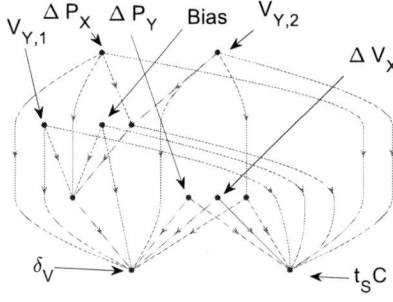

FIGURE 8. Optimized neural network for speed change strategy

Direction Change Results

Figure 9 shows the convergence history using Direction Change strategy.

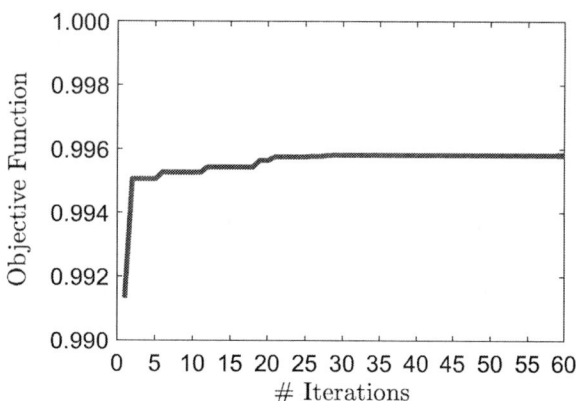

FIGURE 9. Convergence history for direction change strategy

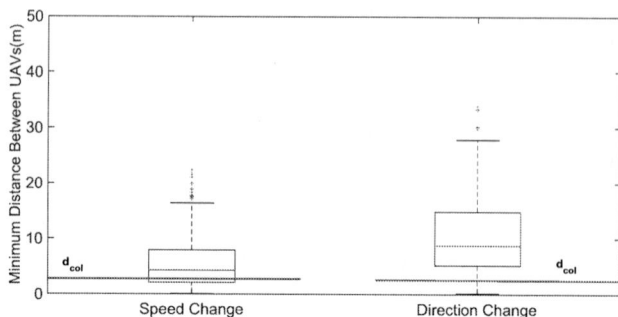

FIGURE 11. Minimum distance distribution for 500 test scenarios; reporting median of Monte Carlo samples

Based on figure 9 it is clear that the Direction Change Strategy has better performance compared to the Speed Change strategy. The initial objective function, as well as the final value, is better for Direction change. It seems that direction change is a more flexible strategy.

Figure 10 shows the structure of the optimized neural network using AGENT.

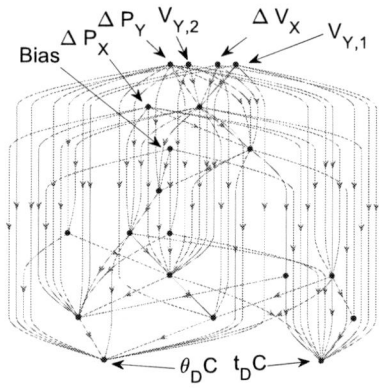

FIGURE 10. Optimized neural network for direction change strategy

The network structure for direction change is more complicated than speed change which may be because for Speed change the optimization could find a more nonlinear transformation and it helped its performance. Usually, a more complex network is associated with a more nonlinear transformation.

High Fidelity Tests

Although the objective function is large,it is not an intuitive parameter. Besides that, the successful performance based on the surrogate model does not necessarily mean that the decision

is valid in a real application. Another possible effect of current formulation is the inability of the system to perform well for the scenarios which it is not trained for.

In order to ensure the validness of the results as well as considering the generalizability of the result, the best network trained for both speed change and direction change was used for 500 completely new scenarios from the same DoE. Each scenario was tested for 20 Monte Carlo samples to illustrate the effect of uncertainty. Figure 11 and 12 (plotted by removing 12 outliers from each plot) Shows the distribution of the "median minimum" distance and total energy consumption of UAVs for both Strategies in new samples. Median is used because compared to mean it is less sensitive to outliers [31].

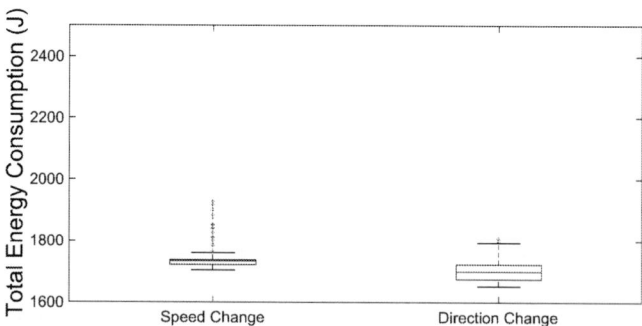

FIGURE 12. Total energy consumption distribution for 500 test scenarios; reporting median of Monte Carlo samples

These results indicate that similar to the trained samples. Direction change strategy generates more distance compared to speed Change strategy while energy consumption is slightly lower. This observation indicates that changing velocity is probably more energy consuming because 1) deceleration reduces the lift force and 2) acceleration increases the drag force which effect in more energy consumption than energy consumption at constant speed.

Besides the median which is a robust choice for studying statistics, a more conservative approach in handling uncertainties is using the worst case scenario. Figure 13 shows the least value over Monte Carlo samples for minimum distance.Similarly, Figure 14 (plotted by removing 25 outliers from Speed Change and 8 outliers from Direction Change) illustrates the worst case scenario of the maximum energy consumption over different Monte Carlo samples.

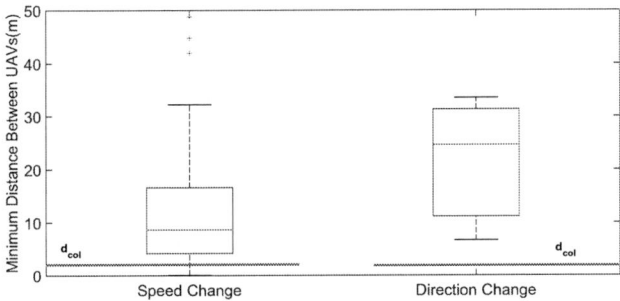

FIGURE 13. Minimum distance distribution for 500 test scenarios; reporting worst of monte carlo samples

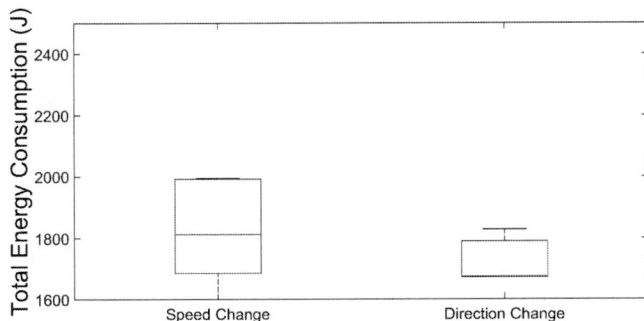

FIGURE 14. Total energy consumption distribution for 500 test scenarios; reporting worst of monte carlo samples

These results show that Direction change strategy is a more trust-able approach compared to Speed Change strategy. It has less energy consumption and more distance. All test scenarios that lead to collision in Direction change strategy are less than speed change strategy. Figure 15 shows the number of scenarios lead to collision out of 500 scenarios for each strategy. In this figure, S.C and D.C. stand for Speed change and Direction change respectively, and M. and W. stand for Median and Worst

case scenarios. The number of collisions in Speed change strategy are 5 for median and 71 for the worst case scenario. These values are 3 and 22 for Direction change strategy.

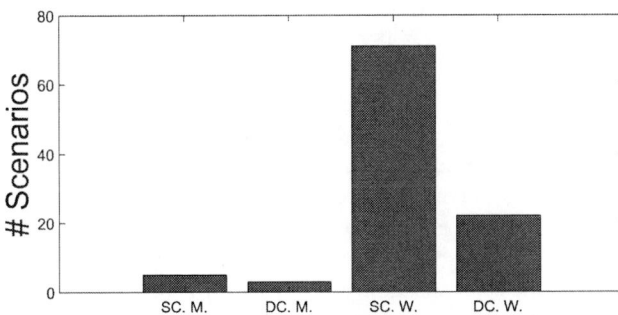

FIGURE 15. Collisions based on median minimum distance and worst case scenario for 500 test scenarios

Finally, calculating minimum distance using trajectory values is a valid idea. The maximum error of the controller in the following trajectory is plotted in the figure 16 (plotted by removing 25 outliers for Speed changes and 8 outliers for Direction Change). This result shows that the error in following trajectory is small for most of the scenarios, i.e., the controller successfully follows the trajectory. Therefore, using trajectory points to estimate the minimum distance is a valid method especially for Direction change.

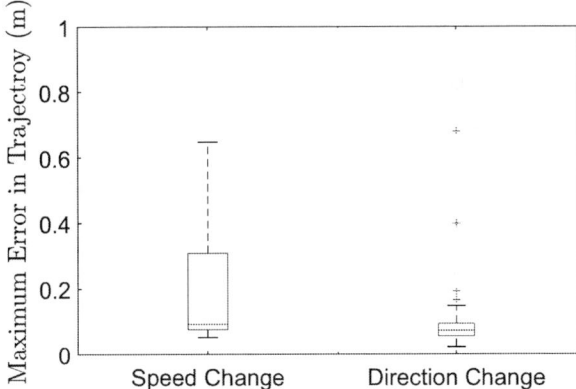

FIGURE 16. Maximum error in following the trajectory for 500 test scenarios

DISCUSSION

The first notable point in the results is the reliability of the strategies. Based on the results, it seems that the Direction

Change strategy is a more reliable compared to Speed change strategy. The first evidence to support this idea is based on low fidelity optimization. The convergence history shows better performance for direction change. On the other hand, because the Direction change had a more complex network, the advantage of Direction change might be the result of better optimization.

In addition, note that Speed change leads to greater energy consumption although it has a shorter path. It seems that operating with undesired speed is a more energy consuming outcome based on the current energy model. On the other hand, the high fidelity results for speed change show larger error, which can be due to control issues. Speed change samples could not be controlled as good as Direction change samples and that lead to high error. Note that the same controller was used for both strategies; separate tuning is not performed. It is conceivable that speed change leads to less stable maneuvers or the decisions from Speed Change neural network are less optimized for the current controller.

Besides control issues, it seems that speed change is a less flexible strategy. This strategy is not useful for head-on collisions, and its abilities shrink when one of the UAVs has very high or low speed. Also, Direction change leads to large, although not necessarily optimal, maneuvers based on its minimum distance. Direction change neural network was able to find a safe, not necessarily energy optimal, decisions to avoid most collision cases through a large path deviation.

CONCLUSION

This paper expands upon a reciprocal collision avoidance mechanism for uniform multi-rotor UAVs. Primary contribution lies in incorporating uncertainty attributed to pose estimation errors, and handling this uncertainty implicitly by training an avoidance action model via neuroevolution.

Two different strategies for collision avoidance are studied. Each strategy is associated with a maneuver to avoid the collision and return to the original velocity and path. In the first strategy, the controlled UAV can change speed, without changing its path. In the second, the controlled UAV changes its direction to avoid a collision. A Monte Carlo sampling approach is used to model the uncertainties during the training process of neural system to be used as the online action model.

In order to decrease the computation time for training each strategy, a surrogate model is constructed to estimate energy consumption, instead of having to simulate the entire control/dynamics for each scenario under the design of experiments and Monte carlo sampling for uncertainty. Also, a classifier is trained to distinguish state/action pairs that lead to trajectories that are not (stably) flyable by the controller. This approach provides additional computational time savings, and facilitates reliability during online application.

The neural networks trained using the neuroevolution ap-

proach called AGENT is tested on high fidelity simulations to ensure their reliability. The results indicate that the direction strategy is particularly capable of achieving safe performance. Future work would involve investigating more computationally frugal approaches to modeling the uncertainty, consideration of uncertainty introduced by wind effects, and finally implementation on hardware platforms to explore the feasibility of this approach when integrated with state of the sensors and onboard computing.

REFERENCES

[1] Pham, H., Smolka, S. A., Stoller, S. D., Phan, D., and Yang, J., 2015. "A survey on unmanned aerial vehicle collision avoidance systems". *arXiv preprint arXiv:1508.07723*.

[2] Tang, S., and Kumar, V., 2016. Translating paths into optimal trajectories for safe coordination of teams of dynamic robots. Robotics: Science and Systems (RSS),Robotics: Science and Systems (RSS), Workshop on On-line Decision-Making in Multi-robot Coordination, June. Ann Arbor, Michigan.

[3] Luo, C., McClean, S. I., Parr, G., Teacy, L., and De Nardi, R., 2013. "Uav position estimation and collision avoidance using the extended kalman filter". *IEEE Transactions on vehicular technology, 62*(6), pp. 2749–2762.

[4] Liao, T. W., 2012. "Uav collision avoidance using a * algorithm".

[5] Zhang, M., Song, J., Huang, L., and Zhang, C., 2016. "Distributed cooperative search with collision avoidance for a team of unmanned aerial vehicles using gradient optimization". *Journal of Aerospace Engineering, 30*(1), p. 04016064.

[6] Gan, S. K., Fitch, R., and Sukkarieh, S., 2012. "Real-time decentralized search with inter-agent collision avoidance". In 2012 IEEE International Conference on Robotics and Automation, IEEE, pp. 504–510.

[7] Richards, A., and How, J. P., 2002. "Aircraft trajectory planning with collision avoidance using mixed integer linear programming". In Proceedings of the 2002 American Control Conference (IEEE Cat. No. CH37301), Vol. 3, IEEE, pp. 1936–1941.

[8] Mcfadyen, A., Durand-Petiteville, A., and Mejias, L., 2014. "Decision strategies for automated visual collision avoidance". In 2014 International Conference on Unmanned Aircraft Systems (ICUAS), IEEE, pp. 715–725.

[9] Yang, X., Alvarez, L. M., and Bruggemann, T., 2013. "A 3d collision avoidance strategy for uavs in a non-cooperative environment". *Journal of Intelligent & Robotic Systems, 70*(1-4), pp. 315–327.

[10] Chang, D. E., Shadden, S. C., Marsden, J. E., and Olfati-Saber, R., 2003. "Collision avoidance for multiple agent systems".

[11] Shim, D. H., Kim, H. J., and Sastry, S., 2003. "Decentralized nonlinear model predictive control of multiple flying robots". In 42nd IEEE International Conference on Decision and Control (IEEE Cat. No. 03CH37475), Vol. 4, IEEE, pp. 3621–3626.

[12] Mueller, E. R., and Kochenderfer, M., 2016. "Multirotor aircraft collision avoidance using partially observable markov decision processes". In AIAA Modeling and Simulation Technologies Conference, p. 3673.

[13] Temizer, S., Kochenderfer, M., Kaelbling, L., Lozano-Pérez, T., and Kuchar, J., 2010. "Collision avoidance for unmanned aircraft using markov decision processes". In AIAA guidance, navigation, and control conference, p. 8040.

[14] Van Den Berg, J., Snape, J., Guy, S. J., and Manocha, D., 2011. "Reciprocal collision avoidance with acceleration-velocity obstacles". In 2011 IEEE International Conference on Robotics and Automation, IEEE, pp. 3475–3482.

[15] Alejo, D., Cobano, J., Heredia, G., and Ollero, A., 2014. "Optimal reciprocal collision avoidance with mobile and static obstacles for multi-uav systems". In 2014 International Conference on Unmanned Aircraft Systems (ICUAS), IEEE, pp. 1259–1266.

[16] Paul, S., 2017. A bio-inspired neural system for energy optimal collision avoidance by unmanned aerial vehicles, ms thesis,.

[17] Shem, A. G., Mazzuchi, T. A., and Sarkani, S., 2008. "Addressing uncertainty in uav navigation decision-making". *IEEE Transactions on Aerospace and Electronic Systems,* *44*(1), January, pp. 295–313.

[18] Rathbun, D., Kragelund, S., Pongpunwattana, A., and Capozzi, B., 2002. "An evolution based path planning algorithm for autonomous motion of a uav through uncertain environments". In Proceedings. The 21st Digital Avionics Systems Conference, Vol. 2, pp. 8D2–8D2.

[19] Prahl, S. A., 1989. "A monte carlo model of light propagation in tissue". In Dosimetry of laser radiation in medicine and biology, Vol. 10305, International Society for Optics and Photonics, p. 1030509.

[20] Stanley, K. O., Clune, J., Lehman, J., and Miikkulainen, R. "Designing neural networks through neuroevolution".

[21] Chidambaran, S., Behjat, A., and Chowdhury, S., 2018. "Multi-criteria evolution of neural network topologies: Balancing experience and performance in autonomous systems". In ASME 2018 International Design Engineering Technical Conferences and Computers and Information in Engineering Conference, American Society of Mechanical Engineers, pp. V02BT03A039–V02BT03A039.

[22] Behjat, A., Chidambaran, S., and Chowdhury, S., 2019. "Adaptive genomic evolution of neural network topologies (agent)for state-to-action mapping in autonomous agents".

[23] Clark, R. A., Punzo, G., MacLeod, C. N., Dobie, G., Summan, R., Bolton, G., Pierce, S. G., and Macdonald, M., 2017. "Autonomous and scalable control for remote inspection with multiple aerial vehicles". *Robotics and Autonomous Systems,* *87*, pp. 258–268.

[24] Ball, Z., Odonkor, P., and Chowdhury, S., 2017. "A swarm-intelligence approach to oil spill mapping using unmanned aerial vehicles". In *AIAA Information Systems-AIAA Infotech@ Aerospace.* p. 1157.

[25] Ferrera, E., Alcntara, A., Capitn, J., Rodrguez Castao, ., Marrn, P., and Ollero, A., 2018. "Decentralized 3d collision avoidance for multiple uavs in outdoor environments". *Sensors,* *18*, 11, p. 4101.

[26] Mellinger, D., and Kumar, V., 2011. "Minimum snap trajectory generation and control for quadrotors". In 2011 IEEE International Conference on Robotics and Automation, IEEE, pp. 2520–2525.

[27] Floreano, D., Dürr, P., and Mattiussi, C., 2008. "Neuroevolution: from architectures to learning". *Evolutionary Intelligence,* *1*(1), pp. 47–62.

[28] Salimans, T., Ho, J., Chen, X., Sidor, S., and Sutskever, I., 2017. "Evolution strategies as a scalable alternative to reinforcement learning". *arXiv preprint arXiv:1703.03864.*

[29] Stanley, K. O., and Miikkulainen, R., 2002. "Evolving neural networks through augmenting topologies". *Evolutionary computation,* *10*(2), pp. 99–127.

[30] Vargas, D. V., and Murata, J., 2017. "Spectrum-diverse neuroevolution with unified neural models". *IEEE transactions on neural networks and learning systems,* *28*(8), pp. 1759–1773.

[31] Leys, C., Ley, C., Klein, O., Bernard, P., and Licata, L., 2013. "Detecting outliers: Do not use standard deviation around the mean, use absolute deviation around the median". *Journal of Experimental Social Psychology,* *49*(4), pp. 764–766.

Proceedings of the ASME 2019
International Design Engineering Technical Conferences
and Computers and Information in Engineering Conference
IDETC/CIE2019
August 18-21, 2019, Anaheim, CA, USA

DETC2019-98027

GAUSSIAN PROCESS EMULATION FOR BIG DATA IN DATA-DRIVEN METAMATERIALS DESIGN

Ramin Bostanabad
Northwestern University
Evanston, IL, USA

Yu-Chin Chan
Northwestern University
Evanston, IL, USA

Liwei Wang
Shanghai Jiao Tong University
Shanghai, China

Ping Zhu
Shanghai Jiao Tong University
Shanghai, China

Wei Chen[1]
Northwestern University
Evanston, IL, USA

ABSTRACT

Our main contribution is to introduce a novel method for Gaussian process (GP) modeling of massive datasets. The key idea is to build an ensemble of independent GPs that use the same hyperparameters but distribute the entire training dataset among themselves. This is motivated by our observation that estimates of the GP hyperparameters change negligibly as the size of the training data exceeds a certain level, which can be found in a systematic way. For inference, the predictions from all GPs in the ensemble are pooled to efficiently exploit the entire training dataset for prediction. We name our modeling approach globally approximate Gaussian process (GAGP), which, unlike most largescale supervised learners such as neural networks and trees, is easy to fit and can interpret the model behavior. These features make it particularly useful in engineering design with big data. We use analytical examples to demonstrate that GAGP achieves very high predictive power that matches or exceeds that of state-of-the-art machine learning methods. We illustrate the application of GAGP in engineering design with a problem on data-driven metamaterials design where it is used to link reduced-dimension geometrical descriptors of unit cells and their properties. Searching for new unit cell designs with desired properties is then accomplished by employing GAGP in inverse optimization.

Keywords: Gaussian processes, Supervised learning, Big data, Metamaterials

NOMENCLATURE

n	Number of training samples
GP	Gaussian process
d	Input dimensionality
$\boldsymbol{x}$	Vector of d inputs
q	Output dimensionality
$\boldsymbol{y}$	Vector of q outputs
$\boldsymbol{R}$	Sample correlation matrix of size $n \times n$
$\boldsymbol{\omega}$	Roughness parameters of the correlation function
MLE	Maximum likelihood estimation
L	Objective function in MLE
δ	Nugget or jitter parameter
n_0	Number of initial random samples
n_s	Number of random samples added to n_0 per iteration
s	Number of times that n_s samples are added to n_0
$\widehat{\boldsymbol{\omega}}_\infty$	Estimate of $\boldsymbol{\omega}$ via MLE with very large training data

1 INTRODUCTION

Fueled by recent advancements in high performance computing as well as data acquisition and storage capabilities (e.g., online repositories), data-driven methods are increasingly employed in engineering design [1-3] to efficiently explore the design space of complex systems by obviating the need for expensive experiments or simulations. For emerging material systems, in particular, large datasets have been successfully leveraged to design heterogeneous materials [4-8] and mechanical metamaterials [9-12].

---

[1] Corresponding Author. Wilson Cook Professor in Engineering Design. Email: weichen@northwestern.edu

Key to data-driven design is to develop supervised learners that can distill as much useful information from massive datasets as possible. However, most large-scale learners such as deep neural networks [13] (NNs) and gradient boosted trees [14] (GBT) are difficult to interpret and hence less suitable for engineering design. Gaussian process (GP) models (aka Kriging) have many attractive features that underpin their widespread use in engineering design. For example, GPs interpolate the data, have a natural and intuitive mechanism to smooth the data to address noise (i.e., avoid interpolation) [15], and are very interpretable (i.e., provide insight into input-output relations) [16, 17]. In addition, they quantify prediction uncertainty and have analytical conditional distributions that enable, e.g., tractable adaptive sampling or Bayesian analysis [18]. However, conventional GPs are not readily applicable to large datasets and have been mostly confined to engineering design with small data. The goal of our work is to bridge the gap between big data and GPs while achieving high predictive accuracy.

The difficulty in fitting GPs to big data is rooted in the repetitive inversion of the sample correlation matrix, R, whose size equals the number of training samples, n. Given the practical features and popularity of GPs, considerable effort has been devoted to resolving their scalability shortcoming. One avenue of research has explored partitioning the input space (and hence the training data) via, e.g., trees [19] or Voronoi cells [20], and fitting an independent GP to each partition. While particularly useful for small to relatively large datasets that exhibit nonstationary behavior, prediction via these methods results in discontinuity (at the partitions' boundaries) and information loss (because the query point is associated with only one partition). Projected process approximation (PPA) [21] is another method where the information from n samples is distilled into $m \ll n$ randomly (or sequentially) selected samples through conditional distributions. PPA is very sensitive to the m selected samples and overestimates the variance [21]. In Bayesian committee machine (BCM) [22], the dataset is partitioned into p mutually exclusive and collectively exhaustive parts with independent GP priors, and then the predictions from all the GPs are pooled together in a Bayesian setting. While theoretically very attractive, BCM does not scale well with the dataset size and is computationally very expensive.

Another avenue of research has pursued subset selection. For example, a simple strategy is to only use $m \ll n$ samples to train a GP [23, 24] where the m samples are selected either randomly or sequentially based on maximizing some criteria such as information gain or differential entropy score. Reduced-rank approximation of R with $m \ll n$ samples is another option for subset selection and has been used in the Nystrom [25] and subset of regressors [26, 27] (SR) methods. The m samples in these methods are chosen randomly or in a greedy fashion to minimize some cost function. While the many variants of subset selection may be useful in some applications, they waste information and are not applicable to very large datasets due to the computational and storage costs. Local methods also use subsets of the data because they fit a stationary GP (for each prediction) to a very small number of training data points that are closest to the query point. Locally approximate Gaussian process [28] (LAGP) is perhaps the most widely recognized local method where the subsets are selected either based on their proximity to the query point or to minimize the predictive variance. Despite being useful for nonstationary and relatively large datasets, local methods also waste some information and can be prohibitively expensive for repetitive use since local samples have to be found and a GP must be fitted for each prediction.

Although the recent works have made significant progress in bridging the gap between GPs and big data, GPs still struggle to achieve the accuracy of the state-of-the-art large-scale supervised learners such as NNs and trees. Motivated by this limitation, we develop a computationally stable and inexpensive approach for GP modeling of massive datasets. The main idea of our approach is to build an ensemble of independent GPs that utilize converged roughness parameters as their hypermeters. This is based on an empirical observation that the estimates of the GP hyperparameters negligibly change as the size of the training data exceeds certain level. While having some common aspects with a few of the abovementioned works, our method is more massively scalable, can leverage multicore or GPU (graphical processing unit) computations [29, 30], and is applicable to very high-dimensional data with or without noise.

As mentioned earlier, big data has enticed new design methods for complex systems such as metamaterials [9-12], which possess superior properties through their hierarchical structure that consists of repeated unit cells. While traditional methods like topology optimization (TO) provide a systematic computational platform to discover metamaterials with unprecedented properties, they have many challenges that are primarily due to the high dimensional design space (i.e., the geometry of unit cells), computational costs, local optimality, and spatial discontinuities across unit cell boundaries (in case multiple unit cells are simultaneously designed). We take a data-driven approach to address these challenges by first building a large training database of many unit cells and their corresponding properties. Unlike previous data-driven works that represent unit cells as signed distance fields [9] or voxels [11], we drastically reduce the input dimension in our dataset by characterizing the unit cells via spectral shape descriptors based on the Laplace-Beltrami (LB) operator. Then, we employ our GP modeling approach to link the geometrical descriptors of unit cells and their properties and, in turn, efficiently discover new unit cells with desired properties.

The rest of the paper is organized as follows. We first review some preliminaries on GP modeling in Sec. 2 and then introduce our novel idea in Sec. 3. In Sec. 4, we validate the accuracy of our approach by comparing its performance against three popular and largescale supervised learning methods on four analytical problems. We demonstrate an application of the GP approach to our data-driven design method for metamaterials in Sec. 5 and conclude the paper in Sec. 6.

2 REVIEW ON GAUSSIAN PROCESS MODELING

Below, we describe how GP emulators (aka surrogates, metamodels, or models) can replace a computer simulator. The procedure is identical if the data is obtained from physical experiments. Let us denote the output and inputs of a computer simulator by, respectively, y and the d dimensional vector $\boldsymbol{x} = [x_{(1)}, x_{(2)}, \ldots, x_{(d)}]^T$ where $\boldsymbol{x} \in \mathbb{R}^d$. Assume the input-output relation is a realization of the random process $\eta(\boldsymbol{x})$:

$$\eta(\boldsymbol{x}) = \sum_{i=1}^{h} \beta_{(i)} f_i(\boldsymbol{x}) + \xi(\boldsymbol{x}), \qquad (1)$$

where $f_i(\boldsymbol{x})$'s are some pre-determined set of basis functions, $\boldsymbol{\beta} = [\beta_{(1)}, \ldots, \beta_{(h)}]^T$ are unknown weights, and $\xi(\boldsymbol{x})$ is a zero-mean GP characterized with its parametric covariance function, $c(\cdot, \cdot)$:

$$cov(\xi(\boldsymbol{x}), \xi(\boldsymbol{x}')) = c(\boldsymbol{x}, \boldsymbol{x}') = \sigma^2 r(\boldsymbol{x}, \boldsymbol{x}'), \qquad (2)$$

where $r(\cdot)$ is the correlation function having the property $r(\boldsymbol{x}, \boldsymbol{x}) = 1$ and σ^2 is the process variance. Various correlation functions have been developed in the literature with the Gaussian correlation function being the most widely used one:

$$r(\boldsymbol{x}, \boldsymbol{x}') = \exp\{-(\boldsymbol{x} - \boldsymbol{x}')^T \boldsymbol{\Omega} (\boldsymbol{x} - \boldsymbol{x}')\}, \qquad (3)$$

where $\boldsymbol{\Omega} = diag(\mathbf{10^{\omega}})$ and $\boldsymbol{\omega} = [\omega_{(1)}, \omega_{(2)}, \ldots, \omega_{(d)}]^T$, $-\infty < \omega_i < \infty$ are the roughness or scale parameters. The collection of σ^2 and $\boldsymbol{\omega}$ are called the hyperparameters.

With the formulation in Eq. (1) and given the n training pairs of $(\boldsymbol{x}_i, y_i)$, GP modeling requires finding a point estimate for $\boldsymbol{\beta}$, $\boldsymbol{\omega}$, and σ^2 via either maximum likelihood estimation (MLE) or cross-validation (CV). Alternatively, Bayes' rule can be employed to find the posterior distributions if there is prior knowledge on these parameters. Herein, we use a constant process mean (i.e., $\sum_{i=1}^{h} \beta_i f_i(\boldsymbol{x}) = \beta$) and employ MLE. These choices are widely practiced because a high predictive power is provided while minimizing the computational costs [28, 31-35].

MLE requires maximizing the multivariate Gaussian likelihood function, or equivalently:

$$[\hat{\beta}, \hat{\sigma}^2, \widehat{\boldsymbol{\omega}}] = \frac{argmin}{\beta, \sigma^2, \boldsymbol{\omega}} \left(\frac{n}{2} log(\sigma^2) + \frac{1}{2} log(|\boldsymbol{R}|) \right.$$
$$\left. + \frac{1}{2\sigma^2} (\boldsymbol{y} - \mathbf{1}\beta)^T \boldsymbol{R}^{-1} (\boldsymbol{y} - \mathbf{1}\beta) \right), \quad (4)$$

where $log(\cdot)$ is the natural logarithm, $\mathbf{1}$ is an $n \times 1$ vector of ones, and $\boldsymbol{R}$ is the $n \times n$ correlation matrix with $(i, j)^{th}$ element $R_{ij} = r(\boldsymbol{x}_i, \boldsymbol{x}_j)$ for $i, j = 1, \ldots, n$. Setting the partial derivatives with respect to β and σ^2 to zero yields:

$$\hat{\beta} = [\mathbf{1}^T \boldsymbol{R}^{-1} \mathbf{1}]^{-1} \mathbf{1}^T \boldsymbol{R}^{-1} \boldsymbol{y}, \qquad (5)$$

$$\hat{\sigma}^2 = \frac{1}{n} (\boldsymbol{y} - \mathbf{1}\hat{\beta})^T \boldsymbol{R}^{-1} (\boldsymbol{y} - \mathbf{1}\hat{\beta}). \qquad (6)$$

Plugging these values into Eq. (4) and eliminating the constants:

$$\widehat{\boldsymbol{\omega}} = \frac{argmin}{\boldsymbol{\omega}} \, nlog(\hat{\sigma}^2) + log(|\boldsymbol{R}|) = \frac{argmin}{\boldsymbol{\omega}} L. \qquad (7)$$

By numerically minimizing L in Eq. (7) one can find $\widehat{\boldsymbol{\omega}}$. Many global optimization methods such as genetic algorithms [36], pattern searches [37, 38], and particle swarm optimization [39] have been employed to solve for $\widehat{\boldsymbol{\omega}}$ in Eq. (7). However, gradient-based optimization techniques are commonly preferred due to their ease of implementation and superior computational efficiency [15, 16, 31]. To guarantee global optimality in this case, the optimization is done numerous times with different initial guesses.

Upon completion of MLE, the following closed-form formula can be used to predict the response at any $\boldsymbol{x}^*$:

$$\hat{y}(\boldsymbol{x}^*) = \hat{\beta} + \boldsymbol{g}^T(\boldsymbol{x}^*) \boldsymbol{V}^{-1} (\boldsymbol{y} - \mathbf{1}\hat{\beta}), \qquad (8)$$

where $\boldsymbol{g}(\boldsymbol{x}^*)$ is an $n \times 1$ vector with i^{th} element $c(\boldsymbol{x}_i, \boldsymbol{x}^*) = \hat{\sigma}^2 r(\boldsymbol{x}_i, \boldsymbol{x}^*)$, $\boldsymbol{V}$ is the covariance matrix with $(i, j)^{th}$ element $\hat{\sigma}^2 r(\boldsymbol{x}_i, \boldsymbol{x}_j)$, and $\boldsymbol{y} = [y_1, \ldots, y_n]^T$ are the responses in the training dataset. The posterior covariance between the responses at the two inputs $\boldsymbol{x}^*$ and $\boldsymbol{x}'$ reads:

$$cov(y^*, y') =$$
$$c(\boldsymbol{x}^*, \boldsymbol{x}') - \boldsymbol{g}^T(\boldsymbol{x}^*) \boldsymbol{V}^{-1} \boldsymbol{g}(\boldsymbol{x}') + \boldsymbol{h}^T (\mathbf{1}^T \boldsymbol{V}^{-1} \mathbf{1})^{-1} \boldsymbol{h}, \quad (9)$$

where $\boldsymbol{h} = (1 - \mathbf{1}^T \boldsymbol{V}^{-1} \boldsymbol{g}(\boldsymbol{x}'))$.

If the training dataset has multiple outputs, one may fit either a single-response GP emulator to each response or a multi-response GP (hereafter denoted by MRGP) to all the responses. We follow [40] and extend the above formulations to simulators with q responses by placing a constant mean for each response (i.e., $\boldsymbol{\beta} = [\beta_{(1)}, \ldots, \beta_{(q)}]^T$) and employing the separable covariance function:

$$cov(\xi(\boldsymbol{x}), \xi(\boldsymbol{x}')) = c(\boldsymbol{x}, \boldsymbol{x}') = \boldsymbol{\Sigma} \otimes r(\boldsymbol{x}, \boldsymbol{x}'), \qquad (10)$$

where $\otimes$ denotes the Kronecker product and $\boldsymbol{\Sigma}$ is the $q \times q$ process covariance matrix with its off-diagonal elements representing the covariance between the corresponding responses at any fixed $\boldsymbol{x}$. The MLE approach described above can also be applied to multi-response datasets in which case σ will be replaced with $\boldsymbol{\Sigma}$ (see [41-44] for the details).

Finally, we note that GPs can address noise and smooth the data (i.e., avoid interpolation) via the so-called nugget or jitter parameter, δ, in which case $\boldsymbol{R}$ is replaced with $\boldsymbol{R}_\delta = \boldsymbol{R} + \delta \boldsymbol{I}_{n \times n}$. If δ is used, the estimated (stationary) noise variance in the data would be $\delta \hat{\sigma}^2$. We have recently developed an automatic method to robustly detect and estimate noise [31].

3 GLOBALLY APPROXIMATE GAUSSIAN PROCESS

Regardless of the optimization method used to solve for $\widehat{\boldsymbol{\omega}}$, each evaluation of L in Eq. (7) requires inverting the $n \times n$

matrix R. For very large n, there are two main challenges associated with this inversion: computational cost of approximately $O(\alpha n^3)$ and singularity of R (since the samples get closer as n increases). To address these issues and enable GP modeling of big data, our essential idea is to build an ensemble of independent GPs that use the same $\hat{\omega}$ and share the training data among themselves. To illustrate, we take the following function over $-2 \leq x \leq 3$:

$$y = x^4 - x^3 - 7x^2 + 3x + 5\sin(5x). \tag{11}$$

The associated likelihood profile (i.e., L) is visualized in **Figure 1** as a function of ω for various values of n. Two interesting phenomena are observed in this figure: (i) With large n, the profile of L does not alter as the training samples change. To observe this, for each n, we generate five independent training samples via Sobol sequence [45, 46] and plot the corresponding L. As illustrated in **Figure 1**, even though a total of 20 curves are plotted, only four are visible since the curves with the same n are indistinguishable. (ii) As n increases, L is minimized at similar ω's.

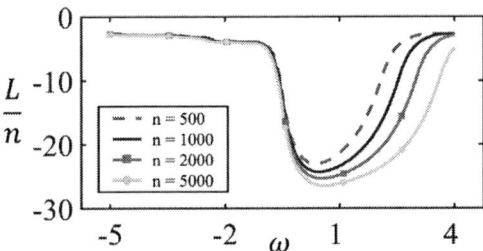

Figure 1 The profile of $\frac{L}{n}$ as a function of ω for various values of n. For each n, five curves are plotted but only four are visible since the curves with the same n are indistinguishable.

While we visualize the above two points with a simple 1D function, our studies indicate that they hold in general (i.e., irrespective of problem dimensionality and the absence or presence of noise) as long as the number of training samples is large. Therefore, we propose the following approach for GP modeling of large datasets.

Assuming a very large training dataset of size n is available, we first randomly select a relatively small subset of size n_0 (e.g., $n_0 = 500$) and estimate $\hat{\omega}_0$ with a gradient-based optimization technique. Then, we add n_s random samples (e.g., $n_s = 250$ or $n_s = 500$) to this subset and estimate $\hat{\omega}_1$ while employing $\hat{\omega}_0$ as the initial guess in the optimization. This process is stopped after s steps when $\hat{\omega}$ does not change noticeably (i.e., $\hat{\omega}_s \cong \hat{\omega}_{s-1}$) as more training data are used. At this point, the latest solution, denoted by $\hat{\omega}_\infty$, is employed to build m GP models each with $n_k \geq n_0 + s \times n_s$ randomly chosen samples (from the entire training data) where $n = \sum_{k=1}^m n_k$. Here, we have assumed that the collection of these GPs (who have $\hat{\omega}_\infty$ as their hyperparameters) approximate a GP that is fitted to the entire training dataset and,

correspondingly, call it globally approximate Gaussian process (GAGP). The algorithm of GAGP is summarized in **Figure 2**.

We point out the following important features regarding GAGP. First, we recommend using gradient-based optimizations throughout the entire process because (i) if n_0 is large enough (e.g., $n_0 > 500$), one would need to select only a few initial guesses to find the global minimizer of Eq. (7), i.e., $\hat{\omega}_0$; and (ii) we want to use $\hat{\omega}_{i-1}$ as the initial guess for the optimization in the ith step. This latter choice ensures fast convergence since the minimizer of L changes slightly as the dataset size increases (see **Figure 1**). To estimate $\hat{\omega}_0$, we recommend the method developed in [31]. Second, for predicting the response, Eq. (8) is used for each of the m GP models and then the results are averaged. In our experience, we have observed very similar prediction results with different averaging schemes (e.g., weighted averaging where the weights are proportional to inverse variance). The advantages of employing an ensemble of models (in our case the m GPs) in prediction is extensively demonstrated in the literature [14, 22]. Third, the predictive power is not sensitive to n_0, s, and n_s so long as large enough values are used for them. For novice users, we recommend starting with $n_0 = 500$, $s = 6$, $n_s = 250$, and equally distributing the samples among the m GPs (we use these parameters in Sec. 5 and for all the examples in Sec. 4). For more experienced users, we provide a systematic way in Sec. 4 to choose these values based on GP's inherent ability to estimate noise by the nugget variance. Lastly, it is pointed out that while GAGP has a high predictive power and is applicable to very large datasets, its implementation is very straightforward because it only entails integrating a GP modeling package such as GPM [31] with the algorithm in **Figure 2**.

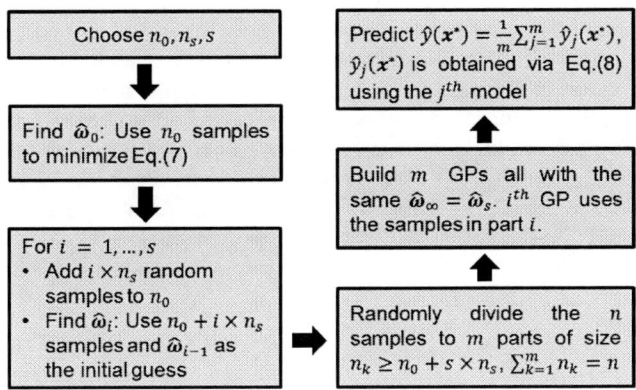

Figure 2 Flowchart of globally approximate Gaussian process, GAGP. It is assumed that a very large training dataset of size n is available.

4 COMPARATIVE STUDIES ON ANALYTICAL EXAMPLES

To validate the performance of GAGP in regression, we compare its predictive power on four examples (Ex1-4) against recognized big data learners: locally approximate Gaussian process (LAGP) [28], gradient boosted trees (XGB) [14], and feed-forward neural networks (NNs) [47]. As shown in Eqs.

(12)-(15), the examples cover a wide range of dimensionality and input-output complexity.

Ex1:
$$y = x^4 - x^3 - 7x^2 + 3x + 5\sin(5x), \quad (12)$$
$$-2 \le x \le 3,$$

Ex2 [48]:
$$y = \frac{\left(1-\exp\left(-\frac{1}{2x_2}\right)\right)\left(x_3 x_1^3 + 1900 x_1^2 + 2092 x_1 + 60\right)}{x_4 x_1^3 + 500 x_1^2 + 4x_1 + 20}, \quad (13)$$
$$\min(\boldsymbol{x}) = [0, 0, 2200, 85]$$
$$\max(\boldsymbol{x}) = [1, 1, 2400, 110],$$

Ex3 [49]:
$$y = 2\pi \sqrt{\frac{x_1}{x_4 + \frac{4x_3 x_4 x_5 x_6}{\left(x_5 x_2 + 19.62 x_1 - \frac{x_4 x_3}{x_2}\right)^2 + 4x_4 x_5 \left(\frac{x_6}{x_7}\right) x_7}}}, \quad (14)$$

$$\min(\boldsymbol{x}) = [30, 0.005, 0.002, 1000, 9 \times 10^4, 290, 340],$$
$$\max(\boldsymbol{x}) = [60, 0.02, 0.01, 5000, 11 \times 10^4, 296, 360],$$

Ex4 [49]:
$$y = \frac{5x_{12}}{1+x_1} + 5(x_4 - x_{20})^2 + x_5 + 40x_{19}^3 - 5x_1 + 0.05x_2$$
$$+ 0.08x_3 - 0.03x_6 + 0.03x_7 - 0.09x_9$$
$$- 0.01x_{10} - 0.07x_{11} + 0.25x_{13}^2 - 0.04x_{14}$$
$$+ 0.06x_{15} - 0.01x_{17} - 0.03x_{18}, \quad (15)$$
$$-0.5 \le x_i \le 0.5 \text{ for } i = 1, \dots 20.$$

For each example, two independent and unique datasets of size 30000 are generated with Sobol sequence [46] where the first one is used for training while the second one for validation. In each example, Gaussian noise is added to both the training and validation outputs. We consider two noise levels to test the sensitivity of the results where the noise standard deviation (SD) is determined based on each example's output range. As we measure performance by root mean squared error (RMSE), the noise SD should be recovered on the validation dataset (i.e., the RMSE would ideally equal noise SD).

We use CV to ensure the best performance is achieved for LAGP, XGB, and NN. For GAGP, we use $n_0 = 500$, $s = 6$, $n_s = 250$, and equally distribute the samples among the $m = \frac{30000}{500+6\times250} = 15$ GPs (i.e., each GP has 2000 samples). The results are summarized in **Table 1** (for small noise SD) and **Table 2** (for large noise SD), and indicate that (i) GAGP consistently outperforms LAGP and XGB, (ii) GAGP and NN both recover the true amount of added noise in the data, and (iii) GAGP achieves very similar results to NN. We note that given the large number of data points, the effect of sample-to-sample randomness on the results is very small and hence not reported.

We highlight that the performance of GAGP in each case could have been improved even further by tuning its parameters via CV (which was done for LAGP, XGB, and NN). Potential parameters include n_0, s, n_s, and $f_i(\boldsymbol{x})$ (see Eq. (1)). However, we intentionally avoid this tuning to demonstrate GAGP's flexibility, generality, and ease-of-use.

Table 1 Root mean squared error (RMSE) with small noise. Smallest errors are in bold.

	Noise SD	LAGP	XGB	NN	GAGP
Ex1 (1D)	0.2	1.271	0.209	**0.200**	**0.200**
Ex2 (4D)	0.1	1.386	0.121	**0.100**	0.103
Ex3 (7D)	0.1	0.129	0.118	**0.100**	**0.100**
Ex4 (20D)	0.1	1.450	0.351	**0.101**	0.103

Table 2 Root mean squared error (RMSE) with large noise. Smallest errors are in bold.

	Noise SD	LAGP	XGB	NN	GAGP
Ex1 (1D)	2	2.270	2.062	**2.000**	**2.000**
Ex2 (4D)	1	1.739	1.123	**1.002**	1.009
Ex3 (7D)	1	1.037	1.098	**1.002**	**1.002**
Ex4 (20D)	1	1.911	1.155	1.011	**1.001**

In engineering design, it is highly desirable to employ interpretable methods and tools that facilitate the knowledge discovery and decision-making process. Contrary to many supervised learning techniques such as NNs and random forests that are black boxes, the structure of GPs can provide qualitative insights. To demonstrate, we rewrite Eq. (3) as $r(\boldsymbol{x}, \boldsymbol{x}') = \exp\left\{-\sum_{i=1}^{d} 10^{\omega_i}\left(x_{(i)} - x'_{(i)}\right)^2\right\}$. If $\omega_i \ll 0$ (e.g., $\omega_i = -10$), then variations along the ith dimension (i.e., $x_{(i)}$) do not contribute to the summation and, subsequently, to the correlation between $\boldsymbol{x}$ and $\boldsymbol{x}'$. This contribution increases as the magnitude of ω_i increases. In a GP with constant mean of β, all the effect of inputs on the output is captured through $r(\boldsymbol{x}, \boldsymbol{x}')$. Hence, as ω_i decreases, the effect of x_i on the output decreases as well. We illustrate this feature with a 2D example as follows. Assume $y = f(x_1, x_2; \alpha) = \sin(2x_1 x_2) + \alpha \cos(\alpha x_1^2)$, $-\pi \le x_1, x_2 \le \pi$ for $\alpha = 2, 4, 6$. Three points regarding f are highlighted:

1. x_1 is more important than x_2 since $\alpha \cos(\alpha x_1^2)$ only depends on x_1 (note that $\alpha \ne 0$) while both inputs affect $\sin(2x_1 x_2)$.
2. As α increases, the relative importance of x_1 (compared to x_2) increases because the amplitude of $\alpha \cos(\alpha x_1^2)$ increases.
3. As α increases, y depends on x_1 with growing nonlinearity because the frequency of $\alpha \cos(\alpha x_1^2)$ increases.

Note that the first two points can be verified by calculating Sobol's total sensitivity indices (SIs) for x_1 and x_2 in f, see **Table 3**. These indices are in $[0, 1]$ range with higher values indicating more sensitivity to the input. Here, SI of x_1 is always 1 but SI of x_2 decreases as α increases. This trend indicates that the *relative* importance of x_1 on y increases as α increases.

We now demonstrate that a GP can distill the above features from a training dataset. To this end, for each α, we fit two GPs;

one with $n = 1000$ training data and the other with $n = 2000$. The hyperparameter estimates are summarized in **Table 3** and indicate that:

- For each α, $\hat{\omega}_1$ is larger than $\hat{\omega}_2$ which implies that x_1 is more important than x_2.
- As α increases, $\hat{\omega}_1$ increases while $\hat{\omega}_2$ only changes insignificantly. This shows that, as α increases, x_1 becomes more important (since $\hat{\omega}_1$ increases), and that the underlying functional relation between x_2 and y does not depend on α (since $\hat{\omega}_2$ does not change).
- For a given α, the estimates change negligibly when n is increased.

Table 3 Effect of sample size and nonlinearity on hyperparameter estimates in the pedagogical 2D example. The Sobol's total sensitivity indices (SIs) are also included.

		$\alpha = 2$	$\alpha = 4$	$\alpha = 6$
$n = 1000$	$\hat{\omega}_1$	2.39	3.11	3.59
	$\hat{\omega}_2$	-1.98	-2.12	-2.11
$n = 2000$	$\hat{\omega}_1$	2.38	3.10	3.54
	$\hat{\omega}_2$	-1.92	-2.18	-2.22
Total SI	x_1	1.00	1.00	1.00
	x_2	0.18	0.05	0.03

To demonstrate the above feature in GAGP, the convergence histories for Ex3 and Ex4 are plotted in **Figure 3** and **Figure 4**, respectively. Similar to **Figure 1**, it is evident that the estimated roughness parameters do not change noticeably as more samples are used in training (only 6 out of the 20 roughness parameters are plotted in **Figure 4** for a clearer illustration). The values of these parameters can determine which inputs (and to what extent) affect the output. For instance, in Ex4, ω_8 is very small so the output must be almost insensitive to x_8. Additionally, since $\omega_4 \cong \omega_{20}$, it is also expected that the corresponding inputs should affect y similarly. These observations completely agree with the analytical relation between x and y in Ex4 where y is independent of x_8 and is symmetric with respect to x_4 and x_{20}.

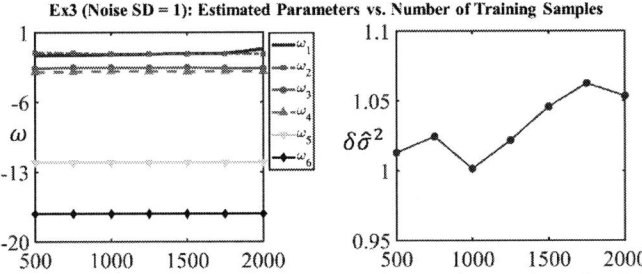

Figure 3 Convergence history in example 3 as the number of training samples is increased from 500 to 2000.

The estimated variance, $\delta\hat{\sigma}^2$, in both examples fluctuates very closely around the true noise variance. $\delta\hat{\sigma}^2$ provides a useful quantitative measure for the expected predictive power

(e.g., RMSE in future uses of the model). Additionally, similar to $\hat{\omega}$, its convergence history helps in determining whether sufficient samples have been used in training. Firstly, the number of training samples should be increased until $\delta\hat{\sigma}^2$ does not fluctuate noticeably. Secondly, via k-fold CV during training, one should ideally recover the noise variance by calculating the RMSE associated with predicting the samples in the ith fold (when fold i is not used in training). If these two values differ significantly, s (or n_s) should be increased. For instance, if the fluctuations on the right panel in **Figure 4** were large, we should have increased s (from 6 to, e.g., 10) or n_s (from 250 to, e.g., 500).

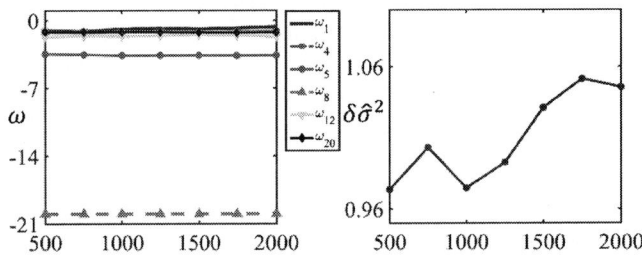

Figure 4 Convergence history in example 4 as the number of training samples is increased from 500 to 2000. For clearer demonstration, only six out of the twenty roughness parameters are plotted.

5 DATA-DRIVEN DESIGN OF METAMATERIALS

To demonstrate the application of GAGP in engineering design, we employ it in a new data-driven method for the optimization of metamaterial unit cells using big data. Although various methods, e.g., TO and genetic algorithm (GA), have been applied to design metamaterials with prescribed properties, these are computationally intensive and suffer from sensitivity to the initial guess as well as disconnected boundaries if multiple unit cells exist. A promising solution is to construct a large database of precomputed unit cells (aka microstructures or building blocks), enabling efficient selection of well-connected unit cells from the database and inexpensive optimization of new unit cells [9-12]. However, with the exception of [12] where unit cells are parameterized via geometric features like beam thickness, research in this area thus far use high-dimensional geometric representations (e.g., signed distance functions [9] or voxels [11]) that increase the memory demand and the complexity of constructing machine learning models that link structures to properties. Reducing the dimension of the unit cell is therefore a crucial step.

In this work, we reduce the dimension of the unit cells in our metamaterial database with spectral shape descriptors based on the LB operator. We then employ GAGP to learn how the effective stiffness tensor of unit cells change as a function of their LB descriptors. After the GAGP model is fitted, we use it to discover unit cells with desired properties through inverse optimization. Furthermore, to present the advantages of a large unit cell database and GAGP, we compare the results to those obtained using a conventional GP model fitted on a smaller database.

5.1 Metamaterials Database Generation

We propose a novel two-stage pipeline method inspired by [11] to generate a large training dataset of unit cells and their corresponding homogenized properties. For demonstration, our primary properties of interest are the effective stiffness tensor components, E_x, E_y, and E_{xy}. As explained below, our method starts by building an initial dataset and then proceeds to better cover the input (geometry) and output (property) spaces.

In stage one, to construct the initial dataset, we select design targets in the property space (the 3D space spanned by E_x, E_y, and E_{xy}). As the bounds of the property space are unknown *a priori*, we sample 1000 points uniformly distributed in $[0,1]^3$. Then, we use the SIMP (Solid Isotropic Material with Penalization) TO method [50] to find the orthotropic unit cells corresponding to each target. This stage generates 358 valid structures, while the remaining 642 points do not result in feasible unit cells, mainly because the uniform sampling places some design targets in theoretically infeasible regions. Moreover, TO may fail to meet the design targets due to sensitivity to the initial shape which is difficult to guess without prior knowledge. These 358 initial structures are shown in **Figure 5** where the Poisson's ratio is used instead of E_{xy} for a better illustration of the space.

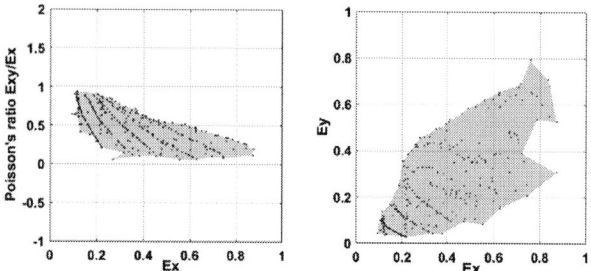

Figure 5 The property space of the initial database with 358 structures.

In stage two, we further populate the initial database via a stochastic shape perturbation algorithm that generates distorted structures with slightly different properties from the original ones. This perturbation technique is performed iteratively to efficiently expand the property space by avoiding expensive TO in achieving prescribed properties. Specifically, the following radial distortion model is used to perturb an existing shape:

$$x_{new} = \begin{cases} x_c + \frac{r_{new}}{r_{old}}(x_{old} - x_c) & if \ r_{old} \leq R_0 \\ x_{old} & if \ r_{old} > R_0 \end{cases}, \quad (16)$$

where x_{new} and x_{old} are the coordinates of the new and original pixel locations, x_c is the coordinate vector of the distortion center, r_{new} and r_{old} are the new and original distances to the distortion center, and R_0 is the outer distortion radius. r_{new} can be expressed as:

$$r_{new} = \begin{cases} \frac{1}{2}R_0\left(1 - cot\left(\frac{\gamma}{2}\right) - \beta\right) & if \ \gamma > 0 \\ \frac{1}{2}R_0\left(1 - cot\left(\frac{\gamma}{2}\right) + \beta\right) & if \ \gamma < 0 \\ r_{old} & otherwise \end{cases}, \quad (17)$$

where

$$\beta = \sqrt{\frac{2}{sin^2(\gamma/2)} - \left(1 + cot\left(\frac{\gamma}{2}\right) - \frac{2r_{old}}{R_0}\right)^2}, \quad (18)$$

and $\gamma \in \left(-\frac{\pi}{2}, \frac{\pi}{2}\right)$ is the angle that controls the amount of distortion. Considering the orthorhombic symmetry of the unit cells, only a quarter of the original structure is distorted and then reassembled to realize the full structure. We adopt the distortion model in Eq. (16) for two reasons. First, its parameters (i.e., R_0, γ, and x_c) have clear interpretations and hence can be easily tuned. In our case, they are all set as random variables with standard uniform distributions to generate a wide range of structures. Second, it preserves the topology of the original unit cell and introduces negligible artifacts (e.g., disconnections and checkerboard patterns) upon perturbation.

To better cover the property space, the database is populated iteratively. In each iteration we first calculate the following score for all the available unit cells:

$$Score = \frac{1}{(d+\varepsilon)2^\rho}, \quad (19)$$

where d is the Euclidean (L2) distance between the stiffness tensor components of each unit cell to the boundaries of the region enclosing the current property space (see **Figure 5** and **Figure 6**), ρ is the number of data points inside the neighborhood within a given radius in the property space (in our experience, sampling is more uniform when $\rho = 0.05$), and $\varepsilon \ll 1$ is used to avoid singularity. Then, we select the N points with the highest scores for stochastic perturbation. The properties of newly generated structures are calculated via numerical homogenization [51] and added to the current property space for the next iteration. After each iteration, newly generated unit cells are checked and discarded if they contain infeasible features such as isolated parts. The perturbation is repeated until the boundary of property space does not expand significantly and the points inside the boundaries are relatively dense. In this second stage, the database is expanded from 358 to 88000 unit cells that cover a wider range of properties (see **Figure 6**).

5.2 Unit Cell Dimension Reduction via Spectral Shape Descriptors

In the previous section, each unit cell in the database is represented by 50×50 pixels. For dimensionality reduction, we use spectral shape descriptors as they retain geometric and physical information. Specifically, we use the LB spectrum, also known as Shape-DNA, which can be directly calculated for any unit cell shape [52, 53].

The LB spectrum is an effective descriptor for the metamaterials database for several reasons: (*i*) It has a powerful discrimination ability and has been successfully applied to shape matching and classification in computer vision, despite being one of the simplest spectral descriptors. (*ii*) All of the complex structures in our orthotropic metamaterials database can be uniquely characterized with the first 10-15 eigenvalues in the LB spectrum. (*iii*) The spectrum embodies some geometrical information, including perimeter, area, and Euler number. This can be beneficial for the construction of the machine learning model as less training data may be required to obtain an accurate model compared to voxel- or point-based representations. (*iv*) Similar shapes have close LB spectrum, which may also help the supervised learning task.

The calculation of the LB spectrum for each unit cell is as follows. For a real-valued function f defined on a Riemannian manifold [52], the Helmholtz equation reads as:

$$\Delta f = -\lambda f, \tag{20}$$

where the Laplace-Beltrami operator Δ is defined as:

$$\Delta := div(grad\ f). \tag{21}$$

The eigenvalues of the Helmholtz equation are the LB spectrum and denoted:

$$0 \le \lambda_1 \le \lambda_2 \le \cdots < \infty. \tag{22}$$

We focus on the LB spectrum of a 2D shape under Dirichlet boundary conditions. In this case, the Helmholtz equation reduces to a Laplacian eigenvalue problem with the Dirichlet boundary condition:

$$\begin{aligned} \frac{\partial^2 f}{\partial x^2} + \frac{\partial^2 f}{\partial y^2} &= -\lambda f \quad in\ \Omega \\ f &= 0 \quad on\ \tau \end{aligned}, \tag{23}$$

where Ω and τ are the interior and boundaries of the domain of interest, respectively.

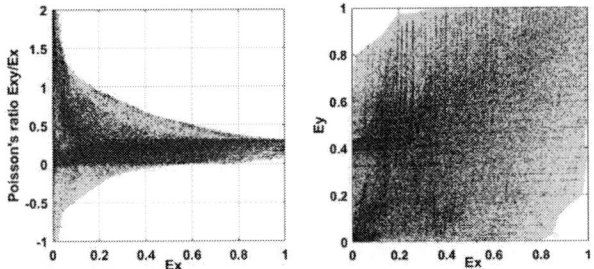

Figure 6 The property space of the expanded database with 88,000 structures.

Finally, the finite element method is employed to obtain the LB spectrum of unit cells [54]; see **Figure 7**. It is noted that our

88000 structures can be uniquely determined with only the first 16 orders of LB spectrum, reducing the input dimension from $50 \times 50 = 2500$ pixels to 16 scalar descriptors. In general, the computation of the LB spectrum takes only a few seconds per unit cell on a single CPU (Intel(R) Xeon(R) Gold 6144 CPU @3.50 GHz). Since this computation is performed once and can be run in parallel, the runtime is small.

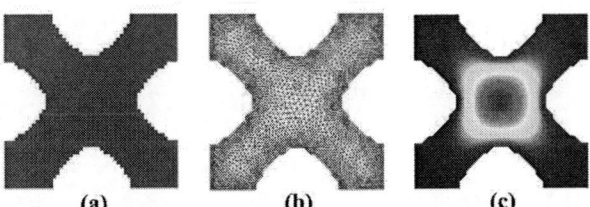

Figure 7 LB spectrum calculation: **(a)** Original structure, **(b)** Finite element mesh, and **(c)** The first eigenfunction.

5.3 Machine Learning - Linking LB Representation to Property via GAGP

Once the dataset is built, we follow the algorithm in **Figure 2** for machine learning, i.e., relating the LB representations of unit cells to their stiffness tensor. We use the same fitting parameters as in Sec. 4 ($n_0 = 500$, $s = 6$, $n_s = 250$), equally distribute the samples among the $m = \frac{88000}{500+6 \times 250} = 44$ GPs, and use Eq. (10) to have a multi-response model that leverages the correlation between the responses to have a higher predictive power. The convergence histories are provided in **Figure 8**, where the trends are consistent with those in Sec. 4. It is observed that the 16 estimated roughness parameters do not change noticeably once more than 1000 samples are used in training. In particular, 3 out of the 16 roughness estimates (corresponding to $\lambda_{14} = \lambda_{15} = \lambda_{16}$) are very small, indicating that the corresponding shape descriptors do not affect the responses; see Eq. (3). The next largest estimate belongs to $\omega_{13} \cong -8$ which corresponds to λ_{13}. The rest of the estimates are all between 2.5 and 3, indicating that the first twelve eigenvalues (shape descriptors) affect the responses similarly and nonlinearly (since large ω_i indicates rough response changes along dimension i). These observations agree well with the fact that the higher order eigenvalues generally explain less variability in the data. The estimated noise variances (one per response) also converge, with E_{xy} having the largest estimated noise variance in the data, which is potentially due to larger numerical errors in property estimation.

To illustrate the effect of expanding the training data from 385 to 88000, we randomly select 28000 samples from the data for validation. Then, we evaluate the mean squared error (MSE) of the following two models on this test set: A conventional GP fitted to the original 385 samples and a GAGP fitted to the rest of the data (i.e., to 60000 samples, resulting in $m = \frac{60000}{500+6 \times 250} = 30$ models). To account for randomness, we repeat this process 20 times. The results are summarized in **Table 4** and demonstrate that: (*i*) Increasing the dataset size (stage two in

Sec. 5.1) creates a supervised learner with a higher predictive power (compare the mean of MSEs for GP and GAGP); (ii) GAGP is more robust to variations than GP (compare the variance of MSEs for GP and GAGP); (iii) With 60000 samples, the predictive power of GAGP is slightly lower than the case where the entire dataset is used in training (compare mean of MSEs for GAGP in **Table 4** with the converged noise estimates in **Figure 8**).

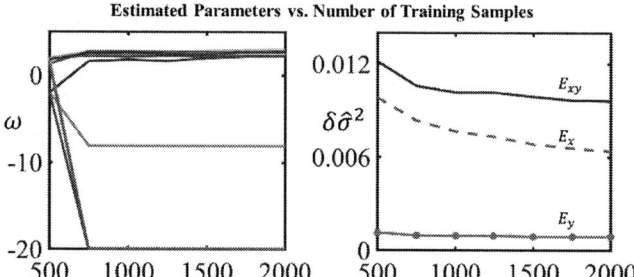

Figure 8 Convergence history as the number of training samples is increased from 500 to 2000. The 16 colored lines in the left panel indicate the histories of the 16 hyperparameters.

Table 4 MSE errors on 28000 random samples. The mean and variance of MSE are calculated over 20 random repetitions.

	Mean of MSE			Variance of MSE ($\times 10^6$)		
	E_x	E_y	E_{xy}	E_x	E_y	E_{xy}
GP	0.048	0.007	0.028	39	5.5	0.45
GAGP	0.008	0.001	0.011	0.12	0.0007	0.04

5.4 Data-Driven Unit Cell Optimization

In this section, we demonstrate how our GAGP model can be employed in an inverse optimization scheme to realize unit cells with target stiffness tensor components. Establishing such an inverse link is highly desirable in structure design as it allows to achieve desired elastic properties efficiently, obviating the tedious and expensive trial and error in TO. Additionally, though not demonstrated in this work, such a link can provide multiple candidate unit cells with the same properties that, in turn, enables tiling different unit cells into a macrostructure while ensuring boundary compatibility.

Our data-driven optimization scheme has two steps: The search for the optimal LB spectrum and the reconstruction of the unit cell given the LB spectrum. Using our GAGP (or GP) model, we directly search for the LB spectrum of the unit cell with the desired properties. We use GA in this search process, which is formulated as:

$$\min_{\lambda} \|E^t - E^p\|_\infty$$
$$s.t. \ \lambda_{i-1} \le \lambda_i$$
$$0.9\lambda_i^0 \le \lambda_i \le 1.1\lambda_i^0, \quad (24)$$

where E^t and E^p are the vectors of, respectively, the target and predicted stiffness tensors, and $\lambda = [\lambda_1, ..., \lambda_{16}]$ and λ^0 are

the LB spectra of the current unit cell and the unit cell closest to the prescribed properties in the property space respectively, with λ_i being the ith order eigenvalue. We choose GA for optimization since the GAGP model is cheap to run and GA ensures global optimality for multivariate and constrained problems. The search space for GA is defined by the LB spectrum of the unit cell in the training dataset whose properties are closest E^t.

After obtaining the optimal LB spectrum, we use a level set method to reconstruct the corresponding unit cell based on [55], with the squared residuals of the LB spectrum as the optimization objective. For faster convergence, the unit cell closest to the optimal LB spectrum in the spectrum space is taken as the initial guess in the reconstruction process.

In the following two examples, the goal is to design structures with desired E_x, E_y, and E_{xy} (see the targets in **Figure 9**). In each example, two unit cells are designed: one with the GP model (fitted to the initial set of 358) and one with the GAGP model (fitted to all 88000 structures). The results are visualized in **Figure 9** and demonstrate that the unit cells identified from GAGP are more geometrically diverse than those obtained via GP. This is likely a direct result of populating the large dataset with perturbed structures and, in turn, providing the GA search process with a wider range of initial seeds. It is also noted that the unit cells designed with GP are similar in shape but different in the size of the center hole, which leads to a significant change in properties.

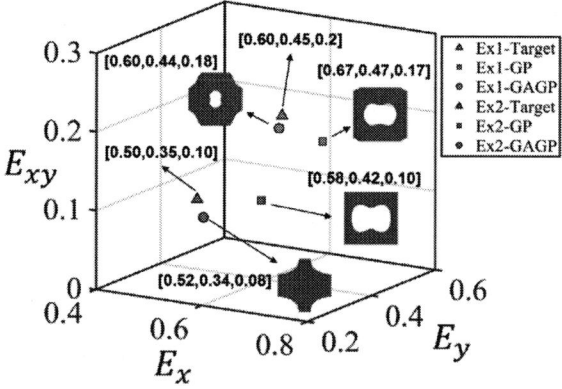

Figure 9 Reconstructed unit cells in the two examples. The results are visualized in the property space, $[E_x, E_y, E_{xy}]$.

From a quantitative point of view, it is observed that our data-driven design method with the large database can, as compared to the small dataset case, discover unit cells with properties that are closer to the target values. For instance, in Ex1, the GAGP result using the large dataset achieves the target E_x, whereas the GP result from the small dataset differs from the target by around 12%. Ex2 shows a similar pattern, with the GAGP and GP results differing from the target E_x by 4% and 16%, respectively. When the small dataset is used, the greater deviations from the target properties can be mainly attributed to insufficient training samples and the relatively small search space. This reinforces the need for a large database of unit cells in the data-driven design of metamaterials, along with an expedient machine learning method for big data.

6 CONCLUSION AND FUTURE WORKS

We propose a novel method to enable Gaussian process modeling of massive datasets. The central idea of our method, named globally approximate Gaussian process (GAGP), is based on the observation that the hyperparameter estimates of a GP converge to some limit values, $\hat{\omega}_\infty$, as more training samples are used. We introduce an intuitive and straightforward method to find $\hat{\omega}_\infty$ and, subsequently, build an ensemble of independent GPs that all use the converged $\hat{\omega}_\infty$ as their hyperparameters. These GPs randomly distribute the entire training dataset among themselves, which allows to make inference based on the entire dataset by pooling the predictions from the individual GPs.

With analytical examples, we demonstrated that GAGP achieves very high predictive power that matches (and in some cases exceeds) that of state-of-the-art machine learning methods such as neural networks and boosted trees. Unlike these latter methods, GAGP is easy to fit and interpret, which makes it particularly useful in engineering design with big data. In our approach, we assume that the noise is stationary with an unknown variance. Considering nonstationary noise variance would be an interesting and useful extension for GAGP. Thrifty sample selection for model refinement (instead of randomly taking subsets of training data) can also improve the predictive power of GAGP and is planned for our future works.

As a case study, we applied GAGP to a data-driven metamaterials unit cell design process that achieves desired elastic properties by transforming the complex material design problem into a parametric one. After mapping reduced-dimensional geometric descriptors (LB spectrum) to properties through GAGP, unit cells with properties close to the target values are discovered by finding the optimal LB spectrum with inverse optimization. This framework provides a springboard for a salient new approach to systematically and efficiently design metamaterials with optimized boundary compatibility, spatially varying properties, and multiple functionalities.

7 ACKNOWLEDGMENTS

The authors are grateful to Prof. K. Svanberg, from Royal Institute of Technology, Sweden, for providing a copy of the MMA code for metamaterial designs. Support from NSF grants (ACI 1640840 and OAC 1835782) and AFOSR FA9550-18-1-0381 are greatly appreciated. Ms. Yu-Chin Chan would like to acknowledge the NSF Graduate Research Fellowship Program under grant DGE-1842165.

8 REFERENCES

[1] Yan, W., Lin, S., Kafka, O. L., Lian, Y., Yu, C., Liu, Z., Yan, J., Wolff, S., Wu, H., and Ndip-Agbor, E., 2018, "Data-driven multi-scale multi-physics models to derive process–structure–property relationships for additive manufacturing," Computational Mechanics, 61(5), pp. 521-541.
[2] Mozaffar, M., Paul, A., Al-Bahrani, R., Wolff, S., Choudhary, A., Agrawal, A., Ehmann, K., and Cao, J., 2018, "Data-driven prediction of the high-dimensional thermal history in directed energy deposition processes via recurrent neural networks," Manufacturing letters, 18, pp. 35-39.
[3] Ghumman, U. F., Iyer, A., Dulal, R., Munshi, J., Wang, A., Chien, T., Balasubramanian, G., and Chen, W., 2018, "A Spectral Density Function Approach for Active Layer Design of Organic Photovoltaic Cells," J Mech Design, 140(11), p. 111408.
[4] Bessa, M. A., Bostanabad, R., Liu, Z., Hu, A., Apley, D. W., Brinson, C., Chen, W., and Liu, Wing K., 2017, "A framework for data-driven analysis of materials under uncertainty: Countering the curse of dimensionality," Comput Method Appl M, 320, pp. 633-667.
[5] Bostanabad, R., Chen, W., and Apley, D. W., 2016, "Characterization and reconstruction of 3D stochastic microstructures via supervised learning," Journal of microscopy, 264(3), pp. 282-297.
[6] Bostanabad, R., Zhang, Y., Li, X., Kearney, T., Brinson, L. C., Apley, D. W., Liu, W. K., and Chen, W., 2018, "Computational microstructure characterization and reconstruction: Review of the state-of-the-art techniques," Prog Mater Sci, 95, pp. 1-41.
[7] Li, X., Yang, Z., Brinson, L. C., Choudhary, A., Agrawal, A., and Chen, W., August 26–29, 2018, "A Deep Adversarial Learning Methodology for Designing Microstructural Material Systems," ASME 2018 International Design Engineering Technical Conferences and Computers and Information in Engineering Conference, American Society of Mechanical Engineers, Quebec City, Quebec, Canada.
[8] Bostanabad, R., Bui, A. T., Xie, W., Apley, D. W., and Chen, W., 2016, "Stochastic microstructure characterization and reconstruction via supervised learning," Acta Materialia, 103, pp. 89-102.
[9] Schumacher, C., Bickel, B., Rys, J., Marschner, S., Daraio, C., and Gross, M., 2015, "Microstructures to control elasticity in 3D printing," ACM Transactions on Graphics, 34(4), p. 136.
[10] Panetta, J., Zhou, Q., Malomo, L., Pietroni, N., Cignoni, P., and Zorin, D., 2015, "Elastic textures for additive fabrication," ACM Transactions on Graphics, 34(4), p. 135.
[11] Zhu, B., Skouras, M., Chen, D., and Matusik, W., 2017, "Two-Scale Topology Optimization with Microstructures," ACM Transactions on Graphics, 36(5), p. 164.
[12] Chen, D., Skouras, M., Zhu, B., and Matusik, W., 2018, "Computational discovery of extremal microstructure families," Science Advances, 4(1), p. eaao7005.
[13] LeCun, Y., Bengio, Y., and Hinton, G., 2015, "Deep learning," Nature, 521(7553), pp. 436-444.
[14] Chen, T., and Guestrin, C., "Xgboost: A scalable tree boosting system," Proc. 22nd ACM SIGKDD International Conference on Knowledge Discovery and Data Mining, ACM, pp. 785-794.
[15] Hassaninia, I., Bostanabad, R., Chen, W., and Mohseni, H., 2017, "Characterization of the Optical Properties of Turbid Media by Supervised Learning of Scattering Patterns," Scientific Reports, 7(1), p. 15259.
[16] Tao, S., Shintani, K., Bostanabad, R., Chan, Y.-C., Yang, G., Meingast, H., and Chen, W., "Enhanced Gaussian Process

Metamodeling and Collaborative Optimization for Vehicle Suspension Design Optimization," Proc. ASME 2017 International Design Engineering Technical Conferences and Computers and Information in Engineering Conference, American Society of Mechanical Engineers.

[17] Bostanabad, R., Liang, B., Gao, J., Liu, W. K., Cao, J., Zeng, D., Su, X., Xu, H., Li, Y., and Chen, W., 2018, "Uncertainty quantification in multiscale simulation of woven fiber composites," Comput Method Appl M, 338, pp. 506-532.

[18] Zhang, W., Bostanabad, R., Liang, B., Su, X., Zeng, D., Bessa, M. A., Wang, Y., Chen, W., and Cao, J., 2019, "A numerical Bayesian-calibrated characterization method for multiscale prepreg preforming simulations with tension-shear coupling," Composites Science and Technology, 170, pp. 15-24.

[19] Gramacy, R. B., and Lee, H. K. H., 2008, "Bayesian treed Gaussian process models with an application to computer modeling," Journal of the American Statistical Association, 103(483), pp. 1119-1130.

[20] Kim, H.-M., Mallick, B. K., and Holmes, C., 2005, "Analyzing nonstationary spatial data using piecewise Gaussian processes," Journal of the American Statistical Association, 100(470), pp. 653-668.

[21] Rasmussen, C. E., 2006, Gaussian processes for machine learning, MIT Press, Cambridge, MA.

[22] Tresp, V., 2000, "A Bayesian committee machine," Neural computation, 12(11), pp. 2719-2741.

[23] Herbrich, R., Lawrence, N. D., and Seeger, M., "Fast sparse Gaussian process methods: The informative vector machine," Proc. Advances in neural information processing systems, pp. 625-632.

[24] Seeger, M., Williams, C., and Lawrence, N., "Fast forward selection to speed up sparse Gaussian process regression," Proc. Artificial Intelligence and Statistics 9.

[25] Williams, C. K., and Seeger, M., "Using the Nyström method to speed up kernel machines," Proc. Advances in neural information processing systems, pp. 682-688.

[26] Rasmussen, C. E., and Quinonero-Candela, J., "Healing the relevance vector machine through augmentation," Proc. 22nd international conference on Machine learning, ACM, pp. 689-696.

[27] Wahba, G., 1990, Spline models for observational data, Siam.

[28] Gramacy, R. B., and Apley, D. W., 2015, "Local Gaussian process approximation for large computer experiments," J Comput Graph Stat, 24(2), pp. 561-578.

[29] Garcia, D. J., Mozaffar, M., Ren, H. Q., Correa, J. E., Ehmann, K., Cao, J., and You, F. Q., 2019, "Sustainable Manufacturing With Cyber-Physical Discrete Manufacturing Networks: Overview and Modeling Framework," Journal of Manufacturing Science and Engineering, 141(2), p. 021013.

[30] Mozaffar, M., Ndip-Agbor, E., Stephen, L., Wagner, G.J., Ehmann, K., and Cao, J., 2019, "Acceleration Strategies for Explicit Finite Element Analysis of Metal Powder-based Additive Manufacturing Processes using Graphical Processing Units," Computational Mechanics.

[31] Bostanabad, R., Kearney, T., Tao, S., Apley, D. W., and Chen, W., 2018, "Leveraging the nugget parameter for efficient Gaussian process modeling," Int J Numer Meth Eng, 114(5), pp. 501-516.

[32] MacDonald, B., Ranjan, P., and Chipman, H., 2015, "GPfit: AnRPackage for Fitting a Gaussian Process Model to Deterministic Simulator Outputs," Journal of Statistical Software, 64(12).

[33] Plumlee, M., and Apley, D. W., 2017, "Lifted Brownian Kriging Models," Technometrics, 59(2), pp. 165-177.

[34] Ranjan, P., Haynes, R., and Karsten, R., 2011, "A computationally stable approach to Gaussian process interpolation of deterministic computer simulation data," Technometrics, 53(4), pp. 366-378.

[35] Sacks, J., Schiller, S. B., and Welch, W. J., 1989, "Designs for Computer Experiments," Technometrics, 31(1), pp. 41-47.

[36] Toal, D. J. J., Bressloff, N. W., and Keane, A. J., 2008, "Kriging hyperparameter tuning strategies," Aiaa Journal, 46(5), pp. 1240-1252.

[37] Audet, C., and Dennis Jr, J. E., 2002, "Analysis of generalized pattern searches," SIAM Journal on optimization, 13(3), pp. 889-903.

[38] Zhao, L., Choi, K., and Lee, I., 2011, "Metamodeling method using dynamic kriging for design optimization," AIAA journal, 49(9), pp. 2034-2046.

[39] Toal, D. J., Bressloff, N., Keane, A., and Holden, C., 2011, "The development of a hybridized particle swarm for kriging hyperparameter tuning," Engineering optimization, 43(6), pp. 675-699.

[40] Conti, S., Gosling, J. P., Oakley, J. E., and O'Hagan, A., 2009, "Gaussian process emulation of dynamic computer codes," Biometrika, 96(3), pp. 663-676.

[41] Arendt, P. D., Apley, D. W., and Chen, W., 2012, "Quantification of Model Uncertainty: Calibration, Model Discrepancy, and Identifiability," Journal of Mechanical Design, 134(10), p. 100908.

[42] Arendt, P. D., Apley, D. W., Chen, W., Lamb, D., and Gorsich, D., 2012, "Improving identifiability in model calibration using multiple responses," J Mech Design, 134(10), p. 100909.

[43] Bayarri, M., Berger, J., Cafeo, J., Garcia-Donato, G., Liu, F., Palomo, J., Parthasarathy, R., Paulo, R., Sacks, J., and Walsh, D., 2007, "Computer model validation with functional output," The Annals of Statistics, pp. 1874-1906.

[44] Conti, S., and O'Hagan, A., 2010, "Bayesian emulation of complex multi-output and dynamic computer models," Journal of statistical planning and inference, 140(3), pp. 640-651.

[45] Sobol, I. M., 1967, "On the distribution of points in a cube and the approximate evaluation of integrals," Zhurnal Vychislitel'noi Matematiki i Matematicheskoi Fiziki, 7(4), pp. 784-802.

[46] Sobol, I. M., 1998, "On quasi-Monte Carlo integrations," Math Comput Simulat, 47(2-5), pp. 103-112.

[47] Hastie, T., Tibshirani, R., Friedman, J., Hastie, T., Friedman, J., and Tibshirani, R., 2009, The elements of statistical learning, Springer.

[48] Bastos, L. S., and O'Hagan, A., 2009, "Diagnostics for Gaussian process emulators," Technometrics, 51(4), pp. 425-438.

[49] Ben-Ari, E. N., and Steinberg, D. M., 2007, "Modeling data from computer experiments: an empirical comparison of kriging with MARS and projection pursuit regression," Quality Engineering, 19(4), pp. 327-338.

[50] Xia, L., and Breitkopf, P., 2015, "Design of materials using topology optimization and energy-based homogenization approach in Matlab," Structural and Multidisciplinary Optimization, 52(6), pp. 1229-1241.

[51] Andreassen, E., and Andreasen, C. S., 2014, "How to determine composite material properties using numerical homogenization," Comp Mater Sci, 83, pp. 488-495.

[52] Reuter, M., Wolter, F.-E., and Peinecke, N., 2006, "Laplace–Beltrami spectra as 'Shape-DNA' of surfaces and solids," Computer-Aided Design, 38, pp. 342-366.

[53] Lian, Z., Godil, A., Bustos, B., Daoudi, M., Hermans, J., Kawamura, S., Kurita, Y., Lavoué, G., Van Nguyen, H., Ohbuchi, R., Ohkita, Y., Ohishi, Y., Porikli, F., Reuter, M., Sipiran, I., Smeets, D., Suetens, P., Tabia, H., and Vandermeulen, D., 2013, "A comparison of methods for non-rigid 3D shape retrieval," Pattern Recognition, 46, pp. 449-461.

[54] Su, S., 2010, "Numerical approaches on shape optimization of elliptic eigenvalue problems and shape study of human brains," The Ohio State University.

[55] Zhu, S., 2018, "Effective shape optimization of Laplace eigenvalue problems using domain expressions of Eulerian derivatives," Journal of Optimization Theory and Applications, 176(1), pp. 17-34.

Proceedings of the ASME 2019
International Design Engineering Technical Conferences
and Computers and Information in Engineering Conference
IDETC/CIE2019
August 18-21, 2019, Anaheim, CA, USA

DETC2019-98115

MULTI-FIDELITY PHYSICS-CONSTRAINED NEURAL NETWORK AND ITS APPLICATION IN MATERIALS MODELING

Dehao Liu and Yan Wang[1]
Woodruff School of Mechanical Engineering
Georgia Institute of Technology
Atlanta, GA 30332, USA

ABSTRACT

Training machine learning tools such as neural networks requires the availability of sizable data, which can be difficult for engineering and scientific applications where experiments or simulations are expensive. In this work, a novel multi-fidelity physics-constrained neural network is proposed to reduce the required amount of training data, where physical knowledge is applied to constrain neural networks, and multi-fidelity networks are constructed to improve training efficiency. A low-cost low-fidelity physics-constrained neural network is used as the baseline model, whereas a limited amount of data from a high-fidelity simulation is used to train a second neural network to predict the difference between the two models. The proposed framework is demonstrated with two-dimensional heat transfer and phase transition problems, which are fundamental in materials modeling. Physics is described by partial differential equations. With the same set of training data, the prediction error of physics-constrained neural network can be one order of magnitude lower than that of a classical artificial neural network without physical constraints. The accuracy of the prediction is comparable to those from direct numerical solutions of equations.

Keywords: machine learning; multi-fidelity model; physics-constrained neural networks; materials modeling; partial differential equations

1. INTRODUCTION

Machine learning (ML) tools, exemplified by the convolutional neural network and its derivatives, have demonstrated success in diverse fields. However, they are very data-hungry during training and can easily fail in applications where data are scarce and expensive to collect. The root cause is the "curse of dimensionality" in training the ML tools. As ML

tools need to capture more detailed patterns or sensitive features, more complex modeling structures need to be introduced with more parameters and degrees of freedom. As a result, training algorithms need to explore and exploit in a very high-dimensional parameter space to search for optimal parameters. When the dimension increases, the volume of parameter spaces increases exponentially, so does the required amount of training data to cover the space and ensure the convergence of training. When the size of the training data set is small, overfitting can occur. That is, the training results in a spurious relationship that looks deceptively good but has low generality outside the labeled data range.

In various engineering and scientific applications, the cost of obtaining a large amount of data from high-fidelity simulations or experiments can be prohibitive. Data sparsity is not helpful to construct meaningful predictive ML models. Therefore, it is still a challenge for current state-of-the-art ML techniques to be applied in the domains of engineering or physical sciences. In the engineering domain, establishing high-dimensional process-structure-property relationships for either product or process design is the essential task. In engineering and scientific communities, human intelligence or knowledge has been embodied as physical laws or models based on centuries of data and knowledge accumulation. Giving up the available physical knowledge and purely relying on data-driven ML tools to identify the cause-effect relationships in physical sciences and engineering can be regarded as reinventing the wheel. Nevertheless, ML provides tools for systematic searching and exploring nonlinear and nonconvex relationships, which is much more efficient than ad hoc discovery. It is believed that training ML tools based on prior knowledge of physics can help navigate the high-dimensional parameter space with a small amount of training data.

---

[1] Contact author: yan.wang@me.gatech.edu

It is envisioned that the efficiency of training ML tools under the constraint of physical knowledge can be improved with small sample sizes. The physical laws or models can guide the searching and optimization procedures [1]. Generally, many physical laws are mathematically described as the relationships between physical quantities in the forms of ordinary differential equations (ODEs) or partial differential equations (PDEs). Some important and useful physical laws include but are not limited to conservation laws, laws of classical mechanics and thermodynamics. These physical laws have become the milestones of knowledge discovery in various scientific and engineering domains. Based upon physical laws or principles, various physics-based modeling and simulation techniques have been proposed to predict the behaviors of physical systems.

Incorporating physical meanings and physical knowledge in artificial neural networks (ANNs) has been studied from different perspectives. The first approach is to customize ANNs and incorporate physical meanings in the architecture. It has been demonstrated that ANN models can be applied to solve some special forms of optimization. For example, quadratic programming problems can be converted to linear complementarity problems and solved iteratively by projection neural networks [2,3]. Some efforts have been made for incorporating prior knowledge into ANNs in order to improve the training efficiency or prediction accuracy. Here, training efficiency means the convergence speed. For instance, prior knowledge can be applied as preprocessing tools to filter training data [4,5], or embedded as some analytical input-output functions in additional layers of ANNs [6], to improve the training efficiency. Prior knowledge can also be expressed as rules and interpreted with weights and basis functions in the ANN architecture, which could be further refined using training data [7,8]. Similarly, finite-element neural networks (FENNs) [9,10] can be constructed by transforming a finite element model to a neural network, where the weights of a FENN have physical meanings of material properties and can be computed in advance without training. FENNs have been used to obtain the solutions of differential equations for both forward and inverse problems. The major challenge of incorporating physical meanings into the ANN architecture is the complexity of customized networks. For instance, the number of weights in FENNs is related to the number of nodes, which could be very large for some high-dimensional problems with complex geometry.

The second approach to incorporate physical knowledge is treating it as constraints so that they can guide the training process. For instance, prior knowledge can be embedded into ANNs as architectural constraints and connection weight constraints to improve the training efficiency [11]. In addition to functional values, the information of derivatives has also been incorporated as prior knowledge for support vector regression [12]. ANNs have been used to approximate the solutions of PDEs. By transforming the original PDEs into their weighted residual forms, the prior knowledge of model forms and boundary values can be incorporated as penalty functions during the training of ANNs [13]. Similarly, the original model forms and boundary conditions, rather than their weighted residual

forms, can be directly embedded as regularization terms into the objective function during the training process [14]. A regularization parameter has been introduced to control the trade-off between data fitting and knowledge-based regularization [15]. It has been shown that regularized ANNs such as multi-layer perceptron (MLP) and radial basis function (RBF) neural networks can help obtain the solutions of ODEs and PDEs with higher accuracy and lower memory requirement than traditional numerical methods [16]. The initial and boundary conditions can also be incorporated as the regularization terms to improve the efficiency of ANN training. For instance, a trial solution is formulated such that it contains the information of both boundary conditions and the model form [17,18]. However, it may be difficult to find trial solutions for boundary value problems that are defined on irregular boundaries. To tackle this problem, a MLP-RBF synergy model [19] was further proposed, where the first part of the trial solution was replaced by the RBF neural network so that the boundary conditions on irregular boundaries can be satisfied. Another way to handle arbitrary irregular boundaries is introducing a length factor [20] into the second part of the trial solution. As a measure of distance from the boundary, the length factor returns zero on the boundary and nonzero inside the boundary so that the first part of the trial solution is unaffected. Similarly, regularized ANNs were applied to approximate the solutions of ODEs [21], and a comparison was conducted between the performance of four different ANNs to solve ODEs [22]. Instead of regularization, information about boundary conditions can be explicitly used as equality constraints between the weights in ANNs such that a constrained backpropagation training can be taken [23–25]. The effectiveness of regularization during the ML training has been demonstrated in the above work. However, the training efficiency is still limited in high-dimensional problems, where the sampling of solutions from PDEs or ODEs can be costly.

In this paper, a multi-fidelity framework for the physics-constrained neural network (PCNN) is proposed to help construct high-dimensional surrogate models more efficiently. Here, PCNNs are constructed to approximate and predict the solutions of PDEs. Some solutions from simulations serve as the training data. The prior knowledge of PDEs, as well as the initial and boundary conditions, are applied to guide the training process of PCNNs with reduced searching space. The multi-fidelity concept is introduced to further reduce the required amount of training data. By combining a low-fidelity physics-constrained neural network (LF-PCNN) and a high-fidelity physics-constrained neural network (HF-PCNN), a multi-fidelity physics-constrained neural network (MF-PCNN) can be created with a lower training cost and higher prediction accuracy. The LF-PCNN is trained with low-fidelity simulation results, whereas the HF-PCNN is trained from high-fidelity simulations. The MF-PCNN is constructed by combining the predictions from the LF-PCNN and the difference between the LF-PCNN and HF-PCNN predictions. The advantage of the MF-PCNN is that the overall amount of training data can be reduced in order to achieve the similar level of accuracy by using the HF-PCNN alone. In this paper, two examples are used to demonstrate the efficiency of the MF-PCNN framework. One example is the prediction of the temperature field in a heat transfer problem, and the other is the prediction of the phase field in a phase transition. It is shown that a MF-PCNN can be constructed with very limited simulation data to achieve a good accuracy of prediction.

In the remainder of this paper, the training of PCNNs, the construction of MF-PCNNs, and the setup of the computational scheme are described in Section 2. The computational results of the examples are shown in Section 3.

2. METHODOLOGY

In MF-PCNNs, the training data for LF-PCNNs and HF-PCNNs can be obtained from the analytical or numerical solutions of PDEs, e.g. from finite-element method (FEM). During the training, the prior knowledge about the form of PDEs or boundary values is added as the regularization terms in the loss function. The knowledge constraints provide guidance to the searching direction for optimization. The MF-PCNN is constructed based on the information from the LF-PCNN as well as the additional information that the HF-PCNN provides. The cost of obtaining high-fidelity information is higher than that of low-fidelity one. Therefore, the allocation of computational resources between high- and low-fidelity simulations can help reduce the overall training cost.

2.1 Training of PCNNs

Generally, a wide range of physical phenomena and dynamics can be described by PDEs, including heat transfer, advection-diffusion process, fluid dynamics, and others. Let us consider a time-dependent parametrized PDE with the general form

$$P\left(u, \frac{\partial u}{\partial t}, \frac{\partial u}{\partial \mathbf{x}}, \frac{\partial^2 u}{\partial t^2}, \frac{\partial^2 u}{\partial \mathbf{x}^2}, \dots\right) = f(t, \mathbf{x}), \; t \in [0, T], \; \mathbf{x} \in \Omega, \quad (1)$$

where $u(t, \mathbf{x})$ is the hidden solution to be found, $f(t, \mathbf{x})$ is a source or sink term, t is the time, $\mathbf{x} = (x_1, x_2, \dots, x_n)$ is the spatial vector, and $\Omega \in \mathbb{R}^n$ denotes the definition domain. This general PDE is subject to initial conditions (ICs)

$$I\left(u, \frac{\partial u}{\partial t}, \frac{\partial^2 u}{\partial t^2}, \dots\right) = g(\mathbf{x}), \; t = 0, \; \mathbf{x} \in \Omega, \quad (2)$$

and boundary conditions (BCs)

$$S\left(u, \frac{\partial u}{\partial t}, \frac{\partial u}{\partial \mathbf{x}}, \frac{\partial^2 u}{\partial t^2}, \frac{\partial^2 u}{\partial \mathbf{x}^2}, \dots\right) = h(t, \mathbf{x}), \; t \in [0, T], \; \mathbf{x} \in \partial\Omega, \quad (3)$$

where $\partial\Omega$ is the boundary of the definition domain. A more compact form of the above initial-boundary value problem can be written as

$$\mathbf{D}[u(t, \mathbf{x})] = f(t, \mathbf{x}), \; t \in [0, T], \; \mathbf{x} \in \Omega, \quad (4)$$
$$\Lambda[u(0, \mathbf{x})] = g(\mathbf{x}), \; t = 0, \; \mathbf{x} \in \Omega, \quad (5)$$
$$\Gamma[u(t, \mathbf{x})] = h(t, \mathbf{x}), \; t \in [0, T], \; \mathbf{x} \in \partial\Omega, \quad (6)$$

where $\mathbf{D}[\cdot]$, $\Lambda[\cdot]$, and $\Gamma[\cdot]$ are differential operators. For example, the three-dimensional (3D) heat equation without the source term corresponds to $\mathbf{D}[u(t, \mathbf{x})] = u_t - \alpha(u_{xx} + u_{yy} + u_{zz}) = 0$, where α is the thermal diffusivity, and the subscripts represent the partial derivatives with respect to either time or space.

In this work, the MLP architecture is used as a demonstration, which includes one input layer for $(t, \mathbf{x})$, multiple hidden layers, and one output layer for $U(t, \mathbf{x})$ to approximate the true solution $u(t, \mathbf{x})$. The neurons are connected with those in the neighbor layers, and the weights represent the strength of connections. The output from the hidden layer to the following layer is calculated as

$$y_i = \varphi\left(\sum w_{ij}\theta_j + b_i\right), \quad (7)$$

where w_{ij} is the weight of the connection between neuron j in the previous layer and neuron i in the current layer, θ_j is the j-th input value from the previous layer, and b_i is the bias for the neuron i in the current layer. φ is a nonlinear activation function, which can be sigmoid, tanh, rectified linear unit, or others.

The weights of a PCNN can be learned by minimizing the mean squared loss or total cost function

$$E = \lambda_T E_T + \lambda_P E_P + \lambda_I E_I + \lambda_s E_s, \quad (8)$$

where

$$E_T = \frac{1}{N_T}\sum_{i=1}^{N_T}|U(t_i^T, \mathbf{x}_i^T) - T(t_i^T, \mathbf{x}_i^T)|^2$$

is the loss caused by the discrepancy between the training data $T(\cdot)$ and the PCNN model prediction $U(\cdot)$, $\{t_i^{(\cdot)}, \mathbf{x}_i^{(\cdot)}\}$ denotes the sampling points in the defined domain, and $N_{(\cdot)}$ denotes the number of sampling points. Similarly,

$$E_P = \frac{1}{N_P}\sum_{i=1}^{N_P}|\mathbf{D}[U(t_i^P, \mathbf{x}_i^P)] - f(t_i^P, \mathbf{x}_i^P)|^2,$$
$$E_I = \frac{1}{N_I}\sum_{i=1}^{N_I}|\Lambda[U(t_i^I, \mathbf{x}_i^I)] - g(\mathbf{x}_i^I)|^2,$$

and

$$E_S = \frac{1}{N_S}\sum_{i=1}^{N_S}|\Gamma[U(t_i^S, \mathbf{x}_i^S)] - h(t_i^S, \mathbf{x}_i^S)|^2$$

are the losses caused by the violations of the model, initial conditions, and boundary conditions as the physical constraints from Eqs. (4)-(6). The constraint on weights of different losses is given as

$$\lambda_T + \lambda_P + \lambda_I + \lambda_s = 1. \quad (9)$$

The relative importance of prior knowledge can be adjusted by changing the weights of physical constraints λ_P, λ_I and λ_S. If the total loss function only includes the training loss E_T, then this is the traditional pure data-driven ANN to solve the initial-boundary value problem. By adding physical losses E_P, E_I and E_S as the regularization terms, the prior physical knowledge can help to reduce the size of searching space and provide guidance for the searching directions in training.

2.2 Construction of MF-PCNNs

The LF-PCNN and HF-PCNN must be trained first before the MF-PCNN is constructed. In this work, the fidelities are determined by the resolutions of FEM simulations given the same density of physical constraints. To be more specific, low-fidelity simulations are used to construct the LF-PCNN during a long time period $t \in [0, T]$, whereas high-resolution simulations are applied for the HF-PCNN during a short time period $t \in [0, T_0]$ $(T_0 < T)$.

After the LF-PCNN and HF-PCNN are trained, the difference between the predictions of the LF-PCNN $U_L(t,\mathbf{x})$ and HF-PCNN $U_H(t,\mathbf{x})$ is calculated as

$$\delta(t,\mathbf{x}) = U_H(t,\mathbf{x}) - U_L(t,\mathbf{x}), \ t \in [0,T_0], \ \mathbf{x} \in \Omega. \quad (10)$$

Then another ANN called difference artificial neural network (DANN) is constructed to predict the difference between the LF-PCNN and HF-PCNN, denoted as $U_\delta(t,\mathbf{x})$, during a longer time period $t \in [0,T]$. The weights of the DANN can be learned by using the observed difference $\delta(t,\mathbf{x})$ as the training data to minimize the mean squared error loss

$$E_\delta = \frac{1}{N_\delta}\sum_{i=1}^{N_\delta}|U_\delta(t_i,\mathbf{x}_i) - \delta(t_i,\mathbf{x}_i)|^2 , \ t \in [0,T_0], \ \mathbf{x} \in \Omega, \quad (11)$$

where N_δ is the number of sampling points for the DANN. It is assumed that the evolution of the difference between the LF-PCNN and HF-PCNN during a longer time period $t \in [0,T]$ can be predicted by the DANN using the observed difference $\delta(t,\mathbf{x})$ as the training data during the short time period $t \in [0,T_0]$. Then the MF-PCNN is a combination of the LF-PCNN and DANN. The prediction from the MF-PCNN during the time period $t \in [0,T]$ is given by

$$U_M(t,\mathbf{x}) = U_L(t,\mathbf{x}) + U_\delta(t,\mathbf{x}), \ t \in [0,T], \ \mathbf{x} \in \Omega. \quad (12)$$

2.3 Experimental setup of the proposed MF-PCNN

The construction and training of the MF-PCNN are accomplished by using Tensorflow [26], which is an open-source Python library for machine learning. The partial derivatives of the ANNs are calculated based on the chain rules using the automatic differentiation [27]. Automatic differentiation is different from the numerical differentiation such as the method of finite difference. By applying the chain rules repeatedly, the derivatives of arbitrary order can be computed automatically and accurately to a working precision.

Two examples are applied to demonstrate the proposed MF-PCNN framework. The first example is a heat transfer problem where the evolution of the two-dimensional (2D) temperature distribution is modeled with the heat equation. The heat transfer example is used to demonstrate the effectiveness of the PCNN and test different weighting schemes of the total loss function. The second example is the phase transition problem where the evolution of the 2D phase field is modeled with the Allen-Cahn equation. The phase transition example is utilized to demonstrate the efficiency of the MF-PCNN framework.

The details of the computational setup for different ML models in the heat transfer and the phase transition examples are listed in TABLE 1 and TABLE 2, respectively. The ANNs, LF-PCNNs, and HF-PCNNs have the same structure of 30-20-30-20. That is, each of the networks has 4 layers. There are 30 neurons in the first and third layer, and 20 neurons in the second and last layer. The structures of the DANNs are simpler to avoid overfitting, which are 5-5-5-5 and 10-10-10-10. Two Gaussian process (GP) surrogate models with the RBF kernel are also constructed to predict the difference between the LF-PCNN and HF-PCNN for comparison purpose. The tanh function is used as the activation function. All of the loss functions of neural networks are minimized by using a gradient-based optimization algorithm called Adam [28] for the consideration of efficiency.

The training data for the ANNs, LF-PCNNs, and HF-PCNNs come from the FEM solutions of COMSOL, whereas the training data for the DANNs and GPs come from the observed differences between the predictions of the LF-PCNNs and HF-PCNNs during the short time period $t \in [0,T_0]$. The sampling strategy is uniform in both temporal and spatial dimensions for the convenience of comparison with the FEM solutions. Other sampling strategies such as random sampling, orthogonal sampling, and Latin Hypercube sampling can also be adopted.

TABLE 1 and TABLE 2 list the sizes of training data sets for the heat transfer and phase transition examples respectively. For instance, the amount of training data for the ANNs is $21\times6\times6$, which means that there are 21 sampling points in the temporal dimension, 6 sampling points in the x direction of the spatial domain, and 6 sampling points in the y direction of the spatial domain. For the PCNNs in the heat transfer example, the number of physical constraints is $41\times11\times11$. That is, there are 41 sampling points in the temporal dimension, 11 sampling points in the x direction and 11 sampling points in the y dimension of the spatial domain. The time period represents the size of the sampling space in the temporal dimension. In the heat transfer example, three different weighting schemes (PCNN1, PCNN2, and PCNN3) are compared. In the phase transition example, two HF-PCNNs (HF-PCNN1 and HF-PCNN2) are trained. The HF-PCNN1 is trained during the time period $t \in [0,0.2]$, whereas the HF-PCNN2 is trained during two time periods, $t \in [0,0.2]$ and $t \in [0.8,1]$. Therefore, the amount of training data and the number of physical constraints for the HF-PCNN2 are twice those of the HF-PCNN1. The observed difference between the predictions of the LF-PCNN and HF-PCNN1 is served as the training data for the DANN1, DANN2, and GP1. Similarly, the observed difference between the predictions of the LF-PCNN and HF-PCNN2 is served as the training data for the DANN3, DANN4, and GP2. For ANNs, LF-PCNNs, and HF-PCNNs, the training of a neural network stops when the total loss E is lower than a threshold value 0.01. Similarly, the training of a DANN stops when the loss function E_δ is below 0.01.

TABLE 1: The setup for different ML models in the heat transfer example

ML model	Structure	Amount of training data ($t\times x\times y$)	Number of physical constraints ($t\times x\times y$)	Time period/s
ANN	30-20-30-20	$21\times6\times6$	0	[0, 1]
PCNN1, PCNN2, PCNN3	30-20-30-20	$21\times6\times6$	$41\times11\times11$	[0, 1]

TABLE 2: The setup for different ML models in the phase transition example

ML model	Structure	Amount of training data ($t \times x \times y$)	Number of physical constraints ($t \times x \times y$)	Time period/s
ANN	30-20-30-20	21×6×6	0	[0, 1]
LF-PCNN	30-20-30-20	21×6×6	21×11×11	[0, 1]
HF-PCNN1	30-20-30-20	9×21×21	5×11×11	[0, 0.2]
HF-PCNN2	30-20-30-20	18×21×21	10×11×11	[0, 0.2], [0.8, 1]
DANN1	5-5-5-5	9×26×26	0	[0, 0.2]
DANN2	10-10-10-10	9×26×26	0	[0, 0.2]
DANN3	5-5-5-5	18×26×26	0	[0, 0.2], [0.8, 1]
DANN4	10-10-10-10	18×26×26	0	[0, 0.2], [0.8, 1]
GP1	RBF kernel	9×26×26	0	[0, 0.2]
GP2	RBF kernel	18×26×26	0	[0, 0.2], [0.8, 1]

3. EXPERIMENTAL RESULTS

In this section, the results for the heat transfer and phase transition examples are shown. The heat transfer example is used to demonstrate the effectiveness of the PCNN and test different weighting schemes of the total loss function. A convergence analysis for the ANN and the PCNN is also conducted. The phase transition problem is to demonstrate the efficiency of the MF-PCNN framework.

3.1 Heat equation

The evolution of temperature distributions can be modeled by parabolic PDEs. The heat equation describes the diffusion process of energy, which is important in modeling microstructure evolution during phase transition. The 2D heat equation with the zero Neumann boundary condition used in this example is

$$
\begin{cases}
u_t - 0.01\big(u_{xx} + u_{yy}\big) = 0, & t,x,y \in [0,1], \\
u(0,x,y) = 0.5[sin(4\pi x) + sin(4\pi y)], \\
u_x(t,0,y) = 0, \\
u_x(t,1,y) = 0, \\
u_y(t,x,0) = 0, \\
u_y(t,x,1) = 0.
\end{cases}
\tag{13}
$$

where u is the 2D temperature field.

The goal of training a neural network is to ensure the prediction $U(t,x,y)$ from the neural network can approximate the true solution $u(t,x,y)$ from FEM simulations with the desired accuracy. Here, the physical loss is

$$
E_P = \frac{1}{N_P}\sum_{i=1}^{N_P}
\left|
\begin{matrix}
U_t(t_i^P, x_i^P, y_i^P) \\
-0.01 U_{xx}(t_i^P, x_i^P, y_i^P) \\
-0.01 U_{yy}(t_i^P, x_i^P, y_i^P)
\end{matrix}
\right|^2 .
\tag{14}
$$

The initial loss is given by

$$
E_I = \frac{1}{N_I}\sum_{i=1}^{N_I}
\left|
\begin{matrix}
U(0, x_i^I, y_i^I) \\
-0.5[sin(4\pi x_i^I) + sin(4\pi y_i^I)]
\end{matrix}
\right|^2 .
\tag{15}
$$

The boundary loss is

$$
E_S = \frac{1}{N_S}\sum_{i=1}^{N_S}
\left[
\begin{matrix}
\left|U_x(t_i^S, 0, y_i^S)\right|^2 + \left|U_x(t_i^S, 1, y_i^S)\right|^2 \\
+\left|U_y(t_i^S, x_i^S, 0)\right|^2 + \left|U_y(t_i^S, x_i^S, 1)\right|^2
\end{matrix}
\right].
\tag{16}
$$

To assess the sensitivity of weights, three weighting schemes of the total loss function are tested and compared with each other. In the PCNN1, the weights are equal and fixed in the total loss function

$$
E = 0.25(E_T + E_P + E_I + E_s).
\tag{17}
$$

In the PCNN2, the weights are unequal and fixed in the total loss function

$$
E = 0.125(E_T + 2E_P + 4E_I + E_s).
\tag{18}
$$

In the PCNN3, the weights are adaptive during the training, which are proportional to the percentages of individual losses in the total loss function

$$
E = \frac{E_T^2 + E_P^2 + E_I^2 + E_S^2}{E_T + E_P + E_I + E_s}.
\tag{19}
$$

Assigning higher weights to the physical constraints indicates that prior knowledge will be more influential in the training process. When the training data is sparse, increasing the number of physical constraints can help improve the training efficiency. In addition, the weights of physical constraints need to be large enough in order to ensure the training efficiency and prediction accuracy. When the weights of physical constraints are assigned, it is also necessary to consider the balance among different losses such that the reduction speeds of the four errors are comparable. The ideal case is that the four losses are reduced at the same speed so that the overall reduction speed of the total loss is maximized.

Here, the training data come from the FEM solutions. FIGURE 1 shows the original FEM solution of the temperature field, as well as the predictions by the traditional ANN, the equally-weighted PCNN1, the unequally-weighted PCNN2, and the adaptively-weighted PCNN3 at $t = 1$, respectively. The errors of the predicted temperature fields compared with the original FEM solution for different neural networks at $t = 1$ are shown in FIGURE 2. Here, the prediction error is the absolute difference between the prediction from a neural network and the FEM solution. The dots in the figures indicate the evaluation points. There are 26×26 evaluation points in the 2D domain. It is seen

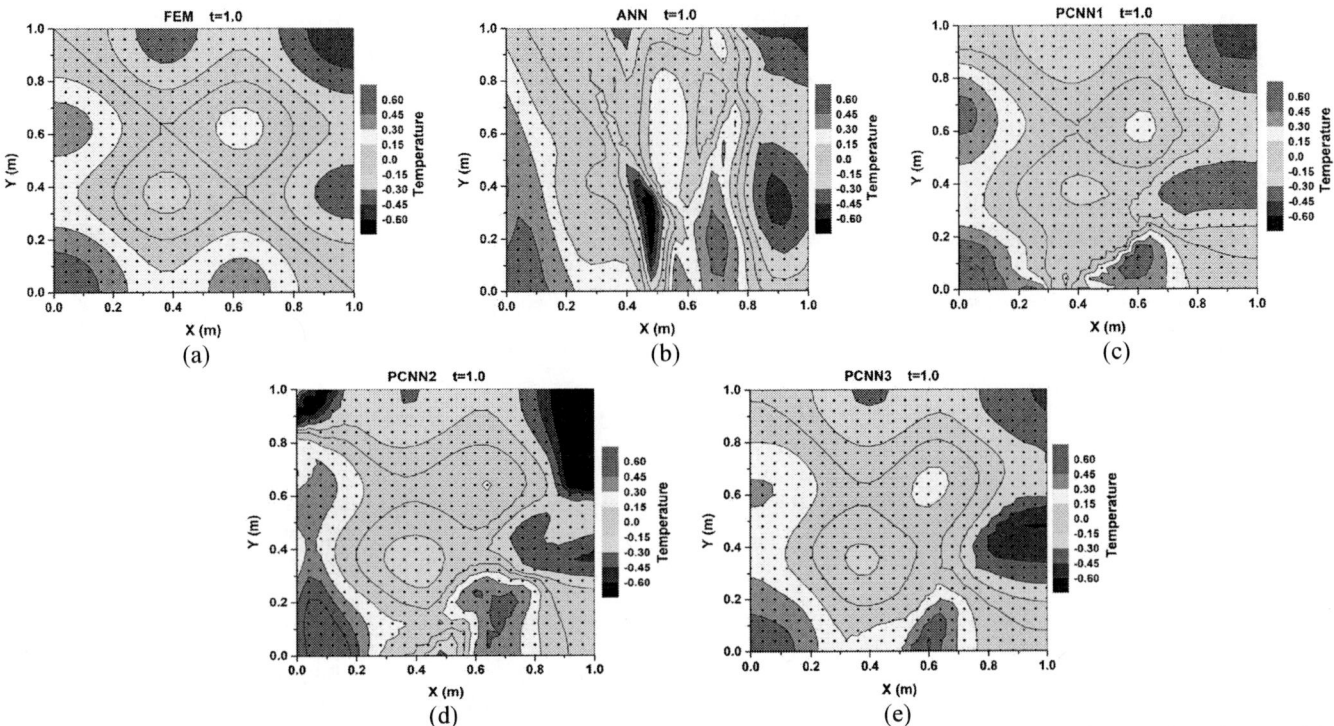

FIGURE 1: The predicted temperature fields from different models at $t = 1$: (a) original FEM solution, (b) traditional ANN, (c) equally-weighted PCNN1, (d) unequally-weighted PCNN2, and (e) adaptively-weighted PCNN3.

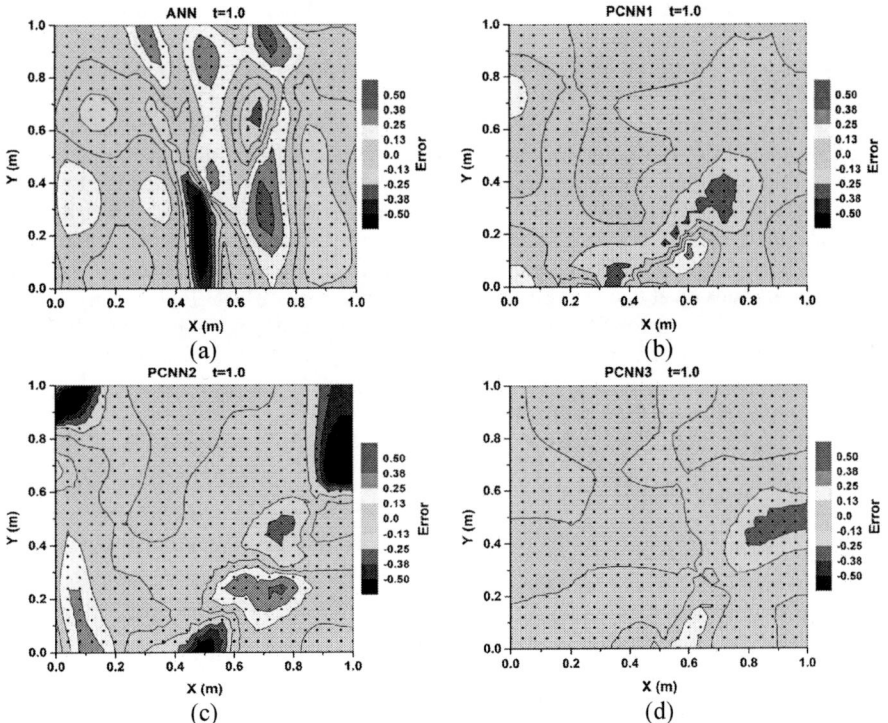

FIGURE 2: The errors of the predicted temperature fields compared to the FEM solution at $t = 1$: (a) traditional ANN, (b) equally-weighted PCNN1, (c) unequally-weighted PCNN2, and (d) adaptively-weighted PCNN3.

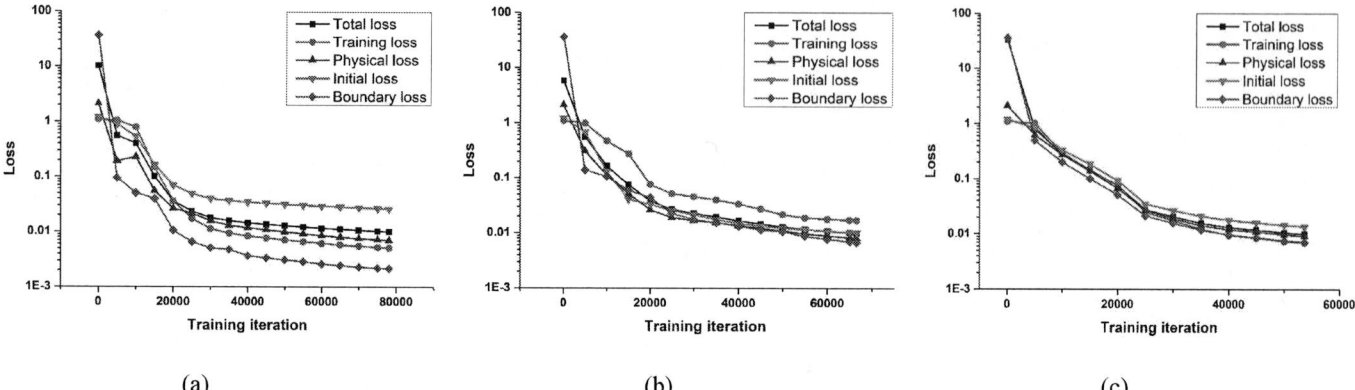

(a)   (b)   (c)

FIGURE 3: The learning curves for different PCNNs: (a) the equally-weighted PCNN1, (b) the unequally-weighted PCNN2, and (c) the adaptively-weighted PCNN3.

that the prediction from the ANN is less accurate than the three PCNNs, because of the small training data set. The error is especially large in the area around saddle points. Notice that the training data for the ANN and PCNNs come from the same LF simulations. With physical constraints added as regularization terms, the prediction errors of the PCNNs are reduced significantly.

The learning curves for different PCNNs are shown in FIGURE 3. For the three different PCNNs, all losses monotonically decrease during the training. However, the difference between the convergence speeds of individual losses varies with the different weighting schemes. For the equally-weighted PCNN1, as shown in FIGURE 3(a), the initial loss is one order of magnitude larger than the boundary loss, meaning that the difference between the convergence speeds of individual losses is large. Therefore, it takes a longer time for the PCNN1 to converge. For the unequally-weighted PCNN2, the weights of physical constraints are higher in order to increase the influence of prior knowledge. As a result, the different losses are within the same order of magnitude, as shown in FIGURE 3(b). As for the adaptively-weighted PCNN3, the weights are dynamically adjusted based on the percentages of individual losses in the total loss function. As shown in FIGURE 3(c), the different losses converged at the same speed and are well-balanced. The training time is the shortest among the three cases.

The quantitative comparison of training time and the mean squared error (MSE) of prediction for four neural networks is listed in TABLE 3. All MSEs of prediction for the PCNN1 and PCNN3 are almost one order of magnitude lower than that for the ANN. As a result of stronger enforcement for the physical constraints, the prediction accuracy of the PCNN2 is higher than that of the PCNN1 at $t = 0$. However, the MSE of prediction at $t = 1$ for the PCNN2 is larger than that of the PCNN1. This could be caused by the in-balance between different losses in the PCNN2. As shown in FIGURE 3(b), the training loss is still larger than the threshold value 0.01 when the training is finished, although the total loss as the weighted average has reached the threshold. The adaptively-weighted PCNN3 has all individual losses well-balanced and has the highest prediction accuracy.

The PCNN3 also has the least training time among the three PCNNs. Notice that the computational time for training the PCNNs is much longer than that for the ANN, because additional information from physical knowledge is used in the training. In engineering applications, simulations, especially high-fidelity ones, are computationally expensive. Therefore, the simulation results as the training data are sparse. However, prior knowledge can be obtained without expensive computation, which can be regarded as the supplemental data for training.

The convergence speeds of the ANN and the adaptively-weighted PCNN3 with respect to the amount of training data are compared in FIGURE 4. It is shown that the required amount of training data to reach certain accuracy level of prediction at time $t = 1$ can be reduced by adding physical constraints. Here, the number of physical constraints of the PCNN3 is $21 \times 6 \times 6 = 756$. The prediction MSEs at $t = 1$ of both ANN and PCNN decrease when the training data size increases. The advantage of PCNN over ANN is obvious when the training data size is small. When the training data size is less than 400, the prediction accuracy can have nearly one order of magnitude difference. To reach the same accuracy level of 0.01, the ANN requires about 900 training data points, whereas the PCNN only needs about 300 training data points. As the training data size increases, the difference of prediction accuracy between the ANN and PCNN gradually reduces.

TABLE 3: Quantitative comparison for different neural networks to solve the heat equation

Neural network	Training time (second)	MSE of prediction at $t = 0$	MSE of prediction at $t = 1$
ANN	8.66	0.1998	0.0293
PCNN1	1475.40	0.0225	0.0079
PCNN2	1259.91	0.0125	0.0350
PCNN3	1019.07	0.0139	0.0055

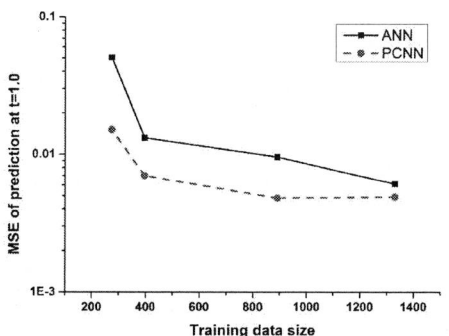

FIGURE 4: Convergence analysis for the ANN and the PCNN3.

3.2 Allen-Cahn equation

The second example is the Allen-Cahn equation, which is a nonlinear reaction-diffusion equation that describes the process of phase transition such as grain growth and spinodal decomposition. It has become the foundational model for the interface diffusion in the phase-field method, which is developed to study phase transitions and interfacial dynamics in materials science. The purpose of this example is to demonstrate the proposed MF-PCNN framework. The Allen-Cahn equation with periodic boundary condition in this example is

$$\begin{cases} u_t - 0.001\left(u_{xx} + u_{yy}\right) = u - u^3, \ t, x, y \in [0,1], \\ u(0, x, y) = 0.5\left[sin(4\pi x) + sin(4\pi y)\right], \\ u(t, 0, y) = u(t, 1, y), \\ u_x(t, 0, y) = u_x(t, 1, y), \\ u(t, x, 0) = u(t, x, 1), \\ u_y(t, x, 0) = u_y(t, x, 1). \end{cases} \tag{20}$$

where a non-conserved variable u is the order parameter or phase field.

Based on the results of the previous example, the weights of the physical constraints are adaptively adjusted as in Eq. (19). The physical loss is given by

$$E_P = \frac{1}{N_P}\sum_{i=1}^{N_P}\left| \begin{matrix} U_t\left(t_i^P, x_i^P, y_i^P\right) \\ -0.001 U_{xx}\left(t_i^P, x_i^P, y_i^P\right) \\ -0.001 U_{yy}\left(t_i^P, x_i^P, y_i^P\right) \\ -U\left(t_i^P, x_i^P, y_i^P\right) + U^3\left(t_i^P, x_i^P, y_i^P\right) \end{matrix} \right|^2. \tag{21}$$

The initial loss is given by

$$E_I = \frac{1}{N_I}\sum_{i=1}^{N_I}\left| \begin{matrix} U\left(0, x_i^I, y_i^I\right) \\ -0.5\left[sin\left(4\pi x_i^I\right) + sin\left(4\pi y_i^I\right)\right] \end{matrix} \right|^2. \tag{22}$$

The boundary loss is given by

$$E_S = \frac{1}{N_S}\sum_{i=1}^{N_S}\left[\begin{matrix} \left|U\left(t_i^S, 0, y_i^S\right) - U\left(t_i^S, 1, y_i^S\right)\right|^2 \\ +\left|U_x\left(t_i^S, 0, y_i^S\right) - U_x\left(t_i^S, 1, y_i^S\right)\right|^2 \\ +\left|U\left(t_i^S, x_i^S, 0\right) - U\left(t_i^S, x_i^S, 1\right)\right|^2 \\ +\left|U_y\left(t_i^S, x_i^S, 0\right) - U_y\left(t_i^S, x_i^S, 1\right)\right|^2 \end{matrix} \right]. \tag{23}$$

As shown in Eq. (12), the prediction of a MF-PCNN is a combination of the LF-PCNN prediction and the difference predicted by a ML model (DANN or GP). First, a low-cost LF-PCNN is trained during the time period $t \in [0, 1]$ and then used as the baseline model. In addition, two high-cost HF-PCNNs (HF-PCNN1 and HF-PCNN2) are constructed. As shown in TABLE 2, the HF-PCNN1 is trained with data for the time period $t \in [0, 0.2]$, whereas the HF-PCNN2 is trained with the data for two time periods, $t \in [0, 0.2]$ and $t \in [0.8, 1]$. Then DANNs and GPs are trained to predict the differences between the LF-PCNN and HP-PCNN predictions during the time period $t \in [0, 1]$. The observed difference between the predictions of the LF-PCNN and HF-PCNN1 during the time period $t \in [0, 0.2]$ serves as the training data for the DANN1, DANN2, and GP1. The network structure of DANN2 is more complex than DANN1. Similarly, the observed difference between the predictions of the LF-PCNN and HF-PCNN2 for two time periods, $t \in [0, 0.2]$ and $t \in [0.8, 1]$, serves as the training data for the DANN3, DANN4, and GP2. Finally, the prediction of the MF-PCNN is the sum of the LF-PCNN prediction and the predicted difference by DANNs or GPs. In this example, the mean square of the difference between the LF simulation and HF simulation is 0.0001 during the time period $t \in [0, 0.2]$. However, since a coarser mesh and larger time step is used in the LF simulation, errors are accumulated over time. Then, the mean square of the difference between the LF simulation and HF simulation becomes 0.0029 during the time period $t \in [0.8, 1.0]$. Therefore, LF simulations are less accurate than HF simulations in the later stage. It is necessary and useful to adopt the MF-PCNN framework to fully utilize the training data with different fidelity.

The predictions of the phase field from different models, including traditional ANN, LF-PCNN, multi-fidelity models (combinations of LF-PCNN and DANNs, as well as LF-PCNN and GPs), at time $t = 0.5$ are shown in FIGURE 5. It is seen that the traditional ANN has larger prediction errors than PCNNs, especially at the saddle points where the true values are zeros. Adding physical constraints can significantly reduce the prediction errors, as in the LF-PCNN. At some saddle points, the phase field predicted by the LF-PCNN is still larger than zero, as shown in FIGURE 5(c). Compared to the LF-PCNN, the prediction errors of MF-PCNNs can be further reduced by adding the prediction of the difference from DANNs or GPs. As shown in FIGURE 5(d-i), the phase field predicted by the MF-PCNNs is almost zero at all saddle points. The difference between the predictions of the LF-PCNN and HF-PCNN can be captured by DANNs or GPs very well.

V02AT03A007-8   Copyright © 2019 ASME

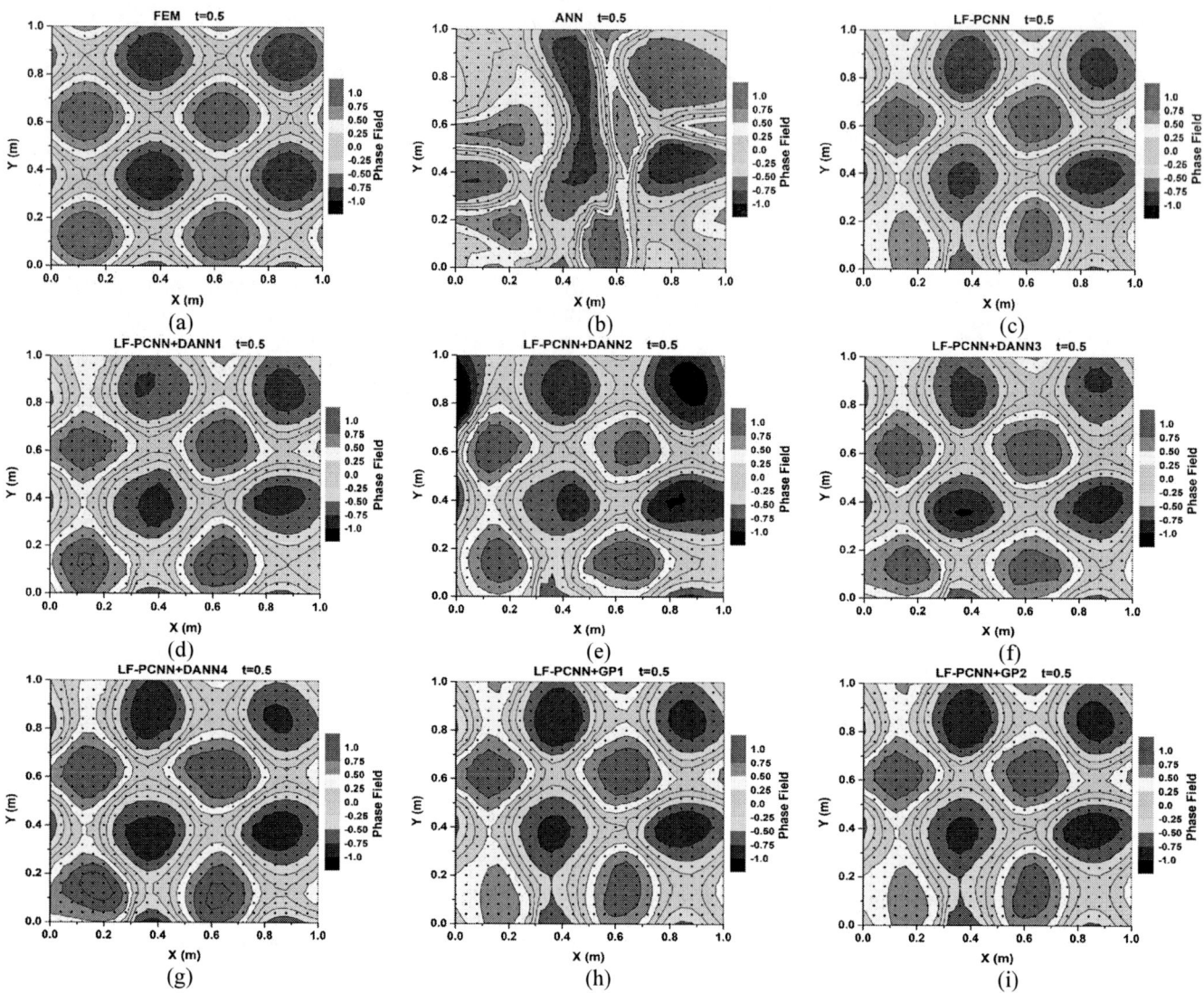

FIGURE 5: The predicted phase fields from different models at $t = 0.5$.

The quantitative comparisons of training time and the MSE of prediction for different ML models to solve the Allen-Cahn equation are listed in TABLE 4, where a MF-PCNN is composed of a LF-PCNN and a ML model to predict the difference. For example, MF-PCNN1 = LF-PCNN+DANN1 means that the MF-PCNN1 is a combination of the LF-PCNN and DANN1. The total training time of a MF-PCNN is the sum of training times for the LF-PCNN, HF-PCNN, and the difference ML model (DANN or GP). It is noted that the prediction of the HF-PCNN1 is used in the training of the MF-PCNN1, MF-PCNN2, and MF-PCNN3, whereas the prediction of the HF-PCNN2 is used in the training of the rest of the MF-PCNNs. Therefore, the training times of the MF-PCNN4, MF-PCNN5, and MF-PCNN6 are longer because of more training data and physical constraints. The training time of the MF-PCNNs with GPs is longer than that of the MF-PCNNs with DANNs because GPs are computationally more expensive.

The MSEs of predictions at different simulated time periods for different ML models are shown in FIGURE 6. In general, the MSE of prediction increases over time for different ML models except the MF-PCNN3 and MF-PCNN6. Since the prediction of the phase field relies on the previous predictions, the error will be accumulated over time. It is noted that the time period $t \in [1, 2]$ is outside the time range $t \in [0, 1]$ of LF training data for the LF-PCNN. Therefore, the error for extrapolation is larger, which is a common issue for most ML models. Nevertheless, the MSEs of extrapolation for the LF-PCNN, MF-PCNN1 and MF-PCNN4 are one order of magnitude lower than that of the ANN.

TABLE 4: Quantitative comparison between different ML models to solve the Allen-Cahn equation

ML model	Training time (second)	MSE of prediction at $t = 0.5$	MSE of prediction at $t = 1.5$
ANN	7.93	0.2215	0.8866
LF-PCNN	774.32	0.0258	0.0684
MF-PCNN1=LF-PCNN+DANN1	774.32+324.37+79.52=1178.21	0.0133	0.0521
MF-PCNN2=LF-PCNN+DANN2	774.32+324.37+25.19=1123.88	0.0753	0.8508
MF-PCNN3=LF-PCNN+GP1	774.32+324.37+1433.66=2532.35	0.0258	0.0684
MF-PCNN4=LF-PCNN+DANN3	774.32+3095.68+100.38=3970.38	0.0114	0.0399
MF-PCNN5=LF-PCNN+DANN4	774.32+3095.68+58.01=3928.01	0.0173	0.1926
MF-PCNN6=LF-PCNN+GP2	774.32+3095.68+10730.40=14600.40	0.0258	0.0684

The MSE of prediction from the MF-PCNN1 is significantly lower than that of the LF-PCNN for $t \in [0, 1]$. The difference between the MSEs however decreases for $t \in [1, 2]$. Furthermore, the MSE of prediction at $t = 0.5$ for the MF-PCNN1 is decreased by about 50%, compared with that of the LF-PCNN. As for the MF-PCNN2, its MSE of prediction is higher than that of LF-PCNN when $t > 0.5$. The MSE of prediction for the MF-PCNN2 is almost the same as that of the ANN when $t > 0.75$. The increased MSE for the MF-PCNN2 is caused by overfitting since the DANN2 has more neurons than the DANN1. The MSE of prediction at $t = 0$ for the MF-PCNN3 is slightly lower than that of the LF-PCNN. However, the MSE of prediction at $t = 0.25$ for the MF-PCNN3 is larger that of the LF-PCNN. The MSE of prediction for the MF-PCNN3 becomes the same as that of the LF-PCNN when $t > 0.5$, which means that the GP1 fails to predict the differences and its output is zero. Notice that $t = 0.5$ is outside the time range $t \in [0, 0.2]$ of the HF training data for the HF-PCNN1. The prediction is based on extrapolation. The errors indicate that DANNs are more robust than GPs for extrapolation.

With more training data and physical constraints, the HF-PCNN2 has two sampling spaces in the temporal dimension, which are [0, 0.2] and [0.8, 1]. The observed difference between the predictions of the LF-PCNN and HF-PCNN2 is served as the training data for the DANN3, DANN4, and GP2. Therefore, the prediction of the difference between the LF-PCNN and HF-PCNN at t = 0.5 has become an interpolation problem. Compared to the MF-PCNN1, the MSE of prediction for the MF-PCNN4 is the lowest among all ML models for the most of the time. With more training data, the MSE of prediction for the MF-PCNN5 is lower than that of the MF-PCNN2. However, the MSE of prediction for the MF-PCNN5 becomes higher than that of the MF-PCNN4 when $t > 0.25$ because of the overfitting. Compared to the MF-PCNN3, the MSE of prediction for the MF-PCNN6 is reduced with more training data when $t \in [0.5, 1.25]$. However, the MSE of prediction for the MF-PCNN6 becomes the same as that of the LF-PCNN when $t > 1.25$, indicating the failure of prediction by GP2.

Among all ML models in this work, the MF-PCNN1 is the best one in comprehensive performance since it has a relatively low training time and very good accuracy. The good generalization of the MF-PCNN1 comes from the simpler neural network structure of the DANN1.

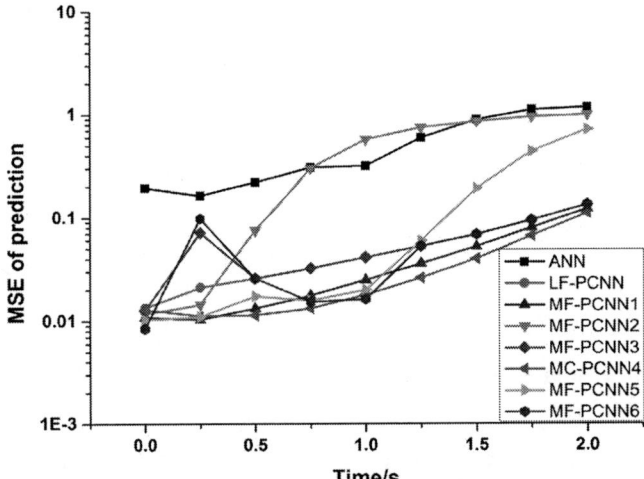

FIGURE 6: The change of MSE of prediction for different ML models.

4. CONCLUSION

In this work, a new scheme of multi-fidelity physics-constrained neural networks is proposed to improve the efficiency of training in neural networks by reducing the required amount of training data and incorporating physical knowledge as constraints. Neural networks with two (or more) levels of fidelities are combined to improve the prediction accuracy. Low-fidelity networks predict the general trend, whereas high-fidelity networks model local details and fluctuations. For the concern of training cost, low-fidelity networks can be trained with low-fidelity data, and the prediction accuracy can be further improved with supplementary high-fidelity data. Thus, the training efficiency is improved from two aspects. The first one is the guidance from the physical knowledge, and the second one is a more cost-effective data collection and sampling strategy.

The physical knowledge can be easily added as the regularization terms into the total loss functions in neural networks. The physical constraints then can help reduce the searching space and guide the searching direction during the training. The proposed formulation is generic and can be extended to other machine learning approaches, where regularization can be similarly applied.

The proposed scheme is demonstrated with two examples of materials modeling. The PCNN is effective for these two different types of PDEs with different boundary conditions. The classical ANN with small training data sets tends to have large prediction errors. By adding physical constraints, the prediction accuracy of the PCNN can be one order of magnitude higher than the one from the classical ANN. Even with limited training data, the prediction of the PCNN is comparable with the original FEM solution. The weights associated with physical constraints can be adjusted to reflect the importance of the prior knowledge. They also affect the prediction accuracy. It is demonstrated that the adaptive weighting scheme results in higher prediction accuracy and shorter training time because the different losses in the total cost function are well balanced and have a similar convergence speed. The convergence analysis shows that the required amount of training data can be reduced by adding more physical constraints. Based on the computational results, DANNs are more capable than GPs to do the extrapolation of the difference between the LF-PCNN and HF-PCNN.

The developed MF-PCNN is an efficient approach to predict unknown relationships by combining the information from physical knowledge and available data. The training efficiency can be significantly improved if the training data from numerical simulations with different fidelities are utilized to construct MF-PCNNs. The training data are not limited to numerical simulation results only. They can also come from experimental measurements. The costs of experimental measurements can also be incorporated into the multi-fidelity scheme, where cost-effective sampling strategies can be taken.

The potential improvement of the current PCNN could be replacing the ANN to be the Recurrent Neural Network (RNN), such as long short-term memory (LSTM) neural network. Unlike feedforward neural networks, RNNs can use their internal state to process sequences of inputs, which may be more appropriate to solve time-dependent problems.

The proposed scheme should not be regarded as the replacement of classical numerical simulation methods (e.g. finite element and spectral methods) for solving partial differential equations. Rather, it enhances the efficiency of engineering design when high-fidelity simulations need to be run repetitively to obtain samples for design optimization. The number of samples for optimization for high-dimensional problem usually is very large. The machine learning approach therefore only shows its advantage for complex problems with high-dimensional searching space with the cost of training justified. The proposed scheme has the potential of making machine learning useful for real-world engineering applications where data sparsity is a common issue.

ACKNOWLEDGEMENTS

This work is supported in part by the George W. Woodruff Faculty Fellowship at Georgia Institute of Technology.

REFERENCES

[1] Karpatne, A., Atluri, G., Faghmous, J. H., Steinbach, M., Banerjee, A., Ganguly, A., Shekhar, S., Samatova, N., and Kumar, V., 2017, "Theory-Guided Data Science: A New Paradigm for Scientific Discovery from Data," IEEE Trans. Knowl. Data Eng., **29**(10), pp. 2318–2331.

[2] Li-Zhi, L., and Hou-Duo, Q., 1999, "A Neural Network for the Linear Complementarity Problem," Math. Comput. Model., **29**(3), pp. 9–18.

[3] Xia, Y., Leung, H., and Wang, J., 2002, "A Projection Neural Network and Its Application to Constrained Optimization Problems," IEEE Trans. Circuits Syst. I Fundam. Theory Appl., **49**(4), pp. 447–458.

[4] Thompson, M. L., and Kramer, M. A., 1994, "Modeling Chemical Processes Using Prior Knowledge and Neural Networks," AIChE J., **40**(8), pp. 1328–1340.

[5] Watson, P. M., Gupta, K. C., and Mahajan, R. L., 1998, "Development of Knowledge Based Artificial Neural Network Models for Microwave Components," 1998 IEEE MTT-S Int. Microw. Symp. Dig. (Cat. No.98CH36192), **1**, pp. 9–12.

[6] Wang, F., 1997, "Knowledge-Based Neural Models for Microwave Design," IEEE Trans. Microw. Theory Tech., **45**(12 PART 2), pp. 2333–2343.

[7] Tresp, V., Hollatz, J., and Ahmad, S., 1993, "Network Structuring and Training Using Rule-Based Knowledge," Adv. Neural Inf. Process. Syst. 5, pp. 871–878.

[8] Towell, G. G., and Shavlik, J. W., 1994, "Knowledge-Based Artificial Neural Networks," Artif. Intell., **70**(1–2), pp. 119–165.

[9] Ramuhalli, P., Udpa, L., and Udpa, S. S., 2005, "Finite-Element Neural Networks for Solving Differential Equations," IEEE Trans. Neural Networks, **16**(6), pp. 1381–1392.

[10] Xu, C., Wang, C., Ji, F., and Yuan, X., 2012, "Finite-Element Neural Network-Based Solving 3-D Differential Equations in Mfl," IEEE Trans. Magn., **48**(12), pp. 4747–4756.

[11] Han, F., and Huang, D. S., 2008, "A New Constrained Learning Algorithm for Function Approximation by Encoding a Priori Information into Feedforward Neural Networks," Neural Comput. Appl., **17**(5–6), pp. 433–439.

[12] Lauer, F., and Bloch, G., 2008, "Incorporating Prior Knowledge in Support Vector Regression," Mach. Learn., **70**(1), pp. 89–118.

[13] Dissanayake, M. W. M. G., and Phan-Thien, N., 1994, "Neural-network-based Approximations for Solving

Partial Differential Equations," Commun. Numer. Methods Eng., **10**(3), pp. 195–201.

[14] Raissi, M., Perdikaris, P., and Karniadakis, G. E., 2018, "Physics-Informed Neural Networks: A Deep Learning Framework for Solving Forward and Inverse Problems Involving Nonlinear Partial Differential Equations," J. Comput. Phys., **378**, pp. 686–707.

[15] de Cursi, J. E. S., and Koscianski, A., 2007, "Physically Constrained Neural Network Models for Simulation," Adv. Innov. Syst. Comput. Sci. Softw. Eng., pp. 567–572.

[16] Shirvany, Y., Hayati, M., and Moradian, R., 2009, "Multilayer Perceptron Neural Networks with Novel Unsupervised Training Method for Numerical Solution of the Partial Differential Equations," Appl. Soft Comput. J., **9**(1), pp. 20–29.

[17] Lagaris, I. E., Likas, A., and Fotiadis, D. I., 1998, "Artificial Neural Networks for Solving Ordinary and Partial Differential Equations," IEEE Trans. Neural Networks, **9**(5), pp. 987–1000.

[18] Shekari Beidokhti, R., and Malek, A., 2009, "Solving Initial-Boundary Value Problems for Systems of Partial Differential Equations Using Neural Networks and Optimization Techniques," J. Franklin Inst., **346**(9), pp. 898–913.

[19] Lagaris, I. E., Likas, A. C., and Papageorgiou, D. G., 2000, "Neural-Network Methods for Boundary Value Problems with Irregular Boundaries," IEEE Trans. Neural Networks, **11**(5), pp. 1041–1049.

[20] McFall, K. S., and Mahan, J. R., 2009, "Artificial Neural Network Method for Solution of Boundary Value Problems With Exact Satisfaction of Arbitrary Boundary Conditions," IEEE Trans. Neural Networks, **20**(8), pp. 1221–1233.

[21] Malek, A., and Shekari Beidokhti, R., 2006, "Numerical Solution for High Order Differential Equations Using a Hybrid Neural Network-Optimization Method," Appl. Math. Comput., **183**(1), pp. 260–271.

[22] Bellamine, F., Almansoori, A., and Elkamel, A., 2015, "Modeling of Complex Dynamic Systems Using Differential Neural Networks with the Incorporation of a Priori Knowledge," Appl. Math. Comput., **266**, pp. 515–526.

[23] Ferrari, S., and Jensenius, M., 2008, "A Constrained Optimization Approach to Preserving Prior Knowledge during Incremental Training," IEEE Trans. Neural Networks, **19**(6), pp. 996–1009.

[24] Di Muro, G., and Ferrari, S., 2008, "A Constrained-Optimization Approach to Training Neural Networks for Smooth Function Approximation and System Identification," Proc. Int. Jt. Conf. Neural Networks, pp. 2353–2359.

[25] Rudd, K., Muro, G. Di, and Ferrari, S., 2014, "A Constrained Backpropagation Approach for the Adaptive Solution of Partial Differential Equations," IEEE Trans. Neural Networks Learn. Syst., **25**(3), pp. 571–584.

[26] M. Abadi, A. Agarwal, P. Barham, E. Brevdo, Z. Chen, C. Citro, G. S., and Corrado, A. Davis, J. Dean, M. Devin, et al., 2016, "TensorFlow: A System for Large-Scale Machine Learning.," 12th USENIX Symp. Oper. Syst. Des. Implement.

[27] Baydin, A. G., Pearlmutter, B. A., Radul, A. A., and Siskind, J. M., 2015, "Automatic Differentiation in Machine Learning: A Survey," J. Mach. Learn. Res., **18**, pp. 1–43.

[28] Lee, D., and Myung, K., 2017, "Read My Lips, Login to the Virtual World," 2017 IEEE Int. Conf. Consum. Electron. ICCE 2017, pp. 434–435.

Proceedings of the ASME 2019
International Design Engineering Technical Conferences
and Computers and Information in Engineering Conference
IDETC/CIE2019
August 18-21, 2019, Anaheim, CA, USA

DETC2019-97777

COMPUTATIONAL DESIGN OF A PERSONALIZED ARTIFICIAL SPINAL DISC WITH A DATA-DRIVEN DESIGN VARAIBLE LINKING HEURISTIC

Zhiyang Yu[1]
Engineering Design and Computing Laboratory
Dept. of Mechanical and Process Engineering
ETH Zürich, Switzerland

Kristina Shea
Engineering Design and Computing Laboratory
Dept. of Mechanical and Process Engineering
ETH Zürich, Switzerland

Tino Stanković
Engineering Design and Computing Laboratory
Dept. of Mechanical and Process Engineering
ETH Zürich, Switzerland

ABSTRACT

A personalized, 3D printed, multi-material artificial spinal disc is expected to not only achieve personalized anatomical fit, but also to restore the natural mechanics of the implanted spinal segment. However, the necessary structure for disc design is not explored and optimizing the design is challenging due to the high-dimensional search space provided by the material distribution precision of multi-material 3D printing as well as necessary nonlinear finite element simulation. Therefore, this study explores the feasibility of two multi-material spinal disc designs and a clustering-based design variable linking method to achieve efficient and effective optimization. The optimization goal is to enable the implant to have natural stiffnesses for five loading cases. The results show that a biomimetic fiber network is necessary for the disc design. Moreover, the optimization performance of the heuristic derived from a clustering-based method is shown to be a good trade-off between the objective function value and the computational time.

Keywords: Total disc replacement; Personalized medicine, Artificial spinal disc; Computational design; Additive manufacturing; Machine Learning; Optimization

1. INTRODUCTION

Total Disc Replacement (TDR) based on replacing natural spinal discs with an artificial spinal disc is considered to be the most promising surgical treatment method so far. In contrast to spinal fusion, TDR is able to preserve the flexibility of the spinal segment after operation [1] and to recover a portion of the patient's original range of motion. Most of the currently existing artificial spinal discs such as activL [2] and SB Charite [3] utilize the design principle of "ball-and-socket" similar to designs for artificial hip and shoulder joint implants. However, this analogy

in design principle sacrifices the functionality for the sake of design simplicity and often yields complications after TDR. The literature reports the main causes for the common complications are the anatomical misfit between the implanted implant and the adjacent vertebrae, as well as the inability of the artificial spinal disc to fully restore physiological range of motion (ROM) of the implanted spinal segment [4].

One strategy to tackle the complications associated with "ball-and-socket" designs is to draw the inspiration for the design of an artificial spinal disc from the natural intervertebral disc which is an anisotropic biological tissue. The natural disc mainly consists of three parts: the nucleus pulposus (NP) in the middle, the surrounding annulus fibrosus (AF) and the cartilaginous endplate [5]. The soft, deformable, water-like NP accounts for 40-50% of the total disc volume, while the AF is composed of 15-25 concentric lamellae that function as the main contributor to the total disc anisotropy. The lamellae consists of a fiber network that has a crisscross pattern embedded in a matrix, while the regional variation of fiber orientation is regarded as the main cause for the AF anisotropy [6]. This is also confirmed by mechanical tests reporting the stiffness variation in the AF ventral-dorsal as well as internal-external direction [7], which is hypothesized to be caused by the adaptation of the disc to the stresses induced by the loads [8, 9].

The anisotropy in natural intervertebral discs inspires the design of artificial spinal discs to incorporate material anisotropy for natural mechanics restoration. The necessity of including material anisotropy in the artificial disc design is further demonstrated by one earlier study [10] that shows the inadequateness of using a single isotropic material to restore the natural disc behavior. The development of AM technology such as multi-material 3D printing, provides a potential solution for

---

[1] Corresponding author: yuz@ethz.ch (Zhiyang Yu)

the realization of personalized implants considering both shape and behavior. Compared to the traditional manufacturing technologies, multi-material AM is able to fabricate products with complex geometries and certain material distribution with great precision, which enables a realization of artificial discs that can not only achieve anatomical fit, but also at the same time restore the natural mechanics of the implanted spinal segment.

In our previous work, a multi-material artificial disc that has natural rotational stiffnesses under four rotational loading conditions is proposed [11]. Although a good stiffness match between the optimized multi-material disc and the natural disc is achieved, there are several limitations in the previous study:

1) The effects of the disc design on the natural mechanics restoration are not explored.

2) The effects of the heuristic for design variable linking on the solution accuracy are not understood.

3) The optimization problem is relaxed by a continuous problem formulation with material properties that do not map to materials available for a certain multi-material 3D printing technique. Moreover, the mechanical failure constraint is not considered. The continuous problem formulation, the possibly unrealizable material range and the potential mechanical failure hinder the fabrication and the usability of the proposed multi-material disc.

4) Only four rotational loading cases are optimized and the critical loading case of compression is neglected.

Based on our previous work and to overcome its limitations, this study explores the feasibility of two disc designs, whose optimization includes a data-driven clustering for obtaining design variable linking heuristic to reduce the search space dimensionality. The artificial disc is aimed to be manufactured using a Stratasys Objet500 Connex3 inkjet-based, multi-material 3D printer and to exhibit natural stiffnesses under five critical loading cases: flexion, extension, right lateral bending, left axial rotation, and compression. The main contributions of this study are summarized in the following points:

1) Exploration of disc designs to overcome the main limitations of current artificial spinal discs by enabling the artificial disc to have personalized geometry and natural stiffness under five loading conditions for natural mechanics restoration.

2) Proposal of a clustering-based design variable linking heuristic that is otherwise difficult to derive manually for search space dimensionality reduction.

3) Proposal of a multi-material disc design that can be manufactured using the Objet500 Connex3 3D printer.

The organization of the paper is as follows: First, a background section discusses the work relevant for this study. In Section 3, the detailed steps of the proposed methodology for artificial disc design are explained. The optimization results of the two proposed disc designs together with the performance comparison of different design variable linking heuristics are presented in Section 4. Section 5 interprets the results of this

work, discusses the algorithm performance as well as the limitations of the method proposed. A conclusion summarizing the findings of this study is presented at the end in Section 6.

2. BACKGROUND

The following sections first introduce the design of a multi-material artificial spinal disc from our previous study. Afterwards, the strategies for design variable linking and the application of clustering in structure optimization are discussed, followed by a brief introduction of the principle of k-means.

2.1 Design of a multi-material artificial spinal disc

The disc proposed in our earlier work (Fig. 1.A) [11] has a biomimetic structure that consists of a matrix and a crisscross fiber network embedded in the matrix. The high resolution by which materials can be distributed within the build space results in a large number of design alternatives, as ideally the material property of each voxel in a 3D object can be regarded as one design variable [10]. Due to the implicit formation of the optimization problem as well as the unreliable and noisy derivatives, a gradient-free Nelder-Mead method was used for the optimization [11]. The high-dimensional space, however, will worsen the performance of the gradient-free optimizer [12]. Therefore, a bio-inspired design variable linking heuristic was implemented in [11] to decrease the dimensionality of the search space. The matrix was divided into four regions, while the material property of each region is defined as one design variable. Further, instead of taking the material property of each fiber as one design variable, the linking heuristic assigns one Young's modulus to a fiber based on its coordinates in the radial and circumferential coordinate system (Fig. 1.B).

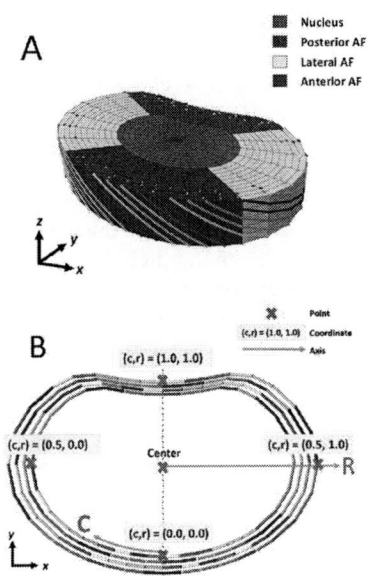

FIGURE 1: THE DESIGN OF A MULT-IMATERIAL ARTIFICIAL SPINAL DISC [11]: DISC DESIGN AND MATRIX SECTIONING (A), FIBER COORDINATE SYSTEM WITH RADIAL [R] AND CIRCUMFERENTIAL [C] AXIS FOR LOCAL FIBER STIFFNESS ASSIGNMENT (B)

2.2 Strategies for design variable linking

The principle of design variable linking in structural optimization is first introduced in 1973 by Schmit and Farshi [13]. It is considered as one global approximation technique and it is targeted at problems that have a large problem size both in terms of design variables, and/or local constraints whose explicit handling affects the efficiency of the optimization process [14]. One study groups structural members that are symmetrical to the y-z and x-z planes together and regards their cross section sizes as one independent design variable [13]. Another study tests different design variable linking strategies such as grouping structural members based on similar cross section areas, and near optimal solutions are obtained from the reduced search space with much less computational time [15, 16]. Different from the abovementioned strategies, Hajela and Sobieski [17] propose an adaptive design variable linking scheme that ranks all the design variables according to their *combined measure of effectiveness* at each optimization iteration, while only the design variables with high effectiveness, i.e. big contribution to the system response, get updated. Instead of linking design variables based on their relative contributions to the objective function, an automatic variable linking method based on enforcing cardinality constraints in the genetic algorithm encoding is used to group trusses [18].

2.3 Clustering and k-means

As an unsupervised learning method in machine learning, clustering is able to group similar data instances together while separating different instances [19]. Thanks to its ability to detect underlying relations between data instances, clustering has been used to group variables with similar densities together after an initial continuous topology optimization, while the grouped elements are considered as one material for subsequent material optimization to achieve thermal compliance [20]. Moreover, promising search regions in a multi-modal search space are identified using clustering [21]. Together with surrogate modeling, the integration of clustering into an evolutionally algorithm is proved to help solving high-dimensional problems with limited computational budget. In the topology optimization, the stress evaluation points that have a similar stress level are clustered together to decrease the number of local constraints and thus to reduce the computational cost [22].

K-means is a clustering algorithm that is widely used in the scientific and industrial fields for partitioning a set of n data points $(\mathbf{y}_1, \mathbf{y}_2, \mathbf{y}_3, \dots, \mathbf{y}_n)$ of dimension d into k disjoint subsets $\mathbf{S} = \{S_1, S_2, S_3, \dots, S_k\}$ by minimizing the sum-of-squares as presented in Eq. (1) where $\boldsymbol{\mu}_i$ is the geometric centroid of the data points in S_i. The optimization process then iteratively updates the centroids $\boldsymbol{\mu}_i$ and then assigns the data points to the new centroids until the assignment does not change [23].

$$\underset{\mathbf{S}}{\arg\min} \sum_{i=1}^{k} \sum_{n \in S_i} |\mathbf{y}_n - \boldsymbol{\mu}_i|^2 \qquad (1)$$

3. METHODOLOGY

Two designs of multi-material artificial spinal discs are proposed and their feasibilities are examined: a one-piece solid disc (Fig. 2. A), and a biomimetic disc (Fig. 2. B) that contains a nucleus and a fiber network.

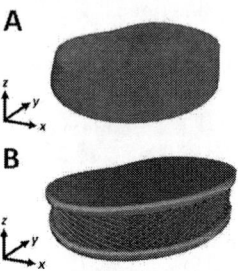

FIGURE 2: TWO DESIGNS OF ARTIFICIAL SPINAL DISC: A SOLID DISC (A), AND A BIOMIMETIC DISC (B)

The overall optimization workflow consists of three steps as shown in Fig. 3. The computational design process starts with the build of a patient-specific finite element disc model with a personalized geometry and an initial FE run to get the values of features $(\mathbf{y}_1, \mathbf{y}_2, \mathbf{y}_3, \dots, \mathbf{y}_n)$ (Step 1), followed by a clustering-based design variable linking technique to cluster structural elements into subsets $\mathbf{S}$ (Step 2). The details of the clustering will be explained in Section 3.2. The final step (Step 3) is a determination of the optimized material distribution in which a Generalized Pattern Search (GPS) method [24] is used to solve a non-linear optimization problem with discrete variables in a reduced search space.

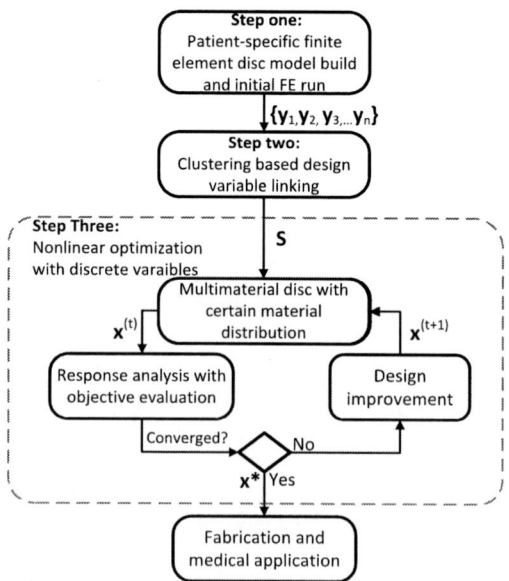

FIGURE 3: THE WORKFLOW OF THE COMPUTATIONAL DESIGN PROCESS OF A PERSONALIZED MULTI-MATERIAL ARTIFICIAL SPINAL DISC

The optimization process in Step 3 is realized using a closed loop: the optimizer gives the design variable vector **x**, which represents the Young's modulus of the structural elements in subsets of **S**. Based on **S**, a Python script then calls ABAQUS to calculate the disc response, i.e. reaction moment or reaction force. The value of the stiffness deviation of the artificial disc is then fed back to the optimizer to update **x** for further optimization.

3.1. Personalized disc finite element model build

The first step in building a personalized finite element (FE) model is to build a patient-specific geometric model of the disc. The key geometric parameters for disc modeling are: anterior-posterior diameter (APD), transverse diameter (TD), height and wedge angle that is defined as the angle formed by the disc top and bottom endplates due to natural spine curvature. Based on the method proposed in [25], the base elliptical cone shape of the disc is formed using a superquadratic equation [26], followed by a superimposition of the Gaussian curves as well as bumps to the superior and inferior surfaces of the disc. The boundary conditions are set to mimic the in-vitro mechanical test environment for an artificial spinal disc. The material candidates for PolyJet 3D printing are modeled as linear elastic materials, which is a realistic assumption based on the small strain within the disc. In the following subsections, the details of the FE model build for both disc designs are explained in detail.

3.1.1 Solid disc FE model build

The solid disc is modeled using a six-node wedge element (C3D6) and an eight-node hexahedron elements (C3D8R) in ABAQUS. A mesh sensitivity analysis then determines the mesh density in disc circumferential, radial, and height-direction. The mesh of the solid disc model (Fig. 4) contains 4086 nodes and 3400 elements that include 3230 hexahedral elements and 170 wedge elements. All the nodes lying on the bottom, or caudal, surface of the disc (Fig. 4.C) are fixed in all degrees of freedom, while the nodes located on the top, or cranial, surface of the disc are coupled to a reference point in six degrees of freedom.

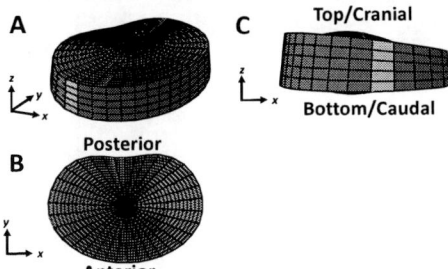

FIGURE 4: FE MODEL OF THE SOLID DISC

Displacement control mode is used in the FE analysis where a rotational angle of 2° under rotational loading conditions and a displacement of 1.5 mm under the compressive loading condition are added to the reference point. The corresponding reaction moments and reaction force of the whole disc are then read out from the reference point. Afterwards, the rotational stiffness and compression stiffness are calculated by dividing either the reaction moment by the rotational angle or the reaction force by the displacement, respectively. This is justified based on an almost linear response of the total disc.

3.1.2 Biomimetic disc FE model build

The biomimetic disc is composed of the following parts: a nucleus, a surrounding fiber network, and two endplates on the top and bottom surfaces (Fig. 5). A function decoupling strategy is used for the biomimetic disc design, as the fiber network is shown to be mainly responsible for flexural and torsional resistance [29]. The two endplates are regarded as rigid bodies and are therefore not modeled.

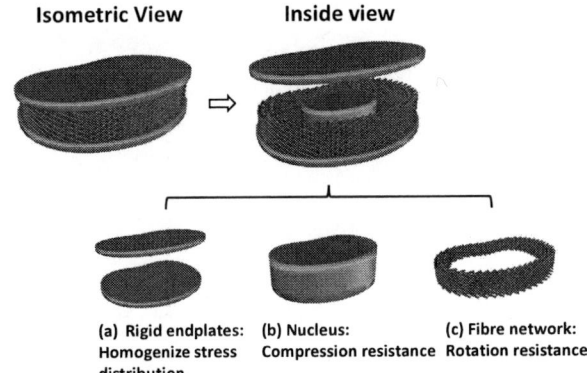

FIGURE 5. THE COMPOSITION AND FUNCTION DECOUPLING OF THE BIOMIMETIC DISC

The nucleus (Fig. 6.A) contains 4180 nodes and 1404 elements, among which 1248 are hexahedral elements and 156 are wedge elements. The fiber network (Fig. 6.B) has four alternative fiber layers with a biomimetic crisscross pattern, and each fiber is modeled with six beam elements, yielding 1472 nodes and 1104 B31 line elements. The boundary conditions for both the nucleus and fiber network are specified in the same way as for the solid disc in Section 3.1.1, i.e. all the nodes lying on the bottom surface are fixed in all degrees of freedom, while the nodes lying on the top surface are coupled in six degrees of freedom to a reference point.

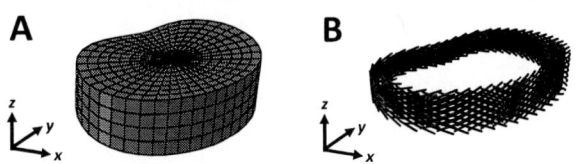

FIGURE 6: FUNCTIONAL UNITS OF THE BIOMIMETIC DISC: A NUCLEUS (A) AND A FIBER NETWORK (B)

3.2. Clustering-based design variable linking

Although the behavior of the disc is reported to be dependent on whether to include a radial or radial-circumferential AF anisotropy [8, 27] the review of the literature shows that there lacks a norm for AF anisotropy modeling, which makes it difficult to manually derive an appropriate design variable linking heuristic to reduce the initial high-dimensional search space. To overcome this, a design variable linking heuristic derived from a data-driven clustering technique is used. Structural elements, i.e. solid elements in the solid disc and fibers in the biomimetic disc, with similar stresses or displacements are grouped together after finite element analysis, which is analogous to the fact that the anisotropic structures of most biological tissues such as bone are the results of their adaption to external stress or strain mechanical stimuli [28]. Table 1 presents three types of structural element features used for clustering in this study: element node displacement (Feature A), element stress (Feature B), and a combination of the previous two (Feature C). The process of the feature calculation from a 3D solid element is illustrated in Fig. 7.

TABLE 1: STRUCTURAL ELEMENT FEATURES

Feature	Feature representation	Notation of feature at 3D solid element
Feature A	Element node displacement	$y_A = \left(y^A_{11}, y^A_{12}, \dots y^A_{ij}, \dots y^A_{pc}\right)$, where p = number of nodes, c = number of loading cases
Feature B	Element stress	$y_B = \left(y^B_{11}, y^B_{12}, \dots y^B_{ij}, \dots y^B_{qc}\right)$, where q = number of elements, c = number of loading cases
Feature C	Combination of Feature A and B	$y_C = (y_A \, y_B)$

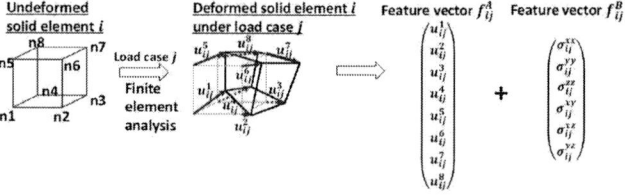

FIGURE 7: FEATURE CALCULATION OF SOLID ELEMENT

As shown in Fig. 7, under a certain loading case j, an FE analysis is conducted and the node displacements as well as the element stresses are read out from a certain structural member i and considered as element features. Afterwards, an extension [30] of the original k-means implementation in Matlab [29] with an integrated Elbow method [31] that determines the number of clusters k is used to cluster n structural members into k clusters ($k \leq n$) based on the features. Feature A, B and C are used for the solid disc optimization under compression and only feature A is used for the other optimization cases. The details will be explained in Section 4. The elbow method defines the number of clusters at the turning point as the optimal value of k. The turning point is defined as the point above which the optimization process for clustering according to the generic model shown in Eq. (1) only has minor gain with further increase of number of clusters. In this study, a maximum cluster number of 30 is used to define k. Each structural member is then assigned to a certain cluster after clustering. By assigning the same material property to the structural members in a certain cluster, the number of design variables is decreased.

3.3. Optimization problem formulation

A nonlinear FE analysis is performed for the disc response calculation due to the large displacement of the disc. The optimization problem is then formulated as to minimize the deviation of stiffness of an artificial spinal disc to that of a natural disc under five loading cases [30, 31]:

$$\min_{\mathbf{x}} f = \sum_{c=1}^{m} (SA_c(E(\mathbf{x})) - SN_c)^2 \tag{2}$$

s.j.t:
$$\mathbf{K}\big(E(\mathbf{x})\big)\mathbf{u} = \mathbf{F} \tag{3}$$

$$\sum_{j=1}^{l_i} \sigma_j \leq l_i \sigma_y(x_i) \tag{4}$$

$$x_i \in \{1, 2, \dots, z\}, i = 1, 2, 3, \dots k \tag{5}$$

where SA and SN are the stiffness of the artificial disc and natural disc respectively, m represents the number of loading cases, and c with a value of 1 to 5 represents the loading case of compression, flexion (F), extension (E), right lateral bending (right LB), and left axial rotation (left AR), respectively as shown in Fig. 8; E is the function that maps certain discrete values of $\mathbf{x}$ to the linear elastic material property of each structure element that belongs to cluster i; $\mathbf{K}(E(\mathbf{x}))$ is the stiffness matrix of the FE model with a certain material distribution, $\mathbf{u}$ is the prescribed displacement vector that serves as boundary conditions of the FE model, $\mathbf{F}$ is the response of the model; l_i is the number of structural members in cluster i, σ_j is the von Mises stress of a certain structural element j that belongs to cluster i, while $\sigma_y(x_i)$ is the yield stress of the material assigned to cluster i; k is the number of clusters, and z is the number of 3D printable materials. The constraint in Eq. (3) is automatically satisfied by the FE analysis, while the stress constraint in Eq. (4) is added to avoid mechanical failure. The clusters are not updated during the optimization as it has been shown that the convergence is more smooth without re-clustering [22].

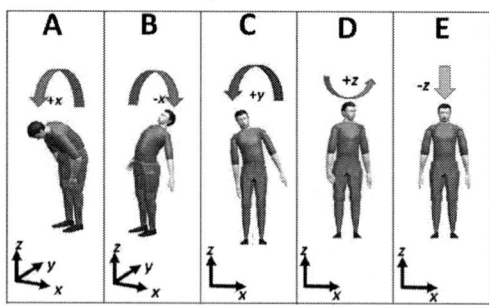

FIGURE 8: THE FIVE LOADING CASES OF THE SPINAL DISC: FLEXION (A), EXTENSION (B), RIGHT LATERAL BENDING (C), LEFT AXIAL ROTATION (D) AND COMPRESSION (E).

4. RESULTS

The results section presents the optimization results of the solid disc in Section 4.1 and the biomimetic disc in Section 4.2 subsequently. The artificial discs with a dimension of a L4-L5 lumbar disc as shown in Table 2 are expected to be fabricated using a Stratasys Objet500 Connex3 3D printer. Nine candidate materials with their material properties are shown in Table 3. An example to illustrate the material mapping process is as follows: suppose the value of x_1 is 3, then the structural elements that belong to the first cluster S_1 are assigned a Young's modulus of 2.65 MPa, representing the digital material F9850.

TABLE 2: DIMENSIONS OF THE ARTIFICIAL SPINAL DISC

Disc Type	Endplate geometry		Height [mm]	Wedge Angle [°]
L4-L5 disc	TD	43.2 mm	10.0	10.0
	APD	32.9 mm		

TABLE 3: THE DIGITAL MATERIALS AVAILABLE [34]

Material mapped code	Material name	Young's modulus [MPa]	Yield Stress [MPa]
1	TangoBlack Plus	0.486	0.85
2	F9840	1.36	1.5
3	F9850	2.65	2.65
4	F9860	22.5	0.45
5	F9870	52.5	1.00
6	F9885	125	2.50
7	F9895	375	7.50
8	RGD8530	2038	49.0
9	VeroWhite Plus	2500	50.0

The whole computational design process is implemented in an in-house Matlab code, and a Generalized Pattern Search (GPS) algorithm [35] within the NOMAD framework [36] is used to optimize the material distribution after clustering according to Fig. 3. The optimization settings include a randomized coordinate search direction and a 0.15 probability of initiating Variable Neighborhood Search (VNS) heuristic [37] to avoid the optimizer to be stuck in local optimum. The constraints are treated using a progressive barrier method [38].

The solid disc is first optimized under compression only and the results are presented in Section 4.1.1, followed by an optimization of the solid disc for five loading cases in Section 4.1.2. The nucleus and the fiber network of the biomimetic disc are optimized for compression and rotational loading cases separately, with Section 4.2.1 presenting the results of the nucleus and Section 4.2.2 the results of the fiber network. Furthermore, Section 4.2.3 serves as a benchmark to compare the performance of different design variable linking heuristics.

4.1 Solid disc optimization

The solid disc is first optimized for compression only using the proposed method in Section 4.1.1, followed by an optimization for five loading cases Section 4.1.2.

4.1.1. Optimization of the solid disc for compression

This section serves to explore the effect of cluster feature type (A, B, and C) on the material distribution and the optimization results. In this experiment only compression is considered to save the computational time. Due to the feature dimension incompatibility of the 6-node wedge elements that lie at the center of the disc and compose only a negligible portion of all the structural members, the clustering is implemented on 8-node solid elements that comprise the majority of the structural members. The stopping criteria of the optimization is set as a

maximum optimization iteration of 500. The initial high-dimensional design space is shown in Fig. 9. The optimization results of using different features are listed in Table 4, while the corresponding clustered pattern and the optimized material distribution are shown in Fig. 10.

FIGURE 9: THE INITIAL DESIGN SPACE OF THE SOLID DISC: ONE CUBE REPRESENTS ONE DESIGN VARAIBLE

TABLE 4: THE OPTIMIZATION RESULTS OF USING DIFFERENT FEATURES

Feature	Resultant No. of clusters	No. of optimization iterations	Optimized compression stiffness [N/mm]	Optimized goal [N/mm]
A	9	171	667.4	663.7
B	9	135	667.4	663.7
C	9	170	658.1	663.7

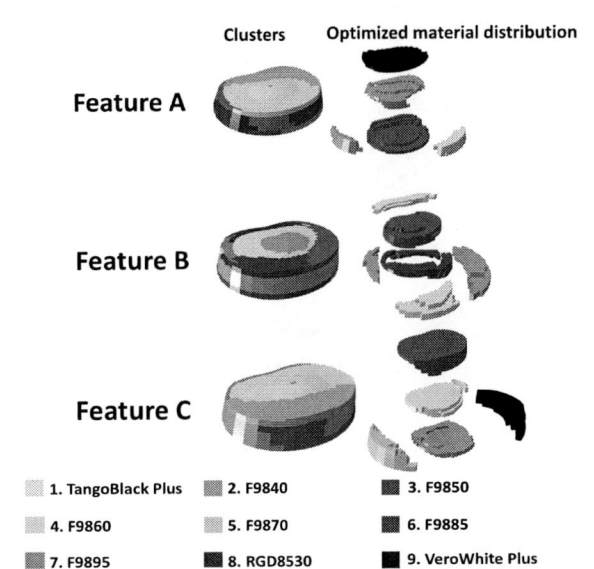

1. TangoBlack Plus	2. F9840	3. F9850
4. F9860	5. F9870	6. F9885
7. F9895	8. RGD8530	9. VeroWhite Plus

FIGURE 10: OPTIMIZED MATERIAL DISTRIBUTION OF SOLID DISC FOR COMPRESSION: THE CLUSTERED PATTERN (LEFT) AND THE OPTIMIZED MATERIAL DISTRIBUTION SHOWN IN AN EXPLODED VIEW FOR CLARITY (RIGHT)

4.1.2 Optimization of the solid disc for five loading cases

The solid disc is then optimized for five loading cases. Feature B is used for the clustering due to its good performance shown in the previous section. The optimized stiffnesses are shown in Table 5 and Fig. 11, while the clustered pattern and the optimized material distribution is shown in Fig. 12.

TABLE 5: THE OPTIMIZED STIFFNESS OF THE SOLID DISC

	Compression stiffness [N/mm]	Rotational Stiffness [Nm/deg]			
		F	E	Right LB	Left AR
Goal	663.7	0.75	0.78	1.23	1.33
Optimized solid disc	272.5	1.02	1.02	0.73	0.60

* F = Flexion, E= Extension, LB = Lateral Bending, AR = Axial Rotation

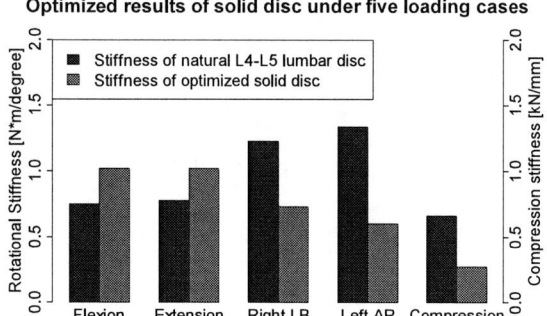

FIGURE 11: COMPARISON OF THE STIFFNESS OF NATURAL DISC AND THE OPTIMIZED SOLID DISC

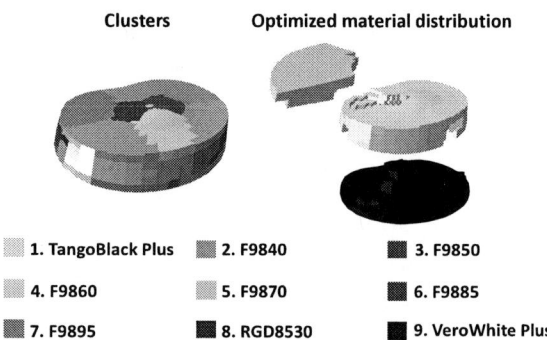

FIGURE 12: OPTIMIZED MATERIAL DISTRIBUTION OF SOLID DISC FOR FIVE LOADING CASES: THE CLUSTERED PATTERN (LEFT) AND OPTIMIZED MATERIAL DISTRIBUTION SHOWN IN AN EXPLODED VIEW FOR CLARITY (RIGHT)

4.2 Optimization of the biomimetic disc

In the following subsections, the optimized result of the nucleus for compression is first shown in Section 4.2.1, followed by a fiber orientation turning and a material optimization in Section 4.2.2.

4.2.1 Optimization of the nucleus for compression

The initial design space of the nucleus is divided into nine concentric circles, and the material property of each concentric circle is defined as a design variable (Fig. 13. A). The optimized material distribution is shown in Fig. 13. B. After optimization for compression, the contribution of the nucleus to the rotational stiffness of the whole disc is calculated.

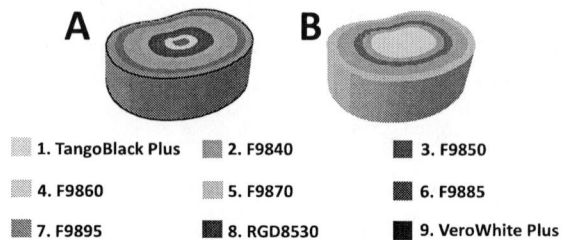

FIGURE 13: NUCLEUS OPTIMIZATION IN THE BIOMIMETIC DISC FOR COMPRESSION: DIVISION OF NUCLEUS (A), OPTIMIZED MATERIAL DISTRIBUTION (B)

4.2.2 Optimization of the fiber network for rotation

A fiber network with a fiber diameter of 0.6 mm is selected so that the number of fibers at each region (posterior, left, anterior, and right region as shown in Fig. 14) enables the stiffness trend of the biomimetic disc to mimic the natural disc. This is achieved when the stiffness magnitudes satisfy the following relation: left AR > right LB > extension > flexion. The final optimized results of the whole disc response are shown in Table 6 and Fig. 15, and the optimized material distribution within the fiber network is shown in Fig. 16.

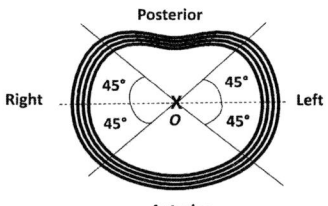

FIGURE 14: DIVISION OF THE FIBER NETWORK: POSTERIOR, LEFT, ANTERIOR, AND RIGHT REGIONS

TABLE 6: OPTIMIZED STIFFNESS OF THE BIOMIMETIC DISC

	Compression stiffness [N/mm]	Rotational Stiffness [Nm/deg]			
		F	E	Right LB	Left AR
Goal	663.7	0.75	0.78	1.23	1.33
Optimized biomimetic disc	663.96	0.70	0.74	1.00	1.60

* F = Flexion, E= Extension, LB = Lateral Bending, AR = Axial Rotation

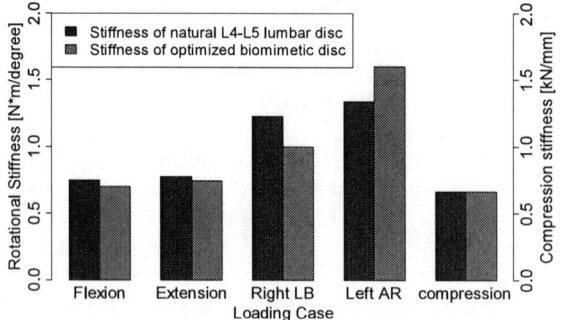

FIGURE 15: COMPARISON OF THE STIFFNESS OF THE NATURAL DISC AND THE OPTIMIZED BIOMIMETIC DISC

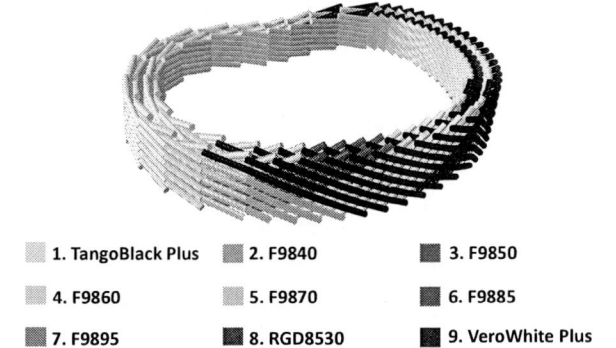

1. TangoBlack Plus	2. F9840	3. F9850
4. F9860	5. F9870	6. F9885
7. F9895	8. RGD8530	9. VeroWhite Plus

FIGURE 16: OPTIMIZED MATERIAL DISTRIBUTION OF THE FIBER NETWORK IN THE BIOMIMETIC DISC

4.3 Comparison of design variable linking heuristics

The performances of three design variable linking heuristics are compared by conducting a continuous optimization of a fiber network for achieving natural stiffness under four rotational loading conditions with different linking heuristics. The fiber network has a total of 212 fibers and a homogenous fiber orientation. The optimization bounds of the design variables are set as from 0.4 MPa to 2 GPa. Strategy one does not include any heuristic and uses 212 design variables directly. In comparison, strategy two, which is used in our previous work, reduces the initial high-dimensional search space to three design variables, i.e. fiber base Young's modulus, fiber circumferential gradient, and fiber radial gradient, as proposed in the previous work [11] and each fiber is assigned a certain stiffness based on its radial and circumferential coordinate. Strategy three uses the clustering-based design variable linking method proposed in this study. Here, in order to compare the performances of different strategies under a certain given computational budget the stopping criteria is according to the default settings of NOMAD framework with an additional rule to limit the maximum optimization iteration to 200. The resultant objective function values are shown in Table 7.

TABLE 7. COMPARISON OF OPTIMIZATION PERFORMANCES

Strategy #	Objective function value
1	0.49
2	0.45
3	0.29

5 DISCUSSION

The discussion section focuses first on the investigation of the necessary structure in an artificial disc design for natural mechanics restoration. Afterwards, the performance of the clustering-based method presented in this study and its comparison with other linking strategies are discussed. Finally, the limitations and the performance of the optimization algorithms close this section.

5.1 Disc Design for natural mechanics restoration

Two artificial disc designs are explored to examine their ability to restore natural stiffness under five critical loading cases. The optimized results of the solid disc model are not ideal: the percentage deviation of the artificial disc's stiffness compared to a natural disc under flexion, extension, right lateral bending, left axial rotation, and compression are 36.0%, 36.0%, 40.6%, 54.9%, 58.9%, respectively (Table 5 and Fig. 11). Moreover, the same stiffness magnitudes relation is observed throughout the optimization process, i.e. extension > flexion > right lateral bending > left axial rotation. The unsatisfying results show that a pure multi-material solid disc design is insufficient for natural mechanics restoration.

The optimization of the biomimetic disc is decoupled as the nucleus and the fiber network are separable in terms of their different roles under five loading cases. As shown in Table 6 and Fig. 15, a good stiffness match to the natural disc is achieved. This can be first explained by the introduction of the crisscross fiber network that has bigger torsion resistance compared to pure solid. In addition, the fiber network introduces additional design flexibility of fiber orientations, which helps to achieve the expected stiffness magnitude relation by tuning the fiber orientation in different regions. In spite of this, the stiffnesses of the optimized disc under lateral bending and axial rotation do not match perfectly to the natural disc, which is mainly due to the limited material candidates. The optimization range is restricted by the nine digital materials available for the Objet500 Connex3 multi-material 3D printer, and the restriction is mostly reflected in the material gap between 375 MPa to 2.038 GPa, i.e.: material 7 and 8 in Table 3. The results show that different optimal material distributions are obtained using different initial disc designs: the solid disc design as shown in Section 4.1 results in a less diverse material distribution and worse objective function value; in comparison, the biomimetic disc design as shown in Section 4.2 shows a more diverse material distribution and better objective function value. Considering the ability of GPS coupled with VNS to thoroughly explore the search space given the small number of design variables, it can be concluded that a biomimetic design is vital for the artificial disc design to restore natural kinematics.

5.2 Clustering based design variable linking

This study uses a data-driven method to obtain the implicit form of the variable linking matrix that relates the high dimensional search space to the lower dimensional search space. The motivation of using a k-means method is the high-dimensional features of the structural members. For example, an 8-node solid element under 5 loading cases yields a 40 dimensional vector for Feature A. The principle of the clustering is similar to a sensitivity test that examines the relative contributions of structural members to the mechanical responses of the whole system, while structural members with similar level of contributions are grouped together.

The feature type is shown to affect the optimization performance. As presented in Table 4, using either node displacement (Feature A) or element stress (Feature B) as features provides a good global approximation of the initial high-dimensional search space and thus a good objective function value is achieved, while using both displacement and stress (Feature C) results in a worse objective function value. A possible explanation is that by using Feature C the structural members are clustered based on strain energy, which is more suitable for a compliance-dominated optimization problems [39]. The advantage of using a clustering based design variable linking heuristic is further proved by comparing the optimization performances of three linking heuristics (Table 7). The main limitation of strategy one is the search complexity due to high-dimensional search space, while a rough approximation is used in Strategy two that manually adds a design variable linking heuristic. Table 8 gives a summary of the three design variable linking heuristics.

TABLE 8. COMPARISON OF THREE DESIGN VARAIBLE LINKING STRATEGIES

#	Heuristic	Cons.	Pros.
1	No heuristic	Large number of design variables and thus long computational time	Good convergence given infinite computational resources
2	Radial and circumferential stiffness gradient	1) Rough approximation 2) Manually set linking strategy with unproven optimality	Faster convergence due to reduced dimensionality
3	Clustering-based design variable linking	The performance is dependent on clustering feature type and number of clusters	1) Trade-off between fast convergence and result 2) No manual work needed

In our previous work, strategy 2 in Table 8 is used to reduce the number of design variables, requiring manually adding a variable linking strategy. In comparison, strategy 3 in Table 8 saves this manual work by automatically linking the design variables. Further, the aim of this study is to highlight the possibility of using a 3D printed multi-material artificial spinal disc to restore the physiological kinematics within a computational framework. However, the clinical application of the proposed design still requires 3D printable materials that are biocompatible.

5.3 Modeling limitations and algorithmic performance

The first limitation is the simplified FE model: only the artificial disc is modeled, while other functional parts of the functional spinal unit such as vertebrae are not included. The simplification is mainly due to the increase in computational cost with the increase of number of elements. Moreover, a perfect fixation of the artificial disc to the adjacent vertebrae is assumed. Furthermore, the cross-sectional areas of the fibers in the biomimetic disc are not optimized, which is due to the fixed small design space and the minimum allowable 3D printing diameter. Nevertheless, challenged by the fact that the

formulated optimization problem is discrete, the clustering-based search space dimensionality reduction technique together with a GPS algorithm perform well in terms of balancing the computational time and the objective function value. To prevent Eq. (4) in accepting infeasible designs, different ways of optimization constraint formulation will be explored in the future work.

6 CONCLUSION

This study explores the feasibility of two multi-material artificial spinal disc designs as well as the optimization performance of using a clustering-based design variable linking heuristic. Results show that the pure multi-material solid disc is inadequate for natural mechanics restoration and the optimized biomimetic disc exhibits a good stiffness match to the natural disc in spite of the limited material candidates, although the clinical implementation of the disc is currently hindered by the biocompatibility of the 3D printing materials. The heuristic derived from clustering is shown to be a good trade-off between the computational time and the objective function value. Future work will involve the exploration of other machine learning techniques to aid in the artificial disc design process as well as the mechanical test verification of the proposed biomimetic disc design.

REFERENCES

[1]   J. Dong et al., "Artificial disc and vertebra system: a novel motion preservation device for cervical spinal disease after vertebral corpectomy," Clinics, vol. 70, no. 7, pp. 493–499, 2015.

[2]   J. J. Yue, R. Garcia, and L. E. Miller, "The activL?? Artificial Disc: A next-generation motion-preserving implant for chronic lumbar discogenic pain," Med. Devices Evid. Res., vol. 9, pp. 75–84, 2016.

[3]   N. Langrana, J. R. Parsons, C. K. Lee, M. Vuono-Hawkins, S. W. Yang, and H. Alexander, "Materials and design concepts for an intervertebral disc spacer. I. fiber-reinforced composite design.," J. Appl. Biomater., vol. 5, no. 2, pp. 125–32, 1994.

[4]   F. Galbusera et al., "Design concepts in lumbar total disc arthroplasty," Eur. Spine J., vol. 17, no. 12, pp. 1635–1650, 2008.

[5]   N. Newell, J. P. Little, A. Christou, M. A. Adams, C. J. Adam, and S. D. Masouros, "Biomechanics of the human intervertebral disc: A review of testing techniques and results," J. Mech. Behav. Biomed. Mater., vol. 69, no. August 2016, pp. 420–434, 2017.

[6]   J. J. Cassidy, A. Hiltner, and E. Baer, "Hierarchical structure of the intervertebral disc," Connect. Tissue Res., vol. 23, no. 1, pp. 75–88, 1989.

[7]   R. Eberlein, G. A. Holzapfel, and M. Fr??hlich, "Multi-segment FEA of the human lumbar spine including the heterogeneity of the annulus fibrosus," Comput. Mech., vol. 34, no. 2, pp. 147–163, 2004.

[8]   G. Marini and S. J. Ferguson, "Modelling the influence of heterogeneous annulus material property distribution on intervertebral disk mechanics," Ann. Biomed. Eng.,

vol. 42, no. 8, pp. 1760–1772, 2014.

[9] P. Adams, D. R. Eyre, and H. Muir, "Biochemical aspects of development and ageing of human lumbar intervertebral discs," *Rheumatology*, vol. 16, no. 1, pp. 22–29, 1977.

[10] T. Belytschko, R. F. Kulak, A. B. Schultz, and J. O. Galante, "Finite element stress analysis of an intervertebral disc," *J. Biomech.*, vol. 7, pp. 277–285, 1974.

[11] Z. Yu, T. Stankovic and K. Shea, "Computational Design of a Personalized Artificial Spinal Disc," *Comput. Inf. Eng. Conf.*, pp. 1–10, 2018.

[12] L. Miguel and R. Nikolaos, "Derivative-free optimization : A review of algorithms and comparison of software implementations."

[13] L. Angeles, "Some Approximation Concepts for Structural Synthesis," vol. 12, no. 5.

[14] U. Kirsch, "Approximate Models for Structural Optimization," vol. 1, pp. 289–325.

[15] J. D. Biedermann and D. E. Grierson, "Training and using neural networks to represent heuristic design knowledge," *Adv. Eng. Softw.*, vol. 27, no. 1–2, pp. 117–128, 1996.

[16] M. F. RUBINSTEIN and R. B. Nelson, "Automated structural synthesis using a reduced number of design coordinates," *Aiaa J.*, vol. 11, no. 4, pp. 489–494, 1973.

[17] P. Hajela, "The Controlled Growth Method — A Tool for Structural Optimization," vol. 20, no. 10, pp. 1440–1441.

[18] H. J. C. Barbosa, A. C. C. Lemonge, and C. C. H. Borges, "A genetic algorithm encoding for cardinality constraints and automatic variable linking in structural optimization," *Eng. Struct.*, vol. 30, no. 12, pp. 3708–3723, 2008.

[19] F. Camastra and A. Vinciarelli, "Clustering methods," *Adv. Inf. Knowl. Process.*, no. 9781447167341, pp. 131–167, 2015.

[20] K. Liu and D. Detwiler, "Optimal Design of Nonlinear Multimaterial Structures for Crashworthiness Using Cluster Analysis," vol. 139, 2017.

[21] H. Dong, B. Song, P. Wang, and Z. Dong, "Surrogate-based optimization with clustering-based space exploration for expensive multimodal problems," *Struct. Multidiscip. Optim.*, vol. 57, no. 4, pp. 1553–1577, 2018.

[22] E. Holmberg, B. Torstenfelt, and A. Klarbring, "Stress constrained topology optimization," *Struct. Multidiscip. Optim.*, vol. 48, no. 1, pp. 33–47, 2013.

[23] P. Berkhin, "A survey of clustering data mining techniques," in *Grouping multidimensional data*, Springer, 2006, pp. 25–71.

[24] C. Audet and J. E. Dennis, "Analysis of Generalized Pattern Searches," *SIAM J. Optim.*, vol. 13, no. 3, pp. 889–903, 2002.

[25] R. Korez, B. Likar, F. Pernuš, and T. Vrtovec, "Parametric modeling of the intervertebral disc space in 3D: Application to CT images of the lumbar spine," *Comput. Med. Imaging Graph.*, vol. 38, no. 7, pp. 596–605, 2014.

[26] A. H.Barr, "Superquadrics and angle-preserving transformations," *IEEE Comput. Graph. Appl.*, vol. 1, no. 1, pp. 11–23, 1981.

[27] J. Noailly, J. A. Planell, and D. Lacroix, "On the collagen criss-cross angles in the annuli fibrosi of lumbar spine finite element models," pp. 203–219, 2011.

[28] A. Brand, M. Stanford, and C. Swan, "How do tissues respond and adapt to stresses around a prosthesis? A primer on finite element stress analysis for orthopaedic surgeons.," *Iowa Orthop. J.*, vol. 23, p. 13, 2003.

[29] D. Arthur and S. Vassilvitskii, *k-means++: The advantages of careful seeding*. Society for Industrial and Applied Mathematics, 2007, pp. 1027–1035.

[30] Sebastien De Landtsheer (2019).kmeans_opt (https://www.mathworks.com/matlabcentral/fileexchange/65823-kmeans_opt),MATLAB Central File Exchange. Retrieved May 3, 2019.

[31] R. L. Thorndike, "Who belongs in the family?," *Psychometrika*, vol. 18, no. 4, pp. 267–276, 1953.

[32] F. Heuer, H. Schmidt, Z. Klezl, L. Claes, and H.-J. Wilke, "Stepwise reduction of functional spinal structures increase range of motion and change lordosis angle," *J. Biomech.*, vol. 40, no. 2, pp. 271–280, 2014.

[33] S. Wang *et al.*, "A combined numerical and experimental technique for estimation of the forces and moments in the lumbar intervertebral disc," *Computer Methods in Biomechanics and Biomedical Engineering*, vol. 16, no. 12. Taylor & Francis, pp. 1278–1286, 2013.

[34] T. Chen, J. Mueller, and K. Shea, "Design and fabrication of a bistable unit actuator with multi-material additive manufacturing," *Annu. Int. Solid Free. Fabr. Symp.* , pp. 2060–2076, 2016.

[35] C. Bogani, M. G. Gasparo, and A. Papini, "Generalized Pattern Search methods for a class of nonsmooth optimization problems with structure," *J. Comput. Appl. Math.*, vol. 229, no. 1, pp. 283–293, 2009.

[36] C. Audet, S. Le~Digabel, and C. Tribes, "{NOMAD} user guide," 2009.

[37] P. Hansen, N. Mladenović, J. Brimberg, and J. A. M. Pérez, "Variable neighborhood search," *Int. Ser. Oper. Res. Manag. Sci.*, vol. 272, no. 1, pp. 57–97, 2019.

[38] P. Controller, "SYNCHRONIZATION OF COUPLED REACTION-DIFFUSION Copyright © by SIAM . Unauthorized reproduction of this article is prohibited .," vol. 51, no. 5, pp. 3486–3510, 2013.

[39] N. Aulig, "Generic Topology Optimization Based on Local State Features," no. November 2016, 1984.

Proceedings of the ASME 2019
International Design Engineering Technical Conferences
and Computers and Information in Engineering Conference
IDETC/CIE2019
August 18-21, 2019, Anaheim, CA, USA

DETC2019-98190

DESIGN AND BIOLOGICAL SIMULATION OF 3D PRINTED LATTICES FOR BIOMEDICAL APPLICATIONS

Paul F. Egan
Texas Tech University
Department of Mechanical Engineering
Lubbock, TX

ABSTRACT

There is great potential for using 3D printed designs fabricated via additive manufacturing processes for diverse biomedical applications. 3D printing offers capabilities for customizing designs for each new fabrication that could leverage automated design processes for personalized patient care, but there are challenges in developing accurate and efficient assessment methods. Here, we conduct a sensitivity analysis for a biological growth simulation for evaluating 3D printed lattices for regenerating bone and then use these simulations to identify performance trends. Four design topologies were compared by generating varied unit cells. Biological growth was modeled in a voxel environment by simulating the advancement of a tissue front by calculating its local curvature. Designs were generated with properties suitable for bone tissue engineering, namely 50%

porosity and microscale pores. The sensitivity analysis determined trade-offs between prediction consistency and computation time, suggesting calculating curvature within a radius of 7.5 voxels is sufficient for most cases. Topologies were compared in bulk with design variations. All topologies had similar tissue growth rates for a given surface-volume ratio, but with differing unit cell sizes. These findings inform future optimization for selecting unit cells based on volume requirements and other criteria, such as mechanical stiffness. A fitted analytical relationship predicted tissue growth rate based on a design's surface-volume ratio, which enables design evaluation without computationally expensive simulations. Lattices were 3D printed with biocompatible materials as proof-of-concepts, demonstrating the feasibility of the approach for future computational design methods for personalized medicine.

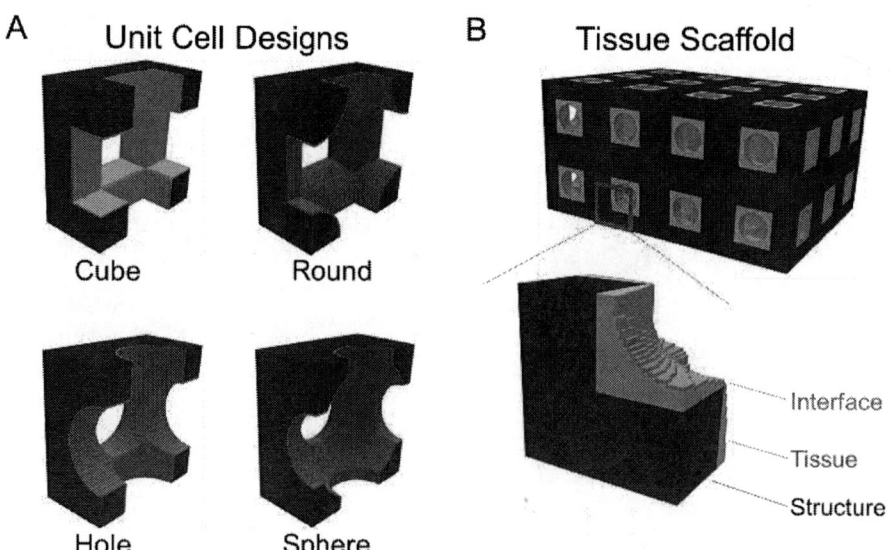

FIGURE 1: VOXEL ENVIRONMENT FOR TISSUE GROWTH SIMULATION WITH DESIGNED UNIT CELLS

1. INTRODUCTION

As 3D printing processes advance and receive widespread use, there is a need for suitable design approaches [1-3]. In biomedical applications, tissue scaffold design requires configuring a structure's stiffness and porosity to support biological tissue growth [4]. Scaffolds repair and promote growth of tissues such as bone [5]. Design enables customization of 3D printable scaffolds for each patient's unique physiology [6]. A challenge is the efficient evaluation of both mechanical and biological scaffold factors [7]. Here, we assess the efficiency and accuracy of tissue growth simulations for 3D printed scaffolds [8]. The approach is novel because it includes biological simulations in the assessment of scaffold functioning, whereas past design approaches focused on mechanics and nutrient transport [9]. Included biological simulations is difficult since they are computationally expensive so only a small portion of the design space is sampled [10]. This study provides foundations for developing efficient design approaches for optimizing scaffolds for personalized medicine.

Scaffold properties influence mechanical and biological functioning and have opposing design requirements [11, 12]. Porosity is the ratio of void volume to total volume and is typically 50% or higher to provide sufficient tissue growth volume and nutrient transport. 3D printed lattices for scaffolds consist of unit cells tuned for desired stiffness at varied porosities [13]. Lattices provide a large design space due to the number of possible decisions in structural configuration [14], therefore motivating the use of computational design [15, 16]. Lattice topology influences both mechanical properties of a structure and how tissue grows based on local geometry [17]. Four unit cell topologies in Figure 1 are informed by a Cube topology from previous research as a control, with three new topologies for comparison [8].

Figure 1A topologies consist of Cube and Round designs built from combining beams to form a lattice. The Round design is similar to the Cube, except it has circular beam cross-sections instead of square beam cross-sections. These differences allow for testing how cross-sectional beam geometry influences tissue growth. The Hole and Sphere are formed by removing cylinder and sphere shapes from a solid structure. These different strategies for forming lattices provide different curvatures and pore sizes that influence tissue growth rate [18, 19]. The Hole and Sphere designs have concave curvatures as opposed to the Cube's open porous beam architecture. The comparison between these topologies has not previously been conducted, in addition to the need to determine how beam cross-sectional area may influence tissue growth rates when comparing the Cube and Round topologies.

The scaffold in Figure 1B is created by patterning a unit cell to form a structure with suitable for medical interventions. Designs were generated in a discretized voxel environment for simulating tissue growth according to the positive curvature of the interface between tissue and void space [20, 21]. The accuracy of the algorithm to predict tissue growth was validated by both *in vitro* and *in vivo* experiments. We therefore focus on applying the algorithm in new situations and improving its efficiency, rather than conducting further validations with costly biological experiments.

In the simulation, structure voxels make up the unit cell and are surrounded by tissue voxels in time step $t = 0$ that represents tissue seeded on the structure's surface. Interface voxels represent volume where tissue would advance upon growing. All other voxels are void. The approach has previously investigated tissue growth in diverse 3D printed structures and found designs with the fastest growing tissue had the least nutrients available to support tissue growth [8]. There is a need to investigate designs not constructed from beams as a basis of comparison, which includes the Hole and Sphere designs in Figure 1. There is also a need to assess the simulation approach's computational efficiency and accuracy, since voxel sizes chosen during discretization influence accuracy and computational time [22].

In this paper, we generate designs using Figure 1 topologies not considered in past work and compare their properties and tissue growth behavior. A sensitivity analysis is conducted by changing the resolution of the discretized environment to find a suitable trade-off between time required for computational evaluation and accuracy. Once a suitable resolution is determined, tissue growth rates are compared across topologies to find trends. This work provides a significant advancement in exploring new design topologies and providing biological assessment methods for predicting tissue growth on complex 3D printed geometries. These steps are necessary step for developing efficient design automation tools.

2. BACKGROUND
2.1 Biological modeling

Biomechanical modeling approaches are informed by the behavior of biological cells that aggregate to form tissues [23]. Cells organize into tissue and interact based on their capabilities to mechanically sense their surrounding environment [24]. Osteoblasts that help construct bone form into three-dimensional tissues according to tension among cell interactions caused by differences in the local curvature [25-27]. Empirical studies that observe three-dimensional tissue growth are used to develop mathematical and computational models for improving understanding of biological phenomena or developing clinical solutions [28-31].

Simulations are used to model tissue growth behavior due to difficulty in predicting how tissue grows on complex surfaces of diverse geometries. Multi-agent based models are used to consider the emergent behavior for how individual cells interact to form a tissue aggregate [32-35]. Due to the need to model individual cell behavior, agent-based approaches are generally inefficient in comparison to other methods for modeling tissue growth at the aggregate level. Numerical and voxel-based approaches that predict growth based on a tissue's curvature have efficiently modeled tissue growth on three-dimensionally complex geometries [8, 20, 36]. These approaches successfully predicted three-dimensional tissue growth on diverse 3D printed scaffold geometries and provide a basis for evaluating designs biologically for design optimization [37].

2.2 Scaffold design

Scaffolds are constructed from a variety of materials that depend on the manufacturing process. For 3D printed scaffolds, polymer and metal materials are common. Recently, polyjet printing has developed polymer scaffolds for spinal cage fusion [1]. Stereolithography is used to achieve microscale lattices [7], with biocompatible materials [38, 39]. Further materials for scaffolds include wood, biomaterials, carbon fibers, and ceramics [40-44]. The most common metal used for scaffolds is titanium [13]. Polymers and metal 3D printing approaches achieve similar resolutions for complex scaffold designs.

There are many properties related to a scaffold's geometry that determine its mechanical and biological functioning [6]. These include the porosity of the structure and pore size, that are altered by tuning unit cell size and density [11, 45]. A higher porosity provides a larger proportion of void volume for tissue growth and nutrient transport. Pore size is related to the surface area of the scaffold, such that smaller pores provide more surface area per volume that leads to a higher volume of seeded tissue. Nutrient transport is influenced by the geometry of pores and scaffold architecture [46], generally larger voids provide greater nutrient transport through the scaffold. Larger pores also reduce scaffold surface area which promotes higher nutrient transport. Other factors are mechanical [47], such as elastic/shear moduli that generally decrease as porosity increases. Scaffolds can have functional gradients, such that varied pore sizes and unit cell sizes alter tissue growth behavior throughout the scaffold [48]. Due to the complexity in individual properties and scaffold architectures, computational design methods are often necessary.

2.3 Optimization

Optimization of biomechanical systems is challenging due to the need for considering a complex design space and the difficulty in evaluating designs efficiently and accurately [49]. Design maps show how one performance output, such as mechanical stiffness, relates to two or more design parameters [13], such as porosity and pore size. Design map approaches have demonstrated a trade-off such that improving tissue growth rates reduces nutrient availability necessary for tissue growth [8]. When these properties are further compared to mechanical stiffness and other design outputs, a complex multi-dimensional design space emerges [50]. These design spaces are typically traversed with topology optimization, response surfaces, and other multivariable optimization approaches suitable for 3D printed structures [17, 51-54]. Generally, scaffolds are designed by considering trade-offs in key properties, or through sampling the design space while controlling for properties [55, 56]. There is a need for efficient computational design approaches such that multiple variables are considered simultaneously. Since biological growth is expensive to compute and limits the efficiency of multivariable optimization, our approach focuses on improving the evaluation of biological growth.

3. METHODS
3.1 Unit cell generation

Unit cells were generated by configuring cubic voxels with length ds in a virtual environment that consist of void voxels, structure voxels, tissue voxels, and interface voxels (Fig. 1). The cubic virtual environment has length $l/2$, where l is the length of a unit cell. Only an eighth of each unit cell is used for simulations, which facilitates modeling efficiency based on symmetry [8].

Cube designs were generated by adding three orthogonal rectangular beams of equal width made up of structure voxels to an environment that consists of all void voxels. Round designs were generated in the same manner, but with beams that had a circular cross-section. Beam widths were adjusted to reach a

specified porosity. Hole designs were generated by adding three cylinders of equal radius of void voxels to an environment filled with structure voxels, therefore cutting out cylindrical holes from a structure. Sphere designs consisted of these same cylinders, plus a large sphere of void voxels in the center of the unit cell. The combination of cylinders and spheres for the Sphere design ensured planar pores existed on the outside of the Sphere unit cell when tuned for varied porosities. If cylinders were not included, many Sphere configurations would consist of a large central void that had no connection to other unit cells in a scaffold that is necessary for nutrient transport and tissue seeding.

After unit cells were generated, tissue voxels were placed adjacent to all structure voxels to seed the scaffold surface. Interface voxels were placed adjacent to all tissue voxels, and represent the boundary between the advancing tissue front and void space.

3.2 Tissue growth simulation

Curvature was calculated for each interface voxel to determine biological growth in the virtual environment. Curvature is calculated using a spherical scanning mask placed on each interface voxel that estimates curvature based on the ratio of voxels representing solid substrate (structure/tissue voxels) to voxels representing void space (interface/void voxels) within the mask. The environment and mask were discretized based on the cellular reach $c_r = 55\mu m$ of tissue to mechanically sense the curvature of the local environment [20]. The reach was mapped to the discretized spherical mask with radius m_r. The voxel length ds in the environment was calculated according to

$$ds = \frac{c_r}{m_r} \qquad (1)$$

which enables tuning of voxel size based on the design input radius for the spherical scanning mask. Voxel sizes are used to map a structure designed with continuous units to a number of discrete voxels. For equation 1, when the mask radius is 5.5voxels, the size of each voxel is 10μm, therefore a unit cell with length of 500μm is mapped to a length of 50voxels. Note, this length is halved to account for only an eighth of the unit cell being simulated.

In each simulation step, the scanning mask calculates the local curvature κ of all interface voxels according to

$$\kappa = \frac{16}{3 \cdot c_r} \left(\frac{m_{on}}{m_{on} + m_{off}} - \frac{1}{2} \right) \qquad (2)$$

where m_{on} refers to all voxels in the scanning mask that are structure or tissue while m_{off} refers to all other voxels in the scanning mask. For a given interface voxel, if the curvature was positive, then that voxel was changed into a tissue voxel for the next step of the simulation and a new interface voxel was placed adjacent to the currently calculated voxel. All new interface voxels placed within a simulation step had their local curvature calculated to determine if it was positive and changes to tissue. These calculations counted newly placed tissue within the same time step as void voxels contributing to the m_{off} total, since this newly placed tissue was not present during that step of the simulation. According to this algorithm, it is possible for a tissue front to advance multiple voxels in regions of high curvature since the algorithm efficiently checks for all potential voxels with positive curvature in the environment each time step, without inefficiently calculating the curvature of every void voxel in the environment.

The positive curvature of all topologies immediately after tissue seeding in a simulation is shown in Figure 2, where all interface voxels with positive curvature are highlighted in red. These designs have a unit cell length of 500μm, a porosity of about 50%, and were evaluated with a mask radius of 9.5voxels. There is positive local curvature in the Cube and Round designs at the beam joints. For the Hole and Sphere designs, there is curvature along the radius of the circular shapes removed from the structure. The Sphere design additionally has circular regions of positive curvature on the surface of its spherical shape.

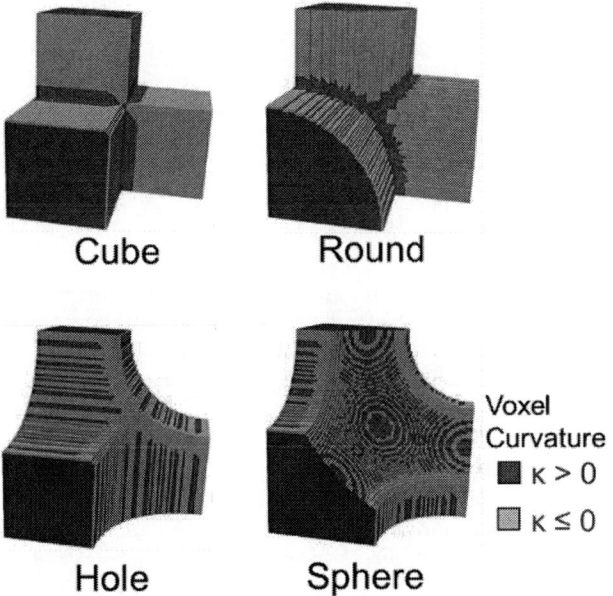

FIGURE 2: UNIT CELL CURVATURE AT $t = 0$

Each simulation step consists of calculating the curvature of voxels that could potentially become tissue (i.e. interface voxels and void voxels near the interface) followed by converting those voxels into tissue and converting void voxels to interface voxels that surround newly formed tissue. Simulations were run until convergence with convergence criteria being the scaffold filling completely with tissue or all interface voxels having zero curvature. Simulations were stopped prior to convergence if their total simulation reached 10,000s; results from these simulations are not reported. All simulations were coded in python and run using one core of an i5-2500k (3.3GHz) processor.

3.3 Design properties

Properties describing the state of the scaffold were calculated each simulation step. Porosity P was calculated by counting the number of structure voxels v_{stru} and tissue voxels v_{tiss} in comparison to the total number of the voxels v_{tot} in the virtual environment. Porosity

$$P = 1 - \frac{v_{stru} + v_{tiss}}{v_{tot}} \qquad (3)$$

describes the proportion of void volume in the scaffold in comparison to total volume.

Pore size p was found by counting the number of void voxels on one side of a structure and multiplying by four, since only a quarter of a planar pore is visible in each unit cell corner. The size is found according to

$$p = ds \cdot \sqrt{4 \cdot v_{pore}} \qquad (4)$$

which allows for comparison of differently shaped pores based on planar area.

The mean positive curvature $\overline{\kappa_+}$ is found by determining the average curvature of all interface voxels with positive curvature using equation 2. The curvature-surface ratio $\overline{\kappa_+}/S$ is

$$\overline{\kappa_+}/S = \frac{\overline{\kappa_+}}{v_{int} \cdot (ds)^2} \qquad (5)$$

where S is determined from the surface area of interface voxels. The surface-volume ratio S/V is

$$S/V = \frac{v_{int}}{v_{tot}} \cdot \frac{1}{ds} \qquad (6)$$

which provides a basis for comparing structures based on density of initial tissue deposition.

3.4 Sensitivity analysis

A sensitivity analysis was conducted by simulating biological growth using masks of varied radii to determine how mask size influenced the calculation of properties from each simulation step for varied designs. An initial analysis was conducted by generating 50% porous Cube and Sphere designs with unit cell lengths of 500μm. These designs were evaluated with masks of 5.5voxels and 9.5 voxels, based on 5.5voxels being a value used in previous studies for Cube scaffolds [8]. The sensitivity of property calculations for the Sphere design was analyzed to a greater extent by varying mask radii from 3.5 to 11.5 voxels in addition to unit cell length from 200μm to 1,000μm.

3.5 Bulk analysis

A bulk analysis was conducted by comparing designs of all topologies by generating 50% porous unit cells of lengths from 200μm to 1,000μm and evaluating with a constant mask radius. Designs were compared by calculating their void filling rate V_{rate} of

$$V_{rate} = P/t_{fill} \qquad (6)$$

where t_{fill} refers to the number of time steps required for a unit cell to completely fill with tissue. This property provides a basis for comparing tissue growth rate performance of varied porosities, which enables comparison to previous studies [8].

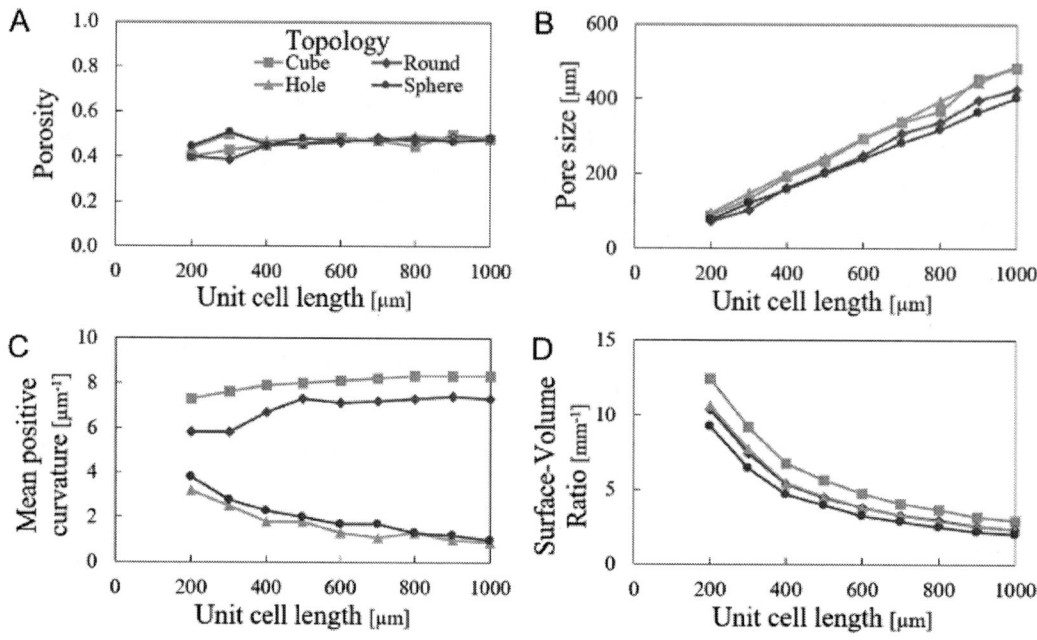

FIGURE 3: PROPERTIES FOR DESIGNED SCAFFOLDS SEEDED WITH TISSUE AT $t = 0$

4. RESULTS
4.1 Design property results

To compare design properties, unit cells for each topology were generated to achieve approximately 50% porosity for unit cell lengths of 1,000μm prior to tissue seeding. Designs were generated in an environment sized for a mask radius of 7.5 voxels. Unit cell lengths were altered in increments of 100μm from 200μm to 1,000μm while maintaining an approximately 50% porosity by scaling added/removed material appropriately [6]. These designs were seeded with tissue and their properties were evaluated in time step $t = 0$ prior to new tissue being added to the scaffold as growth in subsequent steps (Fig. 3).

The porosity (Fig. 3A) for all designs ranged from 39% to 50%, with the biggest fluctuations occurring for unit cells with lengths of 200μm to 300μm. Designs with unit cell lengths of 400μm and greater all had porosity above 45%, which is close to the targeted 50% porosity that is reduced once tissue seeded.

The pore size for (Fig. 3B) all designs ranged from about 80μm to 500μm and generally increased linearly with unit cell length. The Cube and Hole designs had slightly larger pores than the Hole and Sphere designs at a given unit cell length. At a unit cell length of 1,000μm, pore sizes were 484μm for the Cube, 425μm for the Round, 489μm for the Hole, and 403μm for the Sphere designs.

Designs differed in how unit cell length scaled with mean positive curvature. Cube and Round designs had higher mean positive curvature than Channel and Sphere designs for all unit cell lengths (Fig. 3C). Additionally, Cube and Round designs had a slight increase in mean positive curvature as unit cell length increased while Hole and Sphere designs had decreasing mean positive curvature as unit cell length increased.

Surface-Volume ratio (Fig. 3D) for all designs exponentially decayed as unit cell length increased. The scaling of these ratios follows these relationships due to the difference in how each individual property scales with the change in unit cell length. For instance, increasing unit cell length increases surface area according to a square law while volume would increase according to a cubic law.

4.2 Sensitivity results

A sensitivity analysis was conducted to determine how changes in mask radius influenced property calculations over time during the simulation. In previous studies, a mask radius of 5.5voxels was sufficient for beam-based lattices, such as the Cube. Here we investigate whether a 5.5voxel mask radius is sufficient for the Sphere design, or whether a larger 9.5voxel mask influences results (Fig. 4).

In Figure 4A, simulations are shown at time step $t = 60$. This time step corresponds to the point when the Sphere design evaluated with a mask of 5.5voxels converges because it calculates no positive curvature among all interface voxels, thus halting the tissue growth. All simulations reach a similar state at $t = 60$, however, all other simulations besides the Sphere evaluated with mask radius of 5.5voxels continue until the entire void volume is filled with tissue (Fig. 4B). The growth occurs faster for the Cube designs in comparison to the Sphere. When considering the mean positive curvature (Fig. 4C), the mask radius generally models the same growth behavior for both 5.5voxel and 9.5voxel mask radii for the Cube. Both mask sizes model the small peak in mean positive curvature that occurs around $t = 150$ as the Cube's planar pores close before leaving one large central void [8]. The Sphere however, requires the larger mask radius to model tissue growth behavior that fills the entire porous volume. The Sphere has a similar increased in mean positive curvature at around step 240, which corresponds with the closing of its planar pores.

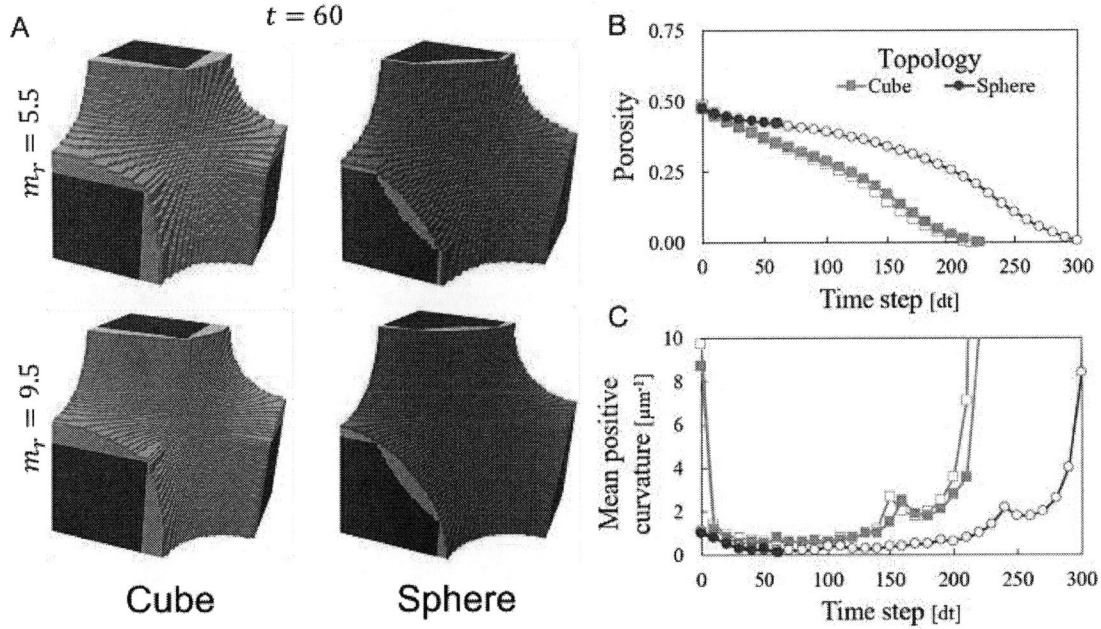

FIGURE 4: GROWTH FOR $l = 500\mu m$ UNIT CELLS WITH MASK RADIUS $m_r = 5.5$ (CLOSED SYMBOLS) OR $m_r = 9.5$ (OPEN)

A second sensitivity analysis was conducted for the Sphere design to determine an appropriate mask that captures the void filling behavior for a broad range designs, with the lowest mask radius possible to reduce computational effort. Sphere designs were generated with 200um to 1,000um unit cell lengths, in increments of 100um. Designs were evaluated with mask radii of 3.5voxels, 5.5voxels, 7.5voxels, 9.5voxels, and 11.5voxels. Time steps required for convergence, computation time at convergence, porosity at convergence, and curvature at $t = 0$ are plotted for each design in Figure 5.

The time steps to convergence increased with unit cell length (Fig. 5A). Simulations with mask radii of 7.5 voxels, 9.5voxels, and 11.5voxels had similar time steps to convergence for each design. Simulations with mask radii of 3.5voxels and 5.5voxels had lower time steps to convergence, which corresponded to simulations that halted before complete void filling.

The increased number of steps to convergence corresponded to an increase in computational time that exponentially increased with unit cell length (Fig. 5B). Time also exponentially increased with mask radius, since larger masks have more voxels used to evaluate curvature locally, therefore increasing the number of computations required for evaluation.

Designs evaluated with masks of radius 7.5voxels or greater required more steps to convergence in comparison to designs evaluated with smaller masks that converged much faster since they converged prior to filling the entire porous volume to reach zero porosity (Fig. 5C). The 300μm length cube evaluated with a mask of radius of 7.5voxels also did not completely fill, which suggests that a mask of at least 9.5voxels is necessary for these smaller designs.

All masks predicted a similar mean positive curvature immediately after tissue seeding for all designs, with the exception of the 3.5voxel radius mask (Fig. 5D). The mean positive curvature generally decreased for designs at a linear rate as unit cell length increased. Results from different mask sizes tended to have higher agreeance as unit cell length increased, which suggests a larger mask is necessary for smaller unit cell lengths.

As a whole, these findings suggest that the mask radius of 7.5voxels is sufficient for predicting Sphere tissue growth with similar accuracy as larger masks in most cases, but with less computation time required.

4.3 Bulk results

Designs of all topology types from Figure 3 were simulated with a mask radius of 7.5voxels until convergence, resulting in all designs having a completely filled void volumes (Fig. 6).

All designs filled at similar rates that increased with unit cell length, ranging from 8 to 329 steps (Fig. 6A). Generally, the Round and Sphere designs required more time to fill than the Cube and Hole designs, with these latter two requiring no more than 250 steps to fill for their largest unit cell length. When designs are plotted according to their surface-volume ratio and void filling rate, all designs follow a similar curve, regardless of topology (Fig. 6B). Designs fill exponentially faster as surface-volume ratio increases. A curve fit to the data with the equation $V_{rate} = 0.0003(S/V)^{2.1463}$ has an $R^2 = 0.9785$ when considering all topologies as a single trend. The high accuracy of this fit suggests it could provide similar evaluations to simulations since it could facilitate efficient calculations for design automation algorithms.

5. DISCUSSION

Four unit cell topologies were used to design scaffolds evaluated with a tissue growth simulation. All designs were generated with a fixed 50% porosity. Cube and Hole designs had higher mean positive curvature (Fig. 3C) in comparison to Round and Sphere designs. The higher mean positive curvature translated to faster tissue growth (Fig. 6A), however, all designs

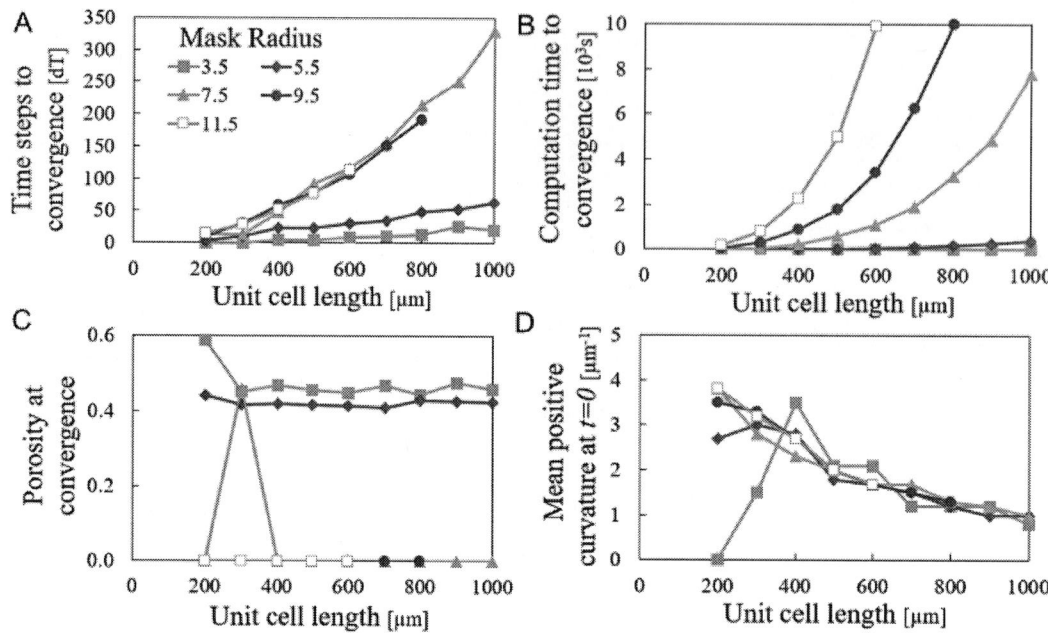

FIGURE 5: SENSITIVY ANALYSIS OF SPHERE DESIGNS EVALUATED WITH MASK RADIUS m_r

only differed by about 30% in how much time they required to fill at the largest unit cell sizes. The square cross-sectional area of beams for the Cube design therefore provided faster tissue growth rate per unit volume in comparison to the circular cross-sectional area of beams for Round designs.

Sensitivity analysis results (Fig. 4) suggest a mask radius of 5.5voxels is sufficient for evaluating Cube designs and 7.5voxels or more is necessary for Sphere designs. As unit cell length is decreased (Fig. 5), a larger mask is preferred. The primary trade-off in choosing a mask size is accuracy vs. computation time. Larger masks require exponentially more time to calculate curvature to evaluate tissue growth but do not predict different behavior in comparison to smaller masks generally, unless smaller masks converge early prior to complete void filling behavior.

The bulk analysis (Fig. 6B) suggests that all topologies have similar trends in void filling rate per surface-volume ratio used to fit a power equation that predicts tissue growth without the

need for further simulations. Such laws have been derived in past studies [8], where a pareto trade-off demonstrated void filling rate and permeability were opposed for a diverse set of scaffold designs. According to this past study validated with *in vitro* data, each step in the simulation corresponds to approximately one tenth of a day in cell culture.

Multiple different 3D printing processes are viable for printing designs considered in this study, including processes for both metals and polymers. In Figure 7, 3D printed samples that were created with a biocompatible material using polyjet printing are presented as proof-of-concepts [7]. The Fig. 7 designs have microscale pores and 50%-70% porosity that is representative of the designs explored in this study.

One limitation in this study was the use of only one evaluation method for calculating local curvature in comparison to the variety of mathematical methods available [57-60]. These approaches could provide different trade-offs in accuracy and computation time if incorporated in a simulation. Computational

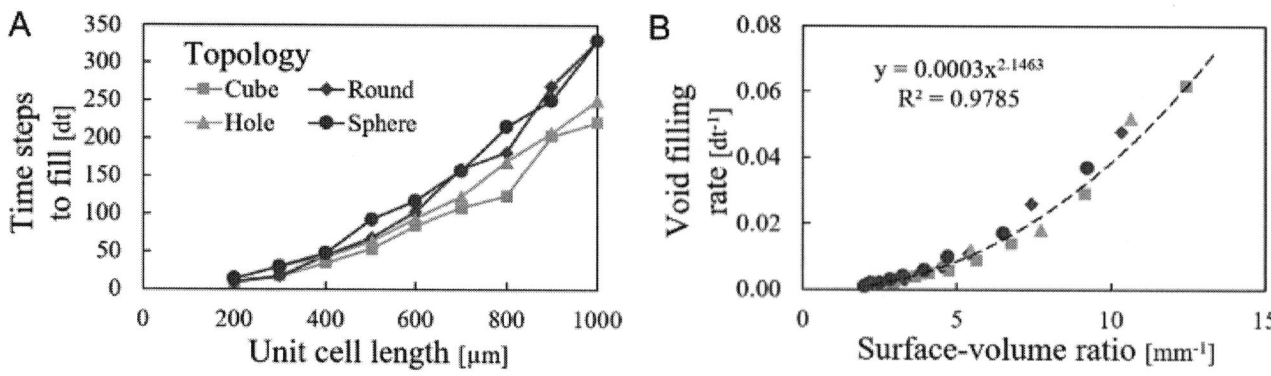

FIGURE 6: BULK SIMULATION RESULTS FOR 50% POROUS DESIGNS

time is improvable by using alternate methods of evaluating curvature on scaffolds that have successfully predicted growth in 3D printed designs [36]. Although the current approach was sufficient for exploring these designs, a more efficient approach would improve evaluation times for larger design spaces in the future.

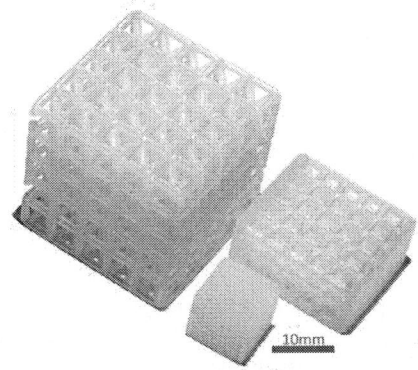

FIGURE 7: PROOF-OF-CONCEPT 3D PRINTED SAMPLES

Future work may focus on developing computational design algorithms for scaffolds that may handle the variety of design properties and behaviors relevant to scaffold performance. These properties may include porosity, pore size, tissue growth rate, and surface-volume ratio included in this study, as well as stiffness, permeability, and other less prominent factors, such as how efficiently cells adhere to a surface upon seeding. The biological growth simulation is improvable by including stochasticity and to adjust growth rates based on nutrient availability. Additionally, further empirical validation is necessary, especially since scaffolds of the same design that are constructed with different materials can significantly influence biological responses. All of these improvements would move improve the biological assessment such that future multivariable optimization processes are more efficient and may explore a larger portion of the design space for comparing biological growth with other design criteria.

6. CONCLUSION

This paper investigated how altering the resolution of a biological simulation influences its capabilities for predicting tissue growth in lattice structures. After determining a suitable resolution, designs of four topologies were compared to find trends that could inform future computational design endeavors for optimizing tissue scaffolds for personalized medicine. These steps provide an efficient means of evaluating biological growth, which is a limiting factor in developing efficient multivariable optimization approaches for scaffold design.

Design topologies consisted of a Cube, Round, Hole, and Sphere topology. Designs of all topologies had similar pore sizes and surface-volume ratios as a function of unit cell length, however, Cube and Hole designs had higher mean positive curvature that translated to faster tissue growth. A sensitivity analysis suggested that using a scanning mask of 7.5 voxel radius is sufficient for modeling tissue growth behavior in most cases. The mask radius consistently simulated growth in all topologies at 50% porosity with unit cell lengths relevant to bone tissue engineering. All topologies were fit to the same curve that

predicted a power law relationship describing how fast tissue fills the void volume as a function of surface-volume ratio.

The discovered power law is useful for future work to efficiently predict biological growth in scaffolds without the need for further simulations, which facilitates faster evaluations for computational design methodologies. Findings suggest unit cell designs are differentiated based on their varied volumes associated with specific surface-volume ratios. Future work may also assess their mechanical properties. This work provides a significant advancement in developing efficient methods for designing scaffolds, which enables the fabrication of customized 3D printed devices for personalized medical treatments.

ACKNOWLEDGEMENTS

3D printed samples in Figure 7 were fabricated and provided by research support at ETH Zurich, Switzerland.

REFERENCES

[1] P. F. Egan, I. Bauer, K. Shea, and S. J. Ferguson, "Mechanics of Three-Dimensional Printed Lattices for Biomedical Devices," *Journal of Mechanical Design,* vol. 141, no. 3, p. 031703, 2019.

[2] M. K. Thompson *et al.,* "Design for Additive Manufacturing: Trends, opportunities, considerations, and constraints," *CIRP Annals-Manufacturing Technology,* vol. 65, no. 2, pp. 737-760, 2016.

[3] X. Xu, C. K. P. Vallabh, A. Krishnan, S. Volk, and C. Cetinkaya, "In-Process Thread Orientation Monitoring in Additive Manufacturing," *3D Printing and Additive Manufacturing,* vol. 6, no. 1, pp. 21-30.

[4] S. J. Hollister *et al.,* "Design control for clinical translation of 3D printed modular scaffolds," *Annals of biomedical engineering,* vol. 43, no. 3, pp. 774-786, 2015.

[5] C.-G. Liu, Y.-T. Zeng, R. Kankala, S.-S. Zhang, A.-Z. Chen, and S.-B. Wang, "Characterization and Preliminary Biological Evaluation of 3D-Printed Porous Scaffolds for Engineering Bone Tissues," *Materials,* vol. 11, no. 10, p. 1832, 2018.

[6] P. F. Egan, V. C. Gonella, M. Engensperger, S. J. Ferguson, and K. Shea, "Computationally designed lattices with tuned properties for tissue engineering using 3D printing," *PloS one,* vol. 12, no. 8, p. e0182902, 2017.

[7] P. Egan, S. Ferguson, and K. Shea, "Design of hierarchical 3D printed scaffolds considering mechanical and biological factors for bone tissue engineering," *Journal of Mechanical Design,* vol. 139, no. 6, 2017.

[8] P. F. Egan, K. A. Shea, and S. J. Ferguson, "Simulated tissue growth for 3D printed scaffolds," *Biomechanics and modeling in mechanobiology,* pp. 1-15, 2018.

[9] H. Kang, S. J. Hollister, F. La Marca, P. Park, and C.-Y. Lin, "Porous biodegradable lumbar interbody fusion cage design and fabrication using integrated global-local topology optimization with laser sintering," *Journal of biomechanical engineering,* vol. 135, no. 10, p. 101013, 2013.

[10] A. Boccaccio, A. E. Uva, M. Fiorentino, L. Lamberti, and G. Monno, "A Mechanobiology-based Algorithm to Optimize the Microstructure Geometry of Bone Tissue Scaffolds," *International journal of biological sciences,* vol. 12, no. 1, p. 1, 2016.

[11] H. Zaharin *et al.,* "Effect of Unit Cell Type and Pore Size on Porosity and Mechanical Behavior of Additively Manufactured Ti6Al4V Scaffolds," *Materials,* vol. 11, no. 12, p. 2402, 2018.

[12] P. Egan, X. Wang, H. Greutert, K. Shea, K. Wuertz-Kozak, and S. Ferguson, "Mechanical and Biological Characterization of 3D Printed Lattices," *3D Printing and Additive Manufacturing,* 2019.

[13] D. Melancon, Z. Bagheri, R. Johnston, L. Liu, M. Tanzer, and D. Pasini, "Mechanical characterization of structurally porous biomaterials built via additive manufacturing: experiments, predictive models, and design maps for load-bearing bone replacement implants," *Acta biomaterialia,* vol. 63, pp. 350-368, 2017.

[14] G. Dong, Y. Tang, and Y. F. Zhao, "A Survey of Modeling of Lattice Structures Fabricated by Additive Manufacturing," *Journal of Mechanical Design,* vol. 139, no. 10, p. 100906, 2017.

[15] S. Zhang, S. Vijayavenkataraman, W. F. Lu, and J. Y. Fuh, "A review on the use of computational methods to characterize, design, and optimize tissue engineering scaffolds, with a potential in 3D printing fabrication," *Journal of Biomedical Materials Research Part B: Applied Biomaterials,* 2018.

[16] M. I. Mohammed and I. Gibson, "Design of three-dimensional, triply periodic unit cell scaffold structures for additive manufacturing," *Journal of Mechanical Design,* vol. 140, no. 7, p. 071701, 2018.

[17] X. Wang *et al.*, "Topological design and additive manufacturing of porous metals for bone scaffolds and orthopaedic implants: A review," *Biomaterials,* vol. 83, pp. 127-141, 2016.

[18] Y. Guyot, I. Papantoniou, F. Luyten, and L. Geris, "Coupling curvature-dependent and shear stress-stimulated neotissue growth in dynamic bioreactor cultures: a 3D computational model of a complete scaffold," *Biomechanics and modeling in mechanobiology,* vol. 15, no. 1, pp. 169-180, 2016.

[19] N. Taniguchi *et al.*, "Effect of pore size on bone ingrowth into porous titanium implants fabricated by additive manufacturing: an in vivo experiment," *Materials Science and Engineering: C,* vol. 59, pp. 690-701, 2016.

[20] C. M. Bidan, F. M. Wang, and J. W. Dunlop, "A three-dimensional model for tissue deposition on complex surfaces," *Computer methods in biomechanics and biomedical engineering,* vol. 16, no. 10, pp. 1056-1070, 2013.

[21] M. Paris *et al.*, "Scaffold curvature-mediated novel biomineralization process originates a continuous soft tissue-to-bone interface," *Acta biomaterialia,* vol. 60, pp. 64-80, 2017.

[22] P. Christen *et al.*, "Voxel size dependency, reproducibility and sensitivity of an in vivo bone loading estimation algorithm," *Journal of The Royal Society Interface,* vol. 13, no. 114, p. 20150991, 2016.

[23] M. Rumpler, A. Woesz, J. W. Dunlop, J. T. van Dongen, and P. Fratzl, "The effect of geometry on three-dimensional tissue growth," *Journal of the Royal Society Interface,* vol. 5, no. 27, pp. 1173-1180, 2008.

[24] I. B. Bischofs and U. S. Schwarz, "Cell organization in soft media due to active mechanosensing," *Proceedings of the National Academy of Sciences,* vol. 100, no. 16, pp. 9274-9279, 2003.

[25] M. Rumpler, A. Woesz, F. Varga, I. Manjubala, K. Klaushofer, and P. Fratzl, "Three-dimensional growth behavior of osteoblasts on biomimetic hydroxylapatite scaffolds," *Journal of Biomedical Materials Research Part A,* vol. 81, no. 1, pp. 40-50, 2007.

[26] C. M. Bidan *et al.*, "How linear tension converts to curvature: geometric control of bone tissue growth," *PloS one,* vol. 7, no. 5, p. e36336, 2012.

[27] J. A. Sanz-Herrera, P. Moreo, J. M. García-Aznar, and M. Doblaré, "On the effect of substrate curvature on cell mechanics," *Biomaterials,* vol. 30, no. 34, pp. 6674-6686, 2009.

[28] J. Sanz-Herrera, J. Garcia-Aznar, and M. Doblare, "A mathematical model for bone tissue regeneration inside a specific type of scaffold," *Biomechanics and modeling in mechanobiology,* vol. 7, no. 5, pp. 355-366, 2008.

[29] A. Carlier, L. Geris, J. Lammens, and H. Van Oosterwyck, "Bringing computational models of bone regeneration to the clinic," *Wiley Interdisciplinary Reviews: Systems Biology and Medicine,* vol. 7, no. 4, pp. 183-194, 2015.

[30] P. R. Buenzli, "Governing equations of tissue modelling and remodelling: A unified generalised description of surface and bulk balance," *PloS one,* vol. 11, no. 4, p. e0152582, 2016.

[31] T. Chen, M. M. McCarthy, H. Guo, R. Warren, and S. A. Maher, "The Scaffold–Articular Cartilage Interface: A Combined In Vitro and In Silico Analysis Under Controlled Loading Conditions," *Journal of biomechanical engineering,* vol. 140, no. 9, p. 091002, 2018.

[32] B. C. Thorne, A. M. Bailey, and S. M. Peirce, "Combining experiments with multi-cell agent-based modeling to study biological tissue patterning," *Briefings in bioinformatics,* vol. 8, no. 4, pp. 245-257, 2007.

[33] N. Garijo, R. Manzano, R. Osta, and M. Perez, "Stochastic cellular automata model of cell migration, proliferation and differentiation: Validation with in vitro cultures of muscle satellite cells," *Journal of theoretical biology,* vol. 314, pp. 1-9, 2012.

[34] J. S. Czarnecki, S. Jolivet, M. E. Blackmore, K. Lafdi, and P. A. Tsonis, "Cellular Automata Simulation of Osteoblast Growth on Microfibrous-Carbon-Based Scaffolds," *Tissue Engineering Part A,* vol. 20, no. 23-24, pp. 3176-3188, 2014.

[35] M. A. Alias and P. R. Buenzli, "Modeling the Effect of Curvature on the Collective Behavior of Cells Growing New Tissue," *Biophysical Journal,* vol. 112, no. 1, pp. 193-204, 2017.

[36] Y. Guyot, I. Papantoniou, Y. C. Chai, S. Van Bael, J. Schrooten, and L. Geris, "A computational model for cell/ECM growth on 3D surfaces using the level set method: a bone tissue engineering case study," *Biomechanics and modeling in mechanobiology,* vol. 13, no. 6, pp. 1361-1371, 2014.

[37] C. M. Bidan, K. P. Kommareddy, M. Rumpler, P. Kollmannsberger, P. Fratzl, and J. W. Dunlop, "Geometry as a factor for tissue growth: towards shape optimization of tissue engineering scaffolds," *Advanced healthcare materials,* vol. 2, no. 1, pp. 186-194, 2013.

[38] C. Kurzmann *et al.*, "Evaluation of resins for stereolithographic 3D-printed surgical guides: the response of L929 cells and human gingival fibroblasts," *BioMed research international,* vol. 2017, 2017.

[39] F. Alifui-Segbaya, J. Bowman, A. R. White, S. Varma, G. J. Lieschke, and R. George, "Toxicological assessment of additively manufactured methacrylates for medical devices in dentistry," *Acta biomaterialia,* vol. 78, pp. 64-77, 2018.

[40] Ž. Kačarević *et al.*, "An Introduction to 3D Bioprinting: Possibilities, Challenges and Future Aspects," *Materials,* vol. 11, no. 11, p. 2199, 2018.

[41] J. Segmehl, V. Studer, T. Keplinger, and I. Burgert, "Characterization of wood derived hierarchical cellulose scaffolds for multifunctional applications," *Materials,* vol. 11, no. 4, p. 517, 2018.

[42] B. Huang, G. Caetano, C. Vyas, J. Blaker, C. Diver, and P. Bártolo, "Polymer-ceramic composite scaffolds: The effect of hydroxyapatite and β-tri-calcium phosphate," *Materials,* vol. 11, no. 1, p. 129, 2018.

[43] J.-B. Lee, W.-Y. Maeng, Y.-H. Koh, and H.-E. Kim, "Porous calcium phosphate ceramic scaffolds with tailored pore orientations and mechanical properties using lithography-based ceramic 3D printing technique," *Materials,* vol. 11, no. 9, p. 1711, 2018.

[44] J. García-Ruíz and A. Diaz Lantada, "3D Printed Structures Filled with Carbon Fibers and Functionalized with Mesenchymal Stem Cell Conditioned Media as In Vitro Cell Niches for Promoting Chondrogenesis," *Materials,* vol. 11, no. 1, p. 23, 2018.

[45] I. Buj-Corral, A. Bagheri, and O. Petit-Rojo, "3D printing of porous scaffolds with controlled porosity and pore size values," *Materials,* vol. 11, no. 9, p. 1532, 2018.

[46] G. Ahn, J. H. Park, T. Kang, J. W. Lee, H.-W. Kang, and D.-W. Cho, "Effect of pore architecture on oxygen diffusion in 3D scaffolds for tissue engineering," *Journal of biomechanical engineering,* vol. 132, no. 10, p. 104506, 2010.

[47] A. Bagheri, I. Buj-Corral, M. Ferrer, M. M. Pastor, and F. Roure, "Determination of the Elasticity Modulus of 3D-Printed Octet-Truss Structures for Use in Porous Prosthesis Implants," *Materials,* vol. 11, no. 12, p. 2420, 2018.

[48] A. Khoda, I. T. Ozbolat, and B. Koc, "Engineered tissue scaffolds with variational porous architecture," *Journal of biomechanical engineering,* vol. 133, no. 1, p. 011001, 2011.

[49] P. Egan, J. Cagan, C. Schunn, F. Chiu, J. Moore, and P. LeDuc, "The D3 Methodology: Bridging Science and Design for Bio-Based Product Development," *Journal of Mechanical Design,* vol. 138, no. 8, 2016.

[50] X. Blasco, J. M. Herrero, J. Sanchis, and M. Martínez, "A new graphical visualization of n-dimensional Pareto front for decision-making in multiobjective optimization," *Information Sciences,* vol. 178, no. 20, pp. 3908-3924, 2008.

[51] A. Anindyajati, P. Boughton, and A. Ruys, "Modelling and optimization of Polycaprolactone ultrafine-fibres electrospinning process using response surface methodology," *Materials,* vol. 11, no. 3, p. 441, 2018.

[52] E. Ulu, R. Huang, L. B. Kara, and K. S. Whitefoot, "Concurrent Structure and Process Optimization for Minimum Cost Metal Additive Manufacturing," *Journal of Mechanical Design,* vol. 141, no. 6, p. 061701, 2019.

[53] H. Kang, C.-Y. Lin, and S. J. Hollister, "Topology optimization of three dimensional tissue engineering scaffold architectures for prescribed bulk modulus and diffusivity," *Structural and Multidisciplinary Optimization,* vol. 42, no. 4, pp. 633-644, 2010.

[54] T. Stanković, J. Mueller, P. Egan, and K. Shea, "A Generalized Optimality Criteria Method for Optimization of Additively Manufactured Multimaterial Lattice Structures," *Journal of Mechanical Design,* vol. 137, no. 11, p. 111705, 2015.

[55] J. M. Sobral, S. G. Caridade, R. A. Sousa, J. F. Mano, and R. L. Reis, "Three-dimensional plotted scaffolds with controlled pore size gradients: effect of scaffold geometry on mechanical performance and cell seeding efficiency," *Acta Biomaterialia,* vol. 7, no. 3, pp. 1009-1018, 2011.

[56] M. De Wild *et al.,* "Osteoconductive lattice microarchitecture for optimized bone regeneration," *3D Printing and Additive Manufacturing,* 2018.

[57] J. Bullard, E. Garboczi, W. Carter, and E. Fuller, "Numerical methods for computing interfacial mean curvature," *Computational materials science,* vol. 4, no. 2, pp. 103-116, 1995.

[58] H. C. Batagelo and S.-T. Wu, "Estimating curvatures and their derivatives on meshes of arbitrary topology from sampling directions," *The Visual Computer,* vol. 23, no. 9-11, pp. 803-812, 2007.

[59] O. I. Frette, G. Virnovsky, and D. Silin, "Estimation of the curvature of an interface from a digital 2D image," *Computational materials science,* vol. 44, no. 3, pp. 867-875, 2009.

[60] M. Kronenberger, O. Wirjadi, J. Freitag, and H. Hagen, "Gaussian curvature using fundamental forms for binary voxel data," *Graphical Models,* vol. 82, pp. 123-136, 2015.

Proceedings of the ASME 2019
International Design Engineering Technical Conferences
and Computers and Information in Engineering Conference
IDETC/CIE2019
August 18-21, 2019, Anaheim, CA, USA

DETC2019-97437

A DIGITAL TWIN-DRIVEN IMPROVED DESIGN APPROACH OF DRAWING BENCH FOR BRAZING MATERIAL

Bingtao Hu
State Key Laboratory of Fluid Power and
Mechatronic Systems, Zhejiang University
Hangzhou, Zhejiang, China
hubingtao@zju.edu.cn

Yixiong Feng
State Key Laboratory of Fluid Power and
Mechatronic Systems, Zhejiang University
Hangzhou, Zhejiang, China
fyxtv@zju.edu.cn

Yicong Gao
State Key Laboratory of Fluid Power
and Mechatronic Systems, Zhejiang
University
Hangzhou, Zhejiang, China
gaoyicong@zju.edu.cn

Hao Zheng
State Key Laboratory of Fluid Power
and Mechatronic Systems, Zhejiang
University
Hangzhou, Zhejiang, China
haozheng@zju.edu.cn

Jianrong Tan
State Key Laboratory of Fluid Power
and Mechatronic Systems, Zhejiang
University
Hangzhou, Zhejiang, China
egi@zju.edu.cn

ABSTRACT

Brazing materials can be made into different shapes to meet the requirements of different scenarios and the welding rod is a very common form. The rough-processed welding rods must be properly finished by the drawing bench to remove the oxide film on the surface and made into a uniform diameter. However, the continuous welding rod often breaks resulting low production efficiency. To reduce the frequency of workers' reconnection operation of broken welding rods, we proposed a digital twin-based approach to improve the design of the structure of the drawing bench. First, we established a full life cycle digital twin model for the welding rod from the formulation stage to the finishing stage. The product ontology of the welding rod was built and the key process parameters were collected. Second, based on the product ontology, the key structural parameter of the drawing bench that affects the internal stress of the welding rod was determined by means of analytic hierarchy process. Third, we modified the key structural parameter in the digital twin model and simulated the finishing process. A near-optimal parameter was found. Last, we improved the structure of the actual drawing bench accordingly and carried out some experiments. The results matched well with the simulation prediction and the frequency of welding rod breaking is significantly reduced, which proved the effectiveness of our proposed improved design approach.

Keywords: Brazing material; Digital twin; Improved design;

1. INTRODUCTION

Brazing is one of the most important methods of precision welding. Generally, the melting point of the brazing material is lower than that of the weldment. The temperature can be heated to the melting point of the brazing material and the liquefied brazing material fills the gap between the solid weldments [1]. Since the brazing process does not destroy the shape and performance of the weldment itself, brazing is widely used [2–3]. Brazing material can be made into different shapes, such as welding rods, welding rings, welding pieces, etc., depending on the application scenario (Figure 1). Welding rod is a very common form. The raw materials of the welding rods need to be matched according to the formulation and put together to be melt into liquid. After the liquid is cooled, it is cast into a metal ingot. Then the metal ingot is rough-processed by a hydraulic machine into the semi-finished products of the wielding rods.

The semi-finished products require further finishing on the drawing bench that will heat and soften them. The semi-finished products will pass through a ring made of harder metal that will rub off the oxide film on the surface and standardize their diameter. However, the continuous welding rod often breaks during the finishing process resulting low production efficiency.

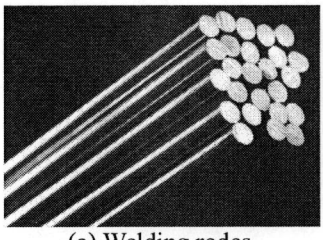
(a) Welding rodes

(b) Welding rings

(c) Welding pieces

(d) Welding strip

Figure 1. Brazing materials in different shapes.

To reduce the frequency of workers' reconnection operation of broken welding rods, we proposed a digital twin-based approach to improve the design of the structure of the drawing bench.

The core idea of digital twin is to build a digital mirror twin of a physical entity. The physical entity can be anything like equipment, products or workshops. The general digital twin consists of three parts, which are the physical entities in physical space, the virtual models in virtual space, and the connected data that tie physical and virtual worlds [4]. Digital twin can collect all data related the physical entity through the new generation information technologies such as sensor technology, Bluetooth technology, and wireless transmission technology, and records them in the database of the virtual model. These data about the entity can be life-cycled, including the design, manufacturing and service phases. In other words, the virtual model in virtual space is a high-fidelity faithful mapping of the entity in physical space [5]. Compared to the general digital prototyping technology, we can monitor the physical entities in real time by means of digital twin, and combine historical data for more realistic simulation, diagnosis and prediction of the performance of the corresponding physical entities. At the same time, these results will be compared and corrected with real-time data to achieve self-correction and evolution of digital twin. Digital twin has aroused widespread interests in both industry and academia, and has been applied to product design, manufacturing and preventive maintenance [6–8]. The application of digital generation enables designers and producers to obtain a large amount of data, so heuristic algorithms and machine learning algorithms for big data are also widely used in data processing. These algorithms are very efficient at processing quantitative and labeled data, but multi-objective decision algorithms are more useful when dealing with ambiguous and unlabeled data [9–12].

In this work, we proposed a digital twin-based approach to improve the design of the structure of the drawing bench. The full life cycle digital twin model of welding rods production was established. In this digital twin model, we measured and recorded the data about the properties of the raw materials, manufacturing processes and frequency of breaking occurrence for the welding rods of different formulations. Based on that, the product ontology of the welding rod was established. Integrating with the product ontology, the analytic hierarchy process (AHP) was used to determine the key structural parameter of the drawing bench that affects the internal stress of the welding rod. Then the key structural parameter was modified in the digital twin model and the historical data were used to drive the simulation of the finishing process. After several rounds of iterative simulation, we found the near-optimal parameter. Last, the actual drawing bench was modified accordingly, based on which some verifying experiments were carried out.

2. THE PROPOSED APPROACH

To reduce the workers' reconnection operation of the broken welding rods and the drawing bench downtime, the proposed digital twin-driven improved design approach can be summarized into four steps: 1) establish a digital twin model, 2) identify the key structural parameter, 3) simulate the improved design based on the data of historical digital twin model and 4) compare the verifying experiment results and simulation predictions (Figure 2). The following is a detailed description of each step.

2.1 Establish a digital twin model

With the wide application of new generation information technologies in industry, the design of mechanical products has entered the data-driven or even big data-driven era. The cloud technology, industrial Internet technology and other data storage, transmission and processing technologies have become important supports. However, the data-driven design for mechanical products mainly places emphasis on the analysis of physical data rather than the virtual models. That is, the performance prediction of the design change or design improvement using physical data to drive the virtual model is less considered, and the verification between the physical experiments and virtual simulations is usually absent. Digital twin, an emerging technology trying to connect the physical and virtual world, provides us with a new paradigm considering the synergy between entities and data.

Unlike general CAD (computer-aided design) technology, which only pays attention to the spatial structure and assembly features of the product, the digital twin model has two important characteristics. One is that the high-fidelity product model comes with a large amount of corresponding data from the full life cycle, and the other is that the data can be collected in real time and can be seamlessly coordinated with the historical data. All data can be obtained not only from the field measurement, but also from the reference books, experience, and simulation and so on.

To clearly describe, store, and organize the large amounts of data from digital twin models, the product ontology is designed in this work to integrate the physical entity and corresponding data of the mechanical product. Product ontology is an extension

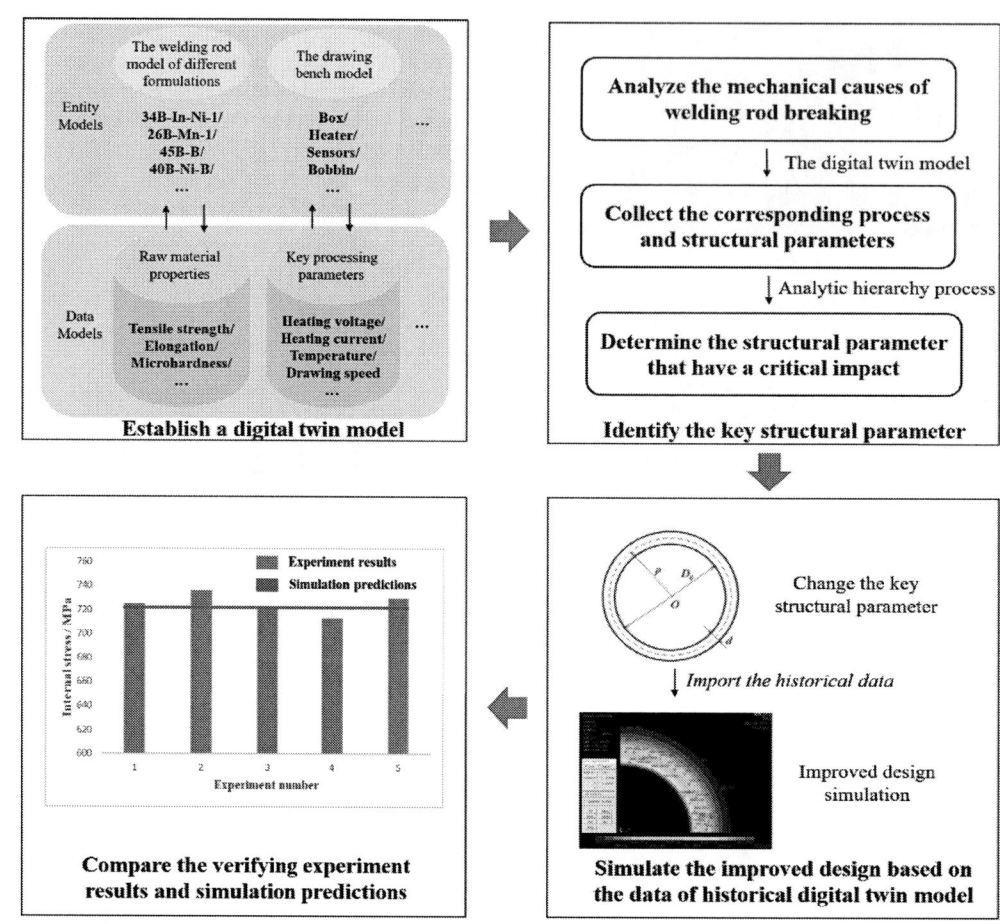

Figure 2. The steps of the proposed digital twin-driven improved design approach

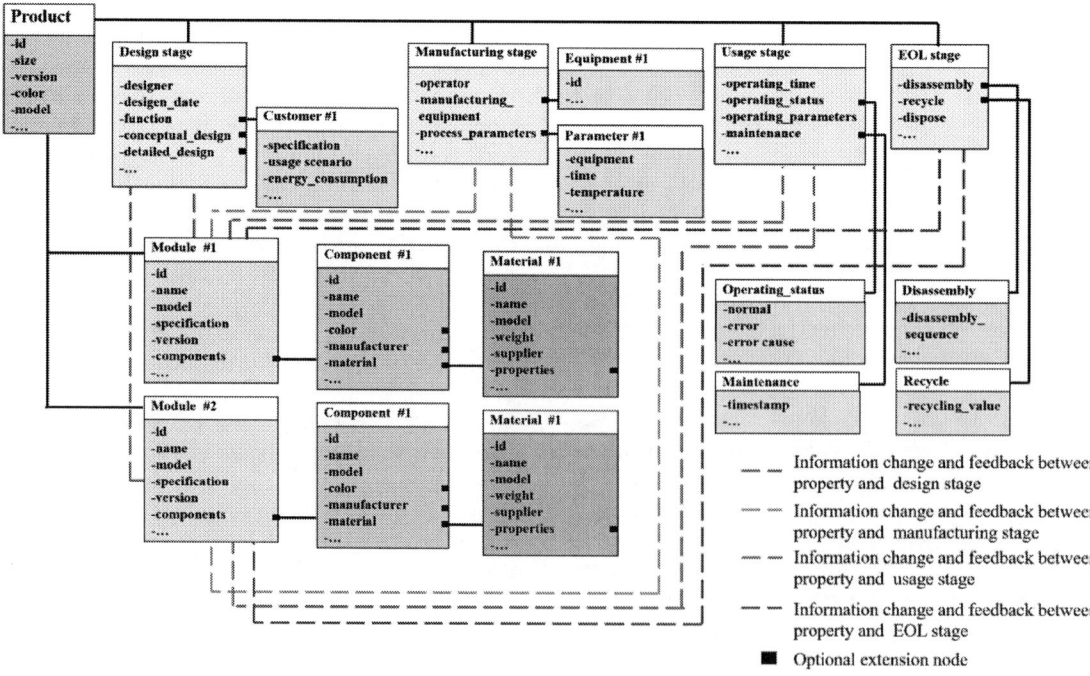

Figure 3. The product ontology of a complex product for its digital twin model.

class with branchable structure. For a digital twin model of a complex product, its product ontology model mainly consists of the data from four stages, i.e., design stage, manufacturing stage, usage stage and EOL (end-of-life) stage, and the property data such as its module data, component data and 3D model data. At each stage, the property data may change accordingly and can be fed back to the iteration of the digital twin model such that they will be connected to the full life cycle of the product. Figure 3 shows a product ontology for its digital twin model.

It should be noted that the product ontology structure of the digital twin model can be added or deleted depending on actual needs. Besides, the each manufacturing equipment in the manufacturing stage can also be considered as a product such that its product ontology can be established. Furthermore, the impact of changes in manufacturing equipment and process parameters on the product can be reflected by the relationship between the two corresponding product ontologies.

The structure of the welding rod is simple and can be defined by several key shape parameters, so the module and component are not included in its digital twin model. On the other hand, the formulations of the electrode can be diverse and the material properties and process parameters are different. Material properties can be measured and recorded by actual experiments, while process parameters need to be acquired from PDM

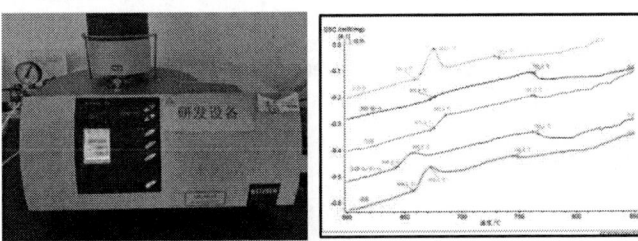

(a)　DSC experimental equipment and result

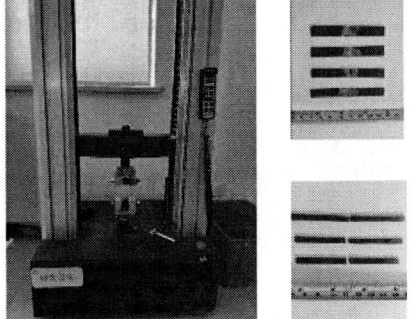

(b)　Tensile experimental equipment and result

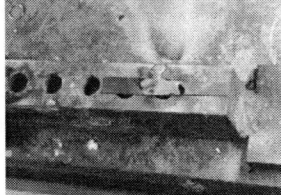

(c)　Welding experimental equipment and result

Figure 4. Part of the experimental equipment and experimental results

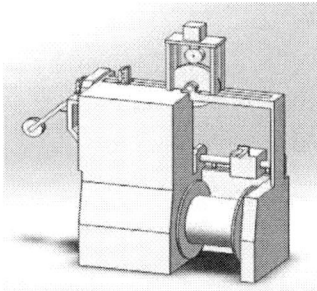

Figure 5. The drawing bench and its digital twin model

(Product Data Management) system or collected by sensors. To establish the product ontology of the welding rods digital twin model, we have done spreading experiment, DSC (Differential Scanning Calorimeter) experiment, tensile experiment, microhardness experiment and welding experiment on 5 different formulations of welding rods, i.e., 26B-Mn-1, 34B-In-Ni-1, 35B, 40B and 45B–B. Part of the experimental equipment and experimental results are shown in Figure 4. In addition, the digital twinning model of the main processing equipment, i.e., drawing bench, was established and key process parameters such as heating voltage, heating current, heating temperature and drawing speed were measured and recorded (Figure 5). Then the product ontology of the digital twin model of welding rod can be completed.

2.2 Identify the key structural parameter

According to the product ontology we established in section 2.1, we discussed with the operators and process designers, and initially determined the factors that may affect the stress inside the electrode as tensile strength (material property), welding rod diameter (product property), heating temperature and drawing speed (process parameters), bobbin diameter (manufacturing equipment property). The reason that the welding rods broke during the finishing process is that its internal stress becomes greater than its tensile strength. Since we want to change the structural parameters of the drawing bench to adjust the processing technology to reduce the frequency of welding rods breaking, we need to find the relevant structural parameters that affect the internal stress of the welding rods during finishing. To determine the key variable that affect the internal stress most, the analytic hierarchy process is employed. AHP refers to a complex decision-making problem as a system, which decomposes the target into multiple subtargets or criteria with several levels of multiple indicators (or criteria, constraints), and calculates the hierarchy through qualitative index fuzzy quantization method [13]. AHP as well as its developed versions like fuzzy AHP and integrated AHP are widely used due to its simplicity, flexibility and availability [14–16].

The general steps for applying AHP to solve the problem are as follows:

(1) Establish a hierarchical model

The decision-making goal, decision-making criteria and the decision-making objects are divided into the highest layer, the middle layer and the lowest layer according to the mutual

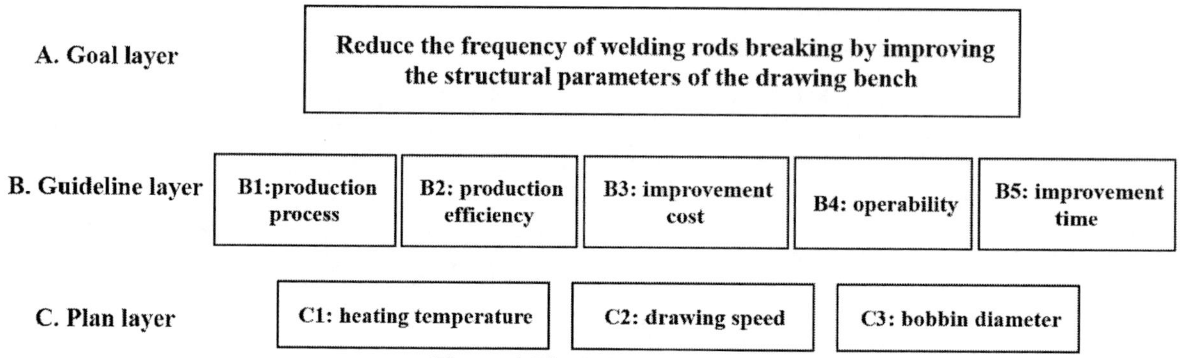

A. Goal layer

> Reduce the frequency of welding rods breaking by improving the structural parameters of the drawing bench

B. Guideline layer

| B1:production process | B2: production efficiency | B3: improvement cost | B4: operability | B5: improvement time |

C. Plan layer

| C1: heating temperature | C2: drawing speed | C3: bobbin diameter |

Figure 6. The AHP structure

relationship between them, and the hierarchical structure diagram is drawn. The highest level refers to the purpose of decision-making and the problem to be solved. The lowest level refers to the alternatives when making decisions. The middle layer refers to the factors considered and the criteria for decision making. For the adjacent two layers, the upper layer is called the target layer, and the lower layer is the factor layer.

(2) Build judgment matrices

When determining the weights between the various factors at each level, if it is only a qualitative result, the importance of the factors may be inconsistent. To reduce the difficulty of comparing factors with different properties and to improve accuracy, not all factors are put together to be compared, but two of them are compared with each other at each time. The matrix formed by the comparison result is called the judgment matrix.

(3) Hierarchical single ordering and consistency check

The feature vector of the largest eigenvalue λ_{max} of the judgement matrix is denoted as W after normalization. The element of W is the ranking weight of the same level factor for the relative importance of the upper level factor. This process is called hierarchical single ordering. The consistency needs to be checked to confirm the hierarchical single order. Supposing the only non-zero eigenvalue of the n-order uniform matrix is n; the largest eigenvalue of the n-th order positive reciprocal matrix A is λ, then $\lambda \geq n$. If and only if $\lambda = n$ t, A can be considered to have passed the consistency test.

(4) Hierarchical total ordering and consistency test

Calculating the weight of all factors at a certain level for the relative importance of the highest level (total target) is called hierarchical total ordering. This process is carried out in order from the highest level to the lowest level.

In this case, our goal is to reduce the frequency of welding rods breaking by improving the structural parameters of the drawing bench. At the same time, we hope the improvement plan has rare impact on the original production process and efficiency. Besides, it should be easy to execute and cost as little time and money as possible. The formulation of the welding rod is generally fixed due to the customer's requirements, so the above-mentioned tensile strength and welding rod diameter cannot be changed. The AHP structure to solve the problem is presented in Figure 6. The comparison matrices of guideline layer and plan

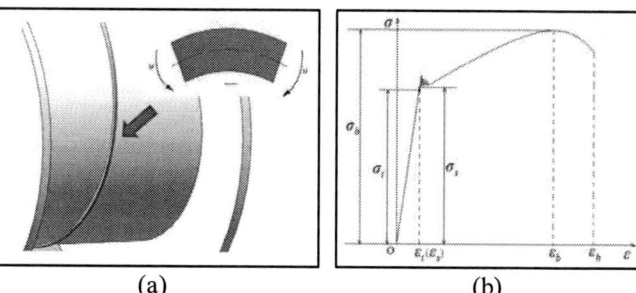

(a) (b)

Figure 7. The illustration of (a) the deformation of the welding rod when it is wound around the bobbin and (b) the general stress-strain curve of the metal during stretching

layer are B and C_1–C_5. Set the matrix consistency threshold to 0.1, it is easy to get all the matrices to meet the consistency requirements by the definition of AHP consistency check. The weight of B to A is

$$W_{B2A} = [0.2636, 0.4758, 0.0538, 0.0981, 0.1087]$$

and the weights of C to B are listed in Table 1.

$$B = \begin{bmatrix} 1 & \frac{1}{2} & 4 & 3 & 3 \\ 2 & 1 & 7 & 5 & 5 \\ \frac{1}{4} & \frac{1}{7} & 1 & \frac{1}{2} & \frac{1}{3} \\ \frac{1}{3} & \frac{1}{5} & 2 & 1 & 1 \\ \frac{1}{3} & \frac{1}{5} & 3 & 1 & 1 \end{bmatrix}$$

$$C_1 = \begin{bmatrix} 1 & 2 & 5 \\ \frac{1}{2} & 1 & 2 \\ \frac{1}{5} & \frac{1}{2} & 1 \end{bmatrix} \quad C_2 = \begin{bmatrix} 1 & \frac{1}{3} & \frac{1}{8} \\ 3 & 1 & \frac{1}{3} \\ 8 & 3 & 1 \end{bmatrix} \quad C_3 = \begin{bmatrix} 1 & 1 & 3 \\ 1 & 1 & 3 \\ \frac{1}{3} & \frac{1}{3} & 1 \end{bmatrix}$$

$$C_4 = \begin{bmatrix} 1 & 3 & 4 \\ \frac{1}{3} & 1 & 1 \\ \frac{1}{4} & 1 & 1 \end{bmatrix} \quad C_5 = \begin{bmatrix} 1 & 1 & \frac{1}{4} \\ 1 & 1 & \frac{1}{4} \\ 4 & 4 & 1 \end{bmatrix}$$

Table 1. The weights of C to B

	B_1	B_2	B_3	B_4	B_5
C_1	0.5954	0.0819	0.4286	0.6337	0.1667
C_2	0.2764	0.2363	0.4286	0.1919	0.1667
C_3	0.1283	0.6817	0.1429	0.1774	0.6667

Table 2. The material properties of one kind of welding rod.

Property	Value
Welding rod diameter d (mm)	3
Elastic limit σ_t (MPa)	545
Elastic modulus E (GPa)	115
Intensity factor λ	0.85
Bending curvature ρ	0.3
Tensile strength σ_b (MPa)	740
Poisson's ratio μ	0.35

Then the final weight of C to A is

$$W_{C2A} = [0.3, 0.245, 0.455]$$

therefore, the structural parameter of the wire drawing bench that needs to be improved is bobbin diameter.

Figure 7 shows the deformation of the welding rod when it is wound around the bobbin and the general stress-strain curve of the metal during stretching. During the welding rod finishing process, the welding rod is continuously twined around the bobbin as the bobbin rotates. As a result, the side of the welding rod away from the center of the bobbin is stretched and becomes easy to break. The smaller the diameter of the bobbin is, the greater the tensile force will be received by the welding rod around it and the easier welding rod is to break. From this perspective, the larger the diameter of bobbin the better. However, if the diameter of the bobbin becomes too large, the weight and volume of the bobbin will become too much, which causes material waste and handover problem. Therefore, a suitable bobbin diameter must be determined.

We measured the material properties of the welding rods of different formulation, one of which is listed in Table 2. When the welding rod is wound around the bobbin, one layer of it is neither affected by tension nor pressure. This layer is called the strain neutral plane. That is, the tensile stress of the strain neutral plane

is 0. The outer edge of the welding rod is subjected to the maximum tensile force, which can be calculated by

$$\sigma_d = \frac{d \times E}{2 \times \rho} \qquad (1)$$

The welding rod can be wound on the bob for many layers, and the layer closest to the bobbin has the smallest radius and therefore the largest strain (Figure 8). The largest strain can be calculated by

$$\varepsilon_d = \frac{d}{d + D_0} \qquad (2)$$

where D_0 is the diameter of bobbin. To ensure that the welding rod does not break, the stress at its outer edge should be less than the tensile strength, i.e., $\sigma_d \le \sigma_b$, that is

$$(1 - \lambda) \times \sigma_t + \frac{d \times \lambda \times E}{d + D_0} \le \sigma_b \qquad (3)$$

therefore

$$D_0 \ge \left(\frac{\lambda \times E}{\sigma_b - (1 - \lambda) \times \sigma_t} - 1 \right) \times d \qquad (4)$$

Substituting the values of the variables we can obtain

$$D_{0min} = 442mm$$

According to our measurement record, the original bobbin diameter is 355mm < 442mm. Therefore, the frequent breakage of the welding rod during finishing process agrees with our analysis.

However, whether the diameter obtained by theoretical calculation alone can meet our goal cannot be finalized. We still need to use the data of the digital twin model integrating with simulation technology for further validation.

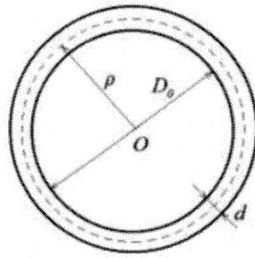

Figure 8. The illustration of the largest strain of the welding rod.

2.3 Simulate the improved design based on the data of historical digital twin model

The digital twin model contains a large amount of data related to the welding rod. These data can be used not only to build the product ontology of the welding rod and to help determine the retrofit solution, but also to support simulation.

We have determined that the theoretical bobbin diameter that can meet the requirement is 442mm. However, considering the friction and drawing speed during actual finishing, whether this theoretical value can reduce the frequency of welding rod breaking still needs to be verified by simulation.

The simulation is executed iteratively as the following steps:
(1) Set property variables and boundary conditions of the simulation environment based on the data in the digital twin model;
(2) Set the bobbin diameter;
(3) Run the simulation to check whether the tensile strength of the welding rod keep always greater than its internal stress during the finishing process;
(4) If so, the bobbin diameter can meet the requirement and will be output, the simulation is over; if not, expand the bobbin diameter according to a certain percentage (in this work the percentage is 1.1), and repeat the steps above.

The initial bobbin diameter is set to D_{0min}.

We run the simulation for the welding rods of 5 different formulations as described above. The simulation of one kind of the welding rod stopped after three iterations, and the final output of the diameter was 535mm. The simulation results are shown in Figure 9.

2.4 Compare the verifying experiment results and simulation predictions

After the simulation driven by digital twin model, we have got the near-optimal bobbin diameter. We do not call it the optimal bobbin diameter because during the optimization process, the diameter is not a continuous value. It is obtained by multiplying by a specific factor as stated before. Next, we modified the bobbin diameter of the drawing bench for 5 different formulations in turn and carried out some manufacturing experiments.

The results show that the frequency of welding rod breaking is significantly reduced for all kinds of welding rods (each machine drops from about 110 times a day to less than 10 times a day). In addition, we have measured the internal stress of the welding rod during finishing. The result of one kind of welding rod is shown in Figure 10, which indicates that the experiment results matched well with our simulation prediction.

It should be pointed out that the conclusion obtained by the AHP method is that the bobbin diameter is the main reason for the rod breaking, but it is not denied that the heating temperature and the drawing speed can also affect the breaking frequency. So after changing the bobbin diameter, we also adjusted the heating temperature and drawing speed accordingly. The results showed that appropriate increase in heating temperature and reduction in drawing speed can slightly reduce the frequency of rod breaking, which can be basically negligible compared to the bobbin diameter.

3. CONCLUSION

In this paper, we have proposed a digital twin-driven improved design approach. Using the drawing bench as a typical case, we have established a digital twin model for the welding rod. Based on the model, the product ontology of the welding rod was built. The data from full life cycle provide solid support for the simulation. The results of the manufacturing experiments indicate that we can dramatically reduce the frequency of

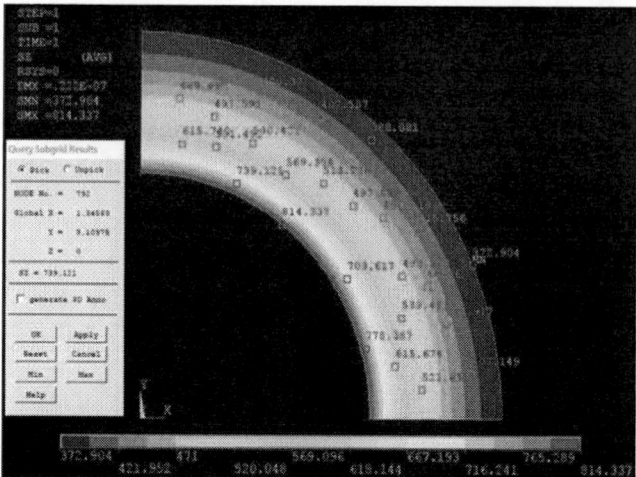

(a) The simulation result of diameter 442mm

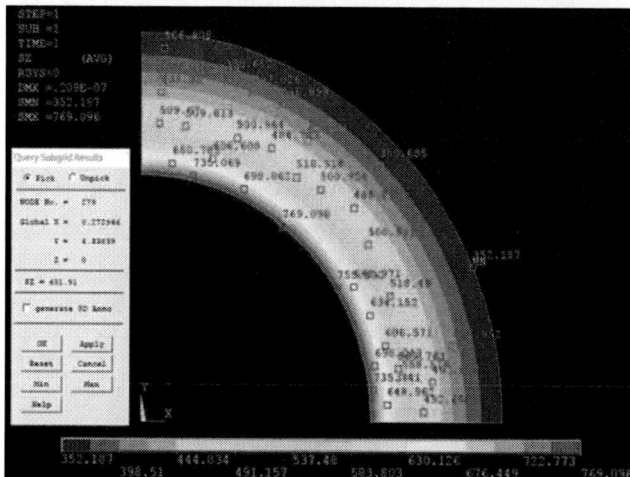

(b) The simulation result of diameter 486mm

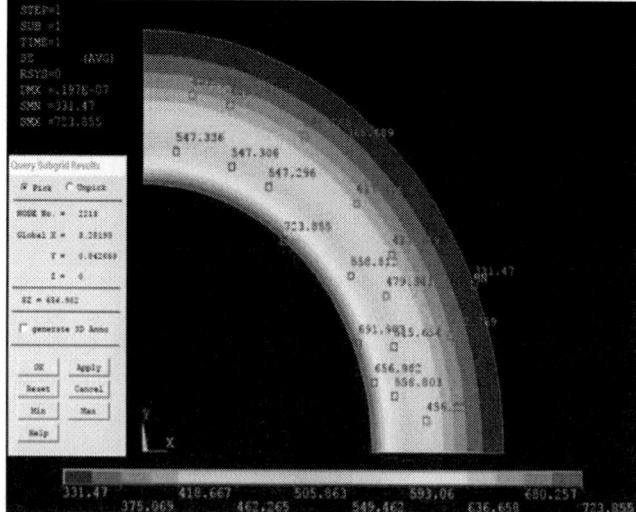

(c) The simulation result of diameter 535mm

Figure 9. The results of the simulation for one kind of welding rod.

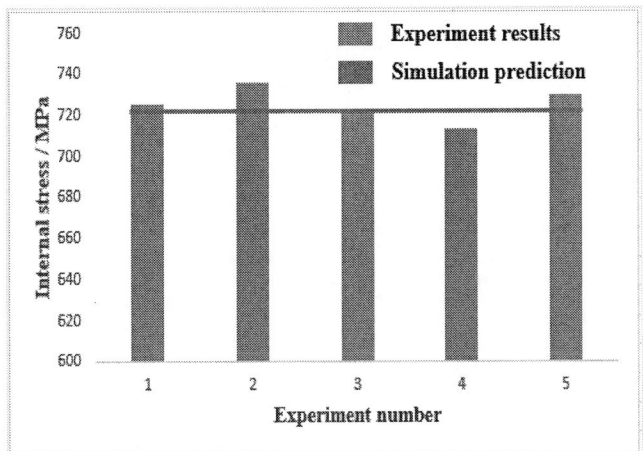

Figure 10. The comparison between experiment results and simulation prediction.

welding rod breaking by changing one single parameter and prove the effectiveness of our proposed approach.

ACKNOWLEDGEMENTS

This work was supported by the National Natural Science Foundation of China (Nos. 51490663, 51821093, 51775489), and Zhejiang Provincial Natural Science Foundation of China (No. LZ18E050001).

REFERENCES

[1] C. Dharmendra, K.P. Rao, J. Wilden, S. Reich. Study on laser welding–brazing of zinc coated steel to aluminum alloy with a zinc based filler, *Materials Science and Engineering: A*, 2011, 528(3): 1497–1503.

[2] Zhihua Song, Kazuhiro Nakata, Aiping Wu, Jinsun Liao. Interfacial microstructure and mechanical property of Ti6Al4V/A6061 dissimilar joint by direct laser brazing without filler metal and groove, *Materials Science and Engineering: A*, 2013, 560: 111–120.

[3] S.Y. Chang, L.C. Tsao, Y.H. Lei, S.M. Mao, C.H. Huang. Brazing of 6061 aluminum alloy/Ti–6Al–4V using Al–Si–Cu–Ge filler metals, *Journal of Materials Processing Technology*, 2012, 212(1): 8–14.

[4] Fei Tao, Fangyuan Sui, Ang Liu, Qinglin Qi, Meng Zhang, Boyang Song, Zirong Guo, Stephen C.-Y. Lu, A. Y. C. Nee. Digital twin-driven product design framework, *International Journal of Production Research*, 2018, 1–19.

[5] Fei Tao, He Zhang, Ang Liu, A. Y. C. Nee. Digital Twin in Industry: State-of-the-Art, *IEEE Transactions on Industrial Informatics*, 2018.

[6] Fei Tao, Jiangfeng Cheng, Qinglin Qi, Meng Zhang, He Zhang, Fangyuan Sui. Digital twin-driven product design, manufacturing and service with big data, *The International Journal of Advanced Manufacturing Technology*, 2018, 94(9–12), 3563–3576.

[7] Jinjiang Wang, Lunkuan Ye, Robert X. Gao, Chen Li, Laibin Zhang. Digital Twin for rotating machinery fault diagnosis in smart manufacturing, *International Journal of Production Research*, 2018: 1–15.

[8] Kannan K, Arunachalam N. A Digital Twin for Grinding Wheel: An Information Sharing Platform for Sustainable Grinding Process, *Journal of Manufacturing Science and Engineering*, 2019, 141(2): 021015.

[9] Yicong Gao, Yixiong Feng, Qirui Wang, Hao Zheng, Jianrong Tan. A multi-objective decision making approach for dealing with uncertainty in EOL product recovery, *Journal of Cleaner Production*, 2018, 204:712–725

[10] Haijun Zhang, Guohui Zhang, Qiong Yan. Digital twin-driven cyber-physical production system towards smart shop-floor, *Journal of Ambient Intelligence and Humanized Computing*, 2018: 1–15.

[11] Xiaomin Li, Jiafu Wan, Hong-Ning Dai, Muhammad Imran, Min Xia, Antonio Celesti. A Hybrid Computing Solution and Resource Scheduling Strategy for Edge Computing in Smart Manufacturing, *IEEE Transactions on Industrial Informatics*, 2019.

[12] Yicong Gao, Yixiong Feng, Zixian Zhang, Jianrong Tan. An optimal dynamic interval preventive maintenance scheduling for series systems. *Reliability Engineering & System Safety*, 2015, 142: 19–30.

[13] Ho William, Ma Xin. The state-of-the-art integrations and applications of the analytic hierarchy process, *European Journal of Operational Research*, 2018, 267(2): 399–414.

[14] Somayeh Nazari, Mohammad Fallah, Hamed Kazemipoor, Amir Salehipour. A fuzzy inference-fuzzy analytic hierarchy process-based clinical decision support system for diagnosis of heart diseases, *Expert Systems with Applications*, 2018, 95: 261–271.

[15] T.R. Ayodele, A.S.O. Ogunjuyigbe, O. Odigie, J.L. Munda. A multi-criteria GIS based model for wind farm site selection using interval type-2 fuzzy analytic hierarchy process: The case study of Nigeria, *Applied energy*, 2018, 228: 1853–1869.

[16] Zhiming Zhang , Witold Pedrycz. Intuitionistic multiplicative group analytic hierarchy process and its use in multicriteria group decision-making, *IEEE transactions on cybernetics*, 2018, 48(7): 1950–1962.

Proceedings of the ASME 2019
International Design Engineering Technical Conferences
and Computers and Information in Engineering Conference
IDETC/CIE2019
August 18-21, 2019, Anaheim, CA, USA

DETC2019-97587

COMPUTER-AIDED DESIGN IDEATION USING INNOGPS

Jianxi Luo[1], Serhad Sarica, Kristin L. Wood
Engineering Product Development Pillar & SUTD-MIT International Design Centre
Singapore University of Technology and Design

ABSTRACT

Traditionally, the ideation of design opportunities and new concepts relies on human expertise or intuition and is faced with high uncertainty. Inexperienced or specialized designers often fail to explore ideas broadly and become fixed on specific ideas early in the design process. Recent data-driven design methods provide external design stimuli beyond one's own knowledge, but their uses in rapid ideation are still limited. Intuitive and directed ideation techniques, such as brainstorming, mind mapping, Design-by-Analogy, SCAMPER, TRIZ and Design Heuristics may empower designers in rapid ideation but are limited in the designer's own knowledge base. Herein, we harness data-driven design and rapid ideation techniques to introduce a data-driven computer-aided rapid ideation process using the cloud-based InnoGPS system. InnoGPS integrates an empirical network map of all technology domains based on the international patent classification which are connected according to knowledge distance based on patent data, with a few map-based functions to position technologies, explore neighborhoods, and retrieve knowledge, concepts and solutions in the near or far fields for design analogies and syntheses. The functions of InnoGPS fuse design science, network science, data science and interactive visualization and make the design ideation process data-driven, theoretically-grounded, visually-inspiring, and rapid. We demonstrate the procedures of using InnoGPS as a data-driven rapid ideation tool to generate new rolling toy design concepts.

1. INTRODUCTION

"What products, services, or systems shall we design next?" is an evergreen question of designers, teams and organizations that aspire to innovate and continually explore new design concepts departing away from their prior designs. Intuitive and directed ideation methods, such as brainstorming, mind mapping and design heuristics [1–3], may empower creative thinking of designers, whereas design ideation is also conditioned on the knowledge and experience of the designers. Design thinking process, market research and user studies may

reveal design opportunities-problems, and technology scanning and expert panels may contribute design solutions [4–7], but such processes are not suitable for rapid ideation. Recently data-driven design methods have been proposed to provide broad external inspirational design stimuli and shown promise for design ideation [8–13], but their adoption in rapid design concept generation are still limited as well.

Here we aim to harness the data-driven design stimulation and directed ideation techniques to introduce a computer-aided rapid design ideation process using the cloud-based InnoGPS system. InnoGPS (www.innogps.com) integrates an empirical network map of all technology domains (based on the international patent classification) which are connected according to knowledge proximity based on patent data, with a few map-based functions to position technologies, explore technology neighborhoods, and retrieve design knowledge, concepts and solutions in near or far fields for new design analogies and synthesis [9,14–17]. An early version of InnoGPS had been used for the analyses of domain-level diversification, innovation and competition of technology firms [18,19], but did not have the functionality for design concept generation.

In this research, we have further developed the system to include the data and capabilities for rapid data-driven design concept generation. In particular, in contrast to the prior data-driven methods that have focused on specialized or sampled design repositories and the retrieval of design stimuli at the document level, InnoGPS can provide the guided retrievals of design information at the concept level in all known domains of technology according to their knowledge proximity or distance, and thus provide rapid and directed inspiration. We will demonstrate the functions and use process of InnoGPS for computer-aided rapid ideation via a case study to generate new design concepts that extend existing rolling toys.

2. RELATED WORK

In established industrial companies, design engineers or product managers often conduct user studies, market research, expert panels, brainstorming sessions and so on to seek inspiration on

---

[1] Contact author: luo@sutd.edu.sg

design concepts for new products, systems and services [5,6]. In the past decade, design thinking process has become popular in innovation practice [4]. Such processes or procedures require time and resources. For rapid ideation of individual designers, structured and directed ideation techniques, such as morphological analysis, SCAMPER and design heuristics, may enhance creative thinking for new design concepts [1–3,17,20]. The efficacy of these techniques is conditioned by the user's knowledge base. Meanwhile, as we enter the big data era, there have been growing efforts in data-driven design methods and tools to provide external design stimuli and inspiration beyond one's own knowledge and also show potentials to support rapid design ideation.

For instance, recent data-driven design efforts have curated design repositories, e.g., asknature.org and moreinspiration.com, which are the databases to store and organize prior innovative designs or nature existences, for designers to retrieve for design inspiration [10,21]. Goucher-Lambert and Cagan [22] recently showed that crowdsourced ideas from novices can be used as design stimuli for expert designers. Siddharth and Chakrabarti [13] developed a web-based tool, Idea-Inspire 4.0, to support the search for analogical stimuli from a manually-populated database of 60 biological concepts and 83 engineered concepts and represent them using function model, text, image, video, and audio. Mukherjea et al. [23] extracted biological terms and their semantic associations from biomedical patent abstracts to build the BioMedical Patent Semantic Web to support knowledge discovery in the specialized biomedical domain.

In contrast to manually curated design repositories, the patent database is a natural and probably the largest design repository to date and it organically accumulates over time as inventors continually file patents for their inventions [24]. Patent documents contain rich design information of the functions, components, structures and working principles of prior designs and are organized by the international patent classification system, which covers all thinkable domains of engineering and technology [8]. Computational methods have been developed to retrieve patents as external design stimuli to support TRIZ [25,26] or Design-by-Analogy [9,27,28].

Recent studies have developed nuanced understanding of the effects of using far-field versus near-field patents as design stimuli. Fu et al. [8] found near-field patents can effectively stimulate new ideas despite limited novelty and far-field patents are ineffective to provide analogical inspiration. They quantified the functional analogy distance between different patented inventions based on latent semantic analysis of patent documents. Srinivansan et al. [12] also found that patent precedents in near fields are more likely to inspire and provide high-quality inspiration, whereas precedents in far fields provide more novel inspiration. They measured the generally-defined knowledge distance between different patent classes to discern near and far fields. By analyzing the cross-domain patenting behaviors of 2 million inventors, Alstott et al. [29] found that inventors are more likely and more productive to exploit fields nearby their prior inventions, but inventions in far fields are more valuable.

These findings based on using patents as design ideation stimuli are aligned with the design creativity literature in

suggesting that designers are more likely to find inspiration from near fields owning to the ease of perception [15,30,31]. Design stimuli from "far" sources contribute to novelty [32–34], but it is difficult for designers to conceive the distant relevance [12,29,30,35] and to derive high comprehensiveness or quality in the conceived new concepts [12,36]. Fu et al. [35] posited that stimuli across a moderate analogical distance between source and target domains are most favorable. He and Luo [37] also found that the most valuable patents are primarily based on prior arts of moderate combination distance.

Such understandings of the multifaceted effects of the knowledge proximity or distance between the original design problem and the source of inspirations on ideation behaviors and outcomes can be utilized to guide the retrieval of patent documents for design inspiration. However, there remain practical challenges to patent data-driven design ideation. First, the size and complexity of the patent database makes it difficult to retrieve the most relevant and inspirational technological knowledge for analogy and synthesis. Second, prior studies were focused on document-level retrievals, but patent documents are difficult for designers to read and comprehend rapidly and thus unsuitable for rapid ideation uses. In this study, we address these challenges and present InnoGPS as a tool to retrieve design information at both patent document and semantic concept levels in different domains according to inter-domain knowledge distance to support rapid design concept generation via analogy and synthesis (as summarized in Figure 1).

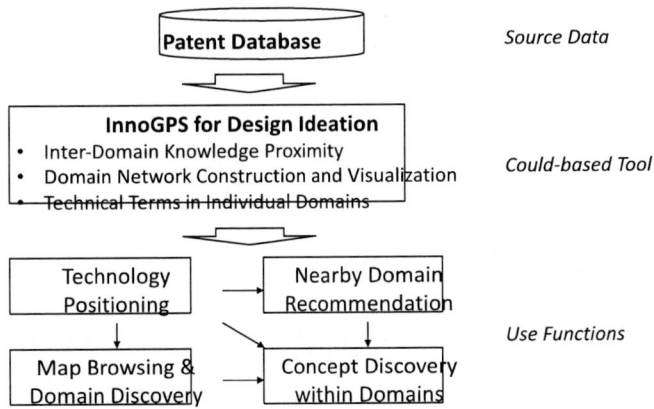

FIGURE 1: OVERVIEW OF THE INNOGPS SYSTEM

3. INNOGPS: DATA, FEATURES AND FUNCTIONS

Luo et al. recently showed that designers may simply browse a network map of technologies defined at the domain level to identify innovation directions [18] or conceive new design opportunities [38] by relating technologies across domains. However, manual map browsing was unguided and design opportunities conceived this way were raw and macro. On that basis, the computer-based InnoGPS has been developed to enable interactive digital map browsing, domain exploration and on-demand design information retrieval within or across domains. Herein, our focus is on the use of InnoGPS for the computer-aided ideation of finer-grained design concepts, with the stimuli at the concept level. For this purpose, the technical

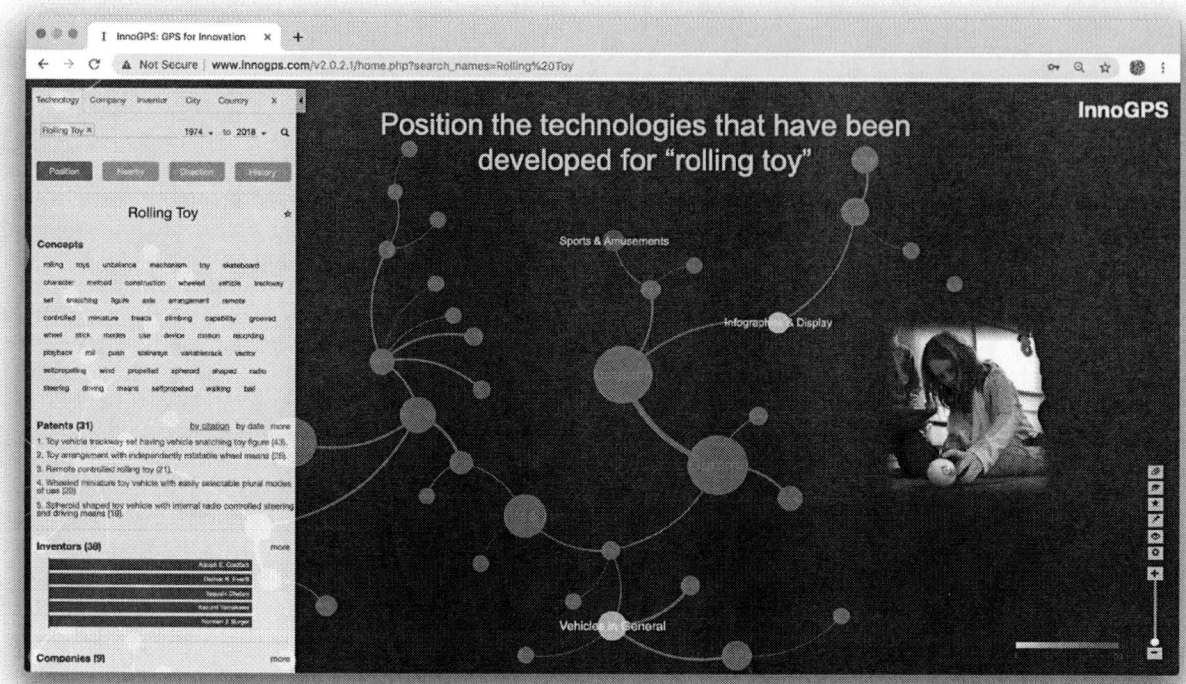

FIGURE 2: POSITION TECHNOLOGIES IN THE PRIOR ROLLING TOY DESIGNS

terms that represent generic engineering concepts (such as functions, components, structures and working principles) are extracted from the patent data in individual domains. In turn, the network map structure serves as a macro guide to direct the retrieval of micro design concepts to inform and inspire design ideation.

The first core component of InnoGPS is the technology space map in a network form. The nodes are all international patent classes that represent different technology domains, e.g., additive manufacturing, biochemistry. The current map in InnoGPS is based on either 3- or 4-digit patent classes of a user's choice, while patent classes can be further decomposed to finer-grained 5 - to 7- digit ones. Patent classes are linked and positioned according to knowledge proximity among represented technology domains. Here the knowledge proximity is calculated as the Jaccard index, i.e., count of shared patent references, normalized by the total count of all unique references of patents in a pair of patent classes.

$$Jaccard\ index = \frac{\left|C_i \cap C_j\right|}{\left|C_i \cup C_j\right|} \qquad (1)$$

where C_i and C_j are the sets of the references of the patents in patent classes i and j; $\left|C_i \cap C_j\right|$ is the number of unique patents in the intersection of the two sets, and $\left|C_i \cup C_j\right|$ is the number of unique patents in the union of the two sets. The references of the patents in a technology domain approximate the knowledge base of the domain. The proximity value in the range of [0,1] indicates the overlapping of the knowledge bases of two domains. Despite the existence of other knowledge proximity measures [39,40], this metric is chosen for its high explanatory power on inventor's historical exploration across domains based on our statistical analysis of all inventors in the entire patent database [41].

To ensure statistical significance, the complete USPTO (United States Patent & Trademark Office) granted utility patent database is used to calculate the knowledge proximity between each pair of domains. Because of the complete coverage of all domains of technology by using the international patent classes and using all patent records in estimating inter-domain knowledge proximity, the resulting network map approximates the total technology space and provides a systematic picture of various technology domains and the structural information of their relative distances or proximities. Then several map-based descriptive, prescriptive and predictive functions in the InnoGPS system may support design ideation.

The interactive "browsing" function allows one to navigate the map to discover the technology domains for conceiving possible relevance to one's original design or interests. The "position" function locates the technology positions of a design domain, i.e., rolling toy, in the total technology space, based on the classifications of the relevant patents retrieved from the total patent database. The position is represented as a vector of the distribution of the patents in respective patent classes (i.e., different nodes in the network). The "information" panel reports the inventor, company and patent information within these map positions. In particular, the information panel also retrieves and reports the semantic-level engineering design concepts in such map positions or a given technology domain.

We consider the provision of semantic-level engineering concepts in each domain the second core component of InnoGPS. They are extracted as the technical terms from the titles of top 10,000 most cited patents classified in each domain,

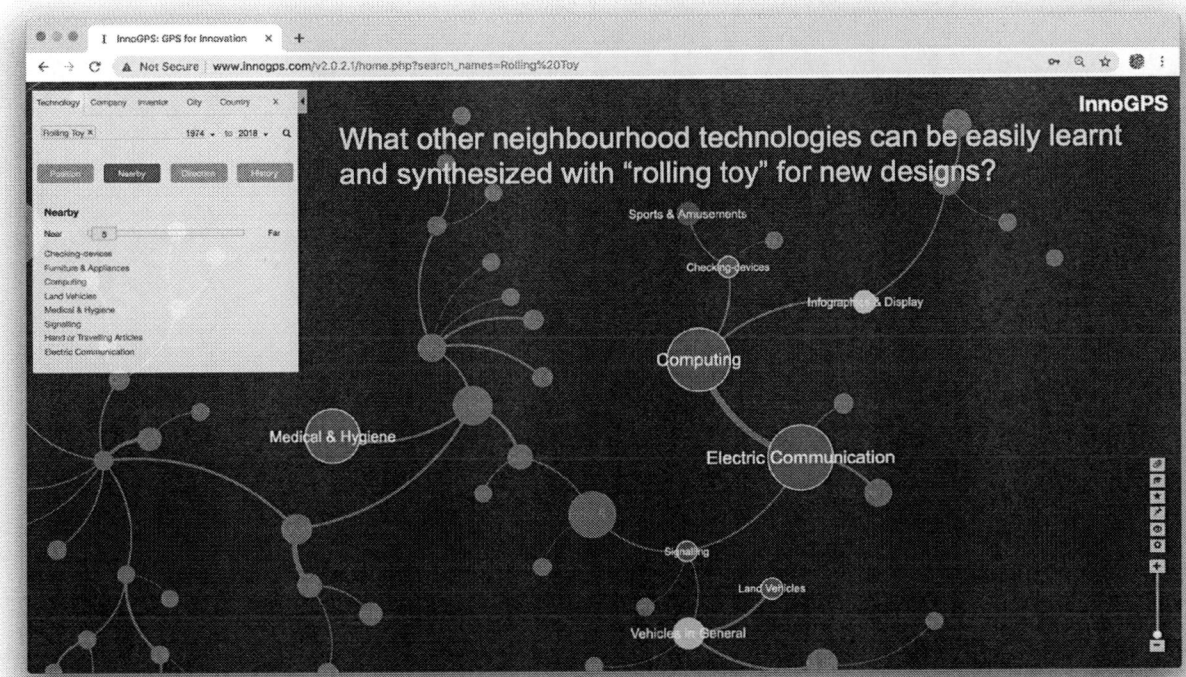

FIGURE 3: NEARBY DOMAINS TO ROLLING TOY DESIGNS

excluding the main texts and legal claims that are often written in disguised and ambiguous terms. Such terms are the most characteristic and reprehensive of the design concepts in respective domains. The extraction of engineering terms uses a sequence of natural language processing (NLP) techniques. The first is to identify phrases by statistically finding words that frequently co-occur and infrequently appear separately [42]. The second step is to detect and remove noisy phrasing formations and stop-words. The third step is to lemmatize multiple inflected forms of words and phrases [43,44]. The fourth step removes the terms that are not included in the large Technology Knowledge Graph – TKG (http://www.tkgraph.net) of more than 4 million technical terms [45,46], which we previously extracted from the entire USPTO granted utility patent database and vectorized using word embedding models. Finally, the remaining terms in each domain are further filtered for their relevancy to the domain by 3 researchers who have 10 to 22 years of experience in engineering. A term would be removed if all three researchers consider it irrelevant to the domain. Then the remaining terms in each patent class are ordered by their occurrence frequencies in the class and represent the most characteristic engineering design concepts in each corresponding domain. Based on our observations, the extracted concepts cover generic functions, components, structures or working principles, in addition to high-level product or system concepts.

In turn, one can discover these generic engineering concepts as domain-specific stimuli during the navigation of technology domains on the map. To direct the map exploration, the "nearby" function recommends the most proximate unexplored domains in the white space to already positioned focal technologies or domains. Such nearby domains host the engineering concepts that can be most easily learnt and comprehended with those in the current positions for design analogies and syntheses, enabled by high knowledge proximity. The nearby domains can be identified in alternative ways. For instance, one can focus on the most proximate domain to each of the current positions, or the top N unexplored domains based on weighted average proximity with all current positions, as follows

$$Weighted\ average\ proximity = \frac{\sum_{i \neq j} \varphi_{ij} x_i}{\sum_{i \neq j} x_i} \qquad (2)$$

where i belongs to all the current positions; j represents each of the unexplored domains in the network; x_i is the count of retrieved patents in domain i; φ_{ij} is the knowledge proximity between domains i and j.

Figure 1 summarizes the source data, key features and functions of InnoGPS, which enable the guided retrievals of diverse design information at the semantic and document levels from the patent database according to inter-domain knowledge distance, for rapid design concept generation.

4. COMPUTER-AIDED RAPID IDEATION FOR NEW "ROLLING TOY" DESIGN CONCEPTS

In the following, we present a case study of using InnoGPS to aid the rapid generation of new design concepts that extend existing spherical rolling toys, as a way to demonstrate the process of using InnoGPS for rapid computer-aided ideation.

Assume we are "rolling toy" designers and need to quickly generate new rolling toy design concepts, facing time and resource constraints. Traditionally, design engineers would conduct expert panels, user studies, market research, patent

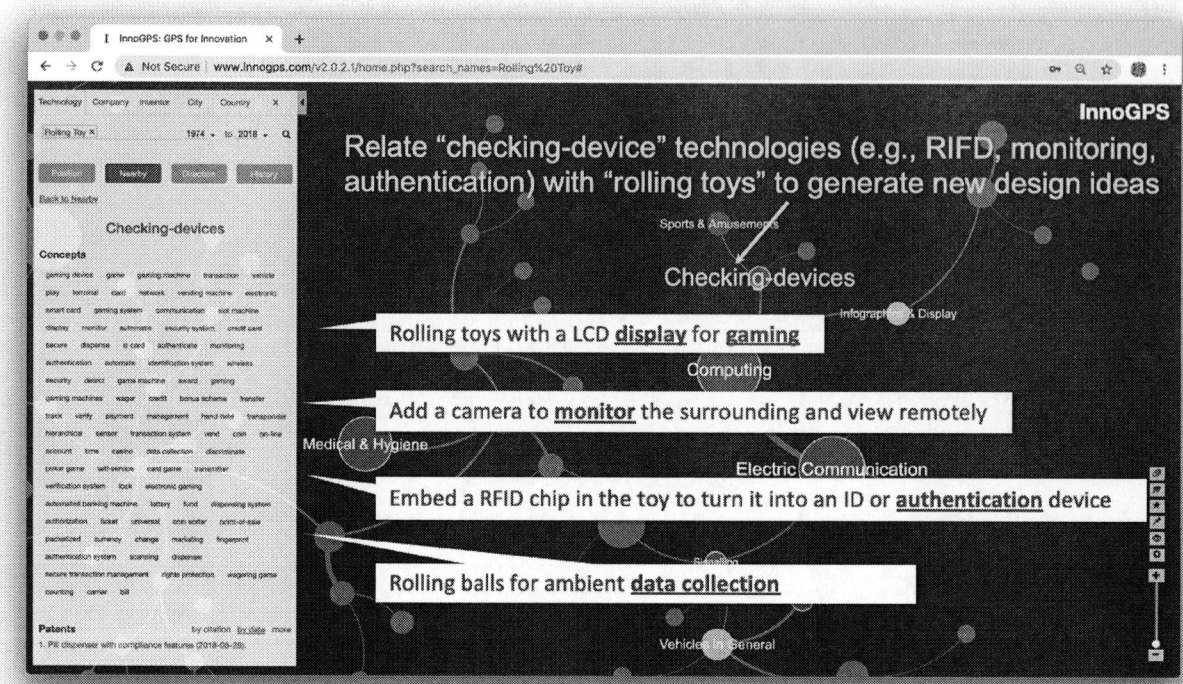

FIGURE 4: LEVERAGE CONCEPTS IN THE CHECKING-DEVICE DOMAIN FOR ROLLING TOY DESIGNS

search, brainstorming sessions and so on to generate new design concepts. Such processes could lead to new concepts but are generally slow and time-consuming. Herein, we use InnoGPS to aid in design concept generation. The first step is to "position" the technologies that have been developed for rolling toys on the map as shown in Figure 2, using the upper left search bar of the interface. "Sports & Amusement", "Infographic & Display", and "Vehicle in General" are highlighted in red color because "rolling toy" patents are found in these domains. The intensity of the red color of the nodes corresponds to rolling toy patent occurrences in respective patent classes. The size of a node corresponds to the total number of patents in the corresponding patent class in the total history. The "grey" domains have no rolling toy designs being found, i.e., no patents found in corresponding classes.

In the information panel on the left-hand side, we can quickly discover and learn the technical terms that are extracted from the titles of rolling toy patents and represent the elementary design concepts for rolling toys, the actual patent documents sorted according to their grant dates or citations, as well as leading inventors and companies that have contributed to the prior designs of rolling toys. Such information informs us of other concepts, designs and technologies that have been adopted in rolling toy designs. So, we can make more informed decisions regarding either adopting them for our own products or differentiating our own designs from them to avoid competition.

InnoGPS is most useful for the exploration of the technology whitespace, i.e., the grey domains on the map, for innovation. The "nearby" function, based on Equation (2), quantitatively identifies and visually highlights the most proximate unexplored domains in the whitespace to the current positions of rolling toy technologies, i.e., the red domains. As shown in Figure 3, one can move the slider bar in the information panel to increase the knowledge distance of the exploration into the whitespace domains from the nearest to the farthest relative to the positions of the technologies already used in prior rolling toy designs. Because of the knowledge proximity, prior design concepts in the nearest domains can be most easily learnt and feasibly synthesized with rolling toy designs. Particularly, the "Checking-Device" domain appears to be the most proximate whitespace domain to rolling toys, based on Equation (2).

We click the node "Checking-Device" on the map to activate the information panel for this specific domain. The panel retrieves the leading inventors and companies, the most cited and newest patents relevant to "checking device". Most importantly, the information panel reports the most characteristic design concepts in the checking device domain (Figure 4). These concepts, which might represent generic functions, components, structures and working mechanisms, have already been extracted from the patent title texts in the domain. Therefore, instead of reading the patent documents one by one for design inspiration, we can quickly go through the design concepts in the "checking device" domain to conceive possible analogies and syntheses with rolling toys.

For instance, "gaming" and "display" inspired us for a new design of rolling toys with an LCD display for kids to play interactive visual games on the rolling ball. The concept "monitor" inspired the inclusion of a camera on the rolling toy to monitor the surrounding remotely. The "authentication" function can be synthesized with a rolling toy by embedding a RIFD chip inside for it to be used as an authentication device. Kids may use the playful rolling balls as their identity keys in the

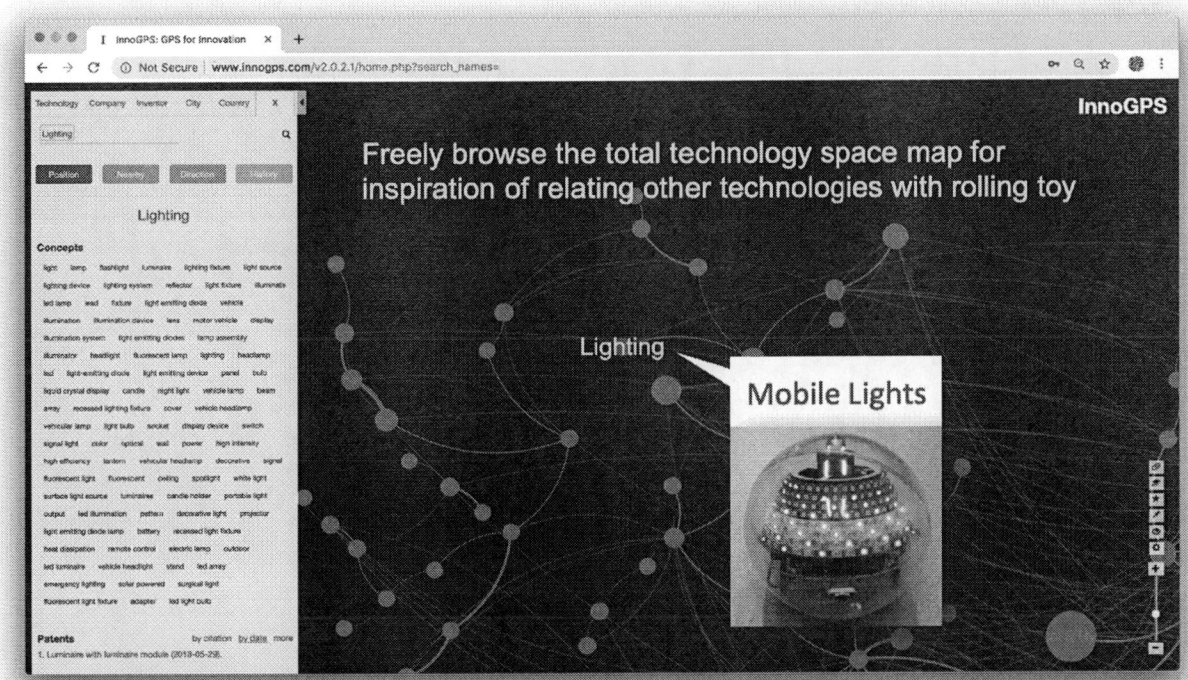

FIGURE 5: FREE MAP NAVIGATION FOR DESIGN INSPIRATION FOR ROLLING TOY DESIGNS

kindergarten. Furthermore, the terms "sensors" for "data collection" inspired us to conceive the use of rolling balls to cruise and collect environment data in difficult-to-visit places, such as underground sewers.

In addition to these most characteristic terms, one can also directly read the patent titles for rapid inspiration. For example, the patent title "dual-mode vehicular controller (2017-11-28)" stimulated us to conceive the design concept of "dual-mode" rolling toys that have a mobile-controlled mode and an autonomous mode for rolling. The patent title "tamper resistant rugged keypad (2017-11-28)" inspired us to conceive the design concept of rolling toys with "rugged" shells, while the surfaces of the existing rolling toys are normally smooth. Without reading the lengthy patent full texts and images, we have been able to be inspired to rapidly generate various divergent design concepts that we would not be able to conceive without the prompts of the term or sentence level stimuli from InnoGPS. These new design concepts are readily porotype-able as they are stimulated by the design information at the semantic level.

Meanwhile, the patent documents are also retrieved and listed in the information panel for each domain. Reading them may provide more detailed inspiration, but requires time and dedication and is unsuitable for rapid wide exploration of the design concept space and the generation of divergent concepts. These patent documents might be more useful for a later phase. After a diverse set of design ideas have been rapidly generated, one can read the source patent documents of an idea's stimulating concepts to exploit the details for inspiration to fine-tune and nuance the idea.

In addition to nearby domains that provide near-field stimuli, one can also freely browse the map regardless of knowledge distance, to discover concepts, technologies, inventors and companies in any domains on the map for design analogies and syntheses. For instance, "lighting" technologies can be synthesized with rolling toys to design moving lights at home or in public spaces (Figure 5). Rolling toy designs and bomb design concepts in the "weapons" domain can be synthesized to design rolling bombs. "Mechanical vibration" technologies can be adopted to either suppress unwanted vibrations or generate useful vibrations for rolling toys. Such simulation is at the domain level, e.g. weapons, lighting. Once the designer conceives a high-level relevance of a domain during the map browsing, he or she can click the node and activate the information panel of the domain that is founded inspirational to discover the generic design concepts for more specific and divergent design stimulation, like those illustrated in Figure 4.

While random map browsing may stimulate one to generate diverse design concepts with the inspiration from different domains of the total technology space, the knowledge distance from the design stimuli to the original design or home design problem, e.g., rolling toy in this case, would condition the feasibility of a potential design project to realize the concepts. As suggested by prior design stimulation studies [12,27,30], the design concepts that synthesize near-field technologies might be more feasible, whereas far-field stimuli will contribute to novelty. With the understanding of such tradeoffs, as shown in Figure 6, the InnoGPS-based visual and quantitative information of the relative distances or proximities between the design stimuli and the original design may further suggest the feasibility and novelty of different design concepts that have been generated in the heuristic map-aided ideation process.

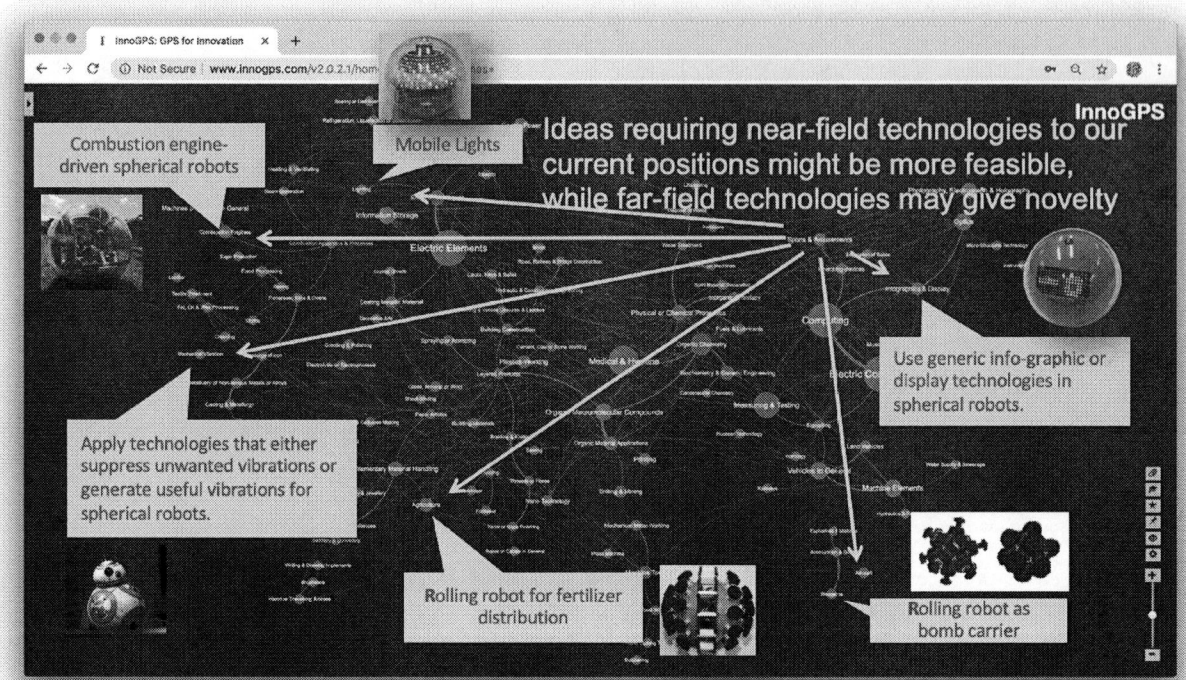

FIGURE 6: EXEMPLAR DESIGN CONCEPTS WITH THE STIMULI OF DIFFERENT DISTANCES TO THE ORIGINAL DESIGN

As a matter of fact, many of the foregoing examples of concept generation were based on analogical thinking, e.g., "design rolling toys with dual modes just like a dual-mode car with controlled and auto-pilot modes", or other design heuristics, "apply rolling toy designs to design rolling bombs". That is, the intuitive design heuristics for rapid concept generation are now supported and empowered by the computer-based tool, which provides term-level design stimuli from the total patent database and directs a designer to retrieve them according to domains and their knowledge distance.

It is particularly noteworthy that the novelty and patentability of the new concepts are naturally ensured when they are generated by synthesizing stimuli from the whitespace to the home design. Additionally, the system also provides rapid visual and quantitative evaluation of the relative novelty of different new design ideas based on the varied knowledge distances crossed by the analogies or syntheses to yield them.

5. DISCUSSION

Now we have demonstrated the functions and use process of InnoGPS for rapid design concept generation. These functions, e.g., "position" and "nearby", are analogous with those of "Google Maps", whereas here they are specific to the technology space and centered on the analysis of knowledge distance. By guide concept and document level design information retrieval across domains by inter-domain knowledge distance, InnoGPS has addressed two major limitations in the prior studies of retrieving patent data as stimuli for rapid concept generation: 1) the retrieval was done at the patent document level, and patent documents are difficult to read and comprehend rapidly

[8,9,12,27,35]; 2) the patent database is vast, so from where in the database to retrieve the patents is always a question [8,28].

On one hand, enabled by design concept retrieval at the semantic level, design inspiration is easier and faster to obtain than from patent documents [8], and the conceived concepts are also more nuanced and specific than those from the static map of technology domains at a macro level [38]. On the other hand, InnoGPS still presents the macro domains and uses the domain network structure to guide the retrieval of concept- and document-level design stimuli by domains. Taken together, the semantic-level information of the design concepts (e.g., functions, components, structures, working principles) as well as design documents in all domains of engineering can be discovered and retrieved according to the knowledge distance among domains for learning and design stimulation. In the other words, InnoGPS provides design stimulation at the concept and document levels, in a holistically integrated fashion organized by the quantified knowledge distance between domains.

The focus on knowledge proximity to organize and retrieve prior design information for stimulating creative ideas follows the established design creativity theories. One's ability to discover, learn, adopt and combine existing but previously-unknown technologies to create new ones is conditioned by the knowledge distance between those unknown but existing technologies and the ones that he/she has already mastered [12,27,30]. One may find it easier and more effective to discover, learn and synthesize technologies nearby or within his/her original designs or specialized positions, but more distant domains and technologies may offer more radical innovation opportunities [32–34,47].

Following these theoretical bases, InnoGPS' functions (e.g., nearby) and design features (i.e. the network map) naturally ensure the novelty of the concepts when they are generated with stimuli from the whitespace to home design problems, and also suggest the relative novelty and feasibility of different concepts according to the knowledge distance between the home designs and the stimuli in the network map. These would ease the later concept evaluation and selection efforts. In contrast, traditional ideation techniques, when used by inexperienced designers without extensive knowledge, are more likely to generate concepts that have already existed elsewhere or are highly infeasible, because the generation process does not necessarily ensure novelty and provide indications on potential design outcomes. Such noisy concepts will need to be evaluated and screened during later concept selection of the design process, indicating additional workloads.

6. LIMITATIONS AND FUTURE WORK

InnoGPS as a computer-aided rapid ideation tool is still limited. First of all, the knowledge distance quantification is based on one singled metric but used for all cases. Despite its statistical superiority over alternatives suggested by prior work [41], in some specific cases other knowledge distance metrics might provide better results. The system and analytics can be improved by allowing the automatic selection of an optimal knowledge proximity metric and resulting map based on the historical data of the case on demand, e.g., rolling toy. Second, currently InnoGPS only implements one network visualization layout for the 3-digit and 4-digit IPC maps. Future research should experiment different visual layouts to possibly enhance user experience and cognition during the interactions with the map. Furthermore, the nearby recommendation is only implemented at the domain level (e.g., IPCs). In the future, we also plan to enable the nearby recommendation at the document (i.e., patents) and concept level (i.e., terms).

In particular, InnoGPS currently only stores and retrieves technical terms from patent data. Not all engineering designs and inventions are disclosed in patent documents; meanwhile, additional design information may be extracted from proprietary enterprise data or public data sources, such as web articles, technical papers, reports, course descriptions, product descriptions on the Internet. Future research may mine the unstructured text data on technologies, extract technical terms from those sources, and organize them in respective domains in the network map for potential retrievals according to knowledge distance to support design analogy or synthesis.

On the basis of the current system design, artificial intelligence capabilities that synthesize machine learning and creativity theories can be added to further enhance the human ideation experience with InnoGPS or carry out more computer ideation beyond the knowledge retrieval and representation. Such capabilities may include but be not limited to the following:

1) Adopt machine learning techniques to learn and understand a user's latent preferences and aptitudes from his or her use behaviors and inputs;
2) On this basis, develop intelligent recommendation of domains, documents and specific design concepts according to the profiles and preferences of users with varied aptitudes for novelty and feasibility;
3) Use natural language generation techniques to generate sentence descriptions of new design ideas (like those we generated ourselves with the aids of InnoGPS) which semantically synthesize design heuristics (e.g., "add a function to an existing design"), the stimulating terms (e.g., "monitor") and the original design (e.g., "rolling toy").

Lastly, the paper has primarily focused on demonstrating the procedures and effectiveness of using InnoGPS to aid in design ideation. We do so in one single case step by step. To further validate the efficacy of the tool and understand its influences on ideation performances, our planned future works include:

4) More advanced use cases of InnoGPS for ideation that move downstream in the design process to produce actual designs, porotypes and inventions for users and experts to rate, test and choose, instead of word ideas.
5) Controlled experiments and in-depth studies of InnoGPS with designers (students and professionals), measuring the ideation performances regarding concept quantity, novelty, quality and variety using the established metrics from the design creativity literature [48-50], and also collecting qualitative feedback from the designers via surveys.
6) Comparisons with Design Heuristics, TRIZ, Mind Mapping, and other Design-By-Analogy techniques, such as AskNature, WordTree and SCAMPER.

7. CONCLUSIONS

This paper presents the process of computer-aided rapid design ideation using the InnoGPS tool. It aims to augment the traditionally intuitive human ideation process by making it more data-driven, theoretically grounded, visually-inspiring, and rapid. It is most useful to empower conceptual design for which computer aids are still underdeveloped, in contrast the existence of computer-aided drafting and analysis at the later stages of the design process. The development of InnoGPS is in line with the efforts to develop methods and tools to search for technologies across domains or disciplines for design, e.g., infused design [51], Interdisciplinary Engineering Knowledge Genome [52], and the efforts to develop data-driven methods and tools to retrieve patents as design stimuli [25–27,35]. We anticipate the adoption of the computer-aided ideation and the InnoGPS tool as part of the integrated Design Innovation process and of a suite of ideation or concept generation techniques, e.g., Brainstorming, Mind Mapping, Design Heuristics. Despite the limitations and ongoing development, interested readers are invited to test use InnoGPS at http://www.innogps.com.

ACKNOWLEDGEMENTS

This research is supported by SUTD-MIT International Design Centre (IDC, idc.sutd.edu.sg). The opinions, findings, and conclusions in this work are the views of the authors and do not necessary reflect the views of the sponsors.

REFERENCES

[1] White, C., Wood, K., and Jensen, D., 2012, "From Brainstorming to C-Sketch to Principles of Historical Innovators: Ideation Techniques to Enhance Student

Creativity," *Journal of STEM Education*, 13, pp. 12–25.

[2] Daly, S. R., Yilmaz, S., Christian, J.L., Seifert, C.M., and Gonzalez, R., 2012, "Design Heuristics in Engineering Concept Generation," *Journal of Engineering Education*, 101, pp. 601–629.

[3] Yilmaz, S., Daly, S.R., Seifert, C.M. & Gonzalez, R., 2016, "Evidence-Based Design Heuristics for Idea Generation," *Design Studies*, 46, pp. 95–124.

[4] Brown, T., 2008, "Design Thinking," *Harvard Business Review*, 86, pp. 84–92.

[5] Chen, W., Hoyle, C., and Wassenaar, H.J., 2013, Decision-Based Design, 1st ed., Springer London.

[6] Cagan, J., and Vogel, C.M., 2002, Creating Breakthrough Products: Innovation from Product Planning to Program Approval, FT Press, Upper Saddle River, NJ.

[7] Camburn, B.A., Auernhammer, J.A., Sng, K.H.E., Mignone, P.J., Arlitt, R.M., Perez, K.B., Huang, Z., Basnet, S., Blessing, L.T. & Wood, K.L., 2017, "Design Innovation : A Study of Integrated Practice," ASME Int. Des. Eng. Tech. Conf. Comput. Inf. Eng. Conf.

[8] Fu, K., Cagan, J., Kotovsky, K., and Wood, K., 2013, "Discovering Structure in Design Databases Through Functional and Surface Based Mapping," *Journal of Mechanical Design*, 135, p. 031006.

[9] Murphy, J., Fu, K., Otto, K., Yang, M., Jensen, D., and Wood, K., 2014, "Function Based Design-by-Analogy: A Functional Vector Approach to Analogical Search," *Journal of Mechanical Design*, 136, p. 101102.

[10] Bohm, M.R., Vucovich, J.P., and Stone, R.B, 2017, "Using a Design Repository to Drive Concept Generation," *Journal of Computing and Information Science in Engineering*, 8, p. 014502

[11] Song, B., Luo, J., Mohan, R.E., and Wood, K.L., 2018, "Data-Driven Platform Design: Patent Data and Function Network Analysis," *Journal of Mechanical Design*, 141, p. 021101.

[12] Srinivasan, V., Song, B., Luo, J., Subburaj, K., Elara, M.R., Blessing, L., and Wood, K., 2018, "Does Analogical Distance Affect Performance of Ideation?" *Journal of Mechanical Design*, 140, p. 071101.

[13] Siddharth, L., and Chakrabarti, A., 2018, "Evaluating the impact of Idea-Inspire 4.0 on analogical transfer of concepts," *Artificial Intelligence for Engineering Design, Analysis and Manufacturing,* 32, pp. 431–448.

[14] Dahl, D.W., and Moreau, P., 2002, "The Influence and Value of Analogical Thinking During New Product Ideation," *J. Mark. Res.*, 39, 47–60

[15] Weisberg, R.W., 2006, Creativity: Understanding Innovation in Problem Solving, Science, Invention, and the Arts, John Wiley & Sons Inc, Hoboken, NJ, US.

[16] Taura, T., and Nagai, Y., 2012, Concept Generation for Design Creativity: A Systematized Theory and Methodology, Springer.

[17] Moreno, D.P., Blessing, L.T., Yang, M.C., Hernández, A.A., and Wood, K.L., 2016, "Overcoming Design Fixation: Design by Analogy Studies and Nonintuitive Findings," *Artificial Intelligence for Engineering Design, Analysis and Manufacturing*, 30, pp. 185–199.

[18] Luo, J., Yan, B., and Wood, K., 2017, "InnoGPS for Data-Driven Exploration of Design Opportunities and Directions: The Case of Google Driverless Car Project." *Journal of Mechanical Design*, 139, p. 111416.

[19] Sarica, S., Yan, B., Bulato, G., Jaipurkar, P., and Luo, J., 2019, "Data-Driven Network Visualization for Innovation and Competitive Intelligence," 52nd Hawaii Int. Conf. Syst. Sci., pp. 127–135.

[20] Otto, K.N., and Wood, K.L., 2001, Product Design: Techniques in Reverse Engineering and New Product Development, Prentice Hall, Upper Saddle River, NJ.

[21] Szykman, S., Sriram, R.D., Bochenek, C., Racz, J.W. & Senfaute, J., 2000, "Design Repositories : Engineering Design's New Knowledge Base," *IEEE Intelligent Systems and Their Applications*, 15, pp. 48–55.

[22] Goucher-Lambert, K., & Cagan, J., 2019, "Crowdsourcing Inspiration: Using Crowd Generated Inspirational Stimuli to Support Designer Ideation," *Design Studies*, 61, pp. 1–29.

[23] Mukherjea, S., Bamba, B., and Kankar, P., 2005, "Information Retrieval and Knowledge Discovery Utilizing a BioMedical Patent Semantic Web," *IEEE Transactions on Knowledge and Data Engineering*, 17, pp. 1099–1110.

[24] Luo, J., and Wood, K.L., 2007, "The Growing Complexity in Invention Process," *Research in Engineering Design*, 28, pp. 421–435.

[25] Cascini, G., and Russo, D., 2007, "Computer-aided Analysis of Patents and Search for TRIZ Contradictions," *Int'l Journal of Product Development*, 4, pp. 52.

[26] Li, Z., Tate, D., Lane, C., and Adams, C., 2012, "A Framework for Automatic TRIZ Level of Invention Estimation of Patents Using Natural Language Processing, Knowledge-Transfer and Patent Citation Metrics," *Computer Aided Design*, 44, pp. 987–1010.

[27] Fu, K., Murphy, J., Yang, M., Otto, K., Jensen, D., and Wood, K., 2014, "Design-by-Analogy: Experimental Evaluation of a Functional Analogy Search Methodology for Concept Generation Improvement," *Research in Engineering Design*, 26, pp. 77–95.

[28] Song, B., Srinivasan, V., and Luo, J., 2017, "Patent Stimuli Search and Its Influence on Ideation Outcomes," *Design Science*, 3, p. e25.

[29] Alstott, J., Triulzi, G., Yan, B., Luo, J., 2017, "Mapping Technology Space by Normalizing Patent Networks," *Scientometrics*, 110, pp. 443–479.

[30] Chan, J., Dow, S.P., and Schunn, C.D., 2015, "Do the Best Design Ideas (Really) Come from Conceptually Distant Sources of Inspiration?" *Des. Studies*, 36, pp. 31–58.

[31] Gick, M.L., and Holyoak, K.J., 1980, "Analogical Problem Solving," *Cognitive Psych.*, 12, pp. 306–355.

[32] Gentner, D., & Markman, A.B., 1997, "Structure Mapping in Analogy and Similarity," *American Psychologist*, 52, pp. 45–56.

[33] Ward, T.B., 1998, "Analogical Distance and Purpose in

Creative Thought: Mental Leaps versus Mental Hops," *Advances in Analogy Research: Integration of Theory and Data from the Cognitive, Comput. and Neural Sciences*, pp.221–230.

[34] Tseng, I., Moss, J., Cagan, J., and Kotovsky, K., 2008, "The Role of Timing and Analogical Similarity in the Stimulation of Idea Generation in Design," *Design Studies*, 29, pp. 203–221.

[35] Fu, K., Chan, J., Cagan, J., Kotovsky, K., Schunn, C., and Wood, K., 2013, "The Meaning of "Near" and "Far": The Impact of Structuring Design Databases and the Effect of Distance of Analogy on Design Output," *Journal of Mechanical Design*, 135, p. 021007.

[36] Keshwani, S., and Chakrabarti, A., 2017, "Influence of Analogical Domains and Comprehensiveness in Explanation of Analogy on the Novelty of Designs," *Res. in Eng. Des.*, 28, pp. 381–410.

[37] He, Y., and Luo, J., 2017, "The Novelty 'Sweet Spot' of Invention," *Design Science*, 3, p. e21.

[38] Luo, J., Song, B., Blessing, L., and Wood, K., 2018, "Design Opportunity Conception Using the Total Technology Space Map," *Artificial Intelligence for Engineering Design, Analysis and Manufacturing,* 32, pp. 449–461.

[39] Alstott, J., Triulzi, G., Yan, B., and Luo, J., 2017, "Inventors' Explorations across Technology Domains," *Design Science*, 3, p. e20.

[40] Yan, B., and Luo, J., 2017, "Filtering Patent Maps for Visualization of Diversification Paths of Inventors and Organizations," *J. of the Association for Information Science and Technology*, 68, pp. 1551–1563.

[41] Yan, B., and Luo, J., 2019, "The Superior Knowledge Proximity Measure for Patent Mapping," http://arxiv.org/abs/1901.03925.

[42] Mikolov, T., Chen, K., Corrado, G., and Dean, J., 2013, "Distributed Representations of Words and Phrases and their Compositionality," in: Adv. Neural Inf. Process. Syst., 26, pp. 1–9.

[43] Nobécourt, J., 2000, "A Method to Build Formal Ontologies from Texts," Proceedings of EKAW, Juan-Les-Pins, Paris, France, pp. 21-27.

[44] Alfonseca, E., and Manandhar, S., 2002, "An Unsupervised Method for General Named Entity Recognition and Automated Concept Discovery," in: 1st Int. Conf. Gen. WordNet.

[45] Sarica, S., Song, B., Low, E., Luo, J., 2019, "Engineering Knowledge Graph for Keyword Discovery in Patent Search," International Conference on Engineering Design (ICED19), Delft, Netherlands.

[46] Sarica, S., Song, B., Luo, J., 2019, "Technology Knowledge Graph for Design Exploration: Application to Designing the Future of Flying Cars," IDETC/CIE - Computers and Information in Engineering Conference, Anaheim, CA.

[47] Chan, J., Fu, K., Schunn, C., Cagan, J., Wood, K., and Kotovsky, K, 2011, "On the Benefits and Pitfalls of Analogies for Innovative Design: Ideation Performance Based on Analogical Distance, Commonness, and Modality of Examples," *Journal of Mechanical Design*, 133, p. 081004.

[48] Shah, J.J., Vargas-Hernandez, N., Smith, S.M., 2003, "Metrics for Measuring Ideation Effectiveness," *Design Studies*, 24, pp. 111-134.

[49] Sarkar, P., Chakrabarti, A., 2011, "Assessing Design Creativity," *Design Studies*, 32(4), pp.348–383.

[50] Oman, S.K., Tumer, I.Y., Wood, K., Seepersad, C., 2013, "A Comparison of Creativity and Innovation Metrics and Sample Validation through In-Class Design Projects," Res. Eng. Des., 24, pp. 65–92.

[51] Shai, O., and Reich, Y., 2004, "Infused Design. I. Theory," *Res. in Eng. Des.*, 15(2), 93- 107.

[52] Reich, Y., and Shai, O., 2012, "The Interdisciplinary Engineering Knowledge Genome," 23(3), pp. 251–264.

**Proceedings of the ASME 2019
International Design Engineering Technical Conferences
and Computers and Information in Engineering Conference
IDETC/CIE2019
August 18-21, 2019, Anaheim, CA, USA**

DETC2019-97691

PRODUCT SERVICE SYSTEM DESIGN IN NEW SITUATIONS: PREDICTION OF DEMAND SURFACES FROM ENVIRONMENT

Bryan C Watson[1], Cassandra Telenko
Georgia Institute of Technology
Atlanta, Georgia

ABSTRACT

Product service systems (PSS), such as DVD rental stations or the subway, face a unique problem slowing their adoption and growth: they are uniquely dependent upon timely or expensive user data for system planning, yet user datasets are only accurate for a small part of the entire PSS. Thus, methods to use the available data effectively and use data collected in one portion of a PSS for system design in another portion could transform PSS design. PSS allow customers to purchase use of a product rather than the product itself, resulting in improved environmental sustainability. The central question examined by this work is: how can designers compensate for situations where the design environment has changed and limited user data is available to inform demand estimations? Our hypothesis is that publicly available socio-demographic and environmental variables can be used to estimate the demand outside of the boundaries previously constrained by available user data. This approach was validated by applying multivariable regressions to a major Bike Share System (BSS) Expansion, outperforming the methods utilized by the BSS operators. The approach is tested in four different design scenarios. When examining all 174 stations added in 2015, our approach shows a moderate correlation with the ideal ordering (Rho=.566, Stations =174, p< .01), while the implemented operator ordering was only weakly correlated (Rho=.334, Stations =174, p< .01). This work demonstrates a partial solution to the problem of transforming available user data into demand for new situations.

Keywords: Demand Estimation, Revealed Preference, Multivariable Regression, Product Service System

NOMENCLATURE

n	User population size
p	User population product affinity
$E[x]$	Expected hourly utilization
μ	Observed hourly utilization.

d_i	Ranking error for calculating Spearman's Rho
s	Number of datapoints evaluated for calculating Spearman's Rho
t_i	Number of rankling ties for calculating Spearman's Rho

1. INTRODUCTION

Product service systems (PSS) face a unique problem slowing their adoption and growth: they are uniquely dependent upon timely or expensive user data for system planning, yet user datasets are only accurate for a small part of the entire PSS. Thus, methods to use the available data effectively and use data collected in one portion of a PSS for system design in another portion could transform PSS design. PSSs, such as DVD rental stations or the subway, allow customers to purchase use of a product rather than the product [1], [2]. PSSs have improved environmental sustainability and lower user cost [1], [2]. For example, one model estimates that PSS implementation of washing machines could lower CO_2 emissions by approximately 10%, reduce fluctuations in supply chain demand, and reduce the number of machines in service[3]. The designer faces a unique challenge when using system data to estimate PSS demand. PSS demand varies throughout the area serviced by the PSS and PSS are often introduced in situations where user data is not available. Additionally, even when user data is available, demand estimation approaches were designed for traditional products, not PSS.

Stated Intention (SI), Stated Preference (SP), and Revealed Preference (RP) demand estimation methods have significant challenges for product service systems. SP and SI approaches can require time-consuming or expensive surveys. SP and SI methods may be inadequate due to user's inability to accurately forecast their demand for a new service. SI results tend to overestimate demand due to self-selectivity bias, non-

---

[1] Contact author: BCWatson@gatech.edu

commitment, and exaggerations intentions to drive the overall results of the survey[4].

To avoid SI and SP, RP demand estimations may be created by analyzing comparable products. For completely novel products, designers may consider creating a test market to observe users; however this may not be practical for large PSS[5]. Without SI, SP, or RP, demand estimation methods such as local knowledge, expert guidance, heuristics, or "gut-feel of the decision maker" may be employed [5]–[7]. As a result, PSS design decisions may not be quantifiably repeatable or built from evidence[8]. For example, when 29 firms estimated the demand for wireless technology in the early 1980s, estimates varied by as much as 650%[5]. These challenges highlighted the need for a new RP approach for PSS demand estimation.

Previous research has used a variety of approaches have been to attempt to overcome these challenges and improve PSS design. Reviews of 80 emerging market case studies were used to generate 9 decision making heuristics for PSS design in developing countries [9]. Beyond general heuristics, another challenge is translating imprecise customer responses into PSS design characteristics. Different individuals may describe the desired PSS differently due to different priorities or speech patterns. One approach to this problem, Song et. al. introduced Industrial Customer Activity Cycle (I-CAC) Analysis, providing an improved method to translate customer requirements into a PSS design [10]. A second approach used Supervised Machine Learning to estimate PSS design configuration from data, achieving an impressive classification accuracy of 93.3% [11]. These studies however, look only at overall PSS design, not utilization of local demand variance within the PSS to inform design. Once a general PSS framework has been chosen, an approach is needed to inform the implementation of the PSS which is sensitive to local demand fluctuations.

In [12], we presented an approach to predict the various level of user demand throughout an existing PSS service area [12]. Our previous work provided a PSS demand estimation starting point. This approach successfully predicted PSS demand in new situations. We estimated binomial distribution parameters n (user population size) and p (user population product affinity) from historical user data throughout the area serviced by the PSS. These parameters provided a reliable prediction of future demand, but only in areas with available user data.

In this scenario, information is available about some users in a given context or scenario; however, the designer must derive an approach to allow him/her to apply the known user information in a new scenario. Insights into this type of problem could allow designers to better design new systems, utilizing available limited data. For example, a framework to allow planners in New York of a new telecommunications system to utilize user demand from other cities to frame their approach could limit demand estimation inaccuracies. While our previous work demonstrated an approach to estimate user demand when planning a service expansion[12], an approach is needed to transform the limited available user data into insights to inform the design of the remainder of the PSS.

Therefore, the central question raised by this investigation is: *how can designers compensate for situations where the design environment has changed and limited user data is available to create SI, SP, or RP demand estimations?* Our hypothesis is that publicly available socio-demographic and environmental variables can be used to estimate the n and p Demand Surfaces outside of the boundaries previously constrained by available user data.

We investigate this question via multivariable regression. Our goal is to infer PSS demand in new areas from environmental and socio-demographic variables alone. User Socio-demographic variables include factors such as income, race, and age. Environmental variables include factors such as restaurants, terrain, and infrastructure conditions[7], [8], [13]–[18].

The result of this approach will be two types of demand surfaces. First, Localized Demand Surfaces (n and p) will be estimated [12]. These surfaces exist *within the boundaries* of available user data. Next regressions are used to discover the relationship between socio-demographic variables, environmental variables and the magnitude of n or p (the Localized Demand Surfaces). Once these relationships are determined, the designer can estimate the values n and p from environmental and socio-demographic variables alone. These new estimates create Regressed Global Demand Surfaces.

This approach was validated by applying these regressions to a major Bike Share System Expansion, outperforming the currently employed methods utilized by the BSS operators. The case study successfully demonstrates an approach for PSS design when designers do not have available user data for new locations.

The contributions of this paper are as follows:
1. We provide a framework to transform geographically limited available PSS user data into design insights for the portion of the system without user data.
2. This provides a second validation of the value of n and p estimations for PSS planning as proposed in our last study[12].
3. We provide a tool for BSS operators planning a system expansion.
4. We identify environmental and socio-demographic variables that correlate with higher BSS use.

2. BACKGROUND

2.1 Previous Work

Our previous efforts presented an RP demand estimation method for a PSS during growth[12]. We utilized user data to estimate binomial parameters, providing insight into user demand in new situations. We proposed modeling user demand as a Binomial Distribution. This distribution outputs the number of successes (x), determined by two parameters: n and p.

$$f(x|p) = \binom{n}{x} p^x (1-p)^{n-x} \qquad (1)$$

The number of trials is n and the probability of success for each trial is p. We utilized a combination of the method of moments and point-estimation calculations to estimate n and p throughout the area serviced by the PSS. Once estimated, the

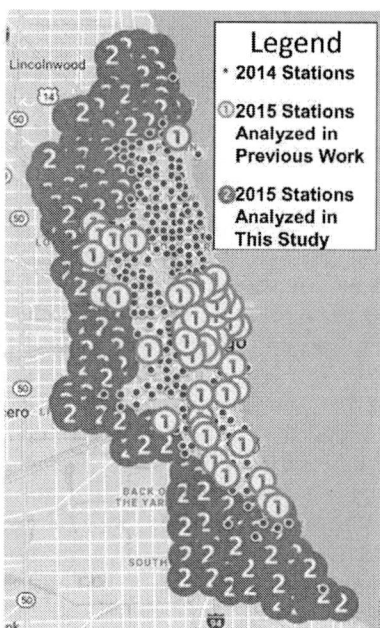

Figure 1: Stations added to DIVVY during 2015 Expansion

binomial parameters create two surfaces allowing estimations of design decisions during a distributed system expansion.

A case study that analyzed the 2015 Chicago Bike Share System (BSS) expansion validated this approach by comparing the implemented stations sizing, the sizing recommended by our approach, and the ideal sizing. BSSs place docks of various sizes with rental bikes around a city. BSS Operators must predict rider demand when selecting the sizes of the new docks. The Divvy BSS had 299 stations in 2014 with an average dock size of 17.4 bicycles per docks (maximum 43, minimum 11).

In 2015, Divvy added 174 stations, 46 were within the boundaries of the existing service area and analyzed in the previous research, while 128 were added outside the boundaries of the existing system (Figure 1) [19]. Station locations were determined with crowdsourcing, input from local government, and interactions with the local community[20].

We used the 2014 subscriber ridership data, the small dots on Figure 1, to create the n and p Localized Demand Surfaces for Chicago. We then tested the predictive power of our approach by predicting the demand for the stations added within the boundaries of the available user data, the one-circles on Figure 1. Demand was quantified by the sizing of the various docks. Algorithm, operator, and ideal dock sizing was compared with Spearman's Rho. The operator size ordering had a moderate correlation with the ideal ordering (rho=.60, stations=46, p<.01). Our approach outperformed the operator ordering, with a very strong correlation with the ideal ordering (rho=.83, stations=46, p<.01). Additionally, the algorithm recommended 94 docks (10% of those added), or about 5,640 pounds of steel, in more optimal locations than the operators actually did.

Although our previous work demonstrated the ability to derive n and p surfaces from existing user data, this approach was limited due to only being applicable to the new PSS

locations within the current boundaries of the existing BSS. We define this type of n and p surfaces as a Localized Demand Surfaces. When this regionally dependent approach was applied to the 128 stations existing outside the 2014 BSS boundaries, Spearman's Rho dropped from .830 (n=46) to .324 (n=128) and the R^2 value for E[x] vs μ dropped from .503 to .003. Although this was an improvement over the implemented operator ordering (rho=.146, n=128), extending the n and p surface beyond the boundaries of the dataset used to build the n and p surface significantly limited this approach's effectiveness.

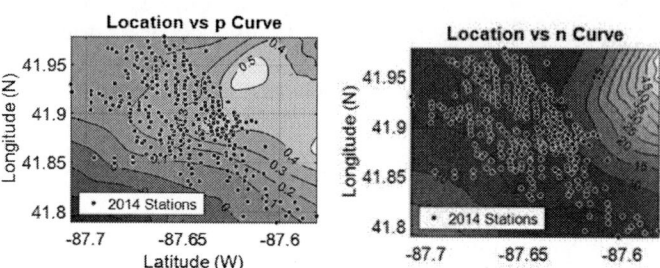

Figure 2: n and p Demand Surfaces derived during Previous Study

2.2 The Problem of User Demand Estimation in New Situations: Insights from Disruptive Innovations

If we seek to incorporate anticipated user demand into PSS design, then we face a unique problem when user data is not available for a new market or situation. Although one could collect a small amount of data and re-perform the analysis of our previous research, this results in opportunity cost and may not be effective[21]. Much previous work has focused on Human-Computer Interaction or software development. This field easily allows for near final prototype testing and rapid design changes of the final product[22], not always possible for mechanical systems. Current applications include active utilization of user context for information delivery and game usability testing[22], [23]. Methods such as user interviews, role playing, or user interactions with prototypes are used to guide the design process[24].

Insights into the investigation of design situations with limited user data question can be found in the current research on the adoption and spread of a disruptive innovation. Disruptive innovations introduce a new technological advancement, dramatically affecting market demand by replacing current technologies[21], [25]. Examples include Micro-Electrical-Mechanical devices (MEMS), Electric-bikes, the personal computer, cloud computing, advanced 3-D printers, and cellular telephones [21], [25]–[28].

A variety of approaches have been proposed to predict demand of disruptive innovations. At the most basic, expert opinions are used. Linton refined this approach by using expert opinions about projected supply and demand to inform Monte-Carlo simulations, allowing inclusion of expert uncertainty, into his models[26]. Application of this approach, yields a probability distribution for expected demand[27]. Diffusion models have also been used to predict demand, however without data to estimate model parameters for the market being examined, accuracy suffers[21]. Useful data, however, is often only available after the disruption process has begun [25].

Identifying disruptive innovations before they shift the market place could provide dramatic benefits to companies [26]. The potential benefit is so large that authors note that even though additional algorithmic work is required, any advance in this field could be leveraged for significant profit [21], [25]–[27]. As a result, many efforts have been made to simply pre-identify disruptive innovations, rather than predict the demand magnitude. This includes analysis of patent data, hazard function, text mining, and web mining [28]–[30].

Previous disruptive innovation research provides insight to our investigation of demand estimation beyond the boundaries of available user data. Disruptive Innovation demand estimation research provides a precedent for using past behavior to predict future demand[26]. Disruptive Innovation profitability requires consideration of each market, analogous to socio-demographic and environmental variables in our work[21], [26]. This provides a basis for our approach of using user population and environmental variables to directly estimate PSS demand in new situations.

Researchers do caution, however, against applying a model from one marketplace to another. Evidence indicates it may be possible to adjust for differences within environments [21]. Thus in this study we utilize regression techniques and in our case study apply our estimations within the same marketplace (different parts of the same city). Both approaches are made more difficult due to the dearth of previous user data[21]. Conducive environments such as bike paths or cultural considerations can be shown to increase the adoption of disruptive technologies[31], providing justification for including environmental variables into our study.

Although similar, the previous work on disruptive innovations differs from the focus of our research in several key areas. First, we examine the situation where some market data is available, while disruptive innovation research attempts to identify if an innovation is disruptive and then predict demand with no market data. User data may not be available because even though the technology is identified as disruptive, it may not be developed enough for market entry. When applying SI, SP, and RP demand estimation, this uncertainty can yield wide variance in demand estimation. For example, expert estimates for the MEMs market size in 2000 ranged from 2 billion to almost 30 billion dollars annually[27]. This is a fundamental difference between identification of disruptive products (disruptive innovation research) and successful implementation of products with limited user data (a design question examined by our research). Secondly, disruptive innovation demand prediction often focuses on dynamic demand growth[26], while this research only examines final steady state demand. Third, disruptive innovation research often examines factors beyond the scope of this investigation, such as the role of friendly government legislature to the innovation[31].

Finally, examining an Innovative PSS may provide insights or overcome traditional challenges faced by disruptive innovation research. The challenge of accurately forecasting multiple sales to a single individual is not applicable to PSS[21].

For a PSS, we are concerned about infrastructure use, and the usage per individual is not a vital statistic as for traditional products. Of note, this gap in repurchasing demand estimation may provide insight into why traditional SI, SP, and RP methods struggle to predict PSS demand.

2.3 Case Study: BSS Demand Estimation in New Situations

Various methods have been historically employed to estimate areas of high demand when planning an initial BSS expansion or installation[8]. There are two distinct approaches to planning BSS station locations. First, experts generated a list of desirable environmental characteristics, analyzed their density within their target cities, and used them to create a BSS potential heat-map. A discussion of BSS planning in Philadelphia and Boise illustrates this first approach. These studies were used by city planners to determine new station location and sizing. Philadelphia planners reviewed the lessons from Montreal, Lyon, and Paris when identifying the nine variables to consider in 2010[14]. Environmental features were given a weighting and were summed within 500m of potential station locations. Philadelphia planners, however, did not consider station sizing; the authors recommended combining this approach with surveys and other qualitative methods[14].

Table 1: Pertinent Characteristics of Previous Regressions

City	Dependent Variable	Environmental Radius	Type of Model	R^2
Boise [67]	Points	820ft and 1640ft	Points for each possible location based on amenities within two distances.	N/A
Philadelphia [66]	Points	500M	Points for each possible location based on amenities	N/A
Brisbane [68]	Frequency of Station Usage	400M	Linear Regression	0.43
Paris [50]	None	Per Hectare	Performance profiles are related to environmental variables, no prediction or R^2 provided.	N/A
Minneapolis and St Paul [69]	Total Station Activity (arrivals and Departures)	400M (200M for food)	Log Linear OLS and Negative Binomial Regression.	.847 and .863
Nice[11]	Car Station Performance	500M	Linear Regression	"fairly robust and have reasonable measures of fitness"
Minneapolis and St Paul [69]	12hr rider counts (not BSS)	Census Block Group	OLS and Negative Binomial Regression	.381 and .476
DC, Denver, Minneapolis and St Paul [51]	Natural Log of Monthly Rentals	400M	Multivariate Linear Regressions	.802, .754, and .801

Increasing in refinement from the approach taken in Philadelphia, Boise planners in 2012 considered 11 variables and two stations sizes when creating their BSS heat-map[13]. Of note, due to Boise being generally flat, the researchers did not include topography considerations but this study incorporated bicycle accidents, reasoning that a high number of accidents implies a large volume of traffic. Thus, unexpectedly, bicycle accidents are positively correlated with ridership. The limitation of these two studies are that without available ridership data, no analysis was performed to assess the accuracy of these approaches

Once ridership data became available, a second generation of BSS environmental studies utilized observed ridership data validate regressions conducted with environmental characteristics. As shown in Table 1 and Figure 3, these studies resulted in a wide variety of accuracies, independent variables, and approaches. A Brisbane study examined the role that infrastructure and environmental features such as elevation played in station usage. The resulting linear regression found an R^2 of .43 when considering these factors alone, highlighting the importance of a conducive cycling environment to BSS demand[17]. An analysis of pedestrian and bicycle traffic in Minneapolis incorporated additional socio-economic factors[8]. This study utilized through-traffic counts of all bicycle traffic and did not focus on BSS usage. Insights include the importance of appropriately varying the radius of influence for environmental variables when considering different transportation modes[8]. A related study analyzed car-share station placements in Nice, France incorporated additional socio-economic factors[7]. Finally, a correlation analysis at the end of an autonomous usage profile study provided evidence that BSS usage is correlated with population, jobs, services, and shops[18].

The previous studies were able to demonstrate moderate correlations between subsets of environmental variables and BSS usage. Studies that combined both environmental and socio-demographic variables resulted in very strong correlations. This approach was demonstrated in both Minneapolis St. Paul, Washington D.C., and Denver with R^2 greater than .8[15], [16]. These studies however, utilized monthly rentals or total demand, rather than n and p as done in this study. The models developed in our approach uses significantly fewer variables than these two studies but with comparable accuracy.

3. METHODOLOGY

This methodology section provides the general framework of our approach and discusses the specifics of the validation of our case study including the environmental data collected, the regression calculation approach, and outline the four tests conducted to validate this approach.

3.1 General Approach for Demand Estimation in New Situations

The following is our overall approach for utilizing binomial parameter estimation with environmental regressions for estimating PSS demand outside the boundary of existing user data. Major steps and considerations are as follows.

1. Gather Demand Data: User data is required for the initial estimation of the Localized Demand Surfaces. Considerations include appropriate data discretization and minimizing the possible values of n to increase estimation accuracy.
2. Create Localized Demand Surfaces (Figure 2). These surfaces provide designer insight for increasing the density of PSS.

Additional details and examples of steps 1 and 2 are provided in ref [12].

3. Gather Environmental and socio-demographic variables: Once Localized Demand Surfaces are generated, the next step is to gather potentially relevant variables including both socio-demographic (income, race, gender) and physical environment (infrastructure, climate, terrain). We recommend these data sources be selected after review of current demand estimation approaches and consultation of system experts.
4. Calculate Multiple Linear Regressions: To create Regressed Global Demand Surfaces, multiple linear regressions are calculated to relate the Localized Demand Surfaces to the environmental and socio-demographic variables gathered in step 3. The purpose of this step is twofold. First, it enables determination of which variables are relevant. Secondly, the final regressions are used to create the Regressed Global Demand Surfaces. In our case study we utilized a step up approach with multi-collinearity checks.
5. Generate Regressed Global User Characteristic Surfaces to predict demand at new locations: Once the regressions are determined in step 4, the environmental data outside the Localized Demand Surfaces is used to calculate n and p for new locations throughout the PSS. n and p then allow calculation of the expected hourly utilization, $E[x]$. Recall, the number of trials is n and the probability of success for each trial is p.

$$E[x] = n * p \qquad (2)$$

Next, steps 3-5 are illustrated through the analysis of the DIVVY BSS expansion case study.

3.2 Step 3: Gather Environmental and Socio-demographic Variable Data

An initial investigation of n and p from Localized Demand Surfaces revealed wide fluctuations of n and p within the same census tract. Thus, two types of input variables were desired. First, variables that continuously varied through the census tract and were not derived from census data. This includes variables such as bike path length, retail, and restaurants. Secondly, data that aggregated several census tracks within a defined radius of the potential BSS station locations. This includes socio-demographic variables such as race and income.

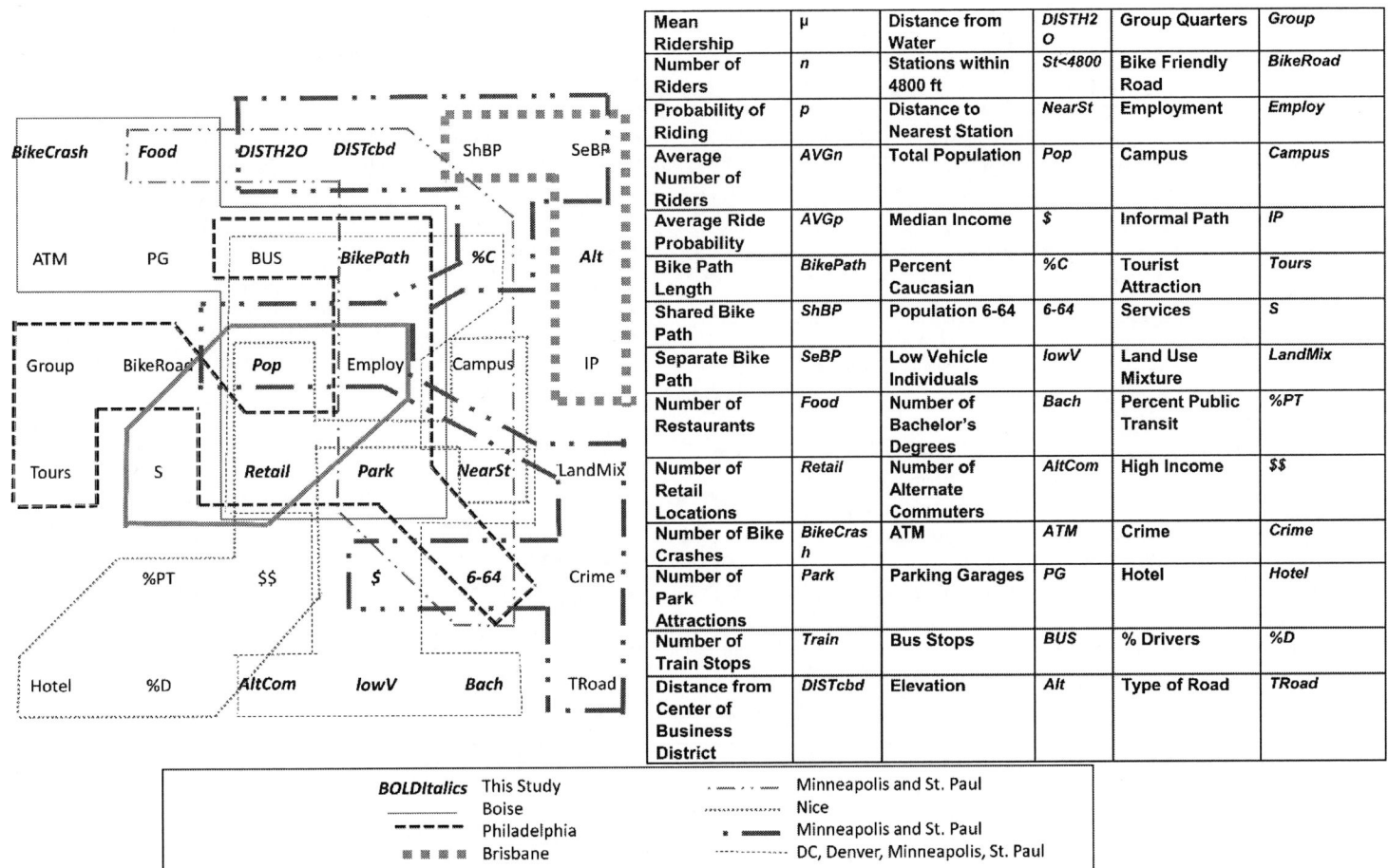

*Figure 3: Previous Studies that Predicted Ridership incorporated a wide variety of independent variables
Note: The most frequently used independent variables are in the center of the grid.*

Table 2: Socio-demographic Variables Evaluated

Variable	Mean and Range	Literature Precedent
Total Population	3,923 (808.5-12,468)	[7], [8], [13], [16]–[18]
Median Income	75,355 (18,535-138,250)	[8], [16]
Percent Caucasian	66.78 (.46-93.88)	[8], [15], [16]
Population 6-64	286 (12.43-1,587)	[8], [14], [15]
Number of Low Vehicle Individuals	700 (23-4981)	[16]
Number of Bachelor's Degrees	797.5 (24-3,094)	[16]
Number of Alternate Commuters.	1,295 (116.1-5,214)	[16]

Table 2 and 3 outline the socio-demographic and environmental variables examined in this paper. Socio-demographic variables describe the population around a station (age, income, race). Environmental variables describe the setting around a station (infrastructure, business density, terrain).

The mean values in Table 2 are for the 300 stations used to create the Localized Demand Surfaces. Two influence radiuses were utilized. 1.6 miles was used for crashes, parks, and paths. 1.6 miles is the average ride length and these features were more likely to be enjoyed or experienced while riding [33]. One-quarter mile was used for restaurants, retail locations, and train stops to be consistent with last-mile literature [34]. Park

attractions were used instead of square footage or number of parks because it should more closely correlate with usage. A small park with many attractions may generate more BSS demand than a large, empty park. Attractions include features such as playgrounds, basketball courts, and restrooms. Bus stops were considered, but not analyzed due to the high density within the 2014 BSS station area. All stations would have had approximately the same number of bus stops within their influence radius. The center of the business district and line approximating the shore of Lake Michigan were estimated by the researcher.

The second type of data collected for analysis was aggregated Census Data. Due to many of the BSS stations existing on the edge of a census track, the census values were weighted and averaged within a .25 mile radius. The raw data was received from www.AmericanFactFinder.com's 2014 American Community Survey with 5 year look ahead (reports B01003, B02001, S1903, B08201, S0601, S0801). The data analysis was performed in ArcGIS 10.3.1. This resulted in an additional seven independent variables for analysis. Number of Caucasians was also considered instead of percent Caucasian, but no significant correlation was discovered.

3.2 Step 4: Calculate Multiple Linear Regression

Once the independent and dependent variable are collected into a single dataset, they are analyzed to determine appropriate linear regressions. Pairwise linear correlation coefficients and associated p-values are presented in Appendix 1. The variables in Tables 2 and 3 were normalized by dividing each datapoint by the mean of their dataset. To identify

Table 3: Environmental Variables Evaluated

Variable	Mean and Range	Source	Description	Literature Precedent
Bike Path Length (Miles)	.0192 (.024-.160)	Chicago Data Portal(CDP) [32]	Total Length of Paths within Radius	[8], [13]–[17]
Number of Restaurants	34.77 (0-169)	CDP 2014 Food Inspections	Total number of Restaurants within radius	[13], [15]
Number of Retail Locations	26.42 (0-80)	CDP 2014 Annual Taxpayer Location List	Total number of Stores within radius	[7], [13], [14], [18]
Number of Bike and Pedestrian Crashes	21.33 (0-55)	CDP	Number of Pedestrian and Bikes hit by cars.	[13]
Number of Park Attractions	170.61 (85-298)	CDP	Total Number of Attractions	[13]–[16]
Number of Train Stops	1.51 (0-16)	CDP	Total number of Train stops.	
Distance from Center of Business District (Miles)	2.76 (.12-7.15)	Calculated	Center of Business District Defined as -41.89N/87.63W.	[8], [15]
Distance from Lake Michigan (Miles)	2.30 (.103-6.42)	Calculated	Lake Michigan between 41.999N/-87.654W and 41.765N/-87.559W	[8], [15]
Number of Stations within 4800 ft	23.37 (2-54)	Measured	Number of Stations within 4800 ft	[16]
Distance to Nearest Station (Miles)	.2377 (.08-.84)	Measured	Distance to Nearest Station	[7], [15]

appropriate dependent-independent variable matching, .4 correlation was chosen for μ and p, while .3 was chosen for n. The n cutoff was lower, due to the weaker general correlation all independent variable showed with n. Variable were examined for multi-collinearity by utilizing the variance inflation factor (VIF), with VIF greater than ten being eliminated from consideration[35]. Next, a step-up approach was used to create the model and ensure all relevant independent variables were included. Finally, a second multi-collinearity check was performed to ensure the final model was accurate.

3.3 CASE STUDY: Procedure to Validate Results

Once the regression is completed, Regressed Global Demand Surfaces are generated. We tested the efficiency of this approach in four different scenarios. Comparison is conducted with Spearman's Rho (ρ), or rank correlation. This allows for ranking comparison [36].

$$\rho = 1 - \frac{6 \sum d_{i\,i}^2}{s^3 - s} \tag{3}$$

d_i is the ranking error and s is the number of stations evaluated.

Spearman's Rho also enables a correction factor for ties. The dock sizes implemented by the operators have ties due to the operators utilizing discrete station configurations. Eq. (9) and Eq. (10) allows equivalent comparisons between rankings with and without ties [36].

$$\rho = 1 - \frac{6(\sum d_i - \sum t_x)}{s^3 - s} \tag{4}$$

$\sum t_x$ is

$$\sum t_x = \frac{\sum(t_i^3 - t_i)}{12} \tag{5}$$

and t_i is the number of rankling ties.

Four tests were designed to validate this approach, providing different scenarios to test the efficacy of the Regressed Global Demand Surfaces to predicting PSS demand.

3.3.1 Testing the 46 Stations Within the Boundaries of the Available User Data

This area (shown with circle 1's on Figure 1) is expected to have higher ridership than those outside the boundary of the Localized Demand Surfaces. Thus, accuracy in this area may dominate overall performance. Additionally, this test is contained within the boundary of the Localized Demand Surfaces used to perform the regression that is the basis for the Regressed Global Demand Surfaces. The predictions from the Regressed Global Demand Surfaces are compared to the Localized Demand Surfaces (the results of our previous research) and the implemented operator ordering.

3.3.2 Testing the 128 Stations Outside the Boundaries of the Available User Data

This section, circle-2 on Figure 1, is expected to have lower ridership per station than those within the boundaries of available user data. Additionally, this test is not contained within the boundary of the Localized Demand Surfaces used to perform the regression that is the basis for the Regressed Global Demand Surfaces. This is the key test of the hypothesis that publicly available socio-demographic and environmental variables can be used to estimate Demand Surfaces outside of the boundaries constrained by user data.

3.3.3 Testing all 174 Stations with the Global User Characteristic Curve

All stations added in the 2015 Divvy BSS Expansion, circle-1 and circle-2 on Figure 1, are tested with the Global User Characteristic Curve to assess the overall implementation in areas of both high and low ridership. The accuracy of the Environment-Derived User Characteristic Curve is compared to the implemented operator ordering.

3.3.4 Testing all 174 Stations with both types of User Characteristic Curves

Finally, all stations added in the 2015 Divvy BSS Expansion are tested with the curve best suited to that area in an attempt to maximize algorithm accuracy. The Localized Demand Surfaces are used to predict ridership within the boundaries of available user user data (circle-1 on Figure 1), while the Global User Characteristic Curve is used to predict the ridership at the stations marked circle-2 on Figure 1. This test combines the results of our previous research and this effort as a best practice for BSS operators considering an expansion

4. RESULTS AND DISCUSSION

4.1 Regression Results

Table 4 summarizes the independent variables that exceeded the correlation and multi-collinearity thresholds for *n, p,* and μ. .4 correlation was chosen for μ and p, while .3 was chosen for *n*.

Each regression coefficient was examined to ensure correlation matched intuition. This was true, except for the park amenities variable. There is a negative correlation of park amenities to *p* and μ (-.845 and -1.58). Previous regressions utilized distance to park vice park amenities, resulting in .061, -,486, and -.485 [15], [16]. The negative coefficient might be because although more individuals choose to ride through the parks, the existence of large parks reduces the population density in those areas, resulting in an overall negative relationship. Additionally, docked bikeshare may be used more for commuting, thus recreational amenities have relative lower appeal, resulting in an overall negative correlation.

Although the final regression for p yielded a strong correlation ($R^2 = .827$), the n regression failed to provide more than a weak correlation ($R^2 = .178$). Although there was evidence of correlation between the predicted n values and the evaluated independent variables, strong predictors were not discovered within the 17 evaluated variables. Number of Train Stops, Total Population, Population 6-64, number of low vehicle households, number of bachelor's degrees, and number of alternate commuters only showed an influence at high values by forcing n to a stable value. Although we attempted to include the influence of these variables by incorporating heuristics into the prediction algorithm, overall model results did not appreciably change, thus they were omitted to minimize model complexity. It is possible that these variables only appear to drive n to a stable value as an artifact of the size of the dataset evaluated.

4.2 Test Results

Summary statistics and overall test results are presented in Table 5. A discussion of these results is presented in the following sections.

4.2.1 Test One Results: The 46 Stations Within the Boundaries of User Data

The Regressed Global Demand Surface was used to predict the ridership for the 46 stations existing within the boundaries of the 2014 existing stations. These are the same stations predicted in our previous work. Algorithm performance slightly degraded from the Localized Demand Surface (Rho = .828, stations=46, p < .01 to Rho= .756, stations=46, p <.01). As expected,

observational user data creates superior performing Demand Surfaces than regressions, but both approaches outperformed the implemented operator ordering's moderate correlation (rho=.590, stations=46, p<.01).

To provide a qualitative assessment of algorithm accuracy, instances where algorithm or operator ordering varied from ideal ordering by more than nine places (1/4 of the sample size) were assessed. This occurred fifteen times in the algorithm ranking and eighteen times for the operator ranking. Instances where

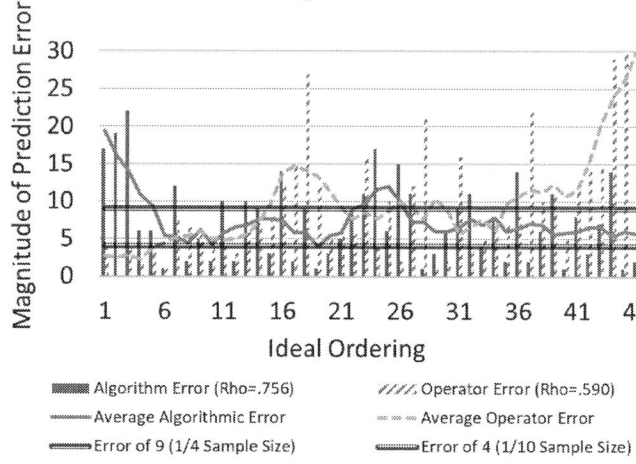

Figure 4: Running Average and Individual Prediction Error for the GUCS and the Implemented Operator Ordering for 46 Stations

Table 4: Resulting n ,p, and µ regressions

Variable	Coefficient			Standard Error			P Value		
	N	P	µ	N	P	µ	N	P	µ
Intercept	.480	2.43	2.89	.0757	~	.227	<.01	<.01	<.01
Median Income	.262	~	~	.0694	~	~	<.01	~	~
St<4800	.096	-.335	~	.0554	.0586	~	<.1	<.01	<.01
Distance from H2O	.162	-.517	-.532	.0891	.0315	.077	<.1	<.01	~
Crash	~	.405	~	~	.0404	~	~	<.01	~
Distance from CBD	~	-.195	~	~	.0539	~	~	<.01	~
Park	~	-.845	-1.58	~	.0956	.201	~	<.01	<.01
Food	~	.0599	.223	~	.0218	.0431	~	<.01	<.01
R2	.178	.827	.457						

Table 5: Summary of Results for Tests 1-4

Test	One			Two		Three		Four	
Predictor	LDS	RGDS	OP	RGDS	OP	GUCF	OP	COMBO	OP
Rho	.828	.756	.590	.295	.143	.559	.334	.566	.334

Legend:
LDS- Localized Demand Surfaces
RGDS- Regressed Global Demand Surfaces
OP- Actual Operator Choices
COMBO- Utilizing both LDS and RGDS.

predicted ordering was within four (1/10 of the sample size) of ideal ordering were also counted. This occurred sixteen times for the algorithm ranking and thirteen times for operator rankings. For an detailed analysis of the Localized Demand Surfaces, refer to our previous work[12].

As in our previous work, we observed that the Implemented Operator Ordering is superior at identifying the highest ranked stations, indicating that combining this approach with current methods might be the best practice for decision makers. Figure 4 shows the error for each prediction from the operators and the algorithms vs the ideal ordering. A running average (every 5 predictions) is plotted to allow easier detection of overall trends. 1/4 and 1/10 sample-size error lines are overlaid to provide a sense of scale for the accuracy of the predictions.

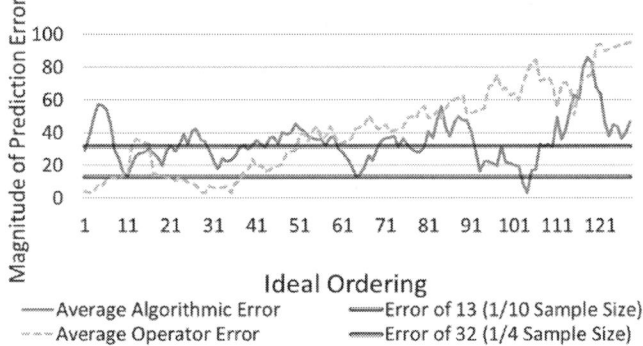

Figure 5: Algorithm and Operator Prediction Error vs Ideal Ranking for 128 stations outside the boundary of available user data.

4.2.2 Test Two Results: the 128 Stations Outside the Boundaries of Available User Data

Next, the utilization of the remaining stations added in 2015 were predicted using the Regressed Global Demand Surfaces. The algorithm ordering showed a weak correlation to the ideal ordering (Rho=.259, stations=128, p< .01), but outperformed the implemented operator ordering, which was only very weakly correlated (Rho=.143, stations=128, p<.1). The μ regression results degraded to Rho=.101.

Figure 5 shows that the implemented operator orderings are generally superior in estimating the demand of the highest performing stations, while the Algorithm performance does not appreciably degrade with ideal ordering. To maintain graph readability, only the running average is plotted.

4.2.3 Test Three Results: all 174 Stations with Regressed Global Demand Surfaces

Next, all 174 BSS stations added during the 2015 BSS expansion were tested and a station ordering was generated from

the Regressed Global Demand Surfaces. The algorithm ordering showed a moderate correlation to the ideal ordering (Rho=.559, stations=174, p< .01) while the implemented operator ordering was weakly correlated (Rho=.334, stations=174, p<.01).

Figure 6 shows that the implemented operator ordering methods are generally superior in estimating the demand of the highest performing stations, while the algorithm performance does not appreciably degrade with ideal ordering. Figure 6 summarizes the accuracy of the algorithm and operator predictions.

4.2.4 Test Four Results: all 174 Stations with both types of User Characteristic Curves

The results of our previous work and tests 1-3 of this case study indicate that the optimal choice for BSS operators considering an expansion would be to use observational data to create Localized Demand Surfaces to estimate new demand for stations within the boundaries of the observational data (our previous work). Then Global User Characteristic Curves should be created and used to estimate station demand outside the boundaries of the observational data.

When applied to this case study, the resulting multi-approach algorithm ordering shows a moderate correlation with the ideal ordering (Rho=.566, Stations =174, p< .01), while the implemented operator ordering was only weakly correlated (Rho=.334, Stations =174, p< .01). Incorporating the Localized Demand Surface for the stations within the boundary of available user data only increased spearman's rho by .007. Excitedly, this indicates that if Regressed Global Demand Surfaces can be expanded further (from city to city vice simply different areas within a city), it might prove to be equally as powerful as Localized Demand Surfaces. This could allow BSS operators to omit costly or time-consuming data collection.

When assessing the correlation between the predicted hourly station utilization and observed hourly utilization, this

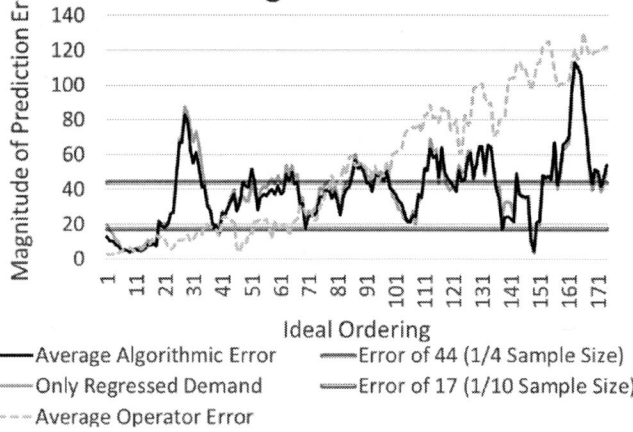

Figure 6: Running Average Algorithm and Operator Prediction Error vs Ideal Ranking for all 174 stations

approach results in a stronger correlation (R^2=.580, stations=174, p<.01) than the implemented station capacities (R^2=.377, stations=174, p<.01).

5. CONCLUSIONS

Through the analysis of four tests of this case study, we have demonstrated a partial solution to our research question: how do you transform available user data into demand in new situations? Through the analysis of the 2015 Divvy BSS expansion, Regressed Global Demand Surfaces were used to better estimate demand for 128 stations existing outside the boundaries of the Localized Demand Surfaces than traditional demand estimation methods. This approach improved total BSS expansion algorithm Spearman's Rho from .333 to .566 when using the Regressed Global Demand Surfaces.

A key gap still exists, however. Our efforts in this research have examined demand estimation in a market with enough user data available to estimate Localized Demand Surfaces. In the BSS case study, the current results can only be applied to a BSS expansion. What about scenarios where only user data from a completely different environment is available? Can the results of this study be used to help BSS operators plan a system installation in a city that currently has no BSS? Does this approach provide a tool for design across markets? These questions will be examined in future work. Additional future work could involve refining a method to combine operator ranking with the results of the Regressed Global Demand Surfaces and investigation of additional predictive environmental and population variables.

ACKNOWLEDGEMENTS

The authors gratefully acknowledge the support of the Georgia Institute of Technology and the three anonymous reviewers for their feedback.

REFERENCES

[1] O. K. Mont, "Clarifying the concept of product-service system," *J. Clean. Prod.*, vol. 10, pp. 237–245, 2002.

[2] T. S. Baines *et al.*, "State-of-the-art in product-service systems," *Proc. Inst. Mech. Eng. Part B J. Eng. Manuf.*, vol. 221, no. 10, pp. 1543–1552, 2007.

[3] P. Wangphanich, "Simulation model for quantifying the environmental impact and demand amplification of a product-service system (PSS)," in *2011 International Conference on Management Science and Industrial Engineering, MSIE 2011*, 2011.

[4] T. Fowkes and J. Preston, "Novel approaches to forecasting the demand for new local rail services," *Transp. Res. Part A Gen.*, vol. 25, no. 4, pp. 209–218, 1991.

[5] R. J. Thomas, "Problems in Demand Estimation For a New Technology," *J. Prod. Innov. Manag.*, vol. 2, no. 3, pp. 145–157, 1985.

[6] ITDP, "The Bike-Sharing Planning Guide," *Institute for Transportation & Development Policy*, 2013. [Online].

Available: https://www.itdp.org/wp-content/uploads/2014/07/ITDP_Bike_Share_Planning_Guide.pdf. [Accessed: 19-Feb-2018].

[7] V. P. Kumar and M. Bierlaire, "Optimizing Locations for a Vehicle Sharing System," in *Swiss Transport Research Conference (STRC) (Ascona, Switzerland)*, 2012.

[8] S. Hankey *et al.*, "Estimating use of non-motorized infrastructure: Models of bicycle and pedestrian traffic in Minneapolis, MN," *Landsc. Urban Plan.*, vol. 107, no. 3, pp. 307–316, 2012.

[9] C. Schafer, R. Parks, and R. Rai, "Design for Emerging Bottom of the Pyramid Markets: A Product Service System (PSS) Based Approach," 2012.

[10] W. Song, X. Ming, Y. Han, and Z. Wu, "A rough set approach for evaluating vague customer requirement of industrial product-service system," *Int. J. Prod. Res.*, 2013.

[11] H. J. Long, L. Y. Wang, J. Shen, M. X. Wu, and Z. B. Jiang, "Product service system configuration based on support vector machine considering customer perception," *Int. J. Prod. Res.*, 2013.

[12] B. C. Watson and C. Telenko, "Binomial Parameter Determination and Mapping for Demand Prediction: A Case Study of Bike Sharing Station Expansion Design," in *Proceedings of the ASME 2018 International Mechanical Engineering Congress and Exposition*, 2018, pp. 1–12.

[13] T. Wuerzer, S. Mason, and R. Youngerman, "Boise Bike Share Location Analysis," Boise, 2012.

[14] G. Krykewycz, C. Puchalsky, J. Rocks, B. Bonnette, and F. Jaskiewicz, "Defining a Primary Market and Estimating Demand for Major Bicycle-Sharing Program in Philadelphia, Pennsylvania," *Transp. Res. Rec. J. Transp. Res. Board*, vol. 2143, pp. 117–124, 2010.

[15] X. Wang, G. Lindsey, J. E. Schoner, and A. Harrison, "Modeling Bike Share Station Activity: Effects of Nearby Businesses and Jobs on Trips to and from Stations," *J. Urban Plan. Dev.*, vol. 142, no. 1, p. 04015001, 2016.

[16] R. Rixey, "Station-Level Forecasting of Bikesharing Ridership," *Transp. Res. Rec. J. Transp. Res. Board*, vol. 2387, pp. 46–55, 2013.

[17] I. Mateo-Babiano, R. Bean, J. Corcoran, and D. Pojani, "How does our natural and built environment affect the use of bicycle sharing?," *Transp. Res. Part A Policy Pract.*, vol. 94, pp. 295–307, 2016.

[18] C. Ome and O. Latifa, "Model-Based Count Series Clustering for Bike Sharing System Usage Mining : A Case Study with the Vélib System of Paris," *ACM Trans. Intell. Syst. Technol.*, vol. 5, no. 3, pp. 1–21, 2014.

[19] J. Zhang, X. Pan, M. Li, and P. S. Yu, "Bicycle-sharing system analysis and trip prediction," in *Proceedings - IEEE International Conference on Mobile Data Management*, 2016, vol. 2016-July, pp. 174–179.

[20] A. Smith, "Crowdsourcing for Active Transportation," *ITE J.*, vol. 85, no. 5, pp. 30–35, 2015.

[21] J. D. Linton, "Forecasting the market diffusion of disruptive and discontinuous innovation," *IEEE Trans. Eng. Manag.*, vol. 49, no. 4, pp. 365–374, 2002.

[22] P. Moreno-Ger, J. Torrente, Y. G. Hsieh, and W. T. Lester, "Usability testing for serious games: Making informed design decisions with user data," *Adv. Human-Computer Interact.*, vol. 2012, pp. 1–13, 2012.

[23] G. Fischer, "Context-Aware Systems: The 'Right' Information, at the 'Right' Time, in the 'Right' Place, in the 'Right' Way, to the 'Right' Person," *Adv. Vis. Interfaces Int. Work. Conf.*, 2012.

[24] H. E. McLoone, M. Jacobson, C. Hegg, and P. W. Johnson, "User-centered design," *Work*, vol. 37, no. 4, pp. 445–456, 2010.

[25] D. Nagy, J. Schuessler, and A. Dubinsky, "Defining and identifying disruptive innovations," *Ind. Mark. Manag.*, vol. 57, pp. 119–126, 2016.

[26] J. D. Linton, "Determining demand, supply, and pricing for emerging markets based on disruptive process technologies," *Technol. Forecast. Soc. Change*, vol. 71, no. 1–2, pp. 105–120, 2004.

[27] J. D. Linton and S. T. Walsh, "Forecasting micro electro mechanical systems: A disruptive innovation," in *PICMET '01. Portland International Conference on Management of Engineering and Technology*, 2001.

[28] I. Bildosola, R. M. Río-Bélver, G. Garechana, and E. Cilleruelo, "TeknoRoadmap, an approach for depicting emerging technologies," *Technol. Forecast. Soc. Change*, vol. 117, pp. 25–37, 2017.

[29] S. Altuntas, T. Dereli, and A. Kusiak, "Forecasting technology success based on patent data," *Technol. Forecast. Soc. Change*, vol. 96, pp. 202–214, 2015.

[30] Y. Jeong, I. Park, and B. Yoon, "Forecasting technology substitution based on hazard function," *Technol. Forecast. Soc. Change*, vol. 104, pp. 259–272, 2016.

[31] Y. Ruan, C. C. Hang, and Y. M. Wang, "Government's role in disruptive innovation and industry emergence: The case of the electric bike in China," *Technovation*, vol. 34, no. 12, pp. 785–796, 2014.

[32] "Chicago Data Portal," *Chicago Data Portal*. [Online]. Available: https://data.cityofchicago.org/. [Accessed: 24-Feb-2019].

[33] E. Fishman, S. Washington, and N. Haworth, "Bike Share: A Synthesis of the Literature," *Transp. Rev.*, vol. 33, no. 2, pp. 148–165, 2013.

[34] H. Dittmar and G. Ohland, "The New Transit Town: Best Practices in Transit-Oriented Development," *Transportation (Amst).*, vol. 71, no. i, p. 253, 2004.

[35] D. A. Belsley, E. Kuh, and R. E. Welsch, *Regression diagnostics*. Hoboken: John Wiley and Sons, 1980.

[36] W. L. Taylor, "Correcting the Average Rank Correlation Coefficient for Ties in Rankings," *J. Am. Stat. Assoc.*, vol. 59, no. 307, pp. 872–876, 1964.

Proceedings of the ASME 2019
International Design Engineering Technical Conferences
and Computers and Information in Engineering Conference
IDETC/CIE2019
August 18-21, 2019, Anaheim, CA, USA

DETC2019-97881

PSEUDO-RIGID BODY DYNAMIC MODELING OF COMPLIANT MEMBERS FOR DESIGN

Vedant
University of Illinois at Urbana-Champaign
Aerospace Engineering
Urbana, IL 61801
Email: vedant2@illinois.edu

James T. Allison
University of Illinois at Urbana-Champaign
Industrial & Enterprise Systems Engineering
Urbana, IL 61801
Email: jtalliso@illinois.edu

ABSTRACT

Movement in compliant mechanisms is achieved, at least in part, via deformable flexible members, rather than using articulating joints. These flexible members are traditionally modeled using Finite Element Models (FEMs). In this article, an alternative strategy for modeling compliant cantilever beams is developed with the objectives of reducing computational expense, and providing accuracy with respect to design optimization solutions. The method involves approximating the response of a flexible beam with an n-link/m-joint Pseudo-Rigid Body Dynamic Model (PRBDM). Traditionally, PRBDM models have shown an approximation of compliant elements using 2 or 3 revolute joints (2R/3R-PRBDM). In this study, a more general nR-PRBDM model is developed. The first n resonant frequencies of the PRBDM are matched to exact or FEM solutions to approximate the response of the compliant system. These models can be used for co-design studies of flexible structural members, and are capable of modeling higher deflection of compliant elements.

1 Introduction

Many complex engineering systems are comprised of multiple mechanical members to attain the desired functionality and performance. Recently, it has been shown that elastic compliance in individual members within a system can be exploited to reduce system complexity [1]. This can be achieved in part by utilizing compliance to create multifunctional components, which can help reduce the required number of discrete components. In addition, compliant mechanisms can help reduce overall volume, improve mechanical precision, and reduce wear.

Modeling such systems is a challenging task. There are several methods of varying fidelity to model the compliant members; some of the well-known methods include: Finite Element Analysis (FEA), lumped-parameter models, and Pseudo Rigid Body Models (PRBMs) [1]. Each of these methods has been shown to have individual strengths and weaknesses. Since many compliant structures undergo large deflection, techniques that approximate the performance of the compliant structure well for large deflection are desirable. PRBMs approximate the performance of a compliant member by modeling them as a series of rigid bodies linked to each other using torsional spring joints. PRBMs have been shown to model properties such as bistability/tristability [2, 3], dynamic behaviors [4, 5], and pose workspace [6].

Pseudo Rigid Body Dynamic Models (PRBDMs) are a variation of PRBMs, where the dynamic response of the model is matched to the expected response from the complaint system. The matching of system response can be performed using several metrics. Some of the existing approaches have utilized the deflection of a compliant member (e.g., a cantilever beam) under constant structural loads. Most initial studies explored the spring stiffness and the position of a single revolute joint (1R-PRBM) to approximate the dynamics [1]. the 1R-PRBM uses a characteristic pivot along the beam, used to approximate the response of any compliant member. Subsequent studies have explored more accurate approximation of the compliant members that undergo larger deflection levels using two revolute joint (2R) [7] and 3R PRBM models [8]. These models focused on mapping the deflection of a compliant member accurately. Most PRBMs are load dependent, where the spring stiffness and joint position depend on the value and type of load applied. This is undesirable for a general model approximation. A survey of multiple PRBMs is provided in Ref. [9]

A comparison of the lumped parameter model against the FEA results for compliant members was performed previously, and it was discovered that the lumped model was more accurate in approximating deflection, whereas the FEA model was more effective at modeling resonant frequencies [10].

Co-design is a class of dynamic system design problems and methods of growing importance that aims to produce system-optimal designs by considering both physical and control system design decisions in an integrated manner [11–13]. Successful application of co-design methods requires the creation of low- to medium-fidelity models that predict the effect of changes both to physical and control system design decisions. Medium-fidelity models that do not depend on computationally expensive steps (e.g. re-meshing) are very desirable for co-design applications, such as compliant mechanisms and intelligent structures. In this study, a method of modeling a non-uniform cantilever beam using nR-link PRBDM models is introduced. The realized PRBDM models will have the same first 'n' resonant modes as the original beam. The natural frequency for this study is obtained using COMSOL[14] eigenvalue analysis, and the eigenvalue of the PRBDM equation is matched using an optimization scheme to find the system parameters that minimize the difference in the eigenvalues of the two systems. Alternate methods to obtain the resonant frequencies include use of analytical methods [15] and simple machine learning methods.

2 Problem Formulation

The PRBDM modeling method accuracy depends on obtaining appropriate values for spring stiffness, the number of joints, and the distance between joints. For design purposes, these model parameters must be estimated based on independent physical design variables, such as geometric parameters. This article investigates the utility of several strategies for mapping design variables to PRBDM parameters. Quantitative comparison of mapping strategies is based on candidate beams whose specifications are generated randomly, and where the first n eigenvalues are approximated using a variety of methods. These approximated eigenvalues are compared against "truth" values obtained for test beams using either analytical methods similar to those presented in Ref. [15], or an FEM eigenanalysis with a coarse mesh.

The dynamical equations for a nR-PRBDM is then formulated. Numerical optimization is used to match the eigenvalues for the PRBDM to the truth values. The cantilever beam, which is constrained to move in a 2D plane, can be approximated as an n-revolute joint rigid multi-body arm, as depicted in Fig 1. This study quantitatively compares several mapping strategies to de-

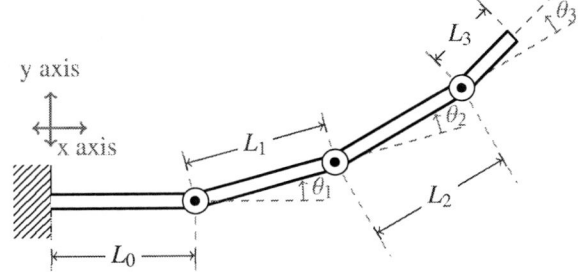

FIGURE 1: Illustration of 4-link/3R PRBDM

termine the choice of spring stiffness and node distance to best match the eigenvalues. Here it is assumed that the panels have a maximum limit of 1 meter for length and width, while the thickness was limited to a minimum of 10 mm.

2.1 Generating random candidates

The mapping strategies are tested by applying them to randomly generated cantilever beams. The candidate beams are generated by first declaring the number of sections in the panel, denoted as an integer p. The generation steps include: choose p random numbers between 0 and 1, order these in ascending order, and append 0 and 1 to this list. These will serve as the X coordinates for the polygon representing distributed beam geometry (similar to the piecewise-linear design description used in Ref. [16]). The next step is to choose $p+2$ additional random real numbers between 0 and 1; these will serve as the Y coordinates for the polygon. Then, define a polygon where its starting point is the origin, its end point is point $(0,1)$, and the start and end points are connected by a piecewise linear curve defined by vertices with positions specified by the ordered list of X and Y values. Reflecting the curve about the X-axis generates the closed polygon representing the planform geometry of

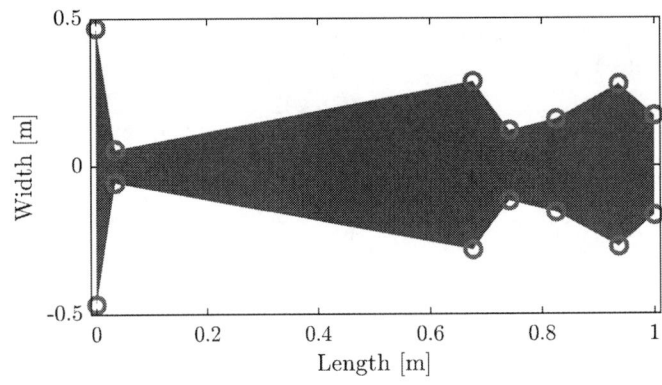

FIGURE 2: Visualization of a randomly generated beam, symmetric about X-axis (black line)

a candidate beam. Extruding this polygon to thickness t (in the Z-direction, out of the page) yields the complete description of a randomly generated test beam. All beams used in the tests here each have uniform thickness in the X-Y plane. One of the candidate panels is shown in Fig 2.

2.2 PRBDM Model

The dynamical model for an n-link arm in general can be represented by Eqn. (1). Assuming the beam is in a gravity-free environment, the contribution of $G(\theta, \psi)$ is defined in Eqn. (2).

$$M(\theta,\psi)\ddot{\theta} + C(\theta,\dot{\theta},\psi)\dot{\theta} + G(\theta,\psi) = \tau \quad (1)$$

$$G(\theta,\psi) = K\theta = \begin{bmatrix} K_{11} & \dots & K_{1n} \\ \vdots & \ddots & \vdots \\ K_{n1} & \dots & K_{nn} \end{bmatrix} \theta \quad (2)$$

In the above equations, θ and $\dot{\theta}$ are the local relative angular positions and velocities for each link. The quantities q and $\dot{q}$ correspond to the angular orientation and velocity of each link with respect to the global/world frame. The vector τ indicates the torque applied at each joint, and ψ is the vector of design parameters for all links. Figure 1 shows an example of a 4 link/3R-PRBDM model. Here, θ_n is the relative angle of the nth link with respect to the $(n-1)$th link. The relationship between θ and q is shown in Eqn. (3). The first link is always aligned to the world X axis, as quantified in Eqn. (4).

$$\theta_j = q_j - q_{j-1}, \ \forall \ j \in [1,2,\dots,n] \quad (3)$$
$$q_0 = 0 \quad (4)$$

The matrix $C(\theta,\dot{\theta},\psi)$ in Eqn. (1) represents the Coriolis effect for the system, but an artificial damping term is added to the system to render the model more tractable for simulation. The modified $C_m(\theta,\dot{\theta},\psi)$ with damping is defined in Eqn. (5):

$$C_m(\theta,\dot{\theta},\psi) = C(\theta,\dot{\theta},\psi) + 0.1 \times \begin{bmatrix} 1 & 0 & 0 \\ 0 & 1 & 0 \\ 0 & 0 & 1 \end{bmatrix}. \quad (5)$$

The eigenvalues (λ) of the PRDBM model can calculated using Eqn. (6):

$$\lambda^2 = (M(\theta,\dot{\theta},\psi))^{-1} K. \quad (6)$$

2.3 Mapping Strategy

The eigenfrequencies for the randomly-generated beams are obtained using COMSOL eigenvalue analysis. The eigenvalues and the mass participation factors are saved. The eigenvalues that have the highest mass participation in the Y direction are filtered and arranged in ascending order.

Numerical optimization is used to match the eigenvalues of an $(n+1)$-link/nR-PRBDM model to the truth values. The optimizer chooses the distance between the nodes L_j and each joint stiffness K_j, $j = 1,\dots,n$, where n is the number of joints. A core

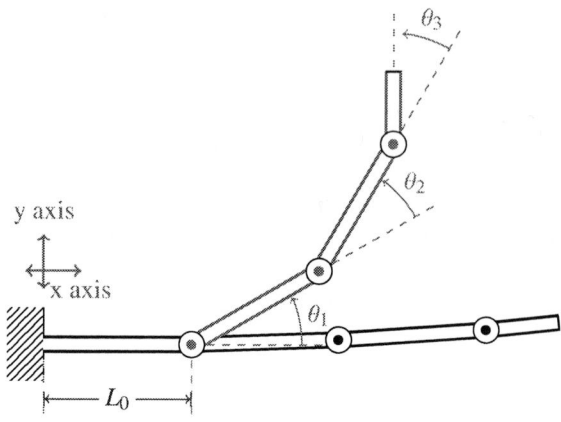

FIGURE 3: A 3R PRBDM at 2 different poses, the grey pose is the position when $\theta = \xi + \zeta\epsilon \times \text{ones}(1,3)$, and the black pose is for $\theta = \xi \times \text{ones}(1,3)$

challenge in this modeling problem is that the joint stiffnesses and eigenvalues depend on position via mass matrix dependence on pose. As a strategy to manage this variation in eigenvalues, we calculate for each design the eigenvalues across a range of poses. We generate this set of poses by sweeping across a range of joint position values θ_i from small angles (ϵ) to $\pi/2$, as depicted in Eqn. (7):

$$\hat{\theta}_i = A_{i,<\cdot>} \quad (7)$$

where $\hat{\theta}_i$ is the vector of all joint angles for pose i, and $A_{i,j}$ is an $m \times n$ matrix where the ith row corresponds to $\hat{\theta}_i$. Specifically:

$$A_{i,<\cdot>} = [\xi + i\epsilon, \dots, \xi + i\epsilon], \quad (8)$$

where ξ is small scalar offset value to prevent singular M (discussed below), and ϵ is small fixed angular value. Here ϵ is chosen such that at the mth pose, the end link world angle q_n is close to (but not exceeding) $\pi/2$. Specifically, in the implementation here, $\epsilon = \pi/2n$. Other definitions of ϵ are possible.

The model given in Eqn. (1) was not tested for any cases where $\theta = 0$, as this would result in a singular mass matrix (M). The pose for the PRBDM for the two boundary cases of $A_{0,<\cdot>}$ and $A_{m,<\cdot>}$ is shown in Fig. 3.

In the most general case, the stiffness of each joint is a continuous function of pose θ. To approximate this relationship in a discrete way, we define m different stiffness values for each joint. This results in $n \times m$ joint stiffness values required to specify a design.

In this study, a reduced-dimension stiffness representation is employed where a single stiffness correction parameter is found that allows us to define a single independent stiffness parameter for each joint, which then maps to m unique stiffness values for each joint for each pose. The rationale for this approach is that the stiffness variation on pose is modeled as a material property.

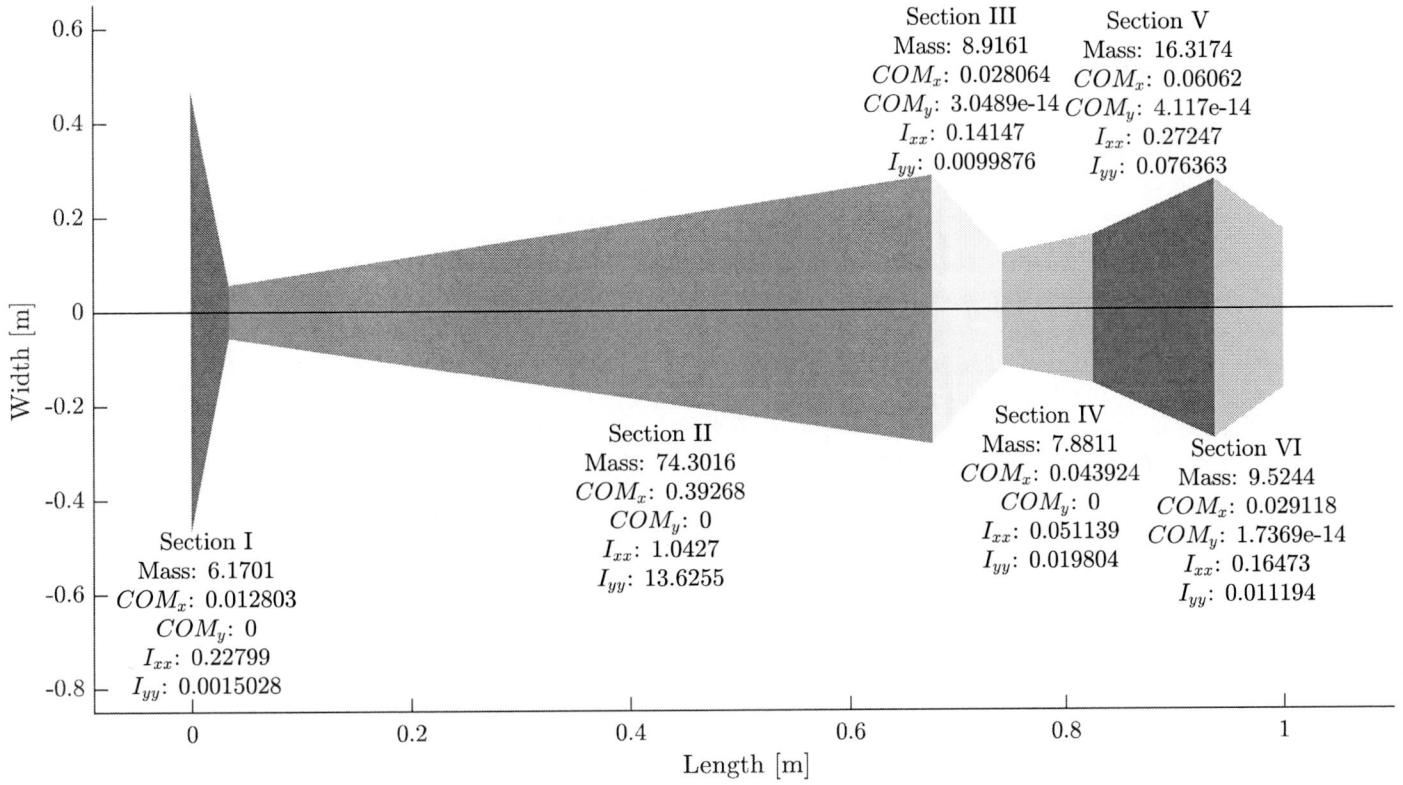

FIGURE 4: Plot of randomly generated beam, with physical properties for each section (link)

More specifically, the stiffness matrix in Eqn. (2) is assumed to be a diagonal matrix, as shown in Eqn. (9):

$$K = \begin{bmatrix} K_{11} & 0 & \dots & 0 \\ 0 & K_{22} & \ddots & \vdots \\ \vdots & \ddots & \ddots & 0 \\ 0 & \dots & 0 & K_{nn} \end{bmatrix}. \qquad (9)$$

To define an initial design for optimization, a unique K is assumed for each unique pose i, of the same form as Eqn. (9). This yields m stiffness matrices; the elements of these m matrices are defined according to Eqn. (10):

$$\hat{K}_i = \begin{bmatrix} K_{11}(i) & 0 & \dots & 0 \\ 0 & K_{22}(i) & \ddots & \vdots \\ \vdots & \ddots & \ddots & 0 \\ 0 & \dots & 0 & K_{nn}(i) \end{bmatrix}. \qquad (10)$$

$\hat{K}_i$ quantifies the stiffness for pose i, where the spring stiffness for each joint j and pose i, $K_{jj}(i)$, is approximated using a linear scaling to reduce the design representation dimension, according to Eqn. (11):

$$K_{jj}(i) = k_i \times \sqrt{\frac{EI_j}{\hat{m}_j}}, \qquad (11)$$

where k_i is defined as a scaling parameter, also referred as stiffness correction factor, that is used as the independent stiffness design variable approximates how joint stiffness depends on pose. E is the material modulus of elasticity, I_j is the area moment of inertia for link j, and $\hat{m}_j$ is the mass of link j. This linear mapping strategy uses m independent design variable values for k_i, along with link mechanical properties, to generate $n \times m$ unique stiffness variables needed to define $\hat{K}_i$, $i = 1, \dots, m$. Eqn. (9) is based on the assumption that the stiffness correction factor is a characteristic of the material and should not depend on the properties unique to each section.

The optimizer reduces the difference between the sorted eigenvalues from both systems (PRBDM and truth values, either FEM or exact) by minimizing the ℓ_2 norm between the sorted eigenvalue vectors, as defined in Eqn. (12):

$$\|\lambda_{error}(\theta_i, \dot{\theta}_i, \psi)\|_2 = \|\lambda_{FEA} - \lambda_{PRBDM}(\theta_i, \dot{\theta}_i, \psi)\|_2. \qquad (12)$$

Here it is assumed that $\dot{\theta}_i = 0$. The optimization problem formulation for eigenvalue matching is defined in Eqn. (13):

$$\min_{L_j, k_i} \sum_{i=1}^{m} \| \lambda_{error}(\theta_i, 0, \psi) \|_2 \tag{13a}$$

$$\text{subject to:} \sum_{j=0}^{n} L_j = 1 \tag{13b}$$

$$L_j \geq 0.025, \tag{13c}$$

$$L_j \leq (1 - 0.025), \tag{13d}$$

where L_j is the length of the jth link (these values are independent of pose). This problem was solved using the MATLAB[1] function fmincon with MultiStart to improve the probability of finding globally-optimal PRBDM parameter values.

The compliant beam model parameter vector ψ is a vector of length $(5 + m)n$. Eqn. (14) details the components of ψ:

$$\psi = [\gamma, k, \hat{m}, J, X_{com}, Y_{com}]^T, \tag{14}$$

where γ is a vector (length n) of normalized link length values:

$$\gamma_i = L_i / L_{total}, \quad L_{total} = \sum_i L_i = 1, \tag{15}$$

k is a vector (length m) of stiffness correction factors for each pose, $\hat{m}$ is a vector (length n) containing the mass of each link (assuming constant cross-section prismatic geometry), J is the vector of rotational mass moments of inertia for each link, and X_{com} and Y_{com} are the center-of-mass locations in the $n-1$th joint frame. Due to the symmetry assumption used here, all values for the vector Y_{com} are zero (at least with respect to numerical tolerances).

Once the number of joints has been chosen for the PRBDM to approximate the compliant member, independent components of ψ can be specified:

$$\psi_{ind} = [\gamma, k]^T. \tag{16}$$

These are the $(n + m)$ independent beam design optimization variables based on the above PRBDM approximation and linear stiffness correction strategy. The next subsection describes how the remaining design parameters in ψ are calculated from ψ_{ind}.

2.4 Design Parameter Calculation

Once ψ_{ind} is specified, the optimization problem given in Eqn. (13) is solved using a multi-start approach with a gradient-based optimization algorithm. The set of start points are generated using a custom strategy, defined in Eqn. (17):

$$L = \text{random}(1, 100, n - 1) \tag{17a}$$

$$\gamma = \left[L, 1 - \sum L \right]^T / 100 \tag{17b}$$

$$k = \text{random}(m) \tag{17c}$$

where $L = [L_1, \ldots, L_{n-1}]$ is the vector of link lengths, random($\cdot$) is a uniform random number generator without replacement, γ is

---

[1]The MATLAB code for these test problems is available in Ref. [17].

Algorithm	fmincon, MultiStart [19–21]
Constraint Tolerance	1e-6
Step Tolerance	1e-6
Max Iterations	10000
Optimality Tolerance	1e-6
Max Function Evaluations	5000

TABLE 1: Optimizer conditions

a normalized length vector based on the random length vector, and k is a randomly generated set of stiffness correction factors (length m). The numerical scaling defined in Eqn. (11) eases numerical solution difficulty due to differing orders of magnitude in the unscaled parameter space. In Eqn. (17a), the random function chooses $n - 1$ unique random integers between 1 and 100. To sample a random uniform distribution under the simplex condition (Eqn. (13b)), the method described in Ref. [18] is used, as seen in Eqn. (17b).

Once the values of γ and k are known, the values of $\hat{m}$, J, X_{com}, and Y_{com} can be estimated using the first and second area moments of the sections of the polygons. During the optimization, for each new link length design γ, the physical properties for each link is calculated, and an updated ψ is obtained. The physical properties for each section is shown in Fig. 4 for a sample design.

3 Test Problems

In this section, two cantilever beams are modeled using the PRBDM defined above. The first example models a uniform rectangular cantilever beam, while the second example corresponds to a randomly-generated beam. The material for the beam is Steel ANSI 4340, available in COMSOL [14].

The PRBDM model for 4-link and 5-link cases are used as candidate models to match the first 4 (and 5) modes, respectively. The optimal stiffness correction factor k_i and joint separation L_j is obtained by solving the optimization problem in Eqn. 13 using the algorithm settings shown in Table 1.

3.1 Test Problem 1: Uniform Rectangular Beam

A uniform rectangular beam is considered for validation of the methods. The test beam used the parameters listed in Table 2.

The PRBDM method is used to find the response of the model and compare the result against the analytical solutions for a uniform beam. The analytical solutions are obtained us-

Length[m]	Width[m]	Thickness[m]
1	0.3624	0.040532

TABLE 2: Uniform beam parameters

ing the Euler Bernoulli beam theory for cantilever plates, based on Eqn. (18), from Ref. [22]. An illustration of a simple beam is shown in Fig 5, and the natural frequency for the nth bending mode is defined as:

$$\omega_n = \alpha_n^2 \sqrt{\frac{EI}{\rho A L^4}} = \alpha_n^2 \sqrt{\frac{E\left(\frac{bh^3}{12}\right)}{\rho(bh)L^4}} \qquad (18)$$

where E is the material modulus of elasticity, I is the beam area moment of inertia about the root, ρ is the beam material density, A is the uniform beam cross-sectional, L is the beam length, and α_n is the coefficient for the nth resonance mode. The value of α_n is obtained by solving Eqn. (19). The first three solutions to this equation are listed in Eqn. (20), where the first value corresponds to the first bending mode, the second value to the second mode, and so on. The first two mode shapes are illustrated in Fig 5, labeled as α_1 and α_2.

$$-1 = \cos(\alpha_n)\cosh(\alpha_n) \qquad (19)$$

$$\alpha_n = [1.875, 4.694, 7.855, \dots] \qquad (20)$$

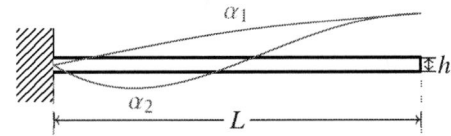

FIGURE 5: Side-view of a uniform beam of length L, thickness h, and width b. The red curves show the first 2 bending modes of the cantilever beam

Maximum element size	0.03 m
Minimum element size	0.001 m
Maximum element growth rate	1.5
Curvature factor	0.3
Resolutions of narrow regions	0.9

TABLE 3: Mesh generation parameters [14].

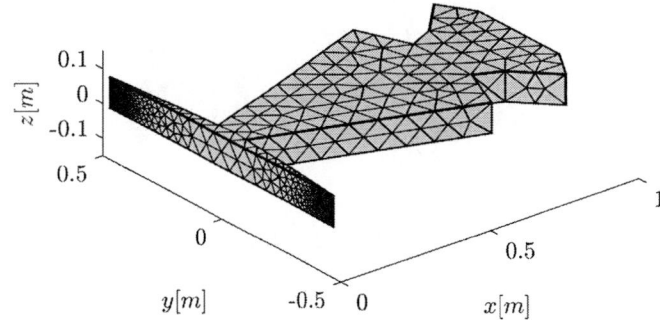

FIGURE 6: Mesh for random panel used for FEM analysis

3.2 Test Problem 2: Random Beam

A random beam is generated using the procedure stated in Sec. 2, for $p = 10$. The randomly generated beam used for the studies presented here is shown in Fig. 2. COMSOL was used to estimate the first 40 natural frequencies for the beam along with the mass participation factors (MPF) for each mode. A 'fine' mesh is used to estimate the resonant frequencies, as shown in Fig. 6, according to the parameters mentioned in Table 3 . The coarse mesh supports computationally-efficient estimation of the resonant frequencies. The resonant frequencies are then filtered according to the MPF to obtain the modes that exhibit bending along the desired axis; Fig. 13 shows the top 12 filtered modes. The vector of 5 eigenvalues obtained for the panel from Fig. 2, using the mesh shown in Fig. 6, is presented in Eqn. (21).

$$\lambda_{panel} = \begin{bmatrix} 55.037, 414.59, 1112.2, 2218.9, 3641.5 \end{bmatrix}^{\mathsf{T}} \qquad (21)$$

4 Results and Discussions

In this section, the results for the two test cases are discussed. The first case maps the resonant values for a uniform cantilever beam to a 3R-PRBDM and 4R-PRBDM, while mapping the first three eigenvalues. The second case discusses one of the randomly generated random beams. The PRBDM matching code, which is available at Ref. [17], was solved for 50,000 different model beams; the beam discussed here is beam number 17,572[2]. The beam was modeled using a 4R-PRBDM and a 5R-PRBDM for different number of eigenvalues.

4.1 Uniform beam

This case uses the MATLAB fmincon function with MultiStart. The problem was solved based on the k_i dimension-reduction strategy discussed above. One of the initial points used for the optimizer was based on uniform space, as

---

[2]The remaining data will be available upon final publication via an archival data repository.

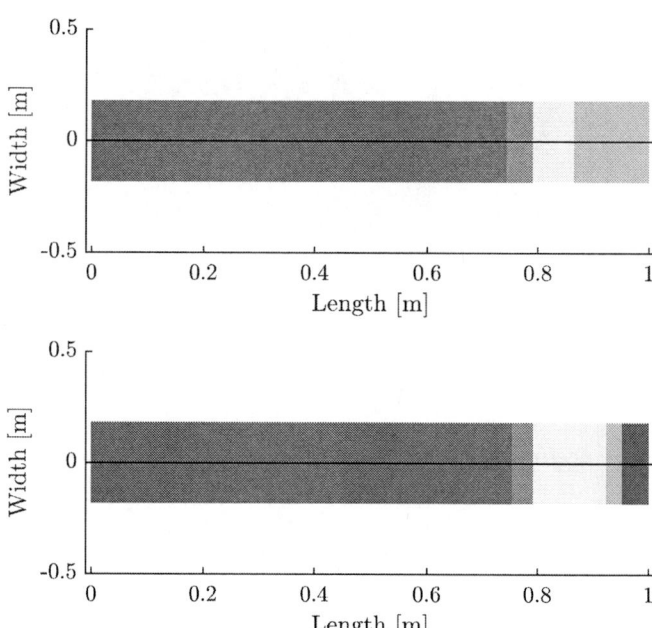

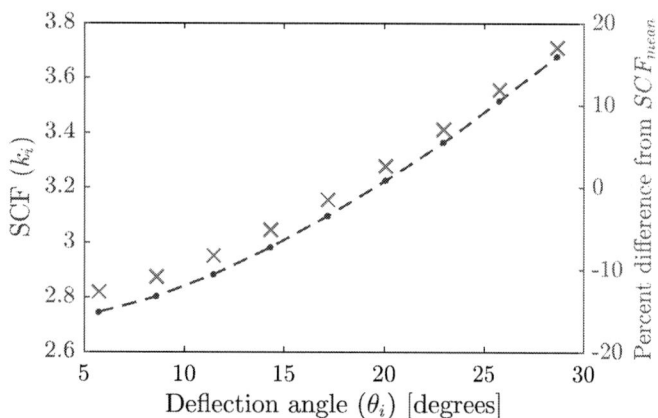

FIGURE 7: Plot of uniform beam, with optimal joint separation. Shaded regions indicate distinct rigid links.

FIGURE 8: Plot of the spring Stiffness Correction Factor (SCF) vs. deflection angle (in degrees), for the 4R-PRBDM uniform beam. The right-hand axis corresponds to the relative change from mean values (Eqn. 24).

defined in Eqn. (22):

$$\gamma_i = \frac{L}{n}, \tag{22a}$$

$$k_i = 1. \tag{22b}$$

The results for the 3R-PRBDM and 4R-PRBDM can be seen in Fig. 7. An interesting observation here is that the length fraction

Joints	Frequencies mapped	Stiffness correction factor	Joint distance[m]	Loss [%]
3	3	$\begin{bmatrix} 2.7441 \\ 2.8032 \\ 2.8830 \\ 2.9815 \\ 3.0960 \\ 3.2246 \\ 3.3652 \\ 3.5164 \\ 3.6776 \end{bmatrix}$	$\begin{bmatrix} 0.7447 \\ 0.0466 \\ 0.0725 \\ 0.1362 \end{bmatrix}$	0.0168
4	3	$\begin{bmatrix} 3.2301 \\ 3.2812 \\ 3.3504 \\ 3.4356 \\ 3.5349 \\ 3.6462 \end{bmatrix}$	$\begin{bmatrix} 0.7555 \\ 0.0369 \\ 0.1323 \\ 0.0281 \\ 0.0472 \end{bmatrix}$	0.0168

TABLE 4: Results for uniform beam

for the first joint γ_1 is relatively similar for both cases. This is similar to the "characteristic length" for a 1R-PRDBM described in [9].

The variation of the spring Stiffness Correction Factor (SCF) with respect to different poses for the 3R-PRBDM case can be seen in Fig. 8. The right-hand side y-axis shows the percent difference between the SCF for the given joint angles to the mean SCF. The mean SCF is defined by Eqn. 23. The variation of the spring stiffness is small, but shows a close to linear trend.

$$\text{SCF}_{\text{mean}} = \frac{1}{m} \sum_{i=0}^{m} k_i \tag{23}$$

$$P_i = \frac{k_i - \text{SCF}_{\text{mean}}}{\text{SCF}_{\text{mean}}} 100 \tag{24}$$

Both the 3R-PRDBM and 4R-PRBDM models were used to map the first three resonant modes of the beam to the PRBDM. The details of the solution obtained for both cases can be seen in Table. 4. The "Loss" column shoes the value of the objective function Eqn. 12.

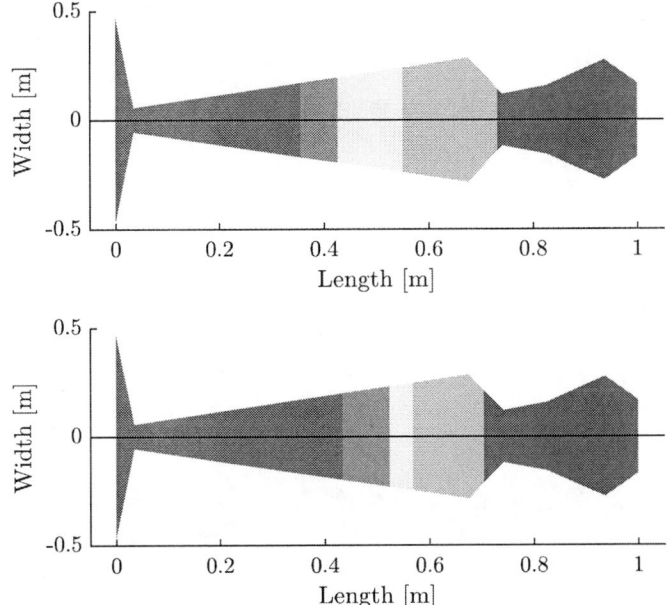

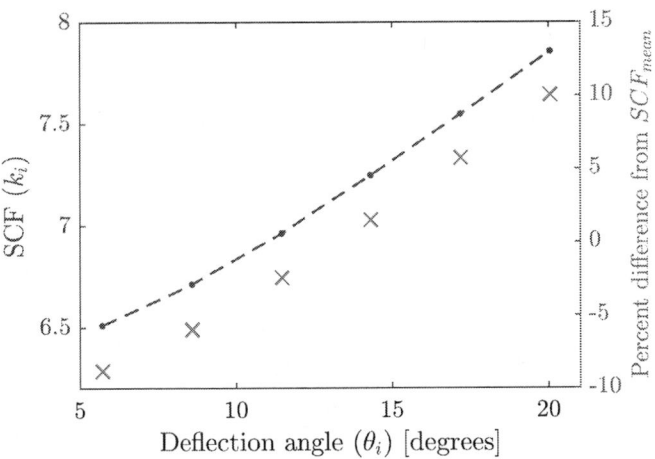

FIGURE 10: Plot of the spring Stiffness Correction Factor (SCF) vs. deflection angle (in degrees), for the 4R-PRBDM random beam case. The right-hand axis corresponds to the relative change from mean values (Eqn. 24).

FIGURE 9: Plot of randomly general beam, with physical properties for each section optimized for eigenvalue matching (4R-PRBDM)

4.2 Random Beam

The random beam was solved for both the 4R-PRBDM and the 5R-PRBDM cases. Both cases were solved to map the first three eigenfrequencies as well as the maximum number of eigenfrequencies that the models allow. The results for the random beam are thus divided into two sections, one for the 4R-PRBDM, and the latter for the 5R-PRBDM.

4.2.1 4R-PRBDM Approximation
The optimization problem to solve for the joint distribution and spring stiffness was initialized using 1,024 unique initial points, and solved using the MultiStart algorithm in MATLAB. One of the initial beam designs is illustrated in Fig 4. The ϵ value chosen for this study was 0.1, and ξ is 0.05, for Eqn. (7).

The optimizer solves for both the link lengths and SCFs simultaneously to best match the eigenvalues. For the 4R-PRBDM model, the optimizer converged to the solution shown in Fig 9. The obtained PRBDM model has eigenvalues which differ by 0.12% of the eigenfrequencies obtained using COMSOL. Similar to the uniform beam case, the length fraction for both cases of the 4R-PRBDM (γ_1) are indicative of a "characteristic pivot" for this non-uniform and higher-dimensional (4R-PRBDM) case.

The SCF for each deflection case is shown in Fig 10. A linear trend is observed for the correction factor with respect to the angle of deflection between each link. A summary of solution parameters can is provided in Table 5.

When the SCF is applied according to Eqn. (11), the spring stiffness is obtained for different deflection angles. The spring stiffness values for each deflection case can be seen in Eqn. (25). The spring stiffness values for all tests performed for the random panel can be seen in Appendix (Sec. 7).

$$K_{4R}^{q1} = 10^5 \times \begin{bmatrix} 1.2089 & 2.1401 & 2.9362 & 4.8353 \end{bmatrix} \tag{25a}$$

$$K_{4R}^{q2} = 10^5 \times \begin{bmatrix} 1.2462 & 2.2062 & 3.0269 & 4.9846 \end{bmatrix} \tag{25b}$$

$$K_{4R}^{q3} = 10^5 \times \begin{bmatrix} 1.2930 & 2.2890 & 3.1405 & 5.1717 \end{bmatrix} \tag{25c}$$

$$K_{4R}^{q4} = 10^5 \times \begin{bmatrix} 1.3456 & 2.3822 & 3.2684 & 5.3823 \end{bmatrix} \tag{25d}$$

$$K_{4R}^{q5} = 10^5 \times \begin{bmatrix} 1.4015 & 2.4810 & 3.4040 & 5.6057 \end{bmatrix} \tag{25e}$$

$$K_{4R}^{q6} = 10^5 \times \begin{bmatrix} 1.4587 & 2.5824 & 3.5431 & 5.8346 \end{bmatrix} \tag{25f}$$

4.2.2 5R-PRBDM Approximation
For the 5R-PRBDM model, the optimizer converged to the solution shown in Fig. 11. The obtained PRBDM model has eigenvalues which differ by 0.2277% of the eigenfrequencies obtained using COMSOL. This is higher than the value for a 4R-PRBDM model, but it must be noted that the 5R-PRBDM model maps 5 eigenfrequencies instead of 4. When 4 eigenfrequencies are mapped using a 5R-PRBDM the accuracy of the model is within 0.012%. The SCF for each deflection case is shown in Fig 12. Although a similar linear trend is observed for the correction factor with respect to the angle of deflection between each link, as with the 4R case, the variation is significantly smaller. A summary of solution parameters can be seen in Table 6. The unscaled spring stiffness for mapping the first 5 eigenfrequencies

can be seen in Eqn. (26).

$$K_{5R}^{q1} = 10^4 \times \begin{bmatrix} 9.3878 & 3.8301 & 2.4465 & 2.2160 & 1.2217 \end{bmatrix} \quad (26a)$$

$$K_{5R}^{q2} = 10^4 \times \begin{bmatrix} 9.4145 & 3.8410 & 2.4535 & 2.2223 & 1.2252 \end{bmatrix} \quad (26b)$$

$$K_{5R}^{q3} = 10^4 \times \begin{bmatrix} 9.4515 & 3.8561 & 2.4631 & 2.2310 & 1.2300 \end{bmatrix} \quad (26c)$$

$$K_{5R}^{q4} = 10^4 \times \begin{bmatrix} 9.4988 & 3.8754 & 2.4754 & 2.2422 & 1.2361 \end{bmatrix} \quad (26d)$$

$$K_{5R}^{q5} = 10^4 \times \begin{bmatrix} 9.5560 & 3.8988 & 2.4904 & 2.2557 & 1.2436 \end{bmatrix} \quad (26e)$$

5 Conclusion and Future Work

In this article, a strategy to approximate the dynamic behavior of compliant members using a reduced-order model. These models are particularly useful for design optimization, as computational expense can be reduced significantly while maintaining reasonable accuracy. Quantification of this tradeoff with respect to design solution accuracy is a topic for future work. Two test cases are presented, but the method has been validated for a large variety of beams. These results and the associated code are available at Ref. [17]. Some preliminary correlation of the spring stiffness is seen with the deflection angles, and the stiffness scaling was sufficient for a linear correlation between the stiffness correction factor and deflection angles. Some evidence of characteristic pivot distances has also been observed for both the uniform beam and the random beam cases. The correlation

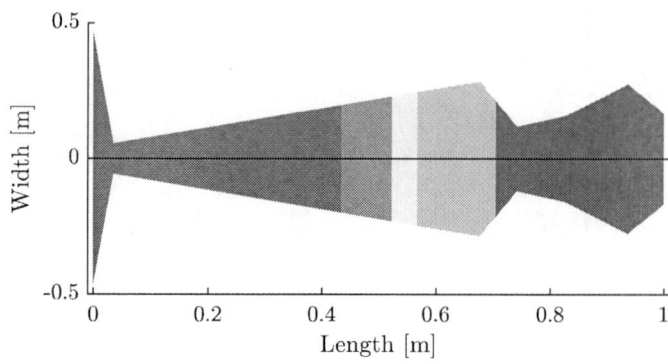

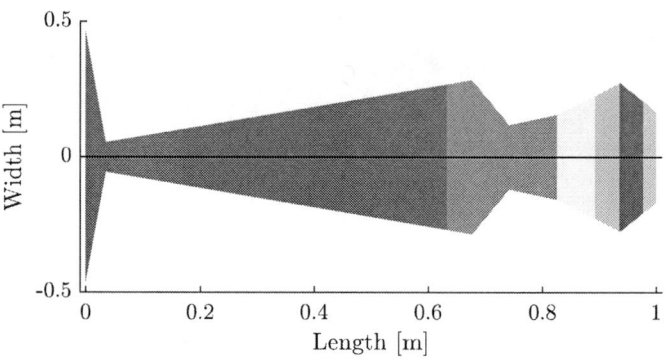

FIGURE 11: Plot of randomly general beam, with physical properties for each optimal section (5R-PRBDM)

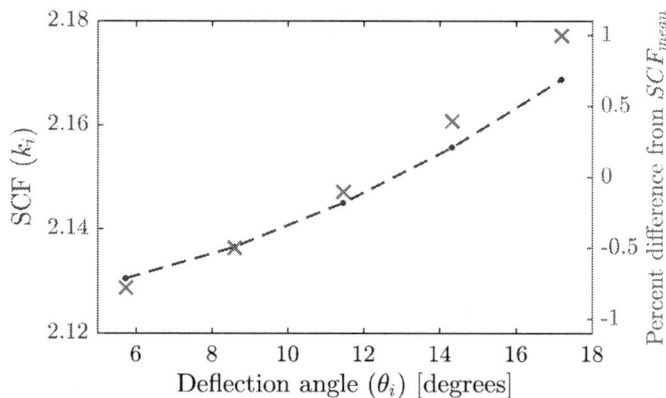

FIGURE 12: Plot of Spring Stiffness Correction Factor (SCF) vs. deflection angle (in degrees), for the 5R-PRBDM random beam case

of the joint distances with the beam shape needs further investigation to show dependence.

More extensive studies that optimize the reduced-order PRBDM models for an enumerated distribution of joint separation could help generate empirically-driven analytic rules to de-

Frequencies mapped	Stiffness correction factor	Join distance[m]	Loss [%]
4	$\begin{bmatrix} 6.5099 \\ 6.7109 \\ 6.9627 \\ 7.2464 \\ 7.5471 \\ 7.8553 \end{bmatrix}$	$\begin{bmatrix} 0.3553 \\ 0.0707 \\ 0.1242 \\ 0.1819 \\ 0.2678 \end{bmatrix}$	0.1001
3	$\begin{bmatrix} 5.0093 \\ 5.2555 \\ 5.5703 \\ 5.9349 \\ 6.3344 \\ 6.7601 \end{bmatrix}$	$\begin{bmatrix} 0.4345 \\ 0.0890 \\ 0.0447 \\ 0.1366 \\ 0.2952 \end{bmatrix}$	0.0536

TABLE 5: Results for 4R-PRBDM, random beam

Frequencies mapped	Stiffness correction factor	Join distance[m]	Loss [%]
5	$\begin{bmatrix} 2.1305 \\ 2.1365 \\ 2.1450 \\ 2.1557 \\ 2.1687 \end{bmatrix}$	$\begin{bmatrix} 0.6323 \\ 0.1930 \\ 0.0667 \\ 0.0435 \\ 0.0418 \\ 0.0228 \end{bmatrix}$	0.2277
3	$\begin{bmatrix} 3.3366 \\ 3.4041 \\ 3.4950 \\ 3.6062 \\ 3.7352 \end{bmatrix}$	$\begin{bmatrix} 0.6632 \\ 0.0165 \\ 0.0844 \\ 0.1595 \\ 0.0093 \\ 0.0671 \end{bmatrix}$	0.0119

TABLE 6: Results for 5R-PRBDM for random beam

termine link lengths and spring stiffness factors. If validated, this would eliminate the need to solve an optimization problem based on truth data, speeding up the implementation of effective PRBDMs for design studies. An efficient method for estimating the resonant frequencies could be obtained using a machine learning framework. Additionally, a reinforcement learning method could be used to train a neural network to provide the joint separation and spring stiffness for PRBDMs across a range of designs.

6 Acknowledgements

This material is based upon work partially supported by the National Science Foundation under Grant No. CMMI-1653118 and partially through the NASA SBIR in collaboration with CU aerospace Contract No. NNX17CA25P.

The authors would like to thank Daniel R. Herber for his guidance with this study.

References

[1] Howell, L. L., 2001. *Compliant mechanisms*. John Wiley & Sons.

[2] Jensen, B. D., and Howell, L. L., 2003. "Identification of compliant pseudo-rigid-body four-link mechanism configurations resulting in bistable behavior". *Journal of Mechanical Design*, **125**(4), pp. 701–708.

[3] Chen, G., Wilcox, D. L., and Howell, L. L., 2009. "Fully compliant double tensural tristable micromechanisms (dttm)". *Journal of Micromechanics and Microengineering*, **19**(2), p. 025011.

[4] Lyon, S., Erickson, P., Evans, M., and Howell, L., 1999. "Prediction of the first modal frequency of compliant mechanisms using the pseudo-rigid-body model". *Journal of Mechanical Design*, **121**(2), pp. 309–313.

[5] Yu, Y.-Q., Howell, L. L., Lusk, C., Yue, Y., and He, M.-G., 2005. "Dynamic modeling of compliant mechanisms based on the pseudo-rigid-body model". *Journal of Mechanical Design*, **127**(4), pp. 760–765.

[6] Midha, A., Howell, L. L., and Norton, T. W., 2000. "Limit positions of compliant mechanisms using the pseudo-rigid-body model concept". *Mechanism and Machine Theory*, **35**(1), pp. 99–115.

[7] Kimball, C., and Tsai, L.-W., 2002. "Modeling of flexural beams subjected to arbitrary end loads". *Transactions-american Society of Mechanical Engineers Journal of Mechanical Design*, **124**(2), pp. 223–235.

[8] Su, H.-J., 2009. "A pseudorigid-body 3R model for determining large deflection of cantilever beams subject to tip loads". *Journal of Mechanisms and Robotics*, **1**(2), p. 021008.

[9] Chen, G., Xiong, B., and Huang, X., 2011. "Finding the optimal characteristic parameters for 3r pseudo-rigid-body model using an improved particle swarm optimizer". *Precision Engineering*, **35**(3), pp. 505–511.

[10] Chudnovsky, V., Mukherjee, A., Wendlandt, J., and Kennedy, D., 2006. "Modeling flexible bodies in simmechanics". *MatLab Digest*, **14**(3).

[11] Fathy, H. K., Reyer, J. A., Papalambros, P. Y., and Ulsov, A., 2001. "On the coupling between the plant and controller optimization problems". In Proceedings of the 2001 American Control Conference.(Cat. No. 01CH37148), Vol. 3, IEEE, pp. 1864–1869.

[12] Allison, J. T., and Herber, D. R., 2014. "Multidisciplinary design optimization of dynamic engineering systems". *AIAA Journal*, **52**(4), Apr., pp. 691–710.

[13] Herber, D. R., and Allison, J. T., 2019. "Nested and simultaneous solution strategies for general combined plant and control design problems". *ASME Journal of Mechanical Design*, **141**(1), Jan., p. 011402.

[14] AB, C., 1998. Comsol multiphysics v. 5.3. [Online]. Available: www.comsol.com.

[15] Herrera-May, A. L., Aguilera-Cortés, L. A., Plascencia-Mora, H., Rodríguez-Morales, Á. L., and Lu, J., 2011. "Analytical modeling for the bending resonant frequency of multilayered microresonators with variable cross-section". *Sensors*, **11**(9), pp. 8203–8226.

[16] Chilan, C. M., Herber, D. R., Nakka, Y. K., Chung, S.-J., Allison, J. T., Aldrich, J. B., and Alvarez-Salazar, O. S., 2017. "Co-design of strain-actuated solar arrays for spacecraft precision pointing and jitter reduction". *AIAA Journal*, **55**(9), Sept., pp. 3180–3195.

[17] Vedant. PRBDM code repository. [Online]. Available: https://github.com/VedantFNO/PRDBM_IDETC2019_97881.

[18] Smith, N. A., and Tromble, R. W., 2004. "Sampling uniformly from the unit simplex". *Johns Hopkins University, Tech. Rep*, **29**.

[19] MultiStart. [Online]. Available: https://www.mathworks.com/help/gads/multistart.html.

[20] Ugray, Z., Lasdon, L., Plummer, J., Glover, F., Kelly, J., and Martí, R., 2007. "Scatter search and local nlp solvers: A multistart framework for global optimization". *INFORMS Journal on Computing*, **19**(3), pp. 328–340.

[21] Glover, F., 1998. "A template for scatter search and path relinking". *Lecture notes in computer science*, **1363**, pp. 13–54.

[22] Plunkett, R., 1963. "Natural frequencies of uniform and non-uniform rectangular cantilever plates". *Journal of Mechanical Engineering Science*, **5**(2), pp. 146–156.

7 Appendix

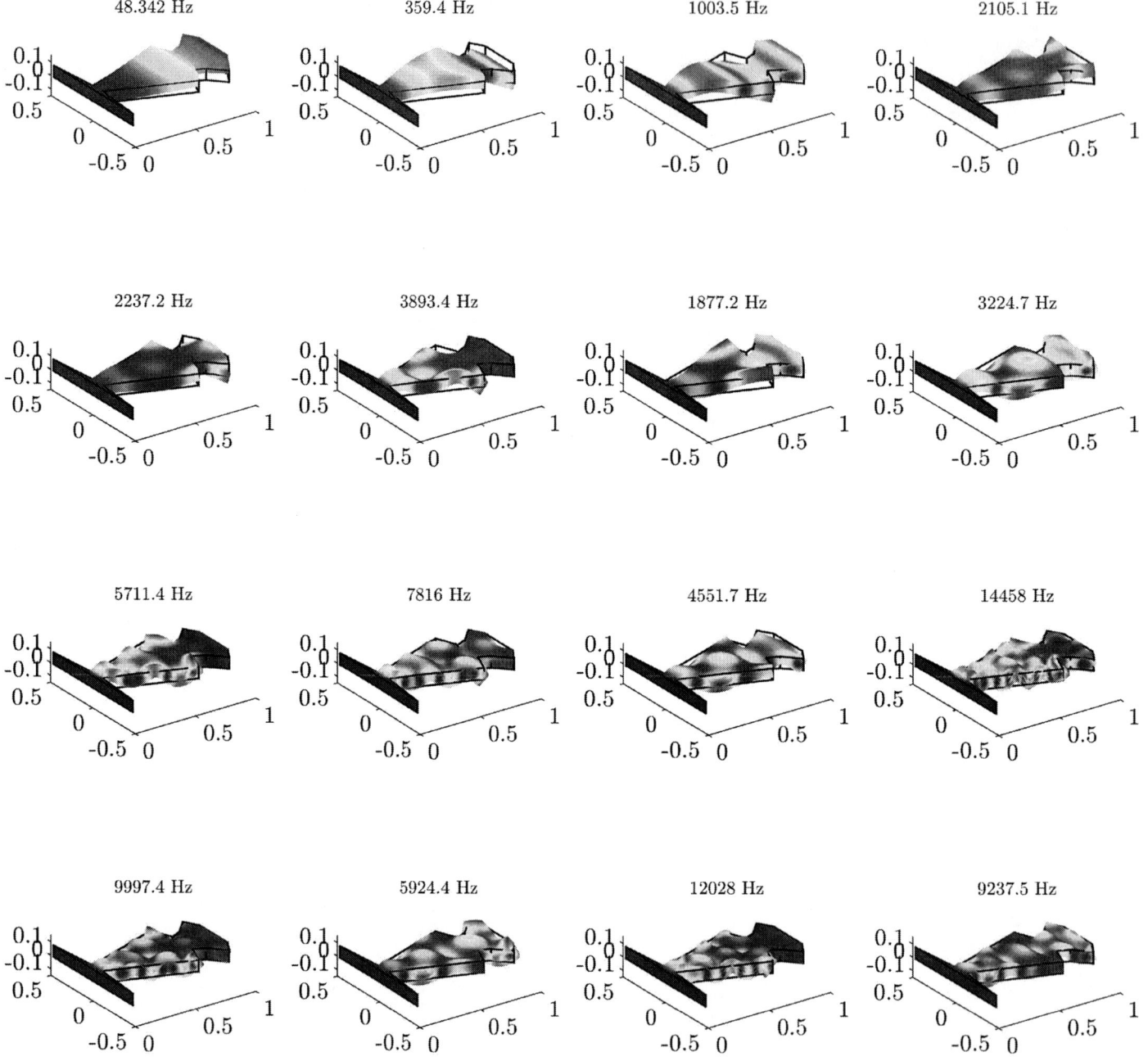

FIGURE 13: Plot of bending modes for 16 filtered eigenvalues, the title for each plot shows the resonant frequency, all axis lengths are in meters

Proceedings of the ASME 2019
International Design Engineering Technical Conferences
and Computers and Information in Engineering Conference
IDETC/CIE2019
August 18-21, 2019, Anaheim, CA, USA

DETC2019-98000

CHECKING THE AUTOMATED CONSTRUCTION OF FINITE ELEMENT SIMULATIONS FROM DIRICHLET BOUNDARY CONDITIONS

Kevin N. Chiu
Dept. of Mechanical Engineering
University of Maryland
College Park, Maryland

Mark D. Fuge
Dept. of Mechanical Engineering
University of Maryland
College Park, Maryland

ABSTRACT

From engineering analysis and topology optimization to generative design and machine learning, many modern computational design approaches require either large amounts of data or a method to generate that data. This paper addresses key issues with automatically generating such data through automating the construction of Finite Element Method (FEM) simulations from Dirichlet boundary conditions. Most past work on automating FEM assumes prior knowledge of the physics to be run or is limited to a small number of governing equations.

In contrast, we propose three improvements to current methods of automating the FEM: (1) completeness labels that guarantee viability of a simulation under specific conditions, (2) type-based labels for solution fields that robustly generate and identify solution fields, and (3) type-based labels for variational forms of governing equations that map the three components of a simulation set—specifically, boundary conditions, solution fields, and a variational form—to each other to form a viable FEM simulation. We implement these improvements using the FEniCS library as an example case. We show that our improvements increase the percent of viable simulations that are run automatically from a given list of boundary conditions. This paper's procedures ultimately allow for the automatic—i.e., fully computer-controlled—construction of FEM multi-physics simulations and data collection required to run data-driven models of physics phenomena or automate the exploration of topology optimization under many physics.

1 INTRODUCTION

Modern advances in computation provide pathways for data-intensive processes in mechanical design. For example, current state of the art approaches to engineering analysis, topological optimization, generative design, and machine learning can benefit from large amounts of training data or methods to generate that data. Generating this data manually—*e.g.*, by having an employee manually run various test cases—is infeasible due to the large quantity needed, so a method to automate the generation of such data is crucial to these data-driven methods. To acquire this data, we wish to automate the simulation of physical phenomena using the Finite Element Method (FEM) to solve the partial differential equations (PDEs) that govern many physical phenomena. From these simulations, we can calculate, *e.g.*, forces acting on objects from fluid stress fields or differences in electrostatic potential. To build these simulations manually is time-consuming and impractical for the large number of simulations needed in machine learning model training, so this paper proposes a method to check that an automatically-generated simulation will run given Dirichlet boundary conditions that satisfy certain properties.

As an illustrative example, consider a fluid flowing through a pipe in the presence of an electrostatic field (*e.g.*, as is common in biological microfluidic devices) shown in Fig. 1. The fluid in the pipe is governed by the Navier-Stokes equations, whereas the electrostatic field is governed by the Poisson equation. If the fluid and electric field are not coupled, then two simulations can be run separately. However, if they are coupled, then both

Copyright © 2019 ASME

fluid and electrostatic boundary conditions are needed to run a fully coupled simulation; if not enough boundary conditions are provided, the coupled simulation cannot run, but some subset of uncoupled simulations may. If an algorithm wanted to automatically construct each of the possible subsets of simulations in this example, it would need to overcome several crucial problems:

1. Choosing a variational equation that models the problem, defining appropriate solution fields to solve the simulation, and applying the correct boundary conditions to ensure convergence of the solution.
2. Guaranteeing the viability of a simulation given the boundary conditions, solution fields, and variational form.
3. Ensuring robustness of simulations to permutations in variational equations and boundary conditions—*i.e.*, so long as there is sufficient information to construct the simulation, the order of that information should not matter.

Specifically, the key contributions of this paper are:

1. We discuss properties of simulation sets necessary to run FEM simulations, differentiating them into what we call *complete*, *partially complete*, and *incomplete* simulations. Under certain conditions, these labels guarantee simulation viability.
2. We propose the use of type-based labeling to generate appropriate solution fields from Dirichlet boundary conditions without knowing ahead of time the types (scalar, vector, or tensor) or dimensions of the boundary conditions. This contrasts previous implementations where we have prior knowledge of the boundary condition types and manually build solution fields based on that knowledge.
3. We apply type-based labeling in variational forms to improve robustness of simulations to ordering permutations of solution fields, boundary conditions, and variational form terms. This gives us the ability to construct simulations more robustly than previous implementations in which, *e.g.*, the order of boundary conditions affects whether a simulation runs or not.

To demonstrate our above contributions, we specifically use the FEniCS library [1] to implement the FEM; however, these contributions are applicable to other implementations of the FEM as well.

2   RELATED WORK

Past approaches to automating the construction of FEM models falls roughly into three camps: (1) Automating Mesh Generation, (2) Knowledge-Based Engineering, and (3) Languages for general purpose, multi-physics solvers.

Much of the previous work in automating FEM focuses on the mesh generation process. For example, [2] and [3] describe basic methods of automating the meshing process, leading to a variety of applications, e.g., from bio-medical [4] to environmental [5]. Such approaches have led to major advances in how to discretize FEM problems, assuming that the variational form of the governing equation and boundary conditions are known. In contrast, our paper focuses on how to assemble all parts of the FEM solution including the solution fields (for a given mesh), the variational form, and the boundary conditions. As such, this past work in automated mesh generation is complementary to this paper, and we assume an appropriate quality mesh is provided (or can be computed using past methods) because our concerns lie with guaranteeing simulation viability and setting up the FEM problem definition.

Above the mesh-level, approaches from knowledge-based engineering (KBE) abstract the mid- and high-level components of a design by capturing rules and hierarchies between levels of different components [6]. By implementing high-level primitives, such as airplane wings or fuselages, KBE allows a designer to consider more detailed improvements by providing a shortcut to methods of analysis early in the design process [7]. When applied to geometry as in [8], KBE allows a simplified method of generating designs by adjusting input parameters to the program. Additionally, KBE can be extended to multi-disciplinary optimization [9] and, with machine learning, to predicting, *e.g.*, manufacturing cycle times [10]. While KBE promotes the use of high-level abstractions in design, such abstractions are largely manually constructed and cannot automatically generate new multi-physics FEM simulations needed for generating data or optimizations in data-driven design.

The last area of related work lies in programmatic languages for writing and simulating PDEs using FEM. Specifically, our work extends the work of [11] and [12], which laid the framework for the FEniCS library [1]—a programmatic language for writing and simulating FEM solutions to PDEs, such as the Unified Form Language [13]. [14], a similar package, also provides a language for implementing complex FEM simulations, acting either as a black-box solver or a framework for custom implementations. While such approaches have helped standardize how one can describe and implement various types of single and multi-physics analysis (*e.g.*, [15]) they still require researchers to write customized code for implementing specific physics, particularly in coupled models. The closest work to this paper is the work of Xia [16], who implements a multi-physics solver that can generate FEniCS code automatically by assuming that boundary conditions exist for every portion of the domain. Our approach differs from [16] in that we make no assumptions regarding the types or number of boundary conditions given, and can automatically generate and check for multiple subsets of physics models that are consistent with the boundary conditions.

3 EXPERIMENTAL SETUP

In this paper, we assume that only *physically-realizable* boundary conditions are given as inputs. For example, fluid flowing through a pipe can have one inlet velocity Boundary Condition (*BC*) with the other end as a pressure *BC*.[1]

For clarity throughout the rest of this paper, we use our previous example of a fluid flowing through a pipe in an electrostatic potential field to demonstrate the core contributions. This example provides us with two different types of physics, namely fluid flows and electrostatic potential fields, governed individually by the Navier-Stokes and Poisson equations, respectively, and by a coupled form in [17]. For the purposes of this paper, we do not consider the fully-coupled form of the equations; instead, we choose a "synthetic" variational form that requires the same types of *BCs* and solution fields as the fully-coupled form to make computation simpler and faster to replicate.

Specifically, we use the following three variational forms:

1. Quasi-static Incompressible Navier-Stokes (Pressure, Velocity)

$$F = \langle u, v \rangle + \langle \nabla u, \nabla v \rangle + \langle \nabla p, v \rangle + \langle \nabla \cdot u, q \rangle \qquad (1)$$

2. Poisson

$$F = \langle \nabla u, \nabla v \rangle - f \cdot v - g \cdot v \qquad (2)$$

3. "Synthetic" (Pressure, Velocity, Poisson)

$$F = \langle u, v \rangle + \langle p, q \rangle + \langle s, t \rangle + f \cdot t + g \cdot t \qquad (3)$$

In the Navier-Stokes equations, u and p are the velocity and pressure, respectively, with v and q the respective test functions. In the Poisson equation, u is the variable of interest with v as the test function. In the Synthetic variational form, u, p, and s are velocity, pressure, and Poisson terms, respectively, with v, q, and t the respective test functions. In both the Poisson and Synthetic variational forms, f and g are internal and boundary source terms, respectively. dx signifies integration over the domain's interior, while ds signifies integration over the boundary.

While the Navier-Stokes and Poisson equations are fairly standard, the "Synthetic" variational form we use is constructed

for the purposes of this paper. This form has no physical meaning; rather, it is used merely as a test equation that requires multiple and different *BCs* and combines a pressure field, a velocity field, and a Poisson field. The only criteria of the Synthetic variational form are that it requires different *BCs* (and subsequently solution fields) and that it converges to a solution, regardless of the physical interpretation of that solution.

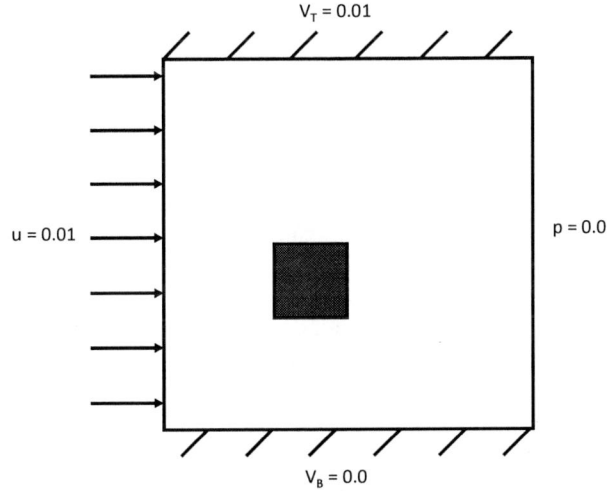

FIGURE 1. Example problem of fluid in a pipe in an electrostatic field. All boundary conditions are non-dimensionalized Dirichlet boundary conditions. The black square is an "obstacle" on which we want to calculate the forces from the fluid flow and the electrostatic potential field.

We also use three types of Dirichlet boundary conditions: fluid pressure, fluid velocity, and Poisson. We separate them into the following sets (denoted $\{BCs\}$) for our experiments:

0. {}
1. {Velocity L}
2. {Pressure R}
3. {Poisson T}
4. {Poisson T, Poisson B}
5. {Poisson T, Pressure R}
6. {Velocity L, Pressure R, Poisson T}
7. {Velocity T, Velocity L, Poisson B}
8. {Velocity B, Velocity T, Velocity L, Pressure R}
9. {Velocity L, Pressure R, Poisson T, Poisson B}
10. {Velocity B, Velocity T, Velocity L, Pressure R, Poisson T}
11. {Velocity B, Velocity T, Velocity L, Poisson T, Poisson B}
12. {Velocity B, Velocity T, Velocity L, Pressure R, Poisson T, Poisson B}

---

[1] In contrast, if both ends mandate velocities leaving the domain with no inputs, then fluid must be generated somewhere within the pipe to conserve mass (*i.e.*, maintain a divergence of 0). If there is no source of fluid, then the simulation will break due to non-physicality, not necessarily because the FEM solution to the PDE is incorrect. (That is, solutions might be mathematically possible but not physically possible.) Assuring that a given set of *BCs* is physically-realizable, while important, is not within the scope of this paper.

Here, "B", "T", "L", and "R" represent "Bottom," "Top," "Left," and "Right" respectively, referring to the edge on the example domain on which the *BC* is applied. As seen in Figure 1, Velocity L is a small, positive, uniform, horizontal flow, while Velocity B and Velocity T are both 0-magnitude flows (*i.e.*, the "no-slip" condition). Pressure R is 0 pressure. Poisson T and Poisson B are uniform positive and zero values, respectively.

These sets of *BC*s are heuristically chosen by the authors to maximize diversity in the following traits in the aforementioned sample problem:

1. Number of boundary conditions —how many boundary conditions are applied
2. Types of boundary conditions —fluid velocity, fluid pressure, and Poisson
3. Combinations of types —*e.g.*, fluid velocity with Poisson, fluid pressure by itself, *etc.*

Although many other possible test problems are possible, we combinatorically choose among these sets to simplify the method of testing our approaches. While a more exhaustive test set would include all 2^6 possible combinations of 6 *BC*s, many of these tests would be redundant for testing our approach (*e.g.*, two sets of *BC*s where the only difference between them is removing one no-slip condition and adding in the other).

This paper refers only to simulations with Dirichlet boundary conditions. While Neumann and Robin boundary conditions are important, they are not considered in this work due to the differences in their implementation in FEM, specifically because they are added as terms in the variational form. In this work, we assume the variational form is already given. It is expected that this paper is compatible with Neumann and Robin boundary conditions with minimal conceptual changes, though the specific implementation may differ if there are no Dirichlet boundary conditions acting on the same field(s).

Our experiment in §4.3 explores the accuracy with which we can automatically label a simulation set given the Variational Form of the governing equation (*VF*) and set of *BC*s while assuming the associated Solution Field (*SF*) is correct. We use the same sets of *BC*s in §5.3, testing different methods of generating these *SF*s. With the now fully-defined and complete simulation sets, we use the *VF*s from above and the same sets of *BC*s to build and run viable simulations to test the implementation of our approach compared to standard methods in §6.4.

4 COMPLETENESS OF VIABLE SIMULATION SETS

Each simulation requires what we call a *simulation set* of $\{BCs\}$, *SF*s, and a *VF*. To solve a PDE, boundary conditions are applied on *SF*s, whose values are manipulated to satisfy the *VF*. Because of these interwoven relations, a viable simulation has a simulation set with two important properties between its different members.

First, we define a *viable* simulation set as a simulation set whose members fully define a simulation in FEniCS (or any other FEM package) that: (1) runs to completion (*e.g.*, no shape mismatches or undefined terms) and (2) converges to a solution (*e.g.*, not a singular matrix, no infinite or NAN values).

In this section, we assume that, for a given *VF*, the *SF*s in the simulation set match the *VF* exactly; we discuss generating these *SF*s in §5. We define a simulation set with at least one *BC* for every variable type in the *VF* without extra as *complete*; if there are unused/extra *BC*s, that simulation set is called *partially complete*, and extra *BC*s can be trimmed off to make a complete simulation set. In both of these cases, the simulation set can define a viable simulation.

If the *VF* contains variables whose types do not have corresponding *BC*s, then that *VF* is considered *incomplete*, leading to a nonviable simulation set.

With this definition , we can decide whether a simulation set is viable or not by making several observations about its members and their relationships to each other.

4.1 Claim 1: A viable simulation must have a bijective function between *SF* and *VF*.

Define $S_f(v_{type})$ as a function that maps a type of variable field in a *VF* to its corresponding *SF*. To prove that $f = S_f(v_{type})$ is bijective, we want to show that f is both surjective and injective.

4.1.1 Claim 1.1: f is surjective.
Assume f is not surjective. According to the definition of "surjective," there must be some *SF* to which f does not map, *i.e.*, *SF* is unused in the *VF*. Because *SF* is unused, then the coefficients of the corresponding row(s) in the stiffness matrix are 0. Because at least one row of the stiffness matrix is all 0s, the stiffness matrix is singular and cannot be inverted, thus leading to no solution to the equation $\mathbf{A}x = b$ where $\mathbf{A}$ is the stiffness matrix. This contradicts our assumption that a simulation converges to a solution, so our initial assumption that f is not surjective must be false. Therefore, f is surjective.

4.1.2 Claim 1.2: f is injective.
Assume f is not injective. According to the definition of "injective," there must be two types v_1 and v_2 where

$$f(v_1) = S_f(v_1) = S_f(v_2) = f(v_2)$$

but $v_1 \neq v_2$. Because

$$S_f(v_1) = S_f(v_2)$$

v_1 and v_2 map to the same variable and are manipulated in exactly the same way in the VF (e.g., they have the same $\{BCs\}$ applied). This implies that the same values satisfy the VF, leading to the same output values in the variable field to which they map. Because the outputs of the simulation are the same, then $v_1 = v_2$, which contradicts our assumption that f is not injective. Therefore, f is injective.

Since we have shown that f is both surjective and injective, f is bijective by definition. Thus, the function that maps a type of variable field to its corresponding SF is a bijective function, so we can treat a variable type and its SF as one unit. For the rest of this section, we assume a VF has its corresponding SFs and refer to them interchangeably unless otherwise specified. In §5, we show how to derive the appropriate SFs for a given VF.

4.2 Claim 2: A viable simulation set must have a surjective function from BC to SF.

Define $apply(bc)$ as a function that maps a BC to its corresponding SF. Thus, we impose a condition such that $g = apply(bc)$ exists and we want to prove g is a surjective function from BC to SF.

4.2.1 Condition 2.1: A viable simulation set does not contain unused BCs.
Assume we have a viable simulation set. An unused BC can be removed without changing the viability of the simulation set, so we remove that BC and redefine the simulation set without changing its solution.

4.2.2 Claim 2.2: g is surjective.
Assume g is not surjective. By definition, there must be some SF on which no BCs are applied. By [18], the stiffness matrix being solved in this simulation is singular, rendering the equation unsolvable. This contradicts our assumption that the simulation set is viable, so our assumption that g is not surjective is incorrect. Thus, g must be surjective.

4.3 Experiments

To ensure completeness labels can be generated automatically, we compare automatically-generated completeness labels with the ground truth labels manually assigned by the author. For each run, we take a single $\{BCs\}$ from §3 and consider its completeness with the three VFs also from §3. Table 1 shows the results of our experiments.

As can be seen from Table 1, our approach correctly labels the simulation sets for every test case we attempted, regardless of the type of VF or its completeness. That there are no partially complete Synthetic tests is due to the Synthetic VF requiring all 3 possible types of BCs. The total number of tests matches the number of unique $\{BCs\}$.

TABLE 1. Number of labels based on VF and ground truth labels. Our approach, row "# Correct," correctly labels 13 out of 13 cases for each VF.

	Poisson	Navier-Stokes	Synthetic
Incomplete	4	8	9
Part. Complete	7	4	0
Complete	2	1	4
# Correct	**13**	**13**	**13**
Total	13	13	13

5 TYPE-BASED INDEXING FOR SOLUTION FIELDS

We previously mentioned that we must define SFs that match the requirements of the VF. In most FEM implementations, one knows the type of physics or equation being solved ahead of time and, consequently, can build SFs that match the VF exactly. However, this approach is limited in that each code is specific to a single type of physics; to simulate a different type of physics, an entirely new simulation must be coded from scratch. The goal of this paper is to develop a generalizable system that can check that multiple single- and multi-physics FEM simulations can run without the need of checking each simulation manually, eventually to generate large datasets of simulations completely automatically.

To this end, we propose the use of type-based labels to derive the corresponding SFs for a simulation set from a list of BCs. We discuss baseline methods of deriving SFs before describing our implementation and experimental comparison of different approaches.

5.1 Baseline Methods for Solution Field Generation

Arbitrary The most naïve approach to generate SFs is to initialize some arbitrarily large number of SFs of multiple dimensions and types (scalar, vector, and tensor). This approach assumes $\{BCs\}$ provides no information about the required SFs. However, this is computationally expensive and makes no guarantees on the viability of the simulation (if there are too many SFs, or alternatively too few). In essence, this can produce too few or too many SFs, and we cannot know which without more information.

Minimal-Maximum As a small improvement on this naïve approach, we can use the number of BCs in $\{BCs\}$ as an upper bound on the number of SFs. In this case, if n is the number of BCs, n scalar fields, n vector fields, and n tensor fields would be initialized. This would, of course, be many more SFs than are needed.

Unique BC Rather than gathering no information from $\{BC\}$, we can assign each BC in $\{BCs\}$ to its own SF. This imposes an upper limit on the number of SFs that can possibly be required in a specified VF because, as stated in §4.1.1, each SF requires at least one BC for convergence. However, this approach does not link variables of the same type together, *e.g.*, if two fluid velocity BCs are given for a two-inlet pipe system, then two unrelated SFs would be generated, leading to ambiguity in which fluid velocity field the VF "should" use and resulting in an inaccurate simulation.

Unique Dimension A third approach could be to build a SF for each BC of a different *dimension*. For example, this method would work for the Navier-Stokes equations as pressure BCs, applied to a scalar SF, and velocity BCs, applied to a 2-D vector SF, would have separate fields. Unfortunately, this approach breaks down when different BCs require SFs of the same dimension, *e.g.*, an electrostatics-coupled Navier-Stokes simulation [17]. Specifically in this example, electrostatic potential is a scalar quantity and requires a scalar SF, and while pressure is also a scalar quantity and requires a scalar SF, electrostatic potential and pressure should not be treated as the same variable. This approach gives a lower bound of the number of SFs required because BCs of different dimensions must have unique SFs.

Unique Names Instead of extracting information from $\{BCs\}$, we can assume that a given $\{BCs\}$ completes the VF of interest and try to derive necessary SFs from the unique variable names in the VF. In fact, this approach correctly derives the number of SFs needed for a specific VF. However, naming each variable in a variational equation is somewhat arbitrary, and a simple difference in convention (*e.g.*, ϕ, electrostatic potential, vs. u, temperature in the heat equation, though both are applications of the Poisson equation) should not necessarily imply a different simulation type entirely (*e.g.*, a general Poisson simulation vs. electrostatic potential or heat simulations), only a different physical interpretation of the solution. Additionally, while there is some sort of convention for naming variables, ordering the SFs that correspond to those variables is loosely alphabetical at best and entirely random at worst. Finally, augmenting a single-physics model to multi-physics, *e.g.*, from Navier-Stokes to electrostatic-coupled Navier-Stokes, requires manually naming the additional SF(s). This ambiguity with both naming and ordering leads to uncertainties in which BCs and which SFs correspond.

Indexed Names A slight modification to the name-based labels stated above replaces names with numerical indices. For example, an electrostatic-coupled Navier-Stokes equation, which requires three SFs, could assign, *e.g.*, velocity to SF_0, pressure to SF_1, and electrostatic potential to SF_2. Numerical indices are easily extensible to an (almost) arbitrarily large number of SFs, but the same issue arises with ordering the SFs in some unam-

biguous manner, *e.g.*, the indices of the velocity and electrostatic potential SFs should not change the viability of the simulation set. This is portrayed in more detail in Table 3.

5.2 Proposed Implementation of Type-Based Labels for Solution Fields

We propose type-based labeling to mitigate the disadvantages of name-based or index-based labels. While this approach has some overhead in encoding more information into the VF, it provides an unambiguous label for each unique type of BC in $\{BCs\}$, which provides more robust checks for completeness, and removes the issue of ordering the SFs as they are referred to only by their type, rather than an arbitrary name or index.

To build the SFs, we use the *builder* class, described in Algorithm 1 and below.

Algorithm 1 Builder

1: Initialize dictionary sf
2: **for** BC in $\{BCs\}$ **do**
3:     Add BC to sf with the key type(BC)
4: **end for**
5: Initialize empty mixed element *mix*
6: **for** key_{type} in sf **do**
7:     Add dimensioned element to *mix*
8: **end for**
9: Build the function space from *mix*
10: **for** key_{type} in sf **do**
11:     **for** BC in key_{type} **do**
12:         $fieldToUse = key_{type}$
13:         $value = BC.value$
14:         $location = BC.location$
15:         Create FDBC and append to list of FDBCs
16:     **end for**
17: **end for**
18: Return mixed function space and FDBCs

A list of BCObjects is the input to this algorithm. We start with a dictionary (line 1) to store each unique *type* (line 3) since it only matters that each *type* has a SF rather than each BC. Once we have extracted the types, we add appropriate elements to a mixed element placeholder (line 7). Each of the added elements has dimensions according to the *type* it represents (*e.g.*, fluid pressure fields are scalars and use scalar elements, whereas fluid velocity fields in 2-D have 2-D vector elements). These dimensions can be derived from the *value* property of a BC object. With this mixed element *mix*, FEniCS can build a function space (line 9) that contains SFs corresponding to the different elements in the mixed element.

Finally, the algorithm creates FEniCS Dirichlet Boundary Condition objects (*FDBCs*), which require information about the solution field to be used, value, and location of each *BC* (lines 10-17). The mixed function space and list of *FDBCs* are returned as outputs. The *FDBCs* are specific to the FEniCS library, but any implementation-specific object or function to apply Dirichlet boundary conditions (*DBCs*) on the stiffness matrix can be used instead of *FDBCs*.

We would like to draw special attention to lines 3 and 12. In line 3, we use a key that, rather than a number or a name, is the *type* of *BC* being applied. This ensures that the function mapping from the *BCs* to the *SFs* is surjective, thus meeting one of the criteria for a complete simulation set from section §4.

In line 12, we again refer to key_{type}. Referring to the solution field by its type is fairly straightforward: we find the *SF* whose key matches the *BC*'s type[2] and apply the *BC* onto that *SF*.

Also of note are lines 6 and 10, where we iterate over the keys matching each type of *BC*, thus ensuring *BCs* of the same types are kept on the same *SF* and those of different types are separated.

5.3 Experiments

We compare our approach of type-based indices with the baseline approaches. We use the sets of *BCs* mentioned in §3. For every ordering permutation of that {*BCs*}, we use each baseline and the proposed approach to generate *SFs* and compare the generated fields to those we generate manually. The ratio of correctly generated[3] *SFs* to total generated is the accuracy of the method. Table 2 shows the results of our experiment.

From Table 2, we note that as we move to more sophisticated methods, we see an increase in the number of {*BCs*} that may have correctly generated *SFs*. We also note that every method is generalizable[4] except method "Unique Names," which uses names of variables to differentiate fields. Even so, methods "Unique Names," "Indexed Names," and "Types" all generate at least some correct *SFs*. With the switch to indexed names, method "Indexed Names" seems to generalize method "Unique Names." However, these two methods perform poorly as the possible number of orderings increases.

To test robustness to *BC* ordering, we run this experiment with every order permutation of the *BCs* in each {*BCs*}. Table 2 includes the results of these permuted ("shuffled") boundary conditions.

From Table 2, the "Unique BC" and "Unique Dimension" approaches correctly generate *SFs* regardless of *BC* ordering, but in general, these approaches are not robust to the combinations of *BCs* that may be encountered. While the "Unique Names" and "Indexed Names" approaches generated the correct *SFs* at least sometimes in all of our test cases, Table 2 shows they lack of robustness when the *BCs* are ordered differently, especially as the number and types of *BCs* increase. Our type-indexed approach "Types" provides the correct references to the corresponding *SFs* regardless of the order of the *BCs* in all of our test cases.

6 TYPE-BASED INDEXING FOR VARIATIONAL FORMS

Similar to the previous section, we use type-based indexing to differentiate between variables in the *VFs*. This consistent labeling allows easy mapping of *SFs* to the corresponding variables in the *VF*. This also facilitates robustness to the order of terms in the *VF*.

6.1 Baseline Methods for Variational Form Encoding

Named Variables As previously mentioned, in many implementations of the Navier-Stokes (e.g., [19]) and Poisson (e.g., [19]) solvers, *SFs* are named with relatively intuitive or conventional letters. For example, in the Navier-Stokes *VF*, the velocity *SF* is often denoted *u*, and the pressure *SF* as *p*, as seen in Table 4, with the test functions *v* and *q* (not shown), respectively. Similarly, if somewhat less interestingly, the Poisson equation's *VF* in Table 4 uses *u* for its single *SF* with a test function of *v*.

Such naming conventions are common in existing state-of-the-art programmatic FEM solvers (such as when using the Unified Form Language [13]), and it assumes knowledge of the *VF* to be used beforehand. In contrast, this paper explores methods for constructing such FEM simulations in FEniCS automatically. Thus, hard-coding the names of *SFs* into the *VF* is not useful.

Indexed Variables Rather than hard-coding variable names, we again consider indexing *SFs* numerically. Table 4 gives an example of this implementation. A simple substitution of *vars*[0] and *vars*[1] for *u* and *p*, respectively, and their counterparts of *testVars*[0] and *testVars*[1] (not shown), allows these *SFs* to be referenced without hard-coded names, if somewhat less compactly. As an added bonus, any number of *SFs* can be referenced merely by increasing the index number. However, we run into the same issues of ambiguous ordering and references as stated in Section 5.

Consider the example in Table 3. In this 2-D Navier-Stokes fluid example, we apply two types of *BCs*, fluid velocity ("V") and fluid pressure ("P"). A velocity *SF* is a 2-D vector field, whereas a pressure *SF* is a scalar field, as seen in columns 2 and

---

[2]Our implementation actually uses an intermediate index-type dictionary, but because this dictionary is bijectively defined, we refer to the index and the type interchangeably.

[3]"Correctly generated," here, means that the algorithm builds the correct number of *SFs* of the correct size (scalar, vector, or tensor) and dimensionality. Additionally, all order permutations are compared to the default sets in §3, *i.e.*, permuted lists must generate the same *SFs* and refer to them in the same manner to be considered "correct."

[4]Generalizable, meaning extendable to different *VFs* without additional adjustments.

TABLE 2. Results of order-permuted *SF* generation using baseline and the proposed type-based approaches. Numbers are ratios of successfully-generated *SF*s to total generated *SF*s. Dashes indicate 0 successes and are used to make the table more readable.

BC Set	Arbitrary	Min-Max	Unique BC	Unique Dimension	Unique Names	Indexed Names	Types
0	-	1.0	1.0	1.0	1.0	1.0	1.0
1	-	-	1.0	1.0	1.0	1.0	1.0
2	-	-	1.0	1.0	1.0	1.0	1.0
3	-	-	1.0	1.0	1.0	1.0	1.0
4	-	-	-	1.0	1.0	1.0	1.0
5	-	-	1.0	-	0.5	0.5	1.0
6	-	-	1.0	-	0.167	0.167	1.0
7	-	-	-	1.0	0.333	0.333	1.0
8	-	-	-	1.0	0.75	0.75	1.0
9	-	-	-	-	0.083	0.083	1.0
10	-	-	-	-	0.3	0.3	1.0
11	-	-	-	1.0	0.3	0.3	1.0
12	-	-	-	-	0.0083	0.0083	1.0
Average	0.0	0.077	0.462	0.615	0.572	0.572	1.0
Generalizable	✓	✓	✓	✓	-	✓	✓

TABLE 3. Different orderings of *BC*s can result in inconsistently-referenced solution fields, even if the same type of simulation is run

BCs Applied	vars[0] Expected	vars[0] Actual	vars[1] Expected	vars[1] Actual
{V, P}	2-D Vector	2-D Vector	Scalar	Scalar
{V, V, P}	2-D Vector	2-D Vector	Scalar	Scalar
{V, P, V}	2-D Vector	2-D Vector	Scalar	Scalar
{P, V}	2-D Vector	Scalar	Scalar	2-D Vector

4 of Table 3. In the first three cases, where V is first, a 2-D vector field is labeled as vars[0], and a scalar field is vars[1]; when P comes before V as in the last row of Table 3, vars[0] now becomes a scalar field, and attempting to perform vector operations on scalar values results in a failed simulation. This required ordering severely limits the number of simulations that can be generated and run automatically as the order of *BC*s should not affect the solution to a viable simulation.

6.2 Proposed Implementation of Type-Based Labels for Variational Forms

We convert the numerical indices in the baseline approach into type-based indices[5]. These types are encoded into the *VF* as *needed types* to make completeness labeling easier. Table 4 provides an example of the (standard) name-based approach, indexed names, and our typed indices. Section 3 gives the full *VF*s we test.

---

[5]Our implementation actually uses an additional dictionary-type object to map indices to types of boundary conditions; however, because mapping between indices and *BC*s is bijectively defined, we can treat them as referring to the same object.

TABLE 4. Several possible references to *VF* variables

	Poisson	Navier-Stokes
Unique Names	**u**	**u, p**
Indexed Names	vars[0]	vars[0], vars[1]
Typed Index	vars["Poisson"]	vars["FluidVelocity"], vars["FluidPressure"]

6.3 Selection of Applicable Boundary Conditions for Viable Simulation Set Generation

With these labeled simulation sets and robust *SF*s, we can choose the sets which are complete and know that the simulation will run. However, partially complete simulation sets still contain extra *BC*s that need to be removed. For this, we can apply our type-based approach to select only the *BC*s that are applicable to our current *VF*. We implement this in Algorithm 2, which returns the set of *BC*s that are applicable to the simulation set. Lines 2 and 9 use our completeness labeling from Section 4 to check whether the there is a need to strip out the applicable *BC*s. Line 4 uses type-checking to determine if a given *BC* is required for a given *VF*.

Algorithm 2 Selector

1: Inputs: *VF*, *BCList*
2: **if** isPartiallyComplete(*VF*, *BCList*) **then**
3:     **for** *BC* in {*BCList*} **do**
4:         **if** *type*(*BC*) is in the required types of *VF* **then**
5:             Add *BC* to the list of applicable *BC*s
6:         **end if**
7:     **end for**
8: **else**
9:     **if** isComplete(*VF*, *BCList*) **then**
10:         All *BC*s are applicable
11:     **else**
12:         Simulation set is incomplete
13:     **end if**
14: **end if**
15: Return applicable *BC*s

6.4 Experiments

We input into our Algorithm 3 objects: a mesh, on which we want to solve the simulation; a set of *BC*s, BCList, which we want to apply; and a *VF*. We take the BCList and VF and label these two, assuming the *SF* corresponding to the *VF* is appropriate. If the result is incomplete, we stop the simulation; if it is partially complete, we use Algorithm 2 (the "Selector") to turn this into a complete simulation. Once the (partial) simula-

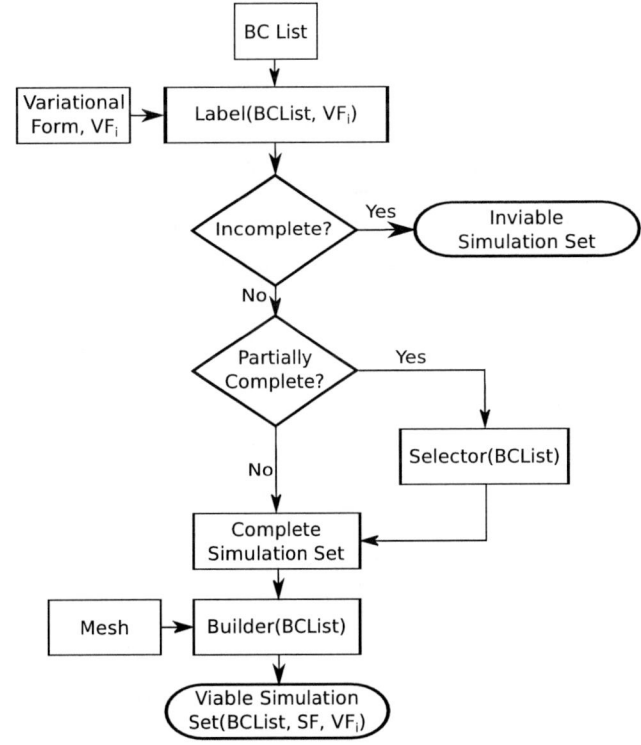

FIGURE 2. Flowchart of Experiments in Section 6.4

tion set is complete, we send it into Algorithm 1 (the "Builder"). This generates the actual *SF*s for the simulation set, with which the FEM can be solved. Figure 2 illustrates a flowchart of the process.

Following this process, we run two related experiments, the first to test the viability of a given order of *BC*s, and the second to test robustness to permutations of *BC* order.

6.4.1 Use of Completeness Labels to Determine Simulation Viability

With our fully defined and viable simulation sets, we now want to determine whether knowing these simulation sets' labels is useful in deciding which simulation sets to run. We run three different groups of simulation sets: "Complete," "Viable" (*i.e.*, both Complete and Partially Complete), and "All" sets. We tabulate the number of attempted simulations

TABLE 5. Results of non-shuffled simulation testing with and without completeness labels. Here, "Complete" refers to a too-restrictive class of simulation sets that are exactly complete; "Viable" refers to the class of simulation sets that are both complete and partially complete; and "All" refers to all of the simulation sets being tested, regardless of completeness.

	Attempted	Success	% Success
Complete	7	7	100
Viable	18	18	100
All	39	8	20.5

TABLE 6. Results of shuffled simulation testing with and without completeness labels. This is an extension of the previous experiment to include different BC orderings.

	# Attempted	# Succeeded	% Succeeded
Complete	897	897	100
Viable	**2765**	**2765**	**100**
All	3084	921	29.9

and the number of successful simulations to calculate the success percent of simulations each group attempts. With 3 variational forms and 13 sets of bcs, we expect $3 \times 13 = 39$ total simulations for each group. Table 5 shows the results of this experiment.

We expect every Complete or Partially Complete simulation to be viable, as is confirmed by the 100% running success rate of those simulation. From Table 5, we also see that, as "Viable" is a superset of "Complete," there are 11 more sets in "Viable" than there are in "Complete." Similarly, "All" is a superset of "Viable," resulting in 21 more attempted simulations.

6.4.2 Robustness of Type-Based Indices to BC Order Permutations

We run a similar setup as Section 6.4.1, but we permute the order of BCs in every combination possible. Again, we calculate the number of simulations attempted by each method and compare it to the number of simulations that were run to completion. These results are in Table 6.

From Table 6, we can see that both "Complete" and "Viable" simulations run with 100% success in our tested cases. However, fewer simulations (in this case, less than 1/3) are run when the "Complete" criterion is used as compared to the "Viable" criterion. "All" simulations encompass 3084 possible cases, of which only 921 (29.9%) are able to run to completion. In this case, "Viable" covers 89.7% of "All" simulations while running to completion more than triple the number of simulations. From this, we can see that the "Viable" set attempts fewer simulations but

runs more successful simulations than previous implementations, thus saving in computational time wasted on failed simulation attempts.

7 DISCUSSION

From the results of our experiment shown in Table 5, we note there are 10 simulations that "Viable" runs successfully but "All" does not. While at first glance, this seems odd, this behavior is expected. Because some of the sets in "All" are partially complete, there are extra BCs that violate condition 4.2.1 from §4. "Viable" sets, on the other hand, contain this completeness label, which informs whether extraneous BCs should be stripped before constructing a simulation. Thus, when an "All" simulation set is used to construct a simulation, extraneous BCs cause the construction to fail, whereas "Viable" simulation sets have extra BCs stripped before construction.

Even taking out Partially Complete simulation sets, one should expect that "All" runs the same number of successful simulations as "Complete" does since both should run only Complete sets. However, Table 5 shows this is not the case: "All" contains 8 successfully-run sets, whereas "Complete" only contains 7 sets. This extra viable simulation set actually comes from BC set 8 ($BC8$), which contains 3 fluid velocity and 1 fluid pressure BC. Specifically, in the "Complete" pass through the BCs, our approach checks $BC8$ against the requirements of the Poisson VF. Finding that there is no Poisson BC, this simulation set (of fluid velocity/pressure BCs and Poisson VF) does not make a viable simulation. However, the "All" pass does not check the BC against the VF; thus, when the "All" algorithm encounters a set of fluid velocity and pressure conditions, a Navier-Stokes simulation is performed, regardless of the intended Poisson simulation. Because the Navier-Stokes VF is technically viable for $BC8$, the simulation completes without any problems, despite the fact that an entirely different set of equations is solved. A similar reasoning applies to the 897 "Complete" vs. 921 "All" simulations in Table 6.

Our results in Table 6 suggest that our approach allows both a larger number of simulations to be run without additional manual processing and that more of the attempted simulations run successfully, especially when combined with completeness labels. While it may seem that many of the simulations being run are redundant or repetitions of other simulations (*e.g.*, $BC3$ and $BC5$ differ in the addition of a Poisson BC), many of these cases include BCs whose addition does not fully-define other types of simulation sets, so running simulations with each of these BCs *should* result in the same exact simulations. In addition, our type-based indices allow the simulations to run without regard to the order of the BCs, providing the framework for automated simulation construction from BCs. Because our approach allows more successful runs with fewer attempts, we claim that implementing the completeness labels and type-based indices successfully allows for robust automatic simulation checking.

However, our work is limited in several aspects. We previously mentioned that each set of *BC*s is assumed to be physically realizable; however, we do not check for whether a given set of *BC*s is physically realizable in this paper. For example, given two velocity inlets to a pipe, our implemented approach will consider that set as viable, even though it is physically impossible for a pipe to have only inlets and no outlets assuming that mass is conserved.

Secondly, we assume that *VF*s are given. The derivation of these equations can be done manually, but the process is difficult to automate while keeping the scope of this paper reasonably limited. We are investigating this in our future work.

Third, we restricted our tests to only 3 types of *VF*s and 3 types of *DBC*s to make experimentally testing our claims more straight-forward. Implementing more types of physics and *VF*s would extend the practical functionality of this approach, though that is separate from the intellectual contributions of the paper.

Fourth, our work only discusses Dirichlet *BC*s (DBCs) without Neumann (NBCs) or Robin *BC*s (RBCs). NBCs and RBCs are implemented differently than DBCs in that they are additional terms in the *VF*, while DBCs are enforced by manipulating values in the stiffness matrix. However, simulation sets would still need to be defined appropriately to create viable simulations, even in cases with pure NBCs or simulations that do not require DBCs, *e.g.*, those using inertial relief methods.

Finally, we have used only boundary value problems with no time dependence in this work. Initial value problems can be solved with similar setups of simulation sets, *e.g.*, assuming each time step in the simulation is quasi-static but depends on the previous state to determine the next state (*e.g.*, the Runge-Kutta method).

Although our initial goal in this work was the autonomous generation of meaningful FEM simulations, we realized that several critical issues arose. This paper is in response to one of those issues, namely that ensuring the viability of a simulation is critical to generating such simulations autonomously.

8 CONCLUSION

In this paper, we proposed three improvements to the algorithmic construction of FEM simulations. First, we discussed conditions between different aspects of a simulation to ensure viability. Second, we proposed a type-based indexing method to generate the correct solution fields for a simulation automatically. Third, we used a similar type-based indexing method to connect boundary conditions and solution fields to a variational form robustly. We showed that our implementation of completeness labels guarantees simulation viability, and when combined with type-based indices for solution fields and variational forms, provides the framework needed for constructing FEM simulations automatically from boundary conditions.

With our contributions, we are able to generate and check for viable physics simulations automatically from collections of physically-realizable boundary conditions. With extension to a larger number of PDEs, we lay the foundations of automatically constructing single- and multi-physics FEM models, allowing future implementations to build and run simulations autonomously with a higher degree of certainty that the simulation will run than current, manually-constructed methods. These automatic simulations can then lead to physics- and data-driven design optimization, machine learning, analysis, and topology optimization, allowing us to apply new computational methods to mechanical design and to provide a path forward for future data-driven design methods.

ACKNOWLEDGMENT

We acknowledge the funding and support provided by DARPA through their Fundamentals of Design program (#HR0011-18-9-0009). The views, opinions, and/or findings contained in this article are those of the author and should not be interpreted as representing the official views or policies, either expressed or implied, of the Defense Advanced Research Projects Agency or the Department of Defense.

REFERENCES

[1] Alnæs, M. S., Blechta, J., Hake, J., Johansson, A., Kehlet, B., Logg, A., Richardson, C., Ring, J., Rognes, M. E., and Wells, G. N., 2015. "The FEniCS project version 1.5". *Archive of Numerical Software,* **3**(100).

[2] Ho-Le, K., 1988. "Finite element mesh generation methods: a review and classification". *Computer-Aided Design,* **20**(1), pp. 27 – 38.

[3] Finnigan, P. M., Kela, A., and Davis, J. E., 1989. "Geometry as a basis for finite element automation". *Engineering with Computers,* **5**(3), Jun, pp. 147–160.

[4] Viceconti, M., Davinelli, M., Taddei, F., and Cappello, A., 2004. "Automatic generation of accurate subject-specific bone finite element models to be used in clinical studies". *Journal of Biomechanics,* **37**(10), Oct, pp. 1597–1605.

[5] Sun, L., Zhao, G., and Yeh, G.-T., 2018. "An automatic quadrilateral mesh generation algorithm applied to 2-d overland flow simulations". *Computational Geosciences,* **22**(5), Oct, pp. 1283–1303.

[6] Rocca, G. L., 2012. "Knowledge based engineering: Between AI and CAD. Review of a language based technology to support engineering design". *Advanced Engineering Informatics,* **26**(2), pp. 159 – 179.

[7] van Tooren, M. J. L., La Rocca, G., Krakers, L., and Beukers, A., 2003. "Design and Technology in Aerospace. Parametric Modelling of Complex Structure Systems Including

Active Components". In 13th International Conference on Composite Materials, San Diego, CA.

[8] van der Laan, A. H., and van Tooren, M. J. L., 2005. "Parametric modeling of movables for structural analysis". *Journal of Aircraft,* **42**(6), November-December, pp. 1605–1613.

[9] La Rocca, G., and van Tooren, M. J. L., 2006. "A modular reconfigurable software modelling tool to support distributed multidisciplinary design and optimisation of complex products". In 16th CIRP International Design Seminar, Kananaskis, Alberta, Canada, 16-19 July.

[10] Quintana-Amate, S., Bermell-Garcia, P., Tiwari, A., and Turner, C., 2017. "A new knowledge sourcing framework for knowledge-based engineering: An aerospace industry case study". *Computers & Industrial Engineering,* **104**, pp. 35 – 50.

[11] Logg, A., 2007. "Automating the finite element method". *Archives of Computational Methods in Engineering,* **14**(2), Jun, pp. 93–138.

[12] Logg, A., and Wells, G. N., 2011. "DOLFIN: automated finite element computing". *CoRR,* **abs/1103.6248**.

[13] Alnæs, M. S., Logg, A., Ølgaard, K. B., Rognes, M. E., and Wells, G. N., 2014. "Unified form language: A domain-specific language for weak formulations of partial differential equations". *ACM Transactions on Mathematical Software (TOMS),* **40**(2), p. 9.

[14] Cimrman, R., 2014. "SfePy - write your own FE application". In Proceedings of the 6th European Conference on Python in Science (EuroSciPy 2013), P. de Buyl and N. Varoquaux, eds., pp. 65–70. http://arxiv.org/abs/1404.6391.

[15] , 2017. Python FEM and Multiphysics Simulations with FEniCS and FEATool. On the WWW, June. URL `www.featool.com/`.

[16] Xia, Q., 2017. "Automated Mechanical Engineering Design using Open Source CAE Software Packages". In FEniCS '18, Oxford, UK, June. URL `github.com/qingfengxia/FenicsSolver`.

[17] Emamy, N., Karcher, M., Mousavi, R., and Oberlack, M., 2015. "A high-order fully coupled electro-fluid-dynamics solver for multiphase flow simulations". In Proceedings of the VI international conference on coupled problems in science and engineering, pp. 753–9.

[18] Logan, D. L., 2014. *A First Course in the Finite Element Method.* Cengage Learning, Boston, MA.

[19] Langtangen, H. P., and Logg, A., 2016. *Extensions: Improving the Poisson Solver.* Springer International Publishing, Cham, pp. 109–141.

Proceedings of the ASME 2019
International Design Engineering Technical Conferences
and Computers and Information in Engineering Conference
IDETC/CIE2019
August 18-21, 2019, Anaheim, CA, USA

DETC2019-98111

USING BAYESIAN OPTIMIZATION WITH KNOWLEDGE TRANSFER FOR HIGH COMPUTATIONAL COST DESIGN: A CASE STUDY IN PHOTOVOLTAICS

Mine Kaya, Shima Hajimirza[1]
Texas A&M University
College Station, TX

ABSTRACT

Engineering design is usually an iterative procedure where many different configurations are tested to yield a desirable end performance. When the design objective can only be measured by costly operations such as experiments or cumbersome computer simulations, a thorough design procedure can be limited. The design problem in these cases is a high cost optimization problem. Meta model-based approaches (e.g. Bayesian optimization) and transfer optimization are methods that can be used to facilitate more efficient designs. Transfer optimization is a technique that enables using previous design knowledge instead of starting from scratch in a new task. In this work, we study a transfer optimization framework based on Bayesian optimization using Gaussian Processes. The similarity among the tasks is determined via a similarity metric. The framework is applied to a particular design problem of thin film solar cells. Planar multilayer solar cells with different sets of materials are optimized to obtain the best opto-electrical efficiency. Solar cells with amorphous silicon and organic absorber layers are studied and the results are presented.

Keywords: transfer optimization, Bayesian optimization, thin film solar cells, external quantum efficiency

NOMENCLATURE

D	dataset
L_D	diffusion length
N	number of data in the training set
n	size of a sample
N_i	number of incident photons on a solar cell
N_p	number of absorbed photons by a solar cell
n_p	carrier generation rate
$\mathbb{P}_c$	collection probability
t	thickness of the solar cell layers
$\mathcal{X}$	input space
$\mathcal{Y}$	output space

w	weight
X	input set
y	output
Greek letters	
η_e	external quantum efficiency
η_i	internal quantum efficiency
λ	wavelength
μ	mean
κ	tradeoff parameter
σ	standard deviation
φ	similarity metric
Subscripts	
k	task
T	training
$*$	new point

1. INTRODUCTION

Engineering design is often an iterative process in which many successive experiments and computer simulations are performed to select the best configuration. Engineers working on a design task gain experience and intuition about similar tasks and may transfer their knowledge to a new problem. However, relying on the human knowledge can be misleading as the human mind can be biased and subjective. Moreover, continuation and documentation of the knowledge can be intractable. Nevertheless, knowledge transfer among tasks can lead to smarter design practices. This is known as transfer learning which has proven to be a promising concept in data science.

Transfer learning has attracted the attention of data scientists for taking advantage for addressing the cold start issue in machine learning problems [1]. For example, in image classification, a large data size is required for training the classifier for every task which can be extremely slow. This was addressed via transfer learning [2,3]. Similarly, web document [4,5], brain-computer interface [6,7], music [8] and emotion [9] classification operations are accelerated and improved using knowledge transfer. Although regression transfer has received

---

1. Contact author: shima.hm@tamu.edu (S. Hajimirza)

Copyright © 2019 ASME

less attention compared to classification transfer [10], it has also been studied to address several problems, such as configurable software performance prediction [11], shape model matching in medical applications [12] and visual tracking [13].

Despite the wide range of applications in machine learning applications, transfer learning has not been evaluated thoroughly in the fields of design and optimization. Recently, a few studies including authors' have proposed using transfer learning to prevent the cold start issue in design optimization methods. In [14] a transfer leaning surrogate-based optimization was proposed for thin film solar cell design where the knowledge transfer is achieved by neural network layers. Li et al. [15] proposed a transfer learning based design space exploration method for microprocessor design. Min et al. [16] investigated the use of transfer learning in aircraft design problems and demonstrated the effectiveness of the proposed algorithms. Gupta et al. [1] reviewed the recent progress in transfer learning in optimization problems and categorized them as sequential, multitasking and multiform transfer optimizations.

Transfer learning can be particularly useful in optimization of black box functions with high computational costs. In these cases, a useful technique to use surrogate (meta) models, which are regression/machine learning tools. Optimizations are performed on this lower fidelity model of the original function in order to reduce the computational cost without sacrificing the accuracy. Surrogates can also be used to create a bridge among different tasks in ways similar to machine learning tasks. Assume that one is dealing with optimization of the same physical objective function but with different input settings. The settings can be different boundary or initial conditions, different sets of physical characteristics, different environments or materials, or any other practical variation. An example can be designing an airfoil with optimal aerodynamic properties under different settings of speed, altitude, allowable material types, etc. Every design problem is unique, but the objective functions are correlated as they pertain to the same underlying physical function. These correlations can be captured and used across various settings. Once a surrogate model is fit and learned for one objective function, it is expected that the knowledge can be transferred to more accurately or more efficiently fit a surrogate model for another similar function. If the process improves the efficiency and accuracy of the surrogate fitting, then it is expected that the black-box optimization is improved in general.

One of the areas where the simulations are computationally costly is material design problems. For example, optical properties of the solar cells can be obtained by solving Maxwell's electromagnetic equations due to the subwavelength features of the solar cells. In such problems, time consuming numerical solvers, such as finite difference time domain (FDTD) are employed which limits exhaustive search [17-19]. In particular, we are interested in a material design problem at nano-scale related to multilayer thin film solar cells. For a fixed set of materials, the objective is to choose the dimensions of the layers. These dimensions affect the photo-electric properties of the solar cell in complex ways that are hard to analytically express or anticipate [20,21]. Recently, we have shown that surrogate-based

optimization methods can be used to solve optimization problems of this nature. We have established their efficacy of those methods in several thin film design problems [22-24]. Consequently, the computational costs for completing optimizations were significantly reduced compared to traditional optimization methods. In the present study, we aim to further demonstrate that both accuracy and speed can be improved in similar problems using transfer learning.

This paper proposes a framework for knowledge transfer in design problems to decrease the computational cost of exhaustive search algorithms without sacrificing accuracy. The proposed method establishes the knowledge transfer among tasks based on the weighted common response surface. We assume that at least one design task has been completed. The aim is to repeat the optimization for a structure with different material choices. This study has the following novelties: first, the Gaussian Process Bayesian optimization (GP-BO) algorithm combined with transfer optimization has never been used in a design problem; second, the similarity among the tasks is determined by a metric calculated as a function of the deviation between tasks. Knowledge transfer among tasks is initiated using the *similarity weight* accordingly.

The remainder of the paper is organized as follows: first, GP-BO based transfer optimization is explained. Then the physical structure of the problem is described and the optimization results are presented and discussed in section 3.

2. MATERIALS AND METHODS

2.1 Bayesian Optimization

Bayesian optimization [25] is a global optimization method that searches the optimum of a function by using a surrogate based on a Gaussian Process. The most important feature of this method is to use the exploration and exploitation of the design space. Therefore a suitable surrogate must provide accurate point estimates as well as uncertainty in the new predictions. Surrogate functions can be selected from a wide range of possibilities and Gaussian Process is one of the most widely used methods in the Bayesian optimization framework. Bayesian optimization targets an acquisition function instead of the original objective. There are a variety of acquisition functions such as probability of improvement, entropy search, upper confidence bound (UCB) and the expected improvement (EI). Specifically, UCB is calculated as follows:

$$UCB(\mathbf{x}) = \mu_y(\mathbf{x}) + \kappa\, \sigma_y(\mathbf{x}) \qquad (1)$$

where $\mu_y(\mathbf{x})$ and $\sigma_y(\mathbf{x})$ are the mean and standard deviation of the prediction at input point $\mathbf{x}$ calculated from using Gaussian Process regression. κ is the exploration-exploitation tradeoff parameter. In this study, κ is set to 2.

2.2 Gaussian Process

Gaussian Process (GP) is a regression tool in which the prior knowledge on a data set is used to make new predictions. The joint distribution of the observations and test values under prior is expressed via the joint normal distribution with zero mean and a variance.

$$\begin{bmatrix} \mathbf{y_T} \\ y_* \end{bmatrix} \sim \mathcal{N} \left(0, \begin{bmatrix} K(X_T, X_T) + \sigma^2 I & K(X_T, \mathbf{x}_*) \\ K(\mathbf{x}_*, X_T) & K(\mathbf{x}_*, \mathbf{x}_*) \end{bmatrix} \right), \quad (2)$$

where X_T and $\mathbf{y_T}$ are the input and output sets for training (i.e. calculated previously). $\mathbf{x}_*$ is the new data point to be predicted (y_*). K is the covariance matrix consisting of kernel functions. Consequently, the conditional distribution of $\mathbf{y}_*$ given $\mathbf{y}$ is:

$$\mathcal{P}(y_*|X_T, \mathbf{y_T}, \mathbf{x}_*) \sim \mathcal{N}\left(\mu_y, \sigma_y^2\right), \quad (3)$$

Then mean prediction of the function μ_y and its variance σ_y^2 are calculated from:

$$\mu_y = K(\mathbf{x}_*, X)[K(X, X)]^{-1}\mathbf{y_T}$$
$$\sigma_y^2 = K(\mathbf{x}_*, \mathbf{x}_*) - K(\mathbf{x}_*, X)[K(X, X)]^{-1}K(X, \mathbf{x}_*) \quad (4)$$

The Bayesian optimization algorithm then maximizes UCB using μ_y and σ_y (see equation 1**Error! Reference source not found.**). At every iteration, $\{X, \mathbf{y}\}$ is updated based on the result of BO and the GP is retrained.

2.3 Transfer Learning in Design Optimization

Engineering design is an iterative procedure where the design space is modified by the gained knowledge along the process. Experience is somewhat natural for human brain to acquire but hard to quantify for computers. This challenge in computational sciences is generally referred to as the cold start issue. Transfer learning or knowledge transfer concept is used to address this.

Let $\mathcal{X}_k$ be the input space of the kth task. $X_k = \{\mathbf{x_1}, \mathbf{x_2}, ..., \mathbf{x_{N_k}}\} \in \mathcal{X}_k$ is the $N_k \times d$ input array of the training data where N_k is the number of training data in the kth dataset and d is the number of features in the problem. The output space is $\mathcal{Y}_k$ and $Y_k = \{y_1, y_2, ..., y_{N_k}\} \in \mathcal{Y}$ is the output array in the dataset obtained from the black-box function. Thus the kth dataset is $D_k = \{\{\mathbf{x_i}, y_i\}, \forall i = 1, ... N_k\} = [X_k \ Y_k]$. In general, we have a predictive model (meta-model) to estimate the most probable value of the function at a new point, $\mathbf{x}^*$ given the dataset and the trained hyperparameters θ of the metamodel.

$$p(y^*|\mathbf{x}^*, D) = \int p(y^*|\mathbf{x}^*, \theta)\, p(\theta|D)\, d\theta \quad (5)$$

One of the solutions to transfer knowledge from experiences to the current task is to assume a joint probabilistic model $p(f_1, f_2, ...)$ and to use a product covariance function [1,26]. Another approach is to assume a common response surface [26,27]:

$$\tilde{y} = \frac{y - \mu_{common}}{\sigma_{common}} \quad (6)$$

This approach can be meaningful when the input and output spaces are exactly the same. However it can be misleading in different spaces. On the other hand, common response surface approach can be generalized to be used between different spaces by considering rank correlations between data, such as Spearman rank correlation coefficient. At each transfer, we can rank the previous tasks according to their correlations with the new one, and transfer can be made between more similar tasks. For example, the following weighted average can be used to obtain a common mean and a standard deviation:

$$\mu_{common} = \frac{\sum_{k=1}^{K} w_k N_k \mu_k}{\sum_{k=1}^{K} w_k N_k}$$
$$\sigma_{common}^2 = \frac{\sum_{k=1}^{K} (w_k N_k)^2 \sigma_k^2}{\sum_{k=1}^{K} (w_k N_k)^2} \quad (7)$$

where w_k is the weight which is a function of similarity among the current task and the task knowledge is transferred from. This similarity can be calculated by the deviation of the responses of the current task (y) and the kth previous task (y_k) to the same input set (X_{init}). The similarity $\varphi(y, y_k)$ can be calculated as $(1 - deviation)$:

$$\varphi(y, y_k) = 1 - \frac{1}{C}\sqrt{\frac{1}{n}\sum_{i=1}^{n}(y_i - y_{k,i})^2} \quad (8)$$

where n is the size of the initial sample. C is the normalization constant which represent the maximum possible deviation. In the extreme case, $y = 0$ and $y_k = 1$ (or vice-versa) which translates to a deviation of 1 thus $C = 1$. However, C can be selected according to the desired allowable deviation. Additionally, it is safer to transfer knowledge from the tasks with high similarity only by establishing a threshold to prevent the initial sample being misleading. Now we can calculate the weight in equation (7).

$$w_k = \begin{cases} \varphi(y, y_k) & \varphi(y, y_k) > 0.5 \\ 0, & \varphi(y, y_k) < 0.5 \end{cases} \quad (9)$$

2.4 Description of the Physical Model

We use the above-mentioned methodology in a design optimization for multi-layer thin film solar cells. For this purpose, a simple multilayer solar cell structure is used consisting of an absorber, an antireflective coating, a back-reflector metal layer and interlayers with thicknesses t_A, t_{ARC}, t_M, t_{IL1} and t_{IL} respectively stacked together as shown in Figure 1. The working principle of solar cells is explained by the photovoltaic effect where an electron-hole pair inside a semiconductor is created due to photon absorption. The electron completes the electrical circuit and generates photocurrent. The interaction between the incoming light and the multilayer structure of solar cell is explained by the Maxwell's electromagnetic theory since the characteristic length of thin film solar cells and the operation wavelengths are at the same order of magnitude ($0.1 - 1$ μm).

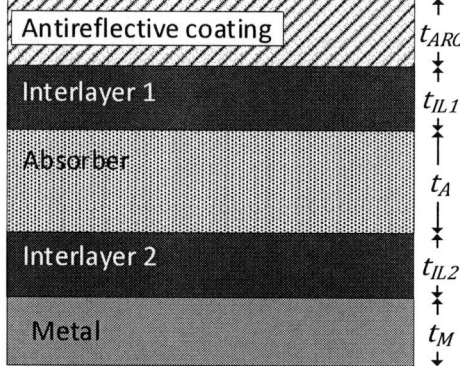

FIGURE 1: SCHEMATIC OF THE MULTILAYER SOLAR CELL

Light-matter interaction in the near field region provides unique properties which strongly depend on the dimensions of the thin film structures. Therefore, a careful optimization of the thin film geometry is required to maximize the solar cell efficiency. The material choice in the solar cell layers greatly affects the optical and electrical properties in addition to the physical dimensions. Nevertheless, including material choices as a design variable makes the optimization problem a mixed integer programming which is known to be computationally costly. Previously, this kind of problems were studied in structure design [28]. Furthermore, for the present problem where the optimizations are done one by one, the optimization study should be repeated ($m_1 \times m_2 \times ... \times m_d$) times for all possible material combinations where d is the input space dimension and m_j ($1 \leq j \leq d$) is the number of choices for the jth input. In this case, knowledge transfer between different material combination tasks is worthwhile, as similar geometries with different material combinations can have similar opto-electrical responses. In general, the initial assumption is that source and target domains are similar [15]. The similarity assumption will be quantified using the metric explained in the previous section to prevent the false similarity and negative transfer [29].

An efficient solar cell must provide desirable optical and electrical properties which can be quantified by the external quantum efficiency (EQE). EQE is defined as the ratio of number of generated electrons (N_e) to the number of incident photons on the solar cell (N_i):

$$\eta_e = \frac{N_e}{N_i} = \frac{N_p}{N_i}\frac{N_e}{N_p} = \eta_A\,\eta_I, \qquad (10)$$

where N_p is the number of absorbed photons and η_A is the absorption efficiency. η_I is called internal quantum efficiency (IQE) which quantifies the electron generation from the

absorbed photons. IQE and EQE are mostly measured experimentally in order to calculate short-circuit current density [30]. There are numerous examples of experimental measurements of IQE and EQE for solar cells. For example, Thouti et al. [31] measured IQE of a textured silicon solar cell to estimate the effective diffusion length due to the presence of resonance effect of silver nanoparticles. There have also been attempts at obtaining analytical expressions for the quantity. Ferrero et al. [32] proposed a method to calculate EQE of photodiodes based on Beer's law for absorption and Shockley–Read–Hall recombination. The structure is divided into front, space-charge (depletion) and rear regions and treated differently regarding recombination. Dibbs et al. [33] approximated the collection model as a step function, which is unity in the depletion region and zero elsewhere. In these two theoretical models, the exact knowledge of the dimensions of depletion zone and p and n layers are required. A probabilistic model of EQE similar in principle to those of Xue et al. [34,35] is used in this work which preserves the dependence of collection probability to absorber thickness and diffusion length. The essence of the model is as follows: In general, in a p-n junction solar cell, $\mathbb{P}_c$ is unity in the depletion region and decreases exponentially as the distance from depletion region increases [32,33].

$$\eta_e(\lambda) = \frac{1}{N_i}\iiint n_p(x,y,z,\lambda)\,\mathbb{P}_c(x,y,z)\,dxdydz, \qquad (11)$$

$$\eta_e = \int \eta_e(\lambda)\,d\lambda \qquad (12)$$

where the first integral is evaluated only within the absorber. $n_p(x,y,z,\lambda)$ is the number of absorbed photons at the given position and wavelength. We assume that the collection probability $\mathbb{P}_c(x,y,z) \equiv \mathbb{P}_c(z)$ so only the vertical position inside the absorber is taken into consideration. Previously [24], the collection probability was modeled as:

$$\mathbb{P}_c(z) = \exp\left(-\frac{|z - t_A/2|}{L_D}\right), \qquad (13)$$

where t_A is the absorber thickness and L_D is the diffusion length of the semiconductor. Readers interested in external quantum efficiency formulation are referred to the literature [24,30-35] including authors' previous studies. After making necessary simplifications and assuming charge generation is uniform along horizontal direction, $n_p(x,y,z,\lambda) \equiv n_p(z,\lambda)$ EQE is expressed as:

$$\eta_e = \frac{N_p}{N_i}\frac{2L_D}{t_A}\left(1 - \exp(-t_A/2L_D)\right) \qquad (14)$$

where N_p and N_i are calculated from:

$$N_p = \frac{1}{hc}\int \lambda\,\alpha(\lambda)\,I(\lambda)d\lambda, \qquad (15)$$

$$N_i = \frac{1}{hc}\int \lambda\,I(\lambda)d\lambda. \qquad (16)$$

L_D is a key material property for an effective carrier collection which depends on the diffusion coefficient and carrier lifetime: $L_D = \sqrt{D\tau}$. When L_D is much longer than the absorber thickness, all the generated carriers contributes the photocurrent and IQE equals to 1 therefore EQE solely depends on the absorption efficiency. However, this is generally not the case for most of the emerging technologies. For example, L_D is ~100 nm for amorphous silicon [36] which restricts the absorber thickness to the same order. On the other hand, the determination of L_D for organic materials is not as straightforward due to the bulk heterojunction (BHJ) structure of organic semiconductors. Although the individual values of L_D are known for donor and acceptor materials, its calculation for bulk heterojunction layers is complicated. Most of the organic materials have very short diffusion lengths ($< 10\ nm$) which significantly limits the possibility of sufficient light absorption. Therefore, bulk heterojunction blend solar cells were proposed in order to limit the distance of all the locations to the donor-acceptor interface to the diffusion length. Therefore, theoretical collection probability is 100% [37-39]. However, this is never the case in practical solar cells due to the microstructure or other internal effects which are out of the scope of the present study. In BHJ organic solar cells, it is better to use the term collection length (L_c) which is defined as the distance that an exciton can travel before reaching the other layer. L_c for organic BHJ corresponds to a value at the order of 100 nm.

Computation of the carrier generation rate requires solving Maxwell's equations which are mostly solved using computational methods except a few available analytical solutions for simple problems. In this study, FDTD method is used where Maxwell's equations are solved on discrete spatial and temporal domain grids, called Yee cells where electric and magnetic field vectors are placed on the edges according to the polarization mode. Also known as Yee's method, FDTD provides accurate and robust as well as fast solution to electromagnetic problems [40].

2.4 Gaussian Quadrature Integration

The use of Gaussian Process is not limited to optimization problems but can be helpful in many areas where uncertainty information is needed. For example, the computation of the integral in equation (14) requires the evaluation of the function many times for only one optimization iteration. This method is similar to sigma-points methods [41], and relies on the evaluation of the objective functions only where most contributions are made. As the uncertainty of the predictions can be obtained in GP predictions, evaluating functions only at high uncertainties will result in an exploration based numerical integration framework. This algorithm is summarized in Table 1.

Table 1. GP-based numerical integration.

Evaluate $I = \int_{x_1}^{x_2} y(x)\,dx$

Input: n: initial number of sampling points, δ: convergence criterion, $N_{iter,max}$: maximum allowed iterations.

1. Sample n initial points: $\mathbf{x}_{1:n}^0 \sim \mathcal{U}(x_1, x_2)$.
2. Evaluate the initial points: $y^0 = y(\mathbf{x}^0)$
3. Fit a $\mathcal{GP}(\mathbf{x}^0, y^0)$
4. Set $\mathbf{x} = \mathbf{x}^0, y = y^0, t = 0$

Until terminate do:

5. Calculate the numerical integral using trapezoidal rule: $I_t = trapz(\mathbf{x}, \mathbf{y})$
6. Find k s.t. $x^{new} = \arg\max_\chi |I_{pred} - I_t|$ where

$I_{pred} = trapz([\mathbf{x}; \chi], [\mathbf{y}; \mu_y(\chi)])$
$\mu_y(\mathbf{x}^{\text{test}}), \sigma_y^2(\mathbf{x}^{\text{test}}) \sim \mathcal{GP}(\chi \,|\, \mathbf{x}, y)$

7. Sample the new point: $x_{new} = \chi$ and calculate $y_{new} = y(\chi)$
8. Set $\mathbf{x} = [\mathbf{x}, x_{new}]^T$ and $\mathbf{y} = [\mathbf{y}, y_{new}]^T$
9. $t = t + 1$
10. Terminate if $\frac{(I_t - I_{t-1})}{I_{t-1}} < \delta$ or $t > N_{iter,max}$ or go to step 5.

2.4 Optimization Procedure

The optimization problem solved in this study is given as:

$$\max_{\mathbf{x}} \eta_e(\mathbf{x}),$$
$$\mathbf{x_L} < \mathbf{x} < \mathbf{x_U}. \tag{17}$$

where $\eta_e(\mathbf{x})$ is given in equation (13). As mentioned earlier, we assume that at least one optimization is already performed (task 0). When we move to a new task, the similarity of the tasks is first evaluated on a small representative batch of data using equation (8). This initial batch is selected as the first n iterations of task 0 so that we make the comparison without spending extra computation. Moreover, this initial batch can also be used for constructing the first GP model. Then the first point which maximizes utility function (1) is found and the GP is updated. The cycle of sampling, evaluation and utility function maximization continues until convergence. Please note that the integral in equation (15) is evaluated using Gaussian Quadrature explained in the previous section. In both GP models, matern 5/2 kernel is preferred. In this study, the infinite-metric GP optimization (IMGPO) algorithm developed by Kawaguchi et al. [42] is used.

The materials used in different layers of the solar cells in different tasks are presented in Table 2.

Table 2. Materials of the layers used in different tasks

#	ARC	IL-1	Absorber	IL-2	Metal
0	ITO	ZnO	P3HT:PCBM	MoO$_3$	Al
1	ITO	SiO$_2$	a-Si	Al$_2$O$_3$	Al
2	Si$_3$N$_4$	PEDOT:PSS	PCPDTBT:PCBM	Al$_2$O$_3$	Al

3. RESULTS AND DISCUSSION

In this section, the optimization results are presented. The similarity metrics among the tasks are shown in Figure 2. Similarity is quantified by the metric in equation (8). Furthermore, Pearson (r) and Spearman (ρ) correlation coefficients are calculated as reference. These coefficients are commonly used to determine correlations among different data sets. As can be seen from the plots and numerical values of r and

ρ, there are strong correlations among almost all of the tasks. However, a strong linear relation can be a weak estimator of similarity as it only shows they increase/decrease with the same change in the inputs. Therefore, the similarity metric is formulated to determine the effect of deviation between the tasks. The numerical values translate to the weights when calculating the mean and standard deviation of the common surface.

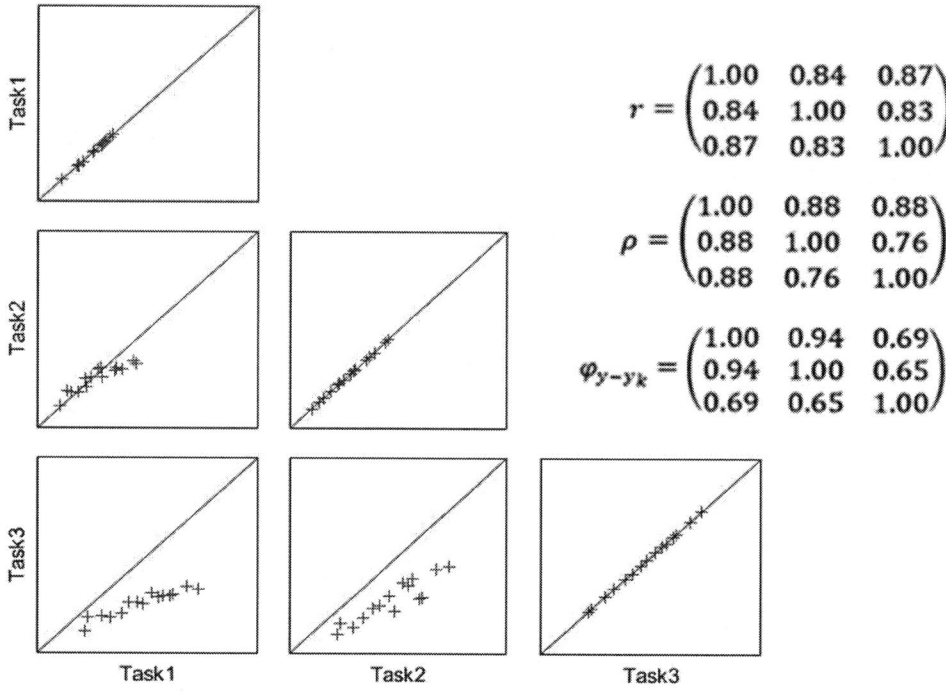

$$r = \begin{pmatrix} 1.00 & 0.84 & 0.87 \\ 0.84 & 1.00 & 0.83 \\ 0.87 & 0.83 & 1.00 \end{pmatrix}$$

$$\rho = \begin{pmatrix} 1.00 & 0.88 & 0.88 \\ 0.88 & 1.00 & 0.76 \\ 0.88 & 0.76 & 1.00 \end{pmatrix}$$

$$\varphi_{y-y_k} = \begin{pmatrix} 1.00 & 0.94 & 0.69 \\ 0.94 & 1.00 & 0.65 \\ 0.69 & 0.65 & 1.00 \end{pmatrix}$$

FIGURE 2: COMPARISON OF THE RESPONSES OF THE TASKS GIVEN THE SAME INPUT SET WITH THE SIMILARITY METRIC (φ), PEARSON (r) AND SPEARMAN (ρ) CORRELATION COEFFICIENCTS

The results of the optimizations are shown in Table 3. Optimal geometry vectors from Bayesian optimization with and without transfer learning are presented. Furthermore, the spectral absorptivity profiles of the optimized designs are presented in Figures 3-5. Note that the black lines are the results of the direct optimizations presented as references. The results of the optimizations show that the final absorptivity profiles are consistent with the direct optimization results.

Table 3. Results of optimizations

Task	t^*_{ARC}	t^*_{IL1}	t^*_{A}	t^*_{IL2}	t^*_{M}	EQE^*
0 no TL	82	15	78	12	108	0.367
1 no TL	26	17	64	26	99	0.368
1 w/ TL	28	20	66	22	102	0.371
2 no TL	40	8	92	8	110	0.355
2 w/ TL	38	12	96	5	100	0.357

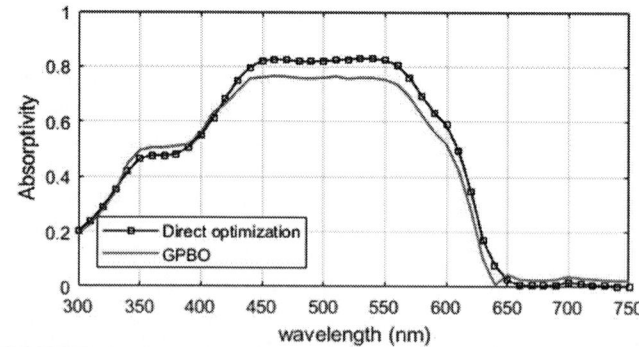

FIGURE 3: ABSORPTIVITY PROFILES OF THE 0TH TASK (BASE) FOR OPTIMIZED GEOMETRIES USING DIRECT OPTIMIZATION AND GAUSSIAN PROCESS BAYESIAN OPTIMIZATION (GPBO)

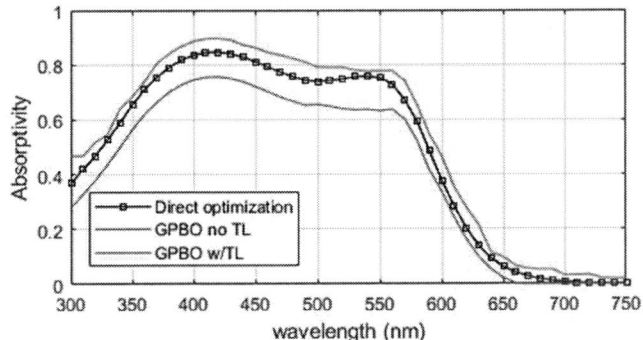

FIGURE 4: ABSORPTIVITY PROFILES OF THE 1ST TASK FOR OPTIMIZED GEOMETRIES USING DIRECT OPTIMIZATION AND GAUSSIAN PROCESS BAYESIAN OPTIMIZATION (GPBO) WITHOUT (NO TL) AND WITH TRANSFER LEARNING (W/ TL).

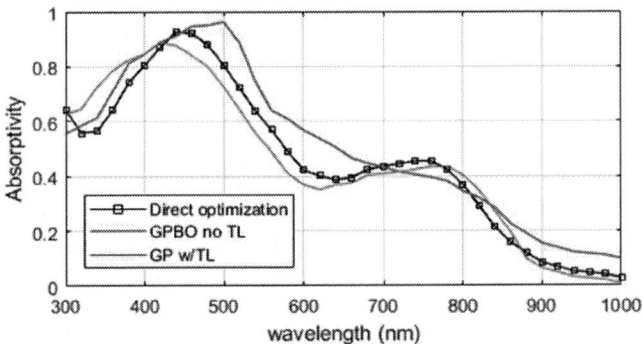

FIGURE 5: ABSORPTIVITY PROFILES OF THE 2ND TASK FOR OPTIMIZED GEOMETRIES USING DIRECT OPTIMIZATION AND GAUSSIAN PROCESS BAYESIAN OPTIMIZATION (GPBO) WITHOUT (NO TL) AND WITH TRANSFER LEARNING (W/ TL).

In all of the cases, direct optimizations were carried out using a heuristic optimization method, i.e. simulated annealing without any approximation methods. Furthermore, it can be seen from Figures 3-5 that GPBO could reach the final absorption profiles close to the ground truth. After GPBO optimizations were performed without transfer learning, the information in Task 0 is transferred to Task 1 and information in Tasks 0 and 1 are transferred to Task 2. The evolution of EQE values during optimizations are also presented in Figures 6-8. One of the conclusions from these results is that transfer learning helps reducing the time of optimization by leading the predictions for the function value improvements to better subspaces in the design space.

The similarity of these cases is mostly originated from the spectral behavior of the absorber materials. Although materials used in this study are not chemically similar, they share some common characteristics. For example, all three materials show peak absorptivity in the visible region. Furthermore, P3HT:PCBM of Task 0 and amorphous silicon of Task 1 have similar band gaps so that the absorption approaches zero at similar wavelengths. These similarities can be observed in the similarity metric.

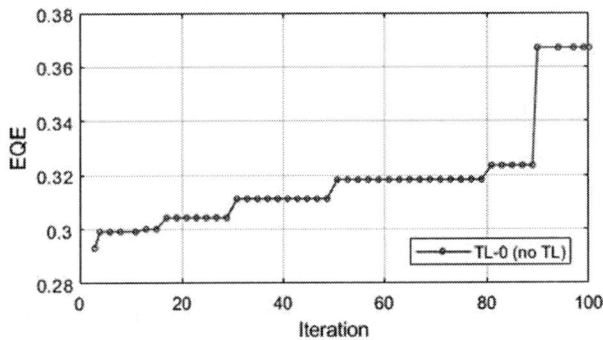

FIGURE 6: EVOLUTION OF THE EQE DURING OPTIMIZATION OF TASK 0 WITHOUT TL.

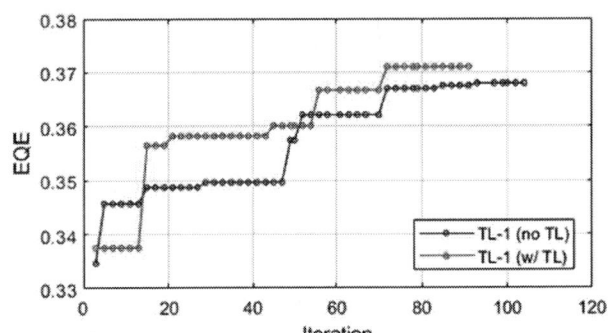

FIGURE 7: EVOLUTION OF THE EQE DURING OPTIMIZATION OF TASK 1 WITHOUT AND WITH TL.

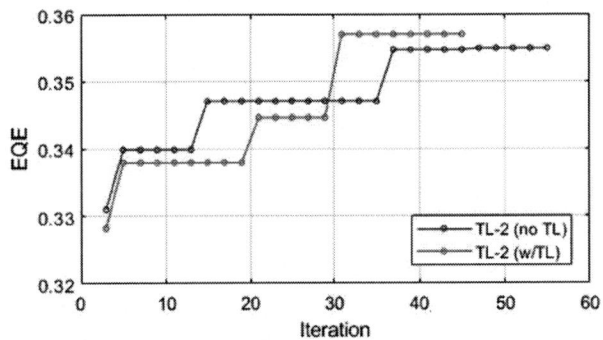

FIGURE 8: EVOLUTION OF THE EQE DURING OPTIMIZATION OF TASK 2 WITHOUT AND WITH TL.

4. CONCLUSION

In this study, a transfer learning based design optimization methodology was presented. The method was used with Bayesian optimization using Gaussian Processes. Knowledge transfer was modeled by means of a common response surface where mean and standard deviations from the previous optimizations are used to increase the accuracy of the Gaussian Process predictions and consequently the speed of the optimization. The common surface is developed using weighted means and standard deviations which enables to transfer knowledge from the most related sources. As a result, 10-20% reduction in the computation time can be achieved.

REFERENCES

[1] Gupta, Abhishek, Ong, Yew-Soon, and Feng, Liang. "Insights on Transfer Optimization: Because Experience is the Best Teacher." *IEEE Transactions on Emerging Topics in Computational Intelligence* Vol. 2 No. 1 (2018): pp. 51-64

[2] Quattoni, Arziadna, Collins, Michael, and Darrell, Trevor. "Transfer learning for image classification with sparse prototype representations." *Proceedings of the IEEE Conference on Computer Vision and Pattern Recognition.* pp. 1-8. 2008.

[3] Raina, Rajat, Battle, Alexis, Lee, Honglak, Packer, Benjamin, and Ng., Andrew Y. "Self-taught learning: transfer learning from unlabeled data." *In Proceedings of the 24th international conference on Machine learning.* pp. 759-766. 2007.

[4] Fung, Gabriel P. C., Yu, Jeffrey X., Lu, Hongjun, and Yu, Philip S.. "Text classification without negative examples revisit." *IEEE Transactions on Knowledge & Data Engineering* Vol. 1 (2006): pp. 6-20.

[5] Dai, Wenyuan, Yang, Qiang, Xue, Gui-Rong, and Yu, Yong. "Boosting for transfer learning." *Proceedings of the 24th International Conference on Machine Learning*, pp. 193–200. 2007.

[6] Zanini, Paolo, Congedo, Marco, Jutten, Christian, Said, Salem, and Berthoumieu, Yannick. "Transfer learning: a Riemannian geometry framework with applications to brain–computer interfaces." *IEEE Transactions on Biomedical Engineering* Vol. 65 No. 5 (2018): pp. 1107-1116.

[7] Waytowich, Nicholas R., Lawhern, Vernon J., Bohannon, Addison W., Ball, Kenneth R., and Lance, Brent J.. "Spectral transfer learning using information geometry for a user-independent brain-computer interface." *Frontiers in neuroscience* Vol. 10 (2016): p. 430.

[8] Choi, Keunwoo, Fazekas, Gyorgy, Sandler, Mark, and Cho, Kyunghyun. "Transfer learning for music classification and regression tasks." Preprint, 2017 http://arxiv.org/abs/1703.09179

[9] Lin, Yuan-Pin, and Jung, Tzyy-Ping. "Improving eeg-based emotion classification using conditional transfer learning." *Frontiers in human neuroscience* Vol. 11 (2017): p. 334.

[10] Pardoe, David, and Stone, Peter. "Boosting for regression transfer." *Proceedings of the 27th International Conference on International Conference on Machine Learning*, pp. 863-870. 2010.

[11] Jamshidi, Pooyan, Velez, Miguel, Kästner, Christian, Siegmund, Norbert, and Kawthekar, Prasad. "Transfer learning for improving model predictions in highly configurable software." *Proceedings of the 12th International Symposium on Software Engineering for Adaptive and Self-Managing Systems*, pp. 31-41. 2017.

[12] Lindner, Claudia, Waring, D., Thiruvenkatachari, B., O'Brien, K., and Cootes, Timothy F.. "Adaptable landmark localisation: applying model transfer learning to a shape model matching system." *Proceedings of the International Conference on Medical Image Computing and Computer-Assisted Intervention*, pp. 144-151. 2017.

[13] Gao, Jin, Ling, Haibin, Hu, Weiming, and Xing, Junliang. "Transfer learning based visual tracking with gaussian processes regression." *Proceedings of the European conference on computer vision*, pp. 188-203. 2014.

[14] Kaya, Mine and Hajimirza, Shima. "Using a Novel Transfer Learning Method for Designing Thin Film Solar Cells with Enhanced Quantum Efficiencies." *Scientific Reports* Vol. 9 No. 1 (2019): p. 5034.

[15] Li, Dandan, Wang, Shenzhang, Yao, Shuzhen, Liu, Yu-Hang, Cheng, Yuanqi, and Sun, Xian-He. "Efficient Design Space Exploration by Knowledge Transfer." *Proceedings of International Conference on Hardware/Software Codesign and System Synthesis (CODES+ISSS)* p.12. 2016.

[16] Min, Alan T. W., Sagarna, Ramon, Gupta, Abhishek Ong, Yew-Soon and Goh, Chi K.. "Knowledge transfer through machine learning in aircraft design." *IEEE Computational Intelligence Magazine* Vol. 12 No. 4 (2017): pp. 48-60.

[17] Hajimirza, Shima, El Hitti, Georges, Heltzel, Alex and Howell, John R.. "Specification of micro-nanoscale radiative patterns using inverse analysis for increasing solar panel efficiency." *Journal of Heat Transfer* Vol. 134 No. 10 (2012): p. 102702.

[18] Hajimirza, Shima, and Howell, John R.. "Design and analysis of spectrally selective patterned thin-film cells." *International Journal of Thermophysics* Vol. 34 No. 10 (2013): pp. 1930-1952.

[19] Hajimirza, Shima. "Expedited Quasi-Updated Gradient Based Optimization Techniques for Energy Conversion Nano-Materials." *Journal of Nanoelectronics and Optoelectronics* Vol. 10 No. 1 (2015): pp. 140-146.

[20] Ferry, Vivian E., Polman, Albert and Atwater, Harry A.. "Modeling light trapping in nanostructured solar cells." *ACS Nano* Vol. 5 No. 12 (2011): pp. 10055-10064.

[21] Atwater, Harry A., and Polman, Albert. "Plasmonics for improved photovoltaic devices." *Nature Materials* Vol. 9 No. 3 (2010): p. 205.

[22] Kaya, Mine and Hajimirza, Shima. "Surrogate based modeling and optimization of plasmonic thin film organic solar cells." *International Journal of Heat and Mass Transfer* Vol. 118 (2018): pp. 1128-1142.

[23] Kaya, Mine and Hajimirza, Shima. "Application of artificial neural network for accelerated optimization of ultra thin organic solar cells." *Solar Energy* Vol. 165 (2018): pp. 159-166.

[24] Kaya, Mine and Hajimirza, Shima. "Rapid Optimization of External Quantum Efficiency of Thin Film Solar Cells Using Surrogate Modeling of Absorptivity." *Scientific Reports* Vol. 8 No. 1 (2018): p. 8170.

[25] Mockus, Jonas. "Application of Bayesian approach to numerical methods of global and stochastic optimization." *Journal of Global Optimization* Vol. 4 No. 4 (1994): pp. 347-365.

[26] Bonilla, Edwin V., Chai, Kian Ming A., and Williams, Christopher K. I.. "Multi-task Gaussian process prediction." *Proceedings of the Advances in neural information processing systems (NIPS)* pp. 153-160. 2008.

[27] Yogatama, Dani and Mann, Gideon. "Efficient transfer learning method for automatic hyperparameter tuning." *Proceedings of the 17th International Conference on Artificial*

Intelligence and Statistics (AISTATS). pp. 1077-1085. Reykjavik, Iceland 2014.

[28] Zhao, Zingxue, Xi, Zhimin, Xu, Hongyi, Yang, and Ren-Jye. "Model bias characterization considering discrete and continuous design variables." *Proceedings of the ASME 2016 International Design Engineering Technical Conferences and Computers and Information in Engineering Conference (IDETC/CIE)* pp. V02BT03A053-V02BT03A053. Charlotte, North Carolina 2016.

[29] Pan, Sinno J., Yang, Qiang, "A Survey on Transfer Learning." *IEEE Transactions on Knowledge and Data Engineering* Vol. 22, pp. 1345-1359 (2009).

[30] Berginski, Michael, Hüpkes, Jürgen, Schulte, Melanie, Schöpe, Gunnar, Stiebig, Helmut, Rech, Bernd, and Wuttig, Matthias. "The effect of front ZnO: Al surface texture and optical transparency on efficient light trapping in silicon thin-film solar cells." *Journal of Applied Physics* Vol. 101 no. 7 (2007): p. 074903.

[31] Thouti, Eshwar, Sharma, Ashok K., Sardana, Sanjay K., and Komarala, Vamsi K.. "Internal quantum efficiency analysis of plasmonic textured silicon solar cells: surface plasmon resonance and off-resonance effects." *Journal of Physics D: Applied Physics* Vol. 47 No. 42 (2014): p. 425101.

[32] Ferrero, Alejandro, Campos, Joaquin, Pons, Alicia, and Corrons, Antonio. "New model for the internal quantum efficiency of photodiodes based on photocurrent analysis." *Applied Optics* Vol. 44 No. 2 (2005): pp. 208-216.

[33] Dibb, George FA, Muth, Mathis-Andreas, Kirchartz, Thomas, Engmann, Sebastian, Hoppe, Harald, Gobsch, Gerhard, Thelakkat, Mukundan et al. "Influence of doping on charge carrier collection in normal and inverted geometry polymer: fullerene solar cells." *Scientific Reports* Vol. 3 (2013): p. 3335.

[34] Xue, Jiangeng, Rand, Barry P., Uchida, Soichi and Forrest, Stephen R.. "A hybrid planar–mixed molecular heterojunction photovoltaic cell." *Advanced Materials* Vol. 17 No. 1 (2005): pp. 66-71.

[35] Xue, Jiangeng, Uchida, Soichi, Rand, Barry P., and Forrest, Stephen R.. "Asymmetric tandem organic photovoltaic cells with hybrid planar-mixed molecular heterojunctions." *Applied Physics Letters* Vol. 85 No. 23 (2004): pp. 5757-5759.

[36] Shah, A. V., Schade, H., Vanecek, M., Meier, J., Vallat-Sauvain, E., Wyrsch, N., Kroll, U., Droz, C., and Bailat, J.. "Thin-film silicon solar cell technology." *Progress in Photovoltaics: Research and Applications* Vol. 12 No. 2-3 (2004): pp. 113-142.

[37] Petoukhoff, Christopher E., Vijapurapu, Divya K., and O'Carroll, Deirdre M.. "Computational comparison of conventional and inverted organic photovoltaic performance parameters with varying metal electrode surface workfunction." *Solar Energy Materials and Solar Cells* Vol. 120 (2014): pp. 572-583.

[38] Nazerdeylami, S., and Dizaji, H. Rezagholipour. "A theoretical study of influence of charge carrier mobility in PTB7: PC71BM bulk heterojunction organic solar cells." *Optical and Quantum Electronics* Vol. 48 No. 11 (2016): p. 506.

[39] Wen, Long, Chen, Qin, Song, Shichao, Yu, Yan, Jin, Lin, and Hu, Xin. "Photon harvesting, coloring and polarizing in photovoltaic cell integrated color filters: efficient energy routing strategies for power-saving displays." *Nanotechnology* Vol. 26, No. 26 (2015): p. 265203.

[40] Schneider, John B. "Understanding the finite-difference time-domain method" (2010), online. http://bbs.hwrf.com.cn/downebd/90845d1368122665-ufdtd.pdf

[41] Karvonen, Toni, and Särkkä, Simo. "Classical quadrature rules via Gaussian processes." *Proceedings of the IEEE 27th International Workshop on Machine Learning for Signal Processing (MLSP)*, pp. 1-6. 2017.

[42] Kawaguchi, Kenji, Kaelbling, Leslie P., and Lozano-Pérez, Tomás. "Bayesian Optimization with Exponential Convergence" *Proceedings of the Advances in Neural Information Processing Systems 28 (NIPS)* pp. 2809-2817. 2015.

Proceedings of the ASME 2019
International Design Engineering Technical Conferences
and Computers and Information in Engineering Conference
IDETC/CIE2019
August 18-21, 2019, Anaheim, CA, USA

DETC2019-98385

DATA-DRIVEN DYNAMIC NETWORK MODELING FOR ANALYZING THE EVOLUTION OF PRODUCT COMPETITIONS

Jian Xie
School of Mechanical Engineering
Beijing Institute of Technology
Beijing, China

Youyi Bi
Integrated Design Automation
Laboratory
Northwestern University
Evanston, IL, USA

Zhenghui Sha
System Integration & Design
Informatics Laboratory
University of Arkansas
Fayetteville, AR, USA

Mingxian Wang, Yan Fu
Ford Analytics
Global Data Insight &
Analytics
Ford Motor Company
Dearborn, MI, USA

Noshir Contractor
Science of Networks in
Communities
Northwestern University
Evanston, IL, USA

Lin Gong
School of Mechanical
Engineering
Beijing Institute of
Technology
Beijing, China

Wei Chen[1]
Integrated Design
Automation Laboratory
Northwestern University
Evanston, IL, USA

ABSTRACT

Understanding the impact of engineering design on product competitions is imperative for product designers to better address customer needs and develop more competitive products. In this paper, we propose a dynamic network based approach to modeling and analyzing the evolution of product competitions using multi-year product survey data. We adopt Separate Temporal Exponential Random Graph Model (STERGM) as the statistical inference framework because it considers the evolution of dynamic networks as two separate processes: formation and dissolution. This treatment allows designers to investigate why two products enter into competition and why a competitive relationship preserves or dissolves over time. In an open market, the available products to customers are continuously changing over the time, posing challenges for conventional modeling methods concerning fixed product input. Consequently, we propose to leverage "structural zeros" in STERGM to tackle the problem of modeling varying product competitors as nodes in dynamic networks. We use China's automotive market as a case study to illustrate the implementation of the proposed approach and its benefits compared to the static network modeling approach based on Exponential Random Graph Model (ERGM). The results show that our approach identifies the driving factors associated with product attributes and current market competition structures for the change of competition in both formation and dissolution processes. The insights gained from this paper can help

designers better interpret the temporal changes of product competition relations and make product design decisions with the aid of dynamic network-based models.

Keywords: engineering design, product competition, dynamic network analysis, ERGM, STERGM

1. INTRODUCTION

The **objective of this paper** is to study the impact of engineering design and existing market competitions on the evolution of product competition relations. To achieve this objective, we develop an approach based on dynamic network modeling using multi-year product survey data. Product competition occurs when customers consider multiple products for evaluation before making the final purchase decision [1]. Therefore, product competition can be indicated by customers' consideration sets [2]. It is therefore of great importance for product designers to gain insights into the factors that can impact product competition relations, such as key design attributes, similarities or differences of design attributes among competing products, design improvement as well as existing market competition structures. These insights can help designers develop more competitive products thus better address customer needs. They can also support companies' strategic decision making, such as branding, product positioning, and marketing.

Our recent studies explored the capability of utilizing network analysis to model product competitions [2,3]. Network-based approaches model products as nodes and the co-

---

[1] Contact author: weichen@northwestern.edu

consideration relations (competition) between products as links. By taking the dependencies among links into consideration, network-based approaches are effective in modeling complex and interdependent relations [4,5]. In our previous research, we developed a unidimensional network-based approach based on Exponential Random Graph Model (ERGM) [6] to study product competitions in the form of product co-considerations [3], and used this model to assess the impact of technological changes (e.g., turbo engine) on product competitions and customers' co-consideration behaviors [1]. Later, we extended the unidimensional network-based approach to a multidimensional network structure where customer social relations were included to study their influence on heterogeneous customer preferences [5]. Despite the strength of these earlier developed network models [1–3,5,7,8], they are static in nature that ignore the dynamic change of product competitions over time. The models are limited in explaining why a product may maintain or lose its competitive advantage over time.

In a highly dynamic market environment, product competition changes over time [9], e.g., due to release of new products, withdrawal of outdated products, technological progress, and social changes. For example, the market share of U.S. brands in the Chinese auto market fell from 12.2% in 2017 to 10.7% in 2018, possibly due to a delayed refresh of the U.S. lineups [10], which demonstrates the negative effect of untimely design improvement on product competition. The change of customer preferences is another potential cause for the evolution of product competition. Fuel-efficient cars have been more desirable since the energy crisis in 1970s as customers become more sensitive to rising gas prices [11]. Since early 2000s, higher fuel efficiency has also contributed to the increasing competitiveness of hybrid vehicles [12]. Therefore, a thorough understanding of the dynamic changes in market competition is of great significance in many engineering design scenarios, such as product feature competition (e.g., whether to upgrade design features of an existing car and by how much) and new product positioning (e.g., whether to develop a new car model to fill a specific market niche).

To address this research need, we propose a dynamic network modeling approach to study the evolution of product competition relations. Specifically, we adopt Separable Temporal ERGM (STERGM) [13,14] as the statistical inference framework for dynamic network modeling. The reasons are in three aspects:

- First, existing degree-based generative dynamic network models (e.g., small world network [15], scale-free network [16], and the dynamic stochastic block model [17]) are limited in modeling the influence of nodal and edge characteristics (e.g., customer demographics and relation strength) on the change of network structures [18]. In contrast, STERGM [13] can study both growing and shrinking dynamics, i.e., how the explanatory factors influence the formation of new links and the dissolution of old links, an important feature to study the impact of product design decisions. Compared to Stochastic Actor-Oriented Model (SAOM) [19], another widely used longitudinal

network model which treats nodes as actors who make active decisions, STERGM is more advantageous for our purpose because it is a link-oriented model that can include nodes which are not actors, e.g., products in our work that cannot be treated as intelligent decision makers [20].

- Second, in the social network literature, STERGM has been applied in modeling various dynamic social relations, such as the evolution of social networks of politicians [21] and international trade networks [22]. Researchers found that both exogenous factors (e.g., economic characteristics of counties) and preexisting network structures (e.g., reciprocity for bilateral trade relations and triadic closure effects for trilateral trade relations) can influence international trade networks over time [22]. Thus, STERGM allows us to investigate how preexisting competition relations would influence product competitions in addition to the impact of product features.

- Third, STERGM can be viewed as an extension to the static ERGM. Therefore, the results from STERGM and ERGM can be compared and connected to further advance our understanding on product competition relations.

The **main contribution** of this work is the development of a dynamic network-based modeling approach rooted in STERGM for studying the evolution of product competition relations using customer survey data from multiple years. On the other hand, in previous work, STERGM only handles the same set of nodes over time in modeling dynamic networks [21,22]. However, for product competition modeling, the sets of products in different years are inconsistent because new products appear in the market and existing products exit the market from time to time. In this research, we leverage the concept of "structural zeros" [23] to tackle the problem (see Section 2.3 for details). So the secondary contribution is the "structure-zero" technique that successfully tackles the inconsistency of product consideration set from year to year in dynamic network modeling. Results from STERGM can be used to explain why two products enter into competition and why a competitive relationship is preserved or dissolved over time. We demonstrate the proposed approach using multiple-year survey data from China's auto market, and show the benefits of our approach compared to the static ERGM.

The remaining of the paper is organized as follows. Section 2 provides the technical background of network analysis, the static network modeling technique (ERGM), the dynamic network modeling technique (STERGM), and the structural zeros method. In Section 3, a general approach for modeling dynamic product competition relations based on STERGM is introduced and illustrated using the data associated with China's SUV market. The process of preparing the dataset, identifying modeling attributes, and handling inconsistent vehicle nodes in dynamic network modeling are explained. The results of STERGM are presented and compared with the results of the static network modeling approach (ERGM). To verify the results, model fit evaluation is performed at both the link level and the network level. Section 4 concludes with closing thoughts, the implications in engineering design and future research opportunities.

2. TECHNICAL BACKGROUND OF DYNAMIC NETWORK ANALYSIS AND MODELING

Network analysis has been recognized as an essential method for analyzing and modeling complex systems in a wide variety of fields such as biology, computer science, social science, and engineering [5,24–26]. With network analysis, the structure of a system is visualized and simplified as a graph, where nodes represent entities in the system and edges/links represent relationships between entities. In this section, two statistical network models for network analysis, i.e., Exponential Random Graph Model (ERGM) and Separable Temporal Exponential Random Graph Model (STERGM), and the structural zeros method are introduced.

2.1 Exponential Random Graph Model (ERGM)

ERGM is a flexible statistical inference framework, which assumes an observed network y as an instance of random networks Y given by the distribution in Eqn. (1),

$$Pr(Y = y) = \frac{exp\left(\theta^T \cdot g(y)\right)}{\kappa(\theta, y)} , \qquad (1)$$

where θ is a vector of corresponding model parameters, and $\kappa(\theta, y)$ is a normalizing constant to guarantee the equation is a proper probability distribution. In the context of this work, network Y captures product competition relations, which are identified based on product co-considerations from the customer survey (see more details in Section 3.1). $g(y)$ is a vector of network statistics of interest including attributes of nodes, attributes of links as well as network structure attributes [5]. For example, the two-star structure in a vehicles competition network shows whether a car is co-considered with other two distant competitors. Fig. 1 provides three exemplary network structures that can be modeled by ERGM. Eqn. (1) suggests that the probability of observing a specific network structure is proportional to the exponent of a weighted combination of network statistics. The estimated ERGM parameters θ indicate the importance of the network statistics to the formation of links in a network. For example, a positive θ of the triangle effect in an ERGM for the vehicle competition network implies that those vehicles involved in three-way competitions are more likely to compete with each other.

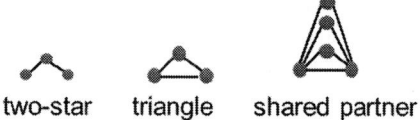

two-star   triangle   shared partner

FIGURE 1: Exemplary network structures in a vehicle competition network. The two-star structure indicates whether a car will have two distant competitors. The triangle structure indicates whether three cars compete with each other (three-way competition). The shared partner structure indicates whether two cars sharing the same set of competitors will be considered.

The estimated ERGM parameters θ can also be used to calculate the log-odds of the formation of certain links, i.e., how likely a link will exist between two nodes given their nodal attributes and the structure of the rest network as shown in Eqn. (2):

$$Logit\, Pr\big(y_{ij} = 1|y_{-ij}\big) = log\,\frac{Pr(y_{ij}=1|y_{-ij})}{Pr(y_{ij}=0|y_{-ij})}$$
$$= \theta^T \cdot \big(g(y|y_{ij} = 1) - g(y|y_{ij} = 0)\big) = \theta^T \cdot \delta_{ij}(y) , \qquad (2)$$

where y_{ij} is the link between node i and j, y_{-ij} is the network excluding the link between node i and j, and $\delta_{ij}(y)$ is the difference of the network statistics between the network where the link between node i and j exists (i.e., $g(y|y_{ij} = 1)$) and the network where the link between node i and j does not exist (i.e., $g(y|y_{ij} = 0)$). If we get positive log-odds for nodes i and j, this indicates that having a link between them is more likely than not having the link and *vice versa*.

2.2 Separable Temporal Exponential Random Graph Model (STERGM)

As an extension of ERGM, STERGM is established to model dynamic networks. As an example shown in Fig. 2, STERGM treats the evolution from the network at time t (Y^t) to the network at time $t + 1$ (Y^{t+1}) as two separate processes: 1) **link formation** in which new links are created following $Pr(Y^+ = y^+|Y^t\theta^+)$, and 2) **link dissolution** in which old links disappear following $Pr(Y^- = y^-|Y^t\theta^-)$. The network at time $t + 1$ is constructed by applying the changes in Y^+ and Y^- to Y^t following $Y^{t+1} = Y^- \cup (Y^+ - Y^t)$. Here θ^+ and θ^- denote the parameters of the formation model (Y^+) and dissolution model (Y^-), respectively.

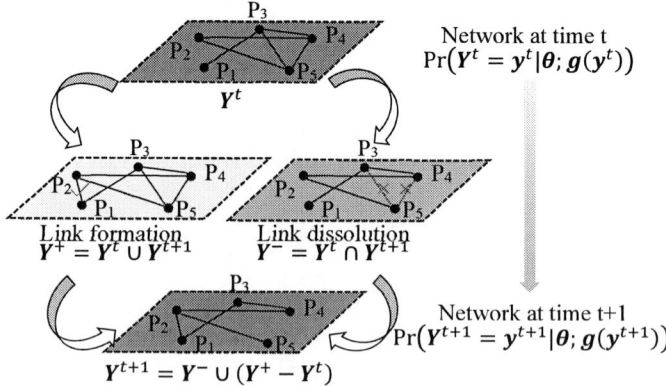

FIGURE 2: Evolution dynamics of product competition network

For each discrete time step, the process of formation and dissolution are independent conditional on the network at time t. This means,

$$Pr(Y^{t+1} = y^{t+1}|Y^t = y^t\theta) = Pr(Y^+ = y^+|Y^t = y^t\theta^+) \cdot Pr(Y^- = y^-|Y^t = y^t\theta^-). \qquad (3)$$

Associating STERGM with ERGM, the probability distribution of formation model and dissolution model are expressed as

$$Pr(Y^+ = y^+ | Y^t = y^t, \theta^+) = \frac{exp\left(\theta^{+T} \cdot g(y^{t+1}, y^t)\right)}{\kappa(\theta^+, y^t)}, \text{ and } \quad (4)$$

$$Pr(Y^- = y^- | Y^t = y^t, \theta^-) = \frac{exp\left(\theta^{-T} \cdot g(y^{t+1}, y^t)\right)}{\kappa(\theta^-, y^t)}. \quad (5)$$

The normalizing denominator $\kappa(\theta^+, y^t)$ and $\kappa(\theta^-, y^t)$ are the sum of network statistics of all possible formation and dissolution networks, respectively. Here these formation and dissolution networks only include possible variations to Y^t (i.e., the additions and subtractions). In contrast, the normalizing denominator of ERGM includes all networks from an empty network to a complete network (i.e., a network in which all nodes are linked with each other).

Similar to ERGM, STERGM can include both exogenous (e.g., nodal attributes) and endogenous (network structures) variables in network modeling. This enables the prediction of products' future competition relations considering the market's present competition structure and the influence from design changes. In addition, not only can STERGM identify the design features contributing to the formation of competition between two products, it can also identify the features influencing the dissolution of competitions. For example, when assessing the influence of certain SUV attributes on vehicles' competitiveness given the current market structure, if the estimated effect of third-row seat feature is positive and statistically significant in the formation model, it implies that improving the design by including third-row seats would make the SUV more likely to be co-considered against its competitors. In the dissolution model, if the estimated parameter of fuel consumption is negative and statistically significant, it means that better fuel economy in the SUV would make its competing relationships more likely to persist (not dissolve).

2.3 Structural zeros method

To address the challenge of modeling changing network nodes in STERGM, we propose a new method based on *structural zeros*. Given a unidimensional network consisting of N nodes, this network can be presented by a $N \times N$ matrix with entries of binary indicator 0 or 1. *Structural zeros*, which also called fixed zeros or necessarily empty cells, are a set of predefined zero inputs in the matrix to restrict the relations among some nodes in the network (i.e., these nodes have no links) [27]. Therefore, the distribution of dynamic network at time $t + 1$ can be written as

$$Pr(Y^{t+1} = y^{t+1} | Y^t = y^t, R^t = r^t, \theta), \quad (6)$$

where $R^t \in \mathbb{R}^{N \times N}$ is binary matrix corresponding to Y^t and r^t is a realization of R^t, such as

$$r = \begin{bmatrix} 0 & 1 & 1 & 0 & 0 \\ 1 & 0 & 1 & 0 & 1 \\ 1 & 1 & 0 & 0 & 1 \\ 0 & 0 & 0 & 0 & 0 \\ 0 & 1 & 1 & 0 & 0 \end{bmatrix}, \quad (7)$$

where blocks denoted as $\mathbf{0}$ are structural zero blocks which enforce y_{15}, y_{x4}, y_{4x} and y_{xx} ($x = 1, 2, \dots, 5$) to be 0 in Y^{t+1}, while the other entries denoted as 1 indicate corresponding relations in Y^{t+1} have no restrictions (i.e., they are free to change to 0 or remain as 1). Fig. 3 shows an exemplary network following this matrix r.

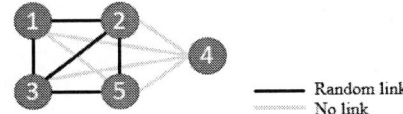

FIGURE 3: A network example based on matrix r

Thus, the dynamic network model based on structural zeros can be represented as

$$Pr(Y^{t+1} = y^{t+1} | Y^t = y^t, R^t = r^t, \theta) = \frac{exp\left(\theta^{+T} \cdot g(y^{t+1}, y^t)\right)}{\kappa(\theta^+, y^t, r^t)} \cdot \frac{exp\left(\theta^{-T} \cdot g(y^{t+1}, y^t)\right)}{\kappa(\theta^-, y^t, r^t)}, \quad (8)$$

where possible networks of Y^+ and Y^- in $\kappa(\theta^+, y^t, r^t)$ and $\kappa(\theta^-, y^t, r^t)$ have been reduced because of r^t.

Several applications of structural zeros can be found in different areas, such as parametric models [28], control system [29] and social network [30,31]. It was used commonly in network modeling since Robins et.al. [30] who used structural zeros to represent the relationship in those ordered pair of actors that are not couple. In addition, Fu et al. [8] and Sha et al. [32] used structural zeros to restrain customers to purchase products outside their consideration sets in a bipartite network-based approach for modeling customer preferences over two stages (consideration and purchase). Snijders et al. [19] used structural zeros to deal with the missing data in friendship networks. In this paper, we employ this concept to handle the issue of varying nodes from one time period to another in dynamic network modeling.

3. UTILIZING STERGM TO MODEL EVOLUTION OF PRODUCT COMPETITIONS
3.1 Overview of the dynamic network analysis approach

Fig. 4 illustrates the procedure of utilizing STERGM for understanding the evolution of product competitions in market. The detailed description of each of the three steps is provided as follows.

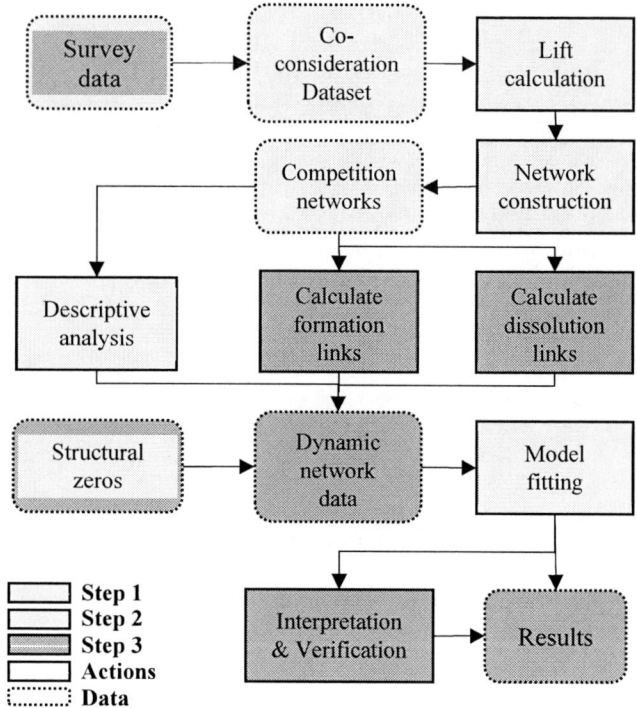

FIGURE 4: Overview of the proposed approach

Step 1-*Network construction and descriptive analysis.* Product competition relationship is identified in this work based on the survey data of considerations, i.e., the products in the same consideration set before a customer makes the final choice. In the formed network, whether the competition relationship exists is determined by the *lift* criterion [7]. The calculation of *lift* and the criterion for link formation are illustrated in Eqn. (9),

$$Lift(i,j) = \frac{Pr(i,j)}{Pr(i) \cdot Pr(j)}, \; y_{ij} = \begin{cases} 1, & if \; lift(i,j) > 1 \\ 0, & otherwise \end{cases}, (9)$$

where $Pr(i,j)$ is the probability of products i and j being co-considered by customers, $Pr(i)$ is the probability of individual product i being co-considered with other products, and y_{ij} is the competition link between product i and product j. The *lift* value measures the likelihood of competition between two products given their respective frequencies of considerations and indicates the dependence of the two products being considered. If the *lift* between product i and product j is greater than a threshold, then the competition link between these two product nodes exists. If two products are completely independent, the *lift* value will be 1 [1]. In this paper, we set the threshold equals to 1 to capture two products that are competed more likely than expected in random [3]. After the *lift* calculation, a unidimensional undirected competition network can be constructed. Fig. 5 provides an illustrative vehicle competition network, in which the numbers represent the *lift* values. A larger *lift* value between two vehicles indicates that they are more frequently co-considered by customers, i.e., there is a stronger

competition relationship between them. Descriptive analysis and visualization of the obtained network can provide an intuitive understanding of the nodal attributes and network structures.

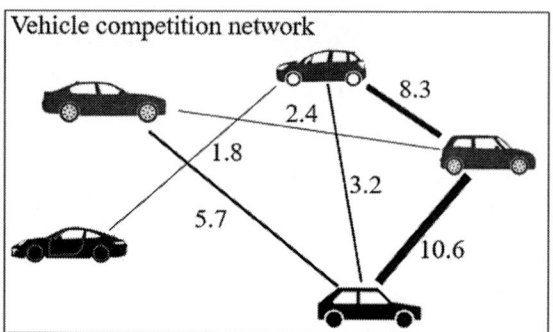

FIGURE 5: An illustrative vehicle competition network

Step 2-*Dynamic network modeling.* After generating the product competition networks in different time periods, we can construct the formation and dissolution networks between each pair of two consecutive networks (see Fig. 2), and use STERGM to investigate the evolution of network structures. The results of descriptive analysis from previous step can help the selection of explanatory factors in network statistics $g(y^{t+1}, y^t)$. To deal with the varying nodes, structural zeros were utilized in STERGM to generate the dynamic network model (see Eqn. 8). Fig. 6 shows how this method works for link formation process by using three networks, represented by adjacency matrices, in time $t-1$, t and $t+1$ (i.e., Y_{t-1}, Y_t, Y_{t+1}) as an example. The formation network between Y_{t-1} and Y_t is noted as Y_{t-1}^+, and the formation network between Y_t and Y_{t+1} is noted as Y_t^+. The *structural zeros* method contains the following steps:

a) Combine all possible formation networks (Y_{t-1}^+, Y_t^+ in this example) in one matrix. If we need to model four-year networks, there will be three formation networks, i.e., three blue regions in Fig. 6 (a). Since STERGM only supports consistent sets of nodes in principle, here the matrices of Y_{t-1}^+ and Y_t^+ include the union set of nodes from Y_{t-1}, Y_t, Y_{t+1}, which means the sizes of these two sub-matrices are the same as shown in the blue regions in Fig. 6 (a). Apparently, the whole matrix is symmetric with zero diagonal cells (self-competitions are not considered).

b) Fill zeros to specific blue regions of Fig. 6 (a) to ensure those new nodes appearing in the other years have no links. For example, the red region Y'^+_{t-1} in Fig. 6 (b) only includes the vehicles appearing in Y_{t-1}, and Y'^+_t only includes the vehicles appearing in Y_t.

c) To ensure that the model captures the changing effect for both preserved and created nodes, the new nodes in the second year of a two-year formation network which are not considered in the red zone in Fig. 6 (b) will be included in the yellow zone in Fig. 6 (c). For example, the difference between the yellow region Y''^+_{t-1} in Fig. 6 (c) and the red region Y'^+_{t-1} in Fig. 6 (b) is the new nodes appearing in Y_t.

When finished, the matrix in Fig. 6 (c) will replace the original matrix in Fig. 6 (a) in fitting the formation model of STERGM. Here we focus on the formation process for illustrative purpose, and the dissolution process follows the same procedures except that the dissolution networks are used.

Step 3-*Interpretation and verification.* Since STERGM is an extension of ERGM with separated effects on formed links and dissolved links, STERGM can be estimated by methods commonly used in ERGM such as maximum likelihood and generalized moments [28]. The estimation results should be verified based on both model fitting and model interpretability. Extracted insights from how product competitions change over time can be used to support product upgrade and new product development strategies in engineering design.

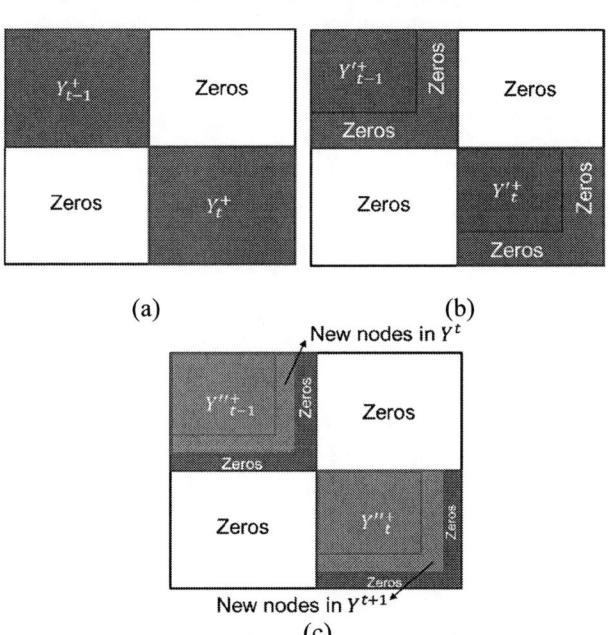

(a) (b)

(c)

Figure 6: Structural zeros used in STERGM for dynamic network with varying nodes

3.2 Dataset for a Case Study on China's Crossover SUV Market

To demonstrate the proposed approach, we carried out a case study using the data from a recognized new car buyer survey from 2013 to 2015 in China's auto market [8]. The dataset in each year consists of 50, 000-70,000 new car buyers' responses including approximately 400 unique car models. Respondents were asked to list up to three vehicles (including final choice) they considered when purchasing the car. Customer demographics, such as income, city of residence and education, and car attributes, such as price, power, and fuel consumption, are reported in the survey and verified by the data company. In each year, the survey was collected every four months. However, we use yearly data in this paper to avoid the seasonal effects on customers' choice behaviors. In this case study, we focus on market competitions among crossover SUVs. A crossover SUV, such as an Audi Q7, BMW X6, or Ford Edge, is a vehicle with

the body and space of an SUV, but the backbone of a sedan. Crossover SUVs inherit the advantages of spacious interiors from vans, the off-road performance of SUVs and the light-weight and fuel economy of sedans. Our interest is in demonstrating why two crossover SUVs enter in a competitive relationship and why the competition ceases.

3.3 Descriptive analysis

Once we obtain the customers' consideration sets from the survey data, the *lift* value for all competitions between car models can be calculated. We constructed three competition networks corresponding to years 2013, 204 and 2015. There are roughly 100 car models from 2013 to 2015 considered in our study (including both crossover SUVs and other conventional or SUV vehicles co-considered with crossover SUVs) and 600 pairs of competitions in each year. Fig. 7 illustratively shows partial competition networks in 2014 and 2015 with 10 vehicle models. The node size is proportional to its degree and the nodes with green color represent selected vehicle models for log-odds calculation in Section 3.5 (this is to verify the STERGM modeling results at link levels). The blue line indicates newly formed competitions in 2015 while the grey lines indicate the dissolved ones. The *lift* value marked on each link shows the strength of competitions. For example, in Fig. 7, Audi FAW Q5 (Q5) competes with three cars - Toyota GAIG Highlander (Highlander), Dongfeng Yulong Luxgen Grand 7 (Luxgen Grand 7), and BMW X6 (X6). For a customer considering the Q5, in 2014 the Highlander is the most likely co-considered car whereas in 2015 the closest competitor changes to X6.

Table 1 provides the mean value and standard deviation (presented in brackets) of vehicle attributes from 2013 to 2015 for all pairs of competing vehicles versus all possible pairs of competing vehicles. The major difference is that the car models in the set of "all possible pairs of competing vehicles" do not necessarily hold a direct competition relation (e.g., car A and car B are not competitors, but they both compete with car C). Based on our previous research in choice modeling and product competitions [2,5,33], three categories of vehicle attributes are considered in this study: 1) regular vehicle attributes such as the price, power and fuel consumption of vehicles; 2) vehicle attribute difference. For example, the price difference between two competed vehicles allows us to investigate whether vehicles with similar or different attributes are more likely to compete with each other; and 3) SUV-relevant attributes, such as seat position. In total, nine vehicle attributes are considered including price, power, fuel consumption (FC), turbo, make origin, all-wheel drive (AWD), seat position, legroom, and third row. Among them, price and power are preprocessed using log2 transformation to handle their non-normal distributions. Turbo, AWD, and third row are binary variables describing whether such a property is available in a car or not. Seat position and legroom are customer satisfaction Likert-scale ratings from 1 (dissatisfy strongly) to 4 (satisfy strongly). It can be seen that the mean values of price, power, fuel consumption, vehicle attributes difference, and third row are lower on average in competing vehicles than that in all possible pairs of competing vehicles.

Some vehicle attributes changed from 2013 to 2015; for example, the mean value of power for all possible pairs of competing vehicles increased from 7.25 to 7.35 and fuel consumption decreased from 10.13 to 9.96.

Table 1. Descriptive analysis of competing vehicle pairs versus all possible pairs of competing vehicles

	Pairs of competing vehicles			All possible pairs of competing vehicles		
	2013	2014	2015	2013	2014	2015
Regular vehicle attributes						
Price	16.95	17.37	17.04	17.43	17.55	17.52
(log2)	(0.94)	(1.22)	(0.99)	(1.22)	(1.37)	(1.32)
Power	7.03	7.20	7.18	7.25	7.28	7.35
(log2)	(0.42)	(0.52)	(0.41)	(0.56)	(0.58)	(0.55)
FC (L per	9.24	9.82	9.36	10.13	10.07	9.96
100 km)	(1.79)	(2.07)	(1.73)	(2.3)	(2.17)	(2.09)
Turbo	0.14	0.28	0.37	0.19	0.28	0.39
	(0.28)	(0.39)	(0.37)	(0.33)	(0.39)	(0.40)
Origin	172	239	138	23	21	18
(US)	(14%)	(16%)	(15%)	(20%)	(15%)	(15%)
Origin	188	366	111	20	36	27
(EU)	(15%)	(25%)	(12%)	(17%)	(26%)	(23%)
Origin (JP)	183	167	107	19	17	18
	(15%)	(11%)	(12%)	(17%)	(12%)	(15%)
Origin	275	128	59	17	14	10
(KR)	(22%)	(9%)	(7%)	(15%)	(10%)	(8%)
Origin	438	566	485	36	51	46
(CN)	(35%)	(39%)	(54%)	(31%)	(37%)	(39%)
Vehicle attribute difference						
Price	0.47	0.49	0.44	1.40	1.57	1.52
difference	(0.39)	(0.43)	(0.36)	(1.01)	(1.13)	(1.09)
Power	0.25	0.25	0.27	0.64	0.67	0.63
difference	(0.21)	(0.22)	(0.21)	(0.47)	(0.49)	(0.46)
FC	1.08	1.08	1.04	2.61	2.48	2.42
differences	(0.94)	(0.94)	(0.86)	(1.95)	(1.80)	(1.71)
SUV-relevant attributes						
AWD	0.14	0.26	0.22	0.23	0.26	0.24
	(0.29)	(0.36)	(0.30)	(0.36)	(0.37)	(0.36)
Seat	3.08	3.08	3.08	3.09	3.07	3.09
position	(0.11)	(0.11)	(0.14)	(0.13)	(0.13)	(0.19)
Legroom	3.20	3.17	3.14	3.21	3.16	3.13
	(0.09)	(0.10)	(0.10)	(0.14)	(0.13)	(0.16)
Third row	0.10	0.14	0.12	0.16	0.16	0.20
	(0.29)	(0.35)	(0.33)	(0.36)	(0.37)	(0.40)

3.4 Results of Dynamic Network Modeling

The R package "tergm" is used to fit the STERGM models [34]. As shown in Table 2, the explanatory variables in our dynamic network models correspond to two types of variables: endogenous variables (i.e., network structures) and exogenous variables consisting of the main effects and homophily effects of vehicle attributes. The main effect measures the impact of the existence or value of a vehicle attribute on the competition link probability, whereas the homophily effect measures the impact of the similarity or difference of the attributes of two vehicles on their competition link probability [2]. In this study, we consider two network structure effects: Geometrically Weighted Edgewise Shared Partner (GWESP), which is referred to as the

shared partner structure in Fig. 1; and Geometrically Weighted Degree (GWD), which measures the centralization effect of the network (i.e., the evenness of degree distribution). In a vehicle competition network, a positive coefficient of GWESP means it is very likely for two cars to compete with each other if they share the same set of competitors. A positive coefficient of GWD means most cars have similar numbers of competitors.

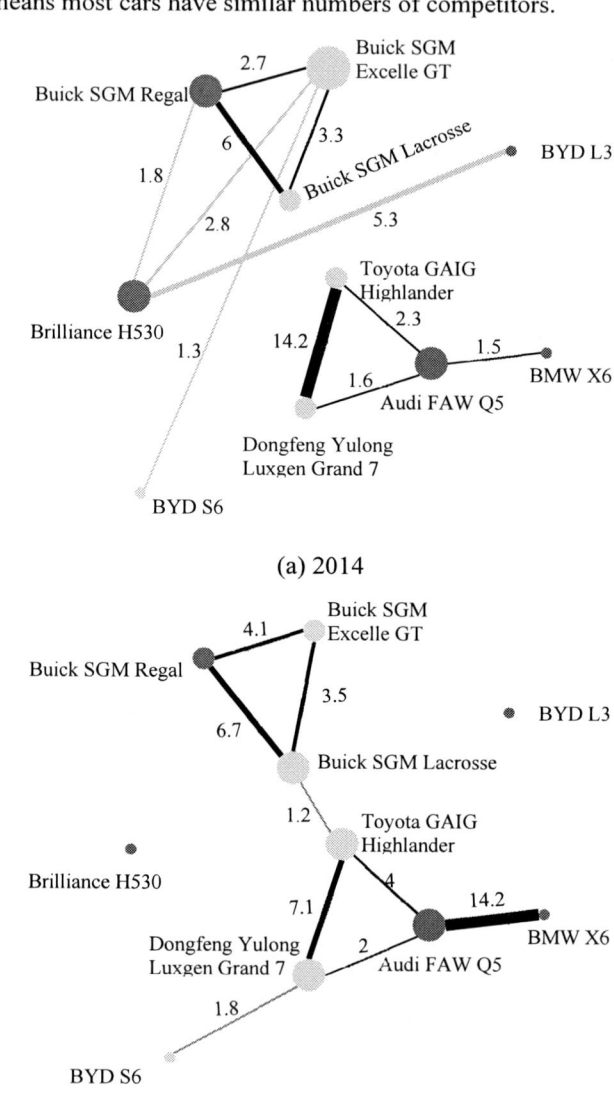

Figure 7: An example of partial vehicle competition networks evolving from 2014 (a) to 2015 (b). Black lines indicate preserved competitions, grey lines indicate dissolved competitions, and blue lines represents new competitions.

Three STERGM models (A, B, C) are created with different model specifications. In Model A, we only consider main effect of regular vehicle attributes and homophily effects, and results are used as the baseline for comparison. Model B includes main effect of regular vehicle attributes, homophily effects, and network effects. In Model C, we consider all endogenous and exogenous variables including main effect of SUV-relevant attributes. The 17% decrease of AIC value from Model A to Model B in Table 2 indicates model improvement due to the introduction of network structure effects. On the other hand, introducing SUV-relevant attributes in Model C leads to slightly worse model fit compared to Model B, and SUV-relevant attributes are not significant in both formation and dissolution models.

When examining the results from the STERGM formation model, the coefficients of all three network structure effects are significant. The positive sign of "closure effect" indicates that a new competition is more likely to form between two vehicles if they shared the same set of competitors previously. The negative sign of "centralization effect" denotes that new competitions are more likely to form between vehicles that have been in competition with many other car models already. Among the regular vehicle attributes, we observe the estimated coefficients of -0.34 and -0.35 for vehicle brands from Japan and Korea, respectively. The negative signs indicate that, compared to domestic vehicles, vehicles from Japan and Korea are less likely to form new competitions from 2013 to 2015. Although the main effects of price, power, FC, and turbo show no significance in the formation model, the homophily effects are significant. The negative coefficient of price difference indicates that vehicles tend to form new competitions with those having similar prices. In contrast, two vehicles with a higher difference in power and FC are more likely to create a new competition between them as time goes by.

When examining the results from the link dissolution model, it is important to note that the estimations measure the persistence of competition, *not* disappearance. As shown in Table 2, the estimates of closure effect, centralization effect and vehicles from Japanese and Korean brands significantly influence the preservation of existing competitions. The interpretation of dissolution model is somewhat similar to the formation model except that one is for the formation of new competitions while the other is for the preservation of old/existing competitions. For example, the estimate of closure effect being 0.41 indicates that existing competitions are more likely to persist over time if two vehicles share the same set of competitors.

To better understand the differences between STERGM and ERGM, we compare the results of the two models in Table 3. It is noted that the statistical significance of the attributes in ERGM is different from one year to another. In many cases, it is also different from the significance obtained from STERGM. These differences can be explained by the fact that ERGM is a static network modeling approach based on single-year data, assuming there are no pre-existing network relations (i.e., no competition at all), whereas STERGM is a dynamic network modeling approach focused on detecting the **changing pattern** that best describes the formation and dissolution of competitions conditional on the pre-existing competitions.

Table 2. Results of the STERGM fitting for dynamic competition networks from 2013 to 2015

Model	Formation			Dissolution		
	A[1]	B[2]	C[3]	A	B	C
Network effects						
Closure		1.15***	1.15***		0.41**	0.41**
Centralization		-2.79***	-2.79***		-1.03***	-1.06***
Edges	3.80*	-3.59**	-3.55*	-2.91	-4.55	-3.18
Main attributes effects						
Price	0.03	0.01	0	-0.24	0.07	0
Power	-0.79***	-0.15	-0.17	0.86	0.17	0.3
FC	0.15***	0.03	0.05*	-0.19	-0.09	-0.13
Turbo	-0.08	0.06	0.07	-0.22	-0.16	-0.19
Origin (US)	0.29***	0	-0.01	0.18	0.05	0.12
Origin (EU)	0.42***	0.07	0.07	0.59*	0.19	0.29
Origin (JP)	-0.39***	-0.34***	-0.34***	-0.53	-0.48*	-0.37
Origin (KR)	-0.26**	-0.35***	-0.33***	-0.39	-0.43**	-0.35*
AWD			-0.04			0.12
Seat position			0.01			0.69
Legroom			0.02			-0.75
Third row			-0.10			0.04
Homophily effects						
Price diff.	-0.33***	-0.22***	-0.22***	-0.14	-0.06	-0.11
Power diff.	0.54***	0.41**	0.42**	0.37	0.41	0.43
FC diff.	0.09**	0.07*	0.06*	0.03	0.05	0.07
AIC	9340	7740	7747	841	787	792

***$p<0.001$, **$p<0.01$, *$p<0.05$.
[1] Model A-only consider regular attributes and homophily effects
[2] Model B-consider regular attributes, homophily effects and network effects
[3] Model C-consider all attributes

The differences in the model coefficients of ERGM over time (2013-2015 as shown in Table 3) imply the change of customer preferences from one year to another. For instance, the coefficient for fuel consumption is insignificant in 2013 but becomes significant in both 2014 and 2015 with a value of 0.07 and 0.10, respectively. On the other hand, a significant coefficient (0.5) for fuel consumption in the STERGM formation model indicates that based on the three-year data, fuel consumption has a positive influence on forming new competitions. It is also found that SUV specific attributes such as AWD are shown to be significant in each year's statistic ERGM modeling but are insignificant in influencing the forming of new competitions over time compared to other main attributes.

Table 4 (continued)

Model name	Price	Power	FC	Make origin
Buick SGM Excelle GT	16.78	6.85	8.67	American
Buick SGM Lacrosse	17.89	7.60	11.07	American
BYD S6	16.71	7.16	10.09	Chinese
Dongfeng Yulong luxury grand 7	17.84	7.46	12.56	Chinese
Toyota GAIG Highlander	18.08	7.56	12.31	Japanese

Table 3. Comparing results of ERGM versus STERGM for competition networks from 2013 to 2015

Coefficients	ERGM			STERGM	
	2013	2014	2015	Formation	Dissolution
Network effects					
Closure	0.77**	0.99**	0.87**	1.15**	0.41**
Centralization	-0.51	0.24	-0.86*	-2.79**	-1.06**
Edges	7.11*	-2.42	2.6	-3.55*	-3.18
Main attributes effects					
Price	-0.30*	-0.15	-0.22	0	0
Power	-0.09	-0.17	-0.06	-0.17	0.3
FC	0.03	0.07*	0.10*	0.05*	-0.13
Turbo	-0.17	0.08	0.07	0.07	-0.19
Origin (US)	0.12	0.13	-0.14	-0.01	0.12
Origin (EU)	0.54**	0.31**	0.13	0.07	0.29
Origin (JP)	0.54**	0.21	-0.08	-0.34**	-0.37
Origin (KR)	0.49**	-0.1	-0.19	-0.33**	-0.35*
AWD	0.56**	0.35**	0.38**	-0.04	0.12
Seat position	-0.02	0.58*	-0.09	0.01	0.69
Legroom	0.22	0.35	0.33	0.02	-0.75
Third row	-0.24*	-0.16*	-0.35**	-0.10	0.04
Homophily effects					
Price diff.	-1.41**	-1.16**	-1.59**	-0.22**	-0.11
Power diff.	-0.56*	-0.42	0.32	0.42**	0.43
FC diff.	-0.10	-0.21**	-0.22**	0.06*	0.07
AIC	2993	3776	2317	7747	792

3.5 Results verification

To further verify the results in Section 3.4, the link-level verification and the goodness-of-fit analysis at the network level are performed.

3.5.1 Link-level verification

Link-level verification compares the log-odds of those hypothetical links (i.e., those links do not exist in observed networks and are solely for testing purposes) to the log-odds of the real links in observed networks. In general, newly formed competitions in observed networks (i.e., real links) are supposed to have higher log-odds than those hypothetical links. The vehicles represented by green nodes in Fig. 7 are selected for the link-level verification. Table 4 provides the value of significant variables in Model C for these vehicles.

Table 4. The value of vehicle attributes for selected vehicles

By inserting the estimated coefficients obtained from Table 2 into Eqn. (2), the log-odds of a link forming conditional on the rest of network can be calculated:

$$Logit\ Pr\big(Y_{ij}^+ = 1\big) = \boldsymbol{\theta}^T \cdot \boldsymbol{\delta}_{ij}^+(y) = -3.55 \times \delta_{Edges} + 0.05 \times \delta_{Fuelconsump} - 0.34 \times \delta_{Origin(JP)} - 0.33 \times \delta_{Origin(KR)} - 0.22 \times \delta_{Price\ diff.} + 0.42 \times \delta_{Power\ diff.} + 0.06 \times \delta_{FC\ diff.} + 1.15 \times \delta_{GWESP} - 2.79 \times \delta_{GWD}\ , \tag{10}$$

Plugging the attribute values from Table 4 into Eqn. (10), we calculate the log-odds of real links compared with a hypothetical link to verify the accuracy of our model; results are illustrated in Fig. 8, using the same set of vehicles shown in Fig. 7. It is observed that the real links (e.g., the competition link between Highlander and Buick SGM Lacrosse) reach higher log-odds than the hypothetical links (e.g., the dash link between Highlander and Buick SGM Excelle GT) in 2015. This indicates that the STERGM results can successfully capture the influence of exogenous variables (e.g., fuel consumption, price difference, power difference, fuel consumption difference, and make origin), and endogenous variables (e.g., centralization and closure effect) on the formation of vehicle competitions shown in Fig. 7.

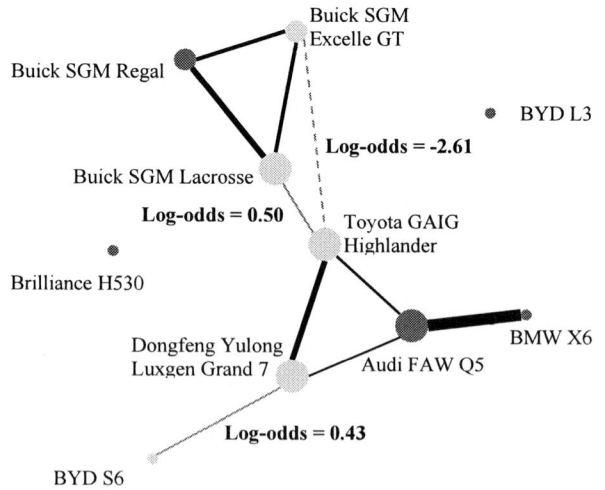

Figure 8: The log-odds results of two newly formed links (solid blue lines) and a hypothetical link (dash blue line) in the competition network from 2014 to 2015

3.5.2 Network level verification

The goodness-of-fit analysis at the network level verifies the model through comparing the simulated networks from the estimated models with observed networks in terms of the distributions of certain endogenous variables such as the degree of nodes and GWESP as well as exogenous statistics. We use competition network in 2014 as the target data for STERGM simulations. Fig. 9 provides the results of 100 simulated competition network of 2014 with STERGM (using competition network in 2013 as the starting network) for examining the explanatory variables. The vertical axis in each plot represents the logit (log-odds) of the relative frequency, the solid line represents the statistics for the observed network, the boxplots indicate the median and interquartile range of the simulated networks, and the light-grey lines represent the range in which 95% of simulated observations fall. We can see most observed value lies in the 95% range of simulated observations which indicates that STERGM performs relatively well in both the formation model and the dissolution model.

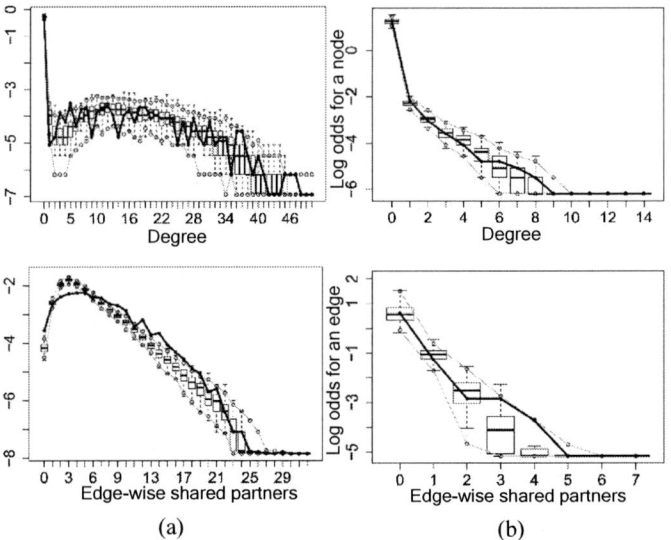

(a)   (b)

Figure 9: Goodness-of-fit plots of STERGM using competition network in 2014 as the target data. (a) Formation model. (b) Dissolution model.

4. CONCLUSION

The major contribution of this study is the development of a dynamic network analysis approach to modeling the evolution of product competition relations. Even though a network-based approach was previously adopted in modeling vehicle competitions to overcome the limitations of DCA [35,36], this is the first attempt to systematically analyze and model product competitions based on longitudinal market data and dynamic network analysis. Different from our previous study on multi-year analysis using cross-sectional network data, this research provides insights into the factors (such as product attributes, homophily effects, and network structure effects) that drive changes of product competitions.

Our proposed STERGM approach models the impact of endogenous variables as well as exogenous variables on the

formation and dissolution of product competitions separately. A three-year customer survey from China's auto market was utilized and three crossover SUV-oriented competition networks were constructed to illustrate the implementation of dynamic network modeling. By utilizing the structural zeros method, we addressed the challenge of longitudinal network modeling with varying sets of nodes from year to year. We observe an obvious improvement of model fit after the network structural effects are introduced into the dynamic model. The results from STERGM formation model indicate that two unrelated vehicles tend to form a competition relation if they are not Japanese or Korean brands, have lower price difference, higher fuel consumption, higher power difference, higher fuel consumption difference, compete with more other cars, or have more shared competitors. The dissolution model results indicate that competing vehicles may lose their competition in the future if they are Korean vehicles, compete with fewer other cars, or have fewer shared competitors. Our verification at both the link level and the network level further demonstrates the model fit.

Our work also illustrates the difference between the static ERGM and the dynamic STERGM. In summary, ERGM is a static network modeling approach assuming there are no pre-existing network relations (i.e., no pre-existing competition at all), whereas STERGM is a dynamic network modeling approach focused on detecting the changing pattern that best describes the formation and dissolution of competitions conditional on the pre-existing competitions.

Our future work will focus on examining the use of STERGM for prediction, given the current competition structure and product design change. For example, we may use the STERGM results obtained in this work to predict the competition network in 2016 based on the competition network in 2015. By studying the impact of improving existing products and releasing new products on future product competitions, this dynamic network modeling approach can support engineering design decisions and companies' strategic decision making.

ACKNOWLEDGEMENTS

The authors gratefully acknowledge the financial support from NSF-CMMI-1436658, Ford-Northwestern Alliance Project, the grant from China Scholarship Council (No. 201706030108) and National Key R&D Program of China (No. 2018YFB1700800).

REFERENCES

[1] Wang, M., Sha, Z., Huang, Y., Contractor, N., Fu, Y., and Chen, W., 2016, "Forecasting Technological Impacts on Customers' Co-Consideration Behaviors: A Data-Driven Network Analysis Approach," *ASME 2016 International Design Engineering Technical Conferences and Computers and Information in Engineering Conference,* Charlotte, NC, USA, August 21-24, 2016, p. V02AT03A040--V02AT03A040.

[2] Wang, M., Sha, Z., Huang, Y., Contractor, N., Fu, Y., and Chen, W., 2018, "Predicting Products' Co-Considerations and Market Competitions for

Technology-Driven Product Design: A Network-Based Approach," Des. Sci. J., **4**.

[3] Sha, Z., Huang, Y., Fu, S., Wang, M., Fu, Y., Contractor, N., and Chen, W., 2018, "A Network-Based Approach to Modeling and Predicting Product Co-Consideration Relations," Complexity, **2018**.

[4] Robins, G., Pattison, P., Kalish, Y., and Lusher, D., 2007, "An Introduction to Exponential Random Graph (P*) Models for Social Networks," Soc. Networks, **29**(2), pp. 173–191.

[5] Wang, M., Chen, W., Huang, Y., Contractor, N. S., and Fu, Y., 2016, "Modeling Customer Preferences Using Multidimensional Network Analysis in Engineering Design," Des. Sci., **2**.

[6] Snijders, T. A. B., Pattison, P. E., Robins, G. L., and Handcock, M. S., 2006, "New Specifications for Exponential Random Graph Models," Sociol. Methodol., **36**(1), pp. 99–153.

[7] Wang, M., Chen, W., Fu, Y., and Yang, Y., 2015, "Analyzing and Predicting Heterogeneous Customer Preferences in China's Auto Market Using Choice Modeling and Network Analysis," SAE Int. J. Mater. Manuf., **8**(2015-1-468), pp. 668–677.

[8] Fu, J. S., Sha, Z., Huang, Y., Wang, M., Fu, Y., and and Chen, W., 2017, "Modeling Customer Choice Preferences in Engineering Design Using Bipartite Network Analysis," *Proceedings of the ASME 2017 International Design Engineering Technical Conferences and Computers and Information in Engineering Conference*, Cleveland, Ohio.

[9] Christensen, C., 1997, "Patterns in the Evolution of Product Competition," Eur. Manag. J., **15**(2), pp. 117–127.

[10] Bloomberg News, 2018, "American Automakers Losing Ground in a Shrinking Chinese Market," Bloomberg.com [Online]. Available: https://www.bloomberg.com/news/articles/2018-09-11/u-s-car-brands-losing-share-of-shrinking-chinese-auto-market.

[11] Turrentine, T. S., and Kurani, K. S., 2007, "Car Buyers and Fuel Economy?," Energy Policy, **35**(2), pp. 1213–1223.

[12] Heffner, R., Kurani, K. S., and Turrentine, T. S., 2007, "Symbolism in Early Markets for Hybrid Electric Vehicles," Tech. Rep. No. UCD-ITS-RR-07-01, Inst. Transp. Stud. Univ. California, Davis, Davis, CA.

[13] Hanneke, S., Fu, W., Xing, E. P., and others, 2010, "Discrete Temporal Models of Social Networks," Electron. J. Stat., **4**, pp. 585–605.

[14] Krivitsky, P. N., and Handcock, M. S., 2014, "A Separable Model for Dynamic Networks," J. R. Stat. Soc. Ser. B (Statistical Methodol., **76**(1), pp. 29–46.

[15] Watts, D. J., and Strogatz, S. H., 1998, "Collective Dynamics of'small-World'networks," Nature, **393**(6684), p. 440.

[16] Barabási, A.-L., and Albert, R., 1999, "Emergence of Scaling in Random Networks," Science, **286**(5439), pp. 509–512.

[17] Xu, K. S., and Hero, A. O., 2013, "Dynamic Stochastic Blockmodels: Statistical Models for Time-Evolving Networks," *International Conference on Social Computing, Behavioral-Cultural Modeling, and Prediction*, Washington, DC, USA, April 2-5, 2013, pp. 201–210.

[18] Block, P., Koskinen, J., Hollway, J., Steglich, C., and Stadtfeld, C., 2018, "Change We Can Believe in: Comparing Longitudinal Network Models on Consistency, Interpretability and Predictive Power," Soc. Networks, **52**, pp. 180–191.

[19] Snijders, T. A. B., de Bunt, G. G., and Steglich, C. E. G., 2010, "Introduction to Stochastic Actor-Based Models for Network Dynamics," Soc. Networks, **32**(1), pp. 44–60.

[20] Leifeld, P., and Cranmer, S. J., 2015, "A Theoretical and Empirical Comparison of the Temporal Exponential Random Graph Model and the Stochastic Actor-Oriented Model," arXiv Prepr. arXiv1506.06696.

[21] Mousavi, R., and Gu, B., 2015, "The Effects of Homophily in Twitter Communication Network of US House Representatives: A Dynamic Network Study," Available SSRN https://ssrn.com/abstract=2666052 or http//dx.doi.org/10.2139/ssrn.2666052.

[22] Lebacher, M., Thurner, P. W., and Kauermann, G., 2018, "International Arms Trade: A Dynamic Separable Network Model With Heterogeneity Components," arXiv Prepr. arXiv1803.02707.

[23] Hunter, D. R., Handcock, M. S., Butts, C. T., Goodreau, S. M., and Morris, M., 2008, "Ergm: A Package to Fit, Simulate and Diagnose Exponential-Family Models for Networks," J. Stat. Softw., **24**(3), p. nihpa54860.

[24] Lusher, D., Koskinen, J., and Robins, G., 2012, *Exponential Random Graph Models for Social Networks: Theory, Methods, and Applications*, Cambridge University Press.

[25] Han, J.-D. J., Bertin, N., Hao, T., Goldberg, D. S., Berriz, G. F., Zhang, L. V, Dupuy, D., Walhout, A. J. M., Cusick, M. E., Roth, F. P., and others, 2004, "Evidence for Dynamically Organized Modularity in the Yeast Protein-Protein Interaction Network," Nature, **430**(6995), p. 88.

[26] Conaldi, G., and Lomi, A., 2013, "The Dual Network Structure of Organizational Problem Solving: A Case Study on Open Source Software Development," Soc. Networks, **35**(2), pp. 237–250.

[27] Tsamardinos, I., and Borboudakis, G., 2010, "Permutation Testing Improves Bayesian Network Learning," *Joint European Conference on Machine Learning and Knowledge Discovery in Databases*, pp. 322–337.

[28] Hu, M.-C., Pavlicova, M., and Nunes, E. V, 2011, "Zero-Inflated and Hurdle Models of Count Data with Extra Zeros: Examples from an HIV-Risk Reduction Intervention Trial," Am. J. Drug Alcohol Abuse, **37**(5),

pp. 367–375.

[29] Blanchini, F., Franco, E., and Giordano, G., 2013, "Structured-LMI Conditions for Stabilizing Network-Decentralized Control," *52nd IEEE Conference on Decision and Control*, pp. 6880–6885.

[30] Robins, G., Pattison, P., and Woolcock, J., 2004, "Missing Data in Networks: Exponential Random Graph (P∗) Models for Networks with Non-Respondents," Soc. Networks, **26**(3), pp. 257–283.

[31] Snijders, T. A. B., 1991, "Enumeration and Simulation Methods for 0--1 Matrices with given Marginals," Psychometrika, **56**(3), pp. 397–417.

[32] Sha, Z., Bi, Y., Wang, M., Stathopoulos, A. Contractor, N., Fu, Y., and Chen, W., 2019, "Comparing Utility-Based and Network-Based Approaches in Estimating Customer Preferences for Engineering Design," *The 21st International Conference on Engineering Design*, Delft, The Netherlands.

[33] Sha, Z., Wang, M., Huang, Y., Contractor, N., Fu, Y., and Chen, W., 2017, "Modeling Product Co-Consideration Relations: A Comparative Study of Two Network Models," *Proceedings of the 21st International Conference on Engineering Design (ICED 17) Vol 6: Design Information and Knowledge, Vancouver, Canada, 21-25.08. 2017.*

[34] Krivitsky, P. N., Handcock, M., and others, 2018, "Tergm: Fit, Simulate and Diagnose Models for Network Evolution Based on Exponential-Family Random Graph Models," Statnet Proj. (http//www. statnet. org). R Packag. version, **3**(2).

[35] Wang, M., Chen, W., Huang, Y., Contractor, N. S., and Fu, Y., 2015, "A Multidimensional Network Approach for Modeling Customer-Product Relations in Engineering Design," *ASME 2015 International Design Engineering Technical Conferences & Computers and Information in Engineering Conference*, Boston, MA.

[36] Wang, M., and Chen, W., 2015, "A Data-Driven Network Analysis Approach to Predicting Customer Choice Sets for Choice Modeling in Engineering Design," J. Mech. Des., **137**(7), p. 71410.

Proceedings of the ASME 2019
International Design Engineering Technical Conferences
and Computers and Information in Engineering Conference
IDETC/CIE2019
August 18-21, 2019, Anaheim, CA, USA

DETC2019-98525

3D SHAPE SYNTHESIS FOR CONCEPTUAL DESIGN AND OPTIMIZATION USING VARIATIONAL AUTOENCODERS

Wentai Zhang   **Zhangsihao Yang**   **Haoliang Jiang**   **Suyash Nigam**
Soji Yamakawa   **Tomotake Furuhata**   **Kenji Shimada**   **Levent Burak Kara**[*]
Carnegie Mellon University, Pittsburgh, PA, USA

ABSTRACT

We propose a data-driven 3D shape design method that can learn a generative model from a corpus of existing designs, and use this model to produce a wide range of new designs. The approach learns an encoding of the samples in the training corpus using an unsupervised variational autoencoder-decoder architecture, without the need for an explicit parametric representation of the original designs. To facilitate the generation of smooth final surfaces, we develop a 3D shape representation based on a distance transformation of the original 3D data, rather than using the commonly utilized binary voxel representation. Once established, the generator maps the latent space representations to the high-dimensional distance transformation fields, which are then automatically surfaced to produce 3D representations amenable to physics simulations or other objective function evaluation modules. We demonstrate our approach for the computational design of gliders that are optimized to attain prescribed performance scores. Our results show that when combined with genetic optimization, the proposed approach can generate a rich set of candidate concept designs that achieve prescribed functional goals, even when the original dataset has only a few or no solutions that achieve these goals.

INTRODUCTION

In engineering design, while design simulation and analysis technologies are well developed and ubiquitous, digital tools to assist the early conceptual design phases are severely limited. Instead, humans still play a critical role in establishing the design

space and the associated parameterizations. However, the heavy reliance on human-driven concept generation and design space exploration make product development particularly challenging for problems in which the geometry/form of the product has a significant impact on performance. As such, the need for digital design tools that support (1) knowledge extraction from configurationally and geometrically different past designs, (2) leveraging this information for large-variance, automatic design synthesis inside and outside of the original design space, and (3) seamless integration into analysis and simulation engines remain a central need in design automation.

In this work, we present a data-driven 3D shape synthesis method to assist human designers in conceptual design. Our approach relies on the observation that past designs may encapsulate useful design information that, if digitally captured, could be used to generate new designs automatically. To this end, we adopt an unsupervised variational autoencoder (VAE) deep learning method that takes input a corpus of 3D designs and extracts a latent design representation. This representation transforms the originally very high-dimensional data into a compact feature vector, where each feature encodes a latent probability distribution function learned over all past designs. Once learned, this representation can be sampled, or elements in this latent space can be interpolated and extrapolated to generate new latent space instances. These new instances can then be projected to the original design space using the decoder of the VAE.

In contrast to the common method of voxelizing 3D data into a 3D binary representation [1–4], we utilize distance transformation maps [5] as the primary input. This involves (1) the conversion of input 3D shapes (commonly acquired in the form

---

[*]Address all correspondences to lkara@cmu.edu

of polygonal models) into real valued distance maps, (2) using the distance map of each original design to train the VAE, (3) automatically converting any synthetically generated new distance map back to a polygonal model for downstream analysis. This allows the synthesized 3D shapes to exhibit much smoother surfaces without suffering from 'pixelatization,' while being amenable to engineering analyses.

We demonstrate the utility of our approach on the 3D outer shape design of gliders. While one approach is to learn a direct mapping from the available past designs to their aerodynamic performance, and use this mapping as a simulator to evaluate new designs, such a mapping would need to be learned for every new engineering objective. Instead, VAE learns a shape generator in an unsupervised manner, where the latent space exploration in the trained architecture allows the newly generated design to be integrated into widely available analysis tools. We demonstrate this in a particular case study, where the study incorporates flight dynamics (albeit simplified) for shape optimization.

Our approach also develops a simple latent space design crossover technique that allows a genetic optimizer to produce a large set of new designs through stochastic latent vector interpolation and extrapolation. While not a requirement in the overall framework, the use of genetic optimization enables a large set of synthetically generated designs to meet the target performance score, even though only a few or no original design solutions could attain the prescribed goals. This, in turn, offers greater conceptual latitude to the human designer in deciding which concepts to further develop.

BACKGROUND

With advances in machine learning, data-driven approaches that model and optimize engineering design problems are becoming increasingly more prevalent. In this section, we will discuss some recent progress in (1) addressing various mechanical problems with data-driven approaches, (2) applications of learning-based methods in design and (3) dimensionality reduction for shape optimization.

Data-driven Methods in Various Mechanical Problems

Sosnovik et al. [6] propose a Convolutional Neural Network that accelerates topology optimization computations. Using the powerful ability of deep learning methods to segment images pixel-wise, the approach predicts final optimal topologies after several iterations of optimization, based upon the initial conditions of the layout.

To address the circuit synthesis in EDA design, Guo et al. [7] propose an active learning strategy for reducing topology evaluation cost for a circuit synthesis problem. They utilize a predictive model with a random forest to approximate true circuit topology performance. Their experimentation reveals that uncertainty and

topology structure may play critical roles in improving the appropriation model accuracy and make a significant contribution to reducing the system evaluation costs.

Data-driven methods have also been used in modeling non-linear physics. Raissi et al. [8] treats the deep neural network as a non-linear function approximator and use the method to identify complex non-linear systems such as Lorenz system and the Glycoltic oscillator model. Umetani et al. [9] develop a data-driven approach to estimate the aerodynamic forces on a glider and its wing shape, and use this for glider design. This enables a user to accurately match the desired trajectory without the aid of costly simulations or experiments.

Learning-based Methods in Design Problems

With a large amount of available data and the advance of hardware technology, data-driven methods have become an increasingly common strategy for the problems that are difficult to approach by the creating physical model or are expensive in computation. Recently, researchers in the mechanical design community have started exploring the machine learning approach to aid the design process.

Fuge et al. [10] devise a framework that relies on collaborative filtering to recommend best design methodologies to solve target design problems, and argue that such approaches can be valuable for novice designers and enhance the overall product development cycle. In order to have an automatic design generator, Chen et al. [11] introduce the BezierGAN, a generative model for synthesizing smooth curves. The model maps a low-dimensional latent representation to a sequence of discrete points sampled from a rational Bezier curve. It is tested on four different design datasets and reveals better capacities in generating realistic 2D smooth shapes when compared with InfoGAN. Similar frameworks can have an impressive performance even for creative hand sketches. Chen et al. [12] propose a model, sketch-pix2seq, based on a sequence-to-sequence variational-auto-encoder (VAE) model called sketch-rnn. With their modification, the model has better performance in learning and generating sketches of multiple categories and shows promising results in creative tasks. However, these works are limited to 2D images or sketches. It is still challenging to extend these works to 3D design representations. Unlike 2d images which can be easily represented as unified RGB/grayscale form with the same size, 3d shapes are usually represented in polygonal meshes, point clouds or implicit surfaces with various lengths and weak correspondence among different shapes.

To have a deeper understanding of the mapping between the shape and the functions of design, Dering et al. [13] propose a deep learning approach based on three-dimensional (3D) convolutions that predict functional quantities of digital design concepts. Testing trained models on novel input yields accuracy as high as 98% for estimating the rank of the functional quantities.

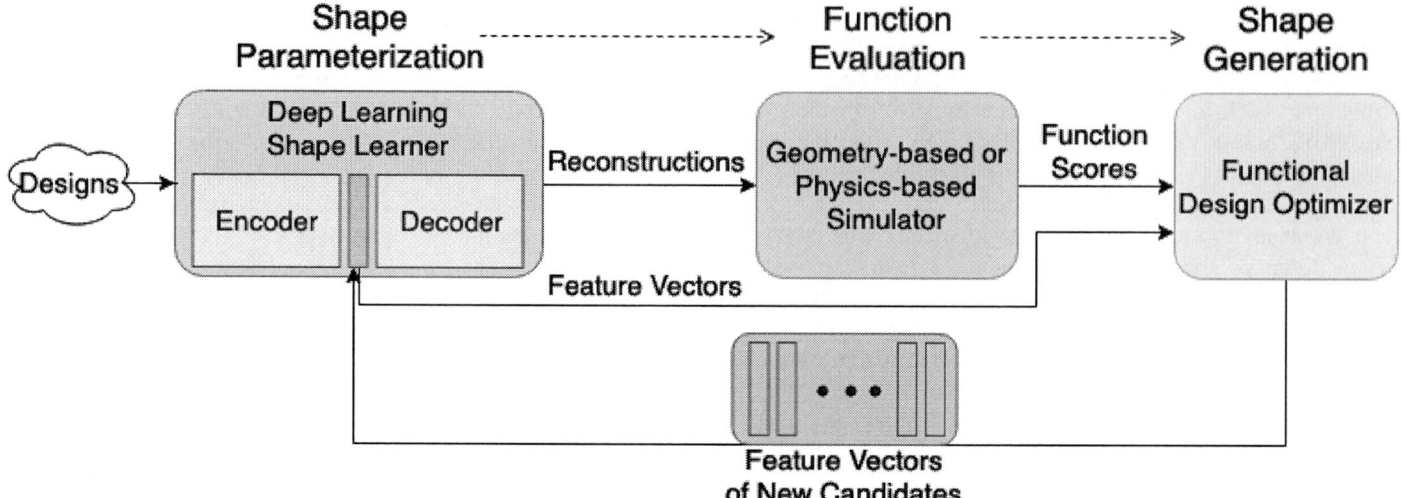

FIGURE 1. The architecture of our proposed data-driven conceptual design pipeline.

This method is also employed to differentiate between decorative and functional headwear. Moreover, Burnap et al. [14] develop a deep learning approach to predict design gaps in the market. Their approach is built on conventions in both quantitative marketing inbounding the heterogeneity of consumer choice preferences, as well as engineering design for bounding the space of possible designs. Raina et al. [15] explore the representation of design strategies as a Hidden Markov Model and their application to engineering design problems. Their results imply the successful transfer of design strategies from human designers to computational agents. They also propose a method to achieve transfer learning in agent-based models through state-based probabilistic models. Burnap et al. [16] use a deep learning based generative model to find the statistical representation of a design space via using a large number of images and design attributes. They test their methods on automobile body design and successfully morph a vehicle into different meaningful body types. In consideration of a sequential design pipeline, Oh et al. [17] manage to combine the generative methods with further topology optimization in automobile wheel design. However, the synthetic designs are still technically immature with one-shot optimization. They claim that an iterative and automatic optimization process can be a better alternative.

Dimensionality Reduction for Shape Optimization

At the heart of a functional shape design process, the idea of reducing the dimension in the original design space plays a critical role in shape optimization. For example, generative topographic mapping has been proposed by Viswanath et al. [18, 19] to tackle with the design problem of 2D aircraft wings. In order to reduce the dimension of design space with a sense of the shape

common features, Proper Orthogonal Decomposition (POD) [20] and Generative Adversarial Networks (GAN) [21] are introduced as effective shape feature extractors. They both validate their design framework in designing a 2D airfoil profile through limited control points.

Although the aforementioned prior researches achieve plausible results when applying a data-driven approach as a tool for the design process in 2D design space, there doesn't exist a close-loop 3D conceptual design pipeline which enhances the design automation and optimization with respect to the functional requirements.

TECHNICAL APPROACH

Fig. 1 shows our proposed framework. The key modules are a deep learned shape encoder-decoder, a geometry or physics-based design simulator, and an optimizer.

Input to our approach is a database of 3D models belonging to the same object category (e.g., aircrafts). These models are most commonly acquired in the form of 3D polygonal surface models. Through unsupervised learning, the variational shape learner extracts a latent feature vector for each of the input designs. The latent space vector has much fewer dimensions than the original shape representation, and thus serves as a dimension reduced encoding of the large design space. This is the first module of our design pipeline (Shape Parameterization).

The geometry- or physics-based simulator is determined by the design performance objectives. It is responsible for testing the design candidates and provide performance scores for the subsequent optimization process (Function Evaluation). The inputs of this module are the reconstructions of some candidate feature vectors using the existing well-trained decoder network.

Through evaluation, we can have all the function scores for these design candidates, which aid the optimizer to generate potential better candidates.

The optimizer utilizes the outcomes of the simulator, together with the latent space representation, to optimize the designs directly in the latent space. New candidates can be generated with a non-gradient based optimization method (Shape Generation). These modules are detailed further in the following sections.

Variational Shape Learner

In this study, we adopt a variational autoencoder model, the Variational Shape Learner (VSL) [22], which builds on the ideas of the Neural Statistician [23] and the volumetric convolutional network [24]. The parameters of the VSL are learned under a variational inference scheme [25]. As shown in Fig. 2, we use a hierarchical VAE (Variational Autoencoder) which consists of an encoder, a decoder and a latent space feature representation.

Distance Maps and Shape Representation: We use a 3D signed distance field (SDF) as the primary design representation in the original space. As opposed to the commonly used binary voxel representation [1–4], this representation allows smooth final surfaces to be constructed over the designs generated by VSL.

SDF is a scalar function of position that defines a closed volume implicitly. The absolute value of the function is the distance from the surface of the solid. A positive value indicates the point is inside of the solid, or negative outside. The boundary is the isosurface where the function value is zero. SDF has been recently used for deep-learned shape completions [5].

In our approach, the SDF is implemented as a tri-linear function defined over a structured lattice. A signed distance is assigned to the lattice nodes and interpolated within the lattice cell. A lattice with $n \times m \times k$ nodes has $(n-1) \times (m-1) \times (k-1)$ cells, and each node is connected to neighboring nodes by multiple edges.

To create the training and testing data, an SDF needs to be generated from a polygonal mesh. The program calculates a distance d from each lattice node to the nearest point on the polygonal mesh. Then $+d$ is assigned to the node if the node is inside of the solid, or $-d$ if outside.

Once an SDF is obtained as an output from the VSL, a polygonal mesh of the boundary needs to be extracted from the SDF for rendering and to facilitate downstream processes.

For this, our approach first places a vertex on each edge with the two nodes having one positive and one negative value. The position of the vertex is smoothly interpolated based on the distance values.

Then the program creates arcs on each rectangle of the lattice cells that has a mix of positive and negative nodes. There are four cases: (1) one negative and three positive, (2) two negative connected and two positive connected, (3) two negative diagonal and the other two positive, and (4) one positive and three negative. Cases (1), (2), and (4) yield one arc, and (3) yields two arcs as shown in Fig. 3.

Finally, polygonal faces in each lattice cell are created by connecting the arcs. The result may become degenerate if a lattice node has exactly zero value, or the node is exactly on the boundary. We append a very small value to the distance values of such nodes to avoid this degeneracy.

Encoder-Decoder Details: For the encoder, the global latent code is directly learned from the input SDF through three convolutional layers with kernel sizes 6, 5, 4, strides 2, 2, 1 and channels 32, 64, 128. The size of the latent vector is 70. Each layer is followed by a ReLU activation layer [26] and a batch normalization layer [27]. Each local latent code is conditioned on the global latent code, the input voxel and the previous latent code (except for the first local latent code) using two fully-connected layers with 100 neurons each. After we learn the global and local latent codes, we concatenate them into a single vector. A 3D deconvolutional neural network with dimensions symmetrical to the encoder of the global latent code is used to decode the learned latent features into an output SDF model.

An element-wise logistic sigmoid [28] is applied to the output layer in order to convert the learned features to occupancy probabilities for each SDF cell. The detailed network architecture can be found in [22].

Physics-based Simulator

The trajectories of the original and synthesized aircrafts are simulated in YS FLIGHT SIMULATOR [29]. While the flight dynamics kernel of the simulator uses a simplified physical model, it is suitable for making a quick estimation of the flight characteristics.

An aircraft in the air is subject to lift, drag, gravity, and thrust, so called the four forces of flight. For this design task thrust is cut to idle, or zero, rendering the designs as gliders. Gravity is $-9.8 m/s^2$ in the y-direction. Lift and drag forces are calculated as:

$$L = \frac{1}{2} C_L \rho v^2 S \qquad (1)$$

$$D = \frac{1}{2} C_D \rho v^2 S, \qquad (2)$$

where L is lift, C_L is lift coefficient, ρ is air density, v is velocity, S is wing area, D is drag, and C_D is drag coefficient. C_L and C_D are functions of α or angle-of-attack. YS FLIGHT SIMULATOR kernel approximates C_L and C_D as a piecewise-linear and a parabolic function respectively.

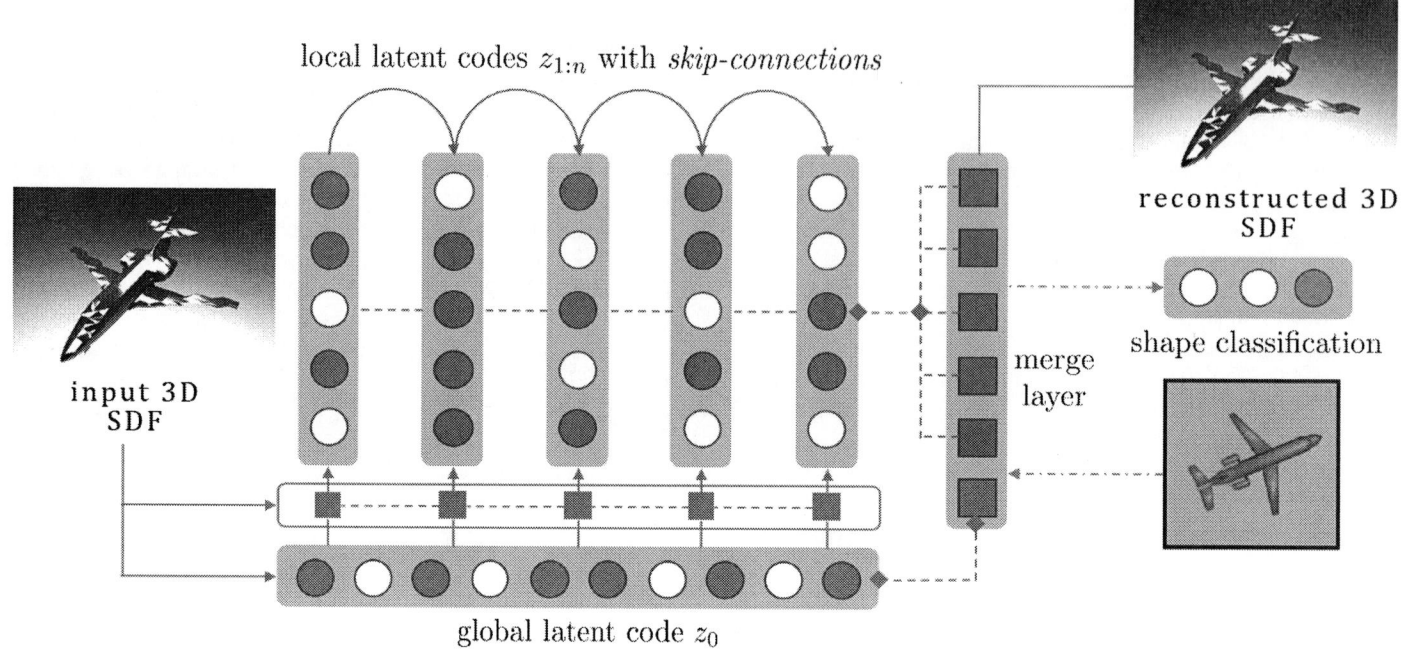

FIGURE 2. A schematic of VSL [22].

The rotation of an aircraft is the hardest to simulate. Unless the moment of inertia and center of gravity is known it is impossible to make an accurate simulation. Instead of estimating the moment of inertia and center of gravity, the simulator kernel approximates a rotation as a second-order system with a stability constant and a maneuverability constant, both of which are empirically specified.

Estimating a reasonable C_L and C_D functions and stability and maneuverability constants for an output design of the network is also a challenge. However, the output design from the network leaves freedom of choosing the airfoil of the wing, moment of inertia, and the center of gravity. It is reasonable to assume that the designs generated by the network can be configured to have a similar characteristic to an existing airplane of a similar geometric signature. We have chosen one from 88 aircraft data in the flight simulator that matches the forward-projection and top-view-projection area ratio the best to the output design from the network and take C_L and C_D functions and stability and maneuverability constants for the characteristic of the designs generated by the network.

Design Optimizer

The main purpose of this module is to explore potentially better design candidates using the compact latent space. We use a gradient-free genetic optimizer following its success in conceptual design [30–32]. Latent feature vectors extracted from various designs are selected and ranked based on their corresponding performance scores. Then, mutation and crossover operations involving interpolation and extrapolation are used to produce subsequent generations.

We observe that conventional crossover renders shapes incomplete. This means that not all the designs in the latent space are valid. To address this problem, we utilize the line crossover operator [33] to generate new offsprings. A child is generated using a linear interpolation between two parents:

$$Child = r \times P_1 + (1 - r) \times P_2, \tag{3}$$

where $r \in [0, 1.2]$ is a random number drawn from a uniform distribution, and P_1 and P_2 are the two parents respectively. The intuition of line crossover is that the linear interpolation or extrapolation of the two parents may provide competitive offsprings [32]. In our case, we expect the line crossover is functionally similar to shape morphing which blends the geometric representation of two models together. The weight r tells the similarity between the child representation and the parent P_1. Sample results of the reconstruction models from the latent feature interpolation and extrapolation are shown in Fig. 4. Note that the voxelized representation is used to show the coarse, binary versions of the designs that are generated from the latent space. The actual output of the network is an SDF.

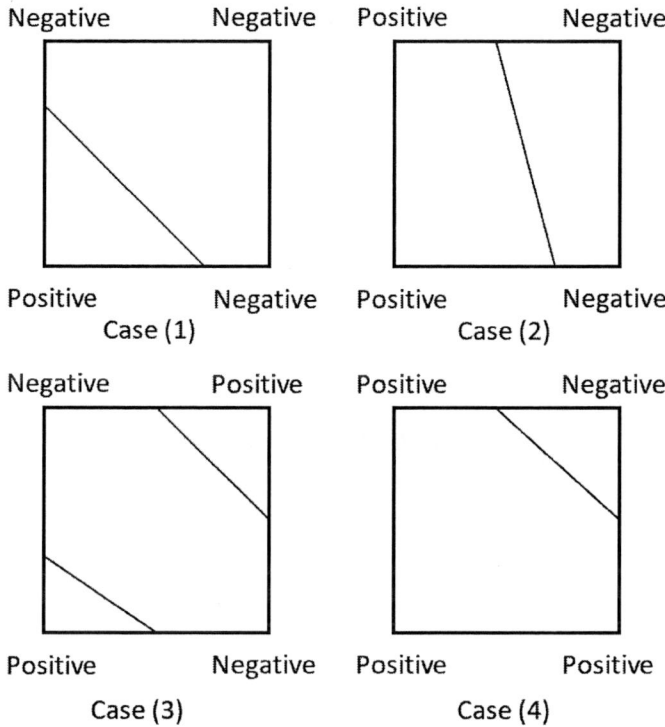

FIGURE 3. Four cases of arc generation in a rectangular face of a lattice cell.

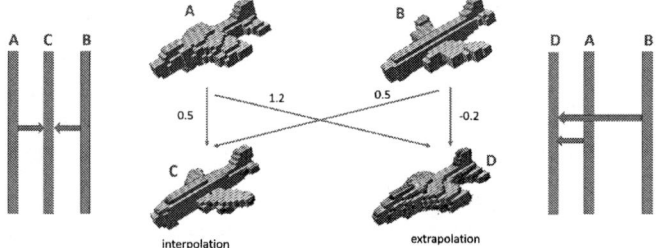

FIGURE 4. Sample reconstructions of the latent vector interpolation and extrapolation.

We first compute the scores for all the models in our dataset. Then, we rank all models and randomly select one model from every (N/n) interval where N is the number of total models, n is the population size (set to 100 in our approach). The probability of crossover and mutation is 0.9 and 0.05 respectively. The optimization goal is to minimize the Mean Square Error:

$$MSE = ||y_t - y_c||_2, \tag{4}$$

where y_t denotes the target functional objective score and y_c denotes the scores of the children generated in each generation.

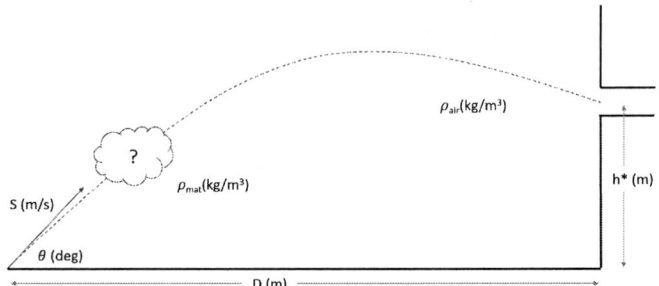

FIGURE 5. A schematic graph of the design task.

CASE STUDY

In this section, we demonstrate our approach on glider aircraft design. All the modules in the pipeline are specified from the suggested options we mentioned above. VSL, YS FLIGHT SIMULATOR and Genetic Optimizer are chosen as the three main modules in our pipeline.

For the data, we use the 4096 airplane models from the ModelNet40 repository [2]. We clean the floating patches in the models and align all the models in the same direction. Then we compute the SDF from the polygonal model to produce the network input $[41 \times 41 \times 41]$ in size for each model. It takes 3 hours to train the network for 1000 epochs on an NVIDIA GeForce 1080 GPU. The learning rate is 5×10^{-3}, and the batch size is 64.

Design Task

As shown in Fig. 5, given the initial launch speed $S(m/s)$, the initial pitch angle $\theta(degree)$, the density of the projectile aircraft $\rho_{mat}(kg/m^3)$ and the density of the air $\rho_{air}(kg/m^3)$ are given, the goal is to design the shape of the glider so that it can go through the gap located at h^*m high from the ground, and Dm horizontally away from the launch point.

In this design task, the constraint is that the projectile candidates should all fit in a $1m \times 1m \times 1m$ box without any propulsion. Among all the parameters, D, ρ_{mat} and ρ_{air} are fixed. S, θ as well as h^* can be set arbitrarily by the user. Designers are usually accustomed to the inverse task: Given the projectile, adjust S and θ to hit the target height, which can be regarded as a tuning process of several parametric design variables. By contrast, in our design problem, the shape of the projectile aircraft cannot be readily parameterized or represented with a limited number of design variables.

RESULTS

In our experiments, we set $\theta = 10^o$, $S = 45.7m/s$, $\rho_{mat} = 1000kg/m^3$, $\rho_{air} = 1.29kg/m^3$ and $D = 100m$. Our main interests lie in the height of the gap $h^*(m)$. We varies the value of h^* to test the performance of our proposed pipeline.

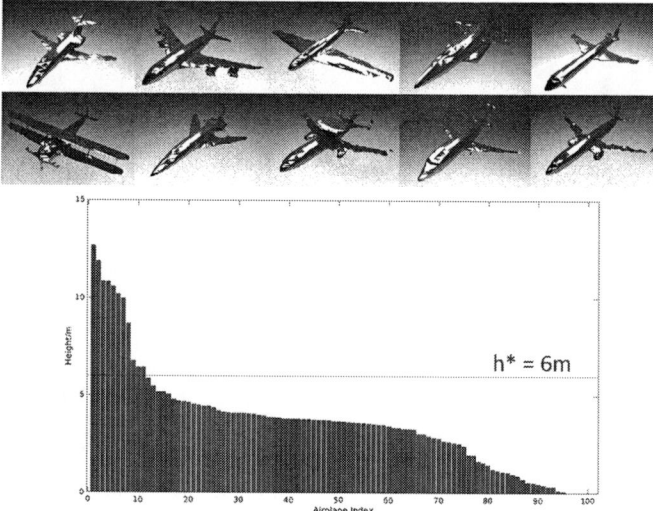

FIGURE 6. **Upper:** 10 randomly selected aircraft SDF models from the initial population. **Lower:** The landing height $h(m)$ distribution of the 100 initial population.

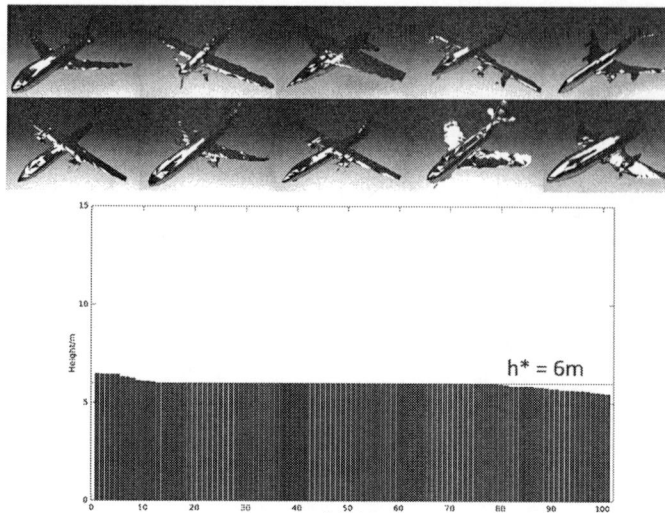

FIGURE 7. **Upper:** 10 randomly selected aircraft SDF models from the optimized population when $h^* = 6m$. **Lower:** The landing height $h(m)$ distribution of the 100 optimized population when $h^* = 6m$.

TABLE 1. PERCENTAGE OF THE DESIGN CANDIDATES WITH THE LANDING HEIGHT h, S.T. $|h - h^*| < \delta$ WHEN $h^* = 6m$.

$\delta(m)$	Initial	Final
0.1	1%	76%
0.5	4%	98%

To have an intuitive sense of a reasonable range for $h*$, we randomly select 100 aircraft models from the dataset. Then we obtain the landing height of each aircraft model using the physics-based simulator we introduce in the CASE STUDY section. The acquired height range is $h \in [0, 12.6]m$. A plot of the sorted height distribution is shown in Fig. 6. Note that zero means the model lands at the ground before $D = 100m$.

First, we set $h^* = 6m$, which is around the mean value of the height range. After about 200 rounds of an iterative optimization process, the distribution of the landing heights for the final design candidates are demonstrated in Fig. 7. Unlike the diverse distribution in Fig. 6, the distribution is much flatter and mostly gather near the target height of $h^* = 6m$. In fact, if the tolerance is 0.1 m, 76% of design candidates in the final population satisfy the design requirement. (See Tab. 1) In the initial population, only 1% of the population can fulfill the same requirements. Fig. 7 also shows the diversity in the final synthetic populations, which will benefit the human designers with more valid options to choose from.

To demonstrate the ability of our pipeline in design space exploration, we intentionally set a design requirement that is originally unfeasible in the existing design candidates. Specifically, we set $h^* = 13.8m$. The height distribution of the suggested design candidates after optimization is shown in Fig. 8. We can observe that all the heights are within $[13.2, 13.9]m$, which exceed the maximum height($12.6m$) the initial designs can reach. According to Tab. 2, although the concentration of heights for final designs is not as impressive as $h^* = 6m$ case, 88% of candidates are able to fulfill the requirement if the tolerance is $0.5m$.

The results also reveal the difficulty of exploration outside the original feasible space. To make the task even more challenging, we set $h^* = 14.5m$ and repeat all the design process. Expectedly, see Fig. 9, the final synthetic candidates cannot reach such a height requirement. But we obtain a design boundary reference in a data-driven manner as well as plentiful promising designs that provide useful suggestions for further manual design exploration.

CONCLUSIONS

This work presents an integrated conceptual design pipeline involving a data-driven shape learner, a function evaluator and a functional design optimizer. The pipeline is verified with a case study on the shape design of projectile aircraft models. When the design objective is set within the objective range of the design exemplars, our algorithm is able to synthesize a large set of design candidates that also satisfy the functional design requirements. It

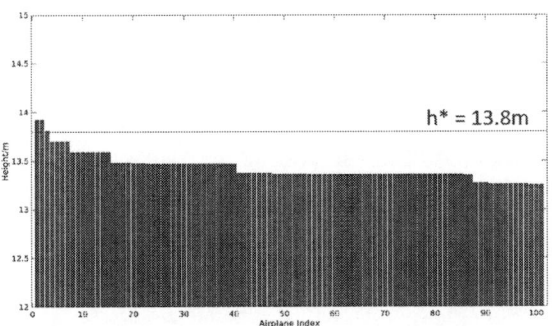

FIGURE 8. The landing height $h(m)$ distribution of the 100 optimized population when $h^* = 13.8m$.

TABLE 2. PERCENTAGE OF THE DESIGN CANDIDATES WITH THE LANDING HEIGHT h, S.T. $|h - h^*| < \delta$ WHEN $h^* = 13.8m$.

$\delta(m)$	Initial	Final
0.1	0%	5%
0.5	0%	88%

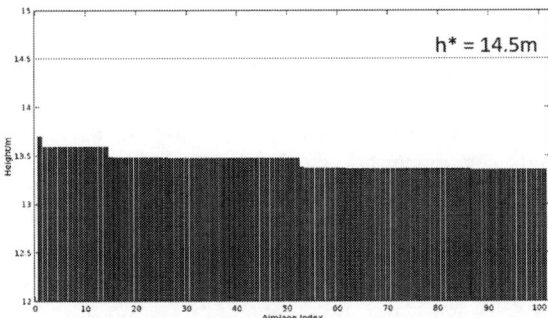

FIGURE 9. The landing height $h(m)$ distribution of the 100 optimized population when $h^* = 14.5m$.

is capable of generating new designs whose performance scores are outside the range of the original models.

Likewise, the same approach can be utilized to generate valid design candidates in other domains when adopting different options for each module. For example, if the design focus is the shape of a 2D beam bridge with load constraints, the shape learner alternative can be a 2D VAE. The physics-based simulator should accordingly be an FEA simulator. The functional design optimizer can adopt PSO.

At the meantime, even for design problems with the same dimension (2D/3D), various data representations can be introduced in the pipeline in terms of the focus area. Depth images are suitable when designing shell structures. While point clouds are beneficial when designing adjacent mesh surfaces. These potential extensions of our pipeline are becoming increasingly promising as several recent learning frameworks like Matterport3D [34] and PointNet [35] have been developed.

LIMITATIONS AND FUTURE WORK

Two time-consuming processes in our pipeline are the training of the shape learner and the geometry-based or physics-based simulation. A potential more efficient method is to directly establish a mapping between the latent feature vector and its corresponding physic properties, which may be a trained network as well. In this manner, designers can completely get rid of the original data format (2D/3D) during the iterative optimization process. The final proposed designs will be reconstructed once the objective is fulfilled.

When manipulating the feature vectors in the functional optimizer module, we don't have a sense of the exact meaning of each dimension in the latent space. This is a common problem which has also problematic for researchers from the machine learning field. Recently, researchers like Wieczorek et al. [36] are exploring learning orthogonal latent features, which may enable the designers to acquire a desired latent space.

REFERENCES

[1] Li, Y., Pirk, S., Su, H., Qi, C. R., and Guibas, L. J., 2016. "FPNN: field probing neural networks for 3d data". In Advances in Neural Information Processing Systems 29: Annual Conference on Neural Information Processing Systems 2016, December 5-10, 2016, Barcelona, Spain, pp. 307–315.

[2] Wu, Z., Song, S., Khosla, A., Yu, F., Zhang, L., Tang, X., and Xiao, J., 2015. "3d shapenets: A deep representation for volumetric shapes". In Proceedings of the IEEE conference on computer vision and pattern recognition, pp. 1912–1920.

[3] Sedaghat, N., Zolfaghari, M., Amiri, E., and Brox, T., 2017. "Orientation-boosted voxel nets for 3d object recognition". In British Machine Vision Conference 2017, BMVC 2017, London, UK, September 4-7, 2017.

[4] Brock, A., Lim, T., Ritchie, J. M., and Weston, N., 2016. "Generative and discriminative voxel modeling with convolutional neural networks". *CoRR, abs/1608.04236*.

[5] Stutz, D., and Geiger, A., 2018. "Learning 3d shape completion from laser scan data with weak supervision". In 2018 IEEE Conference on Computer Vision and Pattern Recognition, CVPR 2018, Salt Lake City, UT, USA, June 18-22, 2018, pp. 1955–1964.

[6] Sosnovik, I., and Oseledets, I. V., 2017. "Neural networks for topology optimization". *CoRR, abs/1709.09578*.

[7] Guo, T., Herber, D., and Allison, J., 2018. "Reducing evaluation cost for circuit synthesis using active learning".

[8] Raissi, M., Perdikaris, P., and Karniadakis, G., 2018. "Multistep neural networks for data-driven discovery of nonlinear dynamical systems".

[9] Umetani, N., Koyama, Y., Schmidt, R., and Igarashi, T., 2014. "Pteromys: Interactive design and optimization of free-formed free-flight model airplanes". *ACM Trans. Graph., 33*(4), July, pp. 65:1–65:10.

[10] Mark Fuge, B. P., and Agogino, A., 2014. "Machine learning algorithms for recommending design methods". *Journal of Mechanical Design, 136*(10), Aug.

[11] Chen, W., and Fuge, M., 2018. "Béziergan: Automatic generation of smooth curves from interpretable low-dimensional parameters". *CoRR, abs/1808.08871*.

[12] Chen, Y., Tu, S., Yi, Y., and Xu, L., 2017. "Sketch-pix2seq: a model to generate sketches of multiple categories". *CoRR, abs/1709.04121*.

[13] Dering, M. L., and Tucker, C. S., 2017. "A convolutional neural network model for predicting a product's function, given its form". *Journal of Mechanical Design, 139*(11), p. 111408.

[14] Burnap, A., and Hauser, J., 2018. "Predicting "Design Gaps" in the Market: Deep Consumer Choice Models under Probabilistic Design Constraints". *arXiv e-prints*, Dec, p. arXiv:1812.11067.

[15] Raina, A., McComb, C., and Cagan, J., 2018. "Design strategy transfer in cognitively-inspired agents". In ASME 2018 International Design Engineering Technical Conferences and Computers and Information in Engineering Conference, American Society of Mechanical Engineers, pp. V02AT03A018–V02AT03A018.

[16] Burnap, A., 2016. "Estimating and exploring the product form design space using deep generative models". In DAC 2016.

[17] Oh, S., Jung, Y., Lee, I., and Kang, N., 2018. "Design automation by integrating generative adversarial networks and topology optimization". In ASME 2018 International Design Engineering Technical Conferences and Computers and Information in Engineering Conference, American Society of Mechanical Engineers, pp. V02AT03A008–V02AT03A008.

[18] Viswanath, A., J. Forrester, A., and Keane, A., 2011. "Dimension reduction for aerodynamic design optimization". *AIAA journal, 49*(6), pp. 1256–1266.

[19] Viswanath, A., Forrester, A., and Keane, A., 2014. "Constrained design optimization using generative topographic mapping". *AIAA journal, 52*(5), pp. 1010–1023.

[20] Cinquegrana, D., and Iuliano, E., 2018. "Investigation of adaptive design variables bounds in dimensionality reduction for aerodynamic shape optimization". *Computers & Fluids, 174*, pp. 89–109.

[21] Chen, W., Chiu, K., and Fuge, M., 2019. "Aerodynamic design optimization and shape exploration using generative adversarial networks". In AIAA SciTech Forum, AIAA.

[22] Liu, S., Giles, C. L., and Ororbia, A., 2018. "Learning a hierarchical latent-variable model of 3d shapes". In 2018 International Conference on 3D Vision, 3DV 2018, Verona, Italy, September 5-8, 2018, pp. 542–551.

[23] Edwards, H., and Storkey, A., 2016. "Towards a neural statistician". *arXiv preprint arXiv:1606.02185*.

[24] Maturana, D., and Scherer, S., 2015. "Voxnet: A 3d convolutional neural network for real-time object recognition". In 2015 IEEE/RSJ International Conference on Intelligent Robots and Systems (IROS), IEEE, pp. 922–928.

[25] Kingma, D. P., and Welling, M., 2013. "Auto-encoding variational bayes". *arXiv preprint arXiv:1312.6114*.

[26] Nair, V., and Hinton, G. E., 2010. "Rectified linear units improve restricted boltzmann machines". In Proceedings of the 27th international conference on machine learning (ICML-10), pp. 807–814.

[27] Ioffe, S., and Szegedy, C., 2015. "Batch normalization: Accelerating deep network training by reducing internal covariate shift". In Proceedings of the 32nd International Conference on Machine Learning, ICML 2015, Lille, France, 6-11 July 2015, pp. 448–456.

[28] Mitchell, T. M., 1997. *Machine Learning*, 1 ed. McGraw-Hill, Inc., New York, NY, USA.

[29] Yamakawa, S., 2019. Ys flight simulator. http://www.ysflight.com. The program is available online since Feburary 2019.

[30] GERO, J. S., and KAZAKOV, V., 2000. "Adaptive enlargement of state spaces in evolutionary designing". *Artificial Intelligence for Engineering Design, Analysis and Manufacturing,* *14*(1), p. 3138.

[31] Cvetkovic, D., and Parmee, I., 1999. "Genetic algorithms based systems for conceptual engineering design".

[32] Rasheed, K., Hirsh, H., and Gelsey, A., 1997. "A genetic algorithm for continuous design space search". *Artificial Intelligence in Engineering,* *11*(3), pp. 295 – 305.

[33] Wright, A., 1999. "Genetic algorithms for real parameter optimization". *Foundations of Genetic Algorithms,* *1*, 06.

[34] Chang, A. X., Dai, A., Funkhouser, T. A., Halber, M., Nießner, M., Savva, M., Song, S., Zeng, A., and Zhang, Y., 2017. "Matterport3d: Learning from RGB-D data in indoor environments". In 2017 International Conference on 3D Vision, 3DV 2017, Qingdao, China, October 10-12, 2017, pp. 667–676.

[35] Qi, C. R., Su, H., Mo, K., and Guibas, L. J., 2017. "Pointnet: Deep learning on point sets for 3d classification and segmentation". In 2017 IEEE Conference on Computer Vision and Pattern Recognition, CVPR 2017, Honolulu, HI, USA, July 21-26, 2017, pp. 77–85.

[36] Wieczorek, A., Wieser, M., Murezzan, D., and Roth, V., 2018. "Learning sparse latent representations with the deep copula information bottleneck". *CoRR,* *abs/1804.06216*.

Proceedings of the ASME 2019
International Design Engineering Technical Conferences
and Computers and Information in Engineering Conference
IDETC/CIE2019
August 18-21, 2019, Anaheim, CA, USA

DETC2019-97301

INVESTIGATING OPTIMAL COMMUNICATION FREQUENCY IN MULTI-DISCIPLINARY ENGINEERING TEAMS USING MULTI-AGENT SIMULATION

Mojtaba Arezoomand *
Global Design Laboratory
Department of Mechanical Engineering
University of Michigan
Ann Arbor, MI 48109
Email: mojiar@umich.edu

Jesse Austin-Breneman
Global Design Laboratory
Department of Mechanical Engineering
University of Michigan
Ann Arbor, MI 48109
Email: jausbren@umich.edu

ABSTRACT

Complex engineering design tasks require teams of engineers with different skills and unique knowledge sets to work together to develop a solution. In these contexts, team communication is critical to successful design outcomes. Previous research has identified effective management of communication frequency as an important dimension of team communication leading to improved design outcomes. Organization research literature has demonstrated a curvilinear relationship in which both frequent and infrequent communication may hamper organizational performance. In contrast, recent work in engineering design research has found an inverse relationship between frequency and technical system performance for simple design tasks. This paper extends this work quantifying the impact of communication frequency on technical system performance by examining multidisciplinary problems. Results from a multi-agent simulation on a six discipline parameter design task for minimizing the weight of a geostationary satellite are presented. Simulation results suggest that the form of relationship between frequency and performance changes significantly depending on the communication pattern. The evidence suggests that for the same design task a planned periodic communication pattern results in a curvilinear relationship, whereas for a stochastic communication pattern a less pronounced monotonic inverse relationship is found.

1 Introduction

New-product development (NPD) processes require multidisciplinary teams to communicate effectively [1, 2]. Communication frequency has been shown by a number of research communities to significantly impact performance [3–12]. In the context of NPD teams, communication has been modeled as the exchange of information between individuals inside and outside of the team [4]. Although there exists a wide range of information created during the design process, models of engineering design teams have defined information as design parameter values [13, 14]. In order to effectively manage communication in the product development process, organizations must ascertain the optimal level of information exchange among team members [11]. However, this is difficult due to the scale and complexity of the socio-technical systems involved.

Previous research on team communication assumed that frequency is monotonically related to performance [15–18]. Some researchers found that this relationship was positively monotonic [15, 16]. However, more recently, empirical and quantitative studies have shown that communication is curvilinearly related to performance. In this context, a curvilinear relation is defined as a non-monotonic relationship with an interior global extremum [10, 11]. These results suggest that there is an optimal communication frequency for a specific team and project [10, 11, 19].

The literature on optimal communication frequency is limited in its treatment of the interactions between process character-

---

*address all correspondence to this author.

istics and the physical system. Current quantitative models focus on either aspects of the process for developing the product [20] or on properties of the physical system [21]. As suggested by socio-technical systems theory [20], these quantitative models could be improved by developing an integrated model of team communications which models both the process aspects as well as the technical and physical aspects of the problem. This study seeks to extend previous work on communication frequency by using a multi-agent simulation which incorporates characteristics of the process and physical system.

2 Background

This study draws upon work on team communication in engineering design research, management, and information.

2.1 Communication in NPD Teams

Information processing theory defines communication in teams as the exchange of information between actors. [22–24]. In this literature organizations performing NPD are regarded as open systems that must process information [25], with limited capacity of doing so [26]. Product development organizations are modeled as processing units that receive input information and turn them into set of outputs, such as a design [4]. For this study, communication in NPD teams is defined as the exchange of information between team members on parameter values to produce a design. This is a subset of the information considered in this literature, as communication could also include information exchange with actors outside the team.

The positive relationship of performance and team communication in the literature has been surveyed by Allen, et al. [27]. Team communication facilitates team processes such as cooperation [28], coordination [29], information processing [30], decision making [31], and creativity [32], [33]. Moreover, communication helps team with work relationships [34] and knowledge sharing [35]. This study draws upon this literature as a theoretical grounding for model creation.

2.2 Communication Frequency

Various dimensions of communication have been investigated in the literature, such as communication frequency, modes, formality, centrality, and rationality [7, 36]. Communication frequency is the first and foremost explanatory variable of team performance discussed in the literature across a wide variety of team types and performance measures due to its importance and ease of management [37].

Early studies found communication frequency to be positively related to performance. Joyce [5] found that powerful project leaders in organization prioritize quantity of internal communication over its quality. In an empirical study of 66 North American and European multinationals, Ghoshal and Bartlett [6] observed more success in the creation, adoption, and diffusion of innovations in subsidiaries with higher levels of interunit communication. Tang et al. reported a positive relationship between Chinese R&D employees' frequency of connection with other team members and external persons with creativity [7]. McDonald et al. [8] found that the frequency of use of communication technologies is higher for better performing global product development teams. Furthermore, Austin [9] observed that frequency of specific types of interactions may lead to better performance.

Other researchers have found results contradicting the above literature using different relationships between frequency and performance. Katz et al. [38] and Fussell et al. [39] have shown that both too little and too much communication can affect performance adversely. Several studies found that higher communication frequency may lead to lower creativity [40–43]. These results suggest that communication frequency and performance may have a more complex relationship than a simple monotonic one. For instance, Hoegl and Leenders [37, 44] found a curvilinear relation between team communication and performance using an empirical study. Specifically, NPD teams performance [44] and teams' creativity [37] may be hindered by both low and high communication frequency.

This study is based upon the subset of this work in NPD which examines the relationship between communication frequency and performance in the context of cross-functional teams (CFT). CFTs consist of team members from different functional areas within an organization bringing their unique "thought worlds" to the team [11, 45] and interacting in every phase of development tasks to design products and processes concurrently [46]. Darawong et al. [47] argued that more frequent communication across functional areas, teams could better utilize external knowledge in NPD tasks. Brown et al. suggested that high internal communication increases the amount and variety of internal information flow which in turn improves development-process performance [1]. Patrashkova et al. found a curvilinear relationship between communication and performance in cross-functional teams using an empirical study of 60 project teams [10]. They observed that the curvilinear relation holds for several performance measures such as goal achievement, project efficiency, and team cohesion and for two communication media namely, face-to-face and e-mail. This paper draws upon this work to form hypotheses regarding the main research questions.

2.3 Simulations of Team Communication

Empirical field studies, such as those cited above, are limited in their ability to isolate the impact of individual factors or phenomena. Simulations of team communication have been widely used to address this problem. These include simulations based on information theory and optimization algorithms.

2.3.1 Information Theory-based Approaches

In the information theory based approach, NPD is modeled as a process through which information in the form of requirements is turned into the information describing the final product [22–24]. Using this method, Patrashkova et al. [11] performed a series of simulations confirming the curvilinear relationship between communication frequency and performance. Building on the Patrashkova et al. model, Kennedy et al. used a Monte Carlo simulation and regression analysis to demonstrate the dependence of optimal communication frequency on the project complexity and communication media [48]. This simulations were limited because they modeled the technical aspects of the design tasks as block boxes. This simplification blocks feedback loops between the technical system and process characteristics modeled in the simulation. For example, McComb et al. [49] related different aspects of the physical system to the optimal team makeup.

2.3.2 Optimization-based Approaches

Optimization algorithms offer another technique for simulating team communication. Yassine used dynamic programming to model the product development process [50]. This work defined optimal decision rules regarding whether to incorporate certain information at a given time in the development process to maximize performance. Their findings suggest that not all available information related to a design activity should be incorporated and teams should stop accumulating information once the information collection exceeds certain value. On the other hand, Lu et al. [21] demonstrated a positive relationship between communication frequency and performance for multi-disciplinary optimization (MDO) problems with no communication cost. Although these optimization approaches provide methods for decomposing the technical aspects of the system and coordinating information exchange or communication between different disciplines, they are limited in their treatment of modeling of the process characteristics of of the socio-technical systems.

2.3.3 Stochastic Optimization-based Approaches

Stochastic optimization approaches have been used to model designer behavior and technical systems in an integrated fashion. Cagan [51] proposed that the process of problem solving starts with a phase of more or less random exploration of space and becomes more deterministic as problem solving progresses. Cagan found the process resembles an accelerated version of the stochastic optimization simulated annealing optimization algorithm [52]. Yu et al. [53] found that high-performing designers completing a parameter design task exhibited problem-solving profiles which resembled a well-tuned simulated annealing optimization algorithm, whereas bottom-ranked subjects used a pseudo random search strategy. Stochastic optimization approaches have also been used to model the interaction between organizational and domain characteristics. Olson et al. proposed a computational framework based on qualitative studies of aerospace organizations in order to examine the impact of informational dependencies on system performance [54].

Similarly, building on the Cagan et al. [51] model, McComb [12] proposed Cognitively-Inspired Simulated Annealing Teams (CISAT) modeling framework. CISAT is a platform based on simulated annealing algorithm for simulating and analyzing design teams which incorporates eight empirically demonstrated cognitive phenomena, such as self-bias, quality bias reduction, operational learning, and satisficing among others. Moreover, a high degree of linear correlation was demonstrated between results of a cognitive study of engineering students designing a structural problem and the results of simulations performed by CISAT. Further work produced a regression model based on the data from a set of test problems to relate the optimal communication frequency and teams size to single disciplinary design problem characteristics [49]. These simulations used in this study are based upon this work using stochastic optimization techniques to model team communication and designer behavior.

2.4 Communication Pattern

The communication pattern is one method for managing communication frequency and effectiveness in organization. Communication patterns which do not involve a system coordinator in fall into two main categories: planned and stochastic. Planned communication pattern represents teams holding schedule meetings periodically throughout the project life-cycle. This can include strategies such as scheduled information passing through interface control documents [55]. In a stochastic communication pattern, information exchange follows a probabilistic pattern and takes place at irregular intervals. For example, this may occur when a team is in an open-plan office and team members interact stochastically [12, 56]. This is an important organizational charasteristic that is modeled in this study.

2.5 Research Gap

Effective management of communication frequency improves system and organizational performance. Determining an optimal frequency for a given design task is difficult given the problem complexity, scale, and coupling between organizational and technical systems. Previous empirical work has examined the relationship between communication frequency and performance but is limited in its ability to investigate how this relationship occurs. Research using computational models has either considered the technical system a black box or ignored the impact of organizational characteristics. A smaller body of work has examined the interaction between the organizational and technical systems. Notably, Olson et al. developed a computational platform to examine structural interdependencies in design [54]. However, the majority of the work on communication frequency addresses design tasks consisting of a single discipline in which team communication may not be as important. This paper seeks to fill this gap by examining the relationship between communication frequency and system performance in multidisciplinary socio-technical systems using a Multi-agent simulation. Given the discrepancies observed in the results from different methods and problem scope, this paper seeks to answer the following research questions:

1. What is the relationship between communication frequency and team performance in multidisciplinary design teams?
2. Does the organizational communication pattern affect the behavior of this relationship?

Given the literature cited above, this work makes the following hypotheses.

H1 There is curvilinear relation between communication frequency and design performance in multidisciplinary design teams.

H2 The relation does not depend on the communication pattern.

3 Methodology

To test the hypotheses a Monte Carlo simulation of a multi-agent model of a geostationary satellite was performed. Consistent with previous work, the multi-agent model is defined in three parts: the technical system formulation, agent decision-making behavior, and the algorithm for coordinating the exchange of information [12, 20, 49]. The technical system is a modified version of the commercial geo-stationary communication satellite in Kannan et al. [57]. It consists of six coupled subsystems representing different disciplines. One agent is assigned to each subsystem optimizing a parameter design problem. Parameter design is typically associated with the detailed stage in engineering design after the design concept is developed [53,58]. The agent's behavior is modeled by a simulated annealing optimization algorithm for their design variables. The communication pattern was varied to test the hypotheses.

3.1 Technical System Formulation

The technical system model consists of six subsystems: payload, attitude determination and control system (ADCS), power, structures, thermal, and propulsion. All of the subsystems have one design variable, except for the payload subsystem which has two. The design variables are distinct, meaning no two disciplines share a design variable. As in many optimization formulations, disciplines are connected through coupling variables [59]. Figure 1 shows the discipline-based DSM which is the same as team-based DSM for each agent is in charge of one discipline. Table 1 shows the design variables for each discipline.

The system has one constraint which requires the composite signal to noise (SNR) ratio of uplink and downlink to be more

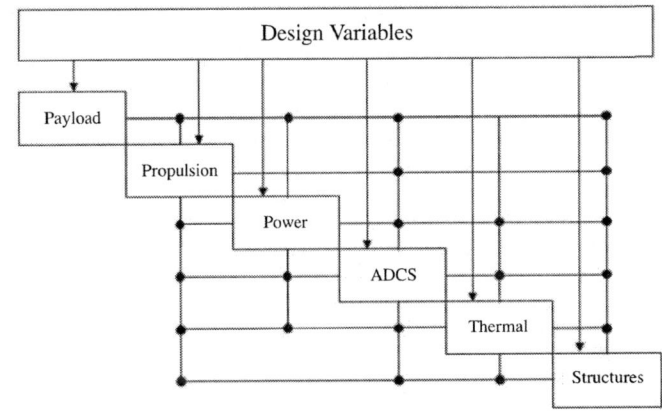

FIGURE 1. Discipline-based Design Structure Matrix (DSM)

TABLE 1. Design variables description

Discipline	Design Variable	Description
Payload	$D_{Sat,trans}$	Satellite transmitting antenna diameter (m)
Payload	$D_{Sat,rec}$	Satellite receiving antenna diameter (m)
ADCS	P_{Sensor}	Power required by the sensors (W)
Power	ε	Battery energy to weight ration
Structures	R	Height to bus ratio of the bus
Thermal	ε_{Rad}	Emissivity of the radiator
Propulsion	I_{sp}	Specific Impulse of the propulsion system (s)

than 40 db in the payload discipline. In the aerospace industry, mass is often chosen as the objective function due to its relationship with cost [57]. Therefore, the objective function of each discipline is set to be the mass of the corresponding subsystem, and the system level objective function is the mass of the satellite, defined as the sum of the subsystems' masses.

3.2 Agent Behavior

Each agent seeks to minimize the mass of the assigned discipline. The design behavior of agents is modeled by simulated annealing [52]. The agents start with a predetermined set of feasible and consistent x_0 which is fixed throughout different simulation scenarios. To get the best performance, different values for the initial temperature T_0, cooling rate γ and number of design iterations were explored. Values of 10,000 and 0.85 were chosen for T_0 and γ, respectively. The algorithm terminates after 100 iterations is reached. Additionally, memory is incorporated into the algorithm to either avoid or prioritize points or directions in the design space [60,61]. This is inline with how humans behave as designers often avoid replacing a better early design for a low performing one found later. Therefore, in this simulation each agent saves a record of points and corresponding objective function values, jumping back to the best point in the list after a fixed number iterations. A range of intervals for memory jumps were examined. Results showed no significant difference in outcome; however, 5 iterations was chosen due to a slightly better performance. Although this method is an ad hoc strategy from design practice with no formal proof of convergence, it was chosen to model engineers' decision-making behavior when encountering low-performing solution sub-spaces.

3.3 Coordination Algorithm

The coordination algorithm draws on established optimization practice from concurrent engineering. The system starts with a set of feasible design variables for which a set of consistent coupling variables exists. After initialization, all the agents perform one iteration of simulated annealing optimization algorithm using a local copy of initial coupling variables. Updated local values are used for the variables generated by the discipline, while initial values are used for those from other disciplines. These initial values are updated through information exchange in two ways: planned communication and stochastic communication.

3.3.1 Planned Communication
In this case, information between disciplines is exchanged periodically. This represents teams scheduled meetings or periodic information passing through interface control documents (ICD) during a project life-cycle [55]. In the algorithm, after a fixed number of iterations, disciplines exchange information on the values of their local coupling variables. The set of local variables is then updated and the agents perform parameter design using the new set of coupling variables. One of the challenges facing distributed optimization methods is the inconsistency of the coupling variables, i.e., for a set of discipline feasible design variables no consistent set of coupling variables may exist. These conflicts are usually handled through negotiations between subsystems or by systems engineers in real-world projects. However, since modeling the agents negotiating or behavior of systems engineers is out of the scope of this paper, two scenarios were incorporated to cope with this problem: simulation with and without redesign.

In the simulation without redesign, right before each information passing session, a separate code is run checking whether there exists a set of consistent coupling variables for the current design variables or not. If such a set exists, the algorithm proceeds to the next step. If not, the algorithm resets all the parameters to the values of the iteration where the most recent consistent set of coupling variables was generated. This includes all the parameters of the simulated annealing algorithm. All the design iterations in between will be discarded. This process may repeat several times until the agents get to a consistent design by trial and error.

Similarly, in simulation with redesign a separate code is run checking for existence of consistent coupling variables. If they exist, the algorithm proceeds. However, if the algorithm is unable to find a consistent set, the same process as simulation without redesign is performed except for two differences. First, each time the algorithm resets the parameters of the simulated annealing to the values of the previous successful information passing session, it divides all the parameters by a factor of γ^k where γ is the cooling rate and k is the number of unsuccessful consecutive information exchange sessions. This means that each time algorithm jumps back to a recent successful session, agents expand their design space exploration by a factor γ^{-1}. If that happens again, the factor turns to γ^{-2}. k will be reset after each successful session. This procedure mimics the human behavior as design failures result in redesigns considering wider design spaces and concepts. The second difference is that all the design iterations leading to a final design will be saved including the trial and error steps.

3.3.2 Stochastic Communication
Team communication can also follow a stochastic pattern taking place at irregular intervals [12, 56]. In this case, each agent decides probabilistically whether to update the coupling variables with the most recent values from the corresponding disciplines or continue using the local values. This decision is made after each iteration and for coupling variables coming from different disciplines separately. As a representation of communication frequency, the probability of interaction is changed and the resulting satellite mass is compared. This assumption does not interfere with the comparisons of different patterns since we are interested in the behaviour of performance relationship with communication frequency; hence any monotonically increasing function of communication frequency on the x-axis is acceptable.

4 Results

Since simulated annealing is stochastic in nature, a Monte Carlo simulation on the algorithm was executed facilitating better comparison of the different scenarios. After 100 iterations, the mean of the satellite final mass of each simulation run was calculated for different communication frequencies. The agents are homogeneous in terms of design behavior. To avoid leaving a design for a worse, after every 5 iterations agents opt for the best design in their design record.

4.1 Planned Communication

10,000 simulation runs were executed for each communication scenario. All the simulations started from the same initial values for design and coupling variables. Six different scenarios were tested: information exchange on every iteration, every 2, 5, 10, 20, and 40 iterations. Figure 2 shows the mean of satel-

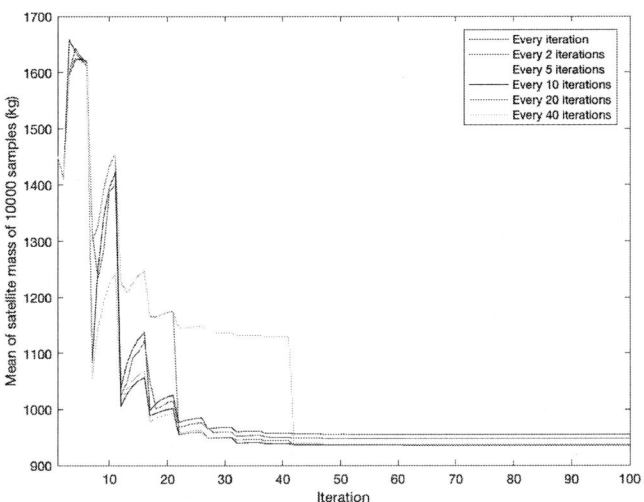

FIGURE 2. Mean of satellite mass versus design iteration for different communication frequencies without redesign

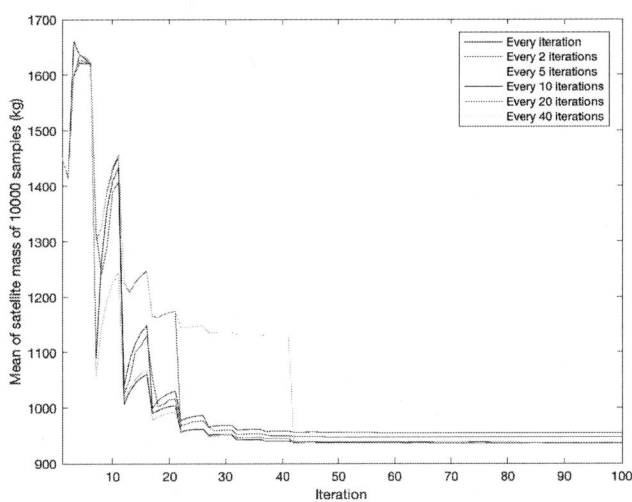

FIGURE 4. Mean of satellite mass versus design iteration for different communication frequencies with redesign

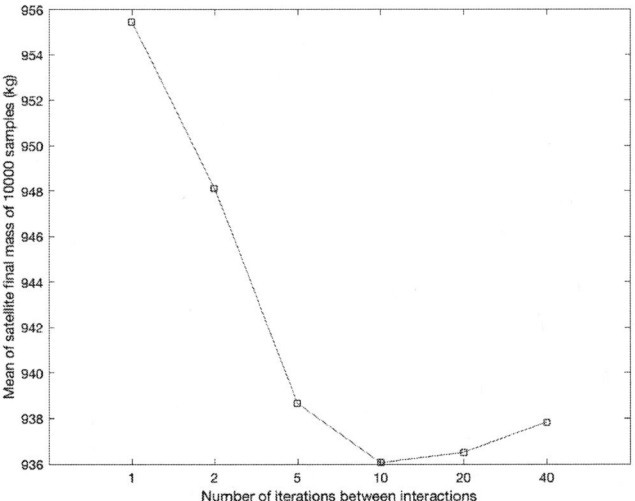

FIGURE 3. Mean of satellite final mass versus communication frequency without redesign

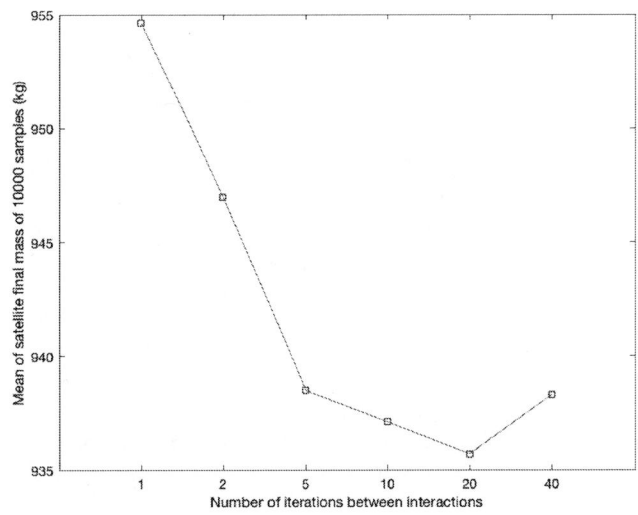

FIGURE 5. Mean of satellite final mass versus communication frequency with redesign

lite mass at each of the 100 design iterations and for different communication frequencies without redesign, and Fig. 3 shows the mean of 10,000 satellite final mass for different frequencies without redesign after 100 iterations. Figure 4 shows the mean of satellite mass at each of the 100 design iterations and for different communication frequencies with redesign, and Fig. 5 shows the mean of 10,000 satellite final mass for different frequencies with redesign after 100 iterations.

4.2 Stochastic Communication

In this case, 5 different scenarios were tested. The probability of information exchange varied between 0 and 1 with 0.25 increments. The interaction happens independently with the predefined probability for all the pair of disciplines that are linked by some coupling variables. Although the problem of finding a multi-disciplinary feasible is rare due to fact that the disciplines update the coupling variables continuously, it still has to be managed for the disciplines interact pairwise rather than col-

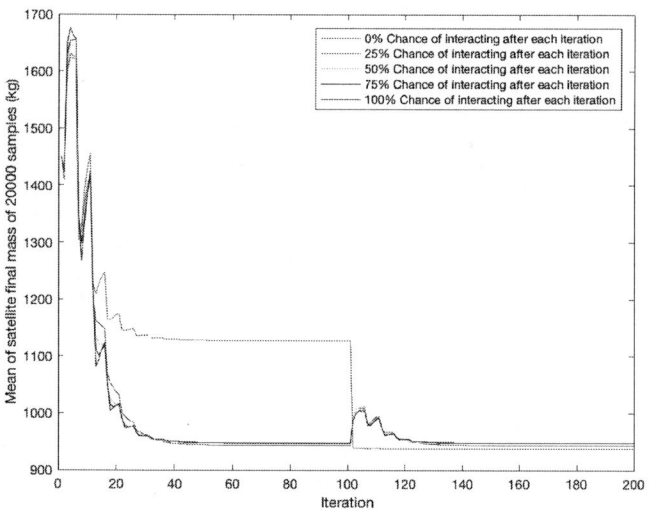

FIGURE 6. Mean of satellite mass versus design iteration for different probabilities of interaction

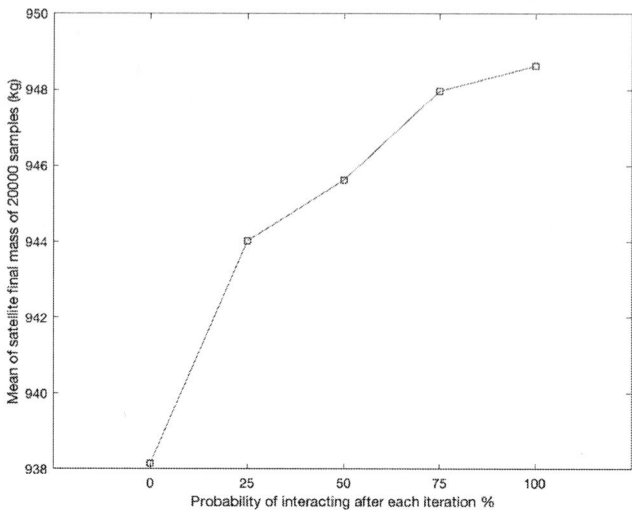

FIGURE 7. Mean of satellite final mass versus probability of interaction

lectively; therefore, the discipline feasible designs may not be multi-discipline feasible. To handle this issue, after the last design iterations, if the system has a consistent solution, the algorithm terminates. Otherwise, the algorithm restarts without discarding previous designs. Since at the last iteration the disciplines try to find a set of consistent coupling variables, there is still a chance that there exists a discrepancy between the consistent coupling variables and the local coupling variables. This discrepancy lead to a jump in satellite mass. Therefore, for this case, 200 iterations were simulated so that the systems converges to a multi-disciplinary feasible solution. Figure 6 shows the average of satellite mass at each of the 100 design iterations and for different probabilities of interaction, and Fig. 7 shows the mean of 20000 satellite final mass for different probabilities of interaction after 200 iterations.

5 Discussion

The notable result from this study is twofold: 1) simulation results suggest that there is an optimal communication frequency for multi-disciplinary design teams when considering the physical system explicitly, and 2) the form of relationship between frequency and performance changes significantly depending on the communication pattern. Based on previous empirical and quantitative works, this study hypothesized a curvilinear relationship between communication frequency and performance in multi-disciplinary design teams. Although the optimal performance happens at different communication frequencies, the two scenarios tested for planned communication demonstrate a curvilinear relationship. In contrast, in stochastic communication scenario

the relationship was inversely monotonic. This suggests that designers should consider their communication pattern when implementing changes to their communication frequency.

Furthermore, Figures 3 and 5 suggest that too much communication can affect performance more adversely than too low communication frequency. If a cost for communication is taken into account, as well as changing the optimal frequency, this negative effect may be amplified. However, as it can be inferred from Fig. 2 and 4 for early design iterations, higher communication frequencies yield better performances. This suggest the possibility that if the project is stopped during early stages due to time overruns or budget shortfalls, there will be a higher chance that the design outcomes for scenarios with higher communication frequencies are superior. Similarly, results from simulations of stochastic communication indicate that more communication results in better performance at early stages, whereas as design progresses the nature of relationship changes to that of Fig. 7.

In the context of CFTs and information theory based modeling, the results of planned communication are in line with the empirical and theoretical findings of Patrashkova et al. [10, 11] which suggest a curvilinear relationship between communication frequency and performance for the planned communication scenario. However, stochastic communication results indicate an inverse relationship between the two being in line with the findings of McComb [49] for some of the test problems. This implies that the structure of communication can indeed define the relation between communication frequency and performance. In terms of process aspect of socio-technical systems, this signifies the importance of the structure of the interaction between individual in a organization. Given the work of [49] on identifying the characteristics of design problems affecting the optimal communication frequency, this work's results necessitate identifying the features of communication structure defining the relation between communication frequency and performance in the context of multi-disciplinary design teams.

The findings of this study are limited by several factors. First, the communication is modeled as a simple exchange of parameter value which does not represent the whole possible communication types and account for differences in the interpretation and perception of information. However, in real-world organizations parameter value exchange is a common type of communication in concurrent design. Second, the behavior of designers is modeled as a completely stochastic process when in fact many of the decisions made by the designers, in particular when exploring the design space, can be based on intuition formed by previous experiences with similar problems. However, previous human studies have shown that simulated annealing algorithm approximates designers behavior well [12, 51, 53]. Third, in real-world problems agents are often part of more than one disciplines. For example, a software engineer may work on the program run by the payload subsystems as well as taking part in coding part of the software required by the ADCS subsystem. By relaxing the one-to-one relation between decomposed technical subsystems and agents, new patterns of behavior may emerge. Finally, previous work has argued that traditional validation methods are less suitable for validating models such as the one presented here, which can be viewed as simplified and isolated representations of real-world phenomena [20, 62, 63]. However, this work utilizes externally validated models proposed in the literature to address these concerns [12, 20].

6 Conclusion & Future Work

This study examines the relationship between communication frequency and performance in multi-disciplinary design teams. Results are presented from a series of multi-agent simulations of a design team engaged in a mass minimization problem of a geostationary satellite. The work compares two different scenarios with communication between team members modeled as either a planned information passing session or a stochastic information exchange. For the respective scenarios, the interval of information passing session and the probability of interaction were varied to answer the following research questions:

1. What is the relationship between communication frequency and team performance in multidisciplinary design teams?

   Results of the Monte Carlo simulation suggest that the hypothesis that there is a curvilinear relation between communication frequency and performance is true for planned communication. As both two scenarios implemented for planned communication, simulation with and without redesign, exhibit a curvilinear relation with the optimum fre-

quency being around one information passing session every 10 to 20 design iterations. For stochastic communication, on the other hand, a monotonic inverse relation between probability of interaction and performance was observed, i.e., best performance was obtained for zero probability of pairwise information exchange.

2. Does the organizational communication pattern affect the behavior of this relationship?

   Results from simulations suggest that the organizational communication pattern does affect the behavior of relationship between communication frequency and team performance. Although this study did not explore all the possible communication patterns, the two different scenarios tested indicate that the communication pattern is in fact a factor impacting the behavior of communication frequency and performance relationship.

This study points to many avenues for future work. As well as addressing the limitations, future work can extend the scope of this work to realistic complex systems design problems by increasing the number of agents, subsystems and complexity of the technical systems. Furthermore, designers' behavior could include more dimensions of human behaviors, such as trust, team cohesion, competency, and motivation. To better understand how the communication pattern can impact the performance, different information exchange architectures should be investigated. For example, the information can be passed through a hub as in systems engineers or team leaders, or it can be exchanged directly by the agents involved. Moreover, further examining of the relation between subsystems physical couplings strength and the optimal pairwise communication frequency gives us more insight into how technical and social interact to bring into existence the emergent behaviors.

REFERENCES

[1] Brown, S. L., and Eisenhardt, K. M., 1995. "Product development: Past research, present findings, and future directions". *Academy of management review,* **20**(2), pp. 343–378.

[2] Denison, D. R., Hart, S. L., and Kahn, J. A., 1996. "From chimneys to cross-functional teams: Developing and validating a diagnostic model". *Academy of Management journal,* **39**(4), pp. 1005–1023.

[3] Dong, A., Hill, A. W., and Agogino, A. M., 2004. "A document analysis method for characterizing design team performance". *Journal of Mechanical Design,* **126**(3), pp. 378–385.

[4] Sosa, M. E., Eppinger, S. D., Pich, M., McKendrick, D. G., and Stout, S. K., 2002. "Factors that influence technical communication in distributed product development: an em-

pirical study in the telecommunications industry". *IEEE transactions on engineering management,* **49**(1), pp. 45–58.

[5] Joyce, W. F., 1986. "Matrix organization: A social experiment". *Academy of Management Journal,* **29**(3), pp. 536–561.

[6] Ghoshal, S., and Bartlett, C. A., 1990. "The multinational corporation as an interorganizational network". *Academy of management review,* **15**(4), pp. 603–626.

[7] Tang, C., 2014. "The impact of connecting with professional virtual forum, team member and external person on r&d employee creativity". *Computers in Human Behavior,* **39**, pp. 204–212.

[8] McDon III, E. F. o., and Kahn, K. B., 1996. "Using hardand softtechnologies for global new product development". *R&D Management,* **26**(3), pp. 241–253.

[9] Austin, J. R., 2000. "Knowing what and whom other people know: Linking transactive memory with external connections in organizational groups.". pp. F1–F6.

[10] Patrashkova-Volzdoska, R. R., McComb, S. A., Green, S. G., and Compton, W. D., 2003. "Examining a curvilinear relationship between communication frequency and team performance in cross-functional project teams". *IEEE Transactions on Engineering Management,* **50**(3), pp. 262–269.

[11] Patrashkova, R. R., and McComb, S. A., 2004. "Exploring why more communication is not better: insights from a computational model of cross-functional teams". *Journal of Engineering and Technology Management,* **21**(1-2), pp. 83–114.

[12] McComb, C., Cagan, J., and Kotovsky, K., 2015. "Lifting the veil: Drawing insights about design teams from a cognitively-inspired computational model". *Design Studies,* **40**, pp. 119–142.

[13] Kalsi, M., Hacker, K., and Lewis, K., 2001. "A comprehensive robust design approach for decision trade-offs in complex systems design". *Journal of Mechanical Design,* **123**(1), pp. 1–10.

[14] Austin-Breneman, J., Yu, B. Y., and Yang, M. C., 2016. "Biased information passing between subsystems over time in complex system design". *Journal of Mechanical Design,* **138**(1), p. 011101.

[15] Allen, T. J., et al., 1984. "Managing the flow of technology: Technology transfer and the dissemination of technological information within the r&d organization". *MIT Press Books,* **1**.

[16] Katz, R., and Tushman, M., 1981. "An investigation into the managerial roles and career paths of gatekeepers and project supervisors in a major r & d facility". *R&D Management,* **11**(3), pp. 103–110.

[17] Ancona, D. G., and Caldwell, D. F., 1991. "Bridging the boundary: external process and performance in organizational teams".

[18] Smith, K. G., Smith, K. A., Olian, J. D., Sims Jr, H. P., O'Bannon, D. P., and Scully, J. A., 1994. "Top management team demography and process: The role of social integration and communication". *Administrative science quarterly,* pp. 412–438.

[19] Hutchins, E., 1995. *Cognition in the Wild.* MIT press.

[20] Crowder, R. M., Robinson, M. A., Hughes, H. P., and Sim, Y.-W., 2012. "The development of an agent-based modeling framework for simulating engineering team work". *IEEE Transactions on Systems, Man, and Cybernetics-Part A: Systems and Humans,* **42**(6), pp. 1425–1439.

[21] Lu, X., Menzel, S., Tang, K., and Yao, X., 2018. "Cooperative co-evolution-based design optimization: A concurrent engineering perspective". *IEEE Transactions on Evolutionary Computation,* **22**(2), pp. 173–188.

[22] Streufert, S., and Streufert, S. C., 1978. *Behavior in the complex environment.* VH Winston & Sons.

[23] Schroder, H. M., Driver, M. J., and Streufert, S., 1967. *Human information processing.* Holt, Rinehart and Winston.

[24] Streufert, S., and Swezey, R. W., 1986. *Complexity, managers, and organizations.* Academic Press.

[25] Thompson, J. D., 1967. *Organizations in action: Social science bases of administration.* New York: McGraw-Hill.

[26] Galbraith, J. R., 1977. *Organization design.* Addison Wesley Publishing Company.

[27] Allen, T. J., 1986. "Organizational structure, information technology, and r&d productivity". *IEEE Transactions on Engineering Management*(4), pp. 212–217.

[28] Pinto, M. B., and Pinto, J. K., 1990. "Project team communication and cross-functional cooperation in new program development". *The Journal of Product Innovation Management,* **7**(3), pp. 200–212.

[29] Hauptman, O., 1990. "The different roles of communication in software development and hardware r&d: Phenomenologic paradox or atheoretical empiricism?". *Journal of Engineering and Technology Management,* **7**(1), pp. 49–71.

[30] Hinsz, V. B., Tindale, R. S., and Vollrath, D. A., 1997. "The emerging conceptualization of groups as information processors.". *Psychological bulletin,* **121**(1), p. 43.

[31] Hirokawa, R. Y., and Poole, M. S., 1996. *Introduction: Communication and group decision making,* Vol. 77. Sage.

[32] Albrecht, T. L., and Ropp, V. A., 1984. "Communicating about innovation in networks of three us organizations". *Journal of Communication,* **34**(3), pp. 78–91.

[33] Nemiro, J. E., 2002. "The creative process in virtual teams". *Communication Research Journal,* **14**(1), pp. 69–83.

[34] Oh, K., Kim, Y.-B., and Lee, J., 1991. "An empirical study of communication patterns, leadership styles, and subordinate satisfaction in r&d project teams in korea". *Journal of*

Engineering and Technology Management, **8**(1), pp. 15–35.

[35] De Vries, R. E., Van den Hooff, B., and de Ridder, J. A., 2006. "Explaining knowledge sharing: The role of team communication styles, job satisfaction, and performance beliefs". *Communication research,* **33**(2), pp. 115–135.

[36] Pemartín, M., Rodríguez-Escudero, A. I., and Munuera-Alemán, J. L., 2018. "Effects of collaborative communication on npd collaboration results: Two routes of influence". *Journal of Product Innovation Management,* **35**(2), pp. 184–208.

[37] Leenders, R. T. A., Van Engelen, J. M., and Kratzer, J., 2003. "Virtuality, communication, and new product team creativity: a social network perspective". *Journal of Engineering and Technology Management,* **20**(1-2), pp. 69–92.

[38] Katz, R., and Allen, T. J., 1982. "Investigating the not invented here (nih) syndrome: A look at the performance, tenure, and communication patterns of 50 r & d project groups". *R&d Management,* **12**(1), pp. 7–20.

[39] Fussell, S. R., Kraut, R. E., Lerch, F. J., Scherlis, W. L., McNally, M. M., and Cadiz, J. J., 1998. "Coordination, overload and team performance: effects of team communication strategies". In Proceedings of the 1998 ACM conference on Computer supported cooperative work, ACM, pp. 275–284.

[40] Lovelace, R., 1986. "Stimulating creativity through managerial intervention". *R&D Management,* **16**(2), pp. 161–174.

[41] Shalley, C. E., 1995. "Effects of coaction, expected evaluation, and goal setting on creativity and productivity". *Academy of Management Journal,* **38**(2), pp. 483–503.

[42] Nicholas, J. M., 1994. "Concurrent engineering: overcoming obstacles to teamwork". *Production and Inventory Management Journal,* **35**(3), p. 18.

[43] Nyström, H., 1979. *Creativity and innovation.* Wiley New York.

[44] Hoegl, M., and Wagner, S. M., 2005. "Buyer-supplier collaboration in product development projects". *Journal of management,* **31**(4), pp. 530–548.

[45] Dougherty, D., 1992. "Interpretive barriers to successful product innovation in large firms". *Organization science,* **3**(2), pp. 179–202.

[46] Chen, S.-J., and Lin, L., 2004. "Modeling team member characteristics for the formation of a multifunctional team in concurrent engineering". *IEEE Transactions on Engineering Management,* **51**(2), pp. 111–124.

[47] Darawong, C., 2015. "The impact of cross-functional communication on absorptive capacity of npd teams at high technology firms in thailand". *The Journal of High Technology Management Research,* **26**(1), pp. 38–44.

[48] Kennedy, D. M., McComb, S. A., and Vozdolska, R. R., 2011. "An investigation of project complexity's influence on team communication using monte carlo simula-

tion". *Journal of Engineering and Technology Management,* **28**(3), pp. 109–127.

[49] McComb, C., Cagan, J., and Kotovsky, K., 2017. "Optimizing design teams based on problem properties: computational team simulations and an applied empirical test". *Journal of Mechanical Design,* **139**(4), p. 041101.

[50] Yassine, A. A., Sreenivas, R. S., and Zhu, J., 2008. "Managing the exchange of information in product development". *European Journal of Operational Research,* **184**(1), pp. 311–326.

[51] Cagan, J., and Kotovsky, K., 1997. "Simulated annealing and the generation of the objective function: a model of learning during problem solving". *Computational Intelligence,* **13**(4), pp. 534–581.

[52] Kirkpatrick, S., Gelatt, C. D., and Vecchi, M. P., 1983. "Optimization by simulated annealing". *science,* **220**(4598), pp. 671–680.

[53] Yu, B. Y., de Weck, O., and Yang, M. C., 2015. "Parameter design strategies: a comparison between human designers and the simulated annealing algorithm". In ASME 2015 International Design Engineering Technical Conferences and Computers and Information in Engineering Conference, American Society of Mechanical Engineers, pp. V007T06A033–V007T06A033.

[54] Olson, J., Cagan, J., and Kotovsky, K., 2009. "Unlocking organizational potential: a computational platform for investigating structural interdependence in design". *Journal of Mechanical Design,* **131**(3), p. 031001.

[55] Walden, D. D., Roedler, G. J., Forsberg, K., Hamelin, R. D., and Shortell, T. M., 2015. *Systems engineering handbook: A guide for system life cycle processes and activities.* John Wiley & Sons.

[56] Stempfle, J., and Badke-Schaub, P., 2002. "Thinking in design teams-an analysis of team communication". *Design studies,* **23**(5), pp. 473–496.

[57] Kannan, H., 2015. "An mdo augmented value-based systems engineering approach to holistic design decision-making: a satellite system case study".

[58] Hirschi, N., and Frey, D., 2002. "Cognition and complexity: an experiment on the effect of coupling in parameter design". *Research in Engineering Design,* **13**(3), pp. 123–131.

[59] Martins, J. R., and Lambe, A. B., 2013. "Multidisciplinary design optimization: a survey of architectures". *AIAA journal,* **51**(9), pp. 2049–2075.

[60] Skaggs, R. L., Mays, L. W., and Vail, L. W., 2001. "Simulated annealing with memory and directional search for ground water remediation design 1". *JAWRA Journal of the American Water Resources Association,* **37**(4), pp. 853–866.

[61] Skaggs, R. L., Mays, L. W., and Vail, L. W., 2001. "Application of enhanced annealing to ground water remedi-

ation design 1". *JAWRA Journal of the American Water Resources Association,* **37**(4), pp. 867–875.

[62] Epstein, J. M., 1999. "Agent-based computational models and generative social science". *Complexity,* **4**(5), pp. 41–60.

[63] Carley, K., 2009. "How to validate an agent-based model. applied agent-based modeling in management research". In Proc. Acad. Manag. Prof. Dev. Workshop.

**Proceedings of the ASME 2019
International Design Engineering Technical Conferences
and Computers and Information in Engineering Conference
IDETC/CIE2019
August 18-21, 2019, Anaheim, CA, USA**

DETC2019-97372

CLASSIFICATION AND EXECUTION OF COUPLED DECISION PROBLEMS IN ENGINEERING DESIGN FOR EXPLORATION OF ROBUST DESIGN SOLUTIONS

Gehendra Sharma
Systems Realization Laboratory @ OU
University of Oklahoma, Norman, OK, USA

Janet K. Allen[1]
Systems Realization Laboratory @ OU
University of Oklahoma, Norman, OK, USA

Farrokh Mistree
Systems Realization Laboratory @ OU
University of Oklahoma, Norman, OK, USA

ABSTRACT

Decision Support Problems (DSPs) are used to model design decisions involving multiple trade-offs. In practice, such design decisions are also coupled, that is, these decisions must be modelled by identifying and addressing the influence they exert on one another. Hence, we need to classify coupled decision problems and to introduce methods for managing uncertainty for such problems. Classification of coupled decision problems allows for the development and execution of decision templates to effect design and to archive design-related knowledge on a computer. Incorporating robustness metrics allows for the exploration of robust design solutions for coupled decision problems by managing uncertainty.

*In this paper, we present a classification scheme for coupled decisions using DSPs, called the **Decision Scenario Matrix** and we illustrate its utility by solving a coupled problem using DSPs. The design of a beam to be used as a fender is used to illustrate the efficacy of the formulation of coupled problems. In the first example, we determine a robust design, that is, determine the dimensions of the fender and simultaneously design the material recognizing that the computational models are incomplete and inaccurate. In the second example, we determine robust design solutions when design decisions are coupled, that is, determine the dimensions of the fender and select the material concurrently. Our focus, in this paper, is on illustrating the efficacy of the method rather than on the results.*

KEYWORDS
Coupled decisions, Decision Scenario Matrix, Managing Uncertainty, Robust Design, Solution Space Exploration

GLOSSARY
Selection Decisions: Decisions that involve making choice among several alternatives considering several attributes.
Compromise Decisions: Decisions that involve determining design variables to satisfy multiple conflicting requirements and attain a certain objective.

NOMENCLATURE
DSP = Decision Support Problem
cDSP = compromise Decision Support Problem
sDSP = selection Decision Support Problem
DCI = Design Capability Index
EMI = Error Margin Index
MSR = Mass to Strength Ratio
AR = Aspect Ratio
ST = Beam Stiffness
Y_{max} = Maximum response in a given design space
Y_{min} = Minimum response in a given design space
Δy = Response deviation
ΔY_{upper} = Upper deviation from a mean response
ΔY_{lower} = Lower deviation from a mean response
URL = Upper Requirement Limit
LRL = Lower Requirement Limit

---

[1]Corresponding author [janet.allen@ou.edu]

1. DESIGN DECISIONS IN ENGINEERING DESIGN

One of the reasons for complexity arising in engineering design is the impact on decisions from various disciplines. For example, the design of a composite structure involves selecting material for the skin and core from various available alternatives and compromise on thickness of skin and core material for obtaining desired properties for the structure. The choice amongst alternatives and determining the thickness of the core and skin material constitutes a coupled decision problem. The problem is strongly coupled since the choice of the material impacts the thicknesses and vice versa. Since, typically, the computational models are abstractions of reality we seek robust solutions, namely, those that are relatively insensitive to uncertainties.

Engineering design incorporates information from several disciplines. This information forms the basis for design decisions. A decision based on the information from one discipline exerts influence on decision based on information from another discipline. This is common in engineering design where decisions are modeled using information from, say, fluid dynamics, thermal science, manufacturing science, economics, material science, etc. We contend that failure to account for the interaction among decisions leads to poor decisions. Further, uncertainty and risks are pervasive and must be managed to effect robust solutions. Hence, our method for robust design involving coupled decisions necessitates:

1. Identifying and accounting for the interactions among decisions.
2. Managing the variability in each decision pertaining to noise and/or any slight changes in design parameters and/or insufficiency/inaccuracy in the mean response function utilized for measuring performance.

A fender is a tubular beam used in marine applications, for example, as a damage mitigator between an oil rig and a supply vessel. The compromise decision involves the determination of beam geometry which includes the mean diameter and the thickness to satisfy conflicting requirements of minimizing mass and aspect ratio while maximizing stiffness. This decision also demands satisfying stress and deflection constraints. The selection decision involves selecting one material from various available alternatives based on attributes such as manufacturability, corrosion resistance, hardness and cost. The determination of design variables (compromise part) requires the information about material and associated properties (selection part). In turn, selection of a material (selection part) requires information about design variables (compromise part) for the determination of material cost, which is an important attribute for ascertaining material selection. The two decisions (selection and compromise) are coupled and must account for the

interaction between them. The fender problem is characterized by one selection goal and three compromise goals; see Table 6.

In this paper, we present a classification scheme for coupled decisions and four types of uncertainty that need to be accounted for in determining robust solutions, that is, solutions that are relatively insensitive to uncertainties arising as a consequence of our ability to model reality exactly.

2. FRAME OF REFERENCE

Smith and co-authors [1] suggest that a decision-based design process involves:

- a series of decisions, some being made concurrently and some sequentially.
- multilevel, multidisciplinary and multidimensional decision-making where interactions occur among subsystems on various levels of the decision tree on one or both directions.

One manifestation of the **Decision-Based Design** (DBD) construct is the **Decision Support Problem Technique** (DSPT) developed to provide support to human designers in exercising judgment in making design decisions [2]. The foundational premise in the DSPT is that "the principal role of an engineer, in the design of an artifact, is to make decisions" [3, 4]. The DSPT is anchored in two principles, namely, the computational models are abstractions of reality and the outcome is one or more *satisficing²* solutions. The axioms needed to characterize "decisions" as Decision Support Problems (DSPs) as proposed and stated in [5] follow.

Axiom 1: Existence of Decisions in the DSPT
"The application of the DSPT results in the identification of decisions associated with the system (and subsystems that may be relevant)."

Axiom 2: Type of Decisions in the DSPT
"All decisions identified in the DSPT are categorized as selection, compromise, or a combination of these."

In the DSPT, the selection decision is defined as, 'the process of making a choice between a number of possibilities considering a number of measures of merit or attributes." Similarly, the compromise decision is defined as, "the decision that requires the 'right' values (or combination) of design variables (or parameters) be determined, such that, the system is feasible with respect to constraints and system performance is maximized." Selection decisions are modelled as selection Decision Support Problems (sDSP) and the compromise decisions are modelled as compromise Decision Support Problems (cDSP). A solution to

---

² "decision makers can satisfice either by finding optimum solutions for a simplified world, or by finding satisfactory solutions for a more realistic world." Herbert Simon.
https://en.wikipedia.org/wiki/Satisficing

the compromise DSP is a satisficing solution. Bannerot and coauthors describe three principal components of DSPT: a design philosophy expressed at present in terms of paradigms, an approach for identifying and formulating DSPs and the software necessary for solution [6].

2.1 Robustness In Coupled Decision Support Problems

2.1.1 Coupled DSPs

The issue in the analysis and synthesis of engineering systems as a single problem is complex [7]. Therefore, it is necessary to decompose the design problem as dependent subsystems and then after solving subsystems recompose them [7]. One schema for partitioning a system into interconnected subsystems is presented in [8].They use a rating scheme to model the strength of interactions. However, the method has been developed for use in designing mechatronic systems, where usually the four types of interactions (structural, energy, signal, material) occur [8]. Our intent in this paper is to partition a system from the standpoint of number of decisions and interactions among these decisions, which makes this independent to the domain of application. The above-mentioned axioms allow us to formulate decisions as either selection or compromise or combination of selection and compromise. In practice, an engineering design problem involves many selection and compromise decisions which depend on one another and exert influence on each other. Such decisions are referred to as coupled decisions. Coupled decisions are modelled by accounting for the interaction between the DSPs as opposed to independent decisions when the individual DSPs do not interact with each other and the decisions can be taken independently. Karandikar and co-authors provide a method for dealing with coupled DSPs [6, 9]. Based on the extent of interaction between DSPs, two forms are proposed: "weak" and "strong" [8]. The two forms result in two DSP based formulations; see Figure 1. The two formulations provide a basis for formulating coupled DSPs based on the strength of interaction between the DSPs. The formulations are as:

1. Weakly coupled formulation
2. Strongly coupled formulation

The weakly coupled formulation is used when there is one-way flow of information between two DSPs. For instance, in a coupled selection-compromise problem, either the influence of selection on compromise or the influence of compromise on selection is missing. In a strongly coupled formulation, there is a two-way flow of information between the constituent DSPs. For instance, in a coupled selection-compromise problem, there exists mutual influence between selection and compromise. A detailed representation, mathematical interpretation and solution approach is included in [1, 10, 11].

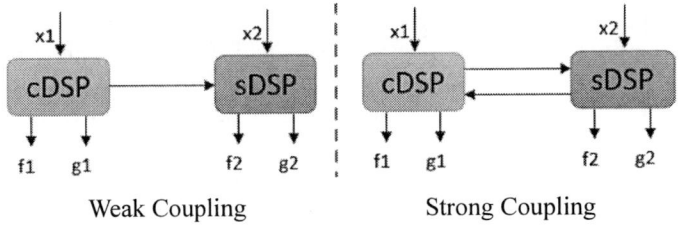

Weak Coupling   Strong Coupling

Figure 1: Illustration of Weakly and Strongly Coupled Decisions

In Figure 1, x1 and x2 = Set of design variables and/or attributes for cDSP and sDSP respectively
f1 and f2 = Constraint functions for cDSP and sDSP respectively
g1 and g2 = Goal functions for cDSP and sDSP respectively

The concise mathematical form for the strong selection – compromise decision problem is shown in Table 1. It is worth noting how the compromise system variables (X) influence selection goal (MF) and how selection alternatives (Y) influence compromise constraints $g_i(X,Y)$ and goals $A_i(X,Y)$.

Table 1: Simplified Mathematical Form for Demonstrating Coupled Selection – Compromise Decision using DSPs

Coupled selection – compromise DSPs	
compromise DSP	selection DSP
Find Compromise System Variables **X**	**Find** Selection System Variables **Y**
Deviation Variables d_i^-, d_i^+	Deviation Variables e_i^-, e_i^+
Satisfy Design Constraints $g_j(X, Y) > 0$	**Satisfy** Selection Constraint $\sum_{i=1}^{n} Y_j = 1$
Constraints on Deviation Variables $d_i^+ \geq 0, d_i^- \geq 0, d_i^+ \cdot d_i^- = 0$	Constraints on Deviation Variables $e_i^+ \geq 0, e_i^- \geq 0, e_i^+ \cdot e_i^- = 0$
Compromise Goals $A_i(X, Y) + d_i^- - d_i^+ = G_i$	**Selection Goal** $MF_j(X) Y_i + e_i^- - e_1^+ = 1$
Bounds B: $X^{(min)} \leq X \leq X^{(max)}$	**Bounds** B: $0 \leq Y_j \leq 1$
Minimize $Z = \{e1^-, \sum_{i=1}^{n} w_i \cdot (di^- + di^+)\}$, $\sum_{i=1}^{3} w_i = 1$	

2.1.2 Robust Decisions – Mitigating the Sensitivity of Variations in Design Performance under Uncertainty

A robust design is a product or process that can be exposed to variations – in manufacturing process and environment, in customer operating and usage conditions, or in the design specifications themselves – without suffering unacceptable performance degradation [12]. The fundamental principles and methods for robust design are anchored in the principles enunciated by Genichi Taguchi [13]. In Taguchi's robust design approach, designed (computational) experiments are carried out to evaluate the effect of control factors (design parameters that are controllable) on nominal response values and sensitivity of responses to variations in noise factors (design parameters that are not controllable or not known). Over the years, there have been several improvements on Taguchi's initial design method. Various design methods have been proposed to incorporate robustness into the model. The categories of robust design methods are based on the source of variation [14]. The four kinds of robust design methods that have been proposed are [15-19]:

1. Type I Robust Design: It is applicable to obtain design solutions that are relatively insensitive to the noise factors.
2. Type II Robust Design: It is applicable to obtain design solutions that are relatively insensitive to the variations in control factors.
3. Type III Robust Design: It is applicable to obtain design solutions that are relatively insensitive to the variations obtained due to different analysis models used in the mathematical formulation of a decision.
4. Type IV Robust Design: It is applicable to obtain design solutions that are relatively insensitive to the propagation of uncertainties across process chains.

In Type I Robust Design we strive to find solutions that are relatively insensitive to noise factors; see Figure 2. In Type II Robust Design we strive to find solutions that are relatively insensitive to variations in the design variables; see Figure 2. In Type III Robust Design we strive to find solutions that are relatively insensitive to variations in the analysis models. This is applicable to problems where more than one analytical form for modeling the response as a function of design variables are available; see Figure 3. In design practice, we require a robust design that can not only be used to address Type I, Type II and Type III robust designs individually but also collectively as shown in Figure 4. In fact, the uncertainty embedded in a system's analysis model cannot be addressed by Type I or Type II Robust Design. This necessitates the need for Type III robust design method to manage uncertainty due to analysis model. In Figure 4, system's model response is illustrated as solid curve and two adjacent dotted curves represent the uncertainty bounds associated with the analysis model.

2.1.3 Robustness in Coupled Decision Support Problems

A class of hierarchical design decision making is introduced in [6, 10]. This class embodies the simultaneous solution of two or more interacting DSPs as a single coupled problem. Bascaran and co-authors present examples to demonstrate coupled selection - selection problem in engineering design [6, 7]. Karandikar and co-authors presents a design problem as a coupled selection – compromise DSPs using two formulation, weak and strong coupling among interacting DSPs [10]. As

discussed earlier, design decisions are open to various sources of uncertainty. The impact of not being able to manage uncertainty at subsystem level becomes large as it propagates from one subsystem to another. Hence, in this paper we showcase an example problem that requires the integration of Type I, Type II and Type III robustness into a coupled selection – compromise DSP.

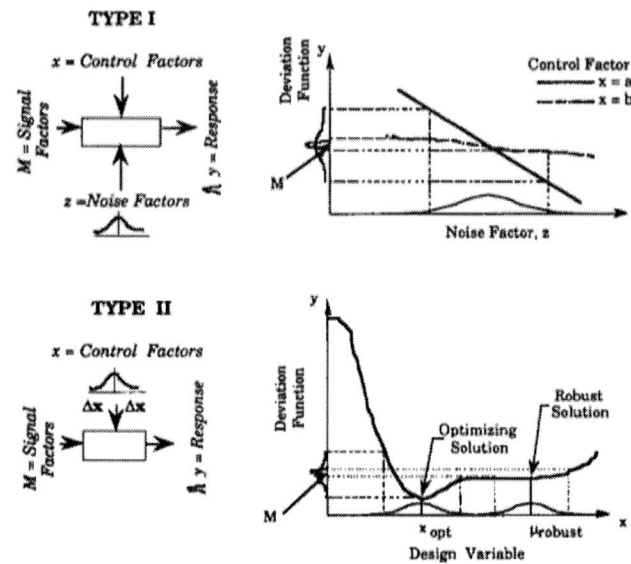

Figure 2: Comparison of Type I and Type II Robust Design [14]

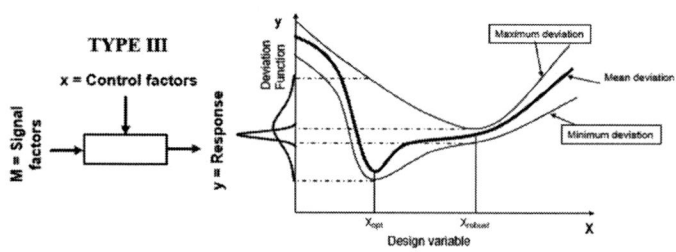

Figure 3: Type III Robust Design

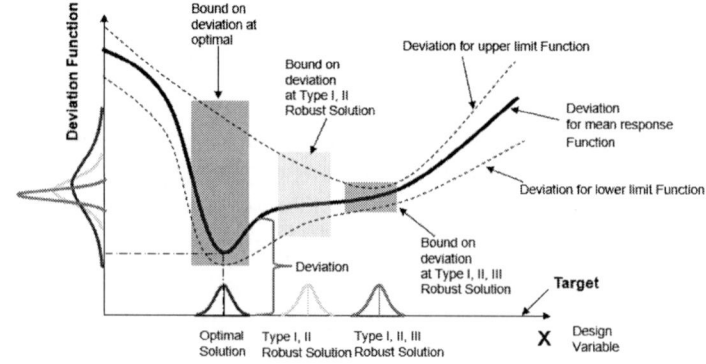

Figure 4: Type I, II and III Robust Design [17]

We propose and present the addition of robustness metrics called Design Capability Index (DCI) and Error Margin Index (EMI) to manage and mitigate the effects to uncertainty propagating in and between interacting DSPs. In Figures 5 and 6 we show the uncertainty bounds due to variations in design variable and model, and the development of mathematical constructs to address such uncertainties; for details see [15, 17].

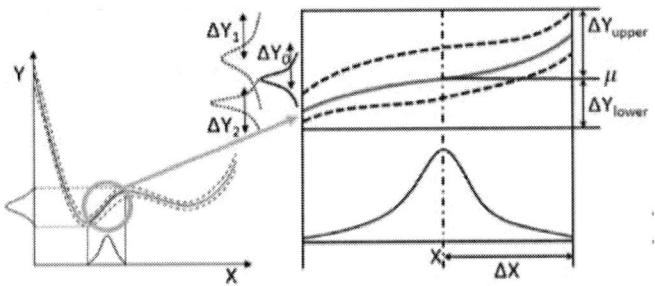

Figure 5: Formulation of Uncertainty Bounds Due to Variations in a Design Variable and a Model [15]

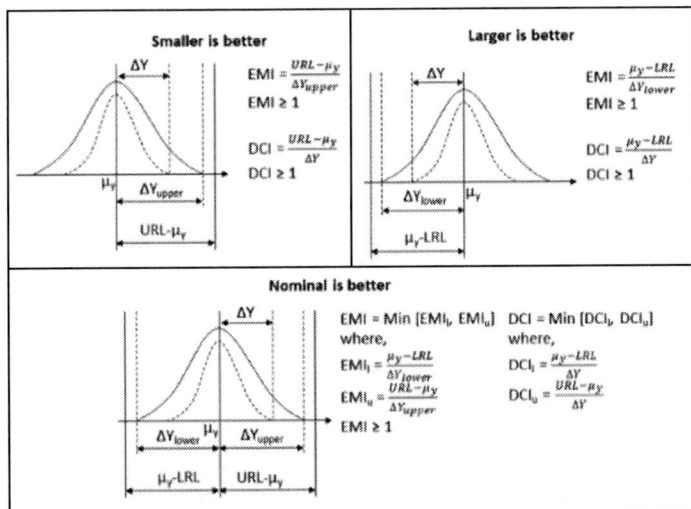

Figure 6: Mathematical Constructs of EMIs and DCIs [15]

In Figure 5, the mean response (μ) for the model is illustrated as a solid red curve and two adjacent dotted curves represent the uncertainty bounds associated with the system model. At x, for the variation of $\pm \Delta x$ in design variable, the expected variation in response given by the mean response model is ΔY_0. Similarly, for the same change in design variable at x, the expected variation in response for the two uncertainty bounds are ΔY_1 and ΔY_2 respectively as shown in the Figure 5. This lets us calculate the maximum expected deviation in response for any given value of x and Δx. In Figure 6, we show the mathematical formulations for implementing EMIs or DCIs as a goal in DSPs. "Smaller is better" signifies that we are looking to minimize the targeted function while "Larger is better" signifies that we are looking to

maximize the targeted function. Further, "Nominal is better" means we are interested in getting a value near to the target set, that is, we want to avoid underachievement as well as overachievement.

Steps for Formulating Goals as DCIs

Step 1: Using a first order Taylor series expansion, estimate the response variation due to variation in the design variable vector x = {x1, x2,……, xn}. The response variation(Δy) for small variations in design variables is

$$\Delta y = \sum_{i=1}^{n} \left| \frac{\partial f}{\partial x_i} \right| . \Delta x_i$$

Step 2: Using the mean response (μ_y) obtained from the mean response model ($f_0(x)$) and the response variation due to variation in design variables (Δy), calculate the DCIs. For a 'Larger is Better' case, the DCI is calculated as

$$DCI = \frac{\mu_y - LRL}{\Delta y}$$

where, LRL is the lower requirement limit. A $DCI \geq 1$ means that the ranged set of design specifications satisfies a ranged set of design requirements and the system is robust against uncertainty in design variables. Higher the value of DCI, higher is the measure of safety against failure due to uncertainty in design variables.

Steps for Formulating Goals as EMIs

Step 1: If a system model has k uncertainty bounds, the response variation (ΔY_j) for each of them for small variation in design variables is calculated as

$$\Delta y_j = \sum_{i=1}^{n} \left| \frac{\partial y_j}{\partial x_i} \right| . \Delta x_i$$

where $j = 0, 1, 2, …, k$ (number of uncertainty bounds).

Step 2: After evaluating the multiple response variations of mean response function and the k uncertainty bound functions for variations in design variables, the minimum and maximum responses by considering the variability in design variables and uncertainty bounds around the mean response are calculated as

$$Y_{max} = Max[f_j(x) + \Delta y_j] \text{ and,}$$
$$Y_{min} = Max[f_j(x) - \Delta y_j]$$

where $j = 0, 1, 2, …, k$ (number of uncertainty bounds), $f_0(x)$ is the mean response function, and $f_1(x)….f_k(x)$ are the uncertainty bound functions.

In Figure 5, we show a mean response function (solid red curve) and two uncertainty bounds (dotted curves in black). At any x, we are able to calculate the value of maximum (Y_{max}), minimum (Y_{min}) and mean response (μ_y) arising due to uncertainty bounds. This will let us calculate the maximum expected deviation in response for any given value of x.

Step 3: Calculate the upper and lower deviation of response at x as

$$\Delta Y_{upper} = Y_{max} - f_o(x) \text{ and}$$

$$\Delta Y_{lower} = f_o(x) - Y_{min}$$

The deviations ΔY_{upper} and ΔY_{lower} is shown in Figure 5.

Step 4: Using the mean response (μ_y) obtained from the mean response model ($f_0(x)$) and the upper and lower deviations (ΔY_{upper} and ΔY_{lower}), the EMIs are calculated as shown in Figure 6. For a 'Larger is Better' case, the EMI is calculated:

$$\text{EMI} = \frac{\mu_y - \text{LRL}}{\Delta Y_{lower}}$$

By applying the concept of robustness metrics, that is, DCI and EMI, the original goals are converted to either EMI or DCI. In Example 2 (Table 6), we have 3 compromise goals. The EMI is applied to our first goal, that is, Mass to Strength Ratio (MSR) because deviation in MSR can occur as a result of variability in design variables and material properties. Our aim here is not to eliminate all possible deviation in MSR but to minimize any such deviation in MSR as a result of variability emerging from change in our expectation about design variables and material properties. Similarly, we have applied DCI to our other two goals, that is, Aspect Ratio (AR) and Stiffness (ST) as we expect deviations in these goals to occur as a result of change in the value of design variables. Understanding of the sources of uncertainty is very important in formulating the goal as either EMI or DCI. By accurately capturing the various sources of uncertainty that are likely to impact performance, we are ensuring the design performance to be robust against such uncertainities.

2.2 Decision Scenario Matrix (DSM)

Given an engineering design problem, it is imperative to understand all the possibilities of formulating the design problem as coupled DSPs. This calls for classifying coupled design problems based on some criteria. The Decision Scenario Matrix (DSM) is created by identifying and classifying decision scenarios based on two criteria: (i) decision types (selection or compromise) and (ii) strength of interaction. Based on the criteria, the decision scenarios or patterns can be identified for a given design problem; see Figure 7.

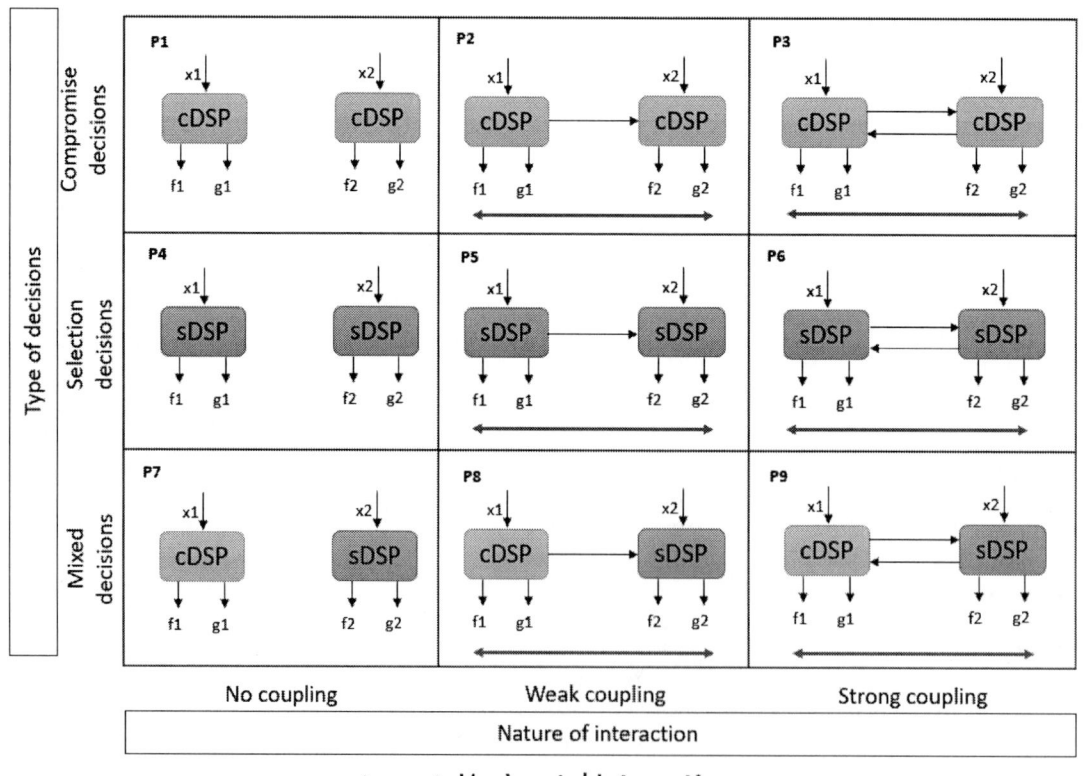

Figure 7: Decision Scenario Matrix (DSM)

Considering the interaction between two design decisions, it is possible to generate 9 different decision scenarios. The Y-axis represent the type of decisions which may take 3 forms:

1. Both design decisions involve compromise
2. Both design decisions involve selection
3. Design decisions involve combination of selection and compromise

Similarly, the X-axis represent strength of interaction among design decisions which may also take 3 forms:

1. There exists no interaction
2. There exists a weak or one-way interaction
3. There exists a strong or two-way interaction

Lack of interaction, referred to as "No coupling" means that the decisions can be modelled as independent decision problems. Weak coupling or one-way interaction means that either of the one decision influences the other. Strong coupling or two-way interaction means that there exists mutual influence among the decisions. The decision interaction between DSPs is referred to as "Horizontal interaction". "Horizontal interaction" indicates information flow within DSPs and exists between decisions that are modelled concurrrently.

Michalek and co-authors present their approach of dealing with the interrelationships in marketing and engineering product design decisions via analytical target cascading [19]. They demonstrate the process of coordinating marketing and engineering design problems to yield the joint optimal solution. The designs obtained by solving such interrelated decisions converge on joint optimality, as such, guarantee better profitability – or any other chosen metric – than the suboptimal solutions achieved by solving them sequentially. However, the ways to deal with uncertainty is not discussed. We explore a similar methodology, but from decision-based design perspective, for addressing interactions in design decisions and uncertainty using coupled DSPs. For modelling interactions, we introduce a classification scheme for addressing decision interactions, which is a foundation for devising decision templates necessary for capturing design knowledge for reuse.

The design decisions in the first example (Table 2) involves only compromise decisions. Hence, this example does not require selection of decision pattern from the Decision Scenario Matrix (DSM). This problem has been solved as a compromise decision problem using the usual DSP approach [2, 3, 20]. However, the second example represents a coupled decision problem. The identified decision pattern for this example is P9 from the DSM and has also been shown in Figure 8.

As the second example (Table 6) involves selection and compromise with mutual influence on one another, the decsion scenario in Figure 2 is an illustrative decision scenario for the example. As discussed in the introduction, this design problem involves influence of compromise decision on selection decision and vice-versa. Identification of pattern are also important in the context of developing knowledge based platform for design exploration, called PDSIDES; for details see [21].

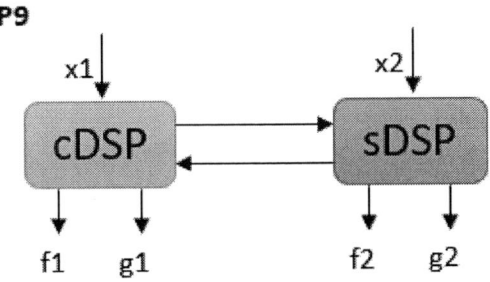

Figure 8: Decision Scenario for Example 2

3. PROCEDURE FOR ROBUST SOLUTIONS EXPLORATION FOR COUPLED DECISION PROBLEMS

In Figure 9, we show the systematic way to approach coupled design problems and to explore potential robust solutions.

Step 1: Problem Statement. In Step 1, the design problem is stated clearly. In Step 1a, the Problem Statement is stated which includes information about design goals, constraints, selection alternatives, attributes and the parameters that are necessary to formulate the design problem. In Step 1b, the control factors, noise factors, responses and their bounds is specified for the given design problem.

Step 2: Model Identification and Selection. In Step 2, the models that are important to capture the functional relationship of the design performance (goals , constraints) with the design variables. The question that needs to be answered is "Are the available models sufficient to help the designer frame the problem?" and "Do the available models provide functional relationships with accuracy that is acceptable to the designer?" An affirmation to these questions signal the end of Step 2. Absence of affirmation to any of these two questions requires us to search for other empirical models or design experiments. As search for empirical models and designing experiments is not a focus and hence, not discussed further.

Step 3: Decision Scenarios. Once the functional relationships are established, the comprehensive nature of the problem becomes more vivid. We are able to answer about the number of decisions, type of decision (compromise or selection or both) and the nature of interaction among decisions (No coupling, Weak Coupling, Strong Coupling). In Step 3, we select the decision scenario/s from the Decision Scenario Matrix (DSM).

Step 4: Coupled Decision Support Problem. Once the functional relationships are established and the decision scenario/s have been selected, the mathematical foundation for

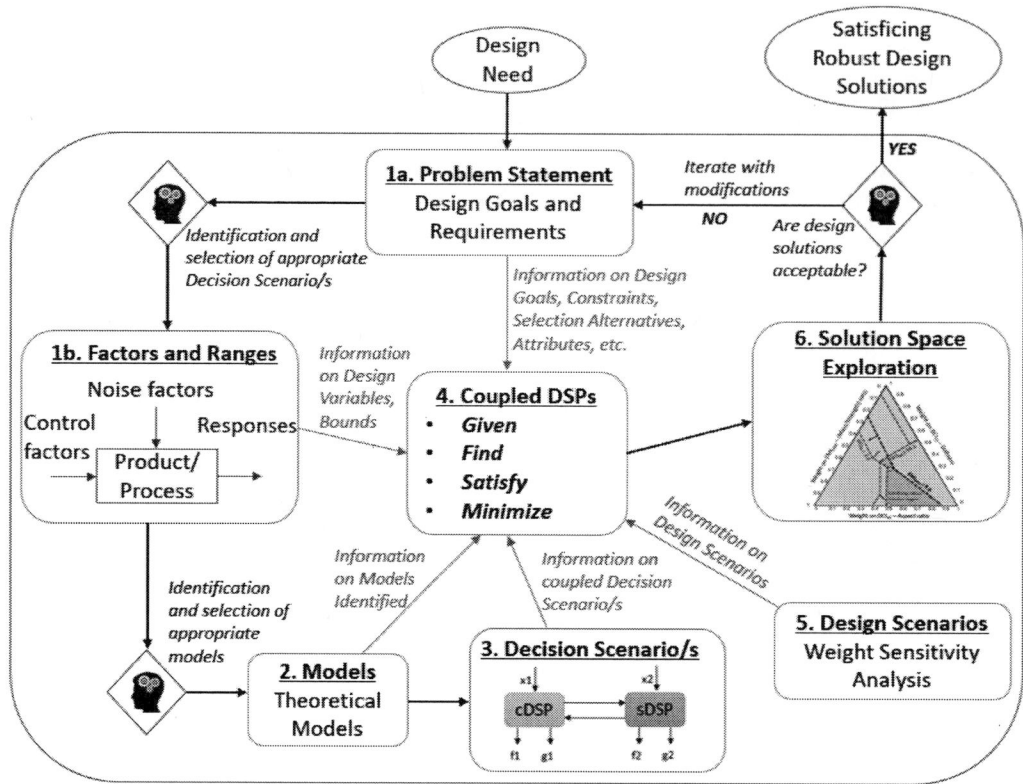

Figure 9: Procedure for Exploring Robust Design Solutions for Coupled Problems

solving coupled DSPs are established by providing detailed mathematical information using the key words – Given, Find, Satisfy, and Minimize. The compromise Decision Support Problem (cDSP) provides the foundational construct for decision support in solving a coupled DSP.

Step 5: Design Scenarios. In Step 5, multiple design scenarios are realized by exercising the coupled DSP formulated in Step 4. These scenarios are selected based on designer's aspiration to effectively capture the design space for the exploration of solution space using different combination of weights on goals. Different weights are assigned to different goals which indicate the designer's interest to achieve the target set to the goals. The identified design scenarios are visualized through ternary plots to evaluate and select design solutions.

Step 6: Solution Space Exploration. In Step 6, design trade-off is carried out for the multiple design scenarios obtained in Step 5, to help designer pick most suitable design solution/s. By stating the acceptable deviation from the target set, an acceptable solution space is outlined in the ternary plots. This is called the satisficing solution space where design solutions are searched. If no design solution exist in the satisficing space, the designer may wish to enlarge the satisficing solution space by accepting larger deviations from target set or to reformulate the design problem.

4. EXPLORING ROBUST DECISIONS – INTEGRATING ROBUSTNESS IN DSP CONSTRUCT AND EXPLORING DESIGN DECISIONS

We demonstrate two example design problems in this paper. The design problem has been bifurcated into two parts. The problem statement contains information that are common to both the examples while the specific problem statements carries information that are distinct to each example.

Problem Statement

The design of a beam, that is to be used as a fender for a floating steel-jacketed platform, is required. This fender must be compatible with the design of floating platform, which specifies a fixed length value L and the specified load P. A tubular cross-section is selected and is characterized by the mean diameter D and the wall thickness t. Restrictions regarding maximum bending stress and deflection on the beam is specified. The quality of the design is measured in terms of design goals which are to be achieved as nearly as possible. Specifically, we need a design that has low weight, stress and aspect ratio while having high beam stiffness. Two important material properties are considered for the design, that is, Young's modulus and yield strength. The design decisions are to be taken to minimize the performance impact from expected variability in design variables and material properties.

Specific Problem Statements	
Example 1 - Robust design with material as design variable: The task is to recommend the value of material properties and the beam dimensions for best performance with respect to the constraints and design quality specified. The material properties are available for selection within the specified bounds.	**Example 2 - Strongly coupled robust design with 3 material alternatives:** The task is to recommend the suitable material and the beam dimensions for best performance with respect to the constraints and design quality specified. There are 3 material alternatives, that is, Cast Iron, Titanium and Copper available for selection.

Table 2: Mathematical Form for Robust Design of a Fender

Example 1: Math formulation of cDSP for Fender Design
Given
System Parameters
Load on the beam (P) = 10,000 lbf
Length of the beam (L) = 100 in
Maximum allowable deflection (δ) = 0.025 in
System Constants
Density of the material (rho) = 0.28 lb/in^3
PI(Π) = 3.142
Find
System Variables
Wall thickness t (in)
Diameter D (in)
Yield Strength of material σy (ksi)
Young's modulus of material E (Mpsi)
Deviation Variables
d_1^+ = Overachievement of EMI_{MSR} goal
d_1^- = Underachievement of EMI_{MSR} goal
d_2^+ = Overachievement of DCI_{AR} goal
d_2^- = Underachievement of DCI_{AR} goal
d_3^+ = Overachievement of DCI_{ST} goal
d_3^- = Underachievement of DCI_{ST} goal
Satisfy
Design Constraints (From the Problem Statement)
Maximum allowable deflection constraint
$$1 - \frac{PL^3}{48EI(\delta)} > 0 \quad \text{(Normalized)}$$
Maximum allowable bending stress constraint
$$1 - \frac{PLD}{8I(\sigma y)} > 0 \quad \text{(Normalized)}$$
System Constraints

Robust solution constraint on EMI_{MSR} goal
$$EMI_{MSR} \geq 1$$
Robust Solution Constraint on DCI_{AR} goal
$$DCI_{AR} \geq 1$$
Robust Solution Constraint on DCI_{ST} goal
$$DCI_{ST} \geq 1$$
Constraints on Deviation Variables
$$d_i^+ \geq 0$$
$$d_i^- \geq 0$$
$$d_i^+ \cdot d_i^- = 0 \quad \text{for } i = 1, 2 \text{ and } 3$$
System Goals
    G1 – Maximize EMI_{MSR} for Mass to Strength Ratio goal

$$\frac{EMI_{MSR}}{EMI_{MSR,Target}} + d1^- - d1^+ = 1$$

    G2 – Maximize DCI_{AR} for Aspect Ratio goal

$$\frac{DCI_{AR}}{DCI_{AR,Target}} + d2^- - d2^+ = 1$$

    G3 - Maximize DCI_{ST} for Stiffness goal

$$\frac{DCI_{ST}}{DCI_{ST,Target}} + d3^- - d3^+ = 1$$

System Bounds
    B1: 0.12 in $\leq$ t $\leq$ 0.75 in
    B2: 3 in $\leq$ D $\leq$ 24 in
    B3: 30 ksi $\leq$ σy $\leq$ 36 ksi
    B4: 27.5 Mpsi $\leq$ E $\leq$ 30.5 Mpsi

Minimize
Deviation Functions
    $Z = \sum_{i=1}^{3} w_i \cdot (d_i^- + d_i^+)$, $\sum_{i=1}^{3} w_i = 1$

The cDSP is solved for 7 different scenarios. These scenarios are selected based on designer's aspiration to effectively capture the design space for the exploration of solution space using different combination of weights on goals. The scenarios and the results are summarized in the Table 3.

Table 3: Goals and Design Variables Achieved for Different Scenarios

Scenarios	Weights			Goals achieved			Design variables			
	w1	w2	w3	EMI_{MSR}	DCI_{AR}	DCI_{ST}	t (in)	D (in)	σy (ksi)	E (Mpsi)
1	1	0	0	14.309	1.099	1.983	0.1337	18.33	35.99	30.5
2	0	1	0	7.586	33.423	2.998	0.4448	13.08	30	27.5
3	0	0	1	0.638	44.438	5.771	0.6822	23.7	30.98	30.48
4	0.5	0.5	0	10.798	30.020	2.881	0.4158	12.83	35.99	30.48
5	0	0.5	0.5	1.632	45.270	5.750	0.6859	23.48	35.92	30.46
6	0.5	0	0.5	15.733	1.804	1.679	0.1413	16.37	35.99	30.49
7	0.34	0.33	0.3	10.787	30.013	2.884	0.4159	12.84	35.99	30.49

These scenarios are selected based on designer's aspiration to effectively capture the design space for the exploration of solution space using different combination of weights on goals. Different weights are assigned to different goals which indicate the designer's interest to achieve the target set to the goals. Assigning 1 (Scenarios 1, 2 and 3) to a goal would mean that designer's interest is to achieve the target set to the goal as closely as possible while ignoring the other goals. For instance, assigning weight $w_1=1$ to EMI_{MSR} (G1 from Table 2) would mean that the designer is inclined to obtain the target set to EMI_{MSR} as closely as possible while not considering the other two goals. Similarly, assigning 0.5 (Scenarios 4, 5 and 6) to two goals mean that the designer is equally interested to achieve the target set to the two goals while not considering the third goal. At last, Scenario 7 means that designer is equally interested to achieve the target set to all three goals. With the solutions obtained for all the scenarios, we are now most interested in exploring the solution space to obtain solutions that are of prime importance to the decision maker, that is, the designer of fender in the present context. With the information tabulated in Table 3, ternary plots for each goal are drawn. The axes of the ternary plots are the weights assigned to each goal and the colored ternary space in the interior indicate the achieved value of that specific goal. For instance, ternary plot for EMI_{MSR} goal shows the value achieved for EMI_{MSR} goal within the ternary space, when different weights are assigned to each goal. Following this, an acceptable region within each ternary plot is identified. Finally, acceptable regions identified from each ternary plot are superimposed into one plot to deduce feasible solution regions considering all 3 goals.

For EMI_{MSR} goal (G1 from Table 2), we are interested in achieving a high value. The solution space in Figure 10 is composed of robust solutions with $EMI_{MSR} \geq 1$ ensuring robustness against model uncertainty as well as parameter uncertainty. The blue region contains the robust solutions that achieves maximum value for EMI_{MSR} goal whereas the red region contains the robust solutions that achieves minimum value for EMI_{MSR} goal. The maximum attained value for EMI_{MSR} goal is 15.730 while the minimum attained value is 1.638. We define an acceptable robust region within the solution space as $EMI_{MSR} \geq 8.6$ identified by the black dashed lines. Any solution points lying within this region is acceptable for us as it satisfies the requirement for mass to strength ratio under model and parameter uncertainty.

For EMI_{AR} goal (G2 from Table 2), we are interested in achieving a high value. The solution space in Figure 11 is composed of robust solutions with $EMI_{AR} \geq 1$ ensuring robustness against parameter uncertainty. The blue region contains the robust solutions that achieves maximum value for EMI_{AR} goal whereas the red region contains the robust solutions that achieves minimum value for EMI_{AR} goal. The maximum attained value for EMI_{AR} goal is 30 while the minimum attained value is 1.090. We define an acceptable robust region within the solution space as $EMI_{AR} \geq 24.2$ identified by the red dashed lines. Any solution points lying within this region is acceptable for us as it satisfies the requirement for aspect ratio under parameter uncertainty.

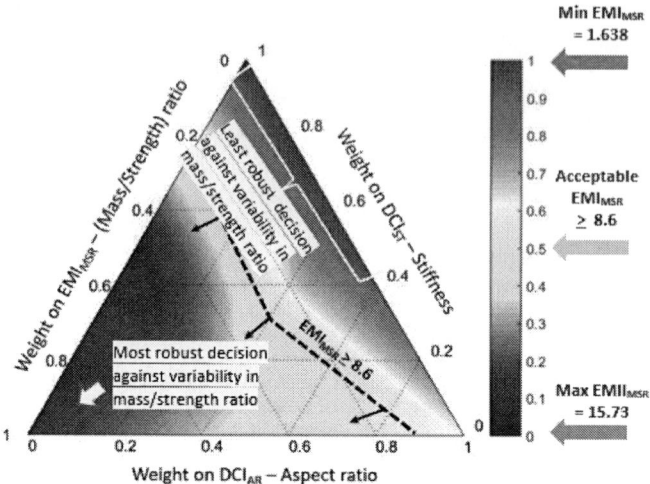

Figure 10: Robust Solution Space for Mass to Strength Ratio

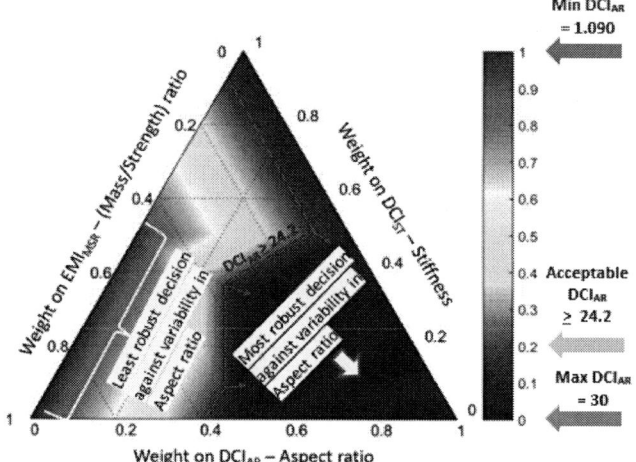

Figure 11: Robust Solution Space for Aspect Ratio

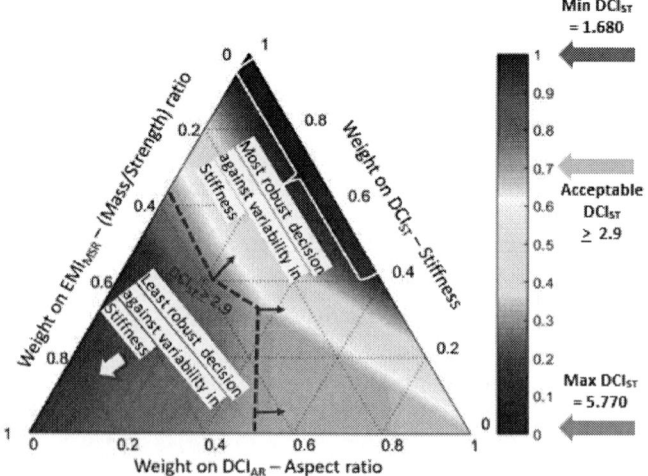

Figure 12: Robust Solution Space for Stiffness

For EMI$_{ST}$ goal (G3 from Table 2), we are interested in achieving a high value. The solution space in Figure 12 is composed of robust solutions with EMI$_{ST}$ ≥ 1 ensuring robustness against parameter uncertainty. The blue region contains the robust solutions that achieves maximum value for EMI$_{ST}$ goal whereas the red region contains the robust solutions that achieves minimum value for EMI$_{ST}$ goal. The maximum attained value for EMI$_{ST}$ goal is 5.770 while the minimum attained value is 1.680. We define an acceptable robust region within the solution space as EMI$_{ST}$ ≥ 2.9 identified by the purple dashed lines. Any solution points lying within this region is acceptable for us as it satisfies the requirement for stiffness under parameter uncertainty.

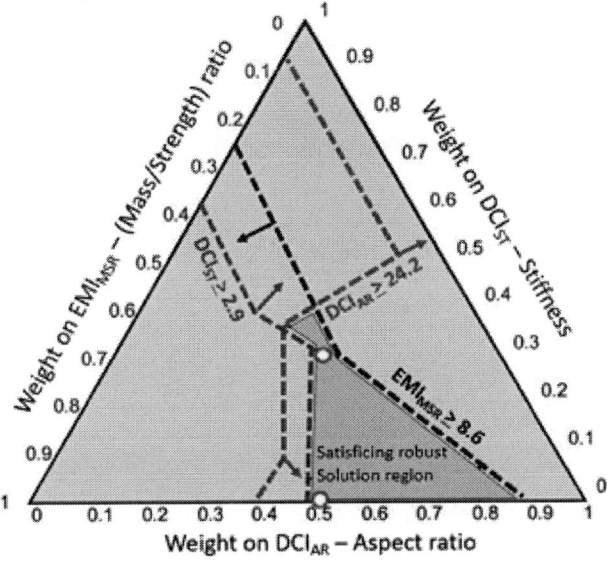

Figure 13: Superimposed Satisficing Robust Solution Space

Following our interest in identifying a satisficing robust solution region against multiple conflicting goals, we derive a superimposed robust solution space as discussed earlier and shown in Figure 13. The green region in Figure 13 is our search space for identifying robust solutions that meets our conflicting need of minimizing mass to strength ratio and aspect ratio while maximizing stiffness. We identify two robust solutions (Scenario 4 and 7) to lie within the green region and are marked by yellow dots with blue edge. The design variables corresponding to these robust solutions are tabulated in Table 4. Although, the results from the two scenarios (4 and 7) change very little, both results are presented as they were obtained while running two different scenarios as can been from Table 3.

Table 4: Robust Solutions Selected

Scenarios	Design Variables			
	t (in)	D (in)	σy (ksi)	E (Mpsi)
4	0.4158	12.83	35.99	30.48
7	0.4159	12.84	35.99	30.49

Before presenting the math formulation for a coupled selection – compromise problem, it is necessary to mathematically formulate coupled selection - compromise design problem using DSPs. The mathematical form for coupled cDSP – sDSP for design of fender is shown in Table 5 and the mathematical form for coupled cDSP – sDSP with robustness is shown in Table 6.

Table 5: Mathematical Form Representing Coupled sDSP-cDSP for Design of a Fender

Design of Fender

Find
MSR, AR, ST, X

Satisfy
$\text{MSR} + d_M^- - d_M^+ = \text{MSR}_{\text{Target}}$
$\text{AR} + d_A^- - d_A^+ = \text{AR}_{\text{Target}}$
$\text{ST} + d_S^- - d_S^+ = \text{ST}_{\text{Target}}$
$\text{MF} + d_{MF}^- - d_{MF}^+ = 1$

Compromise DSP	Selection DSP
Given Length (L)	**Given** Material Alternatives (X_j)
Find Compromise Variables Diameter (**D**) Thickness (**t**) Deviation variables d_i^-, d_i^+	**Find** Selection Variables Material Alternative (**X**) Deviation variables e_i^-, e_i^+
Satisfy Design Constraints Deflection Constraint (DC) DC (D,t,L,**X**) > 0 Bending constraint (BC) BC (D,t,L,**X**) > 0 Constraints on deviation variables $d_i^+ \geq 0, d_i^- \geq 0, d_i^+ \cdot d_i^- = 0$ for i = 1, 2 & 3	**Satisfy** Selection Constraint $\sum_{i=1}^{n} X_j = 1$ Constraints on deviation variables $e_i^+ \geq 0, e_i^- \geq 0, e_i^+ \cdot e_i^- = 0$ for i = 1 **Selection Goal** $MF_j (\mathbf{D}, \mathbf{t}, L) X_i + e_1^- - e_1^+ = 1$ **Bounds** B1: $0 \leq X_j \leq 1$
Compromise Goals $\dfrac{\text{MSR}_{\text{Target}}}{\text{MSR(D,t,L,}\mathbf{X})} + d_1^- - d_1^+ = 1$ $\dfrac{\text{AR}_{\text{Target}}}{\text{AR(D, t, }\mathbf{X})} + d_2^- - d_2^+ = 1$ $\dfrac{\text{ST(D, t, L, }\mathbf{X})}{\text{ST}_{\text{Target}}} + d_3^- - d_3^+ = 1$	

Bounds	
B1: $D^{(min)} \leq D \leq D^{(max)}$	
B2: $t^{(min)} \leq t \leq t^{(max)}$	
Minimize	
$Z = \{e1^-, \sum_{i=1}^{3} w_i \cdot (d_i^- + d_i^+)\}, \sum_{i=1}^{3} w_i = 1$	

Table 6: Mathematical form Representing Coupled sDSP-cDSP with Robustness Metrics for Example 2

Example 2: Math formulation with robustness metrics
Given
Selection System Parameters
Cast Iron yield strength (AS1) = 28 ksi
Titanium yield strength (AS2) = 34.8 ksi
Copper yield strength (AS3) = 27.5 ksi
Cast Iron young's modulus (E1) = 26 Mpsi
Titanium young's modulus (E2) = 15.2 Mpsi
Copper young's modulus (E3) = 19 Mpsi
Cast Iron density (R1) = 0.272 lb/in^3
Titanium density (R2) = 163 lb/in^3
Copper density (R3) = 0.298 lb/in^3
Relative importance of attribute j (I_j)
Normalized rating of alternative i wrt attribute j (R_{ij})
Compromise System Parameters
Load on the beam (P) = 10,000 lbf
Length of the beam (L) = 100 in
Maximum allowable deflection (δ) = 0.025 in
System Constants
PI(Π) = 3.142
Find
Selection System Variables
Cast Iron yield (X1)
Titanium (X2)
Copper (X3)
Compromise System Variables
Wall thickness t (in)
Diameter D (in)
Deviation Variables
e_1^- = Underachievement of MF goal
d_1^+ = Overachievement of EMI_{MSR} goal
d_1^- = Underachievement of EMI_{MSR} goal
d_2^+ = Overachievement of DCI_{AR} goal
d_2^- = Underachievement of DCI_{AR} goal
d_3^+ = Overachievement of DCI_{ST} goal
d_3^- = Underachievement of DCI_{ST} goal
Satisfy
Selection System Constraints
Selection constraint for material alternatives
$\sum_{i=1}^{3} X_i = 1$
Compromise Design Constraints
Maximum allowable deflection constraint
$1 - \frac{PL^3}{48EI(\delta)} > 0$ (Normalized)
Maximum allowable bending stress constraint

$$1 - \frac{PLD}{8I(\sigma y)} > 0 \quad \text{(Normalized)}$$

Compromise System Constraints
    Robust Solution Constraint on EMI_{MSR} goal
        $EMI_{MSR} \geq 1$
    Robust Solution Constraint on DCI_{AR} goal
        $DCI_{AR} \geq 1$
    Robust Solution Constraint on DCI_{ST} goal
        $DCI_{ST} \geq 1$
    Constraints on Deviation Variables
        $d_i^+ \geq 0$
        $d_i^- \geq 0$
        $d_i^+ \cdot d_i^- = 0$ for i = 1, 2 and 3

Coupled selection Goal
    G1 – Maximize Merit Function (MF)
        $MF_i(D,t) X_i + e_1^- - e_1^+ = 1$

Coupled compromise Goals
    G2 – Maximize EMI_{MSR} for mass to strength ratio goal

$$\frac{EMI_{MSR}}{EMI_{MSR,Target}} + d1^- - d1^+ = 1$$

    G3 – Maximize DCI_{AR} for aspect ratio goal

$$\frac{DCI_{AR}}{DCI_{AR,Target}} + d2^- - d2^+ = 1$$

    G4 - Maximize DCI_{ST} for stiffness goal
$$\frac{DCI_{ST}}{DCI_{ST,Target}} + d3^- - d3^+ = 1$$

System Bounds
    B1: 0.12 in $\leq$ t $\leq$ 0.75 in
    B2: 3 in $\leq$ D $\leq$ 24 in
    B3: $0 \leq X1 \leq 1$ (BOOLEAN)
    B4: $0 \leq X2 \leq 1$ (BOOLEAN)
    B5: $0 \leq X3 \leq 1$ (BOOLEAN)

Minimize
$Z = \{e1^-, \sum_{i=1}^{3} w_i \cdot (d_i^- + d_i^+)\}, \sum_{i=1}^{3} w_i = 1$

From the coupled DSP formulation (strong coupling between sDSP and cDSP), 12 different scenarios were implemented in DSIDES whose results are summarized in Table 7.

These scenarios are selected based on designer's aspiration to effectively capture the design space for the exploration of solution space using different combination of weights on goals. All that goals for compromise DSP were assigned equal weights, that is, 0.33 while the weights for attributes in selection DSP are assigned as shown in Table 7. Different weights are assigned to different goals which indicate the designer's interest to achieve the target set to the goals.

Table 7: Goals and Design Variables Achieved for Different Scenarios

Scenarios	Weights				Deviations			Design variables		
	Cost	Corrosion resistance	Machinability	Hardness	EMI_{MSR}	DCI_{AR}	DCI_{ST}	D (in)	t (in)	Material
S1	0.5	0.5	0	0	0.139	0	0.474	18.61	0.516	Titanium
S2	0.5	0	0.5	0	0.8	0.002	0.662	12.98	0.417	Copper
S3	0.5	0	0	0.5	0.762	0	0.643	12.97	0.419	Cast Iron
S4	0	0.5	0.5	0	0.8	0	0.666	12.98	0.42	Copper
S5	0	0.5	0	0.5	0	0	0.738	11.05	0.514	Titanium
S6	0.1	0.3	0	0.6	0.765	0	0.643	12.92	0.424	Cast Iron
S7	0.3	0.3	0.2	0.2	0.362	0	0.39	21.51	0.545	Titanium
S8	0.1	0.2	0.3	0.4	0.814	0	0.663	12.71	0.447	Copper
S9	0.2	0.3	0.1	0.4	0	0	0.509	17.68	0.489	Titanium
S10	0.3	0.2	0.1	0.4	0.761	0	0.643	12.98	0.418	Cast Iron
S11	0.2	0.3	0.2	0.3	0	0	0.552	15.93	0.495	Titanium
S12	0.25	0.25	0.25	0.25	0.885	0	0.569	14.48	0.527	Copper

With the solutions obtained for all the scenarios, we are now most interested in exploring the solution space to obtain solutions that are of prime importance to the decision maker, that is, the designer of fender in the present context. Various scenarios are generated and presented to the designer. The designer then chooses designs that most fit the designer's aspiration. In the present context, designer's wish is to meet the compromise goals ($G2$-EMI_{MSR}, $G3$-DCI_{AR} and $G4$-DCI_{ST} mentioned in Table 6) as closely as possible and select material that can be used to create designs which are corrosion resistant, less expensive and easier to machine.

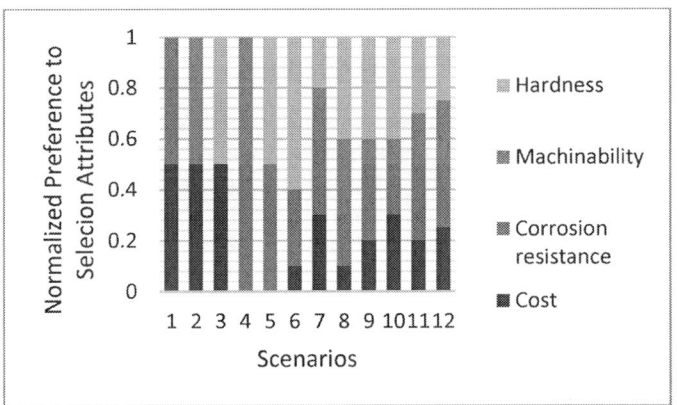

Figure 14: Design Scenarios for Selection Attributes

From Figure 14, we see that in Scenarios 2, 4, 7, 8, 9, 10, 11 and 12 consideration for materials with easier machinability is made. Following from Figure 13, we also see that in Scenarios 1, 4, 5, 6, 7, 8, 9, 10, 11 and 12 consideration for corrosion resistant materials is made. Further from Figure 13, we also see that in Scenarios 1, 2, 3, 6, 7, 8, 9, 10, 11 and 12 consideration for cost is made. As the designer is looking for all three attributes (Machinability, Corrosion Resistance and Cost), Scenarios 7, 8, 9, 10, 11 and 12 are the candidate for potential design solutions. These potential scenarios are to be compared to see which of them satisfy the compromise goals more closely.

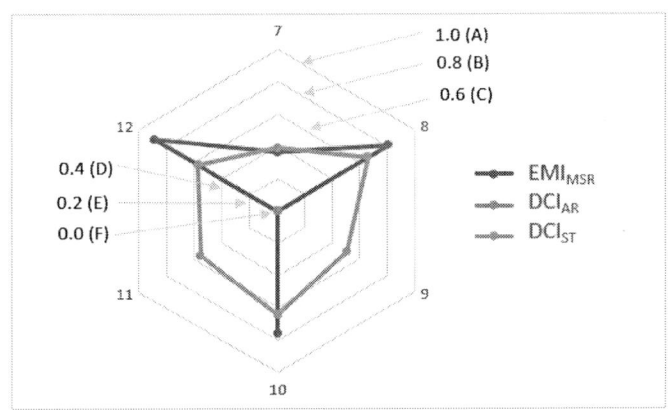

Figure 15: Design Scenarios with Deviations from Compromise Goals

Each corner of the hexagon in Figure 15 represents the six potential design Scenarios 7, 8, 9, 10, 11 and 12. Also, each hexagon represents the normalized deviations from the target set for compromise goals with the outermost hexagon (A) signifying normalized deviation equal to 1. The hexagon second to the outermost hexagon (B) signifies normalized deviation equal to 0.8 and so on with center of the hexagon signifying normalized deviation equal to 0. In Figure 14, as we see that the normalized deviation for DCI_{AR} goal is 0 for all the scenarios, we are now looking for design scenarios that satisfy EMI_{MSR} and DCI_{ST} goals as closely as possible. We do not see any scenarios that have normalized deviation of value equal to 0 for all three compromise goals. We also do not see any scenarios that have normalized deviation within 0.2 for all three compromise goals. However, we find scenario S7 which have normalized deviation within 0.4 for all three compromise goals. Hence, Scenario 7 is the design scenario that closely achieves the three compromise goals.

Based on a designer's aspiration to meet the compromise goals as closely as possible and select material that can be used to create designs with consideration to corrosion resistance, cost and machinability, the robust solution alternative that best satisfies designer's aspiration is shown in Table 8.

Table 8: Robust Solution Selected

	Design Variables		
	Compromise Variables		Selection Variable
Scenario	t (in)	D (in)	Material
7	0.545	21.51	Titanium

Based on the designer's aspiration outlined for this specific design problem, the choice of titanium as a material and dimensions as shown in Table 8 seems appropriate. The intent here is not to suggest the use of Titanium in the design of fender but to demonstrate the solution approach for coupled design problems. The 3 material alternatives (Cast Iron, Copper and Titanium) were chosen as these materials stand out from each other in terms of cost, machinability, corrosion resistance and hardness, allowing us to test if the influence among DSPs are effectively captured. The 12 scenarios tabulated in Table 6 are captured by assigning equal weights to the 3 compromise goals while the weights for attributes in selection DSP are assigned as shown in Table 6. From this table, we can see that as the solutions for selection DSP are changing, the solutions in compromise DSP are also changing and vice-versa. This lets us validate that we have successfully captured the mutual influence among DSPs. Now, by changing the designer's preference allows us to explore for other robust solutions. Further, posing a pool of materials that are more suited for a particular application would allow us to effectively explore robust design solutions. For instance: Gear design problem can be solved by posing material alternatives that are specifically designed to suit gear applications thus, enabling us to compare and make tradeoff study among the available material alternatives.

5. DISCUSSION

In this paper, we present a decision scenario matrix for classifying coupled decisions using DSPs. We also present an approach for addressing the issue of uncertainty involved in coupled design problems. In terms of uncertainty, what has been shown is summarized as follows:

(i) *Robustness against variability in performance due to uncertainty:* Our expectation about how the design should perform becomes more accurate if we can identify and manage sources that alter our expectation. It is crucial to identify the variabilities that can impact design performances. In this context, applying EMI to the performance requirement can be effective in managing uncertainty stemming due to variability in design variables and material properties. For instance: We have applied EMI to Mass to Strength Ratio (MSR) goal (G1 – Table 2 and G2 – Table 6) because deviation in MSR can occur because of variability in design variables and material properties. In the example presented, we have considered the uncertainty in design variables and material properties. Our aim here is not to eliminate all possible deviation in MSR but to minimize any such deviation in MSR because of variability emerging from change in our expectation about design variables and material properties. Similarly, we have applied DCI to our other two goals (G1 and G2 in Table 2; G2 and G3 in Table 6), that is, Aspect Ratio (AR) and Stiffness (ST) as we expect deviations in these goals to occur because of change in the value of design variables.

(ii) *Robust solution exploration by treating material properties as system variable:* One of the biggest benefits that we get by treating material as a variable is that we enlarge our design space to have a larger search space with a possibility of finding better quality designs without compromising the performances. In recent years, tremendous research effort has been put on designing material that empowers us to choose materials with properties beyond the standard set.

(iii) *Robust solution exploration involving concurrent selection – compromise decision:* In case of a material selection the use of selection DSP seems appropriate. Compromise DSPs are more appropriate to determine design variables against multiple conflicting goals. When a decision must been taken when selection and compromise decision are interrelated, coupled DSPs are most appropriate. Here, we present an example demonstrating exploration of robust solutions for a coupled selection-compromise decision.

6. CLOSURE

The principal contributions documented in this paper are highlighted below:

1. Classification of coupled decision problem using Decision Scenario Matrix (DSM)
2. Incorporation of robustness metrics into coupled DSPs
3. Exploration of solution spaces for coupled decision problems

The classification of coupled decision problem allows for the generation of decision scenarios. Given an engineering design problem, the decision matrix provides possible decision scenarios for formulating the design problem as coupled DSPs. Such decision scenarios can support the creation and archival of decision-based templates for coupled decision problems. The modular, executable, reusable decision templates as a means to effect design and to archive design-related knowledge on computer for non-interacting DSPs is described in [21].

Each design decision embodies some degree of uncertainty; our contention is that the impact of uncertainty is more on design solutions when decisions are coupled. In this paper, we illustrate the exploration of robust solutions for coupled selection-compromise decision problem in engineering design. Unlike other coupled design methods, where the goal is limited to finding better design solutions, our focus is to assist designers in the formulation of design problems that involve coupled decisions, exploration of design solution space and preservation of design knowledge for re-use. The latter is particularly important for creating decision templates which can be retrieved and re-used. By identifying and classifying different decision scenarios through Decision Scenario Matrix (DSM), various possible scenarios are identified for any given design problem. This also makes it possible to store pre-designed decision templates that can be particularized with domain dependent information associated with a particular design problem. When a design problem is understood, a particular decision template suitable for the given problem can be retrieved and executed for exploring design solutions. In this paper, we offer method that can be applied to model combinations of selection – compromise decisions for robust solutions. This is very important in engineering design as it involves the decisions associated with different stakeholders (designer, manufacturer, metallurgist, customer, etc.) and coming from various disciplines (thermal science, fluid mechanics, material science, economics, etc.) with some degree of uncertainty as the models derived from these disciplines are the abstraction of reality. We believe that a platform that can model the mutual influence among decisions to carry out robust engineering decisions offers numerous advantages. With the promising benefits it holds, it should find increasing attention and application in engineering design in the days to come.

ACKNOWLEDGMENTS

Gehendra Sharma acknowledges the financial support from the L.A. Comp Chair at the University of Oklahoma. He is thankful to his academic mentors, Professor Janet K. Allen and Professor Farrokh Mistree, who acknowledge financial support from the John and Mary Moore Chair and the L.A. Comp Chair at the University of Oklahoma., Norman.

REFERENCES

[1] Smith, W. F., Kamal, S., and Mistree, F., 1987, "The Influence Of Hierarchical Decisions On Ship Design," *Marine Technology*, 24(2), pp. 131-142.

[2] Mistree, F., Smith, W. F. and Bras, B. A., 1993, "A Decision-Based Approach to Concurrent Engineering" in *Handbook of Concurrent Engineering*, pp. 127-158, (H. R. Parsaei and W. Sullivan, Eds.), New York: Chapman & Hall.

[3] Mistree, F., Smith, W., Bras, B., Allen, J., and Muster, D., 1990, "Decision-Based Design: A Contemporary Paradigm For Ship Design," Transactions, *Society of Naval Architects and Marine Engineers*, 98, pp. 565-597.

[4] Mistree, F., Muster, D., Shupe, J. A. and Allen, J. K., "A Decision-Based Perspective for the Design of Methods for Systems Design," *Recent Experiences in Multidisciplinary Analysis and Optimization*, (Barthelmy, J-F., Ed.), Hampton, Virginia, 1988. NASA CP 3031.

[5] Mistree, F., Smith, W. F., Kamal, S. Z. and Bras, B. A., "Designing Decisions: Axioms, Models and Marine Applications," *Fourth International Marine Systems Design Conference*, Kobe, Japan, pp. 1-24, 1991.

[6] Bascaran, E., Bannerot, R. B., and Mistree, F., 1989, "Hierarchical Selection Decision Support Problems in Conceptual Design," *Engineering Optimization*, 14(3), pp. 207-238.

[7] Bascaran, E., Karandikar, H. M. and Mistree, F., "Modeling Hierarchy in Decision-Based Design: A Conceptual Exposition" in *Design Theory and Methodology 92*, pp. 293-300, (L. A. Stauffer and D. L. Taylor, Eds.), New York: ASME, (1992).

[8] Murphy, T. E., Tsui, K.-L., and Allen, J. K., 2005, "A Review Of Robust Design Methods For Multiple Responses," *Res Eng Des*, 15(4), pp. 201-215.

[9] Karandikar, H., and Mistree, F., 1992, "An Approach For Concurrent And Integrated Material Selection And Dimensional Synthesis," *ASME J Mech Design*, 114(4), pp. 633-641.

[10] Chen, W., Allen, J. K., and Mistree, F., 1993, "Hierarchical Selection In Gas Turbine Maintenance Management," *ASME Advances in Design Automation*, ASME DE, 65, pp. 87-96.

[11] Karandikar, H., and Mistree, F., 1991, "Integration Of Information From Design And Manufacture Through Decision Support Problems," *Applied Mechanics Reviews*, 44(11S), pp. S150-S159.

[12] Murphy, T. E., Tsui, K.-L., and Allen, J. K., 2005, "A Review Of Robust Design Methods For Multiple Responses," *Res Eng Des*, 15(4), pp. 201-215.

[13] Taguchi, G., 1986, "Introduction To Quality Engineering: Designing Quality Into Products And Processes." No. 658.562 T3.

[14] Chen, W., Allen, J. K., Tsui, K.-L., and Mistree, F., 1996, "A Procedure For Robust Design: Minimizing Variations Caused By Noise Factors And Control Factors," *ASME J Mech Design*, 118(4), pp. 478-485.

[15] Choi, H.-J., Austin, R., Allen, J. K., McDowell, D. L., Mistree, F., and Benson, D. J., 2005, "An Approach For Robust Design Of Reactive Power Metal Mixtures Based On Non-Deterministic Micro-Scale Shock Simulation," *Journal of Computer-Aided Materials Design*, 12(1), pp. 57-85.

[16] McDowell D.L., Panchal, J. H., Choi, H.-J., Seepersad C. C., Allen J. K. and Mistree F., *Integrated Design of Multiscale Materials and Products*, Elsevier, New York, (2010). ISBN-13: 978-1-85617-662-0.

[17] Choi, H.-J., 2005, "A Robust Design Method For Model And Propagated Uncertainty," PhD Dissertation, Georgia Institute of Technology, Georgia, Atlanta.

[18] Taguchi, G., 1993, "`Robust Technology Development," *Mechanical Engineering-CIME*, 115(3), pp. 60-63.

[19] Michalek, J. J., Feinberg, F. M., and Papalambros, P. Y., 2005, "Linking Marketing And Engineering Product Design Decisions Via Analytical Target Cascading," *Journal of Product Innovation Management*, 22(1), pp. 42-62.

[20] Mistree, F., Marinopoulos, S., Jackson, D. M., and Shupe, J. A., 1988, "The Design Of Aircraft Using The Decision Support Problem Technique." *NASA contractor report NASA CR-4134*.

[21] Ming, Z., Yan, Y., Wang, G., Panchal, J, Goh, C-H, Allen, J.K. and Mistree, F., 2015, "Ontology-Based Executable Design Decision Template Representation and Reuse," *ASME Design Automation Conference*, Boston, MA, August 2-5. Paper Number DETC2015-46272.

Proceedings of the ASME 2019
International Design Engineering Technical Conferences
and Computers and Information in Engineering Conference
IDETC/CIE2019
August 18-21, 2019, Anaheim, CA, USA

DETC2019-97894

DESIGN OF COMPOSITE STRUCTURES THROUGH DECISION SUPPORT PROBLEM AND MULTISCALE DESIGN APPROACH

Rizwan Khan Pathan [*]
Tata Consultancy Services
TCS Research (TRDDC)
Pune, India

Soban Babu Beemaraj
Tata Consultancy Services
TCS Research (TRDDC)
Pune, India

Amit Salvi
Tata Consultancy Services
TCS Research (TRDDC)
Pune, India

Gehendra Sharma
Systems Realization Laboratory
University of Oklahoma
Norman, OK, USA

Janet K. Allen
Systems Realization Laboratory
University of Oklahoma
Norman, OK, USA

Farrokh Mistree
Systems Realization Laboratory
University of Oklahoma
Norman, OK, USA

ABSTRACT

Composite materials are increasingly being used in load bearing structures due to their high specific stiffness and strength. Designing composite structures involve solving multiple conflicting objectives (e.g weight and deflection) and constraints (e.g failure stress and strain), which is a challenging task. In the absence of an optimal solution, a compromise solution is desired. Concurrent (material selection plus sizing) design approach using Decision Support Problem (DSP) is used to arrive at a compromise solution. In this paper multiscale design approach is proposed, that incorporates the tailoring of material microstructures and sizing to achieve improved compromise solution. The microstructure properties are obtained by using analytical and computational models for various composite materials. These models compute structure-property relations between bulk material properties and their micro-structural constituents. The approach is demonstrated with an example of a sandwich composite cantilever beam subjected to multiple load cases. An efficiency factor (η) is defined to compare the results of concurrent design approach and multiscale design approach.

NOMENCLATURE

B	Breadth of the sandwich beam
BX	Biaxial
cDSP	compromise Decision Support Problem
DSP	Decision Support Problem
d^+, d^-	Deviation variables
E_s	Skin stiffness
G_c	Shear stiffness of the core
h	Honeycomb cell wall length
HC	Honeycomb
L	Sandwich beam length
LCS1	Load case scenario 1
LCS2	Load case scenario 2
LCS3	Load case scenario 3
P	Concentrated point load
q	Uniformly distributed load
sDSP	selection Decision Support Problem
t	Honeycomb foil thickness
t_c	Core thickness

---

[*]Correspondence Author, Tata Consultancy Services, TCS Research (TRDDC), Pune, India, Email: r.pathan@tcs.com.

t_s	Skin thickness
T	Sandwich beam thickness
UD	Unidirectional
V_f	Fiber volume fraction
W	Sandwich beam weight
Z	Deviation function
δ	Tip deflection of the beam
θ	Angle between the inclined wall and ribbon axis
ρ_c	Density of core
ρ_s	Density of skin
ρ_f	Density of fiber
ρ_m	Density of matrix
σ_y	Skin failure strength
τ_y	Core failure strength

1 INTRODUCTION

Designers worldwide are using composite materials to maximize the performance of the products. These materials fulfil functional (e.g. mechanical, physical properties) as well as non-functional (e.g chemical, biological, environmental properties) requirements of the products. Composite materials are combination of two or more materials combined using various manufacturing techniques to yield superior performance. These large number of material alternatives and multiple manufacturing processes coupled with the complex relationships among the different selection parameters often make, material selection and sizing of composites a difficult task. Thus, designers need decision making techniques to assist them to take design decisions in case of multiple objectives and conflicting requirements.

In the preliminary design stage, material selection is carried out for multiple load cases, boundary conditions and design requirements for a given structure. Once the material is selected, topology optimization and sizing is performed to arrive at the final configuration. There have been several attempts to standardize material selection processes. Chiner et al. [1] performed material selection in following stages,

1. Definition of design.
2. Analysis of material properties.
3. Screening of candidate materials, evaluation and decision for optimal solution.

Farag et al. [2] describes the different stages of design and the related activities of the material selection namely initial screening, developing and comparing alternatives, and selecting the optimum solution. Among other methods, Ashby's [3] material selection is more popular in engineering application. Ashby's method employs charts, plotted based on the mechanical and physical properties of the materials available in manufacturer's material databases. Material performance indices are obtained for a specific problem and plotted on the chart. Then,

the materials which lie on these performance indices are selected for a given problem. According to Ashby achieving the match with design requirements involves four fundamental steps.

1. A way for translating design necessities into a requirement for material and process.
2. A method for screening out those that cannot meet the specification, leaving a subset of the original menu.
3. A method for ranking the surviving materials and process, identifying those that have the best potential.
4. An approach of searching for supporting information about the top-ranked candidates.

Ashby's methods are widely used in material selection for mechanical components such as gear, shaft, etc. Aceves et al. [4] considered Ashby's approach for material selection for laminated composite material by utilizing finite element method and constraints on cost, stress and deflection. It employs the Ashby's chart approach to arrive at a solution after eliminating the points in the graph which falls outside the metric lines in the graph. All the methods mentioned above, perform material selection by using discrete materials available in manufacturer's databases. These methods perform material selection and sizing separately. Thus, the methods do not take into account the influence of sizing on material selection and vice versa. The solution achieved by such methods may not be the best combination of material plus sizing for a given application. The best combination is achieved by performing material selection and sizing concurrently. In real world applications, often the design objectives are difficult to achieve due to stringent and challenging requirements. In such cases trade-offs are required, in which a compromise solution is sought.

Decision Support Problem (DSP) facilitates the exploration of satisficing solutions when dealing with engineering design problems. Satisficing solutions are good enough but not necessarily optimal. There exists various DSP techniques used to assist design decisions in case of multiple objectives and conflicting requirements. Coupled DSP is a domain-independent method proposed by Karandikar et al [5] for concurrent design of structures. The coupled DSP combines compromise DSP [6] and selection DSP to solve the design problem where influence of sizing and material selection on each other is taken into account. cDSP is used to take design decisions involving multiple trade-offs in the absence of an optimal solution. Compromise decisions involve determining design variables to achieve multiple conflicting objectives and satisfy multiple constraints. sDSP is used to take decisions involving making a choice among a number of alternatives considering several attributes. The concise mathematical form for selection-compromise decision problem is shown in Table 1. It is worth noting how the compromise system variables X influence selection goal MF and how selection alternatives Y influence compromise constraints $g_i(X,Y)$ and goals $Ai(X,Y)$, detailed description could be found in [5].

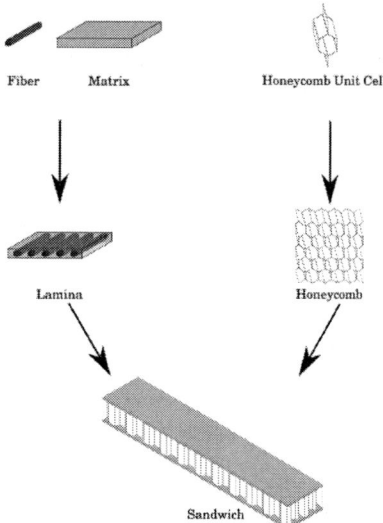

FIGURE 1: MULTISCALE NATURE OF SANDWICH COMPOSITE PANEL.

Composite structures are multiscale (see Fig.1) in nature, its mechanical properties depend on the microstructure, its constituents and their arrangements. These properties are explicitly obtained through structure-property relations as mentioned in Appendix A. These relations are used to obtain suitable material properties by altering the microstructure. *Material selection in concurrent design approach is still carried out using discrete materials mentioned in manufacturer's datasheets. Thus, the approach does not exploit the tailorable nature of composites entirely.* Concurrent design solutions can be further improved upon by including this tailorable nature of composites in the design workflow itself. In this paper, we propose a multiscale design approach to obtain an improvised compromise solution by using multiscale models. These models compute bulk material properties from their microstructures through structure-property relationships. In this approach, based on design goals and constraints, the material microstructure itself evolves to provide better compromise solutions.

In this study, design of a sandwich composite cantilever beam subjected to multiple loads is used as a test problem. The given problem is solved by using two design approaches namely concurrent design approach and multiscale design approach. In the concurrent design approach a coupled DSP technique was used for concurrent material selection and sizing. In the multiscale design approach, the cDSP technique was combined with the multiscale models to achieve the desirable material properties and sizing. These desirable material properties are obtained by designing microstructures through structure-property relationships for each load cases. Multiscale models for skin consist of unidirectional, laminated, braided and woven composites

and for core consist of honeycomb, open and closed cell foams. Design efficiency factors are computed for each load cases. It is defined as ratio of target value to the achieved value.

Section 2 depicts the description and problem formulation of design of sandwich composite beam. Section 3 explains the concurrent design approach to solve the sandwich beam design problem. Section 4 explains the multiscale design approach to solve the test problem. In Section 5, the results are compared and discussed. Structure-property relations for skin and core are mentioned in appendix A.

TABLE 1: COUPLED DSP TEMPLATE.

Coupled selection-compromise DSP	
compromise DSP	selection DSP
Find:	**Find:**
Compromise system variable X	Selection system variable Y
Deviation variables d_i^-, d_i^+	Deviation variables e_i^-, e_i^+
Satisfy:	**Satisfy:**
Design constraints $g_j(X, (\vec{Y})) > 0$	Selection Constraint $\sum\limits_{i=1}^{n} = 1$
Constraint on deviation variables $d_i^+ \geq 0, d_i^- \geq 0, d_i^+ \times d_i^- = 0$	Constraints on Deviation Variables $e_i^+ \geq 0, e_i^- \geq 0, e_i^+ \times e_i^- = 0$
Compromise goals:	**Selection goal:**
$A_i(X, (\vec{Y})) + d_i^+ - d_i^+ = G_i$	$MF_j((\vec{X}))Y_i + e_1^- - e_1^+ = 1$
Bounds:	**Bounds:**
$B : (\vec{X})^{min} \leq \vec{X} \leq (\vec{X})^{max}$	$B : (\vec{Y})^{min} \leq \vec{Y} \leq (\vec{Y})^{max}$
Minimize	
$Z = \{e_1^-, \sum\limits_{i=1}^{n} w_i \times (d_i^- + d_i^+), \sum\limits_{i=1}^{3} w_i = 1\}$	

2 DESIGN OF A SANDWICH COMPOSITE BEAM

The sandwich composite beam consists of a light weight soft core sandwiched between two rigid skin as shown in Fig.2. It increases bending rigidity (see section 2.1) by separating skin from its neutral axis. The sandwich composite beam fulfils functional as well as non-functional requirements. Functional requirements of the sandwich is fulfilled by having stiff skin and soft core. Skin carries bending stresses, while core carries shear stresses. Typically, skins are made up of thin metallic sheets or fiber reinforced composites which offer excellent in plane properties. Cores are usually made up of honeycomb, open and closed cell foams which offer high specific shear stiffness. These functional properties are utilized while designing lightweight load bearing structures. Such panels are widely used in tail section of aircraft to resist high aerodynamic as well as flutter loads. Non-functional requirements of sandwich beam are fulfilled by incor-

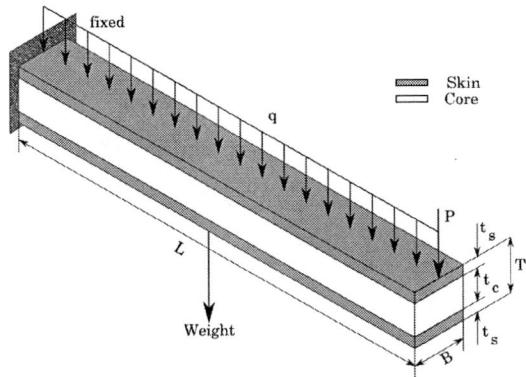

FIGURE 2: SANDWICH COMPOSITE CANTILEVER BEAM.

porating suitable non functional properties for skin and core materials. e.g.- marine panels are made of corrosion resistant skin and a shear stiffened PVC core.

Deflection of sandwich beam depends on the bending and shear rigidity of the beam. High bending rigidity can be achieved by increasing the thicknesses of skin and core materials. But, the thicknesses of skin and core material also dictates the weight of the sandwich beam. Thus, the sandwich beam design is a multiple conflicting objective problem. Also, satisfying non functional requirements along with functional requirements makes the design problem more complex and challenging. Often in such cases no optimal solution is obtained, hence a compromise solution is sought. Description of sandwich design problem is explained in section 2.1.

2.1 DESCRIPTION OF TEST PROBLEM

This section explains the design goals, constraint, loading and boundary condition associated with the sandwich beam test problem. The design goal is to achieve target values of weight (T_w) and tip deflection (T_δ) that are are $14N$ and $10mm$ respectively. The constraints to be satisfied are, stresses in skin and core should not exceed their respective strength value by a factor of 0.5. Sandwich beam is subjected to three different types of load case scenarios (LCS). In LCS1 sandwich composite beam is subjected to uniformly distributed load (q) of 1.5 N/mm and self weight W as shown in Fig. 3a, in LCS2 concentrated point load (P) of 1500 N and self weight as shown in Fig. 3b and in LCS3 uniformly distributed load, concentrated point load and self weight as shown in Fig. 3c. Analytical equations are used to calculate the deflection and weight (see ref [7]). Deflection due to UDL (δ_q), weight (δ_w) and end point load (δ_p) is given by Eqn. (1, 2) and (3) respectively. The deflection of the beam for LCS1, LCS2, and LCS3 is given as $\delta_q + \delta_w$, $\delta_w + \delta_p$ and $\delta_q + \delta_w + \delta_p$ respectively. Total weight of the sandwich beam is given by Eqn. (4).

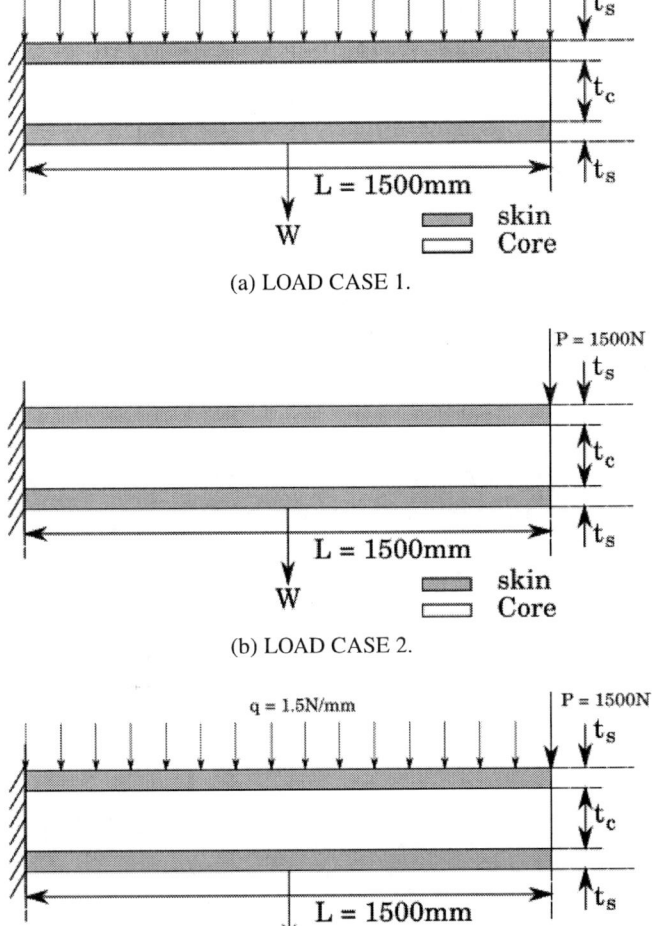

(a) LOAD CASE 1.

(b) LOAD CASE 2.

(c) LOAD CASE 3.

FIGURE 3: LOAD CASE SCENARIOS.

$$\delta_q = \frac{qL^4}{8(EI)_{eff}} + \frac{qL^2}{2(GA)_{eff}} \qquad (1)$$

$$\delta_w = \frac{WL^3}{8(EI)_{eff}} + \frac{WL}{2(GA)_{eff}} \qquad (2)$$

$$\delta_p = \frac{PL^3}{3(EI)eff} + \frac{PL}{(GA)_{eff}} \qquad (3)$$

$$W = 2t_s BL\rho_s g + t_c BL\rho_c g \qquad (4)$$

Where effective bending rigidity $(EI)_{eff}$ and shear rigidity $(GA)_{eff}$ is given by Eqn (5) and (6) respectively.

$$(EI)_{eff} = \frac{E_s B t_s^3}{6} + \frac{E_s B t_c T^2}{2} \qquad (5)$$

$$(GA)_{eff} = \frac{G_c B T^2}{t_c} \qquad (6)$$

Design of sandwich beam incorporating both functional and non-functional properties is carried out using concurrent design method (see section 3) and multiscale design approach (see section 4). The compatibility between skin and core materials and their joining process is also an important material selection criteria but it is not considered in this study.

3 CONCURRENT DESIGN APPROACH

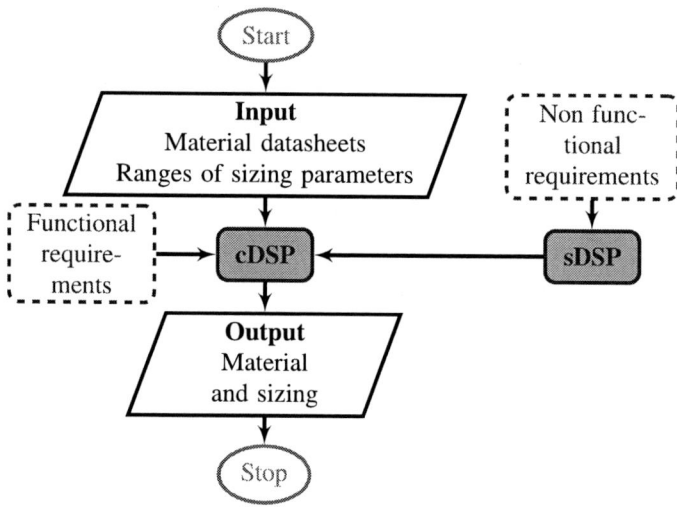

FIGURE 4: CONCURRENT DESIGN APPROACH.

In concurrent design approach, the skin and core materials (chosen from the available material databases) and their corresponding thicknesses are selected concurrently by coupled DSP technique. The compromise decision involves the selection of material of skin and core and their thicknesses against the conflicting requirements of weight and deflection. The approach also takes into account the constraints imposed on stresses in skin and core. The selection decision involves selecting a material from various available alternatives based on non functional attributes such as cost, corrosion resistance, moisture resistant,chemical resistant, fatigue resistant and thermal insulation. The determination of design variables (compromise part) requires information

about material and associated properties (selection part). The two decisions (selection and compromise) are coupled and must account for the impact of the decisions on each other. The flow chart for the process is shown in Fig.4.

The functional and normalized non-functional attributes are mentioned in Table 2. The selection decision is taken by comparing the merit-function values of all the materials. The non-functional attributes of materials are used for merit-function calculations [10], which is based on the relative importance (I) given to these attributes as shown Table 3. These merit functions will be set as goals in the coupled DSP formulation.

The mathematical formulation of coupled DSP to solve the given problem is shown below :-

Given:

$L = 1500mm$.
$B = \dfrac{T}{2}$
The given problem is solved for all the load cases as shown in Fig. 3.
The skin and core materials available for selection are shown in Table 2.

Find:

Material to be used for skin and core.
Skin thickness (t_s), Core thickness (t_c).
Deviation variables

d_1^- Underachievement deviation variable for merit function of skin
d_2^- Underachievement deviation variable for merit function of core
d_3^+ Overachievement deviation variable for weight goal
d_4^+ Overachievement deviation variable for deflection goal

Satisfy:

Constraint:

Strength criteria for skin: $\sigma_s < 0.5\sigma_y$
Strength criteria for core: $\tau_c < 0.5\tau_y$

Goals:

Merit function value for each material (S=skin, C=core); i=1,...,4 and non functional attributes j=1,...,6

$$\sum_{j=1}^{6} R_{ij} I_j S_i + d_1^- - d_1^+ = 1$$

$$\sum_{j=1}^{6} R_{ij} I_j C_i + d_2^- - d_2^+ = 1$$

Meet target deflection(T_δ) = 10mm
Meet target weight(T_W) = 14N

Bounds:

TABLE 2: SKIN AND CORE MATERIALS. [8, 9].

			Functional Attributes			Non-functional Attributes					
								Resistance to			
Material Name	Material family	Type	E(MPa)	G(MPa)	$\rho\,(^{Kg}/_{m^3})$	Moisture	Corrosion	Chemical	Fatigue	Cost	Thermal
ELR0908	E Glass-Epoxy UD	Skin	31830	4910	1800	1.0	0.8	0.6	0.8	0.4	0.6
CL0900	Carbon-Epoxy UD	Skin	108450	4090	1530	0.4	0.6	0.6	0.6	0.8	0.4
CLT0900	Carbon-Epoxy BX	Skin	56380	4270	1530	0.4	0.6	0.6	0.6	0.8	0.4
2024T3	Aluminum	Skin	68900	24950	2700	1.0	0.8	0.6	0.6	0.8	0.4
PCGA-XR23003	Aluminum HC	Core	1020	434	83	1.0	1.0	0.8	0.6	0.4	0.2
DivinycellP150	Polymer Foam	Core	152	40	1507	1.0	0.8	0.6	1.0	0.8	0.8
BaltekSB	Balsa	Core	442.8	362	285	0.2	0.8	0.4	0.8	0.6	0.8
AIREXT92130	Polymer Foam	Core	140	34	135	1.0	0.8	0.8	1.0	1.0	0.8

TABLE 3: RELATIVE IMPORTANCE OF NON-FUNCTIONAL ATTRIBUTES.

Selection Attributes	Relative Importance(I)
Cost	0.5
Resistant to Moisture	0.1
Resistant to Corrosion	0.1
Resistant to Chemical	0.1
Resistant to Fatigue	0.1
Insulation to Thermal	0.1

$$5 \leq t_s(mm) \leq 15$$
$$70 \leq t_c(mm) \leq 90$$
$$80 \leq T(mm) \leq 110$$

Minimize:

$$Z = \{0.25d_1^- + 0.25d_2^- + 0.25d_3^+ + 0.25d_4^+\}$$

x

In this approach, the input values as mentioned in the mathematical formulation are the discrete materials available in databases and ranges of thicknesses of skin and core. The solution is obtained such that it satisfies the strength criteria as mentioned in the constraints of cDSP, while achieving the merit function, weight and deflection goals. The preferences are according to designer's choice, in this case equal preferences were given to the achievement of all the goals. The preferences are given as weights in the deviation function (Z) for each goal. The formulated approach minimizes the deviation function to achieve the goals.

An efficiency factor is defined as ratio of target values to the achieved values $\left(\eta_i = \dfrac{T_i}{i}\right)_{i=W,\delta}$. In the given test case problem, lower values of weight and deflection are always preferred. Thus, $\eta \geq 100\%$ indicates that the achieved values are equal or lower than the specified target values. The solution (skin and core materials and their thicknesses) and efficiencies for all the load cases are shown in Table 4. M-T1, M-T2, M-T3 as shown in Table 4 are the specific material and sizing combinations obtained for load cases LCS1, LCS2 and LCS3 respectively. η_w and η_δ shows the efficiency achieved by the material and sizing combinations for each load cases.

For this design problem, a final solution is obtained by selecting material and sizing combination which provide high efficiencies for all the load cases. The specific material and sizing combinations achieved (M-T1, M-T2, M-T3) are used to obtain the efficiencies in case of all load cases (LCS1, LCS2, LCS3) as mentioned in Table 5. It can be observed that the solution M (Skin CL0900, Core PCGA-XR23003) T3 (t_s=5.16mm, t_c=89.68mm) offers the best efficiency of deflection for all load cases. As the efficiency factor for weight is nearly the same for all the materials, we have selected M-T3 as final solution.

4 MULTISCALE DESIGN APPROACH

The multiscale design approach is investigated to achieve better design efficiencies as compared to the concurrent design approach. The flowchart of the approach is shown in Fig. 5. In this approach initially, the material properties of skin, core and their thicknesses are treated as variables and given appropriate ranges. The appropriate ranges of skin and core

TABLE 4: RESULTS FOR EACH LOAD CASE USING CONCURRENT DESIGN APPROACH.

Test Problems	Selected Materials		Sizing(mm)		$W(N)$	$\delta(mm)$	Efficiency(%)	
							η_W	η_δ
LCS1	M	Skin = CL0900 Core = PCGA-XR23003	$T1$	$t_s = 5.03$ $t_c = 89.16$	14.98	10.01	93.46	99.93
LCS2	M	Skin = CL0900 Core = PCGA-XR23003	$T2$	$t_s = 5.13$ $t_c = 89.75$	15.36	16.58	91.15	60.31
LCS3	M	Skin = CL0900 Core = PCGA-XR23003	$T3$	$t_s = 5.16$ $t_c = 89.68$	15.42	26.06	90.79	38.37

TABLE 5: RESULTS FOR THREE LOAD CASES USING SOLUTION OF EACH LOAD CASE IN CONCURRENT DESIGN APPROACH.

Material	Sizing	Test problem	$W(N)$	$\delta(mm)$	Efficiency(%)	
					η_w	η_δ
M	T1	LCS1	14.98	10.01	93.46	99.93
		LCS2	14.98	17.24	93.46	58.00
		LCS3	14.98	27.18	93.46	36.79
M	T2	LCS1	15.36	9.63	91.15	103.80
		LCS2	15.36	16.58	91.15	60.31
		LCS3	14.98	27.18	93.46	36.79
M	T3	LCS1	15.42	9.60	90.79	104.18
		LCS2	15.42	16.52	90.79	60.53
		LCS3	15.42	26.06	90.79	38.37

material properties are given in Table 6. These ranges are obtained after calculating the minimum and maximum values of skin and core material properties through structure property models (ref appendix A) for variety of input values as shown in Fig. 6 and Fig. 7 respectively. Then, cDSP is used to obtain suitable material properties of skin, core and their thicknesses against the conflicting requirements of weight, deflection and the strength constraints imposed on skin and core materials. In the end, the structure property relationships are used to obtain microstructures that provide these suitable material properties. The microstructure is chosen such that it also satisfies the non functional requirements for given problem.

This approach is formulated as a cDSP problem as shown

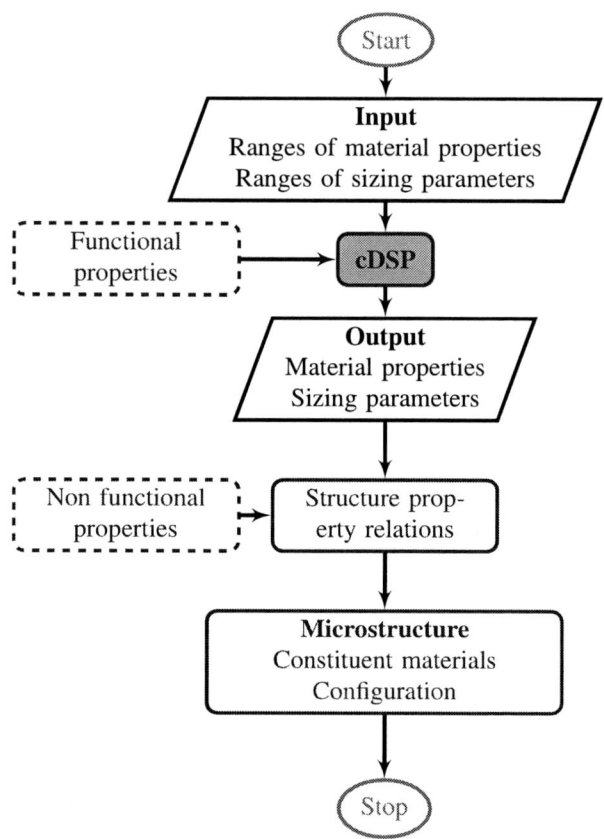

FIGURE 5: MULTISCALE DESIGN APPROACH.

below
Given:

$L = 1500mm.$

$B = \dfrac{T}{2}$

The given problem is solved for all the load cases as shown

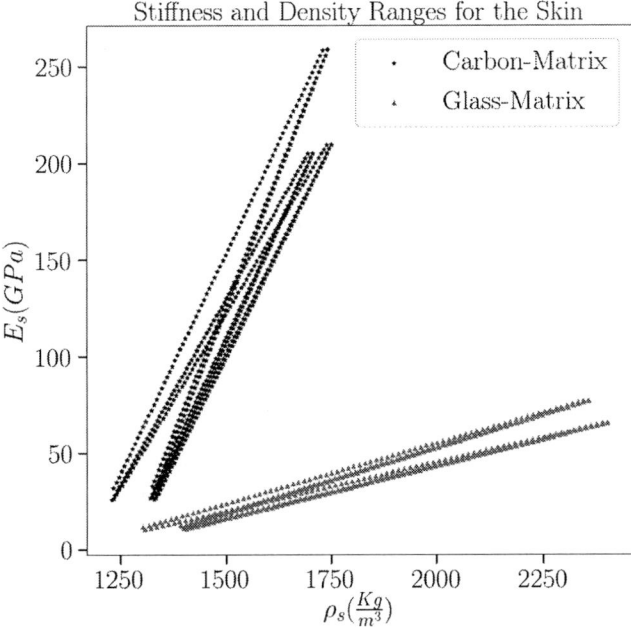

FIGURE 6: ELASTIC MODULUS AND DENSITY RANGES FOR THE SKIN MATERIAL.

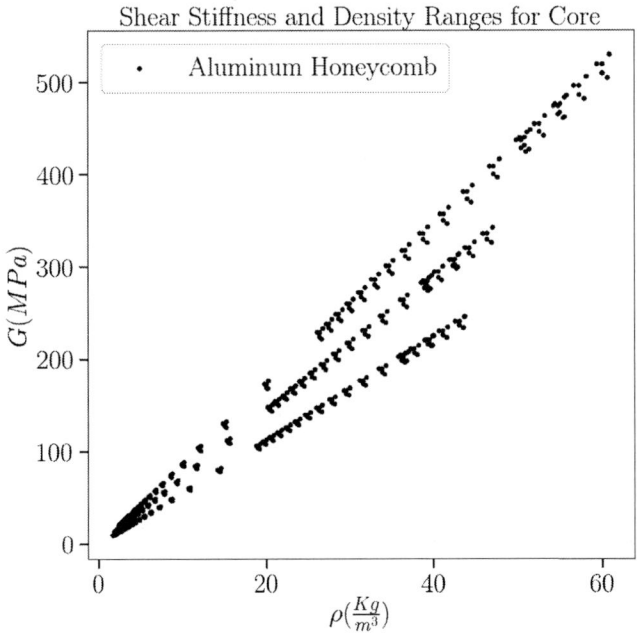

FIGURE 7: SHEAR MODULUS AND DENSITY RANGES FOR THE CORE MATERIAL.

TABLE 6: RANGE FOR MATERIAL PROPERTIES OF SKIN AND CORE.

Material Properties	Type	Min	Max
Elastic modulus (Mpa)	Skin	94060	204310
Density $\left(\dfrac{Kg}{m^3}\right)$	Skin	1406	1651
Shear Modulus(Mpa)	Core	21.6	536.6
Density $\left(\dfrac{Kg}{m^3}\right)$	Core	3.4	86.3

in Fig. 3.

Ranges of material properties for skin and core as shown in Table 6.

Find:

Suitable material properties ($E_s, \rho_s, G_c, \rho_c, t_s, t_c$)
Skin thickness, t_s, Core thickness, t_c
Deviation variables

d_1^+ Overachievement deviation variable for weight goal
d_2^+ Overachievement deviation variable for deflection goal

Satisfy:

System constraint

Strength criteria for skin: $\sigma_s < 0.5\sigma_c$
Strength criteria for core: $\tau_c < 0.5\tau_c$

Goals:

Meet target deflection(T_δ) = 10 mm
Meet target weight(T_W) = 14 N

Bounds:

$5 \leq t_s(mm) \leq 15$
$70 \leq t_c(mm) \leq 90$
$94060 \leq E_s(MPa) \leq 204310$
$1406 \leq \rho_s(\dfrac{Kg}{m^3}) \leq 1651$
$21.6 \leq G_c(MPa) \leq 536.6$

Minimize:

$Z = \{0.5d_1^+ + 0.5d_2^+\}$

In the mathematical formulation, the input values are the material properties and thickness ranges of skin and core. The solution is obtained such that it satisfies the strength criteria as mentioned in the constraints of cDSP, while achieving weight and deflection goals. In this case also equal preferences were

TABLE 7: RESULTS FOR EACH LOAD CASE USING MULTISCALE DESIGN APPROACH.

Test Problem	Material $E, G : MPa, \rho : \frac{Kg}{m^3}$	Sizing(mm)	$W(N)$	$\delta(mm)$	Efficiency(%) η_w	η_δ
LCS1	$M1$ $\begin{cases} E_s = 160250 \\ \rho_s = 1595.48 \\ G_c = 150 \\ \rho_c = 24.17 \end{cases}$	$T1^*\begin{cases} t_s = 6.06 \\ t_c = 79.6 \end{cases}$	14.34	9.11	97.63	109.77
LCS2	$M2$ $\begin{cases} E_s = 202587.34 \\ \rho_s = 1676.17 \\ G_c = 219 \\ \rho_c = 35.29 \end{cases}$	$T2^*\begin{cases} t_s = 5.02 \\ t_c = 89.95 \end{cases}$	14.72	10.08	95.11	99.20
LCS3	$M3$ $\begin{cases} E_s = 204310 \\ \rho_s = 1679.25 \\ G_c = 193 \\ \rho_c = 31.07 \end{cases}$	$T3^*\begin{cases} t_s = 5.36 \\ t_c = 90 \end{cases}$	15.34	15.5	91.26	64.52

assigned to the goals. The formulated approach minimizes the deviation function to achieve these goals.

The solution (skin and core materials and their thicknesses) and efficiencies for all the load cases are shown in Table 7. M1-$T1^*$, M2-$T2^*$ and M3-$T3^*$ are the specific material and sizing combinations obtained for load cases LCS1, LCS2 and LCS3 respectively. The columns η_w and η_δ shows the efficiency achieved by the material and sizing combinations for each load case. A final solution is obtained by selecting a material and sizing combination which provide high efficiencies for all the load cases. The specific material and sizing combinations achieved (M1-$T1^*$, M2-$T2^*$, M3-$T3^*$) are used to obtain the efficiencies in case of all load cases (LCS1, LCS2, LCS3) as mentioned in Table 8. It can be observed that the solution M3 ($E_s = 204310MPa$ $\rho_s = 1679.25^{Kg}/_{m^3}$ $G_c = 193MPa$ $\rho_c = 31.07^{Kg}/_{m^3}$, $T3^*$ ($t_s = 5.36mm$, $t_c = 90mm$) yields the best efficiency of deflection for all load cases. As the efficiency factor for weight is nearly same for all the materials, we have selected $M3 - T3^*$ as final solution. It can be observed that the multiscale approach achieves target goals better as compared to concurrent design approach for all the load cases as shown in Table 9.

The structure property relationships are then used to design microstructures that provide the material properties selected in multiscale design approach (M3). The microstructure is chosen such that it also satisfies the non functional requirements for given problem. The M3 material's mechanical properties are set

TABLE 8: RESULTS FOR THREE LOAD CASES USING SOLUTION OF EACH LOAD CASE IN MULTISCALE DESIGN APPROACH.

Material	Sizing	Test problem	$W(N)$	$\delta(mm)$	Efficiency(%) η_w	η_δ
M1	$T1^*$	LCS1	14.34	9.11	97.63	109.77
		LCS2	14.33	15.12	97.67	66.14
		LCS3	14.34	24.17	97.63	41.37
M2	$T2^*$	LCS1	14.72	6.035	95.11	165.70
		LCS2	14.72	10.08	95.11	99.20
		LCS3	14.72	16.08	95.11	62.19
M3	$T3^*$	LCS1	15.34	5.85	91.25	170.88
		LCS2	15.34	9.68	91.25	103.31
		LCS3	15.34	15.5	91.26	64.52

as target values to obtain the microstructure. Thus, the target values for skin and core namely $E_s^t = 204309MPa$, $\rho_s^t = 1679^{Kg}/_{m^3}$ and $G_c^t = 193MPa$, $\rho_c^t = 31^{Kg}/_{m^3}$ are sought.

TABLE 9: EFFICIENCY FACTOR COMPARISON FOR BOTH APPROACHES.

Selected material		LCS1		LCS2		LCS3	
		η_w	η_δ	η_w	η_δ	η_w	η_δ
Concurrent design approach	M-T3	90.79	104.18	90.79	60.53	90.79	38.37
Multiscale approach	M3-T3*	91.25	170.88	91.25	103.31	91.26	64.52

TABLE 10: FIBER AND MATRIX PROPERTIES.

Material	$E(GPa)$	v	$\rho\left(\dfrac{Kg}{m^3}\right)$	Type
AS-4 Carbon	235	0.2	1810	Fiber
T-300 Carbon	230	0.2	1760	Fiber
IM7 Carbon	290	0.2	1800	Fiber
E-glass	73	0.23	2540	Fiber
S-glass	86	0.23	2490	Fiber
Epoxy 3501-6	4.3	0.35	1270	Matrix
Epoxy (977-3)	3.7	0.35	1280	Matrix
Epoxy (Hy6010)	3.4	0.36	1170	Matrix

4.1 SKIN MICROSTRUCTURE

The skins in the sandwich composites are typically made up of long fiber reinforced polymer composites (e.g, UD, BIAX, woven) consisting of various fibers and matrix as shown in Table 10. The micro-structures that yields $E_s \geq E_s^t$ and $\rho_s \leq \rho_s^t$ are chosen as suitable microstructures. In this problem, the functional and non-functional requirements for skins are only achieved by unidirectional fiber reinforced composites as it yields high longitudinal specific stiffness as compared to the biaxial and woven composite. The structure property relationship for skin as mentioned in appendix A.1 are used to obtain the microstructure yielding the desirable properties. The suitable lamina and it's constituent are shown in Table 11. In case of complicated loading that involves inter-laminar stresses and delamination, a woven or braided composite is more suitable for achieving functional and non functional requirements.

4.2 CORE MICROSTRUCTURE

The core materials are typically made up of honeycomb, open and closed cell foams. The micro-structures that yields $G_c \geq G_c^t$ and $\rho_c \leq \rho_c^t$ are chosen as suitable micro-structures.

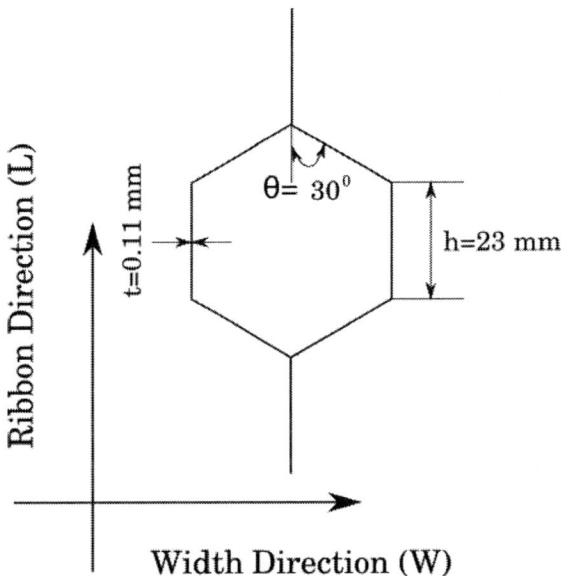

FIGURE 8: ALUMINUM HONEYCOMB MICROSTRUCTURE HAVING BEST $\dfrac{G_c}{\rho_c}$ RATIO.

In this problem, the functional and non-functional requirements for the core are only achieved by aluminum honeycomb as it offers high specific shear stiffness as compared to the open and closed cell foams. The structure property relationship for core as mentioned in appendix A.2 are used to obtain the microstructure yielding the desirable properties. The obtained microstructures are shown in Table 12. The microstructre having best specific shear stiffness $\left(\dfrac{G_c}{\rho_c}\right)$ is shown in Fig 8.

TABLE 11: MICROSTRUCTURE DETAILS FOR THE LAMINA SKIN TO ACHIEVE TARGET MATERIAL PROPERTIES.

$V_f\%$	$E_s(GPa)$	$\rho_s\left(\dfrac{Kg}{m^3}\right)$	$\dfrac{E_s(GPa)}{\rho_s(^{Kg}/m^3)}$	Fiber (Carbon)	Matrix (Epoxy)
70	204	1641	0.125	IM7	3501-6

5 CONCLUSIONS AND FUTURE WORK

In this paper, design of a sandwich composite beam is carried out using concurrent design approach and multiscale design approach. In the former approach, cDSP and sDSP formulation were used to select material for skin and core and their thicknesses for multiple conflicting requirements. Later approach

TABLE 12: MICROSTRUCTURE DETAILS FOR THE CORE TO ACHIEVE TARGET MATERIAL PROPERTIES.

$t(mm)$	$h(mm)$	θ°	$G_c(MPa)$	$\rho_c\left(\dfrac{Kg}{m^3}\right)$	$\dfrac{G_c(MPa)}{\rho_c\left(Kg/m^3\right)}$
0.11	23	30	232	27.68	8.383
0.01	2	30	242	28.94	8.382
0.11	22	30	242	28.94	8.382
0.11	24	30	222	26.53	8.381
0.11	25	30	213	25.47	8.381
0.11	17	45	203	28.95	7.014

combines cDSP and multiscale models to arrive at material properties, material microstructures and sizing for the test problem. Design efficiency (η) showing the achievement of target values are computed for each approach. A unique set of material and thicknesses were selected as final solution that achieves better overall efficiencies for all the load cases. The multiscale approach shows higher design efficiencies as compared to the concurrent design approach. The target material properties in the multiscale approach were achieved by the explicit structure-property relations. The final microstructure for skin and core can be selected from the admissible solutions mentioned in Table 11 and Table 12. The proposed multiscale design approach facilitates the following.

(i) Tailoring multiscale nature of the composite material to achieve the desired target properties by selecting and arranging the constituents.

(ii) Exploring large design space to achieve best performance efficiencies.

(iii) Manufactures can use this methodology to serve designers better by creating new materials, as the former approach has limited selection options.

(iv) In composite structures, failure is governed by local microstructure behavior, this can also be incorporated in the multiscale approach as a design criteria while obtaining the suitable microstructure.

(v) For the combined loadings (e.g, bending and torsion) multiscale approach has a potential to evolve to find the suitable microstructure such as braided composite or laminated composite with varying stacking sequences.

(vi) Functionally graded materials or hybrid composite can also be obtained by multiscale approach based on designer requirements.

(vii) The given approach can solve any design problem in which structure property relations exists between the materials to be used in selection and their microstructures.

Typically, analytical and computational models which are used to compute structure-property relations exhibit uncertainty while predicting actual properties. Manufacturing of polymer composites is highly process dependent and uncertainties in mechanical properties exists due to poor process control. Uncertainty handling methods need to be investigated to incorporate these uncertainties in multiscale design approach .

ACKNOWLEDGMENT

The authors gratefully acknowledge Tata Consultancy Services for their constant guidance and support while undergoing this research work. Farrokh Mistree and Janet K. Allen gratefully acknowledge support from the L A Comp and John and Mary Moore Chair at the University of Oklahoma, Norman.

REFERENCES

[1] M Chiner. Planning of expert systems for materials selection. *Materials & Design*, 9(4):195–203, 1988.

[2] Mahmoud M Farag. Quantitative methods of materials selection. *Handbook of materials selection*, pages 1–24, 2002.

[3] Michael F. Ashby. *Materials Selection in Mechanical Design*. Butterworth-Heinemann, 2011.

[4] C Monroy Aceves, Alexandros A Skordos, and Michael PF Sutcliffe. Design selection methodology for composite structures. *Materials & Design*, 29(2):418–426, 2008.

[5] HM Karandikar and F Mistree. An approach for concurrent and integrated material selection and dimensional synthesis. *Journal of Mechanical Design*, 114(4):633–641, 1992.

[6] Farrokh Mistree, Owen F Hughes, and BA Bras. The compromise decision support problem and the adaptive linear programming algorithm. *Structural Optimization: Status and Promise, MP Kamat, ed., AIAA, Washington, DC*, 1993.

[7] Howard G Allen. *Analysis and design of structural sandwich panels: the commonwealth and international library: structures and solid body mechanics division*. Elsevier, 2013.

[8] Plascore. Technical specification - honeycomb core resource material, plascore,.

[9] Divinycell H. Technical specification h grade, divinycell international ab, laholm, sweden.

[10] N Kuppuraju, S Ganesan, F Mistree, and JS Sobieski. Hierarchical decision making in system design. *Engineering Optimization*, 8(3):223–252, 1985.

[11] Carl T Herakovich. Mechanics of fibrous composites. 1998.

[12] Zvi Hashin and B Walter Rosen. The elastic moduli of fiber-reinforced materials. *Journal of applied mechanics*, 31(2):223–232, 1964.

[13] Shu Ching Quek, Anthony M Waas, Khaled W Shahwan,

and Venkatesh Agaram. Analysis of 2d triaxial flat braided textile composites. *International Journal of Mechanical Sciences*, 45(6-7):1077–1096, 2003.

[14] S Kelsey, RA Gellatly, and BW Clark. The shear modulus of foil honeycomb cores: A theoretical and experimental investigation on cores used in sandwich construction. *Aircraft Engineering and Aerospace Technology*, 30(10):294–302, 1958.

[15] Lorna J Gibson and Michael F Ashby. *Cellular solids: structure and properties*. Cambridge university press, 1999.

[16] Gilmer M Viana and Leif A Carlsson. Mechanical properties and fracture characterization of cross-linked pvc foams. *Journal of Sandwich Structures & Materials*, 4(2):99–113, 2002.

Appendices

A Appendix A
A.1 Structure-Property Relationships for Skin

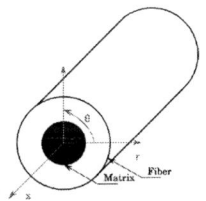

FIGURE 9: CCAM FIBER AND MATRIX [11].

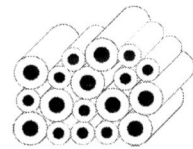

FIGURE 10: ASSEMBLAGE OF CYLINDER [11].

Concentric Cylinder Assemblage Model Micromechanical model of unidirectional laminated composite was given by Concentric Cylinder Assemblage Model (CCAM), which was proposed by Hashin et al [12]. CCAM assumes unidirectional continuous fiber composite is assemblage of fiber core surrounded by a matrix annulus as shown in Fig. 9 and each assemblage is having constant fiber volume fraction see Fig. 10. Density Eqn. (7) and stiffness Eqn. (8) of the assemblage is given in-terms of fiber properties, matrix properties, and fiber volume fraction V_f. Where v is the Poissons ratio; K is the bulk modulus; μ is the shear modulus; f and m are the fiber and matrix, respectively.

$$\rho_s = \rho_f V_f + \rho_m (1 - V_f) \tag{7}$$

$$E_s = V_f E_f + (1 - V_f) E_m + \frac{4 V_f (1 - V_f)(v_f - v_m)^2 \mu_m}{\frac{(1 - V_f)\mu_m}{K_f + \frac{\mu_f}{3}} + \frac{V_f \mu_m}{K_m + \frac{\mu_m}{3}} + 1} \tag{8}$$

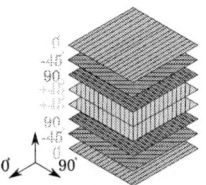

FIGURE 11: STACKED LAMINATE COMPOSITE PLIES.

Classical Laminate Theory Skins could also be made up of laminate as shown in Fig. 11. The laminate consists of several lamina stacked upon each other. These individual laminae have their own thickness and orientation. Based on the lamina properties of all the layers, the effective property of the overall laminate is calculated by using classical laminate theory [11].

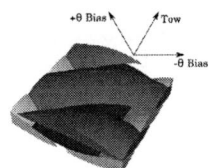

FIGURE 12: 2D TRIAXIAL BRAIDED TEXTILE COMPOSITE.

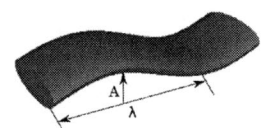

FIGURE 13: WAVELENGTH (λ) AND AMPLITUDE (A) OF BIAS TOW.

2D Triaxial Braided Composite Quek et al. [13] obtained the effective elastic properties of 2D triaxial braided composites in terms of constituents, undulation (see Fig. 12) and wavelength (see Fig. 13).

A.2 Structure-Property Relationships for Core
Hexagonal Honeycomb Kelsey et al. [14] obtained the equation for density and ribbon direction shear modulus of hexagonal honeycomb based on unit deflection method. The density ρ_c is given in Eqn. (9), the ribbon direction shear modulus G_c is given by Eqn. (10). Both the density and shear modulus is function of cell wall angle θ, cell wall length h, cell wall

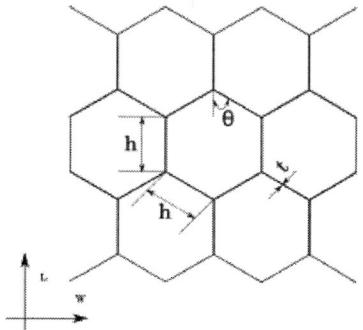

FIGURE 14: HONEYCOMB MICROSTRUCTURE.

thickness t, and cell wall material. Hexagonal honeycomb microstructure is shown in Fig. 14

$$\rho_c = \frac{2}{(1+\cos\theta)\sin\theta}\frac{t}{h}\rho_s \qquad (9)$$

$$G_c = \frac{1+\cos^2\theta}{(1+\cos\theta)\sin\theta}\frac{t}{h}G_s \qquad (10)$$

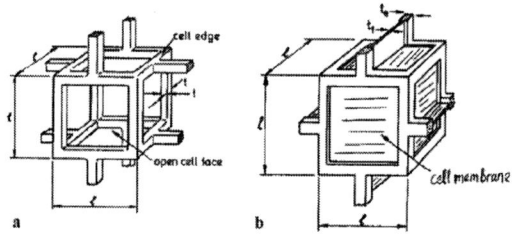

FIGURE 15: OPEN AND CLOSED CELL FOAM [15].

$\phi = 1$ corresponds to open cell foam. Microstructure of open and closed cell foams are shown in Fig. 15.

$$G_c = C_2\phi^2\left(\frac{\rho}{\rho_s}\right)^2 + C_2'(1-\phi)\frac{\rho}{\rho_s}E_s \qquad (11)$$

$$\rho_c = C_3\frac{t}{l}\rho_s \qquad (12)$$

Open and Closed Cell Foam Micromechanical model of closed-cell foams relating foam's mechanical properties to their microstructre based on cell edge bending and face stretching is given by Gibson et al. [15]. The first term in Eqn. (11) corresponds to cell-edge bending and the second term corresponds to cell-wall stretching. C_2 and C_2' are constants and depend on material. For polymeric foam $C_2 = C_2' \approx 3/8$ gives the close approximation, where ϕ is fraction of solid materials at the cell-edge , $(1-\phi)$ is fraction of solid material at cell-wall, ρ/ρ_s is relative density of the foam and E_s is young's modulus of the cell wall material. For a polymeric foam $\phi = 0.5 - 0.6$ gives close relation [16]. Relative density of the foam is given by Eqn. (12). Where C_3 is a constant and depends on the number of edges, faces, etc. Typically $C_3 \approx 1$ is taken for the calculations.

Proceedings of the ASME 2019
International Design Engineering Technical Conferences
and Computers and Information in Engineering Conference
IDETC/CIE2019
August 18-21, 2019, Anaheim, CA, USA

DETC2019-98028

QUANTUM MECHANICAL PERSPECTIVES IN RELIABILITY ENGINEERING AND SYSTEM DESIGN

Vijitashwa Pandey
pandey2@oakland.edu
Industrial & Systems Engineering Department
Oakland University
Rochester, MI 48309

ABSTRACT

Engineering design under uncertainty is an established field. Attempts to extricate the human decision maker from the process generally do not succeed. Surprisingly, even the determination of system parameters and their admissible values needs as many interventional steps from human designers and operators, as the selection of final attributes of the system that the human end user is expected to only interact and be concerned with. In this light, it becomes important to consider the mathematical models that would explain and model the decision making behavior of human beings. Concerningly, this behavior has been seen to violate common sense probability axioms. In this paper, we propose an earnest look at the mathematics of quantum mechanical theory in modeling and manipulating the uncertainties involved in engineering systems. We propose that the state of a system be modeled as a point in an abstract complex vector space as in quantum mechanics. Additionally, at a given point in time it can be interpreted as a superposition of multiple pure states. This change in perspective allows explanation of many commonly observed behaviors, least of which is the inconsistencies in defining what constitutes the failure of a system. We present our approach in the context of reliability engineering as it sees some of the most prevalent use of uncertainty modeling and propagation techniques. However, the implications on design and design theory are also evident. Some motivating examples are provided and directions for future work are identified.

1. INTRODUCTION

A human decision maker is essentially inextricable from the practice of design, manufacture and use of engineering systems. The practice of engineering is subject to uncertainty at almost every step. While there are aspects that lend themselves very well to classical probability based methods, e.g. probability distribution of dimensions produced by a given machine, one routinely encounters problems where the judgment and opinions of human decision makers infiltrate probability assessments, and as a consequence, become subject to the myriad biases, opinions and, quite often, the wisdom of the human decision maker. In such a scenario, tools need to be explored that come to the rescue, either in the form of techniques that attempt to model the decision maker's opinions or reconciliatory methods that accommodate them. In this paper, a quantum probability based approach is proposed for probability calculations regarding engineering system states, and likelihoods of events defined around them.

The purported inadequacy of classical probability methods in design decision making under uncertainty has not evaded researchers. This has not only been limited to the observation of these effects but also reconciliation methods. Fuzzy sets were proposed by Zadeh (1965) wherein one defines degrees of membership for the elements of the set, in part as a way to model relevant uncertainties. More formally, Bayesian analysis has seen extensive application where an expert opinion in terms of a prior probability distribution is updated based on the principles of probability theory. It is well known that the Bayesian method is only as good, or as bad, as the prior used. Dempster-Shaefer theory, sometimes also referred to as evidence theory, makes use of imprecise probabilities to model the lack of knowledge regarding uncertainties (Dempster, 1967, Shaefer, 1976). Imprecise probabilities have also been shown to be a robust method for modeling uncertainty under extremely limited data (Pandey and Nikolaidis, 2008).

On the decision making end of the spectrum, aside from normative decision analysis, many decision making theories also exist, spanning the normative-descriptive continuum. Minimax regret, or the approach to minimize the least-regret from a decision situation, was proposed by Savage (1951). Analytical Hierarchical Processing (AHP) is another method which exploits the ease with which human decision makers make pairwise comparisons (Saaty, 2004). Pugh controlled convergence method is used extensively in engineering design of complex system (Frey, et al., 2009). While all the aforementioned methods have been criticized in some fashion or the other (arbitrary weighting, rank-reversal), the strength of these methods is that they explicitly accommodate and mitigate the difficulties of involving human decision makers in the complicated process that is engineering design.

In light of the above, wherein decision making methods are proposed to *incorporate* a human decision makers' proclivity to deviate from what classical theories mandate, it is important to note that one runs squarely into traps of human limitations in assessing and interpreting *probabilities* of events. Allais and Ellsberg paradoxes are examples of when decision maker deviate from rational choices (Levi, 1986). Walley (1991), Winkler (1966) and de Finetti (1972) experimentally investigated how people assess probabilities of events. Of course, Tversky and Kahneman (1974) explained many errors and biases that creep into probability assessments. And last but not the least, the probability assessments and decision making inconsistencies in groups of decision makers is a well-known phenomenon (Tindale, 1993).

A similar, but arguably more profound, conundrum was confronting physicists in the early 20th century when the peculiarities of how fundamental particles behaved in experiments was noted. Many physicists and mathematicians including Paul Dirac, Neils Bohr and John von Neumann contributed to the formalisms and the formulisms that allowed the description of this behavior (von Neumann, 1932). At the heart of all of this was that a quantum particle's state is best represented as a point in an abstract complex linear vector space. The rules of quantum probability calculations were devised that were able to predict the events that had been observed, to a remarkable degree of accuracy. The ability of quantum physics to predict the world around us is unparalleled.

The use of quantum probability to model human probability assessments, especially in cases where respondents deviate from formal methods, has already seen significant attention in the psychology literature. For example, Busemeyer et al (2006) and Aerts (2009) describe the quantum approach to modeling and understanding human cognition. Further work is also available in explaining the fallacies encountered in considering unions and intersections of events, and also order effects (Trueblood et al., 2011, Busemeyer et al., 2011). However, the applicability of the concepts in engineering design decision making, and within that, the precepts of design under uncertainty has eluded attention. In this paper, we present a perspective, at least for design and reliability analysis that may gain from the concepts of quantum probability. Such an analysis can help explain how system

failures are perceived by decision makers and may even help pose design under uncertainty problems differently.

As mentioned, the application area chosen for this paper is engineering design under uncertainty, which aims at optimizing the expectations of metrics of interest or estimating and subsequently minimizing the failure probability of systems, or both. The research in this field is therefore aimed at finding efficient methods to calculate these probabilities and/or expectations (Mourelatos et al., 2015). Classically, reliability of a component or a system at time t is defined as the probability that it will still be performing its *intended* function at that time under *stated* operating conditions (Kapur and Lamberson, 1977). Clearly, the keywords here are *intended* and *stated*. We hope to show in this paper that the capturing of this intent using inputs from various decision makers is foundational to design under uncertainty, and a quantum approach may be useful in accomplishing that.

As a final note for this section, there also remains the notion of descriptiveness versus normativeness. A method is only as good as the underlying axioms and since QM at the very least modifies the axioms of classical probability theory, we expect this comparison to come to the fore. However, it is clear that design and operation of systems is expressly a human activity. As a consequence, accepting an alternative approach to modeling and predicting how a design decision maker or the end user will act, is only going to enhance our ability to design systems that fulfill their intended function(s).

2. MOTIVATING CONSIDERATIONS
Our motivating example comes from a commonly cited quantum mechanical experiment, the Stern-Gerlach experiment (Gerlach, Stern, 1922). This is followed by a description of some realistic engineering examples later in the section.

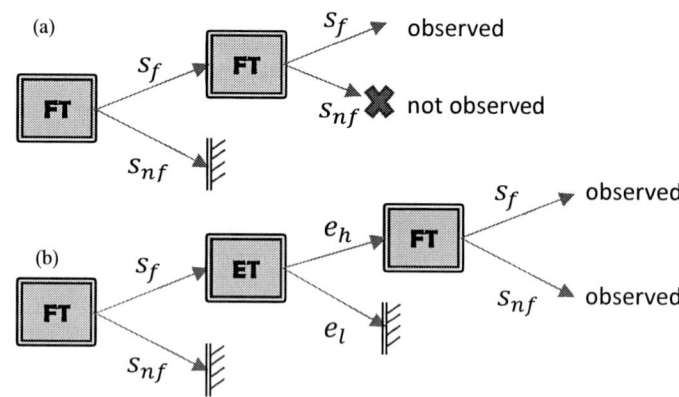

Figure 1: A systems analogue of the Stern-Gerlach experiment.

The Stern-Gerlach Experiment
The Stern-Gerlach experiment is an extremely powerful experiment and exemplar of the peculiarities encountered in the

quantum world. We will try to explain it in an engineering context. Figure 1 details the steps of a hypothetical experiment. Consider a system that can be in two operating states – functional (s_f) and non-functional (s_{nf}) and two efficiency levels, high efficiency (e_h) and low efficiency (e_l). When measured using a test (FT) that determines whether the system is functional or not, the system can be found to be either functional or non-functional. This test is repeatable, in that a functional system is found to be functional if the tests are run in sequence, as many times as needed (in a relatively short period of time and no other tests are performed). Figure 1(a) describes this experimental setup.

If only functional systems are now allowed to proceed for the efficiency analysis test (ET), one sees that they can be divided into high efficiency and low efficiency systems, as shown in figure 1(b). If one were to filter these again based on efficiency one expects to be left with functional systems operating at high efficiency ($s_f + e_h$). Intuitively, running the systems through another FT appears to be moot – we proceed anyway. While we should expect the result to only consist of functional systems, in the quantum world this is not always true! Measurement of the system efficiency levels could destroy any prior information about whether the system is functional or not. What is essentially seen is that as a consequence of this second measurement, the systems revert to a mix of functional and non-functional systems. Intuitively, we do not expect this to happen in the macro world of engineering systems, in this paper, we are positing that if we invoke human judgment and decision making on the determination of system states and efficiencies, such results may very well be encountered. While we may not be able to prove, nor do we intend to, that systems appear to follow quantum rules, decision making regarding them where much of the information is derived from human judgment may well gain an important tool if such a leap were indeed made.

Stern and Gerlach showed the above result with the spin states of silver ions, but the results are essentially the same if FT and ET measure what are called *incompatible observables*. Now, clearly, there are two observations we can make about the similarities and dissimilarities between the proposed consideration of quantum effects in system state determinations and in the original Stern-Gerlach experiment:

1. Before we draw an analogy, we should state clearly that macroscopic systems may not act as in figure 1 on their own. This is expected to essentially be an artifact of a human decision maker's perception and definition of functionality and efficiency. 2. We may consider that in a DM's mind, the system is in a superposition of the pure states of functioning and non-functioning (as well as high and low efficiency). This is further discussed in the later paragraphs with automobile examples.

Despite these apparent challenges, we need not be trepidatious about resorting to quantum mechanical type analysis of reliability of systems. We actually advocate it in this paper, and do so by describing the prevalence of the DM derived probability

assessments and definitions of operating conditions. Departing from simplistic representations of systems and considering, for a moment, a realistic system involving thousands of components, one encounters the need for understanding and incorporating human or human-type judgement. Two examples from the author's experience come to mind.

A large manufacturer of automobiles notices that a small but worrying fraction of the automobiles come back for warranty repairs involving a component (component C). The details of the problem have been intentionally obfuscated to protect the identity of the manufacturer. Since the problems with the automobiles caused by a failed component C are not of a debilitating kind, much of the problems are explained by the customer in a subjective fashion to the dealer who is entrusted with repairing the automobile. These comments are passed on to the manufacturer that accesses the repair logs. Given the design parameters, it may appear possible to predict what is causing the customers to bring the automobile for repairs. Unfortunately, and as expected, statistical analyses of the design parameters that were considered to be contributing to the incidence of warranty did not reveal any meaningful patterns. It was also conjectured that there are other cues unrelated to the component C in the product that the customers were using to determine failure – customers knew more! And finally, as a peculiarity, it was observed that if any of the entertainment systems (e.g. stereo, navigation) in the vehicle stopped functioning, the customer would bring the vehicle back and then request that all the problems be fixed, including component C! There appear to be many cognitive aspects at play. What do the customers consider to be the failure of a system? Are the customers looking at the same parameters to determine failure as the manufacturer? Are failure states of individual components affecting the failure state of the entire automobile in a non-intuitive (may be even irrational) fashion? Is there a time when the customer considers the system to be in a superposition of failed and working states?

Our second example, also in the automobile realm, comes from our personal experience of owning and operating an automobile. There are many partial failures that would contribute to the failure of the automobile system. Clearly, there is a difference between failure of the automobile, the *perception* of failure of an automobile and the *perception* of the reliability of the entire automotive brand that the said automobile belongs to. Take, for example, the battery in the vehicle. From a purely technical standpoint a dead battery makes a vehicle unable to run and should be considered a failure of the system. However, very few people would consider it the failure of the system, let alone hold it against an entire brand, since batteries are expected to have a lifetime lower than most automobiles and are relatively easy to replace. Therefore, while the battery is dead, is the car considered to have failed? There appears to be no clear answer.

In both the above cases– one might be tempted to consider the state of the automobile to be in *a superposition of both failed and working pure states*. When a measurement is made, it collapses into the pure state of either customer deciding the automobile has failed and takes it to the repair shop, or continues to use it considering no failure. The next two paragraphs discuss how this type of decision making permeates system lifecycle and design under uncertainty studies must consider them.

System Lifecycle

A system lifecycle can be decomposed in many different ways. A generic decomposition is presented in figure 2 and consists of four phases. Design of the system, calculation or measurement of system performance characteristics, definition of acceptable levels of the performance characteristics and its use. Customer needs and designer preferences determine what design will be produced. Design engineers and managers determine what operating parameters of the system are, and finally both the designer and customer determine what failure is. Given our earlier discussion, it is evident that multiple decision makers' opinions determine a significant measure of the operating characteristics of a system. It is also entirely plausible that because of the confluence of so many viewpoints, a system is always in a superposition of many pure states and an actual decision made depends on first collapsing the system state into a pure state (eigenstate).

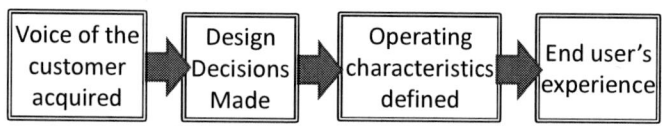

Figure 2: A four stage representation of system lifecycle, depicting multiple opportunities for DM opinions to influence system characteristics.

Design under Uncertainty

Engineering design under uncertainty implies a set of related fields of study (or any combination of these) available in the literature (Nikolaidis et al., 2011). In reliability engineering one is concerned with binary events of success or failure of components and systems. A component's reliability at time t is the probability that it has not failed before t. Mathematically, if t_f is the time when such a component fails, $R(t) = P(t_f \in [0, t])$. If one were to explicitly study the physical processes that lead to said failure, for example, stress exceeding the strength of an element, one is said to consider time dependent reliability of system (Singh, et al., 2009). Component failures propagate to system failures depending on the architecture of the system, usually represented using a topology or a fault tree (Kapur and Lamberson, 1977). One may also generalize reliability analysis into maximization of an objective, subject to minimum reliability requirements as in Reliability Based Design Optimization (Liang et al., 2008). Finally and more generally, one may perform optimization of expectations of suitable metrics of interest. In this paper, the probability of failure of a system embodies both the failures of components (as would be considered in time dependent reliability literature) as well as the architecture of the system – both as determined by designers and

end users. In the next section, we present the formalism of quantum theory and contrast the probability calculations with classical probability calculations. This is followed by an application example followed by the conclusion and discussion section.

3. FORMALISM

Quantum mechanical states are represented by vectors in an abstract complex vector space. When a measurement is performed, a system is forced into an eigenstate of the observable. This is accomplished by an operator (such as the momentum operator) acting on a particle. Here we describe classical probability and contrast it with the quantum approach to probability calculations.

Classical probability is founded on Kolmogorov's axioms which, for the sake of continuity, are presented here. If a measure space (Ω, F, P) is such that $P(\Omega) = 1$ then it can be called a probability space. In such a setup, Ω is the sample space, F is the event space and P is the probability measure. The first Kolmogorov axiom states that the probability of an event is a non-negative real number between 0 and 1. Therefore for all events $E_i \in F$:

$$P(E_i) \in \mathbb{R}, P(E_i) \in [0,1] \qquad \forall E_i \in F$$

The second axiom is that the measure of the entire sample space is 1. This is another way of saying that at least one of the events in the sample space will occur, i.e.:

$$P(\Omega)=1$$

The third axiom is that the probability of the union of mutually exclusive events is equal to the sum of the probabilities of the events. Alternatively, if $E_1, E_2, \dots$ are disjoint subsets of Ω, then:

$$P\left(\bigcup_{i=1}^{\infty} E_i\right) = \sum_{i=1}^{\infty} P(E_i)$$

The far-reaching consequences of the above axioms cannot be overstated. For example, from the above, one can deduce the probability of the null event, intersection and union of events. Most of the results, the readers are undoubtedly already aware of and we would omit them in the interest of brevity.

In quantum mechanics, the state of a physical entity is represented by a ket. We will use the famous Dirac bra-ket notation in this paper. For example, in the earlier section the state of the system can be written as $|s\rangle$. The ket is point in an abstract complex vector space which is distinct from the Euclidean space we are possibly more familiar with. The dimensionality of the space is equal to the number of possibilities after each measurement. Every ket has a corresponding dual, called bra, written as:

$$\langle s| \Leftrightarrow |s\rangle$$

Two kets can be multiplied, both in the form of an inner product and an outer product.

$$\text{Inner product: } \langle s|a\rangle$$
$$\text{Outer product: } |s\rangle\langle a|$$

Notice that unlike the dot product of Euclidean vectors, the inner product is generally not commutative. In fact:

$$\langle s|a\rangle = \langle a|s\rangle^*$$

Where the asterisk signifies the complex conjugate, recall that the kets are points in a *complex* vector space. An observable is represented by an operator that acts on a ket. In the earlier example, if FT is an observable, it will act on the state of the system to generate another ket:

$$FT|s\rangle = |\alpha\rangle$$

Compatible and incompatible observables
In quantum mechanics, an observable is compatible with another if their operators commute. The commutator of two operators, say FT and ET is written as [FT, ET] and is equal to:

$$[FT, ET] = FT \cdot ET - ET \cdot FT$$

If the commutator vanishes then the operators commute otherwise they do not. It must be apparent now to the reader that to be able to get a results as in figure 1, the operators do not commute.

Eigenkets
A ket is considered an eigenket of an operator if the following relationship holds:

$$FT|a_i\rangle = c_i|a_i\rangle$$

Where c_i is simply a scalar and is the eigenvalue associated with eigenket $|a_i\rangle$. A ket can be written as a linear combination of the eigenkets of an operator as:

$$|s\rangle = \sum s_i|a_i\rangle$$

We are now ready to state two postulates of quantum mechanics which are relevant to our discussion:
1. A measurement process throws a system into one of its eigenstates.
2. While it is not possible to predict which eigenstate a system will be thrown into, the probability that the i-th eigenstate will be realized is:

$$P(a_i) = |\langle a_i|s\rangle|^2$$

Clearly, the above two postulates go long way in explaining the seemingly counterintuitive results of quantum experiments. For example, since a measurement throws a system into its eigenstate, with the probability given above, the entire probability of complementary events now is contributed by the subspace spanned by remaining eigenkets. As a result, repeated measurements with the same operator will not change the state of the system, we will get the same eigenket. However, application of another operator will bring the system to its own eigenstate. It is possible to recover the eigenkets of the first operator that were not realized if the first operator was now used again – this explains the results from figure 1. The reader is referred to Sakurai (1985) for a detailed exposition of these concepts and quantum mechanics in general.

Armed with the above results we can define the system's states from our example in figure 1 as points in a complex vector space, whose dimensionality equals the number of outcomes of the experiment. Given an operator, such as *FT*, we can then find the probabilities that the system will be *considered* to be functioning or the *assessed* efficiency level of the system.

Uncertainty Principle Applied to Engineering Systems
Before we show an example, we digress here to discuss the possibility of an uncertainty relationship in systems. Quite expressly, the formalism afforded to us through the use of quantum mechanical postulates demands that we do not end the dalliance without looking at systems and their evolution in time. Readers are undoubtedly familiar with the uncertainty principle in quantum mechanics which relates to the limitations in simultaneously measuring certain quantities. From our earlier discussion these are observables whose operators do not commute. Readers familiar with Fourier transforms and wavelet transforms will know that the uncertainty principle is ubiquitous. This entails that simultaneous precise localization in time and frequency is not possible (Steeb, 2013).

Systems evolve with time, which from an engineering modeling perspective, is the evolution of a metric of interest (*m*). Such modeling and consequent analysis has been studied from many perspectives including the time dependent reliability of systems as well as design evolution and flexibility. As an example, in Pandey and Mourelatos (2015) the stress endured by a hydrokinetic turbine blade is compared in time with the strength of the blade. Clearly the type of the system determines its inertia or the resistance to change, and consequently momentum (*p*). The quantum mechanical analogue of the variation in system metrics with time would be a moving wave packet. And, as such, an uncertainty relation can be derived.

$$\Delta m \Delta p \geq c$$

Where c is a constant that can be derived quantitatively or numerically for a given system. A system may lend itself to a simultaneous understanding of its current state and the rate of change in its state or it may not – depending on the properties of the operators chosen. We intend to address extensions of this and related ideas elsewhere.

4. APPLICATION EXAMPLE

Here we show an example of how quantum theory can be used to model perceptions of system states and calculate probabilities of events. Consider a multicomponent system that can have none, some or all the components functioning at a given time. Figure 3 shows a topology of a simple four component system. When all the components are functioning, the system is considered to be fully functional. If some of the components are working (e.g. components 1, 2 and 4), the system is considered partially functional and finally, if a critical combination of components (e.g. component 4, or components 1 and 3) has failed, the system is considered non-functional. Therefore the system can be in three possible pure states of fully functional (FF), partially functional (PF) and non-functional (NF), or any superposition of them. Notice that there is usually limited consensus between decision makers on system topologies something we intend to address.

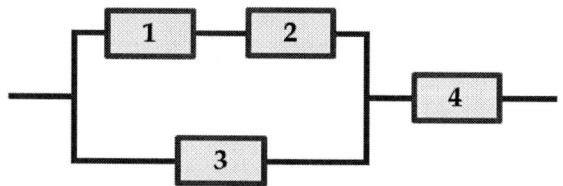

Figure 3: An example four-component system.

Similarly, there are three efficiency levels possible for the system which are, high efficiency (HE), mid-efficiency (ME) and low-efficiency (LE). Readers will notice that this is similar to the example presented in the first section of the paper. Here we provide a worked example of the probability assessments possible through the quantum model.

We demonstrate the forward problem where, given the system state and the eigenkets of the observables under consideration, how one can calculate the relevant probabilities that may account for apparent DM inconsistencies in probability assessments. In practice, the system operators and the corresponding the eigenkets can be fitted using the parameters of the system by back propagating through the equations presented in this section. We will leave the steps of determining these, and the relevant regression steps as future work. Here we show how counterintuitive, and conflicting preferences can be modeled using the mathematics of quantum theory. It is also worth noting, that we are not limited to a few dimensions as our short example demonstrates. The following calculations can be performed for a large system with many states.

As discussed earlier, we look at the basis kets of the vector space that that system state belongs to. Consider three kets, $|a_1\rangle$, $|a_2\rangle$ and $|a_3\rangle$ that correspond to the functionality observable FT. Let

us also assume that their Euclildean representations are as follows:

$$|a_1\rangle = \begin{bmatrix} 1 \\ 0 \\ 0 \end{bmatrix};$$

$$|a_2\rangle = \begin{bmatrix} 0 \\ 1 \\ 0 \end{bmatrix}; \text{ and,}$$

$$|a_3\rangle = \begin{bmatrix} 0 \\ 0 \\ 1 \end{bmatrix}.$$

It is clear that these are orthonormal and of unit length. As mentioned, the kets above are the eigenkets of the FT operator and the concordance between the above kets and system operating status is that:

$$|a_1\rangle \equiv \text{FF}; \ |a_2\rangle \equiv \text{PF}; \text{ and } |a_3\rangle \equiv \text{NF}$$

The above eigenkets form the simplest orthonormal set and there is nothing particularly unique about them otherwise, for the purposes of this discussion. If a system starts out with a state ket given by $|s\rangle = |a_1\rangle$, the FT operator will not modify it and the system will continue to be fully functional until modified in some other way. Similar results hold for the other eigenkets shown.

Similarly to above, we create another set of orthonormal basis kets that correspond to the efficiency observable ET, as:

$$|b_1\rangle = \begin{bmatrix} 0.5773 \\ -0.5773 \\ 0.5773 \end{bmatrix};$$

$$|b_2\rangle = \begin{bmatrix} 0.4082 \\ 0.8165 \\ 0.4082 \end{bmatrix};$$

$$|b_3\rangle = \begin{bmatrix} -0.7071 \\ 0 \\ 0.7071 \end{bmatrix};$$

It is easy to verify that the above kets are unit length and are mutually orthogonal. Again, the above kets have been chosen for demonstration purposes, and have been generated using the Gram-Schmidt algorithm. Starting from an arbitrary set of vectors or kets in any number of dimensions, one can create an orthonormal set by first using the Gram-Schmidt method, and then normalizing the kets. In this basis, the concordance between the kets and the system efficiency levels are:

$$|b_1\rangle \equiv \text{HE}; \ |b_2\rangle \equiv \text{ME}; \text{ and } |b_3\rangle \equiv \text{LE}$$

As before, the eigenkets can be derived and the corresponding observable determined by back propagating from the probabilities described below.

From a given state of the system given by the state ket $|s\rangle$, one can compute the probability of measuring an eigenket (let us say $|a_1\rangle$) as follows.

$$P(|s\rangle \rightarrow |a_1\rangle) = |\langle a_1|s\rangle|^2$$

Where $\langle a_1|s\rangle$ is the inner product in the Dirac bra-ket notation. For Euclidean vectors, this is the inner product given by the well-known sum $\sum_{i=1}^{n} a_{1,i}^* \cdot s_i$. Where the subscript i, denotes the i-th element of the ket and the superscript '*' identifies the complex conjugate. Since the eigenkets chosen do not have complex entries, a simple inner product suffices.

Let us start with the initial state ket given below:

$$|s_0\rangle = \begin{bmatrix} 1 \\ 0 \\ 0 \end{bmatrix}$$

Clearly, the above ket indicates that the system is fully functional. This is similar to the assumption in reliability studies that the reliability functions start with a value of 1 at time, $t = 0$. This can be verified by finding the probability that when the operator FT is used, the state collapses to, or equivalently remains at, the eigenket $|a_1\rangle$:

$$P(|s_0\rangle \rightarrow |a_1\rangle) = |\langle a_1|s_0\rangle|^2 = 1$$

However, if this system is then tested for efficiency we get:

$$P(|s_0\rangle \rightarrow |b_1\rangle) = |\langle b_1|s_0\rangle|^2 = 0.333$$
$$P(|s_0\rangle \rightarrow |b_2\rangle) = |\langle b_2|s_0\rangle|^2 = 0.167; \text{ and}$$
$$P(|s_0\rangle \rightarrow |b_3\rangle) = |\langle b_3|s_0\rangle|^2 = 0.5$$

Which together imply that the system is likely to be operating in high efficiency or a low efficiency mode, and unlikely to be in a mid-efficiency mode. If the system now is allowed to evolve (time evolution or application of other operators) the system let us say, ends up in a state given by the following vector. The vector has been intentionally chosen so that its projection on $|b_1\rangle$ is 0, implying that the system is assumed to have a 0 probability of being in the high-efficiency mode.

$$|s\rangle = \begin{bmatrix} -0.2113 \\ 0.5773 \\ 0.7887 \end{bmatrix}$$

Notice that also from this state it is apparent that the belief that the system is fully functional is low (first entry; sign is irrelevant), while the belief that the system is non-functional is high. This is verified by the magnitude of corresponding entries. We can find exact values of these probabilities as:

$$P(|s\rangle \rightarrow |a_1\rangle) = |\langle a_1|s\rangle|^2 = 0.045$$
$$P(|s\rangle \rightarrow |a_2\rangle) = |\langle a_2|s\rangle|^2 = 0.333; \text{ and}$$
$$P(|s\rangle \rightarrow |a_3\rangle) = |\langle a_3|s\rangle|^2 = 0.622$$

As expected these add up to 1 since when it is tested whether the system is functional or not, one of the above eigenkets has to be realized. Similarly, when we see that

$$P(|s\rangle \rightarrow |b_1\rangle) = |\langle b_1|s\rangle|^2 = 0$$
$$P(|s\rangle \rightarrow |b_2\rangle) = |\langle b_2|s\rangle|^2 = 0.5; \text{ and}$$
$$P(|s\rangle \rightarrow |b_3\rangle) = |\langle b_3|s\rangle|^2 = 0.5$$

This can be interpreted as: the current state of the system indicates that the system is not working in the high efficiency state. If this were an actual system, the implication could also be that since the initial belief indicated that the system is unlikely to be in fully working condition, its probability of being in the high efficiency mode is zero. We now calculate the probability of determining whether the system is operating at high efficiency *after* we test whether the system is working or not and noticing that it is fully functional. We calculate the probability from $|a_1\rangle \rightarrow |b_1\rangle$ and multiply with the probability we calculated earlier, $P(|s\rangle \rightarrow |a_1\rangle)$. The value we get is 0.0148.

If we consider total probability theorem at this point, we see that the probability of encountering a high efficiency system is given by:

$$P(HE) = P(HE|FF)P(FF) + P(HE|PF)P(PF) + P(HE|NF)P(NF)$$

Now since, all the terms in the above equation are positive, we expect, for example that $P(HE) \geq P(HE|FF)P(FF)$. Note that earlier we calculated values of 0 and 0.0148 respectively for the LHS and RHS of the inequality, which clearly is *incompatible*. However, it serves to model any inconsistencies decision makers may exhibit when the system characteristics (operators) are deduced in the reverse analysis i.e. if the system parameters were fit from user responses of assessed probabilities of these events.

An immediate application of the results of the type presented here is in the determination of system topology. For example, a decision maker initially may assume that the system cannot be performing in the high efficiency mode given the state of the system. However, as described earlier, this is found to be inconsistent with if the system is first checked for functionality. Notice, that this reassessment of the system state was one of the motivating examples presented earlier in the paper. One can then use an algorithm similar to that proposed in Pandey and Mourelatos (2012) to derive a system topology. In the paper, a truth table for a system is used, which essentially is a collection of data points of whether, given the states of the components, whether the entire system is functional or not. An evolutionary algorithm is also provided that can iteratively find the topology which can help with further reliability analysis. If a quantum model is used to describe a system's behavior one can find system topologies most consistent with the various stakeholder's opinions. While a topology will not be able to replicate all the quantum effects, it can identify the best fit relationships.

Alternatively, it can model what happens to the topology itself when certain measurements are made.

If we assume, for simplicity, that components 1, 2 and 3 have the same reliability, r, at a given time, while component 4 has a reliability of r_4, we see that a value of $r = 0.382$ and $r_4 = 0.801$ satisfies the probability values calculated for the system functionality state.

$$P(FF) = r_4 r^3 = 0.045$$
$$P(PF) = r_4(3(1-r)r^2 + r(1-r)^2) = 0.333$$
$$P(NF) = 1 - P(FF) - P(PF) = 0.622$$

Notice that $P(NF)$ can also be calculated by exhaustively listing the mutually exclusive events of component failures and adding the associated probabilities. Now, we consider the case where the system efficiency is tested before the system functionality state is determined. Table 1 lists the observations. Clearly, since the probability of the system in a high efficiency state is initially 0, it is not possible for the system to be found in a high efficiency state *and followed by* it being in any of the functionality states. However, it is seen that if the system is first found to be working in a medium or low efficiency mode, the functionality state probabilities change significantly.

Table 1: Combined state probabilities with efficiency test followed by functionality test.

Functionality state	Before ET	Combined probability after ET and found to be:		
		HE	ME	LE
FF	0.045	0	0.0833	0.25
PF	0.333	0	0.333	0
NF	0.622	0	0.0833	0.25

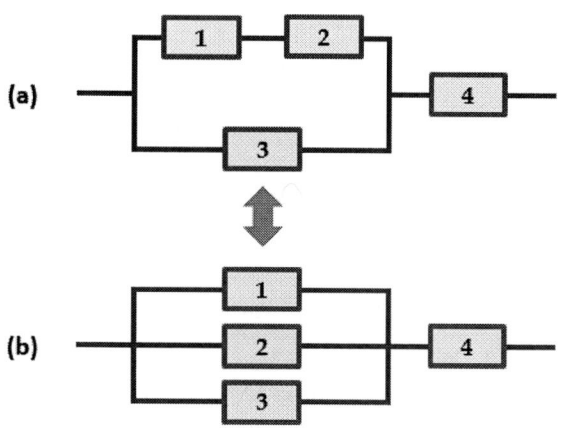

Figure 4: Possible topologies based on DM beliefs.

The component probability values shown earlier cannot always be reconciled with system state probabilities calculated after the efficiency measurement if we insist on the same topology as

figure 3. To do so, one either has to update the reliability calculations or consider a different topology of the system. For example, for the system above, the topology can be modified as shown in the figure 4 below. Notice that the two systems have very similar truth tables, however system (b) is less likely to fail. To draw a parallel with the system discussed above, we see that it is likely that system shown in figure (b) is able to operate at a low efficiency rate even if components 1 and 3 fail. This is not possible with the system in (a).

5. CONCLUSIONS

Human decision makers and experts are such an integral part of engineering practice that no individual step within it, from design to end of lifecycle, is unaffected by it. This really comes to the fore in fields where uncertainty plays a critical role, such as reliability engineering, where one decides what constitutes failure of a system, affecting very deeply the reliability assessments and prediction. Many times, because of lack of knowledge, perspective or even biases, the probability assessments made by humans seem to violate classical probability laws. Quantum probability theory provides a way to reconcile with some of these peculiarities in probability assessments, thereby providing an approach that can be useful in design under uncertainty. In this paper, a case was made for considering systems states, as specified by the designer or perceived by the customer as points in a complex vector space as in quantum theory. This has the potential to allow better modeling of human decision makers' idea of the state of a system, what constitutes failure or success of a system, and how design can proceed in this scenario. An example was provided to show how this approach can be useful in practice.

Future research can extend the proposed work to identify system parameters from the probability assessments given by designers. We will also couple the approach shown with computationally efficient estimation of failure probabilities of systems involving many components. Time dependent reliability of systems can also be investigated using evolution of states, based on results in quantum theory. Finally, implications on both design theory and optimal design can also be investigated.

Acknowledgments
The author would like to thank the anonymous reviewers who helped improve this work.

REFERENCES

1. Aerts, D., 2009. Quantum structure in cognition. Journal of Mathematical Psychology, 53(5), pp.314-348.
2. Busemeyer, J.R., Pothos, E.M., Franco, R. and Trueblood, J.S., 2011. A quantum theoretical explanation for probability judgment errors. *Psychological review, 118*(2), p.193.

3. Busemeyer, J.R., Wang, Z. and Townsend, J.T., 2006. Quantum dynamics of human decision-making. Journal of Mathematical Psychology, 50(3), pp.220-241.

4. De Finetti, B., 1972. Probability, induction, and statistics, Chapter 1 and 2. Wiley London.

5. Dempster, A. P., 1967, "Upper and lower probabilities induced by a multivalued mapping". *The Annals of Mathematical Statistics*. **38** (2): 325–339

6. Frey, D.D., Herder, P.M., Wijnia, Y., Subrahmanian, E., Katsikopoulos, K. and Clausing, D.P., 2009. The Pugh Controlled Convergence method: model-based evaluation and implications for design theory. Research in Engineering Design, 20(1), pp.41-58.

7. Gerlach, W.; Stern, O. (1922). "Der experimentelle Nachweis der Richtungsquantelung im Magnetfeld". Zeitschrift für Physik. 9: 349–352.

8. Kapur, K.C. and Lamberson, L. R., 1977, *Reliability in Engineering Design*, 1st Edition, John Wiley and Sons.

9. L. A. Zadeh (1965) "Fuzzy sets". Information and Control 8 (3) 338–353

10. Levi, I., 1986. The paradoxes of Allais and Ellsberg. Economics & Philosophy, 2(1), pp.23-53.

11. Liang, J., Mourelatos, Z.P. and Tu, J., 2008. A single-loop method for reliability-based design optimisation. International Journal of Product Development, 5(1-2), pp.76-92.

12. Mourelatos, Z.P., Majcher, M., Pandey, V. and Baseski, I., 2015. Time-dependent reliability analysis using the total probability theorem. Journal of Mechanical Design, 137(3), p.031405.

13. Nikolaidis, E., Mourelatos, Z.P. and Pandey, V., 2011. *Design Decisions Under Uncertainty With Limited Information: Structures and Infrastructures Book Series, Vol. 7*. CRC Press.

14. Pandey, V. and Mourelatos, Z.P., 2012, August. Evolutionary System Topology Identification and its Application in Product Reuse. In ASME 2012 International Design Engineering Technical Conferences and Computers and Information in Engineering Conference (pp. 687-701).

15. Pandey, V. and Nikolaidis, E., 2008. Using inexpensive experimental methodology for testing methods for decision under uncertainty. Structure and Infrastructure Engineering, 4(1), pp.1-18.

16. Pandey, V., Mourelatos, Z.P. and Skowronska, A., 2015, August. Flexible Design of Systems Considering Time-Dependent Reliability. In ASME 2015 International Design Engineering Technical Conferences and Computers and Information in Engineering Conference, Boston MA.

17. Saaty, T.L., 2004. Decision making—the analytic hierarchy and network processes (AHP/ANP). Journal of systems science and systems engineering, 13(1), pp.1-35.

18. Sakurai, J.J., 1985. Modern quantum mechanics, The Benjamin/Cummins Publishing Company, Menlo Park, California, 474 pp.

19. Savage, L.J., 1951, "The theory of statistical decision." Journal of the American Statistical Association, vol. 46, pp. 55–67.

20. Shafer, Glenn, 1976, *A Mathematical Theory of Evidence*, Princeton University Press.

21. Singh, A., Mourelatos, Z.P. and Li, J., 2009, January. Design for lifecycle cost using time-dependent reliability. In ASME 2009 International Design Engineering Technical Conferences and Computers and Information in Engineering Conference (pp. 1105-1119). American Society of Mechanical Engineers.

22. Steeb, W.H., 2013. Hilbert spaces, wavelets, generalised functions and modern quantum mechanics (Vol. 451). Springer Science & Business Media.

23. Tindale, R.S., 1993. Decision errors made by individuals and groups. Individual and group decision making: Current issues, pp.109-124.

24. Trueblood, J.S. and Busemeyer, J.R., 2011. A quantum probability account of order effects in inference. *Cognitive science*, *35*(8), pp.1518-1552.

25. Tversky, A. and Kahneman, D., 1974. Judgment under uncertainty: Heuristics and biases. Science, 185(4157), pp.1124-1131.

26. Von Neumann, J., 1932. Mathematical Foundations of Quantum Mechanics. Verlag. J.

27. Walley, P., 1991. *Statistical Reasoning with Imprecise Probabilities*. Chapman and Hall. pp 632-638.

28. Winkler, R.L., 1971. Probabilistic prediction: Some experimental results. Journal of the American Statistical Association, 66(336), pp.675-685.

Proceedings of the ASME 2019
International Design Engineering Technical Conferences
and Computers and Information in Engineering Conference
IDETC/CIE2019
August 18-21, 2019, Anaheim, CA, USA

DETC2019-98035

A PROPOSAL FOR A DECISION SUPPORT FRAMEWORK TO SOLVE DESIGN PROBLEMS IN THE AUTOMOTIVE INDUSTRY

Timothé M. Sissoko[1]
Laboratoire Génie Industriel, CentraleSupélec,
Université Paris-Saclay
Gif-sur-Yvette, France
Renault SAS Technocentre
Guyancourt, France

Marija Jankovic
Laboratoire Génie Industriel, CentraleSupélec,
Université Paris-Saclay
Gif-sur-Yvette, France

Christiaan J. J. Paredis
Department of Automotive Engineering
Clemson University
Greenville, SC, USA

Eric Landel
Renault SAS Technocentre
Guyancourt, France

ABSTRACT

Decision-makers often rely on heuristics and experience to make complex decisions in the industrial context. Often, integrating implicit or expert knowledge as well as uncertainties can lead to decisions that are not necessarily the best ones. Moreover, in engineering design, the decision-making approaches focus on the product itself and do not investigate the necessary effort that is needed to gather additional data in order to devise more precise decision-making models. In our research, we propose to integrate this estimation of additional effort needed for data gathering and decision-making refinement in order to support design teams. This research has been conducted in collaboration with a major car manufacturing company, and in particular in the development process through Modeling and Simulation. The objective is to propose a decision-making model that integrates data-gathering estimation, hence integrating also the estimation of postponing one decision. A decision problem model based upon expected utility combined with the value of information theory is proposed to address this issue. The model has been developed and tested on 4 case studies. We define a decision support framework by integrating the model into a tool and by proposing roles in the decision-making process. We finally present its application on a concrete example.

Keywords: decision-making, normative decision theory, decision support system, value of information, development process, automotive industry

NOMENCLATURE

The nomenclature of the opposites of the following variables and functions can be easily derived and is not expanded in this section.

$pITDC(t)$ — Probability that later or current but not represented changes in the technical definition of the vehicle will impact the outcomes of the current decision.

$P(OK|ITDC)$ — Probability that the vehicle passes the physical test given that changes affected the attributes of the decision.

$P(OK|\overline{ITDC})$ — Probability that the vehicle passes the physical test given that changes did not affect the attributes of the decision.

$P(Fav|OK)$ — True positive rate: the results of the analyses were favorable, and the vehicle passed the physical test.

$Cost_{delay}(t)$ — Cost incurred by the delay in the project: delay propagation on other activities, speeding up, increasing resources to respect the project deadline, likelihood to pay penalties.

Process alternative — In a process-focused approach of the decision, course of actions that include technical modifications, data gathering or delaying other

---

[1] Contact author: timothe.sissoko@centralesupelec.fr

Copyright © 2019 ASME

actions. It directly impacts the planning and resources of the design process.

Artifact alternative In an artifact-focused approach of the decision, course of actions that include technical modifications of the product (the artifact). In a process-focused approach of the decision, artifact alternatives are process alternatives, since modifications impact the planning and resources of the design process.

1. INTRODUCTION

Normative decision theory has been devised in order to support the decision-making. The approach itself is based on the fact that the decision-maker is considered as fully rational and that all data necessary for the decision-making is available. However, in industry, these approaches are used in a smaller scale (computationally challenging, lack of data, conflicting objectives of the decision-makers, etc.). Moreover, in engineering design, decision-making is focused on the product; thus, not necessarily taking into account the time needed for additional data gathering or modelling one decision. In complex system design, decisions are interdependent and can have propagating impacts.

This research is done in a major multinational car manufacturing company and our objective is to support the decision-making process for vehicle development based on modeling and simulation (M&S). Sissoko et al. [1] conducted an empirical study in the automotive industry and emphasized that, in such a context, several reasons cause iterations and delay in the decision-making process. These reasons are a lack of clarity about the situation due to uncertainty about potential changes (because of concurrent design and testing activities) and about cost, quality and delay attributes; a lack of knowledge and trust about the M&S performance and limits; and a miscommunication between decision actors partly due to the fact that complex decisions are *trans-hierarchical* (i.e. decisions escalated at upper hierarchical levels when out of scope information is required, when they involve extra cost, or when they may lead to collateral effects). In addition, decision-makers and stakeholders are sometimes unaware about how their decision influence subsequent decisions and ultimately the profit of the company.

The decision problems that we consider in this research refer to decisions made to solve design issues. We focus on design issues that arise in the vehicle *development phase*. This phase is characterized by design and numerical testing iterations that aim at refining the technical definition of the vehicle to comply to the requirements stated in the *upstream phase*. The vehicle *development phase* lead to a complete and detailed numerical model of the vehicle that is ready to be manufactured in the *industrialization phase*.

We call *issue resolution* the process of defining solutions and gathering information to ultimately incorporate a technical solution. For instance, a design issue can be that a noise, vibration and harshness performance does not meet the requirement. The corresponding decision can be to make a choice between two or more alternatives that will ultimately lead to solving the design issue. The alternatives are courses of actions like "changing the current material X to material Y on the same design" or "analyzing the consequences of incorporating an add-on". Ultimately, the choice of an alternative leads to an irrevocable allocation of resources.

Observations from industry underline the fact that decision-makers (project manager, synthesis architect, etc.) tend to have a process-focused approach. For instance, they consider the fact that the technical definition will be more refined and less likely to change over time, and plan to finalize an issue resolution accordingly. When considering a decision to solve a design issue, they tend to make an intuitive trade-off between the probability that changes in the technical definition will affect their decision about technical solutions and the costs incurred by the delay. However, outcomes related to actions such as data gathering and postponing the issue resolution are not explicit. In this paper, we propose to bridge this gap by proposing to integrate process related data within the decision-making framework entitled IRDS (Issue Resolution Decision Support). IRDS includes a generic model of common decision problems addressing design issues, a customized tool that enables to compute decision problems, a definition of the roles who contribute to the decision-making, and specification of the information flow between these roles. Its decision problem model relies on expected utility maximization and value of information theory. In this paper, we focus on the decision problem model, its structure, variables, and conceptual features.

After a literature review in Section 2, we introduce the IRDS' generic decision problem model in Section 3. We present the cases used to build the model, its structure and principle, as well as its variables, functions, and conceptual features. In Section 4, we describe the roles and data flows involved in the decision-making process supported by IRDS, and an example of the use of IRDS with an industrial example. We discuss the strengths and weaknesses of our proposal in Section 5. Finally, Section 6 draws together the most important observations, highlights the challenges and the opportunities for improvement that need to be addressed in the future work.

2. LITERATURE REVIEW

Decision theory can be split into three approaches: descriptive, normative and prescriptive. The *descriptive* approach focuses on how people actually make decisions in real-world settings. It contributed to the research fields of human factors and cognition. Naturalistic Decision Making (NDM) community emphasized the role of experience in enabling people to rapidly match situations with patterns they have learned and make effective decisions [2]. Researchers observed that people rely on heuristics as opposed to algorithmic strategies and deviate from the principles of optimal performance as defined in the normative approach. The *normative* approach prescribes how people should make decisions, assuming that the decision-maker is fully informed, fully rational and able to compute with perfect accuracy. These assumptions are idealistic and seldom encountered in real-world situations. Indeed, even assuming that decision-makers have bounded rationality [3], they have to deal with uncertainty. This is the purpose of the prescriptive approach [4]. It consists of guiding decision-makers by following principles of rationality and supporting them in their decision-making in practice while dealing with uncertainty and biases. Decision-Based Design has been introduced in the late 80's [5] and defined by Hazelrigg [6] as a normative approach that prescribes a methodology to make unambiguous design alternative selections under uncertainty and risk wherein the design is optimized in terms of the expected utility. Researchers extensively studied this approach and established a strong foundation of the decision theory to design methodology research [7,8]. A formal method and practical application of decision theory is Decision Analysis [9,10]. It consists of decomposing a decision problem and determining the alternative that has the maximum expected utility [11].

Uncertainty has been highlighted in the literature as an important challenge of decision-making [10,12,13]. It can be defined as a lack of information [14], a state of mind characterized by a conscious lack of knowledge [15], or a lack of numerical probabilities of various outcomes [16]. Reducing uncertainty about one's prior beliefs partly involves gathering information and selecting the most valuable information source [17]. Although in the engineering context normative decision theory applications most generally focus on artefact decisions about product features, some researchers stressed that all design decisions are actually process decisions, and therefore the resources spent in the refinement of information should be taken into consideration [18–20]. Hence, actions such as "modify a design specification" or "collect more information about a potential design modification" must coexist in the decision alternatives set and considered when analyzing the decision problem. For alternatives consisting in gathering information, the Value of Information theory helps to determine whether the decision-maker should access an information source or not. Various strategies based on the value of information used with normative decision theory have been discussed [18–20]. In such researches, Bayes' theorem [21,22] is generally used to model how the beliefs are updated when new information is incorporated.

Decision Support Systems (DSS) have been designed to help decision-makers to strive to informed and rational decision-making [23]. They have been subject of prolific research since the concept has been introduced in the early '70s. Either data-driven, communication-driven, document-driven, model-driven, or knowledge-driven [24], they are generally computerized systems that range from systems answering simple queries through data warehouses to systems simulating complex decision situations. Although DSS examples can be easily found

when searching for commercial decision tools software [25,26], examples of applications with real industrial cases often pertain to business confidential information. DSS often rely on formal decision-making methods.

Decision-making methods are commonly used and tested in the engineering design literature. Renzi, Leali and Di Angelo [27], however, emphasize that the proposed methods are applied with illustrative examples streamlining complex design cases in a simpler model to be analyzed. Moreover, to our knowledge, the way decision-making methods are tested in the literature seldom incorporate the value of information and the cost of delaying the finalization of an issue resolution. We believe that such information can be useful in an industrial context. In DBD literature, Thompson and Paredis [20], investigated the decision analysis of design process decisions. They show that the decision analysis of design process decisions provides a more comprehensive model of the problem when multiple sources of information can sequentially be used. This work paves the way of a process-focused approach of decision making through a rigorous and systematic method, namely the decision analysis. We aim to contribute to this research field by extending the design process decisions model through the integration of uncertainties related to product definition evolutions, industrial time constraints, and model characteristics addressing industrial complexity (such as the possibility of concurrent analyses). Hence, we propose to integrate process alternatives within the decision alternatives set of complex systems industrial decision problems.

In the light of this literature review, we assume that the rationality of decision-makers is limited when they face decision problems in the context of our study. We strive to facilitate rational judgment by structuring decision problems and providing valuable information. We propose to use the expected utility maximization derived from the normative approach of decision theory and to handle uncertainty through the value of information and Bayes' theorem. To reduce the modeling effort of the decision problem for each design issue, we propose to build upfront a decision problem model that address common design issues in the vehicle development phase.

3. IRDS' GENERIC DECISION PROBLEM MODEL

Decision-makers struggle with poorly informed decision problems in terms of cost, quality, and time attributes about process alternatives such as: modify design now, analyze a design modification considering the current beliefs about the decision situation, postpone to analyze and modify design at a later time the decision situation). We developed an issue resolution decision support (IRDS) that aims at answering the following general questions: *What are the artifact alternatives (technical solutions)? What if the decision-maker chooses to incorporate a technical solution now? What if he/she decides to collect information about a technical solution? What if he/she postpones the issue resolution finalization?* To do so, IRDS supports the application of decision analysis: an expected utility

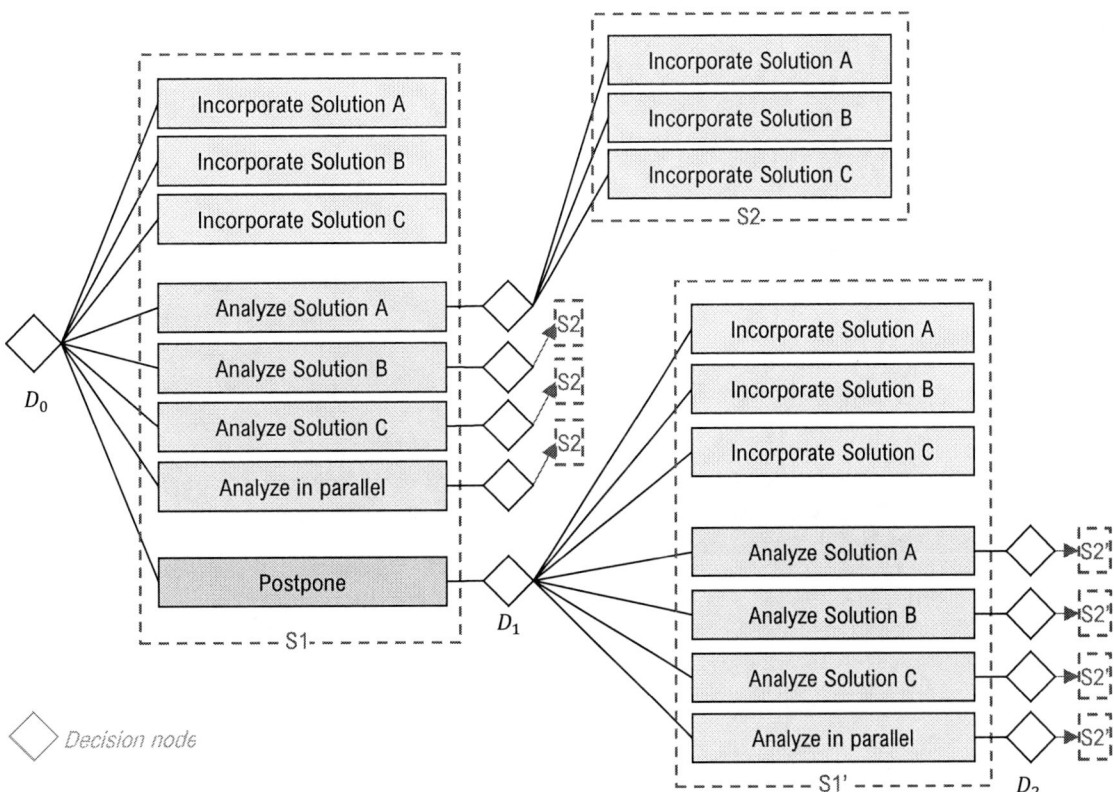

FIGURE 1: STRUCTURE OF IRDS' GENERIC DECISION PROBLEM MODEL

is computed for each decision alternative, and a Bayesian inference is performed to update the beliefs in case of information gathering (for the alternatives consisting in performing an analysis). IRDS is development process oriented: it includes an industrial cost breakdown and takes into consideration the evolution of the design refinement. Depending on the accuracy expected, the data required can be extracted from expert or non-expert estimates; and simulation analyses results.

3.1. Case studies used for model building

In order to propose a generic decision-making model, four case studies have been identified and discussed with industry experts:

1. RC1: Tunnel
2. RC2: Fairing thermal protection
3. TC1: Reinforcer
4. TC2: Analyses for reinforcer

Two of them are real design issues that have been extracted from history and discussions with two experts and two analysts ("*real cases*", RC1 and RC2), and the two others are synthetic design issues ("*toy cases*", TC1 and TC2), that we designed in accordance with experts with the aim to represent common situations.

The two *real cases* were picked at two different stages in the development process. At the upstream edge of the process (RC1), the design maturity is very low and the decision problem information is scarce. In other terms, the technical definition of the vehicle is imprecise and the probability that changes can occur and affect the problem settings is high. Moreover, no numerical simulation results on technical performance are yet available, and the cost estimate is very imprecise. Conversely, at the downstream edge of the process (RC2), closer to the manufacturing milestone, the design maturity is high, and the information is prolific. In other terms, simulation results for technical performance already exist and simulations can be performed with reasonable accuracy. There is also more clarity about the economic and time constraints related to the potential actions.

The two *toy cases* were designed to represent the decision problems encountered between the two edges previously described. One of these *toy cases* helped us to determine the cost breakdown and the influence of the design maturity on the decision (TC1). The other case (TC2) helped us to model the influence of the analyses about the beliefs about the chances of success of one or several technical solutions.

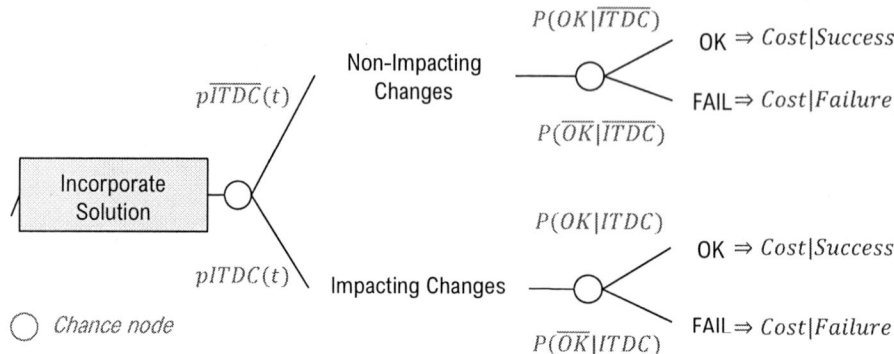

FIGURE 2: STRUCTURE OF "INCORPORATE SOLUTION" ALTERNATIVE

The model was defined and tested progressively. Several feedback loops have been done in order to refine the model and ensure its genericity.

3.2. Structure and principles

IRDS' generic decision problem model can be represented as a pseudo-recursive tree. As shown in Figure 1, to solve a design issue, the decision involves choosing one alternative among three categories:

1. Incorporating a technical solution at the current moment. It corresponds to modifying the design or material of a part or adding a new part, such as a reinforcer. In its broad sense, it also includes sticking to the current technical definition.
2. Analyzing one or several technical solutions concurrently and wait for the results before selecting which technical solution should be incorporated (cf 1.). In other terms, it consists of gathering information about the chances of success of incorporating a technical solution.
3. Postponing the finalization of the issue resolution at a later moment, to then decide whether incorporating a technical solution (cf 1.) or perform analyses (cf 2.). In this case, the decision-maker looks for the most favorable moment to choose to integrate a technical solution. He/she targets a time when the architecture is less likely to evolve and affect the outcomes of the decision.

The current situation (S1) includes the three categories of alternatives. The situation S2 corresponds to the moment when the analysis results are available and when the decision-maker will have to choose which technical solution to incorporate. S1' corresponds the moment when the decision-maker will have to choose between incorporating or performing analyses, after having postponed the finalization of the issue resolution. If in the future, in S1', the decision-maker chooses to perform analyses about one or several technical solutions, he/she will end up to S2'. S2' has the same structure as S2 but has a different time coordinate (and potentially different decision attributes).

At each decision node of the tree, the maximum expected value is used to calculate which is the most profitable alternative.

3.3. Variables, functions and conceptual features

To comply with the company risk policy, we assume that decision-making is risk neutral. Decisions are made to ensure that performance meets the requirement that was defined in a perspective of demand maximization and regulatory compliance in the upstream phase. Also, we consider that in the development phase, a decision addressing a design issue ultimately involves an expense. This expense can result from the implementation of a design change, the performance of analyses or, indirectly, a delay in planning (due to the cost of accelerating subsequent activities to meet the deadline). These two reasons (risk profile and expense-focus) lead us to assimilate the expected utility to an *expected cost*. The preferred alternative is therefore the one that has the minimum expected cost.

As shown in Figure 2, the expected cost of an *Incorporate Solution* alternative is computed with:

- The probability that current or later changes in the technical definition of the vehicle will affect the outcome of the decision, $pITDC(t)$, and its opposite, $p\overline{ITDC}(t)$. This probability is time-dependent as the technical definition of the vehicle becomes more and more detailed as the development phase progresses.
- The probabilities that the vehicle passes the physical tests with the solution implemented whether changes occurred or not $P(OK|ITDC)$ and $P(OK|\overline{ITDC})$, and their opposites, $P(\overline{OK}|ITDC)$ and $P(\overline{OK}|\overline{ITDC})$. Note that during the development phase all the tests are numerical until the vehicle can be manufactured and physically tested in the next phase.
- The cost actually committed in case of success of the solution, composed by the development and analysis costs, the vendor tooling cost, the manufacturing cost (influenced by the number of vehicles manufactured), the supplier engineering cost, and the eventual added value or penalty (e.g. because of addition or subtraction of material weight,

or savings for other vehicle projects that would require the same tooling).

- The cost actually committed in case of failure of the solution. The design issue discovered on physical prototypes is supposed to be solved before the mass production phase. A resolution of this problem may require changes in the manufacturing process. In this case, the manufacturing costs in case of failure of the initial solution replace the manufacturing costs in case of success.

In the case of *Analyze Solution* and *Analyze in parallel* alternatives, the analyses results can be favorable or not regarding one or several technical solutions. "Favorable" means that, considering the accuracy of the analysis, the result ensures that the incorporation of the technical solution will allow the vehicle to pass the physical test. The accuracy of an analysis is broken down by assigning a sensitivity and a specificity. It corresponds respectively to the *true positive rate*, $P(Fav|OK)$ and the *true negative rate*, $P(\overline{Fav}|\overline{OK})$. Whether the analyses results are favorable or not, using sensitivity, specificity and the prior probability about the chance of success, Bayes' rule allows to compute the posterior probability that the vehicle will pass the test. For example, equation (1) represents the posterior probability that the vehicle passes the physical test given that the analysis was favorable. An analysis is performed at a given t_j to which corresponds a $p\overline{ITDC}(t_j)$. Therefore, the analysis allows to update only the belief in the case of $\overline{ITDC}$, i.e. assuming that no changes in the technical definition of the vehicle will have an impact on the results of the decision, $P(OK \cap \overline{ITDC} \cap Fav_A)$. The analysis is based on an already known technical definition and does not yet take into account all possible changes that may occur during the period before the manufacturing phase. The analyses can be performed in parallel for several technical solutions, and the posterior probability that the vehicle passes the test is computed with the product of $P(Fav)$ (and its opposite) for each technical solution, as with the end node $P(OK \cap \overline{ITDC} \cap Fav_C \cap Fav_B \cap Fav_A)$. Finally, a cost of performing analyses must be considered as well as a cost of the delay incurred in the project.

$$P(OK|Fav) = \frac{P(Fav|OK)P(OK)}{P(Fav|OK)P(OK) + P(Fav|\overline{OK})P(\overline{OK})} \quad (1)$$

The *Postpone* alternative corresponds to waiting for the technical definition of the vehicle to be more detailed before deciding on the technical solution to be implemented. Three main reasons motivate decision-makers to postpone the finalization of the issue resolution:

- The belief at any time that the data could be inconsistent, i.e. the simulation results considered at any given time may not correspond to the latest technical definition.
- At the time the decision is considered, at t_0, changes may have occurred; project team already know what are these

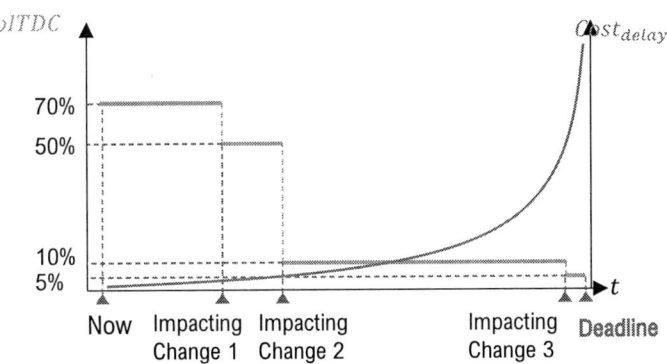

FIGURE 3: NOTIONAL EXAMPLE OF THE RELATIONSHIP BETWEEN TIME, $p\overline{ITDC}$ and $Cost_{delay}$

changes even if they do not appear in the simulation results, or project team do not know what are the changes, but they know that the technical definition has already been affected.

- Changes may occur later in the development process; project team know what they will be (through heuristics, or knowing that process instructions require that the design of certain types of parts be specified only at a given milestone), or they do not know what they will be, but they know that changes will appear.

In all the cases mentioned above, changes in the technical definition of the vehicle may require the design issue to be reworked and resolved late and at a higher cost. When postponing, the decision-maker expect fewer subsequent (or current and unaddressed) changes in the technical definition of the vehicle that could affect the outcomes of the decision under consideration. In general, there are fewer changes over time, hence we propose to consider that the likelihood of reworking on the design issue will decrease and that the expected costs incurred by this rework will also decrease.

It is important to note that we assume that existing technical solutions will be modified according to the evolution of the technical definition of the vehicle. With regard to the *Postpone* alternative, technical solutions should be considered as "types of technical solutions" (which lead to their respective expenses). Considering that the decision situation may change over time, a new and better technical solution may be designed later, the design issue may even disappear because of impacting changes that would be beneficial to meet the required performance. But at the time the decision is considered, at t_0, it is difficult to know. Therefore, we consider that in the worst-case scenario, it is the type of existing technical solution that is most successful that will be incorporated later when the outcomes of the decision will be less likely to be affected by external changes.

For example, if one consider Figure 2 with the following the probabilities: $p\overline{ITDC}(t_0) = 0.3$, $P(OK|\overline{ITDC}) = 0.9$, $P(OK|ITDC) = 0.5$, $P(OK) = P(OK \cap \overline{ITDC}) + P(OK \cap$

$ITDC$), there is 62% chance that the vehicle actually passes the test at the moment the decision is considered, t_0. If the decision-maker chooses to postpone at t_2, $p\overline{ITDC}(t_2) = 0.9$, $P(OK|\overline{ITDC}) = 0.9$, $P(OK|ITDC) = 0.5$, $P(OK) = P(OK \cap \overline{ITDC}) + P(OK \cap ITDC)$. At t_2 there will be 86% chance that the vehicle actually passes the test. Without any consideration of the cost related to the delay in the project caused by a postponement, it is preferable to wait before finalizing the resolution of the problem.

However, a *cost of delay* needs to be taken into account when considering postponing the finalization of the issue resolution. If the project team postpones the finalization of the issue resolution, the technical solution will need to be implemented in the shorter period demanding more resources to respect the deadline of the project. It will also require solving subsequent and depending design issues faster, prioritize these issue resolutions over other activities, and in both cases mobilize engineers. Moreover, some engineers may still be paid whereas they do not provide value to the project while they wait for the decision to be finalized. Finally, the chance to have to pay directly or indirectly delay penalties increases. After discussing with experts and collecting the historical data we propose to model the postponing trade-off with two functions: $p\overline{ITDC}(t)$ that ultimately reflect the evolution of the expected cost committed in case of failure, and $Cost_{delay}(t)$. Consider Figure 3 with the example applied with Figure 2: "Now", at t_0, the decision outcomes have 70% chance to be impacted by changes. After *Impacting Change 2* is made, at t_2, there will be 10% chance of impacts. At that moment, a long time will remain available to implement the decision and carry out the activities that depend on it. Once *Impacting Change 3* will be made, short time will remain available. With regard to project development, closer the to deadline, there is more effort needed to perform the remaining activities on time. Consequently, although postponing can decrease the expected cost of reworking a design issue, it also increases the chance of having to speed up activities to meet the deadline or to pay penalties; in short, to incur expenses. In cases where the project team does not know in advance which change will occur at which time, we propose to define $p\overline{ITDC}(t)$ as a linear function or a simple curve.

Analyzing how the expected cost (or expected value) of the *Postpone* alternative varies when t varies can help to know whether and until when it is profitable to postpone the finalization of the issue resolution. This analysis has been used as a Value of Information analysis related to postponing one's decision. An illustration of this trade-off is presented in Section 4.3.

In industry, it is not always easy to define nominal probabilities and costs used. In this case, we propose to define distributions to represent the uncertainty about the input variables and to use a sampling method to simulate the risk associated with the decision under consideration. In this respect, if being uncertain has a significant impact on the decision or if risk-taking can create high value, then the decision-maker must have the decision analyzed more thoroughly.

The conceptual features we have presented are derived from our observation of industrial issues and practices. Section 4 aims to explain how IRDS it can be integrated into an industrial framework.

4. INDUSTRY APPLICATION
4.1. Proposition of roles in the decision-making process

In order to support the decision making, we propose that the decision analysis is done by someone who can exchange information with project team members and knows to manage a tool based on the generic decision problem model. Hence, in the IRDS framework one can identify 3 different roles: the decision-maker (DM), the decision analyst (DA) and the decision-problem data provider (DaP). DaP are generally numerous (experts, analysts and designers from different disciplines, working on different subsystems), they provide data for decision analysis while responding to DA requests, and provide complementary situational information to the DM. DA gathers decision problem data and *context* information by issuing *data queries*. These data can be *numeric values* obtained with numerical simulations or *verbatims* from experts expressing their beliefs. DA analyses the decision while integrating DM's queries, and provides him/her *decision analysis results* and *suggestions*. DM receives and requests information to DA and DaP, and finally makes the decision to solve the issue. Figure 4 shows what data is conveyed between the 3 roles in the IRDS framework and how the DA interacts with the tool in which IRDS decision problem model is embedded *(IRDS tool)*.

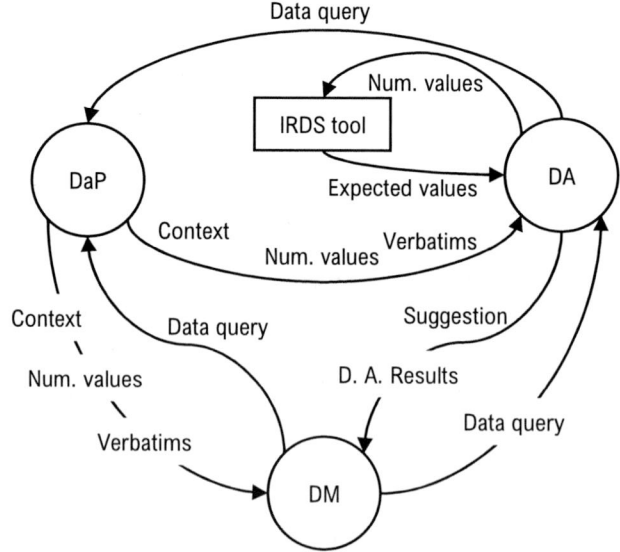

FIGURE 4: DATA FLOW DIAGRAM OF THE DECISION-MAKING PROCESS SUPPORTED WITH IRDS

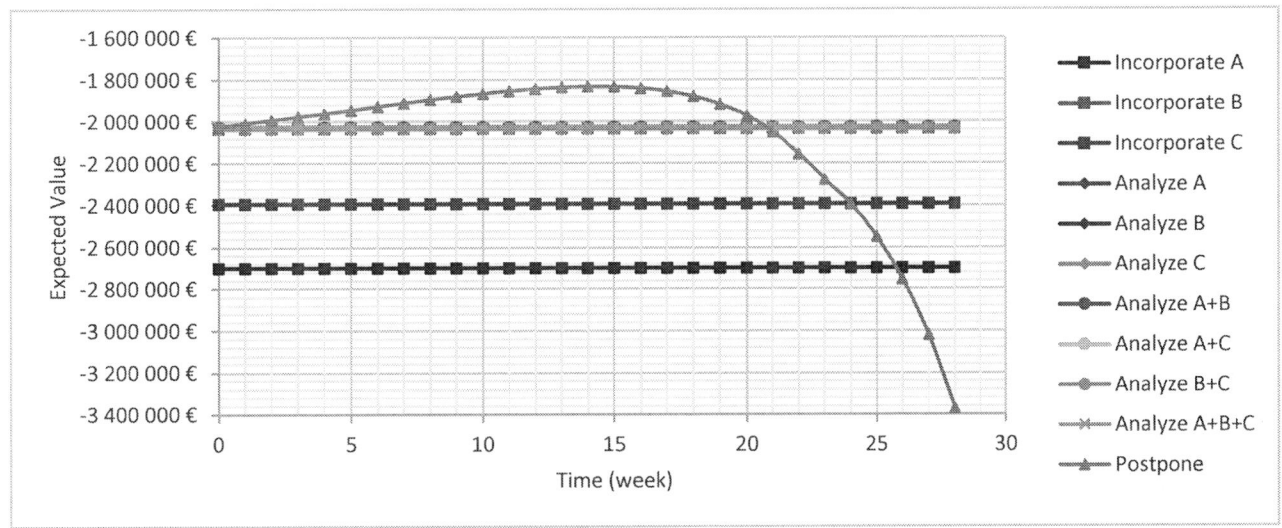

FIGURE 5: EVOLUTION OF THE EXPECTED VALUE OVER TIME WHEN POSTPONING THE ISSUE RESOLUTION COMPARED WITH EXPECTED VALUES OF OTHER DECISION ALTERNATIVES

4.2. Customized toolbox

In order to support the design team, IRDS has been deployed using *Palisade Decision Tools Suite* [28]. With *PrecisionTree 7.5*, a generic decision tree has been developed based upon the model previously discussed. Design team can use automatically defined functions with regard to the type of alternatives selected. For a decision problem considering 3 technical solutions, the decision problem model is constituted of 644 end nodes. We also created tables to define $pITDC(t)$ and $Cost_{delay}(t)$. We propose to use *Palisade's @Risk* [28] to perform sampling distribution simulations. The decision analyst has to define the simulation settings according to the specificities of the decision problems.

Our motivation was to simplify the use of the decision support tool so that the effort is concentrated on the data gathering and analysis rather than on modeling.

4.3. Noise Vibration and Harshness example

To have an example of how to use IRDS, consider an industrial case where a vibration performance does not meet the requirement. Simulation results initially show that two measurement points do not reach the target. Simulation analysts identified the vehicle part involved in this defect and proposed, with designers, two technical solutions (design and material changes). They performed analyses about the efficacity of the technical solutions with regards to vibration performances. As any design change can impact many performances, and passive safety success is a sine qua non condition to accept a design change, simulation analysts also tested passive safety performances.

In order to support the decision making, we deployed IRDS. A DaP initially provided a $P(OK|\overline{ITDC})$ estimate for each

technical solution (10% for Solution A, which is the current technical definition of the vehicle part, 70% for the Solution B, and 90% for the Solution C). To do so he relied on his own experience supported by numerical simulations. The simulations were performed with nominal values for two potential technical solutions. The differences between the 2 proposed technical solutions and the current technical definition in terms of *Manufacturing cost | Success*, and *Weight | Success* were provided in the decision dossier that was already prepared. To use IRDS, DA investigated with DaP to collect missing data.

It is important to note that not all missing data were worth gathering; for instance, development cost difference was supposed to be equal for the two solutions was not necessary to estimate. DA assumed that costs estimates were accurate in case of success and costs estimates in case of failure were rough but plausible according to the experience of DaP. In this case, an order of magnitude was sufficient as input data. However, probabilities varied according to expert's personal representations and whether they were comfortable or not with probabilities. Hence, DA performed sensitivity analyses to estimate to which extent refining beliefs about probabilities was valuable. Depending on the accuracy expected, this data can be easily available for low cost, but it is time consuming since DaP are spread in different services. Sometimes it may require additional effort to estimate costs in the case of failure, since it requires history and/or cost estimates of backup solutions. In this case, DA and DaP assumed that in case of failure, the technical solution that has the best chance to succeed but that is the most expensive will be used as to develop a late solution. It involves extra-costs due to the late work. DA performed analyses with rough approximates of $pITDC(t)$ and $Cost_{delay}(t)$ to assess how sensitive the decision was about these functions, in order to eventually model them more accurately with the help of DaP.

Alternative	Expected Value
Incorporate A	-2 700 000 €
Incorporate B	-2 033 440 €
Incorporate C	-2 392 400 €
Analyze A	-2 035 690 €
Analyze B	-2 036 130 €
Analyze C	-2 036 130 €
Analyze A+B	-2 024 885 €
Analyze A+C	-2 036 690 €
Analyze B+C	-2 034 761 €
Analyze A+B+C	-2 023 792 €

TABLE 1: EXPECTED VALUES OF DECISION ALTERNATIVES

Once the IRDS process led to conclusive results, DA presented a report to DM. As shown in Table 1, excluding the *Postpone* alternative (that is time dependent), the alternative that has the maximum expected value (EV) is *Analyze A+B+C*. Although Solution A (current technical definition of the vehicle part) has only 10% chance of success (prior belief), it costs 1 200 000 € less than Solution B in case of success. Solution B (70% chance of success) costs 1 000 000 € less than Solution C in case of success. These two solutions worth investigating since they would lead to significant savings compared to Solution C if they ever succeed.

Involving $pITDC(t)$ and $Cost_{delay}(t)$, the EV of *Postpone* alternative varies over time, as shown in Figure 5. It is important to remember that the *Postpone* alternative includes a later decision among the other decision alternatives. This later decision will involve a different probability of impacting changes. On Figure 5, we can see that the expected value of *Postpone* increases until week 14. At week 20, its EV is equal to the expected value of choosing *Analyze A+B+C* at week 0. Between week 0 and week 20, an update of the situation will allow to make a better decision and avoid rework. Week 14 is the optimum date to postpone the decision, with a difference of EV of 191 485 €. Postponing until after week 20 will not be profitable, since the EV after week 20 is smaller than choosing *Analyze A+B+C* at week 0. Indeed, any choice after week 20 will be potentially more expensive since costs will be incurred by the delay. In conclusion, the DM should best postpone the decision until week 14.

In the current practice of the company, the decision-maker would not have been provided with such explicit information about the consequences of analyzing technical solutions or postponing the decision. This lack of clarity could have led the decision-maker to iterate more often and mobilize engineers (slowing other activities), or to choose to incorporate a technical solution right away and rework it later at a higher cost. The IRDS framework, and the decision problem model on which it is based, allows to structure a process-focused approach to decision-making and inform the decision-maker accordingly.

5. DISCUSSION

Our proposal presents several strengths: IRDS framework allows to explicitly and systematically identify what the alternatives are, including the action of performing further analyses. Using the decision support framework that we propose is an incentive to bring the team together to discuss the beliefs people have about the inputs of the decision problem. The decision is analyzed with a rigorous mathematical framework. The decision problem model is designed to address most of the decision problems that can be encountered in the development phase. Hence, it allows the project team to not spend resources to build new models for similar problems.

On the other hand, we identified potential disadvantages or weaknesses. Using IRDS requires time and efforts in terms of data gathering and beliefs modeling. In the case where IRDS would lead to the same decision as to the current practice, the use of IRDS would involve unnecessary expenses. In such case, by investigating decision alternatives, the organization can gain insights that can be useful for other purposes (such as improvement of processes and practices). However, this gain is difficult to quantify and out of our current research scope. We acknowledge that IRDS framework does not prevent the users form experiencing biases that can lead to poor decisions. Framing the problem and the actions explicitly can discourage the decision-maker to think broader and reconsider the problem. Moreover, IRDS may not include all the alternatives that are actually available. Another bias is related to data gathering by beliefs elicitation and modeling: if the beliefs deviate too much from the truth, the computation can provide results that are mathematically correct but do not conform to the actual state of reality. Finally, the generic decision problem model at the core of IRDS lies on assumptions that we made. The functional relationships that we impose through our model can also introduce biases as these assumptions may not reflect the beliefs of the people using IRDS. A dialogue on modelling may be necessary between the stakeholders involved in the decision in this case.

6. CONCLUSION AND FUTURE WORK

Decision-making in industry context and in particular in complex system design can be difficult. Often data gathering or additional time necessary for developing a more precise decision-making model is not considered explicitly. In this research, IRDS framework has been proposed in order to integrate process related data in the vehicle Modeling and Simulation based vehicle development process. IRDS allows for process-focused decision-making, incorporating the analysis of the consequences of actions such as information gathering and postponing the choice of artifact alternatives. The proposition of a decision-making model has been defined based upon 4 industry case studies and an additional case of Noise, Vibration and Harshness has been presented as a test example. This model has been designed in accordance with experts to reflect the possibilities and the constraints of the industrial process.

We observed that it is challenging to gather some of the data used for the computation of the decision problem. Engineers struggle to express subjective probabilities when numerical data is scarce or when numerical simulation results contradict their beliefs (e.g. when the numerical model of the system do not take into consideration attributes that are known by the expert). Hence, investigation of methods that allow for data identification that are critical and change the decision (in the light of sensitivity analysis) will be considered as to permit to focus data gathering activities. This is considered by the experts as necessary in order to support decision-making but also future resources consumption.

ACKNOWLEDGEMENTS

We would like to thank Renault's simulation analysts, computer-aided engineering specialists, experts, synthesis architects and vehicle project managers who have brought their insights and feedback to the development of the framework we have presented in this paper.

REFERENCES

[1] Sissoko, T. M., Jankovic, M., Paredis, C. J. J., and Landel, E., 2018, "An Empirical Study of a Decision-Making Process Supported by Simulation in the Automotive Industry," *Proceedings of the ASME 2018 International Design Engineering Technical Conferences & Computers and Information in Engineering Conference.*

[2] Klein, G., Associates, K., and Ara, D., 2008, "Naturalistic Decision Making," 50(3), pp. 456–460.

[3] Adams, E., and Simon, H. A., 1962, "Models of Man, Social and Rational: Mathematical Essays on Rational Human Behavior in a Social Setting.," J. Philos., 59(7), p. 177.

[4] Bell, D. E., Raiffa, H., and Tversky, A., 1988, "Descriptive, Normative, and Prescriptives Interactions in Decision-Making," *Decision Making*, D.E. Bell, H. Raiffa, and A. Tversky, eds., Cambridge University Press, Cambridge, pp. 9–30.

[5] Shupe, J., Allen, J., Muster, D., and Mistree, F., 1988, "Decision-Based Design: Some Concepts and Research Issues," *Expert Systems*, A. Kusiak, ed.

[6] Hazelrigg, G. A., 1998, "A Framework for Decision-Based Engineering Design," J. Mech. Des., 120(4), p. 653.

[7] Hazelrigg, G., 1996, *Systems Engineering: An Approach to Information-Based Design*, Prentice-Hall, Upper Saddle River, NJ.

[8] Chen, W., Hoyle, C., and Wassenaar, H. J., 2013, *Decision-Based Design*, Springer London, London.

[9] Howard, R., 1966, "Decision Analysis: Applied Decision Theory," Proc. Fourth Int. Conf. Oper. Res., pp. 55–71.

[10] Howard, R., and Abbas, A., 2015, *Foundations of Decision Analysis*, Prentice Hall, New-York.

[11] von Neumann, J., and Morgenstern, O., 2007, *Theory of Games and Economic Behavior (60th Anniversary Commemorative Edition)*, Princeton University Press, Princeton.

[12] Dubois, D., and Prade, H., 2010, "Panorama Des Nouvelles Méthodes de Traitement de l'incertitude et de l'imprécision."

[13] Wang, Q., Lee, B. D., Augenbroe, G., and Paredis, C. J. J., 2017, "An Application of Normative Decision Theory to the Valuation of Energy Efficiency Investments under Uncertainty," Autom. Constr., 73, pp. 78–87.

[14] Thiry, M., 2002, "Combining Value and Project Management into an Effective Programme Management Model," Int. J. Proj. Manag., 20(3), pp. 221–227.

[15] Head, G. L., 1967, "An Alternative to Defining Risk as Uncertainty," J. Risk Insur., 34(2), pp. 205–214.

[16] Knight, F., 1964, "Risk, Uncertainty and Profit," Clim. Chang. 2013 - Phys. Sci. Basis, XXXI, pp. 1–30.

[17] Lawrence, D. B., 1999, *The Economic Value of Information*, Springer New York, New York, NY.

[18] Bradley, S. R., and Agogino, A. M., 1994, "An Intelligent Real Time Design Methodology for Component Selection: An Approach to Managing Uncertainty," J. Mech. Des., 116(4), p. 980.

[19] Ling, J. M., Aughenbaugh, J. M., and Paredis, C. J. J., 2006, "Managing the Collection of Information Under Uncertainty Using Information Economics," J. Mech. Des., 128(4), p. 980.

[20] Thompson, S. C., and Paredis, C. J. J., 2010, "An Investigation Into the Decision Analysis of Design Process Decisions," J. Mech. Des., 132(12), p. 121009.

[21] Bayes, M., and Price, M., 1763, "An Essay towards Solving a Problem in the Doctrine of Chances. By the Late Rev. Mr. Bayes, F. R. S. Communicated by Mr. Price, in a Letter to John Canton, A. M. F. R. S.," Philos. Trans. R. Soc. London, 53, pp. 370–418.

[22] Lee, P. M., 2012, *Bayesian Statistics: An Introduction*, Wiley.

[23] Gorry, G. A., and Morton, M., 1971, "A Framework for Management Information Systems," pp. 510–571.

[24] Power, D. J., 2002, *Decision Support Systems: Concepts and Resources for Managers*, Greenwood/Quorum Books, Westport, CT.

[25] Capterra, 2018, "Logiciels d'aide à La Décision (Decision Support Software)" [Online]. Available: https://www.capterra.fr/directory/30544/decision-support/software. [Accessed: 02-Dec-2018].

[26] International Society on MCDM, 2018, "Software Related to MCDM" [Online]. Available: http://www.mcdmsociety.org/content/software-related-mcdm. [Accessed: 02-Dec-2018].

[27] Renzi, C., Leali, F., and Di Angelo, L., 2017, "A Review on Decision-Making Methods in Engineering Design for the Automotive Industry," J. Eng. Des., 28(2), pp. 118–143.

[28] Palisade, 2018, "Products & Services" [Online]. Available: https://www.palisade.com/products.asp.

Proceedings of the ASME 2019
International Design Engineering Technical Conferences
and Computers and Information in Engineering Conference
IDETC/CIE2019
August 18-21, 2019, Anaheim, CA, USA

DETC2019-97171

SUSTAINABLE DESIGN OF RESIDENTIAL NET-ZERO ENERGY BUILDINGS: A MULTI-PHASE AND MULTI-OBJECTIVE OPTIMIZATION APPROACH

Lan Lan,[*] Kristin L. Wood, and Chau Yuen
Singapore University of Technology and Design (SUTD)
8 Somapah Road, Singapore 487372

ABSTRACT

Zero energy building (ZEB) is an important concept for sustainable building design. This paper introduces a holistic design approach for residential net-zero energy buildings (NZEB) by adopting the Triple Bottom Line (TBL) principles: social, environmental, and financial. The proposed approach optimizes social need by maximizing thermal comfort time of natural cooling, and visual comfort time of daylighting. The environmental need is addressed by optimizing energy efficiency, and the financial need is addressed by optimizing life cycle cost (LCC). Multi-objective optimizations are conducted in two phases: the first phase optimizes the utilization rate of natural cooling and daylighting, and the second phase optimizes energy efficiency and LCC. Sensitivity analysis is conducted to identify the most influential variables in the optimization process. The approach is applied to the design of a landed house in a tropical country, Singapore. The results provide a framework and modeled cases for parametric design and trade-off analysis toward sustainable and livable built environment.

NOMENCLATURE

COP   Coefficient of Performance
HDB   Housing and Development Board
IAQ   Indoor Air Quality
LCC   Life Cycle Cost
NSGA-II Non-dominated Sorting Genetic Algorithm

---

[*]Address all correspondence to this author. Email: lan_lan@sutd.edu.sg

NZEB   Net Zero Energy Buildings
PCC   Partial Correlation Coefficient
RES   Renewable Energy System
SBS   Sick Building Syndrome
SHGC   Solar Heat Gain Coefficient
SRRC   Standardized Rank Regression Coefficients
TBL   Three Bottom Line
USPW   Uniform Series Present Worth Factor
VT   Visible Transmittance
WWR   Window to Wall Ratio
ZEB   Zero Energy Building

1 INTRODUCTION

The fast growing global energy use has raised concerns over serious environmental issues, and buildings are one of the major contributors to global energy consumption. Among the efforts of reducing the building energy consumption, zero energy building (ZEB) is an important concept. Significant research has been done to address the challenges in ZEB design. Main focus of the existing work has been put on the optimization of the environmental and the economic impacts of ZEB. However, for a ZEB to be sustainable, it is important to meet all the three requirements indicated by the Triple Bottom Line principles (or the "3Ps"): planet (environmental factor), profit (economic factor), and people (social factor) [1]. The social factor is not well addressed in the existing literature.

Most of the previous work, such as [2–8], simultaneously

optimizes the environmental factor (energy consumption and carbon footprint), and the economic factor (Life Cycle Cost (*LCC*)) using multi-objective optimization. Some literature only consider the economic factor, such as [9, 10]. The social factor is significantly less considered in the existing literature. In [11], social factor (including natural cooling and daylighting) is optimized, but the environmental and the economic factors are not included.

Although it is important to consider all the TBL attributes for a holistic design of NZEB, the research on the combination of all the three TBL principles are lacking. Therefore, this paper introduces a holistic design approach, which aims to optimize all the TBL factors and finds the optimized building performances regarding environmental, economic, and social needs. In addition, most of the existing research were conducted for NZEB in warm and cold climates including various locations across Europe [2, 4, 6–9, 11], east Asia [3, 12], and Middle East and North Africa [5], and there is little research on NZEB in tropical climates. Therefore, this paper evaluates the potential of residential NZEB buildings in a tropical climate.

The remaining sections of this paper are organized as follows. In Section 2, the proposed methodology of residential NZEB design is explained. Section 3 is a case study of residential NZEB design in Singapore. Section 4 includes discussions and limitations. Section 5 includes the conclusions.

2 Methodology
2.1 Triple Bottom Lines
The TBL principles are used to evaluate sustainability using three factors: social, environmental, and economic. For a project to be sustainable, it has to address people's needs and be desired by them, be responsible for the planet and the future of human beings, and bring profit or economic viability so that it can attract business and investment. With the primary aim of being sustainable, NZEBs should align with the TBL principles. However, the social factor is less emphasized in the existing literature. Especially, there is a lack of research which simultaneously takes all the 3Ps into consideration.

While it is well accepted to address the environmental and economic factors in NZEB design, the importance of designing for the social factor should also be recognized [13, 14]. Green buildings should not just help to reduce energy and resource consumption, but also to improve user experience and satisfaction [15]. Buildings are designed, ultimately, for people, so it is important to assist people in terms of health, safety, happiness, and productivity. People's needs towards a building may include human comfort, which consists of thermal comfort, visual comfort, acoustic comfort, and indoor air quality (IAQ). People also have the need of contact with nature in a building. Many studies have shown that access to nature significantly affects human health and productivity [16]. Nature contact may take many

forms in the built environment, such as plants and greenery in buildings, views out of windows, daylighting, and natural wind. In this work, the aim is to optimize selected and critical factors in terms of visual and thermal comfort that is achievable by natural means, i.e. daylighting and natural cooling.

Therefore, this paper proposes a holistic design approach for residential NZEB, which aligns with the TBL principles by optimizing human comfort via natural means, energy efficiency and cost effectiveness.

2.2 Design Optimization Process
In this work, JEPlus+EA [17], a multi-objective optimization tool based on NSGA-II, is coupled with EnergyPlus to explore optimized solutions and to seek insights into the design parametric space that leads to zero energy systems. NSGA-II is one of the most efficient multi-objective evolutionary algorithms [11]. It attempts to perform global optimization by generating offsprings using a specific type of crossover and mutation, and then selecting the next generation according to non-dominated sorting and crowding distance comparison [18].

EnergyPlus [19] is one of the most commonly used whole building simulation software. It has three basic components: a heat and mass balance simulation module, a building systems simulation module, and a simulation manager. EnergyPlus implements detailed building physics algorithms for heat transfer (radiation, convection, and conduction), air and moisture transfer, light distribution, and water flows. It is able to model a broad range of building and mechanical system configurations and conditions.

The proposed process consists of two phases as shown in Fig. 1. The first phase aims to maximize the time when effective daylighting is available and the time when thermal comfort is achievable without air-conditioning. In Phase 1, an EnergyPlus model is first created with basic building geometry, construction information, internal heat gains, airflow network, and daylighting. In JEPlus, passive design variables are defined and linked to the EnergyPlus model. NSGA-II based optimization is performed in JEPlus+EA to minimize the thermal discomfort time (i.e. time when thermal comfort cannot be achieved by natural cooling only) and daylighting unmet time (i.e. time when visual comfort cannot be achieved by daylighting only). The data of the optimization results is imported to Trade Space Visualizer [20] for visualization and analysis of the trade space. From the trade space, best candidates with long effective daylighting time and natural cooling time are selected for further optimization and analysis. In addition, sensitivity analysis is conducted in SimLab [21] to identify the most influential design variables. Based on the sensitivity analysis results, key insights in ZEB can be identified, and highly-sensitive parameters can be adjusted to further improve the building performance.

In the second phase, energy consuming appliances in the

households are taken into consideration and defined as active design variables [22]. The design variables in the second phase comprises selected candidates from Phase 1 and active design variables. Each selected candidate from Phase 1 has respective daylighting unmet time and thermal discomfort time. Therefore, the required light and AC operating schedule can be determined accordingly to estimate the energy consumption of light and AC in EnergyPlus model. Then NSGA-II optimization is conducted to minimize energy consumption and LCC. Preferred designs with least energy consumption and LCC can be identified in Trade Space Visualizer. Most influential variables can be identified using SimLab.

2.3 Objectives

In this study, thermal comfort is determined by the ASHRAE 55 - 2010 Adaptive Model [23], which is suitable for occupants who are naturally acclimatized to the local hot and humid climate conditions, such as in Singapore [24, 25]. The adaptive model relates indoor comfort temperature to prevailing outdoor temperature in naturally ventilated buildings. The model defines two comfort regions: 80% acceptability, and 90% acceptability. The 90% acceptability region is used in this study for a higher requirement of thermal comfort. The acceptable operative temperature ranges with 90% satisfaction rate for naturally conditioned spaces are

$$T_{ot} = 0.31 \times T_o + 17.8 \pm 2.5 \tag{1}$$

where T_{ot} is the operative temperature (°C), calculated as the average of the indoor air dry-bulb temperature and the mean radiant temperature of zone inside surfaces. T_o is the monthly mean outdoor air dry-bulb temperature (°C). Equation (1) is applicable if T_o is between 10 °C and 33.5 °C. The objective to be minimized is the average thermal discomfort rate of all zones when they are occupied, as shown below:

$$O_{P\_1} = 1 - \frac{1}{n} \sum_{i=1}^{n} C_i \tag{2}$$

where n is the number of zones which are evaluated. C_i is the ratio of the time when the satisfaction rate exceeds 90% for the total occupied time of the i_{th} zone.

A lighting level of more than 300 lux is determined to be satisfactory for residential space [26]. In this study, a lighting reference point is set in the centre of the living room at 0.8m above the floor (typical height for measuring light level based on human activity range [27]) to represent the average daylighting level. The object function to be minimized is the average time when daylighting is unavailable or ineffective at the reference point, as shown below.

$$O_{P\_2} = 24 - T/365 \tag{3}$$

where T is the accumulated time when daylight illuminance exceed 300 lux (a light level suggested for general activities in living rooms [28]) at the reference point during a year.

After the optimization of thermal comfort and visual comfort, preferred design candidates are selected for the second optimization of cost effectiveness and energy efficiency. Cost effectiveness is evaluated based on the life cycle cost as shown below:

$$LCC = IC + USPW(N, r_d) \times EC \tag{4}$$

where IC is the initial cost for implementing the design and operating features for both building envelope and building systems. EC is the net annual energy cost required to maintain indoor comfort within the residential building for the selected design and operating features. $USPW$ is the uniform series present worth factor, which converts future recurrent expenses to present costs. USPW depends on the discount rate r_d and lifetime N. The function of $USPW$ is shown below:

$$USPW(N, r_d) = \frac{1 - (1 + r_d)^{-N}}{r_d} \tag{5}$$

Energy efficiency is evaluated based on the total electric energy demand per year, as shown below:

$$E_{total} = E_{AC} + E_L + E_{WH} + E_R \tag{6}$$

where E_{AC}, E_L, E_{WH} and E_R are energy consumptions of air-conditioner, lights, water heater and refrigerator, respectively. The schedules of air-conditioner and lights are determined by the optimization results in the first phase. The schedules of other appliances are based on statistics.

3 Case Study
3.1 Climate of Singapore

Singapore is 1-degree north of the equator and has a tropical rainforest climate, which is characterized by a relatively uniform temperature and pressure, high humidity and abundant rainfall. The daily temperature range has a minimum usually not falling below 23-25°C during the night and maximum not rising above 31-33°C during the day. Therefore, it may be possible to achieve thermal comfort without using air conditioning.

Monthly sunshine hours (with direct irradiance from the sun of at least 120 Watts/m^2) range from 129.6 hours in November to

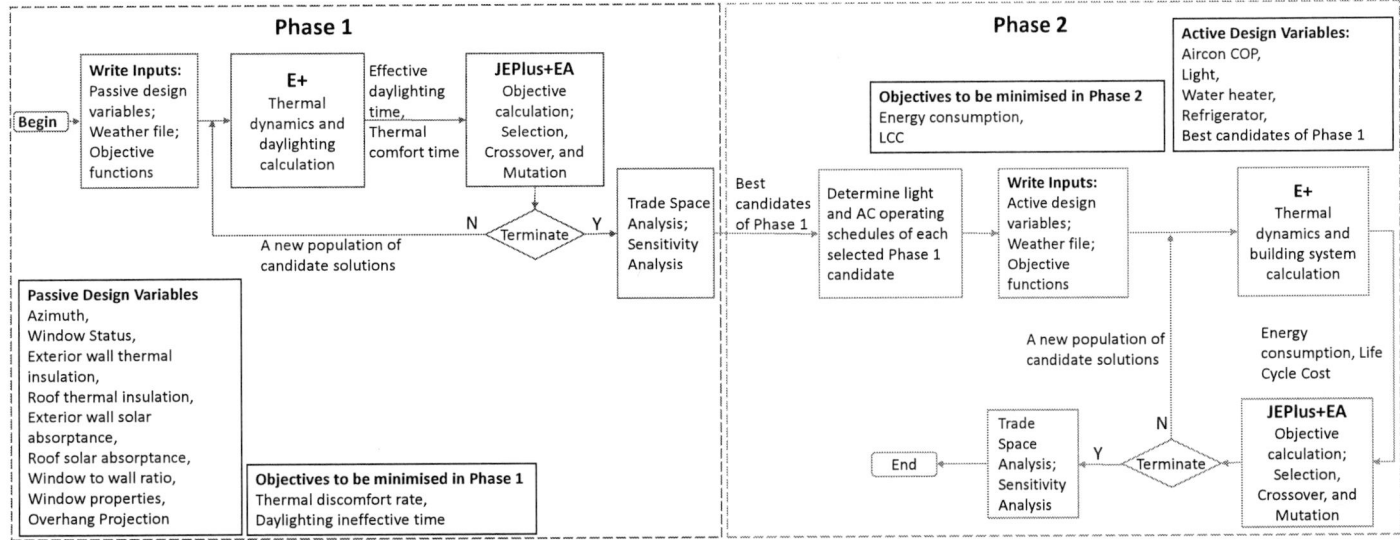

FIGURE 1: DESIGN PROCESS OF NZEB

192.7 hours in March. The long sunshine hours not only provide the potential for utilizing daylighting but also make solar energy the most promising renewable energy source for Singapore when it comes to electricity generation. With an average annual solar irradiance of 1,580 kWh/m²/year and about 50 percent more solar radiation than temperate countries, solar photovoltaic (PV) generation has the greatest potential for wider deployment in Singapore.

3.2 Design Process

3.2.1 Phase 1
Terraced house is the most common landed housing type in Singapore. Most terraced houses are 6 meters wide and 14-20 meters long with 2-3 floors. The floor area is in 200-400m² range. The average household size of landed house is 4.28 people [29].

This case study is based on a hypothetical floor plan of a typical terraced house in Singapore as shown in Fig. 2. The house has three floors with total floor area of 306m² and floor height of 3.5m. It has one living room, four bedrooms, four bathrooms, two study rooms and one kitchen. Exterior windows are located in the front and backside walls. The window in the model is the sliding window type, which is the most commonly used in Singapore. The discharge coefficient is fixed at 0.6, which is a commonly used value for the discharge coefficient of a sharp-rimmed large opening [30]. When fully opened, the effective ventilation ratio of sliding window is 0.5. The side walls are set to be adiabatic as they are shared by two households (terraced house type), and it is assumed to have similar thermal condition and no heat transfer. The internal ceilings are also assumed to be adiabatic. The front and backside exterior walls and the roof

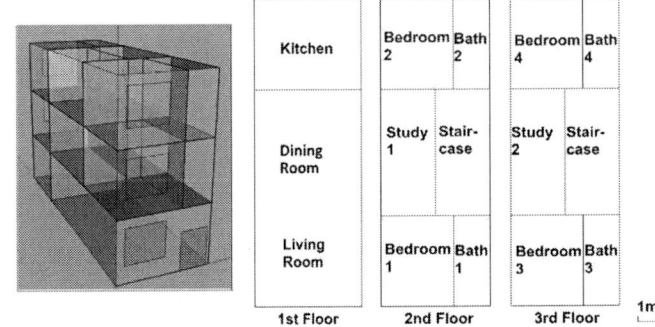

FIGURE 2: GEOMETRY OF LANDED HOUSE

are constructed with 100mm thick brick with 20mm thick cement plasters on both sides. The resulting $U-value$ is 2.76 W/m^2k. Additional insulation is set as a design variable in Phase 1 optimization.

Based on the common occupancy schedule of residential buildings in Singapore [31], the optimization objectives and occupancy schedules of different zones are determined. For bedrooms, only thermal comfort is considered because they are usually occupied at night time (23:30-06:00). For living room, both thermal comfort and daylighting are considered because it can be occupied during both daytime and night time (06:00-23:30). The internal heat source of bedrooms and the living room are mainly from the occupants. The activity levels of people when sleeping and sitting are 72 Watts and 108 Watts, respectively [31]. Based on the average family size of landed house in Singapore,

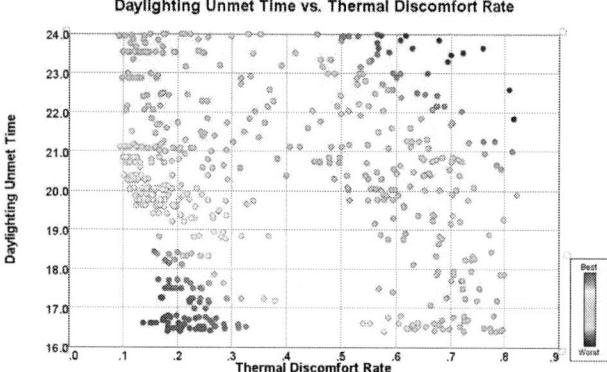

FIGURE 3: OPTIMIZATION RESULT OF PHASE 1

each household is assumed to have 4.28 people. The average number of occupants of each zone is assumed to be 1.07. The design variables with respective options, initial cost and lifespan are shown in Tab. 1. The data is based on existing literatures and publications. There are 9 passive design variables evaluated in Phase 1, resulting in 393,216 combinations. In Tab. 1, azimuth means the building orientation relative to the north. Window status is used to represent the window opening time during a day, i.e. always open, open during daytime only, open during night only, and always closed. Exterior wall/Roof thermal insulation uses polystyrene (conductivity of 0.024 W/mK) with different thickness. Exterior wall/Roof solar absorptance is determined by color. Darker color surface absorbs more solar energy than lighter color surface. Window to Wall Ratio (WWR) is adjusted by varying the window lower edge position while fixing the window width and top edge position. Window types determines the window properties including $U-value$, Solar Heat Gain Coefficient ($SHGC$) and Visible Transmittance (VT). Overhang projection is described by the overhang depth as fraction of window height.

Based on statistics of the most commonly used settings of NSGA-II optimization for NZEB design [11], this study sets the population size as 24, crossover rate as 0.9, mutation rate as 0.355 and max generation as 25. Assuming daylighting and natural cooling is equally important, the 2D scatter plot of the two objectives with preference color shading is shown in Fig. 3. The scatter plot shows all the evaluated design candidates with the best design candidate marked as red color. The minimum daylighting ineffective time is 16.4 hours per day, which means that the longest daylighting hours with illuminance higher than 300 lux is 7.6 hours per day. The minimum thermal discomfort rate is 0.09, but this design candidate is not one of the best as its daylighting unmet time is high. Within the best designs, the best thermal discomfort rate is 0.14.

The sensitivity indexes used are the Partial Correlation Coefficient (PCC) and Standardized Rank Regression Coefficients ($SRRC$). PCC is one of the most suitable indexes for building models considering accuracy and number of inputs [32]. The PCC shows the strength of the correlation between an output Y and an associated input vector X_i, which was cleaned off any effect due to the correlation between the vector X_i and other input vectors. The $SRRC$ is based on regression analysis but is improved by applying the rank transformation of the input and output values (i.e. replacing these values with their ranks). PCC is suitable for linear models while $SRRC$ can be used for nonlinear but monotonic functions among inputs and outputs. The PCC and $SRRC$ tend to supplement each other in measuring sensitivity. Therefore, in this study, both PCC and $SRRC$ are used to provide complex and accurate evaluations.

The sensitivity analysis results are shown in Fig. 4. The PCC and $SRRC$ results show good consistency. It is found that the overhang fraction, window type, and WWR are the most influential variables in daylighting performance. For natural cooling performance, the window status (operating patterns) is found to be the dominant factor over others.

3.2.2 Phase 2 Based on the optimization results in Phase 1, best design candidates with daylighting unmet time less than 18 hours and thermal discomfort rate less than 0.3 are selected for optimization of energy consumption and life cycle cost in Phase 2. A total of 73 candidates, which are the top 13% of all the evaluated candidates in Phase 1, are selected. There are five (5) variables in Phase 2 optimization: the selected candidates from Phase 1, and the four (4) active design variables as shown in Tab. 1.

An air-conditioning system is added to the EnergyPlus model to calculate the cooling energy. The temperature set point of the air-conditioner is assumed to be constant at 24°C. The air-conditioner operating hours are based on the thermal comfort unmet time from Phase 1 simulation. The designed illuminance level is 300 lux for all types of lights. The water heater is assumed to be operating for 0.5 hour each day based on statistics of average shower duration [33]. The fridge is assumed to be always on.

In this study, lifetime N is set to be 25 years (estimated building life), and discount rate r_d is assumed to be 3% [34]. IC is calculated based on the value in Tab. 1. The trade space generated by the optimization is shown in Fig. 5. The minimum energy consumption that can be achieved is 3,167 KWh, and the minimum LCC is 24,673 SGD. A few samples of best designs are listed in Tab. 2.

The PCC and $SRRC$ for sensitivity analysis is shown in Fig. 6. Light type has the most significant impact on both energy consumption and LCC. This may be due to (1) the dramatic performance difference among the different light types, (2) the

TABLE 1: DESIGN VARIABLES

Category	Variable	Description	Options	Initial Cost (S$)	Life Span
Passive Design Variables (Phase 1)	Azimuth	Building orientation relative to the north (°)	Landed: 0, 45, 90, 135, 180, 225, 270, 315 Apartment Building: 0, 45, 90, 135		
	Window Status	window opening time	whole day, daytime only, night-time only, always closed		
	Exterior Wall Thermal Insulation (polystyrene)	Thickness of insulation (mm)	20, 10, 5 , 0	140/m3	30 years
	Exterior Wall Solar Absorptance	determines by color	0.3 (light yellow), 0.5 (green/blue), 0.7 (dark brown), 0.9 (very dark color)		
	Roof Thermal Insulation (polystyrene)	Thickness of insulation (mm)	20, 10, 5 , 0	140/m3	30 years
	Roof Solar Absorptance	determines by color	0.3 (light yellow), 0.5 (green/blue), 0.7 (dark brown), 0.9 (very dark color)		
	WWR	Window-to-Wall ratio	0.2,0.3,0.4,0.5		
	Window Types	$U-value\|SHGC\|VT$	6.7|0.7|0.68, 5.6|0.41|0.42, 4.8|0.59|0.57	38/m2, 220/m2, 200 /m2	30 years
	Overhang Projection	Overhang depth as fraction of window height	0, 0.27 (15 degree), 0.58 (30 degree), 1 (45 degree)		
Active Design Variables (Phase 2)	Air-conditioner Efficiency	Full load COP	3.8, 4.2, 4.6, 5.0	2727, 2925, 3124, 3323	15 years
	Water Heater	Efficiency	0.9 (4000W), 2.7 (heat pump, 1300W)	400, 2000	10 years
	Light type	Efficacy (lm/W) and type	70(LED), 50(CFL), 10(Incadescent)	6/unit, 3/unit, 1.5/unit	25000h, 8000h, 1200h
	Refrigerator	W per 400L	50, 37	905/unit,1050/unit	10 years

optimized natural cooling time, which shortened the AC operating time significantly, and (3) and lighting time is unavoidable at night. Water heater type and air-conditioner operating hours also have relatively big impact on energy consumption and *LCC*.

3.3 Renewable Energy System Requirement

Based on the optimization results, the required capacity of energy generation facility to achieve net-zero energy is estimated. The lowest annual electricity demand of landed house is 3,167kWh. Assuming the electricity consumption of other appliances (e.g. television, fan, kitchen appliance, washing machine, etc.) is 20% of the total household energy consumption [35], the total electricity demand of landed house is 3,959kWh.

To achieve net zero electricity, the required sizes of PV panel for landed building is 16.5m² based on the following calculation.

$$A = \frac{E}{r * H * PR} \quad (7)$$

where A is the required PV panel area. E is the energy demand. r is solar panel efficiency, and 18% efficiency is assumed. H is the annual average solar radiation, and 1,666 kWh/m²/year is used

based on the weather file of Singapore [36]. *PR* is the performance ratio, i.e. the coefficient for losses due to inverter losses, dust, weak radiation, etc.. A *PR* value of 0.8 is assumed. As the rooftop area of landed house is 102 m², it is achievable to build net-zero energy landed house.

4 Discussions and limitations
4.1 Discussions of case study results

Does thermal comfort and visual comfort aims necessarily contradict each other? The results (as shown in Fig. 4) show that among the most influential variables, overhang fraction has positive correlation with thermal comfort while negative correlation with visual comfort. Window type ($U-value\|SHGC\|VT$) has negative correlation with thermal comfort while positive correlation with visual comfort. However, WWR has positive correlation with both thermal comfort and daylighting. Therefore, well balanced overhang fraction and window properties should be chosen by multi-objective optimization. And WWR should be maximized to optimize daylighting and natural cooling.

Does economic and environmental aims necessarily contradict each other? The results (as shown in Fig. 6) shows that lower light efficacy and higher air-conditioner *COP* leads to

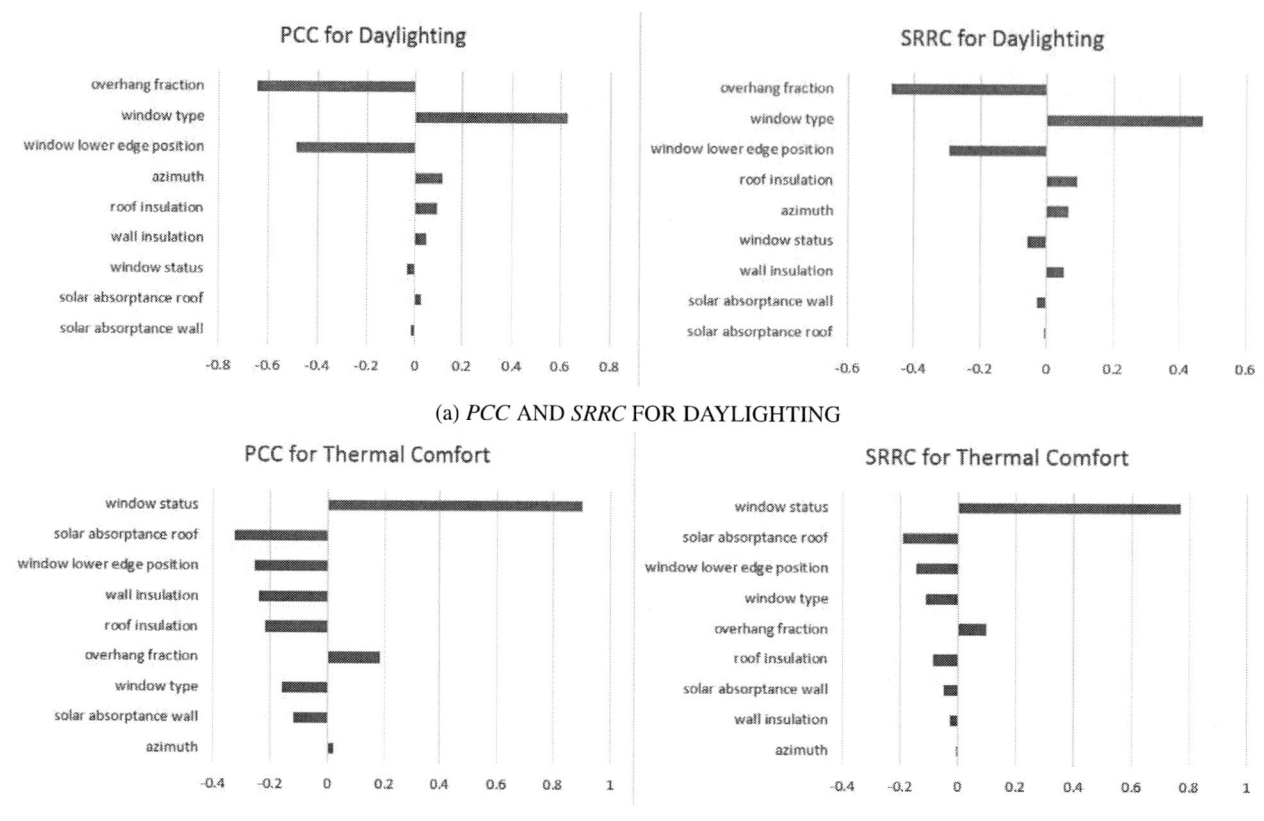

(a) *PCC* AND *SRRC* FOR DAYLIGHTING

(b) *PCC* AND *SRRC* FOR NATURAL COOLING

FIGURE 4: SENSITIVITY ANALYSIS OF PHASE 1

both higher energy efficiency and higher cost effectiveness. This is because the energy savings from the more efficient lights and air-conditioners are able to cover their higher initials costs. However, higher water heater efficiency (heat pump water heater) increases energy efficiency but results in lower cost effectiveness. This is because the initial cost difference is greater than the cost savings from electricity. Thus, high efficiency lights and air-conditioner should be selected for better energy efficiency and cost effectiveness. Water heater type should be selected based on the preference for the environmental factor or the economic factor.

How does Phase 1 result influence Phase 2 results? (Does social aims necessarily contradict environmental and economic aims?) P1 results influence P2 results by determining air-conditioner and light operating rates, and the thermal properties of the building. The objective of P1 is to achieve thermal and visual comfort by natural means, i.e. minimize air-conditioner and light operating rate. This objective contributes to the improvement of energy efficiency and cost effectiveness by reducing cooling and lighting energy consumption in P2. As shown in the P1 sensitivity analysis, natural cooling requires lower solar

absorptance of walls and roof to reduce solar gain, which also benefits the cooling energy consumption in P2. However, natural cooling also requires higher WWR, thinner insulation for internal heat to dissipate outside fast, which may result in higher cooling power in P2. From the sensitivity analysis of P2, lower air-conditioner operating rate (i.e. longer natural cooling hours) leads to higher energy efficiency and cost effectiveness. This means although higher WWR and thinner insulation increases the cooling power of air-conditioner when it is in operation, the energy savings from the shorter operating hours of air-conditioner is able to offset the energy consumption and lead to higher energy efficiency and cost savings. In other words, design for natural cooling increases energy efficiency and cost effectiveness.

4.2 Limitations
The limitations of this study are listed below.

1. The case study is based on the predetermined building shapes and layouts. The building performance (including natural cooling, daylighting, energy and LCC) and solar energy potential should vary as the shape and layout of the

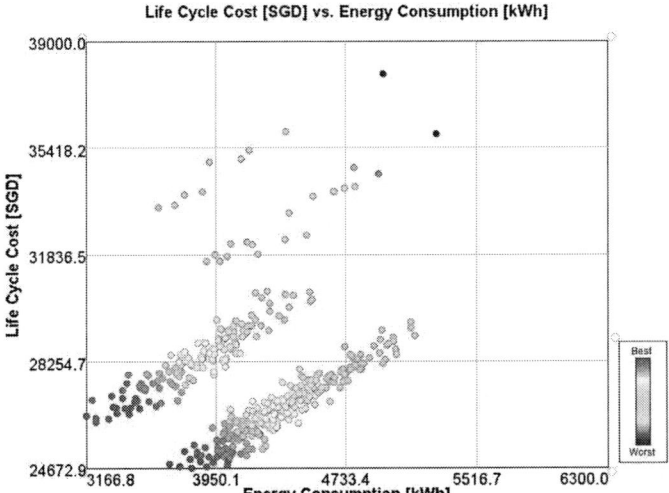

FIGURE 5: OPTIMIZATION RESULT OF PHASE 2

TABLE 2: SAMPLE BEST DESIGNS

Case #	1	2	3	4
Energy Consumption [kWh]	3668	3167	3227	3202
Life Cycle Cost [SGD]	24783	26417	26204	26940
Effective Daylighting Time [hr]	7.40	7.40	7.40	7.40
Natural Cooling Thermal Comfort Rate	0.85	0.85	0.85	0.86
Aircon COP	5	5	4.6	5
Water Heater Type (Power) [W]	4000	1300	1300	1300
Light Type	LED	LED	LED	LED
Refrigerator Type (Power/400L) [W]	37	37	37	37
Azimuth	180	180	180	180
Window Type (U-Value)	6.7	6.7	6.7	6.7
Wall Insulation (Thickness) [m]	0.005	0.005	0	0.02
Window Status Daytime	0	0	0	1
Window Status Night	1	1	1	1
WWR	0.5	0.5	0.5	0.5
Overhang Projection	0.27	0.27	0.27	0.27
Roof Insulation (Thickness) [m]	0	0	0	0
Solar Absorptance Roof	0.3	0.3	0.3	0.3
Solar Absorptance Wall	0.3	0.3	0.3	0.3

building changes.

2. Discount rate and energy price escalation rate are not analyzed as variables in this study. Sensitivity analysis may be conducted to explore the effect of the above factors on the outcome of optimization results.

3. The optimization algorithm applied in this study is selected based on the most used algorithms in the literatures. However, there are many optimization algorithms available. Comparisons between different optimization approaches can be found in [37].

4. It has long been known that occupant behavior has great impact on energy use in buildings [38]. The proposed approach integrates daylighting and natural cooling in the design of NZEB with the aim to improve occupants' satisfaction. The integration of these features also helps to improve building energy efficiency. However, it is important to have occupants' acceptance of utilizing daylighting and natural cooling. Therefore, efforts from other domains, such as behavioral science, are also important to guarantee desired user behavior for NZEB.

5   Conclusions

This paper presents a holistic design approach for residential NZEB. The proposed approach aligns with the TBL principles for sustainable design by optimizing daylighting, natural cooling, energy efficiency and *LCC* of residential NZEB. The optimizations are conducted in two phases: optimizing daylighting and natural cooling in the first phase, and energy efficiency and *LCC* in the second phase. The optimization is performed by coupling JEPlus+EA, a multi-objective optimization tool based on NSGA-II, with EnergyPlus to find the optimized design variables. Sensitivity analysis reveals the most influential design variables on each objective.

This work also evaluates the potential of building residential NZEB in a tropical climate, and a typical landed house in Singapore is evaluated as a case study. The results show that it is achievable to build a net-zero energy landed house. Further discussions on the results of the case study show that although some variables need to be carefully selected to balance daylighting and natural cooling the two objectives do not always contradict each other regarding certain variables, such as *WWR*. Similarly, environmental aims and economic aims do not always contradict each other on certain variables. In fact, more energy efficient appliances tend to result in lower *LCC*. The social aims do not contradict environmental and economic aims as the findings show that designing for daylighting and natural cooling contributes to the improvement of energy efficiency and cost effectiveness.

ACKNOWLEDGMENT

The authors would like to thank Dr. Sumbul Khan for her help. This work is supported by SUTD-MIT International Design Centre (IDC, idc.sutd.edu.sg). Any findings, conclusions,

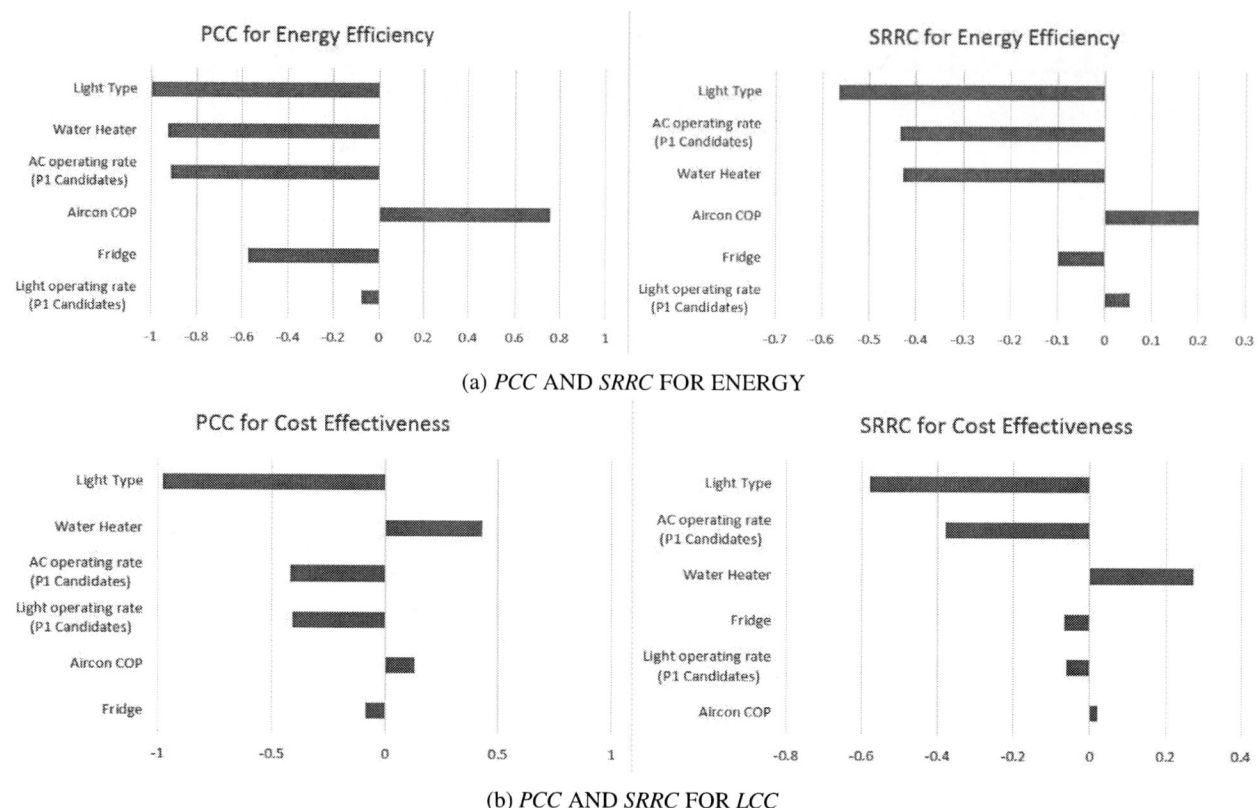

(a) *PCC* AND *SRRC* FOR ENERGY

(b) *PCC* AND *SRRC* FOR *LCC*

FIGURE 6: SENSITIVITY ANALYSIS OF PHASE 2

or opinions expressed in this document are those of the authors and do not necessarily reflect the views of the sponsor.

REFERENCES

[1] T. J. Hall, "The triple bottom line: what is it and how does it work?" *Indiana business review*, vol. 86, no. 1, p. 4, 2011.

[2] M. Hamdy, A. Hasan, and K. Siren, "A multi-stage optimization method for cost-optimal and nearly-zero-energy building solutions in line with the epbd-recast 2010," *Energy and Buildings*, vol. 56, pp. 189 – 203, 2013. [Online]. Available: http://www.sciencedirect.com/science/article/pii/S0378778812004276

[3] Y. Lu, S. Wang, Y. Zhao, and C. Yan, "Renewable energy system optimization of low/zero energy buildings using single-objective and multi-objective optimization methods," *Energy and Buildings*, vol. 89, pp. 61 – 75, 2015. [Online]. Available: http://www.sciencedirect.com/science/article/pii/S0378778814010834

[4] P. M. Congedo, C. Baglivo, D. D'Agostino, and I. Zacà, "Cost-optimal design for nearly zero energy office buildings located in warm climates," *Energy*, vol. 91, pp. 967–982, 2015.

[5] M. Krarti and P. Ihm, "Evaluation of net-zero energy residential buildings in the mena region," *Sustainable Cities and Society*, vol. 22, pp. 116 – 125, 2016. [Online]. Available: http://www.sciencedirect.com/science/article/pii/S2210670716300208

[6] S. K. Pal, A. Takano, K. Alanne, and K. Siren, "A life cycle approach to optimizing carbon footprint and costs of a residential building," *Building and Environment*, vol. 123, pp. 146–162, 2017.

[7] D. D'Agostino and D. Parker, "A framework for the cost-optimal design of nearly zero energy buildings (nzebs) in representative climates across europe," *Energy*, vol. 149, pp. 814–829, 2018.

[8] F. Harkouss, F. Fardoun, and P. H. Biwole, "Multi-objective optimization methodology for net zero energy buildings," *Journal of Building Engineering*, vol. 16, pp. 57–71, 2018.

[9] M. Ferrara, E. Fabrizio, J. Virgone, and M. Filippi, "A simulation-based optimization method for cost-optimal analysis of nearly zero energy buildings," *Energy and Buildings*, vol. 84, pp. 442 – 457, 2014. [Online]. Available: http://www.sciencedirect.com/science/article/

pii/S0378778814006732

[10] A. J. Marszal and P. Heiselberg, "Life cycle cost analysis of a multi-storey residential net zero energy building in denmark," *Energy*, vol. 36, no. 9, pp. 5600–5609, 2011.

[11] S. Carlucci, G. Cattarin, F. Causone, and L. Pagliano, "Multi-objective optimization of a nearly zero-energy building based on thermal and visual discomfort minimization using a non-dominated sorting genetic algorithm (nsga-ii)," *Energy and Buildings*, vol. 104, pp. 378–394, 2015.

[12] H. Li, S. Wang, and H. Cheung, "Sensitivity analysis of design parameters and optimal design for zero/low energy buildings in subtropical regions," *Applied energy*, vol. 228, pp. 1280–1291, 2018.

[13] L. Lan, W. Tushar, K. H. E. Sng, C. Yuen, K. L. Wood, and T. Saha, "Design innovation approaches for sustainable smart energy systems," in *ASME 2018 International Design Engineering Technical Conferences and Computers and Information in Engineering Conference*. American Society of Mechanical Engineers, 2018, pp. V004T05A038–V004T05A038.

[14] B. A. Camburn, J. M. Auernhammer, K. H. E. Sng, P. J. Mignone, R. M. Arlitt, K. B. Perez, Z. Huang, S. Basnet, L. T. Blessing, and K. L. Wood, "Design innovation: A study of integrated practice," in *ASME 2017 International Design Engineering Technical Conferences and Computers and Information in Engineering Conference*. American Society of Mechanical Engineers, 2017, pp. V007T06A031–V007T06A031.

[15] M. Khoshbakht, Z. Gou, Y. Lu, X. Xie, and J. Zhang, "Are green buildings more satisfactory? a review of global evidence," *Habitat International*, 2018.

[16] S. R. Kellert, J. Heerwagen, and M. Mador, *Biophilic design: the theory, science and practice of bringing buildings to life*. John Wiley & Sons, 2011.

[17] Jeplus. [Online]. Available: http://www.jeplus.org/wiki/doku.php

[18] K. Deb, A. Pratap, S. Agarwal, and T. Meyarivan, "A fast and elitist multiobjective genetic algorithm: Nsga-ii," *IEEE transactions on evolutionary computation*, vol. 6, no. 2, pp. 182–197, 2002.

[19] Energyplus. [Online]. Available: https://energyplus.net/

[20] Trade space visualizer. [Online]. Available: http://www.atsv.psu.edu/whatisatsv.html

[21] Simlab. [Online]. Available: https://ec.europa.eu/jrc/en/samo/simlab

[22] C. Withanage, K. Hölttä-Otto, K. Otto, and K. Wood, "Design for sustainable use of appliances: A framework based on user behavior observations," *Journal of Mechanical Design*, vol. 138, no. 10, p. 101102, 2016.

[23] *ANSI/ASHRAE/ASHE Standard 55-2010: Thermal Environmental Conditions for Human Occupancy*, ser.

ASHRAE standard. ASHRAE, 2010. [Online]. Available: https://books.google.com.sg/books?id=SvcrMwEACAAJ

[24] N. Wong, H. Feriadi, P. Lim, K. Tham, C. Sekhar, and K. Cheong, "Thermal comfort evaluation of naturally ventilated public housing in singapore," *Building and Environment*, vol. 37, no. 12, pp. 1267–1277, 2002.

[25] L. Lan, W. Tushar, K. Otto, C. Yuen, and K. L. Wood, "Thermal comfort improvement of naturally ventilated patient wards in singapore," *Energy and Buildings*, vol. 154, pp. 499–512, 2017.

[26] P. Boyce and P. Raynham, *SLL Lighting Handbook*. CIBSE, 2009.

[27] Led light lux levels and light measurement and brightness. [Online]. Available: http://www.liteonled.com.au/buying-guide/understanding-led-lighting/led-lux-levels

[28] J. Yao, "An investigation into the impact of movable solar shades on energy, indoor thermal and visual comfort improvements," *Building and Environment*, vol. 71, pp. 24–32, 2014.

[29] Department of statistics singapore. [Online]. Available: https://www.singstat.gov.sg/

[30] F. Flourentzou, J. Van der Maas, and C.-A. Roulet, "Natural ventilation for passive cooling: measurement of discharge coefficients," *Energy and buildings*, vol. 27, no. 3, pp. 283–292, 1998.

[31] Y. Liu, R. Stouffs, A. Tablada, N. H. Wong, and J. Zhang, "Comparing micro-scale weather data to building energy consumption in singapore," *Energy and Buildings*, vol. 152, pp. 776–791, 2017.

[32] A.-T. Nguyen and S. Reiter, "A performance comparison of sensitivity analysis methods for building energy models," in *Building Simulation*, vol. 8, no. 6. Springer, 2015, pp. 651–664.

[33] Indoor water usage. [Online]. Available: https://www.home-water-works.org/indoor-use/showers

[34] J. J. H. Low, Y. Ko, A. Ilancheran, X. H. Zhang, P. K. Singhal, and S. K. Tay, "Health and economic burden of hpv-related diseases in singapore," *Asian Pacific Journal of Cancer Prevention*, vol. 13, no. 1, pp. 305–308, 2012.

[35] Home energy audit. [Online]. Available: http://www.e2singapore.gov.sg/Households/Saving_Energy_At_Home/HEA.aspx

[36] Weather data. [Online]. Available: https://energyplus.net/weather

[37] M. Hamdy, A.-T. Nguyen, and J. L. Hensen, "A performance comparison of multi-objective optimization algorithms for solving nearly-zero-energy-building design problems," *Energy and Buildings*, vol. 121, pp. 57–71, 2016.

[38] P. C. Stern, "Contributions of psychology to limiting climate change." *American Psychologist*, vol. 66, no. 4, p. 303, 2011.

Proceedings of the ASME 2019
International Design Engineering Technical Conferences
and Computers and Information in Engineering Conference
IDETC/CIE2019
August 18-21, 2019, Anaheim, CA, USA

DETC2019-97190

DESIGNING OPTIMAL ARBITRAGE POLICIES FOR DISTRIBUTED ENERGY SYSTEMS IN BUILDING CLUSTERS USING REINFORCEMENT LEARNING

Philip Odonkor
Graduate Research Assistant
Mechanical & Aerospace Engineering
University at Buffalo
Buffalo, NY 14260
Email: podonkor@buffalo.edu

Kemper Lewis*
Professor
Mechanical & Aerospace Engineering
University at Buffalo
Buffalo, NY 14260
Email: kelewis@buffalo.edu

ABSTRACT

In the wake of increasing proliferation of renewable energy and distributed energy resources (DERs), grid designers and operators alike are faced with several emerging challenges in curbing allocative grid inefficiencies and maintaining operational stability. One such challenge relates to the increased price volatility within real-time electricity markets, a result of the inherent intermittency of renewable energy. With this challenge, however, comes heightened economic interest in exploiting the arbitrage potential of price volatility towards demand-side energy cost savings. To this end, this paper aims to maximize the arbitrage value of electricity through the optimal design of control strategies for DERs. Formulated as an arbitrage maximization problem using design optimization, and solved using reinforcement learning, the proposed approach is applied towards shared DERs within multi-building residential clusters. We demonstrate its feasibility across three unique building cluster demand profiles, observing notable energy cost reductions over baseline values. This highlights a capability for generalized learning across multiple building clusters and the ability to design efficient arbitrage policies towards energy cost minimization. Finally, the approach is shown to be computationally tractable, designing efficient strategies in approximately 5 hours of training over a simulation time horizon of 1 month.

Keywords : Building cluster, operational strategy design, battery storage, reinforcement learning.

------

*Corresponding Author

1 INTRODUCTION

The energy grid is in a state of flux. From solar roofing tiles [1] to battery storage systems [2], distributed energy resources (DER) are diversifying electricity supply, and fundamentally changing the way we access and manage electricity. But as new designs and technologies for renewable DERs become available, grid designers and operators alike are faced with new design challenges in ensuring seamless DER integration, and for maintaining transient and steady-state balances across the grid [3,4]. One notable operational challenge relates to the increased level of volatility across real-time electricity pricing markets, a result of the inherent intermittency of renewable energy [5–7]. Economists have long recognized the adverse impacts of such volatility, which creates allocative inefficiencies across the grid [8]. While price volatility remains a concern, a compelling case can be made for leveraging it, along the optimized control of distributed energy resources to realize significant energy cost savings.

Recent advancements in energy storage systems, demand-side energy management technologies and the smart grid have created multiple avenues for providing grid support as illustrated in Figure 1. With regards to energy cost savings, two applications standout, namely; (i) energy arbitrage - the purchasing and storing of cheap electricity during off-peak times and using or reselling it during peak; and (ii) ancillary services such as frequency regulation and voltage support [10, 11]. The goal of this paper is to design optimal operational strategies that leverage the generation and storage potential of residential DERs to maxi-

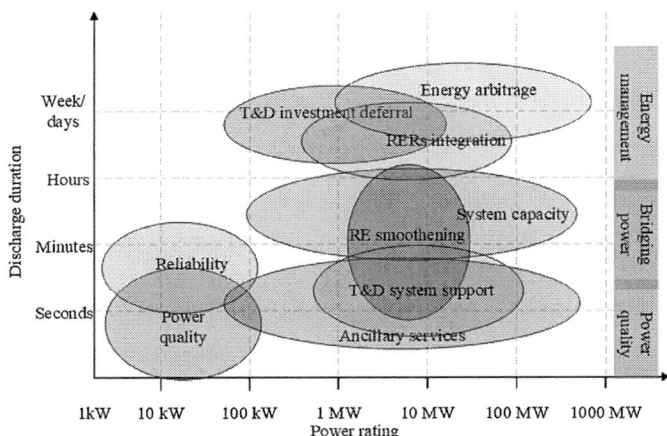

FIGURE 1: *Applications of energy storage systems (ESS) according to duration of discharge and storage capacity [9]*

mize the arbitrage value of electricity, thereby minimizing energy costs. While conceptually easy to understand, the design and subsequent implementation of efficient arbitrage strategies have proved anything but trivial. A notable design challenge stems from the highly dynamic nature of future energy price and demand patterns, making them particularly difficult to forecast [12]. A survey of the literature underlines the difficulties associated with grid-level arbitrage operations, and is discussed in the following section.

2 LITERATURE REVIEW

Arbitrage operations at the grid level has been the focus of several literature findings in recent years. In a study aimed at maximizing revenues from energy arbitrage, Kloess [13] simulated optimal market-based storage operations within the Austrian energy market. This work detailed a notable decline of about 60% in arbitrage revenue starting from the year 2007 to 2011. In a similar vain, Steffen [14] used data captured between 2002-2010 to analyze the profitability of arbitrage operations using pumped-hydro storage (PHS) in Germany. The study found arbitrage profits to be marginal, with the overall results failing to make a strong case for its wholesale adoption in the German energy market. Zafirakis et al. [15] leveraged 5 years worth of historical data to study arbitrage revenues generated within the UK, Spanish, Nordic and Greek energy markets. In addition to observing significant revenue variations across the markets, the study found revenue levels insufficient to justify storage investments.

Other studies such as [16] formulated a linear optimization problem to investigate the economic viability and arbitrage potential of energy storage systems (ESS) in seven U.S markets using 14 different ESS technologies. The study found that the profit-maximizing potential of energy storage systems relied more on their technological characteristics than market price

volatility, which they found to be strongly correlated to the internal rate of return of the ESS. On the other end of the spectrum, Walawalkar et al. [17] detailed a strong economic case of deploying energy arbitrage strategies in tandem with sodium sulfur batteries and flywheel energy storage systems within the New York electricity market.

Beyond grid-wide arbitrage operations, another promising application relates to "small device" energy arbitrage – the ability to buy low and sell high *without* impacting the price of electricity [18]. This type of arbitrage is particularly common in residential applications, and has served as the focus of a number of literature studies [19–22]. These studies typically adopt a "price-taker" model, along with the assumption of perfect knowledge with regards to future energy prices. Zheng et al. [20] for example determined the economic viability of arbitrage in the residential sector using a range of energy storage technologies. The study reported profits as high as 48% of annual electricity costs of a typical household. Building on these findings, Zheng et al. [21] went on to develop peak-shaving demand response (DR) strategies for distributed energy storage systems in residential buildings. The dispatch strategy they used optimized storage capacity and the grid demand limit, allowing for energy cost savings as high as 39% of the households non-DR electricity cost.

With the future of energy generation and storage expected to become increasingly decentralized, there have been studies focused on community-based residential DERs. Hu et al. [23] for example investigated operation strategies in battery equipped building communities (clusters) using memetic algorithms. Odonkor et al. [24, 25] followed this up with specific applications towards Net-Zero energy buildings clusters. Commercially, there is growing interest towards using Artificial Intelligence (AI) to design efficient energy arbitrage strategies. Energy startups such as Upside Energy [26], BeeBryte [27] and Grid-Edge [28] have seen early success in using machine learning towards managing their portfolios of DERs. Despite the early success, the broad and nascent nature of this field of study means that numerous open research challenges still remain, requiring the contribution of many research efforts. Recent literature contributions [12, 29–32] have looked towards addressing some of these research challenges with respect to designing efficient arbitrage policies. In [12], the authors derived an arbitrage policy for storage using reinforcement learning, demonstrating significant performance improvements over existing approaches. The authors in [29] also leveraged reinforcement learning to derive arbitrage policies, but their work varied in terms of scope and control fidelity. Their approach relied on continuous battery control, with applications towards shared battery storage within building clusters. In addition to these works, Elena et al. [31] leveraged reinforcement learning to optimize the scheduling of energy consumption devices within residential homes, realizing notable energy cost and peak minimizations via this approach. Meanwhile, François-Lavet et al. [32] used deep reinforcement learning to control DERs within a microgrid setting. While similar to the work presented in this paper, their model relied on an instantaneous reward signal whereas we actively augment our instantaneous reward signal using historical price data.

As energy supply on the grid becomes diversified, new opportunities exist for extending arbitrage operations to include renewable DERs (i.e., PV panels). Much like grid energy, PV energy has arbitrage value, and can be used towards energy cost savings. Recognizing the overall promise of arbitrage operations using DERs and the learning potential of Reinforcement Learning (RL) algorithms, we demonstrate the feasibility of designing arbitrage policies for PV and battery assets within building communities. To this end, this paper makes three significant contributions over previous work [29]: (i) It extends the reinforcement learning environment to two energy systems, namely a battery storage system and a photo-voltaic system. The additional degrees of freedom this provides allows for a more expansive solution space. This brings about more variations within the learned policies and creates more opportunities for energy cost savings. (ii) It reformulates the the battery reward function to incorporate historical energy price data. This allows the agent to be more exploratory with its policies, allowing it to avoid the premature convergence issues noted in [29]. (iii) Finally, in this work, the time horizon under consideration is expanded from 1 day (24 hours) to 1 month (744 hours). By increasing the time horizon of the study, we increase the agent's ability to approximate arbitrage policies over a wider range of data. This is crucial becomes it marks our progression towards developing a truly generalized learning agent.

To the best of our knowledge, this is the first study to apply reinforcement learning towards the concurrent maximization of arbitrage value of shared distributed energy resources. To achieve this, we consider a single PV and battery storage system shared between two homes. We train a reinforcement learning agent to develop efficient operational strategies to concurrently maximize the arbitrage value of the electricity stored and generated by the battery and PV respectively. Figure 2 provides an overview of the proposed approach.

Figure 2 depicts a total of six numbered control "valves" which orchestrate the flow of electricity within the building cluster. This represents three more valves than were considered in [29]. The three new additions (valves 1, 4 and 6) are related to the photovoltaic system, and control the flow of solar energy within the building cluster. More specifically, valves 1 allows the building cluster to be directly powered by solar energy, valve 4 allows for excess solar energy to be sold to the energy grid, and valve 6 allows the battery storage system to be charged using solar energy. In addition to these new valves, valve 2 allows the cluster to buy electricity directly from the grid, valve 3 allows the cluster to utilize stored battery power, and finally, valve 5 allows the battery to be charged using cheap grid energy. Using reinforcement learning, we train an agent to optimally manage

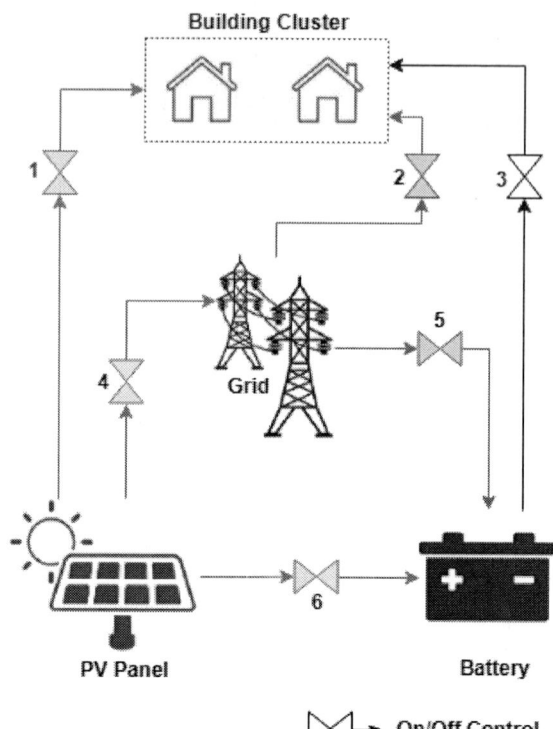

FIGURE 2: *Architecture of the proposed shared DER dispatch problem*

each control valve on an hourly basis to maximize the arbitrage value of electricity flowing within the cluster. The hourly time granularity used was dictated by the granularity of the input data used (energy consumption data and electricity pricing structure), both of which are on an hourly scale. Although this paper applies RL to the arbitrage problem, alternative approaches such as direct and indirect methods are also applicable. Direct methods are particularly interesting because of their ability to scale to large systems and potentially provide continuous solutions for the optimal policy. Their main drawback however stems from an inability to generalize to new situations, an inherent strength of RL. Additionally, they typically require a priori system models and can also have potential convergence issues.

The remainder of the paper details how this was achieved and is structured as follows. Section 3 begins with a brief introduction to Reinforcement Learning. It then details the dynamics of the RL battery and PV environments, along with the learning agent modeled to interact with them. Section 4 integrates the environments with the learning agent and develops a case study which is used to test the viability of the proposed approach. Sections 5 and 6 close the paper with a discussion of our results and concluding remarks respectively.

3 METHODOLOGY

3.1 Background: Reinforcement Learning

Reinforcement learning (RL) is a branch of Machine Learning built on the idea of learning through trial-and-error. RL algorithms are trained from their own experiences, constructing knowledge through their interactions with an environment. Desired behaviors are typically rewarded, while undesired actions are penalized. This feedback mechanism is used to master seemingly complex tasks and has seen successful applications in healthcare [33], robotics [34] and consumer modeling [35].

Reinforcement learning algorithms adopt a standard Markov Decision Process (MDP) formalism [36] defined by the tuple $\langle S, A, R, P, \gamma \rangle$, composed of a set of states S and actions A, a reward function $R(s, a)$, a transition function $P(S'|s, a)$, and a discount factor γ. At time t, an RL agent takes an action ($a_t \in A$) influenced by its current state ($s_t \in S$). The agent then observes a reward feedback ($r(s_t, a_t) \in \mathbb{R}$) released by the environment. It then transitions to a new state (s') and repeats the process. This cyclic process is illustrated in Figure 3.

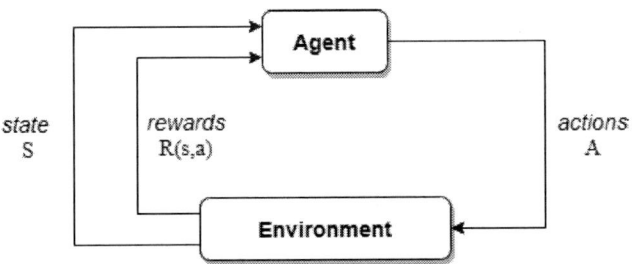

FIGURE 3: *Agent-Environment interaction in reinforcement learning*

A policy, $\pi : S \to A$ dictates an agent's behavior by mapping an observation s_t to an action a. The goal of reinforcement learning is to learn an efficient policy, π, which maximizes the cumulative reward generated by an environment during a given task.

In general, RL can be used to solve problems in which (i) future actions are influenced by the current state of the system; (ii) the goal is to maximize the cumulative reward emitted by the environment; (iii) a fully observable Markov environment is assumed; and (iv) the system is potentially non-stationary. From recent literature [12, 29], we know that the energy storage arbitrage problem aligns well with all four properties. Moreover, with respect to the PV arbitrage problem, all four properties are satisfied in that: (i) varying system states influence future actions with regards to decisions including solar self-consumption and storage; (ii) the operation of the PV aims to maximize total arbitrage profit; (iii) the full state of the PV environment is known at every time step; and (iv) the irradiance, household consumption,

and electricity price profiles are all non-stationary. This supports our decision to use RL in this work.

With a goal of maximizing arbitrage profits, the overall objective function is formulated as follows:

$$\max \quad \sum_{t=1}^{T} \left(\eta_{rt} Bat_t^{dis} - Bat_t^{char} + PV_t^{use} + PV_t^{sell} \right) \quad (1a)$$

$$\text{s.t.} \quad SoC_t = SoC_{t-1} + c_t - d_t \quad \forall t \in \mathcal{T} \quad (1b)$$

$$SoC_{min} \leq SoC_t \leq SoC_{max} \quad \forall t \in \mathcal{T} \quad (1c)$$

$$c_{max} = d_{max} \quad \forall t \in \mathcal{T} \quad (1d)$$

where $Bat_{dis,t}$ and $Bat_{char,t}$ represent the monetary cost of electricity discharged from and charged to the battery respectively. The variable PV_t^{use}, represents the monetary savings of solar self-consumption, while PV_t^{sell} represents the revenue from excess solar energy. The constant η_{rt} represents the round-trip efficiency of the battery. Constraint Eqn. 1b defines the state of charge (SoC) of the battery, where c_t represents the amount of electricity used to charge the battery and d_t represents battery discharge. Constraint Eqn. 1c ensures that the battery cannot be over-charged or over-discharged. Finally, Constraint Eqn. 1d sets the maximum charging and discharging rate of the battery to be equal. This constraint is imposed to reduce the complexity of the battery model.

In order to solve this problem using RL, we need to first adapt it to the general RL framework. From Figure 3, we observe two essential components which need to be identified within our problem: (i) the environment; and (ii) the agent. The environment in our case represents the battery and PV system, while the agent represents the RL algorithm used to solve the objective function. The following subsections provide an in-depth look at how the environments were modeled, along with the RL agent we used.

3.2 RL Environment: Battery Storage

Modeled after Tesla's Powerwall 2 [2], the battery environment is defined by three fundamental parameters, namely: the energy capacity, $E_{rated} = 13.2$ kWh, the power capacity, $P_{rated} = 3.3$ kW, and the round-trip efficiency, $\eta_{rt} = 89\%$. While important, the impact of battery degradation on arbitrage profits is omitted from this study owing to the relatively short time horizon (1 month) considered. The state space of the environment is defined by five time-dependent variables. At any given time, t, the state space is defined by the current time, t, the energy consumption of the building cluster (kW), the state of charge of the battery, SoC_t, the buying price of electricity, E_{price}, and the selling price of electricity, E_{sell}. The electricity consumption profiles for each building was sourced from the Pecan Street Inc. Dataport database [37].

The charge/discharge power capacity of the battery, P_{bat}, is modeled to have a symmetric range bounded by the rated power capacity as formulated in Eqn. 2. The symmetric range allows for positive values to correspond to battery charging and negative values to battery discharge.

$$-P_{rated} \leq P_{bat} \leq P_{rated} \tag{2}$$

To avert the overcharging or discharging of the battery, Eqn. 3 first calculates an unbounded "new charge", (U_{nc}), by adding together the current battery State of charge, $SoC_{(bat,old)}$, with the proposed charge/discharge value P_{bat}.

$$U_{nc} = E_{(bat,old)} + P_{bat} \tag{3}$$

The resulting sum is then compared against E_{rated} using Eqn. 4 to ensure that it does not violate the battery's capacity limit.

$$E_{bat} = max\left(min\left(U_{nc}, E_{rated}\right), 0\right) \tag{4}$$

To account for battery inefficiencies, a loss is incurred every time the battery undergoes a charging cycle. This loss, $loss_{bat}$, is modeled as Eqn. 5.

$$loss_{bat} = (E_{bat} - E_{(bat,old)}) \times (1 - \eta_{rt}) \tag{5}$$

Eqn. 4 is then updated to account for battery inefficiencies. This is presented as Eqn. 6.

$$E_{bat,new} = E_{bat} - loss_{bat} \tag{6}$$

3.2.1 Battery Action Space

The action space encapsulates all possible actions available to the agent within the battery environment. The challenge of handling problems with finite, discrete action spaces has been well-studied in the literature [38, 39] and serves as the basis for us in modeling a discrete action space for our arbitrage operations. Consequently, the battery action space, formulated in Eqn. 7, consists of three discrete actions: *charge*, *discharge* and *dormant*.

$$\mathcal{A} = \{P_{max}, 0, -P_{max}\} \tag{7}$$

The battery environment is assumed to always charge/discharge at its rated power capacity, P_{rated}, hence the action $a = P_{max}$ denotes charging at a rate of P_{rated}, while $a = -P_{max}$ denotes discharging at a rate of $-P_{rated}$.

3.2.2 Battery Reward Function

The reward function generates an appropriate reward signal in response to the agent's charge/discharge actions. The goal of energy arbitrage naturally dictates that this function positively reward charging at low prices and discharging at high prices. Consequently, the function can be formulated as follows:

$$r_t = \begin{cases} P_{max} \times E_{price,t} & \text{if charging} \\ 0 & \text{if dormant} \\ -P_{max} \times E_{price,t} & \text{if discharging} \end{cases} \tag{8}$$

Based on this formulation, a "positive cost" is incurred for charging the battery and a "negative cost" for discharging. If the price of electricity is higher at the time of discharge, then arbitrage value is gained. Despite its simplicity and practicality, this reward function is highly inefficient. Assuming that the battery is initially fully discharged, then the agent has to first incur the temporary costs of charging the battery before it can realize the benefits of arbitrage. While this idea might be intuitive to humans, it is not as obvious to an agent. This often causes the agent to be conservative with its action during training, causing it to converge to sub-optimal policies. Motivated by the work done in [12], we formulated a more effective reward function, one which incorporates historical electricity price data. Specifically, the new reward function is defined as:

$$r_t = \begin{cases} P_{max} \times (E_{price,t} - \overline{E_{avg}}) & \text{if charging} \\ 0 & \text{if dormant} \\ -P_{max} \times (\overline{E_{avg}} - E_{price,t}) & \text{if discharging} \end{cases} \tag{9}$$

where historical average electricity price, $\overline{E_{avg}}$, is calculated in Eqn. 10.

$$\overline{E_{avg}} = (1 - \eta_s)\overline{E_{past}} + \eta_s \overline{E_{price,t}} \tag{10}$$

The variable $\overline{E_{past}}$ represent the average energy cost of all previous energy costs experienced by the agent. The variable η_s serves as a smoothing parameter which allows us to weight the importance between past energy prices and the current energy price. Unlike with Eqn. 8, Eqn. 9 allows for a scenario where charging incurs a "negative" cost. This happens when charging occurs at below the average price, $\overline{E_{avg}}$. This rewards the agent for charging the battery, allowing it to better traverse the solution space. Having explored the battery environment, the next section presents a detailed look at the photo-voltaic environment.

3.3 RL Environment: Photo-Voltaic System

A photo-voltaic (PV) environment allows for the generation of solar energy within the building cluster. The conventional technique for modeling a PV involves describing it using equivalent electrical circuits consisting of current sources, diode(s), and resistors. However, since the focus of this work is primarily concerned with arbitrage strategies, we are not particularly concerned with the inner workings of a PV system. Consequently, rather than modeling a conventional PV environment, we sourced the PV generation data directly from the Pecan Street Inc. Dataport database [37].

3.3.1 PV Action Space
As with the battery environment. The PV action space encapsulates all possible actions available to the agent within the PV environment. The PV action space, formulated in Eqn. 11, consists of four discrete actions: (i) charge battery, PV_{bat}; (ii) power building cluster, PV_{bldg}; (iii) sell to grid, PV_{grid}; and (iv) dormant, PV_{dom}.

$$\mathcal{A} = \{PV_{bat}, PV_{bldg}, PV_{grid}, PV_{dom}\} \tag{11}$$

The variables PV_{bat}, PV_{bldg}, PV_{grid} represent the PV power output applied towards the battery, building, and power grid respectively, measured in kilowatts (kW). The PV system cannot take multiple actions simultaneously, meaning it cannot for example power the building and charge the battery at the same time. Owing to this constraint, should the PV generate more electricity than is required for a given action, the excess electricity is curtailed.

3.3.2 PV Reward Function
As with the reward function for the battery environment, the PV reward function serves to emit a reward signal to the agent in response to actions which impact the PV system. The reward function is formulated as follows:

$$r_{pv,t} = \begin{cases} -PV_{bat} \times (\overline{E_{avg}} - E_{price,t}) & \text{if charging battery} \\ -PV_{bldg} \times (E_{price,t} - \overline{E_{avg}}) & \text{if powering building} \\ -PV_{grid} \times E_{sell,t} & \text{if selling to grid} \end{cases} \tag{12}$$

In this formulation, the variables PV_{bat}, PV_{bldg}, PV_{grid} are negated because they serve to reduce the total energy cost. The agent maximizes its arbitrage potential by charging the battery with PV energy when the difference between the average electricity price and the current electricity price, $E_{price,t}$, is largest. This encourages charging during off-peak times. Furthermore, the reward function encourages the agent to power the building using PV energy during peak times. This reward structure however does not account for curtailment, meaning there is no feedback signal sent to the agent if it allocates excess PV electricity towards an action. To address this issue, a more detailed reward function is formulated. During periods of battery charge, the following reward function is used.

if $SoC_t + PV_{bat,t} > SoC_{max}$ **then**

$$PV_{net} = SoC_{max} - SoC_{t-1}$$

$$PV_{waste} = (SoC_{t-1} + PV_{bat,t}) - SoC_{max}$$

$$r_{pv,t} = (-PV_{net} + PV_{waste}) \times (\overline{E_{avg}} - E_{price,t})$$

else

$$PV_{net} = PV_{bat,t}$$

$$r_{pv,t} = -PV_{net} \times (\overline{E_{avg}} - E_{price,t})$$

end if

where $PV_{bat,t}$ is the PV electricity generated at time t, and SoC_{t-1} is the state of charge of the battery at time $t - 1$. Prior to charging the battery, the agent first ensures that doing so would not exceed SoC_{max}. If it does, it calculates the value needed to fully charge the battery, PV_{net}, and the level of curtailment, PV_{waste}. The reward, $r_{pv,t}$, is calculated as the difference between PV_{net} and PV_{waste}, multiplied by the energy cost. Otherwise, the agent reverts to the original reward function (Eqn. 12). To power the building using solar energy, the following reward logic is used.

if $D_{bldg,t} - PV_{bldg,t} > 0$ **then**

$$r_{pv,t} = -PV_{bldg} \times (E_{price,t} - \overline{E_{avg}})$$

else

$$PV_{net} = \text{demand}_t$$

$$PV_{waste} = PV_{bldg,t} - D_{bldg,t}$$

$$r_{pv,t} = -PV_{net} + PV_{waste} \times E_{sell,t}$$

end if

where $D_{bldg,t}$ is the electricity demand of the cluster at time t, and $PV_{bldg,t}$ is the amount of generated electricity. Prior to using PV energy to power the building, the agent initially checks if demand exceeds generation. If so, it adopts our original reward formulation (Eqn. 12). If not, it calculates the level of curtailment and factors that in when it computes the rewards, $r_{pv,t}$. The variable $E_{sell,t}$ represents the price at which the grid will buy back PV energy, and is set at $0.08/kWh$. It should be noted that this value is different from E_{price}, which is the price at which the grid sells electricity to the cluster. Having explored the battery and PV environments, the next section introduces the learning agent developed to interact with both of these environments.

3.4 Agent: Deep Q-Learning
The artificial agent learns, through trial and error, the best policy, π, to maximize the battery and PV reward functions. To allow for this, we use an RL algorithm known as Q–learning. Q-learning, $Q_\pi(s, a)$, as formalized in Eqn. 13, is an approach for learn-

ing optimal policies. This equation describes the expected discounted return of taking an action, a, in response to an observed state, s, where the discount factor, γ, determines the priority of short-term rewards.

$$Q_\pi(s,a) = \mathbb{E}\Big[\sum_{t=0}^{\infty} \gamma^t r\big(s_t, a_t\big) \Big] \qquad (13)$$

Q-learning is a value-based learning approach in which actions are selected by maximizing the discounted reward at a given state, as formulated in Eqn. 14.

$$\pi(s) = argmax_a Q(s,a) \qquad (14)$$

State-action transition pairs are typically stored within a matrix, however, the large combination of battery and PV state-action pairs for this problem makes it computationally inefficient to use a matrix. To address this issue, we use a Deep Q-Network (DQN) in an approach known as Deep Q-learning to approximate the Q-function. Following the publication of the DQN algorithm, several improvements have been proposed which significantly improve its convergence, stability, and sample efficiency. One of these extensions, known as Double DQN (DDQN) have been adopted and implemented in this work. The intuition behind Double DQN is that the basic DQN formulation has a tendency of non-uniformly overestimating some Q-values, potentially leading to sub-optimal policies. DDQN addresses this shortcoming by decomposing the *max* operation in the target network into two separate action selection and action evaluation tasks.

In the next section, we integrate the DQN agent with the battery and PV environments and highlight some key aspects of our implementation. A case study is also developed to test the proposed approach.

4 PROPOSED MODEL

Figure 4 illustrates the schema used to train the DQN agent. The goal here is to teach the agent, through trial and error, an efficient way to control the PV and battery systems in order to minimize the energy cost incurred by the building cluster. A few key points regarding our implementation are worth nothing. Firstly, the battery and PV environments are developed using the OpenAI Gym. The OpenAI Gym is an open-sourced toolkit for developing and comparing reinforcement learning algorithms. It standardizes several methods related to the mechanics of our PV and battery environments, making it relatively easy to replicate the results of this study. The second point relates to the use of Pytorch as the deep learning framework of choice. Pytorch is a flexible and expressive deep learning Python library developed by Facebook for training neural networks. Being Pythonic in nature, Pytorch is fast and efficient, making it idea for developing

our DQN agent. The "input data" block as shown in Figure 4 encapsulates all external data supplied to the environment. This includes the building demand profiles, hourly energy prices, and the hourly PV production. Finally, a python script is used to initialize the training process, and a visualization module is used to visualize the training results.

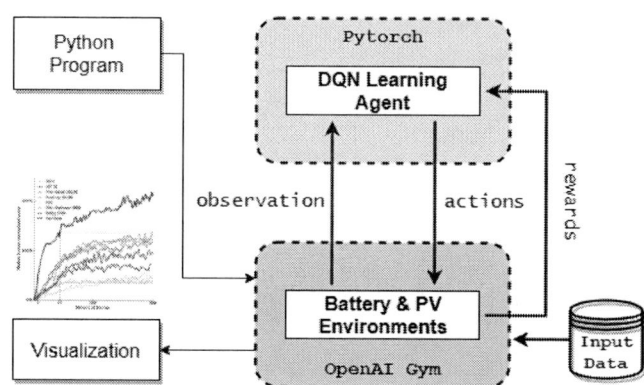

FIGURE 4: *Deep Q-network (DQN) based reinforcement learning for energy arbitrage problem*

Having presented a general overview of the proposed model, the next subsection presents an arbitrage case study developed to test the performance of the learning agent.

4.1 Case Study: DER Energy Arbitrage in Building Cluster

For this case study, we leverage the RL agent to learn efficient arbitrage policies within a 2-building cluster with a shared battery and PV system as originally illustrated in Figure 2. We sampled electricity usage data (see Figure 5) from four randomly selected households in the Austin, Texas area using the Pecan Street Inc. Dataport database [37]. It is worth noting that although the Austin, Texas area is ideal due to its favorable solar exposure, the algorithm is region/climate agnostic, meaning it can theoretically be used in any area. Table 1 provides a brief summary of each building. In each entry, the 'Data ID' identifier corresponds to the numeric identifier used in the Dataport database. The 'Label' column provides the identifiers used in this paper for each building. Finally, the "Demand" and "Generation" columns provide the total electricity consumption and solar energy generation of each home measured for the entire month of July 2018.

Table 2 presents four different building clusters, each of which presents a unique arbitrage problem which we aim to solve with our learning agent. The building clusters are classified according to their total energy consumption. Consequently, Cluster 1 is classified as a low demand cluster, Cluster 2 as a medium

TABLE 1: Basic details of the single-family homes used in this study

Data ID	Label	Total Area (ft^2)	Demand (kW)	Generation (kW)
93	1	2934	1449	647
114	2	1842	1438	464
7017	3	3249	2600	748
8967	4	2001	1916	411

TABLE 3: City of Austin energy pricing schedule [40]

Time of Use	Peak Period	Energy Cost ($/kWh$)
Midnight - 6 AM	Off Peak	0.06
6 AM - 2 PM	Mid Peak	0.10
2 PM - 8 PM	Peak	0.15
8 PM - Midnight	Off Peak	0.06

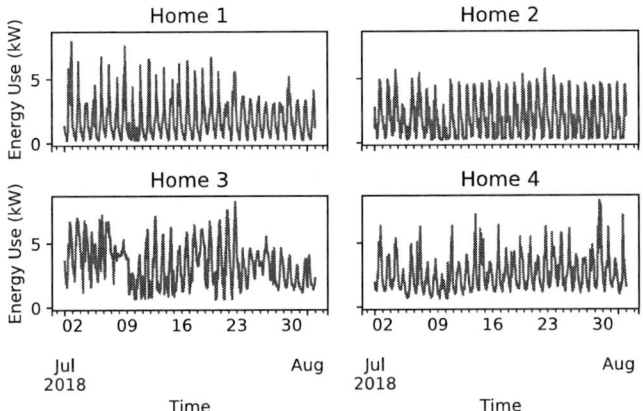

FIGURE 5: *Electricity demand profiles for month of July 2018*

demand, and Cluster 3 as a high demand cluster. These classifications were assigned relative to the demand of the other clusters within the set. By varying the building clusters based on electricity demand, the agent is force to generalize, rather than over fit its policies to suit the demands of a single cluster. Table 3 provides the Time-of-Use pricing structure used to calculate the cost of electricity in this work. The next section presents the results obtained after 6000 episodes of agent learning. Each episode simulates a one month electricity usage cycle, so the agent trains over the equivalent of 500 years of simulated electricity usage. Learning was performed on a 3.50GHz Intel Core i-7 5930k system with 32GB of RAM and a GeForce GTX TITAN graphics card. Each result took on average about 5 hours to run on this system.

TABLE 2: Summary of the building cluster combinations studied

	Homes	Total Demand (kW)	Demand Level
Cluster 1	1 and 2	2887	Low
Cluster 2	2 and 4	3354	Medium
Cluster 3	3 and 4	4516	High

5 RESULTS

Figure 6 illustrates the training results obtained for Building Cluster 1 (labeled C1) across 6000 episodes. We observe a general trend of increasing reward as the number of episodes increases. The blue line represents a smooth moving average using a window of 100 episodes. Readers will recall that the reward function for the battery environment was augmented to allow for learning efficiency (see Eqn. 10). Owing to this, the reward values shown in this figure have no meaning beyond the fact that higher numerical values are preferred.

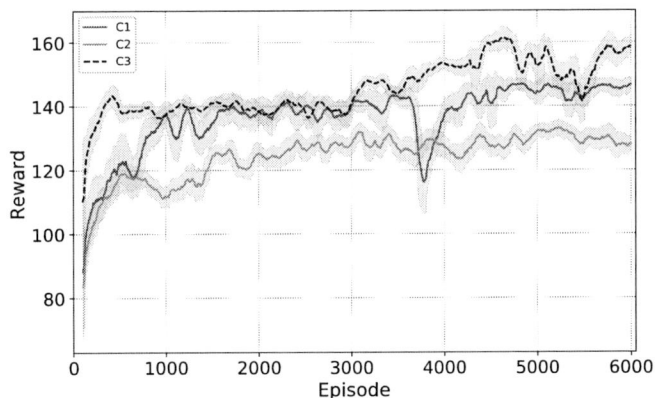

FIGURE 6: *Training results for all three Building Clusters*

To gain more insight, Figure 7 plots the hourly cumulative electricity costs over the course of the 1 month (744 hours) optimization horizon. Three datasets are plotted on this figure. The first, labeled "Electricity Cost" represents the total cost of electricity consumed by the building cluster ($285.40). The second plot, labeled "Original Cost" accounts for solar self-consumption of each building, illustrating the actual electricity cost incurred ($197.63). This value serves as the baseline for our algorithm. Our goal is to improve on this value using arbitrage operations. The final plot, labeled "Arbitrage Cost" shows the performance of our agent's learned policy on the total energy cost of the cluster. We observe a significant improvement over the baseline value. Table 4 provides the actual values associated with each of these plots.

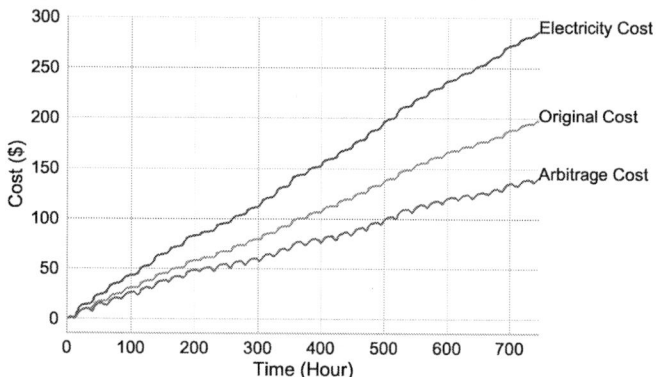

FIGURE 7: *Cumulative electricity costs for Cluster 1*

Figure 6 also illustrates the training results obtained after training the RL agent on data related to building Cluster 2 (labeled C2). From this figure, we again observe an overall trend towards higher rewards over the 6000 episode range.

Figure 8 illustrates the cumulative energy costs incurred by the building cluster. Unlike in Figure 7, on a visual level, the difference between the original and arbitrage cost is not as significant. This can be attributed in part to the the fact that PV generation in Cluster 2 is lower than that of Cluster 1. This is further compounded by the fact that the energy demand of Cluster 2 is larger than that of Cluster 1. Consequently, the agent had a more difficult time learning strategies which significantly improved over the baseline. This however did not seem to impact the learning process too much as Figure 6 does not indicate any significant learning instabilities.

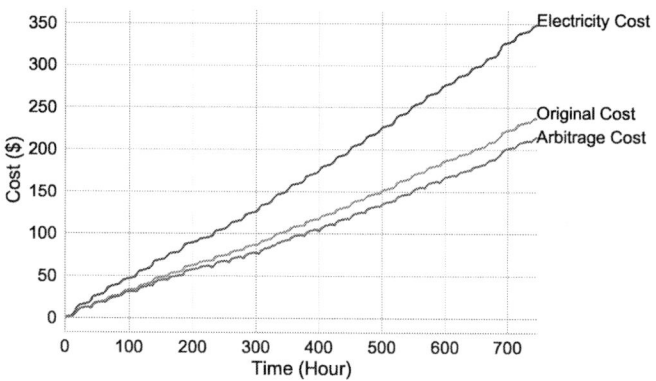

FIGURE 8: *Cumulative electricity costs for Cluster 2*

Figure 6 also details the training progress of the agent with respect to Cluster 3 (labeled C3). Unlike with Clusters 1 and 2, it appears that 6000 episodes was not quite enough to allow the learning to stabilize. We note some erratic learning behavior between episodes 4500 to 6000 as well. It is worth noting that the developed algorithm maintains a small level of stochasticity, even as it converges towards a policy, hence reward perturbations are to be expected, albeit not to the level shown within the noted range. Nevertheless, the data still provided useful insights with regards to the performance of the arbitrage policies developed by the agent for Cluster 3.

In Figure 9, we observe that the learned arbitrage policy outperformed the cluster baseline. Granted, on visual inspection, the difference is not as significant as with the previous clusters, but that is to be expected. This cluster represents the highest energy consumption scenario within the set, and hence presented the biggest challenge to the agent. Table 4 provides a summary of the results obtained. From this table we note a 29.7% energy cost reduction from Cluster 1's baseline value. Similarly, the agent realize 9.65% and 6.48% baseline cost reductions in both Clusters 2 and 3 respectively.

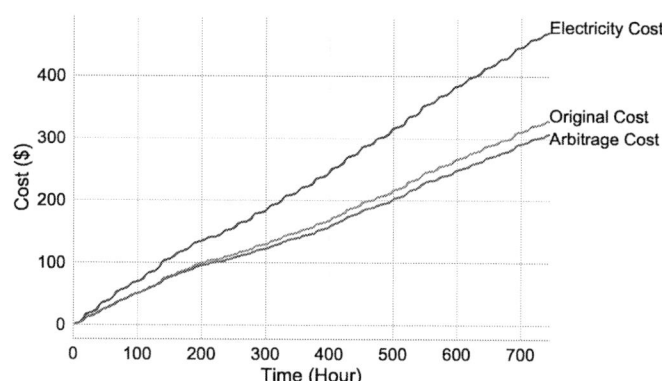

FIGURE 9: *Cumulative electricity costs for Cluster 3*

TABLE 4: Summary of agent performance across all 3 Clusters

	Cluster 1	Cluster 2	Cluster 3
Electricity Cost	$285.40	$348.08	$496.62
Original Cost	$197.63	$237.66	$328.20
Arbitrage Cost	$138.87	$214.71	$306.92
Cost Savings	-29.7 %	-9.65 %	-6.48 %

To further analyze the performance of the algorithm, Figure 10 compares the performance of the algorithm with respect to the agent's reward function. We compare results obtained using Eqn. 8 and Eqn. 9. The two reward functions were evaluated under the same circumstances, with identical system inputs and hyperparameter values. We plot the cumulative profits of the two formulations as they relate to Cluster 1. From Figure 10, we observe that "Reward Func. 2" (Eqn. 9), with a cumulative total of \$138.87, significantly outperforms "Reward Func. 1" (Eqn. 8), which had a cumulative total of \$192.13. Similar results were observed with the other two clusters as well. This demonstrates the effectiveness of our formulated reward function in exploring more of the solution space, allowing it to design more efficient arbitrage strategies.

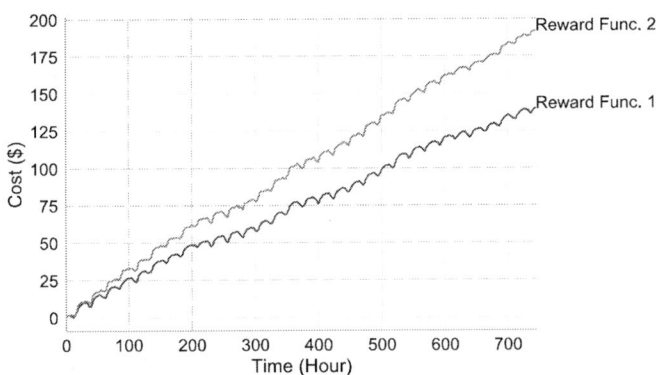

FIGURE 10: *Cumulative rewards of two different reward function formulations applied to Cluster 1*

Lastly, we take a look at the operational strategies designed by the agent to control the PV and battery system. Since displaying all the strategies employed by the agent over the 1 month time horizon is intractable, we show a representative strategy instead. The representative strategy is generated by clustering all agent actions by the time of day. This resulted in 24 unique clusters (24 hours in a day), each containing 31 actions (31 days in July). For each cluster set, the mode was calculated and used as its representative action. This results in 24 representative actions, which when plotted together, produce a single representative 24-hour operational strategy for the battery system (see Figure 11). From this figure, we build a general understanding of the control strategy designed by the agent. We observe that the agent typically charged the battery during the early hours of the morning and discharged it later during peak hours, presumably to maximize arbitrage value. It is worth noting that the "dormant" battery phase represents one of two possible states. It either means the battery is truly dormant, like the name suggests, or that it is being charged with PV energy. PV charging is not represented

as an action here because it is controlled outside the battery environment.

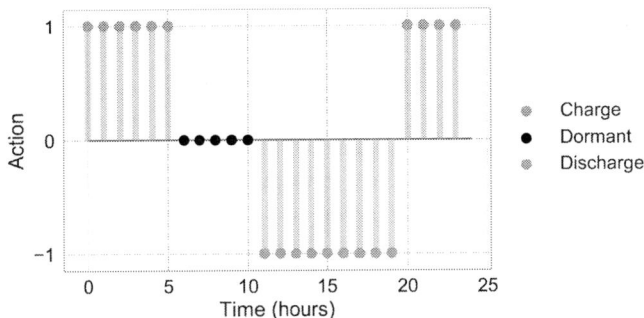

FIGURE 11: *Representative battery operational strategy for Cluster 1*

Similarly, a representative operational strategy was developed for the PV system. Illustrated in Figure 12, we again note a few prominent trends. The agent typically left the PV dormant during the early hours of the day. This was most likely due to the lack of sunlight during this period. Following the dormant period, the agent uses PV energy to charge the battery system. This period aligns well with the "dormant" period shown in Figure 11. The agent then uses PV energy to power the building during the peak time period, before switching back to dormancy. Interesting enough, although the agent has the option to sell PV energy to the grid, it seems this action did not happen frequently enough to leave an impression on the representative strategy.

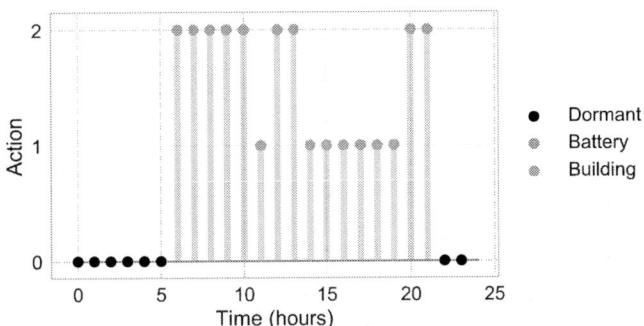

FIGURE 12: *Representative photo-voltaic operational strategy for Cluster 1*

Overall, the agent succeeded in learning and designing efficient arbitrage policies to minimize the energy cost incurred within the building clusters studied. Despite this, there is still room for improvement. In the current formulation, we train a single agent to act on behalf of both buildings within the cluster. It essentially serves as a "super-agent". A more natural decomposition would be to have each building within the cluster serve as an individual agent, resulting in a multi-agent reinforcement learning problem. By employing this approach in future work, we wish to have each agent's action and observation space restricted to only the systems which directly impact them. Doing so not only promises to reduce the dimensionality of agent inputs and outputs, but also effectively increases the amount of training data generated per step of the environment. Additionally, it will allow us to employ multiple pricing structures within the same building cluster, presenting new opportunities for energy cost savings. Finally, switching to a multi-agent RL lends itself well to scaling this approach to multiple buildings and energy systems.

6 CONCLUSION

In this paper we use reinforcement learning to derive temporal arbitrage policies for the operation of shared energy storage and PV systems. The arbitrage problem was modeled as a Markov Decision Process and a Double deep Q-learning agent was trained to derive discrete operational strategies within the battery and PV environments. A reward function which utilizes average price history was designed to promote aggressive exploration of the solution space. This provided significant improvements over more traditional reward functions which focus solely on spot price and cost. Our simulations results show notable cost savings across all three clusters studied, highlighting the agent's ability to adapt to different demand requirements by tailoring it strategies to exploit arbitrage potential.

ACKNOWLEDGEMENTS

The authors would like to acknowledge the National Science Foundation (NSF) (grant CNS-1239093) for their support of some of the early developments in this work. The authors would also like to thank the Department of Mechanical and Aerospace Engineering for additional support of this work.

REFERENCES

[1] Tesla, 2019. Tesla Powerwall. On the WWW, Retrieved Jan 2019. URL https://goo.gl/Z58MPu.

[2] Tesla, 2019. Tesla Powerwall. On the WWW, Retrieved Jan 2019. URL https://goo.gl/6XE63R.

[3] Crabtree, G., 2015. "Perspective: The energy-storage revolution". *Nature, 526*(7575), pp. S92–S92.

[4] Lund, P. D., Lindgren, J., Mikkola, J., and Salpakari, J., 2015. "Review of energy system flexibility measures to enable high levels of variable renewable electricity". *Renewable and Sustainable Energy Reviews, 45*, pp. 785 – 807.

[5] Ketterer, J. C., 2014. "The impact of wind power generation on the electricity price in Germany". *Energy Economics, 44*, pp. 270–280.

[6] Cutler, N. J., Boerema, N. D., MacGill, I. F., and Outhred, H. R., 2011. "High penetration wind generation impacts on spot prices in the australian national electricity market". *Energy Policy, 39*(10), pp. 5939–5949.

[7] Woo, C.-K., Horowitz, I., Moore, J., and Pacheco, A., 2011. "The impact of wind generation on the electricity spot-market price level and variance: The Texas experience". *Energy Policy, 39*(7), pp. 3939–3944.

[8] Allcott, H., 2011. "Rethinking real-time electricity pricing". *Resource and energy economics, 33*(4), pp. 820–842.

[9] Krishan, O., and Suhag, S., 2018. "An updated review of energy storage systems: Classification and applications in distributed generation power systems incorporating renewable energy resources". *International Journal of Energy Research.* doi: 10.1002/er.4285, pp. 1–45.

[10] Byrne, R. H., and Silva-Monroy, C. A., 2012. "Estimating the maximum potential revenue for grid connected electricity storage: Arbitrage and regulation". *Sandia National Laboratories.* SAND2012-3863.

[11] Stoft, S., 2002. "Power system economics: Designing markets for electricity". *Journal of Energy Literature, 8*, pp. 94–99.

[12] Wang, H., and Zhang, B., 2018. "Energy storage arbitrage in real-time markets via reinforcement learning". In Power & Energy Society General Meeting (PESGM), IEEE, pp. 1–5.

[13] Kloess, M., 2012. "Electric storage technologies for the future power systeman economic assessment". In 9th International Conference on European Energy Market (EEM), IEEE, pp. 1–8.

[14] Steffen, B., 2012. "Prospects for pumped-hydro storage in Germany". *Energy Policy, 45*, pp. 420–429.

[15] Zafirakis, D., Chalvatzis, K. J., Baiocchi, G., and Daskalakis, G., 2016. "The value of arbitrage for energy storage: Evidence from european electricity markets". *Applied energy, 184*, pp. 971–986.

[16] Bradbury, K., Pratson, L., and Patiño-Echeverri, D., 2014. "Economic viability of energy storage systems based on price arbitrage potential in real-time us electricity markets". *Applied Energy, 114*, pp. 512–519.

[17] Walawalkar, R., Apt, J., and Mancini, R., 2007. "Economics of electric energy storage for energy arbitrage and regulation in New York". *Energy Policy, 35*(4), pp. 2558–2568.

[18] Sioshansi, R., Denholm, P., Jenkin, T., and Weiss, J., 2009. "Estimating the value of electricity storage in PJM: Arbitrage and some welfare effects". *Energy Economics, 31*(2), pp. 269–277.

[19] Graves, F., Jenkin, T., and Murphy, D., 1999. "Opportunities for electricity storage in deregulating markets". *The*

Electricity Journal, 12(8), pp. 46–56.

[20] Zheng, M., Meinrenken, C. J., and Lackner, K. S., 2014. "Agent-based model for electricity consumption and storage to evaluate economic viability of tariff arbitrage for residential sector demand response". *Applied Energy, 126*, pp. 297–306.

[21] Zheng, M., Meinrenken, C. J., and Lackner, K. S., 2015. "Smart households: Dispatch strategies and economic analysis of distributed energy storage for residential peak shaving". *Applied Energy, 147*, pp. 246–257.

[22] Abdulla, K., De Hoog, J., Muenzel, V., Suits, F., Steer, K., Wirth, A., and Halgamuge, S., 2018. "Optimal operation of energy storage systems considering forecasts and battery degradation". *IEEE Transactions on Smart Grid, 9*(3), pp. 2086–2096.

[23] Hu, M., Weir, J. D., and Wu, T., 2012. "Decentralized operation strategies for an integrated building energy system using a memetic algorithm". *European Journal of Operational Research, 217*(1), pp. 185–197.

[24] Odonkor, P., and Lewis, K., 2016. "Optimization of energy use strategies in building clusters using pareto bands". In ASME 2016 International Design Engineering Technical Conferences and Computers and Information in Engineering Conference, American Society of Mechanical Engineers, Charlotte, NC, DETC2016-59963.

[25] Odonkor, P., Lewis, K., Wen, J., and Wu, T., 2016. "Adaptive energy optimization toward net-zero energy building clusters". *Journal of Mechanical Design, 138*(6). doi:10.1115/1.4033395.

[26] Upside Energy, 2019. Energy is Changing. On the WWW, Retrieved Jan 2019. URL https://goo.gl/bx6ao3.

[27] BeeBryte, 2019. Artificial intelligence for smarter buildings. On the WWW, Retrieved Jan 2019. URL https://goo.gl/PNJcZQ.

[28] GridEdge, 2019. The way we use energy is changing. On the WWW, Retrieved Jan 2019. URL https://goo.gl/PNJcZQ.

[29] Odonkor, P., Lewis, K., Wen, J., and Wu, T., 2018. "Automated design of energy efficient control strategies for building clusters using reinforcement learning". *Journal of Mechanical Design, 141*(2). doi:10.1115/1.4041629.

[30] Odonkor, P., and Lewis, K., 2018. "Control of shared energy storage assets within building clusters using reinforcement learning". In ASME 2018 International Design Engineering Technical Conferences and Computers and Information in Engineering Conference, American Society of Mechanical Engineers, Quebec, Canada, DETC2018-86094.

[31] Mocanu, E., Mocanu, D. C., Nguyen, P. H., Liotta, A., Webber, M. E., Gibescu, M., and Slootweg, J. G., 2018. "On-line building energy optimization using deep reinforcement learning". *IEEE Transactions on Smart Grid*.

[32] François-Lavet, V., Taralla, D., Ernst, D., and Fonteneau, R., 2016. "Deep reinforcement learning solutions for energy microgrids management". In European Workshop on Reinforcement Learning (EWRL).

[33] Liu, Y., Logan, B., Liu, N., Xu, Z., Tang, J., and Wang, Y., 2017. "Deep reinforcement learning for dynamic treatment regimes on medical registry data". In International Conference on Healthcare Informatics (ICHI), IEEE, pp. 380–385.

[34] Kalashnikov, D., Irpan, A., Pastor, P., Ibarz, J., Herzog, A., Jang, E., Quillen, D., Holly, E., Kalakrishnan, M., Vanhoucke, V., and Levine, S., 2018. "QT-Opt: Scalable deep reinforcement learning for vision-based robotic manipulation". *CoRR, abs/1806.10293*.

[35] Lanham, M., 2018. Deep reinforcement learning for the cannabis retail market. On the WWW, Retrieved Jan 2019. URL https://goo.gl/aEEiV2.

[36] Sutton, R. S., and Barto, A. G., 1998. *Reinforcement Learning: An Introduction*, Vol. 1. MIT Press Cambridge, MA.

[37] Pecan Street Inc., 2018. Pecan Street Dataport. On the WWW, Retrieved Mar 2018. URL https://dataport.cloud/.

[38] Duch, W., and Mandziuk, J., 2007. *Challenges for Computational Intelligence*, Vol. 63. Springer Science & Business Media.

[39] Silver, D., Schrittwieser, J., Simonyan, K., Antonoglou, I., Huang, A., Guez, A., Hubert, T., Baker, L., Lai, M., Bolton, A., Chen, Y., Lillicrap, T., Hui, F., Sifre, L., Van den Driessche, G., Graepel, T., and Hassabis, D., 2017. "Mastering the game of go without human knowledge". *Nature, 550*(7676), p. 354.

[40] City of Austin, 2017. Residential electric rate schedules. On the WWW, Retrieved Mar 2018. URL https://goo.gl/Atd66W.

Proceedings of the ASME 2019
International Design Engineering Technical Conferences
and Computers and Information in Engineering Conference
IDETC/CIE2019
August 18-21, 2019, Anaheim, CA, USA

DETC2019-97964

OPTIMIZATION MODEL FOR OWNER-BASED MICROGRIDS USING LSTM PREDICTED DEMAND FOR RURAL DEVELOPMENT

Anosh P. Amaria
anoshama@buffalo.edu
Department of Mechanical
and Aerospace Engineering
University at Buffalo
Buffalo, New York, USA

Ryan Nguyen
ryanguy@buffalo.edu
Department of Mechanical
and Aerospace Engineering
University at Buffalo
Buffalo, New York, USA

Joshua A. Davison
joshuada@buffalo.edu
Department of Mechanical
and Aerospace Engineering
University at Buffalo
Buffalo, New York, USA

Souma Chowdhury
soumacho@buffalo.edu
Department of Mechanical
and Aerospace Engineering
University at Buffalo
Buffalo, New York, USA

John F. Hall
johnhall@buffalo.edu
Department of Mechanical
and Aerospace Engineering
University at Buffalo
Buffalo, New York, USA

ABSTRACT

Over the past several years, microgrids have been setup in remote villages in developing countries such as India, Kenya and China to boost the standards of living of the less privileged citizens, mostly by private companies. However, these systems succumb to increase in demand and maintenance issues over time. A method for scaling the capacity of solar powered microgrids is presented in this paper. The scaling is based on both the needs of the owner and those of the consumers. Data acquired from rural villages characterizes the electrical use with respect to time. Further, it employees a Long-Short Term Memory (LSTM) deep learning model that can help the owner predict future demand trends. This is followed by a model to determine the optimum increase in capacity required to meet the predicted demand. The model is based on empowering the owner to make informed decisions and the equity of energy distribution is the key motivation for this paper. The models are applied to a village in Eastern India to test its applicability. Acknowledging the highly varying nature of demand for electricity and its applications, we propose a rule-based adaptive power management strategy which can be tailored specifically in accordance to the preference of the communities. This will ensure a fair distribution of power for everyone using the system, thereby making it applicable anywhere in the world. We propose to incorporate social and demographic conditions of the user in the optimization to ensure that the profit of the owner does not outweigh the needs of the users.

Keywords: Microgrids, Scalability, Recurrent Neural Networks, LSTM, Adaptive Power Management.

NOMENCLATURE

Cap_0	Current total capacity of the microgrid
Cap_{solar}	Capacity of each solar panel
C_{op}	Current cost of operation
C_{trans}	Cost of transmission upgrade
$C_{upgrade}$	Cost of upgrade (USD)
C_{solar}	Cost of one (1) solar panel
$C_{addition}$	Additional costs (batteries, maintenance, transmission, etc.)
D	Demand (kWh)
T	Time index (varying from 0 until end of life for the system)
I-MR Chart	Individual-Moving Range Chart
LCL	Lower Critical Limit
N_{solar}	Number of solar panels
S	Supply
I	Investment by the owner (USD)
$T_{upgrade}$	Time of upgrade (days)
UCL	Upper Critical Limit
VSAT	Very Small Aperture Terminal
I_r	Clear skies irradiance factor
η	Efficiency of the solar panel
LSTM	Long-Short Term Memory

1. INTRODUCTION

According to the United Nations, 1.2 billion people in the world do not have access to electricity. In today's world it is obvious that electricity plays a vital role in rural development. With uninterrupted electricity, agriculture flourishes, literacy rate increases, social status of the community is improved, and safety is enhanced. With the rapidly increasing world population and the steep decline of fossil fuel reserves, it is imperative to shift to renewable sources of electricity. Populous nations frequently have energy access issues within rural communities. Many of these communities are segregated from centralized power grids due to geographical features such as mountains and rivers. In other cases, the amount of energy generated is simply insufficient for everyone and urban and industrial areas are given preference. In these cases, it has become common to use what is referred to as a "microgrid" as a solution to underdeveloped energy access.

A microgrid is a localized power grid that functions autonomously from the related main power grid. Many of them have made a positive impact on the standard of living of the users. However, this upgrade in social status poses a new challenge for the owners who installed the grid. With the increase in quality of living, the consumption of electricity in these villages consequentially increases. This has resulted in the capacity of many of these microgrids becoming inadequate. Microgrids eventually succumb to other common failure modes such as scalability, maintenance, and power management. Furthermore, these rural communities often have limited resources in terms of capital, technology, and personnel trained in operation and maintenance of the equipment. A systematic approach to power management for these communities would reduce potential for overconsumption of available energy and ensure equity of energy access.

This model is a first step to move towards more reliable decision making in design and working of microgrids for off-grid communities with limited resources. The design must satisfy three essential functional requirements in order to meet the acceptance criteria; it must be scalable and must manage power to provide equity of energy access. Scalability will allow the solution to be effective as the community grows. Equity of energy access will ensure that the scalable resource is available to all demographics within the community. To meet the needs of the community, predictive methods will be used to assess changes within the community and technological dependence as technology becomes increasingly integrated with their day-to-day life.

2. LITERATURE REVIEW

Rural development goes hand in glove with electricity. Currently, most models of energy access are biased towards societies that are industrialized resulting in developing countries being misrepresented. Van Ruijven et al. [1] discussed the importance of energy models which properly represent present-day developing countries. Thiam [4] discussed the growth of energy usage in villages in Africa and their dependence on oil imports. They showed that the country with enough energy has the fuel to develop quickly on a global scale.

Bhattacharyya et. al proposed the use of microgrids to improve electricity access in developing countries. Bhattacharyya further discussed what policies need to be implemented in order to allow microgrid implementation [3]. Shahidehpour et al [18] emphasize that even in large metropolitans like New York, miniature grids will be the need of the hour as the growth of these cities is very fast. Microgrids, as an energy solution, are desirable due to their effective cost of energy (COE) and ability to supply electricity to extremely remote areas independently. However, maintenance and rapid scalability of microgrids provide a major obstacle to effective implementation. Loka et al. [10] presented a real-life implementation of a photovoltaic microgrid system in a rural village with the goal of optimizing power vs. cost. The study presented additional implementation challenges such as the terrain on which pre-fabricated shelters, control rooms, solar panels, and other necessary materials had to traverse to their installation sites.

Kitson et al. [12] simulated a multi-source modular DC microgrid system in which the sources interfaced with the DC grid are managed via droop control. The system combined real energy source and load data from a village in Nepal with short-term and long-term energy modelling to assess DC system control and predict its capacity to provide energy. Zubleta [7] details benefits of DC microgrids over AC microgrids on principles of cost, complexity, and efficiency. Lack of standardization and DC-rated components, however, are present factors impeding the implementation of DC microgrids [7]. Behzadi and Niasati [13] presented issues related to modelling the renewable solar energy-fuel cell-battery system based on various combinations of sizing approaches and power management strategies. They discussed Hybrid Optimization of Multiple Energy Resources (HOMER) software being a powerful and accurate tool to model and analyze the system over an extended simulation time as compared to paper [12]. Puglia et al. [6] found that the usage of battery storage is economically crucial, particularly in areas with low daily electrical consumption and peak loads increasing during time periods where PV arrays have reduced power production. Optimal configuration of a system including battery storage, despite high initial investment cost, presented an economic benefit of 22.2%-25.2% and a 74.7-77% reduction in fuel consumption. While the proposed solution offers long-term economic benefit through reduction of fuel consumption, the high investment cost for the system may be a significant barrier to communities without financial aid. These papers show that there is a need to take the perspective of the owner and the owner collectively needs to be taken into consideration.

MacCarty and Bryden [11] discussed the importance of taking social demographics of the village into consideration while proposing a suitable change in technology for a specific need from the originally used alternative. Barnes and Floor [8] presented an economic perspective on the challenges of supplying energy to rural communities in developing countries. The breakdown of economic cost is applied to the aspects of economic hierarchies Furthermore, discourse was provided on the impact of political and economic policy on the efficacy of micro-grid systems [8]. Alam et al. [5] made an objective comparison on the regulation of microgrids in developing countries. This paper suggests that product delivery of the microgrids does not require formal regulatory supervision but the collective service delivery using local networks requires some form of regulation. Bhattacharyya [2] draws a direct relationship between access to electricity and socioeconomic productivity. Utilizing cross-sectional dependence and panel co-integration, a dependence was identified between the following variables: access to electricity, labor productivity, economic growth, financial development, and gross capital formation.

Wytock et al. [9] presented an analytical model of fail-safe operation of microgrids, also demonstrating the criticality of balancing system fail-safes such that they are not too conservative (decreasing robustness of the system) nor too liberal (increased risk of catastrophic failure). A comprehensive mathematical modelling of individual system parameters using conventional electrical equations was presented by Dash and Bajpai [14]. The modelling aimed to optimize the charging rate of the battery in order to increase its life and prevent frequent starting and stopping of other power sources whilst suggesting that the load demands must be met [14]. A laboratory experiment mentioned by Brka et al. [15] validated the power control algorithms suggested in [13] and [14]. The experiment was comprised of methods to increase the response time of the system to changes in state. Brka et al. proposed predictive power management strategies to preempt change in load and switch modes earlier in order to meet demand [15]. Again, laboratory simulations are very different from real life applications. These systems are sometimes exposed to harsh weather, accidents etc. These conditions cannot be simulated in laboratory environments and therefore cannot fully capture the irregularities in the system parameters.

The literature reviewed shows that models currently used to represent the growth of energy usage in developing countries are inaccurately represented, affecting the way they are treated in terms of power management and scalability [1]. Bhattacharyya [2] suggests that access to electricity has a direct correlation with higher labor rates, presenting the critical nature of proper models to optimize their access to electricity. Williams et. Al [19] throw light upon the challenges owners of these decentralized renewable energy systems face to keep their hands wrapped around these systems. Some of the challenges include, lack of public interest, unclear laws and regulations, marginal expected return on investments etc. Whitefoot et. al [20] highlight the importance of incorporating the cost component in microgrid management by taking a case study of an independent microgrid in an isolated location. They successfully implemented forward looking optimal dispatch of electricity. Their work however did not look into the long term increase in demand of energy.
This work allows owners to make informed decisions regarding scaling microgrid systems. The demand for energy acquired from the system will be used as an input for calculation of the amount of power supplied by the microgrid. The growth of energy demand will be predicted using forecasting machine learning techniques. The resultant data will be used in an optimization model to maximize the profit of the owner. This procedure thus brings the challenges in data acquisition and

data conditioning to make future predictions. Further the proposed adaptive power management will ensure flexibility in power distribution thus making the approach applicable globally.

3. METHOD

The hypothesis is that over time, the demand will gradually increase and the supply will remain constant unless the system is upgraded. In order to validate this hypothesis, we have gathered real time data from Chhotkei, a village in Eastern India where a microgrid is already installed. Chhotkei village is located in the middle of a designated protected land by the Government of India which is why conventional grids cannot be set up in the region. The supply data based on the capacity of the system parameters, weather and geographic conditions is calculated. The supply and demand data is then compared to validate our assumption that eventually the supply will not be able satisfy the demand. Further, using the existing demand data, future demand is predicted with an LTSM to show the trend demand will follow. An optimization model to ensure the profit of the owner of the microgrid throughout its lifespan is proposed and executed for the demand predicted by the LSTM. The paper highlights challenges in remote data acquisition for our analysis which is vital in accurately predicting trends. Acknowledging the fact that frequent upgrade of such systems is not feasible, we propose a power management strategy to ensure equal division of the supply to ensure continuous development of the community in its entirety. Fig. 1 shows the workflow adopted. It should be noted that although this study targets one particular village, the workflow can be adopted for any microgrid.

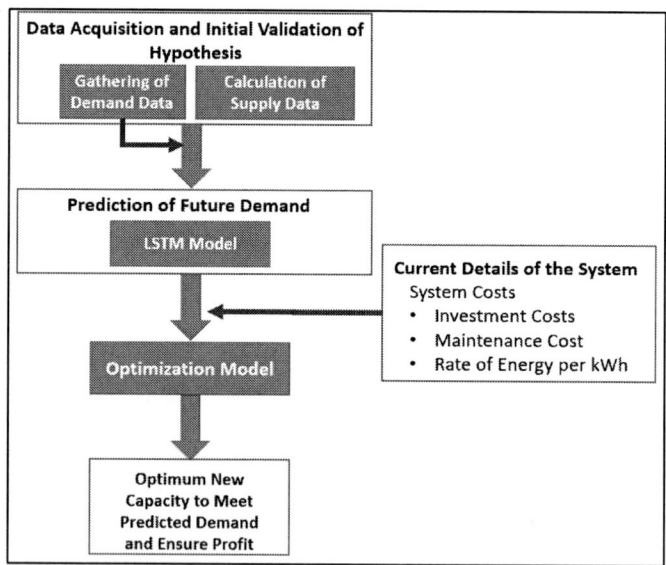

Figure 1: Capacity Prediction Workflow

3.1.1 Gathering the Demand Data: Data was collected using the *Smart Nanogrid*^{TM} System which allows remote access of data for each individual within Chhotkei. The system utilizes a VSAT for internet access to the cloud, providing communication between the microgrid controllers and the cloud network for remote monitoring of up-time and usage data as shown in Fig. 2. The cloud data is available to all remote stakeholders through the Nanosoft Remote Portal for monitoring and timely intervention due to consumer trends, if necessary [16].

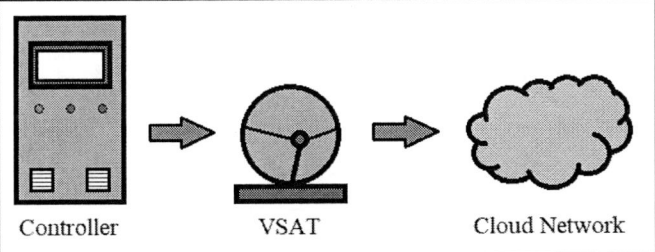

Figure 2: Data Transfer Method, Microgrid to Cloud Network

The dataset chosen for analysis was gathered over the period from February 1st, 2016 through January 31st, 2018, inclusively. This sample was chosen as the data prior to February 1st, 2016 contains many artifacts and anomalies due to communication difficulties between the microgrid controllers and the cloud network. This results in a serious challenge to accurately predict the trend of the demand as will be discussed in later sections. The data of all individual consumers was consolidated for analysis of Chhotkei as a community. Data points are given in kWh/h and, as such, describe the total usage of the microgrid during that hour.

3.1.2: Generation of the Supply Data: Daily energy produced by the microgrid is a function of solar irradiance, the capacity of the solar panels and the hours of sunlight in the day. Since the solar irradiance varies in real-time (depending on weather conditions, cloud cover, etc.), the output of the solar panel changes on a daily basis. The clear skies irradiance coefficient for Chhotkei Village from January 2016 was gathered from the NASA POWER Data Access Viewer [17]. The daily supply is then calculated using the following equation, Eq. 1:

$$S = N_{solar} * Cap_{solar} * I_r * \eta * (Hours\ of\ Sunlight) \qquad (1)$$

This method can be used to predict the supply of any solar powered system in the world whose geographic location and system parameters are known.

3.2 Data Analysis: The first data analysis investigates trends in hourly energy consumption. An I-MR Chart of Total kWh/h for the full dataset as seen in Fig. 3 shows patterns in power load throughout the day. Reference lines visually segregate the data into the following subsets: 12:00 AM – 5:59 AM (Period 1), 6:00 AM – 11:59 AM (Period 2), 12:00 PM – 5:59 PM (Period 3), 6:00 PM – 11:59 PM (Period 4). The Individual Value Chart shows an average hourly consumption of 499 kWh/h, with consumption above the UCL from 8:00 PM through 2:59 AM and consumption below the LCL from 6:00 AM through 12:59 PM. This may be contributed to low energy consumption during Periods 2 and 3 while villagers are working. Energy consumption begins to increase during Period 3 and remains above the UCL during Periods 4 and 1 while the villagers are not working. The Moving Range Chart shows three distinct shifts in energy consumption: 4:00 AM – 5:00 AM when villagers are going to work, 12:00 PM while villagers are on lunch, and 6:00 PM when villagers return home from work. The

analysis shows the hourly changes in demand and the data is therefore useful to analyze system load on an hourly basis. This data can play a vital role in real-time power management, especially in communities where some users consume more power than alotted causing system overload and failure.

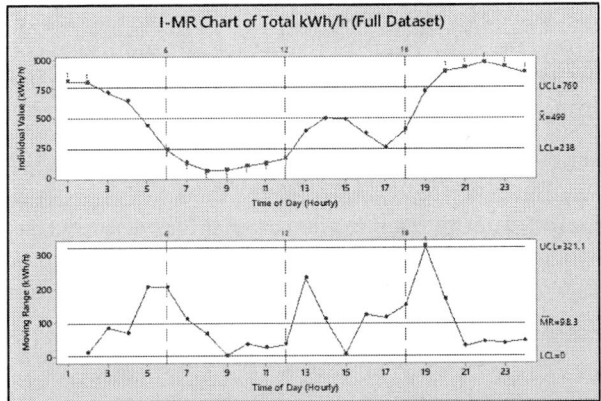

Figure 3: I-MR of Total kWh/h by Time of Day, 2/1/2016 through 1/31/2018. *Y-axis shows hourly sum of kWh/h. X-axis shows hour of the day, starting at 1 = 12:00 AM and ending at 24 = 11:00 PM.*

The second data analysis investigates changes in monthly energy consumption between the time periods of February 2016-January 2017 and February 2017-January 2018. A Line plot of the two time periods in Fig. 4 shows annual energy usage, month to month. Periods of high energy consumption occur from the March through May and from September through November which are likely attributed to seasonal trends in energy needs (farming, heating, air-conditioning, etc.). Monthly data can be used in studies to understand the correlation between energy consumption and seasonal changes in the location. It will give the owner a sense of preparedness as to which time of the year the system would operate at a higher capacity thereby allowing him to make necessary decisions.

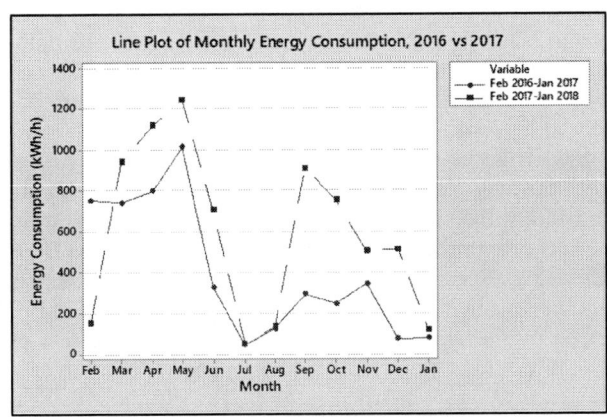

Figure 4: Line Plot of Monthly Energy Consumption, Feb 2016-Jan 2017 vs Feb 2017-Jan 2018.

The third data analysis investigates changes in annual energy consumption between the time periods of February 2016-January 2017 and February 2017-January 2018. A Line Plot of the two time periods shown in Fig. 5 is the annual energy usage, month to month (e.g., March 2016 vs March 2017). For the system under consideration, annual energy consumption has increased from February 2016-January 2017 to February 2017-January 2018. This indicates an increased reliance on energy as Chhotkei becomes accustomed to a readily available energy supply. Monthly energy consumption increased for every month aside from February 2016 and February 2017. This comparison, along with energy consumption statistics from various global organizations validate our assumption that when a community is given access to electricity, their growth and hence energy consumption increases. This results in a need for planned system upgrades to keep up with this demand. The underlying question is by how much should the demand increase over time. In order to answer this question, we utilize an LTSM, a class of Recurrent Neural Network, to predict the future demand using historical demand data details of which are discussed in the following section and then an optimization model to predict the required increase in capacity.

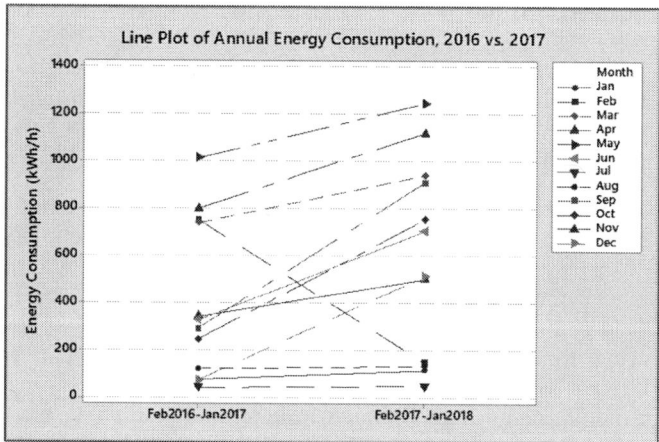

Figure 5: Line Plot of Monthly Energy Consumption, Feb 2016-Jan 2017 vs Feb 2017-Jan 2018.

3.3 The Recurrent Neural Network: A group of 365 Recurrent Neural Networks (RNNs) will take the hourly demand data for Chhotkei Village from February 1, 2016 through January 31, 2018 as the training data. The data will use a shifted window approach in order to develop a more accurate model. The shifted window will consist of ten (10) data points and will be verified by the last point in the data set to validate the models accuracy. The RNN will then predict the first new day of the new year which will be a new point (not validated). The window then moves one data point ahead and then trains a new network for the next data point, using the new ten (10) data points. Fig. 6 is an illustration of the structure of demand prediction using this approach The RNNs prediction will not be skewed by the large discrepancies of energy usage that exist between portions of the data set occurring to due problems in data acquisition mentioned earlier. The demand data is a function of time; therefore, the prediction, is highly sensitive to missing data points. When data relies on time, to adequately capture the trend data must exist for every time step. Unfortunately, this is usually not the case because of failures in the sensors/data collection mechanisms. In our model, there is missing data so the trend is most likely not captured perfectly. This can be improved upon by filling the missing data points with a replica of the previous data point and adding noise to it,

to emulate a real data point and will be performed in future publications.

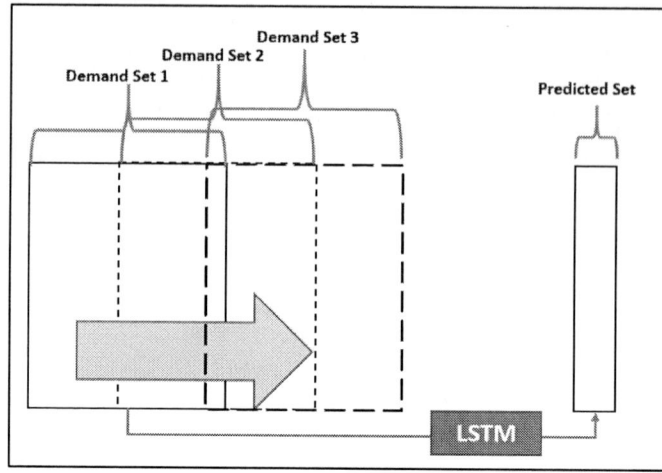

Figure 6: Structure of Shifting Window Approach

The training data will feed forward through the hidden layer and to the outer layer to make a prediction for the next time step. In order to improve accuracy, the type of RNN that will be used will be an LSTM. This is a special case RNN that uses a combination of activation functions within the hidden layer of the network, demonstrated below in Fig. 7 (where σ is the sigmoid function and tanh is the hyperbolic tangent function). The weights are then updated through backpropagation through time (BPTT) using an Adam Optimizer (SGD) to minimize the loss function relative to the weights between each layer. The following diagram in Fig. 7 outlines the LSTM architecture:

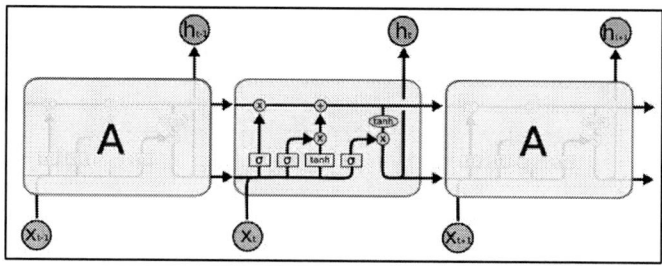

Figure 7: LTSM Structure

Once the LSTM is trained, the network will be used to predict future demand. Now that we have the future demand data for the next year (Fig. 8), we can use the entire data set to find the optimum upgrading capacity required to meet the new increased demand. The upgrading capacity is expressed as a function of the number of solar panels of constant capacity that will be added to the existing system. However, for other cases, variables can be changed based on the degree of knowledge the decision maker has on system parameters. The viewpoint of the owner of the microgrid is thus taken into consideration. The MATLAB function *fmincon* will be used on the generated model to determine how by much the system needs to be upgraded at the end of the two years to meet the increased demand.

V02AT03A025-5   Copyright © 2019 ASME

3.4 Optimization Model for Return on Investment: The optimization model leverages the microgrid investment. It is intended to promote social entrepreneurship by ensuring a rate of return. The model ensures the energy demands of the village are met through scaling. The first term of the function (Eq. 2) defines the selling price of the power generated by the system which is a function of the daily supply and the rate per kWh. The second term is the cost of the system over a particular time span. Equation 2 is the objective function of the optimization model. Due to of lack of cost data available we have only considered the installation cost, the cost of the solar panels, and any additional cost that may be incurred by the owner during the particular time frame. Future studies will further develop the inputs of these terms in greater detail to take into account costs like upgrade of transmission lines, labor costs, inflation rates, battery models etc. to generate more realistic results.

$$Max(N_{solar}) = \left[R * \sum_{T=0}^{T_{upgrade}} S \right] - [I + (N_{solar} * C_{solar}) + C_{addition}] \qquad (2)$$

The optimization model is also subject to the constraint function shown in Eq. 3 that the new capacity must satisfy within at least the 95th percentile of predicted future demand. It should be emphasized that the entire demand should not be the required target as demand varies greatly on a daily basis (Fig. 11). If fulfilment of the entire demand is targeted, it will lead to excessive power supply for most of days of the life of the microgrid. The excess capacity predicted by using the maximum demand in the data set may trigger an unnecessary upgrade in equipment further reducing the profit of the owner for the time.

$$[Cap_0 + (N_{solar} * Cap_{solar})] - (95th\ Percentile * S) \geq 0 \quad (3)$$

NOTE: Each time we infer the optimum time and amount of upgrade for the system, the time variable will be reset and the optimization problem will be resolved independently. The LSTM is used to predict the future demand.

4. APPLICATION & RESULTS

The data acquired from the village of Chhotkei was used for three sub-tasks. First, to study the supply vs. demand trend over the two years of data available. Second, the demand data was used by the LSTM to predict data for another year into the future. Third, the combined dataset (past and predicted) was fed into the optimization model to ensure the constraints are satisfied. Other inputs to the optimization model included the rate per kWh of energy that the owner of the microgrid charges, investment cost, installation cost, and maintenance cost of the microgrid.

4.1 Supply and Demand: This section discusses the results from the conditioned demand data of the microgrid over the past two years superimposed over the calculated supply data for the same time period. The combined plot can be seen below in Fig. 8. We can see from the plot that even over the span of two years, the average demand has increased over time while the average supply is remaining fairly constant. This strongly indicates the need to scale up over time. Further the plot shows

the degree of noise present in the data acquired from the village. As mentioned many days of the data set showed zero demand due to severed connection between the grid and the cloud network. However, in practice this would not have occurred. Those days have been truncated in the figure below leaving only about 420 days of data to work with. Nevertheless, it was this data that was used train the LSTM and run the optimization model. Since this was real data, it can be considered as a much better test case for the robustness of the procedure as compared to randomly simulated results wherein noise would have to be intentionally added.

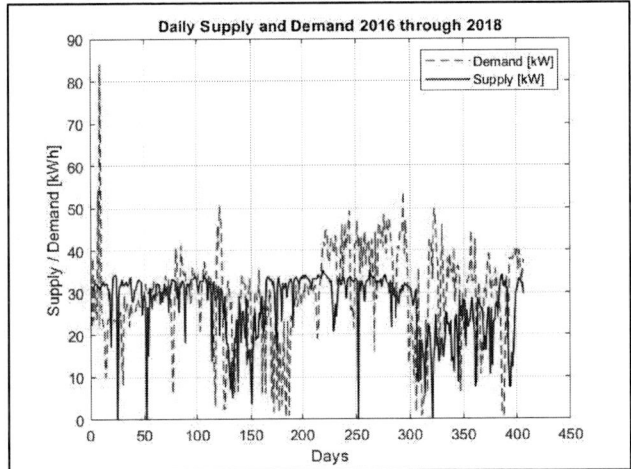

Figure 8: Multivariable Plot of Supply (kWh/Day), 2/1/2016 through 1/31/2018 and Predicted Demand (kWh/Day), LSTM.

4.2 LSTM Results: As discussed in Section 3.3, the LSTM processes one input at a time to produce one output. In implementation, we bin the data points into epochs wherein one epoch is equivalent to one input, with a total of 2000 epochs processed in this study. The output of the Python script of demand predicted by the LSTM is shown in Fig. 9 below.

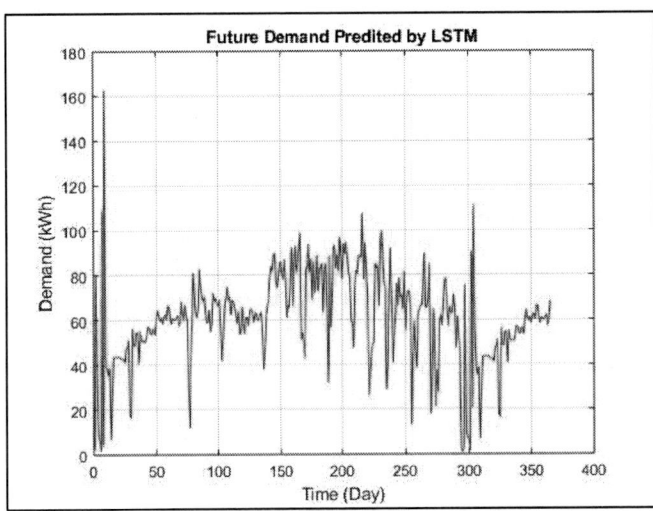

Figure 9: Plot of Predicted Demand (kWh/Day), LSTM

Output values of the LSTM that are at the allowed minimum value (30 kWh/h) are representative of data points in the load data where there are artifacts (e.g., the kWh/h usage data

V02AT03A025-6   Copyright © 2019 ASME

collected show values of 0 for the entire day). While this data was excluded from the analysis performed in Section 3.2, it is imperative to include them for LSTM training as to more accurately predict system output.

The output of the LSTM becomes more accurate as epochs increase sequentially. Each epoch of the LSTM output

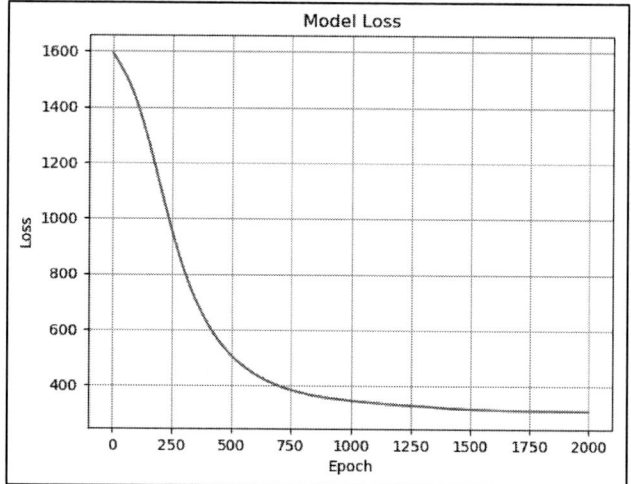

Figure 10: Plot of Loss vs. Epoch, LSTM Output

corresponds to each day of the and demand data. As time progresses, we see that the LSTM more accurately predicts the supply and demand data. Beyond Day 300, the LSTM conservatively estimates supply and demand (e.g., the LSTM output in kWh/h is generally greater than the supply and demand data in kWh/h) which will assist in prevention of energy shortages.

Accuracy of the LSTM was evaluated by measuring loss over sequential epochs as shown in Fig. 10. Starting from the first epoch (x = 1), we begin at a loss greater than an arbitrary magnitude of 2000. As the epochs increase ($x_n > x_{n-1}$), we see that the loss decreases logarithmically; by the last epoch (x = 2000) the loss is essentially a magnitude of 0. This is expected, as the LSTM accuracy should improve as more epochs are processed for algorithm training.

4.2 Results from the Optimization Model: The optimization model shows us that for the owner of the microgrid to remain in profit whilst fulfilling the demand of the village, a total of 110 new solar panels need to be installed to increase the capacity. The cumulative plot of the life of the microgrid as of 2 years of history and 1 year of future prediction is given in Fig. 11. The red plot is the daily demand. The blue line is the average supply of the microgrid operating at the current time. The green line implies what the new capacity will be after 110 new solar panels have been installed.

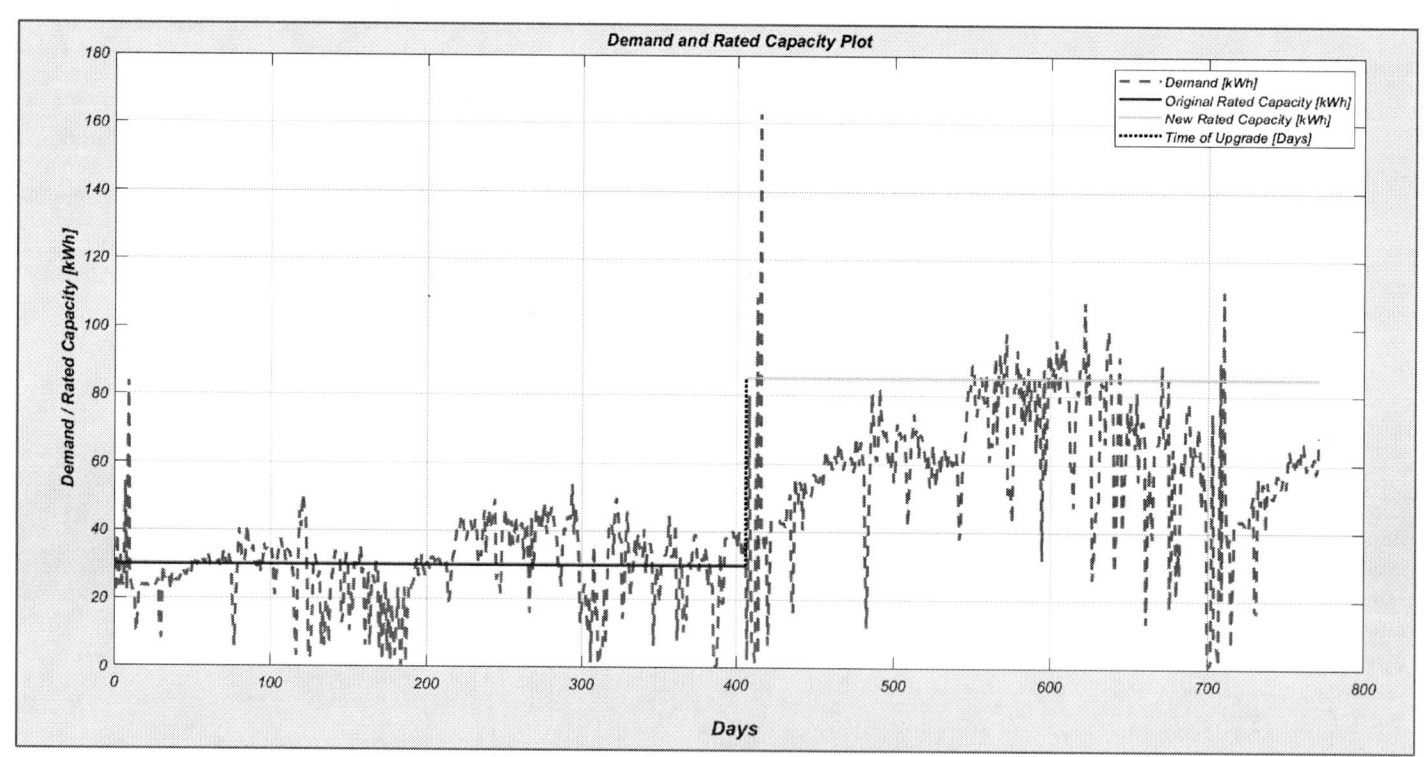

Figure 11: Results from Optimization Model

4.3 Implementation of Adaptive Power Management Strategy: As seen from Fig. 11 and the discussion in Section 3.4 regarding the target capacity of the system, we see that days of unexpectedly high power consumption are not satisfied by the system. This could be attributed to special cases such as

public holidays, festivals or extreme weather conditions. Although these demands cannot be met by the system directly, a control algorithm to ensure equity of power to all patrons of the system can be implemented in real time. Thus, in order to prevent inconvenience to the consumer when supply cannot satisfy demand between two consecutive upgrades, and when the demand goes beyond the 95th percentile, the control algorithm can be activated and ensure fair distribution of power. The basic idea of the control algorithm is shown below in Fig 12.

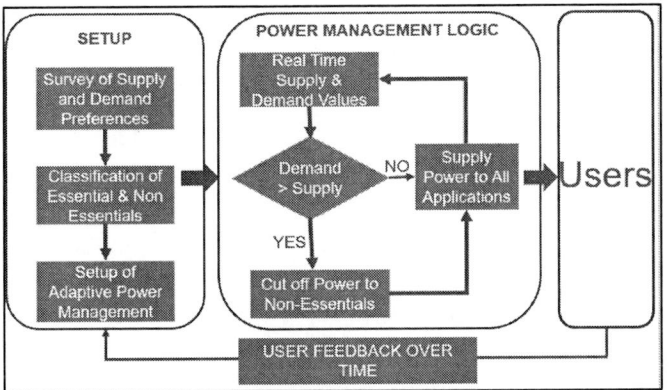

Figure 12: Layout of Adaptive Power Management

The concept is similar to that of load shedding that occurs in many developing countries around the world. Power supply to a region is cyclically cut off every day for a particular amount of time for a particular region the grid is servicing. The difference however is that the entire power supply will not be cut off. In this strategy, all the possible applications of the power in the community will be divided into two subsets: essentials and non-essentials. Essentials may include applications such as street lighting, refrigeration in medical and departmental stores, heating, etc. Non-essentials may include applications such as televisions, computers, etc. Understandably, the non-essentials for a community in one part of the globe may be extremely essential for others. However, for the scale at which microgrids are used in terms of target population and area, it would be feasible to reach a consensus of this division of applications via surveys. The next step is to devise control logic to execute the idea. When supply exceeds demand (supply ≥ demand), all subsets will be provided power and the system will operate normally. When supply cannot satisfy demand (supply < demand), the essentials such as street lighting will take priority while non-essential subsets will be cut-off sequentially. This can easily be implemented in practice using microcontrollers and power electronics. Similarly, a prioritized list can be developed exclusively for those entities that may require to draw more power for special purposes like hospitals or schools. Furthermore, as the community continues to develop and the microgrid is scaled up based on overall demand, we can expect certain non-essentials to be considered converted to essentials. Thus timely feedback from the users is instrumental in smooth and satisfactory functioning of the microgrids.

5. CONCLUSION

From the above discussions we infer that over the life of the microgrid, power demand increases. Our analysis suggests that the original capacity of the system eventually cannot sustain the demand. Future demand can be predicted using machine learning models to a certain degree of accuracy to help the owner of the microgrid to preempt when they will need to scale up the system and by what amount. The accuracy of the prediction however, is highly dependent on the quality and quantity of data. In the present application, the data used is extremely noisy with multiple days of zero demand. This could result in decreased accuracy of the model. This highlights the need to develop methods to compensate for data artifacts. We conclude that a profit and demand fulfillment optimization model can be used to encourage a microgrid owner to continue to invest money into the system while ensuring supply can satisfy user demand and facilitating their development. Optimal increase in capacity required has been presented using the model. Lastly, acknowledging the highly non-linear and dynamic nature of the demands of these developing communities, an adaptive power management strategy with consultation of the users has been proposed to ensure fair usage of power in real time keeping our goal of flexibility and user preference in mind.

6. CLOSING REMARKS

Utilization of microgrids is becoming an increasingly common practice in rural communities which cannot reap the benefits of conventional power grids for multiple reasons. Scalability issues and increasing dependence on energy, however, consistently prove to hinder the ability of these microgrids to fulfill its objective of providing continuous power. The Chhotkei Village microgrid used to implement an LSTM which trends towards 100% accuracy as it is trained and satisfies the essential functional requirements defined for successful provision of energy to the community. The LSTM developed in this paper provides powerful predictive data that allows both the grid owner and other stakeholders to make informed decisions about when to upgrade the microgrid and how much to upgrade it by. Industrial implementation of this LSTM in currently deployed microgrids could greatly contribute to maintenance of operational energy provisions for rural communities whose development depend upon it. It must be understood that unlike conventional grids, the users play an integral role in smooth functioning of these systems. This is especially applicable if microgrids are installed in remote tribal areas across the world wherein the social structure and culture is vastly different from other parts of the world. It therefore can provide rural communities an opportunity to flourish at a faster rate than what they can without energy.

Future work involves employing more detailed LSTM and optimization models. Larger sets of continuous data will also be used in the simulation. Further the predicted demand will be verified by the actual data that we will acquire from the village. This will help engineers to understand the correlation between the location of the microgrid, the quality of life for the users, and the trend in demand for energy using the LSTM. When coupled with the adaptive power management strategy, the LTSM can be expected to boost the reliability of these systems in rural development of all communities irrespective of where they are located in the world. Simulations to verify the functionality of the Adaptive Power Management will be performed. The ultimate goal is to come up with an optimization model taking economics, social demographics and geography into consideration to make the system more robust.

REFERENCES

[1] Van Ruijven, B., Urban, F., Benders, R. M., Moll, H. C., Van Der Sluijs, J. P., De Vries, B., and Van Vuuren, D. P. "Modeling Energy and Development: An Evaluation of Models and Concepts." World Development Vol. 36 No. 12 (2008): pp. 2801-2821.

[2] Bhattacharyya, S. C. "To regulate or not to regulate off-grid electricity access in developing countries." *Energy Policy* Vol. 63 (2013): pp. 494-503.

[3] Bhattacharyya, S. C., and Debajit, P. "Mini-grid based off-grid electrification to enhance electricity access in developing countries: What policies may be required?" Energy Policy Vol. 94 (2016): pp. 166-178.

[4] Thiam, D. R. "Renewable decentralized in developing countries: Appraisal from microgrids project in Senegal." Renewable Energy Vol 35 No. 8 (2010): pp. 1615-1623.

[5] Alam, M. S., Miah, M. D., Hammoudeh, S., Tiwari, A. K. "The nexus between access to electricity and labour productivity in developing countries." Energy Policy Vol 122 (2018): pp. 715-726.

[6] Puglia, G., Moroni, M., Fagnani, R., Comodi, G. "A Design Approach of Off-grid Hybrid Electric Microgrids in Isolated Villages: A Case Study in Uganda." Energy Procedia Vol. 105 (2017): pp. 3089-3094.

[7] Zubleta, L. E. "Are Microgrids the Future of Energy?" IEEE Electrification Magazine Vol. 4 No. 2 (2016): pp. 37-44.

[8] Barnes, D. and Floor, W. "RURAL ENERGY IN DEVELOPING COUNTRIES: A Challenge for Economic Development." Annual Review of Energy and the Environment Vol. 21 (1996): pp. 497-530.

[9] Wytock, M., Salapaka, S., Salapaka, M. "Preventing Cascading Failures in Microgrids with One-sided Support Vector Machines." 53rd IEEE Conference on Decision and Control. 0191-2216: pp. 3252-3258. Los Angeles, CA, December 15-17, 2014.

[10] Loka, P., Moola, S., Polsani, K., Reddy, S., Fulton, S., Skumanich, A. "A case study for micro-grid PV: lessons learned from a rural electrification project in India." Progress in Photovoltaics: Research and Applications Vol. 22 (2014): pp. 733-743.

[11] MacCarty, N., and Bryden, K. M. "Costs and impacts of potential energy strategies for rural households in developing communities." Energy Vol. 138 (2017): pp. 1157-1174.

[12] Kitson, J., et al. "Modelling of an expandable, reconfigurable, renewable DC microgrid for off-grid communities." Energy Vol. 160 (2018): pp. 142-153.

[13] Behzadi, M. S., and Niasati, M. "Comparative performance analysis of a hybrid PV/FC/battery stand-alone system using different power management strategies and sizing approaches." International Journal of Hydrogen Energy Vol. 40 No. 1 (2015): pp. 538-548.

[14] Dash, V., and Bajpai, P. "Power management control strategy for a stand-alone solar photovoltaic-fuel cell–battery hybrid system." Sustainable Energy Technologies and Assessments Vol. 9 (2015): pp. 68-80.

[15] Brka, A., Kothapalli, G., Al-Abdeli, Y. "Predictive power management strategies for stand-alone hydrogen systems: Lab-scale validation." International Journal of Hydrogen Energy Vol. 40 No. 32 (2015): pp. 9907-9916.

[16] Suk, H., Yadav, A., Hall, J. "Scalability Considerations in the Design of Microgrids to Support Socioeconomic Development in Rural Communities." Proceedings of the ASME 2018 International Mechanical Engineering Congress & Exposition.

[17] NASA Prediction of Worldwide Energy Resource, n.d., from https://power.larc.nasa.gov/data-access-viewer.

[18] Shahidehpour, Mohammad, Zhiyi Li, and Mehdi Ganji. "Smart cities for a sustainable urbanization: Illuminating the need for establishing smart urban infrastructures." IEEE Electrification Magazine 6.2 (2018): 16-33.

[19] Williams, Nathaniel J., et al. "Enabling private sector investment in microgrid-based rural electrification in developing countries: A review." Renewable and Sustainable Energy Reviews 52 (2015): 1268-1281.

[20] Whitefoot, J. W., Mechtenberg, A. R., Peters, D. L., & Papalambros, P. Y. (2011, January). Optimal component sizing and forward-looking dispatch of an electrical microgrid for energy storage planning. In ASME 2011 international design engineering technical conferences and computers and information in engineering conference pp. 341-350

Proceedings of the ASME 2019
International Design Engineering Technical Conferences
and Computers and Information in Engineering Conference
IDETC/CIE2019
August 18-21, 2019, Anaheim, CA, USA

DETC2019-98205

SELF-ADAPTING INTELLIGENT BATTERY THERMAL MANAGEMENT SYSTEM VIA ARTIFICIAL NEURAL NETWORK BASED MODEL PREDICTIVE CONTROL

Yuanzhi Liu[*]
The University of Texas at Dallas
Richardson, TX 75080
Email: yuanzhi.liu@utdallas.edu

Jie Zhang[†]
The University of Texas at Dallas
Richardson, TX 75080
Email: jiezhang@utdallas.edu

ABSTRACT

This paper develops a self-adaptive control strategy for a newly-proposed J-type air-based battery thermal management system (BTMS) for electric vehicles (EVs). The structure of the J-type BTMS is first optimized through surrogate-based optimization in conjunction with computational fluid dynamics (CFD) simulations, with the aim of minimizing temperature rise and maximizing temperature uniformity. Based on the optimized J-type BTMS, an artificial neural network (ANN)-based model predictive control (MPC) strategy is set up to perform real-time control of mass flow rate and BTMS mode switch among J-, Z-, and U-mode. The ANN-based MCP strategy is tested with the Urban Dynamometer Driving Schedule (UDDS) driving cycle. With a genetic algorithm optimizer, the control system is able to optimize the mass flow rate by considering several steps ahead. The results show that the ANN-based MPC strategy is able to constrain the battery temperature difference within a narrow range, and to satisfy light-duty daily operations like the UDDS driving cycle for EVs.

Keywords: Battery thermal management system, surrogate-based optimization, artificial neural network, model

predictive control, electric vehicles

1 INTRODUCTION

Lithium-ion batteries (LIB), as the primary traction power source, have been extensively employed in electric vehicles due to its distinguished properties like high energy density, low self-discharging, and low maintenance. However, several critical issues such as gradual aging effects and narrow operating temperature range, still need to be addressed before EVs' next massive expansion. For primary LIB technologies nowadays, studies have suggested that the appropriate operating temperature should be maintained between 15 °C to 45 °C due to its intrinsic chemistry and thermal properties [1, 2]. Otherwise, it may trigger capacity reduction, electrode degradation, or even safety issues under extreme low-temperature and high-temperature environment [3]. Therefore, it is critically important to design and optimally control a battery thermal management system (BTMS) with high efficiency.

A significant amount of research has been conducted in the literature to examine and explore the heat transfer medium coupled with its appropriate structures. State-of-the-art heat transfer mediums include air, fluid, phase change material, heat pipe, and a combination of them [4]. All the mediums other than air and fluid are still under lab experimental stage due to their complexities and unstability. Air-based BTMS is currently widely applied in light-duty EVs due to its unparalleled advantages like light

---

[*]Ph.D. Student, Department of Mechanical Engineering, ASME Student Member.

[†]Assistant Professor, Department of Mechanical Engineering, ASME Professional Member. Address all correspondence to this author.

weight, simple structure, and low cost. Although fluid-based cooling technologies are more preferred to address the newly-rising challenges like fast charging, air-based cooling technologies are still worth to be explored especially together with optimal control strategies because of its excellent performance. For instance, Nissan Leaf 2018 and Volkswagen E-Golf 2017 directly utilize passive air cooling technologies, while Renault Zoe40 2017, Chevrolet Bolt EV, and Toyota Prius Prime use active air cooling strategies. All these models have been successfully updated for several generations, showing that the air-based system is trustworthy and able to meet the market expectations [5].

Existing studies of air-based BTMS have mainly focused on battery pack structure design and optimization [6, 7]. The typical parallel-channel BTMS configuration is usually referred to as a U-type (reverse flow) or a Z-type (parallel flow) structure, depending on the air flow direction. Several modified structures have been proposed with better thermal performance, e.g., uneven battery channel interspacing size [8], tapered manifold configurations [9], and other specific manifold structures that seek to allocate the mass flow rate of each channel uniformly [10, 11]. However, the majority of these studies only optimized the geometry structures for a single stable working condition without adaptively modifying and controlling BTMS under changing working conditions. As a result, when the battery working condition changes, BTMS may fail to work as effectively as its original optimal design or lead to issues like temperature nonuniformity and inefficiency. Several studies have proposed to use classical PID and fuzzy logic systems to control the air-based BTMS at the system level. Because of the intrinsic limitations of U-type, Z-type, or other through-type structures, the thermal performance is impressive but probably far from optimum [12, 13, 14].

This paper seeks to design and adaptively control an air-based BTMS using a model predictive control (MPC) strategy coupled with a newly developed J-type BTMS structure. The J-type air-based BTMS [15] was developed by integrating the benefits of popularly used U- and Z-type structures. The main differences between the novel J-type and traditional U- and Z-type structures are: (i) there are two outlets in the J-type BTMS compared to one outlet in U- and Z-type BTMS; and (ii) two control valves are employed to adjust the openness of each outlet, increasing the flexibility of BTMS to switch from J-mode to either U-mode or Z-mode. To improve the computational efficiency, the channels in the J-type structure are arranged in multiple groups and pre-optimized using surrogate-based optimization. Then the optimized J-type BTMS will employ an MPC strategy along with three control modes, i.e. J-mode, U-mode, and Z-mode, to effectively control the battery temperature.

The remainder of the paper is organized as follows. First, a lithium-ion battery electro-thermal model is developed. Then, the optimization of grouped-channel J-type BTMS is conducted to uniform the battery pack temperature distribution under a benchmark working condition. Thirdly, an artificial neural network (ANN)-based MPC strategy is constructed by considering an urban dynamic driving cycle. Concluding remarks and future work are discussed in the last section.

2 BATTERY ELECTRO-THERMAL AND BTMS MODELING

2.1 Battery Electro-thermal Model

Extensive studies have indicated that the volumetric heat generation rate of LIB is strongly affected by the charging/discharging current, operating temperature, state of charge (SoC), and cycles [16]. A simplified LIB heat generation model can be expressed as Eq. 1 [17].

$$\dot{Q} = I(V - V_{oc}) + IT\frac{\partial V_{oc}}{\partial T} \qquad (1)$$

where $\dot{Q}$ represents the battery heat generation rate, V and V_{oc} denote the open circuit voltage and the cell voltage, respectively. T is the battery cell operation average temperature, $\partial V_{oc}/\partial T$ is named as the entropic heat coefficient, and I is the charging or discharging current.

For simplification, the heat source in Eq. 1 is assumed to be uniform and homogeneous, the radiation heat transfer is neglected here, and the battery thermal behavior can be estimated using the lumped thermal model, as shown in Eq. 2.

$$mC_p\frac{\partial T}{\partial t} = \dot{Q} - hA(T_{cell} - T_\infty) \qquad (2)$$

where m denotes the mass of a battery cell, C_p is the average heat capacity, h represents the convective heat transfer coefficient, A is the effective surface area, and T_∞ is the free stream temperature of the cooling media.

2.2 Battery Equivalent Circuit Model

For the LIB equivalent electric circuit model, several hypotheses have been proposed to characterize the relationships between battery electrical characteristics and its thermal behavior. The first order equivalent circuit is broadly used for traction power source integration, as shown in Fig. 1. The electric model consists of an ideal voltage source, an internal ohmic resistance, and a parallel RC circuit. The RC circuit is utilized here to interpret the dynamic responses. All the parameters are contingent on the SoC, operation temperature, and battery cycle. The mathematical expression of the equivalent circuit is derived and given by:

$$V_t = V_{oc} - I \cdot R_D - I \cdot R_o + (I \cdot R_D)e^{-\frac{t-t_0}{\tau}} \quad (3)$$

where $I \cdot R_D$ is the potential drop on RC circuit, $I \cdot R_o$ is the potential drop on the internal resistance, and $\tau = R_D \cdot C_D$ denotes the time constant.

These parameters in Eq. 3 can be measured using the method of hybrid pulse power characterization (HPPC) test. In this study, graphite/$LiMn_2O_4$ pouch battery cells with a capacity of 1.6 Ah, and a nominal voltage of 3.75 V are used for experiments. The heat generated by the resistances R_o and R_D is considered equal to the internal heat source. The details of experiment setup, parameters analyses, and results validations can be found in Refs. [18, 19]. The extracted data, i.e., current, SoC, and operating temperature, is then utilized to construct the battery thermal model.

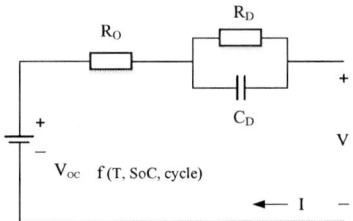

FIGURE 1: First order equivalent circuit model for LIB

2.3 Visualization of Battery Electro-thermal Model

After cross-validation, a Kriging approximation with the second order polynomial regression and exponential error estimation is utilized here to create a battery electro-thermal model based on the experimental data. The model inputs consist of the battery current, SoC, and operating temperature; the model output is the volumetric heat generation rate. The deterministic response $\mathcal{G}(I, SoC, T)$ with three features is formulated as follows:

$$\mathcal{G}(I, SoC, T) = \mathcal{F}(\lambda, I, SoC, T) + \mathcal{R}(\omega, I, SoC, T) \quad (4)$$

where $\mathcal{F}$ is defined as the regression model, and $\mathcal{R}$ is the approximation error. All the Kriging parameters λ, ω are solved by using the generalized least squares estimation method [20].

Figure 2 shows the equivalent volumetric heat generation rate distribution with respect to the operation current,

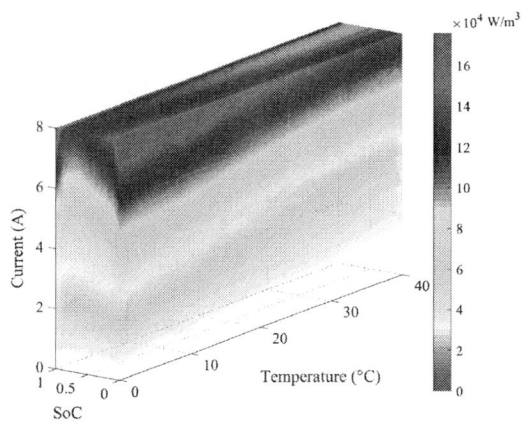

FIGURE 2: The equivalent volumetric heat generation rate distribution with respect to current, SoC, and temperature

SoC, and temperature. It indicates that the internal resistance decreases as the operation temperature increases. This is due to the increased electron mobility and reaction rate at higher temperatures, which will increase heat generation rate. The calorific value is relatively low around 60% SoC, which has also been reported in the literature [16]. Additionally, the overall battery thermal performance is more sensitive to the operating current than other parameters.

2.4 CFD Modeling of J-type BTMS

Ideally, the valve opening of the two control valves on the outlets is expected to be adaptively controlled according to the battery working conditions. However, a large number of simulations are needed to prepare the raw data under various working conditions. So only three modes are considered in this paper for the sake of simplification, which are the J-mode, Z-mode, and U-mode, as presented in Fig. 3. The BTMS prototype consists of ten battery cells, with a geometry size of 151 mm in height, 65 mm in length, and 16 mm in width.

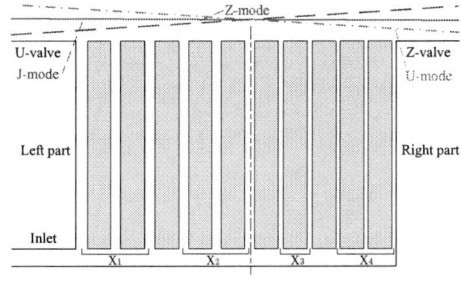

FIGURE 3: Schematic diagram of a J-type air-based BTMS

Figure 4 shows the CFD meshing settings of the J-type BTMS. The total mesh size converges to 1,700,000 ele-

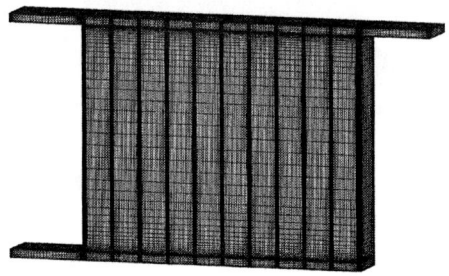

FIGURE 4: The 3D mesh setting of J-type CFD model

ments after grid independence analysis. The inlet and outlet boundary conditions are mass flow rate inlet and pressure outlet, respectively. Radiative heat transfer is neglected because the temperature difference is small. The battery cell is considered to be homogeneous with a uniform heat source.

3 SURROGATE-BASED OPTIMIZATION OF J-TYPE BTMS

Our previous studies have found that the BTMS structure, channel interspacing size in particular, have remarkable influence on the system performance [10]. In this study, surrogate-based optimization is employed to minimize the maximum temperature of battery cells by optimizing the interspacing sizes of grouped-channels. With the optimal structure, BTMS can adaptively adjust the opening degree of the two valves in the J-type structure and the air mass flow rate, seeking to uniformize the temperature among all battery cells.

3.1 Surrogate Model Selection

The Latin hypercube method is employed to perform a design of experiments (DoE). The four grouped-channel interspacing sizes are chosen as the design variables, as illustrated in Fig. 3. All CFD simulations are conducted under the predefined benchmark working condition, in which the mass flow rate is set as 7.1 g/s, and the equivalent heat source is approximately 33,800 W/m^3. The heat source is defined by the battery electrol-thermal model under the benchmark electric condition of 3 C-rate discharging current, 1 SoC, and 300 K environment temperature, and is hooked into Fluent simluations using a user-defined function. A total of 50 simulations are conducted, 75% of which are utilized as training data, and the rest are treated as test data.

An ensemble surrogate model is developed for BTMS modeling. A large pool of surrogate models are first constructed, which consists of five major groups of surrogate models and there are 62 submodels in total regarding dif-

ferent kernel functions or hyper parameters, e.g., Artificial Neural Network (ANN), Kriging/Gaussian Process Regression (GPR), Support Vector Machine (SVM), Radial Basis Functions (RBF), and Polynomial Response Surface (PRS). During the ensemble member model selection and training process, a weighted evaluating criterion is employed here to determine the appropriate models after K-fold cross-validation. Adopted criteria include the normalized maximum absolute error (NMAE) and normalized root mean square error (NRMSE), as given by:

$$NMAE = \frac{1}{n} \sum_{k=1}^{n} \left| \frac{\hat{y}_k - y_k}{y_{max} - y_{min}} \right| \qquad (5)$$

$$NRMSE = \frac{1}{y_{max} - y_{min}} \sqrt{\frac{\sum_{k=1}^{n} (\hat{y}_k - y_k)^2}{n}} \qquad (6)$$

where $\hat{y}$, y, y_{max} and y_{min} denote the corresponding estimated value, actual value, maximum value, and minimum value, respectively. n is the number of test data used in evaluation.

Tabel 1 summarizes the top 6 surrogate models selected from the cross-validation evaluation. These six submodels are integrated together by using the weights calculation method from the Extensive Adaptive Hybrid Function (E-AHF) [21, 22].

TABLE 1: The top 6 surrogate models selected from the cross-validation evaluation

Model-Kernel	NMAE	NRMSE
GPR-Ardmatern32	**3.97**	**5.23**
GPR-Matern32	4.11	5.57
RBF-Cubic	4.14	5.40
PRS-Cubic	4.27	5.59
RBF-TPS	4.37	5.94
KRG-Poly2gauss	4.73	6.41

The estimations and errors of the ensemble surrogate model and the selected member models are shown in Fig. 5. Note that the ensemble model does not necessarily perform the best at every local estimation due to the high nonlinearity of the problem. However, the ensemble model captures the overall trend of the problem and provides the best global accuracy.

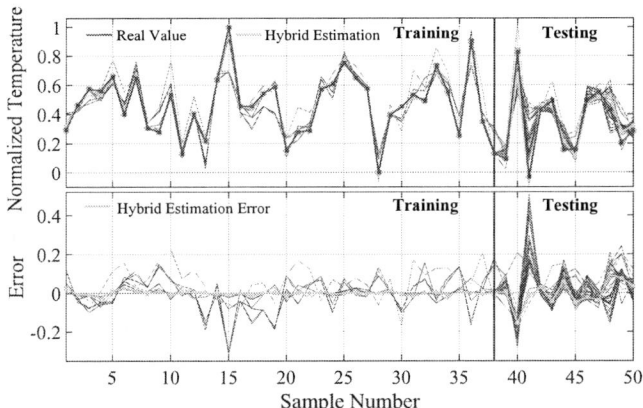

FIGURE 5: Surrogate model estimation and error

3.2 Surrogate-based Optimization and Resampling

The objective is to minimize the maximum temperature $T_{max} = f(x_1, x_2, x_3, x_4)$, and the optimization problem is formulated as:

$$\underset{X}{\arg\min} \quad T_{max} = f(x_1, x_2, x_3, x_4)$$
$$\text{subject to} \quad 2.0 \leq x_i \leq 5.0 \quad i = (1, 2, 3, 4) \tag{7}$$

where $x_1 - x_4$ represent the grouped interspacing size from the left side to the right side.

The genetic algorithm is adopted here to perform the surrogate-based optimization. Figure 6 shows the overall framework of surrogate-based optimization and resampling. Here, a two-stage cluster resampling method is employed, in which the first stage is to cluster the candidate optimal solutions from a surrogate model pool, and the second stage is to generate the adaptive samples using a Gaussian mixed model (GMM). The newly added samples consist of two parts: the selected optimized solutions and the samples generated by GMM. The optimization and resampling process are stopped when the convergence criterion is met, as defined by:

$$BIAS = \left| \frac{Y_k^* - Y_{k-1}^*}{Y_{k-1}^*} \right| \leq 0.001 \tag{8}$$

where Y_k^* and Y_{k-1}^* are the best optimization result of the k-th resampling and $(k-1)$-th resampling, respectively, and 0.001 is a predefined convergence torelance. After two rounds of resampling, as shown in Fig. 7, the resampling results converge to a small design range, where the bias equals to $9.3E\text{-}10^4$. The best sample among all the resampling data is regarded as the optimal solution.

The optimal BTMS design is summarized in Table 2. Compared with the benchmark case with a uniform spacing

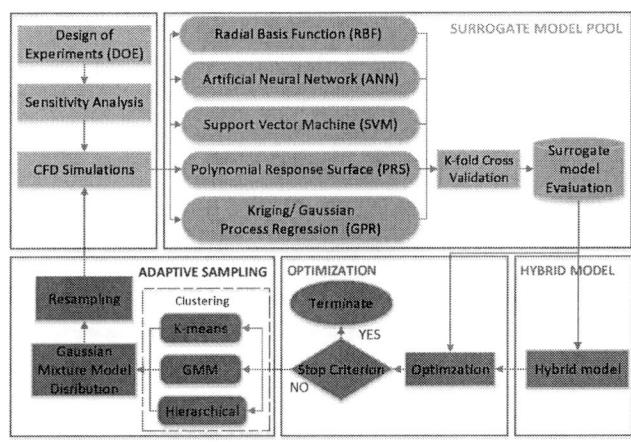

FIGURE 6: The framework of optimization and resampling

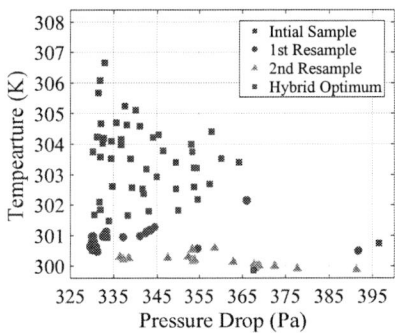

FIGURE 7: The optimization and resampling results

size, the optimal J-type BTMS has a 35.3% reduction in temperature rise, and a 63.4% improvement in temperature uniformity with a cost of 7.5% augment in pressure drop. The CFD simulation result of the optimal J-type BTMS design is illustrated in Fig. 8. It shows a small temperature standard deviation of $0.46\ K$. Though the structure optimization is performed under the benchmark working condition, this optimal structure is also able to uniformize the temperature distribution to some extent under dynamic working conditions, by adaptively controlling the opening degree of the two valves and the air mass flow rate. When the temperature difference between the left section and the right section accumulates to a certain level, the two control valves will change their opening degrees to shift the current cooling mode to another appropriate operation mode.

4 ARTIFICIAL NEURAL NETWORK BASED MODEL PREDICTIVE CONTROL

4.1 ANN Control Model Construction

As mentioned above, in order to prevent unnecessary operations of the two control valves, there are only three operation modes available for the J-type air-based BTMS, i.e., J-mode, U-mode, and Z-mode, as illustrated in Fig.

TABLE 2: The optimal design of J-type BTMS

Parameter	x_1	x_2	x_3	x_4	T_{max}	T_σ	ΔP
Unit	mm	mm	mm	mm	K	K	Pa
J-Benchmark	3	3	3	3	302.15	1.43	347.25
J-Optimized	2.8	2.7	2.2	2	299.98	0.46	391.28

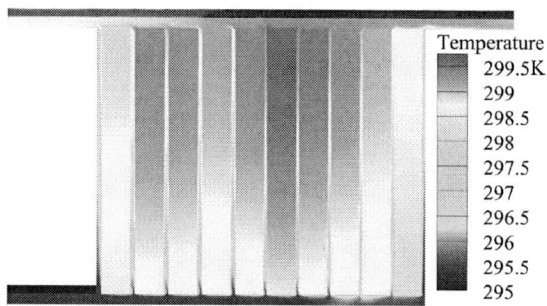

FIGURE 8: CFD simulation validation of the optimal J-type BTMS design

3. A total of 90 CFD simulations are conducted for every mode to characterize the relationship among the mass flow rate, battery temperature, and equivalent heat generation rate. Pressure drop is neglected here, though it is highly related to energy consumption and operating efficiency. Note that the simulations are set in transient mode with a time interval of 5 seconds, which makes the dynamic estimations of different parameters easier by dividing the continuous changing working conditions into discrete steady-state conditions of every 5 seconds and then integrating back to get the final results.

The battery temperature predictor is modeled via surrogate modeling with five inputs, which are the current stage: (i) the maximum temperature at the left side of battery module T_{left_k}, (ii) the maximum temperature at the right side of the battery module T_{right_k}, (iii) operation mode M_k, (iv) mass flow rate $\dot{m}_k$, and (v) heat generation rate $\dot{Q}_k$. There are three outputs of the surrogate model, i.e., $T_{left_{k+1}}$, $T_{right_{k+1}}$, and $\dot{m}_{k+1}$, which are the left side maximum temperature, the right side maximum temperature, and the operation mode of the next stage, respectively, as shown in Fig. 9. The operation mode is designed to switch from one to another when the temperature difference of the battery module exceeds 1 K as given by:

$$
\begin{aligned}
&If\ T_{ref} - T_{right} > 1\ K, \quad then\ switch\ to\ U-mode \\
&If\ T_{right} - T_{left} > 1\ K, \quad then\ switch\ to\ Z-mode
\end{aligned}
\tag{9}
$$

By continuously controlling the operation mode, the temperature difference of the battery module is expected to be fully constrained within a reasonable range. Due to the highly nonlinear nature of the BTMS model, linearized PID control or fuzzy logic control may lose their accuracies under changing control conditions [23]. In this study, an ANN model is employed to establish the control system with a feed-forward structure and two hidden layers. The NRMSE of the ANN model is 11% by cross-validation. Moreover, given the appropriate discharging rate and its corresponding heat generation rate, the mass flow rate is constrained within 0.012 kg/s.

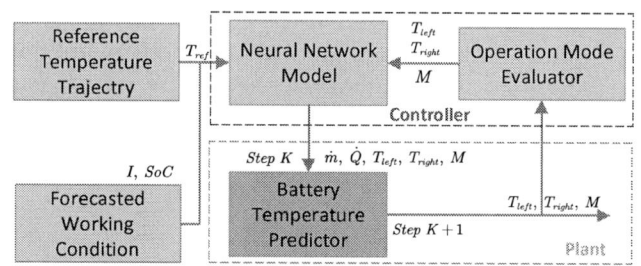

FIGURE 9: The architecture of the ANN controller

4.2 Case Study of ANN Control

Ideally, BTMS as well as battery management system should be controlled adaptively according to driving situations, which could be forecasted based on the driver's personal driving habits and vehicle stream, or transmitted directly from the autopilot system. In this study, the EPA Urban Dynamometer Driving Schedule (UDDS) illustrated in Fig. 10(a) is directly applied to test the thermal management system along with the ANN control strategy. The battery system of the EV consists of 8 series in parallel, and every series has 100 battery cells. The nominal voltage is as high as 375 V.

For a moving vehicle, the equivalent traction power can be estimated using Eq. 10.

$$
P = \frac{V}{1000\eta}\left[mg(\mu cos\alpha + sin\alpha) + \frac{1}{2}\rho A_f C_d V^2 + m\frac{dV}{dT}\right]
\tag{10}
$$

The vehicle parameters and UDDS driving conditions are tabulated in Table 3. Correspondingly, the power profile is shown in Fig. 10(b), where the negative power represents the portion that can be regenerated back into the battery system by a regenerative brake system. Note that the reversible heat is significantly less than the resistance heat, and the charging heat can also be estimated with the

TABLE 3: Specification and driving condition (TESLA Model 3)

Mass	m	1,875 kg	Gradient	α	0
Windward area	A_f	2.22 m^2	Air friction Coeff	C_d	0.24
Standard gravity	g	9.8 m/s^2	Motion efficiency	η	0.98
Air density	ρ	1.16 kg/m^3	Rolling resistance	μ	0.01
Velocity	V	-	Regenerative Coeff	η_r	0.8

electro-thermal model. The equivalent current and the SoC are shown in Fig. 10(c). In this study, the SoC is defined using the Coulomb counting method, as given by:

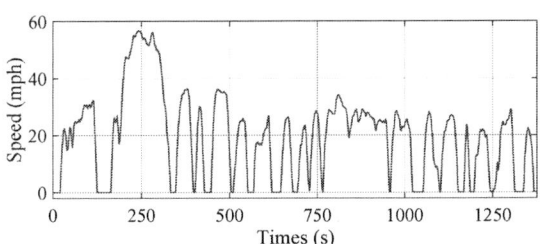

(a) The profile of velocity

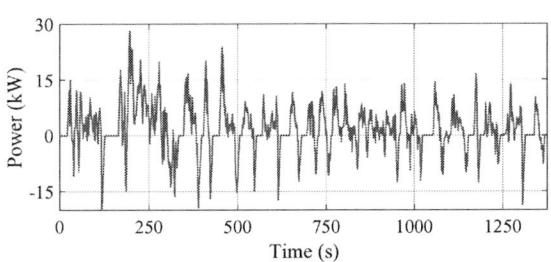

(b) The profile of power

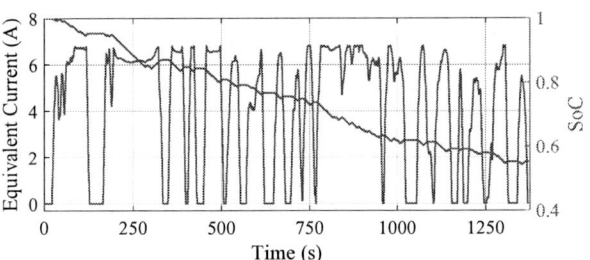

(c) The profile of equivalent current and SoC

FIGURE 10: Profiles of the UDDS driving cycle

$$SoC = SoC_{init} - \frac{\int i\,dt}{Q} \qquad (11)$$

where SoC_{init} and Q denote the SoC of the initial stage and the battery capacity, respectively.

A reference temperature with two portions is predefined based on the UDDS driving cycle: a climbing section that has a temperature rise of 6 K in 20 minutes, and a stationary section of 309 K for the rest of the cycle time. All these parameters are imported into the ANN control system, as shown in Fig. 11. Figure 11(a) shows that the BMTS starts with J-mode, where the left side temperature T_{ref} is dominating. The temperature difference reaches the tipping point in about 11 minutes, and the operation mode switches from J-mode to U-mode. Most of the air flows out of the package from the U-valve, which lowers the battery temperature of the left side and prevents a larger temperature difference. As presented in Fig. 11(b), the mass flow rate is able to adaptively follow the dynamic inputs of heat generation rate under varying working conditions. The results also suggest that the air cooling is capable of dealing with light-duty daily driving cycles like UDDS.

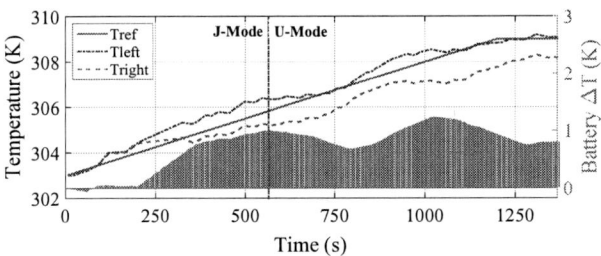

(a) The profile of temperature

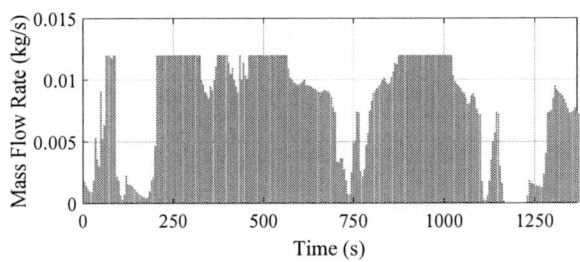

(b) The profile of air mass flow rate

FIGURE 11: ANN control profile of the UDDS driving cycle

4.3 Case Study of ANN-based MPC Strategy

Though the ANN controller is able to successfully control the battery temperature within a narrow range, it is recognized from Fig. 11(b) that the system fails to foresee the rapid increasing temperature, thus resulting in the full load running of the cooling system. As a sequence, the BTMS switches to U-mode that has a relatively lower cooling efficiency. To further improve the cooling efficiency, a MPC control strategy is integrated with the ANN system, as shown in Fig. 12.

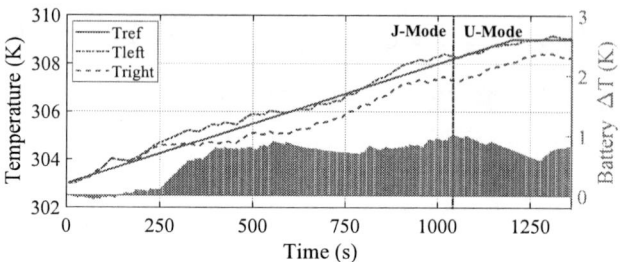

(a) The profile of temperature

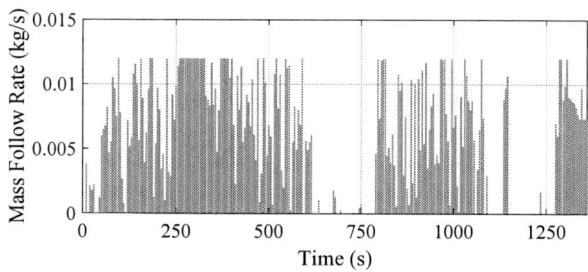

(b) The profile of air mass flow rate

FIGURE 13: ANN-based MPC profile of the UDDS driving cycle

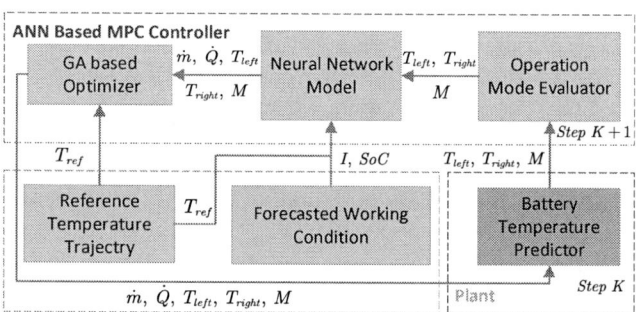

FIGURE 12: The architecture of ANN-based MPC strategy

The genetic algorithm is adopted here to optimize the temperature difference between the reference temperature trajectory and the battery maximum temperature, as given by:

$$\arg\min_{\dot{m}} \quad J = \sum_{k}^{3}(T_{ref_k} - Tmax_k)^2 \tag{12}$$

$$subject\ to \quad 0 \leq \dot{m} \leq 0.012$$

where $\dot{m}$ denotes the mass flow rate, T_{ref} is the reference temperature trajectory, and T_{max} is the maximum temperature of the battery module, defined as $T_{max} = max(T_{left}, T_{right})$. The predictive horizon covers three samples, i.e., 15 seconds in total.

Figure 13 shows the temperature and mass flow rate profiles of the ANN-based MPC strategy. By forecasting the forthcoming load, the control system is able to lower the temperature in advance and provide more control capacity to the system. Compared with the mass flow rate with the ANN control, the ANN-based MPC strategy is more effective, especially in the range between 100 s and 250 s. The NRMSE between the maximum battery temperature and the reference temperature under the ANN-based MPC strategy is 8.5E-4 K compared to 1.3E-3 K

of the ANN control, as illustrated in Fig. 14. It also suggests that the overall temperature difference between the left part and right part of battery module under the ANN-based MPC strategy is much smaller than that of the ANN control case. From the perspective of energy efficiency, we also observe is that the air consumption under ANN-based MPC is much less than that under the ANN control due to different cooling efficiency. Overall, the developed self-adapting intelligent J-type BTMS via ANN-based MPC is capable of controlling the temperature rise as well as the temperature uniformity in a reasonable range.

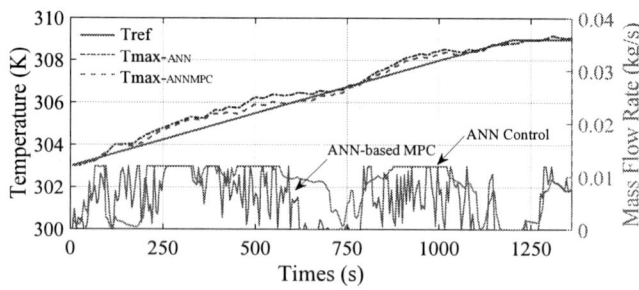

FIGURE 14: A comparison between ANN control and ANN-based MPC strategy

5 CONCLUSION

This paper developed a self-adapting air-based *J*-type battery thermal management system. Based on the newly-established electro-thermal model, surrogate-based optimization was first performed to optimize the structure of the *J*-type BTMS. Results showed that the optimized *J*-type BTMS has a 35.3% reduction in temperature rise, and a 63.4% improvement in temperature uniformity, with a cost of 7.5% augment in pressure drop compared to the benchmark case.

Based on the optimized *J*-type BTMS, an ANN-based MPC model was developed and tested with the UDDS driving cycle. Results showed that the mass flow rate and BTMS operation mode could be adaptively controlled under dynamic working conditions, and the battery temperature difference could be constrained within a narrow range. We also found that the ANN-based MPC strategy performed better than the ANN controller, thereby improving the overall BTMS cooling efficiency. Potential future work will integrate the self-adapting *J*-type battery thermal system together with the operations of other equipments like air conditioner for an optimal battery discharging scheduling.

6 ACKNOWLEDGMENTS

The authors would like to thank Prof. Babak Fahimi and Dr. Zhuo Yang, at the Department of Electrical and Computer Engineering, The University of Texas at Dallas, for providing data on battery electro-thermal modeling.

REFERENCES

[1] Pesaran, A., Santhanagopalan, S., and Kim, G., 2013. Addressing the impact of temperature extremes on large format li-ion batteries for vehicle applications (presentation). Tech. rep., National Renewable Energy Lab.(NREL), Golden, CO (United States).

[2] Ramadass, P., Haran, B., White, R., and Popov, B. N., 2002. "Capacity fade of sony 18650 cells cycled at elevated temperatures: Part i. cycling performance". *Journal of Power Sources, 112*(2), pp. 606–613.

[3] Park, C., and Jaura, A. K., 2003. Dynamic thermal model of li-ion battery for predictive behavior in hybrid and fuel cell vehicles. Tech. rep., SAE Technical Paper.

[4] Xia, G., Cao, L., and Bi, G., 2017. "A review on battery thermal management in electric vehicle application". *Journal of Power Sources, 367*, pp. 90–105.

[5] Grande, L., and He, X. Advanced li-ion beyond li-ion batteries 2018-2028. [Online] Available at: "https://www.idtechex.com/research/reports/advanced-li-ion-and-beyond-li-ion-batteries-2018-2028-000566.asp. [Accessed Jan 2018].

[6] Li, M., Liu, Y., Wang, X., and Zhang, J., 2019. "Modeling and optimization of an enhanced battery thermal management system in electric vehicles". *Frontiers of Mechanical Engineering, 14*(1), pp. 65–75.

[7] Wang, X., Li, M., Liu, Y., Sun, W., Song, X., and Zhang, J., 2017. "Surrogate based multidisciplinary design optimization of lithium-ion battery thermal management system in electric vehicles". *Structural and Multidisciplinary Optimization, 56*(6), pp. 1555–1570.

[8] Xun, J., Liu, R., and Jiao, K., 2013. "Numerical and analytical modeling of lithium ion battery thermal behaviors with different cooling designs". *Journal of Power Sources, 233*, pp. 47–61.

[9] Park, H., 2013. "A design of air flow configuration for cooling lithium ion battery in hybrid electric vehicles". *Journal of power sources, 239*, pp. 30–36.

[10] Kim, J., Oh, J., and Lee, H., 2018. "Review on battery thermal management system for electric vehicles". *Applied Thermal Engineering*.

[11] Shahid, S., and Agelin-Chaab, M., 2017. "Analysis of cooling effectiveness and temperature uniformity in a battery pack for cylindrical batteries". *Energies, 10*(8), p. 1157.

[12] He, F., and Ma, L., 2015. "Thermal management of batteries employing active temperature control and reciprocating cooling flow". *International journal of heat and mass transfer, 83*, pp. 164–172.

[13] Gao, X., Ma, Y., and Chen, H., 2018. "Active thermal control of a battery pack under elevated temperatures". *IFAC-PapersOnLine, 51*(31), pp. 262–267.

[14] Tao, X., and Wagner, J., 2016. "A thermal management system for the battery pack of a hybrid electric vehicle: modeling and control". *Proceedings of the Institution of Mechanical Engineers, Part D: Journal of Automobile Engineering, 230*(2), pp. 190–201.

[15] Liu, Y., Ghassemi, P., Chowdhury, S., and Zhang, J., 2018. "Surrogate based multi-objective optimization of j-type battery thermal management system". In ASME 2018 International Design Engineering Technical Conferences and Computers and Information in Engineering Conference, American Society of Mechanical Engineers, pp. V02BT03A034–V02BT03A034.

[16] Cheng, X.-B., Zhang, R., Zhao, C.-Z., and Zhang, Q., 2017. "Toward safe lithium metal anode in rechargeable batteries: a review". *Chemical reviews, 117*(15), pp. 10403–10473.

[17] Rao, L., and Newman, J., 1997. "Heat-generation rate and general energy balance for insertion battery systems". *Journal of the Electrochemical Society, 144*(8), pp. 2697–2704.

[18] Yang, Z., Patil, D., and Fahimi, B., 2018. "Online estimation of capacity fade and power fade of lithium-ion

batteries based on input–output response technique". *IEEE Transactions on Transportation Electrification,* **4**(1), pp. 147–156.

[19] Yang, Z., Patil, D., and Fahimi, B., 2019. "Electrothermal modeling of lithium-ion batteries for electric vehicles". *IEEE Transactions on Vehicular Technology,* **68**(1), pp. 170–179.

[20] Lophaven, S., Nielsen, H., and Sondergaard, J., 2016. Dace a matlab kriging toolbox [eb/ol].

[21] Zhang, J., Chowdhury, S., and Messac, A., 2012. "An adaptive hybrid surrogate model". *Structural and Multidisciplinary Optimization,* **46**(2), pp. 223–238.

[22] Song, X., Lv, L., Li, J., Sun, W., and Zhang, J., 2018. "An advanced and robust ensemble surrogate model: Extended adaptive hybrid functions". *Journal of Mechanical Design,* **140**(4), p. 041402.

[23] Yan, J., Xu, G., Qian, H., and Xu, Y., 2010. "Battery fast charging strategy based on model predictive control". In 2010 IEEE 72nd Vehicular Technology Conference-Fall, IEEE, pp. 1–8.

Proceedings of the ASME 2019
International Design Engineering Technical Conferences
and Computers and Information in Engineering Conference
IDETC/CIE2019
August 18-21, 2019, Anaheim, CA, USA

DETC2019-97248

STRESS FIELD GUIDED LATTICE STRUCTURE DESIGN BASED ON HEXAHEDRAL MESH

Lingyun Liu, Yizhou Liao, Shuming Gao
State Key Lab of CAD&CG
Zhejiang University
Hangzhou, China

ABSTRACT

Lattice structures are promising for a wide range of applications. The development of additive manufacturing (AM) technology has made it possible to manufacture complex structures. However, designing the optimal lattices of complex solid models efficiently and automatically remains a challenge. Thus, we propose a novel stress-field-guided lattice design method to improve the mechanical properties of a lattice structure. Stress field is used to make the boundary struts of each cell of a lattice structure aligning to the principal stress direction while remaining conformal. Hierarchical cell templates are designed to reduce the computational burden of the cell optimization of a lattice structure. The proposed method is verified experimentally, and the experimental results prove the efficiency and validity of the proposed method.

Keywords: lattice, stress field, hexahedral mesh

1. INTRODUCTION

In recent years, the additive manufacturing (AM) technology has made rapid progress, which has significantly facilitated the manufacturing of complex microstructures. Therefore, the modeling and design of complex structures with special functions and properties (such as lightweight, high strength and stability) has been highly valued by both academia and industry. Accordingly, the lattice structures have become a research hotspot.

Lattice structures have many advantages. First of all, lattice structures have excellent properties such as lightweight, high strength ratio, high stiffness ratio, impact resistance, and high thermal conductivity [1]. The strength of a lattice structure is approximately three times greater than the strength of a foam structure at the relative volumetric density of only 10% [1]. Secondly, the properties of a lattice structure are easy to control, which expands the design space. Recently, many metamaterials have been designed such as structures with a negative Poisson's ratio [1]. Finally, due to the beneficial geometrical characteristics of a lattice structure, the need for a

support structure and difficulty of removing supports during the manufacturing process are far lower than of other microstructures.

There have been numerous studies about automatic design and optimization of high-performance and light-weight lattice structures. Chai et al. [2] proposed a semi-continuous optimization based method that generates the optimal frame structures by solving an optimization problem by minimizing the maximal stress. Chang et al. [3] proposed an improved Size Matching and Scaling (SMS) method, which significantly reduces the time cost of both the design and optimization of lattice structures. Panetta et al. [4] proposed a set of parametric tileable and printable 3D elastic textures achieving a wide range of isotropic elastic material properties which could be used to fabricate objects with a prescribed mechanical behavior. However, their method cannot generate conformal lattice structures. Tang et al. [6] proposed a method to design conformal lattice structures, which calculates each diameter of struts based on the defined project functions and fills hexahedral cells with the diagonal units. Hanks et al. [7] proposed a lattice structure generation method that could customize the hip joint according to the patient's condition. Nessi et al. [8] combined the superformula and tetrahedral meshes to achieve the lightweight design of parts with a complex shape.

Alzahrani et al. [8] proposed an efficient method called the relative density mapping method, which determines the semi-diameters of struts in lattice structures according to the stress field obtained from the topology optimization results of solid models. Chen [10] developed a 3D texture mapping design system, which warps a mesh based on the design requirements such as stress distribution, and then fills the mesh with the lattice units by using the trilinear interpolation. Reinhart [11] proposed a lattice structure design method based on the principal stress lines to overcome the unfavorable load problems such as buckling and shearing. Based on his work, Daynes et al. [12] divided the design space to cells by using the isostatic lines of principal stress, and then filled the cells with the face-centered quadrilateral units to generate the functional graded lattice structures; however, only 2.5D structures could be generated, so that this approach was not suitable for complex 3D models.

The above-mentioned methods can be used to generate lightweight lattice structures, and they have various advantages, but they also have certain shortcomings, which are as follows.

(1) Not fully utilizing the mechanical analysis results to achieve an optimal design of a lattice structure. For instance, the stress field describes well the direction and magnitude of an internal force of a model, but the existing methods based on the stress field take little use of the direction information of the stress field to achieve an optimal design of lattice structures.

(2) It is challenging to generate the conformal lattice structures for complex solid models effectively. Compared to the trimmed lattices, the conforming lattices have better mechanical and surface properties, as well as aesthetic advantage. However, most of the existing methods based on a hexahedral mesh can not generate the conformal lattice structures for complex solid models.

Aiming to overcome the above two shortcomings, in this paper, we propose a novel approach for an optimal lattice design, which generates the conformal high-performance 3D lattice structures by adjusting the hexahedral mesh according to the stress field and infilling the mesh cell with hierarchical template-based structures.

2. OVERVIEW

To achieve an optimal design of 3D lattices, we propose a stress field guided lattice structure design method. Our method follows Michelle's truss structure theory [13]. The basic idea is to reduce the shearing stress of a lattice structure by aligning the boundary struts of each cell of lattice structure with the principal stress direction of the solid model as much as possible, thereby obtaining an optimized lattice. Thus, the key problem is to align the struts of a lattice with the principal stress direction effectively. We solve this problem by using a hexahedral mesh of a solid model as an initial lattice structure and obtain an optimized lattice structure by carrying out multi-stage optimization based on the principal stress field.

For a given 3D solid model and its boundary condition, the procedure of our method can be divided into five stages, as shown in Fig. 1.

i. Stress field extraction. First, discretize the solid model to a hexahedral mesh and calculate the Cauchy's stress tensor field by using the finite element analysis, and then extract the principal stress field and the Von Mises stress field.

ii. Inner geometric optimization of mesh. The objective function of the internal mesh's edges is constructed and optimized by making the optimized edges aligned with the principal stress.

iii. Boundary geometric optimization of mesh. The objective function of edges on a mesh boundary is constructed and optimized under the boundary geometric constraints, making the edges aligned with the principal stress direction while guaranteeing the quality and topology of the mesh boundary.

iv. Template-based cell structure optimization. Construct the hierarchical cell templates of a different complexity following the idea of "ground structure method", and then optimize the lattice cells using the templates based on the cell geometry and stress field.

v. Size optimization. To further reduce the strain energy and improve the mechanical properties of the generated lattices, optimize the diameters of struts by using the Global Response Surface Method (GRSM) [14].

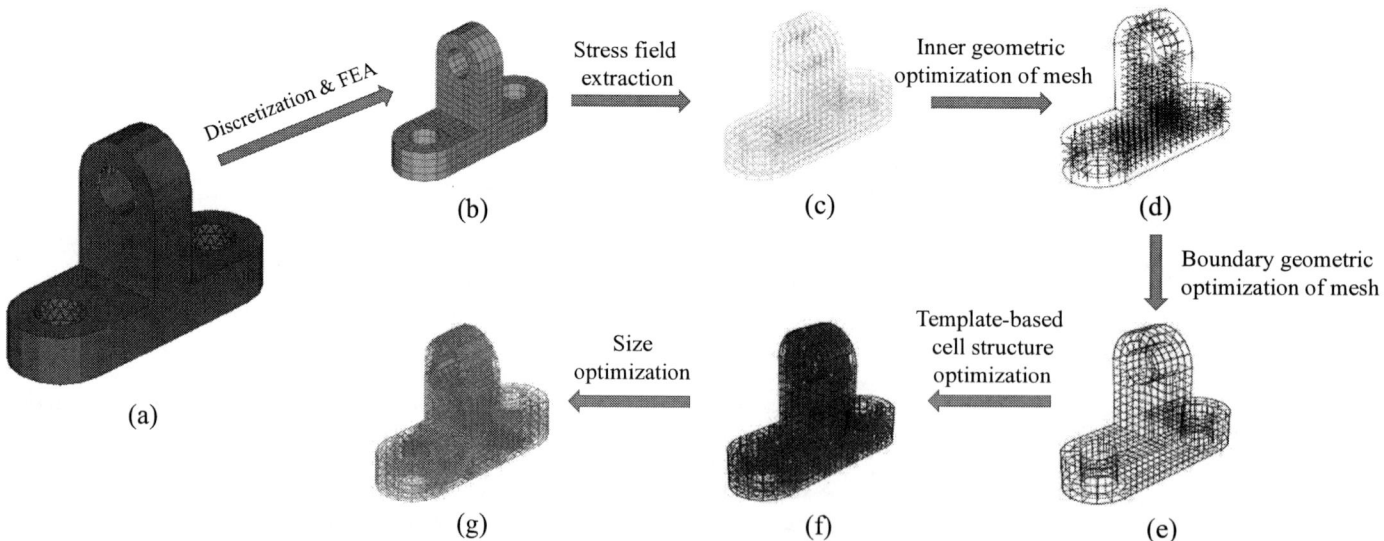

FIGURE 1. THE PROPOSED METHOD WORKFLOW: (a) Model with boundary conditions (b) Hexahedral mesh (c) Stress field (d) Mesh after the inner geometric optimization (e) Mesh after the boundary geometric optimization (f) Lattice after the template-based cell structure optimization (g) Lattice after the size optimization

3. TECHNICAL DETAILS

3.1 Stress field extraction

In order to decrease the shear stress and improve the mechanical properties of a lattice structure, it is necessary to optimize the hexahedral mesh of a solid model based on the stress field. The stress field extraction method is divided into three steps: finite element analysis of solid model, Cauthy's stress tensor field smoothing, and calculation of principal stress field and Von Mises stress field.

3.1.1. Finite element analysis of solid model

For a 3D model with given boundary conditions, the hexahedral mesh generation and finite element analysis are conducted, and isotropic materials and Hooke linear elastic models are used. The Cauchy's stress tensor of each element can be obtained using the finite element analysis results, and it is given by:

$$\sigma = \begin{bmatrix} \sigma_x & \tau_{xy} & \tau_{xz} \\ \tau_{yx} & \sigma_y & \tau_{yz} \\ \tau_{zx} & \tau_{zy} & \sigma_z \end{bmatrix} \quad (1)$$

Thus, the Cauchy's stress tensor is a three-dimensional second-order matrix having six independent parameters, of which σ_x, σ_y, and σ_z denote the normal stresses, and the other matrix elements denote the shear stresses.

3.1.2. Cauchy's stress tensor field smoothing

The principal stress field and Von Mises stress field can be calculated using the Cauchy's stress tensor field, but first, the tensor field needs to be smoothed. Suppose S_i ($i = 0, 1, ..., n$) denotes the sampling points of a point P that satisfies $\|S_i P\| \leq D$, where D is a distance constant that controls the smoothness. The tensor of P after smoothing is given by:

$$\sigma_P = \frac{\sum_0^n \sigma_i}{n} \quad (2)$$

Suppose that R denotes a set of FEM points, and T denotes the Cauchy stress tensor field corresponding to R. The Cauchy's stress tensor for a point X in the model is given by:

$$T_X = \frac{\sum\left(e^{-(\varepsilon\|X - R_i\|)^2} T_i\right)}{\sum e^{-(\varepsilon\|X - R_i\|)^2}} \quad (3)$$

where ε denotes a constant, and in this work, $\varepsilon = 0.5$.

3.1.3. Calculation of principal stress field and Von Mises stress field

Principal Stress Calculation. For a point, according to the Cauchy formula and the reciprocal theorem of shear stresses, Equation (4) can be obtained. By solving its eigenvalues, the principal stresses of R can be obtained; namely, the major principal stress σ_1, the mid-principal stress σ_2, and the minor principal stress σ_3.

$$\begin{vmatrix} \sigma_x - \sigma & \tau_{xy} & \tau_{xz} \\ \tau_{xy} & \sigma_y - \sigma & \tau_{yz} \\ \sigma_{xz} & \tau_{yz} & \sigma_z - \sigma \end{vmatrix} = 0 \quad (4)$$

Von Mises Stress Calculation. The Von Mises stress of a point can be calculated by combining Equation (3) and Equation (1), which results in the following relation:

$$\sigma_v = \frac{1}{\sqrt{2}}\sqrt{\left(\sigma_x - \sigma_y\right)^2 + \left(\sigma_y - \sigma_z\right)^2 + (\sigma_z - \sigma_x)^2 + 6\left(\tau_{xy}^2 + \tau_{yz}^2 + \tau_{zx}^2\right)} \quad (5)$$

Following the maximum distortion energy criterion, the Von Mises stress takes into account the major, mid, and minor principal stresses, which can well describe the stress distribution of a solid model.

3.2 Inner geometric optimization of mesh

Considering the hexahedral mesh structures are widely used in industry, we use a hexahedral mesh of a model as a basis for a lattice structure. In order to improve the mechanical properties of lattice structures generated completely based on the geometry and topology of solid models, it is necessary to use the principal stress field to optimize the hexahedral mesh geometrically so that the direction of the mesh edge can be as uniform as the principal stress field to reduce the shear stress of a lattice structure. Meanwhile, the geometric quality of the hexahedral mesh needs to be guaranteed. Since there is no conformal constraint inside the mesh, first, a more efficient global optimization method is used. The internal mesh optimization can be considered as an unconstrained optimization problem, which is expressed by:

$$\arg\min \left(f_{stress} + \alpha * f_{shape}\right) \quad (6)$$

where f_{stress} and f_{shape} denote the stress term and shape term of a lattice structure, respectively. The specific definition of optimization variables and objective functions is given in the following.

3.2.1 Optimization variables

There are many vertices inside a hexahedral mesh, and the mesh frame field is different from the principal stress field. In order to improve the optimization efficiency, it is impossible to move all the vertices (edges) of the internal mesh. Therefore, some mesh vertices need to be defined for the optimization process. These mesh vertices are used to determine the optimization variables.

Definition: Assume V is a set of all the vertices of a mesh, E is a set of all the edges of a mesh, $V_{boundary}$ and $V_{interior}$ are the boundary and interior vertices of a mesh, and $E_{boundary}$ and $E_{interior}$ are the boundary and interior edges of a mesh respectively. Then, if v_i and v_j are the vertices of e_{ij}, the Cauchy stress tensor of e_{ij} is given by:

$$T_{e_{ij}} = \frac{T_{v_i} + T_{v_j}}{2} \quad (7)$$

Since the axial load capacity of a strut is the largest, according to the first strength theory, we choose a mesh edge whose angle with the direction of the major principal stress $\overrightarrow{\sigma 1_{e_{ij}}}$ is less than θ as a movable edge, which is given by:

$$E_{move} = \left\{ e_{ij} \middle| v_i \in V_{interior} \quad or \quad v_j \in V_{interior} \quad , \langle \overrightarrow{v_i v_j}, \overrightarrow{\sigma 1_{e_{ij}}} \rangle \le \theta \right\} \quad (8)$$

where θ denotes an angle which is always less than 90 degrees, and in this paper, θ is equal to 30 degrees; $E_{adj}(v_i)$ is a set of mesh edges connected to v_i; vertices connected to the movable edges are defined as movable vertices:

$$V_{move} = \left\{ v_i \middle| \exists e_{ij} \in E_{move} \quad and \quad e_{ij} \in E_{adj}(v_i) \right\} \quad (9)$$

3.2.2 Object function

For all movable mesh vertices, an optimal location is obtained by minimizing the global objective function, which is composed of a stress term and a shape term. We assume that the input mesh shape is optimal, so the difference between the optimized mesh and the original mesh is as small as possible. Further, $P_i \in V_{move}$ is the coordinate of a vertex to be optimized, where P_i' is the optimized coordinate, and L is the total number of vertices to be optimized. Then, the mesh shape term is defined as follows:

$$f_{shape} = \sum_{i=1}^{L} \lVert P_i - P_i' \rVert^2 \quad (10)$$

From the above, we can see that it is necessary to minimize the directional difference between the mesh edges and the stress field. Since e_i is a movable edge, f_i is the major principle stress influence factor, and M is the number of edges to be optimized, the stress term is defined as:

$$f_{stress} = \sum_{i=1}^{M} f_i \lVert \overrightarrow{e_i} - \overrightarrow{\sigma 1_{e_i}} \rVert^2 \quad (11)$$

Let α be the penalty factor, then, the objective function is defined by Equation (6), and P' is the variable that needs to be solved. Since the optimization problem represents a minimum solution problem without constraints, a linear system can be constructed to solve P'. The vertices of a mesh M are traversed again, and if they are immovable, then keep the original coordinates; otherwise, update the coordinates to P'.

The comparison of the mesh before and after the optimization is shown in Fig. 2. Using the same volume material, the strain energy and the maximum displacement of a lattice structure corresponding to the original mesh are reduced by about 18% and 17%, respectively. The lowest Jacobian values of the hexahedral cell before and after optimization are 0.654 and 0.605, respectively.

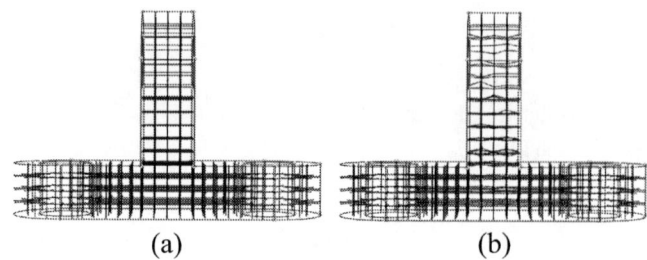

(a)   (b)

FIGURE 2. THE OPTIMIZATION RESULTS: (a) Mesh before the interior optimization (b) Mesh after the interior optimization

3.3 Boundary geometric optimization of mesh

Since the part usually bears forces directly on a surface, the accuracy and the shape quality of a lattice structure boundary are generally considered to be more important. To achieve that the optimized mesh boundary satisfies the geometric constraints of the solid model, we need to ensure that the vertices are still on the boundary curves and surfaces wherein they were before the optimization. Considering the mechanical properties of a lattice structure and the geometric quality of a boundary mesh, the mesh boundary optimization problem can be defined as follows:

$$arg \quad min(\mathbb{f}_{stress} + \beta * \mathbb{f}_{shape}) \quad (12)$$
$$s.t. \quad C_i(P_i') = 0$$
$$S_i(P_i') = 0 \quad (13)$$

where C_i and S_i are the parametric equations of the curves or surfaces wherein P_i was located before the optimization. ; $\mathbb{f}_{stress}$ and $\mathbb{f}_{shape}$ are the stress term and the shape term of a boundary mesh, respectively.

To determine the parameters of the boundary optimization problem, first, according to the principal stress field, the vertices to be optimized on the mesh boundary are obtained by the method presented in the previous section. Then, the objective function of the mesh boundary optimization problem consists of the following two terms: shape term and stress term.

3.3.1. Shape term of mesh boundary

With the aim to guarantee the geometric quality of the boundary of a lattice structure, the shape term [15] of a hexahedral cell is introduced. The average quality of tetrahedrons corresponding to eight vertices of a hexahedral cell is used as a measure of the quality of a hexahedral cell. For a movable vertex, the shape term is defined as:

$$\mathbb{f}_{shape} = \frac{1}{M} \sum_{i=1}^{M} \left(\frac{1}{N} \sum_{k=1}^{N} \frac{|S_{i,k}|^2}{3h(\sigma_{i,k})^{2/3}} \right)^2 \quad (14)$$

3.3.2. Stress term of mesh boundary

With the aim to minimize the difference between the mesh edge and the stress field, for a movable mesh vertex, the stress term is defined as:

$$\mathbb{f}_{stress} = \frac{1}{K}\sum_{i=1}^{K} \mathbb{f}_i \left\| \overrightarrow{e_i} - \overrightarrow{\sigma 1_{e_i}} \right\|^2 \quad (15)$$

$$\mathbb{f}_i = \frac{max(\sigma 1) - min(\sigma 1)}{\sigma 1_{e_i} - min(\sigma 1) + \delta} \quad (16)$$

where K is the number of movable edges adjacent to a vertex p. δ is a small number to avoid dividing zero, and in this work, it equals to 0.01.

After the geometric optimization, the mesh vertices need to be kept on the surfaces and curves wherein they were before the optimization, so the objective function is constructed as follows:

$$P' = \arg min \left(\mathbb{f}_{stress} + \beta * \mathbb{f}_{shape} \right) \quad (17)$$

where β denotes the penalty factor, and P' denotes the coordinates of movable vertices after the optimization. Thus, the parametric equations of curves and surfaces are introduced into the objective function, and the problem is constructed as an unconstrained optimization problem. Finally, the BFGS is used to solve the optimization problem. The optimized hexahedral mesh is shown in Fig. 3. Using the same volume material, the strain energy and the maximum displacement of the lattice structure corresponding to the mesh after optimization are reduced by about 7.1% and 4.9%, respectively. The lowest Jacobian values of the cell before and after the optimization are 0.654 and 0.438, respectively.

FIGURE 3. MESH AFTER THE BOUNDARY OPTIMIZATION

3.4 Template-based cell structure optimization

When the optimized hexahedral mesh is used to construct the lattice structure directly, the resultant lattice structure cannot satisfy the required optimization objective because the hex structure is too simple. Therefore, further structural optimization of the mesh's cells is required. In this work, further optimization is conducted by infilling the mesh cells with the template based structures according to the stress fields.

3.4.1 Template Construction

In order not to exclude the anisotropic mesh cells with good mechanical properties under irregular conditions and to increase the capacity of dealing with complex solid models and boundary conditions, a hierarchical cell template is proposed in this paper. Based on the complexity, the cell template is divided into three levels: L1, L2, and L3. In L1 level, the vertex, the edge midpoint, the face center point, and the body center point of a hexahedron are taken as key points. The L1-level template which has 86 struts is shown in Fig. 4.

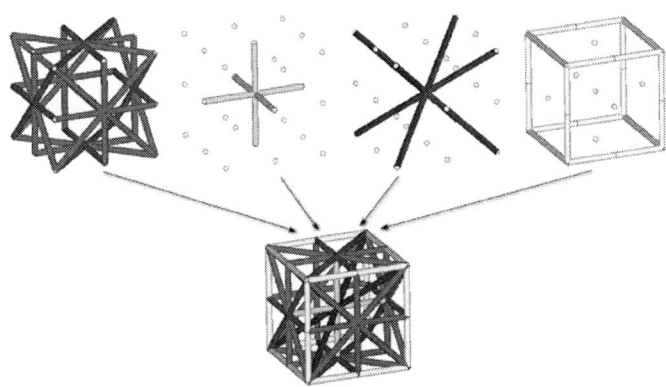

FIGURE 4. L1-LEVEL TEMPLATE AND ITS COMPONENTS

As shown in Fig. 5, the L2-level template can be obtained by using the symmetry of L1-level template in three directions, x-direction, y-direction, and z-direction, and this template has 508 struts in total. Similarly, the L3-level template can be obtained from the L2-level template, and it has 3416 struts in total.

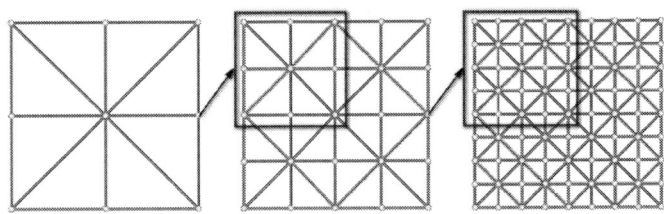

FIGURE 5. SCHEMATIC DIAGRAM OF THE RELATIONSHIP BETWEEN L1-, L2-, AND L3-LEVEL TEMPLATES

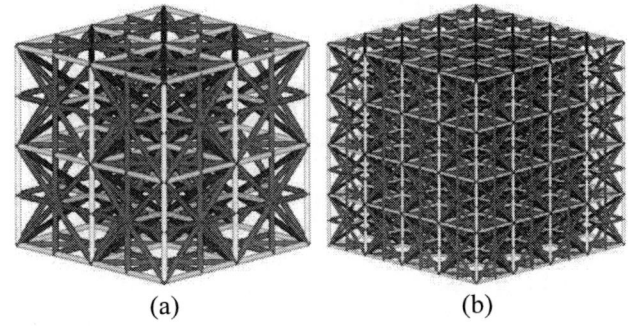

(a)   (b)

FIGURE 6. L2-LEVEL TEMPLATE AND L3-LEVEL TEMPLATE: (a) L2-level template (b) L3- level template

3.4.2 Cell structure optimization

In this paper, an efficient template filling algorithm based on the stress field is proposed to optimize the cell structure, and it is divided into two steps: template matching and connectivity assurance.

1. Template matching

As explained previously, there are three templates with a large difference in the complexity, namely L1-, L2-, and L3-level templates. For instance, the lattice structure generated by L1-level template is too simple, and the lattice structure generated by L3-level template is too complex. Therefore, the mesh cells are classified by the Von Mises stress of the solid model. According to the material distribution theory, we consider that the larger the stress is, the more complex the template should be because a wide design space helps to reduce the hollow ratio of a model and save the amount of materials. Fig. 7 shows the stress distribution of the model.

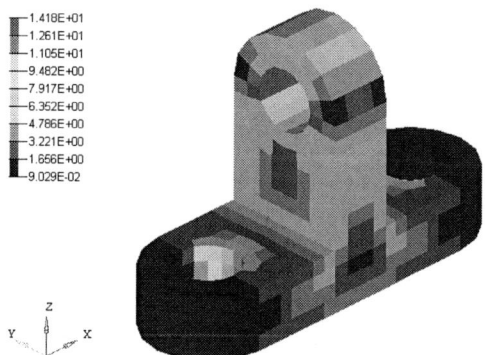

FIGURE 7. THE VON MISES STRESS DISTRIBUTION OF THE MODEL

First, the cell stress is normalized, and then the k-means clustering algorithm is used to classify the cells into three categories according to the stress distribution, corresponding to L1-, L2-, and L3-level templates respectively. As mentioned, the complexity of templates varies greatly; for instance, the number of struts in L1 and L3 levels varies by about 38 times, and the number of key points at the junction in the same levels varies by about 8 times. To ensure the connectivity of adjacent cells, it is specified that the difference between the adjacent cells from different levels can be at most 1, so a cell matched by L3-level template cannot be adjacent to a cell matched by L1-level template. Fig. 8 shows the clustering result after adding the adjacent constraints.

Accordingly, we can determine which template is used to generate the structure of each cell. Since templates are designed using the ground structure method, we need to remove the struts whose performance is relatively poor.

The trilinear interpolation method is used to obtain the coordinates of all the nodes of the current template corresponding to the cell's shape, and then the actual coordinates of all the nodes in the current coordinate system can be obtained.

Then, the internal cell structure is composed of struts whose angle with the direction of the major principal stress is less than θ. The union-find set is used to guarantee the lattice connectivity.

$$element_j = \{beam_k | \langle \vec{f}_{beam_k}, \vec{e}_{beam_k} \rangle \leq \theta\} \qquad (18)$$

In the actual operation, the selection of template and the removal of struts are performed simultaneously. The algorithm is parallel and has high design efficiency. The result is shown in Fig. 11.

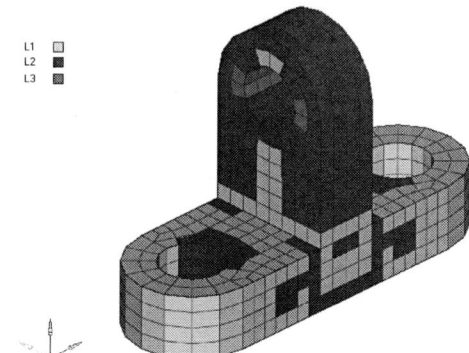

FIGURE 8. CLASSIFICATION OF HEXAHEDRAL CELLS

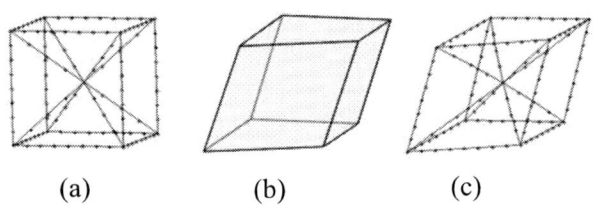

FIGURE 9. TRILINEAR INTERPOLATION OF STRUCTURES: (a) Regular structure (b) Lattice cell (c) Structure after the trilinear interpolation

2. Connectivity Assurance

As can be seen in Fig. 5, the key points of the three templates at the joint surface of the cells are inclusive, so the connectivity between the cells is guaranteed to a certain extent, but there are still some struts which are not stressed and have a node of 1 degree. To reduce the structure complexity, the struts inside the cell are removed directly. On the other hand, to ensure the force transmission between the cells, it is necessary to add some struts at the joint surface of the cell. The struts can be divided into the following two cases.

Case A. Strut with two nodes on the connecting surface. Extend a strut along its direction to a key point whose degree is different from 0. In Fig. 10(a), the blue struts denote the added part.

Case B. Strut with one node on the connecting surface. First, the struts along the orthogonal direction of a node on the connecting surface are complemented. Then, compared with the method a, choose the method that adds fewer struts. In Fig. 10(b), the blue struts denote the added part.

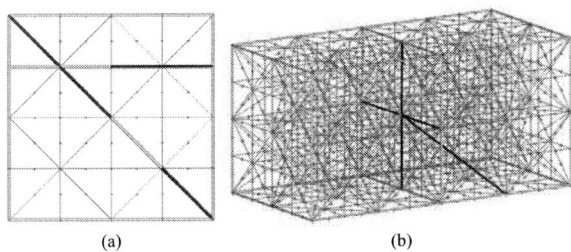

FIGURE 10. CONNECTIVITY ASSURANCE: (a) Case A (b) Case B

TABLE 1. SIZE OPTIMIZATION PROCESS

Given:	P^{BG}, P^F, P^M, i	
Solve:	$D_i \in [D_{min}, D_{max}]$	(19)
Satisfy:	$\sigma_i < \sigma_{max}$	(20)
	$\sum V_i < V_{max}$	(21)
Minimize	$C = \frac{1}{2}U^T K U$	(22)

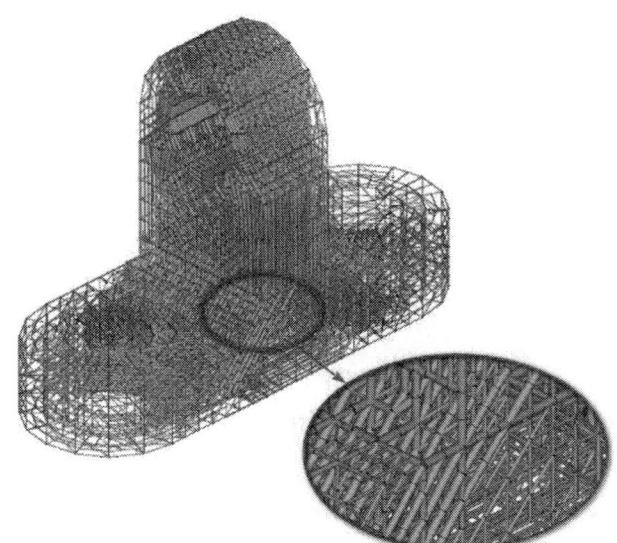

FIGURE 12. LATTICE STRUCTURE AFTER THE SIZE OPTIMIZATION

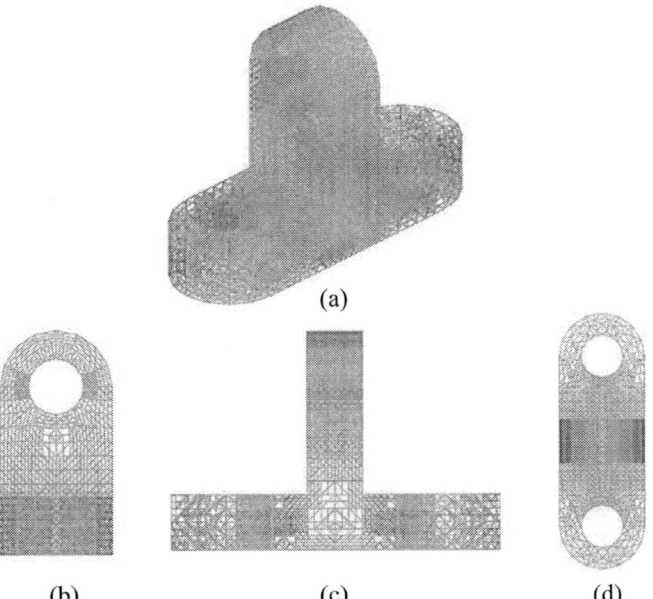

FIGURE 11. LATTICE AFTER THE TEMPLATE-BASED CELL STRUCTURE OPTIMIZATION: (a) Axonometric view (b) End view (c) Front view (d) Vertical view

3.5 Size optimization

With the aim to further improve the mechanical properties of a lattice structure, the size optimization is required. Size optimization of a lattice structure refers to the optimization of the cross-section diameter of struts to minimize the total strain energy of a lattice structure. The formalization of the optimization problem is given in Table 1, where P^{BG}, P^F, and P^M represent the mesh, the given loads and constraints, and the material properties, respectively, D_i denotes the diameter of a $beam_i$ in a lattice structure, and C denotes the total strain energy of a lattice structure.

The FEA of a lattice structure is a non-monotonous and non-linear problem, and the complexity of its solution is proportional to the complexity of a lattice structure. In order to improve the optimization efficiency, the GRSM is selected to optimize the cross-section of struts. Given the volume constraint of 1500 and boundary conditions, the lattice structure after the size optimization is shown in Fig. 12.

4. RESULTS AND DISCUSSION

The proposed method aims to design complex lattice structures with excellent mechanical properties. The proposed method is evaluated experimentally. In the evaluation process, the total strain energy and maximum stress were selected as measures of the mechanical properties of a structure. Chang's method was chosen for comparison with our method. The OPTISTRUCT was used to solve the FEA simulation problem. A desktop computer, having Intel(R) Core(TM) CPU i5-6400 (2.70 GHz) and 16GB of RAM on a 64-Bit Windows operating system was used in all the experiments.

Different loads, shapes, and sizes were used to validate our method. Models used in the experiments and their boundary conditions are shown in Fig. 13. To guarantee the effectiveness of the experiments, both methods used the identical hexahedral meshes as an input. The size of the meshes was 5 mm. The material's Young's modulus was 210000 MPa, and Poisson's ratio was 0.3. The boundary conditions and the obtained results are given in Table 2. The comparison result of our method and Chang's method are given in Table 3. As given in Table 3, our method had less total strain energy, smaller maximum stress, and less maximum displacement. Aligning with the principal directions reduced the shearing stress, which would cause damages to the struts. Thus, the total strain energy, maximum stress, and maximum displacement of the lattice structure were reduced. Meanwhile, our method had fewer struts, which was also an advantage.

TABLE 2. EXPERIMENTAL RESULTS

INDEX / MODEL	Distribued Load (N/mm²)	Material Volume	Opitmization Time(s)	Strut's number
Model 1	0.93	1500	9.44	25480
Model 2	0.72	1500	30.31	12644
Model 3	2.25	7000	13.92	16023
Model 4	1.07	10000	24.47	75648

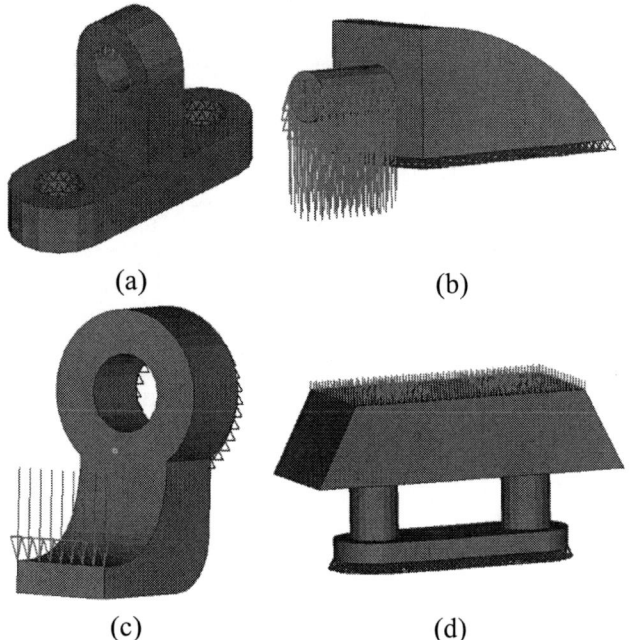

FIGURE 13. TEST MODELS WITH THE BOUNDARY CONDITIONS: (a)Model 1 (b)Model 2 (c)Model 3 (d)Model 4

The stress of the lattices designed by the two methods was normalized, and the corresponding standard deviation was calculated, and it is shown in Fig. 14. The stress distributions of the lattices are shown in Fig. 16, wherein it can be seen that the standard deviation of our method was smaller than of Chang's method, and the stress distribution of our method was more even; thus, it could effectively avoid the local stress to be too large.To further validate our method, we used the mesh size of 10 mm, and compare the two methods under the same conditions. Again, our method had better overall mechanical

properties, and the results were similar to that for the 5-mm mesh. The maximum displacement of model 3 was a little worse than of Chang's method, which probably was caused by the hexahedral mesh generation method.

TABLE 3. COMPARISION RESULTS

INDEX / MODEL	Total strain energy reduction	Total stress reduction	Total displacement reduction	Strut number reduction
Model 1	18.34%	43.98%	19.89%	-3.70%
Model 2	45.00%	39.21%	35.71%	45.16%
Model 3	6.16%	33.13%	9.88%	33.57%
Model 4	23.31%	17.27%	57.75%	-78.58%

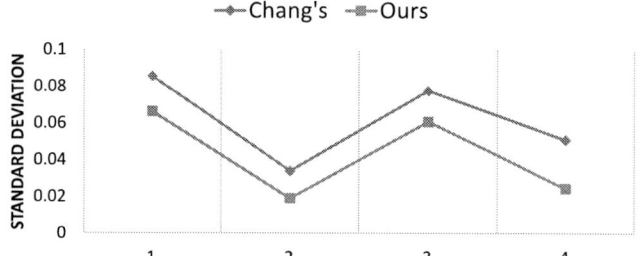

FIGURE 14. STRESS STANDARD DEVIATION OF LATTICES OF THE TWO METHODS

5. CONCLUSION

We propose a novel method to achieve an optimal lattice design by multi-stage optimization of a hexahedral mesh of 3D solid models based on the principal stress field. The proposed method includes: inner vertices geometric optimization of a mesh, boundary vertices geometric optimization of a mesh, template-based cell structure optimization, and size optimization. Our method can effectively design 3D conformal lattices with optimized mechanical properties. The effectiveness of our method is verified experimentally.

However, the proposed method has some shortcomings which we will try to address in our future work. Namely, we will try to improve our method further from the following aspects.

1. We will extend the proposed method to make it able to handle a hybrid hexahedral mesh to support the processing of arbitrarily complex 3D models.

2. We will conduct more comparisons between our method and the existing methods including the non-linear and sequential linear programming methods.

3. We will improve the construction and matching methods of template-based structures so that it can effectively deal with the non-regular hexahedron cells.

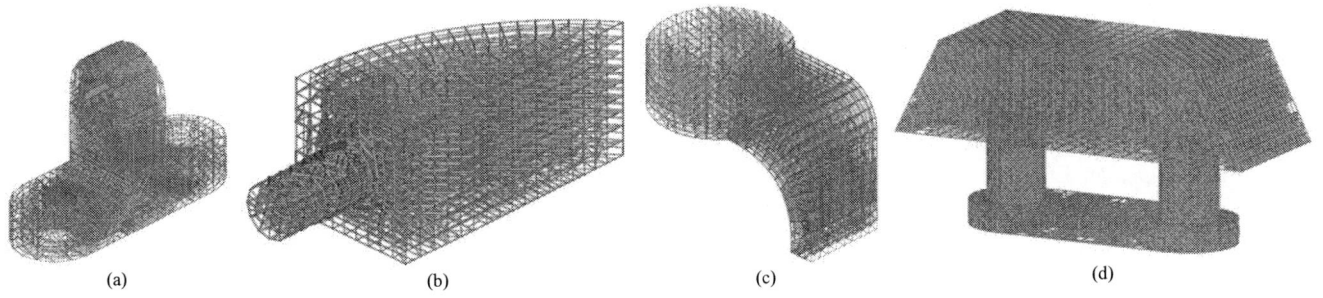

(a)　　　　(b)　　　　(c)　　　　(d)

FIGURE 15. LATTICE STRUCTURES DESIGNED BY OUR METHOD: (a) Lattice of Model 1 (b) Lattice of Model 2 (c) Lattice of Model 3 (d) Lattice of Model 4

Chang's:

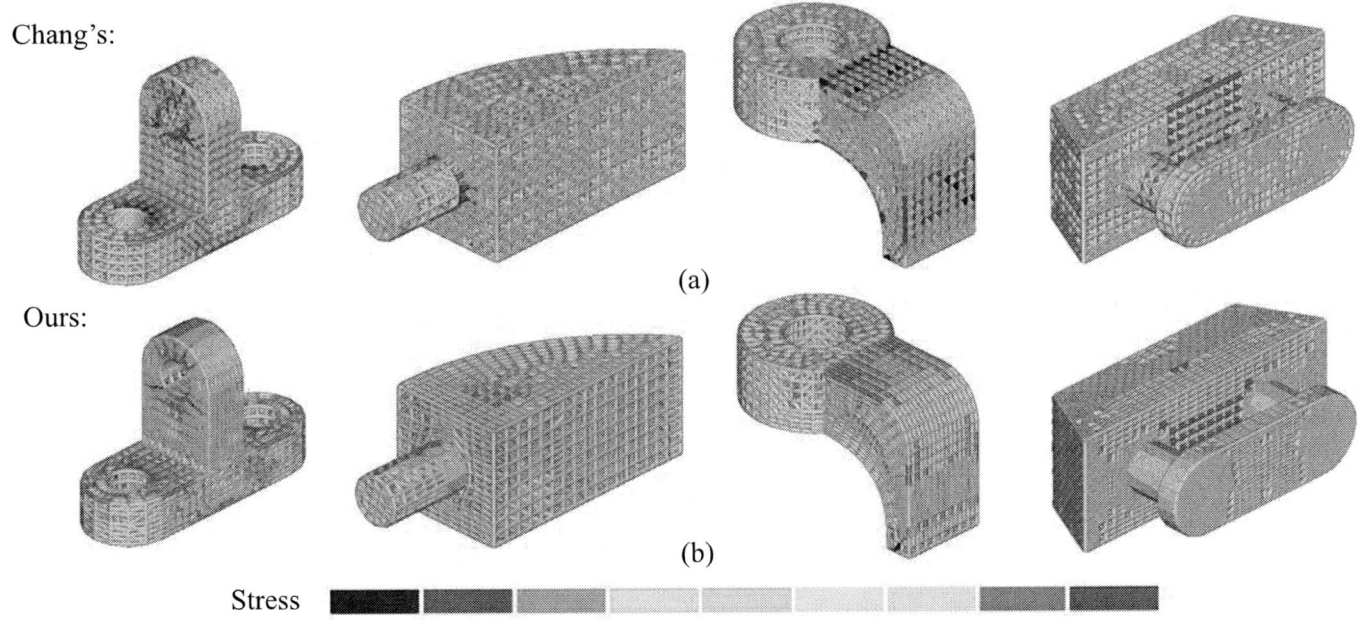

Ours:

Stress

FIGURE 16. STRESS DISTRIBUTION OF LATTICES: (a) Lattice structures designed by Chang's method (b) Lattice structures designed by our method

REFERENCES

[1] Gibson L J, and Ashby M F, *Cellular solids: structure and properties*. Cambridge university press, 1999.

[2] Chai S, Chen B, Ji M, et al, "Stress-oriented structural optimization for frame structures," Graphical Models, 2018, vol.97, pp.80-88.

[3] Chang P S, and Rosen D W, "An improved size, matching, and scaling method for the design of deterministic mesoscale truss structures," *ASME 2011 International Design Engineering Technical Conferences and Computers and Information in Engineering Conference.* American Society of Mechanical Engineers, pp.697-707, 2011.

[4] Panetta J, Zhou Q, Malomo L, et al, "Elastic textures for additive fabrication," *ACM Transactions on Graphics (TOG)*, 34(4): 135, 2015.

[5] Panetta J, Rahimian A, and Zorin D, "Worst-case stress relief for microstructures," *ACM Transactions on Graphics (TOG)*, 36(4): 122, 2017.

[6] Tang Y, Yang S, and Zhao Y F, "Design Method for Conformal Lattice-Skin Structure Fabricated by AM Technologies," *ASME 2016 International Design Engineering Technical Conferences and Computers and Information in Engineering Conference*, American Society of Mechanical Engineers, 2016.

[7] Hanks B, Dinda S, and Joshi S B, "Redesign of the

femoral stem for a total hip arthroplasty for additive manufacturing. In 44th Design Automation Conference," Vol. 2A-2018, American Society of Mechanical Engineers (ASME), 2018

[8] Nessi, Andrea, and Tino Stanković, "Topology, Shape, and Size Optimization of Additively Manufactured Lattice Structures Based on the Superformula," ASME 2018 International Design Engineering Technical Conferences and Computers and Information in Engineering Conference. American Society of Mechanical Engineers, 2018.

[9] Alzahrani M, Choi S K, and Rosen D W, "Design of truss-like cellular structures using relative density mapping method," Materials & Design, 2015, vol. 85, pp. 349-360,2018.

[10] Chen Y, "3D texture mapping for rapid manufacturing," *Computer-Aided Design and Applications*, 4(6): 761-771, 2007.

[11] Reinhart G, and Teufelhart S, "Optimization of mechanical loaded lattice structures by orientating their struts along the flux of force," Procedia CIRP, vol. 12, pp. 175-180, 2013.

[12] Daynes S, Feih S, Lu W F, et al, "Optimisation of functionally graded lattice structures using isostatic lines," *Materials & Design*, vol. 127, pp.215-223, 2017.

[13] Michell A G M. LVIII, "The limits of economy of material in frame-structures," The London, Edinburgh, and Dublin Philosophical Magazine and Journal of Science, 8(47), pp.589-597, 1904.

[14] Pajot J, "Optimal design exploration using Global Response Surface Method: Rail crush, HyperWorks," *Altair Engineering*, 2013.

[15] Knupp P M, "A method for hexahedral mesh shape optimization," *International journal for numerical methods in engineering*, 58(2), pp. 319-332, 2003.

Proceedings of the ASME 2019
International Design Engineering Technical Conferences
and Computers and Information in Engineering Conference
IDETC/CIE2019
August 18-21, 2019, Anaheim, CA, USA

DETC2019-97478

BUT WILL IT PRINT?: ASSESSING STUDENT USE OF DESIGN FOR ADDITIVE MANUFACTURING AND EXPLORING ITS EFFECT ON DESIGN PERFORMANCE AND MANUFACTURABILITY

Rohan Prabhu
Mechanical Engineering
The Pennsylvania State University
University Park, PA, 16802
rohanprabhu@psu.edu

Dr. Scarlett R. Miller
Engineering Design, Industrial Engineering
The Pennsylvania State University
University Park, PA, 16802
scarlettmiller@psu.edu

Dr. Timothy W. Simpson
Mechanical Engineering, Industrial Engineering
The Pennsylvania State University
University Park, Pa, 16802
tws8@psu.edu

Dr. Nicholas A. Meisel[1]
Engineering Design
The Pennsylvania State University
University Park, PA, 16802
nam20@psu.edu

ABSTRACT

Additive manufacturing (AM) enables engineers to improve the functionality and performance of their designs by adding complexity at little to no additional cost. However, AM processes also exhibit certain unique limitations, such as the presence of support material, which must be accounted for to ensure that designs can be manufactured feasibly and cost-effectively. Given these unique process characteristics, it is important for an AM-trained workforce to be able to incorporate both opportunistic and restrictive design for AM (DfAM) considerations into the design process. While AM/DfAM educational interventions have been discussed in the literature, limited research has investigated the effect of these interventions on students' use of DfAM. Furthermore, limited research has explored how DfAM use affects the performance of students' AM designs. This research explores this gap through an experimental study with 123 undergraduate students. Specifically, participants were exposed to either restrictive DfAM or dual DfAM (both opportunistic and restrictive) and then asked to participate in an AM design challenge. The students' final designs were evaluated for (1) performance with respect the design objectives and constraints, and (2) the use of the various aspects of DfAM. The results showed that the use of certain DfAM considerations, such as minimum feature size and support material mass, successfully predicted the performance of the AM designs. Further, while the variations in DfAM education did not influence the performance of the AM designs, it did have an effect on the students' use of certain DfAM concepts in their final designs. These results
highlight the influence of DfAM education in bringing about an increase in students' use of DfAM. Moreover, the results demonstrate the potential influence of DfAM in reducing build time and build material of the students' AM designs, thus improving design performance and manufacturability.

Keywords: design for additive manufacturing, additive manufacturing education, manufacturability

1. INTRODUCTION

Additive manufacturing (AM) defines a set of manufacturing processes that use layer-by-layer deposition of material to build parts [1]. This enables designers and engineers to produce complex parts at little to no additional cost. Here, complexity could be in the geometry of the designs, the features used in their assembly, or the materials used to fabricate them [2]. Companies, such as General Electric, have demonstrated the use of AM capabilities to improve the performance of their products, most notably the nozzle for the GE9X engine [3]. To encourage the use of AM capabilities during design, researchers are constantly exploring novel design methods, tools, and techniques, resulting in the emergence of *opportunistic* design for AM (DfAM). Opportunistic DfAM enables designers to capitalize on the unique capabilities of AM through techniques such as material complexity, multi-material printing, and part consolidation.

In addition to these unique capabilities, AM also introduces certain process limitations. For example, parts manufactured with AM present anisotropic material properties due to the layer-by-layer deposition technique [4]. These limitations, if not

---

[1] Corresponding Author

accounted for, have the potential to decrease the feasibility of AM designs, increase their manufacturing cost, or even lead to build failure. Therefore, to overcome these limitations and reduce build failures, researchers are developing limitation-based DfAM guidelines. These guidelines, known as *restrictive* DfAM, help designers ensure that their designs can be manufactured feasibly, with minimal material waste and build failure. The restrictive DfAM concepts also show similarities to traditional design for manufacturing and assembly (DFMA) guidelines [5] in terms of their focus on the limitations of a specific manufacturing processes. For example, DFMA provides designers with recommendations such as simplifying designs and providing draft angles for sharp corners to improve the manufacturability of their parts with traditional processes.

In addition to the opportunistic and restrictive DfAM concepts, some frameworks [6] suggest the combination of these two aspects of DfAM resulting in dual DfAM. This dual nature of design techniques is unique to AM, and therefore, it is important for engineering design processes to shift from traditional limitation-based DFMA, towards integrating both the opportunistic and restrictive aspects of DfAM. This integration of DfAM in engineering design has the potential to impact the performance of AM designs while ensuring manufacturability.

While several academic institutions have integrated AM and DfAM educational interventions in the engineering curriculum, limited research has explored their effects on the students' incorporation of DfAM considerations into their AM designs. Further, limited research has explored the relationship between DfAM integration and the performance and manufacturability of designs. Understanding this relationship is important as one of the crucial contributions of AM technologies is its ability to improve design performance through added complexity [3,7–9]. Therefore, the present study aims at exploring this gap by evaluating the effects of DfAM education on the participants' DfAM use, and its relationship with the performance and manufacturability of students' AM designs.

2. RELATED WORK

The aim in this research is to explore the effect of DfAM use on the performance and manufacturability of AM designs when introduced in an educational intervention. Therefore, previous research related to the various DfAM guidelines was explored. In addition, current practices in DfAM education were surveyed to help develop the educational intervention. The key findings from the survey of the literature are summarized in this section.

2.1. Design for Additive Manufacturing

The unique characteristics presented by AM has resulted in the emergence of design considerations specifically developed for AM. These DfAM considerations have been applied using several frameworks [6,10–13], of which Laverne, et al. [6] classifies these DfAM considerations into restrictive DfAM and opportunistic DfAM. Restrictive DfAM, as the name suggests, emphasizes on the restrictions or limitations of AM processes and provides design considerations to accommodate them. On the other hand, opportunistic DfAM emphasizes the opportunities or unique capabilities of AM processes and how

best designers can leverage them. A summary of the different opportunistic and restrictive DfAM concepts is seen in Table 1.

Table 1 Summary of DfAM concepts discussed in literature (R: restrictive, O: opportunistic)

	DfAM consideration	Source
R1	Support structure accommodation	[14–18]
R2	Warping due to thermal stresses	[19–22]
R3	Delamination and material anisotropy	[4,23,24]
R4	Stair-stepping and surface roughness	[25–31]
R5	Minimum feature size	[32–35]
O1	Free complexity – geometric and hierarchical	[36–39]
O2	Material complexity and multi-material printing	[40–43]
O3	Part consolidation and printed assemblies	[8,44]
O4	Mass customization	[45–48]
O5	Functional complexity and embedding	[49–52]

Restrictive DfAM is a necessary tool for AM designers as these considerations help reduce build failure and minimize waste of time, cost, and material. An important limitation of AM processes is their limited ability to build overhanging features. This necessitates the use of support material or self-supporting angles and bridging limits to minimize support material [14–18]. Since several AM processes rely on high-temperature melting of solid feed material, parts produced with these processes are prone to warping and cracking due to thermal stresses [19–22]. To minimize warping due to thermal stresses, for instance, designers are encouraged to avoid large flat surfaces or adding thermal walls to their designs to enable better heat dissipation.

The layer-by-layer process used in AM results in the parts having anisotropic material properties [4,23,24]. To avoid delamination between the layers, parts are oriented such that the load-critical features do not bear loads in the build direction. AM processes also result in surface roughness in the build direction due to stair-stepping observed on curves [25–31]. Therefore, parts that have assembly features and need geometric exactness are oriented parallel to the build platform [28]. Finally, given the diverse range of AM processes available, each process has a corresponding minimum feature size and a maximum part size the printer can manufacture. These dimensional limitations affect the accuracy and the number of prints needed to fully manufacture a product [32–35].

Alongside these limitations, AM processes offer new design opportunities for improving part performance. Opportunistic DfAM emphasizes these opportunities offered by AM and helps designers further explore the available design space. One of the most well-known aspects of opportunistic DfAM is the concept of "free complexity" [36]. AM not only provides designers with the freedom to include complex geometries but also extend this complexity at the hierarchical, and functional levels [37–39]. Complexity can also be extended towards the materials available in an AM process, where multiple materials with different characteristics such as rigidity, colour, and transparency can be printed in different combinations [40–43]. Further, AM processes also help minimize assembly time and costs by providing the ability to combine different functional components into one part through part consolidation [8], and design and build assemblies [44] that function with minimal post-processing. The digital manufacturing process followed by AM further permits engineers to manufacture several different parts from the same printer at no additional tooling costs [45]. This enables designers (and consumers) to design and manufacture products that are customized for each user, a concept commonly known as mass customization [46–48]. Finally, AM's unique layer-by-layer process also provides designers with the opportunity to embed external components, such as motors or bearings, by pausing the build at any time [49–52].

Given the uniqueness of DfAM and the growing integration of AM in the industry, several educational institutions have launched initiatives for AM/DfAM education as discussed next.

2.2. DfAM education

While research in AM is constantly refining DfAM methods and providing better tools for engineers and designers, it is also important that future engineers are trained in integrating DfAM in the engineering design process. To meet the growing demand for a workforce skilled in AM, several academic institutions are introducing formal and informal educational interventions focused on both AM and DfAM [36]. Further, a majority of these interventions employ principles of inductive teaching, such as the problem- and project-based learning techniques recommended at the 2013 NSF workshop on AM [36].

An example of a formal AM intervention is the AM course introduced at the University of Texas at Austin and Virginia Tech, where students are introduced to the various AM processes. In addition, students are also exposed to choosing appropriate processes for particular applications and applying their knowledge of AM processes towards solving a design problem [53]. Employing a more self-directed approach, Yang [54] discusses the use of literature reviews to encourage students' exploration of new and ongoing research in AM technologies and its various applications. Similarly, Diegel et al. [55] discuss the use of a problem-based AM educational initiative, where industry participants are exposed to different DfAM concepts in a 4-day hands-on workshop. The use of workshops for AM education has also been demonstrated as a method for addressing the challenges faced by AM education and leveraging the capabilities of AM, particularly in the ideation phases [56,57]. Similarly, Williams et al. [58] demonstrate the use of a project-based intervention as a method for informally introducing DfAM to students. Through the design of remote-controlled ground and air vehicles, students are engaged in exploring the uses of AM and applying DfAM concepts in their designs.

In contrast to these formal initiatives, several academic institutions are constantly working towards providing students access to AM processes to encourage self-learning. For example, the 3D printing vending machine [59] at Virginia Tech allows students to upload their parts for printing and collect it upon completion. A similar service is offered at the maker spaces set up at both, Penn State and Georgia Tech [36,60–62]. Students can utilize these AM services either by uploading their parts online, as in the case of the Penn State's Maker Commons or by directly interacting with the printers. The use of makerspaces for AM education has also been demonstrated through the development of a mobile makerspace that can be transported to remote locations where access to 3D printers is limited [62]. Further, universities such as MIT and Case Western provide students with access to both AM and traditional manufacturing through a network of interconnected makerspaces [63,64]. While these AM services provide students with guidelines for designing AM parts, a majority of these guidelines focus on the restrictive aspects of AM such as warping, support structures, and infill densities. However, limited emphasis has been given to the opportunistic aspects of AM.

A similar emphasis on restrictive DfAM can be seen in the DfAM worksheet developed by Booth et al. [65]. This worksheet helps designers assess their AM designs and has been demonstrated to minimize material wastage by reducing build failure. The DfAM worksheet uses eight factors for assessing the appropriateness of a design to be manufactured using AM, which include: (1) complexity, (2) functionality (load bearing), (3) support material removal, (4) support material accommodation (unsupported features), (5) minimum feature thickness, (6) stress concentrations, (7) tolerances, and (8) geometric accuracy. Of these eight factors, only complexity belongs to the opportunistic DfAM domain, while the remaining fall into the restrictive DfAM domain. This highlights an important issue: *designers are not encouraged enough towards leveraging the capabilities of AM*. Further, their study demonstrates the application of the DFAM worksheet to predict build failure; no information is provided to assess the performance of the AM designs with the worksheet.

In contrast to the restrictive-based DfAM worksheet, Blösch-Paidosh and Shea present the use of opportunistic DfAM-based design heuristics [66]. These heuristics, specifically developed for use in early stages of the design process, emphasize the following opportunistic DfAM concepts: (1) part consolidation, (2) customization, (3) conveying information, (4) material complexity, (5) functional embedding, (6) weight reduction, (7) material distribution and (8) reconfiguration. The study uses qualitative analyses to assess the AM designs for their use of the various heuristics. While this study provides important insights into the participants' use of the various heuristics, little emphasis is given to its effect on the manufacturability and performance of AM designs.

In summary, prior research presents several initiatives that integrate AM and DfAM into the engineering design curriculum. However, limited research has investigated the role of these initiatives on the students' use of DfAM in their designs. Further, limited research has explored the relationship between DfAM use and the manufacturability and performance of the students' AM designs. This is particularly important as integrating DfAM into engineering design has the potential of not only improving design performance through opportunistic DfAM but also ensuring design feasibility through restrictive DfAM. Therefore, the aim in this research is to explore these gaps in research.

3. RESEARCH QUESTIONS

Based on the current state of the literature, this study aims to explore the relationship between students' use of DfAM and the performance of their final designs. To do this, we seek to answer the following research questions:

RQ1: How does the participants' use of DfAM relate to the performance and manufacturability of their final designs? As opportunistic DfAM concepts aim at aiding designers in improving their design performance, we hypothesize that the participants' use of opportunistic DfAM would correlate with lower build material and build time. Further, given the role of restrictive DfAM in improving design feasibility, we hypothesize that students' use of restrictive DfAM would correlate with the generation of designs with better manufacturability.

RQ2: How does the participants' use of DfAM in their designs vary with the content of DfAM education? Since effective learning is demonstrated to correlate with the ability to use the knowledge to solve problems [67,68], we hypothesize that introducing participants to DfAM, either restrictive or dual, would result in greater use of the concepts in their final designs.

RQ 3: How does the performance and manufacturability of participants' designs vary with the content of DfAM education? Given the ability of opportunistic DfAM to improve design performance, we hypothesize that participants who received opportunistic DfAM training will generate ideas with lower build material and build time. Further, the introduction of restrictive DfAM will enable participants to generate designs with better additive manufacturability.

4. METHODOLOGY

To answer these research questions, an experiment consisting of a short intervention lecture followed by a design challenge was conducted. This section discusses the relevant details of the experiment, which was performed as a part of a larger study.

4.1. Participants

The participants (N = 123) in the study were recruited from a junior-level mechanical engineering course at a large public university in the northeastern part of the United States. The course focused on mechanical engineering design methodology, and the experiment was conducted in the fall semester. The participants included juniors (N = 78), and seniors (N = 41), and 5th-year seniors (N = 2) with some participants not reporting their year of study. The participants' previous AM and DfAM experience was collected in a pre-intervention survey and is summarized in Figure 1. As seen in the figure, a majority of the participants had received some formal or informal training in AM. By comparison, fewer participants had received formal or informal training in DfAM.

4.2. Procedure

The experiment was conducted during the second and third weeks of the semester and was broken into two main parts: (1) a DfAM educational intervention and (2) a design challenge. The study was approved by the Institutional Review Board, and informed consent was obtained from the participants before their participation in the study.

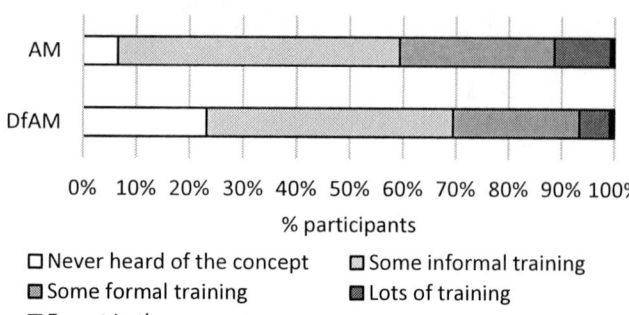

Figure 1 Distribution of participants' previous experience

4.2.1. DfAM educational intervention

Participants consenting to the study were randomly assigned to one of two educational intervention groups: (1) restrictive DfAM (N = 67) or (2) opportunistic and restrictive (dual) DfAM (N = 56). All participants were first given a 20-minute overview lecture on the AM process characteristics. This lecture discussed the material extrusion process available for the design challenge, the contrast between AM and subtractive manufacturing, the digital thread, the Cartesian coordinate system, and filament materials. Next, all participants were given a 20-minute lecture on restrictive DfAM, including build time, minimum feature size, support material, anisotropy, surface finish, and part warping. Finally, the dual DfAM group was given a 20-minute lecture on opportunistic DfAM, which included geometric complexity, mass customization, part consolidation, printed assemblies, multi-material printing, and embedding.

4.2.2. Design challenge

After attending the appropriate DfAM intervention lecture, the participants were asked to complete a design challenge, where they were asked to:

"Design a fully 3D printable free-standing tower for a downscaled wind turbine. The tower must support a motor-blade assembly and must attach to the assembly through a T-slot of given dimensions. The assembly must be able to slide into the slot and stay in place. The motor-blade assembly will include the male side of the t-slot. The objective of the challenge is to minimize the print material and the print time as much as possible while following the constraints listed below. Given the scaling factors of the turbine, the tower must meet the following constraints:
1. The height of the tower must be at least 18 inches (as measured from the ground to the motor).
2. The tower must support the motor (150 grams) assembled with the blades (150 grams).
3. The tower can have a maximum base footprint of 3.5" X 3.5".
4. All components necessary must be completed in one build within the build volume of 11.6" X 7.6" X 6.5"."

The design task was chosen such that minimal domain-specific knowledge, outside of AM, would be required to generate solutions (as suggested by [69]). Further, the wind turbine problem was chosen given the ease with which functional and manufacturing constraints could be placed on the solution space. For example, the task constrains the build volume to 11.6"x7.6"x6.5", though the participants are expected to build a tower 18" tall within this volume.

As part of the design challenge, the participants were first asked to spend 10 minutes individually generating and recording ideas on an idea generation card, with 7 minutes allotted for sketching, and 3 minutes for describing ideas in words. The participants were then given 5 minutes to evaluate their own ideas and note each's strengths and weaknesses. The participants were then given 7 minutes to individually design a final idea with the freedom to redesign, combine, or brainstorm again.

4.2.3. Concept selection and build preparation

After completing the individual concept generation, participants were randomly split into groups of 3 or 4 participants each. Since only the final designs from each group were used for the study, we do not expect the team size to have a major influence on the outcome. Further, while the groups are assigned for a semester-long project within the course, the participants were informed of their groupings for the first time on this day. These groups were formed such that schedule, commute, and commitment levels were matched for similarity, while writing skills, hands-on skills, and shop skills were diversified. This resulted in 44 groups, with 24 groups receiving restrictive DfAM training and 20 groups receiving dual DfAM training.

After being split into groups, each member was given time to present their individual final ideas to the other group members. The team then selected one final idea for the group. Participants were then asked to create a 3D solid model of their group's final idea using Solidworks, prepare a build file using MakerBot Desktop software, and submit it to the university's 3D printing service, which consists of several Makerbot Replicator+ machines. The complete design challenge was conducted within a 3-hour lab session, and participants were not allowed to make any further modifications after submitting their design files. The 3D printed structures, STL files, and .thing (Makerbot build preparation) files were collected from the participants after two weeks. The build files and printed parts were then assessed for their performance using the metrics discussed next.

4.3. Metrics

To assess the performance of the designs, metrics were developed that could evaluate both the performance of the final designs and the participants' use of DfAM. The metrics developed are discussed next.

4.3.1. Manufacturability and Performance of Students' Designs

The manufacturability and performance of the students' designs was assessed with respect to the objectives of the design prompt - minimizing the build material and minimizing the build time. Build time and build material were used as objectives for

the design challenge since these factors have a strong influence on the cost of an AM product [70]. Further, the weight of parts is also an important criterion for assessing design performance in several industries, including aerospace and automotive engineering [71].

The build time and build material were obtained from the build files submitted by the participants. In addition to the objectives of the task, the designs were also evaluated based on their adherence to the design challenge constraints. These constraints were developed based on the general requirements of a wind turbine with the height scaled down to 18". It should be noted that the designs that failed to build successfully, either due to poor design or build preparation, were given zero for all subsequent constraints. A summary of the performance criteria (O = objectives and C = constraints) is shown in Table 2.

Table 2 Metrics used for assessing the manufacturability & performance of the designs (O: objectives, C: constraints)

#	Metric
O1	Build material (g)
O2	Build time (min)
C1	Did it print successfully?
C2	Can it be assembled successfully?
C4	Is the design free standing and can support its own weight?
C5	Does the t-slot attach to the tower?
C6	Does the motor assembly stay in place?
C7	Does the tower bear the motor assembly load?
C8	Is the tower greater than 18" tall?
C9	Is the base footprint within 3.5"x3.5"?
C10	Is the tower built in one build?

4.3.2. Students' Use of DfAM in the design

To assess the students' use of DfAM in their designs, metrics were developed for both opportunistic and restrictive DfAM considerations. Specifically, of the design considerations discussed in Section 2.1, those that were within the scope of the experimental setup were chosen. Specifically, the opportunistic DfAM considerations used were: (1) geometric complexity (2) assembly (functional) complexity, and (3) part consolidation. Meanwhile, the following restrictive DfAM considerations were used: (1) surface roughness and stair-stepping, (2) warping and thermal stresses, (3) support material accommodation, and (4) feature size.

Given the limitations of the open printing facilities available through the university, students would not be able to embed components or use multi-material printing. Furthermore, given the structure and specificity of the task, students have limited scope to generate ideas that can be mass customized, as they are constrained to a specific motor-turbine assembly design. Therefore, these design considerations were excluded from the evaluation. The measurement scales for geometric complexity, feature size, and support material removal were adapted from the DfAM worksheet developed by Booth et al. [65]. The metrics and corresponding DfAM considerations are listed in Table 3. A 3-point scale was used to ensure uniformity across metrics. The results of an analysis based on these metrics is discussed next.

Table 3 Metrics used for assessing the students' use of DfAM in the design challenge and the DfAM consideration associated with each metric.

Metric	Score 1	Score 2	Score 3	DfAM Consideration
Part Complexity	Primitive geometry (ex. square, cylinder)	Complexity/curves that can be machined	Complex/curves that cannot be machined	AM designs can have complex geometries to improve performance as opposed to tradition manufacturing.
Assembly Complexity	Prismatic joint	Prismatic joints with locking features	Unidirectional joints with locking features	AM designs can have complex functional features such as assembly components.
Number of separate parts	--------------------------- Number/value ---------------------------			Designers can reduce part count by combining, thus reducing build time, assembly time and cost.
Part orientation	ZX/ZY (largest dimension in Z-direction)	XZ/YZ (second-largest dimension in Z-direction)	XY/YX (smallest dimension in Z-direction)	AM processes are typically slowest when printing in the z-direction.
Assembly feature orientation	ZX/ZY/XZ/YZ (critical mating features in X or Y planes)		XY/YX (critical mating features in the Z-plane)	The orientation of a part affects its surface finish. Stair stepping is observed when rounded features are printed vertically (along X or Y planes)
Smallest feature size	--------------------------- Value in mm ---------------------------			AM processes have a minimum feature size that can the process can build (~0.5mm for material extrusion [72]).
Smallest tolerance	--------------------------- Value in mm ---------------------------			Adequate tolerances must be given between mating features.
Support material mass	--------------------------- Value in grams ---------------------------			AM designs with overhanging features need support material. Support material mass can be reduced using self-supporting angles and bridging limits. Internal cavities must have access for ease of support material removal.
Support material removal	Internal cavities with support difficult to remove	Easily accessible support material	No support material	
Largest build plate contact	--------------------------- Value in mm^2 ---------------------------			Large flat surfaces are prone to warping due to inadequate heat dissipation and thermal stresses.

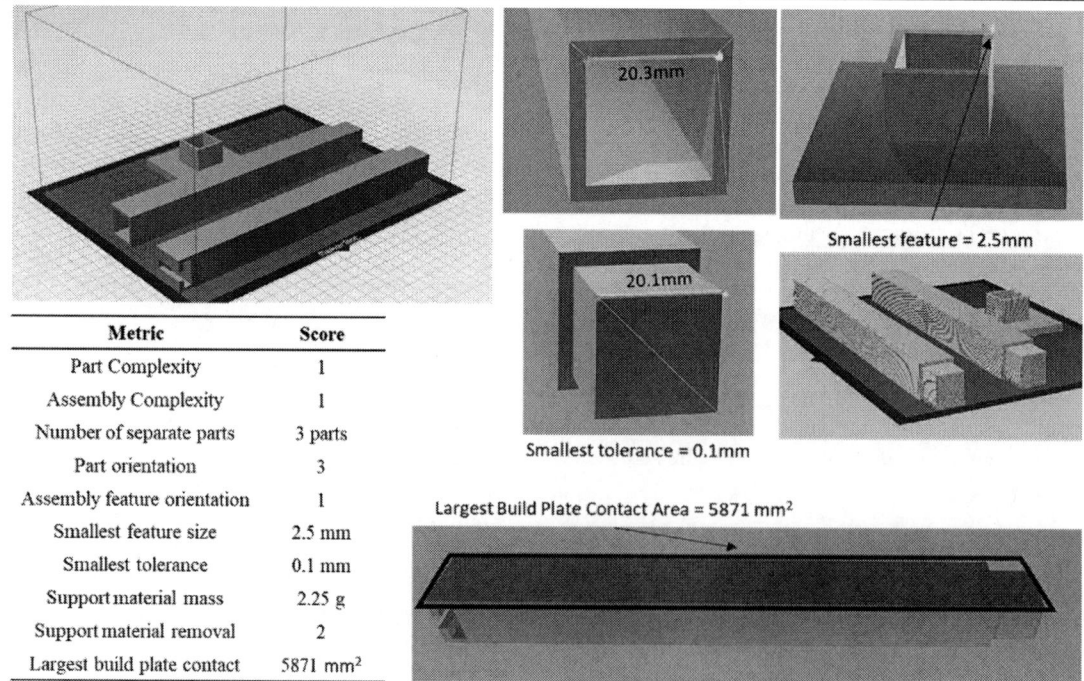

Metric	Score
Part Complexity	1
Assembly Complexity	1
Number of separate parts	3 parts
Part orientation	3
Assembly feature orientation	1
Smallest feature size	2.5 mm
Smallest tolerance	0.1 mm
Support material mass	2.25 g
Support material removal	2
Largest build plate contact	5871 mm^2

Figure 2 Example of assessment of a design using the DfAM metrics

5. DATA ANALYSIS AND RESULTS

To answer the research questions posed in Section 3, we performed statistical analyses with a statistical significance of $\alpha = 0.05$ and a confidence interval of 95%. A sample size of 39 groups was used after accounting for missing data, either due to participants not submitting their built parts or their build files. Among these, 21 groups received restrictive DfAM education and 18 groups received dual DfAM education.

RQ1: How does the participants' use of DfAM relate to the performance and manufacturability of their designs?

The first research question was developed to understand whether participants' use of DfAM had an effect on the performance of their final designs. To answer the research question, first, a multiple linear regression was performed with each objective criteria (i.e., build material and build time) as the dependent variable, and each DfAM criteria (see Table 3) as the independent variable. Before conducting the analysis, all assumptions (e.g., homoscedasticity, normality of residuals) were verified. An outlier was identified based on Cook's distance and the centred leverage values, and the data point was removed from further analysis. The results showed that as a group, the DfAM metrics successfully predicted both build material ($F(10,27) = 4.56$, $p = 0.001$, $R^2 = 0.63$, $R^2_{adj} = 0.49$) and build time ($F(10,27) = 2.74$, $p = 0.02$, $R^2 = 0.50$, $R^2_{adj} = 0.32$). The correlation coefficients, standard errors, and standardized coefficients are as summarized in Table 4.

Table 4 Coefficients for predicting design performance using the DfAM use (significant effects highlighted)

DfAM consideration	Build Material			Build time		
	B	SE$_B$	β	B	SE$_B$	β
Part Complexity	-10.44	19.54	-0.08	58.27	73.73	0.14
Assembly Complexity	17.19	17.43	0.14	13.51	65.75	0.03
Number of parts	9.54	6.89	0.21	**56.68**	**25.99**	**0.38**
Part orientation	0.36	16.04	0.004	-16.82	60.51	-0.05
Assembly feature orientation	-1.12	14.41	-0.01	63.55	54.36	0.22
Smallest feature size	**13.78**	**3.90**	**0.51**	**30.70**	**14.70**	**0.35**
Tolerance	39.48	43.65	0.12	179.86	164.68	0.17
Support material mass	**1.14**	**0.34**	**0.47**	**4.02**	**1.30**	**0.51**
Support material removal	7.81	27.45	0.04	85.34	103.58	0.14
Largest build plate contact area	**0.01**	**0.003**	**0.37**	0.004	0.01	0.06

Bold indicates $p < 0.05$

These results support our hypothesis that the various DfAM concepts influence the performance of AM designs. Specifically, we see that the size of the smallest feature and the support material mass positively correlate with the build time and build material. Furthermore, we see that while the number of parts correlates positively with build time, the maximum build plate contact area correlated positively with build material.

RQ2: How does the participants' use of DfAM in their designs vary with the content of DfAM education?

As seen in the results of RQ1, the participants' use of DfAM influenced the performance of their design. Therefore, the second research question sought to understand the role of DfAM education in bringing about these effects.

To answer the second research question, a series of Mann-Whitney U tests were performed. Specifically, the scores for the designs for each metric discussed in Section 4.3.2 were used as dependent variables, and the educational intervention group was used as the independent variable. The results of the analysis are summarized in Table 5.

Table 5 Comparing DfAM use between the DfAM educational groups (significantly higher values highlighted)

Performance Metric	p	U	z	Mean Rank (Median)	
				Restrictive DfAM	Dual DfAM
Part Complexity	0.57	210	0.61	19.00 (2.00)	21.27 (2.00)
Assembly Complexity	0.13	244.00	1.75	17.38 (1.00)	23.06 (1.75)
Number of separate parts	0.49	214.00	0.77	18.81 (3.00)	21.39 (3.00)
Part orientation	0.44	161.00	-0.93	21.33 (3.00)	18.44 (3.00)
Assembly feature orientation	0.25	147.50	-1.35	21.98 (2.00)	17.69 (1.00)
Smallest feature size	0.13	135.50	-1.51	22.55 (5.00)	17.03 (2.99)
Smallest tolerance	**0.01**	**98.50**	**-2.62**	**24.31 (0.25)**	**14.97 (0.025)**
Support material mass	0.43	217.50	0.80	18.64 (4.0)	21.58 (6.51)
Support material removal	0.86	182.00	-0.22	20.33 (2.00)	19.61 (2.00)
Largest build plate contact	**0.05**	**120.00**	**-1.94**	**23.29 (8174.60)**	**16.17 (6438.9)**

Bold indicates $p < 0.05$

The results show that while there were no significant differences between the educational intervention groups for 8 out of 10 DfAM considerations, the groups did show a significant difference in their use of assembly tolerances and build plate

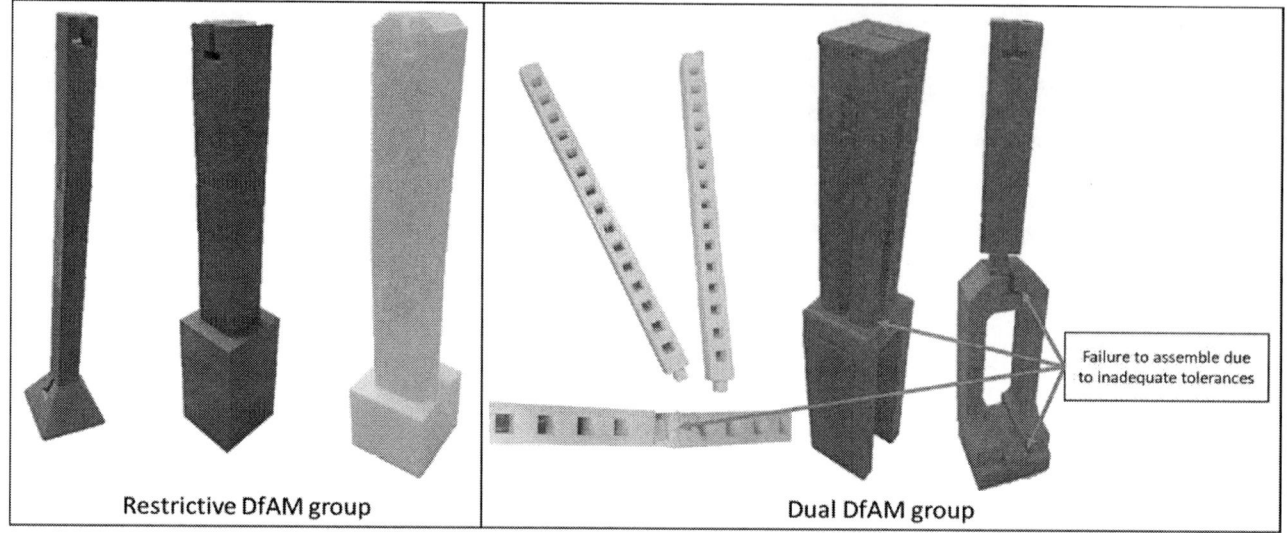

Figure 3 Sample designs: Solid 'blocky' designs with large surfaces by the restrictive DfAM group vs complex designs with poor assembly tolerances by the dual DfAM group

contact area. Specifically, the results showed that the restrictive DfAM group incorporated more appropriate tolerances between their mating features (mean = 0.605 mm, median = 0.25 mm) compared to the dual DfAM group (mean = 0.117 mm, median = 0.025 mm). The tolerances provided by the restrictive DFAM group were closer to the 0.5mm tolerance guideline given during the lecture. Furthermore, the group that received the restrictive DfAM training designed parts with larger build plate contact area (mean = 9622.31 mm^2, median = 8174.6 mm^2) compared to the dual DfAM group (mean = 6651.76 mm^2, median = 6438.9 mm^2). Some representative examples of the designs from each group are shown in Figure 3.

RQ3: How does the performance and manufacturability of the participants' designs vary with the content of DfAM education?

To answer the third research question, first, a one-way analysis of variance (ANOVA) was performed. Specifically, each objective criterion discussed in Section 4.3.1 was used as the dependent variable, and the educational intervention group was used as the between-subjects factor. The data showed no significant outliers or deviations from normality and homogeneity of variances. The results are summarized in Table 6, and we found no statistically significant difference between the two groups for either build material or build time.

Next, Fisher's exact tests [73] were performed for each constraint to check for differences between the educational groups in meeting each constraint. The constraint criteria were used as the dependent variables, and the educational intervention group was used as the between-subjects factor. The results showed that while the group that received only restrictive DfAM education had a greater frequency of success in meeting the constraints, this difference was not statistically significant. In summary, these results refute our hypotheses that the participants who received dual DfAM education would generate designs with lower build time and material.

Table 6 Summary of results comparing the educational intervention groups for their design performance

Performance Metric	p	F	Means (Std. Error)	
			Restrictive DfAM	Dual DfAM
Build material	0.24	1.44	276.53 (18.92)	243.17 (20.43)
Build time	0.57	0.33	896.71 (63.10)	843.67 (68.16)

Performance metric	p	Frequency of success (%)	
		Restrictive DfAM	Dual DfAM
Successful print	0.21	100	88.9
Successful assembly	0.11	76.2	50
Free standing (supports its own weight)	0.11	76.2	50
Attaches to the T-slot	0.09	81	50
Keeps the motor assembly in place	0.34	61.9	44.4
Supports the motor assembly load	0.20	61.9	38.9
Height greater than 18"	0.34	57.1	38.9
Base footprint within 3.5"x3.5"	1.00	90.5	94.4
Built in one build	1.00	95.2	94.4

6. DISCUSSION

The main goal in this study was to understand the effects of DfAM education on the performance and manufacturability of the participants' designs, as well as the influence of DfAM use on these effects. The key findings from the results were:

1. The participants' use of certain DfAM concepts predicted the performance of their designs towards achieving design task objectives.
2. Participants who received only the restrictive DfAM education incorporate more appropriate tolerances, but also have parts with higher build plate contact area than those who received dual DfAM education.
3. Variations in DfAM education do not have a statistically significant effect on the performance of the participants' final designs.

Participants' use of certain DfAM concepts predict the performance and manufacturability of their designs

The first research question was developed to understand whether the performance of the participants' designs could be predicted by their use of the various DfAM concepts. The first key observation was that the smallest feature size in their designs correlated with both the build material and build time. This suggests that designs that tend to have large features tend to take longer to build and use more material, which makes intuitive sense. Therefore, designers must take measures to optimize the size of their features to a minimum, while taking into account the resolution of the chosen AM process and the desired strength of the part. This would enable designers to successfully minimize the time and material consumed by the print.

The second key observation was that the support material required in a design correlated with both the build time and build material. This observation suggests the importance of emphasizing design guidelines such as self-supporting angles and bridging limits. Using these guidelines, designers are able to minimize the amount of support material needed to build their designs. This would help minimize both the time and material used to manufacture the design.

Further, we see that the largest build plate contact area for a component correlated with the build material. This suggests that designs that have large flat surfaces tend to consume more build material throughout the whole design. Therefore, designers must aim at avoiding large flat surfaces in their components, potentially by including complexities at the geometric and functional level. This would help minimize not only the build material but also reduce the risk of warping due to thermal stresses. Finally, we also see that a higher number of components in a design correlated to the time it took to build the design. This further supports the findings of past case studies, where part consolidation has been demonstrated as a technique for improving the manufacturability of designs by reducing build time and build material [3,8].

In summary, these results highlight that integrating the various opportunistic and restrictive DfAM guidelines have a positive influence on the manufacturability of designs. While this is a positive outcome, the dominance of the influence of restrictive DfAM suggests the need for a greater emphasis on applying opportunistic DfAM given its ability to improve design performance by minimizing build time and material.

Variations in DfAM education content affects participants' use of certain DfAM concepts

The second research question was developed to further understand the extent to which variations in DfAM education influenced the use of the various DfAM concepts in the participants' designs. The results showed that participants who received only restrictive DfAM provided *more appropriate tolerances* (closer to the 0.5mm guideline) between assembly mating features compared to those who received dual DfAM education. This suggests a greater emphasis on geometric exactness and interfaces between mating components that could potentially result in their designs being easier to assemble. While this is a positive outcome, given the role of tolerances in improving manufacturability, it also suggests that introducing opportunistic DfAM could potentially reduce the effectiveness of restrictive DfAM education, which supports the findings from previous research [74]. Therefore, educators must ensure that the introduction of DfAM does not dilute students' emphasis on restrictive DfAM. Moreover, this lack of emphasis on restrictive DfAM could be a result of the short duration of the given design challenge. Extending the length of the design activity could potentially provide students with more time and opportunity to apply opportunistic and restrictive aspects of DfAM together.

Finally, the results also show that participants who received only restrictive DfAM education generated designs that had a higher contact area with the build plate. This could potentially lead to a greater risk of build failure due to warping and thermal stresses. While this finding suggests that participants who received restrictive DfAM could have given a lower emphasis on warping and thermal stresses, this outcome could be an effect of the dual DfAM group adding complexity to their AM designs. For example, as seen in Figure 3, participants from the dual DfAM group generated designs with more hollowed out features compared to the restrictive DfAM group where several solid designs were observed. This addition of complexity at the geometric level could have contributed to the reduction in contact area with the build plate without the participants having specifically emphasized this. However, we also observed that despite the added complexity, most designs could still be manufactured using traditional manufacturing processes, thus explaining the lack of difference in the complexity scores between the two educational groups. This could also be attributed to the use of a 3-point scale which might have failed at capturing detailed differences in the complexity of the designs. This inference is further reinforced by the significantly higher contact area among the designs from the restrictive DfAM group, suggesting a potential lack of emphasis on warping by both groups. This finding, therefore, suggests that the current intervention fails to convey the importance of integrating DfAM guidelines for warping and thermal stresses into a design. However, the introduction of opportunistic DfAM, particularly the freedom of complexity, could indirectly help minimize warping.

Variations in DfAM education did not influence the performance and manufacturability of the participants' designs

The third research question was developed to investigate the effect of variations in the content of DfAM education on the performance of the participants' designs. The results showed that the content of the DfAM education did not have a significant effect on the performance of the designs. While the dual DfAM group generated designs with lower mean build time and build material use, as hypothesized, this result was not statistically significant. The results also show that while the designs from the restrictive DfAM groups showed greater success in meeting the constraints, this difference was not significant. These results suggest that the studied DfAM educational intervention did not succeed in bringing about effective learning or application of the various DfAM concepts. This could be attributed to the nature of the lectures where the rapid introduction of the concepts could have affected the students' learning of the concepts. The large amount of information conveyed to the participants in a short time could have limited their ability to absorb and apply all the different opportunistic concepts. Furthermore, the short duration of the design challenge could have limited the time available to apply the various DfAM concepts towards improving the performance of the AM designs.

This could also be attributed to the nature of the design task chosen. The task might not have provided the participants with adequate opportunity to apply some of the DfAM concepts. The lack of differences in the performance of the design outcomes could further be attributed to a relatively low level of incentive among the participants to generate ideas that fully leverage AM capabilities and improve their design performance. Therefore, future research must explore the use of a design challenge with an element of competition (as suggested by [58]) to engage students in generating better design outcomes. Finally, the study primarily focusses on the performance of the designs based on the objectives and constraints of the design challenge. However, there could potentially be differences in the features incorporated in the designs, particularly at the geometry and assembly levels. For example, in terms of material removal, participants from the dual DfAM group employ a variety of strategies such as shell-like designs, trusses, and bulk removal of material. Therefore, future research must explore the assessment of the different features employed in the designs, particularly in terms of their *variety*.

7. CONCLUSION, LIMITATIONS, AND FUTURE WORK

The research and development of AM processes has resulted in an increase in their use in industry, which has consequently developed the need for a workforce skilled in AM and DfAM. Therefore, several academic institutions have undertaken initiatives to integrate AM and DfAM into the undergraduate engineering curriculum. However, limited research has explored the role of DfAM education on the students' use of DfAM in engineering design and its resulting influence on the performance of their design outcomes. The present study explores this gap through an experimental study with undergraduate students consisting of a DfAM educational intervention and a design challenge.

The results of the study show that variations in the content of the educational intervention, namely, restrictive and dual DfAM, does not influence either the build time or the build material of the participants' final designs. However, these variations in the educational content have an influence on the participants' use of assembly tolerances and warping considerations. Finally, the participants' accommodation for feature sizes and support material mass influence the build material as well as the build time. In addition, while the largest build contact area has a small influence on the build material, the number of parts significantly influences the build time. These results, therefore, suggest that the participants' use of the DfAM influences the performance of their design outcomes, thus demonstrating the role of DfAM on improving engineering design outcomes. Further, the results also suggest the low effectiveness of the studied educational intervention, either due to the short length of the lectures and design challenge or due to the choice of the design task.

Although this study demonstrated the effect of DfAM use on the performance and manufacturability of design outcomes, it has several limitations. First, the study was conducted with participants primarily in their junior and senior years of study, with relatively high levels of engineering experience. Future research must compare the effect of engineering experience by comparing students from lower years of study (e.g., freshmen, sophomores). The second limitation of the study is that once the participants were assigned to their groups, they were asked to choose one idea to represent the group; however, the rationale behind the students' selection process is unknown. Future research must explore what factors affect the participants' selection of concepts when engaged in a group design challenge. Such an investigation could not only highlight the participants' emphasis on factors such as manufacturability and creativity but also reflect any biases towards their own ideas. Third, the metrics used in the present study do not provide information on the features used by the participants to manifest the different DfAM concepts. For example, shape complexity could take the form of organic structures, lattice structures, or bulk material removal. This can be seen in the sample designs shown in Figure 3 where the dual DfAM groups employed a variety of strategies such as shells and trusses to introduce complexity to their designs. Future research must explore the design features that are used by the participants to incorporate the different DfAM concepts into their designs. This would help further refine the assessment of their designs.

8. ACKNOWLEDGEMENTS

This research was conducted through the support of the National Science Foundation (Grant No. CMMI-1712234). Any opinions, findings, and conclusions expressed are those of the authors and do not reflect the views of the NSF. We would like to thank Dr. Stephanie Cutler for her guidance and would like to acknowledge the help of Dr. Jason Moore, Dr. Joseph Bartolai, the ME-340 TAs, and members of the Brite Lab and Made by Design Lab for their help with the experiments.

REFERENCES

[1] Crawford, R. H., and Beaman, J. J., 1999, "Solid Freeform Fabrication," IEEE Spectrum, **36**(2), pp. 34–43.

[2] Gibson, I., Rosen, D., and Stucker, B., 2015, *Additive Manufacturing Technologies*.

[3] Smith, H., "3D Printing News and Trends: GE Aviation to Grow Better Fuel Nozzles Using 3D Printing" [Online]. Available: http://3dprintingreviews.blogspot.co.uk/2013/06/ge-aviation-to-grow-better-fuel-nozzles.html. [Accessed: 29-Aug-2017].

[4] Carroll, B. E., Palmer, T. A., and Beese, A. M., 2015, "Anisotropic Tensile Behavior of Ti-6Al-4V Components Fabricated with Directed Energy Deposition Additive Manufacturing," Acta Materialia, **87**, pp. 309–320.

[5] Boothroyd, G., 1994, "Product Design for Manufacture and Assembly," Computer-Aided Design, **26**(7), pp. 505–520.

[6] Laverne, F., Segonds, F., Anwer, N., and Le Coq, M., 2015, "Assembly Based Methods to Support Product Innovation in Design for Additive Manufacturing: An Exploratory Case Study," Journal of Mechanical Design, **137**(12), p. 121701.

[7] Bartolom, N., and Am, R., "Digital Evolution of Cranial Surgery."

[8] Schmelzle, J., Kline, E. V., Dickman, C. J., Reutzel, E. W., Jones, G., and Simpson, T. W., 2015, "(Re)Designing for Part Consolidation: Understanding the Challenges of Metal Additive Manufacturing," Journal of Mechanical Design, **137**(11), p. 111404.

[9] Yang, S., Page, T., and Zhao, Y. F., 2018, "Understanding the Role of Additive Manufacturing Knowledge in Stimulating Design Innovation for Novice Designers," Journal of Mechanical Design, **141**(2), p. 021703.

[10] Atzeni, E., Iuliano, L., Minetola, P., and Salmi, A., 2010, "Redesign and Cost Estimation of Rapid Manufactured Plastic Parts," Rapid Prototyping Journal, **16**(5), pp. 308–317.

[11] Yang, S., and Zhao, Y. F., 2015, "Additive Manufacturing-Enabled Design Theory and Methodology: A Critical Review," International Journal of Advanced Manufacturing Technology, **80**(1–4), pp. 327–342.

[12] Rosen, D. W., 2007, "Design for Additive Manufacturing: A Method to Explore Unexplored Regions of the Design Space," *Eighteenth Annual Solid Freeform Fabrication Symposium*, Austin, Texas, pp. 402–415.

[13] Boyard, N., Rivette, M., Christmann, O., and Richir, S., 2013, "A Design Methodology for Parts Using Additive Manufacturing," High Value Manufacturing: Advanced Research in Virtual and Rapid Prototyping, pp. 399–404.

[14] Hu, K., Jin, S., and Wang, C. C. L., 2015, "Support Slimming for Single Material Based Additive Manufacturing," CAD Computer Aided Design, **65**, pp. 1–10.

[15] Strano, G., Hao, L., Everson, R. M., and Evans, K. E., 2013, "A New Approach to the Design and Optimisation of Support Structures in Additive Manufacturing," International Journal of Advanced Manufacturing Technology, **66**(9–12), pp. 1247–1254.

[16] Kirschman, C., Jara-Almonte, C., Bagchi, A., Dooley, R., and Ogale, A., 1991, "Computer Aided Design of Support Structures for Stereolithographic Components," Proceedings of the 1991 ASME Computers in Engineering Conference, Santa Clara, CA, **91**, pp. 443–448.

[17] Das, P., Chandran, R., Samant, R., and Anand, S., 2015, "Optimum Part Build Orientation in Additive Manufacturing for Minimizing Part Errors and Support Structures," *43rd Proceedings of the North American Manufacturing Research Institution of SME*, Elsevier B.V., pp. 309–330.

[18] "Bridging | Professional 3D Printing Made Accessible | Ultimaker" [Online]. Available: https://ultimaker.com/en/resources/19643-bridging?fbclid=IwAR3r4fC0hDkxPRjFXbD6NylBlhB2q3sIIwgKyPMILF8uHugVCdCkzmcPkkE. [Accessed: 13-Feb-2019].

[19] Zhu, Z., Dhokia, V., Nassehi, A., and Newman, S. T., 2016, "Investigation of Part Distortions as a Result of Hybrid Manufacturing," Robotics and Computer-Integrated Manufacturing, **37**, pp. 23–32.

[20] Nickel, A. H., Barnett, D. M., and Prinz, F. B., 2001, "Thermal Stresses and Deposition Patterns in Layered Manufacturing," Materials Science and Engineering A, **317**(1–2), pp. 59–64.

[21] Li, C., Fu, C. H., Guo, Y. B., and Fang, F. Z., 2015, "A Multiscale Modeling Approach for Fast Prediction of Part Distortion in Selective Laser Melting," Journal of Materials Processing Technology, **229**, pp. 703–712.

[22] Turnbull, A., Maxwell, A. S., and Pillai, S., 1999, "Residual Stress in Polymers - Evaluation of Measurement Techniques," Journal of Materials Science, **34**(3), pp. 451–459.

[23] Ahn, S., Montero, M., Odell, D., Roundy, S., and Wright, P. K., 2002, "Anisotropic Material Properties of Fused Deposition Modeling ABS," Rapid Prototyping Journal, **8**(4), pp. 248–257.

[24] Lee, J., and Huang, A., 2013, "Mechanical Characterization of Parts Fabricated Using Fused Deposition Modeling," Rapid Prototyping Journal, **19**, p. 72.

[25] Boschetto, A., and Bottini, L., 2016, "Design for Manufacturing of Surfaces to Improve Accuracy in Fused Deposition Modeling," Robotics and Computer-Integrated Manufacturing, **37**, pp. 103–114.

[26] Boschetto, A., Bottini, L., and Veniali, F., 2016, "Finishing of Fused Deposition Modeling Parts by CNC Machining," Robotics and Computer-Integrated

Manufacturing, **41**, pp. 92–101.

[27] Campbell, R. I., Martorelli, M., and Lee, H. S., 2002, "Surface Roughness Visualisation for Rapid Prototyping Models R.I.," Computer-Aided Design, **34**, pp. 717–725.

[28] Delfs, P., Tows, M., and Schmid, H. J., 2016, "Optimized Build Orientation of Additive Manufactured Parts for Improved Surface Quality and Build Time," Additive Manufacturing, **12**, pp. 314–320.

[29] Nuñez, P. J., Rivas, A., García-Plaza, E., Beamud, E., and Sanz-Lobera, A., 2015, "Dimensional and Surface Texture Characterization in Fused Deposition Modelling (FDM) with ABS Plus," Procedia Engineering, **132**, pp. 856–863.

[30] Pandey, P. M., Reddy, N. V., and Dhande, S. G., 2003, "Improvement of Surface Finish by Staircase Machining in Fused Deposition Modeling," Journal of Materials Processing Technology, **132**(1–3), pp. 323–331.

[31] Armillotta, A., 2006, "Assessment of Surface Quality on Textured FDM Prototypes," Rapid Prototyping Journal, **12**(1), pp. 35–41.

[32] Fahad, M., and Hopkinson, N., 2012, "A New Benchmarking Part for Evaluating the Accuracy and Repeatability of Additive Manufacturing (AM) Processes," 2nd International Conference on Mechanical, Production, and Automobile Engineering, pp. 234–238.

[33] Moylan, S., Slowinski, J., Cooke, A., Jurrens, K., and Donmez, M. A., 2012, "Proposal for a Standardized Test Artifact for Additive," Proceedings of the 23th International Solid Freeform Fabrication Symposium, pp. 902–920.

[34] Umaras, E., and Tsuzuki, M. S. G., 2017, "Additive Manufacturing - Considerations on Geometric Accuracy and Factors of Influence," IFAC-PapersOnLine, **50**(1), pp. 14940–14945.

[35] Childs, T. H. C., and Juster, N. P., 1994, "Linear and Geometric Accuracies from Layer Manufacturing EOS ~ DTM Dtm Stratasys Sts Helisys Hls," Annals of the CIRP, **43**(2), pp. 163–166.

[36] Simpson, T. W., Williams, C. B., and Hripko, M., 2017, "Preparing Industry for Additive Manufacturing and Its Applications: Summary & Recommendations from a National Science Foundation Workshop," Additive Manufacturing, **13**, pp. 166–178.

[37] Rosen, D. W., 2007, "Computer-Aided Design for Additive Manufacturing of Cellular Structures," Computer-Aided Design and Applications, **4**(1–6), pp. 585–594.

[38] Chu, C., Graf, G., and Rosen, D. W., 2008, "Design for Additive Manufacturing of Cellular Structures," Computer-Aided Design and Applications, **5**(5), pp. 686–696.

[39] Murr, L. E., Gaytan, S. M., Medina, F., Lopez, H., Martinez, E., Machado, B. I., Hernandez, D. H., Martinez, L., Lopez, M. I., Wicker, R. B., and Bracke, J., 2010, "Next-Generation Biomedical Implants Using Additive Manufacturing of Complex, Cellular and Functional Mesh Arrays," Philosophical Transactions of the Royal Society A: Mathematical, Physical and Engineering Sciences, **368**(1917), pp. 1999–2032.

[40] Kaweesa, D. V, Spillane, D. R., and Meisel, N. A., 2017, "Investigating the Impact of Functionally Graded Materials on Fatigue Life of Material Jetted Specimens," Solid Freeform Fabrication Symposium, pp. 578–592.

[41] Garland, A., and Fadel, G., 2015, "Design and Manufacturing Functionally Gradient Material Objects With an Off the Shelf Three-Dimensional Printer: Challenges and Solutions," Journal of Mechanical Design, **137**(11), p. 111407.

[42] Meisel, N., and Williams, C., 2015, "An Investigation of Key Design for Additive Manufacturing Constraints in Multimaterial Three-Dimensional Printing," Journal of Mechanical Design, **137**(11), p. 111406.

[43] Doubrovski, E. L., Tsai, E. Y., Dikovsky, D., Geraedts, J. M. P., Herr, H., and Oxman, N., 2015, "Voxel-Based Fabrication through Material Property Mapping: A Design Method for Bitmap Printing," CAD Computer Aided Design, **60**, pp. 3–13.

[44] Calì, J., Calian, D. A., Amati, C., Kleinberger, R., Steed, A., Kautz, J., and Weyrich, T., 2012, "3D-Printing of Non-Assembly, Articulated Models," ACM Transactions on Graphics, **31**(6), p. 1.

[45] Hopkinson, N., and Dickens, P., 2003, "Analysis of Rapid Manufacturing - Using Layer Manufacturing Processes for Production," Proceedings of the Institution of Mechanical Engineers, Part C: Journal of Mechanical Engineering Science, **217**(1), pp. 31–40.

[46] Pallari, J. H. P., Dalgarno, K. W., and Woodburn, J., 2010, "Mass Customization of Foot Orthoses for Rheumatoid Arthritis Using Selective Laser Sintering," IEEE Transactions on Biomedical Engineering, **57**(7), pp. 1750–1756.

[47] Tuck, C. J., Hague, R. J. M., Ruffo, M., Ransley, M., and Adams, P., 2008, "Rapid Manufacturing Facilitated Customization," International Journal of Computer Integrated Manufacturing, **21**(3), pp. 245–258.

[48] Mohammed, M. I., P. Fitzpatrick, A., and Gibson, I., 2017, "Customised Design of a Patient Specific 3D Printed Whole Mandible Implant," KnE Engineering, **2**(2), p. 104.

[49] De Laurentis, K. J., Kong, F. F., and Mavroidis, C., 2002, "Procedure for Rapid Fabrication of Non-Assembly Mechanisms with Embedded Components," Proceedings of the 2002 ASME Design Engineering Technical Conferences and Computers and Information in Engineering Conference, pp. 1–7.

[50] Aguilera, E., Ramos, J., Espalin, D., Cedillos, F., Muse, D., Wicker, R., and Macdonald, E., 2013, "3D Printing of Electro Mechanical Systems," International Solid Freeform Fabrication Symposium, pp. 950–961.

[51] Lopes, A. J., MacDonald, E., and Wicker, R. B., 2012, "Integrating Stereolithography and Direct Print

Technologies for 3D Structural Electronics Fabrication," Rapid Prototyping Journal, **18**(2), pp. 129–143.

[52] Wicker, R. B., and MacDonald, E. W., 2012, "Multi-Material, Multi-Technology Stereolithography: This Feature Article Covers a Decade of Research into Tackling One of the Major Challenges of the Stereolithography Technique, Which Is Including Multiple Materials in One Construct," Virtual and Physical Prototyping, **7**(3), pp. 181–194.

[53] Williams, C. B., and Seepersad, C. C., 2012, "Design for Additive Manufacturing Curriculum: A Problem-and Project-Based Approach," International Solid Freeform Fabrication Symposium, pp. 81–92.

[54] Yang, L., 2018, "Education of Additive Manufacturing – An Attempt to Inspire Research," *Solid Freeform Fabrication 2018: Proceedings of the 29th Annual International Solid Freeform Fabrication Symposium – An Additive Manufacturing Conference*, pp. 44–54.

[55] Diegel, O., Nordin, A., and Motte, D., 2019, "Teaching Design for Additive Manufacturing Through Problem-Based Learning," *Additive Manufacturing—Developments in Training and Education*, Springer International Publishing AG, pp. 139–149.

[56] Richter, T., Schumacher, F., Watschke, H., and Vietor, T., 2018, "Exploitation of Potentials of Additive Manufacturing in Ideation Workshops," The Fifth International Conference on Design Creativity (ICDC2018), (February), pp. 1–8.

[57] Kumke, M., Watschke, H., Hartogh, P., Bavendiek, A. K., and Vietor, T., 2017, "Methods and Tools for Identifying and Leveraging Additive Manufacturing Design Potentials," International Journal on Interactive Design and Manufacturing, pp. 1–13.

[58] Williams, C. B., Sturm, L., and Wicks, A., 2015, "Advancing Student Learning Of Design for Additive Manufacturing Principles Through An Extracurricular Vehicle Design Competition," Proceedings of the ASME 2015 International Design Engineering Technical Conferences & Computers and Information in Engineering Conference, pp. 1–8.

[59] Meisel, N. A., and Williams, C. B., 2015, "Design and Assessment of a 3D Printing Vending Machine," Rapid Prototyping Journal, **21**(5), pp. 471–481.

[60] "Tips for Designing a 3D Printed Part | Innovation Station" [Online]. Available: https://innovationstation.utexas.edu/tip-design/. [Accessed: 12-Feb-2018].

[61] "Submitting Your 3D Print | Maker Commons" [Online]. Available: https://makercommons.psu.edu/submitting-your-3d-print/. [Accessed: 12-Feb-2018].

[62] Sinha, S., Rieger, K., Knochel, A. D., and Meisel, N. A., 2017, "Design and Preliminary Evaluation of a Deployable Mobile Makerspace for Informal Additive Manufacturing Education," pp. 2801–2815.

[63] "3D Printing Service - MIT Project Manus" [Online]. Available: https://project-manus.mit.edu/3d-printing-

service. [Accessed: 28-Jan-2019].

[64] "3D Printing Services | Case School of Engineering" [Online]. Available: http://engineering.case.edu/sears-thinkbox/use/3d-printing-services. [Accessed: 28-Jan-2019].

[65] Booth, J. W., Alperovich, J., Chawla, P., Ma, J., Reid, T., and Ramani, K., 2017, "The Design for Additive Manufacturing Worksheet," Journal of Mechanical Design, **139**(October 2017), pp. 1–9.

[66] Bloesch-Paidosh, A., and Shea, K., 2018, "Design Heuristics for Additive Manufacturing Validated Through a User Study," Journal of Mechanical Design, (c), pp. 1–40.

[67] Bransford, J. D., Brown, A. L., and Cocking, R. R., 1999, "Learning and Transfer," (1913), pp. 39–66.

[68] Bloom, B. S., 1966, "Taxonomy of Educational Objectives: The Classification of Educational Goals ."

[69] Amabile, T. M., 1996, *Creativity in Context: Update to the Social Psychology of Creativity*, Westview Press.

[70] Niazi, A., Dai, J. S., Balabani, S., and Seneviratne, L., 2006, "Product Cost Estimation: Technique Classification and Methodology Review," Journal of Manufacturing Science and Engineering, **128**(2), p. 563.

[71] Evans, A. G., 2001, "Lightweight Materials and Structures," MRS Bulletin, **26**(10), pp. 790–797.

[72] "Replicator+ 3D Printer - Desktop 3D Printer - Reliable 3D Printing" [Online]. Available: https://www.makerbot.com/3d-printers/replicator/. [Accessed: 13-Feb-2019].

[73] Agresti, A., 2007, *An Introduction to Categorical Data Analysis (2nd Edn)*.

[74] Prabhu, R., Miller, S. R., Simpson, T. W., and Meisel, N. A., 2018, "Teaching Design Freedom: Exploring the Effects of Design for Additive Manufacturing Education on the Cognitive Components of Students' Creativity," Proceedings of the ASME 2018 International Design Engineering Technical Conferences and Computers and Information in Engineering Conference, pp. 1–14.

Proceedings of the ASME 2019
International Design Engineering Technical Conferences
and Computers and Information in Engineering Conference
IDETC/CIE2019
August 18-21, 2019, Anaheim, CA, USA

DETC2019-97480

A COMPARATIVE STUDY OF VIRTUAL REALITY AND COMPUTER-AIDED DESIGN TO EVALUATE PARTS FOR ADDITIVE MANUFACTURING

John K. Ostrander
Engineering Design
The Pennsylvania State University
University Park, PA, 16802
Email: joo5126@psu.edu

Lauren Ryan
Mechanical Engineering
The Pennsylvania State University
University Park, PA, 16802
Email: lur310@psu.edu

Snehal Dhengre
Industrial Engineering
The Pennsylvania State University
University Park, PA, 16802
Email: sud702@psu.edu

Dr. Christopher McComb
Engineering Design
The Pennsylvania State University
University Park, PA, 16802
Email: mccomb@psu.edu

Dr. Timothy W. Simpson
Mechanical Engineering, Industrial Engineering
The Pennsylvania State University
University Park, PA, 16802
Email: tws8@psu.edu

Dr. Nicholas A. Meisel
Engineering Design
The Pennsylvania State University
University Park, PA, 16802
Email: nam20@psu.edu

KEYWORDS
Additive Manufacturing, Design, Virtual Reality, CAD

ABSTRACT

Virtual Reality (VR) has been shown to be an effective assistive tool in the engineering design process, aiding designers in ergonomics studies, data visualization, and manufacturing simulation. Yet there is little research exploring the advantages of VR to assist in the design for the additive manufacturing (DfAM) process. VR may present advantages over traditional computer-aided design (CAD) tools, and these advantages may be more evident as designs become more complex. The following study investigates two types of environments: 1) Immersive Virtual Reality (VR) and 2) Non-Immersive Virtual Reality (CAD) and the advantages that each environment gives to designers to assess parts for additive manufacturing. The two environments are compared to assess potential differences in DfAM decision-making. Participants familiar with DfAM are tasked with evaluating five designs of varying complexity using the Design for Additive Manufacturing Worksheet. Participant scores, evaluation times, and self-reported metrics are recorded and analyzed. Our findings indicate that as part complexity increases, DfAM scores and evaluation times increasingly differ between VR and CAD groups. We found that the VR group
evaluates more complex parts at a faster rate, but with a lower accuracy when compared to the CAD group. In evaluating self-reported metrics, both groups were relatively similar; however, the CAD group reported improved confidence in identifying stress concentrations in DfAM parts. Our findings in this research identify VR as a design evaluation tool that enhances evaluation speed which speaks to its efficiency and usability; however, VR in its current form may not present the resolution necessary to identify smaller details when compared to CAD, the more accurate evaluation tool.

1 BACKGROUND AND MOTIVATION

In 2017, the global Additive Manufacturing (AM) market size was US$ 8,640 million and is expected to reach US$ 76,900 million by the end of 2025 [1]. There are, however, still constraints in AM that affect manufactured parts that must be considered in the design process [2]. Design considerations like support generation, minimum feature size, and stress concentrations define the overall quality of a part and can increase cost and production time. AM also places more responsibility on the designer because the technology is capable of creating more complex designs when compared to traditional manufacturing technologies like casting and machining. AM presents new manufacturing opportunities, but it comes with new restrictions as well; this necessitates a better understanding of Design for Additive Manufacturing (DfAM) knowledge and application for effective decision making by designers.

Currently, parts designed for AM are typically evaluated preemptively using traditional CAD software or specialized build preparation software. Parts may also be evaluated physically, but this requires the component to actually be produced with AM or prototyped through other means. VR may improve the efficiency of DfAM evaluation in that the technology combines the visualization of a CAD environment (evaluating a part before manufacture), with the manipulability and intuitiveness of a physical environment [2,3]. Thus, VR, if effective as a DfAM decision-making tool, may decrease both build failures and the time it takes to evaluate designs.

The following research investigates the extent to which VR benefits designers in DfAM decision-making. Depending on the results, VR or CAD may be proposed as the more effective medium for reducing AM design errors, reducing decision-making times, and enhancing confidence (among other self-reported metrics) in designers. The results of this study may help AM designers make an argument for (or against) incorporating immersive VR into their design process. Furthermore, this research may serve as the base-line for studying the usage of VR for various AM processes to enhance cost-effective production. To simplify terminology, immersive VR is hereafter referred to as 'VR', and non-immersive VR is referred to as 'CAD' for the remainder of this paper.

1.1 Design for Additive Manufacturing

Additive Manufacturing (AM) is an evolving process with extensive support and development through the maker movement as well as growing prevalence in industry [4,5]. A wide range of AM processes allows manufacturing from nanoscale fabrication to large-scale production and appeals to many different makers in research, industry, and education. Unlike traditional, subtractive manufacturing methods, AM utilizes layer-by-layer material deposition to "grow" previously impossible designs from the bottom up through processes like material extrusion, powder bed fusion, vat photopolymerization, material jetting, binder jetting, sheet lamination, and directed energy deposition [4]. This layer-by-layer manufacturing approach enables a variety of unique opportunities to create more complex geometries [6,7], unique material compositions [8], optimized designs [9,10], customized products [11,12], consolidated designs [13], multi-material specimens [14], embedded components [15,16], integrated electronics [17], and functional assemblies [18]. Despite the many opportunities, designers are also faced with many challenges in AM. Problems such as long build times, generated support structures, undesirable surface quality, and problematic anisotropic behavior [5,19]. With its unique opportunities and challenges, AM requires the development of specialized design processes and practices, collectively termed Design for Additive Manufacturing, or DfAM, to ensure AM is reaching its maximum potential [20,21]. DfAM requires a different way of thinking about product design due to increased complexity of parts and process capabilities.

One of the most common forms of DfAM guidance currently available is in the form of design heuristics or guidelines. One such proposed design methodology examines DfAM from either the restrictive elements, the opportunistic elements, or the combined elements of AM processes when designing a part [22]. Restrictive elements, or limitations of the process, may include support structure generation, minimum feature size, post processing considerations, and other elements of the AM process that may restrict a designer. Opportunistic elements, or advantages of the process may include part consolidation, mass customization, free complexity, and other elements of the AM process that give designers more freedom and opportunities with their designs when compared to other means of manufacturing. This methodology identifies the necessary design considerations for whether a component can (and should) be developed using AM [23]. By focusing on these design elements of AM, designers can amplify the opportunities of AM while reducing complications and shortcomings.

Another approach to DfAM involves generalized AM guidelines that can be used to examine specific and common features throughout designs and proposes better practices for AM [5]. For the purposes of this experiment the DfAM guidelines embodied in "The Design for Additive Manufacturing Worksheet"[24] was used by participants to evaluate five different parts selected by AM experts. This DfAM worksheet was designed to assist and educate novice and intermediate designers to improve AM part quality. The worksheet has demonstrated a reduction in "bad parts" designed for AM by requiring designers to consider eight factors [24]: (1) Complexity, (2) Functionality, (3) Material Removal, (4) Unsupported Features, (5) Thin Features, (6) Stress Concentration, (7) Tolerances, and (8) Geometric Exactness. An analysis of print failures was used to successfully validate the worksheet's effectiveness. The worksheet takes about 5 min on average to use, with an 83% drop in number of failed prints after its use [24]. The most commonly reported problems with the designs were functionality and tolerances.

1.2 Design Using Virtual Reality

Virtual Reality (VR) is a broadly used term that is defined in a variety of ways. For this study, we differentiate between VR as Immersive: in which a participant uses a head mounted display (HMD) complete with controllers allowing for full range of motion, or Non-Immersive: where the participant is still interacting with a virtual environment (using computer aided design, or CAD) but is aware of their physical surroundings [25]. As stated earlier, immersive VR is referred to as 'VR', and non-immersive VR is referred to as 'CAD' for the remainder of this paper.

In 1997 it was envisioned that CAD programs would soon converge with VR to enhance product design and quality, reduce fabrication time, and tangentially reduce design costs [26]. Historically, VR use required a vast range of expertise [25] limiting adaptation of the technology; more recently however, high fidelity consumer grade VR has emerged at a low price point [27]. Now that the technology is more accessible, there is a growing interest to incorporate VR to aid with CAD [28].

VR is currently being used as an educational and design tool [29]. VR has proven to be effective in aspects of product design, ergonomics [29], teamwork [30], and confidence [31,32]. VR has also shown to improve perceived learning and enjoyment when compared to traditional 2D displays typically associated with CAD [33]. Additionally, the technology has been shown to shorten manipulation times when compared to working in real-world environments [34] and 2D displays [35]. VR presents advantages over traditional CAD in areas of product dissection, product design and educational lab environments [30], strengthening the argument that VR should be combined with CAD to make for a more powerful design evaluation tool. The advantages of VR have yet to be explored in DfAM, where geometric complexity is a common design feature. We contend that VR offers AM designers a new and immersive environment to evaluate their designs as discussed next.

1.4 Research Objectives

An experiment was designed to explore VR as a decision-making tool in DfAM and uncover advantages (and limitations) the technology has over traditional CAD. Participants were evaluated on their decisions using either VR (Head mounted display with controllers) or CAD (monitor and mouse). The research questions investigated in this study are as follows.

RQ1: Is there a difference between VR and CAD in DfAM scoring with respect to complexity of the evaluated design?

We hypothesize that there will be a growing difference between CAD and VR as design complexity increases. For more complex parts, we hypothesize that the VR group will yield a more accurate evaluation due to the immersive nature of the technology.

RQ2: Is there a difference between VR and CAD in the time it takes participants to make DfAM decisions with respect to complexity of the evaluated design?

Similar to RQ1, we hypothesize that that there will be a growing difference in evaluation time between CAD and VR as complexity increases. Participants were believed to have more prior exposure to CAD than VR; the CAD group was then expected to have faster evaluation times due to familiarity with the software.

RQ3: Is there a difference between VR and CAD in terms of self-efficacy and other self-reported metrics for DfAM?

We hypothesize that there will be significant within-group effects for the VR group, but not the CAD group, as the participants were likely to be more familiar with CAD. We also hypothesized that there would be significant between group effects for self-reported metrics related to DFAM.

2 EXPERIMENT: DFAM DECISION DIFFERENCES BETWEEN VIRTUAL REALITY GROUPS

The following experiment was designed to explore the differences in DfAM evaluation accuracy, evaluation time, and self-reported metrics for DfAM between two VR environments: (1) VR and (2) CAD. This experiment was also designed to explore how varying degrees of shape complexity play a role in these differences. For designs that are fairly simple, it may not matter which environment is used, but as designs become more complex, then one environment may have advantages over the other. The actions performed by both groups (orienting parts, referring to the DfAM Worksheet in the VR environment) were identical to limit potential confounding effects.

2.1 Participants

Participants were recruited from a fourth-year DfAM course from a large university. The DfAM course gave all participants experience in material extrusion, and part evaluation using the DfAM worksheet [24]. A total of 20 students (18 male, 2 female) participated in the study with ages ranging from 20 to 26 years. While most of the participants were in their senior year (17), the study also included one first-year graduate student and two second-year graduate students. The participants' fields of study were Mechanical Engineering (17) and Engineering Design (3). Participant familiarity with AM, CAD, and VR was self-assessed by participants at three levels: Beginner - the participant just started to learn the skill or technology, Intermediate - the participant is in the developing stage to become proficient, and Skilled - the participant is proficient in the domain. The majority of participants had familiarity with AM at an Intermediate level (13), with 6 at the Skilled level and 1 at the Beginner level. The level of participants' familiarity with CAD were Beginner (1), Intermediate (9), and Skilled (10), whereas familiarity with VR was Beginner (17), and Intermediate (3). In short, the majority of participants were familiar with AM, but the participants' familiarity with CAD was higher when compared to VR.

2.2 Procedure

Participants were randomly assigned to one of two groups: (1) the VR group or (2) the CAD group. Both groups were comprised of 10 participants each. The VR group utilized an HTC VIVE [36] comprising a HMD and controllers allowing for full range of arm motion to control an environment rendered using Unity [37]. The CAD group used SolidWorks [38], paired with a mouse and monitor to control the environment (see Figure 1). Although the visualization tools differed, the environments were set up similarly, and both presented participants with the DfAM Worksheet and virtual part renderings. Each participant was engaged in the study privately with a proctor present.

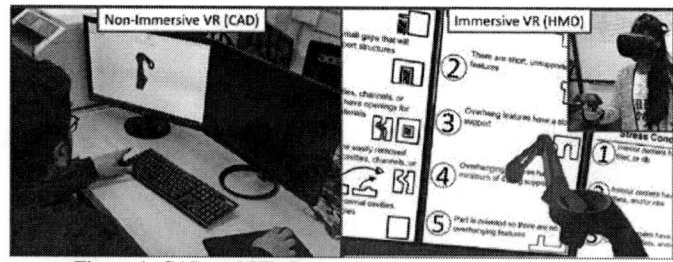

Figure 1: CAD and VR Groups Evaluating A Part for DfAM

After consent for participation was obtained per the approved IRB protocol, each participant completed a pre-survey that comprised a background survey and self-efficacy survey. In this survey, participants filled out their unique identification code comprising of the last two characters of their mothers first name, the last two characters of their birth city, and the two numbers representing their birth month (for example: THIA01). These participant IDs were later used to anonymously identify participants as needed. The background survey asked about familiarity with AM, VR, and CAD software. The survey also assigned each participant an anonymous identification code and identified their area of study and year of study. The participant was then asked to fill out a pre-experiment self-reported metric survey (discussed in detail in Section 3.1). The survey assessed participants' confidence, motivation, anticipated level of success, and anxiety for evaluating DfAM parts. The survey also assessed participants self-reported confidence as it relates to aspects of the DfAM worksheet: evaluating a part for material removal, unsupported features, thin features, stress concentration and geometric exactness.

Following completion of the pre-survey, each participant was randomly assigned to either the VR or CAD environment for DfAM evaluation. Before starting the experiment, each participant was provided an overview of the evaluation tool including where or how to find the parts and worksheet criteria, the necessary controls including how to rotate or move the parts using the mouse or the VR controller, and was instructed in how to communicate DfAM scores to the proctor verbally. Each participant was then tasked with evaluating five similar parts modeled for the Alcoa Airplane Bearing Bracket Challenge on GrabCAD [39]. These parts were chosen by a committee of DfAM experts and varied in levels of complexity from 1 (least complex) to 5 (most complex) as defined by Booth's DfAM worksheet (see Figure 2).

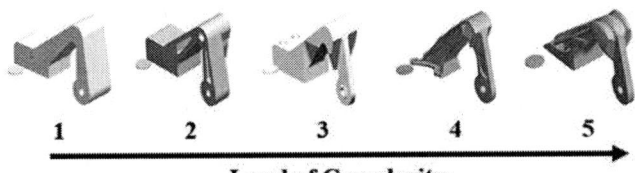

Figure 2: Shape Complexity Levels 1 to 5 of Brackets Designed for the Alcoa Grab CAD Challenge

Participants were informed that these parts were to be manufactured using Material Extrusion. Each participant was independently introduced to one part at a time in a random order to eliminate comparison and ordering effects [40]. Each part rendering also included a model of a penny. Participants were informed the penny was 1.5mm thick to give them a sense of feature scale. For each part, participants were responsible for reporting scores for the restrictive elements as defined in the DfAM Worksheet (Material Removal, Unsupported Features, Thin Features, Stress Concentrations, and Geometric Exactness). Each participant was timed for each part evaluation. After evaluating each of the five parts, participants then took a post-survey consisting of the same self-efficacy questions, and a demographics survey that requested gender, age, and ethnicity.

3 METRICS

Multiple metrics were used to assess the impact of VR and CAD when used as tools for DFAM evaluation. A pre-survey and post-survey were developed to assess self-efficacy among the participants. The participants' DfAM scores were also measured and analyzed along with the evaluation time.

3.1 Self-Efficacy

For this study, a two-part self-efficacy assessment was constructed to measure both the self-efficacy in the DfAM field, in general, as well as self-efficacy of experiment-specific tasks and outcomes. Based on work by Carberry [41], the first part of the survey examines an engineering design process modified for AM-specific design (see Figure 3). The steps were to identify a need to use AM, to develop design solutions using AM, to select the best possible design for AM, to evaluate a part for AM, and to redesign a part for AM. The participants were then asked to assess the following four task-specific concepts of interest for each step: (1) confidence, (2) motivation, (3) anticipated level of success, and (4) anxiety [41]. These four concept categories were chosen to help clarify the more general process steps.

The second part of the survey was developed as an outcome self-efficacy assessment, where participants reported their confidence in evaluating parts for restrictive DfAM considerations. Participants rated their confidence in evaluating parts for material removal, unsupported features, thin features, stress concentration, and geometric exactness (listed in Figure 3). Identically to the Carberry assessment, our survey instructed participants to rate the specific concept of interest for each step or task on a scale of 0 to 100 in increments of 10. This scale was chosen because it gives a more sensitive, increased performance prediction [42]. This self-efficacy survey was given to students pre- and post-experiment to measure any changes in self-efficacy due to participation in the experiment.

Rate your degree of [FILL IN TASK-SPECIFIC CONCEPT OF INTEREST] to perform the following tasks by recording a number from 0 to 100. (0=low; 50=moderate; 100=high)

	0	50	100
Identify a need use for AM			
Develop design solutions using AM			
Select the best possible design for AM			
Evaluate a part for DfAM			
Redesign a part for DfAM			

Rate your degree of confidence to perform the following tasks by recording a number from 0 to 100. (0=low; 50=moderate; 100=high)

Evaluate a part for...	0	50	100
Material Removal			
Unsupported Features			
Thin Features			
Stress Concentration			
Geometric Exactness			

Figure 3: Two-Part Self-Efficacy Survey Tasks

3.2 DfAM Scoring

Participants were instructed to score five different parts using the restrictive aspects of the DfAM worksheet: (1) Material Removal, (2) Unsupported Features, (3) Thin Features, (4) Stress Concentration, (5) Geometric Exactness [24] (see Appendix A1). Complexity was not included in evaluation because it was chosen as one of our independent variables.

To assess how complexity plays a role in DfAM evaluation, AM experts selected five different parts to directly represent the five different levels of complexity as defined by the DfAM Worksheet (as discussed in Section 2.2). Further breakdown of DfAM categories and their use in the experiment are shown in Table 1.

Material Removal and Unsupported Features were scored from one to five while Thin Features, Stress Concentration, and Geometric Exactness were scored from one to three. The total scores for each part for each participant were tallied and later used for data analysis. Two DfAM experts also scored the five parts using these metrics. Cronbach's Alpha was calculated for the two DfAM expert scores to ensure there was no disagreement between scores, and an alpha coefficient of reliability was computed (α=0.75). Expert scores were deemed reliable. The average of the expert score was later compared to participant scores to determine evaluation accuracy in Section 4.

Table 1: Breakdown of DfAM Worksheet categories assessed for own experiment

Category	Used?	Why?	Scale
Complexity	No	Independent variable chosen by AM experts to assess its effects on DfAM.	1 to 5 (5 Steps)
Functionality	No	Five different parts were designed to perform the same task.	1 to 5 (5 Steps)
Material Removal	Yes	Restrictive aspect varying with each part.	1 to 5 (5 Steps)
Unsupported Features	Yes	Restrictive aspect varying with each part.	1 to 5 (5 Steps)
Thin Features	Yes	Restrictive aspect varying with each part.	1 to 5 (3 Steps)
Stress Concentration	Yes	Restrictive aspect varying with each part.	1 to 5 (3 Steps)
Tolerances	No	Five parts were evaluated independently from an assembly.	1 to 5 (3 Steps)
Geometric Exactness	Yes	Restrictive aspect varying with each part.	1 to 5 (3 Steps)

3.3 Evaluation Time

Participants were also timed during the study. When a new part was presented to them, the timer was started, and the timer was not stopped until the participant completed the DfAM Worksheet for the part. The participant was then be presented with a new part, and the process would repeat. Time was recorded in seconds for each part for each participant and was later used in data analysis. Evaluation Time was included as a metric because the implications of one environment being faster than the other could speak to improved efficiency and usability [43]. Task completion time has been used and validated in past research as an effective metric assessing VR effectiveness in developing surgeons skills [44,45], assessing task performance in product dissection [46], and comparing manipulation between VR, CAD and real-world environments [34,35].

4 DATA ANALYSIS

The final DfAM scores and times were tallied and compared using a two-way mixed analysis of variance (ANOVA). The dependent variables were average DfAM score and design evaluation times, the independent variables were design complexity and evaluation environment (VR or CAD). Using the two-way ANOVA, we examined what interactions occur for scores and evaluation time when varying degrees of part complexity are introduced. The self-reported metrics relating to self-efficacy, motivation, outcome expectancy, and anxiety for DfAM were evaluated using independent samples t-tests and

paired samples t-tests. Confidence in using the DfAM Worksheet was also analyzed using the same methods. Results analysis follows.

4.1 RQ1: Is there a difference between VR and CAD in DfAM scoring with respect to complexity of the evaluated design?

Two-way interactions were evaluated based on whether the participant was assigned to the VR or CAD group using a two-way mixed ANOVA. A 95% confidence interval was chosen to determine if results were statistically significant ($p \leq 0.05$). Mauchly's test of sphericity [47] indicated that the assumption of sphericity was met for the two-way interaction for score (p=0.355). This analysis allowed us to evaluate differences in DfAM scoring according to group and part complexity (see Figure 4.) We found that there were no significant interaction effects between evaluation environment and complexity on DfAM scoring (p=0.188) nor were there significant differences found between CAD and VR, in general.

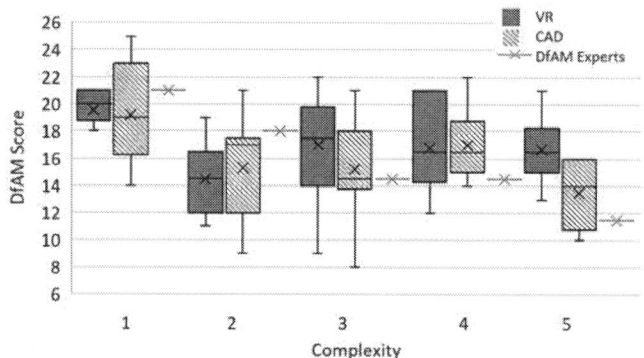

Figure 4: DfAM Scores for Varying Part Complexity

Given that a significant interaction was not found, simple main effects were then evaluated using an independent samples t-test at each level of complexity. The assumption of normality was violated in the VR group for part complexity 1 (p=0.007). The assumption of homogeneity of variances was violated for part complexity 1 (p=0.033) as found using Levene's Test of Equality of Error Variances [48]. The independent samples t-test was still run as parametric tests are robust to deviations from normality and homogeneity of variances violations [49].

Using box and whisker plots, a single outlier was discovered in the VR group for part complexity 1. The t-test was run with and without the outlier, and no significant change in results was found; the outlier was left in the data set for further analysis. A significant difference in DfAM score was found as indicated in Table 2 between VR and CAD groups for the most complex part (p=0.007), where the CAD group scored significantly lower, but was closer to the DfAM expert score.

Table 2: Independent Samples t-test for DfAM Score

Part Complexity	Mean Expert (Score)	Mean VR (Score)	Mean CAD (Score)	Sig.
1	21	19.5	19.2	0.825
2	18	14.5	15.3	0.604
3	14.5	17	15.2	0.303
4	14.5	16.8	17	0.886
5	**11.5**	**16.7**	**13.5**	**0.007**

An independent samples t-test was then performed for each restrictive element of the DfAM Worksheet to determine if there were any differences between CAD and VR for assessing (1) Material Removal, (2) Unsupported Features, (3) Thin Features, (4) Stress Concentration, and (5) Geometric Exactness for the most complex part. Table 3 shows there was a significant difference between CAD and VR evaluation for the restrictive DfAM element Unsupported Features. The VR group had a higher Unsupported Features score, on average, compared to the CAD group; a higher Unsupported Features score means that the part is more likely to require supporting features, or is oriented in a way that minimizes the need for supports. It is important to note that while the data from the DfAM Worksheet is ordinal, the data collected for this study has a sample size greater than 5; therefore, parametric examination methods can be used without the assumption of normality [49].

Table 3: Independent Samples T-Test for Restrictive DfAM Elements Scored for Part Complexity (Level 5)

Part Complexity Level 5	VR Mean	CAD Mean	Sig.
Material Removal	2.5	2.2	0.382
Unsupported Features	**3.4**	**2.1**	**0.006**
Thin Features	3.2	3.0	0.673
Stress Concentration	3.8	3.0	0.087
Geometric Exactness	2.4	2.1	0.306

4.2 RQ2: Is there a difference between VR and CAD in the time it takes participants to make DfAM decisions with respect to complexity of the evaluated design?

Two-way interactions were evaluated based on whether the participant was assigned to the VR or CAD group. This similarly allowed us to explore differences in evaluation time according to group and parts complexity (see Figure 5). Again, we found that there were overall no significant interaction effects between evaluation environment and complexity on evaluation time (p=0.431); however, an increasing mean time difference between VR and CAD evaluation was noted.

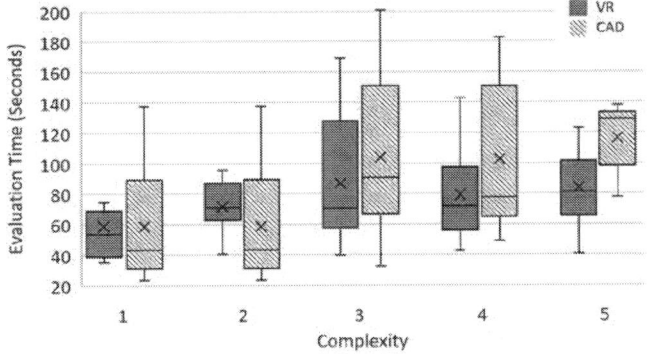

Figure 5: Mean Evaluation Times for Varying Complexities

An independent samples t-test at each level of complexity was performed to evaluate simple main effects. The assumption of normality was met in all cases. The assumption of homogeneity of variances was violated for part complexity 4 (p=0.0019) as determined by Levene's Test of Equality of Error Variances [48]. The independent samples t-test was still run as parametric tests are robust to violations of the homogeneity of variances assumption.

Using box and whisker plots, a single outlier was discovered in the VR group for part complexity 1. The t-test was run with and without the outlier, and no major differences between results were found; therefore, the outlier was left in the data set for further analysis. As shown in Table 4, a significant difference in evaluation time was found between groups for the most complex part (p=0.006) where the VR group evaluated designs at a faster rate.

Table 4: Independent Samples T-Test for Evaluation Time

Part Complexity	Mean VR (Time in Seconds)	Mean CAD (Time in Seconds)	Sig.
1	58.61	58.52	0.995
2	71.66	74.93	0.772
3	86.69	103.99	0.433
4	78.92	102.62	0.206
5	**83.58**	**116.46**	**0.006**

4.3 RQ3: Is there a difference between VR and CAD in terms of self-efficacy and other self-reported metrics for DfAM?

The results for the DfAM self-reported metric pre-surveys and post-surveys are plotted in Figure 6. A difference between groups was observed for anxiety, where the CAD group mean increased by 10.6 points while the VR group mean decreased by 1.2 points. Another difference was seen in anticipated success, where the CAD group mean decreased by 1.2 points, and the VR group mean increased by 4.6 points. The findings for anxiety and anticipated success were not statistically significant.

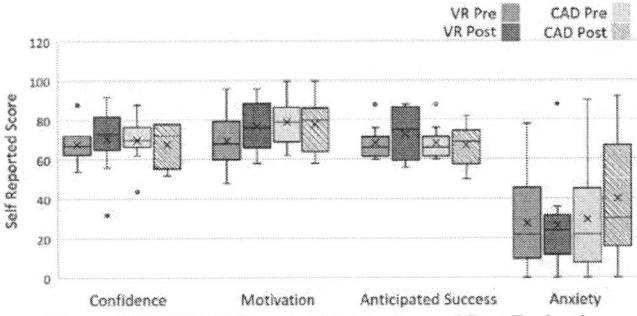

Figure 6: DfAM Self-Reported Metrics Pre and Post Evaluation

An independent samples t-test and paired samples t-test was performed to determine if there were any significant differences within-groups (pre- and post-survey) and between-groups (VR & CAD). The assumption of normality was violated in the VR group for Post-anxiety (p=0.014) and in the CAD group for Post-confidence (p=0.019). The assumption of homogeneity of variances was met for all cases, found using Levene's Test of Equality of Error Variances [48]. The independent samples t-test and paired samples t-test were still run as parametric tests are robust to deviations from normality. A total of six outliers were discovered using box and whisker plots: VR group for Pre-confidence, Post-confidence, and Post-anxiety and CAD group for Pre-confidence, Pre-anxiety, and Pre-Success. The t-tests were run with and without the outliers, and significant changes in results were discovered; therefore, the outliers were removed for further analysis. There was a significant difference found between CAD and VR groups for motivation before participants were exposed to their evaluation tool (VR or CAD) where the VR group had a significantly lower score for motivation (p=0.029). No other significant differences were found.

Table 5: Independent Samples T-Test Between Groups (VR & CAD) and Paired Samples T-Test within Groups (Pre- & Post-Survey) for Confidence, Motivation, Anticipated Success and Anxiety

	Between Group Pre (Sig.)	Between Group Post (Sig.)	Within Group VR (Sig.)	Within Group CAD (Sig.)
Confidence	0.097	0.679	0.182	0.424
Motivation	**0.029**	0.311	0.181	0.598
Success	0.853	0.444	0.380	0.780
Anxiety	0.509	0.411	0.418	0.280

Independent samples t-test and paired samples t-test were also performed for DfAM self-efficacy for the restrictive topics of Material Removal, Unsupported Features, Thin Features, Stress Concentration, and Geometric Exactness. The test was done to try to identify significant between-group effects as well as within-group effects. The assumption of normality was violated in the VR group for Post-Thin Features (p=0.033) and in the CAD group for Post-Stress Concentration (p=0.024). The assumption of homogeneity of variances was met for all cases, found using Levene's Test of Equality of Error Variances [48]. The independent samples t-test and paired samples t-test were still run as parametric tests are robust to deviations from normality. A total of seven outliers were discovered using box and whisker plots: VR group for Pre-Thin Features, Post-Material Removal, and Post-Thin Features and CAD group for Post-Unsupported Features (2), Post-Thin Features, and Post-Stress Concentration. The t-tests were run with and without the outliers, and significant changes in results were discovered. Due to this impact, all of these outliers were removed from analysis. Subsequently, a significant between-group effect was found for Stress Concentration before either group was exposed to their evaluation tool (CAD or VR), the VR group had a significantly higher score than the CAD group (p=0.033). There was also a significant within-group effect in the CAD group for Stress Concentration where the CAD group post-survey scores were significantly higher than pre-survey scores (p=0.001) as seen in Table 6.

Table 6: Independent Samples T-Test Between Groups (VR & CAD) and Paired Samples T-Test Within Groups (Pre & Post) for DfAM Confidence

	Between Group Pre (Sig.)	Between Group Post (Sig.)	Within Group VR (Sig.)	Within Group CAD (Sig.)
Material Removal	0.076	0.345	0.668	0.155
Unsupported Features	0.123	0.786	0.250	0.388
Thin Features	0.096	0.107	0.316	0.051
Stress Concentration	**0.033**	0.829	0.649	**0.001**
Geometric Exactness	0.308	0.837	0.668	0.280

5 DISCUSSION

This experiment was conducted to investigate potential differences in DfAM evaluation as a result of part complexity (as rated by experts) and evaluation environment (VR and CAD). Research questions RQ1, RQ2 and RQ3 are further discussed using the data analysis to support claims.

RQ1: Is there a difference between VR and CAD in DfAM scoring with respect to complexity of the evaluated design?

The different environments do have an effect on DfAM scoring, and complexity does play a role. While we do not find a significant interaction using the two-way mixed ANOVA, we can see increasing differences in scores as complexity increases (see Figure 4). At an part complexity of level 1, both groups report scores with no significant difference between them. Since the part is not very complex in that it is merely a 2D extrusion, there are no intricate features that need be identified. In other words, it does not matter what evaluation environment is used because the part does not warrant any in-depth analysis. In contrast, for the most complex parts, the DfAM scores were found to be significantly different between groups. This finding reveals that if a part is extremely complex the environment used to evaluate that part will have an impact.

It was found that the group using CAD gave significantly lower DfAM scores than the group using VR, and these lower scores were also closer to the DfAM experts' scores. This finding implies that for more complex parts, designers using CAD are more effective in accurately evaluating parts designed for additive manufacturing. It was also found that this significant difference in score was due primarily to the evaluation metric "Unsupported Features", where the VR group had significantly higher scores than the CAD group. In other words, the VR group had a tendency to overestimate the design's ability to be self-supporting.

When evaluating simple parts, it does not matter which evaluation environment is available; the DfAM scores will be the same. With increasing levels of part complexity, however, the evaluation environment will have an impact, and the designer should consider which evaluation environment will give them the advantage. While VR presents a new medium to evaluate parts, CAD may still yield a more accurate evaluation.

RQ2: Is there a difference between VR and CAD in the time it takes participants to make DfAM decisions with respect to complexity of the evaluated design?

Akin to RQ1, there is no significant difference in times for the least complex part; however, as part complexity increases, the overall evaluation time increases, and the evaluation time between groups increases (see Figure 5). A significant difference was found between groups for the most complex parts, where the VR group took less time to evaluate. This contrasts with our initial hypothesis that the CAD group would take less time for more complex parts because of the participants higher reported experience with the environment (see Section 1.4). These results align with published work which found that VR allows for faster manipulation [34,35].

These findings indicate that VR may have an advantage over CAD for quickly evaluating complex DfAM parts. The use of VR to evaluate complex parts will save design team's evaluation time, improving efficiency, and also usability [43]. However, it is important to remember that this fast evaluation time using VR will not be as accurate when compared to CAD. Looking at the mean evaluation time and DfAM score, the VR group was roughly 33 seconds faster than the CAD group in evaluating the most complex object, but the VR group was also 3 points less accurate than the CAD group when compared to the expert score. One metric does not currently have weight over the other; this is dependent on a user's needs during part evaluation. If the user needs to evaluate a large number of complex parts within a short period of time, then evaluation time would be weighted higher. If the user needs high quality evaluations for complex parts and has time to spare, then accuracy of DfAM score would be weighted higher.

RQ3: Is there a difference between VR and CAD in terms of self-efficacy and other self-reported metrics for DfAM?

Overall there were little significant differences in self-reported metrics for both within-group and between group tests. It was noted that the CAD Group had a larger increase in anxiety for evaluating DfAM parts. Those using VR to evaluate DfAM parts may, as a result, feel less anxious in their abilities. It was also noted that the VR Group had a larger increase in anticipated success than the CAD Group; the findings for anxiety and anticipated success were not statistically significant, but they are still in agreement with prior research suggesting that VR increases confidence in task performance [32]. Before evaluation occurred, significant between-group effects were found for DfAM motivation where the CAD group scored significantly higher (see Table 5) and confidence in Evaluating for Stress Concentration where the VR group scored significantly higher (see Table 6).

Before filling out their pre-survey, participants were not informed which group they would be participating in, for this reason it was assumed that participants would have similar pre-survey responses regardless of group. The experiment therefore had no effect on this outcome, and pre-survey differences may be attributed to participants' background knowledge.

Finally, there was a significant within-group effect in which the CAD group scored higher on Stress Concentration confidence after completing their DfAM evaluation. This finding implies that the use of CAD will improve designers' confidence in identifying areas of high or low stress concentration within a part. CAD may have better resolution when compared to VR, allowing designers to identify smaller details. This realization would support findings in RQ2 identifying CAD as a more accurate DfAM evaluation tool.

6 CONCLUSION AND FUTURE WORK

This study was conducted to investigate the potential advantages that VR and CAD present to designers when evaluating parts designed for additive manufacturing. Participants with DfAM experience were tasked with evaluating five parts of varied complexity using the DFAM Worksheet. Participant scores, evaluation times, and self-reported metrics were recorded and analyzed to make conclusions about the two environments.

The experiment results indicated a significant difference in DfAM scores between groups for the most complex parts. It was found that VR had a significantly faster evaluation time for the most complex part but lower accuracy. CAD, however, had a significant increase in Stress Concentration confidence, allowing designers to feel they have improved abilities to identify areas of high and low stress concentration in a part. It is also possible that VR can also improve participants' anticipated success and have less of an effect on Anxiety for DfAM part evaluation, though this was not proven at 95% confidence. Most notably, we found that as part complexity increases, the results between CAD and VR increase, suggesting that this research should be further explored with more complex parts with hard to reach areas.

Future work may also include experiment modifications such as recording build orientations, recording participants "thinking out loud", and higher fidelity emulation. During the experiment build orientation was not specified, nor was orientation preference recorded. This oversight was a limitation to the experiment as orientation has an effect on support generation. Future iterations of this experiment will either record or specify build orientation for more concrete evaluation of DfAM scoring.

This work may also be expanded using a qualitative approach. Participants can be interviewed about their experience with either CAD or VR, and they may also be recorded as they "think aloud" [50] while they are evaluating a part for AM. The recordings may then be coded and analyzed using the inductive approach [51], or conventional content analysis [52] to uncover themes encapsulating participants experience and understanding. A study evaluating a group of more experienced CAD users from an industry setting may further mitigate variability in part feature scoring. Further research using higher fidelity software would also be beneficial. This work limits participants to evaluating the outer shell of a part, however there is a lot of design software in the market today that allows users to visualize supports, estimate build times, visualize build paths, and look at cross sectional areas. Additionally, haptic feedback communicating the weight and feel of each part may also help to better inform participant decisions. A VR emulation with these more advanced design tools may be prove useful and could be validated through similar experimentation.

ACKNOWLEDGMENTS

We would like to thank the members of the Made By Design Lab (MBDL) and the Engineering Design and Optimization Group (EDOG) for their support with the experiment and analysis.

REFERENCES

[1] "3D Printing (Additive Manufacturing) Market - Global Forecast 2025" [Online]. Available: http://heraldkeeper.com/market/3d-printing-additive-manufacturing-market-2018-global-share-trend-segmentation-analysis-forecast-2025-103396.html. [Accessed: 08-Dec-2018].

[2] Bryson, S., 1993, "Virtual Reality in Scientific Visualization," Comput. Graph., 17(6), pp. 679–685.

[3] Gomes De Sá, A., and Zachmann, G., 1999, "Virtual Reality as a Tool for Verification of Assembly and Maintenance Processes," Comput. Graph., 23(3), pp. 389–403.

[4] Gao, W., Zhang, Y., Ramanujan, D., Ramani, K., Chen, Y., Williams, C. B., Wang, C. C. L., Shin, Y. C., Zhang, S., and Zavattieri, P. D., 2015, "The Status, Challenges, and Future of Additive Manufacturing in Engineering," Comput. Des., 69, pp. 65–89.

[5] Williams, C. B., Mistree, F., and Rosen, D. W., 2017, "DETC2008-49353," (Section 2), pp. 1–14.

[6] Brennan-Craddock, J., Brackett, D., Wildman, R., and Hague, R., 2012, "The Design of Impact Absorbing Structures for Additive Manufacture," *Journal of Physics: Conference Series*.

[7] Collins, I. L., Weibel, J. A., Pan, L., and Garimella, S. V., 2019, "A Permeable-Membrane Microchannel Heat Sink Made by Additive Manufacturing," Int. J. Heat Mass Transf.

[8] Loh, G. H., Pei, E., Harrison, D., and Monzón, M. D., 2018, "An Overview of Functionally Graded Additive Manufacturing," Addit. Manuf.

[9] Seifi, H., Rezaee Javan, A., Xu, S., Zhao, Y., and Xie, Y. M., 2018, "Design Optimization and Additive Manufacturing of Nodes in Gridshell Structures," Eng. Struct.

[10] Dede, E. M., Joshi, S. N., and Zhou, F., 2015, "Topology Optimization, Additive Layer Manufacturing, and Experimental Testing of an Air-Cooled Heat Sink," J. Mech. Des.

[11] Ye, H., Chen, Y., Zhou, C., Xu, W., and Kwok, T.-H., 2016, "Mass Customization: Reuse of Digital Slicing for Additive Manufacturing," J. Comput. Inf. Sci. Eng.

[12] Mohammed, M. I., P. Fitzpatrick, A., and Gibson, I., 2017, "Customised Design of a Patient Specific 3D Printed Whole Mandible Implant," KnE Eng.

[13] Jones, G., Kline, E. V., Schmelzle, J., Reutzel, E. W., Simpson, T. W., and Dickman, C. J., 2015, "(Re)Designing for Part Consolidation: Understanding the Challenges of Metal Additive Manufacturing," J. Mech. Des.

[14] Stoner, B., Bartolai, J., Kaweesa, D. V., Meisel, N. A., and Simpson, T. W., 2018, "Achieving Functionally Graded Material Composition Through Bicontinuous Mesostructural Geometry in Material Extrusion Additive Manufacturing," JOM.

[15] Maier, R. R. J., Havermann, D., MacPherson, W. N., and Hand, D. P., 2013, "Embedding Metallic Jacketed Fused Silica Fibres into Stainless Steel Using Additive Layer Manufacturing Technology," *Fifth European Workshop on Optical Fibre Sensors*.

[16] Clark, J. E., Bailey, S. A., Cham, J. G., Cutkosky, M. R., and Full, R. J., 2004, "Fast and Robust: Hexapedal Robots via Shape Deposition Manufacturing," Int. J. Rob. Res.

[17] Espalin, D., Muse, D. W., MacDonald, E., and Wicker, R. B., 2014, "3D Printing Multifunctionality: Structures with Electronics," Int. J. Adv. Manuf. Technol.

[18] Mavroidis, C., DeLaurentis, K. J., Won, J., and Alam, M., 2002, "Fabrication of Non-Assembly Mechanisms and Robotic Systems Using Rapid Prototyping," J. Mech. Des.

[19] Ngo, T. D., Kashani, A., Imbalzano, G., Nguyen, K. T. Q., and Hui, D., 2018, "Additive Manufacturing (3D Printing): A Review of Materials, Methods, Applications and Challenges," Compos. Part B Eng.

[20] Doubrovski, Z., Verlinden, J. C., and Geraedts, J. M. P., 2017, "DETC2011-4," pp. 1–12.

[21] Yang, Z., and Liu, Q., 2007, "Research and Development of Web-Based Virtual Online Classroom," Comput. Educ., 48(2), pp. 171–184.

[22] Laverne, F., Segonds, F., Anwer, N., and Le Coq, M., 2015, "Assembly Based Methods to Support Product Innovation in Design for Additive Manufacturing: An Exploratory Case Study," J. Mech. Des., 137(12), p. 121701.

[23] Kumke, M., Watschke, H., Hartogh, P., Bavendiek, A. K., and Vietor, T., 2018, "Methods and Tools for Identifying and Leveraging Additive Manufacturing Design Potentials," Int. J. Interact. Des. Manuf.

[24] Booth, J. W., Alperovich, J., Chawla, P., Ma, J., Reid, T. N., and Ramani, K., 2017, "The Design for Additive Manufacturing Worksheet."

[25] Mazuryk, T., and Gervautz, M., *Virtual Reality History, Applications, Technology and Future*.

[26] "Virtual Assembly Using Virtual Reality Techniques."

[27] Whitney, D., Rosen, E., Phillips, E. K., Tellex, S., Ullman, D., and Phillips, E., 2018, *ROS Reality: A Virtual Reality Framework Using Consumer-Grade Hardware for ROS-Enabled Robots*.

[28] Kelusak, C., 2018, "How to Make Virtual Reality a Reality in Your Design Firm" [Online]. Available: https://www.autodesk.com/autodesk-university/article/How-Make-Virtual-Reality-Reality-Your-Design-Firm-2018.

[29] Berg, L. P., and Vance, J. M., 2017, "Industry Use of Virtual Reality in Product Design and Manufacturing: A Survey," Virtual Real.

[30] Berg, L. P., and Vance, J. M., 2016, "An Industry Case Study: Investigating Early Design Decision Making in Virtual Reality," J. Comput. Inf. Sci. Eng.

[31] Abulrub, A.-H. G., Attridge, A. N., and Williams, M. A., *The Future of Creative Learning*.

[32] Lynna, J., and Ausburn, F. B., 2008, *Effects of Desktop Virtual Reality on Learner Performance and Confidence in Environment Mastery: Opening a Line of Inquiry.*

[33] Starkey, E. M., Spencer, C., Lesniak, K., Tucker, C., and Miller, S. R., 2017, *DO TECHNOLOGICAL ADVANCEMENTS LEAD TO LEARNING ENHANCEMENTS? AN EXPLORATION IN VIRTUAL PRODUCT DISSECTION.*

[34] Zacharia, Z. C., Olympiou, G., and Papaevripidou, M., 2008, "Effects of Experimenting with Physical and Virtual Manipulatives on Students' Conceptual Understanding in Heat and Temperature," J. Res. Sci. Teach.

[35] Starkey, E. M., McKay, A. S., Hunter, S. T., and Miller, S. R., "LET'S GET PHYSICAL? THE IMPACT OF DISSECTION MODALITY ON ENGINEERING STUDENT DESIGN LEARNING."

[36] "VIVE™ | Discover Virtual Reality Beyond Imagination" [Online]. Available: https://www.vive.com/eu/. [Accessed: 31-Oct-2018].

[37] "Unity" [Online]. Available: https://unity3d.com/. [Accessed: 01-Nov-2018].

[38] "3D CAD Design Software" [Online]. Available: https://www.solidworks.com/. [Accessed: 08-Dec-2018].

[39] "Airplane Bearing Bracket Challenge | Engineering & Design Challenges | GrabCAD" [Online]. Available: https://grabcad.com/challenges/airplane-bearing-bracket-challenge. [Accessed: 08-Dec-2018].

[40] Shaughnessy, J. J., Zechmeister, E. B., and Zechmeister, J. S., 2006, *Research Methods in Psychology*, McGraw Hill, New York.

[41] Adam, R. C., Hee-Sun, L., and Matthew, W. O., 2010, "Measuring Engineering Design Self-Efficacy," J. Eng. Educ., 99(1), p. 71.

[42] G., V., 2001, "Response Format in Self-Efficacy: Greater Discrimination Increases Prediction," Couns. Dev., 33, p. 35.

[43] Hornbæk, K., 2006, "Current Practice in Measuring Usability: Challenges to Usability Studies and Research," Int. J. Hum. Comput. Stud., 64(2), pp. 79–102.

[44] Newmark, J., Dandolu, V., Milner, R., Grewal, H., Harbison, S., and Hernandez, E., 2007, "Correlating Virtual Reality and Box Trainer Tasks in the Assessment of Laparoscopic Surgical Skills," Am. J. Obstet. Gynecol., 197(5), pp. 1–4.

[45] Lendvay, T. S., Brand, T. C., White, L., Kowalewski, T., Jonnadula, S., Mercer, L. D., Khorsand, D., Andros, J., Hannaford, B., and Satava, R. M., 2013, "Virtual Reality Robotic Surgery Warm-up Improves Task Performance in a Dry Laboratory Environment: A Prospective Randomized Controlled Study," J. Am. Coll. Surg., 216(6), pp. 1181–1192.

[46] Bharathi, A. K. B. G., and Tucker, C. S., 2015, "Investigating the Impact of Interactive Immersive Virtual Reality Environments in Enhancing Task Performance in Online Engineering Design Activities," *ASME 2015 International Design Engineering Technical Conferences and Computers and Information in Engineering Conference*, American Society of Mechanical Engineers, p. V003T04A004-V003T04A004.

[47] "Understanding Sphericity - An Introduction to, Testing for, and Interpreting Sphericity | Laerd Statistics" [Online]. Available: https://statistics.laerd.com/statistical-guides/sphericity-statistical-guide.php. [Accessed: 23-Feb-2019].

[48] "Independent T-Test - An Introduction to When to Use This Test and What Are the Variables Required | Laerd Statistics" [Online]. Available: https://statistics.laerd.com/statistical-guides/independent-t-test-statistical-guide.php. [Accessed: 23-Feb-2019].

[49] Norman, G., 2010, "Likert Scales, Levels of Measurement and the 'Laws' of Statistics," Adv. Helth Sci. Educ., 15(1 T), pp. 625–632.

[50] Lundgrén-Laine, H., and Salanterä, S., 2010, "Think-Aloud Technique and Protocol Analysis in Clinical Decision-Making Research," Qual. Health Res.

[51] Elo, S., and Kyngäs, H., 2008, "The Qualitative Content Analysis Process," J. Adv. Nurs., 62(1), pp. 107–115.

[52] Hsieh, H.-F., and Shannon, S. E., 2005, "Three Approaches to Qualitative Content Analysis."

APPENDIX

A1: SCORING SHEET FOR RESTRICTIVE DFAM

Material Removal

1 The part is smaller than or the same size as the required support structure

2 There are small gaps that will require support structures

3 Internal cavities, channels, or holes do not have openings for removing materials

4 Material can be easily removed from internal cavities, channels, or holes

5 There are no internal cavities, channels, or holes

Unsupported Features

1 There are long, unsupported features

2 There are short, unsupported features

3 Overhang features have a slopped support

4 Overhanging features have a minimum of 45deg support

5 Part is oriented so there are no overhanging features

Thin Features

1 Some walls are less than 1/16" (1.5mm) thick

2 Walls are between 1/16" (1.5mm) and 1/8" (3mm) thick

3 Walls are more than 1/8" (3mm thick

Stress Concentration

1 Interior corners have no chamfer, fillet, or rib

2 Interior corners have chamfers, fillets, and/or ribs

3 Interior corners have generous chamfers, fillets, and/or ribs

Geometric Exactness

1 The part has large, flat surfaces or has a form that is important to be exact

2 The part has medium-sized, flat surfaces, or forms that are should be close to exact

3 The part has small or no flat surfaces, or forms that need to be exact

Proceedings of the ASME 2019
International Design Engineering Technical Conferences
and Computers and Information in Engineering Conference
IDETC/CIE2019
August 18-21, 2019, Anaheim, CA, USA

DETC2019-97492

SIMULATION-BASED PROCESS OPTIMIZATION OF METALLIC ADDITIVE MANUFACTURING UNDER UNCERTAINTY

Zhuo Wang
Department of Mechanical Engineering,
Mississippi State university
Starkville, MS, USA

Pengwei Liu
Department of Mechanical Engineering,
Mississippi State University
Starkville, MS, USA

Zhen Hu[1]
Department of Industrial and Manufacturing
systems Engineering
University of Michigan, Dearborn
Dearborn, MI, USA

Lei Chen[1]
Department of Mechanical Engineering
Mississippi State University
Starkville, MS, USA

ABSTRACT

The presence of various uncertainty sources in metal-based additive manufacturing (AM) process prevents producing AM products with consistently high quality. Using electron beam melting (EBM) of Ti-6Al-4V as an example, this paper presents a data-driven framework for process parameters optimization using physics-informed computer simulation models. The goal is to identify a robust manufacturing condition that allows us to constantly obtain equiaxed materials microstructures under uncertainty. To overcome the computational challenge in the robust design optimization under uncertainty, a two-level data-driven surrogate model is constructed based on the simulation data of a validated high-fidelity multi-physics AM simulation model. The robust design result, indicating a combination of low preheating temperature, low beam power and intermediate scanning speed, was acquired enabling the repetitive production of equiaxed-structure products as demonstrated by physics-based simulations. Global sensitivity analysis at the optimal design point indicates that among the studied six noise factors, specific heat capacity and grain growth activation energy have largest impact on the microstructure variation.

1. INTRODUCTION

Additive manufacturing (AM), which creates 3D components layer by layer based on computer aided design (CAD) model, offers great potential for fabricating component with complex geometry in a cost- and time-saving manner [1-3]. One of the major hurdles for the wide application of AM techniques is the variation in the quality of the manufactured parts [4]. That is, various uncertainty sources exist in the AM process and cause variability in the product quality through uncertainty propagation and aggregation. For example, it has been pointed out that the quality and properties of deposits can vary greatly even when all producers used the same materials, processing parameters, and, in some cases, even the same type of AM machine [5].

Uncertainty quantification (UQ) can help resolve the above issue by constructing product variation as functions of the contributing factors and then making effective uncertainty management (UM). One common strategy is to adjust control factors to dampen the variation caused by noise factors, i.e. Type I robust design [6] as illustrated in step 3 in Fig. 1. In the case of AM process, control factors mainly refer to those AM operating settings, such as preheating temperature, electron beam (EB) power and scanning velocity. Operators can purposely manipulate them, namely the process optimization, to minimize variability in product quality caused by random noise factors (uncertainty sources), just like finding out $x = a$ in the simple case shown in Fig. 1. Of course, the robust design in this research is much more complicate than the illustrative case. The product quality has a nonlinear dependence on multiple control factors and noise factors in AM process, of which the later includes the fluctuating power absorption and various uncertain materials properties, as detailed later.

Model-based uncertainty quantification [7-9] especially provides a cheap yet effective way to achieve quality control,

---

[1] Corresponding author: zhennhu@umich.edu; chen@me.msstate.edu

with the development of advanced simulation techniques. Current model-based UQ in AM process is however still at its early stage. Modeling of the entire AM process is a multi-level problem [10]. The product quality is linked to the operating parameters through multiple connected lower-level models (e.g. powder-bed model and heat source model) and upper level models (e.g. melt pool model and solidification model). Nonetheless, most of current UQ analysis are limited to the scope of a single level [5, 11-14], especially the variation in the melt pool caused by uncertainty sources. Thus, yielding a guideline towards direct quality control by optimizing operating parameters is impossible. A systematic UQ and UM spanning multi-level of AM process, i.e. correlating process parameters and uncertain sources arising from different level to the final product quality, is still lacking.

The main reason for the absence of a systematic UQ and UM is that, few reliable multi-level AM simulation models are currently available to support a model-based multi-level study. Various physical phenomena are involved in the entire AM process, hence greatly inhibiting the development of a high-fidelity AM simulation model. Existing multi-level models [15-18] usually failed to take into account, for example, re-melting phenomenon in a layer-wise building and/or undercooling-controlled grain nucleation. But all of them indeed influence the formation of final grain structure [19-21]. Another reason can be related to the multi-level nature, which indicates a large number of uncertain sources across multiple levels. For a powder-bed AM process, there are dozens of factors that may influence the quality of a final manufactured part [22]. Consequently, extensive simulations at numerous points over a high-dimensional space are required during iterations to find the optimal point. This fact tends to make a multi-level UQ and UM computationally intractable, even with a reliable multi-level simulation model in hand.

In this paper, a data-driven process optimization framework is developed, with application to the uncertainty quantification and management in electron beam melting (EBM) of Ti-6Al-4V. It takes advantage of an advanced multi-scale AM simulation model and Kriging surrogate model, allowing effective UQ and UM in a computationally friendly way. Specifically, as shown in Fig. 1, a reasonable number of multi-scale multi-physics simulations were firstly conducted at carefully designed sampling points. This provides training data for training cheap data-driven surrogate models. Through cross-validation, the surrogate models showed great ability in predicting temperature profile and materials microstructure at a much cheaper computational cost. The surrogate models were then utilized to effortlessly carry out uncertainty quantification and robust design optimization. Based on UQ analysis, critical factors in affecting quantity of interest (QoI), here the grain structure described by grain length/width ratio, were successfully identified. In addition, the robust manufacturing condition that makes QoI inert to the variations of noise factors were found, offering manufacturing guidelines to minimize product quality variation.

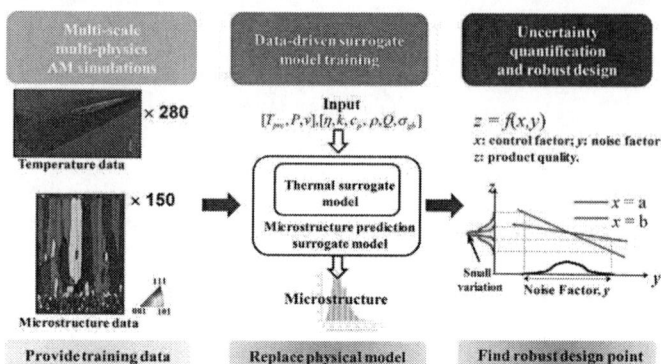

Figure 1. Workflow of the proposed data-driven approach for uncertainty quantification and management

2. PHYSICAL MODEL

The physical model is based on the combination of our recently developed multi-scale multi-physics AM simulation models [23-25]. It spans multi-level by coupling a finite-element (FE) based heat transfer model for simulating temperature field development and grain growth phase field model (PFM) for simulating microstructure evolution. Here we just provide essential aspects of this model, but readers interested in greater detail are referred to Ref. [23-25].

2.1. Finite-element Based Heat Transfer Model

A finite-element based heat transfer model incorporating a moving heat source is utilized to capture the temperature field development during EBM process. The thermal interaction of the moving electron beam with the deposits is described by a Gaussian distribution model [18], as below

$$Q_e\left(x,y,z,t\right)=\eta \cdot P \cdot \frac{4\ln(0.1)}{\pi d^2 z_e} e^{\frac{4\ln(0.1)\left((x-v \cdot t)^2+y^2\right)}{d^2}}\left\{-3\left(\frac{z}{z_e}\right)^2-2\left(\frac{z}{z_e}\right)+1\right\} \quad (1)$$

where η is the absorptive coefficient of the powder, P is the power of electron beam, d is the electron beam diameter, v is the beam velocity, and z_e is the absolute penetration depth of the electron beam.

The initial boundary condition is $T = T_{pre}$ applied to the substrate and deposits, where T_{pre} is the preheating temperature. Convective and radiative hear transfer are employed at the surface.

Note that, the current thermal model enables us to capture instaneous temperature field for every micro time step during a layer-by-layer AM process, but fully coupling two models, namely repeatedly extracting the real-time temperature field and incorporating it into grain growth model, is computationally prohibitive. However, we notice the fact that only the high temperature field enclosed by the isotherm $T = T_{\beta\text{-}transus}$, here denoted as high temperature field $T_\beta(x,z)$ in 2D case, is the useful temperature field for β-grain growth; temperatures below $T_{\beta\text{-}transus}$ are generally believed to have negligible influence on β grain evolution. As shown in Fig. 2, for most of the time, the aforementioned high temperature field $T_\beta(x,z)$ would achieve and remain a relatively steady state, especially in the bulk section that is exactly of the interest in the current study. By ignoring the short-time unsteady state, the temperature field evolvement is equivalent to a steady temperature field moving along with the electron beam. Also, the steady high temperature field $T_\beta(x,z)$ for different layers are assumed to change little under a certain manufacturing condition [26]. As such, once the steady temperature field $T_\beta(x,z)$ is known, we can easily describe the full temperature field evolvement as $T_\beta(x-v\cdot t, z-n\cdot\Delta h)$ and incorporate it into grain growth model. In light of this, a thermal surrogate model, which is able to predict the steady temperature field developed under any condition, could be just trained for a fast approximate of the full thermal process. Correspondingly, the FE-based thermal simulation, in this study, is performed mainly to provide training data of stable temperature fields developed under various conditions. To achieve this, simulating a single-layer building is enough and thus adopted in current thermal simulations.

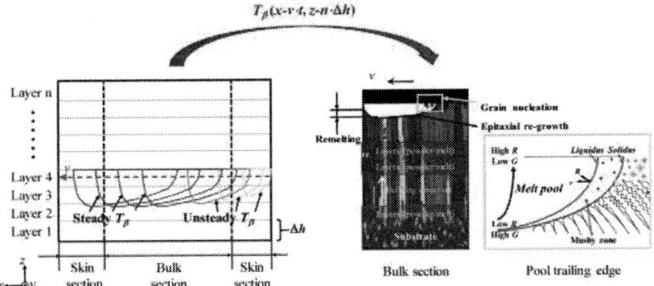

Figure 2. Schematic illustration of the coupled thermal model and grain growth model

2.2. Grain Growth Phase Field Model

We have greatly elaborated the basic grain growth phase field model for applicability in simulating grain evolution during the metal-based AM process. For example, the current PFM uses grain boundary energy, κ_q, that increases with the angle between grain orientation and local thermal gradient ($\kappa_q = \kappa_{q0}\times|\sin\angle(<001>_{axis}, \nabla T)|$), so as to achieve the selective growth of optimally aligned grains essentially due to grains' anisotropy in thermal conductivity, elastic modulus and surface energy. Also, in order to incorporate the temperature effect on grain boundary mobility, a temperature-dependent grain growth kinetic rate coefficient, L_q, using a modified Arrhenius type equation is adopted, as follows [25]

$$L_q(T) = L_0 \cdot \left(\frac{T}{T_a}\right)^b \cdot \exp\left(-\frac{Q}{R_g T}\right) \qquad (2)$$

where T_a is the ambient temperature, R_g is the gas constant of 8.314 J/mol K), L_0 and b (-1<m<1) are constants, and Q is the activation energy for β-grain growth of Ti-6Al-4V alloy.

Besides the aforementioned selective grain growth and temperature-dependent kinetic rate, various physical phenomena involved in the EBM process are also well incorporated in the phase field grain growth simulation; see Fig. 2. First, grain growth simulation starts with a pre-existing grain structure serving as the base for building up the layers, which

essentially mimics the "starter plate" during Ti-6Al-4V EBM fabrication [27]. The new layers are then periodically added to the substrate, presenting as layer-by-layer incremental computational domain in the simulation. A physical phenomenon inherent to the layer-by-layer building fashion is the re-melting of previous layers as the electron beam scans the newly deposited powder layer. In the simulation, grain order parameters are intentionally set to 0 as the front edge of the melt pool (i.e. the isotherm line of $T = T_{liquidus}$) is sweeping through and then re-evolve when the trailing edge arrives. Another important phenomenon is the grain nucleation ahead of the solidification front, which is associated with local thermal conditions (i.e. thermal gradient, G, and solidification rate, R). In general, low G/R values at the pool top will result in large constitutional undercoolings that encourage grain nucleation [28, 29]. The introduction of new grains would directly interrupt the original epitaxial grain growth and even give rise to equiaxed structures in extreme cases. A probabilistic nucleation model [30] describing nucleation probability as a function as local solidification conditions is adopted.

The above grain growth model enables us to get grain microstructure and detailed morphology developed under any condition. However, in this research the grain length/width distribution, i.e. the mean and second moment of grain length/width ratio, is just selected as the microstructure descriptor to facilitate surrogate model training. Therefore, the grain length/width ratio information would be further extracted as training data after microstructure obtainments. This is expected to correspondingly train a microstructure surrogate model that can quickly predicts the grain's length/width ratio distribution and thus the approximate grain structure.

3. PROCESS PARAMETER OPTIMIZATION UNDER UNCERTAINTY

3.1. Uncertainty Sources

Various uncertainty sources or noise factors exist during the metal-based additive manufacturing process. A good example is the fluctuation of power absorption. Power absorption efficiency depends highly on absorbing surface associated with powder packing [31] as well as melt pool dynamics or flow behavior [32, 33], both of which however show very random and unstable nature. This fact brings in the power absorption great uncertainty. For example, the energy absorption coefficient assumed in previous researches exhibits large difference, from 0.6 [27] to 0.9 [34], both for the electron beam melting of Ti-6Al-4V using Arcam® S12 machine. Other uncertainty sources include variation of materials properties of powder particles, natural variability in the temperature boundary condition, etc.

In this study, the following uncertainty sources are considered:

- Fluctuation of power absorption coefficient of powder layers, η,

- and uncertain materials properties of Ti-6Al-4V powders, including the thermal conductivity, k, specific heat capacity, c_p, density, ρ, grain boundary energy, σ_{gb}, and thermal activation energy of grain growth, Q.

Among them the uncertainty sources relevant to grain boundary energy and thermal activation energy of grain growth are introduced at the grain growth level. As shown in Table 1, the reasonable variation range of most factors are determined based on different values used in previous researches. It is noted that, thermal conductivity, specific heat capacity, and density are temperature dependent (usually fitted as cubic polynomial relations $a_0 + a_1T + a_2T^2 + a_3T^3$ [23, 35]), uncertainties of them in this study are thus described by variations in coefficients, a_0 and a_1, of respective polynomial function.

Table 1 Variation of different noise factors

Noise factors	Power absorption, η [27, 34]	Grain boundary energy, σ_{gb} (J/m^2) [36, 37]	Grain growth activation energy, Q (KJ) [38-40]
Lower Bound	0.6	0.8	100
Upper Bound	0.9	0.9	170

Based on the above observations of the variations of different variables, Gaussian distributions are assumed for the coefficients (i.e. a_0 and a_1) of the polynomial functions of the thermal conductivity, specific heat capacity, and density. The mean values are the maximum likelihood estimates (MLE) of the coefficients, and the standard deviations are determined by analyzing the confidence intervals of these coefficients using the data provided in Refs. [23, 41]. Table 2 gives the distribution parameters of these polynomial function coefficients. Following that, Table 3 presents the assumed distributions of η, σ_{gb}, and Q. The distribution parameters of these three parameters are determined according to the variation intervals presented in Table 2. For the sake of illustration, the distribution types are assumed to be Gaussian, Gaussian, and Lognormal, respectively.

Table 2 Distribution parameters of the coefficients of the polynomial functions

Noise factors	Thermal conductivity, k, (W/m K)		Specific heat capacity, c_p (J/Kg K)		Density, ρ (Kg/m^3)	
	a_{10}	a_{11}	a_{20}	a_{21}	a_{30}	a_{31}
Mean	4.7	4.7×10^{-3}	504.6	0.11	4.4×10^3	-0.89
Standard deviation	5.2	5.2×10^{-3}	557.7	0.12	4.9×10^3	-0.99

Table 3 Distribution information of some random variables

Variable	η	σ_{gb}	Q
Distribution Type	Gaussian	Gaussian	Lognormal
Mean	0.75	0.85	130
Standard deviation	0.02	0.01	6

3.2. Design Objective

Materials microstructure is the quantity of interest in this study. In general, the equiaxed grain structure exhibits preferable mechanical performance compared to the columnar grain structure. For example, fine equiaxed microstructures can more easily accommodate strain and promotes tearing [42]. In contrast, large columnar grain structure usually leads to property anisotropy, premature failure under transverse loading and shorter fatigue life [2, 43]. Thus, the design objective in this study is to find out optimal AM process parameters, which enable us to constantly obtain equiaxed grain structure under the existence of various uncertainties.

Specifically, grain structure in this research is described using the length/width ratio of grains, r, which equals 1 for a perfectly equiaxed grain and takes a much larger value for a columnar grain. The objective of design optimization is thus to move the mean value of r closer to one and minimize the standard deviation of r at the same time. Using this objective function, we then formulate the robust design optimization model as below:

$$\min_{\mathbf{d}} w_1(\mu_r(\mathbf{d})-1)^2 + (1-w_1)\sigma_r^2(\mathbf{d})$$
$$s.t.$$
$$\mathbf{d} = [T_{pre}, P, v] \tag{3}$$
$$\mathbf{d}_L \leq \mathbf{d} \leq \mathbf{d}_u$$

where $\mathbf{d}$ is a vector of design variables, $\mathbf{d}_L$ and $\mathbf{d}_u$ are respectively the lower and upper bound of the design variables, T_{pre} is the preheating temperature, P is the EB power, v is EB scanning speed, $\mu_r(\mathbf{d})$ and $\sigma_r(\mathbf{d})$ are respectively the mean and standard deviation of the length to width ratio, in terms of grains of various materials microstructures developed under the manufacturing condition, $\mathbf{d}$, and w_1 is the weight of the mean value in robust design. $w_1 = 0.5$ in this paper since equal weights are assigned to the mean and standard deviation of r.

For given $\mathbf{d}$, $\mu_r(\mathbf{d})$ and $\sigma_r(\mathbf{d})$ are given by

$$\mu_r(\mathbf{d}) = \int_0^\infty r f_{r,\mathbf{d}}(r)dr \tag{4}$$

$$\sigma_r^2(\mathbf{d}) = \int_0^\infty (r - \mu_r(\mathbf{d})) f_{r,\mathbf{d}}(r)dr$$
$$= \int_0^\infty r^2 f_{r,\mathbf{d}}(r)dr - \left(\int_0^\infty r f_{r,\mathbf{d}}(r)dr\right)^2 \tag{5}$$

in which $f_{r,\mathbf{d}}(r)$ is the probability density function (PDF) of r for given $\mathbf{d}$.

3.3. Evaluation of Objective function

Defining the random variables in the manufacturing process as $\boldsymbol{\theta} = [\eta, \sigma_{gb}, Q, a_{10}, a_{11}, a_{20}, a_{21}, a_{30}, a_{31}]$, for given $\mathbf{d}$ and $\boldsymbol{\theta}$, we then have

$$f_{r,\mathbf{d}}(r) = \int f_{r,\mathbf{d}}(r \mid \boldsymbol{\theta}) f_{\boldsymbol{\theta}}(\boldsymbol{\theta}) d\boldsymbol{\theta} \tag{6}$$

and

$$\mu_r(\boldsymbol{\theta}, \mathbf{d}) = \int_0^\infty r f_{r,\mathbf{d}}(r \mid \boldsymbol{\theta}) dr \tag{7}$$

Where $f_{r,\mathbf{d}}(r|\boldsymbol{\theta})$ and $\mu_r(\boldsymbol{\theta},\mathbf{d})$ are the conditional PDF and mean of r for given $\boldsymbol{\theta}$ and $f_{\boldsymbol{\theta}}(\boldsymbol{\theta})$ is the joint PDF of $\boldsymbol{\theta}$.

Plugging Eq. (6) into Eq. (4) yields

$$\mu_r(\mathbf{d}) = \int_0^\infty r \int f_{r,\mathbf{d}}(r \mid \boldsymbol{\theta}) f_{\boldsymbol{\theta}}(\boldsymbol{\theta}) d\boldsymbol{\theta} dr = \int \mu_r(\boldsymbol{\theta}, \mathbf{d}) f_{\boldsymbol{\theta}}(\boldsymbol{\theta}) d\boldsymbol{\theta} \quad (8)$$

The above equation can be approximated using Monte Carlo simulation (MCS) based method as

$$\mu_r(\mathbf{d}) = \int \int_0^\infty r f_{r\mid\boldsymbol{\theta},\mathbf{d}}(r) f_{\boldsymbol{\theta}}(\boldsymbol{\theta}) dr d\boldsymbol{\theta} \approx \frac{1}{n_{MCS}} \sum_{i=1}^{n_{MCS}} \mu_r(\boldsymbol{\theta}^{(i)}, \mathbf{d}) \quad (9)$$

in which n_{MCS} is the number of MCS samples of $\boldsymbol{\theta}$ and $\boldsymbol{\theta}^{(i)}$ is the i-th MCS sample.

Similarly, we have $\sigma_r^2(\mathbf{d})$ as follows

$$\begin{aligned}
\sigma_r^2(\mathbf{d}) &= \int \int_0^\infty r^2 f_{r\mid\boldsymbol{\theta},\mathbf{d}}(r) f_{\boldsymbol{\theta}}(\boldsymbol{\theta}) dr d\boldsymbol{\theta} - \mu_r^2(\mathbf{d}) \\
&= \int \int_0^\infty r^2 f_{r\mid\boldsymbol{\theta},\mathbf{d}}(r) dr f_{\boldsymbol{\theta}}(\boldsymbol{\theta}) d\boldsymbol{\theta} - \mu_r^2(\mathbf{d}) \\
&= \int m_{r2}(\boldsymbol{\theta}, \mathbf{d}) d\boldsymbol{\theta} - \mu_r^2(\mathbf{d})
\end{aligned} \quad (10)$$

where $m_{r2}(\boldsymbol{\theta},\mathbf{d})$ is the second moment of r. Using the MCS samples of $\boldsymbol{\theta}$, Eq. (10) is approximated as

$$\sigma_r^2(\mathbf{d}) \approx \frac{1}{n_{MCS}} \sum_{i=1}^{n_{MCS}} m_{r2}(\boldsymbol{\theta}^{(i)}, \mathbf{d}) - \left(\frac{1}{n_{MCS}} \sum_{i=1}^{n_{MCS}} \mu_r(\boldsymbol{\theta}^{(i)}, \mathbf{d}) \right)^2 \quad (11)$$

Combining Eqs. (9), (11), and (3), the robust design optimization model can be rewritten as

$$\begin{aligned}
\min_{\mathbf{d}} \quad & w_1 \left(\frac{1}{n_{MCS}} \sum_{i=1}^{n_{MCS}} \mu_r(\boldsymbol{\theta}^{(i)}, \mathbf{d}) - 1 \right)^2 + \\
& (1 - w_1) \left[\frac{1}{n_{MCS}} \sum_{i=1}^{n_{MCS}} m_{r2}(\boldsymbol{\theta}^{(i)}, \mathbf{d}) - \left(\frac{1}{n_{MCS}} \sum_{i=1}^{n_{MCS}} \mu_r(\boldsymbol{\theta}^{(i)}, \mathbf{d}) \right)^2 \right]
\end{aligned} \quad (12)$$

$$s.t.$$
$$\mathbf{d} = [T_{pre}, P, v]$$
$$\mathbf{d}_L \le \mathbf{d} \le \mathbf{d}_u$$

The above equation indicates that in order to evaluate the objective function, $\mu_r(\boldsymbol{\theta}^{(i)},\mathbf{d})$ and $m_{r2}(\boldsymbol{\theta},\mathbf{d})$ need to be evaluated repeatedly with $i = 1, 2, \cdots, n_{MCS}$ for given $\mathbf{d}$. A straightforward way of estimating $\mu_r(\boldsymbol{\theta}^{(i)},\mathbf{d})$ and $m_{r2}(\boldsymbol{\theta},\mathbf{d})$ is to perform the multi-scale multi-physics AM simulation (as discussed in Sec. 2) by fixing the random variables $\boldsymbol{\theta}$ at $\boldsymbol{\theta}^{(i)}$. In that case, the computationally expensive AM simulations need to be performed n_{MCS} times to get an evaluation of the objective function for a given $\mathbf{d}$. This makes physics-based robust design of process parameters computationally intractable. In this paper, the data-driven surrogate models are employed to overcome the computational challenges in the robust design optimization.

3.4. Surrogate Modelling of Temperature Field

In this paper, the Kriging surrogate modeling method [44] is employed to build the surrogate models, since Kriging can effectively capture the nonlinearity of the underlying models and can accommodate the noise in the data. For the sake of explanation, we partition the random variables $\boldsymbol{\theta} = [\eta, \sigma_{gb}, Q, a_{10}, a_{11}, a_{20}, a_{21}, a_{30}, a_{31}]$ into two groups, namely $\boldsymbol{\lambda} = [\eta, a_{10}, a_{11}, a_{20}, a_{21}, a_{30}, a_{31}]$, which are the random variables in the melt pool model and $\boldsymbol{\omega} = [\sigma_{gb}, Q]$, which are the random variables in the microstructure model. In what follows, we provide more details of the surrogate modeling of the AM models.

For given $\mathbf{d}$ and $\boldsymbol{\lambda}$, the thermal response is a high-dimensional field response. This makes the surrogate modeling of developed steady temperature field challenging. To overcome this challenge, a singular value decomposition (SVD)-based Kriging surrogate modeling method presented in Ref. [4] is adopted in this paper for the thermal surrogate modeling.

Define the steady temperature field for given $\mathbf{d}$ and $\boldsymbol{\lambda}$ as $T(\mathbf{d}, \boldsymbol{\lambda}, \mathbf{s})$, where $\mathbf{s} \in \Omega_{xyz}$ stands for all the spatial coordinates of the nodes, we first generate N_t training points for $\mathbf{d}$ and $\boldsymbol{\lambda}$ using Latin Hypercube sampling approach [45]. We then perform thermal simulation for each of the training points and obtain the steady temperature field developed under different conditions, $T(\mathbf{d}^{(i)}, \boldsymbol{\lambda}^{(i)}, \mathbf{s})$ $i = 1, 2, \cdots, N_t$. After that, we approximate the original simulation data matrix using SVD as follows [46]

$$T\left(\mathbf{d}^{(i)}, \boldsymbol{\lambda}^{(i)}, \mathbf{s}\right) \approx \sum_{j=1}^m \gamma_j(i) \boldsymbol{\eta}_j(\mathbf{s}), \forall i = 1, 2, \cdots, N_t \quad (13)$$

where $\gamma_j(i)$, $i = 1, 2, \cdots, N_t$; $j = 1, 2, \cdots, m$; are the responses in the latent space, m is the number of important features used in SVD, and $\boldsymbol{\eta}_j(\mathbf{s}), j = 1, 2, \cdots, m$ are the important features.

Using the training points of $\mathbf{d}$ and $\boldsymbol{\lambda}$ and the corresponding latent space response, we then build surrogate models in the latent space as

$$\gamma_j = \hat{g}_j(\mathbf{d}, \boldsymbol{\lambda}), j = 1, 2, \cdots, m \quad (14)$$

For any new prediction point $\mathbf{d}$ and $\boldsymbol{\lambda}$, we have $\gamma_j(i) \approx \hat{g}_j(\mathbf{d}, \boldsymbol{\lambda}^{(i)})$ and thus the surrogate models are used to predict the steady temperature field response at new prediction point as below

$$\hat{T}\left(\mathbf{d}, \boldsymbol{\lambda}, \mathbf{s}\right) \approx \sum_{j=1}^m \mu_{\hat{g}_j}(\mathbf{d}, \boldsymbol{\lambda}) \boldsymbol{\eta}_j(\mathbf{s}) \quad (15)$$

where $\mu_{\hat{g}_j}(\mathbf{d}, \boldsymbol{\lambda})$ is the mean prediction of the j-th latent space surrogate model and $T(\mathbf{d}, \boldsymbol{\lambda}, \mathbf{s})$ is the predicted temperature field response for given $\mathbf{d}$ and $\boldsymbol{\lambda}$.

3.5. Surrogate Modelling of Microstructure

In the grain growth model, there are three types of variables: (1) the shared design variables (i.e. control factors) between the thermal model and the grain growth model, which are denoted as $\mathbf{d}_m$; (2) the inputs from the thermal model which are the latent space responses γ; and (3) the random variables (i.e. noise factors) ω that belong to only the grain growth model. In order to substitute the computationally expensive grain growth model with cheap data-driven surrogate models, we first generate N_{t2} training points for $\mathbf{d}_m$, γ, and ω.

For the i-th training point, we first transform the latent space response $\gamma_1^{(i)}, \gamma_2^{(i)}, \cdots, \gamma_m^{(i)}$ into the original temperature field response as follows

$$\hat{T}\left(i, \mathbf{s}\right) \approx \sum_{j=1}^m \gamma_j^{(i)} \boldsymbol{\eta}_j(\mathbf{s}), \forall i = 1, 2, \cdots, N_{t2} \quad (16)$$

Where $T(i,\mathbf{s})$ is the temperature field response corresponding to i-th training point.

With the input temperature field $T(i,s)$ and the training point $\mathbf{d}_m{}^{(i)}$ and $\boldsymbol{\omega}^{(i)}$, we then perform grain growth simulation and obtain corresponding microstructure statistical moments $\mu_r{}^{(i)}$ and $m_{r2}{}^{(i)}$. Based on the responses of $\mu_r{}^{(i)}$ and $m_{r2}{}^{(i)}$, $i = 1, 2, \cdots, N_{t2}$, we construct surrogate models for μ_r and m_{r2} as follows

$$\mu_r = \hat{g}_{\mu_r}(\mathbf{d}_m, \boldsymbol{\omega}, \boldsymbol{\gamma}) \tag{17}$$

$$m_{r2} = \hat{g}_{m_{r2}}(\mathbf{d}_m, \boldsymbol{\omega}, \boldsymbol{\gamma}) \tag{18}$$

where $\hat{g}$ means the Kriging surrogate model.

After we are able to efficiently evaluate the objective function for given points, we optimize the control factors or design variables by solving the optimization model given in Eq. (12). In this paper, the efficient global optimization method with noise data [61, 62] is employed to solve the optimization model.

4. RESULTS

4.1. Validation of Physical Model

A sound physical model serves as the basis of correct uncertainty quantification and management. Validation of the current multi-scale model in predicting microstructure development is shown in Fig. 3. It is built just by taking advantage of the 150 grain growth simulations for generating training data. A columnar-to-equiaxed transition boundary (the blue curve in Fig. 3 can be clearly delineated, although data scattering presents, e.g. columnar points occasionally observed in the equiaxed region. (Note that, the data scattering should not be mistaken as prediction errors. It happens due to various uncertain factors considered and thus uncertainty in the final materials microstructure in current simulations.) The columnar/equiaxed delineation in Fig. 3 suggests a small equiaxed region in the space of low power and high velocity. This is consistent with previous experimental observation of finer grains with increasing beam speed [47] and lower beam power [48]. It can be explained by the classical nucleation criteria [28]. That is, in this P-v space, correspondingly low thermal gradient and high solidification rate tend to be developed, thus facilitating grain nucleation and equiaxed grain structure formation as indicated by the classical nucleation criteria. The experimentally observed columnar structures in [27] is also successfully located within the as-predicted columnar region. However, fully equiaxed structure that is widely theoretically predicted is still rarely found in experiments as of now. The current P-v map may provide partial explanation for this, since in practice the EBM machine in automatic mode would adopt certain speed functions [49] (namely specific built-in power-velocity combinations instead of arbitrary user-defined ones) to maintain constant and appropriate melt pool cross sectional area (the largest pool area perpendicular to the travel velocity direction). According to the constant pool area P-v map [50], these specific power-velocity combinations would usually be away from the equiaxed region to guarantee sufficiently large pool area. Therefore, besides direct comparisons with experimental results, the ability to explain the gap between previous theoretical and experimental

findings further justifies the P-v map, thus basically validating the microstructure prediction of the adopted grain growth model.

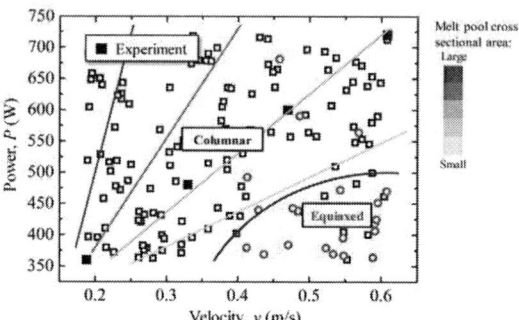

Figure 3. Columnar-to-equiaxed structure transition P-v map based on 150 physical simulations

4.2. Cross-validation of Microstructure Prediction Surrogate model

To train a microstructure prediction surrogate model, we have run 150 grain growth simulations at specific training points. Of them the last 5 microstructure data are set aside for cross-validation; the remainder are training dataset.

Figure 4 shows the cross-validation results of the trained microstructure prediction surrogate model. The comparison shows that the surrogate model successfully captures the grain structure variation (in terms of the mean and second moment of grain length/width ratio) with varying $[\mathbf{d}, \boldsymbol{\theta}]$. In part the acceptable error can be explained by the fact that, the grain evolution is inherently a complex process that includes some probabilistic phenomenon. The structure prediction based on the grain evolution simulation would thus show a little random and stochastic characteristics, while the surrogate model prediction is deterministic.

4.3. Robust Design

Robust design is now readily made with the cheap surrogate models. The design interval are respectively $[T_{pre,L}, T_{pre,U}] = [923K, 1003K]$, $[P_L, P_U] = [360W, 720W]$, $[v_L, v_U] = [0.188 \text{ m/s}, 0.608 \text{ m/s}]$ for the preheating temperature, EB power and scanning speed, based on the accessible manufacturing condition in practice [16, 27]. By solving the objective function Eq. (3), the optimal operating condition is found to be $T_{pre} = 946$ K, $P = 360$ W and $v = 0.42$ m/s. It indicates that equiaxed grain structures could be certainly obtained even with various uncertainty sources, by using a low preheating temperature, low laser power and intermediate scanning speed. The main reason is the introduction of a large number of new grains and highest survival of them under this manufacturing condition, thus greatly interrupting the original columnar grain growth. To be specific, low thermal gradient G that encourages grain nucleation tend to be formed under a low preheating temperature and low laser power. On the other hand, an intermediate scanning speed could result in sufficiently high solidification rate R that allows grain nucleation, at the same time, without causing too much re-melting of new grains in previous layers. A too large speed, in contrary, would usually lead to a small melt pool accompanying with a shallow nucleation layer, thus making equiaxed structure formation vulnerable to uncertainty effects. In conclusion, consistently obtaining equiaxed grain structures during layer-by-layer AM process requires a delicate balance between different process parameters, which not only ensures sufficient grain nucleation, but also guarantees their highest survival from re-melting and competitive growth against existing grains.

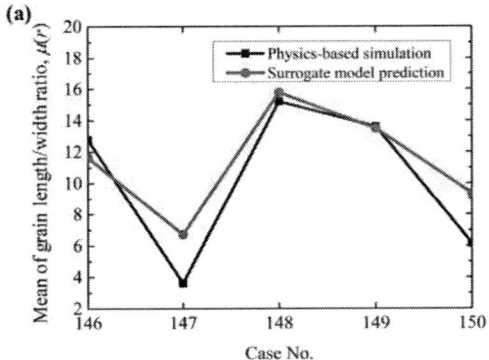

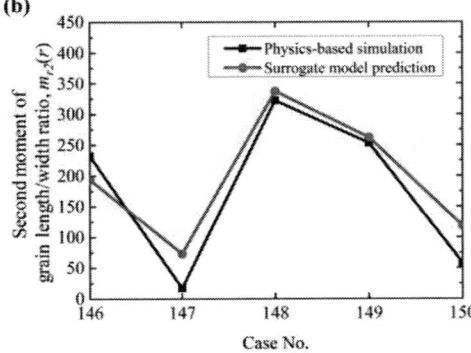

Figure 4. Prediction of the grain length/width ratio distribution as a function of control and noise factors, [**d**, **θ**], by using physics-based AM simulation model and microstructure prediction surrogate model

To verify the robust design result, 20 physical simulations with random noise factors were performed at the robust design point. The first five simulation results are shown in Fig. 5(a). Equiaxed grain structures are indeed constantly produced under the optimal manufacturing condition obtained, although they show some difference due to presence of noise factors or uncertainty sources. For further verification, we have also performed physical simulations at other three test points. Test point 1 is generated by increasing the EB speed a little based on the robust design point, test point 2 is generated by increasing the EB power a little and test point 3 is an arbitrary point. Grain structures developed under those conditions are also shown in Fig. 5(b-d). The grain structure changes greatly under noises

in all three test cases. Meantime, columnar grain structures are usually formed at all test points although small grains can be observed occasionally. The same happens for the other 15 grain structures obtained at each test point. It is thus proved that the robust design point obtained represents the optimal manufacturing condition as compared to probably any of others. The precise obtainment of the robust manufacturing condition, again, demonstrates the capability of the surrogate model in accurately capturing complex relationships between target quantities and process parameters as well as noise factors, besides the cross-validation made earlier.

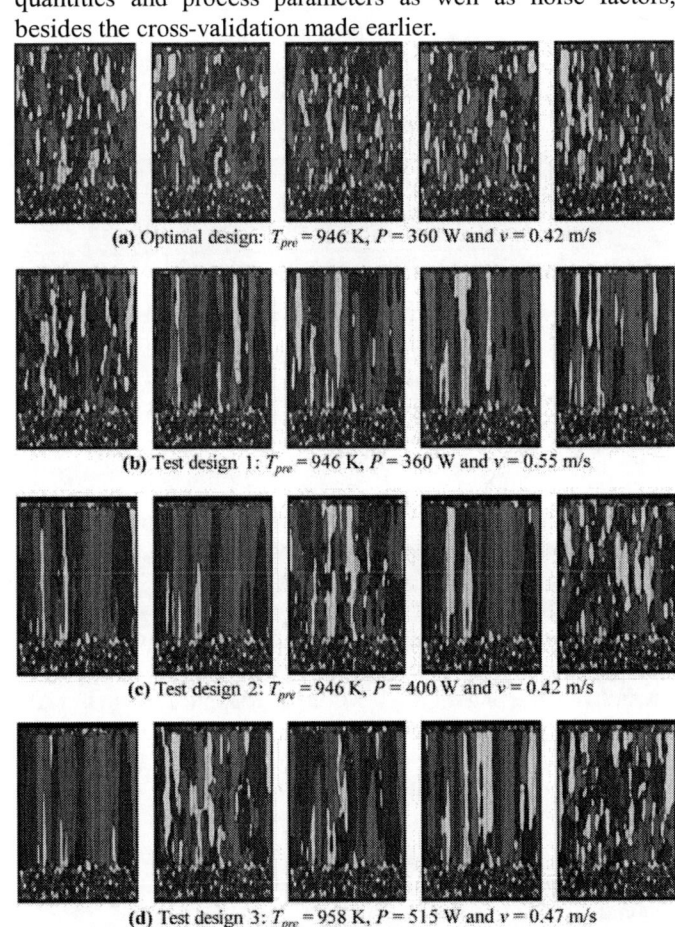

(a) Optimal design: T_{pre} = 946 K, P = 360 W and v = 0.42 m/s

(b) Test design 1: T_{pre} = 946 K, P = 360 W and v = 0.55 m/s

(c) Test design 2: T_{pre} = 946 K, P = 400 W and v = 0.42 m/s

(d) Test design 3: T_{pre} = 958 K, P = 515 W and v = 0.47 m/s

Figure 5. Grain structure developments on the optimal manufacturing condition (robust design point) and test manufacturing conditions under uncertainty

4.4. Sensitivity Analysis

The global sensitivity analysis [51] is also made to reveal the contribution of different uncertainty sources to the variation of materials microstructure. Figure 6 indicates that, the mean of grain length/width ratio is most sensitive to density and grain boundary energy, while the variation of grain length/width ratio shows highest sensitivity to specific heat capacity and grain growth activation energy. Thus, reducing the noise level of density and grain boundary energy could make it more likely to obtain equiaxed grain structures under uncertainty, and reducing the noise level of specific heat capacity and grain growth activation energy would help minimize the grain structure variation among products.

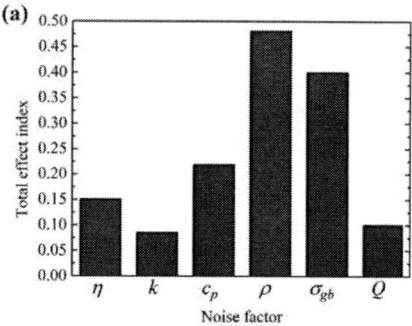

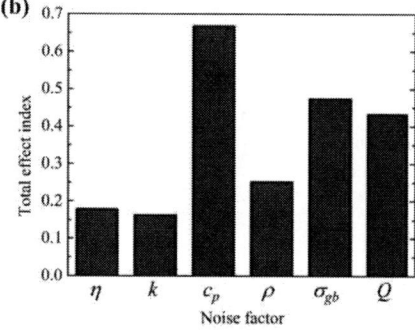

Figure 6. Sensitivity of (a) mean of length/width ratio and (b) variance of length/width ratio to various noise factors.

5. DISCUSSION

In recent years, data-driven approaches have appeared frequently in the engineering field as powerful data analysis and predictive modeling tools. However, their application in the AM study is still rare, if any, with the main function as a simple data fitting tool in terms of low-dimensional inputs and responses [52, 53]. The current study, to authors' best knowledge, presents one of the first efforts to perform multi-level uncertainty quantification and management for metal-based AM process, with the aid of advanced application of the data-driven technique.

A special feature of the current approach is its level-by-level surrogate model building manner, although only the finally obtained microstructure prediction surrogate model is used for uncertainty quantification in this study. This brings about several advantages compared to directly treating all connected levels as a single black-box. Firstly, by level-by-level surrogate model building, validation of corresponding models is able to be made at each level. Hence, incorrect UQ and UM resulting essentially from large prediction errors of either physical or surrogate models at any level can be found easily. Also, benefiting from its modular feature, the individual surrogate model established can be potentially linked to other AM models (either lower or higher levels), thus enabling faster creation of a different multi-level UQ model in future studies. This helps ultimately construct the full network for uncertainty aggregation

in the AM process [10]. For example, the established thermal surrogate model can be connected to lower-level heat source model [31] and/or higher-level lack-of-fusion porosity predictive model [54]. Another advantage offered is that, some noise factors showing little uncertainty propagation could be suspected early from a UQ at lower levels. By eliminating them, the whole UQ procedure would require less physics-based simulations with reduced inputs than a one-step multi-level UQ. This especially alleviates the computational burden for a multi-level UQ that abounds with numerous uncertainty sources from different levels.

Finally, we would like to emphasize that the quantity of interest in this study is materials microstructure. Process optimization is made aiming primarily to improve the microstructure to enhance product quality; other quality metrics like the porosity level have not been accounted for in this research. In this case, the optimal manufacturing condition mainly for obtaining good microstructures may cause other problems, such as insufficient pool overlapping and thus lack-of-fusion porosity. Therefore, this research just selects the materials microstructure as an exemplar target quantity, to bring out the capability of the data-driven technique in facilitating complicate AM process optimizations. Nonetheless, the current approach can be readily extended to, for instance, the process optimization in terms of controlling simultaneously the microstructure and porosity with reasonable modification, like by further coupling the as-mentioned porosity predictive model.

6. CONCLUSION

This paper presents a data-driven AM process optimization approach, backboned by an advanced multi-scale multi-physics AM simulation model and Kriging surrogate model. Based on this approach, multi-level uncertainty quantification and management are successfully performed with respect to the EBM of Ti-6Al-4V. It reveals the highest sensitivity of materials microstructure variation to specific heat capacity and grain growth activation energy. Control factors, here AM process parameters, are optimized through robust design. The optimal combination of process parameters, i.e. low preheating temperature, low laser power and intermediate scanning speed, is suggested to guarantee fabricating products with consistently equiaxed grain structures.

ACKNOWLEDGMENTS

This research was supported by the program of ORAU Ralph E. Powe Junior Faculty Enhancement Award and National Science Foundation under grant CMMI-1662854.

REFERENCES

[1] Brandl, E., Schoberth, A., and Leyens, C., 2012, "Morphology, microstructure, and hardness of titanium (Ti-6Al-4V) blocks deposited by wire-feed additive layer manufacturing (ALM)," Mater. Sci. Eng., A, 532(Supplement C), pp. 295-307.
[2] Donoghue, J., Antonysamy, A., Martina, F., Colegrove, P., Williams, S., and Prangnell, P., 2016, "The effectiveness of

combining rolling deformation with Wire–Arc Additive Manufacture on β-grain refinement and texture modification in Ti–6Al–4V," Mater. Charact., 114, pp. 103-114.

[3] Körner, C., 2016, "Additive manufacturing of metallic components by selective electron beam melting—a review," Int. Mater. Rev., 61(5), pp. 361-377.

[4] Nath, P., Hu, Z., and Mahadevan, S., 2017, "Multi-level uncertainty quantification in additive manufacturing," Solid freeform fabrication, pp. 7-9.

[5] Ma, L., Fong, J., Lane, B., Moylan, S., Filliben, J., Heckert, A., and Levine, L., 2015, "Using design of experiments in finite element modeling to identify critical variables for laser powder bed fusion," International Solid Freeform Fabrication Symposium, Laboratory for Freeform Fabrication and the University of Texas Austin, TX, USA, pp. 219-228.

[6] Chen, W., Allen, J. K., Tsui, K.-L., and Mistree, F., 1996, "A procedure for robust design: minimizing variations caused by noise factors and control factors," J. Mech. Des., 118(4), pp. 478-485.

[7] Chan, S., and Elsheikh, A. H., 2018, "A machine learning approach for efficient uncertainty quantification using multiscale methods," J. Comput. Phys., 354, pp. 493-511.

[8] Zhu, Y., and Zabaras, N., 2018, "Bayesian deep convolutional encoder–decoder networks for surrogate modeling and uncertainty quantification," J. Comput. Phys., 366, pp. 415-447.

[9] Sankararaman, S., Ling, Y., and Mahadevan, S., 2011, "Uncertainty quantification and model validation of fatigue crack growth prediction," Eng. Fract. Mech., 78(7), pp. 1487-1504.

[10] Hu, Z., and Mahadevan, S., 2017, "Uncertainty quantification and management in additive manufacturing: current status, needs, and opportunities," Int. J. Adv. Manuf. Technol., 93(5-8), pp. 2855-2874.

[11] Kamath, C., 2016, "Data mining and statistical inference in selective laser melting," Int. J. Adv. Manuf. Technol., 86(5-8), pp. 1659-1677.

[12] Lopez, F., Witherell, P., and Lane, B., 2016, "Identifying uncertainty in laser powder bed fusion additive manufacturing models," J. Mech. Des., 138(11), p. 114502.

[13] Haines, M., Plotkowski, A., Frederick, C. L., Schwalbach, E. J., and Babu, S. S., 2018, "A sensitivity analysis of the columnar-to-equiaxed transition for Ni-based superalloys in electron beam additive manufacturing," Comput. Mater. Sci., 155, pp. 340-349.

[14] Moser, D., Fish, S., Beaman, J., and Murthy, J., 2014, "Multi-layer computational modeling of selective laser sintering processes," ASME 2014 International Mechanical Engineering Congress and Exposition, American Society of Mechanical Engineers, pp. V02AT02A008-V002AT002A008.

[15] Fallah, V., Amoorezaei, M., Provatas, N., Corbin, S. F., and Khajepour, A., 2012, "Phase-field simulation of solidification morphology in laser powder deposition of Ti–Nb alloys," Acta Mater., 60(4), pp. 1633-1646.

[16] Sahoo, S., and Chou, K., 2016, "Phase-field simulation of microstructure evolution of Ti–6Al–4V in electron beam additive manufacturing process," Addit. Manuf., 9, pp. 14-24.

[17] Acharya, R., Sharon, J. A., and Staroselsky, A., 2017, "Prediction of microstructure in laser powder bed fusion process," Acta Mater., 124, pp. 360-371.

[18] Raghavan, N., Dehoff, R., Pannala, S., Simunovic, S., Kirka, M., Turner, J., Carlson, N., and Babu, S. S., 2016, "Numerical modeling of heat-transfer and the influence of process parameters on tailoring the grain morphology of IN718 in electron beam additive manufacturing," Acta Mater., 112, pp. 303-314.

[19] Gäumann, M., Bezencon, C., Canalis, P., and Kurz, W., 2001, "Single-crystal laser deposition of superalloys: processing–microstructure maps," Acta Mater., 49(6), pp. 1051-1062.

[20] Hunt, J., 1984, "Steady state columnar and equiaxed growth of dendrites and eutectic," Mater. Sci. Eng., 65(1), pp. 75-83.

[21] Gockel, J., Beuth, J., and Taminger, K., 2014, "Integrated control of solidification microstructure and melt pool dimensions in electron beam wire feed additive manufacturing of Ti-6Al-4V," Addit. Manuf., 1, pp. 119-126.

[22] Mani, M., Feng, S., Lane, B., Donmez, A., Moylan, S., and Fesperman, R., 2015, Measurement science needs for real-time control of additive manufacturing powder bed fusion processes, US Department of Commerce, National Institute of Standards and Technology.

[23] Liu, P., Ji, Y., Wang, Z., Qiu, C., Antonysamy, A., Chen, L.-Q., Cui, X., and Chen, L., 2018, "Investigation on evolution mechanisms of site-specific grain structures during metal additive manufacturing," J. Mater. Process. Technol., 257, pp. 191-202.

[24] Liu, P., Cui, X., Deng, J., Li, S., Li, Z., and Chen, L., 2019, "Investigation of thermal responses during metallic additive manufacturing using a "Tri-Prism" finite element method," Int. J. Therm. Sci., 136, pp. 217-229.

[25] Liu, P., Wang, Z., Xiao, Y., Mark, H. F., Cui, X., and Chen, L., 2018, "Insight into the mechanisms of columnar to equiaxed grain transition during metallic additive manufacturing," Addit. Manuf.

[26] Price, S., Lydon, J., Cooper, K., and Chou, K., 2013, "Experimental temperature analysis of powder-based electron beam additive manufacturing," Proceedings of the Solid Freeform Fabrication Symposium, pp. 162-173.

[27] Al-Bermani, S., Blackmore, M., Zhang, W., and Todd, I., 2010, "The origin of microstructural diversity, texture, and mechanical properties in electron beam melted Ti-6Al-4V," Metall. Mater. Trans. A, 41(13), pp. 3422-3434.

[28] Schempp, P., Cross, C., Pittner, A., Oder, G., Neumann, R. S., Rooch, H., Dörfel, I., Österle, W., and Rethmeier, M., 2014, "Solidification of GTA aluminum weld metal: Part 1—Grain morphology dependent upon alloy composition and grain refiner content," Welding journal, 93(2).

[29] Schempp, P., Cross, C., Pittner, A., and Rethmeier, M., 2014, "Solidification of GTA aluminum weld metal: Part 2—thermal conditions and model for columnar-to-equiaxed transition," Welding Journal, 93, pp. 69-77.

[30] Liu, P., Wang, Z., Xiao, Y., Horstemeyer, M. F., Cui, X., and Chen, L., 2019, "Insight into the mechanisms of columnar to equiaxed grain transition during metallic additive manufacturing," Addit. Manuf., 26, pp. 22-29.

[31] Yan, W., Smith, J., Ge, W., Lin, F., and Liu, W. K., 2015, "Multiscale modeling of electron beam and substrate interaction: a new heat source model," Comput. Mech., 56(2), pp. 265-276.

[32] Klassen, A., Bauereiß, A., and Körner, C., 2014, "Modelling of electron beam absorption in complex geometries," J. Phys. D: Appl. Phys., 47(6), p. 065307.

[33] Körner, C., Attar, E., and Heinl, P., 2011, "Mesoscopic simulation of selective beam melting processes," J. Mater. Process. Technol., 211(6), pp. 978-987.

[34] Cheng, B., Price, S., Lydon, J., Cooper, K., and Chou, K., 2014, "On Process Temperature in Powder-Bed Electron Beam Additive Manufacturing: Model Development and Validation," J. Manuf. Sci. Eng., 136(6), p. 061018.

[35] Xiao, Y., Zhan, H., Gu, Y., and Li, Q., 2017, "Modeling heat transfer during friction stir welding using a meshless particle method," Int. J. Heat Mass Transfer, 104, pp. 288-300.

[36] Roth, T. A., and Suppayak, P., 1978, "The Surface and Grain Boundary Free Energies of Pure Titanitunn and the Titanium Alloy Ti-6AI-4 V," Mater. Sci. Eng., 35, pp. 187-196.

[37] Roth, T. A., and Henning, W. D., 1985, "The surface and grain boundary free energies and the self-diffusion coefficient of 5Al · 2.5Sn titanium alloy," Mater. Sci. Eng., 76, pp. 187-194.

[38] Gil, F., and Planell, J., 2000, "Behaviour of normal grain growth kinetics in single phase titanium and titanium alloys," Mater. Sci. Eng., A, 283(1), pp. 17-24.

[39] Ding, R., and Guo, Z. X., 2002, "Microstructural modelling of dynamic recrystallisation using an extended cellular automaton approach," Comput. Mater. Sci., 23(1), pp. 209-218.

[40] Mishra, S., and DebRoy, T., 2004, "Measurements and Monte Carlo simulation of grain growth in the heat-affected zone of Ti–6Al–4V welds," Acta Mater., 52(5), pp. 1183-1192.

[41] Jamshidinia, M., Kong, F., and Kovacevic, R., 2013, "Numerical modeling of heat distribution in the electron beam melting® of Ti-6Al-4V," J. Manuf. Sci. Eng., 135(6), p. 061010.

[42] Martin, J. H., Yahata, B. D., Hundley, J. M., Mayer, J. A., Schaedler, T. A., and Pollock, T. M., 2017, "3D printing of high-strength aluminium alloys," Nature, 549(7672), pp. 365-369.

[43] Donoghue, J., Gholinia, A., Fonseca, J. Q. d., and Prangnell, P., 2015, "In‑Situ High Temperature EBSD Analysis of the Effect of a Deformation Step on the Alpha to Beta Transition in Additive Manufactured Ti‑6Al‑4V," Proceedings of the 13th World Conference on Titanium, Wiley Online Library, pp. 1283-1288.

[44] Hu, Z., and Mahadevan, S., 2016, "Global sensitivity analysis-enhanced surrogate (GSAS) modeling for reliability analysis," Struct. Multidiscip. Optim., 53(3), pp. 501-521.

[45] Helton, J. C., and Davis, F. J., 2003, "Latin hypercube sampling and the propagation of uncertainty in analyses of complex systems," Reliability Engineering & System Safety, 81(1), pp. 23-69.

[46] Hu, Z., Ao, D., and Mahadevan, S., 2017, "Calibration experimental design considering field response and model uncertainty," Comput. Methods Appl. Mech. Eng., 318, pp. 92-119.

[47] Gong, X., Lydon, J., Cooper, K., and Chou, K., 2014, "Beam speed effects on Ti–6Al–4V microstructures in electron beam additive manufacturing," J. Mater. Res., 29(17), pp. 1951-1959.

[48] Parimi, L. L., A, R. G., Clark, D., and Attallah, M. M., 2014, "Microstructural and texture development in direct laser fabricated IN718," Mater. Charact., 89, pp. 102-111.

[49] Narra, S. P., Cunningham, R., Beuth, J., and Rollett, A. D., 2018, "Location specific solidification microstructure control in electron beam melting of Ti-6Al-4V," Addit. Manuf., 19, pp. 160-166.

[50] Gockel, J., and Beuth, J., 2013, "Understanding Ti-6Al-4V microstructure control in additive manufacturing via process maps," Solid Freeform Fabrication Proceedings, Austin, TX, Aug, pp. 12-14.

[51] Hu, Z., and Mahadevan, S., 2018, "Probability Models for Data-Driven Global Sensitivity Analysis," Reliability Engineering & System Safety, https://doi.org/10.1016/j.ress.2018.12.003.

[52] Zhang, W., Mehta, A., Desai, P. S., and III, C. F. H., 2017, "MACHINE LEARNING ENABLED POWDER SPREADING PROCESS MAP FOR METAL ADDITIVE MANUFACTURING (AM)," 2017 Solid Freeform Fabrication Symposium Proceedings.

[53] Tapia, G., Khairallah, S., Matthews, M., King, W. E., and Elwany, A., 2018, "Gaussian process-based surrogate modeling framework for process planning in laser powder-bed fusion additive manufacturing of 316L stainless steel," Int. J. Adv. Manuf. Technol., 94(9-12), pp. 3591-3603.

[54] Tang, M., Pistorius, P. C., and Beuth, J. L., 2017, "Prediction of lack-of-fusion porosity for powder bed fusion," Addit. Manuf.

Proceedings of the ASME 2019
International Design Engineering Technical Conferences
and Computers and Information in Engineering Conference
IDETC/CIE2019
August 18-21, 2019, Anaheim, CA, USA

DETC2019-97607

DIGITAL DESIGN AUTOMATION TO SUPPORT IN-SITU EMBEDDING OF FUNCTIONAL COMPONENTS IN ADDITIVE MANUFACTURING

Manoj Malviya
Mechanical Engineering
The Pennsylvania State University
University Park, PA, 16802
Email: mxm2429@psu.edu

Swapnil Sinha
Mechanical Engineering
The Pennsylvania State University
University Park, PA, 16802
Email: sxs6205@psu.edu

Dr. Nicholas A. Meisel
Engineering Design
The Pennsylvania State University
University Park, PA, 16802
Email: nam20@psu.edu

ABSTRACT

Additive manufacturing (AM) offers access to the entire volume of a printed artifact during the build operation. This makes it possible to embedding foreign components (e.g. sensors, motors, actuators) into AM parts, thus enabling multifunctional products directly from the build tray. However, the process of designing for embedding currently requires extensive designer expertise in AM. Current methods rely on a designer to select an orientation for the embedded component and design a cavity such that the component can be successfully embedded without compromising the print quality of the final part. For irregular geometries, additional design knowledge is required to prepare a shape converter: a secondary piece to ensure a flush deposition surface on top of the embedded component. This research aims to develop a tool to automate these different design decisions for in-situ embedding, thus reducing the need for expert design knowledge. A three-stage process is proposed to 1) find the optimum orientation based on cavity volume and cross-section area, 2) create the necessary cavity geometry to successfully insert the component, and 3) perform a Boolean operation to create the digital design for any requisite shape converter. Performance of the tool is demonstrated with four test cases with varying levels of geometric complexity. These test cases show that the proposed process successfully handles arbitrary embedded geometries, though several limitations are noted for future work.

1 INTRODUCTION AND MOTIVATION

Additive Manufacturing (AM), commonly known as 3D printing, is having a significant impact on the fields of design and manufacturing due to its ability to build products with both extensive geometric and material complexity [1]. User-defined control of material deposition affords unique design opportunities, including in-situ embedding of functional components to create consolidated multifunctional parts. The ability to interrupt the manufacturing process to embed a foreign component into the part allows designers to integrate reliable functional components within the printed structures. This enables the creation of "smart parts" with integrated circuits and sensors directly from the build tray [2]–[5]. Through in-situ embedding, AM can cost-effectively create small batch sizes of multifunctional structures, while also reducing or eliminating the need for post-process assembly [6]. However, as with any other manufacturing process, this application requires a designer to consider a variety of process-related considerations in the design stage to ensure a successful output [7], [8]. For example, decisions on orientation and placement of the embedded component can directly impact the strength and manufacturability of the final multifunctional part. Additionally, the cavities for embedding are often manually designed by a domain expert, making this stage of the process time intensive, especially for embedding components with complex geometry. To assist designers, and thus expedite the process of designing AM parts for multifunctional embedding, this work proposes an automated design methodology. This three-phase method accounts for existing understanding of in-situ embedding design guidelines and uses it to automate the digital design of the requisite printed parts and shape converters. In doing so, the process minimizes the amount of expert domain knowledge required to leverage in-situ embedding in AM.

1.1 Applications of Embedding with AM

Many existing examples of using AM technology to produce multi-functional components center on creating unique designs, where traditional manufacturing techniques could not be used. Won et al [9] attempted fabrication of a robotic hand with careful planning and designing of cylindrical cavities in the design for the cables. Bailey et. al [10] also explored the design process for this technology to create biomimetic mechanisms for embedding complex pneumatic systems and a pressure sensor. They introduced the concept of creating design variations through multiple combinations of component arrangements, made possible through AM.

Actuators and sensors are the most common components embedded into AM parts. Laurentis et. al [11] demonstrated design and manufacturing of non-assembly mechanisms by embedding mechanical components, such as the motor and axles for a set of printed AM gears. Dollar et. al. [12] attempted embedding of cables, magnets, and sensors to create a soft biomimetic robot gripper. The technology was also utilized to create electronic parts with a range of embedded components intended to replace PCBs. This was done by exploring the use of direct writing techniques to create layered electrical interconnects [13]. In-situ embedding has also been explored for creating an electric motor with careful design and process planning for embedding the constituent components: wires, bearings, and actuators [14].

These examples from literature show the wide variety of functional components that researchers are attempting to leverage with in-situ embedding in AM [6]. However, it also showcases the wide variety of geometries that designers must account for in order to ensure successful embedding. As a result, cavity designs in these examples tend to be either 1) oversimplified (i.e., shaped in a primitive form like a cylinder or cuboid, without

conforming to the geometry of the component) or 2) appropriately designed, but done so on an ad-hoc basis by a domain expert. Our research seeks to address the middle gap between these two existing methods, where an automated design approach can appropriately account for design for embedding considerations without requiring significant expert knowledge.

1.2 Design for Embedding Considerations

While not as extensively studied as other applications of design for AM, such as lattice structure design or topology optimization, there is an increasing body of literature dedicated to identifying basic guidelines to drive design decisions in in-situ embedding [15]–[17]. To manufacture multi-functional parts via in-situ embedding, the print process is interrupted once the build operation reaches the top layer of the designed cavity. Any support material in the cavity is removed, the foreign component is placed inside the part, and printing is resumed [18]. The component must be completely encapsulated into the part before resuming the print process to avoid any damage to the print head or the embedded component [19]. Because of this, a component's curvilinear features could demand a need to create a flush surface by inserting a shape converter to cover the cavity before resuming the manufacturing process (Figure 1). The geometry's concavity would be another factor to consider while designing the cavity, as it could challenge the complete insertion of the component.

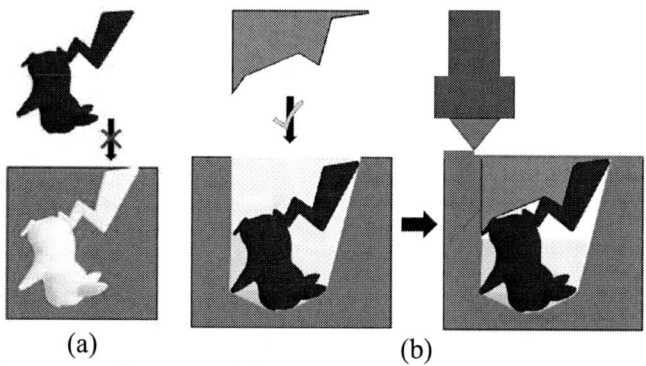

(a)   (b)

Figure 1: Shows (a) failure to insert the complex geometry into the designed cavity at the top layer, and (b) the modified cavity design for feasible geometry insertion, with a shape converter (shown in orange) to create a flush deposition surface when the process is resumed.

The required process interruption, as well as the presence of the cavity, directly affects the material properties of the part [16]. The process interruption required for embedding with these systems cools down the layer, which compromises the cohesion between the resumed and the paused layer [20]. Due to this known weakness at the paused layer, cavity designs could be accordingly designed to minimize the cross-sectional area of the cavity at the paused layer [16]. While these design guidelines provide valuable information for

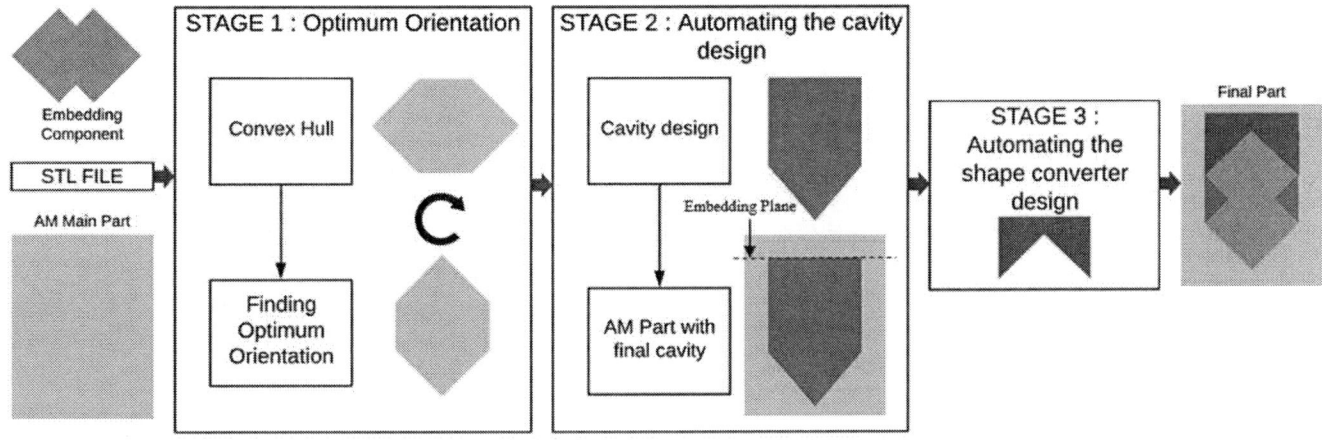

Figure 2: A three-stage methodology tool proposed for automation of digital design for embedding in AM

designers to design for in-situ embedding with material extrusion AM, they currently have to be analytically implemented [16]. Complex curvilinear component geometries pose a challenge for such an analysis.

Currently, the design decisions for in-situ embedding, such as the orientation of the embedded component, the geometry of the embedding cavity, and the shape converter design are made on an ad-hoc basis. This makes in-situ embedding in AM challenging and time-intensive. Therefore, there is a need for an automated design process capable of identifying the ideal component orientation and designing the cavity and shape converter for the part accordingly. This paper presents a functioning framework, informed by design guidelines from prior research [16], [21], to automate this process of design for embedding with AM. The proposed three-stage process includes 1) an embedding orientation selection algorithm that maximizes the cross-sectional area of contact between the paused and resumed layer, 2) a voxel-based approach to automate the cavity design, and 3) a combined shadow-projection/Boolean operation to automate the design of the shape converter.

2 METHODOLOGY

The objective of this research is to automate design for embedding with AM for any arbitrary component geometry through a computational tool. To identify the framework for this tool, the design process was broken into three stages. The cavity design in an AM part for embedding depends on the orientation of the component to be embedded. Therefore, the first stage of the design process is to identify this orientation. Next, the cavity may require modification for successful insertion of the component when the print is paused (as explained in Figure 1). The second stage, therefore, is to modify the cavity design, ensuring the complete insertion of the component. The third stage is to ensure a flush surface before the print is resumed, i.e., to design the shape converter, if required.

The following sub-sections discuss each of the three design stages accompanied by detailed description of the process used to achieve automation; a graphical overview of the overall process is seen in Figure 2. To illustrate each phase of the process, an arbitrarily complex geometry is used throughout this section. The geometric complexity of this object is such that even a domain expert in embedding with AM may encounter difficulty manually identifying the optimal orientation, cavity design, and shape converter design.

2.1 Stage 1: Identifying the Optimum Orientation for Embedding Component

With the understanding of material strength and orientation generated from previous research [16], [21], an embedded component should ideally be oriented so that cross-sectional area of the cavity is minimized at the embedding plane (see Figure 2), which is the cross section of the part at which component is inserted mid-print. Through this consideration, the tensile strength of the part is maximized if loaded perpendicular to the embedding plane. As such, the first stage of the proposed process focuses on obtaining this ideal orientation for any arbitrary component geometry. This first requires special pre-processing of the component geometry to establish the criteria for optimization, followed by the implementation of a suitable optimization technique, elaborated on in the sub-sections below.

2.1.1 Pre-processing the Component Geometry

First, the embedded component's STL file is imported into MATLAB and read by an STL reader [22] for further processing. To facilitate orientation optimization, the cross-section area of the cavity has to be established as a function of orientation. To simplify the cavity design, the component geometry is converted into a convex polyhedron with the least possible vertices. This ensures that concave features of the component geometry, which could restrict insertion of the component in the cavity (see Figure 1) are approximated to a simplified convex form. This also reduces redundant processing of concave features. Several convex-hull algorithms were explored, including the divide and conquer [23], gift wrapping [23], and quick hull [24] approaches. The quick hull algorithm was found to be the most computationally time efficient when compared with the other algorithms [24]. After applying the algorithm to the original component STL file, the resulting mesh, or component hull (as shown in blue in Figure 3), is stored as a separate STL file for further processing.

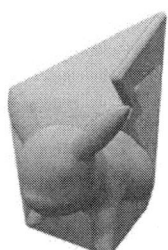

Figure 3: Component hull of the geometry generated through the quick-hull algorithm.

However, a component hull cannot necessarily be used directly toward embedding. If the top layer of the embedding cavity has a cross-sectional area smaller than the largest cross-section of the component hull, then the embedded component will not be able to fit within the cavity. As such, the geometry may need to be further modified for successful component insertion (see Figure 4).

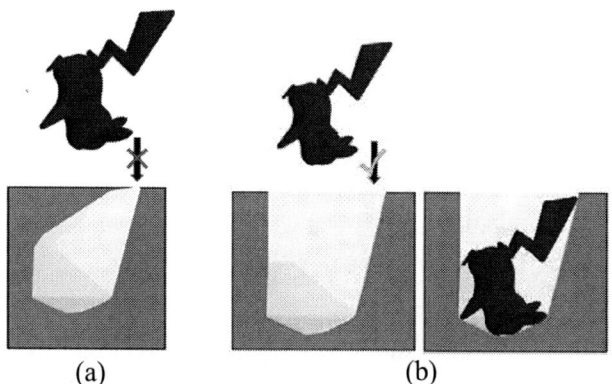

| (a) | (b) |

Figure 4: (a) Component hull as designed does not ensure successful insertion of the component. (b) Modified cavity design ensures the insertion of the component for that orientation.

A shadow projection technique is proposed in this paper to ensure that the cavity is large enough to completely insert the part in a selected orientation. The shadow projection method can be visualized as a shadow on a flat surface by the geometry when parallel light rays are vertically cast over it (see Figure 5).

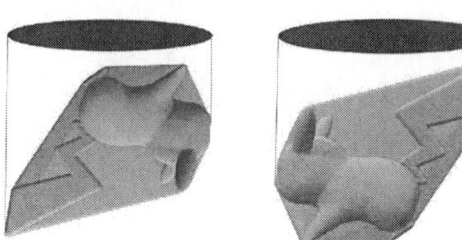

Figure 5: Shows shadow curve or cavity cross section for two different orientations of the case study geometry, to ensure that the geometry can be sucessfully inserted in the respective orienations

Here, all the X-Y coordinates of the vertices on the planes perpendicular to its orientation vector are projected on to the X-Y plane at the top-most plane of that orientation, resulting in a point cloud on the X-Y plane. A 2-dimensional shape that bounds the resulting point cloud is obtained as the cross-section of the cavity at the paused layer, for this orientation (Figure 5). This bound is termed the shadow curve, which is denoted as,

$$S = bound(X, Y) \qquad (2)$$

Where S is the shadow curve, X and Y are all the x and y vertices of the component hull at an arbitrary orientation. The *bound* is the function that obtains the closed curve that bounds all the points. If the cavity was designed with this shadow curve as the cross-section, the component would successfully get inserted in the orientation for that shadow curve.

2.1.2 Orientation Optimization

To obtain the optimum orientation for the component, and thus the embedding cavity, the optimization can be performed by minimizing the cross-section area for the cavity for an orientation x, which is determined by obtaining the area of the closed shadow curve obtained and is denoted as $A_s(x)$. This objective function results in an orientation of the cavity with the maximum strength for the AM part.

The orientation vector of the simplified component hull will act as the design variable in this optimization algorithm. This orientation vector x is defined through an Euler angles transformation [25]. The final orientation axes are visualized (Figure 6) as being achieved in three steps: first, rotating the original x-y-z axes by an angle (α) around z-axis to obtain a new frame (N'-N-z). Next, this frame is rotated by angle (β) around N to obtain a new frame (N'-N-z'). Finally, this frame is rotated by angle (γ) around z' axis to obtain (x'-y'-z').

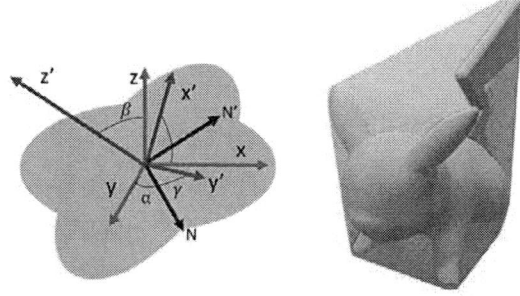

Figure 6: Shows orientation vector for a random orientation x other than the initial orientation.

Relative to the initial orientation, the change in orientation is defined by the angle of rotation from each axis, as shown in Equation (1):

$$x = [\alpha, \beta, \gamma] \qquad (1)$$

It is important to note that, for any arbitrary geometry, two flipped orientations will have the same shadow curve (see Figure 5). This implies that the optimization will give multiple candidate solutions when searching for orientations with minimum $A_s(x)$. While the two flipped orientations need the same cross-section area of the cavity for successful insertion (see Figure 5), the cavity volume, may differ (shown in Figure 7).

Orientations	*Cavity Geometry*

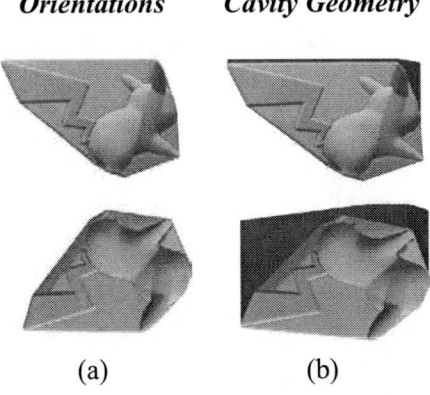

(a)   (b)

Figure 7: Shows two flipped orientations that give same cavity cross section, but different cavity volumes (shown in blue (b)).

Therefore, to address the problem of multiple viable solutions, another objective function is introduced, based on minimizing the total volume of the designed cavity. The volume of the cavity (V_{cv}) is calculated by taking a convex hull of the shadow curve and the component hull of the geometry at the candidate orientation. The shadow curve is fixed as the top surface of the geometry, perpendicular to the candidate orientation (as shown in Figure 8). This produces the overall cavity design for given orientation of the embedded geometry. The volume of the cavity design, V_{cv} is evaluated with divergence theorem [26].

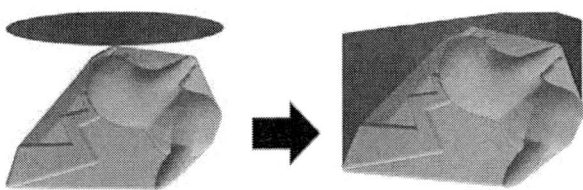

Figure 8: Convex hull of the shadow curve fixed at the top surface of its orientation and component hull, used to determine the volume of the cavity.

With the objectives of minimizing both the cross-sectional area and volume of the embedding cavity, the multi-objective weighted function is formulated in terms of the orientation angles [Equation (3)], as elaborated below:

Objective Function:

$$Min\ f(x) = W_1 A'_s(x) + W_2 V'_c(x) \qquad (3)$$

Where,

$$[0,0,0] \le x = [\alpha, \beta, \gamma] \le [2\pi, 2\pi, 2\pi]$$

$$A'_s(x) = \frac{A_s(x) - A_s(0)}{A_s^{max} - A_s(0)}, \qquad (4)$$

$$V'_c(x) = \frac{V_c(x) - V_c(0)}{V_c^{max} - V_c(0)} \qquad (5)$$

Here, $A_s(x)$ is the cross-section area of the cavity at the paused layer, $V_c(x)$ is the volume of the cavity, x is the orientation vector, W_1 and W_2 are the weights and $f(x)$ is the objective function. W_1 and W_2 defines the contribution of cross section area and volume of cavity in the objective function and is selected as per priority. The cross-sectional area of the cavity at the paused layer and the volume of the cavity are normalized [27] into dimensionless quantities. $V_c(0)$ is the initial volume, and $A_s(0)$ is the initial cross-section area of the cavity at the paused layer, at the initial orientation $x = [0,0,0]$ of the imported component geometry. A_s^{max} and V_c^{max} are the maximum value of the volume of the cavity and cross-section area at the paused layer determined by optimizing each individually.

The cross-sectional area $A_s(x)$, and the volume $V_c(x)$ of the cavity as functions of orientation are too complex to obtain analytically, due to the 3-dimensional nature of the geometry. As such, a stochastic optimization technique is used to identify the optimum orientation for the embedded component. Specifically, a surrogate modelling approach, which acts as an approximation of the objective function and constraints [28], was employed. In this approach, a random subset of orientation angles (x) is initially selected from design space and the objective function (f(x)) is calculated as defined in Equation [3]. In order to determine initial orientations, the Latin hypercube sampling (LHS) method is utilized. LHS method is a popular method for use in a design space with no prior information [29]. Based on this initial data set, a surrogate is constructed to map the values of the design variable, x, to the objective function, f(x). A cubic radial basis function (RBF) method is used in this work to construct the surrogate owing to its relative simplicity and robustness for smaller sample size [30]. A maximum number of 100 evaluations was set as the stopping criteria.

2.2 Stage 2: Automating the Cavity Design

After identifying the optimum orientation for the component through the optimization function demonstrated in Stage 1, the cavity must be translated to a digital design space capable of being sent to an AM system. The cavity design is derived from the processing in Stage 1 (Section 2.1.1, Figure 5), using the shadow projection of the optimum orientation. To create a manipulable cavity in the part, a Boolean operation is performed on a voxelized version of the cavity design and the AM part.

Voxelization enables a geometric object in a continuous surface representation to be converted into a set of discrete cubes, or voxels, that approximate the geometry of the object [31]. Because voxelization captures the entire volume of a geometry, rather than just its surface, it allows for simpler and more robust Boolean operations. While a method to perform Boolean operations with triangulated surfaces exists [32], it was found to be computationally expensive and limited to relatively simple geometries. Instead, a ray intersection method presented by Patil and Ravi [33] is implemented to convert both the AM part's STL and the final cavity mesh into voxels. The unit voxel size set to the minimum printable resolution of the AM system so that the computation time could be reduced [34] without sacrificing fidelity of the final printed structure. On voxelizing, the point cloud data containing facets and vertices information is converted into a 3D logical array that consists of binary data. This pertains to the presence or the absence of the material at each location in 3D space.

The Boolean operation of the two 3D arrays can only be performed for same sized arrays. Since the AM part and the cavity may consist of different numbers of same sized voxels, the *padarray* function in MATLAB [35] is utilized on the cavity design to match their array size (Figure 9). This algorithm creates voxels around the cavity, by deploying dummy voxels in every direction. Additionally, to achieve a successful embedding fit after printing, an added tolerance value is used for the cavity mesh. The

padarray function can be used to provide clearances by scaling the cavity in all direction.

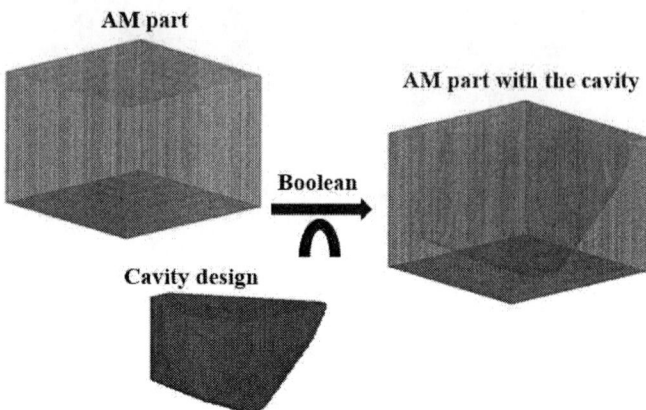

Figure 9: Shows voxelized final AM part with the cavity design, after the Boolean subtraction of the cavity design from the AM part.

2.3 Stage 3: Automation of Shape Converter Design

As discussed in Section 1.2, the primary purpose of the shape converter is to provide the flush surface as a support for the next layer to be deposited (Figure 1(b)).To obtain the shape converter design, a simple Boolean operation can be executed between the voxelized forms of the component hull from Stage 1 and the final cavity design obtained from Stage 2 at the optimum orientation. However, this Boolean operation may result in a discontinuity in the obtained volume due to discrete voxels at the surface (Figure 10 (a)). This could be due to small volume differences between the cavity and the component hull that are approximated by voxelization. This issue of discontinuity in the volume is fixed by adding dummy voxels to the cavity design in all directions through the *padarray* function. This provides a shell thickness to the component hull and results in a continuous volume after the Boolean operation (Figure 10 (b)). The shape converter design is extracted from the shell by trimming the bottom part of the shell at a specified length. The final shape converter design (Figure 10(c)) is thus obtained.

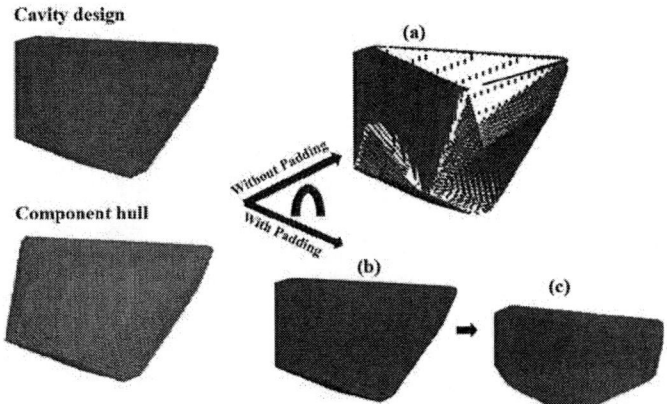

Figure 10: Boolean subtraction of component hull from the cavity design results in (a) discontinuous shape converter design. (b) Padding the cavity design gives a continuous volume of the shape converter, obtained through Boolean operation, which is (c) trimmed to an appropriate height for insertion.

These three stages of the digital design process automatically orient the embedded component's input STL file and designs the cavity and the shape converter geometries according to established design guidelines. Through this automated method, the need for domain expertise is eliminated and the time required to prepare parts for in-situ embedding is drastically reduced. To validate the proposed approach, the following section details several case studies, of varying complexity.

3 CASE STUDY EVALUATION AND DISCUSSION

Since the intent of this tool is to automate the design process for embedding, its performance was evaluated for four different specimens of varying levels of geometric complexity. Orientation was determined and cavity/shape converter designs were created for multiple geometries with different complex features using the proposed process. Computational time was also recorded for each stage of the process. The final designed cavities and shape converters for each of these test cases were printed with a desktop polymer material extrusion system to showcase the feasibility of the proposed design process. The next section shows the results of these evaluations, followed by a discussion on tool performance. All case study geometries were processed through MATLAB 2018b on a PC with a 8 core 3.3 GHz processor and 16GB RAM.

3.1 Adaptability of Proposed Process for Components with Varying Levels of Complexity

To evaluate how the proposed process' ability to account for components of different levels of complexity, four different case study geometries were selected. A simple cuboidal part was used as the base in which the cavity and shape converter were designed. For computation, both weights (W_1 and W_2) in Equation [3] are taken to be 0.5. The design automation tool was used for all four components, and the final part and shape converter design were printed as shown in Figure 10.

All four geometries were successfully processed by the tool, including their cavity and shape converter designs. Success was irrespective of the part's geometric complexity. Importantly, the computed optimal orientations for the snowman, sensor, and motor geometries align with expert intuition. This helps to validate the accuracy of the tool when compared with expert domain knowledge. Unlike these three test cases, the optimum orientation for the squirrel geometry cannot be analytically estimated due to its more complex nature.

Case Study	Initial	Stage1		Stage 2	Stage 3	Final	Printed Parts
		Convex Hull	Orientation	Cavity Design	Shape Converter		
Snowman							
Sensor							
Motor							
Squirrel							

Figure 11: Four case studies conducted for evaluating the performance of the automated tool.

This tests the automated tool for its ability to handle design scenarios where an arbitrary 3D scanned geometry needs to be embedded. As shown in the Figure 10, optimal orientation was obtained, and the cavity and shape converter designs were accordingly generated by the tool

These cases show the effectiveness of the proposed automation tool for finding the optimum orientation, and digitally designing the cavity and the shape converter, despite varied levels of geometric complexities for the embedded components. The next section details the tool's computational performance for each of these four test cases.

3.2 Computational Performance of Proposed Method

A computational time study was conducted to evaluate tool's computational performance, specifically for the most time intensive algorithms in each of the three design stages. These were identified as i) the algorithm to identify the optimum orientation in Stage 1, ii) the shadow projection for the final cavity design in Stage 2, and iii) the Boolean operation for the shape converter design in Stage 3. Observations show that surrogate optimization for finding the optimum orientation took an average of about 63% of the total time of entire process for all test cases. Observations on computation time indicate its high dependence on the features of the test cases. For example, the curvilinear features in the snowman and the squirrel cases lead to a higher overall computation time, possibly due to high numbers of vertices in the component hull geometry for these cases. The sensor and the motor have comparatively lower computation time values in all the stages (See Table 1). However, it is important to note that the time required to prepare the case study geometries for embedding is still significantly lower than the time that would likely be required for even a domain expert to prepare any of the selected geometries for embedding.

Table 1: Computational time for time intensive steps in the design stages.

Test Case	No. of Vertices in Component hull	Orientation Optimization	Shadow Projection	Boolean Operation for Shape Converter
Snowman	13752	582.90s	6.03s	157.02s
Sensor	3042	214.80s	2.84s	104.69s
Motor	1956	157.20s	2.99s	132.70s
Squirrel	5754	244.20s	2.66s	140.42s

4 CONCLUSIONS, LIMITATIONS, AND FUTURE WORK

Design for in-situ embedding of functional components in AM part requires design for manufacturing and part strength considerations. While design guidelines suggest minimizing the cross-sectional area of the cavity for embedding, it is difficult to analytically obtain these cavities for complex geometries. Additionally, currently used digital tools are time intensive for designers to create conformal cavity designs in the part. This paper presents a working framework to automate the process of design for in-situ embedding, by using an optimization algorithm to identify the optimal component orientation and accordingly generate the cavity and the shape converter designs for printing. In case of a predefined orientation for the component, the optimum orientation step can be eliminated, and the cavity and the shape converter design can directly be obtained. Four printed case studies demonstrate the tool's versatility for different component geometries, introducing the possibility to eliminate existing ad-hoc approaches toward design for embedding.

There are some limitations that need to be addressed for complete automation of the design for embedding process. The convex-hull method elaborated in Stage 1 that simplifies the geometry for further processing may create convex hull objects that do not necessarily require geometric changes to be embedded. For example, an inverted "L" shape can be embedded without the need for a convex hull, but the existing algorithm is incapable of recognizing this. Apart from the one-piece shape converter design to create a flush surface, embedded components with concave features may also result in empty spaces in the cavity (shown in Figure 10 final assembly, as dark blue spaces). These spaces may cause damage to embedded components due to available space for free movement. The design tool will be adjusted to expand the shape converter stage into the lateral directions, rather than just to the top embedding surface. Also, this tool only focuses on the orientation of the embedded component, without considering translation of the component in x-y-z space. The *padarray* function can be further augmented to automate the placement for the components, as well as to add tolerances according to the AM system being used. One of the limitations of the surrogate optimization technique used is the stochastic nature of sampling that leads to convergence to one solution, irrespective of multiple available solutions. The optimization process is also computationally expensive, and directly depends on the number of vertices in the component hull of the geometry. Hence, there is a need to explore other optimization techniques.

5 REFERENCES

[1] T. D. Ngo, A. Kashani, G. Imbalzano, K. T. Q. Nguyen, and D. Hui, "Additive manufacturing (3D printing): A review of materials, methods, applications and challenges," *Compos. Part B Eng.*, vol. 143, no. February, pp. 172–196, 2018.

[2] C. Shemelya *et al.*, "Encapsulated copper wire and copper mesh capacitive sensing for 3-D printing applications," *IEEE Sens. J.*, vol. 15, no. 2, pp. 1280–1286, 2015.

[3] B. Stark, B. Stevenson, K. Stow-Parker, and Y. Chen, "Embedded sensors for the health

monitoring of 3D printed unmanned aerial systems," in *2014 International Conference on Unmanned Aircraft Systems, ICUAS*, 2014, pp. 175–180.

[4] S.-Y. Wu, C. Yang, W. Hsu, and L. Lin, "3D-printed microelectronics for integrated circuitry and passive wireless sensors," *Microsystems Nanoeng.*, vol. 1, no. April, pp. 1–9, 2015.

[5] A. Panesar, I. Ashcroft, D. Brackett, R. Wildman, and R. Hague, "Design framework for multifunctional additive manufacturing: Coupled optimization strategy for structures with embedded functional systems," *Addit. Manuf.*, vol. 16, pp. 98–106, 2017.

[6] B.-H. Lu, H.-B. Lan, and H.-Z. Liu, "Additive manufacturing frontier: 3D printing electronics," *Opto-Electronic Adv.*, vol. 1, no. 1, pp. 17000401–17000410, 2018.

[7] N. A. Meisel, "Design for Additive Manufacturing Cosidations for Self-Actuating Compliant Mechanisms Created via Multi-Material PolyJet 3D Printing," *J. Chem. Inf. Model.*, p. 160, 2015.

[8] J. Shang and T. Sueyoshi, "A unified framework for the selection of a Flexible Manufacturing System," *Eur. J. Oper. Res.*, vol. 85, no. 2, pp. 297–315, 1995.

[9] J. Won, K. J. DeLaurentis, and C. Mavroidis, "Fabrication of a Robotic Hand Using Rapid Prototyping," *... ASME Mech. Robot. ...*, pp. 1–7, 2000.

[10] S. A. Bailey, J. G. Cham, M. R. Cutkosky, and R. J. Full, "Biomimetic Robotic Mechanisms via Shape Deposition Manufacturing," in *Robotics Research, The ninth International Symposium*, 2000, pp. 403–410.

[11] K. J. De Laurentis, F. F. Kong, and C. Mavroidis, "PROCEDURE FOR RAPID FABRICATION OF NON-ASSEMBLY MECHANISMS WITH EMBEDDED COMPONENTS," pp. 1–7, 2002.

[12] A. M. Dollar, C. R. Wagner, and R. D. Howe, "Embedded sensors for biomimetic robotics via shape deposition manufacturing," *Proc. First IEEE/RAS-EMBS Int. Conf. Biomed. Robot. Biomechatronics, 2006, BioRob 2006*, vol. 2006, pp. 763–768, 2006.

[13] A. Lopes, M. Navarrete, F. Medina, J. Palmer, E. MacDonald, and R. Wicker, "Expanding Rapid Prototyping for Electronic Systems Integration of Arbitrary Form," *Proc. 2006 Solid Free. Fabr. Symp.*, no. February, pp. 644–655, 2006.

[14] E. Aguilera *et al.*, "3D Printing of Electro Mechanical Systems," in *Proceedings of the 24th Solid Freeform Fabrication Symposium (SFF)*, 2013, pp. 950–961.

[15] S. Sinha and N. A. Meisel, "Influence of Embedding Process on Mechanical Properties of Material Extrusion Parts," *Solid Free. Fabr. 2016*, pp. 847–863, 2016.

[16] S. Sinha and N. A. Meisel, "Quantifying the Effect of Embedded Component Orientation on Flexural Properties in Additively Manufactured Structures," *Solid Free. Fabr. 2018 Proc. 29th Annu. Int.*, pp. 1511–1525, 2018.

[17] N. A. Meisel, A. M. Elliott, and C. B. Williams, "A procedure for creating actuated joints via embedding shape memory alloys in PolyJet 3D printing," *J. Intell. Mater. Syst. Struct.*, vol. 26, no. 12, pp. 1498–1512, 2015.

[18] J. Cham, B. Pruitt, M. R. Cutkosky, M. Binnard, L. E. Weiss, and G. Neplotnik, "Layered manufacturing with embedded components: process planning considerations," in *Proceedings of DETC99: 1999 ASME Design Engineering Technical Conference, DETC/DFM-8910*, 1999, pp. 1–9.

[19] A. Kataria and D. W. Rosen, "Building around inserts: Methods for fabricating complex devices in stereolithography," *Rapid Prototyp. J.*, vol. 7, no. 5, pp. 253–261, 2001.

[20] B. B. Shahriar, C. France, N. Valerie, C. Arthur, and G. Christian, "Toward improvement of the properties of parts manufactured by FFF (fused filament fabrication) through understanding the influence of temperature and rheological behaviour on the coalescence phenomenon," *AIP Conf. Proc.*, vol. 1896, no. April, 2017.

[21] S. Sinha and N. A. Meisel, "Influence of process interruption on mechanical properties of material extrusion parts," *Rapid Prototyp. J.*, vol. 24, no. 5, pp. 821–827, 2018.

[22] E. Johnson, "STL FIle Reader," *MathWorks*, 2011.
.

[23] T. H. Cormen, C. E. Leiserson, R. L. Rivest, and C. Stein, *Introduction to Algorithms*, Second. MIT Press and McGraw-Hill, 2001.

[24] C. B. Barber, D. P. Dobkin, and H. Huhdanpaa, "The quickhull algorithm for convex hulls," *ACM Trans. Math. Softw.*, vol. 22, no. 4, pp. 469–483, 1996.

[25] Weisstein, "Euler Angles," *MathWorld--A Wolfram Resource.* .

[26] M. R. Spiegel, S. Lipschutz, and D. Spellman, *Vector Analysis. Schaum's Outlines*, 2nd ed. McGraw Hill, 2009.

[27] R. T. Marler and J. S. Arora, "Survey of multi-objective optimization methods for engineering," vol. 395, pp. 369–395, 2004.

[28] E. Ulu, E. Korkmaz, K. Yay, O. Burak Ozdoganlar, and L. Burak Kara, "Enhancing the Structural Performance of Additively Manufactured Objects Through Build Orientation Optimization," *J. Mech. Des.*, vol. 137, no. 11, p. 111410, 2015.

[29] A. M. D. Mckay, R. J. Beckman, W. J. Conover, M.

D. Mckay, and R. J. Beckman, "A Comparison of Three Methods for Selecting Values of Input Variables in the Analysis of Output from a Computer Code," *Technometrics*, vol. 21, no. 2, pp. 239–245, 1979.

[30] R. Jin, W. Chen, and T. W. Simpson, "Comparative studies of metamodelling techniques under multiple modelling criteria," no. 1998, pp. 1–13, 2001.

[31] A. Kaufman and S. Brook, "Volume graphics," *Computer (Long. Beach. Calif).*, vol. 26, pp. 51–64, 1993.

[32] G. Mei and J. C. Tipper, "Simple and Robust Boolean Operations for Triangulated Surfaces," pp. 1–18, 2013.

[33] S. Patil and M. T. Student, "Voxel-based Representation , Display and Thickness Analysis of Intricate Shapes," no. December, pp. 7–10, 2005.

[34] B. H. Jared *et al.*, "Additive manufacturing: Toward holistic design," *Scr. Mater.*, vol. 135, pp. 141–147, 2017.

[35] Mathworks, "Image Processing Toolbox." Mathworks, 2006.

**Proceedings of the ASME 2019
International Design Engineering Technical Conferences
and Computers and Information in Engineering Conference
IDETC/CIE2019
August 18-21, 2019, Anaheim, CA, USA**

DETC2019-97649

Optimization of Parts Consolidation for Minimum Production Costs and Time Using Additive Manufacturing

Zhenguo Nie[†], Sangjin Jung[†], Levent Burak Kara[†], Kate S. Whitefoot[*†‡]

[†]Mechanical Engineering, Carnegie Mellon University
[‡]Engineering and Public Policy, Carnegie Mellon University
Pittsburgh, PA, USA

ABSTRACT

This research presents a method of evaluating and optimizing the consolidation of parts in an assembly using metal additive manufacturing (MAM). The method generates candidates for consolidation, filters them for feasibility and structural redundancy, finds the optimal build layout of the parts, and optimizes which parts to consolidate using a genetic algorithm. Optimal results are presented for both minimal production time and minimal production costs, respectively. The production time and cost model considers each step of the manufacturing process, including MAM build, post-processing steps such as support-structure removal, and assembly. It accounts for costs affected by parts consolidation, including machine costs, material, scrap, energy consumption, and labor requirements. We find that developing a closed-loop filter that excludes consolidation candidates with structural redundancy dramatically reduces the number of candidates to consider, thereby significantly reducing convergence time. Results show that, when increasing the number of parts that are consolidated, the production cost and time at first decrease due to reduced assembly steps, and then increase due to additional support structures needed to uphold the larger, consolidated parts. We present a rationale and evidence justifying that this is an inherent tradeoff of parts consolidation that generalizes to most types of assemblies. Subsystems that can be oriented with very little support structures, or have low material costs or fast deposition rates can have an optimum at full consolidation; otherwise, the optimum is likely to be less than 100%. The presented method offers a promising pathway to minimize production time and cost by consolidating parts using MAM. In our test-bed results on an aircraft fairing produced with powder-bed electron-beam melting, the solution for minimizing time is to consolidate 48 components into three discrete parts, which leads to a 33% reduction in unit production time. The solution for minimizing production costs is to consolidate the

components into five discrete parts, leading to a 28% reduction in unit costs.

1. INTRODUCTION

Parts consolidation is a design change in which multiple components that were formerly discrete and assembled together are fabricated as a single part. Through parts consolidation, it is possible to reduce weight and size, minimize assembly operations, improve performance, and prolong service life [1]. Recent research shows that parts consolidation (referred to as consolidation hereafter) has a great potential to improve product or system performance, reduce weight and material usage, and reduce costs. Multiple demonstrations of consolidation in the industry have realized substantial reductions of production or lifecycle costs, weight reductions of up to 60%, and improved reliability [2].

Currently, it is difficult for researchers and manufacturers to identify promising opportunities to redesign products for consolidation using additive manufacturing (AM). Redesign for consolidation is done on an ad-hoc basis without systematically characterizing the effects of consolidating particular parts on assembly operations, production costs and time, or other manufacturer objectives. Complicating matters, determining which parts to consolidate is a combinatorial problem that explodes to large numbers of possible candidates even for assemblies with relatively few parts.

This research develops the first method that optimizes which parts to consolidate in an assembly using AM. Given a user-provided assembly design, the method seeks to minimize costs or time across the full production process consisting of AM setup and build; finishing steps, including support structure removal; and assembly (if needed). Production costs are estimated using a process-based cost model that considers machine, material, and energy inputs; labor; and rejected parts. The method includes six stages to find the optimally consolidated design: generating candidates for consolidation using a connectivity matrix, filtering the candidates based on

---

[*] Corresponding author: kwhitefoot@cmu.edu

feasibility and structural redundancy, optimizing the orientation and layout of parts during build, determining the AM build process parameters, estimating the production costs and time for the design candidates, and finding the optimal design. The optimal design can be obtained for assemblies with a small number of parts by iterating over all candidates. For assemblies with many parts, we develop a genetic algorithm that encodes the part-interfaces that are consolidated in the candidate designs to find a solution with low-costs and production time, respectively.

We demonstrate the methodology on a test-bed assembly selected in collaboration with a company in the aircraft industry. The assembly is a titanium fairing that is produced by electron beam melting (EBM). Results indicate that the solution for minimizing production time is to consolidate 48 components into three parts, which leads to a 33% reduction in production time. The solution for minimizing production costs is to consolidate the components into five parts, leading to a 28% reduction in production costs.

These results illustrate an important tradeoff between the number of consolidated parts and the support structures that are needed during build, which increase production costs and time. For many types of assemblies, it is easier to orient each individual component to reduce support structures than it is to orient consolidated parts because the parts are now larger and have more complex geometry. Consequently, as the number of consolidated parts increases, the total production cost (or time) at first decreases due to the elimination of assembly steps, and then increases due to increased cost (or time) associated with building and removing support structures. Because of these tradeoffs, it is not always optimal to consolidate the entire assembly even when it is feasible to do so.

We provide evidence justifying that this is an inherent tradeoff between consolidation and support structures that apply to most types of assemblies. If the geometry of the assembly is such that it can be oriented with minimal support structures when consolidated, or the material cost is low (e.g., aluminum rather than titanium), or deposition rates are high (e.g., wire-fed direct energy deposition rather than powder-bed fusion), the optimal number of consolidated parts can be 100%. In other cases, the optimal degree of consolidation is less than 100%.

2. LITERATURE REVIEW

Benefits of consolidation. Consolidation can create several advantages in product performance and production, including simplified or eliminated assembly operations, reduced part weight and size, and improved structural performance. The reduction of assembly requirements has tremendous implications, not just for the actual assembly of the components and the consequent cost savings that can be gained, but also from the potential to maximize a design of a product for the purpose in mind and to not have to compromise the design for assembly reasons [3]. Part consolidation in AM has been demonstrated in multiple research case studies [2-3] and industries, including aerospace [4], automotive [5], and energy [6]. Yang et al. [2] studied the part consolidation optimization on a triple clamp. The optimized results showed that the part count reduced from 19

to 7 with less weight by 20% and demonstrated better performance. Schmelzle et al. [6] found that consolidation could reduce a hydraulic manifold's weight by 60% and height by 53% while improving performance and minimizing leak points. General Electric consolidated 230 parts in a compressor bladed disk of a turbine engine to one single part, leading to substantially lower lifecycle costs, 5-10% lower weight, and improved reliability and simplified maintenance [7]. Türk et al. [8] developed a new aircraft instrument panel additively manufactured using Selective Laser Sintering (SLS). They consolidated parts and redesigned the aircraft instrument panel for AM. Compared to the previous design, the number of parts and total weight were reduced by 50% and 41% respectively.

Constraints and tradeoffs with consolidation. Although consolidation has many benefits, it also involves tradeoffs and constraints that can create disadvantages. Consolidation increases coupling among functional requirements, and processing parameters [9]. It also can reduce access for assembly or maintenance [10, 11]. Moreover, it can increase the inputs (e.g., material, labor, or equipment) needed in manufacturing such that production costs actually increase compared to the original (non-consolidated) design [12]. These tradeoffs highlight a need to systematically characterize the effects of consolidation on production costs, time, and other manufacturer objectives and determine optimally consolidated designs.

Parts consolidation optimization. Prior literature has not developed optimization methods for the parts that are chosen for consolidation using AM that consider tradeoffs of associated with the consolidation. In order to optimize AM parts consolidation for reduced production costs or time, the following factors must be considered: how to identify candidate parts for consolidation, optimize the consolidated parts' build orientation, optimize the layout of consolidated and non-consolidated parts into batches, and estimate and minimize the total production costs or time with respect to the consolidated design and AM operational parameters. The remaining literature review focuses on related research dealing with these constituent factors, with an emphasis on approaches applied to AM. TABLE 1 summarizes the factors addressed by this body of literature and the unique contributions of our proposed method.

Identifying candidates for consolidation. Several design guidelines have been proposed to identify feasible candidates for consolidation [13, 14]. For example, Boothroyd et al. [13] proposed some heuristic rules to find potential candidates for consolidation (e.g., during the operation of the product, does the part move relative to all other parts already assembled?). Yang et al. [14] summarized seven feasibility rules for AM-enabled consolidation (e.g., assembly access, material availability), and multiple algorithms based on these rules have been developed that seek feasible consolidation candidates. These algorithms assess the feasibility of different possible combinations of consolidated parts in an assembly and search for the maximum number of parts that can be consolidated that are feasible.

While the literature discussed above has proposed rules for filtering consolidation candidates for feasibility, prior work has not considered filtering candidates for structural redundancy. Depending on the geometry of parts and their interfaces in an assembly, consolidating certain interfaces while leaving others unconsolidated is nonsensical from a structural design standpoint. This is an important consideration for parts consolidation optimization because the number of candidates is combinatorial and can explode to large numbers even for assemblies with relatively few parts. Therefore, efficient filtering of the candidates is necessary to reduce the candidates to consider during optimization.

Build orientation optimization. When consolidating parts, the build orientation of the newly consolidated components must be determined. In AM, build orientation is a crucial process parameter, which affects the surface finish, dimensional accuracy, volumetric error, part strength, production time and cost, and support structures—which are used to uphold internal cavities and overhanging features of a part during build [15, 16]. One key consideration in determining build orientation is to minimize the support structures that are necessary, which directly increase build time, material costs, and the cost and time of post-fabrication steps [17]. The cost effect of build orientation is mainly due to the determination of the best build orientation [18]. Support structure minimization by optimizing build orientation has been an extensively researched area in the AM community [19-23].

Layout configuration. Consolidation also affects the layout configuration of parts that are possible during build. To determine the optimal layout configuration, different types of objective functions have been defined in the literature: (1) fill the build envelope as much as possible [24-30]; (2) minimize build height [27, 28, 30-32]; (3) minimize the the volume of support structures [30-32]. (4) minimize surface roughness [30, 32-34]; (5) minimize staircase error [31, 35]; (6) minimize build time [33-35]. However, optimizing the layout configuration for costs remains an open area of research.

Production cost and time estimation. The integration of AM processes into a production environment requires a cost-model that allows estimation of the production cost per part [17, 36-43]. Rickenbacher et al. [38] developed an integrated cost model, including all pre- and post-processing steps linked to SLM, to optimize build jobs and to manufacture SLM parts more economically by pooling parts from different projects. Ulu et al. [39] proposed a production cost minimization approach for metal AM (MAM) that concurrently optimizes the part structure and process variables, including beam power and velocity. Johnson and Kirchain [37] determined the production costs of parts consolidation in an automotive instrument panel using die-cast magnesium. However, this existing research has not examined the influence of parts consolidation using AM on total production costs or time.

TABLE 1: SUMMARY OF AM CONSOLIDATION RESEARCH.

References	A	B	C	D	E
[14, 44]	×				
[15-23]		×			
Our article	×	×	×	×	×

Annotation:
A: Identify candidates for consolidation
B: Optimize build orientation

C: Optimize layout for build time and/or costs
D: Determine the effects of consolidation on total production costs and/or time
E: Optimize the consolidation of parts

3. METHDOLOGY
As shown in FIGURE 1, the consolidation optimization method consists of six stages: generation of consolidation candidates, selection of consolidation candidates by filters, the configuration of build orientation and layout, determination of processing parameters of MAM, time-based and cost-based modeling of production, and optimization.

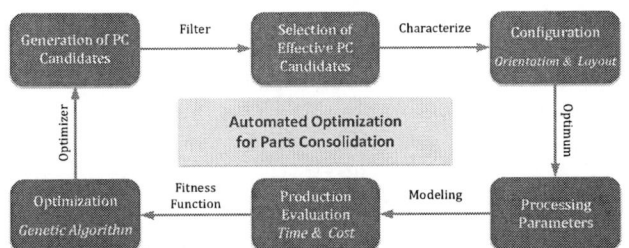

FIGURE 1: SCHEMATIC DIAGRAM OF THE PRODUCTION EVALUATION METHOD.

3.1 Generation of consolidation candidates
The generation of consolidation candidates for a redesign of a given subsystem begins with identifying the components in the original design and the interfaces between them. FIGURE 2 illustrates an example representation of components and interfaces in a network structure of a subsystem consisting of 10 components and 17 interfaces. FIGURE 3 shows an example redesign of the subsystem with consolidation. For the purposes of this paper, we define the *components* as the original discrete parts in the subsystem design, and the *consolidated parts* as the redesigned parts that are made up of one or more components and produced as a monolithic part.

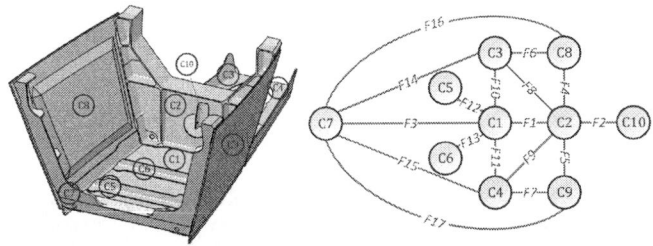

FIGURE 2: A NETWORK STRUCTURE IDENTIFYING DISCRETE COMPONENTS AND THEIR INTERFACES IN AN EXAMPLE SUBSYSTEM DESIGN (1,043×736×692 MM).

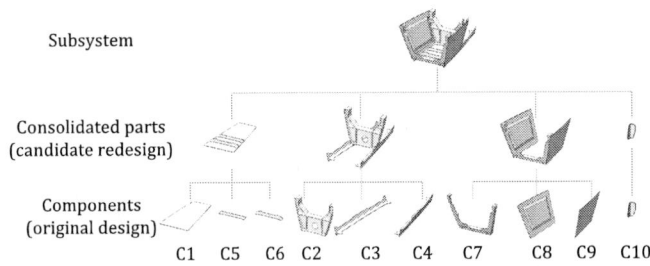

FIGURE 3: TREE STRUCTURE OF PARTS CONSOLIDATION. TEN ORIGINAL COMPONENTS ARE CONSOLIDATED INTO FOUR PARTS THAT ARE ASSEMBLED INTO A SUBSYSTEM.

To depict the topological relationship among all components, the connectivity matrix (symmetric adjacency matrix) of the subsystem is shown in FIGURE 4. The matrix is $C \times C$ where C is the number of components in the original design. Each cell represents a physical connection (interface) between two components. Component pairs that share an interface are in green; and white otherwise. The diagonal cells are in gray and have no meaning (there are no interfaces between a component and itself). Each interface has a binary state to demarcate consolidation: separation (0) or consolidation (1). If the interface has relative motion or material variance, the state is set as a single state: separation. The number of consolidation candidates is 2^F in total, where F is the total number of interfaces with binary states in the original part.

	C1	C2	C3	C4	C5	C6	C7	C8	C9	C10
C1		F1	F10	F11	F12	F13	F3			
C2	F1		F8	F9				F4	F5	F2
C3	F10	F8					F14	F6		
C4	F11	F9					F15		F7	
C5	F12									
C6	F13									
C7	F3		F14	F15				F16	F17	
C8		F4	F6				F16			
C9		F5		F7			F17			
C10		F2								

FIGURE 4: CONNECTIVITY MATRIX OF THE COMPONENTS. GREEN CELLS REPRESENT INTERFACES THAT COULD BE CONSOLIDATED. WHITE CELLS REPRESENT PAIRS OF UNCONNECTED COMPONENTS. GREY CELLS HAVE NO MEANING.

3.2 Structural redundancy and closed-loop filter

As seen in FIGURE 2, the topological structure of the example subsystem includes several strings of three or more components that are all connected together. We call these *ring structures*. For instance, C3-C7-C8 and C1-C4-C7 are two examples of ring structures with three components each, and C1-C3-C8-C2-C0-C4 is a complex ring structure of six components. These ring structures complicate the selection of candidates for consolidation because they can lead to redesigns that do not make sense in practice. To illustrate, consider a subsystem shown in FIGURE 5 where three components are connected end-to-end across three interfaces that are rigidly joined. This part has $2^3 = 8$ consolidation candidates, which are shown in TABLE 2. However, candidates #4, #6 and #7, each has strictly one "0" in their descriptors meaning that all components would be produced monolithically together but with one of the interfaces left for assembly. If the interfaces are rigidly assembled, and there are no performance criteria (e.g., compliance of the interface) requiring an open interface during assembly into the subsystem or during use, a fully consolidated ring structure will yield strictly better performance than one consolidated with a single separated interface. We define such a

consolidated ring-structure, which has one and only separated interface, as *structural redundancy*.

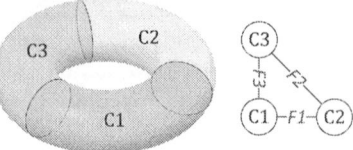

FIGURE 5: SCHEMATIC DIAGRAM OF THE STRUCTURE REDUNDANCY

TABLE 2: CONSOLIDATION CANDIDATES OF A RING STRUCTURE SHOWING STRUCTURAL REDUNDANCY

No.	F1	F2	F3	Redundancy?
1	0	0	0	No
2	0	0	1	No
3	0	1	0	No
4	0	1	1	Yes
5	1	0	0	No
6	1	0	1	Yes
7	1	1	0	Yes
8	1	1	1	No

To reduce the computational complexity of the consolidation optimization, we develop a filter, which we call a *closed-loop filter*, to remove all structurally redundant candidates from the selection. If structural redundancy occurs in a complex ring structure (with more than one loop), the one separated interface must belong to at least one single ring structure (which has one single loop). This means that filtering based on single ring structures alone will simultaneously filter for structural redundancy in complex ring structures. Therefore, an inspection of all single rings is sufficient to confirm whether a candidate is structurally redundant.

We find that the closed-loop filter that we develop dramatically reduces the number of consolidation candidates that need to be considered. Using the example subsystem shown in Fig 2, after the closed-loop filter is applied, the number of candidates decreases significantly from $2^{17} = 131,072$ to 4,920. This means nearly 96.5% of candidates are structurally redundant, which greatly reduces the computational burden imposed by the original search space. In addition, the median of consolidated interfaces reduces from 8 (before filtering) to 5 (after filtering).

3.3 Optimal build orientation to minimize support structure

Build orientation has a great influence on the volume of the support structures required during the build, which accounts for a large proportion of the production cost and time. FIGURE 6 illustrates the volume below a part that requires support structures (represented by green arrows) and how this volume changes with the orientation of the part. The optimal build orientation is defined as the direction in which the support structure is minimized:

$$\vec{v}_{opt} = \underset{\vec{v}}{\arg\min}\, V_{support}(\vec{v}) \qquad (1)$$

where $\vec{v}_{opt}$ is the optimal build orientation, $V_{support}$ is the volume of the space taken up by support structures, and $\vec{v}$ is an arbitrary space vector.

Here we propose a simple voxelization-based method to compute the volume of the support structure $V_{support}$ and obtain the optimal build orientation $\vec{v}_{opt}$ for each part. The voxelization-based method includes four steps: rotation, voxelization, summation, and optimization. The empty domain beneath the geometry is termed as the shadow volume (V_{shadow}). Here we define the support compactness, λ, which is the volume fraction between the support structure volume to the shadow volume, as presented in Equation (2). The support compactness is generally determined by experienced design engineers depending on the material, geometry, and print modality.

$$V_{support} = \lambda \cdot V_{shadow} \qquad (2)$$

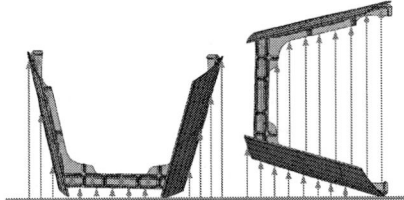

FIGURE 6: BUILD ORIENTATION INFLUENCES THE SUPPORT STRUCTURE. GREEN ARROWS REPRESENT SUPPORT STRUCTURES, WHICH LOCATE IN THE SPACE BENEATH THE GEOMETRY.

(a) Rotation

To minimize support structures, the 3D rotation of a part is parameterized using the three Euler angles as described in [45]. A basic rotation is a rotation around one of the axes in a Cartesian coordinate system. The following three basic rotation matrices rotate vectors (point coordinates) by an angle θ around x-, y-, or z-axis, in three dimensions, using the right-hand rules.

$$R_x(\theta) = \begin{bmatrix} 1 & 0 & 0 \\ 0 & \cos(\theta) & -\sin(\theta) \\ 0 & \sin(\theta) & \cos(\theta) \end{bmatrix} \qquad (3)$$

$$R_y(\theta) = \begin{bmatrix} \cos(\theta) & 0 & \sin(\theta) \\ 0 & 1 & 0 \\ -\sin(\theta) & 0 & \cos(\theta) \end{bmatrix} \qquad (4)$$

$$R_z(\theta) = \begin{bmatrix} \cos(\theta) & -\sin(\theta) & 0 \\ \sin(\theta) & \cos(\theta) & 0 \\ 0 & 0 & 1 \end{bmatrix} \qquad (5)$$

In our research, all the geometric models are written in STL files. In a standard STL file, the geometry is represented by three parts: vertex, facet, and facet normal. For a certain rotation from an arbitrary direction (θ_z, θ_y) to the build orientation, any one of the vertex coordinates and facet normal vector in the STL file will be transformed for update by Equation (6).

$$\begin{bmatrix} x \\ y \\ z \end{bmatrix} = R_y(-\theta_y) \times R_z(-\theta_z) \times \begin{bmatrix} x_0 \\ y_0 \\ z_0 \end{bmatrix} \qquad (6)$$

(b) Voxelization

The rotated part is placed into a minimum bounding cuboid, and then voxelized with a unit length of a. Suppose the cuboid is voxelized into $N_x \times N_y \times N_z$ voxels, so each voxel can be represented by a unique array (i, j, k), where $i = 1, 2, \cdots, N_x$, $j = 1, 2, \cdots, N_y$, and $k = 1, 2, \cdots, N_z$.

As shown in FIGURE 7, each voxel is colored in one of three colors: green (shadow volume), blue (part material), and yellow (empty space).

The positive direction of the Z-axis is defined as up. For a voxel pillar $(i, j, :)$ there are four coloring rules:

(1) If (i, j, k) is in the part or on the surface, then (i, j, k) is part material (blue).

(2) If (i, j, k) is outside the part, and the last one (i, j, N_z) is part material (blue), then (i, j, k) is shadow volume (green).

(3) If (i, j, k) is outside the part, and the last one (i, j, N_z) is not part material (blue), and if the pillar $(i, j, :)$ has at least one voxel that is part material (blue), then the top blue voxel in this pillar is defined as (i, j, K_b), and if $k > K_b$, then (i, j, k) is empty space above the part (yellow), else if $k < K_b$, (i, j, k) is shadow volume (green).

(4) If the whole pillar $(i, j, :)$ has no voxel in the part, then all voxels $(i, j, :)$ are empty space (yellow).

From the figure, it is easy to observe that the part voxels (blue) form define the shadow volume (green) by the space beneath or between part material.

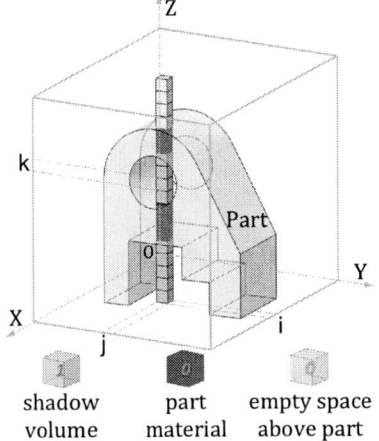

FIGURE 7: SCHEMATIC DIAGRAM OF VOXELIZATION AND STATE FUNCTION DEFINITION. $\Gamma_{ijk} = 1$ FOR GREEN VOXELS, AND $\Gamma_{ijk} = 0$ FOR BLUE AND YELLOW VOXELS.

(c) Summation

We define a state function to identify the shadow volume by:

$$\Gamma_{ijk} = \begin{cases} 0, & \text{part material (blue) or empty space (yellow)} \\ 1, & \text{shadow volume (green)} \end{cases} \qquad (7)$$

The total quantity of voxels in the shadow volume can be written as the summation of Γ_{ijk}, following the expressed in Equation (8). The shadow volume can then be calculated by Equation (9). FIGURE 8 shows the histogram map of the shadow volume for a rotated part. The height at any point (i, j) is $\sum_{k=1}^{N_z} \Gamma_{ijk}$.

$$M = \sum_{i=1}^{N_x} \sum_{j=1}^{N_y} \sum_{k=1}^{N_z} \Gamma_{ijk} \qquad (8)$$

$$V_{shadow} = M a^3 \qquad (9)$$

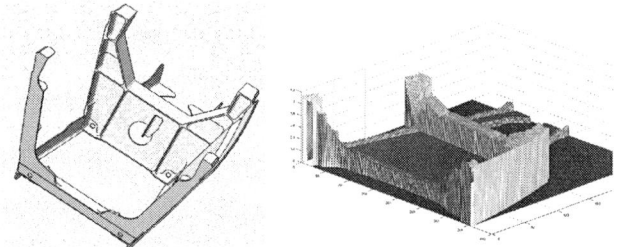

FIGURE 8: A ROTATED PART $\left(\text{DEPICTED AT } \theta_z = \pi/4, \theta_y = \pi/4\right)$ WITH A HISTOGRAM OF THE SHADOW VOLUME AT EACH VALUE OF (θ_z, θ_y).

(d) Orientation optimization

For each part, we compute the shadow volume along the spatial direction (θ_z, θ_y), where $\theta_z \in [0, 2\pi)$ and $\theta_y \in [0, \pi)$ with a step size of $\pi/12$, for a total of 288 directions. The optimal build orientation of the part is the direction with a minimum volume of the support structure. Note that, rotation around the x-axis does not change the shadow volume and so only rotation about the z- and the y-axis is needed to estimate the support structures required for a given part design. (Rotation about the x-axis is considered during the layout of parts into an enclosed volume as discussed in the next section.)

Subsystems with different geometries will vary in the minimum volume of support structures that can be achieved with the optimal orientation. We define a metric that can be used to describe the geometric complexity of a subsystem that contributes to the production costs and time associated with support structures. We call this measure the Shadow Volume Ratio (SVR), which is defined as the ratio between the minimum shadow volume in the optimal build orientation to the total volume of the subsystem.

$$SVR = \frac{(V_{shadow})_{min}}{V_{part}} \qquad (10)$$

where V_{part} is the volume of the part material in the whole subsystem, and $(V_{shadow})_{min}$ is the minimum shadow volume in the optimal build orientation.

3.4 Layout configuration by the bottom-left placement approach

In this work, we employ a coarse voxelization method to represent parts for layout configuration. In voxelization methods, the unit size of voxels influences the computation time of part generation. The coarse voxelization method is more computationally efficient than other methods using fine voxels, and it can guarantee better packing results than using bounding boxes.

We use the bottom-left (BL) placement approach to optimize the packing of parts into the enclosed build envelope. The target of the layout is to pack all the pieces into the bounding box without overlapping to minimize the length required. The advantages of BL are its speed and simplicity when compared with more sophisticated methods that yield better solutions. As the optimal build orientation of each part is determined in Section 3.3, the variables in layout configuration are z-rotation degree and (x, y)-position of each part.

For the purposes of this paper, we use 2-D packing into the build envelope where all parts are placed on the build plate (no stacking of parts). Extension of the methods into 3-D is straight-forward and was considered; however, because the extra support structures that are required for stacking parts contribute significantly to production time and costs, all 3-D packing solutions were dominated by 2-D packing solutions. This is discussed in detail in section 1 of the supporting material (Supplemental Material Section A), available by request.

3.5 Processing parameters of MAM

MAM mainly includes steps as follows: setup, deposition (i.e., build) with the AM machine, support structure removal, post-processing steps, and, if needed, assembly (e.g., riveting, bolting, and welding).

In this research, we consider two different MAM modalities: open platform systems and enclosed volume systems (FIGURE 9). Powder-bed fusion (PBF) using EBM is a classic enclosed volume system, using an electron beam to melt metal powder [46]. Directed energy deposition (DED) using EBM is a classic open platform system [47, 48].

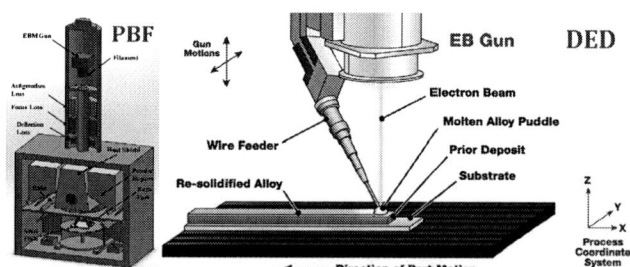

FIGURE 9: ENCLOSED VOLUME SYSTEM WITH PBF EBM (LEFT; REPRINTED WITH PERMISSION FROM ARCAM [49]) AND OPEN PLATFORM SYSTEM WITH DED EBM (RIGHT; REPRINTED WITH PERMISSION FROM SCIAKY [50]).

The process variables of MAM machines not only influence the production time and cost but also determine the process characteristics. The analytical description of the process-solidification map can be derived from the Rosenthal equation [51, 52]. Assuming semi-circular molten pool are, the depth (d) and width (w) of the molten pool can be computed as [39]:

$$d = \frac{w}{2} = \sqrt{\frac{2(a_1 \alpha P + a_2)}{v}} \qquad (11)$$

where a_1 and a_2 are constants that relate to the material, α is the absorption ratio, P is the heat source (i.e., laser or electron beam) power, and v is the travel speed of the heat source.

3.6 Production evaluation

In this section, we describe our model of the production process from two perspectives: the time-based evaluation model and the cost-based evaluation model.

3.6.1 Time-based evaluation model

The total manufacturing time includes four stages: build time on the AM machine, setup time between batches, support removal time and assembly time for joining the parts into the subsystem.

$$T_{total} = T_{build} + T_{setup} + T_{support\_removal} + T_{assembly} \qquad (12)$$

(a) Build time

According to the analysis in Section 3.5, given the heat source power and travel speed, we can obtain the size of the melt pool. The layer thickness is slightly less than the melt pool depth and can be presented by:

$$l = \beta d \qquad (13)$$

where l is the layer thickness, β is a constant ratio. The material deposition rate (MDR) ω can be obtained by:

$$\omega = \frac{\frac{w}{2} l}{1/v} = \frac{2(a_1 \alpha P + a_2) \beta v}{v} \qquad (14)$$

The build time is determined by summing the build time required for the part itself and its required support structures. Therefore, the build time of the whole part is the ratio of total volume and MDR:

$$T_{build} = T_{build-part} + T_{build-support}$$
$$= \sum \frac{(V_{subpart} + V_{support})}{\omega} \qquad (15)$$

where $V_{support}$ is the minimum support volume of the part.

(b) Setup time

As each MAM machine has different requirements for build setup, the time associated with the setup is specified for the particular machine. Total setup time is given by:

$$T_{setup} = t_{setup} \times N_{batch} \qquad (16)$$

where t_{setup} is the unit setup time, and N_{batch} is the average number of batches per product, which can be determined by layout configuration.

(c) Support removal time

Support structure in MAM are usually removed by CNC milling, followed by finishing by grinding or polishing the surface of the part formerly attached to the supports, which depends on the product's surface finish requirements. Therefore, the support removal time includes these two parts:

$$T_{sr} = \sum \left(\frac{V_{support}}{MRR} + \frac{A_{support}}{a_{surface}} \right) \qquad (17)$$

where $V_{support}$ is the support volume of each part, MRR is the estimated material removal rate (mm^3/s) of CNC milling, $A_{support}$ is the area of surface finishing, and $a_{surface}$ is the surface finishing efficiency (mm^2/s).

(d) Assembly time

Assembly includes removing support structure from the part, machining the interfaces, and joining. Equation (18) below specify the assembly time for a subsystem that is assembled through riveting. The first term represents support removal time; the second, machining time of assembly surface; and the third, riveting time:

$$T_{assembly} = T_{sr} + T_{machining} + T_{riveting} \qquad (18)$$

$$T_{machining} = \frac{2}{a_{machining}} \sum_{i=1}^{\mathcal{F}} S_i \qquad (19)$$

$$N_{rivet} = \delta_{rivet} \sum_{i=1}^{F} S_i \qquad (20)$$

$$T_{riveting} = N_{rivet} \times t_{riveting} \qquad (21)$$

where $\mathcal{F}$ is the number of interface pairs after consolidation, S_i is the area of the ith interfaces between parts, $a_{machining}$ is the machining efficiency (mm^2/s) on machining the interfaces, δ_{rivet} is the surface distribution density of rivets, N_{rivet} is the total quantity of rivets, and $t_{riveting}$ is the time consumption of single riveting.

3.6.2 Cost-based evaluation model

The production cost estimates include the input costs (e.g., material, labor, and equipment) associated with MAM setup and build, post-processing steps, and assembly. The part build cost consists of part material cost ($C_{material}$), support material cost ($C_{support}$), machine and maintenance cost ($C_{machine}$), scrap material cost—e.g., powder lost during recovery and recycling (C_{scrap}), and energy consumption cost (C_{energy}).

$$C_{total} = \left(\sum C_{part} \right) + C_{setup} + C_{assembly} \qquad (22)$$

$$C_{part} = C_{material} + C_{support} + C_{machine} + C_{scrap} + C_{energy} \qquad (23)$$

For the production cost model, we follow Ulu et al.[36, 39] in accounting for the cost of the build, setup, material for the part, energy use, lost material during recovery and recycling (scrap), and machine costs. Details are described in Supplemental Material Section B. We add to this model the cost of building and removing support structures as well as the cost of assembly steps as described below.

(a) Support structure cost

The support structure cost contains material cost (C_{sm}) and support removal cost (C_{sr}). The material used for the support structure is usually the same as the part material. Support removal cost (C_{sr}) is given as the product of labor price and the support removal time.

$$C_{support} = C_{sm} + C_{sr} \qquad (24)$$

$$C_{sm} = p_s \rho_s V_{support} \qquad (25)$$

$$C_{sr} = p_{labor} \times T_{sr} \qquad (26)$$

where p_s is the support material price (\$/kg), ρ_m is the support material density (kg/m^3), p_{labor} is labor price (\$/h).

(b) Assembly cost

Assembly cost includes machining cost and riveting cost. The riveting cost contains material (rivet) cost and labor cost. All items are given by:

$$C_{assembly} = C_{machining} + C_{riveting} \qquad (27)$$

$$C_{machining} = p_{machining} \times T_{machining} \qquad (28)$$

$$C_{riveting} = p_{rivet} \times N_{rivet} + p_{labor} \times T_{riveting} \qquad (29)$$

3.7 Optimization by Genetic Algorithm

For subsystems with relatively few parts, the optimal consolidation candidates can be obtained by complete enumeration over all possible candidates. However, as the number of components increases, the candidates increase combinatorially. For example, a subsystem with 20 components has over a million candidates and a subsystem with 30 components has over a trillion. To solve the problem for subsystems with many components, we develop a genetic algorithm (GA) that determines which components to consolidate to reduce production time or costs. FIGURE 10 displays the flow-chart of the algorithm. The input is the connectivity matrix of a subsystem with initial components and interfaces. The closed-loop filter is employed to remove the redundant candidates for each iteration. Production time or cost is used as the fitness function, which depends on which interfaces are consolidated and the optimal orientation and layout of the parts during the build.

In the GA, we encode each consolidation candidate with a chromosome of length equal to the number of interfaces, F, that indicates the interfaces that are consolidated or left separated. According to the analysis in Section 3.1, any consolidation candidate has a one-to-one relationship to an F-digit binary number. Therefore, a binary encoding chromosome is deployed to hold the information on interfaces between connected components, as shown in FIGURE 11. An initial population of consolidation candidates is generated by randomly generating chromosomes, and allowing crossover between candidates and mutation of individual genes to generate subsequent generations.

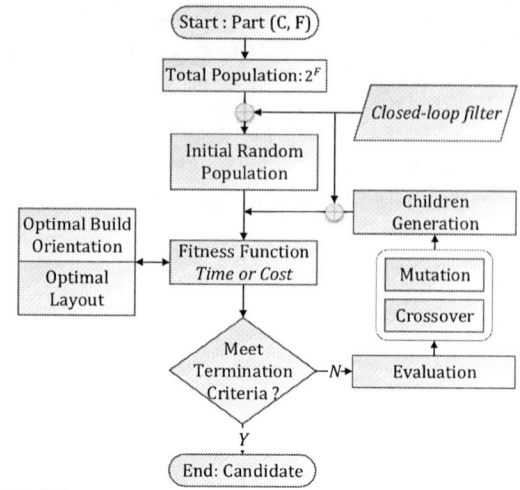

FIGURE 10: FLOWCHART OF GA FOR CONSOLIDATION OPTIMIZATION

Information of PC (Interfaces between neighbor components)

FIGURE 11: A BINARY ENCODED CHROMOSOME HOLDING INFORMATION ON INTERFACES IS DEPLOYED IN GA FOR CONSOLIDATION OPTIMIZATION.

4. OPTIMIZATION RESULTS

We demonstrate the developed method on a test-bed subsystem selected in collaboration with a company in the aircraft industry. The subsystem is an aft fairing, which is composed of 48 parts and 94 interfaces assembled by rivet joints. We simulate the fairing as being produced from Ti-6-Al-4V through both PBF EBM and DED EBM processes, respectively. All the input parameters for these simulations are expressed in detail in **Supplemental Material Section C**. The parameters are derived from literature detailing operating conditions and cost estimates from industrial MAM production facilities [53], as well as equipment suppliers [50]. It is important to note that the cost model does not include costs associated with overhead, management, production plant construction, transportation, or inventory.

Using a subset of the fairing that has a smaller number of components, we first test the GA by comparing results to the global optimum determined by complete enumeration. We then use the GA on the full fairing to find optimal components to consolidate for minimum production costs and time.

4.1 Global optima

The small-scale part (a subsystem of the fairing) has 2^{17} consolidation candidates in total. Using the closed-loop filter, these candidates are reduced to only 4,920. In this case, the optimum for both PBF and DED is the same. The solution is to consolidate nine of the original 10 components (shown in FIGURE 12): C1-C8 and C10 are consolidated into one part, and C9 is produced discretely. The unit production time for PBF is reduced from 77 hours for the original unconsolidated design to 58 hours. For DED, the unit production time is reduced from 54 hours to 43 hours because of the larger input power and travel speed available in this case. For both cases, more than 90% of the production time is required for the build, with 5% or less for assembly.

FIGURE 12: TIME-BASED AND COST-BASED OPTIMALLY CONSOLIDATED PARTS FOR PBF (THE BUILD ORIENTATION IS PERPENDICULAR AND OUTWARD FROM THE PAPER).

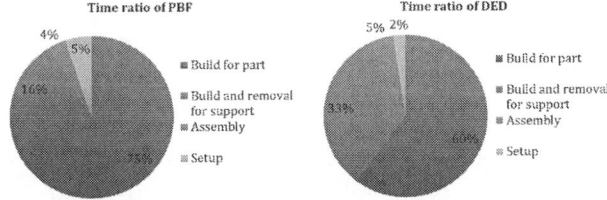

FIGURE 13: PRODUCTION TIME CONTRIBUTORS OF THE OPTIMALLY CONSOLIDATED FAIRING FOR PBF AND DED

The optimal consolidation candidates for minimum production cost are shown in FIGURE 14 for PBF and FIGURE 15 for DED. The solution for DED has three parts: (C2-C10, C1-C3-C4-C5-C6, and C7-C8-C9) with the unit production costs of $9.486K. The PBF solution reduces costs from $20K to $16K per unit. In both cases, over 80% of costs are from part material, manufacturing (machine costs, maintenance, and labor), and support structure material and removal (FIGURE 16). For PBF, the optimally consolidated design has four parts: (C2-C10, C1-C4-C5-C6, C7-C8-C9, and C3). This is because titanium powder, which is used in the PBF case, is significantly more expensive than titanium wire used in the DED case. The increased material price moves the optimum to consolidate fewer parts so that less support structure material is needed to uphold these smaller discrete parts during the build.

FIGURE 14: COST-BASED OPTIMAL CONSOLIDATION CANDIDATE WITH FOUR PARTS FOR PBF (THE BUILD ORIENTATION IS PERPENDICULAR AND OUTWARD FROM THE PAPER).

FIGURE 15: COST-BASED OPTIMAL CONSOLIDATION CANDIDATE WITH TWO PARTS FOR DED (THE BUILD ORIENTATION IS PERPENDICULAR AND OUTWARD FROM THE PAPER).

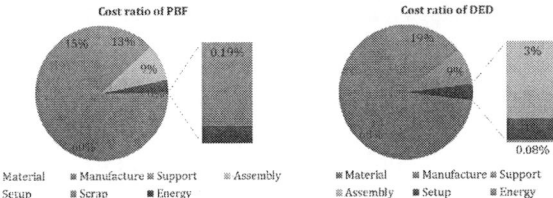

FIGURE 16: PRODUCTION COST CONTRIBUTORS OF THE OPTIMALLY CONSOLIDATED FAIRING FOR PBF AND DED

4.2 Parts consolidation optimization on the small-scale part

Here we test the performance of the developed GA on optimizing the small-scale part (as shown in FIGURE 2) for optimal consolidation. FIGURE 17 displays the estimated PBF unit production cost of each feasible consolidation candidate after filtering, which is computed through complete enumeration. The GA solution is determined with the crossover percentage set to 80%, the mutation percentage set to 30%, and the mutation rate set to 2%. Ten candidates are randomly selected for the initial population, and the algorithm is run until the fifth generation. Five tests repeating this process are conducted to compare solutions to the global optimum. As shown in Table 3, the GA results converge to within 3% of the global optimum in all of these tests.

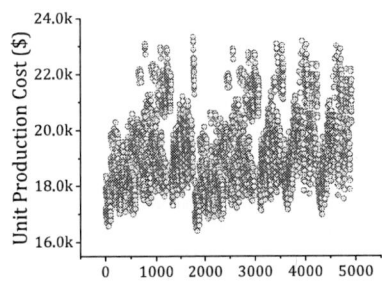

FIGURE 17: COST VALUES OF ALL EFFECTIVE CANDIDATES OF THE SUBSYSTEM BY PBF EBM.

TABLE 3: GROUND TRUTH GLOBAL OPTIMUM AND GA RESULTS.

Items	Target	Test 1	Test 2	Test 3	Test 4	Test 5
Sequence	1838	1808	1783	80	1838	45
Number of parts	4	7	8	6	4	7
Cost (k$)	16.22	16.38	16.56	16.73	16.22	16.68

We also use the algorithm to optimize the consolidation of the full fairing system that consists of 48 components and 94 interfaces, yielding a total of 2^{94} ($\approx 4.95 \times 10^{27}$) total consolidation candidates. Convergence results show that both the time and cost curves flatten at about the 80th generation. Results are described in detail in Supplemental Material Section D.

5. Key tradeoff within the test-bed problems

An important tradeoff was observed in the test-bed problems between the number of parts consolidated, and production costs and time. In this section, we characterize these tradeoffs to understand the main determinants of the optimally consolidated design.

We create a metric to characterize the extent of consolidation in a subsystem and relate the metric to production time and costs. The degree of consolidation (DOC) is defined as the ratio of the number of consolidated interfaces, F_c, in a redesigned system to the number of original interfaces, F:

$$\text{DOC} = \frac{F_c}{F} \qquad (30)$$

The DOC is in the interval [0,1]. F_c ranges from zero to F, where a value of zero represents the original (unconsolidated) subsystem design and $F_c = F$ represents a fully consolidated subsystem that is produced as one monolithic part.

The quantitative tradeoff between the DOC and production cost and time in the fairing subsystem are shown in FIGURE 18. The figure shows each consolidation candidate (in white) as well as the Pareto frontier that minimizes time and cost, respectively, for a particular value of the DOC (in red). As the DOC increases, both time and cost decrease at first and then increase. This illustrates that the optimal DOC is an interior solution.

In FIGURE 19, we plot the constituent factors influencing the tradeoff between the DOC and unit production time and costs. As seen in the figure, the fundamental tradeoff is between reducing assembly (and to a lesser extent setup) steps and increasing support structures. As the fraction of consolidated interfaces increases, the number of assembly and setup steps decreases, reducing the associated production time and costs. However, more support structures are needed to uphold the larger consolidated parts, increasing the build time to construct the support structures and the costs associated with support material and build time. This creates an interior solution of the DOC between 0% and 100%.

We also find that the optimal DOC decreases with increasing support compactness. A detailed characterization of these results is available in the Supplemental Material Section E.

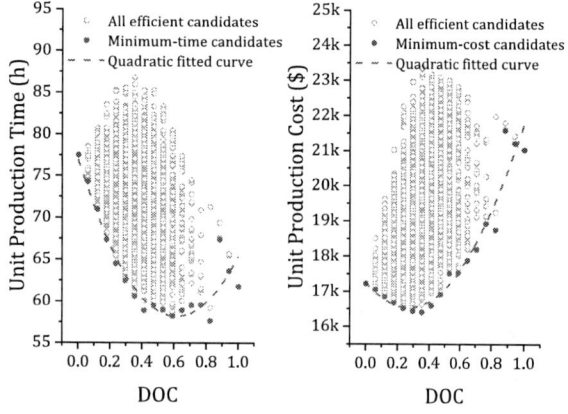

FIGURE 18: UNIT PRODUCTION TIME AND COST VARIATIONS WITH DOC (SUPPORT COMPACTNESS = 0.1, PBF EBM).

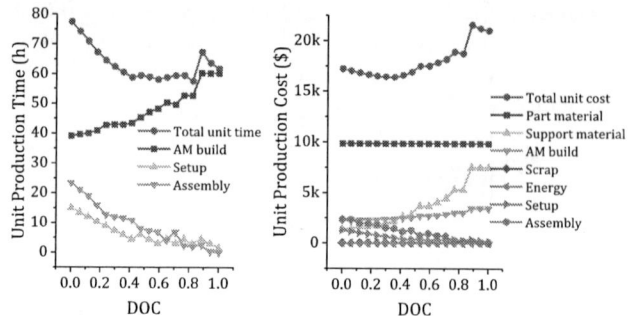

FIGURE 19: MINIMUM TIME AND COST VARIATIONS WITH DOC (SUPPORT COMPACTNESS = 0.1, PBF EBM).

6. Generalization of the identified tradeoffs

In this section, we provide a logical rationale and supporting evidence justifying that the tradeoffs we observe between the optimal degree of consolidation, support structures, and assembly steps are generalizable to many different types of subsystems. We discuss how the magnitude of the tradeoffs depends on the shadow volume ratio of the subsystem, material price and print modality, and the type of joining process used in assembly, and how these factors affect the optimal degree of consolidation. We then find the optimal DOC in a variety of different subsystems to verify that the presented rationale can explain differences in the optima for different geometries, material, print modalities, and assembly requirements.

As we found in the test-bed subsystem (summarized in FIGURE 18 and FIGURE 19), the optimal DOC depends on the tradeoffs between the support structures that are needed to uphold the consolidated parts and the time or cost to assemble the parts. These tradeoffs appear to be inherent to the consolidation of many different types of subsystems using MAM. Part consolidation decreases the number of discrete parts in the subsystem, reducing the time and costs of assembly and build setup. However, as more components are consolidated, they can no longer be individually oriented to minimize support structures; their orientations are now constrained together. As a result, the optimal build orientation for the consolidated part usually requires more support structures than the optimum for the unconsolidated components. Therefore, with the increase in the degree of consolidation, the cost (or time) associated with building and removing the support structures generally increases. The one exception is if the optimal orientation of the consolidated subsystem requires the same amount of support structures as the sum of the support structures needed for the unconsolidated components. For example, this would be the case if the consolidated subsystem could be oriented to have zero support structures, or if the optimal orientation of the consolidated subsystem was the same as the optimal orientation of the discrete components.

FIGURE 20 illustrates how these tradeoffs generalize to different subsystems with different shadow volume ratios, material prices and deposition rates, and assembly requirements.

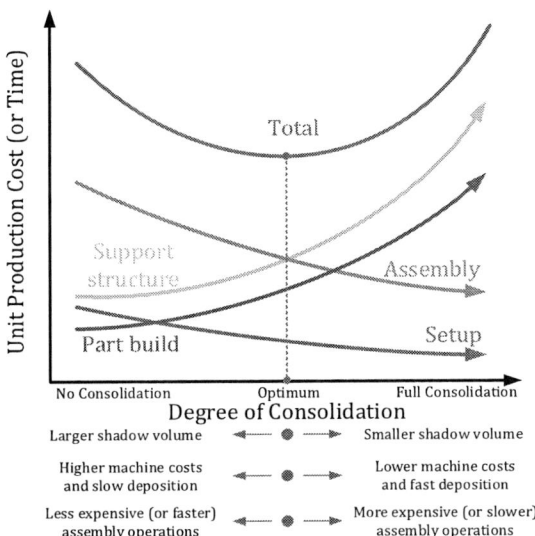

FIGURE 20: KEY TRADEOFF ON THE DEGREE OF CONSOLIDATION

As the figure shows, the optimal DOC for minimizing production cost (or time) moves toward no consolidation if the optimal orientation of the subsystem has a larger shadow volume, and toward full consolidation as the shadow volume decreases. The optimum also depends on material cost and deposition rates as well as the cost (or time) of the required assembly operations. If the material cost is relatively low (e.g., using Aluminum or Stainless Steel instead of Titanium, or wire instead of powder), or deposition rates are faster, the optimum will move toward more consolidation. If the cost (or time) of the assembly operations is relatively lower (e.g., welding instead of riveting), the optimum will move toward less consolidation.

To verify the application of this generalized framework of the key tradeoffs influencing the optimal degree of consolidation, we apply our optimization approach to three very different products in addition to the test-bed subsystem. Specifically, we optimize a puzzle plane, a toy chair, and a heart valve in addition to the aircraft fairing. These products were selected because of their variation in structural complexity in terms of the shadow volume ratio. We examine four different scenarios for each of these products: producing them with Ti6Al4V and Al-6061 material, and using PBF or DED. The plane, toy chair, and heart valve are all assembled with welding whereas the puzzle plane, toy chair, and heart valve are assumed to be assembled with welding whereas the aircraft fairing is riveted together. The size and SVR of each product are shown just below their CAD diagrams in TABLE 4. In the following sections, we describe how the shadow volume ratio, material and print modality, and size influence the optimal DOC results.

(a) Shadow volume effect
As is shown in TABLE 4 the products with a higher SVR, such as the fairing subsystem and the puzzle plane, have an interior optimal DOC. By contrast, the toy chair—which has a smaller SVR—has a local interior minimum but its global minimum is full consolidation. Products with a very small shadow volume, such as the heart valve, have monotonically decreased production time and costs with respect to DOC. The optimal candidate, in this case, is full consolidation. In general, products with a higher SVR *ceterus paribus* will have smaller optimal DOCs for minimum production cost and time.

(b) Material and print modality effects
The aluminum and titanium alloys have different thermophysical properties and material prices. According to Equations (14)-(15), the build time ratio of the two materials can be given by Equation (31). It can be seen that the build time of the aluminum alloy is nearly equal to that of the titanium alloy. However, aluminum has a cheaper material price and is faster to rivet (or weld) than titanium, so unit production cost and time using the aluminum alloy is less than the titanium alloy.

$$\frac{(T_{build})_{Al}}{(T_{build})_{Ti}} = \frac{(a_1 \alpha P + a_2)_{Ti}}{(a_1 \alpha P + a_2)_{Al}} \approx 1.006 \qquad (31)$$

As expected, TABLE 4 shows that when using the titanium alloy, the slope of the production cost and time curves with respect to the DOC decrease. As a result, the optimal DOC for the aircraft fairing and puzzle plane are larger when using aluminum than using titanium. In the toy chair and heart valve cases, the optima remain at 100%.

The same effect of material prices and deposition rates can be seen by comparing PBF and wire-fed DED. PBF has larger material prices, more expensive AM machine costs, and smaller deposition rates than wire-fed DED. As a result, the slope of the production cost and time curves with respect to DOC decrease.

(c) Model size effect
As mentioned in Section 6.1, the enlargement of the shadow volume shifts the optimal DOC toward less consolidation while the enlargement of assembly costs (or time) shifts the optimum point toward more consolidation. These tradeoffs have interesting implications for the relationship between the size of the product and the optimal DOC. The support structure cost (and time) is proportional to the cube of the size, while the assembly cost (and time) is roughly proportional to the square of the size. When the subsystem is scaled up, the support structure cost (and time) increases faster than the assembly cost (and time), and the optimal DOC shifts toward less consolidation.

As shown in TABLE 5, the fairing subsystem model is scaled up and down, respectively, in order to study the tradeoffs between variations of size and optimal DOC. The comparison of the three sizes reveals that scaling up the fairing size to 5 times its original size shifts the DOC to 0%, while scaling down the fairing size by one-fifth shifts it to 100%.

TABLE 4: KEY TRADEOFF VARIATIONS ON GEOMETRY, MATERIAL, AND MAM MODALITY.

Model	Topology	Ti6Al4V	Al-6061
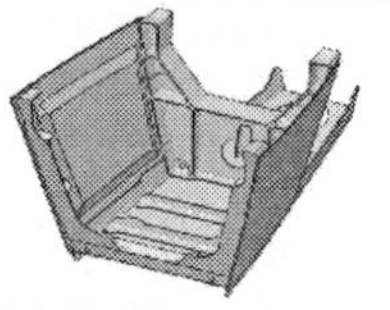 1,043×736×692 mm SUR = 7.59	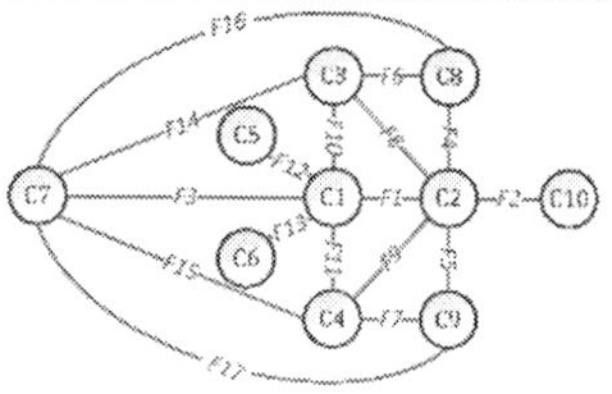	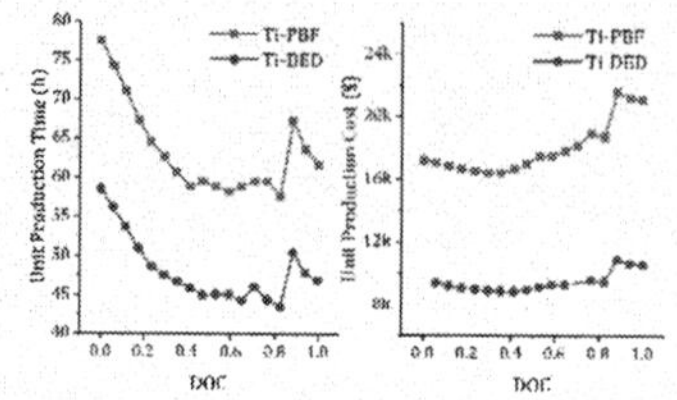	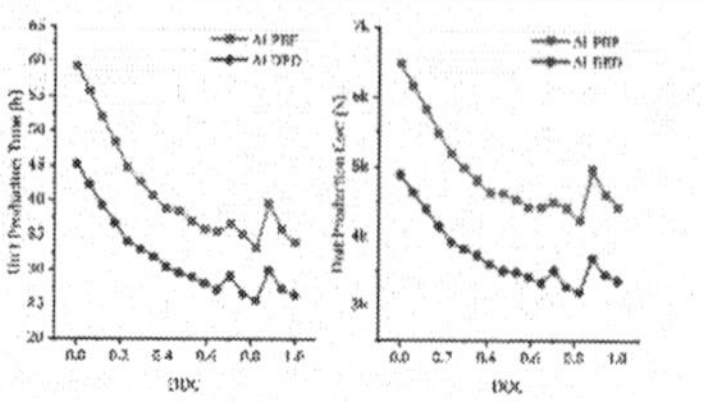
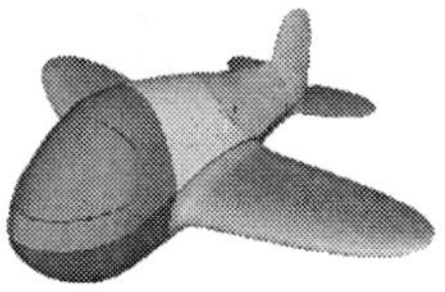 1,058×1,246×415 mm SUR = 0.477	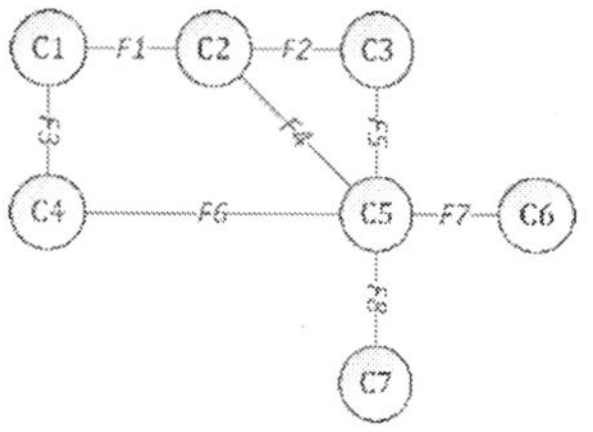	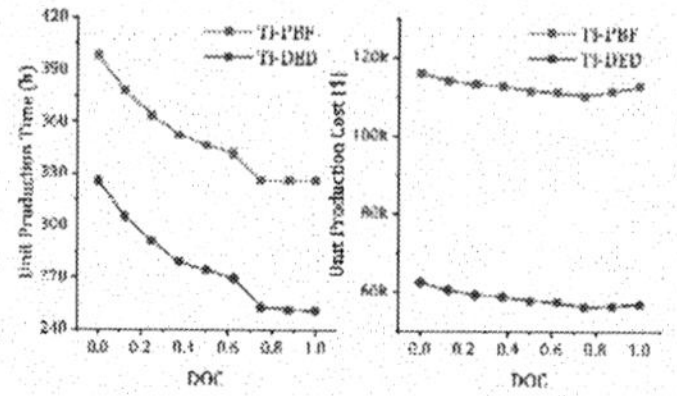	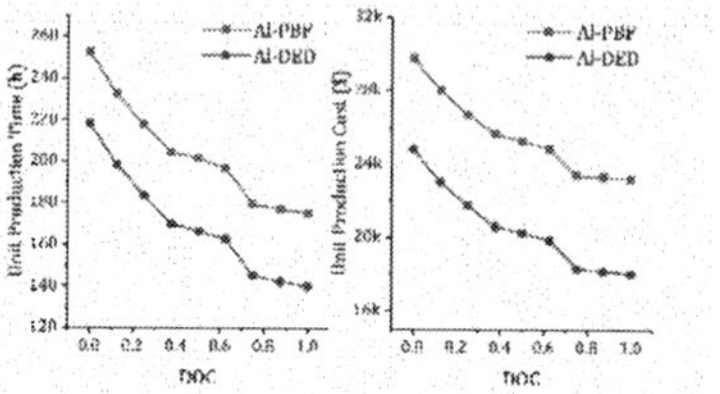
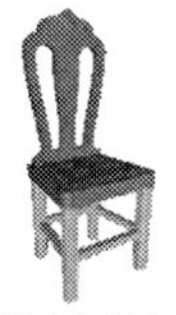 173×61×47 mm SUR = 0.2272	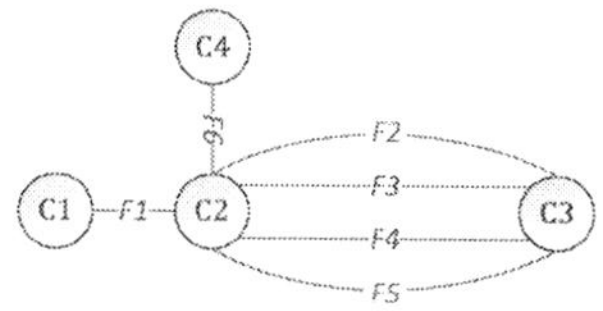	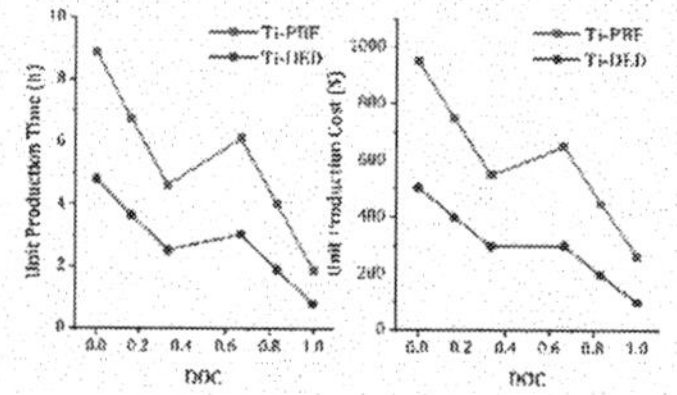	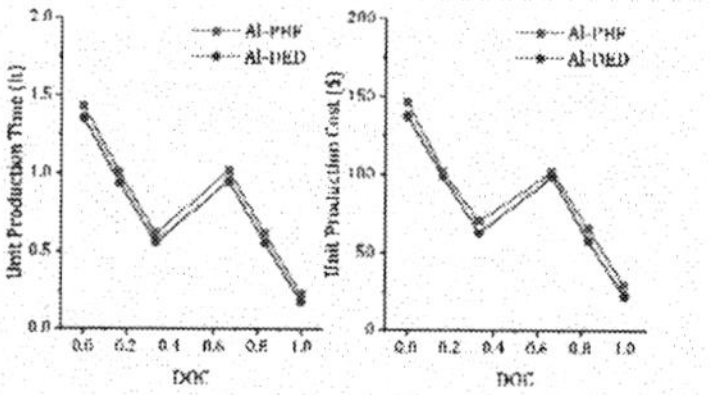
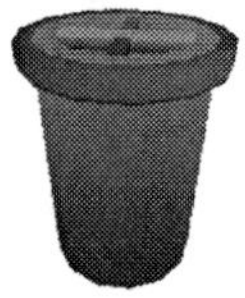 20×8×8 mm SUR = 0	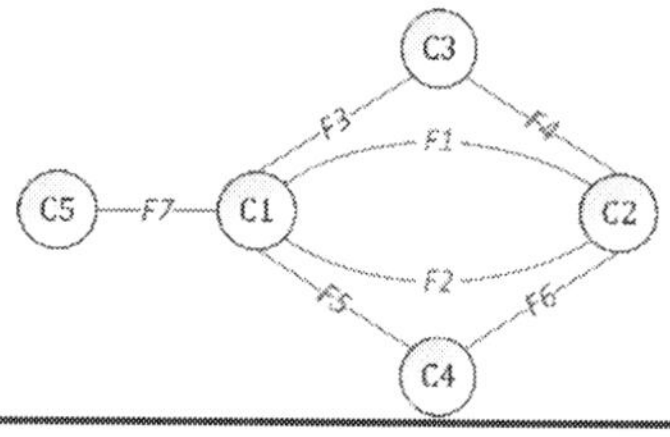	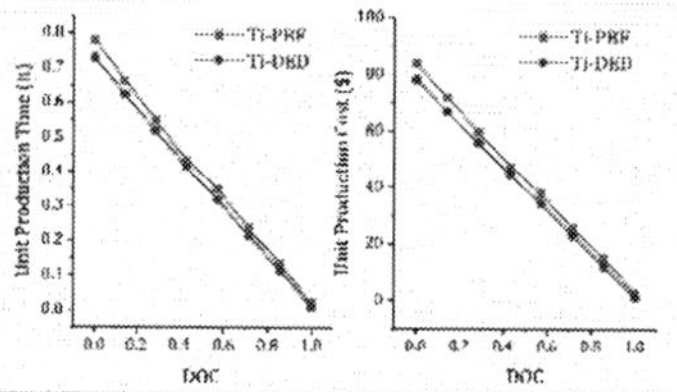	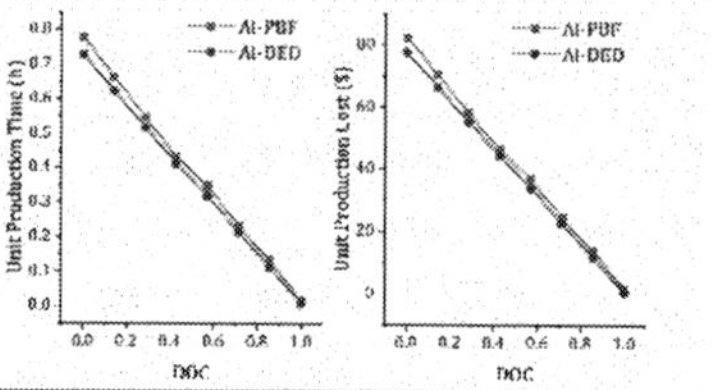

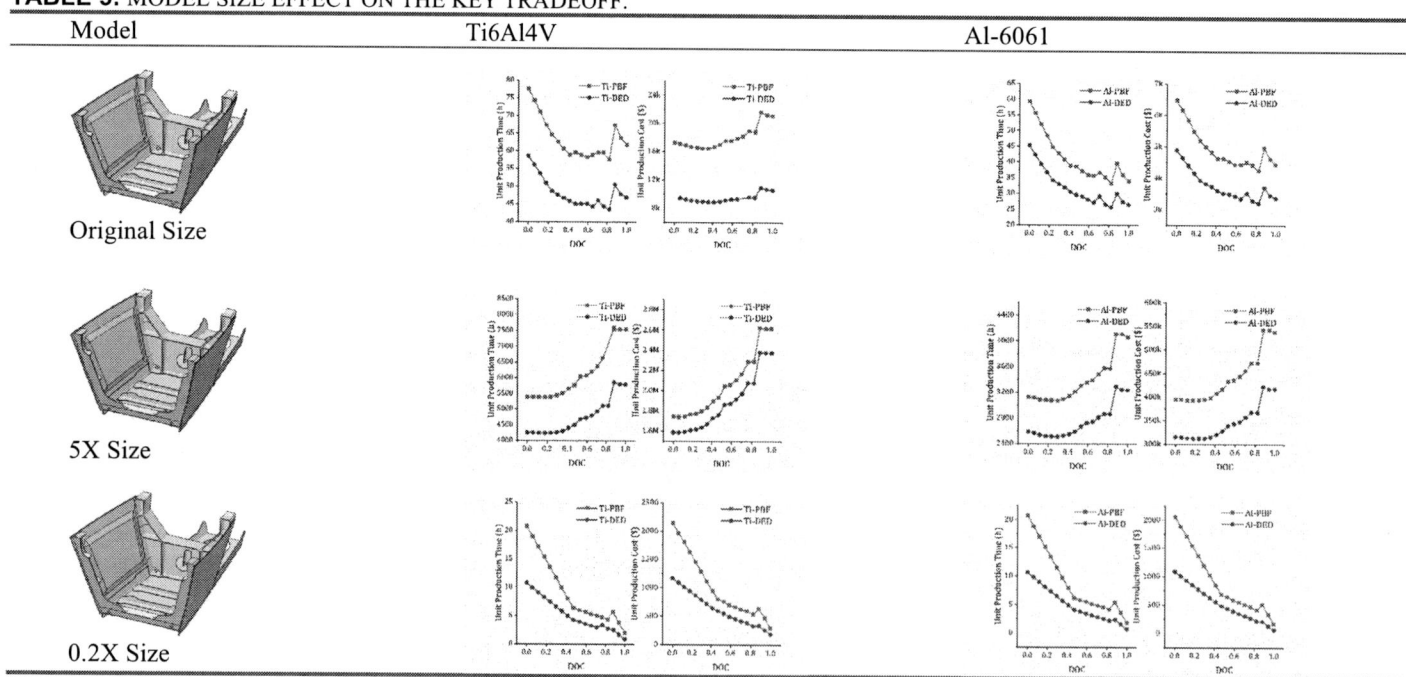

TABLE 5: MODEL SIZE EFFECT ON THE KEY TRADEOFF.

Model	Ti6Al4V	Al-6061
Original Size		
5X Size		
0.2X Size		

7. Conclusions

In this paper, we present the first AM part consolidation optimization method that considers tradeoff between consolidation and manufacturer objectives. Using an objective function of minimizing production costs or time, the method generates and filters consolidation candidates through a connectivity matrix, determines optimal build orientation, layout, and MAM processing parameters, and finds the optimally consolidated design using a GA we develop that encodes each candidate by the interfaces that are consolidated.

Using an aircraft fairing as a test-bed, we compare results for minimum production time and costs for two scenarios: PBF EBM and wire-fed DED EBM. The fundamental tradeoff in both these cases is between reducing assembly (and to a lesser extent setup) steps and increasing support structures. As the fraction of consolidated interfaces increases, the number of assembly and setup steps decreases, reducing the associated production time and costs. However, more support structures are needed to uphold the larger consolidated parts, increasing the build time to construct the support structures and the costs associated with support material and build time. The cost-optimal design for DED has fewer parts than PBF because the metal wire is less expensive than powder, and therefore the additional costs of building support structures are smaller.

We present a rationale and evidence that supports that these tradeoffs generalize to many different types of products. As more interfaces are consolidated, additional support structures are often required for the larger parts, which are more difficult to orient to reduce support structures compared to smaller parts. This increase in support structures drives up production time (due to longer build times) and costs (due to support material and removal costs) which rival reductions in assembly time and costs. We find that products with relatively small shadow volume ratios, lower material costs, faster deposition rates, and more expensive or time-consuming assembly operations will have an optimum closer to 100%.

Acknowledgment

This research was supported in part by Carnegie Mellon University's Manufacturing Futures Institute.

References

[1] Yang, S., Talekar, T., Sulthan, M. A., and Zhao, Y. F., 2017, "A Generic Sustainability Assessment Model towards Consolidated Parts Fabricated by Additive Manufacturing Process," Procedia manufacturing, 10, pp. 831-844.

[2] Yang, S., Tang, Y., and Zhao, Y. F., 2015, "A new part consolidation method to embrace the design freedom of additive manufacturing," Journal of Manufacturing Processes, 20, pp. 444-449.

[3] Hague, R., 2006, "Unlocking the design potential of rapid manufacturing," Rapid manufacturing: an industrial revolution for the digital age.

[4] Uriondo, A., Esperon-Miguez, M., and Perinpanayagam, S., 2015, "The present and future of additive manufacturing in the aerospace sector: A review of important aspects," Proceedings of the Institution of Mechanical Engineers, Part G: Journal of Aerospace Engineering, 229(11), pp. 2132-2147.

[5] Wong, K. V., and Hernandez, A. J. I. M. E., 2012, "A review of additive manufacturing," ISRN Mechanical Engineering, 2012.

[6] Schmelzle, J., Kline, E. V., Dickman, C. J., Reutzel, E. W., Jones, G., and Simpson, T. W., 2015, "(Re) Designing for part consolidation: understanding the challenges of metal additive manufacturing," Journal of Mechanical Design, 137(11), p. 111404.

[7] Frey, D., Palladino, J., Sullivan, J., and Atherton, M., 2007, "Part count and design of robust systems," Systems engineering, 10(3), pp. 203-221.

[8] Türk, D.-A., Kussmaul, R., Zogg, M., Klahn, C., Leutenecker-Twelsiek, B., and Meboldt, M., 2017, "Composites part production with additive manufacturing technologies," Procedia CIRP, 66, pp. 306-311.

[9] Frey, D., Palladino, J., Sullivan, J., and Atherton, M. J. S. e., 2007, "Part count and design of robust systems," 10(3), pp. 203-221.

[10] Booker, J., Swift, K., and Brown, N., 2005, "Designing for assembly quality: strategies, guidelines and techniques," Journal of Engineering Design, 16(3), pp. 279-295.

[11] Boothroyd, G., Dewhurst, P., and Knight, W. A., 2001, Product Design for Manufacture and Assembly, revised and expanded, CRC press.

[12] Combemale, C., Whitefoot, K. S., Ales, L., and Fuchs, E. R., 2018, "Not All Technological Change is Equal: Disentangling Labor Demand Effects of Automation and Parts Consolidation," Available at SSRN 3291686.

[13] Boothroyd, G., Dewhurst, P., and Knight, W. A., 2001, Product Design for Manufacture and Assembly, CRC press.

[14] Yang, S., Santoro, F., and Zhao, Y. F., 2018, "Towards a numerical approach of finding candidates for additive manufacturing-enabled part consolidation," Journal of mechanical design, 140(4), p. 041701.

[15] Taufik, M., and Jain, P. K., 2013, "Role of build orientation in layered manufacturing: a review," International Journal of Manufacturing Technology and Management, 27(1-3), pp. 47-73.

[16] Jibin, Z., "Determination of optimal build orientation based on satisfactory degree theory for RPT," Proc. Computer Aided Design and Computer Graphics, 2005. Ninth International Conference on, IEEE, p. 6 pp.

[17] Thomas, D. S., and Gilbert, S. W., 2014, "Costs and cost effectiveness of additive manufacturing," Special Publication, NIST.

[18] Alexander, P., Allen, S., and Dutta, D., 1998, "Part orientation and build cost determination in layered manufacturing," Computer-Aided Design, 30(5), pp. 343-356.

[19] Langelaar, M., 2016, "Topology optimization of 3D self-supporting structures for additive manufacturing," Additive Manufacturing, 12, pp. 60-70.

[20] Leary, M., Merli, L., Torti, F., Mazur, M., and Brandt, M., 2014, "Optimal topology for additive manufacture: a method for enabling additive manufacture of support-free optimal structures," Materials & Design, 63, pp. 678-690.

[21] Mirzendehdel, A. M., and Suresh, K., 2016, "Support structure constrained topology optimization for additive manufacturing," Computer-Aided Design, 81, pp. 1-13.

[22] Paul, R., and Anand, S., 2015, "Optimization of layered manufacturing process for reducing form errors with minimal support structures," Journal of Manufacturing Systems, 36, pp. 231-243.

[23] Vanek, J., Galicia, J. A. G., and Benes, B., "Clever support: Efficient support structure generation for digital fabrication," Proc. Computer graphics forum, Wiley Online Library, pp. 117-125.

[24] Nyaluke, A., Nasser, B., Leep, H. R., and Parsaei, H. R., 1996, "Rapid prototyping work space optimization," Computers and industrial engineering, 31(1-2), pp. 103-106.

[25] Canellidis, V., Dedoussis, V., Mantzouratos, N., and Sofianopoulou, S., 2006, "Pre-processing methodology for optimizing stereolithography apparatus build performance," Computers in industry, 57(5), pp. 424-436.

[26] Wodziak, J. R., Fadel, G. M., and Kirschman, C., "A genetic algorithm for optimizing multiple part placement to reduce build time," Proc. Proceedings of the Fifth International Conference on Rapid Prototyping, University of Dayton Dayton, OH, pp. 201-210.

[27] Zhang, X., Zhou, B., Zeng, Y., and Gu, P., 2002, "Model layout optimization for solid ground curing rapid prototyping processes," Robotics and Computer-Integrated Manufacturing, 18(1), pp. 41-51.

[28] Hur, S.-M., Choi, K.-H., Lee, S.-H., and Chang, P.-K., 2001, "Determination of fabricating orientation and packing in SLS process," Journal of Materials Processing Technology, 112(2-3), pp. 236-243.

[29] Canellidis, V., Giannatsis, J., and Dedoussis, V., 2013, "Efficient parts nesting schemes for improving stereolithography utilization," Computer-Aided Design, 45(5), pp. 875-886.

[30] Zhang, Y., Gupta, R. K., and Bernard, A., 2016, "Two-dimensional placement optimization for multi-parts production in additive manufacturing," Robotics and Computer-Integrated Manufacturing, 38, pp. 102-117.

[31] Gogate, A., and Pande, S., 2008, "Intelligent layout planning for rapid prototyping," International Journal of Production Research, 46(20), pp. 5607-5631.

[32] Wu, S., Kay, M., King, R., Vila-Parrish, A., and Warsing, D., "Multi-objective optimization of 3D packing problem in additive manufacturing," Proc. IIE Annual Conference. Proceedings, Institute of Industrial and Systems Engineers (IISE), p. 1485.

[33] Pandey, P. M., Thrimurthulu, K., and Reddy, N. V., 2004, "Optimal part deposition orientation in FDM by using a multicriteria genetic algorithm," International Journal of Production Research, 42(19), pp. 4069-4089.

[34] Thrimurthulu, K., Pandey, P. M., and Reddy, N. V., 2004, "Optimum part deposition orientation in fused deposition modeling," International Journal of Machine Tools and Manufacture, 44(6), pp. 585-594.

[35] Phatak, A. M., and Pande, S. S., 2012, "Optimum part orientation in rapid prototyping using genetic algorithm," Journal of manufacturing systems, 31(4), pp. 395-402.

[36] Huang, R., Ulu, E., Kara, L. B., and Whitefoot, K. S., "Cost Minimization in Metal Additive Manufacturing Using Concurrent Structure and Process Optimization," Proc. ASME 2017 International Design Engineering Technical Conferences and Computers and Information in Engineering Conference, American Society of Mechanical Engineers, pp. V02AT03A030-V002AT003A030.

[37] Johnson, M., and Kirchain, R., 2009, "Quantifying the effects of parts consolidation and development costs on material selection decisions: A process-based costing approach," International Journal of Production Economics, 119(1), pp. 174-186.

[38] Rickenbacher, L., Spierings, A., and Wegener, K., 2013, "An integrated cost-model for selective laser melting (SLM)," Rapid Prototyping Journal, 19(3), pp. 208-214.

[39] Ulu, E., Huang, R., Kara, L. B., and Whitefoot, K. S., 2018, "Concurrent Structure and Process Optimization for Minimum Cost Metal Additive Manufacturing," Journal of Mechanical Design.

[40] Baumers, M., Dickens, P., Tuck, C., and Hague, R., 2016, "The cost of additive manufacturing: machine productivity, economies of scale and technology-push," Technological forecasting social change, 102, pp. 193-201.

[41] Dinda, S., Modi, D., Simpson, T. W., Tedia, S., and Williams, C. B., "Expediting Build Time, Material, and Cost Estimation for Material Extrusion Processes to Enable Mobile Applications," Proc. ASME 2017 International Design Engineering Technical Conferences and Computers and Information in Engineering Conference, American Society of Mechanical Engineers, pp. V02AT03A034-V002AT003A034.

[42] Ruffo, M., Tuck, C., and Hague, R., 2006, "Cost estimation for rapid manufacturing-laser sintering production for low to medium volumes," Proceedings of the Institution of Mechanical Engineers, Part B: Journal of Engineering Manufacture, 220(9), pp. 1417-1427.

[43] Yim, S., and Rosen, D., "Build time and cost models for additive manufacturing process selection," Proc. ASME 2012 International Design Engineering Technical Conferences and Computers and Information in Engineering Conference, American Society of Mechanical Engineers, pp. 375-382.

[44] Cardona, C., 2015, "Part Consolidation for Additive Manufacturing Demonstrated in the Design of a 3D-Printed Harmonic Drive," IU Journal of Undergraduate Research, 1(1), pp. 45-49.

[45] Ulu, E., Korkmaz, E., Yay, K., Ozdoganlar, O. B., and Kara, L. B., 2015, "Enhancing the structural performance of additively manufactured objects through build orientation optimization," Journal of Mechanical Design, 137(11), p. 111410.

[46] Gong, H., Rafi, K., Gu, H., Starr, T., and Stucker, B., 2014, "Analysis of defect generation in Ti–6Al–4V parts made using powder bed fusion additive manufacturing processes," Additive Manufacturing, 1, pp. 87-98.

[47] Murr, L. E., Gaytan, S. M., Ramirez, D. A., Martinez, E., Hernandez, J., Amato, K. N., Shindo, P. W., Medina, F. R., and Wicker, R. B., 2012, "Metal fabrication by additive manufacturing using laser and electron beam melting technologies," Journal of Materials Science and Technology, 28(1), pp. 1-14.

[48] Nie, Z., Wang, G., McGuffin-Cawley, J. D., Narayanan, B., Zhang, S., Schwam, D., Kottman, M., and Rong, Y. K., 2016, "Experimental Study and Modeling of H13 Steel Deposition Using Laser Hot-Wire Additive Manufacturing," Journal of Materials Processing Technology, 235, pp. 171-186.

[49] Toh, W. Q., Wang, P., Tan, X., Nai, M. L. S., Liu, E., and Tor, S. B., 2016, "Microstructure and wear properties of electron beam melted Ti-6Al-4V parts: A comparison study against as-cast form," Metals, 6(11), p. 284.

[50] Scaiky Inc., 2015, "Advantages of Wire AM vs. Powder AM," http://www.sciaky.com/additive-manufacturing/wire-am-vs-powder-am.

[51] Rosenthal, D., 1941, "Mathematical theory of heat distribution during welding and cutting," Welding Journal, 20, pp. 220-234.

[52] Tang, M., and Pistorius, P. C., 2017, "Anisotropic mechanical behavior of AlSi10Mg parts produced by selective laser melting," JOM, 69(3), pp. 516-522.

[53] Scaiky Inc., 2016, "EBAM 300 Series," https://www.aniwaa.com/product/3d-printers/sciaky-ebam-300-series/.

Proceedings of the ASME 2019
International Design Engineering Technical Conferences
and Computers and Information in Engineering Conference
IDETC/CIE2019
August 18-21, 2019, Anaheim, CA, USA

DETC2019-97775

COMPUTATIONAL DESIGN OF ACTIVE LATTICE STRUCTURES FOR 4D PRINTED PNEUMATIC SHAPE MORPHING

Cosima du Pasquier
Engineering Design and Computing Laboratory
Dept. of Mechanical and Process Engineering
ETH Zurich, Switzerland
cosimad@ethz.ch

Pascal Koller
Dept. of Mechanical and Process Engineering
ETH Zurich, Switzerland
pkoller@student.ethz.ch

Tino Stankovic
Engineering Design and Computing Laboratory
Dept. of Mechanical and Process Engineering
ETH Zurich, Switzerland
tinos@ethz.ch

Kristina Shea
Engineering Design and Computing Laboratory
Dept. of Mechanical and Process Engineering
ETH Zurich, Switzerland
kshea@ethz.ch

ABSTRACT

With advances in 3D printing and digital fabrication an opportunity is presented to realize highly customized designs whose shape can change and adapt to facilitate their functionality. A computational design method to determine the configuration of 2D pneumatic shape morphing lattices using a direct search method is implemented and assessed. The method is tested using a Kagome unit cell lattice structure, which is particularly well suited for shape morphing. To achieve shape change, beams are replaced by linear actuators such as those found in pneumatic 4D printing, whose number and placement are optimized to replicate a given target shape. The actuator placement and deformation accuracy are given for four main curvature changes: linear, convex, concave and the transition from one to the other. The results are assessed in terms accuracy of deformation and computational effort. It is shown that the method proposed produces structures that can replicate complex shape changes within 1% of the desired shape. Reducing the number of actuators for robustness purposes is shown to affect the results minimally.

Keywords: Shape morphing; Computational design; Optimization; Additive Manufacturing; 4D printing; Pneumatics

1. INTRODUCTION

Whether it is due to a stimulus from within or without, natural systems grow, evolve and adapt to their environment. Although many have tried to mimic this feature for engineering applications, classic design most often resolves the uncertainties of exploitation by keeping its models within a static and rather narrow band of functionality. The field of shape morphing was coined for this purpose exactly; to extend the design space by

enabling structures to deform based on external or internal loading conditions. Rather than aiming to produce only a few states in which the performance is controlled and efficient, shape morphing tries to expand the space of functional and performance optimized behavior during the lifetime of a structure.

Recent research distinguishes two established approaches for shape morphing; reactive load driven systems, in which the actuation is predominantly based on the effects of smart materials interacting with their environment, and proactive deformation driven systems, in which actuation relies on hydraulic or mechanical actuators placed within the system. However, due to the recent advancements in Additive Manufacturing (AM) and digital fabrication, active lattice structures emerge as a feasible solution that allows flexibility in realization considering the two approaches, while at the same time facilitating designs that are also lightweight, load-carrying, and have been shown to conform well with complex surface geometries [1], [2]. The latter especially being the case with Kagome-type cell applications, since they can be statically and kinematically determinate given suitable patch bars on their periphery [3], [4].

One approach to create active lattices facilitated by 4D Printing is to embed pneumatic linear actuators within the structure at certain key locations to achieve a desired shape change. Fig. 1 shows how pneumatic linear actuators can be used to achieve both linear (Fig. 1c) and parabolic (Fig. 1d) displacement. Their design is based on [9] and tuned to achieve the 10% linear deformation prescribed in the numerical model of the work presented here. The structures are printed on a

Connex500 from Stratasys using Verowhite and Agilus Black materials. The modular design of the structure allows the user to change the actuator placement easily. The positioning of actuators within the structure is currently an iterative design process, limited to small lattice samples. A motivation for this paper is to calculate the ideal placement of these pneumatic actuators within a large structure to achieve target deformations. Therefore, this actuator design is the basis for all the dimensions chosen in this paper. However, the fabrication and assembly of the components have an impact on the performance of the structure which is not yet accounted for in the model.

There is a need for a computational method that can automatically identify the optimized arrangement of these actuators given desired deformations. The large number of members, any of which may serve as an actuator, poses a challenging task also given the strong non-linearity of the problem.

To address this challenge the contribution of this work is a computational design method for identification of optimized configurations of pneumatic actuators in a shape morphing lattice structure. The results show that the method converges accurately with respect to the desired shapes. It produces results for a small to medium number of variables in a discrete optimization problem formulation. The relevance of the method proposed to the fields of shape morphing and 4D printing is demonstrated by showing how a small lattice can achieve four main curvature deformations: linear, concave, convex, and the transition between the two.

The structure of this paper is as follows. Existing research and publications concerning shape morphing and 4D printing are summarized within a state-of-the-art. In this context, the focus on pneumatically actuated lattice structures is motivated and a comparison to literature delineates the contributions of the present method to these fields. Then the lattice model that serves as a benchmark to the use case studies and the general optimization procedure are introduced. The optimization method is validated for several lattice sizes, shapes, boundary conditions and by limiting the number of actuators, which gives results more structural integrity. The results are analyzed based on numerical performance, deformation accuracy and actuator layout, the implications of which are then discussed in the context of shape morphing structures.

2. BACKGROUND

This section reviews the related work behind the principle of shape morphing and analyzes how it is applied in the realization of several different engineering applications. The review investigates in depth the two main categories: load driven design and deformation driven design. The distinction between them is explained and illustrated by several application examples that have been explored mainly in the aerospace and the automotive industry. Then, existing 4D printed actuators and their applications are discussed. The section concludes by outlining the motivation for this work.

2.1 Related Work

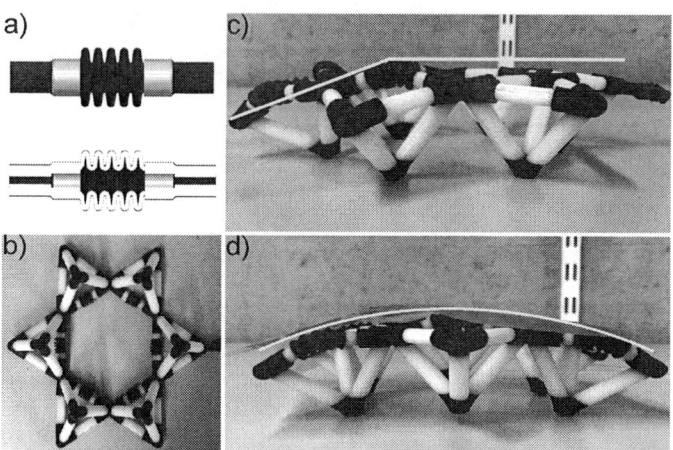

FIGURE 1: A) LINEAR ACTUATOR MODEL B) KAGOME LATTICE TOP VIEW C) LINEAR DEFORMATION OF KAGOME LATTICE D) PARABOLIC DEFORMATION OF KAGOME LATTICE

Due to recent advances in material science and the development of advanced computational methods and tools, the field of shape morphing structures has grown substantially to produce technologies that find applications in the aerospace [1], [6], [7] automotive industries, [8] as well as in soft robotics [9]–[11] and personalized medicine [12], [13]. Creation of morphing, or shape-changing, structures gives designers the ability to enhance the performance and efficiency of a system by varying the structural responses based on load or use cases. By adding the ability to adapt, a usually static design becomes inherently multi-functional. That in turn expands the possible applications and diminishes the need for standardized or niche products. In comparison to the classic designs of compliant structures [14], [15] an additional advantage is the ability to provide adequate structural stiffness while minimizing the actuation energy. By combining the properties of architected materials with computational support tools and additive manufacturing, it is possible to design and realize a structure that is lightweight, robust and adaptable, concepts that are considered contradictory in classic mechanical terms [6], [8], [15].

Structural designs all exhibit a form of reactive morphing: an externally applied load acts as an input and the resulting deformation is the output. Compliance-based topological optimization is often used as a basis for such responses [15], [16]. To achieve reactive morphing, literature focuses on materials with responsive properties such as piezoelectric conductors, shape-memory polymers and alloys or smart composites [15], [17]. The most common application is the design of a morphing wing whose geometry can change given certain inputs [1], [7], [18]. Aerodynamic loading can trigger the transition from one stable state to the next using piezoelectric actuators to enhance in-flight performance [19].

The work from [6] describes the many applications of shape-memory polymers and polymer composites in space, which are currently used for the deployment of antennas, booms and solar-panel arrays. Another application is reported in [1] in which the geometry and placement of lattice structures within a wing are tailored for its stiffness and response to air loading. A car's airfoil and driver's seat with cellular cores have been optimized with a genetic algorithm using prescribed load paths as input [20].

Rather than responding to an external load, the proactive deformation-driven systems are powered from a source within rather than reacting structurally to a load: their input is a strain energy and their output a deformed state. They tend to involve mechanical, electromechanical or hydraulic power supplies [8], [21] which are directly integrated in the design or require external force inputs [22]. The automotive and aerospace industries provide many of the applications reported in the literature. The work from [23] proposes to use an electro-mechanical actuator to improve a helicopter's hover and forward-flight regimes by switching the trailing edge-deflection in a bi-stable design. BMW's morphing car, GINA can control its side panels with an electric and hydraulic control system and a compliant polyurethane skin [8], [24]. MIT's CityCar project equips its car with a foldable system that reduces its car model size from 2.5m to 1.5m for better parking [8].

The systems described above mostly use electromechanical actuators to control their shape morphing. The field of 4D printing has given rise to new types of actuation including thermal, hygroscopic, magnetic and pneumatic actuators. Thermal actuators use the modulation of mechanical properties with respect to temperature in their designs to generate deformation such as bending, folding or contraction. This principle is used in the printing of Liquid Crystal Elastomers (LCEs), where different material concentrations are correlated with actuation temperature. They are printed with Direct Ink Writing (DIW) and can achieve sequential out-of-plane deformation in a small surface [5]. The hygroscopic actuation principle is similar in that it uses different materials' swelling behavior to control the expansion of some elements in a design. [6] uses this to fold up a cube, and [7] achieves complex flower shapes through the control of orientation of cellulose fibrils in printed hydrogels. For magnetic actuation, particles mixed in the ink are oriented with a magnetic field during printing; the subsequent application of a magnetic field induces deformation. This principle was used, for example, in the design of magnetic fasteners [8]. Pneumatic actuation is different as the exact deformations can be controlled and predicted by the geometry of the actuator design The most notable examples involve using linear or bending bellows to power soft grippers [1], [2] and autonomous locomotive systems such as crawlers [3], [4]. Pneumatic actuators are able to sustain overall higher loads predictably while remaining lightweight and using readily available compressed air. These characteristics make them appealing for the design of shape morphing structures and motivate their use in the realization of the designs presented in this paper.

2.2 Motivation

The use of computational methods in 4D printing is limited so far and usually focuses on material placement within a design rather than actuator placement. However, work previously done in shape morphing can be leveraged to develop new approaches in this field. Thus, the motivation for this work is to propose a computational method for the determination of optimized configurations of pneumatic actuators for 4D printed lattice structures to achieve a purposeful range of different shapes. Previous work has handled the complexity of modeling a large, shape changing volume populated by comparatively small unit cells either by using a significantly simplified model or by limiting its design space to statically and kinematically determinate structures [2], [4], [29]. The determinacy criterion alleviates the problem of expensive numerical simulations but simultaneously adds the complex challenge of generating adequate initial structures and thus limiting the flexibility of the approach. Here, we propose an efficient method independent of determinacy that can be used with any lattice unit-cell type.

Determining the configuration of active members within a lattice is a discrete optimization problem. It is most often non-linear due to the analytical complexity of the desired shape change, and is in general not a convex programming problem since a displacement objective can possibly be reached using various actuator combinations. There are several main optimization approaches that can deal with such a complex problem. Meta-Heuristic (MH) approaches, such as Genetic Algorithms and Simulated Annealing, are robust with respect to the problem formulation and offer a pathway to mitigating local optima. However, such approaches require a significant number of function evaluations even for a medium sized problem limiting their applications to mostly truss structures for which the response analysis is not an issue. Another approach would be to solve the problem using the Branch-and-Bound Method (BBM) in such a way that a search space is implicitly enumerated. BBM guarantees to find the global optimum only if the problem is linear or convex, whereas for a general problem a promising search direction might be discarded too soon omitting

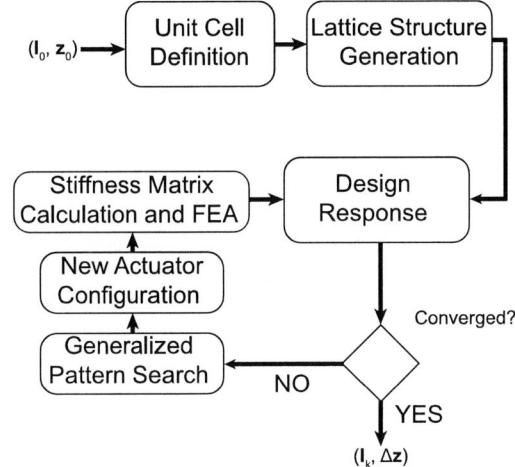

FIGURE 2: WORKFLOW OF THE PROPOSED APPROACH

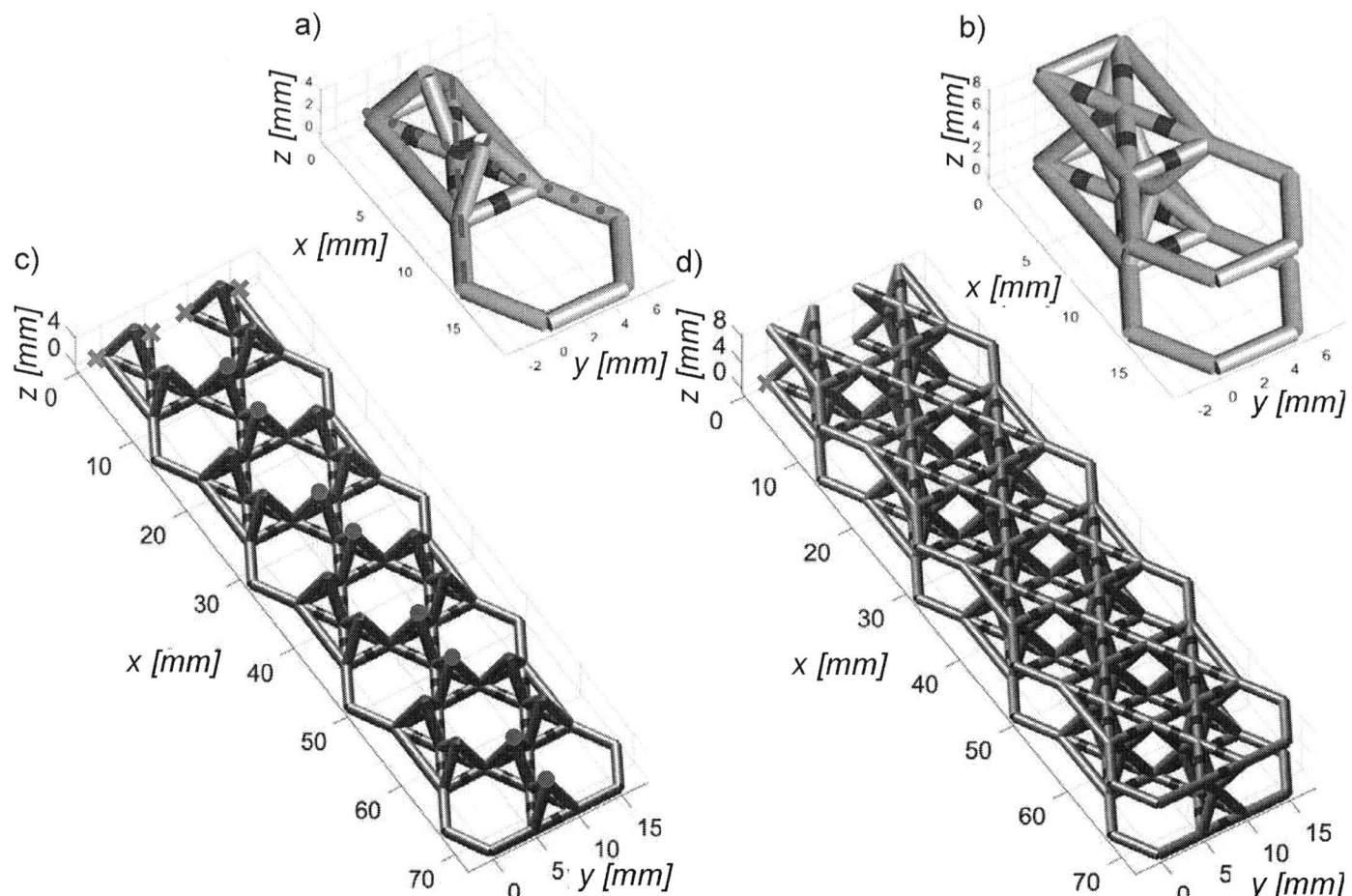

FIGURE 3: A) SINGLE LAYER UNIT-CELL; RED DASEHD LINE: +30°, RED DOTTED LINE: -30°, B) DOUBLE LAYER UNIT CELL C) REFERENCE SINGLE LAYER LATTICE; GREEN MARKINGS: CONTROL NODES POSITIONS, ORANGE CROSSES: BOUNDARY NODES, D) REFERENCE DOUBLE LAYER LATTICE

the global discrete solution. In addition, although the most robust, both SA and BBM are the most time consuming without special heuristics related to the optimization problem [30]. To aim for both efficiency and robustness, we propose the application of a Generalized Pattern Search (GPS) method to solve the non-linear discrete optimization problem involved in this work [31]. GPS delivers accurate results with good efficiency for both discrete and continuous problems and, if coupled with MH methods, it can successfully avoid some local optima.

3. METHOD

In this section, the computational design method for identification of optimized configurations of actuators in a shape morphing lattice structure is defined. Starting with the unit cell, a Kagome cell type design, the generation of the lattice and the actuators is introduced. Then, a series of five case studies according to which the performance of the computational method is evaluated. (A1) is a linear deformation, (A2) a parabolic deformation with different boundary conditions, (A3) a sine shaped deformation, (B) is the benchmark case using two layers of cells instead of one, (C) is the implementation of an

actuator minimization. Finally, the computational method is discussed and the numerical approach to determine the actuator placement is explained.

3.1 Active Lattice Structure Model

3.1.1 Unit Cell Definition. The unit cells used to generate the Kagome lattice for one and two layers are shown in Fig. 3. The members are modeled as 1D Finite Element Beam Elements with 6 DOFs per node and are based on the Euler-Bernoulli beam theory. The blue sections represent the possible actuator placement within a cell.

The members are assembled within regular tetrahedrons whose members are 5 mm long and have a surface area of 1 mm². Extra support is added along the x-direction with the addition of inextensible beams that are 8.66 mm long. A single layered cell is 4.33 mm high and a double-layered cell is 8.66 mm high.

Actuators can replace members that are either at +30°, -30° and 90° of the x-axis. The +30° orientation is depicted by the red dashed line in Fig. 3a and the -30° by the red dotted line. Since the desired deformations are implemented as displacements in z as a function of x, actuators placed at 90° (along the y-axis) contribute less to the shape change and are not considered in the results.

The unit cell member parameters including material properties are summarized in Table 1. These parameters are based on the physical sample shown in Fig. 1.

TABLE 1: MODEL PARAMETERS

Designation	Value
Length	$L = 5$ [mm]
Cross-sectional area	$A_{cross} = 1$ [mm^2]
Beam elongation (linear actuation)	$\Delta d = 0.5$ [mm]
Poisson's ratio	$v = 0.3$
Young's Modulus	$E = 2743.8$ [MPa]

3.1.2 Lattice Structure Generation. To generate the base structure, the unit-cells are tessellated twice in the y-direction and four times in the x-direction. This size is chosen because it has enough cells to replicate complex deformations but the computational effort inferred from the number of variables allows to test for multiple optimization parameters within computational times of 100 to 4000 seconds. The single-layered and double-layered structures contain 168 and 336 beams respectively, out of which 72 (144) can be actuated.

3.1.3 Boundary Conditions. The boundary conditions are shown in Fig. 3c and d, marked by orange crosses. The six Degrees of Freedom (DOFs) of the boundary nodes are all fixed, since it best matches the boundary conditions necessary during FEM, with the exception of the boundary variation case (A2), in which both ends of the structure are pinned, i.e. the three translational DOFs are fixed.

3.1.4 Actuator Implementation. Each beam that can be an actuator is given an additional binary variable, in which $x_i = 0$ and $x_i = 1$ represent the undeformed and deformed state respectively. Linear actuation is assumed that can be implemented through, for example, the pneumatic actuators as shown in Fig. 1 [5]. The states of the individual actuators are determined through the optimization, which minimizes the position difference between the deformed and the desired state. A deformed member is 10% longer, and thus 0.5 mm longer, than a static one. The values are stored within a binary vector **l**, which is the main variable vector of the objective function.

In the model, the actuator placement is defined set by set; each set of actuators represents a row of the lattice in the y-direction, so they are grouped based on their distance to the boundary nodes. The deformation is activated first in the set closest to the boundary ($x = 0$), and then incrementally until the other edge is reached ($x = L_{\text{lattice}}$). The order of actuation is important because the model is based on displacement loads. If the displacement load is directed towards the constrained nodes, it will simply result in an elevated stress within the structure instead of a displacement.

3.2 Case Studies

3.2.1 Benchmarking. To evaluate the results of the five use cases presented in this paper, a reference problem is first set up as a benchmark to all further results. The desired deformation for this problem is defined as:

$$z = \left(\frac{x}{20}\right)^2 + 4 \tag{1}$$

with no additional constraints. This benchmark was chosen as a reference because it proved to be the deformation for which the algorithm converged the best in a preliminary study.

The actuator placement is optimized using Eq. (1) and the results are analyzed based on the following criteria:

- **Normalized error:** $\varepsilon_N = \dfrac{\sqrt{\min_l f(l)}}{n}$, with n: number of nodes, 24 and 53 for one and two layers respectively. The normalization enables comparison of structures of different sizes.
- **Accuracy in %:** $acc. = 100 - \dfrac{\varepsilon_N}{h_{\text{lattice}}}$, with $h_{\text{lattice}} = 4.33$ mm and 8.66 mm for one and two layers respectively. It represents how close the results are to the target shape.
- **Computational effort:** number of iterations, equivalent to the number of objective function evaluations. This metric estimates the impact of shape complexity on the algorithm's performance.
- **Actuator placement:** analysis of the final distribution. This allows to determine the physical validity of the results.

3.2.3 Shape Variation (A). To determine how adaptable the method is, it is applied to four different shape morphing modes: exponential (reference), linear (A1), parabolic (A2) and sine shaped (A3) deformation. The desired displacements are defined as follows:

(A1)   Linear:
$$z = \frac{x}{8} + 4 \tag{2}$$

(A2)   Parabolic:
$$z = -\left(\frac{x - 36}{12}\right)^2 + 13 \tag{3}$$

(A3)   Sinusoidal:
$$z = 4\sin\left(\frac{\pi x}{50} - \frac{\pi}{2}\right) + 8 \tag{4}$$

The objective function is adapted accordingly and the quality of the results is analyzed based on the criteria enumerated above.

3.2.4 Layer Variation (B). To add to the deformation potential of a structure, lattices usually have several layers. In this section, the computational method is tested on a case analog to the benchmark, but with double the layers. This allows the model to take on both convex and concave curvatures.

The results' analysis criteria are then compared to the benchmark to verify whether there is a significant loss of accuracy or increase in computational effort due to the stark increase in variables.

3.2.5 Constrained Actuator Number (C). Every time an actuator replaces a beam within the structure, it compromises slightly the structural integrity of the lattice. In this section, the optimization formulation is defined so as to minimize the number of actuators given a maximum error:

$$\min_{\mathbf{l}} f(\mathbf{l}) = \sum_{k=1}^{n} l_k \qquad (5)$$

where

$$\varepsilon_{max} \leq \sum_{k=1}^{n} [z_{\text{eff}}(x, y, l_k) - z_{\text{des}}(x, y)]^2 \qquad (6)$$

$$\varepsilon_{max} \in [0,1]$$
$$l_i \in \{0,1\}$$

The target deformation is set as a constraint. Only solutions that are within the accuracy of previously calculated target deformations without actuator constraint are accepted.

3.3 Optimization Process

After having introduced the generation and implementation of the lattice model, this section defines how its response is evaluated and updated during the optimization routine. The performance of the method is determined on the basis of case studies that span multiple variable configurations, deformations and actuator constraints.

3.3.1 Problem Formulation.
The deformation of our structure is understood as the displacement of the nodes within the system with the largest z-value, which are marked in Fig. c by green dots. The objective function is defined as the square of the sum of the difference between the effective z-coordinates and the desired z-coordinates:

$$\min_{\mathbf{l}} f(\mathbf{l}) = \sum_{k=1}^{n} [z_{\text{eff}}(x, y, l_k) - z_{\text{des}}(x, y)]^2 \qquad (7)$$

Where:

$$l_i \in \{0,1\}$$

3.3.2 Optimization Method.
The optimization of the configuration of linear actuators within a lattice structures poses several challenges: it is a discrete and non-linear problem, which contains a significant number of variables (71 or 142 in our reference problems). A Generalized Pattern Search (GPS) method is used and the flow diagram of the process shown in Fig. 2.

The particular implementation of GPS used in this work is from NOMAD [34], [35], a derivative-free optimization environment that also offers the possibility to include MH characteristics in the search using the Variable Neighborhood Search (VNS) method to avoid local optima [36]. The NOMAD environment interfaces with MATLAB using the OPTI Toolbox. The results obtained from GPS within the NOMAD environment depend on the seed it uses, which determines the pseudo-random number sequence that control its steps. NOMAD is run using the default seed exclusively, which means the method is deterministic and thus repeatable.

We apply GPS with default settings, meaning that the algorithm can select from $2n$ different coordinate search directions where n is the number of variables. To limit the computational effort, the algorithm stops once the delta of desired to existing shape is smaller than 0.04 mm. The maximum number of iterations is set to 500, except for the sine shape variation (A3), which is set to 2500. The probability for VNS is set to 0.75 and it is only applied to use case (A3).

The finite element analysis involving modeling of lattice structure members using 1D finite element beam elements with 6 DOF per node is carried out using an in-house FEM in MATLAB.

4   RESULTS

In this section, the results from the implementation of the GPS for the optimized configurations of actuators within a shape morphing lattice structure are presented. Its application to five use cases, three shape deformations (A1, A2, A3), an increase in sample size and variables (B) and a minimization of the number of actuators (C), is compared to the benchmark in terms of convergence speed and accuracy and actuator topology. The sensitivity of the method to the choice of starting point is analyzed and, since the problem is non-convex, its convergence is compared to an analogous case where MH are used to avoid local optima.

4.1 Shape Variations (A)

All the results concerning accuracy and performance of the case studies are shown in Table 2.

A linear deformation, use case (A1), is achieved with 46% of the iterations and the computational time necessary for to optimize the benchmark with only an 8% decrease in accuracy. It converges with a relatively high percentage of actuated members, whose directions are less evenly distributed than the benchmark.

In the case of parabolic deformation, use case (A2), the boundary nodes are fixed on either side of the structure, but only in three DOF. Compared to the benchmark deformation, the error increases only by 3% while the optimization converges with merely 13% of the iterations necessary and 14% of the computational effort. Compared to the benchmark deformation, the number of actuators increases by 16% and they are perfectly distributed in the $\pm 30°$ directions. They are also perfectly distributed along the center of the structure.

The sinusoidal deformation, use case (A3), is best achieved using a double layered lattice, so its performance is compared to its corresponding benchmark which is presented in the next section. The calculations for this case are stopped after five times the usual number of iterations allowed to achieve the benchmark target shape. Its error triples, although it still achieves a deformation accurate to 97.4% of the given objective. It needs 40% more actuators, which are interestingly distributed along the center of the structure ($x = L_{lattice}/2$). Of the 19 actuators in the half closest to the constrained nodes, 13 are on the bottom layer, and of the 28 in the half closest to the free edge, 24 are on the top layer. Actuators are fairly evenly distributed among the three directions, as they were for the benchmark double layered lattice.

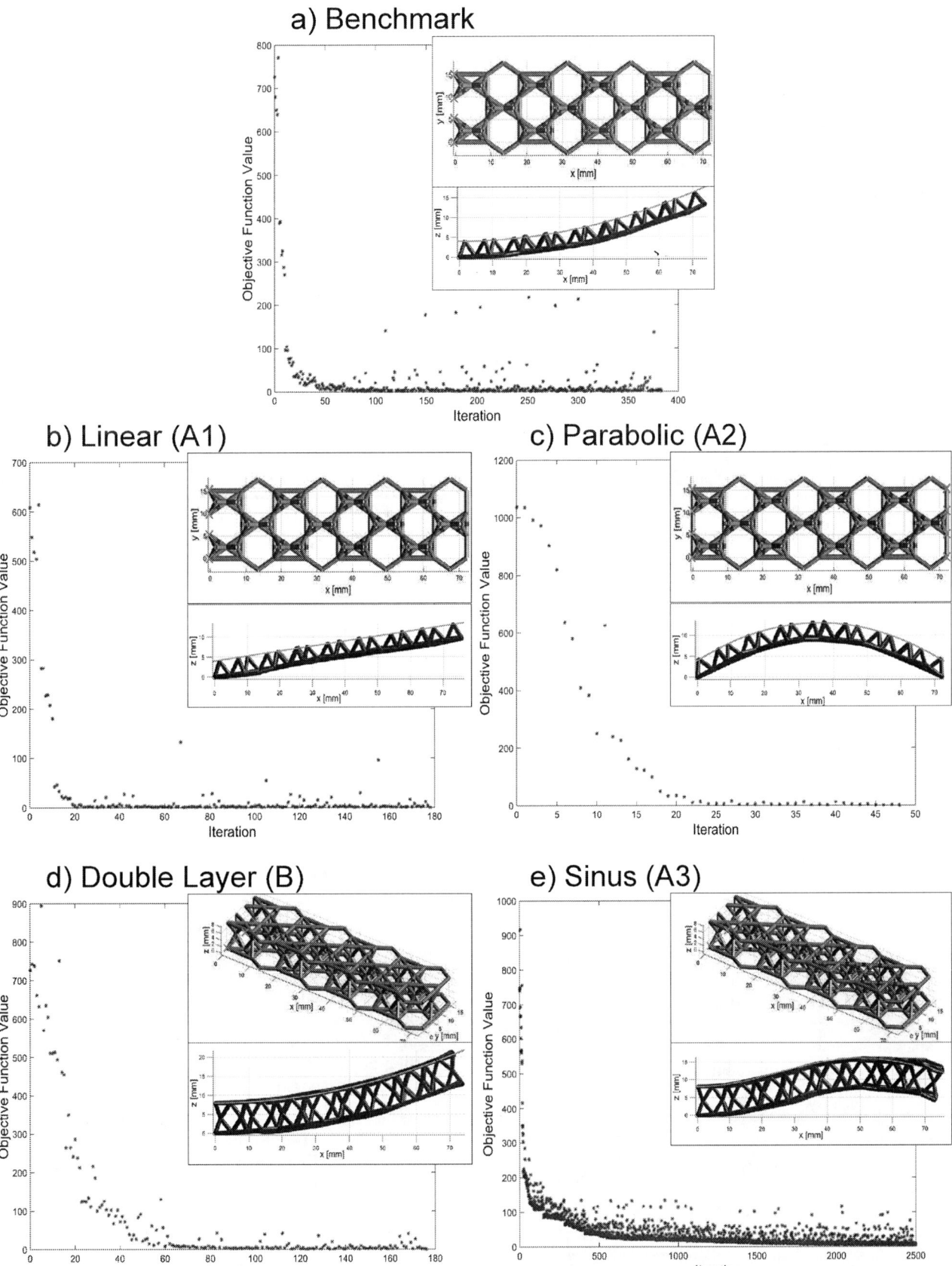

FIGURE 4: CONVERGENCE BEHAVIOR, ACTUATOR PLACEMENT AND DEFORMED STRUCTURE FOR A) BENCHMARK, B) LINEAR DEFORMATION, C) PARABOLIC DEFORMATION, D) DOUBLE-LAYERED STRUCTURE, E) SINUSOIDADEFORMATION; DESIRED DISPLACEMENT IN LIGHT BLUE

TABLE 2: OPTIMIZATION PERFORMANCE

Shape	Normalized error [mm]	Accuracy [%]	Number of iterations
Benchmark	$3.6 \cdot 10^{-2}$	99.2	385
(A1) Linear	$3.89 \cdot 10^{-2}$	99.1	179
(A2) Parabolic	$3.71 \cdot 10^{-2}$	99.1	49
(A3) Sinusoidal	$9.80 \cdot 10^{-2}$	97.4	2500
(B) Double L.	$3.49 \cdot 10^{-2}$	99.2	177
(C) Min. Act.	$3.58 \cdot 10^{-2}$	99.2	1419

TABLE 3: ACTUATOR DISTRIBUTIONS

Shape	Total Number of Actuators	% of Actuated Members	$+30°$	$-30°$
Benchmark	24	33.8	8	9
(A1) Linear	31	43.1	14	10
(A2) Parabolic	28	38.9	13	13
(A3) Sinusoidal	47	33.1	18	12
(B) Double L.	28	19.7	8	10
(C) Min. Act.	10	13.9	4	3

4.2 Layer Variations (B)

Doubling the number of variables for the same target deformation as the benchmark does not slow the algorithm's performance. On the contrary, the larger structure converges within 46% of the iterations the benchmark needs and with 3% more accuracy.

As shown in Table 3, the double layered structure needs a significantly lower actuator concentration to achieve the same deformation as the benchmark, although their repartition is very similar.

4.3 Constrained Actuator Number (C)

Minimizing the number of actuators while constraining the error comes at a computational cost: as shown in Table 2, the algorithm takes approximately four times longer to converge. However, the accuracy actually increases by 1.00% while the number of actuators is decreased by 57.1% and they are similarly distributed in both cases.

4.4 Starting Point Variation

To determine the sensitivity of the computational method to the choice of starting point, the benchmark problem is calculated for the two extreme states: in the first state all actuators are at their original length and in the second state all actuators are elongated.

$$\bar{l}_0 = \begin{pmatrix} 0 \\ \vdots \\ 0 \end{pmatrix} \text{ and } \bar{l}_0 = \begin{pmatrix} 1 \\ \vdots \\ 1 \end{pmatrix}$$

The results are assessed in terms of deformation accuracy and computational effort.

A Variable Neighborhood Search (VNS) option is implemented to assess whether and how much it helps avoid local optima for the fully actuated starting point since it converged with 38% less accuracy than the fully static. The results in Table 4 show that the Meta-Heuristic does decrease the accuracy variation by more than half.

Since the fully static starting point performs the best, it is used for the benchmark and all case studies.

TABLE 4: STARTING POINT VARIATION RESULTS

Case	Fully static l_0	Fully actuated l_0	Fully actuated l_0 with VNS
Normalized error [mm]	$1.63 \cdot 10^{-2}$	$2.25 \cdot 10^{-2}$	$1.82 \cdot 10^{-2}$
Accuracy (% of cell height)	99.6	99.5	99.6

5 DISCUSSION

The discussion section focuses on the optimization performance, the results interpretation and the overall contributions of the method on computational design of shape morphing lattice structures.

The application of the GPS for the optimization of actuator configuration within a lattice shows accurate convergence, as seen in Table, with the exception of the most extreme sinusoidal deformation mode (A3) shown in Fig. 4e. The algorithm consistently converges with less than 1% deviation from the desired shape within 10^2 to 10^3 iterations (as shown in Table 2), which shows good efficiency for a heuristic search method.

The preliminary study of the impact of the starting point on the performance of the GPS is an indicator of the sensitivity of the method to the starting point and non-convexity of the solution space. The initial position and the implementation of a MH component varied the response, although all three remained within 1% deviation from the desired shape. This implies that the starting point could be used to further tune the method, but the improvements would be limited.

The performance of the algorithm on the various use cases shown in Table 3 is uniform and consistent. As one might expect, adding complexity to the desired shape can lead to an increase in error and computational effort. We stipulate however that this could be mitigated with an increase of the lattice size, which would give it more freedom to adapt to different geometries.

The five case studies presented in the Results section demonstrate the flexibility of the method proposed. They show that a relatively small lattice sample can achieve hinging, both convex and concave curvatures as well as a curvature transition.

Comparing the exponential deformation from the benchmark and the linear case (A1), it is shown that one sample lattice can vary its response significantly given the exact same boundaries. This is done with the same level of accuracy and the same level of computational effort. As shown in Fig. 4b, the concentration of elongated members increases regularly, as opposed to the benchmark where there is a concentration at the lifting edge. Both of these patterns are consistent with the deformation increase required of the lattice.

Analyzing the impact of boundary conditions in (A2) makes it clear that constraining additional DOFs influences far more the efficiency of the computational method than constraining more nodes. As shown in Fig. 4d and Table 2, doubling the amount of boundary nodes while only limiting three DOF instead of six results in an eight times faster convergence than the benchmark. This shows that the boundary conditions should be very carefully set with respect to the desired application and are a strong tuning parameter to optimize for an efficient convergence.

The actuator placement to achieve a sine shape, use case (A3), is a clear testament to how convex, respectively concave curvatures are achieved: of the 19 actuators near the boundary nodes, 13 are on the bottom layer, and of the 28 near the free edge, 24 are on top (Fig. 4e). It is clear that the bottom layer promotes convexity and the top concavity, which explains why the actuator positioning is so radically different before and after the curvature transition. This indication as to how the structure best complies with certain deformations could help simplify and scale up the topology optimization. By identifying specific curvature zones, one can tailor the variable set to the actuators which best support the given shape change.

The 'double-layered' lattice (B) shows that an increased number in variables allows to accommodate for deformations more easily. All actuators placed at an angle, and which contribute most to the displacement in the z-direction, are located on the bottom layer. As is shown in the sinusoidal example, the additional variables from having two layers are at disposition to comply with more advanced modes of deformation. The added complexity comes, however, at the cost of optimization accuracy and efficiency.

By minimizing the number of actuators within the lattice (C), we are in fact setting its load bearing capacity as the objective. That this comes at virtually no cost in accuracy and converges with comparable computational effort to the benchmark implies that our method can be generally implemented for various shape morphing applications.

Setting the results back into the context of shape morphing and 4D printing, the method proposed is discussed in four main way: its adaptability, its scalability, its efficiency and structural integrity.

First, the paper shows that it is possible to activate a small lattice sample so as to achieve the four planar curvatures: linear displacement, convex, concave, and the transition between the two. This can be done using any type of cell structure and a varying amount of lattice layers and variables. The option to consider the type of unit cell itself as a variable is still open, which could lead to an even greater range of applications and complexity.

Adding variables however could lead to a compromise in the efficiency of the method. Although the doubling of the lattice size isn't problematic here, this is mainly because the desired shape is the same as the benchmark. Increasing the size of the model while also demanding a more complex deformation would be a challenge for the computational method, as the decrease in accuracy of the sinusoidal case study indicates. Further improvements should focus on improving its scalability.

The examples presented here are relatively small lattices, and real applications call for far larger structures. To accommodate for the increase, the implementation should be tuned to run considerably faster.

As mentioned above, computational efficiency and performance are tightly linked to the size of the problem. The options to subdivide a large structure based on its desired response type, avoiding the excessive use of boundary conditions and minimizing the number of actuators are all parameters that can be tuned and customized for given application requirements.

Considering the size and complexity of active lattice structures as well as the load range necessary in shape morphing applications, we support that pneumatically actuated 4D printing is the appropriate method of fabrication for our designs. The work reported in [3], [37] uses electro-mechanic actuators and a combination of hydraulic and mechanic actuators respectively for realization of active lattices. Such approaches increase significantly the complexity of prototypes and limit the physical validation of the results to few cells and actuators. With the 4D printed design approach shown in Fig. 1, it is straight forward to test different actuator placement configurations with comparatively low cost and low time, while conserving the sturdiness and load range provided by pneumatic actuation. However, printing and assembly introduce irregularities and constraints, especially at the nodes and at the junction between structure and actuator, which are not yet considered in the computational model presented in this paper. Further work is necessary to ensure the correct correlation between the numerical results and the experimental deformations.

6 CONCLUSION

This work proposes a computational method for the determination of optimized configurations of actuators for active lattice structures tested for a range of different shapes. It shows a successful application of the direct search GPS method available within the NOMAD optimization environment to solve a non-linear, discrete programming problem. The GPS method converges efficiently and with acceptable levels of accuracy. Further work should focus on the FEM evaluation of the structure performed at every iteration to improve the efficiency of the method.

The paper demonstrates how the actuator placement can be chosen to achieve multiple deformation modes. The choice of control nodes to achieve our desired shape is completely free. The switch from one layer to a 3D system is straightforward and produces accurate results in both cases. With five use cases, the paper illustrates the flexibility of the method proposed. The examples included demonstrate how to optimize the number of actuators required, which in turn increases the load bearing behavior and is central in shape morphing applications. The structures demonstrated can be 4D printed using pneumatic actuators, thus streamlining the fabrication process for active lattice structures that can support a range of morphed shapes. Future work will focus on fabrication and experimental validation of the results.

REFERENCES

[1] B. Jenett *et al.*, "Digital Morphing Wing: Active Wing Shaping Concept Using Composite Lattice-Based Cellular Structures," *Soft Robot.*, vol. 4, no. 1, pp. 33–48, Mar. 2017.

[2] S. Hyun and S. Torquato, "Optimal and manufacturable two-dimensional, Kagomé-like cellular solids," 2002.

[3] S. L. dos Santos e Lucato, J. Wang, P. Maxwell, R. M. McMeeking, and A. G. Evans, "Design and demonstration of a high authority shape morphing structure," *Int. J. Solids Struct.*, vol. 41, pp. 3521–3543, 2004.

[4] R. G. Hutschinson, "Mechanics of Lattice Materials," University of Cambridge, 2004.

[5] C. du Pasquier, "Modular Pneumatic Toolkit: an Application of 4D-Printing," ETH Zurich, 2017.

[6] Y. Liu, H. Du, L. Liu, and J. Leng, "Shape memory polymers and their composites in aerospace applications: a review," *Smart Mater. Struct.*, vol. 23, no. 2, p. 023001, Feb. 2014.

[7] F. Previtali, G. Molinari, A. F. Arrieta, M. Guillaume, and P. Ermanni, "Design and experimental characterisation of a morphing wing with enhanced corrugated skin," *J. Intell. Mater. Syst. Struct.*, vol. 27, no. 2, pp. 278–292, Jan. 2016.

[8] S. Daynes and P. M. Weaver, "Review of shape-morphing automobile structures: concepts and outlook," *Proc. Inst. Mech. Eng. Part D J. Automob. Eng.*, vol. 227, no. 11, pp. 1603–1622, Nov. 2013.

[9] P. Walters and D. Mcgoran, "Digital fabrication of " smart " structures and mechanisms - creative applications in art and design," *Soc. Imaging Sci. Technol.*, pp. 185–188, 2011.

[10] S. Kim, C. Laschi, and B. Trimmer, "Soft robotics: a bioinspired evolution in robotics," *Trends Biotechnol.*, vol. 31, no. 5, pp. 287–294, 2013.

[11] C. Majidi, "Soft Robotics: A Perspective—Current Trends and Prospects for the Future," *Soft Robot.*, vol. 1, no. 1, pp. 5–11, Mar. 2014.

[12] J. Rossiter, P. Walters, and B. Stoimenov, "Printing 3D dielectric elastomer actuators for soft robotics," 2009, p. 72870H.

[13] B. Gao, Q. Yang, X. Zhao, G. Jin, Y. Ma, and F. Xu, "4D Bioprinting for Biomedical Applications," *Trends Biotechnol.*, vol. 34, no. 9, pp. 746–756, 2016.

[14] J. Wu *et al.*, "Multi-shape active composites by 3D printing of digital shape memory polymers.," *Sci. Rep.*, vol. 6, p. 24224, Apr. 2016.

[15] I. K. Kuder, A. F. Arrieta, W. E. Raither, and P. Ermanni, "Variable stiffness material and structural concepts for morphing applications," *Prog. Aerosp. Sci.*, vol. 63, pp. 33–55, Nov. 2013.

[16] M. Eichenhofer, J. I. Maldonado, F. Klunker, and P. Ermanni, "ANALYSIS OF PROCESSING CONDITIONS FOR A NOVEL 3D-COMPOSITE PRODUCTION TECHNIQUE," 2015.

[17] Q. Ge, H. J. Qi, and M. L. Dunn, "Active materials by four-dimension printing," *Appl. Phys. Lett.*, vol. 103, no. 13, p. 131901, Sep. 2013.

[18] A. F. Arrieta, O. Bilgen, M. I. Friswell, and P. Ermanni, "Modelling and configuration control of wing-shaped bi-stable piezoelectric composites under aerodynamic loads," *Aerosp. Sci. Technol.*, vol. 29, no. 1, pp. 453–461, Aug. 2013.

[19] P. Moseley, J. M. Florez, H. A. Sonar, G. Agarwal, W. Curtin, and J. Paik, "Modeling, Design, and Development of Soft Pneumatic Actuators with Finite Element Method," *Adv. Eng. Mater.*, vol. 18, no. 6, pp. 978–988, Jun. 2016.

[20] K.-J. Lu and S. Kota, "An Effective Method of Synthesizing Compliant Adaptive Structures using Load Path Representation," *J. Intell. Mater. Syst. Struct.*, vol. 16, no. 4, pp. 307–317, Apr. 2005.

[21] S.-H. Ahn, K.-T. Lee, H.-J. Kim, R. Wu, J.-S. Kim, and S.-H. Song, "Smart soft composite: An integrated 3D soft morphing structure using bend-twist coupling of anisotropic materials," *Int. J. Precis. Eng. Manuf.*, vol. 13, no. 4, pp. 631–634, Apr. 2012.

[22] T. Chen and K. Shea, "Design and Fabrication of Hierarchical Multi-Stable Structures Through Multi-Material Additive Manufacturing," in *Volume 2A: 42nd Design Automation Conference*, 2016, p. V02AT03A032.

[23] S. Daynes, P. M. Weaver, and K. D. Potter, "Aeroelastic Study of Bistable Composite Airfoils," *J. Aircr.*, vol. 46, no. 6, pp. 2169–2174, Nov. 2009.

[24] "BMW Group - The Next 100 Years - Brand Visions." [Online]. Available: https://www.bmwgroup.com/en/next100/brandvisions.html. [Accessed: 18-Dec-2018].

[25] R. F. Shepherd, A. A. Stokes, R. M. D. Nunes, and G. M. Whitesides, "Soft Machines That are Resistant to Puncture and That Self Seal," *Adv. Mater.*, vol. 25, no. 46, pp. 6709–6713, Dec. 2013.

[26] A. M. Nasab, A. Sabzehzar, M. Tatari, C. Majidi, and W. Shan, "A Soft Gripper with Rigidity Tunable Elastomer Strips as Ligaments," *Soft Robot.*, 2016.

[27] R. F. Shepherd *et al.*, "Multigait soft robot.," *Proc. Natl. Acad. Sci. U. S. A.*, vol. 108, no. 51, pp. 20400–3, Dec. 2011.

[28] M. Wehner *et al.*, "An integrated design and fabrication strategy for entirely soft, autonomous robots," *Nat. Publ. Gr.*, vol. 536, 2016.

[29] X. Guo, J. Jiang -, S. L. dos Santos Lucato, R. M. McMeeking, and A. G. Evans, "Actuator placement optimization in a Kagome based high authority shape morphing structure," 2005.

[30] J. S. Arora, M. W. Huang, and C. C. Hsieh, "Review Papers Methods for optimization variables: a review of nonlinear problems with discrete," Springer-Verlag, 1994.

[31] V. Torczon and S. J. Optim, "ON THE CONVERGENCE OF PATTERN SEARCH ALGORITHMS *," 1997.

[32] N. Wicks, J. H.-J. of, and undefined 2004, "Sandwich plates actuated by a Kagome planar truss," *... .asmedigitalcollection.asme.org.*

[33] R. MacCurdy, R. Katzschmann, K. Youbin, and D. Rus, "Printable hydraulics: A method for fabricating robots by 3D co-printing solids and liquids," in *2016 IEEE International Conference on Robotics and Automation (ICRA)*, 2016, pp. 3878–3885.

[34] C. Audet and W. Hare, "Derivative-Free and Blackbox Optimization," *Springer Ser. Oper. Res. Financ. Eng.*, p. 302, 2017.

[35] C. Audet and J. E. Dennis, "Mesh Adaptive Direct Search Algorithms for Constrained Optimization," *SIAM J. Optim.*, vol. 17, no. 1, pp. 188–217, Jan. 2006.

[36] N. Mladenović and P. Hansen, "Variable neighborhood search," *Comput. Oper. Res.*, vol. 24, no. 11, pp. 1097–1100, Nov. 1997.

[37] D. Baker and M. I. Friswell, "Determinate Structures for Wing Camber Control," *Smart Mater. Struct.*, vol. 18, 2009.

Proceedings of the ASME 2019
International Design Engineering Technical Conferences
and Computers and Information in Engineering Conference
IDETC/CIE2019
August 18-21, 2019, Anaheim, CA, USA

DETC2019-97840

A DESIGN MODIFICATION SYSTEM FOR ADDITIVE MANUFACTURING: TOWARDS FEASIBLE GEOMETRY DEVELOPMENT

Seyedeh Elaheh Ghiasian, Prakhar Jaiswal, Rahul Rai, & Kemper Lewis[*]

Department of Mechanical and Aerospace Engineering
University at Buffalo
Buffalo, NY 14260, USA
Email: {seyedehe, prakharj, rahulrai, kelewis}@buffalo.edu

ABSTRACT

The substantial role of additive manufacturing (AM) in fabricating unique geometries is undeniable in the domain of design and manufacturing. However, the successful implementation of AM technologies requires a consistency between the geometric specifications of a component and AM manufacturability capabilities and constraints. Otherwise, AM could result in failed prints and a wasteful use of resources. The goal of this research is to provide geometrically feasible designs for AM processes by rectifying the potentially infeasible geometries. To this end, a novel design modification system is presented that addresses the problematic areas of an AM-infeasible component using appropriate redesign solutions. This system also includes a geometric assessment algorithm which identifies the potential problematic part features using a comprehensive evaluation. Based on the obtained manufacturability feedback, the detected problematic features are then modified through a holistic design modification system. The functionality of the presented system is illustrated using a case study, and the effectiveness of the implemented modification approaches is also demonstrated through an experiment.

---
[*]Corresponding Author.

1 INTRODUCTION

Additive Manufacturing (AM) is opening new horizons and shifting many design and manufacturing paradigms by removing many constraints imposed by conventional manufacturing [1]. While AM is billed as an advanced and effective fabrication method, it also includes important limitations that must be considered in a design process including challenges with printing certain geometries, shortages in the number of materials available, and relatively poor dimensional accuracy [2, 3]. To successfully implement AM technologies, the design specifications of products must be compatible with both the capabilities and limitations of AM while meeting all the functional requirements of the product.

Taking advantage of the potential of AM can be achieved through two possible approaches: 1) In an original design process, developing a plan to use AM as the sole fabrication process for a part, or 2) Modifying an already designed component and ensuring it is feasible for the chosen AM process. The first approach, which is an a priori approach, starts by establishing an appropriate design for AM from the conceptual design steps considering both functional specifications and AM process characteristics concurrently [3,4]. This may be used when designing a new product module that consolidates a number of components into an innovative new geometry, previously too costly or complex to fabricate. However, the second approach is an a posteriori approach and focuses on modifying a current design to both take advantage of and to avoid the limitations of the chosen AM method. This may be used to address geometric areas that are problematic to fabricate additively such as thin walls. In this study, we propose a posteriori method for rectifying the problematic geometric features of a current part, previously fabricated using traditional methods. By presenting an automated design modification methodology, we provide feasible designs for successful transition from conventional to additive manufacturing.

For a model to be 3D printable, it must meet a set of topological requirements. This is because AM processes may struggle in creating certain geometries due to limitations in the process type or machine resolution [5]. These problematic geometric features, denoted as *critical areas* in this work, could result in failed fabrication processes [6, 7], a lack of manufacturing precision [8], poor quality products, or an ineffective use of resources (i.e. energy, material, time, cost, human resource). Therefore, these *critical areas* must be corrected in order to achieve a feasible geometry for AM. By addressing the existing *critical areas*, the presented modification system can suggest appropriate redesign solutions using a set of automatic modification processes. This system is integrated with an assessment algorithm that evaluates the geometric feasibility of a given part. The combination of the geometric assessment algorithm and the modification approaches provide an effective detection-modification system for AM design analysis. The major contributions of this paper are:

1. A 3D morphological operations-based modification system to improve the manufacturability of 3D geometries.
2. Fully automated implementation of the strategy which works directly on 3D models, rather than 2D slices.
3. Flexibility in the modification approach allowing the user to express their preferences in the modification criteria.

In this work, we use a case study to illustrate the details of the design modification system. After providing an overview of our AM design modification system in Section 3, the details are explored using a manifold case study in Section 4. A summary of this study and the potential future work are presented in Section 5. The presented redesign system represents a unique contribution to the literature as discussed in the next section.

2 LITERATURE REVIEW

In this section, we discuss the existing related work in the following areas: 1) Design for Additive Manufacturing (DfAM) and the relevant guidelines and frameworks, 2) AM manufacturability and feasibility analysis, and 3) AM redesign methodologies.

According to the work in [3], there exists significant demand for new DfAM methods. These methods, which are classified as a priori design approaches in our study, involve the consideration of AM options from the beginning of a design process. A number of methods have been developed for DfAM which have been thoroughly reviewed in previous work [9–12]. These studies classify these methods based on various criteria, present their advantages and disadvantages, and discuss the remaining opportunities and challenges of DfAM. As one example of this kind of approach, Vayre et al. [7] develop a general design methodology for AM which accounts for manufacturing constraints directly. Also, Ponche et al. [3] describe the detailed stages of a promising design procedure which integrates a global DfAM approach. In addition, a set of new design rules for part consolidation for DfAM purposes is developed by Yang et al. [13]. This study also introduces a numerically assisted approach to automatically implement the new proposed design rules.

While these methods approach DfAM holistically, other works have developed specific sets of design guidelines and frameworks for particular AM technologies [2, 6, 14–20]. These guidelines and frameworks facilitate and provide design guidance for different additive fabrication techniques.

While effective when considering AM from the beginning of a design process, these methods do not address the assessment and modification of already designed components. Applying such a posteriori design methods could either modify the geometry of existing components or adjust AM manufacturing parameters to handle the current geometry. To implement an a posteriori design approach, an assessment first needs to be performed to evaluate the feasibility of the part to be printed additively. For example, Conner et al. [21] present a reference system that evaluates the suitability of printing components using AM from three perspectives: complexity, customization and production volume. Tedia and Williams [22] perform manufacturability analysis for AM focusing on a geometric assessment and an estimation of support material and build time. In [23], we broaden this analysis by presenting a more comprehensive assessment of geometric evaluation identifying seven types of critical feature morphologies while also considering all fabrication times and costs. In another recent study, an interactive tool for AM manufacturability analysis which is capable of predicting achievable tolerances for additively manufactured parts is presented by Budinoff et al. [24]. While these a posteriori methods provide effective foundations for AM manufacturability analysis, they do not apply the modifications necessary to convert infeasible geometries to suitable candidates for additive processes.

To address the geometry modification challenge, Cali et al. [25] propose a semi-automatic redesign method which converts a solid model into a printable and functional model with joints and internal friction. This model can directly be fabricated by an AM process with no need for assembly. A framework for the decomposition of a large 3D model into smaller parts is developed by Luo et al. [26]. The partitioning of the model allows for the printing of multiple smaller parts whose sizes are consistent with the chamber sizes of 3D printers. Both [25] and [26] apply changes in the geometries of current components to provide feasible designs for additive processes by considering AM manufacturability constraints such as the chamber sizes. However, they do not account for the potentially problematic printing features that either may exist in the original model or may be generated through the part's partitioning procedures.

Lindemann et al. [27] present a methodology to screen parts through three phases: information, assessment, and decision. Their methodology also suggests possible redesign approaches and helps to estimate the economic implications of AM. Although this work presents a new economic-based methodology for screening or redesigning a part, it only provides general descriptions of the required screening and redesign processes. In another study, Stava et al. [5] present a correction system for AM which suggests redesign modifications for 3D models based on the structural stress analysis. Despite the utility of this redesign approach, implementing a finite element analysis is usually computationally expensive and may cause an increase in the computational time. To propose balanced shapes for 3D fabrication, Prvost et al. [28] introduce an approach that modifies the objects' volume through a balance optimization process focusing solely on improving its stability. Nelaturi et al. [8] provide geometric manufacturability feedback and model correction for AM by formalizing the measurement of the geometric deviation between the designed and manufactured shapes. Jaiswal and Rai [29] propose a reasoning framework that identifies and corrects the areas of a part's geometry that would be problematic for additive processes. The identified printing areas are rectified using a physics-based mesh deformation scheme. The approaches proposed by [8] and [29] are effective redesign methods which are used for improving the printability; however, they apply localized corrections to 2D slices instead of a 3D model directly. In addition to these research studies, there exist some AM software packages that enable minor component modification; however, the accuracy and utility of these approaches are difficult to assess due to the lack of documentation of the implemented methodologies and also the limited functionality to evaluate the print quality of a 3D model [29].

While these collective works represent significant contributions to the field of DfAM, a number of limitations still exist including automatic detection of different types of 3D printability issues, limited applicability of the methods for 3D models, significantly large computational needs, and limited generalizability for modifying any geometry. The system presented in the next section addresses a number of these issues using a design modification system that operates on voxelized representations of a component. The system automatically identifies and locally modifies different types of *critical areas* of a given part using a purely geometric based approach reducing the computational requirements relative to approaches that use additional analysis such as structural stress analysis. The modification approaches, using the voxelized representation, can also be adjusted to different additive processes and because it is automated would allow

for more large-scale application to a database of part geometry files.

3 DESIGN MODIFICATION SYSTEM

A schematic of the design modification system is shown in Fig. 1. The core functionality is based on voxelized model representations which are developed by segmenting a 3D continuous object to an equivalent 3D discrete model using regular grids of voxels [30]. This technique is capable of storing geometric information in a discrete and efficient format and has wide applications in the domains of computer graphics and image processing [31–33]. It has also been used as a practical tool for studying and simulating the AM process [22, 23, 34] given the similarities between voxelized representations and the layer by layer approach of many additive technologies. The voxelization process implemented in this work is based upon a ray intersection algorithm [35] and is performed in MATLAB.

The overall detection-modification approach is initiated, as shown in Fig. 1, using a STL file which is converted into a voxelized model. Then the AM feasibility assessment, presented in [23], is applied to the voxelized model in order to identify the potential *critical areas* of the part that may be difficult to fabricate using AM (e.g., thin walls, tight/sharp corners, and edges). *Critical areas*, more specifically, refer to the local regions of a part whose corresponding features (e.g., thicknesses or diameters) are less than the minimum printable feature size in the AM process. As a result, these *critical areas* will have defective features, impacting their functionality and handling. Two fundamental groups of *critical areas* have been identified using sets of morphological operations in [23]. Morphological operations are highly effective tools mostly used in image processing, computer vision, and shape recognition [36, 37]. They provide set theoretic and integral formulations for analyzing geometric structures in terms of other (typically simpler) shapes [8]. These operations are highly compatible with 3D voxelized models because of the discretized surfaces, allowing for effective detection of geometric features of a component.

Within the two groups, seven types of *critical areas* are identified through the detection analysis: thin walls, tight/sharp outer corners and edges, tiny bores, small gaps, and tight/sharp inner corners and edges. The first three and the last four are classified into two distinct groups: (Φ) and (Ψ), respectively, based upon the morphological operations used to detect each of them [23]. Once detection analysis is accomplished and if *critical areas* are detected, the modification analysis starts with Group Φ and then moves to Group Ψ as described in the following steps.

Step 1: Modification of Thin Walls from the First Group of Critical Areas (Φ)

As the first step in our design modification approach, the thin walls from Group Φ are modified. Thin walls provide a challenge

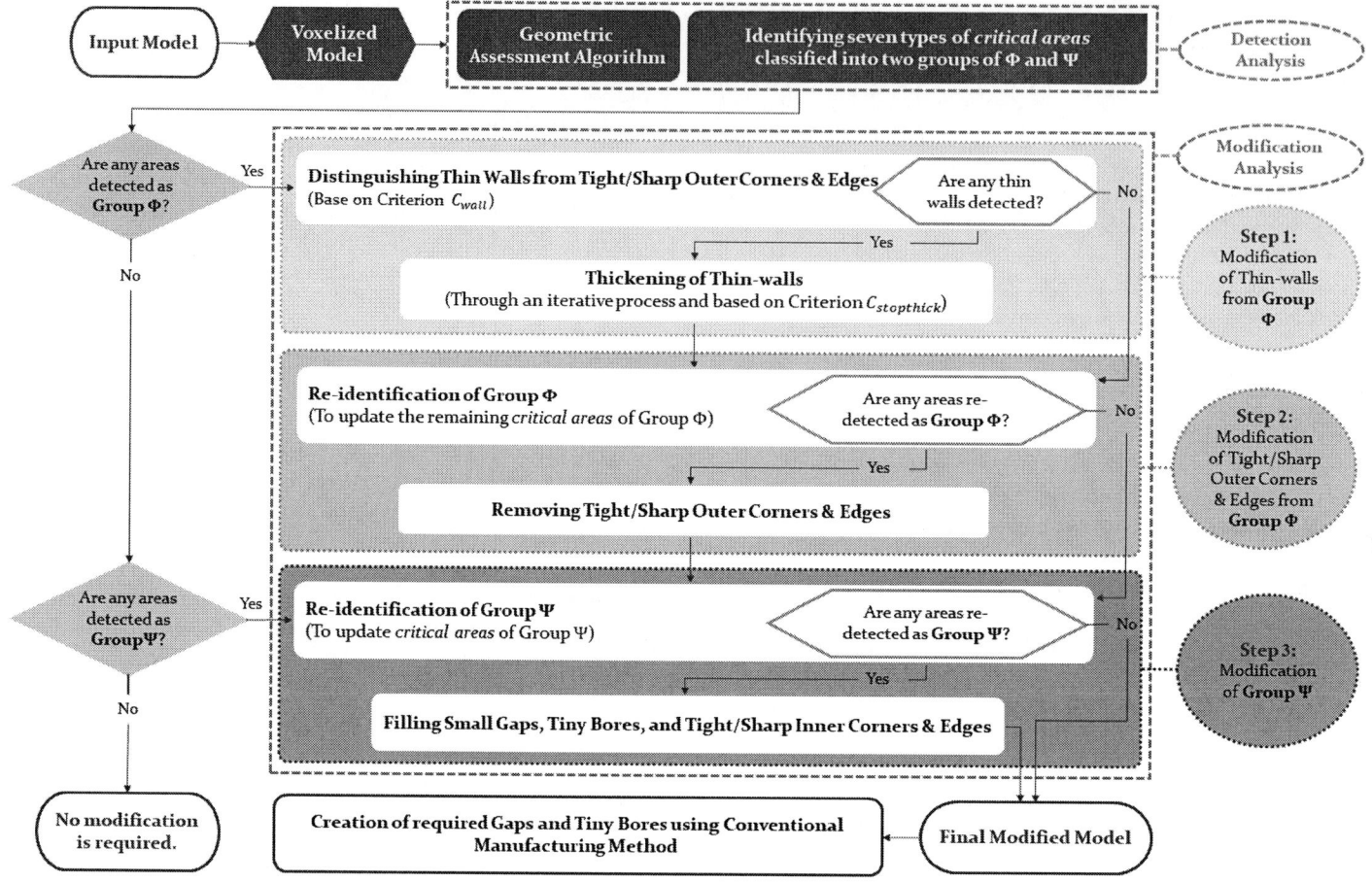

FIGURE 1: Design modification system for AM

for 3D printing, particularly if they compose large regions of the part. These areas may fail to be printed accurately and even if they are printed, undesirable deformations could occur in their locations because of high residual stress caused by part cooling after the fabrication process [16]. Similar to [5, 29], a thickening modification approach is developed for locally correcting thin walls in this study. By providing additional material, the thin walls could reach sufficient thicknesses, allowing for useful part prints.

As Fig. 1 shows, before applying the thickening modification, the thin walls must be distinguished from the tight/sharp outer corners and edges.

Distinguishing Thin Walls from Tight/Sharp Outer Corners and Edges

In the feasibility assessment of the part, thin walls and tight/sharp outer corners and edges are identified together as part of Group Φ since the same morphological operations are used to detect them [23]. However, geometrically these areas require different modifications as thin walls need to be thickened while tight or sharp outer corners and edges must be rounded. To distinguish between these *critical areas*, we define a criterion of C_{wall} in Eqn. (1):

$$C_{wall} : \ v \ (c\tau)^3 \qquad (1)$$

where v is the volume of a *critical area* belonging to Group Φ and τ is the minimum feature size determined according to the chosen AM process. The parameter c is a heuristic parameter and is determined experimentally.

Criterion C_{wall} differentiates between thin walls and tight/sharp outer corners and edges based on the constituent volumes of their associated 3D voxel arrays. For instance, tight/sharp outer corners and edges are composed from small voxel groups and consequently have small volumes while thin walls contain larger volumes due to their geometric structures.

V02AT03A034-4

According to Eqn. (1), the areas with voxel array volumes greater than the value of $(c\tau)^3$ are labeled as thin walls. Otherwise, the area is considered as a tight/sharp outer corner or edge. The voxel arrays associated with thin walls are extracted and used in the next sub-step for further modification.

In this study, the value of 3 is selected for the heuristic parameter of c based on a set of experiments on various components. In these experiments, 20 arbitrary manifold and fan components from GrabCAD.com are evaluated by our feasibility assessment algorithm. Each component includes several *critical areas* detected as Group Φ. For each component, the voxel array volume associated to each *critical area* is calculated and saved. The voxel array volumes associated with the thin walls of the explored components are much greater than the value of $(3\tau)^3$; hence, the value of Parameter c is set to 3 in this work. While this value can help distinguish between thin walls and tight/sharp outer corners and edges for these components, more investigation is required to establish the value of parameter c precisely.

Fig. 2 illustrates a graphical representation of the difference between thin walls and sharp outer edges. The gray regions of the fan represent the areas with no printing challenges while the blue areas represent the *critical areas* whose dimensions are less than 0.12 mm ($\tau = 0.12$ mm) and are therefore detected as part of Group Φ. As Fig. 2 illustrates, the fan part includes three thin walls and several sharp outer edges. Using Eqn. (1) and setting the parameter $c = 3$, these areas can be distinguished from each other since the voxel array volume of each thin wall is large and much greater than the value of $(3\tau)^3$. However, the voxel array volume of each sharp edge is less than $(3\tau)^3$.

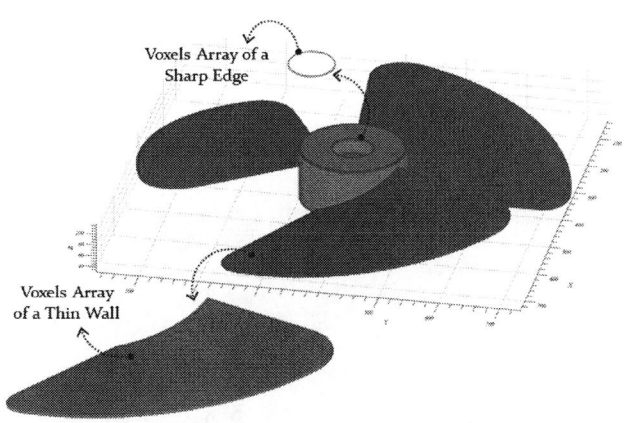

FIGURE 2: A fan with three thin walls and several sharp outer edges. Different colors represent different areas: ▦ 3D voxels array of the part with no printing issues, ▦ 3D voxels arrays of *critical areas* detected as Group Φ. $\Phi \leq 0.12$ mm

If no thin walls are identified, then the next sub-step is skipped, and the process proceeds to Step 2 to modify the tight/sharp outer corners and edges.

Thickening of Thin Walls

After identifying the thin walls, an iterative approach is implemented to thicken each one. The thickening modification is achieved using the *dilation* operation of mathematical morphology. Applying this operation directly on the 3D voxelized model allows for the uniform thickening of the part in the X, Y, and Z directions. To determine whether or not a thin wall is sufficiently thickened, a criterion of $C_{stopthick}$ is defined as represented in Eqn. (2). Based on this criterion, the thin wall is evaluated by the feasibility assessment algorithm after each thickening step in order to identify the remaining thin portions. To do so, the total number of voxels placed in the thin portion, NV_{thin}, is calculated and then divided by the total number of voxels that constitute the relatively thickened wall, NV_{tot}. If the value of this ratio becomes less than the value of the parameter of T_e, the condition of Criterion $C_{stopthick}$ is satisfied, indicating that the thin wall is thickened sufficiently.

$$C_{stopthick}: \quad \frac{NV_{thin}}{NV_{tot}} \leq T_e \tag{2}$$

where, $0 \leq T_e \leq 1$.

Based on the user's preference, both full and partial thickening are possible by adjusting the value of T_e before the modification analysis starts. A T_e value of 0 leads to a fully thickened part (i.e., NV_{thin} becomes 0), while a value of 1 means no thickening modification is required. Any value between 0 and 1 represents partial thickening. The maximum thickness allowed for a thin wall is set to the minimum feature size required for a feasible 3D print. In this way, we not only ensure a feasible thin wall, but we also prevent over-thickening of the wall which could cause material conglomeration and subsequent internal stresses in the wall [27]. Furthermore, it helps avoid major changes in the part's topology as only locally small design modifications are targeted in this study.

Step 2: Modification of Tight/Sharp Outer Corners and Edges

In Step 2 of the design modification approach, the remaining *critical areas* in Group Φ, the tight/sharp outer corners and edges, are modified. Tight/sharp outer corners and edges can cause printing defects, stress concentrations, or hot spots [6]. These areas need to be replaced by curved surfaces in order to minimize acceleration and deceleration of the nozzle movement [7]. Before these areas are modified, we evaluate the modified geometry from Step 1 to assess whether any additional tight/sharp outer corners and edges are created from the geometric modifications to correct the thin walls.

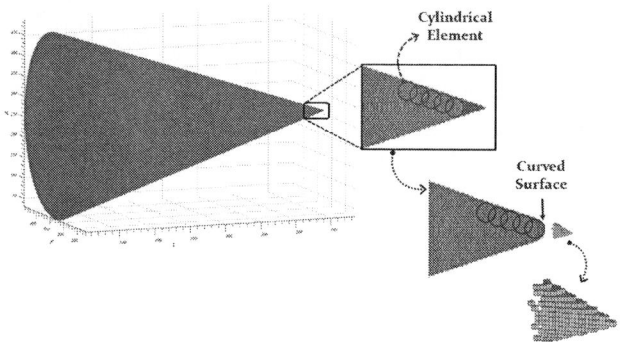

FIGURE 3: The modification of the sharp outer corner of a cone part. Different colors represent different areas: ▨ 3D voxels array of the part with no printing issues, ▨ 3D voxels array of the removed sharp outer corner

For all tight/sharp outer corners and edges, we implement the *opening* filtering operation of mathematical morphology for applying the required modification. Let Ω and Ω' represent respectively the 3D voxels array of the initial model and the modified model obtained from the previous modification, where $\Omega, \Omega' \subseteq \mathbb{R}$. Using the *opening* operation, the 3D voxels array of the updated model obtained from the previous modification (Ω') is firstly eroded by amount of τ, and then the eroded array is dilated by the same amount. The τ parameter is the diameter of a cylindrical structuring element used in the morphological *erosion* and *dilation* filters, and it is set to the minimum feature size required for Group Φ. This process results in a new modified model, Ω'', where the tight/sharp outer corners and edges from Ω' are replaced. By using the cylindrical structuring element in the *opening* morphological operation, the 3D voxels array of the modified part (Ω'') with appropriate filleted corners and edges is obtained as shown in Eqn. (3):

$$\Omega'' = O\left(\Omega', \tau\right) = \delta\left(\varepsilon\left(\Omega', \tau\right), \tau\right) \qquad (3)$$

where, $O\left(\Omega', \tau\right)$ represents the morphological *opening* operation performed on Ω'. In addition, δ and ε indicate the morphological *dilation* and *erosion* operations.

A representation of the creation of appropriately curved surfaces to replace the sharp outer corner of a cone part is presented in Fig. 3. In this figure, the blue colored area is eliminated by the morphological *opening* operation with the cylindrical structuring element of diameter $\tau = 0.12$ mm. By removing this area from the 3D voxels array of the part, the modified cone with appropriate filleted curved surface is obtained.

Step 3: Modification of the Second Group of Critical Areas (Ψ)

This step focuses on modifying the *critical areas* in Group Ψ including tiny bores, small gaps, and tight/sharp inner corners and edges. These regions would be difficult to print due to their inaccessibility or potential powder adhesion in these locations [16, 18].

As discussed in Step 2, the corrected component topology may create new *critical areas*. As such, the modified component geometry from Step 2 is re-evaluated to identify any new areas in Group Ψ.

For the identifies areas, we implement a hybrid modification approach given the common printing challenges of the *critical areas* in Group Ψ. First, these areas are modified geometrically and their locations are filled with excess materials. Then, certain geometries such as tiny bores and necessary gaps can be created using cost effective conventional manufacturing techniques such as drilling.

For the geometric modification of Group Ψ, we apply the *closing* filtering operation of mathematical morphology directly on the 3D voxels array of the modified model obtained from the previous modification (Ω''). Through this operation, the 3D voxels array Ω'' is firstly dilated with a cylindrical structuring element of diameter τ', and then the dilated array is eroded by the same amount. The value of τ' is set to the minimum feature size required for Group Ψ. Using these morphological operations, the existing tiny bores and small gaps are filled, and tight/sharp inner corners and edges are appropriately filleted. The final modified

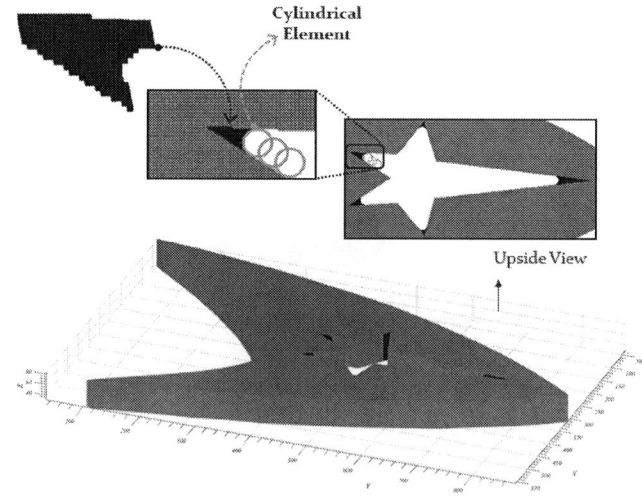

FIGURE 4: The modification of the sharp inner corners of a part. Different colors represent different areas: ▨ 3D voxels array of the part with no printing issues of Group Ψ, ▨ 3D voxels arrays of the areas added the inner sharp corners

3D model, $\overline{\Omega}$, is obtained as shown Eqn. (4):

$$\overline{\Omega} = C\,(\Omega'',\,\tau')\,=\,\varepsilon\,(\,\delta\,(\Omega'',\,\tau'),\,\tau')\qquad(4)$$

where, $C\,(\Omega'',\tau')$ represents the morphological *closing* operation performed on Ω''.

A representation of correcting the sharp inner corners of a part is presented in Fig. 4. In this figure, the red areas represent the material added to the sharp inner corners by the morphological *closing* operation using the cylindrical structuring element of diameter $\tau' = 0.2$ mm. The output of this modification process is an updated part with filleted surfaces in the original locations of the initial sharp inner corners.

As the last step of the modification procedure, the required gaps and tiny bores existing in the initial part are created using conventional manufacturing processes such as machining. This hybrid design modification approach utilizes the advantages of both AM and conventional manufacturing.

It should be noted that if a filled hole or a gap is located in the interior or inaccessible regions of the 3D model, it would be difficult to recreate it with machining. We aim to solve this challenge in our ongoing research by distinguishing the interior and exterior *critical areas* of Group Ψ and only filling those areas that are located in accessible regions of the part.

Together the three main steps and their associated subdivisions of our presented design modification approaches combine to deliver an effective design modification system which can be applied to any type of AM process. In the next section, we illustrate the process for modifying 3D parts using a case study.

4 DESIGN MODIFICATION CASE STUDY

To illustrate the workflow and utility of the system presented in Section 3, we present the design modification of a manifold part shown in Fig. 5. The dimension of the manifold is scaled so that the most dominant diagonal axis of the part's Axis-Aligned Bounding Box (AABB) is set equal to 20 mm. This results in a dimensional size of $16.95 \times 5.87 \times 8.85$ mm. For voxelization, a voxel size of 0.02 mm is used.

4.1 Detection Analysis

The manifold part is evaluated by the feasibility assessment algorithm in order to determine the feasibility of its shape for an AM process. The output of this assessment is illustrated in Fig. 6. The blue and red regions represent the two groups of *critical areas*, respectively: 1) Group Φ: Thin walls and tight/sharp outer corners and edges 2) Group Ψ: Tiny bores, small gaps and tight/sharp inner corners and edges. The dimensions of the blue and red features are less than 0.12 mm and 0.2 mm in any X, Y, Z direction.

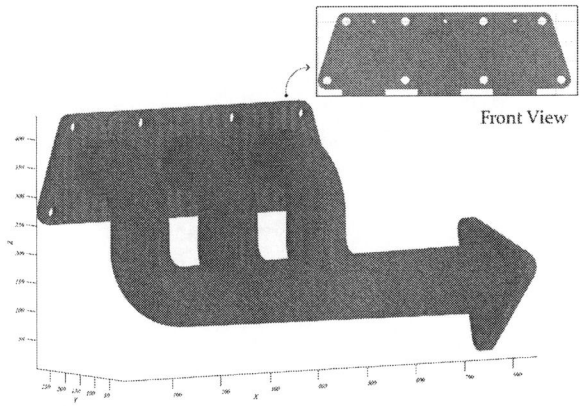

FIGURE 5: The general view of the studied manifold

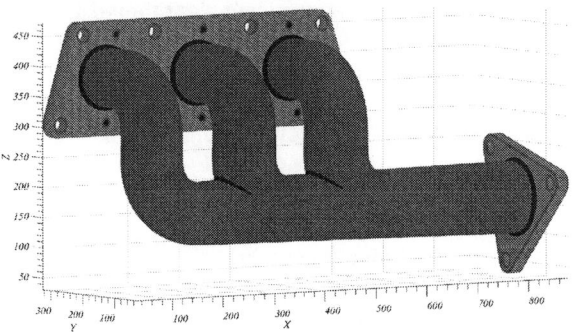

FIGURE 6: All types of *critical areas* of the studied manifold. Different colors represent different areas: ■ Areas with no printing issues, ■ Thin walls and tight/sharp outer edges (Φ), ■ Tiny bores, small gaps and tight/sharp inner corners and edges (Ψ). $\Phi \le 0.12$ mm & $\Psi \le 0.2$ mm

In Fig. 6, large portions of the main body are detected as being a thin wall. In addition, various small gaps, tiny bores, and tight/sharp inner and outer corners and edges are identified. Due to the existence of these *critical areas*, the part's current shape is deemed infeasible for AM, and in need of geometric modification.

4.2 Modification Analysis

Step 1: Modification of Thin Walls from Group (Φ)

As the geometric assessment determines several regions of the manifold as Group Φ of *critical areas* (shown in the blue areas in Fig. 6), the design modification process begins with the correction of this group.

Distinguishing Thin Walls from Tight/Sharp Outer Corners and Edges

Before applying any design corrections, the system explores whether or not there are any thin walls in the manifold using Eqn. (1). The parameter τ used in Eqn. (1) is set to the minimum feature size of Group Φ which is 0.12 mm. Based on the C_{wall} criterion, the process detects one thin wall in the manifold which is represented in Fig. 7. To modify this feature, its associated voxels array is extracted and used in the next sub-step.

Thickening of Thin Walls

The detected thin wall of the manifold shown in Fig. 7 is thickened through an iterative process. As we aim to achieve the full thickening of the thin wall, the parameter of T_e used in Eqn. (2) is set to 0, prior to the start of the modification process. After each thickening step, the geometry of the thin wall is evaluated by our feasibility assessment algorithm to determine whether it is thickened sufficiently. The outputs of the geometric assessment of the thin wall after four iterations of the thickening process are represented in Fig. 8. The ratio of NV_{thin}/NV_{tot} associated with each thickening step is also specified at the bottom of each assessment output. In this figure, the dark gray and blue regions represent the thickened and the remaining thin regions of the relatively thickened wall in each iteration, respectively. The thickening procedure continues until the ratio of NV_{thin}/NV_{tot} equals 0 satisfying Eqn. (2). Fig. 9 represents the modified manifold after thickening the thin wall.

Step 2: Modification of Tight/Sharp Outer Corners and Edges

In this step of the design modification process, the tight/sharp outer corners and edges from Group Φ are modified.

Re-identification of Tight/Sharp Outer Corners and Edges

The geometry of the modified manifold is assessed to determine if new tight/sharp outer corners and edges are created after the thickening modification and the result is shown in Fig. 10. While no tight/sharp outer corners are identified for the manifold, it includes several sharp outer edges. As expected, no thin walls are identified any longer.

Removing Sharp Outer Edges

To modify the sharp outer edges, the morphological *opening* operation is applied to the 3D voxelized manifold. The diameter value of the cylindrical structural elements implemented in the *opening* morphology is set to 0.12 mm, which is the minimum feature size required for Group Φ of *critical areas*. By applying this operation, the voxels arrays of the sharp edges are removed from the 3D binary voxels array of the modified manifold, and consequently filleted edges are created in the original locations

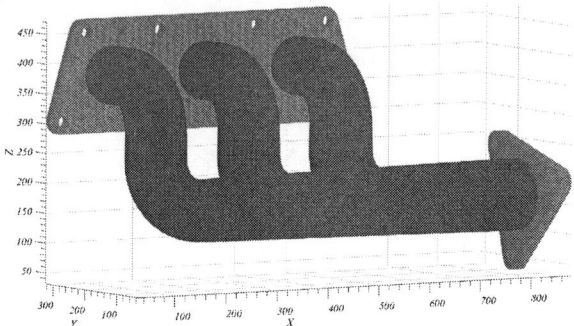

FIGURE 7: The detected thin wall of the manifold. ■ Thin wall, ■ The rest of the main body

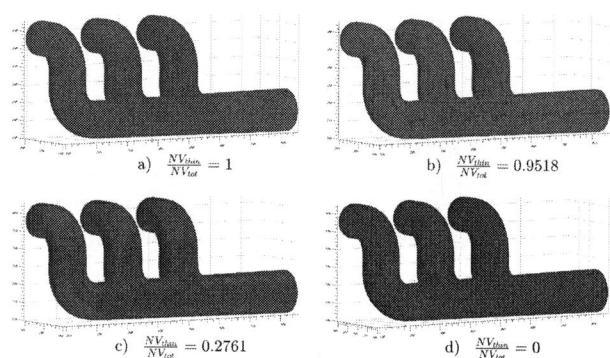

FIGURE 8: The geometric assessment outputs of the thin wall after four thickening steps. ■ Thickened areas, ■ Remaining thin regions of the relatively thickened wall
a) Step 1, b) Step 3, c) Step 6, d) Step 8 of the thickening procedure

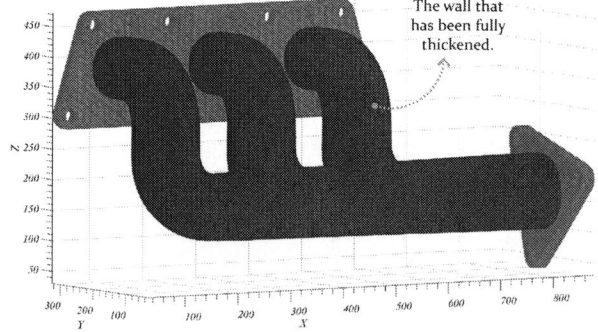

FIGURE 9: The modified manifold after thickening of the initial thin wall. ■ Modified area with no longer printing issues, ■ The rest of the main body

of the initial sharp outer edges. This results in the modified manifold shown in Fig. 11.

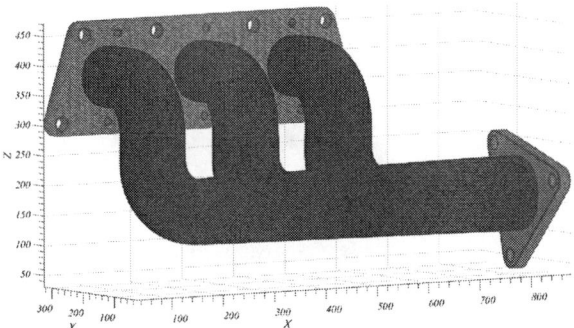

FIGURE 10: The updated Group Φ of *critical areas* of the modified manifold. ■ Modified areas after the thickening modification, ■ Remaining sharp outer edges (Φ ≤ 0.12 mm), ■ The remainder of the main body

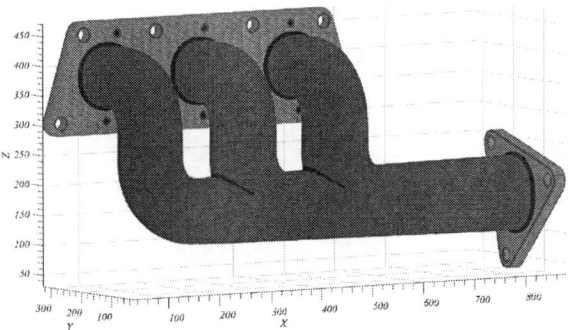

FIGURE 12: The updated Group Ψ of *critical areas* of the modified manifold. ■ Modified areas after correcting Group Φ, ■ Tiny bores, small gaps and tight/sharp inner corners and edges (Ψ ≤ 0.2 mm), ■ Areas with no initial printing issues

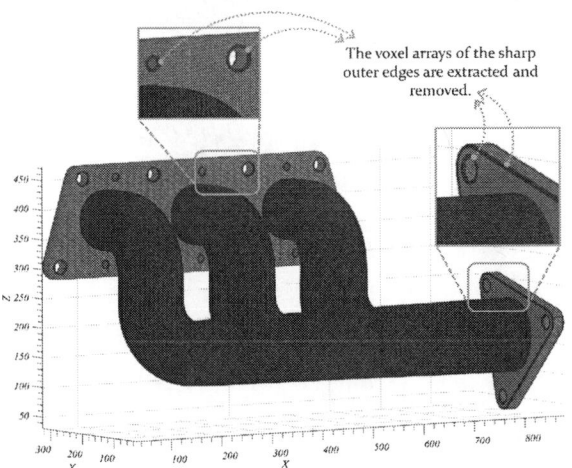

FIGURE 11: The modified manifold after removing sharp outer edges. ■ Modified areas after the thickening modification and the removal of sharp outer edges, ■ The remainder of the main body

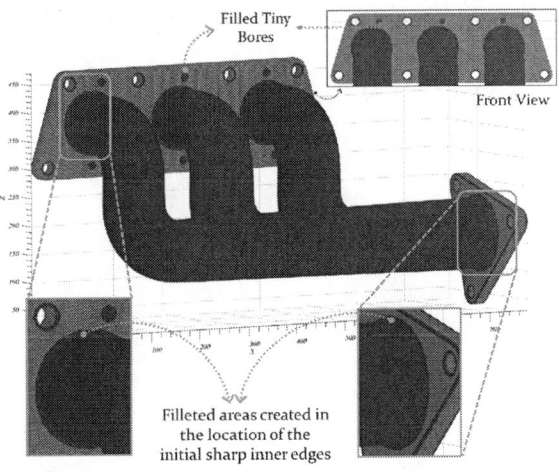

FIGURE 13: The final modified model after the correction of *critical areas* in Group Ψ. ■ Modified areas after correcting Group Φ and Group Ψ, ■ Areas with no initial printing issues

Step 3: Modification of the Second Group of Critical Areas (Ψ)

After modifying the thin wall and tight/sharp outer corners and edges, the *critical areas* in Group Ψ are corrected in this step.

Re-identification of the Second Group of Critical Areas

The geometry of the modified manifold resulting is again re-assessed in order identify any new *critical areas* in Group Ψ. The output of the geometric assessment of this step is represented in Fig. 12. As several regions of the manifold are identified, the next sub-steps are needed to modify these areas.

Modification of Small Gaps, Tiny Bores, and Tight/Sharp Inner Corners and Edges

The detected tiny bores, small gaps and tight/sharp inner corners and edges of the manifold (Fig. 12) are updated using a hybrid modification approach. For the geometric modification, the system applies the *closing* filter to the 3D voxelized manifold. The diameter value of the cylindrical structural elements implemented in the *closing* filter is set to the minimum feature size required for Group Ψ, which is 0.2 mm. In this way, the sufficient fillet sizes are created in the locations of the detected tiny bores, small gaps, and tight/sharp inner corners and edges. By applying this design correction, a final modified manifold is obtained as illustrated in Fig. 13.

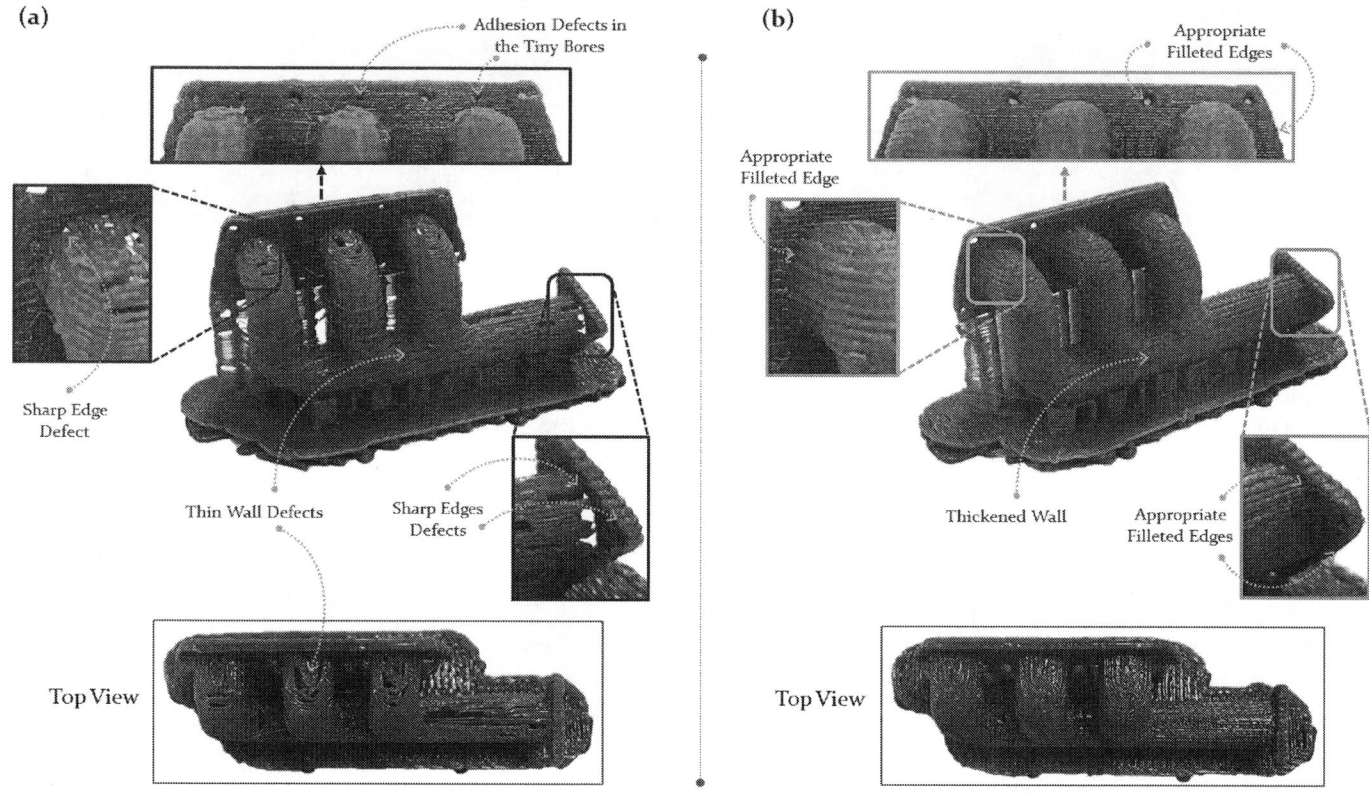

FIGURE 14: The manifold part (a) before and (b) after geometric modifications

Creation of Required Gaps and Tiny Bores

Once the geometric modification of the part is completed, certain required geometries such as tiny bores and necessary gaps can then be created using machining processes.

5 Experimental Verification

In order to illustrate the utility of our approach, the manifold case study was 3D printed before any assessment and modification and then after applying our system. For better visualization, larger scale versions of the original and modified model are printed. The dimensions of the printed parts, which are represented in Fig. 14, are equal to $38.45 \times 13.32 \times 20.07$ mm. We use a *FDM MakerBot* machine, Replicator (*5th generation*) with machine resolution 0.2 mm and material *Polylactic* acid or *PLA* for the fabrication. In Fig. 14(a), the *critical areas* in the original model before modification are shown, and in Fig. 14(b), these same areas are shown after detection and correction.

As shown, the original manifold model included a number of features and areas that were challenging for the 3D printing. This kind of print would result in wasted resources with little value. However, after using our detection-modification ap-

proach, the problematic regions are appropriately modified and feasibly printed, effectively rectifying the geometry without imposing significant changes in the part's topology.

6 CONCLUSIONS AND FUTURE WORK

The presented design modification system addresses a substantial need in design for AM processes. The printable feasibility of any part can be explored prior to printing using a voxelized model, and if the part is detected as an AM-infeasible candidate, automatic modifications are made to ensure that the part's geometry will be compatible with the AM capabilities. A set of visual outputs provided by the system after each step of modification also gives the user a level of understanding of the potential geometric changes caused by each modification. By converting infeasible geometries to feasible ones, engineers can take more advantage of more opportunities in AM. More broadly, the presented system could be adapted for implementation in an open source environment, leveraging the inherent power of large data sets to adapt to the best practices in AM design.

Although this paper provides a foundation for developing geometry re-designs for AM, there are many areas for further development. Our current modification analysis is applicable to the geometric correction of a single component. If a component is part of a larger multi-component module, investigating the acceptability of the final modified design with respect to the presence of the adjacent components is currently up to the user. However, our process could be updated to include geometric constraints imposed by adjacent components which could inform the modification limitations.

Another long-term objective for this work is to add more capabilities to the current design modification methodology by integrating build orientation. Build orientation can impact both *critical areas* and the required design modifications for additive processes. Occasionally, certain geometric modifications should be applied after the determination of the build orientation. For example, the conversion of round internal passages to drop-shape passages (to reduce the support structure amount and inaccessibility regions) is a required correction that needs to be performed after having the build orientation determined. This capability would add increased functionality to the presented modification system.

ACKNOWLEDGMENTS

The authors gratefully acknowledge the New York State Center of Excellence Materials Informatics at the University at Buffalo for the funding of this research. We are also grateful for our industrial research partner for their collaboration on this project. We also acknowledge the Department of Mechanical and Aerospace Engineering at the University at Buffalo for the support of this project. Lastly, we are grateful for the use of supporting CAD files from GrabCAD.com.

REFERENCES

[1] Huang, Y., Leu, M. C., Mazumder, J., and Donmez, A., 2015. "Additive manufacturing: Current state, future potential, gaps and needs". *Journal of Manufacturing Science and Engineering,* *137*(1), p. 014001. DOI:10.1115/1.4028725.

[2] Meisel, N., and Williams, C., 2015. "An investigation of key design for additive manufacturing constraints in multi-material three-dimensional printing". *Journal of Mechanical Design, 137*(11), p. 111406. DOI: 10.1115/1.4030991.

[3] Ponche, R., Hascot, J. Y., Kerbrat, O., and Mognol, P., 2012. "A new global approach to design for additive manufacturing: A method to obtain a design that meets specifications while optimizing a given additive manufacturing process is presented in this paper". *Virtual and Physical Prototyping, 7*(2), pp. 93–105. DOI: 10.1080/17452759.2012.679499.

[4] Ponche, R., Kerbrat, O., Mognol, P., and Hascoet, J. Y., 2014. "A novel methodology of design for additive manufacturing applied to additive laser manufacturing process". *Robotics and Computer-Integrated Manufacturing, 30*(4), pp. 389–98. DOI: 10.1016/j.rcim.2013.12.001.

[5] Stava, O., Vanek, J., Benes, B., Carr, N., and Mch, R., 2012. "Stress relief: Improving structural strength of 3D printable objects". *ACM Transactions on Graphics (TOG), 31*(4), p. 48. DOI: 10.1145/2185520.2185544.

[6] Kumke, M., Watschke, H., and Vietor, T., 2016. "A new methodological framework for design for additive manufacturing". *Virtual and Physical Prototyping, 11*(1), pp. 3–19. DOI:10.1080/17452759.2016.1139377.

[7] Vayre, B., Vignat, F., and F., V., 2012. "Designing for additive manufacturing". *Procedia CIrP, 3*, pp. 632–7. DOI: 10.1016/j.procir.2012.07.108.

[8] Nelaturi, S., Kim, W., and Kurtoglu, T., 2015. "Manufacturability feedback and model correction for additive manufacturing". *Journal of Manufacturing Science and Engineering, 137*(2), p. 021015. DOI: 10.1115/1.4029374.

[9] Doubrovski, Z., Verlinden, J. C., and Geraedts, J. M., 2011. "Optimal design for additive manufacturing: Opportunities and challenges". In International Design Engineering Technical Conferences and Computers and Information in Engineering Conference, American Society of Mechanical Engineers, Washington DC, DETC2011-48131.

[10] Yang, S., and Zhao, Y., 2015. "Additive manufacturing-enabled design theory and methodology: A critical review". *International Journal of Advanced Manufacturing Technology, 80*(1-4), pp. 327–342. DOI:10.1007/s00170-015-6994-5.

[11] Tang, Y., Tang, Y., Zhao, Y. F., and Zhao, Y. F., 2016. "A survey of the design methods for additive manufacturing to improve functional performance". *Rapid Prototyping Journal, 22*(3), pp. 569–590. DOI:10.1108/RPJ-01-2015-0011.

[12] Bikas, H., Stavropoulos, P., and Chryssolouris, G., 2016. "Additive manufacturing methods and modelling approaches: A critical review". *The International Journal of Advanced Manufacturing Technology, 83*(1-4), pp. 389–405. DOI:10.1007/s00170-015-7576-2.

[13] Yang, S., Santoro, F., and Zhao, Y. F., 2018. "Towards a numerical approach of finding candidates for additive manufacturing-enabled part consolidation". *Journal of mechanical design, 140*(4), p. 041701. DOI: 10.1115/1.4038923.

[14] Schmelzle, J., Kline, E. V., Dickman, C. J., Reutzel, E. W., Jones, G., and Simpson, T. W., 2015. "(Re) designing for part consolidation: Understanding the challenges of metal additive manufacturing". *Journal of Mechanical Design, 137*(11), p. 111404. DOI: 10.1115/1.4031156.

[15] Barnard, L., 2008. "Designing for laser sintering". *Journal for New Generation Sciences, 6*(2), pp. 47–59.

[16] Seepersad, C. C., Govett, T., Kim, K., Lundin, M., and Pinero, D., 2012. "A designer's guide for dimensioning

and tolerancing SLS parts". In Solid Freeform Fabrication Symposium, Austin, TX, pp. 921–931.

[17] Adam, G. A., and Zimmer, D., 2014. "Design for additive manufacturing – element transitions and aggregated structures". *CIRP Journal of Manufacturing Science and Technology*, **7**(1), pp. 20–28. DOI:10.1016/j.cirpj.2013.10.001.

[18] Kranz, J., Herzog, D., and Emmelmann, C., 2015. "Design guidelines for laser additive manufacturing of lightweight structures in Tial6v4". *Journal of Laser Applications*, **27**(S1), p. S14001. DOI:10.2351/1.4885235.

[19] Booth, J. W., Alperovich, J., Chawla, P., Ma, J., Reid, T. N., and Ramani, K., 2017. "The design for additive manufacturing worksheet". *Journal of Mechanical Design*, **139**(10), p. 100904. DOI:10.1115/1.4037251.

[20] Mokhtarian, H., Coatana, E., Paris, H., Mbow, M. M., Pourroy, F., Marin, R. R., Vihinen, J., and Ellman, A. A., 2018. "A conceptual design and modeling framework for integrated additive manufacturing". *Journal of Mechanical Design*, **140**(8), p. 081101. DOI:10.1115/1.4040163.

[21] Conner, B. P., Manogharan, G. P., Martof, A. N., Rodomsky, L. M., Rodomsky, C. M., Jordan, D. C., and Limperos, J. W., 2014. "Making sense of 3D-printing: Creating a map of additive manufacturing products and services". *Additive Manufacturing*, **1**, pp. 64–76. DOI:10.1016/j.addma.2014.08.005.

[22] Tedia, S., and Williams, C. B., 2016. "Manufacturability analysis tool for additive manufacturing using voxel-based geometric modeling". In 27th Annual International Solid Freeform Fabrication (SFF) Symposium, pp. 3–22.

[23] Ghiasian, S. E., Jaiswal, P., Rai, R., and Lewis, K., 2018. "From conventional to additive manufacturing: Determining component fabrication feasibility". In International Design Engineering Technical Conferences and Computers and Information in Engineering Conference, American Society of Mechanical Engineers, Quebec City, Canada, DETC2018-86238.

[24] Budinoff, H. D., McMains, S., and Rinaldi, A., 2018. "An interactive manufacturability analysis and tolerance allocation tool for additive manufacturing". In International Design Engineering Technical Conferences and Computers and Information in Engineering Conference, American Society of Mechanical Engineers, Quebec City, Canada, DETC2018-86344.

[25] Cal, J., Calian, D. A., Amati, C., Kleinberger, R., Steed, A., Kautz, J., and Weyrich, T., 2012. "3D-printing of non-assembly, articulated models". *ACM Transactions on Graphics (TOG)*, **31**(6), p. 130. DOI: 10.1145/2366145.2366149.

[26] Luo, L., Baran, I., Rusinkiewicz, S., and Matusik, W., 2012. "Chopper: Partitioning models into 3D-printable parts". *ACM Transactions on Graphics*, **31**(6). DOI: 10.1145/2366145.2366148.

[27] Lindemann, C., Reiher, T., Jahnke, U., and Koch, R., 2015. "Towards a sustainable and economic selection of part candidates for additive manufacturing". *Rapid Prototyping Journal*, **21**(2), pp. 216–27. DOI:10.1108/RPJ-12-2014-0179.

[28] R, R. P., Whiting, E., Lefebvre, S., and Sorkine-Hornung, O., 2013. "Make it stand: Balancing shapes for 3D fabrication". *ACM Transactions on Graphics (TOG)*, **32**(4). DOI: 10.1145/2461912.2461957.

[29] Jaiswal, P., and Rai, R., 2018. "A geometric reasoning approach for additive manufacturing print quality assessment and automated model correction". *Computer-Aided Design*, **109**, pp. 1–11. DOI: 10.1016/j.cad.2018.12.001.

[30] Cohen-Or, D., and Kaufman, A., 1995. "Fundamentals of surface voxelization". *Graphical models and image processing*, **57**(6), pp. 453–461. DOI:doi.org/10.1006/gmip.1995.1039.

[31] Hart, J. C., Sandin, D. J., and Kauffman, L. H., 1989. "Ray tracing deterministic 3D fractals". In ACM SIGGRAPH Computer Graphics, Vol. 23, ACM, New York, NY, pp. 289–296.

[32] Dong, Z., Chen, W., Bao, H., Zhang, H., and Peng, Q., 2004. "Real-time voxelization for complex polygonal models". In 12th Pacific Conference in Computer Graphics and Applications, IEEE, Seoul, South Korea, pp. 43–50.

[33] Greene, N., 1989. "Voxel space automata: Modeling with stochastic growth processes in voxel space". In ACM SIGGRAPH Computer Graphics, Vol. 23, ACM, New York, NY, pp. 175–184.

[34] Telea, A., and Jalba, A., 2011. "Voxel-based assessment of printability of 3D shapes". In International Symposium on Mathematical Morphology and its Applications to Signal and Image Processing, Vol. 6671, Springer, Berlin, Heidelberg, pp. 393–404.

[35] Forstmann, S., and Ohya, J., 2010. "Efficient, high-quality, gpu-based visualization of voxelized surface data with fine and complicated structures". *IEICE Transactions on Information and Systems*, **93**(11), pp. 3088–3099. DOI:10.1587/transinf.E93.D.3088.

[36] Maragos, P., 1996. "Differential morphology and image processing". *IEEE Transactions on Image Processing*, **5**(6), pp. 922–937.

[37] Brockett, R. W., and Maragos, P., 1994. "Evolution equations for continuous-scale morphological filtering". *IEEE transactions on Signal Processing*, **42**(12), pp. 3377–3386. DOI:10.1109/78.340774.

Proceedings of the ASME 2019
International Design Engineering Technical Conferences
and Computers and Information in Engineering Conference
IDETC/CIE2019
August 18-21, 2019, Anaheim, CA, USA

DETC2019-97863

LATTICE STRUCTURE DESIGN FOR ADDITIVE MANUFACTURING: UNIT CELL TOPOLOGY OPTIMIZATION

Bradley Hanks[1]
Pennsylvania State University
University Park, PA

Mary Frecker
Pennsylvania State University
University Park, PA

ABSTRACT

Additive manufacturing is a developing technology that enhances design freedom at multiple length scales, from the macroscale, or bulk geometry, to the mesoscale, such as lattice structures, and even down to tailored microstructure. At the mesoscale, lattice structures are often used to replace solid sections of material and are typically patterned after generic topologies. The mechanical properties and performance of generic unit cell topologies are being explored by many researchers but there is a lack of development of custom lattice structures, optimized for their application, with considerations for design for additive manufacturing. This work proposes a ground structure topology optimization method for systematic unit cell optimization. Two case studies are presented to demonstrate the approach. Case Study 1 results in a range of unit cell designs that transition from maximum thermal conductivity to minimization of compliance. Case Study 2 shows the opportunity for constitutive matching of the bulk lattice properties to a target constitutive matrix. Future work will include validation of unit cell modeling, testing of optimized solutions, and further development of the approach through expansion to 3D and refinement of objective, penalty, and constraint functions.

Keywords: Additive Manufacturing, Topology Optimization, Metamaterial Design

1. INTRODUCTION

Additive manufacturing (AM) is a rapidly expanding area of interest for industry and the scientific community. The layer-by-layer approach to manufacturing a component, using a variety of different methods, leads to vastly increased design freedom [1]. The ability to control the structure and composition at every location throughout the component, without a direct increase in cost, presents a unique opportunity for the design optimization community. However, there remain certain limitations and restrictions to the manufacturing process that must be accounted for through design for additive manufacturing (DfAM) [2]. Through DfAM optimization, topology, cellular, and metamaterial designs can now be feasibly manufactured into functional components [3].

Topology design of cellular materials has been ongoing for over thirty years, dating back to the significant contributions of Bendsoe and Sigmund [4,5]. In these works, the mesostructure or cellular structure of a component is designed to match specific constitutive properties through homogenization. Through their work, they were able to demonstrate mesostructures with extreme mechanical properties, such as designs with Poisson's ratios in the range (-1,1) [6]. At the time of this early research, the limited manufacturing options were recognized as a primary challenge and rapid prototyping is mentioned as a potential prospect. Later, major contributions to the area of mesostructure design, especially in the area of multifunctional structures, were published by Seepersad [7,8]. With recent advances in AM, there is renewed interest in design of cellular structures, or lattice structures.

Lattice structure is a modern term in AM that is used to describe a 3D meso-, or cellular, structure in a component that often replaces a solid section. Lattices are patterned structures, similar to a truss, with a series of members in a specific arrangement [9]. The topology of the most basic repeating structure is called a unit cell, often consisting of a variety of members connected at nodes. These unit cells are the similar to the cellular structures that were designed in early work by Sigmund [6]. The topology of the unit cell defines the bulk properties of the patterned lattice structure.

A variety of researchers are exploring methods to optimize components with a patterned lattice structures by changing the

---

[1] Contact author: bbh5108@psu.edu

unit cell type and resizing the members. Contributions to the optimization of patterned lattice structures have been made by Rosen and members of his research group [10,11]. In many cases, the optimization is applied to the entire structure, producing a functionally graded lattice structure from a simplified library of unit cells formed by merging truss members to form a single cell. On the unit cell optimization side, a number of groups continue to develop work similar to the cellular materials that Sigmund proposed [12–18]. Other work on lattice structures includes intermixing unit cell types to be damage tolerant [19]. While a variety of groups continue to research lattice structures, little work is being done directly on designing unit cell topologies accounting for DfAM principles.

Unit cells are frequently chosen from a list of common structures, without firm understanding of which unit cell topology is best for the application. Tamburrino et al. describe one method for unit cell topology selection based on stretch versus bend dominated unit cells [20]. In commercially available AM software, there is often a list of infill patterns or lattice structures that can be selected in a component, but little is known about which unit cell topology is best. Similarly, in lattice optimization research, a unit cell topology is typically chosen prior to the optimization and then the truss members are resized in order to optimize the component [10,21–24].

In other work, hybrid methods of combining lattice structures through macroscale topology optimization are being applied [25,26]. The superformula has also been used to develop lattice structures [27]. Satterfield et al. demonstrate an example of unit cell synthesis based on basic geometries pieced together [28]. Another example of unit cell generation and optimization is presented by Sharpe and Seepersad [29]. Using a constrained Bayesian optimization, unit cells were optimized to meet certain homogenized constitutive matrices. However, in [29] Sharpe notes that the Bayesian approach is limited by the number of design variables. In other work, Watts and Tortorelli use a geometric projection method with inverse homogenization to generate unit cells [30]. The geometric projection method allows for a continuous mesh. While these are various approaches to optimizing the lattice, there has been little work aimed at generating optimized unit cell topologies specifically for AM. With the exception of periodic minimal surfaces, such as the gyroid, unit cell topologies typically come from common primitive shapes or mimic a crystalline structure. While these unit cells present a variety of options, the majority of these cells were not designed using DfAM principles.

In this paper, a systematic multi-objective unit cell topology optimization method is presented. The unit cells are optimized based on a set of application requirements and subjected to DfAM constraints. In section 2, the discrete topology optimization approach is presented along with the constraint, penalty, and objective functions. Section 3 contains the description and results of two cases studies. Section 4 is a discussion of the case study results, limitations of the current approach, and future work on the project. Finally, section 5 concludes with the major outcomes of the work.

2. METHODS

The basis of the systematic unit cell generation is a discrete topology optimization, and the purpose is to be applicable across a broad range of applications. To this end, the optimization approach is described generally and a variety of constraint, penalty, and objective functions are presented that may be used to tailor the optimization. A summary of DfAM considerations is included in section 2.3.

2.1 Ground Structure Topology Optimization

Ground Structure Topology Optimization (GSTO) is also commonly referred to structural optimization, discrete topology optimization, or truss topology optimization. Contrary to topology optimization that uses a continuum of elements, similar to a typical finite element analysis, GSTO first discretizes the design space into a set of nodes. Next, truss, beam, or frame elements are connected between the nodes to form the structure. Figure 1 shows an example of ground structure topology optimization process. The ground structure method was selected because it is well-suited to model lattice structures consisting of connected truss-like elements.

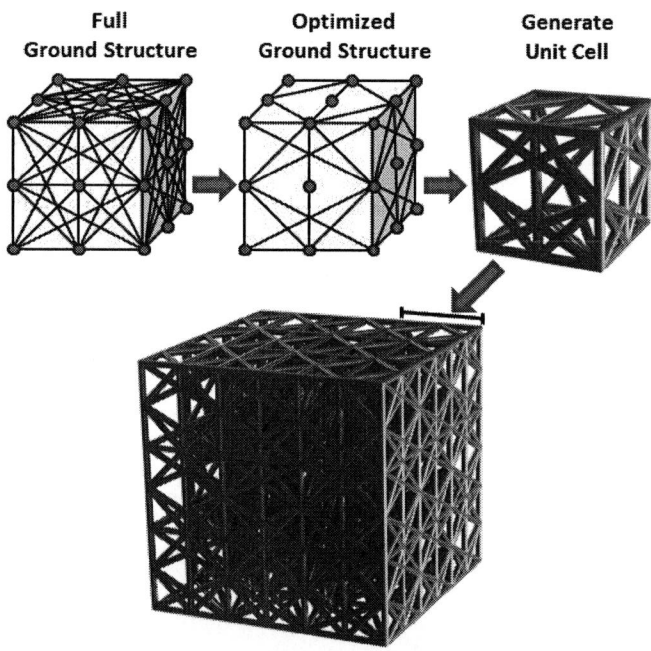

FIGURE 1: ILLUSTRATION OF UNIT CELL OPTIMIZATION APPROACH SHOWING THE FULLY POPULATED INITIAL GROUND STRUCTURE, A REDUCED AND OPTIMIZED STRUCTURE, THE RESULTING SINGLE UNIT CELL, AND A LATTICE FROM THE PATTERNED UNIT CELL.

Generally, in GSTO with truss elements the cross-sectional areas of each element are the design variables and the elements are resized in order to meet the optimization constraint, penalty, and objective functions. As opposed to simply sizing optimization, the area of elements can become so small that their stiffness is irrelevant, allowing for changes in topology.

In this work, all elements are assumed to have circular cross sections. A relevant diameter limit, d_{rel}, is set such that any element that has a cross sectional area A_e with the diameter less than d_{rel}, or $A_e < A_{rel}$, is set to the minimum area, or effectively negligible stiffness. For DfAM, d_{rel}, is chosen based on the machine and build process and represents the minimum feature size.

In this work, the GSTO problem formulation is developed in MATLAB R2018b and solved using the optimization algorithm *fmincon*, a gradient-based nonlinear programming solver that is suitable for structural topology optimization problems. The finite element solution of the structural response of the GSTO uses a linear material model with truss elements with two degrees of freedom per node. The formulation in this paper is implemented for 2D structures, but can be extended to 3D structures. A formal description of the optimization is given in equations (1)-(7). In Eqns. (1)-(7), the goal is to minimize the objective, f, which is defined as the weighted sum of all the sub-objective functions, W, multiplied by the product of all the penalty factors, P. The number of objective and penalty functions are n and np, respectively. The weighting factors, c_i, sum to one. The optimization is subject to constraints on the upper and lower limits of the cross-sectional area of each element, A_e.

Minimize $\qquad\qquad f(W, P)$

Subject to
$$A_e - A_{lower} \geq 0 \qquad (1)$$
$$A_e - A_{upper} \leq 0 \qquad (2)$$
$$A_e = A_{lower} \; for \; A_e < A_{rel} \qquad (3)$$

Where
$$f = (W)(P) \qquad (4)$$

$$W = \sum_{1}^{n} c_i w_i(\boldsymbol{A}, \boldsymbol{l}) , for \; i = 1 \dots n \qquad (5)$$

$$1 = \sum_{1}^{n} c_i , for \; i = 1 \dots n \qquad (6)$$

$$P = p_1(\boldsymbol{A}, \boldsymbol{l})p_2(\boldsymbol{A}, \boldsymbol{l}) \dots p_{np}(\boldsymbol{A}, \boldsymbol{l}) , \\ for \; j = 1 \dots np \qquad (7)$$

2.2 Objective, Penalty, and Constraint Functions

In addition to the standard optimization description presented in Eqns. (1)-(7), a more detailed description of the penalty and objective functions available as well as additional constraint functions are included in this section. The constraints are defined as equalities or inequalities that may be applied to identify feasible and infeasible designs. The penalty functions are defined as a means to negatively impact the objective functions of uneconomical designs.

2.2.1 Objective Functions

Four objective functions are proposed for the optimization: minimization of compliance, unit cell thermal conductance,

homogenized (effective) lattice properties, and a specified volume fraction. The selection of the objective function(s) depends on the design requirements of the unit cell. Consideration of multiple objective functions enables design of multifunctional unit cells.

Minimization of compliance is a common objective function in structural topology optimization. For minimization of compliance, or equivalently the strain energy of the deformed structure under a specified loading condition, is calculated using Eqn. (8),

$$SE = \frac{1}{2} \boldsymbol{u}^T \widetilde{\boldsymbol{K}} \boldsymbol{u} \qquad (8)$$

where SE is the calculated strain energy, $\boldsymbol{u}$ is the global displacement vector, and $\widetilde{\boldsymbol{K}}$ is the global stiffness matrix of the ground structure. Strain energy increases with increased deformation. By minimizing the strain energy, the compliance of the system decreases, maximizing the stiffness of the structure against the applied load(s). In this case, where minimization of compliance is applied to the unit cell, the intent is not to compute the strain energy of the patterned lattice. The purpose is to maximize stiffness in the unit cell against the loading applied.

The minimization of compliance objective function is normalized using Eqn. (9)

$$w_1 = \frac{(SE_{min} - SE)}{(SE_{min} - SE_{max})} \qquad (9)$$

where SE_{min} and SE_{max} are the strain energy calculated with all the element areas set to A_{upper} and A_{lower}, respectively.

Thermal conductance is defined in terms of conductance from one node, or set of nodes, N_{in}, to another node, or set of nodes, N_{out}, and may be minimized or maximized. Temperatures, T_{in} and T_{out}, are assigned to the N_{in} and N_{out}, respectively. The thermal resistance of each truss element within the unit cell is calculated using Eqn. (10),

$$R_e = \frac{L_e}{k A_e} \qquad (10)$$

where R_e is the element thermal resistance, L_e is the element length, A_e is the element area, and k is the material thermal conductivity.

Next, using an analog to Kirchhoff's voltage and current laws for solving an electrical circuit, a linear system of equations is developed to calculate the temperature at each node and the heat flux through each element using R_e, T_{in}, T_{out}, and the connectivity of the elements in the unit cell. By calculating the heat flux through each element, the effective heat flux, q_{eff}, across the unit cell is calculated from the total heat flux leaving N_{in} or entering N_{out}. Finally, the effective thermal conductance is calculated using Eqn. (11),

$$TC = \frac{q_{eff}}{Tin - Tout} \tag{11}$$

where TC and q_{eff} are the effective thermal conductance and heat flux, respectively, from N_{in} to N_{out}.

The thermal conductance objective function is normalized using Eqn. (12),

$$w_2 = \frac{(TC_{min} - TC)}{(TC_{min} - TC_{max})} \tag{12}$$

where TC_{min} and TC_{max} are calculated at with all the element areas set to A_{lower} and A_{upper}, respectively.

Homogenization is a method of calculating the bulk, or effective, mechanical properties of the patterned unit cell by using a set of test strain cases. Through homogenization, the effective constitutive matrix for the bulk lattice properties is calculated following the method described by Sigmund [6]. A summary of that formulation is given in this work, for additional details, see also [5].

In this formulation, the solution is restricted to 2D orthotropic materials, resulting in a [3x3] symmetric constitutive matrix defined by two elastic moduli, two Poisson's ratios, and the shear modulus shown in Eqns. (13)-(14), where $\widetilde{E}$ is the constitutive matrix, σ is the stress, ϵ is the strain, E_1 and E_2 are elastic moduli, v_{12} and v_{21} are Poission's ratios, and G_{12} is the shear modulus for a plane stress condition.

$$\sigma = \widetilde{E}\epsilon \tag{13}$$

$$\widetilde{E} = \begin{bmatrix} \dfrac{E_{xx}}{1 - v_{xy}v_{yx}} & \dfrac{v_{yx}E_{xx}}{1 - v_{xy}v_{yx}} & 0 \\[2ex] \dfrac{v_{xy}E_{yy}}{1 - v_{xy}v_{yx}} & \dfrac{E_{yy}}{1 - v_{xy}v_{yx}} & 0 \\[2ex] 0 & 0 & G_{xy} \end{bmatrix} \tag{14}$$

Using symmetry, the terms in the E_{12} and E_{21} positions are equal, resulting in Eqn. (15), where α is the ratio of the moduli (E_{xx}/E_{yy}) and Poisson's ratios (v_{xy}/v_{yx}), resulting in a constitutive matrix defined by four parameters $(E_{xx}, v_{xy}, G_{xy}, \alpha)$.

$$\widetilde{E} = \frac{1}{\alpha - v_{xy}^2} \begin{bmatrix} \alpha E_{xx} & v_{xy}E_{xx} & 0 \\ & E_{xx} & 0 \\ sym & & (\alpha - v_{xy}^2)G_{xy} \end{bmatrix} \tag{15}$$

For homogenization, three test strain cases are applied to the unit cell, which consist of two axial and one shear case for a 2D unit cell. Based on the deformations and element stiffnesses, averaged mechanical properties are calculated for the unit cell from the element mutual energies using Eqns. (16)-(17),

$$E_{kl}^H = \sum_{e=1}^{NE} Q_{kl}^e \tag{16}$$

$$Q_{kl}^e = \{d_e^k\}^T [k^e] \{d_e^l\}^T, for\ k, l = 1,2,3\ , e = 1 \dots NE \tag{17}$$

where $\widetilde{E}^H$ is the homogenized constitutive matrix, $\widetilde{Q}^e$ is the element mutual energy, $\{d_e^k\}$ is the element displacement vector from the each of the three test strain cases $(k, l = 1,2,3)$, k^e is the element stiffness matrix, and NE is the number of elements.

After calculating the constitutive matrix, the homogenization objective function is calculated using Eqn. (18),

$$w_3 = \sum \frac{(E_{kl}^* - E_{kl}^H)}{E_{kl}^*} \text{ for } k, l = 1, 2, 3 \tag{18}$$

where $\widetilde{E}^*$ is the desired or target homogenized constitutive matrix. This objective function, Eqn. (18), is the percent error between the target and calculated homogenized constitutive matrices.

The fourth objective function is calculated based on the volume fraction, V_f, of the unit cell. The volume fraction is defined using Eqns. (19)-(21),

$$V_{cur} = A^T L \tag{19}$$

$$V_{max} = A_{upper}^T L \tag{20}$$

$$V_f = \frac{V_{cur}}{V_{max}} \tag{21}$$

where V_{cur} is the volume of material, V_{max} is the maximum volume possible, and V_f is the volume fraction. The volume fraction represents the percent of the maximum possible material used in the unit cell, and the desired or target volume fraction, V_f^*, is set by the designer.

The volume objective function is normalized using Eqn. (22).

$$w_4 = \frac{|(V_f^* V_{max}) - V_{cur}|}{(V_f^* V_{max})} \tag{22}$$

2.2.2 Penalty Function

A penalty function is introduced to reduce the complexity of the final topology by penalizing large number of elements in a unit cell. By reducing the number of elements in a unit cell, the unit cell becomes more manufacturable because there would be fewer elements that could overlap and solidify together during printing. The penalty function is defined by Eqns. (23)-(27),

$$ps_e = m\left(\frac{A_{rel}}{A_e}\right) + b\ , for\ A_e < a_2 A_{rel} \tag{23}$$

$$ps_e = 1, \qquad\qquad for\ A_e \geq a_2 A_{rel} \tag{24}$$

$$m = \frac{a_2(a_1 - 1)}{a_2 - 1} \tag{25}$$

$$b = \frac{a_2 - a_1}{a_2 - 1} \tag{26}$$

$$p_i = (ps_1 * ps_2 * \dots * ps_e)^\gamma\ , for\ e = 1 \dots ne_{rel} \tag{27}$$

where A_{rel} is the circular cross-sectional area limit, ne_{rel} is the number of elements with $A_e > A_{rel}$, ps_e is the penalty associated with each relevant element, γ is used to change the importance of the penalty, and (a_1, a_2, m, b) are parameters to define the penalty function values and range. The relevant area limit, A_{rel}, is the circular cross-sectional area at the relevant diameter, d_{rel}, which is based on the minimum feature size of the build process and machine.

By using the penalty function, elements with an area close to A_{rel} become less economical in improving the objective function, encouraging few elements with larger areas as opposed many elements with small areas. An example of the penalty function is shown in Figure 2. In this case (a_1, a_2, m, b) are set to $(2, 100, \frac{100}{99}, \frac{98}{99})$, respectively. The result is that for elements at A_{rel}, $ps_e = 2$. The value of ps_e decreases as A_e increases until $A_e = a_2 A_{rel}$, at which point $ps_e = 1$.

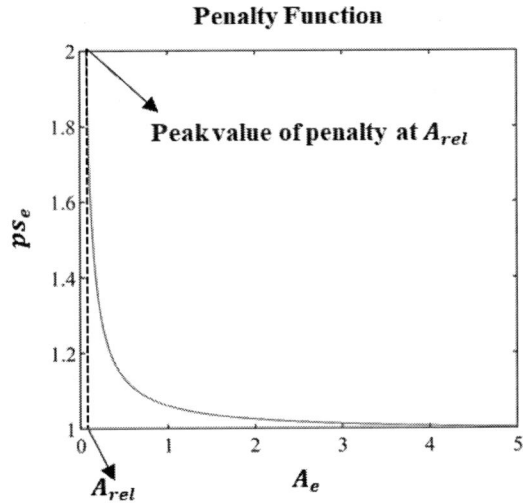

FIGURE 2: GRAPH OF PENALTY FUNCTION THAT ENCOURAGES AN INCREASE IN THE AREA OF EACH ELEMENT AND REDUCES NON-CRITICAL ELEMENTS.

2.2.3 Constraint Functions

Two other constraint functions may be included in the optimization: a limit on total volume and a limit on build angle. These constraint functions are discussed separately from the formal optimization problem definition as optional constraints because they are not required for all objective functions. The limit on total volume is defined by Eqn. (28),

$$A^T L \leq V_f^* V_{max} \qquad (28)$$

where A is a vector of the element cross-sectional areas, L is a vector of the element lengths, V_f^* is the target volume fraction, and V_{max} is the upper limit on the volume.

The build angle constraint is applied to restrict geometries to support free unit cells. It is defined by Eqn. (29),

$$A_e = A_{lower}, for\ 0 < \theta_e < \theta^* \qquad (29)$$

where θ_e is the build angle of each element and θ^* is the upper limit on the build angle based on the printing process. Horizontal members, $\theta_e = 0$, have not been restricted. Based on our preliminary studies of lattice structures, the metal powder bed fusion process can create horizontal members for bridging. Bridging refers to horizontal members connected at each end, forming a bridge between two members. Figure 3 shows several examples of horizontal elements printed in lattice structures built with metal powder bed fusion including 3x3x3 arrays of 4mm and 7mm octet-truss unit cells and 4mm and 6mm cubic unit cells. Note that the horizontal members do print, however, for high element aspect ratios (length/diameter), or in the case of cantilevers, distortion would occur, resulting in poor print quality and increased likelihood of connection failures in the lattice.

FIGURE 3: HORIZONTAL LATTICE MEMBERS ARE SHOWN IN TWO LATTICE EXAMPLES, 3X3X3 ARRAYS OF OCTET-TRUSS UNIT CELLS (4MM, 7MM) AND CUBIC UNIT CELLS (4MM, 6MM).

2.3 DfAM Considerations

DfAM considerations have been included in both constraint and penalty functions for the optimization. First, in the general formulation of the optimization, the minimum feature size is considered using a relevant diameter, d_{rel}, or relevant cross-sectional area, A_{rel}. In GSTO's, element deletion is typically circumvented to avoid singular matrices by considering an element irrelevant when it approaches the lower limit of the cross-sectional area, A_{lower}. Near A_{lower}, the element provides minimal relevance to the overall performance of the unit cell, provided that A_{lower} is several orders of magnitudes smaller a reasonable size. For this formulation the A_{lower} corresponds to a diameter of 1 micron. However, there is a threshold at which members become irrelevant to the performance the unit cell because it falls below the minimum feature resolution of the machine or build process because the element is no longer printable. For elements that fall below the minimum feature size or relevant area, set by the designer, the area is set to A_{lower}.

The second consideration of DfAM for unit cell generation is the build angle limit. Cantilevered lattice elements or low build angles are likely to cause build failures due to the nature of AM processes. As a result, these elements have been restricted by setting their cross-sectional area to A_{lower}, rendering them essentially irrelevant for the unit cell performance. As described in section 2.2.3, the build angle limit has been has not restricted horizontal members because of the capability of bridging.

The penalty function described in section 2.2.2 is used to reduce complexity of the unit cell to also account for minimum feature resolution of AM. For highly complex cells, the unit cell cannot be resolved because elements will blend or mix together, changing the performance as calculated by the optimization. For this purpose, simple cells with fewer elements is desirable. The penalty function accomplishes this by discouraging elements near A_{rel}. As a result, it simultaneously improves print quality and reduces the complexity because it encourages fewer thick members over many thin members.

3. RESULTS AND DISCUSSION

Two case studies are presented to demonstrate the unit cell generation method. Case Study 1 is a multi-objective optimization to show tradeoffs that can exist in optimizing lattice structures for compliance and thermal conduction. Case Study 2 demonstrates a single objective unit cell homogenization approach while considering various DfAM penalty functions.

3.1 Case Study 1

Case Study 1 demonstrates the benefits of multi-objective optimization by considering different weighting factors to show the tradeoffs between objective functions. In this example, the two objective functions are the minimization of compliance and maximization of thermal conductance.

For the minimization of compliance, the load case is shown in Figure 4. Vertical downward loads are applied to the four nodes at the top of the design domain. The purpose of the first objective function, w_1, is to minimize compliance due to loading in the negative Y-direction. While these conditions do not represent periodic boundary conditions that would be present in a unit cell, the purpose is to illustrate the minimization of compliance of a lattice structure under the compressive loading. These loading and boundary conditions may easily be varied by a designer depending on the application. The purpose of the second objective function, w_2, is to maximize thermal conductance in the X-direction, from the left-edge nodes to the right-edge nodes, see Figure 4. A temperature difference between the nodes on the left and right edges of the design domain is applied. The unit cell optimization is solved using a 4x4 array of nodes and assumes symmetry across the vertical and horizontal midplane, shown by the dashed lines in Figure 4.

Weighting values $(c_1 = 0, ..., 1)$ and $(c_2 = 1, ..., 0)$, with increments of 0.01, are applied to each objective function, resulting in a set of 101 solutions. In order to normalize the compliance objective function to be in a similar order of magnitude as the thermal conductance objective function, a scale factor of $1x10^{11}$ is applied to w_1. Also, the formulation is written to minimize each objective, therefore Eqn. (11) is multiplied by -1 to maximize thermal conductance.

DfAM considerations for the lattice structure optimization are the minimum build angle constraint and the relevant diameter limit. A summary of the parameters chosen for the study are shown in Table 1, DfAM considerations are in boldface.

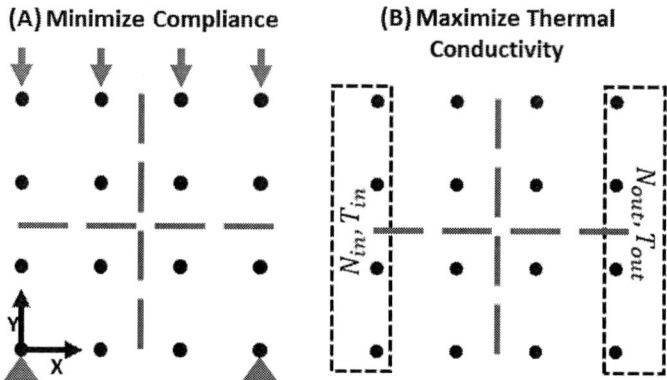

FIGURE 4: CASE STUDY 1: (A) MINIMIZE COMPLIANCE AND (B) MAXIMIZE THERMAL CONDUCTANCE. BLUE DASHED LINES DENOTE SYMMETRY PLANES.

TABLE 1: SUMMARY OF CASE STUDY 1 PARAMETERS. DFAM CONSIDERATIONS ARE SHOWN IN BOLDFACE.

Unit Cell Size	5mm x 5mm
Ground Structure Size	4nodes x 4nodes
Volume Fraction Constraint (V_f^*)	$0.10 * V_{max}$
Elastic Modulus	200 GPa
Thermal Conductivity	54 (W/m·K)
Coefficient of thermal expansion	$1.2x10^{-5}$(m/ m·K)
Diameter Limits	$1x10^{-6}$mm to 2mm
Structural Load	4x: 1000N
T_{in} , T_{out}	100°C, 25°C
Objective 2 scaled	$w_1 = 1x10^{10}w_1$
Weight Values	$(c_1 = 0, 0.01, 0.02, ..., 1)$ $(c_2 = 1, 0.99, 0.98, ..., 0)$
Initial Guess	$A_e = A_{upper}$
DfAM Build Angle Constraint	**35°**
DfAM Relevant Diameter Limit	$d_{rel} = 0.3mm$

The result of the optimization is a set of 101 solutions showing the tradeoffs that exist between the two objective functions. The computation wall time is 2-3 hours using 6 computing cores. Figure 5 illustrates 11 of the solutions with weight value increments of 0.1, showing the range solutions possible. In each plot, the widths of the elements are scaled based on its cross-sectional area. Elements with a diameter less than d_{rel} are excluded from the plots.

From Figure 5, it is shown that the optimal solution for maximizing the thermal conductance is comprised of primarily horizontal bars while the optimal solution for minimization of compliance is comprised of primarily vertical bars. The result is that for a unit cell that has both high thermal conductance and low compliance, there must be a tradeoff.

A graph of the objective function values for the 101 solutions generated are shown in Figure 6. Ideal solutions would simultaneously maximize the thermal conductance and minimize the strain energy. In other words, the optimization searches for designs near the upper left corner of the plot. This figure

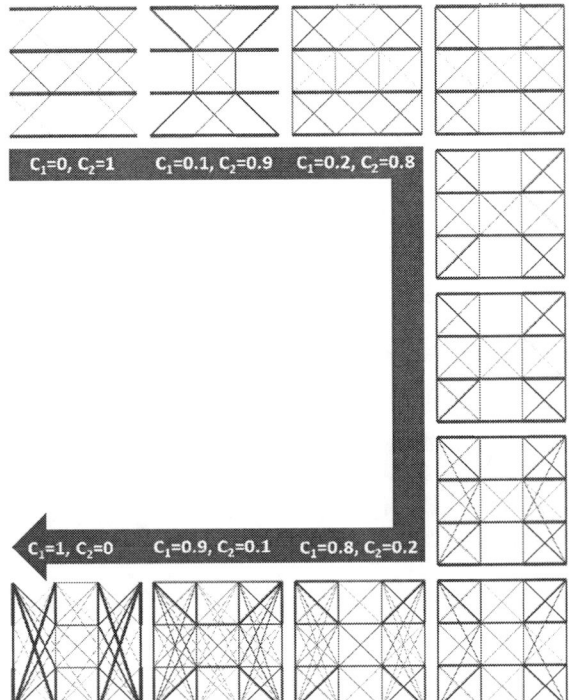

FIGURE 5: CASE STUDY 1 RESULTS SHOWING THE TRADEOFFS BETWEEN MINIMIZATION OF COMPLIANCE AND MAXIMIZATION OF THERMAL CONDUCTANCE.

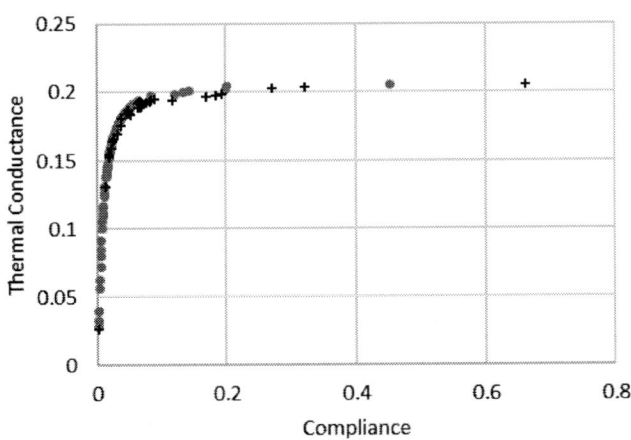

FIGURE 6: COMPARISON OF THE OBJECTIVE FUNCTION VALUES FOR MINIMIZATION OF COMPLIANCE AND MAXIMIZATION OF THERMAL CONDUCTIVITY. THE PARETO-FRONT SHOWS THE TRADEOFF BETWEEN OBJECTIVES.

represents the designs obtained by incrementing the weighting factors and shows the Pareto front that develops. The Pareto front is the set of designs for which no design is better than another in both objectives, or non-dominated designs. These non-dominated solutions form the Pareto front and are all considered

optimal. As one non-dominant design slightly improves in one objective, it gets slightly worse in the other. Dominated designs are worse in both objectives than at least one other solution. In Figure 6, non-dominated solutions are shown in grey circles and dominated designs are shown as black '+'.

3.2 Case Study 2

Case Study 2 is an example of generating unit cells to match an effective constitutive matrix. This objective function is useful when designing a lattice structure for effective orthotropic stiffnesses, for example. In this optimization, the target constitutive matrix is defined with a constraint on the volume fraction. The target constitutive matrix is selected so that the effective modulus of unit cell will be three times larger in the X-direction than the Y-direction. In addition to the objective function and constraints, the penalty function described in section 2.2.2, which encourages fewer and thicker members, and the relevant diameter limit are explored.

Depending on the overall size of the unit cell, high complexity within the unit cell could result in trapped powder or fused members. In addition to the penalty factor to reduce complexity, increasing the relevant diameter limit forces fewer members due to the volume constraint. This effectively reduces the total number of elements that can be used to form the unit cell. In addition, because the optimization algorithm is likely to arrive at local minima, one possibility to find better solutions is to use start from different initial guesses in order to better search the design space.

DfAM considerations applied to this case study are the build angle limit, relevant diameter limit, and penalty function to reduce overall unit cell complexity. A summary of optimization parameters is given in Table 2 with the DfAM considerations shown in boldface.

TABLE 2: SUMMARY OF CASE STUDY 2 PARAMETERS

Unit Cell Size	5mm x 5mm
Ground Structure Size	4nodes x 4nodes
Volume Fraction (V_f^*)	0.15
Diameter Bounds	1×10^{-6}mm to 2mm
Material Modulus Ratio (α)	3
Poisson's Ratio (ν_{12})	0.6
Shear Modulus (G_{12})	$\dfrac{E_{xx}}{3(1+\nu_{xy})}$
Target Constitutive Matrix ($\tilde{E}^*$)	$E_{xx}\begin{bmatrix} 1.136 & 0.227 & 0 \\ & 0.379 & 0 \\ sym & & 0.208 \end{bmatrix}$
DfAM Penalty function parameters (a_1, a_2, m, b) to reduce complexity by penalizing elements near the relevant diameter limit	$(2, 100, \frac{100}{99}, \frac{98}{99})$
DfAM Penalty function levels to adjust the strength of the impact on the objective function	$\gamma = 0, \frac{1}{6}, \frac{1}{3}, \frac{1}{2}, \frac{2}{3}$
DfAM Relevant Diameter Limit	$d_{rel} = 0.3\text{mm}, 0.5\text{mm}, 1.0\text{mm}$
DfAM Build Angle Constraint	35°

For each penalty function level $\left(\gamma = 0, \frac{1}{6}, \frac{1}{3}, \frac{1}{2}, \frac{2}{3}\right)$ and relevant diameter limit $(d_{rel} = 0.3, 0.5, 1.0)$ the optimization was initialized from 20 different randomly selected initial guesses. The range of penalty function levels allows for varied impact from the penalty function. For $\gamma = 0$, there is no penalty factor and with each increasing level the impact of the penalty factor increases, providing a range of solutions. The computation wall time is approximately 3 hours using 6 cores. Figure 7 shows the best solutions from 20 randomized initial guesses, each for a different penalty factor level and relevant diameter limit. Though there are similar features between many of these designs, this shows that there are a variety of solutions that closely match the constitutive matrix. The penalty function and relevant diameter may be used to further refine the solution set and promote or penalize designs with characteristics that are beneficial or challenging for AM processes.

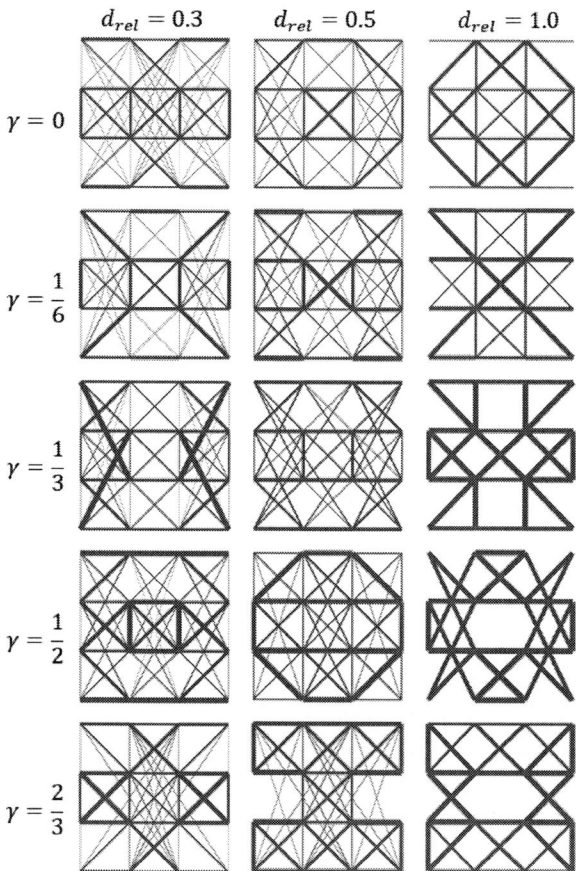

FIGURE 7: LOWEST OPTIMIZATION SCORES FOR THE CONSTITUTIVE MATCHING OBJECTIVE FUNCTION.

In this case, both the penalty function and relevant diameter are used to encourage simpler solutions by encouraging members with larger area, thereby reducing the total number of elements above A_{rel}. While both the penalty and relevant diameter were successful in reducing the complexity of the unit

cell topology, this comes at a cost of error in the homogenized constitutive matrix because there are fewer elements in a unit cell and tighter limits on the diameters of each element. For example, of the 20 random starting positions for $\gamma = 0$, $d_{rel} = 0.3$, 95% of the designs had a total error in the constitutive matrix of less than 10%. However, for $\gamma = \frac{2}{3}$, $d_{rel} = 0.3$ only 25% of designs had an error of less than 10%. For $d_{rel} = 1.0$, only 3 solutions out of out all 5 penalty function levels with 20 initial guesses each (100 optimizations in total) were able to achieve a total error of less than 10%. Therefore, reducing the complexity of the unit cell is desirable but it may come at the expense of less closely matching the constitutive matrix.

Another method to reduce cell complexity is to reduce the number of elements based on the node configuration. For example, setting the area of elements connected to the corners to A_{lower} forces those elements to be inefficient. Figure 8 shows an example of a unit cell generated with $d_{rel} = 0.5$ and $\gamma = \frac{1}{3}$ and no elements connected to the corners. The deformed shape of the unit cell is shown under the same compression load applied horizontally Figure 8A and vertically Figure 8B. As shown in the figure, the deformation is three times larger in the vertical direction as it is in the horizontal direction, demonstrating the target constitutive matrix. An example of the patterned lattice structure based on the unit cell topology shown in Figure 8 is shown in Figure 9.

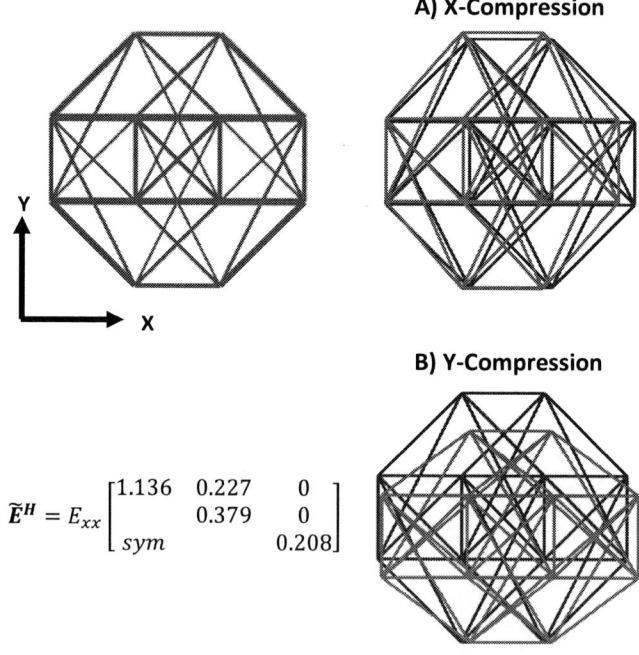

$$\tilde{E}^H = E_{xx} \begin{bmatrix} 1.136 & 0.227 & 0 \\ & 0.379 & 0 \\ sym & & 0.208 \end{bmatrix}$$

FIGURE 8: OPTIMIZED UNIT CELL TOPOLOGY WITH THE AREA OF THE ELEMENTS CONNECTED TO THE CORNERS SET TO THE LOWER LIMIT. COMPRESSION IN X AND Y IS ALSO SHOWN, DEMONSTRATING THE TARGET CONSTITUTIVE MATRIX. TOTAL ERROR IN THE CONSTITUTIVE MATRIX IS 3×10^{-6}%.

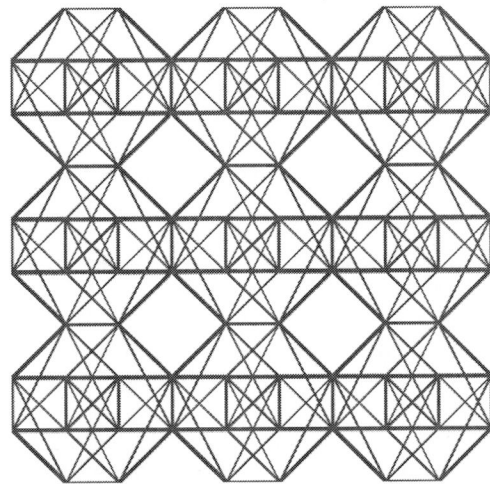

FIGURE 9: EXAMPLE OF A PATTERNED LATTICE STRUCTURE USING A PENALTY FUNCTION LEVEL ($\gamma = \frac{1}{3}$).

4. DISCUSSION

Various limitations exist in the current work and will be described first in general to the GSTO and then with reference to each case study.

In general, GSTOs are limited by the non-continuum design space and the solution is dependent on the selection of node locations. In addition, in the current formulation, truss elements are utilized and all connections at the nodes are pin connections, meaning that each element can only carry axial loads. The interaction between elements that intersect or overlap is also not accounted for in this formulation. While this formulation is simplistic in this regard, it allows for a fast optimization using a reduced set of design variables, especially important for considering the extension of the approach from 2D to 3D. Also, the ground structure model lends itself well to lattice structures that are often based on a set of nodes connected by truss-like elements.

In Case Study 1, one challenge was appropriately scaling each objective function before the weighted sum is calculated for the final objective value. If one objective function is significantly smaller, or larger, than the other than a single objective function is dominant and overshadows the other. Initially, each objective function was normalized based on the range of the objective function value. However, the minimization of compliance objective has such a large range that the normalized value was substantially smaller than the normalized thermal conductance value. In order to rectify this difference, the minimization of compliance objective was scaled by a factor of 1×10^{10}. The scale factor applied places each objective function value on a similar order of magnitude however, this scale factor selection remains case-dependent and somewhat arbitrary. This same issue exists for all optimization where multiple objectives are combined into a single objective optimization.

In Case Study 2, a major challenge continues to be the ability to reduce complexity. With such a variety of solutions that are able to closely match the constitutive matrix, there are numerous local minima where the optimization can become trapped. The penalty function was effective in reducing complexity of the unit cell, however, only to an extent due to the tradeoff with the constitutive matrix error.

The DfAM considerations resulted in unit cells that may be printed without support structures. In all examples shown, the minimum build angle is 35 degrees, with allowance for horizontal members that can be formed by bridging, see Figure 3. The complexity of the cell is a concern for printability. The penalty functions applied showed varied effectiveness in reducing the complexity of the cells to allow clearly defined structures. Overall, the DfAM considerations did improve the printability of the cell, however, future work will continue to refine the optimization and DfAM considerations.

5. CONCLUSION

This work presents a systematic unit cell optimization method to generate lattice structures for AM with specified bulk material properties or characteristics. Using a GSTO approach, two case studies are presented that demonstrate a range of different unit cell design solutions for AM. There are a variety of areas for future work including extension to 3D space and refinement of current objective and penalty functions. With all the penalty and objective functions feeding into a single objective for the optimization, there are a number of challenges with scaling. Because of this challenge, other optimization methods, such as a multi-objective genetic algorithm, will be explored. Through further expansion to 3D space, this approach will be able to generate optimized unit cells, designed for AM, that take advantage of the design freedom allowed.

ACKNOWLEDGEMENTS

The authors would like to acknowledge the support of Dr. Tim Simpson and staff at the Center for Innovative Materials Processing through Direct Digital Deposition (CIMP-3D) for their assistance in printing lattice structures for exploration of manufacturability.

NOMENCLATURE

a_1	Penalty function parameter
a_2	Penalty function parameter
A_e	Element area, mm^2
A_{lower}	Lower limit of element area, mm^2
A_{rel}	Relevant area limit, mm^2
A_{upper}	Upper limit of element area, mm^2
b	Penalty function parameter
c_i	Sub-objective weighting factors
d_e	Element displacement vector, mm
d_{rel}	Relevant diameter limit, mm
$\widetilde{E}$	Constitutive matrix, N/mm^2
$\widetilde{E}^H$	Homogenized constitutive matrix, N/mm^2
$\widetilde{E}^*$	Target homogenized constitutive matrix
E_{xx}	Elastic modulus in X-direction, N/mm^2
E_{yy}	Elastic modulus in Y-direction, N/mm^2
f	Overall optimization objective

G_{xy}	Shear Modulus, N/mm^2
k	Material thermal conductivity, W/m·K
k_e	Element stiffness matrix, N/mm
$\widetilde{K}$	Global stiffness matrix, N/mm
L_e	Element length, mm
m	Penalty function parameter
n	Number of objective functions
ne_{rel}	Number of elements with $A_e > A_{rel}$
np	Number of penalty functions
N_{in}	Node(s) from which TC is measured
N_{out}	Node(s) to which TC is measured
NE	Number of elements
p_i	Penalty function value
ps_e	Element penalty value
P	Product of all penalty factors
q_{eff}	Effective heat flux, W
$\widetilde{Q}^e$	Element mutual energies, N/mm^2
R_e	Element thermal resistance, K/W
SE	Strain energy, N·mm
SE_{min}	Strain energy at A_{upper}, N·mm
SE_{max}	Strain energy at A_{lower}, N·mm
T_{in}	Temperature of N_{in} nodes, °C
T_{out}	Temperature of N_{out} nodes, °C
TC	Thermal conductance, W/K
TC_{min}	Thermal conductance at A_{lower}, W/K
TC_{max}	Thermal conductance at A_{upper}, W/K
$\boldsymbol{u}$	Global displacement vector, mm
V_{cur}	Volume of material, mm^3
V_f	Volume fraction, mm^3
V_f^*	Target volume fraction, mm^3
V_{max}	Maximum possible volume, mm^3
w_i	Sub-objective function values
W	Weighted sum of sub-objectives

Greek Letters

α	Ratio E_{xx}/E_{yy} and ν_{xy}/ν_{yx}
γ	Penalty function importance level
ϵ	Strain
θ^*	Minimum build angle, radians
θ_e	Element build angle, radians
ν_{xy}	Poisson's ratio in XY-direction
ν_{yx}	Poisson's ratio in YX-direction
σ	Stress, N/mm^2

Abbreviations

AM	Additive manufacturing
DfAM	Design for additive manufacturing
GSTO	Ground structure topology optimization

REFERENCES

[1] Gao, W., Zhang, Y., Ramanujan, D., Ramani, K., Chen, Y., Williams, C. B., Wang, C. C. L., Shin, Y. C., Zhang, S., and Zavattieri, P. D., 2015, "The Status, Challenges, and Future of Additive Manufacturing in Engineering," Comput. Des., **69**, pp. 65–89.

[2] Thompson, M. K., Moroni, G., Vaneker, T., Fadel, G., Campbell, R. I., Gibson, I., Bernard, A., Schulz, J., Graf, P., Ahuja, B., and Martina, F., 2016, "Design for Additive Manufacturing: Trends, Opportunities, Considerations, and Constraints," CIRP Ann. - Manuf. Technol., **65**(2), pp. 737–760.

[3] Liu, J., Gaynor, A. T., Chen, S., Kang, Z., Suresh, K., Takezawa, A., Li, L., Kato, J., Tang, J., Wang, C. C. L., Cheng, L., Liang, X., and To, A. C., 2018, "Current and Future Trends in Topology Optimization for Additive Manufacturing," Struct. Multidiscip. Optim., pp. 2457–2483.

[4] Bendsøe, M. P., and Kikuchi, N., 1988, "Generating Optimal Topologies in Structural Design Using a Homogenization Method," Comput. Methods Appl. Mech. Eng., **71**(2), pp. 197–224.

[5] Sigmund, O., 1994, "Materials With Prescribed Constitutive Parameters: An Inverse Homogenization Problem," Int. J. Solids Struct., **31**(17), pp. 2313–2329.

[6] Sigmund, O., 1995, "Tailoring Materials with Prescribed Elastic Properties," Mech. Mater., **20**(4), pp. 351–368.

[7] Seepersad, C. C., Allen, J. K., McDowell, D. L., and Mistree, F., 2006, "Robust Design of Cellular Materials With Topological and Dimensional Imperfections," J. Mech. Des., **128**(6), p. 1285.

[8] Seepersad, C. C., Allen, J. K., McDowell, D. L., and Mistree, F., 2008, "Multifunctional Topology Design of Cellular Material Structures," J. Mech. Des., **130**.

[9] Tao, W., and Leu, M. C., 2016, "Design of Lattice Structures for Additive Manufacturing," 2016 Int. Symp. Flex. Autom., pp. 1–3.

[10] Chang, P. S., and Rosen, D. W., 2013, "The Size Matching and Scaling Method: A Synthesis Method for the Design of Mesoscale Cellular Structures," Int. J. Comput. Integr. Manuf., **26**(10), pp. 907–927.

[11] Graf, G. C., Chu, J., Engelbrecht, S., and Rosen, D. W., 2009, "Synthesis Methods for Lightweight Lattice Structures," *ASME International Design Engineering Technical Conferences and Computers and Information in Engineering Conference*, pp. 1–11.

[12] Liu, L., Yan, J., and Cheng, G., 2008, "Optimum Structure with Homogeneous Optimum Truss-like Material," Comput. Struct., **86**(13–14), pp. 1417–1425.

[13] Huang, X., Zhou, S. W., Xie, Y. M., and Li, Q., 2013, "Topology Optimization of Microstructures of Cellular Materials and Composites for Macrostructures," Comput. Mater. Sci., **67**, pp. 397–407.

[14] Guth, D. C., Luersen, M. A., and Muñoz-Rojas, P. A., 2012, "Optimization of Periodic Truss Materials Including Constitutive Symmetry Constraints," Materwiss. Werksttech., **43**(5), pp. 447–456.

[15] Guth, D. C., Luersen, M. A., and Muñoz-Rojas, P. A., 2015, "Optimization of Three-Dimensional Truss-like Periodic Materials Considering Isotropy Constraints,"

Struct. Multidiscip. Optim., **52**(5), pp. 889–901.

[16] Neves, M. M., Sigmund, O., and Bendsøe, M. P., 2002, "Topology Optimization of Periodic Microstructures with a Penalization of Highly Localized Buckling Modes," Int. J. Numer. Methods Eng., **54**(6), pp. 809–834.

[17] Messner, M. C., 2016, "Optimal Lattice-Structured Materials," J. Mech. Phys. Solids, **96**, pp. 162–183.

[18] Xia, L., and Breitkopf, P., 2015, "Design of Materials Using Topology Optimization and Energy-Based Homogenization Approach in Matlab," Struct. Multidiscip. Optim., **52**(6), pp. 1229–1241.

[19] Pham, M.-S., Liu, C., Todd, I., and Lertthanasarn, J., 2019, "Damage-Tolerant Architected Materials Inspired by Crystal Microstructure," Nature.

[20] Tamburrino, F., Graziosi, S., and Bordegoni, M., 2018, "The Design Process of Additively Manufactured Mesoscale Lattice Structures: A Review," J. Comput. Inf. Sci. Eng., **18**(4), p. 040801.

[21] Gorguluarslan, R. M., Gandhi, U. N., Mandapati, R., and Choi, S.-K., 2015, "A Design and Fabrication Framework for Periodic Lattice-Based Cellular Structures in Additive Manufacturing," *ASME International Design Engineering Technical Conferences & Computers and Information in Engineering Conference*.

[22] Tang, T. L. E., Liu, Y., Lu, D., Arisoy, E. B., and Musuvathy, S., 2017, "Lattice Structure Design Advisor for Additive Manufacturing Using Gaussian Process," *ASME International Design Engineering Technical Conferences and Computers and Information in Engineering Conference*, pp. 1–10.

[23] Stankovic, T., Mueller, J., Egan, P., and Shea, K., 2015, "A Generalized Optimality Criteria Method for Optimization of Additively Manufactured Multimaterial Lattice Structures," J. Mech. Des., **137**(11), p. 111405.

[24] Cheng, L., Zhang, P., Biyikli, E., Bai, J., Robbins, J., and To, A., 2017, "Efficient Design Optimization of Variable-Density Cellular Structures for Additive Manufacturing: Theory and Experimental Validation," Rapid Prototyp. J., **23**(4), pp. 660–677.

[25] Han, Y., and Lu, W. F., 2018, "Optimization Design of Nonuniform Cellular Structures for Additive Manufacturing," Vol. 1 Addit. Manuf. Bio Sustain. Manuf., p. V001T01A033.

[26] Wang, Y., Zhang, L., Daynes, S., Zhang, H., Feih, S., and Wang, M. Y., 2018, "Design of Graded Lattice Structure with Optimized Mesostructures for Additive Manufacturing," Mater. Des., **142**, pp. 114–123.

[27] Nessi, A., and Stankovic, T., 2018, "Topology, Shape, and Size Optimization of Additively Manufactured Lattice Structures Based on the Superformula," *ASME International Design Engineering Technical Conferences & Computers and Information in Engineering*.

[28] Satterfield, Z., Kulkarni, N., Fadel, G., Li, G., Coutris, N., and Castanier, M. P., 2017, "Unit Cell Synthesis for Design of Materials With Targeted Nonlinear Deformation Response," J. Mech. Des., **139**(12), p. 121401.

[29] Sharpe, C., Seepersad, C. C., Watts, S., and Tortorelli, D., 2018, "Design of Mechanical Metamaterials Via Constrained Bayesian," *ASME International Design Engineering Technical Conferences & Computers and Information in Engineering*, pp. 1–11.

[30] Watts, S., and Tortorelli, D. A., 2017, "A Geometric Projection Method for Designing Three-Dimensional Open Lattices with Inverse Homogenization," Int. J. Numer. Methods Eng., **112**(11), pp. 1564–1588.

Proceedings of the ASME 2019
International Design Engineering Technical Conferences
and Computers and Information in Engineering Conference
IDETC/CIE2019
August 18-21, 2019, Anaheim, CA, USA

DETC2019-97865

EVALUATING THE POTENTIAL OF DESIGN FOR ADDITIVE MANUFACTURING HEURISTIC CARDS TO STIMULATE NOVEL PRODUCT REDESIGNS

Alexandra Blösch-Paidosh
Engineering Design and Computing Lab
Department of Mechanical Engineering
ETH Zürich
Zürich, Switzerland

Prof. Saeema Ahmed-Kristensen
Design Products
School of Design
Royal College of Art
London, UK

Prof. Kristina Shea
Engineering Design and Computing Lab
Department of Mechanical Engineering
ETH Zürich
Zürich, Switzerland

ABSTRACT

Additive manufacturing (AM) affords those who wield it correctly the benefits of shape, material, hierarchical, and functional complexity. However, many engineers and designers lack the training and experience necessary to take full advantage of these benefits. They require training, tools, and methods to assist them in gaining the enhanced design freedom made possible by additive manufacturing. This work, which is an extension of the authors' previous work, explores if design heuristics for AM, presented in a card-based format, are an effective mechanism for helping designers achieve the design freedoms enabled by AM. The effectiveness of these design heuristic cards is demonstrated in an experiment with 27 product design students, by showing that there is an increase in the number of *unique capabilities of AM* being utilized, an increase in the *AM novelty,* and an increase in the *AM flexibility* of the generated concepts, when given access to the cards. Additionally, similar to the previous work, an increase in the number of *interpreted heuristics* and *AM modifications* present in the participants' designs when they are provided with the heuristic cards is shown. Comparisons are also made between 8-heuristic and 29-heuristic experiments, but no conclusive statements regarding these comparisons can be drawn. Further user studies are planned to confirm the efficacy of this format at enhancing the design freedoms achieved in group and team design scenarios.

INTRODUCTION

Additive manufacturing (AM) affords those who wield it correctly the benefits of shape, material, hierarchical, and functional complexity [1]. These unique capabilities have sparked the interest of designers and engineers the world over, many of which are eager to incorporate the benefits of AM into their designs. However, they lack the training and experience regarding the processes' capabilities and limitations that exist for other more traditional forms of manufacturing. As a result, prospective AM designers and engineers require training, tools, and methods to assist them in their new ventures and to help them break out of their traditional manufacturing mindsets [2].

In response to this need, many researchers, hobbyists, and industry workers are developing design for additive manufacturing (DfAM) methods and aids to address these training gaps. These DfAM methods and aids fall broadly into two categories: opportunistic and restrictive DfAM [3]. Restrictive methods consist mainly of design and printing guidelines (e.g. [4, 5]) and are predominately useful in the later phases of the product development process [6] , such as *Detailed Design* and *Production Ramp-Up* [7]. On the other hand, the opportunistic methods assist engineers and designers in expanding the limits of their knowledge with regard to AM and improving their design to take better advantage of the unique capabilities that AM affords them. Prior to the past few years, the development of DfAM methods has been dominated by restrictive and opportunistic topology optimization methods [8-

10]. However, recently there has been a lot of interest in developing opportunistic DfAM methods that help people during the *Conceptual Design* phase [11-16], including previous work by the authors in developing design heuristics for additive manufacturing derived from a synthesis of the key AM features of 275 AM artifacts sourced from academia, industry, the media, and hobbyists [15]. These heuristics have already been validated, but they lack a fixed delivery format and their ability to stimulate the inclusion of the four unique capabilities of AM and increase the AM novelty and AM flexibility of the designs has not been evaluated. Therefore, as an extension of this previous work, this paper presents design heuristic cards that correspond to each of the previously developed design heuristics for AM and determines if these design heuristics for AM, presented in a card-based format, are an effective mechanism for helping designers achieve the design freedoms enabled by AM. This is measured through the *unique capabilities of AM* found in designs and the effect they have on the *AM flexibility*, *AM novelty*, and *fluency* of the design solutions.

This paper first offers background information on relevant topics before presenting the design heuristics for additive manufacturing cards and describing the experiment and analysis procedure in the method section. Next, the results of the experiment are reported and discussed. Finally, the paper concludes with a summary and discussion of future work.

BACKGROUND

It has been found that designers find opportunistic additive manufacturing knowledge (AMK) much more useful than restrictive AMK in the early phases of the design process [7]. With this in mind, several research groups have been working on tackling the transfer of opportunistic AMK to designers in the early phases of the design process. For example, [14] looked at using a small AMK booklet presenting ten AM potentials with the goal of stimulating radical and architectural innovation, and [11] investigated combining existing, established design methods with AMK to assist designers with DfAM in the early phases of the design process. [12] provided information sheets about various AM concepts to designers during the *Conceptual Design* phase, and [16] explored the use of Mood Boards for visual inspiration, a checklist of the most important AM design rules, an interactive system for AM design potentials, and 3D models for AM design potentials. Similar to the authors of this paper, [13] developed a list of design principles for AM based on the analysis of a quantity of artifacts from the hobby AM website, Thingiverse[1]. Unlike the design heuristics for AM derived by the authors of this paper, in [15], the design principles included both opportunistic and restrictive principles.

One thing that all these methods have in common is that they all use text to describe each principle/heuristic/potential. Most of the methods also combine these descriptive texts with images in the delivery format. There has also been some investigation into what elements users find functional, practical, and easy to understand when it comes to DfAM aids [7]. They asked

designers to evaluate four different formats: text, physical artifacts, videos, and pictures. Videos, artifacts, and pictures were all highly rated by the participants (5-point Likert scale), but opinions were mixed regarding the text-based formats. However, they did not investigate any combined formats. In the end, they chose to build a video- and picture-based AMK tool, which also includes limited text, presumably, because it is not possible to fully explain concepts and transfer standard nomenclature without at least some text.

All of the previously mentioned studies used some form of verification or validation of their methods. This ranged from comparison with literature [13] to user surveys [7, 11, 16] to third-party evaluation of the designs generated in controlled user studies [12, 14, 15]. Those that used user surveys mainly focused on the opinions of the users as to whether or not they thought the method was useful or easy to use. However, in such cases, objective evaluation of the method in terms of key performance indicators was missing. Several studies examined the effect of the previous AM knowledge of the participant on the assessment of the method [12, 16]. Additionally, almost all of these studies mentioned the importance of the innovative effect that incorporating AM into the early phases of the design process can have. However, only one of these studies objectively assessed this effect [14], and none of the studies looked at the effect of the method on the prevalence of the four unique capabilities of AM in the results, namely shape, material, hierarchical, and functional complexity [1].

Based on the above-mentioned AMK transfer methods, we propose to develop physical cards to accompany each of the previously derived design heuristics for AM, containing a mixture of both text and images so that the relevant AMK is fully transferred and easy to understand. Their ability to effectively assist designers in achieving the design freedoms enabled by AM must also be evaluated, because an AMK transfer method must be shown to objectively and effectively transfer AMK in a controlled experiment and not just receive positive feedback from the users, before it can be deemed a worthwhile tool for designers.

METHOD
Design Heuristic Cards for Additive Manufacturing

As previously stated, physical design heuristics for AM cards are developed that use both text and images to convey the relevant content to the user. Each card contains seven different pieces of information:

1. Design Heuristic
2. Design Heuristic Number
3. Description of the Design Heuristic
4. Design Heuristic Category
5. Image of an Abstract Example
6. Image of an Example from Industry or Literature
7. Short Description of the Real-World Example

An example card with each of these pieces of information labelled is available in Figure 1.

---

[1] https://www.thingiverse.com/

Table 1: Categorized Design Heuristics for AM for which Cards are designed. Superscript numbers indicate to which card set each heuristic belongs. "E" indicates the example card.

Part Consolidation

1 Consolidate parts for better functional performance[3]

2 Consolidate parts to reduce assembly time[3]

3 Consolidate parts to increase robustness[2]

4 Consolidate parts to achieve multiple functions[1]

Customization

5 Customize geometry to use case[1]

6 Customize user interface to use case[3]

7 Customize artifact with decoration[2]

Convey Information

8 Convey information with color[1]

9 Convey information with geometry[3]

10 Convey information with haptics[2]

11 Convey information with light[E]

Material

12 Use single material to achieve recyclability[1]

13 Use metamaterial to achieve unique and graded material properties[3]

14 Use multiple materials to achieve unique and graded material properties[2]

Material Distribution

15 Absorb energy with small interconnected parts[1]

16 Allow movement with small interconnected parts[1]

17 Use material distribution to achieve desired behavior[2]

18 Remove material to provide function[2]

19 Optimize structural topology or geometry[3]

Embed-Enclose

20 Embed functional material[3]

21 Embed functional component[1]

22 Use enclosed, functional parts[2]

Lightweight

23 Replace internal structure with lightweight lattice structure[3]

24 Hollow out artifact to reduce weight[2]

Reconfiguration

25 Create multi-functional artifact with reconfigurable structures[1]

The layout of and content for these cards is inspired by the general design heuristic cards developed by [17], with which the authors have had much success using with students during the last few years. The authors find that the abstract example on the front of the cards helps prevent them from fixating on a specific example, but the real-world example solutions on the back help them understand the heuristic in a real context. Additionally, because the examples are sketched instead of being photographs, the relevant parts of the example can be highlighted and distracting aspects can be lowlighted. However, one aspect of those cards that has proved problematic is that they are not categorized by the relevant design phase, something that makes using them with novices difficult because novices sometimes like to get ahead of themselves while designing. As the design heuristics for AM are largely relevant in the early phases of the design process, the need is not seen to classify them by design phase, but they are categorized into similar groups, so as to assist the user in evaluating multiple possibilities for solving the same problem or providing them with different but related heuristics.

Following the previous work [15], all 29 heuristics are re-examined to eliminate duplicates, confirm their relevancy as a design heuristic for AM, and clarify their wording. Twice, two very similar heuristics are reduced to one, and two heuristics are eliminated. Table 1 contains the final list of the 25 design heuristics for which cards are created. They are each sorted into one of eight AM categories: *part consolidation, customization, information communication, materials, material distribution, embed-enclose, lightweight,* or *reconfiguration*. The full set of

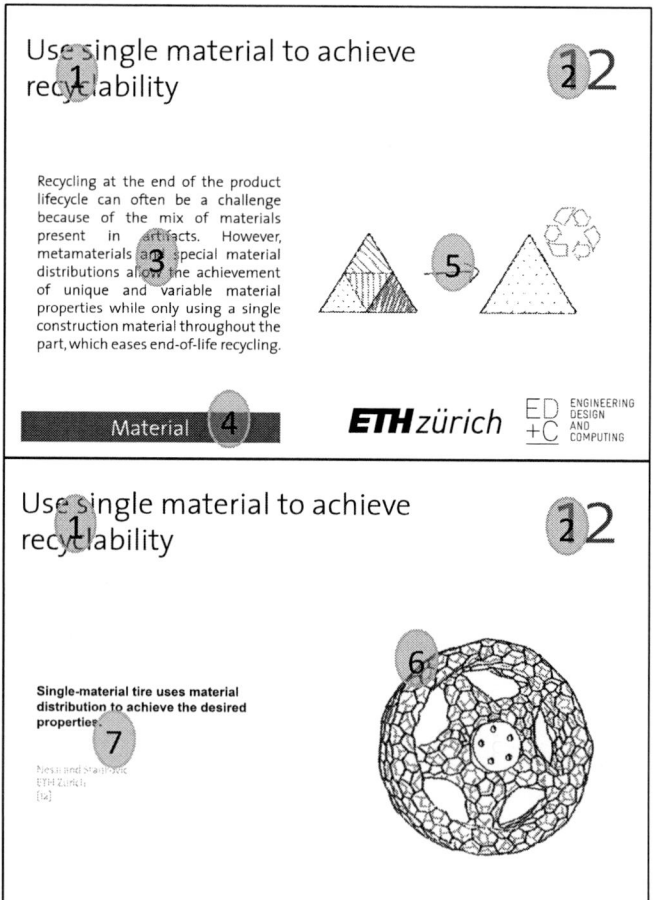

Figure 1. Example design heuristics for AM card. The front of the card is at the top and the back of the card is on the bottom.

The company you are working for is looking at switching over from traditional manufacturing to additive manufacturing. To test out the idea, they would like you to **redesign the Plantui to take advantage of the 4 unique capabilities of additive manufacturing** (i.e. redesign it for additive manufacturing).

Four Unique Capabilities of Additive Manufacturing
- Shape complexity: *ability to build almost any shape*
- Hierarchical complexity: *ability to build something with shape complexity across multiple size scales*
- Functional complexity: *functional artifacts can be produced in single or limited builds*
- Material complexity: *material is variable at every point in the build*

Design Criteria
1. Takes advantage of additive manufacturing
2. Grows plants hydroponically
3. Stores water
4. Provides plants with nutrients
5. Provides plants with water
6. Provides plants with light during growth

Please document **only one concept per page**. Clearly document your concept(s) with sketches and annotations so that they can be understood without you being there to explain it. Please write and sketch clearly. If you need more paper, please raise your hand.

At the end of this portion of the workshop, **please place all your sheets of paper back in your envelope**.

Figure 2. Design Task

cards is available for download from the lab's website[2].

Experiment Design

After the development of the design heuristics for AM cards, their effectiveness at helping designers achieve the design freedoms enabled by AM must be assessed. This is accomplished via a controlled experiment with 27 product design master students at a UK University. Participation in the study is purely voluntary, and the participants are enticed to take part in the study by the opportunity to learn more about AM and DfAM. Each participant also receives a digital copy of the design heuristics for AM cards following the study. The students are not otherwise remunerated for their participation.

The experiment design is described below and additionally in Table 2. Prior to the start of the experiment, the participants are the recipients of an introductory lecture on DfAM and basic AM processes, and the unique capabilities of AM are explained to them. These topics are also interspersed with various AM examples from industry and literature. They are also introduced to the concept of design heuristics and how to use them, and the various portions of the design heuristics for AM cards are explained to them using one of the 25 cards. Following the lecture, they were instructed to fill out an AM Experience Survey, which surveyed their general design skills and DfAM

Table 2. Experiment Design

Order of Activites	Group	
	Control n = 15	Experiment n = 12
1. Lecture on AM (20 Min.)	x	x
2. AM Experience Survey	x	x
3. Recieves Design Heuristics for AM Cards		x
4. Completes Design Task (45 Min.)	x	x

skills and knowledge. It also surveys their degree of awareness of various DfAM topics (both opportunistic and restrictive) in the same manner as [3]. The awareness is surveyed using the same 6-point scale as [3], which is then converted into a linear scale for analysis (0 to 5).

After the participants finish filling out the survey[3], they are randomly split into two groups. The functions of the Plantui[4], a small-scale, commercially available automatic, indoor, hydroponic urban farming device, are described to them during a live demonstration with the device. Then, they are each asked to redesign the Plantui for AM (see Figure 2 for the exact design task). The Plantui is chosen as the basis for the experiment for two reasons. First, it hits a good balance of simplicity and complexity, as it is a relatively small and simple system, but also has multiple functions and parts to which changes can be applied. Second, the authors have previously employed this case study successfully, and it thus allows comparison of data [15].

One group (Control group, 15 participants) receives the design task, an exploded view of the Plantui, a description of the function of each part, and some blank A5[5] pieces of paper on which to draw each of their concepts. The second group (Experimental group, 12 participants) receives everything the control group receives plus one subset of the design heuristics for AM cards. The cards are divided into three subsets of eight cards each. The cards are divided into three subsets because there is concern that the participants may suffer from an information overload for the allotted time if they are provided with all 25 cards, something the authors suspect happened in a previous, failed experiment. They are distributed in such a way so as to as equally as possible distribute the heuristics from each of the eight categories among the three subsets. The heuristic card that is used as the explanatory example is excluded from the study i.e. no subset receives this card. Care is taken during the distribution of the experiment materials to equally distribute the subsets of heuristic cards among the participants in the experimental group.

After the distribution of the materials, the participants are instructed that they have 45 minutes to complete the design task. They are also instructed to use one A5[5] page per concept, although they may include multiple AM capabilities in any one

---

[2] http://www.edac.ethz.ch/Research/Design-Heurestics-AM.html
[3] Only 24 of the 27 participants successfully and completely filled-out the electronic pre-survey. For this reason, when cross-referencing survey data during the analysis, the n-values are lower than for the experimental data.

[4] www.plantui.com
[5] A standard paper size in Europe

concept. Finally, they are instructed that this is an individual task, and that all forms of communication with other participants or the outside world are prohibited during the experiment.

Data Analysis

The designs generated by the participants are evaluated using three different primary indices: *interpreted heuristics*, *AM design modifications*, and *unique capabilities of AM*. Some of these are later combined to derive secondary indices to help evaluate the effectiveness of the cards at enhancing the creativity of the generated designs, namely through *AM novelty*, *design fluency*, and *AM flexibility*. The index value *interpreted heuristics* is derived by analyzing each concept generated by a participant and determining which of the 25 design heuristics one could mine from the design. This is computed both cumulatively (i.e. the sum of all occurrences per concept) and count-wise (i.e. the number of different design heuristics applied per concept). The index value *AM design modifications* is determined by counting the number of AM-relevant changes made to the original Plantui design. If the redesign is too radical to make connections between the design elements, it is excluded from evaluation of this, and only this, index value. Both of these indices were previously used by the authors during an earlier experiment and are derived in the same manner as in the previous work [15]. The final index value *unique capabilities of AM*, however, is new and is derived by evaluating each concept to determine each instance of occurrence of each of the four unique capabilities of AM as described by [1], namely *shape*, *material*, *hierarchical*, and *functional complexity*. This index value is also computed cumulatively and count-wise.

The *AM novelty*, or how unusually the design fulfills the design task with regard to AM, of each concept is assessed in two ways. First, the *degree of AM novelty* is assessed by summing together each occurrence of each *interpreted heuristic* and adding it to the sum of each occurrence of each *unique capability of AM* for each concept. Second, the *breadth of AM novelty* is assessed by summing together a count of the number of different *interpreted heuristics* and a count of the number of different *unique capabilities of AM* for each concept. Designs that employ more heuristics and unique capabilities of AM are considered more novel from an AM perspective because they employ a wider range of AM applications.

The *fluency*, or the number of concepts produced in each group during the allotted time, is also evaluated. This is evaluated to determine if those using the design heuristic cards are less prolific in producing designs than those not using the cards.

To assess the *AM flexibility* of the concepts, or how well the designs are suited to fulfilling the design task through different AM means, the primary indices *interpreted heuristics* and *unique capabilities of AM* are assessed at the *group*-level to determine if some heuristics or unique capabilities are only present in one group or the other. If, for example, some heuristics are only present in the experimental group, then it can be said that the heuristics help produce a more flexible array of designs. Designs that contain a wider variety of heuristics or capabilities are

considered more AM-flexible because they employ AM in different ways and are therefore more likely to have a feasible AM implementation with the final, chosen AM process.

The designs are also evaluated at two different levels: *participant* and *group*. They are examined at the *group*-level so as to separate the individual participant from their concepts and instead be able to compare all of the work generated by the experiment group and compare it to the control group. When evaluated at the *group*-level, each sheet is considered one concept, and each concept is directly evaluated based on the above-discussed three different indices. Assuming that each participant creates at least one concept (i.e. one sheet), the sample size is at least as large as the number of participants in each group. However, when examined only at the *group*-level, any differences discovered in the pre-experiment AM-knowledge survey cannot be controlled for during the analysis. Therefore, they are also examined at the *participant*-level. When examined at the *participant*-level, all of the index values generated for each concept associated with one participant are aggregated together, and the total number of concepts generated by that participant is noted.

Since two coders are not available to evaluate the entire data set, an intercoder reliability analysis is performed on a random sample (10%) of the data for the index value *unique capabilities of AM* using Cohen's Kappa [18] to show intercoder exchangeability. After the establishment of the intercoder exchangeability, the primary coder codes the remaining data. The kappa value for this index value is 0.871, which is considered sufficient to establish intercoder exchangeability [19]. An intercoder reliability analysis is not performed for the other two primary indices because intercoder exchangeability for these indices was already established in previous work by the authors using the same coders and design task [15].

After the index values for each concept are determined and analyzed, the results are compared and contrasted to draw conclusions. All statistical operations are performed using SPSS.

RESULTS

The results comparing the three primary indices at both the *group*- and *participant*-level are visually summarized in Figure 3 and Figure 4, respectively. Throughout the analysis, at the *participant*-level, the primary indices are normalized by dividing the index value by the number of concepts generated by that participant. This is to prevent the particularly prolific designers from dominating the results. Due to space restrictions, only the summarized numerical results are presented here. The full data is available upon request to the authors.

It is found that access to the design heuristic cards affects the number of *AM design modifications* produced by participants at both the *group*- and *participant*-level. Those who have access to the cards produce more *AM design modifications* than those who do not have access to the cards. The same effect is seen on the other primary indices at both levels of analysis and both cumulatively and count-wise. Table 3 summarizes the relevant statistical data for each of these relationships according to the APA standard [20], namely including the *degrees of freedom* (df),

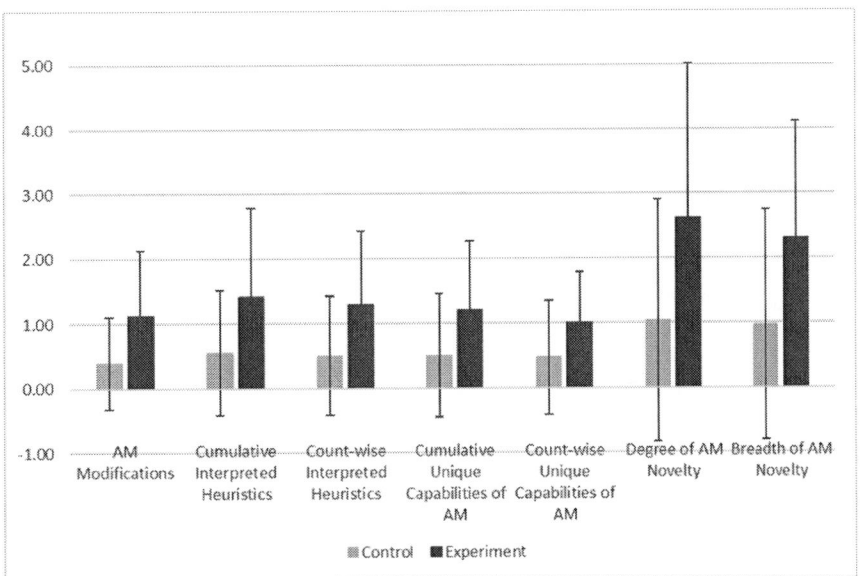

Figure 3. The means of various indices at the *group*-level. The error bars are ± 1 SD of the mean. Control, n = 54. Experiment, n = 34.

t-value (t(df)), the *p-value* (p), and the *effect size* (d). This is also the standard used for reporting the rest of the results in this paper.

At the *participant*-level, it is also found that there is a significant difference between the two groups' mean levels of self-reported *AM design knowledge*, although their mean levels of *DfAM mean topic awareness* show no significant difference (see Figure 4 and Table 3 for details). To determine if this difference in self-reported data affects the results, we run a MANCOVA with the self-reported *AM design knowledge* as the covariate. No significant effect is found (p > 0.05).

The percentage of count-wise *interpreted heuristics* occurring in any given concept that belong to the same card-set

as provided to the participant is analyzed in two MANOVAs (one for each analysis level, experiment group only). A significant effect is found at the *group*-level for the *percentage of heuristics in Set 1* (F(2,31) = 11.54, p < 0.001, η^2 = 0.43), the *percentage of heuristics in Set 2* (F(2, 31) = 3.97, p < 0.03, η^2 = .20), and the *percentage of heuristics in Set 3* (F(2,31) = 39.62, p < 0.001, η^2 = 0.72), but no significant effect is found at the *participant*-level (p > 0.05). Tukey's *post hoc* is used to compare the differences between groups (p < 0.05). The significant relations are available in Table 4, and Figure 5 visually summarizes the data. The indices indicated in Table 3 (plus the *unique capabilities of AM* broken down to the four categories) are also evaluated against

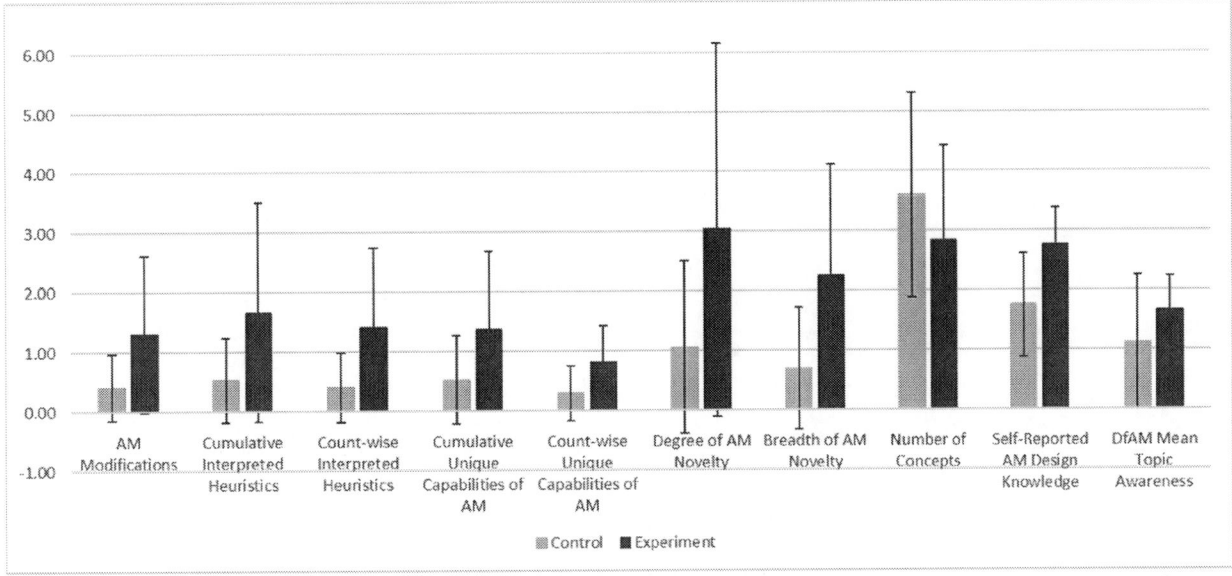

Figure 4. The means of various indices at the *participant*-level. The error bars are ± 1 SD of the mean. Control, n = 15 (Except for self-reported AM design knowledge and mean AM awareness, n=12). Experiment, n = 12. Except for the final three indices, the values are normalized by the number of concepts produced by the participants.

Table 3. Statistical data for the mean comparisons. Non-significant relations are shaded.

Index Value	Group -Level				Participant -Level (normalized)			
	df	t(df)	p	d	df	t(df)	p	d
AM Modifications	86	-3.97	0.000	0.87	25	-2.37	0.026	0.92
Cumulative Interpreted Heuristics	86	-3.51	0.001	0.77	25	-2.20	0.037	0.85
Count-wise Interpreted Heuristics	86	-3.57	0.001	0.78	25	-2.67	0.013	1.03
Cumulative Unique Capabilities of AM	86	-3.24	0.002	0.71	25	-2.11	0.045	0.82
Count-wise Unique Capabilities of AM	86	-2.90	0.005	0.64	25	-2.58	0.016	1.00
Degree of AM Novelty	86	-3.46	0.001	0.76	25	-2.17	0.039	0.84
Breadth of AM Novelty	86	-3.39	0.001	0.74	25	-2.72	0.012	1.05
Number of Concepts	-	-	-	-	25	1.19	0.246	0.46
Self-Reported AM Design Knowledge	-	-	-	-	22	-3.25	0.004	1.33
Mean AM Awareness	-	-	-	-	22	-1.90	0.071	0.78

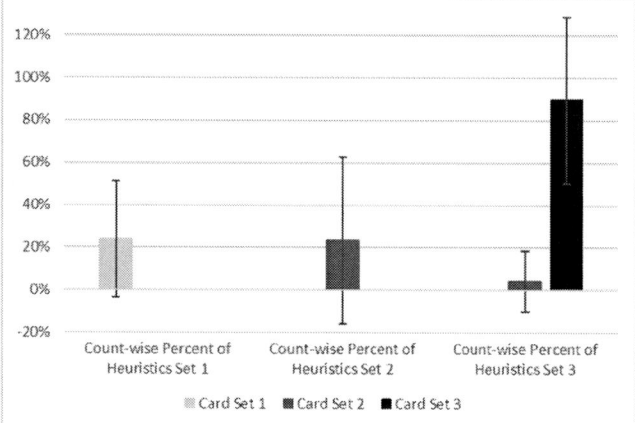

Figure 5. Compares the means of the percentage of heuristics per set at the *group*-level. The error bars are ± 1 SD of the mean. Card Set 1, n = 6. Card Set 2, n = 12. Card Set 3, n = 16.

Table 4. Significant relations between card sets (*group*-level)

Index Value	Significant Relation
Count-wise Percent of Heuristics Set 1	Card Set 1 > Card Set 2
	Card Set 1 > Card Set 2
Count-wise Percent of Heuristics Set 2	Card Set 2 > Card Set 3
Count-wise Percent of Heuristics Set 3	Card Set 3 > Card Set 1
	Card Set 3 > Card Set 2

the different sets of heuristic cards in a MANOVA at the *group*-level, but no significant relations are found (p > 0.05).

The relationship between the relevant *interpreted heuristics* and the participants' self-reported awareness of certain DfAM topics is analyzed to determine if there is any correlation between the number of occurrences of the relevant heuristics and the participants' existing knowledge. No correlations are found.

The effect of the cards on the creativity of the designs through enhanced design freedom is measured by *AM flexibility*, *AM novelty*, and *fluency*. It is found that the pool of concepts generated by the experiment group has a higher degree of *AM flexibility* than the concepts generated by the control group. The categories of *customization* and *lightweight* only occur in experimental group, and outside of these categories, heuristics #3 *Consolidate parts to increase robustness*, #12 *Use single material to achieve recyclability*, #18 *Remove material to provide function*, #19 *Optimize structural topology or geometry*, and #21 *Embed functional component* only occur in the experimental group. Two heuristics only occur in the control group, #11 *Convey information with light* and #22 *Use enclosed functional parts*. The first served as the example heuristic in the pre-experiment brief, and the second occurs only once. In examining the *unique capabilities of AM*, no capability is present in only one group or the other, although the capabilities are more prevalent in the experiment group.

Next, in examining both measures of *AM novelty*, access to the design heuristics affects both of them at both analysis levels. In both cases, the measures of *degree of AM novelty* and *breadth of AM novelty* are higher when the participants have access to the design heuristics for AM cards. The relevant statistical information is in available in Figure 3, Figure 4, and Table 3.

Finally, the *fluency* of the two groups is also examined. The control group produces a total of 54 concepts and the experiment group 34. However, no significant difference is found between the mean number of concepts produced per participant in the control and experiment groups.

These results are compared, where possible, to those results from the authors' previous experiment with the design heuristics for AM [15]. This compares the use of 29 heuristics per person to 8 heuristics per person. There are significant differences between the control groups and the experiment groups. A visual summary of the differences is available in Figure 6 and the statistical data is available in Table 5. The 29-heuristic experiment produces higher levels of *AM modifications* and *interpreted heuristics*, in both the control and experiment groups. However, the fluency of the 8-heuristic experiment is higher.

DISCUSSION

As the *degree of AM novelty* and the *breadth of AM novelty* increase in the presence of the heuristic cards, it can be

Table 5. Statistical data for the inter-experiment mean comparisons at the *participant*-level (not normalized)

Index Value	Control				Experiment			
	df	t(df)	p	d	df	t(df)	p	d
AM Modifications	28	2.57	0.016	0.94	22	3.64	0.001	1.49
Cumulative Interpreted Heuristics	28	2.26	0.032	0.82	19.32	4.49	0.000	1.77
Count-wise Interpreted Heuristics	28	2.64	0.013	0.96	24	4.12	0.000	1.62
Number of Concepts	14.00	-5.84	0.000	2.13	11.00	-4.00	0.002	1.58

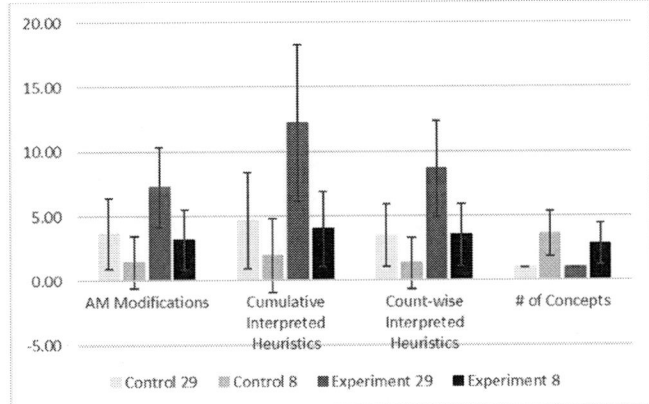

Figure 6. The means of various indices at the *participant*-level comparing a 29 heuristics experiment to a 8 heuristic experiment. The error bars are ± 1 SD of the mean. Control 29, n = 15. Control 8, n = 15. Experiment 29, n = 14 (Except for *AM modifications*, n = 12). Experiment 8, n = 12.

concluded that the cards help produce concepts that are not only more novel from an AM perspective but are also employing a wider range of the novel aspects of AM. This is seen at both the *group*- and *participant*-level, which demonstrates that the effect is not isolated to single concepts, but also affects participants.

As all of the primary indices are higher when the participants have access to the cards, it can be concluded that the cards cause the users to increase the number of AM-relevant modifications that they make. Furthermore, they increase not only the number of heuristics present in their designs but also the number of unique capabilities of AM present in their designs, which indicates that use of the heuristic cards promotes not just the use of specific AM heuristics, but also promotes integration of the four unique capabilities of AM into the designs of AM-novices. As the increases are seen in both the cumulative and count-wise analyses, it can be seen that they are not only increasing the overall quantity but also employing a wider variety of heuristics and capabilities when using the cards. As this behavior is mirrored in both the *group*- and *participant*-level analyses, we see that this effect is seen not only in the groups as a whole, but also for individuals.

A significant difference in the self-reported *AM design knowledge* level between the experiment groups is found, but it is found to have no significant effect on the results. Before the start of the experiment, participants are asked to assess their *AM design knowledge level* as one of the following: *none, fundamental awareness, novice, intermediate, advanced,* or *expert*. As no context-specific definitions are provided for each

of these levels, it is believed that this difference stems from different perceptions of the participants as to what each of these categories mean. Additionally, the survey does not specify if the knowledge is restrictive or opportunistic. As this design task focuses on the opportunistic aspects of AM, those who rate themselves higher in this respect may be thinking about their restrictive knowledge, which is most likely not very useful for this design task. To mitigate this, they are also asked to rate their awareness of several restrictive and opportunistic DfAM topics using a much more descriptive scale, which can then be aggregated to determine their *DfAM mean topic awareness*. The topics covered by the questionnaire skew more heavily toward opportunistic topics, so their mean *DfAM topic awareness* rating also skews toward their opportunistic DfAM awareness. This value is typically lower than their self-reported *AM design knowledge* level and no significant difference between the mean topic awareness levels are found between the groups. Additionally, the individual topic awareness ratings are cross-referenced with the corresponding *interpreted heuristics*, and no significant correlations are found. For these three reasons, it is not believed that the prior knowledge of the participants has any effect on the results, and therefore the effects seen come solely from the use of the cards.

To add to the above conclusion, it is seen that the card set that the participant receives effects which *interpreted heuristics* are most likely to appear in their designs, namely those in the same heuristic card set, which also indicates that the effects seen are due to the cards and not due to the participants' knowledge. One can also assume that the three sets of cards equally promote *AM modifications* and each of the four *unique capabilities of AM*. There is no similar significant effect at the *participant*-level, but it is believed that this is due to the low sample sizes at the *participant*-level.

Related to this, the cards also lead to more flexible designs, as more heuristic categories and heuristics are present in the experiment group than in the control group. Similar to the previous related work [15], the categories of *customization* and *lightweight* only occur in the experimental group. These are categories, which those who are abreast of AM applications know are classic AM applications. Another classic application of AM that is only found in the experimental group is topology and geometry optimization. From this, one can conclude that the heuristic cards are helping the user come up-to-speed regarding AM applications.

A difference in *fluency* between the two groups is to be expected, as one group is not only designing, but also reading

V02AT03A036-8   Copyright © 2019 ASME

and using the cards. This is an effect also seen in other similar experiments [17, 21]. However, it is surprising that there is no significant difference between the two groups when they are analyzed at the *participant*-level. This is believed to be due to a low sample size. However, the difference in *fluency* at the *group*-level may also partially stem from the difference in the number of participants in each group (15 in the control group vs. 12 in the experiment group).

Fluency differences are also observed between the 29- and 8-heuristic experiments. This is believed to be due to the type of participants in each experiment. The 29-heuristic experiment uses engineers-in-training [15] and the 8-heuristic experiment uses product designers-in-training. The engineers produced very detailed concepts containing many heuristics, but only one concept each. The product designers produced many more concepts on average, but most of them were much-less detailed, incorporating perhaps one or two heuristics. This could be a result of their respective trainings. However, in the case of the 29-heuristic experiment group, it is possible that the participants' *fluency* was hampered by the large number of heuristics they had at their disposal. More investigation of this phenomenon would need to be undertaken to make any definite statements.

The count-wise number of *interpreted heuristics* is also an interesting comparison between the two experiments. In the 29-heuristics experiment, the most unique heuristics utilized by one participant is 15 of 29 (52%). In the 8-heuristics experiment, that number is 7 of 8 (88%). This seems to indicate that reducing the number of heuristics supplied to the participants in a fixed-length design task may allow them to apply proportionally more of them. However, this conclusion is also clouded by the fact that, although similar in age, the participants of the two experiments have different training backgrounds and different levels of familiarity with the design task. A more focused study would need to be completed to draw any hard conclusions.

As evidenced by the above results and discussion, use of the heuristic cards by the participants leads to an overall increase in the enabled design freedom of AM. This is measured through the *unique capabilities of AM, fluency, AM flexibility*, and *AM novelty*. Significant differences are found between the two experimental groups in three of the four metrics in favor of the experiment group, and these differences are shown to be a direct result of access to the cards and not the design skills or the existing AMK of the participants. Thus, one can conclude that the design heuristic for AM cards are an effective mechanism for helping designers achieve the design freedoms enabled by AM.

Following this experiment, the number and content of the heuristics are shown to be in a stable state. The cards also function effectively, and general feedback from the participants is neutral to positive about the cards in general and specific aspects of them. Interestingly, different people find different aspects of the card helpful. Some focus-in on the short text, others on the long text, and still others on the images and examples. This informal feedback indicates that the different aspects of the cards are helpful for different people, which

---

[6] A standard paper size in Europe

---

indicates that the mix of text and images is a good decision. It may also be useful to add an additional example in the future. The multi-format aspect of the cards proved to be an asset to the users of the cards. Therefore, in the future, the design heuristics for AM cards will continued to be used in their current format.

One aspect of the cards that does not work so well is the size. They are designed to be printed on A5[6] paper (5.8 x 8.3 inches), but it was found while observing the participants work with them, that they are large and clumsy in the hand, making it difficult to quickly tab through them or examine two at the same time (some of the particular benefits of using a physical-based system). Therefore, in the future, the size of the cards will be reduced, possibly to an A6[6] (4.1 x 5.8 inches).

The effect of the design heuristics for AM cards has been assessed in terms of both *AM flexibility* and *AM novelty*, but there is one aspect of the heuristics' efficacy that has not yet been investigated: *innovativeness*. As this is a classic measure of design creativity, this is something that will be examined in the future.

Thus far, the effect of the design heuristics for AM has only been assessed during individual design scenarios. In the future, they will be studied in group- and team-design scenarios.

CONCLUSION

In this work, we present an extension of our previous work of developing design heuristics for AM [15]. The card-based format for communicating 25 previously derived design heuristics for AM is demonstrated to promote AM design freedom in an experiment with 27 product design students, by showing that there is an increase in the number of *unique capabilities of AM* being utilized, an increase in the *AM novelty,* and an increase in the *AM flexibility* of the generated concepts, when given access to the cards. Additionally, similar to the previous work [15], an increase in the number of *interpreted heuristics* and *AM modifications* present in the participants' designs when they are provided with the heuristic cards is shown. Comparisons are also made between 8-heuristic and 29-heuristic experiments [15], but no conclusive statements regarding these comparisons can be drawn. Further user studies are planned to confirm the efficacy of this format at enhancing the design freedoms achieved in group and team design scenarios.

ACKNOWLEDGMENTS

The authors would like to thank those members of the Design Products Degree Program at the Royal College of Art (Autumn 2018) who participated in this experiment.

REFERENCES

[1] Gibson, I., Rosen, D., and Stucker, B., 2015, "Design for Additive Manufacturing," pp. 399-435. DOI: 10.1007/978-1-4939-2113-3_17

[2] Seepersad, C. C., 2014, "Challenges and Opportunities in Design for Additive Manufacturing," 3D Printing and Additive Manufacturing, 1(1), pp. 10-13. DOI: 10.1089/3dp.2013.0006

[3] Prabhu, R., Miller, S. R., Simpson, T. W., and Meisel, N. A., 2018, "The Earlier the Better? Investigating the Importance of Timing on Effectiveness of Design for Additive Manufacturing Education," Proceedings of the ASME 2018 International Design Engineering Technical Conferences and Computers and Information in Engineering Conference (IDETC/CIE 2018))Québec City, Québec, Canada, 26-29 August 2018.

[4] Adam, G. A. O., and Zimmer, D., 2014, "Design for Additive Manufacturing—Element transitions and aggregated structures," CIRP Journal of Manufacturing Science and Technology, 7(1), pp. 20-28. DOI: 10.1016/j.cirpj.2013.10.001

[5] Stöckli, F., Modica, F., and Shea, K., 2016, "Designing passive dynamic walking robots for additive manufacture," Rapid Prototyping Journal, 22(5), pp. 842-847. DOI: 10.1108/rpj-11-2015-0170

[6] Ulrich, K. T., and Eppinger, S. D., 2008, Product Design and Development, McGraw Hill, Boston, Massachusetts, USA. ISBN: 978-007-125947-7

[7] Laverne, F., Segonds, F., D'Antonio, G., and Le Coq, M., 2016, "Enriching design with X through tailored additive manufacturing knowledge: a methodological proposal," International Journal on Interactive Design and Manufacturing (IJIDeM), 11(2), pp. 279-288. DOI: 10.1007/s12008-016-0314-7

[8] Seepersad, C. C., Allison, J., and Sharpe, C., 2017, "The Need for Effective Design Guides in Additive Manufacturing," In: Proceedings of the 21st International Conference on Engineering Design (ICED17), Vol. 5: Design for X, Design to X, Vancouver, Canada, 21-25 August 2017, pp. 309-316.

[9] Bourell, D. L., Rosen, D. W., and Leu, M. C., 2014, "The Roadmap for Additive Manufacturing and Its Impact," 3D Printing and Additive Manufacturing, 1(1), pp. 6-9. DOI: 10.1089/3dp.2013.0002

[10] Thompson, M. K., Moroni, G., Vaneker, T., Fadel, G., Campbell, R. I., Gibson, I., Bernard, A., Schulz, J., Graf, P., Ahuja, B., and Martina, F., 2016, "Design for Additive Manufacturing: Trends, opportunities, considerations, and constraints," CIRP Annals - Manufacturing Technology, 65(2), pp. 737-760. DOI: 10.1016/j.cirp.2016.05.004

[11] Watschke, H., Bavendiek, A.-K., Giannakos, A., and Vietor, T., 2017, "A Methodical Approach to Support Ideation for Additive Manufacturing in Design Education," In: Proceedings of the 21st International Conference on Engineering Design (ICED17), Vol. 5: Design for X, Design to X, Vancouver, Canada, 21-25 August 2017, pp. 41-50.

[12] Laverne, F., Segonds, F., Anwer, N., and Le Coq, M., 2015, "Assembly Based Methods to Support Product Innovation in Design for Additive Manufacturing: An Exploratory Case Study," Journal of Mechanical Design, 137(12), p. 121701. DOI: 10.1115/1.4031589

[13] Perez, K. B., Anderson, D. S., Höltta-Otto, K., and Wood, K. L., "Crowdsourced Design Principles for Leveraging the Capabilities of Additive Manufacturing," Proc. In: Proceedings of the 20th International Conference on Engineering Design (ICED15), Vol. 4: Design for X, Design to X, Milan, Italy, 27.-30.07.2014.

[14] Yang, S., Page, T., and Zhao, Y. F., 2018, "Understanding the Role of Additive Manufacturing Knowledge in Stimulating Design Innovation for Novice Designers," Journal of Mechanical Design, 141(2). DOI: 10.1115/1.4041928

[15] Blösch-Paidosh, A., and Shea, K., 2019, "Design Heuristics for Additive Manufacturing Validated Through a User Study," Journal of Mechanical Design, 141(4). DOI: 10.1115/1.4041051

[16] Kumke, M., Watschke, H., Hartogh, P., Bavendiek, A.-K., and Vietor, T., 2017, "Methods and tools for identifying and leveraging additive manufacturing design potentials," International Journal on Interactive Design and Manufacturing (IJIDeM), 12(2), pp. 481-493. DOI: 10.1007/s12008-017-0399-7

[17] Murphy, L., Daly, S., McKilligan, S., and Seifert, C. M., 2017, "Supporting Novice Engineers in Idea Generation using Design heuristics," 2017 Asee Annual Conference & Exposition, ASEE, Columbus, Ohio, USA, p. 19008.

[18] Cohen, J., 1960, "A Coefficient of Agreement for Nominal Scales," Educational and Psychological Measurement, 20(1), pp. 37-46. DOI: 10.1177/001316446002000104

[19] Hayes, A. F., 2005, "Statistical Methods For Communication Science," Lawrence Erlbaum Associates, Inc., Mahwah, New Jersey, USA. ISBN: 9780805854879

[20] Schwartz, B. M., Wilson, J. H., and Goff, D. M., 2018, An EasyGuide to Research Design & SPSS, SAGE Publications. ISBN: 9781506385464

[21] Keshwani, S., Lenau, T. A., Ahmed-Kristensen, S., and Chakrabarti, A., 2017, "Comparing novelty of designs from biological-inspiration with those from brainstorming," Journal of Engineering Design, 28(10-12), pp. 654-680. DOI: 10.1080/09544828.2017.1393504

Proceedings of the ASME 2019
International Design Engineering Technical Conferences
and Computers and Information in Engineering Conference
IDETC/CIE2019
August 18-21, 2019, Anaheim, CA, USA

DETC2019-97913

AN OPTIMAL QUANTITY OF SCHEDULING MODEL FOR MASS CUSTOMIZATION-BASED ADDITIVE MANUFACTURING

Yosep Oh
Graduate Research Assistant
Industrial and Systems Engineering
University at Buffalo, The State University of New York
Buffalo, NY 14260
yosepoh@buffalo.edu

Sara Behdad
Assistant Professor
Mechanical and Aerospace Engineering
Industrial and Systems Engineering
University at Buffalo, The State University of New York
Buffalo, NY 14260
sarabehd@buffalo.edu

ABSTRACT

The purpose of this study is to optimize production planning decisions in *additive manufacturing for mass customization (AMMC)* systems in which customer demands are highly variable. The main research question is to find the optimal quantity of products for scheduling, the *economic scheduling quantity (ESQ)*. If the scheduling quantity is too large, the time to collect customer orders increases and a penalty cost occurs due to the delay in responding to consumer demands. On the other hand, if the scheduling quantity is too small, the number of parts per jobs decreases and parts are not efficiently packed within a workspace and consequently the build process cost increases. An experiment is provided for the case of *stereolithography (SLA)* and *2D packing* to demonstrate how the build time per part increases as the scheduling quantity decreases. In addition, a mathematical framework based on ESQ is provided to evaluate the production capacity in satisfying the market demand.

Keywords: additive manufacturing, production planning, economic scheduling quantity, mass customization

NOMENCLATURE

λ The arrival rate of customer demands (units / h)
P The production rate of scheduled parts (units / h)
Q The number of parts for scheduling (units)
T_c Cycle time for scheduling (h)

T_p Completion time for production (h)
T_b Total build time (h)
c_t Penalty cost per hour ($/h)
c_p Build process cost per hour ($/h)
c_m Unit material cost ($/mm^3)
$\bar{v}$ The average volume per part (mm^3/unit)
M The number of AM machines (units)

1. INTRODUCTION

Recently, on-demand production in 3D printing plants, or *3D printing farms,* has been developed as a promising business model [1]–[4]. In this newly developed business model, manufacturers install multiple 3D printers and run them simultaneously to produce hundreds and thousands of parts in a very short period of time. Customers place their orders through e-commerce websites by registering 3D models of what they need. Consumer orders are produced by multiple 3D printers and are shipped to consumers. In these types of business models, consumer orders are coming one by one and are highly variable in terms of features and characteristics. Products and design blueprints are personalized based on each customer order.

The current business models show that Additive Manufacturing (AM) is spotlighted for both mass production and customization. The focus of this study is on a production system for a 3D printing plant with mass customization capabilities named as *AM for Mass Customization (AMMC)*.

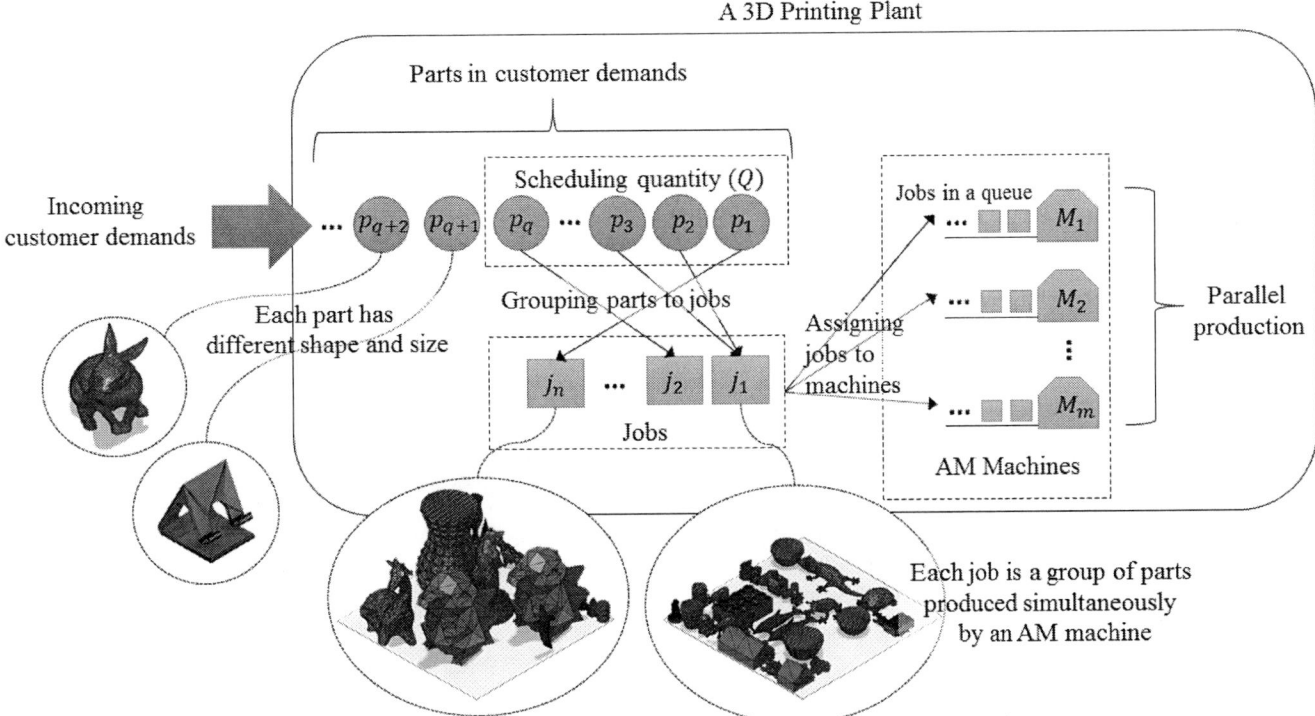

Figure 1: A production system in Additive Manufacturing for Mass Customization (AMMC)

AMMC adopts the *make-to-order (MTO)* production strategy [5]. Once a customer demand arrives, the plant begins the production. In MTO for AMMC, manufacturers have to deal with extremely customized parts, with different size, geometry and shape for individual customers.

Figure 1 represents how the AMMC system works. Customer demands arrive at a 3D printing plant. Each order has its own characteristics. Once the number of incoming orders reaches to Q, known as scheduling quantity, the manufacturer starts defining a production plan to take care of customer orders. Then, parts are grouped into jobs (or builds [6]). A job is a group of parts produced simultaneously by an AM machine (a 3D printer). Then, jobs are assigned to AM machines. Each AM machine has a queue and jobs in a queue are processed by their own AM machine. This is a type of parallel production in which multiple AM machines work on their assigned jobs independently.

The main question in this study is to decide about the scheduling quantity, Q. Defining a plan with a high number of parts (large Q) has a positive effect on minimizing the total build time, since it increases the number of parts per job and consequently improves the packing utilization. However, waiting to collect a high number of orders and then start the production may result in higher lead time and delay in responding to consumer demands. This study develops a method to find the optimal quantity of parts, the *economic scheduling quantity (ESQ)*, Q^*, with the aim of

minimizing the production cost that also includes the penalty cost of delaying in addressing consumer orders.

2. LITERATURE REVIEW

2.1 Production planning for AM

Studies on production planning for AM have considered how to group parts into jobs and how to assign jobs to AM machines. Grouping parts into a single job has usually been focused on using different packing algorithms including 2D packing [7] and 3D packing [8]. In the case of grouping parts into multiple jobs, the approaches for the traditional bin packing problems are adopted for AM [6]. In addition to packing, build orientation determination is another aspect that should be addressed since the packing utilization depends on part orientations. Griffiths et al (2018) provided a heuristic method dealing with both bin packing problem and build orientation problem [9]. However, considering only the grouping issue is not sufficient for managing production systems at a higher level. This is where the planning and scheduling for AM with multiple 3D printers become important.

For assigning jobs or parts to AM machines, most of the previous studies have been based on classical scheduling problems. To name a few studies, Li et al. (2017) proposed heuristic algorithms and mathematical models to minimize the average production cost per volume of material [10]. Kim et al. (2017) suggested a *genetic algorithm (GA)* to match parts to 3D printers in order to minimize makespan [11]. Ransikarbum et al. (2018) solved a part-to-printer assignment problem by using multi-objective optimization [12]. However, previous studies have focused on planning and scheduling without considering the packing. To consider the details of the packing in the production planning level, mathematical models become too complicated and heuristic models result in computational inefficiencies. To overcome this issue, the current study considers the packing issue indirectly by using a result function extracted from a packing simulation experiment.

2.2 The economic order quantity model

Economic order quantity (EOQ) is the order inventory quantity to minimize the total holding and ordering costs proposed by Harris (1913) [13]. The EOQ model is one of the oldest classical production scheduling and inventory control models [14]. It has been extended in many ways including the *economic production quantity (EPQ)* model [15], the reorder point [16], and the stochastic EOQ [17]. EOQ has still being studied as a solid mathematical model for inventory lot sizing [18].

Although the classical EOQ model is based on inventory management [19], the current study focuses on a different production planning problem emphasizing more on customer demands. In order to find the optimal quantity of scheduling for AM, it only adopts the mathematical approach of the EOQ model for identifying the optimal quantity between the trade-off factors.

3. APPROACH

This section describes the proposed mathematical framework for calculating the total build time and the ESQ model. The following assumptions have been used:

- The customer demand arrival rate is known and is constant over time.
- Each customer demand has only one part.
- To make a production plan, Q is consistent for each cycle.
- All AM machines have the same process parameters. The size of the workspace and the model of AM machine are the same.
- Setup time for each job is not considered since it is negligible compared to its build time.

3.1 The relation between build time and scheduling quantity

Each job is completed through a build process by an AM machine, which means that each job has its own build time. The total build time of the jobs, T_b, is the sum of build time for all jobs that result from part grouping by using Q. Once T_b is divided by Q, the build time per part is obtained.

In the case of handling hundreds or thousands of parts using *stereolithography (SLA)* and *2D packing*, it is likely

that the build time per part decreases as Q increases. This is mainly due to the increase in packing utilization.

To validate this, an experiment with the following conditions has been conducted. The build time estimation model and the 2D packing algorithm developed by Oh et al. (2017) are modified and adopted for this experiment [20]. 1000 input parts are randomly generated to simulate customer orders for AMMC. To generate random inputs, parts are arbitrarily chosen from a set of 100 different geometries from Thingiverse.com [21] and, after normalizing their size, the size is re-scaled by multiplying a value from a uniform distribution (1, 10). Without changing the build orientation of parts, the initial orientation is used for part placement. To avoid the undesired case that the shape and size of the generated 1000 parts are biased, the experiment is repeated three times. The width, length, and height of a workspace of an AM machine are 200×200×200 mm, respectively.

To produce 1000 parts based on AMMC, Q is set from 30 (representing a small number of parts) to 1000 (a large number). Table 1 shows T_b based on Q and its build time per part. Given the dataset of Q and the build time per part, the relation can be represented by a non-linear curve model as shown in Equation (1). In this equation, α is mostly affected by the size of workspace and the average volume of parts while β usually indicates the type of AM processes.

$$Build\ time\ per\ part = \alpha + \frac{\beta}{Q} \qquad (1)$$

CurveExpert Professional 2.6.5 is used to identify two parameters, α and β, of the curve model, which are 0.3480 and 3.5095, respectively. Figure 2 represents the curve model fitting the dataset. As shown in the figure, the build time per part decreases and converges as Q increases.

Table 1: T_b to produce 1000 parts depending on Q

Q	T_b (hours)	Build time per part[*] (hours)
30	14.18	0.4727
100	35.96	0.3596
200	74.96	0.3748
300	110.62	0.3687
500	178.78	0.3576
1000	352.80	0.3528
Build time per part[*] $= T_b/Q$		

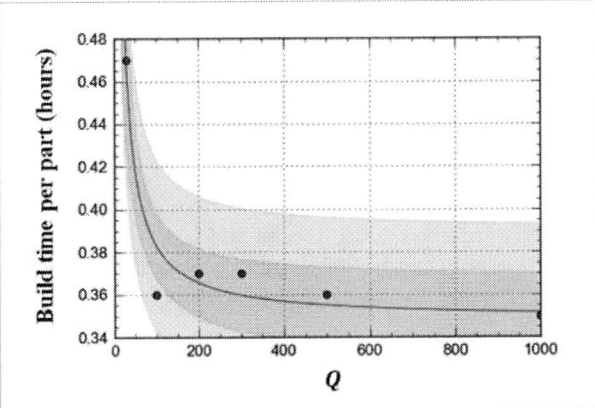

Figure 2: The build time per part depending on Q

3.2 The economic scheduling quantity (ESQ)

Equation (1) is used to estimate T_b based on Q by multiplying Q to both sides of the formula. It is shown as follow:

$$T_b = \alpha Q + \beta \qquad (2)$$

In parallel production, jobs are assigned to multiple AM machines and the machines run simultaneously and independently. Therefore, if a manufacturer has more AM machines, the completion time, T_p of Q units will be smaller. Given this relation, T_p is calculated as follows:

$$T_p = \frac{T_b}{M} = \frac{\alpha Q + \beta}{M} \qquad (3)$$

Figure 3 presents the number of scheduled and not scheduled customer demands over time. With T_c, the solid line that represents the not scheduled customer demands repeats with the same cycle. At the point of scheduling, Q units of customer demands are scheduled by grouping parts into jobs. Then, they are sent to queues of AM machines and the production begins. Therefore, at this point, while the number of not scheduled customer demands becomes zero, the number of scheduled orders goes to Q.

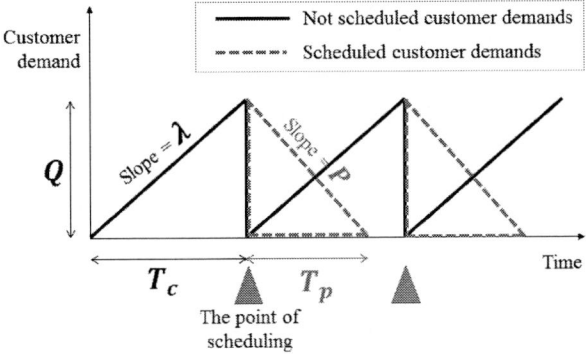

Figure 3: Customer demand level according to time

Since Q units are scheduled for each cycle, with the arrival rate of λ, the time T_c, can be calculated by Equation (4). T_c means the waiting time to make a production plan until Q units of consumer orders arrive.

$$T_c = \frac{Q}{\lambda} \qquad (4)$$

In a similar way, Q units are produced during T_p, therefore, the production rate, P, is as follows:

$$P = \frac{Q}{T_p} \qquad (5)$$

Since the customer demand level increases linearly or decreases between 0 and Q as shown in Figure 3, the average customer demand level can be calculated by Equation (6). Since the customer demand level is periodically repeated, this can be used as the average customer demand level over a time horizon.

The average customer demand level

$$= \frac{Q(T_c + T_p)}{2T_c} \qquad (6)$$

The average penalty cost, $C_t(Q)$, is obtained by multiplying the average customer demand level with the penalty cost per part, c_t. In addition, T_c and T_p are replaced by Equations (4) and (3), respectively.

$$C_t(Q) = \frac{c_t Q(T_c + T_p)}{2T_c} = \frac{c_t Q}{2} + \frac{\alpha \lambda c_t Q}{2M} + \frac{\beta \lambda c_t}{2M} \qquad (7)$$

For each cycle, the cost of build processes is calculated by multiplying T_b with a unit build process cost per hour, c_p.

$$c_p T_b = \alpha c_p Q + \beta c_p \qquad (8)$$

In addition, the material cost to produce Q units is computed as follows:

$$\text{Material cost} = Q \bar{v} c_m \qquad (9)$$

The production cost for each cycle is the build process cost plus the material cost. In order to obtain the production cost per unit time, the cost is divided by the length of cycle time, T_c. Therefore, the annual cost, $G(Q)$, consists of the build process cost, the material cost, and the penalty cost as shown in Equation (10).

$$G(Q) = \frac{\left(\alpha c_p Q + \beta c_p + Q \bar{v} c_m \right)}{T_c} + C_t(Q)$$

$$= \alpha \lambda c_p + \frac{\beta \lambda c_p}{Q} + \lambda \bar{v} c_m + \frac{c_t Q}{2} + \frac{\alpha \lambda c_t Q}{2M} + \frac{\beta \lambda c_t}{2M} \qquad (10)$$

We wish to find Q^* to minimize $G(Q)$. The derivative of $G(Q)$ with respect to Q is obtained as follows:

$$G'(Q) = -\frac{\beta \lambda c_p}{Q^2} + \frac{c_t}{2} + \frac{\alpha \lambda c_t}{2M} \qquad (11)$$

According to Equation (12), $G''(Q) > 0$. Therefore, $G(Q)$ is a convex function of Q and we can get the optimum (minimum) value.

$$G''(Q) = \frac{2\beta\lambda c_p}{Q^3} > 0 \tag{12}$$

The optimal value of Q, ESQ, is obtained from $G'(Q) = 0$.

$$Q^* = \sqrt{\frac{2\beta\lambda M c_p}{c_t(M + \alpha\lambda)}} \tag{13}$$

Some terms in Equation (10) are not functions of the scheduling size, Q, so we have put them in C, as shown in Equation (14). Therefore, the global average annual cost, $G(Q)$, is re-described by the partial average annual cost, $R(Q)$, and C as shown in Equation (15). $R(Q)$, is defined as the sum of $B(Q)$ and $E(Q)$, the buildup cost and penalty cost as functions of Q.

$$C = \alpha\lambda c_p + \lambda\bar{v} c_m + \frac{\beta\lambda c_t}{2M} \tag{14}$$

$$G(Q) = R(Q) + C \tag{15}$$

$$R(Q) = B(Q) + E(Q) \tag{16}$$

$$B(Q) = \frac{\beta\lambda c_p}{Q} \tag{17}$$

$$E(Q) = \left(\frac{c_t}{2} + \frac{\alpha\lambda c_t}{2M}\right)Q \tag{18}$$

In Figure 4, the curves represent $R(Q)$, $B(Q)$ and $E(Q)$. If $B(Q) = E(Q)$ is solved for Q, the ESQ formula, Equation (13), is obtained. This means the minimum Q^* is occurring at the intersection of the two curves, $B(Q)$ and $E(Q)$. This is the point of minimizing $R(Q)$ as well as $G(Q)$.

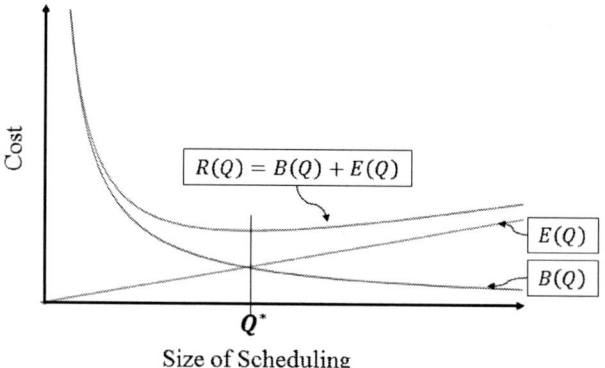

Figure 4: The partial average annual cost, $R(Q)$, consisting of the buildup and penalty costs with only the terms involved Q

3.3 Diagnosis for production status by ESQ

The production efficiency of a 3D printing plant can be analyzed by using ESQ. ESQ can help identify whether the current production capacity is sufficient to deal with the arrival rate of customer demands. This can be achieved by comparing T_c^* and T_p^*. Given Q^*, T_c^* and T_p^* are computed by Equations (4) and (3).

Figure 4 compares the two cases. In Case 1, if $T_c^* > T_p^*$, all scheduled parts are produced before the next cycle. In other words, the production capacity is sufficient to satisfy incoming orders. In Case 2, if $T_c^* < T_p^*$, all scheduled parts cannot be completed before the next cycle. Therefore, some parts that were scheduled in the previous cycle are still in production at the point of arrival of the new scheduled parts. In other words, production capacity is not sufficient to produce incoming parts. This could result in the stacking of customer demands and high penalty cost.

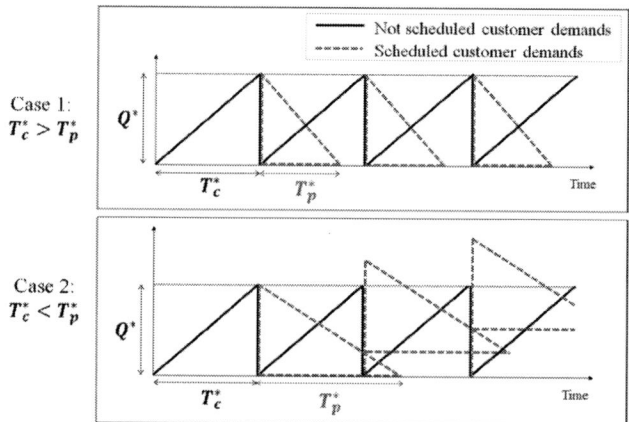

Figure 5: Diagnosis of production status

To fully handle incoming parts, a sufficient number of AM machines are needed. However, investing in too many AM machines may bring unnecessary costs to the system. Therefore, it is important to identify the minimum number of AM machines needed to satisfy consumer orders.

Equation (3) can be represented for M. T_p is replaced by T_c since the machine number is minimized when $T_p = T_c$. Then, T_c is substituted by Q/λ according to Equation (4). Given Q^*, the minimum number of AM machines is calculated by Equation (19). Since the machine number is a positive integer value, the equation has a ceiling function.

$$M^* = \left\lceil \frac{\alpha Q^* + \beta}{T_p} \right\rceil = \left\lceil \frac{\alpha Q^* + \beta}{T_c} \right\rceil = \left\lceil \lambda\left(\alpha + \frac{\beta}{Q^*}\right) \right\rceil \tag{19}$$

4. SENSITIVITY ANALYSIS

This section investigates the way that Q influences the average annual cost. The first sub-section describes the effect of Q on the global cost, $G(Q)$, and the second sub-section shows the impact of Q on the partial cost, $R(Q)$.

4.1 The impact of parameters λ and c_t on G(Q)

In this study, the decision variable Q is determined to minimize the total cost, $G(Q)$. Since $G(Q)$ consists of a variable part, $R(Q)$, and a constant part, C, determining Q is important if $R(Q)$ takes up the large part of $G(Q)$. To identify what conditions increase the effect of $R(Q)$ in $G(Q)$, two parameters, λ and c_t, are investigated. To do this, real estimates are used as shown in Table 2. α, β and $\bar{v}$ come from the experiment in Section 3.1. c_m is based on the material cost of SLA.

Table 2: Real values for major parameters

α	0.3480	β	3.5095
M	10 machines	c_p	10 \$/hour
$\bar{v}$	37928 mm³	c_m	0.00009 \$/mm³

Tables 3 and 4 represent $R(Q)$, C, and $G(Q)$ based on λ and c_t, respectively. Figure 6 shows the $R(Q)/G(Q)$ fraction based on λ and c_t. As λ decreases and c_t increases, $R(Q)$ is getting more portion in $G(Q)$, meaning that the global cost is getting more affected by Q. However, when λ is a large number, both $R(Q)$ and C increase to take care of many customer orders. In this case, even though $R(Q)$ is relatively small compared to C, it is sufficiently large that cannot be negligible.

Table 3: $R(Q)$, C and $G(Q)$ depending on λ ($c_t = 3$)

λ	$R(Q)$	C	$G(Q)$
1	14.76	7.42	22.18
5	35.16	37.10	72.26
10	53.28	74.20	127.48
20	84.51	148.40	232.91
30	113.63	222.60	336.23
60	197.52	445.20	642.72

Table 4: $R(Q)$, C and $G(Q)$ depending on c_t ($\lambda = 10$)

c_t	$R(Q)$	C	$G(Q)$
0.1	9.73	69.11	78.84
0.5	21.75	69.81	91.56
1	30.76	70.69	101.45
2	43.50	72.44	115.95
4	61.52	75.95	137.47
8	87.00	82.97	169.97

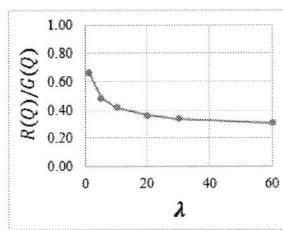 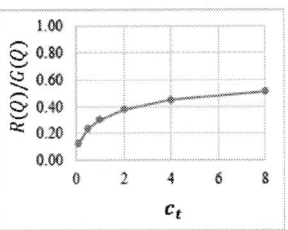

Figure 6: The $R(Q)$ portion of $G(Q)$ depending on λ and c_t

4.2 Sensitivity analysis of Q on R(Q)

Given $\lambda = 20$, $c_t = 1$ and the numbers in Table 2, ESQ is obtained as shown in Equation (20).

$$Q^* = \sqrt{\frac{2\beta\lambda M c_p}{c_t(M + \alpha\lambda)}} = 28.77 \qquad (20)$$

Based on the Q^*, $R(Q^*)$ is calculated as follow:

$$R(Q^*) = \frac{\beta\lambda c_p}{Q^*} + \left(\frac{c_t}{2} + \frac{\alpha\lambda c_t}{2M}\right)Q^* = 48.79 \qquad (21)$$

However, if a manufacturer does not follow ESQ, Q could be different from Q^*. For example, if $Q = 15$, then $R(Q)$ becomes \$59.51. The cost ratio of $R(Q)/R(Q^*)$ is 1.22.

The way that $R(Q)$ is sensitive to Q can be generally expressed by a formula. Suppose R^* is the partial cost at Q^* then it is expressed as Equation (22).

$$R^* = B(Q^*) + C(Q^*)$$

$$= \frac{\beta\lambda c_p}{Q^*} + \left(\frac{c_t}{2} + \frac{\alpha\lambda c_t}{2M}\right)Q^*$$

$$= \beta\lambda c_p \sqrt{\frac{c_t(M + \alpha\lambda)}{2\beta\lambda M c_p}}$$

$$+ \frac{c_t(M + \alpha\lambda c_t)}{2M}\sqrt{\frac{2\beta\lambda M c_p}{c_t(M + \alpha\lambda)}} \qquad (22)$$

$$= \sqrt{\frac{\beta\lambda c_p c_t(M + \alpha\lambda)}{2M}} + \sqrt{\frac{\beta\lambda c_p c_t(M + \alpha\lambda)}{2M}}$$

$$= \sqrt{\frac{2\beta\lambda c_p c_t(M + \alpha\lambda)}{M}}$$

It follows Equation (23) for any Q.

$$\frac{R(Q)}{R^*} = \frac{1}{2Q}\sqrt{\frac{2\beta\lambda c_p M}{c_t(M + \alpha\lambda)}} + \frac{Q}{2}\sqrt{\frac{c_t(M + \alpha\lambda)}{2\beta\lambda c_p M}} \qquad (23)$$

$$= \frac{Q^*}{2Q} + \frac{Q}{2Q^*} = \frac{1}{2}\left(\frac{Q^*}{Q} + \frac{Q}{Q^*}\right)$$

Therefore, if $Q = 15$ and $Q^* = 28.77$, $R(Q)/R^*$ is 1.22 from Equation (23) even though Q^*/Q is 1.92. This shows that the partial cost is relatively insensitive to errors of Q. This point is similar to the concept of EOQ models in the inventory planning literature.

5. CONCLUSION

This paper investigates the concept of production planning in mass customization-based additive manufacturing systems. Specifically, it provides a mathematical method for obtaining the optimal quantity for production planning, that is the *economic scheduling quantity (ESQ)*. In addition, a mathematical framework is provided to analyze the capacity planning in such production systems. Several sensitivity analyses have been conducted to show the impact of the model parameters on the total cost of the system.

The research can be extended in several ways. First, more accurate functions for calculating the build time per part can be extracted from practical experiments. Since the provided function is based on SLA and 2D packing, other conditions and AM processes can be studied to affect the function. Second, when estimating the build time per part other factors such as build orientation should be considered. In addition, the classical EOQ model could be combined with the ESQ model to simultaneously consider inventory control issues for parts as well as production planning. Lastly, the application of the model can be shown in practice for a real case study.

ACKNOWLEDGEMENTS

This material is based upon work supported by the National Science Foundation–USA under grant # CMMI-1727190. Any opinions, findings, and conclusions or recommendations expressed in this material are those of the authors and do not necessarily reflect the views of the National Science Foundation.

REFERENCES

[1] "Voodoo Manufacturing," *Voodoo Manufacturing*. [Online]. Available: https://voodoomfg.com/. [Accessed: 23-Feb-2018].

[2] "Protolabs." [Online]. Available: https://www.protolabs.com/. [Accessed: 19-Feb-2019].

[3] "Markforged." [Online]. Available: https://markforged.com/. [Accessed: 19-Feb-2019].

[4] "Prusa 3D printers." [Online]. Available: https://www.prusaprinters.org/. [Accessed: 23-Feb-2018].

[5] L. J. Krajewski and L. P. Ritzman, *Operations management: processes and value chains*. Pearson/Prentice Hall, 2005.

[6] L. J. P. Araújo, E. Özcan, J. A. D. Atkin, and M. Baumers, "Analysis of irregular three-dimensional packing problems in additive manufacturing: a new taxonomy and dataset," *International Journal of Production Research*, vol. 0, no. 0, pp. 1–15, Oct. 2018.

[7] Y. Zhang, R. K. Gupta, and A. Bernard, "Two-dimensional placement optimization for multi-parts production in additive manufacturing," *Robotics and Computer-Integrated Manufacturing*, vol. 38, pp. 102–117, Apr. 2016.

[8] A. S. Gogate and S. S. Pande, "Intelligent layout planning for rapid prototyping," *International Journal of Production Research*, vol. 46, no. 20, pp. 5607–5631, Oct. 2008.

[9] V. Griffiths, J. P. Scanlan, M. H. Eres, A. Martinez Sykora, and P. Chinchapatnam, "Cost-driven build orientation and bin packing of parts in Selective Laser Melting (SLM)," *European Journal of Operational Research*, Jan. 2018.

[10] Q. Li, I. Kucukkoc, and D. Z. Zhang, "Production planning in additive manufacturing and 3D printing," *Computers & Operations Research*, vol. 83, pp. 157–172, Jul. 2017.

[11] J. Kim, S. Park, and H. Kim, "Scheduling 3D printers with multiple printing alternatives," in *2017 13th IEEE Conference on Automation Science and Engineering (CASE)*, 2017, pp. 488–493.

[12] K. Ransikarbum, S. Ha, J. Ma, and N. Kim, "Multi-objective optimization analysis for part-to-Printer assignment in a network of 3D fused deposition modeling," *Journal of Manufacturing Systems*, vol. 43, pp. 35–46, Apr. 2017.

[13] F. W. Harris, "How Many Parts to Make at Once," *Operations Research*, vol. 38, no. 6, pp. 947–950, Dec. 1990.

[14] W. J. Hopp and M. L. Spearman, *Factory Physics*, 3 edition. Long Grove, Ill.: Waveland Pr Inc, 2011.

[15] E. W. Taft, "The most economical production lot," *The Iron Age*, vol. 101, pp. 1410–1412, 1918.

[16] R. H. Wilson, "A Scientific Routine for Stock Control," *Harvard Business Review*, vol. 13, no. 1, pp. 116–128, 1934.

[17] T. M. Whitin, "Inventory Control Research: A Survey," *Management Science*, vol. 1, no. 1, pp. 32–40, Oct. 1954.

[18] A. Guiffrida, M. A. Bushuev, M. Y. Jaber, and M. Khan, "A review of inventory lot sizing review papers," *Management Research Review*, vol. 38, no. 3, pp. 283–298, Mar. 2015.

[19] S. Nahmias, *Production and operations analysis*, 6th ed. McGraw-Hill/Irwin, 2008.

[20] Y. Oh, C. Zhou, and S. Behdad, "Production Planning for Mass Customization in Additive Manufacturing: Build Orientation Determination, 2D Packing, and Scheduling," in *Proceedings of the ASME 2018 International Design Engineering Technical Conferences & Computers and Information in Engineering Conference*, Quebec, Canada, 2018.

[21] Thingiverse.com, "Thingiverse - Digital Designs for Physical Objects." [Online]. Available: https://www.thingiverse.com/. [Accessed: 29-Sep-2018].

Proceedings of the ASME 2019
International Design Engineering Technical Conferences
and Computers and Information in Engineering Conference
IDETC/CIE2019
August 18-21, 2019, Anaheim, CA, USA

DETC2019-97915

DESIGN FOR ADDITIVE MANUFACTURING USING A MASTER MODEL APPROACH

Anton Wiberg, Johan Persson, Johan Ölvander
Linköping University
Linköping, Sweden

ABSTRACT

The introduction of Additive Manufacturing opens up possibilities for creating lighter, better and customized products. However, to take advantage of the possibilities of Additive Manufacturing, the design engineer is challenged. In this paper, a general design process for the creation of complex products is proposed and evaluated. The proposed method aims to aid a design process in which Topology Optimization (TO) is used for concept development, and the result is then interpreted into a Master Model (MM) supporting design evaluations during detailed design. At the same time as the MM is created, information regarding manufacturing is saved in a database. This makes it possible to automatically generate and export models for manufacturing or CAE analyses. A tool that uses Knowledge-Based Engineering (KBE) to realize the presented methodology has been developed. The tool is specialized for the creation of structural components that connect to other components in an assembly. A case study, part of an aircraft door, has been used for evaluation of the tool. The study shows that the repetitive work when interpreting the topology-optimized design could be reduced. The result comes in the form of a parametric CAD model which allows fast changes and the coupled database enables the export of models for various purposes.

1. INTRODUCTION

Additive Manufacturing (AM) is a group of manufacturing methods that have the similarity of successively building objects based on 3D CAD data, usually using a layer-by-layer approach [1]. AM generates several benefits compared to traditional manufacturing methods. From a component perspective, AM allows the creation of complex shapes that may not be possible, or at best will be expensive, using other manufacturing methods. These complex shapes could be used for the creation of lighter designs or designs with integrated functionality. From a development perspective, AM allows a product development

process that includes faster production of prototypes, the possibility to make changes with relatively small effects on the manufacturing, less need to create drawings and other manufacturing data, and manufacturing with fewer steps than conventional processes [2].

The creation of geometries that fully leverage the advantages of AM is challenging. As stated by Simpson [3], today's design tools and methods need to be improved in order to fully support the design process for AM components. To support this process, the subject Design for Additive Manufacturing (DfAM) has been introduced with the aim of introducing new design methods and tools that aid the design process for AM. One of the technologies that is often mentioned together with AM is Topology Optimization (TO). TO is a generic structural optimization method that uses a finite element-based approach to find the best overall distribution of material in a structure [4]. By using TO, complex shapes with optimal structural design can be achieved, which the freeform manufacturing capabilities of AM can then be used to manufacture. However, it may not be economical or technically feasible to manufacture all kinds of shapes achieved by TO. Therefore, efforts have been directed into the development of new TO formulations and filters, aiming to create designs adopted for AM. This includes formulations that consider support structure minimization, anisotropic material behaviour and more [5]. Despite the advancements in TO, direct manufacturing of the TO shape is in most cases not favourable; instead, some remodelling or transformation to CAD geometry is necessary [6].

This paper contributes to the field of DfAM with the presentation of a design method that uses a Master Model (MM) approach to save knowledge from various stages in the design process in order to reduce the gap between TO, CAD, computer simulations, AM, and post-processing. In the MM, the result of TO is interpreted in a parametric CAD model, which allows easy changes in the geometry. The MM also enables the automatic export of models for different purposes in the design process. This is utilized using Knowledge-Based Engineering (KBE) to reuse existing knowledge and to automate repetitive tasks. In order to support the presented method, a tool aimed at creating models in a structural assembly has been developed. Both the method and tool are demonstrated in a case study.

The paper is divided into six sections: *State of the art, Proposed method, Tool implementation, Case study, Discussion,* and *Conclusion*. In *State of the art,* relevant research and information for the paper is presented. *Proposed method* presents the method, and the section *Tool implementation* shows how the method is implemented in a computer tool. In *Case study,* an example of the usage of the tool is presented. In the *Discussion* section, both the proposed method and the implementation are discussed before the work is wrapped up in the *Conclusion*.

2. STATE OF THE ART

2.1 Design for Additive Manufacturing

The research efforts in the area of DfAM have increased significantly over the last few years [7], [8]. The focus of this research is split and involves everything from choice of components suitable for AM to optimization of manufacturing settings. A wide area of applications, involving everything from bio printing via the automotive and aerospace industries to architecture, has also split the research focus [9]. What these different research areas have in common is that they are all striving to create the best possible components for AM. This means leveraging the advantages of AM while also taking into account the limitations and challenges of the technology [10].

Gibson and Rosen [2] describe a generic AM process containing eight steps: CAD, conversion to STL, transfer to AM machine and STL manipulation, machine setup, build, removal from machine, post-processing, and application of the manufactured component. Two points in the design process are highlighted in order to benefit from the advantages of AM; the first is to consolidate parts in order to reduce the number of parts in a component; the second is to reduce the mass of parts with complex structures.

Consolidation of parts in an assembly could be realized by combining DfAM with Design for Assembly (DfA), whereby methods from DfA are used to identify components that could be integrated into one part [11]. Two different methods specialized for AM are presented by Rodrigue and Rivette [12], and Yang and Zhao [13].

The most common method for creating complex lightweight structures is Topology Optimization (TO). Salonatis and Zarban [14] propose a method in which TO is combined with multi-criteria decision-making between several design alternatives. These alternatives should be evaluated based on several objectives. Vayre et al. [15] have similar ideas but expand the method to include shape optimization with several objectives. Kumke et al. [16] propose a multi-level design method based on one conceptual phase and one detailed design phase. Within the conceptual phase, different concepts are created and evaluated

based on their function, and their technical and economic feasibility. During the detailed design phase, first one or several specific optimizations are performed, including topology, acoustic or flow optimization. Later, the design rules of AM and other perspectives are used to modify the designs. Finally, the design is evaluated using CAE tools to evaluate structural (and other) requirements together with manufacturing evaluation.

Regardless of how the interpretation process of the TO design is performed, there are several basic design rules developed from physical tests. Both Thomas [17] and Adam and Zimmer [18] have compiled lists of the design rules that could be used. Kranz et al. [19] have performed similar experiments but focusing on lightweight structures.

In earlier work [8], an extensive review of the field of DfAM was performed, resulting in a general design methodology with existing tools and methods coupled to each step. The presented methodology consists of 10 steps, from choosing a component to verifying that it could be manufactured. These steps are divided into 3 levels; *System design* is focused on a system or product level, the *Part level* focuses on designing single parts while considering geometrical design rules in the verification of the design step, and in *Process design* manufacturing details are evaluated and implemented for the design. An overview of all the steps in the design methodology is presented in Figure 1.

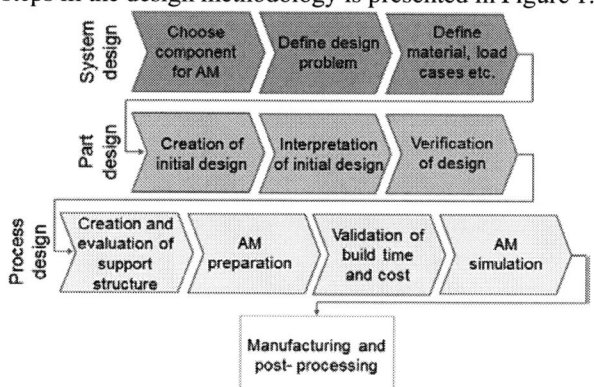

FIGURE 1. A GENERAL DESIGN FOR AM METHODOLOGY, ADOPTED FROM [8].

2.2 Structural Optimization

A general optimization consists of one objective function (f), one set of inequality constraints (g), and one set of equality constraints (h). All of these are affected by changes in a set of variables (x), where one upper and one lower variable limit is set for each variable [21], see equation 1.

$$\min_{x} f(x)$$

Subject to:
$$g_j(x) \leq 0 \quad j = 1...m \quad (1)$$
$$h_j(x) \leq 0 \quad j = 1...p$$
$$x^u \leq x_i \leq x^l$$

Optimization of structural components could be divided into the three categories of size, shape and topology optimization [22]. The difference between these three is how the geometry and the design variables are handled during the optimization. In size optimization, the variables are coupled to geometrical entities such as thickness, width and length. Shape optimization uses variables that control splines or other geometrical shapes. Topology optimization uses a mesh and the material density of each mesh element as design variables [22].

In size optimization, the size of bars or similar is changed, while in shape optimization the geometry is changed in more advanced ways, often based on a parametric CAD model. Topology optimization could be seen as the most general optimization method because it only requires a design space [22]. The advantage of shape optimization based on a parametric CAD model is that the optimized result comes in the form of a CAD model that could be used directly for different analyses. Topology optimization results come in the form of a mesh with various densities, which makes it difficult to use directly and in most cases it needs to be interpreted into a standard CAD model [6].

Different approaches to semi or fully automated methods for the creation of a CAD model based on the results of TO exist in both research and commercial software. In this paper, they are divided into the three categories of *Skeleton models*, *Section based*, and *Surface smoothening*. In the skeleton approach, the TO structure is interpreted as a skeleton of splines which is then solidified using trusses (e.g. [23]). In the section-based method, different cross-sections of the TO shape are extracted, which then build up the CAD model (e.g. [24]). The surface smoothening method builds up the model by using surfaces that are fitted to the TO surface (e.g. [25], [26]).

2.3 Computer Aided Design

If a parametric CAD model is to be used for shape optimization, it is necessary for the model to be flexible in order to take different shapes but still be robust so that it does not fail for certain parts of the design space [27]. Xu [28] makes a comparison with building a house of cards. If the foundation of the house is not well built, the whole house will collapse if something at the bottom is changed. Therefore, it is necessary to use a good, well-structured modelling strategy [29]. Camba et al. [30] have compared three different modelling strategies: Delphi's horizontal modelling, explicit reference modelling and resilient modelling. The best and easiest to use according to Camba is the resilient modelling strategy. This is based on a hierarchy with six levels, in which all the CAD features from level one should be performed before creating the features in level two, and so on. An overview of this hierarchy is shown in Table 1.

Parametric CAD modelling can be split into morphological and topological geometry changes. In topological changes, new geometries are added based on templates in the CAD system. These are controlled by references which are decided when an instantiation of the template is created. Morphological changes control the shape of an already instantiated geometry [27].

TABLE 1. FEATURES OF DIFFERENT LEVELS IN THE RESILIENT MODELLING STRATEGY [30].

Group	Description	Typical features	Notes	Links
1 - Ref	All "Reference" entities are first, making them available to all features	Ref. Bodies, Images, Sketches, Ref Planes, Coord. systems	No Solids	Acceptable to link to, if it appears in the background
2 - Construction	Construction features that will be used to define complex solid features	Surfaces, Project, 3D Curves, Trim, Split	No Solids	Acceptable to link to, if it appears in the background
3 - Core	Core features that determine the model's shape, extents and orientation	Extrude, Sweep, Revolve, Shell	Add material	Acceptable to link to, if it appears in the background
4 - Detail	Detail features complete the shape by only linking to the core group	Extrude, Sweep, Hole, Revolve, Thread	Remove material	Links are acceptable, but no links within the detail group are allowed
5 - Modify	Tilt faces and replicate features, then add any final features	Draft, Pattern, Mirror, Final Features		Acceptable to link if it appears in the background
6 - Quarantine	Volatile features that should only be children	Chamfer, Fillet, Blend	Largest first	Acceptable to link if it appears in the background

3. PROPOSED METHOD

To fully utilize the strengths of topology and shape optimization, a product development methodology that combines the two is proposed. This methodology is inspired by Ulrich and Eppinger's [31] classic product development process and methods for design optimization. Based on the initial product specification, TO is used to identify different concepts. Several different formulations (maximization of stiffness with volume constraint, minimization of mass with stress constraint, manufacturing constraints, etc.) could be used to create different optimal solutions. Detailed design is achieved by interpreting the concepts into parametric CAD models. These models are then used during the Verification and Refinement phases to automatically create and evaluate several designs in a Multi-Disciplinary Analysis (MDA), which is used to create data to support a design decision, see Figure 2.

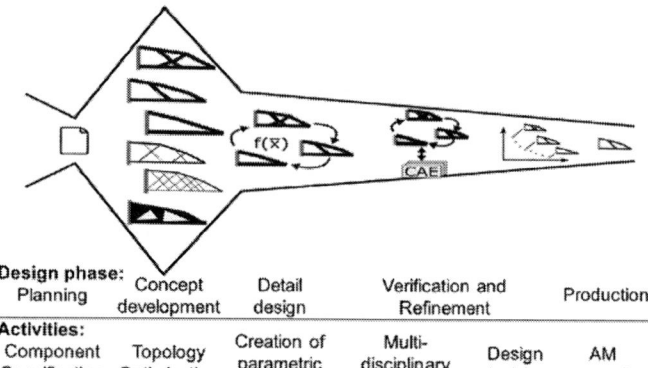

FIGURE 2. GENERAL METHODOLOGY FOR COMBINING TOPOLOGY AND MULTIDISCIPLINARY SHAPE OPTIMIZATION.

In order to achieve an automated process from the detailed design phase all the way through to the design decision, it is proposed to use a Master Model (MM) approach [32]. In an MM approach, one model (a Master Model) is created which contains all the information needed to automatically generate models for various purposes, including simulation and manufacturing. In the MM, the initial TO result is interpreted into a parametric CAD model, including details such as interfaces with other components. The MM approach has two benefits; firstly, it is easy to change the geometry based on parametric values, and secondly, there is a possibility to automatically export different models.

As described in section 2.1, a geometry for manufacturing using AM goes through several steps on its way from CAD to final component. These include an STL representation of the manufacturing geometry. In a build setup where support structure is added, the model is sliced and a g-code for the manufacturing is generated in an AM-prepared geometry. A roughly manufactured geometry direct from the printer is post-processed in different ways before the final geometry is received. During this process, a manufacturing analysis could also be performed in order to verify that manufacturing is possible. As a parallel information track, the geometry is evaluated from other perspectives. This track consists of one or several CAE simulations which mimic the final geometry from their respective functions. An overview of the information flow, with one track for CAE evaluations and another for manufacturing, is shown in Figure 3.

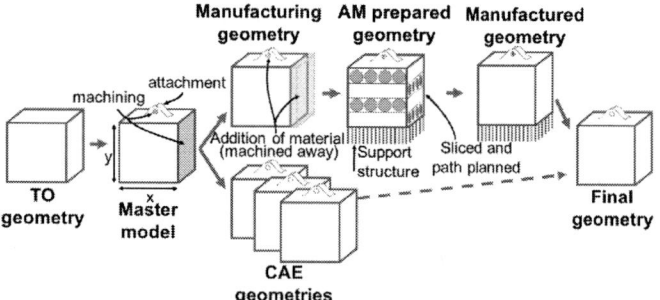

FIGURE 3. OVERVIEW OF PROPOSED INFORMATION PROCESS.

3.1 Master Model representation

Knowledge-Based Engineering (KBE) is used to support the creation of the MM and to store information within it. KBE is an area in which different concepts from object-oriented programming, artificial intelligence, and Computer Aided Design (CAD) are mixed together, with the goal of reducing the repetitive work in engineering design [33]. Within a KBE framework, a generic reuse of rules and design elements could be created by using a formal object-oriented representation of objects within the MM [34]. Object-oriented programming is based on classes, where each class describes a type of objects with a predefined set of attributes and methods. Attributes describe the state of an instance of the class (an object). Methods refers to a set of tasks that describe what you can do with the object. Two things differentiate a KBE framework from ordinary object-oriented programming; the first is a more formal definition of the attributes and methods of the different classes, and the second is a direct coupling between the objects and the geometry [34]. A connection between objects and geometry is resolved by templates in a CAD system, where each class corresponds to a template. The CAD template is controlled by the attributes, which are divided into three categories: *Referencing*, *Relational*, and *Information*. The *Referencing* attributes create the connections between objects and control the references when an object is instantiated in the CAD system. *Relational* attributes are coupled to lengths, sizes, etc. and control morphological CAD changes. This allows rapid changes in the model based on manufacturing or CAE feedback. The *Information* attributes contain information necessary for the export of the CAE and manufacturing models (e.g. post-process operations).

Each class is created using methods that implement the possibility to create, modify, or delete the object. In addition to these three methods, the class could also contain methods that, based on the *Information* attributes, implement changes to the geometry when exporting models for manufacturing or CAE evaluations. An example of how two different objects in a class could appear is shown in Figure 4.

Instance 1	Instance 2		
Ball1	Ball2	*Name*	— Identity
position = pos1	position = pos2	*Reference*	
size = 15.3	size = 12.5	*Relation*	—Attributes
colour = "Red"	colour = "Blue"	*Information*	
create()	create()		
modify()	modify()		
delete()	delete()		— Methods
exportManufacturing ()	exportManufacturing ()		
exportCAE1()	exportCAE1()		

FIGURE 4. EXAMPLE OF TWO INSTANCES (OBJECTS) OF THE CLASS

In order to identify relevant classes to build the MM, it is proposed to use the function-mean modelling method [35]. By breaking down a component into sub-functions and means, thus solving the function, a framework with reusable programming classes could be created. The created classes are realizations of means and these then create a connection between the function of the product and its geometry. The classes could also be divided into sub-classes that build up a nested hierarchy between the different classes. Classes further down in the hierarchy act as *referencing attributes* for the classes above. By following the resilient CAD modelling strategy [36] when identifying the references for the class, a robust CAD model is achieved. An overview of how different classes in an object-oriented programming language could be used to represent a component is shown in Figure 5.

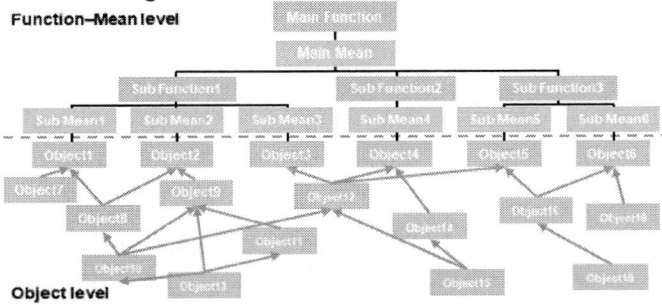

FIGURE 5. OVERVIEW OF HOW A COMPONENT COULD BE REPRESENTED USING OBJECT-ORIENTED PROGRAMMING.

3.2 Information storage in the Master Model

Information about a created component is stored in two parallel ways within the MM: a CAD geometry and a relational database. In a relational database, several tables are used to store the information. It is proposed to use four different tables, where Table A (within Table 2) consists of an identification number and the name of the instance. Table B connects the instance of an object with a class type. In Table C, the available attributes for a specific class are presented. Finally, Table D saves the value of each attribute of the instantiation. What the database would look like for the objects in Figure 4 is presented in Table 2.

TABLE 2. EXAMPLE OF THE DATABASE.

Table A		Table B		Table C		Table D		
ID	Name	ID	Class	Class	Attribute	ID	Attribute	Value
1	Ball1	1	Ball	Ball	Position	1	Position	Pos1
2	Ball2	2	Ball	Ball	Size	1	Size	15.3
				Ball	Colour	1	Colour	Red
						2	Position	Pos2
						2	Size	12.5
						2	Colour	Blue

4. TOOL IMPLEMENTATION

A tool that implements the proposed method has been created using Visual Basic for Applications (VBA). Using an interface in Excel, this tool allows connection to CATIA for the automatic creation of CAD features and parallel saving of information in a database. The goal of the tool is to support the creation of structural components based on the TO results. The functionality of the tool is (so far) limited to the creation of components based on a few basic means.

Function-Mean decomposition was used to identify means of solving the function of a structural component. The main function is to transfer loads between components in an assembly, sub-functions are to transfer loads within the component and to attach to other components. Transformation of loads could be

solved, as defined in Section 2.2, by means of the truss structure, arbitrary shape, and lattice structure. Some of the means for solving the function "attach to other components" are rivets, bolts, and glue.

In the tool implementation, one means to transfer load (truss) and two means to attach (rivets and bolts) have been implemented. The classes for solving the means are built up using other classes in a hierarchy. The resilient modelling strategy [36] is used to identify the hierarchy of the classes. This means, for example, that the class point is lowest in the hierarchy, which then build up splines, and sketches. The top classes, which correspond to the means, are realized in the form of surface-based CAD features. An overview of the function-mean decomposition and the different classes of the studied component is shown in Figure 6.

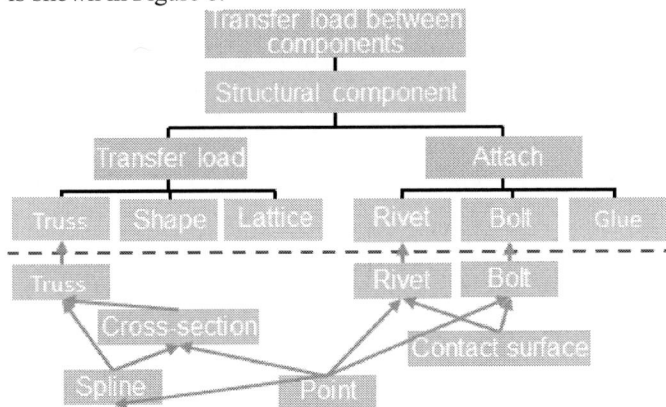

FIGURE 6. FUNCTIONAL DECOMPOSTION OF A STRUCTURAL COMPONENT WITH CLASSES IN AN OBJECT-ORIENTED PROGRAMMING LANGUAGE THAT REALIZES THE COMPONENT.

The Excel interface of the tool is based on four different blocks, each containing a set of buttons coupled to popup windows guiding the design process. In the first block, the process is prepared by the creation of a new CATIA part, the import of a TO shape and an assembly of surrounding components. Blocks two and three are dedicated to creating and handling the geometry and are divided into the two sub-functions identified in the function-mean decomposition. The creation of some basic features, such as points and extraction of surfaces, needs to be done manually in CATIA, guided by instructions from the tool. Features higher in the hierarchy are created with popups that allow the user to add properties to the feature, which are then automatically added into CATIA and the database. The fourth and final block supports the automatic export of models for FE-analysis and manufacturing evaluation. A picture of the interface, and some of its instruction pop-up windows, can be seen in Figure 7.

4.1 CAD

One of the attachment types supported in the tool is rivets. The rivet class is supported by the sub-classes contact surface and referencing point (see Figure 6). An attachment surface is extracted from a surrounding component while the reference point is created on the surface. These two elements need to be manually created in CATIA but with guidance from the tool. Based on these two referencing attributes, the geometry connected to the rivet is automatically created. The rivet class contains attributes that control the placement of the rivet's centre point in relation to the referencing point, the size of the rivet, how thick the material needs to be, and how much material needs to surround the rivet. Information attributes also contain information describing which surfaces need to be post-processed. How two rivets could be placed in a CAD environment and what the referencing and relational attributes describe are shown in Figure 8.

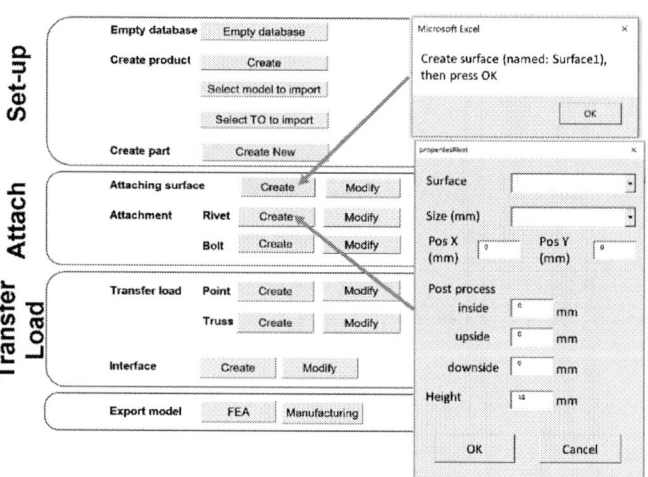

FIGURE 7. INTERFACE OF THE CREATED TOOL. TWO OF THE POPUPS FOR THE CREATION OF A RIVET ARE SHOWN.

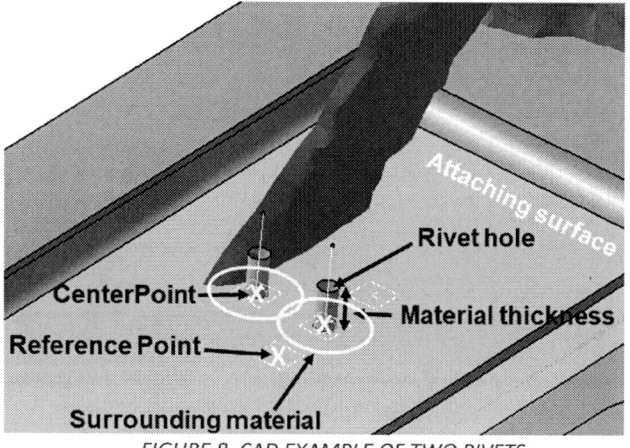

FIGURE 8. CAD EXAMPLE OF TWO RIVETS.

To solve the function "transfer of loads between the attachments", the creation of a truss structure is implemented in the tool. A truss structure is built up of points that support the creation of splines, see Figure 9. The points and splines are used as references for cross-sections. A cross-section consists of an ellipse with two radii and an angle to rotate the ellipse. Several cross-sections, together with a spline, are used to create a truss

section, see Figure 10. In the developed tool, it is necessary to manually create the points and define which of them create a truss. Then, based on this, the spline, cross-sections and truss surface are created automatically.

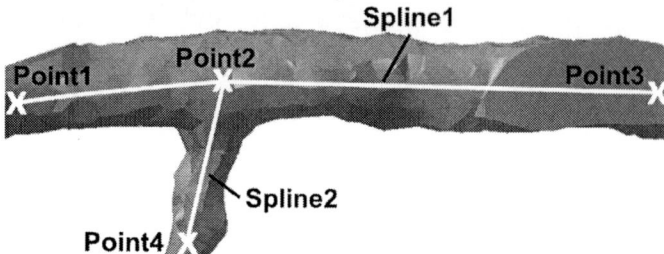

FIGURE 9. POINTS AND SPLINES USED TO BUILD UP TRUSS STRUCTURES BASED ON TO GEOMETRY.

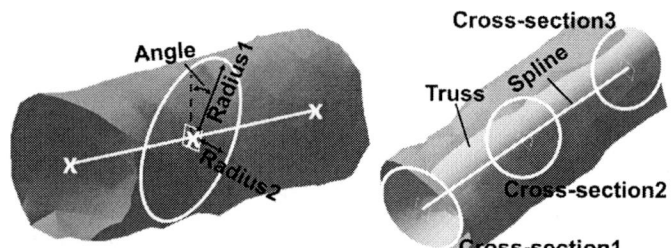

FIGURE 10. LEFT: CROSS-SECTION IN THE FORM OF AN ELLIPSE. RIGHT: TRUSS-SECTION BASED ON SEVERAL CROSS-SECTIONS AND A SPLINE.

4.2 Export of CAE models

The final functionality of the created tool, which is also one of the main purposes of the tool, is the automatic export of models and/or information for CAE evaluations and manufacturing. The created tool supports the export of two different models. One of these is the geometry for manufacturing. Compared to the original geometry, material is added at surfaces that need a higher surface finish than is achieved by AM and hence will be milled afterwards. Information regarding how much material needs to be added and on which surfaces is saved in the informational attributes. The file format for export of a manufacturing geometry is STL. The second model intended for export to CAE analysis is a structural model. This aims to simulate the exact shape of the final geometry and comes in the form of a solid CAD model.

5. CASE STUDY

A case study that covered part of the proposed DfAM process was used to evaluate the tool created and the method behind it. This case study covered a design process from functional requirement to multi-disciplinary analysis. The component in the case study was a lift fitting for an aircraft cargo door. A lift fitting is placed between the door structure and an actuator used for opening and closing the door. Based on the opening and closing cycles, together with other requirements (such as wind loading and mechanical constraints), a set of structural requirements was compiled. These structural requirements took the form of several load cases based on how far the door is opened and also different kinds of jamming situations and other errors that could occur. A stress constraint with a varying safety factor was coupled to each load case. The goal of the study was to design the lift fitting for manufacturing using an additive manufacturing powder-bed fusion technique. The objective of the design was to reduce the weight of the structure, where a certain goal: "weight of the component" was given. A design space where it was permitted to place material was created based on information regarding surrounding structures. An overview of the door geometry, the design space, and the forces on the door (the direction of the forces varied between the load cases), is shown in Figure 11.

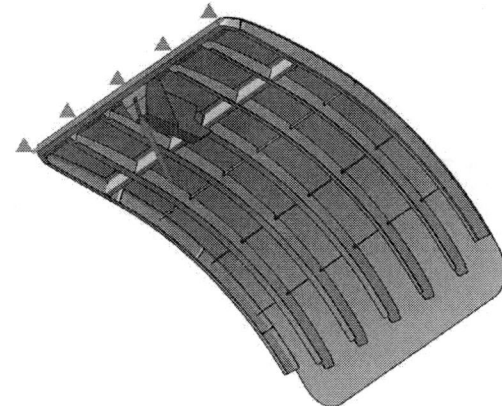

FIGURE 11. SETUP OF CASE STUDY WITH DOOR STRUCTURE IN GREY AND DESIGN SPACE IN YELLOW. THE RED ARROW REPRESENTS THE FORCE FROM THE ACTUATOR AND THE BLUE LINE IS THE HINGE.

5.1 Concept development

As presented in the method section, Topology Optimization was used to identify different concepts. The geometrical setup was like the one presented in Figure 11. The worst load conditions with forces in different directions were used. Several different optimization formulations were employed in order to identify a set of concepts. Formulations including the minimization of mass with stress constraint and the minimization of compliance with volume constraint were both considered.

Based on the weight goal and other requirements, two concepts with similar properties in terms of stiffness and weight were available for the final decision. These are shown in Figure 12. Out of these two concepts, concept 1 would require attachments passing through the skin of the door, which was decided not to be beneficial; therefore, concept 2 was chosen for further development.

FIGURE 12. TWO DIFFERENT CONCEPTS IDENTIFIED USING TOPOLOGY OPTIMIZATION.

5.2 Detailed design

The component was designed in detail using the developed tool, whereby the chosen TO concept and the surrounding components were imported to a common assembly. Then the attachments were added, and as a third step a structure that connects the attachments was created.

The types, number, and placement of attachments were decided together with experts. For connection to the door, rivets were chosen and for connection to the actuator, a bolt was chosen. Using the developed tool, surfaces for attachment were extracted, the placement of the attachments were set, and based on this the geometrical features for the attachments were automatically created. What the model with added attachment looked like is presented in Figure 13.

FIGURE 13. ATTACHMENTS ADDED TO THE COMPONENT USING THE DEVELOPED TOOL.

With the aid of the tool, a truss structure which connects the attachments was created. As presented in Figure 14, several points, splines, and cross-sections were used to build up the structure.

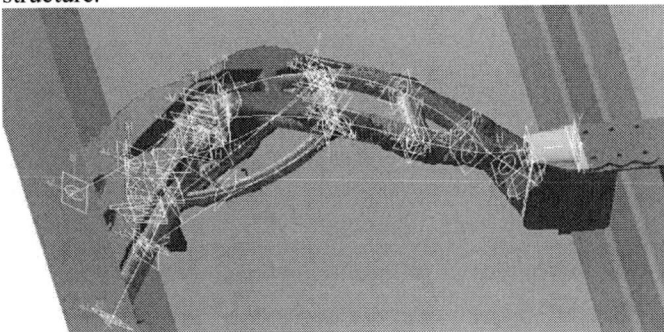

FIGURE 14. COMPONENT BUILT UP USING A TRUSS STRUCTURE.

5.3 Verification and refinement

The final design of the component was achieved by combining the structure and the attachments. A solid model based on the surfaces is automatically obtained from the developed tool. The final design of the developed component is shown in Figure 15.

Two geometries for verification were exported from the Master Model. The first geometry extracted was aimed at manufacturing by AM. As shown in Figure 16, the holes for the attachment will be drilled after the AM process and are therefore removed from the geometry. The second model exported for CAE verification is a model for FE analysis. As shown in Figure 17, an analysis to verify that the structural requirements were fulfilled was performed by an FE analysis.

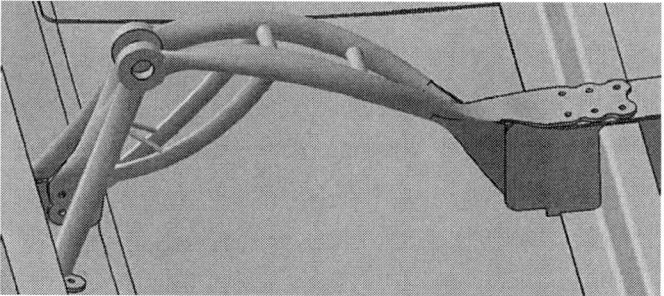

FIGURE 15. FINAL DESIGN OF THE LIFT FITTING STRUCTURE.

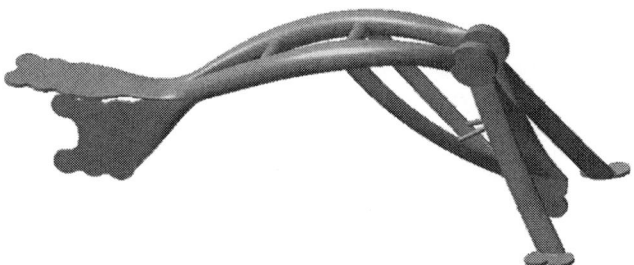

FIGURE 16. GEOMETRY EXTRACTED FOR MANUFACTURING.

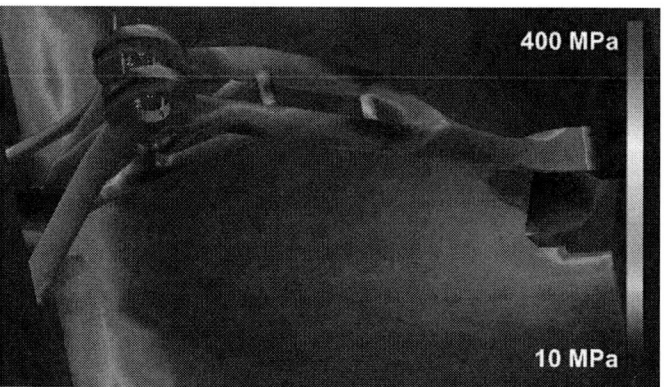

FIGURE 17. STRUCTURAL ANALYSIS OF THE COMPONENT.

5.4 Evaluation
The model created by the developed tool contained some areas of sharp corners and lacked proper blending. These could eventually cause problems during manufacturing or may be initial areas for crack propagation. Therefore, some time was spent to manually edit the model in order to achieve proper blending and a smoother model. Examples of these areas are shown in Figure 18, and the same areas in the modified model are shown in Figure 19.

It is estimated that it took around two hours to create the first model in the developed tool (for the main author). This does not include several discussions and changes in the placement of the interfaces. Another two hours were spent on the manual editing

of the model. Most of the areas could easily be fixed with functions for fillets in the CAD software, while others were more complex and needed to be built up by surfaces (e.g. the left-hand one in Figure 19). The areas that were most difficult to model were the ones connecting the truss structure with the interfaces.

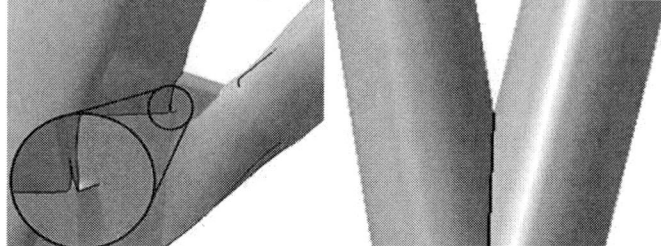

FIGURE 18. TWO EXAMPLES OF PROBLEMATIC AREAS IN THE CAD MODEL DIRECTLY FROM THE DEVELOPED TOOL.

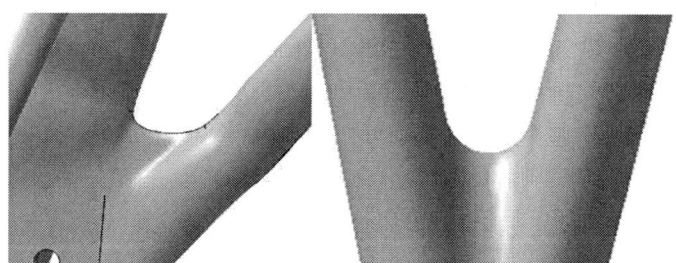

FIGURE 19. THE SAME AREAS AS IN FIGURE 18 AFTER MANUAL MODIFICATION.

The weight of the three models (Topology Optimized, direct from tool, and modified geometry) were compared. The geometry directly from the developed tool was 9% lighter than the TO geometry and the modified geometry was 5% lighter than the TO geometry.

All three geometries were evaluated from a structural perspective for three opening cycles with different jamming and error conditions; in total, 51 different load cases were evaluated. Load cases 1–17 correspond to one opening cycle, load cases 18–34 to the second, and load cases 35–51 to the third. A normalized comparison of the maximum Von Mises stress in the structure during the opening cycles for the three designs are shown in Figure 20. The model with no modifications from the developed tool showed up to 80% higher stress than the geometry direct from the TO. The modified geometry showed up to 35% higher stress for some load cases, but the stress was also up to 9% lower than the TO for other load cases.

It was stated that, for certain load cases, local stress concentrations arose in the model direct from the tool. The manually modified model has more material in certain parts (right-hand part in Figure 16) compared to the TO model; however, at the other end (left-hand part of Figure 16) it has less material. During the opening of the door, different parts of the constructions become load carriers, leading to the varying stresses seen in Figure 20.

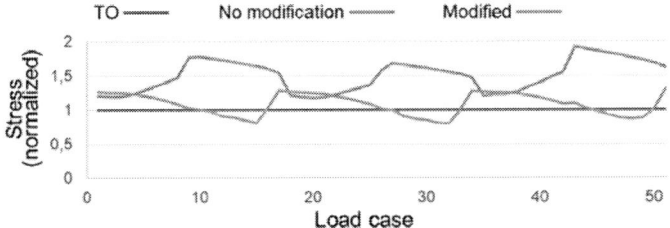

FIGURE 20. A STRESS COMPARSION BETWEEN THE THREE MODELS DURING AN OPENING CYCLE.

6. DISCUSSION

The proposed design methodology is based on existing technologies and assumes that direct manufacturing of the TO result is not suitable in most cases, as many requirements could not be fulfilled by TO today. Research supporting the direct manufacturing of TO results is something that is developing rapidly; however, the authors' opinion is that there is a long way to go before TO can achieve results where no further modifications are necessary. The development of TO formulations for AM is, in the authors' opinion, a good complement to the proposed methodology and adds possibilities at the development stage. Compared to the direct manufacturing of a TO design, a methodology where the TO result is interpreted into a parametric CAD model enables easy changes of the geometry based on feedback from functional evaluation, manufacturing, or post-processing. If the parametric CAD model is combined with an analysis that evaluates the manufacturing cost and feasibility, a Multidisciplinary Design Optimization process could be used to create components that are both structurally optimal and truly feasible. This requires more work on the integration of CAE tools for manufacturing evaluation and the definition of optimization objectives and constraints.

The developed tool is based on quite simple shapes and, so far, has only been developed for structural components. The design automation framework with Function-Mean modelling and coupling to a database is, however, formulated in a generic fashion, and in theory expansion of the tool is easy. By using a function-mean modelling approach, a clear structure within the model could be achieved. This creates a natural division of the problem into sub-problems that are easier to solve, making the task of implementing new features easier. For future work, the function-mean approach also opens up the possibility of including other kinds of functions in the tool, which could be used to incorporate solutions for integrated structures.

In the case study, it was shown that the developed tool did not really fulfil the demand for the creation of smooth CAD surfaces. For the tool to be usable without the need for any modification of the model, it is necessary to develop the ability to create good intersections between structure and interfaces. These problems were solved in the modified geometry, which showed good potential for usage in a real case. The maximum stress in the modified geometry varied a bit from those in the TO structures, and this is probably due to the simplification of the structures. In some regions, the models from the tool and the modified geometry contained more material, but in other regions there was less material. This explains how lower stress could be achieved in the modified geometry compared to the TO geometry. It was also seen that both of the geometries based on the developed tool had a lower mass than the TO geometry. If some of the parameters controlling the cross-sections were fine-tuned, it would probably be possible to reduce the gap between the TO shape and the other two without going above the mass of the TO geometry. Although TO achieved an interesting-looking organic shape, there was not that big a difference between the shape emerging from TO and the much simpler shape involving only trusses.

The estimation of time spent on the case study is based on modelling by the main author of this paper, who knew all the functions within the tool and had a clear idea of how the component was going to be modelled. For someone with less knowledge about the tool and no idea how to model the component, it would certainly take more time. For the modelling of one single geometry, there probably exists software that could do it faster and end up with a smoother model directly. The advantage of using the tool developed in this study is the ease of making changes to the geometry due to the parametric build of the model, and the automatic export of models for manufacturing and FE evaluation. The use of the developed tool is justified in an iterative design process involving several design iterations.

In the developed tool, basic entities such as placement of points and cross-section sizes need to be manually defined. For future studies, it would be interesting to automate this process; possible solutions could be, for example, to use machine learning to identify certain geometrical features within the TO geometry.

7. CONCLUSIONS

This paper proposes a method of design for additive manufacturing that combines topology optimization for concept development and multi-disciplinary analysis for the detailed design and verification of a component. This is enabled using a Master Model approach in which Knowledge-Based Engineering is used to save and reuse information and knowledge in an object-oriented programming language. By introducing a direct coupling between CAD and a database with stored information about the component, repetitive tasks could be reduced. This includes tasks when creating the initial model and when creating models for manufacturing or CAE evaluation.

A tool that implements the proposed method for the creation of structural components was developed. This tool was created in VBA and uses an MS Excel interface for the creation of features in CATIA. The tool was evaluated on a case study consisting of a lift fitting for an aircraft cargo door manufactured using AM. By implementing the methodology on the case study, it was shown that the proposed methodology has the potential to improve the method of working with design for AM. The developed tool also showed potential, but it needs development in order to create smooth models for AM.

Several improvements and ideas for future work were discussed. These include, but are not limited to: automatic integration of CAE software, Multidisciplinary Optimization based on the exported models, improvements off the tool, and the addition of functions to create components with integrated functions.

ACKNOWLEDGMENTS

This work was performed within the AddMan project, which is part of the Clean Sky 2 Joint Undertaking under the European Union's Horizon 2020 research and innovation programme under grant agreement No. 738002.

REFERENCES

[1] ASTM, "ASTM standard f2792, Standard Terminology for Additive Manufacturing Technologies," 2013.

[2] I. Gibson, D. Rosen, and B. (Brent) Stucker, *Additive manufacturing technologies: 3D printing, rapid prototyping, and direct digital manufacturing*, 2nd ed. 20. 2015.

[3] T. W. Simpson, "Trade-offs with AM," *Modern Machine Shop Magazine*, no. June, 2017.

[4] D. W. Rosen, "A review of synthesis methods for additive manufacturing," *Virtual Phys. Prototyp.*, vol. 11:4, pp. 305–317, 2016.

[5] J. Liu *et al.*, "Current and future trends in topology optimization for additive manufacturing," *Struct. Multidiscip. Optim.*, vol. 57, no. 6, pp. 2457–2483, Jun. 2018.

[6] T. Zegard and G. H. Paulino, "Bridging topology optimization and additive manufacturing," *Struct. Multidiscip. Optim.*, vol. 53, no. 1, pp. 175–192, 2016.

[7] S. Yang and Y. F. Zhao, "Additive manufacturing-enabled design theory and methodology: a critical review," *Int. J. Adv. Manuf. Technol.*, pp. 327–342, 2015.

[8] J. Wiberg, A. Persson, J. Ölvander, "Design for additive manufacturing – A review of available design methods and software," *Manuscr. Submitt. Publ.*, 2019.

[9] W. Gao *et al.*, "The status, challenges, and future of additive manufacturing in engineering," *Comput. Des.*, vol. 69, pp. 65–89, 2015.

[10] M. K. Thompson *et al.*, "Design for Additive Manufacturing: Trends, opportunities, considerations, and constraints," *CIRP Ann. - Manuf. Technol.*, vol. 65, no. 2, pp. 737–760, 2016.

[11] O. C. & S. R. N. Boyard, M. Rivette, "A design methodology for parts using Additive Manufacturing," *High Value Manuf. Adv. Res. Virtual Rapid Prototyp.*, 2014.

[12] H. Rodrigue and M. Rivette, "An Assembly-Level Design for Additive Manufacturing Methodology," *IDMME - Virtual Concept*, pp. 1–9, 2010.

[13] S. Yang and Y. F. Zhao, "Additive Manufacturing-Enabled Part Count Reduction: A Lifecycle Perspective," *J. Mech. Des.*, vol. 140, no. 3, pp. 31702–31712, Jan. 2018.

[14] K. Salonitis and S. Al Zarban, "Redesign optimization for manufacturing using additive layer techniques," *Procedia CIRP*, vol. 36, pp. 193–198, 2015.

[15] B. Vayre, F. Vignat, and F. Villeneuve, "Designing for additive manufacturing," *Procedia CIRP*, vol. 3, no. 1, pp. 632–637, 2012.

[16] M. Kumke, H. Watschke, and T. Vietor, "A new methodological framework for design for additive manufacturing," *Virtual Phys. Prototyp.*, vol. 11, no. 1, p. 3, 2016.

[17] D. Thomas, "The Development of Design Rules for Selective Laser Melting," University of Wales, 2009.

[18] G. A. O. Adam and D. Zimmer, "Design for Additive Manufacturing-Element transitions and aggregated structures," *CIRP J. Manuf. Sci. Technol.*, vol. 7, no. 1, pp. 20–28, 2013.

[19] J. Kranz, D. Herzog, and C. Emmelmann, "Design guidelines for laser additive manufacturing of lightweight structures in TiAl6V4," *J. Laser Appl.*, vol. 27, no. S1, p. S14001, 2015.

[20] T. Stankovic, J. Mueller, and K. Shea, "The effect of anisotropy on the optimization of additively manufactured lattice structures," *Addit. Manuf.*, vol. 17, pp. 67–76, 2017.

[21] J. Sobieszczanski-Sobieski, A. Morris, and M. J. L. Van Tooren, *Multidisciplinary Design Optimization supported by Knowledge Based Engineering*. West Sussex: John Wiley & Sons, Ltd., 2015.

[22] P. W. Christensen and A. Klarbring, *An introduction to structural optimization*. Linköping: Springer, 2009.

[23] J.-C. Cuillière, V. François, and A. Nana, "Automatic construction of structural CAD models from 3D topology optimization," *Comput. Aided. Des. Appl.*, vol. 4360, no. October, pp. 1–15, 2017.

[24] S. Larsen and C. G. Jensen, "Converting topology optimization results into parametric CAD models," *Comput. Aided. Des. Appl.*, vol. 6, no. 3, pp. 407–418, 2009.

[25] Y.-L. Hsu, M.-S. Hsu, and C.-T. Chen, "Interpreting results from topology optimization using density contours," *Comput. Struct.*, vol. 79, no. 10, pp. 1049–1058, 2001.

[26] M. H. Hsu and Y. L. Hsu, "Interpreting three-

dimensional structural topology optimization results," *Comput. Struct.*, vol. 83, no. 4–5, pp. 327–337, 2005.

[27] K. Amadori, M. Tarkian, J. Ölvander, and P. Krus, "Flexible and robust CAD models for design automation," *Adv. Eng. Informatics*, vol. 26, no. 2, pp. 180–195, 2012.

[28] X. Xu, *Integrating advanced computer-aided design, manufacturing, and numerical control: principles and implementations, by X. Xu*, vol. 49, no. 11. 2011.

[29] V. S. Douzloo, "An integrated approach to parametric associative design for powertrain components in the automotive industry," *Action Res.*, 2011.

[30] J. D. Camba, M. Contero, and P. Company, "Parametric CAD modeling: An analysis of strategies for design reusability," *CAD Comput. Aided Des.*, vol. 74, pp. 18–31, 2016.

[31] K. T. Ulrich and S. D. Eppinger, *Product design and development*. McGraw-Hill/Irwin, 2012.

[32] G. La Rocca and M. J. L. Van Tooren, "Enabling distributed multi-disciplinary design of complex products: a knowledge based engineering approach," *J. Des. Res.*, vol. 5(3), pp. 333–352, 2007.

[33] W. J. C. Verhagen, P. Bermell-Garcia, R. E. C. Van Dijk, and R. Curran, "A critical review of Knowledge-Based Engineering: An identification of research challenges," *Adv. Eng. Informatics*, vol. 26, no. 1, pp. 5–15, 2012.

[34] G. La Rocca, "Knowledge based engineering: Between AI and CAD. Review of a language based technology to support engineering design," *Adv. Eng. Informatics*, vol. 26, no. 2, pp. 159–179, Apr. 2012.

[35] J. Jiao and M. M. Tseng, "Fundamentals of product family architecture," *Integr. Manuf. Syst.*, vol. 11/7, no. December 2000, pp. 469–483, 1999.

[36] G. Richard, "A resiliant modeling strategy," 2013, pp. 6–7.

Proceedings of the ASME 2019
International Design Engineering Technical Conferences
and Computers and Information in Engineering Conference
IDETC/CIE2019
August 18-21, 2019, Anaheim, CA, USA

DETC2019-98024

RULE OF MIXTURES MODEL TO DETERMINE ELASTIC MODULUS AND TENSILE STRENGTH OF 3D PRINTED CARBON FIBER REINFORCED NYLON

Kaiyue Deng
kaiyuede@buffalo.edu
University at Buffalo
Buffalo, NY, USA

Hamid Khakpour Nejadkhaki
hamidkha@buffalo.edu
University at Buffalo
Buffalo, NY, USA

Felipe M. Pasquali
felipeme@buffalo.edu
University at Buffalo
Buffalo, NY, USA

Anosh P. Amaria
anoshama@buffalo.edu
University at Buffalo
Buffalo, NY, USA

Jason N. Armstrong
jna4@buffalo.edu
University at Buffalo
Buffalo, NY, USA

John F. Hall
johnhall@buffalo.edu
University at Buffalo
Buffalo, NY, USA

ABSTRACT

A model to compute the elastic modulus and tensile properties of 3D printed Carbon Fiber Reinforced Polymers (CFRP) is presented. The material under consideration is Carbon Fiber Reinforced Nylon (CFRN) produced in a Fused Deposition Modeling (FDM) process. A relationship between the nylon raster in each layer and the carbon fiber volume fraction was devised with the help of a scanning electron microscope (SEM). Thirteen groups with different layer configurations and carbon-fiber percentages were formulated and tested to obtain the elastic modulus and tensile strength. This study focused only on the properties along the printed fiber direction. The results from these tests were analyzed within the rule of mixtures framework. The results suggest that the rule of mixtures can be successfully applied to unidirectional CFRP fabricated using additive manufacturing.

Keywords: 3D Printing; Tensile Testing; Continuous Carbon Fiber; Nylon; Rule of Mixture Model

NOMENCLATURE

AM	Additive Manufacturing
CFRN	Carbon Fiber Reinforced Nylon
CFRP	Carbon Fiber Reinforced Polymers
FDM	Fused Deposition Modeling
PLA	Polylactic Acid
SEM	Scanning Electron Microscope
$A_{cf\ per\ layer}$	Carbon fiber area in one layer of the composite sample from cross-sectional view

$A_{cf,\%}$	Area percentage of carbon fiber in the cross-section
D_{fiber}	Diameter of carbon fibers
E_{cf}	Elastic modulus of the carbon fiber
E_{nylon}	Elastic modulus of the nylon matrix
E_{sample}	Elastic modulus of the composite sample
N_{cf}	Number of carbon fiber layers
N_{fiber}	Number of fibers in one filament
N_{floor}	Number of floor layers
N_{path}	Number of printed paths in the raster
N_{roof}	Number of roof layers
$V_{cf,\%}$	Volume percentage of the carbon fiber
$V_{nylon,\%}$	Volume percentage of the nylon matrix
σ_{cf}	Tensile strength of the carbon fiber
σ_{sample}	Tensile strength of the composite sample
σ'_{nylon}	Tensile strength of the nylon matrix at the maximum CFRN elongation

1. INTRODUCTION

Additive Manufacturing (AM) has become extremely popular in recent times. Its inherent ability to manufacture complex parts with ease and little waste have made it a prime choice for many industries including automobile, aerospace, biomedical, and others [1]. Combined with topology optimization, additive manufacturing has the capacity to manufacture highly reliable, application-ready products.

When designing for additive manufacturing it is imperative to understand the mechanical properties of the materials.

Numerous authors have characterized the mechanical properties of 3D printed materials [2–8]. Tian et al. [7] investigated the mechanical properties of carbon fiber reinforced PLA samples prepared by FDM. By varying fabrication parameters, the authors found an optimal fiber content percentage for maximum flexural strength and tensile modulus. Parandoush et al. [8] performed structural tests on carbon fiber printed laminated sheets by changing the fiber orientation and producing crossed ply structures. Ning et al. [9] explored the effects of carbon fiber length and content in FDM of carbon fiber reinforced polymers (CFRP). The authors included carbon fiber powder to plastic pellets to create CFRP filaments and obtained 30% increase in mechanical properties. Van de Werken et al. [10] explored the thermo-mechanical response and fracture mechanics of continuous CFRP. This study found that the strength of the carbon fiber is influenced by the fiber curvature and interface between reinforced and non-reinforced regions. A study by Van der Klift [11] investigated tensile strength of CFRP printed in the Markforged, Mark One, 3D printer. The study discovered that discontinuities originating from the printing process lead to premature failure in their specimens. Furthermore, they measured the fiber content and compared the results obtained to the rule of mixtures. The results indicated that the rule of mixtures was not suitable for modeling the CFRP. However, only two data points were considered. Naranjo-Lozada et al. [12] studied the influence of geometric parameters in the tensile properties. Based on the material properties given by the manufacturer they compared the results of the tensile tests with the rule of mixtures. They concluded that the rule of mixtures is only valid for volume fractions smaller than 11%.

Another group of authors has used the characterized data to model and design parts to be produced with this technology [10,13,14]. In 3D printed CFRP the mechanical properties are largely influenced by the amount of carbon fiber reinforcement. To design for additive manufacturing, knowing accurate mechanical properties is of fundamental importance.

The rule of mixtures is widely used to determine material properties of traditionally manufactured composites. This work in this paper aims to investigate if the rule of mixtures can be used to determine the mechanical properties of continuous carbon fiber reinforced nylon manufactured, produced by FDM process and manufactured in the Markforged Mark X printer.

2. METHODS

Several key elements are needed to reach the objective of this study. First the quantity of carbon fiber in each layer is determined; discussed in Section 2.1. With that information, a set of specimens with different volume fractions is generated and tested; covered in Section 2.2. The properties of the nylon matrix are also obtained, as described in Section 2.3. Finally, the results of the different configurations are discussed within the Rule of Mixtures framework.

2.1. Volume Fraction

In the cross-section area of the printed carbon fiber-reinforced nylon samples, the carbon fiber amount in one layer can be calculated using Eq. (1).

$$A_{cf\ per\ layer} = N_{path} \times N_{fiber} \times \pi \times \left(\frac{D_{fiber}}{2}\right)^2 \quad (1)$$

where, N_{path} is the number of printed paths in the raster, D_{fiber} is the average diameter of the fibers, and N_{fiber} is the number of fibers in a path, which is equal to the average number of fibers in an unprinted filament.

Next, the methods to determine D_{fiber}, N_{fiber}, and N_{path} are described. Before printing the samples using Markforged X printer, multiple short strands of carbon fiber reinforced nylon were cut using the extruder of the printer. After subsequent polishing, the cross section of the filament was visualized using a scanning electron microscope (SEM), as shown in Figure 1.

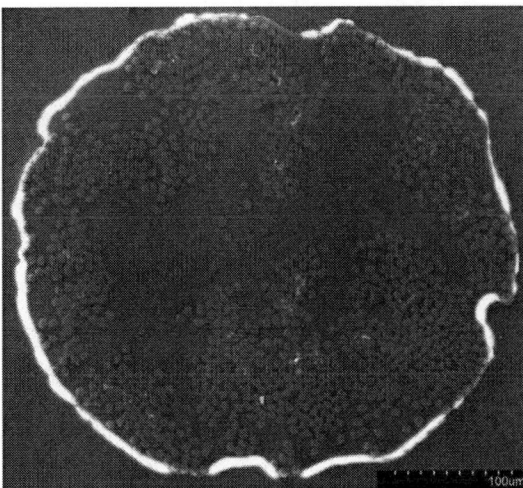

Figure 1. SEM image of Carbon Fiber Filament Cross Section

By measuring 119 fibers three times on different locations, as shown in Figure 2, the mean diameter of the carbon fibers, D_{fiber}, was found to be 7.30 µm, with an uncertainty of 0.05 µm. The number of fibers in one filament, N_{fiber}, were counted and the average value across multiple sections was determined to be 1040 ± 5 fibers.

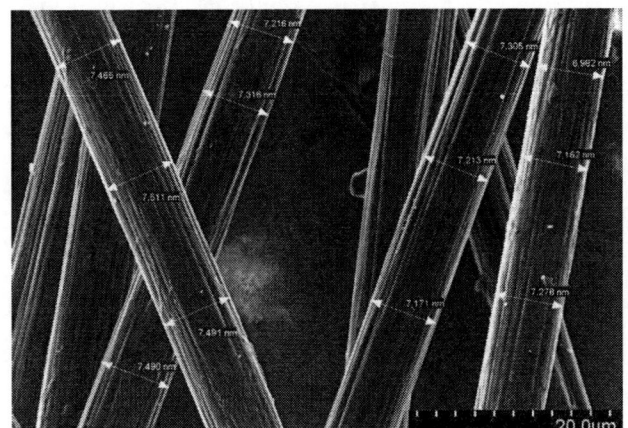

Figure 2. SEM images of Carbon Fibers and their measurement

During the printing of the carbon fiber reinforced nylon samples, it was observed that there were 13 paths of the continuously extruded carbon fiber filament in each fiber layer.

The 13 paths were also confirmed in the raster shown in the Eiger software by Markforged. All of the carbon fiber reinforced nylon samples from section 2.2 have the same number of paths because they have the same sample width. Thus, N_{path}, is 13. Finally, the $A_{cf\ per\ layer}$ of 5.6586 x 10^5 μm^2 was obtained.

2.2. Tensile Testing of Carbon Reinforced Nylon Specimens

The samples were designed in accordance with ASTM D3039-17 [15] and prepared in the Eiger software. Figure 3 shows the toolpath, length and width of the specimen, and stacking of the layers.

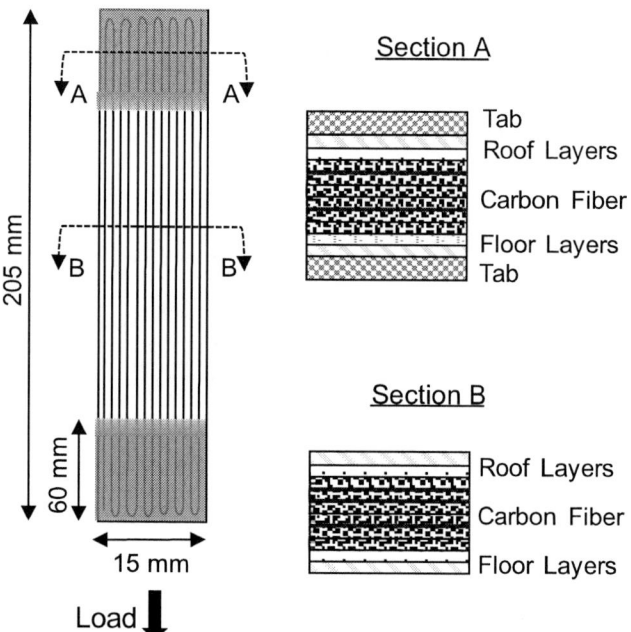

Figure 3. Specimen of Carbon Fiber Reinforced Nylon Sample

The specimens were placed in the X direction of the build plate as defined by ISO/ASTM 52921-13 [16]. The fiber orientation was set as 0 degrees so that the fibers were aligned with the load, as shown in Figure 3. As a standard in the Eiger software, the number of roof and floor layers are set to be equal. Also, the continuous carbon fiber cannot be placed in the roof and floor layers. The remaining middle layers were printed from the carbon fiber filament.

The next step was to produce tabs, which were fabricated from Garolite G-10 sheets. The final tab geometry was 60 mm long and 15 mm wide, with a bevel angle 60° , which is within the suggested range of ASTM D3039-17 [15]. Four tabs were glued onto each carbon fiber sample using 3M Scotch-Weld™ PR Gel adhesive. The specimen thickness is determined by the number of layers, described in Table 1, and the layer thickness that was 0.125mm.

To obtain a sample space with various carbon fiber percentages, the number of carbon fiber layers was used in conjunction with Eq. (1) and 13 configurations were devised. These configurations are shown in Table 1 with its respective calculated carbon fiber area. N_{floor}, N_{cf}, and N_{roof} are the number of floor layers, number of carbon fiber layers and number of roof layers, respectively. From Table 1, the maximum average carbon fiber area percentage appears in Sample 6(4) as 17.5545%, while the minimum is 1.8407% for Sample 15(1).

Table 1. Carbon Fiber Area Percentages

Layer Configuration	$N_{floor}/N_{cf}/N_{roof}$	Carbon Fiber Area %
6(2)	2/2/2	8.7537
6(4)	1/4/1	17.5545
8(2)	3/2/3	6.6091
9(1)	4/1/4	2.9255
9(3)	3/3/3	8.8296
9(5)	2/5/2	14.8321
10(4)	3/4/3	10.7114
12(2)	5/2/5	4.5853
12(4)	4/4/4	8.9239
12(6)	3/6/3	13.3872
15(1)	7/1/7	1.8407
15(3)	6/3/6	5.4739
15(5)	5/5/5	9.2371

Five samples of each of the 13 configurations were printed, 65 samples in total, and tensile testing was performed to determine the elastic modulus and tensile strength. The thickness and width of each specimen was measured using a digital caliper. The tests were conducted in a United SSTM-20kN universal testing machine. The experimental setup is shown in Figure 4.

Figure 4. 3D Printed Carbon Fiber Reinforced Nylon Sample at Failure

The United EZ-2-2 extensometer, mounted in the center of the sample, was used to measure the strain for determining the elastic modulus. Each elastic measurement was repeated three times to get an average value, with a maximum force set to 20 pounds per layer of carbon fiber, so that the strain across all samples was comparable.

2.3 Tensile Testing of Nylon Specimens

The nylon samples were designed in accordance with the ASTM D638-14 Type I [17] specimen and then prepared for printing in Markforged Eiger software. The specimens were printed with Markforged Nylon filament in a Mark X printer.

The test objective was again to obtain the elastic modulus and tensile strength. The United EZ-2-2 extensometer, mounted in the center of the sample, was used to compute measure the strain. The maximum load of 25 lbs. was used to ensure elastic deformation. The test was conducted in a United SSTM-20kN universal testing machine. The experimental setup is shown in Figure 5. After three tests were performed for the elastic modulus measurements, the extensometer was removed and the sample was tested until failure to determine the tensile strength.

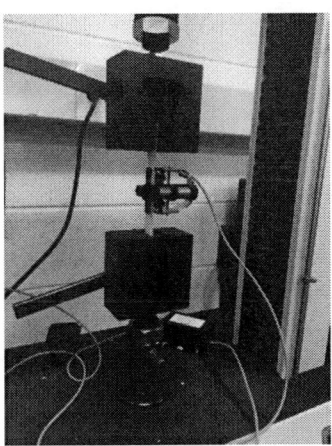

Figure 5. Setup for Elastic Testing of Pure Nylon Sample

3. RESULTS

3.1. Pure Nylon Elastic Modulus Test

Figure 6 shows the stress-strain curves of a pure nylon sample that was elastically deformed; three separate tests are displayed. The elastic modulus was determined using a linear regression of the data.

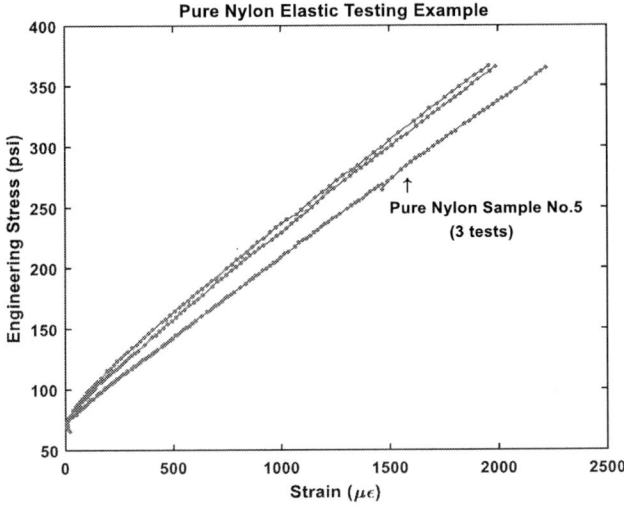

Figure 6. Stress vs Strain Curve-Elastic Testing of Pure Nylon

The average modulus of elasticity of pure nylon was found to be 0.1339 Mpsi (923.21 MPa) with a standard deviation of 0.0076

Mpsi (52.40 MPa) and a coefficient of variation of 5.69%; these values were calculated from 15 data points, with five samples being tested three times each. This value for the elastic modulus of nylon is within 5% of the value specified in the manufacturer's datasheet [18].

3.2. CFRN Elastic Modulus Test

The stress-strain curves from the elastic measurements on four select CFRN samples are shown in Figure 7.

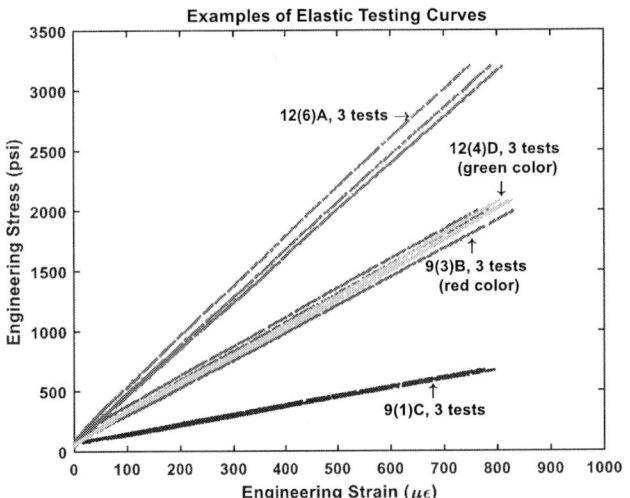

Figure 7. Stress-Strain Curves for CFR Samples

Sample 9(1)C has the lowest elastic modulus which can be attributed to the smaller volume of carbon fiber in the sample. Two samples with similar layer ratio 9(3)B and 12(4)D have comparable slopes indicating similar values for the modulus of elasticity. Sample 12(6)A has the highest slope due to its relatively large carbon fiber content. Table 2 lists the measurement results for all CFRN samples, including the average elastic modulus and corresponding standard deviation and coefficient of variation.

Table 2. Results of Elastic Testing for 13 CFRN Configurations

Layer Configuration	Average Elastic Modulus (Mpsi)	Standard Deviation (Mpsi)	Coefficient of Variation (%)
6(2)	2.3125	0.2460	10.64
6(4)	4.5817	0.2045	4.46
8(2)	1.6986	0.1278	7.52
9(1)	0.8151	0.1104	13.55
9(3)	2.4030	0.2148	8.94
9(5)	3.9044	0.4098	10.50
10(4)	3.0218	0.1816	6.01
12(2)	1.2013	0.1507	12.55
12(4)	2.4646	0.1311	5.32
12(6)	3.6525	0.2943	8.06
15(1)	0.5511	0.0396	7.19
15(3)	1.6436	0.1419	8.63
15(5)	2.5583	0.1284	5.02

Figure 8 shows the elastic modulus measurements as a function of the carbon fiber area percentage. The elastic modulus has a clear linearly increasing trend as the carbon fiber percentage increases, which is the expected result, due to the higher stiffness of the carbon fiber.

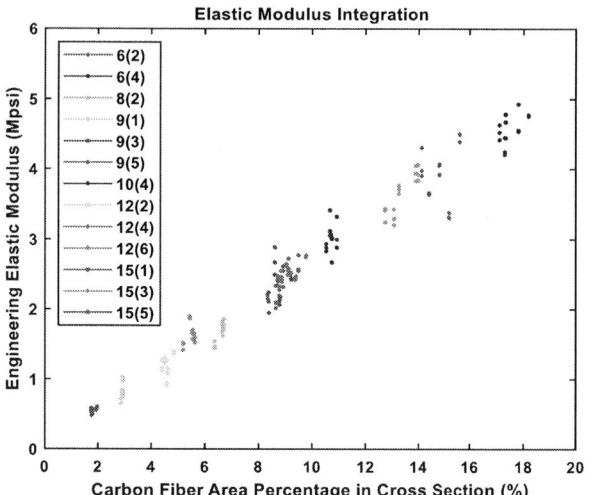

Figure 8. Relation between Elastic Modulus and Carbon Fiber Area %

3.3. Rule of Mixtures for Elastic Modulus

The data obtained in section 3.2 was fitted with a linear model to obtain the relationship of the elastic modulus vs. carbon fiber content. The intercept was constrained to be the pure nylon elastic modulus obtained in section 3.1, which is an implicit requirement for the rule of mixtures. Figure 9 depicts the linear model and its equation, overlaying the experimental data from all elastic measurements; the resulting slope is 25.6895 Mpsi. The corresponding R-squared value is 0.9731, indicating the appropriateness of a linear fit for the data.

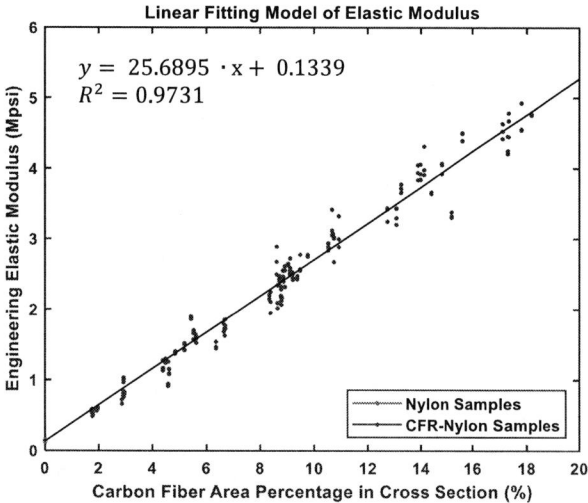

Figure 9. Linear Fit of Elastic Modulus vs Carbon Fiber Area Percentage

The rule of mixtures from Callister, Jr. [19] is described in Eq. (2).

$$E_{sample} = E_{nylon} \times V_{nylon,\%} + E_{cf} \times V_{cf,\%} \qquad (2)$$

E_{sample} is the composite elastic modulus, E_{nylon} and $V_{nylon,\%}$ are the elastic modulus and the volume percentage of the nylon matrix, respectively, and E_{cf} and $V_{cf,\%}$ are the carbon fiber elastic modulus and the carbon fiber volume percentage. Since the cross-section of the specimens is constant lengthwise, Eq. (2) was converted to Eq. (3); from volume percentage into the area percentage. In Eq.(3), $A_{cf,\%}$ is the percentage of the total area that is carbon fiber.

$$E_{sample} = E_{nylon} \times (1 - A_{cf,\%}) + E_{cf} \times A_{cf,\%} \qquad (3)$$

The slope and intercept from the linear fit model in Figure 9 were rearranged to fit Eq.(3), forming Eq.(4):

$$E_{sample} = 0.1339 \times (1 - A_{cf,\%}) \\ + 25.6875 \times A_{cf,\%} \qquad (4)$$

In Eq. (4) the value of the modulus of carbon fiber, E_{cf}, is 25.6875 Mpsi (177.1 GPa). This number falls within the range previously reported by Naito et. al., which varied from 5.95 Mpsi (41 GPa) to 136.3 Mpsi (940 GPa)[20]. Therefore, the rule of mixtures model in Eq.(4) appears well suited to determine the elastic modulus of 3D printed CFRN.

3.4. Pure Nylon Tensile Strength Test

The stress vs. displacement curves of the pure nylon samples are shown in Figure 10. The material exhibits high elongation before failure, which is typical of nylon.

The average tensile strength, determined from five individual samples of pure nylon, was 8.1241e+03 psi (56.01 MPa); the standard deviation was 340.6824 psi (2.35 MPa) and the COV was 4.19 %. This value is in close agreement with the tensile strength of 54 MPa listed in the manufacturer's datasheet [18].

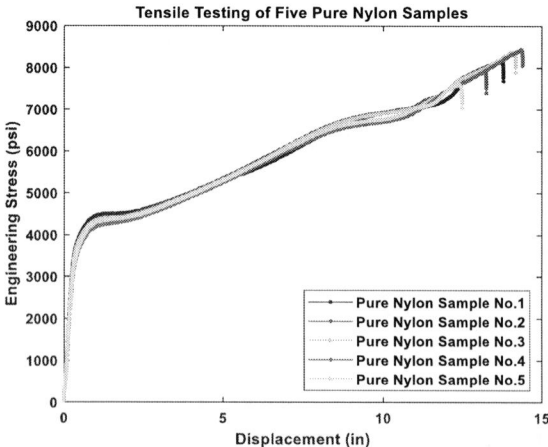

Figure 10. Stress vs. displacement curve for the pure nylon samples

3.5. CFRN Tensile Strength Test

The results of engineering stress at failure for all groups are shown in Figure 11. It is clear that the stress at failure increases linearly with fiber content, similar to the results obtained in section 3.2, and also in agreement with the trend expected by the rule of mixtures.

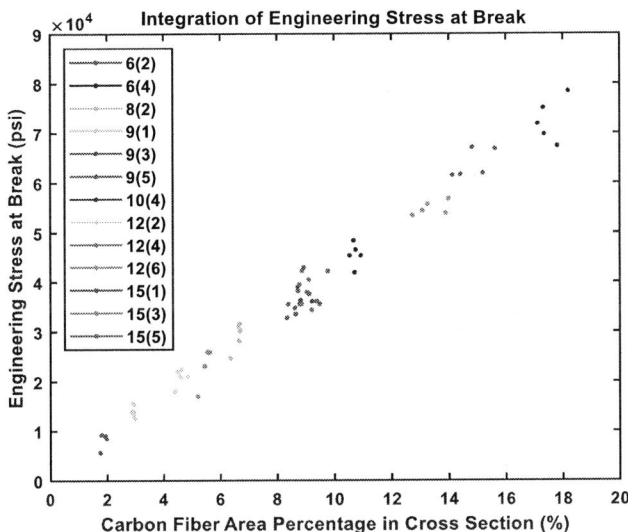

Figure 11. Engineering Stress at Break for CFRN Samples

Table 3 contains the average tensile strength for each configuration across the five tested samples. The layer configuration with the lowest amount of carbon fiber, 15(1), has the highest coefficient of variation. That is because when some samples of this configuration are tested, the carbon fiber breaks while the surrounding nylon remains. The coefficient of variation observed for these tests is comparable to the values observed by Van der Klift et al. [11].

Table 3. Engineering Stress at Break for CFRN

Layer Configuration	Average Engineering Stress at Break (psi)	Standard Deviation (psi)	Coefficient of Variation (%)
6(2)	3.6061e+04	4.0131e+03	11.13
6(4)	7.2521e+04	4.3689e+03	6.02
8(2)	2.9168e+04	2.8263e+03	9.69
9(1)	1.3913e+04	1.0179e+03	7.32
9(3)	3.6361e+04	1.6109e+03	4.43
9(5)	6.3808e+04	2.8997e+03	4.54
10(4)	4.5530e+04	2.3737e+03	5.21
12(2)	2.0851e+04	1.6922e+03	8.12
12(4)	3.9145e+04	2.5441e+03	6.50
12(6)	5.4849e+04	1.3838e+03	2.52
15(1)	8.0969e+03	1.6538e+03	20.43
15(3)	2.3575e+04	3.8751e+03	16.44
15(5)	3.7717e+04	2.9028e+03	7.70

Five samples from different groups were selected to demonstrate the tensile testing curves. Figure 12 shows the plot of stress vs displacement for the five representative samples.

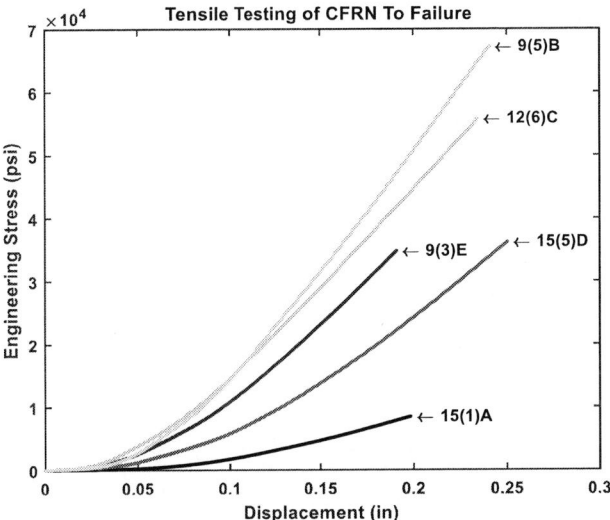

Figure 12. Stress vs. Displacement Curves

Figure 13 depicts the displacement at failure for all the CFRN samples in relation to the carbon fiber area percentage in each of the samples.

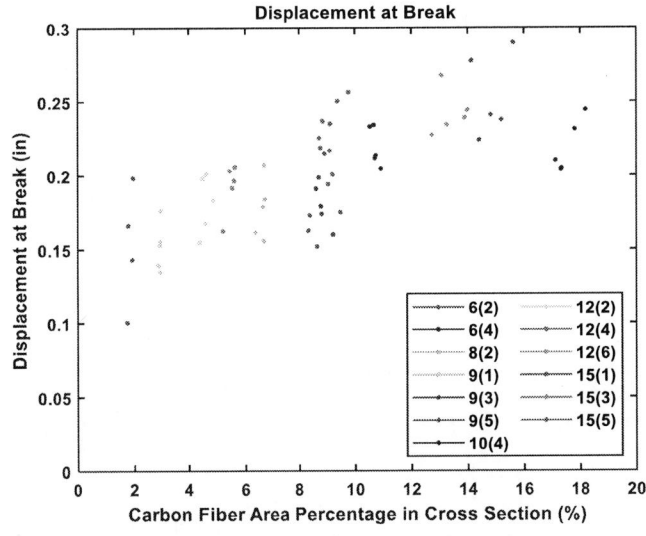

Figure 13. Displacement at Break for CFRN Samples.

The average displacements of each group at break are arranged in Table 4. These results will be used when discussing the results with respect to the rule of mixtures.

Table 4. Displacement at Break among Five Groups of Samples

Layer Configuration	Average Displacement at Break (in)	Standard Deviation (in)	Coefficient of Variation (%)
6(2)	0.1755	0.0239	13.61
6(4)	0.2190	0.0181	8.25
8(2)	0.1773	0.0203	11.45
9(1)	0.1517	0.0164	10.78
9(3)	0.1805	0.0151	8.34
9(5)	0.2541	0.0282	11.11
10(4)	0.2194	0.0133	6.08
12(2)	0.1809	0.0197	10.88
12(4)	0.2206	0.0176	8.00
12(6)	0.2426	0.0153	6.31
15(1)	0.1520	0.0412	27.10
15(3)	0.1918	0.0174	9.05
15(5)	0.2335	0.0221	9.46

3.6. Rule of Mixtures for Tensile Strength

This section develops a rule of mixtures to model the tensile strength of CFRN, similar to the approach used in section 3.3. The tensile strength data from section 3.5 was used to construct a linear model shown in Figure 14.

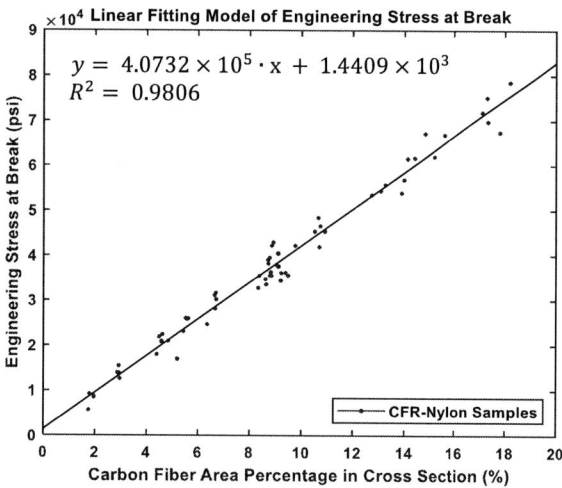

Figure 14 Linear Fitted Model for Tensile Stress vs Carbon Fiber Area Percentage

The rule of mixtures for tensile strength was also adopted from Callister [19] and is shown in Eq. (5).

$$\sigma_{sample} = \sigma'_{nylon} \times (1 - A_{cf,\%}) + \sigma_{cf} \times A_{cf,\%} \quad (5)$$

Where σ_{sample} is the tensile strength of the composite, and $A_{cf,\%}$ is the percentage of carbon fiber in the cross-section. Additionally, σ_{cf} is the tensile strength of the carbon fiber, and σ'_{nylon} is the tensile strength of nylon at the maximum carbon fiber elongation.

The slope and intercept found from Figure 14 were rearranged and inputted in Eq. (5) to obtain Eq. (6).

$$\sigma_{sample} = 1.44 \times 10^3 \times (1 - A_{cf,\%}) + 4.07 \times 10^5 \times A_{cf,\%} \quad (6)$$

By comparison with Eq. 5, it is determined that σ'_{nylon} is equal to 1440 psi (9.93 MPa) and σ_{cf} is 4.07×10^5 psi (2.81 GPa). To investigate the validity of these values, σ'_{nylon} is shown in Figure 15 along with the elastic stress-displacement curves of the pure nylon samples. While the precise elongation is unknown, the maximum displacement can be used for comparison. From Figure 13, the range for maximum displacement is 0.1 in. to 0.2 in. for lower fiber content; the displacement to failure likely increases for higher fiber content due to the larger tab deflections under the higher loading for those samples. From the dashed lines in Figure 15, the stress values of approximately 1220 psi and 2580 psi correspond to displacements of 0.1 in. and 0.2 in., respectively. The value of σ'_{nylon} determined from Eq. (6) falls within this range. Finally, the σ_{cf} obtained in Eq.(6) was also compared to tensile strength values reported by Naito et. al. [20], which range from 1.10 GPa (1.59×10^5 psi) to 5.69 GPa (8.25×10^5 psi). The value of 2.81 GPa determined from Eq. 6 is near the middle of this range. Therefore, the model developed from the rule of mixtures has reasonable values for both σ'_{nylon} and σ_{cf}.

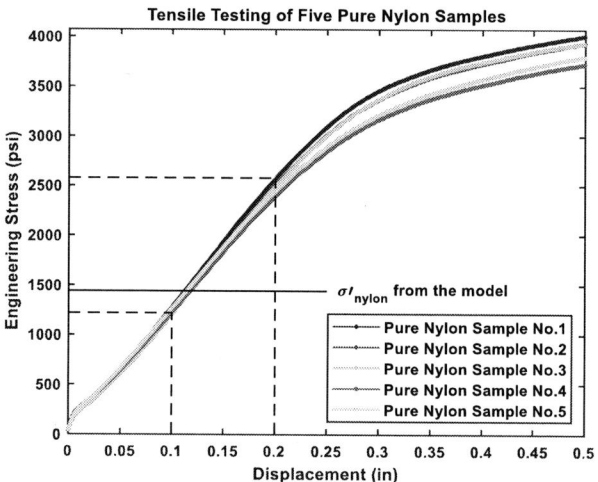

Figure 15. Displacement at Break for CFRN Samples.

4. CONCLUSIONS

This study created a model for determining the elastic modulus and tensile strength for a class of 3D printed CFRP. The model is based on the number of layers and number of carbon fiber filament paths in the specimen, where fibers are oriented unidirectionally along the specimen length. Tensile tests, with loading applied in the axial direction of the fibers, were conducted for 13 different configurations and a linear relationship between the fiber content and the behavior was observed. The model will be valuable for designers to accurately compute the modulus and tensile strength of the components. It will also be useful in optimizing the required level of carbon fiber for this class of 3D printed nylon composite.

This work is a first step in establishing a model that designers can use to predict the mechanical behavior for 3D printed materials. In this study, only one printing direction was studied. Moreover, the direction of the load was applied in the axial direction of the fibers. The nature of the 3D printing process causes anisotropic behavior in the parts. Designers are also in need of models that can predict the behavior for those cases in which loading is not aligned with the fibers. As future work, the authors are studying the tensile strength and modulus of both the fibers and printing that occurs in other directions.

ACKNOWLEDGMENTS

Thanks to University at Buffalo Energy Systems Design Research Group, Digital Manufacturing Laboratory and Materials Testing Laboratory for technical support.

REFERENCES

[1] Guo, N., and Leu, M. C., 2013, "Additive Manufacturing: Technology, Applications and Research Needs," Front. Mech. Eng., 8(3), pp. 215–243.

[2] Costa, C. A., Linzmaier, P. R., and Pasquali, F. M., 2013, "Rapid Prototyping Material Degradation: A Study of Mechanical Properties," IFAC Proc. Vol., 46(24), pp. 350–355.

[3] Steuben, J. C., Iliopoulos, A. P., and Michopoulos, J. G., 2018, "Open Uniaxial Test Machine (OpenUTM): Part 1 — A Low-Cost Electrohydraulic Test Frame for Additive Manufacturing Part Qualification," Proceedings of the ASME IDETC/CIE, DETC2018-86015, Quebec City, Quebec, Canada.

[4] Pham, K. D., O'Brien, W. F., and Case, S. W., 2018, "Characterizing Static and Dynamic Mechanical Properties for Additive Manufactured ULTEM 9085 Used to Construct Flow Control Devices for Turbomachinery Applications," Proceedings of the ASME Turbo Expo, GT2018-75430, Oslo, Norway.

[5] Liverani, E., Lutey, A. H. A., Fortunato, A., and Ascari, A., 2017, "Characterization of Lattice Structures for Additive Manufacturing of Lightweight Mechanical Components," Proceedings of the International Manufacturing Science and Engineering Conference, MSEC2017-2835, Los Angeles, California, USA.

[6] Dizon, J. R. C., Espera, A. H., Chen, Q., and Advincula, R. C., 2018, "Mechanical Characterization of 3D-Printed Polymers," Addit. Manuf., 20, pp. 44–67.

[7] Tian, X., Liu, T., Yang, C., Wang, Q., and Li, D., 2016, "Interface and Performance of 3D Printed Continuous Carbon Fiber Reinforced PLA Composites," Compos. Part Appl. Sci. Manuf., 88, pp. 198–205.

[8] Parandoush, P., Zhou, C., and Lin, D., 2018, "3D Printing of Ultrahigh Strength Continuous Carbon Fiber Composites," Adv. Eng. Mater., p. 1800622.

[9] Ning, F., Cong, W., Wei, J., Wang, S., and Zhang, M., 2015, "Additive Manufacturing of CFRP Composites Using Fused Deposition Modeling: Effects of Carbon Fiber Content and Length," Proceedings of the IMSEC, MSEC2015-9436, Charlotte, North Carolina, USA.

[10] van de Werken, N., Hurley, J., Khanbolouki, P., Sarvestani, A. N., Tamijani, A. Y., and Tehrani, M., 2019, "Design Considerations and Modeling of Fiber Reinforced 3D Printed Parts," Compos. Part B Eng., 160, pp. 684–692.

[11] Van Der Klift, F., Koga, Y., Todoroki, A., Ueda, M., Hirano, Y., and Matsuzaki, R., 2016, "3D Printing of Continuous Carbon Fibre Reinforced Thermo-Plastic (CFRTP) Tensile Test Specimens," Open J. Compos. Mater., 6(1), pp. 18–27.

[12] Naranjo-Lozada, J., Ahuett-Garza, H., Orta-Castañón, P., Verbeeten, W. M. H., and Sáiz-González, D., 2019, "Tensile Properties and Failure Behavior of Chopped and Continuous Carbon Fiber Composites Produced by Additive Manufacturing," Addit. Manuf., 26, pp. 227–241.

[13] Sarvestani, A. N., Werken, N. van de, Khanbolouki, P., and Tehrani, M., 2017, "3D Printed Composites With Continuous Carbon Fiber Reinforcements," Proceedings of the ASME IMECE, IMECE2017-72041, Tampa, FL,USA.

[14] Lee, J.-E., Im, Y.-E., and Park, K., 2016, "Finite Element Analysis of a Customized Eyeglass Frame Fabricated by 3D Printing," Trans. Korean Soc. Mech. Eng. A, 40(1), pp. 65–71.

[15] ASTM International, 2017, "Standard Test Method for Tensile Properties of Polymer Matrix Composite Materials," ASTM D3039/D3039M-17, West Conshohocken, PA, USA

[16] ASTM International, 2013, "Standard Terminology for Additive Manufacturing-Coordinate Systems and Test Methodologies," ISO/ASTM 52921-13, West Conshohocken, PA, USA

[17] ASTM International, 2014, "Standard Test Method for Tensile Properties of Plastics," ASTM D638-14

[18] Marforged, "Mechanical Properties of Continuous Fibers" [Online].Available:https://static.markforged.com/markforged_composites_datasheet.pdf. [Accessed: 25-Feb-2019].

[19] Callister, W. D., and Rethwisch, D. G., 2007, Materials Science and Engineering: An Introduction, John Wiley & Sons New York.

[20] Naito, K., Tanaka, Y., Yang, J.-M., and Kagawa, Y., 2008, "Tensile Properties of Ultrahigh Strength PAN-Based, Ultrahigh Modulus Pitch-Based and High Ductility Pitch-Based Carbon Fibers," Carbon, 46(2), pp. 189–195.

Proceedings of the ASME 2019
International Design Engineering Technical Conferences
and Computers and Information in Engineering Conference
IDETC/CIE2019
August 18-21, 2019, Anaheim, CA, USA

DETC2019-98068

SUBSTRATE OPTIMIZATION FOR HYBRID MANUFACTURING

Brandon R. Massoni
Oregon State University
Corvallis, Oregon, United States

Matthew I. Campbell
Oregon State University
Corvallis, Oregon, United States

ABSTRACT

While advances in metals additive manufacturing continue to make additive a viable option in more scenarios, these processes are generally slower and more expensive than subtractive methods, like machining. The combination of both additive and subtractive, often called hybrid manufacturing, can be used to get the benefits of both processes, while reducing cost. However, dividing a part into the most cost effective additive and subtractive features is often time-consuming and non-intuitive. In this paper, we present a new approach that optimizes the type, size, and position of a substrate within a part. The resulting hybrid manufacturing configuration enables engineers to reach the most cost-effective compromise between additive and machining. A fully implemented method has been developed and tested on several realistic engineering parts. The results are intuitively useful and push the state-of-the-art forward in generating hybrid manufacturing process plans.

1 INTRODUCTION

Milling complex parts down from large billets or forgings has long been the best way to produce complex metallic parts. These mature technologies have been around for years; they are reliable, fast, and relatively inexpensive when compared to metals additive manufacturing (AM). For these reasons, AM has found it difficult to find a place in modern metals manufacturing.

There are certain applications where AM already does provide great benefits in producing new parts – cases where assemblies can be reduced to a single part, or where impossible to machine geometry can be used to greatly reduce weight (e.g., generative design). However, most mechanical parts can be machined on a 5-axis mill, and doing so is usually much less expensive and time consuming than building parts with AM.

Using directed energy deposition (DED) and milling together is an idea referred to as hybrid manufacturing (HM). The advantage of HM is that it can make use of the benefits of both additive and subtractive manufacturing, without being affected by as many of the negatives of either. The goal of using HM is to reduce manufacturing costs and potentially even production time. This can be accomplished by starting with an optimally sized substrate and then machining with a computer numerical control (CNC) mill and printing features with DED to form the final shape (Figure 1).

HM can be accomplished on one machine or on separate dedicated machines. One company has even produced a kit to convert regular CNC milling machines into hybrid machines [1]. Higher production volume is still more effective on dedicated machines because of their faster rates, but these multi-tasking machines are ideal for smaller machine shops and can handle more complex geometry by iteratively changing tools. In both cases, 5-axis additive manufacturing is possible, which has been shown to reduce or even eliminate the need for supports [2,3].

Metals can be deposited additively with a variety of different technologies, but DED paired with CNC machining is the most abundant hybrid manufacturing approach in academic research [4] and industry [5]. Within DED, three popular options are: a laser power source with material delivered as powder, a laser with material delivered as wire, and an electron beam power source with wire. Of the two energy sources, electron beams have a faster deposition rate, but lower accuracy than lasers [6,7]. In our approach, any of these three methods can be considered

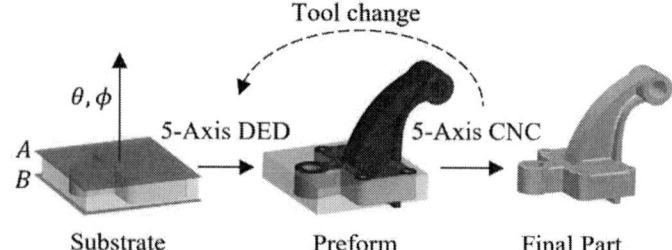

Figure 1: Hybrid manufacturing plan starting with a thick substrate, material is deposited with DED and milled on a CNC. The process can be strictly sequential, or a hybrid machine could iteratively change tools multiple times.

by altering the deposition rate, machine cost, layer height, dimensional tolerance, and material utilization.

The purpose of our research is to aid engineers in deciding when and how to apply HM effectively. Since substrate placement greatly effects production cost and time, it is a key consideration. However, determining the optimal substrate can be very time consuming, because there are so many variables and calculations to be considered. Therefore, we present a method to automatically optimize the substrate type, location, and size, given a cost model and manufacturing parameter inputs.

We formulate this problem as a minimization of manufacturing cost, where the manufacturing plan is determined by four variables (Eqn. 1). The two angles (ϕ and θ) are the azimuth and polar angles of a spherical coordinate system that we collectively refer to as the orientation. A and B are distances along the orientation for two parallel planes. As shown in Figure 1, these planes determine the top and bottom of the substrate within the part.

$$minCost_{direction} = minimize(\theta, \phi, A, B) \qquad (1)$$

Our search method for finding this substrate is presented in Sections 3 and 4, following a description of related research in Section 2. Experiments on a variety of engineering parts is shown in Section 5 to verify the accuracy of the method and validate its usefulness in planning the manufacture of such parts.

2 RELATED WORK

The idea of "hybrid" manufacturing is that a combination of additive and milling processes could be used to produce parts with the high complexity and material utilization of additive and the superior surface finishes of milling. While this idea has some application to plastics, it is primarily beneficial for metals. In addition, it is most applicable to faster deposition rate machines, with lower dimensional tolerances. For these reasons, this work is focused specifically on directed energy deposition, not powder bed fusion or plastics additive.

One of the major challenges in considering HM, especially for multi-axis deposition, is developing an efficient process plan. This includes determining where to place the substrate, selecting an optimal build sequence, and minimizing waste material using multi-axis machines. The individual features should ideally be segmented in a way that collisions are avoided and minimal support material is required. Many approaches have been presented to solve one or both of these issues, but few have taken into consideration the position and sizing of the substrate.

In 2010, Ren et al. [8] presented a method that uses 5-axis hybrid manufacturing to build parts with minimal support material. To accomplish this, they used the centroidal axis to decompose the part into subcomponents, which then informs the slicing orientation for each subcomponent and collision avoidance. In 2017, Chen et al. [9] developed a method to build previously impossible to manufacture structures with 3-axis hybrid manufacturing. They accomplish this by cleverly segmenting the build order, switching between additive and subtractive a minimal number of times throughout the build. In

2018, Kapil et al. [10] presented a method to reduce the need for support structures by tilting the substrate during metal deposition. Additionally, Dai et al. [2] presented an approach that uses 5-axis fused deposition modeling (FDM) to deposit material along curvilinear toolpaths to reduce the need for support material. The layers are deposited on the surfaces of the convex hull to avoid collisions. Lastly, Behandish et al. [11] presented an approach that decomposes the 3D space into atomic cells and applies finite Boolean functions to optimize the build sequence in 3-axis hybrid manufacturing. These approaches consider a breadth of challenges with 3 and 5-axis hybrid manufacturing, but they all assume the substrate is pre-determined.

In DED, the substrate can be somewhere within the part to reduce the production time and cost. Determining where to segment the part for the substrate in hybrid manufacturing has not been explored by many researchers. In 2011, Kerbrat et al. [12] presented a method to segment parts into additive and subtractive subparts by considering machinability. This approach assumes that the additive subpart will be printed separately and then attached by fasteners to the other subpart, and it does not consider the effect of production costs. In 2018, Eldakroury et al. [13] presented a method to optimize the substrate size and location for rectangular blocks and cylindrical substrates. Eldakroury et al. find the maximum inscribed shape along the predefined orientation, which is then translated and scaled to minimize a weighted fitness evaluation. In addition, they present a method to check for potential collisions based on the mean curvature skeleton generated by CGAL, and using Unreal Engine to calculate collisions.

We present a new approach to optimize the substrate placement and size, that builds upon the ideas presented by Eldakroury et al. [13]. Rather than incrementally translating and scaling the substrate, which requires a full objective function (i.e., manufacturing plan) evaluation at each instance, we scan two parallel planes along chosen directions to define stock geometry and cost information. This scanning approach enables us to pre-populate most of the information necessary to generate a solution. Then, we apply a multi-start direct search method to find the best manufacturing plan. As a result, our approach finds a better solution for the same test part from [13] within 0.4 seconds (compared to 63 seconds). This speed improvement allows the consideration of smaller step sizes and more detailed geometry.

Additionally, our method is based on minimizing manufacturing and material costs directly, rather than fitness evaluations. To this end, we use step sizes that are an equal thickness to the additive *layer height*. Lastly, we consider common plate sizes when determining the substrate thickness, which is not included in prior work. Exploration of the search space has revealed that the effect of plate size, especially when reaching the maximum, can significantly impact the location of the optimum. While we have included various new technologies, we have not implemented or improved upon the collision detection evaluation in [13], and as in [13], build sequence planning is beyond the scope of this paper.

3 METHODOLOGY

To find the best substrate type, size, and location for hybrid DED/milling, we move two parallel planes within the part along selected orientations. The substrate is defined as the subsection of the part between the two planes, and the remaining material on top and bottom are defined as sections produced by DED. Milling is performed either after or interchangeably with DED to produce the final shape (Figure 1). This manufacturing plan assumes that the 5-axis DED method can produce the additive features without support material, and that all the part features can be machined with milling at some point in the build process. This does not consider the build sequence, potential collisions, fixturing, or machinability.

3.1 Orientations

The substrate region is defined by four variables (ϕ, θ, A, B), which are split into two sub-problems: finding good orientations (ϕ, θ) and finding optimal distances (A, B). To find good orientations, two heuristics are applied. First, the orientations may be taken from the minimum bounding box (MBB) for the entire part. These orientations often align with major features of the part, but may produce poor results if the MBB is skewed, as in Figure 2. To balance this, we also use three normals – corresponding to the highest collective areas of faces – as the primary axis for aligned minimum bounding boxes (AMBB). This produces up to 12 unique directions (three from the MBB and three from each of the three AMBBs), but often many directions are redundant. In Figure 2, the three face normal directions (N1-3) create the same AMBBs and N1 aligns with the MBB, so five unique orientations are chosen.

The optimization is Section 4.3 is performed along each chosen orientation, such that the total search time scales linearly with the number of orientations. While many other orientations could be considered or the entire space optimized, we have restricted the search space with the MBB and AMBB heuristics to speed up the optimization process. To speed up the evaluations in the distance optimization (Section 4.3), a pre-processing step is performed; along each chosen orientation, the cross sections and two-dimensional convex hulls of those cross sections are identified for each step index. These cross sections and convex hulls are used to define the substrate (Section 3.2) and additive geometry (Section 3.3).

3.2 Substrate

The ideal substrate is the least expensive stock material type and shape that fits the subpart between separation plane A and B. In our approach, we consider rectangular plates, circular bars, and forged rectangular bars. Specifically for plates, we consider sizes between 1/2 to 1-3/4 inches, in 1/8 inch increments and up to 4 inches in 1/4 inch increments. Once each stock type is created for the substrate section in question, the least expensive option is chosen.

The geometry of the substrate is determined with the two-dimensional convex hull for each cross-sections between the two separation planes. For rectangular stock shapes, the bounding rectangle [14] is then found. This bounding rectangle is used to define the substrate's bounding box (Figure 3 top left). If this bounding box has at least one dimension less than the maximum plate size, rectangular plate is considered; otherwise, forged rectangular bar is considered. The thickness of the plate is determined to be along the shortest dimension, but the actual chosen thickness will vary due to discretization in plate sizes.

The central axis for circular bar stock is defined along the largest dimension of the substrate's bounding box. This allows circular bar stock to be defined along the chosen orientation or perpendicular to it, as shown in Figure 3. Its radius is determined by projecting all the cross-sections between A and B onto a plane along the central axis, and then finding the minimum bounding circle (MBC) on this set of points. If the substrate is perpendicular to the orientation, then it slightly over-estimates the additive volume/cost due to using flat planes. This error could be eliminated by subtracting the cylinder from the additive features, but we have not yet implemented such a correction.

3.3 Additive Features

The material above the top separation plane, A, and below the bottom separation plane, B, is to be produced with 5-axis DED, assuming zero support material. To efficiently calculate the additive manufacturing cost using layers, we discretize the search space using a step size equal to the DED *layer height*.

Before optimization, the polygonal cross section at each distance along the orientation is enlarged by an offset of the

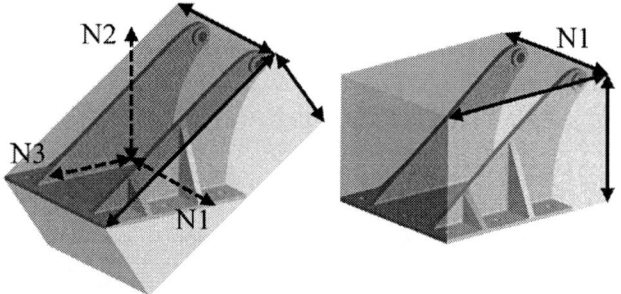

Figure 2: MBB (left) with dashed arrows indicating three highest area face normals (N1-3). AMBB (right) for N1.

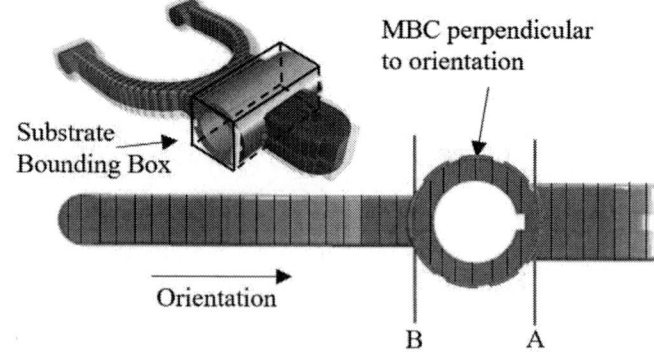

Figure 3: Part with cross sections shown along the orientation (bottom). The MBC (dashed circle) is along the largest dimension of the substrate bounding box (top left).

dimensional tolerance of the DED method, and then stored by step index. To ensure full coverage on all the solid model surfaces with a tolerance of a single layer height, each cross-section is unioned with the next cross section and the prior two. Notice the upper right side of Figure 4, where 0 is shown. The four cross sections at +1, 0, -1, and -2 are unioned to form the polygon at 0, which is then extruded down to form the layer. This is repeated for all cross sections. The volume of each layer is summed together to calculate the additive volume above and below each index. Storing these volumes by index before the optimization allows fast retrieval of the additive volume above A and below B.

Of course, the model in Figure 4 is not manufacturable due to the amount of overhang. However, with DED on 5-axix machines, this can be realized by the illustration in Figure 5, where one portion is printed in the vertical direction without

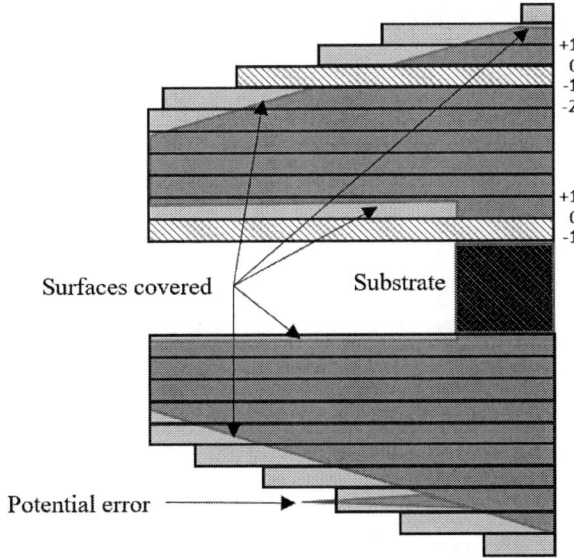

Figure 4: Union operations to estimate additive material. The index offsets indicate which cross sections influence the diagonal patterned layers. Layers do not show horizontal offset for the dimensional tolerance.

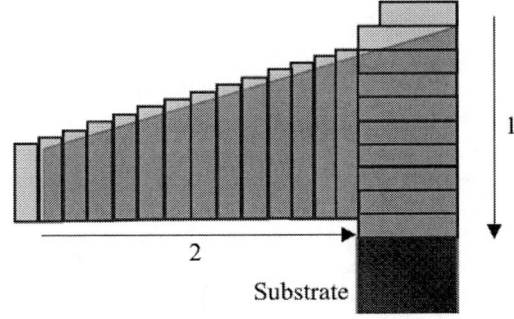

Figure 5: Possible decomposition to make the part printable without support material. We do not currently perform this segmentation.

overhang. The part is then rotated 90° clockwise for printing the "wing" shape on the left. A versatile bed in a 5-axis mill can orient the part so that gravitational effects are no longer significant issues for overhangs. The plan for when the bed is to be rotated is investigated in [8,10], but the effect on volume, and therefore cost, is minimal. Therefore, we are considering these 3D printed sections to be printed from a single direction without much loss in accuracy in the cost function.

4   OPTIMIZATION

4.1   Search Space

With the orientations (ϕ, θ) selected as described above, the search space for the best manufacturing plan is reduced to two dimensions (A, B), which can be discretized by *layer height*. Furthermore, A and B can be restricted, such that B is always less than A. The resulting two-dimensional search space is neither smooth nor monotonic. In most cases, it is multimodal and has discontinuities introduced by the sudden changes in substrate plate sizes and types.

For example, Figure 6 shows the contours of the search space for a part that is symmetric along the chosen orientation. The value for each combination of A and B is the normalized manufacturing cost; the axes represent the distance from the start of the part to its length − in step indices − along the chosen orientation. A candidate solution in the bottom right corner represents when no additive is used, because − at this position − A is at its maximum and B is at its minimum; this describes the largest possible substrate size. The bottom left corner, top right corner, and constraint line (i.e., B = A − 1) represent a minimum substrate size, where nearly the entire part is additively manufactured. From a given candidate point, if B is decreased (moving downwards) or A is increased (moving right), this represents an increase in substrate thickness, whereas any diagonal parallel to the constraint line represents a move that does not change the substrate thickness.

Discontinuities in the search space are common due to: discrete jumps in substrate plate sizes, changes in the type of substrate used, and sudden changes in the part's geometry. Consider the example in Figure 6, where the solution indicated by the arrow is right before a discontinuity in A to the right and in B below. These discontinuities are created by the geometry of the part, because if A was to increase or B decrease, the middle substrate would suddenly need to fill in all the empty space between plane A and B.

Since the search space is triangular, the total number of options is given by Equation 2. The number of indices (n), is determined by the length along the chosen orientation and the specified *layer height* (Eqn. 3).

$$size_{\theta,\phi} = \frac{(n-1)^2}{2} \qquad (2)$$

$$n = \frac{length_{\theta,\phi}}{layer\ height} \qquad (3)$$

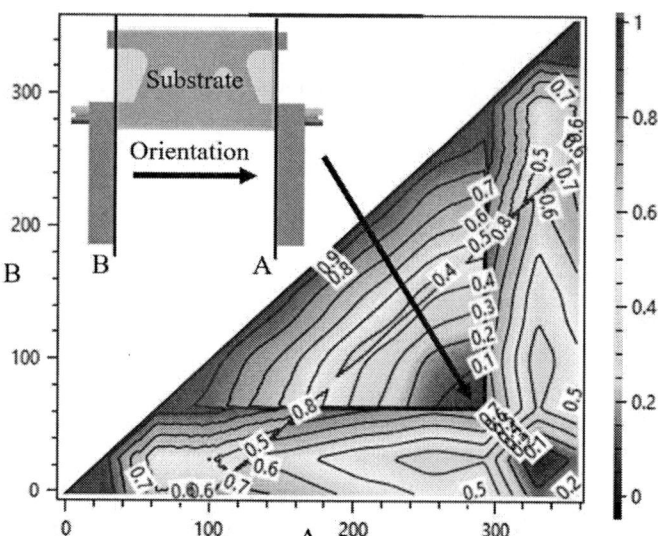

Figure 6: Two-dimensional search space with the optimal solution (top left) indicated by a solid arrow. The values are normalized between the optimal (0.0) and highest cost (1.0) manufacturing plans.

4.2 Objective Function

Assuming large production volumes and mostly automated procedures (i.e., low labor), machine depreciation and material can account for over 90% of the costs in metals additive manufacturing [15,16]. These studies show that machine depreciation is the most significant cost factor when considering low deposition rates (<0.5 lbs/hr), such as with selective laser melting. For faster deposition rates, such as with laser metal deposition (6 lbs/hr) and electron beam additive manufacturing (25 lbs/hr) [6], material price will have a larger impact. There are many other possible factors (e.g., labor rate, amortized fixture cost, machining tools, coolant, maintenance, and heat treatment), but according to these studies, they have a much lower impact. Therefore, we have chosen to simplify manufacturing cost by focusing on the machine depreciation and material cost (Eqn. 4).

Machine depreciation, maintenance, and operating expenses are lumped together in a single cost rate ($CR_{Machine}$). Since milling consumables are ignored, the same rate can be applied to the machine whether it is milling or depositing material. If, however, dedicated machines are used, they would need different rates based on their unique capital costs and operating expenses.

$$C_{Part} = C_{Material} + T_{Total} * CR_{Machine} \qquad (4)$$

The total processing time (Eqn. 5) for a single part is the sum of the time spent depositing material (Eqn. 6), milling (Eqn. 7), and loading/unloading. The lot size is assumed to be large enough that the machine setup time is negligible once divided over the number of parts being produced; however, each part does require time for loading/unloading and fixturing into the machine. $T_{Load/Unload}$ is tiered based on the mass of the final part (<10 lbs = 20 min, 10-50 lbs = 40 minutes, > 50 lbs = 60

minutes). The deposition time is based on a simple rate of deposition and the total additive volume being deposited. While it does not include complexities, such as layer dwell or scanning paths, the approach could be extended to include them because it already has the layer cross sections and the number of layers.

Machining is split into roughing and finishing, each with its own material removal rate (MRR). V_{Rough} is the sum of all extra material that is greater than a defined finishing thickness. This roughing volume is from the substrate and additive material. V_{Finish} is the sum of the remaining excess material. In future work, finishing could apply to specific surfaces as defined by the user beforehand or through the specification of tolerances.

$$T_{Total} = T_{Milling} + T_{Deposition} + T_{Load/Unload} \qquad (5)$$

$$T_{Deposition} = V_{Additive\ Material} * R_{Deposition} \qquad (6)$$

$$T_{Milling} = V_{Rough}/MRR_{Rough} + V_{Finish}/MRR_{Finish} \qquad (7)$$

The cost of material (Eqn. 8) is split between the substrate and the additive sections of the part, as shown in the upper left of Figure 6. The price per unit mass of the substrate is dependent on the stock type, and the mass is determined by the stock size. The remaining sections of the part are formed with DED. The waste (w) is the percentage of additive material that does not join to the substrate and cannot be reused. This is only a concern for laser-powder deposition, since wire DED utilizes all the material.

$$C_{Material} = M_{Substrate} * P_{Billet} + \frac{M_{Additive} * P_{Additive}}{1 - w} \qquad (8)$$

4.3 Multi-Start Hooke and Jeeves

Hooke-and-Jeeves [17] is a direct search method that can be applied to discrete problems. It makes use of exploratory and pattern moves (Figure 7) to quickly converge on local optima. For our specific problem, moves along the diagonal line where slope ($\Delta A/\Delta B$) equals one line are of special interest because they represent a constant substrate thickness. Therefore, the convergence of HJ is improved if we add a special move to make use of this knowledge. Since HJ is a direct search, we need to restart it to find a global optimal in this multimodal space. This approach is called multi-start Hooke-and-Jeeves (MSHJ). Therefore, points (i.e., A and B values) are randomly selected within a grid to ensure that the entire space is explored with a smaller population. Figure 8 shows an example of this distribution and the identified local minima for a max population of 80. Since the grid is triangular, the actual population size will be the highest value in the triangular number sequence (i.e., 1, 3, 6, 10, 15, 21, 28, 36, 45...) below the user-defined maximum.

Prior to deciding on MSHJ, we also considered performing two one-dimensional nested optimizations – a search for the best B for every value of A. This was not efficient or robust for these discontinuous and multimodal spaces. Likewise, an exhaustive search is impractical with complex cost models and small step sizes. Stochastic methods, such as particle swarm optimization

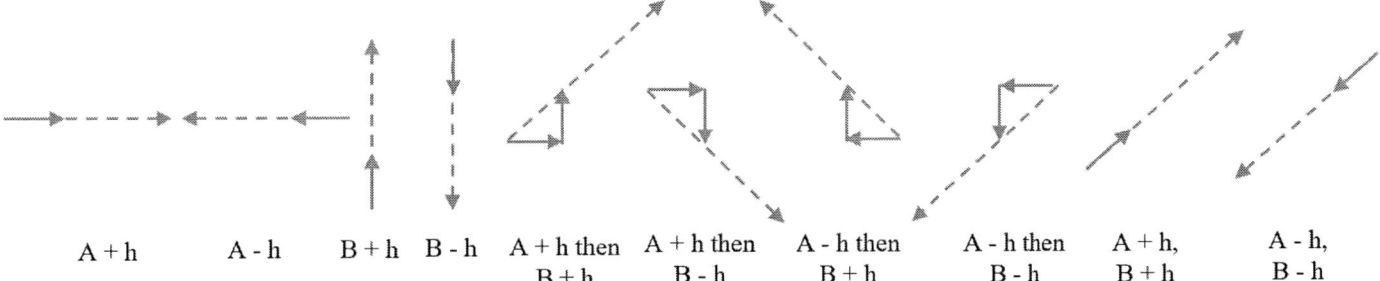

| A + h | A - h | B + h | B - h | A + h then B + h | A + h then B - h | A - h then B + h | A - h then B - h | A + h, B + h | A - h, B - h |

Figure 7: All possible moves from modified Hooke and Jeeves after one iteration. The exploratory moves and diagonal moves are solid arrows, while the pattern moves are dashed arrows. The difference between the exploratory moves and the diagonal moves, is that the exploratory moves will not get to A + h, B + h unless both individual moves were beneficial.

(PSO) [18], were implemented, but their stochastic nature required many more function evaluations to converge than MSHJ. Such stochastic methods are over-engineered for search spaces like this one, since it has relatively few local minima – each of which tends to have a large basin of attraction. MSHJ lacks the stochastic ability to move between exploitation and exploration exhibited in PSO and Simulated Annealing. Instead, the random starting points represent the extent of the exploration. So long as one of these random starts is initiated within the basin of attraction of the global minimum, MSHJ is likely to find the minimizer efficiently. If the space were even simpler, multi-start random hill climbing may be sufficient. The advantage of Hooke-and-Jeeves over random hill climbing is its adjustable step size (h).

Initialization

For a given orientation (ϕ, θ), the search space is discretized into n indices (Eqn. 3), and then split into a grid based on the desired population size. Next, a random point (A, B) within each grid section is evaluated with Equation 4. Modified HJ (steps 1 through 4 below) is then started at each of these random points. The initial step size, h, should be some fraction of the maximum dimension. In our implementation, we chose h to be 10% of the part's length along the chosen orientation.

Step 1: Exploratory Move

For each dimension (A and B), the positive move is checked first (A+h). If the positive move results in a better objective function value, then it is accepted as the new candidate point and the negative move is not considered; otherwise, the negative move (A-h) is checked and adopted if it is better than the current. If neither choice produces a better solution, no move is taken in the A dimension. This is then repeated for the next dimension, B.

Step 2: Diagonal Move

If the exploratory move was unsuccessful, the search moves along the diagonal, where $\Delta A/\Delta B$ equals one. If A+h, B+h is better than the current, the resulting point is accepted. If not, A-h, B-h is checked and – if it is better – it is accepted.

Step 3: Shrink

If no moves in this iteration were an improvement, the process is checked for convergence. The search has converged if the step size (h) drops below a user-defined minimum step size. Once converged, the best solution is returned. Given the fact that the space is discretized to the *layer height*, this serves as an ideal value for the minimum step size. If convergence is not met, then the current step size is reduced according to Equation 9, where r is some cooling rate (r) between zero and one (1/2 is commonly used). Then, the algorithm returns to step 1.

$$h = RoundDown(h * r) \qquad (9)$$

Step 4: Pattern Move

If a move was successful within this iteration, then a larger move along the same direction is considered. Generally, and in our case, this move is twice as big (2*h) as a regular move. If the new candidate has a lower objective function, it is accepted. The search process then continues by returning to step 1.

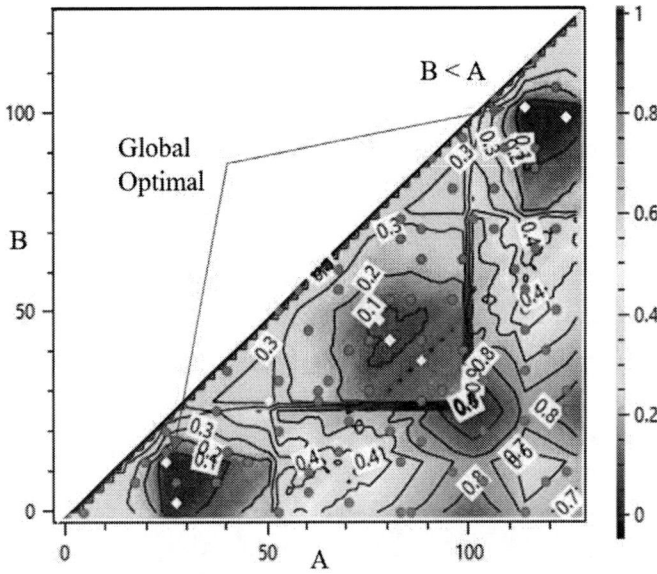

Figure 8: Two-dimensional search space with start locations indicated by magenta circles and local optima indicated by white diamonds.

V02AT03A040-6

Copyright © 2019 ASME

5 RESULTS

In this study, we aim to show whether hybrid manufacturing should be considered for various parts that would traditionally be manufactured with CNC milling. For comparison, we also consider a fully additive approach. In the experiments, the milling parameters and substrate stock material prices are fixed (Table 1). Since the cost of additive and hybrid manufacturing will change significantly as the processes mature over the next few years, we have created two experiments of additive manufacturing input settings to see the effects; one is motivated by high accuracy (Setup 1 in Table 2) and the other by high speed manufacturing (Setup 2 in Table 2). These inputs do not represent specific machines, but aim to show differences that could result for different inputs. Values are informed from our past work in modeling cost for titanium manufacturing. Titanium was chosen because it is a high cost, difficult to machine metal, which makes it more cost effective for additive and hybrid manufacturing than other materials.

Five test parts were chosen for their range of sizes and complexity. Table 3 shows details for each of these parts: the minimum bounding box dimensions, the ratio of stock material mass to final part mass (also known as the "buy to fly" or BTF ratio), number of chosen orientations, and the total runtime to evaluate the part with MSHJ for Setup 1 and Setup 2. This time primarily depends on the physical size of the part, layer height, and the number of orientations considered. The MSHJ algorithm and cost evaluations were programmed in C# and run on a standard desktop computer. The program takes a tessellated solid input (e.g., STL, AMF, 3MF) and outputs both the cost and visual representation of the manufacturing plan (Figures 9 -13).

Table 1: Milling and Substrate Input Parameters – Titanium

$MRR_{Roughing}$	3 in³/min
$MRR_{Finishing}$	0.3 in³/min
Plate Price	$20/lb
Circular Bar Price	$25/lb
Forged Bar Price	$25/lb

Table 2: Additive Material Input Parameters – Titanium

	Setup 1	Setup 2
Cost Rate of Machine	$200/hr	$100/hr
Deposition Rate	0.5 lbs/hr	2 lbs/hr
Layer Height	0.01"	0.1"
Dimensional Accuracy	0.005"	0.05"
Additive Wire Price	$100/lb	$50/lb

Table 3: Part Information

#	MBB Dimensions	BTF	Num. Orient.	Runtime (sec) 1	2
1	5.0" x 7.7" x 2.5"	8.1	3	20.4	0.7
2	2.4" x 2.4" x 6.9"	4.9	3	1.3	0.1
3	29.5" x 11.7" x 20.0"	15	5	18.4	1.1
4	6.3" x 8.1" x 2.0"	5.1	5	22.1	0.4
5	16.3" x 6.5" x 17.8"	9.2	12	65.4	3.5

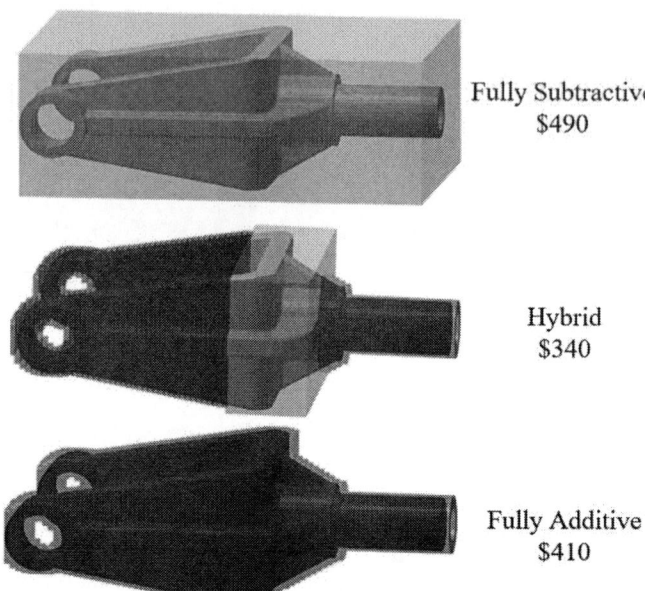

Figure 9 – Part 1: Optimal tradeoff (middle) between fully subtractive (top) and fully additive (bottom) for Setup 2.

Figure 9 shows the comparison of fully subtractive, fully additive, and hybrid solutions for Setup 2. The fully subtractive solution uses the same machine cost rate as additive and hybrid manufacturing. The fully additive solution incurs cost for the material, deposition, and finish machining; however, it does not incur any cost for the substrate (i.e., we assume the part could be sawed off and the substrate reused). With these conditions, the hybrid solution is significantly less expensive on this part. For the remainder of the test parts, the cost breakdown of these extremes are shown in Figure 15 and 16.

Figure 10 shows an example of a part with a low buy-to-fly ratio of 4.9. In this case, Setup 1 fails to find an improvement to fully subtractive manufacturing ($320). However, Setup 2 is able to reduce the cost by 13% when compared with fully subtractive (Figure 10 left). The hybrid solution (Figure 10 right) places the substrate on the plate section of the part, which is clearly better than additively building the full part onto a separate substrate ($270). While this solution is obvious to a designer, that is not always the case. Consider the previous hybrid result in Figure 9; the exact placement and size of that plate is non-intuitive because it does not line up with part features.

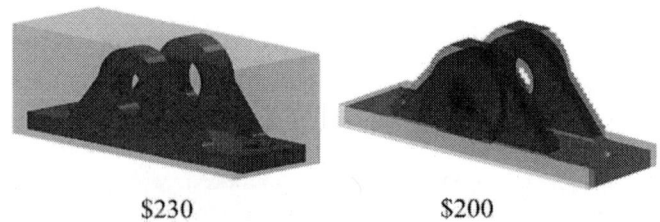

Figure 10 – Part 2: Fully subtractive (left), and optimal hybrid (right) for Setup 2. The Setup 1 best solution is fully subtractive (same as left image, but cost is $320).

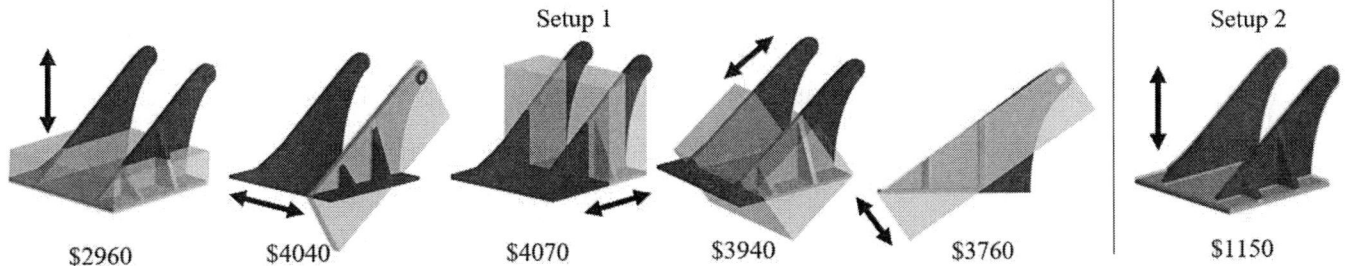

Setup 1   Setup 2

$2960 $4040 $4070 $3940 $3760 $1150

Figure 11 – Part 3: Optimal results for various orientations (indicated by solid arrows) using Setup 1 and Setup 2.

Figure 11 shows the optimal solution for five unique orientations using Setup 1 (left) and the global optimal solution for Setup 2 (right). The total cost of the part could be reduced by 40% and 72%, respectively, when compared to fully subtractive. This is primarily because this part has a high buy-to-fly ratio of 15. In this case, the vertical direction proved to be better than the non-intuitive orientations, but the determination of the optimal plate size is non-trivial for Setup 1 (far left).

Figure 12 shows an example of circular bar stock chosen for the substrate material. For Setup 1, the best solution is still to use a rectangular plate, however, Setup 2 chooses a circular bar. In this case, the circular bar is defined perpendicular to the chosen orientation (black arrows). This example part was recreated from the images in Eldakroury, et al. [13], and the solution for Setup 2 agrees with their results.

Figure 13 shows the method applied to a more complicated part, with 12 orientations (12 unique ϕ and θ pairs) of potential interest. The two rows correspond to the two setups and show the optimal solutions for four unique orientations. The best solution for both setups (leftmost column) is along the same orientation, but Setup 2 has a smaller plate and a second optima (right bottom). Also, the Setup 2 global optimal (left bottom) may have issues with collisions when building the large overhang near the substrate. If collisions are considered, then the second optima would be preferred.

Figure 14 shows the efficiency of MSHJ searching the design space, where the fraction is the number of objective function calls over the size of the search space (all possible

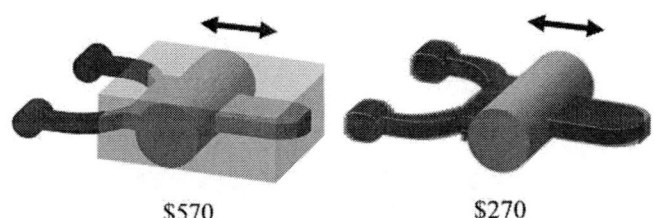

$570 $270

Figure 12 – Part 4: The best result for Setup 1 (left) and Setup 2 (right).

combinations of A and B for a given orientation), as calculated in Equation 2. For large step sizes, such as $s = 0.1$ inches, MSHJ only decreases the number of function calls by 60% to 90%. For smaller discretization, the method is much more necessary, as it reduces the number of objective function calls by over 99% in most cases.

Figure 15 shows the cost breakdowns for Setup 1 for each of the test parts. The costs are shown as percentages of the fully subtractive option, since it is considered the baseline. For most of the parts, a fully additive solution would double the cost. For Part 1, 2, and 4, the best hybrid solution results in cost savings of less than 10%, but for Part 3 and 5 it is much more lucrative. It is also worth noting that while the cost of the additive deposition (i.e., machine cost) dominates the fully additive solutions, it is a much smaller percentage when considering hybrid. This is because hybrid makes use of the lower cost of stock material, as opposed to additive wire.

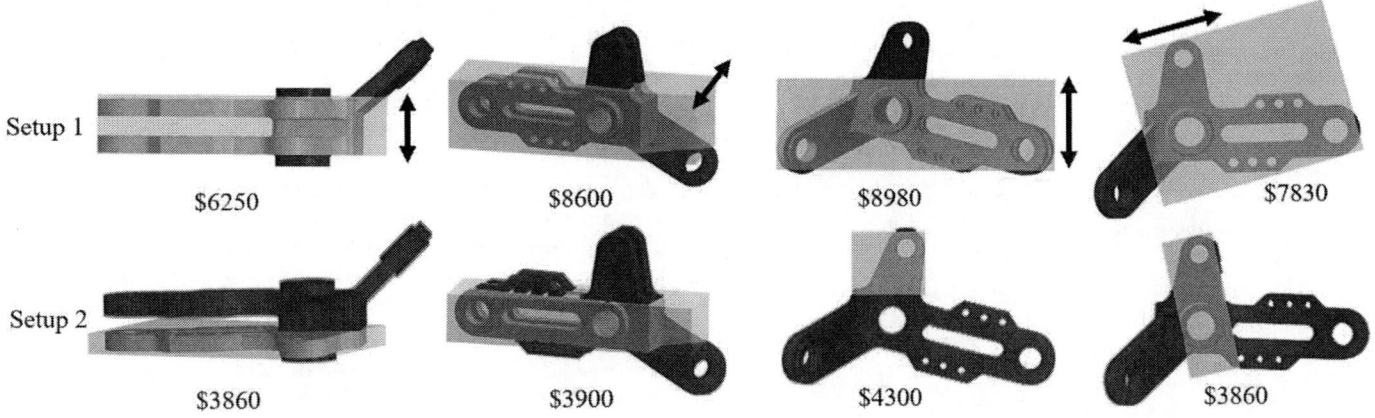

Setup 1

$6250 $8600 $8980 $7830

Setup 2

$3860 $3900 $4300 $3860

Figure 13 – Part 5: Optimal results for various orientations using Setup 1 (top row) and Setup 2 (bottom row).

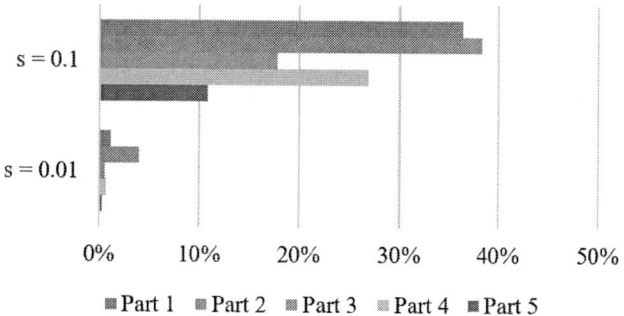

Figure 14: Percent of objective functions evaluated

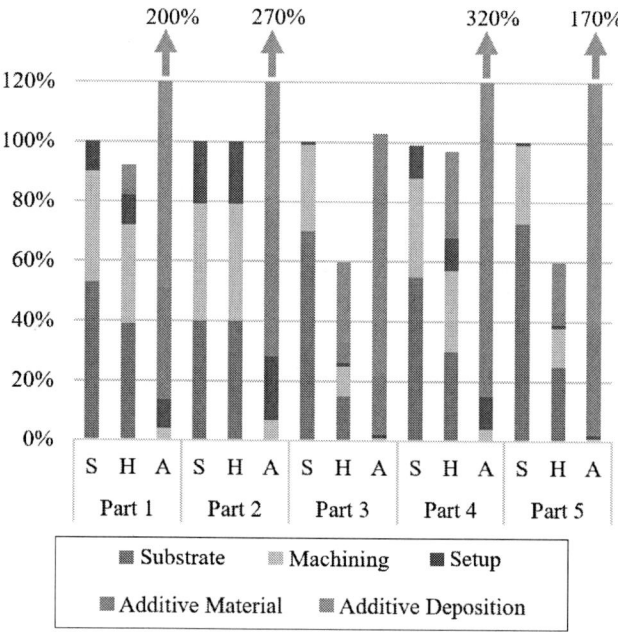

Figure 15: Cost breakdown for fully subtractive (S), the best hybrid (H), and fully additive (A) as percentages of the fully subtractive total cost for Setup 1.

Figure 16 shows the cost breakdowns for Setup 2. For these parts, the fully additive solution is much more competitive, since we are using lower cost titanium wire, a faster deposition rate, and a less expensive machine. For all five parts, the best hybrid option is always better than the either extreme.

6   DISCUSSION

The results in the previous section show the possible evolution of additive manufacturing (AM) and its effects on the cost of various mechanical parts. Currently, AM may not be cost effective in some scenarios, but this will likely change in the near future. The cost of additive material will decrease as it becomes more widely available, and the rate of deposition will continue to increase. Furthermore, the test parts used in this study were not designed to take advantage of the shape complexities available in AM. If the shapes were more organically shaped, as

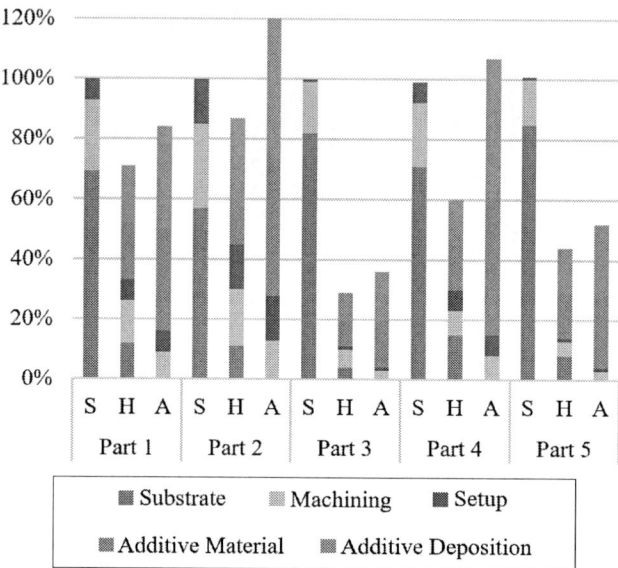

Figure 16: Cost Breakdown of fully subtractive (S), optimal hybrid (H), and fully additive (A) as percentage of the fully subtractive total cost for Setup 2.

opposed to simple prismatic features, additive may have been more beneficial.

While our method is able to estimate cost and can easily be expanded to include more complex cost models, it currently ignores four feasibility concerns. The primary reason not to include these feasibilities is because the optimization process makes hundreds to thousands of manufacturing plans for each candidate orientation pair of angles, ϕ and θ. First, it does not consider material properties or distortion, which may be a particular issue when printing onto thin plates. These properties could be incorporated as constraints within the optimization, using geometric design rules to prevent infeasibilities. Second, our process does not consider build sequence or collisions for 5-axis deposition. To consider the build sequence, planning could be performed at the start to inform the optimization. Otherwise, a post processing step could be applied on each of the local minima until a feasible global solution is found.

Third, we assume that the additively manufactured features could be produced without support material. When complex 5-axis toolpaths are considered, this assumption is valid for most geometries and negligible in some cases where only a little support material is necessary. However, there are certain geometries that will require support material, even with 5-axis deposition. In addition, there are some geometries that are not manufacturable due to collisions. Therefore, future work could analyze the additively manufacturing features in more detail to ensure more accurate costs and feasibility.

Fourth, we assume that all the features are 5-axis machinable, whereas there may be sections of the part that must be produced with additive manufacturing. Future work could also consider 5-axis machinability when placing the substrate material and ordering the additive/subtractive processing steps.

In addition to feasibility limitations, tolerances are not considered for stock sizes. Plate thickness, the saw cutting of plate and circular bar, and forging of rectangular bar are often specified with plus tolerances, guaranteeing that there will be sufficient material to machine down to the final geometry. Since this extra material has only a minor effect on cost, we chose not to include tolerances in the stock geometry. Instead, we assume that the substrate will be purchased by its expected weight.

Lastly, titanium is the only material considered, because of its high material price and low material removal rate. Other materials may also be cost effective for hybrid manufacturing. To consider other materials in our models, the input values in Table 1 and Table 2 would need to be adjusted.

7 CONCLUSION

In this paper, we present a new method to optimize the type, placement, and size of the substrate in hybrid manufacturing, based on a given cost model. A specialized optimization method is developed to identify a single region within a given part that is best made from subtractive machining operations, as opposed to additive operations. This region provides the substrate upon which the remaining features are additively defined. Given discontinuities in the search space due to discrete stock material sizes and features in the part, it is not practical to expect the engineer to find the best substrate size and position. Due to the efficiency of the search process, an engineer can invoke the search to find alternatives quickly regardless of part complexity. The results of the developed algorithm show hybrid manufacturing can reduce costs significantly in many cases – when compared to purely subtractive or additive solutions.

ACKNOWLEDGMENTS

The authors would like to thank the Boeing Company and the Oregon Metal Initiative (OMI) for their financial support on work leading up to this project. Views expressed in this paper are those of the authors and do not necessarily reflect the opinions of these supporters.

REFERENCES

[1]   Jones, J. B., 2014, "The Synergies of Hybridizing CNC and Additive Manufacturing," *RAPID 2014 and 3D Imaging Conferences & Exposition*, Detroit, MI, USA, pp. 1–8.

[2]   Dai, C., Wang, C. C. L., Wu, C., Lefebvre, S., Fang, G., and Liu, Y.-J., 2018, "Support-Free Volume Printing by Multi-Axis Motion," ACM Trans. Graph., **37**(4), p. Article 1.

[3]   Coupek, D., Friedrich, J., Battran, D., and Riedel, O., 2018, "Reduction of Support Structures and Building Time by Optimized Path Planning Algorithms in Multi-Axis Additive Manufacturing," *11th CIRP Conference on Intelligent Computation in Manufacturing Engineering*, pp. 221–226.

[4]   Flynn, J. M., Shokrani, A., Newman, S. T., and Dhokia, V., 2016, "Hybrid Additive and Subtractive Machine Tools - Research and Industrial Developments," Int. J.

Mach. Tools Manuf., **101**, pp. 79–101.

[5]   Lorenz, K. A., Jones, J. B., Wimpenny, D. I., and Jackson, M. R., 2015, "A Review of Hibrid Manufacturing," *26th Solid Freeform Fabrication Conference*, Austin, TX, USA, pp. 96–108.

[6]   Sciaky, 2015, "Wire AM vs. Powder AM" [Online]. Available: http://www.sciaky.com/additive-manufacturing/wire-am-vs-powder-am.

[7]   Ding, D., Pan, Z., Cuiuri, D., and Li, H., 2015, "Wire-Feed Additive Manufacturing of Metal Components: Technologies, Developments and Future Interests," Int. J. Adv. Manuf. Technol., **81**(1–4), pp. 465–481.

[8]   Ren, L., Sparks, T., Ruan, J., and Liou, F., 2010, "Integrated Process Planning for a Multiaxis Hybrid Manufacturing System," J. Manuf. Sci. Eng., **132**(2), p. 021006.

[9]   Chen, L., Xu, K., and Tang, K., 2018, "Optimized Sequence Planning for Multi-Axis Hybrid Machining of Complex Geometries," Comput. Graph., **70**, pp. 176–187.

[10]   Kapil, S., Joshi, P., Kulkarni, P. M., Negi, S., Kumar, R., and Karunakaran, K. P., 2018, "Elimination of Support Mechanism in Additive Manufacturing through Substrate Tilting," Rapid Prototyp. J., **24**(7), pp. 1155–1165.

[11]   Behandish, M., Nelaturi, S., and Kleer, J. de, 2018, "Automated Process Planning for Hybrid Manufacturing," Comput. Des., **102**, pp. 115–127.

[12]   Kerbrat, O., Mognol, P., and Hascoët, J.-Y., 2011, "A New DFM Approach to Combine Machining and Additive Manufacturing," Comput. Ind., **62**(7), pp. 684–692.

[13]   Eldakroury, M. A., Chen, N., and Frank, M. C., 2018, "A New Method for Locating Candidate Substrates for Multi Axis Hybrid Manufacturing Systems," Rapid Prototyp. J., **24**(2), pp. 237–248.

[14]   Toussaint, G. T., 2014, "The Rotating Calipers: An Efficient, Multipurpose, Computational Tool," *Proceedings of the International Conference on Computing Technology and Information Management*, The Society of Digital Information and Wireless Communication, Dubai, UAE, pp. 215–225.

[15]   Lindemann, C., Jahnke, U., Moi, M., and Koch, R., 2012, "Analyzing Product Lifecycle Costs for a Better Understanding of Cost Drivers in Additive Manufacturing," *Solid Freeform Fabrication Symposium*, Austin, TX, USA, pp. 177–188.

[16]   Atzeni, E., and Salmi, A., 2012, "Economics of Additive Manufacturing for End-Usable Metal Parts," Int. J. Adv. Manuf. Technol., **62**(9–12), pp. 1147–1155.

[17]   Hooke, R., and Jeeves, T. A., 1961, "Direct Search Solution of Numerical and Statistical Problems," J. ACM, **8**(2), pp. 212–229.

[18]   Kennedy, J., and Eberhart, R., "Particle Swarm Optimization," *Proceedings of ICNN'95 - International Conference on Neural Networks*, IEEE, pp. 1942–1948.

Proceedings of the ASME 2019
International Design Engineering Technical Conferences
and Computers and Information in Engineering Conference
IDETC/CIE2019
August 18-21, 2019, Anaheim, CA, USA

DETC2019-98103

VOXEL-BASED CAD FRAMEWORK FOR PLANNING FUNCTIONALLY GRADED AND MULTI-STEP RAPID FABRICATION PROCESSES

Cole Brauer
Student
Ira A. Fulton Schools of Engineering
Arizona State University
Tempe, Arizona
Email: cbrauer@asu.edu

Daniel M. Aukes
Assistant Professor
Ira A. Fulton Schools of Engineering
Arizona State University
Tempe, Arizona
Email: danaukes@asu.edu

ABSTRACT

In this paper we describe a new framework for planning functionally graded and multi-step fabrication processes for use in rapid prototyping applications. This framework is contributing to software tools that will simplify planning multi-material manufacturing processes and thereby make this type of manufacturing more accessible. We introduce the material description itself, low-level operations which can be used to combine one or more geometries together, and algorithms which assist the designer in computing manufacturing-compatible sequences. We then apply these tools to several example scenarios. First, we demonstrate the use of a Gaussian blur to add graded material transitions to a model which can then be produced using a multimaterial 3D printing process. Our second example highlights our solution to the problem of inserting a discrete, off-the-shelf part into a 3D printed model during the printing sequence. Finally, we implement this second example and manufacture two example components. The results show that the framework can be used to effectively generate the files needed to produce specific classes of parts.

1 Introduction

Multi-material manufacturing combines multiple fabrication processes to produce individual parts that can be made up of several different materials. These processes can include both additive and subtractive manufacturing methods as well as embedding other components during manufacturing. This yields opportunities for creating single parts that can take the place of an assembly of parts produced using conventional techniques. Some example applications of multi-material manufacturing include parts that are produced using one process then machined to tolerance using another, parts with integrated flexible joints, or parts that contain discrete embedded components such as reinforcing materials or electronics.

Multi-material manufacturing has applications in robotics because, with it, mechanisms can be built into a design without adding additional moving parts. This allows for robot designs that are both robust and low cost, making it a particularly attractive method for education or research. 3D printing is of particular interest in this area because it is low cost, readily available, and capable of easily producing complicated part geometries. Some machines are also capable of depositing multiple materials during a single process.

Current 3D printers are typically used in conjunction with CAM software that takes a CAD-generated file and "slices" it into a layer-based manufacturing solution based on a combination of user preferences and machine capabilities. These programs then output machine code which can be read by the printer. Due to the work-flow of this process, however, these programs typically only take into account the surface geometry of CAD designs. Furthermore, most of the widely-used parametric solid modeling tools used in engineering disciplines to design robots or other devices do not permit users to specify graded or multi-material parts natively. This makes it difficult to assign material properties during the design process and have those properties become directly available for use in manufacturing planning software or with tools like FEA to understand the stiffness of a design.

Thus, up to this point, planning the process to create a part using multi-material manufacturing has been done manually, requiring specialized knowledge of the tools used. The difficulty of this planning procedure can prevent many students and researchers from using multi-material manufacturing.

We envision that due to the capabilities afforded by current and anticipated manufacturing technology, future design workflows will need to consider both the shape and material composition of parts. We anticipate this approach would permit a designer to specify the mass or strength distribution of 3D printed parts, as well as the location and size of material transitions. Such geometries could be used to compute how to split a design into manufacturing sequences occurring on multiple machines, where each sequence considers the capabilities and limitations of each sub-process. Finally, we envision that knowledge of how something is made could lead to automatic generation of support geometry for orienting and fixturing devices in each machine. These concepts apply to a number of manufacturing processes, including additive processes such as single material 3D printing, multi-material 3D printing, and resin casting, and subtractive processes such as laser cutting and CNC milling.

In this paper we present first steps toward a computational framework for processing 3D models and automatically generating viable manufacturing processes. Using a voxel-based descriptor, we then illustrate how this framework could be used in the development of a multi-material component and a component with embedded electronics. Section 2 describes the history and state of the art of multi-material CAD and manufacturing. Section 3 describes the methods by which we represent and compute multi-material geometries. Section 4 describes the algorithms we have defined for producing general manufacturing features and two exemplar algorithms that use these features to plan for specific manufacturing processes. Section 5 describes our implementation of one of these examples to create a multi-material component.

2 Background

The versatility of multi-material manufacturing is illustrated by a number of examples, including the addition of embedded reinforcement [1], integrated joints [2], embedded sensors [3], and embedded actuators [4]. The robotic hand presented by Dollar and Ma demonstrates the use of several of these techniques to produce a robust and low cost device that uses significantly less components that a similar device produced using conventional techniques [5, 6]. The hexapedal robot presented by Cham also demonstrates the use of these techniques to achieve similar

benefits [7].

Kumar has presented a mathematical approach for representing and performing solid geometry operations on a multi-material part [8]. This work describes several possible methods for representing material data in a three-dimensional space and discusses basic material-aware solid geometry operations.

Various structures have been analyzed for computer storage and processing of geometry data combined with material data, including voxels, surfaces, and finite element meshes [9]. Within the area of voxel-based structures, the computational benefits and limitations have been considered and it is noted that although voxel-based models can be computationally intensive, they allow for conceptually simple geometric operations and correspond well to layer based methods for capturing and producing 3D objects [10, 11].

A number of methods for representing graded material properties in a multi-material part have been proposed [9, 12], and methods for generating material gradients based on functional requirements have been discussed [13–17]. Implementations of parts with graded properties have been approached through methods including voxel geometries with multiple discrete materials [4], adjustment of metallurgical properties in metal powder based additive processes [18] and the use of additive processes capable of adjusting the amount of deposited binder material [19].

Work towards the creation of manufacturing process planning algorithms has been done in the areas of SDM manufacturing [20–22], stereolithography [23], assembly sequences [24], and laminate-based robot construction [25]. Validation of generated subtractive manufacturing toolpaths using voxel-based simulation has also been considered [26].

The work on planning for laminate-based construction has been used in the development of a CAD tool, PopupCAD, that allows users to model laminate linkages and generate the cut files needed to produce them. Another software tool, Foundry, has been recently created to aid users in designing multi-material parts that can be produced using multi-material 3D printing [27–29]. This software provides a user-friendly interface that allows for the generation of various features and the modeling of a variety of material types.

This project seeks to combine elements of the prior research described above to create a new planning framework that is capable of automatically identifying the manufacturing steps and generating the required files to produce a user specified part using a series of different materials and processes. This framework will include a model representation supporting graded materials, a mathematical computation framework, and algorithms for process planning and validation. It is intended to apply to a variety of general processes, rather than being limited to a single process type. Finally, it is our intent to develop a user-friendly tool around this framework to make it accessible to a wide range of users.

3 Mathematical Framework

To enable process planning algorithms to be created, a mathematical framework for representing models and performing basic geometric and material operations on model data must first be defined. In addition, steps must be defined to check that the results of operations meet the established constraints of the representation to ensure that they have a valid physical meaning.

3.1 Model Representation

To process multi-material parts, a data structure is needed that can store both the 3D structure of an object as well as its material composition. For this, a voxel-based approach has been selected because it allows for simple definitions of solid geometry operations and can easily represent different materials within a part. A voxel model, represented here by V, has been defined as a four-dimensional array where the first three dimensions represent the 3D space occupied by the model. The fourth dimension contains a vector representing the materials present at that voxel. Each material vector can be defined by

$$V_{\langle i,j,k \rangle} = \langle a, m_0, ..., m_n \rangle, \qquad (1)$$

where a is a boolean value indicating the presence or absence of material at the corresponding voxel, and $m_{0...n}$ are an arbitrary number of material channels that contain floating point values representing the percentages of materials present. The m_0 value represents a null material and is used to express voxels containing a material density of less than 100 percent. For example, a voxel with $a = 1$, $m_0 = 0.5$, $m_1 = 0.5$ will contain material 1 at 50 percent density. For an object containing n materials, these values are constrained by:

$$a \in \{0, 1\} \qquad (2)$$

$$m_{i \in \{0..n\}} \in [0, 1] \qquad (3)$$

$$\sum_{i \in \{0..n\}} m_i = a \qquad (4)$$

Additionally, for the remainder of this paper the following assumptions are made:

1. All voxel models are of size $x \times y \times z \times (n+2)$
2. V^a refers to the $x \times y \times z$ boolean array containing the a value for each voxel in V
3. V^m refers to the $x \times y \times z \times (n+1)$ array containing the $m_{0...n}$ values for each voxel in V
4. V_i^m refers to the $x \times y \times z$ array containing the m_i value for each voxel in V

3.2 Constructive Solid Geometry Emulating Operations

To process model data, basic solid geometry operations for voxels must first be defined. These operations are based on the work described in [8] with the addition of material-specific and morphological operations.

As defined below, these operations are restricted to acting on model arrays of the same dimension and size, represented here by A and B, and must account for both the geometry and material of voxels. If two models with different physical sizes are used, empty voxels must be added around the models as needed to make their array sizes match. Material property vectors are required to be the same length.

The difference of two models is found by first using logical AND ($\wedge$) and NOT ($\neg$) operations on A^a and B^a to obtain a boolean array of the voxels which will contain material in the output. Element-wise multiplication, represented by $\circ$, is then performed between this array and each material channel to determine the values of the material channels in the output model. The resulting model is then found by augmenting these two matrices. This operation can be defined as follows:

$$A \setminus B := \left[A^a \wedge \neg B^a \,\middle|\, \left(A_i^m \circ (A^a \wedge \neg B^a) \right)_{i \in \{0..n\}} \right] \qquad (5)$$

The intersection operation can be defined in a similar manner. In this case, $\cap^L$ and $\cap^R$ denote whether the material of the output is taken from the first or second input model and are defined by:

$$A \cap^L B := \left[A^a \wedge B^a \,\middle|\, \left(A_i^m \circ (A^a \wedge B^a) \right)_{i \in \{0..n\}} \right] \qquad (6)$$

$$A \cap^R B := \left[A^a \wedge B^a \,\middle|\, \left(B_i^m \circ (A^a \wedge B^a) \right)_{i \in \{0..n\}} \right] \qquad (7)$$

The union operation can be defined using the symmetric difference combined with an intersection operation. The material used in overlapping regions is determined by the use of $\cap^L$ or $\cap^R$. The definition using $\cap^L$ is as follows:

$$A \cup^L B := (A \setminus B) + (B \setminus A) + (A \cap^L B) \qquad (8)$$

These operations use the material from either model A or model B. To compute mixtures of materials, operations for addition, subtraction, and multiplication are used. These are defined by applying a logical OR ($\vee$) operation on A^a and B^a, and array operations on A^m and B^m:

$$A +^* B := \left[A^a \vee B^a \,\middle|\, A^m + B^m \right] \tag{9}$$

$$A -^* B := \left[A^a \vee B^a \,\middle|\, A^m - B^m \right] \tag{10}$$

$$cA := \left[A^a \,\middle|\, cA^m \right] \tag{11}$$

The final two operations required are dilation and erosion. These are defined by applying the corresponding binary and grayscale morphological operations across each of the material channels in A using a three-dimensional structuring element S, as illustrated below with dilation:

$$A \oplus S := \left[A^a \oplus_{bin} S \,\middle|\, (A_i^m \oplus_{gray} S)_{i \in \{0..n\}} \right] \tag{12}$$

Erosion $(A \ominus S)$ can be defined in a similar manner.

These operations are used in the creation of manufacturing feature generation algorithms and process generation procedures as discussed in Section 4. It should be noted that some functions may result in voxels that do not meet the constraints established in Eqn. (3) and Eqn. (4). This must be corrected before using the result to generate final manufacturing steps with a cleanup process, as discussed in Section 3.4.

3.3 Material Interface Modification

The basic operations described in the Section 3.2 will maintain a defined boundary between materials. This behavior is consistent with most common manufacturing processes. However, some processes such as 3D printing provide opportunities for creating a precise gradient of material properties within a part via variable infill density or multi-material 3D printing. To permit these capabilities to be leveraged at design time, a blur operation can be defined using the convolution of a three-dimensional Gaussian kernel K and each of the material channels in A:

$$Blur(A, K) := \left[A^a \,\middle|\, \left((A_i^m * K) \circ A^a \right)_{i \in \{0..n\}} \right] \tag{13}$$

The radius of blurring is determined by the size of K. When applied in this manner, the blur operation will affect all voxels that contain material and that are near a boundary between two materials or between a material and empty space. The blurring effect can be restricted to a region R using

$$BlurRegion(A, K, R) := \left(Blur(A, K) \cap^L R \right) \cup^L A, \tag{14}$$

where R is a voxel model such that R^a indicates which voxels blurring should be applied to. This model must have material vectors present, but the values in them will be ignored. As with the functions in the Section 3.2, the blurring operations may result in voxels that do not meet the constraints established in Eqn. (3) and Eqn. (4). This must be corrected before using the result to generate final manufacturing steps, as described below.

3.4 Cleanup Steps

The geometric output from some solid geometry and blurring operations can include voxels with $m_{0...n}$ elements that do not meet the constraints of being in the interval $[0, 1]$ and having a sum equal to a as established in Eqn. (3) and Eqn. (4). To make all voxels follow these constraints, the following steps must be taken for each voxel in the model:

1. Remove negative m values
2. Update a values
3. Scale m values

Negative material values can be removed from a voxel as follows:

$$m_i' = \begin{cases} m_i & m_i > 0 \\ 0 & m_i \leq 0 \end{cases} \tag{15}$$

The new a value for the voxel can then be found using

$$a' = \begin{cases} 1 & \Sigma m \neq 0 \\ 0 & \Sigma m = 0 \end{cases}, \text{ where:} \tag{16}$$

$$\Sigma m = \sum_{i \in \{0..n\}} m_i. \tag{17}$$

Two approaches can be taken to scale the material values in order to satisfy Eqn. (3) and Eqn. (4). The first is to maintain the ratio between all materials including the null material. New $m_{0...n}$ values can be found for a voxel using

$$m_i' = \begin{cases} \frac{m_i}{\Sigma m} & a = 1 \\ 0 & a = 0 \end{cases}. \tag{18}$$

The second option is to adjust the value of the null material without changing any other material values. This can be done using

$$m_0' = \begin{cases} 1 - (\Sigma m - m_0) & a = 1 \\ 0 & a = 0 \end{cases}. \tag{19}$$

Note that the second option only works for voxels with $(\Sigma m - m_0) <= 1$.

4 Manufacturing Algorithms

Using the solid geometry operations discussed in the previous section, algorithms can be created to generate various geometric features needed to plan for manufacturing. These features can then be used in combination with the solid geometry operations to create procedures to generate files for specific materials and manufacturing processes.

4.1 Manufacturing Feature Generation Algorithms

A number of general-purpose features can be defined that are useful in the planning of manufacturing steps, including keep-out, clearance, support, and web. These features are influenced by a number of factors, including the ability of a process to control cutting depth, the tool head size, and the length/kerf/orientation of the cutting tool. As a result, different versions of these operations must be defined for various manufacturing processes depending on constraints and limitations of each process. These features were selected based on the work described in [25] and have been adapted to a voxel-based model representation.

The algorithms presented in this section support four general types of processes: milling, laser machining, FDM 3D printing, and discrete component insertion. In addition, it is assumed that any addition or subtraction of material occurs along the Z axis and that there are no head size or kerf constraints. These algorithms return a region that does not directly correspond to a specific material. As such, occupied voxels in the returned models contain a vector of ones for ease of implementation.

To illustrate the function of the presented algorithms, the sample object shown in Fig. 1 is used. This object includes multiple overhangs and a center cavity. The origin point is considered to be at the front-bottom-left corner of the model and the coordinate axes are aligned as indicated in the figure.

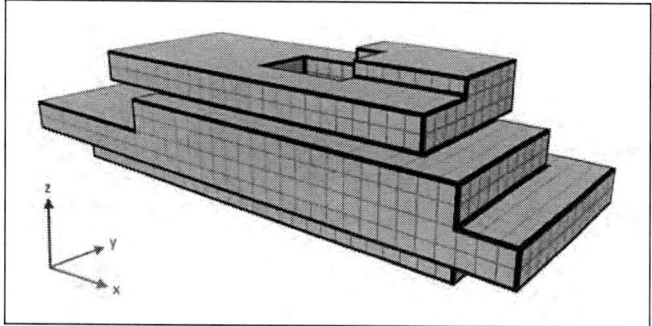

FIGURE 1: Sample object to illustrate the function of various manufacturing feature generation algorithms

4.1.1 Projection To aid in the definition of various manufacturing features, a projection function is first defined in Algorithm 1. This function returns a model that includes all voxels within the workspace that contain material or that lie in the specified direction with respect to a voxel that contains material. Note that a Z value of 1 corresponds to the bottom layer of the model.

Algorithm 1 Model Projection

1: **function** PROJECTZ(V, $direction$)
2:     **for all** $\langle i, j, k \rangle \in \langle \{1..x\}, \{1..y\}, \{1..z\} \rangle$ **do**

3:       $\Sigma_{col} \leftarrow \begin{cases} \Sigma_{K \in \{1..k\}} V_{\langle i,j,K,1 \rangle} & direction = up \\ \Sigma_{K \in \{k..z\}} V_{\langle i,j,K,1 \rangle} & direction = down \\ \Sigma_{K \in \{1..z\}} V_{\langle i,j,K,1 \rangle} & direction = both \end{cases}$

4:       $P_{\langle i,j,k \rangle} \leftarrow \begin{cases} 1_{(n+2)} & \Sigma_{col} \neq 0 \\ 0_{(n+2)} & \Sigma_{col} = 0 \end{cases}$

5:     **end for**
6:     **return** P
7: **end function**

4.1.2 Keep-out The keep-out region for a given process and part represents material which the process may not modify while creating the part. This feature primarily applies to subtractive processes. It includes material that will be present in the final part and regions of the workspace that cannot be accessed without affecting this material. In general, additive processes will have no keep-out region because they deposit material from the bottom up. Methods to find keep-out are defined in Algorithm 2. The results from applying this algorithm to the example model are shown in Fig. 2.

Algorithm 2 Keep-out Generation

1: **function** KEEPOUT(V, $method$)
2:     **if** $method = laser$ **then**
3:       $K^L \leftarrow$ PROJECT(V, $both$)
4:     **else if** $method = mill$ **then**
5:       $K^L \leftarrow$ PROJECT(V, $down$)
6:     **end if**
7:     **return** K
8: **end function**

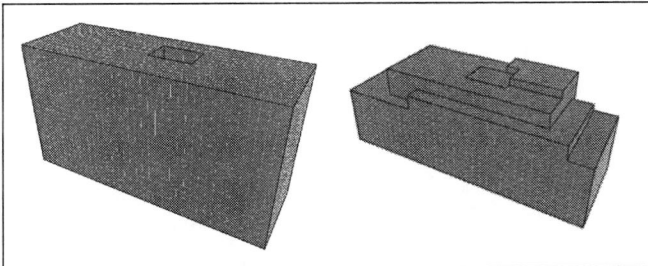

FIGURE 2: Keep-out regions for laser machining (left) and milling (right)

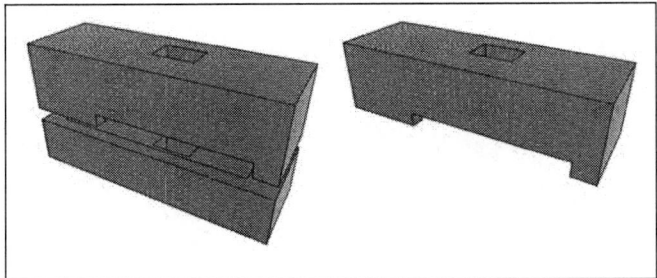

FIGURE 3: Clearance regions for laser machining (left) and milling (right). 3D printing will have a clearance region similar to that for milling, but that also includes regions under overhangs.

4.1.3 Clearance

The clearance region for a given process and part represents regions that will be affected by the process acting on the part. Methods to find clearance are defined in Algorithm 3. The results from these functions are shown in Fig. 3.

Clearance can be used to identify regions of a model A that conflict with the manufacturing of another model B using

$$A^{conflict} = A \cap^L \text{CLEARANCE}(B). \qquad (20)$$

If desired, clearance can be used to modify A to prevent the conflict from occurring. The modified model, A', is found using

$$A' = A \setminus \text{CLEARANCE}(B). \qquad (21)$$

4.1.4 Support

For some parts, support material is needed to hold the part in place during manufacturing. The regions where support material may be added to an object are characterized by the process that is used to remove the supports. Once these regions are determined, they can be used to generate valid supports based on a desired support design. Methods to find the

---

Algorithm 3 Clearance Generation

1: **function** CLEARANCE($V, method$)
2:     $P^U \leftarrow$ PROJECT(V, up)
3:     $P^D \leftarrow$ PROJECT($V, down$)
4:     $P^B \leftarrow$ PROJECT($V, both$)
5:     **if** $method = laser$ **then**
6:         $C \leftarrow P^B \setminus V$
7:     **else if** $method = mill$ **then**
8:         $C \leftarrow P^B \setminus P^D$
9:     **else if** $(method = 3dp) \vee (method = ins)$ **then**
10:        $C \leftarrow P^U$
11:     **end if**
12:     **return** C
13: **end function**

---

supportable regions for planar support structures and to generate support based on a user supplied support model are defined in Algorithm 4. The parameter r_1 is used to determine areas where support is ineffective based on proximity to empty regions that are inaccessible to the removal process. The parameter r_2 represents the desired thickness of the support material. Example results from these functions are shown in Fig. 4.

Algorithm 4 Support Generation

1: **function** SUPPORT($V, method, r_1, r_2$)
2:     $d_1 \leftarrow 2r_1 + 1$
3:     $d_2 \leftarrow 2r_2 + 1$
4:     $S_1 \leftarrow J_{d_1, d_1, 1}$ ▷ Square structuring elements
5:     $S_2 \leftarrow J_{d_2, d_2, 1}$ ▷ for dilation in X and Y
6:     $A \leftarrow$ KEEPOUT($V, method$)
7:     $B \leftarrow A \setminus V$
8:     $C \leftarrow B \oplus S_1$ ▷ Regions where support is ineffective
9:     $D \leftarrow A \oplus S_2$
10:     $E \leftarrow D \setminus A$ ▷ Accessible region of thickness r_1
11:     $S \leftarrow E \setminus C$
12:     **return** S
13: **end function**
14: **function** USERSUPPORT($V, U, method, r_1, r_2$)
15:     $S \leftarrow$ SUPPORT($V, method, r_1, r_2$)
16:     $U' \leftarrow U \cap^L S$
17:     **return** U'
18: **end function**

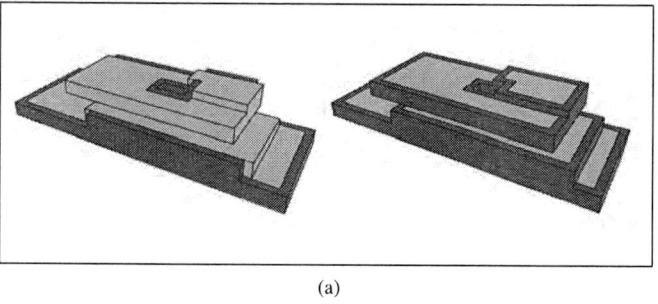
(a)

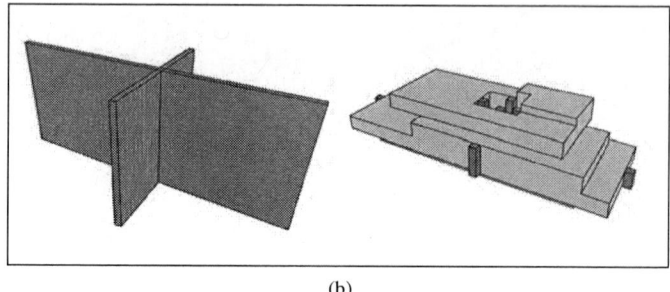
(b)

FIGURE 4: (a) Supportable regions (shown in dark gray) for removal by laser machining (left) and milling (right). (b) User-designed support region (left) and supports generated from the design (right).

4.1.5 Web Web represents scrap material that surrounds a device. It can be used in the creation of supports or layer alignment fixtures. The web for a given removal process can be found using the process defined in Algorithm 5. The parameters r_1 and r_2 represent the distance from the surface of the part to the inside of the web and the width of the web, respectively. An example result is shown in Fig. 5.

Algorithm 5 Web Generation

1: **function** WEB(V, $method$, r_1, r_2)
2:     $d_1 \leftarrow 2r_1 + 1$
3:     $d_2 \leftarrow 2r_2 + 1$
4:     $S_1 \leftarrow J_{d_1,d_1,1}$ ▷ Square structuring elements
5:     $S_2 \leftarrow J_{d_2,d_2,1}$ ▷ for dilation in X and Y

6:     $A \leftarrow$ KEEPOUT(V, $method$)
7:     $B \leftarrow A \oplus S_1$
8:     $C \leftarrow B \oplus S_2$
9:     $D \leftarrow$ (minimum bounding box of C)
10:    $W \leftarrow D \setminus B$

11:    **return** W
12: **end function**

4.2 Exemplar Process Generation Procedures

Using the algorithms defined in the previous sections, procedures can be created to generate the files needed to manufacture a multi-material part. This is demonstrated through manually created procedures for multi-material 3D printing with a blurred region and for 3D printing with a discrete component inserted during the printing process. These procedures are representative of two simple applications and show the utility of the framework. They only make use of a portion of the above operations and algorithms. The remaining operations and algorithms have ap-

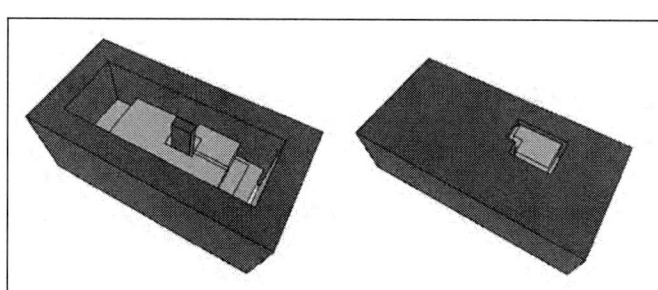

FIGURE 5: Web (shown in dark gray) for removal by laser machining (left) and milling (right)

plications for planning processes for other classes of parts, such as those that make use of subtractive processes or support structures.

4.2.1 Multi-material 3D Printing with a Blurred Region Multi-material 3d printing allows for the creation of models that contain a gradient of materials. The operations discussed in Section 3.3 can be used to automatically generate gradient material transitions given an input model containing two distinct materials. The procedure shown in Algorithm 6 will modify the input model with blurring along any material interfaces, reduce the number of generated materials by grouping materials with similar properties, then export a series of stl files for each distinct generated material. An example input model and the corresponding output are shown in Fig. 6. To create this part, the user would execute the following process:

1. Load generated stl files in multi-material 3D printer software
2. Assign correct material properties to each file
3. Print part
4. Remove part from printer

To create this part manually, a user would be required to individually create a separate model for each material step in the graded area. For a model containing graded areas with a large number of steps or gradients along complex boundaries, this manual workflow would not be practical.

Algorithm 6 Generation of Model and Files for 3D Printing with a Blurred Region

---

1: $V \leftarrow$ User created model
2: $K \leftarrow$ Gaussian kernel of radius r
3: $V' \leftarrow Blur(V, K)$
4: $V'' \leftarrow$ Reduce material count
5: $L \leftarrow$ List of distinct materials in V''

6: **for all** $i \in \{1..|L|\}$ **do**
7: $\quad A_i \leftarrow$ Voxels in V'' containing material L_i
8: $\quad$ Export(A_i)
9: **end for**

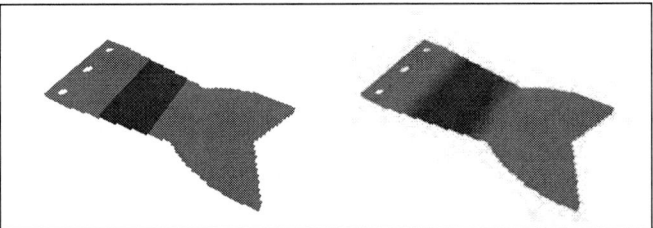

FIGURE 6: Example input model (left) and output model including blurred material transitions (right)

4.2.2 3D Printing with an Inserted Component

One unique capability of multi-material manufacturing is the insertion of a discrete component into a 3D printed object partway though the printing process. Parts added in this way have the advantage of being securely mounted and fully protected. To generate the files to produce this type of part, the procedure shown in Algorithm 7 can be used. This procedure will add any required clearances to the part, generate printer g-code, then add pauses to the g-code at layers where a part must be inserted. To create the part, the user would execute the following process:

1. Load generated g-code file on 3D printer
2. Start printing
3. When printing pauses, insert correct component
4. Resume printing
5. Repeat 3-4 until printing is complete
6. Remove part from printer

The results of this process are discussed further in Section 5.

Algorithm 7 Generation of Model and Files for 3D Printing with an Inserted Component

---

1: $V \leftarrow$ User created model
2: $I \leftarrow$ Inserted components in V
3: $P \leftarrow$ Printed components in V
4: $L \leftarrow \{\}$ $\qquad\qquad$ ▷ Empty list to store pause layers

5: $I^S \leftarrow I \oplus J_{3,3,3}$ $\qquad\qquad$ ▷ Add spacing around I
6: $I^C \leftarrow$ CLEARANCE(I^S, ins) $\qquad$ ▷ Add path to insert I

7: **for all** $k \in \{1..z\}$ **do**
8: $\quad$ **if** $\Sigma I^S_{(:,:,k,:)} = 0$ **then** $\qquad$ ▷ If I^S is not in layer k
9: $\quad\quad I^C_{(:,:,k,:)} =$ Zeros $\qquad$ ▷ Remove layer k from I^C
10: $\quad\quad$ **if** $\Sigma I^S_{(:,:,(k-1),:)} > 0$ **then** $\quad$ ▷ If I^S is in layer $k-1$
11: $\quad\quad\quad$ Add $(k-1)$ to L $\quad$ ▷ Save as a top layer of part
12: $\quad\quad$ **end if**
13: $\quad$ **end if**
14: **end for**

15: $P' \leftarrow P \setminus I^C$ $\qquad$ ▷ Apply spacing and clearance to P

16: $G \leftarrow$ Slice(P') $\qquad\qquad$ ▷ Generate g-code
17: **for all** $l \in L$ **do**
18: $\quad$ AddPause(G, l) $\qquad$ ▷ Add pauses at top layers of I
19: **end for**

20: Export(G) $\qquad\qquad$ ▷ Export final g-code

---

5 Implementation

To illustrate the utility of the framework introduced in this paper, the operations and algorithms discussed in the previous sections were implemented using Python. Two sample parts that used discrete components embedded in 3D printed objects were then created based the process presented in Section 4.2.2.

5.1 Workflow

The workflow used to execute Algorithm 7 is shown in Fig. 7. In addition to the Python implementation of the framework presented in this paper, two additional software packages are required by this workflow. The first, represented by P1, is a program to generate models in a voxel file format and the second, represented by P3, is a slicing program for the 3D printer used. MagicaVoxel [30] and Cura [31] were used for these steps. Various Python libraries were used to facilitate the import and export of the data types required.

5.2 Fabrication of Example Components

Figure 8 shows two exemplar parts that were selected for fabrication. These models were created in MagicaVoxel using a voxel size of 1 mm. They were then processed using Algorithm 7 and created using the steps in Section 4.2.2. The manufacturing

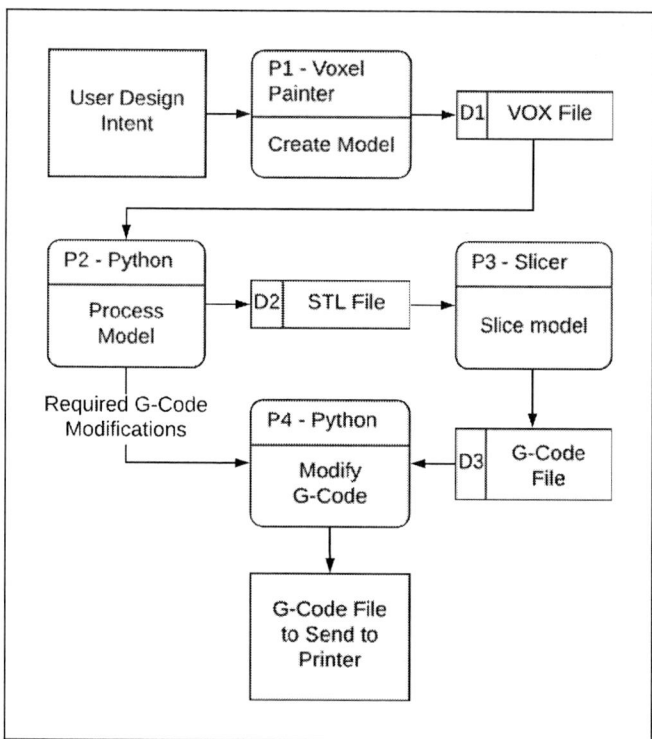

FIGURE 7: Data flow diagram for processing a 3D printed body with an embedded component

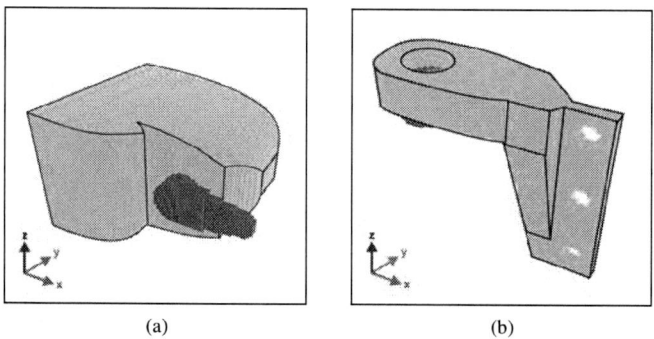

FIGURE 8: Example model designs consisting of (a) a servo motor embedded in a robot body and (b) a servo horn embedded in a mounting arm

process types used by these steps include 3D printing, for which a Prusa i3 MK2S [32] was used, and discrete component insertion. The execution of these steps for the two components is shown in Fig. 9a.

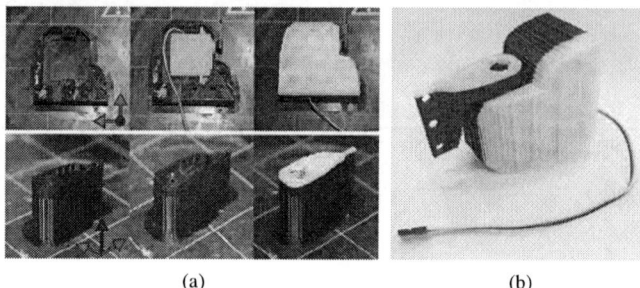

FIGURE 9: (a) Steps taken in the execution of the fabrication process (from left to right): print until pause, insert component, finish printing. (b) Completed and assembled components. The black and white materials were used to highlight the first and second printing stages and are not representative of a true multimaterial computation.

5.3 Implementation Results

The final manufactured parts functioned as intended and are shown assembled in Fig. 9b. Since these parts fall in the same class of components, the same processing algorithm can be used for both with no modifications. To produce these parts manually, the user would have been required to design in all necessary clearances, identify the exact printing layers where the embedded components needed to be inserted, and then modify the g-code by hand to pause at these layers.

A voxel size of 1 mm was identified as providing a good balance between model resolution and processing time. However, due to the precision limitations when rounding dimensions to the nearest millimeter, some cavities in the printed parts were larger than expected. This caused the inserted components to have a loose fit inside of the printed parts. It was also observed that inserted components without a flat top surface, such as the tapered servo horn, could cause the model generation algorithm to produce a model that allowed the component to move or that was missing part of the top layer used to hold the component in place.

Loose fitting parts could be addressed by the use of nonuniform voxel sizes to allow smaller voxels to be placed around inserted components, or by the generation of support features that the user can trim down until a snug fit is achieved. The generation errors for the part without a flat top surface were corrected for this example by manually rotating the part to a more suitable orientation. This could be addressed more generally by the addition of an algorithm that automatically attempts to find the optimal part orientation.

6 Conclusion and Future Work

This project is seeking to make multi-material manufacturing more accessible through the creation of a framework that will allow for the automatic generation of manufacturing steps and files. This paper describes the first steps towards the creation of this framework and illustrates its utility through an example implementation. The results from implementing the framework and using it to create a multi-material component show that the framework can be used effectively to process 3D models and generate the files needed to produce them.

Future work will be towards expanding the framework with additional algorithms and using it in the creation of a software tool that will automate the planning of multi-material manufacturing processes. This tool is intended to automatically determine the best process for producing a part and then provide the user with the process steps in addition to the required files. It should also be able to identify designs that cannot be produced and notify the user of the problem areas. Other areas for future work include adding the ability to import models from standard CAD file formats and optimizing the implementation to permit higher model detail and reduced processing time.

REFERENCES

[1] Belter, J. T., and Dollar, A. M., 2015. "Strengthening of 3d printed fused deposition manufactured parts using the fill compositing technique". *PLOS ONE,* **10**(4), 04, pp. 1–19.

[2] Ma, R., Belter, J. T., and Dollar, A. M., 2015. "Hybrid deposition manufacturing: Design strategies for multimaterial mechanisms via three-dimensional printing and material deposition". *Journal of Mechanisms and Robotics,* **7**, 05, p. 021002.

[3] Dollar, A. M., Wagner, C. R., and Howe, R. D., 2006. "Embedded sensors for biomimetic robotics via shape deposition manufacturing". In The First IEEE/RAS-EMBS International Conference on Biomedical Robotics and Biomechatronics, 2006. BioRob 2006., pp. 763–768.

[4] Hiller, J., and Lipson, H., 2012. "Automatic design and manufacture of soft robots". *IEEE Transactions on Robotics,* **28**(2), pp. 457–466.

[5] Dollar, A. M., and Howe, R. D., 2010. "The Highly Adaptive SDM Hand: Design and Performance Evaluation". *The International Journal of Robotics Research,* **29**(5), feb, pp. 585–597.

[6] Ma, R. R., Odhner, L. U., and Dollar, A. M., 2013. "A Modular, Open-Source 3D Printed Underactuated Hand". In 2013 IEEE International Conference on Robotics and Automation(preprint), IEEE.

[7] Cham, J. G., Bailey, S. A., Clark, J. E., Full, R. J., and Cutkosky, M. R., 2002. "Fast and Robust: Hexapedal Robots via Shape Deposition Manufacturing". *The International Journal of Robotics Research,* **21**(10-11), oct, pp. 869–882.

[8] Kumar, V., and Dutta, D., 1997. "An approach to modeling multi-material objects". *Proceedings of the fourth ACM symposium on Solid modeling and applications - SMA '97,* pp. 336–345.

[9] Jackson, T. D., 2000. "Analysis of FGM Object Representation Methods". PhD thesis, Massachusetts Institute of Technology.

[10] Jense, G., 1989. "Voxel-based methods for cad". *Computer-Aided Design,* **21**, 10, pp. 528–533.

[11] Chandru, V., Manohar, S., and Prakash, C. E., 1995. "Voxel-based modeling for layered manufacturing". *IEEE Computer Graphics and Applications,* **15**(6), Nov, pp. 42–47.

[12] Kou, X. Y., and Tan, S. T., 2007. "Heterogeneous object modeling: A review". *CAD Computer Aided Design,* **39**(4), pp. 284–301.

[13] Qian, X., and Dutta, D., 2004. "Feature-based design for heterogeneous objects". *CAD Computer Aided Design,* **36**(12), pp. 1263–1278.

[14] Samanta, K., and Koc, B., 2005. "Feature-based design and material blending for free-form heterogeneous object modeling". *CAD Computer Aided Design,* **37**(3), pp. 287–305.

[15] Wang, S., Chen, N., Chen, C. S., and Zhu, X., 2009. "Finite element-based approach to modeling heterogeneous objects". *Finite Elements in Analysis and Design,* **45**(8-9), pp. 592–596.

[16] Gupta, V., Kasana, K. S., and Tandon, P., 2010. "Computer Aided Design Modeling for Heterogeneous Objects". *IJCSI International Journal of Computer Science Issues,* **7**(2), pp. 31–38.

[17] Hiller, J. D., and Lipson, H., 2009. "Design Automation for Multi-Material Printing". *20th Annual International Solid Freeform Fabrication Symposium,* pp. 279–287.

[18] Kieback, B., Neubrand, A., and Riedel, H., 2003. "Processing techniques for functionally graded materials". *Materials Science and Engineering A,* **362**(1-2), pp. 81–105.

[19] Chiu, W. K., and Yu, K. M., 2008. "Direct digital manufacturing of three-dimensional functionally graded material objects". *CAD Computer Aided Design,* **40**(12), pp. 1080–1093.

[20] Ramaswami, K., and Prinz, F., 1997. "Process planning for shape deposition manufacturing".

[21] Binnard, M., 1999. *Design by Composition for Rapid Prototyping.* Springer US, Boston, MA.

[22] Hatanaka, M., and Cutkosky, M., 2003. "Process planning for embedding flexible materials in multi-material prototypes".

[23] Kim, H., Choi, J., and Wicker, R., 2010. "Scheduling and

process planning for multiple material stereolithography". *Rapid Prototyping Journal,* **16**(4), pp. 232–240.

[24] Jiménez, P., 2013. "Survey on assembly sequencing: a combinatorial and geometrical perspective". *Journal of Intelligent Manufacturing,* **24**(2), apr, pp. 235–250.

[25] Aukes, D. M., Goldberg, B., Cutkosky, M. R., and Wood, R. J., 2014. "An analytic framework for developing inherently-manufacturable pop-up laminate devices". *Smart Materials and Structures,* **23**(9), aug, p. 094013.

[26] Jang, D., Kim, K., and Jung, J., 2000. "Voxel-based virtual multi-axis machining".

[27] Chen, D., Levin, D. I. W., Didyk, P., Sitthi-amorn, P., and Matusik, W., 2013. "Spec2fab: a reducer-tuner model for translating specifications to 3d prints". *ACM Trans. Graph.,* **32**, pp. 135:1–135:10.

[28] Wang, S.-P., Ragan-Kelley, J., Matusik, W., and Vidimce, K., 2014. "Openfab: A programmable pipeline for multi-material fabrication". *ACM Transactions on Graphics,* **32**, 09.

[29] Vidimce, K., Kaspar, A., Wang, Y., and Matusik, W., 2016. "Foundry : Hierarchical Material Design for Multi-Material Fabrication". *Proceedings of the 29th Annual Symposium on User Interface Software and Technology - UIST '16*, p. 12.

[30] ephtracy. MagicaVoxel. `https://ephtracy.github.io`. Accessed: 2019-05-22.

[31] Ultimaker BV. Ultimaker Cura software. `https://ultimaker.com/en/products/ultimaker-cura-software`. Accessed: 2019-05-22.

[32] Prusa Research s.r.o. Original Prusa i3 MK2S Kit. `https://shop.prusa3d.com/en/3d-printers/59-original-prusa-i3-mk2-kit.html`. Accessed: 2019-05-23.

**Proceedings of the ASME 2019
International Design Engineering Technical Conferences
and Computers and Information in Engineering Conference
IDETC/CIE2019
August 18-21, 2019, Anaheim, CA, USA**

DETC2019-97379

DESIGNING THE CUSTOMER ORDER DECOUPLING POINT TO FACILITATE MASS CUSTOMIZATION

Lin Guo
The Systems Realization Laboratory @ OU
University of Oklahoma, Norman, OK, USA

Suhao Chen
School of Industrial Engineering and Management
Oklahoma State University, Stillwater, OK, USA

Janet K. Allen[1]
John and Mary Moore Chair and Professor
The Systems Realization Laboratory @ OU
University of Oklahoma, Norman, OK, USA

Farrokh Mistree
L.A. Comp Chair and Professor
The Systems Realization Laboratory @ OU
University of Oklahoma, Norman, OK, USA

ABSTRACT

As globalization continues, manufacturing enterprises need to do mass customization with a short lead-time, to satisfy evolving market demands in different regions. One challenge of mass customization is to fulfill orders swiftly at an acceptable cost, meanwhile maintaining the service quality. To do this, the customer order decoupling point – CODP, where the value-adding activities take place, should be designed and adapted to the changing market demands.

In this paper, we propose a Formulation-Exploration method to make decisions on CODP positioning and improve the supply chain to support mass customization. A test problem of auto parts manufacturing is used to establish the efficacy of our method. The Formulation-Exploration method can be used to design supply chains to manage mass customization of products, especially when information is incomplete and inaccurate, goals conflict and multiple types of uncertainty add complexity. In this paper, we focus on the method rather than the results per se.

KEYWORDS

Mass Customization, Customer Order Decoupling Point, Formulation-Exploration, Compromise Decision Support Problem.

GLOSSARY

cDSP	compromise Decision Support Problem
CODP	Customer order decoupling point
DCI	Design capability index
EMI	Error margin index
ESS	Exploration of the solution space
MTS	Make-to-stock production mode
MTO	Make-to-order production mode
PLC	Product life cycle

1 FRAME OF REFERENCE
1.1 What is CODP?

Mass customization greatly enhances the emotional interaction between the designers and the customers [1]. Mass customization is proposed by Naylor et al. [2] as to combine the agile and lean as a new strategy – *leagile*, and it is defined and broadly accepted as "postponing the task of differentiating a product for a specific customer until the latest possible point in the supply network" [3]. Accordingly, being able to determine the differentiation point of a product to adapt to the rapidly changing market is of great importance in supply chain design. This point is normally defined as the customer order decoupling point (CODP)[2]. At this point make-to-stock (MTS) production ends, the value-adding activities take place [5], and the make-to-order (MTO) production starts.

---

[1] Corresponding author: janet.allen@ou.edu
[2] In the literature, the CODP is sometimes called Order Penetration Point (OPP), Decoupling Point or Customer Order Point (COP) [4] Shidpour, H., Da Cunha, C., and Bernard, A., 2014, "Analyzing Single and Multiple Customer Order Decoupling Point Positioning Based on Customer Value: A Multi-Objective Approach," Procedia Cirp, 17, pp. 669-674..

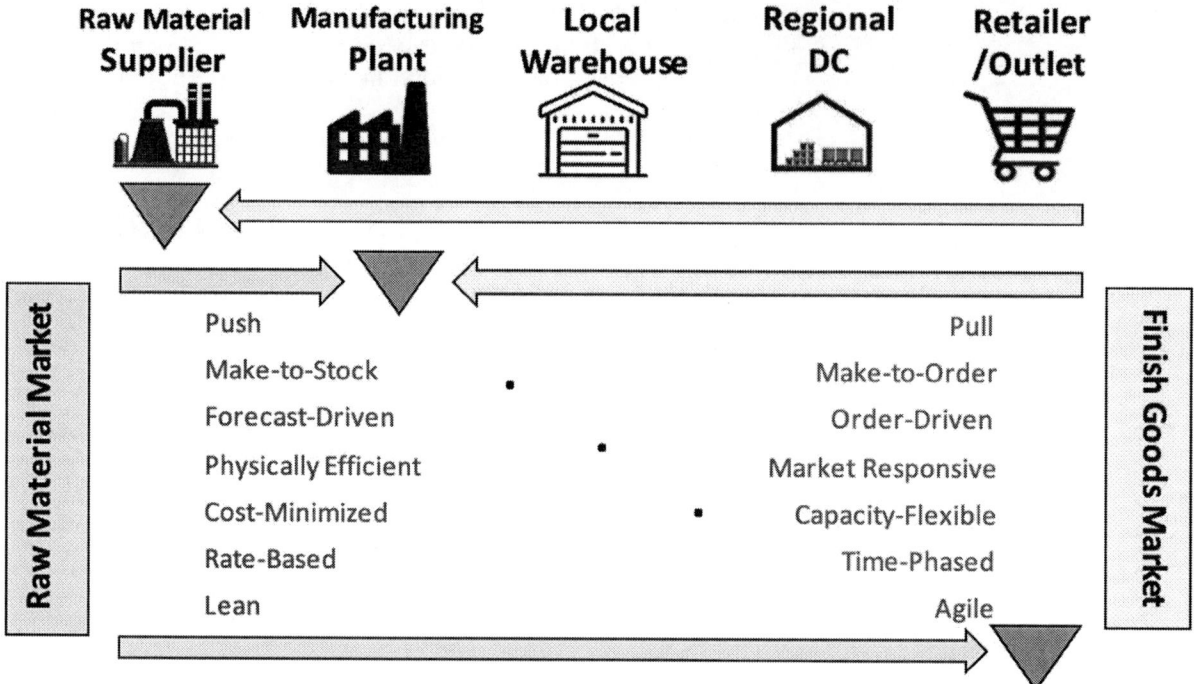

FIGURE 1 THE POSSIBLE POSITION OF CUSTOMER ORDER DECOUPLING POINT IN A SUPPLY CHAIN

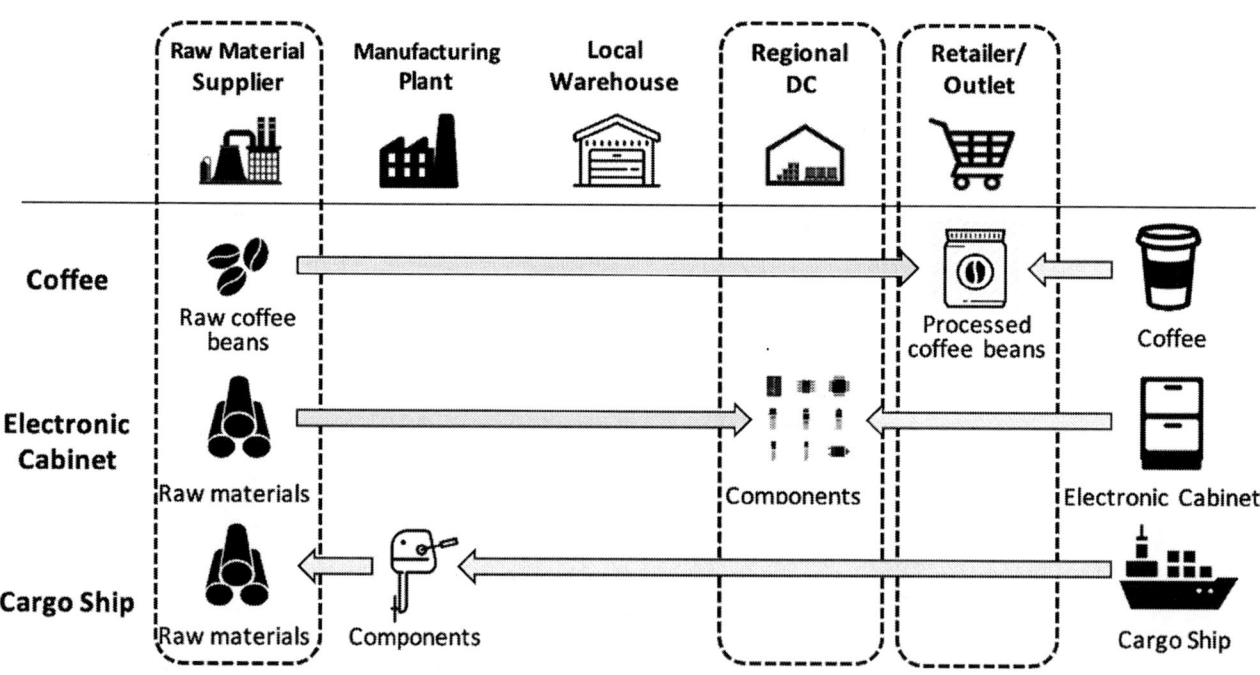

FIGURE 2 THE CUSTOMER ORDER DECOUPLING POINT OF DIFFERENT PRODUCTS

In Figure 1, we illustrate different characteristics of production before (upstream) and after (downstream) the CODP of a supply chain. In MTS production (before CODP), manufacturers make production plans based on demand forecast and pursue physical efficiency. MTS is often used in upstream supply chains in labor-intensive industries, where low-cost and on-time delivery are prioritized. On the contrary, in MTO production, manufacturers produce goods to meet customers' orders and emphasize market response and capacity flexibility. MTO is often used in the downstream supply chain of high-tech industries, or after-market items [6], where differentiation and personalization are valued.

Theoretically, the CODP of a product can be anywhere in the supply chain. Figure 2, we show some typical CODPs. In the coffee industry, the CODP is the raw material of the retailer. When an end customer orders a cup of coffee, the retailer starts making it. In the electronic cabinet supply chain, MTS starts from the raw materials and goes to the regional distribution center (RDC) where stores the standard components. After receiving orders, the assembly plant gets the customers' needs on exact models and the corresponding parts, then starts assembling and shipping. As to the shipbuilding industry, e.g., cargo ships, the whole supply chain is MTO due to its highly customized nature. There are common features of the CODP for different goods: convenient storage, commonality and short lead-time requirement of the goods upstream from the CODP, whereas multiple consumption channels, affordable production shifting cost and relatively long but acceptable lead-time of the goods downstream from CODP, etc.

The aim of mass customization is providing products to satisfy individual customer's requirements with near mass production efficiency [7]. Usually, the efficiency of a supply chain varies significantly as the CODP changes. For a single product, the CODP should vary in different phases of its life cycle to support mass customization. The rapid product upgrades and replacements require fast CODP switching. From the demand side, the users are continuously putting forward new needs, which force the CODP to go upwards along the supply chain. To maintain a near mass production efficiency in a supply chain with fast CODP switching is a challenge, because low cost and product differentiation are contradictory. Therefore, a method that facilitates the exploration of the tradeoffs between cost and flexibility is needed.

1.2 Some Methods on CODP Positioning in Literature

There are two main approaches to identify the CODP positions. One is the qualitative approach, through which the researchers make general rules on CODP locations for different types of products. For example, CODP should be closer to the designer/manufacturer for personalized products, luxury goods or services. The other approach involves quantitative methods – designers position the CODP by applying mathematical modeling. The latter is mainstream in recent years. We focus on the quantitative approach in the literature review. In Appendix A, we list some representative publications on CODP positioning.

Queuing model. Teimoury et al. [9] use a queuing model and the matrix-geometric method to determine CODP in a multi-product supply chain. Customer orders are assumed to arrive according to the Poisson process. The objective is to minimize total cost with constraints of warehouse capacity and service level. The effect of an impatient customer for the arrival of the product on product costs is considered. The assumption of the Poisson customer-order arrival may cause errors, since for different products, or in different phases of the product life cycle, the customer-order arrival vary a lot. Zhou et al. [10] use a two-stage queueing model with customer orders modeled as Markov

processes and find that the variance and correlation of the demand increase the total operation cost. The effect of the CODP position on the unit inventory holding cost, the lead-time quotation policy and the penalty cost for tardiness are considered. However, the single objective is simplified as minimizing the total cost.

Time scheduling model. Liu et al. [18] propose a time scheduling model to examine the dynamic positioning of CODP. Minimizing total cost and maximizing customer satisfaction are the two objectives while the incremental cost of a new order, lead-time and changing values of objectives are constraints. However, the authors do not scale the two objectives with different units – scaling is necessary because the importance and utility of the two objectives may change with circumstances. Jeong [14] proposes a dynamic model based on product life cycle theory to position the CODP and make a production-inventory plan. The author applies optimal control theory to with a quadratic objective function. However, the author does not consider factors as time-varying penalty costs and the delivery lead-time.

Stochastic programming, multi-criteria optimization, dynamic programming, etc. Some authors apply programming methods with stochasticity or multiple criteria. Ghalehkhondabi et al. [11,12] leverage a stochastic programming model to identify CODP by minimizing total cost or maximizing total profit. Constraints of working time, order satisfied percentage and manufacturing capacity are considered. Customer orders are assumed to follow Poisson distribution. Shidpour et al. [5] propose a multi-objective programming model based on manufacturer's profit and customer perceived value to analyze the impacts of single-CODP and multi-CODP while considering service time constraint. They conclude that multi-CODP is preferred than single-CODP for a product portfolio, benefiting both the manufacturer and the customers. Ahmadi et al. [13] build a dynamic programming model to optimize CODP by minimizing holding cost, delivery delay and number of modules not ready to be assigned. The authors assume that the customer demand is deterministic. Liu et al. [15] construct a linear programming model with discrete variables in the logistics industry to position CODP by minimizing total cost of logistics service integrator. Procedure constraints and the lead-time constraints are considered in the model. The authors conclude that the optimal CODP is not affected by decreasing order-transferring and waiting cost but driven to the last procedure when order processing cost decreases. Other methods such as an Analytic Hierarchy Process model [10,19], value network simulation model [16,17] and tandem forecast-driven and order-driven simulation model [21] are also explored to identify the optimal CODP.

In summary, recent researchers in CODP positioning try to address the issue by constructing models with multiple objectives and constraints. There are employed traditional optimization methods such as stochastic programming, dynamic programming and selection model (Analytic Hierarchy Process, AHP), as well as methods "borrowed" from other domains such as value network, tandem circuit design and polychromatic set theory. Researchers find various metaphors for CODP and supply chain design, so as to determine interesting concepts and methods – such as viewing the MTS and MTO production mode as two circuits and looking for an appropriate way to connect them. This phenomenon implies that CODP design is a field combining science and art. With different metaphors, the authors have diverse perspectives and foci. Their observations are interesting and somehow useful to an extent in some situations. However, there are some limitations and we summarize them as follows.

Limitation 1 – There are assumptions in the distribution of non-deterministic methods [8-11]. The authors assume that there is enough data to support decision-making on the distribution of customer-order arrivals, demand forecast accuracy, customer satisfaction quantification, etc. These decisions or assumptions can be wrong and irreversible once the design stage is over, and no adjustments can be done during operations.

Limitation 2 – The optimization methods [4, 8-14] are used to find optimal solutions, that are solutions on the boundary of the feasible solution space (for deterministic approach) or distributed close to the boundary of the fussy feasible solution space (for stochastic approach). As models are incomplete, inaccurate, and embody different levels of fidelity, the solution may be optimal to the model but may not be optimal to the real problem which is way more complicated than the model.

Facing the challenges in the supply chains and mass customization, we need to overcome the limitations. The assumptions on data adequacy can be replaced with the information we obtain from post-solution analysis; typical uncertainties need to be managed in the design and operation stage; Strategies regarding the CODP positioning should be developed for different market environments or different phases in the product life cycle.

In Section 2, we describe the challenges of supply chain and CODP design and pose a research question and hypothesis; in Section 3, we introduce the Formulation-Exploration method with regard to answering the research question; In Section 4, we use a test problem in the automobile industry to validate our method. In Section 5, we summarize the contributions of this paper and suggest possible directions for future work.

2 PROBLEM DESCRIPTION
In this section, we analyze the characteristics and challenges of supply chains designed to deliver products in the global market.

2.1 Challenges in Supply Chains
A large number of goals that may conflict with each other need to be managed (Figure 3). Their priorities may evolve as circumstances change. Since mathematical models used to define supply chains are typically abstractions of reality, we are guided by Simon's maxim of looking for *satisficing* (good enough – not optimum) solutions [22].

According to George Box, "all models are wrong but some are useful" [15], the mathematical models that constitute a supply chain are incomplete and inaccurate [16] and of different fidelity. Decision model do not represent the physical world perfectly [17]. Hence, a designer should work with decision models that embody incompleteness and errors [18].

FIGURE 3 MULTIPLE CONFLICTING GOALS IN SUPPLY CHAINS

Multiple types of uncertainties in decision models affect the performance of the supply chain [19]. In Table 1, we illustrate the four types of robust design in association with the four types of uncertainty, which are defined, extended and summarized by Taguchi [20, 21], Chen et al. [22], Isukapalli [23], Choi et al. [24-26] and interpreted by Nellippallil et al. [27]. In this paper, we tie this to the supply chain. The four types of uncertainty in mechanical engineering have equivalents in the supply chain (Table 1).

Type I robust design is relatively insensitive to noise factors – parameters that cannot be controlled. An example is the uncertainty in the demand side. MTS production is planned based on the demand forecast which may be different from the real demand. MTO production is planned based on the in-taken orders which may be more urgent than usual or withdrawn later. Hence, demand is a parameter with uncertainty that cannot be controlled by the decision maker.

Type II robust design is relatively insensitive to variations in design variables. An example is the uncertainty on the supply side. The actual output of production can be different from the expected volume, because of overuse of the machines, overtime of the labor, etc. The decision maker may set the production volume to a certain level, but the actual output can be different.

TABLE 1 FOUR TYPES OF ROBUST DESIGN AND THE INTERPRETATIONS IN SUPPLY CHAIN

Types of robust design	Types of uncertainty	Example in supply chain	Quantification
I	Parameter uncertainty	Uncertainty in the demand side, such as stochasticity of customers' order	Type I, II: Error Margin Index (EMI), Monte Carlo Simulation, Latin Hyper Cube, First / Second Moment Method, etc.
II	Variable uncertainty	Uncertainty in the supply side, such as variation in productivity	
III	Model structural uncertainty	The number of decision variables, constraints, or mathematical relation between variables change due to machine failure, customer loss, impatient customer or natural disaster	Type III: Design Capability Index (DCI), Variance Function Estimation, Prediction Interval Approach, etc.
IV	Uncertainty created in process chain	The bullwhip effect	Type IV: *Satisficing*, Exploring the Solution Space (ESS), etc.

The Error Margin Index (EMI) is used to deal with Type I and Type II uncertainty by bringing the mean and minimizing the variance of the performances. EMI is a mathematical construct indicating the location of mean system performance and the spread of the performance considering variability in design variables [25].

Type III robust design is relatively insensitive to uncertainties embedded within the model structure, such as dimensionality, constraints or the relation between variables due to unexpected events, such as overcapacity, infrastructure damage, political environment changes. The Design Capability Index (DCI) is used to handle Type III uncertainty. DCI is a mathematical construct for efficiently determining whether a ranged design specification is capable of satisfying a ranged set of design requirements [24]. Using a DCI, the designer identifies a range of design variable with the minimized design specification variation and an acceptable performance variation. Within this range, the impact of the model structural change on the model performance is minimized, meanwhile the performance is good enough and with relatively small variance. Details of using EMI and DCI can be found in [19, 22, 24, 26, 28].

In Type IV robust design, the propagated uncertainty in process chains and the interaction among the three types of uncertainty are managed, for example, the bullwhip effect. We propose to explore the solution space (ESS) to improve the insensitivity of the design considering various situations. Using ESS, we explore design preference, improve design capacity, and boost the system potential. The method is introduced in detail in Section 3.

2.2 A Question and Hypothesis
Based on the limitations in the literature and challenges in CODP design, we pose a question and hypothesis that we address in this paper. By answering the question and validate the hypothesis, we bridge the gap.

Question – *How can we improve the robustness (insensitivity) of the CODP design without sacrificing the performance of the supply chain?*

Hypothesis – *We propose a Formulation-Exploration method to explore the solution space with respect to the design capacity under typical uncertainties. With insight obtained from solution space exploration, we remove some errors and heuristics in the initial design and boost the potential of the system.*

3 THE FORMULATION-EXPLORATION METHOD
To obtain *satisficing* solutions insensitive to multiple types of uncertainty, the enterprises need to apply some systematic methods to make in-time decisions supporting mass customization. We propose the Formulation-Exploration method (Figure 4). The method consists of two stages, namely, problem formulation and solutions space exploration. To converge on a *satisficing* solution, we need to iterate between the two stages.

Compromise Decision Support Problem (cDSP). In the problem formulation stage, the problem is formulated as a cDSP and then solved using the Adaptive Linear Programming algorithm [29]. The mathematical form of a cDSP and why it delivers *satisficing* solutions are stated in Appendix B. Due to the complexity of the supply chain with mass customization, the cDSP usually contain non-convexity and nonlinearity features, so one way of managing the complexity is to approximate the model to a classic linear model and solve the problem using simplex algorithm. The model structural uncertainty (Type III) and process chain added uncertainty (Type IV) are tackled in the problem formulation session.

In the second stage the solution space is explored to find satisficing solutions associated with each design preference (scenario), in different phases in the product life cycle. Type I and II uncertainty are managed in the exploration of the solution space (ESS). In this paper, when we refer to design preferences, we particularly focus on the importance of the different goals.

Weight sensitivity analysis – exploration of the design preferences. We use weight sensitivity analysis to explore how the weight changing affect the system performance regarding the achievement of the goals (Figure 4, "weight sensitivity analysis" [30].

System capacity analysis – identification and management of the sensitive segment and bottleneck. Given that we identify a *satisficing* range of the weights, we need to overcome the capacity limitation of constraints or bounds. Therefore, we use the "system capacity analysis" method to identify the sensitive segments and bottlenecks (Table 2). If an inequality constraint has zero or tiny surplus or slack compared with its right-hand-side value, we define it to be an active constraint. The solution is on or close to the boundary of the active constraint, so the solution is sensitive to the uncertainty happens to the active constraint. If the shadow price of an active constraint is lower than other active constraints, relaxing such an active constraint may not get considerable improvement of the fulfillment of the goals, then we define such a constraint as a "sensitive segment." We then bring the solution away from the sensitive segment by restricting the active constraint, which means adding a buffer to the mathematical model to prevent the solution reaching the physical boundary. If the shadow price of an active constraint is the largest comparing with to other constraints, relaxing the constraint can result in the greatest improvement of the fulfillment of the goals, then we define such an active constraint as a "bottleneck". Then we need to find ways of relaxing the bottleneck in the physical system to boost the system potential, and once there is no more potential of physically relaxing, we bring the solution away from the newly-relaxed boundary by restricting the constraint in the mathematical model – add a buffer to the physical boundary. In such a way, we boost the potential of the physical system while improving its robustness, therefore we neither sacrifice system robustness for a better fulfillment of the goals nor do the opposite.

We hypothesize that using the Formulation-Exploration method, the CODP can be designed/redesigned to obtain satisficing solutions, which support mass customization. A successful company should be able to do repeated design based on the information attained from the previous cycle, so its supply chain can be flexible for mass customization. The Formulation-Exploration method facilitates us to answer the research question and validate the hypothesis we pose in Section 2.2.

In Section 4, we use a CODP design problem in auto industry to examine our proposed methods.

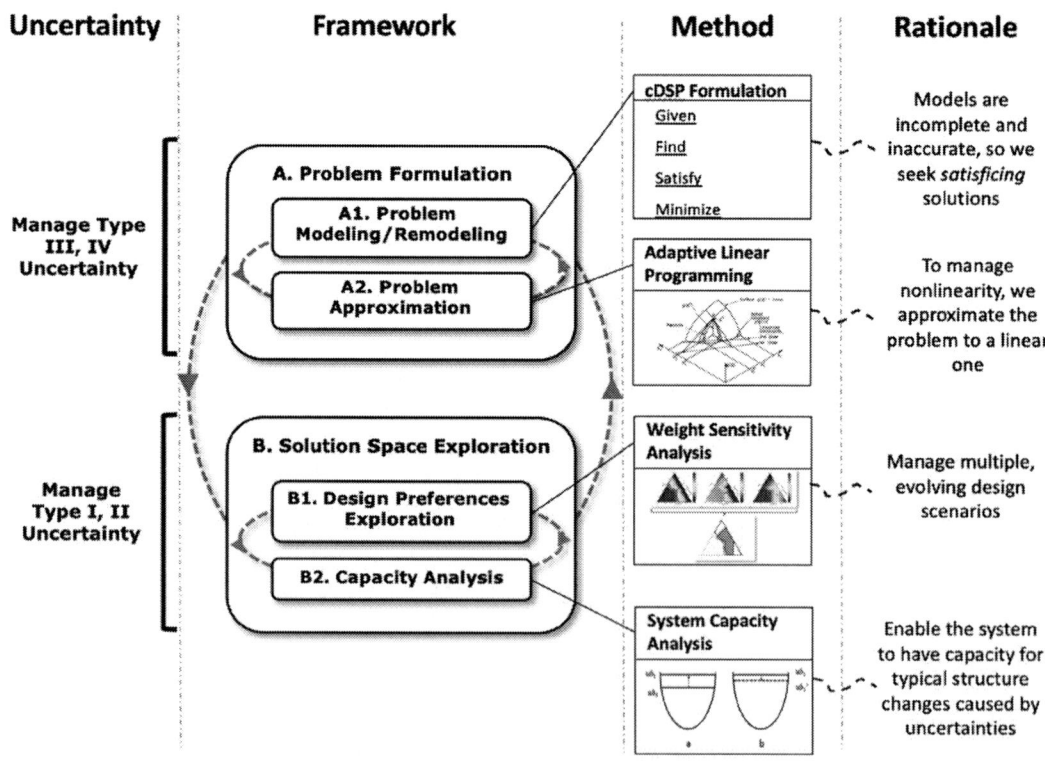

FIGURE 4 FORMULATION-EXPLORATION FRAMEWORK

TABLE 2 ALGORITHM FOR SYSTEM CAPACITY ANALYSIS

---

a. Identify n scenarios of parameters/bounds with uncertainties – ISs.

Begin Iterations of Exploration of the Solution Space

b. Use the latest model to identify the feasible area of weights, and identify m weight scenarios within the feasible area of weights that represent different design preferences – WSs.

c. Plug n ISs and m WSs into the latest model to get x solutions.

d. Plug x solutions into the model in the first iteration and into the model in the current iteration

e. Identify active constraints/bounds and sensitive segment using the model in the first iteration

f. Identify bottleneck using the model in the current iteration

g. if no sensitive segments or bottleneck

  Go to l.

 else

  Continue with h.

h. For each sensitive segment

  Explore the practicality of restricting their right-hand-side values (RHSs) in the model

i. For each bottleneck

  Explore the practicality of relaxing it in the physical system

j. Make model improvement plans (restricting model or relaxing physical system) based on the conclusion in h and i.

k. Improve the model based on the improvement plans in g and **go to** b.

l. The latest model is relatively insensitive to uncertainties and has no potential to achieve a better solution.

End the iteration

---

4 A TEST PROBLEM

To examine our hypotheses, a CODP design problem is used. The baseline model is from [31]. This supply chain is designed for a single product with customized options. To use the test problem to examine our method, we simplify the supply network in of the baseline model to a three-echelon supply chain, formulate it into a cDSP, and position CODP, see Appendix C.

4.1 Problem Formulation

The test problem is a three-echelon supply chain in automobile industry with three players – a supplier, a manufacturer, and a retailer (Figure 5). Our tasks are to position the CODP, determine the production (or purchasing) volume for each player, and to determine the appropriate reliability of each process including the operation for each player and the transportation between two players. The constraints are cost and capacity in each process. We formulate the problem in a compromise Decision Support Problem (cDSP). See Appendix C. We have three goals – profit, service level and variance of reliability of players. We have two types of system variables, the continuous variables and the binary variables.

The production volume of the supplier and the manufacture and the purchasing volume of the retailer are continuous system variables, whereas the CODP of each candidate locations are

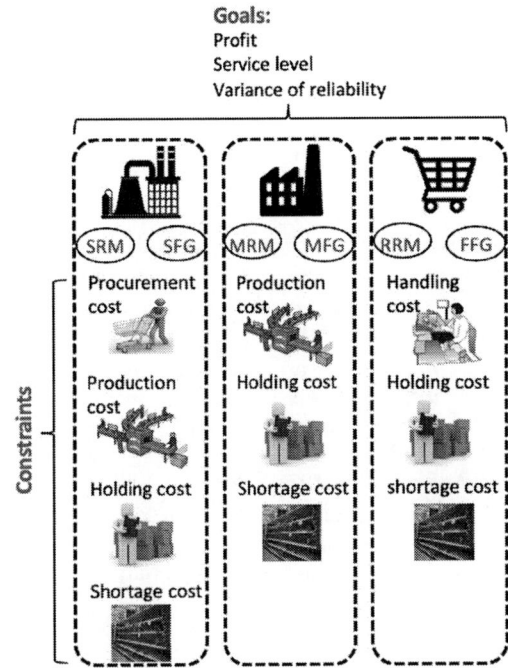

FIGURE 5 A THREE-ECHELON SUPPLY CHAIN

binary variables. Each player has two candidate positions that may store the goods, raw material or finished goods, so there are six candidate locations for CODP in this supply chain, namely, supplier's raw material (SRM), supplier's finished goods (SFG), manufacturer's raw material (MRM), manufacturer's finished goods (MFG), retailer's raw material (RRM), and retailer's finished goods (RFG). In each design scenario that represents a certain design preference, one of the six candidate locations should be the CODP. As the product enters different phases of its life cycle – introduction, growth, maturity, and decline, the preferences on the three goals vary, the CODP may move up and down accordingly.

4.2 Apply the Formulation-Exploration to Improve the Model and Obtain *Satisficing* Solutions

Weight sensitivity analysis. In design preferences exploration, we use a number of weight scenarios (WSs) to represent typical design preferences. We apply the method in [32] to determine the WSs. For a M-goal problem, if we want to divide the weight range for each goal – we define the whole range is [0, 1] – into p pieces, we will have $H = \binom{M + p - 1}{p}$ number of WSs. In this test problem, since we have three goals, $M = 3$. We set $p=2$, so $H=6$. We want one more scenario that assigning equal weights (1/3, 1/3, 1/3) to three goals, thereby we have all together seven WSs (Table 3). Plugging each of the seven WSs into our cDSP, we have seven sub-problems. Solving them one by one, we obtain the results listed in Table 4.

TABLE 3 THE SEVEN WEIGHT SCENARIOS

WS	w1	w2	w3
1	1	0	0
2	0	1	0
3	0	0	1
4	0.5	0.5	0
5	0.5	0	0.5
6	0	0.5	0.5
7	0.33	0.33	0.33

TABLE 4 RESULTS OF THE FIRST-ITERATION

WSc	CODP	Profit	Service Level	Variance of Reliability
1	MRM	30501	99.88%	0.430%
2	RFG	32732	99.89%	0.444%
3	RFG	32764	99.87%	0.408%
4	MRM	30500	99.88%	0.414%
5	MRM	30500	99.88%	0.414%
6	RFG	32807	99.89%	0.428%
7	MRM	30500	99.88%	0.414%

We use Figure 6 to show the achievement of the goals associated with the CODP results. From Figure 6, we conclude that when profit is less important than service level and variance of variability, the CODP should be at the finished goods of the retailer (RFG); if the profit has some importance, the CODP should be at the raw material of the manufacturer (MRM).

System capacity analysis. We identify the holding cost of the manufacturer and retailer have limited capacity but a relatively large shadow price, which means this constraint is a bottleneck. So, if we can reduce the holding cost physically by some means (e.g., applying RFID technology, or using a third-party vender with lower unit cost), the achieved value of the goals can be improved. We assume that the operational team takes action for cost reduction, so we reformulate the problem accordingly using the algorithm in Table 2. Then we run the second iteration of Formulation-Exploration. After three iterations of formulation-exploration, we get the results in Table 5. The final results are better than the first-iteration results (Table 4) in terms of the achieved value for the goals. In the final model, we still have the active constraint on transportation, but as we assume that no further improvements can be physically made, we realize that at this time we have boosted the system potential, and stop iterating.

In this case, we observe that when the weight of the profit goal is zero, which means when in some situations that the enterprise does not set profit as a major goal (in the introduction phase, or decline phase of the product life cycle, PLC), the CODP should be at MRM, the raw material of the manufacturer. In such a situation, from the raw material of the supplier to the raw material of the manufacturer, the production accords to the demand forecast; from the raw material of the manufacturer to the finished goods at the retailer, the production or goods preparation are based on customers' orders. Standard components are stored at MRM, and customization takes place right after MRM. In other words, the production before MRM is for a single product and after MRM is for a product family. The profit is not at a high level, whereas the achievement of service level (the higher the better) and the variation of the reliability (the lower the better) is not always in at a high level either. This indicates that in the introduction or decline phase of a product, we need to sacrifice the performance of the supply chain for a better customized product and a relatively more flexible supply chain.

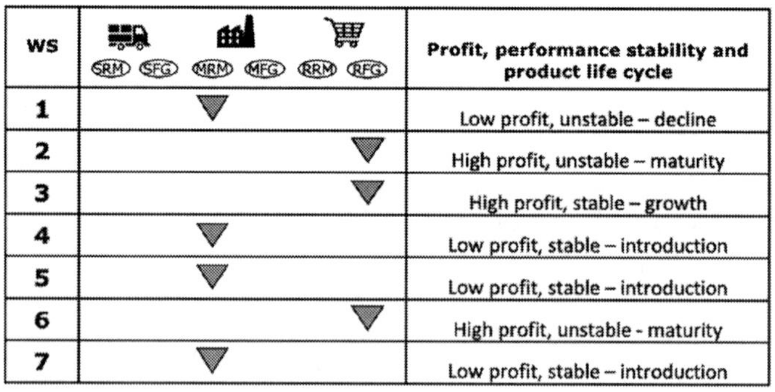

WS	SRM SFG MRM MFG RRM RFG	Profit, performance stability and product life cycle
1	MRM	Low profit, unstable – decline
2	RFG	High profit, unstable – maturity
3	RFG	High profit, stable – growth
4	MRM	Low profit, stable – introduction
5	MRM	Low profit, stable – introduction
6	RRM	High profit, unstable - maturity
7	MRM	Low profit, stable – introduction

FIGURE 6 CODP OF DIFFERENT WEIGHT SCENARIOS

When the profit is the highest priority for the enterprise, which usually happens in the growth phase or the maturity phase of the PLC, the CODP should be at the last candidate node of the supply chain, that is the finished goods of the retailer. In this way, the whole supply chain is an MTS one.

TABLE 5 RESULTS OF THE SEVEN DESIGN SCENARIOS IN THE THIRD ITERATION

WSc	CODP	Profit	Service Level	Variance of Reliability
1	MRM	30501	99.89%	0.398%
2	RFG	33813	99.91%	0.425%
3	RFG	33786	99.89%	0.392%
4	MRM	30500	99.89%	0.408%
5	MRM	32389	99.89%	0.409%
6	RFG	33813	99.91%	0.417%
7	MRM	30500	99.90%	0.407%

The relatively stable and strong demand from customers make the demand forecast more reliable, so the mass customization can be done via the production through the manufacturers to the retailers based on relatively accurate forecast due to market feedback. The whole supply chain is now for a product family. In this case, the mass customization is done with a mass production cost. We tie the results in seven WSs with the different phases in PLC (Figure 7). This indicates that even for a newly, high-tech and highly personalized product, as it enters the growth phase when the demand is relatively stable and predictable, the customization can be done in advance to the customers' order-intake, until the product goes to the decline phase. Therefore, even for a high-tech, highly-personalized product, we do not have to adopt the MTO production mode all the time, because in the growth and maturity phases, we can finish the production of the mass-customized product ahead of the orders with relatively low risk.

4.3 Verification and Discussion
To verify our conclusions and attain insight on the relation between CODP candidate variables and the achieved value of the goals, we do sensitivity analysis of the six CODP candidate locations. In each weight scenario, we fix the CODP in each of the six candidate locations and obtain the achieved value of the goals. We show the comparison of the results with CODP fixed at each point with the *satisficing* results (Table 5) in Table 6. Under each WS, we show the *satisficing* result with an asterisk "*" in the column named "*satisficing*." The percentage numbers in those *non-satisficing* lines are comparison with the *satisficing* ones. "-10%" means that it is 10% worse than the *satisficing* result under the same WS. In some scenarios, if we assign a candidate location as CODP, we cannot obtain a feasible solution because the constraints are always violated. For example, in WS 1, if we set SRM as the CODP, which means the whole supply chain is in MTO production mode, we cannot satisfy all constraints. So, if we set profit as the only goal, given the resource that can be acquired in this supply chain, a whole MTO production mode is not recommend for this product.

From Table 6, we observe that we can never position CODP at two locations because we cannot obtain any feasible solutions. Therefore, the product is not fit for a push pull strategy. In WS

1, 4, 5, and 7, RFG as CODP cannot give feasible solutions, whereas in all the others WSs – WS 2, 3, and 6 – RFG is the *satisficing* solution. This means RFG is a sensitive node, it either gives the best or looses it all. This proves that doing weight sensitivity analysis is necessary so that we only pick up a point as CODP when it gives qualified result in certain design preferences.

TABLE 6 COMPARISON OF *SATISFICING* RESULTS WITH RESULTS FROM OTHER CODP CANDIDATE LOCATIONS

WS	Satisficing	CODP	Profit	Service Level	Variance of Reliability
1		SRM		Infeasible	
1		SFG		Infeasible	
1	*	MRM	30501	99.89%	0.398%
1		MFG	-5.6%	0%	-1.5%
1		RRM	-3.8%	0%	-1.5%
1		RFG		Infeasible	
2		SRM		Infeasible	
2		SFG		Infeasible	
2		MRM		Infeasible	
2		MFG		Infeasible	
2		RRM		Infeasible	
2	*	RFG	33813	99.91%	0.425%
3		SRM		Infeasible	
3		SFG		Infeasible	
3		MRM	-9.7%	0%	-1.5%
3		MFG	-14.8%	0%	-3.1%
3		RRM	-13.1%	0%	-3.1%
3	*	RFG	33786	99.89%	0.392%
4		SRM		Infeasible	
4		SFG		Infeasible	
4	*	MRM	30500	99.89%	0.408%
4		MFG	-5.6%	0%	-1.5%
4		RRM	-3.8%	0%	-1.6%
4		RFG		Infeasible	
5		SRM		Infeasible	
5		SFG		Infeasible	
5	*	MRM	32389	99.89%	0.409%
5		MFG	-11%	0%	-1.2%
5		RRM	-9.1%	-1.0%	-1.1%
5		RFG		Infeasible	
6		SRM		Infeasible	
6		SFG		Infeasible	
6		MRM		Infeasible	
6		MFG		Infeasible	
6		RRM		Infeasible	
6	*	RFG	32807	99.91%	0.417%
7		SRM		Infeasible	
7		SFG		Infeasible	
7	*	MRM	30500	99.90%	0.407%
7		MFG	-5.9%	-1%	+0.3%
7		RRM	-4.3%	-1%	+0.8%
7		FRG		Infeasible	

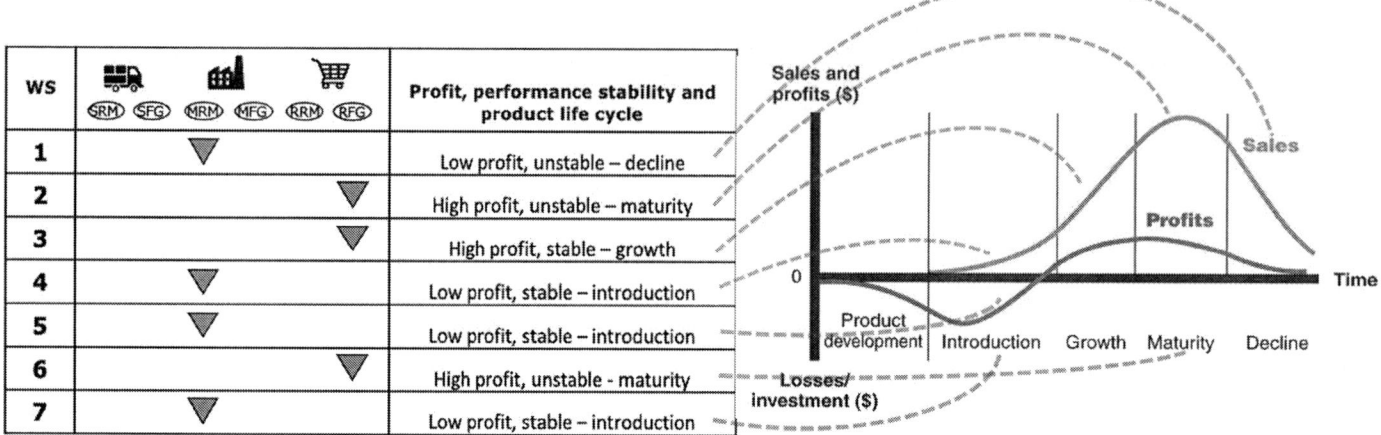

FIGURE 7 CODP, ACHIEVED VALUE OF GOALS IN DIFFERENT PHASES OF A PRODUCT LIFE CYCLE

Among the three goals, profit is relatively sensitive to CODP migration because when we move CODP to *non-satisficing* nodes, the achieved values of profit decrease the most. Service level does not change much as CODP migrates, nor does it change much as WS changes. In this sense, to reduce the computational complexity, we can remove the service level goal.

Among the six candidate locations, only two can be CODP in all seven WSs – MRM and RFG, so it simplifies the decision makers' tasks. They need to focus on these two locations instead of all six locations. The responses of the preference change are all about switching CODP between MRM and RFG. Decision makers do not need to do a one-time investment on fixed-assets to make other locations CODP.

Using the Formulation-Exploration method, when changes take place in the supply chain, such as design preferences unexpect-edly evolve (Type III uncertainty), cost budget cutting-down due to crises (Type II uncertainty), production/ transportation cost or capacity change due to natural disasters (Type I or IV uncer-tainty), etc., we can reformulate the problem and output *satisfic-ing* solutions in a real-time manner. The switching of the phases of the PLC can be captured in time and the production strategy can be updated accordingly.

5 CLOSURE

In this paper, we propose the Formulation-Exploration method for positioning the customer order decoupling point (CODP) to facilitate mass customization. We review the literature on CODP positioning and identify two major limitations – the assumptions about stochastic parameters and relying on model accuracy. We analyze the challenges in supply chain regarding CODP positioning. We define the concept of the robust design to supply chain and categorize the uncertainties in the supply chain into four types. Based on the limitations in the literature and the challenges in the CODP design, we raise a question and

hypothesis. To answer the question, we propose the Formulation-Exploration method to explore the solution space to exploit the potential of the supply chain, using a test problem in automobile industry.

Question – *How can we improve the robustness (insensitivity) of the CODP design without sacrificing the performance of the supply chain to facilitate mass customization?*

Using the Formulation-Exploration method, we explore the solution space by using representative design scenarios and explore the potential of boosting the performance of the system by adopting physical means, such as applying a new technology to cut down a certain cost. By doing this in iterations, we exploit all means that lead us reaching the most desirable *satisficing* solutiongiven the resources on hand. With the Formulation-Exploration method, we can iteratively determine and update the CODP, which facilitate providing products to satisfy individual customer's requirements with near mass production efficiency. We define that the formulation-exploration framework is a dynamic, open meta-design scaffold, on which new, customized modules can be added to manage both domain-dependent and domain-independent design problems. The current four incorporated methods – cDSP, adaptive linear programming, weight sensitivity analysis, and system capacity analysis are examples to launch our practice for *satisficing* design.

We identify several directions for future work. In this paper, we focus on the Formulation-Exploration method instead of the supply-chain design results and the customization performance. In future work, we can improve the model formulation in multiple ways – by developing key performance index for customization, managing more goals, incorporating more players in a supply network, tackling the sourcing problem, etc.

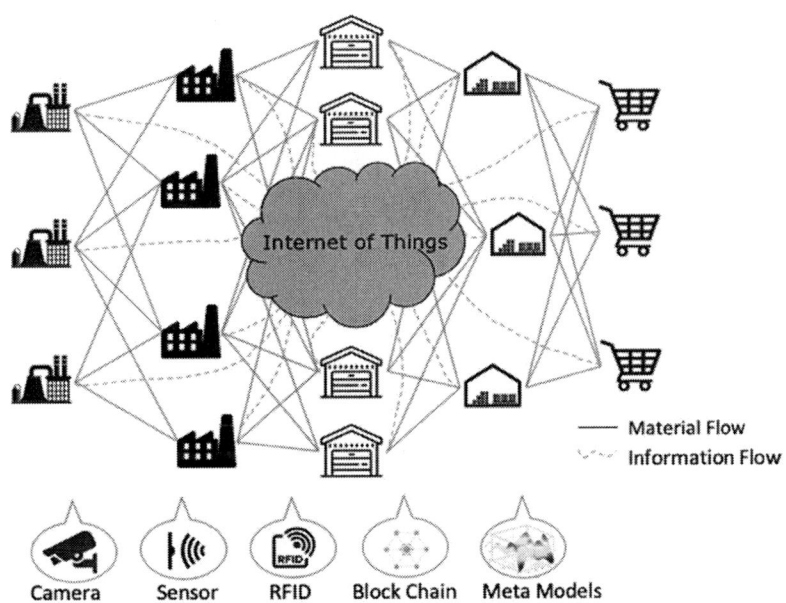

FIGURE 8 USING NEW TECHNOLOGY TO FACILITATE REAL-TIME FORMULATION-EXPLORATION

In addition, as digitalization continues, gathering real-time data from the customers operations along the supply chain can make the design more flexible and adaptive (Figure 8). By analyzing the data we can capture emergent properties of the supply chain and new technologies enrich our framework.

ACKNOWLEDGMENTS

Lin Guo acknowledges funding from the LA Comp Chair at University of Oklahoma. Suhao Chen acknowledges funding from the School of Industrial Engineering and Management at Oklahoma State University. Janet K. Allen and Farrokh Mistree acknowledge financial support from the John and Mary Moore and LA Comp Chairs at the University of Oklahoma.

REFERENCES

[1]   Parker, C. J., 2016, "Human Acceptance of 3D Printing in Fashion Paradox: Is Mass Customization A Bridge Too Far?," *WIT Transactions on Engineering Sciences*, 113, pp. 373-380.

[2]   Naylor, J. B., Naim, M. M., and Berry, D., 1999, "Leagility: Integrating the Lean and Agile Manufacturing Paradigms in the Total Supply Chain," *International Journal of Production Economics*, 62(1-2), pp. 107-118.

[3]   Jacobs, F. R., Chase, R. B., and Aquilano, N., 2004, "Operations Management for Competitive Advantage," *Boston: Mc-Graw Hill*, 64, p. 70.

[4]   Shidpour, H., Da Cunha, C., and Bernard, A., 2014, "Analyzing Single and Multiple Customer Order Decoupling Point Positioning based on Customer Value: A Multi-Objective Approach," *Procedia Cirp*, 17, pp. 669-674.

[5]   Rudberg, M., and Wikner, J., 2004, "Mass Customization in terms of the Customer Order Decoupling Point," *Production Planning and Control*, 15(4), pp. 445-458.

[6]   Iravani, S. M., Liu, T., and Simchi-Levi, D., 2012, "Optimal Production and Admission Policies in Make-To-Stock/Make-To-Order Manufacturing Systems," *Production and Operations Management*, 21(2), pp. 224-235.

[7]   Salvendy, G., 2001, *Handbook of Industrial Engineering: Technology and Operations Management*, John Wiley and Sons.

[8]   Teimoury, E., Modarres, M., Khondabi, I., and Fathi, M., 2012, "A Queuing Approach for Making Decisions about Order Penetration Point in Multiechelon Supply Chains," *The International Journal of Advanced Manufacturing Technology*, 63(1-4), pp. 359-371.

[9]   Zhou, W., Huang, W., and Zhang, R., 2014, "A Two-Stage Queueing Network on Form Postponement Supply Chain with Correlated Demands," *Applied Mathematical Modelling*, 38(11-12), pp. 2734-2743.

[10] Ghalehkhondabi, I., Ardjmand, E., and Weckman, G., 2017, "Integrated Decision Making Model for Pricing and Locating the Customer Order Decoupling Point of a Newsvendor Supply Chain," *Opsearch*, 54(2), pp. 417-439.

[11] Ghalehkhondabi, I., Sormaz, D., and Weckman, G., 2016, "Multiple Customer Order Decoupling Points within a Hybrid MTS/MTO Manufacturing Supply Chain with Uncertain Demands in Two Consecutive Echelons," *Opsearch*, 53(4), pp. 976-997.

[12] Ahmadi, M., and Teimouri, E., 2008, "Determining the Order Penetration Point in Auto Export Supply Chain by

the Use of Dynamic Programming," *Journal of Applied Sciences*, 8(18), pp. 3214-3220.

[13] Jeong, I.-J., 2011, "A Dynamic Model for the Optimization of Decoupling Point and Production Planning in a Supply Chain," *International Journal of Production Economics*, 131(2), pp. 561-567.

[14] Liu, W., Mo, Y., Yang, Y., and Ye, Z., 2015, "Decision Model of Customer Order Decoupling Point on Multiple Customer Demands in Logistics Service Supply Chain," *Production Planning and Control*, 26(3), pp. 178-202.

[15] Box, G. E., and Draper, N. R., 1987, *Empirical Model-Building and Response Surfaces*, Wiley New York.

[16] Talvitie, A., "Inaccurate or Incomplete Data as a Source of Uncertainty in Econometric or Attitudinal Models of Travel Behavior. In: New Horizons in Travel-Behavior Research," *Proc. The Fourth International Conference on Behavioral Travel Modeling*, pp. 559-575.

[17] Norman, G., 1990, "Production and Operation Management: A Problem–Solving and Decision–Making Approach," *USA.: Dryden Press*.

[18] Simon, H. A., 1996, The Sciences of the Artificial, *MIT Press*.

[19] Nellippallil, A. B., Mohan, P., Allen, J.K., Mistree, F., 2018, "Robust Concept Exploration of Materials, Products and Associated Manufacturing Processes," *ASME Design Automation Conference*, Quebec City, Canada, pp. V02BT03A010-V002BT003A010.

[20] Taguchi, G., and Clausing, D., 1990, "Robust Quality," *Harvard Business Review*, 68(1), pp. 65-75.

[21] Taguchi, G., Plains W., 1993, "Taguchi on Robust Technology Development: Bringing Quality Engineering Upstream," *ASME Press Series on International Advances in Design Productivity*, 191, pp. 1-191.

[22] Chen, W., Allen, J. K., Tsui, K.-L., and Mistree, F., 1996, "A Procedure for Robust Design: Minimizing Variations Caused by Noise Factors and Control Factors," *Journal of Mechanical Design*, 118(4), pp. 478-485.

[23] Isukapalli, S., Roy, A., and Georgopoulos, P., 1998, "Stochastic Response Surface Methods (SRSMs) for Uncertainty Propagation: Application to Environmental and Biological Systems," *Risk analysis*, 18(3), pp. 351-363.

[24] Choi, H.-J., Austin, R., Allen, J. K., McDowell, D. L., Mistree, F., and Benson, D. J., 2005, "An Approach for Robust Design of Reactive Power Metal Mixtures based on Non-Deterministic Micro-Scale Shock Simulation," *Journal of Computer-Aided Materials Design*, 12(1), pp. 57-85.

[25] Choi, H.-J., Austin, R., Shepherd, J., Allen, J. K., McDowell, D., Mistree, F. and Benson, D.,"An Approach for Robust Micro-Scale Materials Design under Unparameterizable Variability," *Proc. 10th AIAA/ISSMO Multidisciplinary Analysis and Optimization Conf.*, p. 4331.

[26] Choi, H.-J., Mcdowell, D. L., Allen, J. K., and Mistree, F., 2008, "An Inductive Design Exploration Method for Hierarchical Systems Design under Uncertainty," *Engineering Optimization*, 40(4), pp. 287-307.

[27] Nellippallil, A. B., Song, K. N., Goh, C.-H., Zagade, P., Gautham, B., Allen, J. K., and Mistree, F., 2017, "A Goal-Oriented, Sequential, Inverse Design Method for the Horizontal Integration of a Multistage Hot Rod Rolling System," *Journal of Mechanical Design*, 139(3), p. 031403.

[28] Choi, H., McDowell, D. L., Allen, J. K., Rosen, D., and Mistree, F., 2008, "An Inductive Design Exploration Method for Robust Multiscale Materials Design," *Journal of Mechanical Design*, 130(3), p. 031402.

[29] Mistree, F., Hughes, O. F., and Bras, B., 1993, "The Compromise Decision Support Problem and the Adaptive Linear Programming Algorithm," *Structural Optimization: Status and Promise*, MP Kamat, ed., AIAA, Washington, DC.

[30] Ahmed, S., Goh, C.-H., Allen, J. K., Mistree, F., Zagade, P., and Gautham, B., "Hot Forging of Automobile Steel Gear Blanks: An Exploration of the Solution Space," *Proc. ASME Design Automation Conference*, American Society of Mechanical Engineers, pp. V02BT03A003-V002BT003A003.

[31] Rezapour, S., Farahani, R. Z., and Pourakbar, M., 2017, "Resilient Supply Chain Network Design under Competition: A Case Study," *European Journal of Operational Research*, 259(3), pp. 1017-1035.

[32] Seada, H., and Deb, K., 2014, "U-NSGA-III: A Unified Evolutionary Algorithm for Single, Multiple, and Many-Objective Optimization," *COIN Report*, 2014022.

[33] Liu, W., Wu, R., Liang, Z., and Zhu, D., 2018, "Decision Model for the Customer Order Decoupling Point Considering Order Insertion Scheduling with Capacity and Time Constraints in Logistics Service Supply Chain," *Applied Mathematical Modelling*, 54, pp. 112-135.

[34] Liu, D., Wang, W., and Fu, W., "CODP Position of Leagile Supply Chain Based on Polychromatic Sets Theory," *Proc. IEEE International Conference on Automation and Logistics* (ICAL'09) IEEE, pp. 432-437.

[35] Xu, X., and Liang, Z., "CODP Positioning Based on Extension Superiority Evaluation Model," *Proc. 2011 International Conference on Electronic and Mechanical Engineering and Information Technology (EMEIT)*, IEEE, pp. 4041-4047.

[36] Daaboul, J., Da Cunha, C., Le Duigou, J., Novak, B., and Bernard, A., 2015, "Differentiation and Customer Decoupling Points: An Integrated Design Approach for Mass Customization," *Concurrent Engineering*, 23(4), pp. 284-295.

[37] Daaboul, J., Laroche, F., and Bernard, A., "Determining the CODP Position by Value Network Modeling and Simulation," *Proc. Technology Management Conference (ICE)*, 2010 IEEE International, IEEE, pp. 1-10.

[38] Wikner, J., Naim, M. M., Spiegler, V. L., and Lin, J., 2017, "IOBPCS based Models and Decoupling Thinking," *Intl Journal of Production Economics*, 194, pp. 153-166.

[39] Simon, H. A., and Kadane, J. B., 1975, "Optimal Problem-Solving Search: All-or-None Solutions," *Artificial Intelligence*, 6(3), pp. 235-247.

APPENDIX A SOME REPRESENTATIVE LITERATURE ON CODP POSITIONING – PART I

Appendix A is referenced in Section 1.2. In the literature of CODP positioning, optimization methods as well as the methods "borrowed" from other domains such as value network, tandem circuit design, and polychromatic set theory are used.

Author	Model	Solving method	Parameter — Dynamic / stochastic	Parameter — Static	Decision variables	Objective	Constraints	Observations or conclusions	Limitations or problematic assumptions
Teimoury et al., 2012 [8]	Queuing model	matrix geometric method; direct search		*	CODP position, inventory, product transporting vehicle	Single objective: Minimize total cost	Warehouse capacity, service level	Impatient customers for product arrival. The cost increases as demand arrival rate increases, and it is concave in increasing completed semi-finished goods percentage; capacity growing results in semi-finished percentage ascending.	Assume customer arrival follows Poisson distribution; do not consider the capacity limitation in customer queue
Zhou et al., 2014 [9]	Queuing model	matrix geometric method; grid search	*		CODP position, base-stock level	Single objective: Minimize total cost	Customer arrival distribution, queue length distribution, inventory capacity, customer order fill rate	Consider the effect of CODP on holding cost. Involve lead-time quotation policy and penalty cost for tardiness; the variance and correlation of demand increase the operation cost of the form postponement...	Assume customer orders arriving according to a Markovian arrival process
Jeong et al., 2011 [13]	General dynamic model	Optimal control theory	*		CODP position, production-inventory plan	Single objective: Minimize total cost of the deviation between actual and target production rate and inventory	Manufacturing capacity, relation among production rate, inventory level and demand rate	Overestimating demand ensures optimal decoupling; underestimating demand ensures zero inventory	Assume constrained demand of a product life cycle; do not consider time-vary penalty costs or delivery lead time
Shidpour et al., 2014 [4]	Multi-objective programming	ε-constraint method, simplex		*	CODP position	Bi-objective: Maximize total profit, maximize customer satisfaction	Time service, minimum value of profit objective $-1\ \varepsilon_h$	Multi-CODP is preferred to single-CODP for a product portfolio benefiting both the manufacturer and the	The two objectives are not normalized
Liu et al., 2018 [33]	Time scheduling model	generic algorithm		Having three degrees of indicators	CODP position after inserting a new order	Bi-objective: Minimize cost, maximize satisfaction	Incremental cost of new order, lead-time, changing value of the two objectives	Consider order insertion scheduling; CODP may change as new order comes	CODP is not in manufacturers; Linear weighting is used without scaling the objectives dynamically
Liu et al., 2009 [34]	Polychromatic set theory	color contour evaluation	*		Positioning indicators, influencing factors about members	Multi-level objective: Level 1: competitiveness of supply chain, Level 2: customer satisfaction, cost	Separation of resource, capacity of members	Information decoupling point and logistics decoupling point can be added into CODP model	A huge amount of information is required to process the modeling and reasoning

APPENDIX A SOME REPRESENTATIVE LITERATURE ON CODP POSITIONING – PART II

Author	Model	Solving method	Parameter: Dynamic / stochastic	Parameter: Static	Decision variables	Objective	Constraints	Observations conclusions	Limitations or problematic assumptions
Xu et al., 2011 [9, 35]	CODP selection model based on extension superiority evaluation model	Analytic Hierarchy Process (AHP)		*	CODP position, base-stock level	Single objective: Overall competitiveness (considering market, product, production and resource factors)	Correlation function, correlation degree, standardized correlation degree, priority degree	If CODP is moved towards downstream, generic phase is prolonged	Quantify the correlation degree, priority degree with heuristics
Liu et al., 2015 [14]	Non-conventional linear programming with discrete variables	Matrix transformation, solving multivariate equation set		*	CODP position, amount of logistics service	Single objective: Minimize total cost of logistics service integrator	Procedure constraints, lead-time constraints	Decreasing order-transferring and waiting cost do not switch CODP; order processing cost is decreased as CODP is moved to the last procedure, lead-time	Focus on static condition; customers' satisfaction is oversimplified
Ahmadi et al., 2008 [12]	Dynamic programming model	A four-step algorithm developed by the authors	Static in each stage but dynamic between stages		CODP position, module preparing time; assigned modules, early-arrival	Multi objective: Phase 1: Minimize holding cost, delivery delay; Phase 2: minimize number of modules not ready to be		Holding cost may occur to save transportation cost; inventory batch sizes before CODP should be larger than that after CODP	Assume demand is deterministic and certain; only consider goods exported from one country
Daaboul et al., 2010, 2015 [36, 37]	Value network model	Simulation SimulValor	Static but using scenarios		CODP positions, differentiation point position (PDP)	Single objective: Maximize overall value (cost, quality, delay, subjective and objective view, relevant parties)	Four types of value: financial and image of the enterprise, financial and service of customers	CODP influences delay, inventory and four types of value; PDP and CODP should be integrated in the decision making	No capturing the interrelations between the four types of value
Ghalehkhondabi et al. 2016, 2017 [10, 11]	Stochastic programming model	Optimization method and sensitivity analysis	*		CODP position, inventory, price	Single objective: Minimize total cost, or maximize total profit	Working time, order satisfied percentage, manufacturing capacity.	Variations in demand, unit waiting cost, holding cost affect CODP; insight on decision making is obtained in sensitivity	Assume Poisson-distributed demands; heuristics in price-demand
Wikner et al [38]	Tandem model forecast-driven and order-driven model	Simulation	Parameters with different scenarios		Inventory	Single objective: Minimize deviation between actual inventory and target inventory	Conservation of mass and materials, quantity-based and time-based measures	There is less feedback information than feed-forward information in the supply chain	Assume demand is deterministic without uncertainty

We identify two major limitations in the literature:
- Limitation 1 – There are assumptions in the distribution of non-deterministic methods [8-11].
- Limitation 2 – The optimization methods in [4, 8-14] are used to find optimal solutions, which are sensitive to uncertainties.

APENDIX B THE MATHEMATICAL FORM OF cDSP AND WHY IT DELIVERS SATISFICING SOLUTIONS

Appendix C is referenced in Section 3.

As models are incomplete and inaccurate, instead of searching for optimal solutions, we seek *satisficing* solutions [39], that is, solutions that are relatively insensitive to uncertainties associated with the models and the market demands.

To obtain a *satisficing* solution, we propose using the compromise Decision Support Problems (cDSP) [29]. We identify scenarios and exercise the cDSP to generate *satisficing* solutions and explore the solution space (ESS) to identify the best *satisficing* solution. We introduce the formulation method cDSP in Section 3.1 and the exploration method ESS in Section 3.2.

The mathematical form of the cDSP is shown as follows.

Given
System parameters
A_1: p × n matrix //The coefficients of p equality constraints
A_2: q × n matrix //The coefficient of q inequality constraints
b_1: p-dimension vector //The RHS of the p equality constraints
b_2: q-dimension vector //The RHS of the q inequality constraints
LB: n-dimension vector //The lower bound of the n variables
UB: n-dimension vector //The upper bound of the n variables
C_k: m-dimension vector //The coefficient of the kth goal, $k \in K$
X: n-dimension vector //n system variables
$G_k = C_k \times X$ //The achieved value of Goal k, as a function of system variables X
T_k: A number //The target value of the kth goal, $k \in K$
w_k: A number //The weight of the kth goal,
    $k \in K, 0 \leq w_k \leq 1, \sum_{k \in K} w_k = 1.$

Find
System variables
X: n-dimension vector //n system variables
Deviation variables
d_k^+, d_k^- //Over-achievement and under-achievement of Goal k, $k \in K$

Satisfy
System constraints
$EC_P = A_1 \cdot X - b_1 = 0$ //p equality constraints
$IC_q = A_2 \cdot X - b_2 \geq 0$ //q inequality constraints

System goals
$GOAL_k = \frac{G_k}{T_k} + d_k^- - d_k^+ - 1 = 0$ //Goal k, $k \in K$

Bounds
X ≥ LB //The lower bound inequality functions of X
−X ≥ −UB //The upper bound inequality functions of X

$d_k^- \geq 0, \forall k \in K$
$d_k^+ \geq 0, \forall k \in K$
$d_k^- \cdot d_k^+ = 0, \forall k \in K$ //A goal is either over-achieved or under-achieved
$\sum_{k \in K} w_k = 1$ //The sum of the weights is 1

Minimize
The deviation function
$Z = \sum_{k \in K} w_k \cdot (d_k^- + d_k^+)$ //The weighted sum of deviation variables

In the compromise DSP we minimize a deviation function Z, that is the distance between what we can achieve and what we wish to achieve. The deviation function is a linear sum of the deviation variables and is a measure of the extent to which we are able to achieve the goals that are functions of the system and deviation variables. By minimizing the deviation function we obtain a *satisficing* solution. *Satisficing* solutions contain the optimal solutions. *Satisficing* solutions meet the necessary Kuhn-Tucker condition but may not satisfy the sufficient Kuhn-Tucker conditions because they are not optimal. Optimal solutions meet both necessary and sufficient Kuhn-Tucker conditions (Figure 9). We introduce the ramifications of not meeting the sufficient Kuhn-Tucker conditions as follows.

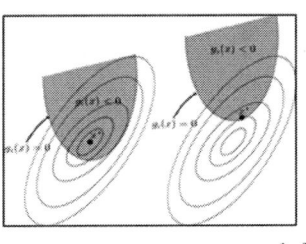

Optimizing
Both the necessary and sufficient Kuhn-Tucker conditions are met at the optimal solution.

(a)

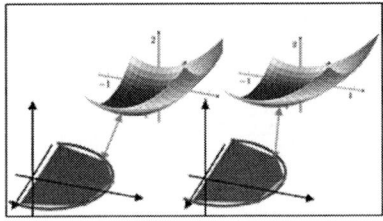

Satisficing
Only the necessary Kuhn-Tucker condition is met at a *satisficing* solution.

(b)

FIGURE 9 SATISFICING SOLUTIONS MEET THE NECESSARY KUHN-TUCKER CONDITION

In the cDSP, in addition to decision variables (system variables) $X = [x_1, x_2, \dots x_n]^T$, we use deviation variables $D = [d_1^-, d_1^+, d_2^-, d_2^+, \dots d_k^-, d_k^+]^T$ to measure the distance between the achieved value and the target value of the goals. D is a vector of variables that independent with X. We use $\mathbb{R}^n$ to represent the n-dimensional space of X, and use $\mathbb{R}^{n+2k}$ to represent the (n+2k)-dimension Space of $[X^T, D^T]^T$. The solution to a cDSP, $[X^{*T}, D^{*T}]^T$, is optimal in $\mathbb{R}^{n+2k}$. If we pick the system variables of the optimal solution, X^*, it is a *satisficing* solution in $\mathbb{R}^n$.

When the necessary Kuhn-Tucker condition is met, the gradient of the merit function is represented as the linear combination of the gradient of all constraints. When the sufficient Kuhn-Tucker conditions are not met, the second order sufficient conditions (SOSC) are not necessarily met.

There are ramifications of not meeting sufficient Kuhn-Tucker conditions. The deviation function Z is the weighted sum of deviation variables, therefore Z is only consists of D: $Z = \sum_{k \in K} w_k \cdot (d_k^- + d_k^+)$. Using the cDSP allows us to avoid the gradient varying as X within its range. In solution space $\mathbb{R}^n$, $\nabla Z(X) \equiv 0$. This means, if designers seek the *satisficing* solution, when any uncertainty happens to a model and the SOSC are lost, as long as the necessary conditions are still met (which is often likely to be so as $\nabla Z(X) \equiv 0$), the *satisficing* solution is still *satisficing*. Hence, the *satisficing* solutions are relatively insensitive to uncertainties.

How does *satisficing* solutions benefit supply chain? Since in supply chain design, the law of diminishing marginal utility applies to the profit versus the production volume increasing. If we seek the optimal solution, with the total profit P as an objective of the optimization model and the with production volume x_v as a design variable, then the objective $P(x_v)$ is a nonlinear function with x_v (Figure 10). So, $\frac{\partial P(x_v)}{\partial x_v}$ is a function of x_v. As the value of x_v changes (Type II uncertainty), $\frac{\partial P(x_v)}{\partial x_v}$ varies continuously so the optimal solution is not optimal anymore. Hence the optimal solution is sensitive to uncertainty. On the contrary, if we seek the *satisficing* solution, with the profit as a goal of the cDSP $\sim \frac{P(x_v)}{T_{profit}} + d_{profit}^- - d_{profit}^+ - 1 = 0$, the deviation function $\sim Z = d_{profit}^- + d_{profit}^+$, and with the production volume x_v as a system variable, then $\frac{\partial Z}{\partial x_v} \equiv 0$, and $\frac{\partial Z}{\partial d_{profit}^{\pm}}$ is either $\{\frac{\partial Z}{\partial d_{profit}^-} = 1, \frac{\partial Z}{\partial d_{profit}^+} = 0\}$ (See Figure 11-a) or $\{\frac{\partial Z}{\partial d_{profit}^-} = 0, \frac{\partial Z}{\partial d_{profit}^+} = 1\}$ (See Figure 11-b). The *satisficing* solution only alters once as x_v changes from $\frac{P(x_v)}{T_{profit}} \leq 1$ to $\frac{P(x_v)}{T_{profit}} \geq 1$. Hence the *satisficing* solution is relatively insensitive to uncertainty.

The nonlinearity due to diminishing profit margin and the uncertainty in productivity are simple examples to illustrate the relatively insensitivity of *satisficing* solutions. However, there are other nonlinearities and uncertainties – diminishing cost

margin, ladder price of transportation, corporate strategy adjustment, etc., which make it rather critical to eye on *satisficing*.

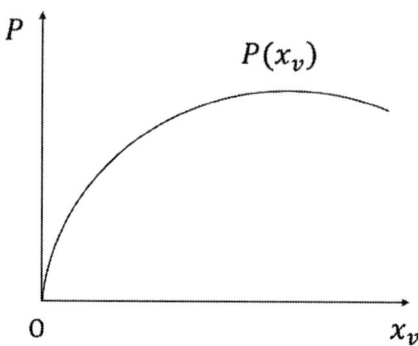

FIGURE 10 THE OBJECTIVE FUNCTION OF AN OPTIMIZATION PROBLEM

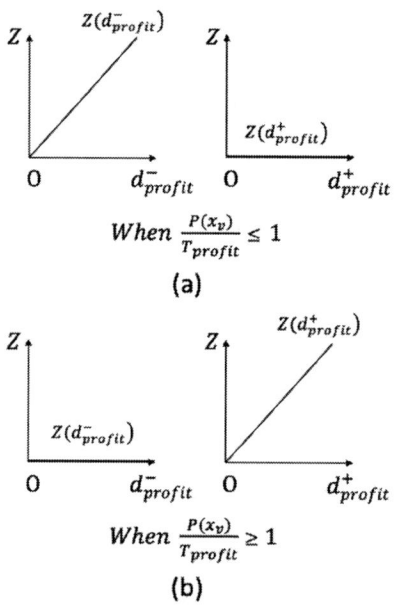

FIGURE 11 THE DEVIATION FUNCTION OF A CDSP

APPENDIX C CODP DESIGN MODEL

Appendix C is referenced in Section 4. It is the mathematical form of cDSP of the test problem of CODP design.

Given

rls	reliability of production of supplier
rlsm	reliability of transportation from supplier to manufacturer
rlm	reliability of production of manufacturer
rlmr	reliability of transportation from manufacturer to retailer
rlr	reliability of operation of retailer
rlrc	reliability of transportation from retailer to customers

$0.997 \leq rls, rlsm, rlm, rlmr, rlr, rlrc \leq 0.999997$
//All reliabilities are between 3σ and 6σ

SRM	CODP at raw material of supplier
SFG	CODP at finished goods of supplier
MRM	CODP at raw material of manufacturer
MFG	CODP at finished goods of manufacturer
RRM	CODP at raw material of retailer
RFG	CODP at finished goods of retailer
ts	production time in supplier
tsm	transportation time from supplier to manufacturer
tm	production time in manufacturer
tmr	transportation time from manufacturer to retailer
tr	operation time in retailer
trc	transportation time from retailer to customer
t	delivery time of this supply chain, $t = f_t(\text{CODP, ts, tsm, tm, tmr, tr, trc})$
VS	production volume of supplier $VS = f_{sm}(rls, rlsm, ts, tsm, SRM, SFG)$
VM	production volume of manufacturer $VM = f_{mr}(rlm, rlmr, tm, tmr, MRM, MFG)$
VR	preparation volume of retailer $VR = f_{rc}(rlr, rlrc, tr, trc, RRM, RFG)$
CS	capacity limit of supplier
CM	capacity limit of manufacturer
CR	capacity limit of retailer
tave	average delivery time of the industry
rlt	reliability of delivery time, $rlt = f_{rlt}(t, tave)$
sl	service level, $sl = rls \cdot rlsm \cdot rlm \cdot rlmr \cdot rlr \cdot rlrc \cdot rlt$
slmin	minimum service level of the industry
pmin	minimum unit price of the industry
pmax	maximum unit price of the industry
pave	average unit price of the industry
p	unit price of retailer's finished goods, $p = f_p(sl, slave, slmin, pave)$
D	market demand of the retailer's finished goods, $D = f_D(sl, slave, slmin, pave)$
hcsrm/scsrm/hcsfg/scsfg/hcmrm/scmrm/hcmfg/scmfg/hcrrm/scrrm/hcrfg/scrfg	unit holding / shortage cost in supplier's manufacturer's retailer's raw material / finished goods

pcs/pcm/pcr	unit production / operation cost of supplier manufacturer retailer
tcsm/tcmr/tcrc	unit transportation cost from supplier to manufacturer manufacturer to retailer retailer to customers
FAS/FAM/FAR	forecast accuracy of supplier manufacturer retailer
FSFG/FMFG/FRFG	forecast of the demand for finished goods at supplier manufacturer retailer (if it is MTS)

$FSFG = f_{SFG}(\text{CODP, FAS, rls, rlsm, p})$
$FMFG = f_{MFG}(\text{CODP, FAM, rlm, rlmr, p})$
$FRFG = f_{RFG}(\text{CODP, FAR, rlr, rlrc, p})$

Find
System Variables

VS	production volume of supplier
VM	production volume of manufacturer
VR	preparation volume of retailer
SRM	CODP at raw material of supplier
SFG	CODP at finished goods of supplier
MRM	CODP at raw material of manufacturer
MFG	CODP at finished goods of manufacturer
RRM	CODP at raw material of retailer
RFG	CODP at finished goods of retailer (SRM, SFG, MRM, MFG, RRM, RFG are binary variables)

Deviation Variables

d_i^-	under-achievement of Goal i
d_i^+	over-achievement of Goal i

Satisfy
System Constraints

SRM+SFG+MRM+MFG+RRM+RFG=1　　　C1
//There is one CODP in this supply chain

sl ≥ slmin　　　C2
//Service level is greater than or equal to the industry minimum service level

PTOT ≥ PTmin　　　C3
//Total profit is greater than or equal to the industry minimum profit

t ≤ tave　　　C4
//Delivery time of this supply chain is no more than the average of the industry

p ≤ pave　　　C5

//Unit price of retailer's finished goods is no more than the average of the industry

Systems Goals

$$\frac{PTOT}{T1} + d_1^- - d_1^+ = 1 \qquad \text{G1}$$

//To reach profit target

$$\frac{SL}{T2} + d_2^- - d_2^+ = 1 \qquad \text{G2}$$

//To reach service level target

$$V + d_3^- - d_3^+ = 0 \qquad \text{G3}$$

Bounds

$$SRM, SFG, MRM, MFG, RRM, RFG \in \{0, 1\}$$

//The CODP variables are Boolean variables.

$$0 \leq VS \leq CS$$

//Production volume of supplier cannot exceed its capacity limit

$$0 \leq VM \leq CM$$

//Production volume of manufacturer cannot exceed its capacity limit

$$0 \leq VR \leq CR$$

//Preparation volume of retailer cannot exceed its capacity limit

Minimize

$$Z = \sum_{i=1}^{3} w_i \cdot (d_i^- + d_i^+)$$

//Minimize the weighted sum of the deviation variables

We explore the design preferences and design capacity in iterations using the algorithm in Table 2. After three iterations, the design is improved to overcome the original bottleneck and the solutions are relatively insensitive to the migration of PLC phases and the changes in the market demand.

**Proceedings of the ASME 2019
International Design Engineering Technical Conferences
and Computers and Information in Engineering Conference
IDETC/CIE2019
August 18-21, 2019, Anaheim, CA, USA**

DETC2019-97680

WORD-OF-MOUTH RECOMMENDATIONS IN AN AUTOMOBILE MARKET SYSTEM

Amineh Zadbood*
School of Systems and Enterprises
Stevens Institute of Technology
Hoboken, New Jersey 07030
Email: azadbood@stevens.edu

Nicholas Russo
School of Systems and Enterprises
Stevens Institute of Technology
Hoboken, New Jersey 07030
Email: nrusso1@stevens.edu

Steven Hoffenson
School of Systems and Enterprises
Stevens Institute of Technology
Hoboken, New Jersey 07030
Email: shoffens@stevens.edu

ABSTRACT

Improving design in the context of market systems requires an understanding of how consumers learn about and evaluate competing products. Marketing models frequently assume that consumers choose the product with the highest utility, which provides businesses insights into how to design and price their products to maximize profits. While recent research has shown the impacts of consumer interactions within social networks on their purchasing decisions, they typically model market systems using a top-down approach. This paper applies an agent-based modeling approach with social network models to investigate the extent to which word-of-mouth (WOM) communications are influential in changing consumer preferences and producer market performance. Using a random network, we study the effects of the number of referrals for a product and the degrees of similarity between the senders and receivers of referrals on purchase decisions. In addition, the eigenvector centrality metric is used to analyze the spread of WOM referrals. The simulation results show that the most influential consumers in the network can create significant shifts in the market share, and a statistical analysis reveals a significant change in the system-level metrics of interest for the competing firms when WOM recommendations are included. The findings incentivize producers to invest in supporting their product development efforts with rigorous social networks analysis so as to increase their market success.

INTRODUCTION

A firm's success in maximizing their profits from a new product depends not only on their design and pricing decisions but also on heterogeneous consumer decision-making [1]. It is therefore critical for firms to study the factors that influence consumer decisions and gain insights on how to update their strategies to increase their market share and profits. Among the most influential factors is the effect that social networks have on purchase decisions. In fact, it has been shown that beyond the early stages of a product or service's life, the effects of marketing efforts such as advertising quickly diminish, and personal recommendations become the main force spreading product information [2, 3]. Many consumers informally communicate their product experiences to others without any incentive by the firms in question [3]. This is referred to as word-of-mouth (WOM) in the literature. WOM can be offline or online through social media and, depending on the experience with the product or service, can be positive, negative, or neutral.

The engineering design literature reports a growing attention to accounting for product information diffusion among consumers to support producer design decisions [4–6]. Recent work by Wang et al. proposed a multidimensional customer-product network for modeling customer preferences that overcomes the shortcomings of previous preference modeling approaches [5]. While their socio-technical system is comprised of interacting human actors as consumers, the framework does not allow for embodying adaptiveness and learning behavior of consumers and producers while they interact in the market. It also does not allow to trace back

---

*Address all correspondence to this author.

system-level behaviors to the individual decisions that generate them. Modeling market systems with multiple heterogeneous stakeholders, who learn over time from their networks, personal experiences, and feedback from the environment, calls for an approach capable of accommodating such complexities. Agent-based modeling (ABM) enables autonomous agents to explicitly adjust their behaviors based on the interactions and feedback, providing researchers opportunities to analyze the emergence of system-level patterns through micro-level modeling [1, 7]. ABM is a decentralized, bottom-up approach to simulate the behavior and decision-making of individuals, whose interactions with one another and the environment govern the overall system behavior [8].

ABM has several capabilities in modeling the influences of individuals on one another in social networks [9]. Zhang et al. models consumers, manufacturers, and policy agencies in an ABM framework to study the factors that influence the diffusion of alternative fuel vehicles (AFV) [6]. Their results indicate that WOM has a positive impact on the willingness of consumers to purchase AFVs. Hamill and Gilbert propose a framework to create agent-based social network models by considering several network properties and utilizing the concept of social circles [10]. The studies mentioned above support the use of ABM to study social influences, and the current study highlights the advantages of using an ABM framework to simulate how WOM recommendations governed by homophily influence producer market performance and design decisions. It is anticipated that such modeling can provide useful insights for designers and producers to estimate demand and derive strategies to gain higher profits.

This paper begins with a review of the relevant literature in design for market systems, word-of-mouth recommendations, network topologies, and social impacts influencing consumer utility. Next, an agent-based modeling approach is proposed to integrate these concepts into a single market systems model. This is examined using an automobile market system model, and the results show how adding WOM effects can change the predicted outcomes.

BACKGROUND
Design for market systems

Recognizing product development as a series of decision-making processes, recent research in the engineering design community seeks to support market-driven design decisions and improve product success [11, 12]. The field of design for market systems (DMS) emerged to incorporate not only consumer and firm perspectives in studying the implications of engineering design decisions in market systems, but also forces from the environment such as policies, regulations, and competition [13]. It addresses how engineering designs and market systems have impacts on one another and on

decision-making by consumers, firms, and governments [13]. DMS studies often include consumer choice models for demand analysis [14, 15], economic game theory models to simulate producer strategies [16, 17], and policy analyses to estimate the implications of their actions on the system [18, 19]. Recently, some studies suggest that complex customer-product relations should be viewed as a socio-technical system and analyzed using social network theory and techniques [4, 5]. Even though the key objective of DMS studies is to analyze these complex relations and support engineering design decisions, there are often multiple stakeholders such as businesses and policy makers who should be taken into account individually while understanding how their micro-level relations lead to macro-level behavior changes in the overall market system. Implementing this requires a systems modeling approach with modeling capabilities for a variety of individuals. This study shows how word-of-mouth communications affect the decision-making of consumers in an agent-based modeling framework, laying the foundation for future work to add policy makers and other stakeholders that may benefit from analyzing consumer relations in their social networks.

Network structures and characteristics

A social network is comprised of individual agents as nodes that are connected to one another by ties or links that represent their social interactions [20]. Determining the shape, structure, and characteristics of networks is a critical step that influences how accurate real-world social interactions among millions of people can be replicated by a hypothetical network structure. Several topologies for social networks have been proposed in the literature, such as: regular lattice, random, small world, and preferential attachment, also called scale-free. Figure 1 depicts these four basic models [21, 22]. The regular lattice model, which has the simplest structure, represents a network of limited size through which each node is connected to its immediate neighbors [20]. The number of neighbors a node is connected to via links is referred to as the node degree. Barbasi and Albert used the preferential attachment features and developed a scale-free network in which new nodes are drawn to create links with those nodes with high degrees [23]. This makes a few nodes have high degrees, while the vast majority of nodes have only a degree of one [21]. However, people are not always aware of who in the social network has a high degree, and, even if they know that, they might not be linked to it [21]. The theory of random networks was first introduced by the mathematicians Erdōs and Rényi [24]. In the network, each node is randomly connected to a number of neighbors, and different pairs of nodes are linked to one another with independent probabilities [24].

Comparing real and random networks yields insights into discovering social elements which stem from factors apart from randomness [25]. Examining the common properties of random lattice networks and their limitations, Watts and Strogatz noticed that there are many real-world networks that cannot be explained by those structures as they have high clustering and short paths [26]. This established the small-world network that bears similarities to lattice and random networks and overcomes their limitations [27]. Due to the prevalence of using random topology as a benchmark for analyzing consumer purchase decisions [28, 29], and its common features with small-world networks [30], this structure is employed in the present study. It allows for increasing our understanding of the effects of WOM referrals on consumer purchase decisions so as to support producer design decisions. This study lays the groundwork for future comparison studies using different structures in DMS problems.

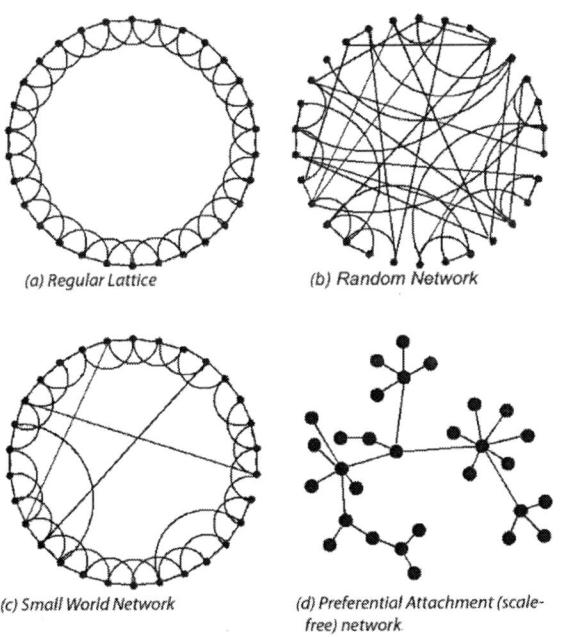

(a) Regular Lattice (b) Random Network

(c) Small World Network (d) Preferential Attachment (scale-free) network

FIGURE 1. Examples of Four Basic Models of Networks with 30 Nodes [21,22].

Given a network structure, social network analysis requires measuring various elements known as network characteristics [25]. One common characteristic is the centrality measure, which is used to identify the nodes of more influence, importance, or power [20, 25]. The key metrics to measure centrality are degree centrality, closeness centrality, betweenness centrality, and Eigenvector centrality [25, 31, 32]. Degree centrality reports the degree of each node but fails to account for the role of a node's position in the network on its high or low connectivity to others [25]. Closeness centrality measures how close a node is to every other node in the network. Betweenness centrality measures the amount of influence a

node has in connecting all the other nodes in the network. Eigenvector centrality is based on the idea that a node's influence is related to how influential its neighbors are. Since the social network analysis in this study revolves around the interactions between each node and its immediate neighbors, metrics which account for the distances between all nodes, such as betweenness centrality and closeness centrality, are much less effective than eigenvector centrality. As such, we use the established eigenvector centrality metric to analyze the impacts of the spread of offline WOM recommendations in consumer social networks on producer design decisions.

Word-Of-Mouth Recommendation

The majority of research related to the impact of social networks on market-based systems pertains to WOM diffusion of products and services. Researchers have studied various factors that influence whether a consumer purchase decision will be affected by referrals from sources within her social network [3]. The most prominent of these include homophily [33, 34], the effect of the strength of relational ties [33], customer satisfaction level of potential senders [3], and confusion within the consumer [35]. Homophily is the degree of similarity between a potential sender and receivers personal attributes [36], and ties between similar people tend to have higher likelihoods of activation as potential product referrers [34]. Brown and Reingren studied referral behaviors by taking into account the impacts of the strength of tie and homophily [33]. Their results indicate that the strength of ties has a significant role in WOM referrals. They discuss that even though their hypothesis regarding the overall impact of homophily was not well supported, future research is needed to understand the impact of homophily on WOM referrals. The present study investigates the influences of homophily on consumer preferences to understand how product purchases are impacted by WOM referrals, and how that information can be used in the context of DMS.

The analysis of WOM referral behaviors is typically used to build diffusion models, which have many applications in different contexts such as epidemiology and viral marketing [37]. The structural models that are formed arise from patterns of contacts within a social system and examine these complex relations where diffusion can be fostered through factors such as node centrality or the bridging of ties. Early models implicitly focused on micro level diffusion in small social groups, whereas newer proposed models tackle the macro level diffusion that occurs across the entire social structure as well. Up until the late 1990's, research was dominated by threshold and critical mass models that attempt to contextualize the spread of influence through both small and widespread networks [30]. On the other end of the continuum of diffusion models are the SIR (susceptible, infected, recovered) contagion models, which have been explored as an alternative to threshold models [38].

More recent research has led to the development of cascade models that are used alongside threshold models to become the standard in modeling complex social networks [2, 39]. The cascade model assumes that whenever a neighbor v of node u adopts a new innovation, then node u will adopt with a probability of $P(u,v)$ [37]. More recent research using empirical data to form networks (rather than assumed relationships) has found traditional linear and cascade models to ineffectively model the nature of true recommendations similar to that of SIR models. They incorrectly assume either a constant probability of converting every time a node interacts with a referrer or that one will convert once the fraction of their contacts who are infected exceed a threshold [37]. This realization led to the development of the acclaimed decreasing cascade model, which uses the greedy algorithm concept and assumes that the probability $P(u,v)$ of being influenced by a recommendation decreases over time [40]. The point of diminishing returns proves the expected number of influenced nodes to be a monotonic and submodular function of the targeted set, which is consistent with observed social network relations [40]. In this study, the diffusion of WOM recommendations is modeled though a similar approach to the decreasing cascade model [40].

Preference and social influence

The growing recognition that consumer decisions are not only influenced by product features but also by social influences has inspired considerable research in recent years [4,41]. Iyengar et al. investigate whether friends influence product purchases in an online social networking site [41]. They utilize empirical purchase data of 200 users and build a model using Bayesian methods. They assume the utility function for each user depends upon three influential factors: the indegree effect, or the number of connections from a node to others, the outdegree effect, or the number of connections directed to a node, and the social influence, or the exposure to the purchases of other contacts. Incorporating social influences in consumer choice models enabled the authors to test different hypotheses about how users are influenced and which users are more influenced. The results provide a classification of users in terms of their degree of influence and reveal a significant and positive relationship between the impacts of friends and purchase probability.

A similar approach was used by He et al. in which social influences within the network are captured and included in the consumer choice models to study the adoption of green vehicles [4]. Two network structures, small-world and scale-free network, are used and the link formation is based on the concept of homophily. They state that it is more likely to have a link between two nodes with shorter distance in their locations. The deterministic part of their utility function is shown in Eqn. (1).

$$U_{ik,t} = U(\beta : A_{ik}, S_i, E_i, N_{i,t}) \qquad (1)$$

Here, $U_{ik,t}$ is the utility of individual i for alternative k in the time period t. A_{ik} denotes desired product attributes. E_i captures the usage context attributes, and S_i is the profile attributes of consumer i. The social influence attribute is included by $N_{i,t}$. In this study, the utility function of each consumer is initially a function of the product attributes and the corresponding part-worth utilities, which later becomes updated by homophily-driven WOM referrals from neighbors as well as the number of referrals.

METHODOLOGY

In this study, an ABM is developed to simulate how producer market performance and design decisions are influenced by the spread of homophily-driven WOM recommendations in consumer social networks. To this end, two research questions are framed: (1) How do word-of-mouth recommendations governed by homophily influence producer market performance? and (2) How do word-of-mouth recommendations governed by homophily influence producer design decisions?

By market performance, we specifically refer to yearly profit and total sales in an automobile market, and the design decisions constitute fuel economy, performance, and price updates. The results are compared with that of a baseline scenario in which consumers are assumed to make purchase decisions independently [42]. Also, since offline means of communication have been found to constitute the majority of all WOM [43–45], we focus on person-to-person recommendations, not accounting for impacts from online sources such as social media.

Model structure

The model consists of two groups of agents: profit-maximizing producers and utility-maximizing consumers. The simulation begins in the year 2019 and runs for 30 years, enabling producers to analyze WOM impacts on their market success in the course of time. Three hypothetical producer agents initially set their first-year specifications for three vehicle attributes: fuel economy, performance (0-60 acceleration time), and price as shown in Tab. 1. In addition to the three producers, there are consumer agents who can generate or consume WOM referrals [46]. A random topology is used to generate consumers' social networks, where connected pairs have social contact with one another. Presently, the probability that any two consumers are connected is $\xi = 0.35$. This was selected after running the simulation several times varying the value of ξ, which did not reveal significant changes in the results. In

TABLE 1. Initial vehicle specifications.

Producers	Fuel economy (mpg)	Performance (0-60 mph time in sec)	Price (USD)
Producer 1	24	10	$18,000
Producer 2	18	6	$22,000
Producer 3	30	8	$26,000

this study, the structure of the social network is assumed to be stable, and the links are not added or deleted over time. All consumer agents are drawn from a sample of 184 individuals surveyed through a previous choice-based conjoint (CBC) study and utility estimation using Hierarchical-Bayes analysis [47]. This information includes individual part worth utilities for each consumer across the three vehicle attributes. Using conjoint analysis, a preference model is assigned to each consumer that maps product attributes to an individual utility in the form shown in Eqn. (2) [48,49].

$$U_{ij} = \sum_k \beta_{ik} x_{jk} \qquad (2)$$

Here, U_{ij} represents the initial utility of consumer i for product j, and β_{ik} denotes a parameter corresponding with consumer i's part-worth utility for attribute k, obtained from the survey data [47]. The parameter x_{jk} is a measure of attribute k in product j. In this case, the attributes represented by k include fuel economy, performance, and price. The distinction between the agent-based framework in this study and nearly all of the approaches in the DMS literature lies in the step following the preference modeling. The majority of the related work employs discrete choice modeling to build a choice probability from the utility function, and then demand is computed by multiplying the choice probability by the size of the market. Our ABM overcomes the limitation imposed by aggregating heterogeneous consumers and suggests counting the number of consumers that purchased a specific vehicle to find demand (quantity sold). This provides researchers much more flexibility to analyze the system-level behavior emerged from heterogeneous consumer interactions.

At the beginning of the simulation in 2019, consumers assess their perceived utility for the three vehicle alternatives on the market and make a purchase decision. Producer profit, π, is calculated as shown in Eqn. (3).

$$\pi = (p - c_t(f,a)) \cdot q(f,a,p) \qquad (3)$$

Here, f, a, and p represent fuel economy, performance, and price per unit, respectively. The total cost per unit, c_t, is a function of the design attributes and is shown in Eqn. (4). Here, the constants c_a and c_f are the costs associated with achieving a particular performance or fuel economy, which is defined for each producer and updated yearly. The total number of consumer purchases from the previous year is denoted by q and referred to as the quantity sold. The notion of a function of design attributes and price implies that the total purchases of a vehicle in the ABM framework depend directly on the perceived value of the level of each of the attributes from the consumer perspective.

$$c_t = c_f f + \frac{c_a}{a} \qquad (4)$$

Producers have a profit-seeking strategy to find the best combinations of the three attributes that maximize their profits. The way consumers change their purchase decisions provides insights for producers to modify their design and pricing strategies according to the yearly total sales and profit they earned at the end of a year. In this study, all three producers have the same strategy and consider three alternatives to update their attributes: They can effect a 2 percent improvement in either fuel economy, price, or performance in the next year. They calculate the expected profit earned in each situation and select the one that results in the highest forecasted profit. Following this, each producer adjusts its cost function by improving either or all of the cost function parameters. In cases where fuel economy yields the highest profit, c_f is improved by 2 percent. If performance is selected to improve next year, a 2 percent improvement is effected on c_a, and an improved price leads to a 1 percent improvement on both c_f and c_a.

As time passes, consumers gain experience with vehicles and WOM recommendations are spread into the social network, which have the effect of updating the utility function from Eqn. (2).

WOM-driven preference update

Over the years, consumers replace their vehicles with new ones. In this study, we assume a vehicle's age is the key driver of the re-purchase intention, and vehicle age enters the utility function with a negative coefficient. The decision-making process of a consumer willing to re-purchase a vehicle is influenced by several factors. Having an initial individual additive utility function in the form of Eqn. (2), she may be influenced by the referrals from her social network. When her current vehicle is of lower utility compared to the new vehicles in the market, her social connections recommend the vehicles they currently own, and she assesses the recommendations based on the number of referrals and the similarity between herself and the sender of WOM. This similarity is captured by a homophily effect governing the impacts of WOM referral on her purchase decision. Consistent with the literature, it is assumed that consumers put more trust in recommendations from people with whom they share more similarities [50–52]. Since the survey data include demographic information about the age and income bracket of agents, the age and income differences are calculated to create a continuous function of homophily, H_{in}, which measures the degree of similarity between consumer i and neighbor n. The homophily value ranges from 0, with the lowest level of similarity, to 1, having identical ages and income brackets. A_{ik} and I_{ik} in Eqn. (5) measure the absolute differences in age and income, respectively, between consumer i and neighbor k, who is connected to consumer i with a probability $\xi = 0.35$. The maximum age difference is 37 years and the maximum income difference is 14 levels. The updated utility function of consumer i for product j, $\hat{U}_{ij}$, after receiving the WOM referral is obtained by Eqn. (6).

$$H_{in} = 1 - (A_{in}/74 + I_{in}/28) \qquad (5)$$

$$\hat{U}_{ij} = U_{ij} + \sum_n H_{in} U_{nj} \gamma \qquad (6)$$

Here, U_{ij} and U_{nj} come from Eqn. (2) for the utility of an individual i and her neighbor n for product j, respectively. The summation on the right-hand side of Eqn. (6) is taken across all neighbors, and γ is a scaling factor reflecting the diminishing returns from the decreasing cascade model, where the change in the adoption rate decreases as more neighbors suggest it [40]. Recommendations may either reinforce a consumer's preference to purchase again from the company producing her current vehicle or raise her preference for alternative vehicles.

Figure 2 summarizes the proposed framework by which the effects of WOM referrals on consumer purchase decisions can be modeled. Smaller differences between the age and income level of the sender and receiver of WOM lead to a higher likelihood that a consumer's utility changes enough to result in a different purchase decision.

RESULTS

A simulation model was developed in Python 3 [53] to represent vehicle purchasing behavior of consumers impacted by their social network, linked with producer strategies to maximize profit and market share. Figures 3, 4, 5, 6, and 7 depict a comparison of sample simulation results when the models are executed for 30 years. In these plots, different colors distinguish the three producers, where Producer 1 enters the market offering the least expensive vehicle, Producer 2 the fastest, and Producer 3 designs the most fuel efficient one. In the baseline scenario, consumer agents are purchasing independently based on their utility in Eqn. (2) [42], which influence producer design decisions and market performance as illustrated in the figures with solid lines. The lines with markers show the effects after incorporating WOM impacts into the model, using Eqn. (6).

Comparing the profit and total sales improvements against the baseline scenario reveals how consumers, when presented with WOM referrals, become more willing to purchase the most fuel efficient vehicle, designed by Producer 3. This resulted in a high jump in the total sales and profits for Producer 3. An analysis of the ways the design decisions are updated over time sheds more light on the success of Producer 3 in the market, which, in this case, is a result of their persistence in improving either of the three attributes. As shown in Figure 5, the producers all increase their fuel economy in a similar fashion as when there were no WOM effects. However, in the presence of WOM effects, Producer 3 puts more efforts into reaching a high fuel economy and lets it remain stable as they find consumers are more willing to pay for a more fuel efficient vehicle when influenced by WOM. Producer 3 has also been more determined in improving its performance over time when observing the influence of WOM on consumer purchase decisions, depicted in Figure 6. Interestingly, the scenario with WOM referrals shows a substantial reduction in producer decisions to lower the prices of their vehicles, as shown in Figure 7.

The change in the dominant producer in the market after accounting for WOM recommendations in the model, as seen in Figures 3 and 4, calls for an analysis of the social network to realize which influential nodes are spreading their WOM into the network. To this end, we use the eigenvector centrality metric, measured for each node in the network by a value ranging from 0 to 1. A node with a value closer to 1 can spread her referral more efficiently, while the centrality decreases as the value is closer to 0. We identified the node with the highest eigenvector centrality across 50 independent runs of the simulation. The consumer with the highest eigenvector centrality owns a vehicle from Producer 3 the most frequently (58%), followed by Producer 2 (28%) and Producer 1 (14%). This reveals the efficiency and abundance of the spread of WOM referrals supporting Producer 3, which explains their success in increasing their market share.

Since the two models, with and without WOM, are stochastic in the generation of the random network and the assignment of age coefficients to individuals, a deeper investigation into the effects of WOM recommendations on consumer purchase decisions requires a statistical analysis across many executions. To achieve this, we performed 50 independent runs of each model to compare the differences in total sales of the producers. Then, in each execution, the two differences are paired to conduct a paired t-test for comparing two related

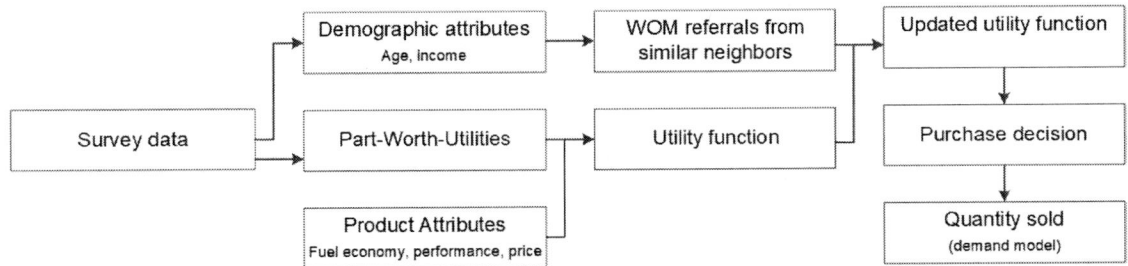

FIGURE 2. WOM referral effects in an ABM framework.

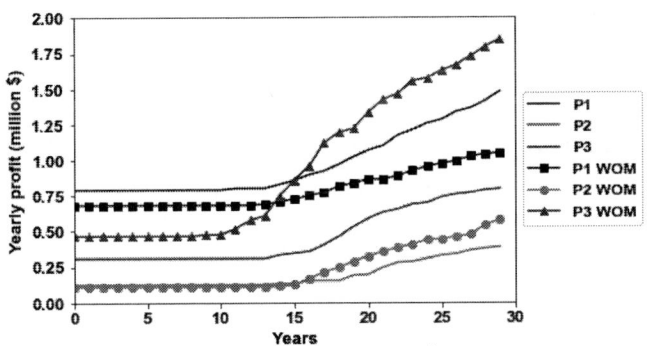

FIGURE 3. A comparison of the profit results for the three producers (P1-P3), with and without word-of-mouth (WOM) effects.

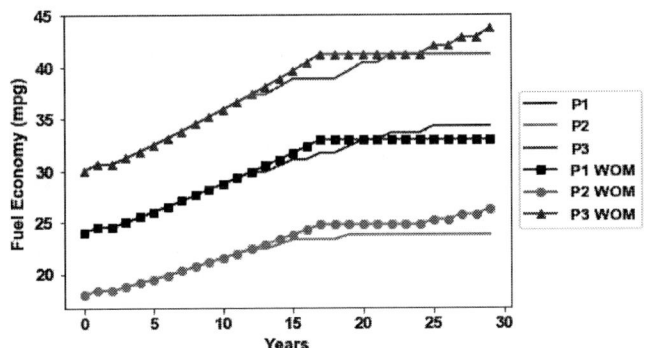

FIGURE 5. A comparison of the fuel economy results for the three producers (P1-P3), with and without word-of-mouth (WOM) effects.

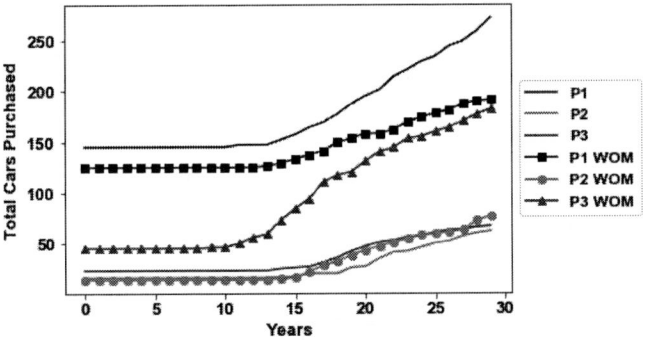

FIGURE 4. A comparison of the market share results for the three producers (P1-P3), with and without word-of-mouth (WOM) effects.

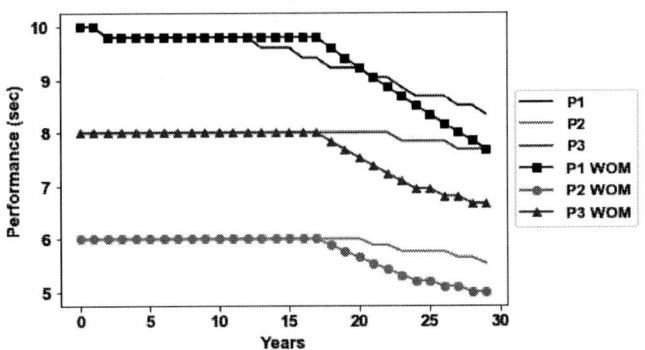

FIGURE 6. A comparison of the acceleration performance (seconds to go from 0 to 60 mph) results for the three producers (P1-P3), with and without word-of-mouth (WOM) effects.

system configurations, with and without WOM effects. The comparison metric subtracts the differences in the total sales of the model without WOM referrals from the one with WOM effects. The statistical analysis was conducted for the differences in the total sales of Producers 1 and 2, 1 and 3, and 2 and 3. The p-values calculated for the three comparisons were all below 0.001, indicating a statistically significant difference between the two models, despite the random network. The large

positive numbers for the mean of the comparison metric shows that the difference in the total sales of producers in the model without WOM has been larger than that of the model with WOM. This highlights the importance of including the effects of WOM when analyzing the impacts of consumer preferences in producer design and pricing strategies.

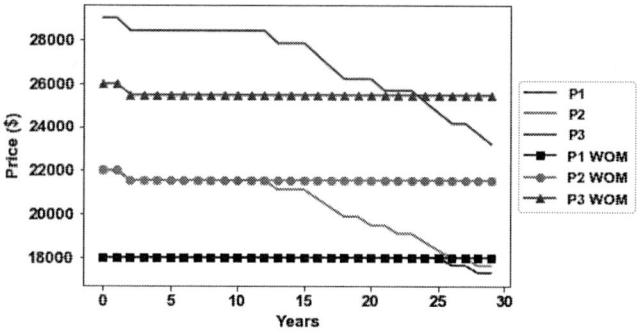

FIGURE 7. A comparison of the price results for the three producers (P1-P3), with and without word-of-mouth (WOM) effects.

DISCUSSION

This study proposes and tests an ABM framework to investigate the impacts of homophily-weighted WOM referrals on the preferences of automobile consumers and producer design decisions and market performance in the context of DMS studies. The bottom-up modeling approach of ABM provides the opportunity to study the decision-making and behavior of individuals while considering their interactions with other agents and the environment. This structure can lead to a better understanding of how the emergent behavior of the system has been influenced by the interactions between individuals, tracing back system-level phenomena to individual behaviors and decisions. The results of executing the stochastic ABM model with the random network 50 times show how the vehicles purchased by the more influential nodes are more successful than in a simulation that does not consider WOM influences.

This work opens many new possibilities for multi-disciplinary modeling to support engineering design and business decision-making. Two particular directions to extend this work include advancements in the ABM structure and the modeling of WOM. The ABM can be made more accurate by conducting new survey experiments that specifically elicit information about social interactions. The ABM can also be applied in the context of economic game theory models, which may more accurately represent how producers' strategic decisions are impacted by their competitors and how an understanding of consumer behaviors help them to improve their performance. Moreover, the current framework with a profit-seeking strategy for each producer demonstrates a simplification of the real-world situation in which producers solve a profit maximization problem every year to know how to update each of the attributes for next year. This step is in progress, where the current profit formulation is the objective function and the design attributes and price are variables.

A comparison of these results with simulations that use other network topologies such as scale-free networks would shed more light on the importance of incorporating social network analysis into market systems analyses. Also, in addition to the homophily metric influencing WOM and consumer purchase decisions, additional influencing factors may be considered, such as:

- Brand loyalty,
- How good or bad experiences with a product affect positive/negative referrer behavior,
- Consumers having different desires/likelihoods to follow the crowd or stand out as different, and
- The strength and weakness of ties.

This can also be accomplished with new survey experiments that elicit this information.

CONCLUSION

This paper presents the addition of word-of-mouth recommendations to an agent-based modeling framework for analyzing design for market systems problems. In particular, it shows the effects of homophily, or the level of similarity between the sender and receiver of a recommendation, on the market system, and it shows how these influences in a social network affect the system-level outcomes of market share and firm profits. The findings show that the most influential consumers can create substantial market share shifts through WOM, regardless of the random network seeding. The agent-based modeling framework enables users to view the emergent behavior of the entire system and trace the patterns to specific individuals in the complex network of consumers. By using this model, producers will be able to obtain insights to update their design and pricing strategies by understanding the impacts of WOM referrals on consumer preferences.

REFERENCES

[1] Wang, Z., Azarm, S., and Kannan, P. K., 2011. "Strategic Design Decisions for Uncertain Market Systems Using an Agent Based Approach". *Journal of Mechanical Design,* *133*(4), p. 041003.

[2] Goldenberg, J., Libai, B., and Muller, E., 2001. "Talk of the network: A complex systems look at the underlying process of word-of-mouth". *Marketing letters,* *12*(3), pp. 211–223.

[3] Anderson, W. E., 1998. "Customer Satisfaction and Word of Mouth". *Anderson EW. Customer satisfaction and word of mouth. Journal of service research. Aug;1(1):5-17.,* *1*(1), pp. 5–17.

[4] He, L., Wang, M., Chen, W., and Conzelmann, G., 2014. "Incorporating social impact on new product adoption in choice modeling: A case study in green vehicles". *Transportation Research Part D: Transport and Environment,* *32*, pp. 421–434.

[5] Wang, M., Chen, W., Huang, Y., Contractor, N. S., and Fu, Y., 2016. "Modeling customer preferences using multidimensional network analysis in engineering design". *Design Science,* **2**.

[6] Zhang, T., Gensler, S., and Garcia, R., 2011. "A study of the diffusion of alternative fuel vehicles: An agent-based modeling approach". *Journal of Product Innovation Management,* **28**(2), pp. 152–168.

[7] Borshchev, A., and Filippov, A., 2004. "From system dynamics and discrete event to practical agent based modeling: reasons, techniques, tools". In Proceedings of the 22nd international conference of the system dynamics society, Vol. 22, Citeseer.

[8] Wellman, M. P., 2016. "Putting the agent in agent-based modeling". *Autonomous Agents and Multi-Agent Systems,* **30**(6), pp. 1175–1189.

[9] Hamill, L., and Gilbert, N., 2015. *Agent-based modelling in economics.* John Wiley & Sons.

[10] Hamill, L., and Gilbert, N., 2009. "Social circles: a sample structure for agent-based social network models". *Journal of Artificial Societies and Social Simulation,* **12**(2), p. 3.

[11] Hazelrigg, G. A., 1998. "A framework for decision-based engineering design". *Journal of mechanical design,* **120**(4), pp. 653–658.

[12] Chen, W., Hoyle, C., and Wassenaar, H. J., 2012. *Decision-based design: Integrating consumer preferences into engineering design.* Springer Science & Business Media.

[13] Michalek, J. J., 2008. "Design for market systems". *Mechanical Engineering,* **130**(11), p. 32.

[14] Li, H., and Azarm, S., 2000. "Product Design Selection Under Uncertainty and With Competitive Advantage". *Journal of Mechnical Design,* **122**(December), pp. 411–418.

[15] Wassenaar, H. J., and Chen, W., 2003. "An approach to decision-based design with discrete choice analysis for demand modeling". *Journal of Mechanical Design,* **125**(3), pp. 490–497.

[16] Shiau, C.-S., and Michalek, J. J., 2008. "Should designers worry about market systems?". In ASME 2008 International Design Engineering Technical Conferences and Computers and Information in Engineering Conference, American Society of Mechanical Engineers, pp. 377–391.

[17] Shiau, C.-S., and Michalek, J., 2007. "a Game-Theoretic Approach To Finding Market Equilibria for Automotive Design Under Environmental Regulation". *ASME 2007 International Design Engineering Technical Conferences & Computers and Information in Engineering Conference,* pp. 1–9.

[18] Hawthorne, B. D., and Panchal, J. H., 2012. "Policy design for sustainable energy systems considering multiple objectives and incomplete preferences". In ASME 2012 International Design Engineering Technical Conferences and Computers and Information in Engineering Conference, American Society of Mechanical Engineers, pp. 253–267.

[19] Hoffenson, S., and Söderberg, R., 2015. "Taxation and Transparency: How Policy Decisions Impact Product Quality and Sustainability". *Journal of Mechanical Design,* **137**(10), p. 101702.

[20] Easley, D., Kleinberg, J., et al., 2010. *Networks, crowds, and markets,* Vol. 8. Cambridge university press Cambridge.

[21] Hamill, L., and Gilbert, N., 2010. "Simulating large social networks in agent-based models: A social circle model". *Emergence: Complexity and Organization,* **12**(4), pp. 78–94.

[22] Wilensky, U., 2005. "Netlogo preferential attachment model". *Center for Connected Learning and Computer-Based Modeling, Northwestern University, Evanston, Illinois, http://ccl. northwestern. edu/netlogo/models/PreferentialAttachment.*

[23] Barabasi, A.-L., and Albert, R., 1999. "Emergence of scaling in random networks". *Science (New York, N.Y.),* **286**(5439), oct, pp. 509–12.

[24] Erdös, and Rényi, A., 1959. "On random graphs, I". *Publicationes Mathematicae (Debrecen),* **6**.

[25] Jackson, M. O., 2008. *Social and economic networks.* Princeton University Press.

[26] Watts, D. J., and Strogatz, S. H., 1998. "Collective dynamics of small-world' networks". *Nature,* **393**(6684), jun, pp. 440–442.

[27] Bohlmann, J. D., Calantone, R. J., and Zhao, M., 2010. "The effects of market network heterogeneity on innovation diffusion: An agent-based modeling approach". *Journal of Product Innovation Management,* **27**(5), pp. 741–760.

[28] Kempe, D., Kleinberg, J., and Tardos, É., 2003. "Maximizing the spread of influence through a social network". In Proceedings of the ninth ACM SIGKDD international conference on Knowledge discovery and data mining - KDD '03, ACM Press, p. 137.

[29] Bohlmann, J. D., Calantone, R. J., and Zhao, M., 2010. "The Effects of Market Network Heterogeneity on Innovation Diffusion: An Agent-Based Modeling Approach". *Journal of Product Innovation Management,* **27**(5), jul, pp. 741–760.

[30] Valente, T., 1996. "Network models of the diffusion of innovations". *Computational and Mathematical Organization Theory,* **2**(2), pp. 163–164.

[31] Wasserman, S., and Faust, K., 1994. *Social network analysis : methods and applications.* Cambridge University Press.

[32] Borgatti, S. P., 2005. "Centrality and network flow". *Social*

Networks, **27**(1), jan, pp. 55–71.

[33] Brown, Jacqueline Johnson, and Reingen, P. H., 1987. "Social ties and word-of-mouth referral behavior". *ournal of Consumer research,* **14**(3), pp. 350–362.

[34] Mcpherson, M., Smith-Lovin, L., Cook, J. M., Mcpherson ', M., Smith-Lovin ', L., and Cook2, J. M., 2001. "Birds of a Feather: Homophily in Social Networks". *Source: Annual Review of Sociology,* **27**, pp. 415–444.

[35] Walsh, G., and Mitchell, 2010. "The effect of consumer confusion proneness on word of mouth, trust, and customer satisfaction". *European Journal of Marketing,* **44**(12), pp. 838–859.

[36] Rogers, E. M., 1983. *Diffusion of innovations.* NewYork.

[37] Leskovec, J., Adamic, L. A., and Huberman, B. A., 2007. "The dynamics of viral marketing". *ACM Transactions on the Web (TWEB),* **1**(1), p. 5.

[38] Watts, D. J., and Dodds, P. S., 2007. "Influentials, Networks, and Public Opinion Formation". *Journal of Consumer Research JOURNAL OF CONSUMER RESEARCH, Inc. ,* **34**(34), pp. 441–458.

[39] Kempe, D., Kleinberg, J., Tardos, É., and Tardos, E., 2015. "Maximizing the Spread of Influence through a Social Network". *ACM Classification: F,* **112**(423), pp. 105–147.

[40] Kempe, D., Kleinberg, J., and Tardos, É., 2005. "Influential nodes in a diffusion model for social networks". In International Colloquium on Automata, Languages, and Programming, Springer, pp. 1127–1138.

[41] Iyengar, R., Han, S., and Gupta, S., 2009. "Do friends influence purchases in a social network?". *Harvard Business School Marketing Unit Working Paper*(09-123).

[42] Zadbood, A., and Hoffenson, S., 2017. "Agent-based modeling of automobile producer and consumer behavior to support design for market systems analysis". In ASME 2017 International Design Engineering Technical Conferences and Computers and Information in Engineering Conference, American Society of Mechanical Engineers, pp. V02AT03A041–V02AT03A041.

[43] Lang, B., and Hyde, K. F., 2013. "Word of mouth: What we know and what we have yet to learn.". *Journal of Consumer Satisfaction, Dissatisfaction & Complaining Behavior,* **26**.

[44] Keller, E., 2007. "Unleashing the Power of Word of Mouth: Creating Brand Advocacy to Drive Growth". *Journal of Advertising Research,* **47**(4), dec, pp. 448–452.

[45] Keller, E., and Fay, B., 2009. "The Role of Advertising in Word of Mouth". *Journal of Advertising Research,* **49**(2), jun, pp. 154–158.

[46] Yang, S., Hu, M., Winer, R. S., Assael, H., and Chen, X., 2012. "An empirical study of word-of-mouth generation and consumption". *Marketing Science,* **31**(6), pp. 952–963.

[47] Hoffenson, S., Frischknecht, B. D., and Papalambros, P. Y., 2013. "A market systems analysis of the us sport utility vehicle market considering frontal crash safety technology and policy". *Accident Analysis & Prevention,* **50**, pp. 943–954.

[48] Green, P. E., and Srinivasan, V., 1978. "Conjoint analysis in consumer research: issues and outlook". *Journal of consumer research,* **5**(2), pp. 103–123.

[49] Michalek, J. J., 2005. "Preference coordination in engineering design decision-making". PhD thesis, The University of Michigan.

[50] Brown, J., Broderick, A. J., and Lee, N., 2007. "Word of mouth communication within online communities: Conceptualizing the online social network". *Journal of Interactive Marketing,* **21**(3), jan, pp. 2–20.

[51] Ruef, M., Aldrich, H. E., and Carter, N. M., 2003. "The Structure of Founding Teams: Homophily, Strong Ties, and Isolation among U.S. Entrepreneurs". *American Sociological Review,* **68**(2), apr, p. 195.

[52] Tang, J., Gao, H., Hu, X., and Liu, H., 2013. "Exploiting homophily effect for trust prediction". In Proceedings of the sixth ACM international conference on Web search and data mining, ACM, pp. 53–62.

[53] Van Rossum, G., and Drake Jr, F., 2003. "The python language reference manual: Network theory ltd, 2003". *URL http://www. network-theory. co. uk/python/language.*

**Proceedings of the ASME 2019
International Design Engineering Technical Conferences
and Computers and Information in Engineering Conference
IDETC/CIE2019
August 18-21, 2019, Anaheim, CA, USA**

DETC2019-97835

INFLUENCE OF OMITTED VARIABLES IN CONSUMER CHOICE MODELS ON ENGINEERING DESIGN OPTIMIZATION SOLUTIONS

Waleed Gowharji
Department of Mechanical Engineering
Carnegie Mellon University
Pittsburgh, PA, USA

Kate S. Whitefoot
Department of Mechanical Engineering
Department of Engineering & Public Policy
Carnegie Mellon University
Pittsburgh, PA, USA

ABSTRACT

This paper examines the impact of Omitted Variable Bias (OVB) within consumer choice models on engineering design optimization solutions. Engineering products often have a multitude of attributes that influence consumers' purchasing decisions, many of which are difficult to include in revealed-preference models due to a lack of data. Correlations among these omitted variables and product attributes included in the model can bias demand parameter estimates. However, engineering design optimization studies typically do not account for this bias. We examine the influence consumer-choice OVB can have on design optimization results. We first mathematically derive how OVB propagates into biased optimal design solutions and characterize properties of optimization problems that affect the magnitude of this bias. We then demonstrate the impact of OVB on optimal designs using a more-realistic engineering optimization case study of automotive powertrain design. In the demonstration, we estimate two sets of choice models: one using only "typically observed" vehicle attributes commonly found in the literature, and one with an additional set of "typically unobserved" attributes gathered from Edmunds.com. We find that the model with omitted variables leads to, in some scenarios, substantial bias in parameter estimates (5-143%) which propagates up to 21% bias in the optimal engine size.

INTRODUCTION

Engineering design research, particularly decision-based design (DBD) and design for market systems (DMS) research, incorporate consumer utility models into design optimization studies to determine profit-optimal designs for new and redesigned products [1–9]. Both stated-preference [10–17] and revealed-preference models using historical purchase data [1,4,6,7,18–21] are commonly used. There is also a recent growing body of work that leverages datamining of purchases to estimate consumer preferences [22–24]. One concern with stated

preference models is that they can substantially over- or under-predict consumer utility parameters due to behavioral biases such as social desirability [25,26]. The latter approaches avoid these behavioral biases by relying on purchase data, but may result in biased estimates from other sources. One key source is omitted variable bias (OVB).

OVB occurs when a product attribute of interest to a modeler is correlated with attributes that are not observed by the modeler because of missing data. Consider, for example, a vehicle choice model that includes the following attributes: fuel economy (a_1) and 0-60 mph acceleration (a_2). The estimates from such a model are biased if there exists an attribute that influences choice and is correlated with either of these attributes but is not accounted for in the model. For example, consumers may value leather seats (a_3). If vehicles that typically have leather seats also tend to have faster acceleration and lower fuel economy (e.g., sport packages), and leather seats are omitted from the model, the estimated parameters associated with acceleration and fuel economy will be biased. This can potentially lead to designs that favor attributes that do not reflect consumers' true preferences (e.g., larger engines with faster acceleration than is optimal in this case).

Because many products have numerous attributes that are typically unobserved, estimating demand models without these attributes can lead to significant impacts on engineering design decisions. For example, in vehicle choice models, not only are leather seats typically unobserved, but also attributes such as climate control, backup cameras, and turning radius [18,27]—all of which could be potentially correlated with attributes of interest to engineering designers (e.g., acceleration performance, fuel economy, dimensions). This is a cause of concern to engineering designers because it makes it very difficult to correctly estimate consumer preference parameters, and more importantly, these biases can propagate through to the optimal design solution.

Copyright © 2019 ASME

It is important to note that OVB is a distinct concept from the goodness-of-fit of a model. Whereas goodness-of-fit of a choice model is concerned with how well the model's choice predictions match observed data, OVB is concerned with properly estimating the influence of variables-of-interest (e.g., product design variables) on consumer choice. The issue with OVB is that it mis-attributes the effect of missing variables to other variables included in the model. The resulting biased parameters can mislead design optimization problems by over- or under-estimating the effect of a variable on consumer choice.

This paper examines the propagation of OVB in consumer choice models to design optimization solutions. First, we analytically derive how OVB influences optimal design solutions in a generic profit maximization problem, and characterize properties that affect the magnitude of the propagated bias on the optimal solution. We then demonstrate the influence of OVB in a case study of a redesigned light-duty automobile as it is a frequently studied product in the DMS literature [1,5,18,19]. It is also an example of an engineered product with many attributes that influence consumer choices that are difficult for researchers to collect data on. In this study, we collect data on typically observed and unobserved vehicle attributes in consumer utility models in the engineering design literature. While the list of attributes we include is more extensive than existing literature, we anticipate that there are additional vehicle attributes that influence consumer choice that we are unable to collect data for (e.g., sunroofs). In order to ensure that we know the "true" parameter values from which we can measure the magnitude of the OVB caused by the "typically unobserved" variables, without confounding any additional biases, we generate synthetic choice data where the "true" values of utility parameters are known. Each choice model—with and without OVB—is then used separately to find optimal powertrain variables that maximize profit.

We find that using the typically observed attributes in the literature leads to a bias that varies from 3% to 21% in the engine size optimization results, which leads to between approximately 6% and 14% lower profits than if the engine was designed to the unbiased "true" optimum. We further find that in our data sample for model-year 2016 vehicles, cargo capacity and turning circle have a large influence on OVB and the optimal engine size, as they are correlated with acceleration, which is a key attribute influenced by engine size. In addition to the vehicle-specific case study, we mathematically characterize how OVB impacts design optimization solutions for more general application to other products. We find that the magnitude of the propagated bias to the optimal solution depends not only on the correlation of the omitted variables with consumer choice and attributes that are influenced by design variables, but also the derivative of the cost function. For a cost function with a convex functional form, the smaller the derivative with respect to the design variable, the larger effect OVB will have on the optimal solution.

LITERATURE REVIEW

A notable challenge when determining profit-optimal engineering design variables is that, for many engineered products, there are a wide variety of attributes that are important for consumer choice but are difficult or impossible to obtain data on [6,18,19,27]. As an illustrative example, Table 1 lists the attributes that are included and those that are omitted in optimal design studies with revealed-preference choice models using sales data [1,5–7,27–30]. As the table demonstrates, many attributes are omitted from these models, even though they are identified among the most desirable attributes consumers consider when purchasing vehicles from surveys conducted by research firms such as Consumer Reports [31–33].

Table 1: Included and omitted attributes in a selection of engineering design literature using revealed-preference consumer choice models.

Attribute	[6]	[27]	[28]	[7]	[30]	[1]	[5]	[29]
Price	✓	✓	✓	✓	✓	✓	✓	✓
Fuel Economy	✓	✓	✓	✓	✓	✓	✓	✓
0-30 Acceleration	Omitted	Omitted	Omitted	Omitted	✓	Omitted	Omitted	Omitted
0-60 Acceleration	✓	✓	✓	✓	Omitted	✓	Omitted	Omitted
Horsepower	Omitted	Omitted	✓	Omitted	Omitted	Omitted	✓	Omitted
Weight	Omitted	Omitted	✓	✓	Omitted	Omitted	✓	Omitted
# Cylinders	Omitted	Omitted	✓	Omitted	Omitted	Omitted	Omitted	Omitted
Torque	Omitted	Omitted	✓	Omitted	Omitted	Omitted	Omitted	Omitted
# Valves	Omitted	Omitted	✓	Omitted	Omitted	Omitted	Omitted	Omitted
Footprint	✓	✓	Omitted	✓	Omitted	Omitted	Omitted	Omitted
Classification	Omitted	✓	Omitted	Omitted	Omitted	Omitted	✓	Omitted
Luxury	Omitted	✓	Omitted	Omitted	Omitted	Omitted	Omitted	Omitted
Size	✓	Omitted	Omitted	Omitted	Omitted	Omitted	✓	Omitted
Turning circle	Omitted	Omitted	Omitted	Omitted	Omitted	Omitted	Omitted	Omitted
wheelbase	Omitted	Omitted	Omitted	Omitted	Omitted	Omitted	Omitted	Omitted
premium leather	Omitted	Omitted	Omitted	Omitted	Omitted	Omitted	Omitted	Omitted
Maximum cargo capacity	Omitted	Omitted	Omitted	Omitted	Omitted	Omitted	Omitted	Omitted

The influence of OVB on parameter estimation is extensively covered in the statistics and econometrics literature [34–37] for both parametric and non-parametric models [38,39]. This literature has developed strategies for addressing OVB, including using instrumental-variables strategies [35,40–42]. The engineering design literature, however, has largely not addressed OVB in choice model estimation or design optimization. Recently, Frischknecht et al. [5] sought to address OVB with respect to only the coefficient on price, and Whitefoot et al. [6] sought to address OVB with respect to price, fuel economy, and acceleration.

Existing literature has found that omitted variables in experimental tests can substantially affect technical performance evaluations. For example, evaluations of how much running an "optimizing compiler" speeds up the execution of a computer program has been found to be biased by up to 10% by omitting variables such as environment size and link order [43]. This is enough to conclude that one optimizing compiler is better than another when it is not [43].Existing literature, however has not investigated the propagation of OVB into product design optimization solutions.

In order to address the gap in the literature, this paper studies OVB in a design optimization context. The paper's results contribute to the literature by (1) characterizing how OVB propagates to optimization solutions, which can allow researchers to *a priori* anticipate which variables may cause propagated bias and the direction of the bias, and (2) quantifying in a case study the impact of OVB on optimization solutions.

CHARACTERIZATION OF PROPAGATED BIAS

This section reviews how omitted variables influence discrete choice parameter estimates and describes how this bias propagates to design optimization solutions. We focus on a multinomial logit utility model estimated using maximum likelihood (MLE), as this is a common approach in the engineering design literature [1,6,29,44]. To derive a closed-form solution to the optimal design problem, we use a simple case where there is one design variable and two product attributes (one continuous and one discrete), no constraints, and consumers are modeled as choosing between the single designed product and an outside good. These assumptions are relaxed in the following sections where we characterize the bias based on properties of the optimization problem and investigate a case study with a large set of discrete and continuous attributes subject to engineering constraints, and with many competing product alternatives.

Let consumer utility for a product be expressed as: $U_{ij} = \beta x_j + \alpha \omega_j + \varepsilon_{ij}$, where x_j is a continuous product attribute, ω_j is a binary discrete product attribute, β and α are utility coefficients, and ε_{ij} is the error term, which is assumed to be type 1 extreme value distribution [3,45]. Let y be the outcome variable indicating whether a consumer chooses product j where y=1 if and only if $U_{ij} \geq U_{ik} \ \forall \ k \neq j$. If ω_j is omitted from the estimation due to lack of data or convenience, its contribution to utility will be absorbed into the error term, such that the error becomes: $\tilde{\eta} = \alpha \omega_j + \varepsilon_j$ [46]. This results in a truncated form of utility: $\tilde{U}_j = \tilde{\beta} x_j + \tilde{\eta}$. In the case where the choice set consists of two products, $j = \{1,0\}$, the unbiased MLE estimate of β is defined by *equation 1*:

$$\hat{\beta} = \frac{\ln(n_1/n_0)}{x_1} \tag{1}$$

where n_1 is the number of times the product $j = 1$ is chosen, and n_0 is the number of times the outside good is chosen. The biased coefficient $\tilde{\beta}$ can be described following Lee [34] as:

$$\tilde{\beta} = \frac{\ln\left(\frac{n_1}{n_0}\right)}{x_1} + \ln\left(\frac{1+e^\psi}{1+e^{\psi+\alpha}}\right) - \ln\left(\frac{1+e^{\psi+\delta}}{1+e^{\psi+\delta+\alpha}}\right) \tag{2}$$

We define b as the bias, which can be seen from the difference between *equation 1* and *2*:

$$b \equiv \ln\left(\frac{1+e^\psi}{1+e^{\psi+\alpha}}\right) - \ln\left(\frac{1+e^{\psi+\delta}}{1+e^{\psi+\delta+\alpha}}\right) \tag{3}$$

The parameters ψ, and δ in the above equations are defined per Lee [34]:

$$\ln\frac{P(\omega=1|x,y)}{P(\omega=0|x,y)} = \psi + x\delta + y\alpha \tag{4}$$

Equation 4 represents the log odds of the probability that the omitted variable is equal to 1 or 0 conditional on the observed

values of x and y. ψ is the intercept; δ is the coefficient on x, the magnitude of which increases with the correlation between x and z; and α is the coefficient on y, the magnitude of which increases with the correlation between y and z. The bias term b goes to zero if, conditional on the response variable y, x and ω are independent [34,47], causing no propagated bias into the optimal design solution. This is the case whether x is continuous, discrete, dichotomous, or polychotomous [34]. Here we focus on a continuous variable.

We wish to see how estimates $\hat{\beta}$ and $\tilde{\beta}$ from *equations 1* and *2* propagate into design optimization solutions. Here, we derive the optimum of a profit function for a single product with a fixed price and a cost function that depends on our design variable of interest, x. For profit maximizing firms, the optimization is specified in *equations 5* through *8*:

$$\max_{\text{wrt } x} \pi = MP(x)\big(p - c(x)\big) \tag{5}$$

$$P(x) = \frac{e^{\beta x + \mu p}}{1 + e^{\beta x + \mu p}} \tag{6}$$

where, π are the firm profits, which is a function of the design variable z, the manufacturing costs, $c(x)$, price of the product, p, and the consumer choice probability, $P(x)$. The biased form of firm profits are as follows:

$$\max_{\text{wrt } x} \tilde{\pi} = M\tilde{P}(x)\big(p - c(x)\big) \tag{7}$$

$$\tilde{P}(x) = \frac{e^{(\beta+b)x + \mu p}}{1 + e^{(\beta+b)x + \mu p}} \tag{8}$$

The first order necessary condition (FONC) of the optimal solution, x^*, is defined by *equation 9*:

$$FOC: \begin{cases} P_j^* = -\left(\frac{\partial P}{\partial p}\right)^{-1} PM + c(x)M \\ \dfrac{\partial c}{\partial x} = -\left(\dfrac{\partial P}{\partial p}\right)^{-1} \dfrac{\partial p}{\partial x} \end{cases} \tag{9}$$

For example, for the cost function $c(x) = hx^2$, the optimal design solution is $x^* = \beta/2h\mu$. If the estimator of β is biased, this bias propagates into the necessary condition for the optimal solution as described by *equation 10*:

$$\tilde{x}^* = \frac{\beta + b}{2h\mu} = \frac{\beta + \ln\left(\frac{1+e^\psi}{1+e^{\psi+\alpha}}\right) - \ln\left(\frac{1+e^{\psi+\delta}}{1+e^{\psi+\delta+\alpha}}\right)}{2h\mu} \tag{10}$$

For the cost function $c(x) = e^{kx}$, the optimal design solution is $x^* = 1/k \ln(\beta/\mu)$. If the estimator of β is biased, this bias propagates into the necessary condition for the optimal solution as described by *equation 11*:

$$\tilde{x}^* = 1/k \ln\left(\frac{\beta + \ln\left(\frac{1+e^\psi}{1+e^{\psi+\alpha}}\right) - \ln\left(\frac{1+e^{\psi+\delta}}{1+e^{\psi+\delta+\alpha}}\right)}{\mu}\right) \tag{11}$$

These properties can be useful for modelers to gain *a priori* intuition about the magnitude and direction of the propagated bias. In the following section, we characterize properties of the propagated bias depending on the parameter values in *equation 4*.

Direction of Propagated Bias

From *equation 10*, we would expect the value of $\tilde{x}^*$ to increase as b increases. This also suggests that an increase in both δ *and* α leads to an increase in $\tilde{x}^*$. In the case where the α and δ terms are both negative or both positive, then the propagated bias will be upward ($\tilde{x}^* > x^*$) pushing the optimal design variable to be larger than the true optimum. If these terms have opposite signs, then the propagated bias will be downward ($\tilde{x}^* < x^*$). The implication of this relationship is that (if $\omega = 1$ is preferred by consumers over $\omega = 0$), the more the omitted attribute is correlated with the design variable, the more the optimization solution will be a larger magnitude than the true optimum. And, conversely, the more anticorrelated the omitted attribute and design variable are, the more the optimization solution will be smaller than the true optimum. (If consumers prefer $\omega = 0$ over $\omega = 1$, then the directions flip.)

Magnitude of Propagated Bias

Equation 11 implies that, intuitively, the larger the absolute value of β is relative to the absolute value of the bias, the smaller the magnitude of the propagated bias will be. The magnitude of the propagated bias also depends on the cost function. Consider a convex functional form of a cost function with a constant second derivative with respect to the design variable (as is the case with $c(x) = hx^2$). The smaller the second derivative, the larger the magnitude of the propagated bias will be. This is because the bias propagates into the model's estimate of how much consumers are willing to pay for an increase in the design variable. If increasing the design variable incurs relatively little cost, then the bias will mislead designers to larger values of the design variable than if the cost change were greater.

Influence of Constraints

The presence of constraints in a design optimization problem will also affect how OVB is propagated into the design solutions. Here, we illustrate the effects constraints have on the propagated bias in simple, idealized cases to gain an intuition of basic properties of these effects. The next section presents a case study of a more-realistic engineering design optimization, which shows that these properties can help to explain results seen in non-idealized cases.

In cases where OVB exists, modelers only observe the biased optimal design solution because, by definition, they do not have the omitted variables that are needed to determine the true optimum. However, as we previously discussed, modelers can anticipate the direction of the propagated bias if they know (or correctly guess) the sign of the correlation between the omitted variable and the design variable, and the correlation between the omitted variable and consumer choice. If no constraints exist between the biased solution and the opposite direction, then the constraints will not affect the propagated bias. In a single dimensional design optimization problem, as discussed above, it is easy to see that if a constraint exists in between the unbiased and the biased solution, then the constraint can only have the effect of reducing the propagated bias into the optimal design solution. However, we will show that for multi-dimensional problems, this is not necessarily the case.

Figure 1 shows a multidimensional contour diagram of an unbiased "true" profit objective functions (left), and one with propagated OVB (right). We define an angle θ that indicates the angle between the direction of translation of the unbiased optimum to the biased optimum and the engineering constraint. For convenience, let the right side of the constraint be infeasible so that when θ is between (0,180), the biased minimum is infeasible and when θ is between (180,360), the unbiased minimum is infeasible.

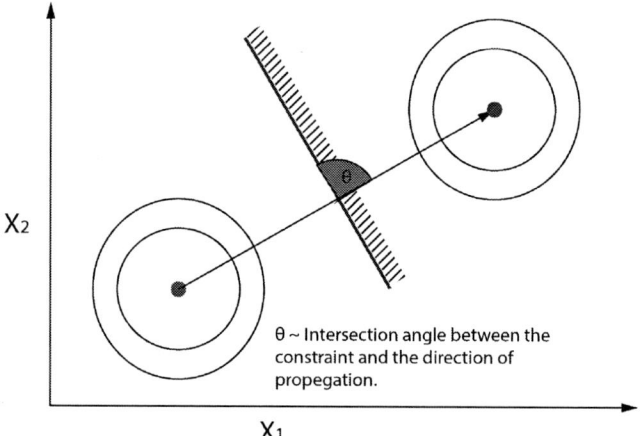

Figure 1: The angle θ, as the intersection angle between the direction of translation due to OVB and the constraint. Depending on this angel, the propagation of bias to the optimum design solution can either be exaggerated or suppressed.

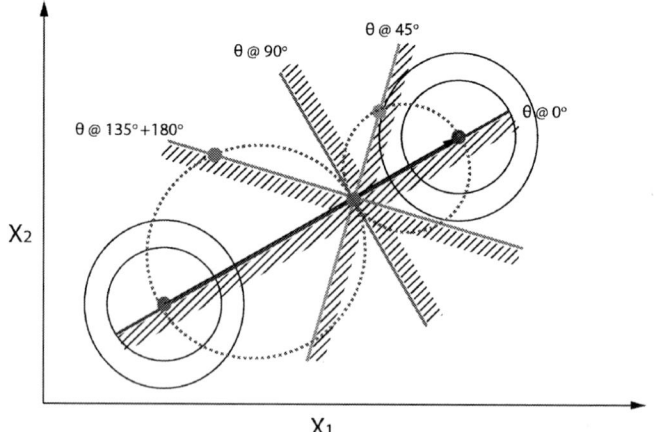

Figure 2: The change in the biased optimum follows a circular path from 0 to 180 degrees when the biased region is actively constrained. Beyond 180 degrees, the unbiased region becomes active, and the unbiased optimum follows a separate circular path.

Now, we look at the case were the direction of the propagated bias has an associated angle, as depicted in Figure 2. In the stylistic case where the contours of the profit functions are concentric circles, the optima follow circular paths as θ increases. In this case, we note that when the intersection angle, θ, is either 0 degrees or 90 degrees, there is no change in the value of x_2^*, so $\tilde{x}_2^* = x_2^*$. Similarly, when the intersection angle is $0 < \theta < 45$ or $135 < \theta < 180$, we find that the value of $|\tilde{x}_2^* - x_2^*|$ increases, while $\tilde{x}_2^* > x_2^*$. When the intersection angle is $90 < \theta < 135$ or $45 < \theta < 90$, we find that the value of $|\tilde{x}_2^* - x_2^*|$ decreases, while also $\tilde{x}_2^* > x_2^*$. Figure 3 depicts the trajectory of the difference in the optimal engineering solution $|\tilde{x}_2^* - x_2^*|$ as a function of the intersection angle θ.

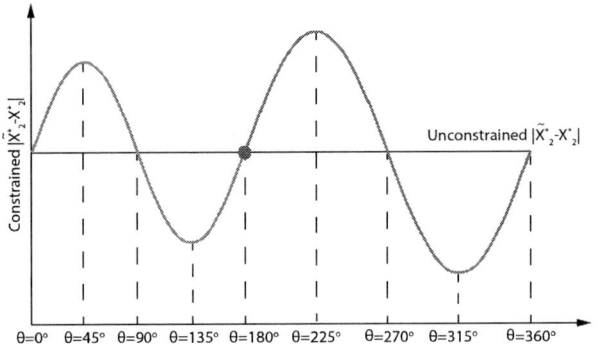

Figure 3: Changing the intersection angle θ from 0 to 360 will cause a cyclical change in the constrained difference of $\tilde{x}_2^* - x_2^*$ around the unconstrained difference $|\tilde{x}_2^* - x_2^*|$.

OPTIMIZATION CASE STUDY

To reveal the impact of OVB in a more realistic engineering design optimization problem, we conduct a powertrain redesign study where a manufacturer maximizes profits with respect to the Engine Displacement (ED) and Final Drive Ratio (FDR) in the transmission of one of its vehicles. In this optimization, we include a consumer choice utility model with two functional forms, one without OVB and one with OVB:

1. The first functional form includes vehicle attributes that are typically observed in the literature (x_j) in addition to attributes that are *typically unobserved* in the literature (ω_j):

$$u_{ij} = \boldsymbol{\beta} \, x_j + \boldsymbol{\alpha} \, \omega_j + \varepsilon_{ij} \quad (12)$$

2. The second functional form contains only the typically observed attributes (x_j). This form is specifically constructed to include OVB:

$$u_{ij} = \boldsymbol{\beta} \, x_j + \varepsilon_{ij} \quad (13)$$

In both *equations 12* and *13*, utility takes on a multinomial logit form where ε_{ij} are independent and distributed according to the type I extreme value distribution. We use synthetically generated consumer choice data to estimate the preference parameters based on willingness to pay estimates from the literature rather than using real-world data. This approach creates a controlled experiment where we know the "true" utility model, and the "true" profit-maximizing design variables from which biases may be measured. Doing so allows the first functional form discussed above to be the "true" form, whereas the estimates from the second functional form indicate the magnitude of the OVB without confounding other sources of bias such as functional form misspecification.

Data

For the case study, we rely on model-year 2016 vehicle attribute data that we gather form *Edmunds.com* [48], which is a website where consumers can shop for vehicles. The attributes we use in the case study are shown in Table 2. Figure 4 depicts the correlation coefficients among all attributes. Positive (in red) coefficients are strongly correlated, and negative ones (in blue) are strongly anti-correlated.

Table 2: List of attributes used in the optimization case study using central, upper, and lower limits of WTP gathered from the literature [31,32].

Attribute	Author	Data Type	Central WTP	WTP Units	Coeff. in this study.
Manufacturer Sales Retail Price (MSRP)	Whitefoot et.al	market data	-	$	0.80
Acceleration	Whitefoot et al.	market data	-800	$/Seconds	-0.13
Combined MPG	Haaf et al.	market data	800	$/MPG	0.05
Cargo capacity, all seats in place	McCarthy and Tay	market data	1100	$/Cubic Feet	0.14
Curb weight	Klier and Lim	market data	-830	$/Pounds	-0.08
wheelbase	Train and Winston	RP survey	6386.98	$/inches	0.57
Turning circle	Espey and Nair	RP Survey	-901	$/ft	-0.09
Width	haaf et al.	Market Data	15573	$/inches	1.56
Height	Liu	RP Survey	250	$/inches	0.03
Length	Consumer Reports	Survey	54	$/inches	0.01
front headroom	Consumer Reports	Survey	165	$/inches	0.02
Basic Warranty	Consumer Reports	Survey	1500	$/years	0.10
dual front side-mounted airbags	Consumer Reports	Survey	585	$	-0.03
post-collision safety system	Consumer Reports	Survey	1545	$	0.15
xenon high intensity discharge headlamp	Consumer Reports	Survey	820	$	0.08
front and rear view camera	Consumer Reports	Survey	3250	$	0.33
keyless ignition	Consumer Reports	Survey	500	$	0.05
leather steering wheel	Consumer Reports	Survey	100	$	0.01
backup camera	Consumer Reports	Survey	865	$	0.09
8-way power driver seat	Consumer Reports	Survey	1360	$	0.14
8-way power passenger seat	Consumer Reports	Survey	1360	$	0.14
heated passenger seat	Consumer Reports	Survey	500	$	0.05
premium leather	Consumer Reports	Survey	930	$	0.09
Driver seat with power adjustable lumbar support	Consumer Reports	Survey	865	$	0.09
Passenger seat with power adjustable lumbar support	Consumer Reports	Survey	865	$	0.09
USB connection	Consumer Reports	Survey	320	$	0.03

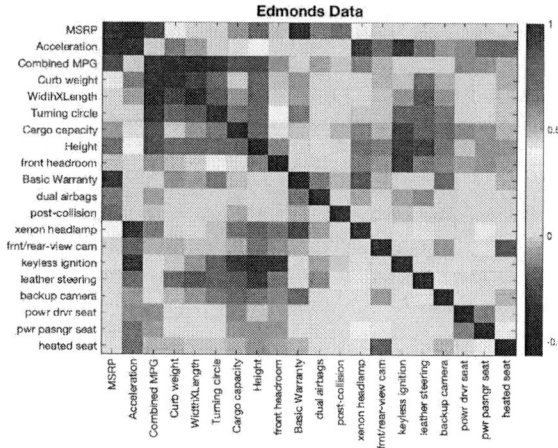

Figure 4: Correlation Coefficients for 2016 Edmunds vehicle data.

Simulated Choice Shares

Simulated choice shares are generated based on a review of estimates of willingness to pay (WTP) in the literature [31]. We generate choices of 500,000 agents according to functional forms based on [5,6,27]. Here, we assume a linear utility model with respect to each of the attributes listed in Table 2. The multinomial logit utility parameters are then estimated from these choice data in order to measure the parameter biases due to omitted variables. We use estimate the parameters using MLE.

Powertrain Optimization

The powertrain optimization is formulated as follows, where the manufacturer seeks to maximize profits with respect to engine displacement, ed, final drive ratio, fdr, and price, p, of an SUV, as shown in *equation 14*:

$$\max_{\substack{wrt \\ \{ed, fdr, p\} \forall j \in J}} \pi = \sum_{j \in J} q_j \left(p_j - (C_E(x) + C_o) \right) - C_I \qquad (14)$$

Following [1,27], a fixed cost, C_I, is included representing the investment costs based on average figures for new production lines, $500,000,000 [1], and marginal vehicle costs are divided into two components: the cost of the engine (C_E), which is a function of the engine size; and a component representing the remaining portion of the costs to produce each vehicle (C_o), which is specified as $7500 following [1]. *Equation 15* reports the cost component associated with the engine is specified as follows, where Pwr is the power of the engine in kW:

$$C_E(x) = 670.51(e^{0.0063*Pwr}) \qquad (15)$$

where the quantity of redesigned vehicles sold, q_j, is determined by multiplying the market share of that vehicle by the total size of the market, M, which is taken to be 1,570,000, as per [1]: $q_i = M \frac{e^{v_i}}{\sum_{j \in J} e^{v_j}}$, where v_i, represents the utility of the redesigned vehicle, and v_J, represents the utility of a vehicle in the choice set (including i). The set of vehicles in the market, J, include 920

vehicles representing a sample of vehicles across all body styles (e.g., small cars, large cars, SUVs, pickup trucks) in the 2016 model-year.

We consider a manufacturer (re)designing a medium size SUV modeled after the Ford Explorer. The relationships between the design variables—engine displacement size and final drive ratio—and vehicle attributes of interest to consumers—namely, 0-60 mph acceleration time and fuel economy—were constructed from performance simulations. Specifically, we used the driveline simulation software *AVL Cruise* [49] to simulate acceleration tests and the U.S. fuel economy test cycles for a large range of feasible values of the design variables. The simulations use a range of vehicle curb weights in addition to simulation parameters such as drag coefficients and frontal area fixed to those of the 2006 Ford Explorer. We then constructed surrogate models from the simulation results that predict 0-60 mph acceleration, fuel economy, and horsepower for given values of the design variables. The following forms for 0-60 acceleration and fuel economy chosen by fitting linear models to the simulation results we obtain from *AVL Cruise*, where ed is engine displacement, fdr is final drive ratio, and wt is weight as shown in Table 3 and 5:

$$Acceleration = t_0 + t_1 * ed + t_2 * wt + t_3 * fdr \\ + t_4 * ed * wt + t_5 * ed * fdr + t_6 * fdr \\ * wt + t_7 * ed^2 + t_8 * wt^2 + t_9 * fdr^2$$

Table 3: Coefficients of the predictive model for acceleration performance using engine displacement, final drive ratio, and weigh.

Coefficient	t_0	t_1	t_2	t_3	t_4
Value	12.4	-2.1e-3	1.4e-3	-8e-1	-2.7e-7
	t_5	t_6	t_7	t_8	t_9
	5.8e-5	-4.4e-5	2.5e-7	7.5e-8	7.2e-2

$$Fuel\ Economy = m_0 + m_1 * ed + m_2 * wt + m_3 * fdr \\ + m_4 * ed * wt + m_5 * ed * fdr + m_6 \\ * fdr * wt + m_7 * ed^2 + m_8 * wt^2 + m_9 \\ * fdr^2$$

Table 4: Coefficients of the predictive model for fuel economy using engine displacement, final drive ratio, and weigh.

Coefficient	m_0	m_1	m_2	m_3	m_4
Value	30.5	-8.3e-4	-1.8e-3	-1.9e-1	2.5e-7
	m_5	m_6	m	m_8	m_9
	-9.8e-6	1.8e-5	-1.9e-7	-1.9e-9	-6.1e-2

Similarly, we also model an active climbing performance constraint which is defined as the maximum ascending gradients a vehicle must surmount while accounting for towing capacity, tow, as shown in Table 5:

$$Climbing\ Performance = s_0 + s_1 * ed + s_2 * wt + s_3 \\ * fdr + s_4 * tow + s_5 * ed * wt + s_6 * ed \\ * fdr \\ + s_7 * ed * tow + s_8 * fdr * wt \\ + s_9 * wt * tow + s_{10} * ed * tow + s_{11} \\ * ed^2 + s_{12} * wt^2 + s_{13} * fdr^2 + s_{14} * tow^2$$

Table 5: Coefficients of the climbing performance constraint

Coeff.	s_0	s_1	s_2	s_3
Value	278	7.6e-3	4.7e-3	-1
	s_4	s_5	s_6	s_7
	-3.7e-5	-6.7e-7	4e-5	2.2e-9
	s_8	s_9	s_{10}	s_{11}
	-2.8e-5	- 6e-9	8.4e-6	7.8e-7
	s_{12}	s_{13}	s_{14}	
	1.1e-7	9.7e-2	2e-9	

RESULTS

In total, we run 8 different simulations to observe the impact OVB has on the optimal design solutions using the Edmunds 2016 dataset. The optimization case study is run for two cases: one in which we control for OVB (using *typically observed* and *unobserved* attributes); and one case where we do not control for OVB (using only the *typically observed* attributes). These two cases are run in each of four scenarios demonstrating the impacts of constraints and consumer willingness to pay (WTP) for key attributes:

1. Base Scenario: unconstrained optimization (choices generated using central WTP for all attributes per [31])
2. Scenario 2: constrained optimization (choices generated using central WTP for all attributes)
3. Scenario 3: unconstrained optimization (choices generated using upper WTP for cargo capacity)
4. Scenario 4: unconstrained optimization (choices generated using lower WTP for cargo capacity)

Scenario 1

Figures 5 and 6 show the choice model estimates for Scenario 1. Figure 5 illustrates that the estimates for the case with no OVB converge to the "true" preference parameters, as expected, with small 95% confidence intervals. This is always the case for each of the scenarios we run, as we control for OVB when generating the choice shares for the unbiased cases. As Figure 6 illustrates, using only the typically observed attributes leads to large biases in the estimates. In this instant, the biases are large enough to cause a change in the sign of the coefficients on acceleration, fuel economy, and curb weight. Substantial biases are also found in Scenario 3 (generated choice shares with upper WTP valuse of cargo capacity), while for Scenario 4 (generated choise shares with lower WTP value of cargo capacity), the bias is limited and the coefficient signs are preserved.

Figure 7 shows how the bias propagates into the profit objective function and the optimal design solutions in Scenario 1. The optimal design results are tabulated in Table 6. These results show that the optimal engine size solution in the biased case is 5% larger than the "true" optimum. The change in final drive ratio is unsubstantial. We also find that the optimal solution for price increases by $766 in the OVB case. In order to gauge how meaningful the 5% propagated bias in the optimal engine size solution is, we simulate a situation where the firm believes the solution is the true optimum without recognizing there is OVB. Specifically, we fix the design of the vehicle at the optimal

solution with OVB (5.9 L, 4.36 final drive, $21,603) and simulate the market share and expected profits using the "true" consumer demand parameters (without OVB). Whereas the firm would have expected profits of $31,847,868 from the optimal design solutions, the true expected profits are $29,858,800, creating $1.95 million after-the-fact losses in anticipated profits because of the unrecognized OVB.

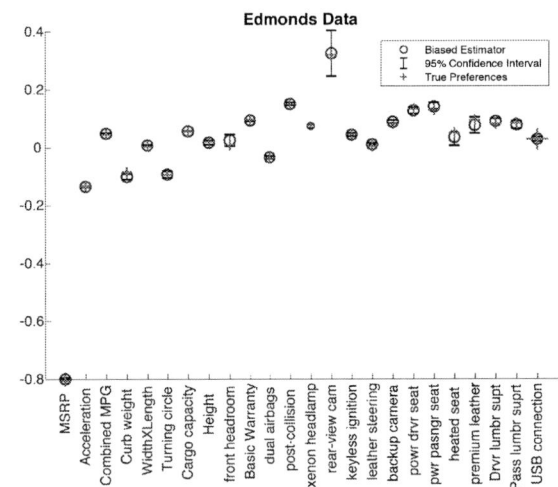

Figure 5: Coefficient estimates (blue circles) using MLE with the fully specified (both typically observed and unobserved attributes). The true parameters are plotted (red crosses) for reference.

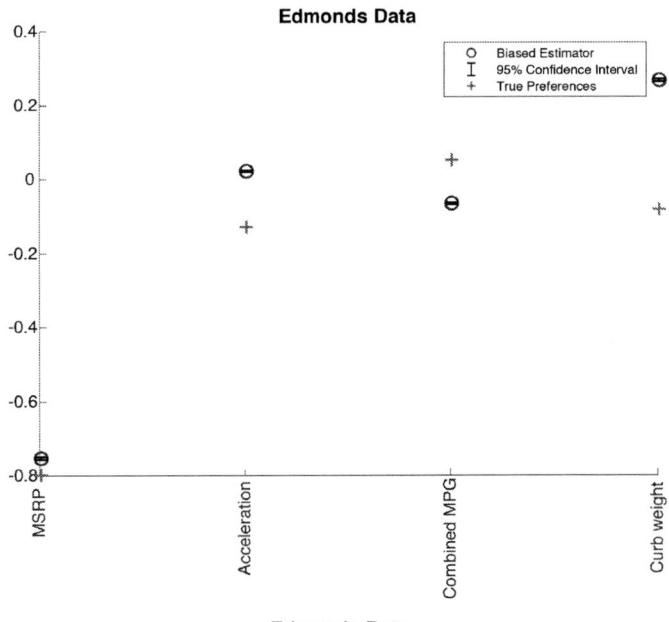

Figure 6:Coefficient estimates (blue circles) using MLE with the misspecified (only typically observed). The true parameters are plotted (red crosses) for reference.

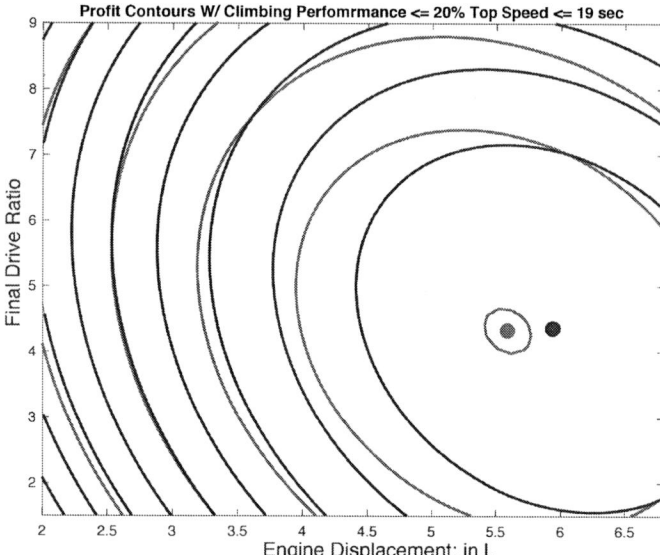

Figure 7: Optimal design unbiased solutions from the fully specified model (green contours and optimum), and the biased solutions from misspecified model (blue contours and optimum).

Table 6: Results from the analysis of Scenario 1. The unbiased, true results are to the left, while what we would expect to see from a biased model are to the right.

	Without OVB	With OVB
Profit	$30,374,261	$31,847,868
ED	5.6 L	5.9 L
FDR	4.32	4.36
MSRP	$20,837	$21,603

In this analysis, we ran several sensitivity tests iteratively including typically unobserved attributes to determine which attributes have the largest influence on the propagated bias. We found that the propagated bias shrinks to close to zero (0.73%) once cargo capacity and the remaining attributes are included. This is intuitive, because the cargo capacity has a relatively large WTP ($1,100), which has a large impact on the magnitude in the α term of *equation 3,* and also is correlation with acceleration (0.3) and fuel economy (-0.25) as high-performance, luxury SUVs tend to offer more cargo space as well as larger engines with higher acceleration performance than small cars. Because of this, we expect that δ from *equation 3* to also be large.

Scenario 2

In Scenario 2, we add a performance constraint to the optimization problem that the vehicle must be able to surmount a minimum ascending gradient of 15% at 60 mph, which actively constrains the solution. Results are shown in Figure 8 and Table 7. Here, we see that the propagated bias into the optimal engine size (+0.2 L) is slightly smaller than in the unconstrained scenario (+0.3 L); however, the propagated bias into the final drive ratio is much larger. The "true" optimum has a final drive ratio of 5.12, whereas the biased optimal solution is 4.84. In contrast, the unconstrained problem changed by only 0.04. This corresponds with the characterization of the propagated bias we

developed in the Characterization Section. Depending on the angle between an active constraint and the direction of the translation of the objective function, the constraint can cause an increase in the propagated bias into an optimal design variable.

Table 7: Results from the analysis of Scenario 2. The unbiased, true results are to the left, while what we would expect to see from a biased model are to the right.

	Without OVB	With OVB
Profit	$29,838,422	$31,764,103
ED	5.8 L	6.0 L
FDR	5.12	4.84
MSRP	$20,933	$21,689

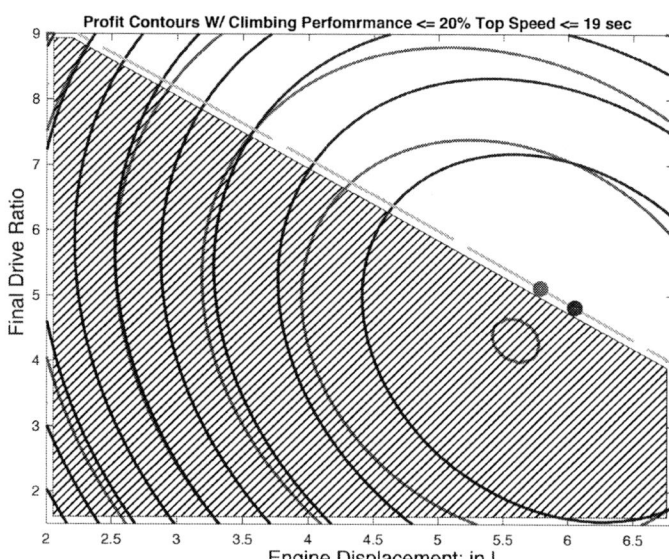

Figure 8: Scenario 2 optimal design unbiased solutions from the fully specified model (green contours and optimum), and the biased solutions from misspecified model (blue contours and optimum). Note that this scenario is Scenario 1 with an active constraint.

Scenario 3

As described above, we find that cargo capacity is correlated with acceleration and fuel economy, and is a considerable source of OVB when omitted from the choice model. In Scenarios 3 and 4, we generate results by varying WTP for cargo capacity in our choice data to the upper bound and lower bound, respectively, of values found in the literature [31]. This allows us to examine how sensitive the propagated bias is to consumer preference for cargo capacity on the optimal design solutions.

The results for the Scenario where consumers' are willing to pay the upper bound of $2,300 for cargo capacity are shown in Table and Figure 9. As we can see, the optimal engine size solutions have changed dramatically. The unbiased model shows an optimal engine displacement of 5.6 L, while the biased model is 6.8 L. There is also a nontrivial difference in the final drive ratio from 4.32 to 4.77. If the firm did not recognize the propagated bias and followed the optimization solution with OVB, the expected profits from the optimization would be $30,900,655, but the true expected profits would be

$24,144,919$, creating $6.76 million after-the-fact losses in anticipated profits.

Table 8: Results from the analysis of Scenario 3. The unbiased, true results are to the left, while what we would expect to see from a biased model are to the right.

	Without OVB	With OVB
Profit	$27,125,529	$30,900,655
ED	5.6 L	6.8 L
FDR	4.32	4.77
MSRP	$20,791	$21,683

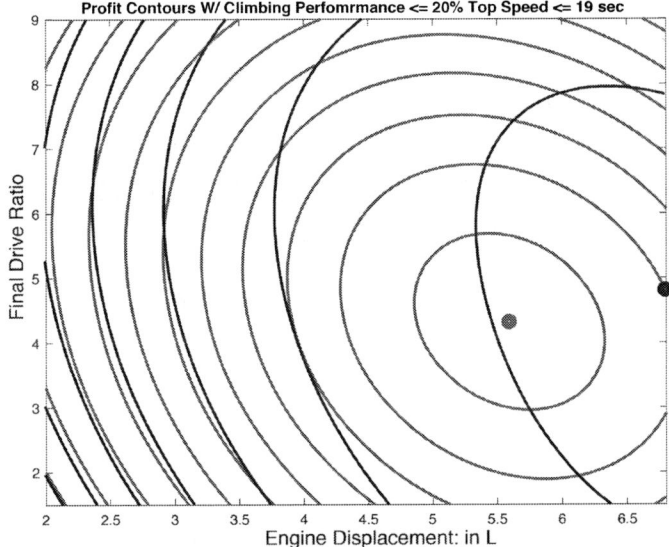

Figure 9: Scenario 3 optimal design unbiased solutions from the fully specified model (green contours and optimum), and the biased solutions from misspecified model (blue contours and optimum). Using a larger WTP value for an attribute correlated with both acceleration and fuel economy leads to large changes in the optimum design variables.

Scenario 4

In Scenario 4, we run the same model using the lower bound of consumers' willingness to pay for cargo capacity to generate choice data. Results are shown in Figure 10 and Table 9: Results from the analysis of Scenario 4 where consumers' willingness to pay for cargo capacity is at the lower bound in the literature.

	Without OVB	With OVB
Profit	$32,929,884	$32,102,717
ED	5.6 L	5.7 L
FDR	4.32	4.33
MSRP	$20,830	$22,367

It is important to note that the attributes that drive the propagated bias may change across years and vehicle choice sets that are available in different regions. From the dataset we use for model-year 2016 vehicles, we find that the propagated bias is sensitive to the inclusion of cargo capacity and consumer WTP for this attribute. In order to test whether this is the case across time, we also collected data for model-year 2018 vehicles. However, we found that correlations of cargo capacity with fuel economy and acceleration were weaker than those found in Figure 4. This points to the difficulty that researchers have in anticipating

which attributes are important to collect data on without *a priori* knowing the structure of the data in terms of the approximate correlations of omitted variables with attributes of interest. Thus, it underscores the importance of using estimation strategies that address OVB when data collection is expensive.

In this scenario, we can see that the propagated bias is very small. The difference in the optimal engine displacement is only 0.07 L, and the difference in the optimal final drive ratio is only 0.01. These results show how sensitive the propagated bias is to cargo capacity. When consumers' willingness to pay for this attribute decreases, the propagated bias vanishes to near zero.

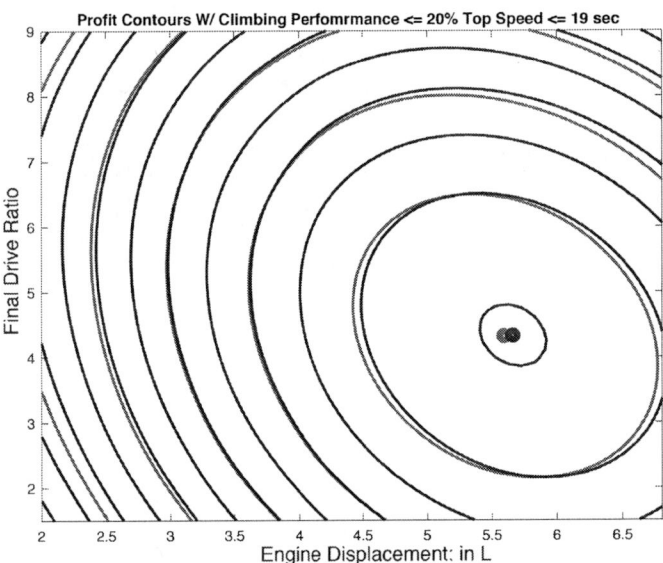

Figure 10: Optimal design solutions in the scenario where consumers' willingness to pay for cargo capacity is at the lower bound in the literature. The true solution is shown in green, and the biased solution is shown in blue.

Table 9: Results from the analysis of Scenario 4 where consumers' willingness to pay for cargo capacity is at the lower bound in the literature.

	Without OVB	With OVB
Profit	$32,929,884	$32,102,717
ED	5.6 L	5.7 L
FDR	4.32	4.33
MSRP	$20,830	$22,367

It is important to note that the attributes that drive the propagated bias may change across years and vehicle choice sets that are available in different regions. From the dataset we use for model-year 2016 vehicles, we find that the propagated bias is sensitive to the inclusion of cargo capacity and consumer WTP for this attribute. In order to test whether this is the case across time, we also collected data for model-year 2018 vehicles. However, we found that correlations of cargo capacity with fuel economy and acceleration were weaker than those found in Figure 4. This points to the difficulty that researchers have in anticipating which attributes are important to collect data on without *a priori* knowing the structure of the data in terms of the approximate correlations of omitted variables with attributes of interest. Thus, it underscores the importance of using estimation strategies that address OVB when data collection is expensive.

CONCLUSION & FUTURE WORK

In this research, we characterize how omitted variables in consumer choice models propagates through to optimal design solutions. Using a simplified problem form, we derive a closed form expression for the propagated bias. The magnitude and direction of the propagated bias is dependent on the correlation of the omitted variable with the design variable, and with consumer choice; the derivative of cost with respect to the design variable; and the willingness to pay for the design variable. We also show that the presence of active constraints can, under certain conditions, lead to a larger propagated bias in a design variable than an unconstrained case.

We demonstrate the impact of OVB on optimal design results in a vehicle powertrain design study. We find that optimal engine size is 5% larger than the true optimum and the difference in final drive ratio is negligible in an unconstrained case. Imposing an active constraint on climbing performance decreases the propagated bias in the engine size solutions but increases it for the final drive ratio to 5%. Results show that designing the powertrain using the biased optimum would lead to significant after-the-fact shortfall in expected profits because of the unrecognized OVB. Based on the specific sample of model-year 2016 vehicles in our simulations, we find that cargo capacity has a particularly large influence on the propagated bias because of its correlations with acceleration and fuel economy. When the lower-bound of consumers' willingness to pay for cargo capacity is used to generate choice data, the propagated bias shrinks to almost zero. When the upper-bound is used, the biased engine size solution is 21% larger than the true optimum.

A note worth mentioning is that while the characterization of OVB derived herein is generalizable for simple cost functions; the manner in which the objective function interacts with the constraints is not. As a result, the insights we demonstrate in our case-study analysis with regards to changes in the optimal design variables (engine displacement & final drive ratio) as well as profits are case specific. We also find that the structure of the correlations matrix can change significantly based on the model-year and how the data sample is gathered, implying that the significance of cargo capacity compared to other omitted variables is likely to vary over years and regions with different choice sets.

With this work, we shed some light on how omitted variables influence propagated bias in optimal design solutions as well as expected profits. The characterization of bias we present give researchers insights into the direction and possible magnitude of the propagated bias they would expect based on *a priori* conjectures of consumer willingness to pay and the strength of correlation with attributes of interest. These insights can be used across a wide-range of product design contexts in real-world applications.

The derivations in the characterization section is presented for the multinomial logit model. This assumes a population with (bounded) homogeneous taste preferences. Extension of these derivations to other choice model forms and to multi-product cases is left for future research.

ACKNOWLEDGMENTS

Funding for this research is provided by the following organizations: the National Science Foundation (NSF) under Grant no. CMMI-1630096, King Abdulaziz City for Science and Technology (KACST), and Carnegie Mellon University (CMU). The authors would like to thank these organizations for their generous support.

REFERENCES

[1] Michalek, J. J., Papalambros, P. Y., and Skerlos, S. J., 2004, "A Study of Fuel Efficiency and Emission Policy Impact on Optimal Vehicle Design Decisions," Journal Of Mechanical Design, **128**, pp. 1062–1070.

[2] Shiau, C.-S., and Michalek, J., 2007, "A Game-Theoretic Approach to Finding Market Equilibria for Automotive Design Under Environmental Regulation," *ASME*, Journal of Mechanical Design, Las Vegas, Nevada, p. 9.

[3] Wassenaar, H. J., Chen, W., Cheng, J., and Sudjianto, A., 2004, "Enhancing Discrete Choice Demand Modeling for Decision-Based Design," Journal of Mechanical Design, **127**(4), pp. 514–523.

[4] Frischknecht, B., and Papalambros, P. Y., 2008, "A Pareto Approach to Aligning Public and Private Objectives in Vehicle Design," *ASME*, Journal of Mechanical Design, Brooklyn, New York.

[5] Frischknecht, B. D., Whitefoot, K., and Papalambros, P. Y., 2010, "On the Suitability of Econometric Demand Models in Design for Market Systems," Journal of Mechanical Design, **132**(12), p. 11.

[6] Whitefoot, K. S., and Skerlos, S. J., 2012, "Design Incentives to Increase Vehicle Size Created from the U.S. Footprint-Based Fuel Economy Standards," Energy Policy, **41**, pp. 402–411.

[7] Ross Morrow, W., Long, M., and MacDonald, E. F., 2014, "Market-System Design Optimization With Consider-Then-Choose Models," Journal of Mechanical Design, **136**(3), p. 031003.

[8] Besharati, B., Luo, L., Azarm, S., and Kannan, P. K., 2006, "Multi-Objective Single Product Robust Optimization: An Integrated Design and Marketing Approach," Journal of Mechanical Design, **128**(4), p. 884.

[9] Williams, N., Azarm, S., and Kannan, P. K., 2008, "Engineering Product Design Optimization for Retail Channel Acceptance," Journal of Mechanical Design, **130**(6), p. 061402.

[10] Wassenaar, H. J., and Chen, W., 2003, "An Approach to Decision-Based Design with Discrete Choice Analysis

for Demand Modeling," Journal of Mechanical Design, **125**, pp. 490–497.

[11] He, L., Chen, W., Hoyle, C., and Yannou, B., 2012, "Choice Modeling for Usage Context-Based Design," Journal of Mechanical Design, **134**(3), p. 031007.

[12] Kroes, E. P., and Sheldon, R. J., 1988, "Stated Preference Methods: An Introduction," Journal of Transport Economics and Policy, **22**(1), pp. 11–25.

[13] McCarthy, P. S., 1996, "Market Price and Income Elasticities of New Vehicles Demand," The Review of Economics and Statistics, **78**(3), pp. 543–547.

[14] Brownstone, D., and Train, K., 1999, "Forcasting New Product Penetration with Flexible Substitution Patterns," Journal of Econometrics, **89**, pp. 109–129.

[15] Dagsvik, J. K., Wetterwald, D. G., and Aaberge, R., 1996, "Potential Demand for Alternative Fuel Vehicles", 165, Statistics Norway Research Department.

[16] Hoyle, C., Chen, W., Ankenman, B., and Wang, N., 2009, "Optimal Experimental Design of Human Appraisals for Modeling Consumer Preferences in Engineering Design," Journal of Mechanical Design, **131**(7), p. 071008.

[17] Michalek, J. J., Feinberg, F. M., and Papalambros, P. Y., 2005, "Linking Marketing and Engineering Product Design Decisions via Analytical Target Cascading," Journal of Product Innovation Management, **22**(1), pp. 42–62.

[18] Haaf, C. G., Morrow, W. R., Azevedo, I. M. L., Feit, E. M., and Michalek, J. J., 2016, "Forecasting Light-Duty Vehicle Demand Using Alternative-Specific Constants for Endogeneity Correction versus Calibration," Transportation Research Part B, **84**, pp. 182–210.

[19] Haaf, C. G., Michalek, J. J., Morrow, W. R., and Liu, Y., 2014, "Sensitivity of Vehicle Market Share Predictions to Discrete Choice Model Specification," Journal of Mechanical Design, **136**, pp. 121402/1–9.

[20] Shiau, C.-S. N., and Michalek, J. J., 2009, "Optimal Product Design Under Price Competition," Journal of Mechanical Design, **131**(7), p. 071003.

[21] Hoyle, C., Chen, W., Wang, N., and Gomez-Levi, G., 2011, "Understanding and Modelling Heterogeneity of Human Preferences for Engineering Design," Journal of Engineering Design, **22**(8), pp. 583–601.

[22] Ma, J., and Kim, H. M., 2014, "Continuous Preference Trend Mining for Optimal Product Design With Multiple Profit Cycles," Journal of Mechanical Design, **136**(6), p. 061002.

[23] Chen, H. Q., Honda, T., and Yang, M. C., 2013, "Approaches for Identifying Consumer Preferences for the Design of Technology Products: A Case Study of Residential Solar Panels," Journal of Mechanical Design, **135**(6), p. 061007.

[24] Ferguson, C.-J., Lees, B., MacArthur, E., and Irgens, C., 1998, "An Application of Data Mining for Product

Design," IEE Colloquium on Knowledge Discovery and Data Mining.

[25] MacDonald, E. F., Gonzalez, R., and Papalambros, P. Y., 2009, "Preference Inconsistency in Multidisciplinary Design Decision Making," Journal of Mechanical Design, **131**, pp. 031009/ 1–13.

[26] Fitzsimons, G. J., Hutchison, W., and Williams, P., 2002, "Non-Conscious Influences on Consumer Choice,"

[27] Whitefoot, K. S., Fowlie, M. L., and Skerlos, S. J., 2017, "Compliance by Design: Influence of Acceleration Trade-Offs on CO_2 Emissions and Costs of Fuel Economy and Greenhouse Gas Regulations," Environmental Science & Technology, **51**(18), pp. 10307–10315.

[28] Klier, T., and Linn, J., 2012, "New-Vehicle Characteristics and the Cost of the Corporate Average Fuel Economy Standard," The RAND Journal of Economics, **43**(1), pp. 186–213.

[29] Shiau, C.-S. N., and Michalek, J. J., 2009, "Should Designers Worry About Market Systems?" Journal of Mechanical Design, **131**(1), p. 011011.

[30] Shiau, C.-S. N., Michalek, J. J., and Hendrickson, C. T., 2009, "A Structural Analysis of Vehicle Design Responses to Corporate Average Fuel Economy Policy," Transportation Research Part A: Policy and Practice, **43**(9–10), pp. 814–828.

[31] Greene, D., Hossain, A., and Beach, R., 2016, "Consumer Willingness to Pay for Vehicle Attributes: What Is the Current State of Knowledge?," 0213244.004.011, RTI International, Ann Arbor, Michigan.

[32] Consumer Reports, 2018, "Must-Have Car Features: Don't Drive Home Without Them."

[33] Gorzelany, J., 2015, "Forbes," The Most-Wanted New-Car Features.

[34] Lee, L.-F., 1982, "Specification Error in Multinomial Logit Models: Analysis of the Omitted Variable Bias," Journal of Econometrics, **20**(2), pp. 197–209.

[35] Wooldridge, J. M., 2010, "Econometric Analysis of Cross Section and Panel Data", MIT Press, Cambridge, Mass.

[36] Lütkepohl, H., 1982, "Non-Causality Due to Omitted Variables," Journal of Econometrics, **19**(2–3), pp. 367–378.

[37] Marais, M. L., and Wecker, W. E., 1998, "Correcting for Omitted-Variables and Measurement-Error Bias in Regression with an Application to the Effect of Lead on IQ: Rejoinder," Journal of the American Statistical Association, **93**(442), p. 515.

[38] 2013, "On the Testability of Identification in Some Nonparametric Models with Endogeneity," Econometrica: Journal of the econometric society, **81**(6), pp. 2535–2559.

[39] Fan, Y., and Li, Q., 1996, "Consistent Model Specification Tests: Omitted Variables and Semiparametric Functional

Forms," Econometrica: Journal of the econometric society, pp. 865−890.

[40] Steven Berry, Levinsohn, J., and Pakes, A., 1995, "Automobile Prices in Market Equilibrium," Econometrica: Journal of the econometric society, **63**(4), pp. 841−890.

[41] Staiger, D., and Stock, J., 1997, "Instrumental Variables Regression with Weak Instruments," Econometrica: Journal of the Econometric Society, **Vol. 65**(No. 3), pp. 557−586.

[42] Stock, J., and Yogo, M., 2002, "Testing for Weak Instruments in Linear IV Regression," t0284, National Bureau of Economic Research, Cambridge, MA.

[43] Mytkowicz, T., Diwan, A., Hauswirth, M., and Sweeney, P., 2010, "The Effect of Omitted-Variable Bias on the Evaluation of Compiler Optimizations," Computer, **43**(9), pp. 62−67.

[44] Frischknecht, B., Whitefoot, K., and Papalambros, P., 2009, "Methods for Evaluating Suitability of Econometric Demand Models In Design for Market Systems," *ASME*, Journal of Mechanical Design, San Diego, California, p. 10.

[45] Chen, W., Hoyle, C., and Wassenaar, H. J., 2013, "Decision-Based Design: Integrating Consumer Preferences into Engineering Design," Springer, London.

[46] Cameron, A. C., and Trivedi, P., 2005, "Mecroeconometrics: Methods and Applications," Cambridge University Press.

[47] Yatchew, A., and Griliches, Z., 1985, "Specification Error in Probit Models," The Review of Economics and Statistics, **67**(1), p. 134.

[48] Edmunds. 2016. "New Cars," www.edmunds.com/new-cars/

[49] Cruise Vehicle System Analysis, 2009, *AVL Cruise − Vehicle System and Driveline Analysis*.

Proceedings of the ASME 2019
International Design Engineering Technical Conferences
and Computers and Information in Engineering Conference
IDETC/CIE2019
August 18-21, 2019, Anaheim, CA, USA

DETC2019-98114

IMPLICATIONS OF COMPETITOR REPRESENTATION ON OPTIMAL DESIGN

Arthur H. C. Yip, Jeremy J. Michalek, Kate S. Whitefoot[1]
Carnegie Mellon University
Pittsburgh, PA, USA

ABSTRACT

We investigate the effect of competitor product representation on optimal design results in profit-maximization studies. Specifically, we study the implications of replacing a large set of product alternatives available in the marketplace with a reduced set of selected competitors or with composite alternatives, as is common in the literature. We derive first-order optimality conditions and show that optimal design (but not price) is independent of competitors under the logit and nested logit models (where preference coefficients are homogeneous), but optimal design results may depend on competitor representation in latent class and mixed logit models (where preference coefficients are heterogeneous). In a case study of automotive powertrain design using mixed logit demand, we find some change in the optimal acceleration performance value when competitors are modeled using a small set of alternatives rather than the larger set. The magnitude of this change depends on the specific form and parameters of the cost and demand functions assumed, ranging from 0% to 3% in our case study. We find that the magnitude of the change in optimal design variables induced by competitor representation in our case study increases with the heterogeneity of preference coefficients across consumers and changes with the curvature of the cost function.

Keywords: optimal design, competitor, logit, nested logit, mixed logit, composite, elemental, heterogeneity

1. INTRODUCTION

The engineering literature on design for market systems integrates consumer choice models within optimal design problems to determine the most profitable product designs and positioning among competing product offerings [1–7]. The optimal design outcomes in these models can be sensitive to choice model specification generally [2,6–9]. However, the implications of competitor representation on optimal design has not been systematically studied.

In optimal design studies, competing products are represented with different practices and at varying levels of detail. Table 1 and its expanded version in the appendix summarize examples of competitor representation in the design for market systems literature. In some studies, competing products are specified at a granular level, and their attributes correspond to those of the near-complete range of real-world products. For example, Choi et al. [10] represent the market of pain relievers with 14 existing brands and types. Morrow et al. [11] include 443 specific automotive design variants at the make-model-engine-option level representing the new car market in the US. In other studies, competing products are represented with a select number of hypothetical products meant to span the options available in the market. For example, Kwak and Kim [12] specify three competing products: a high-, mid-, and low-spec computer; Shiau and Michalek [3] assume four competing products in the weight scales market; Besharati et al. [13] assume that there are three competitive products in the angle grinder market; and Shin and Ferguson [14] assume three cars and three MP3 players that compete with the product under design. In some studies, only a subset of real-world competing alternatives is modeled, based on product segmentation, popularity in the market, and/or proximity with the product under design. For example, in both Shiau et al. [15] and Wassenaar et al. [16], competitors for a mid-size car under design were specified by a choice set with 10-12 other specific mid-size cars of different brands and designs. The choice model formulation in these two studies exclude options in different size segments such as compact and SUV, even though survey data shows that many consumers consider vehicles of different size segments when purchasing a vehicle [17]. In these latter two types of studies, there is an implicit recognition that the actual market consists of many products (more than the 3-7 represented in these examples) but that it would be impractical or infeasible to include all of them in the choice model.

---

[1] Corresponding author: kwhitefoot@cmu.edu

TABLE 1: Examples of Competitor Representation in the Design for Market Systems Literature[2]

Study (author, year)	Market	Number of competing alternatives	Type of competitor representation
Shin & Ferguson, 2016 [14]	Cars	3	Hypothetical
Shin & Ferguson, 2016 [14]	MP3 players	3	Hypothetical
Kwak & Kim, 2012 [12]	Computers	3	Generic/representative
Shiau & Michalek, 2009 [3]	Weight scales	4	Hypothetical
Besharati et al., 2006 [13]	Angle grinders	4	Hypothetical
Li & Azarm, 2000 [18]	Cordless screwdrivers	5	Hypothetical
Zhao & Thurston, 2013 [19]	Cell phones	5	Hypothetical
Wang et al., 2011 [20]	Laptops and smartphones	7	Hypothetical
Shiau et al., 2009 [15]	Midsize cars	10	Generic/representative
Wassenaar et al., 2005 [16]	Midsize cars	12	Detailed (market subset)
Choi et al., 1990 [10]	Pain relievers	14	Detailed
Morrow et al., 2014 [11]	Cars	443	Detailed
Frischknecht et al., 2010 [6]	Cars	473	Detailed

However, Yip et al. [21] found that competitor product representation can substantially affect choice model predictions. Specifically, the study examined the practice of using product "composites"—a type of choice set alternative that represents a category or segment of products in a generic way. For example, instead of a specific "elemental" product design variant such as a Ford Fiesta, the competing choice alternative may be a generic "compact car" or "sedan", specified using a function (usually an average) of "elemental" product alternatives in that grouping. The study found that composite representation could significantly affect choice share predictions unless particular correction factors were applied to the model. Because competitor representation affects choice share prediction, it may affect optimal design conditional on choice share prediction.

Prior studies have characterized the implications of demand model assumptions for engineering design [7], including demand model functional form and specification [2,4,6,8,14,19,21,22], consumer and product heterogeneity [2,8,21,23], and market structure and competition [2,4,6,21]. We contribute to this literature by characterizing the implications of competitor representation for optimal design. Specifically, we investigate conditions under which the optimal design is robust versus sensitive to competitor representation. We pose a generic optimal design problem, derive first-order necessary conditions, and identify properties of the optimality conditions for several popular demand model specifications to determine in which cases the optimal design may depend on competitor representation. Then, in an automotive case study, we investigate the magnitude of this effect in one practical application and assess factors that affect the magnitude.

---

[2] Full table in Annex A.
[3] In this problem formulation, we model firm profit from only a single product. Multiple products, such as a product family or line, would cause the choice share of one alternative to affect the choice share of all other products, whether they

2. INFLUENCE OF COMPETITORS ON OPTIMAL DESIGN

We examine the first-order conditions of the profit-maximizing design and pricing problem under different classes of discrete choice models to understand how competitors affect the optimal design solution. We first show that with logit and nested logit model representations of demand, the optimal design does not depend on any information about competitors—neither the number of alternatives, nor the values of their attributes. We then show how optimal design can be dependent on competitors when consumer preference parameters are heterogeneous (mixed logit and latent-class logit).

Following a common formulation in the literature [1,13,14,24,25], we define a single-period profit-maximization problem where a firm seeks to maximize profits π_j from a single[3] product j with respect to price p_j and a vector of product attributes $\mathbf{x}_j$:

$$\max \pi_j = (p_j - c_j) q_j \qquad (1)$$
$$\text{w.r.t. } p_j, \mathbf{x}_j$$
$$\text{where}$$
$$c_j = f_c(\mathbf{x}_j)$$
$$q_j = f_q(p_j, \mathbf{x}_j, p_k, \mathbf{x}_k \ \forall k \in \mathcal{J} \backslash j)$$

where unit-cost c_j is a function of the design $\mathbf{x}$ of the focal product j, quantity demanded q_j is a function of the design $\mathbf{x}$ and price p of the focal product j and its competitors $k \in \mathcal{J} \backslash j$ ($\mathcal{J}$ being the set of all products in the market). This formulation ignores fixed costs without loss of generality.[4] It also ignores unit-costs that vary with volume and excludes constraints for simplicity. Following common assumptions in the design for market systems literature, we assume that competing products are considered fixed and do not respond to the focal firm's decisions. This is in contrast to the econometric literature and several recent engineering design studies [2,11,26], which determine the Nash equilibrium of competing firms. However, the formulation above where competitors are considered fixed is consistent with each firm's optimality conditions in equilibrium when there are no leading and following firms. We discuss generalizations of this basic model, including extensions to multi-product firms, in the appendix.

Assuming continuous functions, the first-order necessary conditions for this optimization problem are:[5]

$$\frac{\partial \pi_j}{\partial p_j} = \frac{\partial q_j}{\partial p_j}(p_j - c_j) + q_j = 0 \qquad (2)$$

$$\frac{\partial \pi_j}{\partial \mathbf{x}_j} = \frac{\partial q_j}{\partial \mathbf{x}_j}(p_j - c_j) - q_j \frac{\partial c_j}{\partial \mathbf{x}_j} = \mathbf{0} \qquad (3)$$

---

are being (re-)designed or not. How these internal competitors are represented may affect the optimal product design.
[4] Adding a constant to the objective function will not change the design solution.
[5] We use matrix calculus notation here, where the partial derivative of a scalar with respect to a vector indicates a gradient, and gradients are row vectors.

Eq. (2) can be re-arranged as:

$$p_j = c_j - \left(\frac{\partial q_j}{\partial p_j}\right)^{-1} q_j \qquad (4)$$

which says that price at the solution is equal to cost plus a margin that depends on demand and the sensitivity of demand to price. This relationship has some expected properties: If the sensitivity of demand to price in a choice model were lowered toward zero, the price solution would tend to infinity. If the sensitivity of demand to price were raised toward infinity, the difference between price and cost at the solution tends to zero.

Eq. (4) can be substituted into Eq. (3) and then simplified as:

$$\frac{\partial c_j}{\partial \mathbf{x}_j} = -\left(\frac{\partial q_j}{\partial p_j}\right)^{-1} \frac{\partial q_j}{\partial \mathbf{x}_j} \qquad (5)$$

which says that, at a solution, the marginal unit-cost of a design change[6] is equal to the population's aggregated marginal willingness-to-pay for that design change. Willingness-to-pay is the term used in the choice literature for the iso-utility or iso-demand price equivalence of a design or feature change – i.e.: the price change needed per unit change in the design attribute to maintain constant demand.[7,8]

For compact and intuitive notation, we can define

$$\left.\frac{\partial p_j}{\partial \mathbf{x}_j}\right|_{\partial q=0} \equiv -\left(\frac{\partial q_j}{\partial p_j}\right)^{-1} \frac{\partial q_j}{\partial \mathbf{x}_j} \qquad (6)$$

as the population's aggregated marginal willingness-to-pay for a design change (the iso-demand price equivalence of a design change)[9]. We refer to this as the aggregated willingness-to-pay (WTP) from hereon. We can then rewrite the necessary conditions in Eq. (2-3) as:

$$p_j = c_j - \left(\frac{\partial q_j}{\partial p_j}\right)^{-1} q_j \qquad (7)$$

$$\frac{\partial c_j}{\partial \mathbf{x}_j} = \left.\frac{\partial p_j}{\partial \mathbf{x}_j}\right|_{\partial q=0} \qquad (8)$$

In the following sections, we examine the properties of these necessary conditions for the choice model classes used most frequently in the literature [1,8,24,27]. We begin with models

that have homogeneous preference parameters (logit and nested logit) and then examine models with heterogeneous preference parameters (mixed logit and latent-class logit).

2.1. HOMOGENEOUS PREFERENCE PARAMETERS

In this section, we show that when logit and nested logit models are used with linear utility functions to represent demand, the optimal design solution is independent of competing products. Following derivations shown in Shiau and Michalek [15], and Besanko et al. [28], we derive the conditions for first-order optimality in the case of multinomial logit and explicitly show that the optimality conditions reduce to functions of consumer preference parameters and cost, and the profit-maximizing design does not depend on any information about competitors. We then extend these results to the case of nested logit and show that its optimality condition and design solutions are also independent of competitor representation.

Logit

Consider a logit model with a linear utility specification:

$$q_j = m \frac{\exp(\alpha p_j + \boldsymbol{\beta}^\mathsf{T} \mathbf{x}_j)}{\exp(\alpha p_j + \boldsymbol{\beta}^\mathsf{T} \mathbf{x}_j) + \theta} \qquad (9)$$

where m is the market size, α and β are the coefficients for price and the design attribute, respectively; $\theta = \sum_{k \in \mathcal{J} \setminus j} \exp(\alpha p_k + \boldsymbol{\beta}^\mathsf{T} \mathbf{x}_k)$ is a fixed parameter equal to the sum of the exponential of utility of all of product j's competitors; and $\mathcal{J}$ is the choice set of product alternatives available in the market. The partial derivatives of demand with respect to price p and attributes $\mathbf{x}$ are then[10]:

$$\frac{\partial q_j}{\partial p_j} = \alpha q_j \left(1 - \frac{q_j}{m}\right) \qquad (10)$$

$$\frac{\partial q_j}{\partial \mathbf{x}_j} = \boldsymbol{\beta}^\mathsf{T} q_j \left(1 - \frac{q_j}{m}\right) \qquad (11)$$

Substituting Eq. (10-11) into Eq. (7-8), we obtain:

$$p_j = c_j - \frac{1}{\left(1 - \frac{q_j}{m}\right)\alpha} \qquad (12)$$

$$\frac{\partial c_j}{\partial \mathbf{x}_j} = -\frac{\boldsymbol{\beta}^\mathsf{T}}{\alpha} \qquad (13)$$

---

[6] Marginal unit-cost of a design change $(\partial c / \partial \mathbf{x})$ should not be confused with marginal cost of increasing production volume $(\partial C / \partial q$, where C is total cost). Unit-cost c is already "marginal" in the production volume dimension – i.e.: the production cost per incremental unit ignoring fixed costs.
[7] This concept is related to marginal rates of substitution, but it is applied to a population substituting design attributes of a product for price, rather than to an individual substituting one good or service for another.
[8] In the context of a population modeled with heterogeneous consumer preference parameters, the iso-utility framing does not apply (design changes would affect

utility for different consumers differently). We refer to the population's aggregated WTP here as the iso-demand price equivalence of a design change.
[9] Setting $\partial q = 0$, we obtain $\partial q = \frac{\partial q}{\partial p} \partial p + \frac{\partial q}{\partial \mathbf{x}} \partial \mathbf{x} = 0$ and $\therefore \partial p = -\left(\frac{\partial q}{\partial p}\right)^{-1} \frac{\partial q}{\partial \mathbf{x}} \partial \mathbf{x}$ is the marginal price change needed to accompany a marginal design change $\partial \mathbf{x}$ in order to maintain constant demand $(\partial q = 0)$.
[10] Full derivations available in Annex B.

Eq. (12) says that the optimal markup $(p_j - c_j)$ is equal to the negative reciprocal of the choice share for competitors $\left(1 - \frac{q_j}{m}\right)$ multiplied by the price coefficient. Because the choice share of competitors may depend on how they are represented in the model, the profit-maximizing price can depend on competitor representation. However, Eq. (13) shows that, at the optimal solution, the marginal cost of an attribute change is equal to the ratio of the coefficient for the attribute and the coefficient for price – the willingness-to-pay for the design change, which is independent of competitors. Because our cost function depends on only design attribute variables $\mathbf{x}$ and not other variables[11], Eq. (13) implies that the profit-maximizing attribute vector $\mathbf{x}^*$ has a solution that does not depend on variables p, q, or θ. Therefore, the optimal design solution does not depend on the representation of competitors in this case[12]. This finding can be extended to the case of multi-product firms and the case of utility functions that are non-linear in non-price attributes, as detailed in the appendix.

Nested Logit

In the case of nested logit, we also find the first-order optimality condition to be independent of competitor information. Generalizing Eq. (9) to a nested logit specification:

$$q_j = m \left(\frac{\exp\left(\frac{\alpha p_j + \boldsymbol{\beta}^\top \mathbf{x}_j}{\lambda_k}\right)}{\exp\left(\frac{\alpha p_j + \boldsymbol{\beta}^\top \mathbf{x}_j}{\lambda_k}\right) + \rho_k} \right) \times$$
$$\left(\frac{\exp\left(\lambda_k \ln\left(\exp\left(\frac{\alpha p_j + \boldsymbol{\beta}^\top \mathbf{x}_j}{\lambda_k}\right) + \rho_k\right)\right)}{\exp\left(\lambda_k \ln\left(\exp\left(\frac{\alpha p_j + \boldsymbol{\beta}^\top \mathbf{x}_j}{\lambda_k}\right) + \rho_k\right)\right) + \phi_k} \right) \quad (14)$$

where $\rho_k = \sum_{\hat{j} \in \mathcal{J}_k \setminus j} \exp\left(\frac{\alpha p_{\hat{j}} + \boldsymbol{\beta}^\top \mathbf{x}_{\hat{j}}}{\lambda_k}\right)$ and $\phi_k = \sum_{\hat{k} \in \mathcal{K} \setminus k} \sum_{\hat{j} \in \mathcal{J}_k} \exp\left(\frac{\alpha p_{\hat{j}} + \boldsymbol{\beta}^\top \mathbf{x}_{\hat{j}}}{\lambda_k}\right)$ are fixed parameters equal to the sum of exponentiated utility of all of product j's competitors within nest k and outside nest k, respectively; $\mathcal{J}_k$ is the choice set of alternatives within nest k; $\mathcal{K}$ is the set of all nests of products in the market; and λ_k is the nesting parameter for each nest k. Eq. (14) reduces to Eq. (9) when $\lambda_k = 1$.

From Eq. (14), we can find the partial derivatives of demand with respect to price p and attributes $\mathbf{x}$.[13] Substituting the functions of $\partial q_j / \partial p_j$ and $\partial q_j / \partial \mathbf{x}_j$ under nested logit into Eq. (5), we find the second first-order optimality condition also reduces to the ratio of attribute and price preference parameters, which again is only a function of consumer preference

[11] An exception may be where unit-costs vary with market share or sales volume, if they are at a magnitude where economies of scale matter.
[12] Assuming a unique solution exists in the feasible domain. There may be degenerate cases where there are no solutions or multiple solutions.
[13] The full equations can be found in Annex B.

parameters and not a function of variables such as the optimal price or the utility from competitors of the same nest or of other nests:

$$\frac{\partial c_j}{\partial \mathbf{x}_j} = -\frac{\boldsymbol{\beta}^\top}{\alpha} \quad (15)$$

2.2. HETEROGENEOUS PREFERENCE PARAMETERS

In this section, we show that in the case of a logit demand model specified with random coefficients, such as latent-class logit or mixed logit, the optimality conditions do not, in general, establish design variables $\mathbf{x}$ independently of competitors, and competitor representation may affect the optimal design. For example, suppose consumers are modeled as heterogeneous in their preference for attribute x (some prefer more while others prefer less). The optimal solution for attribute x for one firm may depend on whether competing firms are targeting consumers who prefer more of x or those who prefer less of x.

With random coefficients specified using a discrete distribution with groups[14] indexed by i, such as in a latent-class logit model,

$$q_{ij} = m \frac{\exp\left(\alpha_i p_j + \boldsymbol{\beta}_i^\top \mathbf{x}_j\right)}{\exp\left(\alpha_i p_j + \boldsymbol{\beta}_i^\top \mathbf{x}_j\right) + \theta_{ij}} \quad (16)$$

$$q_j = \sum_i q_{ij} \quad (17)$$

where q_{ij} is the share of consumer group i that would choose product j; α_i is the price coefficient for consumer group i; $\boldsymbol{\beta}_i$ is the vector of coefficients for attributes $\mathbf{x}$ for consumer group i; q_{ij} is the share of group i choosing product j; and θ_{ij} is a fixed parameter equal to the sum of exponentiated utilities for group i of product j's competitors. The first-order optimality conditions from Eq. (7-8) become (see Annex B for derivation):

$$p_j = c_j - \frac{q_j}{\sum_i \alpha_i \left(q_{ij}\left(1 - \frac{q_{ij}}{m}\right)\right)} \quad (18)$$

$$\frac{\partial c_j}{\partial \mathbf{x}_j} = -\frac{\sum_i \boldsymbol{\beta}_i^\top \left(q_{ij}\left(1 - \frac{q_{ij}}{m}\right)\right)}{\sum_i \alpha_i \left(q_{ij}\left(1 - \frac{q_{ij}}{m}\right)\right)} \quad (19)$$

Because mixed logit is generally implemented using a finite set of draws to approximate the distribution, these equations apply to the numerically calculated mixed logit case as well.

Eq. (19) indicates that the marginal unit-cost of design change is equal to the ratio of the sum of coefficients $\boldsymbol{\beta}_i$ and α_i of each group, each weighted by $q_{ij}(1 - q_{ij}/m)$. Because

[14] In this formulation, the groups represent equally-sized groups modeled with homogenous consumer preference parameters within the group, also known as latent classes. This is generalizable to groups of different sizes – i.e.: problem could be formulated with multiple equally-sized groups with the same consumer preferences.

the right-hand side of Eq. (19) depends on q_{ij}, which depends on θ_{ij} as well as p_j, the optimal design $\mathbf{x}_j$ is not generally independent of competitor representation.

Although the optimal design may vary depending on competitor representation, it is unclear in general how large this effect may be in practical cases. We use a case study of the US automotive market to demonstrate and test the issue of competitor representation on optimal design.

3. CASE STUDY: RE-DESIGN OF AN AUTOMOBILE

We construct a case study based on a model of the automotive market from the literature [26]. In this case study, we investigate whether and to what extent the optimal design of a single vehicle changes under heterogeneous consumer preferences for different fixed competitor representations. We use data for the 2006 US new car market. On the supply side, we use equations derived from engineering simulations in Whitefoot et al. [26] to define relationships between attributes and costs. We take as decision variables the vehicle's price p and acceleration x, measured as time to accelerate from 0 to 60 mph in seconds. Fuel consumption is calculated as a function of acceleration using their estimation of the Pareto frontier developed through simulation and design of experiments (equivalent to treating fuel consumption as a free variable and constraining the solution to lie on the Pareto frontier). In this model, cost decreases with increasing acceleration time (i.e. smaller engine), with diminishing returns, and is specified by Eq. (20-21):

$$c = 1000\gamma_1 + 1000\gamma_2 \exp\left(-\frac{x}{10}\right) + \gamma_3 w + \frac{\gamma_4 w x}{10} \quad (20)$$

$$\frac{\partial c}{\partial x} = -100\gamma_2 \exp\left(-\frac{x}{10}\right) + \frac{\gamma_4 w}{10} \quad (21)$$

where c is unit-cost in $ per unit, x is acceleration time (0-60 mph) in s, w is weight in lb, and the γ terms are parameters fit to engineering simulation results from [26]. For demand, we use a mixed logit specification and parameters from [26]. The consumer utility is specified by Eq. (22):

$$v_{ij} = \alpha_i p_j + \boldsymbol{\beta}_i^\top \mathbf{x}_j \quad (22)$$

where v is the consumer utility from observed attributes, $\mathbf{x}$ includes the attributes of fuel economy (inverse of fuel consumption), acceleration time, and vehicle area (length times width). The $\boldsymbol{\beta}$ coefficients are specified as independently and normally distributed. Consumer characteristics are ignored.

We hold all competitor products fixed in price and attributes (summarized by the θ parameter), and the firm solves the profit-maximization problem for price and acceleration time of the focal product, with other attributes fixed (except for fuel economy, which is specified as a function of acceleration, as described). We first solve the problem using competitors represented as 470 elemental alternatives, and then we re-solve with competitors represented by 3 composite alternatives (one compact car, one midsize car, and one large car) intended to

represent the 470 elemental alternatives. Each composite is specified by the weighted average price and attributes of their subsumed elemental alternatives – i.e.: the compact car composite's price is the average price of all compact cars, weighted by sales fraction. This type of composite using weighted averages is typical in the literature, though other specifications may include composite correction factors and alternative-specific constants and could lead to different results [21]. We use the *L-BFGS* algorithm implemented in the *R nloptr* package to solve the design optimization problem.

We first show the optimal solution in the base scenario with default parameterization. Then, we vary the heterogeneity in preferences in different scenarios. In a recent review of the automotive demand literature, Greene at al. [27] show that estimates of both mean and the standard deviation of consumer willingness-to-pay for vehicle attributes cover a wide range. We simulate a wide range of preference heterogeneity by applying a multiplier to the standard deviation of the normally distributed preferences found in Whitefoot et al. [26] from 0x to 10x, which is within the range found in Greene at al. [27]. For fuel economy, this represents a fixed $600/mpg at 0x, a heterogeneous range of $472 to $728/mpg (mean +/- 2 standard deviations, covering 95% of values) at 1x (the default parameter values), a range of -$40 to $1240/mpg at 5x, and a range of -$680 to $1880/mpg at 10x. We also test several cases of different cost functions and vehicle size classes to assess their effect on the influence of competitor representation on optimal design solutions.

4. CASE STUDY RESULTS

Figure 1 shows the difference in the optimal decision variables when competitors are represented as 3 composite alternatives rather than 470 elemental alternatives under a range of consumer-preference heterogeneity multipliers. In the base scenario, shown in Fig. 1a, a compact car is redesigned. Using the default preference heterogeneity (1x), we find that competitor representation affects the optimal acceleration time by -0.03% (and optimal price by 4.6%). When the heterogeneity is scaled down to 0x (no preference heterogeneity), the optimal acceleration time is identical regardless of the competitor representation, as expected from our derivation in Section 2.1. With heterogeneity scaled up by 10x, competitor representation affects optimal acceleration time by -2.6%

We repeat these tests for the case of re-designing a large car, which has different consumer utility parameters as estimated by Whitefoot et al. [26]. The optimal design results, in Fig. 1b, show larger differences than the case of the compact car. With a preference heterogeneity multiplier at 10x, competitor representation affected optimal acceleration time by -3.6%. In addition to the results shown in Fig. 1, we ran a series of cases varying the distributions and input parameter values across a wide range, including uniform and bimodal distributions. In most cases where parameter values were in a realistic range, the effect of competitor representation on optimal design is between 0% and -5%. However, in some cases, we found that the optimal design changed up to 20%. Changes in optimal price were larger:

typically between 5% and 15%, and in some cases as large as 30%.

In Fig. 1c and Fig. 1d, we assume a cost function with a steeper (2x) marginal unit-cost of design change, which affects the magnitude of the impact of competitor representation on optimal acceleration time slightly (-2.3% for small car and -3.4% for large car at 10x).

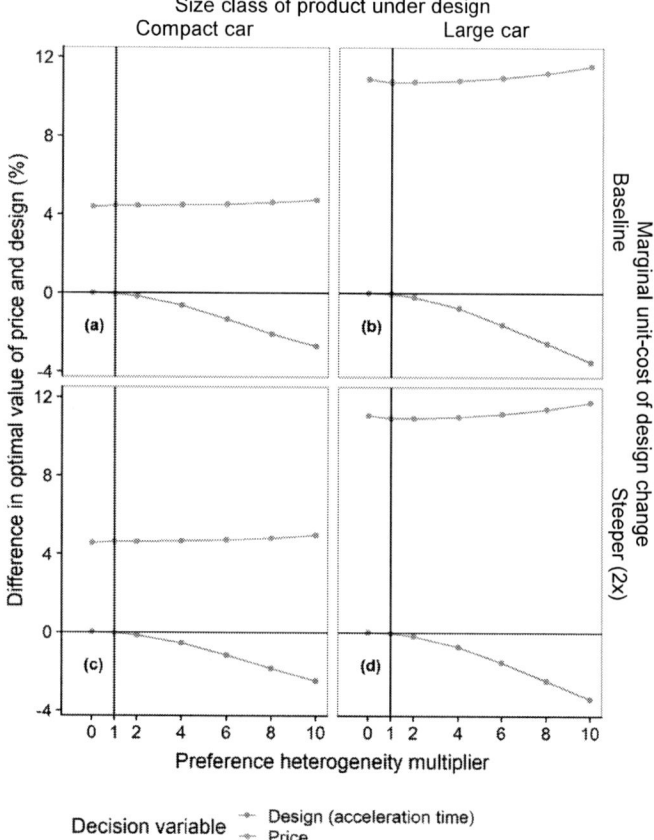

FIGURE 1: Differences in optimal values of price and design (acceleration time) when competitors are represented by elemental vs. composite alternatives

As shown in Section 2.2, the first-order optimality conditions stipulate that at an optimal solution, the marginal unit-cost of a design change is equal to the population's aggregated marginal WTP for that design change (Eq. 8). For the mixed logit case, the aggregated WTP is the ratio of the sum of coefficients β_i and α_i of each consumer (each draw from the distribution), weighted by $q_{ij}(1 - q_{ij}/m)$ (Eq. 19). To help explain how and why both the degree of preference heterogeneity and the curvature of the cost function influence the effect of competitor representation on optimal designs, we plot the marginal unit-cost of design change (MC for short, representing the left-hand side

of Eq. 19) and the aggregated WTP for design change (WTP for short, representing the right-hand side of Eq. 19) in Fig. 2. Both of these values are negative because increasing acceleration time reduces both marginal unit-cost and willingness-to-pay.[15] We see that the optimal solutions for each simulated case lie at each intersection of the MC and WTP curves, where the marginal unit-cost change from increasing acceleration match the iso-demand price change, as expected.

The MC curve is independent of demand assumptions, consumer heterogeneity, or the representation of competitor products because the marginal cost of production volume is independent of production volume in our formulation (Eq. 1). However, WTP is a function of demand and its parameters, and for latent-class logit and mixed logit models, this includes competitor products. As such, when competitors are represented by composites instead of elemental alternatives, the WTP curve shifts, changing the optimal solution. In this case study, the composite representation (solid lines; optima at circles) causes the WTP curve to shift to the left relative to the elemental representation (dashed lines; optima at crosses), and the magnitude of the shift increases with the degree of preference heterogeneity (shown by shades of gray of the pairs of WTP lines, shifting right as the preference heterogeneity multiplier goes from 0x to 10x).

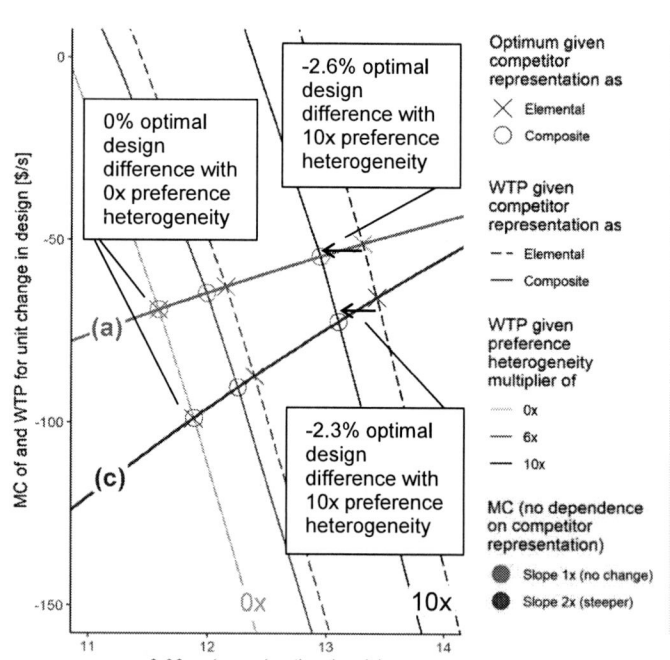

FIGURE 2: Marginal unit-cost of a design change (MC) and the population's aggregated marginal willingness-to-pay for a design change (WTP) as a function of the design variable (acceleration time) for the compact car redesign case

---

[15] For lower values of acceleration time, the population's aggregated marginal WTP for increasing acceleration time is positive because in this model, fuel

economy is also treated as a function of acceleration, following the Pareto frontier.

We also observe that a steeper MC curve (the blue line, noted by (c)) leads to a somewhat smaller effect of competitor representation on optimal design, corresponding to the result in Fig. 1c. We repeat this analysis for the large car case in Figure 3 and obtain similar findings. We summarize the numerical results in Table 2.

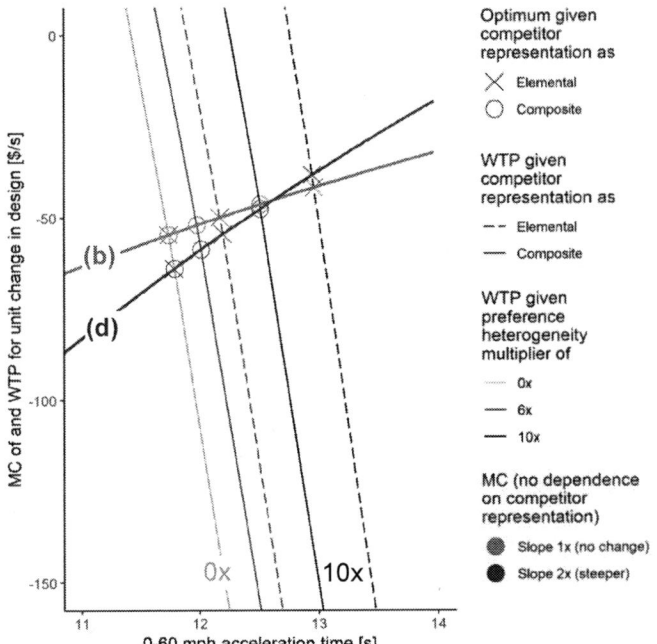

FIGURE 3: Marginal unit-cost of a design change (MC) and the population's aggregated marginal willingness-to-pay for a design change (WTP) as a function of the design variable (acceleration time) for the large car redesign case

TABLE 2: Summary of numerical results in automotive case study - differences in optimal design variable (acceleration time) due to composite representation (vs. elemental representation) of competitor products

Steepness of MC curve	Preference heterogeneity multiplier	Size class of product (car) under design	
		Compact	Large
1x	0x	0 s 0%	0 s 0%
	1x	-0.004 s -0.03%	-0.005 s -0.04%
	10x	-0.36 s -2.6%	-0.45 s -3.6%
2x	0x	0 s 0%	0 s 0%
	1x	-0.005 s -0.04%	-0.005 s -0.04%
	10x	-0.33 s -2.3%	-0.44 s -3.4%

5. CONCLUSION

We derive first-order optimality conditions for profit maximizing design and price and determine that competitor representation does not affect optimal design when demand is modeled using logit or nested logit models (homogeneous consumer preference parameters), but competitor representation may affect optimal design when demand is modeled using latent-class or mixed logit models (heterogeneous consumer preference parameters). Competitor representation may affect optimal price under all demand models. These findings hold for utility functions that are linear in price and cost functions where marginal unit-cost is independent of production volume.

In a case study of automotive design under mixed logit demand, we find that the optimal design (0-60 mph acceleration time) changes when competitors are modeled using a small set of composite alternatives to represent a larger set of vehicles available on the market. The magnitude of this effect depends on the specific form and parameters of the cost and consumer utility functions. In our case study, the magnitude of the change increases with preference heterogeneity and decreases with the steepness of the marginal unit-cost curve.

ACKNOWLEDGMENTS

This research was supported by the National Science Foundation under Grant no. CMMI-1630096, the Natural Sciences and Engineering Research Council of Canada Postgraduate Scholarship, and Carnegie Mellon University.

REFERENCES

[1] Wassenaar, H. J., and Chen, W., 2003, "An Approach to Decision-Based Design With Discrete Choice Analysis for Demand Modeling," J. Mech. Des., **125**(3), p. 490.

[2] Shiau, C.-S. N., and Michalek, J. J., 2008, "Should Designers Worry About Market Systems?," J. Mech. Des., **131**(1), p. 011011.

[3] Shiau, C.-S. N., and Michalek, J. J., 2009, "Optimal Product Design Under Price Competition," J. Mech. Des., **131**(7), p. 071003.

[4] Morrow, W. R., Long, M., and MacDonald, E. F., 2014, "Market-System Design Optimization With Consider-Then-Choose Models," J. Mech. Des., **136**(3), p. 031003.

[5] Chen, W., Hoyle, C., and Wassenaar, H. J., 2013, *Decision-Based Design: Integrating Consumer Preferences into Engineering Design*.

[6] Frischknecht, B. D., Whitefoot, K. S., and Papalambros, P. Y., 2010, "On the Suitability of Econometric Demand Models in Design for Market Systems," J. Mech. Des., **132**(12), p. 121007.

[7] Donndelinger, J., and Ferguson, S. M., 2017, "Design for Marketing Mix: The Past, Present, and Future of Market-Driven Product Design," *IDETC Proceedings*, pp. 1–16.

[8] Haaf, C. G., Michalek, J. J., Morrow, W. R., and Liu, Y., 2014, "Sensitivity of Vehicle Market Share Predictions to Discrete Choice Model Specification," J. Mech. Des.,

136(12), p. 121402.

[9] Donndelinger, J. A., Robinson, J. A., and Wissmann, L. A., 2008, "Choice Model Specification in Market-Based Engineering Design," *IDETC Proceedings, Volume 1: 34th Design Automation Conference, Parts A and B*, ASME, pp. 447–459.

[10] Choi, S. C., Desarbo, W. S., and Harker, P. T., 1990, "Product Positioning Under Price Competition," Manage. Sci., **36**(2), pp. 175–199.

[11] Morrow, W. R., Mineroff, J., and Whitefoot, K. S., 2014, "Numerically Stable Design Optimization With Price Competition," J. Mech. Des., **136**(8), p. 081002.

[12] Kwak, M., and Kim, H., 2012, "Market Positioning of Remanufactured Products With Optimal Planning for Part Upgrades," J. Mech. Des., **135**(1), p. 011007.

[13] Besharati, B., Luo, L., Azarm, S., and Kannan, P. K., 2006, "Multi-Objective Single Product Robust Optimization: An Integrated Design and Marketing Approach," J. Mech. Des., **128**(4), p. 884.

[14] Shin, J., and Ferguson, S., 2016, "Exploring Product Solution Differences Due to Choice Model Selection in the Presence of Noncompensatory Decisions With Conjunctive Screening Rules," J. Mech. Des., **139**(2), p. 021402.

[15] Shiau, C.-S. N., Michalek, J. J., and Hendrickson, C. T., 2009, "A Structural Analysis of Vehicle Design Responses to Corporate Average Fuel Economy Policy," Transp. Res. Part A Policy Pract., **43**(9–10), pp. 814–828.

[16] Wassenaar, H. J., Chen, W., Cheng, J., and Sudjianto, A., 2005, "Enhancing Discrete Choice Demand Modeling for Decision-Based Design," J. Mech. Des., **127**(4), p. 514.

[17] Maritz Research, 2006, *New Vehicle Consumer Survey*.

[18] Li, H., and Azarm, S., 2000, "Product Design Selection Under Uncertainty and With Competitive Advantage," J. Mech. Des., **122**(4), p. 411.

[19] Zhao, Y., and Thurston, D., 2013, "Maximizing Profits From End-of-Life and Initial Sales With Heterogeneous Consumer Demand," J. Mech. Des., **135**(4), p. 041001.

[20] Wang, Z., Kannan, P. K., and Azarm, S., 2011, "Customer-Driven Optimal Design for Convergence Products," J. Mech. Des., **133**(10), p. 101010.

[21] Yip, A. H. C., Michalek, J. J., and Whitefoot, K. S., 2018, "On the Implications of Using Composite Vehicles in Choice Model Prediction," Transp. Res. Part B Methodol., **116**, pp. 163–188.

[22] Haaf, C. G., Morrow, W. R., Azevedo, I. M. L., Feit, E. M., and Michalek, J. J., 2016, "Forecasting Light-Duty Vehicle Demand Using Alternative-Specific Constants for Endogeneity Correction versus Calibration," Transp. Res. Part B Methodol., **84**, pp. 182–210.

[23] Sullivan, E., Ferguson, S., and Donndelinger, J., 2011, "Exploring Differences in Preference Heterogeneity Representation and Their Influence in Product Family Design," *Volume 5: 37th Design Automation Conference, Parts A and B*, ASME, pp. 81–92.

[24] Michalek, J. J., Feinberg, F. M., and Papalambros, P. Y., 2005, "Linking Marketing and Engineering Product Design Decisions via Analytical Target Cascading*," J. Prod. Innov. Manag., **22**(1), pp. 42–62.

[25] Williams, N., Azarm, S., and Kannan, P. K., 2008, "Engineering Product Design Optimization for Retail Channel Acceptance," J. Mech. Des., **130**(6), p. 061402.

[26] Whitefoot, K. S., Fowlie, M. L., and Skerlos, S. J., 2017, "Compliance by Design: Influence of Acceleration Trade-Offs on $CO2$ Emissions and Costs of Fuel Economy and Greenhouse Gas Regulations," Environ. Sci. Technol., **51**(18), pp. 10307–10315.

[27] Greene, D., Hossain, A., Hofmann, J., Helfand, G., and Beach, R., 2018, "Consumer Willingness to Pay for Vehicle Attributes: What Do We Know?," Transp. Res. Part A Policy Pract., **118**, pp. 258–279.

[28] Besanko, D., Gupta, S., and Jain, D., 1998, "Logit Demand Estimation Under Competitive Pricing Behavior: An Equilibrium Framework," Manage. Sci., **44**(11-part-1), pp. 1533–1547.

ANNEX A

TABLE 1A: EXTENDED VERSION OF TABLE 1 WITH EXAMPLES OF COMPETITOR REPRESENTATION IN THE DESIGN FOR MARKET SYSTEMS LITERATURE

Study (author, year)	Market	Number of competing alternatives	Type of competitor representation	Reasoning behind competitor representation
Shin & Ferguson, 2016 [14]	Cars	3	Hypothetical	No reason given
Shin & Ferguson, 2016 [14]	MP3 players	3	Hypothetical	No reason given
Kwak & Kim, 2012 [12]	Computers	3	Generic/ representative	"There are three competing products on the market (i.e., high-spec, mid-spec, and low-spec), and they differ from each other in terms of part specifications and selling price."
Shiau & Michalek, 2009 [3]	Weight scales	4	Hypothetical	"Table 6 shows the specifications of four competing products C1, R2, S3, and T4 in the market, where each product has a unique combination of product characteristics."
Besharati et al., 2006 [13]	Angle grinders	4	Hypothetical	"We assume that in the market for this power tool, there are three competitive products."
Li & Azarm, 2000 [18]	Cordless screwdrivers	5	Hypothetical	No reason given
Zhao & Thurston, 2013 [19]	Cell phones	5	Hypothetical	"a hypothetical market is assumed with five product competitors and their attributes as shown in Table 1. These data were collected from several real products in the current cell phone market."
Wang et al., 2011 [20]	Laptops and smartphones	7	Hypothetical	No reason given
Shiau et al., 2009 [15]	Midsize cars	10	Generic/ representative	Computational reasons. Explanation: Case study: midsize vehicles, 10 generic domestic manufacturers competing in the market. "assuming each manufacturer has a single representative vehicle j in its fleet". "We simulate 10 generic domestic manufacturers competing in the market", with the "assumption of a single vehicle design per producer"
Wassenaar et al., 2005 [16]	Midsize cars	12	Detailed (market subset)	7 models, 12 trims, to represent the midsize segment. "Our implementation is subject to the assumption that customers only consider the 12 vehicle trims when purchasing a vehicle."
Choi et al., 1990 [10]	Pain relievers	14	Detailed	14 existing brands. "small experimental preference data set on 14 over-the-counter analgesic pain relievers" "The reader should note that this example is not meant to be a thorough study of the market, but is simply meant to illustrate the properties of the proposed model and algorithms described in the paper."
Morrow et al., 2014 [11]	Cars	443	Detailed	"This model is not intended to be a high-fidelity model of vehicle design; our intended application of this model is a comparison of numerical methods in a large market."
Frischknecht et al., 2010 [6]	Cars	473	Detailed	No reason given

ANNEX B

EXTENDED DERIVATIONS OF PARTIAL DERIVATIVES AND FIRST-ORDER OPTIMALITY CONDITIONS

The derivations below describe how the first-order optimality conditions reduce to certain forms given competitor representation. These derivations can be extended to firms offering multiple products in the same market and more generic utility functions, to be shown in future work. They can also be extended to problems with constraints by adding Lagrange multipliers and using KKT conditions.

Logit

Given $q_j = m \frac{\exp(\alpha p_j + \boldsymbol{\beta}^\mathsf{T} \mathbf{x}_j)}{\exp(\alpha p_j + \boldsymbol{\beta}^\mathsf{T} \mathbf{x}_j) + \theta}$ and recognizing that the logit choice share of all competitors is defined as $\frac{\theta}{\exp(\alpha p_j + \boldsymbol{\beta}^\mathsf{T} \mathbf{x}_j) + \theta} = \frac{\sum_{k \in \mathcal{J} \setminus j} q_k}{m} = 1 - \frac{q_j}{m}$:

Partial derivative and gradient of quantity demanded with respect to price and design attributes

$$\frac{\partial q_j}{\partial p_j} = \frac{m\left[\alpha \exp(\alpha p_j + \boldsymbol{\beta}^\mathsf{T} \mathbf{x}_j) * \left[\exp(\alpha p_j + \boldsymbol{\beta}^\mathsf{T} \mathbf{x}_j) + \theta_j\right] - \exp(\alpha p_j + \boldsymbol{\beta}^\mathsf{T} \mathbf{x}_j) * \alpha\left[\exp(\alpha p_j + \boldsymbol{\beta}^\mathsf{T} \mathbf{x}_j)\right]\right]}{\left(\exp(\alpha p_j + \boldsymbol{\beta}^\mathsf{T} \mathbf{x}_j) + \theta_j\right)^2} = \frac{m\alpha\theta_j \exp(\alpha p_j + \boldsymbol{\beta}^\mathsf{T} \mathbf{x}_j)}{\left(\exp(\alpha p_j + \boldsymbol{\beta}^\mathsf{T} \mathbf{x}_j) + \theta_j\right)^2} = \alpha q_j \left(1 - \frac{q_j}{m}\right)$$

$$\frac{\partial q_j}{\partial \mathbf{x}_j} = \frac{m\left[\boldsymbol{\beta} \exp(\alpha p_j + \boldsymbol{\beta}^\mathsf{T} \mathbf{x}_j) * \left[\exp(\alpha p_j + \boldsymbol{\beta}^\mathsf{T} \mathbf{x}_j) + \theta_j\right] - \exp(\alpha p_j + \boldsymbol{\beta}^\mathsf{T} \mathbf{x}_j) * \boldsymbol{\beta}\left[\exp(\alpha p_j + \boldsymbol{\beta}^\mathsf{T} \mathbf{x}_j)\right]\right]}{\left(\exp(\alpha p_j + \boldsymbol{\beta}^\mathsf{T} \mathbf{x}_j) + \theta_j\right)^2} = \frac{m\boldsymbol{\beta}\theta_j \exp(\alpha p_j + \boldsymbol{\beta}^\mathsf{T} \mathbf{x}_j)}{\left(\exp(\alpha p_j + \boldsymbol{\beta}^\mathsf{T} \mathbf{x}_j) + \theta_j\right)^2} = \boldsymbol{\beta} q_j \left(1 - \frac{q_j}{m}\right)$$

First-order optimality conditions

$$p_j = c_j - \left[\frac{\partial q_j}{\partial p_j}\right]^{-1} q_j = c_j - \frac{1}{\alpha\left(1 - \frac{q_j}{m}\right)}$$

$$\frac{\partial c_j}{\partial \mathbf{x}_j} = -\left[\frac{\partial q_j}{\partial p_j}\right]^{-1}\left[\frac{\partial q_j}{\partial \mathbf{x}_j}\right] = -\frac{\boldsymbol{\beta}^\mathsf{T}}{\alpha}$$

Nested Logit

Given nested logit choice share:

$$q_j = m \left(\frac{\exp\left(\frac{\alpha p_j + \boldsymbol{\beta}^\mathsf{T} \mathbf{x}_j}{\lambda_k}\right)}{\exp\left(\frac{\alpha p_j + \boldsymbol{\beta}^\mathsf{T} \mathbf{x}_j}{\lambda_k}\right) + \rho_k}\right) \times \left(\frac{\exp\left(\lambda_k \ln\left(\exp\left(\frac{\alpha p_j + \boldsymbol{\beta}^\mathsf{T} \mathbf{x}_j}{\lambda_k}\right) + \rho_k\right)\right)}{\exp\left(\lambda_k \ln\left(\exp\left(\frac{\alpha p_j + \boldsymbol{\beta}^\mathsf{T} \mathbf{x}_j}{\lambda_k}\right) + \rho_k\right)\right) + \phi_k}\right)$$

Partial derivative and gradient of quantity demanded with respect to price and design attributes

$$\frac{\partial q_j}{\partial p_j} = \frac{m\alpha \exp\left(\frac{\alpha p_j + \beta x_j}{\lambda_k}\right)\left(\exp\left(\frac{\alpha p_j + \beta x_j}{\lambda_k}\right) + \rho_k\right)^{\lambda_k - 1}\left(\rho_k\left(\exp\left(\frac{\alpha p_j + \beta x_j}{\lambda_k}\right) + \rho_k\right)^{\lambda_k - 1} + \phi_k\right)}{\lambda_k\left(\phi_k + \left(\exp\left(\frac{\alpha p_j + \beta x_j}{\lambda_k}\right) + \rho_k\right)^{\lambda_k}\right)^2}$$

$$\frac{\partial q_j}{\partial \mathbf{x}_j} = \frac{m\beta \exp\left(\frac{\alpha p_j + \beta x_j}{\lambda_k}\right)\left(\exp\left(\frac{\alpha p_j + \beta x_j}{\lambda_k}\right) + \rho_k\right)^{\lambda_k - 1}\left(\rho_k\left(\exp\left(\frac{\alpha p_j + \beta x_j}{\lambda_k}\right) + \rho_k\right)^{\lambda_k - 1} + \phi_k\right)}{\lambda_k\left(\phi_k + \left(\exp\left(\frac{\alpha p_j + \beta x_j}{\lambda_k}\right) + \rho_k\right)^{\lambda_k}\right)^2}$$

First-order optimality conditions

$$p_j = c_j - \left[\frac{\partial q_j}{\partial p_j}\right]^{-1} q_j = c_j - \frac{\lambda_k\left(\phi_k + \left(\exp\left(\frac{\alpha p_j + \beta x_j}{\lambda_k}\right) + \rho_k\right)^{\lambda_k}\right)^2}{m\alpha \exp\left(\frac{\alpha p_j + \beta x_j}{\lambda_k}\right)\left(\exp\left(\frac{\alpha p_j + \beta x_j}{\lambda_k}\right) + \rho_k\right)^{\lambda_k - 1}\left(\rho_k\left(\exp\left(\frac{\alpha p_j + \beta x_j}{\lambda_k}\right) + \rho_k\right)^{\lambda_k - 1} + \phi_k\right)}$$

$$* m \left(\frac{\exp\left(\frac{\alpha p_j + \boldsymbol{\beta}^\mathsf{T} \mathbf{x}_j}{\lambda_k}\right)}{\exp\left(\frac{\alpha p_j + \boldsymbol{\beta}^\mathsf{T} \mathbf{x}_j}{\lambda_k}\right) + \rho_k}\right) \times \left(\frac{\exp\left(\lambda_k \ln\left(\exp\left(\frac{\alpha p_j + \boldsymbol{\beta}^\mathsf{T} \mathbf{x}_j}{\lambda_k}\right) + \rho_k\right)\right)}{\exp\left(\lambda_k \ln\left(\exp\left(\frac{\alpha p_j + \boldsymbol{\beta}^\mathsf{T} \mathbf{x}_j}{\lambda_k}\right) + \rho_k\right)\right) + \phi_k}\right)$$

$$\frac{\partial c_j}{\partial \mathbf{x}_j} = -\left[\frac{\partial q_j}{\partial p_j}\right]^{-1}\left[\frac{\partial q_j}{\partial \mathbf{x}_j}\right] = -\frac{\boldsymbol{\beta}^\mathsf{T}}{\alpha}$$

Random Coefficients Logit

The case of two equally-sized groups (latent-classes) is presented below, but the derivation is generalizable to n groups of any size, as discussed in footnote 14. Given logit choice share of consumer group i for product j:

$$q_{ij} = m \frac{\exp(\alpha_i p_j + \boldsymbol{\beta}_i^\top \mathbf{x}_j)}{\exp(\alpha_i p_j + \boldsymbol{\beta}_i^\top \mathbf{x}_j) + \theta_{ij}}$$

$$q_j = \sum_{1,2} q_{ij}$$

Partial derivative and gradient of quantity demanded with respect to price and design attributes

$$\frac{dq_j}{dp_j} = m \left(\frac{\alpha_1 \exp(\alpha_1 p_j + \boldsymbol{\beta}_1 \mathbf{x}_j)}{2(\exp(\alpha_1 p_j + \boldsymbol{\beta}_1 \mathbf{x}_j) + \theta_{1j})} - \frac{\alpha_1 \exp(\alpha_1 p_j + \boldsymbol{\beta}_1 \mathbf{x}_j)^2}{2(\exp(\alpha_1 p_j + \boldsymbol{\beta}_1 \mathbf{x}_j) + \theta_{1j})^2} + \frac{\alpha_2 \exp(\alpha_2 p_j + \boldsymbol{\beta}_2 \mathbf{x}_j)}{2(\exp(\alpha_2 p_j + \boldsymbol{\beta}_2 \mathbf{x}_j) + \theta_{2j})} - \frac{\alpha_2 \exp(\alpha_2 p_j + \boldsymbol{\beta}_2 \mathbf{x}_j)^2}{2(\exp(\alpha_2 p_j + \boldsymbol{\beta}_2 \mathbf{x}_j) + \theta_{2j})^2} \right)$$

$$= \frac{\alpha_1 q_{1j} - \alpha_1 q_{1j}^2/m + \alpha_2 q_{2j} - \alpha_2 q_{2j}^2/m}{2} = \frac{\alpha_1 q_{1j}(1 - q_{1j}/m) + \alpha_2 q_{2j}(1 - q_{2j}/m)}{2}$$

$$\frac{\partial q_j}{\partial \mathbf{x}_j} = m \left(\frac{\boldsymbol{\beta}_1 \exp(\alpha_1 p_j + \boldsymbol{\beta}_1 \mathbf{x}_j)}{2(\exp(\alpha_1 p_j + \boldsymbol{\beta}_1 \mathbf{x}_j) + \theta_{1j})} - \frac{\boldsymbol{\beta}_1 \exp(\alpha_1 p_j + \boldsymbol{\beta}_1 \mathbf{x}_j)^2}{2(\exp(\alpha_1 p_j + \boldsymbol{\beta}_1 \mathbf{x}_j) + \theta_{1j})^2} + \frac{\boldsymbol{\beta}_2 \exp(\alpha_2 p_j + \boldsymbol{\beta}_2 \mathbf{x}_j)}{2(\exp(\alpha_2 p_j + \boldsymbol{\beta}_2 \mathbf{x}_j) + \theta_{2j})} - \frac{\boldsymbol{\beta}_2 \exp(\alpha_2 p_j + \boldsymbol{\beta}_2 \mathbf{x}_j)^2}{2(\exp(\alpha_2 p_j + \boldsymbol{\beta}_2 \mathbf{x}_j) + \theta_{2j})^2} \right)$$

$$= \frac{\boldsymbol{\beta}_1 q_{1j} - \boldsymbol{\beta}_1 q_{1j}^2/m + \boldsymbol{\beta}_2 q_{2j} - \boldsymbol{\beta}_2 q_{2j}^2/m}{2} = \frac{\boldsymbol{\beta}_1 q_{1j}(1 - q_{1j}/m) + \boldsymbol{\beta}_2 q_{2j}(1 - q_{2j}/m)}{2}$$

First-order optimality conditions

$$p_j = c_j - \left[\frac{\partial q_j}{\partial p_j} \right]^{-1} q_j = c_j - \frac{q_{1j} + q_{2j}}{\alpha_1 \left(q_{1j}(1 - q_{1j}/m) \right) + \alpha_2 \left(q_{2j}(1 - q_{2j}/m) \right)}$$

$$\frac{\partial c_j}{\partial \mathbf{x}_j} = -\left[\frac{\partial q_j}{\partial p_j} \right]^{-1} \left[\frac{\partial q_j}{\partial \mathbf{x}_j} \right] = -\frac{\boldsymbol{\beta}_1 \left(q_{1j}(1 - q_{1j}/m) \right) + \boldsymbol{\beta}_2 \left(q_{2j}(1 - q_{2j}/m) \right)}{\alpha_1 \left(q_{1j}(1 - q_{1j}/m) \right) + \alpha_2 \left(q_{2j}(1 - q_{2j}/m) \right)}$$

Proceedings of the ASME 2019
International Design Engineering Technical Conferences
and Computers and Information in Engineering Conference
IDETC/CIE2019
August 18-21, 2019, Anaheim, CA, USA

DETC2019-98219

THE IMPACT OF CONSUMER PREFERENCE DISTRIBUTIONS ON DYNAMIC ELECTRICITY PRICING FOR RESIDENTIAL DEMAND RESPONSE

Samuel Dunbar
Graduate Research Assistant
North Carolina State University
Raleigh, NC, United States
scdunbar@ncsu.edu

Scott Ferguson
Associate Professor
North Carolina State University
Raleigh, NC, United States
scott_ferguson@ncsu.edu

ABSTRACT

Demand Response (DR) is the adjustment of consumer electricity demand through the deployment of one or more strategies, e.g. direct load control, policy implementation, dynamic pricing, or other economic incentives. Widespread implementation of DR is a promising solution for addressing energy challenges such as the integration of intermittent renewable energy resources, reducing capacity cost, and improving grid reliability. Understanding residential consumer preferences for shifting product usage and how these preferences are distributed amongst a population are key to predicting the effectiveness of different DR strategies. In addition, there is a need for a better understanding of how different DR programs, system level objectives, and preference distributions will impact different segments of consumers within a population. Specifically, the impacts on their product use behavior and electricity bill. To address this challenge, a product based approach to modeling consumer decisions about altering their electricity consumption is proposed, which links consumer value to their products, instead of directly to the amount of electricity they consume. This model is then used to demonstrate how population level preference distributions for altering product use impact system level objectives.

NOMENCLATURE

i	Electricity consumer index
I	Number of residential electricity consumers
j	Product index
J_i	Number of products owned by consumer i
k	Product use index
K_{ij}	Number of uses of product j by consumer i
t	Continuous time
T	Time period under consideration
h	Hour index
H	Number of hours in time period T
ϕ_{ijk}^t	Consumer dissatisfaction
λ_{ijk}^u	Consumer-product dissatisfaction coefficient
λ_{ijk}^b	Backward shift dissatisfaction coefficient
λ_{ijk}^f	Forward shift dissatisfaction coefficient
t_{ijk}^u	Time of initiation of product use
t_{ijk}^n	Nominal time of initiation of product use
D_{ijk}^t	Dissatisfaction cost
β_i	Dissatisfaction cost coefficient
C_{ijk}^E	Energy cost of product use
C_e	Instantaneous electricity price
P_{ij}	Product power consumption
t_{ijk}^e	End time of product use
d_{ij}	Duration of product use
θ_i	Consumer cost equivalent dissatisfaction
t_{ijk}^m	Mean time of product use
x	Vector of hourly electricity prices
x_h	Electricity price in hour h
D_e	Instantaneous electricity demand

L_h	Hourly electricity load
a_h	Hourly electricity generation cost coefficient
C_T^A	Aggregator cost in period T
R_T^A	Aggregator revenue in period T
$s_{\min}$	Minimum population sensitivity to electricity price
$s_{\max}$	Maximum population sensitivity to electricity price
r	Uniformly distributed random number between 0 and 1

1. INTRODUCTION

In the context of decision-based design (DBD), price is defined as a function of time and should be chosen so that it maximizes utility [1], as shown in Figure 1. With respect to power systems, the price of electricity has been an underutilized tool in influencing consumer demand. Maximizing the utility of an electricity grid operator is a unique case, in that additional demand at a fixed price does not always translate to an increase in utility. The ideal condition for a power generator, in absence of renewable resources, is constant demand where no generation capacity goes unutilized at any period in time. This allows systems to continuously operate at peak efficiency. Instead, electricity demand peaks occasionally throughout the year and system operators must have enough capacity for this peak demand, although most of the time this excess capacity remains unused. Considering the integration of intermittent renewable energy resources, a system operator would want to shift consumer demand from its nominal peak times to times when there is exceptional renewable resource generation to reduce system costs.

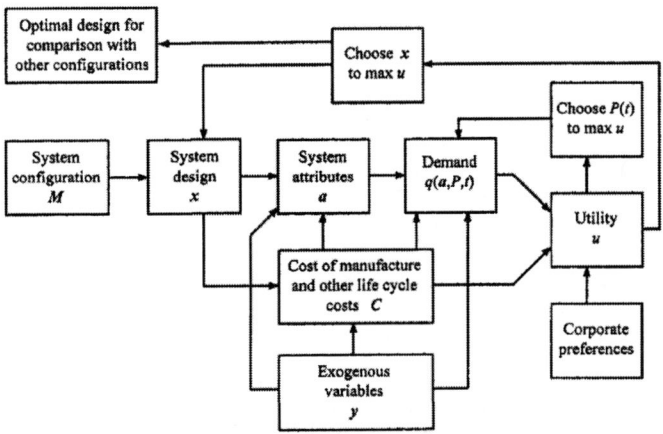

FIGURE 1. A Framework for Decision-Based Engineering Design [1]

Demand Response (DR) is the adjustment of consumer electricity demand through the deployment of one or more strategies, e.g. direct load control, policy implementation, dynamic pricing, or other economic incentives [2]. Using DR, system operators can meet peak demand while reducing their generation capacity. Other challenges can also be addressed, including the integration of intermittent renewable energy resources. Demand Response helps system operators better manage their generation resources and improve system efficiency. Other objectives for the system operator (seen as corporate preferences in the DBD framework) may include things such as minimizing consumer dissatisfaction, increasing system reliability, or reducing consumer energy burden. These preferences are a function of the type of utility providing the electricity. For example, public power utilities that are not-for-profit may more heavily weight system attributes associated with responsiveness and reliability [3].

The analysis of different dynamic pricing strategies has been done mostly in the context of optimization or game-theoretic frameworks. Within these frameworks, utility maximization is commonly used for modeling consumer decisions about electricity usage [4–8]. Utility maximization models the tradeoff between the benefit received from consuming electricity and the financial cost of it. For industrial or commercial consumers, representing their utility functions is fairly straight forward, as their objectives are primarily economic in nature. Residential Demand Response (RDR) is a branch of demand response where the objectives of the consumers are more ambiguous. Chapman, Verbic, and Hill [9] state that many models for RDR use objectives originally developed for industrial or commercial consumers. Furthermore, the majority of these objectives are built on the assumption that consumers value electricity itself instead of understanding the relationship between value and the product that is consuming electricity. By this we mean that the consumers' perceived value of product use is not directly dependent on the amount of electricity the product consumes.

Consumer control over their electrical consumption has often been modeled as continuous, implying they are in constant control of their consumption and it is infinitely adjustable [9]. This formulation does not accurately represent the majority of consumer products which consume hybrid or discrete amounts of electricity. Even if these assumptions are valid in aggregate for a large enough population, they provide little insight into consumer response at the individual level. Further, they do not account for the wide range of heterogeneity in a population's product use. If utilities optimize pricing at the aggregate level, optimization results can result in significant energy burden for populations with large wealth disparity. Where energy burden is defined as the ratio between a household's or individual's income to their electricity bill, over the same time period [10]. Lower income segments may experience larger energy burdens because of an inability to shift their usage to less expensive time periods. A better understanding of how households make tradeoffs between product use and electricity cost could better capture the market systems aspect of DR, resulting in optimization solutions that may better align with more realistic electricity consumption behavior. Stochastic activity-based approaches to residential electricity use have proved valuable for estimating electricity consumption for heterogeneous households over longer periods of time [11,12]. These models provide valuable insight into how different segments of residential electricity consumers use their electricity budget amongst their set of products. Other stochastic models have been developed for application to direct load control based RDR [13] and dynamic pricing based DR [14]. Consumer electricity demand is altered directly by the utility in the case of direct load control, so there is no need for a consumer response model in those scenarios. Stochastic dynamic pricing based DR models do provide additional insight into how consumers may shift their activities, but don't fully represent the consumers decision process.

Modeling of consumers at the individual/household level is critical to substantiating the population level assumptions and determining for what population size these assumptions are valid. It would also allow for the analysis of RDR strategies for smaller populations and the examination of the impacts on different segments of the population.

In this paper a consumer product based dissatisfaction model is proposed to represent the consumer's decision process when considering altering their product use because of dynamic electricity pricing. Specifically, we model the tradeoff consumers make between the dissatisfaction they experience by shifting their time of use and the cost of their electricity consumption. Unlike models that typically consider a consumer's deviation from their normal household level energy consumption during a specific time period, this model is based on deviations from normal times that consumer products would be used. This formulation is similar to a smart-product based dissatisfaction model developed by Yu and Hong [15]. However, Yu and Hong used an exponential dissatisfaction model to represent product dissatisfaction with its current level of energy demand, as compared to its median level of energy demand, for continuous use products. This model was also adapted to larger interruptible discrete loads, such as a Plug-In Hybrid Electric Vehicle and washing machine. It was used in conjunction with an autonomous Home Energy Management System (HEMS), to manage product usage in response to pricing changes. The model was not extrapolated to the system level and didn't model how consumers make decisions about the usage of time-critical appliances in absence of automation.

The dissatisfaction model proposed in this paper can be used to simulate residential consumer response to dynamic electricity pricing, in the absence of automated assistance, such as a HEMS. It links consumer objectives to their appliances, not directly to their energy consumption, and to their product usage behavior, through the implementation of a stochastic product use model. The lack of a HEMS also allows for insight into how automation systems could better assist the consumer decision process and the simulation of the impacts of partial automation on a population of consumers. Other DR programs have been implemented without automated assistance, and this model would be applicable in those cases, since it does not rely on any fundamental technology for implementation. If automation is introduced, this model can be used to represent consumer inputs to those systems as well.

2. CONNECTION TO MARKET SYSTEMS

Market systems research has established the importance of understanding the impacts of heterogenous consumer preferences on product-related design decisions [16,17]. In the context of a dynamic pricing DR strategy, consumer preference is defined as their willingness to shift or curtail certain electrical loads. Understanding what products different segments of consumers are willing to alter usage of and the portion of the population each segment comprises, is critical to predicting the aggregate level response and the impacts different segments will have on one another.

The structure of different market systems also plays an important role in determining how engineering design decisions are made [18,19]. Different DR frameworks implement different levels of automation, control, and connectivity. For example, Mohsenian-Rad et al. [20] proposes a scheduling game based on a full interconnected smart meter network, while Srinivasan et al. [4] makes no such assumptions about any underlying technology. The structure of the market also plays an important role, as some markets have intermediaries between the aggregator and consumer [6], while others do not [9]. The structure of the market can impact objectives of different agents substantially, leading to different model behavior and optimal action policies.

Consumers making decisions about shifting their electricity use will likely do so based on heuristics, rather than a carefully considered tradeoff, without the assistance of automation. How consumers consider their choices has the potential to substantially impact the success of a given RDR program. Noncompensatory decision making has been shown to play an important role in modeling the consumer decision process when purchasing a product [21,22] and has substantial implications for how products are designed. Consumers could also be noncompensatory when making decisions about their electricity usage, in that they could be indifferent to changes in electricity price with respect to one or more products in their set. This could have substantial impacts on the effectiveness of different RDR programs, depending on the power consumption of what products consumers are and are not willing to shift use of.

When considering product design, consumer preferences are usually modeled with respect to product attributes, as drivers of demand. Then an optimization is carried out to maximize market share, revenue, profit, and/or other objectives based on these preference models. In this case, modeling consumer decisions about electricity usage, can't be done effectively with previously developed attribute based preference models. The underlying consumer behavior is different from that of a purchase, although money is still being exchanged so a consumer can satisfy a need. To address this difference in consumer behavior, a different model form is proposed to capture consumers' decision process when making decisions about their product use within a dynamic pricing RDR framework.

3. CONSUMER MODEL DEVELOPMENT

A Taguchi loss function [23] is a quadratic approximation for the loss in quality of a product. It is a representation of the idea that deviation from the nominal value of a design variable results in a negative outcome; a loss in quality and customer dissatisfaction. In this work, a piecewise quadratic function is used for modeling a consumer's loss in satisfaction when they shift the time at which they use a product. As a consumer moves their time of use away from the original (nominal) intended use time, they become increasing dissatisfied. The piecewise characteristic of the function captures a consumer's unequal dissatisfaction with shifting their usage earlier or later. The dissatisfaction function used in this work is shown in Equation 1, where i references a specific consumer and j references a specific product, owned by consumer i. The subscript k indicates the product use event number, throughout the time period in consideration. If a product is used five times for example, k will take on values of one through five.

$$\phi_{ijk}^{t} = \lambda_{ijk}^{u} \left(t_{ijk}^{u} - t_{ijk}^{n} \right)^{2} \qquad (1)$$

Where

$$\lambda_{ijk}^{u} = \begin{cases} \lambda_{ijk}^{b}, & t_{ijk}^{u} - t_{ijk}^{n} \leq 0 \\ \lambda_{ijk}^{f}, & t_{ijk}^{u} - t_{ijk}^{n} > 0 \end{cases} \qquad (2)$$

In Equation 1, t_{ijk}^{u} is the actual time that the use of product j is initiated by consumer i. t_{ijk}^{n} is the nominal time of use that would have occurred in absence of dynamic electricity pricing, λ_{ijk}^{u} is the dissatisfaction coefficient, specific to a single consumer-product pair. Note that this coefficient is represented by λ_{ijk}^{b} when the product is used prior to the nominal time of use, and by λ_{ijk}^{f} when the product is used after the nominal time of use, as shown in Equation 2. Where b denotes a shift backward in the time of product use, and f denotes a shift forward in the time of product use, with respect to the nominal time of use. By creating a piecewise function of this nature, we can represent the difference in the consumer's dissatisfaction sensitivity.

Consumers can be financially incentivized to accept some level of dissatisfaction, represented here as a dissatisfaction cost. This cost reflects the amount of money a consumer would accept as a trade for a corresponding amount of dissatisfaction. For this model, a linear relationship is assumed between a consumer's dissatisfaction and their dissatisfaction cost (D_{ijk}^{t}), as in Equation 3.

$$D_{ijk}^{t} = \beta_{i} \phi_{ijk}^{t} \qquad (3)$$

Where β_{i} is consumer i's dissatisfaction cost coefficient, which is defined as being solely a consumer characteristic, independent of product. This coefficient is repetitive of the amount of money a consumer would be willing to take to incur a proportional amount of dissatisfaction. A consumer's total cost, when considering the start of product use under dynamic electricity pricing, can then be represented by the sum of their dissatisfaction cost and the electricity cost incurred from using the product. For each product use event k, this formulation creates a consumer's disutility function as in Equation 4.

$$C_{ijk} = D_{ijk}^{t} + C_{ijk}^{E} \qquad (4)$$

Where the consumer's energy cost for product use is defined in Equation 5, as the integral of the derivative of the instantaneous electricity price C_{e} multiplied by the instantaneous power consumption P_{ij} of the product.

$$C_{ijk}^{E} = \int_{t_{ijk}^{u}}^{t_{ijk}^{e}} \frac{dC_{e}}{dt} P_{ij} dt \qquad (5)$$

Operating under perfect conditions, consumers would always minimize their total cost. This minimization for each product use becomes an objective function, as shown in Equation 6.

Minimize :

$$C_{ijk}\left(t_{ijk}^{u} \right) = \beta_{i} \lambda_{ijk}^{u} \left(t_{ijk}^{u} - t_{ijk}^{n} \right)^{2} + \int_{t_{ijk}^{u}}^{t_{ijk}^{e}} \frac{dC_{e}}{dt} P_{ij} dt \qquad (6)$$

Where t_{ij}^{e} represents the end of the product use period, defined by Equation 7.

$$t_{ijk}^{e} = t_{ijk}^{u} + d_{ijk} \qquad (7)$$

d_{ijk} is the duration of product use. This duration is a fixed value ($d_{ijk} = d_{ij}$) for products where specific outcomes or cycles are achieved, such as a dishwasher. However, this value can be impacted by variations in electricity pricing for products with variable use durations, such as televisions. Note that there are no constraints placed on the consumer's objective, making it appear easily solvable with standard one dimensional optimization approaches, however it can have multiple local minima, depending on the daily electricity price distribution.

In the absence of automation consumers will likely not know the exact energy cost of their product use at every given moment. Consumers use heuristics to estimate the cost of their consumption, typically over estimating the cost of low power products and under estimating the cost of high power products [24]. This could lead to a substantial difference between their heuristically estimated and true optimums, but for the purposes of this model, consumer heuristics will not be incorporated.

Another consideration of this model is the impacts of dynamic pricing on consumer energy burden. This is represented as a consumer's dissatisfaction with the total cost of their energy consumption, defined by Equation 8.

$$\theta_i = \frac{\sum_{j=1}^{J_i} \sum_{k=0}^{K_{ij}} C_{ijk}^E}{\beta_i} \tag{8}$$

This equation produces units of dissatisfaction but can be used as a surrogate for consumer energy burden. It is defined as the ratio between the cost of their energy consumption and their dissatisfaction cost coefficient. It is assumed that the dissatisfaction cost coefficient is correlated with income, as lower income consumers would be likely to incur higher levels of dissatisfaction, for smaller monetary benefit.

In addition to altering the time at which product use occurs, consumers can also reduce their total usage, reflected in a reduction in the frequency with which they use their products. For the purposes of this study, reduction in total energy usage by the consumers in not considered. Additionally, the price elasticity of demand of electricity tends to be relatively inelastic for the residential sector [25], indicating that this neglection may not drastically impact the results.

For continuous use products, consumers can alter the setting(s) at which the product operates. Note, this is only realistic, provided the alteration of these settings does not inhibit them from satisfying the need they intend to by using the product. For example, cooking a meal with a stove may require a certain temperature, so a consumer would not purposely undercook their food to reduce their electricity bill. Alternatively, consumers might adjust their thermostat periodically and alter the setting based on their expected cost, subject to some comfort threshold.

Other products, typically referred to as white goods, are defined as products for which use is not time critical, meaning they offer a period of time during which the consumer experiences no dissatisfaction in delaying use. An efficient DR strategy should be able to leverage consumer flexibility, before imposing increased costs or any level of dissatisfaction onto consumers.

4. MODEL CALIBRATION

The model should capture the behavior of consumers who are extremely sensitive to changes in electricity pricing and those who are completely indifferent. Indifferent electricity consumers could be modeled by setting the product of their dissatisfaction (λ_{ijk}^u) and dissatisfaction cost (β_i) coefficients equal to infinity, to achieve an optimum of $t_{ijk}^{u*} = t_{ijk}^n$. This is impractical for modeling purposes as dissatisfaction coefficients would be distributed between zero and infinity. This challenge is overcome using a coefficient calibration technique. Coefficient values are maximized subject to a predetermined amount of acceptable error, between the actual and nominal times of use. This calibration method requires two additional

inputs for each consumer product; predetermined error (ε_j) and a set maximum amount of time consumers would be willing to delay product use (Δt_j^c).

The calibration method is based on a linearly decreasing electricity price, from maximum at $t_j^u - t_j^n = 0$ to zero at $t_j^u - t_j^n = \Delta t_j^c$. The electricity price function for a specific product j is displayed in Equation 9.

$$C_e = \max\left[-\left(\frac{C_{e,\max}}{\Delta t_j^c} \right) t + C_{e,\max}, 0 \right] \tag{9}$$

Note that there cannot be a negative electricity price, hence the maximum function. The electricity cost from operating a product under this pricing scheme is given by Equation 10.

$$C_j^E = \int_{t_j^u}^{t_j^e} \max\left[-\left(\frac{P_j C_{e,\max}}{\Delta t_j^c} \right) t + P_j C_{e,\max}, 0 \right] dt \tag{10}$$

For cases where the power consumption from the product can be approximated as constant over the time, solving the integral presented in Equation 8 yields Equation 11.

$$C_j^E = P_j C_{e,\max} \left(\frac{\left(t_j^u \right)^2 - \left(t_j^e \right)^2}{2\Delta t_j^c} + t_j^e - t_j^u \right) \tag{11}$$

$$\forall\ \Delta t_j^c \geq t_j^e - t_j^n$$

Note the constraint placed on this function, allows for the maximum function to be removed, making it easier to deal with. The constraint restricts the functions validity to cases where the maximum time of use window is larger than the difference between the nominal time of use and the end of product use time. Setting up the consumer objective in a similar manner to Equation 4 yields Equation 12.

$$C = \beta \lambda_j^u \left(t_j^u - t_j^u \right)^2 + $$

$$P_j C_{e,\max} \left(\frac{\left(t_j^u \right)^2 - \left(t_j^e \right)^2}{2\Delta t_j^c} + t_j^e - t_j^u \right) \tag{12}$$

For a consumer completely indifferent to changes in electricity pricing, the minimum of this function is located where the actual time of use is the same as the nominal time of use. This requires the dissatisfaction and dissatisfaction cost coefficients to equal infinity. Instead, let the difference between the time of use and the nominal time of use equal the predefined error value. This sets a limit on the maximum coefficient values. While this doesn't completely capture the behavior of indifferent consumers, most dynamic pricing schemes utilize step functions for electricity price, not continuously changing functions. This should result in a smaller error value when subject to step function pricing, unless the nominal time of use is within the predefined error to the next pricing period. Updating the consumer objective with the error term and product duration yields Equation 13.

$$C = \beta \lambda_j^u \varepsilon_j^2 + P_j C_{e,\max}\left(d_j - \frac{d_j^2 + 2\varepsilon_j d_j}{2\Delta t_j^c} \right) \quad (13)$$

The minimum of Equation 12 will exist at the point at which the dissatisfaction cost and energy cost terms are equivalent to one another. Rearranging the equation and solving for the dissatisfaction and dissatisfaction cost coefficients, yields Equation 14.

$$\left[\beta \lambda_j^u \right]_{\max} = \frac{P_j C_{e,\max}}{\varepsilon_j^2}\left(d_j - \frac{d_j^2 + 2\varepsilon_j d_j}{2\Delta t_j^c} \right) \quad (14)$$

$$\forall\ \Delta t_j^c \geq \varepsilon_j + d_j$$

5. SYSTEM MODEL

The system comprises a single RDR aggregator responsible for purchasing and distributing electricity to a community of consumers. The aggregator can set the price of the electricity at each hour h throughout time period T, prior to the start of the period. The aggregator then communicates this information to the household agents within the model, so all consumers know the hourly electricity prices before the period begins. The community has an energy budget comprised of all the energy consumer by each of the individual consumers. The consumers distribute their energy usage across their set of products, at different times throughout the day. This model is displayed in Figure 2.

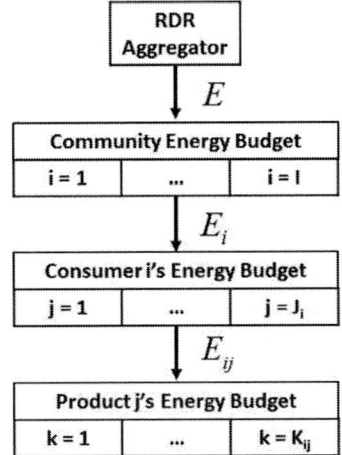

FIGURE 2. Hierarchical Energy Aggregation Model

The hourly electricity cost model for the aggregator was adopted from [20] is given by Equation 15.

$$C_h^A = a_h L_h^2 \quad (15)$$

Where L_h is the aggregator's hourly electricity load; the aggregation of the consumers' demand during hour h and a_h

is the hourly generation cost coefficient. The total electricity generation cost for the RDR aggregator then becomes Equation 16.

$$C_T^A = \sum_{h=1}^{H} C_h^A \quad (16)$$

Where H is the total number of hours in time period T. The revenue generated by the aggregator during period T is equivalent to the aggregation of all the individual consumer energy costs, given by Equation 17.

$$R_T^A = \sum_{i=1}^{I}\sum_{j=1}^{J_i}\sum_{k=0}^{K_{ij}} C_{ijk}^E \quad (17)$$

The RDR aggregator is given three objectives to minimize; the daily electricity demand variance (f_1), hourly pricing variance (f_2), and difference between actual profit and a profit target (f_3), as represented by Equation 18. In the case presented here, the profit margin can be considered zero or as accounted for in the generation cost, given by Equation 15. This reflects that RDR aggregators will be regulated like a traditional utility company where profit margins are set at a fixed level over extended periods of time. Constraints were added to limit the maximum hourly electricity price (Equation 19) and ensure that the aggregator recovered its total energy cost during the time period (Equation 20).

Minimize:
$$\{f_1(x), f_2(x), f_3(x)\} \quad (18)$$

Subject to:
$$g_1(x) = \max[x] - C_{e,\max} \leq 0 \quad (19)$$
$$g_2(x) = C_T^A(x) - R_T^A(x) \leq 0 \quad (20)$$

Where x is the hourly electricity price vector. The daily demand variance is given by Equation 21.

$$f_1(x) = \frac{1}{T}\int_{t=0}^{T}\left(D_e(t) - \bar{D}_e \right)^2 \quad (21)$$

Where $D_e(t)$ is the continuous electricity demand throughout the time period T and $\bar{D}_e$ is the mean demand during the same time period. The hourly pricing variance was calculated using Equation 22.

$$f_2(x) = \frac{1}{H-1}\sum_{h=1}^{H}\left(x_h - \bar{x} \right)^2 \quad (22)$$

Where x_h is the price of electricity in hour h and $\bar{x}$ is the average electricity price in time period T. The hourly pricing variance was selected as an objective since large fluctuations in electricity price may lead to a larger reduction in demand variance but cause consumers more dissatisfaction as they shift more of their product uses in response to these fluctuations. Consumers have also been shown to prefer pricing strategies with less variations in price throughout the day [26], which would be important to a public power utility. The third objective assigned to the aggregator was to minimize the square of the difference between the cost and revenue generated over time period under consideration. This is expressed in Equation 23.

$$f_3(x) = \left(C_T^A(x) - R_T^A(x) \right)^2 \qquad (23)$$

6. METHODOLOGY

In this section, a methodology is presented for demonstrating the impacts of consumer preference distributions on the optimal solutions for a dynamic pricing RDR program. Impacts on aggregator objectives, daily electricity demand variance, consumer cost, and consumer energy burden are explored. This is accomplished through manipulation of the population level sensitivity to electricity pricing, which is used to bound the distributions of dissatisfaction coefficients. As population sensitivity to electricity price decreases, the RDR aggregator increases the price differential between hours, encouraging more consumers to shift their use. While this is effective for reducing demand variance, it results in greater levels of dissatisfaction amongst the community and has the potential to substantially increase the energy burden of a portion of the population.

To demonstrate the impact of dissatisfaction coefficient distributions on the optimal hourly pricing rates, all households within the community were assigned a homogeneous set of products and product usage probabilities at each step in time. The products, and product characteristics, assigned to each of the households are displayed in Table 1.

TABLE 1. Product Characteristics: Power and Duration [13]

Product	Power (W)	Duration (hr)
Stove and Oven	588	0.7
Microwave Oven	800	0.1
Coffee Maker	239	0.4
Dishwasher	922	0.9
Washer and Dryer	1,873	1.9
Television	105	1.0
Second Television	75	1.0
Computer	125	1.0

The product attribute values were modified from [13], to have constant power consumption, while maintaining the original durations and total energy consumption per use. This was done to simplify coefficient calibration and reduce computational time. Nominal product use schedules were generated for all households within the community. For every hour within the time period and every product within each consumer's set, a uniformly distributed random number was generated and compared with the probability of product use during that hour. If the randomly generated number was less than the probability of product use, the consumer was set to use that product at a uniformly distributed random time within that hour. Each using the modified weekday product use frequencies, displayed in Table 2 and hourly weekday product use probabilities from Paatero and Lund [13].

TABLE 2. Weekday Product Use Frequency [13]

Product	Frequency of Use (Events per Day)
Stove and Oven	0.56
Microwave Oven	0.98
Coffee Maker	0.98
Dishwasher	1.16
Washer and Dryer	0.30
Television	1.95
Second Television	0.28
Computer	0.70

For the purpose of this simulation, the RDR aggregator is assumed to have perfect knowledge of the population. While this is obviously not the case in reality, tools like machine learning coupled with smart meter data could provide accurate models for the aggregator. Here, the aggregator's objectives are at the population level, so a model for each household is not required unless further insight into specific impacts on different individual segments or members of the population is desired.

The aggregator's optimization problem was solved using a Multiobjective Genetic Algorithm (MOGA). Binary encoding was used for representing hourly electricity prices to one decimal place. Tournament selection was used to choose parents from the population at each iteration, initially drawing two vectors of hourly prices (x) from the population at random. The design with the better fitness was selected to be the first parent and this process was repeated to select the second parent. The criteria used to determine fitness, in the order in which they were tested, lower number of constraint violations, lower domination count, and larger average normalized distance to the four nearest neighboring designs. Uniform crossover was performed on the two parent strings to generate two children, with a 50% probability of each bit crossing over between parents. Bit-wise mutation was then performed on the children with a 5% probability of occurrence. At each iteration, the population was doubled, and the fitter half was allowed to survive until the next generation.

At each price vector evaluation, the population of household agents each solves their own local optimization problems corresponding to all of their scheduled product use events. A bounding phase algorithm in combination with golden section was used to minimize each household's objective for each product use event. To address the possibility of multiple local minima existing for the consumer objectives, 11 different initial starting locations were used, evenly distributed between three hours before and after the nominal time of product use.

The best solution found from the different starting locations was then set as the consumer's actual time of product use.

The simulation was executed with five different populations of household agents; each having a different dissatisfaction coefficient distribution. The dissatisfaction coefficients were uniformly distributed amongst the agents, between a different upper and lower bound for each population. This was done twice for each consumer product pair, for each the backward (b) and forward (f) shift dissatisfaction coefficients. This is displayed in Equation 24.

$$\beta_i \lambda_{ij}^u = \left((1 - s_{max}) + r(s_{max} - s_{min}) \right) \times \left[\beta \lambda_j^u \right]_{max} \quad (24)$$
$$u = b, f$$

Where s_{min} and s_{max} are the maximum and minimum bounds on the dissatisfaction coefficients, as a portion of the maximum value for each product. These terms represent the population level sensitivity to changes in electricity price. Higher values correspond to a population more willing to deviate from their nominal times of use, and vice-versa. This is used to generate different populations by distributing their dissatisfaction coefficients over different ranges. Note that a uniform distribution was assumed for all, indicated by r, which is a uniformly distributed number between zero and one.

Generating the agents' dissatisfaction coefficients using Equation 24, assumes that these coefficients are independent of each other for a single consumer, i.e. the coefficients assigned to a consumer for one product, do not influence the coefficients assigned for any other products in their set. We view this as a simplifying assumption for this study and will further explore this correlation in future work. Equation 25 was then applied to separate the dissatisfaction and dissatisfaction cost coefficients for each household level agent.

$$\beta_i = \min_j \left[\beta_i \lambda_{ij}^u \right], \quad u = b, f \quad (25)$$

For this analysis, the magnitude of the dissatisfaction cost coefficient is not important, but its relative magnitude still provides insight into consumer characteristics. The lower bound and upper bound of each populations dissatisfaction coefficient distribution is displayed in Table 3.

TABLE 3. Minimum and Maximum Electricity Pricing Sensitivities for the Homogeneous Populations

Scenario	s_{min}	s_{max}
1	0.0	0.2
2	0.2	0.4
3	0.4	0.6
4	0.6	0.8
5	0.8	1.0

Prior to each simulation, 100 agents were generated; assigned a product use schedule and dissatisfaction coefficients. Five different product use schedules were generated for all 100 agents and used for each of the five populations, resulting in a total of 25 different simulations. The maximum amount of time consumer would consider shifting their product usage (Δt_j^c) was taken as 3 hours and the predetermined error (ε_j) was assigned a value of five minutes, for all products in the consumers' set. The maximum hourly electricity price ($C_{e,max}$) was set to 30 cents per kilowatt hour and the hourly electricity generation cost model coefficient (α_h) was assigned a value of 0.3 cents per kilowatt hour squared [20].

7. RESULTS

Results in this section explore the impact of consumer sensitivity to electricity pricing on the RDR aggregator's objective functions. Specifically, we investigate how the Pareto frontier shifts towards lower levels of daily electricity demand variation when the population's sensitivity to electricity price is increased. We compare the characteristics of the nominal demand curve under flat rate pricing with the demand curve produced when dynamic pricing is used. This allows for an investigation of how optimal hourly prices impact the daily electricity demand curve when price sensitivity of the population is increased. Further, potential impacts on consumers are discussed.

Characteristics of five product use schedules generated for the consumers are displayed in Table 4. The flat price is the constant electricity price charged by the RDR aggregator so that costs of generation for the day could be met.

TABLE 4. Product Use Schedule Characteristics

Schedule	Total Load (kWh)	Peak Demand (kW)	Demand Variance (kW²)	Flat Price (cents)
1	298.7	28.5	87.8	5.7
2	367.7	34.5	130.1	7.0
3	299.8	31.5	102.8	6.1
4	336.3	36.3	115.5	6.5
5	294.2	29.4	93.4	5.8

7.1 Pareto Frontier Analysis

The Pareto optimal solutions for the optimization problem posed from the perspective of the RDR aggregator are shown in Figures 3, 4, and 5. Recall, the variance in electricity demand is reduced through the process of consumers shifting their product use times, not through reduction the amount of electricity consumed. In this paper, we assume the total energy consumption as fixed, and will revisit this assumption in future work. The aggregator's cost difference is presented in cents squared, so a value of five million would correspond to slightly above 22 dollars, or an average of 22 cents per home. Color in these plots differentiates the five different scenarios, described by the population electricity pricing sensitivities, displayed in Table 3.

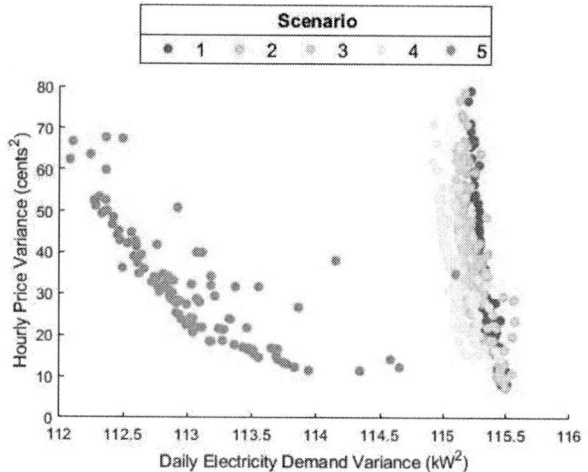

FIGURE 3. Aggregator Pareto Optimal Solutions for Objectives 1 and 2, Scenarios 1-5, Schedule 4

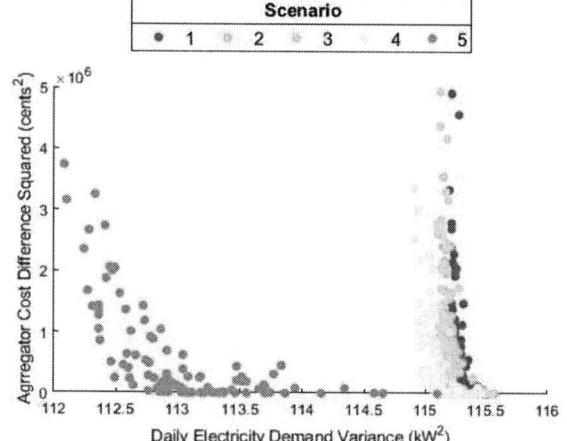

FIGURE 4. Aggregator Pareto Optimal Solutions for Objectives 1 and 3 Scenarios 1-5, Schedule 4

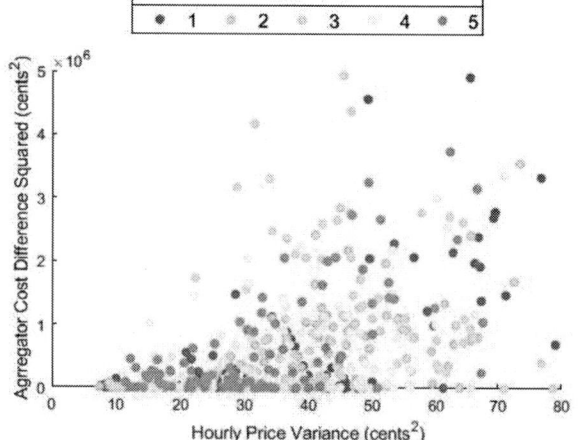

FIGURE 5. Aggregator Pareto Optimal Solutions for Objectives 2 and 3, Scenarios 1-5, Schedule 4

The population of agents in Scenario 5 has the highest sensitivity to electricity prices, giving them the lowest dissatisfaction coefficient values, which means that they are more willing to accept a shift in product usage in response to higher electricity prices. This is reflected by the substantially lower achievable demand variation for this scenario. The distance between the clusters also indicates that the relationship between population sensitivity and minimum achievable demand variation may be nonlinear.

Table 5 displays daily electricity demand variance and the reduction in demand variance from the nominal case, for the minimum demand variance and compromise solutions, averaged across all five usage schedules. Scenario 5 resulted in the largest reductions in demand variance, by nearly an order of magnitude as compared to Scenarios one through four. Scenarios 1 and 2 resulted in almost no change in demand variation as compared to the nominal case. Results for the comprise solution are also shown in Table 5, averaged across all five usage schedules. Here, the 'compromise solution' is the solution from the Pareto frontier with the shortest normalized Euclidian distance to the utopia point. Again, Scenario 5 resulted in the largest reductions in demand variance, by nearly an order of magnitude as compared to Scenarios one through four.

Table 5. Optimal Pricing Solution Impacts on Demand Variance Averaged Over Schedules 1-5

Scenario	Minimum Demand Variance Solution		Compromise Solution	
	Average Demand Variance (kW^2)	Average Variance Reduction (%)	Average Demand Variance (kW^2)	Average Variance Reduction (%)
1	105.7	0.24	105.9	0.00
2	105.7	0.26	105.9	0.01
3	105.5	0.41	105.9	0.06
4	105.3	0.61	105.8	0.09
5	102.9	2.78	104.9	0.94

7.2 Daily Electricity Demand Impacts

The pricing solution for Scenario 1 that resulted in the lowest daily demand variation for a fixed product use schedule is displayed in Figure 6. The large variation in electricity price during periods of high demand appears because consumers are responding to the price difference between hours and the magnitude of the price at the current time. Increasing price alone does not alter consumer behavior in this simulation because curtailments are not considered, and no incentives exist. Fundamentally, the electricity price differential between adjacent hours is the driver of consumers shifting their product use, as opposed to the electricity price magnitude alone.

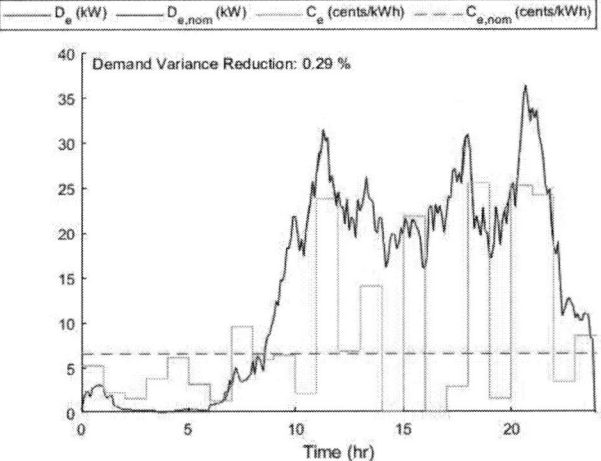

FIGURE 6. Minimum Demand Variation Pricing Solution for Scenario 1, Schedule 4

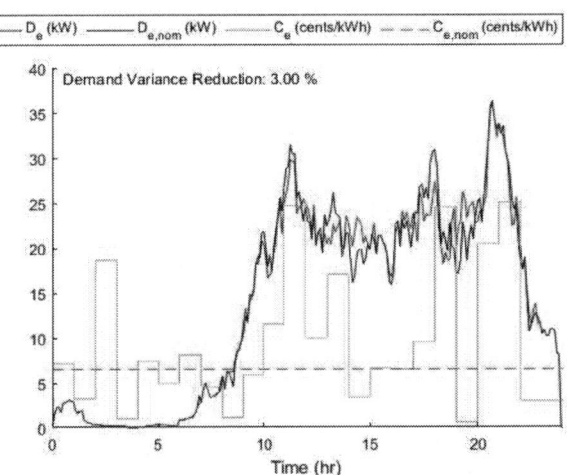

FIGURE 7. Minimum Demand Variation Pricing Solution for Scenario 5, Schedule 4

There is minimal difference between the nominal (blue) and actual (red) demand curves because the population in Scenario1 is least sensitive to variations in electricity pricing. Initially one may expect that during periods of low demand, such as between hour 1 and hour 10 in this example, the electricity price would be near zero. In this simulation, the cost of generating electricity for a given hour and the price of electricity during that same hour are decoupled, so electricity price does not correspond to generation cost on an hourly basis. The higher electricity prices during these low demand periods exist so that the aggregator can meet their profit target. This occurs because larger fluctuations in use are not heavily penalized because they have a small impact on daily demand variance.

The population in Scenario 1 only altered their use by a small amount, even when presented with drastic pricing differences between adjacent hours. It is likely that real consumers will have sensitivities much higher than those reflected in Scenario 1. Scenario 5 has the highest population sensitivity, meaning that they are most willing to alter their product use schedules. They do this because they are most interested in reducing their electricity costs. The pricing solution that resulted in the lowest daily demand variation for Scenario 5 is displayed in Figure 7.

The population of agents in Scenario 5, being the most sensitive to changes in electricity pricing, shifts their use considerably more than Scenario 1 when subjected to large hourly pricing differentials. The hourly electricity prices approximately follow the shape of the demand curve, especially during periods of higher demand from approximately hour 10 to hour 24. Peaks in demand correspond to peaks in prices, and lower prices correspond to demand valleys. Incentivizing consumers to shift their usage to periods of lower demand helps flatten the demand curve. The compromise pricing solution for Scenario 5 is shown in Figure 8. As before, the 'compromise solution' is the solution from the Pareto frontier with the shortest normalized Euclidian distance to the utopia point. The utopia point in this analysis is zero for all three of the RDR aggregator's objectives.

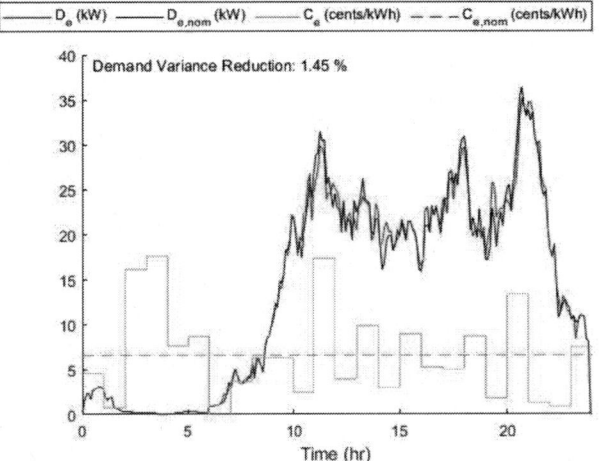

FIGURE 8. Compromise Pricing Solution Demand for Scenario 5, Schedule 4

While demand variance is not reduced as greatly in the compromise solution, there is less variation in the hourly electricity prices throughout the day. This solution also fully minimizes the aggregator's third objective, meeting the aggregator's profit target. The pattern of higher prices corresponding to demand peaks and lower prices corresponding to demand valleys is also seen in Figure 8.

The maximum reduction in the daily demand variance as a function of the mean population level sensitivity to electricity pricing is shown in Figure 9. Here, each of the five product use schedules and scenarios are represented. The maximum reductions in daily demand variance for each of the populations and use schedules, correspond to the minimum demand variance pricing solutions.

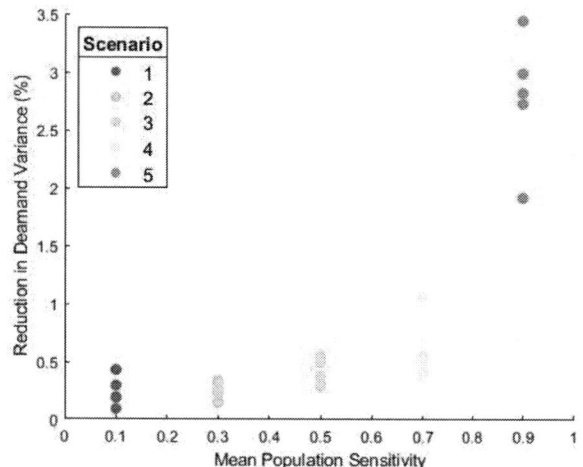

FIGURE 9. Maximum Reduction in Daily Electricity Demand Variance as a Function of Mean Population Sensitivity to Electricity Price

As population level sensitivity to price increases, the reduction in the daily electricity demand increases substantially. The population of agents in Scenario 5 offers substantially more reduction in electricity demand variance, as compared to the others. This large differential implies that the reduction in daily demand variance likely does not correspond linearly with population level sensitivity. Table 6 displays the daily electricity cost and cost dissatisfaction equivalent for each of the five populations, averaged over the five different usage schedules. These values were calculated using the compromise solution found for each population and usage schedule.

TABLE 6. Average Daily Cost and Cost Dissatisfaction Equivalent for Each Scenario (Compromise Solution)

Scenario	Average Consumer Energy Cost ($)	Average Cost Equivalent Dissatisfaction (Diss.)
1	0.20	0.09
2	0.20	0.12
3	0.20	0.17
4	0.20	0.30
5	0.20	17.19

Note that the average consumer energy cost stays the same, as the difference between the profit target and actual profit was approximately zero for all the compromise solutions found. The average cost dissatisfaction equivalent increases at a relatively slow rate from Scenario 1 to Scenario 4, then increases dramatically from Scenario 4 to Scenario 5. This is due to several members within Scenario 5 having near zero dissatisfaction cost coefficients (β_i) and larger daily energy costs. This results in dramatically higher cost equivalent dissatisfaction values. While the cost equivalent dissatisfaction values are primarily a result of the way in which agents are generated, observing these values for an actual population can provide important insight to an RDR aggregator.

8. DISCUSSION

The distribution of consumer preferences amongst the population is important to understand when designing a dynamic pricing strategy for RDR. Different preference distributions can produce drastically different results with respect to the daily electricity demand variance. Understanding what level of variation reduction is achievable is imperative for grid planning and energy generation resource management. If the population tends to be more sensitive to dynamic pricing and lower levels of demand variance are achievable, a system operator may be able to forego adding additional capacity to the system. It would also make the integration of renewable energy resources more appealing if demand could be shifted to periods of higher power generation.

On the consumer side, wide variability in the distribution of consumer preferences was shown to potentially impact more sensitive portions of the total population in a negative way. While the large disparity in electricity cost dissatisfaction equivalent is shown in Table 6 is hypothetical, it does indicate that there is a substantial tradeoff between demand variance reduction and consumer energy burden. To further explore this tradeoff, more information about consumer heterogeneity would need to be known, like what other consumer behaviors correspond to lower dissatisfaction cost coefficients.

Without considering curtailment in this model, hourly electricity price differential became the driver of shifting consumer product use. While the differentials in the minimum demand variance solution were relatively large, they did not produce a reduction in demand variance above 3.5%. This indicates that the consumer preference distributions chosen for this analysis may be more conservative than those used when modeling a real population. These results also indicate that reduction in demand variance and cost dissatisfaction equivalent are likely nonlinear, with respect to the population sensitivity to electricity price.

Determining the coefficients for this model for an actual population of consumers can be accomplished through the use of survey data or smart meter data gathered directly from the home. Smart meter data would be required from both periods of flat pricing and dynamic pricing, to gather information about the consumers' behavior unperturbed by dynamic pricing and their behavior after they are subjected to it. Survey questions should inquire about consumers' general electricity usage and product characteristics, such as when they ordinary use certain products, what products they own, the power consumption of those products, and the duration they use them. In addition to basic product information, the survey should seek to quantify what financial incentives consumers would be willing to accept for a given delay in product use and what is the maximum amount of time they would be willing to delay use. Surveys could be used for estimating a population's response to dynamic pricing in advance of it being implemented. This would allow better initial pricing strategies to be developed before any data about the population's response is known.

9. CONCLUSION

In this paper a product-based model for the simulating consumer response to dynamic electricity pricing was presented. This model was used to explore the impacts of population level sensitivity to electricity pricing on the optimal hourly pricing solutions in an RDR framework. The model links consumer willingness to shift their energy consumption with their products, instead of with their total household level energy consumption. Further, it demonstrates that the population's sensitivity to electricity pricing does not correspond linearly to consumer energy burden or the reduction in daily electricity demand variance. In absence of curtailment, modeling consumers as shifting their electricity loads with respect to time as opposed to altering the total amount of electricity consumed, demonstrated that the price differential between adjacent hours is the primary driver for reducing daily electricity demand variation, as opposed to the magnitude of the prices themselves. This leads to some counter intuitive solutions, where prices during low demand periods may be increased to meet profit margin, while allowing for a larger achievable price differential during peak hours.

This model can be seeded with survey data or smart meter data to fit a particular population of interest. This can be used to simulate the response of different consumer populations, before different pricing strategies are implemented, which is important to understanding the potential impacts of dynamic pricing on individual households.

REFERENCES

[1] G. A. Hazelrigg, "A Framework for Decision-Based Engineering Design," *J. Mech. Des.*, vol. 120, no. December, pp. 653–658, 1998.

[2] M. H. Albadi and E. F. El-Saadany, "Demand response in electricity markets: An overview," *2007 IEEE Power Eng. Soc. Gen. Meet. PES*, pp. 1–5, 2007.

[3] American Municipal Power Inc., "What is Public Power?"

[4] D. Srinivasan, S. Rajgarhia, B. M. Radhakrishnan, A. Sharma, and H. P. Khincha, "Game-Theory based dynamic pricing strategies for demand side management in smart grids," *Energy*, vol. 126, pp. 132–143, 2017.

[5] P. Yang, G. Tang, and A. Nehorai, "A game-theoretic approach for optimal time-of-use electricity pricing," *IEEE Trans. Power Syst.*, vol. 28, no. 2, pp. 884–892, 2013.

[6] M. Yu and S. H. Hong, "Incentive-based demand response considering hierarchical electricity market: A Stackelberg game approach," *Appl. Energy*, vol. 203, pp. 267–279, 2017.

[7] P. Samadi, H. Mohsenian-Rad, R. Schober, and V. W. S. Wong, "Advanced Demand Side Management for the Future Smart Grid Using Mechanism Design," *IEEE Trans. Smart Grid*, vol. 3, no. 3, pp. 1170–1180, 2012.

[8] A. J. Conejo, J. M. Morales, and L. Baringo, "Real-time demand response model," *IEEE Trans. Smart Grid*, vol. 1, no. 3, pp. 236–242, 2010.

[9] A. C. Chapman, G. Verbic, and D. J. Hill, "A healthy dose of reality for game-theoretic approaches to residential demand response," *Proc. IREP Symp. Bulk Power Syst. Dyn. Control - IX Optim. Secur. Control Emerg. Power Grid, IREP 2013*, pp. 1–13, 2013.

[10] A. Drehobl and L. Ross, "Lifting the High Energy Burden in America's Largest Cities: How Energy Efficiency Can Improve Low-Income and Underserved Communities," *Am. Counc. an Energy-Efficient Econ.*, no. April, p. 56, 2016.

[11] Y. Leroy, E. Chapotot, B. Yannou, T. Zaraket, and S. Minel, "A Stochastic Activity-Based Approach for Forecasting Occupant-Related Energy Consumption in Residential Buildings," no. August, p. V004T06A034, 2015.

[12] Y. Leroy, S. Minel, T. Zaraket, E. Chapotot, and B. Yannou, "An Occupant-Based Energy Consumption Model for User-Focused Design of Residential Buildings," *J. Mech. Des.*, vol. 137, no. 7, p. 074501, 2015.

[13] J. V. Paatero and P. D. Lund, "A model for generating household electricity load profiles," *Int. J. Energy Res.*, vol. 30, no. 5, pp. 273–290, 2006.

[14] A. Chrysopoulos, C. Diou, A. L. Symeonidis, and P. A. Mitkas, "Response modeling of small-scale energy consumers for effective demand response applications,"

Electr. Power Syst. Res., vol. 132, pp. 78–93, 2016.

[15]   M. Yu and S. H. Hong, "A Real-Time Demand-Response Algorithm for Smart Grids: A Stackelberg Game Approach," *IEEE Trans. Smart Grid*, vol. 7, no. 2, pp. 879–888, 2016.

[16]   C. Turner, S. Ferguson, and J. Donndelinger, "Exploring Heterogeneity of Customer Preference to Balance Commonality and Market Coverage," pp. 1–13, 2011.

[17]   E. Sullivan, S. Ferguson, and J. Donndelinger, "Exploring Differences in Preference Heterogeneity Representation and Their Influence in Product Family Design," 2012, pp. 81–92.

[18]   C.-S. N. Shiau and J. J. Michalek, "Should Designers Worry About Market Systems?," *J. Mech. Des.*, vol. 131, no. 1, p. 011011, 2008.

[19]   N. Williams, S. Azarm, and P. K. Kannan, "Engineering Product Design Optimization for Retail Channel Acceptance," *J. Mech. Des.*, vol. 130, no. 6, p. 061402, 2008.

[20]   A. H. Mohsenian-Rad, V. W. S. Wong, J. Jatskevich, R. Schober, and A. Leon-Garcia, "Autonomous demand-side management based on game-theoretic energy consumption scheduling for the future smart grid,"

IEEE Trans. Smart Grid, vol. 1, no. 3, pp. 320–331, 2010.

[21]   W. Ross Morrow, M. Long, and E. F. MacDonald, "Market-System Design Optimization With Consider-Then-Choose Models," *J. Mech. Des.*, vol. 136, no. 3, p. 031003, 2013.

[22]   J. Shin and S. Ferguson, "Exploring Product Solution Differences Due to Choice Model Selection in the Presence of Noncompensatory Decisions With Conjunctive Screening Rules," *J. Mech. Des.*, vol. 139, no. 2, p. 021402, 2016.

[23]   M. Phadke, *Quality Engineering Using Robust Design.* Prentice Hall, 1989.

[24]   S. Z. Attari, M. L. DeKay, C. I. Davidson, and W. Bruine de Bruin, "Public perceptions of energy consumption and savings," *Proc. Natl. Acad. Sci.*, vol. 107, no. 37, pp. 16054–16059, 2010.

[25]   S. Gyamfi, S. Krumdieck, and T. Urmee, "Residential peak electricity demand response - Highlights of some behavioural issues," *Renew. Sustain. Energy Rev.*, vol. 25, pp. 71–77, 2013.

[26]   Smart Grid Consumer Collaborative, "The Empowered Consumer," 2016.

Proceedings of the ASME 2019
International Design Engineering Technical Conferences
and Computers and Information in Engineering Conference
IDETC/CIE2019
August 18-21, 2019, Anaheim, CA, USA

DETC2019-97456

INTEGRATED SYSTEM DESIGN AND CONTROL OPTIMIZATION OF HYBRID ELECTRIC PROPULSION SYSTEM USING A BI-LEVEL, NESTED APPROACH

Li Chen[a],[1], Huachao Dong[a],[b], Zuomin Dong[a]
a. University of Victoria, B.C., Canada
b. Northwestern Polytechnical University, Xi'an, China

ABSTRACT

Hybrid electric powertrain systems present as effective alternatives to traditional vehicle and marine propulsion means with improved fuel efficiency, as well as reduced greenhouse gas (GHG) emissions and air pollutants. In this study, a new integrated, model-based design and optimization method for hybrid electric propulsion system of a marine vessel (harbor tugboat) has been introduced. The sizes of key hybrid powertrain components, especially the Li-ion battery energy storage system (ESS), which can greatly affect the ship's life-cycle cost (LCC), have been optimized using the fuel efficiency, emission and lifecycle cost model of the hybrid powertrain system. Moreover, the control strategies for the hybrid system, which is essential for achieving the minimum fuel consumption and extending battery life, are optimized. For a given powertrain architecture, the optimal design of a hybrid marine propulsion system involves two critical aspects: the optimal sizing of key powertrain components, and the optimal power control and energy management. In this work, a bi-level, nested optimization framework was proposed to address these two intricate problems jointly. The upper level optimization aims at component size optimization, while the lower level optimization carries out optimal operation control through dynamic programming (DP) to achieve the globally minimum fuel consumption and battery degradation for a given vessel load profile. The optimized Latin hypercube sampling (OLHS), Kriging and the widely used Expected Improvement (EI) online sampling criterion are used to carry out "small data" driven global optimization to solve this nested optimization problem. The obtained results showed significant reduction of the vessel LCC with the optimized hybrid electric powertrain system design and controls. Reduced engine size and operation time, as well as improved operation efficiency of the hybrid system also greatly decreased the GHG emissions compared to traditional mechanical propulsion.

Keywords: hybrid electric propulsion, powertrain optimization, optimal control, energy management, battery performance degradation, life-cycle cost model, global optimization.

1. INTRODUCTION

Greenhouse gas (GHG) emissions from the transportation sector have been one of the major sources of national pollutant inventory in North America. As shown in Fig. 1, both American and Canada are under urgent pressure to reduce CO_2 from transportation. Hybridization and electrification are considered effective methods to improve system efficiency and reduce emissions from both land- and water- based transportation [1]. Hybrid electric powertrain systems normally include an internal combustion engine (ICE), a large battery energy storage system (ESS), electric machines (generators and motors) and power electronics. Commonly used hybrid powertrain configurations can be classified into series, parallel, and power-split systems. Depending on the hybrid architecture, gear sets and gear reductions may also be needed.

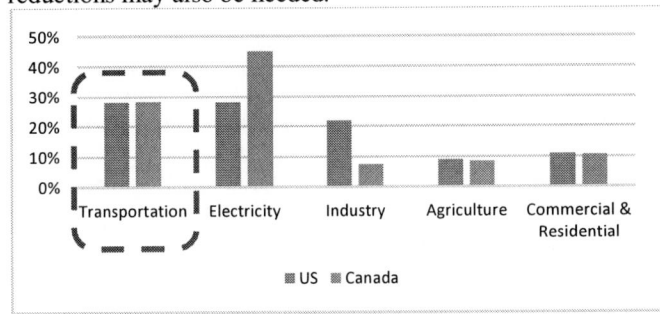

Figure 1: CO_2 Equivalent Emissions by Economic Sector from U.S. and Canada (2016) [2, 3].

Li-ion battery is becoming the main type of rechargeable battery used as ESS in hybrid propulsion systems with

---

[1] Contact author: chenli@uvic.ca

outstanding performance of energy and power density. Recent study showed that average price of Li-ion battery ESS is about $300/kWh for electrified vehicles [4]. However, it is still the most expensive component in the medium and full hybrid powertrain systems. Moreover, the battery deteriorates fast under heavy charging/discharging cycles, and it is very difficult to project the degradation process due to the complicated electrochemical reactions. In general, a Li-ion battery ESS tends to have much lower lifetime than other components (such as ICE, electric machines) in hybrid powertrain.

The Li-ion battery ESS, being the most expensive component with the shortest lifetime, must be appropriately sized and used in the hybrid powertrain system. Battery performance degradation and life heavily depends upon the actual use pattern and operating temperature. The control algorithm of the hybrid systems plays an important role in extending battery lifetime by avoiding harsh charging and discharging during usage.

Most published research on optimal design of hybrid electric systems have generally focused on optimal component sizing with energy demands [5]. Optimal control algorithms for hybrid energy management can be obtained based on selected (not necessarily optimized) power components [6]. Patil [7] combined design and control optimization for a series PHEV, yet the total life-cycle cost were not included due to the lack of battery degradation model. Fathy, et al. [8] examined the plant and controller optimization problems on the coupled conditions, however there needs more sophisticated algorithms to deal with complex dynamic hybrid powertrain system.

Given these circumstances, this paper introduced a model-based design and optimization method to find the optimal design and optimal control algorithm for the hybrid electric propulsion system through a bi-level, nested approach. With the top level optimizing key component sizes and bottom level searching for the optimal control logic, this method can solve the problem in an integrated framework. The overall objective function of this nested problem is to minimize the total life-cycle cost (LCC) of the hybrid propulsion system in 20 years lifetime. A data-driven global optimization algorithm was adopted to solve the formulated optimization problem.

The paper is organized as follows. Section 2 builds the hybrid electric marine propulsion system model for a tugboat operated in a Canadian Harbor as a study case, with detailed battery performance degradation model and LCC model. Section 3 formulates a bi-level nested optimization problem and solves the problem through the new data-driven global optimization framework. Section 4 draws the conclusion of this work.

2. MODELING OF HYBRID ELECTRIC MARINE PROPULSION SYSTEM

2.1 Model of Tugboat Operation Cycle

A typical harbor tugboat usually has very dynamic load profiles when assisting large ships onto and off a berth through pushing and pulling. It also sometimes has long time sailing and waiting in the sea, during which only low constant power

requirement is needed. It is a perfect case for applying a hybrid electric propulsion system, which can gain significant emission reductions by closing engines under low power requirement and using pure electric propulsion.

The traditional tugboat has purely mechanical propulsion with two marine diesel engines, and each has a maximum 2,300 kW power. The designed maximum bollard pull of studied tugboat is 75 tonne. Two additional diesel generator sets are needed for auxiliary and machinery loads. The traditional tug design usually takes generic load data as an input (as shown in Fig. 2). The choice of engine capacity mainly depends on required maximum bollard pull. However it can result in high fuel consumption and air pollutants when the engine works in the off-design area.

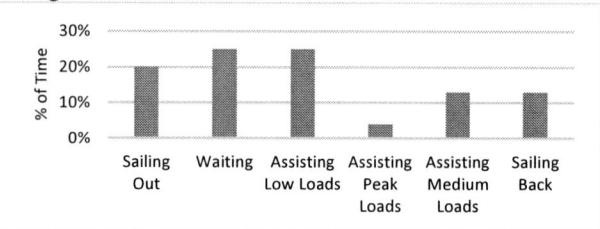

Figure 2: Generic Operation Profile of a Harbor Tugboat.

The hybrid propulsion of tugboat creates more control variables to achieve better efficiency for engines. Batteries can kick in when the power requests are low, and the engine can be either shut off or operated at a higher load in a more efficient area to charge the battery. The decision to turn on or off the engine must rely on the dynamic driving profile. Therefore, a dynamic load profile with timescale was developed, based on the previous generic operation file (as shown in Fig. 3). Basically, it follows a general work sequence by sailing out and waiting for orders, then giving pushes and pulls, finally finishing the job and returning in 2.5 hours.

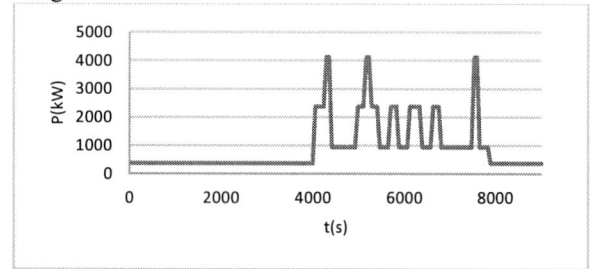

Figure 3: Generated Tugboat Dynamic Operation Profile with Timescale.

2.2 Model of Hybrid Electric Propulsion System and Its Key Components

Power transfer from diesel (or gasoline) engines to ship propellers in traditional marine propulsion systems are either done using mechanical links with gear reductions or through a diesel electric system with diesel generators and electric propulsion motors. The series hybrid electric marine propulsion system is formed by adding a battery ESS and hybrid system control to provide operating flexibility as well as a more environmentally friendly solution. It works as a fully electrified system as all the power from ICE will be transformed into electric energy (Fig. 4). The rechargeable ESS acts as a buffer to store and supply energy. Through optimized powertrain system control, it can be expected to have lower fuel consumption and emissions compared to conventional one.

Hybrid electric systems applied on ships construct a micro power grid on board to provide electricity for electrical consumers. With the fast development of power electronics, both alternating current (AC) and/or direct current (DC) distribution can be used onboard. In this study, a DC power grid was chosen for the integrated hybrid electric propulsion system. As presented in Fig. 4, the arrows indicate energy flow directions in the hybrid system.

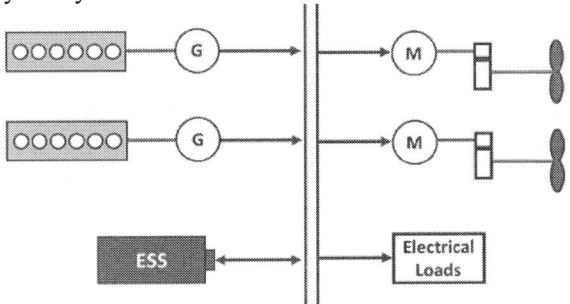

Figure 4: Integrated Hybrid Electric Propulsion System.

The key components designed in the hybrid electric propulsion system were the engine and Li-ion battery ESS. Other components such as electrical generator/motors and power electronics were considered as power loss models with specific efficiency maps in the hybrid system model. Diesel engines were modeled with engine efficiency and emission maps acquired from suppliers and transformed into lookup tables. The mass of fuel consumption was calculated using the engine specific fuel consumption map.

$$m_{fuel} = \int_{t_0}^{t_f} \left(P_{eng,t} \times BSFC_P \right) dt \qquad (1)$$

where t is the time step from t_0 to t_f; $P_{eng,t}$ (kW) is the power output at time t; $BSFC_P$ (g/kWh) is the brake specific fuel consumption at corresponded power P.

For each ship call, engine emissions were also calculated using corresponded emission factors [9].

$$E = \int_{t_0}^{t_f} \left(P_{eng,t} \times EF_P \right) dt \qquad (2)$$

where EF_P (g/kWh) is the emission factor at power P.

The total equivalent CO_2 (CO_2e) was calculated to weight the global warming impacts of different types of GHG with respect to the effects of CO_2. The 100-year Global Warming Potential (GWP) was adopted according to suggestions from U.S. Environmental Protection Agency (EPA). Specifically, the GWP for CH_4 and N_2O are 25 and 298, respectively. CO_2e can be calculated by multiplying GWP values to each GHG.

The Li-ion battery performance was modeled as an electric circuit with a voltage source and passive elements. Two resistor-capacitor (RC) circuits were used to simulate the concentration polarization and electrochemical polarization process. This equivalent circuit model is displayed in Fig. 5, with an open circuit voltage source V_{oc}, an internal resistor R_i, and two RC circuits with R_1, C_1 and R_2, C_2.

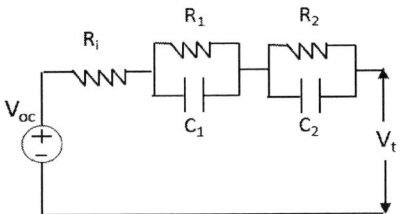

Figure 5: Equivenlent Circuit Model of Li-ion Battery.

The voltage dropped by inner resistance is determined by current, I, and resistor, R_i.

$$V_i = IR_i \qquad (3)$$

The dynamic voltage changes during charge/discharge processes are determined by two RC circuits, and their voltages were denoted as V_1 and V_1.

$$\dot{V}_1 = -\frac{V_1}{R_1 C_1} + \frac{I}{C_1} \qquad (4)$$

$$\dot{V}_2 = -\frac{V_2}{R_2 C_2} + \frac{I}{C_2} \qquad (5)$$

Battery terminal voltage is determined by Kirchhoff laws:

$$V_t = V_{oc} - V_i - V_1 - V_2 \qquad (6)$$

Battery state of charge (SOC) can be calculated using the Coulomb counting method, denoted as the integral of the ration of current variation and capacity Q over a period of time from t_o to t_f (assuming positive current is discharging).

$$SOC = -\int_{t_0}^{t_f} \frac{I}{Q} dt \qquad (7)$$

Performance degradation in the battery's lifetime is an inevitable process. Temperature, current, and state of charge (SOC) are all relevant factors. In general, degradation mechanisms are different between battery anode and cathode. The main reasons for carbon-based anode aging are: solid electrolyte interface (SEI) formation and growth, loss of available lithium due to side reactions, lithium metal plating at low temperature and high rate current, etc. As for lithium metal oxide cathodes, wearing of active materials, compound structure changing, and electrolyte dissolving are relevant to performance decay[10]. In conclusion, battery aging rate can be affected by the charge and discharge current rate (C_{rate}), the state of charge (SOC), the change of SOC (ΔSOC) in usage, temperature (T), operating time (t), etc. The capacity losses (Q_{loss}) can be expressed as:

$$Q_{loss} = f(C_{rate}, SOC, \Delta SOC, T, t) \qquad (8)$$

Bloom, et al. [11] presented the calendar life and power fade of Li-ion battery follow the Arrhenius kinetics with a $(t)^{1/2}$ through accelerated lifetime study, which can be expressed as:

$$P = Ae^{\left(-\frac{E_a}{RT}\right)}t^{0.5} \tag{9}$$

where P is power, t is time, A is the pre-exponential factor, E_a is the activation energy, R is the ideal gas constant, and T is the temperature in K.

Battery cycle life prediction is a key factor that determines the replacing frequency of the ESS during the tugboat lifetime. In other words, the total investment cost calculation using a battery ESS in the hybrid propulsion is highly affected by the estimation of battery operation life. In this work, it was assumed that battery temperature could be well controlled using advanced battery thermal management. Battery lifetime is predicted based on the Arrhenius model, where the capacity loss is a function of C-rate and ΔSOC [12]:

$$Q_{loss} = Ae^{\left(\frac{-E_a + B \cdot C_{rate}}{RT}\right)}(A_h)^z \tag{10}$$

where, A is a pre-determined coefficient, B is the coefficient of C-rate, A_h is the total throughput capacity, and z is the power of A_h which is around 0.5 in most of cases [12].

A_h calculates the total energy flow and influenced by battery cycling numbers N_{bat}, ΔSOC and normal capacity.

$$A_h = N_{bat} \cdot \Delta SOC \cdot Q \tag{11}$$

N_{bat} is the total cycling numbers within the lifetime. In this paper, it was assumed that battery is dead when $Q_{loss} = 20\%$. Therefore, the calculation of cycling number can be derived based on the previous function:

$$N_{bat} = \left(\frac{Q_{loss}}{Ae^{\left(\frac{-E_a + B \cdot C_{rate}}{RT}\right)}}\right)^{\frac{1}{z}} \frac{1}{Q \cdot \Delta SOC} \tag{12}$$

This model was implemented in MATLAB/Simulink. The results of cycling number for 20Ah capacity LiFePO₄ as a function of C-rate and ΔSOC was plotted in Fig. 6.

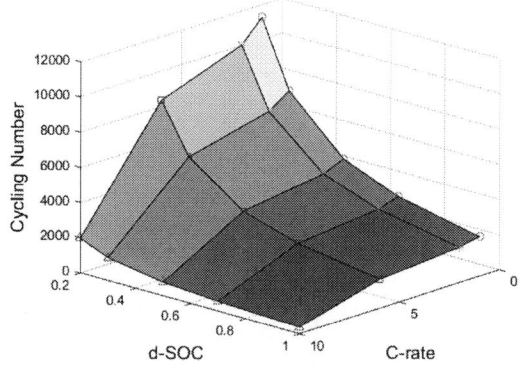

Figure 6: Cycling Number Variation with C-rate and ΔSOC.

2.3 Life-Cycle Cost (LCC) Model of Propulsion System

The life-cycle cost (LCC) model was built in this study to serve as part of the optimization objectives for hybrid electric propulsion system, including the cost of main powertrain components (i.e. power sources, electric machines and power converters). The costs of other ship elements such as hull and propellers have been excluded in this study, and they can be easily added into the results if needed. All costs in this study are given in Canadian dollars.

Several traditional LCC calculation methods are used in industry, such as the LCC method from NORSOK standards (NORSOK originally stands for "the Norwegian shelf's competitive position") [13]. However, these existing LCC models are not suitable for complex hybrid marine propulsion system design. Due to the cost-intensive Li-ion battery ESS, it is important to evaluate the battery replacement cost and residual cost during the total lifetime. A new LCC model is therefore proposed in this work. The main elements in the LCC model include capital cost (C_{cap}), operation cost (C_{ope}), and residual cost (C_{resd}).

$$LCC = C_{cap} + C_{ope} + C_{resd} \tag{13}$$

Net Present Value (NPV) analysis was used in the LCC model to calculate the present value of cash flows over the entire lifetime. After setting the base year of costs, the future costs are discounted back to the base year so that present value of the cost flow can be represented.

$$NPV = \sum_{t=0}^{N_t} \frac{C_t}{(1+r)^t} \tag{14}$$

where C_t is the net cost in year t, which can be assumed equal for all the years; N_t is the total lifetime in year to be evaluated; and r is the annual discount rate/inflation rate. The lifetime of the tugboat in this study is assumed to be 20 years (i.e., $N_t =$ 20). Only the costs directly related with powertrain components were considered.

Capital costs include initial purchase costs for all the components in the propulsion systems, as well as reinvestment cost for Li-ion battery ESS.

$$C_{cap} = C_{eng} + C_{hyb} + C_{ess} + C_{rin} \tag{15}$$

Hybridization cost (C_{hyb}) accounts for electric motors, generators and power converters. Engine and ESS cost (C_{eng} and C_{ess}) are related to component sizes.

$$C_{eng} = p_{eng} P_{eng} \tag{16}$$

$$C_{ess} = p_{ess} E_{ess} \tag{17}$$

where P_{eng} and E_{ess} are the engine power (kW) and battery ESS energy (kWh), and p_{eng} and p_{ess} are the prices for engine ($/kW) and ESS ($/kWh).

C_{rin} is the reinvestment cost, and it only counted for replacement cost of battery ESS in this work, since the battery usually has lower lifetime than other components. The operation life of the battery ESS (L_{bat}) is calculated as a function of designed ESS capacity, and this operation life is also influenced by the power control and energy management strategies of the hybrid electric powertrain system.

$$C_{rin} = \sum_{t=0}^{N_t} \frac{C_{ess}}{(1+r)^t} k_t \tag{18}$$

where k_t is the replacement frequency, which is a function of the battery lifetime (L_{bat}). r is the annual inflation rate. L_{bat} is the key parameter that determines the reinvestment capital costs. The optimal result of L_{bat} must be determined at the system level considering both engine and ESS operation conditions.

$$k_t = f(L_{bat}) = \begin{cases} 1, & t = n \cdot L_{bat}, t \leq N \\ 0, & otherwise \end{cases} \quad (19)$$

When the battery needs to be replaced in year t, then $k_t = 1$, otherwise, k_t is 0.

Operational cost includes the total fuel consumption cost (C_{energy}) and engine maintenance cost (C_{maint}) during lifetime operation.

$$C_{ope\_tot} = \sum_{i=0}^{N} \frac{C_{energy} + C_{maint}}{(1+r)^i} \quad (20)$$

where r is the annual inflation rate. $C_{energy} = p_{fuel} m_{fuel}$ determined by fuel price (\$/kg) and total mass of fuel consumption (kg). In this study the yearly average price of diesel oil was adopted. C_{maint} is highly affected by engine operational time, engine size and many other factors. Specific maintaining cost for studied tugboat engine were estimated based on acquired information from ship company and engine manufacturer.

The residual costs (or salvage costs) of replaced Li-ion batteries is nontrivial for this expensive component. Retired batteries from hybrid transportation vehicles can be reused for residential energy storage and load leveling in a smart grid application. Research has shown that the second use of a Li-ion battery could be provided at a relatively low price based on the techno-economic analysis [14].

In this study, residual costs are the remaining value in the replaced battery ESS, which is also determined by the replacement times and residual price.

$$C_{resd} = \sum_{i=0}^{N} \frac{p_r Q_r}{(1+r)^i} k_i \quad (21)$$

where p_r is the price for the remaining value (\$/kWh), Q_r is the remaining capacity (kWh), and r is the annual inflation rate.

3. NESTED OPTIMIZATION OF HYBRID ELECTRIC PROPULSION SYSTEM DESIGN

The model-based design of a hybrid propulsion system has to consider both design and control optimization at the same time to acquire the optimum result. In this section, a bi-level, nested optimization (BLNO) problem was formed to find the optimal design of the hybrid propulsion system. The upper level aims at finding the optimal key component sizes under an optimized hybrid control strategy, where the lower level can provide the optimal control under a given driving profile and fixed component design.

Generally, a BLNO problem can be described as below,

$$\min_{x_u \in X_{up}} F_{up}(x, c_l^*)$$
$$s.t.$$
$$c_l^* = \underset{c_l \in C_{low}}{argmin}\left(f_{low}(x_l, s_l) \,|\, x_l \in X_{low}, s_l \in S\right) \quad (22)$$
$$G_u(x) \leq 0$$
$$g_l(x) \leq 0$$

where F_{up} and f_{low} are upper- and lower-level objective functions; and G_u and g_l represent the upper and lower constraints, respectively. In Eq. (19), the upper design space is X_{up} and the lower design space is referred to X_{low}. The control variable space in the lower level is referred to as C_{low}. In the optimization, the design variables are $x = (x_u \in X_{up}, x_l \in X_{low}) \in X_w$, the control variables are $c_l \in C$, and the system state variables are $s_l \in S$. Each cycle the upper-level optimization needs a complete optimization process of the lower-level problem, imposing high computational cost especially when the lower-level optimization is time-consuming.

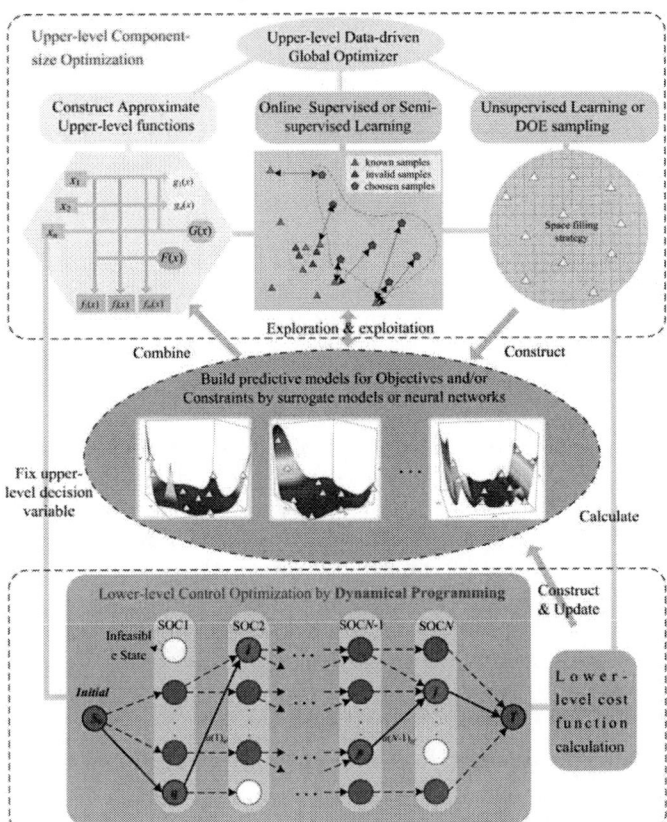

Figure 7: Demonstration of the BLNO Framework.

To ensure the lower level search can acquire the global optimal control sequence, dynamic programming (DP), also called smart enumeration, has been employed. Although DP can provide optimal control over the entire "trip", it also leads to long computational time. For the upper level, data-driven global optimization techniques are then essential to reduce the computation cost in completing the entire integrated optimization process. Figure 7 summarizes the presented optimization framework and the required techniques, including unsupervised sampling, supervised learning, global optimization, approximate modeling, DP, and optimization integration. In the upper level, optimized Latin hypercube sampling (OLHS) [15], Kriging and the widely used Expected Improvement (EI) [16] online sampling criterion were used to carry out "small data"-driven global optimization.

3.1 Optimal Sizing of Key Components – Upper Level

The sizing of key powertrain components in a hybrid system is critical to the total system costs while maintaining superb performance. In the previously discussed hybrid electric propulsion system, the key components are the engines and Li-ion battery ESS, whereas other components are mainly related to these two. Therefore, the upper level design variables are chosen as the maximum engine power ($x_1 = P_{eng}$) and battery ESS capacity ($x_2 = E_{ess}$).

$$x_{up} = [x_1 \quad x_2]' \tag{23}$$

The main objective for the top-level optimization is to minimize the total LCC of hybrid propulsion system over 20-year ship operation. This is a non-convex problem subject to a number of inequality constraints which can be formed as

$$\min_{x_u \in X_{up}} F_{up} = LCC(x_{up}, c_l^*)$$
$$s.t. P_{eng,min} \leq x_1 \leq P_{eng,max}$$
$$E_{ess,min} \leq x_2 \leq E_{ess,max} \tag{24}$$
$$c_l^* = \underset{c_l \in C_{low}}{\mathrm{argmin}} \left(f_{low}(x_l, s_l) \,\middle|\, x_l \in X_{low}, s_l \in S \right)$$

Before getting the LCC value, the whole lower-level DP process needs to be finished to find the optimal pathway, which can be regarded as an expensive black-box model. Kriging is a robust approximation method that is good at predicting nonlinear model, and the EI sampling criterion can guide the surrogate-assisted global optimization. Therefore, Kriging is used to predict LCC, and "maximizing EI" is used to update the Kriging model, and to perform the data-driven global optimization. Specifically, EI is summarized as below,

$$EI(\mathbf{x}) = \begin{cases} I \cdot \Phi\left(\dfrac{I}{s}\right) + s \cdot \Phi\left(\dfrac{I}{s}\right) & \text{if } s > 0 \\ 0 & \text{if } s = 0 \end{cases} \tag{25}$$
$$I(\mathbf{x}) = y_{\min} - \hat{y}(\mathbf{x})$$

where s refers to the estimated mean square error at a to-be-tested point $\mathbf{x}$, $y_{\min}$ denotes the present best LCC value, $\hat{y}$ is the predictive LCC value from Kriging, and $I(\mathbf{x})$ is the estimated improvement. Figure 8 shows the demonstration of EI. The blue area denotes the "Probability Improvement" (PI) that reflects the size of EI values.

In addition, in each search cycle, the grey wolf optimization (GWO) algorithm [17] is used to capture the maximum EI value whose corresponding sample point will be supplemented to the database for the subsequent update of Kriging.

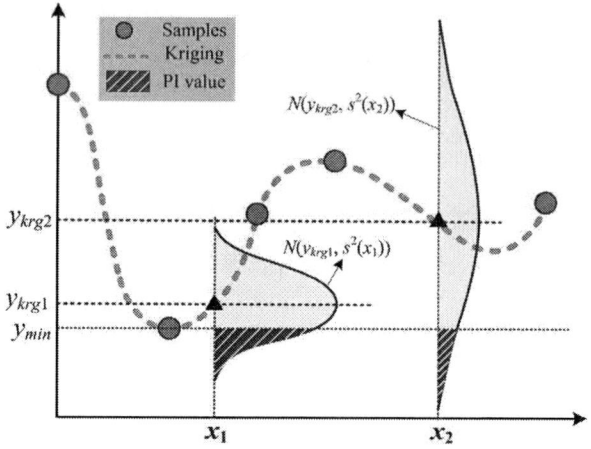

Figure 8: Demonstration of EI.

3.2 Optimal Control of Hybrid Energy System – Lower Level

The lower level optimization, in this problem, deals with the hybrid energy management strategy under a given driving profile. Regardless of the propulsion component design, the essence of hybrid energy management is to find the optimal control policy to achieve the minimum objectives, which are to:

- minimize the total mass of engine fuel consumption (m_f)
- reduce the battery performance degradation (Q_d)

For this multi-objective optimization problem, it can be formulated as an integral over the time horizon from t_0 to t_f.

$$\min_{c \in C} f_{low} = \int_{t_0}^{t_f} \left(a_1 \cdot k_{mf} + a_2 \cdot k_{bat} \right) dt \tag{26}$$

where k_{mf} and k_{bat} represent the coefficients that scaled fuel mass consumption and battery degradation into the same order of magnitude. a_1 and a_2 are weighting factors to adjust the two objectives.

The lower level design variable is the depth-of-discharge (DOD) of the Li-ion battery. Since the usage of the battery can heavily affect battery performance and degradation, it is therefore must be designed and constrained within a certain range.

$$DOD_{min} \leq x_l \leq DOD_{max} \tag{27}$$

The system state in this dynamic system is affected by both state variables ($s(t)$) and control variables ($c(t)$).

$$\dot{s}(t) = f(s(t), c(t), t) \tag{28}$$

where $s \in S$ is state variable and $c \in C$ is control variable.

This optimization problem is subject to constraints related to physical limitations and energy conservation. In general, the SOC of a hybrid propulsion system must meet the criterion at the end of the trip. Specifically, the final SOC must be the same as the initial SOC for a hybrid system without the ability to recharge the battery ESS from the power grid. The charging/discharging current rate (C_{rate}) must be controlled within a certain range according to the battery specification. The torque and speed from the engine and electric motor must also meet certain constraints.

DP is a numerical method to solve control problems by providing the optimal solution for a given time period [6]. It is an off-line method as it requires a priori knowledge about the entire optimization horizon to give the optimal trajectory. To implement DP in the hybrid system control algorithm, the state variable is discretized with time step δt. The length of the control vector is T, where $T = \frac{t_f}{\delta t}$. State space variable is also discretized accordingly.

The purpose of DP is to find the optimal control policy $c^* = [c_1\ c_2\ ...\ c_T]'$ in the discretized time zone which can lead to the minimum value of cost function. To achieve that, the cost-to-go function at each step $t = k$ ($k = 1,2,...,T.$) must be modeled first.

$$f(s_k, k) = L_k(s_k, c_k) + \sum_{i=k}^{T} L_i(s_i, c_i) \quad (29)$$

Starting from the final time step when $k = T$, the final cost function is $L_T(s_T, c_T)$ and it calculates all feasible states as

$$L_T = f(s_T, c_T) = a_1 \cdot k_{mf,T} + a_2 \cdot k_{bat,T} \quad (30)$$

When it progresses backward, the cost-to-go function at $t = k$ captures all possible pathways from k to T for all feasible states at each computational node ($k = 1,2,...,T-1$). The optimal control state at each node can be obtained by finding the minimum value of the cost-to-go function. This process is repeated until reaching to the initial step $k = 1$.

3.3 Optimal Results of BLNO Problem

The bi-level, nested optimization problem for the hybrid electric marine propulsion system is formulated as:

$$\min_{x \in X_w} F_{up} = LCC(x_{up}, c_l)$$

$$s.t.\ c_l = \underset{c_l \in C_{low}}{\arg\min} f_{low}(x_l, s_l)$$

$$\begin{cases} P_{eng,min} \leq x_1 \leq P_{eng,max} \\ E_{ess,min} \leq x_2 \leq E_{ess,max} \\ DOD_{min} \leq x_3 \leq DOD_{max} \\ s_T = SOC_f \\ s_k = f(s_{k-1}, c_{k-1}) + s_{k-1} \\ SOC_{min} \leq s_k \leq SOC_{max} \\ C_{rate,min} \leq C_{rate,k} \leq C_{rate,max} \\ P_{bat,min} \leq P_{bat,k} \leq P_{bat,max} \\ T_{eng,min} \leq T_{eng,k} \leq T_{eng,max} \\ \omega_{eng,min} \leq \omega_{eng,k} \leq \omega_{eng,max} \\ T_{mtr,min} \leq T_{mtr,k} \leq T_{mtr,max} \\ \omega_{mtr,min} \leq \omega_{mtr,k} \leq \omega_{mrt,max} \end{cases} \quad (31)$$

This problem was solved by aforementioned data-driven global optimization method. A surrogate model was constructed through a design of experiments (DOE) and online supervised sampling and the search results are presented in Fig. 8. The LCC value of optimized hybrid electric propulsion system is 4.72 million Canadian dollars. Compared to the traditional tugboat using mechanical propulsion, the total LCC is reduced by 36.86% in 20 years operation time.

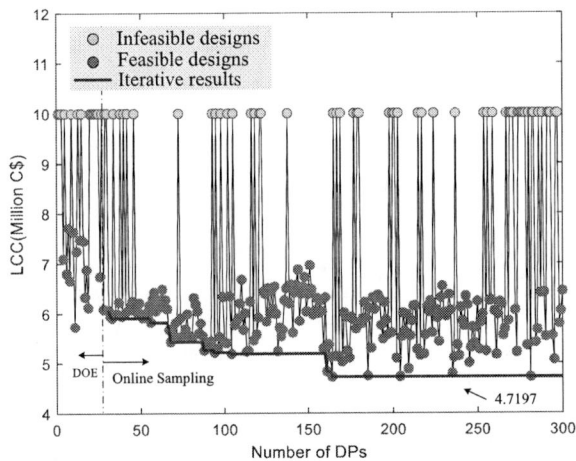

Figure 8: Optimization Results of the BLNO Problem.

The upper-level component sizing optimal results are $x_{up} = [1797,513]$, and the lower level result is $x_{low} = 0.22$. The power distribution between the engine and Li-ion battery ESS in each driving cycle was determined by DP in the lower level optimization program. The optimal results are presented in Fig.9. The engine has been turned down when power requirements are low to avoid inefficient engine operation and instead uses energy from the battery ESS to power the ship. Engine is employed for high power requirements, while the battery ESS shaved the peak power and allowed the engine to operate efficiently. The surplus energy from the engine was used to charge the battery, and allowed it to maintain a certain SOC.

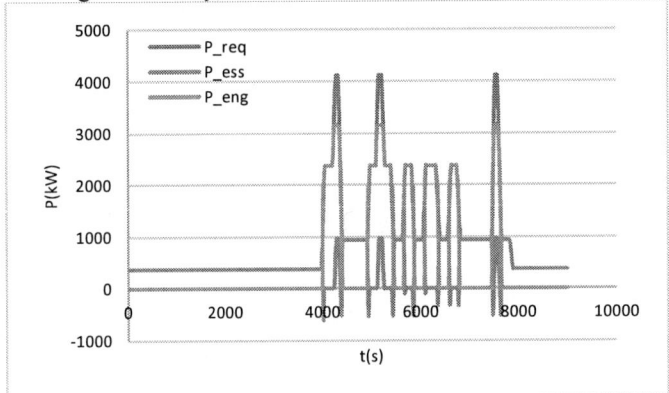

Figure 9: Power Distribution between Engine and ESS.

Hybridization is an effective means to downsize the engine, leading to lower capital cost, as well as less fuel consumption and emissions. The optimized engine size is about 21.87% smaller than the original mechanical propulsion system. Even though the initial investment cost is increased by adding a Li-ion battery ESS, the total operational cost can be greatly reduced. Moreover, engine operation has been reduced by 80.31%, which can save considerable engine maintenance and operational costs.

The yearly CO_2e emission has also been greatly reduced in the hybrid electric system by 34.86%, compared to the conventional mechanical propulsion system.

4. CONCLUSION

This study has presented a new and an effective method to solve the interrelated, optimal powertrain component sizing and optimal control problems in developing hybrid electric propulsion systems. The new integrated approach for minimizing fuel consumption, emissions and LCC jointly has been formulated as a bi-level, nested optimization (BLNO) problem. The upper-level optimization identifies the optimal powertrain component sizes with minimum fuel consumption and the least battery degradation, while the lower-level optimization ensures the optimal powertrain control/energy management solutions of each feasible powertrain system design under a given power load pattern. The complex and computationally-intensive optimization problem is solved using new and innovative surrogate model-based global optimization methods. The integrated powertrain component sizing and control optimization of the hybrid electric propulsion system led to improved fuel efficiency, emissions and life-cycle costs. Continuing work will be focused on comparing different optimization algorithms in solving the new BLNO problem to acquire faster and more accurate results.

ACKNOWLEDGEMENTS

Supports from Natural Science and Engineering Research Council of Canada, Transport Canada, Seaspan, and Dennis and Phyllis Washington Foundation, as well as from National Natural Science Foundation of China (Grant No. 51805436) are gratefully acknowledged. The authors also like to thank the support from Vince den Hertog, Brendan Smoker and Robin Stapleton at Robert Allan Ltd., and the feedbacks from Anthony Truelove at UVic Clean Transportation Lab.

REFERENCES

[1] K. Koprubasi, "Modeling and control of a hybrid-electric vehicle for drivability and fuel economy improvements," Doctor of Philosophy, Mechanical Engineering, The Ohio State University, 2008.

[2] "National Inventory Report 1990-2016: Greenhouse Gas Sources and Sinks in Canada," *Environment and Climate Change Canada,* 2018.

[3] "Inventory of U.S. Greenhouse Gas Emissions and Sinks: 1990–2016.," *United States Environmental Protection Agency,* 2018.

[4] P. A. Nelson, K. G. Gallagher, I. D. Bloom, and D. W. Dees, "Modeling the performance and cost of lithium-ion batteries for electric-drive vehicles," Argonne National Laboratory (ANL)2012.

[5] J. Shen, S. Dusmez, and A. Khaligh, "Optimization of sizing and battery cycle life in battery/ultracapacitor hybrid energy storage systems for electric vehicle applications,"

IEEE Transactions on industrial informatics, vol. 10, pp. 2112-2121, 2014.

[6] O. Sundström, D. Ambühl, and L. Guzzella, "On implementation of dynamic programming for optimal control problems with final state constraints," *Oil & Gas Science and Technology–Revue de l'Institut Français du Pétrole,* vol. 65, pp. 91-102, 2010.

[7] R. M. Patil, "Combined Design and Control Optimization: Application to Optimal PHEV Design and Control for Multiple Objectives," 2012.

[8] H. K. Fathy, J. A. Reyer, P. Y. Papalambros, and A. Ulsov, "On the coupling between the plant and controller optimization problems," in *Proceedings of the 2001 American Control Conference.(Cat. No. 01CH37148),* 2001, pp. 1864-1869.

[9] L. Browning and K. Bailey, "Current methodologies and best practices for preparing port emission inventories," *ICF Consulting report to Environmental Protection Agency,* 2006.

[10] M. Ecker, N. Nieto, S. Käbitz, J. Schmalstieg, H. Blanke, A. Warnecke, *et al.,* "Calendar and cycle life study of Li(NiMnCo)O$_2$-based 18650 lithium-ion batteries," *Journal of Power Sources,* vol. 248, pp. 839-851, 2014.

[11] I. Bloom, B. Cole, J. Sohn, S. Jones, E. Polzin, V. Battaglia, *et al.,* "An accelerated calendar and cycle life study of Li-ion cells," *Journal of Power Sources,* vol. 101, pp. 238-247, 2001.

[12] J. Wang, P. Liu, J. Hicks-Garner, E. Sherman, S. Soukiazian, M. Verbrugge, *et al.,* "Cycle-life model for graphite-LiFePO4 cells," *Journal of Power Sources,* vol. 196, pp. 3942-3948, 2011.

[13] "Life Cycle Cost for Systems and Equipment," *NORSOK Standard,* April, 1996.

[14] C. Heymans, S. B. Walker, S. B. Young, and M. Fowler, "Economic analysis of second use electric vehicle batteries for residential energy storage and load-levelling," *Energy Policy,* vol. 71, pp. 22-30, 2014.

[15] D. R. Jones, M. Schonlau, and W. J. Welch, "Efficient global optimization of expensive black-box functions," *Journal of Global optimization,* vol. 13, pp. 455-492, 1998.

[16] R. Jin, W. Chen, and A. Sudjianto, "An efficient algorithm for constructing optimal design of computer experiments," in *ASME 2003 International Design Engineering Technical Conferences and Computers and Information in Engineering Conference,* 2003, pp. 545-554.

[17] S. Mirjalili, S. M. Mirjalili, and A. Lewis, "Grey wolf optimizer," *Advances in Engineering Software,* vol. 69, pp. 46-61, 2014.

Proceedings of the ASME 2019
International Design Engineering Technical Conferences
and Computers and Information in Engineering Conference
IDETC/CIE2019
August 18-21, 2019, Anaheim, CA, USA

DETC2019-98393

STRUCTURAL CONSEQUENCE ANALYSIS: TOWARDS THE QUANTIFICATION OF COMPONENT CONSEQUENTIAL IMPORTANCE IN SYSTEM ARCHITECTURE DESIGN

Hannah S. Walsh
School of Mechanical, Industrial,
and Manufacturing Engineering
Oregon State University
Corvallis, OR, USA

Mohammad Hejase
Intelligent Systems Division
NASA Ames Research Center
Moffet Field, CA, USA

Daniel Hulse
School of Mechanical, Industrial,
and Manufacturing Engineering
Oregon State University
Corvallis, OR, USA

Guillaume Brat
Intelligent Systems Division
NASA Ames Research Center
Moffet Field, CA, USA

Irem Y. Tumer
School of Mechanical, Industrial,
and Manufacturing Engineering
Oregon State University
Corvallis, OR, USA

ABSTRACT

There is a major push in safety-critical systems to consider system risk early in the design process in order to avoid costly redesign later on. However, existing techniques, which may be labor-intensive and be subject to many sources of uncertainty, rely on failure mode and failure rate data, which can only be estimated in the early design phase. This paper proposes a network-based technique for assessing the consequential importance of a particular component to enable designers to consider hazards in the design of the system architecture without the use of estimated failure rates. Structural consequence analysis represents connectivity between components with a network and provides an explicit representation of risk prevention and mitigation techniques, such as redundancy. The network is augmented with a measure of the consequence of the failure of the "end" components, or sinks, which can be backpropagated through the network to compute the consequence associated with the failure of all components. Based on this consequence, designers can consider mitigation strategies, such as redundancy or increased
component reliability. The approach is demonstrated in the design of an electric system to control an aileron of an unmanned aircraft system (UAS). It is found that structural consequence analysis can identify potentially important components without failure rate data, allowing designers to proactively design for risk earlier in the design process.

1 INTRODUCTION

Risk prevention and mitigation often come at the cost of design elegance. For example, component redundancy is a typical method to meeting safety requirements, but can result in increased operational complexity, up-front design cost, and decreased performance in engineered systems. This is exemplified by the space shuttle, which had multiple computers that "voted" on the correct computation [1], adding complexity. Another example of redundancy adding complexity is in the case of cold standby redundancies, in which switching to the redundant system requires detecting a failure of the main system and switch-

This work was authored in part by a U.S. Government employee in the scope of his/her employment.
ASME disclaims all interest in the U.S. Government's contribution.

Copyright © 2019 ASME

ing over to the redundant system [2]. This additional complexity can increase development costs and time [3], and redundancies add cost and weight, the latter being of particular concern for aerospace systems. For these reasons it is important to consider a trade-off between risk design strategies and complexity, cost, and weight, as well as the consequence of failure. As a result, it is preferable to design inherent safety into the product, if possible, which eliminates the sources of risk, before considering costly protective design features [4, Section 7.3.3], and to allocate redundancy between components in a smart, efficient way.

In the risk-based design of engineered systems, it is often helpful to consider each individual component's contribution to overall risk. Two classical metrics that have been extensively used in literature to express a component's importance are the Statistical Importance and the Structural Importance. However, these metrics are based on the fault tree and not a full network mapping causes to consequences, which is required to fully represent the safety architecture of engineered systems [5].

In this research, an alternative approach, structural consequence analysis (SCA), is proposed for quantifying component importance based on its position in the causal network, as opposed to its direct reliability in a fault tree. The approach utilizes a directed network representation of the engineered system to propagate the consequences of sink nodes, which are components that do not relay information to any other components within the architecture, to quantify the consequence of failure of the other nodes in the network and thus obtain a measure of component importance. There are a few benefits to this approach. First, SCA combines information about differing consequences of failure for different sink nodes. Second, the network representation allows for an explicit representation of redundancy. Third, SCA provides greater expressiveness about common sources of hazards and component interactions that lead to failure than fault trees. Fourth, SCA allows designers to represent causal mechanisms of failure within engineered systems to better consider direct inherent safety in the system. A measure of structural importance is further provided with component consequence consequent metrics, enabling designers to consider changes to system structure in early design.

2 LITERATURE REVIEW

Before introducing the methodology, the literature will be reviewed. First, component importance measures will be reviewed, with an emphasis on fault tree metrics. Second, the network literature will be reviewed, with an emphasis on applications to system design. Third, a review is provided for representations of risk in engineered systems.

2.1 Component Importance Measures

Fault tree analysis is a commonly used method for risk assessment in engineering design. It shows various paths leading to a "top-event", an undesirable system state [6]. Fault trees, which are constructed using expert analysis and historical data, are used to assess whether a design meets risk requirements. Various fault tree metrics can be used to compute the importance of each component within the system architecture. Minimal cut sets, which are combinations of events that cause the top event, and the top event probability can be computed from the fault tree as well. This probability can be compared to risk requirements. If it is found that the top event probability does not meet risk requirements, the design can be updated by adding redundancy. Adding redundancy lowers the top event probability. The selection of the particular component to which redundancy is added to reduce this top event probability is of interest in this research.

Traditionally, the decision of where to add redundancy is made with the help of component importance metrics such as statistical importance and Birnbaum importance. Statistical importance, sometimes called Fussell-Vesely importance, is a measure of the contribution of cut sets containing the component of interest to the total system risk.

There are two forms of Birnbaum importance. One, the reliability importance, uses failure rates of components and one, the structural importance, does not use the actual failure rates of components (it instead assumes equal values for all component failure rates) [7, 8]. Neither version takes into account the failure rate of the component of interest itself.

Recent research has advanced and expanded upon fault tree analysis in various ways. Fault tree analysis has been augmented with fuzzy set theory to account for imprecision in event probabilities used to calculate the top event probability [9]. Sinnamon and Andrews proposed a technique for more accurately computing the top event probability in the case that there is a large number of minimal cutsets [10]. Dynamic fault trees have been used to study how temporal information about the failure of multiple components affects risk [11]. To implement a dynamic fault tree, Boudali et. al. and Dugan et. al. both used Markov chains [12, 13]. Others including Liant et. al. used a Monte Carlo approach [14]. Zhang et. al. used a Petri net based approach [15]. Binary decision diagrams have been used to efficiently compute importance measures [16]. Other fault tree extensions include dependent events [17] and repairable fault trees [18].

2.2 Network Representations of Engineered Systems

A network is a collection of nodes and edges. Nodes, sometimes called vertices, are the elements that make up a network. These nodes are connected by edges, sometimes called links. For the purposes of this paper, the terms network and graph are interchangeable. The difference between the two is that in a network, nodes represent something physical, whereas a graph is simply a mathematical representation. Networks have been used to study the world-wide web [19], various aspects of biology [20], and social networks [21, 22], among others.

Euler is credited with initiating the study of graphs by using a graph to solve the famous Königsberg bridge problem [23]. Erdős and Rényi later studied random graphs, in which all nodes have equal probability of connections [24]. Random graphs, however, are not found so often in the real world as compared to the scale-free networks studied by Barabási [23]. Graphs and networks are often analyzed using their degree distribution plot, which is a histogram of the degree, or number of connections, of all nodes; in a random graph, the degree distribution is a bell-curve shape while in a scale-free network, the degree distribution follows a power law [23].

More recently, networks have been applied to system design. Specifically, they has been used to identify critical design parameters [25–28], analyze system modularity [29], analyze the effect of design changes [30], study collaboration in design [31], predict customer preferences [32, 33], and model system architecture [34, 35]. The next section will cover in more detail how this research uses networks to model engineered systems.

2.3 Representations of Risk in Engineered Systems

Many tools have been developed in the literature to represent the causes, mechanisms, and consequences of failure in engineered systems. Failure Modes and Effects Analysis (FMEA) is most commonly used in design, to determine the components with the highest overall risk priority to direct design and operational decision-making towards the minimization of faults which occur in those components [4, Section 10.4] [36, Section 11.6]. FMEA is performed in a table, and as a result represents causality in a component as set of listed causes which lead to a mode (a specific change of state or behavior of the component) which in turn results in a set of listed effects.

Fault Tree Analysis (as discussed earlier) is another common approach to design for risk, and is used to identify all of the fault events which would lead to a single top-event [4, Section 10.3] [36, Section 11.6]. As a *tree structure*, Fault Tree Analysis is capable of representing an entire chain of causality, allowing one to deductively trace top-level system failure effects to individual failure mechanisms in failure mechanisms which occur in individual components or component interactions. Fault trees are often used for risk assessment to determine the overall probability of failure, given probabilities for each of the causes of failure [37, Chapter 12]. Reliability Block Diagrams (RBDs) are also used to calculate the overall reliability of a system given individual component reliabilities in a system architecture, often to represent different forms of redundancy in a system [37, Chapter 9]. As with FTA, the limitation of the RBD approach is that only one generic effect of component failure is considered–system failure.

Event Tree Analysis (ETA) is another tree structure that instead shows all of the effects of a given event, and is similarly capable of representing an entire set of consequences that occur as a result of an event. ETA is often used to determine the probabilities of sets of consequences, given there are a set of protective safety features (e.g. alarms, seat belts, etc.) to mitigate the consequences of an event which each have a given effectiveness. [37, Chapter 12] Bow tie diagrams are structures which combine the fault tree and event tree for a single fault. These diagrams are used to provide a comprehensive understanding of the causes and effects of a hazardous event, and are used in the final development of a safety case for a system to show that the overall risk associated with that event is minimized [38].

Model-based approaches also exist which represent the propagation of risk through a system, and a variety of modelling approaches exist to perform this [39]. Design approaches, such as [40], focus on the propagation through the function structure of the design, since this is thought to be a representation of the design that represents the flow of tasks in a system while allowing a significant amount of design freedom [4]. A general optimization approach to designing with these fault models is provided in Hulse et al. [41], while Arlitt and Van Bossuyt [42] envision a designer-oriented fault visualization approach to minimizing risk. While these approaches exist, much of the work of model-building is siloed away as model assumptions and few visualization techniques exist (other than those listed above) to give designers the insight needed to understand and act on the model.

The limitation of these approaches is that the representation of the causal network which links root causes to system-level effects is at least partially hidden from the designer, either concatenating the true causal mechanisms and effects into lists (FMEA), focusing solely on the causes or effects of a single event (FTA/ETA), or focusing on the causes and effects of a single event (Bow Ties), rather than the set of causes leading to multiple events leading to multiple effects. In this work, we seek to represent this overall flow of causality with a directed acyclic network (as recommended by Dezfuli et al. [43]), based on the recognition of Denney et. al. [5] that a full representation of risk in a system requires a view of the entire Safety Architecture– the comprehensive mapping of risk sources to risk consequences through an intermediary structure of events.

3 NETWORK MODELS OF SYSTEM ARCHITECTURE

In SCA, system architecture is represented as a network. This representation allows for the modeling and analysis of the connectivity between the components as well as the explicit representation of redundancy. Networks are a natural representation of system architecture due to their close association with the commonly-used design structure matrix (DSM). A DSM represents the connections between components via a matrix, which is equivalent to the adjacency matrix which represents the connections between nodes in a network. A network, then, is simply a graphical representation of the connections of a DSM.

3.1 Network Modeling Methodology

In SCA, a network is used to represent system architecture. In general, a network consists of nodes (sometimes called vertices) connected by edges (sometimes called links). Here, the nodes represent components, and the edges are *directed*, meaning they indicate that information flows from a trailing edge to a leading edge. There are also *undirected* networks, where no direction of information flow is indicated. Whether directed or undirected, an edge between nodes represents some sort of interaction between those nodes. In the case of a directed network of components, the connections represent a physical connection and a direction of flow of information through the system architecture (e.g. a signal or electric current). A simple example is given in Fig. 1.

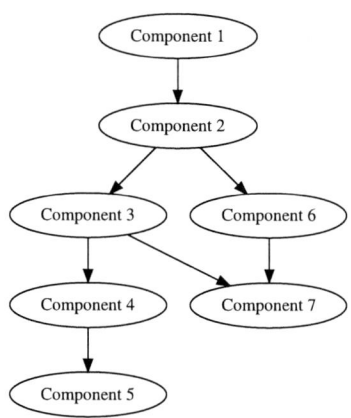

FIGURE 1: SIMPLE EXAMPLE OF AN ARCHITECTURAL REPRESENTATION OF SYSTEM ARCHITECTURE.

The nodes of a directed network can be categorized as either a *source*, *middle node*, or *sink*. A source, such as Component 1 in Fig. 1, has outgoing edges but no incoming edges. A middle node, such as Component 2, has both incoming and outgoing edges. A sink node, such as Component 5, has incoming edges but no outgoing edges. In a system architecture network, the source is likely to be a power source, while the sink could be a control surface or propeller.

3.2 Representing Redundancy

The network based representation of system architecture allows for the explicit representation of redundancy. In this representation, a component that is redundant in parallel to another

component shares the same incoming and outgoing edges. In the example given in Fig. 2, Component 2 Redundant is redundant to Component 2. This representation does not itself distinguish between active and passive redundancy.

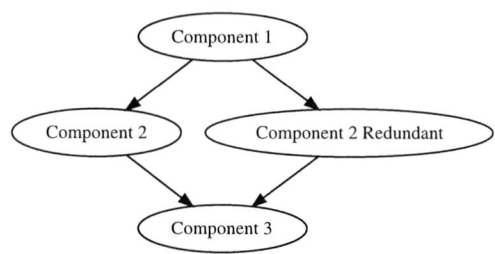

FIGURE 2: REPRESENTATION OF REDUNDANCY, WHERE THE REDUNDANT COMPONENT IS CALLED COMPONENT 2 REDUNDANT.

3.3 Eliminating Algebraic Loops

The algorithm for computing consequence severity involves propagating consequence information from one end of the network to the other. For the causal structure to be consistent with the methods used in this paper, it must be represented as a directed acyclic graph. For this reason, when generating a graph of consequences from a system diagram for preliminary analysis, it is important to eliminate algebraic loops. There are two main situations in which an algebraic loop may form: (1) when the system includes feedback, or (2) when a redundant component is shared between two other components.

To eliminate algebraic loops caused by feedback, the connection between the component providing information and the component receiving information required for feedback is removed. For example, consider a sensor that is sensing motor voltage. The connection in the network between the sensor and motor will be removed. This means that, in this case, the sensor will be treated as a source node. Removal of that edge can be justified by reasoning via the example that there is no interest in propagating a motor malfunction to the sensor since a sensor failure would lead to a motor malfunction in the worst case (an event that has already occurred).

There are also situations in which the addition of a shared redundant component leads to the creation of internal loops. For instance, consider the network in Fig. 3. In its current form in Fig. 3, BuR is redundant to Bu1 only. However, since Bu1 and Bu2 are the same type of component, BuR could be used as a redundancy for both. This form is shown in Fig. 4.

The problem with this representation of the cross redundancy is that it creates a loop that does not physically exist between Co1 and BuR. The loop does not physically exist because

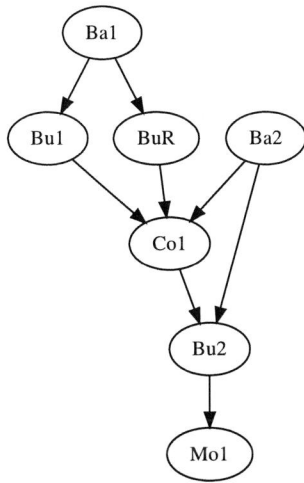

FIGURE 3: NETWORK WITH A REDUNDANT COMPONENT, BuR, WHICH IS REDUNDANT TO Bu1.

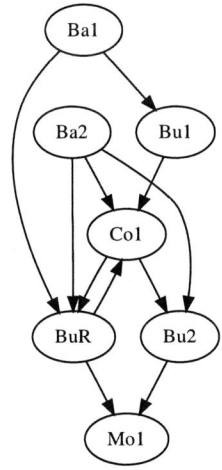

FIGURE 4: BuR IS REDUNDANT TO Bu1 and Bu2. THERE IS A LOOP BETWEEN Co1 AND BuR.

BuR cannot be redundant to both components at the same time. The solution is to create two nodes for BuR: BuRA and BuRB, as shown in Fig. 5.

Not only does this solution eliminate the algebraic loop, but it also correctly implies that BuR can only be redundant to either Bu1 or Bu2 at one point in time, even though it can be activated for either of the two components.

4 STRUCTURAL CONSEQUENCE

The consequence metric proposed in this paper quantifies the consequence of component failures within an architecture

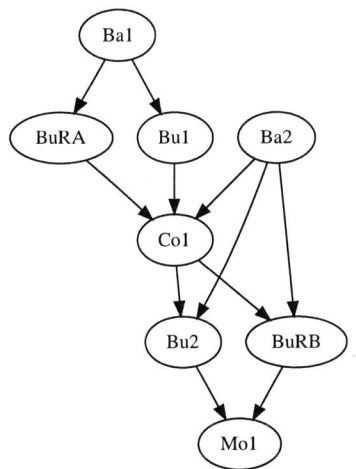

FIGURE 5: ALGEBRAIC LOOP ELIMINATED BY SPLITTING BuR INTO BuRA AND BuRB.

based on the layout of components and the flow of information/connections. The metric falls within the range of $[0,1]$, where 0 corresponds to the absence of consequences and 1 to the most severe consequences. This metric is computed through initialization of sink nodes (end components of a network) with user-defined consequences and back-propagating the consequences throughout the network via a propagation algorithm (presented in Section 4.5) and set operations.

4.1 Construction of the Consequence Space

A consequence space Ω is first constructed by aggregating all distinct and possible sources of cost and lumping them into a single set. The set of all defined consequences make up the consequence space and the overall cost of the entire consequence space is set to 1. Note that consequence in this paper refers to an undesirable event that occurs due to malfunctions. Within the consequence space, failure scenarios constituting different proportions of the total cost are represented. Intersections within the space represent consequences that are common to different failure scenarios. A metric S can then be associated with these consequences to quantify the ordinal severity (e.g. 1-10 or cost) to allow for set operations to be performed.

In the subsections that follow, let C_{k_1} represent the consequence of a node/component k_1 in the network, $E_{k_1:k_2}$ represent the consequence of a link/edge connecting C_{k_1} to C_{k_2}, and $S(C_{k_1})$, $S(E_{k_1:k_2})$ represent the consequential severity (cost) of a component k_1 and an edge $E_{k_1:k_2}$ respectively. Note that the severity of the entire consequence space is 1 ($S(\Omega) = 1$).

4.2 Scoring the Sink Nodes of the Network

The sink nodes of the network represent components at the endpoints of an architecture, meaning they do not have any outgoing edges. Components represented by sink nodes should typically be ones for which consequences can be estimated based on available data such as expert opinion, or field data. For example, let the sink node represent an aircraft aileron; although the electrical architecture operating an aileron might be novel, a cost estimate can still be deduced based on previous instances of aileron failures. Figure 6 shows an example of how consequences and costs are assigned in Ω for the sink nodes.

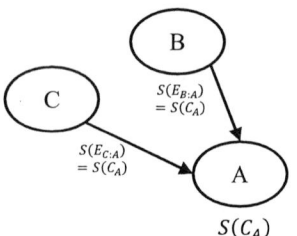

FIGURE 7: EXAMPLE OF SCORING EDGES WITH NO REDUNDANCIES.

4.3.2 Scoring Edges with Redundancies

Although in practice redundancy is expressed in terms of components being redundant to one another, this paper represents redundancy via the edges of the network. Two edges are redundant if (i) their tails are connected to different components, (ii) their heads are connected to the same component and (iii) they share the same function.

Failure of a set of redundant edges leads to the failure of the component they feed. This means that the union of n redundant edges $E_{i_1:j} \cdots E_{i_n:j}$ bears the same consequence as the component C_j the edges all feed.

$$S(E_{i_1:j} \cup E_{i_2:j} \cdots E_{i_n:j}) = S(C_j) \qquad (2)$$

Property 2: The union of a collection of redundant edges bears the same consequence as the component they feed.

Assumption 1: All redundant edges are assumed to have equivalent consequences, where edge consequences represent mutually exclusive subsets of the component consequence with equivalent severities.

This means that,

$$S(E_{i_1:j}) = S(E_{i_2:j}) = \cdots = S(E_{i_n:j}) \qquad (3a)$$

$$E_{i_k:j} \cap E_{i_m:j} = \emptyset \quad k,m \in \{1,\ldots,n\} | k \neq m \qquad (3b)$$

Using *Property 1* and *Assumption 1*,

$$S(E_{i_1:j}) = S(E_{i_2:j}) = \cdots = S(E_{i_n:j}) = \frac{S(C_j)}{n} \qquad (4)$$

Assumption 1 means that each redundant component equally shares the "responsibility" of the total consequences of all redundant edges. Let the overall consequence C_j be broken down into a set of shared and mutually exclusive consequences $C_j^{i_1:j}, C_j^{i_2:j}, \cdots C_j^{i_n:j}$ corresponding to subsets of the overall consequence shared by the redundant edges $E_{i_1:j} \cdots E_{i_n:j}$ (such that $C_j^{i_k:j} = E_{i_k:j}$). This notation is introduced in order to allow proper

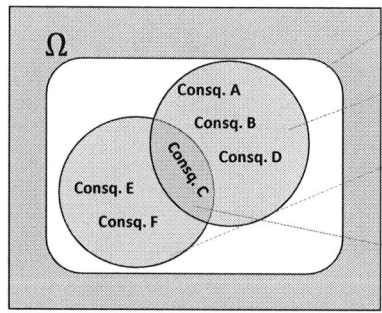

FIGURE 6: CONSEQUENCE SPACE VISUALIZATION.

4.3 Scoring the Edges of the Network

All edges included in the network are assumed to have functions necessary for the operation of the end components (sink nodes). In Section 4.3.1 the scoring scheme for edges with no redundancies is presented, and in Section 4.3.2, the scoring scheme for edges with redundancies is presented.

4.3.1 Scoring Edges with No Redundancies

Failure of an edge for which no redundancies or protective factors exist leads to the potential failure of the component it feeds (at the head of the edge). This means that the failure or loss of the edge bears the same consequence as the loss of the component it feeds.

$$S(E_{i:j}) = S(C_j) \qquad (1)$$

An example of this can be seen in Fig. 7. Loss of either of the edges connecting B to A, or A leads to loss of A.

Property 1: An edge for which no redundancies or protective features exist bears the same consequence as the component it feeds.

propagation and traceability of consequences in the network. The following properties hold:

$$C_j^{i_l:j} \cap C_j^{i_m:j} = \emptyset \quad l,m \in \{1,\cdots,n\} \; l \neq m$$

$$C_j^{i_1:j} \cup C_j^{i_2:j} \cdots \cup C_j^{i_n:j} = C_j$$

$$C_j^{i_l:j} \cup C_j = C_j$$

$$C_j^{i_l:j} \cap C_j = C_j^{i_l:j}$$

In Fig. 8, the edges connecting D to A, and DR to A are redundant. The consequences of A are equally shared between the two edges.

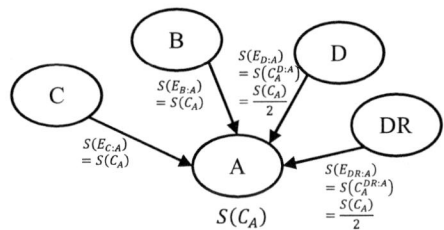

FIGURE 8: EXAMPLE OF SCORING EDGES WITH REDUNDANCIES.

4.4 Scoring the Nodes of a Network

Failure of a component that is not a sink leads to the failure or loss of all the pieces of information and links (edges) originating from that component.

Property 3: A component bears the same consequence as all of the outgoing edges from that component.

$$S(C_i) = S(E_{i:j_1} \cup E_{i:j_2} \cdots \cup E_{i:j_n}) \quad (6)$$

The severity of a component is evaluated in the same manner that the probabilities of set unions are evaluated in probability theory.

It is important to note that set operations are first used in node severity calculations. Once the sets have been fully simplified, severities are then quantified.

Figure 9 illustrates an example of scoring nodes in a network. Each of the nodes bear the same consequences as the union edges that are outgoing from them.

4.5 Propagation Algorithm

The propagation algorithm assumes two characteristics of the network:

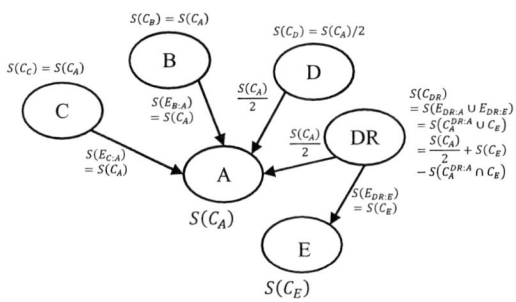

FIGURE 9: EXAMPLE OF SCORING NODES.

1. The network has no undirected edges (all edges have a direction).
2. The network has no internal loops.

If either of these assumptions do not apply, the propogation algorithm will not score nodes accurately and the network model must be revised (see Section 3.3).

The algorithm propagates the consequences of all sinks through the network until all nodes have been scored. This begins with a random sink node. First, the incoming edges associated with sink nodes are scored. Then, the propagation algorithm picks a random node that has all of its outgoing edges scored. This continues until all nodes, including source nodes, are scored. This process is represented visually in the flowchart in Fig. 10.

This algorithm ensures that the current node (the node that the algorithm is currently scoring) will always have all of its outgoing edges scored. This way, the algorithm will not have to "guess" any node scores and the computed node scores will be deterministic.

5 CASE STUDY

The methodology presented in the previous sections is demonstrated using a case study of a simplified electric subsystem of an all-electric fixed-wing unmanned aircraft system (UAS). The electric subsystem consists of several components and provides power to elecro-mechanical actuators used to manipulate an aileron and a flap of the UAS. A block diagram of the base (no redundancy) electric subsystem is given in Fig. 11.

Two different architectures that include redundancies and no single points of failure are presented for the electric subsystem. The first architecture, illustrated in Fig. 12, consists of the addition of a single redundancy in parallel to each of the components in Fig. 11. The second is based on a smart and re-configurable architecture with parallel and cross-redundancies (redundancies shared by different components) and is illustrated in Fig. 13. The motivation of the second architecture is based on the necessity of being able to perform comprehensive analysis on complex

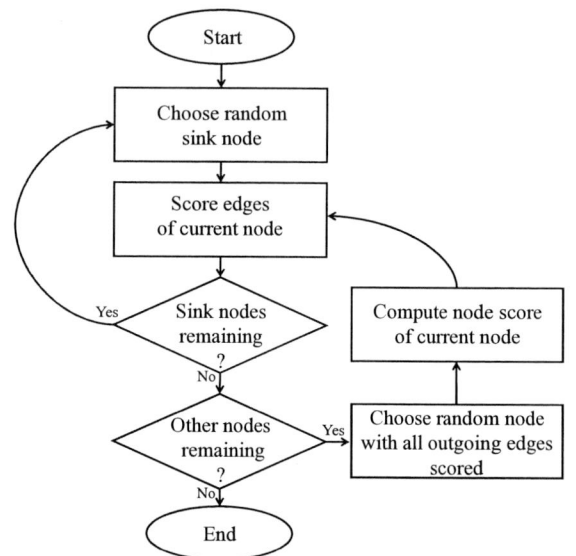

FIGURE 10: FLOWCHART OF PROPAGATION ALGORITHM.

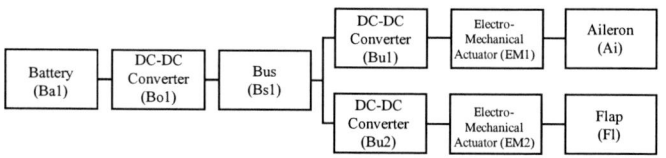

FIGURE 11: ELECTRICAL SUBSYSTEM BLOCK DIAGRAM FOR BASE ARCHITECTURE WITH NO REDUNDANCIES ADDED.

architectural designs for future aircraft that are optimized via re-configuration. The structural consequence metric is used to evaluate and quantify the structural consequence of each component within the two architectures. Note that the architectures are not based on realistic systems, they are hypothetical and developed for the purpose of demonstrating SCA.

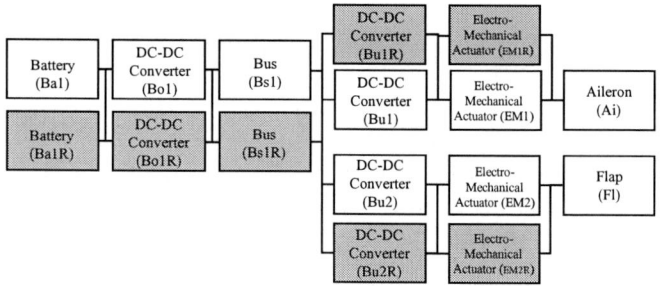

FIGURE 12: ELECTRICAL SUBSYSTEM BLOCK DIAGRAM FOR ARCHITECTURE WITH PARALLEL REDUNDANCIES.

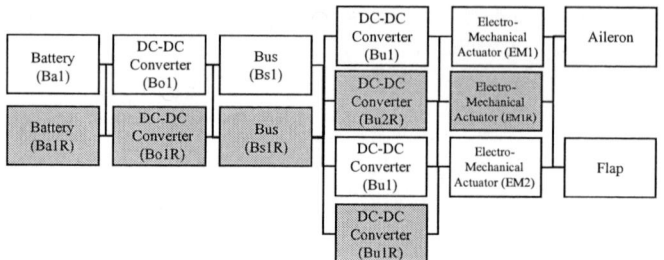

FIGURE 13: ELECTRICAL SUBSYSTEM BLOCK DIAGRAM FOR ARCHITECTURE WITH RE-CONFIGURABLE REDUNDANCIES.

As indicated by the process outlined in Section 4, a consequence space is first constructed and consequences are assigned to both the aileron and the flap. In this case study, consequences with a cumulative cost of $800,000$ are assigned to an aileron failure, consequences with a cumulative cost of $300,000$ are assigned to flap failures, and consequences with a cost of $100,000$ are shared between the two components.

The entire consequence space Ω has an overall cost of $1,000,000$, which is normalized to a severity of 1 ($S(\Omega) = 1$). Based on this normalization, the following serverities are computed:

$$S(C_{Ai}) = 0.8$$
$$S(C_{Fl}) = 0.3$$
$$S(C_{Ai} \cap C_{Fl}) = 0.1$$

It is worth mentioning that a consequence set captures a list of events (with associated costs) that occur as a result of a component failure. Each sink component has a consequence space formed by the list of events that can occur due to loss or failure, where intersections in consequence spaces of different components represent events that are mutual to both the components.

A network-based representation is first constructed for the base architecture, the parallel architecture, and the smart architecture based on Section 3. These network representations are illustrated in Figs.14–16. The white and gray portions of the nodes Bu1R and EM1R represent two re-configurable components with the same connections. Based on the properties and algorithm in Section 4, consequence sets from the sink nodes are back propagated in the three networks, and the severities (costs) of these consequences are quantified at each node based on the consequence space using operations that are analogous of probability measures in sample spaces. The computed edge and node severities for all three architectures are captured in Figs.14–16 and tabulated in Table 1.

A few examples of consequence propagation will be provided based on the parallel redundancies architecture depicted in

Fig. 15. $EM1$ and $EM1R1$ are redundant with respect to $Ai1$. The consequence set C_{Ai1} is therefore symmetrically split (Property 2, and Assumption 1) into two sets $C_{Ai1}^{EM1:Ai1}$ and $C_{Ai1}^{EM1R1:Ai1}$ corresponding to the edges $E_{EM1:Ai1}$ and $E_{EM1R1:Ai1}$ respectively.

The severities of $E_{EM1:Ai1}$ and $E_{EM1R1:Ai1}$ can be computed as:

$$S(E_{EM1:Ai1}) = S(C_{Ai1}^{EM1:Ai1}) = \frac{S(Ai)}{2} = 0.4$$

$$S(E_{EM1R1:Ai1}) = S(C_{Ai1}^{EM1R1:Ai1}) = \frac{S(Ai)}{2} = 0.4$$

The consequences of $EM1$ and $EM1R1$ are then computed using Property 3,

$$S(C_{EM1}) = S(E_{EM1:Ai1}) = 0.4$$
$$S(C_{EM1R1}) = S(E_{EM1R1:Ai1}) = 0.4$$

Similar to $S(E_{EM1:Ai1})$ and $S(E_{EM1R1:Ai1})$, the edges $S(E_{Bu1R:EM1R})$, $S(E_{Bu1:EM1R})$, $S(E_{Bu1R:EM1})$, and $S(E_{Bu1:EM1})$ all have severities of 0.2.

The consequences of $Bu1$ and $Bu1R$ can be computed via Property 3. Note that the union operator is mathematically evaluated in the same manner as probability measures in sample spaces.

$$\begin{aligned} S(C_{Bu1}) &= S(E_{Bu1:EM1} \cup E_{Bu1:EM1R}) \\ &= S(E_{Bu1:EM1}) + S(E_{Bu1:EM1R}) - \\ & \quad S(E_{Bu1:EM1} \cap E_{Bu1:EM1R}) = 0.4 \end{aligned}$$

Similarly, $S(Bu1R) = 0.4$. Note that due to mutual exclusivity of the redundant edges, $S(E_{Bu1:EM1} \cap E_{Bu1:EM1R}) = \emptyset$.

A similar process is followed for the rest of the nodes and edges in between $Ai1, Fl1$ and $Bs1, Bs1R$.

The consequence of $Bs1$ and $Bs1R$ are dependent on consequences originating from both Al and Fl. By employing Properties 1–3, and using the fact that C_{Al} and C_{Fl} do indeed intersect due to shared consequences, $S(C_{Bs1})$ (and similarly $S(C_{Bs1R})$) can be computed as follows,

$$\begin{aligned} S(C_{Bs1}) &= S(E_{Bs1:BuR1} \cup S(E_{Bs1:Bu1} \cup S(E_{Bs1:BuR2} \cup S(E_{Bs1:Bu2}))) \\ &= S(E_{Bs1:BuR1}) + S(E_{Bs1:Bu1}) \\ & \quad + S(E_{Bs1:BuR2}) + S(E_{Bs1:Bu2}) \\ & \quad - S(E_{Bs1:Bu1} \cap E_{Bs1:Bu2}) \\ & \quad - S(E_{Bs1:Bu1} \cap E_{Bs1:Bu2R}) \\ & \quad - S(E_{Bs1:Bu1R} \cap E_{Bs1:Bu2}) \\ & \quad - S(E_{Bs1:Bu1R} \cap E_{Bs1:Bu2R}) = 0.5 \end{aligned}$$

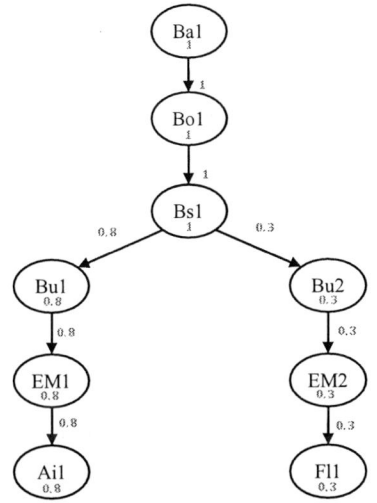

FIGURE 14: NETWORK REPRESENTATION OF THE BASE ARCHITECTURE.

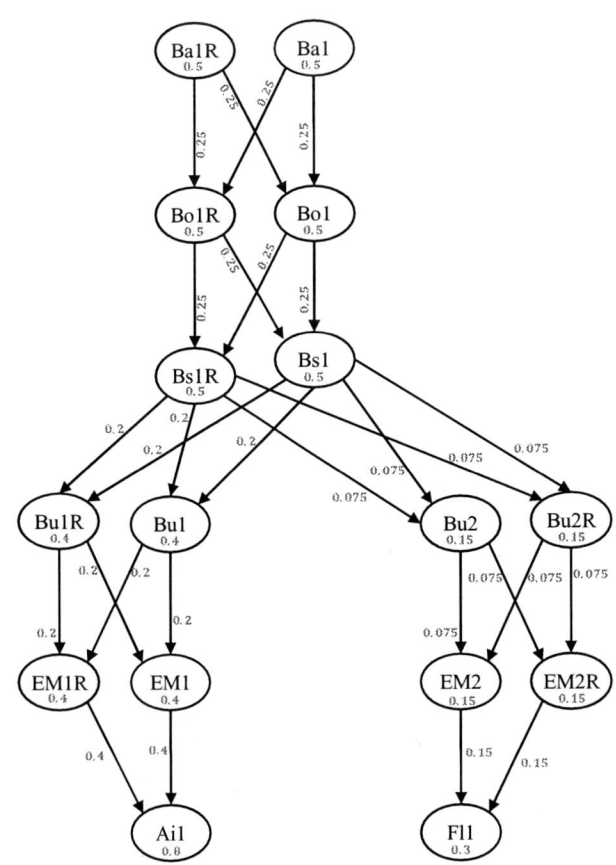

FIGURE 15: NETWORK REPRESENTATION OF THE ARCHITECTURE WITH PARALLEL REDUNDANCIES.

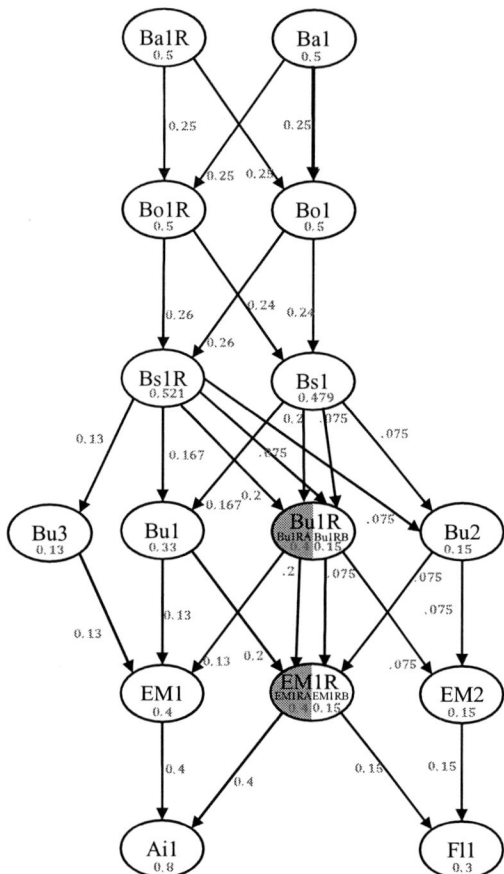

FIGURE 16: NETWORK REPRESENTATION OF THE ARCHITECTURE WITH PARALLEL AND RE-CONFIGURABLE REDUNDANCIES (WHITE & GRAY).

TABLE 1: COMPONENT SCA SEVERITY SCORES FOR BASE, PARALLEL, AND RE-CONFIGURABLE ARCHITECTURES.

Component	Base	Parallel	Re-config.
Ba1	1	0.5	0.5
Ba1R1	N/A	0.5	0.5
Bo1	1	0.5	0.5
Bo1R1	N/A	0.5	0.5
Bs1	1	0.5	0.479
Bs1R1	N/A	0.5	0.521
Bu1	0.8	0.4	0.33
Bu1R1	N/A	0.4	0.4
Bu1R2	N/A	N/A	0.13
Bu2	0.3	0.15	0.15
Bu2R1	N/A	0.15	N/A
EM1	0.8	0.4	0.4
EM1R1	N/A	0.4	0.4
EM2	0.3	0.15	0.15
EM2R1	N/A	0.15	N/A
Ai1	0.8	0.8	0.8
Fl1	0.3	0.3	0.3

The results show that SCA allows the designer to identify nodes for which there are high consequences of failure based on the system architecture. Such information is valuable as it allows for quantification of component failure consequence and early identification of system weak points.

6 DISCUSSION

SCA provides value to the design process by giving designers the ability to represent the chain of causality that leads to various consequences in a system. Unlike fault tree analysis, this view of causality is not limited by single "top event" consequences but by the set of consequences that could occur in a system, consequences that could have different impacts to safety, maintenance and repair, availability, and convenience. This expressiveness allows the designer to assess the true system-level impacts of a failure and make design changes that would lead to different changes in the causal graph that lead to different sets of consequences.

Additionally, as shown in the case study, assessing the structural importance of components from the point of view of the consequences they produce allows the designer to consider a novel set of structural importance metrics to assess redundancy schemes. These metrics, based on the "virtual severity" of component failure, show the designer which redundancies are most consequential in the architecture, and thus which components are the weakest points from the structural standpoint. Table 1 shows the effect of different redundancy strategies on this consequence share. As shown added redundancies reduce the share of severity associated with each component, while allowing reconfiguration changes the amount of "share" each component will take, spreading the amount across several (in the case of Bu1), or concentrating it (in the case of Bs1). This shows how virtual severity can be used as a structural metric to inform design.

The difference between this virtual consequence and the direct consequence of failure should be noted. While direct consequence is defined as the direct consequence that would occur given a node event occurs, there is no direct consequence of a redundant component failing (to functionality, at least). This virtual consequence, the proportion of the failure consequence of the redundancy scheme associated with each component, therefore, requires some technical interpretation beyond direct severity. Nevertheless, it does show the relative "share" of importance of redundant components, which is useful to quantify for developing redundant architectures.

7 CONCLUSIONS

This research has proposed a novel metric for measuring the consequence of components within a system architecture that requires neither a fault tree nor failure rate information. A case study demonstrated how the metric is able to guide early design decisions and provide early estimates for redundancy placement, which can be later revised once more detailed system information becomes available. Use of this metric in early design may help designers from underestimating the costs of adding redundancy and may help them plan for size and weight considerations early in the design process. This can aid in designers' efforts to design safe systems without overly sacrificing performance or cost.

8 FUTURE WORK

Future work should further develop structural consequence analysis into a fully-expressive formalism to analyze the flow of causality in a system and demonstrate the implementation of this tool in software for immediate automated visualization of consequence propagation. Additionally, future work should show the application of this method to consider a more expanded case study, considering the internal causal mechanisms that produce faults in components rather than a simple system structure. Additionally, the effect of protective factors other than redundancy should be considered and factored into this method. This would be complemented by quantifying additional structural metrics to determine how inherently safe or hazardous a system is based on the graph structure.

ACKNOWLEDGMENT

This research is supported by the System-Wide Safety (SWS) project in the Airspace Operations & Safety program (AOSP) in the NASA Aeronautics Research Mission Directorate (ARMD). The research is also partially supported by the National Science Foundation CMMI 1562027, and NASA award number 80NSSC17M0018. Any opinions or findings of this work are the responsibility of the authors and do not necessarily reflect the views of the sponsors or collaborators.

REFERENCES

[1] Sklaroff, J., 1976. "Redundancy management technique for space shuttle computers". *IBM Journal of Research and Development,* **20**(1), February, pp. 20–28.

[2] Coit, D., 2001. "Cold-standby redundancy optimization for nonrepairable systems". *IIE Transactions,* **33**(6), pp. 471–478.

[3] Sinha, K., Suh, E. S., and de Weck, O., 2017. "Correlating integrative complexity with system modularity". In *ASME 2017 International Design Engineering Technical Conferences & Computers and Information in Engineering Conference (IDETC/CIE 2017)*, Vol. 2A: 40th Design Automation Conference. ASME, New York, p. V02AT03A048.

[4] Pahl, G., and Beitz, W., 2007. *Engineering design: a systematic approach.* Springer Science & Business Media.

[5] Denney, E., Pai, G., and Whiteside, I., 2017. "Modeling the safety architecture of uas flight operations". In International Conference on Computer Safety, Reliability, and Security, Springer, pp. 162–178.

[6] Vesely, W. E., Goldberg, F. F., Roberts, N., and Haasi, D. F., 1981. The fault tree handbook. Tech. Rep. NUREG0492.

[7] Barlow, R. E., and Proschan, F., 1975. "Importance of system components and fault tree events". *Stochastic Processes and their Applications,* **3**(2), pp. 153–173.

[8] Birnbaum, Z., 1968. On the importance of different components in a multicomponent system. Tech. Rep. 54.

[9] Yuhua, D., and Datao, Y., 2005. "Estimation of failure probability of oil and gas transmission pipelines by fuzzy fault tree analysis". *Journal of Loss Prevention in the Process Industries,* **18**(2), pp. 83–88.

[10] Sinnamon, R. M., and Andrews, J., 1998. "Improved accuracy in quantitative fault tree analysis". *Quality and Reliability Engineering International,* **13**(5), pp. 285–292.

[11] Ruijters, E., and Stoelinga, M., 2015. "Fault tree analysis: A survey of the state-of-the-art in modeling, analysis and tools". *Computer science review,* **15–16**, pp. 29–62.

[12] Boudali, H., Crouzen, P., and Stoelinga, M. "A compositional semantics for dynamic fault trees in terms of interactive markov chains". *International Symposium on Automated Technology for Verification and Analysis ATVA 2007: Automated Technology for Verification and Analysis,* pp. 441–456.

[13] Dugan, J., Bavuso, S., and Boyd, M., 1990. "Fault trees and sequence dependencies". *Annual Proceedings on Reliability and Maintainability Symposium.*

[14] Liang, X., Yi, H., Zhang, Y., and Li, D., 2009. "A numerical simulation approach for reliability analysis of fault-tolerant repairable system". *Proceedings of the 8th International Conference Reliability, Maintainability and Safety,* pp. 191–196.

[15] Zhang, X., Miao, Q., Fan, X., and Wang, D., 2009. "Dynamic fault tree analysis based on petri nets". *Proceedings*

of the 8th International Conference Reliability, Maintainability and Safety, pp. 138–142.

[16] Dutuit, Y., and Rauzy, A., 2001. "Efficient algorithms to assess component and gate importance in fault tree analysis". *Reliability Engineering System Safety, 72*(2), pp. 213–222.

[17] Buchacker, K., 2000. "Modeling with extended fault trees". *Proceedings of the 5th International Symposium on High Assurance Systems Engineering*, pp. 238–246.

[18] Bobbio, A., and Codetta-Raiteri, D., 2004. "Parametric fault trees with dynamic gates and repair boxes". *Proceedings of Reliability and Maintainability Symposium*, pp. 459–465.

[19] Albert, R., Jeong, H., and Barabsi, A. L. "Diameter of the world-wide web". *Nature, 401*, pp. 130–131.

[20] Mason, O., and Verwoerd, M., 2007. "Graph theory and networks in biology". *IET Systems Biology, 1*(2), pp. 89–119.

[21] Kossinets, G., 2006. "Effects of missing data in social networks". *Social networks, 28*, pp. 247–268.

[22] Freeman, L. C., 1978. "Centrality in social networks conceptual clarification". *Social networks, 1*(3), pp. 215–239.

[23] Barabási, A.-L., 2003. *Linked: The new science of networks*. Plume.

[24] Erdős, P., and Rényi, A., 1959. "On random graphs". *Publicationes Mathematicae Debrecen, 6*, pp. 290–297.

[25] Haley, B., Dong, A., and Tumer, I. Y., 2014. "Creating faultable network models of complex engineered systems". In *ASME 2014 International Design and Engineering Technical Conferences & Computers and Information in Engineering Conference (IDETC/CIE 2014)*, Vol. 2A: 40th Design Automation Conference. ASME, New York, p. V02AT03A051.

[26] Haley, B., Dong, A., and Tumer, I., 2016. "A comparison of network-based metrics of behavioral models of complex engineered systems". *Journal of Mechanical Design, 138*(12), p. 121405.

[27] Walsh, H. S., Dong, A., and Tumer, I. Y., 2017. "The structure of vulnerable nodes in behavioral network models of complex engineered systems". In *ASME 2017 International Design Engineering Technical Conferences & Computers and Information in Engineering Conference (IDETC/CIE 2017)*, Vol. 7: 29th International Conference on Design Theory and Methodology. ASME, New York, p. V007T06A019.

[28] Walsh, H., Dong, A., and Tumer, I., 2018. "The role of bridging nodes in behavioral network models of complex engineered systems". *Design Science, 4*.

[29] Sosa, M., Eppinger, S., and Rowles, C., 2007. "A network approach to define modularity of components in complex products". *Journal of Mechanical Design, 129*(11), pp. 1118–1129.

[30] Ma, S., Jiang, Z., and Liu, W., 2016. "A design change analysis model as a change impact analysis basis for semantic design change management". *Proceedings of the Institution of Mechanical Engineers, Part C: Journal of Mechanical Engineering Science, 231*(13), pp. 2384–2397.

[31] Ball, Z., and Lewis, K., 2018. "Observing network characteristics in mass collaboration design projects". *Design Science, 4*(e4).

[32] Wang, M., Sha, Z., Huang, Y., Contractor, N., Fu, Y., and Chen, W., 2016. "Forecasting technological impacts on customers co-consideration behaviors: A data-driven network analysis approach". In *ASME 2016 International Design Engineering Technical Conferences & Computers and Information in Engineering Conference (IDETC/CIE 2016)*, Vol. 42nd Design Automation Conference. ASME, New York, p. V02AT03A040.

[33] Wang, M., Chen, W., Huang, Y., Contractor, N. S., and Fu, Y., 2016. "Modeling customer preferences using multidimensional network analysis in engineering design". *Design Science, 2*(e11).

[34] Baldwin, C., MacCormack, A., and Rusnak, J., 2014. "Hidden structure: Using network methods to map system architecture". *Research Policy, 43*(8), pp. 1381–1397.

[35] Sosa, M., Mihm, J., and Browning, T., 2011. "Degree distribution and quality in complex engineered systems". *Journal of Mechanical Design, 133*(10), p. 101008.

[36] Ullman, D., 2009. *The mechanical design process*. McGraw-Hill Science/Engineering/Math.

[37] Lewis, E. E., and Lewis, E. E., 1987. *Introduction to reliability engineering*, Vol. 2. Wiley New York et al.

[38] Clothier, R., Denney, E., and Pai, G. J., 2017. "Making a risk informed safety case for small unmanned aircraft system operations". In 17th AIAA Aviation Technology, Integration, and Operations Conference, p. 3275.

[39] Joshi, A., Whalen, M., and Heimdahl, M., 2005. "Model-based safety analysis final report". *NASA Techreport*.

[40] Kurtoglu, T., Tumer, I. Y., and Jensen, D. C., 2010. "A functional failure reasoning methodology for evaluation of conceptual system architectures". *Research in Engineering Design, 21*(4), pp. 209–234.

[41] Hulse, D., Hoyle, C., Goebel, K., and Tumer, I. Y., 2019. "Quantifying the resilience-informed scenario cost sum: A value-driven design approach for functional hazard assessment". *Journal of Mechanical Design, 141*(2), p. 021403.

[42] Arlitt, R., and Van Bossuyt, D. L., 2019. "A generative human-in-the-loop approach for conceptual design exploration using flow failure frequency in functional models". *Journal of Computing and Information Science in Engineering, 19*(3), p. 031001.

[43] Dezfuli, H., Benjamin, A., Everett, C., Feather, M., Rutledge, P., Sen, D., and Youngblood, R., 2015. "NASA system safety handbook. Volume 2: System safety concepts, guidelines, and implementation examples".

Proceedings of the ASME 2019
International Design Engineering Technical Conferences
and Computers and Information in Engineering Conference
IDETC/CIE2019
August 18-21, 2019, Anaheim, CA, USA

DETC2019-98404

AN EXCESS BASED APPROACH TO CHANGE PROPAGATION

Daniel Long
Graduate Research Assistant
North Carolina State University
Raleigh, NC, USA
delong@ncsu.edu

Dr. Scott Ferguson
Associate Professor
North Carolina State University
Raleigh, NC, USA
scott_ferguson@ncsu.edu

ABSTRACT

This research demonstrates how the Decision Based Design (DBD) approach can be used for determining a system's lifecycle value when including excess. Prior research has shown that excess (the degree to which a component or attribute is sized beyond the minimum required to support the initially fielded system) can reduce the cost of changing a system. Theoretically, excess inhibits change propagation within a system and could be strategically added to increase the value of that system. Including excess, however, also adds cost and potentially impacts system performance. Prior research has not quantitatively linked excess as a means of limiting change propagation to system lifecycle value. This work advances the existing literature by considering how excess is imbedded in a system and what impact excess has on the system's total value. After being introduced, the method is demonstrated on a desktop computer example. Results from the study are used to show how decisions about power supply capacity can be optimized by incorporating excess to achieve flexibility.

INTRODUCTION

All engineered systems are designed in a specific context. This context includes the 1) technologies and tools available during design, 2) performance expectations and requirements during operation, and 3) other products the system interacts with. Modern systems are built in, and for, a context where the pace of change is increasingly frenetic. Researchers have advanced design practice by providing tools that model the ramifications of system change, but significant room for progress remains.

The accelerating pace of change increases the uncertainty of assumptions made during the design process. For example, today's automobile manufacturer must consider a myriad of questions: Will autonomous systems be ready for deployment in the next five years? Will a carbon tax raise fuel prices and drive consumer preference towards more efficient or electric vehicles? Will charging stations be prevalent enough to support a plug-in

option for electric vehicles? The assumptions made by designers when answering these questions drive design decisions for both individual system design and the selection of product families.

Complex system design is also made challenging because of the expense associated with designing and building them. One consequence is that complex systems remain in service for extended periods. Barber et al. [2] describes three strategies designers may use when planning for future changes:

1) Design to current requirements only
2) Design to meet predicted end of life requirements
3) Design so that the system may be modified to adapt to new circumstances

The first strategy is useful if the system can be inexpensively replaced (as in simple consumer electronics) and the second strategy is effective when the system modification cost is prohibitive (as in telecommunication satellites). The third strategy, embedding changeability, requires compromise and has subsequently become an active area of design research. While the idea of including flexibility is viewed favorably in literature, significant open questions remain. These questions are highlighted by the ontological debates about the very definition of flexibility [3–6]. Despite the lack of consensus, researchers have proposed useful ideas about what changeability encompasses and have developed tools for exploring and modeling changeability.

This research focuses on the aspect of changeability that enables/allows for system changes after it has been put into operation. Prior flexibility research has focused on characterizing how system architecture affects change propagation throughout a system. In this approach, a system is represented as a network of components connected by design dependencies. This can be done using a Design Structure Matrix (DSM) as in Fricke and Schulz [1] and Clarkson et al. [7]. The cause for change propagation between two components along a specific pathway is frequently derived from expert opinion and is specified as a probability. Once this network has been

Copyright © 2019 ASME

established, a variety of metrics can be calculated to determine the likelihood of change propagation and how components are likely to experience, or contribute to, change propagation.

Despite prior research, there are still aspects of change propagation and flexibility in engineering design research that are poorly understood. One area is understanding the influence that component sizing (design variable selection) and interface definition has on change propagation. Adopting the concept of "excess" from Tackett et al. [8], we can measure the extent by which a component or attribute is sized beyond the minimum required to support the initially fielded system. Absence of excess within a system can result in a loss of expected value and premature system obsolescence.

Including excess in a system raises fundamental design questions. Designers must decide in what systems and situations is excess advantageous, which components or system attributes should have excess, and how much excess should be included. Existing design tools inadequately answer these questions.

This research uses the Decision Based Design (DBD) framework developed by Hazelrigg [10] as the foundation by which a new method is crafted for answering questions about including excess. The DBD framework is an overarching guide for how evaluating the selection of component sizing (system design variable selection) can be approached, but the specifics of implementation are left to the designer. The method and example shown in subsequent sections are an embodiment of the DBD framework when considering excess inclusion in support of future system changes.

Specifically, this paper will: 1) introduce an approach for evaluating system changes using the DBD framework, change propagation, and excess, 2) detail an object-oriented model that executes the approach, and 3) demonstrate the approach on a desktop computer test case using historical data to determine optimal component selection and excess inclusion.

The following section will review existing research necessary to be combined and/or extended to support this effort.

BACKGROUND

To date, system flexibility research has no accepted objective measure of goodness. Saleh et al. [6] compares the state of changeability research to safety research decades ago as "vague and difficult to improve, yet critical to competitiveness". This section will cover approaches foundational to either the study of flexibility and how this leads to work in changeability and excess, or the present research specifically.

Change Propagation Method (CPM)

A large body of existing research dedicated to the study of flexibility focuses on system architecture. System architecture is defined by Ulrich [11] as: 1) the arrangement of function elements, 2) the mapping from functional elements to physical components and 3) the specification of the interfaces among interacting physical components. Flexibility research has therefore been defined in terms of these three elements.

The Change Propagation Method (CPM) is one of the most widely referenced methods for understanding how change to one component in a system can necessitate changes to other connected components. CPM is the result of a pair of papers published after design researchers examined the changes made during the design of the Westland EH101 helicopter [7,12].

CMP uses Design Structure Matrices (DSMs) [13] to capture the change relationships which exist between components. In CPM two DSMs are created: one represents the likelihood change propagating from one component to another and the other represents the anticipated scale of the impact of the change. The value assigned to each pair of components is a probability between zero and one. The initial matrices are generated using expert opinion and only attempts to capture the direct risk of change. CMP has been extended and modified to include a variety of type of connections [14,15], the impact of change on requirements, performance, and function [16,17], more detail on how indirect propagation effects design and initiating components [18], and for use in developing product families [19]. This method provides some insight into how change may propagate throughout a system in a way that informs designers about what components should be focused on but does not provide much insight into how this guidance should be used.

Real options analysis

Real options analysis is a tool rooted in the desire to estimate value under uncertainty in financial markets. A financial option is the right, but not the obligation, to buy or sell shares for a fixed price at some point in the future [20]. The appeal of an option is the hedge it provides against uncertainty. By purchasing an option, the holder spends money now to protect against future uncertainty. Real options analysis was successfully employed in the financial field and was later expanded to include determining the value of options on non-financial (or real) investments. Engineering design researchers have advanced the idea of real options to include the value of real options "in" a system. Real options in a system are options that provide flexibility though some attribute or design feature. This could be preparing the foundation of a bridge during initial construction to add a second deck later if needed [21] or a parking garage with initial construction made to support the addition of more levels [22].

The value of real options analysis is that is encourages designers to consider alternatives to a point design. This includes traditional options like the option to delay a project or the option to stage project growth. Real options can also be expanded to consider what impact system margin may have on the future system value in the presence of uncertainty, a part of this research.

Excess and system evolvability

Excess (or reserve margin) is defined as *the quantity of usable surplus of a system resource*. This is a slightly modified version of the definition of "excess" found in Allen [23] and Tackett [8] and is distinct from safety margin. Margin can result from system flows [24] (like maximum electrical power through a component), or from aggregate system properties like total weight, internal volume, or stability [8]. Margin is typically a measure of how far a system parameter is from a maximum value as defined by a requirement or performance target. An example found in Tackett et al. [8] is electric power generation on a nuclear power aircraft carrier. If the reactor has a maximum output of 200 MW and the system's components only use 150 MW there is an excess of 50 MW that can be used to support future changes.

Several existing studies have highlighted the need for margin to support changeability. In one study, Tilstra [25] performed a survey of patents and consumer products identified as flexible to discover common design guidelines. Guidelines 23 "Controlling the tuning of design parameters" and 24 "Providing the capability for excess energy storage or importation" both identify the need for margin to support flexibility.

Tackett et al. developed a method for quantifying the changeability (or evolvability) of a system based on excess using the principle of elastic potential energy [8]. In their model they proposed a model that quantifies a system's ability to change to a new configuration (E) as a function of a system's excess capacity (X) and a "gain per unit excess" (g_x). This formulation for a system's evolvability is shown in Equation 1:

$$E = \int_{x_l}^{x_u} g_x * X * dX \qquad (1)$$

where x_l and x_u are the upper and lower bounds of usable excess. A system's ability to evolve can then be measured by its available excess and the value gained from the excess integrated.

Allen used the results of Tacket et al. when quantifying the usability of available margin for a system [23]. This method determined the evolvability of an automated assembly station. Watson then developed a method linking the quantity and usability of margin with value [26].

In closest proximity to the present research, excess was modeled at the component level in Cansler et al. [24]. This work addressed the need of exploring margins within a system by mapping functional flows between components and identifying the limiting components. White and Ferguson refined this approach and compared the evolvability of different architectures that performed the same functions [27]. The work in this paper builds on these efforts and quantitively maps system flows as a means for identifying and tracking excess within a system. Since these flows are what connect components, it is logical that properly modeling them can reveal which flows are important to future changes.

Model-based Systems Engineering

Finally, this research takes advantage modeling approaches like Model Based System Engineering (MBSE) or Object Oriented Design. The premise is that instead of documents as the paper of record for a system, the design is instead stored digitally in a series of interconnected objects. These objects can represent system connections, specific analysis, requirement capture, and any other aspect of design. The interconnected nature means that it is easier to track what impact one block may have on other in the system. Delligatti [28] provides a good introduction to a popular language for implementing MBSE called SysML. This research leans on the concept that different aspect of a system can be modeled as distinct objects connected to one-another to form a complete system.

WHAT IS MISSING?

There are four gaps in the literature (as outlined in the previous section) that drive this work. First, while excess and increased system flexibility are intuitively linked, the placement of excess in a system and how it enables flexibility are not understood. Second, the propagation of change through a requirement, constraint, or a performance parameter is not well modeled with existing tools. Change drivers often begin in one of these three categories. Requirements, constraints, and system performance must therefore be included in a change propagation model. Third, existing research does not adequately model how multiple sequential changes may alter the system's architecture and change its ability to incorporate additional changes. A system that seems flexible after initial design may become highly inflexible after one or two changes are made. Finally, to determine the value (costs and benefits) of flexibility, a framework must include explicit models of lifecycle costs and benefits. Without at least an attempt to evaluate long-term impacts, one can only get a relative comparison of flexibility between different system options.

OVERVIEW OF THE APPROACH

This section describes a broad overview of the approach used for addressing the questions posed in the introduction. Figure 1 shows the modified version of the DBD framework used as a guide in this approach. Each block in Figure 1 represents a portion of the approach that is covered in greater detail in the following section.

The **system configuration** block involves assigning functions to components and specifying flows between components. As discussed in the background, there are several methods from the literature that guide this process. During system configuration each component is treated as a black box with flows entering and exiting. The configuration specifies the way that functional elements are encapsulated in modules or components.

The **system design** portion of Figure 1 is where the designer must select or design the mechanisms within each component block. This is the step that design variables (component sizing) decisions are made for each component. The design variables are subject to constraints and requirements from basic physics (e.g. energy must be conserved), interactions between components (e.g. interfaces must conform to connected components), and external requirements. The design variables are inputs to system attributes.

Exogenous variables exist outside the system and designers typically have little or no influence over these variables. Exogenous variables are often the primary agents driving future needs or changes in system requirements.

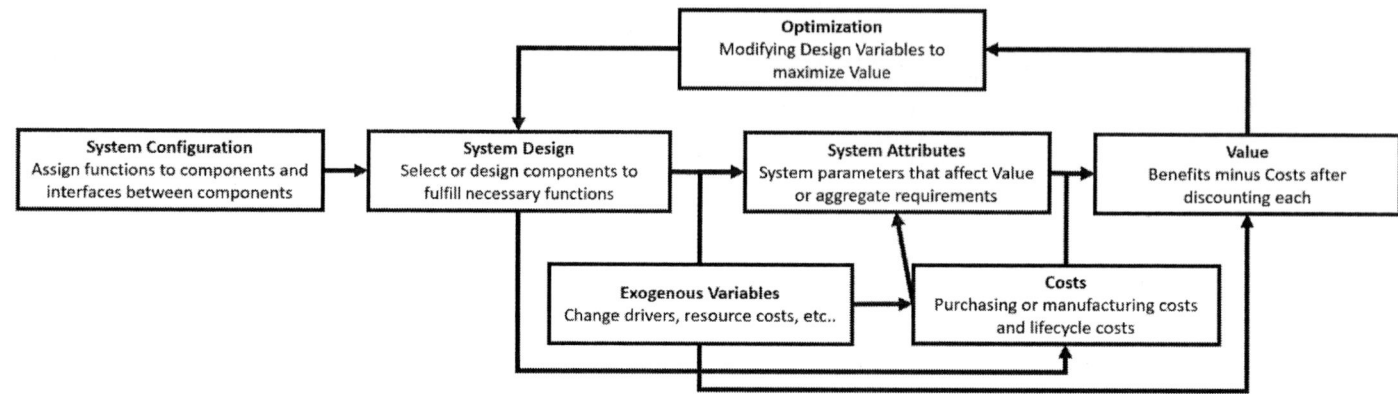

Figure 1: Modified DBD framework adapted from Hazelrigg [10].

The premise of excess based flexibility is that spending more in the present increases the overall value of the system throughout its lifecycle. Therefore, discussing the net lifecycle value of a system requires considering the *costs* of building and operating the system. The costs and value boxes are the external conditions that determine the overall "goodness" of the system. Decisions made during configuration and design influence these properties, but they are also subject to the uncertainties imposed by the exogenous variables.

The *value* of the system is the tally of all the costs and benefits provided by the system over the course of its lifecycle. Value provides the metric by which decisions about system design and configuration can be made. While it may be an arduous task to collect the necessary information to estimate value, it enables the optimization of early design decisions. This approach seeks specifically to use lifecycle value to inform the inclusion of excess within a system.

The next section describes in detail the methods used to implement each part of the approach. It should be noted that existing techniques also can be said to follow a similar approach to the one discussed in this section, but none cover each aspect in Figure 1 to inform design decisions about excess placement.

Real Options research has been used to evaluate the inclusion of specific change options but is limited in its focus to a small number of change pathways (typically one). It does not consider system architecture beyond its influence on that small subset of changes. Change propagation methods provide significant guidance in optimizing system configuration but largely ignores the system design block, and therefore cannot be used to inform design variable specification. MDO is used for design variable selection but most problem statements do not fully model the value of the system over the full lifecycle. Even existing excess research has modeled systems at the system attribute level and has not included change propagation when making changes to a system. The VDD and DBD frameworks provide guidance for what should be done but implementation is left to the designer. Implementing change propagation within the context of a DBD model is an original contribution of this research.

The following detailed methodology section walks through each block in Figure 1 with enough detail to understand how each block was implemented. Sections where tools already exist are kept brief so that the focus is on the contributions made by this research. At the end of the detailed methodology the reader should understand the process used well enough to follow how it is applied to the case study.

DETAILED METHODOLOGY

The detailed methodology steps through how the approach is implemented for a generic system. For consistency with the case study, a desktop computer is used as the example. Each of the following sections implement a portion of Figure 1 and provides implementation detail with either existing tools or as developed for this approach. Figure 2 shows a roadmap for how each block of Figure 1 is implemented. The change propagation modeling is sufficiently complex to warrant discussion on its own after the other aspects of the methodology in this section.

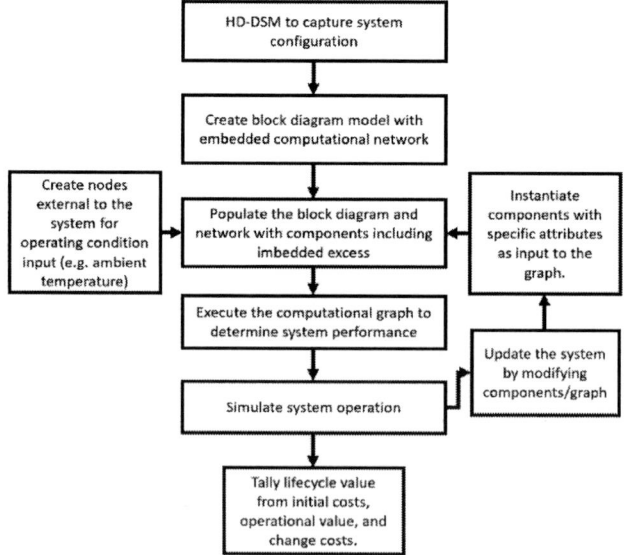

Figure 2: Outline of the methodology.

Initial System Configuration

The system configuration uses existing tools to establish components and their connections. The work done by Tilstra et al. [29] on the High-Definition Design Structure Matrix (HD-DSM) provides an excellent method for representing the initial system configuration. The HD-DSM represents the components and their connections either as a network or as a series of DSM matrices with one for each interaction type. Figure 3 shows an example of a DSM for the Electrical Energy layer of for the desktop computer example.

	External	Power Supply	Hard Drive	GPU	Motherboard	CPU	RAM	Cooling System
External	-	x						
Power Supply	x	-	x	x	x			x
Hard Drive		x	-					
GPU		x		-				x
Motherboard		x			-	x	x	x
CPU					x	-		
RAM					x		-	
Cooling System		x		x	x			-

Figure 3: Electrical layer for Computer HD-DSM. The x's show where there are interactions between components.

Block Diagram with Computational Network

The system design is built upon the System Configuration as a template. Many existing change propagation methods use expert opinion or a strength of interaction metric as an abstraction to capture design variables in their analysis. However, quantification of flows including how they are created, modified, and consumed is required for this analysis.

The system model must be sufficiently detailed that performance may be calculated based solely on external inputs and system configuration. This representation must also be amenable to change by an algorithm and updatable so that design variables, exogenous inputs, and value propositions can be modified and new performance values calculated automatically. To accomplish this, computational networks are embedded in blocks that may be added or removed from the system.

Using block diagrams with embedded computational graphs is an original contribution of this research.

Flow Tracking via Computation Network

Quantitative flow modeling requires the algorithmic manipulation of flows so that changes to them are appropriately propagated. We propose that flows are modeled using computational graphs. Computational graphs are commonly used in Deep Learning Artificial Neural Networks to simplify the backpropagation steps required for gradient ascent. Figure 4 shows a simple example.

Within each calculation block is the assignment of a variable or the execution of a computation on existing variables. For example, if the values presented in Figure 4 represent electric

power then blocks a, b, and d are consuming the power provided by block e. Since block e is a calculation, and the direction for calculations is specified, the total power for the system can be quickly and automatically calculated.

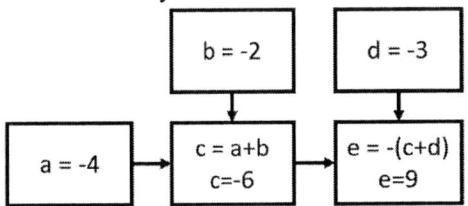

Figure 4: Computation graph example showing how calculations flow along edges.

The calculations necessary for determining how flows are altered are contained within the blocks of the graph which are in turn embedded within the appropriate components from System Configuration. Figure 5 shows an example for calculating the power input required by the system power supply. The grey blocks are *calculations* that require input, perform a calculation, and return output while the white box is an *attribute* box of the power supply and serves only as input to other nodes. This allows flows for the entire system to be calculated regardless of how they are distributed among the constituent components. Figure 5 shows how the total power required by the power supply in the computer is determined based solely on attributes from other components and the efficiency of the supply itself.

The distribution of the calculation blocks within each component also provides a convenient way to express how component sizing affects the architecture for the system. Excess is imbedded in the system by modifying these attributes to exceed the minimum required for initial system functionality. Families of components have similar interfaces and computational graph structures enabling an algorithm to efficiently swap one component for another.

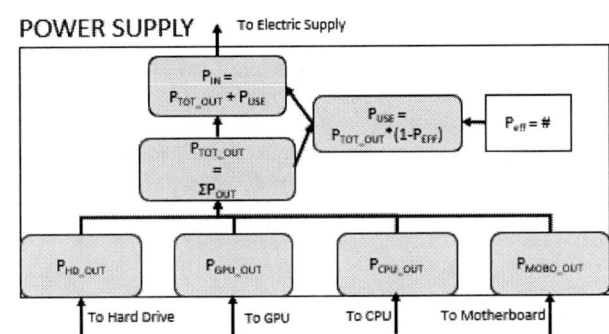

Figure 5: Example of a computational graph with attributes and calculations imbedded in a block.

Functions of arbitrary complexity can be contained within each block providing a flexible method for manipulating flows. This allows for the use of more advanced computational models like finite element analysis packages. The specification of flow values provides necessary data for designers by ensuring the system operates within an allowable design space by incorporating constraints and requirements and estimating the system's attributes that are pertinent for value generation.

Requirements Tracking

Once flow types are quantitatively modeled, a mechanism for representing requirements and constraints is added. These are logical statements that use flow values as input to indicate whether the system is in an allowable state.

The output of a requirement is a binary value dependent on requirement satisfaction. Requirement violations indicate that some aspect of a system is in a non-allowable state and a change is required. One advantage of incorporating requirements into the system flows is that the edges (connections between nodes) connecting flows to the requirement can be traced back to reveal where changes may return the system to an allowable state.

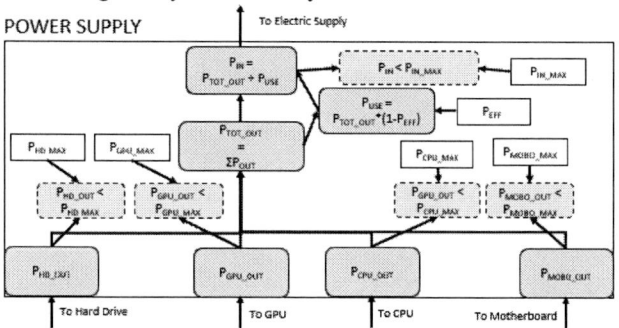

Figure 6: Example block graph showing how requirements are implemented in the computational graph.

Figure 6 shows an example where requirement blocks have been added to the electrical power layer of the Power Supply. The component attributes box shows the attributes pertinent to this diagram; which in this case is the maximum power allowed out to each attached component and in from the external power source. The power needs for other components flow to the power supply which then updates how much power is required from the supply. Each requirement is updated as either satisfied or not dependent on the flow value and the component attribute.

A complication is that the flows may be traced back an arbitrary distance resulting in a large space of potential changes. This mirrors the reality in which a designer has many different options of change propagation pathways. There are many potential inputs that may impact the decision of which set of changes should be used, so this issue will be revisited after additional elements of the DBD framework have been included.

Exogenous Variables

This section discusses two categories of exogenous variables that are included in the model.

Technology

Technology has two direct influences on the system. The first is that technology drives the composition and connection of elements within a system's architecture. The second impact technology has is determining what *attributes* a component is

allowed to have with the given state of technology. These are each discussed below.

The first impact that technology has is determining component connections during initial system configuration. The technology types used specify the functions fulfilled and the flow connections required by each component. The inclusion of specific technologies can drive two systems that perform similar functions to be distinct and generally incompatible. While some elements will be similar, the two systems are sufficiently different that it is not worth the effort to convert one into the other. Changes to standard interfaces are where this impact is most significant. Changes in the specifications for a standard interface (like the change in the socket for a CPU) should be considered when selecting initial system components. If a component requires a specific type of connection, that interface can limit maximum future performance without requiring the additional replacement of the interfacing component.

The second impact that technology has on the system is the specification of available attributes and interfaces for each component. The performance values of components are a function of what technology is currently available. For example, when selecting a computer graphics card (GPU) there are sets of attributes (calculation power, memory cache size, memory cache speed, power use, heat generation, cost, etc.) available. Selecting a GPU requires trading between the various attributes to find optimal balance. The set of allowable attributes will fall along a Pareto frontier along which improving one attribute can only be done to the detriment of another. The technology available determines where the frontier is. For example, over time the maximum theoretical computational power for a GPU has improved at the same time that electric power consumption has fallen. There are also now graphics cards that use far less power, designed for mobile computing, that are largely similar in price to the more powerful desktop graphics cards but provide less computing power.

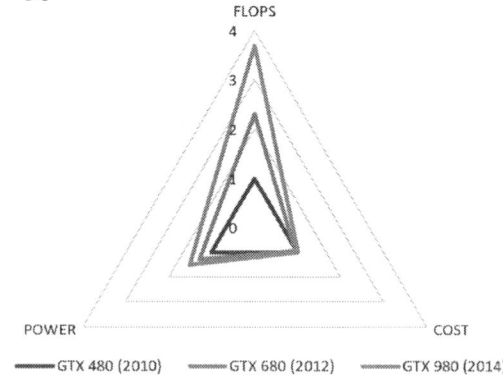

Figure 7: Normalized NVIDIA GPU comparison showing how the fractional improvement in component attributes over time

Figure 7 shows an example of this phenomenon. The performance of three equivalent graphics cards from different years is shown. The fractional improvement in each attribute has been normalized to the 2010 value and cost has been adjusted for inflation. Each year for roughly the same cost the power required goes down while the computational power goes up. When designing a system, a designer can only select from the available sets of attributes, but as time passes the available sets shift. This shifting is exactly the kind of phenomenon a well-designed system can capitalize on. Having the proper type of excess can allow the computer can be upgraded to use the more advanced components without resorting to replacing the entire system.

Environmental and Systems of Systems Variables

Systems exist in, and must interface with, the real world. There are therefore variables outside the boundary of the system that affect its performance. This includes interfaces with external systems and the environment in which the system operates. These variables must be accounted for when composing the change propagation model as they frequently drive system changes.

Figure 8 shows two ways that external variables are included in the system model. The first is through the interface with electrical system connection. Each interface is labeled with its type as a reference. The interface between the power cord and the external power source is NEMA 1-15 Grounded and the electric power flow must pass through this interface. The interface has a maximum power rating which is accounted for by a requirement inside the interface.

The temperature of the air external to the system also impacts system performance. The external air acts as the ultimate heat sink for heat generated by the computer. The cooler the outside air, the less work is required to keep operating temperatures below their maximum. The power supply in Figure 8 is be cooled directly with external air. By including the external environment, any properties that are important to system performance can be included at the appropriate level of detail.

Specifying a standard interface type can simplify the representation of compatible interfaces. All the requirements for a standard type are satisfied by the "Interface Match" requirement. This allows the geometry and positioning aspects of the interface to be abstracted.

System Attributes, Costs, and Value

A convenient way to conceptualize a system's lifecycle is through the use of epochs as introduced by Ross and Rhodes [30]. The epoch is a discrete period in which the system context is substantially the same. When a change in the system or system's context occurs, the value proposition is modified, and a new epoch begins.

As described in Hazelrigg [31], the costs for a system are broken into two categories: recurring and non-recurring. The value generated by a system can be considered the same way. Recurring costs and benefits are those that exist between epoch shifts. These are distributed in time across an epoch. This includes power costs and generated income. To model these, the system attributes are used in a time-dependent simulation that takes system attributes as input and returns dates and values created during the epoch to tabulate the system's overall value. When an epoch shift occurs the attributes, external model, or both may change, and the new values and model are used.

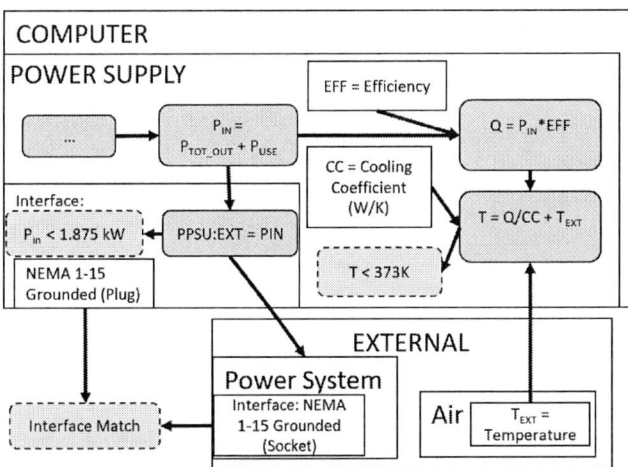

Figure 8: System model demonstrating how standard interface matching abstracts geometry detail and how attributes of the external system included.

Non-recurring costs are those that happen only at discrete points like during initial system development, at the end of system life, and at the boundaries between epochs. Non-recurring costs include the costs of initial system design and construction, making changes to a system, and system retirement. Holding the system configuration constant allows the non-recurring costs to solely be a function of the constituent components. The costs incurred by building or modifying a system are determined by the costs of the components plus the labor and resource costs to connect each interface.

The first step to determine non-recurring costs is select initial components and tally their costs. A set of components are selected with attributes determined by current available technologies. Cost is one of the variables for a component. Components are selected from the set of alternatives available during initial design. During epoch shifts a new set of components, with updated performance values reflecting newly available technology, is available. The cost to purchase (or design and build) the component is recorded each time a component is added to the system.

The labor and resource costs associated with initial system assembly and disassembly during a system modification are also determined for each modification made. These costs are important as significant change costs will push system designers away from flexibility and towards strategies 1 and 2 from Barber et al.[2].

Including assembly and disassembly costs is relatively straightforward using the block diagram approach. The labor and resource requirements to replace components are estimated during initial system configuration. Additional costs, like those associated with post-modification verification and validation work, can be added to the blocks incurring those costs.

Figure 9 shows an example of adding a new CPU to a computer. There are three aspects of the diagram of interest. The first is the prerequisite assembly/disassembly interface blocks. These indicate interfaces that must be connected or disconnected

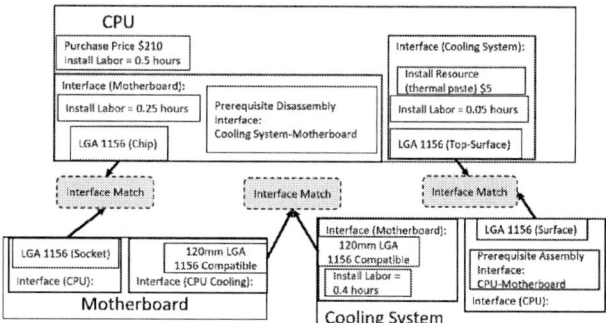

Figure 9: Assembly diagram showing how component, labor, and resource costs are represented in the model.

(incurring the associated costs) before the containing interface can be modified. For example, the prerequisite disassembly for the CPU-Motherboard interface indicates that the cooling system must be disconnected from the motherboard before the CPU can be disconnected from the motherboard.

The next two aspects shown in Figure 9 are costs related to replacing components. The first is the cost associated with the component and not the interfaces. This represents the time and resources required for the component alone. For instance, the labor associated with the CPU is the time required to unpack it and prepare it for use in the system.

The second cost related aspect is associated specifically with the cost to connect or disconnect interfaces. There are three connection interfaces in Figure 9: one to connect the motherboard and CPU, one to connect the cooling system and CPU, and a final one connecting the cooling system with the motherboard. Each interface may have labor costs, resource costs, or both. The motherboard-CPU interface has only a labor cost for placing the CPU into the socket while the cooling system-CPU interface requires the application of a thermal paste for an additional cost.

All-together the costs for making a modification are procedurally generated. The example of adding the CPU in Figure 9 would require the $215 (for the CPU and thermal paste) in resources and 1.2 hours labor for the CPU setup and to connect the three interfaces.

Procedurally computing modification costs is necessary to represent the costs of making changes to a system. The procedural nature of the value and cost calculations, and speed with which simulations can be done, are prerequisite to the ability to evaluate the many possible combinations of components and exogenous variables. Only by sampling the space of potential future contexts in a methodical way can the risks and rewards of flexibility be rigorously understood.

Change Propagation

The framework developed in previous sections is now sufficient to address the topic of how change propagates in this block and flow-based model. The change propagation process requires the steps shown in Figure 10.

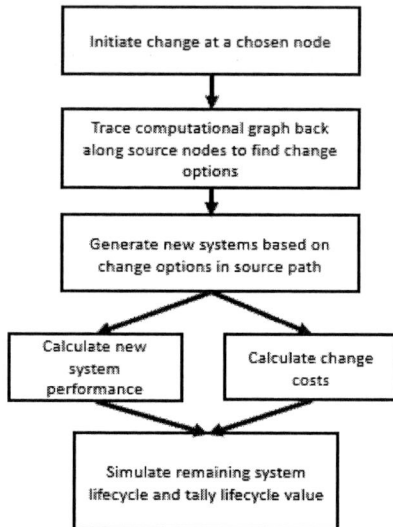

Figure 10: Simulation strategy for assessing change propagation and linking it to lifecycle value.

This method of performing, incorporating, and propagating changes in a systematic way is an original contribution of this research.

Initiate Change at Chosen Node

The source of change is the element of the model that is modified. This can be any element that exists in the system like a requirement, performance element, or a specific component. So long as the element being changed has a node in the system it can be a change source. Example change sources might be a node tied to system performance, and new component, or a new requirement.

Consider an example where a desktop is used as a controller for real-time control of a robotic system. A new control algorithm is implemented which requires more computational power. The controller performance must be improved to implement the new algorithm. Figure 11 shows a diagram containing the pertinent parts of a computer system for this change. The Iterations per Task attribute must increase because of the new algorithm's increased task time. This causes the Task Time requirement to be violated and change propagation is initiated from there.

Generate Change Pathways

After a change initiator has been identified the next task is to determine what changes could be made to bring the system back to an allowable state. This requires the addition of a new object that represents where and how changes may be made to the system called "Change Options". These change options are linked to the nodes they influence and serve as the mechanism by which changes may occur. This can include actions like replacing components (changing the GPU) or exercising options (adding a second GPU to the system). Figure 11 shows two Change Options available in the desktop controller example.

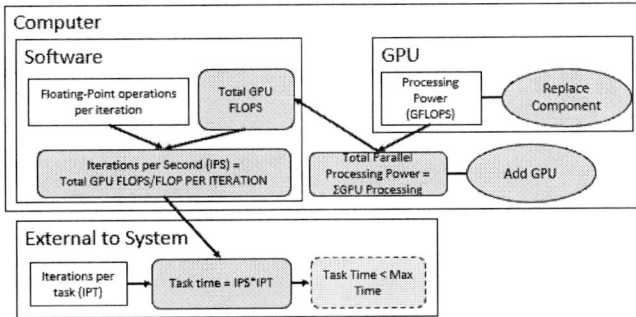

Figure 11: Change option example showing the addition of the "Add GPU" and "Replace Component" change options.

The process for generating change pathways begins at the change initiator. The computational graph is traced backwards along the node dependencies until a Change Option is reached. In the case of the system presented in Figure 11 the violated requirement is traced back to the two available change options: "Replace GPU" and "Add GPU".

Once all applicable Change Options are identified, each possible instantiation of the combination of the attributes is generated. In the example assume that there are three possible graphics cards [a,b,c], and that the motherboard supports up to two graphics cards. Each possible combination of graphics cards and number of cards would be generated. Since paired graphics cards must be of the same type the resulting list would be [(a,-), (a,a), (b,-), (b,b), (c,-), (c,c)]. A system with each combination is created using the combinations from the list.

Each new system is then checked for violated requirements. If any requirement violations are identified, the node for that requirement is used as the change node for a second iteration of the procedure. Components or exercised options that were selected in a prior iteration may not be included in subsequent iterations. For example, if combination (a,a) from above required more power than currently available, they would be excluded from change in the next iteration. Only the options to try different power supplies (to increase supply) or other components in the system (to find lower power alternatives) would be tried. If no change options that have not already been used are found, the system design is discarded as not feasible.

The result is the generation of a list of viable systems that satisfy both the initial change and have no violated requirements. Each candidate on the list is then evaluated for goodness.

Evaluate Change Pathways

Once a list of candidate systems has been generated the options must be evaluated to rank their lifecycle values. The cost of the changes made are compared to the change in value generation over time. The aspects of value discussed in the categories of costs and benefits section are used in the evaluation. These are broken into two categories: recurring and non-recurring costs.

Change Cost Estimation

The first step in evaluating change pathways is to estimate the cost of modifying the system. The change pathway provides a list of components or exercised change options used to meet the new system specifications.

Disassembly costs are first calculated. The labor and resource costs required to disconnect each of the new component's interfaces are determined. Next the costs of the components selected during change pathway generation are added to the tally. Then the assembly costs are tallied by summing the costs to connect each interface associate with new components. Finally, any post assembly costs like system set-up or testing are added to the ledger. The non-recurring costs are assumed to occur instantaneously relative to the overall system lifetime and is therefore discounted as one sum.

System Performance and Lifecycle Simulation

The new system and its attributes are used in the recurring cost model to evaluate how well the new system performs. This model generates the series of values and costs throughout the remainder of system's lifecycle. These include resource costs like electrical power and generated value converted to a dollar amount. These are combined with the original system costs and the costs from the system modification to provide an estimate for the overall value of the system.

This method can be used with a variety of starting components, epoch shifts, and estimates of exogenous variables to characterize how well certain configurations of components perform. If stochastic models are used for exogenous variables, then the entire method can be run repeatedly to characterize how robust the lifecycle value is for a range of future scenarios. The system designer may then select the design that best fits the decision maker objectives.

The detailed methodology described in this section provides a way to implement the DBD framework when making decisions about where to place excess within a system to make them more flexible. It provides a holistic view of the system including costs and lifecycle performance in a representation that is amenable to modification by algorithm enabling a designer to generate full system lifecycle paths (including change and change propagation) for a wide variety of systems in a way not done by prior research. The following case study takes the method outlined in this section and applies it to an example system using historical data to generate realistic lifecycle trajectories.

DESKTOP COMPUTER EXAMPLE STUDY

The test case uses the method outlined in the prior section to model desktop computer lifecycles. Once modeled the system costs and value are simulated for a period of four years. The system is then put through a change scenario with the intent to improve performance by replacing the original components with those available in 2014. The updated computer variants are then simulated for an additional four years. The ability to automatically generate changes and assess their impact on the system lifecycle is not something possible with prior research.

This enables a designer to assess the marginal value of increasing component attributes in a holistic way otherwise not possible.

To the extent possible, the components (performance parameters and costs) used are those found from historical records to realistically capture technology trends. Examples of that historical information like the component specifications and initial system builds are included below, but all historical information can be found at:

https://sdoresearch.wordpress.ncsu.edu/

The results in the analysis section provide lifecycle values while including change propagation. The ability to incorporate change propagation into a lifecycle value analysis can determine both optimal system configuration (analysis 1) and specific component sizing (analysis 2).

System Model

The test case begins with the creation of a generic system configuration. Figure 12 shows the diagram for RAM. This diagram shows how the RAM information (memory size, throughput, and power use) is sent through interfaces to other components within the computer.

The generic guides are then filled in with specific attributes from historical record. Each place in Figure 12 where the word "instance" appears is filled in with specific component details.

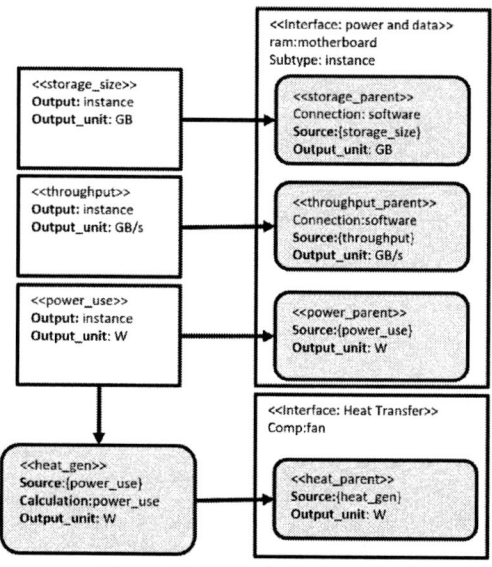

Figure 12: RAM Block Diagram

For example, if the specific type of memory used in the computer is 1 GB of DDR2-800 memory the instance attributes assigned would be: storage_size = 1 GB, throughput = 6400 GB/s, power_use = 5W, and the subtype for the interface would be DDR2. Also included (but not shown) for the component would be a cost for the component ($29) and the connect/disconnect costs for each interface ($2 and $1 for the motherboard).

The same process is repeated for each component within the system. Components for which information was collected include: CPU, Graphics Card (GPU), RAM, motherboard, power supply, and case fan. A balance of systems component was included to represent other parts of the computer for which data was not collected (optical drives, hard drives, ethernet cards, sound cards, etc...). The balance of system component provides a lump energy consumption and heat generation approximately equal that of neglected components.

name	ram: motherboard	throughput (GB/S)	storage_size (GB)	power_use (W)	cost ($)
DDR2-1066-3Gb (2013)	DDR2	8500	3	5	35
DDR2-800-1Gb (2010)	DDR2	6400	1	5	29
DDR3-1066-1Gb (2013)	DDR3	8500	1	4	5
DDR3-1066-3Gb (2013)	DDR3	8500	3	4	15
DDR3-1333-1Gb (2013)	DDR3	10600	1	4	7
DDR3-1333-3Gb (2010)	DDR3	10600	3	4	89
DDR3-1333-3Gb (2013)	DDR3	10600	3	4	21
DDR3-1600-1Gb (2013)	DDR3	12800	1	4	8.3
DDR3-1600-3Gb (2010)	DDR3	12800	3	4	127
DDR3-1600-3Gb (2013)	DDR3	12800	3	4	25
DDR3-1866-1Gb (2013)	DDR3	14900	1	4	9
DDR3-1866-3Gb (2013)	DDR3	14900	3	4	27

Figure 13: RAM Component Information

Figure 13 shows example data collected for RAM. Where possible, information was taken directly from historical records. The cost information includes both values found specifically for the component and values generated using a regression. Similar tables were generated for each component in the system. Tables typically consist of 10-20 distinct instances for each component.

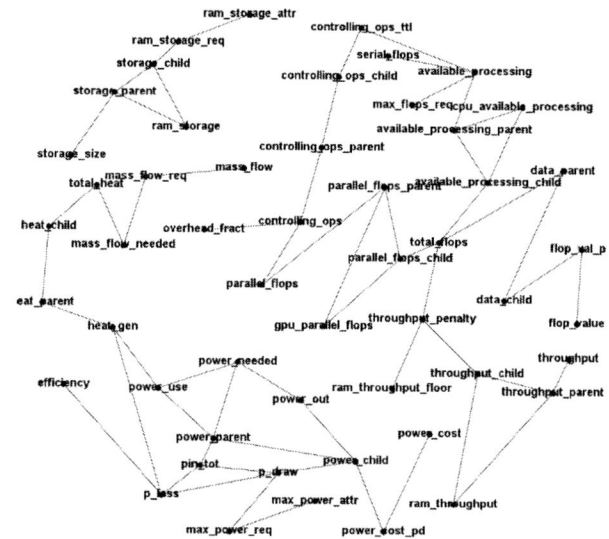

Figure 14: Sample system network map showing each node in the system and the edges connecting them.

The specific components chosen for the initial system were also drawn from historic record. The gaming magazine *PC Gamer* had a repeating article with a build guide for enthusiasts to use when building their own computer. The guide included three different levels of performance (Entry Level, Mid-Range, and Dream System) that offered increasing levels of performance at increasing cost. The builds include specific component recommendations along with the price one should be expected to pay for each component. The December 2010 issue was used to create the baseline systems used in the first part of the analysis.

A network map of one of the systems can be found in Figure 14. This figure shows the computational graph of the system without including the blocks. The graph shows how the attributes, requirements, and computations from each of the constituent components form the cohesive network of the system. Three primary clusters (RAM storage, processing power, and the power/heat removal system) can be observed.

Value Testing and Change Propagation

Once models using specific component attributes were created, they were then used as input for a value simulation. There are many potential propositions for converting computer performance into value. The one selected for this study was a straightforward conversion of computational power into dollars. This is a simplified but realistic proposition as it is the premise of block-chain based digital currency mining. The exchange rate was based on the number of Floating-Point Operations the computer could perform with the value decayed at an exponential rate with a halving time of 24 months (inspired by Moore's law). The value generation was discretized by month and added to a ledger of income and costs.

The recurring cost for the system is the cost of electric energy used by the system. The monthly energy use is tallied and expensed at a rate beginning at \$0.11/kw*hr. The energy cost increases at an annual rate of 3% per year. This value is also calculated on a monthly basis and added to the ledger.

Once the simulation reaches 2014 it is paused, a change process is initiated with the total computation capacity as the change initiator. Candidate change pathways are generated by traveling along each component's dependencies to preceding nodes adding each new branch. For the computer, the components added are the GPU, CPU, and RAM. A list of all possible combinations (between 15,000 and 50,000 depending on the motherboard) of available components of each type is assembled and a system is generated with each. If the hypothetical system has no violated requirements (e.g. too little power from the supply) then it is added to a list for further consideration. If requirement violations are found, the violated requirement becomes the target node for another round of change pathway generation. The procedure is repeated for further requirement violations while one change option can be identified that has not already been exercised. If the requirements can be satisfied, the system is also added to the list for further consideration and otherwise is discarded. The total change cost is determined by which components were changed and the total is added to the ledger on the date of the change.

After each combination of components has been attempted, those without requirement violations are simulated for another 4 years of use. The value generated and power used are tallied for each and stored. After all systems have been simulated the costs on each ledger are all discounted back to 2010 dollars and are summed up as a final measure of the system's overall value.

Analysis

Two types of analysis were performed using the above method. Each of these are discussed in turn in this section.

Analysis 1: Three Initial Builds

In analysis 1, the three configurations suggested by the *PC Gamer* staff were each used as an initial system. Each was put through the process above and the results for how well each system was compared to determine how the different configurations faired. *This test provides insight into the balance between initial investment in the system and the return on that investment.* More powerful systems will provide more ongoing value (while potentially consuming more electric energy) and will theoretically require fewer changes during the change step. However, the more powerful systems also cost significantly more during at the outset. Figure 15 shows the major components for each system.

	Dream	Mid Level	Entry Level
Power Supply	850W 87%	750W 87%	450W 87%
RAM	DDR3-1600-3Gb (2x)	DDR3-1333-3Gb (2x)	DDR2-800-1Gb
Graphics Card	Radeon HD 5870 (2x)	Radeon HD 5870	GeForce GT 220
CPU	Core I7-950	Core I7-920	Athlon-X2
Motherboard	P6T Deluxe V2	P6T	M3A76-CM
Case Fan	75cfm	45cfm	30cfm

Figure 15: Initial system components for each of the three initial system configurations used in analysis 1.

Each system was used as an input to the simulation and the lifecycle values were tallied. The overall results from this experiment can be seen in the violin chart in Figure 15. Each category has a vertical density plot that shows the distribution of lifecycle value returns. The Dream system clearly outperforms the other two and no Entry Level combinations break even.

The conclusion from this figure is that the performance gain from a more expensive build translates into a better lifecycle return on investment. The higher end components generate value quickly enough to offset the initial costs and increased power consumption.

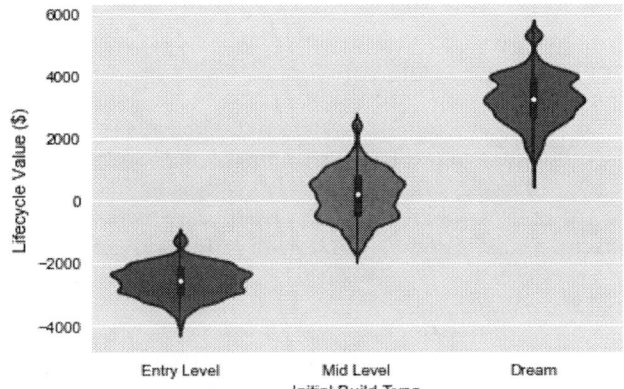

Figure 16: Lifecycle value comparison of initial system builds showing increasing lifecycle value potential for higher performance initial systems.

The top of each distribution has a large tail attributable to a single high-performance GPU. The tops of each distribution are the rebuilds in which the GPU is either one or two of this particular GPU. A gap then exists between those builds and the builds using other cards. While this is likely an artifact of the sample of GPUs used as input, the method clearly identifies systems with those cards as superior with the input used. The best system identified overall used two of this GPU with only two modifications to the initial Dream system. One change was to increase the capacity of the power supply and the other was to remove one of the original RAM components to save on power cost.

The results also suggest that not changing the initial system is better than most other options for change. The unchanged Dream system was in the 95th percentile for value for all attempted configurations while the Mid-Level was in the 81st and the Entry-Level system was in the 79th. A conclusion that can be drawn from this is that the higher performance the initial configuration the less likely changes were to improve system performance.

The more expensive systems also replaced fewer components during the rebuild. The average number of components in common between the initial and final design were 0.33, 1.30, and 1.94 for the Entry-Level, Mid-Level, and Dream systems respectively. This implies that the higher performing systems require fewer additional changes to support an improvement in performance.

This analysis demonstrates that the method can identify and rank systems in order of lifecycle value, but the value of excess is still conflated with other parameters. The Dream system performed best in the simulation, but was the improvement due to better performance during the initial period or was it because fewer components required replacement? The second analysis performed sought to answer this question.

Analysis 2: Power Supply Sizing

The second analysis focuses on the value of excess in a specific component. *This analysis demonstrates that the methodology can identify optimal component sizing for individual components.* The power supply does not contribute directly to system computation performance but does need to supply more power than the system uses. Including a larger power supply in the initial system could potentially reduce the cost of future changes. To test what impact the power supply has on total lifecycle value the entry level system was used as a starting point with five different trials, each with a different capacity power supply. This allows a retrospective look, given the specific value proposition used, to see exactly what size would be optimal.

Figure 17 is a plot showing how the system lifecycle value and average number of common components changes with each power supply. As would be intuitive, as the power supply increases in size, the average number of components in common increases since the larger supply can support more system configurations. There is very little change once power supply exceeds 1000W.

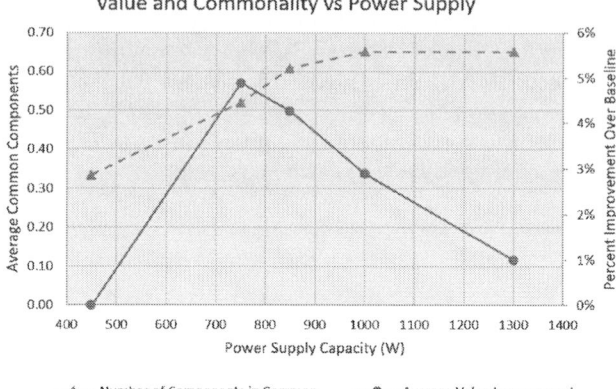

Figure 17: The graph shows that larger initial power supplies increase component commonality but that lifecycle value improvement peaks at 750W.

The lifecycle value of the systems for each power supply shows power supplies that are too large or too small are suboptimal. Too small and the supply must be replaced for must new system configurations. Too large and there is risk that the extra cost incurred does not yield enough benefit in subsequent configurations. Figure 17 supports the notion that there is an optimal selection for the power supply size is 750W.

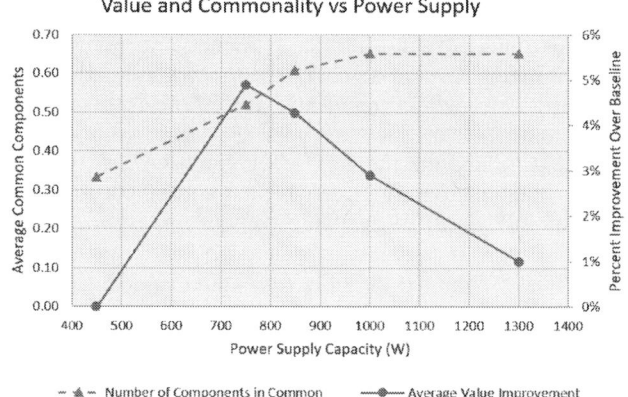

Figure 17: Value and commonality vs power supply showing that larger initial power supplies decrease increases component commonality but that lifecycle value improvement peaks at 750W

CONCLUSIONS

This research demonstrates how the Decision Based Design (DBD) approach can be used for determining a system's lifecycle value when including excess. While the analysis presented is relatively simple and the technology trajectories known with certainty, the combination of excess and change propagation within a DBD based framework holds promise for being a more rigorous way to determine the value of flexibility enabled by excess not currently possible with any other techniques.

We demonstrate in the example study how this approach can be used for finding the sweet spot when including excess in the initial design. We showed in the first analysis that higher performance systems had better lifecycle values. The Dream system performed the best and the highest value systems were those that added a few better components but made fewer modifications overall. The second analysis showed that when considering all possible changes for a system, while holding all components except the power supply constant, the optimal size for maximizing the average lifecycle for system combinations is the 750W sized supply. This enables many good configurations without the risk of not using its excess incurred by larger supplies.

There is significant future work as we continue the development of this approach for complex engineered systems. The first addition is to use the methodology to help select a "best" starting system. This research assumed a set of starting systems to reduce the computational burden for initial development, but ideally the method would test and optimize based on many different initial systems.

This approach provides a foundation for evaluating how and where to include excess by combining system lifecycle evaluation with change propagation to allow simulation of potential changes and lifecycle trajectories. This foundation is enabled by the development of a system representation that can automatically execute both primary changes and subsequent change propagation and then estimate system lifecycle performance including those changes.

REFERENCES

[1] Fricke, E., and Schulz, A. P., 2005, "Design for Changeability (DfC): Principles to Enable Changes in Systems throughout Their Entire Lifecycle," Syst. Eng., **8**(4).

[2] Barber, P., Buxton, I., Stephenson, H., and Ritchey, I., 1999, "Design for Upgradeability: Extending the Life of Large Made to Order Products at the Design Stage," Int. Prod. Dev. Manag. Conf., pp. 75–85.

[3] Ferguson, S., Siddiqi, A., Lewis, K., and de Weck, O. L., 2007, "Flexible and Reconfigurable Systems: Nomenclature and Review," *Volume 6: 33rd Design Automation Conference, Parts A and B*, ASME, pp. 249–263.

[4] Ryan, E. T., Jacques, D. R., and Colombi, J. M., 2013, "An Ontological Framework for Clarifying Flexibility-Related Terminology via Literature Survey," Syst. Eng., **16**(1), pp. 99–110.

[5] Ross, A. M., Rhodes, D. H., and Hastings, D. E., 2008, "Defining Changeability: Reconciling Flexibility, Adaptability, Scalability, Modifiability, and Robustness for Maintaining System Lifecycle Value," **11**(3), pp. 246–262.

[6] Saleh, J. H., Mark, G., and Jordan, N. C., 2009, "Flexibility: A Multi-Disciplinary Literature Review and a Research Agenda for Designing Flexible Engineering Systems," J. Eng. Des., **20**(3), pp. 307–323.

[7] Clarkson, P. J., Simons, C., and Eckert, C., 2004, "Predicting Change Propagation in Complex Design," ASME Des. Eng. Tech. Conf., **136**(August 2004), pp.

788–797.

[8] Tackett, M. W. P., Mattson, C. A., and Ferguson, S. M., 2014, "A Model for Quantifying System Evolvability Based on Excess and Capacity," J. Mech. Des., **136**(5), p. 051002.

[9] Long, D., and Ferguson, S., 2017, "A Case Study of Evolvability and Excess on the B-52 Stratofortress and F/A-18 Hornet," pp. 1–11.

[10] Hazelrigg, G. a., 1998, "A Framework for Decision-Based Engineering Design," J. Mech. Des., **120**(4), p. 653.

[11] Ulrich, K., 1995, "The Role of Product Architecture in the Manufacturing Firm," Res. Policy, **24**(3), pp. 419–440.

[12] Eckert, C., Clarkson, P. J., and Zanker, W., 2004, "Change and Customisation in Complex Engineering Domains," Res. Eng. Des., **15**(1), pp. 1–21.

[13] Browning, T. R., 2001, "Applying the Design Structure Matrix to System Decomposition and Integration Problems: A Review and New Directions," IEEE Trans. Eng. Manag., **48**(3), pp. 292–306.

[14] Pasqual, M. C., and De Weck, O. L., 2012, "Multilayer Network Model for Analysis and Management of Change Propagation," Res. Eng. Des., **23**(4), pp. 305–328.

[15] Morkos, B., Shankar, P., and Summers, J. D., 2012, "Predicting Requirement Change Propagation, Using Higher Order Design Structure Matrices: An Industry Case Study," J. Eng. Des., **23**(July 2012), pp. 902–923.

[16] Koh, E. C. Y., Caldwell, N. H. M., and Clarkson, P. J., 2012, "A Method to Assess the Effects of Engineering Change Propagation," Res. Eng. Des., **23**(4), pp. 329–351.

[17] Hamraz, B., Caldwell, N. H. M., and John Clarkson, P., 2012, "A Multidomain Engineering Change Propagation Model to Support Uncertainty Reduction and Risk Management in Design," J. Mech. Des., **134**(10), p. 100905.

[18] Koh, E. C. Y., Caldwell, N. H. M., and Clarkson, P. J., 2013, "A Technique to Assess the Changeability of Complex Engineering Systems," J. Eng. Des., **24**(7), pp. 477–498.

[19] Simpson, T. W., Bobuk, A., Slingerland, L. A., Brennan, S., Logan, D., and Reichard, K., 2012, "From User Requirements to Commonality Specifications: An Integrated Approach to Product Family Design," Res. Eng. Des., **23**(2), pp. 141–153.

[20] Nembhard, H., and Aktan, M., 2010, *Real Options in Engineering Design, Operations, and Management*, CRC Press.

[21] Kalligeros, K., 2009, "Real Options in Engineering Design," *Real Options in Engineering Design, Operations, and Management*, pp. 127–153.

[22] Neufville, R. De, Asce, L. M., Scholtes, S., and Wang, T., 2005, "Real Options by Spreadsheet: Parking Garage Case Example Valuing Real Options by Spreadsheet:

Parking Garage Case Example," J. Infrastruct. Syst., **12**(3), pp. 1–19.

[23] Allen, J. D., Mattson, C. A., and Ferguson, S. M., 2016, "Evaluation of System Evolvability Based on Usable Excess," J. Mech. Des., **138**(9), p. 091101.

[24] Cansler, E. Z., White, S. B., Ferguson, S. M., and Mattson, C. A., 2016, "Excess Identification and Mapping in Engineered Systems," J. Mech. Des., **138**(8), p. 081103.

[25] Tilstra, A. H., Backlund, P. B., Seepersad, C. C., and Wood, K. L., 2015, "Principles for Designing Products with Flexibility for Future Evolution," Int. J. Mass Cust., **5**(1), pp. 22–54.

[26] Watson, J. D., Allen, J. D., Mattson, C. A., and Ferguson, S. M., 2016, "Optimization of Excess System Capability for Increased Evolvability," Struct. Multidiscip. Optim., **53**(6), pp. 1277–1294.

[27] White, S., and Ferguson, S., 2017, "Exploring Architecture Selection and System Evolvability," *Proceedings of the ASME 2017 International Design Engineering Technical Conferences and Computers and Information in Engineering Conference*, pp. 1–16.

[28] Delligatti, L., 2014, *SysML Distilled: A Brief Guide to the Systems Modeling Language*, Addison-Wesley.

[29] Tilstra, A. H., Seepersad, C. C., and Wood, K. L., 2009, "Analysis of Product Flexibility for Future Evolution Based on Design Guidelinesand a High-Definition Design Structure Matrix," *ASME 2009 International Design Engineering Technical Conferences and Computers and Information in Engineering Conference (IDETC/CIE2009), August 30–September 2, 2009 , San Diego, California, USA*, ASME, pp. 951–964.

[30] Ross, A. M., and Rhodes, D. H., 2008, *Using Natural Value-Centric Time Scales for Conceptualizing System Timelines through Epoch-Era Analysis*.

[31] Hazelrigg, G. A., 2012, *Fundamentals of Decision Making for Engineering Design and Systems Engineering*, Pearson Education Inc.

Proceedings of the ASME 2019
International Design Engineering Technical Conferences
and Computers and Information in Engineering Conference
IDETC/CIE2019
August 18-21, 2019, Anaheim, CA, USA

DETC2019-98428

ESTIMATING THE VALUE OF EXCESS: A CASE STUDY OF GAMING COMPUTERS, CONSOLES AND THE VIDEO GAME INDUSTRY

Darshan Yadav
Graduate Research Assistant
Florida Institute of Technology
Melbourne, Florida, USA
yadavd2015@fit.edu

Daniel Long
Graduate Research Assistant
North Carolina State University
Raleigh, North Carolina, USA
delong@ncsu.edu

Beshoy Morkos[1]
Associate Professor
Florida Institute of Technology
Melbourne, Florida, USA
bmorkos@fit.edu

Scott Ferguson
Associate Professor
North Carolina State University
Raleigh, North Carolina, USA
scott_ferguson@ncsu.edu

ABSTRACT

A widely held belief among design practitioners is that an ideal design solution is the one that meets all the requirements while minimizing surplus cost incurred by exceeding requirements. In this research, we challenge this notion by exploring if providing design "excess", the ability of a solution to exceed certain requirements, can increase the value of a solution to its end users. A case study is performed in the video game industry to explore if design excess is prevalent and its impact on the industry. This study is performed by examining various PC builds (budget, mid-range, and high-end dream) and gaming consoles (Microsoft Xbox and Sony PlayStation) over an 18-year period. Based on a thorough investigation of video game requirements and capacity of different hardware, we find that design excess has existed in computer hardware and is intentionally used as a design property. The results indicate that mid-range solution provide the greatest value to its customers. Further, PC excess based value is adjusted during years when consoles are released. Using measurements of excess, this study also reveals a shift in technology push versus pull that occurs during the mid-2000s and is observable through the lens of system excess.

KEYWORDS: *design excess, value of excess, video games, requirements push and pull*

1. INTRODUCTION

The goal of any designer is developing solutions that meet all requirements in the most economically efficient manner. Design variables are first defined, and components are then designed/selected to fulfill this condition. However, for many reasons, requirements change during and after the design process [1]. The change in requirements and their propagation to other requirements is not yet fully understood [2]. Sometimes change in one requirement may cause a change in 10 other requirements; other times, no other requirements may be affected by these changes [3]. Because of this uncertainty, it is difficult for designers to generate ideal solutions for any problem. This factor of uncertainty is often compensated for by adding a margin (or a buffer) by which the system exceeds its initial requirements. Various margin types have been identified as a means of compensating for different sources of uncertainty. Eckert and Isaksson [4] differentiate these as safety and design margins. Safety margins are added as a guard against specific safety related risks. Design margins are added to accommodate for unexpected changes during a design process and/or to "future proof" the system after its deployment. The latter is referred to herein as "excess" and is the focus of the present research. Differentiating between design excess and overdesign is important; excess in this research is considered as an additional feature of a design solution and something that needs to be appropriately engineered.

Excess serves as a hedge against uncertain future requirements. In the modern world, the pace of requirements

---

[1] Contact author: bmorkos@fit.edu

change for a system is particularly frantic. Fricke and Schulz [5] identified three categories that drive this particular phenomenon - technology evolution, changing customer preferences, and changes in the interfaces between systems. A designer may therefore design a system that meets present day requirements, but that system may experience rapid requirement changes that result in early obsolescence.

There are three strategies a designer may employ when coping with requirements changes discussed by Barber et al [6]. The first is designing a system such that it meets the present-day requirements only. This strategy is most advantageous when the system is inexpensive, and the pace of requirements change is very fast. The second strategy is designing a system that can fulfill projected requirements through the end of the system's planned life. These strategies are useful in completely opposite scenarios as the latter is relevant for an expensive system and a slower pace for requirements change. The final strategy is orthogonal to the other two and allows for system changes during its lifecycle when satisfying emerging or changing requirements. The inclusion of flexibility often comes at an additional cost or reduced initial performance. This research focuses on evaluating the tradeoff between the first two strategies.

1.1. Motivation

There are known instances when excess – though unnecessary at the time – is intentionally incorporated in the system. This excess, at the onset, often comes at a cost but does not provide the end user with additional utility. However, system excess has become a strategic property of systems as consumers specifically consider the longevity of the system with respect to its excess. Currently, little is known about excess and the value it provides to the users over time. If understood properly, excess could be a strategic feature of a system that designers can control early in the design process and incorporate in potential solutions.

In this paper, we explore the relationship between changing system requirements, excess, and a perception of the end user's product value. We explore this relationship so that designers can make educated design decisions when incorporating excess. We look in the current consumer market to see if this concept of design excess is real. And by real, we don't just mean it exists. We rather mean that it is there, and that it has been implemented strategically by designers in industry practice.

2. BACKGROUND

Allen and Mattson [7] performed a systematic analysis to demonstrate benefits of providing design excess in consumer market. This was done by determining Net Present Values (NPV) from both the manufacturer's (defense contractors, automobile manufacturers, pharmaceuticals and electronics manufacturers) as well as consumer's perspectives. Based on various parameters such as changes in requirement after launch of a product, discount rates, R&D costs, etc., it was concluded that providing excess to products can result in higher NPVs for manufacturers and consumers.

The fundamental reason to consider including excess in a system is that the extra resources allocated today could increase the overall value of the system when accounting for the entire lifecycle. Testing this notion requires use of a lifecycle-centric

framework that provides a model linking the choices made during design to future system performance.

Two candidates for such a framework are Value-Driven Design (VDD) proposed by Collopy and Hollingsworth [7] and Decision-Based Design (DBD) proposed by Hazelrigg [8,9]. A representation of the DBD framework can be found in Figure 1. DBD considers a design from the manufacturer's perspective, but by combining the Demand and Utility boxes into a single box. Utility can be defined as a product's value to its intended user. The premise of each framework is that decisions made during the System Configuration and System Design steps are made under uncertainty with regards to the future. The most significant sources of uncertainty are typically the Exogenous Variables.

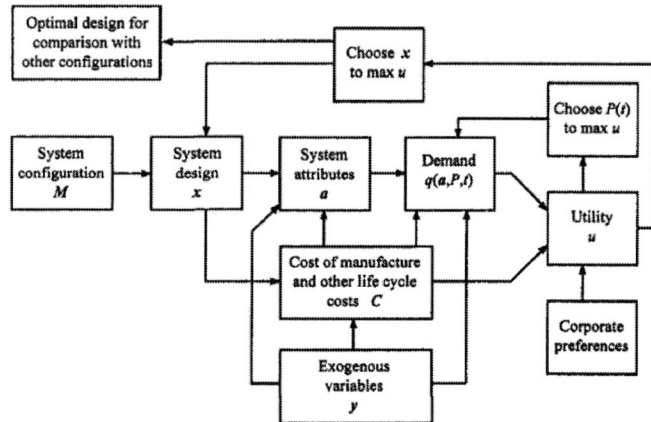

FIGURE 1 ILLUSTRATION OF DBD FRAMEWORK *[8]*

While the DBD framework provides a template for what should be considered, VDD specifically looks to model the connection between system configuration/design choices with the uncertain exogenous variables and the resulting expected value. This is accomplished by mapping system attributes to a specific monetary value. This value can be used as feedback to system designers to push the design toward optimality.

The difficulty inherent in VDD lies in modeling exactly how system attributes impact the value of the system. Prior research has focused on how system configuration (and to a lesser degree system design) increase future value by enabling system changes. This body of research focuses on system architecture, defined by Ulrich [10], and limiting the cost of making changes by incorporating modularity to limit change propagation [11–13]. An excellent overview of this research can be found in [14] by Jarratt et al.

There have been three branches of research that have more applicability to evaluating the sizing decisions made during system design:

The first is the Multi-Attribute Tradespace Exploration (MATE) process discussed by Ross et al. in [15]. MATE is a process for explicitly mapping stakeholder preferences and system attributes into a utility score that represents the value under uncertainty. The utility score is then fed back into the trade space exploration for the system and used to inform design choices. This early instantiation provided a starting point for incorporating VDD into a design process.

The second branch of research is Real Options research as discussed by de Neufville [16]. Real Options uses options analysis, developed in finance to determine the value of a financial option, to determine the value of a real option in a system. The concepts are relatively analogous. In finance an option is purchased by expending resources in the present to give the option in the future to buy or sell an asset at price different from the then current value. In Real Options the expenditure is instead used to build in some aspect of a system that can be exercised to modify system variables to increase system utility to better match system performance with exogenous variables. Real Options have been used to model architecture options in architecture [17], aircraft design [18], and modularity in general [19], but concerns have also be raised as to the applicability of the assumptions made when transferring the approach from the finance domain to the engineering domain by Borison [20].

The third branch is Excess or Evolvability research, introduced by Tackett et al. [21], and focuses exclusively on component sizing as a means for enabling future changes to the system. Excess is defined as in the introduction as design margin is added to accommodate for unexpected changes during a design process and/or to "future proof" the system after it has been placed into service. Evolvability is the ability of a system to be changed while in service and generally uses up excess to make the transition to the new state [22]. Watson et al. [23] explicitly tied excess to a systems ability to evolve and the benefit realized by that evolution.

Long and Ferguson [24] performed a qualitative study suggested that excess played a role in the ability for military aircraft to evolve to new circumstances providing some historical validation for the value of including excess. Cansler and Ferguson [25] suggested using functional flow information to first map and then quantify the design excess in mechanical systems by eliminating non-critical flows. By doing so it is relatively simpler to assign quantitative values to critical qualities of a system.

Real Options, MATE, and Excess based Evolvability research have provided tools for evaluating the benefits of excess as related to making changes to a system but little research has been done to assess how excess can be evaluated for use in preserving system value without presuming the system will be changed after being placed in service. This research contributes to the study of excess by providing a quantitative analysis of excess and its impact on system utility using the uniquely detailed historical records of various consoles, computer builds and video game requirements.

3. METHODOLOGY

The study in this paper is motivated by the following questions:

- What is the value of excess from the perspective of the end user, and is excess worth the extra up-front expense that must be accepted during initial purchase?
- Should designers provide solutions and products with built-in excess? If so, how much of this excess should

there be, and where should it strategically be placed within the system?

Understanding the relationship between excess and value is important because consumers may find the purchase price of products with high excess prohibitive. Conversely, we hypothesize that if minimal (or no) excess is included in a system, consumers may lose interest in products that have limited longevity in the presence of changing requirements. Given these extremes, it is expected that there is an optimal tradeoff between product price (dictated by costs) and the excess that is provided. However, a major challenge associated with studying this tradeoff is tracking and recording requirements changes. One can find how a product's performance specifications change over time, but it is difficult to record and keep track of the way requirements change and why they change. These changes are often either poorly documented or proprietary and/or not well understood. To exacerbate this challenge, requirements are constantly redefined in how they are elicited as developers use various methods (depending on the technological and competitive landscape) for identifying requirements [26,27].

This study explores whether industry is currently designing with excess in mind. We look specifically at the video game industry because game developers communicate the necessary performance capabilities a computer or console must have with potential customers. These system requirements must be met so that the product (the game) can be used. This information must also be stated simply and with enough clarity so that customers can understand if their system meets a minimum threshold.

We use the system requirements of multiple video games per year, over a period of multiple years, as a description of how hardware requirements change for a computer or console. Users that purchase a system often expect that they will play newer games on these systems for multiple years. For desktop computers especially, consumers are presented with multiple variants that can be purchased. Consumers typically upgrade their computers every 4-6 years while console users upgrade every around the 6-year mark [28,29]. In this paper, we question whether buying a high-end computer offers more value over a span of multiple years than a low-end or mid-range system. Additionally, we explore if excess exists within computers and consoles and propose an approach for measuring how that excess is consumed over time.

Public repositories listing the system requirements for video games go back at least 25 years [30,31]. This information provides an ideal record for understanding how requirements changes across a significant period of time. These requirements frequently have both "minimum" (satisficing) and "recommended" (optimal) settings that distinguish between the required hardware capable of simply just running the game versus what is needed for optimal performance. An example of this can be found in Table 1 and Table 2.

Additionally, many venues have published records listing the recommended computer hardware consumers should purchase (or build if they are so inclined) for the best possible experience. These publications list the individual components that must be purchased with the current market price at the time of publication. Most importantly, and of particular interest in this paper, is that these build guides often suggest build combinations that meet various price segments. For example, the PC Gamer

TABLE 1 HARDWARE REQUIREMENTS FOR A CONSOLE GAME

Console	Version	Processor			Memory	Video Card		
		# of Cores	Clock Speed (MHz)	MAD (GFLOPS)	Size (MB)	Clock Speed (MHz)	Size (MB)	MAD (GFLOPS)
PlayStation	PS4	8	1600	25.6	8192	800	Shared	1843
Xbox	Xbox One	8	1750	28	8192	853	8192	1310

magazine lists their recommended systems yearly, classifying the different builds as a budget, mid-range, and dream build [32] (exhibit Table 3). Conversely, consoles often come with limited build combinations. Often the limited customization options involve available storage space or the initial amount of memory. Benchmarked performance data of computer and console hardware is also readily available from the manufacturers or can easily be calculated. Thus, making it easier to correlate the cost of different components to their performance and value to consumers.

Given the wealth of data available, this paper reports on a video game and computer/console study that explores the excess provided in different gaming platforms: PCs and consoles (PlayStation and Xbox). Distinguishing between PCs and consoles is important when conducting our analyses. In the PC market, hardware performance requirements increase every year as new games are developed. This could be considered a requirements pull scenario, as the new games that enter the market may require the customer to purchase a new system so that they can play the latest games. For consoles, manufacturers establish the boundaries of hardware performance for a certain time period (often described in generations of the console – such as PlayStation, PlayStation 2, etc.), and this reflects a requirements push scenario as game developers must adjust accordingly for these different platforms. Two graduate students worked independently in collecting the data required for this study. The authors identified at least 10 performance intensive video games each year, from 2001 to 2018 that could be played on PCs as well as consoles. In total we gathered the minimum and recommended hardware requirements of over 190 video

games for both the PC and consoles (Appendix-A). A sample list of video games from the year 2001 is presented in Table 4

These requirements were then paired with three different PC builds (budget, mid-range and dream build) and consoles for every subsequent year. For PCs, we compared how long each build could run games introduced in the years after that computer was constructed. This was done by calculating the performance capabilities of the available hardware and then comparing these capabilities to the video game requirements. We utilized GFLOPS floating point measurement method to compute the capabilities of CPUs and GPUs. GFLOPS measurement accounts for a processors' number of cores and clock speed and therefore, is a good indicator of their performance.

Each build was then given a rating of 0 or 1 depending on whether it could run at least 50% of the games of a particular year. Minimum and recommended hardware requirements were assessed as individual calculations. The total number of years a build was deemed usable for new games was considered equivalent to the excess within the system. For consoles, on the other hand, the number of years between launches of the console's next generation was considered as proportional to the amount of remaining excess. That is, the remaining excess in the PS3 for the year 2010 would be three as PS4 was launched in 2013. We measure excess in this way for consoles as games designed for the next generation console did not run on the current (or prior) generations of the console. However, the games are often backwards compatible - the Xbox One can play Xbox 360 games, for example.

.

TABLE 2 HARDWARE REQUIREMENT FOR A PC GAME

Game Settings	Processor				Memory	Video Card				
	Intel Model	# of Cores	Clock speed (GHz)	MAD (GFLOPS)	Size (GB)	NVIDIA Model	# of Cores	Clock speed (MHz)	Size (MB)	MAD (GFLOPS)
Minimum	i5-2500k	4	3.3	26.4	6	GTX 660	960	980	2048	1881.6
Recommended	i7-3770	4	3.4	27.2	8	GTX 770	1536	1046	2048	3213.3

Table 3 PC Build Recommendations for a Particular Year

Computer Build-2017	Processor				Video Card				Cost	
	Model #	# of Cores	Clock Speed (GHz)	MAD (GFLOPS)	Model #	# of Cores	Clock Speed (MHz)	MAD (GFLOPS)	Cost	Normalized w.r.t. 2001
Budget	Ryzen 5 1400	4	3.4	27.2	GTX 1050 Ti	768	1392	2138.1	$584.9	$420.8
Mid-Range	Ryzen 5 1600	6	3.6	43.2	RX 580	2304	1340	6174.7	$880.3	$633.3
Dream	i7-8700K	6	4.7	56.4	GTX 1080	2560	1733	8873.0	$1,426.9	$1,049.2

TABLE 4 VIDEO GAMES LIST FOR THE YEAR 2001

Command & Conquer: Red Alert 2
Rainbow Six: Covert Ops
American McGee's Alice
Sudden Strike
Myst III: Exile
X-COM Enforcer
Eurofighter Typhoon
Baulder's Gate II: Throne of Bhaal
Legends of Might and Magic
Microsoft Train Simulator

An illustration of the research method employed is shown in Figure 2. As seen in the figure, during each year of the exploration, video games are identified, and their respective requirements are documented. The video game requirements are documented based on minimum and recommended settings. The hardware (PCs, Consoles) are analyzed to determine their capability to play the said games at that period. This is determined by calculating the required CPU/GPU, and memory capabilities. This data is used to measure the excess in each hardware based on how many years after hardware release it can still play new game releases. When a system cannot play any more games (does not meet minimum settings), the excess is considered to be fully consumed.

Table 5 shows an example of how excess is consumed for all three PC build types constructed in the year 2001. The rows in the table indicate the fraction of games a particular build can run from that year on the recommended settings. For example, the 2001 dream build can run 28.5% of games released in 2003 on recommended settings. Since this is below our established 50% threshold, the 2001 dream-build is considered not useable by the user in 2003. Across all three build types, the excess-based life for all three build types of 2001 is 2 years (rating of total 2). The average excess-based life for each build type is determined by the average of the number of years the machine is viable for both minimum and recommended settings.

Another angle to view this analysis is consumption rate. Since it is unknown how much excess is initially provided to a PC, the only way to determine this is to consider the rate at which this excess is used during the span of the system – a term we coin

consumption rate. The equation for consumption rate is shown in Equation [1. Consumption rate is defined as the percentage of excess consumed per year where following the release of the system.

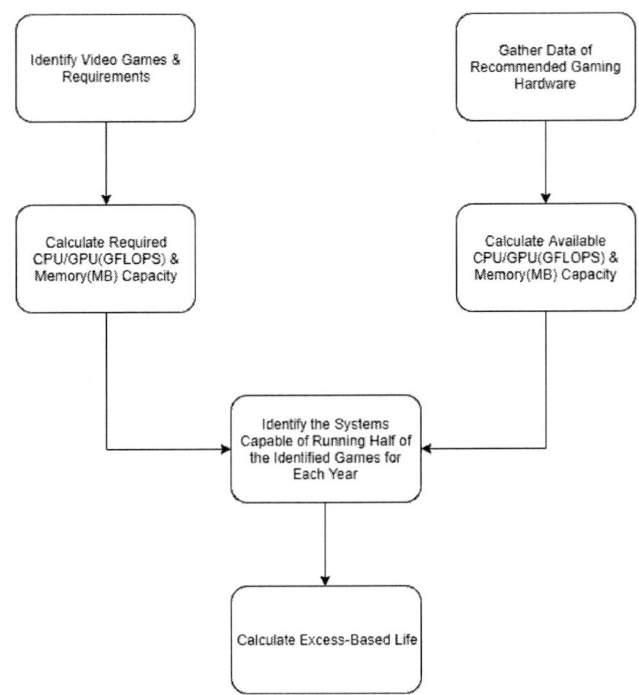

FIGURE 2 RESEARCH METHOD FLOWCHART

TABLE 5 REPRESENTATIVE EXCESS-BASED LIFE CALCULATION FOR COMPUTERS BUILT IN 2001 WHEN USING THE RECOMMENDED SETTINGS

Year	2001		
	Budget	Mid-Range	Dream
2001	1	1	1
2002	0.571	1	1
2003	0.142	0.143	0.285
2004	0	0	0

$$\dot{C} = \frac{\sum_{i=1}^{n} \sum_{t=1}^{l} e_i}{n} \qquad [1]$$

Where

$\dot{C} = consumption\ rate$

n = number of model years used in analysis

l = length of years where excess is greater than zero for particular PC model and build type

e_i = total excess for particular year model PC

For instance, if a PC is released in 2001 and can play 75% of the games released in 2002, then it's consumption rate during that span is 25% (25% of excess from previous year was consumed). If in 2003, the same system is able to play only 10% of the games, then its consumption rate during that year is (86% of excess from previous year was consumed) giving it an average consumption rate of 55% during the two-year span. Table 6 provides the excess consumption rates for select years of the study (the complete list is provided in Appendix-B).

As illustrated in Figure 3**Error! Reference source not found.**, the consumption rate for PC Model years is segmented into two recognized regimes. Prior to 2010, there are no drastic changes in excess consumption rate between PC years as technology pushed what types of games should and are available. However, after 2010, we recognize the start of a technology pull regime where games dictated the type of technology needed and

graphic cards had to respond with drastic changes. Drastic changes included new releases that accommodated the gaming needs. Further, such new releases were developed to last a specific amount of time (hence the oscillations in consumption rates as high excess solutions would be released every two years. The phasing of the budget and mid-range solutions is due to the graphics card solution being economical the following year. This pull regime caused all PC build types experience major oscillations in excess consumption rate

TABLE 6 EXCESS CONSUMPTION RATE FOR SELECT YEARS

	2001	2005	2010	2014
Budget	54.4	43.3	50	28.5
Mid-Range	46.4	48.3	15.4	6.6
Dream	42.8	30.5	33.3	7.7

4. RESULTS

Excess-based life was calculated for budget, mid-range, and dream computer builds and the two consoles - PlayStation and Xbox – over the years 2001 to 2018. We then analyzed this against the normalized price of required components for each year's different PC builds. These results permit the comparison of excess-based life and its value for all three PC builds and the two gaming platforms (PC, PlayStation and Xbox).

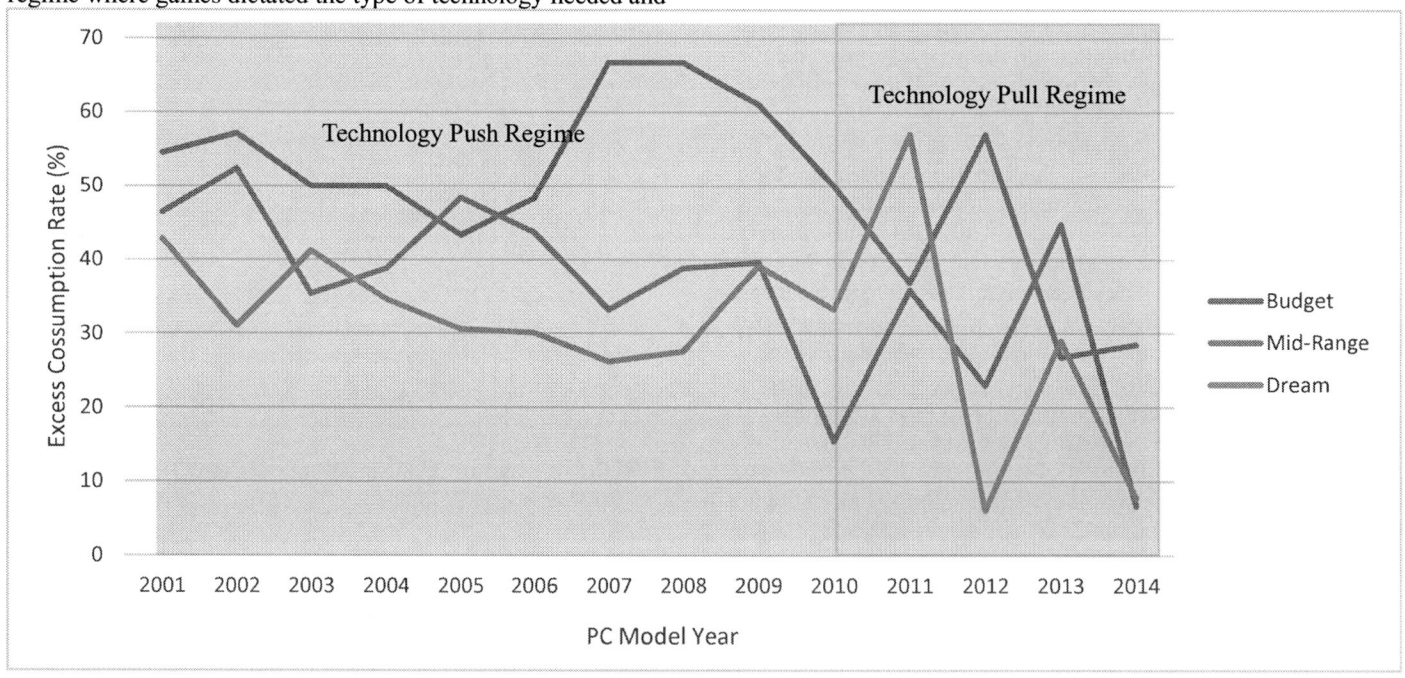

Figure 3 EXCESS CONSUMPTION RATE ACROSS STUDY YEARS

TABLE 7 GENERATIONAL EPOCHS FOR EXCESS-BASED LIFE COMPARISON

Generational Epoch	Time Period	PlayStation Version	Xbox Version
1	2000-2005	PS2	Xbox
2	2006-2012	PS3	Xbox 360
3	2013-2018	PS4	Xbox One

Figure 4 shows the amount of excess-based life present in each generational epoch (refer Table 7) across all three gaming platforms. The generational epochs coincide with the launch of a new generation of consoles. i.e. 2000-05, 2006-12 and 2013-18. The excess-based life of PCs is determined by taking an average of excess-based life of all three build types for both minimum and recommended settings.

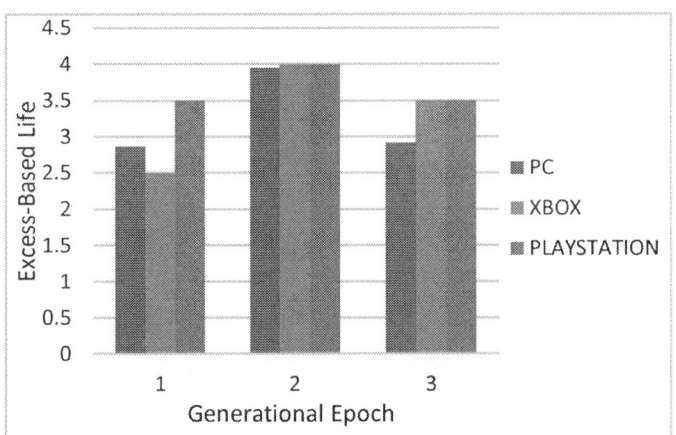

FIGURE 4 COMPARISON OF GENERATIONAL EXCESS-BASED LIFE

The peak value of excess-based life across all systems is around 4 years. The ANOVA analysis for data in Figure 4 indicated that there is not any significant difference in excess-based life between the three gaming platforms (p>0.88).

TABLE 8 ANOVA OF EXCESS-BASED LIFE (YEARS) FOR ALL THREE PLATFORMS

Groups	$\bar{x} \pm s$
Recommended Average	2.5 ± 0.709
Minimum Average	4.1 ± 1.804
Playstation	3.5 ± 3.438
Xbox	3.5 ± 3.514

A comparison of excess-based life (years) between PCs and consoles is shown in Figure 5. In creating this figure, we average the excess-based life across the budget, mid-range, and dream builds. We then determine excess-based life using either the minimum specifications or the recommended game specifications. There is a 50% increase in excess-based life (from 4 years to 6 years) when running games on minimum settings

starting from the year 2001 to 2013. The excess-based life when running games on recommended settings has also increased by the same amount (2 years) from the year 2001 to 2014.

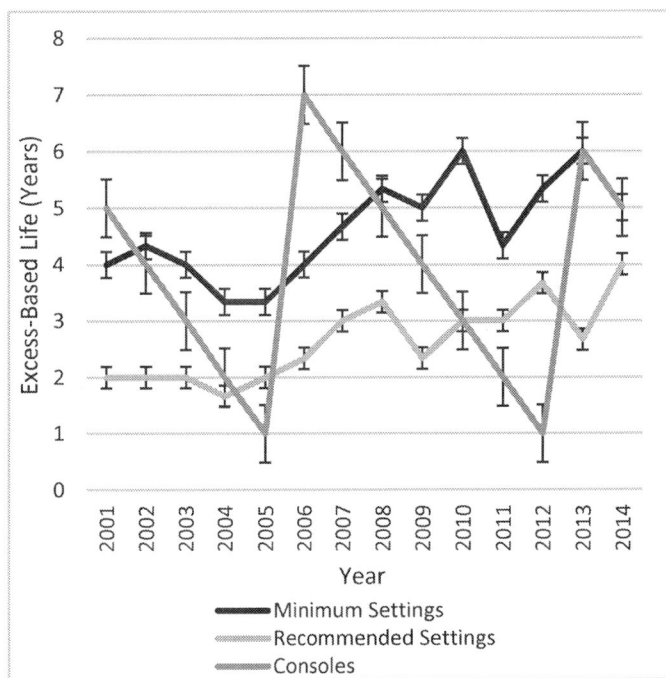

FIGURE 5 EXCESS COMPARISON B/W CONSOLES AND PCS FOR DIFFERENT SETTINGS

There is a systematic decline in excess-based life of gaming platforms from 2014 onwards. This might be due to unavailability of future data and may not necessarily imply reduction in excess. The data pertaining to years 2015-18 is presented in Table 10. ANOVA analysis in Table 9 shows that the excess-based life in consoles and PCs for minimum and recommended requirements are statistically different. This is determined by performing a mean comparison between the excess based life on the various PC build types and consoles.

TABLE 9 ANOVA ANALYSIS OF PC AND CONSOLE EXCESS

	SS	df	MS	F	p-value	F crit value
Between Groups	24.1	3	8.06	3.4	0.02	2.74
Within Groups	157.4	67	2.349			
Total	181.5	70				

We can see the different values of excess-based life for different PC build types in Figure 6. Mid-range builds provide the highest value for money to consumers over the 14-year period analyzed. A more detailed analysis of the value associated with excess-based life is discussed in the subsequent section of this paper.

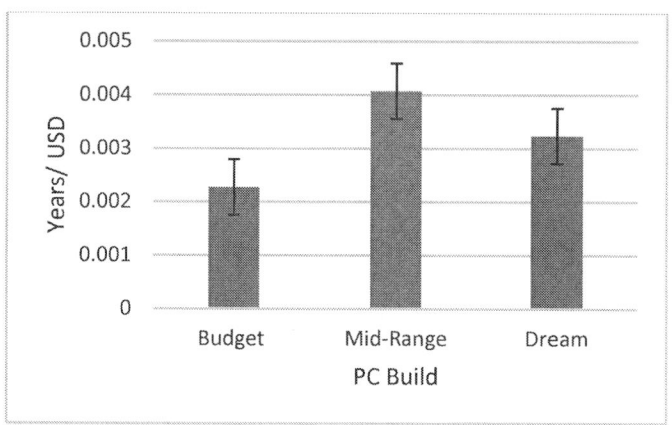

FIGURE 6 VALUE FOR DESIGN EXCESS IN PC

5. DISCUSSION

The results of this study indicate some noteworthy discoveries that may shed light on how excess should be implemented in systems and how excess-based life changes generationally. In the study presented here, we investigate hardware in both PCs and consoles with respect to the gaming industry. The hardware requirements of gaming platforms are becoming more demanding generationally as developers attempt to improve on various facets of their products such as game engines, game mechanics, graphics, etc. This increase in demand is caused by game developers trying to satisfy their consumers by providing improved gaming experience while maintaining game costs [33]. However, the PC (where we see requirements pull) and console (where we see requirements push) hardware manufactures address this increase in demand differently.

We can see from Figure 4 how the hardware technology has evolved over the last two decades in the PC and console product lines. The consoles were technologically superior to PCs in early 2000s and provided more excess-based life to the users (3.5 years of excess). However, PC hardware have been more competitive with their console counterpart for the past 10-12 years by providing increased excess-based life. Here, the absolute values of this excess-based life need to be considered with a pinch of salt. While inferring from this information, we must remember that the for the time period studied, all the consoles were rated for performance at 30 FPS (frames per second). On the other hand, PCs were rated to run games at 60 FPS, meaning they had more demanding hardware requirements to run the same software. So, it's important to note that this difference in excess-based life is not significant. A potentially more important finding in this exploration is that the consoles and PC exhibited similar excess trends. During the second and third generation, all excess increased then decreased respectively compared to the prior generation.

The analysis in Figure 5 provides detailed insight into the evolution of excess-based life. In consoles, we see a see-saw type behavior of excess-based life as initially expected. This is due to the sudden introduction of significantly improved hardware in the beginning which over time shows decay of excess. We observe that there is significantly higher excess-based life in PCs when running games on minimum settings compared to recommended settings. This was also hypothesized in the beginning. Interestingly, the difference in excess-based life for running the games on these settings is almost the same throughout. This was unexpected because the peak performance requirements have increased at a higher rate than minimum performance requirements. This suggests that the hardware manufacturers have been able to keep up with the increased demand most of the time except when we see drops in excess a few times.

TABLE 10 FRACTION OF GAMES PLAYED BY DIFFERENT BUILDS FROM 2015 TO 2018

		2015			2016			2017			2018		
		Budget	Mid	Dream	Budget	Mid	Dream	Budget	Mid	Dream	Budget	Mid	Dream
2015	Minimum	1	1	1	0	0	0	0	0	0	0	0	0
	Recommended	0.7	0.9	1	0	0	0	0	0	0	0	0	0
2016	Minimum	1	1	1	0.54	1	1	0	0	0	0	0	0
	Recommended	0.545	1	1	0.09	1	1	0	0	0	0	0	0
2017	Minimum	1	1	1	0.3	1	1	0.8	1	1	0	0	0
	Recommended	0.5	0.9	0.7	0.1	0.9	0.9	0	1	1	0	0	0
2018	Minimum	0.9	1	1	0.3	1	1	0.6	1	1	1	1	1
	Recommended	0.4	0.6	0.7	0.1	0.6	0.7	0	1	1	0.7	1	1

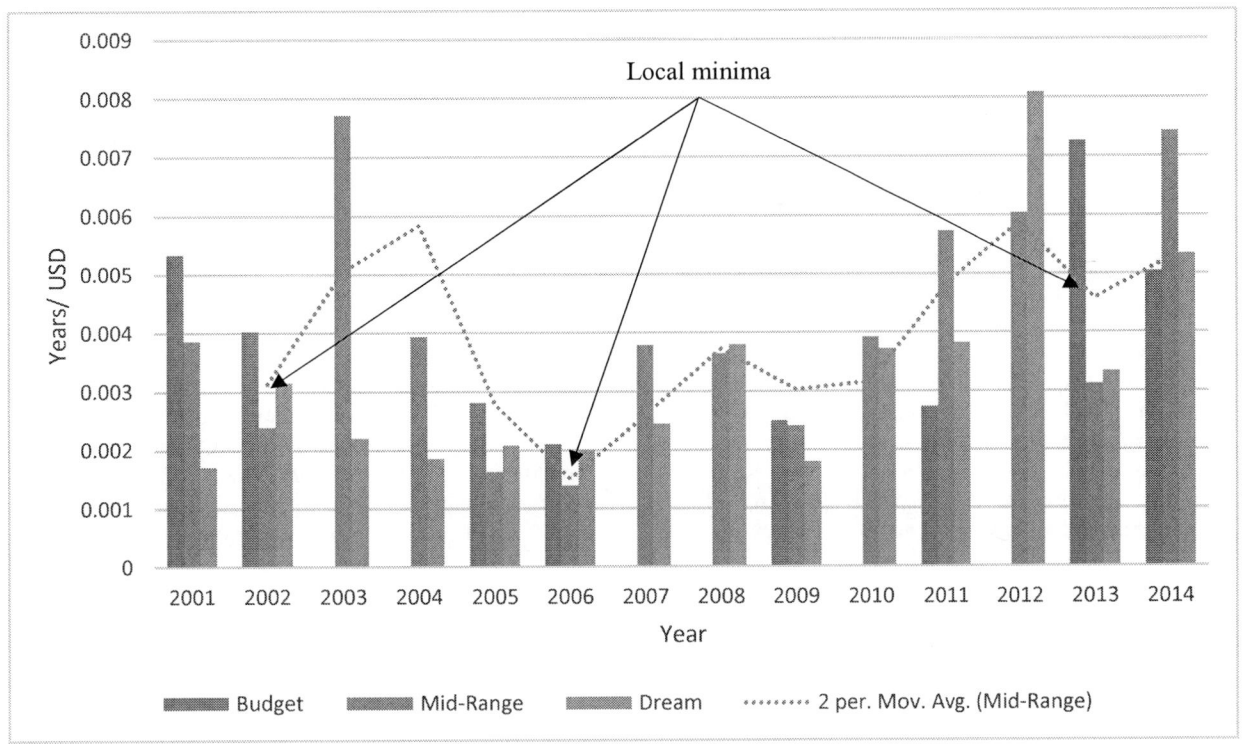

Figure 7 Value of Excess-Based Life in the Different PC Systems from 2001-2014

The most interesting finding of this work is the trend of excess-based life value for consumers over the years of the study. It was expected that the dream builds would provide the highest excess-based life per dollar as they are superior in performance to both the budget and mid-range builds. This led us to believe that it would be worthwhile for consumers to spend more money on the dream build which would last longer. Figure 6 shows that mid-range builds provide the highest excess-based life per dollar over the years. Intriguingly from Figure 7 we can also observe that local minima on a 2-year moving average of excess-based life per dollar exist systematically. The excess-based life per dollar in PCs drops every time a new generation of consoles is released. i.e. year 2002-launch of Xbox, 2006- launch of PS3 and Xbox 360, and 2013-launch of PS4 and Xbox One. This is fascinating as it shows that despite PC (requirements pull) hardware being superior in performance, is driven by needing to adjust to changes in console technology (requirements push).

We would like to clarify that all the discoveries presented in the paper are pertaining to excess-based life, which may not necessarily be excess of a system. It is difficult to say how excess is first, being allocated by the designers and second, being consumed over time by the users.

5.1 Limitations

One notable limitation of the approach of the study is that we only used GFLOPS measurement unit to analyze performance of the CPUs and GPUs. When performing this analysis, we do not account for various other processor performance parameters such as hyperthreading, texture rate, memory bandwidth, etc. These parameters will have some impact on the excess and excess-based life of certain hardware but are unaccounted for in this study due to its initial scope.

As discussed earlier, we find that the mid-range PC builds provide the most value of design excess to consumers. But we lack the data pool of mixture of numerous nonhomogeneous products to generalize the discoveries. Without studying the correlation between the higher value of excess and customer response to this excess, it is extremely difficult to conclude any significant findings. If the consumers, knowingly or unknowingly, tend to buy products with higher excess then this work can have substantial impact on how we design products.

6. CONCLUSIONS

In this paper, a case study is performed to examine excess in PC and console gaming systems. The study explores how excess is utilized in this market and if its phenomenon can be explained. Over a span of 18 years, various PC builds (budget, mid-range, and high-end dream) and consoles (Microsoft Xbox and Sony PlayStation) are compared against the minimum and recommended settings of several games during each respective year. Using measurements of excess, findings are observed that suggest excess can be strategically controlled depending on the type of build, market situation, and state of technology.

This study reveals observations that provide valuable insight on the concept of excess as a system property. By system property, we suggest that systems intentionally and purposefully employ excess and it should not be treated as a byproduct of attempts to improve robustness and/or resilience. Rather, excess is intentionally and perhaps a design property that is used to corroborate between game and graphic card designers.

Of the major findings is the value of excess in such systems. It is observed that budget solutions do not provide the greatest value. Further, the high-end "dream" solutions also do not provide the greatest value. In repeated scenarios, the mid-range solution provides users with the greatest value. This leads to questioning why users would purchase a high-end dream solution. We hypothesize there are social phenomenon that play into that decision making. As this study also considers gaming consoles, it is interesting to observe the response of PC systems to gaming consoles. When considering the analysis of value of excess-based life, the PC build types all experience lows during the time consoles are released. It could by hypothesized that consoles would naturally exceed the capabilities of the PC and the PC would concede defeat during those years. If this is the case, this further reinforces the belief that excess is strategically used as a design property.

A technology push to pull regime is witnessed in the data during 2010. During the push regime, there are no major changes in excess consumption rate between PC years as technology pushed what types of games should and are available. During the pull regime, games dictated the type of technology needed and pushed the bounds of what the graphics card could do. The 2010 transition point where we see significant cyclical variations in excess consumption rates for desktop computers may be the result of increased collaboration between video game designers and hardware manufacturers. This coordination is especially significant if it occurs with graphics card designers, as modern video games often take advantage of increased computational capabilities, power available, and offer new features (interactive environments, shading, texturing, etc.). NVIDIA specifically has followed a 24-month release cycle, suggesting a correlation between the release of new technology (graphics card) and new AAA-games that leverage these new capabilities. It is also important that the number of major video games manufacturers has consolidated during the period of this study. This has occurred because of mergers and the increased cost associated with developing new games.

From a market systems perspective, coordination between hardware and software providers could have significant impact on the end user (the buyer). This coordination would influence the value proposition of purchasing hardware with a large amount (or any) excess. However, we also see that starting in 2008 the average excess consumption rate has been decreasing. This decrease may imply that recent hardware advancements are being tied with marketing strategies rather than the continual releases of games that were seen in the early 2000s. Additionally, we also see greater synergy of game titles between PCs and their peer consoles, implying that there is greater coordination across the segmentation grid for video game developers. This creates the cyclical nature of excess consumption rate after 2010. Further understanding the relationship and interactions between hardware and software developers is a research question for the design of complex systems. This relationship has not been comprehensively explored by the engineering design community.

6.1 Future Work

An immediate area of improvement here would be to increase the video game sample size from each year. This would portray a more accurate picture of excess within the PC hardware. The sample size of 10-12 games per year was sufficient for exploring excess within different systems. However, a larger sample size is required to draw strong conclusions.

In this research we work with over 60 unique products to study the existence of design excess phenomena in consumer products. Although early results are promising and confirms its existence, we still do not fully understand if this excess is provided by designs on purpose or is it a byproduct of some other requirements. This question requires further and a more thorough investigation before it can be answered. It would also be invaluable to get a large manufacturer on board to co-operate in this research aspect to validate any potential findings.

Current product pool only includes one tiny segment of the consumer market (gaming platforms). In order to check the robustness of the findings and to make any general claims, more consumer market areas and products need to be studied for design excess. One key objective of any future research in this area would be to develop a generalized value function for design excess. Such a function would greatly help designers to predict future requirements with more confidence and incorporate design excess early in the design process to create products with higher values for consumers.

REFERENCES

[1] Hein, P. H., Voris, N., and Morkos, B., 2017, "Predicting requirement change propagation through investigation of physical and functional domains," Res. Eng. Des., pp. 1–20.

[2] Morkos, B., and Summers, D., 2010, "Requirement Change Propagation Prediction Approach: Results from an Industry Case Study," International Design Engineering Technical Conferences and Computer & Information in Engineering Conference, Quebec, Canada.

[3] Morkos, B., 2012, "Computational representation and reasoning support for requirements change management in complex system design."

[4] Eckert, C., and Isaksson, O., 2017, "Safety Margins and Design Margins: A Differentiation between Interconnected Concepts," Procedia CIRP, **60**, pp. 267–272.

[5] Fricke, E., and Schulz, A. P., 2005, "Design for changeability (DfC): Principles to enable changes in systems throughout their entire lifecycle," Syst. Eng., **8**(4).

[6] Barber, P., Buxton, I., Stephenson, H., and Ritchey, I., 1999, "Design for upgradeability: extending the life of large made to order products at the design stage," Int. Prod. Dev. Manag. Conf., pp. 75–85.

[7] Collopy, P. D., and Hollingsworth, P. M., 2011, "Value-Driven Design," J. Aircr., **48**(3), pp. 749–759.

[8] Hazelrigg, G. a., 1998, "A Framework for Decision-Based Engineering Design," J. Mech. Des., **120**(4), p. 653.

[9] Hazelrigg, G. A., 2012, Fundamentals of Decision Making for Engineering Design and Systems Engineering, Pearson Education Inc.

[10] Ulrich, K., 1995, "The role of product architecture in the manufacturing firm," Res. Policy, **24**(3), pp. 419–440.

[11] Martin, M. V, and Ishii, K., 2002, "Design for variety: developing standardized and modularized product platform architectures," Res. Eng. Des., **13**, pp. 213–235.

[12] Clarkson, P. J., Simons, C., and Eckert, C., 2004, "Predicting Change Propagation in Complex Design," J. Mech. Des., **126**(5), pp. 788–797.

[13] Engel, A., and Browning, T. R., 2016, "Designing Products for Adaptability : Insights from Four Industrial Cases Designing Products for Adaptability : Insights from Four Industrial Cases."

[14] Jarratt, T. A. W., Eckert, C. M., Caldwell, N. H. M., and Clarkson, P. J., 2011, "Engineering change: an overview and perspective on the literature," Res. Eng. Des., **22**(2), pp. 103–124.

[15] Ross, A. M., Hastings, D. E., Warmkessel, J. M., and Diller, N. P., 2004, "Multi-Attribute Tradespace Exploration as Front End for Effective Space System Design," J. Spacecr. Rockets, **41**(1).

[16] deNeufville, R., 2003, "Real Options: Dealing With Uncertainty in Systems Planning and Design," Integr. Assess., **4**(1), pp. 26–34.

[17] Sharman, D. M., and Yassine, a. a., 2007, "Architectural Valuation using the Design Structure Matrix and Real Options Theory," Concurr. Eng., **15**(2), pp. 157–173.

[18] Mathews, S., Datar, V., and Johnson, B., 2007, "A Practical Method for Valuing Real Options: The Boeing Approach," J. Appl. Corp. Financ., **19**(2).

[19] Gamba, A., and Fusari, N., 2009, "Valuing Modularity as a Real Option," Manage. Sci., **55**(11), pp. 1877–1896.

[20] Borison, A., 2005, "Real Options Analysis: Where Are the Emperor's," J. Appl. Corp. Financ., **17**(2), pp. 17–31.

[21] Tackett, M. W. P., Mattson, C. A., and Ferguson, S. M., 2014, "A Model for Quantifying System Evolvability Based on Excess and Capacity," J. Mech. Des., **136**(5), p. 051002.

[22] Allen, J. D., Mattson, C. A., and Ferguson, S. M., 2016, "Evaluation of System Evolvability Based on Usable Excess," J. Mech. Des., **138**(9), p. 091101.

[23] Watson, J. D., Allen, J. D., Mattson, C. A., and Ferguson, S. M., 2016, "Optimization of excess system capability for increased evolvability," Struct. Multidiscip. Optim., **53**(6), pp. 1277–1294.

[24] Long, D., and Ferguson, S., 2017, "A Case Study of Evolvability and Excess on the B-52 Stratofortress and F/A-18 Hornet," Proceedings of the ASME 2017 International Design Engineering Technical Conferences and Computers and Information in Engineering Conference, pp. 1–11.

[25] Cansler, E. Z., White, S. B., Ferguson, S. M., and Mattson, C. A., 2016, "Excess Identification and Mapping in Engineered Systems," J. Mech. Des., **138**(8), pp. 81103–81111.

[26] Morkos, B., Joshi, S., Summers, J. D., and Mocko, G. M., 2010, "Requirements and Data Content Evaluation of Industry In-House Data Management System," International Design Engineering Technical Conferences and Computers and Information in Engineering Conference, ASME, Montreal, Canada, pp. DETC2010-28548.

[27] Morkos, B. W., and Summers, J. D., 2009, "Elicitation and Development of Requirements Through Integrated Methods," International Design Engineering Technical Conferences and Computer & Information in Engineering Conference, ASME, San Diego, CA.

[28] Hale, J. L., 2017, "When is it time to buy a new computer? These are the tell-tale signs you need to say goodbye," Bustle.

[29] Shah, A., 2016, "The PC upgrade cycle slows to every five to six years, Intel's CEO says," IDG News Serv.

[30] Viki, "Game System Requirements."

[31] Sutton, J., "Game-Debate."

[32] Vederman, G., 2001, "Hard Stuff Trinity," PC Gamer, p. 130.

[33] 2018, Essential Facts About the Computer and Video Game Industry.

APPENDIX
APPENDIX-A: LIST OF VIDEO GAMES ANALYZED

Year	Game Title
2018	Red Dead Redemption 2
	Fallout 76
	Battlefield V
	Hitman 2
	Call of Duty: Black Ops 4
	Just Cause 4
	FIFA 19
	Far Cry 5
	Assassin's Creed Odyssey
	Shadow of the Tomb Raider
2017	Mass Effect: Andromeda
	Injustice 2
	Sniper Elite 4
	FIFA 18
	Middle-Earth: Shadow of War
	Call of Duty: WWII
	Wolfenstein 2: The New Colossus
	Star Wars: Battlefront II
	Assassin's Creed Origins
	Need for Speed Payback
2016	Titanfall 2
	Deus Ex: Mankind Divided
	Watch Dogs 2
	Dark Souls III
	Call of Duty: Infinite Warfare
	Hitman
	Doom
	Battlefield 1
	FIFA 17
	Dishonored 2
	Far Cry Primal
2015	Rise of the Tomb Raider
	Call of Duty: Black Ops III
	FIFA 16
	Assassin's Creed: Syndicate
	Fallout 4

Year	Game Title
2015	The Witcher 3: Wild Hunt
	Star Wars: Battlefront
	Just Cause 3
	Rainbow Six: Siege
	Battlefield Hardline
2014	Assassin's Creed: Unity
	FIFA 15
	Call of Duty: Advanced Warfare
	Dark Souls II
	Middle-Earth: Shadow of Mordor
	Titanfall
	Dragon Age: Inquisition
	Wolfenstein: The New Order
	Far Cry 4
	Watch Dogs
	Borderlands: The Pre-Sequel
2013	Need for Speed Rivals
	Assassin's Creed IV: Black Flag
	Call of Duty: Ghosts
	FIFA 14
	Battlefield 4
	Injustice: Gods Among Us
	Tomb Raider
	Grand Theft Auto V
	BioShock Infinite
	Crysis 3
	Metro: Last Light
	Dead Space 3
	Saints Row IV
2012	Dishonored
	Mass Effect 3
	Far Cry 3
	Assassin's Creed III
	Call of Duty: Black Ops II
	FIFA 13
	Prototype 2
	Max Payne 3
	Counter Strike: Global Offensive
	Need for Speed: Most Wanted
	Borderlands 2

Year	Game
2011	Dark Souls
	Dragon Age II
	Crysis 2
	The Elder Scrolls V: Skyrim
	Star Wars: The Old Republic
	Assassin's Creed: Revelations
	Call of Duty: Modern Warfare 3
	FIFA 12
	Battlefield 3
	Saints Row: The Third
	Dead Space 2
	Need for Speed Shift 2 Unleashed
	Serious Sam 3: BFE
	Deus Ex: Human Revolution
	The Witcher 2: Assassins of Kings
2010	Fallout: New Vegas
	Mass Effect 2
	Assassin's Creed: Brotherhood
	Call of Duty: Black Ops
	FIFA 11
	Battlefield: Bad Company 2
	BioShock 2
	Metro 2033
	Just Cause 2
	Need for Speed: Hot Pursuit
	Lara Croft and the Guardian of Light
	Prince of Persia: The Forgotten Sands
	Red Dead Redemption
2009	Dragon Age: Origins
	Wolfenstein
	Assassin's Creed II
	Call of Duty: Modern Warfare 2
	FIFA 10
	Battlefield 1943
	Borderlands
	Prototype
	Left 4 Dead 2
	Need for Speed: Shift
2008	Fallout 3
	Far Cry 2
	Call of Duty: World at War

Year	Game
2006	Grand Theft Auto IV
	FIFA 09
	Battlefield: Bad Company
	Saints Row 2
	Left 4 Dead
	Tomb Raider: Underworld
	Dead Space
	Devil May Cry 4
	Rainbow Six: Vegas 2
	Prince of Persia
	Need for Speed: Undercover
2007	Mass Effect
	Assassin's Creed
	Call of Duty 4: Modern Warfare
	FIFA 08
	Need for Speed: ProStreet
	Spider-Man 3
	Crysis
	Team Fortress 2
	BioShock
	Tomb Raider: Anniversary
	The Witcher
2006	Call of Duty 3
	Need for Speed: Carbon
	Rainbow Six: Vegas
	Saints Row
	Just Cause
	The Elder Scrolls IV: Oblivion
	Tomb Raider: Legend
	FIFA 07
	Battlefield 2142
	Rainbow Six: Critical Hour
2005	Call of Duty 2
	Devil May Cry 3: Dante's Awakening
	FIFA 06
	Battlefield 2
	Need for Speed: Most Wanted
	Ultimate Spider-Man
	Serious Sam 2
	Rainbow Six : Lockdown
	Prince of Persia: The Two Thrones

	Far Cry
	Grand Theft Auto: San Andreas
	FIFA Football 2005
	Battlefield Vietnam
	Need for Speed: Underground 2
2004	Spider-Man 2
	Half-Life 2
	Prince of Persia: Warrior Within
	Counter Strike: Condition Zero
	Counter Strike: Source
	Doom 3
	Call of Duty
	FIFA Football 2004
2003	Max Payne 2
	Deus Ex: Invisible War
	Need for Speed: Underground

	Tomb Raider: The Angel of Darkness
	Rainbow Six 3: Raven Shield
	Prince of Persia: The Sands of Time
	Star Wars: Knights of the Old Republic
	Devil May Cry 2
	The Elder Scrolls III: Morrowind
	Grand Theft Auto: Vice City
	FIFA Football 2003
	Battlefield 1942
2002	Need for Speed: Hot Pursuit 2
	Spider-Man
	Hitman 2: Silent Assassin
	Serious Sam: The Second Encounter
	Rainbow Six: Lone Wolf
	Star Wars Jedi Knight II: Jedi Outcast

APPENDIX-B: AVERAGE EXCESS CONSUMPTION RATE

	2001	2002	2003	2004	2005	2006	2007	2008	2009	2010	2011	2012	2013	2014
Budget	54.4	57.1	50	50	43.3	48.3	66.6	66.6	61	50	36.8	57	26.7	28.5
Mid-Range	46.4	52.3	35.4	38.7	48.3	43.7	33.2	38.8	39.6	15.4	35.9	22.9	44.8	6.6
Dream	42.8	31	41.2	34.7	30.5	30.1	26.1	27.5	39.1	33.3	57	6	29	7.7

Proceedings of the ASME 2019
International Design Engineering Technical Conferences
and Computers and Information in Engineering Conference
IDETC/CIE2019
August 18-21, 2019, Anaheim, CA, USA

DETC2019-98429

USING SEMANTIC FLUENCY MODELS IMPROVES NETWORK RECONSTRUCTION ACCURACY OF TACIT ENGINEERING KNOWLEDGE

Thurston Sexton[*]
Systems Integration Division
Engineering Laboratory
National Institute of Standards and Technology
Gaithersburg, Maryland 20871
Email: thurston.sexton@nist.gov

Mark Fuge
Dept. of Mechanical Engineering
University of Maryland
College Park, Maryland 20742
Email: fuge@umd.edu

ABSTRACT

Human- or expert-generated records that describe the behavior of engineered systems over a period of time can be useful for statistical learning techniques like pattern detection or output prediction. However, such data often assumes familiarity of a reader with the relationships between entities within the system—that is, knowledge of the system's structure. This required, but unrecorded "tacit" knowledge makes it difficult to reliably learn patterns of system behavior using statistical modeling techniques on these written records. Part of this difficulty stems from a lack of good models for how engineers generate written records of a system, given their expertise, since they often create such records under time pressure using shorthand notation or internal jargon. In this paper, we model the process of maintenance work order creation as a modified semantic fluency task, to build a probabilistic generative model that can uncover underlying relationships between entities referenced within a complex system. Compared to more traditional similarity-metric-based methods for structure recovery, we directly model a possible cognitive process by which technicians may record work-orders. Mathematically, we represent this as a censored local random walk over a latent network structure representing tacit engineering knowledge. This allows us to recover implied engineering knowledge about system structure by processing written records. Additionally, we show that our model leads to improved generative capabilities for synthesizing plausible data.

1 INTRODUCTION

Due in part to an explosion of interest in statistical modeling techniques, specifically machine learning (ML), much recent effort has been devoted to using various forms of engineering data for training these models. These models, trained on historical engineering data to detect patterns of classification, fault detection, performance estimates, etc., promise to reliably automate many of these labor-intensive tasks, freeing the time of designers and maintainers for more high-level decisions. However, in technical fields like engineering, the available historical data is often difficult to use directly — the experts creating it in the past generally assumed it would be read and adapted by colleagues or experts in their own field. This causes analysts to, quite often, lack the information needed to appropriately represent and process this data. One cannot simply use, *e.g.*, written lab notebooks, technical reports, or maintenance work-orders (MWOs) as is, taking them at face value: words and concepts with more general meaning to the layman will have domain-specific technical application that must be accounted for if a statistical model is to learn a robust representation of the semantic space. In this paper, our goal is to infer how the original data creators/experts structure their own knowledge about the problem at hand. This "structured knowledge" can then be used create more reliable models for engineering learning tasks.

This paper presents initial techniques to automatically infer key parts of this tacit structured knowledge, and explores a mechanism to extract it from observations/historical records written

---

[*]Address all correspondence to this author.

This work was authored in part by a U.S. Government employee in the scope of his/her employment.
ASME disclaims all interest in the U.S. Government's contribution.

Copyright © 2019 ASME

by human experts. To do this, we frame the act of recording engineering events as a type of memory recall, which we assume occurs within a broader "network" of system relationships that structure the expert's knowledge about a system's behaviors (but that we do not have direct access to and thus must infer through examples). Specifically, we show that:

1. By explicitly modeling work-order generation as non-Markovian memory recall over learned object relationships, we can more accurately recover those relationships than by using more traditional token similarity measures, and subsequently,
2. learning such relationships provides a generative model of each object's conditional relationships in the form of a graph, for which performing a random walk from points of interest (*e.g.*, a Failed part) will synthesize more realistic new data.

We demonstrate this on two examples of maintenance work orders: (1) synthetically generated work orders from real-world engineering systems with a known ground-truth structures; and (2) actual maintenance work orders from an excavator. In both cases, we show that by building a probabilistic model that accounts for (and subsequently learns) how experts structure their implicit knowledge of a domain, one can often achieve significantly better performance (as measured by standard information retrieval metrics) than existing methods of structure recovery.

2 RELATED WORK

Using data to infer the underlying structure of a complex system is a long-standing goal within both systems engineering and other domains that depend upon accurate network recovery, such as: biological systems and disease transmission vector modeling [1,2]; uncovering economic interactions and social networks [3,4]; inferring physical models by learning governing equations [5,6]; or even description generation in computer vision, and quantifying how humans reason about belonging and causality in ambiguous images or contexts [7,8]. For written (text-based) documents specifically, we can group major methods to perform structure recovery from unstructured written documents into roughly three camps: (1) prescriptive rule definition, (2) training statistical models (NLP), and (3) "folksonomies" and tag-based crowdsourcing.

2.1 Prescriptive Rules

The most straight-forward way to make tacit knowledge computable is to explicitly design the relationships as they are assumed to exist. An expert (or set of experts) define what objects are allowed to exist in the domain, and how those concepts relate to each other. These rules are then mapped onto the observed data, similar to constructing a thesaurus. This manually

constructed rule-set can take the form of ontologies, *e.g.*, but they are always structured representations formed from from mixtures of domain expertise and example data, which can then be used to parse remaining data, and restrict the format of future data. For example, ISO-15926 defines a data model [9,10] with which one can constrain engineering records to have precise, unambigious meanings, and later work built on the standard construct ontologies with which to reason over these meanings and their relationships [9,11–13].

In practice, however, a particular domain or data-set will not have existing, applicable ontologies or data structures, and time investment needed to create them for sufficiently generalized usage is commonly out-of-scope for analysts to dedicate. Some work has been done to automate this process [14], but such techniques generally require us to rely on language-specific syntactical rules (*i.e.*, grammar). Data-entry errors and shorthand are ubiquitous in technical records, where grammar is often low-priority if system-familiarity is assumed. In these cases, sophisticated systems of rules are still often developed, potentially with reduced formalism or scope, taking the form of keyword recognition and filtering rules to find a priori "useful" patterns for analysis [15,16] In engineering design, similar manually-created rule-sets that define concept relationships are involved in constructing Design Structure Matrices (DSMs), which are often derived from expert input or technical/project documents [17–19]. Regardless, this paper assumes that the need for low-cost, low effort estimates of a system's "rules" is not met by requiring a designer to manually intervene.

2.2 Natural Language Processing

Rather than build patterns manually, natural language processing (NLP) often deals with the use of significant quantities of text to discover latent patterns automatically. This requires finding mathematical representations of text, like "bag-of-words" weightings [20], topic models [21,22], or semantic vector embeddings [23,24]. These transformations enable the use of text-based documents in statistical models that can, for instance, train a classifier to automate labeling of work orders [25]. The success of this approach is fundamentally linked to the notion that supervised ML models use *labeled training data* to learn these patterns, and the quality of the model increases with the amount of available labeled data-points — using this approach with few labels presents a problem of diminishing returns. Time saved by automating document classification scales with the amount of time spent labeling document classifications.

This trade-off is problematic in highly technical and jargon-filled domains, where existing models from more generalized training sets cannot be easily or reliably transitioned. In addition, the actual patterns being "learned" are quite often difficult to interpret and use for humans [26], stemming from the so-called "black box" nature of these models, despite an inherent need to justify our engineering decisions with evidence-based reasoning. This paper puts forward an unsupervised model, able to function with few training samples, but only as a stepping stone in the process toward encoding the types of prescriptive knowledge that can be used to communicate and train future operators/designers.

2.3 Folksonomies

In contexts where dedicated annotation labor can be difficult to secure, significant research has been done to present less restrictions to casual annotators, and understand how natural classification and labeling schemes arise in social communities, *e.g.* online tagging efforts [27]. Tags, a form of multi-label classification, allow concepts to be derived freely in the course of work, where repeated and cross-contextual usage leads to a naturally-arising set of useful, domain-specific concepts; this is commonly referred to as a *folksonomy*, a portmanteau of "folk" and "taxonomy" [28]. Because folksonomies generally ask users to determine minimal representative labels rather than strict classifications (*i.e.*, tags), each label can be seen in multiple contexts. The predominant way to analyze these tags, then, is by their co-occurrences with each other: intuitively, highly co-occurring tags are considered "similar." [29, 30]. A basic, but commonly used measure of this co-occurrence is the *cosine similarity*: if, over a set of C documents, tag t_k has binary vector $u_k = \{\mathbf{1}_c(t_k) : c \in C\}$, then the cosine similarity s between the binary occurrence vectors of the tags t_1, t_2 is defined as:

$$s(t_1, t_2) = \frac{u_1 \cdot u_2}{\|u_1\| \|u_2\|} \qquad (1)$$

This measure has seen consistent usage in folskonometric methods to structuring relationships between tagged concepts in useful ways [31–33]. Because various annotators will perceive the importance and relevance of each tag differently in each context, these ambiguities are typically overcome through crowdsourcing, by having large numbers of users tag. This allows a statistical "smoothing" over differences in expertise. However, in the case of technical tags from a few experts, this benefit from large numbers of annotators is not something that we can count on. Additionally, the types of relationship information we might want is not purely statistical/distributional similarity, as experts creating documents will have several core views about what "being related" in their system entails. Consequently, we believe it is important to exploit potential cognitive processes by which these tags might be produced, to enforce a greater degree of information *precision* than typical similarity measures might allow for in their desire for increased information *recall*.

3 MODELING WORK ORDER CREATION

As discussed above, common techniques for discovering structure in human-annotated or natural-language data primarily rely on frequency and co-occurrence information of discrete objects/concepts. These are powerful and easy-to-apply models of speech or the written word, but can miss key causal links implied in the original text, which are difficult to extract this way without significant amounts of data or relevant pre-training. Instead, this paper proposes that by explicitly modeling the conditional dynamics of how humans recall concepts within this data—which, for the purposes of this work will be limited to MWO's—we can extract the conditional relationships between the mentioned objects or concepts that best match what was recorded by the experts.

This section first describes the concept of *semantic fluency*—a existing psychological theory of concept recall—and how that theory relates to the construction of written engineering documents, specifically MWOs in this paper. We then describe a computational method to implement the concept of semantic fluency using Initial-Visit Emitting Random Walks (INVITE) [34]—a probabilistic model of graph walks that is non-Markovian.

3.1 Semantic Fluency

When a technician begins to record a MWO, they try to search their memory for words that represent concepts relevant to the MWO itself. These consist of items, problems that were encountered with some items, and how other items were used to solve these problems [35]. The exact psychological mechanisms by which a person searches through their memory is still an active area of research and has been modeled in various ways. Some recent studies [36] propose that concepts are recalled sequentially by foraging in "semantic patches"—in brief, that humans sequentially recall concepts that are "near" each other in some person-specific semantic space built through experience.

Specifically, these patches are thought of as existing in a high-dimensional concept-space,[1] and the likelihood that some concept is recalled next is based on combining both associative and categorical knowledge into a similarity measure between the current recalled entity and the next. The classic psychological experiment for this model is the Semantic (or, Verbal) Fluency test:

1. Recall and record an object (*e.g.*, an animal);
2. Record the next object of this type you think of;
3. Continue recording for the remaining time

The reader is encouraged to try this process out for themself. One advantage of this test lies in not restricting (or having

---

[1]Though less applicable in technical or domain-specific corpuses where examples are too few and far between, this is the intuition that leads to the success of vector-based semantic embeddings like `gloVe` or `word2vec` [23, 24].

to specify apriori) the relationship between objects required to record subsequent ones. For example:

$$\text{dog} \rightarrow \text{cat} \rightarrow \text{tiger} \rightarrow \text{lion} \rightarrow \text{elephant} \rightarrow \text{wolf} \cdots$$

For example, it is common for animal-based semantic fluency lists to start with household pets, potentially switching to entirely unrelated categories like "large cats," for further exploration, before either retracing back to a previous category (*e.g.*, canines to "wolf" via "dog") or onward via new similarities (*e.g.*, African animals to "elephant" via "lion"). Different people can create different fluency lists, owing to differences in how they psychologically structure relationships between concepts.[2]

The key contribution of this work is to propose that explicitly modeling this process lends itself well to recovering engineering knowledge from text-based technical records. While, technicians are not purely sampling arbitrary system concepts, as you might a list of animals, we nevertheless assume that each subsequent concept written in a MWO is directly conditional on what was written previously. [3] Then, an MWO consists of "jumps" between concepts that depend upon previously "visited" concepts. This assumption allows us to infer relationships between concepts given examples of MWOs. This boils down to two key components of the technician's cognitive task when recalling relevant information to write down MWO's:

- A technician records concepts sequentially, as he or she recalls unique defining characteristics of the MWO.
- They recall these characteristics by remembering links between them, and any recently recalled characteristics.

This differs from a standard Bag of Words model—where all entities occurring in a document are assumed to be linked through co-occurrence—and from n^{th}-order language models—where relations are limited to the nearest (or, previous) n entities. In technical shorthand (like MWOs), objects listed later on may be linked to any of the previously mentioned objects, not strictly those directly adjacent to it. For instance, the MWO *"Leaking hydraulic valve; cleaned oil spill and replaced O-ring"* consists of a sequence of concepts (*leak→hydraulic→valve...*), not all of which share the same causal structure: perhaps "hydraulic", "valve", and "leak" are all potentially subsets of a hydraulic "system", but "replace", "clean", and "oil" all have potential to span subsystems. Similarly, in this MWO, "oil" would likely be considered as linked with "leak", more than it would to the *closer* entity "replace". This illustrates nicely the trade-off

---

[2]*e.g.*

$$\text{dog} \rightarrow \text{walk} \rightarrow \text{run} \rightarrow \text{gym} \rightarrow \cdots \quad \text{vs}$$
$$\text{dog} \rightarrow \text{home} \rightarrow \text{family} \rightarrow \text{meal} \rightarrow \cdots$$

[3]This is standard practice in the language modeling domain [37].

---

between categorical and associative memory foraging that [36] discusses at length, and is precisely the feature of MWOs we exploit to extract a more sparse representation of system relationships through the Initial-Visit Emitting Random Walks semantic fluency model, which we detail next.

3.2 Initial-Visit Emitting Random Walks

Based on the above modelling assumptions, we demonstrate the application of an Initial-Visit Emitting Random Walks (INVITE) as initially described by [34], on recovering system structures from MWOs. Say the set of components or concepts in our system is denoted by the node-set n. A set of T tags [4] can be denoted as a Random Walk (RW) trajectory $\mathbf{t} = \{t_1, t_2, t_3, \cdots t_T\}$, where $T \leq n$. However, this limit on the size of T assumes tags are a set of unique entries: any transitions between previously visited nodes in $\mathbf{t}$ will not be directly observed, making the transitions observed in $\mathbf{t}$ strictly non-Markovian, and allowing for a potentially infinite number of possible paths to arrive at the next tag.

Instead of directly computing over this intractable model for generating $\mathbf{t}$, the key insight from the original INVITE paper comes from partitioning $\mathbf{t}$ into $T - 1$ Markov chains with absorbing states, where previously visited nodes are transient states, and unseen nodes are absorbing. It is then possible to calculate the absorption probability into the kth transition ($t_k \rightarrow t_{k+1}$) using the *fundamental matrix* of each partition. If the partitions at this jump consist of q transient states with transition matrix amongst themselves $\mathbf{Q}^{(k)}_{q \times q}$, and r absorbing states with transitions into them from q as $\mathbf{R}^{(k)}_{q \times r}$, the Markov chain $\mathbf{M}^{(k)}_{n \times n}$ has the form

$$\mathbf{M}^{(k)} = \begin{pmatrix} \mathbf{Q}^{(k)} & \mathbf{R}^{(k)} \\ \mathbf{0} & \mathbf{I} \end{pmatrix} \tag{2}$$

where $\mathbf{0}$, $\mathbf{I}$ represent lack of transition between/from absorbing states. It follows from [38] that the probability P of a chain starting at t_k being absorbed into state $k + 1$, letting $\mathbf{N} = (\mathbf{I} - \mathbf{Q})^{-1}$, is given as

$$P(t_{k+1} | t_{1:k}, \mathbf{M}) = \mathbf{N}^{(k)} R^{(k)} \Big|_{q,1} \tag{3}$$

The probability of being absorbed at $k + 1$ conditioned on jumps $1 : k$ is thus equivalent to the probability of observing the

---

[4]While traditional application of "tagging" assumes the set of labels to be strictly un-ordered (as in multi-label classification), we follow [15,35] by assuming tags are generated directly from text by keyword recognition. It is thereby trivial to reverse the process, assigning each tag a position as the first time its corresponding keyword was recognized in the original text.

$k+1$ INVITE tag. If we approximate an a priori distribution of tag probabilities to initialize our chain as $t_1 \sim \text{Cat}(n, \theta)$ (which could be empirically derived or simulated), then the likelihood of our observed tag chain $\mathbf{t}$, given a Markov chain, is

$$\mathscr{L}(\mathbf{t}\,|\,\theta\,;\mathbf{M}) = \theta(t_1) \prod_{k=1}^{T-1} P(t_{k+1}\,|\,t_{1:k}\,;\mathbf{M}) \qquad (4)$$

Finally, if we observe a corpus of tag lists $\mathbf{C} = \{\mathbf{t}_1, \mathbf{t}_2, \cdots, \mathbf{t}_c\}$, and assume θ can be estimated separately from $\mathbf{M}$, then we can finally frame the problem as minimizing our loss function, the negative log-likelihood of our corpus over $\mathbf{M}$:

$$\mathbf{M}^* = \underset{\mathbf{M}}{\arg\min} \quad \sum_{i=1}^{C} \sum_{k=1}^{T_i-1} -\log P\left(t_{k+1}^{(i)}\,\Big|\,t_{1:k}^{(i)}, \mathbf{M}\right) \qquad (5)$$

3.3 Implementation

As stated in Eq. 5, the optimization is constrained; in addition to requiring row-stochasticity, the matrix N is only guaranteed to exist if self-transitions are disallowed, as proved in [34]. Similar to that implementation, we introduce a softmax re-parameterization of $\mathbf{M}$ that allows the optimization to be unconstrained in $\mathbb{R}^{n \times n}$, and guaranteeing row-stochasticity.

$$M_{i,j} \leftarrow \frac{\exp(M_{i,j})}{\left[\sum_j \exp(\mathbf{M}_i)\right]_j}$$

However, we introduce several modifications to this reparameterization:

Edge Weights Because it is important for our purposes to estimate the weight (*i.e.*, importance) of each relationship, we to not require (as in [39]) that the structure of $\mathbf{M}$ is un-weighted—in this case each relationship would either exists or not exist. However, sparsity of $\mathbf{M}$ is still desirable, so we apply an L_1-penalty to the loss function, adding an $(\alpha/T) \cdot \|\mathbf{M}\|_1$ term to Eq. 5. The parameter α should generally be tuned via cross-validation where possible, but to demonstrate effectiveness in an unsupervised setting (as is expected to be the case when no "true" $\mathbf{M}$ is yet known), we use $\alpha = 0.01$, which was found to be robust to sensitivity trials for one log-factor in either direction.

Edge Direction In addition, Eq. 5 implies that $\mathbf{M}$ represents a *directed* graph. Though we model each tag as being generated conditional on preceding tags alone, we wish to preserve the intuition that relationships between tags are still assumed to be bi-directional, while not strictly enforcing $\mathbf{M}$ to be symmetric (undirected), as in [39]. Put simply, one-directional relationships can be useful when they are strictly the case (*e.g.*, oil→leak), but we may not wish to encourage one-directional relations that are quirks of imbalanced data and how people talk (gear 1 ↔ gear 2) To ensure the recovered weights in each direction are meaningful, and to speed-up recovery of what we assume is a "symmetry-dominant" $\mathbf{M}$, we bias it toward symmetry via an update to each entry prior to softmax:

$$M_{i,j} \leftarrow \max\left\{M_{i,j}, M_{j,i}\right\}$$

Because of these alterations, the analytic gradient for the INVITE loss function described in [34] no longer applies; instead, we make use of automatic differentiation as a means to ensure accurate gradient calculations under the above modifications [40]. The package `autograd` [41] was used to exploit a number of convenience functions for doing so, in the Python programming language.

4 EXPERIMENTS

The first experiment demonstrates the tractability of the INVITE model in the context of MWO-type data by generating synthetic MWOs from real engineering systems as described in [42]. We use these synthetic MWOs to (1) measure the network recovery accuracy of the INVITE model, (2) compute the sample efficiency of the INVITE model, and (3) compare the INVITE model to co-occurrence similarity thresholding models currently used in the state-of-the-art.

Second, we apply our proposed method to a corpus of real excavator MWOs, for which a "true" underlying structure does not yet exist. We compare the plausibility of work orders sampled from our network estimation to the original dataset and benchmark our model with respect to purely associative sampling.

Due to the high dimensionality of Eq. 5, and the noisy nature of observations, we use Stochastic Gradient Descent (SGD) to perform optimization of $\mathbf{M}$. Specifically, we use the ADAM algorithm [43], which modifies the gradient estimation for each iteration with first- and second-order momentum estimates from previous iterations, improving convergence behavior. Because each tag transition is considered a reliable observation, and the underlying structure of $\mathbf{M}$ is generally sparse relative to the complete adjacency graph between the set of all tags, a learning rate of 0.9 was used, but with minibatches of 5 censored lists each. Exponentially-weighted learning-rate decay was used, along with time-discounted averaging of $\mathbf{M}$, with settings as suggested by [44].

4.1 Exp. 1: Recovering Known Engineering Networks

To validate the ability of our method to accurately reconstruct engineering networks under varying data quantities, we

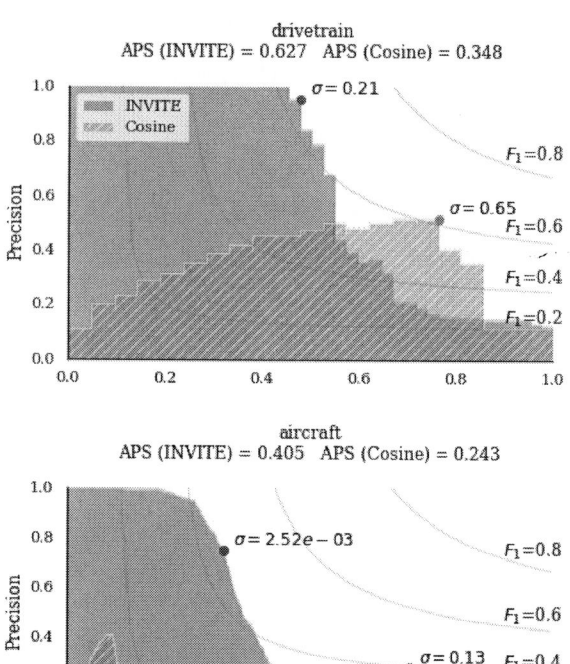

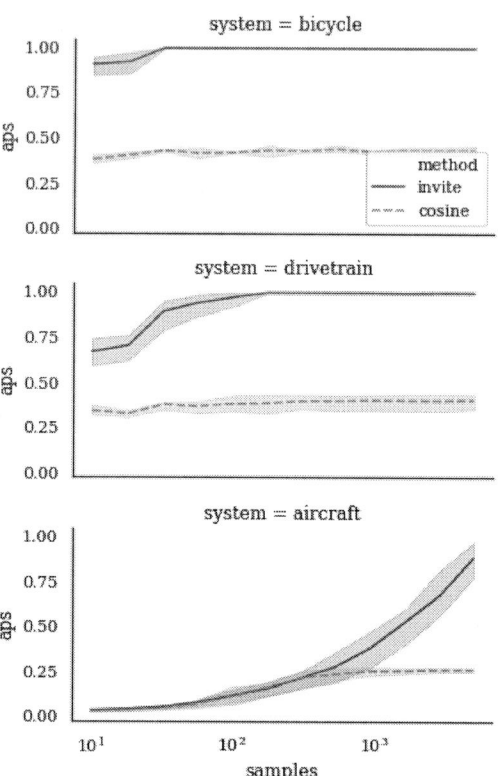

FIGURE 1: Comparing INVITE and Cosine-Similarity thresholding performance for recovering true network structure. **Top**: Recovery performance (precision vs. recall) for the *drivetrain* network. Trained on 18 samples ("work orders"), at 3 tags each. **Bottom**: Same comparison for the more complex *aircraft* network, trained on 1634 samples at 5 tags each. Also shown are the F_1-score iso-lines, along with F_1-optimal thresholds (σ) for each model setting.

FIGURE 2: Mean average precision score (APS) for the three system networks of [42], shown with mean APS over sample lengths $T \in \{3, 4, 5\}$, and a 1000-bootstrap-sample 95% confidence interval. INVITE consistently outperforms similarity thresholding in low-data, low-complexity scenarios. In complex networks, performance is comparable until a significant number of samples are available, after which a lack of sparsity causes the cosine method's performance to plateau.

first synthesize censored tag lists from true component networks described in [42, 46]: a bicycle ($n = 10$), an automotive drivetrain ($n = 18$), and an aircraft ($n = 375$). Drawn layouts for each network are provided for reference in the appendix. For each network, censored random walks were generated by performing a random walk over the nodes until either 100 transitions or all nodes have been visited. The first unique visit to each node was recorded to simulate censoring, and the lists were clipped to the first 3,4, or 5 node visits, to reflect the typical number of tags seen in real MWO datasets (see Exp. 2, below, for an example). The number of censored lists used to train the models was evenly sampled at 11 intervals on a log-scale from 10 - 5000 lists, for a total $3 \times 11 \times 3 = 108$ trials.

Because the original networks are relatively sparse (See Table 1), the classification of edges as "existing" or "not existing"

can be framed as a class-imbalanced information retrieval problem. Given some measure of node similarities (entries in the recovered adjacency matrix), we wish to threshold **M** such that, for a given threshold value $\sigma \in [0, 1]$, the entries of a thresholded adjacency matrix $\mathbf{M}^\sigma$ are given by:

$$M_{i,j}^\sigma = \begin{cases} 1, & \text{if } M_{i,j}^* \geq \sigma \\ 0, & \text{otherwise} \end{cases}$$

Prior to selecting a specific threshold, it is useful to recognize how robustly each model performs under varying threshold values, since the underlying "true" networks are available. For class-imbalanced learning problems like this, precision-recall (P-R) curves can elucidate model robustness under varying threshold sensitivities [47]. In Fig. 1, the precision-recall curves for

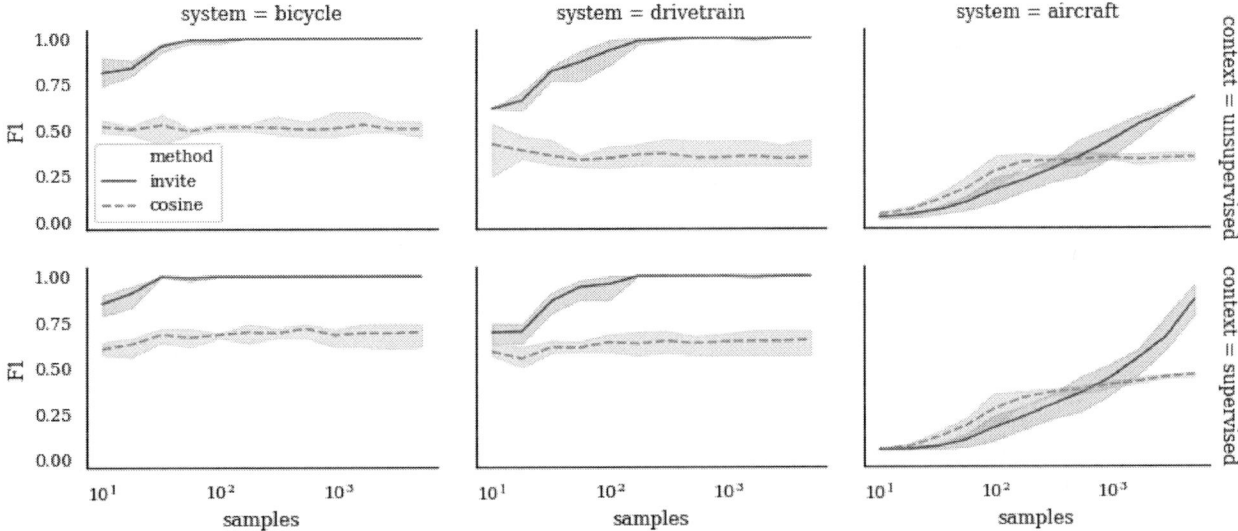

FIGURE 3: Mean F_1 reconstruction scores for the three system networks of [42], shown with mean score over sample lengths $T \in \{3,4,5\}$, and a 1000-bootstrap-sample 95% confidence interval. In an unsupervised context (top row), thresholds for node similarity were selected using a knee-finding heuristic [45], for where the EDF of edge-weights showed maximum curvature. In a supervised context (bottom row), the optimal threshold was selected as one that maximized the model's F_1-score. The INVITE method significantly outperforms pure co-occurrence similarity thresholds as the number of samples increases, and because the EDF is much more spread-out for cosine similarities, picking a "good" threshold is much more difficult than the sparsity-inducing INVITE models.

TABLE 1: Engineering component network summary for Experiment (1). Network models adapted from [42].

Model	Nodes	Sparsity
bicycle	10	80.0%
drivetrain	18	88.4%
aircraft	375	97.5%

the drivetrain model demonstrate that INVITE can quickly recover simpler networks with under 20 observations at relatively few "tags" each, while Cosine Similarity only robustly captures the global structure of the network—precision is relatively invariant over wide ranges of recall. This is even more pronounced for more complex networks, with the INVITE model capable of achieving either high precision or high recall, while the Cosine Similarity threshold has difficulty improving it under any circumstance. One way to summarize this robustness under varying threshold is calculate the average the precision score (APS) gained by each threshold's increase of recall R:

$$\text{APS} = \sum_{\sigma} [R(\sigma_i) - R(\sigma_{i-1})] P(\sigma_i) \qquad (6)$$

The APS score will not give a "good" σ, but instead summarizes the total "goodness" of each model across possible σ. APS scores for the INVITE and Cosine Similarity models are shown against training set size in Fig. 2. APS eventually plateaus for the cosine model in every case. INVITE can perfectly recover the bicycle and drivetrain structures after around 100 samples. For the aircraft network, while INVITE has nearly identical performance to cosine similarity below 500 samples, INVITE's APS almost reaches 1.0 with 5000 samples.

In practice, selecting the value for σ will depend on whether training examples are available: if not, a heuristic threshold such as knee-finding can be applied; if examples are available, it is possible to use performance measures appropriate for imbalanced learning problems (*e.g.* the F_1-score), and optimize the threshold for this value. In the common case that no training labels are available (no "true" structures are known), a common heuristic for thresholding values posits that diminishing returns occur for the retrieval function after the point of maximum curvature on the empirical distribution function (EDF) of values to threshold at the point of diminishing returns—*e.g.*, using a so-called "knee-finding" algorithm. To test the performance of both the cosine-similarity (bag-of-words) and INVITE recovered networks with respect to the originals, we apply the `kneedle` algorithm [45] to calculate a threshold σ. The F_1-score can then be calculated for $\mathbf{M}^{\sigma}$ as for each training-set size (see unsuper-

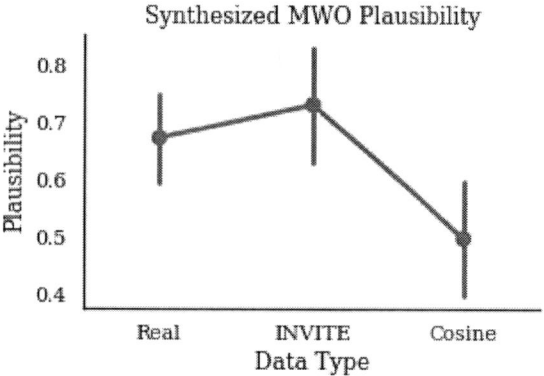

FIGURE 4: Mean plausibility ratio for 100 MWOs, both real and synthesized by sampling $\mathbf{M}^\sigma$ recovered from INVITE and Cosine methods. Confidence intervals show the inter-quartile range of 1000 bootstrap samples.

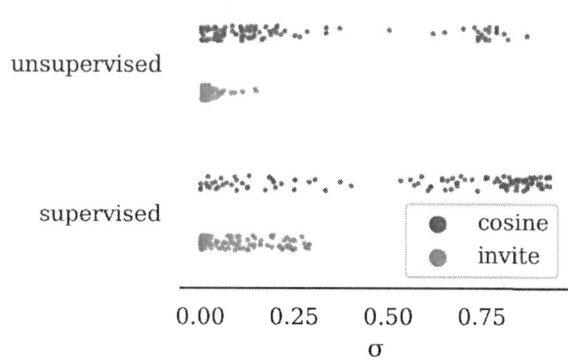

FIGURE 5: Thresholds selected by knee-finding heuristic (unsupervised) and optimal F_1-score (supervised). The cosine similarity performance was more sensitive to σ selection than INVITE.

vised context in the top row of Fig. 3). If parts of the underlying structure are known a priori, it is possible to tune σ so that the F_1-score of $\mathbf{M}^\sigma$ vs. the "test-set" (the true $\mathbf{M}$) is maximized. These can likewise be found on the bottom of Fig. 3

4.2 Exp. 2: Real-World Excavator MWOs

To assess the applicability of INVITE to real-world scenarios, we apply our model to tags annotated for a mining dataset (8264 MWOs) pertaining to 8 similarly-sized excavators at various sites across Australia [15, 48]. The tags were created by a subject-matter expert spending 1 hour of time in the annotation assistance tool `nestor` [49], using a methodology outlined in [50]. The tag annotations were limited to objects (bolt, motor, fan, *etc.*), problems (leak, missing, cracked, *etc.*), and solutions (replace, repair, stick, *etc.*) that occurred at least 50 times each in the original corpus, for a total of 77 unique tags. Subsequently, the same settings for solving Eq. 5 were used as in the previous experiment, though the optimization was initialized with the cosine similarity matrix to speed convergence.

To test whether the INVITE model was able to learn a robust representation of the system structure, we perform blind tests of the generative capability of each recovered network. First, the starting tag probability θ was set as the observed distribution of first tags in the original dataset. Then, censored random walks of length $T = 5$ were sampled from both an INVITE and a cosine-similarity recovered network, without thresholding. This is intended to preserve weighted relationships between tags, for the purposes of data synthesis. The expert was then given a list of 100 randomly mixed MWOs, made from 40 real work-orders,[5] 30 INVITE censored lists, and 30 cosine-similarity cen-

sored lists. The resulting lists were filtered to only contain lists of tags not explicitly found in the original data. The expert was then asked to blindly classify each MWO as being "plausible" or "not plausible", such that an MWO would or would not reasonably occur based only on the tags in each. The fraction of real, INVITE, and cosine-generated work orders marked as "plausible" can be found in Fig. 4. Both the real and INVITE-synthesized MWOs are within a similar plausibility range, between $60\% - 80\%$, while the cosine similarity MWOs are between $40\% - 60\%$ plausible, overall.

5 DISCUSSION

To enable optimization in continuous space, our model does not enforce un-weighted, un-diriected graphs, as in Zemla & Austerweil's extension of the INVITE model that optimizes in discrete space [39]. Their version is intended to induce sparsity without artificially introducing new tuning parameters, as we have in introducing σ and α. In practice, however, it is reasonable to select a low-valued positive $\sigma \ll 1$, or use a knee-finding heuristic on the EDF of edge weights. This is because the L_1 penalty on edge weights, plus the tendency of INVITE to route random walks through commonly-visited chokepoint nodes, naturally drives "unnecessary" edge weights to near-zero probability. As seen in Fig. 5, the "knee" in the edge-weight EDF for INVITE was nearly always near-zero, for all experiments. Even when oracle information was allowed to tune σ, the best F_1-score was still to be found strictly below $\sigma = 0.3$, with a majority below 0.1. In this sense, the signal-to-noise ratio of our INVITE model is quite high, making the selection of a good σ much less

---

[5]The process in [49] displays extracted concepts in order of their statistical

importance, for the purposes of keyword recognition. As such, the annotator does not interact with the original work-orders directly during tagging.

difficult.

In truth, this thresholding does not completely solve the problem of knowledge extraction. If the goal of automatically extracting knowledge graphs is to suggest whether causal relationships *exist* between tags, and not primarily to synthesize data, weights are not needed in communicating these links, and may obfuscate important understanding below vastly more obvious relationships.

Another issue related to thresholding is how the row-stochastic constraint affects edge-value distribution: technically, each row in our model will be re-normalized every iteration, independent of other rows. This means that the row-normalization inherently de-symmetrizes $\mathbf{M}$. In reality, though we might model some asymmetry in node relations[6], the modality of direction being discovered by sentence-structure (ordering of the written tags) is not equivalent to the types of directionality we might want to discover. In memory, it could be beneficial to assume that the probability of transitioning between tags should be bi-directional, and allow desirable directionality to be proxied by local tree-like structures that reduce centrality of tags farther out. This bi-directional assumption implies making $\mathbf{M}$ a doubly-stochastic matrix. This has the added benefit of placing $\mathbf{M}$ on a simplex, *i.e.*, the space of permutation-invariant matrices belonging to the Birkhoff Polytope. There are recent developments [51, 52] in this space that could prove highly useful at reducing the state-space we search over.

Finally, our method is not intended to serve as a complete, end-to-end processing of natural language text into structured knowledge. Ultimately, the final structuring will need to be performed by humans. Instead, we believe the most efficient tools to assist in knowledge recovery will pose annotation questions in a lower-dimensional state-space, easier for a human to verify or edit *quickly*. Figure 6 illustrates how we believe IN-VITE takes steps toward this goal. Common structure recovery techniques, like cosine-similarity thresholding, tend to discover global structure quite well, but over-estimate the connectivity of local communities where hierarchical relationships are unknown-yet-assumed by the data. By definition, co-occurrence (bag-of-words) metrics are treating these work-orders more like un-censored random walks, starting from any tag and transitioning to any other in the list. Consequently, the local resolution of the structure it approximates is going to be fundamentally limited for tree-like communities, more reflecting a 2nd- or 3rd-order power graph[7] of the true, underlying structure.

In contrast, INVITE tends to concentrate edges to nodes that are highly central, forming "chokepoints" where global-scale transition mechanisms are unknown, but largely preserving local tree-like structures in outer communities. From an active learning perspective, humans tend to be quite good at verifying global-scale connections as viable or not—editing spurious connections in every over-dense local community is a much more difficult task for us than recognizing spurious individual connections to a small set of highly abstract concepts.

We believe this feature can be exploited to create better knowledge-structuring assistance tools in an active learning context. Such a tool could additionally benefit from a recent explosion in interest for preserving hierarchical relationships in vector space, *e.g.*, via Poincaré embeddings [54]. Additional care must be taken to allow flexible annotation of different kinds of relationships,[8] and allow for multiple (potentially disagreeing) annotators, subsequently suggesting relationship types for review. We envision a type of "topic model" over the space of knowledge graphs [55], or potentially a set of independent "graph components" that maximally explain the distribution of edge types in a community [56].

6 CONCLUSIONS AND FUTURE WORK

This paper presented a method to recover a structured representation of engineering knowledge from unstructured written documents (specifically, Manufacturing Work Orders), based on initial-visit emitting random walks (INVITE). Compared to previous methods, our technique preserves local connectivity structures, even in tree-like communities. This can lead to (1) better generative capability for synthesizing plausible documents (such as work-orders) in a simulation context; and (2) allowing us to cast the knowledge-structuring problem in probabilistic context that is potentially amendable to active-learning; this can minimize the number of local-scale edits needed relative to global-scale, abstract connections that humans can easily spot and correct.

Overall, the model we describe here can enable experts and novices alike to benefit from tacit system knowledge contained within frequently unused mountains of technical work-orders, by quickly prototyping computable representations of this knowledge for downstream usage in analysis pipelines. We believe that by explicitly incorporating cognitive theories into our modeling assumptions about how technicians might represent and then recall their knowledge in maintenance work-orders, we can accelerate the training and use of unsupervised data-driven expert systems in engineering design.

ACKNOWLEDGMENT

Thanks to Dr. Michael Brundage (NIST) for his efforts annotating and rating MWOs, and to Dr. Melinda Hodkiewicz (Univ. Western Australia) for providing the excavator data, and for many enjoyable and enlightening discussions on this topic.

---

[6] *e.g.* in hierarchies: "gear-1" may be a member of "gearbox," making the link gear-1 $\rightarrow$ gearbox a stronger link in a technician's head than the other direction.

[7] The graph G's n^{th}-order powergraph $P(G, n)$ has an edge between any two nodes if the minimum path length between those nodes in G is at most n.

---

[8] *e.g.*, Walsh *et al.* actually construct three types of structured system representations: functional, parametric, and component (which we use here)

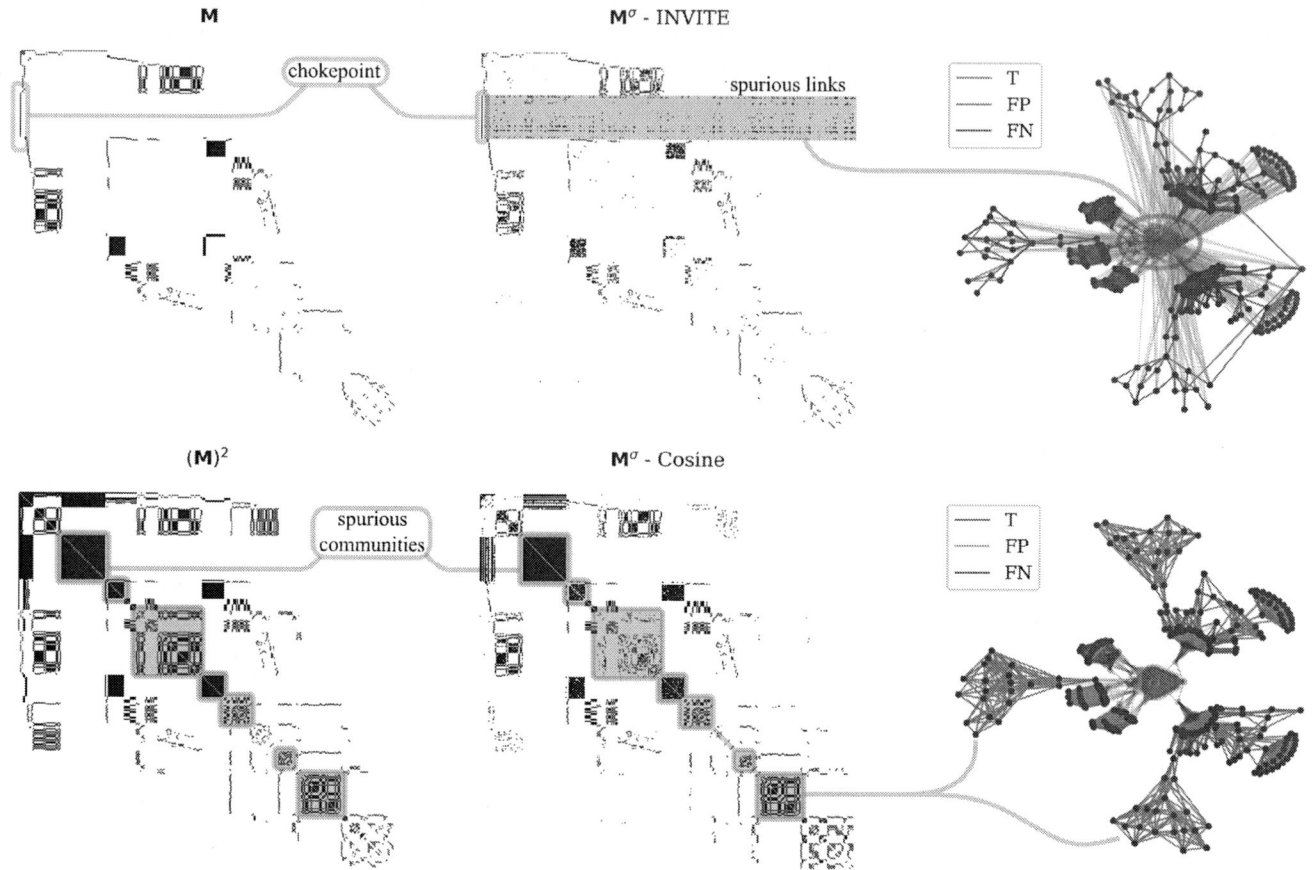

FIGURE 6: Comparing aircraft model network reconstruction for F_1-optimal INVITE ($F_1 = 0.75$) and Cosine Similarity ($F_1 = 0.49$) methods. Shown are the original "true" adjacency matrix $\mathbf{M}$, its 2nd-order power graph $(\mathbf{M})^2$, and the thresholded adjacency matrices $\mathbf{M}^\sigma$ for both INVITE and Cosine Similarity. For visualization, the matrix rows/columns are sorted by the closeness centrality of each node [53], better indicating which nodes form core/integral components in the system (upper-left) and which are more likely a part of localized "edge" communities (bottom and right). These edge communities are highlighted in the graph layouts on the right, where nodes in the top 25th percent most central (and their edges) are transparent. INVITE directly estimates the underlying structure of $\mathbf{M}$ by accounting for node censoring in observations, concentrating uncertain edges into a few highly-connected "chokepoint" nodes. Cosine similarity mistakes co-occurrence of components in a sample for direct relationships, forming dense, spurious communities throughout the graph that are reflective of higher-order powers of $\mathbf{M}$, as shown here. Concentrating the false-positive relationships (FP) in a few highly central nodes makes INVITE a viable candidate for querying human experts for annotation/critique in an active-learning context.

7 DISCLAIMER

The use of any products described in this paper does not imply recommendation or endorsement by the National Institute of Standards and Technology, nor does it imply that products are necessarily the best available for the purpose.

REFERENCES

[1] Guimerà, R., and Sales-Pardo, M., 2009. "Missing and spurious interactions and the reconstruction of complex networks". *Proceedings of the National Academy of Sciences,* *106*(52), pp. 22073–22078.

[2] Gomez-Rodriguez, M., Leskovec, J., and Krause, A., 2012. "Inferring networks of diffusion and influence". *ACM Transactions on Knowledge Discovery from Data (TKDD),* *5*(4), p. 21.

[3] Linderman, S., and Adams, R., 2014. "Discovering latent network structure in point process data". In International Conference on Machine Learning, pp. 1413–1421.

[4] De Paula, Á., Rasul, I., and Souza, P., 2018. "Recovering social networks from panel data: identification, simulations and an application".

[5] Battaglia, P. W., Hamrick, J. B., Bapst, V., Sanchez-Gonzalez, A., Zambaldi, V., Malinowski, M., Tacchetti, A., Raposo, D., Santoro, A., Faulkner, R., et al., 2018. "Relational inductive biases, deep learning, and graph networks". *arXiv preprint arXiv:1806.01261*.

[6] Raissi, M., Perdikaris, P., and Karniadakis, G. E., 2017. "Machine learning of linear differential equations using gaussian processes". *Journal of Computational Physics, 348*, pp. 683–693.

[7] Krishna, R., Zhu, Y., Groth, O., Johnson, J., Hata, K., Kravitz, J., Chen, S., Kalantidis, Y., Li, L.-J., Shamma, D. A., et al., 2017. "Visual genome: Connecting language and vision using crowdsourced dense image annotations". *International Journal of Computer Vision, 123*(1), pp. 32–73.

[8] Speer, R., Chin, J., and Havasi, C., 2017. "Conceptnet 5.5: An open multilingual graph of general knowledge". In Thirty-First AAAI Conference on Artificial Intelligence.

[9] ISO 15926-1:2004, 2004. Industrial automation systems and integration – Integration of life-cycle data for process plants including oil and gas production facilities – Part 1: Overview and fundamental principles. Standard, International Organization for Standardization, Geneva, CH, July.

[10] Leal, D., 2005. "ISO 15926" life cycle data for process plant": An overview". *Oil & gas science and technology, 60*(4), pp. 629–637.

[11] ISO/TS 15926-8:2011, 2011. Industrial automation systems and integration – Integration of life-cycle data for process plants including oil and gas production facilities – Part 8: Implementation methods for the integration of distributed systems: Web Ontology Language (OWL) implementation. Standard, International Organization for Standardization, Geneva, CH, Oct.

[12] Batres, R., West, M., Leal, D., Price, D., Masaki, K., Shimada, Y., Fuchino, T., and Naka, Y., 2007. "An upper ontology based on ISO 15926". *Computers & Chemical Engineering, 31*(5-6), pp. 519–534.

[13] Klüwer, J. W., Skjæveland, M. G., and Valen-Sendstad, M., 2008. "Iso 15926 templates and the semantic web". In Position paper for W3C Workshop on Semantic Web in Energy Industries; Part I: Oil and Gas.

[14] Kumar, N., Kumar, M., and Singh, M., 2016. "Automated ontology generation from a plain text using statistical and nlp techniques". *International Journal of System Assurance Engineering and Management, 7*(1), pp. 282–293.

[15] Hodkiewicz, M., and Ho, M. T.-W., 2016. "Cleaning historical maintenance work order data for reliability analysis". *Journal of Quality in Maintenance Engineering, 22*(2), pp. 146–163.

[16] Ho, M., 2015. "A shared reliability database for mobile mining equipment". PhD thesis, University of Western Australia.

[17] Eppinger, S. D., and Browning, T. R., 2012. *Design structure matrix methods and applications*. MIT press.

[18] Browning, T. R., 2016. "Design structure matrix extensions and innovations: a survey and new opportunities". *IEEE Transactions on Engineering Management, 63*(1), pp. 27–52.

[19] Ellinas, C., Allan, N., Durugbo, C., and Johansson, A., 2015. "How robust is your project? from local failures to global catastrophes: A complex networks approach to project systemic risk". *PloS one, 10*(11), p. e0142469.

[20] Robertson, S., 2004. "Understanding inverse document frequency: on theoretical arguments for idf". *Journal of documentation, 60*(5), pp. 503–520.

[21] Steyvers, M., and Griffiths, T., 2007. "Probabilistic topic models". *Handbook of latent semantic analysis, 427*(7), pp. 424–440.

[22] Blei, D. M., Griffiths, T. L., and Jordan, M. I., 2010. "The nested chinese restaurant process and bayesian nonparametric inference of topic hierarchies". *Journal of the ACM (JACM), 57*(2), p. 7.

[23] Mikolov, T., Chen, K., Corrado, G., and Dean, J., 2013. "Efficient estimation of word representations in vector space". *arXiv preprint arXiv:1301.3781*.

[24] Pennington, J., Socher, R., and Manning, C., 2014. "Glove: Global vectors for word representation". In Proceedings of the 2014 conference on empirical methods in natural language processing (EMNLP), pp. 1532–1543.

[25] Sharp, M., Sexton, T., and Brundage, M. P., 2017. "Toward semi-autonomous information". In IFIP International Conference on Advances in Production Management Systems, Springer, pp. 425–432.

[26] Chang, J., Gerrish, S., Wang, C., Boyd-Graber, J. L., and Blei, D. M., 2009. "Reading tea leaves: How humans interpret topic models". In Advances in neural information processing systems, pp. 288–296.

[27] Strohmaier, M., Körner, C., and Kern, R., 2012. "Understanding why users tag: A survey of tagging motivation literature and results from an empirical study". *Web Semantics: Science, Services and Agents on the World Wide Web, 17*, pp. 1–11.

[28] Vander Wal, T., 2007. Folksonomy. http://vanderwal.net/folksonomy.html.

[29] Specia, L., and Motta, E., 2007. "Integrating folksonomies with the semantic web". In European semantic web conference, Springer, pp. 624–639.

[30] Mousselly-Sergieh, H., Egyed-Zsigmond, E., Gianini, G.,

Döller, M., Kosch, H., and Pinon, J.-M., 2013. "Tag similarity in folksonomies". In INFORSID, Vol. 29, Inforsid, pp. 319–334.

[31] Heymann, P., and Garcia-Molina, H., 2006. Collaborative creation of communal hierarchical taxonomies in social tagging systems. Tech. rep., Stanford.

[32] Henschel, A., Woon, W. L., Wachter, T., and Madnick, S., 2009. "Comparison of generality based algorithm variants for automatic taxonomy generation". In Innovations in Information Technology, 2009. IIT'09. International Conference on, IEEE, pp. 160–164.

[33] Schvaneveldt, R. W., Durso, F. T., and Dearholt, D. W., 1989. "Network structures in proximity data". In *Psychology of learning and motivation*, Vol. 24. Elsevier, pp. 249–284.

[34] Jun, K.-S., Zhu, J., Rogers, T. T., Yang, Z., et al., 2015. "Human memory search as initial-visit emitting random walk". In Advances in neural information processing systems, pp. 1072–1080.

[35] Sexton, T., Brundage, M. P., Hoffman, M., and Morris, K. C., 2017. "Hybrid datafication of maintenance logs from ai-assisted human tags". In 2017 IEEE International Conference on Big Data (Big Data), IEEE, pp. 1769–1777.

[36] Hills, T. T., Todd, P. M., and Jones, M. N., 2015. "Foraging in semantic fields: How we search through memory". *Topics in Cognitive Science, 7*(3), pp. 513–534.

[37] Lv, Y., and Zhai, C., 2009. "Positional language models for information retrieval". In Proceedings of the 32nd international ACM SIGIR conference on Research and development in information retrieval, ACM, pp. 299–306.

[38] Doyle, P. G., and Snell, J. L., 2000. "Random walks and electric networks". *arXiv preprint math/0001057*.

[39] Zemla, J. C., and Austerweil, J. L., 2018. "Estimating semantic networks of groups and individuals from fluency data". *Computational Brain & Behavior, 1*(1), pp. 36–58.

[40] Baydin, A. G., Pearlmutter, B. A., Radul, A. A., and Siskind, J. M., 2018. "Automatic differentiation in machine learning: a survey". *Journal of Marchine Learning Research, 18*, pp. 1–43.

[41] Maclaurin, D., 2016. "Modeling, inference and optimization with composable differentiable procedures". PhD thesis.

[42] Walsh, H. S., Dong, A., and Tumer, I. Y., 2019. "An analysis of modularity as a design rule using network theory". *Journal of Mechanical Design, 141*(3), p. 031102.

[43] Kingma, D. P., and Ba, J., 2014. "Adam: A method for stochastic optimization". *arXiv preprint arXiv:1412.6980*.

[44] Bottou, L., 2012. "Stochastic gradient descent tricks". In *Neural networks: Tricks of the trade*. Springer, pp. 421–436.

[45] Satopaa, V., Albrecht, J., Irwin, D., and Raghavan, B., 2011. "Finding a" kneedle" in a haystack: Detecting knee points in system behavior". In 2011 31st International Conference on Distributed Computing Systems Workshops, IEEE, pp. 166–171.

[46] Haley, B. M., Dong, A., and Tumer, I. Y., 2016. "A comparison of network-based metrics of behavioral degradation in complex engineered systems". *Journal of Mechanical Design, 138*(12), p. 121405.

[47] Saito, T., and Rehmsmeier, M., 2015. "The precision-recall plot is more informative than the roc plot when evaluating binary classifiers on imbalanced datasets". *PloS one, 10*(3), p. e0118432. An optional note.

[48] Hodkiewicz, M. R., Batsioudis, Z., Radomiljac, T., and Ho, M. T., 2017. "Why autonomous assets are good for reliability–the impact of 'operator-related component'failures on heavy mobile equipment reliability". In Annual Conference of the Prognostics and Health Management Society 2017.

[49] Madhusudanan Navinchandran, F., Bones, L., Brundage, M., Hoffman, M., Moccozet, S., and Sexton, T., 2018. Nestor: a toolkit for quantifying tacit maintenance knowledge, for investigatory analysis in smart manufacturing.

[50] Sexton, T., Hodkiewicz, M., Brundage, M. P., and Smoker, T., 2018. "Benchmarking for keyword extraction methodologies in maintenance work orders". In PHM Society Conference, Vol. 10.

[51] Adams, R. P., and Zemel, R. S., 2011. "Ranking via sinkhorn propagation". *arXiv preprint arXiv:1106.1925*.

[52] Linderman, S. W., Mena, G. E., Cooper, H., Paninski, L., and Cunningham, J. P., 2017. "Reparameterizing the birkhoff polytope for variational permutation inference". *arXiv preprint arXiv:1710.09508*.

[53] Sabidussi, G., 1966. "The centrality index of a graph". *Psychometrika, 31*(4), pp. 581–603.

[54] Nickel, M., and Kiela, D., 2017. "Poincaré embeddings for learning hierarchical representations". In Advances in neural information processing systems, pp. 6338–6347.

[55] Gerlach, M., Peixoto, T. P., and Altmann, E. G., 2018. "A network approach to topic models". *Science advances, 4*(7), p. eaaq1360.

[56] Park, B., Kim, D.-S., and Park, H.-J., 2014. "Graph independent component analysis reveals repertoires of intrinsic network components in the human brain". *PloS one, 9*(1), p. e82873.

A APPENDIX - WALSH *ET AL.* NETWORKS

bicycle

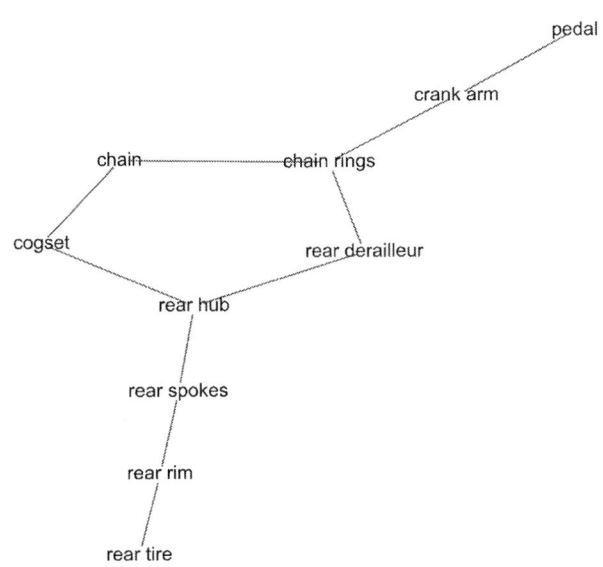

drivetrain

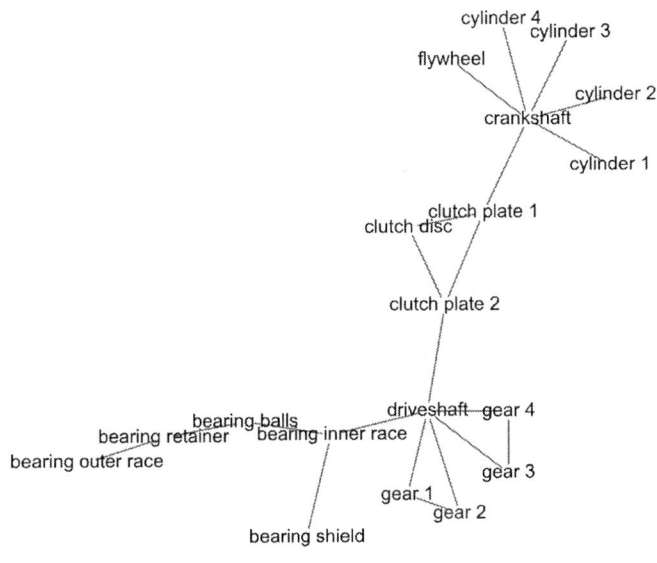

aircraft

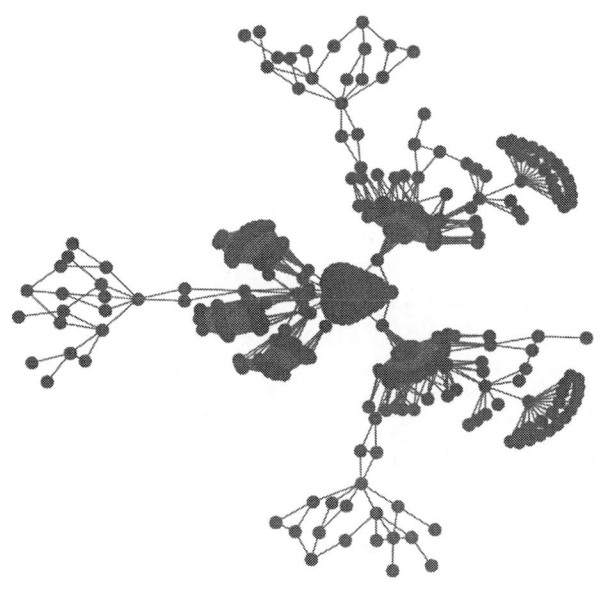

Proceedings of the ASME 2019
International Design Engineering Technical Conferences
and Computers and Information in Engineering Conference
IDETC/CIE2019
August 18-21, 2019, Anaheim, CA, USA

DETC2019-97370

Topology Optimization of Multi-Material Lattices for Maximal Bulk Modulus

Hesaneh Kazemi
Dept. of Mechanical Eng.
University of Connecticut
191 Auditorium Road, U-3139
Storrs, Connecticut 06269
e-mail: hesaneh.kazemi@uconn.edu

Ashkan Vaziri
Dept. of Mechanical and Industrial Eng.
Northeastern University
254 Richards Hall, 360 Huntington Ave.,
Boston, Massachusetts 02115
e-mail: vaziri@coe.neu.edu

Julián A. Norato[*]
Dept. of Mechanical Eng.
University of Connecticut
191 Auditorium Road, U-3139
Storrs, Connecticut, 06269
e-mail: norato@engr.uconn.edu

In this paper, we present a method for multi-material topology optimization of lattice structures for maximum bulk modulus. Unlike ground structure approaches that employ 1-d finite elements such as bars and beams to design periodic lattices, we employ a 3-d representation where each lattice bar is described as a cylinder. To accommodate the 3-d bars, we employ the geometry projection method, whereby a high-level parametric description of the bars is smoothly mapped onto a density field over a fixed analysis grid. In addition to the geometric parameters, we assign a size variable per material to each bar. By imposing suitable constraints in the optimization, we ensure that each bar is either made exclusively of one of a set of a multiple available materials or completely removed from the design. These optimization constraints, together with the material interpolation used in our formulation, make it easy to consider any number of available materials. Another advantage of our method over ground structure approaches with 1-d elements is that the bars in our method need not be connected at all times (i.e., they can 'float' within the design region), which makes it easier to find good designs with relatively few design variables. We illustrate the effectiveness of our method with numerical examples of bulk modulus maximization for two-material lattices with orthotropic symmetry, and for two- and three-material lattices with cubic symmetry.

1 Introduction

Topology optimization aids in the design of structural concepts by finding the optimal layout of material within a prescribed design space. In addition to design of the struc-

tural components, it has been utilized to design architected materials. Architected materials are those for which the material distribution at one or more length scales lower than the (macro) scale of the structure where they will be employed is controlled to obtain desired effective properties. Among these materials, lattice structures with a periodic unit cell have been among the most studied due to the relative simplicity of their geometric representation.

There exist different methods to obtain the optimal topology of the unit cell, herein referred to as the microstructure. The design of microstructures was first demonstrated with density-based topology optimization approaches. An inverse homogenization method to design 2D microstructures that minimize weight subject to specified elastic properties was presented in [1], and later extended to 3D structures [2]. This and similar procedures have been employed in various applications, cf., for example, [3–9] and the recent review [10]. Other topology optimization approaches have also been used for design of architected materials, such as the level-set method (e.g., [11–14]) and evolutionary methods [15–18]. In addition to designing the microstructure, other methods have been proposed that simultaneously design the topology of the micro and macrostructures, for instance, [19–27]. In this work, however, we restrict ourselves to the design of the microstructure.

The density-based topology optimization of microstructures with multiple phases was first introduced in [28] for the design of materials with extreme thermal expansion. A similar interpolation scheme was used in [29] to optimize the effective bulk modulus. The method in [30, 31] designs multi-phase materials for maximal thermal conductivity using the interpolation schemes in [32, 33].

---

[*]Corresponding author.

Copyright © 2019 ASME

Despite the increasing availability of multi-material additive manufacturing techniques, the multi-material designs provided by the foregoing methods are not always easy to manufacture. For instance, these organic designs may have closed cavities that cannot be readily printed without supports and would not allow for easy support removal. Multi-material lattices that have struts made of one out of several available materials may thus facilitate manufacturing. Researchers have employed various techniques to design structures made of distinct geometric primitives while retaining the ability to employ a fixed mesh for the finite element analysis. One of these techniques is the geometry projection method [34–36], wherein a high-level parametric description of the geometric primitives is smoothly mapped onto a continuous density field discretized with the fixed analysis grid. An important characteristic of this method is the assignment of a size variable to each geometry primitive that is penalized in the spirit of density-based topology optimization techniques, so that a value of zero signifies the entire removal of that primitive from the design. The method in [37] designs multi-material lattice structures in 3D by combining the geometry projection with an adaptation of the multi-material interpolation scheme of [28]. The effectiveness of this method was demonstrated by designing two-material lattices for maximal bulk modulus and for minimal Poisson's ratio. As indicated in [38], however, the extension of the interpolation formula of [28] to designing with more than two materials is not straightforward.

The method in [39] introduces a new interpolation scheme to accommodate geometric primitives made of different materials. A size variable per each of the available materials is assigned to each of the primitives, and an adaptation of the discrete material optimization (DMO) method [38] is used to ensure that each of those variables is zero or unity, and that at most one material size variable is unity for each primitive. This penalization scheme readily accommodates any number of materials. In this paper we present an extension of this method to design multi-material 3D lattice structures. We employ inverse homogenization to design the base cell to extremize effective properties of the macrostructure. To obtain macrostructures with desired symmetries, we employ reflections of the geometry projection to enforce symmetries on the lattice design with respect to an arbitrary number of symmetry planes. We demonstrate the efficacy of our method by designing multi-material lattices with maximal bulk modulus given a weight constraint.

The remainder of this paper is organized as follows. In Section 2 we discuss the geometry projection method and the multi-material interpolation scheme. Section 3 describes the homogenization method employed to calculate effective properties of the macrostructure. Section 4 introduces the use of reflection matrices to apply symmetries on the design of the base cell. We describe the optimization problem in Section 5. Section 6 presents numerical examples of our method, and we draw conclusions of our work in Section 7.

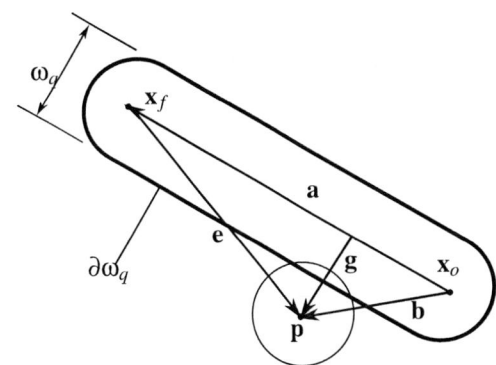

Fig. 1: Bar geometry

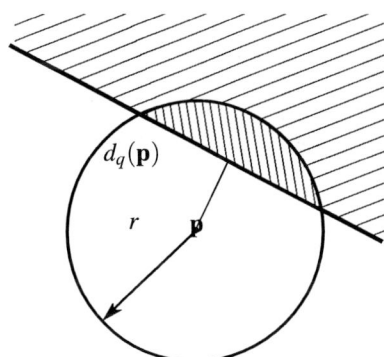

Fig. 2: Sample window for calculation of projected density

2 Geometry Projection

The geometry projection method is a topology optimization technique wherein a high-level parametric description of a structure made of geometry primitives is smoothly mapped onto a density field. This density is subsequently discretized via a fixed finite element mesh and an ersatz material is used to perform the analysis. In this work, a bar q is modeled by a cylinder of diameter w and semispherical ends, parameterized by the positions of the endpoints of its medial axis, $\mathbf{x}_{q_o}$ and $\mathbf{x}_{q_f}$ (cf. Fig. 1). The density assigned to a point $\mathbf{p}$ in space is calculated as the volume ratio of the portion of a sample window $\mathbf{B}_{\mathbf{p}}^r := \{\mathbf{x}|\ \|\mathbf{p} - \mathbf{x}\| \le r\}$ that intersects the solid structure ω:

$$\rho(\mathbf{x}, r) := \frac{|\mathbf{B}_{\mathbf{p}}^r \cap \omega|}{\mathbf{B}_{\mathbf{p}}^r} \tag{1}$$

We approximate the foregoing projected density as the volume ratio of the circular cap of height $r - \phi_q$ (cf. Fig. 2), i.e.:

$$\rho_q(\phi_q, r) = \begin{cases} 0 & \text{if } \phi_q > r \\ \frac{1}{2} + \frac{\phi_q^3}{4r^3} - \frac{3\phi_q}{4r} & \text{if } -r \le \phi_q \le r \\ 1 & \text{if } \phi_q < -r \end{cases} \tag{2}$$

where $\phi_q(\mathbf{p})$ is the signed distance from $\mathbf{p}$ to bar q, which can be computed as:

$$\phi_q(d_q, w) := d_q(\mathbf{x}_{q_o}, \mathbf{x}_{q_f}, \mathbf{p}) - \frac{w}{2} \quad (3)$$

In the expression above, d_q is the distance from $\mathbf{p}$ to the medial axis $\overline{\mathbf{x}_{q_o} \mathbf{x}_{q_f}}$ of bar q, given by:

$$d_q(\mathbf{x}_{q_o}, \mathbf{x}_{q_f}, \mathbf{p}) = \begin{cases} \|\mathbf{b}\| & \text{if } \mathbf{a} \cdot \mathbf{b} \leq 0 \\ \|\mathbf{g}\| & \text{if } 0 < \mathbf{a} \cdot \mathbf{b} < \mathbf{a} \cdot \mathbf{a} \\ \|\mathbf{e}\| & \text{if } \mathbf{a} \cdot \mathbf{b} > \mathbf{a} \cdot \mathbf{a} \end{cases} \quad (4)$$

where

$$\mathbf{a} := \mathbf{x}_{q_f} - \mathbf{x}_{q_o} \quad (5)$$

$$\mathbf{b} := \mathbf{p} - \mathbf{x}_{q_o} \quad (6)$$

$$\mathbf{e} := \mathbf{p} - \mathbf{x}_{q_f} \quad (7)$$

$$\mathbf{P}_\mathbf{a}^\perp := \mathbf{I} - \frac{1}{\|\mathbf{a}\|^2} \mathbf{a} \otimes \mathbf{a} \quad (8)$$

$$\mathbf{g} := \mathbf{P}_\mathbf{a}^\perp \mathbf{b} \quad (9)$$

with $\mathbf{P}_a^\perp$ denoting the perpendicular projector on $\mathbf{a}$.

The foregoing projected density corresponds to a single bar. In the original geometry projection method [34, 35], which considers only primitives made of the same isotropic material, the resulting projected density was computed using a p-norm as a smooth approximation of the maximum function, which is equivalent to a Boolean union of the implicit, density-based representations of the individual primitives. To accommodate the intersection of primitives made of dissimilar materials, a different aggregation scheme was presented in [39]. In this scheme, an effective density corresponding to material i is defined at each point as

$$\rho_{eff}^i(\mathbf{z}, \mathbf{p}) = \frac{\sum_{q=1}^{N_b} \tilde{H}_\varepsilon(-\phi_q(\mathbf{z}, \mathbf{p}))\rho_q w_i^q(\mathbf{z})}{A + B} \quad (10)$$

where

$$A = \sum_{q=1}^{N_b} \left(\tilde{H}_\varepsilon(-\phi_q(\mathbf{z}, \mathbf{p})) \sum_{i=1}^{N_m} \alpha_i^q \right) \quad (11)$$

$$B = 1 - KS_q(\tilde{H}_\varepsilon(-\phi_q(\mathbf{z}, \mathbf{p})) \sum_{i=1}^{N_m} \alpha_i^q) \quad (12)$$

$$KS_i(\mathbf{x}) := \frac{1}{k} ln \left(\sum_i e^{k x_i} \right) \quad (13)$$

In these expressions, N_b denotes the number of bars and N_m the number of available materials; A approximates the number of solid bars which contain $\mathbf{p}$; B renders unity if $\mathbf{p}$ is in

the void region and does not intersect any solid bar in order to avoid a division by zero; and KS is the Kreisselmeier-Steinhauser smooth approximation to the maximum function. $\tilde{H}_\varepsilon$ is a smooth relaxation of the Heaviside function given by

$$\tilde{H}_\varepsilon(x) = \begin{cases} 0 & \text{if } x < -\varepsilon \\ \left[\frac{1}{2} + \frac{x}{2\varepsilon} + \frac{1}{2\pi} \sin(\frac{\pi x}{\varepsilon})\right]^p & \text{if } -\varepsilon \leq x \leq \varepsilon \\ 1 & \text{if } x > \varepsilon \end{cases} \quad (14)$$

where the parameter p indicates the sharpness of the approximation. Finally, the w_i^q denote weights that determine the relative contribution of bar q to the effective density for material i. These weights are defined in a similar manner to the DMO method (although using the bar size variables instead of element densities) as:

$$w_i^q = (\alpha_i^q) \prod_{j=1}^{N_m} (1 - (\alpha_{j \neq i}^q)) \quad (15)$$

Once we obtain the effective densities ρ_{eff}^i at point $\mathbf{p}$ for all materials $i \in \{1, \cdots, N_m\}$, the effective elastic tensor at point $\mathbf{p}$ is computed as:

$$\mathbb{C}(\mathbf{z}, \mathbf{p}) = \mathbb{C}_{min} + \sum_{i=1}^{N_m} (\mathbb{C}_i - \mathbb{C}_{min}) \rho_{eff}^i(\mathbf{z}, \mathbf{p}) \quad (16)$$

with $\mathbb{C}_i$ the elasticity tensor for material i and $\mathbb{C}_{min}$ the elasticity tensor of a weak isotropic material that ensures the analysis is well-posed. We note that the interpolation of material properties of Eq. 16, and the fact that the penalization of intermediate values of the material size variables and the mutual material exclusion is done through separate constraints (described in Sections 5.1 and 5.2), makes it easy to accommodate any number of materials in the design. This is in contrast to multi-material interpolation schemes that couple the material interpolation with the penalization and mutual material exclusion, making it more difficult to extend those schemes to more than two materials.

3 Homogenization

To compute the effective properties of the periodic lattice structure, we employ numerical homogenization (cf., [40–43]). The components of the effective elastic tensor $\mathbb{C}^H$ are computed as

$$C_{ijkl}^H = \frac{1}{|Y|} \int_Y C_{pqrs} (\varepsilon_{pq}^{0(ij)} - \varepsilon_{pq}^{*(ij)})(\varepsilon_{rs}^{0(kl)} - \varepsilon_{rd}^{*(kl)}) \, dy \quad (17)$$

where Y is the domain of the base cell, C_{pqrs} denotes the components of the elasticity tensor of Eq. 16, $\boldsymbol{\varepsilon}^{0(kl)} = \mathbf{e}_k \otimes \mathbf{e}_l$ correspond to six applied unit strains, and

$$\varepsilon_{pq}^{*(kl)} = \frac{1}{2} \left(\frac{\partial \chi_p^{(kl)}}{\partial y_q} + \frac{\partial \chi_q^{(kl)}}{\partial y_p} \right) \quad (18)$$

with $\boldsymbol{\chi}^{(kl)} \in \mathcal{U}_{adm}$ being the solutions to the six problems

$$\int_Y C_{ijpq} \frac{\partial \chi_p^{(kl)}}{\partial y_q} \frac{\partial v_i}{\partial y_j} \, dy = \int_Y C_{ijkl} \frac{\partial v_i}{\partial y_j} \, dy, \forall \mathbf{v} \in \mathcal{U}_{adm} \quad (19)$$

where $\mathbf{v}$ denotes the test function and $\mathcal{U}_{adm} := \{\mathbf{u} | \mathbf{u} \in H^1(Y), \mathbf{u} \text{ is } Y\text{-periodic}\}$ is the set of admissible solutions.

4 Symmetry

To attain desired material symmetries on the effective macrostructure, we allow the imposition of an arbitrary number of symmetry planes in the base cell. These planes define a reference region on which we define the bars that make up the lattice. Then, to compute the projected density at a point in a different region of the base cell, we reflect that point with respect to the appropriate symmetry planes so that its reflection ends up in the reference region, and thereafter we perform the geometry projection as usual. This strategy, which was employed in [39], is similar to the one introduced in [37]. We accordingly modify the sensitivities to account for these reflections. In the case when all symmetry planes pass through the origin of the coordinate system, we can reflect any point to the base section by multiplying all the reflection matrices corresponding to the symmetry planes that divide the base cell such that the original and reflected points lie within different sections. The reflected point is thus computed as

$$\hat{\mathbf{p}} := \prod_{s=1}^{N_s} \mathbf{R}_s \mathbf{p} \quad (20)$$

where N_s is the number of symmetry planes and $\mathbf{R}_s$ is the reflection matrix corresponding to symmetry plane s.

5 Optimization

In this paper we maximize the effective bulk modulus of the macrostructure subject to a resource constraint. Instead of using a volume fraction constraint and specifying an arbitrary volume fraction limit for each individual material, we impose a weight constraint that takes into account the physical densities of the materials as in [39]. To ensure that a) all size variables are either zero or unity, and b) at most one material in each bar has a size variable value of unity, we impose discreteness and mutual material exclusion constraints respectively. This is the strategy employed in [39] to design multi-material structures, and we briefly describe it in the next sections for completeness.

5.1 Discreteness Constraint

To facilitate manufacturing, we desire size variables to be either zero or unity, and therefore intermediate values

should be avoided. To this end, we impose the equality constraint

$$g_d(\mathbf{z}) := 4 \underset{i,q}{KS}(\boldsymbol{\alpha}^T(1-\boldsymbol{\alpha})) = 0 \quad (21)$$

where $KS(\mathbf{x})$ is the function defined in Eq. 13, and $\boldsymbol{\alpha} = [\boldsymbol{\alpha}_1^T \, \boldsymbol{\alpha}_2^T \dots \boldsymbol{\alpha}_{N_b}^T]^T$ is the vector of all size variables, with $\boldsymbol{\alpha}_q$ the vector of size variables of bar q for all N_m materials. Since the optimizer we employ in this work cannot incorporate equality constraints, we replace the constraint of Eq. 21 with the inequality constraint

$$g_d(\mathbf{z}) \le \varepsilon_d \ll 1 \quad (22)$$

A continuation strategy is employed to prevent the design from quickly getting stuck in poor local minima. This is done by gradually decreasing an initial value $\varepsilon_d^{(0)}$ by a step $\Delta\varepsilon_d$. We only start this decrease after the relative change in the objective function in consecutive iterations is less than a specified value Δf^*, i.e.:

$$\text{If } \Delta f^{(I+1)} \le \Delta f^* \text{ then } \varepsilon_d^{(I+1)} \leftarrow \max(\varepsilon_d^{(I)} - \Delta\varepsilon_d, \varepsilon_d^*) \quad (23)$$

with ε_d^* the final constraint limit and $\Delta f^{(I+1)} := \frac{|f^{(I+1)} - f^{(I)}|}{f^{(I)}}$ the relative change in the objective function at iteration $I+1$.

5.2 Mutual Material Exclusion Constraint

We also desire bars to be exclusively made of a single material (or to be void). To ensure a bar is made of at most one material, we impose the constraint

$$g_m(\mathbf{z}) := \underset{q}{KS}\left(\sum_{i=1}^{N_m} \alpha_i^q\right) - 1 \le 0 \quad (24)$$

We employ a similar continuation strategy to the one afore described for the discreteness constraint.

5.3 Optimization Problem

The bulk maximization problem is given by:

$$\max_{\mathbf{z}} K(\mathbf{z}) := \frac{1}{9}(C_{1111} + C_{1122} + C_{1133} + C_{2211} + C_{2222}$$

$$+ C_{2233} + C_{3311} + C_{3322} + C_{3333}) \qquad (25)$$

subject to

$$w_f := \frac{1}{\gamma_{ref}|\Omega|} \sum_{i=1}^{N_m} \gamma_i \int_{\Omega} \rho_{eff}^i(\mathbf{z}, \mathbf{p}) dv \leq w_f^* \qquad (26)$$

$$a(\mathbf{u}^{(kl)}(\mathbf{z}), \mathbf{v}) = l(\mathbf{v}, \boldsymbol{\varepsilon}^{0(kl)}), \forall \mathbf{v} \in \mathcal{U}_0, \mathbf{u}^{(kl)} \in \mathcal{U} \qquad (27)$$

$$g_d(\mathbf{z}) \leq \varepsilon_d^{(I)} \qquad (28)$$

$$g_m(\mathbf{z}) \leq \varepsilon_m^{(I)} \qquad (29)$$

$$\mathbf{x}_{q_0}, \mathbf{x}_{q_f} \in \Omega \qquad (30)$$

$$0.0 \leq \alpha_i^q \leq 1.0 \qquad (31)$$

where Ω is the region occupied by the the design envelope, and $\omega \subseteq \Omega$ is the region occupied by the design. γ_i is the physical density for material i, γ_{ref} is a reference density chosen to be unity, and w_f is the weight fraction. The displacement $\mathbf{u}^{(kl)}$ for the applied unit strain $\boldsymbol{\varepsilon}^{0(kl)}$, $k, l = 1, \cdots, 3$, is obtained as the solution to the analysis problem of Eq. 27, with $\mathbf{v}$ denoting the test function. The admissible sets for trial and test functions are defined as $\mathcal{U} := \{\mathbf{u} | \mathbf{u} \in H^1(\Omega), \mathbf{u} \text{ is } Y\text{-periodic}, \mathbf{u}(\mathbf{c}) = \mathbf{0}\}$ and $\mathcal{U}_0 := \{\mathbf{v} | \mathbf{v} \in H^1(\Omega), \mathbf{v}|_{\Gamma} = \mathbf{0}, \mathbf{v}(\mathbf{c}) = \mathbf{0}\}$, respectively. We impose zero displacements at the center of the unit cell $\mathbf{c}$ to prevent rigid-body motions. Finally, a and l in Eq. 27 are the energy bilinear and load linear forms respectively given by:

$$a(\mathbf{u}, \mathbf{v}) := \int_{\Omega} \nabla \mathbf{v} \cdot \mathbb{C}(\mathbf{z}, \mathbf{p}) \nabla \mathbf{u} \, dv \qquad (32)$$

$$l(\mathbf{v}, \boldsymbol{\varepsilon}) := \int_{\Omega} \nabla \mathbf{v} \cdot \mathbb{C}(\mathbf{z}, \mathbf{p}) \boldsymbol{\varepsilon} \, dv \qquad (33)$$

As in our previous works [35, 36, 39], we scale the design variables so that each scaled variable $\hat{z}$ falls within the range $[0, 1]$, and impose a move limit m on each design variable at each optimization iteration I as:

$$\max(0, \hat{z}^{(I-1)} - m) \leq \hat{z}^{(I)} \leq \min(1, \hat{z}^{(I-1)} + m) \qquad (34)$$

6 Examples

We present numerical examples to demonstrate the effectiveness of our method. We implement it in C++ using the deal.II library [44, 45], which provides significant functionality for the finite element analysis as well as interfaces for parallel linear algebra libraries and for post-processing. We consider homogeneous, isotropic, and linearly elastic materials with Poisson's ratio $\nu = 0.3$, and with different material densities and elastic moduli. In all examples, we use hexahedral, trilinear elements.

We use the method of moving asymptotes (MMA) [46,47] as our optimizer, using the default parameters in [47]. In particular, we employ the MPI implementation of MMA in [48]. We set $\varepsilon_d^0 = 1.0$ and $\varepsilon_d^* = 0.01$ for the discreteness constraint, and $\varepsilon_m^0 = 0.3$ and $\varepsilon_m^* = 0.01$ for the mutual material exclusion constraint. The stopping criterion is the relative change in compliance between consecutive iterations, $\Delta f^* \leq 10^{-3}$. The power $p = 2$ is used in the smooth Heaviside approximation of Eq. 14. We use $k = 25$ for the KS approximation of Eq. 13. We use a move limit $m = 0.1$ in Eq. 34. All examples are run using six compute nodes with 24 Intel Haswell cores each, CPU speed of 2.59 GHz and 128 GB of memory per node.

6.1 Two-material Lattices with Maximal Bulk Modulus and Orthotropic Symmetry

In this example we consider three orthogonal planes of symmetry to obtain an orthotropic design. These planes are the ones that pass through the center of the unit cell and are perpendicular to the faces. The initial design consists of bars of width $w = 0.1$ and with almost zero length, which become essentially spheres (a zero length is not used because it causes the sensitivities to not be defined, as demonstrated in [35]). Fig. 3a shows the initial design after applying symmetry, which consists of 125 spheres. Two materials with Young's moduli $E_1 = 10$ and $E_2 = 5$, and physical densities $\gamma_1 = 0.9$ and $\gamma_2 = 0.45$ are available for the design. For all bars, the size variables corresponding to these two materials are set to $\alpha_1^q = \alpha_2^q = 0.5$ in the initial design. The unit cell is meshed with $64 \times 64 \times 64$ elements. We perform the op-

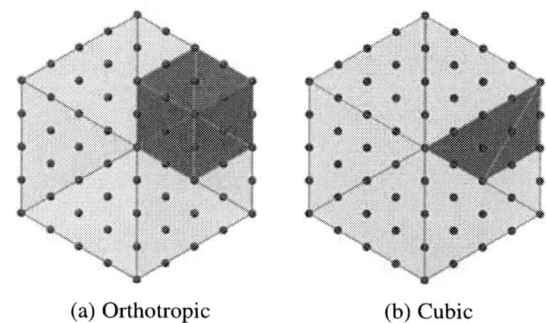

(a) Orthotropic   (b) Cubic

Fig. 3: Unit cell with initial design (spheres) after reflection with respect to material symmetry planes. Red regions indicate regions that are reflected.

timization for weight fraction limits of $0.01, 0.02, \cdots, 0.05$, and present the results in Table 1. For this problem and these weight fraction limits, the optimization produces two-material designs. We also observe that small changes in the weight fraction limit may lead to entirely different designs. For instance, we present results for a narrower weight fraction limit range of $w_f^* \in \{0.020, 0.021, \cdots, 0.030\}$ in Table 2. As this table shows, even those smaller changes may produce different unit cell designs. We also observe that, as expected, as we increase the weight fraction limit the optimal bulk modulus increases. There are three exceptions to this trend (the designs with $w_f^* = 0.024$, $w_f^* = 0.026$ and $w_f^* = 0.029$), which we attribute to premature convergence to a suboptimal local minimum. To escape entrapment from these poor minima, we could employ the tunneling technique recently proposed in [49]; however, this is outside the scope of this paper and deferred to future work. Besides the fact that the geometry projection technique can get entrapped in a suboptimal local minimum, it is also apparent from these examples that this problem has many local minima.

6.2 Two- and Three-material Lattices with Maximal Bulk Modulus and Cubic Symmetry

In this section we investigate maximal bulk modulus designs for lattices made of two and three materials, and with cubic symmetry. To ensure the cubic symmetry, we define nine symmetry planes, which correspond to the three planes for orthotropic symmetry previously described, plus the six planes that partition the cube into two equal triangular prisms. The bar dimensions and initial design are the same as in the previous example. For the two-material designs, the elastic moduli and physical densities are also the same. We again use a $64 \times 64 \times 64$ elements mesh for the analysis.

We perform the optimization for weight fraction limits $w_f^* \in \{0.01, 0.02, \cdots, 0.1\}$ and present the corresponding results in the left-hand section of Table 3. To show that, as in the previous example, very different designs are obtained with small changes of the weight fractions, we also perform the optimization for weight fraction limits in the range $w_f^* \in [0.021 - -0.039]$, shown in the center and right-hand sections of 3. As before, the overall trend is that as the weight fraction limit increases, the effective bulk modulus increases as well. There are several exceptions to this trend, which we again attribute to convergence to different local minima. And once again, we note that this problem seems to exhibit many local minima, as small perturbations to the weight fraction limit lead to substantially different designs.

Another worthwhile observation is that our formulation to impose the material symmetries does not prevent the possibility that a bar is cut by the boundaries of the unit cell or the symmetry planes in such a way that the reflection of bars in the reference region produces a material distribution that is not distinctly made of bars. For example, the design corresponding to the weight fraction limit $w_f^* = 0.05$ in Table 3 cuts the bars made of material 1 on the faces of the unit cell in such a way that they form a thin wall; in fact, this particular result is interesting in that, by cutting the bars in this way, the optimizer produces a near-closed-cell structure. In light of this, it would be advantageous to introduce some mechanism in our formulation to ensure that struts remain whole upon reflection. A technique to achieve this by imposing constraints on the shortest distance between a bar and its reflections is demonstrated in [37]. This, however, is outside of the scope of the current work and we defer it to the future.

To demonstrate the flexibility of our method in incorporating multiple materials, we repeat the previous problem but add a third material with elastic modulus $E_3 = 7.5$ and physical density $\gamma_3 = 0.675$. We perform the optimization for four values of the weight fraction limit, $w_f^* = \{0.04, 0.044, 0.047, 0.05\}$. We choose these values because other values in the range $[0.04, 0.05]$ that we experimented with rendered designs made of only two materials. The resulting designs are shown in Table 4.

For the design with $w_f^* = 0.05$, we also show the history of the design in Table 5. The three rightmost columns show the design at different iterations during the optimization, with the colors indicating the size variables values for each of the three available materials. This plot is interesting in that it illustrates two aspects of this particular problem. First, due to the reflections to ensure cubic symmetry, the ten bars in the reference region result in 125 bars upon reflection, which, as seen in Table 5, can lead to complex lattice structures. Second, as seen in the last row of Table 5 (corresponding to the last iteration in the optimization), there are some bars that attain a near-zero value for α_i^q, $i = 1, 2, 3$, which means they have been entirely removed from the design. The ability to entirely remove bars from the design is an advantage of the geometry projection formulation that facilitates the removal of unneeded bars.

The convergence plots for this problem are provided in Fig. 4. Fig. 4a shows the objective function iteration history. The abrupt increase in the objective function occurs when the continuation strategy starts and begins to tighten the limits on the constraints. Although the objective function value for the final design is higher than some of the values achieved in preceding iterations, these designs are infeasible as they do not satisfy the final tight limits on the discreteness and mutual material exclusion constraints. The iteration history of the constraints is shown in Fig. 4b, where we observe all three constraints being satisfied by the final design.

7 Conclusions

This work presented a topology optimization method for the maximum bulk modulus design of multi-material lattice structures. Each lattice strut can be made of exclusively one material out of a set of available materials with different elastic moduli and physical densities, or be altogether removed. The numerical examples demonstrate the effectiveness of the method. The three-material design examples also show that the proposed formulation facilitates the incorporation of more than two available materials. Although the designs produced by our method still pose some fabrication challenges, the fact that the lattice is an open-cell structure arguably makes it less difficult to manufacture than some of the organic multi-material designs rendered by free-form (density-based and level set) topology optimization techniques, which may produce fully enclosed cavities and complex boundaries between material phases. Future work will be devoted to incorporating geometric considerations to overcome fabrication challenges.

K	w_f^*	iso	side	top	front
0.02172	0.01				
0.03992	0.02				
0.05835	0.03				
0.07905	0.04				
0.09840	0.05				

Table 1: Maximal bulk modulus designs for orthotropic two-material lattices and weight fraction limits $w_f^* \in \{0.01, 0.02, \cdots, 0.05\}$. Red bars are made of material 1, blue bars are made of material 2, and bars that have been removed from the design (i.e., with $\alpha_1^q, \alpha_2^q \approx 0$) are not shown.

K	w_f^*	iso	side	top	front
0.03930	0.021				
0.04435	0.022				
0.04816	0.023				
0.04587	0.024				
0.04951	0.025				
0.03881	0.026				
0.05270	0.027				
0.05672	0.028				
0.05496	0.029				

Table 2: Maximal bulk modulus designs for orthotropic two-material lattices and weight fraction limits $w_f^* \in \{0.020, 0.021, \cdots, 0.030\}$. Red bars are made of material 1, blue bars are made of material 2, and bars that have been removed from the design (i.e., with $\alpha_1^q, \alpha_2^q \approx 0$) are not shown.

K	w_f^*	K	w_f^*	K	w_f^*
0.05485	0.031	0.04338	0.021	0.01041	0.01
0.06281	0.032	0.04623	0.022	0.04075	0.02
0.05600	0.033	0.04915	0.023	0.07492	0.03
0.07518	0.034	0.05709	0.024	0.07850	0.04
0.09180	0.035	0.05347	0.025	0.15103	0.05
0.07802	0.036	0.05691	0.026	0.13232	0.06
0.07820	0.037	0.05498	0.027	0.17170	0.07
0.07847	0.038	0.05816	0.028	0.18664	0.08
0.07404	0.039	0.04986	0.029	0.15468	0.09

Table 3: Maximal bulk modulus designs for cubic two-material lattices with different weight fraction limits. Red bars are made of material 1, blue bars are made of material 2, and bars that have been removed from the design (i.e., with $\alpha_1^q, \alpha_2^q \approx 0$) are not shown.

Acknowledgements

Support from the National Science Foundation, award CMMI-1634563 to conduct this work is gratefully acknowledged. This work was also partially supported by a fellowship grant from GE?s Industrial Solutions Business Unit under a GE-UConn partnership agreement. The views and conclusions contained in this document are those of the authors and should not be interpreted as necessarily representing the official policies, either expressed or implied, of Industrial Solutions or UConn. We also thank Prof. Krister Svanberg for kindly providing his MMA Matlab optimizer to perform the optimization.

References

[1] O. Sigmund, Materials with prescribed constitutive parameters: an inverse homogenization problem, International Journal of Solids and Structures 31 (17) (1994) 2313–2329.

[2] O. Sigmund, Tailoring materials with prescribed elastic properties, Mechanics of Materials 20 (4) (1995) 351–368.

[3] M. Neves, H. Rodrigues, J. M. Guedes, Optimal design of periodic linear elastic microstructures, Computers & Structures 76 (1-3) (2000) 421–429.

[4] O. Sigmund, S. Torquato, I. A. Aksay, On the design of 1–3 piezocomposites using topology optimization, Journal of materials research 13 (4) (1998) 1038–1048.

[5] N. de Kruijf, S. Zhou, Q. Li, Y.-W. Mai, Topological design of structures and composite materials with multiobjectives, International Journal of Solids and Structures 44 (22-23) (2007) 7092–7109.

[6] S. J. Cox, D. C. Dobson, Band structure optimization of two-dimensional photonic crystals in h-polarization, Journal of Computational Physics 158 (2) (2000) 214–224.

[7] J. K. Guest, J. H. Prévost, Design of maximum permeability material structures, Computer Methods in Applied Mechanics and Engineering 196 (4-6) (2007) 1006–1017.

[8] J. K. Guest, J. H. Prévost, Optimizing multifunctional materials: design of microstructures for maximized stiffness and fluid permeability, International Journal of Solids and Structures 43 (22-23) (2006) 7028–7047.

[9] S. Torquato, S. Hyun, A. Donev, Multifunctional composites: optimizing microstructures for simultaneous transport of heat and electricity, Physical review letters 89 (26) (2002) 266601.

[10] M. Osanov, J. K. Guest, Topology optimization for architected materials design, Annual Review of Materials Research 46 (2016) 211–233.

[11] S. Zhou, W. Li, G. Sun, Q. Li, A level-set procedure for the design of electromagnetic metamaterials, Optics express 18 (7) (2010) 6693–6702.

[12] V. Challis, A. Roberts, A. Wilkins, Design of three dimensional isotropic microstructures for maximized stiffness and conductivity, International Journal of Solids and Structures 45 (14-15) (2008) 4130–4146.

[13] M. Otomori, T. Yamada, K. Izui, S. Nishiwaki, N. Kogiso, Level set-based topology optimization for the design of light-trapping structures, IEEE Transactions on Magnetics 50 (2) (2014) 729–732.

[14] R. Picelli, R. Sivapuram, S. Townsend, H. A. Kim, Stress topology optimisation for architected material using the level set method, in: World Congress of Structural and Multidisciplinary Optimisation, Springer, 2017, pp. 1254–1269.

[15] X. Huang, Y. Xie, Optimal design of periodic structures using evolutionary topology optimization, Structural and Multidisciplinary Optimization 36 (6) (2008) 597–606.

[16] X. Huang, A. Radman, Y. Xie, Topological design of microstructures of cellular materials for maximum bulk or shear modulus, Computational Materials Science 50 (6) (2011) 1861–1870.

[17] X. Huang, Y. M. Xie, B. Jia, Q. Li, S. Zhou, Evolutionary topology optimization of periodic composites for extremal magnetic permeability and electrical permittivity, Structural and Multidisciplinary Optimization 46 (3) (2012) 385–398.

[18] A. Radman, X. Huang, Y. Xie, Topological optimization for the design of microstructures of isotropic cellular materials, Engineering optimization 45 (11) (2013) 1331–1348.

[19] H. Rodrigues, J. M. Guedes, M. Bendsoe, Hierarchical optimization of material and structure, Structural and Multidisciplinary Optimization 24 (1) (2002) 1–10.

[20] J. Guedes, E. Lubrano, H. Rodrigues, S. Turteltaub, Hierarchical optimization of material and structure for thermal transient problems, in: IUTAM Symposium on Topological Design Optimization of Structures, Machines and Materials, Springer, 2006, pp. 527–536.

[21] P. Coelho, P. Fernandes, J. Guedes, H. Rodrigues, A hierarchical model for concurrent material and topology optimisation of three-dimensional structures, Structural and Multidisciplinary Optimization 35 (2) (2008) 107–115.

[22] L. Liu, J. Yan, G. Cheng, Optimum structure with homogeneous optimum truss-like material, Computers & Structures 86 (13-14) (2008) 1417–1425.

[23] B. Niu, J. Yan, G. Cheng, Optimum structure with homogeneous optimum cellular material for maximum fundamental frequency, Structural and Multidisciplinary Optimization 39 (2) (2009) 115.

[24] J. Yan, G.-d. Cheng, L. Liu, A uniform optimum material based model for concurrent optimization of thermoelastic structures and materials, International Journal for Simulation and Multidisciplinary Design Optimization 2 (4) (2008) 259–266.

[25] J. Deng, J. Yan, G. Cheng, Multi-objective concurrent topology optimization of thermoelastic structures composed of homogeneous porous material, Structural and Multidisciplinary Optimization 47 (4) (2013) 583–597.

[26] X. Huang, S. Zhou, Y. Xie, Q. Li, Topology optimization of microstructures of cellular materials and composites for macrostructures, Computational Materials

K	w_f^*	iso	side
0.087540	0.04		
0.082621	0.044		
0.095888	0.047		
0.11972	0.05		

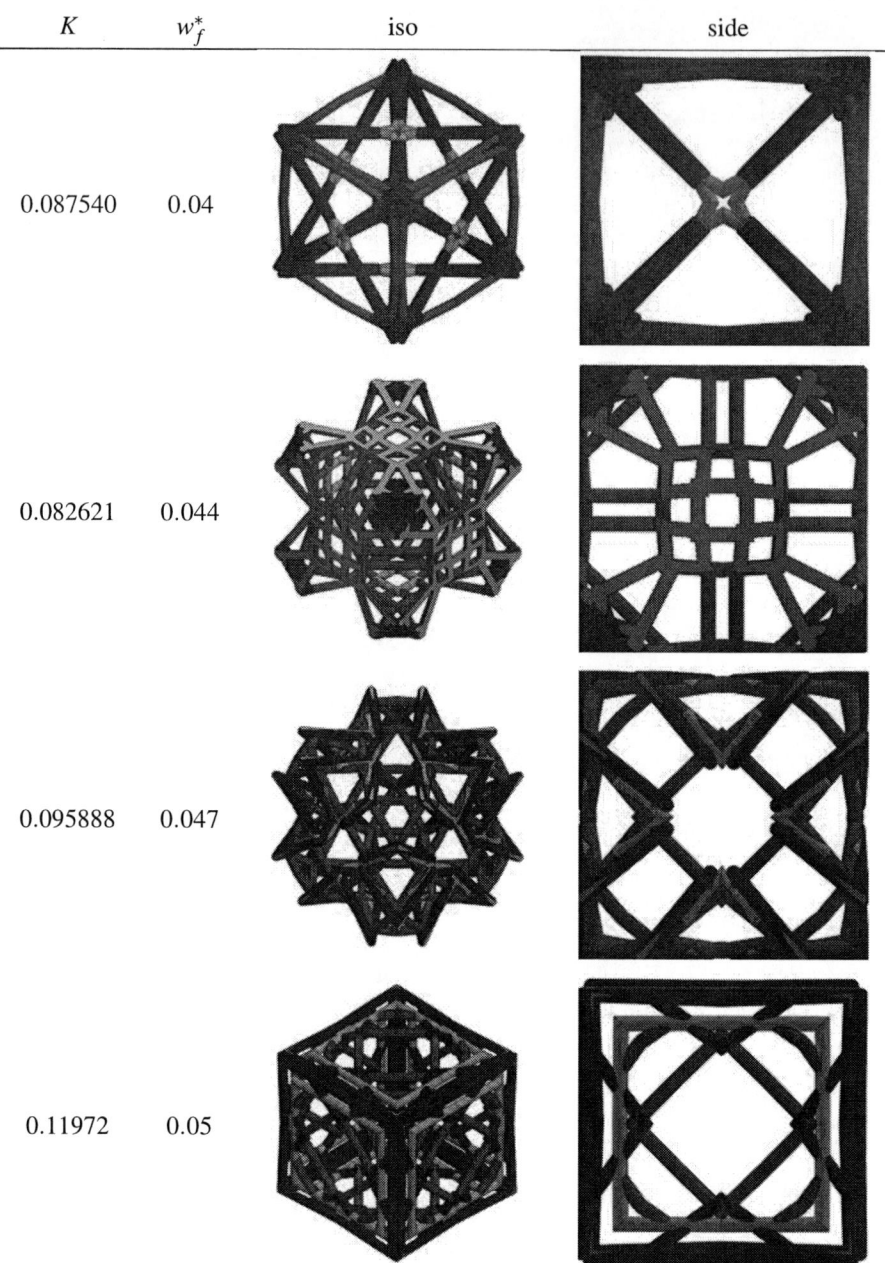

Table 4: Maximal bulk modulus designs for cubic three-material lattices with different weight fraction limits. Red bars are made of material 1, green bars are made of material 2, blue bars are made of material 3, and bars that have been removed from the design (i.e., with $\alpha_1^q, \alpha_2^q, \alpha_3^q \approx 0$) are not shown.

Science 67 (2013) 397–407.

[27] X. Yan, X. Huang, Y. Zha, Y. Xie, Concurrent topology optimization of structures and their composite microstructures, Computers & Structures 133 (2014) 103–110.

[28] O. Sigmund, S. Torquato, Design of materials with extreme thermal expansion using a three-phase topology optimization method, Journal of the Mechanics and Physics of Solids 45 (6) (1997) 1037–1067.

[29] L. V. Gibiansky, O. Sigmund, Multiphase composites with extremal bulk modulus, Journal of the Mechanics

and Physics of Solids 48 (3) (2000) 461–498.

[30] S. Zhou, Q. Li, The relation of constant mean curvature surfaces to multiphase composites with extremal thermal conductivity, Journal of Physics D: Applied Physics 40 (19) (2007) 6083.

[31] S. Zhou, Q. Li, Computational design of multi-phase microstructural materials for extremal conductivity, Computational Materials Science 43 (3) (2008) 549–564.

[32] M. P. Bendsøe, O. Sigmund, Material interpolation schemes in topology optimization, Archive of applied

Iteration	material 1	material 2	material 3
0			
10			
25			
50			
100			
150			
236			

Table 5: Some design iterations for the maximal bulk modulus design of a cubic three-material lattice with weight fraction limit $w_f^* = 0.05$. The three rightmost columns show the same geometric description of the bars at the corresponding iterations; the color scale in those columns, however, indicates the value of the size variable α_i^q, with $i = 1, 2, 3$ denoting materials 1 (red), 2 (green) and 3 (blue) respectively. In the last iteration, bars that do not have a solid color (red, green or blue) have $\alpha_1^q, \alpha_2^q, \alpha_3^q \approx 0$ and have been removed from the design.

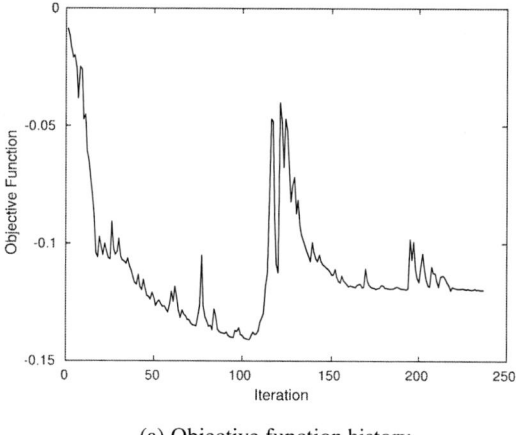 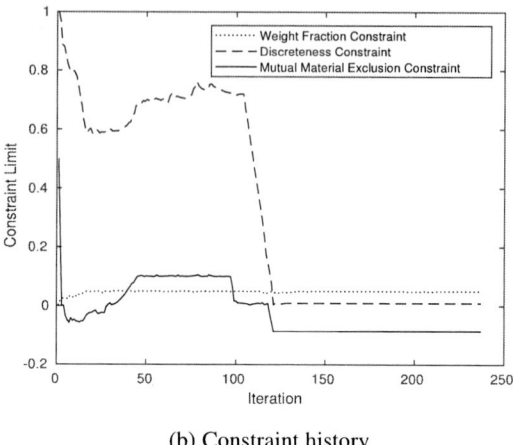

(a) Objective function history (b) Constraint history

Fig. 4: Iteration history for the objective and constraint functions for the maximal bulk modulus design of a cubic three-material lattice with weight fraction limit $w_f^* = 0.05$.

mechanics 69 (9-10) (1999) 635–654.

[33] M. Stolpe, K. Svanberg, An alternative interpolation scheme for minimum compliance topology optimization, Structural and Multidisciplinary Optimization 22 (2) (2001) 116–124.

[34] B. Bell, J. Norato, D. Tortorelli, A geometry projection method for continuum-based topology optimization of structures, in: 12th AIAA Aviation Technology, Integration, and Operations (ATIO) Conference and 14th AIAA/ISSMO Multidisciplinary Analysis and Optimization Conference, 2012, p. 5485.

[35] J. Norato, B. Bell, D. Tortorelli, A geometry projection method for continuum-based topology optimization with discrete elements, Computer Methods in Applied Mechanics and Engineering 293 (2015) 306–327.

[36] S. Zhang, J. A. Norato, A. L. Gain, N. Lyu, A geometry projection method for the topology optimization of plate structures, Structural and Multidisciplinary Optimization 54 (5) (2016) 1173–1190.

[37] S. Watts, D. A. Tortorelli, A geometric projection method for designing three-dimensional open lattices with inverse homogenization, International Journal for Numerical Methods in Engineering.

[38] J. Stegmann, E. Lund, Discrete material optimization of general composite shell structures, International Journal for Numerical Methods in Engineering 62 (14) (2005) 2009–2027.

[39] H. Kazemi, A. Vaziri, J. A. Norato, Topology optimization of structures made of discrete geometric components with different materials, Journal of Mechanical Design 140 (11) (2018) 111401.

[40] J. Guedes, N. Kikuchi, Preprocessing and postprocessing for materials based on the homogenization method with adaptive finite element methods, Computer methods in applied mechanics and engineering 83 (2) (1990) 143–198.

[41] B. Hassani, E. Hinton, A review of homogenization and topology optimization i—homogenization theory for media with periodic structure, Computers & Structures 69 (6) (1998) 707–717.

[42] B. Hassani, E. Hinton, A review of homogenization and topology opimization ii—analytical and numerical solution of homogenization equations, Computers & structures 69 (6) (1998) 719–738.

[43] B. Hassani, E. Hinton, A review of homogenization and topology optimization iii—topology optimization using optimality criteria, Computers & structures 69 (6) (1998) 739–756.

[44] W. Bangerth, R. Hartmann, G. Kanschat, deal.II – a general purpose object oriented finite element library, ACM Trans. Math. Softw. 33 (4) (2007) 24/1–24/27.

[45] G. Alzetta, D. Arndt, W. Bangerth, V. Boddu, B. Brands, D. Davydov, R. Gassmoeller, T. Heister, L. Heltai, K. Kormann, M. Kronbichler, M. Maier, J.-P. Pelteret, B. Turcksin, D. Wells, The `deal.II` library, version 9.0, Journal of Numerical Mathematics 26 (4) (2018) 173–183. `doi:10.1515/jnma-2018-0054`.

[46] K. Svanberg, A class of globally convergent optimization methods based on conservative convex separable approximations, SIAM Journal on Optimization 12 (2) (2002) 555–573.

[47] K. Svanberg, MMA and GCMMA, versions september 2007, Optimization and Systems Theory (2007) 104.

[48] N. Aage, E. Andreassen, B. S. Lazarov, O. Sigmund, Giga-voxel computational morphogenesis for structural design, Nature 550 (7674) (2017) 84.

[49] S. Zhang, J. A. Norato, Finding better local optima in topology optimization via tunneling, in: ASME 2018 International Design Engineering Technical Conferences and Computers and Information in Engineering Conference, American Society of Mechanical Engineers, 2018, pp. V02BT03A014–V02BT03A014.

Proceedings of the ASME 2019
International Design Engineering Technical Conferences
and Computers and Information in Engineering Conference
IDETC/CIE2019
August 18-21, 2019, Anaheim, CA, USA

DETC2019-97390

INVERSE THERMO-MECHANICAL PROCESSING (ITMP) DESIGN OF A STEEL ROD DURING HOT ROLLING PROCESS

Anand Balu Nellippallil[1], Pranav Mohan[2], Janet K. Allen[2*], Farrokh Mistree[2]

[1]Center for Advanced Vehicular Systems
Mississippi State University, Starkville, MS, USA

[2]The Systems Realization Laboratory @ OU
The University of Oklahoma, Norman, OK, USA

ABSTRACT

The production of steel products involves a series of manufacturing processes. The material Thermo-Mechanical Processing (TMP) history at each process affects the final properties and performances of the product. Experiments and plant trials to predict these properties and performance of steel products are expensive and time consuming. This has resulted in the need for computational design methods and tools that support a human designer in realizing such complex systems involving the material, product and manufacturing processes from a simulation-based design perspective.

In this paper, we present a Goal-oriented Inverse Design method to achieve the integrated design exploration of materials, products and manufacturing processes. The key functionality offered is the capability to carry out a microstructure-mediated design satisficing specific processing requirements and performance goals of the product. Given models to establish the information flow chain, a designer can use the method for the decision-based design exploration of material microstructure and processing paths to realize products in a manufacturing process chain. The efficacy of the method is tested using an industry-inspired hot rolling problem to inversely design the thermo-mechanical processing of a steel rod. The focus here is the method and associated design constructs which are generic and support the formulation and decision-based design of similar problems involving materials, products and associated manufacturing processes.

Keywords: Integrated Design of Materials, Products and Manufacturing Processes

1. FRAME OF REFERENCE

Steel research is a growing field as there is a constant drive towards producing newer grades of steel with improved properties and performance. Hot rolling is one major manufacturing process in the steel manufacturing process chain for the production of semi-products like rods and sheets. A round rod produced after hot rolling is processed further to produce automotive components like gears. Careful controlling and managing of the thermal and deformation conditions in a metal forming process to obtain desired microstructures and mechanical properties is termed as Thermo-Mechanical Processing (TMP) [1]. The material Thermo-Mechanical Processing (TMP) history carried out at each process influences the final properties and performances of the end product. Plant trials are one way of predicting these properties and performances of products. However, these plant trials are typically expensive and time consuming. Further, they do not support incorporating design changes at early stages of design. Due to these limitations, there is an increasing demand to address this problem from a simulation-based design perspective with the idea of incorporating design changes at early stages of design. From a simulation-based design perspective, there are two types of approaches in realizing a material-product system. The first, is a bottom-up cause and effect approach where material scientists focus on accurately predicting the chemistry-processing-microstructure-property-performance relationships to realize the end product. The major challenge in this focus is the modeling and linking of material phenomena at multiple length and time scales using domain specific tools, defined as multiscale modeling [2]. The second, is a systems-based goals/means approach where systems designers focus on achieving system-level goals by developing domain independent

---

* Corresponding Author, E-mail: janet.allen@ou.edu

Copyright © 2019 ASME

design methods and tools that support the integrated "design exploration" of the materials, product and manufacturing processes [3]. To fully realize a complex material-product system, an integration of both these approaches must happen – which is the interface of design and engineering sciences. We focus on the second approach and view design as a goal-oriented synthesis activity in which a human designer aims at identifying material chemistry, structures and processing paths that satisfice certain required product-level properties and performances. The role of the human designer here is that of a decision maker and is required to make decisions given the information available. Material models are typically incomplete, inaccurate and not of equal fidelity. Added to this is the randomness associated with material microstructure. All these are sources of uncertainty under which the decision maker/designer must make decisions. Due to these reasons, we focus on identifying 'satisficing' design solutions that show good performance under the presence of uncertainty rather than optimum solutions that hold for narrow conditions, while failing under the presence of uncertainty. As part of the *satisficing* community our intention from the design side is to satisfice a set of conflicting goals [4].

In this paper, we present a Goal-oriented Inverse Design (GoID) method to achieve the integrated design exploration of materials, products and manufacturing processes. The key functionality offered via this method is the capability to carry out a microstructure-mediated design satisficing specific processing requirements and performance goals of the product. Given models to establish the information flow chain, the method supports the decision-based design exploration of material microstructure and processing paths to realize products in a manufacturing process chain. The mathematical construct used in the method to support a designer in formulating the design problem and generating satisficing solutions under uncertainty is the compromise Decision Support Problem (cDSP). The formulation and exercising of the cDSP to generate satisficing design solutions is done in a systematic manner using the Concept Exploration Framework (CEF). The efficacy of the method is tested using an industry-inspired hot rolling problem to design a rod with target microstructure. The method and the associated design constructs are generic and supports the formulation and decision-based design exploration of similar problems involving materials, products and associated manufacturing processes.

The outline of the paper is as follows. In Section 2, we describe the hot rod rolling problem addressed in this paper. In Section 3, we describe the Concept Exploration Framework and the cDSP construct. In Section 4, we present the Goal-oriented Inverse Method and its application to hot rolling of rod. The mathematical models used for the hot rolling design problem are described in Section 5. In Section 6, the mathematical formulation of the cDSP for the problem is presented. The exploration of the solution space using ternary plots is covered in Section 7. We end the paper with our closing remarks in Section 8.

2. PROBLEM DESCRIPTION – INVERSE THERMOMECHANICAL PROCESSING (ITMP) DESIGN

One major application of TMP in hot rolling is to obtain fine grains of austenite during hot deformation. The fine grains of austenite formed during hot rolling support the formation of fine grains of ferrite after phase transformation during the cooling process that follows. These refined grains of ferrite formed from fine austenite grains support improved mechanical properties of hot rolled products like strength and toughness. Our interest here lies in solving an inverse problem, that is, given the target end mechanical properties and material microstructures that the hot rolled product needs to satisfy, how to inversely design the thermo-mechanical processing during hot rolling so as to identify the austenite grain sizes and the hot rolling processing conditions that meets these end targets. We term this as the Inverse Thermo-Mechanical Processing (ITMP) design problem. The specific industry-inspired problem that we are addressing is described next.

The big picture of the industry-inspired problem that we are addressing is described in detail in [5] and is not repeated here. The end goals are to minimize the effects of microsegregates by managing banded microstructures of ferrite and pearlite and achieve specified target values of the mechanical properties like yield strength, tensile strength, hardness and toughness for the end rod product. Based on our inverse analysis of this problem from a goal-oriented perspective in [6], we conclude that to achieve the target mechanical properties and manage banded microstructure, a specific microstructure with a high ferrite fraction, low ferrite grain size, low pearlite interlamellar spacing is preferred after cooling stage. Further the design of microstructure after rolling is carried out and we conclude that to achieve the specific requirements for a low ferrite grain size and satisfy the mechanical property requirements, a low final austenite grain size is preferred after rolling. In this paper, we address the problem dealing with hot rolling thermo-mechanical processing where the end goal is to achieve the minimum final austenite grain size (AGS) after rolling.

In order to predict the final AGS after rolling several phenomena involved in the restoration of the microstructure need to be considered. These include dynamic recovery, dynamic recrystallization, metadynamic recrystallization, static recovery, static recrystallization, and grain growth after recrystallization [7]. To frame a boundary for the problem we focus on static recrystallization and grain growth that happens in a hot rolled product in the final rolling pass. We assume that the strain values are less than the critical values for initiating dynamic or metadynamic recrystallization in this study. Once the grains are fully recrystallized, grain growth happens and defines the final austenite grain size after rolling. Hence there is a requirement for recrystallized fraction to achieve close to 100% to initiate grain growth with a minimum requirement of at least 95% to satisfy full recrystallization.

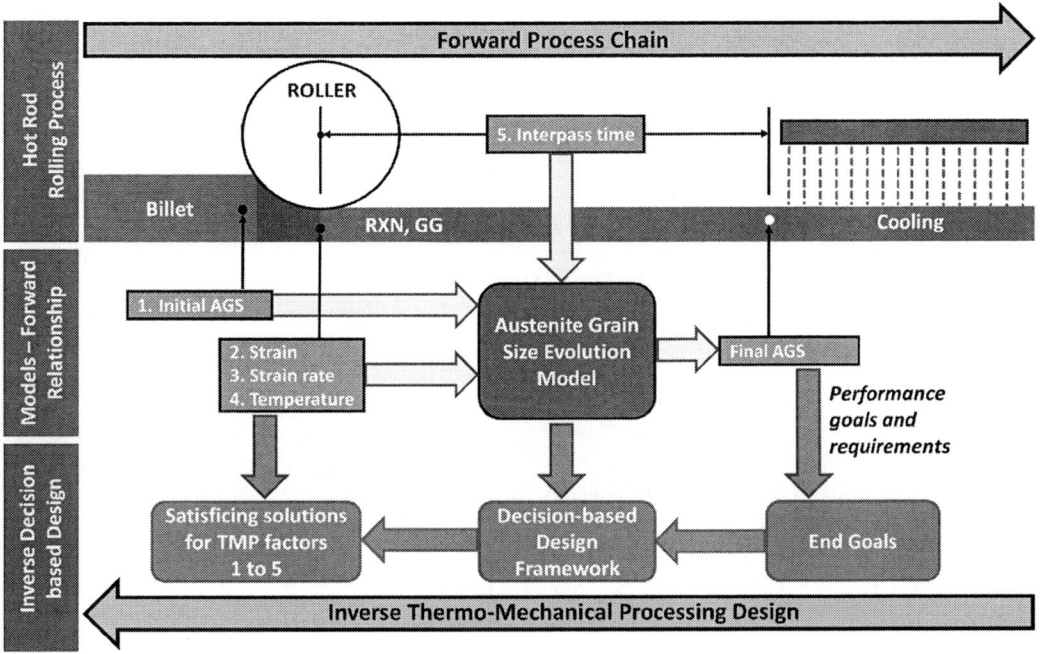

FIGURE 1: INVERSE THERMO-MECHANICAL PROCESSING (ITMP) DESIGN PROBLEM

Another important factor contributing to the microstructural phenomena during hot rolling is the grain surface to grain volume ratio (S_v). Increase in S_v tends to decrease ferrite grain size and thereby support ferrite refinement [8]. The requirement therefore based on literature study is to achieve a target S_v value of 30 mm^{-1} with a desire to explore designs that support higher values of S_v, if possible. In Figure 1, we illustrate the key aspects of the Inverse Thermo-Mechanical Processing design problem. Three aspects of the design problem are shown in the figure. In the upper portion of Figure 1, we pictorially represent the hot rod rolling process. We represent the input material to the rolling unit as the billet which is typically the output of a casting unit. However, the input material for our problem could also be the thermo-mechanically processed semi-rod from a previous rolling unit if we focus on just the rolling processes. To simplify and bound our design problem, we focus on just one rolling pass with the assumption that the input is a billet with an initial Austenite Grain Size (AGS) that is further hot rolled to a rod and sent to the cooling unit where phase transformation of austenite to ferrite, pearlite, etc. occurs. The key thermo-mechanical processing factors (TMP factors) influencing AGS evolution and recrystallization are identified as strain, strain rate, initial AGS, temperature and interpass time [9]. Mathematical models that establish relationships between the identified factors and the problem specific requirements for final AGS, recrystallized fraction and grain surface to volume ratio are identified. Using these models and their integration, specific problem-dependent information is generated and passed across the hot rolling process chain – defined as vertical and horizontal integration of models to establish forward material workflow, see [6]. We represent the forward model-based material workflow in the middle section of Figure 1. Based on the integrated forward models and problem specific information, we formulate the problem statement to be addressed in this paper. The *Inverse Thermo-Mechanical Processing (ITMP)* design problem statement follows: *Given the required microstructural targets and models to establish microstructural evolution, what should the input thermo-mechanical and microstructural parameters be for the model-based realization of the material (steel), product (rod), and the manufacturing processes (multi-stage hot rod rolling)?* Our goal therefore is to present a design method that supports a designer/decision-maker in inversely identifying satisficing solutions for the thermo-mechanical processing factors strain, strain rate, temperature, initial austenite grain size and interpass time while meeting the conflicting microstructure and property/performance requirements set across each stage of the multi-stage hot rod rolling process chain, pictorially represented in the lower section of Figure 1.

In Section 3, we describe the procedure for design and solution space exploration using an adapted Concept Exploration Framework (CEF). The adapted CEF is used to systematically design manufacturing processes, materials and products where already developed empirical models to capture problem specific information are available.

3. THE ADAPTED CEF FOR DESIGN AND SOLUTION SPACE EXPLORATION

The Concept Exploration Framework (CEF) is a generic Decision-Based Design framework that supports systematic problem formulation, and design and solution space exploration of satisficing, robust design solutions [6]. In Figure 2, we show the adapted CEF to carry out systematic problem formulation and solution space exploration of the hot rod rolling problem addressed. The design and solution space exploration procedure are divided into five steps as shown in Figure 2.

Step 1: Design Task Clarification. In Step 1, the design task related to the hot rod rolling problem is clarified. This includes defining the goals, constraints and problem specific requirements (Step 1a). The design space is also defined by identifying the design factors (control factors, noise factors) and the responses along with their possible ranges (Step 1b).

Step 2: Model Identification and Selection. Manufacturing processes such as hot rod rolling have been widely studied over the years through experiments and simulations. These have led to the development of several empirical models that capture the material and manufacturing phenomena during the process. These models are validated for specific classes of materials and are used extensively, see [9-11]. An important requirement for framing decision support problems using CEF is the availability of such models to establish the functional relationships. If models are not available or the available models are computationally expensive, then there is a need to carry out simulation-based design of experiments to develop surrogate models or meta-models that capture the relationships in an approximate manner with relatively good accuracy. However, we do not address such a situation in the problem discussed in this paper. Empirical models that establish the microstructural evolution during hot rolling of rod are identified and selected systematically according to the needs of the problem in Step 2.

Step 3: The compromise Decision Support Problem (cDSP). The cDSP is the foundational mathematical construct for design decision support in the CEF. The cDSP is a hybrid construct that incorporates features from both goal programming and traditional mathematical programming. The foundational assumption underlying the cDSP based decision support is that models are typically incomplete, inaccurate and not of equal fidelity and therefore have associated uncertainty. Given these uncertain models, the designer using the cDSP seek to explore multiple possible solutions that satisfice the conflicting goals that are present. The problem specific information is captured using the four keywords in the cDSP – Given, Find, Satisfy and Minimize. The details regarding the formulation and execution of the cDSP are available in [12] and are not repeated here. In Step 3, as depicted in Figure 2, all the information gathered to formulate the problem are communicated to the cDSP. The designer captures all the information using the four cDSP keywords and formulates the design problem of interest.

Step 4: Design Scenarios. In Step 4, the cDSP formulated is exercised for different sets of design scenarios. These scenarios are identified by assigning different weight preferences to the conflicting goals present. Enough design scenarios to capture and visualize the solution space for further exploration are identified. A ternary solution space is generated after exercising the cDSP for each of the identified scenarios.

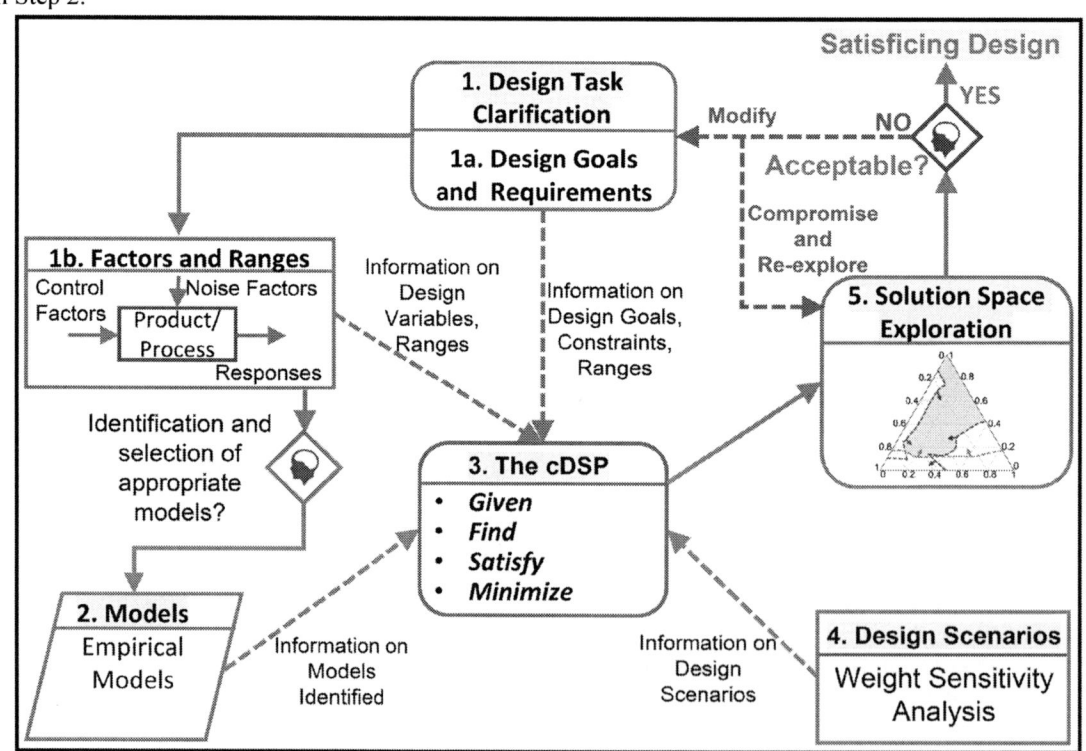

FIGURE 2: PROCEDURE FOR PROBLEM FORMULATION AND SOLUTION SPACE EXPLORATION USING CEF

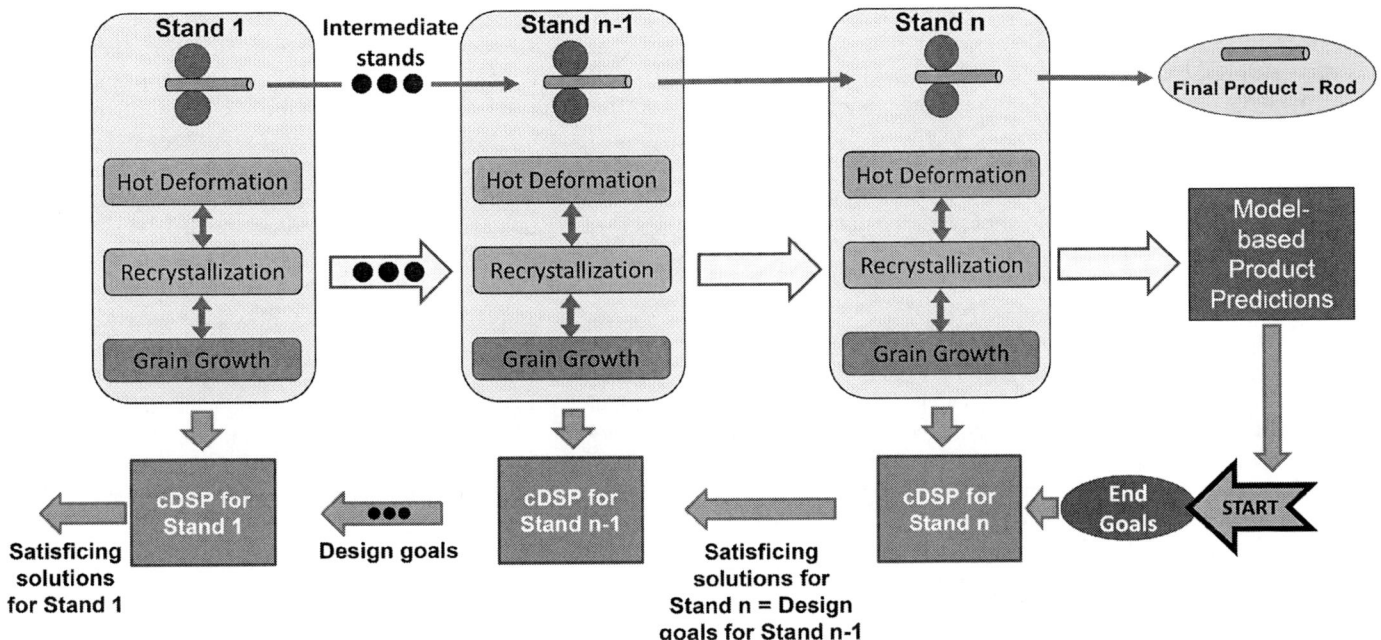

FIGURE 3: GOAL-ORIENTED INVERSE DESIGN METHOD FOR HOT ROD ROLLING PROBLEM

Step 5: Solution Space Exploration. In Step 5, a solution space is generated in a ternary space using the results from the exercised cDSPs. The solution space is explored, and the designer identifies satisficing solution regions by carrying out design trade-offs. Each solution inside the satisficing solution space is evaluated and the designer who is the decision maker finally identifies solution/s that satisfices his/her design requirements in the best possible manner. If a satisficing region is not identified, the designer is left with two options as shown with the red dashed arrows in Figure 2. The designer can either opt to explore further by compromising on the design goals to identify whether a satisficing solution region exists or can modify the overall design goals and requirements to formulate a new problem design space.

The adapted CEF along with the associated features are incorporated into a Goal-oriented Inverse Design (GoID) method that facilitate the integrated design exploration of the manufacturing process chain involving the material and the product. In Section 4, we describe the Goal-oriented Inverse Design method from the standpoint of the hot rolling problem discussed in this paper.

4. GOAL-ORIENTED INVERSE DESIGN METHOD FOR ITMP DESIGN

The Goal-oriented Inverse Design (GoID) method is a generic decision-based design method that supports a designer in identifying satisficing design solutions across a manufacturing process chain meeting specific design goals and requirements set at each stage of the process chain. The method has been

successfully used in the past to design the material microstructure and processing paths during the cooling process after rolling, see [6]. The key functionality offered by the method involves the designer's capability to inversely pass design solutions and set design goals across the process chain. The whole method is goal-oriented as the designer starts with the end goals defined for the material, product or process and designs the system and subsystems to meet these goals. The design decisions made for a stage are communicated to the previous stages in an inverse manner using the cDSP construct in the GoID method. In Figure 3, we show the schematic of the GoID method with application to the hot rod rolling process. The rolling process chain is represented in the form of separate rolling stands (Stand 1 to Stand n) to produce the final rod product. As the material passes through each stand, it is thermo-mechanically processed to a specific geometry and microstructure. The GoID method applied to the hot rod rolling problem involves two phases as described below.

Phase 1: Forward modeling. The first phase in the method is the forward modeling of the manufacturing process chain in order to generate the material and product information workflow. For the hot rod rolling problem, we focus on three major aspects – hot deformation of the material during rolling, the recrystallization of austenite grains and the grain growth that follows recrystallization, see Figure 3. In Phase 1, model and simulation-based analysis is carried out to generate information workflow regarding the material and product across all the stages of rolling process (rolling stands). In the problem discussed, Phase 1 is realized by developing a finite element-based hot rod rolling model to predict the stress, strain, strain rate and temperature evolution as the material passes through the rolling stands. The microstructural phenomena are predicted by using an Austenite Grain Size evolution model that predicts the recrystallized fraction, austenite grain size and grain growth depending on the rolling conditions and parameters. On completion of Phase 1 across the manufacturing process chain, we are able to develop the model-based information workflow for the final product produced, illustrated in Figure 3. The forward modeling phase for the hot rod rolling problem is discussed in detail in Section 5.

Phase 2: Goal-oriented Inverse Design. In this phase of the design method is the process of carrying out decision-based design exploration across the manufacturing process chain starting from the end goals and requirements defined for the problem. In the lower section of Figure 3, we show a schematic representation of the design exploration process. The cDSP for the rolling stand n is formulated by incorporating specific microstructural and property/performance goals for the rod produced. The cDSP is formulated and explored following the procedural steps of the adapted CEF discussed in Section 3. The mathematical models identified in Phase 1 for forward modeling is used to formulate the design problem in Phase 2. For the problem discussed, we look at three end goals, namely minimize final austenite grain size, maximize static recrystallized fraction and achieve target value for grain surface to volume ratio with a desire to explore larger values if possible. These goals and requirements are discussed in detail in Section 2. On solving the cDSP for stand n and exploring the solution space, the designer is able to identify the combinations of the input factors initial austenite grain size, strain, strain rate, temperature and interpass time that satisfices the conflicting goals present. These combinations are then passed as goals for the next cDSP for stand n-1. The whole process of problem formulation and solution space exploration is repeated for rolling stand n-1 and satisficing solutions are identified. The process is continued until the entire problem boundary is covered depending on how this is defined with respect to the manufacturing process chain. The method is generic and can be used to design similar manufacturing process chains where there is sequential information sharing. The method supports design information coordination and goal-oriented decision support for a complex system involving sequential flow of information/energy/material.

5. MATHEMATICAL MODELS FOR HOT ROD ROLLING PROBLEM

In this section, we introduce the mathematical models used to formulate the hot rod rolling problem. These models are used to establish the forward modeling and material information workflow for the problem. In Section 5.1, we discuss the finite element model for the hot deformation during rolling process. In Section 5.2, we discuss the static recrystallization, grain growth and other microstructure specific models for the problem discussed.

5.1 Finite Element Hot Rod Rolling Model

To predict the austenite grain size evolution, the thermo-mechanical processing information in the form of strain, strain rate and temperature is needed. We gather this information using a finite element (FE) hot rod rolling model; the details of this model are discussed briefly here. A four-stand coupled temperature-displacement FE hot rod rolling model is developed in ABAQUS. A simulated initial billet of square cross section is passed through rollers and it is sequentially deformed to oval and circular cross sections to finally produce a rod of definite diameter. The billet is modeled as a deformable body using the thermally coupled brick element C3D8RT. Material properties for steel as a function of temperature are assigned to the block. Yield stress values for steel at different plastic strains are used to model the material's plastic deformation behavior. Analytical models are used to model the rollers and the surface profile of roller grooves, [13]. The rollers are modeled and meshed as discrete right bodies with R3D4 elements. The billet is assigned an initial temperature before rolling. Further details about the hot rod rolling model are available in [14]. In this study, we focus only on the last rolling pass. In Figure 4, we show the temperature contours for the rod after the last rolling pass. The strain, strain rate and temperature information from the last rolling pass from the FE model are used for subsequent analysis in an Austenite Grain Size (AGS) evolution model, discussed next.

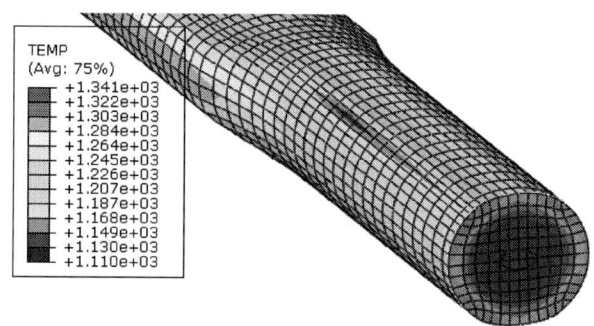

FIGURE 4: FINITE ELEMENT HOT ROD ROLLING MODEL

5.2 Microstructure Evolution Models

During hot rolling, recrystallization occurs which is basically formation of new grains from old grains. Recrystallization is typically followed by grain growth where grains coarsens and arrange themselves in a uniform manner. Modeling recrystallization and grain growth have been studied over the years and several researchers have developed empirical relationships to capture these phenomena, see [10] for a comprehensive list. In this paper, we use the AGS evolution model proposed by Hodgson and Gibbs [9]. This model is used to consider static and metadynamic recrystallization, and grain growth. In the design study carried out in this paper, we assume that the strain (ε) is less than the critical strain (ε_c) for dynamic or metadynamic recrystallization to be initiated.

TABLE 1: MODELS SPECIFIC TO MICROSTRUCTURAL EVOLUTION FOR THE HOT ROLLING PROBLEM

Model	Equation	Reference
Critical Strain	$\varepsilon_c = 5.6 * 10^{-4} * D_0^{0.3} * Z^{0.17}$	Hodgson and Gibbs [9]
Static Recrystallized Fraction	$X = 1 - exp\left[-0.693\left(\dfrac{t_{ip}}{t_{0.5}}\right)\right]$	
Time for 50% recrystallization	$t_{0.5} = 2.3 * 10^{15} * \varepsilon^{-2.5} * D_0^2 * exp\left(\dfrac{230000}{RT}\right)$ R is the universal gas constant	
Zener-Hollomon Parameter	$Z = \dot{\varepsilon} * exp\left(\dfrac{300000}{RT}\right)$	
Static Recrystallized Grain Size	$D_{SRX} = 343 * \varepsilon^{-0.5} * D_0^{0.4} \, exp\left(-\dfrac{45000}{RT}\right)$	
Grain Growth	$D_{GG}^7 = D_{SRX}^7 + 1.5 * 10^{27}\left(t_{ip} - 4.32 * t_{0.5}\right) * exp\left(-\dfrac{400000}{RT}\right)$	
Grain Surface to Volume Ratio	$S_v = \dfrac{24\varepsilon}{\pi D_0}[0.491\, exp(\varepsilon) + 0.155\, exp(-\varepsilon) + 0.1433\, exp(-3\varepsilon)\,]$	Suehiro and co-authors [15]

Hence, we use only the static recrystallization and grain growth models proposed by Hodgson and Gibbs. The grain surface area to volume ratio is defined using the empirical model developed by Suehiro and co-authors [15]. In Table 1, we include the models used in this paper to predict the microstructural phenomenon during hot rod rolling process from the problem perspective. These are standard empirical models applicable to hot rolling of steels. For details on the applicability of these models, see [10]. The procedure for calculating the final austenite grain size value using the AGS evolution model developed by Hodgson and Gibbs is provided by Kwon and co-authors [13] and is not repeated here. These microstructural models are dependent on the independent TMP variables of the design study, namely, initial austenite grain size (D_0), strain (ε), strain rate ($\dot{\varepsilon}$), temperature (T) and interpass time (t_{ip}). Further variables are defined in the cDSP. Given these models and the information flow chain, we are able to formulate the cDSPs starting from end goals following the GoID method described in Section 4. In Section 6, we discuss the cDSP formulated for the last rolling stand n and the associated solution space exploration.

6. cDSP FORMULATION USING CEF FOR ROLL STAND n

Once the models are identified and integrated to establish the forward information workflow, the designer starts the inverse decision-based design exploration starting from end goals. The cDSPs are formulated starting from the last rolling pass. The formulation and exploration procedures are as described using CEF in Section 3. In this Section, we discuss the cDSP for the last rolling stand formulated to achieve the end product goals for the rod. The product goals for final austenite grain size, static recrystallized fraction and grain surface to volume ratio are defined as a function of the Thermo-Mechanical Processing factors initial austenite grain size, strain, temperature, interpass time and strain rate. The requirement for minimizing final AGS is defined as a minimization goal for the austenite grain size after grain growth. The requirement for achieving 100% static

recrystallization is defined as a maximization goal with a constraint defined to ensure 95% of the grains are recrystallized to initiate grain growth. The grain growth constraint is defined to ensure that the grain size after grain growth is always greater than the grain after static recrystallization. A critical strain constraint is defined to ensure that the critical strain is always greater than the strain in the material deformed. This is defined to ensure the non-initiation of dynamic and metadynamic recrystallization in the material. An interpass time constraint is defined to ensure that there is sufficient time for recrystallization and grain growth to happen. The third goal for surface to volume ratio is defined in the minimization format to achieve a target value of 0.03 µm⁻¹. However, the desire is to explore larger values of this goal if such solutions exist in the solution space generated. On exercising this cDSP and carrying out solution space exploration, the designer is able to inversely identify the combination of thermo-mechanical processing (TMP) factors that best satisfices the end goals set for the material, product and process. The TMP factors thus identified are communicated as goals and requirements to the cDSP for the rolling stand that precedes, and the process is repeated to cover the problem boundary defined using the Goal-oriented Inverse Design method proposed. The cDSP reads as follows.

Given:

(1) End requirements identified for the rod rolling process

➢ Minimize final austenite grain size (D_{GG})

➢ Maximise recrystallized grain fraction (X)

➢ Achieve target grain surface area to volume ratio (S_v)

➢ Target value for austenite grain size, D_{GG} = 25 µm

➢ Target value for recrystallized grain fraction, X = 1.0

➢ Target value for grain surface area to volume ratio, S_v = 0.03 µm⁻¹

➢ Minimum limit of recrystallized grain fraction, X = 0.95

➢ Maximum limit of recrystallized grain fraction, X = 1.0

➢ Grain surface area to volume ratio requirement, S_v: Explore whether larger values are possible

(2) The empirical correlations of microstructural evolutions
➤ Critical strain to decide whether SRX or MDRX will take place

$$\varepsilon_c = 5.6 * 10^{-4} * D_0^{0.3} * Z^{0.17} \quad (1)$$

➤ Zener-Hollomon Parameter

$$Z = \dot{\varepsilon} * exp\left(\frac{300000}{RT}\right) \quad (2)$$

➤ Recrystallized grain fraction

$$X = 1 - exp\left[-0.693\left(\frac{t_{ip}}{t_{0.5}}\right)\right] \quad (3)$$

➤ Time taken for half of the grain to recrystallize

$$t_{0.5} = 2.3 * 10^{15} * \varepsilon^{-2.5} * D_0^2 * exp\left(\frac{230000}{RT}\right) \quad (4)$$

➤ Grain size after recrystallization

$$D_{SRX} = 343 * \varepsilon^{-0.5} * D_0^{0.4} \, exp\left(-\frac{45000}{RT}\right) \quad (5)$$

➤ Grain size after grain growth

$$D_{GG}^7 = D_{SRX}^7 + 1.5 * 10^{27}\left(t_{ip} - 4.32 * t_{0.5}\right) * exp\left(-\frac{400000}{RT}\right) \quad (6)$$

➤ Grain surface area to volume ratio

$$S_v = \frac{24\varepsilon}{\pi D_0}[0.491 \, exp(\varepsilon) + 0.155 \, exp(-\varepsilon) + 0.1433 \, exp(-3\varepsilon)] \quad (7)$$

(3) Variability in system variables
The system variables are provided in the Find and Satisfy section.

Find:

System Variables

X_1 initial austenite grain size before the billet is rolled (D_0) – Range: 40 to 100μm

X_2 strain when the rod passes under the roller (ε) – Range: 0.03 to 0.5

X_3 temperature during the rolling process (T) – Range: 1173 to 1473K

X_4 interpass time (t_{ip}) – Range: 3 to 13 seconds

X_5 strain rate ($\dot{\varepsilon}$) – Range: 5 to 10 s^{-1}

Satisfy:

System Constraints

➤ Critical strain constraint

$$\varepsilon_c - \varepsilon \geq 0 \quad (8)$$

➤ Grain growth constriant

$$D_{GG} - D_{REX} \geq 0 \quad (9)$$

➤ Interpass time constraint

$$t_{ip} - 4.32 * t_{0.5} \geq 0 \quad (10)$$

➤ Minimum recrystallized fraction constraint

$$X - 0.95 \geq 0 \quad (11)$$

➤ Maximum recrystallized fraction constraint

$$1.0 - X \geq 0 \quad (12)$$

System Goals

Goal 1:

➤ Minimize austenite grain size after recrystallization and grain growth

$$\frac{D_{GG,Target}}{D_{GG}(X_i)} - d_1^- + d_1^+ = 1 \quad (13)$$

Goal 2:

➤ Maximize recrystallized grain fraction

$$\frac{D_X(X_i)}{D_{GG,Target}} + d_2^- - d_2^+ = 1 \quad (14)$$

Goal 3:

➤ Achieve grain surface to volume ratio target

$$\frac{S_{v,Target}}{S_v(X_i)} - d_3^- + d_3^+ = 1 \quad (15)$$

Variable Bounds

X_1 – 40 to 100μm

X_2 – 0.03 to 0.5

X_3 – 1173 to 1473K

X_4 – 3 to 13 seconds

X_5 – 5 to 10 seconds^{-1}

Bounds on deviation variables

$$d_i^-, d_i^+ \geq 0 \; and \; d_i^- * d_i^+ = 0, i = 1,2,3 \quad (16)$$

Minimize:

We minimize the deviation function. Represented as *DF* in this paper.

$$DF = \sum_{i=1}^{3} W_i(d_i^- + d_i^+); \sum_{i=1}^{3} W_i = 1 \quad (17)$$

From the cDSP, we explore the solution space by exercising the cDSP for different design scenarios, discussed next in Section 7.

7. SOLUTION SPACE EXPLORATION

We select 19 different design scenarios (DS) based on judgement to capture and visualize the solution space for exploration, Table 2. Different weight preferences are assigned to each goal for these 19 design scenarios. Scenarios 1-3 denote situations where the designer's interest is to achieve the target close to one of the goals while the other goals are not considered. Scenarios 4-6 denote situations where a designer gives equal preferences to two goals while not considering the third goal. Scenarios 7-12 denote situations where a designer gives higher preference to one goal, lesser preference to a second and zero preference to a third goal. In Scenario 13, the designer assigns goal preferences equally. Scenarios 14-19 are situations where all goals have equal preferences to two goals. These are used to capture the interior points of the ternary space generated. From these scenarios, we exercise the cDSP for each of these scenarios and develop ternary spaces for exploration. The three axes of ternary plots denote the weights assigned for the three goals. The interior color contour represents the achieved values of the goal for which the ternary plot is made. The corresponding values of control factors are identified from the plot. Next, we discuss the solution space exploration for each goal defined for the hot rod rolling problem.

TABLE 2: DESIGN SCENARIOS AND WEIGHTS

DS	W_1	W_2	W_3
1	1	0	0
2	0	1	0
3	0	0	1
4	0.5	0.5	0
5	0.5	0	0.5
6	0	0.5	0.5
7	0.25	0.75	0
8	0.25	0	0.75
9	0.75	0	0.25
10	0.75	0.25	0
11	0	0.75	0.25
12	0	0.25	0.75
13	0.33	0.34	0.33
14	0.2	0.2	0.6
15	0.4	0.2	0.4
16	0.2	0.4	0.4
17	0.6	0.2	0.2
18	0.4	0.4	0.2
19	0.2	0.6	0.2

Different weight preferences are assigned to each goal in these 19 design scenarios. Scenarios 1 to 3 denote a situation where the designer's interest is to achieve the target close to one of the goals while the other goals are not considered. Scenario 4-6 denote situations where designer gives equal preferences to two goals while not considering the third goal. Scenarios 7-12 denote situations where designer gives higher preference to one goal, lesser preference to second and zero preference to third goal. In Scenario 13, the designer assigns equal preferences to all the three goals. Scenarios 14 to 19 are situations where all goals are given preferences with equal preferences for any two goals. These are used to capture the interior points of the ternary space generated. Given these 19 scenarios, we exercise the cDSP for each of these scenarios and develop ternary spaces for exploration. The three axes of ternary plots denote the weights assigned for the three goals. The interior color contour represents the achieved values of the goal for which the ternary plot is made. The corresponding values of control factors are identified from the plot. Next, we discuss the solution space exploration for each goal defined for the hot rod rolling problem.

The ternary space for Goal 1 is shown in Figure 5. For Goal 1, the interest of the process designer is to achieve a minimum final austenite grain size after rolling. A minimum austenite grain size supports the formation of fine ferrite grains that further improves the mechanical properties of the rod after cooling. A target value of 25 μm is defined for this goal. Based on our previous study in [6], we have identified a value of 30 μm to be acceptable to satisfy the requirements from cooling stage after rolling. The solution region with final AGS value less than 30 μm is identified using the red dashes lines in Figure 5. We see in Figure 5 that the solution space thus identified lie mostly in the region with higher weight preferences assigned to Goal 1, weights between 0.5 and 1. Any combination of weights on goals from this region satisfies the AGS goal.

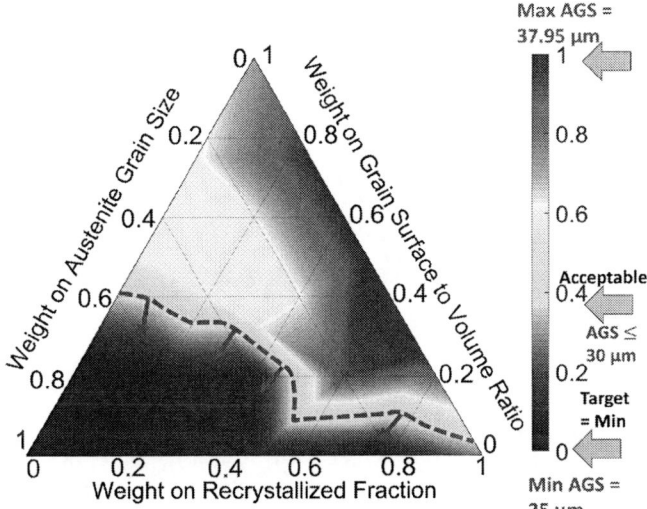

FIGURE 5: TERNARY PLOT FOR GOAL 1 – MINIMIZE AUSTENITE GRAIN SIZE

For Goal 2, see Figure 6, the interest of the process designer is to maximize the static recrystallized fraction to 100%. A minimum recrystallized fraction of 95% is required and this requirement is defined as a constraint in the cDSP. On analyzing the ternary space shown in Figure 6, we observe that the red contour region satisfies the requirement for 100% recrystallized fraction All the other solution regions satisfies a recrystallized fraction of 99.8% and above. Since these high values of recrystallized fraction are satisfied, all the solution points are acceptable to us. This also ensures that grain growth phenomena will happen following recrystallization and therefore Goal 1 is valid from the problem perspective. We therefore accept all the solution points as shown using the violet arrows on the color contour in Figure 6.

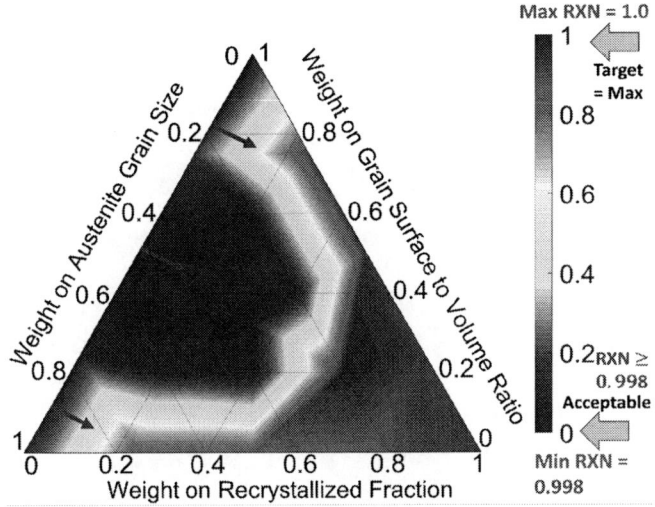

FIGURE 6: TERNARY PLOT FOR GOAL 2 – MAXIMIZE RECRYSTALLIZED FRACTION

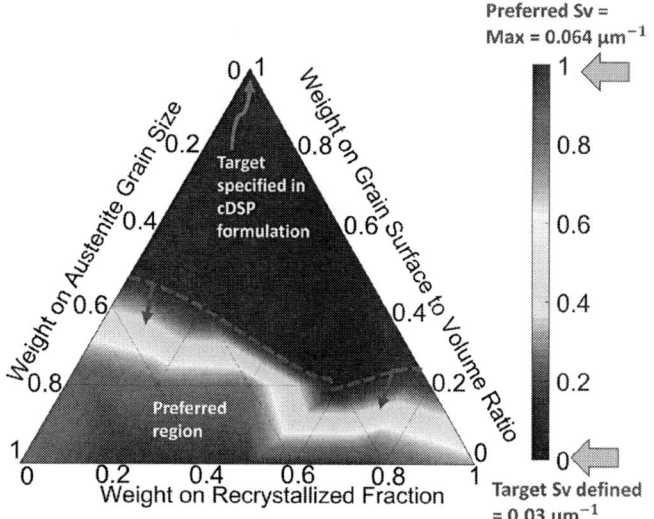

FIGURE 7: TERNARY PLOT FOR GOAL 3 - ACHIEVE GRAIN SURFACE TO VOLUME RATIO TARGET

For Goal 3, see Figure 7, the interest of the process designer is to achieve a target value of 0.03 μm^{-1} for the grain surface to volume ratio goal. On analyzing Figure 7 we see that higher values for this goal is achieved in regions where austenite grain size is minimized (red color region in Figure 7). Achieving higher values of S_v is preferred as it supports the minimization of ferrite grain size and its refinement. Also, the need to minimize grain coarsening is supported by larger values of S_v. For these reasons the regions with larger values of S_v is preferred and are identified using the dashed brown lines in Figure 7.

To make a decision and identify a satisficing region for all the three goals, we superpose the ternary regions, Figure 8. The colored region satisfices all the three goals in the best possible manner. The design scenarios (DS) listed in Table 2 that lie in this superposed region are in Table 3.

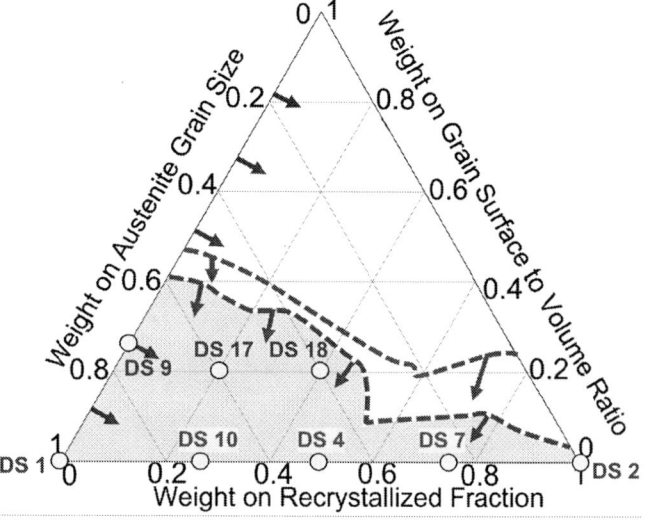

FIGURE 8: SUPERPOSED TERNARY PLOT

TABLE 3: EVALUATION OF SOLUTIONS

DS	Thermo-Mechanical Processing (TMP) Factors					Goals Defined for Hot Rolled Rod		
	D_0 μm	ε	T K	t_{ip} s	$\dot{\varepsilon}$ s^{-1}	D_{GG} μm	X	S_v μm^{-1}
1	40	0.353	1173	3.824	10	25	0.998	0.057
2	50	0.4	1173	8	10	28.57	1.0	0.053
4	40	0.382	1173	3.792	10	25	0.999	0.064
7	40	0.382	1173	3.792	10	25	0.999	0.064
9	40	0.352	1173	3.862	10	25	0.998	0.057
10	40	0.382	1173	3.792	10	25	0.999	0.064
17	40	0.352	1173	3.826	10	25	0.998	0.057
18	40	0.347	1173	3.948	10	25.16	0.997	0.056

On evaluating the design solutions in Table 3, we observe that all the solutions are relatively close and produce responses that are relatively insensitive to changes. Based on the analysis of the results, we observe the following. The final austenite grain size tends to decrease when the initial austenite grain size is low. Thus, one requirement that can be placed on previous rolling stages is to minimize the austenite grain sizes of the rod at each rolling pass. An increase in grain surface to volume ratio is observed at those regions with low initial austenite grain size. An increased grain surface to volume ratio increases the number of nucleation sites resulting in more nuclei and thereby decreasing the ferrite grain sizes. This in turn improves the final mechanical properties of the rod produced after rolling and subsequent cooling processes.

From the solution points, we pick solution points from design scenarios 4, 7 and 10 that best satisfice the goals defined. The associated Thermo-Mechanical Processing (TMP) factor values are identified and selected to achieve the system level goals set for the problem. These include an initial austenite grain size value of 40 μm strain of 0.382, temperature of 1173 K, interpass time of 3.792 seconds and strain rate of 10 seconds^{-1}. The generic method proposed can be used to design the previous rolling stands to meet the requirements set by the cDSP explored and the identified TMP factors.

8. CLOSING REMARKS

In this paper, we present a Goal-oriented Inverse Design (GoID) method to achieve the integrated design exploration of materials, products and associated manufacturing processes. Given the forward material (information) workflow using models, the proposed method can be used to support the goal-oriented inverse design exploration of material microstructures and processing paths to meet specific end properties and performances desired for the material and product. We demostrate the utility of the method using an industry-inspired hot rolling example problem to inversely design the thermo-mechanical processing of a steel rod. The key functionality tested using this example problem is the designer's capability to carry out a microstructure-mediated thermo-mechanical processing design satisfying specific performance goals of the material (steel) and the end product (rod). The method and associated design constructs are generic and has the potential to be applied to co-ordinate information flow and human decision-making across processes/levels in order to realize an end goal – a key functionality that allows the application of this method to other complex system problems involving sequential model-based flow of information across processes/levels.

ACKNOWLEDGMENTS

Anand Balu Nellippallil thanks the Center for Advanced Vehicular Systems (CAVS), Mississippi State University for supporting this work. The authors thank TRDDC, Tata Consultancy Services, Pune for supporting this work (Grant No. 105-373200). Janet K. Allen and Farrokh Mistree gratefully acknowledge financial support from the John and Mary Moore Chair and L.A. Comp Chair respectively at the University of Oklahoma.

REFERENCES

[1] Dieter, G. E., Kuhn, H. A., and Semiatin, S. L., 2003, "Handbook of Workability and Process Design," *ASM International*, Materials Park, OH.

[2] Horstemeyer, M. F., 2012, "Integrated Computational Materials Engineering (ICME) for Metals: Using Multiscale Modeling to Invigorate Engineering Design with Science," John Wiley & Sons, Hoboken, NJ.

[3] McDowell, D. L., Panchal, J., Choi, H.-J., Seepersad, C., Allen, J. K., and Mistree, F., 2009, "Integrated Design of Multiscale, Multifunctional Materials and Products," Butterworth-Heinemann, Waltham, MA.

[4] Mistree, F., and Allen, J. K., 1997, "Position Paper Optimization in Decision-Based Design," *Optimization in Industry*, Palm Coast, FL, Mar, pp. 23-27.

[5] Nellippallil, A. B., 2018, "The Integrated Realization of Materials, Products and Associated Manufacturing Processes," Doctoral Dissertation, School of Aerospace and Mechanical Engineering, University of Oklahoma, Norman.

[6] Nellippallil, A. B., Rangaraj, V., Gautham, B. P., Singh, A. K., Allen, J. K., and Mistree, F., 2018, "An Inverse, Decision-Based Design Method for Integrated Design Exploration of Materials, Products and Manufacturing Processes," *Journal of Mechanical Design*, vol. 140, no. 11, pp. 111403.

[7] Kuziak, R., Cheng, Y.-W., Glowacki, M., and Pietrzyk, M., 1997, "Modeling of the Microstructure and Mechanical Properties of Steels during Thermomechanical Processing," *NIST Technical Note (USA)*, vol. 1393, pp. 72.

[8] Jägle, E., 2007, "Modelling of Microstructural Banding During Transformations in Steel'," Master of Philosophy Dissertation, Department of Materials Science & Metallurgy, University of Cambridge.

[9] Hodgson, P., and Gibbs, R., 1992, "A Mathematical Model to Predict the Mechanical Properties of Hot Rolled C-Mn and Microalloyed Steels," *ISIJ International*, vol. 32, no. 12, pp. 1329-1338.

[10] Pietrzyk, M., Cser, L., and Lenard, J., 1999, "Mathematical and Physical Simulation of the Properties of Hot Rolled Products," Elsevier, Kidlington, Oxford.

[11] Majta, J., Kuziak, R., Pietrzyk, M., and Krzton, H., 1996, "Use of the Computer Simulation to Predict Mechanical Properties of C-Mn Steel after Thermomechanical Processing," *Journal of Materials Processing Technology*, vol. 60, no. 1-4, pp. 581-588.

[12] Mistree, F., Hughes, O. F., and Bras, B., 1993, "Compromise Decision Support Problem and the Adaptive Linear Programming Algorithm," *Progress in Astronautics and Aeronautics*, vol. 150, pp. 251-290.

[13] Kwon, H.-C., Lee, Y., Kim, S.-Y., Woo, J.-S., and Im, Y.-T., 2003, "Numerical Prediction of Austenite Grain Size in Round-Oval-Round Bar Rolling," *ISIJ International*, vol. 43, no. 5, pp. 676-683.

[14] Nellippallil, A. B., Song, K. N., Goh, C.-H., Zagade, P., Gautham, B., Allen, J. K., and Mistree, F., 2017, "A Goal-Oriented, Sequential, Inverse Design Method for the Horizontal Integration of a Multistage Hot Rod Rolling System," *Journal of Mechanical Design*, vol. 139, no. 3, pp. 031403.

[15] Suehiro, M., Sato, K., Tsukano, Y., Yada, H., Senuma, T., and Matsumura, Y., 1987, "Computer Modeling of Microstructural Change and Strength of Low Carbon Steel in Hot Strip Rolling," *Transactions of the Iron and Steel Institute of Japan*, vol. 27, no. 6, pp. 439-445.

Proceedings of the ASME 2019
International Design Engineering Technical Conferences
and Computers and Information in Engineering Conference
IDETC/CIE2019
August 18-21, 2019, Anaheim, CA, USA

DETC2019-97547

GAUSSIAN PROCESS BASED CRACK INITIATION MODELING FOR DESIGN OF BATTERY ANODE MATERIALS

Zhuoyuan Zheng[1]

University of Illinois at Urbana-Champaign
Urbana, IL 61801, USA

Yanwen Xu[2]

University of Illinois at Urbana-Champaign
Urbana, IL 61801, USA

Bo Chen[3]

University of Illinois at Urbana-Champaign
Urbana, IL 61801, USA

Pingfeng Wang[4]

University of Illinois at Urbana-Champaign
Urbana, IL 61801, USA

ABSTRACT

Silicon-based anode is one of the promising candidates for the next generation lithium ion batteries (LIBs) to achieve high power/energy density. However, the major drawback limiting the practical application of Si anode is that Si experiences significant volume change during its lithiation/de-lithiation cycles, which induces high stress and causes degradation and pulverization of the anode. This study focuses on the crack initiation performances of Si anode during the de-lithiation process. A multi-physics based finite element (FE) model is built to simulate the electrochemical process and crack generation during de-lithiation. On top of that, a Gaussian Processes (GP) based surrogate model is developed to assist the exploration of the crack initiation performances within the anode design space. It is found that, the thickness of the Si coating layer T_{Si}, the yield strength σ_{Fc} of Si material, the cohesive strength between Si and substrate σ_{Fs}, and the curvature of the substrate ρ have large impacts on the cracking behavior of Si. This coupled FE simulation-GP surrogate model framework is also applicable to other types of LIB electrodes.

Key words: Silicon anode, Crack initiation, Multi-physics modeling, GP based surrogate model

1. INTRODUCTION

The lithium ion batteries (LIBs), as one of the secondary (rechargeable) batteries, outperform other battery systems because of their high energy/power density, and good cycle life performances [1, 2], and have become the key component of energy storage devices for portable electronic devices such as smart phones and tablet PCs, as well as electric vehicles (EVs). Commercially available LIBs commonly use lithium metal oxide or phosphates ($LiCoO_2$, $LiMn_2O_4$) as the cathode electrode and graphite as the anode. However, these LIBs are gradually approaching their performance limits, and cannot meet the growing demands for higher volumetric and gravimetric energy/power densities, longer cycle life, and reduced cost and improved safety [3], especially for the emerging electric transportation systems.

To address these challenges, numerous efforts have been made to develop new electrodes with enhanced lithium storage capacity. Silicon is a promising candidate for anode active material in LIBs. Different from the intercalation mechanism of graphite for Li ions storage, i.e. Li ions diffusing and residing in the interstitial sites within the host lattice, Si will react with lithium, leading to the bonds breaking between Si atoms, and the formation of Li_xSi alloy, accompanied by the dramatic structural changes in the process [4]. With no constrains from the atomic framework of the host material, Si is able to store much more lithium comparing with the intercalation electrodes [5]. For instance, Si has the highest known theoretical specific capacity of 4200 mAh g^{-1} for $Li_{22}Si_5$ alloy phase; meanwhile,

---

[1] Postdoctoral Research Associate, zhuoyuan@illinois.edu.
[2] Graduate Research Assistant, yanwenx2@illinois.edu.
[3] Graduate Research Assistant, bochen3@illinois.edu.
[4] Associate Professor, pingfeng@illnois.edu. Corresponding Author.

the capacity for graphite is ~370 mAh g^{-1} (with a fully-lithiated graphite phase of LiC$_6$) [6].

However, due to the "alloy" lithium storage mechanism, the volume of silicon changes significantly (up to 300%) during its lithiation/de-lithiation cycles, which creates large internal stresses, and leads to the pulverization of anode and reduction of capacity [7]. Various works have been conducted to study the lithiation-induced stress and its influences on the fracture and mechanical degradation of Si anode. H. Yang et al. [8] and X. Liu et al. [9] used a 3D finite element, small-strain model to simulate the morphological change and stress generation in crystal Si nanowires. Anisotropic swellings and fractures were found depending on different axial crystallographic orientations, which provided the insight of crack initiation during the first lithiation/de-lithiation cycle. J. Li et al. [10] developed a modified spring-block model to capture the essential features of cracking patterns in amorphous Si thin film electrodes. Besides, dimensionless calculation based studies were also conducted to study the stress generated within a simple spherical Si particle during lithiation [11, 12]. However, most of the previous works considered the Si anode with relatively straightforward structures, e.g. particles, nanowires etc. With the improvements of electrode fabrication methods, a number of novel structured Si anodes are proposed [13-15]. The needs for developing new methods to study the electrodes with complicated structures are growing.

Aside from the physics based simulation methods, data-driven techniques [16] are also widely used for electrode design optimization. Different from physics-based models, which are built upon certain physics of failure and can be used to explore the failure mechanism of a certain battery, data-driven techniques can effectively and efficiently capture the general changing trends of the electrode properties affected by various design parameters. The data-driven models, however, require a lot of experimental results as training data, which can be expensive to acquire; meanwhile, these models are often not applicable to investigate the underlying physical mechanisms. This study presents a mixed model, based on both physics-of-failure simulation and Gaussian Processes (GP) surrogate modeling technique, to investigate the lithiation induced stress on the Si anode and provide the optimal design accordingly. GP surrogate modelling technique has been extensively utilized in engineering applications because of its accuracy and efficiency [17, 18].

In this study, a multi-physics-based finite element model is built to investigate the crack initiation performances of Si anode and the critical design parameters, including thickness of Si layer T$_{Si}$, the cohesive strength between the Si layer and the substrate, and the curvature of the substrate. Afterwards, a GP based surrogate model is developed, with the simulated results as training data, to predict the cracking behaviors of the anodes. The GP surrogate model results are then used to analyze the general trends of the Si anode performances. The coupled FE simulation-GP surrogate model framework can be used for the exploration of the influences of the substrate configurations.

2. FINITE ELEMENT MODELING APPROACH

A multi-physics-based 3D finite element model is established via COMSOL Multiphysics to simulate the crack initiation of Si anodes with complex structures during the de-lithiation process. The design parameters of the anode are used as input, and the area of the portion of the Si layer that does not crack apart, i.e. Si island, is the output of the model. The model consists of two sub-models. In the first part, the electrochemical dynamics of the Si anode during de-lithiation, which is the lithium diffusion process, is simulated; subsequently, the lithium concentration profile is acquired and to be used to evaluate the crack initiation performances based on a standard cohesive zone modeling (CZM) technique.

2.1 The Governing Equations

Insertion of lithium into the amorphous Si (a-Si) host is considered as a single-phase reaction in this model. The lithium concentration profile of Si during the de-lithiation process is determined by Fick's law (1). When considering thin Si layer as host material, the hydrostatic stress gradient effects have limited impact on the diffusion flux J and lithium concentration change [19], and is neglected in this study:

$$J = -D\nabla c \tag{1}$$

where, c is the concentration of lithium ion, D is the diffusion coefficient, ∇ is gradient operator. Further combining Fick's law (1) with the mass conservation equation, $\partial c/\partial t + \nabla \cdot J = 0$, we have

$$\partial c/\partial t + \nabla\left(-D\nabla c\right) = 0. \tag{2}$$

The boundary condition of Eq. (2) at the Si-electrolyte surface is controlled by the electric current density i_n,

$$J \cdot n = i_n/F, \tag{3}$$

where, n is the normal vector of the anode-electrolyte surface, F is the Faraday constant. Meanwhile, periodic boundary condition is applied to other surfaces for approximating a large Si anode structure.

Afterwards, lithium concentration profile is used to simulate the de-lithiation induced volume reduction and crack initiation process. The volumetric shrinkage of Si caused by de-lithiation is analogous to the 3D thermal expansion process [20-22], so that the stress and deformation of Si during de-lithiation are correspondingly treated as the thermal contraction of a material surrounded by a low temperature media. An isotropic volume contraction of amorphous Si is assumed. Therefore, the total deformation of Si consists of two parts, elastic deformation that obeys the Hooke's law and an additional diffusion-induced term:

$$\varepsilon_{ij} = \varepsilon_{el} + \varepsilon_{Li} = \frac{1}{E}\left[(1+v)\sigma_{ij} - v\sigma_{kk}\delta_{ij}\right] + \frac{c\Omega}{3}\delta_{ij}, \quad (4)$$

where, ε_{ij} and σ_{ij} are the components of strain and stress tensors, respectively. δ_{ij} is the Kronecker delta, ε_{el} and ε_{Li} are the elastic and lithiation induced strain, respectively. Ω is the partial molar volume (m³/mol). E and v are the Young's modulus and Poisson's ratio, respectively. The lithium diffusion equation (2) and the stress-strain equation (4) are coupled through lithium concentration term c.

Cohesive zone modeling (CZM) technique is utilized to investigate the crack initiation of Si during volume shrinkage. Recently, CZM methods are widely used to analyze the crack initiation and growth in metals [23, 24] and polymers [25, 26]. In the following, the basic features of the model are outlined.

Fig. 1 shows the schematic of the cohesive model. The CZM presets a cohesive surface Γc that permits the material planes adjacent to the surface to separate, and uses a rate-independent, history-dependent irreversible traction-separation law that corelates the cohesive traction vector T and the displacement jump vector Δ along the cohesive surface to evaluate the resistance to crack nucleation and growth. In the tensile (mode I) separation mode, which is the focus of the work, this cohesive model has the form

$$T = \frac{\zeta}{1-\zeta}\frac{\Delta}{\Delta_C}\sigma_{max} \quad (5)$$

where, T and Δ are the cohesive traction and crack opening displacement vectors, respectively. σ_{max} is the tensile cohesive failure strength. Δ_c is the critical value of the opening displacement jump, beyond which complete delamination is regulated. Δ_c is codetermined by σ_{max} and the fracture toughness G_{1C} of the cohesive surface by

$$G_{1C} = \frac{1}{2}\sigma_{max}\Delta_C. \quad (6)$$

ζ is a coefficient lain in [0, 1), and is updated every timestep, and is defined as

$$\zeta = \min\left(\zeta_p, \langle 1 - \Delta/\Delta_C \rangle\right), \quad (7)$$

where, ζ_p denotes ζ value from the previous timestep. $\langle a \rangle$ represents a function that $\langle a \rangle = a$ if $a > 0$, $= 0$ otherwise. ζ monotonically decreases over time and is used to quantify the real-time damage state and prevent the healing of the de-bonded/failed interfaces during the loading/unloading cycles. The initial value of ζ is set at 0.98 (around 1).

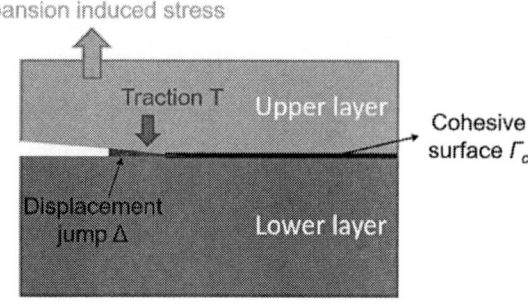

FIGURE 1. Schematic of cohesive zone modeling.

2.2 Finite Element Model Implementation

The governing equations are then implemented in the 3D continuum space FE model (COMSOL Multiphysics). The multi-physics simulation model consists of two sub-modules, the electrochemical (lithium battery) module and the mechanics (crack initiation) module, which are solved sequentially. The electrochemical sub-model is to calculate the lithium concentration c with respect to the de-lithiation process. Then, c is extracted from the simulation results and imported to the crack initiation model. The model simulates the Si half-cell battery, using pure lithium metal and Si as two electrodes. 1M lithium hexafluorophosphate (LiPF₆) in 1:1 Ethylene Carbonate : Diethyl Carbonate (EC:DEC), one of the most commonly used liquid electrolytes, is used in the model. The lithiation-induced expansion is simulated analogously to the thermal expansion process with lithium concentration c substituting temperature in the model to acquire stresses and deformations of the anode. Table 1 lists the electrochemical parameters used in the simulation. The lithium diffusion coefficients in electrolyte and Silicon are kept constant. Silicon active material has a maximum capacity c^{max} of 3.367×10^5 mol/m³ based on the theoretical specific capacity 3579 mAh/g of the Li₃.₇₅Si [27], which is considered as fully lithiated state in this study. The open circuit voltages (OCV) of Si versus lithium as a function of the state of charge is measured from the low-rate discharging of the half cell batteries.

TABLE 1. Electrochemical parameters used in the model

PARAMETERS	VALUE
Li⁺ diffusion coefficient in electrolyte D_{Li^+}	2.6×10^{-10} m²/s [28]
Li⁺ transference number t^+	0.363 [28]
Li⁺ diffusion coefficient in amorphous Si $D_{Li^+\text{-}Si}$	1×10^{-16} m²/s [29]
Si maximum capacity c^{max}	3.367×10^5 mol/m³ [27]

To study the initiated crack patterns, a few assumptions are made. A Si thin layer is coated on a metal support substrate. The substrate is also used as a current collector of the electrode to transport electrons (e⁻) to the external electric field. The cracks are considered to be generated only during the de-lithiation process due to the volume shrinkage of Si layer and are propagated vertically (perpendicular to substrate). Fig. 2 illustrates the schematic of the crack initiation model with a flat substrate at the bottom. Note that, the substrate can be changed to a curved one with designated curvature depending on specific circumstances. The coated Si layer is divided into small cubes with a length and width of 10×10 nm, and the height of the cubes represents the thickness of the Si coating layer and is treated as a controllable input parameter. Each of the individual cube is bonded to its adjacent ones obeying CZM. Besides, Si layer is also adhered to the substrate. Periodic boundary condition is applied to all four vertical surfaces to reflect the bulk Si. During the de-lithiation process, cracks can be formed inside Si along the cohesive surfaces Γ_{Si}; meanwhile, a delamination may also occur at the Si-substrate interface $\Gamma_{Interface}$, resulting in a slipping of Si cubes. Therefore, the overall crack initiation and propagation performances are the competition results of the two crucial forces (fracture criteria), cracking force of Si F_c and slipping force of Si-substrate interface F_s [30]: if F_c is exceeded by the de-lithiation-induced stress at certain cohesive surface of Γ_{Si}, a crack is initiated; correspondingly, if F_s is exceeded at $\Gamma_{Interface}$, Si layer will delaminate from substrate and move/slip along with surrounding cubes. In CZM model, F_c and F_s are evaluated via the cohesive strength σ_{max} and fracture toughness G_{1C}, where the σ_{Fc} and $G_{1C\text{-}Fc}$ of cracking force are those of the yield strength and toughness of Si material; and σ_{Fs} and $G_{1C\text{-}Fs}$ of slipping force reflect the properties of the Si-substrate interface. The ratio $k = \sigma_{Fc}/\sigma_{Fs}$ is also considered as a controllable variable in this study to investigate the crack initiation properties. After full de-lithiation, cracks may be generated along either Γ_{Si} or $\Gamma_{Interface}$ to form several Si islands. The areas of the Si islands separated by cracks are used to analyze the crack initiation and propagation performances.

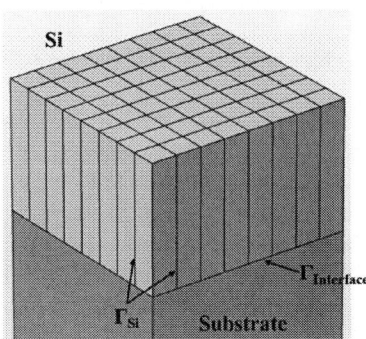

FIGURE 2. Schematic of the crack initiation model.

Annealed nickel is chosen as the current collector substrate. Ni substrate endures elastoplastic deformation. The mechanical properties of nickel (annealed) are obtained from online resources (MatWeb) with Young's modulus E_{Ni} = 200 GPa, Poisson's ratio v_{Ni} = 0.30 and yield strength σ_{Ni} = 600 MPa, (assuming perfect plasticity). As for silicon active material, the modulus decreases in a linear manner with increasing lithium concentration leading to significant elastic softening. The non-constant Young's modulus and Poisson's ratio of silicon with respect to the lithiation state are shown in Fig. 3 [31]. Silicon also bears elastoplastic deformation, with the yield strength σ_y = σ_{Fc} =170 MPa, and toughness G = $G_{1C\text{-}Fc}$ =15 J/m². The Si layer and Ni substrate are in contact with no inter-penetration.

Note that, by calibrating the mechanical parameters of the active material and substrate with experiments or theoretical calculation, the model can be widely applied to thin film LIB electrodes with other active materials.

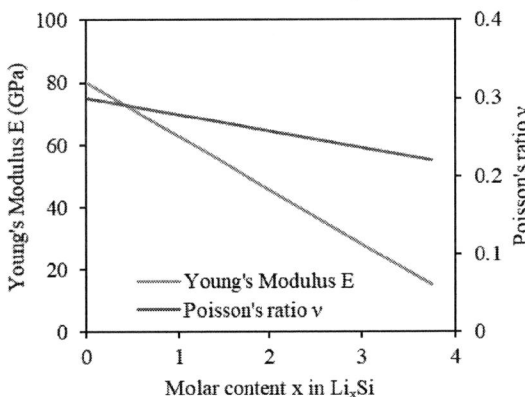

FIGURE 3. Non-constant young's modulus and poisson's ratio of silicon as a function of molar content x in Li$_x$Si from 0 to 3.75.

3. GAUSSIAN PROCESS SURROGATE MODELING

The finite element multi-physics model introduced above is usually computationally expensive, especially when the design space is high dimensional. To further improve the efficiency to explore the electrode design space without losing accuracy, a GP based surrogate model is introduced. GP surrogate model is a statistical procedure that generates an estimated surface from a scattered set of points via a covariance governed Gaussian process interpolation method. It requires training samples for model construction and predicting responses of new sample points [32].

Assume the true performance function (the area of Si island in this case) G(x) has a form of

$$G(\mathbf{x}) \approx G_{GP}(\mathbf{x}) = \boldsymbol{f}^T(\mathbf{x}) \cdot \boldsymbol{\alpha} + S(\mathbf{x}), \qquad (6)$$

where, $f^T(\mathbf{x}) = \left[f_1(\mathbf{x}), ..., f_b(\mathbf{x}) \right]$ is a basis function, $\boldsymbol{\alpha} = \left[\alpha_1, ..., \alpha_b \right]$ is a regression coefficient vector, and $S(\mathbf{x})$ is a Gaussian stochastic process with zero mean and certain covariance matrix. The covariance function between two input x_i and x_j is expressed as

$$\mathbf{Cov}_{(i,j)} = \sigma^2 \cdot \exp\left[-\sum_{p=1}^{k} a_p \left| x_i^p - x_j^p \right|^{b_p} \right], \quad (7)$$

where, a_p and b_p are parameters of the GP model, and k is the number of the input parameters $\mathbf{x}$. The GP model is then trained with the imported simulation results $(\mathbf{X_S}, \mathbf{G_S})$ to generate the initial battery performance function $G(\mathbf{x})$. For any new point $\mathbf{x'}$, its performance can be estimated via maximizing likelihood function [33] as a normal distributed variable with mean $\hat{G}(\mathbf{x'})$ and variance $\hat{e}(\mathbf{x'})$:

$$\hat{G}(\mathbf{x'}) = \mu + \mathbf{r}^T \mathbf{R}^{-1}(\mathbf{G} - \mathbf{A}\mu), \quad (8)$$

$$\hat{e}(\mathbf{x'}) = \sigma^2 \left[1 - \mathbf{r}^T \mathbf{R}^{-1} \mathbf{r} + \left(1 - \mathbf{A}^T \mathbf{R}^{-1} \mathbf{r} \right)^2 \Big/ \mathbf{A}^T \mathbf{R}^{-1} \mathbf{A} \right], \quad (9)$$

where, $\mathbf{r}$ is the correlation vector between $\mathbf{x'}$ and the input parameters $\mathbf{X}$.

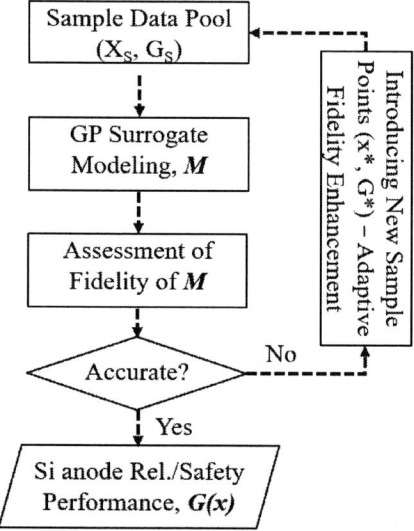

FIGURE 4. Flow chart of developing the GP model with enhanced adaptive fidelity.

Influenced by the sampling situations of the data pool, the initial GP model may potentially have low fidelity in some regions of the electrode design space. Hence, the maximum expected information value (MEIV)-based adaptive sampling method is further used to improve the model fidelity. Fig. 4 illustrates the flow chart of developing the GP model with enhanced adaptive fidelity. In general, three consecutive steps are involved in the approach: (1) an initial set of sample points $(\mathbf{X_S}, \mathbf{G_S})$ are generated by FEA model to develop the initial GP model, where $\mathbf{X_S}$ is the model input parameters, and $\mathbf{G_S}$ is the corresponding output performance. In this study, random sampling strategy is used to select the input $\mathbf{X_S}$. (2) In the next step, the simulation results $(\mathbf{X_S}, \mathbf{G_S})$ are stored in a sample data pool and used to construct the predictive models M employing the GP based surrogate modeling technique [33-35]. (3) The most important sample points $\mathbf{x^*}$ are then selected using the MEIV-based adaptive sampling technique. The performances of these points $\mathbf{G^*}$ are evaluated with the FEA model, and the new sample points $(\mathbf{x^*}, \mathbf{G^*})$ are imported into the sample data pool to upgrade the surrogate model. (4) Cumulative confidence level (CCL) [33] is used to qualify the accuracy of reliability of the surrogate model. Eventually, the updated GP model acquires enhanced fidelity, which can be used for electrode performance prediction and optimization design.

Table 2 lists the main design variables (inputs) and their ranges considered in this study. The ratio $k = \sigma_{Fc}/\sigma_{Fs}$ controls the two failure mechanisms of the anode structure, i.e. whether the Si layer will break and generate cracks (F_c), or the coated Si material will delaminate from the substrate (F_s). With σ_{Fc} as an intrinsic property of Si material and kept as constant (170 MPa), k varies along with σ_{Fs} from 0.1 to 2.0, where smaller k indicates that Si is more vulnerable to fracture, and larger k implies the anode has lower resistance to delamination. The influence of the thickness of Si coating layer on crack initiation performance is also explored from 20 to 650 nm. Last but not least, the impact of flat and curved substrates is investigated, with the mean curvature of the substrate spanning from -0.6 to 0.6.

TABLE 2. Main design variables of the silicon anode and their ranges

Design Variable	Initial Design	Design Range
$k = F_c/F_s$	0.5	$0.1 \sim 2.0$
T_{Si} (nm)	200	$20 \sim 650$
Mean curvature (μm^{-1})	0 (Flat)	$-0.6 \sim 0.6$

4. RESULTS AND DISCUSSTION

4.1 Electrochemical Modeling Results

The change profile of lithium concentration during the de-lithiation process is simulated via the electrochemical sub-model. The de-lithiation process is assumed to be conducted under the constant current (CC) protocol with a very slow de-lithiation rate (C/5, i.e. the discharge current will discharge the entire battery in 5 hours) to eliminate the non-uniform lithium concentration in the Si active material. Fig. 5a illustrates the calculated state of charge (SoC) of the Si anode during de-lithiation process, where SoC is a measure of the present battery capacity as a percentage of its maximum capacity. Under the CC protocol, the lithium concentration decreases in a linear manner with respect to time. The SoC of Si along the thickness direction of the thin anode layer after de-lithiation are illustrated in Fig. 5b, where the thickness $\hat{T}_{Si}$ is a normalized value. $\hat{T}_{Si} = 0$ represents the Si-substrate contact interface $\Gamma_{Interface}$, and $\hat{T}_{Si} = 1$ denotes the Si-electrolyte surface. It can be seen that, the lithium concentration gradient along the thin anode layer is rather small. This can be mainly attributed to the small lithium diffusion length along the thin Si layer (at nano-scale), relatively large diffusivity of lithium in Si, and slow discharging rate. In the following, the distribution of lithium in the Si layer is treated as homogeneous at all time.

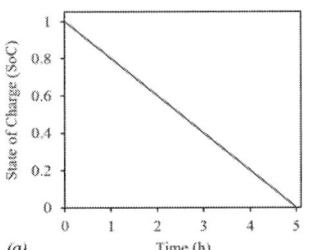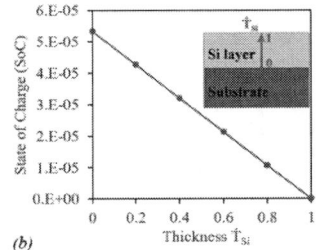

FIGURE 5. (a) Simulated state of charge (SoC) of the Si anode during de-lithiation; (b) SoC along the thickness direction of the Si layer after fully de-lithiated. The inset illustrates the normalized thickness axis, where $\hat{t}_{Si} = 0$ represents the Si-Ni interface and $\hat{t}_{Si} = 1$ is the Si-electrolyte surface.

4.2 Crack Initiation on Flat Substrate

Fig. 6 illustrates the representative simulation results of the crack patterns (top view) of Si layers with different thicknesses coated on a flat substrate after de-lithiation/volume shrinkage. The red dash line circled regions are the portion of the Si layer that does not crack apart, i.e. the Si islands. Although, due to the modeling assumptions, the Si island has a square shape which is not accurate comparing with experiments, the model can still capture the general trend of crack initiation behaviors, and thus be used to analyze the impacts of critical design parameters. It is observed that, with the increase of the thickness of Si layer, the area of the Si island increases. Fig. 7a shows the gap between adjacent Si cubes upon the de-lithiation process, where the cubes are adhered to the substrate and shrink (ΔV, labelled as green arrow). It can be observed, the gap distance at the top surface is ~ 2.3 nm and keeps decreasing downwards; when reaching the bottom, the gap reduces to 0, meaning the adjacent cubes are still combined at this moment and the crack is not completely formed yet. This result suggests the process of the formation of cracks, as illustrate in Fig. 7b: (1) during de-lithiation, the Si layer undergoes a tensile stress induced by the volume reduction; (2) in the beginning, when the deformation starts to grow, the tensile stress $\sigma_{Tensile}$ increases correspondingly (Eq. 4); (3) when it exceeding the yield strength of Si $\sigma_y = \sigma_{Fc}$, the crack is initiated from the top surface; (4) with the crack gradually approaching the bottom, the propagation of the gap is retarded by the adhered interface σ_{Fs}; (5) eventually, when deformation-induced stress $\sigma_{Tensile}$ overcomes the synergistic effect of σ_{Fc} and σ_{Fs}, the Si island is thoroughly separated by the crack.

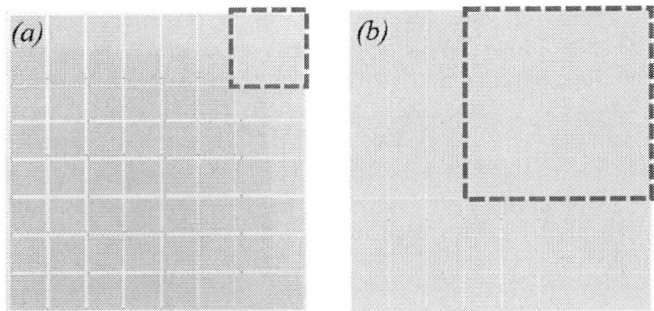

FIGURE 6. Top view of the Si layers with different thicknesses T_{Si} coated on a flat substrate after full de-lithiation: (a) 75 nm, (b) 400 nm.

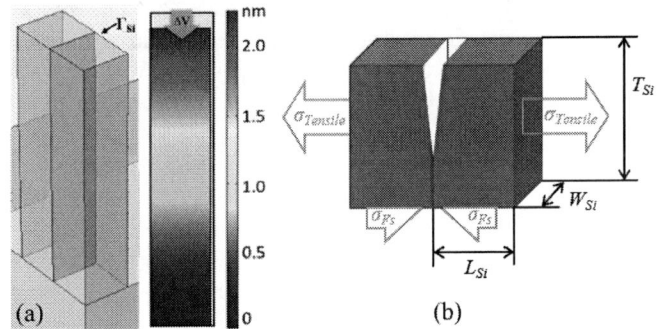

FIGURE 7. (a) Gap distance between two adjacent Si cubes during the de-lithiation process; (b) Schematic of the crack formation process.

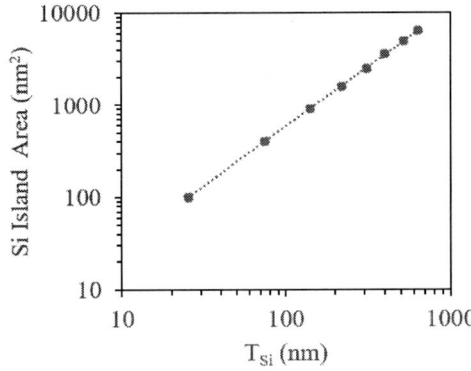

FIGURE 8. A scaling relationship between the area of Si island and the film thickness T_{Si}.

Fig. 8 plots the area of Si island with respect to the thickness of the coating layer T_{Si}, where the fracture criteria ratio k is fixed as 0.5. A scaling relationship is observed, indicating the area A is in an exponential manner with T_{Si}: $A \propto T_{Si}^n$. n is approximated as 1.3 in this study. The exponential relationship between A and T_{Si} is also reported somewhere else [10]. The causes of this phenomenon can be explained as follows. With very slow deformation rate (discharge rate C/5), the crack propagation can be considered as static. By applying the static force equilibrium condition (Fig. 7b), we have $\sigma_{Tensile} \times (T_{Si} \times W_{Si}) = \sigma_{Fs} \times (L_{Si} \times W_{Si})$, where T_{Si}, W_{Si} and L_{Si} are the thickness, width and length of the Si island, respectively. Therefore, along with the increase of T_{Si}, L_{Si} also grows, i.e. the area of Si island increases. Besides, the power n is largely influenced by the mechanical data assigned in the model, including modulus of the active material, yield strength σ_y, toughness of Si, and the adhesive properties at Si-substrate interfaces. Hence, with changes of these parameter values, the value n also varies.

In addition to T_{Si}, the impact of the adhesive properties of the Si-substrate interfaces is also investigated. Fig. 9 shows the relationship between Si island area and the fracture criteria ratio k, where the thickness of Si layer is fixed at 200 nm, and $k = \sigma_{Fc}/\sigma_{Fs}$ is controlled by changing the cohesive strength σ_{Fs}. A second order polynomial relationship is found between Si island area A and ratio k. The area of Si island decreases along with the ratio k. This is because, smaller k suggests that the cohesive strength σ_{Fs} is enhanced, so that the Si layer becomes more tightly adhered on the substrate; in other word, the Si layer becomes relatively vulnerable to cracking. As a result, to absorb the same amount of strain energy caused by the Si volume shrinkage, systems with small k need to form more cracks, leading to the reduction of the Si island area.

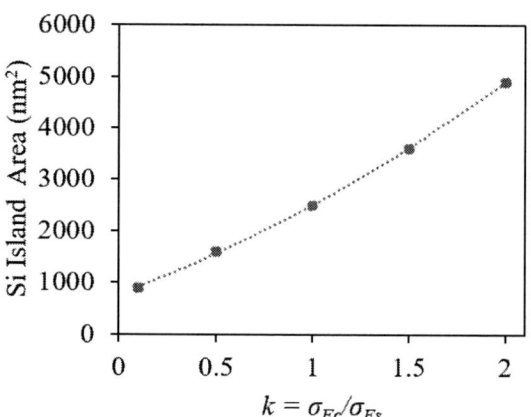

FIGURE 9. Si island area as a function of the fracture criteria ratio k.

4.3 The Impact of Curved Substrate

As mentioned above, with the improvement of electrode fabrication technology, much more sophisticated Si anode structures are proposed in the recent years. One of the common features among the novel electrodes is that, the current collector is usually elaborately designed and fabricated into a 3 dimensional bi-continuous supporting scaffold, so that a thin layer of Si active material can be conformally coated on the scaffold [13, 36]. The major advantages of these kind of structures are that these architectures achieve high power density by simultaneously reducing the ion and electron transport lengths; meanwhile, they provide large porosity that allows Si material to expand without causing pulverization. Among these electrodes, however, Si layer often needs to be coated on a curved substrate rather than a simple flat one. Therefore, the influences of the curvature of the substrate are studied using the 3D FE model.

Fig. 10 illustrates the area of Si island with respect to the curvature of the substrate. The two insets are examples of the Si anodes with negative (-0.5 μm^{-1}) and positive curvatures (0.5 μm^{-1}), respectively. The thickness of the Si layer is fixed at 200 nm, and ratio k is kept as 0.5. Substrate with positive curvatures leads to a reduction of Si island area, suggesting an exacerbated fracture scenario. On the contrary, when the curvature is negative and decreases, a larger area of Si island is found, indicating that substrates with negative curvature tend to prevent the formation of cracks. This is because, as shown in Fig. 11, for the negatively curved substrate, the Si cube has a reduced area from the bottom up. In this case, the deformation-induced stress, especially in the beginning state of de-lithiation process, is very small and even become negative/compressive (-σ). Consequently, the initiation of crack is restrained on the Si surface and the area of the Si island is largely increased. On the other hand, the area of Si cube coated on a positively curved substrate gradually increases from the bottom up, which leads to an intensified tensile deformation stress and an aggravated crack initiation. What's more, according to our previous works, it is found that, with the same Si-substrate contact properties, such as the same cohesive strength and fracture toughness, the contact interface with a negative curvature tends to be more vulnerable to delamination than flat substrate, which benefits the integrality of the Si layer indirectly.

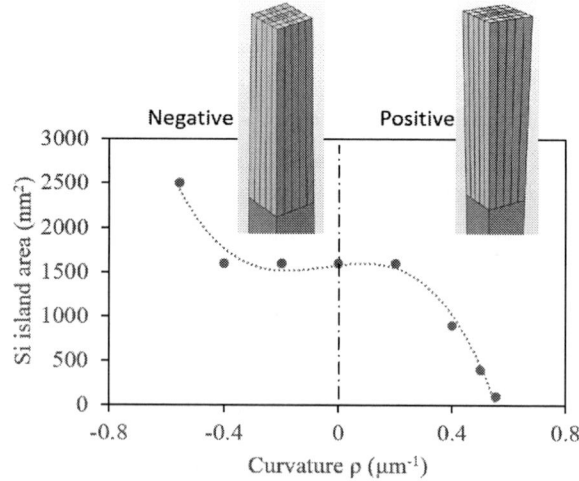

FIGURE 10. The area of Si island with respect to the curvature of the substrate. The insets are the Si anodes with curvatures of -0.5 and 0.5 μm^{-1}, respectively.

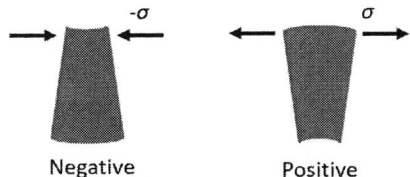

FIGURE 11. Schematics of the stress analysis of the Si cubes during the beginning of de-lithiation.

4.4 GP Surrogate Model Predictions

GP based surrogate model is used to predict the Si island area within the design space. FE based simulation results with different Si anode design parameters, i.e. fracture ratio k, Si layer thickness T_{Si} and curvature of the substrate ρ, are used as training data to develop the surrogate model. And the cumulative confidence level (CCL) [33] is adopted to qualify the accuracy of reliability of the surrogate model. In order to have a clear comparison, the Si cracking performance is analyzed using two performance surfaces, as illustrated in Fig. 12-13. Fig. 12 shows the predicted cracking areas of the Si layer on a flat substrate (curvature $\rho = 0$) with respect to T_{Si} and ratio k. Within the whole parameter space, a monotonically increasing surface is generated from (20 nm, 0.1) to (600 nm, 2.0) corner. Similarly, as illustrated in Fig. 13, a relatively smooth performance surface is observed. By increasing T_{Si} or curvature ρ, the Si island area also increases.

Fig. 14 illustrates the reliability (CCL value) of the GP surrogate model with increasing of data points used to train the model. It can be seen that, the reliability of the model is gradually improved with the increase of numbers of the sample points and reaches as high as 0.997 when 20 training data points are used. This demonstrates that the developed GP surrogate model can rather precisely predict the performance of the Si anode within the design space range and can be further used for design optimization.

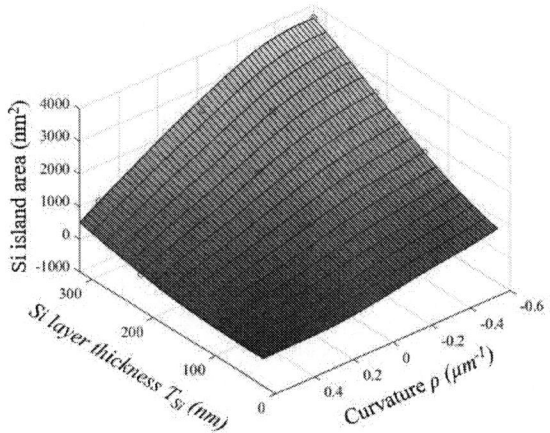

FIGURE 13. Si island area surface w.r.t. the Si layer thickness T_{Si} and curvature of the substrate ρ. The red scatters are the simulated results.

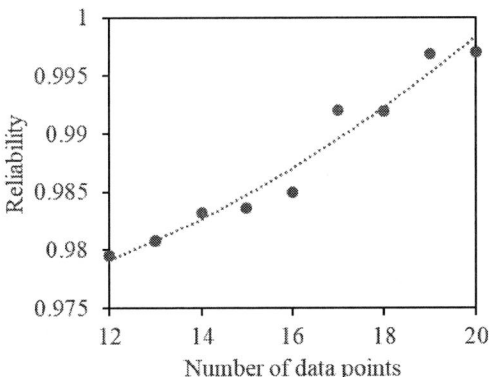

FIGURE 14. Reliability of the GP based surrogate model with increasing of data points.

5. CONSLUSION

In this work, we studied the crack initiation performances of Si anode during the de-lithiation process. Multi-physics-based simulations coupled with GP based surrogate model are utilized to investigate the formation of cracks on Si layer and the critical factors that largely influence the cracking behaviors. By applying a constant current discharging profile with a low de-lithiation rate, the Li$^+$ concentration in Si active material is decreasing linearly, and the concentration gradient along the Si layer thickness direction is rather small, which is thus neglected by assuming a uniform Li$^+$ concentration distribution in Si. The thickness of the Si coating layer T_{Si}, the yield strength σ_{Fc} of Si material, the cohesive strength between Si and substrate σ_{Fs}, and the curvature of the substrate ρ are studied as the main factors that affect the formation of cracks. The area of the Si island has a scaling relationship with T_{Si}; meanwhile the increase of cohesive strength σ_{Fs} of Si-substrate interface could lead to larger Si island area. Furthermore, the curvature of the substrate plays an important role in the crack initiation behaviors of Si.

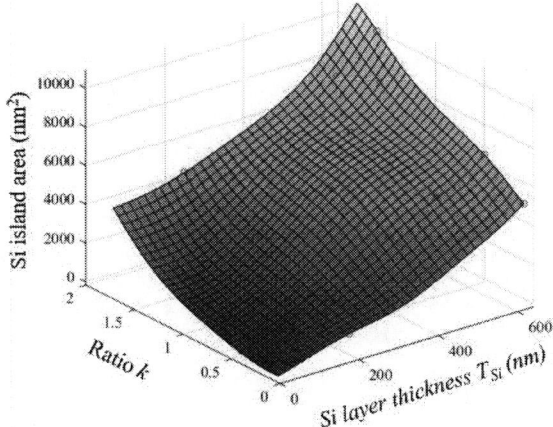

FIGURE 12. Si island area performance surface w.r.t. the Si layer thickness T_{Si} and fracture ratio k. The red scatters are the simulated results.

Substrate with negative mean curvature could prevent the generation of crack; positively curved substrate, on the contrary, exacerbates the cracking formation. The GP based surrogate model is further implemented to predict the cracking properties in the design space. GP surrogate model is found to be a powerful tool to effectively and efficiently explore the design space and assist further optimization design. It is worth mentioning that, by adjusting and calibrating the parameters of the models, this finite element simulation coupled GP surrogate model framework can be extensively used to analyze the properties of the electrodes of other active materials and with complicated geometry structures.

ACKNOWLEDGMENTS

This research is partially supported by the National Science Foundation through the Faculty Early Career Development award (CMMI-1813111), the National Science Foundation Engineering Research Center for Power Optimization of Electro Thermal Systems (POETS) with cooperative agreement EEC-1449548, and the Office of Naval Research (ONR) through the Defense University Research-to-Adoption (DURA) Initiative (N00014-18-S-F004).

REFERENCES

[1] Tarascon, J. M. and Armand, M. "Issues and challenges facing rechargeable lithium batteries." *Nature* VOL. 414 (2001) pp.359-367.

[2] McDowell, M. T., Lee, S. W., Nix, W. D. and Cui, Y. "25th Anniversary Article: Understanding the Lithiation of Silicon and Other Alloying Anodes for Lithium-Ion Batteries." *Advanced Materials* VOL. 25 (2013) pp.4966-4984.

[3] Armand, M. and Tarascon, J. M. "Building better batteries." *Nature* VOL. 451 (2008) pp.652-7.

[4] Limthongkul, P., Jang, Y.-I., Dudney, N. J. and Chiang, Y.-M. "Electrochemically-driven solid-state amorphization in lithium-silicon alloys and implications for lithium storage." *J Acta Materialia* VOL. 51 (2003) pp.1103-1113.

[5] Kasavajjula, U., Wang, C. and Appleby, A. J. "Nano-and bulk-silicon-based insertion anodes for lithium-ion secondary cells." *J Journal of Power Sources* VOL. 163 (2007) pp.1003-1039.

[6] Boukamp, B. A., Lesh, G. C. and Huggins, R. A. "All‐Solid Lithium Electrodes with Mixed‐Conductor Matrix." *J. Electrochem. Soc.* VOL. 128 (1981) pp.725-729.

[7] Kasavajjula, U., Wang, C. and Appleby, A. J. "Nano- and bulk-silicon-based insertion anodes for lithium-ion secondary cells." *Journal of Power Sources* VOL. 163 (2007) pp.1003-1039.

[8] Yang, H., Fan, F., Liang, W., Guo, X., Zhu, T. and Zhang, S. "A chemo-mechanical model of lithiation in silicon." *Journal of the Mechanics and Physics of Solids* VOL. 70 (2014) pp.349-361.

[9] Liu, X. H., Zheng, H., Zhong, L., Huang, S., Karki, K., Zhang, L. Q., Liu, Y., Kushima, A., Liang, W. T. and Wang, J. W. "Anisotropic swelling and fracture of silicon nanowires during lithiation." *Nano letters* VOL. 11 (2011) pp.3312-3318.

[10] Li, J., Dozier, A. K., Li, Y., Yang, F. and Cheng, Y.-T. "Crack pattern formation in thin film lithium-ion battery electrodes." *Journal of The Electrochemical Society* VOL. 158 (2011) pp.A689-A694.

[11] Purkayastha, R. and McMeeking, R. "A parameter study of intercalation of lithium into storage particles in a lithium-ion battery." *Computational Materials Science* VOL. 80 (2013) pp.2-14.

[12] Christensen, J. and Newman, J. "Stress generation and fracture in lithium insertion materials." *Journal of Solid State Electrochemistry* VOL. 10 (2006) pp.293-319.

[13] Zhang, H. G. and Braun, P. V. "Three-Dimensional Metal Scaffold Supported Bicontinuous Silicon Battery Anodes." *Nano Letters* VOL. 12 (2012) pp.2778-2783.

[14] Li, J., Wang, J., Yang, J., Ma, X. and Lu, S. "Scalable synthesis of a novel structured graphite/silicon/pyrolyzed-carbon composite as anode material for high-performance lithium-ion batteries." *J Journal of Alloys Compounds* VOL. 688 (2016) pp.1072-1079.

[15] Yang, J., Wang, Y.-X., Chou, S.-L., Zhang, R., Xu, Y., Fan, J., Zhang, W.-x., Liu, H. K., Zhao, D. and Dou, S. X. J. N. E. "Yolk-shell silicon-mesoporous carbon anode with compact solid electrolyte interphase film for superior lithium-ion batteries." VOL. 18 (2015) pp.133-142.

[16] Ng, S. S. Y., Xing, Y. and Tsui, K. L. "A naive Bayes model for robust remaining useful life prediction of lithium-ion battery." *Applied Energy* VOL. 118 (2014) pp.114-123.

[17] Meng, Z., Zhang, D., Liu, Z. and Li, G. "An Adaptive Directional Boundary Sampling Method for Efficient Reliability-Based Design Optimization." *Journal of Mechanical Design* VOL. 140 (2018) pp.121406-121406-12.

[18] Fan, X., Wang, P. and Hao, F. "Reliability-based design optimization of crane bridges using Kriging-based surrogate models." *J Structural Multidisciplinary Optimization* VOL. (2019) pp.1-13.

[19] Cho, H. H., Glazer, M. P. B., Xu, Q., Han, H. N. and Dunand, D. C. "Numerical and experimental investigation of (de)lithiation-induced strains in bicontinuous silicon-coated nickel inverse opal anodes." *Acta Materialia* VOL. 107 (2016) pp.289-297.

[20] Park, J., Lu, W. and Sastry, A. M. "Numerical Simulation of Stress Evolution in Lithium Manganese Dioxide Particles due to Coupled Phase Transition and Intercalation." *Journal of the Electrochemical Society* VOL. 158 (2011) pp.A201-A206.

[21] Verbrugge, M. W. and Cheng, Y. T. "Stress and Strain-Energy Distributions within Diffusion-Controlled Insertion-Electrode Particles Subjected to Periodic Potential

Excitations." *Journal of the Electrochemical Society* VOL. 156 (2009) pp.A927-A937.

[22] Cheng, Y. T. and Verbrugge, M. W. "Evolution of stress within a spherical insertion electrode particle under potentiostatic and galvanostatic operation." *Journal of Power Sources* VOL. 190 (2009) pp.453-460.

[23] De-Andrés, A., Pérez, J. and Ortiz, M. "Elastoplastic finite element analysis of three-dimensional fatigue crack growth in aluminum shafts subjected to axial loading." *J International Journal of Solids Structures* VOL. 36 (1999) pp.2231-2258.

[24] Deshpande, V. S., Needleman, A. and Van der Giessen, E. "A discrete dislocation analysis of near-threshold fatigue crack growth." *Acta Materialia* VOL. 49 (2001) pp.3189-3203.

[25] Maiti, S. and Geubelle, P. H. "Cohesive modeling of fatigue crack retardation in polymers: Crack closure effect." *Engineering Fracture Mechanics* VOL. 73 (2006) pp.22-41.

[26] Bower, A. F. and Guduru, P. R. "A simple finite element model of diffusion, finite deformation, plasticity and fracture in lithium ion insertion electrode materials." *Modelling and Simulation in Materials Science and Engineering* VOL. 20 (2012)

[27] Cho, Y. H., Booh, S., Cho, E., Lee, H. and Shin, J. "Theoretical prediction of fracture conditions for delithiation in silicon anode of lithium ion battery." *Apl Materials* VOL. 5 (2017)

[28] Xu, K. "Nonaqueous liquid electrolytes for lithium-based rechargeable batteries." *Chemical Reviews* VOL. 104 (2004) pp.4303-4417.

[29] Ding, N., Xu, J., Yao, Y. X., Wegner, G., Fang, X., Chen, C. H. and Lieberwirth, I. "Determination of the diffusion coefficient of lithium ions in nano-Si." *Solid State Ionics* VOL. 180 (2009) pp.222-225.

[30] Li, J. C., Dozier, A. K., Li, Y. C., Yang, F. Q. and Cheng, Y. T. "Crack Pattern Formation in Thin Film Lithium-Ion Battery Electrodes." *Journal of the Electrochemical Society* VOL. 158 (2011) pp.A689-A694.

[31] Feifei, F., Shan, H., Hui, Y., Muralikrishna, R., Dibakar, D., Vivek, B. S., Adri, C. T. v. D., Sulin, Z. and Ting, Z. "Mechanical properties of amorphous Li x Si alloys: a reactive force field study." *Modelling and Simulation in Materials Science and Engineering* VOL. 21 (2013) pp.074002.

[32] Rubinstein, R. Y. and Kroese, D. P., *Simulation and the Monte Carlo method*. John Wiley & Sons (2016).

[33] Wang, Z. Q. and Wang, P. F. "A Maximum Confidence Enhancement Based Sequential Sampling Scheme for Simulation-Based Design." *Journal of Mechanical Design* VOL. 136 (2014)

[34] Wang, P. F., Wang, Z. Q. and Almaktoom, A. T. "Dynamic reliability-based robust design optimization with time-variant probabilistic constraints." *Engineering Optimization* VOL. 46 (2014) pp.784-809.

[35] Wang, Z. Q. and Wang, P. F. "A double-loop adaptive sampling approach for sensitivity-free dynamic reliability analysis." *Reliability Engineering & System Safety* VOL. 142 (2015) pp.346-356.

[36] Pikul, J. H., Braun, P. V. and King, W. P. "Performance Modeling and Design of Ultra-High Power Microbatteries." *Journal of the Electrochemical Society* VOL. 164 (2017) pp.E3122-E3131.

Proceedings of the ASME 2019
International Design Engineering Technical Conferences
and Computers and Information in Engineering Conference
IDETC/CIE2019
August 18-21, 2019, Anaheim, CA, USA

DETC2019-97617

GENERATIVE DESIGN OF MULTI-MATERIAL HIERARCHICAL STRUCTURES VIA CONCURRENT TOPOLOGY OPTIMIZATION & CONFORMAL GEOMETRY METHOD

Long Jiang[1]
[1]Department of Mechanical Engineering
State University of New York
at Stony Brook
Stony Brook, NY 11794, USA
Email: long.jiang@stonybrook.edu

Shikui Chen[1,*]
Department of Mechanical Engineering
State University of New York
at Stony Brook
Stony Brook, NY 11794, USA
Email: shikui.chen@stonybrook.edu

Xianfeng David Gu[2]
[2]Department of Computer Science
State University of New York
at Stony Brook
Stony Brook, NY 11794, USA
Email: gu@cs.stonybrook.edu

ABSTRACT

Topology optimization has been proved to be an automatic, efficient and powerful tool for structural designs. In recent years, the focus of structural topology optimization has evolved from mono-scale, single material structural designs to hierarchical multimaterial structural designs. In this research, the multi-material structural design is carried out in a concurrent parametric level set framework so that the structural topologies in the macroscale and the corresponding material properties in mesoscale can be optimized simultaneously. The constructed cardinal basis function (CBF) is utilized to parameterize the level set function. With CBF, the upper and lower bounds of the design variables can be identified explicitly, compared with the trial and error approach when the radial basis function (RBF) is used. In the macroscale, the 'color' level set is employed to model the multiple material phases, where different materials are represented using combined level set functions like mixing colors from primary colors. At the end of this optimization, the optimal material properties for different constructing materials will be identified. By using those optimal values as targets, a second structural topology optimization is carried out to determine the exact mesoscale metamaterial structural layout. In both the macroscale and the mesoscale structural topology optimization, an energy functional is utilized to regularize the level set function to be a distance-regularized level set function, where the level set function is maintained as a signed distance function along the

design boundary and kept flat elsewhere. The signed distance slopes can ensure a steady and accurate material property interpolation from the level set model to the physical model. The flat surfaces can make it easier for the level set function to penetrate its zero level to create new holes. After obtaining both the macroscale structural layouts and the mesoscale metamaterial layouts, the hierarchical multimaterial structure is finalized via a local-shape-preserving conformal mapping to preserve the designed material properties. Unlike the conventional conformal mapping using the Ricci flow method where only four control points are utilized, in this research, a multi-control-point conformal mapping is utilized to be more flexible and adaptive in handling complex geometries. The conformally mapped multi-material hierarchical structure models can be directly used for additive manufacturing, concluding the entire process of designing, mapping, and manufacturing.

1 Introduction

With the rapid development of the structural design methodology and the modern additive manufacturing technology, in recent years, building multiscale structures with space-varying metamaterials has become possible [1, 2, 3]. On the other hand, by introducing different construction materials to be redistributed inside the design domain properly, the final multimaterial structure can be expected to have a performance boost [4]. Besides, introducing multiple constructing materials can make

---

*Address all correspondence to this author.

some design objectives easier to be achieved, compared with using only one material [5]. This paper is aiming at designing hierarchical multimaterial structures

Topology optimization is a powerful and advanced tool for designing structures with high performances. There are a number of different approaches in this field and a comparative review between can be found in [6]. Among those approaches, the level set methods stand out for its flexibility in handling topological changes and in generating clear design boundaries [7, 8, 9]. For more details about the level set methods, the readers can be referred to [10] for more information. Generally, by using the zero level of the level set function to model the structural design boundary [11], one level set function can separate the design domain into two sub-domains: the void domain and the material domain. However, when two or more material phases have to be included, this implicit boundary representation has to be modified accordingly. The 'color' level set method [12], the piecewise constant level set method [13] and the reconciled level set method [14] are several widely used schemes for modeling multiple material phases within the level set framework. In this paper, the 'color' level set method is utilized. Similar to mixing colors from three primary colors, the 'color' level set method uses n different level set functions to represent up to 2^n different material domains. Therefore, the overall effective material property can be interpolated by assembling the separate constitutive material together with the help of the Heaviside function of each level set function.

On the other hand, structures with multiple scales possess fine-tuned mesoscale filling properties with a low overall density [15, 16, 17]. Thus, they can be found in a wide range of engineering applications [18]. In multiscale or hierarchical structure designing, the key issue is to find the constructing metamaterials with tailored properties. Within the level set framework, the metamaterial designs have covered the topics of designing the negative permeability metamaterails [19], the negative Poisson's ratio metamaterials [20, 21], the electromagnetic metamaterials [22], the zero/negative thermal expansion metamaterials [23] and so on. However, only designing the metamaterial to be filled inside the macroscale structure can not fully explore the potential of the multiscale structure. Ideally, the mesoscale structures and the macroscale structures should be designed in a concurrent manner since the macroscopic loading and boundary conditions will affect not only the macroscale overall structural layout, but also the optimal mesoscale metamaterial properties. Sivapuram et al. [24] developed the concurrent structural topology optimization for multiscale structures where the mesoscale metamaterials have fixed pre-defined locations. Wang et al. [25] proposed the concurrent design of multiscale structures filled with spatially-varying graded microstructures to ensure the connectivity between adjacent mesoscale units. Both Sivapuram's work with pre-defined metamaterial locations and Wang's work with similar topological feature metamaterials are all meant to alleviate the heavy computational cost of introducing too many different types of metamaterials in different locations. Another approach for the concurrent design of multiscale structure is proposed by Li et al. [26], where the density based method is employed in the macroscale and the level set approach is employed in the mesoscale. The mesoscale metamaterial is determined by the intermediate density generated from the macroscale optimization. This combination converts the undesired intermediate densities of the density based approach into an advantage of the entire design methodology and the numerical examples have verified the effectiveness of this process.

Conventionally, the level set methods use a virtual velocity field derived from the sensitivity analysis to evolve the design boundary [27, 28, 7] and solve the Hamilton-Jacobi partial differential equation (PDE) to describe the dynamics of the boundary motion [29]. However, this level set framework faces some drawbacks [30, 31]. A promising solution is to employ the parameterized level set method [32]. Combined with mathematical programming and the gradient-based optimizer MMA [33], introducing multiple design constraints can become straightforward and the numerical efficiency can be improved. The optimal structure generated via parametric level set approach has also been reported having the advantage of requiring less prefabrication time for additive manufacturing [34], which is a preferred feature for combining designing with applications. However, when the radial basis function (RBF) is used in the conventional parametric level set method (PLSM) [30, 35], the upper and lower bounds for the design variable can not be explicitly identified. Therefore, in our previous research, a cardinal basis function (CBF) is proposed [36] to replace the RBF kernel function. With CBFs, the design variable bounds can be explicitly set to be the lower and upper bounds of the corresponding level set function, avoiding the trial and error approach when the RBF kernel function is used. Moreover, by introducing a distance regularization energy functional [37] to regularize the shape of the level set function throughout the optimization process, the level set function can be maintained a distance-regularized shape for an accurate material property interpolation. The flat surfaces of the distance-regularized level set function can stabilize the optimization process and can help to create of new holes [36].

This paper further develops our previous concurrent topology optimization of multiscale structures [38, 39] from single material to multiple materials. The 'color' level set method is utilized for the multimaterial representation. On the macroscale, the overall structural topology and the optimal corresponding mesoscale metamaterial properties are optimized simultaneously. By using the optimal metamaterial properties as targets, a second optimization process is carried out to find out the structural layout with an isotropy constraint [40, 41, 42]. The structural topology optimization on both scales are all carried out within the CBF-based parametric level set method framework. The distance regularization energy functional is minimized during the optimization process to maintain the distance-regularized shape of the level set functions. After achieving the structural layouts for both scales, the angle-preserving conformal mapping is employed to finalize the multiscale structure. The angle-preserving mapping characteristic and the isotropy of the mesoscale metamaterials can mathematically ensure the consistency of the material properties after the mapping of the multiscale structure.

The remaining paper is organized as follows. In section 2, the 'color' level set method for designing multimaterial structure is introduced. The CBF kernel functions and the concurrent parametric level set function are detailed in section 3. Section 4 presents the mesoscale metamaterial design and the multi-control-point conformal mapping technique used for multiscale structure finalization. The numerical examples are listed in section 5. The conclusions are drawn in section 6.

2 'Color' Level Set Model for Multimaterial Representation

In the conventional level set representation with a single level set function, the design boundary is implicitly described as the zero level of the one dimensional higher level set function [8, 7]. However, when multiple structure phases are introduced, the number of the level set functions has to increase as well. With the 'color' level set representation, n level set functions can divide the design domain into up to 2^n different regions. The level set functions used can be described as:

$$\begin{cases} \Phi_k(\boldsymbol{x}) > 0, & (\boldsymbol{x} \in \Omega_k \setminus \Gamma_k) \\ \Phi_k(\boldsymbol{x}) = 0, & (\boldsymbol{x} \in \Gamma_k), \quad k = 1,...,n \\ \Phi_k(\boldsymbol{x}) < 0, & (\boldsymbol{x} \in D \setminus \Omega_k \cup \Gamma_k) \end{cases} \quad (1)$$

In Eq. 1, Φ_k denotes the kth level set function and Ω_k, Γ_k represent the region where the kth level set function has positive value and its corresponding boundary, respectively. D represents the design domain. The example of identifying different regions inside the design domain by the sign of the level set function is illustrated in Figure 1.

With this 'color' level set representation, each material phase can be represented by combining different level set functions together. For example, the elastic tensor of a two-material structure at a given point $\boldsymbol{x}$ can be expressed as:

$$\boldsymbol{D}^{(2)}(\boldsymbol{x}, \Phi) = H(\Phi_1)\{[1 - H(\Phi_2)]\boldsymbol{D}_1 + H(\Phi_2)\boldsymbol{D}_2\} + [1 - H(\Phi_1)]\boldsymbol{D}_0. \quad (2)$$

In Eq.2, the $\boldsymbol{D}^{(2)}$ represents the interpolated elastic tensor with two materials. The $H(\Phi)$ is the Heaviside function of the level set function Φ. $\boldsymbol{D}_1$ and $\boldsymbol{D}_2$ are the elastic tensors of two different constructing materials. The $\boldsymbol{D}_0$ is a dummy elastic tensor with a small positive value to avoid singularities. Generally, as

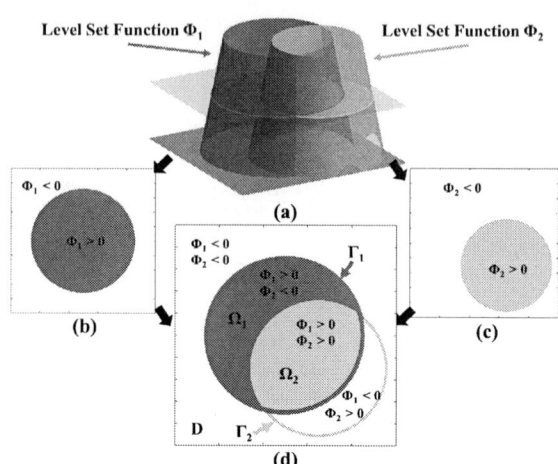

FIGURE 1. The 'color' level set representation. (a) The two level set functions. (b) The sign of level set function 1. (c) The sign of level set function 2. (d) The combination of two level set functions. In this research Ω_1 represents the 1st material phase with boundary Γ_1 and Ω_2 represents the 2nd material phase with boundary Γ_2. The reset regions are considered as void inside the design domain D.

can be seen from Figure 1, the level set function Φ_1 is used to distinguish the material region from the void region. Next, inside the material region, the level set function Φ_2 is used to determine whether the given region should have the material property 1 or 2. With this 'color' level set, the material property inside the design domain can be calculated. Although the whole process is similar to the multimaterial representation of the density-based approach, the 'color' level set scheme can retain the clear design boundary advantage of the level set methods.

3 Concurrent Macro-scale and Meso-scale Optimization with CBF-based Parametric Level Set Method (PLSM)

With a given kernel function at the jth node as Ψ_j, the kth level set functions for multimaterial representation can be parameterized into the following form:

$$\Phi_k(\boldsymbol{x}) = \sum_{j=1}^{m} \Psi_j(\boldsymbol{x})\mu_{kj}, \quad k = 1,...n. \quad (3)$$

In conventional parametric level set method, the kernel function is commonly selected as the RBF. However, with a given support radius, the neighbouring RBF kernel functions will overlap with each other. Therefore, the corresponding weights, namely μ_{kj} in Eq.3, do not have a clear upper and lower bounds. As the design variables [30], those bounds should be passed to the optimizer explicitly. This issue can be solved by constructing the cardinal basis function (CBF) as the kernel function for the level set function parameterization [36]. The CBF has the Kronecker delta property as:

$$\Psi_j(x_i) = \begin{cases} 1, & (if \ i = j) \\ 0, & (if \ i \neq j) \end{cases} \quad j = 1,...,m. \quad (4)$$

When the CBF is used for the level set function parameterization, the corresponding weights will be the upper and lower bounds of the level set function itself. This explicit bounds can maintain the numerical stability of the optimizer and avoid the trial and error approach for guessing those bounds.

Generally, a multimaterial optimization for minimal mean structural compliance can be formulated as:

$$
\begin{aligned}
Min: \quad J \quad &= \int_D \boldsymbol{\varepsilon}(\boldsymbol{u}) : \boldsymbol{D}^* : \boldsymbol{\varepsilon}(\boldsymbol{u}) d\Omega \\
s.t.: \quad a \quad &(\boldsymbol{u},\boldsymbol{v},\Phi) = l(\boldsymbol{v},\Phi) \\
&Vol_k \leq Vol_k^t, \quad (k = 1,...,n) \\
&\mu_{kj}^L \leq \mu_{kj} \leq \mu_{kj}^U, \quad (j = 1,...,m).
\end{aligned} \quad (5)
$$

In Eq.5, the $\boldsymbol{\varepsilon}(\boldsymbol{u}) : \boldsymbol{D}^* : \boldsymbol{\varepsilon}(\boldsymbol{u})$ represents the stain energy density of the structure with a elastic tensor $\boldsymbol{D}^*$ calculated from the aforementioned 'color' level set multimaterial representation. $\boldsymbol{u}$ is the displacement field and $\boldsymbol{v}$ is the test function. The kth material has the volume of Vol_k that is constrained by its volume target Vol_k^t. At most, the number of different material phases can reach to 2^n. However, in this paper, this is simplified to only n different phases. The lower and upper bounds for the design variable μ_{kj} can be easily get from the upper and lower bound of the corresponding level set function. The energy bilinear form $a(\boldsymbol{u},\boldsymbol{v},\Psi)$ and the load linear form $l(\boldsymbol{v},\Psi)$ are detailed as:

$$
\begin{aligned}
a \ (\boldsymbol{u},\boldsymbol{v},\Phi) &= \int_D \boldsymbol{\varepsilon}(\boldsymbol{u}) : \boldsymbol{D}^* : \boldsymbol{\varepsilon}(\boldsymbol{u}) d\Omega \\
l \ (\boldsymbol{v},\Phi) &= \int_\Gamma \boldsymbol{t} \cdot \boldsymbol{v} d\Gamma.
\end{aligned} \quad (6)
$$

To calculate the volume for each material phase, the following equation can be formulated:

$$Vol_k = \int_D \prod_{i=1}^{k} H(\Phi_i) d\Omega, \quad i = 1,...,n. \quad (7)$$

This volume fraction formulation can be understood in a more intuitive way. For example, when 2 level set functions are used, the total number of potential material phases can reach up

to $2^2 = 4$. However, in this research for simplification, 2 level set functions are used to only represent 2 material phases:

$$Vol_1 = \int_D H(\Phi_1) d\Omega \quad , \quad Vol_2 = \int_D H(\Phi_1) H(\Phi_2) d\Omega. \quad (8)$$

Here, Vol_1 is utilized to calculate the total volume and Vol_2 is used to calculate one of the two separate material volumes. The other material volume can be easily achieved by finding the difference between these two.

When combined with the gradient-based optimizer MMA, the derivatives of the objective function and the constrains with respect to the design variable have to be calculated through the sensitivity analysis [43, 44]. By given a pseudo time interval t and considering a two-material structure, the derivative of J in Eq.5 can be expressed as [45]:

$$\left. \frac{dJ}{dt} \right|_{\Phi_k} = \sum_{j=1}^{m} \left(\int_D \beta^k \Psi_j d\Omega \frac{d\mu_{kj}(t)}{dt} \right), \quad k = 1, 2, \quad (9)$$

where β^k takes the form:

$$\beta^k = -\boldsymbol{\varepsilon}(\boldsymbol{u}) : \frac{\partial \boldsymbol{D}(\boldsymbol{x},\Phi)}{\partial \Phi_k} : \boldsymbol{\varepsilon}(\boldsymbol{u}) \quad (10)$$

By applying the chain rule, the derivative of the objective function with respect to the design variables can be expressed as:

$$\frac{\partial J}{\partial \mu_j^k} = \int_D \beta^k \Psi_j d\Omega, \quad (j = 1,...,m; \quad k = 1,2) \quad (11)$$

Similarly, the derivatives for the volume constraints can be achieved in a similar manner:

$$\frac{dVol_1}{d\mu_{1j}} = \int_D \delta(\Phi_1)) \Psi_j d\Omega \quad j = 1,...,m \quad (12)$$

$$\frac{dVol_1}{d\mu_{2j}} = \int_D \delta(\Phi_1) H(\Phi_2) \Psi_j d\Omega \quad j = 1,...,m \quad (13)$$

$$\frac{dVol_2}{d\mu_{2j}} = \int_D \delta(\Phi_2) H(\Phi_1) \Psi_j d\Omega \quad j = 1,...,m \quad (14)$$

To optimize the material properties of the two constructing material along with the topology of the entire structure, the material properties can be treated as design variables [39]. The corresponding derivatives are achieved via the forward finite difference scheme. In a general form, the derivative of function f at point $\boldsymbol{x}$ can be defined as:

$$f'(x) = \lim_{h \to 0} \frac{f(x+h) - f(x)}{h}. \tag{15}$$

An energy functional can be minimized along with the objective function to maintain the distance-regularized level set function. The readers can be referred to our previous works [30, 36] regarding the energy functional formulation and derivatives.

4 Formulations for Optimization and Conformal Mapping of Isotropic Metamaterials

With the optimal material properties achieved in the previous stage, in this section, the structural topology optimization of the metamaterial [39] is introduced. To ensure the isotropy of the metamaterial, an extra isotropy constraint is included in the least square optimization statement:

$$\text{Min:} \quad J_{meta} = \frac{1}{2} \sum_{ijkl}^{n} (C_{ijkl}^{H} - C_{ijkl}^{*})^2$$

$$\text{s.t.:} \quad \int_{D} H(\Phi) d\Omega \leq Vol_t \tag{16}$$

$$a(\boldsymbol{u}, \boldsymbol{v}, \Phi) = l(\boldsymbol{v}, \Phi)$$

$$C_{1212}^{*} = (C_{1111}^{H} + C_{2222}^{H})/4 - C_{1122}^{H}/2$$

Here the Φ is the corresponding level set function for the metamaterial design. The C_{ijkl}^{H} is the homogenized elasticity tensors with the targets at C_{ijkl}^{*}. The volume of the metamaterial is constrained by the volume target Vol_t. The structural isotropy [46] is ensured when the condition of $C_{1212}^{*} = (C_{1111}^{H} + C_{2222}^{H})/4 - C_{1122}^{H}/2$ is satisfied. With the strain energy method, the derivatives can be calculated by using the strain energy under different loading scenarios. The details of this process can be found in [39].

With the optimal overall layout of the multimaterial structure and the detailed layout of the metamaterials, the multimaterial, multiscale structure can be formulated by the local shape-preserving conformal mapping [47]. Some similar works can be found in [48]. The angle-preserving effect of conformal mapping can be illustrated as follows. As can be seen from Figure 2, the pattern on the human face freeform surface in Figure 2(b) still keeps the angles of the checkerboard from Figure 2(a). That is to say the properties can be preserved after the mapping when the boxes are considered as metamaterial unit cells. Below are some basic information about the multiple control point conformal mapping. Let $\omega = f(z) : \mathbb{C} \to \mathbb{C}$ be a complex function on the plane. Denote:

$$\frac{\partial}{\partial \bar{z}} := \frac{1}{2} \left(\frac{\partial}{\partial x} + i \frac{\partial}{\partial y} \right), \tag{17}$$

where i is the unit imaginary root. Then f is said to be *conformal* if

$$\frac{\partial f}{\partial \bar{z}} = 0. \tag{18}$$

Under discrete settings, conformal mappings can be computed by discrete Ricci flow method [49, 50, 51]. Some more discrete Ricci flow algorithms regarding efficiency and adaptivity improvements are reported in [52, 53]. For further conformal mapping algorithms, the readers are refereed to [54] for more information..

The isotropy of the metamaterails will be insensitive to the rotation caused by the mapping and the angle preserving characteristic will ensure the consistency of the designed metamaterial properties. When a highly distorted area has to be mapped, the sharp corners need to be taken care of. By introducing multiple control points, the adaptivity of the mapping to the area distortions can be improved.

Given a triangular mesh $\Sigma = (V, F, E)$, a face element is denoted with corner vertex v_i, v_j and v_k by f_{ijk}, and the angle between rays are denoted as $\overrightarrow{ij}$ and $\overrightarrow{ik}$ by θ_i^{jk}. Then the discrete Gaussian curvature at vertex v_i is defined by

$$K_i = \begin{cases} 2\pi - \sum_{f_{ijk} \in F} \theta_i^{jk} & \text{if } v_i \notin \partial\Sigma, \\ \pi - \sum_{f_{ijk} \in F} \theta_i^{jk} & \text{if } v_i \in \partial\Sigma, \end{cases} \tag{19}$$

where $\partial\Sigma$ is the boundary of mesh Σ. Now the discrete Ricci flow is defined as follows. Given a circle packing metric to Σ, i.e on each vertex v_i a positive real number is defined as γ_i, then the edge length between vertices v_i and v_j is $l_{ij} = \gamma_i + \gamma_j$. With those parameters, all angles can be calculated in Σ. Denote $u_i = \log \gamma_i$, then the discrete Ricci flow is defined as

$$\frac{du_i(t)}{dt} = (\bar{K}_i - K_i), \tag{20}$$

where $\bar{\boldsymbol{k}} = (\bar{K}_1, \bar{K}_2, \ldots, \bar{K}_n)^T$ is the user defined target curvature.

In our case, a conformal mapping from a irregular planar region to a polygonal region needs to be found such that the inner angles are either $\pi/2$ or $3\pi/2$. The polygonal region is filled with regular metamaterial structures, and then they are mapped back with the inverse of the computed conformal mapping, which is also a conformal mapping that preserves local shapes. To realize this, on the boundary of input mesh $\partial\Sigma$, multiple control points $W := \{w_1, w_2, \ldots, w_k\} \subseteq \partial\Sigma$ can be selected based on the need.

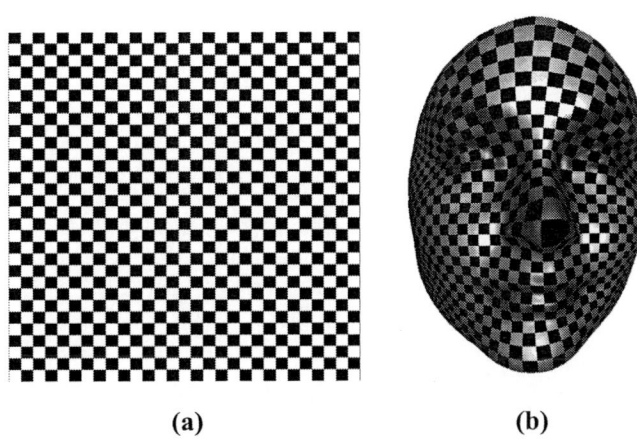

(a)   (b)

FIGURE 2. The conventional conformal mapping Ricci flow method. (a) The Checkerboard Pattern. (b) The Freeform Surface.

Then the target curvature on each vertex are defined by:

$$\bar{K}_i = \begin{cases} 0 & \text{if } v_i \notin \partial\Sigma, \\ 0 & \text{if } v_i \in \partial\Sigma \setminus W, \\ -\pi/2 \text{ or } \pi/2 & \text{if } v_i \in W. \end{cases} \quad (21)$$

Here $\bar{K}_i = \pi/2$ is chosen if the target polygonal region has a outward right angle and $\bar{K}_i = -\pi/2$ is chosen if the target polygonal region has a inward right angle at point v_i. Compared to Ricci flow method, as shown in Figure 2, which will map the input region to a freeform surface via four control points, the proposed multi-control-point method provides more flexibility with the benefit of lower area distortion. The detailed information and the mapping work flow can be found in our recent paper [39].

5 Numerical Results
5.1 MBB Beam

In this section, a multimaterial MBB beam structure is designed under the CBF-based concurrent PLSM framework. The macroscale topology optimization boundary condition is illustrated in Figure 3. A $F = 1$ force is applied at the lower center of a 2-by-1 domain with fixed lower corners. The domain is discretized into 100×50 elements. The Young's modulus of the soft material is given the range from 0.05 to 0.1 and the hard one is between 0.15 and 0.2. The initial Young's modulus values of the soft material and the hard material are given 0.075 and 0.175, respectively. The overall material volume is constrained at 60% and the hard material volume is constrained at 30%. The con-

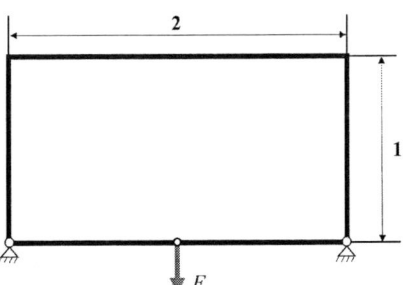

FIGURE 3. The Boundary Condition of the MBB Beam Structure Example

vergence history of the optimization process is shown in Figure 4. The total and the hard material volume for the final design is 59.995% and 29.995%, respectively. The Poisson's ratio for all materials are set to be 0.3. The optimal Young's modulus for the soft metamaterial is 0.1 and the hard one is 0.2, respectively. By using these two values as targets, the second topology optimization is carried out to get the isotropic metamaterial layouts. The Young's modulus for constructing both metamaterials is the same at 1. The volume is 30% and 40% for the soft and hard metamaterial, respectively. A bounding box is introduced to the metamaterials to ensure the connectivity of the adjacent metamaterial unit cells. An isotropy polar plot scheme, as shown in Figure 5, is utilized to illustrate the isotropy of the designed metamaterials, since it is not practical to hit all the targets in the least square objective function in Eq.16. The details of this plot can be found in [39]. By using the conformal mapping, the mapped MBB beam structure is illustrated in Figure 6. By exporting the conformally-mapped structure as a CAD model and send it into FEA package, the actual structural performance can be calculated as 86.29, compared with 72.3939 in Figure 4. For more details about exporting the CAD model and the FEA analysis, the readers are referred to our further journal version of this paper.

5.2 Michell-type Structure

In this section, the Michell-type structure is designed with the boundary condition shown in Figure 7. With the same optimization setting, the structure topology optimization is carried out with the total volume constraint of 80% and the hard material volume constraint of 40%. The evolution of the design is shown in Figure 8, with the optimal Young's modulus of 0.1 and 0.2 for the soft and hard metamaterial, respectively. The final structure has the total volume of 80% and the hard material volume is 39.992%. The corresponding conformally mapped multiscale structure is shown in Figure 9. The corresponding FEA verification calculates the actual total strain energy as 121.42, compared with 102.5102 shown in Figure 8.

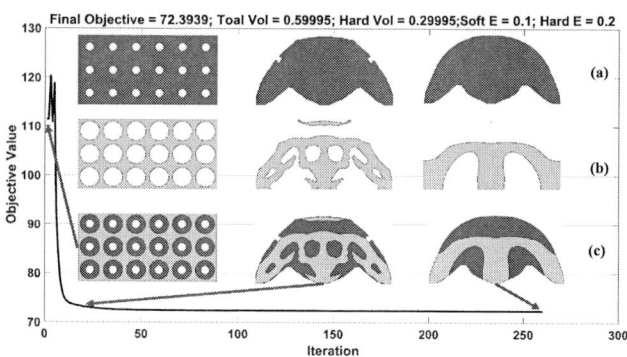

FIGURE 4. The Evolution History of the MBB Beam Structure Example. (a) The Evolution History of the First Zero-Level Set Function. (b) The Evolution History of the Second Zero-Level Set Function. (c) The Evolution History of the Actual Multimaterial Structure. Red: Soft Material. Green: Hard Material.

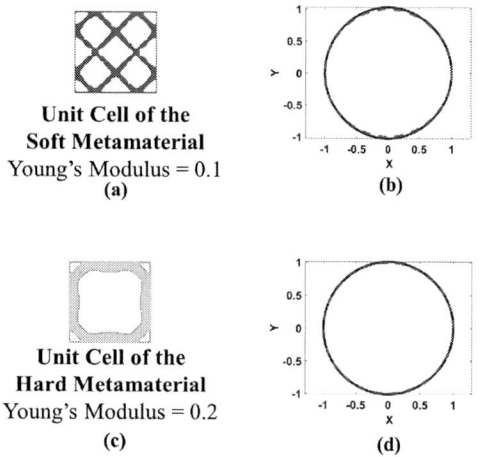

FIGURE 5. Metamaterials and their isotropy plot. (a) The Soft Metamaterial. (b) The Isotropy Polar Plot of the Soft Metamaterial. (c) The Hard Metamaterial. (d) The Isotropy Polar Plot of the Hard Metamaterial. (Red: Reference Standard Circle. Blue: Isotropy Polar Plot of the Current Metamaterial.)

5.3 Short Cantilever Beam

In this section, the short cantilever beam structure is designed within a 1-by-1 design domain discretized into 50×50 elements. The boundary conditions are shown in Figure 10. The material property settings and the topology optimization settings are kept the same as the previous ones. The total volume constraint is 80% and the hard material volume constraint is 40%.

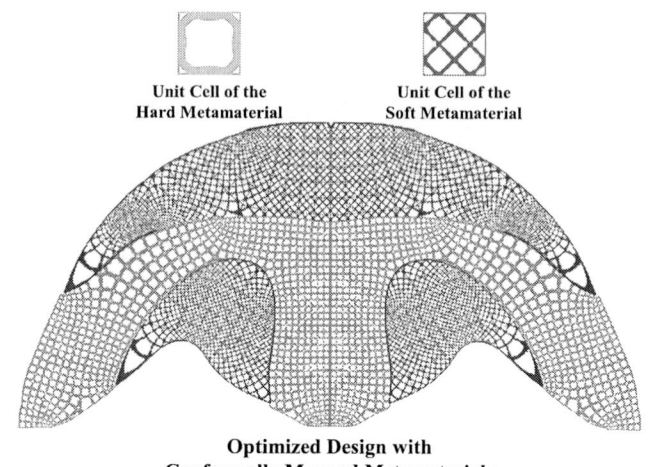

FIGURE 6. The Optimized MBB Beam Structure with Conformally Mapped Hard (Red) and Soft (Green) Metamaterials.

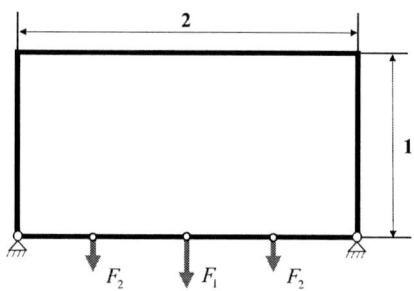

FIGURE 7. The Boundary Condition of the Michell-type Structure Example

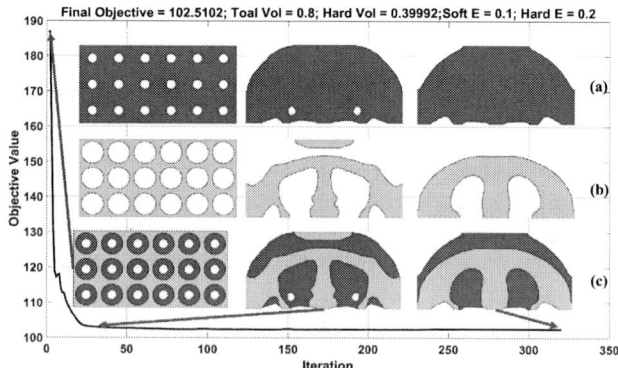

FIGURE 8. The Evolution History of the Michell-Type Beam Structure Example. (a) The Evolution History of the First Zero-Level Set Function. (b) The Evolution History of the Second Zero-Level Set Function. (c) The Evolution History of the Actual Multimaterial Structure. Red: Soft Material. Green: Hard Material.

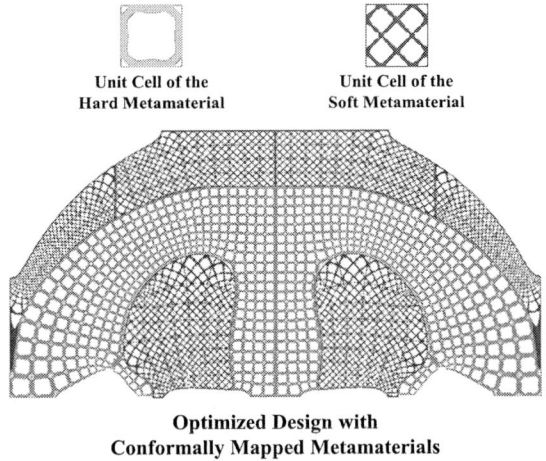

Unit Cell of the Hard Metamaterial

Unit Cell of the Soft Metamaterial

Optimized Design with Conformally Mapped Metamaterials

FIGURE 9. The Michell-type Structure with Conformally Mapped Hard (Green) and Soft (Red) Metamaterials.

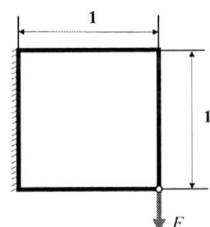

FIGURE 10. The Boundary Condition of the Short Cantilever Beam

The evolution of the design is shown in Figure 11. The final design has the total volume of 79.972% and the hard material volume of 39.989%, together with the optimized Young's modulus of 0.1 and 0.2 for the soft and hard metamaterial, respectively. The corresponding conformally mapped multiscale structure is shown in Figure 12. The corresponding total strain energy for the conformally-mapped structure is calculated as 71.12 in FEA, compared with 67.7091 as shown in Figure 11.

6 Conclusions

In this paper, a concurrent CBF-based PLSM topology optimization framework is proposed to design multimaterial hierarchical structures. With the 'color' level set representation, multiple material phases can be discriminated inside the design domain. By using the CBF kernel function, the explicit design variable bounds can be passed to MMA. The proposed approach can handle multiple constraints in a straightforward manner. With the help of the local shape-preserving conformal mapping with multiple control points, the designed metamaterial properties can be mathematically preserved after the mapping when the isotropic metamaterials are used. The FEA analysis results verify the

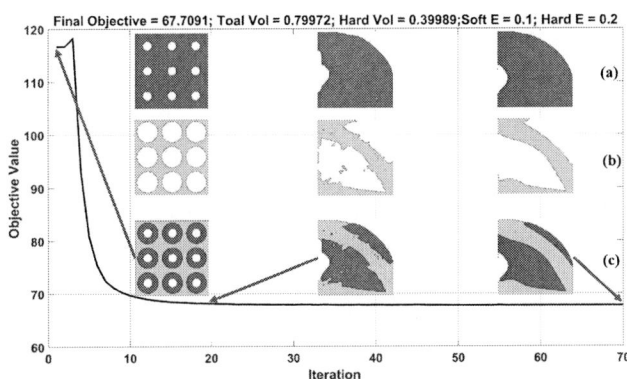

FIGURE 11. The Evolution History of the Short Cantilever Beam Structure Example. (a) The Evolution History of the First Zero-Level Set Function. (b) The Evolution History of the Second Zero-Level Set Function. (c) The Evolution History of the Actual Multimaterial Structure. Red: Soft Material. Green: Hard Material.

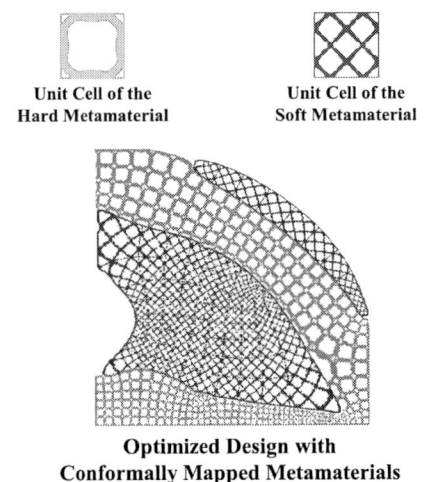

Unit Cell of the Hard Metamaterial

Unit Cell of the Soft Metamaterial

Optimized Design with Conformally Mapped Metamaterials

FIGURE 12. The Short Cantilever Beam Structure with Conformally Mapped Hard (Green) and Soft (Red) Metamaterials

high fidelity of the proposed mapping scheme. Therefore, the designing-mapping-manufacturing process is concluded.

ACKNOWLEDGMENT

The authors acknowledge the support from the National Science Foundation of the United States (Grants No. CMMI1462270 and CMMI1762287), Ford University Research Program (URP), and the start-up fund from the State University of New York at Stony Brook. The authors would also like to thank Mr. Yang Guo from the Department of Computer Science at Stony Brook University for his help with the implementation of the conformal mapping method in this paper.

REFERENCES

[1] Fullwood, D. T., Niezgoda, S. R., Adams, B. L., and Kalidindi, S. R., 2010. "Microstructure sensitive design for performance optimization". *Progress in Materials Science,* **55**(6), pp. 477–562.

[2] Deng, J., Yan, J., and Cheng, G., 2013. "Multi-objective concurrent topology optimization of thermoelastic structures composed of homogeneous porous material". *Structural and Multidisciplinary Optimization,* **47**(4), pp. 583–597.

[3] Vlasea, M., Shanjani, Y., Bothe, A., Kandel, R., and Toyserkani, E., 2013. "A combined additive manufacturing and micro-syringe deposition technique for realization of bioceramic structures with micro-scale channels". *The International Journal of Advanced Manufacturing Technology,* **68**(9-12), pp. 2261–2269.

[4] Sigmund, O., 2001. "Design of multiphysics actuators using topology optimization–part ii: Two-material structures". *Computer methods in applied mechanics and engineering,* **190**(49-50), pp. 6605–6627.

[5] Sigmund, O., and Torquato, S., 1996. "Composites with extremal thermal expansion coefficients". *Applied Physics Letters,* **69**(21), pp. 3203–3205.

[6] Sigmund, O., and Maute, K., 2013. "Topology optimization approaches". *Structural and Multidisciplinary Optimization,* **48**(6), pp. 1031–1055.

[7] Allaire, G., Jouve, F., and Toader, A.-M., 2002. "A level-set method for shape optimization". *Comptes Rendus Mathematique,* **334**(12), pp. 1125–1130.

[8] Wang, M. Y., Wang, X., and Guo, D., 2003. "A level set method for structural topology optimization". *Computer methods in applied mechanics and engineering,* **192**(1-2), pp. 227–246.

[9] Sethian, J. A., and Wiegmann, A., 2000. "Structural boundary design via level set and immersed interface methods". *Journal of computational physics,* **163**(2), pp. 489–528.

[10] van Dijk, N. P., Maute, K., Langelaar, M., and Van Keulen, F., 2013. "Level-set methods for structural topology optimization: a review". *Structural and Multidisciplinary Optimization,* **48**(3), pp. 437–472.

[11] Sethian, J. A., 1996. "Theory, algorithms, and applications of level set methods for propagating interfaces". *Acta numerica,* **5**, pp. 309–395.

[12] Wang, M. Y., and Wang, X., 2004. ""color" level sets: a multi-phase method for structural topology optimization with multiple materials". *Computer Methods in Applied Mechanics and Engineering,* **193**(6-8), pp. 469–496.

[13] Wei, P., and Wang, M. Y., 2009. "Piecewise constant level set method for structural topology optimization". *International Journal for Numerical Methods in Engineering,* **78**(4), pp. 379–402.

[14] Merriman, B., Bence, J. K., and Osher, S. J., 1994. "Motion of multiple junctions: A level set approach". *Journal of Computational Physics,* **112**(2), pp. 334–363.

[15] Gibson, L. J., and Ashby, M. F., 1999. *Cellular solids: structure and properties.* Cambridge university press.

[16] Christensen, R. M., 2000. "Mechanics of cellular and other low-density materials". *International Journal of Solids and Structures,* **37**(1-2), pp. 93–104.

[17] Valdevit, L., Jacobsen, A. J., Greer, J. R., and Carter, W. B., 2011. "Protocols for the optimal design of multifunctional cellular structures: From hypersonics to micro-architected materials". *Journal of the American Ceramic Society,* **94**(s1).

[18] Han, S. C., Lee, J. W., and Kang, K., 2015. "A new type of low density material: shellular". *Advanced Materials,* **27**(37), pp. 5506–5511.

[19] Zhou, S., Li, W., Chen, Y., Sun, G., and Li, Q., 2011. "Topology optimization for negative permeability metamaterials using level-set algorithm". *Acta Materialia,* **59**(7), pp. 2624–2636.

[20] Wang, Y., Luo, Z., Zhang, N., and Kang, Z., 2014. "Topological shape optimization of microstructural metamaterials using a level set method". *Computational Materials Science,* **87**, pp. 178–186.

[21] Vogiatzis, P., Chen, S., Wang, X., Li, T., and Wang, L., 2017. "Topology optimization of multi-material negative poisson's ratio metamaterials using a reconciled level set method". *Computer-Aided Design,* **83**, pp. 15–32.

[22] Zhou, S., Li, W., Sun, G., and Li, Q., 2010. "A level-set procedure for the design of electromagnetic metamaterials". *Optics express,* **18**(7), pp. 6693–6702.

[23] Wang, Y., Gao, J., Luo, Z., Brown, T., and Zhang, N., 2017. "Level-set topology optimization for multimaterial and multifunctional mechanical metamaterials". *Engineering Optimization,* **49**(1), pp. 22–42.

[24] Sivapuram, R., Dunning, P. D., and Kim, H. A., 2016. "Simultaneous material and structural optimization by multiscale topology optimization". *Structural and multidisciplinary optimization,* **54**(5), pp. 1267–1281.

[25] Wang, Y., Chen, F., and Wang, M. Y., 2017. "Concurrent design with connectable graded microstructures". *Computer Methods in Applied Mechanics and Engineering,* **317**, pp. 84–101.

[26] Li, H., Luo, Z., Gao, L., and Qin, Q., 2018. "Topology optimization for concurrent design of structures with multi-patch microstructures by level sets". *Computer Methods in Applied Mechanics and Engineering,* **331**, pp. 536–561.

[27] Allaire, G., Jouve, F., and Toader, A.-M., 2004. "Structural optimization using sensitivity analysis and a level-set method". *Journal of computational physics,* **194**(1), pp. 363–393.

[28] Osher, S. J., and Santosa, F., 2001. "Level set methods for optimization problems involving geometry and constraints:

I. frequencies of a two-density inhomogeneous drum". *Journal of Computational Physics, 171*(1), pp. 272–288.

[29] Osher, S., and Sethian, J. A., 1988. "Fronts propagating with curvature-dependent speed: algorithms based on hamilton-jacobi formulations". *Journal of computational physics, 79*(1), pp. 12–49.

[30] Jiang, L., and Chen, S., 2017. "Parametric structural shape & topology optimization with a variational distance-regularized level set method". *Computer Methods in Applied Mechanics and Engineering, 321*, pp. 316–336.

[31] Luo, Z., Tong, L., Wang, M. Y., and Wang, S., 2007. "Shape and topology optimization of compliant mechanisms using a parameterization level set method". *Journal of Computational Physics, 227*(1), pp. 680–705.

[32] Wang, S., and Wang, M. Y., 2006. "Radial basis functions and level set method for structural topology optimization". *International journal for numerical methods in engineering, 65*(12), pp. 2060–2090.

[33] Svanberg, K., 2007. "Mma and gcmma-two methods for nonlinear optimization". *vol, 1*, pp. 1–15.

[34] Jiang, L., Ye, H., Zhou, C., Chen, S., and Xu, W., 2017. "Parametric topology optimization toward rational design and efficient prefabrication for additive manufacturing". In ASME 2017 12th International Manufacturing Science and Engineering Conference collocated with the JSME/ASME 2017 6th International Conference on Materials and Processing, American Society of Mechanical Engineers, pp. V004T05A006–V004T05A006.

[35] Luo, Z., Wang, M. Y., Wang, S., and Wei, P., 2008. "A level set-based parameterization method for structural shape and topology optimization". *International Journal for Numerical Methods in Engineering, 76*(1), pp. 1–26.

[36] Jiang, L., Chen, S., and Jiao, X., 2018. "Parametric shape and topology optimization: A new level set approach based on cardinal basis functions". *International Journal for Numerical Methods in Engineering, 114*(1), pp. 66–87.

[37] Li, C., Xu, C., Gui, C., and Fox, M. D., 2010. "Distance regularized level set evolution and its application to image segmentation". *IEEE transactions on image processing, 19*(12), pp. 3243–3254.

[38] Jiang, L., Chen, S., and Wei, P., 2018. "Concurrent optimization of structure topology and infill properties with a cardinal-function-based parametric level set method". In ASME 2018 International Design Engineering Technical Conferences and Computers and Information in Engineering Conference, American Society of Mechanical Engineers, pp. V02BT03A006–V02BT03A006.

[39] Jiang, L., Guo, Y., Chen, S., Wei, P., Lei, N., and Gu, X. D., 2019. "Concurrent optimization of structural topology and infill properties with a cbf-based level set method". *Frontiers of Mechanical Engineering*, pp. 1–19.

[40] Challis, V., Roberts, A., and Wilkins, A., 2008. "Design of three dimensional isotropic microstructures for maximized stiffness and conductivity". *International Journal of Solids and Structures, 45*(14-15), pp. 4130–4146.

[41] Radman, A., Huang, X., and Xie, Y., 2013. "Topological optimization for the design of microstructures of isotropic cellular materials". *Engineering optimization, 45*(11), pp. 1331–1348.

[42] Guth, D., Luersen, M., and Muñoz-Rojas, P., 2015. "Optimization of three-dimensional truss-like periodic materials considering isotropy constraints". *Structural and Multidisciplinary Optimization, 52*(5), pp. 889–901.

[43] Van Keulen, F., Haftka, R., and Kim, N., 2005. "Review of options for structural design sensitivity analysis. part 1: Linear systems". *Computer methods in applied mechanics and engineering, 194*(30-33), pp. 3213–3243.

[44] Choi, K. K., and Kim, N.-H., 2006. *Structural sensitivity analysis and optimization 1: linear systems*. Springer Science & Business Media.

[45] Wang, Y., Luo, Z., Kang, Z., and Zhang, N., 2015. "A multi-material level set-based topology and shape optimization method". *Computer Methods in Applied Mechanics and Engineering, 283*, pp. 1570–1586.

[46] Neves, M., Rodrigues, H., and Guedes, J. M., 2000. "Optimal design of periodic linear elastic microstructures". *Computers & Structures, 76*(1-3), pp. 421–429.

[47] Ahlfors, L. V., 2010. *Conformal invariants: topics in geometric function theory*, Vol. 371. American Mathematical Soc.

[48] Wang, H., Chen, Y., and Rosen, D. W., 2005. "A hybrid geometric modeling method for large scale conformal cellular structures". In ASME Computers and Information in Engineering Conference, Long Beach, CA, Sept, pp. 24–28.

[49] Jin, M., Kim, J., Luo, F., and Gu, X., 2008. "Discrete surface ricci flow". *IEEE Transactions on Visualization and Computer Graphics, 14*(5), pp. 1030–1043.

[50] Chow, B., Luo, F., et al., 2003. "Combinatorial ricci flows on surfaces". *Journal of Differential Geometry, 63*(1), pp. 97–129.

[51] Zeng, W., and Gu, X. D., 2013. *Ricci flow for shape analysis and surface registration: theories, algorithms and applications*. Springer Science & Business Media.

[52] Gu, X., He, Y., Jin, M., Luo, F., Qin, H., and Yau, S.-T., 2008. "Manifold splines with a single extraordinary point". *Computer-Aided Design, 40*(6), pp. 676–690.

[53] Jin, M., Luo, F., and Gu, X., 2006. "Computing surface hyperbolic structure and real projective structure". In Proceedings of the 2006 ACM symposium on Solid and physical modeling, ACM, pp. 105–116.

[54] Gu, X. D., Zeng, W., Luo, F., and Yau, S.-T., 2012. "Numerical computation of surface conformal mappings". *Computational Methods and Function Theory, 11*(2), pp. 747–787.

Proceedings of the ASME 2019
International Design Engineering Technical Conferences
and Computers and Information in Engineering Conference
IDETC/CIE2019
August 18-21, 2019, Anaheim, CA, USA

DETC2019-97628

THERMOMECHANICAL TOPOLOGY OPTIMIZATION OF LATTICE HEAT TRANSFER STRUCTURE INCLUDING NATURAL CONVECTION AND DESIGN DEPENDENT HEAT SOURCE

Tong Wu
Dept. of Mechanical Engineering
Purdue University
West Lafayette, Indiana, USA
wu616@purdue.edu

Joel C. Najmon
Dept. of Mechanical Engineering
Purdue University
West Lafayette, Indiana, USA
jnajmon@purdue.edu

Andres Tovar
Dept. of MechC and Energy Engr.
Purdue School of Engr. and Technology
IUPUI, Indianapolis, Indiana, USA
tovara@iupui.edu

ABSTRACT

Lattice Heat Transfer (LHT) structures provide superior structural support while improving the heat transfer coefficient through their high surface-to-volume ratios. By using current Additive Manufacturing (AM) technologies, LHT with highly complex structures is possible. In this study, the design concept of LHT is further improved by implementing a thermomechanical topology optimization method. With utilization of design-dependent heat source, the method can be applied to generate stiffer LHT structures under mechanical and thermomechanical loads, without decreasing their thermal performance; relative to a design made of a uniform LHT having the same mass fraction. Two numerical examples are presented to illustrate how to use the proposed approach to design LHT sections. The results show that the mechanical performance can be improved more than 50% compared to a uniform LHT with the same mass fraction, without decreasing the thermal performance. The method does not require a fluid mechanics model, thus it is computational effective and particularly suitable for the conceptual design stage. The resulting optimized lattice is made possible by utilizing additive manufacturing technologies.

Keywords: Lattice heat transfer; thermomechanical; design-dependent heat source; topology optimization.

1 Introduction

Lattice Heat Transfer (LHT) structures provide superior structural support while improving the heat transfer coefficient through their high surface-to-volume ratios [1–3] (Fig. 1). The accelerated development of additive Manufacturing (AM/3D Printing) technologies enables the design and production of intricate lattice structures, offering significant cost savings, particularly in designs having high geometric complexity [4–6]. Recently, the multifunctional advantages of lattice structures has made them popular in applications which simultaneously require mechanical high strength and heat transfer rate [7,8], such as gas turbine blades and injection mold cooling systems. In gas turbine blades, inserting lattice structures as a cooling layer maintains sufficient structural blade strength, while improving the heat transfer rate two to three times compared to that of a smooth channel [9–11]. Lattice layers have been implemented as cooling system for injection molding, leading to a 20% reduction in cooling time, compared to the design having non-lattice conformal cooling [12–14]. Current LHT structures are mainly composed of uniformly distributed unit cells each having the same randomly generated foam-like porous medium, as well as structures similar to fins and pillar arrays (Fig. 1). A common way to find an optimized LHT structure is through analysis of simulation (recorded after designs are generated) and experimental data (recorded after the structures are produced) [1–3,7,8]. However, this approach limits design freedom, utilizing sample data methods that do not ensure optimality of the structure. To overcome

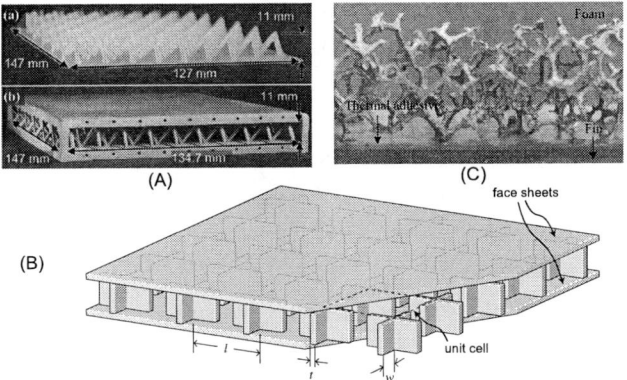

Ref: (A) K N Son, et al., 2017; (B) D T Queheillalt et al., 2008; (C) SS Feng et al., 2014

FIGURE 1. MULTIFUNCTIONAL LATTICE HEAT SINK WITH (A) TETRAHEDRAL LATTICE [8] AND (B) TRUNCATED SQUARE [7], AND (C) METAL FOAM [1].

these drawbacks, in this study, a topology optimization approach is proposed to attain flexible and complex lattice structures that significantly improve the design optimality.

Since a LHT structure is required to transfer heat while withstanding mechanical and thermal stresses induced by the loads and the temperature gradient, a thermomechanical topology optimization should be incorporated to optimize a LHT structure. Traditional thermomechanical topology optimization has been employed to create thermal actuators [15–17], a thermal management device for spacecraft [18–20], and injection molds [21, 22]. However, in these approaches convective heat transfer on the structure's boundary surfaces are not considered and remain scarce in literature. The thermomechanical topology optimization scheme employed in this study considers the convective heat transfer on these surfaces.

Convective heat transfer models that have been discussed in the context of topology optimization theory are primarily composed of a thermal-fluid (conjugate) heat transfer model, or a solid-structure heat transfer model. In thermal-fluid model based topology optimization, natural and forced convective heat transfer are affected by a fluid field resulting from Navier-Stokes equation or Darcy equation [23–25]. Consequently it can be time-consuming to couple with the thermomechanical model and unattractive for early stage conceptual design studies. An alternative approach is to employ a design-dependent heat source in the topology optimization of heat transfer problem without a fluid model. The design-dependent heat source varies with the element states or material itself. This was originally used to solve heat conduction topology optimization problems [26]. Further, design-dependent heat source implementation has been done to analyze heat transfer models that consider heat conduction, convection, and internal heat generation [27–29].

In this study, convective heat transfer and design-dependent heat sources are coupled with a thermomechanical model, resulting in a novel topology optimization method. The method is specifically tailored to the design of multifunctional lattice heat transfer (LHT) structures requiring adequate thermal, mechanical and thermomechanical performance. In this method, it is assumed the design-dependent heat source is only located at the fluid phase, and the optimized fluid and solid phase distribution is available as a result of thermomechanical topology optimization. The paper is organized as follows: In section two, the finite element analysis of thermomechanical model is briefly presented. In section three, the proposed thermomechanical topology optimization including sensitivity analysis is illustrated; Two numerical examples are shown in section four. Finally, the summary of this work is presented in section five.

2 Finite element analysis of thermomechanical model

In a coupled thermomechanical model, the purely mechanical and thermomechanical loads caused by non-uniform temperature field should be considered. The overall thermal and mechanical performance is significantly influenced by the heat source distribution. In this study, it is assumed that the heat source is only applied to areas containing fluid. With the application of the proposed method, the optimized shape, location and numbers of areas are obtained. Before the illustration of the method, procedures of thermomechanical finite element analysis (FEA) is briefly described as follows, where natural convection and design-dependent heat source are incorporated.

2.1 Thermal model with convection and design-dependent heat load

In a thermal model, the energy dissipation can be written as

$$Q = \frac{1}{2} \int_\Omega \nabla T^\mathsf{T} \kappa(\theta) \nabla T d\Omega - W_q, \qquad (1)$$

where κ is the thermal conductivity tensor, ∇T indicates temperature gradients, and W is the external work. Discretizing Eq. (1) yields

$$\mathbf{Q} = \mathbf{T}(\boldsymbol{\theta})^\mathsf{T} \mathbf{K}_t(\boldsymbol{\theta}) \mathbf{T}(\boldsymbol{\theta}) - \mathbf{q}^\mathsf{T} \mathbf{T}(\boldsymbol{\theta}), \qquad (2)$$

In a static equilibrium state $\frac{\partial \mathbf{Q}}{\partial \mathbf{T}} = \mathbf{0}$, the Fourier equation in Matrix form is formulated:

$$\mathbf{K}_t(\boldsymbol{\theta}) \mathbf{T}(\boldsymbol{\theta}) = \mathbf{q}(\boldsymbol{\theta}) \qquad (3)$$

where the global stiffness of heat transfer $\mathbf{K}_t$ is composed of stiffness matrix of thermal conduction $\mathbf{K}_{\text{cond}}$ and natural convection

K_{conv}:

$$\mathbf{K}_t(\boldsymbol{\theta}) = \mathbf{K}_{cond}(\boldsymbol{\theta}) + \mathbf{K}_{conv}(\boldsymbol{\theta}). \tag{4}$$

In Eq. (4),

$$\mathbf{K}_{cond}(\boldsymbol{\theta}) = \sum_{e=1}^{n_e} \int_{\Omega} \nabla \mathbf{N}^{\mathsf{T}} \boldsymbol{\kappa}(\theta) \nabla \mathbf{N} \mathrm{d}V, \tag{5}$$

where

$$\boldsymbol{\kappa}(\theta) = \boldsymbol{\kappa}_{\min} + (\boldsymbol{\kappa}_s - \boldsymbol{\kappa}_{\min}) \theta^{p_1}, \tag{6}$$

where p_1 is a penalty number. This material interpolation scheme indicates the thermal conductivity is higher when the solid material containing more volume in an element. Notably, $\boldsymbol{\kappa}_{\min}$ is the minimum thermal conductivity for an element to avoid singularity in matrix computation. The stiffness of natural convection is formulated as

$$\mathbf{K}_{conv} = \sum_{e=1}^{n_e} h \int_{V} \mathbf{N}^{\mathsf{T}} \mathbf{N} \mathrm{d}V, \tag{7}$$

where h is convective heat transfer coefficient. In Eq. (3), $\mathbf{q}$ is the boundary heat source vector which composed of constant heat source $\mathbf{q}_0$ and a design-dependent heat source $\mathbf{q}_0(\boldsymbol{\theta})$.

$$\mathbf{q}(\boldsymbol{\theta}) = \mathbf{q}_0 + \mathbf{q}_0(\boldsymbol{\theta}), \tag{8}$$

where

$$\mathbf{q}_0(\boldsymbol{\theta}) = \sum_{e=1}^{n_e} h_1(\theta) \int_{V} \mathbf{N} \mathrm{d}V. \tag{9}$$

The design-dependency is represented by the following material interpolation function

$$h_1(\theta) = h_f(1 - \theta^{p_2}). \tag{10}$$

The design-dependent heat source is implemented through voids containing variable fluid levels. This material interpolation scheme assumes there is a maximum design-dependent heat source h_f when the element V is filled with fluid, and zero when the element is filled with solid. When the element is filled with fluid, the design dependent heat source $\mathbf{q}_0$ reaches its maximum value $\overline{\mathbf{q}_0}$.

2.2 Thermomechanical model with convection and design-dependent heat load

Solving the heat transfer model will result in a non-uniform temperature field, which induces thermo-elastic strain and stress fields. For a thermomechanical structure, the strain and stress relations can be described by

$$\mathbf{D}(\boldsymbol{\varepsilon} - \boldsymbol{\varepsilon}_{\mathrm{T}}) = \boldsymbol{\sigma} \tag{11}$$

where $\mathbf{D}$ is the elasticity tensor, and $\boldsymbol{\varepsilon}$ is strain due to mechanical load. $\boldsymbol{\varepsilon}_{\mathrm{T}}$ is strain due to thermal-elastic load coupling the temperature field derived from thermal model. The elementwise thermal-elastic strain $\boldsymbol{\varepsilon}_{\mathrm{T}}$ is formulated as

$$\boldsymbol{\varepsilon}_{\mathrm{T}e} = \alpha(\theta)(\mathbf{N}\mathbf{T}_e(\theta) - \mathbf{T}_0)\mathbf{1} \tag{12}$$

where α is thermal expansion coefficient related to proportion of solid phase in the an element θ and penalty number p_3:

$$\alpha(\theta) = \alpha_{\min} + (\alpha_0 - \alpha_{\min})\theta^{p_3}. \tag{13}$$

$\mathbf{T}_e(\theta)$ is the elemental temperature obtained from thermal model. The strain energy density is

$$\Phi = \frac{1}{2} \int_{\Omega} (\boldsymbol{\varepsilon} - \boldsymbol{\varepsilon}_{\mathrm{T}})^{\mathsf{T}} \mathbf{D}(\theta)(\boldsymbol{\varepsilon} - \boldsymbol{\varepsilon}_{\mathrm{T}}) \mathrm{d}\Omega - \mathrm{W}, \tag{14}$$

Discretizing Eq. (14) yields

$$\Phi = \frac{1}{2} \mathbf{u}^{\mathsf{T}} \mathbf{K}_{elast}(\boldsymbol{\theta}) \mathbf{u} - \mathbf{f}_{th}^{\mathsf{T}}(\boldsymbol{\theta}, \mathbf{T}(\boldsymbol{\theta})) \mathbf{u} - \mathbf{f}^{\mathsf{T}} \mathbf{u}, \tag{15}$$

In Eq. (15), the thermo-elastic load $\mathbf{f}_{th}$ is given by

$$\mathbf{f}_{th}(\boldsymbol{\theta}, \mathbf{T}(\boldsymbol{\theta})) = \mathbf{K}_{mt}(\boldsymbol{\theta}) \mathbf{T}(\boldsymbol{\theta}), \tag{16}$$

where $\mathbf{K}_{mt}$ is thermo-mechanical coupling stiffness matrix

$$\mathbf{K}_{mt}(\boldsymbol{\theta}) = \sum_{e=1}^{n_e} \int_{V} \mathbf{B}^{\mathsf{T}} \mathbf{D}(\theta_e) \alpha \mathbf{N} \mathrm{d}V, \tag{17}$$

and nodel temperature $\mathbf{T}(\boldsymbol{\theta})$ is derived from Eq. (15) associated to the thermal model. $\mathbf{B}$ is the matrix representing strain-displacement relation. The stiffness matrix of elasticity, $\mathbf{K}_{elast}$, is

$$\mathbf{K}_{elast}(\boldsymbol{\theta}) = \sum_{e=1}^{n_e} \int_{V} \mathbf{B}^{\mathsf{T}} \mathbf{D}(\theta_e) \mathbf{B} \mathrm{d}V. \tag{18}$$

where $\mathbf{u}$ is the nodal displacement, and $\mathbf{f}$ is the external force.

In a static equilibrium state $\frac{\partial \Phi}{\partial \mathbf{u}} = \mathbf{0}$, Hook's law in matrix form is formulated as

$$\mathbf{K}_{\text{elast}}(\boldsymbol{\theta})\mathbf{u}(\boldsymbol{\theta}) = \left(\mathbf{f}_{\text{th}}\left(\boldsymbol{\theta}, \mathbf{T}(\boldsymbol{\theta})\right) + \mathbf{f}\right)^{\mathsf{T}}\mathbf{u}(\boldsymbol{\theta}) \qquad (19)$$

3 Proposed thermomechanical topology optimization

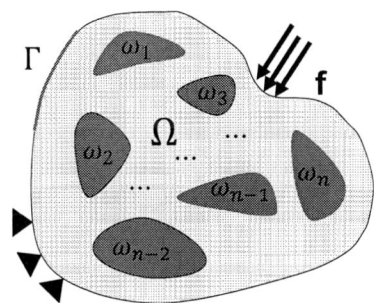

FIGURE 2. An Illustration of design domain, constant and design-dependent heat source for proposed method.

As aforementioned, the heat source can be divided to constant heat source $\mathbf{q}_0$ and design-dependent heat source $\mathbf{q}_0^i(\boldsymbol{\theta})$. In this problem, we assume a constant heat source is applied to the boundary surface Γ in Figure. 2. As described in Eq. (9) and Eq. (10), the heat source is applied to the cavities $\omega_1, \omega_2, ..., \omega_n$ in Figure 2. The shape, location, area, and numbers of these cavities are unknown. These will be defined later with the proposed algorithm. However, the sum of these design-dependent heat sources should have an upper bound $\bar{q}$, if the volume fraction of all the cavities have an upper bound. These heat source conditions can be formulated as:

$$\mathbf{q}_0 \in \Gamma, \quad \mathbf{q}_0^i(\boldsymbol{\theta}) \in \omega_i \quad (i = 1 \ldots n),$$
$$\sum_{i=1}^{n} \mathbf{q}_0^i = \overline{\mathbf{q}} \qquad (20)$$

The heat load could induce a non-uniform temperature field, thus a thermal load field $\mathbf{f}_{\text{th}}\left(\boldsymbol{\theta}, \mathbf{T}(\boldsymbol{\theta})\right)$ is obtained. A superposition of the internal thermal load and external mechanical load $\mathbf{f}$ is applied as the total load to formulate the load of the thermomechanical problem.

In a thermomechanical topology optimization problem, it is desirable to obtain high thermal and mechanical performance. A common metric for thermal performance is heat compliance given by

$$J_t = \mathbf{q}(\boldsymbol{\theta})^{\mathsf{T}}\mathbf{T}(\boldsymbol{\theta}). \qquad (21)$$

For mechanical performance, the mechanical compliance is adopted as

$$J_{\text{m}} = \left(\mathbf{f}_{\text{th}}\left(\boldsymbol{\theta}, \mathbf{T}(\boldsymbol{\theta})\right) + \mathbf{f}\right)^{\mathsf{T}}\mathbf{u}(\boldsymbol{\theta}). \qquad (22)$$

In this study, mechanical compliance is used as the objective function for thermomechanical topology optimization. The thermal compliance is defined as a constraint that should be smaller than the reference value of the initial design, J_t^0:

$$\begin{aligned} \text{minimize} \quad & J_{\text{m}}(\boldsymbol{\theta}) \\ \text{subject to} \quad & J_t(\boldsymbol{\theta}) \leq J_t^0(\boldsymbol{\theta}) \end{aligned} \qquad (23)$$

At the same time, the mass constraints for the structure and elements are given by

$$\begin{aligned} & m(\boldsymbol{\theta}) \leq m(\boldsymbol{\theta}_0) \\ & \theta^{\min} \leq \theta \leq \theta^{\text{solid}}, \end{aligned} \qquad (24)$$

where m is the mass of the structure, $m(\theta)_0$ is a given constant, $\theta^{\min}$ is the minimum allowable density for each element (which is the relative density of fluid phase in this study), and θ^{solid} is the relative density of solid phase, equal to 1. Fourier's law Eq. (3) and Hook's law Eq. (19) must also be satisfied. Finally, the problem statement of the proposed thermomechanical topology optimization is

$$\begin{aligned} \text{find} \quad & \boldsymbol{\theta}^* \in \mathbb{R}^{n_c} \\ \text{minimize} \quad & J_{\text{m}}(\boldsymbol{\theta}) \\ \text{subject to} \quad & J_t(\boldsymbol{\theta}) \leq J_t^0(\boldsymbol{\theta}) \\ & m(\boldsymbol{\theta}) \leq m(\boldsymbol{\theta}_0) \\ & \theta^{\min} \leq \theta \leq \theta^{\text{solid}} \\ & \mathbf{q}_0 \in \Gamma, \quad \mathbf{q}_0^i(\boldsymbol{\theta}) \in \omega_i \quad (i = 1 \ldots n), \\ & \sum_{i=1}^{n} \mathbf{q}_0^i = \overline{\mathbf{q}} \\ \text{satisfying} \quad & \mathbf{q}(\boldsymbol{\theta}) = \mathbf{K}_t(\boldsymbol{\theta})\mathbf{T}(\boldsymbol{\theta}) \\ & \mathbf{f}_{\text{th}}\left(\boldsymbol{\theta}, \mathbf{T}(\boldsymbol{\theta})\right) + \mathbf{f} = \mathbf{K}_{\text{elast}}(\boldsymbol{\theta})\mathbf{u}(\boldsymbol{\theta}). \end{aligned} \qquad (25)$$

3.1 Sensitivity analysis

To analyze the sensitivity of this problem Eq. (25) can be rewritten in the form of Lagrangian function Ł:

$$\text{Ł} = \Big(\mathbf{f}_{\text{th}}\big(\boldsymbol{\theta},\mathbf{T}(\boldsymbol{\theta})\big)+\mathbf{f}\Big)^{\top}\mathbf{u}(\boldsymbol{\theta})+\lambda_J\Big(\mathbf{q}(\boldsymbol{\theta})^{\top}\mathbf{T}(\boldsymbol{\theta})-\beta J_{\text{t}}^{0}(\boldsymbol{\theta})\Big)+$$
$$\boldsymbol{\lambda}_m^{\top}\Big(\mathbf{K}_{\text{elast}}(\boldsymbol{\theta})\mathbf{u}(\boldsymbol{\theta})-\mathbf{f}-\mathbf{f}_{\text{th}}\big(\boldsymbol{\theta},\mathbf{T}(\boldsymbol{\theta})\big)\Big)+$$
$$\boldsymbol{\lambda}_t^{\top}\left(\mathbf{K}_t(\boldsymbol{\theta})\mathbf{T}(\boldsymbol{\theta})-\mathbf{q}(\boldsymbol{\theta})\right) \tag{26}$$

Where $\boldsymbol{\lambda}_m^{\top}$ and $\boldsymbol{\lambda}_t^{\top}$ are adjoint vectors, and λ_J is a penalty which is activated when the thermal compliance is greater than the initial design value, β is a relaxation factor equal to 1.1:

$$\begin{aligned}\text{if}\quad & J_{\text{t}}(\boldsymbol{\theta})\leq\beta J_{\text{t}}^{0}(\boldsymbol{\theta}),\quad \lambda_J=0\\ \text{otherwise}\quad & \lambda_J=1\end{aligned} \tag{27}$$

Since $\mathbf{f}_{\text{th}}$ a function of relative density $\boldsymbol{\theta}$ and temperature $\mathbf{T}(\boldsymbol{\theta})$, $\mathbf{T}(\boldsymbol{\theta})$ is a function of $\boldsymbol{\theta}$, and $\mathbf{q}(\boldsymbol{\theta})$ is a function of $\boldsymbol{\theta}$, the derivatives of the Lagrangian for each element θ are written as

$$\frac{\partial\text{Ł}(\boldsymbol{\theta})}{\partial\theta}=\mathbf{u}(\boldsymbol{\theta})^{\top}\frac{\partial\mathbf{f}_{\text{th}}}{\partial\theta}+\mathbf{u}(\boldsymbol{\theta})^{\top}\frac{\partial\mathbf{f}_{\text{th}}}{\partial\mathbf{T}(\boldsymbol{\theta})}\frac{\partial\mathbf{T}(\boldsymbol{\theta})}{\partial\theta}+(\mathbf{f}+\mathbf{f}_{\text{th}})^{\top}\frac{\partial\mathbf{u}(\boldsymbol{\theta})}{\partial\theta}+$$
$$\lambda_J\mathbf{T}(\boldsymbol{\theta})^{\top}\frac{\partial\mathbf{q}}{\partial\theta}+\lambda_J\mathbf{q}^{\top}\frac{\partial\mathbf{T}(\boldsymbol{\theta})}{\partial\theta}+$$
$$\boldsymbol{\lambda}_m^{\top}\left(\frac{\partial\mathbf{K}_{\text{elast}}(\boldsymbol{\theta})}{\partial\theta}\mathbf{u}(\boldsymbol{\theta})+\mathbf{K}_{\text{elast}}(\boldsymbol{\theta})\frac{\partial\mathbf{u}(\boldsymbol{\theta})}{\partial\theta}\right.$$
$$\left.-\frac{\partial\mathbf{f}_{\text{th}}}{\partial\theta}-\frac{\partial\mathbf{f}_{\text{th}}}{\partial\mathbf{T}(\boldsymbol{\theta})}\frac{\partial\mathbf{T}(\boldsymbol{\theta})}{\partial\theta}\right)+$$
$$\boldsymbol{\lambda}_t^{\top}\left(\frac{\partial\mathbf{K}_t(\boldsymbol{\theta})}{\partial\theta}\mathbf{T}(\boldsymbol{\theta})+\mathbf{K}_t\frac{\partial\mathbf{T}(\boldsymbol{\theta})}{\partial\theta}-\frac{\partial\mathbf{q}(\boldsymbol{\theta})}{\partial\theta}\right) \tag{28}$$

In order to cancel $\frac{\partial\mathbf{u}(\boldsymbol{\theta})}{\partial\theta}$ term and $\frac{\partial\mathbf{T}(\boldsymbol{\theta})}{\partial\theta}$, the value in adjoint vectors can be defined to satisfy

$$\Big((\mathbf{f}+\mathbf{f}_{\text{th}})^{\top}+\boldsymbol{\lambda}_m^{\top}\mathbf{K}_{\text{elast}}(\boldsymbol{\theta})\Big)\frac{\partial\mathbf{u}(\boldsymbol{\theta})}{\partial\theta}=\mathbf{0}$$
$$\Big(\mathbf{u}(\boldsymbol{\theta})^{\top}\frac{\partial\mathbf{f}_{\text{th}}}{\partial\mathbf{T}(\boldsymbol{\theta})}+\boldsymbol{\lambda}_t^{\top}\mathbf{K}_t(\boldsymbol{\theta}) \tag{29}$$
$$-\boldsymbol{\lambda}_m^{\top}\frac{\partial\mathbf{f}_{\text{th}}}{\partial\mathbf{T}(\boldsymbol{\theta})}+\lambda_J\mathbf{q}(\boldsymbol{\theta})^{\top}\Big)\frac{\partial\mathbf{T}(\boldsymbol{\theta})}{\partial\theta}=\mathbf{0},$$

By sequentially solving the above two equations, the sensitivity

is derived as

$$\frac{\partial L(\boldsymbol{\theta})}{\partial\theta_c}=\mathbf{u}(\boldsymbol{\theta})^{\top}\frac{\partial\mathbf{f}_{\text{th}}}{\partial\theta}+\lambda_J\mathbf{T}(\boldsymbol{\theta})^{\top}\frac{\partial\mathbf{q}(\boldsymbol{\theta})}{\partial\theta}+$$
$$\boldsymbol{\lambda}_m^{\top}\left(\frac{\partial\mathbf{K}_{\text{elast}}(\boldsymbol{\theta})}{\partial\theta}\mathbf{u}(\boldsymbol{\theta})-\frac{\partial\mathbf{f}_{\text{th}}}{\partial\theta}\right) \tag{30}$$
$$+\boldsymbol{\lambda}_t^{\top}\left(\frac{\partial\mathbf{K}_t(\boldsymbol{\theta})}{\partial\theta}\mathbf{T}(\boldsymbol{\theta})-\frac{\partial\mathbf{q}(\boldsymbol{\theta})}{\partial\theta}\right).$$

4 Numerical examples

Two numerical examples are shown in this section, namely Design 1 and Design 2. Both of these two designs share the same boudary conditions associated to heat transfer, but their boundary condition for thermomechanical models are different. In Design 1, the bottom edge is fixed, and compressive pressure is imposed on the top edge. In Design 2, the boundary condition and mechanical load locations are following a typical Messerschmitt-Blkow-Blohm (MBB) beam example. Assume the initial design domain is composed of an X-bracing lattice structure having a mass fraction of $m(\boldsymbol{\theta}_0)=0.5$. The X-bracing lattice structure is adopted since it can be produced without requiring additional materials for supporting structure in additive manufacturing.

To reveal the capability of the proposed method, for each design, four scenarios (A-D) are examined. The results of the initial design for each four scenarios are listed as Scenarios a-d. In Scenario A and a, only thermal boundary conditions and supports for thermomechanical model are applied. All the mechanical loads are induced from a non-uniform temperature field derived from the heat transfer model. In Scenario B and b, an additional external constant heat flux is applied. Then, in Scenario C and c, an external pressure or force is imposed on the boundary. Finally, in Scenario D and d, the convective heat transfer coefficient is doubled. The penalization numbers values are $p_1=5$, $p_2=3$, and $p_3=1.2$. The relative density of fluid phase $\theta^{\min}$ is equal to 0.15. For all of these examples, non-dimensional parameters are used.

4.1 Example 1:A LHT section withstanding compressive mechanical load

In the first example, a rectangular design domain having 120×240 elements is fixed at the bottom edge. In the thermal model, the design-dependent heat source is imposed on the entire design domain. The maximum value of the heat source value per element $\overline{q_0}$ is 0.01, convective heat transfer coefficient h is equal to 0.005. The thermal model results in a non-uniform temperature field, inducing thermomechanical loads. In scenarios A and a, only the induced thermomechanical loads are considered in the mechanical model (Fig. 3). In scenarios B and b, a boundary heat source $q_2=0.1$ is applied to each node on the bottom edge (Fig. 4). Then, in scenarios C and c, a compressive pressure

TABLE 1. A SUMMARY OF BOUNDARY CONDITION FOR SCENARIOS.

Scenario	1. Boundary condition of thermal problem	2. Boundary condition of thermomechanical problem
a and A	Volumetric design-dependent heat source	Thermomechanical load
b and B	In addition to volumetric design-dependent heat source, apply a constant heat source to the bottom edge	Thermomechanical load
c and C	In addition to volumetric design-dependent heat source, apply a constant heat source to the bottom edge	In addition to thermomechanical load, apply external mechanical load to the boundary edges
d and D	In addition to volumetric design-dependent heat source, apply a constant heat source to the bottom edge, increasing convective heat transfer coefficient	In addition to thermomechanical load, apply external mechanical load to the boundary edges

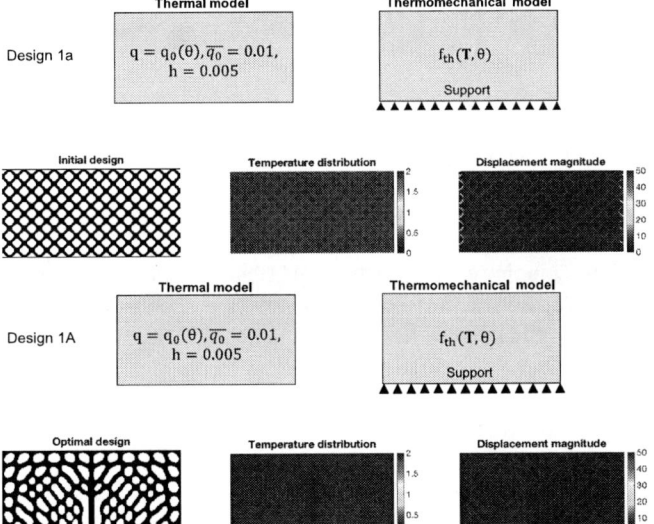

FIGURE 3. TOPOLOGY, TEMPERATURE DISTRIBUTION, AND DISPLACEMENT MAGNITUDE DISTRIBUTION OF INITIAL AND OPTIMAL DESIGN FOR SCENARIO 1a AND 1A.

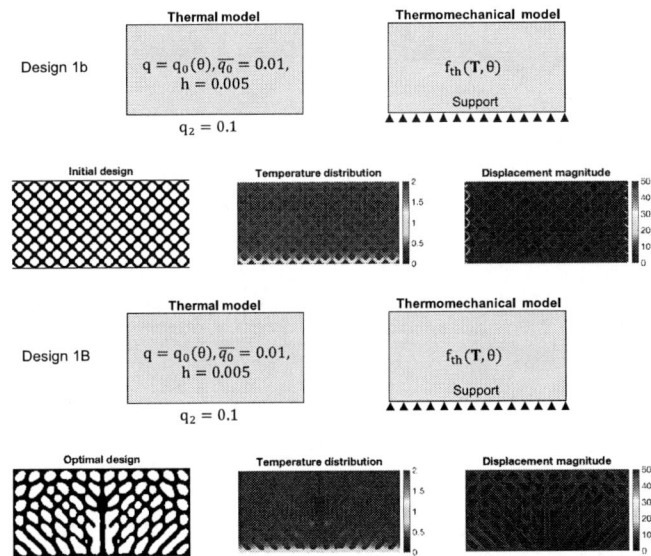

FIGURE 4. TOPOLOGY, TEMPERATURE DISTRIBUTION, AND DISPLACEMENT MAGNITUDE DISTRIBUTION OF INITIAL AND OPTIMAL DESIGN FOR SCENARIO 1b AND 1B.

$p=0.005$ is applied to each node on the top edge of the design domain (Fig. 5). In scenarios D and d, the convective heat transfer h is doubled from 0.005 to 0.01 (Fig. 6).

The topology, temperature distribution and displacement magnitude plot of the first example are shown in Fig. 3 to Fig. 6, and the key results are listed in Table 2. The mechanical compliance of optimal designs are only 40% to 52% of the initial designs, while thermal compliance is kept at 95% to 102% of the initial designs. Additionally, other metrics can also be compared such as mean displacement magnitude $\widehat{\| u \|}$, maximum displacement magnitude $\overline{\| u \|}$, maximum temperature $\widehat{T}$ and mean temperature $\overline{T}$. The magnitude of mean displacement of the optimal designs is 56% to 77% of the initial designs, and the maximum displacement of the optimal designs is only 21% to 27% of the initial designs. The mean temperature of the optimal designs is 91% to 94% of initial designs, which implies a limited compromising of thermal performance to achieve an increased mechanical performance. The maximum temperature of the optimal designs is 72% to 102% of the initial designs, which implies that the overall heat transfer performance is generally maintained, but that the local heat transfer performance is not be guaranteed.

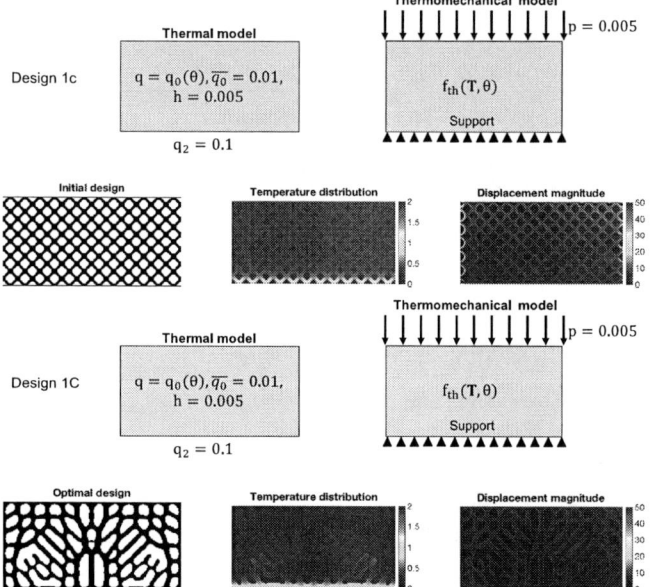

FIGURE 5. TOPOLOGY, TEMPERATURE DISTRIBUTION, AND DISPLACEMENT MAGNITUDE DISTRIBUTION OF INITIAL AND OPTIMAL DESIGN FOR SCENARIO 1c AND 1c.

TABLE 2. KEY RESULTS OF EXAMPLE 1: A LHT SECTION WITHSTANDING COMPRESSIVE MECHANICAL LOAD.

Case	$J_m(\boldsymbol{\theta})$	$J_t(\boldsymbol{\theta})$	$\widehat{\lVert u \rVert}$	$\overline{\lVert u \rVert}$	$\widehat{T}$	$\overline{T}$
Design 1a	285.90	1843.1	2.3133	47.617	0.9331	2.1032
Design 1A	160.38	1736.4	1.7944	10.006	0.9027	2.1518
Design 1b	282.53	1863.1	2.3755	47.718	1.0408	2.7952
Design 1B	152.36	1758.1	1.8206	10.172	0.9477	2.153
Design 1c	301.76	1863.1	4.2062	50.115	1.0408	2.7952
Design 1C	158.8	1798.4	2.672	11.392	0.9571	2.164
Design 1d	83.613	958.25	3.1024	27.8121	0.5212	1.8108
Design 1D	33.704	984.42	1.7487	7.5790	0.4945	1.3121

4.2 Example 2: A LHT section with boundary conditions of a MBB beam

In the second example, the thermal model remains the same as the first example with the exception that the supports are located at left edge and the right bottom corner, in terms of a half MBB beam. In scenarios A and *a*, only the induced thermomechanical loads are considered in the mechanical model (Fig. 7). In scenarios B and *b*, a boundary heat source $q_2=0.1$ is applied to each node on the bottom edge (Fig. 8). Then, in scenarios C and *c*, an external load $p=1.2$ is applied to the top left corner of the design domain (Fig. 9). In scenarios D and *d*, the convective

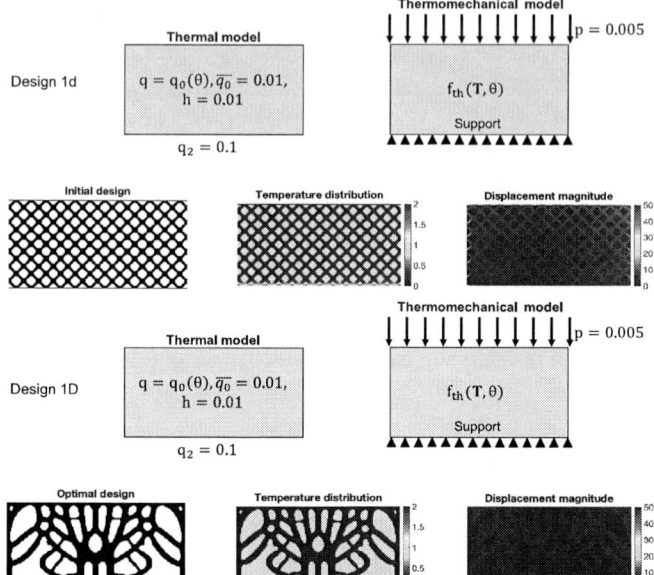

FIGURE 6. TOPOLOGY, TEMPERATURE DISTRIBUTION, AND DISPLACEMENT MAGNITUDE DISTRIBUTION OF INITIAL AND OPTIMAL DESIGN FOR SCENARIO 1d AND 1D.

heat transfer h is doubled from 0.005 to 0.01 (Fig. 10).

The topology, temperature distribution and displacement magnitude plot of the second example are shown in Fig. 3 to Fig. 6, and the key results are listed in Table 3. The results show trends similar to the first example: The mechanical compliance of optimal designs is only 27% to 46% of the initial designs, while thermal compliance is maintained at 94% to 110% of the initial designs. The magnitude of mean displacement of the optimal designs is 26% to 79% of the initial designs, and the maximum displacement of the optimal designs is only 22% to 46% of the initial designs. The mean temperature of the optimal designs is 91% to 100% of initial designs, which again implies a limited compromising of thermal performance to achieve an increased mechanical performance. The maximum temperature of the optimal designs is 73% to 103% of the initial designs, which implies that the overall heat transfer performance is generally maintained, but that the local heat transfer performance is not be guaranteed.

5 Conclusion

This study presents a novel thermomechanical topology optimization method with consideration of convective heat transfer and design-dependent heat sources, that takes advantage of the benefits of a multifunctional LHT structure. The heat source is dependent on the material phase, thus the optimized solid-fluid interface and heat source distribution can be obtained through

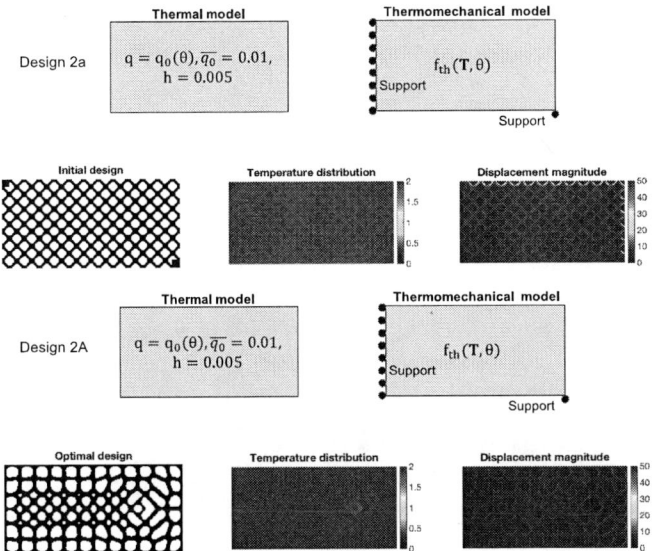

FIGURE 7. TOPOLOGY, TEMPERATURE DISTRIBUTION, AND DISPLACEMENT MAGNITUDE DISTRIBUTION OF INITIAL AND OPTIMAL DESIGN FOR SCENARIO 2a AND 2A.

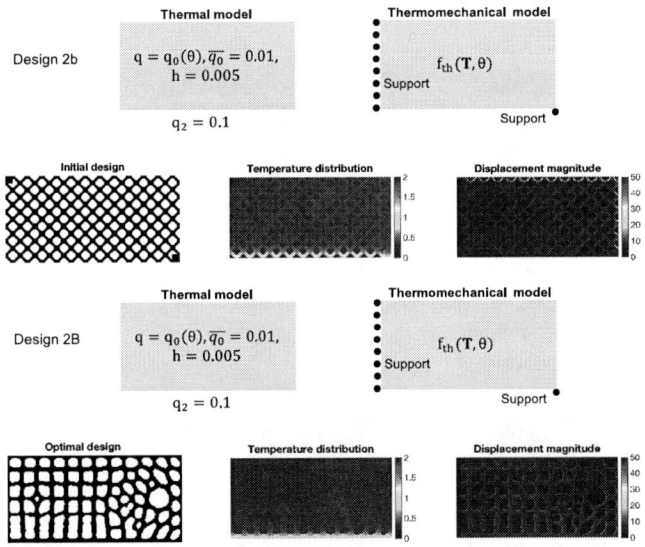

FIGURE 8. TOPOLOGY, TEMPERATURE DISTRIBUTION, AND DISPLACEMENT MAGNITUDE DISTRIBUTION OF INITIAL AND OPTIMAL DESIGN FOR SCENARIO 2b AND 2B.

the method. Since the method does not require a fluid mechanics model, it is computationally efficient and convenient for application in the conceptual design stage. With the application of this method, the mechanical stiffness of the LHT structure due to mechanical and thermomechanical loads is significantly improved,

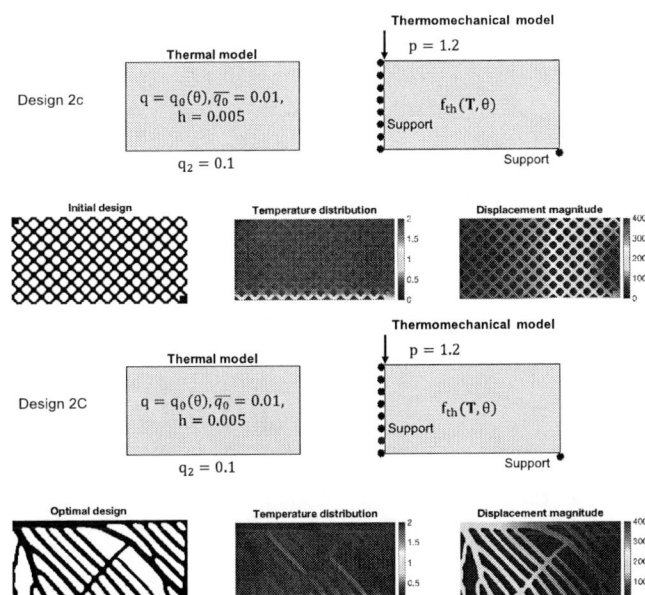

FIGURE 9. TOPOLOGY, TEMPERATURE DISTRIBUTION, AND DISPLACEMENT MAGNITUDE DISTRIBUTION OF INITIAL AND OPTIMAL DESIGN FOR SCENARIO 2c AND 2C.

TABLE 3. KEY RESULTS OF EXAMPLE 2: A LHT SECTION WITH BOUNDARY CONDITIONS OF A MBB BEAM.

Case	$J_m(\boldsymbol{\theta})$	$J_t(\boldsymbol{\theta})$	$\widehat{\|u\|}$	$\overline{\|u\|}$	$\widehat{T}$	$\overline{T}$
Design 2a	350.86	1855.4	3.5224	53.560	0.9981	2.109
Design 2A	164.04	1753.6	2.2592	12.1389	0.9049	2.1697
Design 2b	347.32	1874.7	3.8300	52.242	1.0457	2.7952
Design 2B	156.20	1886.1	3.0133	15.246	0.9923	2.1489
Design 2c	814.94	1874.7	127.36	416.61	1.0457	2.7952
Design 2C	327.53	2032.2	55.0129	192.24	1.0517	2.1649
Design 2d	551.70	963.79	126.97	394.31	0.5236	1.8108
Design 2D	152.30	1058.7	32.516	111.15	0.5128	1.3163

while the overall heat transfer performance is maintained. The final design shows complex lattice structures that can be created with current additive manufacturing technologies.

Finally, limitations of this method and future work are addressed. First, the proposed method uses a design-dependent heat source to replace an accurate fluid mechanics model. It does not contain a velocity field and therefore it could not reflect the temperature gradient caused by forced convection in the fluid. Consequently the method is limited to the investigation of problems where the velocity difference in the fluid is small. Secondly, the results of numerical example show that the method does not

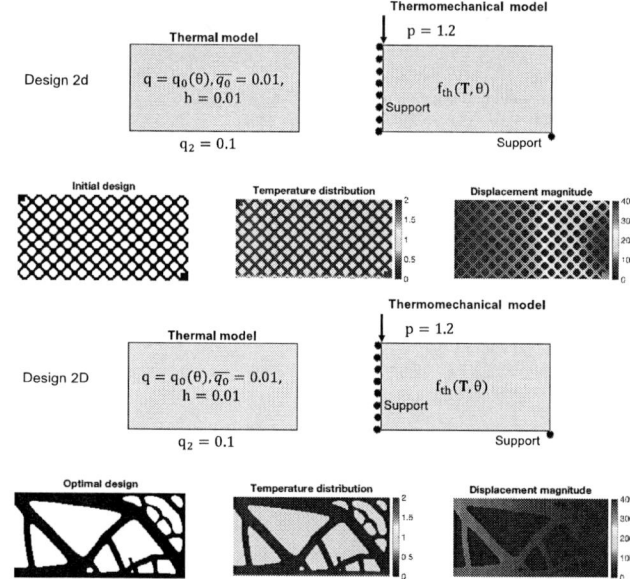

FIGURE 10. TOPOLOGY, TEMPERATURE DISTRIBUTION, AND DISPLACEMENT MAGNITUDE DISTRIBUTION OF INITIAL AND OPTIMAL DESIGN FOR SCENARIO 2d AND 2D.

always maintain the local heat transfer performance. For a detailed design, thermal-fluid-structure coupled simulation and experimental study is also required.

REFERENCES

[1] Feng, S., Kuang, J., Wen, T., Lu, T., and Ichimiya, K., 2014. "An experimental and numerical study of finned metal foam heat sinks under impinging air jet cooling". *International Journal of Heat and Mass Transfer,* **77**, pp. 1063–1074.

[2] Al-Athel, K. S., Aly, S. P., Arif, A. F. M., and Mostaghimi, J., 2017. "3d modeling and analysis of the thermomechanical behavior of metal foam heat sinks". *International Journal of Thermal Sciences,* **116**, pp. 199–213.

[3] Paknezhad, M., Rashidi, A., Yousefi, T., and Saghir, Z., 2017. "Effect of aluminum-foam heat sink on inclined hot surface temperature in the case of free convection heat transfer". *Case studies in thermal engineering,* **10**, pp. 199–206.

[4] Murr, L. E., Gaytan, S. M., Ramirez, D. A., Martinez, E., Hernandez, J., Amato, K. N., Shindo, P. W., Medina, F. R., and Wicker, R. B., 2012. "Metal fabrication by additive manufacturing using laser and electron beam melting technologies". *Journal of Materials Science & Technology,* **28**(1), pp. 1–14.

[5] Hussein, A., Hao, L., Yan, C., Everson, R., and Young, P., 2013. "Advanced lattice support structures for metal additive manufacturing". *Journal of Materials Processing Technology,* **213**(7), pp. 1019–1026.

[6] Beyer, C., and Figueroa, D., 2016. "Design and analysis of lattice structures for additive manufacturing". *Journal of Manufacturing Science and Engineering,* **138**(12), p. 121014.

[7] Queheillalt, D. T., Carbajal, G., Peterson, G., and Wadley, H. N., 2008. "A multifunctional heat pipe sandwich panel structure". *International Journal of Heat and Mass Transfer,* **51**(1-2), pp. 312–326.

[8] Son, K. N., Weibel, J. A., Kumaresan, V., and Garimella, S. V., 2017. "Design of multifunctional lattice-frame materials for compact heat exchangers". *International Journal of Heat and Mass Transfer,* **115**, pp. 619–629.

[9] Bunker, R. S., 2004. "Latticework (vortex) cooling effectiveness: Part 1stationary channel experiments". In ASME Turbo Expo 2004: Power for Land, Sea, and Air, American Society of Mechanical Engineers, pp. 909–918.

[10] Acharya, S., Zhou, F., Lagrone, J., Mahmood, G., and Bunker, R. S., 2005. "Latticework (vortex) cooling effectiveness: Rotating channel experiments". *Journal of turbomachinery,* **127**(3), pp. 471–478.

[11] Rao, Y., and Zang, S., 2014. "Flow and heat transfer characteristics in latticework cooling channels with dimple vortex generators". *Journal of Turbomachinery,* **136**(2), p. 021017.

[12] Au, K., and Yu, K., 2007. "A scaffolding architecture for conformal cooling design in rapid plastic injection moulding". *The International Journal of Advanced Manufacturing Technology,* **34**(5-6), pp. 496–515.

[13] Au, K., and Yu, K., 2011. "Modeling of multi-connected porous passageway for mould cooling". *Computer-Aided Design,* **43**(8), pp. 989–1000.

[14] Brooks, H., and Brigden, K., 2016. "Design of conformal cooling layers with self-supporting lattices for additively manufactured tooling". *Additive Manufacturing,* **11**, pp. 16 – 22.

[15] Sigmund, O., 2001. "Design of multiphysics actuators using topology optimization–part i: One-material structures". *Computer methods in applied mechanics and engineering,* **190**(49), pp. 6577–6604.

[16] Sigmund, O., 2001. "Design of multiphysics actuators using topology optimization–part ii: Two-material structures". *Computer methods in applied mechanics and engineering,* **190**(49-50), pp. 6605–6627.

[17] Du, Y., Luo, Z., Tian, Q., and Chen, L., 2009. "Topology optimization for thermo-mechanical compliant actuators using mesh-free methods". *Engineering Optimization,* **41**(8), pp. 753–772.

[18] Thurier, P. F., 2014. "A two-material topology optimization method for the design of a passive thermal control interface".

[19] Zhang, W., Yang, J., Xu, Y., and Gao, T., 2014. "Topology optimization of thermoelastic structures: mean compliance minimization or elastic strain energy minimization". *Structural and Multidisciplinary Optimization,* *49*(3), pp. 417–429.

[20] Gao, T., Xu, P., and Zhang, W., 2016. "Topology optimization of thermo-elastic structures with multiple materials under mass constraint". *Computers & Structures,* *173*, pp. 150–160.

[21] Wu, T., Liu, K., and Tovar, A., 2017. "Multiphase topology optimization of lattice injection molds". *Computers & Structures,* *192*, pp. 71–82.

[22] Wu, T., and Tovar, A., 2018. "Multiscale, thermomechanical topology optimization of self-supporting cellular structures for porous injection molds". *Rapid Prototyping Journal.*

[23] Alexandersen, J., Aage, N., Andreasen, C. S., and Sigmund, O., 2014. "Topology optimisation for natural convection problems". *International Journal for Numerical Methods in Fluids,* *76*(10), pp. 699–721.

[24] Sato, Y., Yaji, K., Izui, K., Yamada, T., and Nishiwaki, S., 2017. "An optimum design method for a thermal-fluid device incorporating multiobjective topology optimization with an adaptive weighting scheme". *Journal of Mechanical Design.*

[25] Zhao, X., Zhou, M., Sigmund, O., and Andreasen, C., 2018. "A poor man's approach to topology optimization of cooling channels based on a darcy flow model". *International Journal of Heat and Mass Transfer,* *116*, pp. 1108–1123.

[26] Gao, T., Zhang, W., Zhu, J., Xu, Y., and Bassir, D., 2008. "Topology optimization of heat conduction problem involving design-dependent heat load effect". *Finite Elements in Analysis and Design,* *44*(14), pp. 805–813.

[27] Iga, A., Nishiwaki, S., Izui, K., and Yoshimura, M., 2009. "Topology optimization for thermal conductors considering design-dependent effects, including heat conduction and convection". *International Journal of Heat and Mass Transfer,* *52*(11), pp. 2721–2732.

[28] Dede, E. M., Nomura, T., and Lee, J., 2014. *Multiphysics Simulation.* Springer.

[29] Joo, Y., Lee, I., and Kim, S. J., 2017. "Topology optimization of heat sinks in natural convection considering the effect of shape-dependent heat transfer coefficient". *International Journal of Heat and Mass Transfer,* *109*, pp. 123–133.

Proceedings of the ASME 2019
International Design Engineering Technical Conferences
and Computers and Information in Engineering Conference
IDETC/CIE2019
August 18-21, 2019, Anaheim, CA, USA

DETC2019-97659

DESIGN OF GRADIENT NANOTWINNED METAL MATERIALS USING ADAPTIVE GAUSSIAN PROCESS BASED SURROGATE MODELS

Haofei Zhou[1]

Department of Engineering Mechanics, Zhejiang University
Hangzhou, Zhejiang, P.R. China

Xin Chen[2]

University of Illinois at Urbana-Champaign
Urbana, IL 61801, United States

Yumeng Li[3]

University of Illinois at Urbana-Champaign
Urbana, IL 61801, United States

ABSTRACT

Inspired by gradient structures in the nature, Gradient Nanostructured (GNS) metals have emerged as a new class of materials with tunable microstructures. GNS metals can exhibit unique combinations of material properties in terms of ultrahigh strength, good tensile ductility and enhanced strain hardening, superior fatigue and wear resistance. However, it is still challenging to fully understand the fundamental gradient structure-property relationship, which hinders the rational design of GNS metals with optimized target properties. In this paper, we developed an adaptive design framework based on simulation-based surrogate modeling to investigate how the grain size gradient and twin thickness gradient affect the strength of GNS metals. The Gaussian Process (GP) based surrogate modeling technique with adaptive sequential sampling is employed for the development of surrogate models for the gradient structure-property relationship. The proposed adaptive design integrates physics-based simulation, surrogate modeling, uncertainty quantification and optimization, which can efficiently explore the design space and identify the optimized design of GNS metals with maximum strength using limited sampling data generated from high fidelity but computational expensive physics-based simulations.

KEY WORDS

Design, Engineering Materials, Gradient Nanostructured Metals, Gaussian Processes, Surrogate Modeling

1. INTRODUCTION

Recently, gradient nanostructured (GNS) metals, including gradient (GNG) metals and gradient nanotwinned (GNT) metals, have emerged as a new class of materials with tunable microstructures. GNS metals could be designed to exhibit an unusual combination of ultrahigh strength, good tensile ductility, enhanced strain hardening, superior fatigue and wear resistance [1-5]. Using surface mechanical grinding treatment (SMGT), Fang et al. [1] prepared a GNG Cu specimen with a doubled yield strength and a comparable ductility with respect to coarse grained Cu. In GNG interstitial free (IF) steel attained through surface mechanical attrition treatment (SMAT), Wu et al. [2] reported an extra work hardening behavior controlled by a unique combination of non-uniform deformation and multiaxial stresses. In a twinned induced plasticity (TWIP) steel, pre-torsion and subsequent tension can be applied to form a hierarchical nanotwinned structure with a gradient nanotwin density along the radial direction, leading to enhanced strength and ductility [3]. Recently, Cheng et al. [4] used a direct-current electro-deposition method to synthesize GNT Cu samples with combined gradients in grain size and twin thickness. Due to a unique patterning of geometrically necessary dislocations in grain interiors, a sufficiently large structural gradient led to an improved strength and work hardening that can exceed even the strongest component of the gradient microstructure.

Although GNS metals are shown to possess superior

---

[1] Assistant Professor, haofei_zhou@zju.edu.cn.
[2] Graduate Research Assistant, xinc4@illinois.edu
[3] Assistant Professor, yumengl@illinois.edu. Corresponding Author.

mechanical properties as compared with their gradient-free counterparts, it remains largely unknown whether there exists a specific gradient structure that optimizes their mechanical properties [6]. Existing manufacturing approaches, including surface tooling [7, 8] and mechanical treatments [3, 9], generate limited volume fractions of gradient surface layers or negligible degrees of structural gradient along the gradient direction. In addition, most previous studies were focused on only one type of gradient structure, which limits our ability to optimize the mechanical properties of GNS metals. For instance, a typical GNG structure synthesized by surface mechanical treatment possesses a gradient nanograined surface layer and a coarse-grained core. The grain size of the top-most surface layer usually ranges from dozens of nanometers to submicrons, while it increases to microns in the coarse-grained core [1, 2]. Recently, Lin et al. [10] used electrodeposition to prepare GNS Ni samples with a similar change in grain size from 29 nm to 4 μm. The degree of gradient size gradient could be accurately controlled. It is observed that a maximum uniform elongation that exceeds even that of coarse-grained Ni could be achieved at an optimal profile of gradient size distribution, indicating that the ductility of GNS metals can be optimized through control of structural gradient. However, the underlying mechanism for such an optimized mechanical property is still unclear, which is hindering the design and potential application of GNS metals.

The goal of this work is to deepen the understanding of the following fundamental questions: How do structural gradients in grain size and twin thickness affect the strength and plastic deformation mechanisms of GNS metals? And does there exist a specific structural gradient that optimizes their strength? Uniaxial tension simulations have been conducted using molecular dynamics (MD) simulations to evaluate the strength of GNT Cu samples with different gradients in grain size and twin thickness. The collective effects of grain and twin gradients in the simulated platform are investigated for their effects on strength and deformation mechanisms of GNT Cu. An adaptive design framework is developed to explore the interested design space and identify the optimized design of GNT with the target properties through the integration of simulation-based surrogate modeling, uncertainty quantification and optimization. The developed design framework involves Gaussian process (GP) for constructing surrogate model based on physics-based simulations, which is MD simulations in our work, to predict the material property in the design region without experimental and simulation sampling data. Machine learning is emerging as a powerful tool in material science and engineering [11-15] for accelerating the material design. GP process used in our study is a non-parametric Bayesian machine learning technique, which returns robust variance estimates of predictions for uncertainty quantification without extra computational cost [16,17]. Employing an adaptive sampling scheme, a design loop can be realized to explore the design space in an efficient manner with physics-based simulation only performed in critical regions. GNT material system with the optimized strength is identified through the design iteration and validated through physics-based simulation. The present results shed lights on the design of GNS structures for optimal mechanical properties.

NOMENCLATURE

β: GP regression coefficient
Corr: correlation function
d :   grain size
λ :   twin thickness
g_d :   gradient of grain size
g_λ: gradient of twin thickness
$G_{GP}(\mathbf{x})$:   Gaussian process predictor
$\hat{e}(\mathbf{x})$:   mean square error
$\mathbf{R}$:   correlation matrix
Ψ: the likelihood function

2. METHODOLOGY

This section introduces the detailed information about MD simulations and the adaptive design framework we proposed to explore the effects of grain size and twin thickness gradients on the strength of GNT Cu.

2.1. Molecular Dynamics Simulations

Fully three-dimensional GNT Cu samples containing two homogeneous columnar-grained nanotwinned components with $[1\bar{1}\bar{1}]$ out-of-plane texture (NT-A and NT-B) were constructed for atomistic simulations (Fig. 1). Each GNT sample contains ~21,000,000 atoms and has dimensions of ~$100 \times 100 \times 25$ nm^3. The volume fractions of NT-A and NT-B are equal in each GNT sample. The structure of homogeneous component NT-B used to generate the GNT samples was fixed to possess average grain size of $d_B = 50$ nm and twin thickness of $\lambda_B = 6.3$ nm. By varying the grain size (d_A) and twin thickness (λ_A) of the homogeneous component NT-A, GNT samples with various gradients in grain size and twin thickness were generated.

To quantify the gradients in grain size and twin thickness, we define two structural parameters, $g_d = d_B/d_A - 1$ and $g_\lambda = \lambda_B/\lambda_A - 1$. NT-A and NT-B have equivalent grain size and twin thickness when $g_d = 0$ and $g_\lambda = 0$. The system was initially equilibrated at 300K for 500 ps, followed by uniaxial tensile loading along the x-axis to a total strain of 15% at a constant strain rate of 5×10^8 s^{-1}. Throughout the simulation, periodic boundary conditions were applied in all three directions. Constant temperature and zero pressure in the non-stretching directions (i.e., y- and z-axes) were controlled by Nose-Hoover thermostatting and barostatting [18, 19]. The embedded atom method potential [20] was used to compute the interatomic forces, and the integration time-step was fixed at 1 fs. The common neighbor analysis (CNA) method [21] was used to identify defects that emerge during plastic deformation.

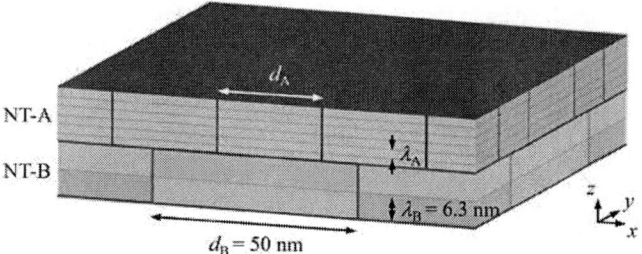

Figure 1. A typical GNT sample in MD simulations composed of two homogeneous nanotwinned layers NT-A and NT-B forming grain size and twin thickness gradients along the z-axis. Grain size and twin thickness are denoted as d_A and λ_A in NT-A, and d_B and λ_B in NT-B. Uniaxial tensile loading is applied along the x-axis.

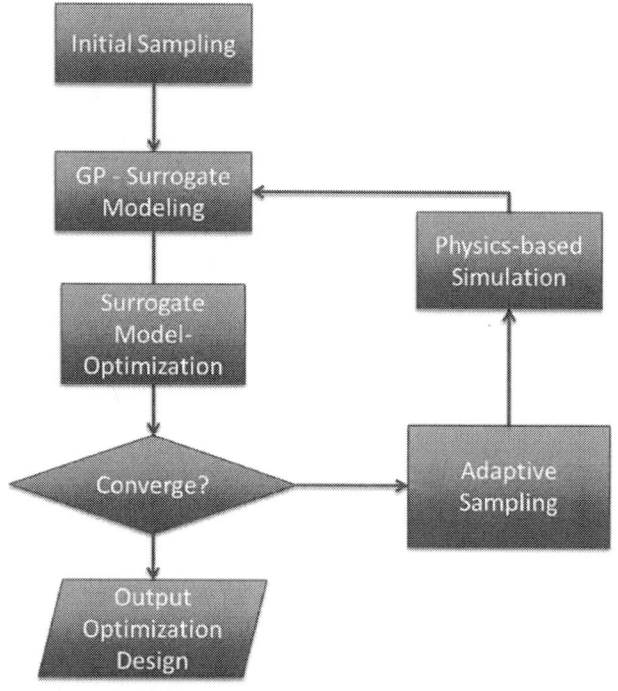

Figure 2. The schematic adaptive design framework based on GP-Surrogate modeling.

2.2. Adaptive Design based on GP Surrogate Model

It is often extremely challenging and costly to dig into the optimization problem of GNS metals using experimental or physics-based computational methods, which is the MD simulation method in the present study. Indeed, our MD simulations have investigated limited design parameters featuring the structural gradients in grain size and twin thickness of GNT, which leaves vast unexplored design spaces. To address the challenges associated with the optimized design of GNT metals, we established an adaptive design framework schematically shown in Fig. 2, which integrates physics-based

simulations, surrogate modeling, uncertainty quantification and optimization. The key components of the developed adaptive design framework include: 1) an initial sparse random training data set from MD simulations; 2) surrogate model that uses the training data set to learn the feature-property relationship; 3) uncertainty quantification of the surrogate model for updating; 4) identifying the critical new sampling points for updating the surrogate model to improve the fidelity; 5) the use of the converged surrogate model to search the unexplored design space for optimized design.

In this paper, a GP based surrogate model is employed to establish how the strength of GNT metals varies with the input features **x** over the whole chosen design domain. While using the GP regression, the material performance is assumed to be generated as

$$G(\mathbf{x}) \approx G_{GP}(\mathbf{x}) = \mathbf{f}^{\mathbf{T}}(\mathbf{x}) \cdot \boldsymbol{\beta} + S(\mathbf{x}) + \varepsilon \qquad (1)$$

where $\mathbf{f}^{\mathbf{T}}(x)=[f1(\mathbf{x}), ..., fb(\mathbf{x})]$ is the base functions, $\boldsymbol{\beta} =[\beta_1, \beta_2...,\beta_b]$ the regression coefficient vector, $S(\mathbf{x})$ the Gaussian stochastic process with zero mean and certain covariance matrix, and ε the uncorrelated noise that follows normal distribution. Generally, $\mathbf{f}^{\mathbf{T}}(\mathbf{x}) \boldsymbol{\beta}$ can be replaced by a constant global mean μ based on the fact that a constant mean in GP model is generally adequate to model the performance function. By assuming noise-free observations, the GP model can be written as:

$$G_{GP}(\mathbf{x}) = \mu + S(\mathbf{x}) \qquad (2)$$

In the GP model shown in Eq. 2, the Gaussian covariance function is adopted for the stochastic process $S(x)$. For two random inputs x_i and x_j, the Gaussian covariance function can be expressed as

$$\mathbf{Cov}_{(i,j)} = \sigma^2 \mathbf{R}_{(i,j)} \qquad (3)$$

where $\mathbf{R}$ is the correlation matrix. The (i, j) entry of the matrix $\mathbf{R}$ is described as

$$\mathbf{R}_{(i,j)} = Corr(\mathbf{x}_i, \mathbf{x}_j) = \exp\left[-\sum_{p=1}^{N} a_p \mid x_p^i - x_p^j \mid^{b_p}\right] \qquad (4)$$

where $Corr$ is the correlation function. a_p and b_p are parameters of the GP model. With n number of observations (X, Y) in which $X=[x_1,...,x_n]$ and $Y =[y(x_1),...,y(x_n)]$, the log likelihood function of the GP model can be given as

$$\Psi = -\frac{1}{2}\left[n \ln(2\pi) + n \ln \sigma^2 + \ln|\mathbf{R}| + \frac{1}{2\sigma^2}(\mathbf{Y} - \mathbf{A}\mu)^T \mathbf{R}^{-1}(\mathbf{Y} - \mathbf{A}\mu)\right] \qquad (5)$$

where $\mathbf{A}$ is an n×1 unit vector. Then μ and σ^2 can be obtained by maximizing the likelihood function as

$$\mu = \left[\mathbf{A}^T \mathbf{R}^{-1} \mathbf{A}\right]^{-1} \mathbf{A}^T \mathbf{R}^{-1} \mathbf{Y} \qquad (6)$$

$$\sigma^2 = \frac{(\mathbf{Y} - \mathbf{A}\mu)^T \mathbf{R}^{-1}(\mathbf{Y} - \mathbf{A}\mu)}{n} \qquad (7)$$

With the GP regression model, the response for any given sample point $\mathbf{x}$ can be estimated as

$$G_{GP}(\mathbf{x}) = \mu + \mathbf{r}^T \mathbf{R}^{-1}(\mathbf{Y} - \mathbf{A}\mu) \qquad (8)$$

where r is the correlation vector between x and the samples $X=[x_1,...,x_n]$, and the i-th element of r is given by $r(i) = Corr(x, x_i)$. The mean square error $\hat{e}(\mathbf{x})$ can be accordingly estimated by

$$\hat{e}(\mathbf{x}') = \sigma^2 \left[1 - \mathbf{r}^T \mathbf{R}^{-1}\mathbf{r} + \frac{(1 - \mathbf{A}^T \mathbf{R}^{-1}\mathbf{r})^2}{\mathbf{A}^T \mathbf{R}^{-1}\mathbf{A}} \right] \qquad (9)$$

It is noted that employing a GP-based surrogate model allows the uncertainty quantification of the prediction over the whole design domain through the estimated mean square error in a natural way.

As the physics-based MD simulations are computationally expensive, we start with a low fidelity GP-based surrogate model based on a samll size of initial random sampling [X_E, Y_E]. Then the GP-based surrogate model is updated gradually with new sampling points to enhance the fidelity. An adaptive sampling scheme is employed to determine the most wanted new sampling points for the sequential updating of the GP-based surrogate model. The adaptive sampling scheme will identify the new sampling point to be where the highest uncertainty (i.e. the highest estimated variance) is observed or where the current possible maximum property is, in order to decrease the overall error of the GP-based surrogate model. True response value Y* of the selected new sample data X* will be calculated using MD simulations. Then [X*, Y*] is added to [X_E, Y_E] and the GP-based surrogate model will be updated accordingly. GP-based surrogate model is updated iteratively by adding new sampling points until the overall error is below a certain threshold. After that, the GP-based surrogate model is used for providing the prediction of material property over the whole design space and identifying the optimized design by combing with optimization.

3. RESULTS AND DISCUSSIONS

This section demonstrates the developed adaptive design framework for finding the optimized structural gradient in GNT Cu with maximum flow stress and some insights about the relationship between the structure gradients and material property from MD simulations.

3.1. Design and Optimization History

As shown in Fig. 3, starting with a small number of sampling data points, the GP-based surrogate model is initially set up with a relatively low fidelity. During the iteration process, new sampling points are added to update the surrogate model, which shows a decreasing error as shown in Fig. 3. Due to the difficulty of preparing simulation models with high resolution,

new sampling points are generated to have a close distance to the desired sampling points determined by the adaptive sampling rules. Three design iterations are performed to start with different initial sampling points. They all exhibits decreasing error during the iteration history with new sampling data added.

All iterations are stopped when the maximum variance reduces below the convergence threshold of 0.01 in our current work. Optimization is complemented using the trained surrogate model with a desired fidelity. In all the iterations, a total of 20 sampling points are used. Five-fold correlation analysis is conducted as shown in Fig. 4. It can be seen that the errors are centered around a low value, which demonstrates a good fidelity of the established surrogate model.

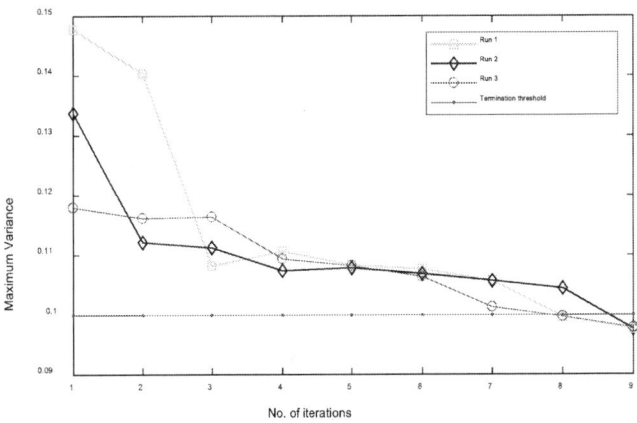

Figure 3. The history of Maximum variance and uncertainty of the current optimization results over the design iteration.

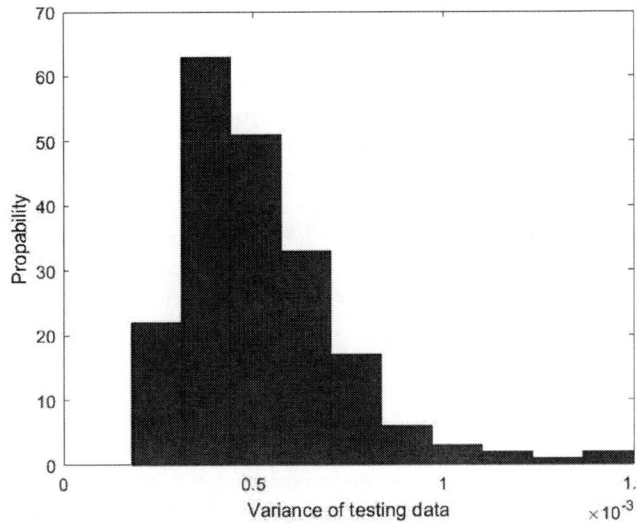

Figure 4. The mean square error distribution for 5 folder cross validation analysis with 100 replicate runs.

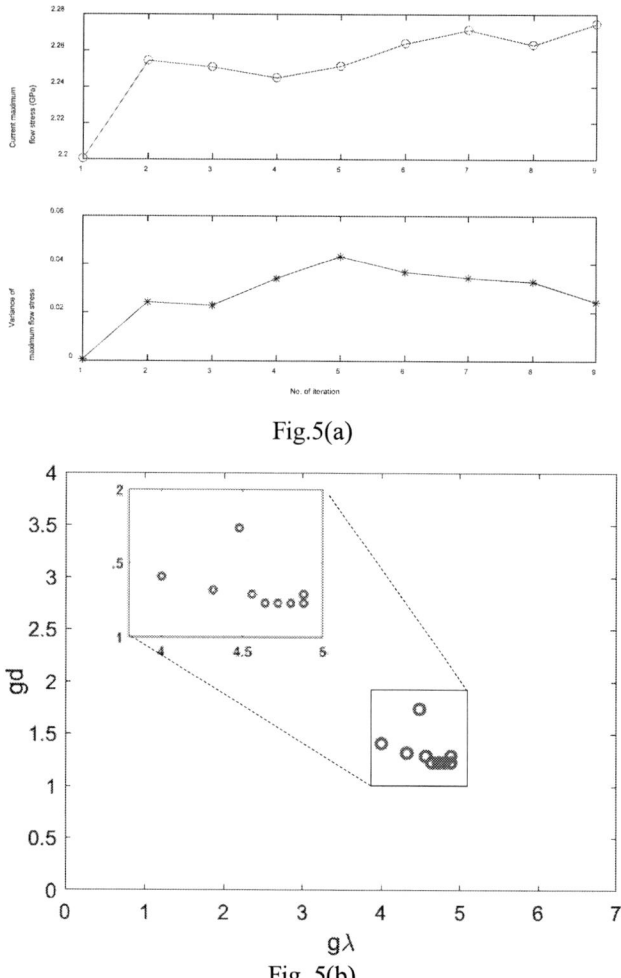

Fig.5(a)

Fig. 5(b)

Figure 5. The adaptive surrogate modeling and design optimization process: (a) the maximum flow stresses and corresponding prediction variances in each iteration, and (b) the predicted optimized design point in the 2D design space with maximum flow stress in each iteration.

The optimized design is identified to be around $g_\lambda = 4.88$ and $g_d = 1.23$ with a prediction of maximum flow stress around 2.274 GPa. To validate our model, we conducted a simulation with $g_\lambda = 5$ and $g_d = 1$ close to the optimized design. The simulated true flow stress is 2.256 GPa, while the predicted value is 2.265 GPa (error=0.0039). This validates the optimized design and demonstrates the high fidelity of the established surrogate model. We plot the current maximum flow stress estimated in each step of one design iteration (top panel) and the corresponding variance (bottom panel) in Fig 5a. Except for the first iteration, all the predicted maximum flow stresses stay in a small region between 2.245 and 2.274 GPa. We also plot the best design in each iteration in the design space (Fig. 5b). Most of the best designs are clustered in a small region.

This demonstrates that, although a highly nonlinear relationship exists between the flow stress and the structural gradients (Fig 6a), high-fidelity surrogate model can still be established based on a limited size of sampling data.

3.2. Insights from the MD Simulations

Tensile stress versus strain curves for GNT samples with four different grain gradients g_d and four different twin gradients g_λ are shown in Fig. 7a. It can be seen in Fig. 7b that, for a given g_λ, the average flow stress first increases with increasing g_d, reaching a maximum at $g_d = 1$, and then decreases with further increasing g_d. On the other hand, for a given g_d, the average flow stress increases monotonically with increasing g_λ (Fig. 7c).

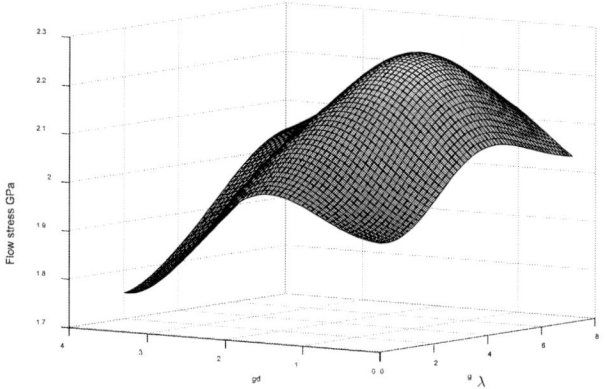

Figure 6. Flow stress over the design space predicted by the GP-based surrogate model.

To explain the gradient effect on average flow stress observed in Fig. 7, deformation patterns of the GNT samples with different structural gradients are examined (Fig. 8). The result for the layer NT-B in GNT samples with $d_B = 50$ nm and twin thickness of $\lambda_B = 6.3$ nm at 7.5% strain is shown in Fig. 8a where plastic deformation is governed by threading (mode II) dislocations. The movement of threading dislocations are confined inside the twin lamellae. Confined threading dislocations are frequently observed in highly oriented nanotwinned metals, nanoscale thin films and multilayered materials [22-24]. The stress required to move a threading dislocation can be expressed by the well-known confined layer slip (CLS) model in the form of [25]

$$\sigma_{CLS} = M \frac{\mu b \sin\varphi}{8\pi\lambda}\left(\frac{4-v}{1-v}\right)\ln\left(\frac{\alpha\lambda}{b\sin\varphi}\right) + \sigma_0 \qquad (10)$$

where μ is the shear modulus, v the Poisson's ratio, M the Taylor factor, φ the angle between slip plane and the layer interface, b the Burgers vector of the dislocation and α a coefficient representing the extent of dislocation core. σ_0 represents the resistance to dislocation motion due to lattice friction, the prestress associated with the twin boundaries, and the interaction between the threading dislocation and a pre-existing array of misfit segments on the neighboring twin boundaries. Eq. 10 shows that the dependence of flow stress on twin thickness scales with $\lambda^{-1}\ln\lambda$. It leads to moderate strengthening when threading dislocations are dominant at twin thicknesses above 5 nm [25, 26]. This may explain the slight increase in yield strength of the GNT samples when the gradient in twin thickness, g_λ, increases from 0 to 1. Since λ multiplies its logarithm in Eq. 10, the CLS model would predict a stress drop when the twin thickness approaches 1 nm [25], and thus fail to fully capture the monotonic increase in yield strength of the GNT samples as g_λ reaches 5 (Fig. 7c).

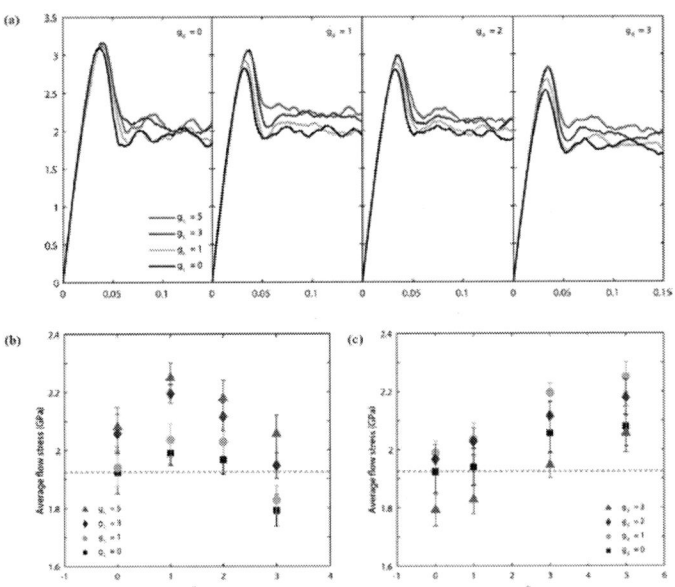

Figure 7. Stress-strain relations and flow stress of the simulated GNT Cu samples. (a) Stress-strain curves of GNT Cu with different values of grain and twin gradients, g_d and g_λ. Average flow stress as a function of g_d (b) and g_λ (c). The average flow stress is measured from a strain of 6-15% for all simulated samples.

Fig. 8b shows the deformation pattern at 7.5% strain for the layer NT-A with $d_A = 50$ nm and $\lambda_A = 1.04$ nm, corresponding to a GNT structure with gradient parameters of $g_\lambda = 5$ and $g_d = 0$. It is observed that plastic deformation is governed by necklace-like dislocations extended over multiple twin boundaries, which was referred to as correlated necklace dislocations (CNDs) or jogged dislocations [4, 25 27]. Each CND consists of multiple short component dislocations in adjacent nanotwins, connected by either constricted or extended unit jogs at the twin boundaries. The glide of CNDs involves the collective movement of the component dislocations as well as the continuous pinning and depinning of the units jogs at the twin boundaries. The critical stress required for the de-pinning of a CND has been proposed as follows [25]:

$$\sigma_{\text{CND}} = M\left(\frac{f_{\text{pin}}}{b\lambda} + \tau_0\right) \tag{11}$$

where τ_0 is the lattice friction stress. Here, f_{pin} represents the pinning force exerted by the twin boundaries on the CND during its motion. Our previous simulations have demonstrated that there exists a transition from threading dislocations to CNDs as the twin thickness is reduced below a critical value around 1 nm [25]. Eq. 11 explicitly shows that the critical stress to drive a CND scales with λ^{-1}, leading to a continuous strengthening behavior with increasing g_λ.

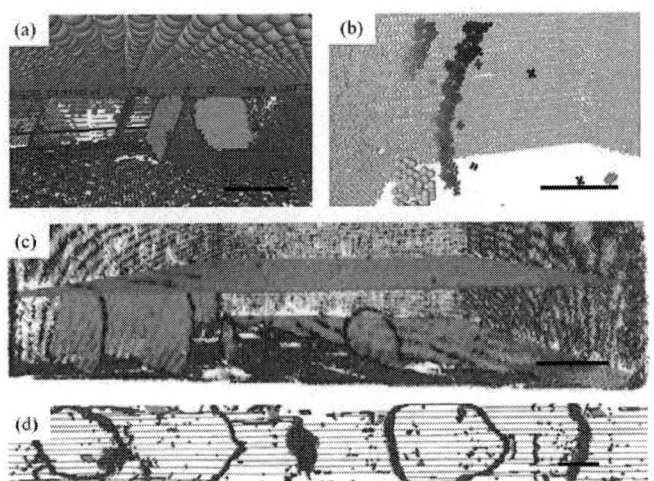

Figure 8. Deformation patterns of GNT samples with different structural gradients. (a) Threading dislocations governs plastic deformation of NT-B layer with $\lambda_B = 6.3$ nm and $d_B = 50$ nm in all GNT samples. (b) Correlated necklace dislocations (CNDs) governs plastic deformation in NT-A layer with $\lambda_A = 1.04$ nm and $d_A = 50$ nm. (c) A combination of threading dislocations and inclined (Hall-Petch type) dislocations observed in NT-B layer of GNT samples with $g_d = 1$ and $g_\lambda = 5$. (d) Detwinning governed by the glide of twinning partial dislocations at twin boundaries in GNT samples with $g_d = 3$ and $g_\lambda = 5$. Colors are assigned to atoms according to their local lattice structure in (a,c,d) and spatial coordinates in (b). The scale bar stands for a length of 5 nm.

As discussed above, the twin gradient effect on strength (Fig. 7c) can be qualitatively captured by either CLS model for moderate g_λ or CND model for large g_λ. To account for the grain gradient effect on strength observed in Fig. 7b, we examined the deformation pattern of the GNT sample with $g_d = 1$ and $g_\lambda = 5$. Fig. 8c shows the deformation pattern at 7.5% strain for the layer NT-B. We observed the nucleation of threading dislocations from the lateral grain boundaries, which is similar to those found in Fig. 8a. In addition, we observed dislocation loops emitting from the interface between the two layers NT-A and NT-B and gliding toward nearby twin boundaries. These are frequently referred to as mode I dislocations in the literature [22, 28, 30]. The activation of a particular dislocation type depends on the local stress state. According to the Schmid factor calculations of dislocations in highly oriented nanotwinned structures [22], mode I dislocations are supposed to dominate plastic deformation when the external loading is perpendicular to the twin boundaries, while mode II dislocations are dominant when loading is parallel to the twin boundaries. The simultaneous activation of mode I and mode II dislocations is thus unexpected and must be related to the change of local stress state induced by the substantial structural gradient [2, 4, 30-32]. Different from the Burgers vectors of threading dislocations that are parallel to the twin boundaries, the Burgers vectors of mode I dislocation are inclined to the twin boundaries. Since twin boundaries provide adequate barriers to the slip of these inclined mode I dislocations, the dependence of yield strength on twin thickness can be written in the form of the classical Hall-Petch (HP) strengthening law, $\sigma_{HP} \propto \lambda^{-1/2}$ [26, 33]. It is known that flow stress controlled by HP strengthening is much higher than that controlled by CLS strengthening [26]. More importantly, it has been revealed that the simultaneous activation of mode I and mode II dislocations can trigger the formation of bundles of concentrated dislocations uniformly distributed in grain interiors, leading to a unique patterning of geometrically necessary dislocation with ultra-high dislocation densities [4]. This is believed to cause the extra strengthening observed for GNT samples with $g_d = 1$ (Fig. 7b).

It is intriguing to observe the maximum strength around a critical grain gradient $g_d = 1$. We will next investigate the physical origin of the stress drop above this critical gradient. Fig 8d shows the deformation pattern at 15% strain for the layer NT-A with $d_A = 12.5$ nm and $\lambda_A = 1.04$ nm, corresponding to a GNT structure with gradient parameters of $g_\lambda = 5$ and $g_d = 3$. Twinning partial dislocations are observed to glide along numerous twin planes, leading to the migration of twin boundaries and the coarsening of the twin lamellae. It is seen that detwinning occurs primarily at the interface between the layers NT-A and NT-B where local stress state is changed by the substantial structural gradient and resolved shear stress is sufficiently high to drive the movement of twinning partial dislocations along the twin boundaries. The nucleation of the twinning partial dislocations should be assisted by the stress concentration at the grain boundaries or the misfit segments at the interface left by the threading dislocations in the layer NT-B. Apparently, the number of nucleation sources of the twinning partial dislocations at the interface increases with increasing grain gradient, i.e., $n_s \propto g_d$. Following the approach developed by Li et al.[34], we derive the critical stress required to activate nucleation-governed detwinning at the interface as follows:

$$\sigma_d = \sigma_a - k \ln\left(\alpha \left(g_d + 1\right) \frac{d_B}{\lambda_A} \right) \tag{12}$$

where σ_a stands for the athermal resistance to dislocation nucleation. The pre-factor k of the second term on the right-hand side of Eq. 12 depends on temperature, local stress concentration and activation volume. The pre-factor in the logarithm, α, is related to Debye frequency and strain rate. For a given d_B and λ_A, it could be seen that the flow stress decreases with increasing grain gradient g_d, consistent with our current simulations results (Fig. 7b). Eq. 12 also suggests that the yield strength decreases with increasing twin thickness, which has been known as the nucleation-governed softening behavior of nanotwinned metals when the twin thickness drops below a critical value [34]. The dependence of flow stress on twin thickness is thus determined by the competition between CND-governed strengthening (Eq. 11) and detwinning-governed softening (Eq. 12). Since the twin thickness is in the logarithmic function in Eq. 12, the magnitude of detwinning-induced stress reduction ($\propto \ln\lambda$) is weaker than that that of CND-induced stress enhancement ($\propto \lambda^{-1}$) as λ decreases. As a result, for a given g_d, the flow stress should increase with increasing g_λ as indicated in Fig. 7c.

Table 1. Gradient effect on strength and deformation mechanism of GNT Cu

Structural gradient	Deformation mechanism	Strength
Twin gradient, g_λ	Threading dislocations	Strengthening
	Correlated necklace dislocations	Strengthening
Grain gradient, g_d	Inclined slip to twin boundaries	Strengthening
	Twinning partial dislocations	Softening

4. CONCLUSIONS

This paper presents an adaptive design framework that integrates MD simulations, GP-based surrogate modeling, adaptive sampling and optimization for investigating the effects of structural gradients on the strength of CNT Cu.

In this work, it has been demonstrated that the GP-based surrogate model can achieve a high fidelity using a limited sampling data to describe a highly nonlinear feature-property relationship, which is critical for material design. The trade-off between the fidelity of surrogate model and the expense of generating new sampling points is balanced by employing the adaptive sampling to identify the critical sampling points for enhancing the fidelity of the surrogate model in an iterative manner. Employing adaptive sampling and surrogate modeling, we address the challenges of exploring the design space with limited sampling data points due to the expensive computational cost of physics-based simulation. Combining physics-based MD simulations and the surrogate modeling, we are able to obtain a number of insights about the relationship between the structural gradients and the strength of CNT Cu.

It has recently been demonstrated that in a GNT Cu subject to external loading parallel to the twin boundaries, the strength increases progressively with an enhanced structural gradient, which is believed to be controlled by the gradients in both grain size and twin thickness [4]. Note that the present study undercovers the individual effects of grain/twin gradients on strength and deformation mechanisms. We demonstrate the importance of creating sufficiently large twin gradients in GNT samples, as the strength is found to increase with increasing twin gradient regardless of the grain gradient involved in the GNT structure. More importantly, our findings point out that it is crucial to carefully design the grain gradient of GNT samples because there exists a critical grain gradient at which the strength is maximized. Above the critical grain gradient, a detwinning-induced softening mechanism would become dominant and the strength will decrease with further increasing grain gradient.

We have summarized the effect of grain/twin gradients on strength and deformation mechanisms of GNT Cu in Table 1 for comparison. In the regime of threading dislocations, the confined layer slip (CLS) model (Eq. 10) can be used to account for the strengthening of GNT samples for moderate g_λ. In the regime of CNDs, the jog-depinning model (Eq. 11) is essential to capture the continuous strengthening behavior of GNT samples for large g_λ, as the CLS model fails when the twin thickness is comparable to the cutoff radius of dislocation cores. As g_d increases, the classical Hall-Petch (HP) model based on the mechanism of dislocation slip inclined to twin boundaries should be incorporated to explain the extra strengthening of GNT samples. However, the reduced grain size also promotes the nucleation of twinning partial dislocations at the interface, leading to a detwinning-induced softening mechanism (Eq. 12) for large g_d. The competition between HP strengthening and detwinning-induced softening leads to a maximum strength at a critical g_d.

ACKNOWLEDGMENTS

H. Zhou acknowledges support by Zhejiang University through "Hundred Talents Program". X. Chen and Y. Li are grateful for the support provided by the University of Illinois at Urbana-Champaign. The authors are grateful to Prof. H.J. Gao for insightful discussions.

REFERENCES

[1]. Fang T H, Li W L, Tao N R, Lu K. Revealing extraordinary intrinsic tensile plasticity in gradient nano-grained copper. Science, 2011, 331(6024): 1587-1590.

[2]. Wu X L, Jiang P, Chen L, Yuan F P, Zhu Y T T. Extraordinary strain hardening by gradient structure. Proceedings of the National Academy of Sciences

U.S.A., 2014, 111(20): 7197-7201.

[3]. Wei Y J, Li Y Q, Zhu L C, Liu Y, Lei X Q, Wang G, Wu Y X, Mi Z L, Liu J B, Wang H T, Gao H J. Evading the strength-ductility trade-off dilemma in steel through gradient hierarchical nanotwins. Nature Communications, 2014, 5: 3580.

[4]. Cheng Z, Zhou H F, Lu Q H, Gao H J, Lu L. Extra strengthening and work hardening in gradient nanotwinned metals. Science, 2018, 362(6414): eaau1925.

[5]. Roland T, Retraint D, Lu K, Lu J. Fatigue life improvement through surface nanostructuring of stainless steel by means of surface mechanical attrition treatment. Scripta Materialia, 2006, 54(11): 1949-1954.

[6]. Lu K. Making strong nanomaterials ductile with gradients. Science, 2014, 345(6203): 1455-1456.

[7]. Lu K, Lu J. Nanostructured surface layer on metallic materials induced by surface mechanical attrition treatment. Materials Science and Engineering a-Structural Materials Properties Microstructure and Processing, 2004, 375: 38-45.

[8]. Wu X L, Jiang P, Chen L, Zhang J F, Yuan F P, Zhu Y T. Synergetic Strengthening by Gradient Structure. Materials Research Letters, 2014, 2(4): 185-191.

[9]. Liu Y, Wei Y J. Gradient driven anomalous reversible plasticity in conventional magnesium alloys. Extreme Mechanics Letters, 2016, 9: 158-164.

[10]. Lin Y, Pan J, Zhou H F, Gao H J, Li Y. Mechanical properties and optimal grain size distribution profile of gradient grained nickel. Acta Materialia, 2018, 153: 279-289.

[11]. Mannodi-Kanakkithodi A, Pilania G, Huan T D et. Al. , Machine Learning Strategy for Accelerated Design of Polymer Dielectrics, Scientific Reports, 2016, 6, 20952.

[12]. Bassman L, Rajak P and Kalia R et. Al., Active Learning for Accelerated Design of Layered Materials, Computational Materials, 2018, 6, 74.

[13]. Artrith N and Urban A , An implementation of artificial neural-network potentials for atomistic materials simulations:Performance for TiO2 , Computational Materials Science 2016(114): 135–150.

[14]. Li Y, Xiao W. and Wang P. , "Uncertainty Quantification of Artificial Neural Network based Machine Learning Potentials", Proceedings of the ASME 2018 International Mechanical Engineering Congress and Exposition, IMECE 2018 (IMECE2018-88071)

[15]. Li Y, Wang P and Xiao W., Uncertainty Quantification of Atomistic Materials Simulation with Machine Learning Potentials, 2018 AIAA Non-Deterministic Approaches Conference, AIAA SciTech Forum (AIAA 2018-2166)

[16]. Rubinstein R.Y. and Kroese D.P., Simulation and the Monte Carlo method. Vol. 707. John Wiley & Sons, 2011.

[17]. Wang, Z., and Wang, P., "A maximum confidence

enhancement based sequential sampling scheme for simulation-based design." Journal of Mechanical Design 136.2 (2014): 021006.

[18]. Hoover W G. Constant-Pressure Equations of Motion. Physical Review A, 1986, 34(3): 2499-2500.

[19]. Nose S. A Unified Formulation of the Constant Temperature Molecular-Dynamics Methods. Journal of Chemical Physics, 1984, 81(1): 511-519.

[20]. Mishin Y, Mehl M J, Papaconstantopoulos D A, Voter A F, Kress J D. Structural stability and lattice defects in copper: Ab initio, tight-binding, and embedded-atom calculations. Physical Review B, 2001, 63(22): 224106.

[21]. Honeycutt J D, Andersen H C. Molecular-Dynamics Study of Melting and Freezing of Small Lennard-Jones Clusters. Journal of Physical Chemistry, 1987, 91(19): 4950-4963.

[22]. Lu Q H, You Z S, Huang X X, Hansen N, Lu L. Dependence of dislocation structure on orientation and slip systems in highly oriented nanotwinned Cu. Acta Materialia, 2017, 127: 85-97.

[23]. Nix W D. Yielding and strain hardening of thin metal films on substrates. Scripta Materialia, 1998, 39(4-5): 545-554.

[24]. Misra A, Hirth J P, Hoagland R G. Length-scale-dependent deformation mechanisms in incoherent metallic multilayered composites. Acta Materialia, 2005, 53(18): 4817-4824.

[25]. Zhou H F, Li X Y, Qu S X, Yang W, Gao H J. A Jogged Dislocation Governed Strengthening Mechanism in Nanotwinned Metals. Nano Letters, 2014, 14(9): 5075-5080.

[26]. You Z S, Li X Y, Gui L J, Lu Q H, Zhu T, Gao H J, Lu L.

Plastic anisotropy and associated deformation mechanisms in nanotwinned metals. Acta Materialia, 2013, 61(1): 217-227.

[27]. Zhou H F, Gao H J. A Plastic Deformation Mechanism by Necklace Dislocations Near Crack-like Defects in Nanotwinned Metals. Journal of Applied Mechanics-Transactions of the Asme, 2015, 82(7): 071015.

[28]. Lu K. Stabilizing nanostructures in metals using grain and twin boundary architectures. Nature Reviews Materials, 2016, 1(5): 16019.

[29]. Zhu T, Gao H J. Plastic deformation mechanism in nanotwinned metals: An insight from molecular dynamics and mechanistic modeling. Scripta Materialia, 2012, 66(11): 843-848.

[30]. Yang M X, Pan Y, Yuan F P, Zhu Y T, Wu X L. Back stress strengthening and strain hardening in gradient structure. Materials Research Letters, 2016, 4(3): 145-151.

[31]. Zeng Z, Li X Y, Xu D S, Lu L, Gao H J, Zhu T. Gradient plasticity in gradient nano-grained metals. Extreme Mechanics Letters, 2016, 8: 213-219.

[32]. Wang Y, Yang G X, Wang W J, Wang X, Li Q, Wei Y J. Optimal stress and deformation partition in gradient materials for better strength and tensile ductility: A numerical investigation. Scientific Reports, 2017, 7: 10954.

[33]. Lu K, Lu L, Suresh S. Strengthening Materials by Engineering Coherent Internal Boundaries at the Nanoscale. Science, 2009, 324(5925): 349-352.

[34]. Li X Y, Wei Y J, Lu L, Lu K, Gao H J. Dislocation nucleation governed softening and maximum strength in nano-twinned metals. Nature, 2010, 464(7290): 877-880.

Proceedings of the ASME 2019
International Design Engineering Technical Conferences
and Computers and Information in Engineering Conference
IDETC/CIE2019
August 18-21, 2019, Anaheim, CA, USA

DETC2019-97672

DISTRIBUTED DESIGN OF TWO-SCALE STRUCTURES WITH UNIT CELLS

Xingchen Liu
International Computer Science Institute
Berkeley, CA
Email: xingchen@icsi.berkeley.edu

ABSTRACT

The use of unit cell structures in mechanical design has seen a steady increase due to their abilities to achieve a wide range of material properties and accommodate multi-functional requirements with a single base material. We propose a novel material property envelope (MPE) that encapsulates the attainable effective material properties of a given family of unit cell structures. The MPE interfaces the coarse and fine scales by constraining the combinations of the competing material properties (e.g., volume fraction, Young's modulus, and Poisson's ratio of isotropic materials) during the design of coarse scale material properties. In this paper, a sampling and reconstruction approach is proposed to represent the MPE of a given family of unit cell structures with the method of moving least squares. The proposed approach enables the analytical derivatives of the MPE, which allows the problem to be solved more accurately and efficiently during the design optimization of the coarse scale effective material property field. The effectiveness of the proposed approach is demonstrated through a two-scale structure design with octet trusses that have cubically symmetric effective stiffness tensors.

1 INTRODUCTION
1.1 Motivation

Additive manufacturing (AM) allows mechanical components with complex shapes and internal structures to be manufactured without significantly increasing the cost or turnaround time. This unique feature of AM leads to a design space that is too vast to be represented by existing design methods. For the first time in an extended period, our ability to design falls short to our ability

to manufacture.

One effective way to approach such complexity is through scales. In the last decade, several methods have been developed for the design and modeling of two-scale structures. For example, periodic lattice structures may be efficiently represented by composing periodic functions with implicitly [1] or parametrically [2] defined shapes. Unit cell approaches decompose the coarse-scale shape into disjoint regions that are filled by unit cells with predefined shapes and material properties [3,4,5,6,7]. Open-cell Voronoi foams [8] has been proposed to generate random structures with an extensive range and gradient of elasticity. The sample-based approach is most versatile among all methods, capable of modeling a wide range of material structures while fulfilling the requirement of effective material properties on the coarse scale [9,10].

Despite the abundance of new methods, the lack of a common mathematical model to compare and categorize these methods leads to the proliferation of ad hoc approaches, which further increase the difficulties in the composition and interoperability with other two-scale modeling methods and classical solids. To address this issue, Liu and Shapiro [11] proposed a framework for modeling multiscale material structures by recursive composition of two-scale material structures. The framework links the scales by establishing an explicit relationship between shape–material properties at the fine scale and material properties at the coarse scale and ensures the interchangeability and interoperability of different representations on different scales via queries.

Albeit successful in generalizing the existing two-scale modeling approaches through the common mathematical model, the framework is not yet complete for designing multiscale structures

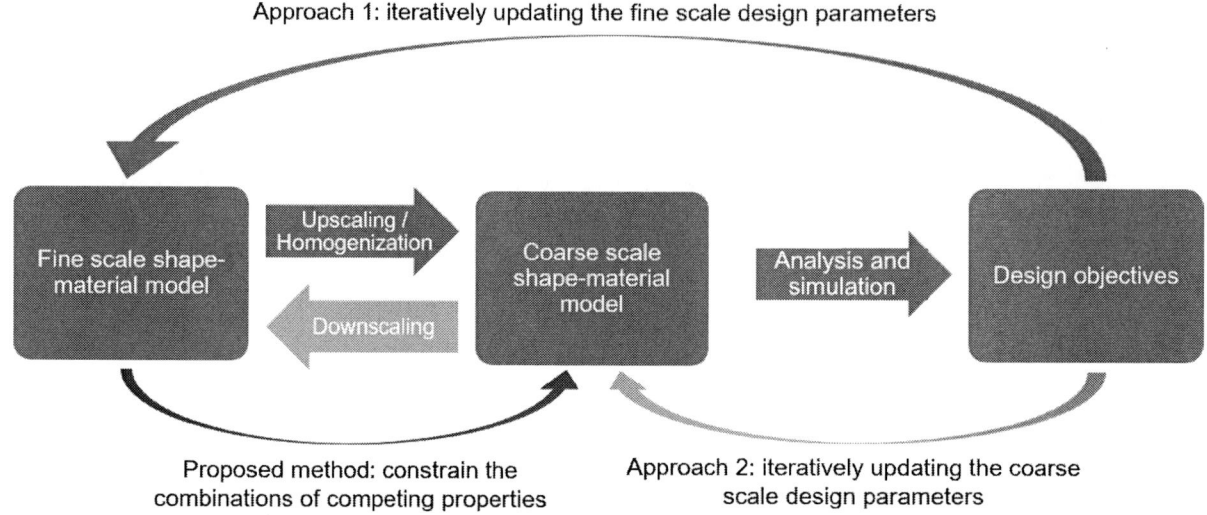

FIGURE 1: Two approaches of multiscale structure design.

in that it lacks a mechanism to constrain the design of the material property on the coarse scale. As a result, the designed coarse scale material property may not be attainable by the given fine scale structures [12], which is also one of the drawbacks that hindered the development of the free material optimization [13]. This issue only becomes more prominent when facing the demand of higher dimensional material property spaces required in designs with anisotropic material behavior, non-linear mechanics, or multi-physics.

1.2 Contributions and outline

In the current work, we propose a novel concept of material property envelope (MPE) that encapsulates all attainable effective material properties of interest by a given family of material structures. The MPE constrains combinations of the competing material properties, e.g., volume fraction vs. Young's modulus, during the design of coarse scale material properties, decoupling design parameters of fine-scale material structures from the coarse scale design and optimization, significantly reducing the complexity which improving the scalability of multiscale design problems.

The concept of material property space already exists in mechanical design with classical engineering materials: Ashby charts is often understood as the two-dimensional projections of a higher dimensional material property space [14]. In studies of architected materials, Ashby charts are often presented to convey the information on ranges of and relationships between different material properties. In studies of metamaterials, the material properties are usually plotted on top of Ashby charts of classical engineering materials, contrasting the novel material properties that are not found in naturally occurring materials. Albeit prevalent in two

dimensions, the direct use of high dimensional material property space in automated mechanical design has not been systematically studied.

The rest of the paper is organized as follows. After comparing two different approaches of integrating effective material properties into the design of multiscale structures among related work in Section 2.1, we review the compositional framework for multiscale structure modeling proposed by Liu and Shapiro [11] in Section 2.2. In Section 3, we start our discussion with the need for a material property envelope to constrain the design of coarse scale material properties as well as the sufficiency and necessity of respecting such constraints. In section 4, we discuss the sampling and reconstruction approach of modeling the MPE of a given family of material structures. In particular, we extend the method of moving least square (MLS) into dimension higher than three to support the high dimension material property envelope. The method of moving least square is used for its ability to provide analytical derivatives of the level set function representing the MPE. This is important because coarse scale material property design is often formulated as an optimization problem, and analytical derivatives allow faster convergence of the optimization, especial with the high dimensional design space. We formulate the problem of coarse scale material property design as an MPE constrained material optimization and demonstrate the effectiveness of the proposed approach with examples in Section 5, where we also discuss the relationship between the proposed approach and SIMP-based topology optimization. Section 6 concludes with discussions.

We note that geometric and topological properties, such as volume fraction and surface area, as well as statistical properties, such as correlation functions, are not considered as material

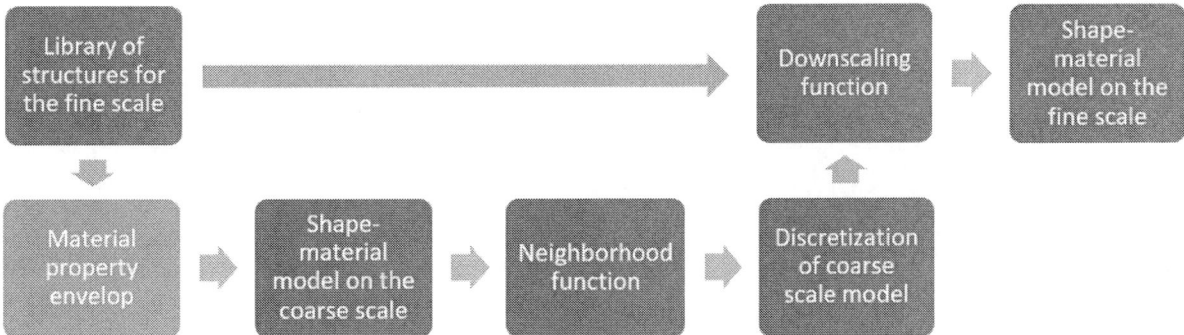

FIGURE 2: Two-scale structure design with multiscale framework.

properties in the classical sense. In the present work, we collectively refer these properties together with physical and mechanical properties, such as stiffness and thermal conductivity, as material properties for the ease of discussion. The proposed work is closely related to the material selection problem in mechanical design formulated by Ashby. Interfacing the scales with material property envelope allows us to combine multiple families of fine-scale structures or intersect material property envelope for an interoperable design. Different material structures, as well as classical engineering materials, can be compared used concurrently through the proposed material property envelope, enabling the interchangeable design among multiple families of material structures. We limit the scope of the current discussion to linear elasticity.

2 BACKGROUND AND RELATED WORK
2.1 Effective material properties in two-scale structure design

Traditionally, existing methods are categorized into groups such as procedural models [8, 15], unit cell approaches [7, 5, 6], functional compositions [16, 1], and sample-based approaches [10, 9, 17], etc. based on the representations of the fine scale structures and their tessellations of the coarse scale shape into two-scale structures. In this review, we choose to focus on the use of effective material properties of the fine scale structures to identify the gaps in existing two-scale structure design methods.

With few exceptions, the effective material property of the fine scale structure is used to reduce the complexity of the design problem. This is often achieved in one of two distinct ways upon close examination. In the first approach, effective material properties are used in place of the detailed fine scale structures for multiscale structural analysis, drastically reducing the complexity of solving the forward problem in the design process. Design methods that adopt this approach often use optimization algorithms to update the design parameters for an automated design iteratively. For example, the density of the Voronoi seeds and

the thickness of the Voronoi edges are varied to create the spatially varying material properties in procedural Voronoi foams method [8]. The densities of the iso, octet, and ORC trusses, which have a one-to-one correspondence to the truss diameters, are optimized directly in the designs of the two-scale cantilever beam and MBB beam [4].

Both of these two studies streamlined the analysis of the designed structure through surrogate models which map the design parameters to the homogenized material property of the fine scale structure. However, the design remains on the fine scale as the design space is completely parameterized by design parameters on the fine scale. Furthermore, the complexity of the design problem scales with the number of design parameters on the fine scale, limiting the use of this approach beyond two scales. This approach prohibits the use of other types of structures (with the same or a broader range of effective material properties) or even the same structures with different parameterizations in place of the existing one without overhauling the entire design process, creating problems in the interchangeability and interoperability of fine-scale structures.

In the second approach, effective material properties are designed directly as part of the shape-material model on the coarse scale and then mapped into the design parameters (or indices) of the fine scale structures after the coarse scale design is finished [5, 6, 9]. In contrary to the first approach, the design space of the two-scale structure is parameterized through the shape-material model on the coarse scale. Different fine scale structures and solid materials with the same material properties can be used interchangeably for the same material property on the coarse scale. The design of the coarse scale material properties could be manual, e.g., material painting [5], or automated, e.g., material optimization [9]. This approach is less rigorous in that the material properties on the coarse scale are rarely constrained beyond simple bounds, resulting in coarse scale material properties that may not be attainable by the given fine-scale structures.

We note that methods focus on discovering new unit cell architectures [18, 4], synthesis of unit cells on the fine-scale focused on achieving extreme material properties, such as negative Poisson ratio [19] and negative or turnable rate of thermal expansions [20] and methods to design two-scale structures with piecewise constant material properties [21] on the coarse scale are outside the scope of current discussion and therefore are not discussed in detail.

2.2 Multiscale shape–material modeling by composition

Liu and Shapiro [11] proposed a framework for modeling multiscale structures by recursive composition of two-scale material structures. The framework includes three key ingredients: (1) single scale shape-material model supported by fundamental single-scale queries, (2) mechanisms to link the coarse and fine scales, such that homogenized shape and material properties of the structures on the fine-scale correspond to those on the coarser scale, and (3) multiscale queries that reduce to single-scale queries performed at appropriate scales.

For a multiscale structure with a total of I scales, $S^i = (\Omega^i, M^i)$ is used to represent the shape and material property on i-th scale, with $i = 1$ being the coarsest scale and $i = I$ being the finest scale. A neighborhood N^i is a set of points and may take up various shapes. For practical purposes, a square is commonly used. A neighborhood function on each scale $N^i(x)$ is therefore defined by positioning neighborhood N^i to location x.

The neighborhood function is an important concept that serves dual purposes. Firstly, shape and material models are intrinsically connected in multiscale models such that

$$\Omega^i = \Omega^{i+1} \bullet N^{i+1} \tag{1a}$$

$$M^i(x) = E(S^{i+1}, N^{i+1}(x)), \quad x \in \Omega^i \ominus -N^{i+1} \tag{1b}$$

where $\Omega^i \ominus -N^{i+1}$ is the set of all locations that N^{i+1} fits inside Ω^i, and E the process of evaluating effective material property. Secondly, the sampling of neighborhood function induces the discretization such that Equation 1 is only satisfied at a finite number locations. The nonconformal grids used in the unit cell approach is one example of such discretization.

Within the framework, the process of generating fine-scale structure is represented as a downscaling function D:

$$D^{i+1} : S^i \mapsto S^{i+1} \tag{2}$$

that satisfies Equation 1. Multiscale material structures are represented (implicitly) by the recursive composition of two-scale material structures:

$$S^I = D^I \circ D^{I-1} \circ \cdots \circ D^2(S^1) \tag{3}$$

The framework is compositional by construction, providing a general and systemically way to reduce multiscale design problems into multiple design problems on a single scale. Interactions between adjacent scales are encapsulated in terms of a small number of standard abstractions, primitives, and computations, which liberates multiscale modeling from dependence on specific representations by allowing "mixing and matching" a variety of standard shape and material representations that support the basic queries. In the next section, we discuss the need for a material property envelope to constrain the design of coarse scale material properties within this framework.

3 MATERIAL PROPERTY ENVELOPE
3.1 MPE in multiscale structure design

In a strict sense, the composition $D^{i+1} \circ D^i$ in Equation 3 can be built only if D^i's image equals D^{i+1}'s domain; in a wider sense it is sufficient that the former is a subset of the latter. However, this design space includes all shape-material models that D^{i+1} can accommodate and is too vast to capture directly. Instead, we aim for a necessary condition of the composability: material properties on the coarse scale should be attainable by the given downscaling function in all neighborhoods. This condition becomes sufficient with disjoint neighborhoods, and the connectivity of the fine-scale structure between adjacent neighborhoods is assumed. Such assumption is commonly used in the unit cell approaches [22, 5] and Voronoi cell approach [8]. The sufficient condition for downscaling functions with overlapping neighborhoods, e.g., sample-based approach [23], is more complicated. For this reason, we limit our focus to downscaling functions with disjoint neighborhoods in the present paper. We interchangeably use downscaling function and family of structures in the rest of the paper as the image of the downscaling function is equivalent to the family of structures.

Let E_i representing the homogenization process for a set of $n \geq 2$ competing material properties $\{p_1, p_2, \cdots, p_n\}$ spaning the material property space P. The material property envelope of a downscaling function D can be defined set-theoretically as:

$$\{(p_1, p_2, \cdots, p_n) \in P | \forall i, \exists S \in im(D) \text{ s.t. } E_i(S) = p_i\} \tag{4}$$

where $im(D)$ represents the resultant structure of the downscaling function D.

Apart from the degenerate cases, the dimension of the envelops generally equals the number of design parameters of the structure or the number of the material properties of interest, whichever is less. For example, octet trusses studied in [24] are parametrized by the diameter of the rods in the truss and the Poisson's ratio of the based material. The cubic symmetry in the effective material properties of octet trusses allows the elasticity tensor to be characterized by three material parameters, commonly as Young's modulus E, Shear modulus G, and Poisson's ratio ν

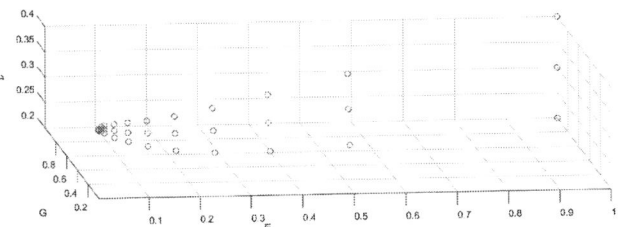

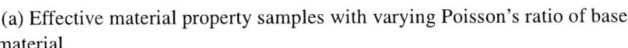

(a) Effective material property samples with varying Poisson's ratio of base material.

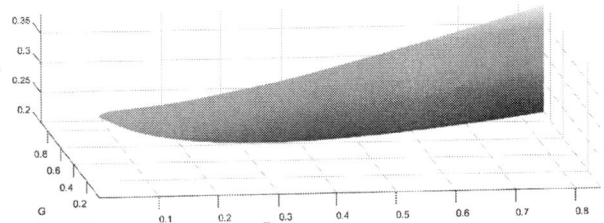

(b) Fitted cubic polynomial surface of the sampling data point.

FIGURE 3: Material property samples and envelopes of Octet truss. Only three dimensions of the four dimensional material space is shown in the plot. The sampling data and polynomial reconstruction are reproduced from [22].

(see Figure 3). Together with the volume fraction of the trusses, the MPE of octet trusses resides in a four-dimensional material property space. However, the effective material properties of the trusses will be concentrated around a two-dimensional surface because only two parameters are used to change the shape and material of the trusses. Knowing the dimension of the MPE *a priori* helps the reconstruction of MPE from sampling data set, which are discussed in the following sections. When such information is not readily available, a quick localized principal component analysis (PCA) could help determine the dimension of data set as well.

The material property envelope is closely related to the material selection approach proposed by Ashby, commonly known as Ashby chart [14]. Ashby charts are two-dimension projections of high dimensional material property space with every individual bubble corresponding to a different class of engineering materials. Different material indices, commonly a ratio between competing material properties (e.g., strength vs. density), are used to select material for different mechanical components, such as rods, beams, and plates. The method is developed to facilitate the material selection of mechanical components manufactured by traditional subtractive manufacturing processes, where the interior of the component is a homogeneous solid. To the best of our knowledge, the material selection scheme is not fully automated and can only be used to design structures with a homogeneous interior. The primary focus of the present work is the design of two-scale structures. However, the proposed method can be seen as an extension of the Ashby chart approach to select the material directly in the high dimension material property space.

4 RECONSTRUCTING MPE FROM SAMPLED MATERIAL PROPERTIES

Analytical relationship between different material properties has been studied in the literature. For example, Voigt and Ruess bounds [25] provide upper and lower limits of the effective material properties given the volume fractions of each base materials.

Hashin-Shtrikman bounds [25] further restrict the bulk and shear moduli for two-phase isotropic materials at different volume fractions. However, these analytical bounds are usually too general to be applied to any specific family of material structures.

Given a material structure, computing its effective material properties usually involves solving a number of partial differential equations with different boundary conditions [26, 27]. As a result, a direct analytical relationship between competing material properties is generally not available. In the current work, we adopt a sampling and reconstruction approach to approximate the material property envelope defined in Equation 4. For practical purposes, the MPE is represented as a level set, which is a scalar field defined over the entire material property space, with a one-to-one correspondence to the set-theoretic definition. The zero and negative level sets correspond to the boundary and the interior of the MPE, respectively.

To approximate the material property envelope, we first populate the space of competing material properties by sampling the effective material properties of the given material structure family. For material structures with procedural or parametrical representations, the material property envelope can be computed through a parameter sweep, where every design parameters of the material structure are sampled at regular or irregular intervals covering the entire range of parameters.

For material structures with non-parametric representations, such as voxel-based unit cells design by topology optimization, we enumerate all available material structures and record their effective material properties. We note that the material properties envelope is inherently discontinuous in this case. However, a continuous approximation may be constructed if a particular optimization algorithm prefers it. In the current work, we implement existing material structures and their effective material properties reported in the literature.

It is often convenient to reconstruct MPEs of low intrinsic dimensions through the fittings of analytical curves and surfaces. For example, the relationships between volume fractions, the Poisson's ratio of the based material and Young's modulus, Shear modulus, or Poisson's ratio has been approximated by cubic polynomials surfaces in [22]. An exponential curve is used in [8] to represent the relationship between the volume fraction and Young's modulus for Voronoi foams, with the assumption that the Poisson's ratio is constant. However, generalizing analytical curves and surfaces fittings to model higher dimensional MPEs is a non-trivial task as the required degree of freedom proliferate with the dimension of the material property envelope and the degree of the polynomials.

When the MPE resembles a two dimensional manifold in a three-dimensional material property space, the reconstruction of MPE is similar to the surface reconstruction from point clouds, a well-studied problem in computer graphics [28]. Over the last two decades, several methods have been proposed including distance field [29], radial basis functions [30], Delaunay triangulation [31], and moving least squares [32], to list a few.

Albeit similar, two significant differences exist. Firstly, the intrinsic dimension of the MPE may be equal to or less than the dimension of the material property space, which is different from the classical surface reconstruction problem that the surface is almost always two dimensions in a three-dimensional space. Secondly, surfaces in point cloud reconstructions usually represent a continuous, closed surface (i.e., of a solid object) while the lower dimensional MPE may be a "non-manifold", which may require additional trimming operations. Besides, analytical derivatives of MPE are often preferred by many off-the-shelf optimization algorithms as it significantly increases the speed of every iteration and solved the optimization problem more accurately (see next section for more details).

We use the method of moving least squares (MLS) to reconstruct the MPE from the sampling data. The method not only is easy to extend to and scales well with high spatial dimensions to accommodate the high dimensional material property space but also provides analytical derivatives w.r.t every material property of the fitted function with correctly chose weighted function. MLS can be understood through the weighted least squares approximation which adds a local weight function θ to the traditional least square method:

$$\min_f \sum_i \theta(\|\bar{x} - x_i\|)\|f(x_i) - f_i\|^2 \qquad (5)$$

where f_i is the function value sampled at point x_i, f is a degree m polynomials in d dimensional space:

$$f(x) = b(x)^T c(\bar{x}) \qquad (6)$$

where $b(x) = [b_1(x), ..., b_k(x)]^T$ is the polynomial basis vector and $c(\bar{x}) = [c_1(\bar{x}), ..., c_k(\bar{x})]^T$ is the vector of unknown coefficients. If θ is locally defined, $f(x)$ is also defined only locally. The unknown coefficients $c(\bar{x})$ is a function of $\bar{x}$ since Equation 5 is

weighted by distance to $\bar{x}$. By setting the partial derivatives of equation 5 w.r.t. $c(\bar{x})$ to zero, $c(\bar{x})$ is solved as

$$c(\bar{x}) = [\sum_i \theta(d_i)b(x_i)b(x_i)^T]^{-1} \sum_i \theta(d_i)b(x_i)f_i \qquad (7)$$

where $d_i = \|\bar{x} - x_i\|)$

The MLS method was proposed by Lancaster and Salkauskas [33] for surface generation from 3D point cloud data. In MLS, $\bar{x}$ is moved over the entire domain and $f(x)$ can be computed for every $\bar{x}$ location. It has been shown that the global function $f(x)$ is continuously differentiable if and only if the weighting function is continuously differentiable [34]. We use a Gaussian function in the current work. MLS is also a local approach: the interpolated value of any point can be computed locally on demand without processing. This is in stark contrast with the distance field approach [7], which is a global approach and require a significant computational resource to construct.

We add a set of augmented data points to the original sampling data set to represent the combinations of the material properties that are unattainable by the given downscaling function. We associate the original data points and the augmented ones with values of opposite signs. A numerical value that loosely approximates the distance to the nearest original data point is assigned to every augmented data point with the method proposed in [35]. Figure 4 shows the reconstruction of MPE by interpolation. The sampling data points and augmented data points are shown in red and blue, respectively. The reconstructed MPE is shown in Figure 4(b) and is used to constrain the design optimization in the next section.

5 COARSE SCALE MATERIAL PROPERTY DESIGN WITH MPE

There exist multiple ways to design coarse scale material property distributions. For example, a large body of work exists that designing the material properties field as one or more vector fields through the inverse distance interpolation [36, 37, 38]. However, design spaces of multiscale structures are often too vast to be captured by designers, especially in designs with multi-functions or multi-objective. Optimization becomes a valuable tool in the conceiving stage of the design process. In this section, we formulate the problem of designing coarse scale material property as a PDE-constrained optimization problem which usually takes the form

$$\min_m J(u, m)$$
$$\text{subject to:} \quad F(u, m) = 0, \quad h(m) = 0, \quad g(m) \le 0 \qquad (8)$$

where the vector m contains the material properties, $F(u, m) = 0$ is a system of PDEs parametrized by m with solution vector u,

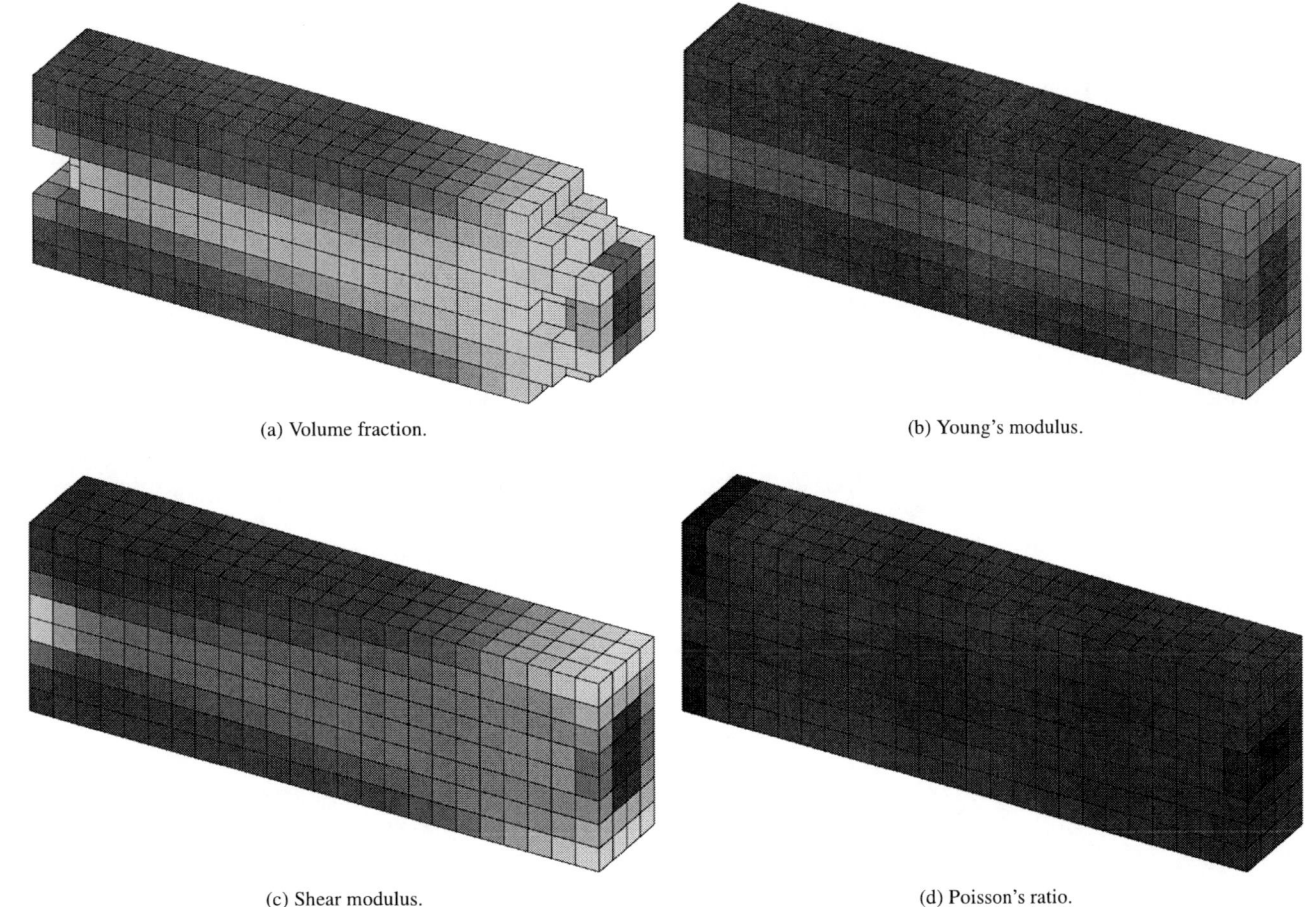

(a) Volume fraction.

(b) Young's modulus.

(c) Shear modulus.

(d) Poisson's ratio.

FIGURE 4: Optimization results of coarse scale mateiral property fields of a cantilever beam. Elements with volume fractions less than 20% are not shown in (a).

and $J(u,m)$ is the scalar-valued objective functional that is to be minimised. The equality and inequality constraints $h(m) = 0$ and $g(m) \leq 0$ enforce additional conditions on the material properties. The displacement field u of a structure defined within a fixed design domain Ω is governed by the equations of static equilibrium:

$$\text{div}\,\sigma + b = 0, \quad x \in \Omega \tag{9}$$

under the assumption of small strains

$$\varepsilon = \frac{1}{2}[\nabla u + (\nabla u)^T] \tag{10}$$

and linear elastic material response

$$\sigma = C\varepsilon \tag{11}$$

The constitutive relation C is a function of material properties vector m. It is possible that only a subset of m is used to parameterize C as m may include properties, such as volume fraction and surface area, which are not part of the constitutive relation.

Depending on the dimensionality, the MPE may take the form of either equality or inequality constraints. The full dimensional MPE ($\Delta D = 0$) is represented by a level set of negative values, therefore can be modeled as the inequality constraint $g(m) \leq 0$. On the other hand, lower dimensional MPE ($\Delta D \geq 1$) is represented by the zero level set, therefore can be modeled as the equality constraint $h(m) = 0$. We note that MPE constrains the combinations of material properties such that it is attainable by the given family of material structures as all locations $x \in \Omega$. This constraint will be enforced over every element in the finite element discretization.

Many algorithms for constrained nonlinear optimization require gradients information of the objective function and constraints. Ordinarily, the gradient information of the constraints is calculated numerically by finite difference approximation. In the next section, we provide the partial derivatives of the constraints analytically, allowing the problem to be solved more accurately and efficiently. This is particularly important as the number of constraints we have is on the same order as the number of finite elements. The derivation of the analytical gradient of the objective function is problem dependent and will be explained through examples.

5.1 Analytical partial derivatives of MPE as constraints

When reconstructed by the method of moving least squares, the scalar value representing the MPE is continuously differentiable with a continuously differentiable weighting function θ. Subsituting Equation 7 into Equation 6, the partial derivative of $f(x)$ w.r.t. x_i:

$$\frac{\partial f(x)}{\partial x_i} = \left(\frac{\partial b(x)^T}{\partial x_i} A^{-1} B - A^{-1} \frac{\partial A}{\partial x_i} A^{-1} B + b(x)^T A^{-1} \frac{\partial B}{\partial x_i} \right) f_i \tag{12}$$

where

$$A = \sum_j \theta(d_j) b(x_j) b(x_j)^T, \qquad \frac{\partial A}{\partial x_i} = \sum_j \frac{\partial \theta(d_j)}{\partial x_i} b(x_j) b(x_j)^T,$$

$$B = \sum_j \theta(d_j) b(x_j), \qquad \frac{\partial B}{\partial x_i} = \sum_j \frac{\partial \theta(d_j)}{\partial x_i} b(x_j), \tag{13}$$

and j is the index of the sampling data points.

5.2 Sensitivity of the objective function

Common objectives of structural design problems include minimizing the compliance or maximize the stiffness of the structure, minimizing the maximum stress within the restructure, matching the deformation profile of the structure to design a compliant mechanism, to list a few. In this section, we focus on the problem of designing the material property field minimizing the compliance of a structure. The designed material property field will be realized through octet truss lattices, whose effective elasticity tensors exhibit cubic symmetry and can be adequately described by Young's modulus E, shear modulus G, and Poisson's ratio v. Also, the design is constrained by the total volume of the material. The elasticity tensor with cubic symmetry take the form:

$$C = \begin{bmatrix} C_{11} & C_{12} & C_{12} & 0 & 0 & 0 \\ C_{12} & C_{11} & C_{12} & 0 & 0 & 0 \\ C_{12} & C_{12} & C_{11} & 0 & 0 & 0 \\ 0 & 0 & 0 & C_{33} & 0 & 0 \\ 0 & 0 & 0 & 0 & C_{33} & 0 \\ 0 & 0 & 0 & 0 & 0 & C_{33} \end{bmatrix}, \tag{14}$$

where

$$C_{11} = \frac{E(1-v)}{(1+v)(1-2v)}, \quad C_{12} = \frac{Ev}{(1+v)(1-2v)}, \quad C_{33} = G \tag{15}$$

It is straightforward to derive $\frac{\partial(C)}{\partial(E)}$, $\frac{\partial(C)}{\partial(G)}$, and $\frac{\partial(C)}{\partial(v)}$. We note that the partial derivative w.r.t to volumen fraction is zero as C is not a function of the volume fraction of the lattice structure. Figure 5 shows the optimized material property fields of a centilever beam. The designed distribution of effective property will be realized by octet truss lattices.

6 CONCLUSION

In the present work, we proposed the novel concept of the material property envelope to facilitate the design of multiscale structures with architected materials. The material property envelope encapsulates the attainable effective material properties of a given material structure family and interfaces the coarse and fine scales by constraining the combinations of the competing material properties. A sampling and reconstruction approach is proposed to model the MPE numerically. When used as constraints in design optimization, the MPE is represented as a continuous scalar function over the material property space of interest and provides the analytical partial derivatives to allow the optimization problem to be solved more accurately and efficiently.

By decoupling design parameters of fine-scale material structures from the coarse scale design and optimization, the novel interface significantly reduces the complexity while improving the scalability of multiscale design problems. The proposed formulation provides a mechanism to compare multiple families of material structures as well as classical engineering materials. Multiple material families can work together to enable a large design space through the union of MPEs. An interchangeable design becomes possible through the intersection of MPEs from multiple material families.

Acknowledgements

The author would like to thank Vadim Shapiro from UW-Madison and International Computer Science Institute, Carolyn Conner Seepersad from UT-Austin, Daniel A Tortorelli and Seth

Watts from LLNL for useful discussions. This research was supported by the Defense Advanced Research Projects Agency's TRAnsformative DESign Program. This work was partially performed under the auspices of the U.S. Department of Energy by Lawrence Livermore Laboratory under contract DE-AC52-07NA27344. The responsibility for errors and omissions lies solely with the author.

REFERENCES

[1] Fryazinov, O., Vilbrandt, T., and Pasko, A., 2013. "Multiscale space-variant FRep cellular structures". *Computer-Aided Design,* **45**(1), jan, pp. 26–34.

[2] Massarwi, F., and Elber, G., 2016. "A B-spline based framework for volumetric object modeling". *CAD Computer Aided Design,* **78**, pp. 36–47.

[3] Osanov, M., and Guest, J. K., 2016. "Topology Optimization for Architected Materials Design". *Annual Review of Materials Research,* **46**(1), pp. 211–233.

[4] Watts, S., and Tortorelli, D. A., 2017. "A geometric projection method for designing three-dimensional open lattices with inverse homogenization". *International Journal for Numerical Methods in Engineering.*

[5] Panetta, J., Zhou, Q., Malomo, L., Pietroni, N., Cignoni, P., and Zorin, D., 2015. "Elastic textures for additive fabrication". *ACM Transactions on Graphics,* **34**(4), pp. 135:1–135:12.

[6] Schumacher, C., and Bickel, B., 2015. "Microstructures to Control Elasticity in 3D Printing". *ACM Transactions on Graphics (ACM SIGGRAPH 2015),* **34**(4).

[7] Zhu, B., Skouras, M., Chen, D., and Matusik, W., 2017. "Two-Scale Topology Optimization with Microstructures".

[8] Martínez, J., Dumas, J., and Lefebvre, S., 2016. "Procedural voronoi foams for additive manufacturing". *ACM Transactions on Graphics,* **35**(4), pp. 1–12.

[9] Liu, X., and Shapiro, V., 2017. "Sample-Based Synthesis of Functionally Graded Material Structures". *Journal of Computing and Information Science in Engineering,* **17**(3), p. 031012.

[10] Liu, X., and Shapiro, V., 2014. "Random Heterogeneous Materials via Texture Synthesis". *Computational Materials Science,* **99**, pp. 177–189.

[11] Liu, X., and Shapiro, V., 2018. "Multiscale shape–material modeling by composition". *CAD Computer Aided Design,* **102**.

[12] Liu, X., and Shapiro, V., 2016. "Sample-based design of functionally graded material structures". *ASME 2016 International Design Engineering Technical Conferences and Computers and Information in Engineering Conference IDETC/CIE 2016,* pp. 1–15.

[13] Kocvara, M., Stingl, M., and Zowe, J., 2008. "Free material optimization: Recent progress". *Optimization,* **57**(1), pp. 79–100.

[14] Ashby, M. F., 1994. "Materials selection in mechanical design". *Metallurgia Italiana.*

[15] Martínez, J., Song, H., Dumas, J., and Lefebvre, S., 2017. "Orthotropic k-nearest foams for additive manufacturing". *ACM Transactions on Graphics,* **36**(4), pp. 1–12.

[16] Rumpf, R. C., Pazos, J. J., Digaum, J. L., and Kuebler, S. M., 2015. "Spatially variant periodic structures in electromagnetics". *Philosophical Transactions of the Royal Society of London A: Mathematical, Physical and Engineering Sciences,* **373**(2049), jul.

[17] Liu, X., and Shapiro, V., 2017. "Sample-based synthesis of two-scale structures with anisotropy". *Computer Aided Design,* **90**, pp. 199–209.

[18] Hopkins, J. B., Shaw, L. A., Weisgraber, T. H., Farquar, G. R., Harvey, C. D., and Spadaccini, C. M., 2015. "Organizing cells within non-periodic microarchitectured materials that achieve graded thermal expansions". *Proceedings of the ASME 2015 International Design Engineering Technical Conferences & Computers and Information in Engineering Conference IDETC/CIE 2015*(310), pp. 1–11.

[19] Larsen, U. D., Sigmund, O., and Bouwstra, S., 1997. "Design and fabrication of compliant micromechanisms and structures with negative Poisson's ratio". *Journal of Microelectromechanical Systems,* **6**(2), pp. 99–106.

[20] Sigmund, O., and Torquato, S., 1999. "Design of smart composite materials using topology optimization". *Smart Materials and Structures,* **8**(3), jun, pp. 365–379.

[21] Du, Z., Zhou, X.-Y., Picelli, R., and Kim, H. A., 2018. "Connecting Microstructures for Multiscale Topology Optimization with Connectivity Index Constraints". *Journal of Mechanical Design.*

[22] Watts, S., and White, D. A. "Simple, accurate surrogate models of the elastic response of three-dimensional open truss micro- architectures with applications to multiscale topology design Simple , accurate surrogate models of the elastic".

[23] Liu, X., 2017. "Sample-Based Material Structure Modeling". PhD thesis, The University of Wisconsin-Madison.

[24] Watts, S., Arrighi, W., Kudo, J., Tortorelli, D. A., and White, D. A., 2019. "Simple, accurate surrogate models of the elastic response of three-dimensional open truss microarchitectures with applications to multiscale topology design". *Structural and Multidisciplinary Optimization,* may.

[25] Hill, R., 1952. "The Elastic Behaviour of a Crystalline Aggregate". *Proceedings of the Physical Society. Section A,* **65**(5), may, pp. 349–354.

[26] Liu, X., and Shapiro, V., 2016. "Homogenization of material properties in additively manufactured structures". *Computer-Aided Design,* **78**, may, pp. 71–82.

[27] Hollister, S. J., and Kikuchi, N., 1994. "Homogenization

theory and digital imaging: A basis for studying the mechanics and design principles of bone tissue.". *Biotechnology and bioengineering,* **43**(7), mar, pp. 586–596.

[28] Berger, M., Alliez, P., Tagliasacchi, A., Seversky, L. M., Silva, C. T., Levine, J. a., and Sharf, A., 2014. "State of the Art in Surface Reconstruction from Point Clouds". *Eurographics 2014, Eurographics STARs.*

[29] Amenta, N., and Kil, Y. J., 2004. "Defining point-set surfaces". *ACM Transactions on Graphics.*

[30] Carr, J. C., Beatson, R. K., Cherrie, J. B., Mitchell, T. J., Fright, W. R., McCallum, B. C., and Evans, T. R., 2001. "Reconstruction and representation of 3D objects with radial basis functions". In Proceedings of the 28th annual conference on Computer graphics and interactive techniques - SIGGRAPH '01.

[31] Cazals, F., and Giesen, J., 2006. "Delaunay triangulation based surface reconstruction". In *Effective Computational Geometry for Curves and Surfaces.*

[32] Alexa, M., Behr, J., Cohen-Or, D., Fleishman, S., Levin, D., and Silva, C. T., 2003. "Computing and rendering point set surfaces". *IEEE Transactions on Visualization and Computer Graphics.*

[33] Lancaster, P., and Salkauskas, K., 1981. "Surfaces generated by moving least squares methods". *Mathematics of Computation.*

[34] Levin, D., 1998. "The approximation power of moving least-squares". *Mathematics of Computation,* **67**(224), pp. 1517–1532.

[35] Zhu, Y., and Bridson, R., 2005. "Animating sand as a fluid". In ACM SIGGRAPH 2005 Papers on - SIGGRAPH '05.

[36] Kou, X., and Tan, S., 2008. "Heterogeneous object design: an integrated CAX perspective". *Heterogeneous objects modelling and applications,* pp. 42–59.

[37] Qian, X., and Dutta, D., 2004. "Feature-based design for heterogeneous objects". *Computer-Aided Design,* **36**(12), pp. 1263–1278.

[38] Biswas, A., Shapiro, V., and Tsukanov, I., 2004. "Heterogeneous material modeling with distance fields". *Computer Aided Geometric Design,* **21**(3), mar, pp. 215–242.

Proceedings of the ASME 2019
International Design Engineering Technical Conferences
and Computers and Information in Engineering Conference
IDETC/CIE2019
August 18-21, 2019, Anaheim, CA, USA

DETC2019-97675

A FRAMEWORK OF MULTI-FIDELITY TOPOLOGY DESIGN AND ITS APPLICATION TO OPTIMUM DESIGN OF FLOW FIELDS IN BATTERY SYSTEMS

Kentaro Yaji[*], Shintaro Yamasaki, Shohji Tsushima, Kikuo Fujita
Department of Mechanical Engineering, Graduate School of Engineering, Osaka University,
2-1, Yamadaoka, Suita, Osaka 565-0871, Japan

ABSTRACT

We propose a novel framework based on multi-fidelity design optimization for indirectly solving computationally hard topology optimization problems. The primary concept of the proposed framework is to divide an original topology optimization problem into two subproblems, i.e., low- and high-fidelity design optimization problems. Hence, artificial design parameters, referred to as seeding parameters, are incorporated into the low-fidelity design optimization problem that is formulated on the basis of a pseudo-topology optimization problem. Meanwhile, the role of high-fidelity design optimization is to obtain a promising initial guess from a dataset comprising topology-optimized design candidates, and subsequently solve a surrogate optimization problem under a restricted design solution space. We apply the proposed framework to a topology optimization problem for the design of flow fields in battery systems, and confirm the efficacy through numerical investigations.

INTRODUCTION

Topology optimization [1] is a powerful design tool for automatically generating novel shapes of structures from blank design domains, without relying on a designer's intuitions. During the past three decades, topology optimization has been applied to fundamental research topics, namely, stiffness maximization problems in structural mechanics, as well as design problems pertaining to various physical phenomena such as heat conduction, fluid flow, and electromagnetic wave. The basic framework and engineering applications can be found in [2]. In addition, re-

cent developments of topology optimization were elegantly summarized in [3, 4].

Meanwhile, it is widely known that topology optimization cannot avoid multimodality. Particularly, instability is often observed when solving highly nonlinear and/or complex multiphysics problems. Consequently, massive computational burden with patient parameter studies is often required to obtain a local minimum in such computationally hard topology optimization problems. Hence, it can be concluded that directly solving such complex topology optimization problems is highly laborious for expanding to practical engineering applications.

To avoid this issue, pseudo-models have been widely utilized in topology optimization. A representative example is the compliant mechanism design problem [2], in which linear deformation is typically assumed; by contrast, the original aim of this design problem is to realize an intended mechanism considering nonlinear large deformations. Recently, pseudo-models for solving topology optimization problems addressing complex fluid systems have been proposed. For instance, Zhao et al. [5] proposed a topology optimization method using a linear model based on the Darcy model for solving flow channel design problems considering turbulent heat transfer. Similarly, Asmussen et al. [6] proposed a simplified model for solving natural convection problems. Zeng et al. [7] proposed a pseudo-two-dimensional (2D) topology optimization model for solving three-dimensional (3D) heat sink design problems. The abovementioned studies focused on indirectly solving complex topology optimizations using pseudo-models. However, it is noteworthy that pseudo-models contain approximation errors in most cases; therefore, a local optimum does not guarantee a satisfactory solution.

---

[*]Corresponding author: yaji@mech.eng.osaka-u.ac.jp

As a method to globally search satisfactory solutions in complex optimization problems, multi-fidelity design optimization has attracted significant attention [8–10]. The basic concept of multi-fidelity design optimization is to decompose an original optimization problem into low- and high-fidelity design optimization problems for realizing an efficient design optimization in reasonable numerical cost. Hitherto, most studies on multi-fidelity design optimization has focused on parametric optimization problems, e.g., airfoil design [11] and electromagnetic device design [12], under a small number of design variables. The majority of multi-fidelity design optimization is based on surrogate optimization, which is gradient-free optimization and therefore requires only forward analyses for constructing surrogate models. Although surrogate optimization is popular for solving complex optimization problems, the usable number of design variables is limited. Hence, it is clear that directly applying surrogate optimization to topology optimization problems is not valid because many design variables, e.g., $O(10^4)$ in 2D and $O(10^6)$ in 3D, must be treated in topology optimization problems. In other words, the *curse of dimensionality* cannot be avoided in these cases. In general, adjoint sensitivity analysis, which is independent of the number of design variables, is essential in topology optimization.

The aim of our research is to construct a framework for solving complex topology optimization problems. Hence, on the basis of multi-fidelity design optimization, we propose a novel framework, named "Multi-Fidelity Topology Design (MFTD)," in which an original topology optimization problem is decomposed into low- and high-fidelity design optimization problems. The role of low-fidelity design optimization is to generate promising design candidates by solving a pseudo-topology optimization problem, which can be solved in reasonable numerical cost using adjoint sensitivity analysis. Hence, we introduce a few design parameters, called "seeding parameters," into the low-fidelity design optimization. The seeding parameters are varied such that various patterns of topology-optimized configurations can be generated. In high-fidelity design optimization, the promising initial guess for the high-fidelity design optimization is estimated by obtaining the best combinations of seeding parameters. In addition, using the promising initial guess, the optimum solution is found under a few design variables using function-based surrogate models such as polynomial regression, Kriging, support vector regression, and radial basis function [13].

We apply MFTD to a flow field design problem for battery systems such as flow batteries [14, 15]. These batteries contain flow channels to enhance battery performances, e.g., charging and discharging performances [16]. Hitherto, various patterns of flow channels have been proposed; however, geometrically simple patterns are typically employed in conventional battery systems [17, 18]. For evaluating battery performances, 3D simulations of electrolyte flow and electrochemical reaction are required, at the least. Although a forward analysis for this simulation is possible, directly applying topology optimization to such flow field design problems is computationally challenging insofar large-scale cluster computers cannot be used [19, 20]. Herein, as a first development of a radical research topic, we address this complex topology optimization problem and aim to reveal the efficacy of MFTD.

MULTI-FIDELITY TOPOLOGY DESIGN

In this section, we discuss the framework of MFTD in general expressions. The concrete expressions for battery system design problems are described in a later section.

Basic Concept of Topology Optimization

Topology optimization is a type of structural optimization, and focuses on optimizing material distribution in a fixed design domain $D \subset \mathbb{R}^d$, where d is the spatial dimension. The material distribution corresponding to $\Omega \subseteq D$ is expressed using the characteristic function $\chi_\Omega : D \to \{0, 1\}$, as follows:

$$\chi_\Omega(\mathbf{x}) = \begin{cases} 1 & \text{if } \mathbf{x} \in \Omega \\ 0 & \text{if } \mathbf{x} \in D \setminus \Omega, \end{cases} \quad (1)$$

where $\forall \mathbf{x} \in D$ represents the spatial position. Owing to the discreteness of χ_Ω, topology optimization requires a regularization or relaxation technique. The most popular is the density approach [2], whose basic concept is to replace the characteristic function with a continuous function, i.e., $\gamma : D \to [0, 1]$, where γ is termed a design field.

Using the design field γ, a topology optimization problem is generally formulated as follows:

$$\begin{aligned} \underset{\gamma}{\text{minimize}} \quad & J(\gamma) \\ \text{subject to} \quad & G_i(\gamma) \leqslant 0, \ i = 1, ..., M, \\ & 0 \leqslant \gamma(\mathbf{x}) \leqslant 1, \ \forall \mathbf{x} \in D, \end{aligned} \quad (2)$$

where J is an objective functional, and G_i are constraints. The optimization problem above can be solved using mathematical programming techniques. Owing to the high degree of freedom of γ, gradient-based optimization with the incorporation of adjoint sensitivity anaylsis is typically used when solving topology optimization problems.

In topology optimization research, most studies focus on directly solving a topology optimization problem, formulated as expressed in (2), even if the optimization problem is computationally hard. We emphasize that the traditional method to directly solve the topology optimization problem is an important research topic; meanwhile, we believe that the development of

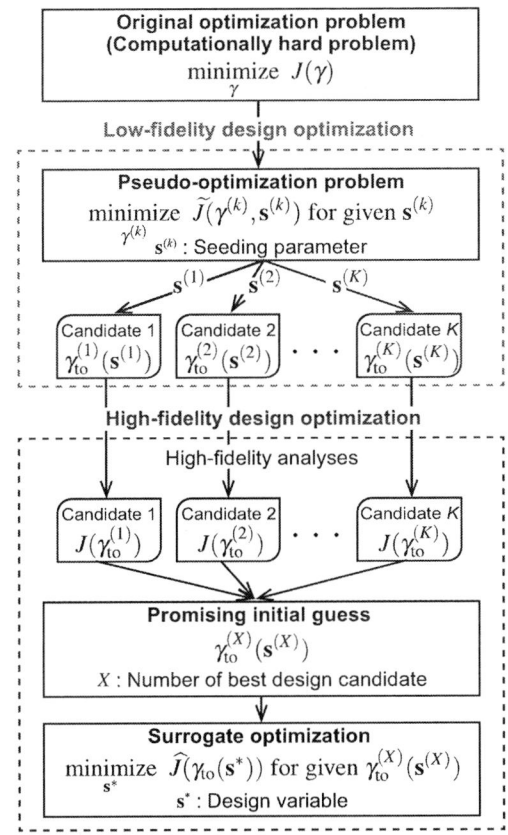

FIGURE 1. PROPOSED FRAMEWORK.

an alternative approach to indirectly solve the topology optimization problem is equally important. This conviction is essential for solving practical design optimization problems by topology optimization.

Proposed Framework

The aim of our study is to obtain a promising design solution in computationally hard topology optimization problems. Hence, we propose a novel framework (termed MFTD) that is based on the concept of multi-fidelity design optimization. A schematic illustration of MFTD is shown in Fig. 1, which expresses the following procedures.

1. The original topology optimization problem is reformulated as two subproblems, i.e., low- and high-fidelity design optimization problems;
2. The low-fidelity design optimization problem aims to generate various patterns of design candidates, which are generated by solving a pseudo-topology optimization problem. Hence, artificial design parameters, i.e., seeding parameters are fed into the low-fidelity design optimization problem;

3. The original objective is evaluated using the design candidates above under the high-fidelity analysis model, and the promising initial guess for the later surrogate optimization problem is estimated by obtaining the best combinations of seeding parameters.
4. Using the promising initial guesses, surrogate optimization in the high-fidelity design optimization problem aims to obtain the optimum solution with respect to a few design variables that are newly defined and/or are chosen from a set of seeding parameters using a function-based surrogate model.

The key concept of MFTD is that the design space of the original topology optimization problem is purposely restricted for solving complex topology optimization problems.

On the basis of the above, the low-fidelity design optimization problem is formulated as follows:

$$
\begin{aligned}
\underset{\gamma^{(k)}}{\text{minimize}} \quad & \widetilde{J}(\gamma^{(k)}, \mathbf{s}^{(k)}) \\
\text{subject to} \quad & \widetilde{G}_i(\gamma^{(k)}, \mathbf{s}^{(k)}) \leqslant 0, \ i = 1, ..., M, \\
& 0 \leqslant \gamma^{(k)}(\mathbf{x}) \leqslant 1, \ \forall \mathbf{x} \in D, \\
\text{for given} \quad & \mathbf{s}^{(k)} \in S, \ k = 1, ..., K,
\end{aligned}
\tag{3}
$$

where $\widetilde{J}$ and $\widetilde{G}_i$ are pseudo-objective and constraint functionals, respectively. It is noteworthy that these pseudo-functionals are assumed to be easily calculated in comparison with the original ones in (2). In addition, $\mathbf{s} = \{s_1 \ s_2 \ ... \ s_{N_{sd}}\}^{\text{T}}$ is the set with respect to N_{sd} types of seeding parameters; $\mathbf{s}^{(k)}$ ($k \in \mathbb{N}^+$) expresses the sample points of $\mathbf{s}$; $S = \{\mathbf{s}^{(1)} \ \mathbf{s}^{(2)} \ ... \ \mathbf{s}^{(K)}\}^{\text{T}}$ is the set of the sample data. We consider that the low-fidelity design optimization problem in (3) is solved for generating the K types of topology-optimized configurations. As mentioned earlier, the obtained design candidates, $\gamma_{\text{to}}^{(k)}(\mathbf{s}^{(k)})$, are evaluated using the high-fidelity analysis model for revealing their exact performance; subsequently, the best combinations of seeding parameters are regarded as the promising initial guess, $\gamma_{\text{to}}^{(X)}(\mathbf{s}^{(X)})$, where X is the number of promising initial guesses for the high-fidelity design optimization problem.

We now consider the formulation of the high-fidelity design optimization problem. The aim of high-fidelity design optimization is to obtain the optimum solution by utilizing a function-based surrogate model. It is noteworthy that the design solution space with respect to $\mathbf{s}$ cannot always be addressed in a surrogate optimization framework. For instance, if the topologies of the sample data are significantly different from each other, the construction of a precise surrogate is then no longer possible. Therefore, the design space of the high-fidelity design optimization problem must be restricted on an admissible design space S_{ad}. In practice, we use $\mathbf{s}^* \in S_{\text{ad}} \subseteq S \times S_{\text{nw}}$, where S_{nw} is the newly defined design variable space, as the design variable in the

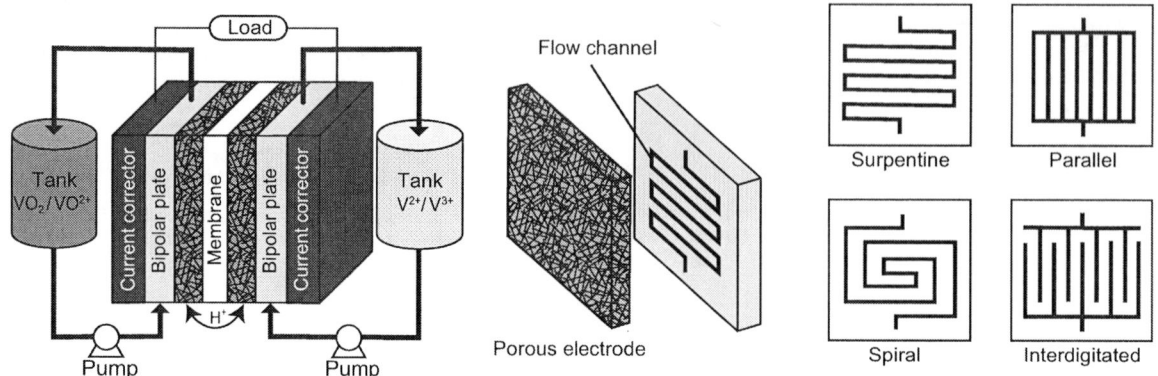

FIGURE 2. FLOW BATTERY SYSTEM WITH INTERNAL FLOW CHANNEL CONFIGURATIONS.

high-fidelity design optimization problem:

$$
\begin{aligned}
&\underset{\mathbf{s}^*}{\text{minimize}} && \widehat{J}(\gamma_{\text{to}}(\mathbf{s}^*)) \\
&\text{subject to} && \widehat{G}_i(\gamma_{\text{to}}(\mathbf{s}^*)) \leqslant 0, \ i = 1, ..., M, \\
& && \underline{s}_j^* \leqslant s_j^* \leqslant \overline{s}_j^*, \ j = 1, ..., N_{\text{sd}}^*, \\
&\text{for given} && \gamma_{\text{to}}^{(X)}(\mathbf{s}^{(X)}).
\end{aligned}
\tag{4}
$$

Here, $\underline{s}_j^*$ and $\overline{s}_j^*$ are the jth lower and upper limits of s_j^*, respectively; $\widehat{J}$ and $\widehat{G}_i$ are the surrogate objective and constraint functions for the original ones, respectively. These surrogates are constructed using the sample data pairs, e.g., $\{(\mathbf{s}^{*(1)}, J^{(1)}), (\mathbf{s}^{*(2)}, J^{(2)}), ..., (\mathbf{s}^{*(L)}, J^{(L)})\}$, where L is the sample number of $\mathbf{s}^*$. It is noteworthy that each response $J^{(l)}$, $l = 1, ..., L$, $(l \in \mathbb{N}^+)$, is evaluated through the original objective functional on the given topology-optimized configuration, namely, $J^{(l)} = J(\gamma_{\text{to}}(\mathbf{s}^{*(l)}))$. In practice, $\mathbf{s}^*$ should be chosen such that the topology of γ_{to} is preserved when solving the high-fidelity design optimization problem in (4) for constructing a precise surrogate model.

FLOW FIELD DESIGNS OF BATTERY SYSTEMS

As a case study of MFTD, we consider a topology optimization problem for the design of flow fields in battery systems. In this study, although we address the flow field of a vanadium redox flow battery [14, 15] as a concrete case study, we remark that the proposed framework can be applied to other battery systems such as other flow batteries as well as fuel cells. It is widely known that the battery performance, e.g., charging/discharging efficiency, depends significantly on the flow channel configuration. Figure 2 shows the schematic illustration of the flow battery system, in which various patterns of flow channels are utilized.

Because the optimum pattern of the flow channel is not obvious, topology optimization is a promising approach for the design of innovative battery systems.

Meanwhile, it is difficult to directly solve such topology optimization problems because the battery systems should be conducted through 3D simulations considering complex physical phenomena, e.g., multiphysics problems pertaining to electrolyte flow and electrochemical kinetics. In fact, although the construction of numerical simulation models for battery systems has attracted attention as a pioneering step for enhancing battery performances, the development of such simulation models is still ongoing owing to its complexity.

Herein, a low-fidelity design optimization problem is formulated on the basis of a previous study [21], in which a simplified electrochemical reaction model was proposed for the design of 2D flow fields of flow batteries through topology optimization. In addition, the high-fidelity design optimization problem is defined as a 3D flow channel design problem. The analysis model that imposes several assumptions, and the concrete governing equations and problem formulations are described hereinafter.

Analysis Model

As shown in Fig. 2, a flow battery is composed of electrodes facing each other across a membrane. In the vanadium redox flow battery system, the electrochemical reactions in the battery cell can be expressed as follows:

$$
\text{Negative electrode: } \text{V}^{2+} \rightleftharpoons \text{V}^{3+} + \text{e}^- \tag{5}
$$

$$
\text{Positive electrode: } \text{VO}_2^+ + 2\text{H}^+ + \text{e}^- \rightleftharpoons \text{VO}^{2+} + \text{H}_2\text{O} \tag{6}
$$

Here, the left and right arrows express charging and discharging, respectively.

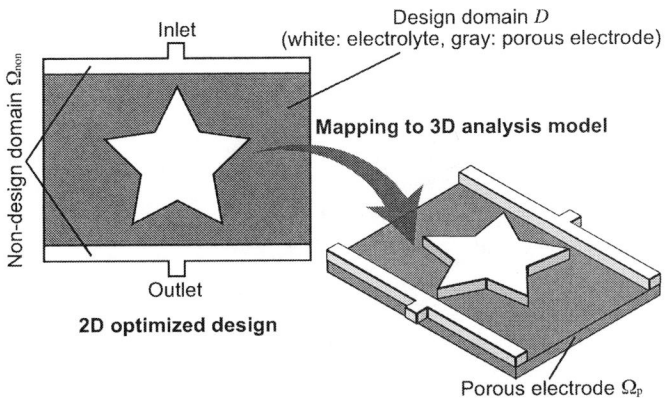

FIGURE 3. ANALYSIS MODEL.

As shown in Fig. 3, we consider the analysis domain $\mathscr{O} = D \cup \Omega_{\text{non}}$, where Ω_{non} is the non-design domain. The fixed design domain D is composed of the electrolyte domain Ω_f and solid domain Ω_s that correspond to Ω and $D \setminus \Omega$ as the general expressions in (1), respectively. The porous electrode domain Ω_p is part of the non-design domain in this analysis model. For simplification, in our study, only the negative electrode is considered, as in a previous study [22], and the battery performance is regarded as the charging efficiency. In addition, the following assumptions are introduced [23]:

1. The electrolyte flow is treated as a steady-state incompressible viscous fluid flow;
2. The battery performance can be improved by enhancing the reaction rate of one type of vanadium species V^{2+};
3. Migration can be ignored in the species transport equation.

Governing Equations

Base on the abovementioned analysis model and assumptions, the governing equations for the electrolyte flow and mass transport are briefly discussed herein. The governing equations are defined in dimensionless forms as in a previous study [21,24].

Because the electrolyte is a steady-state incompressible viscous fluid, the fluid velocity $\mathbf{u} : \mathscr{O} \to \mathbb{R}^d$ and the pressure $p : \mathscr{O} \to \mathbb{R}$ satisfy the continuity and Navier–Stokes equations, as follows:

$$\nabla \cdot \mathbf{u} = 0, \tag{7}$$

$$(\mathbf{u} \cdot \nabla)\mathbf{u} = -\nabla p + \frac{1}{Re}\nabla^2 \mathbf{u} + \mathbf{F}, \tag{8}$$

where Re is the Reynolds number, and $\mathbf{F} : \mathscr{O} \to \mathbb{R}^d$ is the body force. The governing equations in (7) and (8) can be solved under the appropriate boundary conditions [25]. As shown in Fig. 3,

because the analysis model is composed of fluid and porous domains, the Navier–Stokes equation in (8) is slightly modified as the Brinkman equation, in which the body force $\mathbf{F}$ in the porous domain Ω_f is defined as follows:

$$\mathbf{F} = \alpha \mathbf{u}. \tag{9}$$

Here, α is the inverse permeability that can be estimated by the Carman–Kozeny equation. It is noteworthy that $\mathbf{F} = \mathbf{0}$ is satisfied in the fluid domain Ω_f.

The governing equation for the transport of the vanadium species $c : \mathscr{O} \to \mathbb{R}$ is defined as

$$\mathbf{u} \cdot \nabla c = \frac{1}{Pe}\nabla^2 c + S. \tag{10}$$

Here, Pe expresses the Péclet number, and $S : \Omega_p \to \mathbb{R}$ is the reaction term in the porous domain Ω_p. Although the reaction term S is generally estimated using the Butler–Volmer equations, we use the following simplified model [21]:

$$S = \kappa(1 - c), \tag{11}$$

where $\kappa = a|\mathbf{u}|^b$, in which a and b are positive parameters, is the volumetric mass transfer coefficient that depends on the local velocity.

Problem Formulation

In this section, we formulate the optimization problem for the design of flow field in a battery system. Because battery performance is typically correlated with the reaction rate of species, we consider the following optimization problem:

$$\begin{aligned} \underset{\gamma}{\text{maximize}} \quad & J(\gamma) = \int_{\Omega_p} \kappa(1 - c)\,\mathrm{d}\Omega \\ \text{subject to} \quad & 0 \leqslant \gamma(\mathbf{x}) \leqslant 1, \ \forall \mathbf{x} \in D. \end{aligned} \tag{12}$$

Originally, this optimization problem must be solved on a 3D simulation model that requires massive computational cost compared with a 2D model. Furthermore, the nonlinearity of electrolyte flow often renders it difficult to realize computationally stable derivations of the optimized solutions. Hence, on the basis of MFTD, we reformulate the optimization problem in (12) as two subproblems, namely, low- and high-fidelity design optimization problems. As shown in Fig. 3, the flow channel configuration is estimated by solving a 2D topology optimization problem; in addition, the high-fidelity analysis for evaluating the original objective J in (12) is conducted on the 3D analysis model, in which the 2D optimized configuration is excluded

V02AT03A059-5

Copyright © 2019 ASME

for the 3D simulation. The detailed formulations of the low- and high-fidelity design optimization problems are described here-inafter.

Low-Fidelity Design Optimization   On the basis of a previous study [21], we formulated the low-fidelity design optimization problem as follows:

$$
\begin{aligned}
\underset{\gamma^{(k)}}{\text{maximize}} \quad & \widetilde{J}(\gamma^{(k)}, \mathbf{s}^{(k)}) = \int_{\widetilde{D}} \tilde{\kappa}_\gamma (1 - \tilde{c}) \, d\Omega \\
\text{subject to} \quad & 0 \leqslant \gamma^{(k)}(\mathbf{x}) \leqslant 1, \ \forall \mathbf{x} \in D, \\
\text{for given} \quad & \mathbf{s}^{(k)} = \{ s_1^{(k)} \ s_2^{(k)} \}^{\mathrm{T}}, \ k = 1, \ldots, K.
\end{aligned} \tag{13}
$$

Here, the pseudo-objective functional $\widetilde{J}$ is evaluated by solving the following pseudo-governing equations:

$$
\nabla \cdot \tilde{\mathbf{u}} = 0, \tag{14}
$$

$$
(\tilde{\mathbf{u}} \cdot \nabla) \tilde{\mathbf{u}} = -\nabla \tilde{p} + \frac{1}{\widetilde{Re}} \nabla^2 \tilde{\mathbf{u}} - \tilde{\alpha}_\gamma \tilde{\mathbf{u}}, \tag{15}
$$

$$
\tilde{\mathbf{u}} \cdot \nabla \tilde{c} = \frac{1}{\widetilde{Pe}} \nabla^2 \tilde{c} + \tilde{\kappa}_\gamma (1 - \tilde{c}), \tag{16}
$$

where $\tilde{\mathbf{u}}$, $\tilde{p}$, and $\tilde{c}$ are the state variables corresponding to the original ones in (7), (8), and (10), respectively; $\widetilde{Re}$ and $\widetilde{Pe}$ are the Reynolds number and Péclet number for the pseudo-governing equations, respectively. In addition, the design-dependent functions, $\tilde{\alpha}_\gamma$ and $\tilde{\kappa}_\gamma$, are defined as follows:

$$
\tilde{\alpha}_\gamma = f_\gamma \tilde{\alpha}, \quad \tilde{\kappa}_\gamma = f_\gamma \tilde{\kappa}(\tilde{\mathbf{u}}) \quad \text{with} \quad f_\gamma = \frac{q(1-\gamma)}{\gamma + q}, \tag{17}
$$

where $\tilde{\alpha}$ is the inverse permeability for the expression of the solid domain as in a previous study [26]; $\tilde{\kappa} = \tilde{a}|\tilde{\mathbf{u}}|^{\tilde{b}}$, where $\tilde{a}$ and $\tilde{b}$ are positive parameters corresponding to a and b, respectively, is the volumetric mass transfer coefficient; q is a positive parameter for controlling the convexity of f_γ.

As mentioned earlier, we focus on using the seeding parameters, $\mathbf{s}^{(k)}$ for generating a promising initial guess. On the basis of the results in a previous study [21], we redefine $\widetilde{Re}$ and $\tilde{a}$ for generating various patterns of topology-optimized configurations, as follows:

$$
\widetilde{Re}(s_1) = \widetilde{Re}_{\min} + (\widetilde{Re}_{\max} - \widetilde{Re}_{\min}) s_1, \ 0 \leqslant s_1 \leqslant 1, \tag{18}
$$

$$
\tilde{a}(s_2) = \tilde{a}_{\min} + (\tilde{a}_{\max} - \tilde{a}_{\min}) s_2, \qquad 0 \leqslant s_2 \leqslant 1, \tag{19}
$$

where the subscripts 'min' and 'max' express the lower and upper limits, respectively. It is noteworthy that the incorporation of

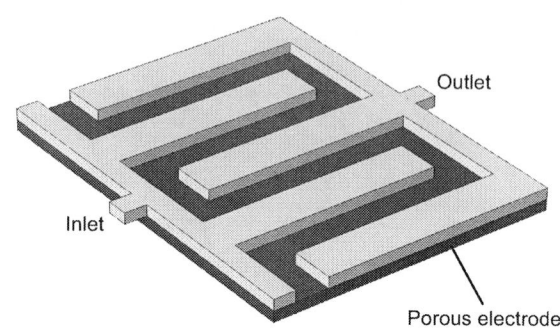

FIGURE 4.   REFERENCE MODEL.

the seeding parameters into the low-fidelity design optimization problem is not unique. In addition, the estimation of the lower and upper limits requires parameter studies in advance.

High-Fidelity Design Optimization   After obtaining the promising initial guess $\gamma_{\text{to}}^{(X)}$ by solving the low-fidelity design optimization problem in (13), we solve the following optimization problem:

$$
\begin{aligned}
\underset{s^*}{\text{maximize}} \quad & \widehat{J}(\gamma_{\text{to}}(s^*)) \\
\text{subject to} \quad & s_{\min}^* \leqslant s^* \leqslant s_{\max}^*, \\
\text{for given} \quad & \gamma_{\text{to}}^{(X)}(\mathbf{s}^{(X)}).
\end{aligned} \tag{20}
$$

Here, s^* is newly defined as the design variable. In this study, we define s^* as an iso-contour value of $\gamma_{\text{to}}^{(X)}$. The range of s^* should be set such that the topology of γ_{to} is preserved. It is noteworthy that the surrogate objective $\widehat{J}$ is constructed by calculating the original one, J, at each sample point.

NUMERICAL EXAMPLES

Next, we discuss the numerical examples in this study. The parameters for the low- and high-fidelity design optimization problems are shown in Table 1 and 2. The analysis model is discretized using the finite element method (FEM). In this study, all simulations for the flow field are conducted using COMSOL Multiphysics, which is commercial finite-element software. In the low-fidelity design optimization, we use a partial differential equation-based filter [27, 28] for ensuring the smoothness of the topology-optimized configuration. In addition, we use sequential linear programming for updating the design variables in the

FIGURE 5. SAMPLE DATASET GENERATED BY LOW-FIDELITY DESIGN OPTIMIZATION (100 DESIGN CANDIDATES).

low-fidelity design optimization. The detailed numerical implementation of the low-fidelity design optimization can be found in [21].

To evaluate the obtained design solution, we introduce a reference model that is based on an interdigitated flow field, as shown in Fig. 4. The analysis model is discretized using a 3.71×10^5 unstructured mesh. Under the same parameter settings in Table 2, the original objective is $J = 1.07 \times 10^{-5}$ on the reference model.

Figure 5 shows all the design candidates generated by solving the low-fidelity design optimization problem, in which the sample number of seeding parameters is set to $K = 100$. Figure 6 shows the heatmap for revealing the design solution space with respect to the seeding parameters s_1 and s_2. As shown in Fig. 6, the promising design candidates that exhibit superior performances compared with the reference model, are primarily distributed in the right region.

For the high-fidelity design optimization, we choose the top three promising candidates, $(s_1, s_2) = (0.0, 0.1), (0.7, 0.2), (0.8, 0.2)$, from the datasets shown in Fig. 5 and 6. The design variable s^*, corresponding to the iso-contour of $\gamma_{to}^{(X)}$ is sampled at five points using equally spaced sampling, i.e., $s^* = \{0.1, 0.3, 0.5, 0.7, 0.9\}^T$; subsequently, we construct the surrogate function $\hat{J}$ using a regressing Kriging surrogate model [29]. Figure 7 shows the surrogate functions of the promising candidates, in which the optimum points are estimated using the efficient global optimization technique [30]. As shown in Fig. 7, a sample point is added in the case of $(s_1, s_2) = (0.7, 0.2), (0.8, 0.2)$ at the vicinity of $s^* = 0.2$. The maximum performance, $J = 1.24 \times 10^{-5}$, is obtained at $(s_1, s_2, s^*) = (0.7, 0.2, 0.18)$, and the obtained result is superior compared with that of the reference model ($J = 1.07 \times 10^{-5}$). Figure 8 shows this best design solution generated by our proposed framework. It can be confirmed that the electrolyte

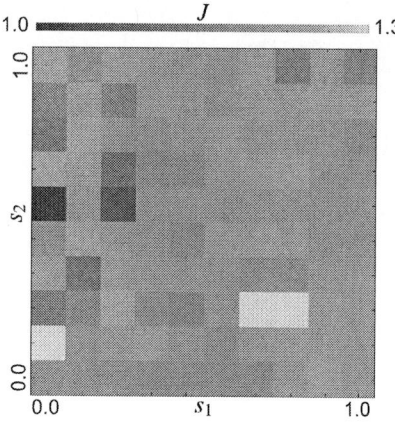

FIGURE 6. DESIGN SPACE OF SEEDING PARAMETERS.

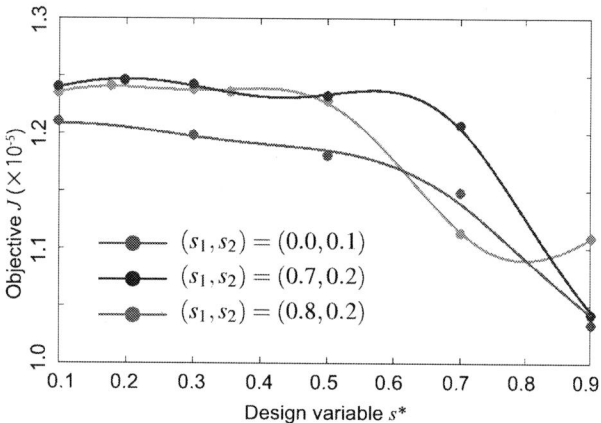

FIGURE 7. SURROGATE MODELS OF DESIGN CANDIDATES.

TABLE 1. PARAMETER SETTINGS FOR THE LOW-FIDELITY ANALYSIS MODEL.

Parameter	Symbol	Value
Péclet number	$\widetilde{Pe}$	1.5×10^{-6}
Tuning parameter	q	0.01
Inverse permeability	$\tilde{\alpha}$	6.0×10^{3}
Reynolds number	$(\widetilde{Re}_{\min}, \widetilde{Re}_{\max})$	(20, 200)
Coefficient parameter of $\tilde{\kappa}$	$(\tilde{a}_{\min}, \tilde{a}_{\max})$	(0.01, 0.1)
Exponent parameter of $\tilde{\kappa}$	$\tilde{b}$	0.7

TABLE 2. PARAMETER SETTINGS FOR HIGH-FIDELITY ANALYSIS MODEL.

Parameter	Symbol	Value
Péclet number	Pe	6.0×10^{-6}
Inverse permeability	α	6.0×10^{2}
Reynolds number	Re	100
Coefficient parameter of κ	a	3.2×10^{-4}
Exponent parameter of κ	b	0.4

is flowing in the porous electrode. Because the objective J in (12) is dependent on the velocity in the porous electrode, the space between the interdigitated flow channels is important for maximizing J. In fact, as shown in Fig. 7, we can confirm that the wider space corresponding to the larger value of s^* performs worse. It is confirmed that the surrogate optimization problem in (20) is important for achieving a satisfactory performance in this numerical example.

CONCLUDING REMARKS

A framework, i.e., MFTD was proposed herein for solving computationally hard topology optimization problems. MFTD is based on multi-fidelity design optimization, whose concept is to decompose an original optimization into low- and high-fidelity design optimization problems. For adapting the concept of multi-fidelity design optimization to topology optimization,

we proposed using seeding parameters to generate various patterns of design candidates by solving low-fidelity topology optimization problems. In the high-fidelity design optimization process, MFTD focused on generating promising initial guesses that were estimated under the high-fidelity analysis model; subsequently, the optimum design solution can be led by constructing surrogate models under a restricted design solution space.

To confirm the efficacy of the proposed framework, we applied MFTD to a topology optimization problem for the design of flow fields in battery systems. In our numerical example, the low-fidelity design optimization problem was formulated as a topology optimization problem based on a previous study [21], in which a simplified analysis model was proposed for the design of 2D flow fields of redox flow batteries. By utilizing the results of a previous study, we chose the Reynolds number and a parameter of the volumetric mass transfer coefficient as the seeding parameters for generating various patterns of topology-optimized design candidates. We generated 100 design candidates and found the promising initial guesses on the heatmap; subsequently, they were used for revealing the design solution space of the seeding

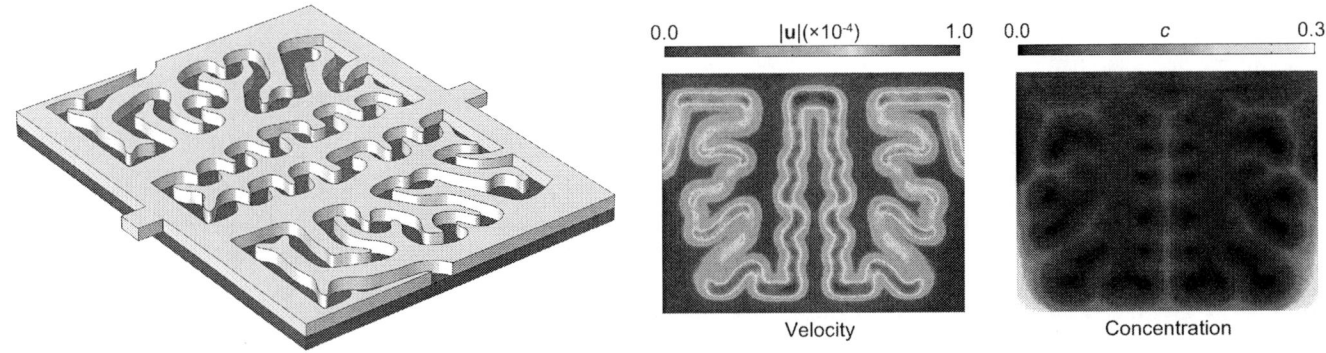

Velocity · Concentration

3D view of best design · Distributions on the middle surface of porous electrode

FIGURE 8. BEST DESIGN GENERATED BY MFTD.

parameters. Using the promising initial guesses, we constructed Kriging surrogate models with respect to the design variable—an iso-contour of the topology-optimized configuration in the low-fidelity design optimization problem—in the surrogate optimization problem. Consequently, we obtained a satisfactory solution under the 3D high-fidelity analysis model, and the obtained result indicated superior performance in comparison with a reference model.

Meanwhile, the proposed framework presented an issue because the appropriate formulation of the low-fidelity design optimization problem was determined heuristically. It is noteworthy that, in general, such issues for determining formulations exist when solving topology optimization problems, as discussed in [31]. We emphasize that obtaining the appropriate formulations for low-fidelity design optimization problems are important. To further verify the results, the efficacy of the proposed framework should be demonstrated in other design optimization problems in the future.

ACKNOWLEDGMENT

This work was partially supported by JSPS KAKENHI Grand Number 18K13674, and The Mazda Foundation.

REFERENCES

[1] Bendsøe, M. P., and Kikuchi, N., 1988. "Generating optimal topologies in structural design using a homogenization method". *Computer Methods in Applied Mechanics and Engineering, 71*(2), pp. 197–224.

[2] Bendsøe, M. P., and Sigmund, O., 2003. *Topology optimization: theory, methods, and applications*. Springer, New York.

[3] Sigmund, O., and Maute, K., 2013. "Topology optimiza-

tion approaches". *Structural and Multidisciplinary Optimization, 48*(6), pp. 1031–1055.

[4] Deaton, J. D., and Grandhi, R. V., 2014. "A survey of structural and multidisciplinary continuum topology optimization: post 2000". *Structural and Multidisciplinary Optimization, 49*(1), pp. 1–38.

[5] Zhao, X., Zhou, M., Sigmund, O., and Andreasen, C. S., 2018. "A "poor man's approach" to topology optimization of cooling channels based on a Darcy flow model". *International Journal of Heat and Mass Transfer, 116*, pp. 1108–1123.

[6] Asmussen, J., Alexandersen, J., Sigmund, O., and Andreasen, C. S., 2019. "A "poor man's" approach to topology optimization of natural convection problems". *Structural and Multidisciplinary Optimization, 59*(4), pp. 1105–1124.

[7] Zeng, S., Kanargi, B., and Lee, P. S., 2018. "Experimental and numerical investigation of a mini channel forced air heat sink designed by topology optimization". *International Journal of Heat and Mass Transfer, 121*, pp. 663–679.

[8] Alexandrov, N. M., and Lewis, R. M., 2001. "An overview of first-order model management for engineering optimization". *Optimization and Engineering, 2*(4), pp. 413–430.

[9] Park, C., Haftka, R. T., and Kim, N. H., 2017. "Remarks on multi-fidelity surrogates". *Structural and Multidisciplinary Optimization, 55*(3), pp. 1029–1050.

[10] Peherstorfer, B., Willcox, K., and Gunzburger, M., 2018. "Survey of multifidelity methods in uncertainty propagation, inference, and optimization". *SIAM Review, 60*(3), pp. 550–591.

[11] Leifsson, L., and Koziel, S., 2010. "Multi-fidelity design optimization of transonic airfoils using physics-based surrogate modeling and shape-preserving response prediction". *Journal of Computational Science, 1*(2), pp. 98–106.

[12] Bandler, J. W., Cheng, Q. S., Nikolova, N. K., and Ismail, M. A., 2004. "Implicit space mapping optimization exploiting preassigned parameters". *IEEE Transactions on Microwave Theory and Techniques, 52*(1), pp. 378–385.

[13] Forrester, A. I. J., Sobester, A., and Keane, A. J., 2008. *Engineering design via surrogate modelling: a practical guide.* John Wiley & Sons.

[14] Bartolozzi, M., 1989. "Development of redox flow batteries. A historical bibliography". *Journal of Power Sources, 27*(3), pp. 219–234.

[15] Cunha, Á., Martins, J., Rodrigues, N., and Brito, F., 2015. "Vanadium redox flow batteries: a technology review". *International Journal of Energy Research, 39*(7), pp. 889–918.

[16] Xu, Q., Zhao, T., and Zhang, C., 2014. "Performance of a vanadium redox flow battery with and without flow fields". *Electrochimica Acta, 142*, pp. 61–67.

[17] Xu, Q., Zhao, T., and Leung, P., 2013. "Numerical investigations of flow field designs for vanadium redox flow batteries". *Applied Energy, 105*, pp. 47–56.

[18] Tsushima, S., and Suzuki, T., 2016. "Modeling and simulation of an interdigitated vanadium redox flow battery with interfacial mass transfer resistance". No. PRTEC-15318, The First Pacific-Rim Thermal Engineering Conference.

[19] Alexandersen, J., Sigmund, O., and Aage, N., 2016. "Large scale three-dimensional topology optimisation of heat sinks cooled by natural convection". *International Journal of Heat and Mass Transfer, 100*, pp. 876–891.

[20] Yaji, K., Ogino, M., Chen, C., and Fujita, K., 2018. "Large-scale topology optimization incorporating local-in-time adjoint-based method for unsteady thermal-fluid problem". *Structural and Multidisciplinary Optimization, 58*(2), pp. 817–822.

[21] Yaji, K., Yamasaki, S., Tsushima, S., Suzuki, T., and Fujita, K., 2018. "Topology optimization for the design of flow fields in a redox flow battery". *Structural and Multidisciplinary Optimization, 57*(2), pp. 535–546.

[22] Ma, X., Zhang, H., and Xing, F., 2011. "A three-dimensional model for negative half cell of the vanadium redox flow battery". *Electrochimica Acta, 58*, pp. 238–246.

[23] You, D., Zhang, H., and Chen, J., 2009. "A simple model for the vanadium redox battery". *Electrochimica Acta, 54*(27), pp. 6827–6836.

[24] Yaji, K., Yamada, T., Kubo, S., Izui, K., and Nishiwaki, S., 2015. "A topology optimization method for a coupled thermal–fluid problem using level set boundary expressions". *International Journal of Heat and Mass Transfer, 81*, pp. 878–888.

[25] Olesen, L. H., Okkels, F., and Bruus, H., 2006. "A high-level programming-language implementation of topology optimization applied to steady-state Navier-Stokes flow". *International Journal for Numerical Methods in Fluids, 65*(7), pp. 975–1001.

[26] Borrvall, T., and Petersson, J., 2003. "Topology optimization of fluids in Stokes flow". *International Journal for Numerical Methods in Fluids, 41*(1), pp. 77–107.

[27] Kawamoto, A., Matsumori, T., Yamasaki, S., Nomura, T., Kondoh, T., and Nishiwaki, S., 2011. "Heaviside projection based topology optimization by a PDE-filtered scalar function". *Structural and Multidisciplinary Optimization, 44*(1), pp. 19–24.

[28] Lazarov, B. S., and Sigmund, O., 2011. "Filters in topology optimization based on Helmholtz-type differential equations". *International Journal for Numerical Methods, 86*(6), pp. 765–781.

[29] Forrester, A. I. J., Keane, A. J., and Bressloff, N. W., 2006. "Design and analysis of "noisy" computer experiments". *AIAA journal, 44*(10), pp. 2331–2339.

[30] Jones, D. R., Schonlau, M., and Welch, W. J., 1998. "Efficient global optimization of expensive black-box functions". *Journal of Global Optimization, 13*(4), pp. 455–492.

[31] Yamasaki, S., Yaji, K., and Fujita, K., 2019. "Knowledge discovery in databases for determining formulation in topology optimization". *Structural and Multidisciplinary Optimization, 59*(2), pp. 595–611.

Proceedings of the ASME 2019
International Design Engineering Technical Conferences
and Computers and Information in Engineering Conference
IDETC/CIE2019
August 18-21, 2019, Anaheim, CA, USA

DETC2019-97722

VISUALIZING AND EVALUATING HIGH-DIMENSIONAL MAPPINGS OF SETS OF HIGH PERFORMANCE DESIGNS

Clinton B. Morris
clintonm@parc.com
Computation for Automation in
Systems Engineering
Palo Alto Research Center
Palo Alto, California, USA

Michael R. Haberman
haberman@arlut.utexas.edu
Mechanical Engineering Department
and Applied Research Laboratories
The University of Texas at Austin
Austin, Texas, USA

Carolyn C. Seepersad
ccseepersad@mail.utexas.edu
Mechanical Engineering Department
The University of Texas at Austin
Austin, Texas, USA

ABSTRACT

Design space exploration can reveal the underlying structure of design problems. In a set-based approach, for example, exploration can map sets of designs or regions of the design space that meet specific performance requirements. For some problems, promising designs may cluster in multiple regions of the input design space, and the boundaries of those clusters may be irregularly shaped and difficult to predict. Visualizing the promising regions can clarify the design space structure, but design spaces are typically high-dimensional, making it difficult to visualize the space in three dimensions. To convey the structure of such high-dimensional design regions, a two-stage approach is proposed to (1) identify and (2) visualize each distinct cluster or region of interest in the input design space. This paper focuses on the visualization stage of the approach. Rather than select a singular technique to map high-dimensional design spaces to low-dimensional, visualizable spaces, a selection procedure is investigated. Metrics are available for comparing different visualizations, but the current metrics either overestimate the quality or favor selection of certain visualizations. Therefore, this work introduces and validates a more objective metric, termed *preservation*, to compare the quality of alternative visualization strategies. Furthermore, a new visualization technique previously unexplored in the design automation community, t-Distributed Neighbor Embedding, is introduced and compared to other visualization strategies. Finally, the new metric and visualization technique are integrated into a two-stage visualization strategy to identify and visualize clusters of high-performance designs for a high-dimensional negative stiffness metamaterials design problem.

1. INTRODUCTION

Engineering design problems are commonly hierarchical and multilevel, which requires precise coordination between models at each level. If the levels are highly coupled and the models are computationally expensive, blackbox, nonconvergent, and/or highly nonlinear, identification of an optimal design may be exceptionally difficult. Therefore, alternatives to optimization-based methods have been investigated to explore and map design spaces. One such alternative approach is the set-based method which classifies and tracks sets or ensembles of satisfactory designs or designs that meet some performance requirement(s) [1]. The regions of interest in the input design space, defined by these ensembles of satisfactory designs, are often disjoint and irregularly shaped because of the inherent complexity and nonlinearity of many design problems such as the metamaterials design problem of interest in this paper [2, 3]. Therefore, set-based design approaches must accurately map these complex, possibly disjoint, and irregularly shaped satisfactory design regions in design spaces that are commonly high-dimensional, i.e., defined by more than three design variables.

Various approaches have been adopted to generate these inverse mappings. The Inductive Design Exploration Method (IDEM) creates flexible mappings of a multilevel design space and supports uncertainty analysis [4]. Malak and Paredis [5] used support vector machines (SVM) to explore feasible design regions. Zeliff *et al.* [6] utilized a Multi-Objective Genetic Algorithm (MOGA) modified by random forest, kNN, and Naïve Bayesian classifiers to identify non-dominated solutions efficiently. Chen and Fuge [7] recently developed a method for identifying feasible designs in the presence of implicit

Copyright © 2019 ASME

constraints by balancing exploration and exploitation with a Gaussian process-based classifier. Shahan and Seepersad adopted an approach based on Bayesian network classifiers (BNC) to capture the arbitrary shape of those promising regions resulting in substantially less classification error when compared to interval-based mappings [8].

While each of these methods have their advantages for mapping design spaces, drawbacks still exist. Examples include the discriminative nature of SVMs, the possible inaccuracies of the modified MOGA, or the computational expense of a Gaussian process with many training points. For all set-based design approaches, however, dimensionality of the design space presents major challenges for interpreting these inverse mappings of the design space. The curse of dimensionality not only increases the computational resources needed to explore the design space but also makes it difficult to visualize interesting regions of the design space identified by set-based methods. Visual mappings provide intangible insight into the physics of the problem and relationships between design variables. For nonlinear problems encountered in materials design, for example, viewing inverse mappings of the design space can provide insights into the problem that could not have been obtained by any other means. To present intuitive mappings when more than three variables are present, a visualization framework tailored to multilevel, set-based design must be developed.

The work by Chen *et al.* provided a framework for visualization and introduced algorithms that are the state of the art for visualization in the field of set-based design. Their visualization framework and subsequent design space exploration was intended, however, for single-level design problems in which the design regions reside on lower dimensional manifolds [9]. Expanding on this previous work, this paper provides a general visualization approach that is applicable to both single and multilevel design problems in which the high-dimensional design spaces may not lie on low dimensional manifolds.

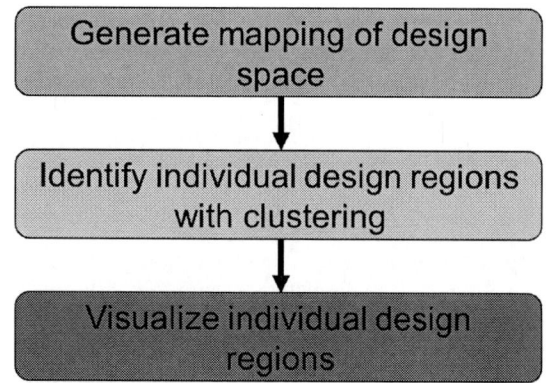

Figure 1: Flowchart of proposed visualization methodology for set-based design approaches.

As shown in Figure 1, a three-step approach is proposed to visualize inverse mappings of design regions of interest. First,

the design space must be mapped into sets of satisfactory designs or regions of interest in the design space. The BNC approach is utilized in this work, but other approaches can be used, as well, as discussed earlier. After the regions of interest have been identified, clustering is utilized to identify individual ensembles of promising designs. Following identification of the ensembles, a dimension reduction/visualization technique is used to visualize each individual ensemble. The visualization technique transforms high dimensional mappings into a low-dimensional space so the mappings can be visualized and interpreted by the designer.

Clustering and visualization are not simple tasks, and a vast body of literature proposes alternative approaches, each with their own problem-dependent strengths and weakness. In previous work, the authors identified the Nyström modified implementation of self-tuning spectral clustering (STSC) as a viable clustering method to identify high-dimensional design regions or ensembles of designs of interest. Unlike many of the other clustering approaches utilized in the design automation community [9-17], it does not need to perform iterations to determine the number of clusters. Furthermore, it is far more computationally efficient than the STSC approach utilized in Chen *et al.* although some confidence is lost in the accuracy of the partitions [18]. For further details in the comparison of the Nyström modified STSC and the STSC, the reader is directed to Morris and Seepersad [18].

After identifying each of the design regions or clusters of interest, the designer must select an appropriate method to visualize each cluster. Therefore, a brief review of suitable visualization strategies for design is provided in Section 2. The review concludes with the introduction of a visualization approach previously unexplored in the design automation community, t-distributed stochastic neighbor embedding (t-SNE). Often, though, it is not clear which approach is the most suitable for a specific design problem. Furthermore, many of these visualizations require the selection of tuning parameters or heuristics that affect the output of the visualization approach. To assist designers in selecting a suitable visualization and appropriate heuristics, metrics have been introduced to compare the quality of visualizations. These metrics discussed in Section 3, however, can erroneously assess the quality of a visualization or favor certain visualizations. Therefore, a new metric, *preservation*, is developed and introduced in Section 4 to quantify the quality of a visualization. This metric provides the designer with an objective measure of how well the visualization preserved the original high dimensional topology in the low dimensional space. In Section 5, the new metric is compared to previously established metrics, and the quality of the t-distributed stochastic neighbor embedding (t-SNE) visualization approach is evaluated for a materials design problem. The overall outcome is a two-stage visualization strategy for identifying and visualizing clusters of high-performance designs for high-dimensional design problems.

In summary, this paper will provide several contributions to the design automation community. First, t-SNE, a visualization strategy, will be introduced and compared to a visualization strategy currently used in the design automation community. Second, a new metric, *perplexity*, will be developed and validated for objectively comparing different visualization strategies . Third, a complete, general approach for identifying and visualizing ensembles of designs will be demonstrated. The goal of the complete approach is to assist designers in visually uncovering relationships between design variables.

2. VISUALIZATION STRATEGIES

The final task of the visualization strategy in Figure 1 is to reduce the dimensionality of the design regions of interest by mapping into a lower dimensional space. Selection of the most appropriate visualization approach is a difficult task because efficacy is often problem dependent. Rather than claim that there is a "best" visualization approach, this work provides a review of visualization approaches with commentary to assist the designer in selecting appropriate approaches for comparison. An approach previously unused in the design automation community, t-SNE, is identified as a prospective visualization approach and demonstrated in the examples, but other approaches can be considered.

A complete review of all visualization techniques available across all disciplines would be too lengthy for this paper. Instead, this review provides a general overview of various visualization techniques, categorized based on their similarities. The reader is encouraged to see the reviews provided by van der Maaten *et al.* [19] or Liu *et al.* [20] if they are interested in obtaining further information. In general, the terms dimension reduction and visualization will be used interchangeably even though they are distinct areas of focus. This work assumes any use of dimension reduction is intended for visualization purposes.

van der Maaten [19] categorized visualization or dimension reduction techniques into two different families: (1) convex techniques or (2) nonconvex techniques. A convex technique minimizes a convex objective function to uncover the low dimensional structure of a high dimensional data set while a nonconvex technique optimizes a nonconvex objective function.

These families can be subdivided further into various other classes of techniques. The convex family of techniques include approaches that rely on spectral analysis to solve the convex objective functions. These spectral techniques are either full spectral or sparse spectral depending on the formulation of the objective function. Full spectral techniques, such as kernel principal component analysis (KPCA) [21], use the eigendecomposition of a high dimensional data set to determine the orthogonal vectors, or principal components, along which most of the variation in the data occurs. One of the limitations of these full spectral approaches is that the computational expense of calculating eigenvalues and eigenvectors scales cubically with the number of points in the dataset [19]. The other member of the convex family of dimension reduction techniques is sparse spectral techniques. The reason these techniques are considered sparse is because they do not consider the global structure of the dataset; rather, they solve a sparse or generalized eigenproblem that considers the local structure of the data. These approaches have found success in many fields, but they commonly collapse regions of high dimensional data into points or rescale the axes to distort the topography [19].

The other family of dimension reduction approaches, nonconvex techniques, is far more general than convex techniques. Some examples of nonconvex approaches are Sammon Mappings [22], Locally Linear Embedding (LLE) [23], Manifold Charting [24], Isomaps [25], t-Distributed Stochastic Neighbor Embedding (t-SNE) [26] and multilayer autoencoders [27]. Multilayer autoencoders are becoming especially popular for visualization because they are based on neural networks that are common across dozens of fields including the design automation community. Neural network based approaches are not utilized in this research, however, because they can require extensive resources for training the networks, and they are highly blackbox, which leads to a lack of intuition about the visualization for the designer.

The nonconvex visualization approach, t-SNE, [26] has been gaining popularity for visualization due to its effective and efficient results. It retains a balance of both global and local structures of high dimensional data sets. To the best knowledge of the authors, this approach has yet to be implemented to visualize design spaces, so part of the focus of this work is to introduce a novel visualization method that can accompany a novel implementation approach. The formulation of t-SNE is largely based on the original stochastic neighbor embedding (SNE) approach but modifies the algorithm slightly to achieve significantly better visualizations.

The t-SNE approach builds on the SNE approach but modifies the formulation to balance both global and local structure to produce better mappings of multiple clusters or regions of interest. The stochastic approach takes a high dimensional datapoint, x_i, and transforms it into a low-dimensional point, y_i through a nonlinear mapping obtained by minimizing the divergence of two conditional probabilities, a nonconvex function.

First, the high-dimensional pairwise distance between two points is converted to a conditional probability, $p_{j|i}$ that represents the similarity of two points, x_i and x_j in the high dimensional space. The conditional probability is given by Equation 1 where $p_{i|i} \equiv 0$ and σ_i is a local tuning parameter.

$$p_{j|i} = exp\left(\frac{\|x_i - x_j\|_2^2}{2\sigma_i^2}\right)\left(\sum_{k \neq i}^{N} exp\left(-\frac{\|x_i - x_k\|_2}{2\sigma_i^2}\right)\right)^{-1} \quad (1)$$

Equation 1 allows for the conditional probability between two points in the high dimensional space, p_{ij}, to be defined as the weighted average of the associated conditional probabilities. Next, the joint probability describing the similarity of low dimensional points, q_{ij}, is described by the student t-distribution with one degree of freedom provided in Equation 2.

$$q_{ji} = \frac{\left(1 + \|y_i - y_j\|_2^2\right)^{-1}}{\sum_{k \neq l}(1 + \|y_k - y_l\|_2^2)^{-1}} \quad (2)$$

With the joint probabilities at each scale established for the t-SNE approach, the cost function given by Equation 3 must be defined for optimization which is the Kullback-Leibler divergence measure.

$$C = \sum_{i=1}^{N} \sum_{j=1}^{N} p_{ij} \log \frac{p_{ij}}{q_{ij}} \quad (3)$$

By minimizing Equation 3, a nonlinear mapping of high dimensional points to low dimensional points is obtained. The form of the joint probabilities results in t-SNE retaining global and local structure better and requiring less computation than SNE. Specifically, the formulation of q_{ij} leads to a heavy-tailed distribution which means that the tails of the distribution are not bounded by negative exponential distributions. It allows moderate separation in the high dimensional space to be modeled more faithfully by larger distances in the low dimensional spaces because there is more "probabilistic mass" distributed into the tails. Also, for large pairwise distances in the low dimensional space, the t-distribution approximates an inverse square law. This feature facilitates better separation of clusters when visualizing multiple clusters while still balancing local structure within clusters—much the same way that the gravitational force on a planet varies with its structure but the interactions between planets assumes they are point masses. Furthermore, the formulation p_{ij} as the average of conditional probabilities reduces the number of terms in the gradient of C and hence the the computation time required to obtain the optimal mapping [26].

The t-SNE approach, though, like many convex and nonconvex visualization approaches, relies on a tuning parameter that influences the results of the visualization. The tuning parameter for t-SNE, perplexity or *Perp*, which is provided by Equation 4 and 5, tunes the conditional probabilities for the high dimensional data.

$$Perp = 2^{H(p_{j|i})} \quad (4)$$
$$\text{where}$$
$$H(p_{j|i}) = \sum_{j=1}^{N} p_{j|i} \log_2 p_{j|i} \quad (5)$$

Effectively, the perplexity is the continuous analog of neighborhood selection. Selection of the value of the perplexity is arbitrary, although the original paper on t-SNE recommended a perplexity value of approximately 50 as a general guideline.

The subjectivity of heuristics coupled with the vastness of available visualization strategies motivates the need for an objective measure of the quality of visualizations. This work proposes a new metric—preservation—that attempts to combine the best features of other metrics while eliminating some of their shortcomings.

3. EVALUATING THE QUALITY OF VISUALIZATIONS

The vast number of visualization and dimension reduction techniques makes it difficult to select the most appropriate approach for a specific application. While this work proposes using t-SNE to visualize satisfactory design regions, other approaches may perform better for specific problems. But what does it mean for a visualization to outperform another visualization technique? Saying one is better than the other is quite subjective, so there have been attempts to quantify the quality of a visualization. The two most common metrics used to objectively measure the quality of a visualization are trustworthiness and continuity, which are typically used together [28].

Informally, trustworthiness quantifies the points that are mapped closely in the low-dimensional space but not in the high-dimensional space while continuity quantifies the points that are close in the high-dimensional space but are not mapped closely in the low-dimensional space. To formally provide a definition of each metric some terms need to be defined. First, N is the total number of points to be visualized, and k is the number of points defined to be in the neighborhood of a point, i.e., the k closest points. Next, $r(i, j)$ is the rank order in terms of distance for the jth point relative to the ith point in the high-dimensional space and $\hat{r}(i, j)$ is the rank order in terms of distance for the jth point relative to the ith point in the low-dimensional space. Finally the set, $U_k(i)$, is the set of points that are in the k-nearest neighborhood (kNN) of the ith point in the low-dimensional space but not in the high-dimensional space and $V_k(i)$, is the set of points that are in the k-nearest neighborhood of the ith point in the high-dimensional space but not in the low-dimensional space.

With these terms established, trustworthiness, $T(k)$, and continuity, $C(k)$, can be evaluated by Equations 6 and 7. Notice that both metrics are dependent on the neighborhood size.

$$T(k) = 1 - \frac{2}{Nk(2N - 3k - 1)} \sum_{i=1}^{N} \sum_{j \in U_k(i)} (r(i,j) - k) \quad (6)$$

$$C(k) = 1 - \frac{2}{Nk(2N - 3k - 1)} \sum_{i=1}^{N} \sum_{j \in V_k(i)} (\hat{r}(i,j) - k) \quad (7)$$

Both metrics are bounded between 0 and 1 where a perfect mapping will have a trustworthiness and continuity of 1 and a low-quality mapping will be lower. It is important to note, though, that a mapping need not be perfect to have a continuity and trustworthiness of 1 because the metrics are not concerned with the exact distances separating points, but with the rank ordering of the distances. This distinction is intentional, so the metric can be used for categorical problems. By using only a rank ordering, non-perfect mappings can cause the metric to be falsely high for continuous spaces.

To elucidate how the metrics can overestimate the quality of a mapping a simple example is provided. Consider a set of points in $\mathbb{R}^2$ that are initially in the L-shaped configuration shown in Figure 2.

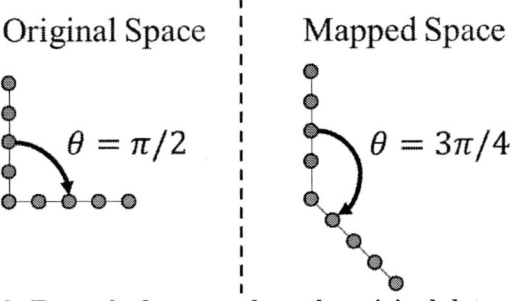

Figure 2: Example data set where the original dataset (left) can be mapped to the dataset (right) by varying the angle, θ.

Define the angle, θ, to be the angle between the two lengths of the L-shaped structure. Initially $\theta = \pi/2$, and a mapping is defined that simply varies θ within the interval $[0, 2\pi]$. Figure 2 also provides a specific mapping of the original space where the angle between arms, $\theta = 3\pi/4$. For each value of θ in the interval, the continuity and trustworthiness between the resultant mapping and original space were determined. Four different values of the neighborhood size were considered, $k = 5, 10, 25$, and 50, for the dataset where $N = 100$. The results of the various mappings are shown in Figure 3 where the horizontal axis is the angle of rotation in radians and the vertical axes are the metrics.

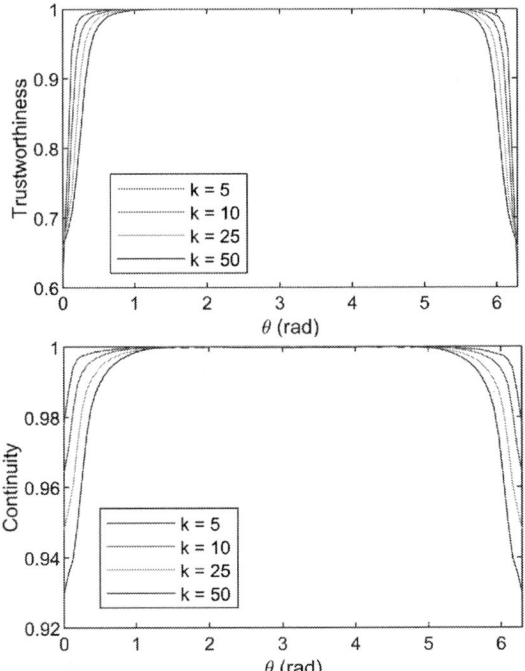

Figure 3: Trustworthiness (left) and continuity (right) as a function of θ for the L-Shaped example structure.

As stated previously, a value of 1 for continuity and trustworthiness should only indicate a perfect mapping, but this is not the case for Figure 3. Within rotational symmetry, the only

perfect mapping should be at odd integer intervals of $\pi/2$, that is $\pi/2, 3\pi/2, 5\pi/2, ...$, but as indicated in the plots of Figure 3, continuity and trustworthiness are simultaneously 1 for nonperfect mappings. This outcome is due to the rank ordering of distances, rather than the consideration of their continuous values in the metrics. Even through various rotations, the ranks of the kth closest points remain the same because, generally, points are closest to other points on the same line segment. If the whole line is rotated, then the rank ordering of distances will be largely unchanged.

Another metric that has been used is the geodesic distance inconsistency (GDI) which was proposed by Chen *et al* [29] in their work on visualizing design spaces. The metric was used to evaluate the quality of three different dimension reduction/visualization techniques, PCA, KPCA, and stacked denoising autoencoder (SdA). The GDI is measuring the correlation between the similarity and geodesic matrices, such that high correlation implies the geodesic distances and Euclidian distances increase and decrease linearly with each other. In contrast to trustworthiness and continuity, a low GDI indicates a quality visualization; therefore, to be consistent, this work will consider only the correlation term to compare the metrics on the same scale. While the GDI metric does not fall into the same pitfalls as trustworthiness and continuity, it is biased towards selecting methods such as Isomaps and Local Linear Embedding (LLE) as the best visualization techniques because they minimize an objective function that is related to the GDI metric [7, 23, 25]. The GDI metric was based on these methods, so it will naturally be biased to select these methods. Therefore, a metric that preserves distances well, conserves the rank ordering of the neighborhood, and is less biased for selection of specific visualization methods should be used. This work proposes such a metric. While slightly more complex to explain than the other metrics, it is hypothesized that the metric, which is called preservation in this work, provides more insight into the quality of the visualization.

4. INTRODUCTION OF THE PRESERVATION METRIC

The proposed metric combines aspects of the previous metrics that lead to preferable evaluation of visualizations. Much of the formulation of preservation, P, is also borrowed from spectral clustering due to its ability to balance both local and global structure. At a high level, the metric generates a kernel matrix on the original cluster in the high-dimensional space, K^O, and a kernel matrix on the mapped cluster in the low-dimensional space, K^M, that can be compared to measure the quality of the dimension-reduction and associated visualization. To generate the high-dimensional kernel matrix for a cluster, Equation 8 is provided, which has a similar form to the kernel used for STSC.

$$K_{ij}^0(k) = \exp\left(-\frac{d^2(x_i, x_j)}{\sigma_i^o(k)\sigma_j^o(k)}\right) \quad (8)$$

The value of the ith and j^{th} component of the kernel matrix is related to the Euclidian distance, d, between the points x_i and

x_j via a self-tuning Gaussian. The smoothing parameters, $\sigma_i^o(k)$ and $\sigma_j^o(k)$, are provided by Equation 9 in which the dummy variable, l, is used to represent the lth point of interest, x_l^k is the kth closest point to point x_l.

$$\sigma_l^o(k) = d\left(x_l, x_l^k\right) \qquad (9)$$

Secondly, to generate the low-dimensional kernel matrix for the mapped cluster, Equation 10 is provided in a similar form to the kernel used for STSC, as well.

$$K_{ij}^M(k) = \exp\left(-\frac{d^2\left(y_i, y_j\right)}{\sigma_i^M(k)\sigma_j^M(k)}\right) \qquad (10)$$

The value of the ith and j^{th} component of the kernel matrix is related to the Euclidian distance, d, between the mapped points y_i and y_j, which correspond to the original points x_i and x_j. The smoothing parameters, $\sigma_i^M(k)$ and $\sigma_j^M(k)$, have a slightly different form than their high-dimensional counterpart, as provided by Equation 11.

$$\sigma_l^M(k) = d\left(y_l, \tilde{y}_l^k\right) \qquad (11)$$

Once again, the dummy variable, l, is used to represent the lth mapped point of interest, x_l, and the subtle, but important difference, lies in the definition of $\tilde{y}_l^k$. Unlike in previous formulations, it is *not* the kth closest point to point y_l, rather it is the point in the low-dimensional space that corresponds to the kth closest point in the original high-dimensional space, x_l^k.

Finally, with the two kernel matrices defined, the preservation, $P(k)$, can be determined via Equation 12 where $\|X\|_F$ is the Frobenius norm of a matrix.

$$P(k) = 1 - \frac{1}{2}\left\| \frac{K^O}{\|K^O\|_F} - \frac{K^M}{\|K^M\|_F} \right\|_F \qquad (12)$$

Preservation, which is conservatively bounded in the range [0,1] by the Cauchy-Schwartz inequality, provides information regarding the mismatch between each weighted pairwise distance across spaces. If the two kernels are equivalent, implying all distances are conserved, then the difference is the zero matrix with Frobenius norm 0 leading to a preservation of 1. The structure of preservation has features that should allow it to be a more reliable measure of visualization quality. By setting the tuning parameters in the low dimensional space based on the mapping of the kth closest points in the high dimensional space, mappings that preserve distance *and* ordering of points are favored. Rather than prefer one or the other as is the case for the other metrics, they were combined into one single metric. Also, all elements in the kernels contribute to preservation ensuring that points that are far away from one another in the high-dimensional space are penalized for being too close to one

another in the low-dimensional space. With preservation introduced, it can be used to compare visualization strategies and to tune their heuristics.

5. VALIDATION OF THE PRESERVATION METRIC

The accuracy of the preservation metric can be investigated by comparing it to trustworthiness and continuity for a variety of examples. The first example is a simple one for which trustworthiness and continuity metrics lead to inaccurate results. Specifically, the preservation metric is applied to the L-Shaped structure of Figure 2. As shown in Figure 4 the preservation metric assumes a value of 1 only at the increments of theta initially identified as "perfect" mappings, which are odd integer multiples of $\pi/2$.

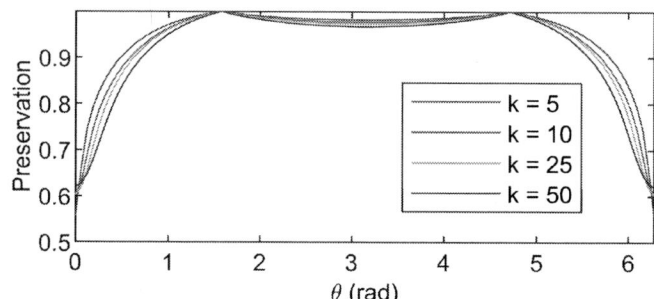

Figure 4: New metric, preservation, as a function of θ for the L-Shaped example structure.

More complex examples, though, are needed to further validate and compare preservation to trustworthiness, continuity, and GDI. This requires the selection of a group of high-dimensional datasets and a visualization technique to reduce the dimensionality. KPCA, a technique previously used to visualize design regions [29], was selected to visualize five high-dimensional datasets. Five different datasets, each consisting of 243 points, were generated in a 24-dimensional space: (1) a spiral or swiss roll, (2) the surface of a cross, (3) a multimodal surface (4) a hyperspherical shell, and (5) a complex parametric geometry. Each of the first 3 clusters truly lies on a 2-dimensional manifold but 24-dimensional noise was added to map the structures to a high-dimensional space. Images are provided of the structure in $\mathbb{R}^3$ in Figure 5 (note that more than 243 points are shown to help clarify the structures).

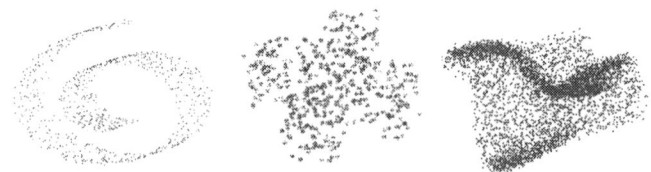

Figure 5: 3-Dimensional projection of the first three clusters, swiss roll (left), cross (center), and twin peaks (right), that were embedded in a 24-Dimensional Space

The first three datasets are described first, followed by the hyperspherical shell and complex parametric geometry.

The first three data sets were selected because of their varying degrees of difficulty for visualization. First, the swiss roll was selected because both global and local structure must be balanced to visualize it accurately. Since it spirals, points can be close based on absolute Euclidian distance, but if you must travel between them on the manifold, the points are far away due to the necessary path. This requires the visualization technique to be sophisticated enough to balance global and local preservation. The cross was selected because of its sharp corners and abrupt transitions in directions along the outer surface. The multimodal surface, or twin peaks is a standard visualization structure that was selected due to its complex surface geometry. Since the first three structures lie on low dimensional manifolds, whatever technique is used to visualize them in $\mathbb{R}^3$ should indicate a metric score that is nearly 1, if not 1.

The next cluster is a 5-Dimensional hyperspherical shell. It is anticipated that the quality metrics will be lower than 1 but not substantially lower than 1 for the hyperspherical shell because it cannot be truly visualized in 3-dimensions. The final cluster to be used for comparisons of metrics is the complex parametric shape described by Equation 13 with parameters, $t_1, t_2, t_3, t_4,$ and t_5. This shape was selected due to its complex geometry and its difficulty for visualization in lower dimensions.

$$\boldsymbol{x}_{difficult} = \cos{(t_1)}\boldsymbol{e_{15}} + \tanh{(3t_2)}\boldsymbol{e_{16}} + (t_1 + t_3)\boldsymbol{e_{17}} + \cdots$$
$$t_4 \sin(t_2)\boldsymbol{e_{18}} + t_5\cos{(t_2)}\boldsymbol{e_{20}} + (t_4 + t_5)\boldsymbol{e_{21}} + \cdots$$
$$t_2\boldsymbol{e_{22}} + t_3 t_4 \boldsymbol{e_{23}} + t_1\boldsymbol{e_{24}} + \epsilon_{noise} \qquad (13)$$

$$where$$

$$t_i \in [0,2\pi]$$

It is anticipated that for such visualizations the metrics should not be able to attain a value close to 1 because it is too complex for significant feature visualization.

KPCA was performed on each of the datasets. The tuning parameter in the form of the Gaussian kernel, σ, was varied at 250 equal increments on the interval [0.001, 3] For each resultant dimension reduction/visualization, the trustworthiness, continuity, GDI, and preservation metrics were evaluated. Also, for trustworthiness, continuity, and preservation the kNN value was set to four different values: 12, 24, 36, and 48, which correspond to flooring 5%, 10%, 15%, and 20% of the number of points in each cluster, respectively. Rather than supply a complete set of results, the characteristic results are included in this text for conciseness. The results of the KPCA study for the first 3 clusters (swiss roll, cross, and twin peaks) are presented in Figure 6 in which the value of the kernel tuning parameter is plotted against the value of the various metrics. The kNN value is 24 for trustworthiness, continuity, and preservation because the results were largely insensitive to its value.

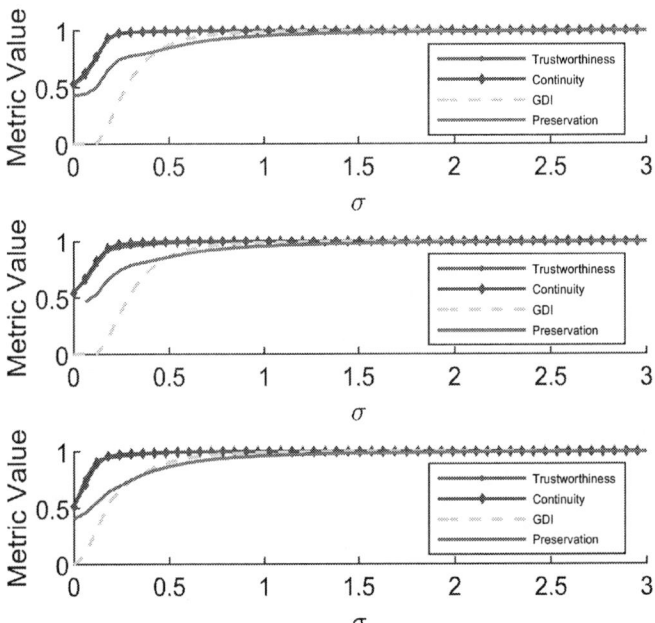

Figure 6: Results of metric comparison study for the KPCA dimension reduction for the (top) swiss roll, (middle) cross, and (bottom) twin peaks for _kNN_ = 24.

Figure 6 indicates that preservation provides similar results to GDI for the 3 clusters. Also, as expected, the metrics indicate that the KPCA dimension reduction/visualization technique extracts the low dimensional structure of the 3 clusters for relatively large values of the tuning parameter. The early convergence to a metric value of 1, indicates that trustworthiness and continuity may overestimate the quality of the dimension reduction because they are concerned more with the rank ordering of the distances rather than their actual values. To support this claim, the visualizations of the swiss roll at $\sigma = 0.35$ and $\sigma = 2$ are presented in Figure 7 to determine visually whether they are quality representations of the high dimensional cluster. The metric values at the two heuristics are provided in Table 1 and demonstrate that there is a discrepancy in relative quality for trustworthiness and continuity compared to GDI and preservation at $\sigma = 0.35$. The metrics agree at $\sigma = 2$.

Metric	$\sigma = 0.35$	$\sigma = 2$
Trustworthiness	0.99	1
Continuity	0.99	1
GDI	0.71	1
Preservation	0.79	0.99

Table 1: Metric values for the KPCA visualizations of the swiss roll structure for two different heuristic values.

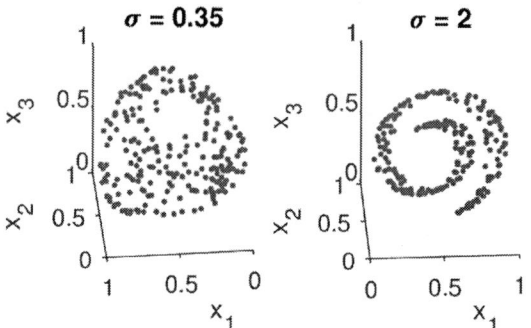

Figure 7: Resulting visualization/dimension reduction of the swiss roll using KPCA for $\sigma = 0.35$ (left) and $\sigma = 2$ (right).

It is immediately clear from inspection of Figure 7 that the visualization on the left does not preserve the structure of the swiss roll as well as the one on the right. As shown in Table 1, trustworthiness and continuity indicate that the inferior visualization was of high quality which it clearly is not. The GDI and preservation metrics, however, do indicate that the visualization on the left was of lesser quality. This overestimation of quality was consistent for each of the first three clusters visualized with KPCA. The remaining two datasets, the hyperspherical shell and the complex parametric form, are not easily mapped to a low-dimensional space. Therefore, a useful metric should indicate that the mappings are not perfect.

KPCA was performed again on the hyperspherical shell and the complex parametric form. The tuning parameter in the form of the Gaussian kernel, σ, was varied at 250 equal increments in the interval [0.001, 3] For each resultant dimension reduction/visualization, the trustworthiness, continuity, GDI, and preservation metrics were evaluated. Also, for trustworthiness, continuity, and preservation the kNN value was set to four different values: 12, 24, 36, and 48, which correspond to flooring 5%, 10%, 15%, and 20% of the number of points in each cluster respectively.

The results of the KPCA study for the datasets are presented in Figure 8 in which the value of the kernel tuning parameter is plotted against the value of the various metrics. The kNN value is 24 for trustworthiness, continuity, and preservation because the results were largely insensitive to its value.

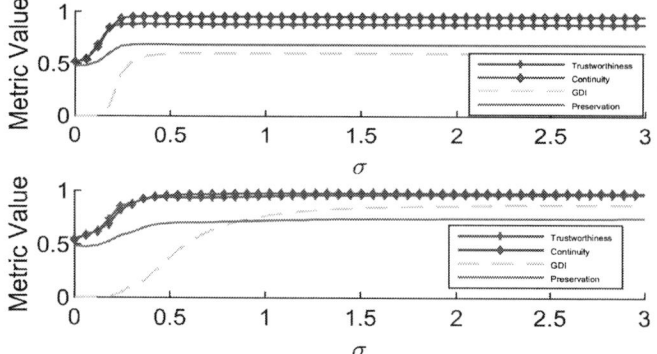

Figure 8: Results of metric comparison study for the KPCA dimension reduction for the hyperspherical shell (top) and complex parametric form (bottom) for $kNN = 24$.

The visualizations of the hyperspherical shell and complex parametric datasets resulted in low preservation and GDI values while the continuity and trustworthiness values were high once again. The lower values of GDI and preservation indicate that these clusters cannot be perfectly mapped to a low-dimensional space which is a key insight for designers when viewing a design space. These results inform the designer that more than three variables strongly affect the performance of a design if the GDI and preservation are below 1. Trustworthiness and continuity are high because they consider only the rank ordering of distances. Typical visualization approaches preserve local distances reasonably well, so the set of the k-closest points in the high dimensional space is likely similar to the set of the k-closest points in the low dimensional space. The exact order may not be preserved; for example, the second closest point in the high-dimensional space may not be the second closest in the low-dimensional space but it is likely to be in the k-closest neighborhood. This will drive the sets U_k and V_k to be small in Equations 1 and 2 and thus the trustworthiness and continuity to be large. This is why even relatively poor representations that do not preserve structure, but do preserve distance, also perform well for these two metrics.

Based on the results of these comparisons, the remainder of this section focuses on GDI and preservation. For KPCA, both the GDI and preservation metrics followed similar trends; as the heuristic increased, the quality stabilized to a constant value less than 1. There is some discrepancy for low heuristic values. For the hyperspherical and complex parametric datasets, GDI has a much lower value than preservation as $\sigma \rightarrow 0$. Therefore, the KPCA visualizations of the hyperspherical shell at $\sigma = 0.05$ and $\sigma = 2$ are provided for inspection in Figure 9.

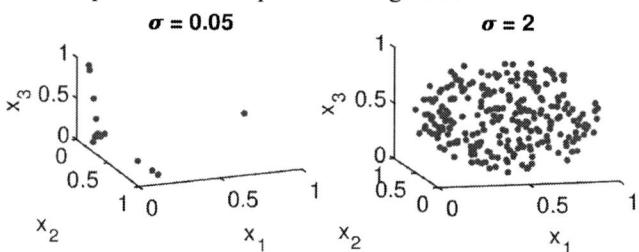

Figure 9: Resulting visualization/dimension reduction of the hyperspherical shell using KPCA for $\sigma = 0.05$ (left) and $\sigma = 2$ (right).

Once again, visual inspection immediately elucidates the quality of the visualizations in which the image on the right of Figure 9 is more spherical (although not a shell) than the image on the left. The left image appears to have collapsed the hyperspherical shell to points on the principal axes demonstrating how drastically heuristics can affect visualizations. The GDI metric provides a better indication than preservation that the right image is of exceptionally poor quality. The GDI is a measure of linear correlation between mappings of distances, so a near zero value of GDI indicates a random mapping with no correlation. Preservation requires a weighted summation in its formulation, which may result in different lower bounds than the conservative limits obtained from the Cauchy-Schwartz inequality. It is hypothesized that the practical lower bounds of the preservation metric may be 0.5 instead of 0, but this must be investigated further. If this is true, the lowest value of each metric was obtained for the poor mapping at $\sigma = 0.05$.

These studies indicate that GDI and preservation are the superior metrics for assessing the quality of a visualization. GDI appears to provide a better assessment of extremely poor visualizations. This may be a consequence of an incorrectly scaled preservation metric such that the lower bound of preservation is actually 0.5. If this is the case, then preservation performs as well as GDI for assessing visualization quality. Furthermore, preservation is exceptionally difficult to optimize due to its formulation which requires not only optimizing the distances between points but also tracking the rank ordering of proximity. GDI or functions closely resembling GDI, though, are optimized for methods such as Isomaps or LLE. They seek to minimize geodesic distances which inherently make GDI biased to indicate Isomaps or LLE as a superior visualizations when compared to other visualization techniques [7, 23, 25]. Preservation does not result in such biases, so for the remainder of this work it will be utilized as the sole metric and integrated into the proposed visualization strategy. In the following section preservation is applied to assess the quality of a new visualization strategy, t-SNE, which will be compared to KPCA for an engineering problem.

6. COMPARISON OF VISUALIZATION TECHNIQUES

Preservation was validated as a viable metric to objectively compare visualization techniques so it will be utilized to compare t-SNE to KPCA within the complete visualization framework proposed in Section 1 and shown in Figure 1. The motivation for comparing t-SNE with KPCA is that KPCA has been used previously to visualize design regions. Therefore, the KPCA approach provides a baseline standard for quality visualizations to which t-SNE can be compared. Furthermore, KPCA is a convex technique while t-SNE is nonconvex and this class discrepancy showcases the generality of the preservation metric.

The complete visualization strategy will be utilized to visualize mappings of design regions for a multilevel materials design problem. Specifically, a set-based design approach generates a mapping of the input design space with multiple clusters or regions of satisfactory designs. These regions are identified through spectral clustering and visualized with both KPCA and t-SNE. The quality of the visualizations is compared using preservation, and the highest quality visualizations are presented for insight into the materials design problem.

The multilevel materials design problem of interest is the design of negative stiffness (NS) metamaterials. In previous work, they have been shown to exhibit simultaneous structural strength and vibration mitigation capabilities [30-32]. They derive these exceptional capabilities by exploiting microstructural mechanisms. The microstructure of interest in this work is the NS inclusion that derives it NS behavior from its curved beams. When the curved beams are subjected to a transverse loading, they exhibit nonlinear stress-strain behavior indicative of NS shown in Figure 10.

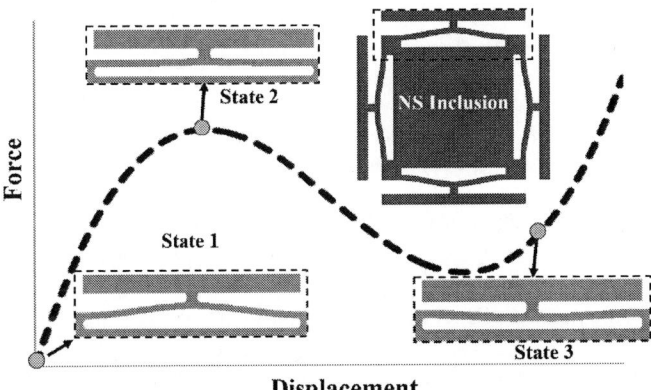

Figure 10: Proposed geometry for a NS inclusion (top right) with associated force-displacement curve for a transversely loaded beam.

The force-displacement curve shown in Figure 10 indicates that the stiffness of the NS inclusions is a function of the transverse displacement. As the inclusion transitions from State 2 to State 3, it traverses a region of negative stiffness in which the structure assists subsequent displacement. By initially prestraining the inclusions to operate in the negative stiffness regime and embedding low volume fractions of NS structures within a viscoelastic material, composites or NS metamaterials with exceptional effective damping and little to no loss in strength can be obtained [32-34]. To obtain a NS metamaterial with these desired properties, an appropriate inclusion geometry and viscoelastic matrix must be properly selected. Identification of the appropriate geometry, initial prestrain, and complimentary viscoelastic matrix is a multilevel design problem. Figure 11 provides an illustration and flowchart describing the multilevel design of the NS metamaterials.

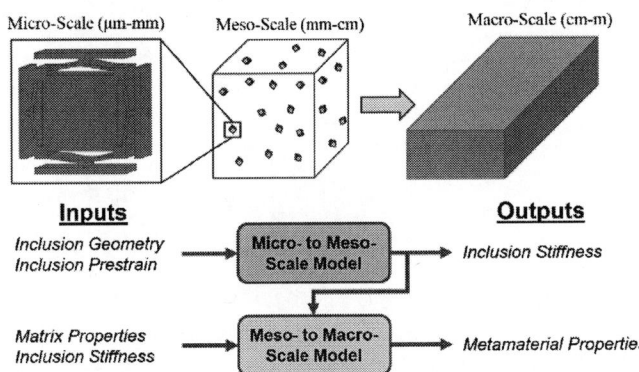

Figure 11: Schematic illustrating the multilevel nature of the design of negative stiffness metamaterials (top). Flowchart illustrating the connectivity of the design and performance spaces (bottom).

In the first design space, or micro scale, the geometry and material properties of the inclusion and its initial prestrain are varied to determine the effective stiffness tensor of the inclusion. The second scale, or the meso scale, is simultaneously a design and performance space in which the stiffness tensor of the inclusion, volume fraction of the inclusions within the matrix, and viscoelastic matrix properties are varied to determine the effective composite properties at the macro scale. To model the relationship between each design and performance space two models—the micro-to-meso and meso-to-macro—were developed. The micro-to-meso model is a finite element, direct energy-based homogenization method that relates the NS inclusion characteristics to its stiffness tensor. The model also evaluates the stiffness as a function of the displacement of the beam. The meso-to-macro model is based on effective medium theory and relates the strain-dependent inclusion stiffness tensor, volume fraction, and matrix properties to the dynamic material properties of the composite or metamaterial [33].

To better define the multilevel design problem, design variables and performance parameters must be defined and the most relevant variables are illustrated in Figure 12. Previously, at the micro scale, the beam thickness, t, apex height, h, and Young's modulus of the NS inclusion, E_{inc}, were varied while the beam length, L, and beam separation, W, were fixed. In this work, a fourth variable is introduced, which is the ratio of beam prestrain, d, to the apex height, $r_d = d/h$. The beams are prestrained in compression, and the geometry of the beam bounds r_d on the interval between zero and two. By introducing a fourth variable, the design space is no longer simple to visualize. At the meso scale the inclusion stiffness tensor, C^{Inc}, is supplied by the micro-to-meso model, the inclusion volume fraction, v_f, is held constant, and the matrix storage modulus, E_m, and loss factor, η_m, are fixed to approximately match the material properties of a polyurethane elastomer [33,35]. Finally, the variables describing the macro scale performance space are the effective storage modulus, E_c^{eff}, normalized by the storage modulus of the matrix, E_m', and the loss factor, η_c^{eff}, of the composite, normalized by the loss factor of the matrix, η_m [8, 14, 18, 27, 36, 37].

The assigned values and ranges at the micro scale are selected to be consistent with two microstereolithography systems. Microstereolithography systems were chosen to manufacture the NS inclusion because they can produce the micron-sized features required for the NS inclusions. The first microstereolithography system uses a photopolymer with a Young's modulus between 0.60 and 0.66 GPa while the second system uses a photopolymer with Young's modulus between 0.72 and 0.77 GPa.

With the variables and parameters defined, the BNC approach is applied to the NS metamaterials design problem to identify regions of the design space likely to lead to satisfactory performance. For details regarding the BNC approach and how it is applied in this work, the reader is directed to Morris *et al.* [18]. A performance threshold is set at the macro-scale, requiring the effective loss factor of the NS metamaterial to be at least 4 times the loss factor of the polymer matrix. This classification is propagated to the micro-scale design space. The micro-scale design space is further classified to restrict the Young's modulus of the inclusion to the bounds of the two photopolymers. This classification based on modulus is forward propagated to the meso and macro-scales.

The BNC classifier was trained to identify the decision boundaries of the satisfactory micro-scale design space(s) in four dimensions. The dimensionality of the design space prevents the results from being visualized directly, but queries of the design space indicated that the promising designs clustered into more than one region of the design space. Therefore, the proposed visualization approach was applied to visualize each of the promising regions of the design space. First, 2000 points were identified that were on or near the decision boundary of each satisfactory design region. A point was considered near the decision boundary if the magnitude of its associated posterior class determinant was less than 0.05. This threshold was arbitrarily selected and smaller values could be used to obtain points even closer to the decision boundary with an accompanying increase in computation time. After identifying points on the decision boundary, the Nyström STSC method was performed (*m = 300*) to identify the number of clusters in each micro-scale design space. Two clusters or regions of promising designs were identified, which most likely corresponded to the different microstereolithgraphy materials used.

Following identification of the design regions using the Nyström modified STSC, t-SNE and KPCA were applied to each of the clusters. For t-SNE the perplexity value was varied from 1 to 2000 and for KPCA σ was varied from 0.01 to 4. The best preservation values for each of the visualizations and clusters is provided in Table 2 with the associated heuristic value. For preservation k was 10% of the number of elements in a cluster.

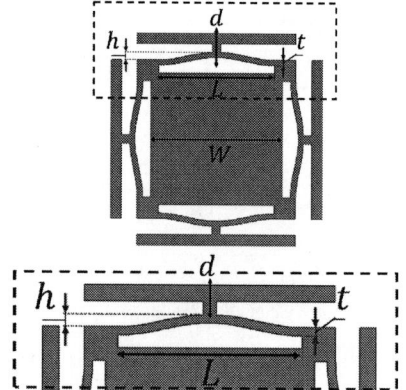

Figure 12: NS Inclusion geometry described parameterized by the shown variables.

	Cluster 1		Cluster 2	
	Heuristic value	Preservation	Heuristic value	Preservation
t-SNE	880	0.78	947	0.81
KPCA	4	0.85	4	0.87

Table 2: Best preservation values and associated heuristics for KPCA and t-SNE visualizations of design regions of a NS metamaterials design problem.

Table 2 provides some insights into the structure of the design space of the NS inclusion. First, it appears that the visualization yielded by t-SNE and KPCA are of similar quality. It could be argued that the KPCA visualization is slightly better than the t-SNE visualization, but the difference is small. Second, the suboptimal value of preservation suggests there are more than 3 design variables that influence the performance of the metamaterial. Based on the author's domain specific knowledge, the result is not unexpected, but if the designer did not have a strong understanding of the design space, these results would provide significant information. To gain more insight into the design space mapping, each cluster was visualized in Figure 13 using the optimal heuristic of t-SNE. Due to the similarity in quality of both visualizations, only the t-SNE visualizations are provided. The t-SNE approach was used to visualize the entire design space, which is also provided in Figure 13. The different colors in Figure 13 are simply used to distinguish the clusters.

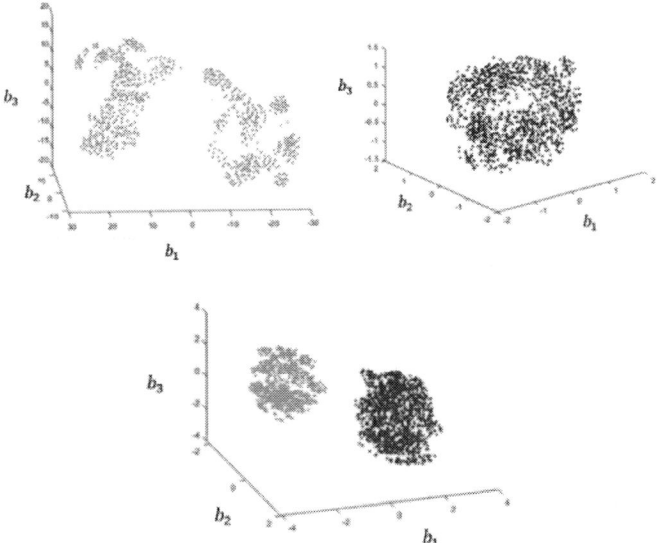

Figure 13: Visualization of each cluster/design region (top left and right) and entire design space (bottom) using t-SNE.

From the global mapping two clusters can be identified in this space, and the local mappings provide high resolution visualizations of each cluster. The structure of the design regions are also quite strange in which the green cluster has a narrowing and the blue cluster has a hole within it. Due to the nonlinear

nature of the mappings, it is unclear how the different topologies correspond to physical geometries. The axes do not represent any physical quantities, rather, they are simply a consequence of the the nonlinear mapping that minimizes the cost function. Furthermore, for the global mapping it is unclear if the different clusters correspond to different material properties.

To help explain what is physically represented by the data, the visualization approach was augmented to allow the designer to select points in each mapping and link them to an image of the associated inclusion. Furthermore, the color of the inclusion indicates the material properties of its base material. A gradient of color from green to blue is used in which blue indicates a lower Young's modulus and green indicates a higher Young's modulus. The value of each of the design variables is also provided for each cluster which allows the designer to see how designs vary across the design space. Figure 14 provides snapshots of the tool illustrating that points on Cluster 1 have higher Young's moduli than Cluster 2.

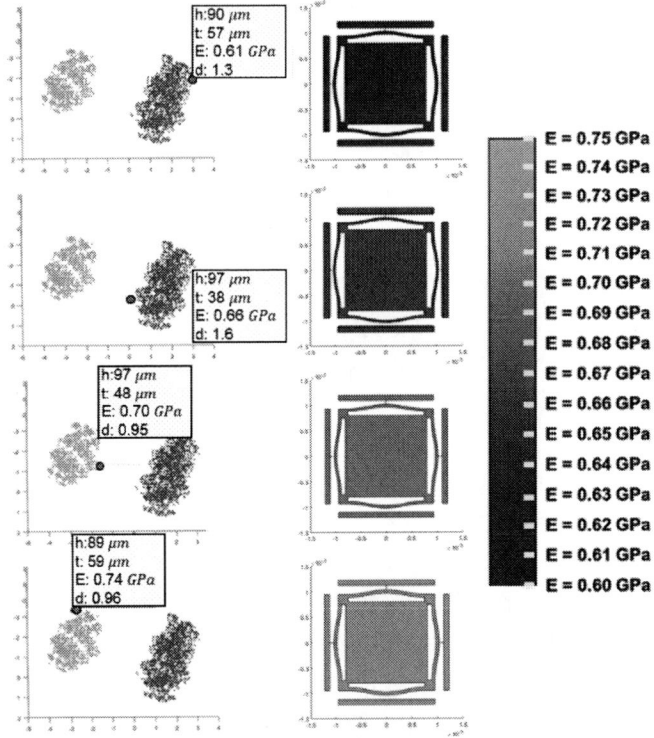

Figure 14: Snapshot of the visualization tool used to clarify how designs vary across nonlinear mappings.

Figure 14 corroborates the notion that the clusters are primarily separated by their Young's modulus due to the color of the inclusion in each cluster. Even more interestingly, the points that are closest to the other cluster have a more similar blue green color. The points farthest from each other are the most different in color. The main limitation of the visualization approach is that t-SNE is a nonlinear mapping, so the meaning of the axes is unclear. Future work could investigate clarifying the meaning of the axes, although this may not always be feasible.

7. CONCLUSIONS AND FUTURE WORK

A visualization strategy was presented with the goal of providing designers with an intuitive low-dimensional visualization of ensembles or sets of design in an inherently high dimensional design space. As part of the strategy, a new metric, preservation, was introduced for assessing the quality of a visualization, and it was compared to other standard metrics. The metric performed at least as well as the standard metrics without being biased towards any particular visualization techniques. As a result, it facilitates comparison of different visualization strategies. A new visualization strategy, t-SNE, was applied to an engineering design problem for the first time.

The newly introduced metric and visualization strategy were integrated into a two-stage visualization approach and applied to a multilevel materials design problem. t-SNE and KPCA performed similarly for the materials design problem. The visualizations provided some insight into the dimensionality and structure of the design space, and highlighted the clustering of promising designs into distinct regions. However, there are some important challenges associated with implementing t-SNE. Primarily, the transformed axes associated with t-SNE do not have a physical meaning which is common for dimension reduction/visualization strategies but poses a major obstacle for designers. Therefore, future work should investigate methods of conveying information regarding nonlinear mappings into easily understandable representations. A first step in this direction was illustrated for the example problem, for which color coding was introduced to overlay physical parameters onto the visualization.

While there are advantages and drawbacks to various nonlinear visualization techniques, the preservation metric is intended to facilitate investigation of a variety of visualization techniques and evaluation of their quality using a single, objective measure. Designers do not need to be experts in visualization to know which approach to utilize if preservation facilitates an informed comparison. However, additional work is needed to investigate optimization of visualization heuristics to maximize preservation. Bayesian optimization could be useful for this purpose.

While this paper was focused on visualizing regions of the design space obtained by set-based design, optimization is still extensively utilized. Therefore, the connection between set-based design and optimization-based methods should be further explored. The mappings obtained by set-based design could potentially provide constraints to optimization-based strategies. Such a hybrid-strategy will require investigation of the computational tradeoffs between sampling the space to explore the decision boundaries and exploiting the space through optimization.

ACKNOWLEDGEMENTS

The authors would like to acknowledge support from the National Science Foundation under Grant No. CMMI -1435548. We would also like to acknowledge Jordan Matthews and Tim Klatt for their past assistance in formulating the metamaterials example problem and our collaborators at Lawrence Livermore National Laboratory, Dr. Christopher Spadaccini and Logan Bekker, for their assistance with microstereolithography. Any opinions, findings and conclusions or recommendations expressed in this material are those of the authors and do not necessarily reflect the views of the sponsor.

REFERENCES

[1] A. Ward, J. K. Liker, J. J. Cristiano and D. K. Sobek, "The Second Toyota Paradox: How Delaying Decisions can Make Better Cars Faster," *Sloan Management Review,* vol. 36, no. 3, p. 43, 1995.

[2] J. Matthews, T. Klatt, C. Morris, C. C. Seepersad, M. R. Haberman and D. W. Shahan, "Hierarchical Design of Negative Stiffness Metamaterials Using a Bayesian Network Classifier," *Journal of Mechanical Design,* vol. 138, no. 4, pp. 1-12, 2016.

[3] C. Morris, L. Bekker, M. R. Haberman and C. C. Seepersad, "Design Exploration of Reliably Manufacturable Materials and Structures With Applications to Negative Stiffness Metamaterials and Microstereolithography," *Journal of Mechanical Design,* vol. 140, no. 11, p. 111415, 2018.

[4] J. H. Panchal, M. G. Fernández, J. J. Christiaan, J. K. Paredis and F. Mistree, "An Interval-based Constraint Satisfaction (IBCS) Method for Decentralized Collaborative Multifunctional Design," *Concurrent Engineering,* vol. 15, no. 3, pp. 309-323, 2007.

[5] R. J. Malak and C. J. Paredis, "Using Support Vector Machines to Formalize the Valid Input Domain of Predictive Models in Systems Design Problems," *Journal of Mechanical Design,* vol. 132, no. 10, p. 101001, 2010.

[6] K. Zeliff, W. Bennette and S. Ferguson, "Zeliff, Kayla, Walter Bennette, and Scott Ferguson. "Benchmarking the Performance of a Machine Learning Classifier Enabled Multiobjective Genetic Algorithm on Six Standard Test Functions," in *Internation Design Engineering Technical Conferences and Computers and Information in Engineering Conference,* Cleveland, Ohio, 2017.

[7] W. Chen and M. Fuge, "Beyond the Known: Detecting Novel Feasible Domains Over an Unbounded Design Space," *Journal of Mechanical Design,* vol. 139, no. 11, p. 11405, 2017.

[8] D. W. Shahan and C. C. Seepersad, "Bayesian Network Classifiers for Set-Based Collaborative Design," *Journal of Mechanical Design,* vol. 134, no. 7, pp. 1-14, 2012.

[9] N. Knerr and D. Selva, "Cityplot: Visualization of High-Dimensional Design Spaces with Multiple Criteria," *Journal of Mechanical Design,* vol. 138, no. 9, p. 091403, 2016.

[10] X. Zhang, T. W. Simpson, M. Frecker and G. Lesieutre, "Supporting Trade Space Exploration of Multi-Dimensional Data with Interactive Multi-Scale Nested Clustering Aggregation," in *International Design Engineering Technical Conference,* San Diego, CA, 2009.

[11] G. Stump, M. Yukish, J. Martin and T. Simpson, "The ARL Trade Space Visualizer: An Engineering Decision-Making Tool," in *Multidisciplinary Analysis and Optimization Conference*, Albany, New York, 2004.

[12] F. V. Paulovich, M. F. Oliveira and R. Minghim, "The Projection Explorer: A Tool for Projection-Based Multidimensional Visualization," in *Computer Graphics and Image Processing*, Minas Gerais, Brazil, 2007.

[13] A. Kusiak and J. Wang, "Decomposition of the Design Process," *Journal of Mechanical Design,* vol. 115, no. 4, pp. 687-695, 1993.

[14] P. J. Newcomb, B. Bras and D. W. Rosen, "Implications of Modularity on Product Design for the Life Cycle," *Journal of Mechanical Design,* vol. 120, no. 3, pp. 483-490, 1998.

[15] J. Jiao and M. M. Tseng, "A Methodology of Developing Product Family Architecture for Mass Custimization," *Journal of Intelligent Manufacturing,* vol. 10, no. 1, pp. 3-20, 1999.

[16] T. Shao and S. Krishnamurty, "A Clustering-Based Surrogate Model Updating Approach to Simulation-Based Engineering Design," *Journal of Mechanical Design,* vol. 130, no. 4, pp. 041101-041114, 2008.

[17] F. Borjesson and K. Höltta-Otto, "Improved Clustering Algorithm for Design Structure Matrix," in *International Design Engineering Technical Conference*, Chicago, 2012.

[18] C. Morris and C. C. Seepersad, "Identification of High Performance Regions of High-Dimensional Design Spaces With Materials Design Applications," in *ASME. International Design Engineering Technical Conferences and Computers and Information in Engineering Conference*, Cleveland, OH, 2017.

[19] L. van der Maaten, E. Postma and J. van den Herik, "Dimensionality Reduction: A Comparitive," *Journal of Machine Learning Research,* vol. 10, pp. 66-71, 2009.

[20] S. Liu, D. Maljovec, B. Wang, P. T. Bremer and V. Pascucci, "Visualizing High-Dimensional Data: Advances in the Past Decade," in *Eurographics Conference on Visualization*, Cagliari, Italy, 2015.

[21] B. Schölkopf, A. Smola and K.-R. Müller, "Kernel principal component analysis," in *International conference on artificial neural networks*, Berlin, Germany, 1997.

[22] J. W. Sammon, "A Nonlinear Mapping for Data Structure Analysis," *IEEE Transaction on Computers,* vol. 100, no. 5, pp. 401-409, 1969.

[23] S. T. Roweis, L. K. Saul and G. E. Hinton, "Glocal Coordination of Local Linear Modals," in *Advances in Neural Information Processing Systems*, Vancouver, 2002.

[24] M. Brand, "Charting a Manifold," in *Advanceing Neural Information Processing Systems*, Vancouver, 2003.

[25] J. B. Tenenbaum, V. De Silva and J. C. Langford, "A Global Geometric Framework for Nonlinear DImensionality Reduction," *Science,* vol. 290, no. 5500, pp. 2319-2323, 2000.

[26] L. van der Maaten and G. Hinton, "Visualizing Data Using t-SNE," *Journal of Machine Learning Research,* vol. 9, pp. 2579-2605, 2008.

[27] G. E. Hinton and R. R. Salakhutdinov, "Reducting the Dimensionality of Data with Neural Networks," *Science,* vol. 313, no. 5786, pp. 504-507, 2006.

[28] J. Venna and S. Kaski, "Local Multidimensional Scaling with Controlled Tradeoff Between Trustworthiness and Continuity," *Proceeding of WSOM,* vol. 5, pp. 695-702, 2005.

[29] W. Chen, M. Fuge and J. Chazan, "Design Manifolds Capture the Intrinsic Complexity and Dimension of Design Spaces," *Journal of Mechanical Design,* vol. 139, no. 5, p. 051102, 2017.

[30] M. R. Haberman, Y. H. Berthelot and M. Cherkaoui, "Micromechanical Modeling of Particulate Composite for Damping of Acoustic Waves," *Journal of Engineering Materials and Technology,* vol. 128, no. 3, pp. 320-329, 2006.

[31] Y. Koutsawa, M. R. Haberman and M. Cherkaoui, "Multiscale Design of a Rectangular Sandwich Plate with Viscoelastic Materials," *International Journal of Mechanics and Materials in Design,* vol. 5, no. 1, pp. 29-44, 2009.

[32] C. S. Wojnar and D. M. Kochmann, "A Negative-Stiffness Phase in Elastic Composites Can Produce Stable Extreme Effective Dynamic but Not Static Stiffness," *Philosophical Magazine,* vol. 94, no. 6, pp. 532-555, 2013.

[33] T. Klatt and M. R. Haberman, "A Nonlinear Negative Stiffness Metamaterial Unit Cell and Small-On-Large Multiscale Material Model," *Journal of Applied Physics,* vol. 114, no. 3, pp. 1-12, 2013.

[34] R. S. Lakes, "Extreme damping in composite materials with a negative stiffness phase," *Physical review letters,* vol. 86, no. 13, p. 2897, 2001.

[35] D. S. Huh and S. L. Cooper, "Dynamic Mechanical Properties of Polyurethane Block Polymers," *Polymer Engineering & Science,* vol. 77, no. 5, pp. 369-376, 1971.

Proceedings of the ASME 2019
International Design Engineering Technical Conferences
and Computers and Information in Engineering Conference
IDETC/CIE2019
August 18-21, 2019, Anaheim, CA, USA

DETC2019-97774

COMPUTATIONAL DESIGN OF 4D PRINTED SHAPE MORPHING MULTI-STATE LATTICE STRUCTURES

Thomas Lumpe, Kristina Shea[1]
Engineering Design and Computing Laboratory
Dept. of Mechanical and Process Engineering
ETH Zurich, Zurich, Switzerland

ABSTRACT

4D printed structures can change their properties and functionalities as a response to a change in the environmental conditions, such as a change in the temperature. A heat stimulus can be used to trigger a transition between two states of a shape memory polymer. Specially designed structures made from these materials can transform into different shapes at different temperatures and can be useful for applications in morphing wings or car panels. Most of these structures, however, are still designed by hand and possess limited load carrying capabilities in at least one of their states. Here, it is shown how complex lightweight structures with multiple stable states can be designed using material modeling and structural optimization methods. By distributing different materials to different parts of the structure, local stiffness gradients are introduced, giving rise to architected global deformations under a single, locally applied load. The shape deformations can be either continuous over the whole structure or discrete only in small regions. The results demonstrate how active materials can be used in a new way to design shape morphing, lightweight lattice structures with different stable states and without sacrificing their structural capabilities.

Keywords: 4D printing; lattice structures; shape morphing; optimization; shape memory polymer

1. INTRODUCTION

Active or smart materials can change their properties in response to external stimuli, such as a change in temperature or pressure. The possibility to 3D print these materials allows to explore the complex geometric design space inherent to additive manufacturing. Combining this with a distinct topological and geometric design enables the development of active structures with properties and functionalities far beyond conventional structures. Recent progress in the design of 4D printed structures includes structures with architected properties such as bistability [1], controlled shape transformations [2]–[4] and self-assembly [5], [6]. These functionalities are conventionally achieved via complex mechanical assemblies of numerous single parts and thus are heavy, expensive and prone to failure. Applications in many fields where weight and cost are important factors for the efficiency of products, e.g. in the aerospace, in the automotive or in the biomedical industry, could benefit from the additional functionalities and unique properties that 4D printed structures offer.

However, most of these structures are still designed by hand and thus are limited by human imagination and experience. To overcome these limitations and to exploit the capabilities that the combination of active materials and geometric freedom via 3D printing offer, computational methods need to be developed. Most of the recent studies focus on optimizing the final shape of structures after thermal or pneumatic actuation by changing the geometry or the material distribution of the initial, as-printed structure [7], [8]. Active structures that exhibit multiple stable states are also investigated [3], but only few studies exist where these structures are computationally designed for multiple stable states [9], [10]. In many cases, the deformations are measured rather qualitatively or the obtained structures consist of compliant materials and do not possess any advanced load carrying capabilities. To be relevant for practical lightweight engineering applications like morphing airfoils or car panels, structures need to be rigid, load-carrying as well as compliant, shape morphing, which is naturally contradictory.

This research seeks to explore the possibilities that active materials offer and combine them with advanced computational design tools to introduce additional functionalities in lightweight

---

[1] Contact author: kshea@ethz.ch

lattice structures allowing for tailoring their shape morphing behavior. More specifically, the contribution of this work is a novel concept that allows rigid lattice structures to temporarily change their shape in a reversible way, while maintaining their structural capabilities. By using established lattice topologies and extensive material knowledge, we ensure that our designs are structurally sound and thus can be applied to engineering problems.

We start by investigating the thermo-mechanical properties of the materials used in this work. From this, we derive a concept that allows a given lattice structure to change its global shape in a predefined way at elevated temperatures. We then formulate the optimization problem that exploits these characteristics. Results for the optimized structures are then presented and discussed.

2. BACKGROUND

Many active structures are based on the combination of materials with different mechanical and thermo-mechanical properties. Multi-material 3D printers allow for designing and printing structures where every voxel (volume pixel) at the microscale can be a different material. The material properties, like strength or stiffness, can vary by orders of magnitude between the different materials. In addition, their response to a change in environmental conditions, like an increase in temperature or humidity, can be different and further extend the range of possible properties. As in classical composite materials, the interaction between different materials can yield unique properties that cannot be achieved by single materials.

One class of material that is often used in active structures are shape memory polymers (SMPs). SMPs exist with different mechanical properties from rigid brittle to rubbery compliant at room temperature and are available for different 3D printing technologies like material jetting, stereolithography and fused deposition modeling. Above the material-dependent glass transition temperature (T_G), the Young's modulus drastically decreases, and large strains can be applied without damaging the material. Cooling the material below the T_G while maintaining the deformations fixes the strains. This process is generally referred to as 'programming'. The material is now in a second stable state, which is different to the initial, as-printed state, due to the programmed deformations. Upon heating the material again above the T_G, the original shape is recovered.

These materials can be used to store energy in deformed structures, which in turn can be used as actuators. Chen et al. [11] combine SMP actuators with bistable mechanisms to obtain structures that can change their shape autonomously on heating. A structure is proposed that can deploy from a flat 2D state into multiple 3D shapes. Since different materials have different actuation temperatures, multiple actuators can be placed in a structure, which then actuate sequentially at different temperatures, yielding multi-state structures or autonomous swimming robots. [12] Each actuator, however, has to be programmed manually every time before the structure can be actuated. For a lattice structure with several hundred or thousand active members, this is not feasible. To avoid this issue,

computational methods can be used to predict the programmed shape of a complex structure at simple loading conditions.

One of the first studies that uses SMPs and computationally designs the shape of structures in the activated state is Maute et al. [9]. Two materials that are at room temperature rigid and compliant respectively, but become both compliant at higher temperatures, are combined in a fiber-matrix arrangement. By applying linear deformations to the composite at higher temperatures, strains are stored and fixed in the originally rigid SMP. At lower temperatures, the mismatch in strains of the two materials leads to a global deformation of the composite structure. Level set topology optimization methods together with a finite element analysis are used to predict the final shape of the composite sheets. This approach enables parabolic, cosine or twisting deformations of the sheets.

More studies focus solely on the shape morphing behavior of structures, not involving any functional materials. Han and Lu [13] investigate how the geometry of lattice structures can be tuned to control their shape morphing behavior. A parametrized reentrant unit cell can have both negative and positive Poisson's ratio, depending on the parametrized angle. A tessellation of this unit cell hence yields local gradients of the Poisson's ratio, which in turn trigger a global deformation on application of an external load. The distribution of the angles in the structure is determined by an evolutionary algorithm. Continuous wave deformations of 2D and 3D structures are achieved. Similar to this, Mirzaali et al. [14] map planar deformations of auxetic and non-auxetic unit cells to predefined regions to obtain elastic deformations of soft cellular materials, which can adapt to complex curvatures of real life objects like a pumpkin or a vase. Both studies use only a single unit cell topology and vary the shape. Since only compliant, elastic materials are used, the morphed state is only temporary and the structures do not carry large loads.

Zhu et al. [15] propose a two stage approach to obtain solid multi-material structures that can morph their shape under external deformations. In a first step, the deformation behavior of unit cell microstructures is precomputed. The microstructures consist of one rigid and one compliant material and are generated with a topology optimization algorithm to cover the biggest possible range of metamaterial properties (Young's modulus and Poisson's ratio). In a second step, the microstructures are mapped to the design domain to obtain prescribed nonlinear deformations subject to an external loading boundary condition. Since the focus of these structures is on the shape morphing behavior, none of them, however, possesses any active properties. Their deformation is purely elastic and their behavior is similar to conventional mechanisms.

An approach that combines both lightweight structures and active shape morphing, is proposed by Weeger et al. [10]. Different materials are assigned to three-dimensional beam structures, where the material can also vary within one beam. This allows for a global nonlinear deformation under specified loading conditions at high temperatures. Using SMPs, this deformation can be fixed by cooling the structure, and the original shape can be recovered upon reheating. However, the structures have only one state with a predictable mechanical response and limited load carrying capabilities due to the use of both rigid and compliant materials. The inhomogeneous material distribution leads to an unpredicted mechanical behavior in the initial state at lower temperatures.

In this study, we aim to design lightweight lattice structures that can change their global shape in a deterministic way under given boundary conditions. Instead of programming single actuators separately as it is shown in many previous works, a simple actuation of the whole structure is desired to facilitate the actuation and reduce the structural complexity. To be applicable to a broad range of applications, a general optimization framework is required that can account for different types of lattice structures. Further, the structures should be load carrying in both states with distinct mechanical behavior and reversibility of the deformations.

3. MATERIALS AND METHODS

This section introduces the different materials used and their thermomechanical characteristics and shows how an optimization framework can be utilized to generate active lattice structures with different deformed shapes.

3.1 Materials and Concept

To overcome the problem of actuating single members in a lattice separately, we introduce a concept that allows the structure to morph into a specific shape upon heating up the whole structure uniformly. We make use of the temperature dependency of the Young's modulus of SMPs and the different glass transition temperatures of different materials of the Objet500 Connex3 3D printer (Stratasys, Eden Prairie, MN, USA). This material jetting process enables printing multiple materials simultaneously. By mixing two base materials in different quantities during printing, twelve digital materials with intermediate properties can be obtained. Figure 1a shows the T_G of the two base materials VeroWhite+ (VW) and Agilus 30 (Agilus), of the twelve digital material obtained by mixing these two base materials, and of a third base material, HighTemp (adapted from [11]).

Four groups can be identified: (1) six materials from Agilus to FLX9885 with a T_G between 0 °C and 10 °C. (2) FLX9895 with a T_G of about 30 °C. (3) Six materials from RGD8530 to VW with a T_G between 55 °C and 60 °C. (4) HighTemp with a T_G of about 70 °C. At room temperature, the materials in the first two groups are already in a rubbery state and are compliant. They all have a low Young's modulus between about 0.5 MPa and 5 MPa at room temperature. The materials in the third and fourth group are rigid at room temperature with Young's moduli between 1500 MPa and 2300 MPa. Figure 1b shows the Young's modulus for VW and HighTemp at temperatures between room temperature (23 °C) and 80 °C. It can be observed that VW and HighTemp have approximately the same Young's modulus at room temperature. Since the T_G is different for both materials, the change in mechanical properties of the two materials also differs with increasing temperature. Until 50 °C, both materials have a similar Young's modulus. At higher temperatures, the Young's modulus of VW significantly decreases, while the

Young's modulus of HighTemp is still orders of magnitude higher due to its higher T_G. The transition between the low and high temperatures is not digital but rather a continuous change. At the respective T_G, both materials are already considerably softer than at room temperature. This yields two materials that have the same Young's modulus at room temperature, but a different stiffness at elevated temperatures.

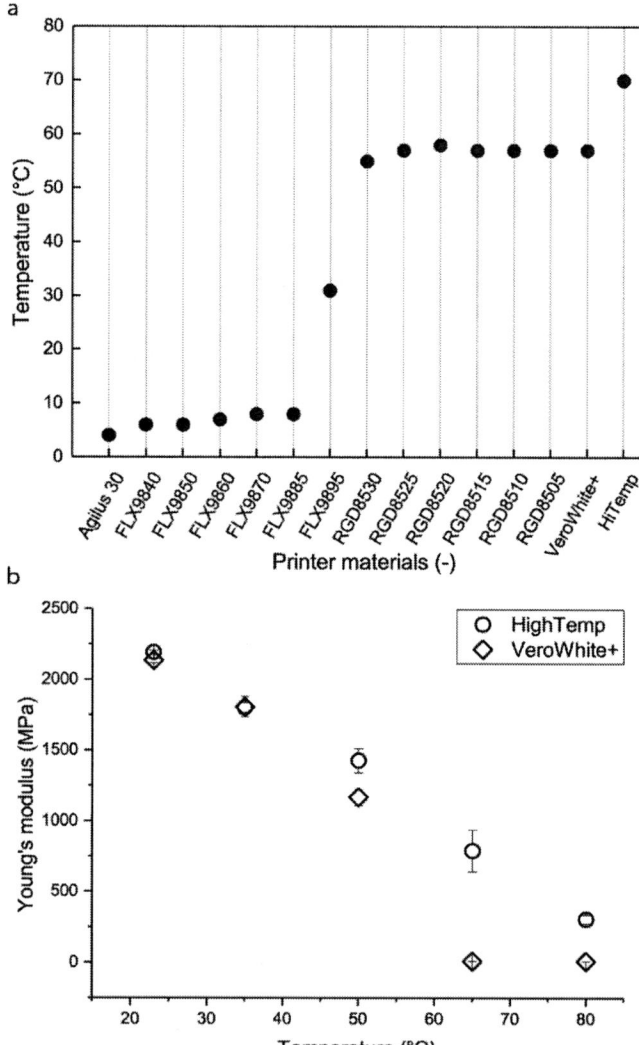

FIGURE 1: (a) Glass transition temperature of materials that can be 3D printed with the Stratasys Objet3 Connex500 3D printer. (Adapted from [11]) (b) Young's modulus of the two materials HighTemp and VeroWhite+ at different temperatures between room temperature (23 °C) and 80 °C.

Instead of using the heating to actuate a structure by recovering the initial state of previously programmed parts [16], we use the different temperature response of the two materials VW and HighTemp to locally reduce the Young's modulus in a structure. Figure 2a shows schematically the topology of a 2D lattice structure. Black lines indicate members with HighTemp material and gray lines refer to members with VW material. At room temperature, both materials have approximately the same Young's modulus. The lattice has a regular structure with a homogeneous stiffness and a clearly defined mechanical response. At 80 °C, the stiffness of the VW members is reduced to ~8 MPa, whereas the HighTemp members have a Young's modulus of ~305 MPa. As members highlighted in black are about 38 times stiffer than the gray shaded members are, the overall mechanical response of the structure is now governed by the material distribution in the structure and can be assumed to be different from the response at room temperature. Thus, the placement of the two materials in the structure can be used to introduce local stiffness gradients on heating the whole structure.

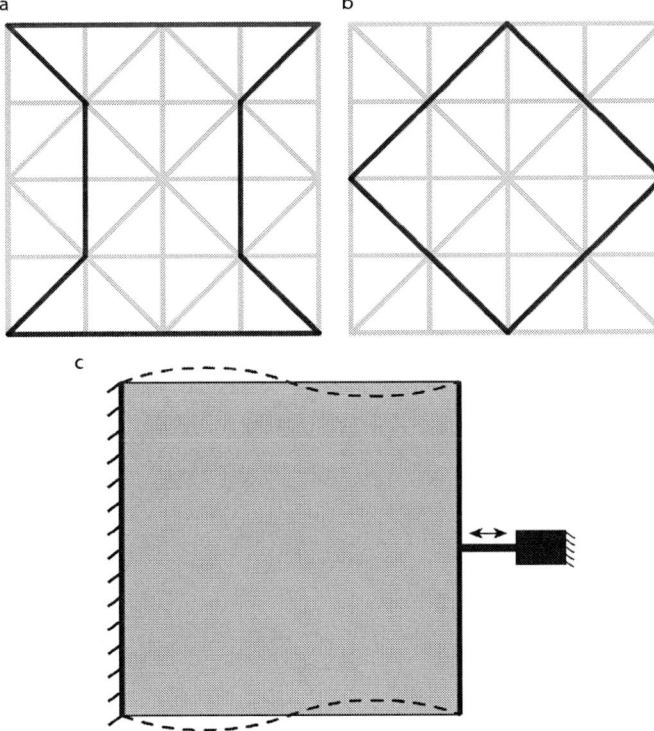

FIGURE 2: (a) A lattice structure with compliant members (grey) and rigid members (black). The rigid members form an auxetic structure with negative Poisson's ratio. (b) The same lattice structure but with different rigid members, which form a structure with positive Poisson's ratio. (c) Schematic of a lattice structure with external actuation. Depending on the internal distribution of compliant and rigid members, a specified global shape, like a sine wave at the top and a negative sine wave at the bottom, can be achieved.

We make use of this concept to design load carrying lattice structures that can morph into a specific shape at elevated temperatures and under a simple external mechanical actuation. Consider the schematic lattice structure in Figure 2a subject to a compressive vertical load. At room temperature, all members have the same Young's modulus and can carry the same load. On heating, the grey members become more compliant than the black members. The structural response is governed by the more rigid, black members. As the black members describe the topology of a 2D reentrant unit cell, the whole structure can be expected to show auxetic behavior (Poisson's ratio $v<0$). Figure 2b shows a different material distribution in the members. At high temperatures, the stiffer members form a rectangular structure that will expand horizontally when compressed in the vertical direction ($v>0$). This simple example shows that it is possible to locally change the Poisson's ratio in a structure on heating by assigning the two materials VW and HighTemp to different members. Figure 2c schematically shows a larger 2D lattice as it could be integrated in a mechanical structure. The grey area is the design space and can be populated by any given lattice topology. The structure is supported at the left boundary and a linear deformation or load can be locally applied at the right boundary, e.g. by a pneumatic actuator. The global deformation at the top and the bottom is governed by the material distribution inside of the structure, i.e. which member consists of which material. Different material distributions yield different global deformations, e.g. the double sine shape indicated by the dashed lines.

As an illustrative example of the whole thermo-mechanical process, Figure 3 shows a discrete lattice structure in all states of the shape morphing cycle. In the first step, the structure is in its initial state at room temperature ($T = 23$ °C). Even though two different materials are used (VW and HighTemp), all members have the same Young's modulus of about 2300 MPa. The structure is rigid and the mechanical behavior is the same as that of a conventional lattice structure with the same shape, size and topology. On heating up the structure to 80 °C in step 2, the Young's moduli of the materials change. Lines drawn in black indicate HighTemp members, which remain stiff, while VW members, indicated by gray lines, become considerably more compliant. In this heated state, the programming deformation is locally applied to the right side of the structure. The non-homogeneous stiffness distribution in the structure at high temperatures causes a non-uniform global deformation. Due to the optimized distribution of materials, this global deformation transforms the structure into the desired shape, e.g. the sine shape seen here in step 3. When the structure is cooled down to room temperature ($T = 23$ °C), the deformations and strains in the material are fixed in step 4. This is enabled by the shape memory properties of the materials, which allow for large programming strains at high temperatures $T > T_\mathrm{G}$ without damaging the material and for fixing these strains on cooling. The structure is now in its second stable state. All members have again the same Young's modulus of about 2300 MPa at room temperature. The overall topology, and thus the load-carrying capabilities, are preserved, only the shape of the lattice changed. Re-heating the structure activates the "memory" of the SMPs and allows for recovering the initial state and shape without any external actuation, making the whole process reversible.

For simple lattices with only few members and simple desired deformations, the material assignment can be done manually (e.g. as in Figure 2a). However, it becomes almost impossible for structures with hundreds or thousands of members and complex desired deformation behavior such as the sine shape in Figure 3. Hence, we introduce an optimization algorithm to find the distribution of the two materials VW and HighTemp for all members in a lattice structure, given the boundary and load conditions and a desired shape in the morphed configuration.

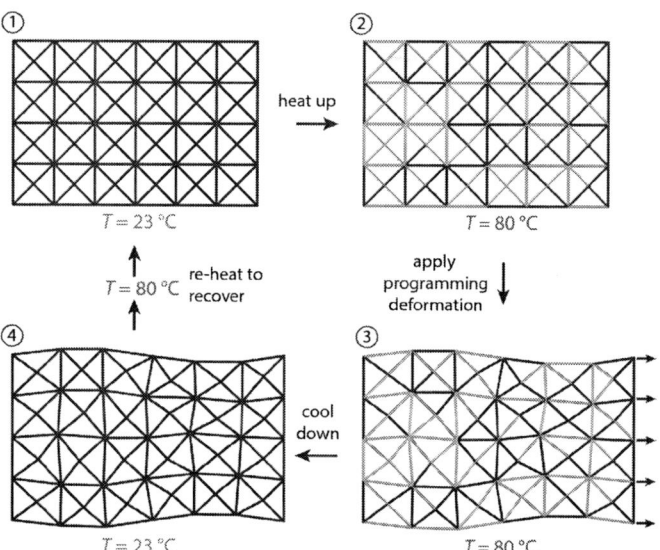

FIGURE 3: All steps of the shape morphing cycle for the "sine" structure. All members have the same Young's modulus at room temperature (1). On heating it up to 80 °C, the stiffness in the members changes according to the different materials (2). The local programming deformation is applied to the right side of the structure and the local stiffness gradients give rise to the global deformation, i.e. the sine shape here (3). The deformations are fixed by cooling down the structure. The structure is now in the second stable, load carrying state. (4) Due to the shape memory properties of the materials, the whole process is reversible and the initial state can be recovered on re-heating.

3.2 Optimization

The general optimization problem is illustrated in Figure 4a. We start with a given lattice structure in the design domain Ω, characterized by the positions of its nodes and the connectivity that describes how the nodes are connected by the members. We use the 2D equivalent of the 3D face centered cubic (FCC) unit cell here as a ground structure. This structure is statically overdetermined, i.e. it has redundant bars in the original state. Choosing an overdetermined topology gives the optimization algorithm the possibility to find a solution for different desired deformations for a single ground structure. Assigning a compliant material to a member relates to removing this member in the context of determinacy, since it elastically releases degrees of freedom at the associated joints. A lattice structure thus can only deform in ways that are "contained" in the original ground structure. In this study, we restrict ourselves to the use of lattices constructed from tessellated unit cells. This makes it straight

forward to illustrate our method and to compare the results, as the topology and the mechanical response are generally more illustrative than in stochastic structures. However, the method is generally applicable to any type of determinant or overdetermined lattice structure.

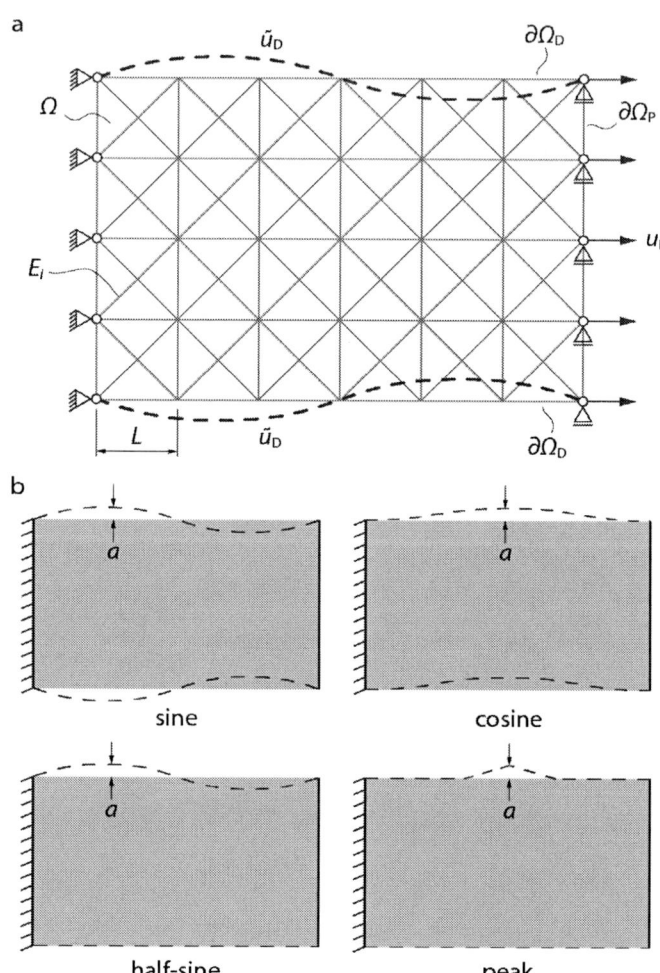

FIGURE 4: (a) Base model of the optimization problem with boundary conditions and desired deformation (dashed line). (b) The four desired deformation modes shown in this study: sine, cosine, half-sine and peak.

The ground structure in Figure 4 consists of 6x4 unit cells with a side length of $L = 17$ mm each. The overall structure has the dimensions 102 mm x 68 mm. After generating the ground structure, we apply a set of boundary conditions to support the structure and suppress any rigid body modes. All nodes at the left are fixed horizontally and vertically, while all nodes at the right still allow horizontal sliding. A prescribed displacement u_p on the nodes in the subset $\partial\Omega_P \subset \Omega$, i.e. all nodes on the right, represents the external actuation. Here, we apply a programming deformation of $u_p = 10$ mm. The desired deformation $\tilde{u}_D$ is defined on the nodes in the subset $\Omega_D \subset \Omega$. To show the functionality of the approach, we consider different desired deformation modes, schematically shown in Figure 4b. All four modes have a maximum desired amplitude of $a = 3$ mm. The design variables are the Young's moduli E_i of all m members, where $i = 1 \ldots m$. As we use the two materials VW and HighTemp, we obtain a discrete problem where each member can only be made of VW or HighTemp, with $E_{VW} = 8$ MPa and $E_{HighTemp} = 305$ MPa respectively. This choice of parameters represents a load case where the structure is actuated at a temperature of $T = 80$ °C $> T_{G,HighTemp} > T_{G,VW}$.

To avoid localized deformations at large stresses and strains in the model, constituting premature failure, a constraint on the maximum strain in each member ε_{max} is enforced. Since the maximum allowable stress of HighTemp at $T = 80$ °C is much higher than the maximum stress of VW, a stress constraint as it is often found in optimization problems would not protect the compliant members in this case. The failure strain, however, is $\varepsilon_{max} = 0.2$ for both materials at $T = 80$ °C, which makes a strain constraint feasible here. In summary, the optimization problem is formulated as:

$$\min_{E_i} f(E_i) = \left(\widetilde{u}_D - u_D(E_i)\right)^2$$
$$s.t. \qquad \varepsilon_i \leq \varepsilon_{max}$$
$$with \qquad E_i \in \left\{E_{VW}, E_{HighTemp}\right\}$$
$$\mathbf{Ku} = \mathbf{f}$$

with u_D as the actual deformation on the boundary Ω_D due to the current choice of design variable and the strain ε_i in the ith member. $\mathbf{Ku} = \mathbf{f}$ is the finite element (FE) equation, which is solved directly and is used to determine the structural response of the lattice. In this study, we use an in-house implementation of a linear-elastic truss FE analysis code written in MATLAB. The chosen parameters yield a nonlinear integer optimization problem with 154 design variables, i.e. the number of trusses, and 154 nonlinear constraints, i.e. one strain constraint for each truss.

An efficient algorithm for discrete design variables is NOMAD, a C++ implementation of the Mesh Adaptive Direct Search algorithm (MADS) under general nonlinear constraints [17]. The NOMAD solver (version 3.8.1) is used via the OPTI toolbox MATLAB interface (version 2.28) [18]. The objective function is evaluated at trial points on a mesh, which is adapted every step. NOMAD was originally designed to solve nonlinear, non-smooth, expensive derivative-free optimization problems with a small number of evaluations. Its possibility to deal with a large number of discrete bounded design variables (up to ~200) efficiently while considering nonlinear constraints makes it suitable to solve the optimization problem in this study. Table 1 provides the non-default parameters used. The option 'ortho 2n' for the search directions is considered the most efficient choice for many problems and is therefore chosen here [19]. This option uses variable orthogonal search directions without randomness. The variable neighborhood search (VNS) metaheuristic perturbs the current iterate and can help to escape local optima. The seed controls the search directions of NOMAD. It is assigned randomly by default. Fixing the value yields a deterministic

search behavior and helps to make the results reproducible. [20]

Table 1: Non-default parameters used for the NOMAD optimization algorithm.

Parameter name	Value
direction_type	ortho 2n
vns_search	0.5
seed	0

The optimization terminates after 30,000 function evaluations or if the change in the objective function between two successive iterations is less than 10^{-7}. The objective function and the constraint function are each evaluated once per iteration. The starting point for each optimization run is chosen randomly. To ensure statistical significance of the results and to account for the randomness in the starting point and the VNS, 10 optimization runs are carried out for every desired deformation mode.

3.3 Prototyping and Testing

For all four desired deformation modes, the structure with the best objective function out of all 10 optimization runs is 3D printed on a Stratasys Objet500 Connex3 material jetting printer. The structures are heated up to $T = 80$ °C in an Instron 3119 environmental chamber and are deformed in the environmental chamber at the same constant temperature with an Instron ElectroPuls E3000 mechanical testing machine. To facilitate the gripping of the specimens, a sheet of HighTemp material with a width of 20.0 mm and a thickness of 2.0 mm is added to the left and right of each lattice. Tensile grips are used to attach the lattices to the testing machine. The deformations in the experiment are measured from images with the image processing software ImageJ2 [21].

4. RESULTS

Figure 5 shows the four optimized structures in the deformed state obtained from the FE analysis. Each plot represents the structure with the best objective function value after 10 consecutive optimization runs. The red lines show the desired deformations with a maximum amplitude of 3 mm. The "sine" structure in Figure 5a is symmetric along the horizontal axis, showing a sine wave at the top and a negative sine wave at the bottom. In the right contracting part of the structure where a positive Poisson's ratio is required, the rigid, black HighTemp members form rhombic structures. In the left part, where a negative Poisson's ratio expands the structure, the black members form partly an auxetic structure.

Figure 5b shows the "cosine" structure, which deforms symmetrically with respect to the vertical axis by shifting upwards at the top and the bottom with a cosine shape. Black members form long diagonal lines from the top right and from the top left to the middle. The "half-sine" structure in Figure 5c remains straight at the bottom and follows a sine shape at the top. Rhombic structures formed by black members can be observed in contracting regions. In contrast to the first three examples, the "peak" structure in Figure 5d does not show a continuous deformation, but a discrete movement of a single node at the top in the middle, while all other nodes do not move upwards or downwards. Below the node in the middle, longer diagonal lines of rigid material can be observed.

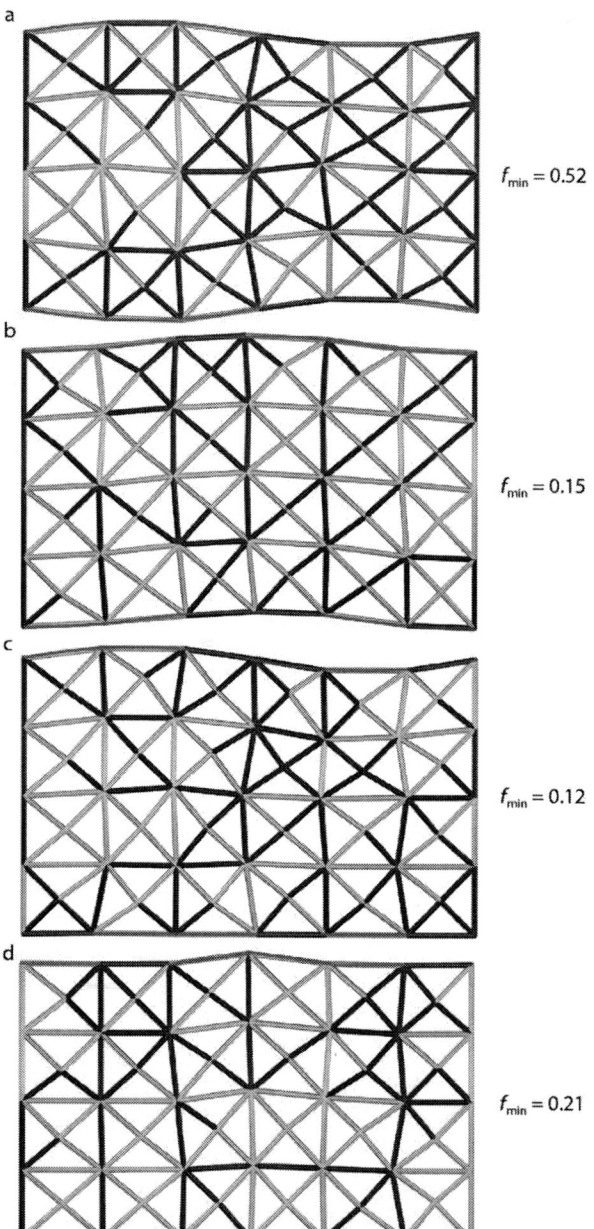

FIGURE 5: (a)-(d) Optimized structures in the deformed state, as predicted with the FE analysis. Black members represent the stiffer HighTemp material and grey members represent the more compliant VW material. Each structure is the one with the best objective function value after 10 consecutive optimization runs.

Figure 6 shows the objective function value over all iterations for the optimized structure in Figure 5a with the minimum objective function value of $f_{min} = 0.52$. Similar convergence behavior is observed for all other structures as well, but with different minimum values of the objective function. The broad band of objective function values shows the explorative character of the search algorithm. At about ~21,500 iterations and ~24,000 iterations, the algorithm detects a minimum and the VNS is activated to escape from this region. This is visible in the two peaks, where the objective function suddenly increases.

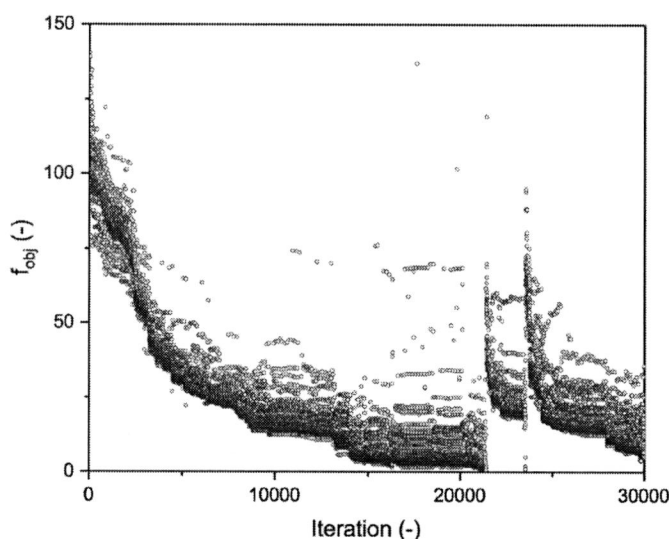

FIGURE 6: Objective function value for 30,000 iterations during the optimization of the structure in Figure 5a.

As the random starting points generate different designs, ten optimization runs are conducted for every desired deformation behavior. Out of the ten runs, the best objective function value (Min.), the worst objective function value (Max.), the arithmetic mean of all ten objective function values and the respective standard deviation are presented in Table 2. All structures achieve a minimum objective function value of $f_{obj} < 1$. The "half-sine" structure has a mean objective function value of less than one, while all other structures have higher mean values. The worst value of the objective function during the ten runs is also considerably lower for the "half-sine" structure than for the other three.

TABLE 2: Statistical results of ten consecutive optimization runs for each structure. The table presents the minimum and maximum value of the objective function that occurred during the ten runs, as well as the arithmetic mean of all ten values and the respective standard deviation.

	Objective function (-)			
	Min.	Max.	Mean	Std. Dev.
sine	0.52	19.63	5.17	5.99
cosine	0.15	16.57	3.44	5.56
half-sine	0.12	3.47	0.93	1.00
peak	0.21	10.96	3.45	3.19

Figure 7 shows the 3D printed optimized structures in the deformed state. The deformed states as predicted by the FE simulation (c.f. Figure 5) are shown transparent in red below the images for comparison. Since both VW and HighTemp have similar colors, the different materials cannot be distinguished visually in the printed structures. The comparison with the simulation shows very good agreement. The desired shapes can be obtained qualitatively with a high accuracy. However, the amplitude of the deformations in the experiments is partly lower than in the simulations.

Table 3 provides the error between the experimental deformations, which are measured visually via ImageJ, and the simulations at the top and bottom nodes. For all four structures, the mean error is below 0.5 mm, with the lowest value for the "peak" structure with 0.17 mm. The maximum error that occurs on a node is significantly higher with up to 1.56 mm for the "cosine" structure. A minimum error of 0.0 in parts of the model, i.e. no visible difference between experiment and simulation, is achieved for all structures. The maximum errors occur locally in regions where the structures expand to the top or bottom, i.e. regions with negative Poisson's ratio, such as the left part of the "sine" in Figure 7a, the top part of the "cosine" in Figure 7b or the top left part of the "half-sine" in Figure 7c. Further, some members that are loaded in compression show buckling in the experiments, which is not observed in the simulations.

TABLE 3: Statistical evaluation of the error between the experimental deformations, measured visually via ImageJ, and the simulated deformations at the top and bottom nodes of the structure.

	Experimental error (mm)			
	Mean	Std. Dev.	Max.	Min.
sine	0.39	0.52	1.55	0.0
cosine	0.42	0.42	1.56	0.0
half-sine	0.37	0.36	1.33	0.0
peak	0.17	0.24	0.86	0.0

5. DISCUSSION

The distribution of the two materials in the structures locally introduce stiffness gradients, which enable the controlled morphing deformations. As seen in Figure 5, the optimization algorithm identifies substructures from rigid HighTemp material that partly resemble known structures with positive or negative Poisson's ratio like the reentrant structure or rhombic topologies. These patterns account for the contraction or expansion in local regions of the model. The material assignment of the other members does not show a distinct pattern but is necessary to further tune the deformation behavior and precisely match the prescribed shapes.

The mean error of maximum 0.42 mm for all four structures confirms the good fit between experiments and simulations. Even though the maximum value that occurs at single nodes is much higher with up to 1.56 mm, the high standard deviations and visual inspections show that these deviations occur only

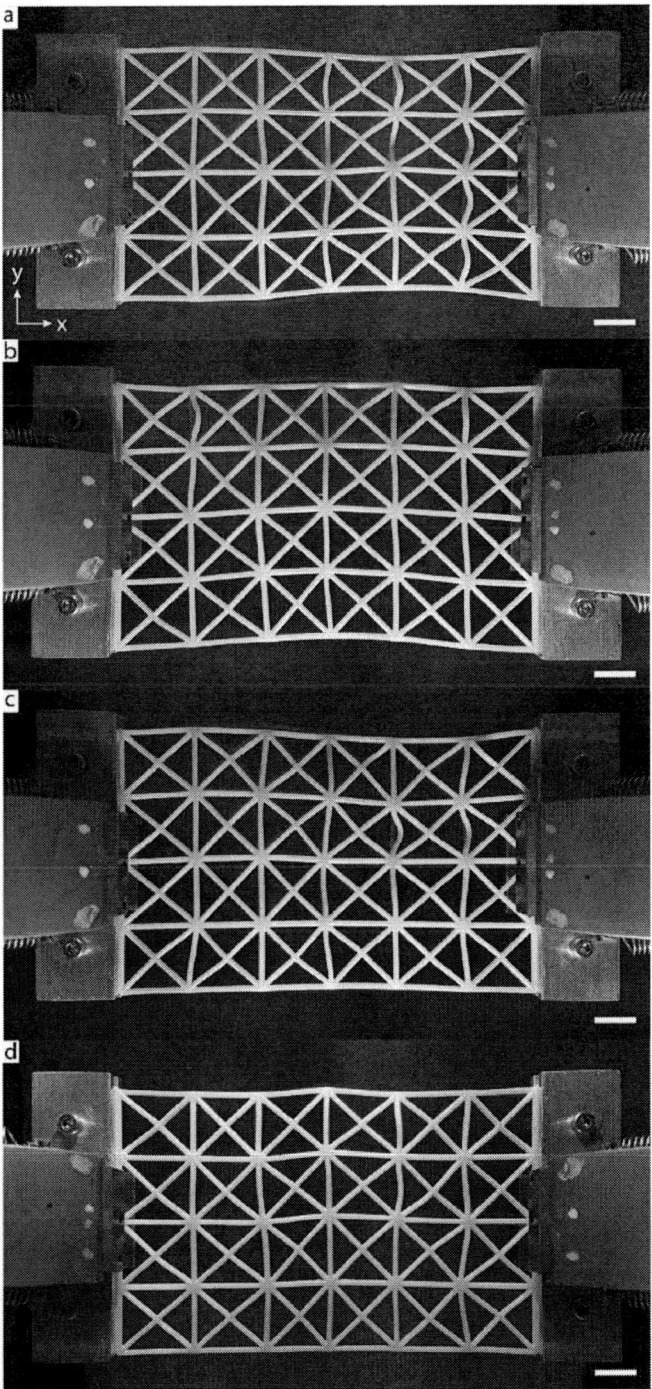

FIGURE 7: (a)-(d) 3D printed optimized structures, deformed at 80 °C in the environmental chamber. The deformed shapes from the FE simulation (as in Figure 5) are shown transparent in red for comparison. The length scales indicated with thick white lines in the bottom right corner of each figure represent 1 cm.

locally in small regions of the structures, whereas other regions show no visible error at all. The large deviations occur mostly in regions where locally a positive Poisson's ratio is required, i.e.

regions that are moving towards the 'outside'. Since these regions need to be 'pushed' outwards, compressive forces act in the members compared to regions that are 'pulled' inwards. Since the model in the experiment is not constrained out of plane in z-direction, manufacturing inaccuracies or small inaccuracies in setting up the experiments can cause the structure to warp out of plane. This is observed during the experiments and can partly account for the deviations between simulations and experiments. Other effects that can influence this discrepancy can be found in the FE modeling of the structures.

The optimization is carried out using a truss FE model that does not account for bending in the members and assumes pin-joints at the vertices. The real structures however are printed with continuous material at the joints, i.e. they are fixed and do not allow rotation, and the members can deform in bending. By limiting our approach to stretch-dominated unit cells only, the modeling can be nonetheless accurate. Members in stretch-dominated cells are mostly loaded in tension and compression, i.e. they do not experience bending moments. Thus, the mechanical behavior can be described by truss elements with sufficient accuracy. Since truss elements are computationally cheaper than beam elements, this choice also enhances the general performance of the optimization routine compared to beam-based methods. The general validity of this modelling approach is confirmed by the overall good agreement between simulation and experiments, as discussed at the beginning of this section.

By assuming a linear-elastic material behavior and small deformations, material nonlinearities and geometric nonlinearities are neglected in this study. The assumption of a linear-elastic material is sufficient since the strains during actuation do not exceed the yield strain of both base materials at 80 °C, as enforced by the strain constraints, and the stress-strain curves show approximately linear behavior in this range. As buckling of members can be observed in the structures, geometric nonlinearities seem to be more important. Comparing Figure 5 and Figure 7, it is visible that only compliant members made from VW buckle. These members have approximately 38x lower stiffness than the stiffer HighTemp members during actuation. Due to this order of magnitude discrepancy in stiffness, the structural response is mostly governed by the stiffer members, only a small contribution of the compliant members can be assumed. As the additional decrease in stiffness after buckling does not change much in the overall stiffness distribution, the nonlinear effects of member buckling can be neglected in the simulation.

Due to the nonlinear objective function and constraints, the optimization problem is inherently non-linear and non-convex, i.e. finding a global optimum is not guaranteed. This further implies that the starting point influences the quality of the solution. This behavior can be observed for this problem. Due to the choice of search direction type and fixed seeding of the VNS, the NOMAD algorithm yields deterministic, repeatable results. The randomness in the starting point generation introduces a stochastic factor to the optimization, resulting in different objective function values and different structures at the end of every optimization run. The low mean value and standard deviation of the objective functions of the half-sine structure indicate that it is easier for the algorithm to find good solutions for problems where only local smooth deformations are required compared to problems where deformations everywhere in the structure (sine/cosine problem) or non-smooth deformations like kinks (peak problem) are required. The low mean values of the objective function for all structures in Table 2 indicate that there exist many local minima with these objective function values, but different distributions of the design variables. The VNS helps the algorithm to escape from these local minima and often finds solutions with lower objective function values. However, only few designs are found with objective function values below $f_{obj} < 0.2$. Even though a global convergence to these low values is often favored, it is not necessary in this case. We consider solutions "sufficient", if the squared error between the desired shape $\tilde{u}_D$ and the actual shape u_D at all nodes is less than 1. Below this error, deviations are almost barely visible anymore and the manufacturing and modeling uncertainties are likely to have a higher influence than the computational error on the final shape of the printed structures. Even though some of the desired deformation modes and the boundary conditions are symmetric, e.g. the "sine" structure in Figure 5a, the distribution of materials in the optimized structure is not exactly symmetric. This further suggests that the solutions found are not global optima, but rather local minima.

A limitation of the current approach is the number of discrete design variables. NOMAD is an efficient algorithm to solve optimization problems with up to about 200 discrete variables but becomes increasingly inefficient for problems with more variables. The simple 2D example presented here already has 154 members. For more complex topologies or 3D structures, the number of design variables increases exponentially. To be able to optimize these types of structures, more efficient modeling and optimization techniques have to be developed.

The advantage of the approach presented in this study over methods where single actuators must be programmed separately lies in the reversibility of the deformation and the simple external actuation. Instead of locally actuating single members within the structure, an external load at a part of the boundary is sufficient to globally deform the structure in a controlled way. The strain constraints avoid solutions where members in a very small area are deformed on actuation but distribute the deformation evenly over the whole design space. This avoids localized stresses, distorts the original cells less and thus preserves the structural integrity even in the deformed, actuated state. The combination of the indirectly actuated shape morphing mechanisms with the functional shape memory materials enables the design of multi-state morphing lightweight structures, which can reversibly deform and are load carrying in both the original and the morphed state.

6. CONCLUSION

In summary, we introduce a novel concept to design lattice structures that are able to actively change their shape while maintaining their structural capabilities. In contrast to conventional, rigid lattices with properties tailored to a specific use case, the structures presented here can deform in a controlled way, changing their shape from one stable state to a second stable state and back. This behavior is achieved by exploiting the difference in thermomechanical properties of materials 3D printed with the material jetting technology. Optimization methods are used to assign different materials to a ground structures, introducing local stiffness gradients at higher temperatures. Upon a global, single actuation load, the inhomogeneous stiffness causes locally varying deformations, which are controlled by the exact placement of the available materials. It is shown that different types of deformations like continuous wave shapes or discrete movements of single nodes can be achieved. The shape memory effect of the SMPs allows for recovery of the initial state, which makes the whole process reversible. Due to the generality of the presented optimization framework, any statically determinant or overdetermined lattice can be used as a ground structure. This makes our approach suitable for all types of high-performance applications where distinct mechanical properties in multiple states of operation are required, e.g. in active airfoils or shape morphing car panels.

ACKNOWLEDGEMENTS

Funding for this research was provided by the ETH Strategic Focus Area Advanced Manufacturing through the project "Sustainable Design of 4D Printed Active Systems".

REFERENCES

[1] T. Chen, J. Mueller, and K. Shea, "Integrated Design and Simulation of Tunable, Multi-State Structures Fabricated Monolithically with Multi-Material 3D Printing," *Scientific Reports*, vol. 7, pp. 1–8, 2017.

[2] M. Wagner, T. Chen, and K. Shea, "Large Shape Transforming 4D Auxetic Structures," *3D Printing and Additive Manufacturing*, p. 3dp.2017.0027, 2017.

[3] J. Wu *et al.*, "Multi-shape active composites by 3D printing of digital shape memory polymers," *Scientific Reports*, vol. 6, no. 1, p. 24224, 2016.

[4] D. Raviv *et al.*, "Active Printed Materials for Complex Self-Evolving Deformations," *Scientific Reports*, vol. 4, no. 1, p. 7422, 2015.

[5] Q. Ge, C. K. Dunn, H. J. Qi, and M. L. Dunn, "Active origami by 4D printing," *Smart Materials and Structures*, vol. 23, no. 9, p. 094007, 2014.

[6] Z. Ding, C. Yuan, X. Peng, T. Wang, H. J. Qi, and M. L. Dunn, "Direct 4D printing via active composite materials," *Science Advances*, vol. 3, no. 4, p. e1602890, 2017.

[7] J. H. Pikul, S. Li, H. Bai, R. T. Hanlon, I. Cohen, and R. F. Shepherd, "Stretchable surfaces with programmable 3D texture morphing for synthetic camouflaging skins,"

Science, vol. 358, no. 6360, pp. 210–214, 2017.

[8] Z. Ding, O. Weeger, H. J. Qi, and M. L. Dunn, "4D rods: 3D structures via programmable 1D composite rods," *Materials and Design*, vol. 137, pp. 256–265, 2018.

[9] K. Maute, A. Tkachuk, J. Wu, H. Jerry Qi, Z. Ding, and M. L. Dunn, "Level Set Topology Optimization of Printed Active Composites," *Journal of Mechanical Design*, vol. 137, no. 11, p. 111402, 2015.

[10] O. Weeger, Y. S. B. Kang, S.-K. Yeung, and M. L. Dunn, "Optimal Design and Manufacture of Active Rod Structures with Spatially Variable Materials," *3D Printing and Additive Manufacturing*, vol. 3, no. 4, pp. 204–215, 2016.

[11] T. Chen and K. Shea, "An Autonomous Programmable Actuator and Shape Reconfigurable Structures using Bistability and Shape Memory Polymers," *3D Printing and Additive Manufacturing*, vol. 5, no. 2, 2018.

[12] T. Chen, O. R. Bilal, K. Shea, and C. Daraio, "Harnessing bistability for directional propulsion of untethered, soft robots," *Proceedings of the National Academy of Sciences*, 2018.

[13] Y. Han and W. Lu, "Evolutionary design of nonuniform cellular structures with optimized Poisson's ratio distribution," *Materials & Design*, vol. 141, pp. 384–394, 2017.

[14] M. J. Mirzaali, S. Janbaz, M. Strano, L. Vergani, and A. A. Zadpoor, "Shape-matching soft mechanical metamaterials," *Scientific Reports*, vol. 8, no. 1, pp. 1–7, 2018.

[15] B. Zhu, M. Skouras, D. Chen, and W. Matusik, "Two-Scale Topology Optimization with Microstructures," *ACM Transactions on Graphics (TOG)*, vol. 36, no. 5, p. 16, 2017.

[16] Q. Ge, A. H. Sakhaei, H. Lee, C. K. Dunn, N. X. Fang, and M. L. Dunn, "Multimaterial 4D Printing with Tailorable Shape Memory Polymers," *Scientific Reports*, vol. 6, no. 1, p. 31110, 2016.

[17] C. Audet, S. Le Digabel, C. Tribes, and V. Rochon Montplaisir, "The NOMAD project." .

[18] J. Currie and D. I. Wilson, "OPTI: Lowering the Barrier Between Open Source Optimizers and the Industrial MATLAB User," *Foundations of Computer-Aided Process Operations*. Georgia, USA, 2012.

[19] M. A. Abramson, C. Audet, J. E. Dennis Jr., and S. Le Digabel, "OrthoMADS: A deterministic MADS instance with orthogonal directions," *SIAM Journal on Optimization*, vol. 20, no. 2, pp. 948–966, 2009.

[20] C. Audet, S. Le Digabel, and C. Tribes, "NOMAD user guide," *Technical Report G-2009-37, Les cahiers du GERAD*, 2009.

[21] C. T. Rueden *et al.*, "ImageJ2: ImageJ for the next generation of scientific image data," *BMC Bioinformatics*, vol. 18, no. 529, pp. 1–26, 2017.

Proceedings of the ASME 2019
International Design Engineering Technical Conferences
and Computers and Information in Engineering Conference
IDETC/CIE2019
August 18-21, 2019, Anaheim, CA, USA

DETC2019-97833

TOPOLOGY DESIGN WITH CONDITIONAL GENERATIVE ADVERSARIAL NETWORKS

Conner Sharpe
c_sharpe@utexas.edu
Mechanical Engineering Department
The University of Texas at Austin
Austin, Texas, USA

Carolyn Conner Seepersad
ccseepersad@mail.utexas.edu
Mechanical Engineering Department
The University of Texas at Austin
Austin, Texas, USA

ABSTRACT

Deep convolutional neural networks have gained significant traction as effective approaches for developing detailed but compact representations of complex structured data. Generative networks in particular have become popular for their ability to mimic data distributions and allow further exploration of them. This attribute can be utilized in engineering design domains, in which the data structures of finite element meshes for analyzing potential designs are well suited to the deep convolutional network approaches that are being developed at a rapid pace in the field of image processing. This paper explores the use of conditional generative adversarial networks (cGANs) as a means of generating a compact latent representation of structures resulting from classical topology optimization techniques. The constraints and contextual factors of a design problem, such as mass fraction, material type, and load location, can then be specified as input conditions to generate potential topologies in a directed fashion. The trained network can be used to aid concept generation, such that engineers can explore a variety of designs relevant to the problem at hand with ease. The latent variables of the generator can also be used as design parameters, and the low dimensionality enables tractable computational design without analytical sensitivities. This paper demonstrates these capabilities and discusses avenues for further developments that would enable the engineering design community to further leverage generative machine learning techniques to their full potential.

INTRODUCTION

Deep learning techniques have exploded in popularity in recent years because of their ability to accurately represent complex data distributions [1]. The deep learning revolution has been motivated in part by computing hardware advances, but algorithmic advances have also been key in pushing the limit of machine learning capabilities. In particular, the development of deep convolutional neural networks has led to rapid progression in the fields of image processing and computer vision [2, 3]. Convolutional neural networks (CNNs) are designed to take advantage of structured data by preserving locality with their operations. This attribute, along with the limited number of parameters compared to fully-connected networks, has enabled deep convolutional networks to represent structured data such as images with unprecedented detail and accuracy. A number of representation learning algorithms have pushed the frontier of image processing in recent years. Variational autoencoders (VAEs) [4] and generative adversarial networks (GANs) [5] provide means of fitting complex high-dimensional data distributions with just a few latent variables. These distributions can then be sampled to generate new unseen results, which is an exciting possibility in the design field.

The structure of image data, in which continuous entities are represented as discretized pixels with distinct intensities, is fundamentally similar to finite element meshes, in which continuous designs are represented as discretized elements with distinct material properties. In particular, topology optimization, in which designers seek to optimize a layout of material distribution within a defined finite element domain relative to some physical performance objective, seems to be a natural fit with convolutional networks. A number of recent publications have sought to leverage the advancements in deep convolutional networks to enhance the topology design process. In [6] the authors utilize CNNs to process intermediate results from the initial iterations of a conventional topology optimization algorithm and predict the final result, thus speeding convergence time without sacrificing significant structural accuracy. In [7], the authors construct a CNN that maps from displacement and loading boundary conditions encoded as image data to a predicted output topology. They also demonstrate a procedure for upscaling the output topology to higher resolution representations. Other recent works have explored the use of deep convolutional networks [8] and fully-connected feed forward neural networks [9] to create mappings to boundary conditions outside of the training set. These publications are valuable enhancements of the existing topology optimization framework, but have slightly different aims than this work.

This paper is focused on the creation of a flexible but compact generative design representation that can facilitate concept generation through visual exploration of designs and enable further design exploration through the manipulation of a small number of latent variables. There have been a few recent works related to this concept. In [10] the authors use a VAE to define a design representation of desired dimensionality. The authors train the model on compliance minimization results, and then perform design in the latent space on a multiobjective heat transfer problem. Their results represent a good illustration of the utility of low-dimensional generative representations in cross-domain design applications, as the authors are able to significantly expand the Pareto frontier utilizing the new generative approach. However, recent studies have shown that the independence assumptions and reconstruction loss functions used in VAEs make them susceptible to blurry or low-quality samples compared to GANs [11]. The work of [12] demonstrates the use of GANs to represent high-resolution material microstructures and perform design in the latent space as well. Our current work is in a similar vein to these approaches, but extends the approach to include prescribed conditions and constraints to focus the design generation process. Specifically, the framework of conditional generative adversarial networks (cGANs) [13] is adopted to encode conditions as inputs that determine the modes of the data distribution that are generated. Ongoing concurrent work from [14] has also explored the use of cGANs to aid topology design, but their results have been limited to a single load case. Here, we consider a wide array of conditions and resulting topologies to evaluate the potential of cGANs in generative design.

The next two sections provide the necessary technical background for cGANs and the topology optimization problem respectively. The following section presents the specifics of the optimization data generation and network training process. Then, results are shown for various settings of prescribed input conditions to demonstrate the topology design tool. A discussion follows to evaluate the results and discuss the many avenues of future work that would further develop the utility of generative machine learning approaches in topology design.

CONDITIONAL GENERATIVE ADVERSARIAL NETWORKS FOR DIRECTED REPRESENTATION LEARNING

Generative adversarial networks [5] are a form of unsupervised representation learning. GANs attempt to learn a mapping from a random distribution of a small prescribed number of latent variables to outputs that match the black box distribution of the training data. Instead of trying to minimize a point-wise reconstruction error such as mean square error, as is done in VAEs, the generator network that performs the mapping instead tries to fool an opposing discriminator network. The

discriminator network takes both authentic training data and "fake" data produced by the generator network as input and tries to determine which inputs are real and which are fakes. The networks are trained in tandem in the minimax game described by Equation 1 where x is the data space, z is the latent variable space, D is the discriminator network, G is the generator network, and E denotes an expected value.

$$
\begin{aligned}
\min_G \max_D V(D, G) \\
= E_{x \sim p_{data}(x)}[\log D(x)] \\
+ E_{z \sim p_z(z)}\Big[\log\big(1 \\
- D(G(z))\big)\Big]
\end{aligned}
\tag{1}
$$

The generator seeks to maximize the discriminator error during training and the discriminator tries to minimize its error until they reach an equilibrium. If the training process has gone well, then at the equilibrium point the generator can produce realistic outputs that match the training data distribution based on random samples of a small number of input latent variables. The probabilistic nature of the mapping yields the ability to generate new unseen outputs that have many of the same characteristics of interest seen in the training set. This attribute could serve as a valuable way for designers to visually explore potential designs using the trained generator.

The original instantiations of GANs utilized fully-connected neural networks. The adversarial training paradigm was later adapted to utilize CNNs in deep convolutional generative adversarial networks (DCGANs) to great effect [3]. CNNs preserve data structure by performing all operations locally. A convolutional filter is placed over a small section of the input data and a nonlinear transformation is applied to produce an output. The filter is then moved by some distance defined by the stride parameter to a new section of the input data, and the procedure is repeated until the entire input has been processed. The parameters of the convolutional layer, including filter size, stride, and data padding, can be utilized to adjust the output size of the layer. In DCGANs, multiple layers are stacked to process inputs and produce the desired output, be it a probabilistic classification for the discriminator or a data sample for the generator. The output of each intermediate convolutional layer can be interpreted as a feature map that represents a different level of abstraction of the data. DCGANs have demonstrated the ability to produce realistic samples of complex images such as human faces or city landscapes.

Conditional GANs expand on traditional GANs by providing additional labels or conditions as inputs to both the generator and discriminator networks as depicted pictorially in Figure 1 [13]. These conditions force the generator to learn to generate samples that mimic the conditional distribution of the data as described mathematically in Equation 2 where y are the conditional variables and all other variables definitions are the same as Equation 1.

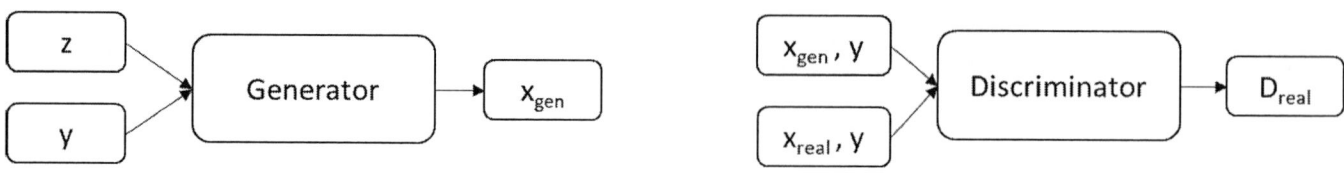

Figure 1. Pictorial of inputs and outputs for the generator and discriminator networks in a cGAN. The latent variables z are combined with conditional variables y as inputs to the generator to output generated data, x_{gen}. The conditional variables, y, are provided along with either actual training data, x_{real} or generated data, x_{gen}, as inputs to the discriminator. The training scheme of Equation 2 forces the networks to learn realistic combinations of data and conditions.

$$\min_{G} \max_{D} V(D, G)$$
$$= E_{x \sim p_{data}(x)}[\log D(x|y)]$$
$$+ E_{z \sim p_z(z)}\left[\log\left(1 - D(G(z|y))\right)\right] \quad (2)$$

As a practical example in the original cGAN paper the authors sought to emulate handwritten digits. The label of the digit was provided as a condition so that the generator would learn to produce samples that look like a 7 when provided with the label corresponding to 7, or a 9 when provided with the label corresponding to 9. In contrast, traditional GANs would learn to generate any realistic digit 0-9 and produce samples consisting of random combinations of digits.

The ability to direct the sample generation process is of potential interest to designers, as relevant physical information such as boundary conditions or material type could then be provided to the network to direct the generator to produce potential designs that pertain to the specific context of the problem. The goal of this paper is to explore this capability by using a set of topology optimized designs as the training set. Structural design for minimum compliance is utilized because of the ease of generating data with this common and computationally inexpensive problem. However, we note that the same process and potential benefits would hold for design outside of the structural domain if sufficient training data were available.

TOPOLOGY OPTIMIZATION OVERVIEW

Topology optimization approaches seek to define a distribution of material that maximizes a performance objective of interest subject to one or more constraints. There are a number of approaches to topology optimization, including level set methods, evolutionary methods, and heuristic methods. Here we focus on gradient-based methods that operate on element-wise approximations of a continuous density field. The Solid Isotropic Material with Penalization (SIMP) method [15] is the most popular version of this approach. The compliance minimization problem is posed in Equations 3-6.

$$\min_{x} c(x) = U^T K U = \sum_{e=1}^{N}(x_e)^p u_e^T k_0 u_e \quad (3)$$

$$\frac{V(x)}{V_0} = vf \quad (4)$$

$$F = KU \quad (5)$$

$$0 < x_{min} \leq x \leq 1 \quad (6)$$

where c is the compliance, x is the material density field, U is the global displacement field, K is the global stiffness matrix, k_0 is the element stiffness matrix, entries with the subscript e represent the elemental versions of their global counterparts, and p is the penalty. The power law approximation of the elemental compliance maintains smooth derivatives in the optimization. The penalty is a positive exponent greater than one to encourage designs to converge to fully solid and void regions by making intermediate volume fractions mechanically inefficient. However, advances in additive manufacturing processes have created the possibility of producing intricate architected materials that can be tiled together to form bulk materials with desirable effective properties [16]. These material architectures, often referred to as mesostructures or unit cells, can be manufactured at a wide range of volume fractions while maintaining mechanical performance. This capability removes the need to artificially penalize intermediate volume fractions; instead one can accurately represent the performance of an architected material throughout a spectrum of relative densities and then let the optimizer choose the density field that is most mechanically efficient. The authors in [17] generate differentiable surrogate models for a number of different unit cell geometries to represent their effective elastic responses. They then perform coarse-scale topology optimization, in which each element in the mesh represents a unit cell of prescribed relative density. They demonstrate optimization results with the same compliance objective function as Equation 3 without the $(x_e)^p$ scaling factor. The primary modification to traditional SIMP algorithms is the incorporation of the density dependence of the constitutive response into the element stiffness matrices as described in Equation 7 where C^h is the homogenized stiffness tensor and G is the finite element shape function gradient matrix.

$$k_e = \int G^T C^h(\rho) G \, dv \quad (7)$$

The results presented in [17] show that optimal coarse-scale topologies vary based on the selected material architecture. Some unit cells benefit from a large number of low and intermediate volume fraction elements that produce a smoother density field to distribute the load over a wider area. Additionally, the paper demonstrates the potential for performance gains when utilizing designed mesostructures as opposed to the fully solid and void designs produced when applying a penalization technique such as SIMP on a coarse-scale mesh of the design domain. The outlined approach is utilized to generate the multiscale topology optimization training data shown in the results section.

METHODOLOGY

In this work, topology optimization data from the standard SIMP approach and the multiscale approach was utilized to train different cGANs. Volume fractions, load locations, and material type (bulk versus architected material) were prescribed as conditional inputs to the network.

The classic 99 line topology optimization code in Matlab from Sigmund [15] was used to generate the training set of SIMP optimization results. The domain was defined as a 2-dimensional rectangular mesh of 32x64 square elements of unit width. The displacements were fixed on the left side to approximate the classic cantilever beam problem with an aspect ratio of 2. The volume fraction was randomly selected from the uniform distribution between 0.2 and 0.4 and stored for network training. A single point load was applied to a randomly selected node in the right half of the beam, yielding 32x33=1056 possible loading locations. The normalized (x,y) coordinates of the load location were stored for inclusion in network training. The x and y magnitudes of the load were randomly selected on the uniform distribution -1 to 1, yielding a random resultant loading direction between 0° and 360°. All other optimization parameters were equivalent to the original paper. A total of 17,500 optimization cases were simulated to produce the training set of topologies, which required about 15 hours of runtime to complete when executed in parallel on a desktop machine with 2 Xeon E5 CPUs with 24 cores each at 3.00GHz speed. The SIMP results were not post-processed to be completely solid or void, which left a small amount of intermediate volume fractions in the designs at the edges of solid features.

Next, the adapted multiscale topology optimization code presented in [17] was used to generate optimization results for an octet truss microarchitecture. The code is intended to perform 3D FEA, so a mesh size of 32x64x1 was used to replicate the SIMP optimization domain shape. The volume fraction, load location, and load direction were varied using the same protocol. A total of 8,400 results were generated, which required approximately 3 days to run on the same machine because of the more computationally expensive 3D FEA.

The topology optimization data was utilized to train cGANs using the Keras API [18]. The architecture of the cGAN generator and discriminator networks is shown in Figure 2. Some of the architecture and training details were adapted from a public implementation of the DCGAN algorithm on MNIST data [19]. The generator consists of an initial dense layer followed by a series of batch normalized convolutional layers with stride 1 and upsampling layers to increase dimensionality until the final shape of the topology is achieved. The discriminator network contains batch normalized convolutional layers of stride 2 to perform downscaling on input topologies. A kernel size of 3x3 is used in every convolutional layer. The output of the series of convolutional layers in the discriminator is flattened and concatenated with the output of a dense layer that processes the prescribed conditions, and the combined data is then mapped through two more dense layers to produce the classification

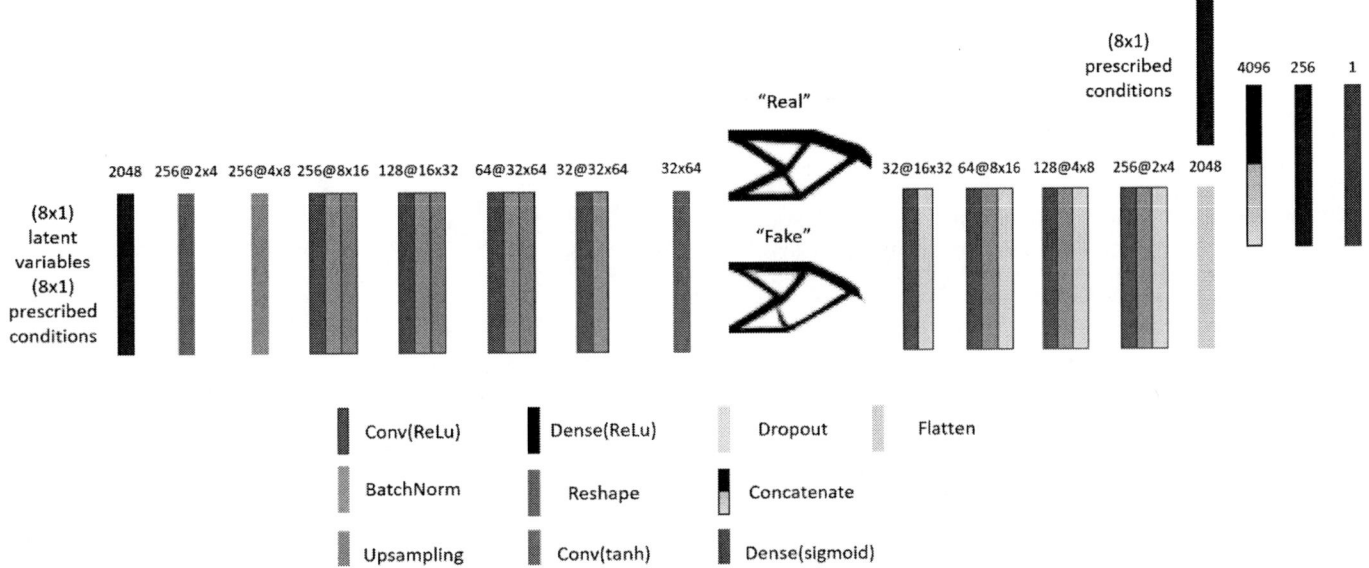

Figure 2. A simple schematic of the implemented cGAN architecture. The output shape of each layer is shown above in the format (# of filters @ filter shape) for the 2D convolutional layers. Activation functions are included in parentheses where appropriate. The generator network takes noise and prescribed condition vectors as input and outputs a topology. The discriminator takes in topologies along with the vector of prescribed conditions and outputs a probability that the topology is from the "real" set of traditional optimization results versus the "fake" set of results produced by the generator network.

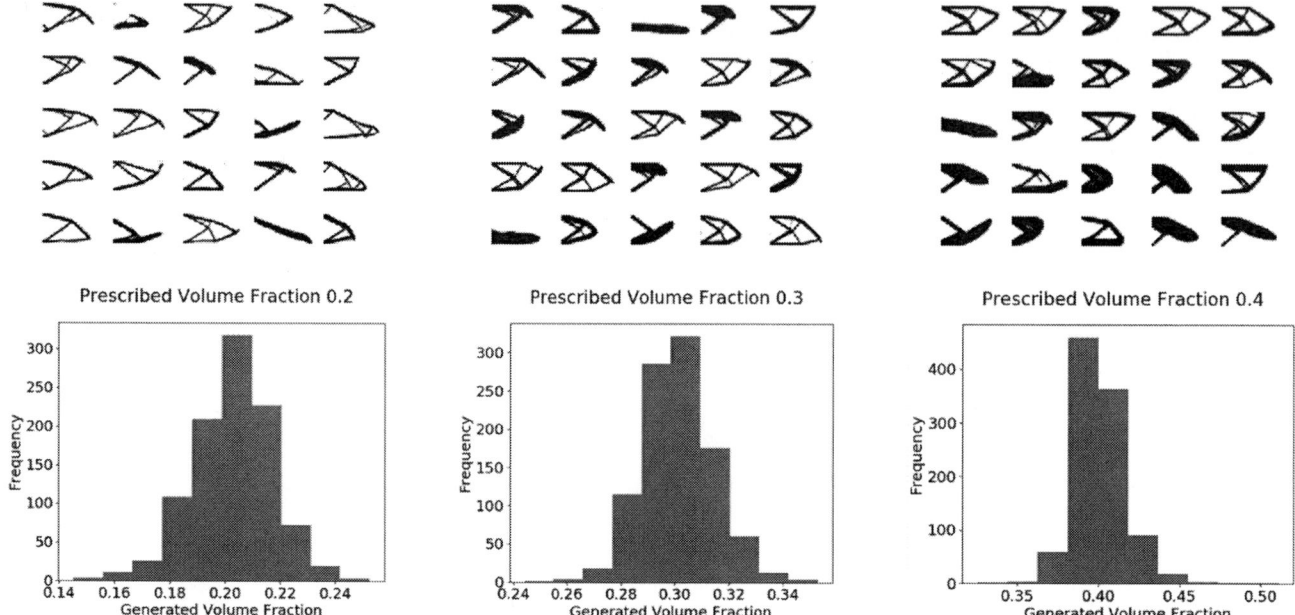

Figure 3. Randomly sampled topologies produced by the generator for different prescribed volume fractions provided as conditional input. The distribution of volume fractions for 1000 randomly generated topologies at each representative volume fraction is shown in the histograms.

output. The dimensionality of the prescribed input conditions is adjusted to match the number of latent variables. For instance, if only a single scalar volume fraction is prescribed the scalar is repeated to form an 8x1 vector of identical entries. The appropriate dimensionality of input conditions warrants further research in the future, but matching dimensions was used in this work because it was found to yield better results empirically. The networks were trained for 20,000 iterations each containing a batch size of 100 topologies. This was an arbitrary stopping point, as developing quantitative metrics for GAN convergence and quality is an active area of research. The output images did not change significantly based on visual inspection from 15,000 to 20,000 iterations, which informed the choice of stopping point. The training process took about 4 hours to complete on an HPC cluster of 4 Intel Xeon Phi 7250 CPUs with 68 cores each at 1.4 GHz.

RESULTS WITH SIMP DATA

An initial test was conducted with only volume fraction included as a real-valued prescribed condition to the networks. Volume fraction is allowed to vary continuously between 0.2 and 0.4. Figure 3 shows random samples from the trained generator for different prescribed volume fractions. A wide variety of topologies are produced, reinforcing the ability of generative networks to quickly sample a space of potentially promising designs to consider. The variety of topologies stems from the range of loading conditions utilized to generate the training data. There is a clear visual correlation between the prescribed volume fractions and the amount of material in the generated topologies.

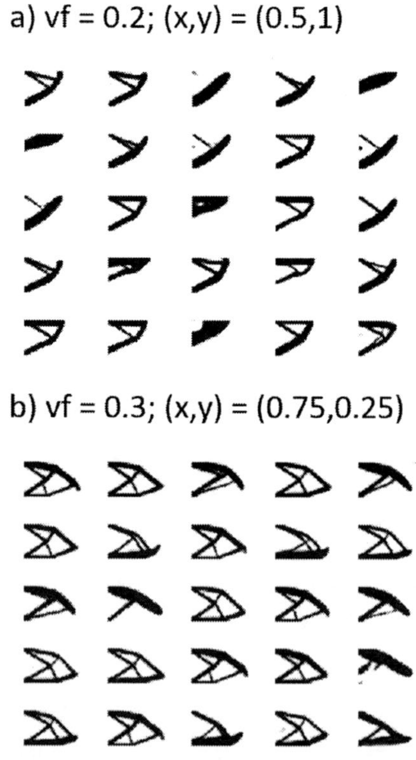

a) vf = 0.2; (x,y) = (0.5,1)

b) vf = 0.3; (x,y) = (0.75,0.25)

Figure 4. Generated results for two representative test cases with volume fraction and load location prescribed as input to the cGAN.

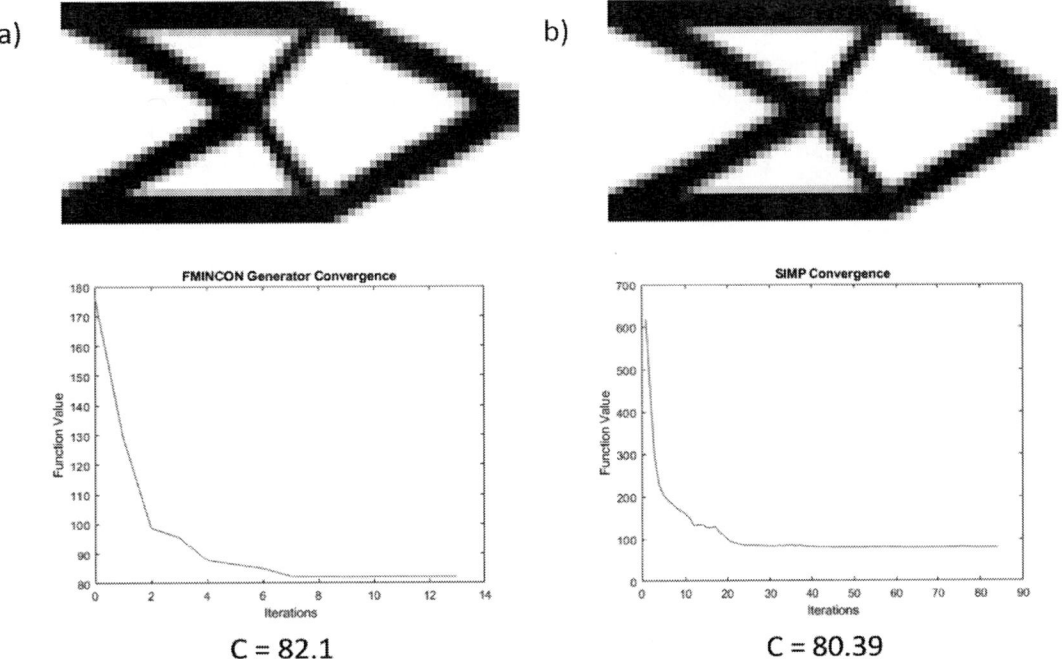

Figure 5. Optimization results for a downward load at the midpoint of the free end of the cantilever beam with a volume fraction of 0.4. Design results and convergence plots are shown for a) design in the 8 dimensional latent space using numerical gradients and b) traditional SIMP design.

To quantify the effect of the volume fraction condition more precisely, histograms were generated from 1000 random samples of the generator at the specified volume fractions. The results in Figure 3 show that the generator produces a distribution centered on the prescribed volume fraction constraint. While the conditional generator does not guarantee results that satisfy the prescribed inequality constraint, it only produces results that are reasonably close to the desired amount of material. The distributions could likely be tightened even further by treating volume fraction as a discrete conditional label with only a few allowable settings during training instead of allowing the constraint to vary freely within the specified range. However, the real-valued condition provides additional freedom in the design phase while still producing results that are clustered around the prescribed condition.

Next, the networks were trained with both volume fraction and load location in normalized (x,y) coordinates provided as real-valued conditions to the networks. The inclusion of load location provides more control over the topology generation process. This trend can be observed in Figure 4, where generated results for two representative cases of different prescribed volume fractions and load locations are shown. The influence of load location is visually clear. There is less variation in topology as compared to Figure 3, as should be expected since location is no longer a degree of freedom. The topologies still vary based on the direction of the resultant load

vector, but some structural load path from the point load to the fixed boundary is present in all of the generated results.

It was hypothesized that the consistent presence of a meaningful load path would allow efficient design convergence in the low-dimensional latent space using numerical techniques. The hypothesis was based on the fact that all changes in generated topology would produce meaningful changes in structural performance due to the change in load path, whereas if the location is not controlled then many completely distinct topologies will perform very poorly due to the lack of a coherent load path for a load applied at a given location. In order to test this, an optimization was conducted with the same compliance minimization objective but using the latent variables of the generator as the design parameters. Equation 8 depicts the problem, where G(z) replaces x as the density field and the rest of the problem is as described previously in Equations 3-6.

$$\min_{z} c\big(G(z)\big) \qquad (8)$$

The fmincon optimization package in Matlab was used to solve the problem with a step size of 1e-6 as the termination criterion. Finite differencing was used to obtain the derivatives of the generator's latent variables. The 8-dimensional input space to the generator thus yielded 9 function calls for each design iteration: 1 for the base function call to evaluate performance of the current design and 8 to obtain numerical derivatives with respect to each of the 8 latent variables. Figure 5 shows the results of the optimization in the latent space for a downward load in the middle of the free end of the beam at a volume fraction of 0.4, along with the SIMP results for comparison. The generator design reached nearly asymptotic error after 7 iterations or 63 function calls. The generator convergence takes fewer iterations but is less efficient overall than the SIMP result, which reaches nearly asymptotic error in about 25 iterations but with only 1 function call per iteration. Less efficient convergence should be expected when comparing with algorithms that utilize analytical gradients. However, the ability to produce an optimal design from only 8 design parameters with a feasible number of function calls could prove more valuable in future design cases of larger scale or of a different class without available sensitivities.

RESULTS WITH SIMP AND MULTISCALE DATA

In order to test the effect of including material type as a conditional input, 10,000 random SIMP results were selected along with 8,400 octet truss results to form a new training set. The cGAN was trained with volume fraction and binary encoded material type (SIMP or octet truss) as conditional inputs. Randomly sampled generated results requested with the SIMP or octet truss label are shown in Figure 6, along with randomly

sampled results from the training set for comparison. The effect of utilizing an architected unit cell is visually clear; the octet truss results contain the same dominant load paths as the SIMP results, but frequently utilize intermediate volume fractions distributed between solid features to smooth the load distribution and improve structural performance. The conditional generator captures this distinction, as results generated with the octet truss label frequently contain this intermediate density smoothing whereas results generated with the SIMP label do not.

DISCUSSION

The results presented here demonstrate the potential of conditional generative network techniques in the engineering design process. The approach can be used to quickly generate large varieties of promising designs that adhere to specified design conditions such as allowable mass fraction or material type. Loading locations can also be specified a priori to further direct the design generation process towards more relevant designs. The cGAN approach is capable of providing flexible yet detailed design representations with single digit design variables for these 2D problems. The low dimensional representation coupled with the ability to provide direct loading condition

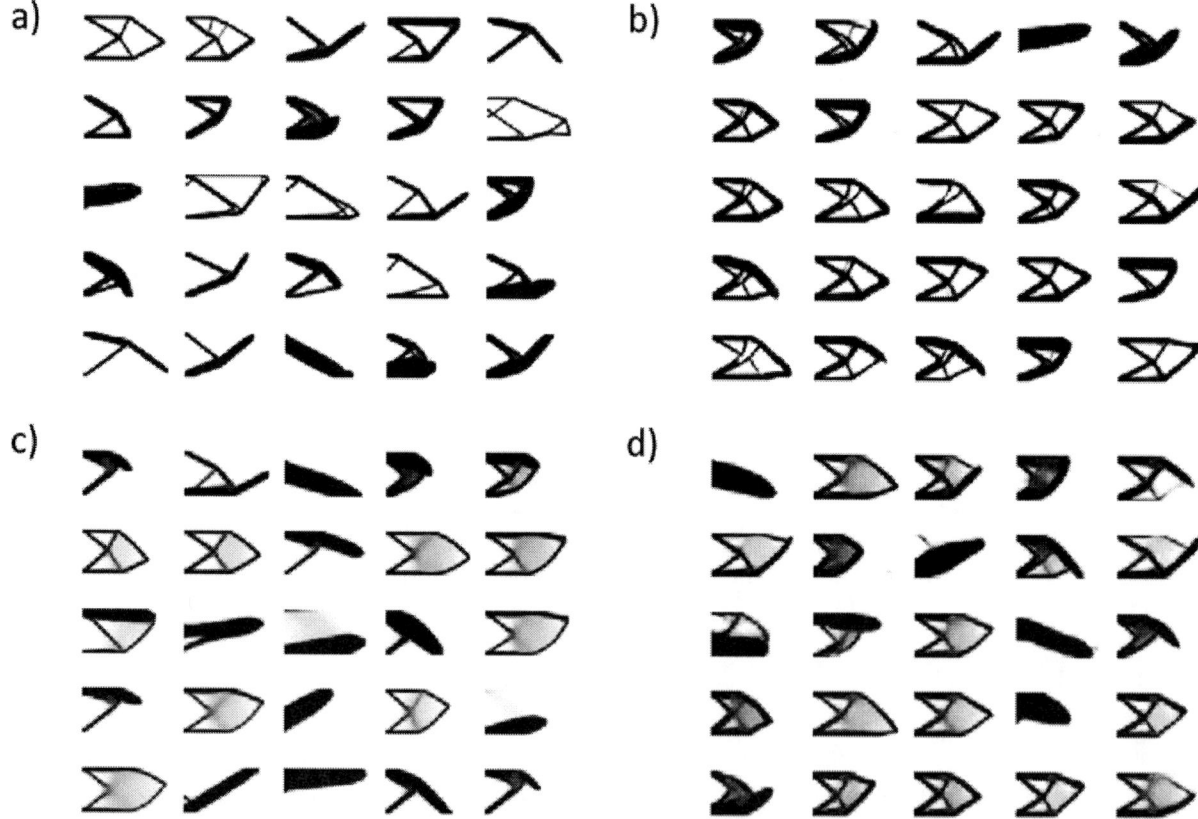

Figure 6. Subfigure a) shows randomly sampled SIMP results, subfigure b) shows random samples of the cGAN with the SIMP label input, subfigure c) shows randomly sampled octet truss optimization results, and subfigure d) shows random samples of the cGAN with the octet truss label input.

information can enable computationally efficient design through numerical techniques without the use of analytical sensitivities. This capability was demonstrated on the same class of structural compliance problems used to train the generative network. In this case the SIMP algorithm was more efficient, but one can envision the generative design parameterization being of greater value for other applications of larger scale or without analytical sensitivities available.

The convergence of the SIMP approach slows considerably for problems with a higher resolution mesh, as the number of design variables grows directly with the number of finite elements in the model. The generative approach does not exhibit this direct scaling, and would thus be of greater utility for more finely resolved design problems. The approach could also be used for multiple related but distinct problem classes. For instance, the generative design tool could potentially be used to efficiently navigate through structurally sound topologies to find designs that are also conducive to desirable heat transfer properties as in [10], or to design structures for electromagnetic performance. The compressed design representation could also be used to design for structural applications with plastic deformation such as impact loading where analytical sensitivities are unavailable; the necessary dominant load paths would automatically be accounted for because of the elastic analysis preceding network training.

Finally, the ability to condition on material type allows designers to quickly visualize how their choice of material would change the resulting topology. The cGAN representation also enables the possibility of efficiently selecting a unit cell type and optimizing the macroscale density field simultaneously with the same generative design tool as opposed to performing separate optimizations for each potential architected material. A computationally efficient multiscale design approach of this nature would provide a valuable step towards realizing the potential performance benefits afforded by modern additive manufacturing. The possibilities enabled by generative deep learning techniques are wide-ranging, but the current state of research still has many shortcomings that must be addressed to realize their full potential.

FUTURE WORK

There are a number of limitations of the generative approach as presented in this work. One issue is that it is very possible for a generator to produce topologies that look visually similar to the training data but contain critical discontinuities in material that cause structural performance to degrade to unusable levels. Incorporating some measure of predicted structural performance into a cGAN network could be a valuable approach to mitigating the occurrence of this issue. Another issue is that the networks are trained on a single domain size of 32x64 elements, and changing this domain size would require further training data and associated computational expense. One approach is to down- or up-scale the mesh resolution using an adversarial approach without any noise to produce a scale mapping as in [7]. However, this requires optimized topologies

at the desired resolution to use for training which implies further optimization expenditures.

The results in this paper are generated on a 2D rectangular design domain. This application is similar to the most common use case of convolutional networks applied to square images, and it is still a significant limitation in engineering design. Volume convolutions can be used to extend the deep convolutional neural network to 3D meshes, as was demonstrated successfully in [6]. Adapting research into irregular meshes to allow generative networks in arbitrary domain shapes would be necessary to move towards more flexible design tools.

Finally, there is a legitimate concern that the significant upfront cost of a large volume of optimizations to produce sufficient training data may be intractable in other design domains with more expensive underlying simulations. In the short term this concern can be partly mitigated by taking advantage of data augmentation techniques to generate cheap additions to the training set. One approach that is intuitive to engineering design is to take advantage of domain symmetry wherever it is available. In this paper symmetry is mirrored over the horizontal midline of the domain, which could be utilized to double the amount of training data by including the mirrored twins for each optimization case. Other translational or rotational symmetries could be utilized depending on the design case. Additionally, augmentation techniques in which small amounts of noise are added have been utilized successfully in previous deep learning image generation work.

In the longer term, the issue of upfront data cost could be mitigated by a concerted effort in the engineering design community to share design data whenever possible. There are hundreds of groups around the world that constantly produce optimization results for problems of different physics, boundary conditions, and mesh resolutions. Creating an organized public repository and encouraging engineers to upload design results and the conditions that produced them would allow these generative techniques to grow in power and applicability as results for problems of interest build over time. Common optimization cases such as those described here could quickly have such a density of examples as to yield further optimizations within the application space largely unnecessary. Robust data channels would also open up the possibility of transfer learning, where pre-trained architectures could be used to quickly build accurate generative design tools for new design applications with lower upfront data generation requirements.

ACKNOWLEDGMENTS

The authors would like to acknowledge Dan Tortorelli, Seth Watts, and Brenda Ng from the Lawrence Livermore National Laboratory for valuable conversation. This work was performed under the auspices of the U.S. Department of Energy by Lawrence Livermore National Laboratory under contract DE-AC52-07NA27344 and with funding from the Defense Advanced Research Projects Agency (DARPA). The views, opinions and/or findings expressed are those of the authors and should not be interpreted as representing the official views or policies of the Department of Defense or the U.S. government.

REFERENCES

[1] Y. LeCun, Y. Bengio and G. Hinton, "Deep Learning," *Nature,* vol. 521, pp. 436-444, 2015.

[2] A. Krizhevsky, I. Sutskever and G. Hinton, "ImageNet Classification with Deep Convolutional Neural Networks," in *NIPS,* Lake Tahoe, Nevada, 2012.

[3] A. Radford, L. Metz and C. Soumith, "Unsupervised Representation Learning with Deep Convolutional Generative Adversarial Networks," in *ICLR,* San Juan, Puerto Rico, 2016.

[4] D. Kingma and M. Wellinig, "Auto-Encoding Variational Bayes," in *ICLR,* Banff, Canada, 2014.

[5] I. Goodfellow et. al., "Generative Adversarial Networks," in *NIPS,* Montreal, Canada, 2014.

[6] S. Banga et. al., "3D Topology Optimization Using Convolutional Neural Networks," *arXiv:1808.07440,* 2018.

[7] Y. Yu, T. Hur and J. Jung, "Deep Learning for Determining a Near-optimal Topological Design without any Iteration," *Structural and Multidisciplinary Optimization,* 2018.

[8] Y. Zhang et. al., "A Deep Convolutional Neural Network for Topology Optimization with Strong Generalization Ability," *arXiv:1901.07761,* 2019.

[9] R. Cang, H. Yao and Y. Ren, "One-shot Generation of Near-optimal Topology through Theory-driven Machine Learning," *Computer-Aided Design,* vol. 109, pp. 12-21, 2019.

[10] T. Guo et. al., "An Indirect Design Representation for Topology Optimization Using Variational Autoencoder and Style Transfer," in *2018 AIAA/ASCE/AHS/ASC Structures, Structural Dynamics, and Materials Conference,* 2018.

[11] V. Dumoulin et. al., "Adversarially Learned Inference," in *ICLR,* Toulon, France, 2017.

[12] Z. Yang et. al., "Microstructural Materials Design Via Deep Adversarial Learning Methodology," *Journal of Mechanical Design,* vol. 140, no. 11, pp. 111416-1-10, 2018.

[13] M. Mirza and S. Osindero, "Conditional Generative Adversarial Nets," *arXiv:1411.1784,* 6 November 2014.

[14] M. Shen and L. Chen, "A New CGAN Technique for Constrained Topology Design Optimization," *arXiv:1901.07675.*

[15] O. Sigmund, "A 99-line Topology Optimization Code Written in Matlab," *Structural and Multidisciplinary Optimization,* vol. 21, pp. 120-127, 2001.

[16] X. Zheng et. al., "Ultralight, Ultrastiff Mechanical Metamaterials," *Science,* vol. 344, no. 6190, pp. 1373-1377, 2014.

[17] S. Watts et. al., "Simple, Accurate Surrogate Models of the Elastic Response of Three-dimensional Open Truss Micro-architectures with Application to Multiscale Topology Design," *Structural and Multidisciplinary Optimization (In Press),* 2019.

[18] F. Chollet, "Keras," GitHub, 2015. [Online]. Available: https://github.com/fchollet/keras. [Accessed January 2019].

[19] E. Linder-Noren, "GitHub," 1 August 2018. [Online]. Available: https://github.com/eriklindernoren/Keras-GAN/blob/master/dcgan/dcgan.py. [Accessed January 2019].

Proceedings of the ASME 2019
International Design Engineering Technical Conferences
and Computers and Information in Engineering Conference
IDETC/CIE2019
August 18-21, 2019, Anaheim, CA, USA

DETC2019-97905

STRESS-CONSTRAINED DESIGN OF FUNCTIONALLY GRADED LATTICE STRUCTURES WITH SPLINE-BASED DIMENSIONALITY REDUCTION

Jenmy Zimi Zhang
jenmy.zhang@autodesk.com
Autodesk Research
Toronto, Ontario, Canada

Conner Sharpe
c_sharpe@utexas.edu
Mechanical Engineering Department
The University of Texas at Austin
Austin, Texas, USA

Carolyn Conner Seepersad
ccseepersad@mail.utexas.edu
Mechanical Engineering Department
The University of Texas at Austin
Austin, Texas, USA

ABSTRACT

This paper presents a computationally tractable approach for designing lattice structures for stiffness and strength. Yielding in the mesostructure is determined by a worst-case stress analysis of the homogenization simulation data. This provides a physically meaningful, generalizable, and conservative way to estimate structural failure in three-dimensional functionally graded lattice structures composed of any unit cell architectures. Computational efficiency of the design framework is ensured by developing surrogate models for the unit cell stiffness and strength as a function of density. The surrogate models are then used in the coarse-scale analysis and synthesis. The proposed methodology further uses a compact representation of the material distribution via B-splines, which reduces the size of the design parameter space while ensuring a smooth density variation that is desirable for manufacturing. The proposed method is demonstrated in compliance minimization studies using two types of unit cells with distinct mechanical properties. The effects of B-spline mesh refinement and the presence of a stress constraint on the optimization results are also investigated.

Keywords: functionally graded lattice, multiscale structures, mesostructural yield, B-splines, stress-constrained optimization

1. INTRODUCTION

Topology optimization of continuum structures and functionally graded materials typically focuses on maximizing stiffness or minimizing compliance, but the design of optimal material distribution with stress constraints has gained momentum in recent years. Existing literature has addressed several inherent challenges in stress-constrained optimization that are not present in purely volume-constrained compliance minimization. Examples include the highly nonlinear and nonconvex nature of the design problem, often associated with singular topologies (cf.[1],[2]), and the locality of the stress constraints, which requires a balance between efficiency and accuracy (cf. [3]-[5]). Most of these investigations are concerned with solid-void topology design of homogeneous structures. When designing functionally graded lattice structures with stress constraints, however, a reliable and physically meaningful prediction of yield is needed for lattice unit cells with intermediate densities between 0 and 1. While homogenization enables the interpretation and realization of intermediate material densities in terms of the effective stiffness of specific mesostructures, the homogenized or macro-scale stress underpredicts the peak stress in the underlying mesostructures. (Here, the term mesostructure refers to the arrangement of material into unit cells that comprise a lattice structure.) The point of initial failure in the architected material depends on the detailed local geometry and material composition of the unit cell, which are not retained during the homogenization process.

This work focuses on stress-constrained multiscale design problems. It differs from existing work in the literature which focus on stress-constrained designs at the mesostructural level (cf. [6]-[8]). These studies aim to design architected materials with reduced stress concentrations to delay the point of failure in the mesostructure given any macroscopic loads. However, they do not enable the efficient prediction of when failure occurs in the mesostructure as a function of the homogenized strains, which is of primary interest to this paper. The ability to do so is needed in the multiscale design of functionally graded lattices composed of these mesostructures.

Various methods have been proposed in the literature to define yielding in materials with intermediate densities for the purpose of structural optimization. In the context of designing solid-void topologies, power-law relationships between stress and material density have been introduced as an interpolation scheme that is designed to alleviate singularities as density approaches zero [2],[4]. Physics-based stress-density relationships have been derived analytically and used to optimize special classes of materials, such as rank-2 sequential laminates in two dimensions (2D) by Duysinx and Bendsøe [9] and then laminates of any rank by Allaire et al. [10]. Lipton and Steubner [11] optimized functionally graded lattices with multiscale stress constraints that connect macro-scale to local-scale stresses for fiber reinforced structures in 2D. Stump et al. [12] obtained stress-constrained designs of functionally graded lattices with arbitrary material compositions using the averaged stress within each phase, but they did not account for the geometry of the material interfaces, which are especially important for architected materials. A generalizable yet accurate way to describe yield at intermediate densities for arbitrary mesostructural architectures has not emerged until very recently. Pasini et al. [13] used on-the-fly homogenization nested within the optimization loop to directly determine the effective stiffness and local peak stresses when designing functionally graded lattices in 2D. A more computationally scalable approach applicable to three-dimensions (3D) was proposed by Cheng et al. [14]. The yield envelopes of metallic mesostructures at intermediate densities were approximated by Hill's criterion. Surrogate models describing the material constants in the Hill's criterion as a function of density were constructed from plastic simulations of the unit cell.

An alternative way to predict yield at the mesoscale is proposed here for the design of functionally graded lattices in 3D with any unit cell architectures consisting of solid and void. The proposed measure of yield differs from [14] in that it can be determined by analyzing the available simulation data for homogenization, thus avoiding additional nonlinear simulations of the unit cell. Furthermore, it is guaranteed to be conservative to the accuracy of the unit cell simulations. It is based on the worst-case stress in the mesostructure, which was introduced in [6] in the context of designing architected materials with improved strength performance. Here, it is used to provide a conservative estimate of meso-scale yield when designing at the macrostructural level.

The stress prediction capability is part of a computationally tractable framework for designing lattice structures for stiffness and strength. Surrogate models capturing the behavior of the mesostructure as a function of density are generated to avoid the difficult problem of analyzing detailed fine-scale features when designing the coarse-scale structure. Furthermore, the coarse-scale density field is parametrized by B-splines. B-splines have been used previously to represent density in topology optimization where the authors demonstrated the ability of splines to provide automatic filtering and length scale control, produce mesh-independent results for multi-resolution design, and handle design dependent loads [15],[16]. The use of splines

also provides control over the smoothness of the density field, which could help mitigate the possibility of large stress concentrations at transitions from nearly solid elements to low volume fraction elements with small strut sizes. Smooth density variation could also aid in manufacturing, where sharp changes in the structure could lead to discontinuities and high temperature gradients in laser-based processes and a corresponding degradation in performance. Finally, operating on spline control points can result in orders of magnitude fewer design parameters than element-wise design approaches. This compact representation enables tractable numerical differentiation for gradient-based optimization, which is of value when using commercial analysis software that makes it difficult to obtain analytical sensitivities.

The procedure for obtaining surrogate models of stiffness and strength is presented in Section 2. The details of the B-spline representation are given in Section 3. Sections 4 and 5 describe the optimization problem formulation and results. Finally, Section 6 concludes with a discussion of the demonstrated results and possibilities for future work to address weaknesses of the proposed method and make further progress in multiscale design.

2. REDUCED-ORDER MODELS FOR THE LATTICE STRUCTURE UNIT CELLS

The development of surrogate or reduced-order models (ROM) to describe the homogenized mechanical response of the mesostructures as a function of mesostructural parameters is necessary for computationally tractable multiscale design. The presence of fine-scale features in a functionally graded lattice requires an exponential increase in meshing and simulation costs if the detailed mesostructural geometries are analyzed directly in traditional approaches to computer-aided design. The approach presented here mitigates the computational burden by precomputing the structural response of defined lattice structure unit cell geometries of interest and capturing the response using simple surrogate models. The constitutive properties for each element in the finite-element mesh of the macroscale lattice can then be defined by querying the cheap surrogate models for the homogenized mechanical properties of the unit cell of interest at the prescribed macroscale relative density. This approach maintains reasonable physical accuracy with much cheaper finite-element simulations.

2.1 Design Representation using Geometric Projection

Training data for the surrogate models is obtained via finite-element analyses of the unit cells, the geometries of which are represented using the geometric projection approach described in [17],[18]. Rod primitives are used to define the struts of the unit cell by the 3D coordinates of each endpoint and the rod diameter, thus providing mesh-independent means of defining the unit cell designs. To analyze the structural performance of the unit cell, the rods are projected into an orphan mesh of 80^3 3D continuum finite elements, yielding a total of 512,000 elements. In the finite element projection, the geometry is described implicitly by assigning appropriate material properties to each element. This is done by sampling around the centroid of each element in a spherical volume that circumscribes the cubic brick element in order to determine the fraction of the element that lies within the boundaries of rod primitives. Elements that lie completely within the rods describing the unit cell geometry are assigned a volume fraction of 1 and the Young's modulus of the constituent material of the rod. Elements that lie completely outside of the geometry are considered to be void space and assigned a Young's modulus many orders of magnitude below the constituent material, although still non-zero to maintain numerical stability of the finite element simulation. Elements that lie partly within the geometry are assigned a Young's modulus that linearly interpolates the Young's modulus of the constituent material proportionally to the volume fraction of the element. The analysis presented here uses the same approximate interpolation approach as [19] where elements are separated into discretized volume fraction bins to make writing the analysis input file in the commercial FEA software ABAQUS more tractable. The accuracy of the design discretization through geometric projection and the resulting finite-element simulations increase with finer mesh resolution, but must be balanced against the associated computational expense. The mesh resolution used here was deemed a reasonable trade-off between accuracy and expense in light of previous geometric projection work such as [18]. However, adaptive mesh refinement and exact interpolation approaches would yield further increases in the accuracy of the analyses and the resulting surrogate models of homogenized unit cell properties. The use of the geometric projection method to represent designs incurs an additional computational cost compared to voxelized design representations, but the drastic reduction in design parameters and enforcement of simple geometric primitives that are known to be manufacturable make it suitable for our application.

2.2 Constitutive Relationship

Surrogate models for the constitutive relationship seek to capture the effective mechanical response of the mesostructures in the context of a macroscale lattice. This work utilizes asymptotic expansion homogenization (AEH) to estimate the homogenized constitutive tensor within a lattice [15]. AEH utilizes the periodicity of cellular lattices to estimate homogenized bulk properties at the lattice scale as shown in Equation 1,

$$\mathbb{C}_H = \sum_{ij} \frac{1}{|\Omega|} \int \mathbb{C}(\nabla \chi_{ij} \otimes E_{ij}) dv, \qquad (1)$$

where the homogenized constitutive tensor $\mathbb{C}_H$ is computed from constituent material properties $\mathbb{C}$, microscale corrector displacements χ_{ij}, and microscale strains E_{ij}. The derivation of this equation is excluded here for brevity, but readers are encouraged to read [20],[21] for further details. Six forward simulations are solved for the six independent unit test strains (three normal strains and three shear strains) to obtain the necessary microscale results for Equation 1. These test strains

utilize periodic boundary conditions that are applied in ABAQUS following the procedure outlined in [22].

Several different rod diameters were sampled to span a range of relative densities from approximately 2% up to 100% or fully solid. The homogenized constitutive tensor was obtained at each relative density. The data was then fit using a weighted least-squares approach where the relative error of each entry was minimized instead of the absolute error to avoid washing out the results at low volume fractions. Each of the three independent elements of the cubically symmetric constitutive tensor for the unit cell geometries of interest were fit using independent third-order polynomial functions. These surrogate models were then included in user-material definitions in the ABAQUS subroutine UMAT to define the material properties for each unit cell geometry.

2.3 Model of Yield Stress

The unit cell is assumed to yield when the peak von Mises stress in the unit cell exceeds the yield stress of the constituent material. The yield surface of a specific unit cell geometry is a function of the relative density as well as the magnitude and direction of the homogenized stress state. An isotropic yield model is proposed here based on the worst-case stress in the microstructure. This approach has the advantage of being simple yet conservative by avoiding the difficulty associated with capturing the complex anisotropic shape of the yield surface in a high-dimensional stress space. At the same time, it can be obtained from analyzing the linear elasticity simulation data available from the homogenization process discussed in Section 2.2.

The worst-case stress, λ, for a mesostructure with a fixed relative density is defined as [6]

$$\lambda(y^e) = \max_{\|\hat{S}_M\|=1} \sigma_m(\hat{S}_M, y^e), \qquad (2)$$

where σ_m is the von Mises stress at a point y^e in the mesostructure. In other words, the worst-case stress is the largest von Mises stress at point y^e given any possible homogenized stress, $\hat{S}_M$, with a norm of 1. Simplifying the subsequent discussions with vector notations for the stress and strain, it can be shown that the worst-case stress in fact points to the largest eigenvalue of a 6-by-6 matrix $\mathbb{T}_v$ given by

$$\mathbb{T}_v(y^e) = \mathbb{C}_H^{-1} \mathbb{G}(y^e)^T \mathbb{V}^T \mathbb{V} \mathbb{G}(y^e) \mathbb{C}_H^{-1}. \qquad (3)$$

where $\mathbb{C}_H$ is the homogenized stiffness tensor (in matrix form) obtained in Section 2.2, and $\mathbb{V}$ is another 6-by-6 matrix defined such that $\sqrt{2/3}\, \mathbb{V}\sigma$ gives the deviatoric stress of any stress vector σ. The tensor $\mathbb{G}(y^e)$ maps from macro strain to the mesostructural stress at y^e, and is the quantity integrated in Equation 1 to obtain $\mathbb{C}_H$ as follows:

$$\mathbb{C}_H = \frac{1}{|\Omega|} \int \mathbb{G}(y^e) dv. \qquad (4)$$

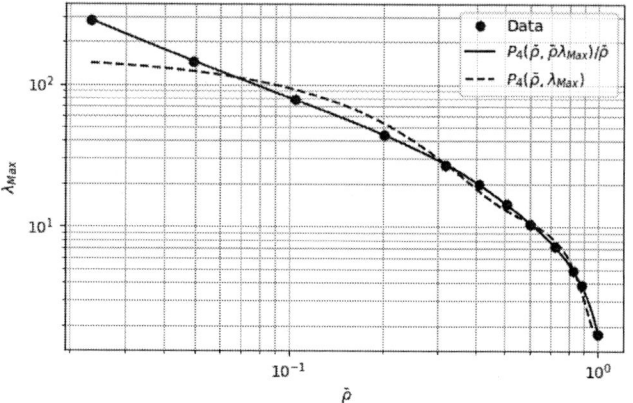

Figure 1 WORST-CASE STRESS ROM FOR THE OCTET TRUSS USING A CUBIC POLYNOMIAL PLUS A $1/\bar{\rho}$ TERM (0.82% MEAN RELATIVE ERROR) AND A FOURTH ORDER POLYNOMIAL (13% MEAN RELATIVE ERROR).

The readers are welcome to reference [6] for additional details. For any S_M whose norm is not 1, Equation 3 defines a quadratic mapping between $S_M = \|S_M\|\hat{S}_M$ and σ_m such that

$$\sigma_m^2(S_M, y^e) = S_M^T \mathbb{T}_v(y^e) S_M = \|S_M\|^2 \hat{S}_M^T \mathbb{T}_v \hat{S}_M. \tag{5}$$

It can further be shown that

$$\max_{S_M} \sigma_m(S_M, y^e) = \|S_M\| \lambda(y^e). \tag{6}$$

This means the local von Mises stress at any point y^e in the mesostructure is at most λ times the magnitude $\|S_M\|$ of the homogenized stress S_M. A single worst-case stress measure is then determined for the entire unit cell by computing λ_{Max} as the largest $\lambda(y^e)$ over all y^e. For a given relative density, λ_{Max} indicates the largest von Mises stress that can possibly be present anywhere in the unit cell as an amplification factor of the homogenized stress magnitude. It is independent of the direction of S_M. The resulting yield model is therefore isotropic. However, the worst-case stress has merit in its simplicity in that it can be described by a scalar function of the relative density. More importantly, it provides a conservative estimate of the homogenized stress magnitude, below which designers can be certain that the peak von Mises stress in the mesostructure is below the yield stress of the constituent material.

To construct a ROM for the worst-case stress, let $\bar{\rho}$ denote the relative density. Then for each $\bar{\rho}_j$ where there is simulation data for the constitutive ROM, the worst-case stress measure is obtained in following manner: the matrix $\mathbb{T}_v(y^e)$ is assembled for every point y^e on the unit cell mesh; the largest eigenvalue $\lambda(y^e)$ of $\mathbb{T}_v(y^e)$ is then computed; $\lambda_{Max}(\bar{\rho}_j)$, which is the maximum $\lambda(y^e)$ over all y^e on the mesostructure, is finally determined. Once all the $(\bar{\rho}_j, \lambda_{Max}(\bar{\rho}_j))$ pairs are gathered, the weighted least-squares procedure from Section 2.2 is used to fit the data. The ROM consists of a third-order polynomial and an additional basis proportional to $1/\bar{\rho}$, which is introduced to capture the rapid growth of λ_{Max} as $\bar{\rho}$ goes to zero. The ROM is obtained by fitting a fourth-order polynomial to $(\bar{\rho}_j, \bar{\rho}_j \lambda_{Max}(\bar{\rho}_j))$ and lowering the degree in each term by 1. Figure 1 shows that the proposed form of the ROM improves the quality of the fit compared to a fourth-order polynomial ROM.

3. MATERIAL DISTRIBUTION VIA B-SPLINE

Tensor-product B-spline surfaces (cf. [23]) are used to distribute the lattice structure over the macro-scale geometry of interest. Although this paper focuses primarily on distributions of $\bar{\rho}$, distributions of other material properties can be parameterized in a similar manner. The value of $\bar{\rho}$ at any point on the domain can be described by the height of a two-dimensional open B-spline surface at the corresponding location, which is given by

$$\bar{\rho}(u, w) = \sum_{i=1}^{m} \sum_{j=1}^{n} \bar{\rho}_{i,j} N_i^k(u) N_j^l(w). \tag{7}$$

In Equation 7, $u, w \in [0,1]$ represent the parametric coordinates of the point where $\bar{\rho}$ is evaluated, m, n indicate the number of B-spline control points, $\bar{\rho}_{ij}$ is member of an $m \times n$ array of control points, and $N_i^k(u)$ and $N_i^l(w)$ are basis functions with degrees k and l, respectively. The parametric coordinates are further determined from the physical coordinates (x, y) as follows:

$$\begin{bmatrix} u \\ w \end{bmatrix} = \begin{bmatrix} x/x_{Max} \\ y/y_{Max} \end{bmatrix}, \tag{8}$$

where x_{Max} and y_{Max} for a rectangular domain are the maximum dimensions in x and y, respectively. For a non-rectangular domain, x_{Max} and y_{Max} will be the dimensions of the smallest bounding box, whose lower left corner is at the origin. The array of control point values dictates the shape of the surface, and they are designated as design parameters for the optimization.

Quadratic B-splines are used for the results in this paper, where $k = l = 3$. To evaluate Equation 7, the knot intervals to which u and w belong are first identified. There are $m - 2$ and $n - 2$ knot intervals, respectively, for the two parametric coordinates, which are distributed uniformly between 0 and 1. Let s and t denote the indices of the intervals containing u and v, and let $u_s, w_t \in [0,1]$ be the local coordinates of u and w within their respective intervals, then Equation 7 is equivalent to the following matrix equation:

$$\bar{\rho}_{st}(u_s, w_t) = \frac{1}{4} \begin{bmatrix} u_s^2 \\ u_s \\ 1 \end{bmatrix}^T \mathbb{M} \mathbb{P}_{st} \mathbb{M}^T \begin{bmatrix} w_t^2 \\ w_t \\ 1 \end{bmatrix}, \tag{9}$$

where

$$\mathbb{M} = \begin{bmatrix} 1 & -2 & 1 \\ -2 & 2 & 0 \\ 1 & 1 & 0 \end{bmatrix}, \tag{10}$$

$$\mathbb{P}_{st} = \begin{bmatrix} \bar{\rho}_{s,t} & \bar{\rho}_{s,t+1} & \bar{\rho}_{s,t+2} \\ \bar{\rho}_{s+1,t} & \bar{\rho}_{s+1,t+1} & \bar{\rho}_{s+1,t+2} \\ \bar{\rho}_{s+2,t} & \bar{\rho}_{s+2,t+1} & \bar{\rho}_{s+2,t+2} \end{bmatrix}.$$

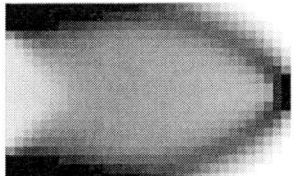

(a) Solid-void distribution via the SIMP penalty, i.e. $E \propto \rho^3$, with an optimized compliance of 52.36.

(b) Smoothly graded distribution using $E \propto \rho$, with an optimized compliance of 39.18.

Figure 2 CANTILEVER BEAM WITH OPTIMIZED MATERIAL DISTRIBUTION WHERE THE DENSITY IN EACH ELEMENT IS VARIED INDEPENDENTLY.

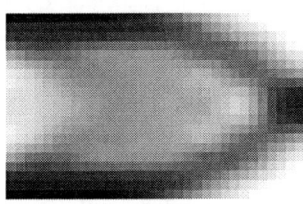
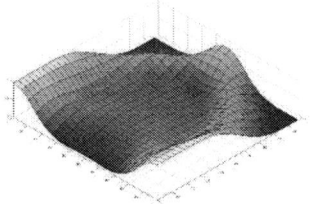

(a) Material distribution using 10×10 control points, with an optimized compliance of 39.85.

(b) The B-spline surface whose height describes the material distribution in (a).

Figure 3 CANTILEVER BEAM WITH OPTIMIZED MATERIAL DISTRIBUTION VIA B-SPLINE.

To motivate the use of B-splines, a simple 2D test was conducted to study the effect of removing the artificial penalty in the traditional Solid Isotropic Material with Penalization (SIMP) approaches to topology optimization. The artificial penalty in SIMP makes intermediate material densities inefficient, thus encouraging designs to converge to solid and void regions. This penalty is imposed because traditional subtractive manufacturing techniques produce only fully solid material. However, advanced additive manufacturing techniques have enabled the production of architected materials that can be manufactured at intermediate densities. By removing the SIMP penalty, the 2D test aimed to investigate the effect of allowing intermediate densities. A simple cantilever beam example was set up with a 20×32 element mesh and a downward load applied in the middle of the free end of the beam. One optimization was conducted with the typical SIMP penalty exponent of 3, and another was conducted with a linear scaling (or a penalty exponent of 1). Figure 2 shows the intuitive results that intermediate material densities are structurally efficient when linear elasticity scaling is used without an artificial penalty. The ability to effectively distribute loads over a wider area while still obeying total mass constraints is one of the primary potential benefits of multiscale design of lattice structures. This result is similar to what has previously been demonstrated in related works on variable thickness sheet design (cf. [25],[26]) and free material design (cf. [27], [28]). The primary distinction between the cited works and the present one lies in the use of surrogate

models for stress-constrained design of lattices, as well as the use of splines for reduced dimensionality of the design problem.

The presence of smoother material distributions is encouraging for the use of B-splines as the design representation since splines can naturally represent smooth topologies with a much smaller number of parameters then element-based approaches. Another study was conducted with an identical loading case to Figure 2(b), but with a 10×10 field of control points as design parameters. Figure 3 illustrates the result of this investigation. The two designs are comparable both visually and in terms of the resulting structural compliance. For the finite-element mesh resolution of 20×32 used here, this translates to a reduction by more than a factor of 6 in the number of design parameters. Even larger dimensionality reduction would be achieved in cases where the analysis requires finer mesh resolution. However, lattice unit cell geometries will all scale differently with relative density. Depending on the unit cell selected and the loading case applied, the optimal design may still have sharp changes in topology. In these cases, increasing the control point resolution for the B-spline surfaces would be necessary to converge towards the true optimal design. A control mesh refinement study is included in Section 5.1 to determine the appropriate number of design parameters.

4. SIMULATION FRAMEWORK

This section describes the computational design workflow for a functionally graded structure filled with lattice unit cells of a given type. The lattice structure is 3D but the material field distribution to be optimized is 2D, meaning that it does not vary through the third dimension of the structure. Two types of objectives are considered in this paper: to minimize the compliance integrated over the structural domain Ω:

$$\min_P \int_\Omega \frac{1}{2} S_M^T \bar{\epsilon} d\Omega, \tag{11a}$$

or to minimize the total volume fraction of material over the domain:

$$\min_P \int_\Omega \bar{\rho} d\Omega. \tag{11b}$$

Each optimization is subject to some or all of the following constraints:

$$0.02 \le \bar{\rho} \le 1.0, \tag{12a}$$

$$\int_\Omega \bar{\rho} \, d\Omega \le V_{Max}, \tag{12b}$$

$$\lambda_{Max} \|S_M\| < \sigma_Y, \tag{12c}$$

$$\nabla \cdot S_M + F = 0, \tag{12d}$$

$$S_M = \mathbb{C}_H \bar{\epsilon}. \tag{12e}$$

The design parameter P is an array of size mn flattened from a matrix of size $m \times n$ containing the values of $\bar{\rho}_{i,j}$ from Equation 7. Equation 12a imposes a physically realistic upper bound of 100% on the relative density and a lower bound of 2% based on manufacturability considerations. Equation 12b defines a material budget for the optimization where V_{Max} is the maximum volume fraction of the full structure. Equation 12c is the stress constraint defined based on the worst-case stress ROM proposed in Section 2.3. Equation 12d is the partial differential equation (PDE) constraint given by the structural equilibrium equation in terms of the homogenized stress and strain. Equation 12e defines the homogenized constitutive relationship based on the ROM in Section 2.2. The readers are reminded that both λ_{Max} and $\mathbb{C}_H$ are spatially varying quantities which are functions of $\bar{\rho}$, which is in turn a function of P.

The optimization problem is solved numerically using the interior-point algorithm embedded within the *fmincon* function in MATLAB. Thanks to the reduced dimensionality of the design problem, gradients of the objective and any nonlinear constraints with respect to P can be computed via finite-difference without incurring a significant cost. This feature simplifies the use of off-the-shelf commercial software in the optimization pipeline. For the results presented here, ABAQUS is used to enforce the PDE constraint in Equation 12d by performing a structural analysis at each design iteration. Equation 12e is implemented in UMAT for user materials. During a finite-element simulation, the values of $\bar{\rho}$ are needed at the Gauss-Quadrature points of each element, where the constitutive relationship is used to compute the stress from the strain. At each quadrature point, $\bar{\rho}$ is first determined from Equation 8 as a function of its location and the B-spline control point values. The constitutive ROM in Section 2.2 is then evaluated to determine the local stress-strain relationship for the given value of $\bar{\rho}$. Likewise, for Equation 12c, the value of $\bar{\rho}$ at each Gauss-Quadrature point is used to evaluate λ_{Max} at the same point according to the worst-case stress ROM from Section 2.3. This evaluation is implemented in the post-processing user subroutine URDFIL, in which the value of $\bar{\rho}$ and the magnitude of the homogenized stress, $\|S_M\|$, are queried for each quadrature point to predict the worst-case stress $\lambda_{Max}\|S_M\|$ in the underlying mesostructure. The subsequent sections provide more detailed discussions on the numerical treatment of the density constraint in Equations 12a and the volume fraction calculation in Equations 11b and 12b, the stress constraint in Equation 12c, and the beam and L-bracket geometries on which the material distribution is optimized.

4.1 Density and Volume Constraints

This section describes the implementation of Equations 12a and 12b as linear inequality constraints. For the density constraints in Equation 12a, simple box constraints on the design parameters P are insufficient to bound the pointwise values of $\bar{\rho}$ due to the implicit relationship between $\bar{\rho}$ and P through the B-spline parameterization. It must therefore be enforced as an additional set of constraints.

To define the linear constraints corresponding to Equation 12a, it can be realized from Equation 7 that although $\bar{\rho}$ varies nonlinearly in space or with respect to the parametric coordinates (u, w), $\bar{\rho}$ depends linearly on P for a fixed (u, w). In other words, let $\bar{\rho}_{uw} \in R^{N \times 1}$ be an array of density values for all quadrature points on the finite-element mesh, it is possible to define a coefficient matrix $N_{uw} \in R^{N \times nm}$ such that

$$\bar{\rho}_{uw} = N_{uw}P. \tag{13}$$

The entries in N_{uw} can be derived analytically from Equations 9 and 10. In the case of this work, they are determined numerically by: (1) setting one entry in P to 1 at a time with all remaining entries equal to 0, and (2) evaluating $\bar{\rho}_{uw}$ via Equation 8 to obtain the entries of N_{uw} in the corresponding column. This procedure is executed offline prior to the optimization and the results stored. Using Equation 13, Equation 12a is enforced discretely by

$$0.02 - N_{uw}P \le 0, \tag{14}$$
$$N_{uw}P - 1.0 \le 0.$$

The left-hand-side of Equation 12b is evaluated via numerical integration in a finite-element simulation. It is a weighted sum of the relative density values in $\bar{\rho}_{uw}$ for all Gauss-Quadrature points on the domain. Given the linear relationship between $\bar{\rho}_{uw}$ and P in Equation 13, the total volume fraction of the structure, V, also depends linearly on P. The linear relationship is given by

$$V = N_V P, \tag{15}$$

where $N_V \in R^{1 \times nm}$ is the weighted sum of each column in N_{uw} or the numerical integral of $\frac{\partial \bar{\rho}}{\partial P}$ over the entire structure. Equation 15 is used to define the objective function in Equation 11b. It is also used to enforce Equation 12b as a single linear inequality constraint given by

$$N_V P - V_{Max} \le 0. \tag{16}$$

4.2 Aggregated Stress Constraints

The stress constraint in Equation 12c must be independently satisfied at each Gauss-Quadrature point in the domain, at which the stresses are computed in a finite-element simulation. This results in a large number of nonlinear constraints, which could hinder the optimization convergence. This issue is addressed by replacing a point-wise evaluation of Equation 12c with an aggregation function which averages the discrete stress values over a small region on the finite-element mesh. For a domain that is divided into N_R regions, there would be N_R stress constraints. Let $j = 1, \dots, N_R$ be the indices of the regions and Q_j be the total number of Gauss-Quadrature points in the jth region, the aggregation function is given by

$$\frac{1}{Q_j} \sum_{i=1}^{Q_j} \mathcal{R}\left(\frac{\lambda_{Max}\|S_{M,i}\|}{\sigma_Y} - 1\right) < S_{tol}, \tag{17}$$

where S_{tol} is a small positive tolerance, $\|S_{M,i}\|$ is the magnitude of the homogenized stress at the ith quadrature point, and $\mathcal{R}$ is a smooth hinge function given by

$$\mathcal{R}(\xi) = \begin{cases} \beta(1 + \xi), & \xi < 0 \\ \beta[\exp(100\alpha\xi) - 100\alpha\xi + \xi], & 0 \le \xi < 0.01, \\ \beta[A(\xi - 0.01) + B] & \xi \ge 1 \end{cases}$$

$$A = 100\alpha \exp \alpha + (1 - 100\alpha), \tag{18}$$
$$B = \exp(\alpha) + 0.01(1 - 100\alpha).$$

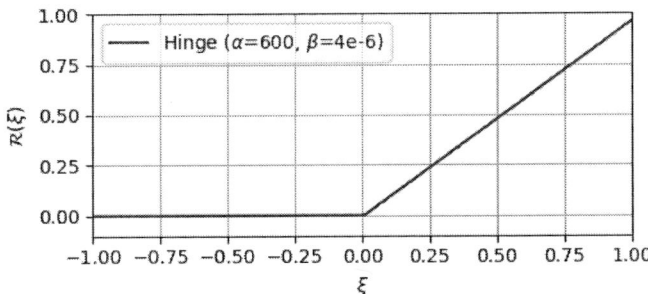

Figure 4 THE SMOOTH HINGE FUNCTION.

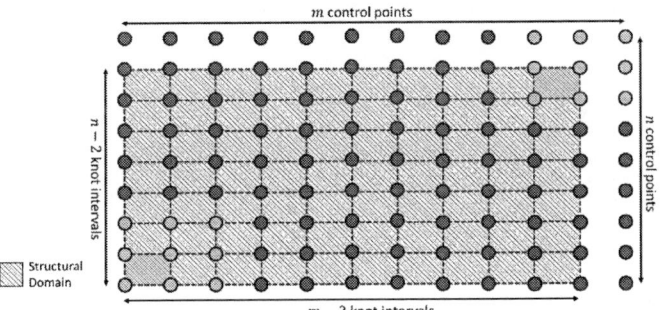

Figure 5 REGIONS FORMED BY THE KNOT INTERVALS OF AN $m \times n$ GRID OF CONTROL POINTS, WHICH ARE ALSO THE N_R REGIONS FOR ENFORCING THE STRESS CONSTRAINT, i.e. $N_R = (m-2)(n-2)$.

The smoothness of $\mathcal{R}$ at the hinge and its slope for $\xi < 0$ are respectively controlled by parameters α and β. The hinge function, as shown in Figure 4, takes on a small value when Equation 12c is satisfied at the point i in region j. As a result, the value of S_{tol} in Equation 17 controls the amount of stress constraint violation in each region. Suitable values of α, β, and S_{tol}, along with the number and size of the N_R regions dividing the domain, determine how tightly the stress constraint is satisfied over the structure. For the results presented in this paper, $\alpha = 600$ and $S_{tol} = \beta = 4 \times 10^{-6}$ have worked well.

For this work, the regions are chosen to be those formed by intersecting the knot intervals of the B-spline representation of the material field. This is illustrated in Figure 5. For the choice of B-spline parameterization described in Section 3 and more specifically Equations 9 and 10, densities everywhere within each region indexed by s, t are determined by the set of 3×3 control points with indices from s to $s + 2$, and from t to $t + 2$. As examples, Figure 5 highlights two regions along with the control points responsible for the density variation within each region. Dividing the structure in this manner makes it easier to satisfy the stress constraint globally during optimization.

4.3 Cantilever Beam Problem

The design workflow is demonstrated with an extruded cantilever beam in 3D. The beam, shown at the top of Figure 6, is of rectangular shape with a dimension of $30 \times 5 \times 2.4$ meters. The constituent material for the graded lattice is assumed to be aluminum, which has a Young's modulus of 68.9GPa, a

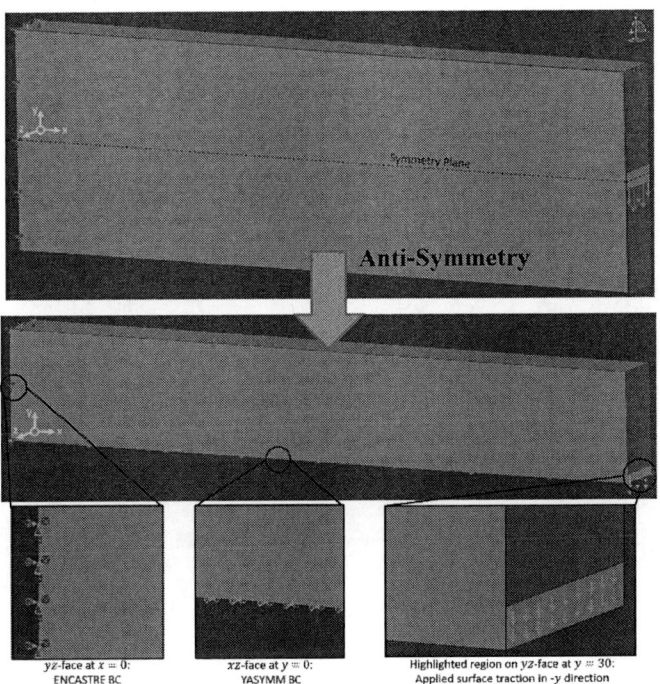

Figure 6 THE FULL CANTILEVER BEAM GEOMETRY AND THE HALVED GEOMETRY AFTER EXPLOITING THE ANTI-SYMMETRY OF THE DOMAIN. DIRECTIONS OF THE COORDINATE AXIS AND THE LOCATION OF THE ORIGIN ARE INDICATED IN YELLOW.

Poisson's ratio of 0.33, and a yield strength of 276MPa. A surface traction of 25MPa is applied over the middle 16% of the beam cross-sectional area, which is highlighted in pink in Figure 6. All optimization studies of the cantilever beam in this paper are subject to a volume constraint, V_{Max}, of 40%.

The domain and the loading exhibit anti-symmetry about the mid-plane in y, which can be leveraged to reduce the costs of the structural simulation and design. Anti-symmetry in y means that the y-displacements are fully symmetric in the top and bottom halves of the domain, whereas the x- and z-displacements are respectively symmetric in magnitudes but opposite in signs. Simulation of the full domain can therefore be reduced to the top half of the geometry above the symmetry plane, as long as appropriate boundary conditions are imposed on the symmetry face. The resulting domain with the applied boundary conditions is shown near the bottom of Figure 6. On the yz-face at $x = 0$, standard clamped boundary conditions (ENCASTRE BC in ABAQUS) are applied where all displacement components are fixed at zero. On the opposite face at the free end of the beam, a downward surface traction is applied over the bottom 8% of the halved beam cross-section. The anti-symmetry condition (YASYMM in ABAQUS) is enforced on the xz-face at the bottom of the halved beam, where the x- and z-components of the displacements are set to zero.

Symmetry in the material field is additionally enforced. Let there be $i = 1, \cdots, m$ control points along the length of the beam

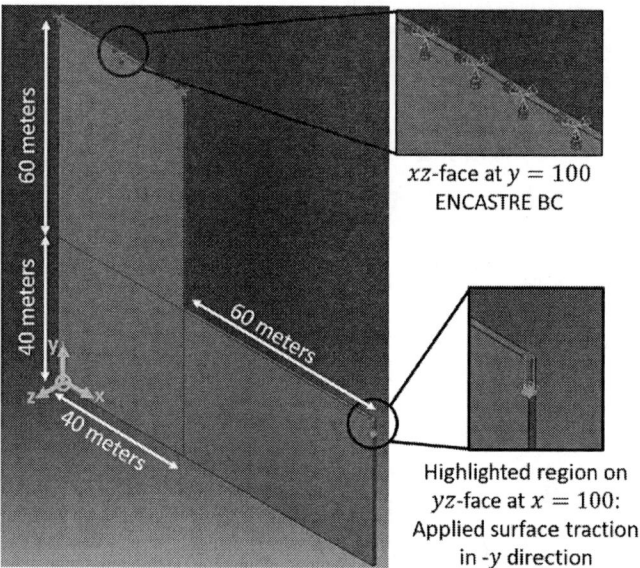

Figure 7 GEOMETRY OF THE L-BRACKET WITH THE APPLIED BOUNDARY CONDITIONS. THE THICKNESS OF THE STRUCTURE IN Z IS 1 METER.

and $j = 1, \cdots, n$ along the height. Let $j = 1,2$ further denote control points closest to the symmetry plane. Then it can be shown by manipulating Equations 9 and 10 that a zero gradient of $\bar{\rho}$ across the symmetry plane can be ensured by setting $\bar{\rho}_{i,1} = \bar{\rho}_{i,2}$ for all $i = 1, \cdots, m$. This assumption also results in a total of $m(n-1)$ design parameters. Section 5.2 validates the anti-symmetry condition on a uniform and a compliance-minimized density distribution, respectively.

4.4 L-Bracket Problem

The present workflow is additionally demonstrated on an L-bracket geometry that is commonly used to benchmark stress-constrained structural topology optimization methods. In this case, the material distribution on the L-bracket is optimized without modifying the geometry, providing a comparison between the present approach and the classical solid-void design approaches. The geometry of the L-bracket is illustrated in Figure 7 along with its dimensions and boundary conditions. The L-bracket is clamped at the top (ENCASTRE BC in ABAQUS). A surface traction of 100MPa in the downward or negative y direction is applied to the area indicated in red which is of 1×1 meter. The geometry is discretized by elements of size $1 \times 1 \times 1$ for structural analyses. The constituent material for the lattice is aluminum as in the cantilever beam problem.

Figure 8 illustrates the B-spline parameterization of the material field. A B-spline grid of resolution 10×10 is used, which overlays the L-bracket geometry. Control points in gray do not contribute to the relative density in any part of the structural domain. Thus, they are excluded from the set of design parameters. Figure 8 also shows in purple the N_R regions for the aggregated stress constraints. The original regions formed by the knot intervals, as shown in Figure 5, are refined in both x and y

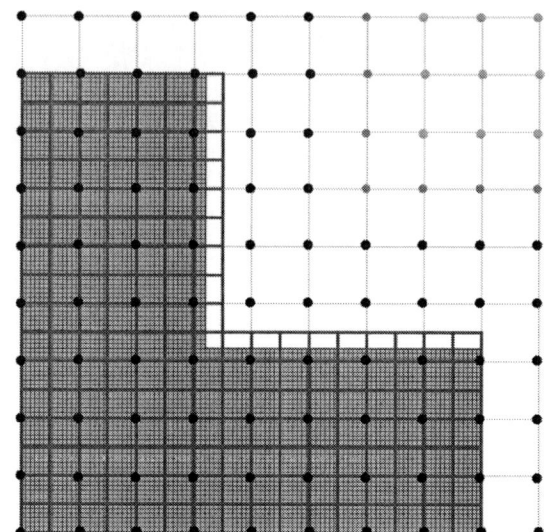

Figure 8 B-SPLINE GRID PARAMETRIZING THE L-BRACKET GEOMETRY SHOWN WITH REGIONS ON THE DOMAIN FOR STRESS-CONSTRAINT AGGREGATION.

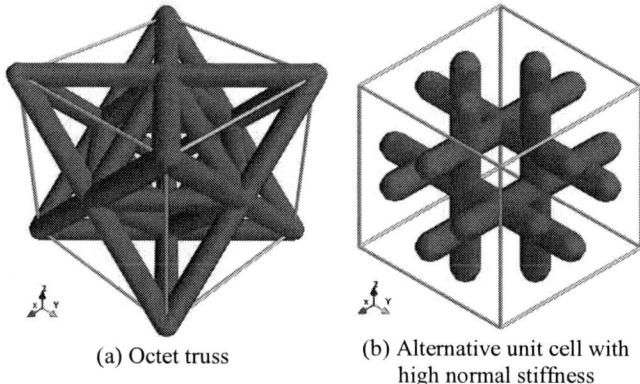

(a) Octet truss   (b) Alternative unit cell with high normal stiffness

Figure 9 TWO TYPES OF UNIT CELL GEOMETRIES USED AS THE MESOSTRUCTURE IN THE FUNCTIONALLY GRADED LATTICE.

to ensure that the size of each region is small enough to achieve tight control of the largest stress in each region. There is a total of 84 design parameters and 175 stress constraints.

5. RESULTS

To demonstrate that the proposed framework can be used to design for different mesostructures, two types of lattice unit cells are considered for the optimization problem outlined in Section 4: the octet truss [29] and an alternative mesostructural architecture designed to have a higher normal stiffness but lower shear stiffness compared to the octet truss [19]. Geometries of the unit cells are illustrated in Figure 9. Expressions for the constitutive and worst-case stress ROMs are listed in Table 1 and Table 2 for the two unit cell types, respectively. Figure 10 further provides a visual comparison between the properties of the two unit cell types by plotting the ROMs from Table 1 and Table 2 with the relative density on the x-axis. In all cases, material

V02AT03A063-8

$$\mathbb{C}_{1111} = 1.5344\bar{\rho}^3 - 0.3635\bar{\rho}^2 + 0.3244\bar{\rho} - 0.0029$$
$$\mathbb{C}_{1122} = 1.0011\bar{\rho}^3 - 0.5136\bar{\rho}^2 + 0.2301\bar{\rho} - 0.0033$$
$$\mathbb{C}_{1212} = 0.0884\bar{\rho}^3 + 0.2102\bar{\rho}^2 + 0.0836\bar{\rho} + 0.0006$$
$$\lambda_{Max} = -33.54\bar{\rho}^3 + 87.26\bar{\rho}^2 - 86.17\bar{\rho} + 28.15 + \frac{6.034}{\bar{\rho}}$$

Table 1 ROM FOR THE OCTET TRUSS

$$\mathbb{C}_{1111} = 0.8786\bar{\rho}^3 + 0.1951\bar{\rho}^2 + 0.4064\bar{\rho} + 0.0003$$
$$\mathbb{C}_{1122} = 0.8754\bar{\rho}^3 - 0.2212\bar{\rho}^2 + 0.0713\bar{\rho} - 0.0021$$
$$\mathbb{C}_{1212} = 0.2445\bar{\rho}^3 + 0.1932\bar{\rho}^2 - 0.0255\bar{\rho} + 0.0001$$
$$\lambda_{Max} = -34.29\bar{\rho}^3 - 8.544\bar{\rho}^2 + 153.4\bar{\rho} - 171.8 + \frac{63.04}{\bar{\rho}}$$

Table 2 ROM FOR THE ALTERNATIVE UNIT CELL TYPE.

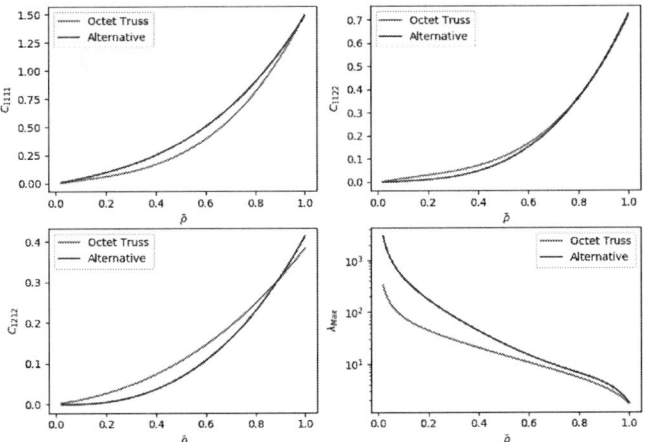

Figure 10 PLOTS OF THE ROM IN Table 1 and Table 2.

constants in the homogenized constitutive relationships are normalized by the Young's modulus of the constituent material. Results in this section are organized into four parts. Section 5.1 investigates the appropriate number of control points to be used in subsequent studies. Section 5.2 verifies the anti-symmetry assumption. Section 5.3 includes the stress-constrained compliance minimization results for the octet truss. These results are then compared with compliance minimized designs using the alternative unit cell type with and without a stress constraint in Section 5.4. Finally, Section 5.5 presents two optimized lattice designs for the L-bracket based on stress-constrained volume minimization and volume-constrained compliance minimization, respectively.

5.1 Control Mesh Refinement

This section investigates the effect of control mesh refinement on the optimized compliance of a functionally graded structure made of octet truss unit cells. The stress constraint is not considered in this study. The optimized material distributions for 3 combinations of control mesh resolution are shown in Figure 11. The control point resolutions are superimposed on the

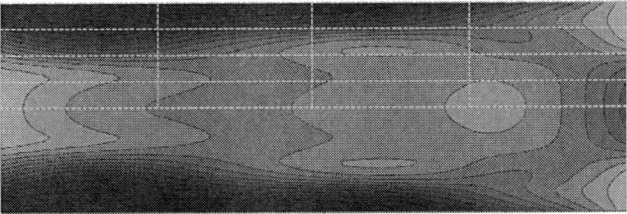

(a) 6 × 6 control points, compliance = 9.838×10^5 Nm

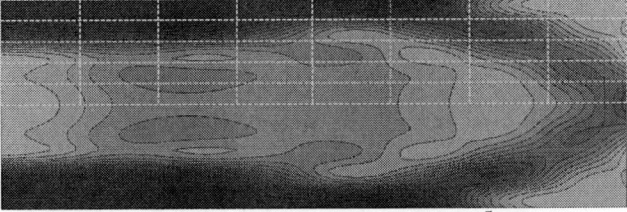

(b) 10 × 7 control points, compliance = 9.267×10^5 Nm

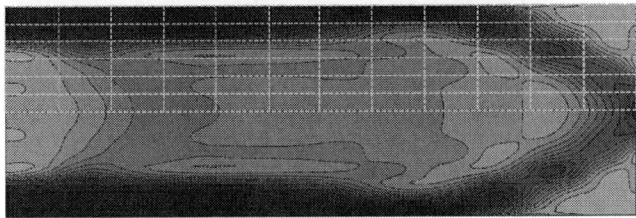

(c) 14 × 8 control points, compliance = 9.125×10^5 Nm

Figure 11 DENSITY DISTRIBUTIONS ON THE FULL DOMAIN FOR COMPLIANCE MINIMIZATION OBTAINED WITH INCREASING CONTROL POINT RESOLUTIONS. THE DOTTED LINES INDICATE THE KNOT INTERVALS AT EACH REFINEMENT LEVEL.

top half of the domain, on which the optimization is performed. Recall that for the quadratic B-spline material representation used in this paper, the number of knot intervals is two less than the number of control points in each direction.

It can be observed that the lowest resolution of 6 × 6 is able to capture the general regions of high and low relative densities, with optimized compliance within 8% of the optimized compliance using 14 × 8 control points. At the intermediate refinement level of 10 × 7 control points, the difference in optimized compliance is less than 1.6% of that achieved with 14 × 8 control points. The compliance of an unoptimized structure with a uniform distribution of $\bar{\rho} = 0.4$ is 3.171×10^6 Nm. Therefore, in all three cases, the optimization has improved the performance of the structure by 70%. On the other hand, a larger number of design parameters results in more optimization iterations due to the need to explore a larger design parameter space. In this case, in which the design gradients are approximated with finite-differencing, increasing the number of design parameters results in a further increase in the computational cost. Using more control points to parametrize the material field thus offers diminishing returns in designing these functionally graded structures. Based on the findings of this investigation, subsequent design studies will be conducted with a control point resolution of 10 × 7 to achieve a smooth

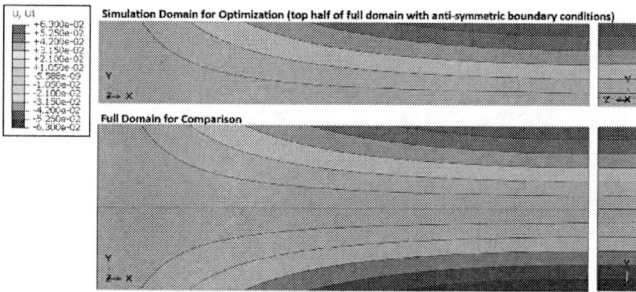

(a) *x*-displacement field on the halved domain and the full domain

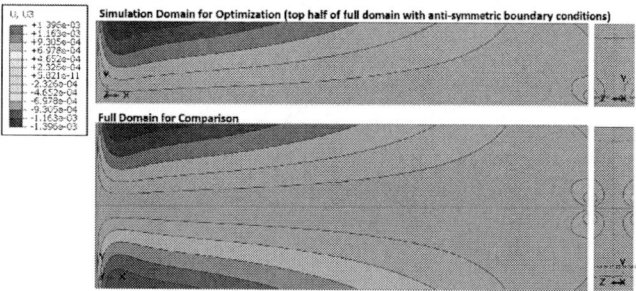

(b) *z*-displacement field on the halved domain and the full domain

Figure 12 VERIFICATION OF THE ANTI-SYMMETRY ASSUMPTION VIA COMPARISON WITH SIMULATION ON THE FULL DOMAIN: UNIFORM DENSITY DISTRIBUTION.

gradation in density with enough fidelity to improve structural performance while keeping the dimensionality of the design space low.

5.2 Verification of the Anti-Symmetry Assumption

This section verifies the anti-symmetry assumption by comparing the *x*- and *z*-displacement fields on the halved domain with those obtained via simulations on the full beam geometry. This study is conducted on a uniform density distribution of 0.4 with results shown in Figure 12, and on the compliance-minimized density distribution with 10×7 control points with results shown in Figure 13. Contours in the first column are displacements on the *xy*-face at $z = 2.4$, whereas contours in the second column are displacements on the *yz*-face at $x = 30$. For both density fields, the *x*- and *z*-displacements are equal in magnitude and opposite in sign, thus validating the anti-symmetry in the solution fields. Furthermore, the zero-contour of the displacement fields on the full domain are aligned with the location of the symmetry plane in all cases. Therefore, the anti-symmetric boundary conditions enforced on the symmetry face of the halved domain are also validated. Finally, a node-by-node comparison of all components of the displacement shows that the simulation results on the full domain can be reproduced using those on the halved domain with appropriate sign changes to numerical precision.

5.3 Stress-Constrained Designs with the Octet Truss

This section investigates various stress-constrained designs using the octet truss unit cell and compares their performance to compliance minimized designs without the stress constraint. The

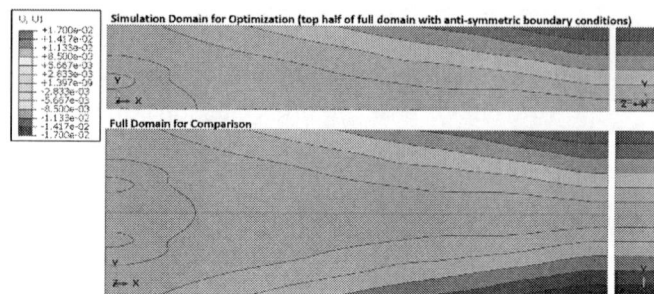

(a) *x*-displacement field on the halved domain and the full domain

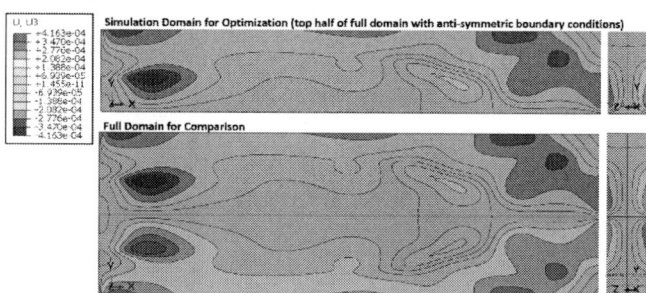

(b) *z*-displacement field on the half domain and the full domain

Figure 13 VERIFICATION OF THE ANTI-SYMMETRY ASSUMPTION VIA COMPARISON WITH SIMULATION ON THE FULL DOMAIN: COMPLIANCE-MINIMIZED DENSITY DISTRIBUTION FROM Figure 11B.

optimal material distribution for minimal compliance is independent of the magnitude of the applied load at the free end of the beam. This is due to the linear scaling between the load and the displacements over the linear elastic domain. In contrast, a stress-constrained design will be load-dependent. For this reason, two different load cases are considered: an end load of 25MPa and 50MPa, respectively. Figure 14 shows the worst-case stress distribution, normalized by the yield stress, on the compliance minimized design from Figure 11b. In Figure 14a, there is a small region of constraint violation where the symmetry plane intersects the free end of the beam and the worst-case stress in the mesostructure is 1.188 times greater than the yield stress of the constituent material. When the applied load is increased to 50MPa, regions near the top and bottom corners of the fixed end are also failing, with the worst-case stress in the mesostructure up to 2.376 times that of the yield stress.

Figure 15a and Figure 16a show the compliance minimized density distributions from problem formulations that include the stress constraint for the two load cases. The corresponding normalized worst-case stress distributions can be found in Figure 15b and Figure 16b. In both cases, the feasibility of the structure has been improved as indicated by the reduction in maximum worst-case stress.

For the 25MPa load case, the largest worst-case stress is reduced to roughly 1.011 times the yield stress of the constituent material. The mild stress constraint violation is due to the choice of aggregation function discussed in Section 4.2. It is possible to enforce the stress constraint more tightly by varying the user-specified parameters. Doing so, however, makes the

V02AT03A063-10

Copyright © 2019 ASME

(b) Applied load = 25MPa; maximum worst-case stress is 1.188 times the yield stress of the constituent material.

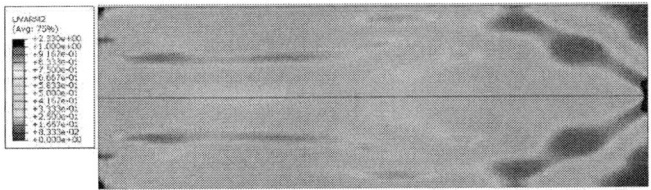

(b) Applied load = 50MPa; maximum worst-case stress is 2.376 times the yield stress of the constituent material.

Figure 14 DISTRIBUTION OF THE WORST-CASE STRESS NORMALIZED BY THE YIELD STRESS ON THE COMPLIANCE-MINIMIZED DESIGN WITH THE OCTET TRUSS. REGIONS VIOLATING THE STRESS CONSTRAINT ARE SHOWN IN BLACK.

optimization problem more nonlinear and more challenging to solve. A constraint violation of 1% is reasonable given that the worst-case stress is a conservative estimate and that a factor of safety is often added in practice. The reduction in the maximum worst-case stress is accompanied by an insignificant 0.02% increase in compliance. This result is expected because only a very small portion of the structure is failing in Figure 14a without the stress constraint. It is therefore not surprising that a small change in the density distribution is sufficient to ensure the structural integrity without sacrificing much of the overall structural stiffness. In the case of the 50MPa load, the optimized material distribution in Figure 16a is more distinctly different from that in Figure 11b. The largest worst-case stress is reduced to roughly 1.017 times the yield stress of the material. This is accompanied by a 13% increase in compliance compared to the non-stress-constrained result. Both load cases demonstrate the importance of considering structural failure in the design of functionally graded lattice structures, as well as the ability of the B-spline material parameterization to produce designs satisfying different functional requirements.

5.4 Lattice Design with an Alternative Unit Cell

This section investigates functionally graded lattice designs with the alternative unit cell type described in Figure 6b and Table 2. For a lattice beam with a uniform density distribution of 0.4, the compliance is 1.956×10^6 Nm, which outperforms an octet truss lattice with uniform density by almost 40%. Figure 17 shows the optimized density and normalized worst-case stress distribution without enforcing the stress constraint. The optimized compliance is 8.9% higher than that of the octet truss shown in Figure 11b, and it is evident that the worst-case stress in many regions of the structure is above the yield stress of the

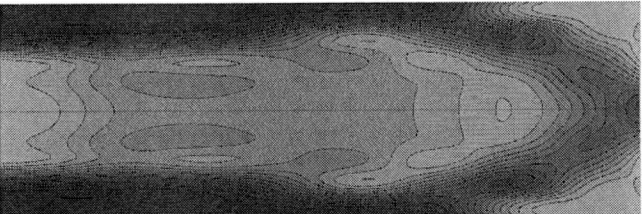

(a) Optimized density distribution with stress constraint. Compliance is 9.269×10^5 Nm.

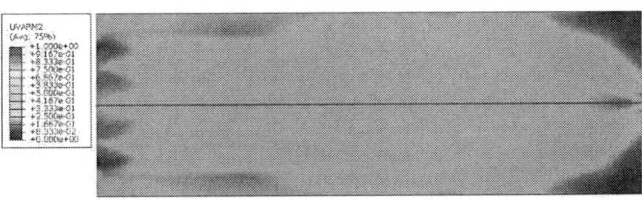

(b) Normalized worst-case stress distribution on the optimized design with maximum at 1.011 times the yield stress of the constituent material.

Figure 15 STRESS-CONSTRAINED DESIGN WITH THE OCTET TRUSS AND AN APPLIED LOAD OF 25MPA.

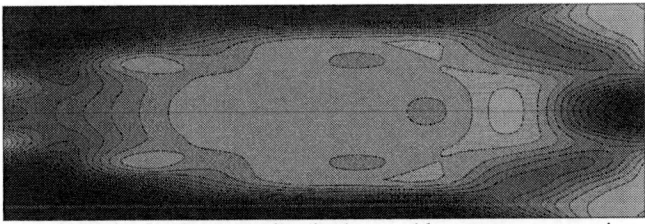

(a) Optimized density distribution with stress constraint. Compliance is 4.203×10^6 Nm.

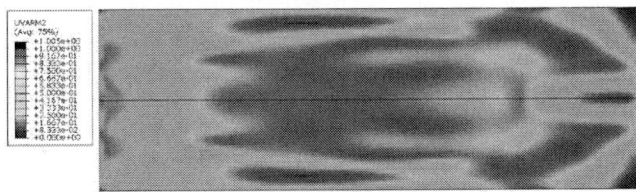

(b) Normalized worst-case stress distribution on the optimized design with maximum at 1.017 times the yield stress of the constituent material.

Figure 16 STRESS-CONSTRAINED DESIGN WITH THE OCTET TRUSS AND AN APPLIED LOAD OF 50MPA.

constituent material. This result is not unexpected given the large bending moment experienced by the high aspect-ratio cantilever beam. Figure 18 illustrates the stress-constrained design using the alternative unit cell in terms of the density and normalized worst-case stress distributions. Figure 18a shows a density distribution that varies more gradually than that in Figure 17a. This change reduces the largest worst-case stress in Figure 18a on the domain to 1.013 times the yield stress. High stress areas indicated in black in Figure 17b have also been substantially reduced. This is accompanied by a 22.5% increase in the structural compliance compared to Figure 17a. A comparison between Figure 11b and Figure 17a, as well as between

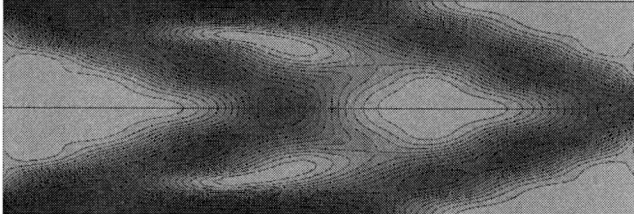

(a) Optimized density distribution without stress constraint; compliance is 1.009×10^6 Nm.

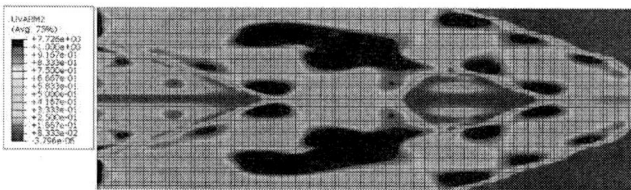

(b) Normalized worst-case stress distribution on the optimized design with maximum stress at 7.94 times the yield stress of the constituent material.

Figure 17 COMPLIANCE MINIMIZED DESIGN WITH THE ALTERNATIVE UNIT CELL TYPE (Figure 9B) AND AN APPLIED LOAD OF 25MPA.

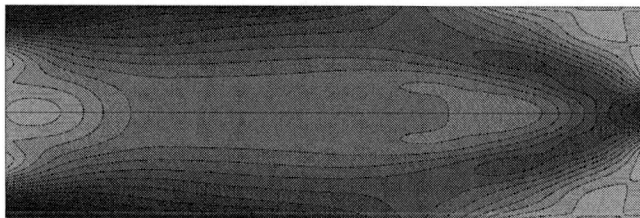

(a) Optimized density distribution with stress constraint; compliance is 1.236×10^6 Nm.

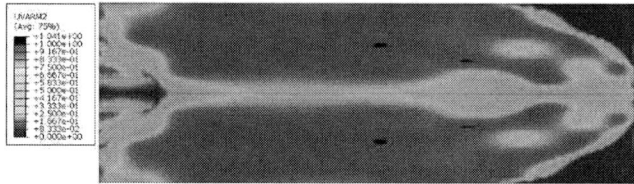

(b) Normalized worst-case stress distribution on the optimized design with maximum stress at 1.013 times the yield stress of the constituent material.

Figure 18 STRESS-CONSTRAINED DESIGN WITH THE ALTERNATIVE UNIT CELL TYPE (FIGURE 7B) AND AN APPLIED LOAD OF 25MPA.

Figure 15a, Figure 16a, and Figure 18a, further shows that the proposed B-spline parameterization is able to produce visually distinct material distributions, and that the proposed design framework can accommodate mesostructures with different mechanical properties.

5.5 Lattice Design on the L-Bracket

Two design studies are conducted on the L-bracket geometry with the octet truss as the mesostructure. The first is a

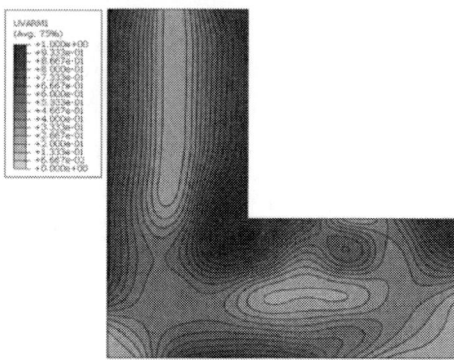

(a) Optimized density distribution for the stress-constrained volume minimization design. Volume fraction is 37.5%, compliance is 3.348×10^7 Nm.

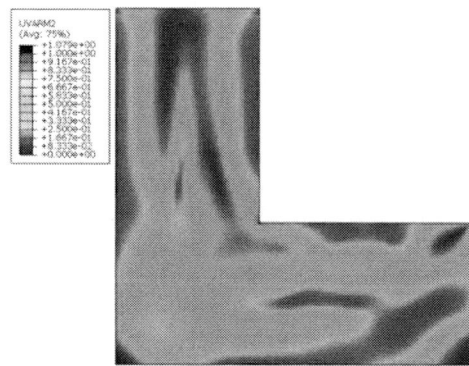

(b) Normalized worst-case stress distribution on the optimized design with maximum stress at 1.0066 times the yield stress of the constituent material.

Figure 19 LATTICE DESIGN FOR STRESS-CONSTRAINED VOLUME MINIMIZATION.

stress-constrained volume minimization problem. Equation 11b is used as the objective, subject to all constraints in Equation 12 except for 12b. The design is initialized with a uniform density distribution of 40%. Figure 19a shows the optimized density field. The resulting volume fraction is 37.5%, the compliance is 3.348×10^7 Nm, and the maximum stress constraint violation is approximately 0.66% of the yield stress. Distribution of the normalized worst-case stress in Figure 19b shows that the stress constraint is satisfied in most parts of the domain. The present study of the L-bracket differs from ones in the literature (cf. [4],[30]) in that changes in topology due to complete removal of materials in a region are not permitted due to the lower bound in the relative density. For this reason, the shape of the re-entrant corner is unaltered and is instead reinforced via adding materials to that region. The use of B-spline parameterization further produces a smoother material field distribution compared to designing with the densities of individual elements. Despite these differences, the overall distribution of materials over the L-bracket as well as the optimized volume fraction are comparable with the results reported by similar studies in the literature.

The second study minimized the compliance with a volume constraint given by Equation 12b, where $V_{Max} = 37.5\%$ is the

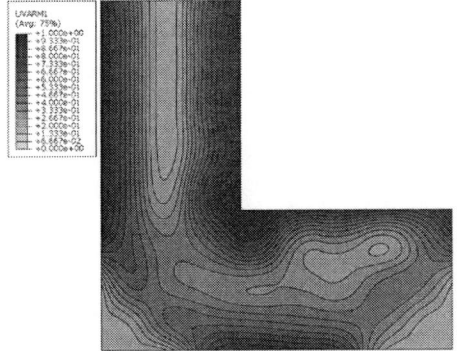

(a) Optimized density distribution for the stress-constrained volume minimization design. Compliance is 3.050×10^7 Nm.

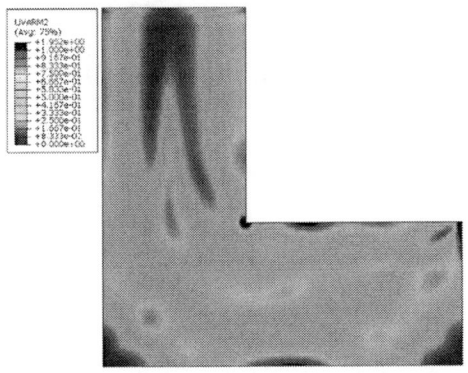

(b) Normalized worst-case stress distribution on the optimized design with maximum stress at 1.819 times the yield stress of the constituent material.

Figure 20 LATTICE DESIGN FOR VOLUME-CONSTRAINED COMPLIANCE MINIMIZATION.

minimized volume fraction from the previous study. In contrast with the first study where there is no consideration of compliance, the second study does not consider the maximum stress in the objective or in the constraints. The optimized density distribution is shown in Figure 20a, accompanied by the normalized worst-case stress distribution in Figure 20b. The optimized compliance is 3.050×10^7 Nm, which is 8.9% lower than the compliance from the first study as expected. The decrease in compliance comes at the cost of multiple regions at risk of structural failure, which are indicated in black in Figure 20b. The largest worst-case stress on the L-bracket is almost twice the yield stress of the constituent material.

The two studies in this section show that the present framework produces designs that are comparable with those from the literature. Furthermore, they demonstrate that the B-spline parameterization is applicable to material field design on non-rectangular domains.

6. DISCUSSIONS AND FUTURE WORK

This paper describes a computational framework for designing functionally graded lattices for stiffness and strength. The lattices are composed of unit cell mesostructures. The analysis and optimization of such multiscale structures are made computationally tractable with the use of surrogate models. More specifically, stress-constrained design optimization is enabled by ROMs characterizing yield at the mesostructural level based on a worst-case stress measure. A compact design representation that also allows for control over the smoothness of the material field is achieved via B-splines. The proposed methodology is demonstrated by optimizing functionally graded lattices made of two types of unit cell structures. It is shown that the present framework allows for the design of structurally efficient lattices able to withstand different design loads. The B-spline parameterization offers sufficient fidelity to achieve significant improvements in structural compliance and to produce visually distinct material distributions for different unit cells and for design domains of different shapes.

In terms of future work, the proposed methodology can be applied to explore multiscale lattice designs with a wider range of unit cell types and in different application settings. The stress prediction capability can be improved by incorporating anisotropy into the yield model such that failure of the mesostructure depends on the direction of the homogenized stress tensor. Accuracy of the yield model may also benefit from taking into account different smoothed joint geometries, which could have an impact on the point of initial failure in the mesostructure. Finally, additional improvements in the structural efficiency of the optimized lattice structures may be possible with a computational framework that concurrently designs the material field and the shape/topology with suitable manufacturability constraints.

ACKNOWLEDGEMENTS

The authors would like to thank Daniel A. Tortorelli from the Lawrence Livermore National Lab for his suggestions throughout this research. This work was performed under the auspices of the U.S. Department of Energy by Lawrence Livermore National Laboratory under contract DE-AC52-07NA27344 and with funding from the Defense Advanced Research Projects Agency (DARPA). The views, opinions and/or findings expressed are those of the authors and should not be interpreted as representing the official views or policies of the Department of Defense or the U.S. government. LLNL-PROC-768641.

REFERENCES

[1] G. D. Cheng and X. Guo. "ϵ-Relaxed Approach in Structural Topology Optimization." *Structural Optimization*, Vol. 13 No. 4 (1997), pp. 258-266. DOI 10.1007/BF01197454.

[2] M. Bruggi. "On an Alternative Approach to Stress Constraints Relaxation in Topology Optimization." *Structural and Multidisciplinary Optimization*, Vol. 36 No. 2 (2008), pp. 125-141. DOI 10.1007/s00158-007-0203-6.

[3] J. París, F. Navarrina, I. Colominas, and M. Casteleiro. "Topology Optimization of Continuum Structures with Local and Global Stress Constraints." *Structural and Multidisciplinary Optimization*, Vol. 40 No. 4 (2009), pp. 419–437. DOI 10.1007/s00158-008-0336-2.

[4] C. Le, J. Norato, T. Bruns, C. Ha, and D. Tortorelli. "Stress-Based Topology Optimization for Continua." *Structural and Multidisciplinary Optimization*, Vol. 41 No. 4 (2010), pp. 605-620. DOI 10.1007/s00158-009-0440-y.

[5] E. Holmberg, B. Torstenfelt, and A. Klarbring. "Stress Constrained Topology Optimization." *Structural and Multidisciplinary Optimization*, Vol. 48 No. 1 (2013), pp. 33-47. DOI 10.1007/s00158-012-0880-7.

[6] J. Panetta, A. Rahimian, and D. Zorin. "Worst-Case Stress Relief for Microstructures." *ACM Transactions on Graphics*, Vol. 36 No. 4 (2017), pp 122:1-122:16, DOI 10.1145/3072959.3073649.

[7] M. Collet, L. Noël, M. Bruggi, and P. Duysinx. "Topology optimization for microstructural design under stress constraints." *Structural and Multidisciplinary Optimization*, Vol. 58 No. 6 (2018), pp 2677–2695, DOI 10.1007/s00158-018-2045-9.

[8] R. Picelli, R. Sivapuram, S. Townsend, H. A. Kim. "Stress Topology Optimisation for Architected Material Using the Level Set Method." *Advances in Structural and Multidisciplinary Optimization WCSMO 2017*, Springer (2017), DOI 10.1007/978-3-319-67988-4_94.

[9] P. Duysinx and M. P. Bendsøe. "Topology Optimization of Continuum Structures with Local Stress Constraints." *International Journal of Numerical Mechanical Engineering*, Vol. 43 No. 8 (1998), pp. 1453-1478.

[10] G. Allaire, F. Jouve and H. Maillot. "Topology Optimization for Minimum Stress Design with the Homogenization Method." *Structural and Multidisciplinary Optimization*, Vol. 28 No. 2-3 (2004), pp. 87-98. DOI 10.1007/s00158-004-0442-8.

[11] R. Lipton and M. Stuebner. "Optimal Design of Composite Structures for Strength and Stiffness: An Inverse Homogenization Approach." *Structural and Multidisciplinary Optimization*, Vol. 33 No. 4-5 (2007), pp. 351–362, DOI 10.1007/s00158-006-0089-8.

[12] F. V. Stump, E. C. N. Silva, and G. H. Paulino. "Optimization of Material Distribution in Functionally Graded Structures with Stress Constraints." *Communications in Numerical Methods in Engineering*, Vol. 23 No. 6 (2007), pp. 535-551, DOI 10.1002/cnm.910.

[13] D. Pasini, A. Moussa, and A. Rahimizadeh. "Stress-Constrained Topology Optimization for Lattice Materials." *Encyclopedia of Continuum Mechanics*, (2018), DOI 10.1007/978-3-662-53605-6_249-1.

[14] L. Cheng, J. Bai, A. C. To. "Functionally Graded Lattice Structure Topology Optimization for the Design of Additive Manufactured Components with Stress Constraints." *Computer Methods in Applied Mechanics and Engineering*, Vol. 344 (2019), pp 334–359. DOI 10.1016/j.cma.2018.10.010.

[15] X. Qian. "Topology Optimization in B-spline Space." *Computer Methods in Applied Mechanics and Engineering*, Vol. 265 (2013), pp 15-35. DOI 10.1016/j.cma.2013.06.001.

[16] Y.-D. Seo, H.-J. Kim, S.-K. Youn. "Isogeometric Topology Optimization Using Trimmed Spline Surfaces." *Computer Methods in Applied Mechanics and Engineering*, Vol. 199 (2010), pp 3270-3296. DOI 10.1016/j.cma.2010.06.033.

[17] J. Norato, B. Bell and D. Tortorelli, "A Geometry Projection Method for Continuum-Based Topology Optimization with Discrete Elements," *Computer Methods in Applied Mechanics and Engineering*, Vol. 293 (2015), pp. 306-327.

[18] S. Watts and D. Tortorelli, "A Geometric Projection Method for Designing Three-dimensional Open Lattices with Inverse Homogenization," *International Journal for Numerical Methods in Engineering*, Vol. 112 No. 11 (2017), pp. 1564-1588.

[19] C. Sharpe et. al., "Design of Mechanical Metamaterials via Constrained Bayesian Optimization" in *Proceedings of the International Design Engineering Technical Conference*, (2018), Paper No: DETC2018-85270.

[20] P. Chung, K. Tamma and R. Namburu, "Asymptotic Expansion Homogenization for Heterogeneous Media: Computational Issues and Applications," *Composites Part A: Applied Science and Manufacturing*, Vol. 32 No. 9 (2001), pp. 1291-1301.

[21] G. Allaire, *Shape Optimization by the Homogenization Method*, Springer (2002).

[22] M. Shahzamanian et. al., "Representative Volume Element Based Modeling of Cementitious Materials," *Journal of Engineering Materials and Technology*, Vol. 136 No. 1 (2013), pp. 1-16.

[23] S. Watts et. al., "Simple, Accurate Surrogate Models of the Elastic Response of Three-Dimensional Open Truss Micro-Architectures with Application to Multiscale Topology Design," *Structural and Multidisciplinary Optimization* (2019) - accepted.

[24] M. E. Mortenson. *Geometric Modeling*. John Wiley & Sons, New York (1997).

[25] M. P. Rossow and J. E. Taylor, "A Finite Element Method for the Optimal Design of Variable Thickness Sheets," *AIAA Journal*, Vol. 11 No. 11 (1973), pp. 1566-1569, DOI 10.2514/3.50631.

[26] J. Petersson, "A Finite Element Analysis of Optimal Variable Thickness Sheets," *SIAM Journal on Numerical Analysis*, Vol. 36 No. 6 (1999), pp. 1759-1778, DOI 10.1137/S0036142996313968.

[27] J. Zowe, M. Kočvara, and M. P. Bendsøe, "Free material optimization via mathematical programming," *Mathematical Programming*, Vol. 79 No. 1 (1997), pp. 445-466, DOI 10.1007/BF02614328.

[28] M. Kočvara and M. Stingl, "Free material optimization for stress constraints," *Structural and Multidisciplinary Optimization*, Vol. 33 No. 4 (2007), pp. 323-335, DOI 10.1007/s00158-007-0095-5.

[29] V. S. Deshpande, N. A. Fleck, and M. F. Ashby. "Effective Properties of the Octet-Truss Lattice Material," *Journal of the Mechanics and Physics of Solids*, Vol. 49 No. 8 (2001), pp. 1747-1769, DOI 10.1016/S0022-5096(01)00010-2.

[30] R. Picelli, S. Townsend, C. Brampton, J. Norato, and H. A. Kim, "Stress-Based Shape and Topology Optimization with the Level Set Method," *Computer Methods in Applied*

Mechanics and Engineering, Vol. 329 (2018), pp. 1-23, DOI
10.1016/j.cma.2017.09.001.

Proceedings of the ASME 2019
International Design Engineering Technical Conferences
and Computers and Information in Engineering Conference
IDETC/CIE2019
August 18-21, 2019, Anaheim, CA, USA

DETC2019-97934

MULTIMATERIAL TOPOLOGY OPTIMIZATION OF THERMOELECTRIC GENERATORS

Xiaoqiang Xu[1]
[1]Department of Mechanical Engineering
State University of New York at Stony Brook
Stony Brook, NY 11794, USA
Email:Xiaoqiang.Xu@stonybrook.edu

Yongjia Wu[2]
[2]Department of Mechanical Engineering
Virginia Tech
Blacksburg, VA 24060, USA
Email: Yjwu2015@vt.edu

Lei Zuo[2]
[2]Department of Mechanical Engineering
Virginia Tech
Blacksburg, VA 24060, USA
Email: Leizuo@vt.edu

Shikui Chen[1,*]
[1]Department of Mechanical Engineering
State University of New York at Stony Brook
Stony Brook, NY 11794, USA
Email: shikui.chen@stonybrook.edu

ABSTRACT

Over 50% of the energy from power plants, vehicles, oil refining, and steel or glass making process is released to the atmosphere as waste heat. As an attempt to deal with the growing energy crisis, the solid-state thermoelectric generator (TEG), which converts the waste heat into electricity using Seebeck phenomenon, has gained increasing popularity. Since the figures of merit of the thermoelectric materials are temperature dependent, it is not feasible to achieve high efficiency of the thermoelectric conversion using only one single thermoelectric material in a wide temperature range. To address this challenge, this paper proposes a method based on topology optimization to optimize the layouts of functional graded TEGs consisting of multiple materials. The objective of the optimization problem is to maximize the output power and conversion efficiency as well. The proposed method is implemented using the Solid Isotropic Material with Penalization (SIMP) method. The proposed method can make the most of the potential of different thermoelectric materials by distributing each material into its optimal working temperature interval. Instead of dummy materials, both the P and N-type

electric conductors are optimally distributed with two different practical thermoelectric materials, namely Bi_2Te_3 & $PbTe$ for P-type, and Bi_2Te_3 & $CoSb_3$ for N-type respectively, with the yielding conversion efficiency around 12.5% in the temperature range $T_c=25°C$ and $T_h=400°C$. In the 2.5D computational simulation, however, the conversion efficiency shows a significant drop. This could be attributed to the mismatch of the external load and internal TEG resistance as well as the grey region from the topology optimization results as discussed in this paper.

1 INTRODUCTION

Since its discovery, the thermoelectric effect, which can convert temperature differences into the electric voltage and vice versa, has aroused a lot of interests among researchers, especially in such an energy sustainable society. One important application of such effect is the thermoelectric generator (TEG), which can transform a tremendous amount of waste heat generated from various sources like home heating, automotive exhaust, and industrial processes into electricity [1]. Such devices occupy unique advantages over other thermal power-generation devices.

---

*Address all correspondence to this author.

For example, it can be designed in flexible size, operate without moving parts thus quite reliable, and very environmentally friendly [2]. However, one major weakness is that the efficiency of a TEG has been relatively low, which, to a certain extent, impedes the broader application of such technology into various fields.

Typically, besides the temperature difference between the hot and cold end, the efficiency of a TEG relies heavily on the figure of merit of the thermoelectric material, $zT = \alpha^2 T/\rho\kappa$, where T is absolute temperature, α is the Seebeck coefficient, ρ is the electrical resistivity and κ is the thermal conductivity, respectively [3]. Over the past several decades, tremendous effort has been made to enhance the zT value of the thermoelectric materials. For example, several breakthroughs have been made in the nanoscale from the standpoint of materials science, such as the all-scale hierarchical architectures to reduce the thermal conductivity [4], preparing thermoelectric materials like Bi_2Te_3 in quantum-well super-lattice structures [5] and band engineering [6, 7] to increase the power factor (α^2/ρ). Due to these advances, the figure of merit of the TE materials could reach 1.8 with the corresponding conversion efficiency increasing to 11%-15% [2].

In addition to the progress from the material science perspective, researchers from the mechanical engineering community also contribute by exploring the full potential of the existing thermoelectric materials. One typical example is the segmented thermoelectric generator [8], in which the P and N-type elements are usually divided into several segments. In each segment, a proper thermoelectric material is employed to achieve the highest efficiency in the temperature interval of that segment. For simple regular geometry, it is convenient to determine the dimensions for each segment. But it is difficult to treat with a TEG with complex and irregular geometry. Therefore, in this paper, we propose a new method based on topology optimization to find the optimal thermoelectric materials distribution to achieve the highest conversion efficiency.

Topology optimization has emerged for nearly 30 years since its introduction by MP Bendse and N Kikuchi in 1988 [9], which aims at finding the proper materials layout in a prescribed domain to obtain the best performance for a certain purpose. One outstanding characteristic of topology optimization lies in that it can give you designs independent on your initial guess. Several major topology optimization approaches have been proposed over the years, including SIMP (Solid Isotropic Material Penalization) [10, 11], level set method [12, 13], phase field [14] and topological derivative [15]. In this paper, SIMP is employed since it is quite straightforward. To implement the SIMP, several relationships between the density variable and the thermoelectric properties need to be established. And to make it more accurate, the temperature-dependence of the thermoelectric materials properties are taken into consideration in the form of cubic spline interpolation functions. The output power and conversion efficiency are treated as the objectives to be optimized.

This paper is organized as follows: in Section 2, the governing equations for thermoelectric phenomena are introduced. The topology optimization formulation is given in Section 3, including computational model description, objective functions and material interpolation scheme. Section 4 details the numerical implementation, followed by the topology optimization results and 2.5D numerical verification in Section 5. Finally, in Section 6, the results are discussed, and some concluding remarks are given.

2 THERMOELECTRIC GOVERNING EQUATIONS

A typical TEG is shown in Figure 1, which usually consists of heat source, heat sink, P and N-type thermoelectric elements, external load, and electrodes (gray regions). It works based on the Seebeck effect, where an electromotive force is built due to the movement of charge carriers in the presence of a temperature gradient between the hot and cold ends. When connected to an external load to build a circuit, there will be a current flowing through to produce electric power. For P-type, the current is carried by holes, while for N-type, it is by electrons. We can imagine the potential application of such a TEG device wherever there is a temperature gradient.

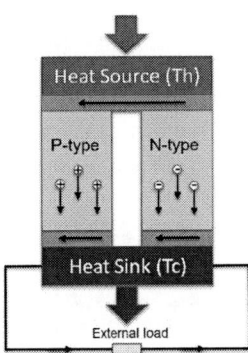

FIGURE 1: The configuration of a typical TEG device

This multi-physics optimization problem involves heat transfer in solids, electron migration, and thermoelectric effect. For simplicity, we only consider the steady state and assume that the thermoelectric materials are isotropic with regards to the thermoelectric properties like Seebeck coefficient, electrical conductivity, and thermal conductivity. Consulting [16, 17], the governing equations for electrical and thermal conductions can be given as follows:

$$\nabla \cdot \boldsymbol{J} = 0 \qquad (1)$$
$$\nabla \cdot \boldsymbol{q} = f \qquad (2)$$

The above equation (1) and (2) are coupled with the following thermoelectric constitutive equations:

$$\boldsymbol{J} = \sigma(\boldsymbol{E} - \alpha \nabla T) \qquad (3)$$
$$\boldsymbol{q} = \beta \boldsymbol{J} - \kappa \nabla T \qquad (4)$$

where:
$\boldsymbol{J}$ = electric current density vector, A/m^2,
$\boldsymbol{q}$ = heat flux density vector, W/m^2,
$f = \boldsymbol{J} \cdot \boldsymbol{E}$ is the heat generation rate per unit volume, W/m^3,
σ = electric conductivity, S/m,
$\boldsymbol{E} = -\nabla V$, electric field intensity vector, V/m; V is electric potential, V,
α = Seebeck coefficient, V/K,
T = absolute temperature field, K,
$\beta = T \cdot \alpha$ denotes the Peltier coefficient, V,
κ = thermal conductivity, W/(m$\cdot$K).

The boundary conditions are:

$$V = V_0, \text{fixed electric potential} \qquad (5a)$$
$$T = T_c, \text{fixed temperature} \qquad (5b)$$
$$T = T_h, \text{fixed temperature} \qquad (5c)$$
$$\boldsymbol{n} \cdot \boldsymbol{J} = 0, \text{electrical insulation} \qquad (5d)$$
$$\boldsymbol{n} \cdot \boldsymbol{q} = 0, \text{thermal insulation} \qquad (5e)$$

By imposing the above boundary conditions, the equation (1) and (2) can be solved with respect to the two state variables T and V, which will be further used to compute the objective functions and constraints.

3 TOPOLOGY OPTIMIZATION FORMULATION

Topology optimization has been a robust tool for finding the optimal materials layout for a particular purpose. In this paper, different thermoelectric materials are to be optimally distributed in the P & N-type thermoelectric elements, i.e., the design domain. For simplicity, in the topology optimization stage, only 2D problem is considered.

3.1 Computational model description

As seen in Figure 2, a 2D computational model is built. Unlike the conventional configuration in Figure 1, this model adopts

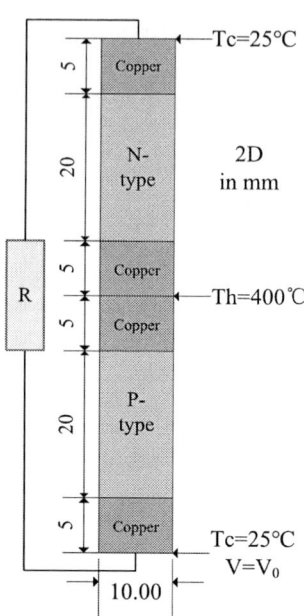

FIGURE 2: Diagram of the computational model

a stack junction, which could reduce the parasitic losses from the connection between the P and N-type electrical conductors [18]. The thickness is 1mm to make it a thin plate structure. The temperature for the hot and cold end is set to be 400 °C and 25 °C respectively. An electrical potential V_0 is assigned to the bottom surface. Copper acts as electrodes. An external load R is connected to form an electrical circuit and produce power. All other outer boundaries of the whole domain are electrically insulated and adiabatic, corresponding to the Neumann boundary conditions in equations (5d) and (5e).

3.2 Objective functions

In this topology optimization problem, we have two objective functions to optimize, namely output power and conversion efficiency. When there is a large amount of heat sources, maximum output power is preferably wanted. While when the heat source is limited, we prefer a maximum conversion efficiency, which is defined as the ratio between output power and the total heat flow from the source. For output power:

$$P_{out} = \int_{\Omega_1} \boldsymbol{J} \cdot \boldsymbol{E} d\Omega \qquad (6)$$

where $\boldsymbol{J}$ is the electric current density vector; $\boldsymbol{E}$ is the electrical field intensity vector as mentioned earlier; Ω_1 is the external resistor domain.

The conversion efficiency is defined as follows:

$$\eta = \frac{P_{out}}{Q} = \frac{\int_{\Omega_1} \boldsymbol{J} \cdot \boldsymbol{E} d\Omega}{|\int_{\Omega_2} \boldsymbol{q} \, d\Omega|} \qquad (7)$$

where Q is the total heat flow from the source; $\boldsymbol{q}$ is the heat flux density vector; Ω_2 is the domain from which the heat flows in.

Thus, the objective of the topology optimization problem is to maximize either the output power or the conversion efficiency, which can be formulated as follows:

$$\text{maximize } J_1 = P_{out}, \qquad (8)$$
$$\text{or maximize } J_2 = \eta, \qquad (9)$$

subject to:

$$\int_{\Omega_{PN}} dx < V_f \cdot A. \qquad (10)$$

In the above equations, Ω_{PN} refers to regions occupied by thermoelectric materials of the final design; V_f is the volume ratio; A is the initial area of the P and N-type elements. In practice, a large number of TEG units would be connected electrically in series and thermally in parallel to produce massive power, requiring a lot of thermoelectric materials. Therefore, it is advisable to impose a volume constraint in the designing process of a TEG unit taking the total cost and weight into consideration.

3.3 Material interpolation scheme

The crux of the SIMP method is that a proper relationship between the design variables and corresponding physical properties must be established. In this paper, both the P and N-type thermoelectric materials are optimized. Following [19], we employ the following material interpolation scheme in this paper:

$$\begin{aligned}
\alpha(\rho_1, \rho_2) &= \rho_1^p (\rho_2^p \alpha_1 + (1 - \rho_2^p)\alpha_2), \\
\kappa(\rho_1, \rho_2) &= \rho_1^p (\rho_2^p \kappa_1 + (1 - \rho_2^p)\kappa_2), \\
\sigma(\rho_1, \rho_2) &= \rho_1^p (\rho_2^p \sigma_1 + (1 - \rho_2^p)\sigma_2),
\end{aligned} \qquad (11)$$

where $0 < \rho_1 < 1, 0 < \rho_2 < 1$; p is the penalty factor and is set to be 3. The symbols $\alpha_i, \kappa_i, \sigma_i, i = 1,2$ denote the Seebeck coefficient, thermal conductivity and electrical conductivity of the two thermoelectric materials respectively. At each point of the design domain, there are two design variables, where ρ_1 is used to determine whether it is material or void and ρ_2 is used to indicate which material it is.

4 NUMERICAL IMPLEMENTATION

The procedures of the topology optimization process are shown in Figure 3. To perform the numerical implementation, the governing equations must be solved correctly to obtain the state fields, i.e., temperature and electrical potential fields in this case, and further to compute the objective functions and constraints. If it is not converged, sensitivity analysis needs to be conducted and later to update the design variables using the Method of Moving Asymptotes (MMA) algorithm [20] in this paper. Once converged, some post-processing steps, like smoothing the boundary to facilitate the manufacturing process, need to be included before the final design is presented. The above procedures are realized with the Comsol multiphysics FEA software.

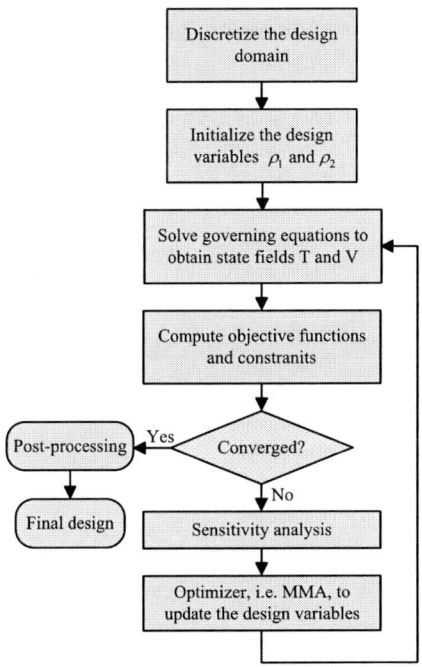

FIGURE 3: The flowchart for the topology optimization process

4.1 Material properties

The properties of the copper electrodes are: $\alpha = 6.5 \times 10^{-6} \text{V/K}, \kappa = 400 \text{W/(m} * \text{K)}, \sigma = 5.998 \times 10^7 \text{S/m}$. Because the temperature ranges from $T_c = 25\,^\circ\text{C}$ to $T_h = 400\,^\circ\text{C}$, two different thermoelectric materials are employed for each of the P and N-type elements, as shown in Table 1 based on the ZT value from [1]. Such selection makes sure that there are a low-temperature and high-temperature thermoelectric material relatively for each type element, making it possible to make the most of the thermo-electric capability of each material.

P type	Bi_2Te_3	PbTe
N type	Bi_2Te_3	$CoSb_3$

TABLE 1: Material selection

The temperature-dependency of thermoelectric materials properties is taken into account and plotted based on the data from [21,22] in Figure 4.

4.2 Finite element formulation

The governing equations are highly nonlinear due to the coupling of electrical and thermal conductions as well as the strong temperature dependence of the thermoelectric materials properties. They are solved in discretized forms by standard finite element method [23].

$$\begin{bmatrix} \mathbf{K}^{VT} & \mathbf{K}^{VV} \\ \mathbf{K}^{TT} & \mathbf{0} \end{bmatrix} \begin{Bmatrix} \mathbf{T}^n \\ \mathbf{V}^n \end{Bmatrix} = \begin{Bmatrix} \mathbf{0} \\ \mathbf{Q}^P + \mathbf{Q}^E \end{Bmatrix} \quad (12)$$

$$\mathbf{K}^{VT} = \sum_1^n \int_{\Omega^e} [B]^T \alpha \sigma [B] d\Omega^e \quad (13a)$$

$$\mathbf{K}^{VV} = \sum_1^n \int_{\Omega^e} [B]^T \sigma [B] d\Omega^e \quad (13b)$$

$$\mathbf{K}^{TT} = \sum_1^n \int_{\Omega^e} [B]^T \lambda [B] d\Omega^e \quad (13c)$$

$$\mathbf{Q}^P = \sum_1^n \int_{\Omega^e} [B]^T \beta [\mathbf{J}] d\Omega^e \quad (13d)$$

$$\mathbf{Q}^E = \sum_1^n \int_{\Omega^e} [N]^T [\mathbf{J}]^T [\mathbf{E}]^T d\Omega^e \quad (13e)$$

where $\mathbf{T}^n, \mathbf{V}^n$ represent the nodal temperature and nodal electrical potential vector respectively. N is the linear shape function and $B = \nabla N$. The finite element analysis is performed using mapped quadrilateral elements.

The above discretized system is obtained following the general procedures: first multiply the original governing equations with test functions, integrate over the whole domain by parts, and utilize the 2D divergence theorem.

4.3 Sensitivity analysis

The sensitivity information, i.e., the total derivative of objective functions with respect to the design variables, must be derived to update the design in each iteration. In this paper, the adjoint method [24] is employed, which starts from the discretized system as shown in Section 4.2. The equation (12) can be rewritten in the following residual form.

$$\mathbf{R}(\rho, \mathbf{U}) = \mathbf{K}(\rho, \mathbf{U}) \cdot \mathbf{U} - \mathbf{F}(\rho, \mathbf{U}) \quad (14)$$

where $\mathbf{R}$ is the residual vector; $\mathbf{K}$ is the global stiffness matrix; $\mathbf{U}$ is the state variable vector, i.e., $\{\mathbf{T}, \mathbf{V}\}$ in this case and $\mathbf{F}$ is the global load vector.

Then the general Lagrangian function can be formulated as:

$$\mathbf{L}(\rho, \mathbf{U}) = J(\rho, \mathbf{U}) + \lambda^T \cdot \mathbf{R}(\rho, \mathbf{U}) \quad (15)$$

λ^T is the Lagragian multiplier. The differentiation of the objective function J in relation to design variable ρ is calculated by:

$$\frac{dJ}{d\rho} = \frac{d\mathbf{L}}{d\rho} = \frac{\partial J}{\partial \rho} + \frac{\partial J}{\partial \mathbf{U}} \cdot \frac{d\mathbf{U}}{d\rho} + \lambda^T \cdot [\frac{\partial \mathbf{R}}{\partial \rho} + \frac{\partial \mathbf{R}}{\partial \mathbf{U}} \cdot \frac{d\mathbf{U}}{d\rho}] \quad (16)$$

The crucial point of the adjoint method in conducting sensitivity analysis is to get rid of the derivative of the state variables with respect to the design variables, which can be done by solving the following adjoint equation for proper λ^T:

$$\lambda^T \cdot \frac{\partial \mathbf{R}}{\partial \mathbf{U}} = -\frac{\partial J}{\partial \mathbf{U}} \quad (17)$$

Once λ^T is obtained, we can get the sensitivity information from equation (16).

5 PRELIMINARY RESULTS
5.1 Topology optimization results

The following topology optimization results in Figure 5 are obtained by setting the volume ratio $V_f = 0.6$ and the external load $R = 0.1\Omega$. The initial design is given such that $\rho_1 = V_f$ and $\rho_2 = 1$. With reference to Figure 4(d), the results make sense in that the relatively high-temperature thermoelectric materials PbTe and $CoSb_3$ are distributed near the hot end, while the relatively low-temperature thermoelectric materials Bi_2Te_3 near the cold end. In other words, each thermoelectric material is placed

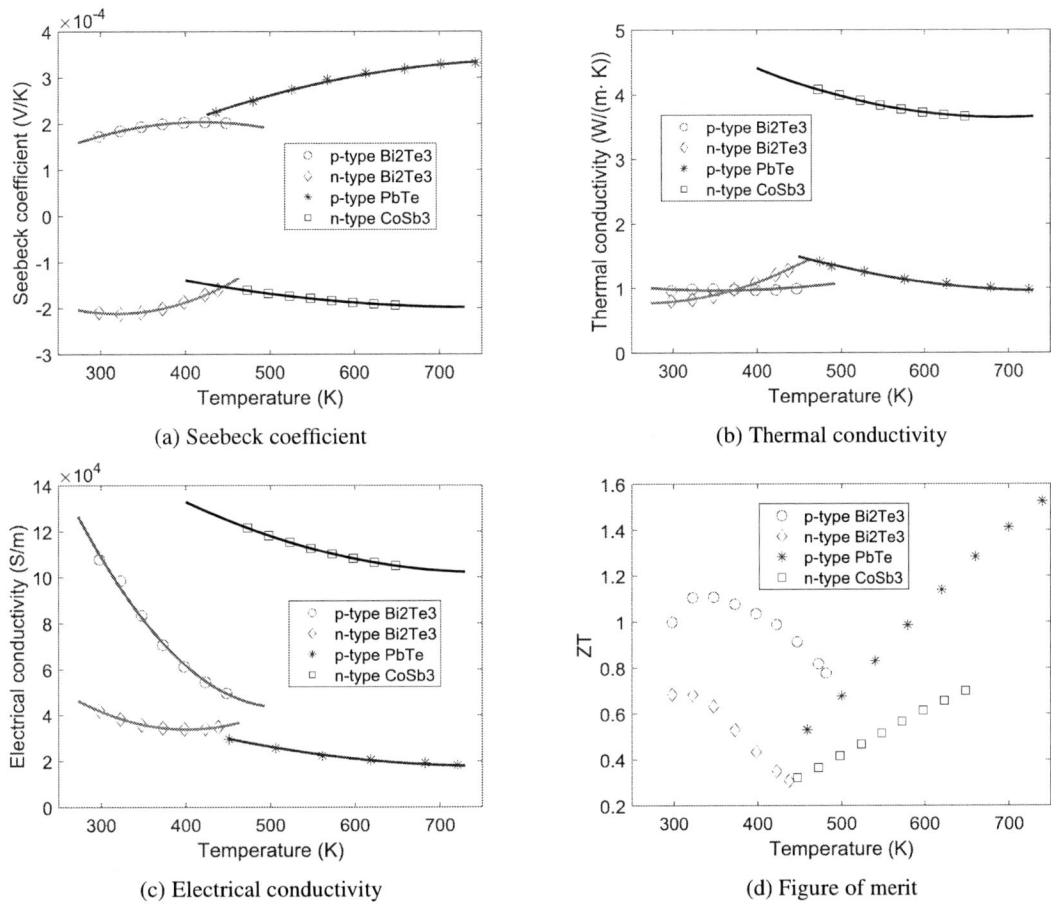

(a) Seebeck coefficient

(b) Thermal conductivity

(c) Electrical conductivity

(d) Figure of merit

FIGURE 4: Thermoelectric materials properties

into its optimal working temperature interval to fully exploit their thermoelectric capability. The conversion efficiency could reach 12.4%.

The convergence curves for conversion efficiency and output power are shown in Figure 6. As expected, both objectives improve significantly compared with the initial designs.

By tuning the volume ratio V_f from 1 to 0.4, we can obtain a series of optimized TEG structure as shown in Figure 7 and Figure 8. The highest conversion efficiency of 4 different volume ratio cases could reach 13.24%. It is noticeable from Figure 8a and 8b that although the initial volume ratio V_f is given as 1 and 0.8, the actual volume ratio of the final design is 0.936 and 0.745 respectively. This means that in terms of conversion efficiency, it is not the more thermoelectric materials, the higher conversion efficiency.

5.2 Numerical Verification

In this section, a 2.5D model is built and simulated based on the optimized TEG structures obtained from the topology optimization algorithm proposed in this paper. As shown in Figure 9 and Figure 10, two prototypes are built corresponding to the results for both output power and conversion efficiency as the objective function with the volume ration $V_f = 0.6$. The overall dimensions of these two models are $60 \times 10 \times 1$ mm, which is in consistence with the 2D geometry with a thickness of 1mm as described in Section 3.1. The same boundary conditions are applied as in Figure 2. That is: the two end surfaces of copper are assigned low temperature with $T_c=25\,^\circ$C, and the middle plane of the central copper electrode is treated as the hot end with $T_h=400\,^\circ$C. A zero electrical potential is placed on the lower end copper surface. And all other surfaces of the domain are prescribed as adiabatic and electrically insulated boundaries. The external resistor is still set to be $R = 0.1\Omega$.

The temperature and electrical potential distribution of the two simulated models are shown in Figure 11, 12, 13 and 14 re-

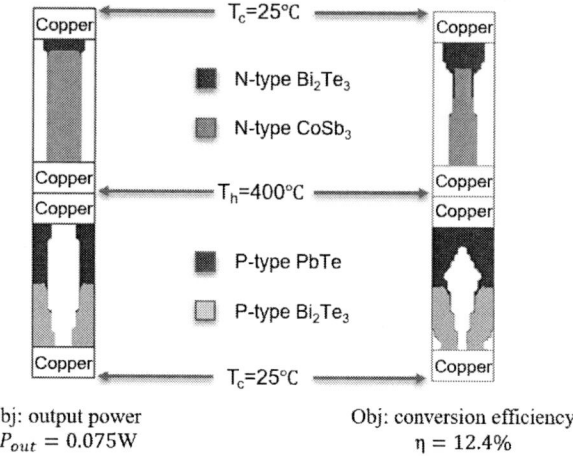

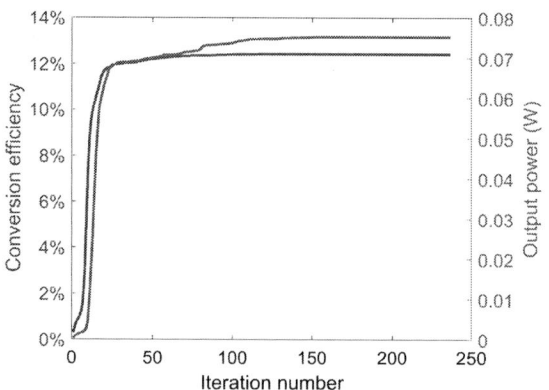

FIGURE 5: Topology optimization results for $V_f = 0.6$

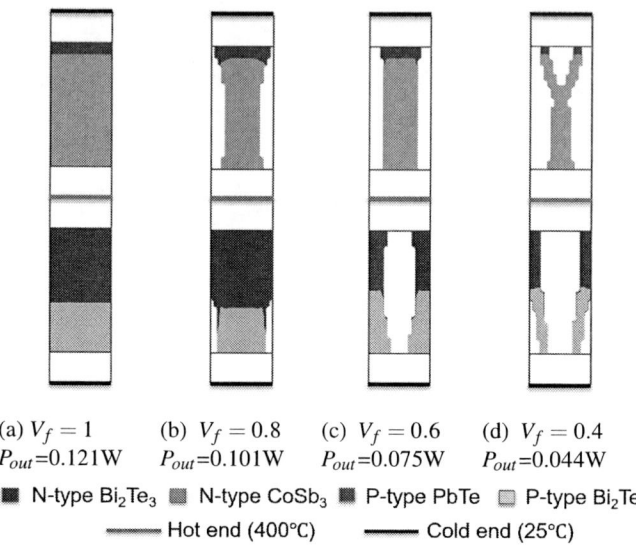

(a) $V_f = 1$	(b) $V_f = 0.8$	(c) $V_f = 0.6$	(d) $V_f = 0.4$
P_{out}=0.121W	P_{out}=0.101W	P_{out}=0.075W	P_{out}=0.044W

■ N-type Bi$_2$Te$_3$ ▨ N-type CoSb$_3$ ■ P-type PbTe ▢ P-type Bi$_2$Te$_3$
━━ Hot end (400°C)   ━━ Cold end (25°C)

FIGURE 7: Different results for output power as the objective

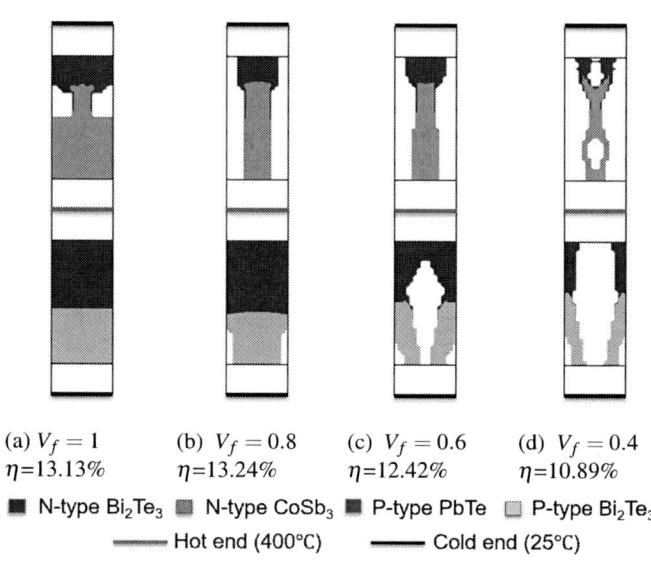

(a) $V_f = 1$	(b) $V_f = 0.8$	(c) $V_f = 0.6$	(d) $V_f = 0.4$
η=13.13%	η=13.24%	η=12.42%	η=10.89%

■ N-type Bi$_2$Te$_3$ ▨ N-type CoSb$_3$ ■ P-type PbTe ▢ P-type Bi$_2$Te$_3$
━━ Hot end (400°C)   ━━ Cold end (25°C)

FIGURE 8: Different results for efficiency as the objective

FIGURE 6: Convergence history for $V_f = 0.6$

spectively. It is worth noticing, however, that the simulated output power $P_{out} = 0.036$W and conversion efficiency $\eta = 7.30\%$ are quite smaller than those from the topology optimization results. The causes of these discrepancies will be discussed in details in Section 6.

6 DISCUSSION AND FUTURE WORK

In this paper, a SIMP-based topology optimization method is applied to the thermoelectric problems with the aim of maximizing the output power and conversion efficiency by optimizing the layouts of different thermoelectric materials in a prescribed domain. With the external load fixed at $R = 0.1\Omega$, the highest output power and conversion efficiency could reach 0.075W and 12.42% respectively for $V_f = 0.6$, which is quite promising compared with the results from [8]. However, these two indexes ob-

tained from the subsequent 2.5D simulation are not that appealing, with each dropping nearly by 50%. This could be attributed to the following aspects.

First, for a certain layout of the TEG structure, the maximum output power, and conversion efficiency can be achieved when the external load value equals to or approach the internal electrical resistance of the TEG [25]. This means that the value of the external resistor needs to be updated in each iteration of the

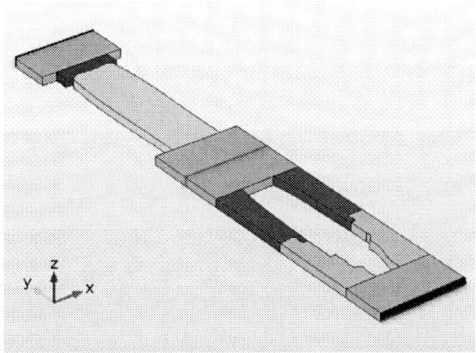

■ N-type Bi$_2$Te$_3$ ■ N-type CoSb$_3$ ■ P-type PbTe ☐ P-type Bi$_2$Te$_3$
—— Hot end (400°C) ☐ Copper —— Cold end (25°C)

FIGURE 9: 2.5D model corresponding to $V_f = 0.6$ and output power as the objective

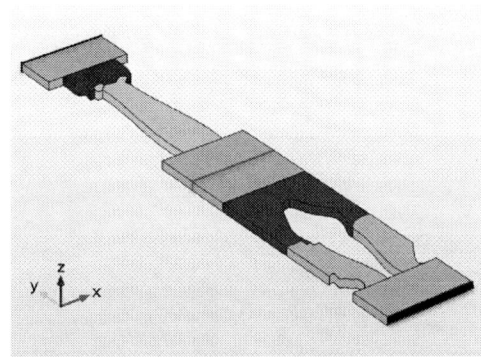

■ N-type Bi$_2$Te$_3$ ■ N-type CoSb$_3$ ■ P-type PbTe ☐ P-type Bi$_2$Te$_3$
—— Hot end (400°C) ☐ Copper —— Cold end (25°C)

FIGURE 10: 2.5D model corresponding to $V_f = 0.6$ and conversion efficiency as the objective

topology optimization process to match that of the TEG functioning like a battery. But in this paper, the external load is kept as a constant for all different volume ratio conditions, which could leave some space for improvement of the two objective functions. Nevertheless, it still makes sense, for example, when you intend to optimize a TEG structure to charge a device whose electrical resistance is a constant. In the 2.5D simulation, however, the internal resistance of the TEG changes due to the removing of TE materials. As a consequence, the ratio between the fixed external load R and the internal resistance changes accordingly, resulting in a lower output power and conversion efficiency than expected. This argument can be favored by the fact that an optimal external load R can be found to achieve the highest value of the two objective functions as shown in Figure 15. The black dash line

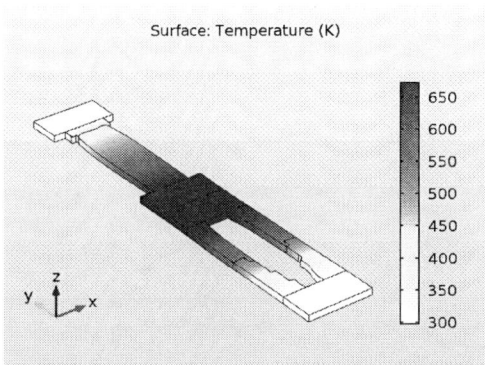

FIGURE 11: Temperature distribution corresponding to $V_f = 0.6$ and output power as the objective

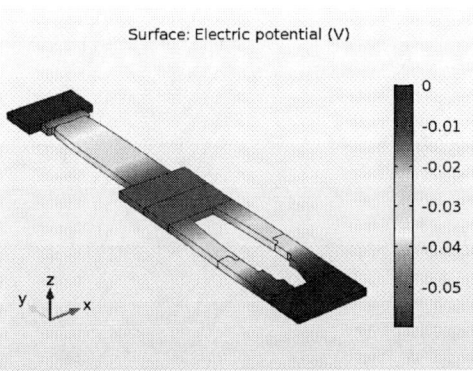

FIGURE 12: Electrical potential distribution corresponding to $V_f = 0.6$ and output power as the objective

corresponds to $R = 0.1\Omega$ as used in the 2.5D simulations.

Second and more important, as a common issue in SIMP-based topology optimization methods, there would exist a grey region representing the transition from one thermoelectric material to the other. Such a grey area corresponds to no physical material. One approach of getting rid of these transition zones is "power-law", in which a penalty factor is introduced in the interpolation scheme to penalize the intermediate density variables and force them to approach 0 or 1. In this paper, although the penalty factor p is set to be 3, the topology optimization results still show a quite large grey area. As shown in Figure 16, the surface for $\rho_1^p * \rho_2^p$ exhibits a quite large transition zone in the bottom P-type element (circled areas). The clear boundary between different thermoelectric materials can be obtained by setting up a threshold for ρ_1 and ρ_2 to distinguish one thermoelectric material from the other. However, this treatment would arouse another problem. The final designs are, to a certain degree, like segmented TEG. The performance of such structures depends not only on combining different thermoelectric materials but also on

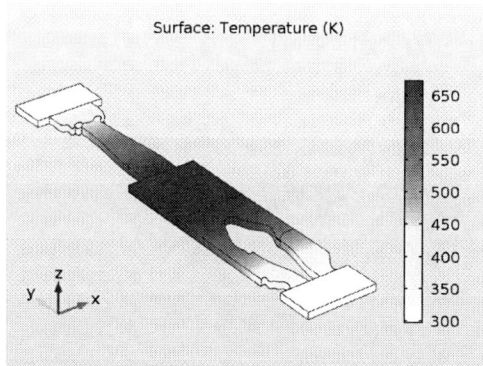

FIGURE 13: Temperature distribution corresponding to $V_f = 0.6$ and conversion efficiency as the objective

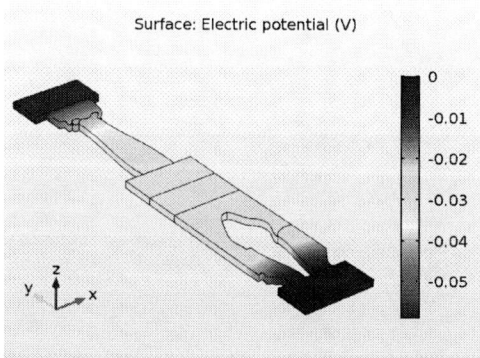

FIGURE 14: Electrical potential distribution corresponding to $V_f = 0.6$ and conversion efficiency as the objective

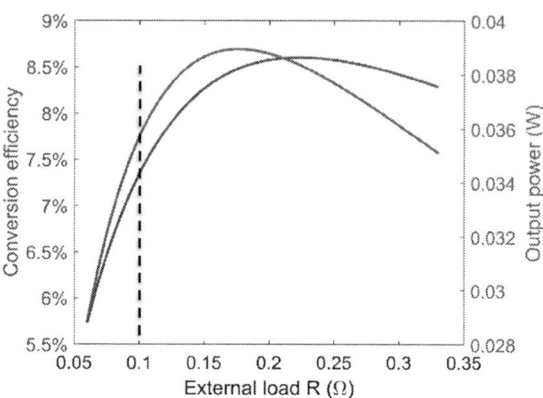

FIGURE 15: Conversion efficiency and output power as a function of external load R in the 2.5D simulations

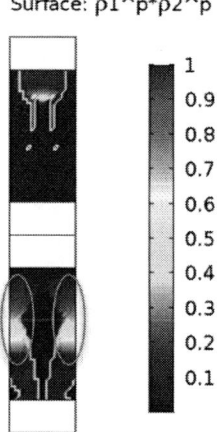

FIGURE 16: Grey region corresponding to $V_f = 0.6$ and conversion efficiency as the objective

the compatibility factor. The compatibility factors must be close enough to ensure high conversion efficiency [26, 27, 28]. Thus, merely giving a threshold to separate one thermoelectric material from the other manually could yield a substantial difference in the compatibility factors and further low output power and conversion efficiency.

Based on the above discussion, several strategies may be adopted in the future to further improve current results: (1) include an inner optimization process to update the value of external load to match that of the TEG to ensure highest output power and conversion efficiency in each iteration; (2) try different interpolation scheme to reduce the grey region; (3) turn to alternative topology optimization methods, e.g. level-set method, which is known for its ability for making clear boundaries, to readdress this problem.

ACKNOWLEDGMENT

The authors acknowledge the support from the National Science Foundation (Grants No. CMMI1462270, CMMI1762287, and ECCS1508862), Ford University Research Program (URP) (Award No. 2017-9198R), and the start-up grant from the State University of New York at Stony Brook.

REFERENCES

[1] Snyder, G. J., and Toberer, E. S. "Complex thermoelectric materials". *Nature materials,* **7**(2), p. 105.

[2] Zhang, X., and Zhao, L.-D. "Thermoelectric materials: Energy conversion between heat and electricity". *Journal of Materiomics,* **1**(2), pp. 92–105.

[3] Zhao, L.-D., Lo, S.-H., Zhang, Y., Sun, H., Tan, G., Uher, C., Wolverton, C., Dravid, V. P., and Kanatzidis, M. G. "Ultralow thermal conductivity and high thermoelectric figure of merit in SnSe crystals". *Nature, 508*(7496), p. 373.

[4] Biswas, K., He, J., Blum, I. D., Wu, C.-I., Hogan, T. P., Seidman, D. N., Dravid, V. P., and Kanatzidis, M. G. "High-performance bulk thermoelectrics with all-scale hierarchical architectures". *Nature, 489*(7416), p. 414.

[5] Hicks, L., and Dresselhaus, M. S. "Effect of quantum-well structures on the thermoelectric figure of merit". *Physical Review B, 47*(19), p. 12727.

[6] Pei, Y., Wang, H., and Snyder, G. J. "Band engineering of thermoelectric materials". *Advanced materials, 24*(46), pp. 6125–6135.

[7] He, J., Kanatzidis, M. G., and Dravid, V. P. "High performance bulk thermoelectrics via a panoscopic approach". *Materials Today, 16*(5), pp. 166–176.

[8] El-Genk, M. S., and Saber, H. H. "High efficiency segmented thermoelectric unicouple for operation between 973 and 300 K". *Energy Conversion and Management, 44*(7), pp. 1069–1088.

[9] Bendsøe, M. P., and Kikuchi, N. "Generating optimal topologies in structural design using a homogenization method". *Computer methods in applied mechanics and engineering, 71*(2), pp. 197–224.

[10] Bendsøe, M. P. "Optimal shape design as a material distribution problem". *Structural optimization, 1*(4), pp. 193–202.

[11] Zhou, M., and Rozvany, G. "The COC algorithm, Part II: topological, geometrical and generalized shape optimization". *Computer Methods in Applied Mechanics and Engineering, 89*(1-3), pp. 309–336.

[12] Allaire, G., Jouve, F., and Toader, A.-M. "Structural optimization using sensitivity analysis and a level-set method". *Journal of computational physics, 194*(1), pp. 363–393.

[13] Wang, M. Y., Wang, X., and Guo, D. "A level set method for structural topology optimization". *Computer methods in applied mechanics and engineering, 192*(1-2), pp. 227–246.

[14] Bourdin, B., and Chambolle, A. "Design-dependent loads in topology optimization". *ESAIM: Control, Optimisation and Calculus of Variations, 9*, pp. 19–48.

[15] Sokolowski, J., and Zochowski, A. "On the topological derivative in shape optimization". *SIAM journal on control and optimization, 37*(4), pp. 1251–1272.

[16] Landau, L. D., Bell, J., Kearsley, M., Pitaevskii, L., Lifshitz, E., and Sykes, J., 2013. *Electrodynamics of continuous media*, Vol. 8. elsevier.

[17] Takezawa, A., and Kitamura, M. "Geometrical design of thermoelectric generators based on topology optimization". *International Journal for Numerical Methods in Engineering, 90*(11), pp. 1363–1392.

[18] Bell, L. E. "Cooling, heating, generating power, and recovering waste heat with thermoelectric systems". *Science, 321*(5895), pp. 1457–1461.

[19] Bendsøe, M. P., and Sigmund, O. Topology optimization: theory, methods and applications. 2003.

[20] Svanberg, K. "The method of moving asymptotesa new method for structural optimization". *International journal for numerical methods in engineering, 24*(2), pp. 359–373.

[21] Kajikawa, T., Rowe, D., and Rowe, D. "Thermoelectric handbook: macro to nano". *CRC/Taylor & Francis, Boca Raton, FL, USA.*

[22] Heremans, J. P., Jovovic, V., Toberer, E. S., Saramat, A., Kurosaki, K., Charoenphakdee, A., Yamanaka, S., and Snyder, G. J. "Enhancement of thermoelectric efficiency in PbTe by distortion of the electronic density of states". *Science, 321*(5888), pp. 554–557.

[23] Zienkiewicz, O. C., Taylor, R. L., Zienkiewicz, O. C., and Taylor, R. L., 1977. *The finite element method*, Vol. 36. McGraw-hill London.

[24] Allaire, G. "A review of adjoint methods for sensitivity analysis, uncertainty quantification and optimization in numerical codes". *Ingénieurs de l'Automobile, 836*, pp. 33–36.

[25] Goldsmid, H. J., 2010. *Introduction to thermoelectricity*, Vol. 121. Springer.

[26] Wu, Y., Yang, J., Chen, S., and Zuo, L. "Thermo-element geometry optimization for high thermoelectric efficiency". *Energy, 147*, pp. 672–680.

[27] Snyder, G. J. "Application of the compatibility factor to the design of segmented and cascaded thermoelectric generators". *Applied physics letters, 84*(13), pp. 2436–2438.

[28] Hadjistassou, C., Kyriakides, E., and Georgiou, J. "Designing high efficiency segmented thermoelectric generators". *Energy conversion and management, 66*, pp. 165–172.

Proceedings of the ASME 2019
International Design Engineering Technical Conferences
and Computers and Information in Engineering Conference
IDETC/CIE2019
August 18-21, 2019, Anaheim, CA, USA

DETC2019-98105

BAYESIAN OPTIMIZATION OF EQUILIBRIUM STATES IN ELASTOMERIC BEAMS

David Yoo, Carson Wiley, Andrew Gillman, Vincent Chen
UES, Inc.
4401 Dayton Xenia Road
Beavercreek OH 45432
david.yoo.1.ctr@us.af.mil,
carson.willey.ctr@us.af.mil, andrew.gillman.2@us.af.mil,
vincent.chen.2.ctr@us.af.mil

Abigail Juhl, Philip Buskohl*
Air Force Research Laboratory
2179 12th Street
Wright-Patterson AFB, OH 45433
abigail.juhl.1@us.af.mil, philip.buskohl.1@us.af.mil
* Corresponding author

ABSTRACT

Architected elastomers have demonstrated great potential for energy absorption, multi-resonant vibration isolation, and multi-bandgap acoustic control, due to the reversibility and programmability of their mechanical instabilities. However, computational design tools are needed to explore the large parameter space that regulates buckling-based mechanical instability behavior. In this study, we develop a machine-learning-based design optimization framework to control equilibrium states of a bistable elastomeric beam, while also regulating the required energy to transition between these configurations. Leveraging symmetry to reduce the design space, the research is performed on a single element, an inclined, slender beam, through a Fourier series parameterization. To evaluate the force-displacement response of the bistable beam, a nonlinear finite element analysis (FEA) with an arc-length continuation method is employed in a commercial FEA tool. Due to the highly non-convex bistability response of the beam in this proposed design space and the computational cost of the FEA analysis, a Bayesian optimization is implemented to promote a better trade-off between the number of function evaluations and rate of convergence. Bayesian optimization depends on several optimization parameters, which are systematically tuned in this study to characterize their role on the optimization process. With the proposed method, the equilibrium displacement and the ratio of output to input energy between stable states can be optimized within a few tens of iterations. A multi-objective optimization is also carried out to

study the trade-off between equilibrium position and the energetics to transition between the bistabilities.

KEYWORDS: *Architected Elastomers, Multi-Resonant Vibration Control, Bistable Beam, Finite Element Analysis, Machine Learning, Bayesian Optimization*

1. INTRODUCTION

Hyperelastic lattice-based architectures have demonstrated potential for energy absorption, and multi-resonant vibration isolation and bandgap controls, due to their inherent buckling instabilities [1-3]. Compared to other traditional acoustic and vibration metamaterials [4-6], tunable vibration control is enabled for architected elastomers through the multi-stable configurations which exist in a single design. Mechanical constraints play a critical role in the accessibility of multi-stable behavior, as demonstrated in a previous study on prismatic, slender beams [1]. Shan et al demonstrated a transition between mono-stable and bi-stable beam buckling by tuning the aspect ratio and angle of the inclined beam. Our study builds on this work by exploring the untapped design space of non-prismatic beam geometries to further tune the equilibrium position and energy ratios of the bi-stable states.

In this study, we carry-out shape optimization on a single slender bistable beam defined with a clamped-roller boundary condition. We consider two objectives: i) minimize displacement at the second stable state and ii) the energy ratio between the stable configurations. In the proposed design space, the thickness of the beam along the length direction is

This work was authored in part by a U.S. Government employee in the scope of his/her employment.
ASME disclaims all interest in the U.S. Government's contribution.

Copyright © 2019 ASME

prescribed using a Fourier series, with the coefficients serving as the design variables. The Fourier series representation guarantees a continuous shape for the beam that is directly manufacturable, without incorporating a post-process thresholding step often required in SIMP-based optimization approaches [7, 8]. For a given periodic beam design, a FEA is carried-out using an arc-length continuation method [10, 11] to determine the nonlinear force-displacement response of beam under large strains. Given the geometric and material nonlinearity of the beam, the response surfaces of the objective functions are highly non-convex.

To navigate this complex design space, a Bayesian optimization approach is implemented to optimize the beam shape with respect to its bistability behavior. Bayesian optimization is a robust global-based optimization method that generally works well for highly non-convex problem, and especially for those in which the response of interest contains a significant amount of noise [12-16]. Bayesian methods have previously been leveraged to classify the attenuation behavior in elastomers loaded with negative stiffness inclusions [17, 18]. The primary advantage of Bayesian optimization is the potential for convergence under fewer function evaluations, which is important due to the required nonlinear FEA. At each iteration of Bayesian optimization, a Gaussian process (GP) surrogate model is estimated, based on training data from the FE model. The surrogate model provides insight into the global behavior of the design space, which can also assist the designer build intuition.

The challenges in employing Bayesian optimization are selection of the best covariance function to predict the objective function using GP, and the determination of a good balance between exploration and exploitation. Once those can be determined, hyper-parameters, i.e., kernel length scale and kernel scaling parameter, are then optimized by maximizing the log-likelihood function. Another challenge is the occurrence of mono-stable solutions in the design space, which create non-defined values for the equilibrium position objective function. In order to still incorporate these solutions in the GP model, a large fictitious equilibrium position is assigned as a penalty to inform the Bayesian model that poor performing candidates exist in this region. The penalty displacement for non-bistable design is also an important parameter tuned in this study. Bayesian optimization is carried-out to obtain target displacement and stability at the second bistable state. The multi-objective optimization is then carried-out to find the compromised design between target displacement and targeted energy ratio at the second stable state.

2. DESIGN PROBLEM SETUP AND ANALYSIS

In this study, the thickness of the beam along the length direction is defined with a set of Fourier series coefficients. The upper contour of the proposed beam design can be defined as

$$f_{top}(x) = \frac{a_{10}}{2} + \sum_{n=1}^{m} r_{1n} \sin\left(\frac{P_{1n}\pi x}{L} + \alpha_{1n}\right)$$

where the magnitude, r_{1n}, periodicity, P_{1n}, and phase, α_{1n}, for n = 1 to m are design variables for the beam geometry. a_{10} is determined to maintain constant area between beam design remains identical and is given as

$$L_o\left(\frac{t_o}{2}\right) - \int_0^L \frac{a_{10}}{2} + r_{11}\sin\left(\frac{P_{11}\pi x}{L} + \alpha_{11}\right) + r_{12}\sin\left(\frac{P_{12}\pi x}{L} + \alpha_{12}\right) + \cdots + r_{1n}\sin\left(\frac{P_{13}\pi x}{L} + \alpha_{13}\right) dx = 0$$

(1)

where t_o and L_o are thickness and length of the original/baseline beam. While the design of baseline beam is defined as rectangular beam with t_o and L_o, change of magnitude, periodicity, and phase of the proposed beam can be varied with r_{1j}, p_{1j}, and α_{1j}, respectively. The geometry of bottom beam contour is similarly defined (a_{20}, r_{2j}, p_{2j}, and α_{2j}). The schematic for the proposed design space with Fourier series coefficients is shown in Fig. 1 (a). Schematic of the boundary condition of the proposed beam design with inclined angle is shown in Fig. 1 (b),

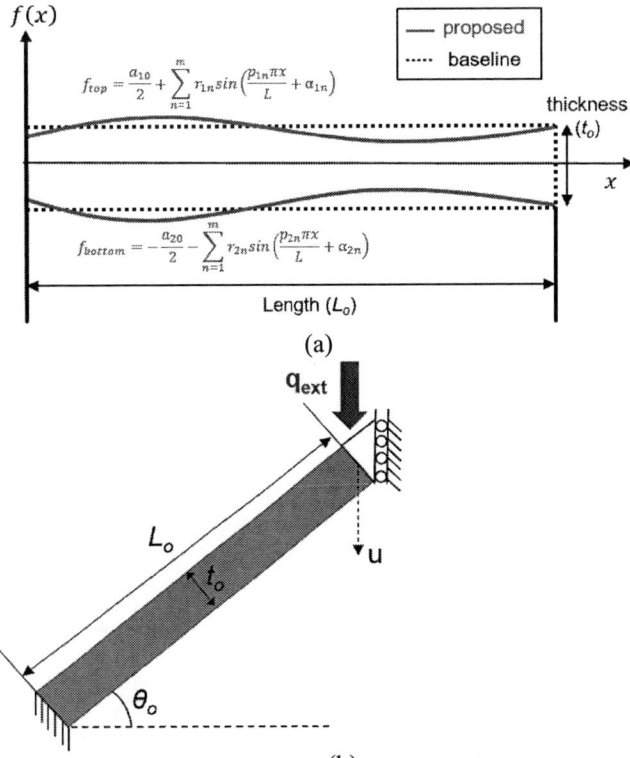

FIGURE 1. Bistable beam design space and boundary conditions. (a) Design space of beam shape is parameterized by a Fourier series expansion. (b) Schematic of beam with inclination angle θ_0 and applied displacement boundary conditions [1].

where q_{ext} is the applied external force, and the u is the corresponding displacement. Vertical displacement constrains the horizontal motions and enables a bistable configuration [1].

The bistable behavior for a given beam design is characterized by a nonlinear force-displacement response, which can be obtained with FEA using a Newton Raphson method or Arc-Length continuation scheme. Here quasi-static FEA is carried-out while neglecting any dynamic/inertial effects. While Newton Raphson method, in general, works efficiently up to a critical point, it usually fails to converge for highly nonlinear response, *e.g.*, response of bistable beam with multiple critical points (Fig. 2). In the Newton Raphson method, the force-displacement relationship is obtained by iteratively applying the constant step incremental force to the structure, and solving the effective change of the displacement. The process is repeated until the full response is obtained. In an arc-length method, the step incremental force is modified at every iteration to prevent non-converging problem [10, 11]. The method is briefly explained in the following section.

The equilibrium condition should be satisfied at every point in force-displacement response in Fig, 2, which means the applied external force and internal reaction force from the structure should balance with each other. The equilibrium condition at any displacement in Fig. 2 can be described as

$$\mathbf{g}(\mathbf{u}, \lambda) = \mathbf{q}_i(\mathbf{u}) - \lambda \mathbf{q}_{ef} = 0, \tag{2}$$

where **u** is the displacement vector of FE model, $\mathbf{q}_i$ is the vector of internal reaction forces from FE model, λ is the parameter that determines magnitude of external force, and $\mathbf{q}_{ef}$ is the vector of the unit of external forces. Here, instead of applying a constant λ, it is adjusted at every step to improve convergence. For the arc-length method, the incremental form of the equation is given as

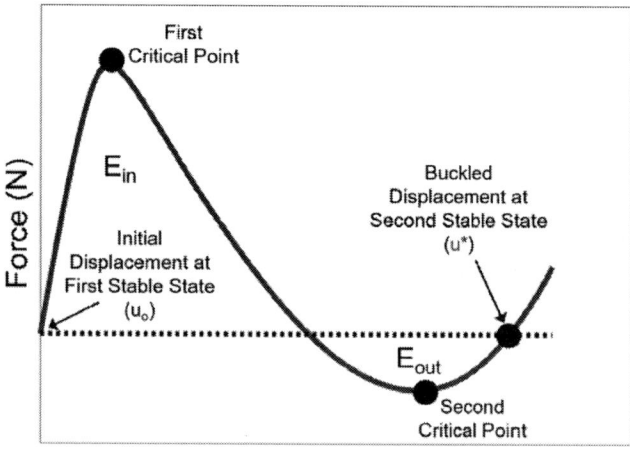

FIGURE 2. Schematic for force-displacement response of bistable beam, which contains two stable equilibrium points (u_o and u^*). Energies required to switch between the equilibrium configurations are labeled E_{in} and E_{out}.

$$a = \left(\Delta \mathbf{u}^T \Delta \mathbf{u} + \Delta \lambda^2 \varphi^2 \mathbf{q}_{ef}^T \mathbf{q}_{ef}\right) - \Delta l^2 = 0, \tag{3}$$

where Δl is the radius of intersection at equilibrium, φ is used for scaling between external force and displacement, and $\Delta \mathbf{u}$ and $\Delta \lambda$ are incremental displacement vector and external force parameter [10, 11]. Equations (2) and (3) can be approximated using Taylor series expansion as

$$\mathbf{g}_n = \mathbf{g}_o + \frac{\partial \mathbf{g}}{\partial \mathbf{u}} \delta \mathbf{u} + \frac{\partial \mathbf{g}}{\partial \lambda} \delta \lambda \tag{4a}$$

$$a_n = a_o + 2\Delta \mathbf{u}^T \delta \mathbf{u} + 2\Delta \lambda \delta \lambda \varphi^2 \mathbf{q}_{ef}^T \mathbf{q}_{ef}, \tag{4b}$$

where subscripts o and n represent old and new, respectively. Solving Eq. (4a) and (4b) by taking both $\mathbf{g}_n$ and a_n as 0 then yields

$$\begin{bmatrix} \delta \mathbf{u} \\ \delta \lambda \end{bmatrix} = - \begin{bmatrix} [K_T] & -\mathbf{q} \\ 2\Delta \mathbf{u}^T & 2\varphi^2 \Delta \lambda (\mathbf{q}^T \cdot \mathbf{q}) \end{bmatrix}^{-1} \begin{bmatrix} \mathbf{g}_o \\ a_o \end{bmatrix}, \tag{5}$$

where K_T is the tangent stiffness matrix. Once Eq. (5) is solved for $\delta \mathbf{u}$ and $\delta \lambda$, $\Delta \mathbf{u}$ and $\Delta \lambda$ can be updated at every iteration, while obtaining force-displacement response.

While the tilted slender beam can contain a geometric nonlinearity with a linear constitutive model (Hooke's Law), a nonlinear one can amplify the nonlinearity. In Ref. [1], a tilted slender beam with PDMS material exhibits strong bistable behavior, and it is demonstrated that it can be very precisely modelled using Neo-Hookean model up to a strain of about 100% [1,7,8], which is given as

$$W = C_1(\bar{I}_1 - 3) + D_1(J - 3), \tag{6}$$

where C_1 and D_1 are material constants; $\bar{I}_1 = J^{-2/3}(\omega_1^2 + \omega_2^2 + \omega_3^2)$, where J is Jacobian matrix and ω_i are the principal stretches of the right Cauchy-Green stretch tensor. In Eq. (6), $C_1 = \frac{\mu}{2}$ and $D_1 = \frac{K}{2}$ where μ and K are shear and bulk moduli, respectively. The shear modulus of the proposed beam in this study is assigned as $\mu = 0.32$MPa. A very small value of K is assigned such that $\frac{K}{\mu} \approx 2500$ to model the nearly incompressibility.

3. GAUSSIAN PROCESS REPRESENTATION OF DESIGN SPACE & ACQUISITION FUCTION
3.1 Gaussian Process

In Bayesian optimization, the decision for design update is made based on a response function modeled with a Gaussian Process (GP) [12-16]. In this study, a GP surrogate is applied to predict the objective function response surface based on initial training data of force-displacement profiles evaluated by FEA. Using the Gaussian process, the joint distribution of the response functions **h**, and the training data $\mathbf{h}^*$ can be obtained based on a prior as [12]

$$\begin{bmatrix} \mathbf{h} \\ \mathbf{h}^* \end{bmatrix} \sim N \left(\mathbf{0}, \begin{bmatrix} K(\mathbf{X},\mathbf{X}) & K(\mathbf{X},\mathbf{X}^*) \\ K(\mathbf{X}^*,\mathbf{X}) & K(\mathbf{X}^*,\mathbf{X}^*) \end{bmatrix} \right), \qquad (7)$$

where $\mathbf{X}$ and $\mathbf{X}^*$ are training inputs and inputs for response prediction; N represents a normal distribution and K is the covariance function. The predictive equations for Gaussian process regression based on observations are given as [12]

$$\mathbf{h}^*|\mathbf{X}^*,\mathbf{X},\mathbf{h} \sim N \left(\begin{matrix} K(\mathbf{X}^*,\mathbf{X})K(\mathbf{X}^*,\mathbf{X})^{-1}\mathbf{h}, \\ K(\mathbf{X}^*,\mathbf{X}^*) - K(\mathbf{X}^*,\mathbf{X})K(\mathbf{X},\mathbf{X})^{-1}K(\mathbf{X}^*,\mathbf{X})^{-1} \end{matrix} \right). \tag{8}$$

As shown in Eq. (8), the predicted response surface highly depends on the type of selected covariance function. By selecting the appropriate covariance function, the response surface can be approximated closely to the true function.

The most well-known covariance function is the squared exponential, whose equation is given as [12]

$$k_{\text{SE}}(r) = \exp\left(-\frac{r^2}{2l^2}\right), \qquad (9)$$

where parameter l is the length scale that determines how far the function can be extrapolated from the training data. This is a relatively smooth covariance function and is infinitely differentiable. However, many physical systems possess discontinuities, which as not well estimated by the intrinsic smoothness of the squared exponential kernel [13]. The transition between mono-stable and bi-stable solutions in our problem creates a discontinuity, which may be effected by this limitation of the exponential kernel. The Matérn class of covariance functions was developed to address this discontinuity issue, and is given by

$$k_{\text{Matérn}}(r) = \frac{2^{1-\nu}}{\Gamma(\nu)}\left(\frac{\sqrt{2\nu}r}{l}\right)^\nu K_\nu\left(\frac{\sqrt{2\nu}r}{l}\right), \qquad (10)$$

where K_ν is the modified Bessel function. As shown in Eq. (10), the covariance function highly depends on the parameter, ν. The most widely used Matérn covariance functions are the ones with $\nu = 0.5, 1.5, 2.5$, with which Eq. (10) is simplified as

$$k_{\nu=1/2}(r) = \exp\left(-\frac{r}{l}\right)$$
$$k_{\nu=3/2}(r) = \left(1 + \frac{\sqrt{3}r}{l}\right)\exp\left(-\frac{\sqrt{3}r}{l}\right)$$
$$k_{\nu=5/2}(r) = \left(1 + \frac{\sqrt{5}r}{l} + \frac{5r^2}{3l^2}\right)\exp\left(-\frac{\sqrt{5}r}{l}\right) \qquad (11)$$

Moreover, it is worth noting that Matérn covariance in Eq. (10) approaches the squared-exponential covariance function in Eq. (9), as ν approaches infinity. In general, the smaller ν is, the less smooth the covariance function becomes. In this research,

different covariance functions in Eq. (9) and (11) are studied to best predict response functions.

3.2 Acquisition Function: Lower Confidence Bound

The Bayesian optimizer makes decisions using the distribution of the response function while balancing *exploration* and *exploitation* [14, 15]. Exploration corresponds to searching the design space where large uncertainties in the response exist, while exploitation can be done by seeking the design space where average response is minimized. If optimization is biased toward exploration, it can become inefficient (requiring more iterations), while if it is biased toward exploitation the algorithm may converge to a local optima. To determine an appropriate balance, the lower confidence bound acquisition function is used in this study, which is given as

$$a_{\text{LCB}}(x;\beta) = \mu(x) - \beta\sigma(x), \qquad (12)$$

where β is the trade-off parameter that controls between exploration and exploitation; $\mu(x)$ and $\sigma(x)$ are mean and variance of the predicted response surface as a function of design space x.

4. BAYESIAN OPTIMIZATION OF BISTABLE BEAM
4.1 Design Optimization Formulation

The general optimization problem for beam design on its bistability behavior can be formulated using a design space with Fourier series coefficients (Fig. 1) as

$$\text{find } (r_{1i}, p_{1i}, \alpha_{1i}, r_{2i}, p_{2i}, \alpha_{2i})$$

$$\text{to minimize } O = z\big(f(r_{1i}, p_{1i}, \alpha_{1i}, r_{2i}, p_{2i}, \alpha_{2i})\big)$$

$$\text{subject to:}$$
$$\text{top beam constraint:} r_{1i}^L \le r_{1i} \le r_{1i}^U, p_{1i}^L \le p_{1i} \le p_{1i}^U, \alpha_{1i}^L$$
$$\le \alpha_{1i} \le \alpha_{1i}^U$$
$$\text{bottom beam constraint:} r_{2i}^L \le r_{2i} \le r_{2i}^U, p_{2i}^L \le p_{2i}$$
$$\le p_{2i}^U, \alpha_{2i}^L \le \alpha_{2i} \le \alpha_{2i}^U$$
$$\tag{13}$$

where superscripts L and U represent lower and upper bounds for each design parameter. Although there are infinitely many design variables in the general Fourier series formulation posed in Eq. (13), only a finite set of coefficients are used as design variables for practical purpose, with which bistability behavior can be still well-controlled in the proposed design space. Once we obtain a force-displacement response based on the arc-length method with FEA in **Section 2** for a given Fourier series beam design $\big(f(r_{1i}, p_{1i}, \alpha_{1i}, r_{2i}, p_{2i}, \alpha_{2i})\big)$, we can then evaluate objectives (O) in Eq. (13). Two objectives for design optimization are considered in this study, one is displacement at second stable state (u* in Fig. 2), and other is energy ratio between stable state (s*), which is defined as

$$s^* = \frac{E_{out}}{E_{in}} \qquad (14)$$

where E_{in} and E_{out} are input and output energies (Fig. 2). In Eq. (14), E_{out} becomes 0 when the beam is not bistable, thus s* is 0. When $E_{out} = 0$, the displacement of the second stable state simply does not exist. In this study, a penalty displacement is applied for non-bistable designs to inform the GP surrogate that this a not a good solution. The tuning of this penalty term is explored in **Section 4.2**.

The interaction between the displacement and the energy ratio objective is also of interest for this study. This can be investigated by solving a multi-objective optimization, for which the objective function in Eq. (13) can be defined as compound objective function, which is given as

$$O = z\big(f(r_{1i}, p_{1i}, \alpha_{1i}, r_{2i}, p_{2i}, \alpha_{2i})\big) = w_1 \frac{u^*}{u_o} - w_2 \frac{s^*}{s_o} \quad (15)$$

where $w_1 + w_2 = 1$, and w_1 and w_2 are weights applied to each objective; u_o and s_o are displacement and stability at second stable state for baseline beam design, and they can be used to normalize scales.

4.2 Optimization Parameter Tuning

Accuracy and efficiency of Bayesian optimization highly depend on optimization parameter tuning. In this section, smoothness of the Matérn kernel ($v = 1/2, 3/2, 5/2$), the trade-off parameter between exploration and exploitation of the acquisition function ($\beta = 1, 3, 5$) and the penalty displacement applied to non-bistable designs are systematically studied. For optimization parameter tuning, two Fourier series coefficients are used as design variables, which are $t/10 \leq r_{11} \leq t/4$ and $0 \leq p_{11} \leq 8$. In addition, the bottom layer is constrained that $r_{21} = r_{11}$ and $p_{21} = p_{11}$, such that it mirrors the top layer. Slenderness ratio (t_o/L_o) and inclination angle (θ_o) for baseline design are 0.12 and 40°, respectively, where t_o is 1.0 mm. 100 training data points are used for GP prediction with squared exponential covariance function to determine the penalty displacement, and 5 initial random training data points are used for Bayesian optimization to determine both optimal values of β and v.

Figure 3 (a) shows a contour plot of the GP predicted surface with only bistable training data (76 of 100) on the displacement at the second stable state (u*) in terms of the magnitude change (r_{11}) and periodicity (p_{11}) of the beam design. The GP surrogate predicts a weakly bistable beam for the design with the smallest displacement, as shown in the force-displacement response of Fig. 3 (e). Interestingly, the GP surface indicates a basin of small displacements between the periodicities of 2 and 6, which intensifies with increasing magnitude. Figure 3 (b) – (d) show the results of the GP predictions, with different values of *penalty displacements* (PD = 5.0 mm, 6.0 mm, and 7.0 mm) applied to the monostable training data. The force-displacement responses of the optimal designs obtained from GP predictions from (b) – (d) are all shown in Fig. 3 (f). With a penalty displacement of 5.0 mm, the obtained optimal design is monostable, indicating the error in the GP estimate. With both penalty displacements of 6.0 mm and 7.0 mm, the obtained optimal design exhibits bistabilities (When the penalty displacement is 6.0 mm, the energy ratio at the second stable is nearly zero (s* $\cong$ 0)). The error of prediction from GP from the FE model with penalty displacement equal to 6.0 mm is found to be 6.1 %.

Observation on training data in Fig. 3 (a) – (d) indicates that u* as a function of magnitude change and periodicity of beam design would be highly nonlinear, as bistable and monostable regions are not clearly divided. Moreover, while little effect of u* to magnitude change is observed on GP prediction without monostable training data (Fig. 3 (a)), GP predicted surfaces significantly vary in terms of magnitude change, when penalty displacements are applied. The penalty approach enables the mono-stable data to influence the GP surrogate and improve the prediction of bistable designs. However, the penalty creates a sharp transitions in the response surface, which may be difficult for the GP surrogate to capture. According to Fig. 3 (c) and (f), the optimal region is effectively localized in the GP predicted surface with the penalty function of 6.0 mm, which will be used in the design case studies.

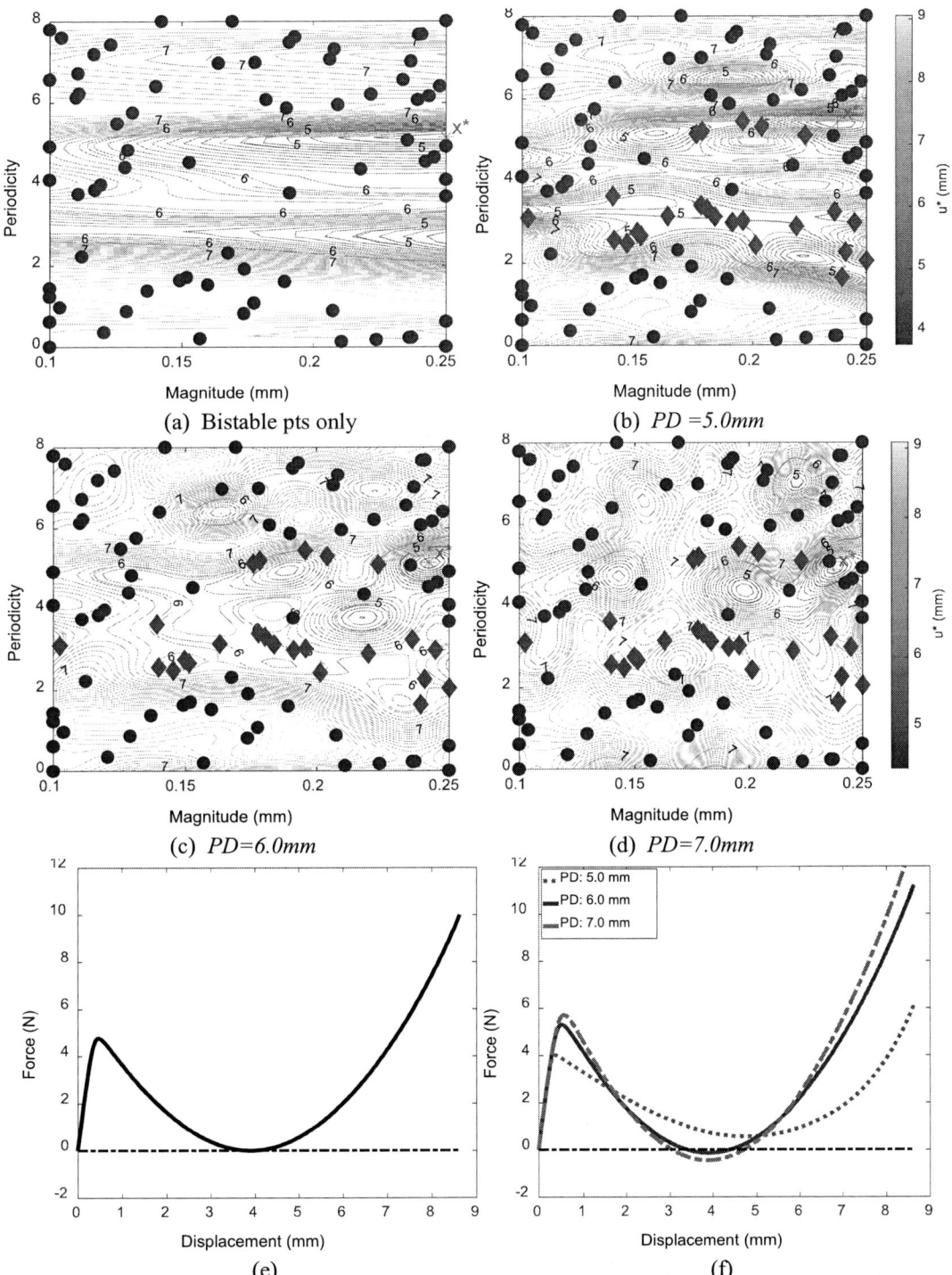

FIGURE 3. Penalty displacement of monostable solutions reshapes GP surrogate to promote bistable designs. (a) - (d): Contour plots of GP predictions with squared exponential covariance functions using 100 training data points. (bistable training data: ●; monostable training data: ◆; x*: predicted optimal design). Only bistable training data is considered for (a), and different penalty displacements of (b) 5.0 mm, (c) 6.0 mm, and (d) 7.0 mm are applied for monostable training data for (b) – (d). With penalty displacement of (c) 6.0 mm, optimal region is effectively localized in GP predicted surface (e) & (f): Force displacement responses of optimal designs obtained from GP predictions of (a) – (d). In (f), u* is minimized with penalty displacement of 6.0 mm.

V02AT03A065-6   Copyright © 2019 ASME

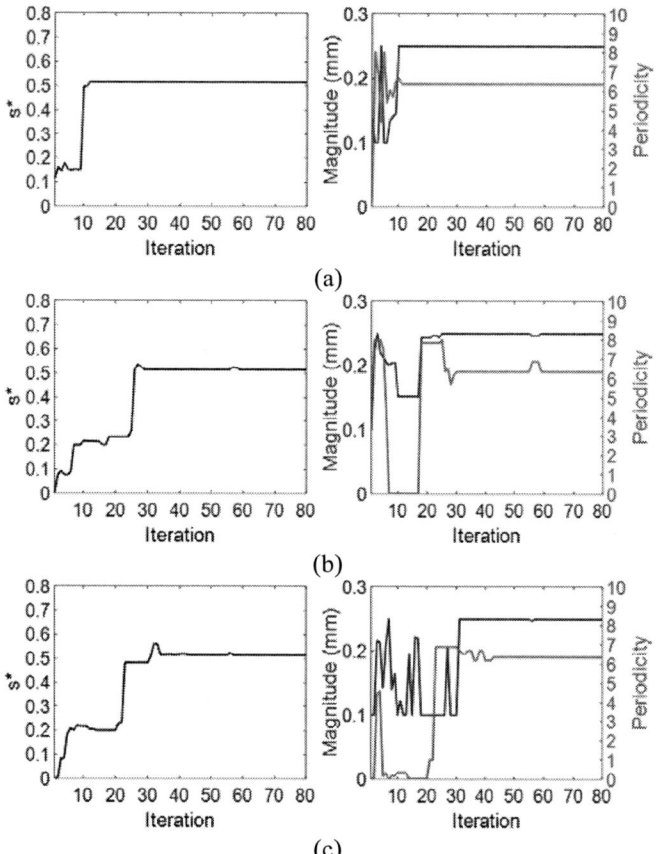

FIGURE 4. Exploration increases iteration count before convergence. Optimal design search history for different balances of exploitation and exploration: (a) $\beta = 1$, (b) $\beta = 3$, and (c) $\beta = 5$. The quickest convergence is observed when $\beta = 1$.

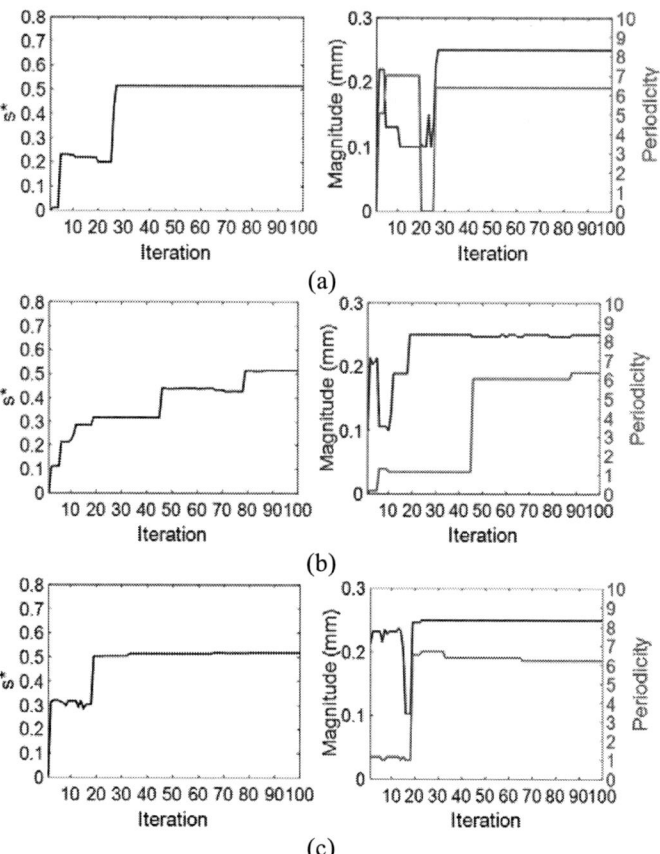

FIGURE 5. Shown are optimal design search history for different types covariance function with different smoothness: (a) $v = 1/2$, (b) $v = 3/2$, and (c) $v = 5/2$

Figure 4 shows the design search histories for magnitude change and periodicity of beam design with different value of trade-off parameter (β) between exploration and exploitation. The objective is to maximize the energy ratio between the equilibrium states (s*). For all values of $\beta = [1, 3, 5]$, it is shown that they are all very quickly converged to the identical optimal designs, within 40 iterations. The Bayesian optimization converged most quickly to an optimal design when $\beta = 1$, highlighting the strength of an exploration biased acquisition function for this design problem.

Figure 5 show design search histories for magnitude change and periodicity of beam design with different value of v, which define the smoothness of the Matérn covariance function. Once again, the optimizer converged to the correct design for all v values. The convergence was fastest with the smallest v value ($v = \frac{1}{2}$), which better captures the non-smoothness of the design space. However, there is not a clear trend in convergence with increasing v values, which could be a consequence of the rather limited training dataset used in this parameter study.

4.3 Design Optimization Case Studies

Based on the optimization parameter tuning in the previous section, the design optimization was performed with the following parameters: $\beta = 3$, $v = \frac{1}{2}$, and penalty displacement of 6.0 mm. The baseline design and design space are all identical to the ones used in **Section 4.2**.

The first design problem is to minimize the location of the stable configuration (u*). The search history to minimize the displacement at the second stable state is shown in Fig. 6 (a) and (b), which shows that it converges very efficiently to the optimal design within 25 iterations. The obtained optimal change of magnitude was 0.25 mm (max allowable), optimal periodicity is 5.1, and the minimized displacement was 4.58 mm. The displacement was reduced 42% from the baseline design (5.1 mm vs 7.9 mm). The force displacement response of the optimal design is shown in Fig. 6 (c), and the geometries of the optimal design at two stable configurations are shown in Fig. 6 (d).

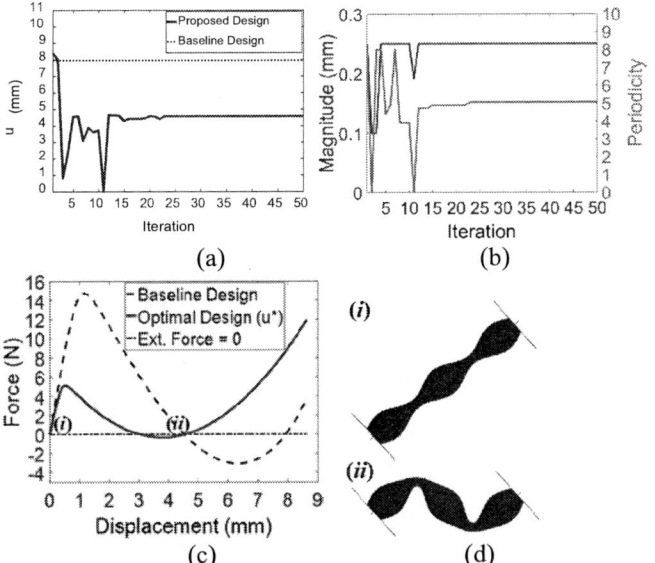

FIGURE 6. Design search history to minimize displacement at second stable state (u^*): (a) objective function vs iteration and (b) the optimal design vs iteration. (c-d) Force-displacement response of the optimal design in the (i) unloaded and (ii) buckled stable states.

The objective of the second optimization problem is to maximize the ratio of output to input energy, s^* (Fig. 7). The design search converges efficiently to the optimal design within 27 iterations. The obtained optimal change of magnitude was 0.25 mm (max allowable), optimal periodicity is 6.4, and the maximized s^* was 0.52. Compared to the baseline beam design, the optimized energy ratio increased 158%.

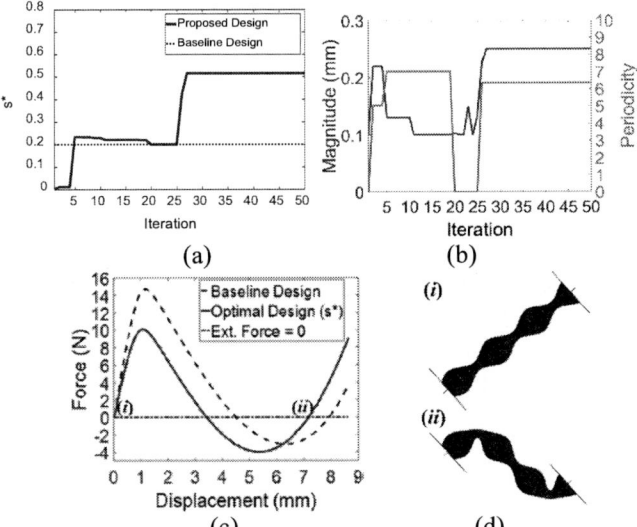

FIGURE 7. Design search history to maximize energy ratio at second stable state ($s^* = E_{out}/E_{in}$): (a) objective function vs iteration and (b) the optimal design vs iteration. (c-d) Force-displacement response of the optimal design in the (i) unloaded and (ii) buckled stable states.

As shown in Fig. 6 and 7, the yielded optimal designs are significantly different depending, on the objective (u^* or s^*). When the objective is to minimize displacement at second stable state, the energy ratio to retain equilibrium state becomes very weak, as indicated by Fig. 6 (c). This can be resolved by carrying-out a multi-objective optimization, where the compound objective is defined in Eq. (15). Since the ultimate goal of the design is to control equilibrium state, the weights biased toward objective u^*, with $w_1 = 0.9$ and $w_2 = 0.1$.

The search history for multi-objective optimization is shown in Fig. 8 (a) and (b), respectively. Force displacement response of the optimal design is shown in Fig. 8 (c), where it can be noticed that the stability of the design is slightly improved compared to Fig. 6 (c). The energy ratio indeed significantly increased from 0.046 to 0.119. The minimized displacement is also slightly increased from 4.58 mm to 4.84 mm. The suggested optimal periodicity of the design here is 4.9, which is only 3.9 % smaller than the one of the optimal design that minimizes u^*, which indicates that s^* is highly sensitive to design variables, and also the non-convexity of the problem, since higher periodicity of the design is suggested for the one that maximizes s^*.

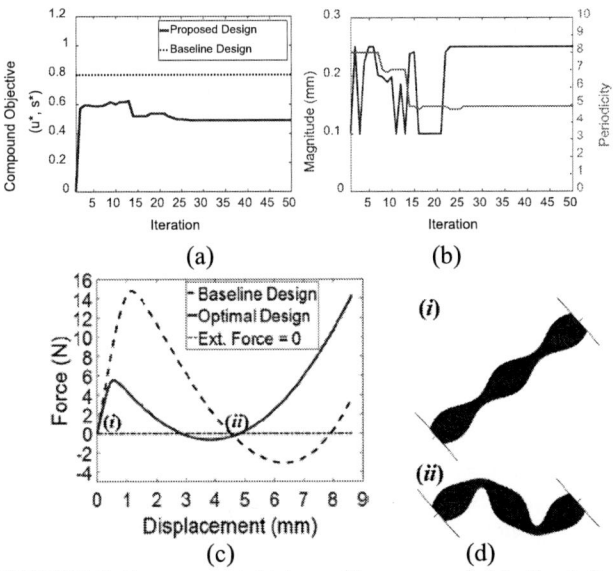

FIGURE 8. Design search history with compound objective (u^*, s^*): (a) objective function vs iteration and (b) the optimal design vs iteration. (c-d) Force-displacement response of the optimal design in the (i) unloaded and (ii) buckled stable states.

5. CONCLUSION

A design optimization method has been developed for bistable elastomeric beam for potential applications in vibration isolation and acoustic bandgap controls. The geometry of the beam was parametrized with a Fourier series, where the coefficients corresponding to magnitude and periodicity served as design variables. Since the response of the bistability behavior in terms of the design variables are highly non-convex, and a nonlinear FEA is required during the optimization, Bayesian optimization was implemented to improve the convergence with respect to the number of required high fidelity function evaluations. Important optimization parameters of Bayesian optimization are systematically studied for both accuracy and efficiency. When the optimizer is biased to exploitation, it converges to the optimal design most quickly. Convergence also improves with the Matérn covariance functions using $v = 0.5$, which indicates the non-smoothness of the design space. We also implemented a penalty scheme to represent the monostable designs in the bistable GP model, which improved the effectiveness of the surrogate model. Based on the studied optimization parameters, the design optimization case studies converged to the optimal designs within 30 iterations. The displacement at the second stable state is reduced by 42%, and the stability is improved by 58%. A competition between the two objectives was investigated thought a weighted multi-objective formulation, demonstrating the potential to balance the performance trade-off. Given the efficient convergence of the Bayesian approach, our future work will leverage this method to create Pareto fronts of these competing design objectives to fully characterize the trade-off.

REFERENCES

[1] Shan, S., Kang, S. H., Raney, J. R., Wang, P., Fang, L., Candido, F., Lewis, J. A., and Bertoldi, K., 2015. Multistable Architected Materials for Trapping Elastic Strain Energy. *Advanced Materials* **27** (29): 4296 – 4301.

[2] Harne, R. and Wang, K. W., 2013. A Review of the Recent Research on Vibration Energy Harvesting via Bistable Systems. *Smart Materials and Structures* **22** (2): 023001.

[3] Pellegrini, S. P., Tolou, N., Schenk, M., and Herder, J. L., 2012. Bistable Vibration Energy Harvesters: A Review. *Intelligent Material Systems and Structures* **24** (11): 1303-1312.

[4] Chen, Y., Li, T., Scarpa, F., and Wang, L., 2017. Lattice Metamaterials with Mechanically Tunable Poisson's Ratio for Vibration Control. *Physical Review Applied* **7** (2): 024012.

[5] Emmanuele, B. and Ruzzene M., 2013. Internally Resonating Lattices for Bandgap Generation and Low-Frequency Vibration Control. *Journal of Sound and Vibration* **332** (25): 6562-6579.

[6] dell'Isola, F., Maurini, C., and Porfiri, M., 2004. Passive Damping of Beam Vibrations through Distributed Electric Networks and Piezoelectric Transducers: prototype design and experimental validation. *Smart Materials and Structures* **13** (2): 299.

[7] James, K. A., and Waisman, H., 2016. Layout Design of a Bi-stable Cardiovascular Stent Using Topology Optimization. *Computer Methods in Applied Mechanics and Engineering* **305**: 869-890.

[8] Bruns, T. E., Sigmund, O., and Tortorelli, D. A., 2002. Numerical Methods for the Topology Optimization of Structures that Exhibit Snap-Through, *International Journal for Numerical Methods in Engineering* **55** (10): 1215-1237.

[9] Qiu, J., Lang, J. H., and Alexander, H. S., 2004. A Curved-Beam Bistable Mechanism. *Journal of Microelectromechanical Systems* **13** (2): 137-146.

[10] Crisfield, M. A., 1983. An Arc-Length Method Including Line Searches and Accelerations. *International Journal for Numerical Methods in Engineering* **19** (9): 1269-1289.

[11] Crisfield, M. A., 1993. "Non-linear Finite Element Analysis of Solids and Structures." *Wiley* 1.

[12] Williams, C. K., and Rasmussen, C. E., 2006. "Gaussian Process for Machine Learning." *The MIT Press* **2**, **3** (4).

[13] Stein, M. L., 1999. Interpolation of Spatial Data, *Springer-Verlag*, Newyork.

[14] Snoek, J., Larochelle, H., and Adams, R. P., 2012. Practical Bayesian Optimization of Machine Learning Algorithms. *In Advances in Neural Information Processing Systems*: 2951-2959.

[15] Calandra, R., Gopalan, N., Seyfarth, A., Peters, J., and Deisenroth, M. P., 2016. Bayesian Optimization for Learning Gait under Uncertainty. *Annals of Mathematics and Artificial Intelligence* **76** (1-2): 5-23.

[16] Shahriari, B., Swersky, K., Wang, Z., Adams, R. P., and De Freitas, N., 2016. Taking the human out of the loop: A review of Bayesian optimization. *Proceedings of the IEEE*, **104** (1): 148-175.

[17] Sharpe, C., Morris, C., Golds berry, B., Seepersad, C. C., Haberman, M., Shahan, D., 2016. Hierarchical Design of Negative Stiffness Metamaterials Using a Bayesian Network Classifier. *Journal of Mechanical Design* **138** (4): 041404.

[18] Sharpe, C., Morris, C., Goldsberry, B., Seepersad, C. C., and Haberman, M. R., 2017. Bayesian Network Structure Optimization for Improved Design Space Mapping for Design Exploration with Materials Design Applications. *In ASME International Design Engineering Technical Conference and Computers and Information in Engineering Conference*: V02BT03A004-V02BT03A004

Proceedings of the ASME 2019
International Design Engineering Technical Conferences
and Computers and Information in Engineering Conference
IDETC/CIE2019
August 18-21, 2019, Anaheim, CA, USA

DETC2019-98222

DATA-CENTRIC MIXED-VARIABLE BAYESIAN OPTIMIZATION FOR MATERIALS DESIGN

Akshay Iyer, Yichi Zhang
Mechanical Engineering
Northwestern University, USA

Aditya Prasad
Material Science & Engineering
Rensselaer Polytechnic Institute, USA

Siyu Tao, Yixing Wang
Mechanical Engineering
Northwestern University, USA

Linda Schadler
Engineering & Mathematical Sciences
The University of Vermont, USA

L Catherine Brinson
Mechanical Engineering & Material Science
Duke University, USA

Wei Chen[1]
Mechanical Engineering
Northwestern University, USA

ABSTRACT

Materials design can be cast as an optimization problem with the goal of achieving desired properties, by varying material composition, microstructure morphology, and processing conditions. Existence of both qualitative and quantitative material design variables leads to disjointed regions in property space, making the search for optimal design challenging. Limited availability of experimental data and the high cost of simulations magnify the challenge. This situation calls for design methodologies that can extract useful information from existing data and guide the search for optimal designs efficiently. To this end, we present a data-centric, mixed-variable Bayesian Optimization framework that integrates data from literature, experiments, and simulations for knowledge discovery and computational materials design. Our framework pivots around the Latent Variable Gaussian Process (LVGP), a novel Gaussian Process technique which projects qualitative variables on a continuous latent space for covariance formulation, as the surrogate model to quantify "lack of data" uncertainty. Expected improvement, an acquisition criterion that balances exploration and exploitation, helps navigate a complex, nonlinear design space to locate the optimum design. The proposed framework is tested through a case study which seeks to concurrently identify the optimal composition and morphology for insulating polymer nanocomposites. We also present an extension of mixed-variable Bayesian Optimization for multiple objectives to identify the Pareto Frontier within tens of iterations. These findings project Bayesian Optimization as a powerful tool for design of engineered material systems.

Keywords: Data-centric Material Design, Latent Variable Gaussian Process, Mixed-variable Bayesian Optimization, Acquisition Functions, Nanocomposites

1. INTRODUCTION

The launch of the Material Genome Initiative (MGI) [1] has revolutionized the way advanced material systems are designed with targeted performance. Moving away from the traditional experiment-based cause-effect approach, MGI strives to elucidate the relationship between Processing-Structure-Property (PSP) [2] domains for material design. It requires development of new methods within each of the three domains and protocols to manage information flow across domains. A holistic design strategy for bi-directional traversal of PSP relations requires us to address certain key issues – cost effective processing techniques, microstructure representation and reconstruction, dimensionality reduction and tractable optimization techniques, to name a few.

The emergence of open-source material databases [3-6] and the increasing popularity of machine learning techniques are accelerating our ability to address some of these challenges using a data-centric approach. NanoMine [5, 6], a nanocomposite material database with in-built data curation, exploration and analysis capabilities, represents this approach in the field of polymer nanocomposites. It collects nanocomposite characteristics reported in the literature and from individual research labs including microstructure, processing conditions, and material properties. An ontology-enabled knowledge graph framework helps NanoMine establish relationships between those characteristics. A collection of module tools for microstructure characterization & reconstruction and simulation software to model bulk nanocomposite material response augments knowledge generated by experimental data. Integrating these different sources of knowledge is critical for materials design. However, generating experimental or simulated data for the vast design space defined by the almost infinite combinations of constituents, microstructure morphology, and processing conditions is impractical. This

---

[1] Corresponding Author email: weichen@northwestern.edu

signifies the need for data-centric methodologies that can effectively interrogate existing data and guide "on-demand" computer simulations and physical experiments to accelerate the search of new high performing materials.

A range of optimization-based techniques have been developed to support the design of microstructural material systems. Among the existing methods developed by the author group, Physical Descriptors based [7, 8] and Spectral Density function [9, 10] based microstructure designs are the most commonly used approaches due to their physically meaningful characterization, relative ease of reconstruction and low dimensional representation. Recently, we have shown deep learning based Generative Adversarial Networks (GAN) to be effective tools for low dimensional microstructure representation and design [11]. Convolutional layers in GAN can capture higher order spatial correlations in complex morphologies, but the requirement of several thousand microstructures for training presents a barrier.

As microstructural design involves expensive microstructure simulations to assess material properties, most of the existing methods rely on building surrogate models (global metamodels) for replacing the physics-based simulations in parametric optimization. However, metamodel based optimization is not well suited for material systems with highly nonlinear behavior, disjoint design space, limited data and high experiment/simulation cost; a common scenario in material science. In contrast, Bayesian Optimization (BO) [12, 13] has emerged as a viable proposition in material design. BO adaptively samples new designs conditioned on information available from existing data, balancing exploration and exploitation to efficiently locate the global optimum. Balachandran et al. [14] used this adaptive sampling strategy to design M_2AX compounds with desired elastic properties using orbital radii of individual components as design variables. Their work highlighted the importance of incorporating prediction uncertainties in the sampling procedure and noted that purely exploitative strategies often result in suboptimal outcomes. Li et al. [15] devised an adaptive experimental optimization (AEO) framework to develop a novel fluid processing platform for synthesis of short polymer fibers. The AEO framework incorporates material and process related parameters to optimize

a set of qualitative and quantitative objectives. All design variables are quantitative and related to process optimization for a single combination of polymer and solvent. Similar examples of accelerated material design using BO have been reported by others [16, 17]. However, all existing applications of BO considered thus far have only quantitative or qualitative variables; while mixed variable problems containing both qualitative and quantitative variables is common in material design. Choice of constituents in any material system can be treated as qualitative variables, while microstructure descriptors, processing and operating parameters (temperature, RPM, wavelength etc.) manifest as quantitative variables. For example, nanocomposite design involves selecting the optimal combination of qualitative (polymer, nanoparticle, surface modification) and quantitative (microstructure descriptors) variables.

In this article, we present a data-centric Bayesian Optimization framework for material design and innovation. The guiding hypothesis is that a Bayesian inference approach can effectively model knowledge contained in an available dataset as a prior and guide "on-demand" computer simulations and physical experiments to accelerate the search of optimal material designs. Therefore, the framework is flexible to incorporate data generated by experiments as well as simulations or machine learning. We demonstrate how Latent Variable Gaussian Processes, a novel GP modelling strategy for mixed variable problems with inbuilt uncertainty estimation, plays a critical role in Bayesian Optimization to efficiently navigate complex, non-linear design space. Then, LVGP based BO is extended to multi-objective optimization, a common scenario in design of multi-functional material systems. The efficacy of the proposed framework is demonstrated through a design case study focused on identifying the optimal composition and microstructure for insulating nanocomposites.

2. DATA-CENTRIC MIXED-VARIABLE BAYESIAN OPTIMIZATION (BO) FRAMEWORK FOR MATERIALS DESIGN

In this section, we present the data-centric material design framework and discuss the two driving concepts of BO – surrogate model and acquisition functions.

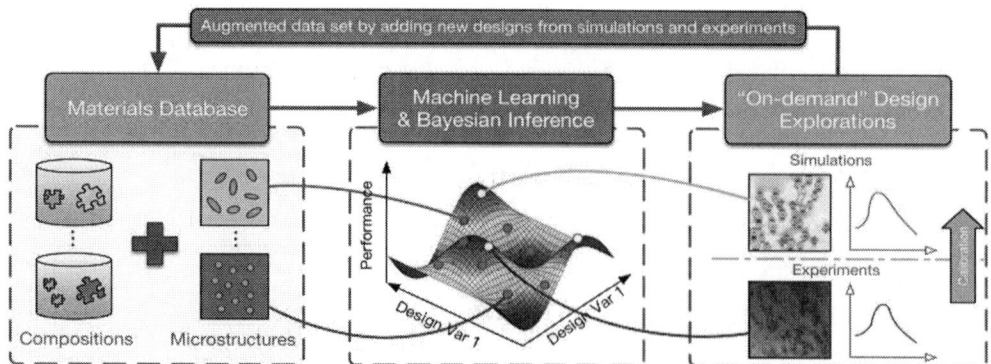

Figure 1: Bayesian optimization approach treats the existing data set as prior knowledge, chooses new samples, and builds machine learning models using curated and new experiment/simulation data to capture p-s-p relations for optimization.

V02AT03A066-2   Copyright © 2019 ASME

2.1 Data-Centric Material Design Framework and Concept of BO

Fig. 1 shows the proposed framework, integrating curated material databases with material property simulations and machine learning. The framework is initiated from a **materials database** of curated experimental and simulated data describing material properties with appropriate attributes. For example, the NanoMine [5] system we developed contains data describing mechanical, electrical and optical properties of nanocomposites – labelled by composition (polymer, nanoparticle, surface chemistry, nanoparticle weight ratio), microstructure, processing conditions and associated metadata (source of data, processing equipment used, etc.). Based on PSP relationships, some of these material characteristics are known to influence material properties and are treated as design variables in BO. These characteristics maybe quantitative (e.g., microstructure descriptors) or qualitative (e.g., type of nanoparticle or polymer, type of surface treatment). Material database can be used to train **machine learning** models which predict material properties from the corresponding design variables. Since the database contains only a small fraction of points in a vast design space, the machine learning model possesses interpolation uncertainty, which varies from point to point and plays an important role in BO-based **"on-demand" design explorations**. Using predictions and uncertainty quantification of a machine learning model, **Bayesian inference** determines the design that shows the most "potential" for improvement in a given material property. There are several metrics, commonly known as **acquisition functions**, for evaluating "potential" improvement. After a promising design is identified by the acquisition function, its corresponding material property is evaluated using either simulation or experiment. Once property evaluation is complete, the sample design is added to the database and the above steps will be repeated until the termination criteria is satisfied based on available computational and experimental resources. It should be noted that as a part of the Bayesian inference framework, experimental data can also be used for calibrating unknown model parameters in simulation models. For example, finite element simulations for prediction of dielectric properties in nanocomposites requires calibration of change in polymer mobility at the polymer-nanoparticle interphase (the region surrounding nanoparticles) as characterized by the shift in dielectric loss peaks [18] obtained from the experimental dielectric spectra.

Let us represent the general global optimization problem as

$$x^* = \underset{x \in X}{argmin}\, f(x), \qquad (1)$$

where x is the input design variable vector in the input space X and $f(\cdot)$ is the objective function. Based on the data collected about $f(\cdot)$, BO uses statistical models to model the true $f(\cdot)$ function and produce fast statistical predictions of $f(x)$ at any given input x. The statistical model must be flexible to tackle situations where x is qualitative, quantitative or a combination of both. We adopt Gaussian process (GP) models as the statistical model in our BO framework as GP models can flexibly model highly nonlinear behavior with a small amount of

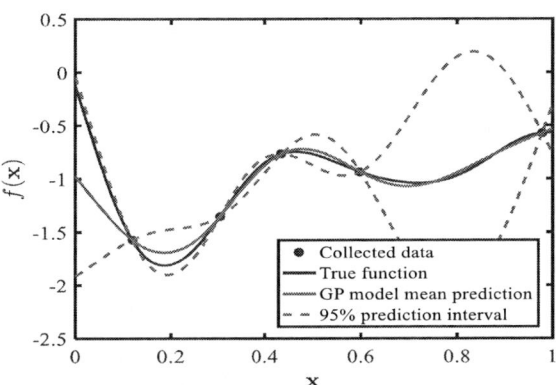

Figure 2: 1D example of a GP model fitted to collected data of $f(\cdot)$

hyperparameters and conveniently provides analytical solutions for its prediction uncertainty.

Fig. 2 is a 1D example of a GP model fitted to the collected data of $f(\cdot)$. At each input x, the output $f(x)$ is regarded as a normally distributed random variable and the GP model predicts its mean and variance. The 95% prediction interval in the figure reflects the confidence bounds of the prediction [19, 20].

The **acquisition functions** (aka utility functions, infill functions) in BO are functions that quantify the benefit of evaluating $f(x)$ at any input x in the next optimization iteration based on the statistical predictions of $f(x)$. Fig. 3 illustrates profiles of a few widely used acquisition functions for the 1D example problem in Fig 2. Plotted acquisition functions are expected improvement (EI) [21], probability of improvement (PI) [22], upper/lower confidence bound (UCB) [23] and knowledge gradient (KG) [24]. In general, an acquisition function aims for either exploitation, exploration, or both of a design space. Here, **exploitation** means evaluating at input x where prediction of $f(x)$ has high cumulative probability of being better than the current best solution from the perspective of optimizing a design objective. On the other hand, **exploration**

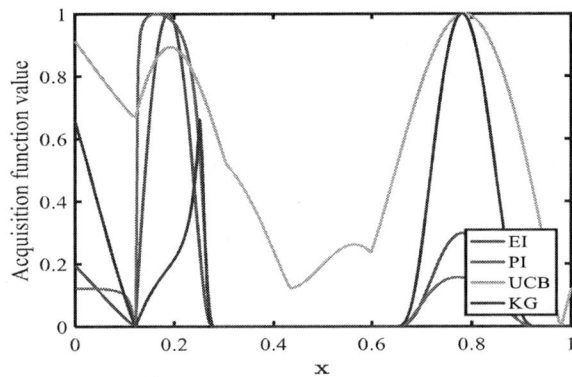

Figure 3: Different acquisition function profiles in a 1D example problem, based on the statistical predictions in Fig.2. The value of the acquisition functions is a measure of the benefit of evaluating f(·) at that input.

implies evaluating at input x where prediction of $f(x)$ has large uncertainty, which warrants the chance that either x is a potential global optimum or the evaluation of $f(x)$ leads to the discovery of the global optimum near x. On many occasions, exploitation and exploration are quite contradictory goals to achieve. Therefore, acquisition functions typically strike a balance between the two.

2.2 Latent Variable GP Modelling for Mixed-Variable Problems

The standard GP methods were developed under the premise that all input variables are quantitative, which does not hold in many real engineering applications. We recently proposed a latent variable Gaussian processes (LVGP) [25] modeling method that maps the levels of the qualitative factor(s) to a set of numerical values for some latent quantitative variable(s). The latent variables transform the underlying high dimensional physical attributes associated with the categorical variables into the latent-variable space and quantify the "distances" between samples. As a by-product, the latent variable mapping of the qualitative factors provides an inherent ordering and structure for the levels of the factor(s), which leads to substantial insight into the effects of the qualitative factors. To avoid misunderstanding, note that the latent variables are only used internally inside LVGP models. When LVGP models are used for predictions, they still take mixed-variable inputs in the original mixed-variable input spaces.

To describe the LVGP approach, the input variables are denoted as $w = (x, t)$, where $x = (x_1, x_2, \dots, x_p)$ represents p quantitative variables and $t = (t_1, t_2, \dots, t_q)$ is the vector of q qualitative variables. With $i = 1, 2, \dots, q$, the qualitative variable t_i has m_i levels $\{l_1^{(i)}, l_2^{(i)}, \dots, l_{m_i}^{(i)}\}$. The output variable is denoted as y, and a set of data points of input-output pairs are noted as $\{(w_1, y_1), \dots, (w_N, y_N)\}$. Consider the GP model

$$Y(\cdot) = \mu + G(\cdot), \qquad (2)$$

where μ is the constant prior mean, and $G(\cdot)$ is a zero-mean GP with covariance function $k(\cdot, \cdot) = \sigma^2 r(\cdot, \cdot | \varphi)$. σ^2 is the prior variance of the GP, and $r(\cdot, \cdot | \varphi)$ is the correlation function parameterized with φ. The true model $y(\cdot)$ is regarded as a realization of the GP $Y(\cdot)$. Once the form of the correlation function $r(\cdot, \cdot | \varphi)$ is specified, the hyperparameters (μ, σ^2, φ) can be estimated through maximum likelihood estimation (MLE) or other principles such as minimizing cross-validation errors. If the independent variables of the correction function $r(\cdot, \cdot | \varphi)$ are only the continuous variables x, one can use the common Gaussian correlation function

$$r(x, x' | \varphi) = exp\left\{-\sum_{i=1}^{p} \varphi_i(x_i - x_i')^2\right\}, \qquad (3)$$

which quantifies the correlation between $G(x)$ and $G(x')$ for any input locations $x = (x_1, \dots, x_p)$ and $x' = (x_1', \dots, x_p')$ based on their 2-norm distance scaled by φ. However, in the mixed-variable problem, it is not straightforward to incorporate the qualitative variable t in such a correlation function, as it makes no sense to compute $t_i - t_i'$ as the difference between levels. The

LVGP method handles this by mapping the qualitative variables t to quantitative ones.

In the LVGP method, the m_i levels of the qualitative variable t_i are mapped to m_i latent numerical vectors $\{z^{(i)}(l_1^{(i)}), \dots, z^{(i)}(l_{m_i}^{(i)})\}$ of a latent variable $z^{(i)} \in \mathbb{R}^d$, where d is the dimensionality of $z^{(i)}$. A modeler is free to choose the value of d as a modeling parameter, although setting $d = 2$ has been shown to be advisable for most problems. The original mixed-type input variables $w = (x, t)$ are thus mapped to purely continuous variables $(x, z^{(1)}(t_1), \dots, z^{(q)}(t_q))$. A correlation function like Eq. 3 can be subsequently constructed as

$$r(w, w' | \varphi, Z) = exp\left\{-\sum_{i=1}^{p} \varphi_i(x_i - x_i')^2 \right.$$
$$\left. -\sum_{i=1}^{q} \|z^{(i)}(t_i) - z^{(i)}(t_i')\|_2^2\right\}, \qquad (4)$$

where Z is the collection of all the latent parameters denoted by $\{z^{(1)}(l_1^{(1)}), \dots, z^{(1)}(l_{m_1}^{(1)}), z^{(2)}(l_1^{(2)}), \dots, z^{(q)}(l_{m_q}^{(q)})\}$.

With the correlation structure in Eq. 4, and following Eq. 2, the log-likelihood for the given dataset $\{(w_1, y_1), \dots, (w_N, y_N)\}$ is

$$L(\mu, \sigma^2, \varphi, Z) = -\frac{N}{2} ln(2\pi\sigma^2) - \frac{1}{2} ln|R(\varphi, Z)| -$$
$$\frac{1}{2\sigma^2}(y - \mu\mathbf{1})^T R(\varphi, Z)^{-1}(y - \mu\mathbf{1}), \qquad (5)$$

where y is the column vector $(y_1, \dots, y_N)^T$, $\mathbf{1}$ is the N-by-1 vector of ones, and $R(\varphi, Z)$ is an N-by-N matrix with $(i, j)^{th}$ element being $R_{ij} = r(w_i, w_j | \varphi, Z)$ for $i, j = 1, \dots, N$. The MLE values of the hyperparameters $(\mu, \sigma^2, \varphi, Z)$ are obtained as

$$[\hat{\mu}, \widehat{\sigma^2}, \widehat{\varphi}, \widehat{Z}] = arg \max_{\mu, \sigma^2, \varphi, Z} L(\mu, \sigma^2, \varphi, Z). \qquad (6)$$

For estimating the hyperparameters, the latent vector $z^{(i)}(l_1^{(i)})$ corresponding to the first level $l_1^{(i)}$ is fixed at zero to avoid indeterminacy issue. This is because the correlation quantified by Eq. 4 only depends on the relative distances

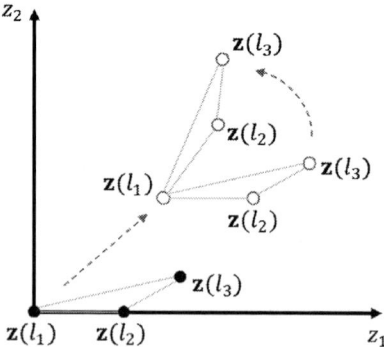

Figure 4: Indeterminacy caused by translation and rotation when $d = 2$. Fixing the continuous variable values, the three different configurations for the mapped latent values $\{z(l_1), z(l_2), z(l_3)\}$ have the same pairwise distances and hence the same correlation values. Index i is omitted for simiplicity.

between $\mathbf{z}(t)$ and $\mathbf{z}(t')$ but not their respective absolute values. When $d \geq 2$, some other elements in the collection of latent vectors $\{\mathbf{z}^{(i)}(l_1^{(i)}), \ldots, \mathbf{z}^{(i)}(l_{m_i}^{(i)})\}$ need to be set to zero to avoid indeterminacy. In the $d = 2$ case, the second element of $\mathbf{z}^{(i)}(l_2^{(i)})$ needs to be set to zero, as is illustrated in Fig 4. The total number of latent parameters that are subject to optimization is thus $\sum_{i=1}^q (2m_i - 3)$.

After obtaining the estimates of the hyperparameters, the mean prediction for $y(\mathbf{w}_{(n)})$ and the associated mean squared error (MSE) (for uncertainty prediction) are

$$\hat{y}(\mathbf{w}_{(n)}) = \hat{\mu} + \mathbf{r}^T(\mathbf{w}_{(n)}) \mathbf{R}^{-1} (\mathbf{y} - \hat{\mu}\mathbf{1}), \qquad (7)$$

$$\begin{aligned} MSE[\hat{y}(\mathbf{w}_{(n)})] = \widehat{\sigma^2} [& r(\mathbf{w}_{(n)}, \mathbf{w}_{(n)}) \\ & - \mathbf{r}^T(\mathbf{w}_{(n)}) \mathbf{R}^{-1} \mathbf{r}(\mathbf{w}_{(n)}) \\ & + W^2 (\mathbf{1}^T \mathbf{R}^{-1}\mathbf{1})^{-1}], \end{aligned} \qquad (8)$$

where $W = 1 - \mathbf{1}^T \mathbf{R}^{-1} \mathbf{r}(\mathbf{w}_{(n)})$ and $\mathbf{r}(\mathbf{w}_{(n)})$ is a N-by-1 vector whose i^{th} element is $r(\mathbf{w}_i, \mathbf{w}_{(n)})$.

2.3 Multi-Objective Bayesian Optimization

A multi-objective BO approach is needed because material design often involves targets for multiple properties. The general multi-objective optimization problem can be formulated as

$$\min_{\mathbf{x} \in X} \{y_1(\mathbf{x}), y_2(\mathbf{x}), \ldots, y_s(\mathbf{x})\}, \qquad (9)$$

where $\mathbf{x}$ is the design input, X is the design space, s is the number of objective functions, and $\{y_1(\cdot), y_2(\cdot), \ldots, y_s(\cdot)\}$ is the set of the objective functions that share the same design inputs. The solution to this problem is a so-called Pareto set (aka Pareto front or Pareto frontier) consisting of design points that achieve Pareto optimality [26]. To identify the Pareto frontier for Eq. 9 numerically, the objective functions are evaluated at a certain number of design inputs. Of all the evaluated design points, one selects the set of design points that are not dominated by any other ones [26]. Here, a design point $\mathbf{x}$ is not dominated by another one $\mathbf{x}'$ if there exists at least one $i \in \{1, 2, \ldots, s\}$ such that $y_i(\mathbf{x}) < y_i(\mathbf{x}')$. This set of design points is regarded as a representation of the true Pareto set.

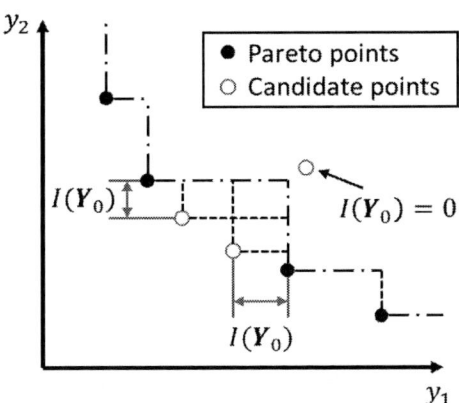

Figure 5: Example of the values of $I(\mathbf{Y}_0(\mathbf{x}_0))$. The depiction is in 2D output space.

Several BO methods have been proposed in the literature for such a multi-objective setting. In our implementation, we choose the widely-used expected maximin improvement (EMI) formulation [27], which we describe as follows. Let the current Pareto set be composed of input set $P_X = \{\mathbf{x}_1, \mathbf{x}_2, \ldots, \mathbf{x}_k\}$ and output set $P_Y = \{\mathbf{y}_1, \mathbf{y}_2, \ldots, \mathbf{y}_k\}$, where k is the number of points in the Pareto set and $\mathbf{y}_i = [y_1(\mathbf{x}_i), y_2(\mathbf{x}_i), \ldots, y_s(\mathbf{x}_i)]^T, i = 1, 2, \ldots, k$. For any given new input $\mathbf{x}_0$, the corresponding output is predicted by the uncertainty quantification model as $\mathbf{Y}_0(\mathbf{x}_0) = [Y_1(\mathbf{x}_0), Y_2(\mathbf{x}_0), \ldots, Y_s(\mathbf{x}_0)]^T$, where $Y_j(\mathbf{x}_0), j = 1, 2, \ldots, s$ is a random variable. To quantify how much the random outputs $\mathbf{Y}_0(\mathbf{x}_0)$ would improve the current Pareto set, we use the minimax improvement metric

$$\begin{aligned} I(\mathbf{Y}_0(\mathbf{x}_0)) = \min_{\mathbf{x}_i \in P_X} \Big\{ & max \left(\{y_j(\mathbf{x}_i) - Y_j(\mathbf{x}_0)\}_{j=1}^s \right. \\ & \left. \cup \{0\} \right) \Big\}. \end{aligned} \qquad (10)$$

The larger the value of $I(\mathbf{Y}_0(\mathbf{x}_0))$ is, the more improvement the output $\mathbf{Y}_0(\mathbf{x}_0)$ is considered to make.

With this formula, if the output $\mathbf{Y}_0(\mathbf{x}_0)$ would be dominated by at least one point in the current Pareto set, then $I(\mathbf{Y}_0(\mathbf{x}_0)) = 0$, which means no improvement. Otherwise, $I(\mathbf{Y}_0(\mathbf{x}_0))$ would be a positive value quantifying the improvement. The value of $I(\mathbf{Y}_0(\mathbf{x}_0))$ is illustrated by a 2D example case in Fig. 5, with one of the candidate points being $I(\mathbf{Y}_0) = 0$ and the other two points with a positive value $I(\mathbf{Y}_0)$. The criterion for choosing the new evaluation input $\mathbf{x}_0^*$ is to maximize the EMI given in Eq. 10, i.e.,

$$\mathbf{x}_0^* = \arg\max_{\mathbf{x}_0 \in X} E(I(\mathbf{Y}_0(\mathbf{x}_0))). \qquad (11)$$

When the original problem (Eq. 9) has mixed-variable input space X, Eq. 11 is a mixed-variable optimization problem. To solve Eq. 11, we use a zero-order optimization strategy, where we generate a large set of candidate points in the input space, and then choose the one with the largest EMI as $\mathbf{x}_0^*$. For evaluating the expectation in Eq. 11, we use Monte Carlo simulation, as the analytical formula for EMI is too complex for $s \geq 3$, which is the case for our application problems.

3. CASE STUDY: CONCURRENT COMPOSITION AND MICROSTRUCTURE DESIGN FOR INSULATING NANOCOMPOSITES

Polymer nanocomposites are ideal candidates for insulating materials with potential application in high voltage rotating machines [28]. Three major electrical properties that determine suitability for this application are breakdown strength, dielectric permittivity and dielectric loss. Breakdown strength (U_d) is the minimum voltage at which current flows through an insulating material. Dielectric permittivity (ϵ) characterizes the degree of electrical polarization experienced by material while dielectric loss ($tan\delta$) is a measure of the amount of heat generated under an alternating electric field. Ideally, one would want high U_d, low ϵ and low $tan\delta$ but tradeoffs between U_d Vs ϵ and ϵ Vs $tan\delta$ have been observed [29, 30]. Automation of polymer nanocomposite design is considerably less developed than that for alloys due to challenges associated with searching the vast design space defined by the almost infinite combinations of polymers, nanoparticle, surface chemistries, microstructure morphology, and processing conditions.

Fig. 6 depicts the mixed-variable BO framework customized for case study, indicating the various modules involved and information flow between them. Our investigations are initiated from a materials database (**Module 1**) comprising nanocomposite samples with varying compositions, corresponding microstructures and measurement of dielectric properties. We consider nanocomposites with silica nanoparticles (aka filler) dispersed in two types of polymers - polystyrene (PS) and polymethylmethacrylate (PMMA). It is a common practice to enhance interphase properties by appending functional groups to the surface of nanoparticles, a process known as surface modification. We consider silica nanoparticles with three surface modifications in this study – Chloro-, Amino- and Octyl-silanes. Samples corresponding to all six polymer and surface modification combinations were prepared. Nanoparticle dispersion, a key microstructure descriptor influenced by choice of polymer, surface modification and processing conditions is quantified from TEM images using the Spectral Density Function (SDF) along with nanoparticle volume fraction. The identified range of these microstructure descriptors will be used as bounds in the design process.

Our database also contains experimental measurements of all three dielectric properties, to be used for calibrating the nanoparticle-polymer interphase parameters in the finite element analysis (FEA) model as well as training machine learning models for the breakdown strength. These properties are known to be influenced by material composition (choice of filler, polymer, surface modification), filler volume fraction and dispersion. Dielectric permittivity ϵ and loss $tan\delta$ are evaluated using FEA (**Module 3**), where interphase properties are characterized by a shift in the nanocomposite properties w.r.t pure polymer properties and obtained by calibration (**Module 2**). In **Module 3** SDF based microstructure reconstruction is used to generate 2D Representative Volume Elements (RVEs) with desired filler volume fraction and dispersion; which is evaluated using FEA. Since microstructure reconstruction using SDF is stochastic and involves uncertainties, we generate three reconstructions and take their average ϵ and $tan\delta$ to calculate the objective function for each design. **Module 3** represents the most computationally intensive module in our framework, requiring several minutes of simulations on a 16 core Intel Xeon® 2.4 GHz CPU with 192 GB RAM. On the other hand, **Module 4** is an empirical machine learning model employing Random Forrest technique, trained on experimental data present in nanocomposite database, to predict the breakdown strength U_d as a function of material design variables.

With bounds for design variables identified and models to predict dielectric properties, our study progresses to BO for designing insulating nanocomposites (**Module 5**). BO is initiated with a small set of training data and adaptively samples subsequent designs to approach the global optimum. We use LVGP with two-dimensional latent variables for each qualitative variable, as the surrogate model. Its in-built uncertainty quantification is leveraged for performing Bayesian Inference using the Expected Improvement acquisition criterion. At each iteration, the LVGP model is updated with a new design whose dielectric properties are evaluated using Modules 3 & 4. We present single and multi-objective optimization formulations for the nanocomposite design problem and discuss performance of LVGP-BO framework in both scenarios. In the following subsections, we describe each module in detail, followed by results.

3.1 Nanocomposite Preparation and Dielectric Property Measurement (Module 1)

Silica nanoparticles (diameter 14 nm) in methyl ethyl ketone were procured from Nissan Inc. The surface of the nanoparticles was modified using three monofunctional silane coupling agents:

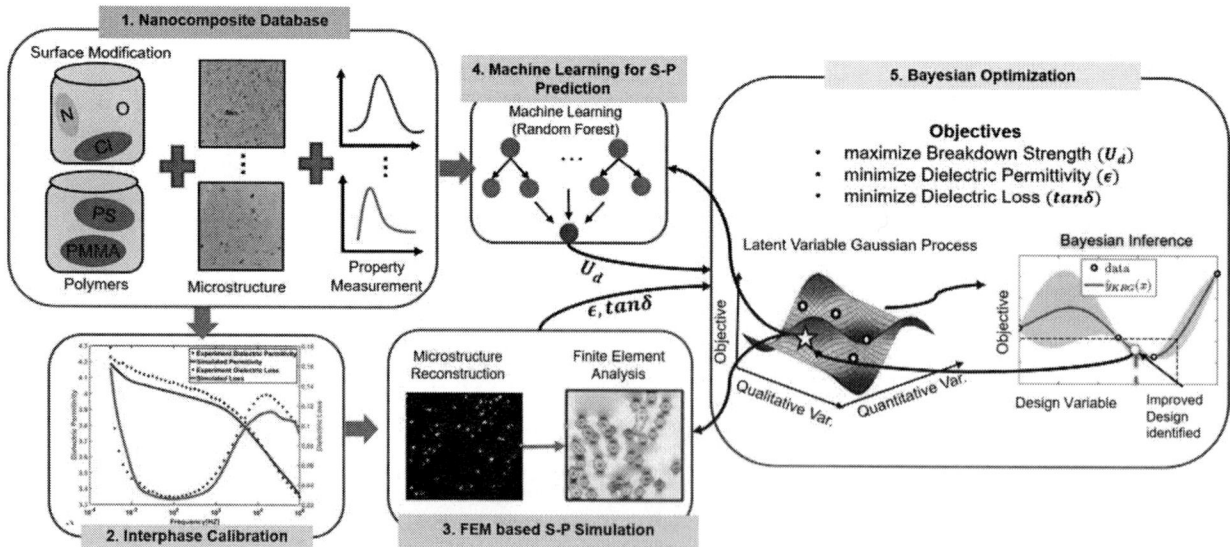

Figure 6: Bayesian Optimization framework for insulating polymer nanocomposites design

aminopropyledimethylethoxysilane (Amino), chloropropyledimethylethoxysilane (Chloro) and octyldimethylmethoxysilane (Octyl), from Gelest Inc. Polystyrene (PS) from Goodfellow Corporation and polymethylmethacrylate (PMMA) from Scientific Polymer Products Incorporated were used as the polymer. Nanocomposites were prepared in a Thermo Haake Minilab, co-rotating twin screw extruder. The mixing was carried out at 200°C for the PMMA nanocomposites and 150°C for the PS nanocomposites. Mixing parameters such as screw speed and specific energy input were varied to obtain a range of different dispersion states. A JEOL 2010 transmission electron microscope (TEM) was used to characterize the dispersion state of the nanocomposites. For each sample, 30 images were collected from different sections to obtain a representation of the overall dispersion state of the nanocomposites. The TEM images were binarized using a Niblack algorithm [31]. Dielectric spectroscopy measurements to ascertain ϵ and $tan\delta$ were carried out in a Novocontrol Alpha Analyzer. All nanocomposite samples were also subjected to dielectric breakdown testing. Hot pressed samples were placed in a ball-plane electrode setup [32]. The testing apparatus was kept in an oil bath to reduce the incidences of flash over and the voltage ramp-rate was fixed at 500 V/s. For each nanocomposite sample, 30 specimens were tested and the voltage at breakdown for each sample was recorded. The observed values were then fit to a Weibull function to obtain the dielectric breakdown strength, which is defined as the electric field at which the probability of breakdown was 63.2%.

3.2 Microstructure Characterization and Reconstruction (Module 1 & 3)

Spectral Density Function (SDF) is a frequency domain microstructure representation capable of capturing spatial correlations of complex heterogeneous materials. Mathematically, SDF $\rho(k)$ can be evaluated as:

$$\rho(\boldsymbol{k}) = |\mathcal{F}\{\mathcal{M}\}|^2, \qquad (12)$$

where $\mathcal{M}$ is the binarized microstructure, $\mathcal{F}(.)$ is the Fourier Transform operator and $\boldsymbol{k}$ is the frequency vector. For isotropic microstructures, SDF can be radially averaged about zero frequency such that the frequency vector $\boldsymbol{k}$ is reduced to a scalar k; making SDF a one-dimensional function of frequency. Although it is known to be the Fourier Transform of Two-point Autocorrelation function and hence encapsulates equivalent morphological information, Yu et al. [9] have shown that SDF is a more convenient representation to parametrize and design microstructures. These features are also evident from the analysis of nanocomposite microstructures. After binarizing TEM images using Niblack Algorithm [31] and assuming isotropy, SDF was evaluated using Eq. 12. We noticed that SDF of all microstructures approximately follows an exponential distribution that can be parametrized with two variables – shape parameter α and scale parameter θ:

$$\rho(k) = \alpha * \exp\left(-\frac{k}{\theta}\right). \qquad (13)$$

A total of 1719 TEM images (approx. 30 images per sample) were characterized using SDF and parameters α and θ were ascertained by curve fitting using Eq. 13. The average R^2 value for fitting was 0.91. Fig. 7 shows three microstructures along with their one-dimensional SDF and curve fitting. Filler dispersion increases from Fig. 7a-c and is reflected in a slower decay rate of SDF which can be quantified by θ. It was noticed that α varies in a very small interval [0.39, 1.84] and has very little influence on SDF profile. On the other hand, scale parameter θ varies between [0.62, 6.55], changing the rate of decay of SDF and consequently characterizing the dispersion of

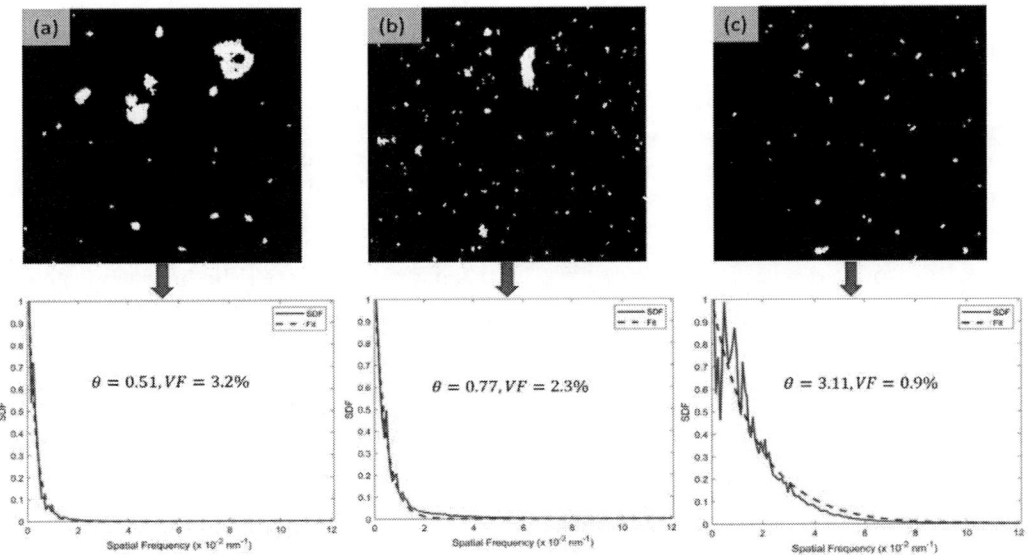

Figure 7: Three representative microstructure images with varying dispersions and their SDF (blue curve) and corresponding curve fit using Eq. 13(red dashed curve). Design Variables VF and θ for each image shown in inset

Table 1: Summary of design variables used in case study

Variable	Type	Range/Levels
Polymer Type (P)	Qualitative	$\{PMMA, PS\}$
Surface Modification Type (S)	Qualitative	$\{Chloro, Octyl, Amino\}$
Filler Volume Fraction (VF)	Quantitative	$[0.7\%, 8.3\%]$
Filler Dispersion (θ)	Quantitative	$[0.62, 6.55]$

filler aggregates. Thus, we will consider θ as a microstructure design variable and fix α to its mean value 1.1. Additionally, filler composition is represented by its volume fraction (VF), computed as the fraction of white pixels in binarized image, and is found to be in $[0.7\%, 8.3\%]$. The range of θ and VF identified here will be used to define bounds for these variables in design formulation.

Microstructure reconstruction is an integral part of the material design framework, since at each iteration of optimization, the material properties must be evaluated for the new microstructure. Here we adopt the analytical Cahn's Scheme [33] for reconstruction.

3.3 Interphase Calibration and Finite Element Analysis for Dielectric Permittivity and Loss (Module 2 & 3)

Finite element (FE) models have been developed to simulate the continuum level properties for polymer nanocomposites [34-36]. One of the challenges in structure-property relationship of nanocomposites is modeling the interphase. Since direct measurement of interphase properties, especially in situ, is limited experimentally, one approach is to calculate the interphase properties inversely. Using a FEA model with known nanoparticle and polymer properties, the interphase properties can be tuned until the predicted properties match the experimental measurement of the bulk composite properties [37, 38]. We developed an automated optimization-based method to solve this inverse calibration problem, details of which can be found in Wang et al. [18]. This method helps identify the dielectric shifting factors that will optimally match the experimental data through adaptive optimization. For the case study, interphase properties for all six polymer-surface modification combinations are calibrated using this adaptive optimization strategy (Module 2) and later used in FEA model (Module 3).

3.4 Machine Learning for Breakdown Strength Prediction (Module 4)

Dielectric breakdown of nanocomposites is a complex phenomenon and requires atomistic scale simulations to decode the complex interactions occurring in the interphase. Our investigation in this direction are ongoing. Therefore, in this case study, we use a random forest [39] model trained on experimental data for rapid evaluation of U_d as a function of material design variables during optimization. Random forest technique was chosen due to its ability of handling mixed variables, superior computational efficiency and minimal

possibility of overfitting. Training data comprised U_d measurement (expressed in kV/mm) of 52 samples at 60 Hz. The features used for predicting U_d are the two qualitative (polymer type, surface modification type) two quantitative (VF and θ) design variables. The trained random forest with 1000 trees has $R^2 = 0.9$.

3.5 Material Design Formulation for Mixed variable BO (Module 5)

Our goal is to identify nanocomposites with high U_d, low ϵ and low $tan\delta$ suitable for electrical insulation. The design space consists of four variables, two qualitative and two quantitative, as summarized in Table 1. Choice of polymer and surface modification are qualitative variables with two (PS, PMMA) and three (Octyl, Chloro, Amino) levels respectively. Dispersion and volume fraction are quantitative variables with their bounds identified using SDF in Section 3.2. We present both single and multi-objective BO strategies for this case study, using the same set of design variables with different objective formulations. For single objective BO, we formulate an objective function that weighs all three normalized properties (indicated by *) equally and adds (subtracts) each property depending on whether it needs to be minimized (maximized):

$$\min_{s \in S, p \in P, m \in M} tan\delta^* + \epsilon^* - U_d^*$$
$$S: \{Chloro, Octyl, Amino\}$$
$$P: \{PMMA, PS\} \qquad (14)$$
$$M: microstructures \; with \; 0.7\% \leq VF$$
$$\leq 8.3\%, 0.62 \leq \theta \leq 6.55,$$

where objective is to be minimized over a design space consisting of all possible combinations of surface modification (S), polymers (P) and microstructures (M). In contrast, multi-objective aims to find candidate designs lying on the Pareto Frontier – a characteristic boundary comprising designs where no objective can be improved without deterioration of others. With three dielectric properties of interest, we define a multi-objective optimization problem as follows:

$$\min_{s \in S, p \in P, m \in M} tan\delta, \epsilon, -U_d,$$
$$S: \{Chloro, Octyl, Amino\}$$
$$P: \{PMMA, PS\} \qquad (15)$$
$$M: microstructures \; with \; 0.7\% \leq VF$$
$$\leq 8.3\%, 0.62 \leq \theta \leq 6.55,$$

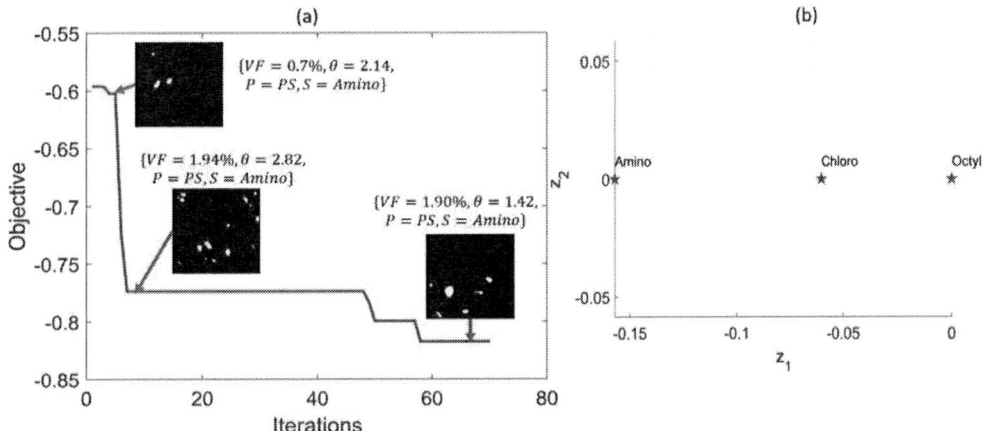

Figure 8: (a)Optimization history for single objective BO that converged to objective = -0.817 along with three designs evaluated in the process (b) Latent variable space for surface modification (**S**) variable

In this case, we do not normalize dielectric properties and all variables have their usual meaning.

3.6 Results and Discussion

We performed 70 & 120 iterations of BO for single and multi-objective formulations respectively, as specified by Eq. 14-15. Each BO is initiated with 30 random initial samples where quantitative variables $\{VF, \theta\}$ are generated by Latin Hypercube Design and qualitative variables polymer & surface modification type are sampled uniformly.

3.6.1 Results from Single Objective Bayesian Optimization

We performed ten replicates of single objective BO; each replicate initiated with 30 random samples. We observed that all replicates converge to optimal design with objective value of -0.798 ± 0.009. The best solution identified was -0.817, which corresponds to design $\{VF = 1.9\%, \theta = 1.42, P = PS, S = Amino\}$ with material properties $tan\delta = 0.0013$, $\epsilon = 2.108$ and $U_d = 132.421 \frac{kV}{mm}$. Fig. 8a shows optimization history for one replicate and depicts evolution of design during optimization. We observe that amino-modified Silica nanoparticles in PS with low filler volume fraction and dispersion is ideal to meet our requirements of high U_d, low $tan\delta$ and ϵ. These findings are consistent with our previous investigations that found $tan\delta$ and ϵ increase with dispersion. An added benefit of using LVGP is the ability to visualize latent space and draw insights about behavior of various levels for each qualitative variable. Fig. 8b shows the latent variable space for **S**- the three-level qualitative variable representing surface modification. We notice that amino surface modification is positioned relatively further away from other two – suggesting that its effect on dielectric properties is different. Also, we notice in Fig. 8b that all three levels are positioned along the z_1 axis with little variation in z_2; suggesting that the three surface modifications depend on a single latent factor and the use of two-dimensional latent variable representation is sufficient. We

believe this latent factor may be the interfacial energy descriptor described by Hassinger et al. [40]. The ordering of these three levels, Octyl-Chloro-Amino (or vice-versa), also matches the order of their corresponding interfacial energy descriptor.

To demonstrate the efficacy of BO in identifying optimal designs for problems with limited computational budget, we compare its performance against Genetic Algorithm (GA) [41]. MATLAB's implementation of GA for mixed integer optimization was used in this study and applied to problem formulation defined by Eq. 14. For a fair comparison with BO, GA was configured to terminate after 100 objective function evaluations (10 generations with population size of nine). Fig. 9 compares the optimal designs identified by 10 replicates of GA versus BO. Each BO replicate was initiated with 30 random samples. We see that regardless of initial samples provided, BO can consistently converge to the optimum design with low variability whilst GA is highly susceptible to the initial population that's usually generated randomly. This shows that the BO strategy of utilizing GP model uncertainty quantification to intelligently select new designs for evaluation and improve surrogate LVGP model makes it robust & faster at approaching

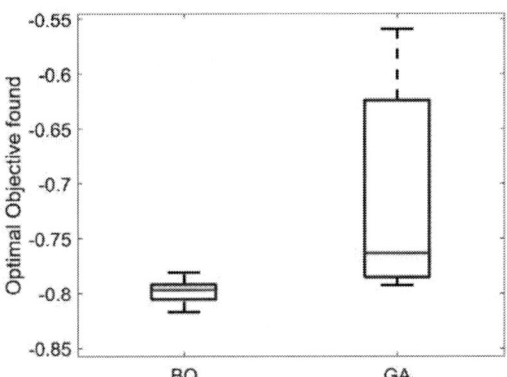

Figure 9: Comparison of 10 replicates of BO and GA for single objective optimization

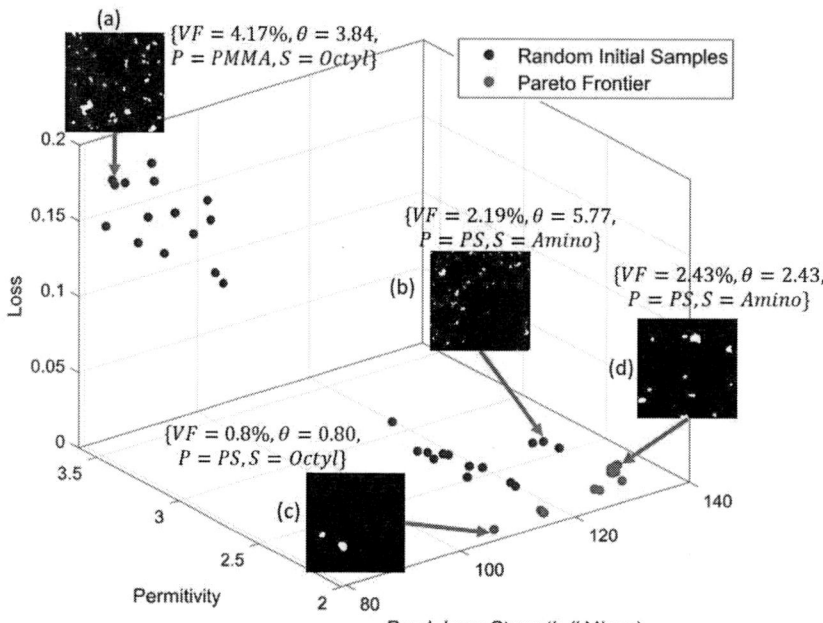

Figure 10: Summary of 120 iterations of MOBO

global optimum compared to other algorithms that ignore this information.

3.6.2 Results from Multi-Objective Bayesian Optimization (MOBO)

120 iterations of MOBO were performed starting with 30 random initial samples. Fig. 10 plots the random samples and 11 designs that were identified on the Pareto front. A noticeable feature in this plot is that the initial samples create two clusters corresponding to two polymers under consideration. The cluster located in the low U_d, high $tan\delta$ and ϵ region (top left corner in Fig. 10) exclusively contains PMMA based samples. This is consistent with our knowledge of dielectric behavior of PMMA. On the other hand, PS based samples have higher U_d, lower $tan\delta$ and ϵ; suggesting that they are better suited for electrical insulation application as compared to PMMA samples. This is

also reflected in the fact that all 11 designs identified on the Pareto Front are exclusively PS based. Notice that Pareto Front obtained by MOBO shows significant improvement w.r.t random initial samples and thus underlines the capability of uncertainty driven MOBO to locate improved designs.

The designs on the Pareto Frontier show large variations for U_d as compared to those of $tan\delta$ and ϵ. For example, designs (c) and (d) in Fig. 10 has properties $\{U_d = 107.83 \frac{kV}{mm}, \epsilon = 2.07 \& tan\delta = 0.0012\}$ and $\{U_d = 135.05 \frac{kV}{mm}, \epsilon = 2.26 \& tan\delta = 0.0020\}$ respectively. Table 2 lists all Pareto designs and their dielectric properties. It suggests that designs with octyl surface modification have lower $tan\delta$, ϵ and U_d while amino surface modification induces higher $tan\delta$, ϵ and U_d. The behavior of designs with chloro surface modification is intermediate; as suggested by its relative

Table 2: Pareto points obtained by MOBO

P	S	VF(%)	θ	U_d (kV/mm)	ϵ	$tan\delta$
PS	Octyl	0.884	0.803	107.835	2.072	0.0012
PS	Octyl	1.13	1.417	117.165	2.107	0.0014
PS	Chloro	2.539	1.504	132.914	2.229	0.0012
PS	Chloro	2.017	2.287	133.814	2.26	0.0013
PS	Chloro	0.981	1.124	117.069	2.083	0.0011
PS	Chloro	1.671	1.325	127.64	2.147	0.0011
PS	Amino	2.623	1.058	132.384	2.143	0.0014
PS	Amino	2.854	1.904	133.492	2.22	0.0018
PS	Amino	2.57	1.777	132.922	2.215	0.0018
PS	Amino	2.429	2.425	135.049	2.262	0.0020
PS	Amino	2.105	2.477	134.355	2.259	0.0020
PS	Amino	1.633	1.356	127.898	2.125	0.0014

positioning in latent space in Fig. 8b. Similarly, high (low) filler VF and θ leads to high (low) $tan\delta$, ϵ and U_d. Thus, we see a tradeoff between the three properties of interest. Which point in the Pareto front should be chosen as the design solution will depend on the designer's preference function, to be created based on the application, how the material is deployed, and device level performance.

Once the optimal design is identified, the corresponding processing condition can be obtained by mapping the optimized design variables to processing energy using the PS relationship established in our previous work [40]:

$$\bar{I}_{filler} = f(matrix)\sinh^2(2\,W_{PF}/W_{FF} - 1)\log(E_\gamma + 1) + C_0, \quad (16)$$

where $\bar{I}_{filler}$ is the normalized interphase area, f(matrix) and C_0 are polymer (aka matrix) dependent constants, W_{PF}/W_{FF} is the filler-matrix compatibility descriptors and E_γ is the processing energy descriptor that we seek. For illustration, we choose the design (d) in Fig. 10, favoring high breakdown strength of 135.05 kV/mm, as our optimal solution. Microstructure reconstruction corresponding to $VF = 2.43\%$, $\theta = 2.43$ is performed and $\bar{I}_{filler}$ is found to be 0.185. For PS, f(matrix) and C_0 are 0.00995 and 0.08798 respectively. For amino-modified Silica nanoparticles dispersed in PS, $W_{PF}/W_{FF} = 0.95$. Plugging these values in Eq. 16 leads to $E_\gamma = 10.47\ kJ/g$. Thus, we can identify designs satisfying application specific material properties and deduce processing parameter necessary for manufacturing.

4. CONCLUSIONS

In this article, we presented a data-centric Bayesian Optimization framework for materials design and innovation. Our framework pivots around LVGP, a novel machine learning method that projects qualitative variables onto a continuous latent space for covariance formulation. Uncertainty quantification by LVGP helps in navigating a complex design space using expected improvement acquisition function that balances exploration and exploitation. Visualization of latent space estimated by LVGP provides insight into behavior of levels for each qualitative variable. Generalization of LVGP based BO for multi-objective problems is developed using the expected minimax improvement acquisition criterion. The efficacy of the proposed framework is demonstrated using a case study centered on designing novel insulating nanocomposites with optimal choice of polymer, surface modification and morphology. We considered silica nanoparticles with Amino, Chloro and Octyl surface modifications dispersed in PS, PMMA polymers. Our implementation integrates empirical data with state-of-the-art techniques in interphase modelling, SDF based MCR for dimensionality reduction, and FEA-based structure-property simulations. Design formulation for single and multi-objective Bayesian optimization was presented using two qualitative (polymer and surface modification) and two quantitative (filler volume fraction and dispersion) variables. Both formulations led to similar optimal designs comprising Amino modified silica nanoparticles dispersed in PS matrix with low volume fraction and dispersion. The relative positioning of surface modification levels in latent space corroborates existing knowledge about their behavior. Processing energy required for fabrication of optimal design was evaluated using processing to structure mapping, to complete the bi-directional traversal across PSP paradigms and demonstrate the material genome approach to material design. While LVGP based BO and MOBO are applicable for any engineering design problem, their ability to facilitate concurrent optimization of composition and microstructure w.r.t. one or more properties, makes them a powerful tool for design of functional materials.

For future work, developing accurate simulation models based on Molecular Dynamics and Density Functional Theory is necessary for understanding & evaluating material properties such as Dielectric breakdown strength and interphase behavior. Additionally, we are working on continuously expanding the capabilities of the NanoMine that allows the exploration of curated data gathered from literature and the exploitation of data in the proposed BO framework.

ACKNOWLEDGEMENT

Support from NSF grants (ACI 1640840, CMMI 1729452, CMMI 1818574, CMMI 1729743, CMMI 1537641, OAC 1835782) and Center for Hierarchical Materials Design (ChiMaD NIST 70NANB14H012) are greatly appreciated.

REFERENCES

[1]   J. P. Holdren, "Materials genome initiative for global competitiveness," *National Science and technology council OSTP. Washington, USA,* 2011.

[2]   G. B. Olson, "Computational design of hierarchically structured materials," *Science,* vol. 277, no. 5330, pp. 1237-1242, 1997.

[3]   A. Jain *et al.*, "Commentary: The Materials Project: A materials genome approach to accelerating materials innovation," *Apl Materials,* vol. 1, no. 1, p. 011002, 2013.

[4]   J. E. Saal, S. Kirklin, M. Aykol, B. Meredig, and C. Wolverton, "Materials design and discovery with high-throughput density functional theory: the open quantum materials database (OQMD)," *Jom,* vol. 65, no. 11, pp. 1501-1509, 2013.

[5]   H. Zhao, X. Li, Y. Zhang, L. S. Schadler, W. Chen, and L. C. Brinson, "Perspective: NanoMine: A material genome approach for polymer nanocomposites analysis and design," *APL Materials,* vol. 4, no. 5, p. 053204, 2016.

[6]   H. Zhao *et al.*, "NanoMine schema: An extensible data representation for polymer nanocomposites," *APL Materials,* vol. 6, no. 11, p. 111108, 2018.

[7]   H. Xu, D. A. Dikin, C. Burkhart, and W. Chen, "Descriptor-based methodology for statistical characterization and 3D reconstruction of microstructural materials," *Computational Materials Science,* vol. 85, pp. 206-216, 2014.

[8]   H. Xu, Y. Li, C. Brinson, and W. Chen, "A descriptor-based design methodology for developing heterogeneous microstructural materials system," *Journal of Mechanical Design,* vol. 136, no. 5, p. 051007, 2014.

[9]   S. C. Yu *et al.*, "Characterization and Design of Functional Quasi-Random Nanostructured Materials Using Spectral Density Function," *Journal of Mechanical Design, 139(7), 071401.*

https://doi.org/10.1115/1.4036582, vol. 139, no. July, pp. 135-145, 2016.

[10]   U. Farooq Ghumman *et al.*, "A Spectral Density Function Approach for Active Layer Design of Organic Photovoltaic Cells," *Journal of Mechanical Design,* vol. 140, no. 11, pp. 111408-111408-14, 2018.

[11]   Z. Yang, X. Li, L. Catherine Brinson, A. N. Choudhary, W. Chen, and A. Agrawal, "Microstructural Materials Design Via Deep Adversarial Learning Methodology," *Journal of Mechanical Design,* vol. 140, no. 11, pp. 111416-111416-10, 2018.

[12]   B. Shahriari, K. Swersky, Z. Wang, R. P. Adams, and N. De Freitas, "Taking the human out of the loop: A review of bayesian optimization," *Proceedings of the IEEE,* vol. 104, no. 1, pp. 148-175, 2016.

[13]   D. R. Jones, M. Schonlau, and W. J. Welch, "Efficient global optimization of expensive black-box functions," *Journal of Global optimization,* vol. 13, no. 4, pp. 455-492, 1998.

[14]   P. V. Balachandran, D. Xue, J. Theiler, J. Hogden, and T. Lookman, "Adaptive strategies for materials design using uncertainties," *Scientific reports,* vol. 6, p. 19660, 2016.

[15]   C. Li *et al.*, "Rapid Bayesian optimisation for synthesis of short polymer fiber materials," *Scientific Reports,* vol. 7, no. 1, p. 5683, 2017/07/18 2017.

[16]   A. M. Gopakumar, P. V. Balachandran, D. Xue, J. E. Gubernatis, and T. Lookman, "Multi-objective Optimization for Materials Discovery via Adaptive Design," *Scientific reports,* vol. 8, no. 1, p. 3738, 2018.

[17]   D. Xue, P. V. Balachandran, J. Hogden, J. Theiler, D. Xue, and T. Lookman, "Accelerated search for materials with targeted properties by adaptive design," *Nature communications,* vol. 7, p. 11241, 2016.

[18]   Y. Wang *et al.*, "Identifying interphase properties in polymer nanocomposites using adaptive optimization," *Composites Science and Technology,* vol. 162, pp. 146-155, 2018.

[19]   S. Tao *et al.*, "Enhanced Gaussian Process Metamodeling and Collaborative Optimization for Vehicle Suspension Design Optimization," presented at the International Design Engineering Technical Conferences and Computers and Information in Engineering Conference, 2017. Available: http://dx.doi.org/10.1115/DETC2017-67976

[20]   R. Bostanabad, T. Kearney, S. Tao, D. W. Apley, and W. Chen, "Leveraging the nugget parameter for efficient Gaussian process modeling," *International Journal for Numerical Methods in Engineering,* vol. 114, no. 5, pp. 501-516, 2018.

[21]   J. Mockus, V. Tiesis, and A. Zilinskas, "The application of Bayesian methods for seeking the extremum," *Towards global optimization,* vol. 2, no. 117-129, p. 2, 1978.

[22]   H. J. Kushner, "A new method of locating the maximum point of an arbitrary multipeak curve in the presence of noise," *Journal of Basic Engineering,* vol. 86, no. 1, pp. 97-106, 1964.

[23]   N. Srinivas, A. Krause, S. M. Kakade, and M. Seeger, "Gaussian process optimization in the bandit setting: No regret and experimental design," *arXiv preprint arXiv:0912.3995,* 2009.

[24]   W. Scott, P. Frazier, and W. Powell, "The correlated knowledge gradient for simulation optimization of continuous parameters using gaussian process regression," *SIAM Journal on Optimization,* vol. 21, no. 3, pp. 996-1026, 2011.

[25]   Y. Zhang, S. Tao, W. Chen, and D. W. Apley, "A Latent Variable Approach to Gaussian Process Modeling with Qualitative and Quantitative Factors," *arXiv preprint arXiv:1806.07504,* 2018.

[26]   Y. Censor, "Pareto optimality in multiobjective problems," *Applied Mathematics and Optimization,* journal article vol. 4, no. 1, pp. 41-59, March 01 1977.

[27]   D. C. T. Bautista, "A sequential design for approximating the pareto front using the expected pareto improvement function," The Ohio State University, 2009.

[28]   J. R. Weidner, F. Pohlmann, P. Gröppel, and T. Hildinger, "Nanotechnology in high voltage insulation systems for turbine generators-First results," *17th ISH, Hannover, Germany,* 2011.

[29]   J. W. McPherson, J. Kim, A. Shanware, H. Mogul, and J. Rodriguez, "Trends in the ultimate breakdown strength of high dielectric-constant materials," *IEEE transactions on electron devices,* vol. 50, no. 8, pp. 1771-1778, 2003.

[30]   Wei Chen *et al.*, "Materials Informatics and Data System for Polymer Nanocomposites Analysis and Design," in *Big, Deep, and Smart Data in the Physical Sciences*, 2018.

[31]   W. Niblack, *An Introduction to Image Processing*. Englewood Cliffs, NJ: Prentice-Hall, 1986, pp. 115-116.

[32]   T. Krentz *et al.*, "Morphologically dependent alternating-current and direct-current breakdown strength in silica–polypropylene nanocomposites," *Journal of Applied Polymer Science,* vol. 134, no. 1, 2017.

[33]   J. W. Cahn, "Phase separation by spinodal decomposition in isotropic systems," *The Journal of Chemical Physics,* vol. 42, no. 1, pp. 93-99, 1965.

[34]   Y. Huang *et al.*, "Predicting the breakdown strength and lifetime of nanocomposites using a multi-scale modeling approach," *Journal of Applied Physics,* vol. 122, no. 6, p. 065101, 2017.

[35]   X. Li *et al.*, "Rethinking Interphase Representations for Modeling Viscoelastic Properties for Polymer Nanocomposites," *arXiv preprint arXiv:1811.06238,* 2018.

[36]   H. Zhao *et al.*, "Dielectric spectroscopy analysis using viscoelasticity-inspired relaxation theory with finite element modeling," *IEEE Transactions on Dielectrics and Electrical Insulation,* vol. 24, no. 6, pp. 3776-3785, 2017.

[37]   M. G. Todd and F. G. Shi, "Validation of a novel dielectric constant simulation model and the determination of its physical parameters," *Microelectronics journal,* vol. 33, no. 8, pp. 627-632, 2002.

[38]   P. Maity, N. Gupta, V. Parameswaran, and S. Basu, "On the size and dielectric properties of the interphase in epoxy-alumina nanocomposite," *IEEE Transactions on Dielectrics and Electrical Insulation,* vol. 17, no. 6, 2010.

[39]   L. Breiman, "Random forests," *Machine learning,* vol. 45, no. 1, pp. 5-32, 2001.

[40]   I. Hassinger *et al.*, "Toward the development of a quantitative tool for predicting dispersion of nanocomposites under non-equilibrium processing conditions," *Journal of Materials Science,* vol. 51, no. 9, pp. 4238-4249, May 2016.

[41]   D. E. Goldberg, *Genetic algorithms*. Pearson Education India, 2006.

Proceedings of the ASME 2019
International Design Engineering Technical Conferences
and Computers and Information in Engineering Conference
IDETC/CIE2019
August 18-21, 2019, Anaheim, CA, USA

DETC2019-98341

MULTI-SCALE DESIGN OF META-MATERIALS WITH OFFSET PERIODICITY

Rushabh Sadiwala *

Department of Mechanical Engineering
Clemson University
Clemson, SC 29634, USA
Email: rsadiwa@clemson.edu

Georges Fadel

Department of Mechanical Engineering
Clemson University
Clemson, SC 29634, USA
Email: fgeorge@clemson.edu

ABSTRACT

Meta-materials are a class of artificial materials with a wide range of bulk properties that are different from the base material they are made of. The term meta-material in the context of this research refers to a continuous, heterogeneous structure with prescribed elastic properties. Such meta-materials are designed using Topology Optimization (TO). Tools like SIMP interpolation, mesh filtering and continuation methods are used to address the numerical issues with Topology Optimization.

In a previous research [1], by offsetting meta-material layers by a half-width of the Unit Cell, an auxetic honeycomb-like geometry was obtained. This was the first time such a shape was observed as the result of Topology Optimization targeting the effective shear modulus using square Unit Cells. This was obtained while designing the shear beam of a non-pneumatic wheel.

This research studies the design of meta-materials using offsets other than zero or half-widths. The same problem [1] was solved for different values of offset, and the obtained geometries and volume fractions are studied. It is concluded that it may be beneficial for designers to consider offsetting meta-material layers with offsets other than half-width, to design novel, potentially better performing structures.

NOMENCLATURE

ρ Density of material at any given point in the domain.
C_0 Young's Modulus of the base material of the meta-material.

C_e Young's Modulus of th e^{th} element in the meta-material.
p   SIMP penalty.
E_{ij}^* Target elastic properties for optimization.
E_{ij}^M Effective elastic properties of the meta-material.
E_{ii}^U Effective normal elastic moduli in the ith direction, calculated using Volume Averaging
G_{ij}^U Effective shear moduli, calculated using Volume Averaging
b   Introduced offset value in meta-material layers.
h   Size of elements
r_{min} Size of mesh filtering used for Topology Optimization.

INTRODUCTION

Topology Optimization (TO) is a structural optimization tool used to find the optimal distribution of material in a design space subjected to certain stimuli (such as structural loading, heat input, electromagnetic field etc.). Topology Optimization helps to solve a broad set of design problems from different engineering disciplines.

The most common problems are minimum compliance and minimum weight problems. Other problems include maximum heat transfer rate, maximum porosity, etc.

Meta-material design is one such problem that can be solved using topology optimization. Meta-materials are a class of artificial materials which are structurally designed to achieve a set of properties which are different from the base material they are made of. Such meta-materials can also achieve properties which are not found in any traditional, homogeneous material. In the

---

*Address all correspondence to this author.

Copyright © 2019 ASME

context of this research, meta-materials are heterogeneous, periodic structures, designed to achieve a specific elastic behavior.

First the system in which the meta-material is to be used is optimized, considering a homogeneous material with properties E_{ij}^*. The properties E_{ij}^* are optimized to achieve the objectives and constraints of the system optimization problem.

Then, topology optimization is used to find the optimal distribution of the base material with properties E_{ij}^0 subject to E_{ij}^* as the effective elastic property of the resultant structure. This research primarily focuses on the second step of the meta-material design process.

Meta-material design typically uses the mathematical Asymptotic Homogenization theory to evaluate the effective elastic property (E_{ij}^M) of a given meta-material. The assumption of the mathematical tool requires the meta-material to be Y-periodic, where the domain Y, is far smaller in size than the global domain. This domain (Y) is known as the Unit Cell (UC). Because of this assumption, the applicability of asymptotic homogenization in the design of certain meta-material components is limited.

Field homogenization, also called Volume Averaging is another analytical tool that can be used to evaluate the elastic properties of a material distribution. Volume averaging was originally developed to evaluate the global properties of composite materials. It was previously demonstrated by Chris Czech that this tool can be used to solve meta-material design problems where asymptotic homogenization could not be used [2], such as when the domain is not Y-periodic (non-simple or offset geometry), and when designing a meta-material outside the scaling limit of asymptotic homogenization.

The Volume Averaging method evaluates the meta-material using the RVE, rather than the RUC (Repeated Unit Cell). This eliminates the sizing constraints on the UC that homogenization theory requires. If the system of the meta-material is small enough in one or more dimensions, a meta-material with UCs bound by homogenization's requirements would be too small to be manufactured [3] [4].

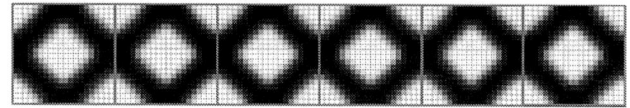

FIGURE 1: A meta-material which violates the homogenization requirements in the vertical direction(only one cell)

The design domain Y is said to be Y-periodic if some physical property of the domain follows the periodicity equation within the domain:

$$f(x + nY_1 + mY_2) = f(x) \qquad (1)$$

Where,
x is the position of any given point in the design domain,
n, m are arbitrary integers, and
Y is the domain bound by the vectors Y_1 and Y_2

For meta-materials, the periodic function is the material tensors $C_{ijkl}(x)$ at a given point x. This means C_{ijkl} at any point in the UC is equivalent to C_{ijkl} at the corresponding point in any other cell in the global design domain.

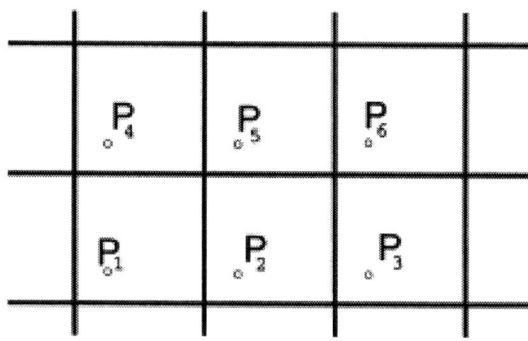

FIGURE 2: Periodicity conditions require functions to be identical at each point P in the domain [5].

The global design domain is discretized into smaller, periodic unit cells with the same material distribution. Each UC is further discretized into a number of finite elements. This discretization serves two purposes: to be used for any finite element analysis required for the evaluation of the meta-material, and to quantify the material distribution for topology optimization.

A density function, $\rho(x)$, is used to quantify the material distribution. For simplicity, the density function is approximated as being constant within each element e. The constant densities ρ_e are then chosen as the design variables for topology optimization. ρ_e ranges from 0(void) to 1(material). ρ_e can be a discrete (black or white) or a continuous (gray) variable. A discrete design variable will require a discrete optimization such as integer programming, which is computationally costly and is NP-complete (no guaranteed convergence). Intermediate values of the design variables lack any physical interpretation. A material interpolation scheme is required to translate the "artificial" variables to a real material property. In this research, the densities ρ_e are translated into the elastic tensor C_e using Solid Isotropic Material with Penalization (SIMP) interpolation [6].

The elastic tensor of element e, is given by:

$$C_e = \rho_e^p . C_0 \qquad (2)$$

The penalty, p, is an arbitrary constant that penalizes intermediate densities to make them tend to a black or white (1-0) solution. It is usually chosen between 2 and 4. Penalties less than 2 do not provide enough penalization, while those larger than 4 have a lower convergence rate [6]. After optimization, elements with densities larger than a certain value (usually, $\rho_i > 0.3$) are interpreted as elements with material, and the remaining as elements without material. The solution is called SIMP-convergent if the interpreted topology has a behavior close to the optimized artificial topology [1].

The optimum structure is the one which has the least volume fraction while having an elastic behavior equal to the desired target. Hence, the optimization problem is formulated as:

$$
\begin{aligned}
\underset{\rho}{\text{Minimize}} \quad & V(\rho) \\
\text{Subject To:} \quad & E_{ij}^M(\rho) - E_{ij}^* = 0 \\
& 0 \le \rho_e \le 1, \ e = 1, \dots, n
\end{aligned} \qquad (3)
$$

Field Homogenization (Volume Averaging)

Field Homogenization, also called Volume Averaging is an evaluation tool, originally developed for calculating the global elastic properties of conventional fiber composites by Drago et. al. [3]

Unlike asymptotic homogenization, the evaluation is performed on the Representative Volume Element (RVE) of the global domain, and not the RUC. The RVE is defined as the volume which accurately represents the properties of the global material. The smallest RVE which can be analyzed to accurately evaluate the global properties of the material is called the RVE limit. In the context of meta-materials, RVE limit can either be a single UC or a collection of UCs. [3] [4]

There are two sets of boundary conditions which can be applied to the design domain, homogeneous traction or homogeneous displacements. Homogeneous traction boundary conditions apply a uniform traction force, σ_{ij}^0, on the boundary of the domain. Homogeneous displacement boundary conditions enforce a uniform boundary deformation on the domain. [3] [4]. The homogeneous traction conditions do not work properly with Topology Optimization, as a finite traction on an element with 0 density will induce an infinite strain. So, only homogeneous displacement conditions are considered. The different Boundary Conditions that can be applied on the RVE boundaries, S [3] [4] are:
Displacement BCs for Transverse Normal loading - For evaluation of E_{22}^U (or E_{33}^U)

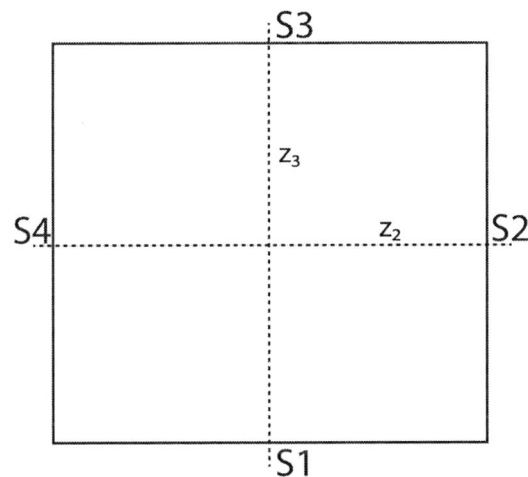

FIGURE 3: Representative Volume Element (RVE) boundaries

$$z \in S: \quad u_2(z) = \varepsilon_{22}^0 z_2 \quad u_3(z) = -\varepsilon_{33}^0 z_3$$
The unknown ε_{33}^0 is determined subject to the integral constraint, $\bar{\sigma}_{33} = 0$ on S1 and S3

Displacement BCs for Transverse Shear loading - For evaluation of G_{23}^U

$$z \in S: \quad u_2(z) = \varepsilon_{23}^0 z_3 \quad u_3(z) = \varepsilon_{23}^0 z_2$$

Using finite element stiffness matrix, K, for solid 2-D, linear, isotropic elasticity, the displacement field $u(z_2, z_3)$ is calculated as the solution for the linear system [3] [4]:

$$[K]\{u\} = \{F\} \qquad (4)$$

The effective modulus of the periodic structure, E_{ij}^M is then calculated as the solution to:

$$\bar{\sigma}_i = E_{ij}^M \bar{\varepsilon}_j \qquad (5)$$

Where, $\bar{\varepsilon}$ is the average strain, and $\bar{\sigma}$ is the average stress induced in the RVE due to the displacement field $u(z_2, z_3)$. The average stresses and strains are calculated using the average stress and strain theorems [3] [4].

META-MATERIALS WITH OFFSET PERIODICITY

A typical meta-material is a regular periodic structure with square or rectangular unit cells. In principle, meta-material design using regular meso-structures can obtain the full spectrum of topologies, as Topology Optimization creates and shapes holes in the meso-structure as required. However, meta-materials with shapes like honeycombs or auxetic honeycombs cannot be obtained by the standard Topology Optimization process targeting elastic or shear modulus, as observed in the literature. [1]

Diaz et al. used a parallelogram unit cell with regular periodicity to design meta-materials [7]. By using such parallelogram unit cells, with varying angles, they could obtain honeycomb shapes as the optimal design of the meta-material design process. A change in the parallelogram angle changes the shape of the global and local design domains. This requires re-discretization of the finite elements every time the angle changes.

Chris Czech proposed a way to obtain honeycomb or auxetic honeycomb shapes from the meta-material design process by using square unit cells with irregular periodicity [1]. Square or rectangular unit cells are tessellated in a staggered arrangement as shown in Figure 5. Changes in the staggered arrangement do not change the shape of the global and local design domains. Hence, such meso-structures have an advantage over parallelogram meso-structures, as they do not require re-discretization for every change in the topological connectivity. It was demonstrated that such irregular periodicities can lead to a broader class of meta-materials such as ones with auxetic honeycomb geometries [1].

Chris Czech successfully obtained an auxetic honeycomb meta-material by using square unit cells with layers offset by a half-width of the unit cell. However, no research exists in the literature that uses arrangements with offsets other than zero and half-width of the unit cell. This research will investigate the meta-material shapes obtained by using tessellations with varying offsets.

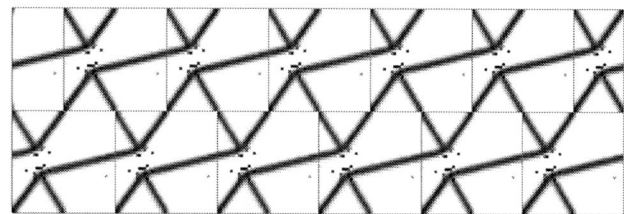

FIGURE 4: Meta-material designed using square unit cells offset by their half-widths [1]

Consider the UC geometry in Figure 5. Layers in y_2 are shifted by b relative to the previous layer. For simplicity and uniformity of the discretization, the offset parameter b is limited to be an integer multiple of h_1 (the element size in y_1). If b is continuous instead of discrete, the finite element discretization will not be uniform. An additional node per element would be required at the interface of two meta-material layers. This non-uniformity also means that the domain has to be re-discretized for every change in the offset.

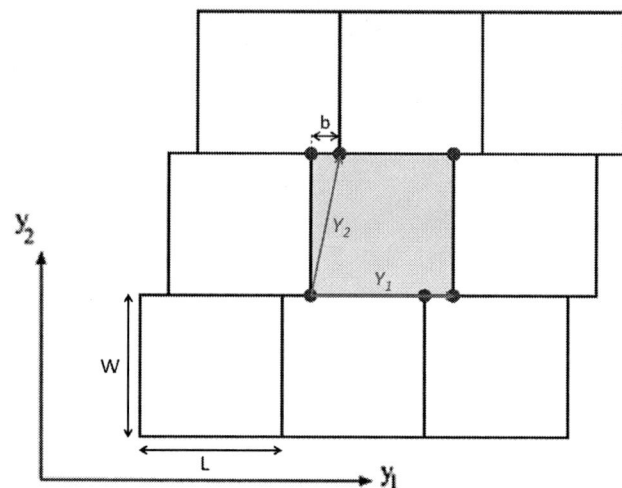

FIGURE 5: Square UC w/ offset tessellation

Introducing offset between the layers of UC also results in violation of the Y-periodicity condition. In other words, for orthogonal Unit Cells, the meta-material has to be orthogonally periodic. Polygonal Unit Cells must be used for non-orthogonally periodic meta-materials for evaluation by homogenization [7]. For the Unit Cell in Figure 5, $Y_1 = [L, 0]^T$ and $Y_2 = [b, W]^T$, where L, W, and b are the lengths, widths and the offsets of the UC respectively.

Because of this violation, asymptotic homogenization cannot be used to evaluate offseted meta-materials.

Since Volume Averaging is applied on the RVE and not on a single UC, the analysis is not affected by the change in periodicity.

OPTIMIZATION OF THE SHEAR BEAM OF A NON-PNEUMATIC WHEEL

The design of the shear beam of a non-pneumatic wheel consists of two steps. The top-level, or the system-level optimization is solved by Thyagaraja et. al. [8] The goal of this step is to identify the target moduli of the next step of optimization, aiming to reduce the hysteresis losses in the beam, by using a meta-material that is designed to replace rubber. Sensitivity analysis on the design variables identified the shear beam thickness, sl_{Thk} and shear modulus G_{23} as the most influential variables for the system-level optimization. The solution to the optimization problem was found to lie on the curve:

$$G_{23}.sl_{Thk} = 67MPa.mm \qquad (6)$$

Top Level: Non-pneumatic wheel optimization

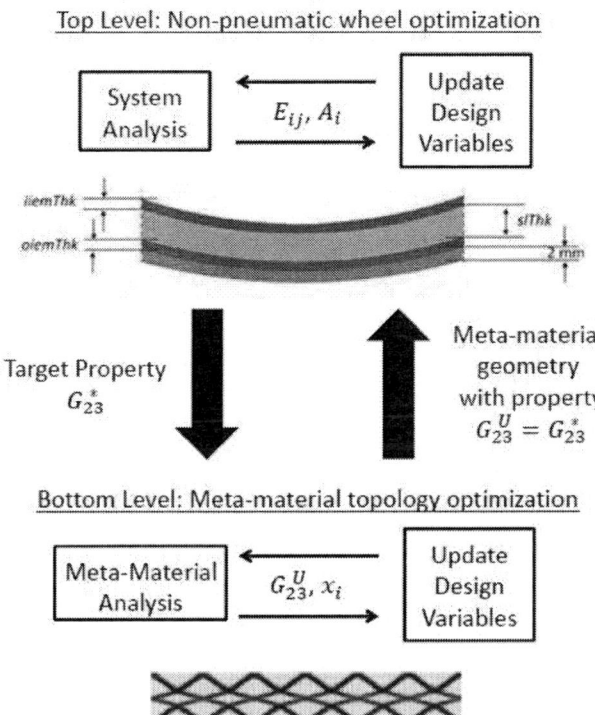

FIGURE 6: Two-level optimization process used for meta-material design of the shear beam of the non-pneumatic wheel [8]

The solution to the top-level optimization establishes the targets of the bottom-level optimization (i.e. Topology Optimization). The bottom-level optimization was solved by Dr. Czech, for meta-materials with half-width offset periodicity [4]. The base material chosen was Poly-Carbonate (PC) with E=2.7 GPa and v=0.42. Since the shear beam's radial dimension was in the order of 1 cm, no more than 2 or three cells could exist in the radial direction for the meta-material structure to be manufacturable. Such a meta-material violates the homogenization scaling limit, hence volume averaging analysis was required to calculate the effective meta-material moduli. A high-resolution mesh of 40x40 elements in each UC was chosen. Four different initial points were considered as shown in Figure 8. Optimization was performed on all integer values of sl_{Thk} between 5 and 12 mm. [4]

The optimization problem was formulated as:

$$\underset{\rho}{\text{Minimize}} \quad V(\rho)$$
$$\text{Subject To:} \quad G_{23}^U(\rho) - G_{23}^* = 0 \qquad (7)$$
$$0 \le \rho_e \le 1, \ e = 1,\dots,n$$

All Topology Optimization problems solved by Dr. Czech

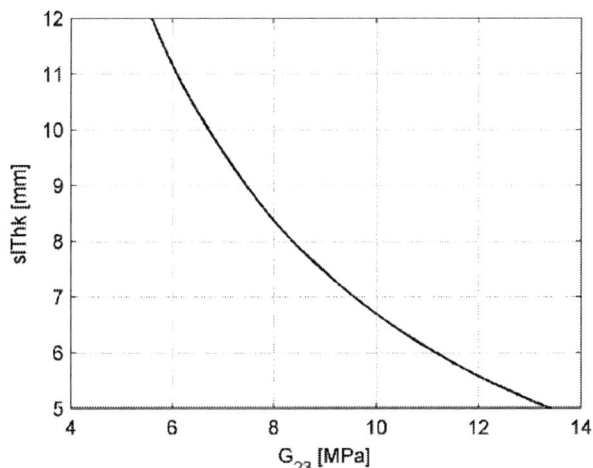

FIGURE 7: Solution to the top-level optimization for design of the shear beam of the non-pneumatic wheel [8] [4]

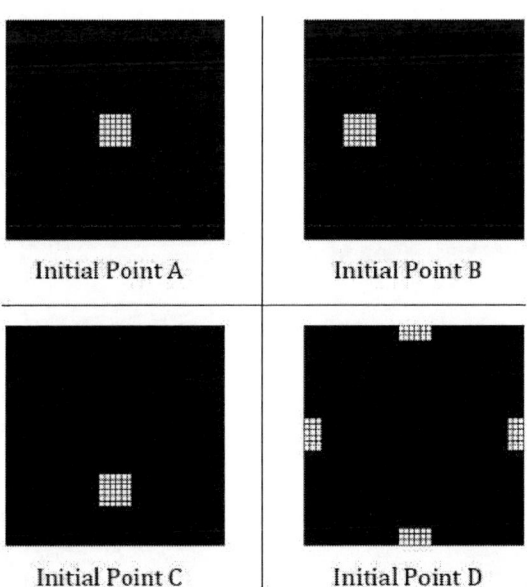

FIGURE 8: Different initial points used for the Topology Optimization of the shear beam of the non-pneumatic wheel [4]

[4] are repeated while using unit cells with offset periodicity. The offsets range from 0%(0 element) to 50%(20 element) in 5%(2 element) increments. Selected results to this problem are shown in Table 1.

Discussion on Topology Optimization Results Selected results to the Topology Optimization problem are shown

V02AT03A067-5

TABLE 1: Selected Solutions to Topology Optimization of the shear beam of the non-pneumatic wheel as obtained by Czech et. al. [1]

sl_{Thk}(mm)	G_{23}^U(MPa)	V	Meta-Material
Initial Point A			
6	11.7	0.117	
7	9.57	0.102	
12	5.58	0.054	
Initial Point B			
5	13.40	0.08	
7	9.57	0.121	
8	8.42	0.064	
12	5.58	0.104	
Initial Point C			
5	13.39	0.072	
8	8.38	0.115	

TABLE 2: Optimization results for sl_{Thk} = 12 mm, G_{23}^U = 5.58 MPa, initial point A.

Offset	V	Meta-Material
0	0.0859	
2	0.0529	
4	0.0521	
6	0.0539	
8	0.0514	
10	0.0509	
12	0.0509	
14	0.0890	
16	0.0881	
18	0.0886	
20	0.1039	

in Table 2 to Table 4.

It can be observed that for problems with zero offset, the result obtained are meta-materials with a cross X-like geometry. It is a well known fact that such structures are the stiffest in shear when minimizing the volume fraction [9].

Another class of meta-material is obtained for intermediate offsets (from >0 elements to 12 elements). These meta-materials have a bristle /-like geometry. These meta-materials have the lowest volume among all the different classes of meta-materials obtained, however, such meta-materials can buckle when loaded in compression. An interesting observation is that the angle of bristles change from a near-45 degree angle, becoming steeper as offset increases.

For offsets >12 elements, the meta-materials have a staggered x-like geometry, which consists of alternating layers of auxetic hexagons and crosses. These geometries have a lower volume fraction than pure x-like geometries, without any obvious buckling issue.

TABLE 3: Optimization results for $sl_{Thk} = 7$ mm, $G_{23}^U = 9.57$ MPa, initial point B.

Offset	V	Meta-Material
0	0.0941	
2	0.0632	
4	0.0624	
6	0.0616	
8	0.0610	
10	0.0608	
12	0.0604	
14	0.1022	
16	0.0996	
18	0.1002	
20	0.1066	

TABLE 4: Optimization results for $sl_{Thk} = 5$ mm, $G_{23}^U = 13.38$ MPa, initial point C.

Offset	V	Meta-Material
0	0.1078	
2	0.0707	
4	0.0697	
6	0.0694	
8	0.0684	
10	0.0683	
12	0.0687	
14	0.0702	
16	0.1120	
18	0.0997	
20	0.1341	

Some problems yield a near auxetic honeycomb structure (Table 4, for offset equal to 18 elements). These meta-materials are quite close to the auxetic honeycomb obtained by Dr. Czech [4] as seen in Table 1.

It was also observed that it is not necessary that extreme offsets of zero or half-width will converge to lower volume fractions than intermediate ones.

The results obtained by Dr. Czech for an offset of 20 elements are not reproduced. However, one must note that the volume fractions of the results produced are sometimes lower than the volume fractions of the meta-materials produced by Dr. Czech. Also, if optimization is attempted with the meta-material produced by Dr. Czech as the initial point, it returns the same meta-material without performing any iterations. These are the characteristics of a local minimum. Such local minimum issues are quite common for a Topology Optimization problem [10]. Local minima for this problem could be caused by TO parameters like the filter size r_{min}, the SIMP parameter and/or the numerical optimization algorithm used.

To prevent such local minima issues, continuation methods can be used [10]. However, it should be noted that such methods do not guarantee a global minimum, but the resultant geometry can be quite close to it [10]. All optimization results shown in Tables 2 to 4 are obtained using a continuation method on the mesh filter size [10], r_{min} (gradually decreased from 3 to 1.5 elements).

All optimization problems in this research deals only with the effective properties of the meta-materials. The designed meta-materials, especially the ones with low volume fractions and/or with a bristle /-like geometry, are susceptible to failure due to Von-Mises stresses or buckling stresses. A constraint can be added to the optimization formulations to prevent such failures.

CONCLUSION

This research was posed as an extension to Dr. Czech's research in half-width offset meta-materials. That research was the first time an auxetic honeycomb shaped meta-material was observed as the solution of Topology Optimization targeting effective shear modulus, using square UCs. It demonstrated that half-width offsets can lead Topology Optimization to output a different classes of meta-materials.

The same optimization problem (design of the shear beam of a non-pneumatic wheel) was solved for different offsets. Based on the optimization results it is observed that it is indeed possible to converge on differently shaped meta-materials for different offsets.

Another observation is that lower volume fractions may be obtained at intermediate (non-zero, non-half width) offsets. Therefore it is beneficial to consider designing meta-materials across every possible offset values.

REFERENCES

[1] Christopher Czech, Paolo Guarneri, and Georges Fadel. Meta-Material Design of the Shear Layer of a Non-Pneumatic Wheel Using Topology Optimization. In *Volume 3: 38th Design Automation Conference, Parts A and B*, page 893. ASME, 8 2012.

[2] Christopher Czech, Paolo Guarneri, James Gibert, and Georges Fadel. On the accurate analysis of linear elastic meta-material properties for use in design optimization problems. *Composites Science and Technology*, 72:580–586, 2012.

[3] Anthony Drago and Marek Jerzy Pindera. Micro-macromechanical analysis of heterogeneous materials: Macroscopically homogeneous vs periodic microstructures. *Composites Science and Technology*, 67(6):1243–1263, 2007.

[4] Christopher Czech. Design of Meta-Materials Outside the Homogenization Limit Using Multiscale Analysis and Topology Optimization. *Ph.D. Dissertation, Clemson University*, pages 10–13, 2012.

[5] B Hassani and E Hinton. A review of homogenization and topology optimization Ihomogenization theory for media with periodic structure. *Computers & Structures*, 69(6):707–717, 1998.

[6] M. P. Bendsøe and O Sigmund. Material interpolation schemes in topology optimization. *Archive of Applied Mechanics (Ingenieur Archiv)*, 69(9-10):635–654, 1999.

[7] Alejandro R Diaz and Andre Benard. Designing materials with prescribed elastic properties using polygonal cells. *INTERNATIONAL JOURNAL FOR NUMERICAL METHODS IN ENGINEERING Int. J. Numer. Meth. Engng*, 57:301–314, 2003.

[8] Niranjan Thyagaraja, Prabhu Shankar, Georges Fadel, and Paolo Guarneri. Optimizing the shear beam of a non-pneumatic wheel for low rolling resistance. In *Volume 5: 37th Design Automation Conference, Parts A and B*, pages 33–42. ASME, 1 2011.

[9] Weihong Zhang, Gaoming Dai, Fengwen Wang, Shiping Sun, and Hicham Bassir. Using strain energy-based prediction of effective elastic properties in topology optimization of material microstructures. *Acta Mechanica Sinica/Lixue Xuebao*, 23(1):77–89, 2007.

[10] O. Sigmund and J. Petersson. Numerical instabilities in topology optimization: A survey on procedures dealing with checkerboards, mesh-dependencies and local minima. *Structural Optimization*, 16(1):68–75, 8 1998.

**Proceedings of the ASME 2019
International Design Engineering Technical Conferences
and Computers and Information in Engineering Conference
IDETC/CIE2019
August 18-21, 2019, Anaheim, CA, USA**

DETC2019-98386

OPTIMIZING TOPOLOGY AND FIBER ORIENTATIONS WITH MINIMUM LENGTH SCALE CONTROL IN LAMINATED COMPOSITES

Chuan Luo, James K. Guest[1]
Department of Civil Engineering, Johns Hopkins University, Baltimore, Maryland, 21218, United States

ABSTRACT

Discrete Material Optimization (DMO) has proven to be an effective framework for optimizing the orientation of orthotropic laminate composite panels across a structural design domain. The typical design problem is one of maximizing stiffness by assigning a fiber orientation to each subdomain, where the orientation must be selected from a set of discrete magnitudes (e.g., 0°, ±45°, 90°). The DMO approach converts this discrete problem into a continuous formulation where a design variable is introduced for each candidate orientation. Local constraints and SIMP style penalization are then used to ensure each subdomain is assigned a single orientation in the final solution. The subdomain over which orientation is constant is typically defined as a finite element, ultimately leading to complex orientation layouts that may be difficult to manufacture. Recent literature has introduced threshold projections, originally developed for density-based topology optimization, into the DMO approach in order to influence the manufacturability of solutions. This work takes this idea one step further and utilizes the Heaviside Projection Method within DMO to provide formal control over the minimum length scale of structural features, holes, and patches of uniform orientation. The proposed approach is demonstrated on benchmark maximum stiffness design problems in terms of objective function, solution discreteness, and manufacturability. Numerical results suggest that projection-based methods can play an important role in controlling the manufacturability of optimized material distributions in optimized design and that solutions are near-discrete with performance properties comparable to designs without manufacturing considerations.

Keywords: Topology optimization, Discrete Material Optimization, Manufacturability, Heaviside Projection, Composite materials, Minimum length scale, Density Filters

1. INTRODUCTION

Topology optimization has gained wide acceptance as a powerful design tool for identifying highly efficient structural solutions. The goal is to optimize the distribution of a given material across a design domain with the resulting connectivity defining the structural topology [1,2]. The vast majority of research in topology optimization has assumed the material used to manufacture the resulting structural designs is homogenous and isotropic. Although appropriate for many materials and fabrication processes, the assumption is generally invalid for the case of laminated composites. A typical fiber-reinforced laminate composite is composed of a stack of anisotropic laminate sheets that are bonded together. A common goal in laminate design is thus to determine the optimal orientation of each laminate sheet within a stack (the stacking sequence) and across the structural domain.

Ideally laminate orientations would be tailored at each point in space within the structural domain, allowing use of classical homogenization-based topology optimization methods that include an orientation angle as an independent design variable [3]. Taking such an approach, however, would likely result in optimized orientation distributions being incredibly complex and detailed, making them either impossible or expensive to manufacture in practice. At a larger scale, one can consider the case of precise arrangement of continuous fibers using modern advanced fiber placement machines, motivating the development of associated continuous fiber placement design methods [4,5]. On the other hand, engineers in aerospace and wind turbine industries often prescribe a discrete set of allowable orientations for each laminate sheet (e.g., 0°, ±45°, 90°) and a minimum panel size over which this orientation is held constant. The corresponding design optimization problem is then a large scale,

---

[1] Contact author: jkguest@jhu.edu

discrete-linked variable problem, a challenging class of problems to solve.

The Discrete Material Optimization (DMO) method [6] has arisen as an effective approach to handling the discrete material selection problem in topology optimization. The DMO approach relaxes the discrete condition and introduces a continuous independent design variable for each candidate material orientation at each location the material can be placed (typically each finite element in the discretized problem). Elemental stiffness is then expressed as a combination of the stiffnesses corresponding to candidate orientations, with SIMP penalization used to drive design variables to 0-1 magnitudes, and local constraints on each element to ensure only one orientation is used at a location [6]. Voids can also be accounted for by considering them as a candidate orientation having minimal stiffness [7].

As the resulting DMO problem is continuous, gradient-based optimizers can be used to identify optimized solutions as in classical topology optimization. However, optimizing orientation at the finite element level can lead to rapidly varying orientation distributions that may still be challenging to manufacture, as well as introduce issues of solution mesh dependency, both of which are well known challenges in topology optimization [8]. The original DMO approach circumvented these issues by predefining patches composed of multiple finite elements, each required to have the same orientation. These patch discretizations are fixed during the course of the optimization, meaning the designer is required to define the patch sizes, shape and locations a priori. Noting this limitation, Sørensen and Lund [9] recently proposed implementing projection filters within the DMO framework, essentially allowing the optimizer to influence these decisions.

Projection-based methods in topology optimization essentially express physical design variables, typically relative densities, as a continuous function of independent optimization variables [10]. This function is defined for each finite element over a geometric space, called the neighborhood, using some form of a regularized Heaviside step function. This step function is key to achieving clear (near-discrete) distributions of relative density, offering a distinct advantage over linear filtering techniques [11,12,13] that lead to a blurry topological boundaries or interfaces between adjacent materials. It is the combination of the neighborhood geometry and regularization function that provide the designer with control over geometric features, which have been manipulated to achieve minimum length of designed structural features [8], void features [14,15], or both [10,14,15, 16], machining constraints [17], overhang constraints [18], and even discrete objects such as fibers [19,20], among others. An important property of the regularized Heaviside functions in these works is that all have either positive or negative curvatures through the entire variable space, an important detail that guarantees the geometric control being pursued is rigorous and quantifiable.

An alternative to these Heaviside projection formulations is so-called thresholding projection functions, referred to as volume preserving filters in the original work [21]. These filters typically use hyperbolic tangent functions as the approximation to the step function and tend to produce clear (near-binary) representations of topology, as in standard projection. The hyperbolic tangent functions are also quite convenient for perturbing a given topology, such as in robust topology optimization approaches [22]. Sørensen and Lund [9] successfully implemented the threshold projection methodology in the context of DMO and demonstrated that orientations tended to become grouped together using this approach, without the need to predefine patches of constant orientation. They also observed that the length scale of the designed features were less than the projection radius, and that mesh independence of solutions was not guaranteed. These observations are not unique to DMO implementations, and are ultimately a result of the properties of thresholding projection. Despite its advantages, thresholding projection does not guarantee strict geometric control, but rather may only be said to influence geometric properties. Although often numerical results show a relation between the radius of the projection neighborhood and the length scale of designed features, it can be readily shown that features at the scale of a single element can be designed, regardless of the projection radius, when using only the thresholding projection approach.

Herein we look to extend the novel projection-based DMO approach presented in Reference [9] by replacing the thresholding projection scheme with the original projection step function of Reference [10], with the aim of guaranteeing the minimum length scale of orientation distributions in a physically meaningful and convenient manner. Specifically, the designer may specify the minimum patch size, defined as a physical dimension, as well as the corresponding shape of this minimum patch. These patches would then be assembled using adhesives or fasteners, or in some cases could be achieved with tow steering. In this work, we consider projected shapes that are square instead of the widely used radial projection domains, thereby allowing tight packing of the projected orientation patches. Additionally, by using different projection radii for each orientation, we demonstrate that the designer may prescribe different minimum patch sizes for voids and/or for different orientations. Finally, we note that guaranteeing minimum length scale also provides a path to ensuring DMO solutions are mesh independent, as observed in classical topology optimization [8,10].

The original [6] and updated DMO formulations and algorithms are presented in Section 2, followed by results found when solving the benchmark minimum compliance design problems previously studied in [9,27]. As the resulting optimization problems are continuous, all problems utilize SIMP formulations, are solved using the Method of Moving Asymptotes gradient-based optimizer [23] with sensitivities computed using the adjoint method.

2. MATERIALS AND METHODS
We consider the case of optimizing the distribution of orthotropic materials with n^c different possible fiber orientations across a structural design domain. When the void

phase is also considered, the problem becomes one of simultaneously optimizing topology and fiber orientations across the topology.

2.1 Discrete Material Optimization parameterization

The material constitutive properties for the lth layer at eth element can be expressed as an assembly of constitutive tensors C_i from candidate material i as follows:

$$C^{el} = \sum_{i=1}^{n^c} x_{iel} C_i = x_{1el} C_1 + \cdots + x_{n^c el} C_{n^c}; \tag{1}$$

$$x_{iel} \in \{0; 1\}; \quad \forall\, (i,e,l) \tag{2}$$

$$\sum_{i=1}^{n^c} x_{iel} = 1; \quad \forall\, (e,l) \tag{3}$$

where x_{iel} is the material selection variable indicating whether candidate material i is present in layer l at element e. If the ith candidate material is chosen for the lth layer at eth element, then $x_{iel} = 1$; otherwise, $x_{iel} = 0$. Equations (2) and (3) ensure that only one candidate material is selected for each element. It should be noted that the candidate materials can either be isotropic, anisotropic, or void, although herein we consider only orthotropic materials.

Following the DMO approach, the binary condition (2) is relaxed and the SIMP method [1] is used to create continuous formulation such that gradient-based optimization algorithms could be applied. Equation (2) is thus replaced with

$$0 \le x_{iel} \le 1; \quad \forall\, (i,e,l) \tag{4}$$

and equation (1) with the following SIMP representation:

$$C^{el}(x) = E_0 + \sum_{i=1}^{n^c} x_{iel}{}^p (C_i - E_0); \ p \ge 1; \ \forall\, (e,l) \tag{5}$$

where E_0 is a small positive definite matrix to avoid singularity of the stiffness matrix. For SIMP penalty $p > 1$, intermediate elemental design variables become unfavorable as stiffness is reduced.

For simultaneously optimizing topology and fiber orientation, an additional variable controlling the topology could be used, as in typical multi-material topology optimization problems [11,24]. Alternatively, the linear equality constraints (3) could be relaxed to linear inequality constraints to allow no candidate to be chosen [25]. Here we consider the void phase as one of the candidate materials [7] such that the projection algorithm proposed in the paper could conveniently and effectively control the minimum length scale of both solid material phases and the void phase.

2.2 Projection formulation

The Heaviside Projection Method [10] is implemented to control the minimum length scale of patches of like orientation. Nodal design variables ϕ_{iml} located in the lth layer at node m of the finite element mesh are introduced as independent design variables. These nodal design variables are projected onto the finite element space, with mapping done separately for each candidate material phase of the element by computing the average of the nodal design variables in the neighborhood set of N^{ie} within the filtering region shown in Fig.1(b). Here we use the same in-plane neighborhood for each layer of the laminate composite, although this need not be the case. The weighted average μ_{iel} is computed using linear filtering [13] as

$$\mu_{iel} = \frac{\sum_{m \in N^{ie}} w(x_{im} - \bar{x}^e) \phi_{iml}}{\sum_{m \in N^{ie}} w(x_{im} - \bar{x}^e)} \tag{6}$$

where $w(x_{im} - \bar{x}^e)$ is the weighting function relating node m to the element e, often chosen as a proximity-based function defined as

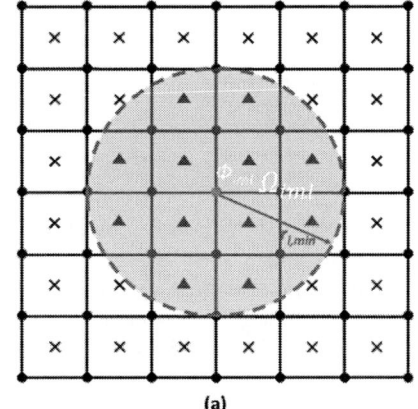 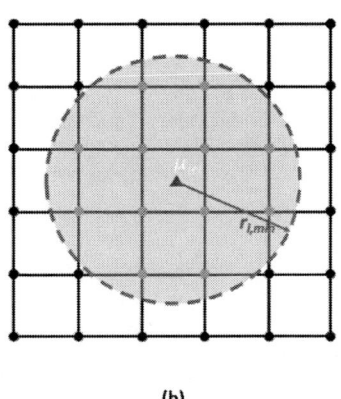

(a)	(b)

FIGURE 1: (a) the active nodal design variable ϕ_{iml} (green dot) is projected onto the element space over the distance $r_{i,min}$. The upward triangles indicate elements affected by the active nodal design variable and the symbol × indicates elements not affected by it. (b) From the elemental perspective, the element design variable μ_{iel} (upward triangle) receives the projected contributions from nodal design variables (green dots) within the distance of $r_{i,min}$. The black dots indicate nodal design variables that have no effect on the element.

$$w(x_{im} - \bar{x}^e) = \begin{cases} 1 - \dfrac{||x_{im} - \bar{x}^e||}{r_{min}}; & \text{if } im \in N^{ie} \\ 0; & \text{otherwise} \end{cases} \quad (7)$$

or uniform weighting function defined as

$$w(x_{im} - \bar{x}^e) = \begin{cases} 1; & \text{if } im \in N^{ie} \\ 0; & \text{otherwise} \end{cases} \quad (8)$$

It is generally well-known from studies in classical topology optimization that the uniform weighting function is capable of obtaining more discrete solutions than the linear weighting function.

Direct use of the functions (6) as the physical element variable would lead to non-discrete distributions of orientation angles, much like in typical density-based topology optimization [13]. We therefore use projection methods, which express the elemental orientations as a nonlinear function of the filtered variables. As previously mentioned, the use of projection methods within DMO is not new idea. Reference [9] utilized a thresholding projection expressed as a hyperbolic tangent function of the filtered variables (6). Although this led to solutions that were nearly discrete, minimum length scale could not be guaranteed. The nonlinear Heaviside Projection Method (HPM) function [10], on the other hand, allows us to achieve both discrete distributions and guarantees a user-defined length scale for each orientation. The HPM expresses the elemental orientations as a function of the linearly filtered variables μ_{iel} via the following regularized Heaviside function:

$$x_{iel} = 1 - e^{-\beta \mu_{iel}} + \mu_{iel} e^{-\beta} \quad (9)$$

where the parameter $\beta \geq 0$ dictates the curvature of the projection function that approaches the discrete Heaviside step function as β approaches infinity, as shown in Fig. 2. We also note that a different regularization parameter β_i could be used for each orientation projection, although this option is not used for the work herein.

2.3 Definition of minimum length scale

The shape of the projection domain becomes very important when attempting to achieve strict control of minimum length scale with multiple materials or orientations. In order for the approach to be effective, projected material shapes must lie tangent, without gaps between or overlaps of different material projections. This is quite similar and thoroughly explained in past projection works involving multiple materials, including two-phase length scale control [15,16] and discrete object projection [19] in density-based topology optimization. The widely used radial projection scheme, where the neighborhood set is the circle shown in Figure 1, for example, will dramatically limit topological complexity as only simple topologies will allow many different circles to lay tangent. To facilitate tangency, as well as utilize a manufacturing primitive that is more consistent

with composite panels, we utilize a square projected shape of side length $2r_{min}$ as shown in Fig. 3.

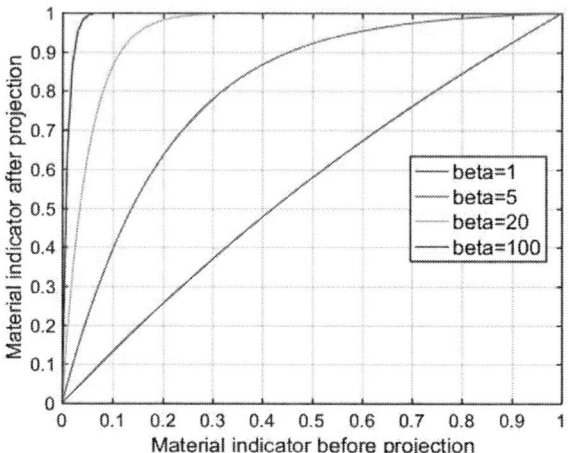

FIGURE 2: Regularized Heaviside function shown with various magnitudes of β

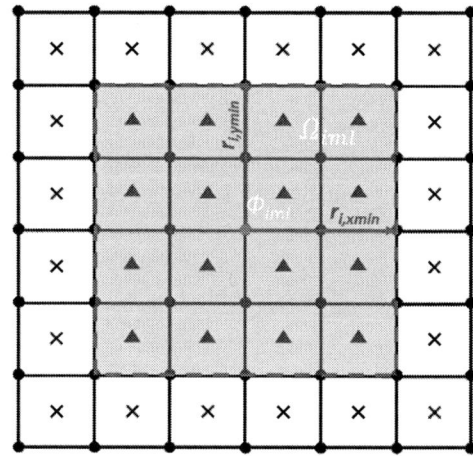

FIGURE 3: The active nodal design variable (green dot) is projected onto the element space within a distance of r_{xmin} and r_{ymin} in the rectangular filter

To ensure minimum length scale through the entire design area, special care should be taken when using the projection method at the boundaries of the design domain. As in [19], the independent nodal design variables are removed or set to inactive near the boundaries of the design domain within a distance of r_{min}, while the elemental design variables in these regions do receive the projection from nodal design variables in other regions.

2.4 Optimization formulation

The proposed method is demonstrated on benchmark maximum stiffness design problems [7] under a structural volume constraint. Simultaneous optimization of fiber orientation and topology optimization is considered by introducing the void phase as one of the candidate materials. The optimization formulation is given as

$$
\begin{aligned}
\min_{\phi} \quad & c(\phi) = f^T u(\phi) + \frac{M}{nel} \sum_{i,e,l} x_{iel} (1 - x_{iel}) \\
\text{s.t.} \quad & K(\phi)u = f \\
& \sum_{i,e,l} x_{iel} V_{el} \le \bar{V} \\
& \sum_{i=1}^{n^c} x_{iel} = 1; \ \forall \, e, l \\
& \sum_{i=1}^{n^c} \phi_{iml} \le 1; \ \forall \, m, l \\
& 0 \le \phi_{iml} \le 1; \ \forall \, i, m, l
\end{aligned}
\tag{10}
$$

where c is the compliance, f are the applied nodal forces, u are nodal displacements, and K is the global stiffness matrix. The quadratic penalization term is introduced in the objective function to gradually force the elemental design variables to distinct solutions and M increases from 1 to a large number during the optimization process. The variable V_{el} denotes the volume of the element e in layer l, $\bar{V}$ is the allowable volume of material within the design domain, $i = 1, \dots, n^c$ and n^c is the number of candidate materials (orientations), $e = 1, \dots, nel$ and nel is the number of elements, $l = 1, \dots, nl$ and nl is the number of layers, and $m = 1, \dots, nn$ and nn is the number of nodal design variables for each material phase.

2.5 Sensitivity analysis

The sensitivities of the objective function with respect to the independent design variables are calculated using the chain rule as follows

$$
\frac{\partial c}{\partial \phi_{iml}} = \sum_{e \in \Omega_{iml}} \frac{\partial c}{\partial x_{iel}} \frac{\partial x_{iel}}{\partial \mu_{iel}} \frac{\partial \mu_{iel}}{\partial \phi_{iml}}
\tag{11}
$$

where Ω_{iml} is the set of elements that are a function of ϕ_{iml} and the sensitivities with respect to the elemental material variables $\frac{\partial c}{\partial x_{ij}}$ are computed using the adjoint method as

$$
\frac{\partial c}{\partial x_{iel}} = f^T \frac{\partial u}{\partial x_{iel}} = -d^T \frac{\partial K}{\partial x_{iel}} d = -d_e^T \frac{\partial K^e}{\partial x_{iel}} d_e
\tag{12}
$$

Here d_e and K^e are the nodal displacements and element stiffness matrix corresponding to element e. By differentiating (6) and (9) the following expression is obtained.

$$
\frac{\partial x_{iel}}{\partial \mu_{iel}} = e^{-\beta} + \beta e^{-\beta \mu_{iel}}
\tag{13}
$$

$$
\frac{\partial \mu_{iel}}{\partial \phi_{iml}} = \frac{w(x_m - \bar{x}^e)}{\sum_{m \in N^e} w(x_m - \bar{x}^e)}
\tag{14}
$$

2.6 Optimizer

All problems are solved using the Method of Moving Asymptotes (MMA) gradient-based optimizer [23], which solves a sequence of sub-problems that are convex approximations of the objective functions and constraints.

A continuation scheme is introduced for both the SIMP penalty coefficient p and the Heaviside Projection parameter β. The continuation method is developed to avoid convergence to undesirable local optimum. For our problems, the initial value of p is set as 1.5 and an increment of $\Delta p = 0.5$ is applied every 50 iterations. The initial value of M is 1 and it increases by five times every 50 iterations. The maximum number of iterations is 250. For the Heaviside parameter β, we used $\Delta \beta = 0.1$ and increase β from 1 to 25. It is suggested in Reference [26] that initial asymptotes be conservative and expressed as a function of β, and this is followed here:

$$
s_0 = \frac{0.5}{\beta + 1}
\tag{16}
$$

2.7 Measure of solution discreteness

The measure of design non-discreteness [14] is extended to the multi-material topology optimization to quantify the discreteness of the final solutions:

$$
M_{nd} = \frac{\sum_{i,e,l} x_{iel}(1 - x_{iel})}{nl * nel * n^c * \frac{1}{n^c} * \left(1 - \frac{1}{n^c}\right)} \cdot 100\%
\tag{17}
$$

The measure is normalized such that $M_{nd} = 100\%$ if the elemental material design variables have the value of $1/n^c$ and $M_{nd} = 0\%$ when all the elemental material design variables have the value of 0 or 1, i.e., achieving the fully discrete solutions.

3. RESULTS AND DISCUSSION

3.1 Benchmark examples

The proposed method is demonstrated on benchmark maximum stiffness design problems. Two benchmark problems P.1 and P.2 are considered here for material orientation selection optimization problems and simultaneous topology and material selection optimization, respectively. It should be noted that the benchmark problems considered here have only one layer of material, but extension of the method to consider multiple layers is straightforward. The in-plane material properties used for all problems are shown in Table 1, and include Young's moduli in the principal directions (E_{11} and E_{22}), shear modulus G, and Poisson's ratio v.

Problem P.1 is a short cantilever plate that has L=60, H=20, t=1, and a uniformly distributed load P=1 applied on the top edge of the plate (Fig. 4(a)). The material is assumed orthotropic with 6 different allowable fiber orientations of [0°, 30°, -30°, 60°,

-60°, 90°] degrees. The finite element analysis is solved using 4 node quadrilateral plane stress elements.

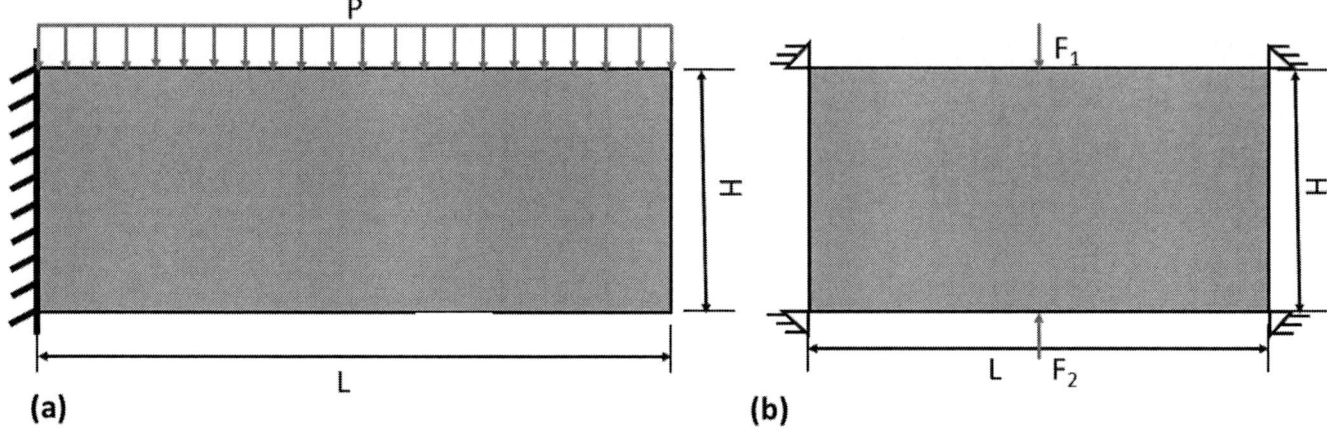

FIGURE 4: Benchmark examples: (a) a short cantilever plate under distributed load and (b) a pinned plate subjected to two independent load cases

Problem P.2 is a pinned plate subjected to two independent load cases (Fig. 4(b)). The dimension of the plate is L=4, H=2, and t=0.5·10⁻³. The applied loads F1 and F2 are two independent unit forces acting in opposite directions as shown. The objective function is an averaged compliance of the structure under the two independent load cases. In this case, the void phase and volume constraints of the solid materials are taken into consideration and the topology and orientation within the topology are simultaneously optimized. The volume constraint for this case is set as 70% of the full design domain. The candidate materials are orthotropic with 4 different allowable orientations of [0°, 45°, -45°, 90°] degrees, as well as a void phase. The finite element analysis is also solved using 4 node quadrilateral plane stress elements.

Table 1: Material properties of the candidate materials for the two benchmark problems.

EXAMPLE NO.	E_{11} (GPa)	E_{22} (GPa)	G_{12} (GPa)	ϑ_{12}
P.1 [27]	32	4	1.4	0.3
P.2 [9]	34	8.2	4.5	0.29

3.2 Material continuity and minimum length scale

Figure 5 shows the orientation distributions achieved when solving problems P.1 and P.2 using finite element meshes of 60×20 and 40×20 elements, respectively, with and without projection. This yields a number of elemental design variables totaling 7,200 for P.1 and 4,000 for P.2.

As can be seen from Fig.5(a1) and Fig.5(b1), where no filters are used, the design variable orientations are permitted to vary rapidly from element to element, as orientation at a location is not affected by the neighboring variables. This clearly indicates that a minimum length scale larger than a single element size cannot be controlled. It should be noted that the large area with rapidly varying material orientation in Fig.5(a1) has fibers orientated in -30 and 60 degrees. The explanation for is that these regions have principal stresses with similar magnitudes orientated at -30 and 60 degrees, indicating the two fiber orientations in these locations are of same importance to the stiffness of the structure. When using Heaviside projection, the fiber orientation distributions appear much more regular and the minimum length scales of patches of like orientation are strictly satisfied through the entire design domain as shown in Fig. 5(a2) and Fig. 5(b2). This is observed as the projection square shown in black may fit within each patch of constant material distribution. This results in a design that is more easily assembled from patches of like orientation, which are connected through fasteners or adhesives, or in some cases can be achieved with tow steering. We note, however, that any resulting discrete interfaces are not explicitly captured in the finite element model used here.

By comparing the compliance of these designs, it is shown that the optimized solutions without projection have the lowest compliance. This is expected, as these solutions are found without any geometric restriction imposed. The compliance obtained by the Heaviside projection decreases less than 15% for the two cases. The compliance is also dependent on the size of the projection domains, which are further discussed in the next section.

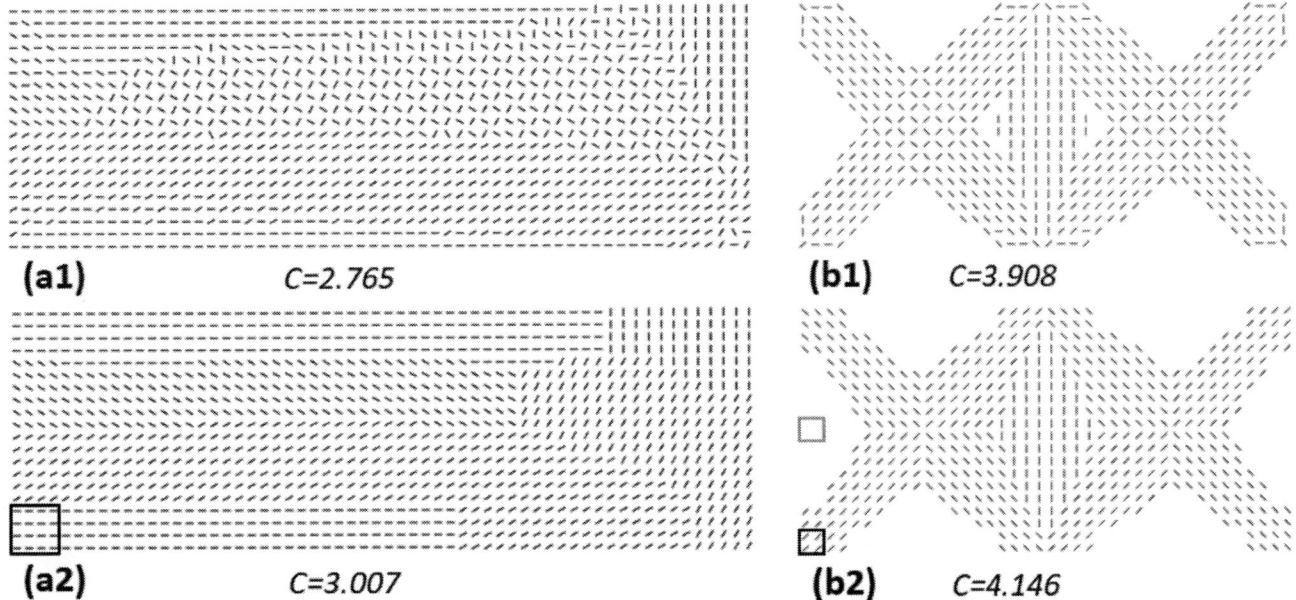

FIGURE 5: Optimized fiber orientation distributions and topology for (a1) problem P.1 without projection and (a2) problem P.1 using Heaviside projection with a 4×4 element rectangular projection, and for (b1) problem P.2 without projection scheme and (b2) problem P.2 using Heaviside projection with a 2×2 element rectangular projection. The dark black rectangle indicates the filter used for solid materials or fiber orientations, and the orange rectangle indicates the filter used for the void phase (where applicable).

3.3 Different length scales for the solid material phase and void phase

Figure 6 shows the effect of different projection length scales for the fiber orientation phases for problem P.1. The mesh for this problem is 60×20. The space-filling fiber orientation distributions are achieved everywhere in the design domain. As the minimum allowable length scale increases, less complex fiber distributions appear and the compliance of the optimized solution increases, an effect of trading reduced complexity for reduced performance. Increasing the minimum allowable feature size may even lead to elimination of some of the fiber orientations. For example, no fiber orientations of 90 degree appear in the optimized case Fig. 6(c).

Figure 7 shows the effect of different minimum length scales for both fiber orientation phases and the void phase for problem P.2. The mesh for the problem is 80×40 and the corresponding number of elemental design variables is 16,000. By comparing Fig.7(a) and Fig.7(b), it is seen that the smaller minimum allowable length scale of voids allows small holes on the interior of the structure, which become fully solid features when using larger magnitudes of the minimum length scale of the void phase. By comparing Fig.7(b) and Fig.7(c), the minimum allowable length scale of the fiber orientation phases may also affect the topology of the void phase since only one candidate materials could be chosen in each element. The minimum length scale is satisfied through the design domain for both material phases and the void phase. This is important as the minimum length feature could not only affect the mechanical performance of the structure but also allow more flexible manufacturing tools and processes.

As can be seen from the distribution of x_{iel} in both Fig. 6 and Fig. 7, the non-discreteness of the optimized design also increases with the length scale of the projection. This is a well understood effect and can be mitigated by increasing the regularization parameter β as described in [26]. However, the non-discreteness of all designs is quite small in magnitude.

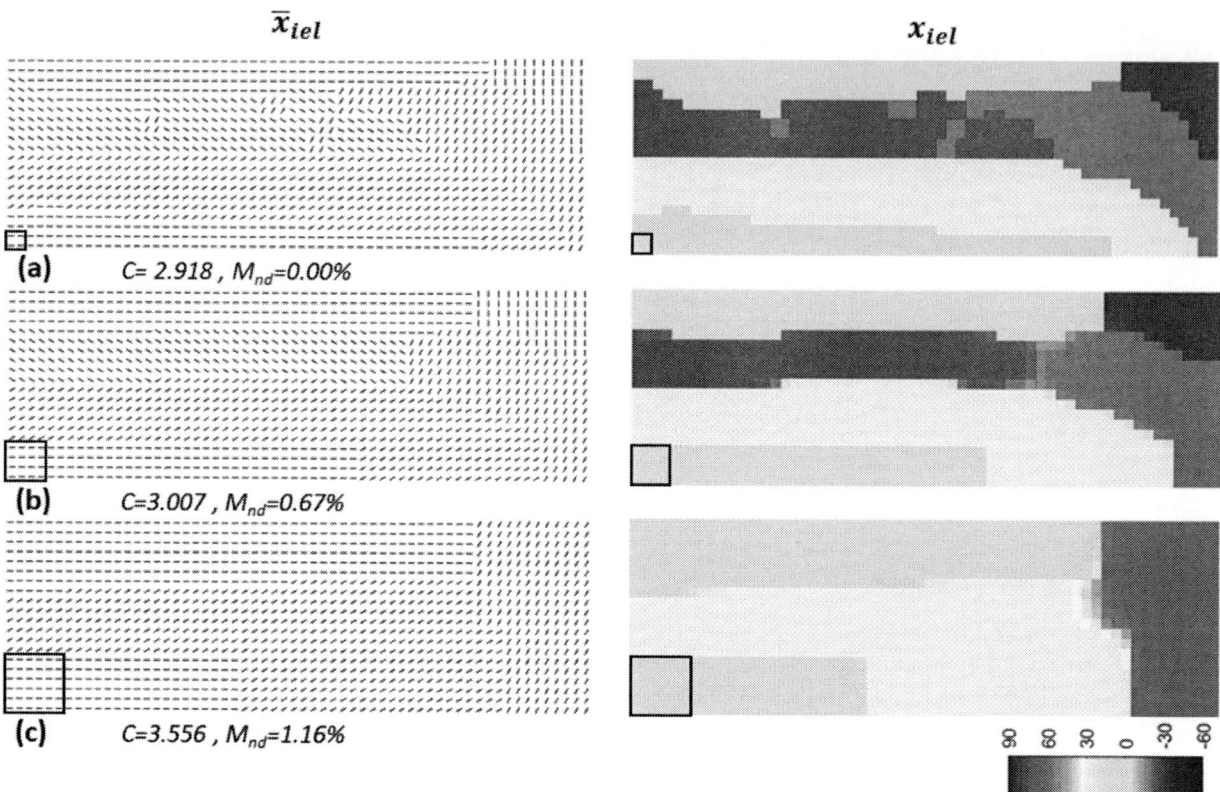

FIGURE 6: Optimized fiber orientation distributions x_{iel} for problem P.1 using different minimum allowable lengths scales of (a) $r_{xmin} = 1$, $r_{ymin} = 1$, (b) $r_{xmin} = 2$, $r_{ymin} = 2$, and (c) $r_{xmin} = 3$, $r_{ymin} = 3$. The black squares indicate the designer prescribed minimum allowable length scale for the fiber orientations. Different colors represent different candidate fiber orientations.

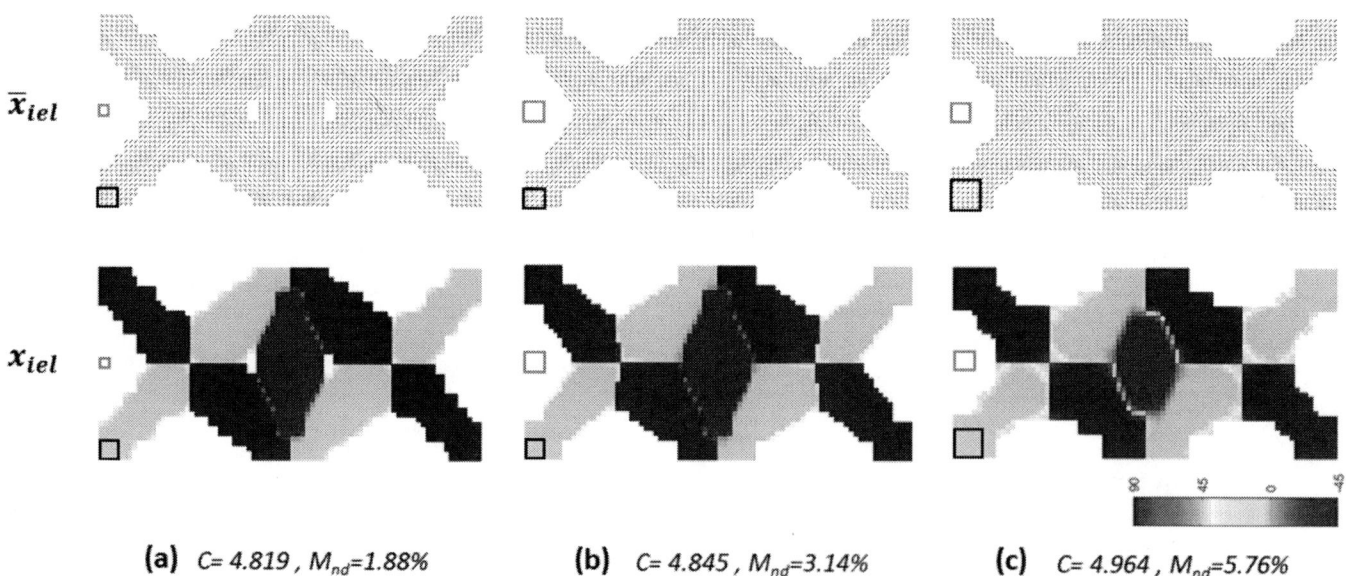

FIGURE 7: Optimized material and topology distributions x_{iel} for problem P.2 using different minimum allowable length scales of (a) $r_{xmin} = 0.1$, $r_{ymin} = 0.1, R_{xmin} = 0.05$, $R_{ymin} = 0.05$ (b) $r_{xmin} = 0.1$, $r_{ymin} = 0.1, R_{xmin} = 0.1$, $R_{ymin} = 0.1$, and (c) $r_{xmin} = 0.15$, $r_{ymin} = 0.5, R_{xmin} = 0.1$, $R_{ymin} = 0.1$. Parameters r_{xmin} and r_{ymin} are the minimum length scale of the fiber orientation phases, shown by the black squares, and R_{xmin} and R_{ymin} are the minimum length scale for the void phase, shown by the orange square. Different colors represent different candidate fiber orientations.

To illustrate the solution discreteness more clearly, we plot the distribution of each candidate material orientation separately using a black and white contour, as is shown in Fig. 8. As can be seen in this figure, each element is clearly assigned only a single orientation, and there is no overlap between the individual orientation figures. It is also observed that the elements are either black or white, without any grey which would indicate a non-discrete orientation selection.

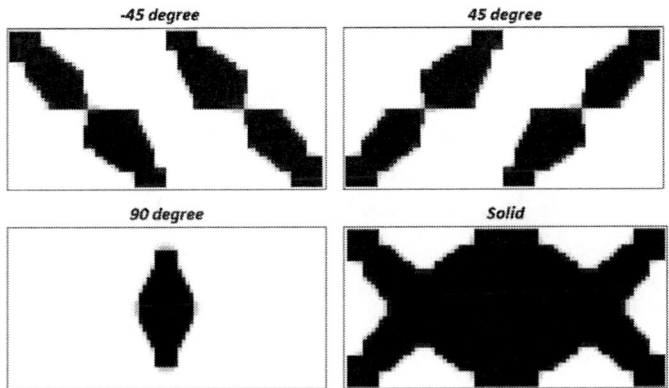

FIGURE 8: Distribution of candidate material orientations with -45°, 45°, and 90° for problem P.2 in Fig.7(b). The 0° solution is entirely zero and not shown here. On the bottom right corner is the structural design obtained by superimposing the distribution of the three non-zero fiber orientations.

The convergence history shown in Fig. 9 illustrates how the normalized compliance and discreteness evolve during the optimization process. The jumps in compliance are due to the continuation steps where the SIMP penalty and quadratic penalty term are increased every 50 iterations. The non-discreteness measure of the design consistently decreases from the initial guess, eventually converging to a minimal magnitude, suggesting the SIMP approach with projection is effective for achieving clear distributions of material and orientation.

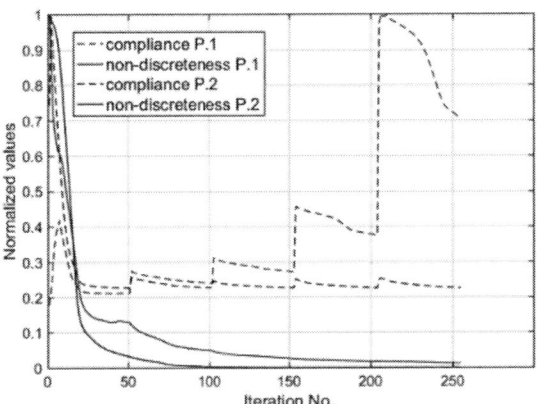

FIGURE 9: Iteration history of normalized objective function value (compliance) and non-discreteness of the elemental design variables. The corresponding curves are shown in Fig.6(a) and Fig.7(a).

3.4 Mesh independence

As the approach enforces a minimum length scale on designed features, including void, structural and orientation, the approach should be capable of circumventing the fundamental instability of solution mesh dependence. This is in contrast to using the thresholding filter, which does not provide mesh independence, as noted in [9]. Coarse and fine meshes are therefore used for problems P.1 and P.2 to test this hypothesis. As is shown in Fig. 10, for problem P.1, the fiber orientation distribution obtained using two different meshes are very similar for the material selection optimization problem. For problem P.2, the fiber orientation distribution and topology of the structure are also very similar for different meshes. Although minor deviations are observed, it is likely these differences are due to the algorithm converging to different local minima, as the underlying problem is nonconvex. However, additional investigation is required before mesh independence can be said to be achieved. Finally, we note that as the mesh becomes finer, smoother boundaries between different material phases are obtained. The compliance and non-discreteness measure for these additional problems are summarized in Table. 2.

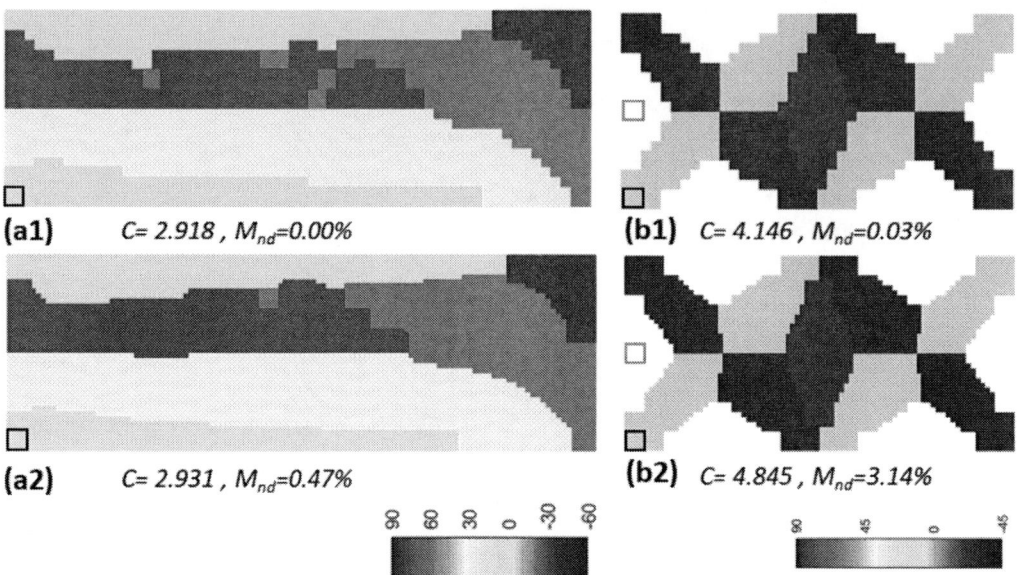

FIGURE 10: Optimized material distribution for problem P.1 and P.2 with different mesh sizes (a1) mesh 60×20, (a2) mesh 120×40, (b1) mesh 40×20, and (b2) mesh 80×40

Table 2: Results of compliance and non-discreteness of the optimized design for problems P.1 and P.2 using different meshes and minimum allowable length scales

Problem	Filters	Mesh 60×20		Mesh 120×40	
		Compliance (C)	Non-Discreteness (M_{nd})	Compliance (C)	Non-Discreteness (M_{nd})
P.1	no projection	2.765	1.93%	2.733	8.92%
	$r_x=r_y=0.5$	-	-	2.918	0.00%
	$r_x=r_y=1$	2.918	0.00%	2.931	0.47%
	$r_x=r_y=1.5$	-	-	2.965	1.02%
	$r_x=r_y=2$	3.007	0.67%	-	-
	$r_x=r_y=3$	3.556	1.16%	-	-
		Mesh 40×20		Mesh 80×40	
P.2	no projection	3.908	5.16%	4.391	2.25%
	$r_x=r_y=0.1 \ R_x=R_y=0.05$	-	-	4.819	1.88%
	$r_x=r_y=0.1 \ R_x=R_y=0.1$	4.146	0.03%	4.845	3.14%
	$r_x=r_y=0.15 \ R_x=R_y=0.1$	-	-	4.964	5.76%
	$r_x=r_y=0.2 \ R_x=R_y=0.1$	4.284	4.33%	-	-

4. CONCLUSION

In this work, the Heaviside Projection Method (HPM) is introduced in the framework of DMO for simultaneously optimizing the structural topology and fiber orientation across the topology in the design of laminated composites. The work builds directly on the work of [9] but replaces a thresholding projection scheme with the original form of the Heaviside Projection Method to achieve strict control over the minimum allowable length scale of patches of consistent orientation. Through this approach, the designer may influence solution complexity and therefore manufacturability, and may prescribe different minimum allowable length scales to the void phase (holes) and/or fiber orientations, providing a flexible implementation. Preliminary numerical results suggest that the projection-based method can play an important role in controlling the manufacturability of optimized fiber orientations in composite laminates, and that solutions are near-discrete with performance properties comparable to the solutions without filtering approach for the considered problems. The approach can be extended to optimizing the stacking sequence [9], as well as to design problems beyond maximum stiffness design, such as considering strength and failure requirements.

ACKNOWLEDGEMENTS

This work was supported by the National Aeronautics and Space Administration (NASA) Space Technology Research Institute (STRI) for Ultra-Strong Composites by Computational Design (US-COMP), grant number NNX17AJ32G. Any opinions, findings, and conclusions or recommendations expressed in this article are those of the author(s) and do not necessarily reflect the views of NASA. The authors also thank Krister Svanberg for kindly providing the MMA optimizer code.

REFERENCES

[1] Bendsøe, M. P., 1989, "Optimal Shape Design as a Material Distribution Problem," Struct. Optim., 1(4), pp. 193-202.

[2] Bendsøe, M. P., and Sigmund, O., 2003, Topology Optimization: Theory, Methods, and Applications.

[3] Bendsøe, M. P., and Kikuchi, N., 1988, "Generating optimal topologies in structural design using a homogenization method," Comput. Methods Appl. Mech. Eng., 71(2), pp.197-224.

[4] Nomura, T., 2015, "General Topology Optimization Method with Continuous and Discrete Orientation Design Using Isoparametric Projection," Int. J. Numer. Methods Eng., 27(148), pp. 148–153.

[5] Zhou, Y., Nomura, T., and Saitou, K., 2018, "Multi-Component Topology and Material Orientation Design of Composite Structures (MTO-C)," Comput. Methods Appl. Mech. Eng., 342, pp. 438–457.

[6] Stegmann, J., and Lund, E., 2005, "Discrete Material Optimization of General Composite Shell Structures," Int. J. Numer. Methods Eng., 62(14), pp. 2009–2027.

[7] Hvejsel, C. F., Lund, E., and Stolpe, M., 2011, "Optimization Strategies for Discrete Multi-Material Stiffness Optimization," Struct. Multidiscip. Optim., 44(2), pp. 149–163.

[8] Sigmund, O., and Petersson, J., 1998, "Numerical Instabilities in Topology Optimization: A Survey on Procedures Dealing with Checkerboards, Mesh-Dependencies and Local Minima," Struct. Optim., 16(1), pp. 68–75.

[9] Sørensen, R., and Lund, E., 2015, "In-Plane Material Filters for the Discrete Material Optimization Method," Struct. Multidiscip. Optim., 52(4), pp. 645–661.

[10] Guest, J. K., Prévost, J. H., and Belytschko, T., 2004, "Achieving Minimum Length Scale in Topology Optimization Using Nodal Design Variables and Projection Functions," Int. J. Numer. Methods Eng., 61(2), pp. 238–254.

[11] Sigmund, O., and Torquato, S., 1997, "Design of Materials with Extreme Thermal Expansion Using a Three-Phase Topology Optimization Method," J. Mech. Phys. Solids, 45(6), pp. 1037–1067.

[12] Bourdin, B., 2001, "Filters in Topology Optimization," Int. J. Numer. Methods Eng., 50(9), pp. 2143–2158.

[13] Bruns, T. E., and Tortorelli, D. A., 2001, "Topology Optimization of Non-Linear Elastic Structures and Compliant Mechanisms," Comput. Methods Appl. Mech. Eng., 190(26–27), pp. 3443–3459.

[14] Sigmund, O., 2007, "Morphology-Based Black and White Filters for Topology Optimization," Struct. Multidiscip. Optim., 33(4–5), pp. 401–424.

[15] Guest, J. K., 2009, "Topology Optimization with Multiple Phase Projection," Comput. Methods Appl. Mech. Eng., 199(1–4), pp. 123–135.

[16] Carstensen, J. V., and Guest, J. K., 2018, "Projection-Based Two-Phase Minimum and Maximum Length Scale Control in Topology Optimization," Struct. Multidiscip. Optim., 58(5), pp. 1845–1860.

[17] Guest, J. K., and Zhu, M., 2013, "Casting and Milling Restrictions in Topology Optimization via Projection-Based Algorithms," Proceedings of the ASME Design Engineering Technical Conference, 3 (PARTS A AND B), pp. 913-920.

[18] Gaynor, A. T., and Guest, J. K., 2016, "Topology Optimization Considering Overhang Constraints: Eliminating Sacrificial Support Material in Additive Manufacturing through Design," Struct. Multidiscip. Optim., 54(5), pp. 1157–1172.

[19] Guest, J. K., 2015, "Optimizing the Layout of Discrete Objects in Structures and Materials: A Projection-Based Topology Optimization Approach," Comput. Methods Appl. Mech. Eng., 283, pp. 330–351.

[20] Guest, J. K., 2014, "Projection-Based Topology Optimization Using Discrete Object Sets," ASME 2014 International Design Engineering Technical Conferences and Computers and Information in Engineering Conference, pp. V02BT03A013-V02BT03A013.

[21] Xu, S., Cai, Y., and Cheng, G., 2010, "Volume Preserving Nonlinear Density Filter Based on Heaviside Functions," Struct. Multidiscip. Optim., 41(4), pp. 495–505.

[22] Schevenels, M., Lazarov, B. S., and Sigmund, O., 2011, "Robust Topology Optimization Accounting for Spatially Varying Manufacturing Errors," Comput. Methods Appl. Mech. Eng., 200(49–52), pp. 3613–3627.

[23] Svanberg, K., 1987, "The Method of Moving Asymptotes-a New Method for Structural Optimization," 24(October 1985), pp. 359–373.

[24] Gibiansky, L. V., and Sigmund, O., 2000, "Multiphase Composites with Extremal Bulk Modulus," J. Mech. Phys. Solids, 48(3), pp. 461–498.

[25] Hvejsel, C. F., and Lund, E., 2011, "Material Interpolation Schemes for Unified Topology and Multi-Material Optimization," Struct. Multidiscip. Optim., 43(6), pp. 811–825.

[26] Guest, J. K., Asadpoure, A., and Ha, S. H., 2011, "Eliminating Beta-Continuation from Heaviside Projection and Density Filter Algorithms," Struct. Multidiscip. Optim., 44(4), pp. 443–453.

[27] Pedersen, P., 2004, "Examples of Density, Orientation, and Shape-Optimal 2D-Design for Stiffness and/or Strength with Orthotropic Materials," Struct. Multidiscip. Optim., 26(1–2), pp. 37–49.

Proceedings of the ASME 2019
International Design Engineering Technical Conferences
and Computers and Information in Engineering Conference
IDETC/CIE2019
August 18-21, 2019, Anaheim, CA, USA

DETC2019-98463

AN ADAPTIVE AND EFFICIENT BOUNDARY APPROACH FOR DENSITY-BASED TOPOLOGY OPTIMIZATION

Reza Behrou[*]
Reza Lotfi
Josephine V. Carstensen
James K. Guest
Department of Civil Engineering, Johns Hopkins University, Baltimore, Maryland 21218
Email: rbehrou@jhu.edu, reza.lotfi@gmail.com, jvcar@mit.edu, jkguest@jhu.edu

ABSTRACT

This paper presents an adaptive nodal boundary condition scheme to systematically enhance the computational efficiency and circumvent numerical instabilities of the finite element analysis in density-based topology optimization problems. The approach revisits the idea originally proposed by Bruns and Tortorelli to eliminate the contribution of void elements from the finite element model and extends this idea to modern projection methods to stabilize the implementation, facilitate reintroduction of material, and consider additional physics. The computational domain is discretized on a fixed finite element mesh and a threshold density is used to determine if an element is sufficiently low relative density to be "removed" from the finite element analysis. By eliminating low-density elements from the design domain, the number of free Degrees-Of-Freedom (DOFs) is reduced, thereby reducing the solution cost of the finite element equations. Perhaps more importantly, it circumvents numerical instabilities such as element distortion when considering large deformations. Unlike traditional solids-only modeling approaches, a key feature of the projection-based scheme is that the design and finite element spaces are separate, allowing the design variable sensitivities in a region to remain active (and potentially non-zero) even if the corresponding analysis elements are removed from the finite element model. This ultimately means material reintroduction is systematic and driven by the design sensitivities. The Solid Isotropic Material with Penalization (SIMP) approach

is used to interpolate material properties and the Heaviside Projection Method (HPM) is used to regularize the optimization problem and facilitate material reintroduction through the gradient-based optimizer. Several benchmark examples in areas of linear and nonlinear structural mechanics are presented to demonstrate the performance of the proposed approach. The resulting optimized designs are consistent with literature and results reveal the performance and efficiency of the developed method in reducing computational costs without numerical instabilities known to be due to modeling near-void elements.

INTRODUCTION

Topology optimization as a design methodology offers a powerful tool to determine the optimal layout of material within the design domain. Although significant progress has recently been made to enhance capability and performance of topology optimization methodologies, improving computational efficiency remains a key goal and circumventing numerical instabilities in nonlinear mechanics remains an open challenge. This paper presents an adaptive and efficient nodal boundary methodology that both systematically enhances the computational efficiency and the numerical stability of the finite element analysis in density-based topology optimization problems.

A long-standing drawback of material distribution approaches is that they require the entire design domain to be modeled, including the void material in structural design problems. Alt-

---

[*]Corresponding author: Reza Behrou (rbehrou@jhu.edu).

hough such regions are structurally insignificant and are removed during manufacturing of the final design, they remain in the structural model during optimization to enable reintroduction of material as the design evolves. It can, therefore, be said that their presence is necessitated for design, and not for analysis, purposes. In fact, elements with negligible volume fraction (density) are quite detrimental to analysis as they (1) amplify computational inefficiency and (2) are susceptible to numerical instabilities under nonlinear mechanics, such as excessive distortions under large deformations. A number of works have therefore focused on combating these two issues in structural problems; see for example [1, 2].

Reducing computational expense has been a major research focus in the topology optimization community. As computational cost is mainly driven by the cost of solving the finite element system of equations, research efforts have typically aimed at reducing the cost of finite element analyses by using advanced solvers and efficient reanalysis techniques [3–6], or reducing the size of the system to be solved. The latter includes re-meshing, such as uniform re-meshing (e.g. [7]) or adaptive design dependent re-meshing in the cases where the independent design and state variables share the same mesh [8–10]. Alternatively, the design and state variables may be defined on different meshes, allowing different discretizations to be applied to the design and analysis spaces [11–15].

The difficult problem of circumventing numerical instabilities associated with low-stiffness modeling under nonlinear mechanics has also drawn the attention of researchers. For problems governed by finite deformations, Bruns and Tortorelli [2] used a hyper-elastic constitutive model, stiffening elements that undergo large deformations to prevent distortion. Buhl et al. [1] proposed removing DOFs associated with low-stiffness elements from Newton-Raphson convergence criteria, thereby preventing oscillations associated with the nodes of void elements [16]. Yoon and Kim [17] proposed an Element Connectivity Parameterization (ECP) approach where the entire design domain is modeled as stiff continuum elements connected with zero-length links. The link stiffnesses are then optimized to determine topology. Void regions are represented by flexible links, and thus distortions are localized in the link elements, not the relatively stiff continuum elements, preventing numerical instabilities.

In this paper, we emphasize separation of the design and analysis spaces and revisit the idea of simply "removing" elements with negligible density from the analysis model—we'll refer to this as single-phase modeling. The presented approach not only prevents numerical instabilities localized in such elements, but also reduces the size of the finite element system of equations to be solved, significantly reducing the computational cost. The idea of single-phase modeling is certainly not new to topology optimization. This has long been a selling point of Evolutionary Structural Optimization (ESO) approaches [18, 19] and discrete implementations of level set methods [20–22], with savings as-

sociated with the latter specifically quantified in [23] for fluids. For material distribution approaches, Maute and Ramm [11] implemented the idea with a formal re-meshing scheme, where relative density iso-lines were computed on the design space and a threshold line was selected to represent the structural boundary. As this boundary evolves, the structural domain is re-meshed. This adaptive scheme was then extended to problems governed by elastoplastic material models [24].

This work closely follows the approach proposed by [25] where a fixed finite element discretization and the element "removal" check at the individual element level were performed. Elements with a relative density, ρ^e, below a threshold, denoted here as ρ_t, are removed from the finite element analysis. In [25], a Gaussian elimination on the full global stiffness matrix was used to highlight degrees of freedom leading to instabilities, which are then stabilized by prescribing deformations. Herein, artificial boundary conditions are used to reduce the number of free DOFs to be solved.

The key feature of the proposed approach is that the design and analysis spaces are separate and discretized on different meshes. The independent design variables are projected onto the finite element space to determine the physical design for analysis. This means that regions of the analysis mesh may be made inactive and thus not to be modeled, yet maintain the ability to reappear provided the corresponding region of the design variable space indicates this is desirable. In [25], a linear density filter was used to relate these spaces. Here, we use nonlinear projection [26]. It is shown, through a simple sensitivity analysis study for a structural problem, that this nonlinearity more rapidly enables the reintroduction of material into void regions. Further, by using the standard optimizer parameter setting found in [27], we also avoid disconnected features that were discussed in [25]. A key point of emphasis is that this material reintroduction is driven purely by the formal sensitivity analysis. This makes the method different than soft-kill methods such as Bi-directional Evolutionary Structural Optimization (BESO) where element reintroduction is driven by heuristic rules, as discussed in detail by [28].

The proposed method is applied to benchmark structural design problems governed by linear and nonlinear mechanics. The density-based topology optimization method is used to relate physical properties to optimization variables. Adjoint sensitivity analysis is used to compute the sensitivities of the objective and constraints with respect to design variables. The resulting optimized designs reveal the performance and efficiency of the method in reducing the computational costs for a variety of engineering applications. No instances of numerical instabilities were detected, including issues of islanding and disconnected load paths reported in [25].

The remainder of the paper is organized as follows. The generalized formulation of the optimization problems and the corresponding sensitivity analysis is first discussed, followed by the discussion of the proposed approach, we then study the main characteristics of the proposed method with numerical examples, followed by concluding remarks.

OPTIMIZATION PROBLEM

Before presenting the new scheme, it is convenient to briefly review the generalized topology optimization problem formulation and corresponding sensitivity analysis. We focus first on topology optimization considering structural mechanics, and provide extensions in later sections.

TOPOLOGY OPTIMIZATION FORMULATION

The generalized form of the topology optimization problem is defined as follows:

$$
\begin{aligned}
&\min_{\phi} \quad f(u(\phi), \phi) \\
&\text{subject to} \begin{cases} H_i(u(\phi), \phi) = 0 & \forall i = 1, ..., N_h \\ G_j(u(\phi), \phi) \leq 0 & \forall j = 1, ..., N_g \\ \phi_{min} \leq \phi_k \leq \phi_{max} & \forall k = 1, ..., N_\phi \end{cases}
\end{aligned} \quad , \quad (1)
$$

where f is the objective function, ϕ is the vector of independent design variables, u is the vector of state variables satisfying the finite element discretized governing equations, $R(u(\phi), \phi)$. At each design iteration, the Nested ANalysis and Design (NAND) approach is used to solve the finite element problem. H_i and G_j are the equality and inequality constraints with N_h and N_g denote the number of equality and inequality constraints, respectively. The independent design variables, ϕ_k, are bounded with ϕ_{min} and ϕ_{max} set herein to magnitudes of zero and one, respectively [27].

SENSITIVITY ANALYSIS

Adjoint sensitivity analysis is used to compute the sensitivities of the objective and constraint functions. The sensitivities of the objective and constraints with respect to independent design variables is performed using the adjoint method. The generalized form of adjoint sensitivities of a criterion, $\mathscr{C}$, the objective or a constraint, with respect to a design variable, ϕ_k, is written as a combination of explicit and implicit terms as follows:

$$
\frac{d\tilde{\mathscr{C}}}{d\phi_k} = \frac{\partial \tilde{\mathscr{C}}}{\partial \phi_k} + \frac{\partial \tilde{\mathscr{C}}}{\partial u} \frac{du}{d\phi_k}. \quad (2)
$$

The implicit relationship between the state and design variables can be eliminated by using the total derivatives of converged re-

sidual equations as follow:

$$
\frac{dR}{d\phi_k} = \frac{\partial R}{\partial \phi_k} + \frac{\partial R}{\partial u} \frac{du}{d\phi_k} = 0. \quad (3)
$$

Solving Eq. (3) for $du/d\phi_k$ and substituting the result into (2) yields:

$$
\frac{d\tilde{\mathscr{C}}}{d\phi_k} = \frac{\partial \tilde{\mathscr{C}}}{\partial \phi_k} + \tilde{\lambda}^T \frac{\partial R}{\partial \phi_k}, \quad (4)
$$

where $\tilde{\lambda}$ is the vector of adjoint solutions, computed by solving the following adjoint problem:

$$
\left(\frac{\partial R}{\partial u}\right)^T \tilde{\lambda} = -\left(\frac{\partial \tilde{\mathscr{C}}}{\partial u}\right)^T. \quad (5)
$$

MATERIAL INTERPOLATION

In density-based topology optimization, a continuous 0-1 relative density transition requires a continuous and differentiable relation for density-based material properties such that both physical response and design sensitivities can be computed. The main objective of density-based approaches such as SIMP [29] is to interpolate material properties with an artificial law, referred as material interpolation, and penalize intermediate density values such that the optimization results converge to 0-1 binary solutions. In this paper, the SIMP approach is used. Using the SIMP approach, the Young's modulus (stiffness) of an element is defined as a function of elemental relative density as follows:

$$
E^e(\rho^e) = (\rho^e_{min} + (\rho^e)^\eta) E^e_0, \quad (6)
$$

where $\eta \geq 1$ is the SIMP exponent penalty term, E^e_0 is the Young's modulus of the pure solid material and ρ^e_{min} is a small positive number (e.g., 10^{-4}) to maintain positive definiteness of the global stiffness matrix. Alternatively, the elemental stiffness can be expressed as:

$$
E^e(\rho^e) = (\rho^e)^\eta E^e_0 + E_{min}, \quad (7)
$$

where E_{min} is a small magnitude of Young's modulus to maintain positive definiteness of the global stiffness matrix (with $\rho_{min} = 0$ in this formulation).

We note the goal of this paper is to eliminate the modeling of elements with low relative density. The parameters ρ_{min} and E_{min} thus become inconsequential as they do not appear in the finite element analysis, and thus may be set to zero.

REGULARIZATION

In density-based topology optimization, it is well-known that numerical instabilities of checkerboard patterns and mesh dependent solutions [30] can readily appear. The HPM provides a means of circumventing these instabilities by imposing a minimum length scale on structural features while enabling a sharp transition of relative densities from solid to void [26]. Using the SIMP approach, the elemental density, ρ^e, is computed at the elemental level, by filtering and projecting the independent design variables, ϕ. These operators eliminate mesh-dependent and checkerboard designs and regularize the optimization problems. In HPM, the independent design variables associated with a material phase are projected onto the finite element space by a regularized Heaviside function [26]. The connection between the design variable space, where the optimization is performed, and the finite element space, where the physical equilibrium is solved, is established through a radial projection. In this method, the projection radius is chosen as the minimum desired feature size, r_{min}. The design variables are mapped onto the elements by computing the weighted average, $\mu^e(\phi)$, using linear filtering scheme as follows [2]:

$$\mu^e(\phi) = \frac{\sum\limits_{j \in N^e} w_j \phi_j}{\sum\limits_{j \in N^e} w_j}, \quad \text{with} \quad w_j = \frac{r_{min} - \| x_j - x^e \|}{r_{min}}, \quad (8)$$

where N^e denotes the neighborhood set containing all design variables located within a distance r_{min} of the centroid of e, x_j is the position of design variable j, and x^e is the central position of element e.

In linear density filtering, the weighted average μ^e is set equal to the relative density:

$$\rho^e(\phi) = \mu^e(\phi). \quad (9)$$

Using linear density filtering, however, creates a transition between the solid and void phases, creating a "blurry" boundary effect. This motivated development of the HPM, where element relative densities are expressed as a nonlinear (regularized) Heaviside function of μ^e as follows:

$$\rho^e(\phi) = 1 - e^{-\beta \mu^e(\phi)} + \frac{\mu^e(\phi)}{\phi_{max}} e^{-\beta \phi_{max}}, \quad (10)$$

where $\beta \geq 0$ dictates the curvature of the regularization which approaches the Heaviside function as $\beta \to \infty$ [26] and guarantees minimum length scale. Accounting for the filtering and projection, the sensitivities of the objective, constraints, and residual equations are computed by applying the chain rule as follows:

$$\frac{\partial \tilde{\mathscr{C}}}{\partial \phi_k} = \frac{\partial \tilde{\mathscr{C}}}{\partial \rho^e} \frac{\partial \rho^e}{\partial \mu^e} \frac{\partial \mu^e}{\partial \phi_k},$$
$$\frac{\partial R}{\partial \phi_k} = \frac{\partial R}{\partial \rho^e} \frac{\partial \rho^e}{\partial \mu^e} \frac{\partial \mu^e}{\partial \phi_k}. \quad (11)$$

GRADIENT BASED OPTIMIZATION ALGORITHM

In all design examples considered herein, the Method of Moving Asymptotes (MMA) optimizer is used to solve the optimization problem (1). The adjoint method is used to compute the gradients of the objective and constraints with respect to independent design variables, ϕ, via Eqs. (4), (5), and (11) as described. The linearized system of governing equations is solved using Newton's method. Direct method is used to solve both the linearized sub-problems within the Newton iterations and the adjoint sensitivity analysis (5).

The optimization problem is considered converged if the relative change of the objective function is less than ε_{MMA} and all constraints are satisfied. The nonlinear problem is considered converged if the relative residuals are less than 10^{-8}.

The projection regularization parameter is selected following [27], and the MMA asymptote's move factors are set to a_{MMA} and b_{MMA}. A continuation methods on the SIMP exponent penalty term, η, is used. This ensures smooth evolution in the design and avoids wiggling in the design history and premature convergence to local minima. To avoid premature transition to the following step, a minimum number of iterations per continuation step, N_{it}^c, is considered.

SENSITIVITY STUDY

In this section, we study and compare the sensitivity analysis for a simple cantilever beam (Figure 1) using three popular approaches, namely sensitivity filtering [31], linear density filtering [2], and HPM [26], to illustrate why the proposed method is effective. The design variable distributions and resulting topologies for the three cases are shown in Figures 1(b) and 1(c), respectively. For the sensitivity filter, the dependent (ρ^e) and independent (ϕ) design variables are identical (i.e. $\rho^e = \phi$). Although the linear sensitivity filter is not capable of achieving a 0-1 binary topology, we use a 0-1 distribution as the initial guess. A direct comparison of properties is difficult as the solid-void interface is sharp in HPM and blurry in the linear density filter. To maintain a fair comparison, we use topologies with the same volume of material and define the structural boundary threshold in linear density filter as $\rho^e = 0.5$, which is the reason the design variable distributions are different.

The derivatives, $df/d\rho^e$, are shown Figure 1(d). For structural problems, these derivatives are simply elemental strain energies scaled by the coefficient $\eta(\rho^e)^{(\eta-1)}$, see Eq. (6). This

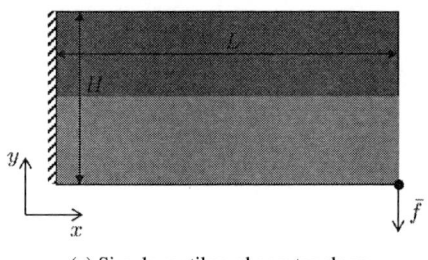

(a) Simple cantilever beam topology

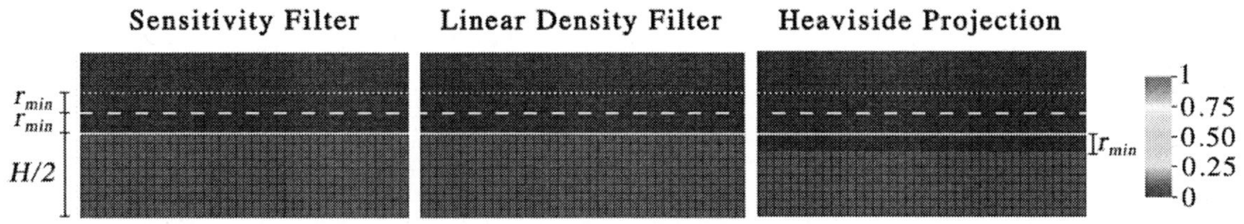

Sensitivity Filter   **Linear Density Filter**   **Heaviside Projection**

(b) Design variable, ϕ

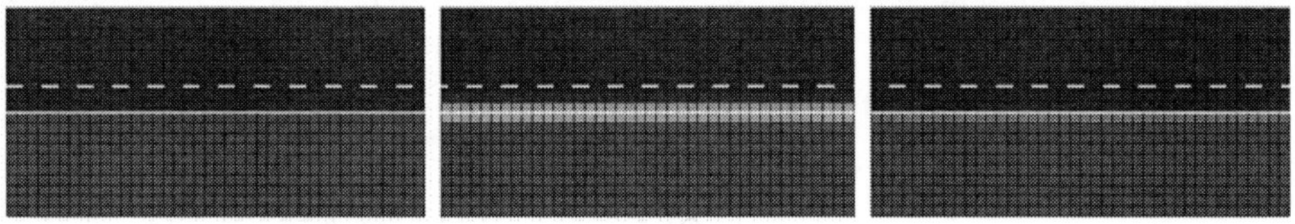

(c) Topology, ρ^e

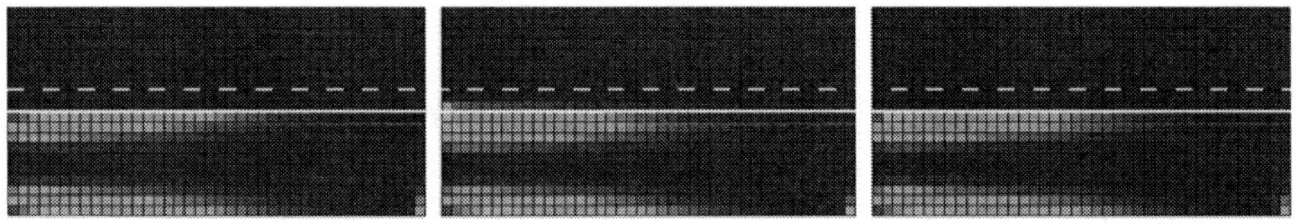

(d) $-df/d\rho^e$ (normalized)

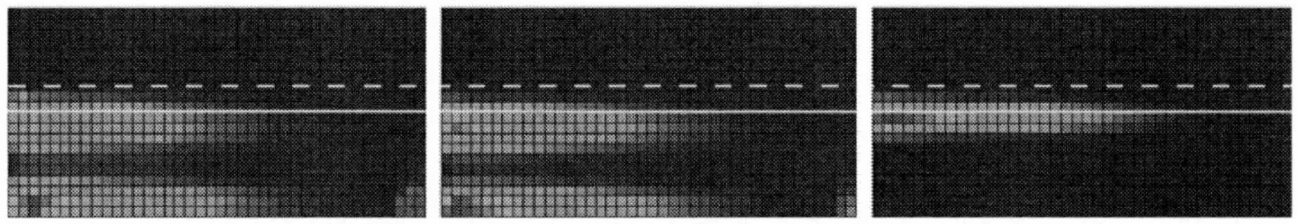

(e) Sensitivity, $-df/d\phi$ (normalized)

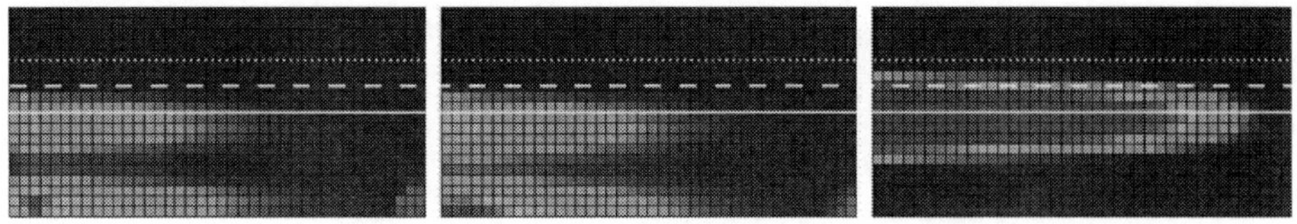

(f) Potential topological influence of sensitivity

FIGURE 1: Sensitivity study of three different approaches

coefficient is (or approaches) zero magnitude for void (or low-density) elements when the SIMP exponent value exceeds one. As shown in Figure 1, for $\eta = 3$, this drives the elemental derivatives to zero regardless of the strain energy in these low-density elements. This supports the logic that low-density elements need not be modeled as their contribution to the design variable sensitivities is negligible.

The design variable sensitivities, $df/d\phi$, are shown in Figure 1(e), where the case of the sensitivity filter represents the filtered sensitivities. The key observation here is that the design variable sensitivities are nonzero in regions extending a distance r_{min} beyond the topological boundary. This implies that design variables located in void space may achieve nonzero sensitivities, and therefore maintain the capability to change magnitude, even if the nearest elements have zero magnitudes of $df/d\rho^e$. This supports the idea that material may reappear in elements even if they are removed from the finite element analysis.

Figure 1(f) illustrates "reach" of these sensitivities, illustrating how they would intend to change topology. These are visualized by simply treating the normalized design variable sensitivities as design variables, and thus in the case of linear density filtering and Heaviside projection passing them through the linear filter and regularized Heaviside functions, respectively. This is not a rigorous estimation as we neglect effect of constraints such as upper bounds, but rather illustrates the geographic extent to which topological changes are possible. In the case of sensitivity filter, topological growth is limited to a distance of r_{min} from the original topological boundary. The reach of the linear density filter and projection methods, however, is $2r_{min}$. Design variables within a distance r_{min} may change, subsequently leading to topological growth a distance r_{min} from these variables. In the case of linear density filter, however, the intensity of this growth falls off rapidly. The largest sensitivities, $df/d\phi$, are located on the interior of the topology. Elements at or beyond a distance r_{min} from the structural boundary see little intensity due to the relatively small weights in the sensitivity calculation, which are then damped further by the small weights in the density filter. In the case of Heaviside projection, however, the sensitivity intensity is the strongest at the topological boundary and the Heaviside function amplifies the potential impact of small sensitivities, leading to a stronger reach at further distances than the linear density filter. This is clearly can be seen in Figure 1(f) where the red spike is centered on the topological boundary and extends to a distance of $2r_{min}$.

To summarize, for a SIMP exponent greater than one, material reintroduction essentially grows from structural boundaries in all cases, enabling void material that is not modeled to be reintroduced. This "growth" is limited to a distance of r_{min} in the case of the sensitivity filter and a distance of $2r_{min}$ in the case of linear density filter and Heaviside projection. A key difference in the latter two is that the potential rate of material reintroduction. The linear filter has a nonzero but negligible effect beyond a dis-

tance of r_{min} from the structural boundary, while the Heaviside projection does dissipate until a distance of $2r_{min}$. This enables reintroduction of material more quickly, as well as the fundamental benefit of achieving 0-1 binary designs.

While the proceeding figures motivate and justify the use of solids-only modeling, the rate of material reintroduction is significantly less in case where intermediate volume fractions are unpenalized (i.e. $\eta = 1$). In such cases, the derivatives, $df/d\rho^e$, are nonzero at all locations and thus neglecting to model the voids causes potentially valuable sensitivity information to be lost. Finally, we note that the plots in Figure 1 assumed a linear distance weighting function as it is the most commonly used. It is widely known that using uniform weighting reduces grey regions at topological boundaries for Heaviside projection [26].

IMPLEMENTATION OF THE ADAPTIVE NODAL BOUNDARY CONDITIONS

The proposed adaptive nodal boundary conditions scheme is applied on a fixed finite element discretization following the original work [25]. Herein we apply this concept to implement adaptive nodal boundary conditions on displacement. The approach consists of detection and elimination steps. The detection step simply identifies elements that are deemed to be "void", as determined by a threshold magnitude ($\rho^e \leq \rho_t$). The elimination step then eliminates DOFs associated with those elements from the finite element model. As shown in Figure 2, this includes any displacement DOFs that are fully surrounded by elements identified as void elements. Displacement DOFs on the void-solid interface are left as free DOFs.

In this approach, eliminating DOFs is a straightforward process and reduces the size of the equilibrium system to be solved by avoiding assembling of void elements into the global stiffness matrix, leading to computational savings and reducing cost. The key point is that the contribution of elements with volume fraction below a threshold magnitude can be "removed" from the analysis model. This "removal" has taken various forms, including approximating artificial boundary conditions numerically [25] or inserting them explicitly [32], or neglecting the behavior of Degree-Of-Freedom (DOF) associated with the element in the nonlinear analysis convergence criteria [16]. The insertion of boundary conditions effectively eliminates strain energies from appearing in the removed elements, and thus the derivatives of stiffness-related metrics with respect to element volume fractions are zero. However, the derivative of these metrics with respect to the independent design variables may be nonzero due to the filtering approach [25] as shown in Figure 1(f).

Of course it must be acknowledged that formally implementing nodal boundary conditions as the design evolves introduces a discontinuity in the otherwise continuous topology optimization problem, and that designs are likely to be dependent on the chosen magnitude of ρ_t. In the case of solid mechanics

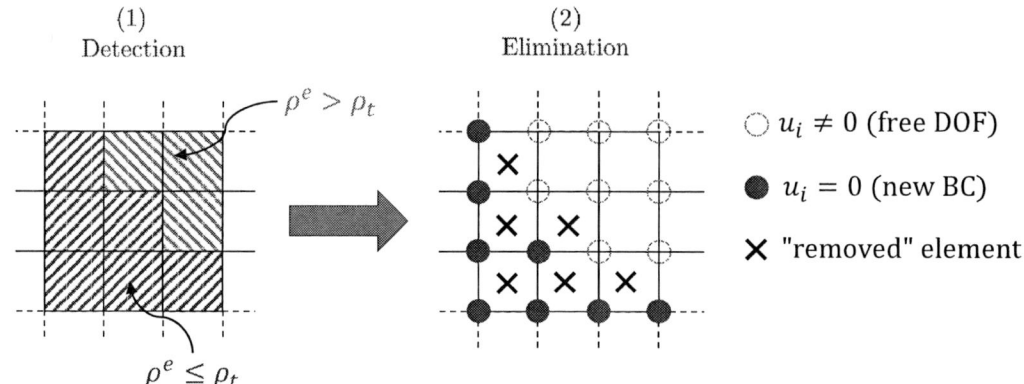

FIGURE 2: Schematic representation of nodal boundary conditions for structural problems

problems, different threshold magnitudes may lead to different solutions, although if chosen small enough these solutions tend to have very similar objective function magnitudes.

Perhaps most importantly, eliminating the need to model low-density elements circumvents numerical instabilities that can localize in such elements. Examples include element distortion that can lead to negative Jacobian errors in problems considering finite deformation, and artificial modes in eigenvalue problems.

ALGORITHMIC SUMMARY

To summarize, the following operations are performed at each design iteration for a given ϕ and corresponding ρ^e: (1) nodal boundary conditions are added to DOFs that are fully surrounded by elements having $\rho^e \leq \rho_t$ (Figure 2), (2) the forward problem is solved, (3) sensitivity analysis is performed, (4) a design change in ϕ is computed by MMA, and (6) ϕ is projected onto ρ^e to update topology. This procedure is repeated until convergence.

NUMERICAL EXAMPLES

In this section, we study characteristics of the proposed adaptive nodal boundary conditions for structural mechanics problems with linear and nonlinear behaviors. Multiple optimization problems are solved considering different values of threshold density, ρ_t. For a given threshold density, the computational efficiency is investigated through comparison of the performance of optimized design and number of equations to be solved, at each design iteration.

CANTILEVER BEAM We first explore the characteristics of the adaptive nodal boundary conditions scheme through a 2D linear elastic cantilever beam subjected to a point load as

shown in Figure 3. The objective is to minimize the strain energy subject to the volume constraint as follows:

$$\min_{\phi} \quad f = \frac{1}{2} \int_{\Omega} \sigma_{ij} \varepsilon_{ij} d\Omega$$

subject to:

$$R(u(\phi), \phi) = 0 \qquad , \tag{12}$$

$$\sum_{e \in \Omega} \rho^e(\phi) v^e \leq V_{max}$$

$$\phi_{min} \leq \phi_k \leq \phi_{max}$$

with the weak form of the governing equation given as:

$$R(u(\phi), \phi) = \int_{\Omega} \nabla w_u \cdot \sigma d\Omega - \int_{\Omega} w_u \cdot b d\Omega - \int_{\Gamma} w_u \sigma \cdot n d\Gamma = 0, \tag{13}$$

where σ is the stress tensor, ε is the strain tensor, w_u is the displacement test function, b is the body force vector, n is the outgoing normal vector, v^e is the elemental volume, and V_{max} is the maximum allowable volume fraction of solid materials. The computational domain is discretized by 400×100 elements. Table 1 summarizes the finite element and design parameters defining the setup of the problem and the solution procedure.

The resulting optimized structures for different values of the threshold density are shown in Figure 4. The optimized designs for different ρ_t are compared with the reference design (design without nodal boundary conditions). The comparison shows similar adaptive topology optimized designs. However, as the threshold increases some features evolve/disappear in the optimized designs, compared to the reference design. This results in small differences in the objective function of the optimized designs. The evolution of the objective function and number of

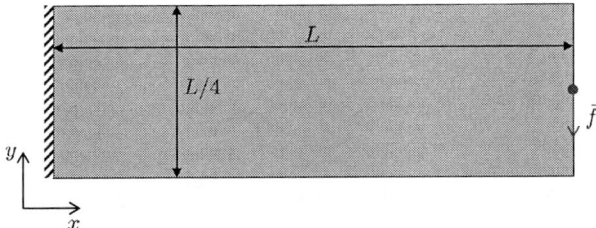

FIGURE 3: Schematic of 2D cantilever beam problem

TABLE 1: Model and optimization parameters for 2D linear elastic cantilever beam

Description	Symbol	Value
beam length	L	1 m
applied force	$\bar{f}$	1.0×10^7 N
Young's modulus	E_0	200 GPa
Poisson's ratio	v	0.33
SIMP exponent	η	2 – 10
		with 0.5@N_{it}^c
number of iterations per step	N_{it}^c	30
small positive number	ρ_{min}^e	1.0×10^{-4}
filter radius	r_{min}	7.625×10^{-3}
curvature of regularization	β	40
prescribed volume fraction	V_{max}	0.5
MMA constraint penalty	-	1000
MMA convergence tolerance	ε_{MMA}	1.0×10^{-6}
MMA asymptote's move factor	a_{MMA}	1.05
MMA asymptote's move factor	b_{MMA}	0.55

free DOFs with design iteration are shown in Figures 5 and 6, respectively. The observed jumps in the objective function are caused by the continuation in the SIMP exponent penalty. Figure 6 shows the nature of the proposed nodal boundary conditions and the capability of removing free DOFs in the system of equations and therefore a consistent reintroduction in computational cost as the design evolves.

Figure 6 shows a monotonically decreasing in number of free DOFs and therefore does not clearly illustrate that material (and associated free DOFs) can be reintroduced into removed void elements. Although it is possible this has occurred, it will be more clearly shown in the following examples.

FORCE INVERTER To challenge the nodal boundary conditions algorithm, we consider the well-known inverter compliant mechanism design problem shown in Figure 7. The objective is to maximize negative displacement at the output point for a given force at the input point. Due to symmetry, only half of the design domain is considered and discretized with 240×120 elements. Following [31], Two springs are added to the model at the input and output points with with stiffness k_{in} and k_{out}, respectively. Table 2 summarizes the finite element and design parameters for the force inverter problem. Considering linear elastic behavior, the objective is defined as:

$$\min_{\phi} \quad \mathrm{f} = L^T u$$

subject to:

$$R(u(\phi), \phi) = 0 \quad , \qquad (14)$$
$$\sum_{e \in \Omega} \rho^e(\phi) v^e \leq V_{max}$$
$$\phi_{min} \leq \phi_k \leq \phi_{max}$$

with the weak form of the governing equation given in Eq. (13) and L is a vector with value zeros except one at the entry associated with the output point displacement degree of freedom.

The resulting optimized designs for the inverter using different values of the threshold density are shown in Figure 8. As can be seen, the optimized topologies are all quite similar to the reference design. Although very slight variations can be seen in the thickness and shape of some members, only the design of Figure 8(e) has a slight variation in topology with a small pore near the top and bottom junction of the right hinge. Even when using an aggressive threshold of 0.1, the algorithm is capable of identifying a high performance inverter mechanism.

The evolution of the objective function and number of free DOFs with design iteration are shown in Figures 9 and 10, respectively. The observed jumps in the objective function are caused by the continuation in the SIMP exponent penalty, but the convergence is otherwise monotonically decreasing. Figure 10 highlights the ability of the algorithm to remove, and then reintroduce, elements and associated DOFs. This is seen by the decreases and subsequent increases in Figure 10. It is well-known that the compliant mechanism design problem can undergo significant changes in topology during the design evolution. This figure confirms that the algorithm can readily reintroduce elements and associated DOFs as these design changes occur. We reiterate that this reintroduction of material is driven entirely by the formal sensitivity analysis and gradient-based optimizer.

ρ^e

0.00 0.20 0.40 0.60 0.80 1.00

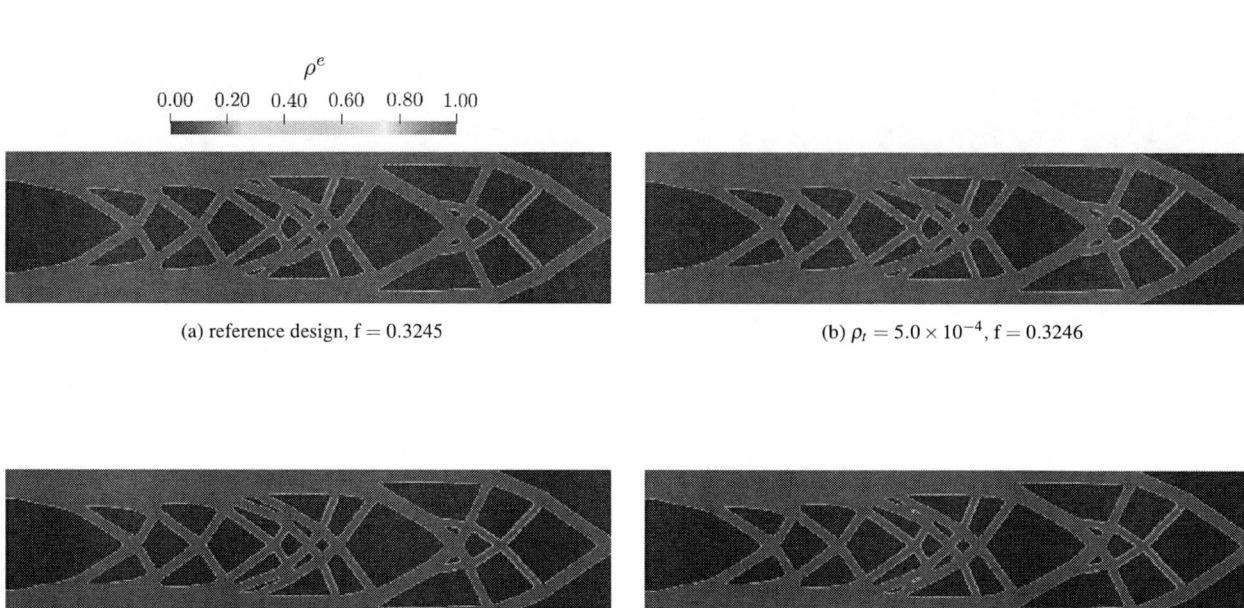

(a) reference design, f = 0.3245

(b) $\rho_t = 5.0 \times 10^{-4}$, f = 0.3246

(c) $\rho_t = 1.0 \times 10^{-3}$, f = 0.3251

(d) $\rho_t = 5.0 \times 10^{-3}$, f = 0.3250

(e) $\rho_t = 1.0 \times 10^{-2}$, f = 0.3251

(f) $\rho_t = 1.0 \times 10^{-1}$, f = 0.3253

FIGURE 4: Topology optimized designs for 2D linear elastic cantilever beam with different values of ρ_t

GEOMETRIC NONLINEARITY   In this section, topology optimization of structures under large deformations is considered. To this end, we consider the optimization problem solved by [1]. The geometrically nonlinear behavior is modeled using the total Lagrangian finite element formulation. The optimization problem is formulated as follows:

$$\min_{\phi} \quad f = F^T U$$

subject to:

$$R(u(\phi), \phi) = 0 \quad , \quad (15)$$
$$\sum_{e \in \Omega} \rho^e(\phi) v^e \leq V_{max}$$
$$\phi_{min} \leq \phi_k \leq \phi_{max}$$

with the weak form of the governing equation given as:

$$R(u(\phi), \phi) = \int_{\Omega} P \cdot \text{Grad}\, \eta \, d\Omega - \int_{\Omega} \bar{B} \cdot \eta \, d\Omega - \int_{\Gamma} \bar{T} \cdot \eta \, d\Gamma = 0,$$
$$(16)$$

where F is the external force vector, U is the displacement solution in the reference configuration, P is the first Piola-Kirchhoff stress tensor, $\bar{B}$ is the body force vector, $\bar{T}$ is the traction in the reference configuration, η is the test function, and Grad is the gradient operator in the reference configuration. For more information about the formulation of finite element and optimization problems, the reader is referred to [1].

The optimized designs for different applied external load and density threshold are given in Figure 14. Comparison of the designs shows that some features are removed/reintroduced by varying the density threshold. These features are more pro-

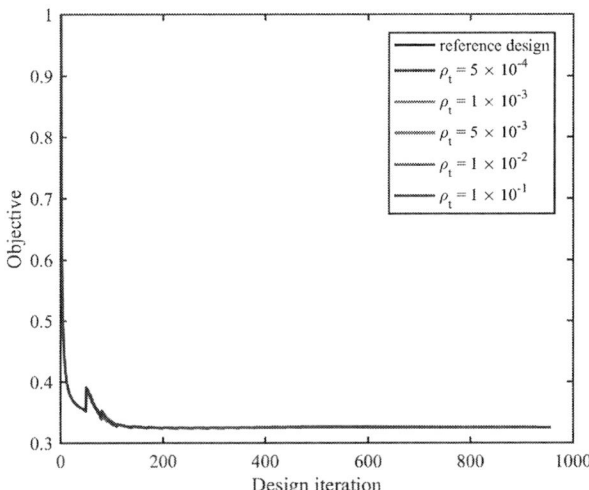

FIGURE 5: Evolution of the objective function for different density thresholds (2D cantilever beam)

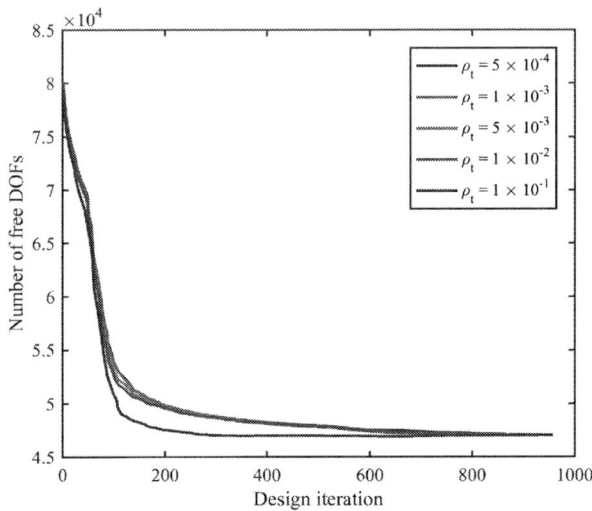

FIGURE 6: Evolution of the number of free DOFs with design iteration (2D cantilever beam)

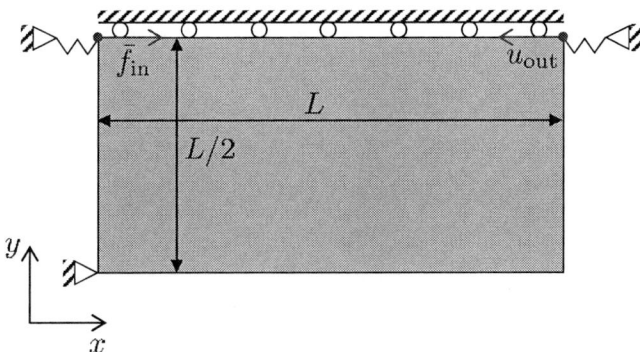

FIGURE 7: Schematic of force inverter design

TABLE 2: Model and optimization parameters for 2D force inverter design problem

Description	Symbol	Value
beam length	L	120
applied force	$\bar{f}_{in}$	1.0
spring stiffness	k_{in}	1.0
spring stiffness	k_{out}	1.0×10^{-3}
Young's modulus	E_0	1.0
Poisson's ratio	ν	0.3
SIMP exponent	η	$2 - 10$
		with $0.5@N_{it}^c$
number of iterations per step	N_{it}^c	50
small positive number	ρ_{min}^e	1.0×10^{-3}
filter radius	r_{min}	2.0
curvature of regularization	β	100
prescribed volume fraction	V_{max}	0.3
MMA constraint penalty	-	1000
MMA convergence tolerance	ε_{MMA}	1.0×10^{-6}
MMA asymptote's move factor	a_{MMA}	1.05
MMA asymptote's move factor	b_{MMA}	0.55

nounced as the external load increases. The evolution of number of free DOFs with design iteration is shown in Figures 11, 12, and 13 for $F_{ext} = 12$, $F_{ext} = 96$, and $F_{ext} = 144$, respectively. The evolution of free DOFs shows a decreasing trend, with frequent jumps, indicating the reintroduction of material. These jumps are more pronounced for smaller magnitudes of ρ_t, indicating it is "easier" to reintroduce material when using small threshold magnitudes. Interestingly, as the external load increases some features appear/disappear in the topology optimized designs. This could be related to the nonlinear behavior of the system, distribution of the strain energy, and local minimum in the system.

We note that the observed reductions in the free DOFs are

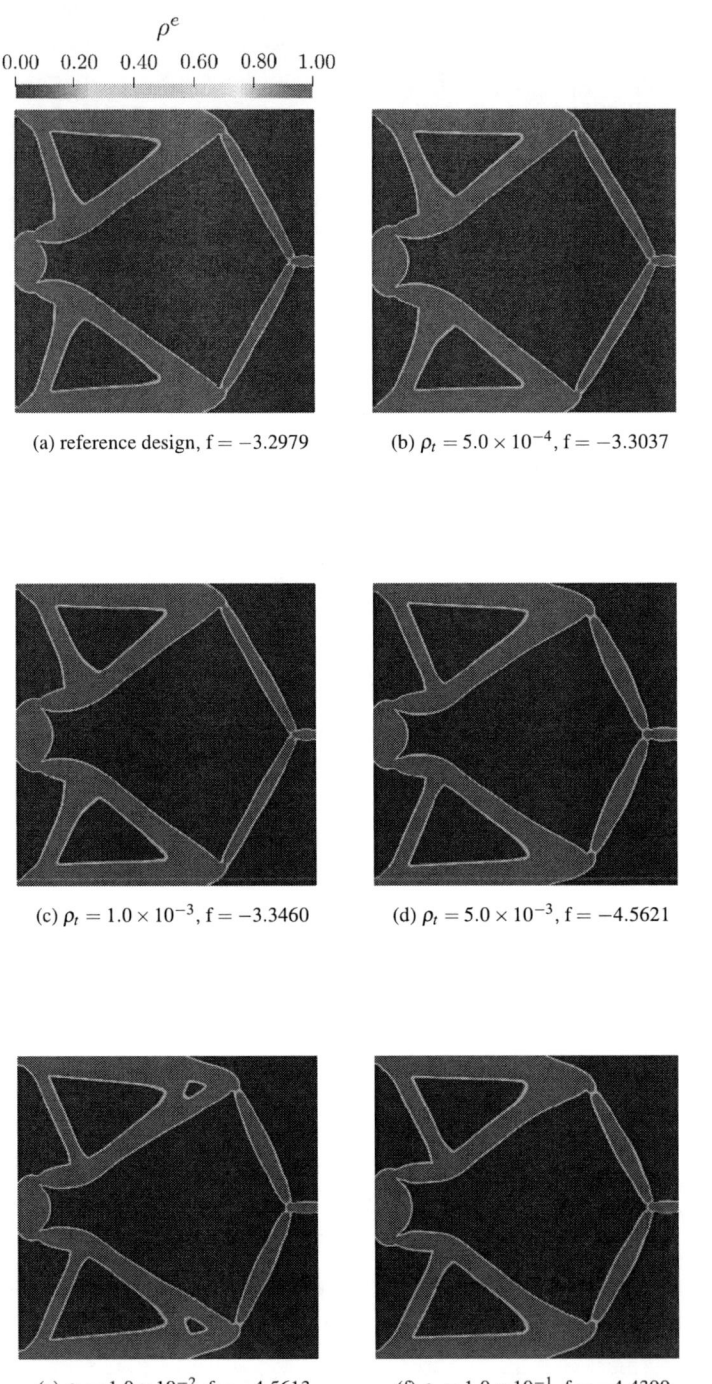

(a) reference design, f = −3.2979

(b) $\rho_t = 5.0 \times 10^{-4}$, f = −3.3037

(c) $\rho_t = 1.0 \times 10^{-3}$, f = −3.3460

(d) $\rho_t = 5.0 \times 10^{-3}$, f = −4.5621

(e) $\rho_t = 1.0 \times 10^{-2}$, f = −4.5613

(f) $\rho_t = 1.0 \times 10^{-1}$, f = −4.4399

FIGURE 8: Topology optimized designs for 2D linear elastic force inverter with different values of ρ_t

particularly beneficial when considering the added cost of non-linear structural analysis. Additionally, using this approach, no

instabilities related to element distortions were encountered. It should be noted that some solutions exhibit one node hinges (Fi-

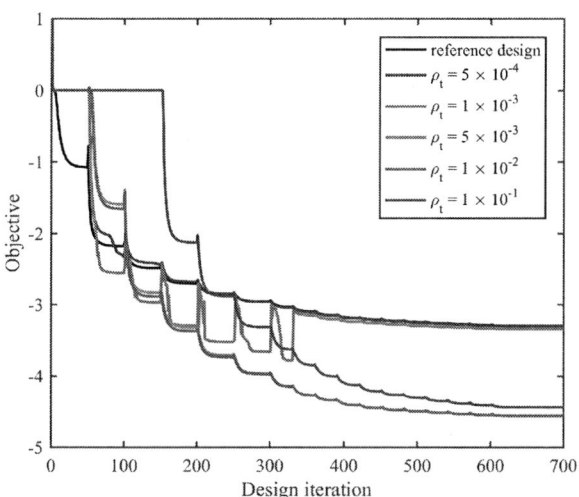

FIGURE 9: Evolution of the objective function for different density thresholds (2D force inverter)

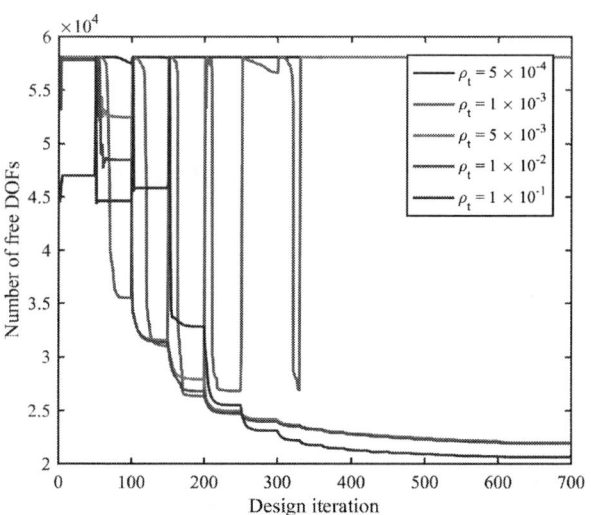

FIGURE 10: Evolution of the number of free DOFs with design iteration (2D force inverter)

gure 14(c), (i), (o)). This is due to definition of minimum length scale, as discussed thoroughly in [26, 33]. One could simply use the robust approach ([34]) to avoid this issue. Finally, it is noted that some solutions have members composed of intermediate volume fractions. These features would likely be eliminated by using the robust approach ([34]) or larger magnitudes of η and/or β.

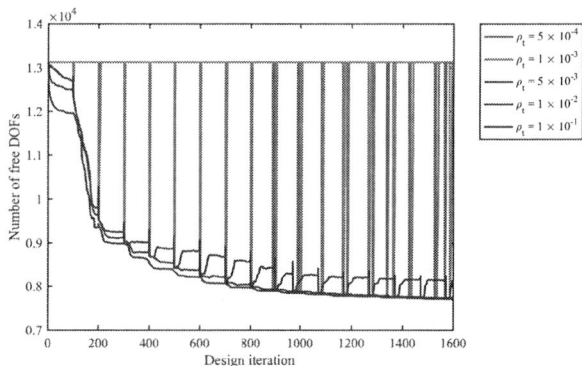

FIGURE 11: Evolution of number of free DOFs with design iteration in the structure with geometric nonlinearity, $F_{ext} = 12$

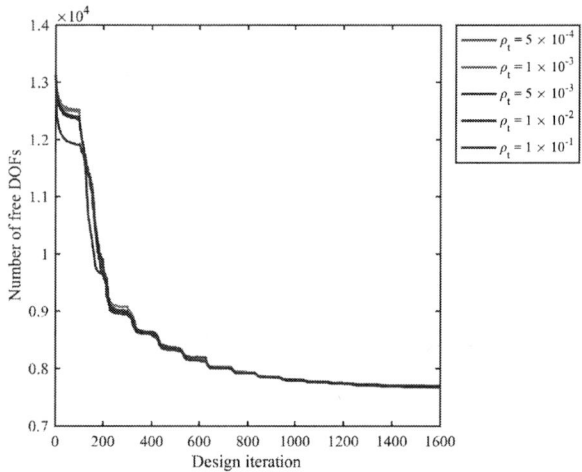

FIGURE 12: Evolution of number of free DOFs with design iteration in the structure with geometric nonlinearity, $F_{ext} = 96$

CONCLUSIONS

We revisit and extend the idea of [25] to adaptively manipulate nodal boundary conditions scheme to systematically enhance computational efficiencies and circumvent numerical instabilities of the finite element analysis in density-based topology optimization problems. An elemental removal check is performed, on a fixed finite element mesh, and elements with density below a threshold density are systematically removed from the finite element analysis. The main feature of the proposed method is that the analysis and design spaces are separate and discretized on different meshes. This indicates while the elemental removal check is performed on the finite element space the design space remains active during the optimization procedure and enables the ability for material reintroduction driven by purely formal sensitivity analysis. Schematics illustrate that the use of nonlinear

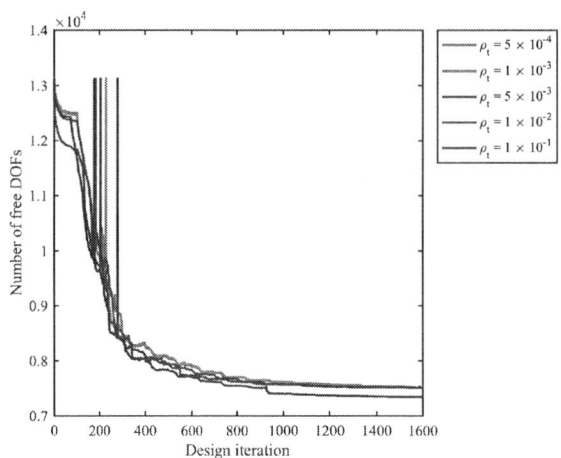

FIGURE 13: Evolution of number of free DOFs with design iteration in the structure with geometric nonlinearity, $F_{\text{ext}} = 144$

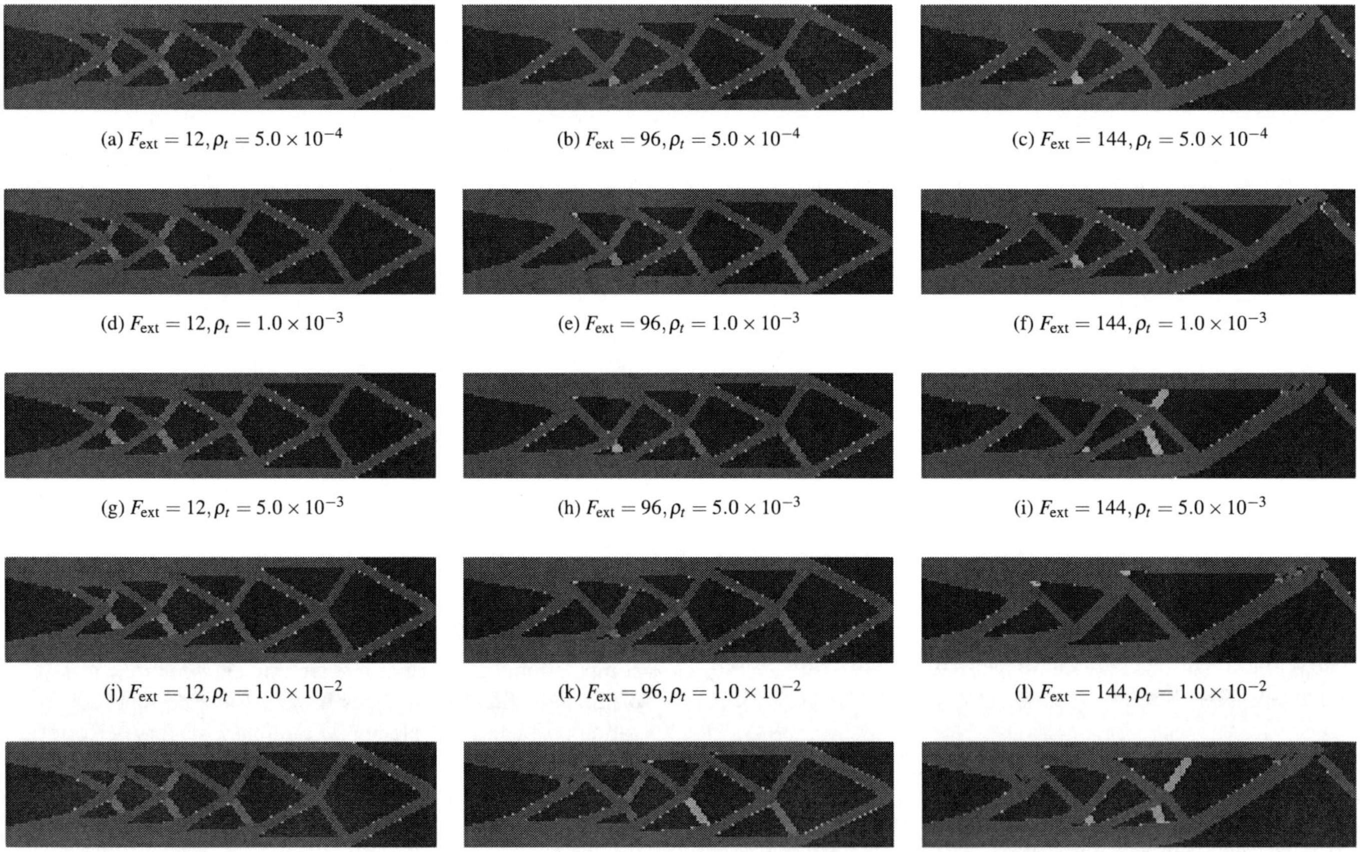

(a) $F_{\text{ext}} = 12, \rho_t = 5.0 \times 10^{-4}$

(b) $F_{\text{ext}} = 96, \rho_t = 5.0 \times 10^{-4}$

(c) $F_{\text{ext}} = 144, \rho_t = 5.0 \times 10^{-4}$

(d) $F_{\text{ext}} = 12, \rho_t = 1.0 \times 10^{-3}$

(e) $F_{\text{ext}} = 96, \rho_t = 1.0 \times 10^{-3}$

(f) $F_{\text{ext}} = 144, \rho_t = 1.0 \times 10^{-3}$

(g) $F_{\text{ext}} = 12, \rho_t = 5.0 \times 10^{-3}$

(h) $F_{\text{ext}} = 96, \rho_t = 5.0 \times 10^{-3}$

(i) $F_{\text{ext}} = 144, \rho_t = 5.0 \times 10^{-3}$

(j) $F_{\text{ext}} = 12, \rho_t = 1.0 \times 10^{-2}$

(k) $F_{\text{ext}} = 96, \rho_t = 1.0 \times 10^{-2}$

(l) $F_{\text{ext}} = 144, \rho_t = 1.0 \times 10^{-2}$

(m) $F_{\text{ext}} = 12, \rho_t = 1.0 \times 10^{-1}$

(n) $F_{\text{ext}} = 96, \rho_t = 1.0 \times 10^{-1}$

(o) $F_{\text{ext}} = 144, \rho_t = 1.0 \times 10^{-1}$

FIGURE 14: Topology optimized designs of structure with geometric nonlinearity and for different values of ρ_t

projection methods allows material to be reintroduced more readily, over a large length scale, even when using linear filters, with the added benefit of creating 0-1 (crisp) topologies.

Numerical benchmark examples in the areas of linear and nonlinear structural mechanics are presented to show the performance of the proposed method which was also recently extended to fluid topology optimization problems [35]. The approach not only prevents numerical instabilities caused by elements with negligible density, but also reduces the size of the finite element system of equations to be solved and eventually reducing the computational cost. Additionally, instances of islanding and disconnected loads, which were reported in the original work [25], were circumvented by using modified MMA parameters of [27]. Interestingly, selection of the threshold relative density ρ_t did not have significant impact on solution quality for the considered cases of $\rho_t \leq 0.1$. Indeed the algorithm performed quite robustly for the considered design problems. The ability of HPM to readily reintroduce material likely plays a critical role in achieving this robust behavior.

ACKNOWLEDGMENTS

The work was supported by the US National Science Foundation under Award Number 120103. Any opinions, findings, and conclusions or recommendations expressed in this article are those of the author(s) and do not necessarily reflect the views of the National Science Foundation.

REFERENCES

[1] Buhl, T., Pedersen, C. B., and Sigmund, O., 2000. "Stiffness design of geometrically nonlinear structures using topology optimization". *Structural and Multidisciplinary Optimization, 19*(2), pp. 93–104.

[2] Bruns, T. E., and Tortorelli, D. A., 2001. "Topology optimization of non-linear elastic structures and compliant mechanisms". *Computer Methods in Applied Mechanics and Engineering, 190*(26), pp. 3443–3459.

[3] de Sturler, E., Paulino, G. H., and Wang, S., 2008. "Topology optimization with adaptive mesh refinement". In Proceeding of the 6th international conference on computational of shell and spatial structures, IASS-IACM, pp. 28–31.

[4] Amir, O., Bendsøe, M. P., and Sigmund, O., 2009. "Approximate reanalysis in topology optimization". *International Journal for Numerical Methods in Engineering, 78*(12), pp. 1474–1491.

[5] Amir, O., Stolpe, M., and Sigmund, O., 2010. "Efficient use of iterative solvers in nested topology optimization". *Structural and Multidisciplinary Optimization, 42*(1), pp. 55–72.

[6] Amir, O., and Sigmund, O., 2011. "On reducing computational effort in topology optimization: how far can

we go?". *Structural and Multidisciplinary Optimization, 44*(1), pp. 25–29.

[7] Kim, I. Y., and De Weck, O., 2005. "Variable chromosome length genetic algorithm for progressive refinement in topology optimization". *Structural and Multidisciplinary Optimization, 29*(6), pp. 445–456.

[8] Stainko, R., 2006. "An adaptive multilevel approach to the minimal compliance problem in topology optimization". *International Journal for Numerical Methods in Biomedical Engineering, 22*(2), pp. 109–118.

[9] de Sturler, E., Paulino, G. H., and Wang, S., 2008. "Topology optimization with adaptive mesh refinement". In Proceeding of the 6th international conference on computational of shell and spatial structures, IASS-IACM, pp. 28–31.

[10] Bruggi, M., and Verani, M., 2011. "A fully adaptive topology optimization algorithm with goal-oriented error control". *Computers & Structures, 89*(15), pp. 1481–1493.

[11] Maute, K., and Ramm, E., 1995. "Adaptive topology optimization". *Structural and Multidisciplinary Optimization, 10*(2), pp. 100–112.

[12] Guest, J. K., 2007. "Reducing dimensionality of design variable space in topology optimization". *Proceedings 7th World Congress on Structural and Multidisciplinary Optimization, Seoul, Korea*, pp. 1799–1806.

[13] Guest, J. K., and Smith Genut, L. C., 2010. "Reducing dimensionality in topology optimization using adaptive design variable fields". *International Journal for Numerical Methods in Engineering, 81*(8), pp. 1019–1045.

[14] Nguyen, T. H., Paulino, G. H., Song, J., and Le, C. H., 2010. "A computational paradigm for multiresolution topology optimization (mtop)". *Structural and Multidisciplinary Optimization, 41*(4), pp. 525–539.

[15] Panesar, A., Brackett, D., Ashcroft, I., Wildman, R., and Hague, R., 2017. "Hierarchical remeshing strategies with mesh mapping for topology optimisation". *International Journal for Numerical Methods in Engineering, 111*(7), pp. 676–700.

[16] Pedersen, C. B., Buhl, T., and Sigmund, O., 2001. "Topology synthesis of large-displacement compliant mechanisms". *International Journal for Numerical Methods in Engineering, 50*(12), pp. 2683–2705.

[17] Yoon, G. H., and Kim, Y. Y., 2007. "Topology optimization of material-nonlinear continuum structures by the element connectivity parameterization". *International Journal for Numerical Methods in Engineering, 69*(10), pp. 2196–2218.

[18] Xie, Y. M., and Steven, G. P., 1997. "Basic evolutionary structural optimization". In *Evolutionary Structural Optimization*. Springer, pp. 12–29.

[19] Querin, O., Steven, G., and Xie, Y., 1998. "Evolutionary structural optimisation (eso) using a bidirectional algorithm". *Engineering Computations, 15*(8), pp. 1031–

1048.

[20] Wang, M. Y., Wang, X., and Guo, D., 2003. "A level set method for structural topology optimization". *Computer Methods in Applied Mechanics and Engineering,* *192*(1), pp. 227–246.

[21] Allaire, G., Jouve, F., and Toader, A.-M., 2004. "Structural optimization using sensitivity analysis and a level-set method". *Journal of Computational Physics,* *194*(1), pp. 363–393.

[22] Challis, V. J., 2010. "A discrete level-set topology optimization code written in matlab". *Structural and Multidisciplinary Optimization,* *41*(3), pp. 453–464.

[23] Challis, V. J., and Guest, J. K., 2009. "Level set topology optimization of fluids in stokes flow". *International Journal for Numerical Methods in Engineering,* *79*(10), pp. 1284–1308.

[24] Maute, K., Schwarz, S., and Ramm, E., 1998. "Adaptive topology optimization of elastoplastic structures". *Structural and Multidisciplinary Optimization,* *15*(2), pp. 81–91.

[25] Bruns, T. E., and Tortorelli, D. A., 2003. "An element removal and reintroduction strategy for the topology optimization of structures and compliant mechanisms". *International Journal for Numerical Methods in Engineering,* *57*(10), pp. 1413–1430.

[26] Guest, J. K., Prévost, J. H., and Belytschko, T., 2004. "Achieving minimum length scale in topology optimization using nodal design variables and projection functions". *International Journal for Numerical Methods in Engineering,* *61*(2), pp. 238–254.

[27] Guest, J. K., Asadpoure, A., and Ha, S.-H., 2011. "Eliminating beta-continuation from heaviside projection and density filter algorithms". *Structural and Multidisciplinary Optimization,* *44*(4), pp. 443–453.

[28] Zhou, M., and Rozvany, G., 2001. "On the validity of eso type methods in topology optimization". *Structural and Multidisciplinary Optimization,* *21*(1), pp. 80–83.

[29] Bendsøe, M. P., 1989. "Optimal shape design as a material distribution problem". *Structural and Multidisciplinary Optimization,* *1*(4), pp. 193–202.

[30] Sigmund, O., and Petersson, J., 1998. "Numerical instabilities in topology optimization: a survey on procedures dealing with checkerboards, mesh-dependencies and local minima". *Structural optimization,* *16*(1), pp. 68–75.

[31] Sigmund, O., 1997. "On the design of compliant mechanisms using topology optimization". *Journal of Structural Mechanics,* *25*(4), pp. 493–524.

[32] Carstensen, J. V., Lotfi, R., Guest, J. K., Chen, W., and Schroers, J., 2015. "Topology optimization of cellular materials with maximized energy absorption". In ASME 2015 International Design Engineering Technical Conferences and Computers and Information in Engineering Conference, American Society of Mechanical Engineers,

pp. V02BT03A014–V02BT03A014.

[33] Carstensen, J. V., and Guest, J. K., 2018. "Projection-based two-phase minimum and maximum length scale control in topology optimization". *Structural and Multidisciplinary Optimization,* *58*(5), pp. 1845–1860.

[34] Sigmund, O., 2009. "Manufacturing tolerant topology optimization". *Acta Mechanica Sinica,* *25*(2), pp. 227–239.

[35] Behrou, R., Ranjan, R., and Guest, J. K., 2019. "Adaptive topology optimization for incompressible laminar flow problems with mass flow constraints". *Computer Methods in Applied Mechanics and Engineering,* *346*, pp. 612–641.

AUTHOR INDEX

Proceedings of the ASME 2019
International Design Engineering Technical Conferences and Computers and Information in Engineering Conference
Volume 2A

Ahmed-Kristensen, Saeema V02AT03A036

Allen, Janet K. V02AT03A019, V02AT03A020

.... V02AT03A042, V02AT03A053

Allison, James T. V02AT03A013

Amaria, Anosh P. V02AT03A025, V02AT03A039

Arezoomand, Mojtaba V02AT03A018

Armstrong, Jason N. V02AT03A039

Aukes, Daniel M. V02AT03A041

Austin-Breneman, Jesse V02AT03A018

Ayoub, Jackie V02AT03A002

Beemaraj, Soban Babu V02AT03A020

Behdad, Sara V02AT03A037

Behjat, Amir V02AT03A005

Behrou, Reza V02AT03A069

Bi, Youyi V02AT03A016

Blosch-Paidosh, Alexandra V02AT03A036

Bostanabad, Ramin V02AT03A006

Brat, Guillaume V02AT03A048

Brauer, Cole V02AT03A041

Brinson, L. Catherine V02AT03A066

Buskohl, Philip V02AT03A065

Cagan, Jonathan V02AT03A001

Campbell, Matthew I. V02AT03A040

Carstensen, Josephine V. V02AT03A069

Chan, Yu-Chin V02AT03A006

Chen, Bo V02AT03A054

Chen, Lei V02AT03A030

Chen, Li V02AT03A047

Chen, Shikui V02AT03A055, V02AT03A064

Chen, Suhao V02AT03A042

Chen, Vincent V02AT03A065

Chen, Wei V02AT03A006, V02AT03A016

.... V02AT03A066

Chen, Xin V02AT03A057

Chiu, Kevin N. V02AT03A014

Chowdhury, Souma V02AT03A005, V02AT03A025

Contractor, Noshir V02AT03A016

Cunningham, James D. V02AT03A003

Davison, Joshua A. V02AT03A025

Deng, Kaiyue V02AT03A039

Dhengre, Snehal V02AT03A029

Dong, Huachao V02AT03A047

Dong, Zuomin V02AT03A047

Du Pasquier, Cosima V02AT03A033

Dunbar, Samuel V02AT03A046

Egan, Paul F. V02AT03A009

Fadel, Georges V02AT03A067

Feng, Yixiong V02AT03A010

Ferguson, Scott V02AT03A046, V02AT03A049

.... V02AT03A050

Frecker, Mary V02AT03A035

Fu, Yan V02AT03A016

Fuge, Mark D. V02AT03A014

Fuge, Mark V02AT03A051

Fujita, Kikuo V02AT03A059

Furuhata, Tomotake V02AT03A017

Gabani, Krushang V02AT03A005

Gao, Shuming V02AT03A027

Gao, Yicong V02AT03A010

Ghiasian, Seyedeh Elaheh V02AT03A034

Gillman, Andrew V02AT03A065

Gong, Lin V02AT03A016

Gowharji, Waleed V02AT03A044

Gu, Xianfeng David V02AT03A055

Guest, James K. V02AT03A068, V02AT03A069

Guo, Lin V02AT03A042

Haberman, Michael R. V02AT03A060

Hajimirza, Shima V02AT03A015

Hall, John F. V02AT03A025, V02AT03A039

Hanks, Bradley V02AT03A035

Hejase, Mohammad V02AT03A048

Hoffenson, Steven V02AT03A043

Hu, Bingtao V02AT03A010

Hu, Zhen V02AT03A030

Hulse, Daniel V02AT03A048

Iyer, Akshay V02AT03A066

Jaiswal, Prakhar V02AT03A034

Jankovic, Marija V02AT03A022

Jiang, Haoliang V02AT03A017

Jiang, Long V02AT03A055

Juhl, Abigail V02AT03A065

Jung, Sangjin V02AT03A032

Kara, Levent Burak V02AT03A017, V02AT03A032

Kaya, Mine V02AT03A015

Kazemi, Hesaneh V02AT03A052

Koller, Pascal V02AT03A033

Lan, Lan V02AT03A023
Landel, Eric V02AT03A022
Lewis, Kemper V02AT03A024, V02AT03A034
Li, Yumeng V02AT03A057
Liao, Yizhou V02AT03A027
Liu, Dehao V02AT03A007
Liu, Lingyun V02AT03A027
Liu, Pengwei V02AT03A030
Liu, Xingchen V02AT03A058
Liu, Yuanzhi V02AT03A026
Long, Daniel V02AT03A049, V02AT03A050
Lotfi, Reza V02AT03A069
Lumpe, Thomas V02AT03A061
Luo, Chuan V02AT03A068
Luo, Jianxi V02AT03A011
Malviya, Manoj V02AT03A031
Massoni, Brandon R. V02AT03A040
McComb, Christopher V02AT03A001, V02AT03A029
Meisel, Nicholas A. V02AT03A028, V02AT03A029
.... V02AT03A031
Michalek, Jeremy J. V02AT03A045
Miller, Scarlett R. V02AT03A028
Miller, Simon W. V02AT03A003
Mistree, Farrokh V02AT03A019, V02AT03A020
.... V02AT03A042, V02AT03A053
Mohan, Pranav V02AT03A053
Morkos, Beshoy V02AT03A050
Morris, Clinton B. V02AT03A060
Najmon, Joel C. V02AT03A056
Nejadkhaki, Hamid Khakpour V02AT03A039
Nellippallil, Anand Balu V02AT03A053
Nguyen, Ryan V02AT03A025
Nie, Zhenguo V02AT03A032
Nigam, Suyash V02AT03A017
Norato, Julian V02AT03A052
Odonkor, Philip V02AT03A024
Oh, Yosep V02AT03A037
Olvander, Johan V02AT03A038
Ostrander, John K. V02AT03A029
Pandey, Vijitashwa V02AT03A021
Paredis, Christiaan J. J. V02AT03A022
Pasquali, Felipe M. V02AT03A039
Pathan, Rizwan Khan V02AT03A020
Persson, Johan V02AT03A038
Prabhu, Rohan V02AT03A028
Prasad, Aditya V02AT03A066
Rai, Rahul V02AT03A034
Raina, Ayush V02AT03A001
Russo, Nicholas V02AT03A043
Ryan, Lauren V02AT03A029
Sadiwala, Rushabh V02AT03A067
Salvi, Amit V02AT03A020

Sarica, Serhad V02AT03A011
Schadler, Linda V02AT03A066
Schwarz, Jonas V02AT03A004
Seepersad, Carolyn C. V02AT03A060
Seepersad, Carolyn Conner V02AT03A062
.... V02AT03A063
Sexton, Thurston V02AT03A051
Sha, Zhenghui V02AT03A016
Sharma, Gehendra V02AT03A019, V02AT03A020
Sharpe, Conner V02AT03A062, V02AT03A063
Shea, Kristina V02AT03A004, V02AT03A008
.... V02AT03A033, V02AT03A036, V02AT03A061
Shimada, Kenji V02AT03A017
Simpson, Timothy W. V02AT03A003, V02AT03A028
.... V02AT03A029
Sinha, Swapnil V02AT03A031
Sissoko, Timothe M. V02AT03A022
Stankovic, Tino V02AT03A008, V02AT03A033
Tan, Jianrong V02AT03A010
Tao, Siyu V02AT03A066
Telenko, Cassandra V02AT03A012
Tovar, Andres V02AT03A056
Tsushima, Shohji V02AT03A059
Tucker, Conrad S. V02AT03A003
Tumer, Irem Y. V02AT03A048
Vaziri, Ashkan V02AT03A052
Vedant V02AT03A013
Walsh, Hannah S. V02AT03A048
Wang, Liwei V02AT03A006
Wang, Mingxian V02AT03A016
Wang, Pingfeng V02AT03A054
Wang, Yan V02AT03A007
Wang, Yixing V02AT03A066
Wang, Zhuo V02AT03A030
Watson, Bryan C. V02AT03A012
Whitefoot, Kate S. V02AT03A032, V02AT03A044
.... V02AT03A045
Wiberg, Anton V02AT03A038
Wiley, Carson V02AT03A065
Wood, Kristin L. V02AT03A011, V02AT03A023
Wu, Tong V02AT03A056
Wu, Yongjia V02AT03A064
Xie, Jian V02AT03A016
Xu, Qianli V02AT03A002
Xu, Xiaoqiang V02AT03A064
Xu, Yanwen V02AT03A054
Yadav, Darshan V02AT03A050
Yaji, Kentaro V02AT03A059
Yamakawa, Soji V02AT03A017
Yamasaki, Shintaro V02AT03A059
Yang, Jessie V02AT03A002
Yang, Zhangsihao V02AT03A017

Yip, Arthur H. C. V02AT03A045
Yoo, David... V02AT03A065
Yu, Zhiyang... V02AT03A008
Yuen, Chau V02AT03A023
Yukish, Michael A.. V02AT03A003
Zadbood, Amineh V02AT03A043
Zhang, Jenmy Zimi V02AT03A063
Zhang, Jie V02AT03A026
Zhang, Wentai V02AT03A017
Zhang, Yichi V02AT03A066
Zheng, Hao V02AT03A010
Zheng, Zhuoyuan V02AT03A054
Zhou, Feng.. V02AT03A002
Zhou, Haofei V02AT03A057
Zhu, Ping V02AT03A006
Zuo, Lei V02AT03A064